REVIEWS IN MINERALOGY AND GEOCHEMISTRY

Volume 88 2022

Diamond: Genesis, Mineralogy and Geochemistry

EDITORS

Karen Smit
University of the Witwatersrand, South Africa

Steve Shirey
Carnegie Institution for Science, USA

Graham Pearson
University of Alberta, Canada

Thomas Stachel
University of Alberta, Canada

Fabrizio Nestola
University of Padova, Italy

Thomas Moses
Gemological Institute of America, USA

Series Editor: ***Ian Swainson***

MINERALOGICAL SOCIETY OF AMERICA
GEOCHEMICAL SOCIETY

COVER ILLUSTRATION

Rough diamond crystals from unspecified African mines. Left: 15.96 ct octahedron, Center: 22.32 ct macle, Right: 10.50 ct octahedron containing a visible orange-red garnet inclusion.

Credit: Sir Oppenheimer Student Collection. Photo by Robert Weldon. © GIA.

Reviews in Mineralogy and Geochemistry, Volume 88

Diamond: Genesis, Mineralogy and Geochemistry

ISSN 1529-6466 (print)
ISSN 1943-2666 (online)
ISBN 978-1-946850-10-2

THE MINERALOGICAL SOCIETY OF AMERICA
3635 CONCORDE PARKWAY, SUITE 500
CHANTILLY, VIRGINIA, 20151-1125, U.S.A.
WWW.MINSOCAM.ORG

Diamond: Genesis, Mineralogy and Geochemistry

88 *Reviews in Mineralogy and Geochemistry* **88**

PREFACE

After more than 80 volumes of Reviews in Mineralogy and Geochemistry (RiMG), we now have a volume on diamond—*Diamond: Genesis, Mineralogy and Geochemistry*. Diamond is the record-setter in many mineralogical properties such as hardness, diffusivity, thermal conductivity, purity, and covalency of bonding. Similarly, diamond, as the premier gemstone of the mantle holds primacy for geological features such as age and depth of origin. Diamond was among the first crystalline structures to be solved by X-ray diffraction and the first materials measured for their Raman spectrum. The second issue of the multi-society journal Elements was devoted to diamonds. At more than 80 billion USD in yearly commercial value, diamond sets the record for the most traded, valuable mineral on the planet. Despite its chemical simplicity, diamond has been the object of more research effort, and had more scientific and popular press pages written about it, than any other mineral. How odd it is then that a RiMG volume on diamond took fifty years to be published! Here we address that shortcoming.

The late emergence of Volume 88 is perhaps like the emergence of a now-famous understudy in a Broadway show who used her time well to prepare for the real performance. After more than half a century, geologists now thoroughly understand many aspects of the mantle-convection-driven process of plate tectonics. This understanding is an essential backdrop to understanding how diamond forms because diamond is the only mantle gem to be directly created by plate tectonics, wherever its carbon has been derived. Over the same half century, many analytical techniques such as mass spectrometry (SIMS, TIMS, LA-ICP-MS), spectroscopy (Raman, IR), microanalysis and sampling (EPMA, STXM, TEM, FIB, XRD) have developed such improved sensitivity and spatial resolution that they now can handle the purity and minute lattice imperfections in diamond and allow the study of its very tiny mineral and fluid inclusions. Similarly, advances in the size, depth, and pressure attainable by experimental apparatus such as piston cylinder, multi-anvil and diamond anvil presses have reproduced and mapped the stability of the mineral inclusions trapped in the diamond structure to reveal the P,T conditions of diamond formation and to allow stringent constraints to be placed on the mineralogy of the mantle host rocks.

The delay has been worth the wait. Through diamond study, we now have a way to see directly and to great depths, the plate tectonic-driven recycling processes, now and in the past, of the most dynamic and active rocky planet in the solar system, our own Earth. We now can evaluate the physical and chemical effects of Earth's geodynamics with the deepest samples available: diamond and its mineral and fluid cargo. Diamond played a central role in the recent decade-long exploration of the deep carbon geochemical cycle by the Deep Carbon Observatory (cf. Hazen et al. Carbon in Earth and Orcutt et al. Deep Carbon). The minerals contained in these rarest, deepest diamonds are among the most special samples we have on Earth. Like lunar rocks and soils, like unique meteorites, like samples returned from comet flybys, or asteroid landers — they are messengers from another time and place that we are not going to get by any other means. Science simply demands that we study them to understand the inner workings of our planet.

1529-6466/22/0088-0000$00.00 (print) http://dx.doi.org/10.2138/rmg.2022.88.00
1943-2666/22/0088-0000$00.00 (online)

Over time, the interior of a planet controls everything: its exterior, its surface features, its geochemical cycles, its atmosphere, its hydrosphere, and its habitability. Earth is the only template we have for direct study of the relationship between the solid mantle/crust and the hydrosphere/atmosphere of a rocky exoplanet. If we want to understand how the atmosphere evolved or life began we have to understand Earth's interior better. Earth is the only place where, now and in the near future, we will be able to obtain enough samples and enough measurements to understand how interior dynamics couple with surface properties. If, for example, we can find unique connections between plate tectonics and certain characteristics of the atmosphere, then we might be able to use exoplanet atmospheric observations to infer something about the solid exoplanet. But this will be impossible to do the other way around.

The purpose and goal of this new volume is to assemble all the chief current knowledge about natural diamond in one body for the use of the Earth Science community. The contents of this volume are wide-ranging with the goal to leave little out so that any scientist could reach for this volume to obtain as much basic diamond knowledge as necessary. RiMG volumes have always served this role. An important feature of *Diamond: Genesis, Mineralogy and Geochemistry* is that the authors of several chapters used new and up-to-date databases that were expressly compiled for the purposes of accuracy in writing their chapters. These databases are available for community use at: https://dataverse.scholarsportal.info/dataverse/diamond.

The authors of this volume have had to work in times of unprecedented difficulties in collaboration due to the COVID-19 pandemic and yet they have managed, in all cases, to produce monumental works that are not only state-of-the-art reviews, but that set the scene for diamond research in the coming decades. We thank every contributor for overcoming these difficulties and collaborating to produce a truly comprehensive volume. Should anyone take on the task up up-dating this volume, in perhaps another 50 years, we truly hope that they will be able to do so in a pandemic-free world that more resembles the environment so conducive to research that we all desire.

Reviewers, beyond the editorial team, worked hard to improve the quality of these long chapters. Our thanks go to Mike Ashfold, Gerhard Brey, Ingrid Chinn, Phil Diggle, Wali Faryad, Yana Fedortchouk, Boris Feygelson, Andrea Giuliani, Jon Goss, Herman Grutter, Ben Harte, Peter Heaney, Juske Horita, Janne Koornneef, Konstantin Litasov, Volker Lorenz, Martin Kunz, Tania Marshall, Sami Mikhail, Tom McCandless, Oded Navon, Lutz Nasdala, George Read, Steve Richardson, Roberta Rudnick, Hans-Peter Schertl, Zachary Sharp, Evan Smith, Andrew Steele, Tatsuki Tsujimori, Daniel Twitchen, Fanus Viljoen, Michelle Wenz, and Dmitry Zedgenizov.

We are grateful to Ian Swainson and Rachel Russell for their patience and dedication to ensuring that the RiMG series remains a product of the highest quality. We thank the Deep Carbon Observatory for bringing the diamond community together at so many meetings, all facilitated by the Sloan Foundation, who part-financed the Open Access for this volume. Thank you also to the Geological Survey of Canada and the Gemological Institute of America who provided further Open Access funds for this volume.

Karen Smit (Johannesburg, South Africa)

Steve Shirey (Washington, DC, USA)

Graham Pearson (Edmonton, AB, Canada)

Thomas Stachel (Edmonton, AB, Canada)

Fabrizio Nestola (Padova, Italy)

Thomas Moses (New York City, NY, USA)

Diamond: Genesis, Mineralogy and Geochemistry

88 *Reviews in Mineralogy and Geochemistry* 88

TABLE OF CONTENTS

1 A Review of the Geology of Global Diamond Mines and Deposits

BA Kjarsgaard, M de Wit, LM Heaman, DG Pearson, J Stiefenhofer, N Janusczcak, SB Shirey

2 Morphology of Monocrystalline Diamond and its Inclusions

JW Harris, KV Smit, Y Fedortchouk, M Moore

3 Polycrystalline Diamonds from Kimberlites: Snapshots of Rapid and Episodic Diamond Formation in the Lithospheric Mantle

DE Jacob, S Mikhail

5 Crystallographic Methods for Non-destructive Characterization of Mineral Inclusions in Diamonds

RJ Angel, M Alvaro, F Nestola

6 Mineral Inclusions in Lithospheric Diamonds

T Stachel, S Aulbach, JW Harris

7 Geochemistry of Silicate and Oxide Inclusions in Sublithospheric Diamonds

MJ Walter, AR Thomson, EM Smith

8 Raman Identification of Inclusions in Diamond

EM Smith, MY Krebs, P-T Genzel, FE Brenker

9 Fluid Inclusions in Fibrous Diamonds

Y Weiss, J Czas, O Navon

10 Pressure and Temperature Data for Diamonds

P Nimis

11 Geochronology of Diamonds

KV Smit, S Timmerman, S Aulbach, SB Shirey, SH Richardson, D Phillips, DG Pearson

12 Diamond Spectroscopy, Defect Centers, Color, and Treatments

BL Green, AT Collins, CM Breeding

13 Synthesis of Diamonds and Their Identification

UFS D'Haenens-Johansson, JE Butler, AN Katrusha

14 Experimental Petrology Applied to Natural Diamond Growth

RW Luth, YN Palyanov, H Bureau

15 Carbon and Nitrogen in Mantle-Derived Diamonds

T Stachel, P Cartigny, T Chacko, DG Pearson

Reviews in Mineralogy & Geochemistry
Vol. 88 pp. 1–118, 2022

A Review of the Geology of Global Diamond Mines and Deposits

Bruce A. Kjarsgaard

Geological Survey of Canada
Natural Resources Canada
Ottawa, Ontario, K1A0E8
Canada

bruce.kjarsgaard@NRCan-RNCan.gc.ca

Mike de Wit

Department of Earth Sciences
Stellenbosch Unviversity
Stellenbosch
South Africa

dewit@icon.co.za

Larry M. Heaman, D. Graham Pearson

Department of Earth and Atmospheric Sciences
University of Alberta
Edmonton, Alberta
T6G 2E3
Canada

gdpearso@ualberta.ca

Johann Stiefenhofer

Anglo American
Johannesburg
South Africa

Johann.Stiefenhofer@angloamerican.com

Nicole Janusczcak

BHP
Toronto, Ontario
Canada

nicole.januszczak@bhp.com

Steven B. Shirey

Earth and Planets Laboratory
Carnegie Institution for Science
5241 Broad Branch Road NW
Washington DC 20015, U.S.A.

sshirey@carnegiescience.edu

1529-6466/22/0088-0001$15.00 (print)
1943-2666/22/0088-0001$15.00 (online)
http://dx.doi.org/10.2138/rmg.2022.88.01

INTRODUCTION

Diamond is not a common rock-forming mineral (an exception being "diamondite"; see Jacob and Mikhail 2022, this volume) nor a common crustal mineral (the exceptions being "UHP diamonds"; see Dobrzhinetskaya et al. 2022, this volume); it is a scarce or minor mantle mineral whose rarity belies its importance in understanding mantle geochemistry and dynamics, as the numerous chapters in this volume attest. A complete understanding of natural diamond formation and the mantle geology it reveals, starts with an understanding of the magmas that deliver diamonds many hundreds of kilometers upward from the mantle to the upper crust. There, at Earth's surface, diamonds are found in primary magmatic deposits of kimberlite and olivine lamproite, or secondary deposits weathered from these primary sources. The goal of this chapter is to summarise the important geological aspects of primary and secondary diamond deposits found at the Earth's surface.

This chapter covers both primary and secondary diamond deposits because both have played an important role in the science of diamond geology, and in the gem diamond trade. A significant scientific research focus since the 1980's has been on lithospheric diamonds, sourced from kimberlite mines; however, some of the first and most important discoveries of sublithospheric diamonds were made from the alluvial diamond fields of Australia (Ororoo), Brazil (Juina) and West Africa (Kan Kan) as described in Stachel et al. (2022, this volume). Africa, where the beneficiation of both primary and secondary diamond deposits has been substantial, provides a ready comparison of the importance of both of these types of deposit. Since the discovery of diamonds in Africa in 1854 until the end of 2019, Africa has produced almost 3.6 billion carats (Bct) of diamonds out of a total global production of some 5.9 Bct (updated from de Wit et al. 2016). Of the 3.6 Bct African production it is estimated that 1.3 Bct have been mined out of placer and paleo-placer (including marine) deposits, which represents 22% of total global production. With the combined production of placers and paleo-placers from South America, India, Australia and Siberia it is estimated that close to 1.5 Bct (or just over 25% of the global production), have come from these deposits. Furthermore, these diamonds are typically better quality having survived erosion from a primary source, subsequent transport and deposition and hence are of higher carat value.

Despite the importance of alluvial diamonds produced with respect to total diamond production in terms of their values, the alluvial deposits themselves have little to teach us directly about the mantle aside from providing isolated diamond research specimens. In this respect, an overview of the magmas that host primary diamond deposits, with specific aspects such as geochemistry, mineralogy, and age of these rocks, along with diamond grade and value data is provided herein for four important reasons.

First, kimberlites, carbonate-rich olivine lamproites (CROL—see *Magmatic Source Rocks: General Background and Historical Perspectives*) and olivine lamproites (see Table 1 for the mineralogy of these three rock types) are samples of magmas and their fluids moving through the lithospheric mantle precisely where the vast majority (>90%) of all diamonds have formed. While the nature of diamonds as xenocrysts imply that they are typically derived from an older generation of melt/fluid than their magmatic host rocks, they also bear some broad similarities to younger generations of diamond-forming fluids. In the case of fibrous diamonds (Harris et al. 2022, this volume), epitaxial overgrowth of a second generation of diamond suggests some kimberlites (or kimberlite-related melts), are diamond-forming magmas.

Second, as most kimberlites are diamond-poor or are barren of diamonds, with only ~1% proving to be economic, understanding the key aspects of the magmatic host rock–diamond relationship and the geological occurrences of these deep-seated magmas are critical factors in establishing that newly discovered kimberlites, CROLs, or olivine lamproites contain sufficient quantities of diamonds of a size and quality that permits economically viable mining.

Table 1. Mineralogy of kimberlite, carbonate-rich olivine lamproite (CROL), and olivine lamproite.

Minerals present	Kimberlite	Carbonate-rich olivine lamproite (CROL)	Olivine lamproite
Megacryst suite (> 1 cm)	Common to absent	Absent, rare	Absent, ?very rare?
	Cognate and antecryst minerals		
Olivine	Common to absent	Rare to absent	Rare to absent
Phlogopite	Minor to absent	Common to minor	Common, minor, rare
Spinel	Common to rare (spinel cores)	Minor to absent (spinel cores)	Minor to absent (spinel cores)
Diopside	Absent	Rare to absent	Rare to absent (Minor)
	Phenocryst /Microphenocrysts (0.2–1.0 mm)		
Olivine phenocrysts	Common	Common	Common
Olivine microphenocrysts	Common	Minor to rare	Minor to rare
Phlogopite phenocrysts	Minor, rare, absent	Common	Common
Phlogopite microphenocrysts	Minor, rare (poikilitic)	Common	Common (poikilitic)
Spinel phenocrysts	Common to minor	Minor to rare	Minor to rare
Diopside microphenocrysts	Absent	Variable: Common to absent	Minor to absent
	Groundmass (<0.2 mm)		
Olivine	Common	Minor to rare	Rare, absent
Phlogopite	Common, laths (poikilitic)	Common (poikilitic)	Common (poikilitic)
Spinel	Common to minor	Minor to rare	Minor to rare
Glass	Absent, exceptionally rare?	Absent (not reported)	Common, minor, rare
Apatite	Common to rare	Common to minor	Minor to rare
Perovskite	Common to minor, rare	Minor to rare	Minor to rare
Carbonate (calcite, dolomite)	Common to minor	Common to minor	Rare, absent
Monticellite	Common to absent	Absent, very rare	Absent
Diopside	Absent	Variable: Minor to absent	Rare to absent
Leucite	Absent	Rare to absent	Minor to rare
Sanidine—K Feldspar	Absent	Rare to absent	Rare to absent
Ti-K-richterite	Absent	Rare to absent	Minor to rare
Ilmenite	Minor, rare to absent	Minor to rare	Rare to absent
Zr-silicates	Rare to absent	Rare to absent	Rare to absent
Hollandite, priderite	Rare to absent	Rare to absent	Rare to absent
Brucite	Rare to absent	Absent	Absent
Periclase	Rare to absent	Absent	Absent

Minerals present	Kimberlite	Carbonate-rich olivine lamproite (CROL)	Olivine lamproite
		Mineral chemistry zoning trends	
Spinel	Trend 1; Trend 2; Trend 1–2; Trend 3	Trend 2; Trend 1–2	Trend 2
Phlogopite	Phlogopite–eastonite; phlogopite–tetraferri-phlogopite	Phlogopite–tetrafer-riphlogopite; phlogopite–Al-poor phlogopite	Phlogopite–Al-poor phlogopite; Al-poor phlogopite–tetraferriphlogopite
		Alteration	
Serpentine	Common to rare	Common to rare	Common to minor
Talc	Rare	Rare	Common to minor
Calcite, dolomite	Common to absent	Common to absent	Rare to absent
Magnetite	Common to absent	Common to absent	Rare to absent
Diopside	Common to absent		

Third, kimberlites, CROLs and olivine lamproites are the dominant magma types that actually deliver the deepest-derived samples of mantle lithologies in the form of xenoliths of peridotite and eclogite. The geologic history of these samples, and studies of diamondiferous xenoliths and diamond inclusions, has become essential knowledge in understanding how diamonds, especially lithospheric diamonds, form.

Fourth, kimberlite-derived diamonds are known to contain sublithospheric mantle mineral inclusions. These mineral inclusions and the kimberlite magmas that carry them inform us about the deep Earth, such as when and how deep slabs devolatilize, how these deep supercritical fluids react with mantle lithologies, what is the nature of this fugitive mantle fluid reservoir, and what is the relationship to mantle convection. There is a substantive quantity of literature on this currently active field of research and the reader is directed to early work (e.g., Sobolev 1977; Nixon 1987), and chapters in the *Treatise on Geochemistry* (Pearson et al. 2003, 2018; Pearson and Wittig 2014), as well as the chapters in this volume and references therein.

Brief historical overview of the key diamond discoveries

There have been numerous descriptions of the history of diamond discovery and mining, from its inception to present day and we shall not attempt to replicate this history. The interested reader is referred to reviews by Balfour (1987), Kirkley et al. (1992), Harlow (1998), Erlich and Hausel (2002) and Wilson et al. (2007a).

The first known mining of diamonds on an economic scale occurred in the Indian alluvial deposits, in the region of the Godavari and Krishna rivers, exploited since ~ 2000 B.C., but first described in western literature by Gaveia de Orta in 1565 and later by Tavernier in 1676 (cited in Balfour 1987). For over 2000 years, the main recognised source of the world's diamonds was from India. The second significant discovery of diamonds that led to economic mining occurred in Brazil in 1725 when Bernardo Francisco Lobo purchased diamonds from locals, prompting a "rush" to exploit alluvial diamonds that established the city now known as Diamantina (Erlich and Hausel 2002). This discovery was followed, in 1844, by the discovery of rich alluvial deposits in Bahia State (Brazil).

The subsequent identification of diamonds in South Africa led to a third global source of diamonds. The first diamond discovery in South Africa, which prompted the exploration rush that ultimately led to the finding of diamond in a primary source rock, took place in 1866 or 1867 when a diamond was discovered on the banks of the Orange River at the farm "De Kalk"

(Janse 1995). After numerous other diamond finds, including an 83.5 carat (ct) alluvial stone in 1869, a 50 ct diamond was extracted by a farmer named De Klerk from what was to be recognised as the Jagersfontein kimberlite pipe (Wilson et al. 2007a). This discovery is usually credited as the first diamond find from a "primary" deposit, although it is now evident that diamonds had been mined in India well before that time, from what was later recognised as the Majhgawan olivine lamproite (Wilson et al. 2007a). Discoveries of primary diamond deposits in South Africa that became major mines include Jagersfontein (1869 or 1870) Dutoitspan and Bultfontein (1869), Koffiefontein (1870), De Beers and Kimberley (1871), Wesselton (1890), and Cullinan (aka, also known as, Premier) in 1903, Letšeng-la-Terai (1957), Finsch, (1963) and lastly Venetia (1979), as documented by Janse (1995) and Wilson et al. (2007a).

Following these initial discoveries, mainly in the 1800s, large scale commercial diamond mining was slow to develop in other areas of the world. A detailed broader history of the development of diamond mining across Africa is given by Janse (1995, 1996) who noted the inception of alluvial mining in Zaire (now the Democratic Republic of Congo, DRC) in 1903 and the Mbuji Mayi area in 1918, and in Angola in 1912. The Colossus kimberlite was discovered in Zimbabwe in 1907, along with the richly diamondiferous kimberlites at Mbuji Mayi (DRC) in 1946, and many diamondiferous kimberlites in the Lunda Norte region of Angola in 1952, with the Catoca kimberlite discovered in 1985 (Janse 1996, 2007). Key discoveries elsewhere in Africa include the Mwadui (Williamson) diamond mine in Tanzania, discovered in 1940, and the Koidu area kimberlites in Sierra Leone in 1948 (Janse 1996).

Though diamonds had been reported in Russia since 1829 from the Ural Mountains, mining primary deposits on a large scale only commenced following the discovery of Siberian (southern Yakutia) kimberlites by Larisa Popugayeva in 1954—a discovery that was to lead to the establishment of the major diamond mines of Mir (1954) and its associated placers (1957); Sytykanskaya (1955); Udachnaya (1957), 23rd Party Congress (1959); Aikhal (1960); Internationalaya (1969) and Jubileinaya (1975), establishing Russia as the world's largest producer for a time (Janse 2007).

In global terms, three other highly significant discoveries of diamonds are of note. Firstly, exploration by De Beers in the Kalahari desert led to the discovery, in 1967, of the vast Orapa kimberlite in Botswana, followed in 1972 by the discovery of the Jwaneng kimberlite (Brook 2012). The development of these two kimberlite-hosted diamond mines alone placed Botswana as the world's leading diamond producer for the decades that followed, with other significant developments being the opening of the Letlhakane, Damtshaa, Lerala mines and Karowe mines (Janse 1995, 1996, 2007; Brook 2012), and more recently the Ghaghoo mine (2014–2017).

Although alluvial diamonds had been found in Australia in 1861, and small-scale extraction from placers occurred in the Copeton area of New South Wales since 1884, large scale production from a primary deposit commenced only with the discovery in 1979 of the richly diamondiferous Argyle olivine lamproite pipe, whose exploration and discovery history is documented by Smith et al. (2018).

The establishment of large-scale commercial diamond mining in North America took much longer. Despite diamonds being found in the eastern USA in the 1840s and in glacial tills in the Great Lakes region since 1876 (Kjarsgaard and Levinson 2002), exploration success was slow to materialise. The Prairie Creek (Arkansas, USA) olivine lamproite was discovered in 1906, although it did not prove to be economic. However, the discovery of diamondiferous kimberlite in the Lac de Gras region of the Northwest Territories in 1990 (Kjarsgaard and Levinson 2002) led to one of the greatest staking rushes in Canadian history and the commencement of large-scale diamond mining at the Ekati Mine in 1998 and Diavik Mine in 2002 (Kjarsgaard and Levinson 2002; Janse 2007).

Classification of mined diamond deposits

Primary magmatic deposits classified by rock type. Three rock types host primary magmatic diamond mines and include kimberlite, olivine lamproite, and carbonate-rich olivine lamproite (CROL; formerly termed micaceous kimberlite, or Group II kimberlite, or orangeite). The mineralogy of these rocks types are defined in Table 1, and described in more detail in this Chapter. Kimberlites occur on stable cratonic continental crust; there are no known kimberlites on oceanic crust or young orogenic belts (Fig. 1). These deep-seated magmas are derived from hundreds of kilometers beneath Earth's surface and due to their affinity with thick continental lithospheric mantle are intimately associated with cratons. Olivine lamproite and carbonate-rich olivine lamproite also occur on stable cratonic continental crust (Fig. 2). All primary magmatic-hosted diamond mines are intimately associated with cratons. Cratons, as defined by Pearson et al. (2021) have thick lithospheric roots (>150 km) and have not been disturbed tectonically since ~1 Ga.

Secondary diamond deposits classified by geologic setting and age. Three categories of placer diamond deposits have been recognised: retained placers, transient placers, and terminal placers (Bluck et al. 2005). These three different types of placer deposits are decribed in more detail in *Review of Secondary Global Diamond Deposits.*

Tiered classification system of primary magmatic-hosted diamond mines

Diamonds mined in primary magmatic source rocks are found in magmatic rocks classified as kimberlite, olivine lamproite and carbonate-rich olivine lamproite (previously referred to as micaceous kimberlite, or Group II kimberlites, or orangeites)—see Table 1 for details. Diamonds in these rocks occur as sparsely dispersed diamond xenocrysts, both micro- and macro-diamonds. A macrodiamond is defined (using Canadian National Instrument 43-101 reporting standards) as diamonds that do not pass through a 0.85 mm square mesh screen. However, it is also relevant to include a definition of the term "macrodiamond", which from a diamond sale point of view includes diamonds that do not pass through a +1 Diamond Trading Company (DTC) screen, which has a round aperture of 1.092 mm.

In total, there are globally ~7,000 known occurrences of these three rock types, dominantly kimberlite (~6,500; Giuliani and Pearson 2019), with subordinate numbers of olivine lamproite (Bergman 1987) and carbonate-rich olivine lamproite (Skinner 1989). However, only ~1000 kimberlites (~15%) contain macrodiamonds (de Wit et al. 2016), and are of potential economic interest. A similar (or lower) percentage of macrodiamond bearing olivine lamproites are also known. In contrast, from the available data, if taken as representative, it appears that significantly more than 15% of CROL are macrodiamond-bearing, given the high number of macrodiamond-bearing CROL in southern Africa (Table 2). In total, there are < 100 producing or past producing diamond mines globally (Table 2)—an amazingly small number that has supplied humanity's demand for diamonds.

Key parameters of global diamond mines include diamond grade (carat/tonne), diamond value (US$/carat), mine size (hectare), and ore value (US$/tonne), as listed in Table 2. Together these parameters can be utilized to determine the potential value and economic viability of the deposit. Note that carats per hundred tonnes (cpht) is also widely used as diamond deposit grades are often less than 1 carat per tonne (Table 2) As noted by Field et al. (2008), these four key parameters do not define "large" versus "small" mines. Instead, the key features that define the "scale" of a mine are the rate of diamond production (carats per year) and the lifespan of the mining operation, the latter not necessarily being directly related to size (area) and/or tonnes of contained ore. Simply stated, the economic viability of a diamond mine, or any mine for that matter, requires that the value of the ore (US$/tonne) needs to exceed the mining, processing and other affiliated costs, while factoring in the initial capital outlay of the project. A number of these costs are a strong function of the geographic location of the mine and the available infrastructure. For instance, the exploration and operating costs for diamond mines in southern

African, northern Russian or Canadian diamond deposits vary greatly. de Wit et al. (2016) refined the "size" concept of diamond mines of Field et al. (2008), introducing a four-Tier diamond mine classification system. In their definition, diamond mines are characterized by their yearly carat production, as well as the in-situ value of diamonds produced over the life of the mine. Here, we further modify the "Tier Structure", defining a total of five tiers, Tier 1 being the most valuable and Tier 5 being the least valuable, using the following rationale:

- **Tier 1.** World Class Mine. US$13B or greater in-situ diamond value produced up to 2019/20. Twelve mines.
- **Tier 2.** Large Mine. Produce >1 million carats/per year for 5 years minimum, but has not exceeded US$13B in value to date. Some Tier 2 mines, with continued (future) mining, may achieve Tier 1 status. Twenty-three mines.
- **Tier 3.** Medium Mine. Produce >0.4–1 million carats/per year for 5 years minimum. Fifteen Mines
- **Tier 4.** Small Mine. Produce 50,000– 400,000 carats/per year for 5 years minimum. Twenty-seven mines.
- **Tier 5.** Very Small Mine. 5,000 –50,000 carats/year/per year for 5 years minimum. Five mines.

The US$B13B cut-off for Tier 1 mines is approximately one-third lower than that suggested by De Beers (2014) and de Wit et al. (2016), who placed the cut-off for Tier 1 diamond mines at US$20B. Our division at US$13B was selected based on a clear-cut break on a cumulative probability plot (Fig. 3; derived from data in Table 2) of mine value. It could be argued that the ~US$50B Jwaneng deposit should be in a Tier of its own, being a clear outlier at the top of the global deposit value list (Fig. 3; Table 2). Importantly, note that the data in Table 2 is derived from a wide variety of public sources and may not be as robust as the confidential or difficult to obtain data held by large diamond mining companies—especially with respect to data for tonnes of ore mined, diamond grade, and diamond value—as determined on a yearly basis. Furthermore, the reader should note there are a number of past-producing diamond mines that are not listed in Table 2, either because they did not meet the 5-year mine life criteria, or simply because, there is limited or no data available to classify them. This is especially true for diamond mines that operated >100 years ago, for example many of the kimberlite or carbonate-rich olivine lamproite (CROL) diamond mines in South Africa. However, these former mines, as well as current advanced diamond exploration projects, are shown on global, and continent or craton-scale maps throughout this Chapter, for completeness. The "Tier Structure" established above underpins the following sections of this Chapter as context for categorising or discussing the different aspects of diamond deposits.

MAGMATIC SOURCE ROCKS

General background and historical perspective

The primary magmatic host rocks for diamond mines include kimberlite, olivine lamproite, and carbonate-rich olivine lamproite (CROL). Additional rock types such as ultramafic lamprophyre (variety aillikite) are known to be diamond-bearing (Hutchison and Frei 2009; Nielsen et al. 2009), as well as some possible kamafugite localities, but these have not as yet been proven to be mineable resources. The nomenclature, terminology and classification of magmatic diamond source rocks, specifically kimberlite, CROL and olivine lamproite, has been the source of much contention, fueling numerous debates and sowing significant confusion. We attempt to provide some clarity by briefly reviewing this issue below, though we acknowledge that in such inherently variable and complex magmatic systems, a degree of subjectivity remains.

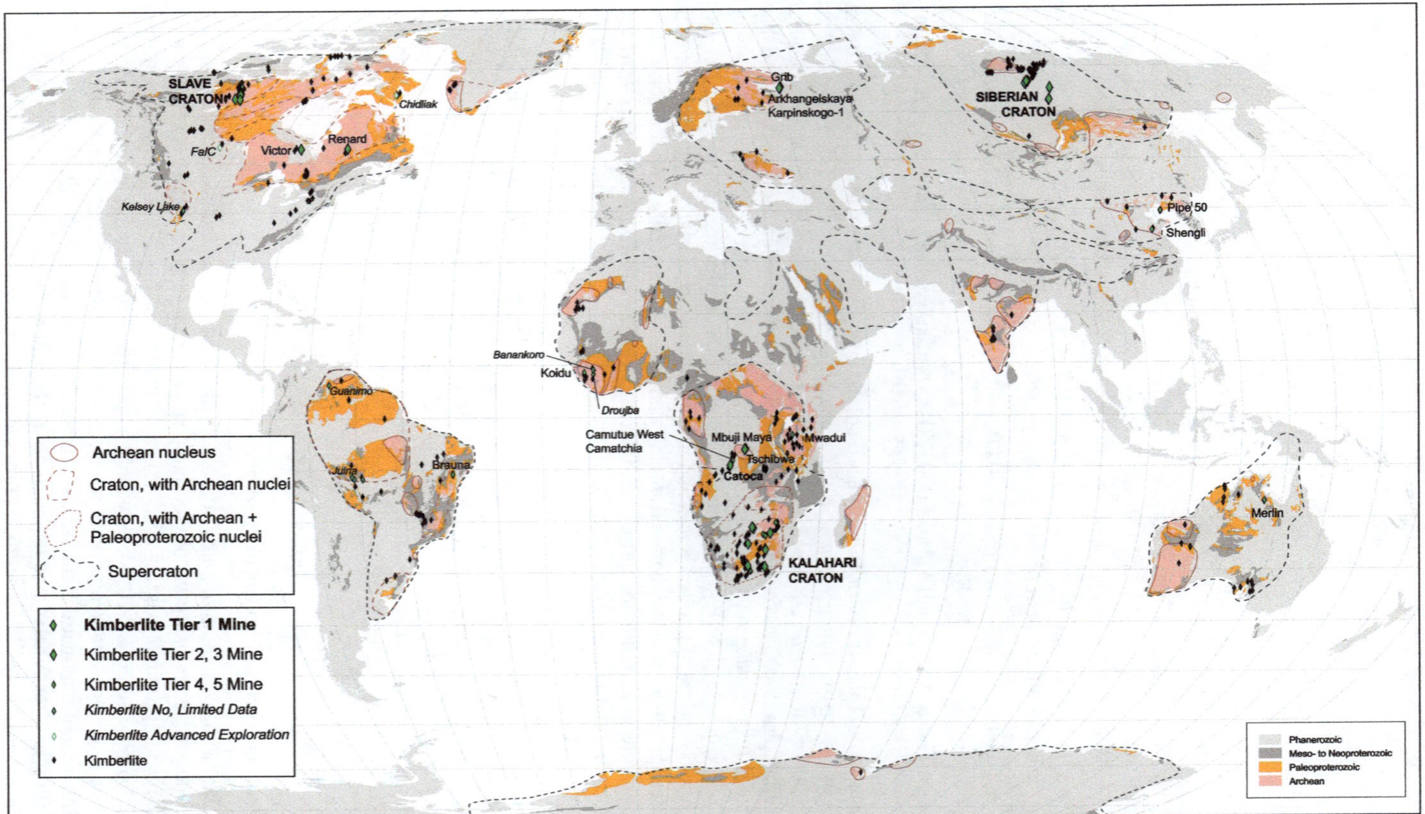

Figure 1. Global distribution of kimberlite and kimberlite-hosted diamond mines, past producing mines and advanced exploration projects. Additional kimberlite locations show the main fields and clusters globally. Bedrock geology backdrop from Chorlton (2007); note the dominant bedrock geology is that at the surface; i.e., Phanerozoic rocks that overlie Archean or Proterozoic rocks are shown on this map as Phanerozoic. Craton outlines from Pearson et al. (2021). Robinson map projection.

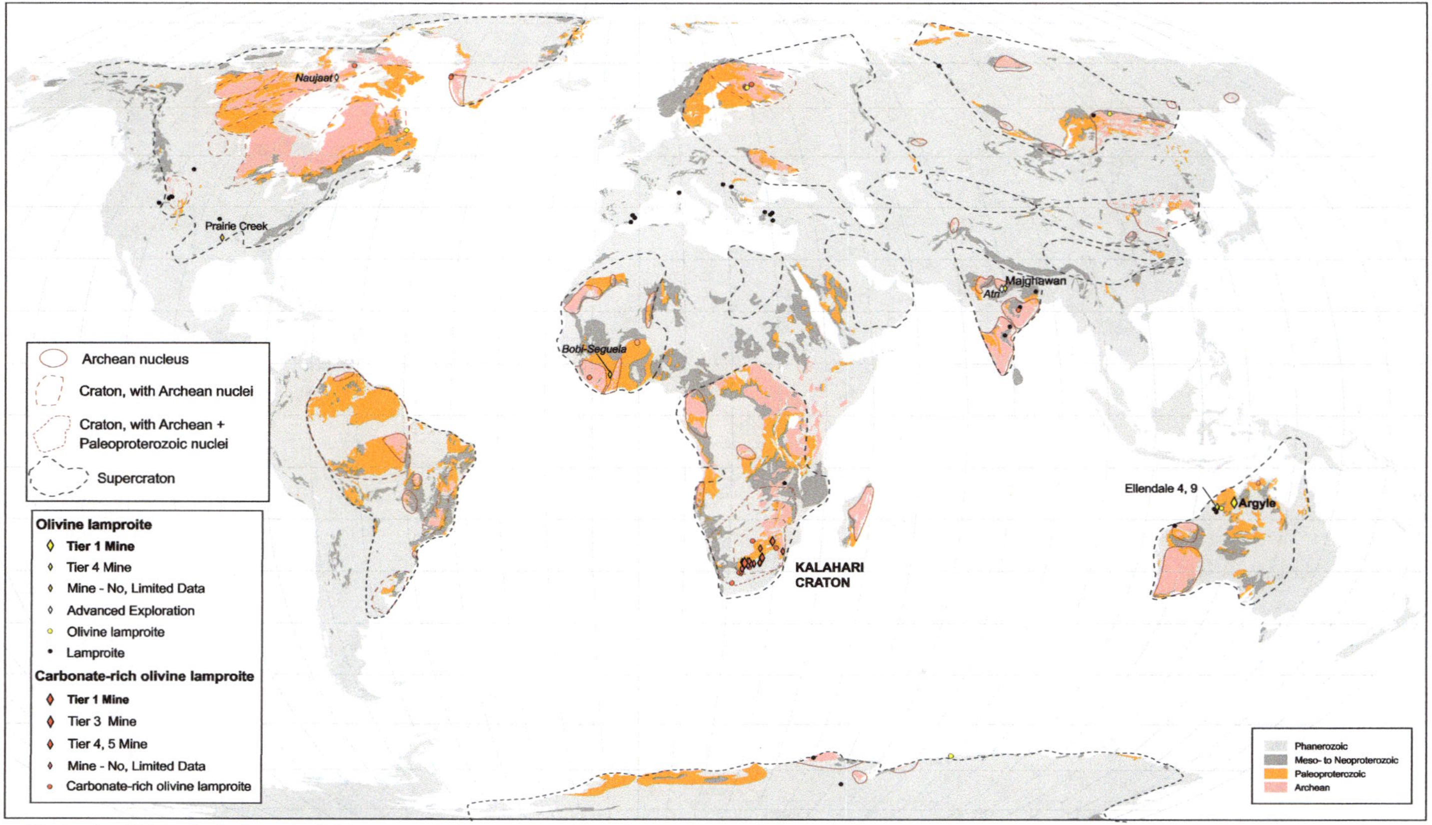

Figure 2. Global distribution of CROL- and olivine lamproite-hosted diamond mines, past producing mines and advanced exploration projects. Additional CROLs, olivine lamproites and lamproites (e.g., leucite lamproites, Mediterranean lamproites) locations show the main fields and clusters globally. Bedrock geology backdrop from Chorlton (2007); note the dominant bedrock geology is that at the surface i.e., Phanerozoic rocks that overlie Archean or Proterozoic rocks are shown on this map as Phanerozoic. Craton outlines from Pearson et al. (2021). Robinson map projection.

Table 2. Eighty-two producing or past producing magmatic-hosted global diamond mines, and their key parameters.

Mine	Field	Country	Rock Type	Grade (ct/t)	Stone Value (US$/ct)	Ore Value (US$/t)	Area (ha)	US$B to 2019/20	Mine Tier	Age (Ma)
Catoca	Lunda South	Angola	K	0.645	86	63.6	63.6	14.654	1	117.9
Camutue West	Lunda North	Angola	K	0.096	59	57.12	9	0.91	4	*120*
Camatchia (Luo)	Lunda North	Angola	K	0.17	200	34	29.4	0.93	4	*120*
Ellendale 9	Ellendale	Australia	OL	0.049	220	10.78	46	0.387	4	22
Ellendale 4	Ellendale	Australia	OL	0.14	220	30.8	76	0.079	4	22
Merlin	EMU	Australia	K	0.2	150	30	small pipes	0.076	4	382
Argyle	East Kimberley	Australia	OL	3.5	20	70	47	17.92	1	1126
Orapa (AK1)	Orapa	Botswana	K	0.95	70	66.5	118	29.54	1	92
Karowe (AK6)	Orapa	Botswana	K	0.153	690	105.57	9.5	0.966	4	93
Letlhakane	Orapa	Botswana	K	0.284	200	56.8	15.2	7.436	2	92.1
Damtshaa	Orapa	Botswana	K	0.215	200	43	13.5	0.86	4	94
Jwaneng	Jwaneng	Botswana	K	1.56	110	171.6	54	52.58	1	235
Lerala	Martin's Drift	Botswana	K	0.255	55	14.03	small pipes	n.d.	4	1364
Brauna	Brauna	Brazil	K	0.29	338	98.02	2	0.358	4	682
Beartooth	Lac de Gras	Canada	K	1.1	100	110	0.5	0.28	3	53
Koala North	Lac de Gras	Canada	K	0.5	143	71.5	0.5	0.24	4	53.3
Panda	Lac de Gras	Canada	K	1.11	168	186.48	3.3	4.03	2	53.3
Koala	Lac de Gras	Canada	K	1.26	200	252	4.5	5.17	2	53.3
Pigeon	Lac de Gras	Canada	K	0.41	150	61.5	1	0.17	3	*54*
Sable	Lac de Gras	Canada	K	0.9	128	115.2	2.4	0	2	*54*
Misery	Lac de Gras	Canada	K	4.17	50	208.5	1.1	2.06	2	*55*
A-154 North	Lac de Gras	Canada	K	3.021	166	501.54	1.2	5.014	2	55.2
A-154 South	Lac de Gras	Canada	K	4.778	126	602.08	2.6	7.003	2	55.5

Mine	Field	Country	Rock Type	Grade (ct/t)	Stone Value (US$/ct)	Ore Value (US$/t)	Area (ha)	US$B to 2019/20	Mine Tier	Age (Ma)
Fox	Lac de Gras	Canada	K	0.37	241	89.17	11	2.14	2	56
Lynx	Lac de Gras	Canada	K	0.8	230	184	0.6	0.07	4	56
A-21	Lac de Gras	Canada	K	2.7	126	340.2	1	1.286	2	56
A-418	Lac de Gras	Canada	K	3.426	90	308.32	1.4	2.957	2	56
Victor	Lac de Gras	Canada	K	0.25	458	114.5	15	3.708	3	177
Snap Lake	SE Slave	Canada	K	1.46	144	210.24	dike	1.21	2	523
Tuzo	SE Slave	Canada	K	1.23	57	70.11	1.4	0	2	530
Hearne	SE Slave	Canada	K	2.08	70	145.6	1.5	0.665	2	534
GK5034	SE Slave	Canada	K	1.84	82	150.88	2	1.51	2	542
Renard-2	Otish Mts	Canada	K	0.84	85	71.4	1	1.836	2	643.4
Pipe 50	Fuxian	China	K	1	20	20	0.9	0.14	4	464
Shengli	Mengyin	China	K	0.5	20	20	0.4	0.14	4	476
Mbuji Mayi M1	Mbuji-Mayi	DRC	K	1.3	13	16.9	22.6	0.71	3	69.9
Tshibwe	Bena-Baya	DRC	K	0.381	16	6.1	60	0.98	2	85
Dokolwayo	Dokolwayo	Eswatini	CROL	0.4	100	40	2.8	0.072	4	203
Majhgawan	Hinota	India	OL	0.13	220	28.6	6.5	0.22	4	1072
Mothae	Lesotho	Lesotho	K	0.027	931	25.14	8.8	0	5	87
Kao	Lesotho	Lesotho	K	0.054	350	18.9	19.8	0.476	4	89
Liqhobong	Lesotho	Lesotho	K	0.27	80	21.6	8.6	0.38	3	91
Letšeng Satellite	Lesotho	Lesotho	K	0.023	2750	63.25	4.7	n.d.	4	91
Letšeng Main	Lesotho	Lesotho	K	0.023	1650	37.95	17.2	2.805	4	94
Sytykanskaya	Alakit	Russia	K	0.6	85	51	6	0.842	3	344

Mine	Field	Country	Rock Type	Grade (ct/t)	Stone Value (US$/ct)	Ore Value (US$/t)	Area (ha)	US$B to 2019/20	Mine Tier	Age (Ma)
Komsomolskaya	Alakit	Russia	K	0.38	20	76	4	1.4	4	357
Jubileinaya	Alakit	Russia	K	0.84	100	84	59	17.227	1	358
Aikhal	Alakit	Russia	K	5.35	100	535	3	14.493	1	378
Udachnaya	Daldyn	Russia	K	1.5	80	120	25	18.055	1	360
Zarnitsa	Daldyn	Russia	K	0.21	185	38.85	21.5	3.511	2	362.3
Internationalnaya	Malo-Botuobiya	Russia	K	8.73	160	1396.8	1.7	26.397	1	360
Mir	Malo-Botuobiya	Russia	K	5.55	130	721.5	6.9	17.639	1	363
23rd Party Congress	Malo-Botuobiya	Russia	K	4	120	480	1.5	2.4	2	433
Botuobinskaya	Nakyn	Russia	K	5.19	55	285.45	2.9	0.213	3	365
Nyurbinskaya	Nakyn	Russia	K	4.13	55	227.15	2.4	3.912	2	366
Grib	Verkhotina	Russia	K	0.7	105	73.5	14.2	2.661	2	374
Arkhangelskaya	Lomonosov	Russia	K	0.86	100	86	19	0.861	2	375
Karpinskogo-1	Lomonosov	Russia	K	1.14	100	114	14	1.368	2	380
Koidu K1	Koidu	Sierra Leone	K	0.67	350	234.5	0.5	0.642	4	143
Koidu K2	Koidu	Sierra Leone	K	0.33	350	115.5	0.5	0.583	4	146
Jagersfontein	Jagersfontein	South Africa	K	0.12	200	24	12	1.9	4	85.8
Dutoitspan	Kimberley	South Africa	K	0.32	140	44.8	10.8	2.8	3	86.5
De Beers	Kimberley	South Africa	K	0.64	100	64	5.1	2.3	3	86.5
Kimberley	Kimberley	South Africa	K	1.5	100	150	3.7	2.2	3	87
Bultfontein	Kimberley	South Africa	K	0.54	75	40.5	9.7	1.838	3	88.3
Wesselton	Kimberley	South Africa	K	0.37	75	27.75	8.7	2.138	3	89.6
Koffiefontein	Koffiefontein	South Africa	K	0.08	510	40.8	11.1	6.12	4	93.9
Finsch	Lime Acres	South Africa	CROL	0.36	99	35.64	17.9	12.989	1	118

Mine	Field	Country	Rock Type	Grade (ct/t)	Stone Value (US$/ct)	Ore Value (US$/t)	Area (ha)	US$B to 2019/20	Mine Tier	Age (Ma)
Frank Smith	Barkley West	South Africa	K	0.05	500	25	7.4	n.d.	5	114
Main, Bobbejaan	Bellsbank	South Africa	CROL	0.5	250	125	dike	n.d.	5	118
Star	Theunissen	South Africa	CROL	0.62	323	200.26	dike	n.d.	5	124
Roberts Victor	Boschof	South Africa	CROL	0.45	100	45	2	0.044	4	126
Voorspoed	Kroonstad	South Africa	CROL	0.24	150	36	12	1.062	3	132
Lace/Crown	Kroonstad	South Africa	CROL	0.41	180	73.8	2.9	0.135	3	133.2
Helam	Swartruggens	South Africa	CROL	2	100	200	dike	n.d.	5	145.1
Marsfontein	Klipspringer	South Africa	CROL	1.73	142	245.66	0.4	0.256	3	155
The Oaks	Marnitz	South Africa	K	0.34	105	35.7	1	0.103	4	505
Venetia	Venetia	South Africa	K	1.22	125	152.5	23.3	19.55	1	521
Cullinan	Pretoria	South Africa	K	0.466	185	86.21	32.2	29.415	1	1153.3
Mwadui	Shinyanga	Tanzania	K	0.07	307	21.49	146	6.831	4	52
Murowa	Murowa	Zimbabwe	K	0.9	65	58.5	4.5	0.215	4	527

Notes: K = kimberlite, OL = olivine lamproite, CROL = carbonate-rich olivine lamproite. Mine area, for example in twin or triple pipes is the total area; for dikes, or mines consisting of multiple small bodies (dike enlargements/blows/diatreme buds) the area is not determined (n.d), or termed small pipes. The emplacement age data for each mine is listed when known; if unknown the geochronology data is inferred and shown in italics, based on data from other bodies in that field (see Geochronology section for further details). Sources of diamond mine data: Janse (1993, 1995, 1996, 2007), De Beers (2004), Kjarsgaard (2007a), Read and Janse (2009) and references therein. Post-2000 data is mainly based on public domain National Instrument 43-101 and JORC compliant reports, and company reports e.g., ALROSA, Anglo American plc, BHP Billiton Limited, Rio Tinto plc, etc.; Roffey et al. (2018) provide a detailed listing of these sources.

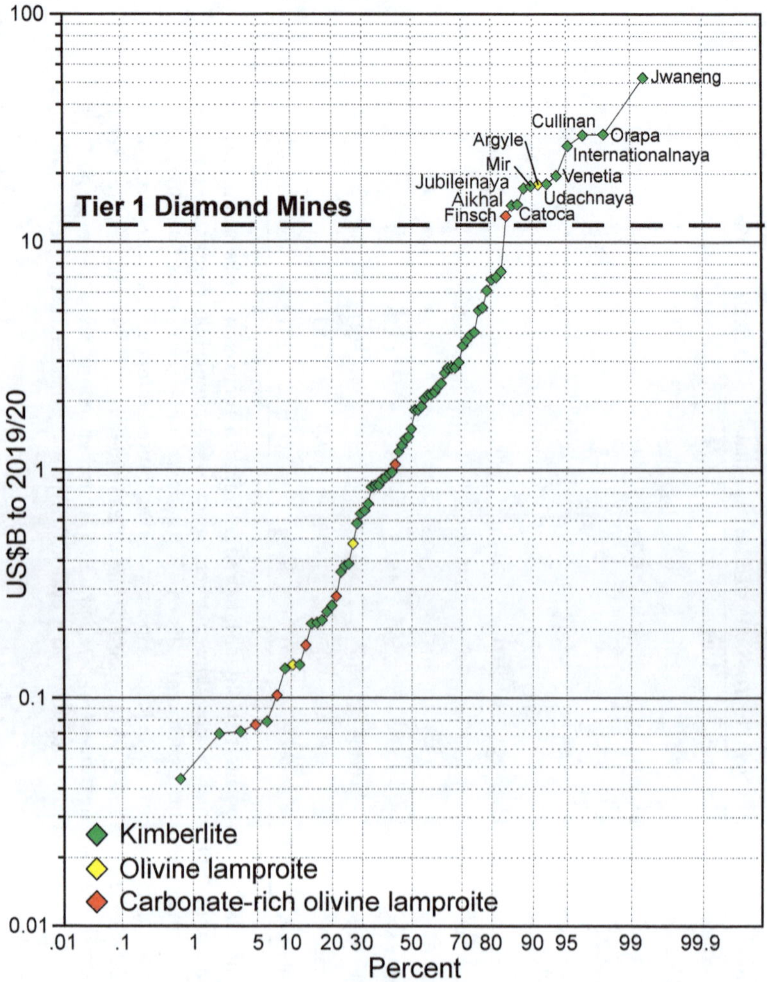

Figure 3. Cumulative probability plot for the value of individual diamonds mined up to 2019/20 (data from Table 2). The Tier 1 versus Tier 2 division is defined by the break between the Finsch Mine (US$13B—Tier 1) and the Letlhakane Mine (US$7.4B—Tier 2). All primary magmatic hosted diamond mines are color coded by rock type (Kimberlite, CROL, olivine lamproite). All Tier 1 mines are labeled by locality.

The term kimberlite (following type-locality nomenclature rules of the day) was introduced by Lewis (1888) for the distinctive volcanic rocks being mined for diamonds at Kimberley, South Africa, which he described as "porphyritic volcanic peridotite of basaltic structure". Importantly, Lewis (1888) recognized distinctions between kimberlite, kimberlite breccia and kimberlite tuff. Subsequently, the landmark monograph of Wagner (1914) subdivided kimberlites into two distinct petrographic types, a "basaltic" variety, being poor in mica and equivalent to the kimberlite described by Lewis (1988), and a "lamprophyric" or "micaceous" variety, rich in phlogopite mica. Wagner (1914) further described the "basaltic" variety as being exceptionally rich in large and small olivine crystals (50–75 modal%), with the "lamprophyric" variety as having up to 50% phlogopite, with less olivine. The terms "basaltic" and "lamprophyric" were used by Wagner to emphasize the structure/texture/habit/fabric of these diamondiferous volcanic and intrusive rocks, but his term "basaltic" did not imply the presence of plagioclase (cf. Dawson 1980; Mitchell 1986, 1995).

The hallmark paper of Smith (1983) on Late Jurassic–Cretaceous age southern African kimberlites utilized Sr–Nd–Pb isotope systematics to subdivide the studied sample suite into two distinct rock types, which he termed Group I (Gp I) and Group II (Gp II) kimberlites. Smith (1983) specifically stated that his Group I and II kimberlites corresponded to the "basaltic" and "micaceous" varieties of kimberlite as defined by Wagner (1914) and also by Dawson (1967, 1980). However, Mitchell (1986; p. 6 and 7) noted that contemporary advances in kimberlite petrology from 1970-1985 resulted in the final overthrow of the Lewis–Wagner classification, in favour of a scheme (Clement and Skinner 1979, 1985) based on the modal abundance of primary kimberlite minerals. However, the publications of Clement and Skinner (1979, 1985), Clement et al. (1984), and subsequently by Mitchell (1986) did not distinguish, but instead lumped all kimberlite types together, including the petrographic (Wagner 1914) and isotopically distinctive (Smith 1983) Group I and Group II kimberlite types.

Dawson (1980, 1987) stated that southern African micaceous/Group II kimberlites are the Kaapvaal craton's variant of globally observed lamproites, based on similarities in mineralogy and geochemistry. Later, Mitchell and Bergmann (1991) suggested a few southern African micaceous/Group II kimberlite localities could be lamproites. Tainton and Browning (1991), Tainton (1992), and Tainton and McKenzie (1994) examined the Kaapvaal lamproite versus micaceous/Group II kimberlite problem in detail, concluding all these rocks are a variety of lamproite. Mitchell (1995) subsequently re-named micaceous/Group II kimberlites with the term "orangeite", following on the original suggestion by Wagner (1928), with these rocks deemed to occur only in southern Africa. However, studies by Mahotkin (1998) and O'Brien and Tyni (1999) described and classified rocks from Finland and western Russia as Gp II kimberlite, and also as olivine lamproite. Sarkar et al. (2018) described and classified kimberlite pipes and orangeite sills, and kimberlite dikes and olivine lamproites from northern and southern Melville Peninsula (Arctic Canada), respectively; the olivine lamproites have similarities in mineralogy, bulk composition and age to the lamprophyric rocks of Sisimiut, Greenland studied by (Scott 1979, 1981) and Thy et al. (1987). However, the Sisimiut rocks are better described as carbonate-bearing olivine lamproites, due to their calcite contents. Carbonate-bearing and carbonate-rich olivine lamproites have also been documented from e.g., Kodomali and Behradih in India (Lehmannn et al. 2010; Rao et al. 2011). Therefore, the rocks variably termed micaceous kimberlite/Gp II kimberlite/orangeite are not globally restricted to southern Africa.

More recently, Pearson et al. (2019) clarified and revised these nomenclature issues by (1) utilizing the term kimberlite instead of Group I kimberlite or archetypal kimberlite; and; (2) introducing the term carbonate-rich olivine lamproite (CROL) for global occurrences of rocks formerly known as micaceous kimberlite or Group II kimberlite or orangeite. Simply put, CROL are carbonate-bearing or carbonate-rich modal variants of olivine lamproite. This terminology is used herein, and key aspects of the mineralogy and geochemistry of kimberlite, carbonate-rich olivine lamproite, and olivine lamproite are described in the following subsections of this chapter.

The diamond-bearing volcano

The volcanology terms employed in this chapter are common to all three magma types (i.e., kimberlite, carbonate-rich olivine lamproite and olivine lamproite), which form intrusive and extrusive rocks that host diamond mines, These different magma types all form volcanic landforms and volcanic structures that are similar to those observed in small-volume alkali basalt systems, as previously suggested for kimberlites (Lorenz 1975, 1987; Kjarsgaard 2003, 2007a,b) and olivine lamproites (e.g., Atkinson et al. 1984; Smith and Lorenz 1989; Boxer et al. 1989; Stachel et al. 1994). These are termed maar-diatreme volcanoes (Wagner 1914; Mannard 1962; Lorenz 1975), or "small volcanoes" (White and Ross 2011). That these three different magmas can generate similar styles of volcanism and volcanic products is not a new idea (e.g., see Harris 1984; Atkinson et al. 1984; Atkinson 1989; Nixon 1995). Kimberlite geologists have been—and to some extent still are—under the mistaken belief that exceptionally volatile-rich "kimberlite" (i.e., kimberlite and carbonate-rich olivine lamproite) magmas are unique and thus do not have to conform to the basic tenets of magma physics and volcanology, such that standard volcanological nomenclature and terminology, (e.g., Fisher 1961, 1966; Schmid 1981; McPhie et al. 1993), have not been applied. Recent, revised kimberlite nomenclature and terminology (compare Cas et al. 2008a,b, 2009 with Scott Smith et al. 2013, 2018) is highly problematic and contradictory and cannot be construed as improvements to this situation; see also Field et al. (2008) for further discussion on this topic. In contrast, studies of olivine lamproite (and lamproite) have followed standard volcanology terminology and nomenclature (e.g., Jaques et al. 1986; Rayner et al. 2018a,b).

Terminology. The key geomorphological/structural elements of maar-diatreme volcanoes associated with diamond mines are illustrated in Figure 4 (modified after S. Kurszlaukis, written communication 2003, 2020; Kjarsgaard 2007a,b; White and Ross 2011). These elements include: (1) the paleo-eruptive land surface (aka pre-eruptive land surface); (2) positive relief, with respect to the paleo-eruptive land surface (and rarely preserved), tuff cone and/or tuff ring; (3) crater, crater wall, crater floor; (4) diatreme (upper and lower), diatreme wall, diatreme fill; (5) root zone, and; (6) feeder dike. Note that root zones are neither defined by their size or shape, nor by their consituent rocks types (see *lithofacies nomenclature)*.

A maar-diatreme volcano is defined by the crater floor lying below the paleo-eruptive land surface and is formed by the eruption of magma (Lorenz 1973); the maar crater is surrounded by a tuff ring when preserved. Tuff cones (aka cinder cones, scoria cones) may overlie a maar, thus obscuring an older maar crater and tuff ring plus underlying diatreme. Tuff rings and tuff cones are differentiated on the basis of their morphometric parameters (Wood 1980; Vespermann and Schminke 2000). These include the crater rim height (HCR), the crater rim diameter (DCR), the HCR/DCR ratio, maximum bedding dips, and the crater floor depth below the paleo-eruptive land surface for tuff rings; tuff cones use the cone height (HCO), the crater rim diameter (DCR), the cone basal diameter (DCO), and HCO/DCO and DCR/DCO ratios. Note that for kimberlite, CROL and olivine lamproite magmatic systems, the volcanic edifice (when observed) is typically consolidated/lithified and hence tuff cone or tuff ring is preferred terminology instead of tephra (i.e., loose, unconsolidated).

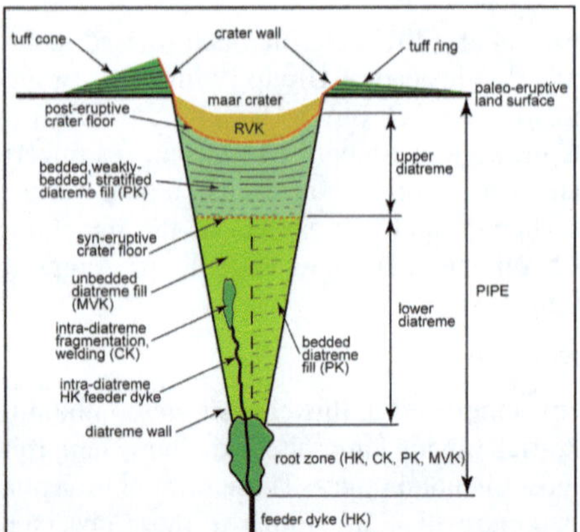

Figure 4. Cross-section (highly schematic composite, no scale implied) of a kimberlite maar-diatreme volcano, at the post eruptive stage in which the maar-crater has been partially in-filled, illustrating the key terms used in this chapter. RVK = resedimented volcaniclastic kimberlite; PK = pyroclastic kimberlite; MVK = massive volcaniclastic kimberlite; HK = hypabyssal kimberlite; CK = coherent kimberlite. The terminology of this non-genetic maar-diatreme volcano as well as the associated lithofacies can be equally applied to olivine lamproite or CROL. Modified after Kurszlaukis (written communication 2003, 2020), Kjarsgaard (2007a,b) and White and Ross (2011).

The root zone plus the diatreme is termed a pipe (alternately, the diatreme structure; White and Ross 2011); importantly, note that a pipe can consist of multiple, discrete and individual diatremes. The root zone and the diatreme can be in-filled with an exceptionally wide array of different rock types (lithofacies), derived from magmas that are extrusive (typically fragmental), or, are intrusions (non-fragmental) such as dikes, sills, and feeder conduits for tephra jets. As well, resedimented material that can include large blocks of wall rock that are not in-situ can constitute part of the pipe in-fill. Upper diatreme deposits are more typically bedded, diffusely bedded or stratified; lower diatreme deposits are more typically massive in nature. However, bedded pyroclastic rocks, or blocks of this material can be observed in the lower diatreme (e.g., Clement 1982). The crater, an open hole with a bottom or base, can be syn-eruptive and in-filled by primary pyroclastic, volcaniclastic and resedimented material, or post-eruptive when volcanic activity has ceased and the crater is open to air, or filled with water, or, infilled by younger resedimented ("epiclastic") material.

Lithofacies nomenclature. A key aspect of the description of rocks in kimberlite (and also CROL and olivine lamproite) volcanoes is the use and application of non-genetic terminology (Kjarsgaard 2003, 2007a,b; Sparks et al. 2006; Field et al. 2008); this is certainly not a new idea with respect to volcanic systems in general (e.g., McPhie et al. 1993). A very simple, non-genetic nomenclature system (Kjarsgaard 2003, 2007a,b; Sparks et al. 2006; Field et al. 2008) to describe rocks from kimberlite (and also from CROL and olivine lamproite) magmatic systems is preferred. Here kimberlite is used as an example of the subdivisions: (1) volcaniclastic kimberlite (VK), i.e., fragmental rocks, and (2) coherent kimberlite (CK), typically, non-fragmental intrusive rocks. With more detailed information, these two fundamental root rock names can be further subdivided: massive volcaniclastic kimberlite (MVK), which were formerly termed tuffisitic kimberlite or tuffisitic kimberlite breccia (TK/TKB; Clement 1982); pyroclastic kimberlite (PK), and; resedimented volcaniclastic kimberlite (RVK), which were formerly and variably termed epiclastic kimberlite. MVK, PK and RVK are all varieties of VK. Introduction of type-locality lithofacies terms such as "Kimberley-type pyroclastic—KPK" (Mitchell et al. 2009, 2012) and "Fort à la Corne-type pyroclastic—FPK" (Scott Smith et al. 2013, 2018), when the fragmentation and depositional processes are not or are very poorly understood, represents the nadir of kimberlite lithofacies classification.

Coherent kimberlites (CK) are typically 1–3 m in width and manifest as either dikes or sills, and more rarely as dike enlargements termed blows, or as small plugs. If the rock is demonstrably an intrusion, the use of the term hypabyssal, as in hypabyssal kimberlite (HK) is perfectly acceptable, for example HK dike or HK sill. These hypabyssal intrusions can be associated with a pipe, and include: precursor (aka antecedent), contemporaneous, and consequent (aka subsequent) dikes and sills (Wagner 1914; Clement 1982). The most important and key application of the term CK is when the rock is not obviously fragmental, nor obviously an intrusion, based on the available information. Examples of bonafide CK are rare, but would include pyroclastic rocks that have welded or agglutinated in the root zone or within the diatreme and appear dike- or sill-like in nature (Sparks et al. 2006). At the Victor Diamond mine (Canada), detailed studies led van Straaten et al. (2011) to interpret that lava fountaining and welding generated clastogenic coherent kimberlite that in appearance resembled intrusive hypabyssal kimberlite sills. Additional possible examples of bonafide CK include rootless lava flows, or lava lakes in which the magma has fragmented (via fire fountaining) and subsequently welded. While kimberlite (and CROL) lava lakes and lava flows are rare (or rarely preserved), they are quite common in olivine lamproite and lamproite magmatic systems (Jaques et al. 1986; Mitchell and Bergman 1991). Furthermore, note that the existence of a lava lake or lava flow, while extrusive, is not synonymous with magma fragmentation and welding, as these rocks could simply be related to flows from very passive fissure eruptions. At the Igwisi Hills, Brown et al. (2012) could find no evidence that the kimberlite lavas formed by welding of spatter.

Pipe formation and preservation. Our understanding of diamondiferous volcanoes is derived from "temporal snapshots" of individual volcanoes, which includes: rocks at the present-day erosional surface, rocks observed in open pit and underground mining exposures, and rocks from drill cores. Hawthorne (1975) astutely observed that South Africa kimberlites had experienced significant erosion, in contrast to kimberlites in Angola, Botswana, Tanzania and the Democratic Republic of Congo (DRC), which preserved RVK and/or "epiclastic kimberlite" that infilled a basin (i.e., a syn- or post-eruptive crater) above the diatreme. The preservation of a kimberlite volcano and its surficial edifice is dependent on the age and degree of cementation of the rocks; however, the regional-scale post-volcanic depositional and/or erosional environment is key. Importantly, variably preserved and intact tuff rings and tuff cones are rare, but known from mid-Cretaceous age kimberlites at Fort à la Corne (Canada) due to post-eruptive sedimentation and burial (e.g., Zonneveld et al. 2004), and the Holocene age Igwisi Hills kimberlites (Tanzania) that are preserved due to minimal erosion (Reid et al. 1975; Dawson 1994; Brown et al. 2012). In contrast, kimberlite pipes in the Kimberley area of South Africa area have been subjected to ~850 m of erosion (Hanson et al. 2006, 2009; Stanley et al. 2015), with original (ca. 1870's) surface exposures at mid-diatreme depths. The observation that near surface hypabyssal dikes, sills, blows, plugs and root zones exist without any fragmental rocks (i.e., a diatreme consisting of fragmental rocks at the same stratigraphic level) does not imply there was significant erosion and removal of the diatreme. Rather, this is simply a reflection of near-surface magma fragmentation (or not!), i.e., how the magma degassed, and/or interacted with ground and/or surface water.

As emphasized by Nixon (1995), no single example of a diamondiferous volcano is known that has an exposed feeder dike, plus root zone, plus diatreme, plus tuff ring or cone, and crater (as per Fig. 4). Nixon (1995) further underscored the notion that no single example of a diamondiferous volcano would be definitive, due to the exceptionally wide variations that are known in pipe geometry and internal morphology, and pipe or crater size and shape (see Figs. 5 through 12). This simple statement of Nixon (1995) is true for kimberlite, CROL and olivine lamproite pipes. The notion that there are three highly distinctive classes/types of kimberlite pipes globally (Field and Scott Smith 1999; Skinner and Marsh 2004; Scott Smith 2006, 2008) is inconsistent with known observations. These tripartite endmembers serve little if any purpose, because each individual model is highly oversimplified, to the point of being misleading, especially from an exploration perspective (Kjarsgaard 2007a,b). Volcanological studies at the Murowa mine (Zimbabwe) led Moss et al. (2013) to arrive at a similar conclusion; he noted that in general, kimberlites cannot be easily or completely described by any single class/type of pipe and its implied emplacement model.

Juvenile magmatic degassing versus phreatomagmatic eruption styles. The volcanology, volcanic architecture and volcanic deposits of olivine lamproites (and lamproites) in terms of their modes of formation are not controversial—there is general agreement that phreatomagmatic processes, in which a hot magma interacts with ground or surface water in a highly explosive manner—is key to understanding magma fragmentation (e.g., Zimanowski et al. 1991, 2015; Büttner et al. 2002). In contrast, two kimberlite pipe formation models are actively debated: the exsolution of juvenile magmatic CO_2 dominant (due to near-surface lowering of the confining pressure) volatile driven pipe formation model, and the phreatomagmatic pipe formation model; compare Scott Smith (1999) with Lorenz et al. (1999), but see also Lorenz and Kurszlaukis (2007), Kurszlaukis and Lorenz (2008), and Field et al. (2008) and their contained references for further discussion. For example, diametrically opposed pipe formation models are described for two nearby (<1 km distance) kimberlite pipes from the Diavik Mine (Lac de Gras, Canada) that are of identical age (within uncertainty). The A418 pipe is considered to be of phreatomagmatic origin (Porritt et al. 2013), while the A154 North and A154 South pipes are considered to be driven by exsolution of magmatic volatiles (Moss et al. 2009, 2018). More recently, the Diavik A154 North and A154 South pipes were reinvestigated by Tovey et al. (2020), who interpreted that groundwater availability and phreatomagmatism controlled kimberlite fragmentation processes.

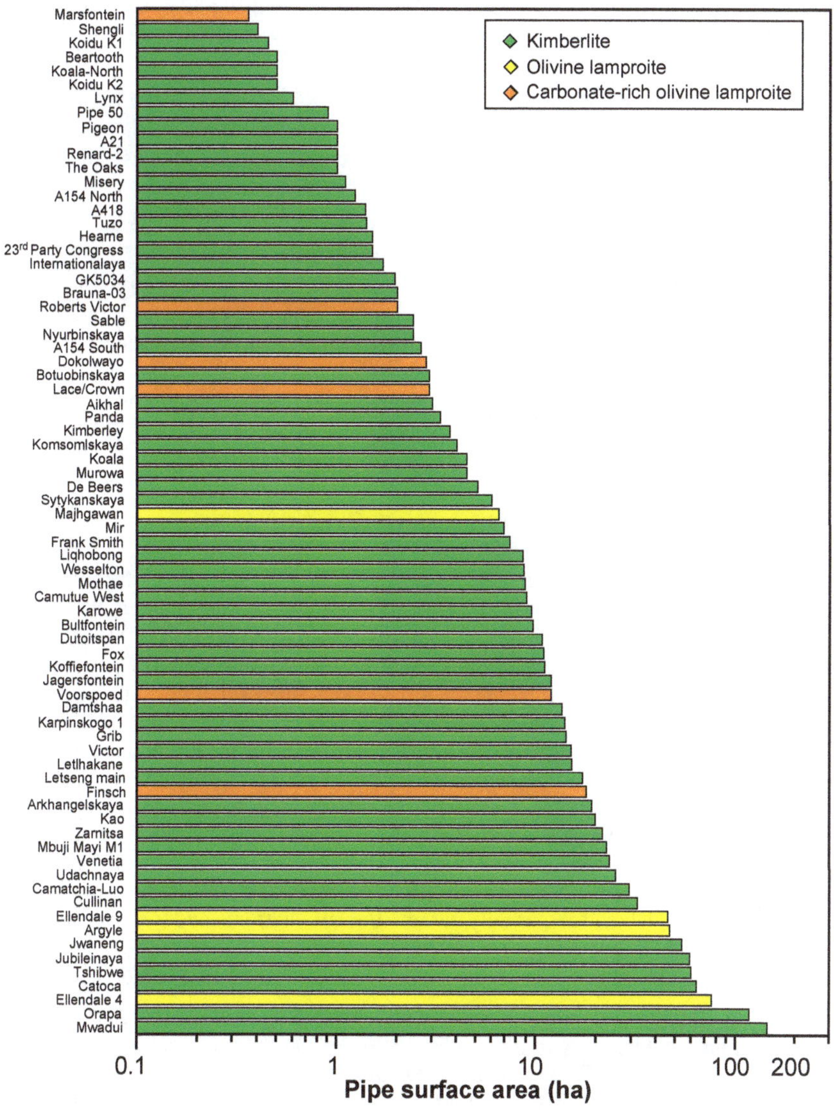

Figure 5. Pipe surface area (ha) of global diamond mines, log histogram plot (data from Table 2). All primary magmatic hosted diamond mines are color coded by rock type (Kimberlite, CROL, olivine lamproite). Note that the area can either represent the surface area of the pipe, or represent the surface area of the volcanic edifice (e.g., the tuff ring or "apron" deposits surrounding the diatreme) at the start of mining.

Given that carbonate-rich olivine lamproites have high whole rock CO_2 concentration levels (see *Whole Rock Geochemistry*), the debate regarding kimberlite pipe formation models is thus equally applicable to CROL pipe formation. Of course, a multitude of pipe formation models, which variably combine aspects of both magmatic and phreatomagmatic processes are entirely feasible. In this respect, the induced phreatomagmatic explosion experiment of a volatile-rich (CO_2 and H_2O) kimberlite melt generated very high energy shock waves, and produced pyroclasts with exceptional ejecta speeds of >400 m/s (Kurszlaukis et al. 1998). Specific features of diamondiferous volcanoes, such as exceptionally large pipes, are perhaps resultant from such processes. One could speculate further that the huge size of the olivine lamproite-hosted Tier 1 Argyle mine resulted from higher magmatic CO_2 contents than is the "norm" for an olivine lamproite magma (see Abersteiner et al. 2022).

Diamond mine volcanoes. Examples of kimberlite, CROL and olivine lamproite diamond mines, advanced exploration projects, and non-economic maar-diatreme volcanoes are utilized to illustrate key basic concepts via plan views, cross sections, isometric views, and 3-D solids images in Figures 6–12.

A variety of variably intact tuff cones, tuff rings and maar craters are known from Fort à la Corne (Canada), and the Igwisi Hills (Tanzania) kimberlites. The ca. 105–95 Ma kimberlites at the Fort à la Corne advanced exploration project have typically well preserved volcanic edifices due to burial by subsequent kimberlite volcanic events, but mainly by rapid deposition of contemporaneous and younger mid-Cretaceous deltaic, and near-shore and deeper marine sediments (Leckie et al. 1997; Kjarsgaard 2003, 2007a,b; Zonneveld et al. 2004; Kjarsgaard et al. 2009). The Fort à la Corne kimberlite volcanoes are typically multiphase, as has been established for many kimberlites, but they are also polygenetic (cf. Fulop and Kurszlaukis 2016; see also *Geochronology*, this chapter).

From drill intersections and 2-D seismic imaging, the early Joli Fou (EJF) eruptive phase at the #169 kimberlite forms a cone-like volcanic edifice with ~100 m of relief above the paleo-eruptive land surface (the JF shale), and overlies an older, small, flared (~45°) diatreme; (Leckie et al. 1997; Kjarsgaard et al. 2007). The top of the EJF tuff cone is flattened and has a reduced height due to marine shoreface reworking and slumping (Leckie et al. 1997). The Pense P2 eruptive phase at the Orion South kimberlite also has ~100 m of relief above the paleo-eruptive land surface (the Pense shale), with a truncated and irregular cone-like form; the top of the P2 kimberlite is in direct contact with Quaternary till, a glacial erosion surface (Kjarsgaard et al. 2009). The P2 edifice overlies a >150 m deep, 500 m wide and flared (~45°) P2 diatreme. At the Star kimberlite, the EJF eruptive phase forms a ~400 m diameter tuff ring with the rim having 44 m of elevation above the paleo-eruptive land surface (the JF shale); the EJF crater floor is ~49 m below the paleo-eruptive surface (Fig. 6a) and together they define a tuff ring–maar crater (Zonneveld et al. 2004). The EJF crater is underlain by a flared (~45°) diatreme that extends ~100 m below the EJF crater floor (Fig. 6a). The EJF crater itself is filled by RVK, and then subsequently by a younger phase of kimberlite volcanism from an adjacent diatreme (the Mid-JF), which has partially cut into the EJF diatreme and tuff ring (see Fig. 10 in Zonneveld et al. 2004). The Tokapal kimberlite (India) has a wide pyroclastic apron around a central feeder vent (Mainkar et al. 2004), similar to some of the Fort à la Corne kimberlites.

For the Igwisi Hills (Tanzania), erosion of these young Holocene, ca. 12,000 to 6,000 years old kimberlites are negligible, with good preservation of the three adjacent volcanic edifices (Sampson 1953; Reid et al. 1975; Dawson 1994; Brown et al. 2012), with the following adapted from Brown et al. (2012). The NW volcano has pyroclastic kimberlite deposits with bedding dips of 8–32° and the crater floor is at or below the paleo-surface, i.e., a tuff ring and maar crater. A significant lava flow breaches the east side of the tuff ring; the crater is interpreted to be filled by a lava lake. The SW volcano is a tuff cone, with beds dipping at 6–31°. A lava flow breaches the east side, fed by a perched lava lake. The central volcano has pyroclastic beds dipping at 6–31°, the volcanic edifice has an angle of repose of 24° and is suggested to be a tuff (scoria) cone, although interpretation is problematic due to the significant lava coulee that obscures the morphology of the east side of the central volcano. The Igwisi Hills and the Fort à la Corne examples each illustrate substantive variations in styles of volcanism and volcanic architecture for kimberlites that are in close proximity within a cluster or field, and are of similar age. These kimberlites have deposits consistent with volcanic formation processes ranging from fire fountaining (magmatic degassing driven) at the Igwisi Hills (Brown et al. 2012) and at Fort à la Corne (welded PK; Leckie et al. 1997), to tephra jets associated with phreatomagmatic processes at Fort à la Corne (McClintock et al. 2009; Kjarsgaard et al. 2009).

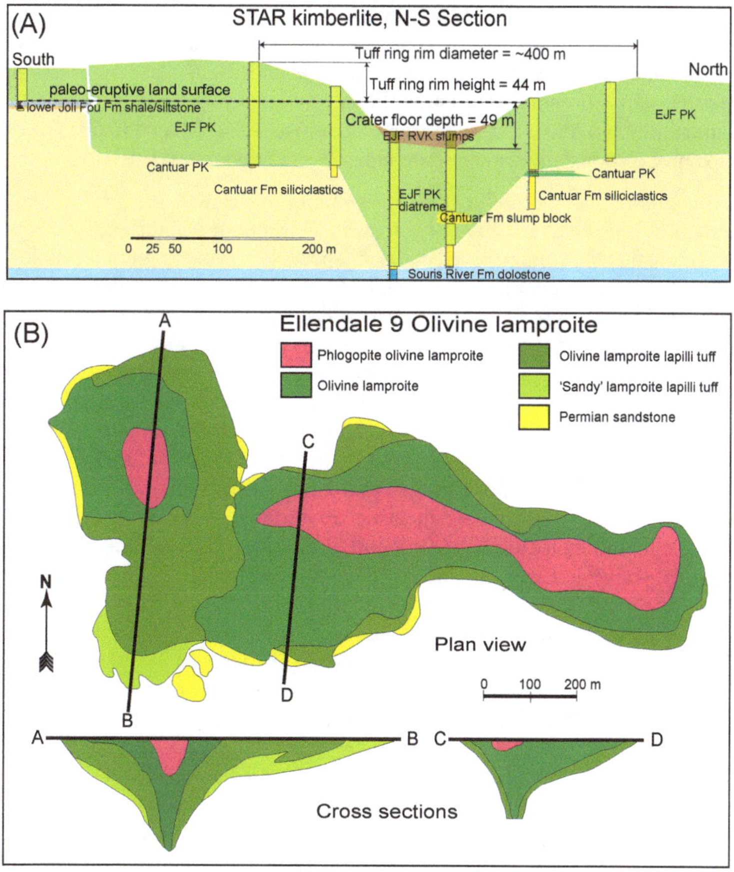

Figure 6. (**A**) North–South simplified cross section (partial, not the full N-S extent) of the Star kimberlite pipe and its host rock geology (Fort à la Corne, SK, Canada), showing only the early Joli Fou (EJF) tuff ring, and the partially in-filled maar crater that overlies the EJF diatreme. Adapted from B.A. Kjarsgaard (unpublished); see also Harvey et al. (2009). (**B**) Simplified plan and section views of the Ellendale 9 (Australia) olivine lamproite pipe. Note the occurrence of at least two distinct diatremes in this pipe. Area of phlogopite olivine lamproite constitutes a lava lake. Adapted from Smith and Lorenz (1989).

A number of other kimberlite pipes actively mined for diamonds (or past producers) are known that resemble the deposits described above, with specific similarities such as pipes with significant PK and RVK deposits that are preserved both external and internal to the pipe/maar crater. These include the 22 ha Tier 2 Mbuji Mayi (M1) mine, DRC (Fieremans 1953; Meyer de Stadelhofen 1963), the 60 ha Tier 2 Tshibwe mine, DRC (de Wit et al. 2016) and possibly quite a number of other examples in Angola (e.g., Camafuca-Camazamba). The olivine lamproite pipes in the Ellendale field (Jaques et al. 1986) also bear some morphological similarities to the kimberlite volcanoes in the DRC and Angola listed above. The geology of the 76 ha Tier 4 Ellendale 4 and 46 ha Tier 4 Ellendale 9 olivine lamproite-hosted past producing diamond mines are well documented by Jaques et al. (1986), Smith and Lorenz (1989) and Stachel et al. (1994). Plan and section views of Ellendale 9 (in which the tuff ring is not preserved due to erosion), are shown in Figure 6b. The E-W elongated Ellendale 9 pipe is interpreted to be the product of two (or more) flared diatremes, each >200 m deep, filled with well-bedded "sandy" quartz-rich olivine lamproite base surge deposits of phreatomagmatic origin, pyroclastic mass flow deposits, and variably welded spatter deposits (Jaques et al. 1986; Smith and Lorenz 1989;

Stachel et al. 1994). Coherent olivine lamproite and coarser-grained madupitic phlogopite olivine lamproite lava lakes overlie these rocks. There are certainly other known olivine lamproites with similarities to Ellendale 4 and 9, with the pipes filled by fragmental lamproite, and then by overlying spatter deposits and/or lava lakes of coherent lamproite. Examples would include the weakly diamondiferous Prairie Creek, Arkansas (Bolivar 1984; Mitchell and Bergmann 1991) and the Kapamba, Zambia olivine lamproite pipes (Scott Smith et al. 1989).

Since the important and influential observations of Hawthorne (1975) on minimally eroded kimberlite pipes that preserve crater in-fill deposits in central and southern Africa, a number of significant new studies have been published. New geology and volcanology data and interpretations have been generated as a result of greater exposures of these kimberlite pipes from decades of diamond mining since the 1970's. The geology of the 118 ha Orapa (Botswana) Tier 1 diamond mine, a kimberlite twin pipe that coalesced at the surface, is described by Field et al. (1997) and Kilham et al. (1998), with more recent studies and overviews in Field and Stiefenhofer (2006), Field et al. (2008), Gernon et al. (2009), de Wit et al (2016) and Kruger and Maphane (2017). The 54 ha Tier 1 Jwaneng (Botswana) diamond mine, a kimberlite triple pipe (Fig. 7) that coalesced at the surface, was noted to be quite different as compared to a typical South African kimberlite, due to the pipes being dominantly in-filled with VK, bedded VK, PK and RVK, but not massive volcaniclastic kimberlite (MVK) as is typical of a Kimberley cluster pipe (Field and Scott Smith 1999; Webb et al. 2003; Skinner and Marsh 2004). More recent geological studies on Jwaneng include the work of Mmualefe (2017) and the overviews presented by Field et al. (2008), and de Wit et al. (2016). A cross section of the geology of the Jwaneng pipe is shown in Figure 7. The Koffiefontein mine kimberlite (South Africa) contains bedded PK at depth in the diatreme, in what was originally described as TK/TKB (Naidoo et al. 2004). The Tier 4 Mwadui (Tanzania) diamond mine is the world's largest (146 ha) kimberlite-hosted diamond mine, but with low diamond grades (Table 2). The geology at Mwadui was initially described by Tremblay (1956), Mannard (1962) and Edwards and Howkins (1966). More recent studies were undertaken by Stiefenhofer and Farrow (2004), who concluded it is an almost perfectly preserved example of a post-eruptive kimberlite crater that is in-filled by a wide variety of types of RVK to ~600 m depth, with the tuff ring surrounding the crater having been eroded away. Interestingly, these RVK crater in-fill deposits are interpreted to be underlain by diatreme-filling PK (and not MVK). The Tier 1 Catoca (Angola) kimberlite-hosted diamond mine is quite large (63.6 ha; Table 2) and has ~250 m of preserved post-eruptive crater in-fill (Pervov et al. 2011). These RVK deposits and their geometry appear to bear a number of similarities to those at Mwadui. Another very large (59 ha; Table 2) and complex multiphase kimberlite is the Tier 1 Jubileinaya (Russia) diamond mine (Kurszlaukis et al. 2009, 2015). Here a number of smaller pre-cursor pipes are cut by a younger and much larger main pipe comprised dominantly of MVK, which itself is overlain by PK, and in turn overlain by a variety of post-eruptive crater in-fill RVK (Kurszlaukis et al. 2009).

More highly eroded examples of diamondiferous volcanoes, with no preserved volcanic edifice or post-eruptive crater in-fill are much more common in the geological record. The Tier 1 Argyle diamond mine (Fig. 8) is a large (47 ha, Table 2) steep sided and elongated olivine lamproite pipe that is interpreted to have formed from several distinct individual diatremes that have coalesced, and are aligned along a fault (Rayner et al. 2018a,b; Fig. 8), with the fault presumably coincident with an olivine lamproite feeder dike. The Argyle mine is interpreted to have formed by numerous phreatomagmatic eruptions (Rayner et al. 2018a,b). At the Argyle mine, the steep-walled individual diatremes (and collectively, as a larger elongated pipe structure) resemble classic steep-walled kimberlite diatremes, as observed in the Kimberley area (Clement 1982; Clement et al. 1986) and at many other global kimberlite localities. The 6.5 ha Tier 4 Majghawan olivine lamproite-hosted mine is also a steep sided diatreme, being circular–elliptical in plan-view (Rao 2007). The Atri (Bunder) twin pipe is also a steep-walled olivine

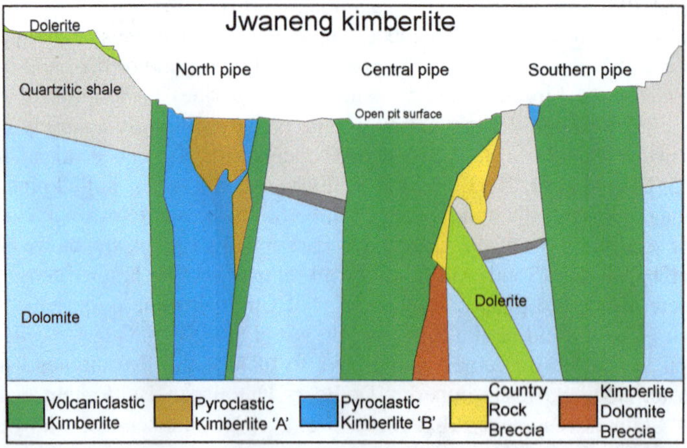

Figure 7. Simplified cross section of the Jwaneng (Botswana) triple kimberlite pipe mine and its host-rock geology. This section shows the three pipes that coalesced at the surface and are currently mined; the fourth pipe (not mined) has been omitted for clarity. Adapted from de Wit et al. (2016).

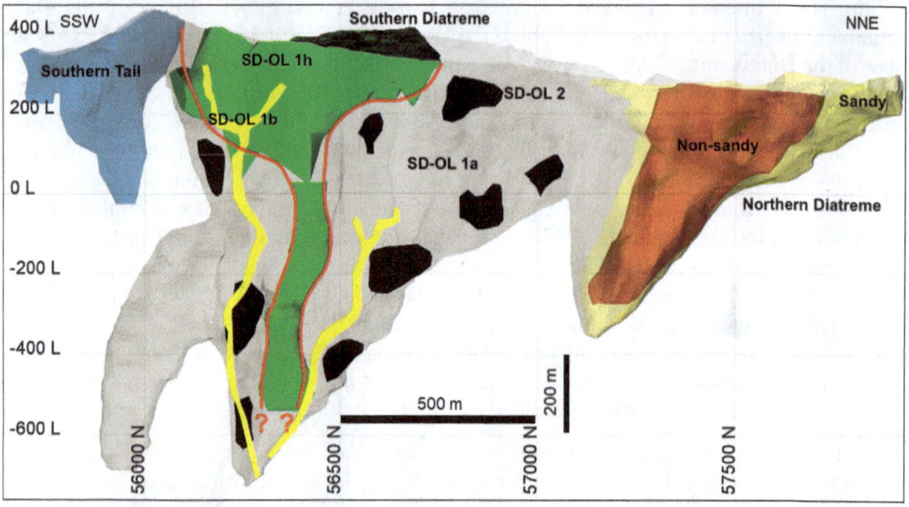

Figure 8. 3-D view of the elongated, steep-walled olivine lamproite pipe that hosts the Argyle Mine, Australia. Note the multiple olivine lamproite diatremes that make up this this pipe. The northern diatreme consists of two main sub-units, a sandy and a non-sandy phase. The southern diatreme consists of unit SD-OL 1a (**grey**), a well-bedded to massive to chaotic pyrclastic olivine lamproite; unit SD-OL 2 (**black**), a massive to crudely bedded olivine and juvenile pyroclast poor olivine lamproite; unit SD-OL 1h (**green**), a bedded to well-bedded pyroclastic olivine lamproite, and; unit SD-OL 1b (**yellow**), a massive to chaotic textured pyroclastic olivine lamproite interpreted as feeder conduits or fragments thereof. After Rayner et al. (2017, 2018b), with permission.

lamproite pipe (Das et al. 2018). The Argyle, Majghawan and Atri olivine lamproite pipes closely resemble steep-walled kimberlite diatremes i.e., they do not have "champagne-bowl shaped" or a flared diatreme (e.g., Mitchell 2020b), as per Ellendale 4 or 9 (see Fig. 6b), or Prairie Creek. The kimberlite-hosted Tier 1 Aikhal (Russia) diamond mine (Kostrovitsky et al. 2015) has a number of similarities in terms of overall pipe morphology with Argyle, including a strongly elongated (in plan-view), steep-walled pipe, which consists of multiple discrete individual diatremes.

The Tier 1, CROL-hosted 18 ha Finsch diamond mine (Fig. 9; Clement 1982), which along with the Tier 4 CROL-hosted12 ha Voorspoed mine, is somewhat unusual as many CROL in southern Africa and globally tend to form smaller-sized pipes, and/or dike and blow systems (Table 2). However, the Finsch mine has numerous similarities to South African kimberlite pipes, having a steep-walled diatreme and the pipe being in-filled by a variety of lithofacies including massive volcaniclastic, hypabyssal, and bedded/diffusely bedded volcaniclastic CROL (Clement 1982; Ekkerd et al. 2003; Field et al. 2008). Another steep-walled pipe is the 25 ha Tier 1 Udachnaya (Russia) diamond mine, a kimberlite twin pipe (Kostrovitsky et al. 2105). Plan and cross section views of the Udachnaya East and West pipes are shown in Figure 10. The 32 ha Tier 1 Cullinan (South Africa) diamond mine is another kimberlite twin pipe, with multiple discrete kimberlite phases. The geology of Cullinan mine is described by Wagner (1914), Bartlett (1998), Field et al. (2008) and de Wit et al. (2016). Other excellent examples of steep-walled kimberlite diatremes would have to include the five diamond mines in the Kimberley cluster, South Africa (Kimberley, De Beers, Dutoitspan, Bultfontein and Wesselton). The studies of Clement (Clement 1982; Clement et al. 1986) are seminal with respect to our fundamental understanding of these pipes, and kimberlite geology in general. More recent work on the Kimberley cluster pipes is summarized in Field et al. (2008) and de Wit et al. (2016).

Much smaller area pipes (<3 ha) that host economic diamond deposits include multiple Tier 2 mines at both the Ekati and Diavik mines (Canada) in the Lac de Gras kimberlite field, and the Tier 1 Internationalaya and Tier 2 23rd Party Congress mines (Russia) in the Malo-Botuoba kimberlite field (Fig. 11b; Table 2). At Lac de Gras, the A154 North and A154 South pipes of the Diavik mine (Fig. 11a) have very small surface areas (1.2 and 2.6 ha, respectively), which increase with depth (Moss et al. 2018). Multiple different kimberlite lithofacies are observed in the Diavik kimberlites, as illustrated in Figure 11a for the A154 North pipe. Moss et al. (2018) demonstrated that A154 South pyroclastic kimberlite (PK) eruptions in-filled the top of the adjacent A154 North crater (PK4-N in Fig. 11a). The observation that a kimberlite eruptive event from a nearby pipe can in-fill a crater in an older adjacent kimberlite was also described by Zonneveld et al. (2004) for the Star kimberlite, Fort à la Corne (Canada). The Internationalaya mine (Fig. 11b) is also a small (1.7 ha) steep-walled, but irregular shaped pipe that expands and contracts with depth (Kostrovitksy et al. 2015). The CROL-hosted 0.4 ha, Tier 3 Marsfontein M1 mine, and the kimberlite-hosted Tier 4 Koidu K1 and K2 pipes (both ~0.5 ha) in Sierra Leone, plus the ~1 ha The Oaks (South Africa) diamond mines are additional examples of very small pipes that have been mined (Table 2). These Tier 3 and 4 examples from Africa are further described by Field et al. (2008) and de Wit et al (2016).

In terms of surface area, the smallest diamond mines are hosted by dikes, sills, and blows, and these are typically Tier 4 or Tier 5 mines. Classic localities include the CROL-hosted Tier 4 Roberts Victor mine (Fig. 12a), which consists of two dikes with three affiliated blows and two small pipes (Wagner 1914; Field et al. 2008; de Wit et al. 2016) and the CROL dikes and blows that host diamond mines in the Bellsbank area (Fig. 12b) of South Africa (Clement 1973; Mitchell 1995; Field et al. 2008; de Wit et al. 2016).

Summary of diamond mine volcanology. Magmatic-hosted diamond deposits are quite variable in surface area (Fig. 5), 3-D geometry, morphology and volcanic architecture (dikes, sills, blows, root zones, pipes, tuff rings, tuff cones, apron deposits; Figs. 6 to 12). A 1:1 correlation does not exist between these parameters and the primary magma type (kimberlite, CROL, olivine lamproite).

Notwithstanding active or past producing diamond mines or advanced exploration projects, a paucity 3-D solids models and ore tonnage data exist for most kimberlites, CROLs and olivine lamproites. Due to the highly variable geometry and geology of these deposits, even when the surface area is known, a significant amount of drilling and drill core is required to generate

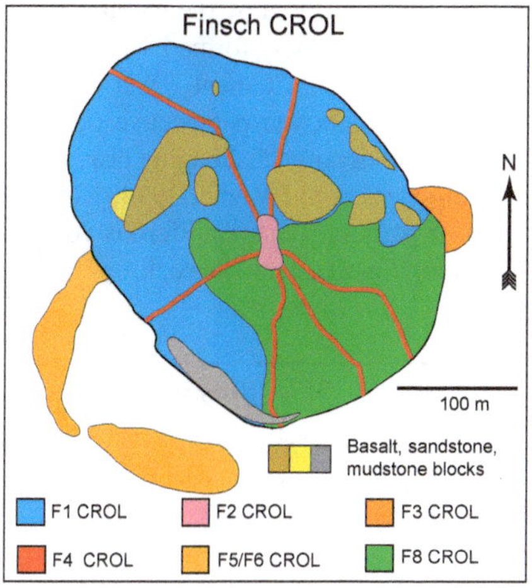

Figure 9. Simplified plan view of the Finsch (South Africa) CROL-hosted diamond mine, at the 610 m level. Note the individual CROL phases in and adjacent to the pipe, as well as the blocks of host-rock basalt, sandstone and mudstone in the pipe. Re-drawn from Field et al. (2008), after Clement (1982) and Ekkerd et al. (2003).

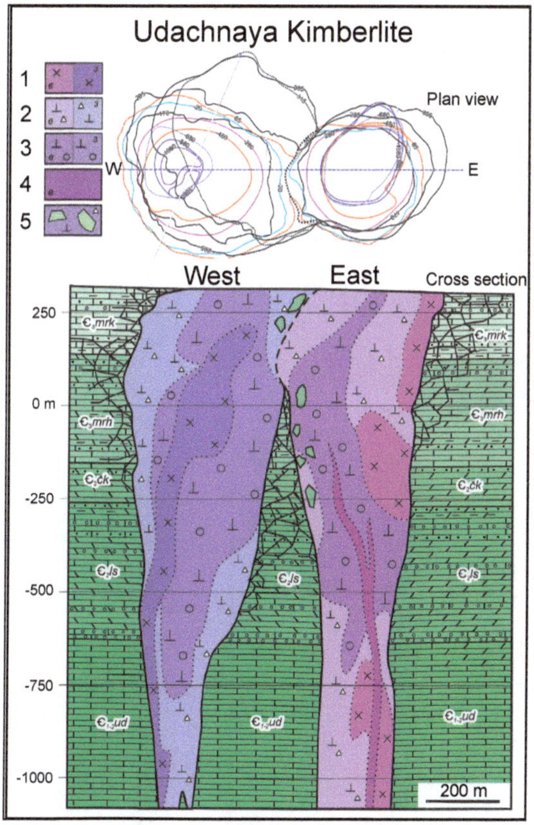

Figure 10. Udachnaya kimberlite twin pipe mine and its host rock geology. Plan and section views. Note the multiple individual phases of kimberlite in each pipe. Geology units as follows: 1 = phase 1, pyroclastic kimberlite; 2 = phase 2, kimberlite breccia; 3 = phase 3, autolithic kimberlite breccia; 4 = phase 4, pyroclastic kimberlite; 5 = sedimentary xenoliths in kimberlite. Note geology legend units 1, 2, and 3 are divided in half with the left side the east diatreme and the right side the west diatreme. After Kostrovitsky et al. (2015), with permission.

these 3-D models and calculate ore tonnages. This is further complicated by the fact that these diamondiferous volcanoes all have distinctive and discrete lithofacies, formed by differing processes that can have exceptionally variable diamond grades (e.g., Clement 1982; Clement et al. 1986; Kjarsgaard 2007a,b; Field et al. 2008). From an economic perspective (i.e., the determination of ore reserve or resource tonnage) this is an exceptionally challenging task, as not all of a diamondiferous volcano may contain economically viable ore (Wagner 1914; Nowicki 2014; Kjarsgaard et al. 2019). In other words, only part or parts of a diamondiferous volcano could be of economic interest, with the rest being low diamond grade, or waste rock.

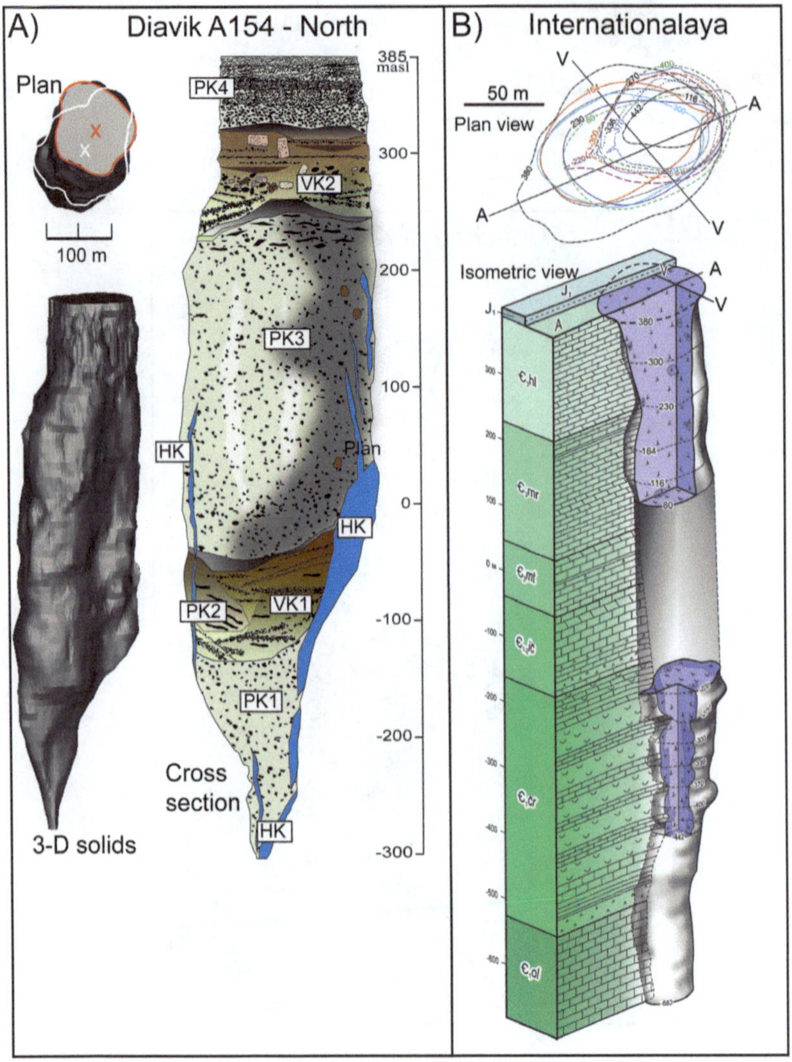

Figure 11. (**A**) Plan and 3-D solids and cross section of the A154—North kimberlite, Diavik Mine, Lac de Gras kimberlite field, NT, Canada. Note the multiple phases of kimberlite in the A154 North cross section. HK = hypabyssal kimberlite; PK1, PK2, PK3 = pyroclastic kimberlite phases 1, 2, 3; VK1, VK2 = volcaniclastic kimberlite phases 1 and 2. After Moss et al. (2018), with permission. (**B**) The steep-walled Internationalaya, Russia kimberlite, plan and isometric views. After Kostrovitsky et al. (2015), with permission.

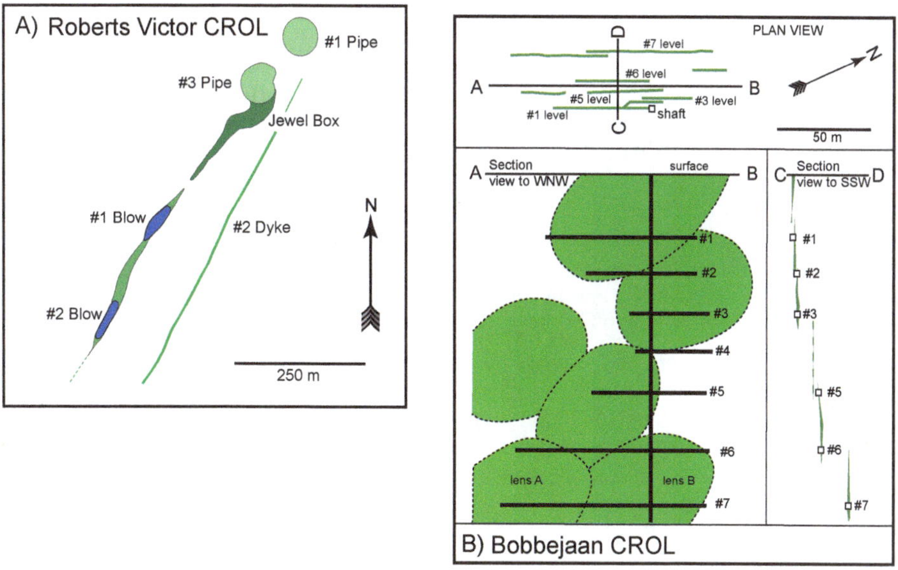

Figure 12. (**A**) Roberts Victor Mine (South Africa) CROL dike—blow—pipe system. Re-drawn after Wagner (1914), Gurney and Kirkley (1996) and de Wit et al. (2016). Bobbejaan Mine (Bellsbank area, South Africa), CROL en-echelon dike system. Adapted and re-drawn from Clement et al. (1973). Note there is diamond grade and rough diamond value variation between the pipes, blows and dikes in (**A**) and the dike lenses shown in (**B**).

Mineralogy and mineral chemistry

Comparative mineralogy of magmatic rocks that host diamond mines. Kimberlite, CROL and olivine lamproite contain a widely variable load of xenocrysts, including macrodiamonds. A summary and comparison of the mineralogy of kimberlite, carbonate-rich olivine lamproite (CROL) and olivine lamproite is listed in Table 1. The petrography of ultramafic lamprophyres (UMLs, variety aillikite), which have some similarities with kimberlite and CROL, is compared to and described and discussed in detail by Tappe et al. (2005). A first order observation from Table 1 is that the mineral assemblages (and the modal abundances of these minerals) for these three rock types can be overlapping, and/or similar (see also Tappe et al. 2005). Olivine, spinel, phlogopite, apatite and perovskite are common minerals to all three rock types. The application of mineral chemistry—specifically mineral zoning trends—has been previously utilized by Mitchell (1986) to classify, distinguish and discriminate "kimberlites" (i.e., historically combined kimberlite plus CROL), lamproites, and UMLs. A number of important observations concerning the mineralogy and mineral chemistry of these "lamprophyre clan rocks" was reviewed by Rock (1988). Subsequently, Mitchell and Bergman (1991) and Mitchell (1995) re-visited these mineralogical and mineral chemistry discriminant classifications with respect to "archetypal kimberlites", "orangeites", lamproites, and UML.

Here, we re-examine the suites of minerals observed in kimberlite, CROL and olivine lamproite, and scrutinise the mineral chemistry zoning trends for groundmass phlogopite and spinel for these three primary diamond-bearing magma types. Our focus is on the primary mineralogy of little altered, hypabyssal or coherent examples of cratonic kimberlite, CROL and olivine lamproite. Note that we do not examine in detail "evolved" CROL, or "evolved" olivine lamproites, or primary leucite lamproites, or Mediterranean lamproites, for which we use the generic term lamproite, even though some of these rocks may be diamond-bearing.

Incorporation of lithospheric mantle and crustal xenoliths and xenocrysts is commonly observed in kimberlite, CROL and olivine lamproite, more so in volcaniclastic (fragmental) rocks. The incorporation of mantle and crustal materials, and their partial or complete digestion or assimilation into the melt can result in the formation of a variety of "new" minerals, or magmatic overgrowths on existing minerals (Caro et al. 2004; Gaudet et al. 2018; Dalton et al. 2019). Due to the volatile-rich nature of kimberlite and CROL (both CO_2- and H_2O-rich) and olivine lamproite (H_2O-dominant) as detailed in *Whole Rock Geochemistry, this chapter*, there can be significant carbohydrothermal and hydrothermal alteration with concomitant precipitation of subsolidus mineral phases and suites of these minerals (e.g. serpentine plus magnetite, serpentine plus magnetite plus calcite, etc) as described in more detail below. The magmatic crystallization of, versus the fluid precipitation of, e.g., calcite or magnetite certainly can complicate the classification of these three rock types.

Kimberlite—General and petrographical aspects. Macrodiamonds in kimberlite are xenocrysts. Hypabyssal kimberlite is typically rich in macrocrysts (large crystals), which includes mineral grains from 0.5–10 mm in size. Note that the 0.5 mm size break is completely arbitrary, as macrocryst minerals can form part of a continuum with similar minerals < 0.5 mm and > 10 mm in size; as an example, see the discussions in Moss et al. (2010) and Moore et al. (2021) regarding olivine grain size in kimberlites. Macrocrysts have varied origins, and can be comprised of mantle xenocrysts such as olivine, Cr-diopside, enstatite and Cr-pyrope garnet derived from the dis-aggregation of peridotite xenoliths (dunite, harzburgite, lherzolite and wehrlite), omphacite and pyrope-almandine garnet derived from dis-aggregated eclogite xenoliths, and Mg-ilmenite and phlogopite plus amphibole xenocrysts from metasomatized peridotites (e.g., Dawson 1980; Nixon 1987). Macrocrysts in kimberlite also include bonafide cognate high pressure kimberlite magma-derived phenocrysts, with olivine being the classic example. More recently, the term antecryst (high pressure phenocrysts crystallized from an earlier-formed and "allied" proto-kimberlite or kimberlite magma) has also been applied to a subset of olivine macrocrysts (Sobolev et al. 2015; Soltys et al. 2020). An additional complexity of kimberlite mineralogy is the occurrence of the megacryst suite of minerals (olivine, phlogopite, Cr–Ti-pyrope, Cr-diopside, enstatite, Mg-ilmenite, zircon, baddelyite), which are large (>1 cm) single grains (or fragments thereof) up to ~30 cm in size. Both a Cr-poor (Boyd and Nixon 1973) and Cr-rich (Eggler et al. 1979) suite are known, but the distinction between them in kimberlites globally is somewhat poorly defined. Megacryst suite mineral(s) occur in variable modal proportions in kimberlite. The Cr-poor suite of megacryst minerals was interpreted as high pressure (cognate) minerals crystallized from a kimberlite or proto-kimberlite magma (e.g., Gurney et al. 1979; Schulze 1987). More recent studies suggest the Cr-rich suite represents greater mantle lithosphere interaction with kimberlite magma, whereas the Cr-poor suite has a less dominant lithosphere interaction with kimberlite magma (e.g., Pivin et al. 2009; Bussweiler et al. 2016, 2018).

In a kimberlite, the coarser-grained minerals (megacrysts, macrocryts, antecrysts and cognate phenocrysts) are set in a finer-grained matrix, which imparts a distinctive inequigranular texture to these rocks—prominent in outcrop, drill core and thin section. Kimberlite matrix mineralogy, often viewed as complex, is in fact relatively simple (Table 1). Olivine, spinel and phlogopite phenocrysts and microphenocrysts (typically < 0.2–1 mm) are observed in quite variable modal proportions; these grains can have xenocrystic or antecrystic cores. Importantly, in kimberlite, phlogopite can be rare or absent (e.g., Nielsen and Sand 2008; Tovey et al. 2021), but may also be very abundant (e.g., Taylor and Kingdom 1999; Howarth and Giuiani 2020). The olivine, spinel and phlogopite (when present) phenocrysts are set in a very fine-grained (< 0.2 mm) groundmass assemblage consisting of a suite of minerals of magmatic origin, including: olivine, spinel, phlogopite–kinoshitalite and phlogopite–tetraferriphlogopite mica, apatite, perovskite, monticellite, calcite, dolomite and ilmenite, with less common rutile and sulfides. Any given kimberlite groundmass contains only a subset (i.e., not all) of these minerals. Additional minor, rare minerals are also observed in kimberlite, but these can often only be identified by electron beam methods via back scattered electron images (BSEI) on a scanning electron microscope (SEM) or electron microprobe (EMP) with energy dispersive spectrometry (EDS) or wavelength dispersive (WDS) analysis, due to their small size (<10 μm). Glass (not a mineral) is present only as exceptionally rare melt inclusions (Howarth and Büttner 2019), but Skinner and Marsh (2004) have suggested that groundmass antigorite in kimberlite represents devitrified glass. These findings require further study, however. Combinations of calcite, dolomite, serpentine, magnetite and rare talc (plus other minerals) are precipitated at sub-solidus temperatures from fluids, which variably overprint the macrocrysts and magmatic crystallized phenocryst/microphenocryst/groundmass mineral phases and can constitute significant modal proportions of a kimberlite. In Lac de Gras, C–O isotopic analyses of precipitated calcite (with serpentine) and dolomite coupled with modelling determined the fluids to be of deuteric origin (Wilson et al. 2007b). Based on analyses of serpentine minerals and modelling, Mitchell (2013) proposed a deuteric origin for serpentinizing fluids. In contrast, Sparks (2013) has advocated a crustal origin for serpentinizing fluids in kimberlites, while Giuliani et al. (2014) has modelled that serpentinizing fluids could have either a meteoric, or a mixed deuteric–meteoric origin.

Crustal contamination of kimberlite can be quite common; granitoid contamination of kimberlite magma results in higher Si-, Al-, alkali-contents and the potential for crystallization of non-diagnostic kimberlite minerals such as diopside (e.g., Fulop et al. 2018; Gaudet 2018). For example, diopside, titanite and amphibole are not observed in uncontaminated kimberlite, as the silica activity is too low in these ultrabasic melts for these minerals to crystallize (Mitchell 1986, 1995, 2020a). Similarly, alkali- and plagioclase feldspars and the feldspathoid minerals kalsilite and nepheline are not observed due to the low Na, K and Al-activities in the kimberlite melt; the absence of melilite is interpreted to be due to low Na- and/or Al-activities (Mitchell 1995; Kjarsgaard et al. 2009; Foley et al. 2019).

Carbonate-Rich Olivine Lamproite (CROL)—General and petrographical aspects. Macrodiamonds in CROL are xenocrysts. Similar to kimberlite, a wide array of macrocryst minerals occur in CROLs, consisting dominantly of xenocrysts derived from a variety of mantle xenoliths (see *Kimberlite—General and petrographical aspects*). As with kimberlites, olivine xenocrysts are observed in variable quantities in CROL. Mantle peridotite and eclogite xenoliths, plus minerals derived from their disaggregation can be very common in some CROL (e.g., Roberts Victor eclogites). Megacryst suite minerals are not characteristic of CROL, but are observed at a few localities (Skinner et al. 1994; Moore and Gurney 1991). In CROL, cognate (high pressure) phlogopite phenocrysts (and thus potentially also antecrysts) can be common. In contrast, phlogopite antecrysts, cognate phenocrysts and phenocrysts in kimberlites are a minor to rare phase, and of potentially have cryptogenic origin (Mitchell 1995). The mineralogy of CROL is characterized by the occurrence of olivine and phlogopite (macrocrysts, phenocrysts, microphenocrysts), and spinel phenocrysts/microphenocrysts set in a groundmass of apatite, calcite (or dolomite), spinel, perovskite and phlogopite. Note that diopside microphenocrysts and groundmass diopside may be present, or absent (Wagner 1914; Mitchell 1995). The occurrence of magmatic calcite in CROL is consistent with the observation that there is a statistical correlation between measured wholerock CaO and CO_2 concentrations (as per kimberlites; see *Whole Rock Geochemistry*). CROL are both CO_2- and H_2O-rich (as per kimberlite), and similar to kimberlite they contain subsolidus low temperature fluid-precipitated minerals that can include calcite, dolomite, serpentine, talc and magnetite. A distinctive subset of "evolved CROL" occur at Pniel and Postmasburg (South Africa) that originally were suggested to be lamproites by Tainton and Browning (1991). These rocks, as well as other "evolved CROL" localities in South Africa identified by Mitchell (1995) have petrographic similarities to lamproites with respect to the presence of sanidine and potassium richterite in the groundmass and more rarely as microphenocryts. In evolved CROL, olivine macrocrysts and phenocrysts may be mantled/overgrown by parallel olivine aggregates ("dog tooth habit") as observed at Postmasburg and Sover North (Mitchell 1995). Plagioclase feldspars, the feldspathoid minerals kalsilite and nepheline, and melilite are absent in CROL.

Olivine Lamproite—General and petrographical aspects. Macrodiamonds in olivine lamproite are xenocrysts; leucite lamproite can be diamondiferous, but these rocks have lower diamond tenor and have not been proven to be economically viable thus far (Mitchell and Bergman 1991; Mitchell 2020b). Similar to kimberlite and CROL, a suite of macrocryst minerals are present in olivine lamproites, that can include, for example, olivine and phlogopite, plus xenocrysts derived from a variety of mantle xenoliths (see *Kimberlite—General and petrographical aspects*). However, mantle xenoliths and xenocrysts derived from their disaggregation are typically rare in olivine lamproite and even less so in lamproite, as compared to kimberlite or CROL. The megacryst suite of minerals (typical of many kimberlites) is completely absent (or possibly exceptionally rare) in olivine lamproite. However, high pressure cognate phenocrysts (and thus potentially also antecrysts) such as phlogopite, diopside, richterite and olivine, as well as cognate xenoliths (e.g., phlogopite clinopyroxenite, glimmerite), are observed in both olivine lamproite and lamproite, but these are perhaps more common in lamproite (Mitchell and Bergman 1991).

The mineralogy of olivine lamproite is characterized by the occurrence of olivine (macrocrysts, phenocrysts, microphenocrysts; Jaques et al. 1986; Jaques and Foley 2018), plus spinel and phlogopite phenocrysts/microphenocrysts that are set in a groundmass of glass, often altered. Fine-grained (<50 μm) apatite, perovskite and leucite are variably observed, but may be rare or absent (Jacques et al. 1986). In coarser-grained coherent or hypabyssal (aka "magmatic") olivine lamproite, poikilitic phlogopite plates can be common, and in the very coarsest-grained examples interstitial or poikilitic K-richterite is also observed (Jaques et al. 1986). Olivine macrocrysts with "dog-tooth habit" overgrowths occur in some olivine lamproites (e.g., Prairie Creek; Scott Smith and Skinner 1984). In leucite lamproite and more evolved lamproites, apatite, perovskite, leucite, sanidine, diopside, wadeite and ilmenite occur in the groundmass (Jaques et al. 1986; Mitchell and Bergman 1991); however, only a subset of these minerals may be present in any given sample. These minerals, if present, are far more common and typically of larger grain size than compared to what is observed in olivine lamproite. Additional accessory minerals of lamproite include rutile, hollandite–priderite, jeppeite, and Ti–Zr garnets, but these are rare or absent in olivine lamproites (Mitchell and Bergman 1991). Crustal contamination can also be quite common in volcaniclastic lamproite, for example the high quartz contents at Argyle and Ellendale (Australia) and Prairie Creek (Arkansas, USA), derived from their siliciclastic host-rock sediments. Alteration products of serpentine and talc are typical of olivine lamproites; secondary carbonate minerals are observed. Minor modal amounts of calcite, interpreted as primary have also been observed in olivine lamproites, providing a potential link between olivine lamproites and CROLs. The feldspathoid minerals kalsilite and nepheline are not observed in olivine lamproites, nor are melilite or plagioclase.

Mineral chemistry—General aspects of kimberlites, CROLs and olivine lamproites. The compositions of the various minerals observed in kimberlites, CROLs and olivine lamproites can be quite distinct as compared to other rock types/magmatic lineages. For example, in olivine lamproite, lamproite and evolved CROL, amphiboles are K-, Ti-rich richterites (e.g., Mitchell 1985, 1995); in lamproites, sanidine (if present) is often iron-bearing, with Fe_2O_3 substituting for Al_2O_3 (e.g., Carmichael 1967). Comparing kimberlites, CROLs and olivine lamproites, a mineral phase can have near-identical, or conversely quite disparate compositions and zoning trends. A significant amount of major- and minor-element mineral chemistry data determined by EMP (+/- trace elements determined by laser ICP-MS) exists for olivine, spinel, phlogopite, diopside, monticellite and more recently for apatite, perovskite and carbonates. Spinel and phlogopite are considered to have unique compositions and zoning trends that are diagnostic with respect to kimberlite, CROL and olivine lamproite/lamproite

(Mitchell 1986, 1995; Mitchell and Bergman 1991; Mitchell et al. 2019). Our understanding of the compositional range of olivine and its paragenetic significance (xenocrystal or magmatic in origin) in kimberlite, CROL and olivine lamproite has advanced significantly, but is still imperfectly understood and debated (e.g., Bussweiler et al. 2015; Giuliani 2018, Jaques and Foley 2018; Howarth and Nembambula 2021; Moore et al. 2021). However, identification of xenocrystic olivine versus magmatic formed olivine, provides insight to lithospheric mantle sampling, and thus potentially macrodiamond sampling.

Spinel zoning trends. Two different plots—the "reduced" and "oxidized" spinel prisms (Irvine 1965; Haggerty 1976; Mitchell 1986)—are typically used to examine compositional trends of spinel in kimberlite, CROL, olivine lamproite, and many other basic or ultrabasic rock types. A projection to the front face of the reduced or oxidized spinel prism is often utilized to generate a bivariate plot. The oxidized prism projection (Fig. 13) has axes of Mg# (using a stoichiometric calculation for Fe^{2+}) versus $Fe^{3+}/(Fe^{3+} + Cr + Al)$, while the reduced prism projection (Fig. 14) has axes of Mg# (using Fe_{total}) versus $Ti/(Ti + Cr + Al)$. The use of Fe_{total} instead of Fe^{2+} to determine Mg# in spinel shifts the Mg# to lower values, as discussed by Pasteris (1982). The reduced and oxidized spinel prism plots (or the projected bivariate plots) have been suggested to provide diagnostic discrimination between spinel from kimberlite, CROL, and olivine lamproite.

Kimberlite spinel "Trend 1" compositions (the magnesian ulvöspinel trend; Mitchell 1986) are defined by spinel evolution at a high, and near constant Mg# (Figs. 13, 14). "Trend 2" spinel compositions (the titanomagnetite trend; Mitchell 1986) observed in olivine lamproite, lamproite, CROL and also some kimberlites are defined by spinel zoning with decreasing Mg#, followed by subsequent evolution at constant and low Mg# (Figs. 13, 14). While many kimberlites have spinel compositions that fall into the Trend 1 field, this however, is an oversimplification. A number of kimberlites have spinel compositions that fall into Trend 2, such as those at the Koidu mine (Tompkins and Haggerty 1984), the Jagersfontein mine (Taylor and Kingdom 1999), the Renard mine (Birkett et al. 2004), De Beers peripheral (Pasteris 1983), Rich (Roeder and Schulze 2008), Buffalo Hills K6 (Eccles et al. 2004), Tunraq and Elwin Bay (Mitchell 1986, 1995) and Zagodachnaya and Marushkaya (Rozova et al. 1982).

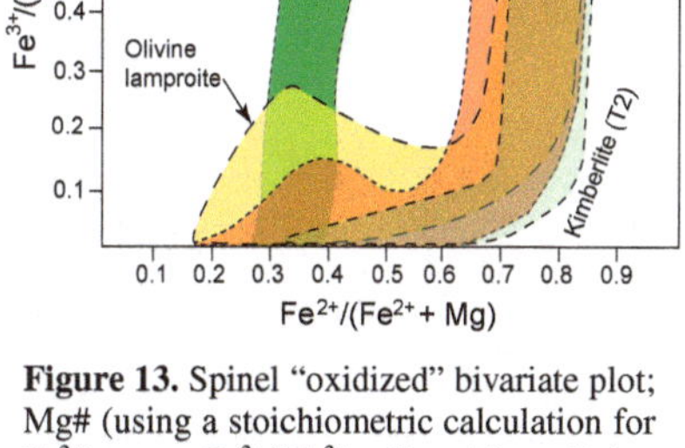

Figure 13. Spinel "oxidized" bivariate plot; Mg# (using a stoichiometric calculation for Fe^{2+}) versus $Fe^{3+}/(Fe^{3+} + Cr + Al)$, outlining the region of kimberlite Trend 1 and Trend 2 spinel compositions, and olivine lamproite and CROL Trend 2 spinel compositions. See text for details and data sources.

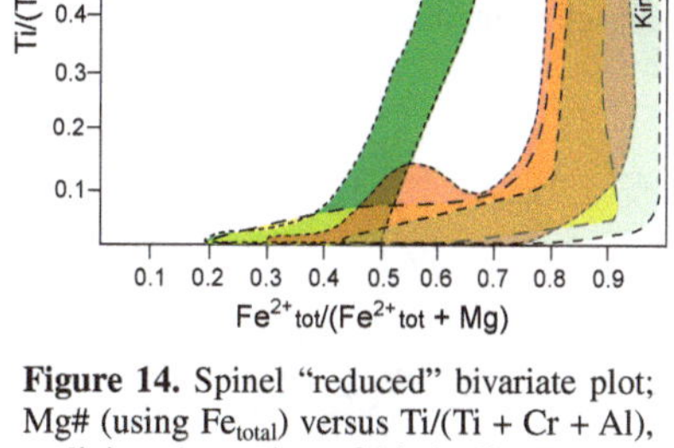

Figure 14. Spinel "reduced" bivariate plot; Mg# (using Fe_{total}) versus $Ti/(Ti + Cr + Al)$, outlining the region of kimberlite Trend 1 and Trend 2 spinel compositions, and olivine lamproite and CROL Trend 2 spinel compositions. See text for details and data sources.

Furthermore, some kimberlites (e.g. Jagersfontein, De Beers, Tunraq, Elwin Bay) have both Trend 1 and Trend 2 spinels, which are observed in distinct phases of kimberlite. The situation is additionally complicated by kimberlites with spinel compositions that lie intermediate between Trend 1 and Trend 2 (T1–2), such as those from the Jericho mine (Roeder and Schulze 2008), Tonguma in Sierra Leone (Howarth and Giuliani 2020) and Maniitsoq in Greenland (Nielsen et al. 2009). The Maniitsoq kimberlite is also of additional interest as it contains spinel compositions with higher Mg# than typical for Trend 1 spinels. Of additional intrigue is the recognition of Trend 1 spinels in some carbonatites (Roeder and Schulze 2008).

CROL and olivine lamproite contain spinel grains with compositions consistent with Trend 2 (Figs. 13, 14); an exception being the New Elands CROL with spinel compositions that are intermediate between Trend 1 and Trend 2 (Roeder and Schulze 2008). Because a variety of non-unique spinel chemistry trends are known, Roeder and Schulze (2008) emphasized the application of a molar Cr^{3+}–Al^{3+}–Fe^{3+} spinel ternary plot (Fig. 15). They described eight spinel zoning trends for kimberlites (and CROLs). The three main types of spinel zoning trends in kimberlites include Trend 1 and Trend 2, together with a third trend, in which spinel zones from chromite to pleonaste. Roeder and Schulze (2008) also noted that pleonaste can further evolve to Trend 1 spinel compositions (see also Abersteiner et al. 2019; Tovey et al. 2020), and that Trend 1 spinel can evolve to pleonaste.

Phlogopite zoning trends. Application of two bivariate plots (Al_2O_3 versus FeO_{tot} and Al_2O_3 versus TiO_2) for micas have been suggested to permit discrimination between kimberlite, CROL, olivine lamproite, and other rock types (Figs. 16, 17; e.g., Mitchell 1995). Compositionally zoned groundmass phlogopite micas in kimberlite have previously been noted to have an Al-enrichment trend (phlogopite-eastonite trend), whereas phlogopite in CROL and olivine lamproite exhibit decreasing Al with increasing Fe and variable (increasing; constant or decreasing) Ti contents (e.g., Mitchell and Bergman 1991; Mitchell 1995; Mitchell et al. 2019). The typical kimberlite groundmass phlogopite zoning trend is one of Al-enrichment (often with Ba-enrichment; Spriggs 1988), with decreasing Ti and constant or decreasing Fe (Figs. 16, 17). However, this phlogopite Al-enrichment trend, while typical of kimberlite does not describe all kimberlite micas. A zoning trend of decreasing Al, and decreasing or constant Ti, with Fe-enrichment is also known (i.e., a phlogopite-tetraferriphlogopite zoning trend; Figs. 16, 17). Moreover, this tetraferriphlogopite (TFP) trend is observed in a number of kimberlites globally, such as: Jagersfontein mine, South Africa (Taylor and Kingdom 1999); Renard mine, Canada (Birkett et al. 2004); Koidu mine and Tonguma, Sierra Leone (Mitchell 1995; Howarth and Giuliani 2020); Ororoo, Australia (Scott Smith et al. 1984); Mayeng, South Africa (Apter et al. 1984); Skinners sill, South Africa (Mitchell 1984); Antochka, Guinea (Mitchell 1995); K4 Buffalo Hills, Canada (Eccles et al. 2004); Porpoise, Lac de Gras, Canada (Armstrong et al. 2004); Guaniamo, Venezuela (Kaminsky et al. 2004) and Adamantin, Quebec, Canada (Barnett and Laroulandie 2017). An Al–Fe–Mg ternary plot is also useful for examining phlogopite mica Al-enrichment and Al-depletion trends (Rock 1988; Mitchell 1995).

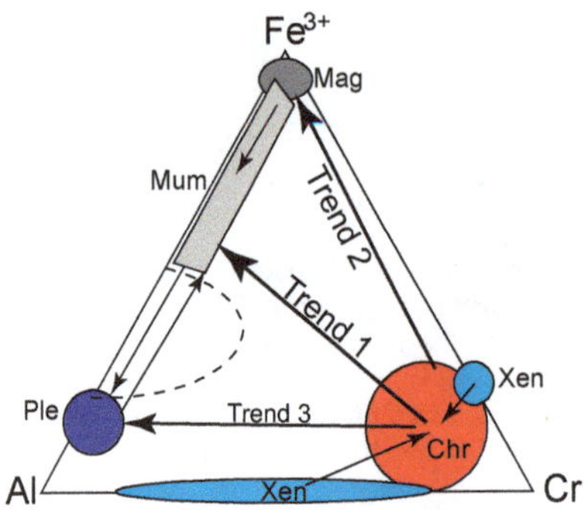

Figure 15. Spinel molar Cr^{3+}–Al^{3+}–Fe^{3+} ternary plot, showing the spinel xenocryst–chromite zoning trends, and the chromite–spinel Trend 1, chromite–spinel Trend 2, and chromite–spinel Trend 3 (pleonaste trend), as well as additional, less common zoning trends. Xen = xenocryst spinel; Chr = chromite; Ple = pleonaste; Mag = magnetite; Mum = magnesian ulvöspinel magnetite. Modified and adapted from Roeder and Schulze (2008).

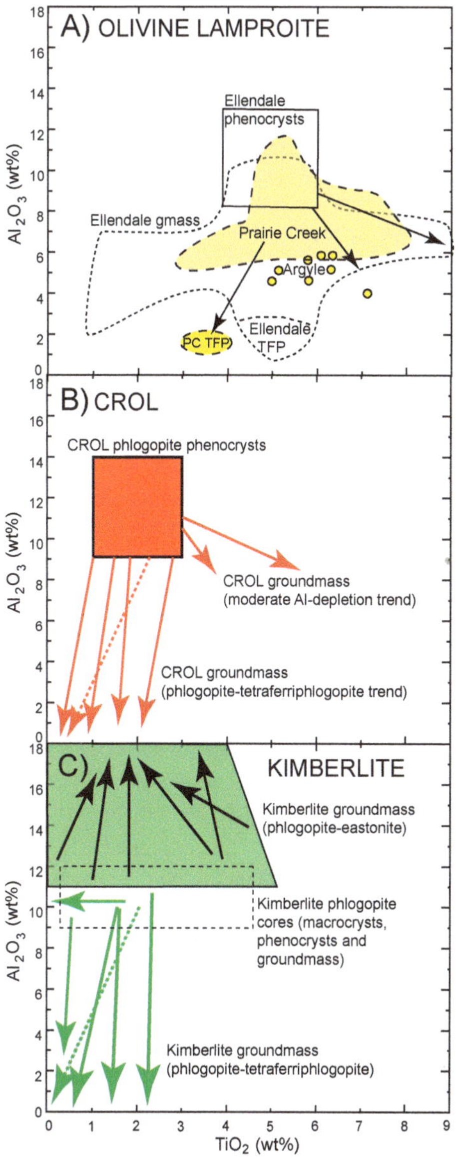

Figure 16. Bivariate plot of Al_2O_3 versus TiO_2 for phlogopite mica from (**A**) olivine lamproites, (**B**) CROL, and (**C**) kimberlites. For olivine lamproites and CROLs, zoning trends are from phenocryst cores to rims and groundmass grains; both these rock types contain phlogopite that exhibit moderate- and/or extreme-Al depletion trends. Kimberlite zoning trends can be observed within groundmass grains, and as phlogopite cores (macrocrysts, phenocrysts, microphenocrysts) with compositional zoning to the margins. In kimberlites, phlogopite Al-enrichment and Al-depletion trends are observed. See text for details and data sources.

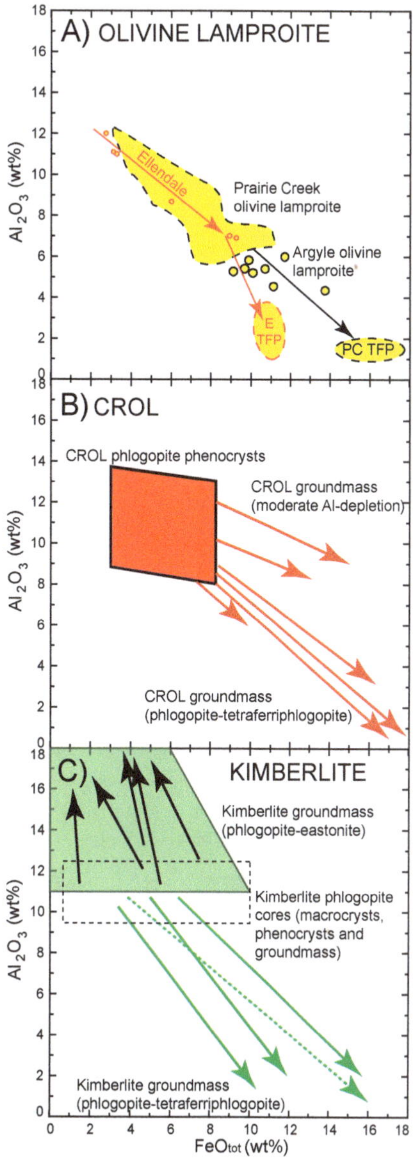

Figure 17. Bivariate plot of Al_2O_3 versus FeO_{total}, for phlogopite mica from (**A**) olivine lamproites, (**B**) CROL, and (**C**) kimberlites. For olivine lamproites and CROLs, zoning trends are from phenocryst cores to rims and groundmass grains; both these rock types exhibit phlogopite with moderate-Al depletion- and extreme-Al depletion (to TFP) trends. Kimberlite zoning trends can be observed within only groundmass grains, and as phlogopite cores (macrocrysts, phenocrysts, microphenocrysts) with compositional zoning to the margins. See text for details and data sources

CROL micas exhibit a trend of phlogopite zoning towards Al-depleted tetraferriphlogopite. These micas typically have an extreme Al-depletion trend, with very low Al_2O_3 concentrations (to <2 wt% Al_2O_3) and concomitant decreasing or constant TiO_2 (< 2 wt%), with FeO-enrichment (Figs. 16, 17). Examples include the Sover, Lace, New Elands, Finsch, Helam (Swartruggens) and Star (Theunissen) CROL-hosted diamond mines (Figs. 16, 17; data of Mitchell 1995). However, a "moderate" phlogopite Al-depletion trend (to 8–9 wt% Al_2O_3) is noted at the Voorspoed CROL-hosted diamond mine, and from a CROL dike in the Postmasburg area (Figs. 16, 17; data of Mitchell 1995). Tetraferriphlogopite is also reported from Sisimiut, Greenland (Thy et al. 1987). A subtle difference between the phlogopite—tetraferriphlogopite zoning trends for CROL and kimberlite is observed: CROL mica tend to have slightly higher FeOt concentrations (14–18 wt%) as compared to kimberlite mica (11–16 wt%) at the lowest Al_2O_3 concentrations (Figs. 16, 17). Thus, these kimberlite TFP micas have comparatively higher Mg# as compared to CROL TFP micas.

Olivine lamproite micas typically exhibit a phenocryst–microphenocryst—groundmass Al-depletion trend that is quite variable depending upon the locality. In general, there are a limited number of studies with detailed core to rim mica zoning trends. At the Argyle olivine lamproite, with increasing Fe, alumina decreases from 6 to 4 wt% Al_2O_3, with titanium concentrations of 5–7 wt% TiO_2 (Jaques et al. 1986). For olivine lamproites from the Ellendale field, coarse groundmass micas exhibit decreasing alumina (12 to 6 wt% Al_2O_3) with decreasing Mg# (90 to 65) and Ti concentrations increasing from ~4 to ~10 wt% TiO_2. Fine groundmass micas exhibit decreasing alumina (8 to 1 wt% Al_2O_3) with decreasing Mg# (90 to 60) and variable titanium concentrations of ~1 to ~9 wt% TiO_2 (Jaques et al. 1986). Prairie Creek micas have 11 to 5 wt% Al_2O_3, with 5 to 9 wt% FeO_{tot} and titanium concentrations of 3–8 wt% TiO_2; tetraferriphlogopite rims are also observed with 1–2 wt% Al_2O_3, 3–4.5 wt% TiO_2 and 14–17 wt% FeO_{tot} (Scott Smith and Skinner 1984; Mitchell and Bergman 1991). Tetraferriphlogopite mica has also been documented from olivine lamproites at the American mine, Arkansas (Mitchell 1985), Atri, India (Das et al. 2018) and at Argyle (Jaques et al. 1986). In olivine lamproite, these TFP micas appear as distinct rims (i.e., phlogopite has a distinctive compositional break, and a TFP rim), in comparison to CROLs in which TFP micas are typically part of a continuum of compositions with phlogopite that is zoned to tetraferriphlogopite (Figs. 16, 17).

Olivine chemistry. A significant number of studies have examined olivine compositions in kimberlite, CROL and olivine lamproite. In kimberlites, the key issue is examination and identification of xenocrystic olivine versus magmatic olivine (with the latter collectively including megacrysts, cognate phenocrysts, antecrysts, phenocrysts and microphenocrysts); this is termed the kimberlite "olivine problem" (see Fig. 18). Two quite disparate views are held, namely that all olivine is xenocrystic, excepting the thin rims and rinds (when present) that are interpreted as magmatic overgrowths (e.g., Kamenetsky et al. 2008; Brett et al. 2009). In contrast, Moore et al. (2021) suggest that the majority of olivine in kimberlites is not xenocrystal, but of magmatic origin. A third point of view is that olivine in kimberlite can be both of xenocrystal origin and magmatic (forming at mantle to crustal pressures) origin (e.g., Boyd and Clement 1977; Kjarsgaard et al. 2010; Giuliani 2018; Soltys et al. 2018; Lim et al. 2018; Mitchell et al. 2019; Soltys et al. 2020). In kimberlite, olivine has a continuum of grain sizes (e.g., Field et al. 2009; Moss et al. 2010; Harvey et al. 2013; Moore et al. 2021), with larger grains tending to be rounded, and smaller grains typically subhedral to euhedral (Mitchell 1995; Mitchell et al. 2019). The absence of a 1:1 compositional correlation (e.g., Mg# and Ni, Mn, Ca) between olivine from granular and sheared peridotite xenoliths and kimberlite macrocrystal olivine (with both olivine types sourced from the same kimberlite or kimberlite field), clearly suggest that not all olivine macrocryst cores are derived from disaggregation of mantle peridotite (e.g., Kjarsgaard et al. 2010; Giuliani 2018; Soltys et al. 2020; Moore et al. 2021; see Fig. 18). Specifically, this would include olivine that are not clearly part of the megacryst suite, which have Mg# ~89–78, (e.g., Boyd and Clement 1977; Lim et al. 2018; Giuliani 2018; Moore et al. 2021).

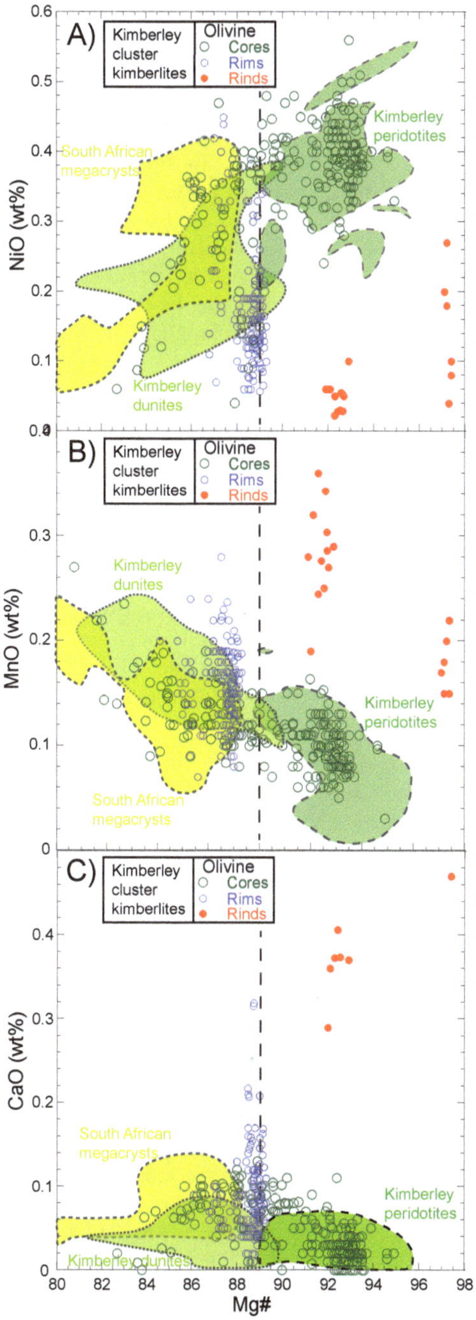

Olivine compositions in CROL are remarkably similar to those in kimberlite, with a wide range in modal abundance and grain size, with larger grains often rounded and smaller grains being rounded, subhedral and euhedral (Mitchell 1995; Moore 1998; Howarth and Nembambula 2021). Euhedral and subhedral olivine phenocrysts in CROLs appear to have a more restricted compositional range with Mg# 89–93, as compared to kimberlites with Mg# 85–93 (Mitchell 1995; Mitchell et al. 2019) However, more data for olivine in CROL (and complimentary mantle xenoliths) would be useful to confirm this notion.

Olivine macrocrysts from both the Ellendale, Australia and Prairie Creek, U.S.A. olivine lamproites are interpreted to contain both xenocrystal olivine and phenocrystal olivine (Jaques and Foley 2018; Scott Smith and Skinner 1984). A detailed study of olivine from the Ellendale olivine lamproites (Jaques and Foley 2018) determined that olivine from mantle xenoliths (Mg# 90.1–92.4), larger (>1 mm) olivine macrocryst grains (Mg# 89.7–92.9), and smaller (<1 mm) olivine cores (Mg# 90–93) all have essentially similar Mg#'s, with a subset of phenocryst cores having slightly lower Mg#'s (87–90). However, olivine phenocryst zoning trends (e.g., Ni, Ca, Mn versus Mg#) and rim compositions on all olivine grains indicate that the >1 mm versus <1 mm size break corresponds remarkably well to the xenocrystal olivine and phenocrystal olivine compositional break (Jacques and Foley 2018). The Wajrakarur (India) P4 and P12 olivine lamproites have olivine xenocryts with Mg# ~91–94. However, olivine phenocrysts have lower Mg# of ~89–85 in P4 and ~87–83 in P12 (Shaikh et al. 2018; Sarkar et al. 2021) i.e., they are more evolved as compared to olivine phenocrysts in Ellendale or Prairie Creek olivine lamproites.

Figure 18. (**A**) Mg# vs NiO (**B**); Mg# vs MnO; (**C**) Mg# vs CaO for Kimberley cluster kimberlite-derived olivine macrocrysts cores, rims and rinds. Composition of olivine from coarse and sheared mantle peridotites, and olivine from dunites (interpreted as magmatic cumulates that are associated with the megacryst suite) from Kimberley cluster kimberlites are shown as fields. The field for southern African olivine megacrysts is also shown. All raw olivine data from Soltys et al. (2020).

Summary of mineral chemistry and the discrimination of kimberlite, CROL and olivine lamproite. The petrography and mineralogy of the three rock types is outlined in Table 1, which details the similarities and differences between them. Some broad generalizations are as follows.

Kimberlites typically have spinel Trend 1 compositions (chromite—magnesian ulvöspinel trend), Importantly however, some kimberlites have spinel Trend 2 compositions (chromite–titanomagnetite trend), as well as spinel compositions that lie intermediate between Trend 1 and Trend 2 (T1–2), or, have higher Mg# than is the "norm" for Trend 1 spinels. Kimberlites also usually have phlogopite micas that exhibit zoning with Al- (and Ba-) enrichment and concomitant Ti- (and Fe-) depletion (the phlogopite-eastonite trend). A second kimberlite mica zoning trend is also known, an Al- and Ti-depletion trend with increasing Fe (i.e., phlogopite to tetraferriphlogopite). Note that kimberlites with atypical spinel or phlogopite mica zoning trends are hosts to diamond mines.

CROLs typically have spinel Trend 2 compositions (chromite–titanomagnetite trend). Phlogopite mica has a characteristic extreme Al- and Ti-depletion with Fe-enrichment (phlogopite to tetraferriphlogopite) trend. A second mica trend has moderate Al-depletion with increasing Fe- and Ti-enrichment. CROL also rarely have spinel compositions intermediate between Trend 1 and Trend 2 (T1–2).

Olivine lamproites are characterized by spinel Trend 2 compositions (chromite–titanomagnetite trend), and phlogopite mica with an Al-depletion and Fe-enrichment trend from phlogopite towards tetraferriphlogopite. The occurrence of tetraferriphlogopite (TFP) in olivine lamproites does not represent a continuous zoning trend from phlogopite, but is discontinuous with a compositional break between Al-poor phlogopite and TFP. Micas in olivine lamproites are Ti-rich as compared to those observed in kimberlites and CROLs.

We note that the compositional trend or trends defined by a single mineral type should not be used in isolation to distinguish one diamond-bearing rock type from another. Compositional trends of spinel and phlogopite, while useful, cannot be construed as diagnostic when considered in isolation (or together), as previously inferred (Mitchell 1986, 1995). We caution that on their own, spinel chemistry trends cannot be used with confidence to distinguish kimberlite (Trend 2) from CROL or olivine lamproite (Trend 2). Phlogopite with Al-depletion and compositional zoning to tetraferriphlogopite is also not diagnostic of kimberlite, or CROL, or olivine lamproite. A sound understanding of the mineralogy is key (Table 1). Spinel and phlogopite discriminant plots should be used along with other criteria (e.g., mineralogy, whole rock geochemistry, isotopic studies, etc.) wherever possible. At present, only an unknown sample with Trend 1 composition spinels and a mica Al-enrichment (phlogopite–eastonite) trend could be construed as a kimberlite.

Any individual or discrete, intrusive or extrusive phase within a given occurrence of kimberlite, CROL or olivine lamproite can have distinctive spinel and phlogopite compositions and zoning trends (e.g., Naidoo et al. 2004; van Straaten et al. 2011). The modal amount of olivine and the relative proportions of xenocrystal olivine to magmatic olivine is exceedingly variable in kimberlite, CROL, and olivine lamproite. Taken together, these observations are very useful for diamond exploration geologists with respect to understanding the geology, and hence furthering understanding of the economic potential of kimberlites, CROLs and olivine lamproites.

Whole rock geochemistry

History of geochemical studies of kimberlite, CROL and olivine lamproite. Kimberlite, the dominant rock type that hosts diamond mines is arguably the most hybridized, diversely altered and xenolith-rich magmatic material to have found their way to Earth's surface. A long history of kimberlite research is only now arriving at a consensus of their definition and classification while their origin remains as contentious as ever—perhaps an indication of the importance of these rocks as geochemical probes of Earth's mantle. This part of the chapter will briefly review the geochemistry of kimberlite, classified on the basis of petrography and mineral chemistry as bonafide kimberlites, and then compare minimally contaminated hypabyssal kimberlite with carbonate-rich olivine lamproite and olivine lamproite. Our aim is to provide the reader with a view of the range of compositions and key geochemical characteristics of the three main magmatic rock types that host diamond deposits. We do not focus on the geochemistry of volcaniclastic (fragmental) examples of these three rock types, except to illustrate their substantive geochemical diversity as compared to minimally contaminated and/or altered hypabyssal samples, and to consider their likely alteration, and mantle and crustal contamination pathways and relationships.

One of the first studies to incorporate whole rock analyses of kimberlites was by Wagner (1914) who applied geochemistry to help distinguish kimberlite from what he thought to be "similar" rocks observed in South Africa, e.g., olivine melilitite. Wagner was perhaps the first author to highlight the lower alumina and alkalis, and higher Mg of what he termed "basaltic" kimberlites compared to these other rock types considered at that time to be "similar", or "related". After a hiatus of research on the topic, the 1970's and early 1980's ushered in new era of whole rock geochemical studies of kimberlites that included significant datasets, e.g. Ilupin and Lutts (1971), Nixon (1973), Gurney and Ebrahim (1973), Fesq et al. (1975), Dawson (1978, 1980) and Clement (1982). In these studies, many of the samples analysed were massive volcaniclastic kimberlite and kimberlite breccia that contained high proportions of entrained crustal and/or mantle xenoliths/xenocrysts. Importantly, these data confirmed the broadly high-Mg and Si-poor geochemical nature of kimberlites determined in earlier studies (e.g., Wagner 1914). The landmark papers by Smith (1983) and Smith et al. (1985) recognized the need for geochemical studies on samples less influenced by crustal input and provided a high-quality dataset for a suite of southern African hypabyssal kimberlites. These studies importantly identified, separated and distinguished Gp Ia (occurring within the Archean nucleus of the craton) and Gp Ib (occurring in the Proterozoic regions of the craton) kimberlites, and Gp II kimberlites (i.e., CROLs) in the Kaapvaal craton. Spriggs (1988) made the first detailed geochemical studies of circum-cratonic kimberlites (from Namibia), focusing on hypabyssal varieties. Taylor et al. (1994) examined whole rock geochemical data from kimberlites from West Africa, making detailed comparison with Kaapvaal kimberlites. This period was followed by a marked proliferation of numerous and noteworthy new geochemical studies, with important contributions by le Roex et al. (2003), Harris et al. (2004), and Becker and le Roex (2006) containing analyses of kimberlites from southern Africa, and by Price et al. (2000) from the Jericho kimberlite, Slave Craton, Canada. Significant bodies of work were also published on the kimberlites of the Lac de Gras field, Slave Craton, Canada (Kjarsgaard et al. 2009; Tappe et al. 2013), West Greenland kimberlites (Tappe et al. 2011), kimberlites from Finland (Dalton 2019) and Na-, Cl-, and CO_2-rich kimberlites from Udachnaya-East, Russia (Kamenetsky et al. 2007a,b).

While the recognition of Group II kimberlite (referred to as CROL here) by Smith (1983) in the Kaapvaal Craton was an immensely significant advance, further studies on these rocks were hindered by the notion that they only occur within southern Africa (Mitchell 1995). However, CROL are now recognized globally (see Fig. 2), with known localities in Finland and adjacent Russia within the Kola composite Craton (e.g., Mahotkin 1998; O'Brien and Tyni 1999; Kargin et al. 2014), in the Man Craton, West Africa (e.g., Skinner et al. 2004; Howarth and Giuliani 2020), in the Bastar Craton, India (e.g., Mainkar and Lehmann 2007; Lehmann et al. 2010; Rao et al. 2011) and in the Rae Craton, Canada (Sarkar et al. 2019). The geochemistry of numerous CROL localities of the Kalahari composite Craton in South Africa, Botswana and Eswatini is well documented (e.g., Wagner 1914; Smith 1983; Smith et al. 1985; Dawson 1987; Fraser and Hawkesworth 1992; Skinner et al. 1994; Tainton and Mckenzie 1994; Becker and le Roex 2006; Coe et al. 2008; Howarth et al. 2011).

Though long recognised (e.g., Cross 1897), the term lamproite was popularised by Niggli (1923). These compositionally unusual rocks with elevated K and Mg contents only became a focus of intense global interest with the discovery of the Tier 1 Argyle diamond mine and the Tier 4 Ellendale 4 and Tier 4 Ellendale 9 diamond mines in western Australia, with all three mines hosted by olivine lamproite (e.g., Jaques et al. 1984, 1986). Several comprehensive studies on olivine lamproites followed soon after (Fraser et al. 1985, 1992; Tainton and McKenzie 1991), in addition to an overview by Mitchel and Bergmann (1991).

The differing rock classification at some specific localities e.g., CROL and also olivine lamproite, or kimberlite and also UML (e.g., West Greenland), remains problematic. The Aries pipe in the Kimberley Block (Australia) is variably interpreted as a kimberlite and also a CROL, compare Edward et al. (1992) with Taylor et al. (1994), or Mitchell (1995), or Downes et al. (2006). The reader is encouraged to read the many papers published on these issues in the extended abstracts (https://ikcabstracts.com /index.php/ikc), and the Proceedings volumes of the International Kimberlite Conference over the last 50 years.

A more recent and comprehensive review of the comparative geochemistry (including major-, trace-element and tracer isotope systematics) of kimberlite, CROL, and olivine lamproite is provided by Pearson et al. (2019). Here, we focus on the most commonly used criteria—major elements—in identifying and characterising these rocks. Major elements have also been used in suggested schemes for evaluating the diamond-potential of kimberlites (e.g., Valsilenko et al. 2012).

Major element database. Numerous examples of kimberlite major element data exist in downloadable databases such as the GEOROC database, which contains >4,500 analyses of rocks reported as kimberlites. We assembled our own kimberlite database using a combination of the GEOROC database and data in the Proceedings of the International Kimberlite Conferences (prior to 2003). This database, with 5,133 major element analyses, labelled as "Kimb–All" in our plots, was assembled to provide an overview on the range of compositions of rocks that have been referred to as kimberlites in the literature, including a number of volcaniclastic kimberlites, kimberlites with large amounts of crustal contamination (e.g., Fulop et al. 2018), or altered kimberlites. We screened this database to eliminate rocks that did not meet the basic petrographic/mineralogical classification of a kimberlite (see below).

To augment the general kimberlite database, we sought to constrain, more accurately, the composition of kimberlite magmatic rocks less affected by significant crustal contamination or alteration. This aim involved the assembly of a much smaller ($n = 519$) database of kimberlite, referred to here as "Kimb–HK", as the samples are exclusively hypabyssal/coherent. In addition, we assembled and scrutinised similar published datasets and databases for carbonate-rich olivine lamproite (n = 305) and olivine lamproite (n = 306) whole rock analyses. The source references of the data are listed in on-line Appendix A. The databases were constructed using the following guidelines:

i) Confident identification and classification of the rocks based on current or previous petrographic/mineral chemistry and/or tracer isotopic studies studies.

ii) Identification of the rocks as hypabyssal rocks (intrusions) rather than fragmental volcaniclastic rocks.

iii) The data are accompanied by a description of analytical methods with co-analysed reference materials, from a reputable laboratory, or the publication uses methods previously published by that laboratory.

iv) We avoided most rocks from regions where low-T alteration problems are known to be severe (e.g., India) while recognising that there likely are numerous reliable analyses from these regions. Time limits did not permit an extensive investigation of this issue on an individual sample basis.

v) We avoid analyses on rocks that are clearly identified as containing significant crustal contamination, or suites of rocks where this was in fact the objective of the study (e.g., the extensive and well characterized Snap Lake dataset of Fulop et al. 2018).

vi) We avoid rocks containing apparently "excessive" MgO, where instrument calibration or olivine accumulation may have been a major problem.

vii) Unlike in the "Kimb–All" database, in the "Kimb–HK" database we avoided studies where many samples had been analysed from the same kimberlite body (e.g., the extensive studies of Russian kimberlites by Vasilenko et al. 2012) to avoid giving undue weight to particular individual locations.

viii) We screened kimberlites using the Clement Contamination Index (CCI) and the ln (Si/Al) versus ln (Mg/Yb) approach of Dowall (2004) and Kjarsgaard et al. (2009), checking that we did not inadvertently screen-out kimberlites with minimal contamination that may be enriched in K-rich phlogopite, which will drive up the CCI value (Clement 1982). Note that these two screening techniques are not directly applicable to CROL or olivine lamproite, as these approaches have not been rigorously tested on these rock types, which typically have higher CCI values than kimberlites (Fig. 19).

In applying these criteria, we recognize that we may have screened out many "useful" kimberlite, CROL and olivine lamproite analyses, simply because the publications do not contain sufficient information with which to judge the reliability and nature (e.g., hypabyssal versus volcaniclastic kimberlite, degree of alteration) of the data (e.g., Lapin et al. 2007) and or are written in a language not easily accessible to the authors (e.g., Ilupin and Lutts 1971). We partition rocks that were formerly known as "micaceous" kimberlite" (Wagner 1914) or "Group II kimberlites" (Smith 1983; Smith et al. 1985) or orangeites (Wagner 1928; Mitchell 1995) into a category of lamproites referred to as "carbonate-rich olivine lamproites" or CROLs (Pearson et al. 2019), and further investigate the validity of this division. The lamproite database is divided into Mg-rich lamproites that are confidently identified as olivine lamproites, and apply a nominal cut-off requirement of MgO >14 wt% for this category. Note that Jaques and Foley (2018) employed a cut-off at 10 wt% MgO.

Crust–mantle contamination: geochemical diversity of kimberlite, CROL and olivine lamproite. Kimberlites especially are notable for their high load of crust and mantle xenoliths and are one of the few rock-types where "contamination" by both crust and mantle is a serious issue, clouding interpretation. This problem can also be a serious issue with CROLs and olivine lamproites. Quantifying the effects of crustal contamination on kimberlite compositions is key to understanding their geochemistry including parental melt compositions, especially for samples that lack obvious petrographic evidence for contamination (Fig. 19a). Though crustal contamination effects are less severe for some of the incompatible elements that are highly enriched in kimberlites, e.g., Nd and Sr, the effects of this process must still be considered carefully.

A variety of approaches have been developed, with an early popular approach being the "Clement's Contamination Index" (Clement 1982), or CCI, where:

$$CCI = (SiO_2 + Al_2O_3 + Na_2O) / (MgO + 2K_2O)$$

Typically, CCI values < 1 are considered uncontaminated (Clement 1982; Mitchell 1986), an approach taken by Nowicki et al. (2008). However, Clement (1982) reported on apparently uncontaminated kimberlites with CCI values up to 1.5, but these are from phlogopite-rich kimberlites, or from samples that are now recognized as CROLs (Fig. 19b). We have adopted a maximum CCI value of 1.3 in our own kimberlite screening. Although the CCI is widely used,

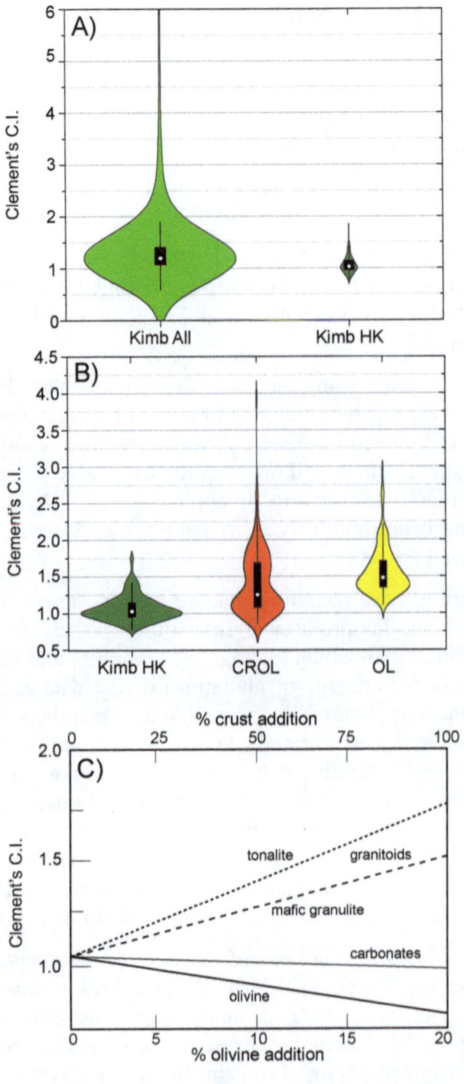

Figure 19. Violin, and box and whisker plots of Clement's Contamination Index (CCI) for: (**A**) Kimb-All and Kimb-HK; note for this plot, the Kimb-HK data was not screened for CCI < 1.3, and; (**B**) Kimb-HK, CROL and olivine lamproite. The box and line plot showing the median (**white circle**) and the interquartile range (50% of the data—the **box, in black**) with the lines representing the upper and lower quartiles of the data; this is combined with a "violin" plot, which is in essence a probability density function of the data. The sources of the data are in the on-line appendix. (**C**) Mass balance for mixing various crustal lithologies and mantle olivine (Mg#92) into a kimberlite with equivalent composition to the Wesselton aphanitic kimberlite (le Roux et al. 2003), as a function of varying CCI. Note the opposing effects of olivine addition versus most crustal lithologies and the insensitivity of this index to carbonate (limestone) contamination. Re-drawn after Dowall (2004).

this approach lacks sensitivity due to the mutually opposing effects of olivine addition from peridotites—a ubiquitous process in kimberlites—and assimilation of typical silicic crust, such as granite (Fig. 19c). Other approaches have suggested that examination of a combination of SiO_2, Al_2O_3, MgO, Pb and HREE can provide more robust screens for crustal contamination;

for example, the application of using Yb, low in "uncontaminated" kimberlites (<0.5 ppm), and typically higher in granitic rocks (~1 ppm) was shown to provide improved sensitivity (le Roex et al. 2003).

There is wide agreement that kimberlites, CROLs and olivine lamproites have incorporated varying proportions of the cratonic lithospheric mantle they have traversed, and that this appropriation of mantle rock is responsible for most of the diamond grade, as >90% of mineral inclusions in diamonds, from most locations, are of lithospheric origin (Stachel and Harris 2008; see Stachel et al. 2022, this volume). Pearson et al. (2019) examined the influence of the local lithospheric mantle composition, and the degree of its incorporation on the bulk composition of kimberlites and suggested there is a strong correlation between the kimberlite Mg/Si ratio and that of entrained lithospheric mantle xenoliths from the same kimberlite or kimberlite field (Fig. 20). This association was strengthened in a follow-up study by Giuliani et al. (2020). These studies demonstrated that incorporation and/or assimilation of orthopyroxene (Mg/Si ~ 0.83) and olivine (Mg/Si ~ 1.64) clearly influences the bulk kimberlite Mg/Si ratio. Kimberlite locations that erupted through olivine-rich, orthopyroxene-poor cratonic mantle, such as Greenland (North Atlantic Craton), have higher Mg/Si than those erupted through the more orthopyroxene-rich cratonic mantle e.g., beneath the Kaapvaal, Siberian and Slave Cratons. Incorporation of lithospheric mantle thus imparts a local flavor to kimberlite compositions (Pilbeam et al. 2013; Pearson et al. 2019; Giuliani et al. 2020), potentially creating differences such as the higher Ti contents of Kaapvaal and Greenland kimberlites relative to Lac de Gras kimberlites (Pearson et al. 2019).

More complex approaches to examine crustal contamination, and combined crustal and mantle contamination were developed by Dowall (2004) and Kjarsgaard et al. (2009), utilising element ratios to combat the "closure effects" inherent in compositional data, an example of which is illustrated in Figure 21 for a suite of hypabyssal kimberlites from Lac de Gras, Slave Craton, Canada. The natural log of Si/Al is a useful screen that is very sensitive to the addition of high Al crustal rocks and neatly divides the Lac de Gras kimberlite sample suite into two distinct groups that are a function of their contamination (Fig. 21). Crustally contaminated rocks that are identified using this measure are also classified as being significantly contaminated on the basis of higher Yb abundances and lower ln Mg/Yb.

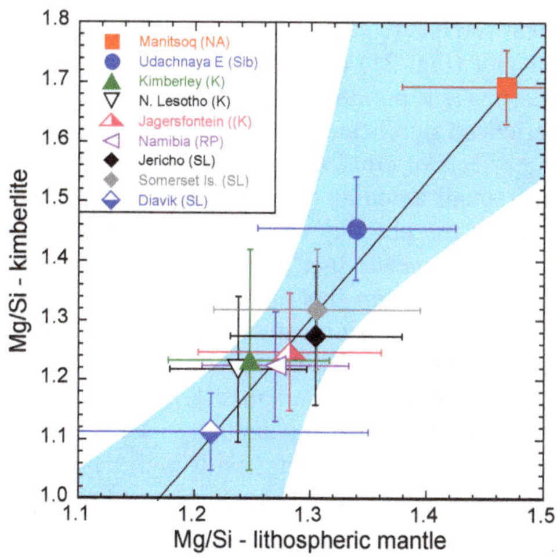

Figure 20. Plot of average molar Mg/Si ratio for kimberlites from nine global localities, versus the molar Mg/Si ratios of coarse peridotite xenoliths from the same kimberlite or kimberlite field. After Pearson et al. (2019), with permission.

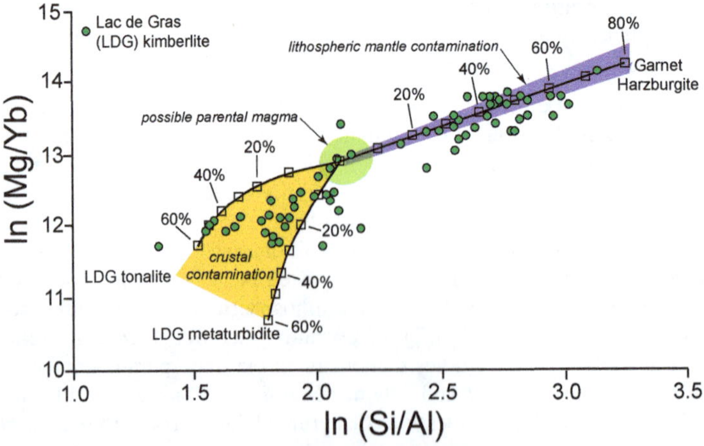

Figure 21. Plot of ln Mg/Yb versus ln Si/Al illustrating the competing effects of crustal contamination versus peridotitic lithospheric mantle assimilation for a "possible parental" Lac de Gras kimberlite magma. Data plotted are the bulk compositions of Lac de Gras hypabyssal kimberlites, unfiltered for the effects of crustal contamination. Endmember metaturbidite and tonalite compositions are from the Lac de Gras area of the central Slave Craton. Kimberlite and crustal whole rock data data are from Kjarsgaard et al. (2009) and Tappe et al. (2013). The garnet harzburgite compositions is a is a typical Slave craton harzburgite.

If, however, we examine the range of CCI values for kimberlites within the two groups of Lac de Gras hypabyssal kimberlites defined from ln Si/Al considerations, they largely overlap in CCI values (Kjarsgaard et al. 2009). Furthermore, a significant number of "non-contaminated" samples, identified on the basis of the ln Si/Al parameter, actually have CCI significantly >1, i.e., they would be flagged as contaminated if we used only this index according to the suggestions of Mitchell (1986) and Nowicki et al. (2008). One reason for this limitation of the CCI is that crustal contamination and peridotite incorporation have mutually opposing effects (see Figs. 19, 21, and also below) that severely limit the sensitivity of this index. On this basis we suggest that the sole use of Clement's C.I. as a screen for identifying crustally contaminated kimberlites be abandoned, or is at least utilized in conjunction with other proposed geochemical screens (e.g., le Roux et al. 2003; Kjarsgaard et al. 2009). The ln Mg/Yb versus ln Si/Al plot (Fig. 21) offers the possibility of visualising the competing effects of crustal assimilation versus peridotite incorporation, and is effective at screening out kimberlites that have experienced significant contamination by continental crust. Nonetheless, the diametrically opposing effects of crust versus mantle incorporation mean that it is likely impossible to guard against small amounts of crustal assimilation in kimberlites apparently free of crustal xenoliths, given the ubiquity of peridotite incorporation. The dilution of Yb by addition of mantle peridotite means that identifying even 5% addition of crust when accompanied by 20% peridotite addition (not an unreasonable mass fraction; e.g., Soltys et al. 2018) remains challenging. Lastly, using parameters such as Si/Al and Yb will not assist in detecting kimberlites containing variable amounts of crustal contamination from platform carbonate sediments, e.g., Udachnaya, Russia, and at Somerset Island, Jericho and Chidliak in Canada. Notwithstanding this limitation, a potentially useful application of the ln Mg/Yb versus ln Si/Al plot (Fig. 21) is that kimberlites with high degrees of lithospheric mantle contamination can be identified—and these may be of interest for their diamond potential.

Key aspects of the geochemistry of kimberlite, CROL and olivine lamproite. Most reviews of kimberlites and kimberlite geochemistry (e.g., Mitchell 1986, 1995) classify these rocks as silica-undersaturated, alkaline, potassic, and MgO- and CO_2-rich rocks. However, Kjarsgaard et al. (2009) and Pearson et al. (2019) noted that, for global kimberlites with minimal crustal contamination they are also H_2O-rich. Furthermore, their relatively low K + Na contents result in them having molar (Na+K)/Al < 1, i.e., they are not alkaline in nature and thus cannot be termed either "sodic" or "potassic" and with K_2O concentrations that are <3 wt%, these rocks certainly cannot be termed "ultrapotassic". In fact, kimberlites typically exhibit K_2O contents similar to mid-ocean ridge basalts (MORB)! In contrast, lamproites (including olivine lamproites and carbonate-rich olivine lamproites) are considered bona-fide alkaline rocks (Mitchell 2020a,b). Herein we utilize a combination of "violin", and box and whiskers statistical plots of the whole rock data to re-examine the similarities and differences between the three main magmatic rock types that host diamond mines, and provide an update on their basic geochemical parameters and ratios.

We start by examining the large, unscreened kimberlite database (Kimb-All; includes fresh and altered samples of both volcaniclastic and hypabyssal kimberlites) and compare this to a hypabyssal kimberlite database (Kimb-HK) to illustrate some of the potential variations in elemental concentration with respect to contamination and alteration (Fig. 22). Wider variations in concentration levels (both higher and lower) for SiO_2 and Al_2O_3 are observed in the Kimb-All data as compared to the Kimb-HK data, with the very high levels of SiO_2 (40–65 wt% SiO_2) and Al_2O_3 (5–18 wt% Al_2O_3) interpreted to result from crustal contamination and/or alteration processes (Figs. 22a,b). In contrast, MgO concentrations are lower in the Kimb-All data as compared to Kimb-HK data, interpreted to result from alteration processes (Figs. 22c, d) and dilution from continental crust. The available and reliable CO_2 data for kimberlites is surprisingly small, with 1487/5133 of the Kim-All samples, while the Kimb-HK data set is improved, with 371 of 519 samples having CO_2 data. Measured CO_2 levels are slightly higher in the Kimb-HK data (median = 5.8 wt%) compared with the Kimb-All data (median value 4.6), perhaps reflecting the more degassed character of the volcaniclastic rocks included in the Kimb-All database. The available and reliable H_2O data for kimberlites is very limited, with 386/5133 of the Kimb-All samples, while the Kimb-HK data set is much improved, with 329 of 519 samples having H_2O data. H_2O levels are similar for both kimberlite categories (6.88 wt% H_2O Kimb-HK versus 6.99 for Kimb-All) (Figs. 22e,f). Molar (Na+K)/Al and K/Al are lower in the Kimb-All data set as compared to the Kimb-HK data, due to a suggested combination of higher Al from crustal contamination, and similar or lower Na and/or K as a result of alteration and de-gassing processes (Figs. 22d, g, h). Based on these observations from comparing the Kimb-All and Kimb-HK data sets, we will subsequently only examine the Kimb-HK data set (screened to CCI values <1.3) for comparative purposes with the CROL and olivine lamproite data sets.

The SiO_2 concentration levels for all three rock types is <52 wt% SiO_2, i.e., they are all basic in composition, and typically, at <45 wt% SiO_2, are ultrabasic in composition (Fig. 23a). TiO_2 concentration levels for all three rock types (Fig. 23b) is quite variable at 0–10 wt% TiO_2, with olivine lamproites typically having higher concentration levels (median 3.1 wt% TiO_2) as compared to kimberlites (median 1.7% TiO_2) and CROLs (median 1.1 % TiO_2). Kimberlites have exceptionally low Al_2O_3 contents (<1–4 wt% Al_2O_3); CROL's and olivine lamproites have similar, but higher concentration levels of ~1–9 wt% Al_2O_3 (Fig. 23c). The highest MgO contents are observed in kimberlites (20–40 wt% MgO, mean 30 wt% MgO); MgO is lower in CROLs and olivine lamproites (mean of 24 and 22 wt% MgO, respectively), but there is quite wide scatter (<10 to >40 wt% MgO) observed in the CROL dataset (Fig. 23d). Sodium contents for all three rock types are <2 wt% Na_2O, notwithstanding a few samples with higher concentrations, which are suggested to be alteration related; kimberlite and CROL are dominated by samples with <0.5 wt% Na_2O, with olivine lamproite exhibiting slightly higher concentrations (Fig. 23e). Many kimberlites have very low Na_2O concentrations, close to or below the instrumental detection limit, often attributed to leaching of Na during alteration, or partitioning of Na into coexisting fluid (e.g., Kamenetsky et al. 2007a,b, 2009; Giuliani et al. 2017; Abersteiner et al. 2017, 2019). Potassium concentrations are lowest in kimberlite (mean ~1 wt% K_2O) and higher in CROL and olivine lamproite, with means of

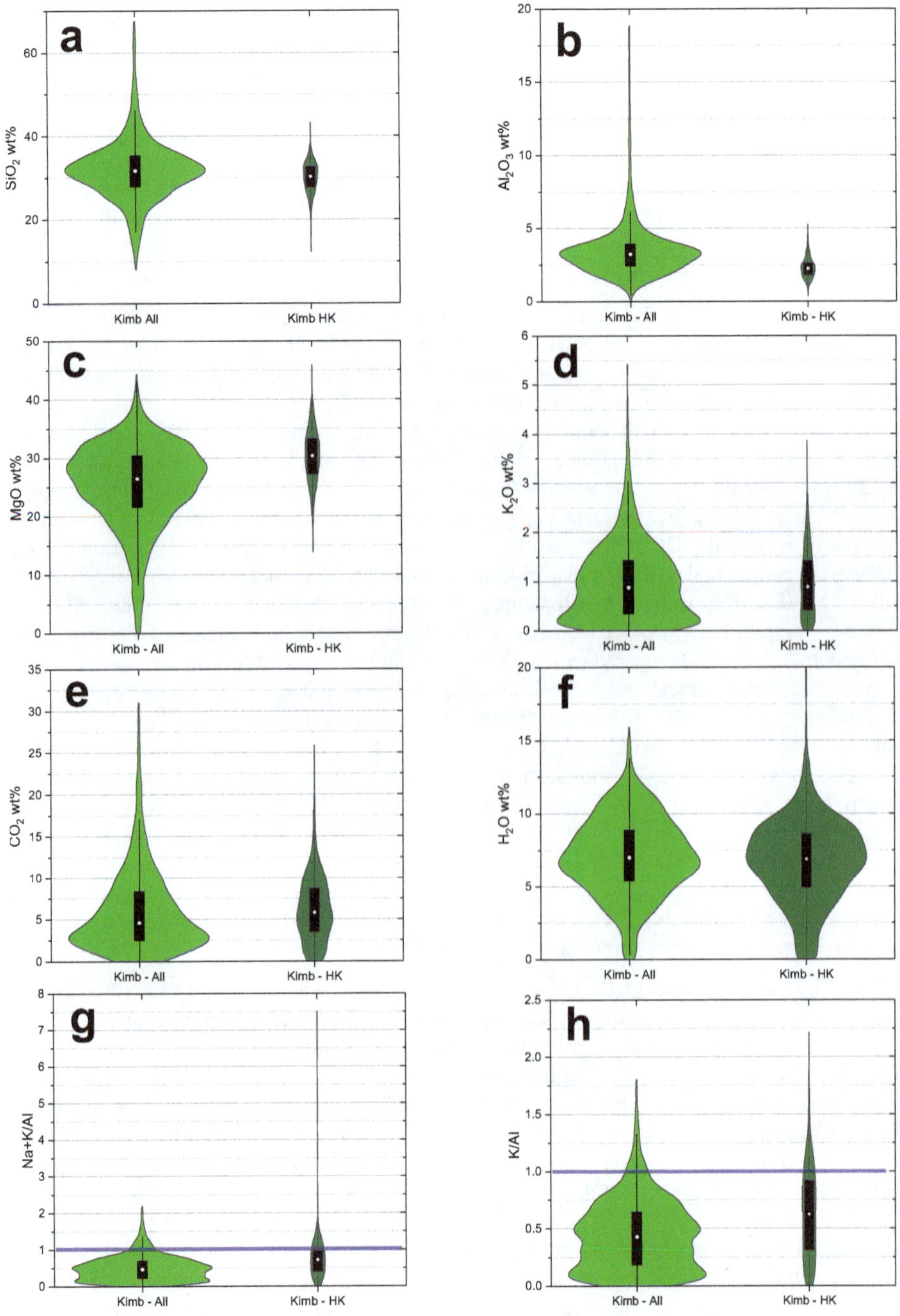

Figure 22. Kimb-All versus Kimb-HK violin plus box plots for (**a**) SiO_2, (**b**) Al_2O_3, (**c**) MgO, (**d**) K_2O, (**e**) CO_2, (**f**) H_2O (**g**) (Na+K)/Al, (**h**) K/Al. The **blue line in (g)** divide alkaline (>1) and subalkaline (<1) samples. The **blue line in (h)** demarcates samples with K/Al > 1 (perpotassic). See text for discussion. The box and line plot showing the median (**white circle**) and the interquartile range (50% of the data—the **box, in black**) with the lines representing the upper and lower quartiles of the data; this is combined with a "violin" plot, which is in essence a probability density function of the data. The sources of the data are in the on-line appendix A.

~2.8 and 3.8 wt% K_2O, respectively (Fig. 23f). Phosphorus concentration levels are similar in kimberlite and CROL (mean of ~0.7 and 0.8 wt% P_2O_5, respectively), but higher in olivine lamproite at ~1.5 wt% P_2O_5 (Fig. 23g).

Kimberlite and CROL have similar CO_2- and H_2O-rich compositions; in comparison, olivine lamproite is typically very CO_2-poor (or absent), with higher H_2O contents (Fig. 23h, i). Moreover, for CROL there is a statistically significant correlation between CaO and CO_2 concentrations (Pearson's $r = 0.87$), indicating that the CO_2 dominantly resides in carbonate (calcite), as is observed in kimberlite (Kjarsgaard et al. 2009). In contrast there is no statistical correlation between CaO and CO_2 for olivine lamproites.

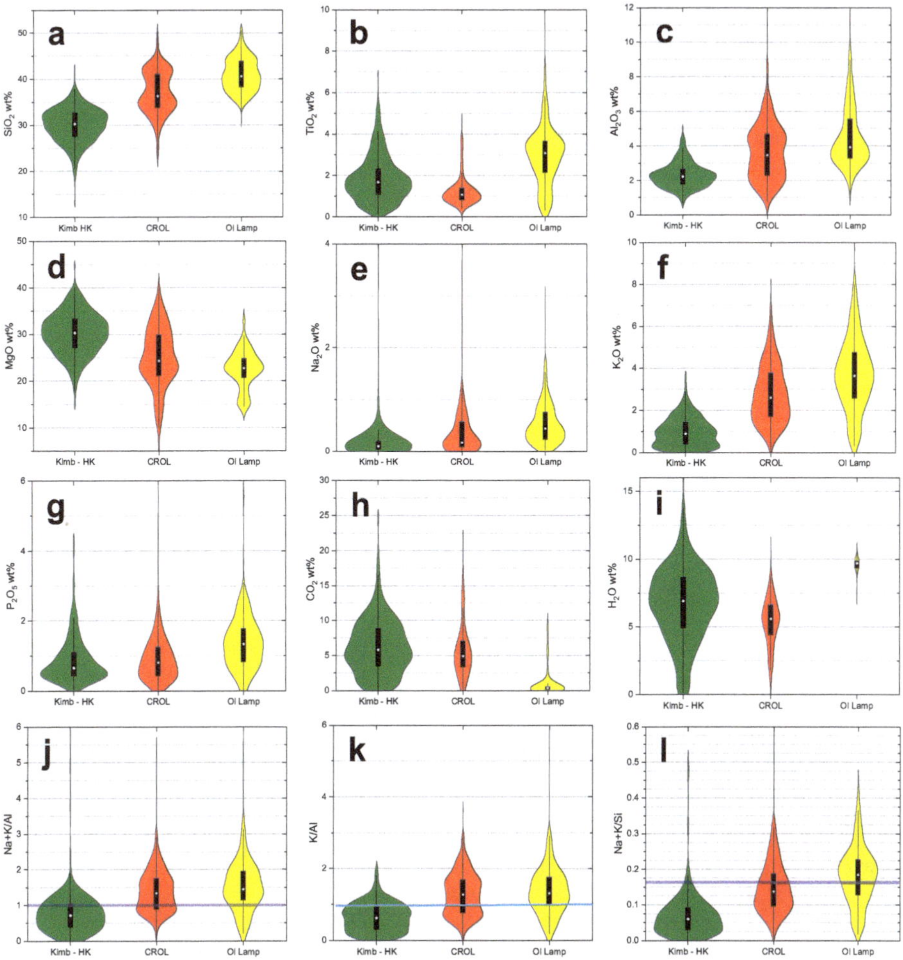

Figure 23. Kimb-HK, CROL, and olivine lamproite violin + box plots for (**a**) SiO_2, (**b**) TiO_2, (**c**) Al_2O_3, (**d**) MgO, (**e**) Na_2O, (**f**) K_2O, (**g**) P_2O_5, (**h**) CO_2, (**i**) H_2O (**j**) (Na+K)/Al, (**k**) K/Al, (**l**) (Na+K)/Si. The **blue lines in (j)** divide alkaline (>1) and subalkaline (<1) samples. The **blue lines in (k)** demarcates samples with K/Al > 1 (perpotassic). The **blue line in (l)** demarcates samples with (Na+K)/Si > 0.1666 that are considered alkaline on the basis of a silica deficiency with respect to alkalis. See text for further discussion. The box and line plot showing the median (**white circle**) and the interquartile range (50% of the data—the **box, in black**) with the lines representing the upper and lower quartiles of the data; this is combined with a "violin" plot, which is in essence a probability density function of the data. The sources of the data are in the on-line appendix A.

Applying the peralkalinity index—molar (Na+K)/Al—clearly illustrates that ~75% of the kimberlite sample suite have values <1, hence kimberlites are not alkaline rocks (Fig. 23j). Because kimberlites are not alkaline, they cannot thus be termed "potassic" even though K/Na is >1. Much of the variation in K/Na for kimberlites may be ascribed to alteration or Na loss to fluids (see references cited above). In contrast >75% of the CROL and olivine lamproites have (Na+K)/Al >1 and are thus alkaline (Fig. 23j), and also potassic, with K>Na. Molar K/Al data for these two rock types is very similar, with ~75% of the CROLs and olivine lamproites in the sample sets being perpotassic with K/Al >1 (Fig. 23k). An additional approach suggested by Mitchell (2020a) is to examine alkalinity by application of the "Shand Index" (molar Na+K:Al:Si = 1:1:6). Shand (1922) stated that "an alkaline rock, then, if names are to mean anything, should be one in which the alkalis are in excess of 1:1:6, either alumina or silica being deficient". Sorenson (1974, p. 6) further noted "this means that contents of alkali feldspar and/or mica do not entitle a rock to be termed alkaline". We have utilized the Shand Index to examine alkali/alumina (i.e., the peralkalinity index; Fig. 23j) and alkali/silica ratios as tests for kimberlite alkalinity and conclude that kimberlites are neither alumina or silica deficient relative to alkalis. This is clearly illustrated by Figure 23l, a plot of (Na+K)/Si, that demonstrates that kimberlites have values less than 1/6 (0.166) and thus cannot be considered alkaline rocks by this criterion. Mitchell (2020a) has stated that crystallization of phlogopite in kimberlite is an expression of their alkaline nature. However, this is manifestly not the case, as per Shand (1922), or Sorenson (1974).

Summary of the geochemistry of kimberlite, CROL and olivine lamproite. From our analysis of the major element composition of the three primary magma types that host diamond deposits, they have some common attributes such as being MgO- and volatile-rich, and ultrabasic (or basic, for a subset of olivine lamproites) in terms of silica contents. Kimberlites can clearly be distinguished from other diamondiferous rocks such as olivine lamproites and their carbonate-rich equivalents, CROL (often referred to as micaceous kimberlite, orangeites, or Group II kimberlite). In contrast to the vast majority of published literature, we re-iterate (Kjarsgaard et al. 2009) that kimberlites, the dominant primary magma type that form diamond mines, cannot be described as alkaline rocks, nor are they potassic. Kimberlites, as sampled at Earth's surface, are better described as silica-poor, ultrabasic rocks (mean MgO ~ 30.2 wt %), that are both CO_2- and H_2O- rich, with relatively low K, Na and Al contents.

PRIMARY DIAMOND MINES

Distribution of primary magmatic-hosted diamond mines

The global distribution of primary magmatic-hosted diamond mines is illustrated in Figure 24. There is a 1:1 correspondence between diamond mines and supercratons. These supercraton regions (Pearson et al. 2021) are "modern" (e.g., utilizing mantle seismic tomography information) outlines of the Earth's cratonic regions, for areas previously termed stable Precambrian cratons in the earlier literature (e.g., Dawson 1980; Janse 1984; Nixon 1995). The age of last major tectonomagmatic disturbance of cratons was originally suggested to be older than 1.5 Ga (Kennedy 1964; Clifford 1966), although Pearson et al. (2021) have revised this age to > ~1 Ga and extended the definition of the craton itself to include the presence of a deep (>150km) lithospheric mantle root. Supercratons are made of two or more composite cratons that are themselves compromised of several cratonic nuclei that are typically Archean in age.

A high proportion of kimberlite-hosted diamond mines lie within Archean cratonic nuclei. Examples include Victor and Renard (Superior Craton), all Slave Craton examples, Kelsey Lake (Wyoming Craton), Mwadui (Tanzanian Craton), Koidu (Man Craton), Mbuji Mayi and Tshibwe (Kasai Craton); all these examples were emplaced through Archean crustal rocks (Fig. 24). However, a number of other kimberlite-hosted diamond mines are associated with

Archean plus Paleoproterozoic crust (and associated mantle lithosphere of the same age span), such as is observed in the Kalahari (e.g., Orapa), Siberian (e.g., Mir, Internationalaya, Udachnaya), and Kola (e.g., Grib, Arkhangelskaya) composite cratons (Fig. 24).

Carbonate-rich olivine lamproites, of quite variable diamond contents, are observed in the Kalahari, Karelian, Indian, Rae-Laurentia and West Greenland cratonic regions (Fig. 2), but CROL-hosted diamond mines are only known from the Kaapvaal Archean nuclei within the Kalahari composite craton (Figs. 2, 25, 26; Table 2). Olivine lamproite-hosted diamond mines (e.g., Argyle, Ellendale 4 and 9, Majghawan), while within supercratons (Figs. 2, 24), all appear to reside in supercraton marginal positions, adjacent to Archean, or Paleo- or Mesoproterozoic terrains.

The majority of magmatic-hosted diamond mines are found in three key areas, the Kalahari (Figs. 24, 25, 26) and the Siberian composite cratons (Figs. 24, 27, 28), and in the Archean Slave Craton (Figs. 24, 29). Important, Tier 1 diamond mines are also found outside these regions, such as the Catoca kimberlite (Angola-South African supercraton) and the Argyle olivine lamproite (Australian supercraton), amongst other significant diamond mines shown on the global maps in Figures 1, 2 and 24. These Tier 1 diamond mines, plus additional key diamond mines and deposits are also shown on the detailed maps that accompany the following section on *Review of Global Secondary Diamond Deposits.*

The Kalahari composite craton provides useful insight to the distribution of the diamond mines with respect to its constituent Archean and Paleoproterozoic terranes (Figs. 25, 26). The Archean Limpopo terrane contains the Tier 1 Venetia mine, the Tier 4 Murowa mine, and the River Ranch deposit; all are kimberlite-hosted. The Paleoproterozoic Magondi fold belt, to the west of the Zimbabwe Archean nuclei, contains the Tier 1 Orapa mine, the Tier 2 Letlhakane mine, and the Tier 4 Karowe and Damtshaa mines. All of these mines are in a restricted geographic area collectively known as the Orapa kimberlite field (Field et al. 2008). The Lerala mine is also within the Magondi belt. Further west of the Magondi belt, the Okwa inlier contains the kimberlite-hosted Ghagoo (Gope-25) mine. The Kaapvaal Archean nucleus, within the Kalahari composite craton, hosts a plethora of kimberlite-hosted diamond mines (Table 2; Figs. 25, 26). These include the Tier 1 Jwaneng and Cullinan mines, the multiple Tier 3 mines that constitute the Kimberly cluster of mines (Kimberley, De Beers, Bultfontein, Dutoitspan, Wesselton), and also multiple Tier 4 (e.g., Koffiefontein, Jagersfontein, Letšeng, Palmietgat, The Oaks) and Tier 5 (e.g., Mothae, Frank Smith) mines. The Kaapvaal Archean nucleus also contains a number of other kimberlite-hosted historic mines and deposits e.g., Leicester, Kamfersdam, Belgravia and Otto's Kopje, for which data are limited or incomplete in terms of diamond grade, rough diamond value, and yearly carats mined.

Within the Archean Kaapvaal nucleus are also a number of important carbonate-rich olivine lamproite hosted diamond mines (Table 2; Fig. 26). These include the Tier 1 Finsch mine, the Tier 3 Marsfontein, Voorspoed and Lace/Crown mines, the Tier 4 Roberts Victor, Helam and Dokolwayo mines, and the Tier 5 Bellsbank fissure mines and Star (Theunissen area) mine. A significant number of additional CROL-hosted historic diamond mines, mainly beneficiated in the late 1800's to early 1900's e.g., Postmas, West End, Peiserton, West Kimberley, Paardeberg (Fig. 26), have no, limited or incomplete data for diamond grade, rough diamond value, and yearly carats mined. A history of mining of the significant diamond deposits in South African within the Kalahari Craton is provided by Wilson et al. (2007a), while Brook (2012) documents the history of the Botswana diamond mines.

The Siberian composite craton contains numerous kimberlites, including a very significant number of important kimberlite-hosted diamond mines (Table 2; Figs. 27, 28). The Daldyn kimberlite field contains the Tier 1 Udachnaya mine, the Tier 2 Zarnitsa mine, and the Dalnayay and Zagadochnaya advanced exploration projects (Table 2; Fig. 28a).

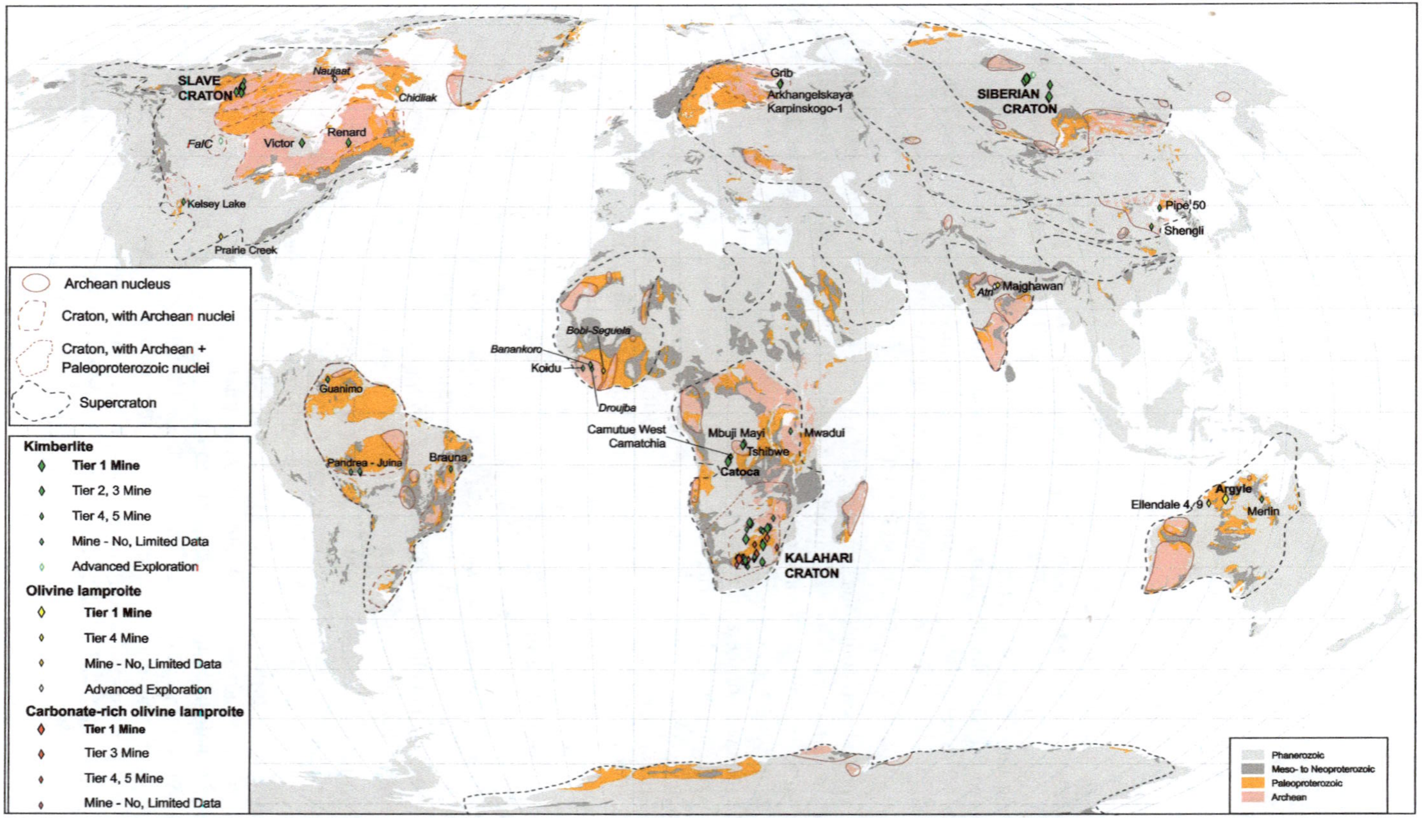

Figure 24. Global diamond mines in primary magmatic source rocks (kimberlite, olivine lamproite, and carbonate-rich olivine lamproite). Advanced exploration projects and former mines with no or limited data are also shown. Tier 1 diamond mines indicated by bold font. Detailed maps of the diamond mines and deposits of the Kalahari Craton, Siberian Craton (southern Yakutia area) and the Slave Craton follow (Figs. 25 to 29). Bedrock geology backdrop from Chorlton (2007); note the dominant bedrock geology is that at the surface i.e., Phanerozoic rocks that overlie Archean or Proterozoic rocks are shown on this map as Phanerozoic. Craton outlines from Pearson et al. (2021). Robinson map projection.

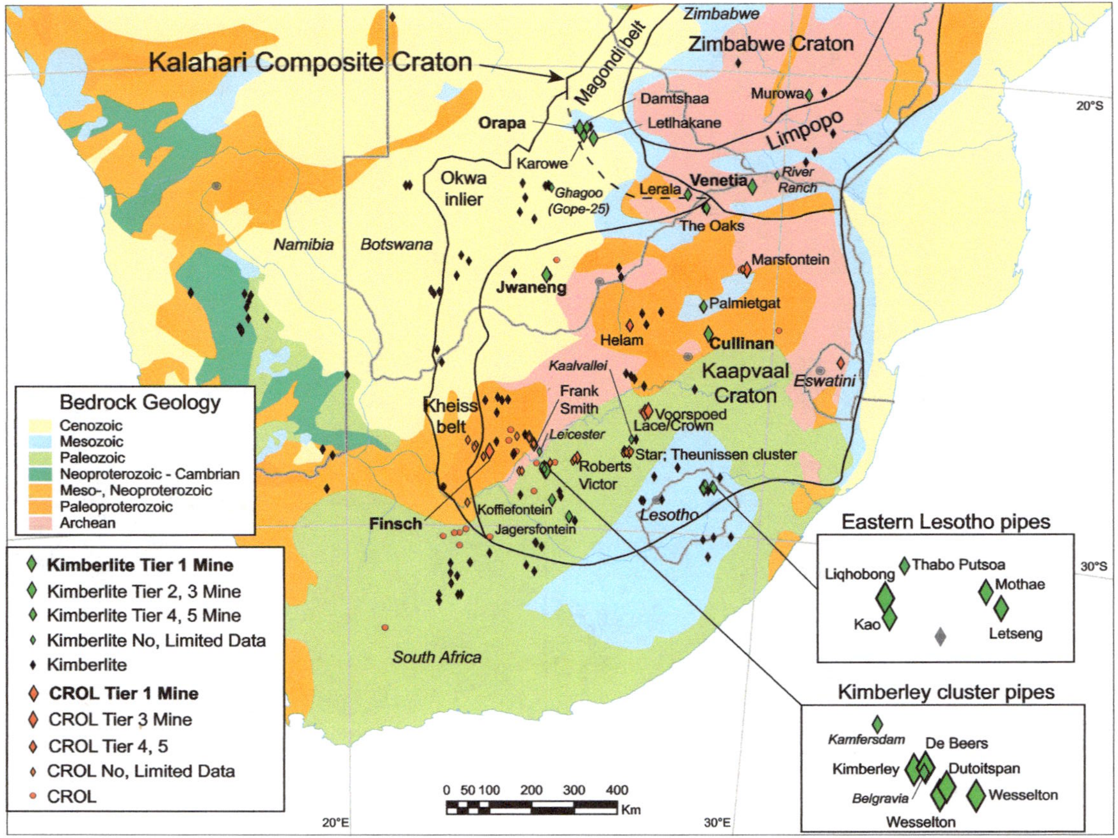

Figure 25. Spatial distribution of Kalahari Craton kimberlites and carbonate-rich olivine lamproites, and their primary magmatic-hosted diamond mines and deposits. Tier 1 diamond mines indicated by bold font. Mine data from Table 2. The subdivisions of the Kalahari composite craton (Archean nuclei, and Paleoproterozoic rocks) are modified after Griffin et al. (2003) and Field et al. (2008). Inset maps show kimberlites, and kimberlite-hosted diamond mines and deposits in the Kimberley and Lesotho kimberlite clusters. Mercator map projection.

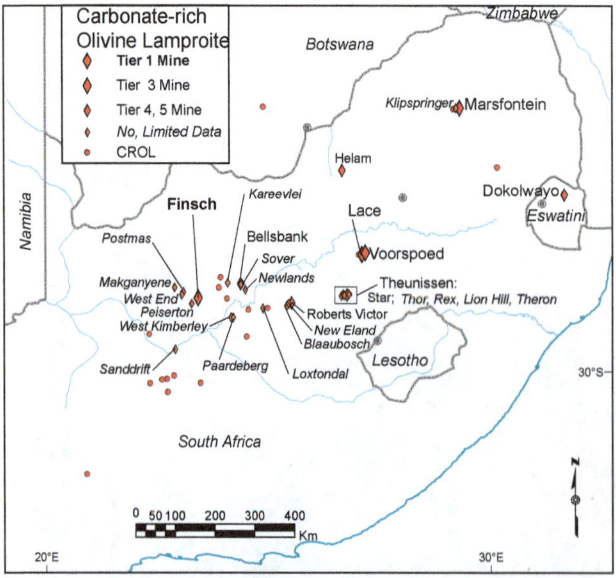

Figure 26. Spatial distribution of carbonate-rich olivine lamproites, and CROL-hosted diamond mines and deposits of the Kaapvaal Craton. Finsch CROL Tier 1 diamond mine indicated by bold font. See text for details. Mercator map projection.

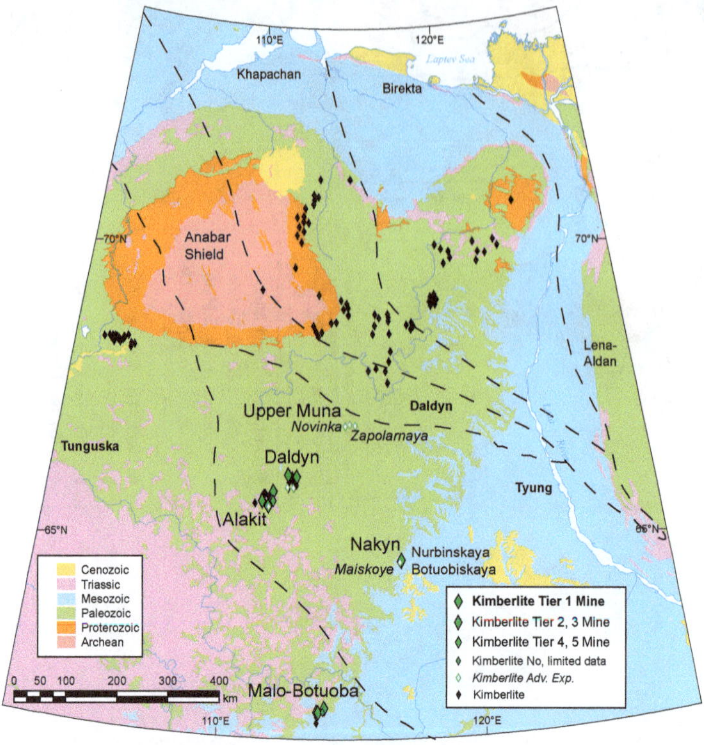

Figure 27. Spatial distribution of Siberian Craton (southern Yakutia) kimberlite-hosted diamond mines, deposits, advanced exploration projects and kimberlites. Tectonic subdivisions after Smelov and Zaitsev (2013). Note that the key kimberlite fields (Daldyn, Alakit-Markha, Nakyn) with diamond mines, as well as the Upper Muna field and its advanced exploration projects all lie in the Tyung domain; the Malo-Botuoba field lies in the Tunguska domain. Mercator map projection.

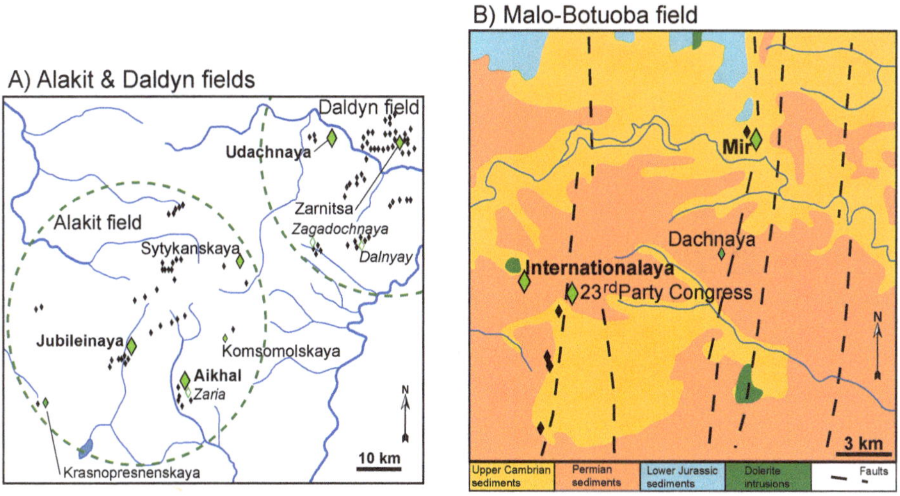

Figure 28. Detail maps of Siberian Craton kimberlites and kimberlite-hosted diamond mines, deposits and advanced exploration projects. Tier 1 diamond mines indicated by bold font. Mine data from Table 2. (**A**) Alakit-Markha and Daldyn kimberlite fields (modified after Ashchepkov ey al. 2015 and Kostrovitsky et al. 2015). (**B**) Malo-Botuoba field (modified after Agashev et al. 2018).

The Alakit-Markha kimberlite field contains the Tier 1 Aikhal and Jubileinaya mines, the Tier 3 Sytykanskaya mine, the Tier 4 Komsomolskaya mine, the Zaria advanced exploration project, and the Krasnopreskenskaya deposit (Table 2; Fig. 28a). The Malo-Botuoba kimberlite field contains the Tier 1 Mir and Internationalaya mines, the Tier 2 23rd Party Congress mine, and the Dachnaya deposit (Table 2; Fig. 28b). The Nakyn kimberlite field contains the Tier 2 Nurbinskaya mine, Tier 3 Botuobiskaya mine, and the Maiskoye advanced exploration project (Table 2; Fig. 27). The Upper Muna kimberlite field contains advanced exploration projects at Novinka and Zapolarnaya, plus at other kimberlites in this area. (Fig. 27).

The Slave Craton, one of the Archean nuclei within the Laurentian supercraton (Figs. 1, 24), contains numerous kimberlites and kimberlite-hosted diamond mines (Table 2; Fig. 29). The Lac de Gras kimberlite field in the central part of the Slave Craton contains two diamond mines, Ekati and Diavik, each comprised of multiple individual mines, as well as advanced exploration projects (Table 2; Fig. 29, inset map). The Diavik mine consists of four Tier 2 Mines (A154 North, A154 South, A418 and A21). The Ekati mine consists of the Tier 2 Koala, Panda, Sable, Fox and Misery mines, the Tier 3 Pigeon and Beartooth mines, and the Tier 4 Koala North and Lynx mines. Advanced exploration projects include the A841, Jay and Leslie kimberlites (Table 2; Fig. 29, inset map). The southeast Slave Craton contains the past producing Snap Lake Mine, and the three Tier 2 Mines (GK5034, Tuzo, Hearne) that constitute the Gahcho Kué mine (Table 2; Fig. 29). To the east of the Gahcho Kué mine area, there is advanced exploration at the Kelvin and Faraday kimberlites (Table 2; Fig. 29). In the north Slave Craton, the Jericho mine was short-lived (<5 years), and limited mine data are available (Fig. 29).

Economic characteristics of global diamond mines

Within a diamond mine, significant grade variations (typically greater than an order of magnitude) are observed between individual intrusive phases and/or extrusive phases and/or resedimented volcaniclastic phases (Clement 1982; Kjarsgaard 2007a). In a diamond mine, this can include kimberlite phase(s) which are of very low grade or very low rough diamond value (and thus very low ore value) and were not mined as ore, such as the "West End" kimberlite in the Kimberley (Big Hole) Mine (Wagner 1914). Furthermore, within an

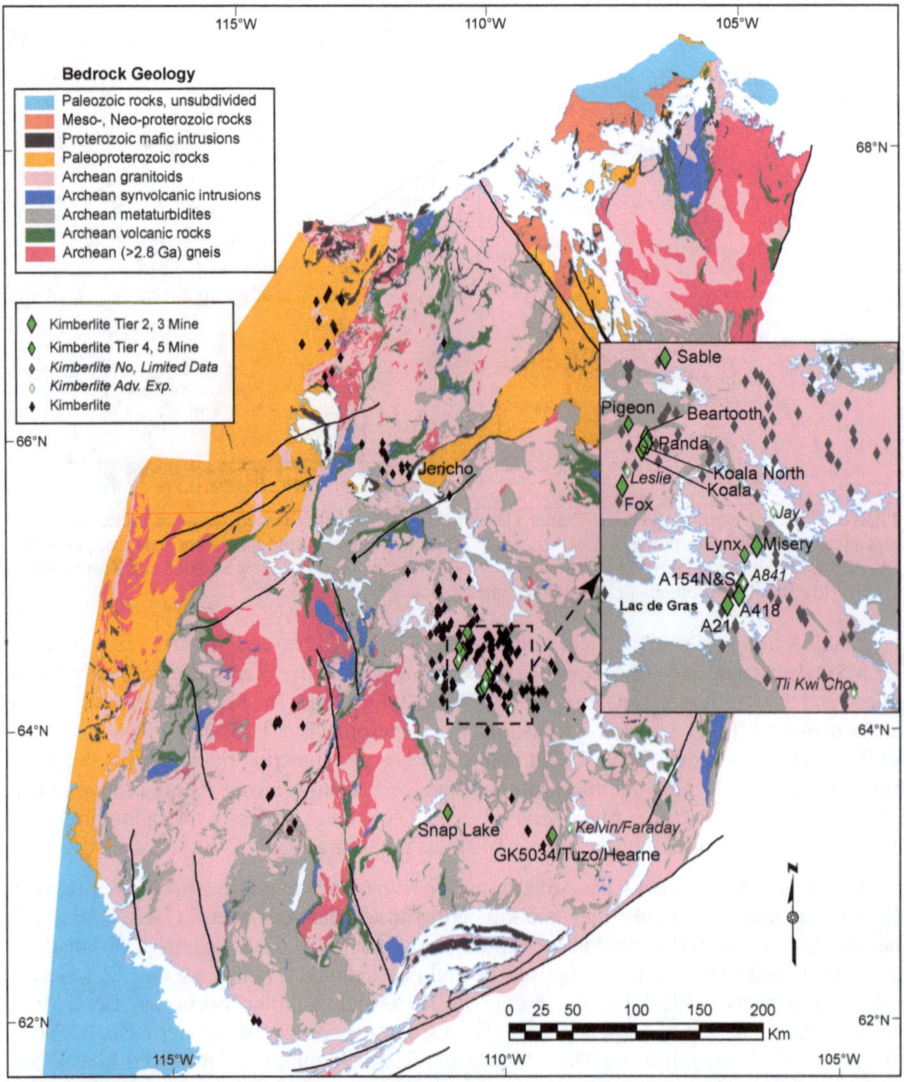

Figure 29. Spatial distribution of Slave Craton kimberlite-hosted diamond mines. Inset map shows mines in the Lac de Gras region of the central Slave Craton. Geology simplified and modified after Stubley and Irwin (2019). Inset map modified after Kjarsgaard et al. (2002).

individual kimberlite phase of a diamond mine, there can also be significant diamond grade variation of more than an order of magnitude. The De Beers kimberlite of the Kimberley cluster of pipes (Fig. 25, inset) provides an excellent example of grade variation between kimberlite phases, and grade variation within a single phase (Clement 1982; Clement et al. 1986; Kjarsgaard 2007a). Hence, the compiled diamond grade data as shown in Table 2 and in Figure 30 is based on interpreted average grades for mined-out deposits, or "head frame" grades for active mines. Also note that many diamond grades in Table 2 are inferred due to a lack of public domain data, or are based on data from exploration or feasibility studies, which may not be a true representation of the run of mine grade due to the smaller size of these preliminary samples. With these caveats in mind, diamond grade for economic diamond deposits typically ranges from ~0.025 to ~2 carats per tonne (carat/tonne; ct/t), equivalent to 0.005–0.4 ppm by mass (Fig. 30). Less commonly, global diamond mines have grades up to 5 ct/t and rarely (e.g., the exceptional Internationalaya mine) to 9 ct/t, equivalent to 1–2 ppm by mass (Fig. 30). Thus, there is more than two orders of magnitude in diamond grade variation for diamond mines globally, from 0.025 to 9 ct/t, as shown on the log histogram plot of diamond grade in Figure 30. This wide variation in diamond grade is observed for kimberlites, compare Letšeng (Lesotho) with the lowest diamond grade ever mined in a large scale mine, and the very high grade Internationalaya mine (Russia). For olivine lamproites compare the low-grade Ellendale 9 mine with the much higher-grade Argyle mine (both in Australia). For carbonate-rich olivine lamproites, compare the lower grade Voorspoed mine with the higher grade Helam (Swartruggens) mine (both in South Africa).

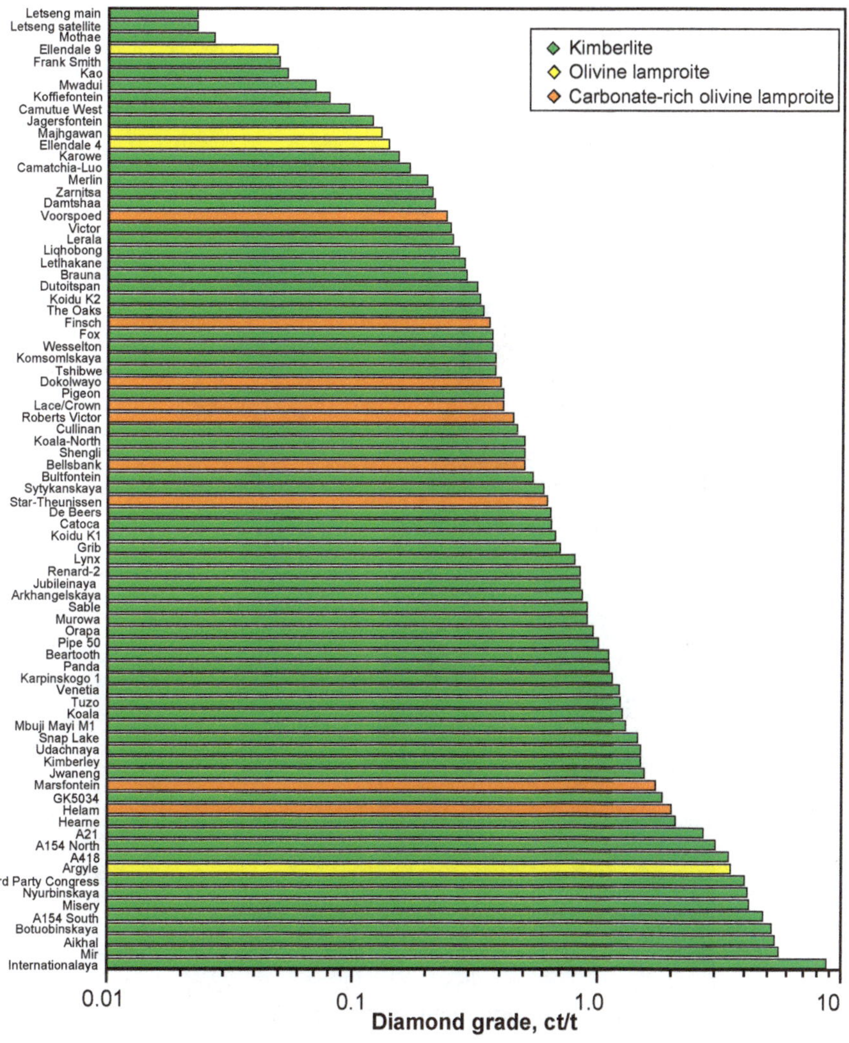

Figure 30. Histogram plot (log scale) illustrating the variation in diamond grade (in carats per tonne; ct/t) for global diamond mines (active and past-producers). Data from Table 2. All primary magmatic hosted diamond mines are color coded by rock type (Kimberlite, CROL, olivine lamproite).

In contrast to the vast majority of commodities (e.g., gold, iron ore, lithium), there is not a single "value" for rough diamonds. Hence, the value of an individual rough diamond can be exceptionally variable, depending upon size (the carat weight), color, and quality. Rough diamonds range from non-gem and near worthless diamonds (boart; <US$1/ct) to exceptional gem quality diamonds (>US$1,000,000/ct). Since a mine typically produces a quite wide range of rough diamonds with respect to value (boart to gem), the average rough diamond value, stated in US$/ct (Fig. 31), is an exceptionally important economic parameter. Note that the caveats previously stated regarding data in Table 2 for diamond grade are also applicable to average rough diamond value; for most active mines, average rough diamond value is considered proprietary data. In this regard, some of the data shown in Table 2 and Figure 31 are probably low, due to use of

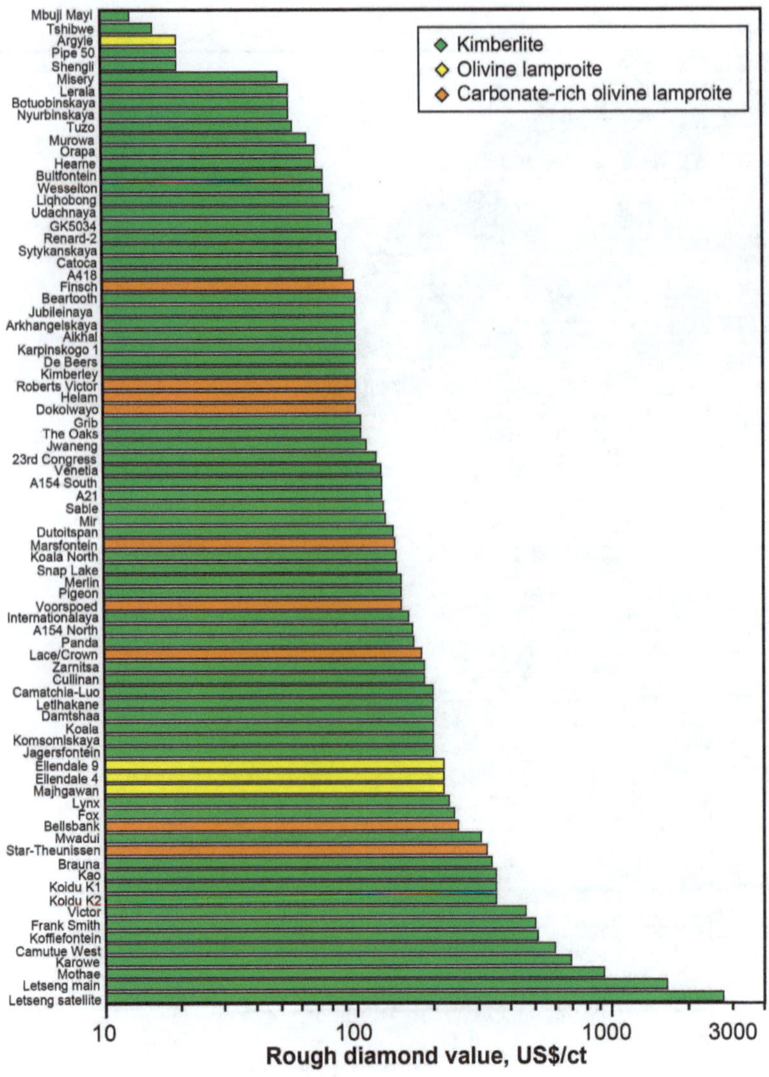

Figure 31. Histogram plot (log scale) illustrating the variation in rough diamond value (in US$ per carat; US$/ct) for global diamond mines (active and past-producers). Data from Table 2. All primary magmatic hosted diamond mines are color coded by rock type (Kimberlite, CROL, olivine lamproite).

exploration or feasibility sampling data. For example, the average value of rough diamonds from the Panda kimberlite (Lac de Gras, Slave Craton, Canada) from the mine feasibility study was US$130/carat, whereas the Panda run of mine rough diamond value increased ~30% to US$168/carat (Kjarsgaard 2007a). This increase can be due to many factors, including lower stone breakage during mining as compared to reverse circulation drilling during exploration sampling, but perhaps more importantly mining typically recovers larger diamonds which can strongly influence average rough diamond values. The more efficient sampling of larger diamonds during mining is due to a statistical under sampling of the larger sieve classes of diamond in exploration or feasibility sampling, even when 3,000 to 10,000 carat parcels of diamond are recovered for evaluation purposes. An additional highly significant problem in comparing average rough diamond values from global mines is the fluctuation in book price over time (years, decades, centuries), due to supply and demand, and general global economic conditions.

The known range for average diamond value from current and past producing mines is quite variable (>2 orders of magnitude), but typically ranges from ~US$13/ct to ~US$300/ct, with exceptional mines having average rough diamond values of ~~US$450/ct to ~US$2,750/ct (Table 2 and in the log histogram plot of rough diamond value in Fig. 31). This wide variation in average rough diamond value for global diamond mines is observed for kimberlites, e.g., compare low value diamonds from Mbuji Mayi or Tshibwe (both in the DRC) and high value diamonds from Karowe (Botswana) or Letšeng (Lesotho) as seen in Figure 31. The variation in average rough diamond value is also observed for olivine lamproites; compare Argyle (lower value diamonds) and Ellendale 4 or 9 (higher value diamonds) mines in Australia Fig. 31). For carbonate-rich olivine lamproites, compare lower value Finsch and higher value Star (Theunissen) average rough diamond values from these two South Africa mines (Fig. 31).

Due to the wide range of diamond grades and rough diamond values observed in individual diamond mines globally, the ore value (Table 2), expressed as US$/t (= grade [ct/t] X rough diamond value [US$/ct]) is a fundamental parameter utilized in assessing the economic potential of a diamond deposit. Global diamond mines have ore values that range from ~US$6/t to ~US$1400/t (log histogram plot, Fig. 32). This wide variation in ore value for global diamond mines is observed for kimberlites, compare the low ore value at Tshibwe (the DRC) with the very high value ore from Internationalaya (Russia). For olivine lamproites, compare lower value Ellendale 9 ore versus higher value ore from Argyle (both in Australia). For carbonate-rich olivine lamproites, compare lower value ore at Finsch with higher value ore at Marsfontein (both in South Africa).

The relationship between diamond grade and rough diamond value is further illustrated on a bivariate log–log plot in Figure 33, which also shows iso-ore value lines. For comparison, the Mbuji Mayi (DRC) and Kao (Lesotho) kimberlites have similar ore values (~17–19 US$/t), but Mbuji Mayi has high diamond grade with low rough diamond value, whereas Kao has low diamond grade, with high rough diamond value (Fig. 33). This general observation regarding the relationship between grade, diamond value and ore value is also applicable to olivine lamproite and carbonate-rich olivine lamproite diamond mines and deposits (Fig. 33).

KIMBERLITE AND LAMPROITE GEOCHRONOLOGY AND THE TEMPORAL DISTRIBUTION OF PRIMARY DIAMOND DEPOSITS

Primary diamond deposits are hosted by kimberlite and a subset of olivine lamproites i.e., olivine lamproite and carbonate-rich olivine lamproite (CROL). Economic diamond-bearing pipes and/or intrusions occur in numerous countries (~15) and typically are found in cratons areas where a relatively thick (>150-km-thick) subcontinental lithospheric mantle is preserved. Globally, the majority of kimberlites and lamproites were emplaced in the last ~25% of Earth history, with 98% younger than 1.2 Ga (Heaman et al. 2019; this study). There is great interest in understanding whether there is any geodynamic control on enhanced periods of kimberlite and lamproite emplacement and whether temporal windows exist for the formation of these primary magmatic diamond deposits. Recent overviews on the nature of primary diamond deposits in southern Africa (Field et al. 2008; de Wit et al. 2016) showed that kimberlites of economic interest were emplaced over a span of ~1 billion years, with major deposits occurring at approximately 1150 Ma (e.g., Cullinan), 520 Ma (e.g., Venetia), 235 Ma (e.g.,

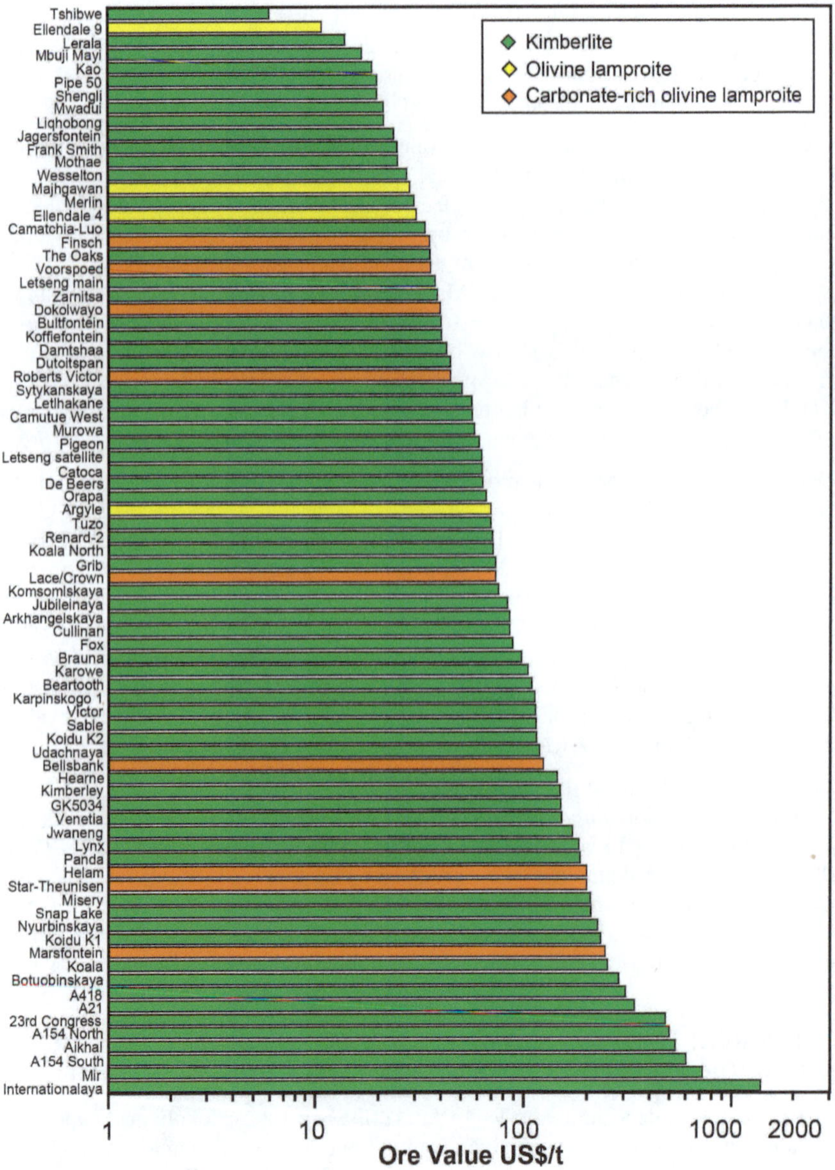

Figure 32. Histogram plot (log scale) illustrating the variation in ore value (in US$ per tonne; US$/t) for global diamond mines (active and past-producers). Data from Table 2. All primary magmatic hosted diamond mines are color coded by rock type (Kimberlite, CROL, olivine lamproite).

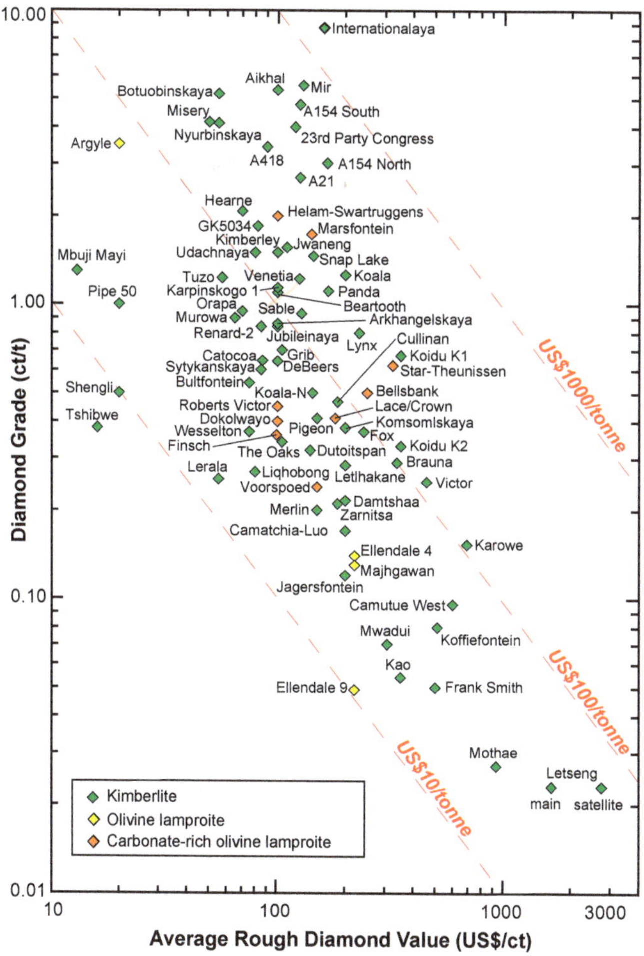

Figure 33. Bivariate log–log plot of diamond grade (ct/t) versus rough diamond value (US$/ct) for global diamond mines (active and past-producers). Data from Table 2. All primary magmatic hosted diamond mines are color coded by rock type (Kimberlite, CROL, olivine lamproite). **Red dashed lines** are iso-ore value lines (US$/t). Note that diamond mines with quite disparate diamond grades and rough diamond values can have very similar ore values; see text for further discussion.

Jwaneng), 95–85 Ma (e.g., Orapa, Kimberley, Letšeng), with CROL of economic interest in the time interval 200–110 Ma (e.g., Dokolwayo, Helam, Voorspoed, Finsch).

In this section we specifically address 1) the importance of applying a higher fidelity screened kimberlite geochronology database, 2) the global age inventory and emplacement history of kimberlites and lamproites, 3) the geochronology of diamond mines and deposits in kimberlite, CROL and olivine lamproite, and 4) the existence of temporal windows that contain a greater proportion of the key economic deposits that have been mined for diamonds. We refer to the "Tier" system of diamond deposits established above (Table 2) that classified the economic value of the various diamond mines that appear as examples in this section.

Veracity of kimberlite and lamproite geochronology

Several kimberlite geochronology databases have been published. Some focus on regions within continents or countries, usually where the majority of known diamond deposits occur e.g., Russia (Smelov and Zaitsev 2013), southern Africa (Jelsma et al. 2004, 2009; Moore et al. 2008; Griffin et al. 2014), and North America (Heaman et al. 2003, 2004). Very few studies have investigated the global pattern of kimberlite emplacement (Heaman et al. 2003, 2019; Tappe et al. 2018a), or the timing of primary magmatic diamond deposit formation. Establishing a robust geochronology database for kimberlites, olivine lamproites and CROLs is challenging because by their nature they are heterogeneous rocks that consist of mixtures of minerals that crystallized from the magma and a plethora of entrained, xenolithic/xenocrystic mantle and crustal material (e.g., Giuliani and Pearson 2019). In some cases, numerous geochronology attempts, by several dating techniques, produced a large range of radiometric dates for a single kimberlite pipe. For example, the Udachnaya East kimberlite has 47 published radiometric dates by four different dating methods yielding a range of dates between 756–326 Ma (most are summarized in Smelov and Zaitsev 2013). In contrast, our approach is different; we use a single Udachnaya East pipe emplacement age of 363.2 ± 1.8 Ma (MSWD = 2.7) based on the agreement of two separate U–Pb perovskite age determinations (Heaman and Mitchell 1995; Kinny et al. 1997). It is certainly possible that there are multiple kimberlite emplacement events within a single pipe that are sufficiently distinct to be resolvable at the precision of current dating methods, as demonstrated for the Renard 2 kimberlite in Quebec in which kimberlite magmatism at this pipe spans ~20 m.y. (Ranger et al. 2018). However, often the large age range reported for a single kimberlite body reflects inclusion of spurious dates that don't accurately record the time of kimberlite emplacement. Spurious dates (too old and too young) can be generated by all geochronology methods and can result from analysing entrained materials, such as crustal or mantle xenocrystic mica or zircon, from a disturbance affecting the dating system that can cause loss or gain of radionuclides (e.g., metamorphism, metasomatism, alteration, weathering), and challenges specific to the analytical methods used, such as incomplete fission track etching (Haggerty et al. 1983) or analyzing mixed mineral fractions. Other limitations with many existing kimberlite geochronology databases, such as 1) many of the dated samples in these databases are not kimberlite—they include dates from "related" and potentially "unrelated" rocks, such as carbonatites, kamafugites, picrites, aillikites, alnöites, melilitites, and various types of lamprophyres, 2) some reported kimberlite dates are inferred, they are not direct determinations but are estimated on the basis of the age of nearby kimberlites that are assumed to be contemporaneous, and 3) the isotopic data for many reported dates are unpublished so the veracity of these cannot be evaluated. We caution against the use of inferred dates because they can be misleading, as many fields contain kimberlites in close proximity that are emplaced over enormous time spans. For example, the central Slave Craton Lac de Gras field (Canada) contains kimberlites erupted over a time span of ~280 m.y. (Sarkar et al. 2015), and in the Siberian craton the Alakit-Markha and Daldyn fields contain kimberlites erupted over 150 m.y. and 200 m.y., respectively.

The kimberlite and lamproite emplacement dates used here are based on a screened geochronology database (Heaman et al. in review) The source references of these dates are listed in on-line Appendix B that contains one date per kimberlite or lamproite, unless there is strong evidence that there are multiple magmatic events with distinct and different age. Where possible, this single date is based on the weighted mean calculation of multiple dating attempts where the determined ages are in agreement within analytical uncertainty. This screened geochronology database of kimberlites and lamproites does not include rock types classified as "related rocks" in previous compilations. The biggest effect by excluding "related rocks" is for Greenland, with 57 emplacement ages that are published (mostly aillikite, ultramafic lamprophyre, and carbonatite) but only seven are kimberlite or lamproite.

Global patterns of kimberlite and lamproite emplacement

Kimberlites and lamproites occur as small intrusions and quite variable but often small volcanic pipes that are often difficult to identify in the field because they are easily weathered, eroded, and/or covered by other deposits. These rock types occur on all continents but their temporal and spatial distribution is uneven. Most known kimberlites pipes and dikes are observed within Archean cratons and it is not surprising that the highest concentration of known kimberlites occur in diamond mining districts where there has been extensive exploration efforts. In this respect, the current global distribution of discovered kimberlites and lamproites, and those for which emplacement dates are available, is biased towards regions with economic diamond deposits. Any kimberlite or lamproite geochronology database will have this inherent bias. Despite this, creating an accurate geochronology database is important for several reasons; it is vital for evaluating the spatio-temporal distribution of kimberlites or lamproites, predicting the location of new clusters/fields, establishing their geodynamic setting, and ultimately understanding their origins. Combining the timing of kimberlite or lamproite magmatism with their diamond tenor enables an evaluation of the possible links between the timing of economic diamond deposits and their geodynamic settings.

A compilation of more than 2350 dates for kimberlites and related rocks was recently summarized in Heaman et al. (2019). Similar to most other compilations, this geochronology database was not filtered, a variety of rock types in addition to kimberlite were included that occur co-spatially with kimberlite clusters/fields. Some kimberlites, such as Udachnaya East discussed above, are over-represented in unfiltered compilations because there is a large number of potentially spurious dates spanning a large age range despite no clear evidence of a protracted emplacement history spanning millions or tens of millions of years. Unfiltered compilations may blur details but can provide a general overview of kimberlite emplacement patterns; the most recent compilation identifies four broad periods of enhanced kimberlite magmatism globally in the last quarter of Earth history at 1,200–1,050 Ma, 600–480 Ma, 400–320 Ma, and 170–50 Ma (Heaman et al. 2019).

Here we have scrutinised the available geochronology data more thoroughly, omitting spurious dates and attempting to avoid over-emphasis on kimberlites with multiple age determinations. We specifically focus on the age distribution of kimberlites, and olivine lamproites and carbonate-rich olivine lamproites, the latter a smaller subset of lamproites (Pearson et al. 2019), because these are the two lamproite sub-types that host economic diamond deposits. Many CROLs reported in this database were previously classified as Group II kimberlites (e.g., Smith 1983). The distribution of dates by continent/country are as follows, with the number of dated occurrences in parentheses; Africa (14 countries; 234), Antarctica (7), Australia (51), Brazil (32), Canada (274), China (2), Finland (14), Greenland (7), India (33), Russia (336), Venezuela (1), and U.S.A. (35).

The total number of compiled dates after screening is 1026; kimberlite (n = 933; 90.9%) and lamproite (n = 93; 9.1%). Despite the uneven abundance of kimberlite and lamproite pipes or intrusions in this database, there is sufficient age data to evaluate whether there are enhanced periods of magmatism and whether there is any overlap in the timing of these magmatic events. We introduce the term epoch to identify globally significant periods of kimberlite magmatism. Epochs are defined here as >30 m.y. periods in Earth history where there is evidence for enhanced kimberlite magmatic activity on a global scale (i.e., in a specific epoch there are occurences in multiple cratons/countries). A kimberlite epoch includes a minimum of 25 known pipes or intrusions and is separated by a >15 m.y. hiatus, or a period of significantly reduced magmatism. We also identify significant temporal periods of kimberlite magmatism that have the potential to become epochs with future geochronology studies on existing, and newly discovered occurrences. For the 93 dated lamproites compiled for this study, we define 6 periods of enhanced lamproite magmatism that each has a minimum of 4 known pipes or intrusions.

Kimberlites. The age distribution of kimberlite magmatism globally is shown in histogram plots (Figs. 34, 35, 36a,b) at increasingly more detailed scale. Greater than 50% of the kimberlite dates are Mesozoic (252–66 Ma) or younger so we highlight their age distribution at the most detailed scale (Figs. 36a,b). A first-order observation from these screened compilation plots is that a larger number of discrete kimberlite epochs are identified compared to the unscreened compilations (e.g., Heaman et al. 2003, 2019; Jelsma et al. 2009; Tappe et al. 2018a). This higher age resolution database emphasizes the value of only reporting high fidelity age determinations (i.e., omitting spurious dates), not over-representing the number of dates reported for a single locality, and being selective about which lithologies are compiled. Although globally it appears that kimberlite magmatism is broadly continuous since ~700 Ma (no substantial emplacement hiatus during this time), eight discrete epochs of enhanced kimberlite magmatism can be discerned from ~1,200–30 Ma, based on frequency distribution plots (Figs. 34, 35). This is a significant increase from the four broad periods of kimberlite/lamproite magmatism based on unfiltered age compilations (e.g., Tappe et al. 2018a; Heaman et al. 2019). These kimberlite epochs with the number of pipes or intrusions per epoch (denoted in parentheses) are:

1) 1,155–1,075 Ma ($n = 26$)
2) 650–585 Ma ($n = 29$)
3) 562–505 Ma ($n = 29$)
4) 440–408 Ma ($n = 39$)
5) 382–340 Ma ($n = 112$)
6) 250–214 Ma ($n = 69$)
7) 197–139 Ma ($n = 247$)
8) 109–45 Ma ($n = 254$)

The duration of most kimberlite epochs spans between 30–80 million years. The epochs generally encompass multiple kimberlite fields that individually have emplacement ages spanning <~20–30 m.y. (e.g., Heaman and Kjarsgaard 2000; Heaman et al. 2015). These eight kimberlite epochs capture ~85% of the dated kimberlites. Very young kimberlites with emplacement dates younger than 45 Ma ($n = 3$), include those in the Kundelungu field, DR Congo (29–33 Ma; Batumike et al. 2008) and the Igwisi Hills volcanoes (Tanzania), that preserve the youngest known pyroclastic rocks and lava flows, with ages of 12,100–6, 000 yr (Brown et al. 2012). There are eight dated kimberlites older than 1,155 Ma, including those in the 1.8–1.6 Ga Kuruman cluster, South Africa (Donnelly et al. 2012), and the 2,128 Ma Turkey

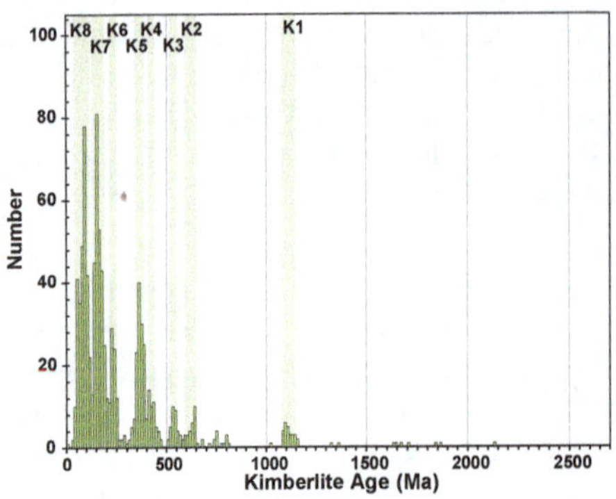

Figure 34. Distribution of kimberlite emplacement dates between 2,500–0 Ma ($n = 934$). The **green shaded bars** represent global epochs of kimberlite magmatism (K1–K8). Bin width is 10 Ma, 180 bins.

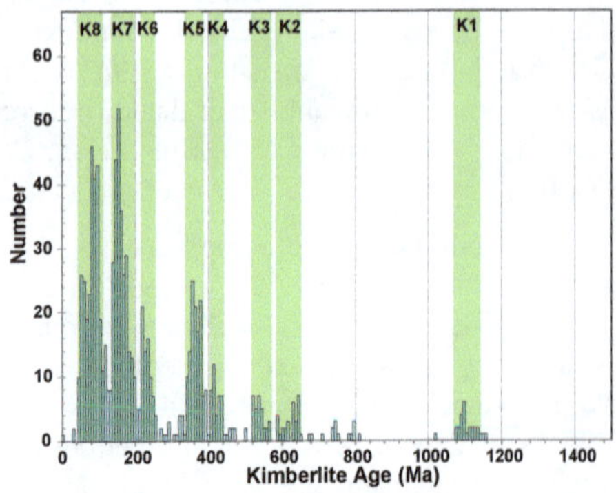

Figure 35. Distribution of kimberlite emplacement dates between 1,200–0 Ma ($n = 924$) highlighting the 8 identified global epochs (K1–K8) of kimberlite magmatism (**green shaded bars**). Bin width is 7 Ma (larger than the typical 95% confidence limit uncertainty on most kimberlite dates), 180 bins.

Well intrusion, Australia (Jourdan et al. 2012), the latter we consider to be the oldest known kimberlite reliably identified and dated. Thus far, only one kimberlite with sufficient deposit value to be considered economic has been identified that is older than 1.2 Ga, the 1,364 Ma Lerala (Martin's Drift) mine in Botswana.

More than half the kimberlites in the database are Mesozoic or younger and we show a detailed evaluation of the age distribution for these occurrences in Figures 36a,b. Two kimberlite epochs (7 and 8) are currently recognized in this most prolific time of kimberlite magmatism (Fig. 36a). In Figure 36b we show that there is potentially more kimberlite emplacement age structure during the Mesozoic, the light green shaded fields highlighting enhanced sub-periods of magmatism. Epoch 7 (197–139 Ma) can be further subdivided into two significant pulses, at 180–168 Ma and 160–139 Ma. Epoch 8 (109–45 Ma) can be further subdivided into 4 significant pulses at 103–77 Ma, 75–67 Ma, 65–51 Ma and 48–44 Ma (Fig. 36b). A substantial reduction of kimberlite magmatism occurred globally between ~210–180 Ma and ~140–100

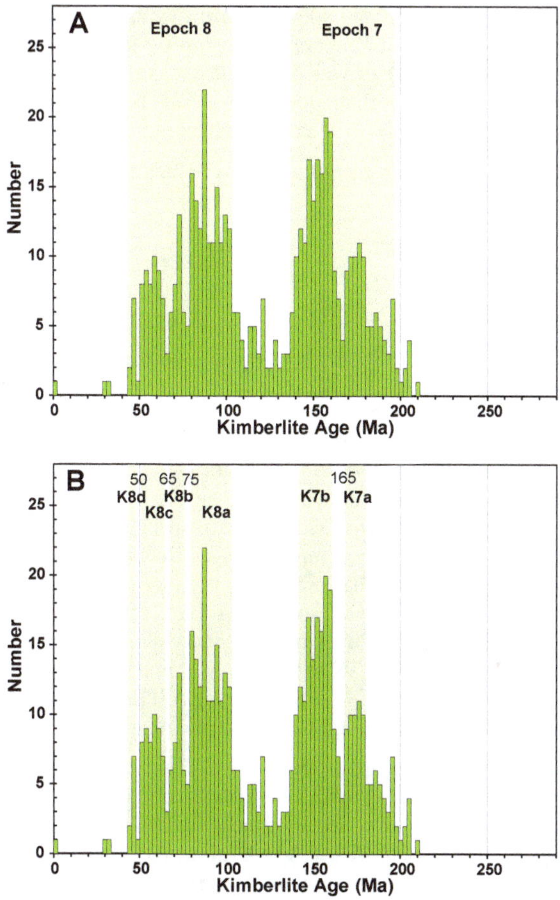

Figure 36. (**A**) Age histogram for Mesozoic and Cenozoic (<210 Ma) kimberlites (n = 553). More than 50% of all dated kimberlites are represented. Shaded green fields (K7 and K8) denote the two youngest kimberlite epochs. Bin width is 2 Ma, 100 bins. (**B**) Detailed evaluation of kimberlite geochronology epochs K7 and K8, identifying up to 6 sub-periods of enhanced magmatism with potential small "time breaks" at ~165, ~125, ~75, ~65, and ~50 Ma. Bin width is 2 Ma, 100 bins.

Ma, in contrast to a greater frequency during earlier Jurassic (180—142 Ma), and subsequent Late Cretaceous (< 100 Ma) times (Figs. 36a,b). High-resolution kimberlite geochronology in the future will reveal whether epochs 7 and 8 can be further refined into additional epochs and whether there are true hiatuses in kimberlite magmatism at ~ 165, 75, 65, and 50 Ma, and lower frequency at 210–180 and 140–100 Ma (Fig. 36b).

Lamproites. Although less abundant than kimberlites, lamproites (dated localities $n = 93$) occur sporadically over the last quarter of Earth history. Unlike kimberlites, a high proportion (~45%) of all lamproite types on Earth are younger than 35 Ma. In this overview we have omitted, and do not discuss a significant number of dated non-diamondiferous lamproite occurrences, such as the leucite lamproites at the Leucite Hills in the U.S.A. ($n = 18$), or the Mediterranean lamproites in Europe ($n = 42$), the latter occurring in a region dominated by Neotectonics far from cratonic regions. Furthermore, in the compilation we have distinguished lamproites (L; $n = 31$), olivine lamproites (OL; $n = 14$) such as the past producing Tier 1 Argyle Mine, and carbonate-rich olivine lamproites (CROL; $n = 48$), such as the Tier 1 Finsch Mine. A high proportion of the lamproites (~65%) occur in three countries/regions, Australia (16), India (11), and southern Africa (33). The oldest lamproites are the Paleoproterozoic 1,719 Ma Por'ya Guba (Russia) and 1,682 Ma Yangare-02 (Australia) intrusions. The youngest is the ~56,000 yr Gaussberg (Antarctica) olivine lamproite. A small number of lamproites are temporally coincident with the kimberlite epochs described above, indicating that a variety of magmas of deep-seated origin can occur in each epoch.

There are six periods where four or more lamproite magmatic events occur (L1 to L6 in Figs. 37a,b):

1) 1,230–1,130 Ma ($n = 10$)
2) 1,080–1,040 Ma ($n = 6$)
3) 550–500 Ma ($n = 5$)
4) 235–200 Ma ($n = 4$)
5) 135–110 Ma ($n = 32$)
6) 22–17 Ma ($n = 14$)

The two youngest time periods of significantly enhanced lamproite magmatism, do not qualify as epochs based on the definition above for kimberlites. Furthermore, two of these periods (L3; 550–500 Ma, and L4; 235–200 Ma) are synchronous with kimberlite epochs K3 and K6, respectively, as described above. However, the two periods of Mesoproterozoic lamproite magmatism do not coincide with Mesoproterozoic kimberlite epoch K1 (Figs. 37a,b). A more detailed histogram of Mesozoic and younger lamproite magmatism is shown in Figure 38, colour coded by country, that highlights the fact that most periods of lamproite magmatism are specific to certain countries/cratons (i.e., most are not global in extent). The 135–110 Ma period (L5 in Fig. 37b) is further subdivided in Figure 38 into two discrete enhanced sub-periods between 135–132 Ma (L5a; $n = 9$; 2 in Russia, 6 CROL in South Africa) and 125–115 Ma (L5b; $n = 20$; all CROL in South Africa). The whole of the 135–110 Ma L5 lamproite period coincides with a lull in global kimberlite magmatism.

The older Mesoproterozoic magmatic period (~1,230–1,130 Ma) is the only lamproite activity that is truly global in extent, with occurrences documented in Australia, Finland, Greenland, India, and Russia (Fig. 39). Some of these lamproites are highly diamondiferous, such as the 1,126 Ma past producing Tier 1 Argyle mine, Australia. A temporally equivalent Mesoproterozoic cluster with CROLs and olivine lamproites occurs in the Karelian craton of Finland and Russia (O'Brien et al. 2007; Phillips et al. 2017; Dalton et al. 2020) and includes the 1,180 Ma Kostomuksha and 1,204 Ma Lentiira intrusions. The younger (1,080–1,040 Ma) Mesoproterozoic lamproite occurrences are all from India (Das et al. 2018). Some of these are diamondiferous, such as the 1,079 Ma Atri deposit and 1,072 Ma Majhgawan mine (Bundelkhand Craton), but have much lower diamond grades than at Argyle. A new finding

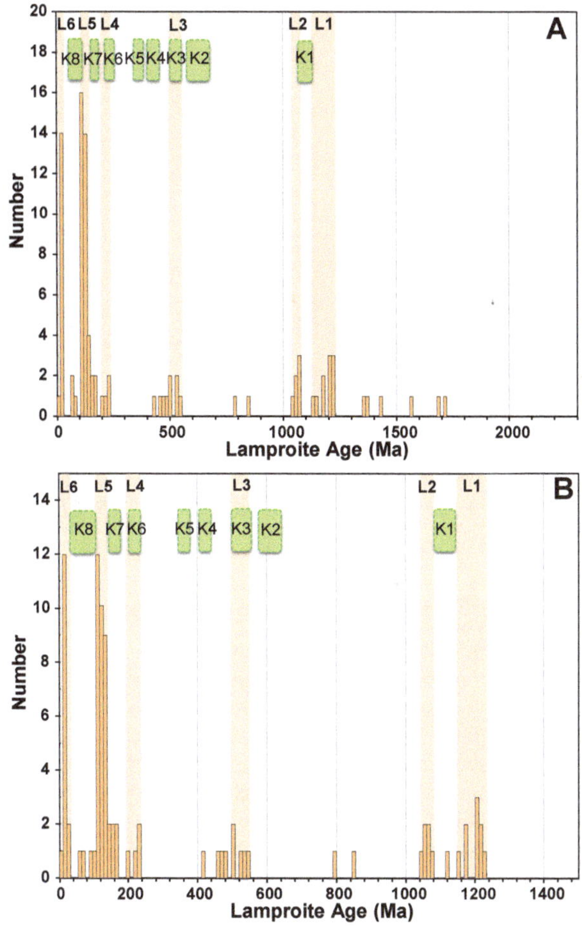

Figure 37. (**A**) Age histogram (2,300–0 Ma) for lamproites (n = 93). **Shaded orange fields** denote enhanced periods of magmatism. Kimberlite epochs are denoted by **green fields with green dashed outlines**. Bin width is 16 Ma. Note—lamproite age data does not include Mediterranean lamproites. (**B**) Age histogram (1,500–0 Ma) for lamproites (n = 87). **Shaded orange fields** denote enhanced periods of lamproite magmatism. Kimberlite epochs are denoted by **green fields with green dashed outlines**. Bin width is 10 Ma. Lamproite age data do not include Mediterranean lamproites.

from this compilation is the existence of coeval Cambrian kimberlite and lamproite magmatism (Fig. 40). The majority of known Cambrian lamproites occur in the newly discovered Aviat (CROL) and Qilalugaq-Naujaat clusters, northern Rae craton, Canada (Sarkar et al. 2018). One other Cambrian lamproite is reported from Priestly Peak, Antarctica (Black and James 1983).

The youngest period of lamproite magmatism is 22–17 Ma (n = 14) and is exclusive to NW Australia. It is curious that there are only three dated olivine lamproites and zero CROLs that are younger than 35 Ma; two of the former localities are in NW Australia, including the two Tier 4 past-producing diamond mines, the 22 Ma Ellendale 4 and 9 pipes. The majority (~65%; 60/93) of dated lamproites are younger than 250 Ma (Fi. 37b). Their most prolific period of emplacement, CROL dominated, was between 135–115 Ma (Fig. 38). Also noteworthy is the conspicuous absence of lamproites from some well-known diamond-producing regions, such as the Slave Craton in Canada.

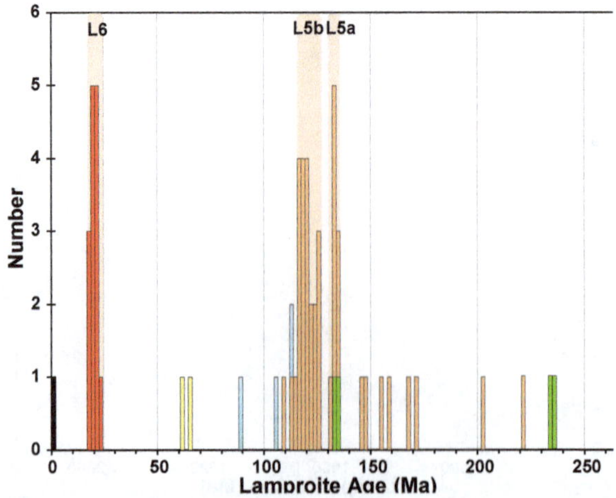

Figure 38. Age histogram of Mesozoic and younger lamproites ($n = 60$). Columns are colour-coded by country: **Gray**-Antarctica, **Red**-Australia, **Yellow**-India, **Blue**-USA, **Orange**-Southern Africa, **Green**-Russia. Enhanced periods of lamproite magmatism denoted by light orange shaded fields. Bin width is 2 Ma.

Summary of global kimberlite and lamproite age distributions

There are two distinct emplacement patterns that describe the timing and distribution of kimberlite and lamproite magmatism. For kimberlites this includes, 1) epochs that are widespread globally and occur in multiple countries/cratons, and 2) epochs that are largely restricted to a specific country or craton. Kimberlites from both of these emplacement patterns are known to contain economic diamond deposits. Using a much smaller geochronology dataset, Heaman et al. (2003) identified four periods of kimberlite magmatism that they considered to be global in extent because contemporaneous kimberlite magmatism was recognized in multiple countries; 60–45 Ma (Canada, Russia, Tanzania), 95–100 Ma (Canada, Russia, South Africa), 140–160 Ma (Canada, Russia, South Africa), and 215–240 Ma (Botswana, Canada, Russia). Based on limited geochronology data available at the time, they also proposed a number of additional potential "global periods" of magmatism, such as a Cambrian kimberlite event recognized in both the Slave and Kaapvaal cratons. Here we propose a slight revision to these conclusions since the few 60–45 Ma age kimberlites previously identified in Russia (e.g., Beta) are ultramafic lamprophyres and not kimberlites senso stricto. Heaman et al. (2003) also identified one period of late Devonian to early Carboniferous kimberlite magmatism (350–370 Ma) that was not global in extent and largely occurred in the Siberian (southern Yakutia) Craton, Russia. These periods of kimberlite magmatism identified previously have a slightly more restricted duration in the current screened compilation, but largely overlap kimberlite epochs 1–5 described here. Lamproite geochronology was not evaluated by Heaman et al. (2003).

The new screened, and hence more objective geochronology database supports the majority of these previous findings but also amplifies our understanding of the number and extent of kimberlite epochs. Ten discrete periods of magmatism are now recognized, eight kimberlite epochs and two periods of enhanced lamproite magmatism. It highlights that some periods of kimberlite magmatism are more prolific and widespread than previously thought. For example, the largely Jurassic epoch 7 kimberlite magmatism (197–139 Ma) was previously recognized in three countries (Canada, Russia and South Africa), but now consists of 248 kimberlites and also 6 lamproites, which have been identified from 12 countries (Angola, Australia, Botswana, Canada, Greenland, Guinea, Lesotho, Russia, Sierra Leone, South Africa, Tanzania, U.S.A.), but are relatively rare or absent in other countries (Brazil, China, and India). Another example of a global kimberlite, and now also a lamproite event, that is more evident from the new data compilation is the late Neoproterozoic to early Cambrian magmatic event (32 intrusions or pipes; Fig. 40) now recognized in 8 countries (Botswana, Canada, Finland, Greenland, Namibia, Russia, South Africa, Zimbabwe).

The geochronology of primary diamond deposit formation

The age of diamond formation in the mantle (Smit et al. 2022, this volume), and the emplacement age of primary magmatic rocks that host diamond deposits emplaced in the crust can differ by billions of years. Diamond formation ages are commonly ancient, as they can reside in the mantle for long periods of time before being entrained in younger kimberlite/lamproite intrusions or pipes. Mesoarchean, Neoarchean and Paleoproterozoic diamonds are known from the Kalahari Craton (Richardson 1986; Richardson et al. 1990, 1999, 2009; summary in Smit et al. 2022, this volume). Mesoarchean diamonds have also been reported from the Slave Craton, Canada. For example, a Re-Os dating study of peridotitic sulfide inclusions in diamond was interpreted to indicate 3.52 Ga growth of diamonds entrained in the Panda kimberlite (Westerlund et al. 2006), whereas the Panda kimberlite pipe that contains the diamonds was emplaced at 53 Ma (Creaser et al. 2004). The processes that form macrodiamonds in the mantle are quite distinct from the processes that form the kimberlite magmas. Diamond formation in the NW Slave craton (e.g., the 173 Ma past-producing Jericho mine) likely occurred during periods of mantle metasomatism related to east-dipping Paleoproterozoic subduction of oceanic crust and the local thermal/metasomatic influence of the 1.27 Ga Mackenzie mantle plume (Heaman et al. 2006; Heaman and Pearson 2010). In contrast, some diamond formation ages correspond to the emplacement age of the magmas that entrained the diamonds and brought them to the surface. Examples include some gem diamonds at Koffiefontein (Pearson et al. 1998), Jwaneng (Gress et al. 2021), Cullinan (Richardson 1986), and fibrous diamonds from a number of localities (Timmerman et al. 2019).

A total of 81 diamond mines (active and past-producers; Table 2), 10 advanced exploration projects and 10 former mines with essentially no diamond production data available, are examined. Of these, 84 are kimberlite and 17 are OL/CROL. Diamond mines and deposits occur in 18 countries; Angola (5), Australia (4), Botswana (8), Brazil (2), Canada (25), China (2), Democratic Republic of Congo (2), Eswatini (1), Guinea (2), India (2), Ivory Coast (1), Lesotho (5), Russia (15), Sierra Leone (2), South Africa (20), Tanzania (1), U.S.A. (2), and Zimbabwe (2). Twelve of these deposits are classified as Tier 1 based on their mined diamond value (to 2019/20) of >13 US$B; Aikhal, Argyle, Catoca, Cullinan, Finsch, Internationalnaya, Jubileinaya, Jwaneng, Mir, Orapa, Udachnaya, and Venetia. Tier 1 deposits occur in most kimberlite epochs except K2 (650–585 Ma) and K7 (197–135 Ma). Two Tier 1 deposits of the same age, 118 Ma Finsch (CROL) and 118 Ma Catoca (kimberlite), do not fall within a proposed kimberlite epoch but do coincide with an enhanced period of lamproite magmatism (L5 = 135–110 Ma), which is prolific in southern Africa. The kimberlites or lamproites for most of these diamonds deposits have accurate emplacement dates, however the age for 13 mines or deposits is unknown (or unpublished) and often inferred based on the age of kimberlites in their vicinity, highlighting the need for additional high-precision geochronology. For example, of the diamond mines listed in Table 2 from Angola, only the 118 Ma Tier 1 Catoca kimberlite has been dated.

The temporal pattern of global diamond deposit formation is similar to the global pattern of kimberlite/lamproite emplacement (Fig. 41). The oldest (1,432 Ma Bobi, Côte d'Ivoire) and youngest (22 Ma Ellendale, Australia) deposits are both hosted in olivine lamproite. There is at least one diamond mine that occurs in each of the eight kimberlite epochs (green-shaded fields in Fig. 41) and the two periods of enhanced lamproite activity that do not overlap any of the kimberlite epochs. The majority of diamond deposits mined so far (51/79; ~65%) are found in kimberlites or lamproites younger than 252 Ma; this is not surprising because the majority of known kimberlites and lamproites are Mesozoic in age.

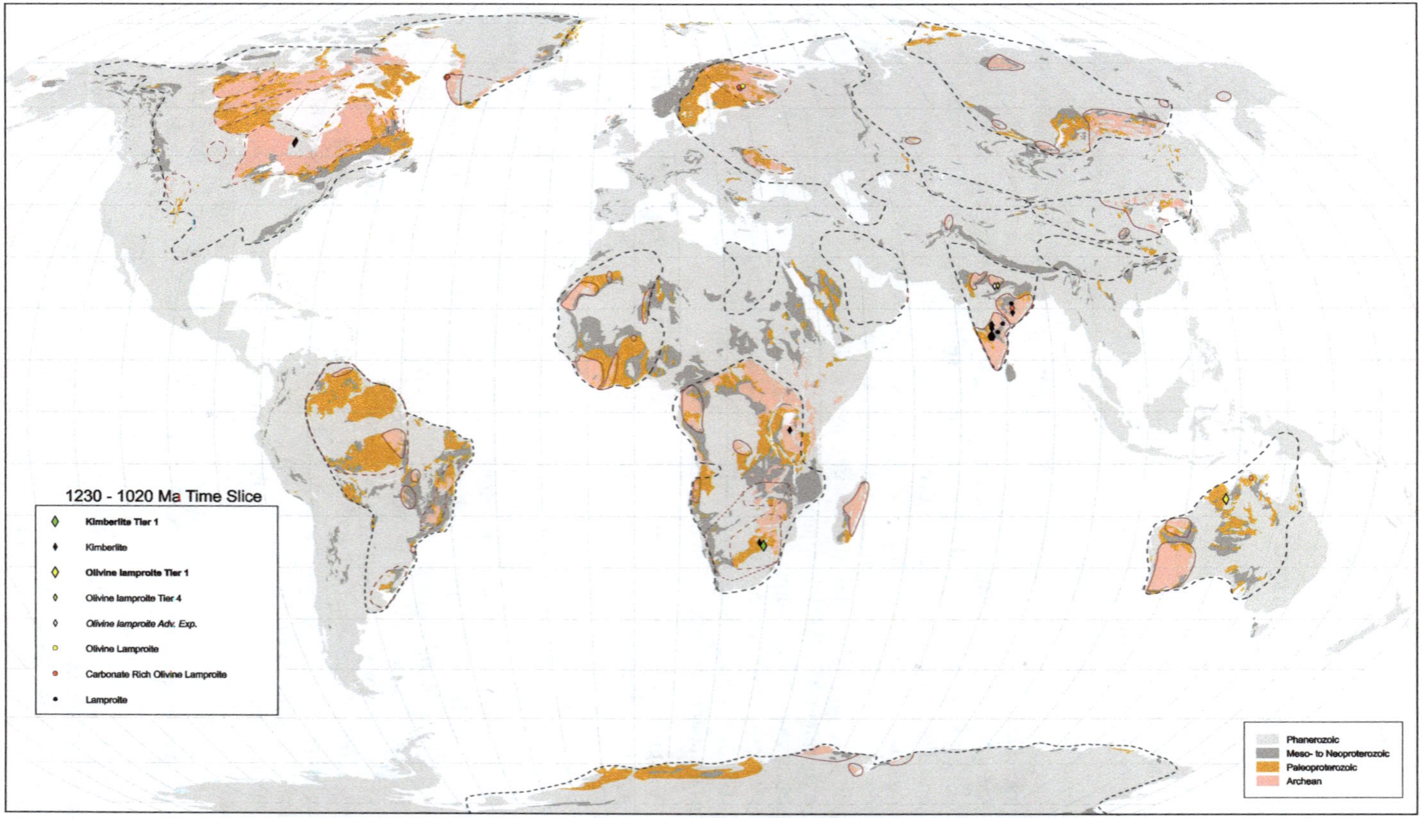

Figure 39. The global distribution of dated kimberlites, olivine lamproites and carbonate rich olivine lamproites during the time interval 1,230–1,020 Ma. **Red solid lines** outline areas of Archean dominated bedrock; **red dashed lines** outline areas of Archean–Paleoproterozoic cratonic regions; **black dashed lines** outline cratonic areas based on mantle geophysics and bedrock > 1 Ga (modified after Pearson et al. 2021). Robinson map projection.

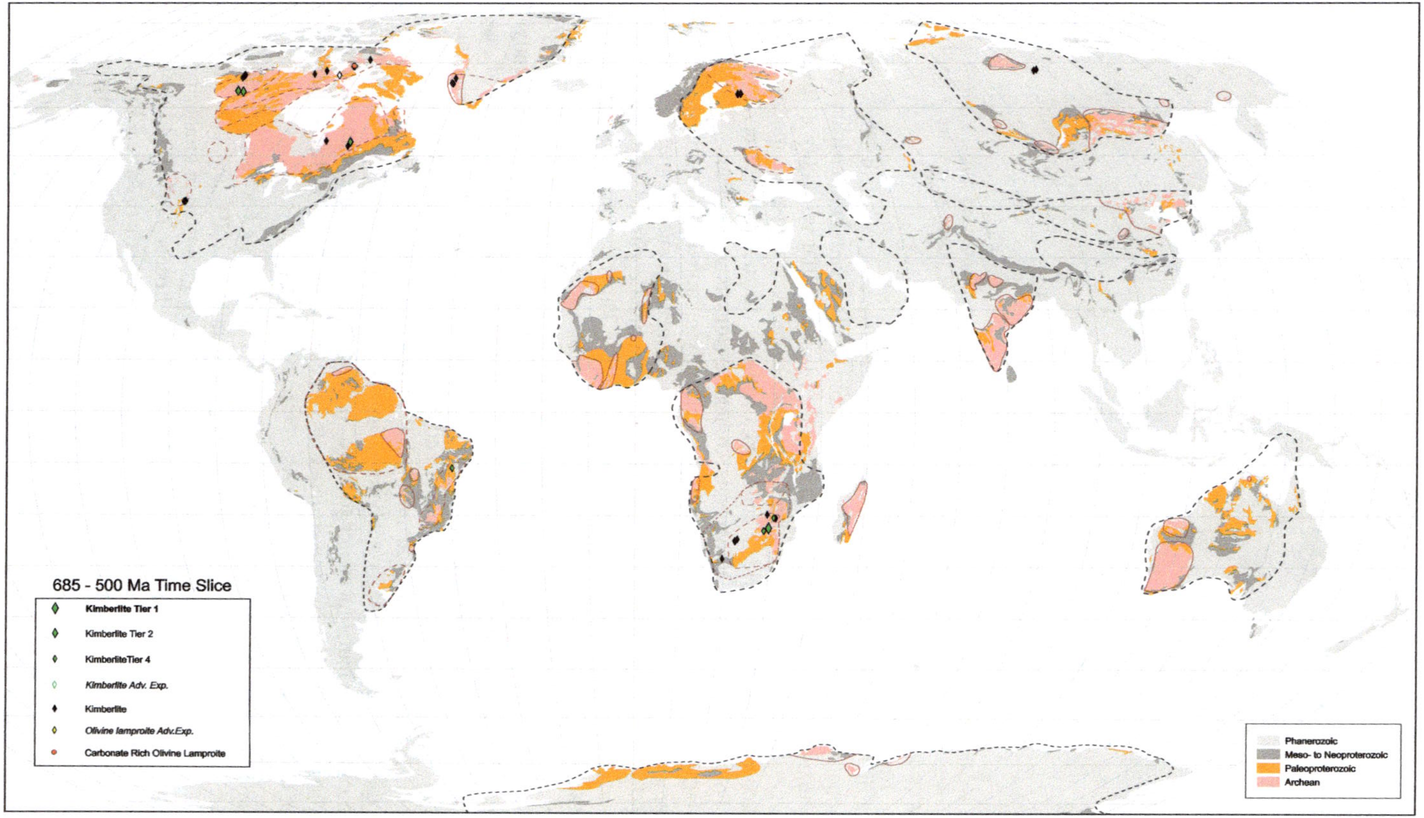

Figure 40. The global distribution of kimberlites, olivine lamproites and carbonate rich olivine lamproites during the time interval 685–500 Ma. **Red solid lines** outline areas of Archean dominated bedrock; **red dashed lines** outline areas of Archean–Paleoproterozoic cratonic regions; **black dashed lines** outline cratonic areas based on mantle geophysics and bedrock > 1 Ga (modified after Pearson et al. 2021). Robinson map projection.

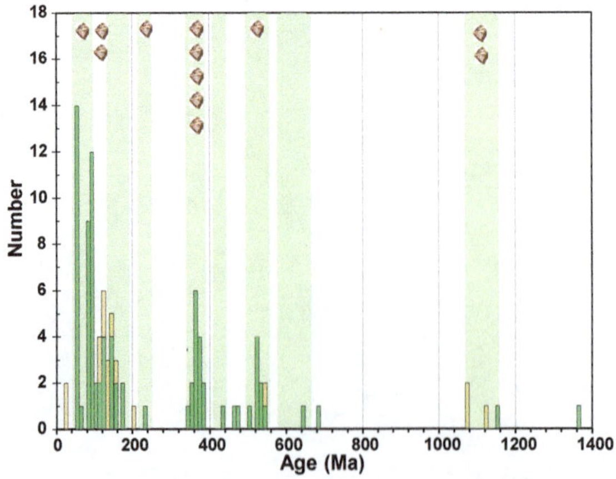

Figure 41. Age histogram (1,400–0 Ma) of worldwide primary diamond mines and deposits (**green** = kimberlite-hosted, **orange** = OL/CROL-hosted); $n = 98$. **Diamond symbols** denote the number of Tier 1 mines in each kimberlite epoch. Bin width is 10 Ma, 120 bins.

From a diamond exploration perspective, it is instructive to know if there are times when diamondiferous kimberlites are more prolific and whether certain kimberlite epochs are more economic than others. A possible reason for this potential variability is that the mantle sources for diamonds may change over time. Local or craton-scale modification of the subcontinental lithospheric mantle (SCLM) can impact the preservation of diamonds. For example, the SCLM beneath Archean cratons is a well-known source of diamonds, but parts of this lithosphere can be chemically modified and/or removed by a variety of processes (Pearson et al. 2021, and references therein), such as metasomatism, thermal erosion during the interaction with one or more mantle plumes (e.g., Liu et al. 2021), or lithosphere delamination. Hence kimberlite or lamproite magmas emplaced prior to lithosphere modification or removal could host economic diamond deposits. Conversely, post-disturbance kimberlite magmatism, such as beneath an area of thinned/hotter lithosphere, is an unlikely source of diamonds, unless the lithosphere subsequently cooled (and thickened) sufficiently to return back into the diamond stability field, as proposed for the Victor Mine, Attawapiskat field, Canada (Smit et al. 2014; Stachel et al. 2018). Griffin et al. (2014) showed that the Sr and Nd isotopic composition of Jurassic and Cretaceous kimberlite magmas erupted into the Kalahari Craton, South Africa become more homogeneous after about 100 Ma, possibly reflecting the removal of ancient metasomatized SCLM at this time, or a temporal switch to more "primitive" deeper magma sources at this time (Woodhead et al. 2019). Both interpretations can explain the paucity of post-100 Ma CROLs, and CROL-hosted diamond deposits in southern Africa.

In a previous study we evaluated whether there is a correlation between the timing of kimberlite emplacement and diamond potential (Heaman et al. 2003) and showed that the majority of economic diamond deposits in various cratons are hosted in kimberlites of different age; Devonian–Carboniferous kimberlites in Russia, Eocene kimberlites in North America, and Cretaceous, Permian and Cambrian kimberlites in southern Africa. Based on this it was concluded that diamond potential is not specifically related to global patterns of kimberlite emplacement. However, in the nearly 20 years since that study there have been numerous new diamondiferous kimberlite discoveries worldwide and more than double the number of kimberlite dates reported, allowing a re-evaluation of this tenet.

Figures 42a,b are plots that show the age of economic diamond deposits in relation to in-situ mined value in US billion dollars (to 2019/20; log scale), and their diamond grade (carat/

tonne), respectively. The deposits are separated into those in kimberlites (green circles) and those in OL/CROLs (orange circles). Also shown for comparison are the 8 epochs of global kimberlite magmatism (light green shaded fields) and 2 distinct enhanced periods of lamproite magmatism (light orange shaded fields) as identified in Figures 34 and 37. Diamond deposits occur in all main epochs of kimberlite, and enhanced periods of lamproite magmatism, but their temporal distribution and diamond grades vary significantly. A first order observation is that there is no correlation between in-situ value and age (Fig. 42a), with Tier 1 (highest value) deposits spanning the range of geological time examined. Using a 1 US$B cut-off, there are a similar number of higher and lower value diamond mines over a 1.2 b.y. time-frame. A second observation from these data is that, apart from Tier 1 Argyle (~18 US$B) and Tier 1 Finsch (~13 US$B), most OL/CROL-hosted deposits are typically lower in mined value (<1 US$B), due to their small size (tonnage), and/or lower diamond grade, and/or lower stone value (see Figs. 5, 30–33).

Two kimberlite epochs (109–45 Ma and 382–340 Ma) contain a substantial endowment (~65%) of known economic primary diamond deposits. As noted in previous studies (Heaman et al. 2003, 2019) some periods of diamondiferous kimberlite are specific to certain cratons. In the 109–45 Ma epoch, 93% (13/14) of the 65–45 Ma diamond mines occur in the Slave craton, Canada, with Mwadui, Tanzania being the exception. For the 110–70 Ma diamond mines, 79% (19/24) occur in the Kalahari Craton, southern Africa (Botswana, Eswatini, Lesotho, South Africa, Zimbabwe). In the 382–340 Ma epoch, all diamond mines are in Russia, with 87% (11/14) in southern Yakutia (Siberian Craton) and 3/14 in the Arkhangelsk area (Kola Craton). However, some diamond-prospective kimberlite epochs occur across multiple cratons; the Neoproterozoic to earliest Cambrian kimberlite-hosted diamond mines are known from both the Slave and Kalahari Cratons. In contrast, there are three kimberlite epochs where economic diamond deposits are scarce, and contain only a small number of diamond mines. There are two mines in the 650–585 Ma epoch, including the 643 Ma Tier 2 Renard 2 (Canada) and the 682 Ma Tier 4, Brauna (Brazil, which is slightly older than epoch K2); one mine in the 440–405 Ma epoch, the Tier 2, 23rd Party Congress (Russia), and; one mine in the 250–210 Ma epoch. However, the latter importantly includes the richest diamond mine worldwide, the 235 Ma Tier 1 Jwaneng kimberlite, Botswana with an in-situ value of 53 US$B.

The majority (10 of 12) of high-value Tier 1 mines (≥13 US$B) are hosted by kimberlite. The exceptions are the 1,126 Ma Tier 1 Argyle olivine lamproite in Australia and the 118 Ma Tier 1 Finsch CROL in South Africa. The number of Tier 1 mines in each kimberlite epoch (shown with diamond symbols in Fig. 41) is as follows; 1,155–1,075 Ma (2; Cullinan, and the Argyle olivine lamproite), 650–585 Ma (0), 562–505 Ma (1; Venetia), 440–408 Ma (0), 382–340 Ma (5), 250–214 Ma (1; Jwaneng), 197–139 Ma (0), and 109–45 Ma (1; Orapa). The high abundance of Tier 1 diamond mines in the 382–340 Ma epoch (i.e., Aikhal, Udachnaya, Jubileinaya, Mir and Internationalaya) is a spatial clustering, related entirely to the Devonian-Carboniferous kimberlites of southern Yakutia, Siberian Craton. As noted above, two Tier 1 mines occur in the 135–110 Ma period of enhanced lamproite magmatism, the 118 Ma Finsch CROL (13.0 US$B), South Africa, and, the 118 Ma Catoca kimberlite (14.7 US$B), Angola.

In Figure 42b there is also no obvious correlation between diamond mine emplacement age and diamond deposit grade (acknowledging that grade can vary widely in a single deposit). Most diamond mines have diamond grades between 0.1 and 1.0 ct/t. Higher grade (>1 ct/t) diamond mines occur in the Mesoproterozoic (1,126 Ma Argyle olivine lamproite, Australia), at Devonian–Carboniferous time in southern Yakutian kimberlites, through to Eocene (56–53 Ma Ekati and Diavik mine kimberlites, Slave Craton, Canada). Mesoproterozoic OL/CROL, of quite variable diamond grades occur in Australia (1,126 Ma Argyle) and India (1,079 Ma Atri and 1,072 Ma Majhagawan).

In Figure 42c we highlight the distribution of diamond deposits by country/region, Canada (red circles), Russia (orange circles), and southern Africa (Botswana, Eswatini,

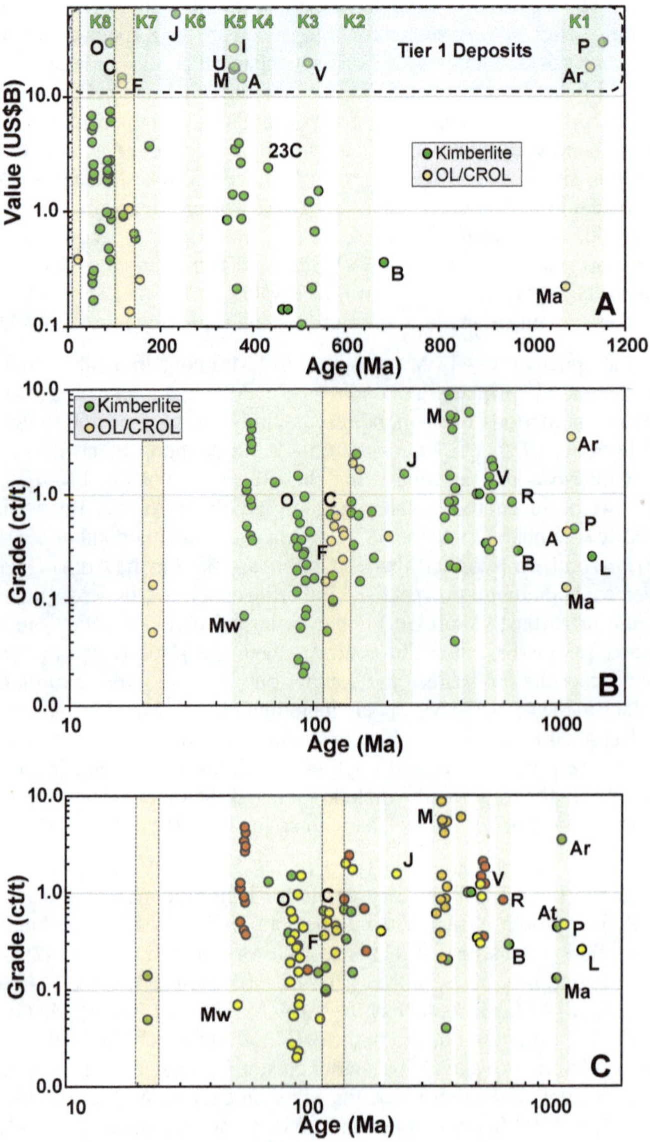

Figure 42. (**A**) Diamondiferous kimberlite and lamproite emplacement age versus diamond mine/deposit in-situ value (US$B log scale). Kimberlites are denoted with **green circles**, lamproites denoted with **orange circles**. Major periods of kimberlite and OL/CROL magmatism shown with **green and orange shaded fields**, respectively. Tier 1 deposits have mine-life values of >13 US$B (**blue shaded field**). (**B**) Diamondiferous kimberlite and lamproite emplacement age versus diamond grade (carat/tonne), log–log plot. (**C**) Diamondiferous kimberlite and lamproite emplacement age versus diamond grade (carat/tonne) highlighting deposit distribution by country: Canada (**red circles**), Russia (**orange circles**), and southern Africa (Botswana, Eswatini, Lesotho, South Africa, Zimbabwe; **yellow circles**), log–log plot. Main kimberlite (**green**) and OL/CROL periods (**orange**) shown as shaded bars. **Green circles** represent other diamond mines/deposits not located in southern Africa, Russia, or Canada. Individual deposits denoted in 42B and 42C as follows: A—Aikhal; Ar—Argyle (Tier 1); B—Brauna (Tier 4); C—Catoca (Tier 1); F—Finsch (Tier 1); J—Jwaneng (Tier 1); M—Mir (Tier 1); Ma—Majhagawan (Tier 4); Mw—Mwadui; O—Orapa (Tier 1); R—Renard (Tier 2); U—Udachnaya; V—Venetia (Tier 1); 23C—23rd Party Congress.

Lesotho, South Africa, Zimbabwe; yellow circles). Note that in the following text, the names of *"Advanced Diamond Exploration Projects"* are shown in *italics*. In some countries the diamond mines occur in specific kimberlite/lamproite epochs. The Majhagawan mine and *Atri advanced exploration project* (India) occur in ~1.1 Ga olivine lamproite. Across Russia (Arkhangelsk and southern Yakutia), >90% (14/15) of the kimberlite diamond mines formed in the 382–340 Ma kimberlite epoch, and most of these (9/15) were emplaced in an even more restricted time period between 356–366 Ma (all in southern Yakutia). In contrast, the Canadian and southern African diamond deposits are distributed between multiple kimberlite epochs. In Canada diamond mines and advanced exploration projects occur during 4 epochs: 56–51 Ma Lac de Gras and 103 Ma *Fort à la Corne* in the 109–45 Ma epoch; 177 Ma Victor, 172 Ma Jericho and 150 Ma *Chidliak* in the 197–139 Ma epoch; 523 Ma Snap Lake, 541 Ma Hearne, 542 Ma GK5034, 542 Ma Tuzo and 546–531 Ma *Qilalugaq-Naujaat* in the 565–505 Ma epoch; 643 Ma Renard 2 in the 650–610 Ma epoch. In southern Africa, diamond deposits formed during six kimberlite epochs and 1 OL/CROL period. Most southern Africa diamond mines occur in Late Cretaceous (< 100 Ma) kimberlite pipes ($n = 26$; ~70%), but importantly include the 250–210 Ma (Jwaneng), 570–520 Ma (Venetia, Murowa, The Oaks, *River Ranch*), and 1155–1075 Ma (Cullinan) mines in kimberlite pipes. In addition, at 180–140 Ma are the Marsfontein and Helam (Swartruggens area) diamond mines in CROL dikes or pipes.

An important new finding in this study is that ~80% of the 32 Cambrian pipes or intrusions worldwide are kimberlite and a relatively high proportion (~25%) are economic (Figure 42a), being either past producers, or are currently mined for diamonds, or are advanced exploration projects. In Canada, this includes Snap Lake, GK5034, Hearne, Tuzo, Qilalugaq-Naujaat, Kelvin, and Faraday, and in southern Africa the Venetia and Murowa mines. These new dating results clearly indicate that economically important Cambrian kimberlites and lamproites are more widespread than previously thought and that this epoch of magmatism is of considerable interest to diamond explorers.

Diamond deposit temporal windows

There are a number of factors that contribute to the generation of primary diamond deposits, including: 1) those that contribute to the formation, preservation and, in some cases, destruction of diamonds in the mantle, and; 2) those that are responsible for entraining and transporting diamonds to the Earth's surface. It is reasoned herein that the nature (magma composition, oxidation state, diamond entrainment capacity, and ascent rate), location (depth of magma generation, ascent path relative to diamond-rich mantle regions) and timing of kimberlite magmatism are key factors in forming primary diamond deposits. For example, if certain periods of kimberlite magmatism are corrosive to diamonds or they don't ascend through diamond-enriched mantle then such magmas would be barren of diamonds. Regarding the importance of kimberlite timing, in mantle sections where diamond formation is young and close in age to periods of broadly synchronous kimberlite formation (e.g., Koffiefontein), older kimberlites in the field cannot entrain diamonds that have not yet formed. In some locations mantle processes occurring after diamond formation, such as metasomatism, delamination, or thermal perturbations, could destroy or modify the quality of diamonds. For example, some diamond deposits have a high abundance of diamonds with graphite coatings that could form during subsequent mantle metasomatism. This could explain why, in certain fields where kimberlite magmatism occurs periodically over a long period of time, diamond mines occur in specific kimberlite epochs.

Below we focus on three examples of diamond deposit temporal windows that occur in some of the most productive primary magmatic sourced diamond mining areas on Earth, the Cretaceous Kalahari Craton deposits in southern Africa (Botswana, Eswatini, Lesotho, South Africa, Zimbabwe), the Devonian-Carboniferous Siberian Craton deposits in Yakutia, Russia (plus temporally equivalent kimberlite-hosted deposits in the Kola Craton), and the Paleocene (Eocene) central Slave Craton deposits in the Northwest Territories, Canada.

Kalahari Craton, southern Africa. The majority of diamond mines (*n* = 33) in southern Africa (Botswana, Eswatini, Lesotho, South Africa, Zimbabwe) are associated with Mesozoic age kimberlites and CROLs. A summary of the timing of this Mesozoic kimberlite (green) and OL/CROL (orange) magmatism is shown in Figure 43 (bin size is ~4 m.y.). Also shown in Figure 43 is the timing of economic diamond deposits (denoted by diamond symbols). There are two dominant periods of magmatism (135–105 Ma and 95–70 Ma) and both contain important diamond producers/past producers (20 and 13 mines, respectively). In the time period between 150–100 Ma, ~65% (33/50) of the dated localities in southern Africa are CROL. The majority of CROL magmatism (135–115 Ma) overlaps with but is generally older than southern Africa kimberlite magmatism (55–120 Ma). Two peaks in CROL magmatism can be discerned in Figure 43, the main peak is between 125–113 Ma containing the notable Tier 1 Finsch mine and a narrow secondary peak between 135–132 Ma (*n* = 7; Blaaubosch, Besterskraal, Lace/Crown, Roberts Victor, Saltpetrepan, Stieniesrus-Rex Mine, and Voorspoed). Economic diamond deposits occur in both periods of CROL magmatism.

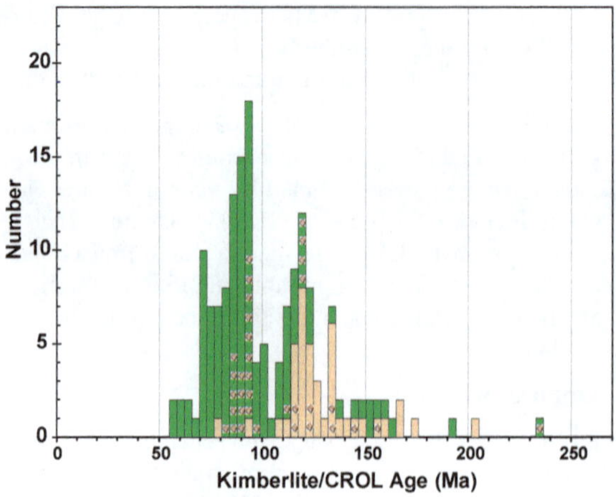

Figure 43. Histogram of Mesozoic and Cenozoic kimberlite (**green**) and CROL (**orange**) emplacement ages from southern Africa (Botswana, Eswatini, Lesotho, South Africa, Zimbabwe). Emplacement age of primary magmatic diamond deposits are shown with a diamond symbol. Bin width = 3.5 Ma.

A more detailed summary of the emplacement ages for Late Cretaceous (100–70 Ma) southern Africa kimberlites is shown in Figure 44. The producing/past producing diamond mines are labelled. A significant proportion (36%; 16/44) of kimberlite-hosted diamond mines in southern Africa occur within a narrow (~10 m.y.) time window between ~95–86 Ma in the Late Cretaceous.

Siberian and Kola Cratons, Russia. A very high proportion of diamond mines in Russia are observed in Paleozoic age kimberlites (Fig. 45). These kimberlites are colour coded to distinguish those that occur in the Arkhangelsk field (dark green circles) in the Kola Craton, from those that occur in the southern Yakutia kimberlite fields of the Siberian Craton (white and light green circles). The Yakutia kimberlites are further distinguished, to identify those that occur within kimberlite fields (Alakit, Daldyn, Nakyn, Malo-Botuobiya) where there is active diamond mining (light green circles), in contrast to kimberlite in fields with no active diamond mines (white circles). The locality names of the kimberlite diamond mines are also labelled. Diamond mines occur in Paleozoic kimberlites that have a large range in emplacement age (~90 m.y.). The oldest Paleozoic diamond mine is the 433 Ma 23rd Party Congress kimberlite, the youngest is the 344 Ma Sytykanskaya kimberlite. However, the most striking feature of this plot is that 60%

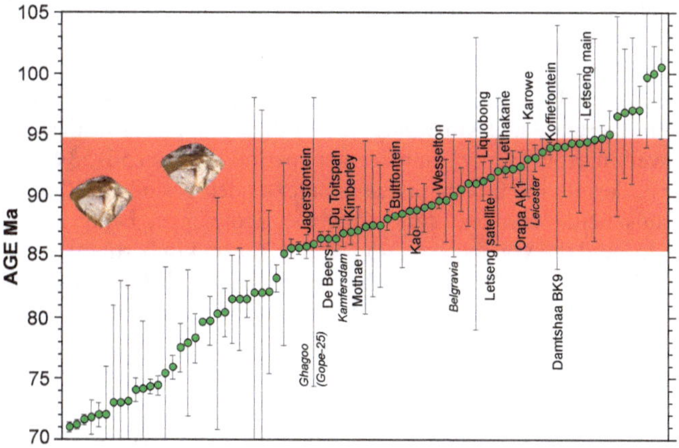

Figure 44. Compilation of kimberlite emplacement dates in southern Africa within the age range of 100–70 Ma. The 17 Cretaceous age diamond mines are denoted by name; all are within the narrow (~10 m.y.) temporal window between 95–85 Ma.

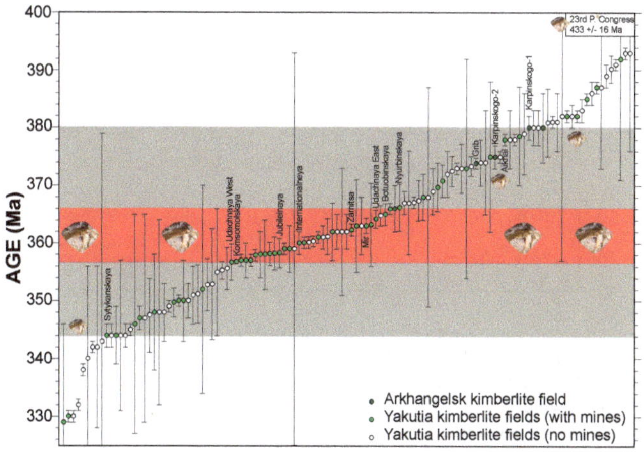

Figure 45. Age distribution of Paleozoic kimberlites in Russia. The kimberlites are subdivided into those from the Arkhangelskaya (**dark green**) and southern Yakutia (**white and light green circles**) kimberlite fields. Yakutian kimberlites that occur within fields that host diamond mines and deposits are denoted with **light green circles**. Individual diamond mines are identified. The **red shaded region** denotes the proposed temporal window for 75% of the Yakutian diamond mines (9 of the 12), except the Aikhal, Sytykanskaya and 23rd Party Congress mines. The broader 380–344 Ma timespan encompasses 14 of the 15 Russian diamond mines.

(9/15) of Russian kimberlite-hosted diamond mines were emplaced within the narrow (~10 m.y.) period between 366–356 Ma (red shaded field), and 14/15 Russian diamond mines (Yakutia and Arkhangelsk regions) over ~45 m.y. within the time span 380–344 Ma (grey shaded field). As noted previously, this group of Paleozoic kimberlites is exceptional, with five Tier-1 diamond mines, more than within any other epoch, or within a single craton.

Central Slave Craton, Canada. The Lac de Gras kimberlite field in the central Slave Craton, Canada consists of more than 250 kimberlites, 53 of which have well-established emplacement dates (Fig. 46). An Eocene diamond mine window for the Lac de Gras field, of 52–56 Ma (~50–58 Ma including age uncertainties) based on seven kimberlites from the Ekati and Diavik mines was first reported on by Kjarsgaard et al. (2002). The occurrence of dated diamond-bearing kimberlites in this field was further delineated by Sarkar et al. (2015; ~45–55 Ma) and Heaman et al. (2019; ~45–56 Ma). In this study we specifically re-investigate the age pattern of the kimberlite diamond mines. The ages of Lac de Gras kimberlites are shown in Figure 46. In this evaluation, the ten dated kimberlite-hosted diamond mines in the Lac de Gras field (Fig. 46) have a very narrow (3 m.y.) emplacement period between 56–53 Ma (51–61 Ma with age uncertainties), despite the fact that there is an extended period (~280 m.y.) of kimberlite magmatism in this field, with one kimberlite, the Eddie pipe, being as old as ~325 Ma (Sarkar et al. 2015) and the youngest at ~45 Ma (the Aaron kimberlite; Heaman et al. 2004). Furthermore, of all the 56–53 Ma dated kimberlites in the Lac de Gras field, a high proportion (10/15) contain economic diamond deposits and support the contention that diamond deposit temporal windows are real and significant.

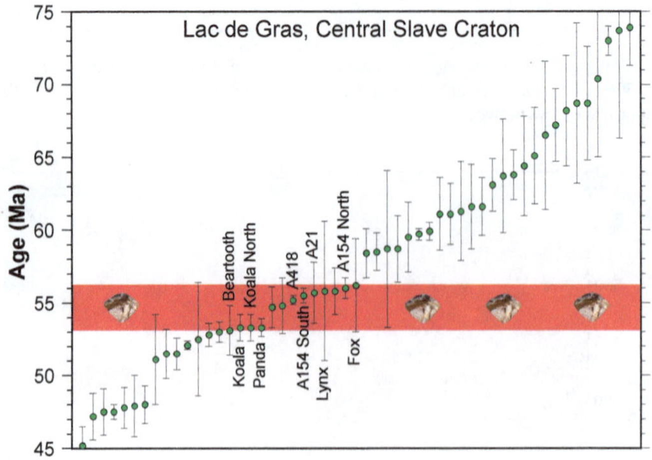

Figure 46. Compilation of Eocene-Cretaceous kimberlite emplacement ages from the Lac de Gras kimberlite field, central Slave Craton, NT, Canada. Ten of the thirteen kimberlite mines/deposits listed in Table 2 have precise emplacement dates and their pipe names are denoted. The **red band** highlights a remarkably short temporal diamond window (56–53 Ma) for this field. It should be noted that a 325 Ma kimberlite also exists in this field, that is not economic.

The examples outlined above demonstrate that a high proportion of diamond mines in arguably the most productive diamond-mining regions on the planet (Russia, southern Africa, Canada) are in kimberlites that were emplaced within extremely narrow temporal windows (≤10 m.y.), within restricted geographic areas. It is noteworthy that these three temporal windows are specific to each craton/country (366–356 Ma, 95–86 Ma, 56–53 Ma, respectively).

Origins of kimberlite diamond deposit temporal windows

The various kimberlite provinces and fields that have detailed geochronology available, as examined above, make a strong case that kimberlites erupting within specific time windows are more likely to be of higher economic value than others. There are clear exceptions, such as Jwaneng, but in fields where many kimberlites have been dated, a case for temporal clustering can be made. The time windows for all or the majority of economic kimberlites vary from ~ 10 m.y. for Yakutia and for the Kalahari Craton, to ~3 m.y. for the Lac de Gras field. This relationship has made kimberlite geochronology a critical part of diamond exploration in some regions. The cause of the apparent correlation between kimberlite age and the likelihood of economic viability, i.e., diamond deposit temporal windows, is unknown.

In many respects, this apparent temporal relationship runs contrary to "conventional wisdom" that diamondiferous kimberlites represent a random event with respect to kimberlite sampling of un-related diamond source rocks in the mantle (e.g., Gurney et al. 2005).

In order to better understand the origin of diamond deposit temporal windows it is important to have a clear understanding of the abundance, distribution, and origin of the diamonds recovered in these deposits together with salient features of the kimberlite magmas that transport the diamonds to the surface; such as their depth of formation, ascent pathway through the mantle and crust, and carrying capacity for transporting mantle material. Some insight into the origin of diamonds can be gleaned from the nature of their mineral inclusions, although only a fraction of mined diamonds contains such inclusions (Stachel et al. 2022, this volume). Previous diamond mineral inclusion studies (Stachel and Harris 2008) have shown that diamonds occur throughout the Earth's mantle below the diamond–carbon stability transition at > ~140 km depth. Certain diamonds (~8%) originate from ultra-deep mantle sources (e.g., >410 km depth in the transition zone or lower mantle; Stachel et al. 2000; Tappert et al. 2005; Pearson et al. 2014; Walter et al. 2022, this volume) based on the preservation of unusual mineral inclusions that are only stable at great depths in the mantle (e.g., majorite, jeffbenite, ringwoodite). If a large inventory of super-deep diamonds exists in the transition zone or lower mantle, then they will only be sampled by kimberlite magmas that form at equally great depths, unless mantle convection acts as an intermediary, somehow bringing these diamonds to asthenospheric depths, or to the base of the lithosphere.

Accepting the former assertion of a deep source region, one end-member explanation for a temporal restriction in diamond deposits is that the Tier 1 diamond mines are in kimberlites that have these very deep mantle sources and have entrained a large cargo of super-deep plus lithospheric diamonds. However, few diamond mines listed in Table 2 are known to contain super-deep diamonds (exceptions include Argyle, Cullinan, Diavik, Ekati, Jagersfontein, Karowe, and Letšeng). Unless there are a large number of undetected super-deep diamonds occurring in mines that formed during a distinct kimberlite diamond mine temporal window, we conclude that there is no correlation between very deep kimberlite magma formation and the formation of diamond deposit temporal windows. A possible association between depth of kimberlite magma generation and the economic value of a deposit is the presence of very large, highly valuable Type II diamonds, which only occur in kimberlites, but not in olivine lamproite- or CROL-hosted diamond mines. It is possible that Rayleigh–Taylor instabilities at a deep Earth boundary layer, such as the lower mantle / upper mantle boundary, or the core / mantle boundary may produce short temporal bursts of kimberlite magmatism that are capable of transporting these deeply derived, very high value diamonds that make some low-grade deposits (e.g., Letšeng, Lesotho), economically viable. However, such a hypothesis would have to invoke different depths of origin for other kimberlites of similar or different age, that did not contain these high value gems.

Because diamond inclusion studies indicate that most inclusion-bearing diamonds (92%) originate from relatively shallow (140–250 km depth) SCLM lithologies, primarily peridotite and eclogite (Stachel and Harris 2008), other explanations must be sought for the temporal window association of diamond deposits. The oldest diamonds (>2.7 Ga) typically occur in harzburgite lithologies whereas eclogitic and lherzolitic diamonds are typically younger than 2.7 Ga and are predominantly of Proterozoic age (Shirey et al. 2013; Smit et al. 2022, this volume). There is a growing list of deposits (Cullinan, Koffiefontein, Damtshaa, Finsch, Jwaneng, Orapa, Letlhakane, Venetia, Mir, 23rd Party Congress) with diamonds that formed during multiple distinct diamond forming events, often spanning a large time-frame (>2 billion years; see Smit et al. 2022, this volume for summary). An important feature of some diamond deposits listed in Table 2 is that there are multiple discrete periods of diamond formation at a single mine (Smith et al. 1991; Pearson et al. 1998; Richardson et al. 2004; Koornneef et al. 2017). In an innovative study of diamonds recovered from the 363 Ma Mir kimberlite, part of the Siberian Paleozoic diamond deposit temporal window, it was shown that sulphide inclusions extracted from diamond core and rim growth zones yield distinct Re–Os dates of 2.1 and 1.0 Ga, respectively (Wiggers de Vries et al. 2013). The 92 Ma Orapa and Letlhakane mines both recover diamonds that formed at distinct time periods (Timmerman et al. 2017; Gress et al. 2021). In fact, it seems that the more diamonds that are dated from a given deposit, the higher the likelihood of finding multiple generations of diamond growth in the underlying mantle lithosphere that was subsequently sampled by the kimberlite. These diamond growth events can often be linked to large-scale (global) mantle processes, such as the formation of Large Igneous Provinces, periods of continental rifting, and enhanced periods of oceanic lithosphere subduction. We conclude that a pre-requisite for the formation of economic primary diamond deposits is the existence of mantle regions that are rich in diamonds, produced by multiple diamond forming events, which are subsequently sampled by passing kimberlite/lamproite magmas during their ascent. However, there is no obvious connection between the generation of diamond mineralization in the mantle and the formation of diamond deposit temporal windows, since some of the Tier 1 deposits (e.g., Orapa) that contain multiple diamond growth ages occur within a temporal window, and some (e.g., Jwaneng, Catoca) do not occur in a diamond deposit temporal window.

Another possible explanation of why certain temporal periods of kimberlite magma formation could be more favourable to containing diamond deposits is that magma generation occurred at sufficient depth within the diamond stability field (>140 km) and the source region was relatively spatially focused. During ascent these magmas passed through a diamond-rich region in the mantle and were successful at entraining a large cargo of mantle material from this region. As the magmas continued to ascend through to the crust they followed local structures, forming dikes, sills and pipes at the surface. This can be described as the common magma source hypothesis for synchronous diamond deposit formation. If correct, one prediction of this hypothesis is that the diamond mines or deposits in kimberlites (or lamproites) should have similar ages, geochemistry, and radiogenic isotopic compositions that reflect the nature of their common source. Although there are some published Lac de Gras kimberlite Sr–Nd–Hf isotopic data (Dowall 2004; Tappe et al. 2013; Woodhead et al. 2019; Tovey et al. 2021), there are insufficient radiogenic isotope data for the Eocene age kimberlite diamond mines in the Lac de Gras field, NT, Canada (those listed in Table 2) at present to test this hypothesis. However, in southern Africa, radiogenic isotope studies of the Cretaceous age kimberlite-hosted diamond mines (Woodhead et al. 2009; Griffin et al. 2014) indicate two possible sources for these kimberlite magmas based on their perovskite initial strontium isotope compositions; a lower $^{87}Sr/^{86}Sr_i$ group (0.7032–0.7036; Kao, Letšeng, Liquobong, Lethlakane, Orapa) and a higher $^{87}Sr/^{86}Sr_i$ group (0.7039–0.7046; Jagersfontein, De Beers, Dutoitspan, Matsoku, Wesselton, Koffiefontein). These perovskite Sr isotope results either point to two different kimberlite magma sources, or there is a single magma source ($^{87}Sr/^{86}Sr_i$ ~0.7032) that was variably contaminated with more radiogenic lithosphere. We conclude that a single magma hypothesis could explain the existence of some diamond formation temporal windows, but that more research is required.

REVIEW OF GLOBAL SECONDARY DIAMOND DEPOSITS

Background

Secondary diamond deposits are formed by a range of sedimentary and diagenetic processes that operate under different climatic and geomorphological conditions of which the fluvial and marine environments have shown to have been the most effective. Placers specifically are accumulations of valuable minerals (e.g., diamond) formed by gravity separation during sedimentary processes. Diamond has a specific gravity higher than most common minerals (SG = 3.52) and hence, much like any other heavy mineral, there is sufficient density contrast to separate it from the worthless gangue (non-ore) minerals, and also most rock types. Only if the deposit is winnowed in this way can the minerals be concentrated to economic levels. The importance of trap-sites and other ingredients required to develop alluvial deposits has been covered in contributions such as Sutherland (1982), Muggeridge (1995), Marshall and Baxter-Brown (1995) and Jacob et al. (1999).

Diamond is also the hardest mineral (hardness of 10) and is therefore tough, but it can break along fractures by interparticle collision. Hence diamonds do break during transport especially those that are weakened by fractures and inclusions, and as a result the percentage of durable "cleaner" and superior quality stones increases during sedimentary cycle(s). Diamonds are also resistant to chemical weathering and can therefore be concentrated during diagenesis and/or pedogenesis of the host material when this is broken down and removed.

To assess a complete source-to-sink model, a robust understanding of stratigraphy, geomorphology and climatic conditions is required. Identification of hiatal periods when there is neither erosion nor sedimentation, is crucial in the stratigraphic record. New technologies have been applied to improve the source-to-sink resolution such as, single grain geochronology, cosmogenic dating, low-temperature thermochronological estimates for long-term erosion rates, off-shore seismic studies to quantify sediment fluxes from the catchment areas (sources) to off-shore basins (sinks). Finally, uplift and exhumation events, critical in the development of secondary deposits, and timing thereof are now integrated with landscape evolution and depositional models. Not many source-to-sink systems are complete and we may only see part of the journey of diamonds that have been eroded out of kimberlites or lamproites to their final resting place.

Three categories of placer deposits have been recognized: *retained* (such as in the kimberlite craters, or deposits associated with, but external to kimberlites; diamonds trapped in karst sediments, or in older intra-cratonic sediments); *transient* (diamond-bearing sediment accumulations along dispersal routes or within the active drainage basin); and *terminal* placers (at the terminal end of the transport system usually along basin margins subjected to marine reworking processes) (Bluck et al. 2005).

Archean

South Africa (Gauteng Province): Witwatersrand supergroup—Transient/terminal placer (Mesoarchean). The oldest reported alluvial diamonds are from the conglomerates of the Witwatersrand Supergroup gold placer (Fig. 47) where several hundred carats were recovered from the Turffontein Subgroup (previously the Elsburg Series) of the Central Rand Group, particularly on the West Rand, but some also from the East Rand (Raal 1969). Single zircon U/Pb dating results show that sedimentation in the basin occurred between 3,074 and 2,714 Ma, a period of some 360 m.y. (Robb and Robb 1998). The Elsburg Reef is between 2,894 Ma and 2,780 Ma in age, and the diamonds from here are between 1 and 2 ct in size with one 8 ct reported stone (Young 1914). However, Smart et al. (2016) suggest that diamonds were also recovered from the base of the Witwatersrand Supergroup (West Rand Group) deposited between 3.1 and 2.9 Ga.

Deposition in the Witwatersrand basin commenced in a shallow marine environment, forming the lower West Rand Group. During the subsequent Central Rand Group regression stage (Tucker et al. 2016), sedimentation occurred on coalescing lobes on extensive braid deltas or alluvial fans that spread along the margins of the basin with typical braided stream processes (Minter and Leon 1991). Since there is evidence of localized marine incursions, fluvial deposits may in part have been upgraded by marine reworking (Tucker et al. 2016). The headwaters of the rivers, based on paleocurrent analysis along the northern margin of the basin, were to the north and northwest. Diamonds have been described ranging from cadmium-yellow to yellow-green and dark green (Williams 1932). The color is only skin deep and is due to radiation damage from the uranium that is present in the same conglomerates (Raal 1969). Raal (1969) also mentioned that most of the Wits diamonds are Type I, and Denny (1897) described most of them as rhombic dodecahedra.

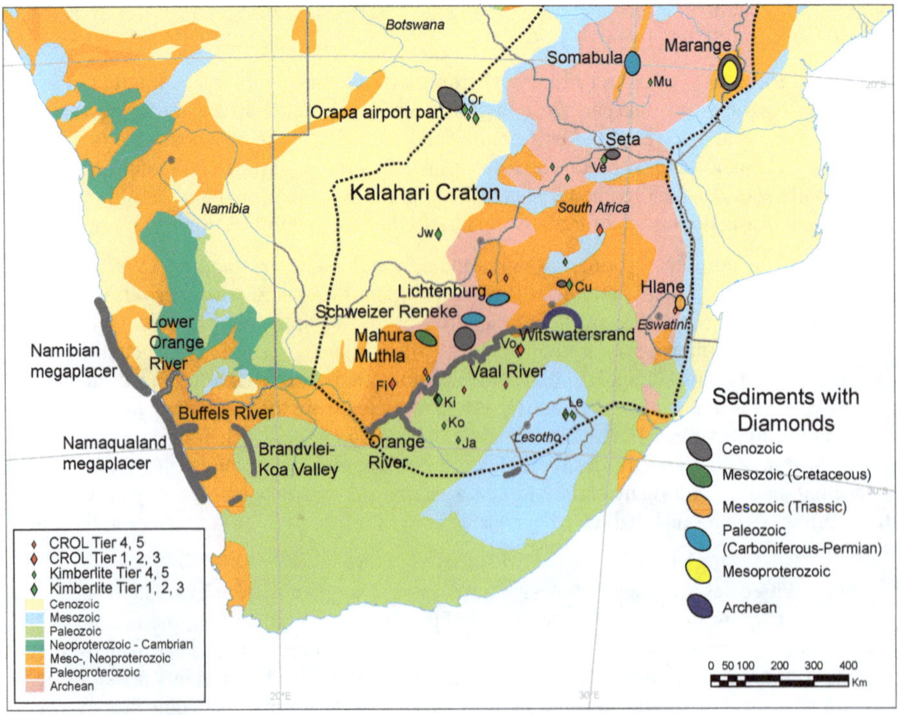

Figure 47. Southern Africa placer deposits, as labeled on the map (Marange, Somabula, Lichtenburg/Ventersdorp, Schweizer Reneke, Vaal River, Orange River, Brandvlei/Koa valley, Lower Orange River, Buffels River, Namaqualand megaplacer, Namibian megaplacer, Mahura Muthla, Witwatersrand, Hlane, Orapa Airport pan and Seta). Primary magmatic kimberlite and CROL mines/deposits as follows: Mu, Murowa; Or, Orapa; Ve, Venetia; Jw, Jwaneng; Cu, Cullinan; Vo, Voorspoed; Fi, Finsch; Ki, Kimberley pipes; Ko, Koffiefontein; Ja, Jagersfontein; Le, Letšeng. Mercator map projection.

Canada (Nunavut): Tree River area—Terminal placer (Mesoarchean). Three diamonds have recently been recovered from a Mesoarchean gold-bearing conglomerate (2.96 Ga) on the northern part of the Slave Craton (Fig. 48) in the Tree River area (Timmerman et al. 2020). These quartz-pebble conglomerates are at the base of the Central Slave Cover Group and have been interpreted as continental margin deposits followed by a marine transgression (Haugaard et al. 2021). This indicates that diamonds were released from deeply derived volcanic rocks and concentrated in Archean marine settings not only in South Africa but also in Canada.

Canada (Ontario): Wawa area—Retained and transient/terminal placers (Neoarchean). It has been shown that erosion rates of the central shield areas in Canada have generally been low (Flowers et al. 2006). Cretaceous fluvial networks in northern Canada drained eastwards to Hudson Bay and the Labrador Sea (MacMillan 1973), however, minimal uplift of the shield areas since the Eocene prevented the development of extensive alluvial deposits in Canada. Further, any weathered material would have been eroded and transported by the Quaternary Laurentide ice sheet from the Canadian shield towards the south where diamonds have been found in glacial till in parts of the United States, such as Wisconsin and Indiana (Kjarsgaard and Levinson 2002).

There is however one reported Canadian paleo-placer in the Wawa area on the Superior Craton (Fig. 48). This is spatially associated with a diamondiferous matrix- and clast-supported volcaniclastic breccias of the Catfish assemblage within the Archean Michipicoten Greenstone Belt. This crudely bedded and poorly-sorted volcaniclastic unit is a "retained"

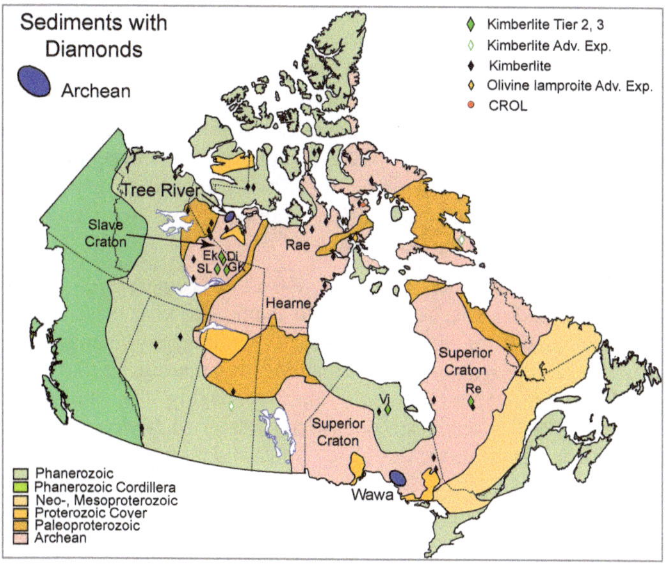

Figure 48. Canada placer deposits, as labeled on the map (Tree River and Wawa). Primary magmatic kimberlite and CROL mines/deposits as follows: Ek, Ekati; Di, Diavik, SL, Snap Lake; GK, Gahcho Kué; Vi, Victor; Re, Renard.

placer that has been interpreted as debris flows (Lefebvre et al. 2003). The breccias and associated lamprophyric dikes, both diamond-bearing, resulted possibly from a subduction-related Archean eruption (Lefebvre et al. 2003; Stachel et al. 2006), and both have been dated at between 2,744 Ma and 2,618 (Kopylova et al. 2010).

Unconformably overlying these volcanic rocks, but still part of the Michipicoten Greenstone Belt, are the Leadbetter Conglomerates (Wendland et al. 2012). These 2,701–2,697 Ma (Miller et al. 2012) conglomerates are dominated by debris flows at the base, followed by channel bars and lags representing a braided fluvial system on a sub-aerial alluvial fan which was terminated by a marine transgression. The best concentration of the diamonds is associated with braid-plain deposition and occurs in the middle of the upward-fining sequence (Wendland et al. 2012). The forecasted grades of the latter range from 0.11 to 0.29 ct/t (Ryder et al. 2008). However, based on diamond studies, the stones from the conglomerates seem to have a different source compared to those from the volcaniclastic rocks and dikes (Kopylova et al. 2010). Wendland et al. (2012) suggest that the main reason for the existence of this placer is its proximity to a primary source(s) and not through hydrodynamic concentration of diamonds from a distal source.

Proterozoic

Brazil (Amapá State): Vila Nova—Retained placer? (Paleoproterozoic). Diamonds in Brazil were discovered on the banks of the Jequitinhonha River in 1725 near the village of Arraial do Tijuco (later renamed Diamantina) Minas Gerais State, while artisanal miners were searching for alluvial gold. Brazil has produced close to 50 Mct since the discovery of its first diamond and most of these have been derived from conglomerates (Romaria, Minas Gerais State), glacial (paleo-diamictite) rocks (Tibagi, Parana), metasedimentary rocks (Diamantina, Minas Gerais State and Chapada, Bahia State) and recent alluvials. However, diamonds have only been mined successfully out of the Espinhaço Supergroup (Mesoproterozoic) and the Capacete Formation of the Late Cretaceous Mata da Corda Group (palaeo-placers; Pereira et al. 2017). The rest have come from Quaternary deposits.

The oldest known occurrence of diamonds in Brazil are the Palaeoproterozoic Vila Nova Group metavolcanic-sedimentary rocks (Fig. 49) located in the central southeast part of the Amapá State (Spier and Ferreira Filho 1999). Mafic and ultramafic rocks are covered by this volcanic–sedimentary sequence and both are metamorphosed to upper amphibolite facies. Diamond-bearing fine- and coarse-grained clastic meta-sediments including the meta-conglomerates have been dated at 2,264 Ma and suggested to represent a Palaeoproterozoic greenstone belt broadly comparable to the Guyana Shield greenstone belts (Spier and Ferreira 1999). The Vila Nova occurrences are also contemporaneous with the Birrimian meta-sediments of West Africa (Tompkins and Gonzaga 1989).

Brazil (Roraima State) and Guyana: Roraima Supergroup—Transient fluvial placer (Paleoproterozoic). Late Palaeoproterozoic diamonds have been associated with the Roraima Supergroup which forms a continuous area of chemically mature sedimentary rocks (orthoquartzite, quartz arenite and polymictic conglomerate) that can be traced over large tracks of the Guiana Shield in parts of Venezuela, Brazil and Guyana (Fig. 49), with outliers in Suriname (Grubb and McCallum 1991). This thick sequence of relatively flat-lying unaltered Precambrian clastic sediments was deposited unconformably on the underlying basement complex roughly between 1.95 and 1.78 Ga (Reis et al. 2017). No diamonds have been reported coming directly from the Roraima units, but almost all alluvial deposits, whether terrace or recent river gravels that have been exploited for diamonds, are associated with the rivers that flow across the basal Arai Formation of the Roraima Group (Meyer and McCallum 1993). Fleischer (1998) believes that the diamonds are restricted to lithologies of the middle member of the Roraima Group. The sediments of the Arai Formation have been interpreted as braided fluvial facies with paleo-terraces (Reis et al. 2017). Interestingly the diamonds from Boa Vista (Roraima State) show almost no sign of abrasion and have similar green spots that are also found on the diamonds from the Guaniamo kimberlite sills (Fig. 49) and it has been suggested that the Boa Vista diamonds have been derived from a nearby kimberlite source (Tappert et al. 2006).

Phanerozoic-aged uplift of the Guyana Shield combined with climatic fluctuations has eroded and upgraded these siliciclastic rocks into colluvial-type deposits proximal to exposed outcrop of the Roraima rocks (Watkins 2009). These and two Pleistocene terraces with braided river type alluvium and a Holocene age gravel-rich valley fill in the surrounding rivers have all been mined (Grubb and McCallum 1998). The grade and diamond quality improve in successively younger terraces and are best in the recent channel fills. The diamonds are either gem or near gem with only 10% boart (Meyer and McCallum 1993), and Grubb and McCallum (1998) report up to 65% gem-quality in the recent alluvium. The main sources of sediments of the Roriama Group were rocks of Trans-Amazonian age showing common sources from the greenstone belts in the northernmost portion of Guyana and Venezuela, with paleo-flow direction coming from the north and northeast (Reis et al. 2017). Reid (1974) has suggested that the diamonds were derived from Proterozoic kimberlites in West Africa (Ivory Coast, Ghana or Liberia).

Diamonds in Guyana were discovered in 1890, with most of the deposits occurring within 25 km of the Roraima escarpment in recent fluvial sediments of the Puruni and Mazaruni valleys, with the highest grades at the bedrock contact. In more distal cases they are associated with the Cenozoic White Sand Series, which are thought to be derived from the erosion of the Roraima sediments. Some 60% of the diamonds are gem quality and the average size range is between 10 and 15 stones/carat (st/ct) (Guyana Office for Investment 2019). It is estimated that Guyana has produced some 10 Mct since discovery.

Diamonds in Venezuela came from many localities scattered over the Guyana Highlands of the Bolívar province, but mainly below the steep rises of the highlands. Many of these alluvial deposits occur in tributaries of the Orinoco River, such as the Paragua, Caroni and Cuyon, as terraces and recent gravels in channels in the navigable parts of the rivers, which were exploited by mechanical dredges.

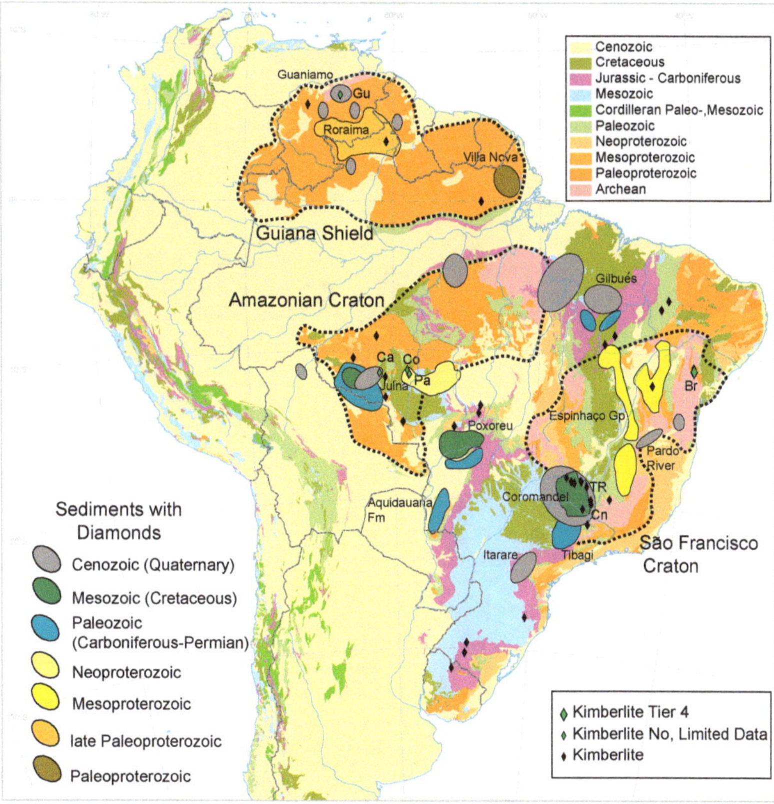

Figure 49. South America placer deposits, as labeled on the map (Villa Nova, Roraima, Espinhaço Group, Tibagi, Itarare, Coromandel, Poxoreu, Guaniamo, Aquidauana Formation, Juína, Gilbués, Pardo River). Primary magmatic kimberlite mines/deposits as follows: Gu, Guaniamo; Br, Brauna; Ca, Carolina; Co, Collier; Pa, Pandrea; Br, Brauna; TR, Tres Ranchos; Cn Canastra. **Black dotted lines** outline the Shield/Cratons, as labeled. Mercator map projection.

Brazil (Minas Gerais, Bahia States): Espinhaço Supergroup—Transient fluvial/glacial placer (Mesoproterozoic). Diamonds have been mined from the Mesoproterozoic meta-conglomerates and breccias of the Espinhaço Supergroup (Minas Gerais and Bahia States) that stretches as a north-south belt across the São Francisco Craton (Fig. 49). Diamonds occur in two formations. Firstly, in the Sopa-Brumadinho Formation (Diamantina Group; Fig. 50) around the town of Diamantina in central Minas Gerais State that has a minimum age of 1,715 Ma (Pedreira and de Waele 2008). Secondly, to the north in the Chapada Diamantina area in Bahia State, diamonds are found in the slightly younger Tombador (Figs. 51, 52) and Moto do Chapéu Formations that are both part of the Chapada Diamantina Group, which has a maximum age of 1,515 Ma (Pedreira and de Weale 2008). It is here that significant amounts of carbonado are also found, especially close to the town of Lençois (Fig. 52).

In the Diamantina area the meta-conglomerates have been variably interpreted as alluvial fans, glaciofluvial sediments, coastal marine facies, and breccias associated with debris flows in a tidal flat environment (Chaves et al. 2001), or as possible volcanic breccias (Fleischer 1998). The meta-conglomerates have grades between 0.04 and 0.08 ct/m³ and some 5 Mct have been estimated to have been produced from the Diamantina district alone (Karfunkel et al. 1994). The diamonds are generally small, well-sorted in the 0.6 to 0.9 ct/st range and of good quality (Chaves and Uhlein 1991), which suggests periods of reworking and upgrading.

Figure 50. Poorly sorted breccias of the Sopa-Brumadinho Formation, at Campo Sampio west of Diamantina, Brazil.

Figure 51. Conglomerates of the Tombador Formation near Lencois, Brazil.

Figure 52. Diamonds from the Tombador conglomerates, Brazil. Note the presence of carbonado (black).

Some have green and brown surface colors. Diamonds from the Bahia State area are colorless, brown and grey. There are no kimberlitic indicator minerals in these rocks and no primary sources have yet been linked to these diamonds. Sources could be unknown, eroded or buried distal kimberlites located in the São Francisco Craton. Another idea is a proximal source located within the confines of the diamondiferous Proterozoic sedimentary basin (Battilani et al. 2007). Recently it has been shown that metamorphosed igneous rocks intruded into the lower Tombador Formation, and these contain microdiamonds. Erosion and reworking of these intrusive rocks could be the source of diamonds which tend to be concentrated in the upper levels of the Tombador and Moto do Chapéu Formations (Battilani et al. 2007).

The drainage basin of the Jequitinhonha River (Minas Gerais State), which rises in the Espinhaço Mountains, was responsible for most of Brazil's historic diamond production, and these were mainly derived from weathering of the Mesoproterozoic Espinhaço Supergroup. Grades in the recent alluvium in this river were up to 0.6 ct/m^3 of mainly gem quality diamonds (Dupont 1991). Mining was initially mainly by hand although later junior companies introduced dredges to mine the river gravels.

Zimbabwe (Mutare/Chimanimani Districts): Chiadzwa (Marange)—Terminal placer (Mesoproterozoic). Diamonds were discovered in outcropping Mesoproterozoic conglomerate in the Marange area of eastern Zimbabwe in 2001 (Fig. 47). These basal sediments of the Umkondo Group (>1.11 Ga) comprise thin (10 cm–2m) compositionally mature, quartz pebble and cobble conglomerates (Fig. 53). The deposit represents a craton margin, basin edge deposit of the Umkondo Foreland Basin fed by an easterly directed drainage from the Zimbabwe Craton by comparatively small rivers (Ward et al. 2013). Gravel and sand, including diamond, were introduced to the Umkondo Basin shoreline at Chiadzwa in the east (the Makodzi anomaly), where considerable reworking and concentration took place by wave action and tidal energy variations during an initial transgression of the basin (Zhou 2015). In the absence of longshore drift, diamond concentration falls sharply over a short distance from north to south (Ward et al. 2013). It produced extremely coarse (5 to 7 carat stones), but poor-quality diamonds that become finer-grained further eastwards into the basin where at Chimanimani the average size is between 0.5–1 ct/st (Ward et al. 2013). Diamond quality improves from the Makodzi area towards the east into the Umkondo Foreland Basin. Furthermore, the high degree of abrasion displayed by many of the diamonds is a reflection of the reworking processes which also contributed to the higher grades. Another unique feature of the Marange diamonds is the brownish to black coating indicative of radiation damage and as a result it is estimated that only some 10% are of gem quality (Fig. 54).

Erosion of these Mesoproterozoic conglomerates has also produced diamondiferous Quaternary alluvium comprised of surface lag, talus (scree slope), sheetwash and ephemeral riverine sediments and these have been the primary mining targets. Together, these conglomerates and recent sediments have reportedly produced ~65 Mct to date, although in reality it is estimated that this could be as much as ~100 Mct (de Wit et al. 2016).

Figure 53. Metaproterozoic conglomerates from the basal Umkondo Group at Marange, eastern Zimbabwe. Note the well rounded quartz clasts dominate the pebble population.

Figure 54. Poor quality but large diamond present in the Marange conglomerates, eastern Zimbabwe (photo courtesy J.D. Ward).

India (Andhra Pradesh State): Banganapalle—Terminal placer (Meso-, Neoproterozoic). India and Indonesia are historically the first countries where diamonds were found, with the Krishna River (described later) having produced some of the world's famous diamonds. There are two main areas with palaeoplacers in India, the Banganapalle conglomerates that are associated with the Dharwar Craton in the south, and the Panna diamond belt on the northern margin of the Vindhyan Basin on the Bundelkhand Craton central-north (Fig. 55), supplemented by Quaternary diamondiferous gravels.

The late Meso- to early Neoproterozic Banganapalle Formation occurs at the base of the Kurnool Group and can be followed for 225 km in the Cuddapah Basin (Bertram 2010). The thin quartzitic conglomerate beds vary in thickness from 0.01 to 3 m, are flat lying and have been interpreted as a beach gravel developed during a transgressive phase (Fareeduddin and Mitchell 2012). The average grade is between 2–3 ct/100t with some 75% gem quality diamonds (Joy et al. 2012). Diamonds from these conglomerates are distinct from those derived from the alluvials of the Krishna River; they have no radiation damage nor any abrasion features on their surfaces and are believed to have been derived from proximal source(s) (Ravi et al. 2012). The proximal occurring Chelima lamproites, of 1,370–1,287 Ma age (Joy et al. 2012), are now believed to be the source for the diamonds in this terminal placer, since the mineral chemistry of mantle minerals in the conglomerates eliminates kimberlites occurring to the SW as its potential source (Joy et al. 2012). Also, Type IIa diamonds, which occur among the Krishna River diamonds, are not seen in the Banganapalle conglomerates.

India (Madhya Pradesh State): Panna diamond belt—Retained/terminal placer (Meso-, Neoproterozoic). In the Panna belt (Fig. 55) diamond-bearing conglomerates occur in the Meso- to Neoproterozoic Vindhyan Supergroup. Sedimentation of the upper part of this supergroup, the Rewa Group, started around 1,100 Ma and continued until 650 Ma (Ray 2006) and include the diamondiferous Itwa (Lower Rewa Group), and Jhiri and Gahadra (both Upper Rewa Group) conglomerates, which are separated by shale horizons. The lower Itwa conglomerates are poorly sorted and have up to boulder size sub-rounded to sub-angular clasts that are dominated by footwall lithologies, in a sandy matrix. Its thickness varies from 0.05 to 2 m (Soni et al. 2002). Diamond grades of as high as 27.9 ct/100 t have been recorded and there is also a high percentage of gem quality diamonds (Fareeduddin and Mitchell 2012). These deposits formed during uplift resulting in high gradient rivers that accelerate erosion and rapid incision bringing into the basin poorly-sorted coarse-grained sediments enriched in diamonds that were reworked out of pre-incision erosion surfaces or retained placers (Gupta et al. 2003). The diamondiferous Jhiri conglomerate (Fig. 56), a terminal placer which is between 0.02 to 0.70 m thick with isolated channels, form banket sheets of clast- and matrix-supported lenses. It is well-sorted and composed of well-rounded pebble size clasts and these lenses have been interpreted as shoreline lags (Gupta et al. 2003). The Gahadra conglomerate, containing well-rounded pebbles composed mainly of local sandstone (Soni et al. 2002; Rau 2007), are reported to have grades of up to 28 ct/100t (Fareeduddin and Mitchell 2012). Based on evaluation work done by Geological Survey of India, some 22.8 Mt of indicated resource has been established for the Jhiri and Itwa conglomerates (Rau et al. 2012). The local 1,040 Ma old Majhgawan (average grade 14 cpht) and Hinota diamond-bearing olivine lamproites, are interpreted as the likely sources of the diamonds in these conglomerates (Fareeduddin and Mitchell 2012).

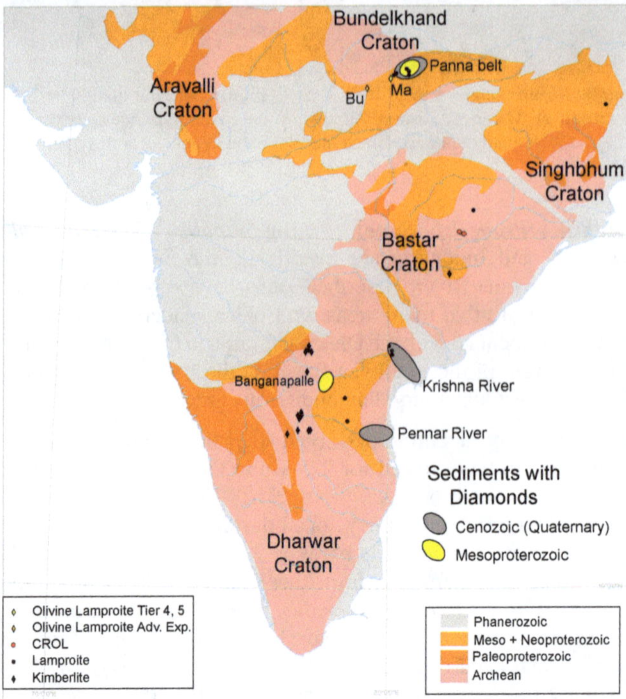

Figure 55. India placer deposits, as labeled on the map (Banganapalle, Panna Belt, Krishna and Pennar Rivers). Primary magmatic olivine lamproite mines/deposits as follows: Ma, Majhgawan; Bu, Bunder (Atri). Mercator map projection.

Diamonds released by erosion of these Proterozoic conglomerates are mined in recent gravels of the Baghain and Ranj Rivers. The former is the most important, with gravels returning grades of up to 26 ct/100t (Fareeduddin and Mitchell 2012), and has a Proven Diamond Resource of 1.1 Mct (Rau et al. 2012). Gravels along the Ranj and other tributaries have yielded considerably lower grades. Remnants of lateritic gravel, which occur as a thin deflation layer on the Baghain plateau often trapped in joints, have been mined by artisanal miners with variable success.

Figure 56. Diamonds in fine-grained Jhiri conglomerates from the Vindhyan Supergroup in the Panna Belt, northern India.

Paleozoic

Russia: Urals—Reworked terminal placer (Lower–Middle–Upper Devonian). The occurrence of alluvial diamonds along the western edge of the middle and northern Ural Mountains has been known since 1829. The diamonds are found in two main areas; near Perm in the middle and northern Ural Mountains (Fig. 57) where diamonds occur in the Lower Devonian (Emsian stage) Takaty Formation (Laiginhas et al. 2009), and in the Timan areas further north (Fig. 57) where diamonds are derived from the basal conglomerates of the Middle Devonian Travyanka, and Upper Devonian Nadezhda and Kumushka Formations (Konstantinovskii 2003). Sources for these Devonian placer diamonds are not known but Sobolev et al. (2019) noted, based on the chemical composition of certain diamond inclusions as well as the high eclogitic/peridotitic inclusion ratios, that the Ural diamonds show striking similarities to those from the northeast Siberian placers (except for their carbon isotope compositions), and in part to those from the Arkhangelsk kimberlite province to the west/northwest. However, the eruption age of its primary magmatic source rocks, based on $^{40}Ar/^{39}Ar$ dating of clinopyroxene inclusions in some Urals diamonds is 472 Ma (Laiginhas et al. 2009), which is older than the *ca.* 380—374 Ma Lomonosov and Grib diamond mines in the Arkhangelsk kimberlite province and precludes them as a source.

Ural Mountain diamonds are concentrated in 1 to 5 m thick basal conglomerates, composed of alluvial sediment, that were partially reworked within a coastal-marine zone. The diamonds, mainly rounded dodecahedra, were thus concentrated in terminal placers along a shoreline on an eastern active margin of the East European Craton during the Devonian (Laiginhas et al. 2009). Only rare pyrope garnets are associated with these conglomerates. Erosion and reworking of these Devonian sediments have produced numerous Quaternary transient placers in the form of terraces and recent river gravels preserved in Mesozoic–Cenozoic karst depressions, which have been mined. The diamonds, many of which have green and brown spots, are well-sorted, and also abraded (Konstantinovskii 2003).

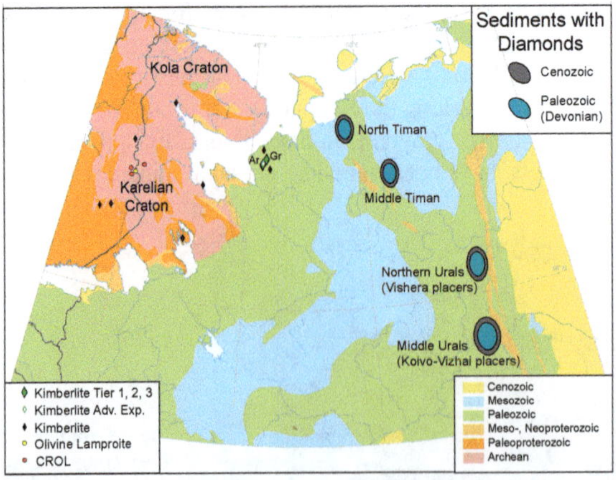

Figure 57. Western Russia (East European Platform) placer deposits, as labeled on the map (Northern Timan placers, Middle Timan placers, Northern Urals—Vishera placers Middle Urals/Perm—Koivo/Vizhai placers). Primary magmatic kimberlite mines/deposits as follows: Gr, Grib; Ar, Arkhangelskaya. Mercator map projection.

Russia (Republic of Sakha): central Siberian Shield—Retained placers (Carboniferous and Permian). There are some 1,070 kimberlites known in Yakutia in 15 Palaeozoic and Mesozoic fields. Of these 195 kimberlites occur in the central Siberian Craton and are predominantly Palaeozoic in age, which include most of the economic kimberlites such as Mir and Udachnaya. Some 875 kimberlites occur in the northern Siberian Craton (Anabar-Olenek subprovince) almost 200 km north of Udachnaya, and are almost barren or at best are weakly diamondiferous (Sobolev et al. 2018). Many of these Triassic and Jurassic–Cretaceous kimberlites are eroded; they have been described as hypabyssal kimberlite (Sun et al. 2014; Sobolev et al. 2018).

There are two types of placer deposits on the Siberian platform. Retained placers which are closely associated with the known kimberlite mines in the central Siberian Craton, and transient placers found in the north that are associated with the Anabar-Olenek drainage basins (also referred to as the Anabar, Lower Olenek and Lena alluvial deposits, discussed under the *Cenozoic* section).

Poorly-sorted retained placers, close to the kimberlite mines/deposits (Fig. 58, Mir, Nyurbinskaya, Botuobinskaya, Udachnaya, Verkne-Munskoye), occur as Lower and Middle Carboniferous, Lower Permian and in Triassic–Jurassic proximal alluvials fans. These have an abundance of coarse kimberlitic minerals which supports a proximal source (Kedrova et al. 2022; Konstantinovskii 2003).

The Vostochnaya (Fig. 59) and Solur retained placers are two spatially adjacent alluvial paleo-placers of Upper Palaeozoic and Mesozoic age, respectively, some 25 km northwest of the Mir mine, between the Irelyakh and Chuonalyr Rivers. The poorly sorted sediments are described as clayey conglomerates and fine breccias interpreted as alluvial fan deposits which were controlled by repeated tectonic movement along faults resulting in the formation and burial of these placers (Konstantinovskii 2003). The Carboniferous (Serpukhovian stage) Vostochanaya deposit stretches over 5 km and is on average 0.7 m thick (Petrov et al. 2016). The Lower Jurassic Solur deposit has an average thickness of 2.4 m and is mainly made up of pebbly conglomerate with interbedded sands, siltstones, clays and coals (Micon International Co Ltd 2016). The grades of the Vostochnaya deposit is on average much higher than that of the Solur deposit.

The Vodorazdelnye Galechniki deposit is a buried retained alluvial placer close to the Mir mine under some 2 m of waste, and sourced diamonds from both the Mir and Sputnik

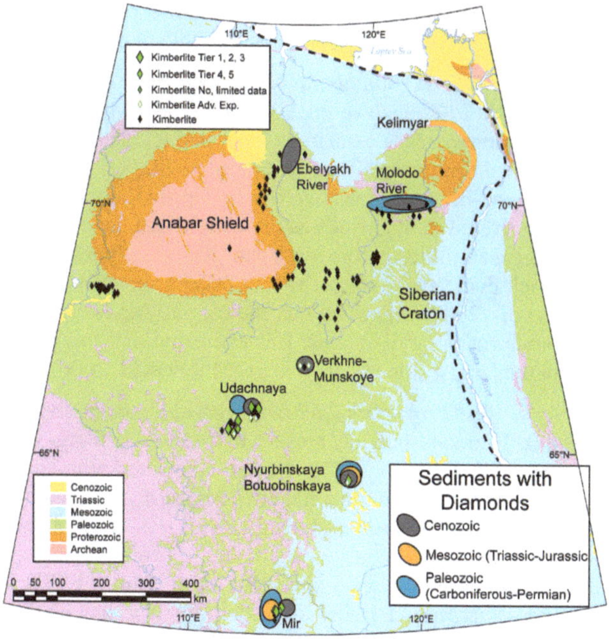

Figure 58. Russia (Siberia) placer deposits, as labeled on the map (Ebelyakh River, Molodo River, Kelimyar, Mir, Udachnaya, Verkhne-Munskoye, Nyurbinskaya-Botuobinskaya). Mercator map projection.

Figure 59. Section of Carboniferous Vostochanaya retained placer near the Mir diamond mine in central Siberia, Russia (photo courtesy J.D. Bristow).

kimberlites. It is a Late Triassic proximal alluvial fan that was upgraded by marine processes at the interface of the fan (Konstantinovskii 2003). The diamonds, which are concentrated in the basal gravel, show no sign of wear (Afanasiev and Pokhilenko 2013).

Zimbabwe (Gweru District): Somabula—Residual placer (Permo-Carboniferous). Diamonds were first reported in Zimbabwe in 1903 (Moore and Moore 2006) and came from deflated glaciofluvial conglomerates at Somabula in central Zimbabwe (Fig. 47). The linear, southeast–northwest aligned Somabula Karoo outlier, preserved on Archean basement, has a discontinuous basal diamond-bearing gravel overlain by Upper Karoo sediments (Moore et al. 2009). The basal braided river system gravel was interpreted as a lag deposit formed by fluvial winnowing of former Permian tillites before the Upper Karoo sediments were deposited. The paleo-current and isopach evidence indicate that the flow of the Somabula River was to the northwest and Moore et al. (2009) suggested that the ca. 520 Ma Murowa kimberlites to the south might be the source for these diamonds. It is estimated that this deposit has produced close to 20,000 ct with the largest stone a 50 ct boart (Moore et al. 2006).

South Africa (North West Province): Lichtenburg/Ventersdorp area—Residual placer (Permo-Carboniferous). Diamond-bearing gravels of the Lichtenburg-Ventersdorp area of the North West Province in South Africa (Fig. 47) are associated with sinuous North–South orientated "runs" that occur exclusively on a flat erosional surface comprised almost entirely of Transvaal dolomites. Some of the coarse-grained gravels (texturally diamictites) that make up the runs, occur within sinkholes directly or indirectly linked to the runs. The gravels have been mined since 1926 and have produced some 12 Mct of diamonds. The runs are narrow, elongated, positive ridges that meander across the dolomite surface. They are up to 30 km long, between 80 and 300 m wide and up to 20 m high (Fig. 60) New data, including field evidence, geomorphological studies, age-dating of diamond inclusions and zircons, indicate that the runs are Permian paleo-eskers deposits that formed during final deglaciation of the Dwyka continental ice sheet, rather than post-Gondwana drainage deposits linked to southward flowing late Cenozoic river systems, as has previously been assumed (de Wit 2016).

Figure 60. Permo-Carboniferous and diamond-bearing sinuous eskers on the Transvaal dolomites near Lichtenburg, South Africa.

Brazil (Mato Grosso do Sul State): Paranaíba and Parana Basins—Transient glacial placer (Permo–Carboniferous). The Carboniferous glacial outwash plains of the Aquidauana Formation (Mato Grosso do Sul State; Fig. 49) and the fluvio-deltaic facies of the Poti Formation (Piauí State) further north, are related to the Carboniferous glaciation of the Paranaíba Basin, and have been reported to contain diamonds (Tompkins and Gonzaga 1989). In addition, mineralized parts of recent drainages in the Parana Basin are intimately associated with glaciofluvial conglomerates and reworked tillites of the Itararé Group. It has been suggested that the glacial centres were located in southern Africa (Dos Santos et al. 1996), and based on glacial groove direction, pebble orientations, and paleocurrent patterns, that these glaciogenic deposits (and hence their diamonds) are interpreted to be derived from the Kaapvaal Craton (Perdoncini and Soares 1999).

Australia (South Australia State): Springfield Basin—Transient glacial placer (Permo-Carboniferous). Within the Springfield Basin (Fig. 61) diamonds occur exclusively in basal conglomerate, which are most likely glaciofluvial sedimentary rocks of Permo–Carboniferous age, and that in part were subsequently reworked into younger sediments (Tappert et al. 2009). Many of the diamonds from the younger sediments resemble those derived from the nearby Jurassic kimberlites at Eurelia. However, the diamonds found in gold mining operations SE of Adelaide in the Echunga area are abraded and have green and brown radiation spots, and Tappert et al. (2008) suggested that these placer diamonds and the "older" diamonds from the Springfield Basin were transported to south Australia by Permo-Carboniferous glaciers from the eastern part of Antarctica.

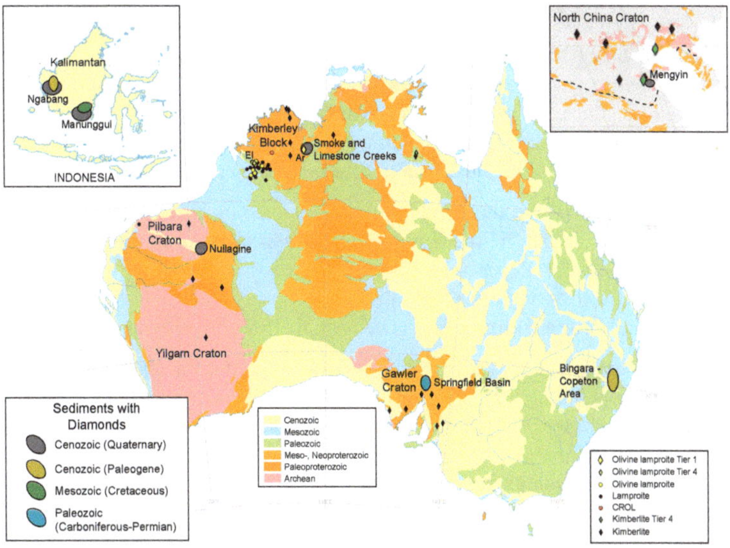

Figure 61. Australia placer deposits as labeled on the map (Springfield Basin, Bingara-Copeton, Smoke and Limestone Creeks; Nullagine). Primary magmatic olivine lamproite mines/deposits as follows: Ar, Argyle; El, Ellendale. Mercator map projection. Indonesia placer deposits as labeled on the **top left inset** map (Ngabang, Manunggul); China placer deposit (Mengyin) as labeled on the **top right inset** map.

South Africa (North West Province): Schweizer-Reneke area—Terminal placer (Permian). Diamondiferous gravels in the Schweizer-Reneke area, south of Lichtenburg (Fig. 47), have been subdivided into a derived gravel unit and four younger alluvial units (Marshall 1990). The derived unit is the oldest and formed as a result of the landscape lowering by weathering and deflation. These gravels are preserved as remnants on a pre-Karoo surface and have often been referred to as "Rooikoppie" gravels (de Wit et al. 2000). Recent research has shown that these deflated gravels may be closely linked to the Permian paleo-shoreline of the Ecca Group, which has reworked some of the diamondiferous glacial sediments of the Dwyka Group (Ward et al. in prep) that can be linked to the residual placers of the Lichtenburg/ Ventersdorp area (de Wit 2016). The younger alluvial units occur at respective lower elevations towards the Vaal River and have reworked the Permian paleo-placers. Some 2.5 Mct have been recovered from this area, with diamond values much higher than those from Lichtenburg and Ventersdorp (de Wit 2016).

Mesozoic

Eswatini (Lubombo Region): Hlane—Transient fluvial placer (Upper Triassic). In northeast Eswatini (formerly Swaziland), diamonds were discovered in Upper Triassic grits and conglomerates (Elliot Formation) at Hlane (Fig. 47). These sediments represent erosionally based, clast-supported channel-fills and host a well-sorted population with 95% of the diamonds that fall in the 0.5 to 2 mm fraction (Turner and Minter 1985). This transient placer formed as a result of a single phase of tectonic uplift and denudation accompanied by climatic change (Turner and Minter 1985). Sedimentological studies of these grits led directly to the discovery of the primary source, the Dokolwayo (Hawthorne et al. 1982) carbonate-rich olivine lamproite.

Russia (Republic of Sakha): Kelimyar River region, northern Siberian Shield—Terminal marine placers (Upper Triassic and Jurassic). Diamond placers were discovered on the northern Siberian platform in 1965. Several types have been identified: terminal near-shore marine placers of Triassic and Jurassic ages, early Jurassic deltaic deposits and Neogene to Quaternary transient alluvial deposits (Prokopchuk 1972). The mineralized Upper Triassic basal conglomerates, which were found to host diamonds and kimberlitic indicator minerals, represent a Mesozoic terminal placer (Grakhanov et al. 2010, 2015; Sobolev et al. 2013; Nikolenko et al. 2018). These Triassic (Rhaetian Stage) basal marine shoreline facies of between 0.1 to 1 m thick, can be traced from the lower reaches of the Lena River (reference section along the Kelimyar River) westwards along the Laptev Sea as it wraps around the Anabar and Olenek blocks (Fig. 58). These sediments formed during transgressions across paleo-surfaces that represent major breaks in sedimentation after the Middle Devonian uplift of the Anabar and Olenek blocks.

Detrital zircons from these sediments returned an age range of 245–230 Ma. Comparable aged zircons are also found in the Jurassic marine rocks (Grakhanov et al. 2015) and in the Cenozoic Ebelyakh River (Fig. 58) placer (Sobolev 2018), which illustrates the level of reworking towards younger formations. These ages are also similar to the weakly diamondiferous local Triassic kimberlites (Davis et al. 1980; Smelov and Zaitsev 2013).

The coastal and deltaic placers are generally lenticular in shape with low diamond concentration (Prokopchuk 1972). The Upper Triassic deposits are thought to represent an entire second-order transgressive–regressive cycle (Egorov and Mørk 2000) while the Lower Jurassic deposits rest on an erosion surface above these formations (Ilyina and Egorov 2008). These Upper Triassic sediments have indicator minerals derived from the Triassic kimberlite bodies, but no indicator minerals derived from the Palaeozoic kimberlite-hosted diamond mines much further to the south (Grakhanov et al. 2010).

Cretaceous

Indonesia (Kalimantan Barat, Kalimantan Selatan Provinces)—Transient fluvial placer (Upper Cretaceous–Lower Paleogene). Diamonds were found in west Kalimantan (Barat) around 600 AD associated with the Landak and Kapuas Rivers near Ngabang (Fig. 61, inset map). Much later, in the 1600's, diamonds were recovered from the Martapura drainage basin in the southeast (Selatan) at Cempaka, along the Apukan, Riam Kanan and Riam Kiwa rivers (Fig. 61, inset map). Other small occurrences of diamonds have been reported from the more central part of Kalimantan. In total just under 2 Mct have been produced and at its peak the annual production for the island was between 20,000 and 30,000 ct of generally small (average of 0.3 ct/st) but gem-quality diamonds (Spencer et al. 1988). The largest diamond recorded is a 166.8 ct stone from southeast Kalimantan.

In the southeast, diamonds occur in clastic rocks of the Upper Cretaceous to Lower Paleogene Manunggul Formation (Fig. 62) that crop out in the Meratus Mountains, and in Pleistocene fanglomerates and alluvial fans or sheet wash sediments. Diamonds also occur in Holocene alluvials (Fig. 63) and recent alluvium and river terraces (White et al. 2016), but in the more distal areas around Cempaka, a large part of this region is covered by swamps.

Figure 62. Cretaceous Manunggul conglomerates in southeast Kalimantan, Indonesia

Figure 63. Younger Holocene gravels being concentrated using wooden cone-shaped "pans', around Cempaka, Indonesia.

The Manunggul Formation is believed to be the source for the Cempaka diamonds. Continuous uplift of the Meratus ophiolite and erosion of the Manunggul conglomerates was the source of the diamonds to the younger sediments. Although the Manunggul Formation has been mined, grades in the Pleistocene fanglomerates and the alluvial gravels of the younger paleochannels were higher as a result of reworking processes. Interestingly Spencer et al. (1998) noted that the upper gravels proved to have higher grades than the basal gravels. Around Ngabang in the west of Kalimantan, diamonds occur in Eocene stream channels, and in derived recent stream beds that drain exposed areas of these ancient stream channels, usually near the flanks of uplands (Spencer et al. 1998).

White et al. (2016) recognized two groups of Cempaka diamonds, based on the study of the heavy mineral suite from the Cempaka alluvial deposit. One group, supported by evidence of long-distance transport and/or multiple recycled stages, indicates that these were either already present in Borneo at least by the Early Cretaceous and were reworked several times, or they were transported from NW Australia to Borneo before it rifted from Gondwana in the Late Jurassic. A second, more proximal sourced group was interpreted to have come from local diamondiferous diatremes emplaced during the Neogene tectonic events (White et al. 2016). Smith et al. (2009), based on the abrasion features and the surface occurrence of radiation damage brown and green spots on many of the diamonds, concluded that the diamonds have gone through multiple cycles of fluvial transportation. This is supported by Kueter et al. (2016) who, based on zircon provenance and diamond morphologies, indicate that the great majority of the Kalimantan and other SE Asia diamonds have northern Australian source(s).

South Africa (Northern Cape Province): Mahura Muthla—Transient placer (Upper Cretaceous). Remnants of diamond-bearing fluvial gravels of Cretaceous age have been preserved on the dolomites of the Ghaap Plateau, northwest of Kimberley (Fig. 47). This north-westerly orientated paleo-channel was directly linked to the Kalahari River drainage basin that drained the northern part of the Northern Cape and southern Botswana via the paleo-Molopo and lower Orange River. By the Early Cenozoic the latter had captured the Vaal and the upper and middle Orange Rivers that had previously been part of the Karoo River during the Cretaceous. The gravels, which are highly calcretized, have yielded over 3,500 carats, the largest diamond was an 8 ct stone; these diamonds were probably sourced from some of the carbonate-rich olivine lamproites known from the Ghaap Plateau to the south and southeast (de Wit et al. 2009).

Angola/DRC (Lunda Norte/Kasai Provinces): Calonda/Kwango Formations—Retained placer (Upper Cretaceous). In Angola, with the exception of the recent alluvial diamond deposits around the Cretaceous diamond-bearing kimberlites, all diamond placers (Fig. 64) are directly related to the presence of the Upper Cretaceous Calonda Formation in northern Angola (Pereira et al. 2003) and the equivalent Upper Kwango Group in southern DRC (Roberts et al. 2015). Many of the diamond-bearing kimberlites in northeast Angola have been dated at between 113 and 145 Ma, including Catoca at 118 Ma (Robles-Cruz et al. 2012). A group of older, low-grade, Triassic kimberlites on the west side of the Kasai Craton are the sources for small-scale diamond operations in recent stream sediments in the Lubia area (Jelsma et al. 2013).

The Cretaceous period of kimberlite emplacement is followed by a phase of subsidence of the Congo Basin from 110 to 60 Ma (Linol et al. 2015) which triggered the deposition of the Upper Kwango Group, also referred to as C3 and C4 (Calonda Formation) of Late Cretaceous age (Roberts et al. 2015). This formation, which is highly weathered, consists of a series of sandstones and conglomerates (Fig. 65), and locally contains diamonds. The basal gravels are generally upward-fining with siliceous resistant and non-resistant clasts up to boulder-size.

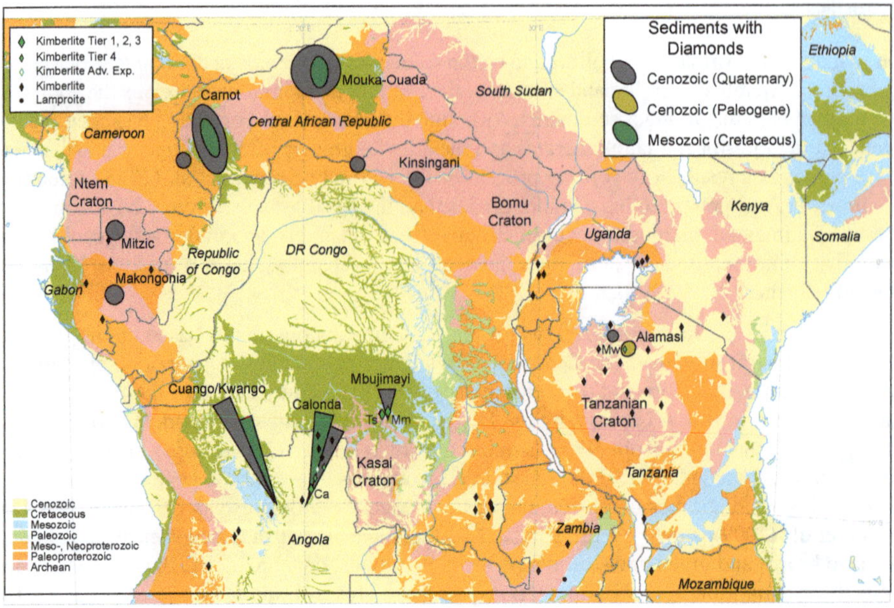

Figure 64. Central Africa placer deposits as labeled on the map (Mouka-Ouadda, Carnot, Mitzic, Makongonio, Cuango/Kwango River, Calonda, Mbujimayi, Alamasi, Kinsangani). Primary magmatic kimberlite mines/deposits as follows: Ca, Catoca; Ts, Tschibwe; Mm, Mbuji Mayi, Mw, Mwadui. Mercator map projection.

The thickness can be up to 250 m in structurally controlled grabens, but in the diamondiferous areas the thickness is generally less than 50 m.

In Angola the basal conglomerates of the Calonda Formation represent the first cycle of diamond capture after release from the proximal kimberlites. This diamondiferous formation occurs in channels and runs from the Lucapa area in a broad subsided part of the western Kasai Craton towards the northeast to cover the region of the Luachimo (Longatshimo) and Chicapa (Tshikapa) Rivers (Ambroise 1991). This northward orientated Cretaceous river system entered the DRC south of Tshikapa. The diamonds here are generally smaller and believed to have been derived from the erosion of the Angolan kimberlites and Calonda sediments, but some economically viable deposits still exist up to 600 km from source (Sutherland 1982). Reworking of the intensely chemically weathered Calonda sediments has resulted in upgraded younger Miocene, Plio–Pleistocene and recent deposits on floodplains, terraces (with an elevation of 1 to 40 m above the present river level) and slope eluvium (directly derived from Calonda conglomerates), but particularly in the Late Quaternary alluvial deposits (Spaggiari and de Wit 2021). The Cenozoic onset of erosion of the Calonda, which has been recorded by younger alluvial deposits, is closely linked to the Oligocene (30 to 35 Ma) and post-Pliocene uplifts of the Angolan Highlands (White et al. 2009). These uplifts initiated alluvial cycles that have upgraded the diamonds to a higher percentage of gem quality stones. The alluvial deposits in the Tshikapa triangle have produced over 90 Mct and are part of the largest diamond megaplacer in central Africa, stretching from Lucapa in northeast Angola to Tshikapa in the southern DRC that has contributed in excess of 200 Mct (de Wit et al. 2016).

Brazil (Minas Gerais State): Capacete, Coromandel area—Retained placer (Upper Cretaceous). Fernandes et al. (2014) have shown that the basal polymictic conglomerates of the Upper Cretaceous Capacete Formation, uppermost in the Mata da Corda Group, contain diamonds (Fig. 49). Diamonds were mined from lithified conglomerates at the Romaria mine (western Minas Gerais State) on the northeastern side of the Paraná Basin in a tributary of the Bagagem River at Água Suja (Svisero et al. 2017). The Tauá conglomerate, which contains the diamonds (Fleischer 1998), forms the base of the Upper Cretaceous Uberaba Formation, which is an equivalent to the Capacete Formation conglomerates. These conglomerates have been interpreted as a debris flow or alluvial fan deposit consisting of materials from surrounding Neoproterozoic meta-sediments (Araxá Group) and Lower Cretaceous Botucatu sandstone and basalt (Areado Group) (Pereira et al. 2017). The diamonds are concentrated in a 6 m thick conglomerate deposit which has grades of 0.05 to 0.12 ct/m³ (Pereira et al. 2017). The diamond suite is dominated by rhombododecahedral shapes. They are generally colorless with 70% gem quality (Svisero et al. 2017) and are similar to diamonds from the proximal Cretaceous diamondiferous Três Ranchos cluster kimberlites (99–81 Ma) (Pereira et al. 2017).

The source of these alluvial diamonds, however, has been debated. Some suggest that erosion of the many local kimberlites released diamonds to be incorporated into the Capacete conglomerates (Karfunkel et al. 2014) and were subsequently distributed throughout the Coromandel region (Fernandes et al. 2014). Others propose that the diamonds have been transported by ice from the São Francisco Craton by Neoproterozoic (Jequitái, Ibiá glaciations) and Paleozoic (Santa-Fé de Minas tillites) glacial events (Tomkins and Gonzaga 1989, Gonzaga et al. 1994). Read et al. (2004) on the other hand, based on stratigraphic relationships and mantle-derived indicator mineral suites, advocate that the thickness of the underlying lithospheric mantle of the Quiricó Basin and surrounding areas of the Coromandel thinned dramatically during the onset of the Upper Cretaceous Mata da Corda Group. The Mata de Corda Group sediments covered the Lower Cretaceous kimberlites, which were emplaced when the mantle was still cool enough to form and preserve diamond to be transported by kimberlites to the surface.

Fernandes et al. (2014) interpreted the basal Capacete conglomerates as being deposited by alluvial fans and braided rivers as a result of local Late Cretaceous uplift. These probably formed the main diamond source rock in the Santo Antônio do Bonito, Santo Inácio and Douradinho river alluviums. The possibility that the Neoproterozoic and Palaeozoic diamictites as well as the Cretaceous kimberlites, have all contributed to the mix at certain localities has not been excluded (Fernandes et al. 2014). However, most of the mined alluvial deposits are of Late Cenozoic age and mainly from terraces and floodplain gravels.

Brazil (Mato Grosso State): Poxoréu—Retained placer (Upper Cretaceous). Rivers such as the Coité, São João, Poxoréu and Pomba in the northwestern part of the Paraná Basin (Fig. 49) have been producing diamonds since the 1930s with grades up to 0.05 ct/m^3 and with some 27% being gem-quality (Souza 1991). All these diamonds are recovered from sedimentary rocks of the Upper Cretaceous Bauru Group (Uberaba Formation), which is associated with a half graben (Svisero 1995; Souza 1991). Diamonds are also recovered in various outliers of residual gravel on the high plateaus to the W and NW of Poxoréu (Fleischer 1998), which are thought to be related to conglomerates of the same formation (Weska 1996). Although the Capacete and Uberaba Formations are time equivalents, Quintão et al. (2017) have shown that the diamonds from the Uberaba Formation are not related to those from the Romaria mine in the Capacete conglomerates.

Cenozoic

Tanzania (Shinyanga Region): Alamasi—Retained "epiclastic" placer (Paleogene). The first diamonds found in Tanzania were from alluvial gravels at Mabuki in 1917, which were mined from 1925 onwards, and led to the discovery of Tanzania's first kimberlite, the Mabuki pipe. The Mwadui mine (Fig. 64), the world's largest kimberlite-hosted diamond mine, was found thereafter and mining here initially concentrated on surficial deposits adjacent to and over the pipe. This residual placer of eluvial gravels is mainly lag deposits that resulted from the weathering of the tuff ring surrounding the pipe and the uppermost deposits within the kimberlite crater. Limited down wasting resulted in tremendous increase of the diamond grades in these lag deposits, particularly adjacent to the pipe—the Alamasi deposit (Fig. 66). A regional geomorphological study suggested that very little erosion had occurred since the eruption of the kimberlite at 52 Ma (Stiefenhofer and Farrow 2004), and hence there has been limited dispersion of diamonds and indicator minerals <1 km away from the pipe (Edwards and Howkins 1966). Field (2010) makes the point that these deposits are technically not "alluvials" as there has not been any recognizable fluvial transport and are therefore explained as Eocene residual remnants of proximal reworked eluvial material.

Australia (New South Wales State): Bingara and Copeton—Transient placer (Paleogene). There are two main areas of alluvial mining in Australia; one along the east coast and the other associated with the Argyle olivine lamproite in the Kimberley Block of northwest Australia (Fig. 61). Some small occurrences of alluvial diamonds occur in southern Australia as described earlier.

Alluvial diamonds are found in eastern Australia from Queensland in the north to Tasmania in the south, but have only been mined in the New South Wales area (since 1867). Historical production from the Bingara and Copeton areas (Fig. 61) up to 1973 was 34,000 and 167,000 carats, respectively; the total production for these two areas has been estimated to be 500,000 carats (Davies et al. 1999). The average stone size in Bingara is 0.2 ct and the largest reported stone 2.6 ct; the Copeton area has an average stone size of 0.25 ct with the largest stone being a 12 ct piece of boart (MacNevin 1977).

The diamonds occur in upper Cenozoic gravels, sands and silts that are beneath Miocene basalts. The gravels occur in channels preserved as remnants of a palaeo-river system. Reconstruction of the palaeo-drainage system in the Bathurst region suggest that the general flow direction was towards the north, from the late Mesozoic and through to the Cenozoic (Cham 1998). In some areas the basal planar cross-bedded conglomerates, interpreted as

Figure 65. Calonda Formation gravels of Cretaceous age at Chitotolo, Angola. Note the weathered and "bleached" state of the locally derived clasts.

Figure 66. Poorly sorted and coarse-grained conglomerates at the outer edge of the Mwadui kimberlite, Tanzania.

lateral accreting channels and typically associated with fluvial processes, has been the target for the miners. In other areas, such as the Craddocks claim, breccias are overlain by upward-coarsening sandstone beds and are believed to be associated with a sequence of volcaniclastic rocks into which the diamonds were incorporated, in processes resembling debris flows or even via primary volcaniclastic processes (Fig. 67). Pillow lavas, basalt-filled valleys, intercalated sediments between basalt flows and almost "tuffaceous" textures in the fluvial sediments all indicate that volcanic activity interrupted the fluvial system transporting the diamonds.

The diamonds can be grouped into two distinct populations. One group, similar to those found in kimberlites and lamproites, has abrasion features and radiation damage

Figure 67. Poorly-sorted diamondiferous breccias overlain by volcaniclastic rocks in the Callas Hill audit, near Bingara, southeast Australia.

suggesting that they represent diamonds recycled from older sedimentary deposits. These could have been transported northwards by Permian glaciers that affected most of southern Australia and derived from primary source(s) as far afield as Antarctica (Davies et al. 2002). This interpretation is supported by inclusion studies that suggest these diamonds are associated with ~340 Ma magmatism (Burgess et al. 1998). The second group, characterized by unusual heavy C isotope compositions and Ca-rich eclogitic inclusions (Sobolev et al. 1984), has been linked to a young subduction model (Davies et al. 1999). There are no typical diamond indicator minerals associated with these sediments.

South Africa (North West, Northern Cape Provinces): Orange–Vaal River terraces—Transient placers (Paleogene to Quaternary). The Vaal–Orange River system (Fig. 47) is the main route along which the diamonds have been transported from the Kaapvaal Craton to the southwest African coast. There were two main ancestral rivers, flowing westwards since break up of west Gondwana: the Kalahari River, with its lower part following the present Lower Orange River and the Molopo River out of Botswana, and; the Karoo River that drained most of the Vaal and middle and upper Orange Rivers before it entered the Atlantic Ocean much further to the south than currently (de Wit 1999). River capture by the end of the Cretaceous resulted in the present-day Orange–Vaal drainage configuration. This interpretation is supported by diamond inclusion geochronology studies which suggests that many of the diamonds along the west coast of South Africa (Namaqualand) are sourced from kimberlites or carbonate-rich olivine lamproites with eruption ages of between 115 and 300 Ma via the Karoo River (Phillips and Harris 2018). Using the same techniques, Phillips et al. (2009) were able to show that most of the diamonds along the west coast of Namibia, north of the Orange River mouth, are derived from younger (<90 Ma) kimberlites and also older carbonate-rich olivine lamproites, and were transported to the coast by the present Orange–Vaal River configuration. Longshore drift has carried diamond populations from both of these rivers northwards along the coast.

Examples of retained placers in this system include the diamond deposit on Nooitgedacht close to Kimberley with diamonds from the Kimberley mines (Fig. 47), before it became incorporated into transient Vaal River placers (de Wit 2004), as well as deposits adjacent to some of the Lesotho kimberlites, such as the Letšeng diamond mine (Fig. 47).

The oldest (Eocene) Orange River deposit (Bluck et al. 2005) has a low grade (<0.5 ct/100t) and low average stone size (<0.4 ct). The Oligocene pre-Proto Orange River terrace deposits have grades up to 35 ct/100t and an average stone size of 2 ct, these are followed by the Miocene Proto Orange River terraces with grades of 1 to 5 ct/100t and average stones sizes of 1 to 2 carats, and then by the Meso Orange terraces (Plio-Pleistocene; Figs. 68, 69) with low

Figure 68. Coarse-grained Plio–Pleistocene gravels along the Vaal River at Gong Gong near Barkly West, South Africa. Note imbrication of the oversize boulders.

Figure 69. Gravels being mined in the town of Windsorton along the Vaal River, north of Kimberley, South Africa.

grades <0.5 ct/100t. This suggests that there was an early entry of fine diamonds in the Eocene, followed by the main flush in the pre-Proto time and since then a declining diamond grade; however, big stones are associated with the younger terraces (Bluck et al. 2005).

Terraces along the middle Orange River, where terrace elevation varies from some 270 m (Late Cretaceous?) to some 20 m (Pliocene) above the river, have a quite notable large diamond population, with several +200 ct stones and at least 22 +100 ct stones recovered (De Meillon 2019; Norton et al. 2007). This area produces a mixture of diamonds sourced from the Cretaceous kimberlite in Lesotho (e.g., Letšeng), the Free State and Northern Cape Province (Kimberley, Koffiefontein, Jagersfontein, Finsch etc.), Permo-Carboniferous sources, as well as the reworked deposits in the Northwest Province and the Witwatersrand basin. In addition to the terrace stratigraphy along the Vaal and Orange Rivers, there are large areas within the river valleys covered by what is termed "Rooikoppie" or derived gravel. These are thin but chemically mature deflation deposits from older alluvial gravel or tillite units, consisting mainly of well-rounded and often polished siliceous pebbles in a loose red colored sand that are often highly profitable (Marshall 2004).

The economic sections of the upper Orange are restricted to residual placers next to the Lesotho kimberlites. From Lesotho to Hopetown the Orange River flows over horizontally bedded sandstones and shales of the Karoo Supergroup with only Jurassic dolerite dikes as potential trap-sites and these are not conducive to development of economic placers (Fig. 47). Only from Hopetown to Prieska, where the river cuts through the Dwyka tillites at the base of the Karoo, diamonds are trapped in the coarse gravel fabrics developed as a result of the breakdown of the tillites such that viable deposits are formed (de Wit 1996). From Prieska to Noordoewer there are only a few isolated and small remnant alluvial deposits, but the final 270 km stretch of the lower Orange River from Noordoewer to the mouth at Alexander Bay (Fig. 70) is where the lower Orange River economic deposits occur.

Figure 70. The Orange River and its delta, with aggressive wave action and long-shore drift transporting diamonds northwards. Looking northwards along Namibia shoreline.

Some of the large-scale trap-sites that have formed are splay deposits were the river exits narrow gorges, cut into bedrock, and develops coarse grained terraced fans. This is seen along the Vaal (e.g., Windsorton, Barkly West) and the Orange Rivers (de Wit 1996). These are often controlled by faults or due to changes in basement lithologies. Trap-sites on a smaller scale are coarse cobble-boulder size basal gravels close or on bedrock, with the development of scour pools, push bars and bedrock highs where turbulence created by boundary conditions at these fixed bedrock sites, are conducive for stable growth of gravel that retains the concentrated diamonds (Jacob et al. 1999).

South Africa/Namibia (Northern Cape Province/Karas Region): West Coast—Terminal placers (Paleogene to Quaternary). Epeirogenic uplift of southern Africa in the Cretaceous (de Wit 2007) has been the driving force in the development of the alluvial placers along the Orange River but also the creation of the spectacular marine deposits along the coast of southwest Africa (Figs. 47, 70). The diamonds in this terminal marine setting are hosted in fluvial, marine, deflation and aeolian placer types with temporal ranges from at least the end of the Cretaceous to the modern day. This large deposit, which stretches over 1,500 km from the Namaqualand coast of South Africa northwards to the Skeleton Coast of Namibia (Hallam 1964; Corbett 1996; de Wit 1996; Bluck et al. 2005), is separated into two distinct terminal placers, defined as the Namaqualand and Namibian megaplacers (Bluck et al. 2005), which from their initial discovery until 2019, have produced 53 and 112 Mct, respectively. The bulk of the diamonds were derived from the Namibian deposits along the Sperrgebiet coast, including almost 16 Mct from unique deflation and aeolian deposits (de Wit et al. 2016).

Offshore mining in this area was pioneered by a Texas oilman, Sammy Collins in the early 1960s but it was only in the 1970s that the significant deposits in water depths of 120–140 m were discovered. These deposits consist of mostly thin accumulations of coarse clastic submerged beach gravel, overlying an Eocene and Cretaceous clay footwall (de Wit et al. 2016). Following a long period of prospecting and technical development, mining of these marine deposits started in earnest in the early 1990s, and approximately 25 Mct have been recovered from the offshore areas. The greatest remaining potential along the West Coast is located in these deposits, with indicated/measured and inferred resources at 12.3 and 70.6 Mct and grades of 0.08 and 0.07 ct/m^2 respectively for the midwater and marine placers combined (Anglo American Ore reserves and mineral resources report 2019).

Russia (Republic of Sakha): Ebelyakh (Anabar River basin), northern Siberian Platform—Transient fluvial placers (Neogene to Quaternary). Although diamonds are present in the Mesozoic terminal placers (see above), these are uneconomic to mine and act only as a source to the viable Quaternary transient alluvial placers that are associated with tributaries of the Anabar River (Fig. 58). These occur towards the northeastern edge of the Siberian platform, east of the Archean Anabar shield, west of the Palaeoproterozoic Olenek uplift and some 450 km north-northeast of the Udachnaya and Zarnitsa mines. These alluvials are lenticular shaped deposits following valley profiles and are enriched in areas where they directly erode the older diamond bearing coastal and deltaic deposits. Other placers in northern Siberia, are found along the Molodo River (Fig. 58), a tributary of the Lena River, some 500 km northeast of the Udachnaya and Zarnitsa mines. This region is considered to be one of the world's largest secondary diamond fields (Micon 2016); combined these transient alluvial deposits have produced between 4.8 and 5.2 Mct/year between 2014 and 2016 with resource grades of between 0.77 to 1.33 ct/t (Alrosa 2017).

The Ebelyakh placer is underlain by Paleozoic and Mesozoic sediments and has a cover of unconsolidated Cenozoic sediments. Much of this placer is diamondiferous, mainly as the results of transgressive-regressive cycles (Ilyina and Egorov 2008). Paleo-weathering surfaces developed at various stratigraphic levels during periods of regression from the end of the Carboniferous to the Neogene, concentrating diamonds as a result of chemical and physical deflation. Reworked remnants of these diamond-bearing weathering surfaces and older sediments form the economic Neogene to Quaternary age transient diamond placers along several tributaries of the Anabar River, such as the Ebelyakh River. The Ebelyakh River deposits cover 83 km from its confluence with the Anabar River and is on average some 80 m wide, and include deposits along different branches of the Ebalyakh River. All the alluvial formations, the riverbed, the floodplain deposits and terraces, as well as the reworked weathering crusts in the river valley are diamond bearing.

The next major tributary to the Anabar River is the Maiat River some 25 km to the north, whose economic alluvial deposits occur mainly in branches in its upper reaches. Between the Ebelyakh and Maiat rivers is the Billiakh River and associated alluvial deposits. The Ebelyakh, Maiat and Billiakh rivers are right tributaries of the Anabar River, draining areas to the east. Other placers occur on left tributaries to the Anabar and include the Khara Mas stream, almost opposite the Billiakh River, and the Bolshaya Kuonamka (B-K) River some 20 km south of and opposite to the Ebelyakh River. The B-K River drains near-shore marine deposits of Meso-, and Neoproterozoic (Ediacaran) and early Cambrian age that are intruded by barren or low-grade Triassic and Jurassic kimberlites. Over 95% of the B-K reserves are comprised of Quaternary upper and lower floodplain (river placer) deposits, the rest being contained in three terraces that are of lower volume—but higher grades. The diamonds from the latter deposits are mainly colorless and significantly abraded (Afanas'ev et al. 2009). Indicator minerals derived from Upper Palaeozoic sediments also occur in these younger sediments (Grakhanov et al 2010), emphasizing the importance of the process of reworking and upgrading of older marine deposits to form the transient placer diamond deposits. This is also illustrated by placer gold recovered from some of these rivers, which has likewise been derived from the reworking of older sequences (Gerasimov 2019). By comparing diamonds from Palaeozoic, Mesozoic and Quaternary alluvial deposits, which are dominated (>85%) by eclogitic stones (Shatsky et al. 2019), and Triassic kimberlites of the same district, it was found that these kimberlites have only contributed to a part of the Anabar placer diamond population (Sobolev et al. 1999). Since the diamonds from the northern placer deposits are different to those from the southern Yakutian Palaeozoic kimberlite-hosted diamond mines (Grakhanov et al. 2010; Sobolev et al. 2018), the source(s) of these diamonds is still not known (Olev 1973; Sobolev et al. 1999).

Australia (Western Australia State): Smoke and Limestone Creeks of Argyle—Retained placers (Neogene to Quaternary). The discovery of diamonds in Smoke Creek in Western Australia (Fig. 61) led directly to the AK1 Argyle olivine lamproite pipe (Rayner et al. 2018; Smith et al. 2018), which occurs within the Halls Creek mobile belt, east of the Kimberley Craton (Shigley et al. 2001). Alluvial diamond deposits occur along Smoke Creek and Limestone Creek and these retained placers developed because of the proximity to a high-grade primary source (average grade 3 ct/t) that is part of a major positive topographic feature, with streams flowing to the north (Smoke Creek) and to the south (Limestone Creek). Local Devonian-aged conglomerates provided coarse clasts to the alluvial fabric necessary to concentrate diamonds, and finally the older alluvial terraces were upgraded by chemical deflation during periods of laterization.

Both creeks have been mined extensively and a total of 44.3 Mct and 6.8 Mct have been mined out of Smoke Creek and Limestone Creek, respectively (B. Janse pers. comm. 2020). Smoke Creek is mineralized over its entire length from the Argyle lamproite pipe to where it enters Lake Argyle, a distance of some 35 km (Fig. 71). The creek has three terrace levels according to elevation and the difference in the proximal reach between the oldest and youngest terraces is roughly 10 m. Pits dug in the 1 to 2 m thick oldest terrace show typical coarsening-upwards sequences indicative of an alluvial braidplain or alluvial fan system with gravel bars that have well-developed bar-top pavements. The grade of the older terrace was on average 3 to 4 ct/t reaching up to 15 ct/t in more proximal sites. The younger terraces are texturally less mature and become sandier downstream suggesting an increase in aggradation with distance. This is reflected in the reduced grades of between 1 to 2 ct/t.

The Limestone Creek alluvials have a very similar stratigraphy with three terraces, possibly from Pliocene or even Miocene age through to the Late Pleistocene for the youngest terrace, above the Holocene present-day river deposits. The oldest terrace deposit has been interpreted as a veneer of residual lag gravel on limestone where karstification of the carbonates in the form of sinkholes contributed to its preservation. Both Limestone and Smoke creeks are comparatively minor low-energy alluvial systems but with increased sediment transport capabilities during the wet season.

Figure 71. Aerial view looking south, of Argyle mine (Australia) in the gap of the Ragged Range hills with Smoke Creek draining to the north. Note the terraces levels in the valley.

Russia (Republic of Sakha): Molodo River, northeast Siberian platform—Transient fluvial placers (Quaternary). The Molodo placer is on a tributary to the Lena River south of the Olenek high (Fig. 58). Diamonds are mined from Quaternary alluvial sediments in the midstream section of the Molodo. The river has cut through Permian coal, Lower Triassic volcanics and Jurassic sandy-silty strata and the bedrock substrate is weathered Lower to Middle Cambrian carbonates and siliciclastics. The economic part of the deposit is in the present river bed gravels and in the five terraces of the lower and upper floodplain which are Pleistocene in age. Almost 40% of the diamond reserves are held in the upper floodplain and another 48% in the recent alluvium (Micon 2016). A high concentration of diamonds was also found near the mouth of the Bulkur River, a left tributary of Lena River and to the north of the Molodo River.

Siberia (Republic of Sakha): Mir, Udachnaya, Nyurbinskaya and Botuobinskaya alluvials, central Siberian Craton—Retained placers (Quaternary). There are two recent and proximal retained placers close to the Mir kimberlite, associated with the Irelyakh River (Fig. 58). These are sandy to pebbly river and floodplain gravels that include four associated terraces which are up to 6.4 m thick and covered by some 2 m of overburden. These have been mined by open pit methods. The Gornoye placer, 26 km southeast of Mir, features river gravels that stretch for almost 6 km, with a width of between 150 and 1,650 m along the Irelyakh and Malaya Botuobiya rivers. Both have been dredged and generally produce good quality diamonds.

Other retained secondary deposits that are also proximal and intimately related to the primary kimberlite sources include the Kluch Piropovy (av. grade 0.55 ct/m^3) and Zakonturnaya placers (grade from 0 to 1.24 ct/m^3) on the Piropovy stream that has eroded kimberlite and diamonds from the Udachnaya mine (Fig. 58) to produce proximal clay-rich flood deposits. The scattered Verkhne-Munskoye retained placers (Fig. 58) formed from weathering and erosion of the five diamondiferous kimberlites (including the Zapolarny and Magnitny kimberlites) of the upper Muna field, some 160 km northeast of the Udachnaya mine. The Nyurbinskaya and Botuobinskaya placers (Fig. 58), are gravels that are adjacent to, or partially overly the kimberlite-hosted diamond mines with the same names (Micon 2016). Although the diamond grade in all these examples drops off quickly away from the primary source, a 214.6 ct stone was recovered from the Nyurbinskaya placer close to the pipe (Alrosa press release 2016).

Venezuela (Bolivar State): Guaniamo—Transient fluvial placers (Quaternary). In 1969 newly discovered deposits at Guaniamo in Venezuela (Fig. 49) resulted in a large increase in diamond production from alluvial terraces and recent river sediments in that country, particularly from the Cuchivero River in the Orinoco River Basin. Initially it was thought that the diamonds had been derived from the Roraima sediments. However, in 1982 the Guaniamo kimberlites sills and dikes were found in the Venezuelan portion of the Guyana Shield (Baxter-Brown and Baker 1991; Meyer and McCallum 1993). These were dated at 712 Ma (Kaminsky et al. 2004) and clearly indicated that Guaniamo kimberlite and Roraima diamond populations are not related. This also suggests that the Palaeoproterozoic primary sources that supplied diamonds to the base of the Roraima Supergroup might still exist. Venezuela is estimated to have produced some 16 Mct by 2010 of which some 14.5 Mct have likely come out of alluvial deposits (B. Janse pers. comm. 2020).

Brazil (Paraná State): Tibagi and Itararé—Transient reworked placers (Quaternary). Although diamonds occur in many stratigraphic sequences, in most instances it is only the reworking of the weathering products that are upgraded sufficiently to permit economic exploitation. Hence there are many areas in Brazil that are, or have been, mined from Late Cenozoic terraces, floodplains and present river channels. Much of the terrace development along the major rivers in the Paraná Basin is Quaternary in age (Oliveira et al. 2019).

Tibagi and Itararé are the most important diamondiferous sites of the eastern Paraná Basin (Fig. 49) and occur in high and low terraces and in active stream sediments of the Tibagi, Cinzas, Peixe and Verde Rivers. These deposits, which have grades of up to 0.62 ct/m^3, were originally discovered by gold prospectors. The diamonds are generally small with an average stone size of between 0.1 to 0.3 ct/st but of high value (90% gem quality; Liccardo and Chieregati 2013). No kimberlites have ever been found in the region and there are no kimberlite indicators associated with these placers. The alluvial diamonds are spatially associated with the glacial-marine and glaciofluvial facies deposits of the Permo–Carboniferous Itararé Group (Liccardo and Chieragti 2013); mineralized parts of recent drainages are often located over the glaciofluvial conglomerates of the Itararé Group.

Brazil (Minas Gerais State): Coromandel area—Transient placer (Quaternary). The Coromandel area and specifically the Alto Paranaíba province (Fig. 49) of the southern São Francisco River basin (Minas Gerais State), is occupied by the Quiricó basin, which is underlain by Neoproterozoic Bambuí Group sediments and contains Lower (Areado Group) and Upper (Mata da Corda Group) Cretaceous sediments and volcaniclastics (Read et al. 2004). Rivers draining this area are a major source area for large high-value diamonds (Karfunkel et al. 2014), such as the 726.6 ct Presidente Vargas, 455 ct Darci Vargas and 400.7 ct Coromandel from the northward draining Santo Antônio do Bonito River, and the 262 ct Star of the South from the Bagagem River. High value diamonds are also found east of the head waters of the Santo Antônio do Bonito River, in the upper and north-eastward flowing Abaeté drainage basin (Read et al. 2004), which has produced a 452 ct stone.

Although the source of the diamonds has been discussed previously, the large diamonds could also have been transported from the São Francisco Craton by two glacial events: the Neoproterozoic Jequitaí, and Cambrian Santa-Fé glaciations (Tompkins and Gonzaga 1989; Gonzaga et al. 1994), although no diamonds have so far been recovered from these glacial sediments and the primary sources of the large diamonds have never been found.

Brazil (Mato Grosso State): Juína—Transient placer (Quaternary). This area is associated with the southern Amazonian Craton where the alluvium of the Juína Mirim, Vinte e Um de Abril and Cinta Larga rivers have been producing diamonds that are likely derived from Cretaceous sediments of the Parecis Basin. In the 1970-80's the rich Juina placer deposits in Mato Grosso State (Fig. 49) produced up to 5-6 million carats annually (Kaminsky et al. 2009), with diamond grades varying from 7 (Cinta Larga River basin) to 0.6 ct/m^3 (Juína Mirim River Basin) (Tremblay 2005), and the major producers had mined some 40 Mct by 2014. Although some of the stones proved to be superdeep diamonds (sub-lithospheric) and mainly Type IIa stones (e.g., Harte et al. 1999; Kaminsky et al. 2009), most diamonds are of industrial quality (85%). The biggest diamond found was 452 ct and at least 15 plus 100 ct stones have been reported with eight over 200 ct (Nannini et al. 2017). In 2006 a group of diamondiferous kimberlitic pipes (Pandrea 1 to 7) were found at the head of a drainage system and some 30 km upstream from the alluvial placers. Diamonds from these kimberlites and the placer deposits proved to be from the same population, which suggests that these kimberlites were probably the source for most of the alluvial diamonds. The kimberlites are Late Cretaceous, Cenomanian/Turonian in age (*ca.* 100–90 Ma), and although the basal conglomerates of the younger sedimentary Cretaceous Parecis Formation contain diamonds, they are not economic. Only the recent river alluvial material at Juína, that has been upgraded from the primary sources via the Upper Cretaceous conglomerates, are mined economically.

Brazil (Piauí, Mato Grosso, Bahia states): other deposits—Mainly transient placers (Quaternary). Diamonds, generally small but good quality, were discovered in 1946 southwest of Gilbués (Piauí State; Fig. 49) along the Riachão stream, an area known for significant historical alluvial diamond production (Junior Mining Network, May 2017). Gravels were mined from the river alluvium and terrace deposits, and it was subsequently the area where the first kimberlite, Redondão, was found in Brazil (Svisero et al. 1977). It has been suggested that the diamonds have been derived from the Lower Palaeozoic (Poti Formation), Upper Palaeozoic (Santa Fé Group, Piaui Formation; Gomes et al. 1972), and/or Cretaceous sediments (Fleischer 1998). This area seems to be off-craton and judging from the generally small sizes but good quality of diamonds, these deposits are likely to have been reworked from older sediments and not related to the nearby Cretaceous kimberlites.

Other deposits include Arenapolis (Mato Grosso State) where alluvials have been mined in Quaternary sediments associated with the headwaters of the Paraguay River and where the diamonds were interpreted to be derived from the Cretaceous conglomerates of the Parecis Formation. This is supported by the high degree of abrasion of these stones (Tappert et al. 2006).

Diamonds were found in recent alluvial sediments in the Lower Pardo River Basin (Bahia State), between the São Francisco and Jequitinhonha rivers (Fig. 49) and some 60 km inland from the coast, and it is estimated that some 980,000 ct were recovered between 1881 and 1890 (Pan American Union 1907, p. 582). Since the rivers cut through the Salobro Formation, a sequence of Lower Cambrian meta-conglomerates and pebbly meta-greywackes, it has been assumed to have been the source of the diamonds. However, no diamonds were ever found in the Salobro Formation and it is therefore more likely that these diamonds had ultimately been derived from the older Espinhaço Supergroup.

India (Andhra Pradesh State): Krishna and Pennar Rivers—Transient placer (Quaternary). In India diamonds were discovered over four thousand years ago. The country has produced some of the world's famous diamonds, such as the Koh-i-noor (1st cut 186 ct), Nizamz (440 ct), Great Moghul (787 ct), Hope (115 ct when purchased in 1666 by Jean Baptiste Tavenier, now 45.52 ct and on display in the Smithonian), Regent (410 ct) and others, all from the Krishna River alluvials (Ravi et al. 2013).

The Quaternary gravels along the Krishna and Pennar Rivers (Fig. 55) have long been known to be a source of gem quality diamonds. The greatest intensity of mining was along a 60 km stretch of the Krishna River between Kollur and Paritala, where diamond bearing gravel is some 30 cm thick. Old workings are also present along the banks of the Pennar River in the area south of the Krishna River. The Krishna diamonds have an average weight of 0.83 ct and nearly 40% of the stones are over 0.5 ct (Ravi et al. 2012). The alluvial diamonds display abrasion features such as percussion marks as well as radiation damage (green and brown spots) suggesting that there has been recycling through older sediments (Ravi et al. 2012). Furthermore, small diamonds have been reported from terminal beach placers of the Krishna–Godavari delta region, but are not considered to be of economic value (Subrahmanyam et al. 2005).

Angola/DRC (Malanje/Kwango Provinces): Cuango/Kwango River—Transient placer (Quaternary). Diamonds are associated with alluvial sediments within the Cuango/Kwango River over a distance of some 700 km (Fig. 64). This river and its tributaries tap into diamondiferous kimberlites and Calonda Formation sediments in its headwaters, to develop very rich deposits in floodplain, terraces and recent river gravels. The maximum age for the Kwango terraces is between 0.3 and 0.5 Ma (de Wit and Thorose 2015). The highest diamond concentrations were found where the river cut the basement, forming rich deposits that have been exposed in river diversions at Luzamba, Cafunfo and Luremo in Angola; in the 1980s these areas were producing some 100,000 ct/month. There is a significant decrease in stone size in the Cuango Valley, from an average stone size of 1.9 ct (Modern Mining, March 2017) in the small Caculio tributary in the upper reaches, to less than 10 st/ct in typical terrace and river deposits along the middle and lower reaches of the Kwango River along the border with the DRC (de Wit and Thorose 2015).

Central African Republic (Mambéré-Kadéï, Lobaye and Haute-Koko Prefectures): Carnot and Mouka-Oudda Formations—Transient placers (Quaternary). Diamonds from the Central African Republic (CAR) are exclusively derived from alluvial deposits (Fig. 64). To date, the country has produced some 25 Mct, of which ~75% are gem quality (Censier 1996). The alluvial deposits show an obvious spatial relationship with two horizontally-bedded Mesozoic fluvial formations: the Mouka-Ouadda Formation in the northeast (Haute-Koko) and the Carnot Formation in the west of the country (Mambéré-Kadéï, Lobaye). Flow directions of these Late Cretaceous palaeodrainage systems are to the north into basins of the Central African Rift system. These two formations unconformably overly Palaeozoic tillites and glaciofluvial deposits on an outwash plain (Censier 1996).

Most of the diamonds have been mined out of recent alluvial sediments that occur in streams on basement proximal to the Mesozoic formations and in streams flowing over these formations. The following alluvial placer deposit types are present: alluvial flat or floodplain, channel deposits and low and high terrace deposits (Chirico et al. 2010). The most prospective alluvial deposits occur on hard basement with fixed trap-sites such as scours, sluits, waterfalls and rapids, but always in proximity of the Mesozoic formations. The terraces may be lateritized, especially if it they are covered by colluvium. One restricted digging operation, in the northeast and upstream from the village of N'Zako, comprises a major, 60–80 m thick, benched, red excavation, and is known for its larger stones (Fig. 72). This digging regularly produces +50 ct stones and Bardet (1974) reports a +300 ct diamond from this deposit.

Figure 72. Diggings at N'zako where some of the largest diamonds have been produced in the Central African Republic.

There are no kimberlite indicator minerals reported from any of the CAR alluvial deposits, but the heavy mineral suite of the diamondiferous gravels and the Mesozoic sandstones are very similar (Censier 1996), suggesting that the Late Cretaceous formations are the source of the diamonds (Censier 1996). Because the clasts in the conglomerates of the Mesozoic formations are believed to have been derived from the Palaeozoic tillites, it is possible that the diamonds could also have been reworked out of the latter. Since the transport direction for both the Mesozoic fluvial as well as the Palaeozoic glaciofluvial systems was from the south, it is reasonable to suggest that the source(s) are located in the northern DRC although the Proterozoic metasediments overlying much of the CAR basement have also been highlighted as a possible, but more unlikely source rock (Chirico et al. 2010).

Diamonds have been produced north of Kisangani in the DRC (Fig. 64) mainly from recent alluvium and are of a similar population to those from the CAR. Most of the diamonds are rhombododecahedral (Censier and Toureng 1995), with a high proportion of the diamonds showing percussion marks, edge abrasions, network patterns and general surface corrosion (Harris 1985).

Sierra Leone (Eastern Province): Kono and Tongo—Retained/Transient placers (Quaternary). The pre-2008 production of diamonds in West Africa was mainly from alluvial sources, but in Sierra Leone the Koidu kimberlite-hosted diamond mine came on stream in 2012 and this increased significantly diamond output. There are two main alluvial mining areas in eastern Sierra Leone, Kono, close to the Koidu kimberlites, and Tongo, near the Tongo kimberlite dikes (Fig. 73). These areas are associated with recent sediments of the Sewa and Moa rivers, respectively. The Kono area in particular is known for the recovery of some very large stones such as the 969.8 ct Star of Sierra Leone, a 770 ct stone and the 709.5 ct Peace Diamond. In some potholes from the Sewa River, extremely high grades of over 1,000 ct/t have been achieved (Marshall et al. 2013). Diamond production from the Tongo area has been exclusively from alluvial sources with relatively high grades and high-quality stones, though in general the stone size at Tongo is smaller with no reported occurrence of +100 ct stones. The alluvials are present in river channels, flood plains and terraces as part of a Late Pleistocene to recent laterally migrating channels with isolated bedrock trapsites often associated with cross-cutting dikes.

The Mano River (Fig. 73) further east has also been targeted for alluvial mining, but it is a less prolific producer. Since the Cretaceous, diamonds have been transported along the Sewa, Moa and Mano Rivers to coastal areas, a distance of some 200 km. These Atlantic-directed rivers were rejuvenated as a result the subsiding continental margin that uplifted watershed areas (Sutherland 1982) following rifting at about 185 Ma.

Liberia (Lofa County): Mano, Lofa and Yambasi rivers—Transient placers (Quaternary). Production in Liberia of around 0.4 Mct/a, has been exclusively from alluvial sources and primarily from the west of the country close to the border with Sierra Leone in the Mano, Lofa and Yambasi River basins (Fig. 73). The alluvial deposits are similar to those in other parts of West and Central Africa and occur in high- and low-level terrace gravels, alluvial river flats and deep plunge pools within the active rivers and in the current river bed. Large gem-quality diamonds up to 170 ct as well as industrial diamonds have been recovered in diggings in west Liberia (Gunn et al. 2018). At Camp Alpha in the northwest, diamonds exceeding 300 ct have been found in alluvial gravels (S. Haggerty pers. comm. 2014).

Ivory Coast (Woroba, Vallée du Bandama Districts): Séguéla and Tortiya—Retained placers (Quaternary). Diamond production in the Ivory Coast, which has amounted to some 8.5 Mct, has mainly been from the alluvial fields at Tortiya in the north-central part of the country, and at Séguéla some 150 km to the southwest (Fig. 73). At Séguéla there are meta-

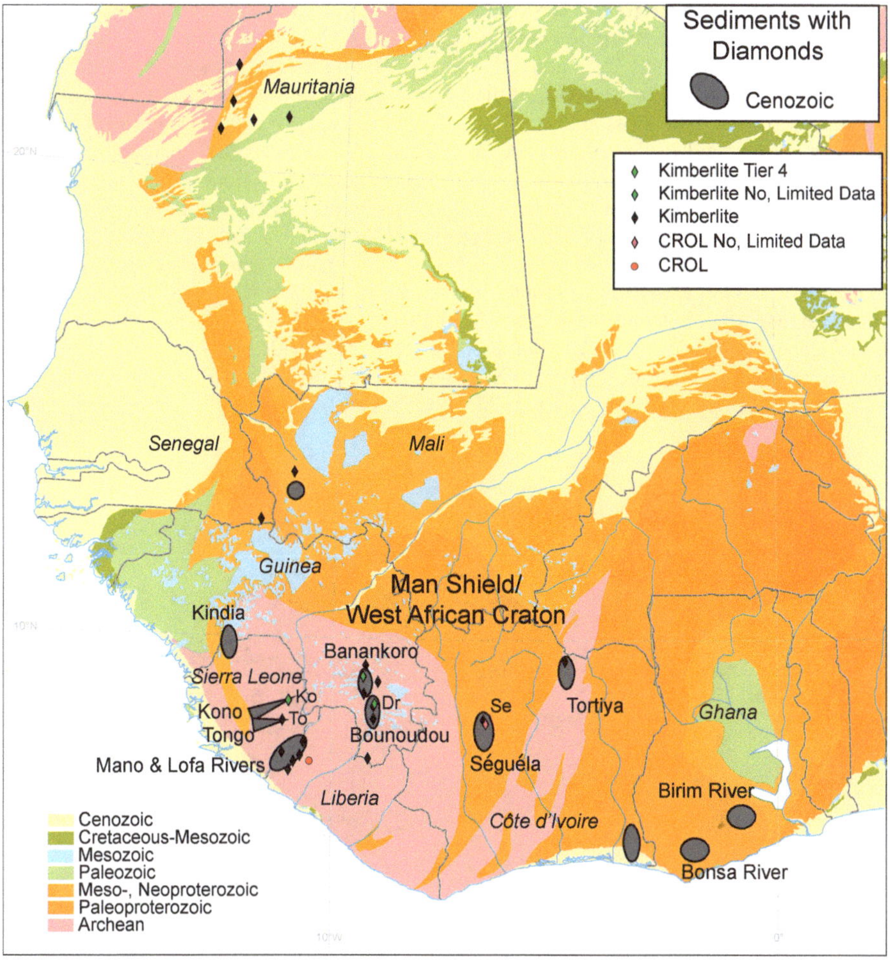

Figure 73. West Africa placer deposits as labeled on the map (Kono, Tongo, Mano and Lofa Rivers, Tortiya, Séguéla, Banankoro, Bounoudou, Kindia, Birim River, Bonsa River). Primary magmatic kimberlite and CROL mines/deposits as follows: Ko, Koidu; To, Tongo; Dr, Droujba; Se, Séguéla.

kimberlite (or metamorphosed carbonate-rich olivine lamproite) dikes dated at between 1,145 and 1,429 Ma (Bardet 1974), but exploitation here is restricted to small-scale mining of the associated recent colluvial and alluvial sediments proximal to these dikes. Diamonds are small, some 4 to 5 st/ct, and only one third is gem-quality.

The diamonds at Tortiya along the Bou River are sourced from a Proterozoic meta-volcanic-sedimentary complex trending NNE. Bardet (1974) believes the diamonds occur in conglomerates of the Birimian Supergroup and these were probably introduced during sedimentation at ~2.3 Ga, prior the Eburnian Orogeny (Milesi et al. 1992). After the break-up of Gondwana, the only major erosional events that would have affected Tortiya are sea level and climatic changes associated with Quaternary glaciations. The bulk of the diamonds, which are generally small and typically between 0.2–2.5 mm in size, occur in Quaternary colluvial deposits that are less than 10 m thick. These were formed by the weathering, erosion, and re-concentration of earlier deposits (Teeuw 2002) and the highest grades are found towards the base of what appears to be ferruginized mudflows.

Guinea (Kankan, Nzèrékoré, Kindia Regions): Banankoro, Bounoudou, Kindia areas—Retained and transient placers (Quaternary). In Guinea all of the ~20 Mct diamond production to date has come from alluvial deposits. There are four known kimberlite clusters in Guinea, all located in the southeast of the country and all have associated alluvial diamond fields (Figs. 73, 74). The most prolific of these is the Banankoro cluster where over 20 pipes and dikes were found in the 1950s. This area has seen extensive alluvial mining since the 1930's and has produced the most diamonds, including some large stones from the large scale commercial alluvial mining at the Aredor Mine at Kérouané along the Baoulé River in the 1990's to mid-2000's. Here the grades range from 5 to 9 ct/100t, and a resource of almost 1 Mct was declared (de Wit et al. 2016). Terraces west of the main mining area, which are between 20 and 30 m above the present river level, are extensively lateritised and fine-grained, with a granular- to pebble-size basal gravel. Both the retained placers, in the form of eluvial-alluvial deposits close to primary sources (and associated with small tributaries), and the transient alluvial placers that represent reworked gravels with rich deposits in flats and lower terraces of the larger rivers, are present (Chirico et al. 2012).

Figure 74. Chemically mature conglomerates at Kindia, Guinea, with well-rounded predominately siliceous clasts.

The watersheds of the Atlantic and Niger directed drainages had a major influence in the dispersal, size and quality of the alluvial deposits. The topography of rivers of the northeast directed Niger River Basin, such as the Baoulé, have a more mature relief with broad valleys and well-developed terraces. Limited down cutting has retained the diamonds close to its sources and little reworking has resulted in low to moderate diamond concentration in the main trunk rivers (Chirico et al. 2012). The Atlantic-directed drainages have been rejuvenated and the main concentration occurs in the large channels reworked out of older terraces (Sutherland 1993).

Some 70 km to the southeast of Banakoro alluvial mining has occurred in the Bounoudou region. For many years this was the largest diamond producing area of Guinea. Extremely high grades of up to 2.60 ct/t have been recorded next to the Droujba kimberlite, although grades of 0.25 ct/t are more common. A resource of 1.5 Mct at a value of 30 US\$/ct (2016 prices) was established for the flats of this part of the Diani River basin (de Wit et al. 2016).

Diamonds have also been mined on a small scale in western Guinea in the Kindia and Forécariab areas. One-meter thick gravels occupy paleochannels and low terraces along the Atlantic-draining Konkouré River. These are structurally controlled and clearly follow bedrock fractures. These deposits are located on the edge of Palaeozoic Bove Basin. The bedrock varies from Late Archean and Proterozoic gneisses to Ordovician sandstones and conglomerates, but since no kimberlitic minerals are associated with these gravels it has been suggested they have been reworked from white conglomeratic sandstone and diamictite of the Ordovician Pita Group.

Ghana (Eastern and Western Regions): Birim and Bonsa Rivers—Transient placers (Quaternary). Diamonds in Ghana are concentrated in two main areas, one near the Birim River close to Akwatia and another field in the Bonsa River further the southwest, referred to as the Birim and Bonsa diamond fields (Fig. 73). These are exclusively alluvial deposits and no diamond bearing kimberlites have been found in this part of West Africa. However, diamonds have been recovered from actinolite—tremolite schists that have been interpreted as "syn-eruptive volcaniclastic mega-turbidites derived from a diamond-bearing komatiite" similar to some of the diamond bearing rocks that have been described in French Guiana in South America (Canales 2005). Field evidence suggests that these komatiitic rocks are coeval with the meta-sediments indicating an age of 2,155 –2,085 Ma for the diamond bearing schists.

The concentration of the recent alluvial deposits is the result of the extensive weathering of these Proterozoic diamond-bearing phyllites, and subsequent redistribution into the local river systems. However, these diamond-bearing phyllites and schists are not mined. The regolith of weathered bedrock on the interfluve areas is covered by a laterite crust and erosion of these have supplied diamonds to the river terraces and floodplain gravels. Eroded quartz veins have provided large amounts of gravel clasts, that with bedrock traps (e.g., the sudden widening of the channel), are important constituents that help to retain the diamonds in the trap. Grades within these alluvial deposits are highest for the river flats, decreasing in the lower and higher terraces respectively, from 43, to 24 to 16 ct/100t for the Birim deposits, and from 27, to 14 to 7 ct/100t for the Bonsa deposits (Chirico et al. 2010).

The diamonds are in general small with average stone sizes of 0.033 ct/st for Birim and 0.025 ct/st for Bonsa and both fields have a high percentage (~35% and ~45%, respectively) of industrial diamonds (Chirico et al. 2010). Most of the diamonds are macles and dodecahedral crystals, and abrasion features appear more prominent on those from Bonsa area.

It is estimated that of the 120 Mct that have been mined in Ghana, 108 Mct are from the Birim area and 12 Mct from Bonsa, with 89 Mct and 2.6 Mct of diamonds still remaining in the Birim and Bonsa fields, respectively (de Wit et al. 2016). However, due to the patchy nature and lower grades of the remaining deposits, most of these would not be economically viable except as a by-product of gold dredging operations in the Bonsa region, and as artisanal operations in both areas.

Finally, isolated and less significant diamondiferous sediments have been reported from Gabon (Fig. 64), Mali (Fig. 73), China (Figs. 61, 75), and in California (associated with alluvial

Figure 75. Lower Cenozoic, upwards-coarsening diamond-bearing gravels at Mengyin, Shandong, China

gold recovery). Alluvial diamonds in Mali are not associated with the Kéniéba kimberlites, which have been shown to be barren.

Summary of economic diamond placers

This review has highlighted a few critical components for the development of economic placers. Most important is probably some form of epeirogenic uplift (e.g., Espinhaço range, Roraima basin, southern Africa, Anabar, West Africa etc.) to initiate erosion, transport and concentration of heavy minerals. The presence of retained placers in these areas of uplift would add a dimension of reworking and upgrading which is illustrated by the concentration of placers in the more recent transport systems during the Cenozoic. Hence recycling produces large diamond placers, such as the Roraima (Brazil) and Ebelyakh (Russia) deposits, but reworking in terminal settings, where there is upgrading of transient placers by marine processes, produces the largest diamond placers such as the Namibian, Namaqualand and Marange deposits. In a Gondwana context it is interesting to note how many diamonds are linked to having been distributed by the Permo-Carboniferous glaciation affecting deposits in Africa, Brazil and Australia.

ACKNOWLEDGEMENTS

This chapter is dedicated to Professor Peter H. Nixon who, since discovering the Letšeng la Terae deposit in 1957 has made a lifetime of important contributions to our field, inspiring many along the way. We thank the volume editors for the patience with us during the compilation of this chapter. Nick Pokhilenko and Sergie Kostrovitsy kindly granted permission to use images from their book for Figures 10 and 11b. Murray Rayner and Stephen Moss kindly provided images used for Figures 8 and 11a, respectively, and permission to reproduce them. Evelina Wieleba's GIS expertise in producing the global and craton/country scale maps is greatly appreciated. We thank reviewers George Read and Volker Lorenz for tackling a rather lengthy manuscript, as well as additional reviews specifically on the mineralogy and geochemistry section by Andrea Giuliani, and on the secondary diamond deposit section by Tania Marshall. Thomas Stachel and Stephan Kurszlaukis provided highly useful commentary and information on the volcanology section. BAK acknowledges funding from the GSC and the Geomapping for Energy and Minerals (GEM) program.

REFERENCES

Abersteiner A, Golovin A, Chayka I, Kamenetsky VS, Goemann K, Ehrig K, Rodemann T (2022) Carbon compounds in west Kimberley diamondiferous lamproites (Australia): Insights from melt and fluid inclusions. Gond Res (accepted)

Abersteiner A, Kamenetsky VS, Goemann K, Giuliani A, Howarth GH, Castillo-Oliver M, Thompson J, Kamenetsky M, Cherry A (2019) Composition and emplacement of the Benfontein kimberlite sill complex (Kimberley, South Africa): Textural, petrographic and melt inclusion constraints. Lithos 324:297–314

Afanas'ev VP, Zinchuk NN, Logvinova AM (2009) Distribution of placer diamonds related to Precambrian sources. Geol Ore Deposits 51:675–683

Afanasiev VP, Pokhilenko NP (2013) Wear of diamonds: An experimental study and field evidence. *In:*Proceedings of the 10th International Kimberlite Conference, DG Pearson, HS Grutter, JW Harris, BA Kjarsgaard, H O'Brien, NVC Rao, S Sparks (eds)1:317–321

ALROSA (2017) IR-release: Q4 and FY 2016 Operational Overview, 30 January 2017

Ambroise M (1991) Geology of the NE Angolan kimberlite region. 5th International Kimberlite Conference, Extended Abstracts, 6–9

Apter DB, Harper FJ, Wyatt BA, Smith BHS (1984) The geology of the Mayeng kimberlite sill complex, South Africa. *In:*Kimberlites and Related Rocks, Proceedings of the Third International Kimberlite Conference. Kornprobst J (ed) Elsevier Science Publishers, Clermont Ferrand, France 1:43–57

Armstrong JP, Wilson M, Barnett RL, Nowicki T, Kjarsgaard BA (2004) Mineralogy of primary carbonate-bearing hypabyssal kimberlite, Lac de Gras, Slave Province, Northwest Territories, Canada. Lithos 76:415–433

Ashchepkov I, Logvinova AM, Reimers LF, Ntaflos T, Spetsius ZV, Vladykin NV, Downes H, Yudin DS, Travin AV, Makovchuk IV, Palesskiy VS, Khmel'nikova OS (2015) The Sytykanskaya kimberlite pipe: Evidence from deep-seated xenoliths and xenocrysts for the evolution of the mantle beneath Alakit, Yakutia, Russia. Geosci Front 6:687–714

Atkinson WJ (1989) Diamond exploration philosophy, practice, and promises: a review *In:*Kimberlites and Related Rocks, Proceedings from the Fourth International Kimberlite Conference. Ross J (ed) GSA Spec Publ 14, Blackwell, Perth, Australia 2:1075–1107

Atkinson WJ, Hughes FE, Smith CB (1984) A review of the kimberlitic rocks of Western Australia. *In:* Kimberlites and Related Rocks, Proceedings of the Third International Kimberlite Conference. Vol 1. Kornprobst J, (ed) Elsevier Science Publishers, Clermont Ferrand, France, p 195–224

Balfour I (1987) Famous Diamonds. William Collins & Co., London

Bardet MG (1974) Geologie du Diamant. Gisements de Diamant d'Afrique. Memoire 83, Vol 2, B.R.G.M., Paris

Bartlett PJ (1998) Premier Mine. 7th International Kimberlite Conference, Large Mines Field Excursion, University of Cape Town, South Africa, p 39–49

Battilani GA, Gomes NS, Guerra WJ (2007) The occurrence of microdiamonds in Mesoproterozoic Chapada Diamantina intrusive rocks—Bahia/ Brazil. Anais da Academia Brasileira de Ciências 79: 321–332

Batumike JM, Griffin WL, Belousova EA, Pearson NJ, O'Reilly SY, Shee, SR (2008) LAM-ICPMS U–Pb dating of kimberlitic perovskite: Eocene–Oligocene kimberlites from the Kundelungu Plateau, D.R. Congo. Earth Planet Sci Lett 267, 609–619

Baxter-Brown R, Baker NR (1991) Discover of diamond deposits in the Quebrada Grande catchment Venezuela. International Kimberlite Conference, Extended Abstracts, v5

Becker M, Le Roex AP (2006) Geochemistry of South African on- and off-craton, Group I and Group II kimberlites: Petrogenesis and source region evolution. J Petrol 47:673–703

Bergman SC (1987) Lamproites and other potassium-rich igneous rocks: a review of their occurrence, mineralogy and geochemistry. *In:* Alkaline Igneous Rocks. Vol 30. Fitton JG, Upton BGJ, (eds) Geol Soc London, Spec Publ, p 103–190

Bertram CN (2010) Sedimentology, Age and Stable Isotope Evolution of the Kurnool Group, Cuddapah Basin. BSc(Hons) thesis (unpubl.), University of Adelaide, Australia

Black LP, James PR (1983) Geological history of the Napier Complex of Enderby Land. Antarctic Earth Science, International Symposium 4, RL Oliver et al. (ed) Canberra, Austral Acad Sci, p 11–15

Bluck BJ, Ward JD, de Wit MCJ (2005) Diamond mega-placers: southern Africa and the Kaapvaal craton in a global context. *In:*Mineral Deposits and Earth Evolution. I McDonald et al. (eds) Geol Soc London, Spec Publ 248:213–245

Bolivar SL (1984) An overview of the Prairie Creek intrusion, Arkansas. AIME Fall Meeting, October 1984, Preprint 84–346

Boxer GL, Lorenz V, Smith CB (1989) The geology and volcanology of the Argyle (AK1) lamproite diatreme, Western Australia. *In:* Kimberlites and Related Rocks, Proceedings from the Fourth International Kimberlite Conference. Ross J (ed) GSA Spec Publ 14, Blackwell, Perth, Australia, 1:140–152

Boyd F, Clement C (1977) Compositional zoning of olivines in kimberlite from the De Beers Mine, Kimberley, South Africa. International Kimberlite Conference Extended Abstracts 2:39–41

Boyd F, Nixon P (1973) Origin of the discrete nodules in the kimberlites of northern Lesotho. *In:* International Kimberlite Conference: Extended Abstracts 1:51–54

Brett RC, Russell JK, Moss S (2009) Origin of olivine in kimberlite: Phenocryst or impostor? Lithos 112:201–212

Brook MC (2012) Botswana's Diamonds. Longpack Co Ltd, China

Brown RJ, Manya S, Buisman I, Fontana G, Field M, MacNiocaill C, Sparks RSJ, Stuart FM (2012) Eruption of kimberlite magmas: physical volcanology, geomorphology and age of the youngest kimberlitic volcanoes known on earth (the Upper Pleistocene/Holocene Igwisi Hills volcanoes, Tanzania) Bull Volcanol 74:1621–1643

Burgess R, Phillips D, Harris JW, Robinson DN (1998) Antarctic diamonds in South-eastern Australia? Hints from $^{40}Ar/^{39}Ar$ Laser probe dating of clinopyroxene inclusions from Copeton diamonds. Extended Abstract 7th Int. Kimberlite Conference, Cape Town, South Africa

Bussweiler Y, Stone RS, Pearson DG, Luth RW, Stachel T, Kjarsgaard BA, Menzies A (2016) The evolution of calcite-bearing kimberlites by melt-rock reaction: evidence from polymineralic inclusions within clinopyroxene and garnet megacrysts from Lac de Gras kimberlites, Canada. Contrib Mineral Petrol 171:1–25

Bussweiler Y, Pearson DG, Stachel T, Kjarsgaard BA (2018) Cr-rich megacrysts of clinopyroxene and garnet from Lac de Gras Kimberlites, Slave Craton, Canada—Implications for the origin of clinopyroxene and garnet in cratonic lherzolites. Mineral Petrol 112 Suppl 2:S583–S596

Büttner R, Dellino P, La Volpe L, Lorenz,V, Zimanowski B (2002) Thermo-hydraulic explosions in phreatomagmatic eruptions as evidenced by the comparison between pyroclasts and products from Molten Fuel Coolant Interaction experiments. J Geophys Res Solid Earth 107(B11):2277

Carmichael IS (1967) The mineralogy and petrology of the volcanic rocks from the Leucite Hills, Wyoming. Contrib Mineral Petrol 15:24–66

Caro G, Kopylova MG, Creaser RA (2004) The hypabyssal 5034 kimberlite of the Gahcho Kue cluster, southeastern Slave craton, Northwest Territories, Canada: a granite-contaminated Group-I kimberlite. Can Mineral 42:183–207

Cas R, Porritt L, Pittari A, Hayman P (2008a) A new approach to kimberlite facies ten-ninology using a revised general approach to the nomenclature of all volcanic rocks and deposits: Descriptive to genetic. J Volcanol Geotherm Res 174:226–240

Cas RAF, Hayman P, Pittari A, Porritt L (2008b) Some major problems with existing models and terminology associated with kimberlite pipes from a volcanological perspective, and some suggestions. J Volcanol Geotherm Res 174:209–225

Cas RAF, Porritt L, Pittari A, Hayman PC (2009) A practical guide to terminology for kimberlite facies: A systematic progression from descriptive to genetic, including a pocket guide. Lithos 112:183–190

Cham RA (1998) Palaeodrainage and its significance to mineral exploration in the Bathurst region, NSW. Cooperative Research Centre for Landscape Evolution and Mineral Exploration, Monograph 3, p 38–54

Chaves MLCS, Uhlein A (1991) Depósitos diamantiferos da região do Alto/Médio Rio Jequitinhonha, Minas Gerais. *In:* Principals Depósitos Minerais do Brasil, DNPM and CPRM, Brasilia. Vol. IV, p 117–138

Chirico PG, Barthélémy F, Ngbokoto FA (2010) Alluvial Diamond Resource Potential and Production Capacity Assessment of the Central African Republic. US Geol Surv Scientific Investigations Report 2010–5043

Chirico PG, Malpeli KC, Van Bockstael M, Diaby M, Cissé K, Amadou Diallo T, Sano M (2012) Alluvial Diamond Resource Potential and Production Capacity Assessment of Guinea. US Geol Surv Scientific Investigations Report 2012–5256

Chorlton L (2007) Generalized geology of the world: bedrock geology and major faults in GIS format. Geol Surv Can Open File 5259

Clement CR (1982) A comparative geological study of some major kimberlite pipes in the Northern Cape and Orange Free State. Ph.D. thesis, University of Cape Town, Cape Town, South Africa

Clement CR, Skinner EMW (1979) A textural-genetic classification of kimberlite rocks. Kimberlite Symposium II, Cambridge, U.K., Extended Abstracts

Clement CR, Skinner EMW (1985) A textural-genetic classification of kimberlites. Trans Geol Soc S Afr 88:403–409

Clement CR, Dawson JB, Geringer GJ, Gurney JJ, Hawthorne JB, Krol L, Kleinjan L, van Zyl AA (1973) Guide for the first field excursion. 1st International Kimberlite Conference, University of Cape Town, South Africa

Clement CR, Skinner EMW, Scott Smith BH (1984) Kimberlite redefined. J Geol 92:223–228

Clement CR, Harris JW, Robinson DN, Hawthorne JB (1986) The De Beers kimberlite pipe—a historic South African diamond mine. *In:*Mineral Resources of Southern Africa. Anhaeusser CR, Maske S (eds) Geol Soc S Afr, Johannesburg, p 2193–2214

Clifford TN (1966) Tectono-metallogenetic units and metallogenetic provinces of Africa. Earth Planet Sci Lett 1:421–434

Coe N, Le Roex A, Gurney J, Pearson DG, Nowell G (2008) Petrogenesis of the Swartruggens and Star Group II kimberlite dyke swarms, South Africa: constraints from whole rock geochemistry. Contrib Mineral Petrol 156:627–652

Corbett IB (1996) A review of diamondiferous marine deposits of western southern Africa. Afr Geosci Rev 3:157–174

Creaser RA, Grutter H, Carlson J, Crawford B (2004) Macrocrystal phlogopite Rb-Sr dates for the Ekati property kimberlites, Slave Province, Canada: evidence for multiple intrusive episodes in the Paleocene and Eocene. Lithos 76:399–414

Cross CW (1897) The igneous rocks of the Leucite Hills and Pilot Butte, Wyoming. Am J Sci 4:115–141

Dalton H, Giuliani A, O'Brien H, Phillips D, Hergt J, Maas R (2019) Petrogenesis of a hybrid cluster of evolved kimberlites and ultramafic lamprophyres in the Kuusamo area, Finland. J Petrol 60:2025-

Dalton H, Giuliani A, Phillips D, Hergt J, Maas R, Matchan E, Woodhead J, O'Brien H (2020) A comparison of geochronological methods commonly applied to kimberlites and related rocks: three case studies from Finland. Chem Geol 558:119899

Das H, Kobussen AF, Webb KJ, Phillips D, Maas R, Soltys A, Rayner MJ, Howell D (2018) The Bunder Diamond Project, India: Geology, geochemistry, and age of the Saptarshi lamproite Pipes. *In:* Geoscience and Exploration of the Argyle, Bunder, Diavik, and Murowa Diamond Deposits. Davy AT, Smith CB, Helmstaedt HH, Jaques AL, Gurney JJ (Eds.) Allen Press Inc., Kansas, p 201–222

Davies RM, O'Reilly SY, Griffin WL (1999) Growth Structures and nitrogen characteristics of Group B alluvial diamonds from Bingara and Wellington, Eastern Australia. *In:* Proc 7th International Kimberlite Conference. JJ Gurney, JL Gurney, MD Pascoe, SH Richardson (eds) Cape Town, South Africa 1:156–163

Davies RM, O'Reilly SY, Griffin WL (2002) Multiple origins of alluvial diamonds from New South Wales, Australia. Econ Geol 97:109–123

Dawson JB (1967) A review of the geology of kimberlite. *In:*Ultramafic and Related Rocks. Wyllie PJ (ed) Wiley, New York, p 241–251

Dawson JB (1980) Kimberlites and Their Xenoliths. Springer, Berlin

Dawson JB (1987) The kimberlite clan: relationship with olivine and leucite lamproites, and inferences for upper mantle metasomatism. *In:* Alkaline Igneous Rocks. Fitton JG, Upton BGJ, (eds) Geol Soc London, Spec Publ 30, London, p 95–101

Dawson JB (1994) Quaternary kimberlitic volcanism on the Tanzania craton. Contrib Mineral Petrol 116:473–485

De Beers (2014) Diamond exploration. The Insight Report, p 46–49

de Wit MCJ (1996) The distribution and stratigraphy of inland alluvial diamond deposits in South Africa. Afr Geosci Rev 3:175–189 de Wit MCJ (2004) The diamondiferous sediments on the farm Nooitgedacht (66), Kimberley, South Africa. S Afr J Geol 107:477–488

de Wit MJ (2007) The Kalahari Epeirogeny and climate change: differentiating cause and effect from core to space. Inkaba yeAfrica Spec Vol. S Afr J Geol 110:367–392

de Wit MCJ (2016) Early Permian diamond-bearing proximal eskers in the Lichtenburg/Ventersdorp area of the North West Province, South Africa. S Afr J Geol 119:585–606

de Wit MCJ, Marshall TR, Partidge TC (2000) Fluvial deposits and drainage evolution. *In:* The Cenozoic of Southern Africa. TC Partridge, RR Maud (eds) Oxford Monographs on Geology and Geophysics 40:55–72

de Wit MCJ, Thorose E (2015) Diamond-bearing gravels along the Lower Kwango River DRC. Ch 16 *In:*Geology and Resource Potential of the Congo Basin, Regional Geology Reviews. MJ de Wit et al. (eds) Springer, Berlin, p 341–360

de Wit MCJ, Bhebhe Z, Davidson J, Haggerty SE, Hundt P, Jacob J, Lynn M, Marshall T, Skinner C, Smithson K, Stiefenhofer J, Robert M, Revitt A, Spaggiari R, Ward J (2016) Overview of diamond resources in Africa. Episodes 39:199–238

de Wit MCJ, Ward JD, Bamford MK, Roberts MJ (2009) The significance of the cretaceous diamondiferous gravel deposit at Mahura Muthla, northern cape province, South Africa. S Afr J Geol 112:89–108

Dobrzhinetskaya LF, O'Bannon EF III, Sumino H (2022) Non-cratonic diamonds from UHP metamorphic terranes, ophiolites and volcanic sources. Rev Mineral Geochem 88:191–256

Donnelly CL, Griffin WL, Yang J-H, O'Reilly SY, Li Q-L, Pearson NJ, Li X-H (2012) In situ U–Pb dating and Sr–Nd isotopic analysis of perovskite: constraints on the age and petrogenesis of the Kuruman Kimberlite Province, Kaapvaal Craton, South Africa. J Petrol 53:2497–2522

Dos Santos PR, Rocha-Campos AC, Canuto JR (1996) Patterns of late Palaeozoic deglaciation in the Paraná Basin, Brazil. Palaeogeogr Palaeoclimatol Palaeoecol 125:165–184

Dowall DP (2004) Elemental and Isotopic Geochemistry of Kimberlites from the Lac de Gras field, Northwest Territories, Canada. Ph.D. thesis, Durham University

Downes PJ, Wartho JA, Griffin BJ (2006) Magmatic evolution and ascent history of the Aries micaceous kimberlite, Central Kimberley Basin, Western Australia: Evidence from zoned phlogopite phenocrysts, and UV laser Ar-40/Ar-39 analysis of phlogopite–biotite. J Petrol 47:1751–1783

Eccles DR, Heaman LM, Luth RW, Creaser RA (2004) Petrogenesis of the Late Cretaceous northern Alberta kimberlite province. Lithos 76:435–459

Edwards CB, Howkins JB (1966) Kimberlites in Tanganyika with special reference to the Mwadui occurrence. Econ Geol 61:537–554

Edward D, Rock NMS, Taylor WR, Griffin BJ, Ramsay RR (1992) Mineralogy and petrology of the Aries diamondiferous kimberlite pipe. J Petrol 3:1157–1191

Eggler DH, McCallum ME, Smith CB (1979) Megacryst Assemblages in kimberlites from northern Colorado and southern Wyoming: Petrology, geothermometry–barometry, and areal distribution. *In:* The Second International Kimberlite Conference. Boyd FR, Meyer HOA (eds) Am Geophys Union, Santa Fe, USA 2:213–226

Ekkerd J, Stiefenhofer J, Field M, Lawless P (2003) The geology of Finsch mine, Northern Cape Province, South Africa. International Kimberlite Conference: Extended Abstracts, Vol 8

Erlich EI, Hausel WD (2002) Diamond Deposits: Origin, Exploration, and History of Discovery. Society for Mining Metallurgy and Exploration Inc

Fernandes AF, Karfunkel J, Hoover DB, Sgarbi PBdA, Sgarbi GNC, Oliveira GD, de Souza Pereira Gomes JC, Kambrock K (2014) The Basal Conglomerate of the Capacete Formation (Mata da Corda Group) and its Relation to Diamond Distributions in Coromandel, Minas Gerais State, Brazil. Braz J Geol 44:91–103

Fesq HW, Bibby DM, Erasmus CS, Kable EJD, Sellschop JPF (1975) A comparative trace element study of diamonds from Premier, Finsch and Jagerfontein mines, South Africa. *In:* Proceedings of the First International Kimberlite Conference, Phys Chem Earth, Vol 9. Ahrens LH, Dawson JB, Duncan AR, Erlank AJ (eds) Pergamon Press, Cape Town South Africa, p 817–836

Field M (2010) An assessment of the diamond potential of the Itilima property, Tanzania. Report prepared for Midlands Minerals Inc, Diakim Consulting Ltd

Field M, Scott Smith BH (1999) Contrasting geology and near-surface emplacement of kimberlite pipes in Southern Africa and Canada. *In:* The JB Dawson Volume, Proceedings of the VIIth International Kimberlite Conference. Gurney JJ, Gurney JL, Pascoe MD, Richardson SH (eds) Reed Roof Design, Cape Town, p 214–237

Field M, Steifenhofer J (2006) Orapa as a Type 1 kimberlite pipe. 8th International Kimberlite Conference, Kimberlite Emplacement Workshop, Extended Abstracts, Saskatoon, Canada

Field M, Gibson JG, Wilkes TA, Gababotse J, Khutjwe P (1997) The geology of the Orapa A/K1 kimberlite Botswana: further insight into the emplacement of kimberlite pipes. *In:* Kimberlites, Related Rocks and Mantle Xenoliths, Proceedings of the Sixth International Kimberlite Conference. Vol 1. Dobretsov NL, Mitchell RH (eds) Russ Geol Geophys Vol 38, Novosibirsk, Russia, p 24–39

Field M, Stiefenhofer J, Robey J, Kurszlaukis S (2008) Kimberlite-hosted diamond deposits of southern Africa: A review. Ore Geol Rev 34:33–75

Field M, Gernon TM, Mock A, Walters A, Sparks RSJ, Jerram DA (2009) Variations of olivine abundance and grain size in the Snap Lake kimberlite intrusion, Northwest Territories, Canada: A possible proxy for diamonds. Lithos 112:23–35

Fieremans C (1953) Geologie en geochemie der Diamantvelden van Belgisch-Congo. Technisch-Wetenschappelijk Tijdschrift, Jaargang 22, no 4–5, 1–16. Forminière 1956. Ouvrage publié par la Société Internationale Forestière et Minière du Congo a l'occassion du 50e anniversaire de sa création. Édition M L Cuypers, Bruxelles, 15 juin 1956

Fisher RV (1961) Proposed classification of volcaniclastic sediments and rocks. Geol Soc Am Bull 72:1409–1414

Fisher RV (1966) Rocks composed of volcanic fragments and their classification. Earth Sci Rev 1:287–298

Fleischer R (1998) A rift model for the sedimentary diamond deposits of Brazil. Mineralium Deposita 33:238–254

Flowers RM, Bowring SA, Reiners PW (2006) Low long-term erosion rates and extreme continental stability documented by ancient (U–Th)/He dates. Geology 34:925–928

Foley SF, Yaxley GM, Kjarsgaard BA (2019) Kimberlites from source to surface: insights from experiments. Elements: 15:393–398

Fraser KJ, Hawkesworth CJ (1992) The petrogenesis of Group-2 ultrapotassic kimberlites from Finsch Mine, South-Africa. Lithos 28:327–345

Fraser KJ, Hawkesworth CJ, Erlank AJ, Mitchell RH, Scott Smith BH (1985) Sr, Nd and Pb isotope and minor element geochemistry of lamproites and kimberlites. Earth Planet Sci Lett 76:57–70

Fulop A, Kurszlaukis S (2016) Monogenetic v. polygenetic kimberlite volcanism: in-depth examination of the Tango Extension Super Structure, Attawapiskat kimberlite field, Ontario, Canada. *In:*Monogenetic Volcanism. Nemeth K, Carrasco-Nunez G, Aranda-Gomez JJ, Smith IEM (eds) Geol Soc London, Spec Publ 446:205–224

Fulop A, Kopylova M, Kurszlaukis S, Hilchie L, Ellemers P, Squibb C (2018) Petrography of Snap Lake Kimberlite Dyke (Northwest Territories, Canada) and its interaction with country rock granitoids. J Petrol 59:2493–2518

Gaudet M, Kopylova M, Muntener C, Zhuk V, Nathwani C (2018) Geology of the Renard 65 kimberlite pipe, Québec, Canada. Mineral Petrol 112:433–445

Gerasimov B (2019) Typomorphism of fine placer gold and potential primary sources of the Anabar mineral subprovince (North-eastern Siberian Platform) IOP Conf Ser Earth Environ Sci 362:012039

Gernon TM, Fontana G, Field M, Sparks RSJ, Brown RJ, Mac Niocaill C (2009) Pyroclastic flow deposits from a kimberlite eruption: The Orapa South Crater, Botswana. Lithos 112:566–578

Giuliani A (2018) Insights into kimberlite petrogenesis and mantle metasomatism from a review of the compositional zoning of olivine in kimberlites worldwide. Lithos 312:322–342

Giuliani A, Pearson DG (2019) Kimberlites: from deep earth to diamond mines. Elements 15:377–380

Giuliani A, Phillips D, Kamenetsky VS, Fiorentini ML, Farquhar J, Kendrick MA (2014) Stable isotope (C, O, S) compositions of volatile-rich minerals in kimberlites: a review. Chem Geol 374:61–83

Giuliani A, Soltys A, Phillips D, Kamenetsky VS, Maas R, Goemann K, Woodhead JD, Drysdale RN, Griffin WL (2017) The final stages of kimberlite petrogenesis: petrography, mineral chemistry, melt inclusions and Sr–CO isotope geochemistry of the Bultfontein kimberlite (Kimberley, South Africa) Chem Geol 455:342–356

Gomes HA, Silva EFA, Quinh JS, Lopes JI (1972) Projecto Gilbués, Relatóriò Final, Vol 3 CPRM-DNPM, Recife

Gonzaga GM, Teixeira NA, Gaspar JC (1994) The origin of diamonds in western Minas Gerais, Brazil. Mineralium Deposita 29:414–421

Grakhanov SA, Malanin YuA, Pavlov VI, Afanas'ev VP, Pokhilenko NP, Gerasimchuk AV, Lipashova AN (2010) Rhaetian diamond placers in Siberia. Russ Geol Geophys 51:127–135

Grakhanov SA, Zinchuk NN, Sobolev NV (2015) The age of predictable primary diamond sources in the northeastern Siberian Platform. Dokl Earth Sci 465:1297–1301

Gress MU, Pearson DG, Chinn IL, Thomassot, E, Davies, GR (2021) Mesozoic to Paleoproterozoic diamond growth beneath Botswana recorded by Re–Os ages from individual eclogitic and websteritic inclusions. Lithos 388–389:106058

Griffin WL, O'Reill SY, Natapov LM, Ryan CG (2003) The evolution of lithospheric mantle beneath the Kalahari Craton and its margins. Lithos 71:215–251

Griffin WL, Batumike JM, Greau, Y, Pearson NJ, Shee SR, O'Reilly, SY (2014) Emplacement ages and sources of kimberlites and related rocks in southern Africa: U–Pb ages and Sr–Nd isotopes of groundmass perovskite. Contrib Mineral Petrol 168:10–32

Grubb MD, McCallum ME (1991) Genesis of diamond placers on the Guiana Shield, South America. Fifth International Kimberlite Conference, Brazil. Extended Abstracts, p 151–153

Gunn AG, Dorbor JK, Mankelow JM, Lusty PAJ, Deady EA, Shaw RA, Goodenough KM (2018) A review of the mineral potential of Liberia. Ore Geol Rev 101:413–431

Gupta S, Jain KC, Srivastava VC, Mehrotra (2003) Depositional environment and tectonism during the sedimentation of the Semri and Kaimur Groups of rocks, Vindhyan Basin. J Pal Soc India 48:181–190

Gurney JJ, Ebrahim S (1973) Chemical composition of Lesotho kimberlites. *In:* Lesotho Kimberlites. Nixon PH (ed) Maseru National Development Board, Maseru, p 280–284

Gurney JJ, Kirkley MB (1996) Kimberlite dyke mining in South Africa. Afr Geosci Rev 3:191–201

Gurney JJ, Helmstaedt HH, le Roex AP, Nowicki TE, Richardson SH, Westerlund, KJ (2005) Diamonds: crustal distribution and formation processes in time and space and an integrated deposit model. Econ Geol 100:143–177

Haggerty S (1976) Oxide minerals. Rev Mineral 3:6–88
Haggerty SE, Raber, E, Naeser, CW (1983) Fission track dating of kimberlitic zircons. Earth Planet Sci Lett 63:41–50
Hallam CD (1964) The geology of the coastal diamond deposits of southern Africa. *In:* The Geology of Some Ore Deposits in Southern Africa. SH Haughton (ed) Geol Soc S Afr 2:671–728
Hanson EK, Moore JM, Robey J, Bordy EM, Marsh JS (2006) Re-estimation of erosion levels in Group I and II kimberlites between Lesotho, Kimberley and Victoria West, South Africa. 8th International Kimberlite Conference, Kimberlite Emplacement Workshop, Saskatoon, Canada, Extended Abstracts
Hanson EK, Moore JM, Bordy EM, Marsh JS, Howarth G, Robey JVA (2009) Cretaceous erosion in central South Africa: evidence from upper-crystal xenoliths in kimberlite diatremes. S Afr J Geol 112:125–140
Harlow GE (1998) The Nature of Diamonds. Cambridge University Press, Cambridge
Harris JW, Smit KV, Fedortchouk Y, Moore M (2022) Morphology of monocrystalline diamond and its inclusions. Rev Mineral Geochem 88:119–166
Harris M, le Roex A, Class C (2004) Geochemistry of the Uintjiesberg kimberlite, South Africa: petrogenesis of an off-craton, group I, kimberlite. Lithos 74:149–165
Harris PG (1984) Kimberlite volcanism. *In:* Kimberlite Occurrence and Origin: A Basis for Conceptual Models in Exploration. Glover JE, Harris PG (eds) Uni West Aust Spec Publ 8:125–142
Harvey S, Kjarsgaard B, McClintock M, Shimell M, Fourie L, Du Plessis P, Read G (2009) Geology and evaluation strategy of the Star and Orion South kimberlites, Fort a la Corne, Canada. Lithos 112:47–60
Haugaard R, Waterton P, Ootes L, Pearson DG, Luo Y, Konhauser K (2021) Detrital chromites reveal Slave cratons missing komatiite. Geology 49:1079–1083
Hawthorne B (1975) Model of a kimberlite pipe. Phys Chem Earth 9:1–15
Hawthorne JB, Carrington AJ, Clement CR, Skinner EMW (1982) Geology of the Dokolwayo kimberlite and associated palaeo-alluvial diamond deposits. *In:*Kimberlites, Diatremes, and Diamonds: Their Geology, Petrology, and Geochemistry. HOA Meyer, FR Boyd (eds) Am Geophys Union, Washington, DC
Heaman LM, Mitchell RH (1995) Constraints on the emplacement age of Yakutian province kimberlites from U–Pb perovskite dating. 6th International Kimberlite Conference, Novosibirsk, Russia, 6IKC Extended Abstract, p 233
Heaman LM, Kjarsgaard BA (2000) Timing of eastern North American kimberlite magmatism: continental extension of the Great Meteor hotspot track? Earth Planet Sci Lett 178:253–268
Heaman LM, Pearson, DG (2010) Nature and evolution of the Slave Subcontinental Lithospheric Mantle. Lithoprobe Synthesis Volume II—Parameters, Processes and the Evolution of a Continent. Can J Earth Sci 47:369–388
Heaman LM, Kjarsgaard BA, Creaser RA (2003) The timing of kimberlite magmatism in North America: implications for global kimberlite genesis and diamond exploration. Lithos 71:153–184
Heaman LM, Kjarsgaard BA, Creaser RA (2004) The temporal evolution of north American kimberlites. Lithos 76:377–397
Heaman LM, Creaser RA, Cookenboo HO, Chacko T (2006) Multi-stage modification of the Northern Slave mantle lithosphere: evidence from zircon-and diamond-bearing eclogite xenoliths entrained in Jericho kimberlite, Canada. J Petrol 47:821–858
Heaman LM, Pell J, Grütter HS, Creaser RA (2015) U–Pb geochronology and Sr/Nd isotope compositions of groundmass perovskite from the newly discovered Jurassic Chidliak kimberlite field, Baffin Island, Canada. Earth Planet Sci Lett 415:183–199
Heaman LM, Phillips, D, Pearson DG (2019) Dating kimberlites: methods and emplacement patterns through time. Elements 15:399–404
Howarth GH, Büttner SH (2019) New constraints on archetypal South African kimberlite petrogenesis from quenched glass-rich melt inclusions in olivine megacrysts. Gondwana Res 68:116–126
Howarth GH, Giuliani A (2020) Contrasting types of micaceous kimberlite-lamproite magmatism from the Man Craton (West Africa): New insights from petrography and mineral chemistry. Lithos 362:19
Howarth GH, Nembambula T (2021) Petrogenesis of Kaapvaal lamproites (aka orangeites) constrained by the composition of olivine and similarities with kimberlites and other diamondiferous lamproites. Lithos 406:106499
Howarth GH, Michael E, Skinner W, Prevec SA (2011) Petrology of the hypabyssal kimberlite of the Kroonstad group II kimberlite (orangeite) cluster, South Africa: Evolution of the magma within the cluster. Lithos 125:795–808
Hutchison MT, Frei D (2009) Kimberlite and related rocks from Garnet Lake, West Greenland, including their mantle constituents, diamond occurrence, age and provenance. Lithos 112:318–333
Ilupin IP, Lutts BG (1971) The geochemical composition of kimberlite and questions on the origin of kimberlite magma. Sovietskaya Geologiya 6:61–73
Ilyina N, Egorov AY (2008) Upper Triassic of northern Middle Siberia: stratigraphy and paynology. Polar Res 27:372–392
Irvine TN (1965) Chromian spinel as a petrogenetic indicator: Part 1. Theory. Can J Earth Sci 2:648–672
Janse AJA (1984) Kimberlites—where and when. *In:* Kimberlite Occurrence and Origin: A basis for conceptual models in exploration, Glover JE and Harris PG (eds), Uni West Aust Spec Pub 8: 19–62
Janse AJA (1993) The aims and economic parameters of diamond exploration. *In:*Diamonds: Exploration, Sampling and Evaluation. Sheahan P, Chater A (eds) Short Course Proceedings, Prospectors and Developers Association of Canada, Toronto, p 173–184

Janse AJA (1995) A history of diamond sources in Africa: Part I. Gems Gemol 31:228–255

Janse AJA (1996) A history of diamonds in Africa: Part II. Gems Gemol 32:2–30

Janse AJA (2007) Global rough diamond production since 1870. Gems Gemol 43:98–119

Jaques AL, Foley SF (2018) Insights into the petrogenesis of the West Kimberley lamproites from trace elements in olivine. Mineral Petrol 112:519–537

Jaques AL, Lewis JD, Smith CB (1986) The kimberlites and lamproites of Western Australia. Geol Surv West Austral, Perth

Jaques AL, Lewis JD, Smith CB, Gregory GP, Ferguson J, Chappell BW, McCulloch MT (1984) The diamond-bearing ultrapotassic (lamproitic) rocks of the west Kimberley region, Western Australia. *In:*Kimberlites and Related Rocks, Proceedings of the Third International Kimberlite Conference. Vol 1. Kornprobst J (ed) Elsevier Science Publishers, Clermont Ferrand, France, p 225–254

Jacob DE, Mikhail S (2022) Polycrystalline diamonds from kimberlites: Snapshots of rapid and episodic diamond formation in the lithospheric mantle. Rev Mineral Geochem 88:167–190

Jacob RJ, Bluck BJ, Ward JD (1999) Tertiary-age diamondiferous fluvial deposits of the Lower Orange River Valley, southwestern Africa. Econ Geol 94:749–758

Jelsma HA, de Wit MJ, Thiart, C, Dirks, PHGM, Viola, G, Basson IJ, Anckar E (2004) Preferential distribution along transcontinental corridors of kimberlites and related rocks of Southern Africa. S Afr J Geol 107:301–324

Jelsma H, Barnett W, Richards S, Lister G (2009) Tectonic setting of kimberlites. Lithos 112:155–165

Jourdan F, Thern E, Wilde SA, Frewer L (2012) $^{40}Ar/^{39}Ar$ dating of unusual minerals (tourmaline, K-richterite, yimengite, wadeite and priderite) and applicability to the geological record. EGU Conference Abstract

Joy S, Jelsma HA, Preston RF, Kota S (2012) Geology and diamond provenance of the Proterozoic Banganapalle conglomerates, Kurnool Group, India. *In:*Palaeoproterozoic of India. Mazumder R, Saha D (eds) Geol Soc London Spec Publ 365:197–218

Kaminsky FV, Sablukov SA, Sablukova LI, Channer D (2004) Neoproterozoic "anomalous" kimberlites of Guaniamo, Venezuela: mica kimberlites of "isotopic transitional" type. Lithos 76:565–590

Kaminsky FV, Khachatryan GK, Andreazza P, Araujo D, Griffin WL (2009) Super-deep diamonds from kimberlites in the Juina area, Mato Grosso State, Brazil. *In:* Proc. 9th International Kimberlite conference. SF Foley, S Aulbach, GP Brey, HS Grütter, HE Höfer, DE Jacob, V Lorenz, T Stachel, AB Woodland (eds) Lithos 112S: 833–842

Kamenetsky VS, Kamenetsky MB, Sharygin VV, Golovin V (2007a) Carbonate-chloride enrichment in fresh kimberlites of the Udachnaya-East pipe, Siberia: A clue to physical properties of kimberlite magmas? Geophys Res Lett 34:101029

Kamenetsky VS, Kamenetsky MB, Sharygin VV, Faure KA, Golovin V (2007b) Chloride and carbonate immiscible liquids at the closure of the kimberlite magma evolution (Udachnaya-East kimberlite, Siberia) Chem Geol 237:384–400

Kamenetsky VS, Kamenetsky MB, Sobolev AV, Golovin AV, Demouchy S, Faure K, Sharygin VV, Kuzmin DV (2008) Olivine in the Udachnaya-East kimberlite (Yakutia, Russia): Types, compositions and origins. J Petrol 49:823–839

Karfunkel J, Chaves MLSC, Svisero DP, Meyer HOA (1994) Diamonds from Minas Gerais, Brazil: an update on sources, origin and production. Int Geol Rev 36:1019–1032

Karfunkel J, Hoover D, Fernandes AF, Sgarbi GNC, Kambrock K, Oliveira GD (2014) Diamonds from the Coromandel area, west Minas Gerais State, Brazil: an update and new data on surface sources and origin. Braz J Geol 44:325–338

Kargin AV, Nosova AA, Larionova YO, Kononova VA, Borisovsky SE, Koval'chuk EV, Griboedova IG (2014) Mesoproterozoic orangeites (Kimberlites II) of West Karelia: Mineralogy, geochemistry, and Sr–Nd isotope composition. Petrology 22:151–183

Kedrova TV, Bogush IN, Zinchuk NW, Bhardukhinov LD, Lipashova AN, Saltykova VP (2022) Diamond placers of the Nakyn kimberlite field. Russ Geol Geophys 63:245–254

Kennedy W (1964) The structural differentiation of Africa in the Pan-African (± 500 m.y.) tectonic episode. 8th Annual Report of the Research Institute of African Geology, University Leeds, p 48–49

Kilham J, Field M, Steifenhofer J (1998) Orapa and Lethlakane Mines. *In:*7th International Kimberlite Conference, Large Mines Field Excursion, University of Cape Town, South Africa, p 11–21

Kirkley MB, Gurney JJ, Levinson A (1992) Age, origin and emplacement of diamonds: A review of scientific advances in the last decade. CIM Bull 85:48–57

Kjarsgaard, BA (2003) Volcanology of kimberlite. *In:*Diamonds Short Course Notes. Tosdal R (ed) Cordilleran Round-Up, Vancouver BC, January 29–30

Kjarsgaard BA (2007a) Kimberlite diamond deposits. *In:*Mineral Deposits of Canada: A Synthesis of Major Deposit-Types, District Metallogeny, the Evolution of Geological Provinces, and Exploration Methods. Vol Special Publication No. 5. Goodfellow WD (ed) Geological Association of Canada, Ottawa, p 245–272

Kjarsgaard BA (2007b) Kimberlite pipe models: Significance for exploration. *In:*Proceedings of Exploration 2007. Milkereit B (ed) PDAC, Toronto p 667–677

Kjarsgaard BA, Levinson AL (2002) Diamonds in Canada. Gems Gemol 38:208–238

Kjarsgaard BA, Wilkinson L, Armstrong J (2002) Geology, Lac de Gras Kimberlite Field, Central Slave Province, Northwest Territories—Nunavut. Geol Surv Can Open File 3228, 1:250,00 scale, with descriptive notes and figures

Kjarsgaard BA, Leckie DA, Zonneveld JP (2007) Discussion of "Geology and diamond distribution of the 140/141 kimberlite, Fort a la Corne, central Saskatchewan, Canada", by A Berryman BH Scott Smith and BC Jellicoe. Lithos 97:422–428

Kjarsgaard BA, Pearson DG, Tappe S, Nowell GM, Dowall DP (2009) Geochemistry of hypabyssal kimberlites from Lac de Gras, Canada: Comparisons to a global database and applications to the parent magma problem. Lithos 112:236–248

Kjarsgaard BA, Pearson DG, Malarkey J (2010) The kimberlite olivine phenocryst/macrocryst/xenocryst problem, re-visited. GeoCanada 2010, Calgary, May 2010

Kjarsgaard BA, Januszczak N, Stiefenhofer J (2019) Diamond exploration and resource evaluation of kimberlites. Elements 15:411–416

Kinny PD, Griffin BJ, Heaman LM, Brakhfogel FF, Spetsius ZV (1997) SHRIMP U–Pb ages of perovskite from Yakutian kimberlites. Russ Geol Geophys 38:97–105

Konstantinovskii AA (2003) Epochs of diamond placer formation in the Precambrian and Phanerozoic. Lithol Min Res 38:530–546

Koornneef JM, Gress MU, Chinn IL, Jelsma HA, Harris JW, Davies, GR (2017) Archaean and Proterozoic diamond growth from contrasting styles of large-scale magmatism. Nat Commun 8:648

Kopylova M, Bruce L, Ryder J (2010) Diamonds in an Archaean greenstone belt: diamond suites in unconventional rocks of Wawa, northern Ontario (Canada) Geophys Res Abstracts 12:EGU2010-6835

Kostrovitsky ZV, Spetsius DA, Fon-der-Flaas GS, Suvorova LF, Bogush IN (2015) Atlas of Primary diamond deposits of Yakutian kimberlite province. Pokhilenko NP (ed), Scientific Council of NIGP ALROSA, Mirny

Kruger K, Maphane K (2017) Orapa, Lethlakane and Damtshaa Mines field guide. *In:* Desert Gems Botswana's Major Mines, 11th International Kimberlite Conference Field Guide, Gabarone, Botswana

Kueter N, Soesilo J, Fedortchouk Y, Nestola F, Belluco L, Juliana Troch J, Wälle M, Guillong M, Von Quadt A, Driesner T (2016) Tracing the depositional history of Kalimantan diamonds by zircon provenance and diamond morphology studies. Lithos 265:159–176

Kurszlaukis S, Lorenz V (2008) Fomation of "Tuffisitic Kimberlites" by phreatomagmatic processes. J Volcanol Geoth Res 174:68–80

Kurszlaukis S, Büttner R, Zimanowski B, Lorenz V (1998) On the first experimental phreatomagmatic explosion of a kimberlite melt. J Volcanol Geotherm Res 80:323–326

Kurszlaukis S, Mahotkin I, Rotman AY, Kolesnikov GV, Makovchuk IV (2009) Syn- and post-eruptive volcanic processes in the Yubileinaya kimberlite pipe, Yakutia, Russia, and implications for the emplacement of South African-style kimberlite pipes. Lithos 112:579–591

Laiginhas F, Pearson DG, Phillips D, Burgess R, Harris JW (2009) Re–Os and $^{40}Ar/^{39}Ar$ istopoe measurements of inclusions in alluvial diamonds from the Ural mountains: constrains on diamond genesis and eruption ages. Lithos 112S:714–723

Lapin AV, Tolstov AV, Vasilenko VB (2007) Petrogeochemical characteristics of the kimberlites from the Middle Markha region with application to the problem of the geochemical heterogeneity of kimberlites. Geochem Int 45:1197–1209

le Roex AP, Bell DR, Davis P (2003) Petrogenesis of Group I Kimberlites from Kimberley, South Africa: Evidence from Bulk-rock Geochemistry. J Petrol 44:2261–2286

Leckie DA, Kjarsgaard BA, Bloch J, McIntyre D, McNeil D, Stasiuk L, Heaman L (1997) Emplacement and reworking of Cretaceous, diamond-bearing, crater facies kimberlite of central Saskatchewan, Canada. Geol Soc Am Bull 109:1000–1020

Lefebvre N, Kopylova M, Kivi K, Barnett R (2003) Diamondiferous volcaniclastic debris flows of Wawa, Ontario, Canada. 8th International Kimberlite Conference, Long abstracts, FLA 0289, 5 pp

Lehmannn B, Burgess R, Frei D, Belyatsky B, Mainkar D, Rao NVC, Heaman LM (2010) Diamondiferous kimberlites in central India synchronous with Deccan flood basalts. Earth Planet Sci Lett 290:142–149

Lewis HC (1888) The matrix of diamond. Geol Mag 5:129–131

Liccardo A, Chieregati LA (2013) A extração de diamantes na história geológica e mineral no Paraná. Boletim de geociências paranaense 70:166–179

Lim E, Giuliani A, Phillips D, Goemann K (2018) Origin of complex zoning in olivine from diverse, diamondiferous kimberlites and tectonic settings: Ekati (Canada), Alto Paranaiba (Brazil) and Kaalvallei (South Africa) Mineral Petrol 112:539–554

Linol B, de Wit MJ, Barton E, Guillocheau F, de Wit MCJ, Colin JP (2015) Facies analyses, chronostratigraphy and paleo-environmental reconstructions of Jurassic to Cretaceous Sequences of the Congo Basin. Ch 8 *In:*Geology and Resource Potential of the Congo Basin, Regional Geology Reviews. MJ de Wit et al. (eds) Springer-Verlag, p 135–161

Liu J, Pearson DG, Wang LH, Mather KA, Kjarsgaard BA, Schaeffer AJ, Irvine GJ, Kopylova MG, Armstrong JP (2021) Plume-driven re-cratonisation of deep continental lithospheric mantle. Nature 592:732–736

Lorenz V (1975) Formation of phreatomagmatic maar-diatreme volcanoes and its relevance to kimberlite diatremes. *In:* Proceedings of the First International Kimberlite Conference, Phys Chem Earth, Vol 9. Ahrens LH, Dawson JB, Duncan AR, Erlank AJ (eds) Pergamon Press, Cape Town, South Africa, p 17–27

Lorenz V (1987) Phreatomagmatism and its relevance. Chem Geol 68:149–156

Lorenz V, Kurszlaukis S (2007) Root zone processes in the phreatomagmatic pipe emplacement model and consequences for the evolution of maar-diatreme volcanoes. J Volcanol Geotherm Res 159:4–32

MacNevin AA (1977) Diamonds in New South Wales. Geol Surv NS Wales, Mineral Resources No. 42

Mahotkin IL (1998) Petrology of the Group 2 kimberlite–olivine lamproite (K2L) series from the Kostomksha area, Karelia, NW Russia. 7th International Kimberlite Conference, Cape Town, South Africa, p 529–531

Mahotkin IL, Skinner EMW (1998) Kimberlites from the Archangelsk region—a rock type transitional between kimberlites, melnoites and lamproites. 7th International Kimberlite Conference, Cape Town, South Africa, p 532–534

Mainkar D, Lehmannn B (2007) The diamondiferous Behradih kimberlite pipe, Mainpur kimberlite field, Chhattisgarh, India: Reconnaissance petrography and geochemistry. J Geol Soc India 69:547–552

Mainkar D, Lehmannn B, Haggerty SE (2004) The crater-facies kimberlite system of Tokapal, Bastar district, Chhattisgarh, India. Lithos 76:201–217

Mannard GW (1962) The Singida Kimberlite Pipes. PhD thesis, McGill University, Montreal, Canada

Marshall T (1990) The Nature, Origin and Evolution of the Diamondiferous Gravels of the South Western Transvaal. Ph.D thesis (unpubl.), University of the Witwatersrand, Johannesburg

Marshall T (2004) Rooikoppie deposits in South Africa. Rough Diamond Rev 6:21–25

Marshall TR, Baxter-Brown R (1995) Basic principles of alluvial diamond exploration. J Geochem Explor 53:277–272

Marshall TR, Ward JD, de Wit MCJ (2013) African Alluvial diamond deposits. Presentation Diamond Course Oct. 2013, University of Pretoria

McClintock M, Ross PS, White JDL (2009) The importance of the transport system in shaping the growth and form of kimberlite volcanoes. Lithos 112:465–472

McPhie J, Doyle M, Allen RL (1993) Volcanic Textures. University of Tasmania

Meyer de Stadelhofen C (1963) Les breches kimberlitiques du Territoire de Bakwanga (Congo) Arch Sci Geneve 16:8–143

Meyer HOA, McCallum ME (1993) Diamonds and their sources in the Venezuelan portion of the Guyana Shield. Econ Geol 88:989–998

Micon International Co Ltd (2016) Alrosa Group of companies: Independent expert report on the reserves and resources of the diamond assets (JORC Code), Russian Federation

Milesi JP, Ledru P, Feybesse JL, Dommanget A, Marcoux E (1992) Early Proterozoic ore deposits and tectonics of the Birimian orogenic belt, West Africa. Precambrian Res 58:305–344

Miller CE, Kopylova MG, Ryder J (2012) Vanished diamondiferous cratonic root beneath the southern Superior province: evidence from diamond inclusions in the Wawa meta conglomerate. Contrib Mineral Petrol 164:697–714

Mitchell R (1984) Mineralogy and origin of carbonate-rich segregations in a composite kimberlite sill. Neues Jahrb Mineral Abh 150:185–197

Mitchell R (1985) A review of the mineralogy of lamproites. Trans Geol Soc S Afr 88:411–437

Mitchell RH (1986) Kimberlites: Mineralogy, Geochemistry, and Petrology. Plenum Press, New York

Mitchell RH (1995) Kimberlites, Orangeites and Related Rocks. Plenum Press, New York

Mitchell RH (2013) Paragenesis and oxygen isotopic studies of serpentine in kimberlite. Paragenesis and oxygen isotopic studies of serpentine in kimberlite. *In:* Proceedings of the 10th International Kimberlite Conference, Vol 1, Pearson DG, et al. (eds) Springer, p 1–12

Mitchell RH (2020a) Potassic Alkaline Rocks: Leucitites, Lamproites, and Kimberlites. Encycl Geol, 2nd edition

Mitchell RH (2020b) Igneous Rock Associations 26. Lamproites, Exotic Potassic Alkaline Rocks: A Review of their Nomenclature, Characterization and Origins. Geosci Can 47:119–142

Mitchell RH, Bergman SC (1991) Petrology of Lamproites. Springer

Mitchell RH, Skinner EMW, Scott Smith BH (2009) Tuffisitic kimberlite from the Wesselton Mine, South Africa: Mineralogical characteristics relevant to their formation. Lithos 112 S1:452–464

Mitchell RH, Scott Smith BH, Skinner EMW (2012) Mineralogy of magmaclasts and interclast matrices of Kimberley-type pyroclastic kimberlites from Kao, Letseng-la-Terae, Lethlakane and Premier kimberlite pipes, southern Africa. International Kimberlite Conference Extended Abstracts 10:10IKC-094

Mitchell RH, Giuliani A, O'Brien H (2019) What is a kimberlite? Petrology and mineralogy of hypabyssal kimberlites. Elements 15:381–386

Mmualefe MK (2017) Jwaneng diamond mine, Botswana: History, geology and mining. *In:* Desert Gems, Botswana's Major Mines, 11th International Kimberlite Conference Field Guide, Gabarone, Botswana, p 1–31

Moore R, Gurney J (1991) Garnet megacrysts from Group II kimberlites in southern Africa. International Kimberlite Conference: Extended Abstracts 5, p 298–300

Moore A, Moore J (2006) A glacial ancestry for the Somabula diamond-bearing alluvial deposit, Central Zimbabwe. S Afr J Geol 109:625–636

Moore A, Costin G, Proyer A (2021) Cognate versus xenocrystic olivines in kimberlites–A review. Earth Sci Rev 221:103771

Moore A, Blenkinsop T, Cotterill F (2008) Controls on post-Gondwana alkaline volcanism in Southern Africa. Earth Planet Sci Lett 268:151–164

Moore AE, Cotterill FPD, Broderick T, Plower D (2009) Landscape evolution in Zimbabwe from the Permian to present with implications for kimberlite prospecting. S Afr J Geol 112:65–88

Moss S, Russell JK, Brett RC, Andrews GDM (2009) Spatial and temporal evolution of kimberlite magma at A154N, Diavik, Northwest Territories, Canada. Lithos 112 S1:541–552

Moss S, Russell JK, Smith BHS, Brett RC (2010) Olivine crystal size distributions in kimberlite. Am Mineral 95:527–536

Moss S, Webb, K, Hetman, C, Manyumbu A (2013) Geology of the K1 and K2 kimberlite pipes at Murowa, Zimbabwe. *In:* Proc 10th International Kimberlite Conference, J Geol Soc India, Spec Issue 2, p 35–50

Moss S, Porritt L, Pollock K, Fomradas G, Stubley M, Eichenberg D, Cutts J (2018) Geology, mineral chemistry, and structure of the kimberlites at diavik diamond mine: indicators of cluster-scale cross-fertilization, mantle provenance, and pipe morphology. *In:* Geoscience and Exploration of the Argyle, Bunder, Diavik and Murowa diamond deposits. Davy AT, Smith CB, Helmstaedt H, Jaques AL, Gurney JJ (eds) Soc Econ Geol Spec Publ 20, p 287–318

Muggeridge MT (1995) Pathfinder sampling techniques for locating primary sources of diamond: Recovery of indicator minerals, diamonds and geochemical signatures. J Geochem Explor 53:183–204

Naidoo P, Stiefenhofer J, Field M, Dobbe R (2004) Recent advances in the geology of Koffiefontein Mine, Free State Province, South Africa. Lithos 76:161–182

Nannini F, Neto IC, Silveira FV, Cunha LM, de Oliveira RG, Weska RK (2017) Áreas kimberlíticas e diamantíferas do estado do Mato Grosso. Projeto diamante Brasil, Serviço Geológico do Brasil—CPRM, Série Pedras Preciosas, No 12

Nielsen TF, Sand KK (2008) The Majuagaa kimberlite dike, Maniitsoq region, West Greenland: constraints on an Mg-rich silicocarbonatitic melt composition from groundmass mineralogy and bulk compositions. Can Mineral 46:1043–1061

Nielsen TFD, Jensen SM, Secher K, Sand KK (2009) Distribution of kimberlite and aillikite in the Diamond Province of southern West Greenland: A regional perspective based on groundmass mineral chemistry and bulk compositions. Lithos 112:358–371

Niggli P (1923) Gesteins und Mineralprovinzen: Verlag Gebrüder Borntraegar, Berlin

Nikolenko EI, Logvinova AM, Izokh AE, Afanasiev VP, Oleynikov OB, Biller Aya. (2018) Polyphase inclusions in chrome spinels from Upper Triassic gravelites from the northeastern part of the Siberian Platform. Dokl Earth Sci 480:656–660

Nixon PH (1973) Lesotho Kimberlites. Lesotho National Development Corporation, Maseru, Lesotho

Nixon PH (1987) Mantle Xenoliths. John Wiley & Sons, Chichester

Nixon PH (1995) The morphology and nature of primary diamondiferous occurrences. J Geochem Explor 53:41–71

Norton G, Bristow J, van Wyk H (2007) Alluvial deposits and diamonds of the Lower Vaal and Middle Orange River, northern Cape Province RSA (2007) S Afr Inst Min Metal Diamonds—Source to Use. Extended Abstracts

Nowicki T (2014) On the value of geology in kimberlite evaluation: are we oversampling? Kimberley Diamond Symposium and Trade Show, Kimberley, September 11–13, Proc Geol Soc S Afr

O'Brien H, Tyni M (1999) Mineralogy and geochemistry of kimberlites and related rocks from Finland. *In:* Proc 7th International Kimberlite Conference, Gurney JJ, Gurney JL, Pascoe MD, Richardson SH (eds) Red Roof Design, Cape Town, p 625–636

O'Brien H, Phillips D, Spencer R (2007) Isotopic ages of Lentiira–Kuhmo–Kostomuksha olivine lamproite–Group II kimberlites. Bull Geol Soc Finland 79:203–215

Oliveira SC, Pupim FN, Stevaux JC, Assine ML (2019) Luminescence chronology of terrace development in the Upper Paraná River, southeast Brazil. Front Earth Sci 7:200

Pan American Union (1907) Monthly International Bureau of the American Republics, Jan—Jun, Washington, 14

Pasteris JD (1982) Representation of compositions in complex titanian spinels and application to the De Beers kimberlite. Am Mineral 67:244–250

Pasteris JD (1983) Spinel zonation in the De Beers kimberlite, South Africa: possible role of phlogopite. Can Mineral 21:41–58

Pearson D, Wittig N (2014) The formation and evolution of cratonic mantle lithosphere: Evidence from mantle xenoliths. Ch 3.6 *In:* Treatise on Geochemistry. Carlson RW (ed) Elsevier, p 255–292

Pearson DG, Shirey SB, Harris JW, Carlson, RW (1998) Sulphide inclusions in diamonds from the Koffiefontein kimberlite, S. Africa: constraints on diamond ages and mantle Re–Os systematics. Earth Planet Sci Lett 160:311–326

Pearson DG, Canil D, Shirey SB (2003) Mantle samples included in volcanic rocks: xenoliths and diamonds. *In:* Treatise on Geochemistry. Carlson RW (ed) Elsevier. The mantle and core 2:171–276

Pearson DG, Brenker FE, Nestola F, McNeill J, Nasdala L, Hutchison MT, Matveev S, Mather K, Silversmit G, Schmitz S, Vekemans B (2014) Hydrous mantle transition zone indicated by ringwoodite included in diamond. Nature 507:221–224

Pearson D, Liu J, Smith C, Mather K, Krebs M, Bulanova G, Kobussen A (2018) Characteristics and origin of the mantle root beneath the Murowa diamond mine: Implications for craton and diamond formation. *In:* Geoscience and Exploration of the Argyle, Bunder, Diavik and Murowa Diamond Deposits. Vol 20. Spec Publ Soc Econ Geol, p 403–424

Pearson DG, Woodhead J, Janney PE (2019) Kimberlites as geochemical probes of earth's mantle. Elements 15:387–392

Pearson DG, Scott JM, Liu J, Schaeffer A, Wang LH, van Hunen J, Szilas K, Chacko T, Kelemen PB (2021) Deep continental roots and cratons. Nature 596:199–210

Pedreira AJ, de Waele B (2008) Contemporaneous evolution of the Paleoproterozoic–Mesoproterozoic sedimentary basins of the São Francisco-Congo Craton. *In:* West Gondwana Pre-Cenozoic Correlations Across the Southern Atlantic Region. Pankhurst RJ et al. (eds) Geol Soc London Spec Publ 294:33–48

Perdoncini LC, Soares PC (1999) O diamante na bacia do rio Santa Rosa, Tibagi (pr) Revista Brasileira de Geociências 29:299–306

Pereira E, Rodrigues J, Reis B (2003) Synopsis of Luanda geology, NE Angola: Implications for diamond exploration. Comun Inst Geol e Mineiro 90:189–212

Pereira RS, Fuck RA, França OS, Leite AA (2017) Evidence of young, proximal and primary (YPP) diamond source occurring in alluviums in the Santo Antônio do Bonito, Santo Inácio and Douradinho rivers in Coromandel region, Minas Gerais. Braz J Geol 47:383–401

Petrov OV, Morozov AF, Shatov VV, Molchanov AV, Terekhov AV, Lukyanova LI, Artem'ev DS, Belova VN, Khalenev VO (2016) Russia, *In:* Mineral Resources in the Arctic. Geol Surv Norway Spec Publ Ch 9, p 374–482

Pervov VA, Somov SV, Korshunov AV, Dulapchii EV, Felix JT (2011) The Catoca kimberlite pipe, Republic of Angola: A Paleovolcanological model. Geol Ore Dep 53:295–308

Phillips D, Harris JW (2009) Diamond provenance studies from $^{40}Ar/^{39}Ar$ dating of clinopyroxene inclusions: An example from the west coast of Namibia. Lithos 112S:793–805

Phillips D, Zhong D, Matchan EL Maas R, Farr H, O'Brien H, Giuliani A (2017) A Comparison of Geochronology Methods Applied to Kimberlites and Related Rocks from the Karelian Craton, Finland. 11th International Kimberlite Conference, Botswana, Extended Abstract, 11IKC-4880

Phillips D, Harris JW, de Wit MCJ, Matchan EL (2018) Provenance history of detrital diamond deposits, West Coast of Namaqualand, South Africa. Mineral Petrol 112:S259–S273

Pilbeam L, Nielsen T, Waight T (2013) Digestion fractional crystallization (DFC): an important process in the genesis of kimberlites. Evidence from olivine in the Majuagaa kimberlite, southern West Greenland. J Petrol 54:1399–1425

Pivin M, Femenias O, Demaiffe D (2009) Metasomatic mantle origin for Mbuji-Mayi and Kundelungu garnet and clinopyroxene megacrysts (Democratic Republic of Congo) Lithos 112:951–960,

Porritt LA, Russell JK, McLean H, Fomradas G, Eichenberg D (2013) A Phreatomagmatic kimberlite: The A418 kimberlite pipe, Northwest Territories, Canada. *In:* Proc 10th International Kimberlite Conference, J Geol Soc India, Spec Issue 2:97–107

Price S, Russell J, Kopylova M (2000) Primitive magma from the Jericho Pipe, NWT, Canada: constraints on primary kimberlite melt chemistry. J Petrol 41:789–808

Prokopchuk BI (1972) Zoning in the distribution of diamond placers on old platforms. Dokl Akad Nauk SSSR 212:100–103

Quintão DA, Caxito FdA, Karfunkel J, Vieira FR, Seer HJ, de Moraes LC, Ribeiro LCB, Pendrosa-Soares AC (2017) Geochemistry and sedimentary provenance of the Upper Cretaceous Uberaba Formation (Southeastern Triângulo Mineiro MG, Brazil). Braz J Geol 47:159–182

Raal FA (1969) A study of some gold mine diamonds. Am Mineral 54:292–296

Ranger I, Heaman LM, Pearson DG, Muntener C, Zhuk V (2018) Punctuated, long-lived emplacement history of the Renard 2 kimberlite, Canada, revealed by new high precision U–Pb groundmass perovskite dating. Mineral Petrol 112:639–651

Rao NVC, Lehmannn B, Mainkar D, Belyatsky B (2011) Petrogenesis of the end-Cretaceous diamondiferous Behradih orangeite pipe: implication for mantle plume–lithosphere interaction in the Bastar craton, Central India. Contrib Mineral Petrol 161:721–74

Rau TK (2007) Panna diamond Belt, Madhya Pradesh—A critical review. J Geol Soc India 69:513–521

Rau TK, Wadhwa R, Mishra V, Fareeduddin (2012) Field guide to central Indian lamproite field. Guidebook prepared for the 10th International Kimberlite Conference 12016 Feb 2012

Ravi S, Sufija MV, Patel SC, Sheikh JM, Sridhar M, Kaminsky FV, Khachatryan GK, Nayak SS, Bhaskara Rao KS (2012) Diamond potential of the Eastern Dharwar Craton, India: their physical and infrared characteristics. Short Abstract 10th International Kimberlite Conference, p 282–283

Ravi S, Sufija MV, Patel SC, Sheikh JM, Sridhar M, Kaminsky FV, Khachatryan GK, Nayak SS, Bhaskara Rao KS (2013) Diamond potential of the Eastern Dharwar Craton, Southern India, and a reconnaissance study of physical and infrared characteristics of the diamonds. *In:*Proceedings of the 10th International Kimberlite Conference, DG Pearson, HS Grutter, JW Harris, BA Kjarsgaard, HO'Brien, NV Chalapathi Rao, S Sparks (eds) 1:335–348

Ray JS (2006) Age of the Vindhyan Supergroup: A review of recent findings. J Earth System Sci 115:149–160

Rayner MJ, Moss SW, Jaques AL, Lorenz V, Boxer GL, Smith CB, Webb K (2017) New insights into volcanic processes and diamond grades from deep mining at Argyle. IKC extended abstracts 11

Rayner MJ, Moss SW, Lorenz V, Jaques AL, Boxer GL, Smith CB, Webb K (2018a) New insights into volcanic processes from deep mining of the southern diatreme within the Argyle lamproite pipe, Western Australia. Mineral Petrol 112:351–363

Rayner MJ, Jaques A, Boxer G, Smith C, Lorenz V, Moss SW, Webb K, Ford D (2018b) The geology of the Argyle (AK1) diamond deposit, Western Australia. *In:* Geoscience and Exploration of the Argyle, Bunder, Diavik, and Murowa Diamond Deposits. Society of Economic Geologists

Read GH, Janse AJA (2009) Diamonds: Exploration, mines and marketing. Lithos 112:1–9

Reid AR (1974) Proposed origin of Guianan diamonds. Geology 2:67–68

Reid AM, Donaldson CH, Dawson JB, Brown RW (1975) The Igwisi Hills extrusive "kimberlites". *In:* Proceedings of the First International Kimberlite Conference, Phys Chem Earth, Vol 9. Ahrens LH, Dawson JB, Duncan AR, Erlank AJ, (eds) Pergamon Press, Cape Town, South Africa, p 199–218

Reis NJ, Nadeau S, Fraga LM, Betiollo LM, Faraco MTL, Reece J, Lachhman D, Ault R (2017) Stratigraphy of the Roraima Supergroup along the Brazil–Guyana border in the Guiana shield, Northern Amazonian Craton—Results of the Brazil–Guyana Geology and Geodiversity Mapping Project. Braz J Geol 47:43–57

Richardson, SH (1986) Latter-day origin of diamonds of eclogitic paragenesis. Nature 322:623–626

Richardson SH, Erlank AJ, Harris JW, Hart SR (1990) Eclogitic diamonds of Proterozoic age from Cretaceous kimberlites. Nature 346:54–56

Richardson SH, Chinn IL, Harris, JW (1999) Age and origin of eclogitic diamonds from the Jwaneng Kimberlite, Botswana. Proc 7th International Kimberlite Conference 2:709–713

Richardson SH, Shirey SB, Harris, JW (2004) Episodic diamond genesis at Jwaneng, Botswana, and implications for Kaapvaal Craton evolution. Lithos 77:143–154

Richardson SH, Poml, P, Shirey SB, Harris, JW (2009) Age and origin of peridotitic diamonds from Venetia, Limpopo Belt, Kaapvaal-Zimbabwe craton. Lithos 112:785–792

Robles-Cruz SE, Escayola M, Jackson S, Galí S, Pervov V, Watangua M, Gonçalves A, Melgarejo JC (2012) U–Pb SHRIMP geochronology of zircon from the Catoca kimberlite, Angola: Implications for diamond exploration. Chem Geol 310–311:137–147

Roberts EM, Jelsma H, Hegna (2015) Mesozoic sedimentary cover sequences of the Congo Basin in the Kasai Region, Democratic Republic of Congo. Ch 9 *In:*Geology and Resource Potential of the Congo Basin, Regional Geology Reviews, MJ de Wit et al. (eds) Springer-Verlag, p 163–191

Rock NM (1988) Lamprophyres. Springer

Roeder P, Schulze D (2008) Crystallization of groundmass spinel in kimberlite. J Petrol 49:1473–1495

Roffey S, Rayner MJ, Davy AT, Platell RW (2018) Evaluation of the AK1 diamond deposit, Western Australia. *In:* Geoscience and Exploration of the Argyle, Bunder, Diavik and Murowa Diamond Deposits. Davy AT, Smith CB, Helmstaedt H, Jaques AL, Gurney JG (eds) Soc Econ Geol Spec Publ 20, p 65–88

Rozova YV, Frantsesson EV, Pleshakov AP, Botova MM, Filipova LP (1982) High iron chrome spinels in kimberlites of Yakutia. Int Geol Rev 24:1417–1425

Sampson DH (1953) The volcanic hills of Igwisi. Rec Geol Surv Tanganika 3:47–53

Sarkar C, Heaman LM, Pearson DG (2015) Duration and periodicity of kimberlite volcanic activity in the Lac de Gras kimberlite field, Canada and some recommendations for kimberlite geochronology. Lithos 218–219:155–166

Sarkar C, Kjarsgaard BA, Pearson DG, Heaman LM, Locock A, Armstrong JP (2018) Geochronology, classification and mantle source characteristics of kimberlites and related rocks from the Rae Craton, Melville Peninsula, Nunavut, Canada. Mineral Petrol 112:653–672

Sarkar S, Giuliani A, Ghosh S, Phillips D (2021) Petrogenesis of coeval lamproites and kimberlites from the Wajrakarur field, Southern India: New insights from olivine compositions. Lithos 406:106524

Schmid R (1981) Descriptive nomenclature and classification of pyroclastic deposits and fragments: Recommendations of the IUGS Subcommission on the Systematics of Igneous Rocks. Geology 9:41–43

Schulze DJ (1987) Megacrysts from alkalic volcanic rocks. *In:* Mantle Xenoliths. Nixon PH (ed) John Wiley & Sons Ltd., Chichester, p 433–451

Scott BH (1979) Petrogenesis of kimberlites and associated potassic lamprophyres from central West Greenland. *In:* The Second International Kimberlite Conference. Vol 1. Boyd FR, Meyer HOA (eds) Am Geophys Union, Santa Fe, USA, p 190–205

Scott BH (1981) Kimberlite and lamproite dykes from Holsteinborg, Greenland. Medd Gronland Geosci 4:3–24

Scott Smith BH (2006) Canadian kimberlites: geological characteristics relevant to emplacement. 8th International Kimberlite Conference, Kimberlite Emplacement Workshop, Saskatoon, Canada, Extended Abstracts

Scott Smith BH (2008) Canadian kimberlites: geological characteristics relevant to emplacement. J Volc Geotherm Res 174:9–19

Scott Smith BH, Skinner EMW (1984) A new look at Prairie Creek, Arkansas. *In:*Kimberlites 1: Kimberlites and related rocks. Kornprobst J (ed) Elsevier, Amsterdam, p 255–283

Scott Smith BH, Danchin RV, Harris JW, Stracke KJ (1984) Kimberlites near Orroroo, South Australia. *In:* Kimberlites and Related Rocks, Proceedings of the Third International Kimberlite Conference. Kornprobst J (ed) Elsevier Science Publishers, Clermont Ferrand, France, 1:121–142

Scott Smith BH, Skinner EMW, Loney PE (1989) The Kapamba lamproites of the Luangwa Valley, Eastern Zambia. *In:* Kimberlites and Related Rocks, Proceedings from the Fourth International Kimberlite Conference. Ross J (ed) GSA Spec Publ No.14, Blackwell, Perth, Australia, 1:189–205

Scott Smith BH, Nowicki TE, Russell JK, Webb KJ, Mitchell RH, Hetman CM, Harder M, Skinner EMW, Robey JVA (2013) Kimberlite Terminology and Classification. *In:* Proceedings of the 10th International Kimberlite Conference, Pearson DG, et al. (eds) Spec Pub J Geol Soc India 2:1–17

Scott Smith BH, Nowicki TE, Russell JK, Webb K, Mitchell RH, Hetman C, Robey JVA (2018) A Glossary of Kimberlite and Related Terms. Scott Smith Petrology Incorporated, North Vancouver

Shaikh AM, Kumar SP, Patel SC, Thakur SS, Ravi S, Behera D (2018) The P3 kimberlite and P4 lamproite, Wajrakarur kimberlite field, India: mineralogy, and major and minor element compositions of olivines as records of their phenocrystic vs xenocrystic origin. Mineral Petrol 112:609–624

Shand SJ (1922) The problem of the alkaline rocks. Proc Geol Soc S Afr 25:xix–xxxii

Shatsky V, Zedgenizov D, Ragozin A, Kalinina V (2019) Silicate melt inclusions in diamonds of eclogite parageneis from placers on the northeastern Siberian craton. Minerals 9:412

Shigley JE, Chapman J, Ellison RK (2001) Discovery and mining of the Argyle diamond deposit, Australia. Gems Gemol 37:26–41

Shirey SB, Cartigny P, Frost DJ, Keshav S, Nestola F, Nimis P, Pearson DG, Sobolev NV, Walter MJ (2013) Diamonds and the geology of mantle carbon. Rev Mineral Geochem 75:355–421

Skinner EMW (1989) Contrasting Group I and Group II kimberlite petrology: towards a genetic model for kimberlites. *In:*Kimberlites and Related Rocks, Proceedings from the Fourth International Kimberlite Conference. Ross J (ed) GSA Spec Publ 14, Blackwell, Perth, Australia, 1:528–544

Skinner EMW, Viljoen KS, Clark TC, Smith CB (1994) The petrography, tectonic setting and emplacement ages of kimberlites in the southwestern border region of the Kaapvaal craton, Prieska area. *In:* Proceedings of the 5th International Kimberlite Conference, CPRM, Brazil. Meyer HOA, Leonardos OH (eds) 1:80–97

Skinner E, Marsh JS (2004) Distinct kimberlite pipe classes with contrasting eruption processes. Lithos 76:183–200

Skinner EMW, Apter DB, Morelli C, Smithson NK (2004) Kimberlites of the Man Craton, West Africa. Lithos 76:233–260

Smart KA, Tappe S, Stern RA, Webb SJ, Ashwal LD (2016) Early Archaean tectonics and mantle redox recorded in Witwatersrand diamonds. Nat Geosci 9:255–259

Smelov AP, Zaitsev AI (2013) The age and localization of kimberlite magmatism in the Yakutian Kimberlite Province: Constraints from isotope geochronology—An overview. *In:* Proceedings of 10th International Kimberlite Conference, DG Pearson et al. (eds.) Spec Issue J Geol Soc India 1:225–234

Smit KV, Pearson DG, Stachel, T, Seller, M (2014) Peridotites from Attawapiskat, Canada: Mesoproterozoic reworking of Palaeoarchaean lithospheric mantle beneath the Northern Superior superterrane. J Petrol 55:1829–1863

Smit KV, Timmerman S, Aulbach S, Shirey SB, Richardson SH, Phillips D, Pearson DG (2022) Geochronology of diamonds. Rev Mineral Geochem 88:567–636

Smith C (1983) Pb, Sr and Nd isotopic evidence for sources of southern African Cretaceous kimberlites. Nature 304:51–54

Smith CB, Lorenz V (1989) Volcanology of the Ellendale lamproite pipes, Western Australia. *In:* Kimberlites and Related Rocks, Proceedings from the Fourth International Kimberlite Conference. Ross J (ed) GSA Spec Publ 14, Blackwell, Perth, Australia, 1:505–520

Smith C, Gurney J, Ebrahim N, Skinner E, Clement C (1985) Geochemical character of southern African kimberlites: a new approach based on isotopic constraints. Trans Geol Soc S Afr 88:267–280

Smith C, Gurney JJ, Harris JW, Otter ML, Kirkley MB, Jagoutz, E (1991) Neodymium and strontium isotope systematics of eclogite and websterite paragenesis inclusions from single diamonds, Finsch and Kimberley Pool, RSA. Geochim Cosmochim Acta 55:2579–2590

Smith CB, Bulanova GP, Kohn SC, Milledge HJ, Hall AE, Griffin BJ, Pearson DG (2009) Nature and genesis of Kalimantan diamonds. Lithos 112S:822–832

Smith CB, Atkinson WJ, Tyler EW, Hall AE, Macdonald I (2018) The discovery of the Argyle pipe, western Australia: The world's first lamproite hosted diamond mine. *In:* Geoscience and exploration of the Argyle, Bunder, Diavik and Murowa diamond deposits. Vol 20. AT Davy CB Smith, H Helmstaedt, AL Jaques, JJ Gurney (eds) Soc Econ Geol, p 49–64

Sobolev NV (1977) Deep-seated inclusions in kimberlites and the problem of the composition of the upper mantle. (Translated from the Russian edition 1974) Am Geophys Union, Washington

Sobolev NV, Yefimova ES, Lavrent'yev YG, Sobolev VSD (1984) Dominant calcsilicate association of crystalline inclusions in placer diamonds from southeastern Australia. Dokl Akad Nauk SSSR 274:172 5

Sobolev NV, Yefimova ES, Koptil VI (1999) Mineral inclusions in Diamonds in the Northeast of the Yakutian Diamondiferous Province. *In:* JJ Gurney, JL Gurney, MD Pascoe, SH Richardson (eds) Proceedings 7th International Kimberlite Conference, Cape Town 2: 816–822

Sobolev NV, Logvinova AM, Nikolenko EI, Lobanov SS (2013) Mineralogical criteria for the diamond potential of Upper Triassic placers of the northeastern margin of the Siberian Platform. Russ Geol Geophys 54:903–916

Sobolev NV, Sobolev AV, Tomilenko AA, Kovyazin SV, Batanova VG, Kuz'min DV (eds) (2015) Paragenesis and complex zoning of olivine macrocrysts from unaltered kimberlites of teh Udachnaya-East pipe,Yakutia: relationship with the kimberlite formation conditions and evolution. Russ Geol Geophys 56:260–279

Sobolev NV, Sobolev AV, Tomilenko AA, Kuz'min DV, Grakhanov SA, Batanova VG, Logvinova AM, Bul'bak TA, Kostrovitskii SA, Yakovlev DA, Fedorova EN, Anastasenko GF, Nikolenko EI, Tolstov AV, Reutskii VN (2018) Prospects of search for diamondiferous kimberlites in the northeastern Siberian Platform. Russ Geol Geophys 59:1365–1379

Sobolev NV, Logvinova AM, Tomilenko AA, Wirth R, Bul'bak TA, Luk'yanova Li, Fedorova EN, Reutsky VN, Efimova ES (2019) Mineral and fluid inclusions in diamonds from the Urals placers, Russia: Evidence for solid molecular N_2 and hydrocarbons in fluid inclusions. Geochim Cosmochim Acta 266:197–219

Soltys A, Giuliani A, Phillips D (2018) A new approach to reconstructing the composition and evolution of kimberlite melts: a case study of the archetypal Bultfontein kimberlite (Kimberley, South Africa). Lithos 304:1–15

Soltys A, Giuliani A, Phillips D, Kamenetsky VS (2020) Kimberlite metasomatism of the lithosphere and the evolution of olivine in carbonate-rich melts—evidence from the Kimberley kimberlites (South Africa) J Petrol 61:egaa062

Soni MK, Jha DK, Rao TK, Mukul Tiwari (2002) Proterozoic diamondiferous placers, Panna diamond belt M.P. (2002) International Conference on Diamond and Gemstones, Raipur, India. Extended abstracts, p 19–21

Sorenson H (1974) Introduction. *In:*The Alkaline Rocks, Sorenson H (ed) Wiley, New York, p 1–15

Sparks RSJ (2013) Kimberlite volcanism. Annu Rev Earth Planet Sci 41:497–528

Sparks RSJ, Baker L, Brown RJ, Field M, Schumacher J, Stripp G, Walters A (2006) Dynamical constraints on kimberlite volcanism. J Volcanol Geotherm Res 155:18–48

Spencer LK, Dikinis SD, Keller PC, Kane RE (1988) The Diamond deposits of Kalimantan, Borneo. Gems Gemol 24:67–80

Spier CA, Ferreira Filho CF (1999) Geologia, estratigrafia e depósitos minerais do projeto Vila Nova, escudo das Guianas, Amapá, Brasil. Revista Brasileira de Geociências 29:173–178

Spriggs A (1988) An Isotopic and Geochemical Study of Kimberlites and Associated Alkaline Rocks from Southern Namibia. PhD Dissertation, University of Leeds, UK

Stachel T, Harris JW (2008) The origin of cratonic diamonds—Constraints from mineral inclusions. Ore Geol Rev 34:5–32

Stachel T, Lorenz V, Smith CB, Jaques AL (1994) Evolution of four individual lamproite pipes, Ellendale volcanics field (Western Australia) *In:*Kimberlites, Related Rocks and Mantle Xenoltihs: Proceedings of the Fifth International Kimberlite Conference. Vol 1. Meyer HOA, Leonardos OH (eds) CPRM, Araxa, Brazil, p 177–194

Stachel, T, Harris JW, Brey GP, Joswig W (2000) Kankan diamonds: lower mantle inclusion parageneses. Contrib Mineral Petrol 140:16–27

Stachel T, Banas A, Aulbach S, Smit KV, Wescott P, Chinn IL, Kong J (2018) Victor Mine: Neoproterozoic lherzolitic diamonds from a thermally modified cratonic root. Mineral Petrol 112S:311–324

Stachel T, Aulbach S, Harris JW (2022) Mineral inclusions in lithospheric diamonds. Rev Mineral Geochem 88:307–392

Stanley JR, Flowers RM, Bell DR (2015) Erosion patterns and mantle sources of topographic change across the southern African Plateau derived from shallow and deep records of kimberlites. Geochem Geophys Geosyst 16:3235–3256

Stiefenhofer J, Farrow DJ (2004) Geology of the Mwadui kimberlite, Shinyanga district, Tanzania. Lithos 76:139–160

Stubley MP, Irwin D (2019) Bedrock geology of the Slave Craton, Northwest Territories and Nunavut. Northwest Territories Geological Survey, Open File Report 2019–01

Subrahmanyam AV, Anil Kumar V, Deshmukh RD, Viswanathn G (2005) Discovery of microdiamonds in beachplacers of east coast, Andhra Pradesh, India Curr Sci 88:1227–1228

Sun J, Liu C-Z, Tappe S, Kostrovitsky SI, Wu F-Y, Yakovlev D, Yang Y-H, Yang J-H (2014) Repeated kimberlite magmatism beneath Yakutia and its relationship to Siberian flood volcanism: insights from ins situ U–Pb and Sr–Nd perovskite isotope analysis. Earth Planet Sci Lett 404:283–295

Sutherland DG (1982) The transport and sorting of diamonds by fluvial and marine processes. Econ Geol 77:1613–1620

Sutherland DG (1993) Drainage Basin evolution in southeast Guinea and the development of diamondiferous placer deposits. Econ Geol 88:44–54

Svisero DP, Meyer HOA, Tsai HM (1977) Kimberlite minerals from Vargem (Minas Gerais) and Redondão (Piauí) diatremes, Brazil: and garnet lherzolite xenolith from Redondão diatreme. Revista Brasileira de Geosciências, 7:1–13

Svisero DP, Shigley JE, Weldon R (2017) Brazilion Diamonds: a historical and recent perspective. Gems Gemol 53:2–33

Tainton KM (1992) The Petrogenesis of Group-2 Kimberlites and Lamproites from the Northern Cape Province, South Africa. University of Cambridge

Tainton KM, Browning P (1991) The Group-2-kimberlite–lamproite connection: Some constraints from the Barkly–West District, Northern Cape Province, South Africa. International Kimberlite Conference, Extended Abstracts, v5

Tainton KM, McKenzie D (1994) The generation of kimberlites, lamproites, and their source rocks. J Petrol 35:787–817

Tappe S, Foley SF, Jenner GA and Kjarsgaard BA (2005) Integrating ultramafic lamprophyre into the IUGS classification of igneous rocks: Rationale and implications. J Petrol 46:1893–1900

Tappe S, Pearson DG, Kjarsgaard BA, Nowell G, Dowall D (2013) Mantle transition zone input to kimberlite magmatism near a subduction zone: origin of anomalous Nd–Hf isotope systematics at Lac de Gras. Earth Planet Sci Lett 371–372:235–251

Tappe S, Smart K, Torsvik T, Massuyeau M, de Wit M (2018) Geodynamics of kimberlites on a cooling Earth: Clues to plate tectonic evolution and deep volatile cycles. Earth Planet Sci Lett 484:1–14

Tappert R, Stachel T, Harris JW, Muehlenbachs K, Ludwig T, Brey GP (2005) Diamonds from Jagersfontein (South Africa): Messengers from the sublithospheric mantle. Contrib Mineral Petrol 150:505–522

Tappert R, Stachel T, Harris JW, Muehlenbachs K, Brey GP (2006) Placer diamonds from Brazil: Indicators of the composition of the earth's mantle and the distance to their kimberlitic sources. Econ Geol 101:453–470

Tappert R, Foden J, Muehlenbachs K, Stachel T, Goryniuk M, Wills K (2008) The diamonds of South Australia. 9th International Kimberlite Conference, Extended abstract, 9IKC-A-00209

Tappert R, Foden J, Stachel T, Muehlenbachs K, Tappert M, Wills K (2009) The diamonds of south Australia. Lithos 112S:806–821

Taylor WR, Kingdom L (1999) Mineralogy of the Jagersfontein kimberlite—An unusual Group I micaceous kimberlite—and a comment on the robustness of the mineralogical definition of "Orangite". *In:* Proceedings of the Seventh International Kimberlite Conference. Gurney JJ, Gurney JL, Pascoe MD, Richardson SH (eds) Red Roof Design, Cape Town, South Africa 2:861–866

Taylor WR, Tompkins LA, Haggerty SE (1994) Comparative geochemistry of West-African Kimberlites —Evidence for a micaceous kimberlite endmember of sublithospheric origin. Geochim Cosmochim Acta 58:4017–4037

Teeuw RM (2002) Regolith and diamond deposits around Tortiya, Ivory Coast, West Africa. Catena 49:111–127

Thy P, Stecher O, Korstgard JA (1987) Mineral chemistry and crystallization sequences in kimberlite and lamproite dikes from the Sisimiut area, central West Greenland. Lithos 20:391–417

Timmerman S, Koornneef JM, Chinn IL, Davies GR (2017) Dated eclogitic diamond growth zones reveal variable recycling of crustal carbon through time. Earth Planet Sci Lett 463:178–188

Timmerman S, Yeow H, Honda M, Howell D, Jaques AL, Krebs MY, Woodland S, Pearson DG, Ávila JN, Ireland TR (2019) U–Th/he systematics of fluid-rich "fibrous" diamonds–evidence for pre-and syn-kimberlite eruption ages. Chem Geol 515:22–36

Timmerman S, Nestola F, Stachel T, Reimink JR, Ielpi A, Jackson V, Pearson G, Banas A, Stern RA, Vezinet A, Mircea C (2020) Diamond-bearing metasediments point to thick, cool lithospheric root established by Measoarchean beneath parts of the Slave Craton. AGU Fall Meeting, Abstract paper 732477

Tompkins LA, Gonzaga GM (1989) Diamonds in Brazil and a proposed model for the origin and distribution of diamonds in the Coromandel Region, Minas Gerais, Brazil. Econ Geol 84:591–602

Tovey M, Giuliani A, Phillips D, Moss S (2020) Controls on the explosive emplacement of diamondiferous kimberlites: New insights from hypabyssal and pyroclastic units in the Diavik mine, Canada. Lithos 360:105410

Tovey M, Giuliani A, Phillips, D, Pearson DG, Sarkar C, Nowicki T, Carlson J (2021) The spatial and temporal evolution of primitive melt compositions within the Lac de Gras kimberlite field, Canada: Source evolution vs lithospheric mantle assimilation. Lithos 392–393:106142

Tremblay M (1956) Geology of the Williamson Diamond Mine, Mwadui, Tanzania. PhD thesis, McGill University, Montreal, Canada

Turner BR, Minter WEL (1985) Diamond-bearing upper Karoo fluvial sediments in NE Swaziland. J Geol Soc London 142:765–776

van Straaten BI, Kopylova M, Russell J, Smith BS (2011) A rare occurrence of a crater-filling clastogenic extrusive coherent kimberlite, Victor Northwest (Ontario, Canada). Bull Volcanol 73:1047–1062

Vasilenko V, Kuznetsova L, Tolstov A, Minin V (2012) Evaluating the diamondiferous potential of unaltered kimberlites by the population models of their composition. Geochem Int 50:988–1006

Vespermann D, Schminke H-U (2000) Scoria cones and tuff rings. *In:*Encyclopedia of Volcanoes. Sigurdsson H, Houghton B, Rymer H, Stix J, McNutt S (Eds), Academic Press, New York, p 653–697

Wagner PA (1914) The diamond fields of southern Africa. The Transvaal Leader, Johannesburg

Wagner PA (1928) The evidence of kimberlite pipes on the constitution of the outer part of the Earth. S Afr J Sci 25:127–148

Walter MJ, Thomson AR, Smith EM (2022) Geochemistry of silicate and oxide inclusions in sublithospheric diamonds. Rev Mineral Geochem 88:393–450

Ward JD, de Wit MCJ, Revitt AW, Abson JP (2013) Geological and economic aspects of the Proterozoic Umkondo Group Diamond placer near Marange, Zimbabwe. Abstract, Geoforum, Johannesburg

Ward JD, de Wit MCJ, de Jager P (in prep) Placers of the Southern Alluvial Diamond Field, North West Province: a case for Palaeozoic marine influence in the "Rooikoppie Gravels"

Watkins JM (2009) Diamonds in South America. Presentation PROEXPLO Congress Lima, Peru

Webb K, Stiefenhofer J, Field M (2003) Overview of the geology and emplacement of the Jwaneng DK2 kimberlite, southern Botswana. 8th International Kimberlite Conference Extended Abstracts

Wendland C, Fralick P, Hollings P (2012) Diamondiferous, Neoarchean fan-delta deposits, western Superior Province, Canada: sedimentology and provenance. Precambrian Res 196:46–60

Westerlund KJ, Shirey SB, Richardson SH, Carlson RW, Gurney JJ, Harris JW (2006) A subduction wedge origin for Paleoarchean peridotitic diamonds and harzburgites from the Panda Kimberlite, Slave Craton; evidence from Re–Os isotope systematics. Contrib Mineral Petrol 152:275–294

White JD, Ross P-S (2011) Maar-diatreme volcanoes: A review. J Volcanol Geotherm Res 201:1–29

White N, Al-Hajri Y, Fishwick S (2009) Scales of transient convective support beneath West Africa. Geology 37:883–886

White LT, Graham I, Tanner D, Hall R, Armstrong RA, Yaxley G, Barron L, Spencer L, van Leeuwen TM (2016) The provenance of Borneo's enigmatic alluvial diamonds: A case study from Cempaka, SE Kalimantan. Gondwana Res 38:251–272

Wiggers de Vries DF, Pearson DG, Bulanova GP, Smelov AP, Pavlushin AD, Davies GR (2013) Re–Os dating of sulphide inclusions zonally distributed in single Yakutian diamonds: Evidence for multiple episodes of Proterozoic formation and protracted timescales of diamond growth. Geochim Cosmochim Acta 120:363–394

Wilson MGC, McKenna N, Lynn MD (2007a) The Occurrence of Diamonds in South Africa. Council for Geoscience, Pretoria

Wilson MR, Kjarsgaard BA, Taylor BE (2007b) Stable isotopic composition of magmatic and deuteric carbonate phases in hypabyssal kimberlite, Lac de Gras field, Northwest Territories, Canada. Chem Geol 242:435–454

Wood CA (1980) Morphometric evolution of cinder cones. J Volcanol Geotherm Res 7:387–413

Woodhead J, Hergt J, Phillips D, Paton C (2009) African kimberlites revisited: in situ Sr–isotope analysis of groundmass perovskite. Lithos 112:311–317

Woodhead J, Hergt J, Giuliani A, Maas R, Phillips D, Pearson DG, Nowell GM (2019) Kimberlites reveal 2 billion years of evolution of a deep isolated mantle reservoir. Nature 573:578–581

Zhou T (2015) Geology, regional diamond exploration and diamond provenance of the Proterozoic diamondiferous Umkondo conglomerates, Umkondo group, Eastern Zimbabwe. MSc thesis (unpubl), Rhodes University, Grahamstown, South Africa

Zimanowski B, Frohlich G, Lorenz V (1991) Quantitative experiments on phreatomagmatic explosions. J Volcanol Geotherm Res 48:341–358

Zimanowski B, Büttner R, Dellino P, White JDL, Wohletz KH (2015) Magma–water interaction and phreatomagmatic fragmentation. *In:*Encyclopedia of Volcanoes. Sigurdsson H, Houghton B, Rymer H, Stix J, McNutt S (Eds), Academic Press, New York, p 473–484

Zonneveld J-P, Kjarsgaard BA, Harvey SE, Heaman LM, McNeil DH, Marcia KY (2004) Sedimentologic and stratigraphic constraints on emplacement of the Star Kimberlite, east-central Saskatchewan. Lithos 76:115–138

Reviews in Mineralogy & Geochemistry
Vol. 88 pp. 119-166, 2022

Morphology of Monocrystalline Diamond and its Inclusions

Jeff W. Harris

School of Geographical and Earth Sciences
Gregory Building, University of Glasgow
Glasgow G12 8QQ, U.K.

jeff.harris@glasgow.ac.uk

Karen V. Smit*

Gemological Institute of America
New York City, NY 10128, U.S.A.

Department of Earth and Atmospheric Sciences
University of Alberta
Edmonton T6G 2E3 Alberta, Canada

*Now at the University of the Witwatersrand: *karen.smit@wits.ac.za*

Yana Fedortchouk

Department of Earth and Environmental Sciences,
Dalhousie University, Halifax, Canada

yana@dal.ca

Moreton Moore

Royal Holloway University of London,
Egham, Surrey TW20 0EX, U.K.

m.moore@rhul.ac.uk

INTRODUCTION

Since first being discovered in India over 2000 years ago, the most striking and universally recognized shape of an uncut diamond is the octahedron. This shape was revered from early times, in part because of the hardness of the crystal, but principally because of the ability of that crystal to display a spectrum of color in its uncut state. To ancient people, this property gave the crystal mystical powers and a diamond octahedron soon became a talisman believed to protect the wearer from pestilence or ensure survival in battle. The octahedron reached Europe at the beginning of the Roman Empire (27 BC–AD 476) usually set in gold rings and always in an uncut form. For centuries thereafter diamond was owned exclusively by the nobility and broader ownership developed only after new discoveries in Brazil and South Africa in the 18th and 19th Centuries respectively. Since that time, the octahedron along with other important shapes such as the macle (a twin form), the dodecahedron (a secondary form), and even irregular diamonds, have helped create a world-wide diamond jewelry industry with retail sales worth USD 78.13 billion in 2018 (Tacy 2018).

1529-6466/22/0088-0002$05.00 (print)
1943-2666/22/0088-0002$05.00 (online)
http://dx.doi.org/10.2138/rmg.2022.88.02

For some diamonds, it was not always an easy journey. Once released from their primary rock, diamonds are affected by processes such as dissolution that can change their morphology (e.g., to rounded dodecahedral shapes) and/or impose new surface features. Additionally, about one percent of them had the responsibility of carrying synchronously formed mineral inclusions, subsequently so beloved by diamond researchers. Diamond is, however, an extremely resilient mineral and can survive considerable physical and chemical challenges.

Rough diamonds have a variety of morphologies, surface features, and mineral or fluid inclusions that mean no two rough diamonds are exactly alike. All these features can tell an interesting story of diamond's geological history deep within Earth—both in the mantle rocks where diamonds grew and during their subsequent volcanic transport.

Based on diamond morphology, distinctions can often be made between suites of rough diamonds from different kimberlites (e.g., Harris et al. 1975, 1979). This likely reflects specific mantle regions being sampled during kimberlite eruption, combined with the volatile composition of the kimberlite resulting in specific secondary morphologies. Also, surface features of diamonds are heavily influenced by the composition and temperature of the dissolving fluid/melt, and are another tool that mantle geochemists can use to study carbon-bearing fluids/melts in the deep Earth. Investigating the differences in diamond morphologies and surface features between different kimberlites, therefore, can potentially help geologists understand the compositional variety of kimberlite magmas, getting a step closer to the primary melt composition, which has been a longstanding enigma in mantle geochemistry.

This chapter is divided into four sections and leads the reader through the external and internal morphologies, the surface features and finally the characteristics of the mineral inclusion content of this amazing mineral.

PRIMARY GROWTH MORPHOLOGY

In this section we discuss the morphologies of the principal growth forms of diamond, which can be monocrystalline, twinned, aggregated, irregular, fibrous or coated. These morphologies are illustrated in classic texts, such as Fersmann and Goldschmidt (1911), Sutton (1928), Williams (1932), and Orlov (1977). For information on the morphological characteristics of specific diamond types such as framesite, carbonado and ballas, all usually grouped under the title 'bort', the reader is referred to Jacob and Mikhail (2022, this volume).

Octahedral diamond

The ideal diamond octahedron has eight triangular faces of roughly equal size. Perfect octahedra are relatively rare, and more often diamond octahedra are slightly distorted, elongated or flattened.

Considering its internal crystallographic structure, the octahedron is the expected growth form for diamond. When a crystal grows, atoms adhere to its surface in a regular fashion; and several different kinds of face may be formed. The faces that eventually dominate are those which grow at the slowest rate. This is because fast-growing faces will grow out of the crystal. The slowest-growing faces are usually those that require the greatest number of atoms per unit area (Wulff 1901). For the case of the diamond structure, these are the octahedral {111} planes. Based on this simple model, one might therefore expect all diamond crystals to be octahedra, but this is not so. There are other considerations to take into account: the growing conditions, the local availability of carbon, and the effects of incorporated defects, such as dislocations enhancing growth rates. There are indeed countless natural diamonds that are octahedral, but there are many other shapes, to be discussed below. Some typical octahedral diamonds are shown in Figure 1.

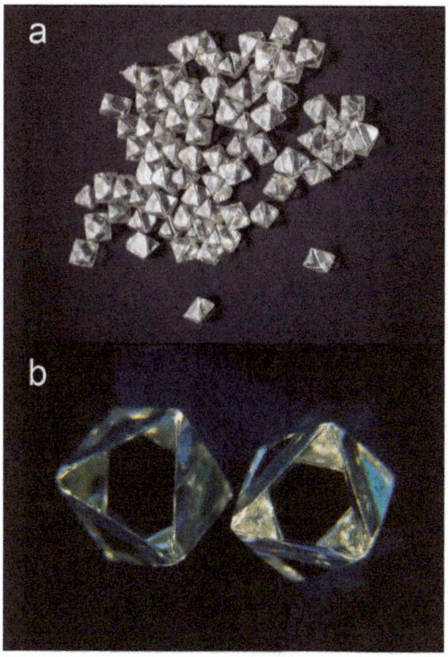

Figure 1. Octahedral diamond crystals. (**a**) Rough octahedral diamonds from Jwaneng, all in the 3 ct range. (**b**) Two unusually sharp-edged and equant 1 ct octahedral diamonds from Angola. A 1 carat octahedron has an edge length of approximately 5 mm. [Image in (a) reproduced from Weldon and Shor (2014), Gems and Gemology. Used with permission. Image in (b) courtesy of Thomas Hunn Co.]

Diamonds of cubic habit

The second monocrystalline primary form of diamond is the cube whose occurrence is rarer than the octahedron but is not uncommon. Cubic diamonds are found in production from Brazil, Siberia, Sierra Leone, the Democratic Republic of the Congo, South Africa and Botswana.

Gem cubes in the size range 2–6 mm from Botswana are not usually well-shaped, nor are their surfaces flat. In many instances these surfaces are associated with cube edges which are distinctly rounded, but with well-developed striations set at right-angles to these edges: see Fersmann and Goldschmidt (1911) for examples. In the 2–4 mm size range for diamonds from Siberia (Fig. 2; Zedgenizov et al. 2016) shapes are similar but striated edges appear to be absent.

Figure 2. Irregular orange and yellow cuboid alluvial diamonds from NE Siberia. [Images by Dmitry Zedgenizov (Sobolev Institute of Geology and Mineralogy).]

As curved cube faces can be inclined up to 30° from true cubic directions (Lang 1974a; Suzuki and Lang 1976) and because of all their other characteristics, diamonds with this habit are better termed 'cuboid'. Within the smaller diamond sizes (1–2 mm) flat faced and more regular shaped cubes exist and these can have translucent or opaque diamond coats. They are found particularly in the Mbuji Mayi production from the Democratic Republic of Congo (DRC) (Fig. 3), but also in diamonds from Sierra Leone (Grantham and Allen 1960) and Wawa, Canada (Stachel et al. 2006). With the larger gem diamonds their characteristics contrast with the planar cube faces often found on synthetic diamonds, especially in combination with octahedral faces in the latter. The relative development of the two forms, {100} and {111} can be varied by changing the temperature (Bovenkerk 1961; Yamaoka et al. 1977), and cooler conditions favor growth of cubic faces. The temperature can be adjusted to create synthetic diamonds in the shape of a cubo-octahedron (Moore 1985), which is a 'blocky' morphology chosen for industrial sawing applications. A review contrasting the morphologies of natural and synthetic diamonds has been written by Sunagawa (1995).

A second cube form of diamond is the result of fibrous octahedral growth. The resultant opaque cube shape can be regular with flat faces, edges being slightly rounded. For details on how both types of cuboid diamond grow, see Moore and Lang (1972) and the *Internal Morphology* section below.

Figure 3. 70 carats of rough cubic diamonds from the Democratic Republic of Congo, with an interpenetrant twin in the center. [Image courtesy of Thomas Hunn Co.]

Mixed-habit diamonds with cubo-octahedral morphology

There is clear evidence that some more moderately sized diamonds have shapes that are the result of a mutual growth between octahedron and cube. Invariably in these cases, octahedral growth is faceted whilst that of the cube is curved and non-faceted. If one direction is progressively dominant over the other, then the resulting shape may be an octahedron or a cube, but the internal structure, as seen by cathodoluminescence, identifies the two contributions (see *Internal Morphology*). If one of the directions is distinctly faster, gem cubes, for example, may have prominent octahedral corners or have pronounced edges giving the diamond a distinct skeletal appearance, sometimes referred to as re-entrant cubes (Tappert and Tappert 2011).

In shape classification studies on Botswanan diamonds in the 2–6 mm size range, a small but noticeable proportion of mixed-habit diamonds were recorded in the non-industrial diamond fraction (Harris et al. 1986). At Jwaneng they took the form of cubo-octahedra and cubo-dodecahedra, whilst at Orapa they were cubo-dodecahedra only, this shape in both localities being a consequence of dissolution. In similar work on South African diamonds (Harris et al. 1979), cubo-octahedra were occasionally recorded from the 1 mm fraction at the Cullinan (formerly Premier) mine.

Coated stones

When fibrous growth (characterized by fibers branching in <111> directions) occurs from the surface of an octahedral diamond, already formed by faceted {111} growth, then a 'coated stone' is the result (Kamiya and Lang 1965). Why a growing diamond should change from faceted growth to fibrous growth is still a topic for discussion, but clearly it indicates a marked change in growth conditions. Impurities incorporated during growth could be partially responsible for the change in growth directions. These impurities attach themselves to the growing surface and prevent the lateral spread of diamond growth parallel to {111} planes and the formation of smooth faces. In some specimens, X-ray topographs revealed not only the sharp interface between the good-quality octahedral 'core' and the 'coat' but also several successive epochs of fibrous growth (Fig. 4). Since the fibers grow perpendicularly from these octahedral faces, the resultant faces on the coated stone are also triangles: maintaining the same size, shape and orientation as the triangular faces on the 'core' diamond. The size of the 'core' may therefore be simply determined by measuring the slant edge lengths of the coated stone, which are equal to the edge-lengths of the octahedron (Machado et al. 1985). An interested jeweler can cleave off the imperfect coat material on {111} planes to reveal the gem inside!

In addition to octahedral faces, a coated stone may also exhibit cube faces and dodecahedral faces, ideally forming a rhombicuboctahedron (Machado et al. 1985). The six cube faces grow from the six corners of the octahedron by fibrous growth in four octahedral directions. The twelve dodecahedral faces have grown from the twelve edges of the octahedron, each in just two directions. For example, the (110) face is made from fibers running in [111] and [$11\bar{1}$] directions. Close examination of these dodecahedral faces by scanning electron microscopy showed that they are not flat but finely corrugated, with the sloping sides parallel to {111}. Some sides have been seen to have trigons upon them, confirming their assignment to {111}. The dodecahedral form {110} is therefore not a true growth form for natural diamond (although it is sometimes reported for synthetic diamond). The faces, being made up of fine-scale {111} surfaces, may be described as 'pseudo-dodecahedral'.

Coated diamonds are principally found in the production from the DRC (Kamiya and Lang 1965; Boyd et al. 1987; Akagi and Masuda 1988; Navon et al. 1988), although noticeable amounts are recovered from Sierra Leone and Canada (Grantham and Allen 1960; Gurney et al. 2004; Yelisseyev et al. 2004; Petts et al. 2016).

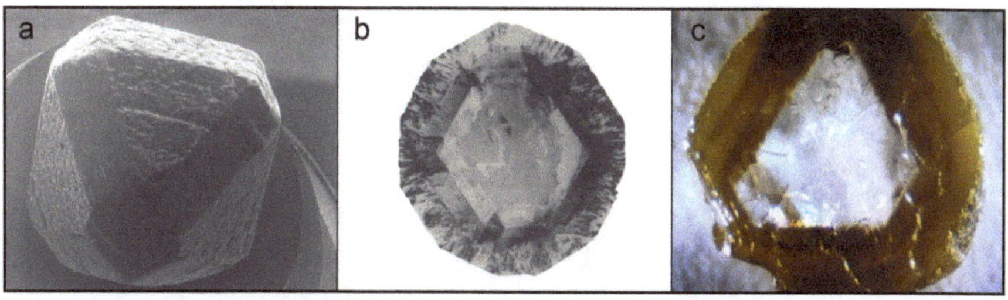

Figure 4. Faceted growth followed by fibrous growth. (**a**) Secondary electron image of a cubo-octa-dodec diamond with 8 mm diameter (**b**) Section X-ray topograph of the same diamond showing multiple layers of fibrous growth. (**c**) A coated stone from Sierra Leone (4 mm across) showing a clear center and distinct growth layering within the coat. [Images in (a,b) reprinted from Machado et al. (1985). Image in (c) by Jeff Harris.]

Tetrahedral diamond

Ever since Bragg and Bragg (1913, 1914) determined the crystal structure of diamond, most scientists have accepted that its point group symmetry is $\frac{4}{m}\bar{3}\frac{2}{m}$ (that is, cubic holosymmetry). Very occasionally natural diamonds have been found that have the appearance of being tetrahedra and these are illustrated in the classic works of Fersmann and Goldschmidt (1911), Sutton (1928) and Williams (1932). If primary, the existence of tetrahedral diamond changes the point group symmetry to the hemihedral class $\bar{4}3m$; and therefore to a class that does not have a center of symmetry. For this to be true, the tetrahedral habit would have to be a growth form arising from unimpeded growth from a center, resulting from internal inter-atomic forces; rather than acquired by external events such as fracture, cleavage along {111}, plastic deformation or dissolution.

Seager (1979) made a careful study of one such diamond and noted that whereas three of the four 'tetrahedral faces' were similar in appearance, the fourth was quite different. Using high-resolution X-ray topography to probe the interiors of 11 diamonds, Yacoot and Moore (1993) concluded that the tetrahedral diamonds studied were {111} cleavage fragments from larger stones, sometimes together with twinning from the origin on one, or two non-parallel, {111} planes. No diamonds have been found that actually grew as flat-faced tetrahedra, and thus the diamond structure remains firmly in the category of the cubic holosymmetric point group $\frac{4}{m}\bar{3}\frac{2}{m}$.

Diamond twins and aggregates

Twinned crystals. Instead of single-crystals, natural diamond can also grow as pairs of twins that are aligned along crystallographic planes. There are two main varieties: contact ('spinel-type') twins for faceted octahedral growth and interpenetrant ('fluorite-type') cubes for fibrous growth. Twinning of diamonds formed by cuboid growth has not been reported.

Because diamond possesses full cubic symmetry, twinning in diamond by a 180° rotation about a [111] direction is the same as twinning by a reflection in a (111) plane. *Contact twins* therefore develop by faceted growth on {111} planes and are known as 'macles' in the diamond trade (Fig. 5). They are often of a thin triangular shape with small re-entrants at the corners; and the twin plane is not always planar but often shows some inter-growth (Yacoot et al. 1998). These authors also found that the origin of contact-twinning is nearly always at the center of the diamond and this is often marked by a visible inclusion (see *Internal Morphology* section below). *Interpenetrant cubes* are rare, and they are usually colored yellow or brown. In the few specimens examined, twinning appears to have started from the center (Machado et al. 1998).

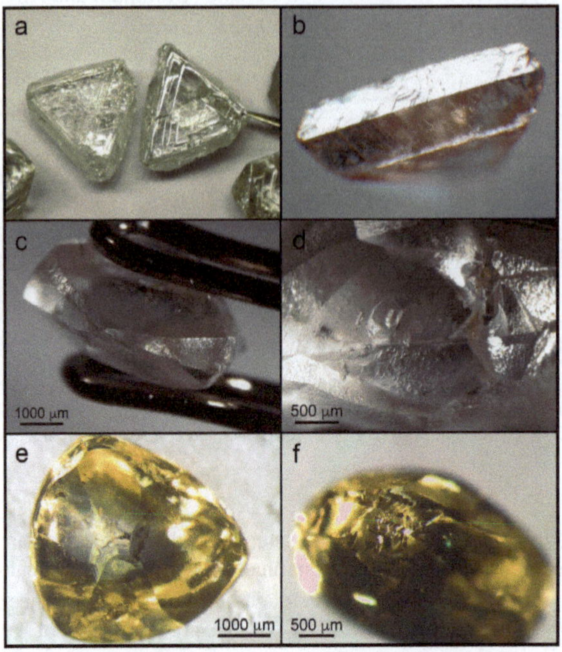

Figure 5. Features of typical macles. (**a**) A set of slightly dissolved (resorbed) macles with the arm of tweezers for scale. (**b**) A single macle showing a straight twin plane with characteristic herring-bone, with scale the same as in (a). (**c**) A dodecahedral macle showing the truncation of the dodecahedroid faces by the macle line. (**d**) A stepped macle plane. (**e**) A dissolved macle with an obvious medial line indicating a tetrahexahedroid (THH) crystal. (**f**) A thicker THH macle. [Images (a, b) used with permission of Anetta Banas (University of Alberta). Images (c-f) by Karen Smit.]

Multiple twinning on {111} planes can produce interesting and beautiful morphologies in natural diamond. A 'Star of David' morphology can form by the combined growth of two macles, the flat faces of which are in contact and the triangular outlines being oriented in opposite directions (Fig. 6). The angular relationship between these two macles is a rotation of 180° about the common [111] normal to the parallel twin planes.

Figure 6. 'Star of David' diamonds—six-pointed star-shaped contact twins. Left diamond is 0.92 ct (6.15 × 6.84 × 2.66 mm); right diamond is 0.98 ct (6.48 × 6.66 × 2.45 mm). [Image by Robert Weldon (GIA). Diamonds courtesy of Trillion Diamond Company, NY.]

When twinning occurs on non-parallel planes, a cyclic twin (or 'star twin') can result. Such twins are not uncommon among synthetic diamonds, but they are rather rare in natural stones. The angle between two {111} planes is approximately 70.5°. This means that five twinned crystals can grow around a common [110] axis, with some room to spare: 5 × 70.5 = 352.5°, leaving a narrow diamond wedge of angular thickness of approximately 7.5° (to complete the full circle of 360°), see Fersman (1955).

Diamond aggregates. Two or more intergrown diamonds can also be aggregate crystals (Fig. 7). They can be distinguished from twinned diamonds if there is a lack of crystallographic alignment between crystals. However, there are some examples where (now) octahedral diamonds started as aggregates showing crystallographic alignment (Fig. 8).

Figure 7. Aggregate of octahedral crystals. [Image by Karen Smit.]

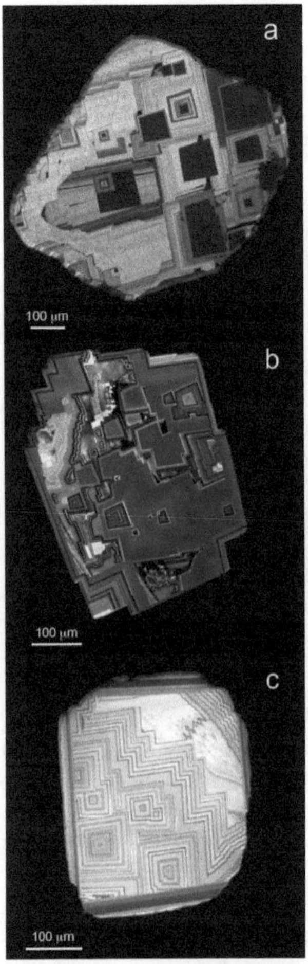

Figure 8. Cathodoluminescence of the internal growth features show that these (now octahedral) Ekati diamonds probably started as aggregates, and show that aggregates may have crystallographic alignment. [Images in (a,b) by Zhihai Zhang (Dalhousie University). Image in (c) from Zhang (2016).]

Irregular diamonds

Some diamonds show little or no discernible morphology. If the shape is fracture controlled then such diamonds can be divided into two groups: those that have broken in the kimberlite or lamproite and whose edges have been rounded by later dissolution (old breaks), or those whose edges are still sharp (new breaks) and which probably resulted from damage during diamond recovery or in an alluvial environment.

At high temperatures above 900 °C, diamond deformation is typically ductile, so just how diamonds fracture in the igneous environment is uncertain, but Clement et al. (1986) showed that explosive or implosive brecciation occurred directly around parts of the root zone of the De Beers mine in Kimberley. The authors considered the angular breccia to be a consequence of *in situ* shattering of the country rock by a precursor gas phase. If this kind of event occurred more generally during or just after eruption, when the kimberlite was cooling, then shock waves from such explosions might fracture a proportion of the diamond content, over a range of near surface depths.

If an irregular diamond is without obvious fractures, then the shape is principally controlled by dissolution. Diamonds in this category are not uncommon and many large sub-lithospheric diamonds have irregular shapes, such as the newly-recognized CLIPPIR diamonds (Cullinan-like Large Inclusion-Poor, relatively Pure, Irregularly shaped, Resorbed diamonds; (Smith et al. 2016b, 2017). A specific source for these large diamonds is the Letšeng-la-terae mine in Lesotho (now commonly shortened and called the Letšeng mine, Shor et al. 2015). The irregular nature of sub-lithospheric diamonds could be due to sequences of breakage and dissolution in the deep Earth. Deep Focus Earthquakes (Zhan 2020), show that brittle deformation is indeed possible at such depths. Sub-lithospheric diamonds often have rounded, irregular and complex surfaces (Fig. 9).

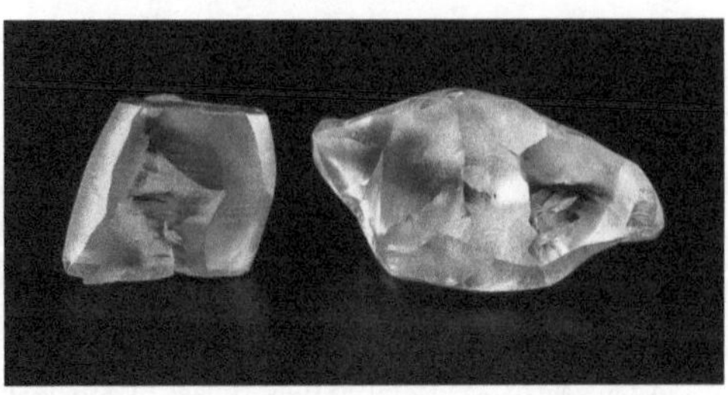

Figure 9. Faint pink (3.72 ct) and greyish blue (9.55 ct) irregular diamonds from the Letšeng mine in Lesotho. [Reprinted from Shor et al. (2015) Gems and Gemology. Used with permission.]

SECONDARY MORPHOLOGIES OF DIAMOND

When diamonds are recovered from mantle xenoliths, those that are completely enclosed in the xenolith are invariably sharp-edged. Robinson (1979) considered this to be particularly true for diamonds in eclogitic xenoliths. Although the majority of diamonds recovered directly from xenoliths are octahedra or macles, cube-shaped microdiamonds (0.5 mm) have been recovered from eclogitic xenoliths from Udachnaya (Spetsius 1998). Once released from xenoliths, these diamond shapes can remain pristine and show very little attack (or dissolution) by fluids either during mantle storage or during kimberlite transport. In other instances, diamonds are transformed from their primary shapes into secondary shapes by fluid dissolution. Changes induced by dissolution are progressive: for octahedra, the results of this process are recognized as dodecahedroids or tetrahexahedroids (THH) and for macles, the end-products are pin-cushion macles (Fig. 10).

A century-old controversy as to whether the dodecahedral shape was a growth form or the result of dissolution from larger stones was settled by Moore and Lang (1974). They used X-ray topography (XRT) of the internal growth surfaces within dozens of dodecahedra to show that the dissolution of octahedral diamonds was responsible for turning the twelve edges of the octahedron into the twelve rounded pseudo-faces of the rhombic dodecahedron. The concentric 'growth horizons' within each diamond were one-time external surfaces of the crystal; and their octahedral shape resulted from good faceted growth. They are visible in XRT because strain-producing particles have altered the lattice parameter by a few parts per million. These strain-producing particles may be the result of impurities incorporated during crystal growth.

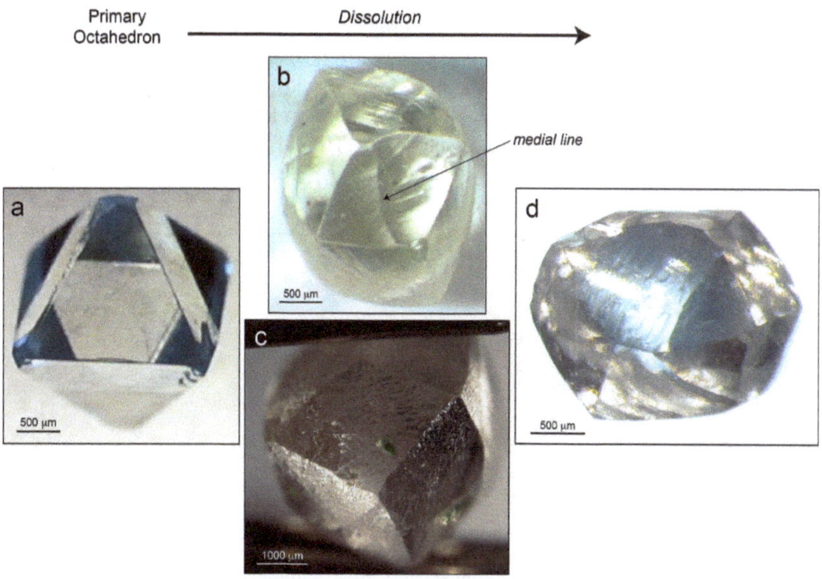

Figure 10. The sequence of dissolution from an octahedron (**a**) to a tetrahexahedroid (THH) (**d**). During dissolution, each octahedral face is reduced to a vertex surrounded by three rhombi to form a dodecahedroid. Across each short diagonal of a rhombus is a sinuous medial line. If this medial line is regarded as a divide between two faces then the resulting polyhedron is best described as a tetrahexahedron {*hk*0} (**b**). If that line is not regarded as a face edge but just a minor feature, and the rhombi are flatter, then the polyhedron is a dodecahedron (**c**). For both crystals, further dissolution creates more rounded THHs (**d**). Neither of these secondary shapes has true crystal faces, but rather curved dissolved surfaces that only approximate similar crystallographic directions to crystal faces, and accordingly the crystal names end with "-oid". [Images in (a,b) by Jeff Harris. Image in (c) by Karen Smit. Image in (d) by Yana Fedortchouk.]

The outermost growth horizons are truncated by the present external surface of each diamond, unequivocally demonstrating that the crystal was once bigger than it is now; and that the rhombic dodecahedral form is a dissolution shape. Furthermore, differential etching of growth horizons with different impurity concentrations has caused many of the diamonds to have terraces surrounding their triad axes (see *Surface Features* section).

The manner by which dissolution proceeds can be described on an atomic scale by the motion of kinks along stable monomolecular steps parallel to <110> directions. Each rhombic 'face' is often comprised of two parts, with a common boundary running along the minor diagonal of the rhombus. Unlike the rhombus edge, this minor diagonal (known as the medial line) is usually sinuous and rarely intersects the triple points at either end of the short diagonals of the rhombus. The line may be regarded as the intersection of two solution fronts starting from two four-fold apices of the original octahedron.

If this medial line is not regarded as a face edge but just a minor feature, then the polyhedron is a dodecahedron. If the medial line is regarded as a divide between two faces then the resulting polyhedron is best described as a tetrahexahedron $\{hk0\}$. For easier visualization, these forms can be considered as a theoretical morphological spectrum with the dodecahedron and cube as end-points, *although it is emphasized that cubes do not form in this way.* The dodecahedron is bounded by 12 rhombic (lozenge-shaped) faces, each of which may be considered as being composed of two triangles divided by the minor diagonal. Considering the two triangles to bend about the minor diagonal and the angle between their normals to increase, the morphology transforms from dodecahedron with zero angle to dodecahedroid, then THH, and finally a cube.

Neither of these secondary shapes has true crystal faces, but rather curved dissolved surfaces that approximate similar crystallographic directions to crystal faces. To account for the curved surfaces for both these shapes, they are termed hereafter as tetrahexahedroids (THHs—with 24 dissolved surfaces) and dodecahedroids (with 12 dissolved surfaces). When remnants of octahedral faces are present on these rounded diamonds, they show trigonal shapes on dodecahedroids and ditrigonal shapes on THHs. Examples of dodecahedroids and THHs are shown in Figures 11 and 12.

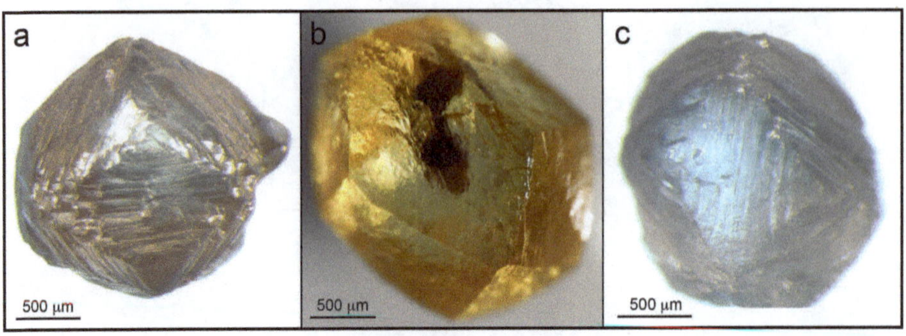

Figure 11. Further examples of dodecahedroid crystals. [Image in (a,c) by Yana Fedortchouk. Image in (b) by Karen Smit.]

An ongoing debate concerns how much importance is placed on the medial line. For example, Orlov (1977) used dodecahedroid in his descriptions of Russian diamonds. In shape classifications of diamonds from the major mines in South Africa and Botswana, Harris et al. (1979, 1986) used dodecahedron, whilst Robinson et al. (1989) favored THH. Recent surface feature work on Canadian diamonds also used THH (Fedortchouk 2019).

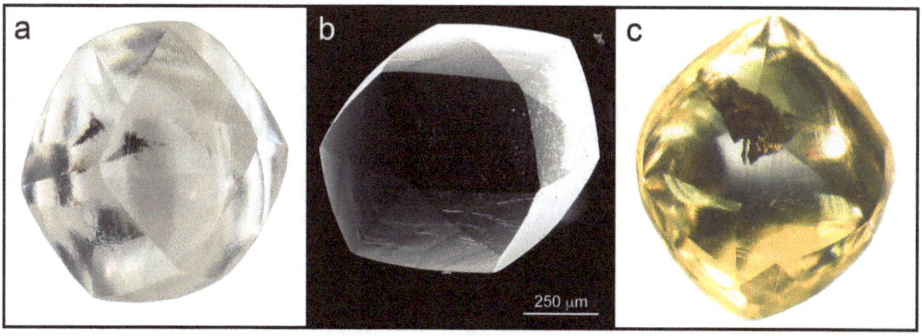

Figure 12. Further examples of tetrahexahedroid (THH) crystals. [Image in (a) by Jian Xin (Jae) Liao (GIA). Image in (b) by Yana Fedortchouk. Image in (c) by Karen Smit.]

Macles undergo the same dissolution as the octahedron, but the process is constrained by the twin plane. Either side of that plane, the original octahedral surfaces eventually divide into three curved faces with medial lines that are all closed off by the twin divide. Nevertheless, the resulting surfaces exhibit all the features found on dodecahedroid surfaces. Figure 5c–f shows some typical macles that have undergone dissolution.

Dodecahedroid and tetrahexahedroid (THH) formation conditions

Early work by Chepurov et al. (1985) on how mafic silicate melts dissolved diamonds at high pressure, was followed by further studies mostly in the noughties (2000–2009). Most of the new work involved varying amounts of H_2O and/or CO_2 in starting materials of natural kimberlite or lamproite (Kozai and Arima 2005), carbonate and silicate melts (Khokhryakov and Pal'yanov 2007, 2010), or H_2O- and CO_2-fluids (Fedortchouk et al. 2007). Pressure conditions for these experiments covered the range for diamond occurrence in the sub-cratonic lithospheric mantle—between 1 and 7.5 GPa and temperatures between 1150 and 1750 °C.

Invariably the diamond products were referred to as dodecahedroids or THHs, but images show that some of the resultant diamond shapes were difficult to clearly define, although authors attempted to do so. Khokhryakov and Pal'yanov (2007) and Fedortchouk et al. (2007), however, show dissolved diamonds where the 'medial line' is straight and connected to the triple points either side of the short diagonal of a rhombus. In both sets of experiments, the THHs formed by dissolution in a carbonate/water mix.

The experiments by Khokhryakov and Pal'yanov (2007, 2010) were conducted mostly at 5.7 GPa and 1400 °C. These temperatures are somewhat higher than the average diamond formation conditions in the cratonic lithosphere (Stachel and Luth 2015) and they argued strongly that all dissolution of diamond took place in the mantle, principally because late-stage changes would not be compatible with the short eruption times from mantle depths. On the other hand, the experiments by Fedortchouk et al. (2007) showed that serious dissolution took place within hours at 1 GPa and 1150–1500 °C, suggesting dissolution of diamond during the late stages of kimberlite eruption was possible. Fedortchouk and Zhang (2011) considered that dodecahedroid formation occurred in the mantle, whereas THHs were late-stage alteration products of dodecahedroids. Dissolution and formation of THHs occurred during kimberlite ascent, and it was severe enough to leave very few dodecahedroids.

Earlier studies of diamond shape from southern African sources mostly reported dodecahedroids. A possible solution to the dilemma of why THHs were not commonly found may relate to differences in the diamond sizes studied: the southern Africa studies (Harris et al. 1979, 1983) were on 2–6 mm diamonds whereas those that reported THHs were on ~1 mm Canadian microdiamonds (Fedortchouk and Zhang 2011). Late-stage dissolution on a small dodecahedroid would have a greater ability to change the shape into a THH than a similar level of dissolution on a larger similar diamond.

Figure 13 is an example of a so-called pseudohemimorphic diamond (named by Robinson 1979) that appears to be in two parts, a dodecahedroid top and an octahedral bottom. However, the diamond is a single crystal that shares the same axes and has identical color between the two halves. Pseudohemimorphic diamonds are not common, although they were noticeable in the production from the Venetia mine in South Africa (Jeff Harris, personal observation). These diamonds result from partial dissolution of an original octahedron, where portions of the diamond are protected from dissolution by kimberlite fluids, by being partly embedded in another mineral or xenolith.

Figure 13. A 2–3 mm partially dissolved (resorbed) pseudohemimorphic South African diamond displaying a dodecahedroid top and pristine octahedral lower portion. [Image by Jeff Harris.]

INTERNAL FEATURES OF DIAMOND

In general, diamond has a complex internal growth history. In the early days, the internal growth features were examined after etching with potassium nitrate at temperatures around 575 °C or using X-ray topographic techniques. For details of the diverse X-ray techniques that can be used to examine the interior of diamonds see Moore (2009) and references therein. The most common modern technique to investigate diamond growth history is through cathodoluminescence (CL) imaging, that is created by luminescence instrumentation attached to optical microscopes, and instruments such as scanning electron microscopes (SEMs). Instruments such as the DiamondView™ that use UV rather than electron excitation can also be used to create luminescence images (Welbourn et al. 1996).

Initially the study of diamond plates by some of the above methods was the realm of physicists who wished to determine and locate defects, such as dislocations or stacking faults in diamond crystals, particularly synthetic ones. The patterns exposed by these techniques however, were also used to elucidate the way natural diamond grows, much of the initial work being completed in the second half of the 20th century. Notable amongst this work was that of Seal (1965) who studied 86 diamonds from mines in seven countries and discovered virtually all the main types of growth patterns.

Influence of nitrogen content on internal features

An important factor that influences many of the internal features we see in diamond is the presence or absence of nitrogen, where nitrogen-containing (>5 at. ppm) diamonds are termed Type I, and nitrogen-free (<5 at. ppm) diamonds are Type II. The absence of nitrogen in Type IIa (nitrogen- and boron-free) and IIb (boron-containing) diamonds lessens the yield stress by a factor of two, relative to Type I diamonds (Evans and Wild 1965). Consequently, Type II diamonds plastically deform more easily than Type I diamonds.

The influence of nitrogen on the internal strength of diamond was demonstrated for half-colorless half-brown diamonds from the Argyle mine in Australia (Nailer et al. 2007). Colorless zones had 1.6× more nitrogen than brown zones, yet clearly the two halves of each diamond would have experienced exactly the same pressure, temperature, and stress conditions. The presence of nitrogen, therefore, stiffened the diamond against natural plastic deformation, and the nitrogen-rich sectors remained colorless whereas the nitrogen-poor sectors became brown due to deformation (Fig. 14).

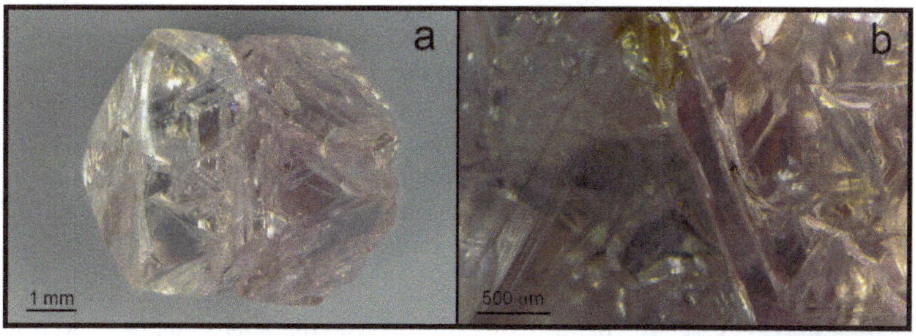

Figure 14. Half-colorless, half-pink aggregate diamond from Argyle (Australia). Higher nitrogen content in the colorless diamond is thought to stiffen the diamond against natural plastic deformation, also observed in other Argyle diamonds by Nailer et al. (2007). [Images by Sally Eaton-Magaña (GIA).]

The relative resilience of Type I diamond to stress fields does nevertheless cause some plastic deformation, frequently observed as deformation slip planes or deformation twins, though the original growth patterns are preserved (see *Surface Textures* section below). On the contrary, Type IIa and IIb diamonds show almost ubiquitous deformation. Deformation starts through the introduction of lattice dislocations, and with subsequent prolonged mantle annealing, these dislocations migrate through the crystal to form dislocation loops which polygonize to form mosaic networks (Hanley et al. 1977; Sumida and Lang 1981). These networks are visible in CL (Fig. 15) due to their frequent association with a luminescent source known as Band-A emission (Hanley et al. 1977; Graham et al. 1991), and are the most commonly observed internal feature in Type IIa and IIb diamond (e.g., Smith et al. 2016b, 2018). The lack of any observable growth zones most likely relates to the lack of nitrogen-related defects in these diamonds, or otherwise to deformation overprinting any growth features. Regardless, the distinct defect characteristics of Type I and Type II diamonds means that there may be no necessity to invoke different external stress regimes in the mantle for the different internal patterns observed.

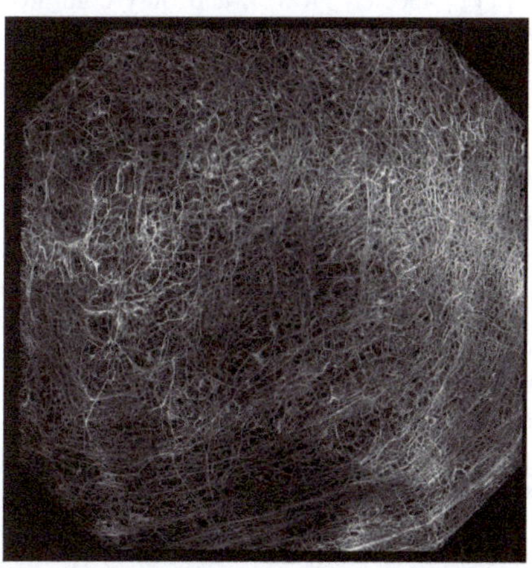

Figure 15. Cathodoluminescence image of a polygonized dislocation network on the table of a faceted Type IIb diamond. 4.8 mm × 4.8 mm. [Image by Karen Smit.]

Internal features of Type I (nitrogen-containing) diamonds

In the majority of diamonds, the growth of Type I single diamond octahedra and cubes is not continuous, but consists of distinct growth episodes. This growth zonation gives a distinct stratigraphy to the diamond when viewed on either (100) or (110) polished sections under CL. Diamonds are not normally examined on {111} surfaces because as a growth surface, no pattern may be seen and additionally {111} surfaces are difficult to polish. Perfect octahedral patterns from the crystal center to the edges are only rarely seen—for example in a Russian octahedron (Bulanova et al. 1999) and in an elongated octahedron from Ghana (Seal 1965) (Fig. 16).

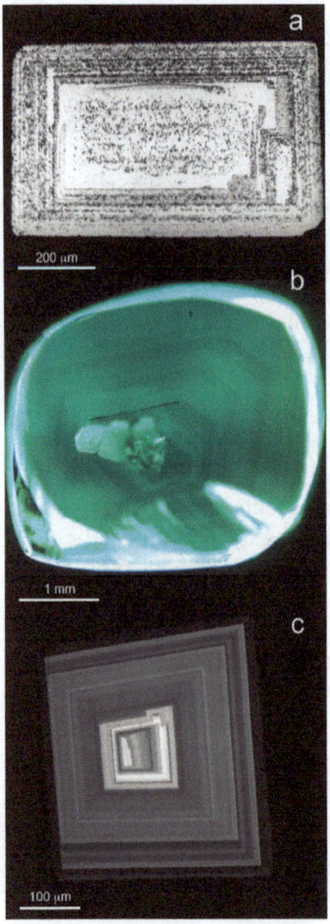

Figure 16. (**a**) Regular octahedral growth patterns observed in an etched (100) surface of a slightly elongate octahedral diamond from Ghana. (**b, c**) Regular octahedral growth patterns observed in CL images of diamonds from Ellendale (Australia) and Ekati (Canada). [Image in (a) reprinted from Seal (1965) with permission from the Mineralogical Society of America. Image in (b) from Smit (2008). Image in (c) from Zhang (2016).]

In some cases, growth zones that are visually distinct in CL between the core area and the periphery show sharp variations in isotopic characteristics. Such observations indicate that there have been at least two distinct and separate growth events and age dating of inclusions indicated that these different zones grew billions of years apart (Wiggers de Vries et al. 2013; Bulanova et al. 2014).

Not all Type I diamonds have a definable growth pattern. Figure 17 shows diamonds that have poorly defined central regions with structural order towards the edge. In the Bultfontein diamond, the central area might be part of a cuboid growth zone before octahedral surfaces developed. The darker growth zones, with a low CL response are thought to be Type II (low nitrogen) diamond.

Using X-ray topography, Moore and Lang (1972) studied diamonds of cubic habit and recognized two very distinct growth patterns—cuboid and fibrous. *Cuboid growth* (Fig. 18a) consists of a hill-and-valley structure (in the scientific literature variously described as 'hummocky', 'crinkly' or 'cobbled') of average orientation in one of the cube directions <100>. *Fibrous cubes* (Fig. 18b) have a fibrous (or 'columnar') internal structure, starting from the center of the stone and running, not in <100> directions, but in the <111> octahedral directions towards the cube corners. The sectors in between these principal directions are filled by branched fiber growth in four directions. The incorporation of impurities (which also causes the color) prevents the lateral growth of layers in the more usual {111} faceted growth, and only columns are formed (Moore and Lang 1972). The roughness of cube 'faces' arises from fiber-branching. For example, the (001) surface of the cube can be reached by taking steps parallel to [111], $[\bar{1}11]$, $[\bar{1}\bar{1}1]$ and $[1\bar{1}1]$ in a multiplicity of paths of equal length. The cubic shape is a natural consequence of such growth. Some post-growth dissolution can cause the already-rough surface to be covered in square-shaped pyramidal pits called tetragons (Sutton 1928).

Figure 19 shows a diamond that exhibits partial cubic patterns, which are sometimes blocky and riddled with nitrogen-free Type II diamond. A principal conclusion after examining such complex surfaces is that we still have a lot to learn about how some diamonds grow. See also Figure 8.

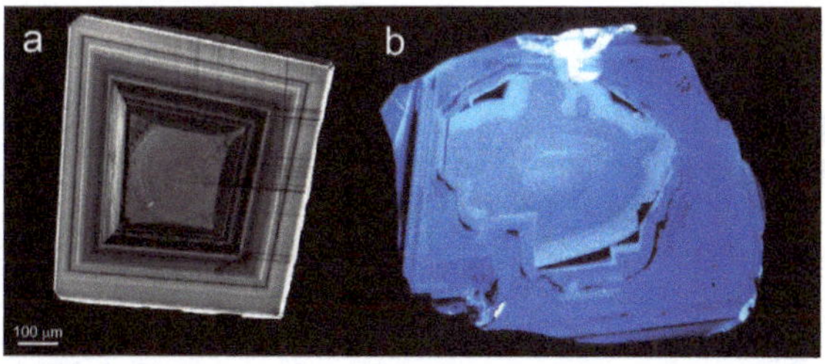

Figure 17. Cathodoluminescence images of diamonds that have poorly defined central regions with more orderly octahedral growth towards the edge. (**a**) Diamond from Grizzly kimberlite (Ekati mine) and (**b**) Diamond from Bultfontein (South Africa). [Image in (a) by Zhihai Zhang (Dalhousie University) and (b) by Jeff Harris.]

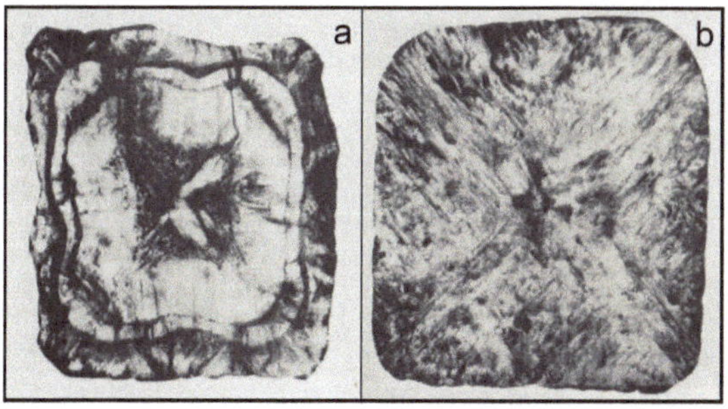

Figure 18. (**a**) Mid-section X-ray topograph of a colorless diamond cube of width 1.8 mm, showing truncated cuboid growth in its outermost regions. The growth rate of an octahedral face (top right) has been increased by dislocations, and consequently this face (or 'truncation') has grown out by the time it reaches the next visible outline. (**b**) Mid-section X-ray topograph of a brown diamond cube of width 1.9 mm, showing fibrous growth in <111> directions. [Figs 1 and 2a from Moreton Moore and A. R. Lang (1972) On the internal structure of natural diamonds of cubic habit, The Philosophical Magazine: A Journal of Theoretical Experimental and Applied Physics, 26:6, reprinted by permission of the publisher Taylor & Francis Ltd.]

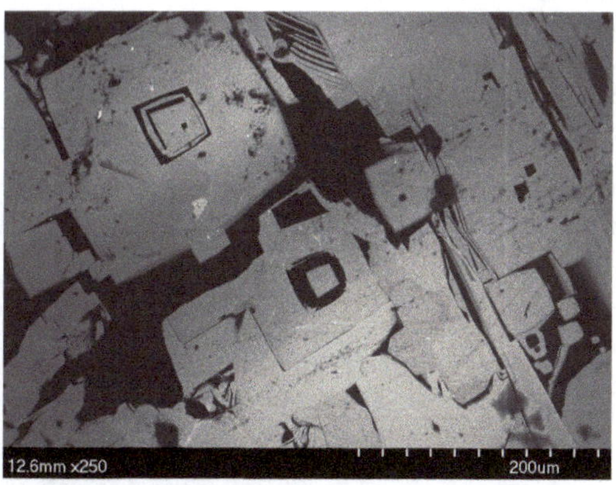

Figure 19. A (100) surface of a Bingara diamond (SE Australia) showing a complex but basically cuboid growth pattern, with Type I and Type II (dark) diamond intimately associated. [Image used with permission of Rondi Davies (Queensborough Community College, City University New York).]

Mixed-habit diamonds with cubo-octahedral growth

Harrison and Tolansky (1964) cut a clear 2.4 mm good quality, but slightly yellow, coated octahedron into eleven slices and etched them to examine in detail the growth history patterns of what became known as center-cross diamonds (Fig. 20). The center-cross pattern results from mixed-habit growth of flat octahedral planes and more hummocky cuboid surfaces (Lang 1979). The cuboid growth sectors in mixed-habit diamonds often contain tiny particles (1–5 µm), which if in sufficient concentration give rise to the visible petal arrays seen in some diamonds (Lang 1979; Rakovan et al. 2014). Cuboid zones often contain higher concentrations of nitrogen, nickel, hydrogen, methane and graphite (Bulanova et al. 2002; Lang et al. 2007; Rondeau et al. 2007; Howell et al. 2013; Smit et al. 2016b) (Fig. 21).

Figure 20. A clear 2.4 mm coated octahedron with an internal growth patterns of what became known as center-cross diamonds, now known as mixed-habit or cubo-octahedral diamonds. [Republished with permission of The Royal Society (U.K.), from Growth history of a natural octahedral diamond. Harrison ER, Tolansky S (1964) Proc R Soc London A 279:490–496; permission conveyed through Copyright Clearance Center, Inc.]

Mixed-habit diamonds can show a wide range of growth patterns that depend on whether cuboid or octahedral growth is dominant. Figure 22a shows a crystal developed from a cube core to an octahedral outer shape and Figure 22b the reverse, a possible controlling condition being mantle temperature. An example where directional growth is less perfect is shown in Figure 22c, giving rise to a repetitive pattern of cube or octahedral dominance. The mechanisms of how mixed-habit crystals can develop from a cubic core to an octahedral outer shape are detailed in Zedgenizov et al. (2006) and Howell et al. (2013).

There are only a limited number of studies on mixed-habit diamonds from specific locations. Welbourn et al. (1989) studied the variation of nitrogen content and other spectroscopic features between cuboid and octahedral sectors in mixed-habit Jwaneng diamonds. Bulanova (1995) and Bulanova et al. (2002) documented inclusion parageneses and carbon and nitrogen isotopic characteristics in mixed-habit Yakutian diamonds. Kaminsky and Khachatryan (2004) examined

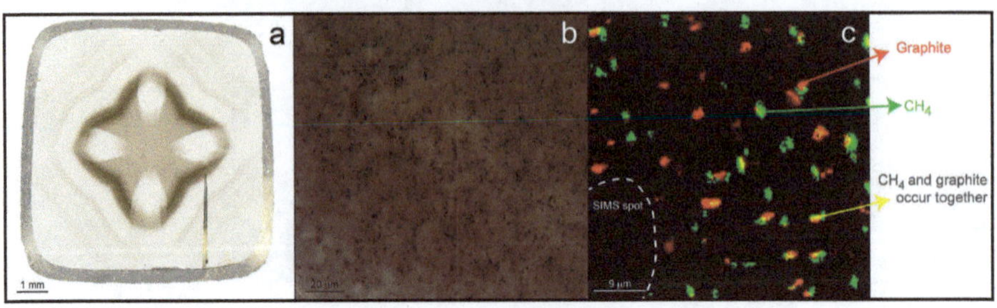

Figure 21. **(a,b)** In some cubo-octahedral diamonds, the cuboid sectors are visible in plain light because they often contain tiny micron-sized particles. **(c)** In Marange, and other diamonds of unknown origin, these particles have been identified as graphite by Raman spectroscopy (Howell et al. 2013; Smit et al. 2016b). [Images in (a,b) by Karen Smit. Raman map in (c) reprinted from Smit et al. (2016b).]

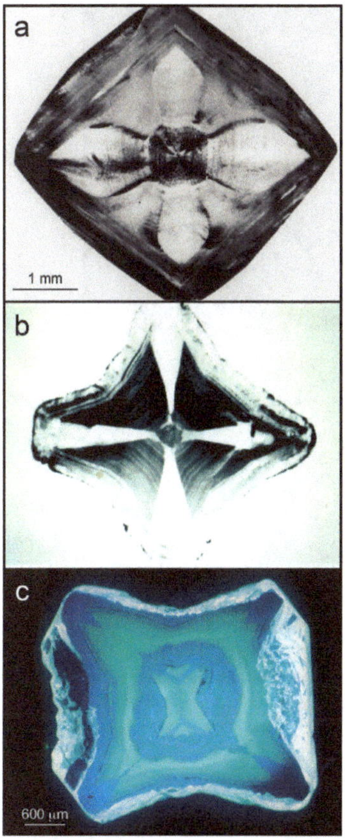

Figure 22. (**a**) Mixed-habit cubo-octahedral diamond in which there has been a smooth variation in ratio and rate of growth between cuboid surfaces and {111} faces. (**b**) A Jwaneng diamond (2–3 mm across) showing smooth variation between octahedral and cubic growth, resulting in a re-entrant cube. (**c**) A re-entrant Jwaneng cube showing repetition of octahedral and cuboid growth zones. [Image in (a) reproduced from Lang (1974b) with permission from Elsevier. Image in (b) reproduced from Welbourn et al. (1989) with permission from Elsevier. Image in (c) courtesy of De Beers.]

the variation of nitrogen content and aggregation state in Yakutian, Urals and Arkhangelsk mixed-habit diamonds. Davies et al. (2004) and Aulbach et al. (2009) both document cubo-octahedral diamonds in eclogitic and peridotitic diamonds from Lac de Gras (Canada). Smit et al. (2016b) studied mixed-habit diamonds from Marange (Zimbabwe), and used fluid inclusion content combined with isotopic characteristics of the diamonds to model diamond growth from hydrous fluids.

Features of Type Ib diamonds

Type Ib diamonds contain abundant neutrally charged isolated substitutional nitrogen (N_S^0, known as C centers) and account for less than 0.1% of natural diamonds (Harlow 1998), although this percentage is not well constrained. This rarity is because over geological time and at temperatures typical of the cratonic lithosphere, isolated nitrogen aggregates to A centers (N pairs) and B centers (4N around a vacancy). For further details on these and other defects in diamonds, see Green et al. (2022, this volume). Preservation of N_S^0 in natural diamonds is normally explained through either short mantle residency times or storage at $T < 700$ °C (Taylor et al. 1996), whereas most cratonic diamonds are billions of years old and resided at $T > 1000$ °C (Stachel and Harris 2008; Shirey and Richardson 2011).

Very few natural Type Ib diamonds contain 100% C centers, and the majority of these diamonds have undergone some aggregation to A centers. Aside from age and mantle temperature, the extent of Type Ib-IaA aggregation also depends on the amount of nitrogen initially incorporated into the diamond lattice during growth. If total nitrogen content is low, nitrogen aggregation may remain low since during annealing and diffusion, one nitrogen atom does not meet another.

In recent years, there have been several studies that document the internal features of rare natural Ib diamonds from French Guiana, eastern Sierra Leone, NE Siberia, and Botswana (Smith et al. 2016a; Zedgenizov et al. 2016; Smit et al. 2018b; Timmerman et al. 2018). These diamonds are mostly yellow-orange due to the presence of isolated nitrogen.

Diamonds recovered from highly altered rocks interpreted to be komatiites in French Guiana exhibit a mixture of growth signatures that are blocky, sectorial, granular or polycentric, and associated with intense plastic deformation (Fig. 23a,b; Smit et al. (2016b). At Zimmi (Sierra Leone), brownish- or greenish-yellow alluvial dodecahedroid-THH diamonds have complex internal patterns (Fig. 23c,d; Smit et al. (2018b). In part the diamond appears to be like a calving ice-sheet, the darker diamond being a second generation of diamond infiltrating the original. The origin of these patterns is unclear, although they do not appear to be related to *plastic deformation lines/planes* commonly seen in these diamonds.

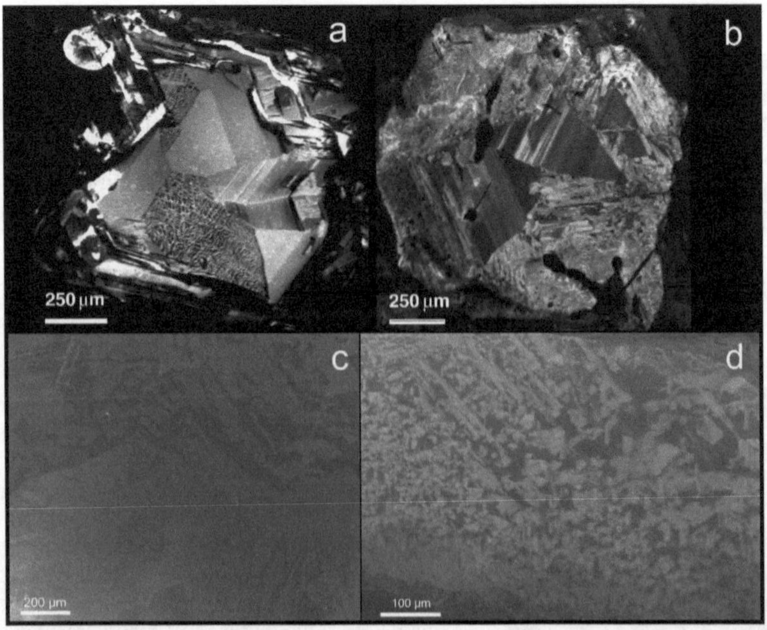

Figure 23. (**a**) and (**b**) Internal patterns of two Dachine Type Ib diamonds with blocky polycentric growth. (**c**) and (**d**) Internal patterns of unknown origin visible in CL images of Type Ib diamonds from Zimmi. [Images in (a, b) from Smith et al. (2016a) with permission from Elsevier. Images in (c,d) reprinted from Smit et al. (2018), Mineralogy and Petrology.]

Some cuboid alluvial diamonds from NE Siberia exhibit faint concentric growth patterns (Fig. 24; Zedgenizov et al. (2016), whereas other diamonds from this locality had heterogeneous patterns and high levels of plastic deformation. In some diamonds, growth zoning correlates with heterogeneous yellow color distribution. In Botswana, the Orapa mine has produced some irregular cubo-dodecahedral diamonds, with some color differences between the different growth sectors (Timmerman et al. 2018).

At typical mantle temperatures, the first stage of nitrogen aggregation in diamond (isolated nitrogen to nitrogen pairs) has been shown to occur very quickly (Taylor et al. 1996). Thus the preservation of Type Ib diamonds needs special geological environments shortly after their formation that are distinct to those of more common Type Ia diamonds. The presence of deformation features in Type Ib diamonds may indicate a rapid rise to cooler parts of the mantle (e.g., Smit et al. 2016a, 2018b), however this may not be true for all localities and

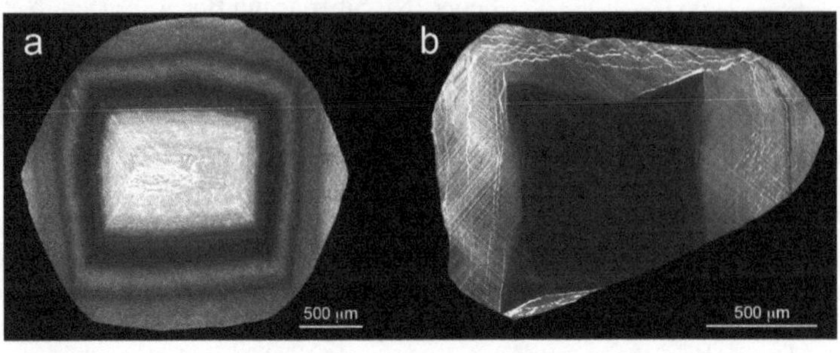

Figure 24. Cathodoluminesence images showing the growth patterns of cuboid alluvial diamonds from NE Siberia: (**a**) Dodecahedroid diamond with cuboid growth zones. (**b**) Rounded cuboid diamond with central cuboid domain. [Reprinted from Zedgenizov et al. (2016), with permission from Elsevier.]

Type Ib diamonds may also survive due to the diamonds' long residence in low temperature portions of the diamond stability field. Additionally, the Siberian diamonds could have formed close to the time of kimberlite eruption. One unifying aspect of these rare diamonds is that at present, they all belong to the eclogitic paragenesis.

Twinned diamonds

The internal growth structures for both contact and interpenetrant twins have been studied principally by X-ray topographic techniques (Machado et al. 1998; Yacoot et al. 1998). For contact twins, the twin plane was found not to be always planar but with twin interpenetrant components. As mentioned above, the initial contact twinning was shown to be nearly always at the center of the diamond and this was sometimes marked by a visible inclusion (Fig. 25). The second group consisted of fibrous diamonds. X-ray topographs of this type of crystal showed that fibrous growth is in octahedral directions (see Machado et al. 1998) occurring without the need of branching as happens with single fibrous cubes and again, nucleation was near the crystal center.

Figure 25d shows a section across a typical macle twin plane, one side of which exhibits a complex growth pattern, the other having no pattern at all. As this surface is generated by a potassium nitrate etch, the absence of a pattern would signify a Type II diamond adjacent to a Type I. Seal (1965) suggested that the stark difference may be due to crystal orientation effects. If that proposal should not be the case and Type I and II diamond exist across a twin plane, then some serious thinking about diamond formation is required.

Concluding remarks

Within the sizes of diamonds usually studied—microdiamonds up to 5 mm, with rare examples of larger stones—the main growth patterns of diamond have probably now been identified. A thorough understanding of the internal growth structure of a diamond—and core-to-rim growth directions—is important since the outward variation of carbon and nitrogen isotopes can sometimes be related to the redox conditions of the diamond source fluid.

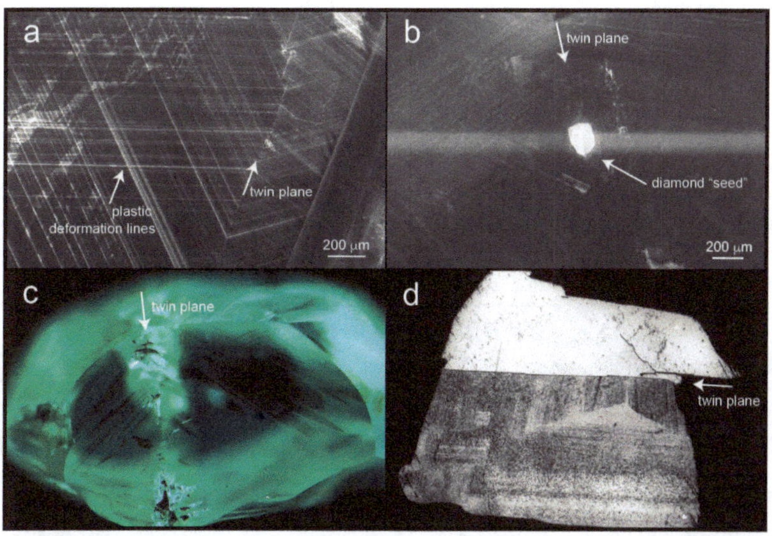

Figure 25. Examples of twin planes on the internal surfaces of diamonds: (**a**) Twin plane + plastic deformation lines. (**b**) Diamond inclusion visible along a twin plane in a Zimmi diamond. (**c**) Etching along a twin plane in an Ellendale diamond. (**d**) twinned diamond from Ghana (2 mm length) showing distinct growth characteristics either side of the twin plane. [Images in (a,b,c) by Karen Smit. Image in (d) reproduced from Seal (1965) with permission from the Mineralogical Society of America.]

SURFACE TEXTURES

The surface of natural diamond shows relief features of varying degrees of complexity, which are the result of either crystal growth or subsequent partial dissolution of growth layers during metasomatic events in the mantle and/or during transportation by the deep-seated mantle magmas (kimberlites and lamproites) to Earth's surface. Any post-emplacement alluvial history may add further traces of mechanical wear on diamond surfaces. The shapes of the surface features created during metasomatic events invariably reflect the crystal symmetry of the primary diamond shapes of octahedra and cubes (and their twin equivalents). In the case of the dissolved dodecahedroids and THHs, features are either crystallographically approximate to the symmetries of those crystals, or independent of such symmetry. With alluvial diamonds initial impact damage is not governed by crystallography, although the development of that damage may have a crystallographic component.

There is a great variety of surface textures observed on all the principal shapes of natural diamond. Some have a very common general occurrence whereas others are extremely rare. In almost any diamond population there are diamond crystals with unusual surface features, whose origins are still difficult to interpret. Many surface features have been reproduced in controlled experiments at high pressures and temperatures, providing constraints on the conditions of their formation within the Earth.

Linking surface features to growth or dissolution events in the mantle requires knowledge on (1) how to discriminate growth features from dissolution features, and (2) how to discriminate dissolution features formed during mantle metasomatism from those induced by interaction with kimberlite magma. Finally, as the surface features of micro-diamonds can be linked back to specific kimberlite lithologies and a depth interval within the kimberlite pipe, a better understanding of the relationship between diamond dissolution style and kimberlite type and lithology can be established. As surface features can also be reproduced in controlled experiments, these relationships help to reconstruct the emplacement mechanisms of different kimberlite bodies.

In this section we first describe features that are seen on diamonds from kimberlites or lamproites and then consider a similar assessment of alluvial diamonds.

Octahedral growth features

Perfectly shaped flat-faced sharp-edged octahedra are not common in most diamond productions. Invariably diamond growth is imperfect, which gives rise to three principal surface features: triangular plates, imbricated surfaces, and pyramids.

Triangular plates. These have an outline that follows the shape of diamond's octahedral face, can be single but are usually multiple, and with each consequent plate they decrease in size (Fig. 26). The thickness of the plates can vary but usually it is possible to clearly

Figure 26. (**a**) Triangular plates on an Orapa diamond. (**b**) Triangular plates on a Diavik diamond (7.97 × 4.73 × 3.58 mm). (**c**) A striated octahedron (3 mm across) resulting from diminishing step growth. [Image in (a) by Yana Fedortchouk. Image in (b) by Jian Xin (Jae) Liao (GIA). Image in (c) by Jeff Harris.]

distinguish one step-face (plate) from another. The edges and corners of the growth step-faces can be sharp or slightly rounded by later dissolution. Diamonds with growth step-faces are fairly common. Experimental work by Yamaoka et al. (1977) shows that this growth shape is confined to a narrow growth region close to the diamond/graphite equilibrium line. The growth origin of these faces is evident from the stepped growth patterns on CL images of sections through the diamonds with this surface feature (Fig. 27).

Imbricated surfaces. If the incomplete growth of triangular plates prevents an octahedral vertex from forming and the edges of particular plates overlap, then the result is an *imbricated surface*. In some instances imbrication creates a pseudo-cube face made up of the triangular vertices of the plates (Fig. 28).

Pyramids. These are positive-relief triangles, with their base oriented parallel to the edges of the octahedral face and cover only part of an octahedral face. The more obvious pyramids develop at the octahedral vertices giving rise to a negative triangular depression in the middle of that face.

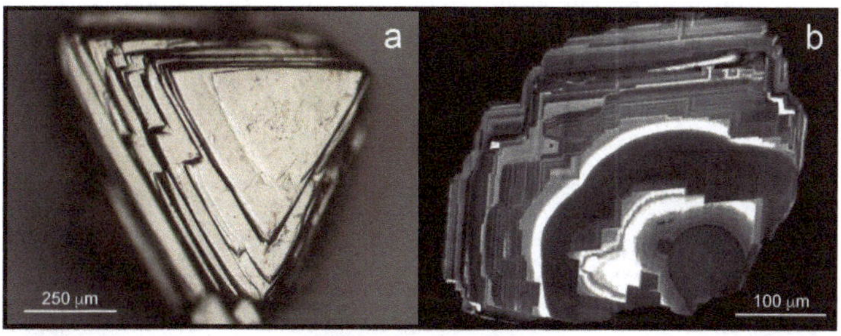

Figure 27. (**a**) An octahedral diamond from Misery (Ekati mine) showing well-defined thick triangular plates with straight sides and slight rounding due to dissolution. (**b**) CL image showing that triangular plates follow the stepped growth pattern of the diamond confirming their growth origin. [Image in (a) by Yana Fedortchouk. Image in (b) by Zhihai Zhang (Dalhousie University).]

Figure 28. A South African octahedron exhibiting imbrication. [Image by Jeff Harris.]

Octahedral dissolution features

Dissolution features considered in this instance are those that are associated with major octahedral to dodecahedroid (or THH) conversion (Fig. 10), but dissolution still leaves a recognizable octahedral shape. These features also apply to primary octahedral macle surfaces.

Shield-shaped and serrate laminae are consequences of the dissolution of triangular plates on octahedral surfaces, or the octahedral surface itself. Dissolution commences at the corners and edges and moves across a face creating a triangular convex shape resembling a shield (Fig. 29). In the case of an octahedral face the shield is formed from the partial exposure

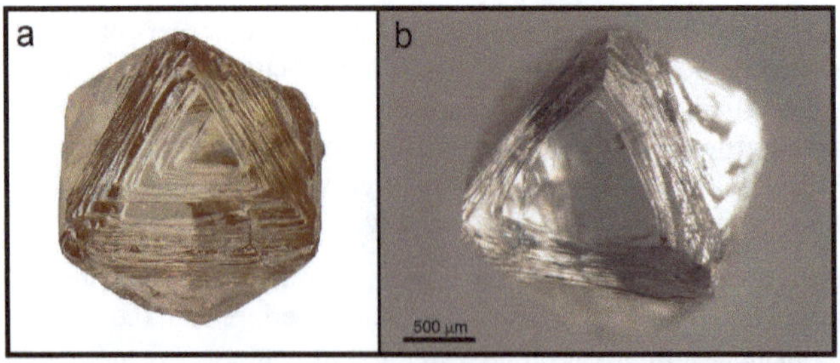

Figure 29. Shield-shaped laminae in octahedral diamonds from (**a**) Diavik (Canada) with 5.26 × 4.92 × 6.17 mm dimensions and (**b**) Koffiefontein (South Africa). [Image in (a) by Jian Xin (Jae) Liao (GIA). Image in (b) by Nicole Meyer (University of Alberta).]

of the original growth layers of the diamond. Serrate laminae only partially cover a face often having a zig-zag outline to their edges (Fig. 30). They result from dissolution of an imbricated surface or on diamonds with thin internal growth zoning at mantle conditions. They also often appear to be accompanied by numerous trigonal etch pits (see below). *Shield-shaped and serrate laminae* very similar in form to those found on natural diamonds have been produced at mantle pressures of 5–7 GPa during diamond dissolution experiments in silicate-carbonate melts (Khokhryakov and Pal'yanov 2007). Laminae shapes are influenced by variations in temperature and also amount of carbonate (Fedortchouk et al. 2019).

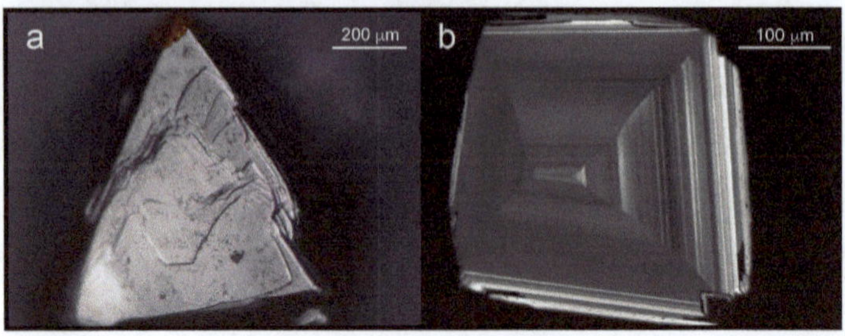

Figure 30. Octahedral diamond with serrate laminae caused by dissolution (Misery kimberlite, Ekati Mine). (**a**) Microphotograph showing serrate laminae. (**b**) Cathodoluminescence image showing thin growth zones, ideal for serrate laminae formation through dissolution. [Images by Zhihai Zhang (Dalhousie University).]

Trigons. Trigons are triangular features that are defined as being either 'positive' or 'negative' (Frank et al. 1958). A positive trigon has the same orientation as the triangular octahedral face of the diamond; whereas a negative trigon has the opposite orientation.

Negative trigons are a very common feature on octahedral surfaces. They can vary in size, depth and shape, even on the same diamond. Trigons range from about 5 μm in diameter to ~100 μm. Two principal trigon morphologies can be distinguished (1) plain flat-bottomed (f/b) trigons (Fig. 31a,b) and (2) deeper stepped-faced and point-bottomed (p/b) trigons (Fig. 31c,d) that are sometimes associated with hexagons (see below). Flat-bottomed trigons have even been observed on diamonds recovered from inside an eclogite xenolith (Larry Taylor, pers. comm.) When f/b and p/b trigons occur on the same diamond, they are typically grouped by size ranges. In some kimberlites, diamonds with small f/b trigons (less than about 5 μm in diameter) evolve into p/b as they grow larger whereas in other kimberlites the opposite trend is noted (Fedortchouk 2015 and unpublished data). Also, f/b and p/b trigons have different relationships between their diameter and depth. For example, on the same diamond f/b trigons often have similar diameters but a large range of depths, whereas p/b trigons show a strong positive correlation between their diameter and their depth (Fedortchouk 2015).

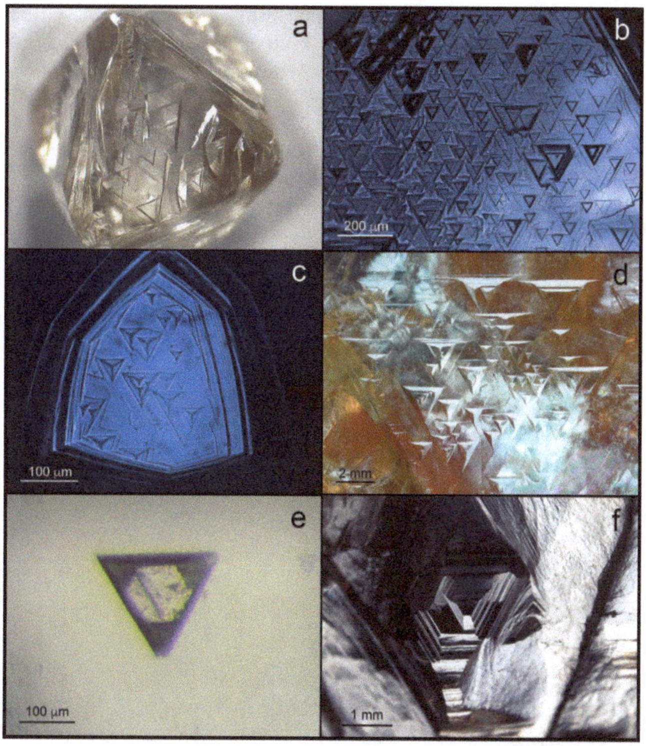

Figure 31. Negative trigons. (**a,b**) Flat-bottomed (f/b) trigons on diamonds from Gahcho Kué and Snap Lake. (**c,d**) Point-bottomed (p/b) trigons on a Misery kimberlitic diamond and Sierra Leone alluvial diamond. (**e,f**) Truncated trigons combining walls of positive and negative orientation (Orapa cluster AK15 and Letšeng). [Images in (a,b) courtesy of Ingrid Chinn (De Beers Exploration). Images in (c,e) by Yana Fedortchouk. Images in (d,f) by Evan Smith (GIA).]

Experiments conducted at 1–3 GPa and 1150–1350°C suggest that higher CO_2 content in kimberlitic fluids correlates with (1) increasing size and abundance of p/b trigons at the expense of f/b trigons (Fedortchouk 2015; Zhang et al. 2015) and also (2) the development of positively-oriented walls inside negative trigons (resulting in truncated trigons).

Their formation on octahedral surfaces often seems random, but Lang (1964) and Frank and Lang (1965) established by X-ray topography that point-bottomed (pyramidal) negative trigons were etch-pits resulting from preferential removal of material from locations where dislocations outcropped on the external surface of the crystal. Frank and Lang (1965) recorded that no dislocations outcrop beneath flat-bottomed trigons. For the theoretical grounds explaining the association of point- and flat-bottomed etch pits on minerals with different types of defects see Lüttge (2006). Sometimes lines of negative trigons are observed running across octahedral faces parallel to octahedral edges, and trigon development exploits the weakness created at the surface by *plastic deformation lines/planes*.

Positive trigons are very rare on octahedral surfaces of natural diamond and are usually found only on the smallest sized stones. At Snap Lake (Canada), positive trigons were found on microdiamonds of size ≤ 300 µm to 1.7 mm (Li et al. 2018). During diamond classification studies by Harris in the 1970s (unpublished data), *positive trigons* were seen in the ~1 mm-sized diamonds from Finsch (South Africa). As present-day cut-offs for diamond recovery are now usually set at ~2 mm, diamonds showing this surface feature have probably become particularly rare. In addition, at the Ekati kimberlite cluster (Canada) *positive trigons* occur on microdiamonds recovered from coherent hypabyssal kimberlite in the Grizzly pipe but are lacking on microdiamonds from volcaniclastic kimberlites (Yana Fedortchouk, unpublished data). Thus, it may be that positive trigons are a feature of diamond dissolution in hypabyssal conditions, where high-temperature fluids are present at near-surface pressures.

Figure 32 shows f/b positive trigons on a Finsch mine diamond and one from Snap Lake. Those from Finsch have edges of ~20 µm whilst Snap Lake edges are 1–3 µm. At both localities *positive trigons* overprint earlier dissolution features such as f/b negative trigons and may not be evenly distributed over an octahedral face. At Snap Lake some very small positive trigons were found on the rounded edges of octahedra (Fig. 6a in Li et al. 2018). In South Africa this trigon type has also been noted with two other rare surface features known as *transverse hillocks* and *imbricate wedges* (Robinson 1979) (see Fig. 1c,d in Li et al. 2018). Experiments show that low pressure, high oxygen fugacity (fO_2), and fluid with > 90 mol% of CO_2 promote positive trigon development (Evans and Sauter 1961; Yamaoka et al. 1980; Zhang 2016). These conditions suggest that etching took place after kimberlite emplacement, representing very late-stage dissolution by kimberlitic volatiles.

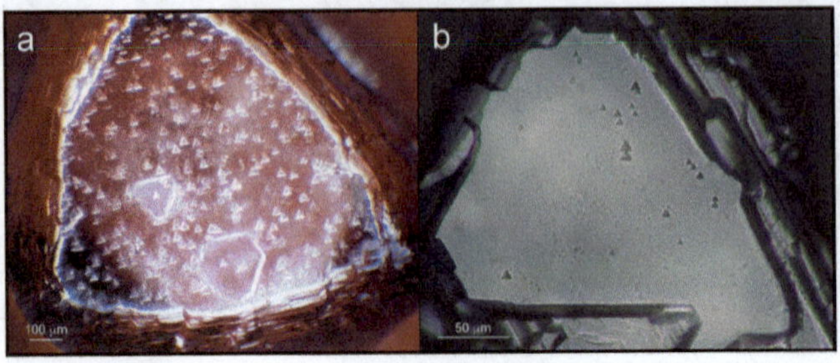

Figure 32. (**a**) A Finsch octahedron exhibiting positive flat-bottomed trigons in part overprinting earlier larger flat-bottomed truncated negative trigons. (**b**) Flat-bottomed (f/b) positive trigons on a Snap Lake diamond. [Image in (a) by Jeff Harris. Image in (b) by Yana Fedortchouk.]

Hexagons are six-sided pits that develop on octahedral faces and combine the components of positive and negative trigons (Fig. 33). Negatively oriented walls are steeper and flatter than those with positive orientation, which are shallower and show *striations*, *tetragonal or circular pits*. Hexagons are usually larger and deeper than trigons, their base being flat or rugged, the former often having *small trigons* on the bottom. *Hexagons* are less common than *trigons*, but diamond populations from almost any kimberlite have some of this type of mantle dissolution. They were relatively common on diamonds in hypabyssal kimberlite from Snap Lake (Li et al. 2018) and the lamproitic Argyle Mine in Western Australia (Hall and Smith 1985). Figure 33a shows a typical diamond with hexagons from Argyle.

As negative and positive trigons have been shown to be two separate dissolution events, the formation of hexagons is most likely the result of dissolution conditions being close to the boundary region for these dissolution mechanisms. At Snap Lake this relationship was studied at the 1–10 µm scale (Li et al. 2018) where there appeared to be an association, positive trigons developing inside negative pits making them hexagons (Fig. 33d). Experiments indicate that the larger hexagon formation is promoted by high CO_2 content (> 90 mol%) of magmatic fluid (Fedortchouk et al. 2007; Zhang 2016) and by dissolution in carbonate melts (Fedortchouk et al. 2019).

Cubic growth features

Flat-faced sharp-edged gem quality cube diamonds are much more rare than octahedra with equivalent characteristics. Perhaps the nearest to cube perfection are some of the diamonds from Wawa in Canada (see Fig. 2 in Stachel et al. 2006).

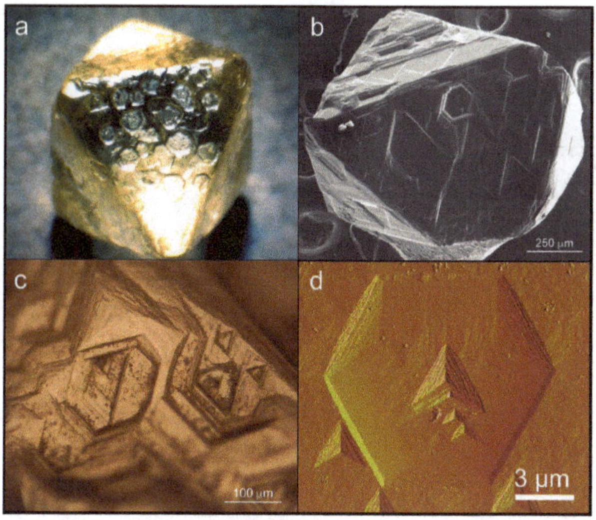

Figure 33. (**a**) A 5 mm Argyle octahedron exhibiting hexagons, a common surface feature of Argyle diamonds (Australia). Hexagons on diamonds from (**b,c**) Leslie kimberlite, Ekati Mine (Canada) and (d) Snap Lake Mine (Canada). [Image in (a) reprinted from Hall and Smith (1984) with permission from the University of Western Australia. Images in (b–d) by Yana Fedortchouk.]

The original growth of the larger gem diamonds may have created the smooth hummocky surfaces observed, but the association with tetragons (see below), probably indicates that original growth features have been modified or obliterated by dissolution. The smaller and more regular gem and opaque flat-faced cube diamonds have surfaces which are finely pockmarked, but whether or not this is a growth feature is uncertain.

Cubic dissolution features

Dissolution causes the rounding of cube edges creating a small modification to the original shape. Cubic dissolution features however are not constrained only to diamond cubes. They can also develop on any {100} plane exposed to dissolution (e.g., an octahedron broken along a (001) plane with tetragonal pits and tetragons on that surface—Tom McCandless, pers. comm.).

Development of a single face along the cubic edge transforms cubes towards dodecahedroids while development of two faces one on either side of the original edge results in THH morphology (Khokhryakov and Pal'yanov 2007). These authors also show through dissolution experiments that transformation of diamond cubes into THHs requires twice the weight loss than octahedral transformations. The other very obvious and dominant cubic feature comprises crystallographically controlled etching consisting of tetragons, a feature like trigons, whose orientation is dependent on temperature and fO_2 (Fedortchouk et al. 2009).

Tetragons are tetragonal pits set at 45° to the cube edge (in the case of the usual negative tetragons) and which may cover all {100} faces (Fig. 34). The surfaces of some coated cubes from the Democratic Republic of the Congo have distinct, but incomplete cube patterns which parallel the cube edges. Whether this observation indicates an irregular growth of the diamond coat or a subsequent partial dissolution is uncertain (Jeff Harris, unpublished data).

In an extensive study of South African diamonds only one was found to have positive tetragons (with sides parallel to the cube edges) (Robinson 1979). Small *tetragons* may appear on four-fold surfaces of other diamond crystals, at crystal apices and positively oriented walls of hexagonal pits. The tetragon base can be deep point-bottomed or very shallow flat-bottomed

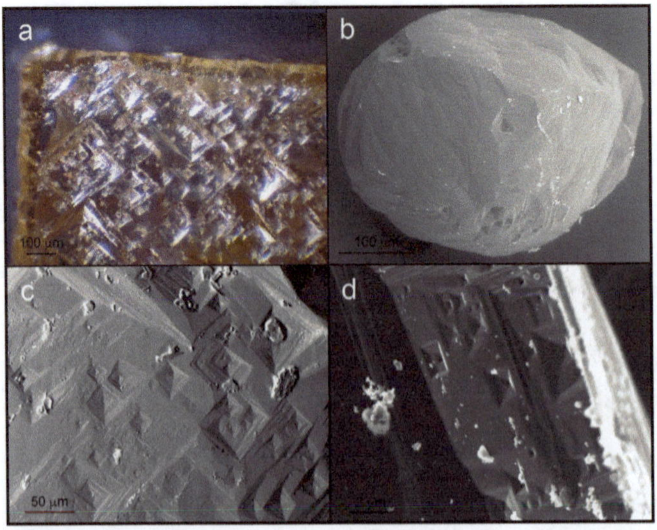

Figure 34. (**a**) Tetragons set in the negative orientation covering the whole of a cube face on a diamond from the DRC. (**b,c**) Dodecahedral diamond showing trigon dissolution pits in [111] direction (centre of the crystal) and point-bottomed (p/b) tetragons in [100] direction (upper left and lower areas of the crystal). (**d**) Small point-bottomed (p/b) tetragons formed on positively-oriented walls of hexagonal pits. [Image in (a) by Jeff Harris. Images in (b–d) by Yana Fedortchouk.]

and both types usually show terraced (stepped) walls. Robinson (1979) suggested that tetragon pits could represent the incipient forms of the *circular pits* sometimes found on diamonds and formed during late stages of dissolution by kimberlitic fluid.

Dodecahedroid and THH features

Two main surface features develop during dissolution of diamond octahedra to dodecahedroids and THHs, during ascent in kimberlite/lamproite magma: (1) low relief features and *plastic deformation lines/planes* linked to diamond crystallography, and (2) features which are independent of crystallographic control.

Low-relief features. The curved surface of a dodecahedroid face is sometimes smooth, featureless and glossy, but more commonly consists of an overall or partial *terracing*. The terraces arise because dissolution exposes the intermittent growth layers of the original octahedral surfaces, where terrace shapes are centered around <111> directions (Fig. 35).

Commonly associated with terracing are *hillocks*, which form on individual dodecahedroid surfaces normally elongated parallel to the long diagonal of the face. *Hillocks* can form as isolated features, but usually occur in groups or cover all faces of a diamond crystal. *Hillocks* form a complete array of shapes from smooth drop-shaped to pyramidal with relatively sharp and pointy apices. Their clear link to diamond crystallography can be seen, for example, through their alignment around three-fold axes and the way they radiate from four-fold axes (Fig. 36). Observational evidence indicates that there is a general uniformity of hillock type on dodecahedroid diamonds from the same kimberlite phase. This infers that the dissolution conditions in kimberlite magmas are the determining factor for hillock shape (Fedortchouk et al. 2017), however, in some cases, hillock shape could be influenced by minor cuboid growth and plastic deformation (Ingrid Chinn pers. comm.). A study of diamond plates cut through dodecahedroid diamonds from Udachnaya and Yubileynaya (Yakutia) showed that there was a strong association between hillock development and areas of intense plastic deformation (Khokhryakov and Pal'yanov 2009).

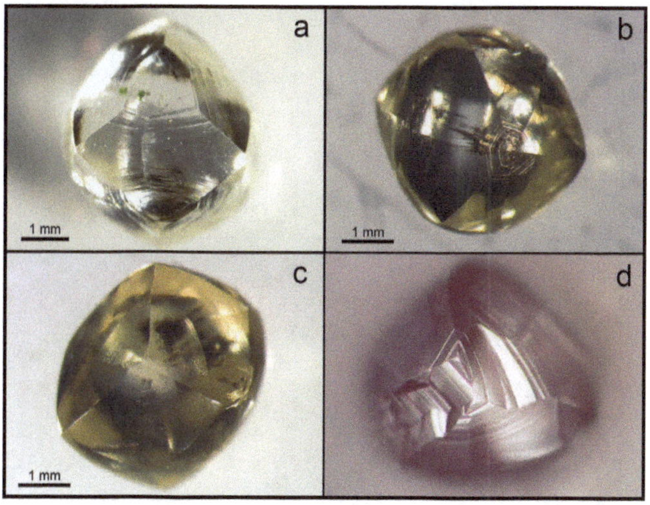

Figure 35. Examples of terraces on THH diamonds from volcaniclastic kimberlites. [Images in (a,b,c) by Karen Smit. Image in (d) by Yana Fedortchouk.]

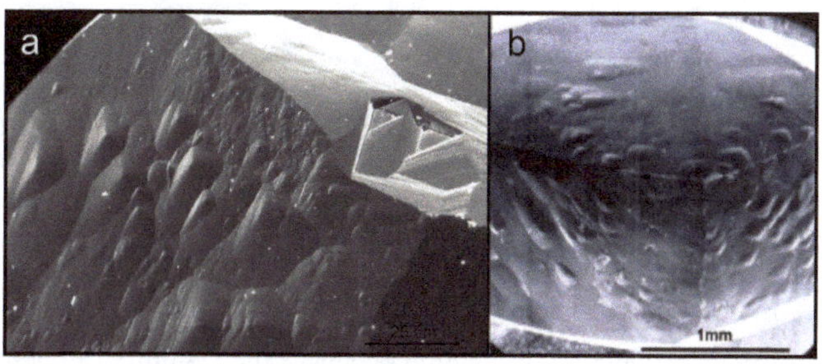

Figure 36. Hillocks are linked to diamond crystallography: in their alignment around three-fold axes, and the way they radiate from four-fold axes. [Image in (a) by Yana Fedortchouk. Image in (b) reproduced from Robinson (1979) with permission from the author.]

These low relief features have been reproduced in dissolution experiments in the presence of C–O–H fluid at 1 GPa and 1250–1350 °C (Kozai and Arima 2005; Fedortchouk et al. 2007) and often have a dominating presence in microdiamond populations from volcaniclastic facies kimberlite (Fedortchouk et al. 2017) both of which suggest that they develop when diamonds are carried by volatile-rich kimberlite magmas.

Plastic deformation lines/planes. The presence of variably spaced lines/planes running across a diamond (Fig. 37) is evidence that the diamond has been plastically deformed along its cleavage planes during residence at high temperatures in the mantle. The conditions for the transition from brittle to ductile (plastic) deformation in diamond shows that above 1000 °C, and in the pressure range expected for the sub-continental lithospheric mantle (1 to 6 GPa), ductile behavior is not very dependent on pressure (DeVries 1975).

More than one of the four equivalent sets of {111} cleavage planes may be involved when a diamond is plastically deformed. In these cases, the intersections of the *plastic deformation lines/planes* sometimes give rise to a *shagreen texture* over the dodecahedroid surface. Although this feature can be recognized on octahedral diamonds (see above), they

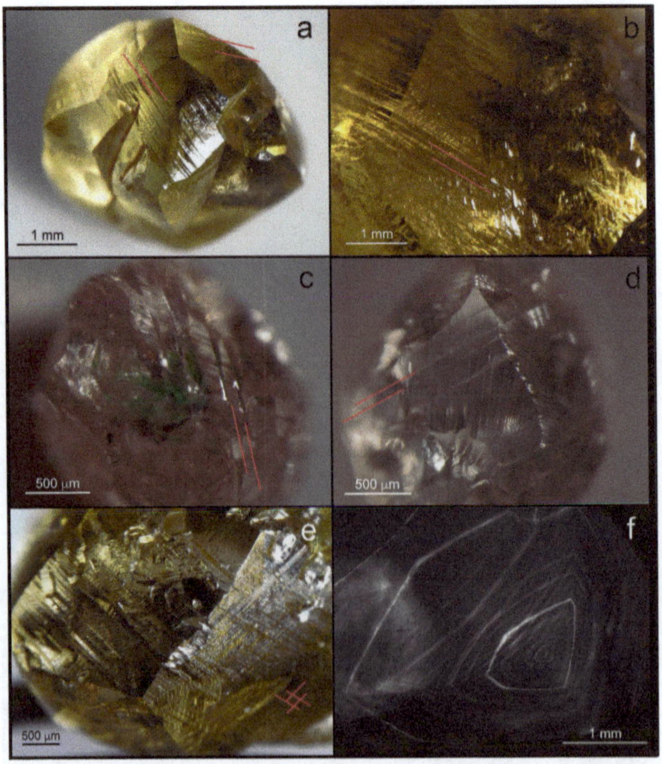

Figure 37. Plastic deformation lines/planes, indicated with **red lines** for clarity in (**a**) to (**e**). (**f**) Cathodoluminescence image that shows bright defects decorating the octahedral planes along which deformation occurred. [Images in (a,b,e,f) by Karen Smit. Images in (c,d) by Nicole Meyer (University of Alberta).]

are described here because diamond dissolution enhances the visibility of *deformation lines/ planes* on dodecahedroid surfaces. The appearance of deformation lines/planes is variable, and can appear wavy (e.g., Argyle diamonds) or as straight planes through the diamond; and this distinction is probably due to differences in nitrogen content and aggregation state (Gaillou et al. 2010, 2012; Howell et al. 2015). Straight *deformation lines/planes* in 'non-Argyle-type' pink diamonds have been found to be due to deformation micro-twinning (Mineeva et al. 2007; Gaillou et al. 2010; Titkov et al. 2012).

The terms *lamination line* and *deformation lamellae* are also used to describe this surface feature, and to a lesser extent, *glide planes, slip planes* or *graining*. Historically, they were also known as *Tanganyika lines* because of their common occurrence in diamonds from the Williamson mine in Tanzania.

A measure of the proportion of diamonds deformed in a particular size fraction in a general diamond production can be gained by assessing the proportion of pink and brown diamonds present, as these colors are linked to plastic deformation (Collins 1982; Harris et al. 1983; Robinson et al. 1986; Fisher et al. 2009; Gaillou et al. 2010, 2012; Titkov et al. 2012; Howell et al. 2015). Not all deformed diamonds are brown, however, and experiments indicate that the color can be annealed out leaving just the deformation lines (Moses et al. 1999; Collins et al. 2000) a process that can occur naturally during high temperature storage in the mantle. For further details on the relationship between plastic deformation and pink and brown coloration in diamonds see Green et al. (2022, this volume).

Non-crystallographic features. *Corrosion sculpture* occurs on dodecahedroid surfaces as individual cavities of ~50 μm depth which are elliptical to irregular in shape with maximum diameters from 50 to 300 μm (Fig. 38a,b). The bases of the cavities are either smooth or striated and the cavities usually cover all the faces and edges of a crystal. *Shallow depressions* are generally restricted to dodecahedroid surfaces. These are equally irregular but larger in area, much shallower, and tend to occur in clusters. They are noticeable because they give a light frosting to the diamond surface (Fig. 38c).

Robinson et al. (1986) showed that in the 2.9 to 3.4 mm diamond size range for the major South African mines, high proportions of *corrosion sculpture* in a production were often matched by a smaller but significant proportion of *shallow depressions*, which might suggest similar conditions for their formation. Additionally, at the Wesselton mine in South Africa this relationship was kimberlite specific.

A more recent study by Fedortchouk et al. (2017) of 1 to 4.5 mm diamonds from the BK1 kimberlite (Orapa cluster) showed that *corrosion sculptures* were abundant in hypabyssal facies; and *shallow depressions* on diamonds in transitional partially fragmented hypabyssal facies. *Corrosion sculptures* are common on 1 to 3.5 mm diamonds in hypabyssal Gahcho Kué kimberlite, Northwest Territories, Canada (1 to 3.5 mm diamond size: Ingrid Chinn, pers. comm.). Experimentally, corrosion features and cavities develop on diamonds in volatile-undersaturated alkaline basalt, synthetic kimberlite, and carbonate melts at $P < 3$ GPa (Kozai and Arima 2005; Fedortchouk et al. 2007; Arima and Kozai 2008), but are not reported at pressures between 5 and 7 GPa (Khokhryakov and Pal'yanov 2007, 2010; Fedortchouk et al. 2019). Thus their occurrence in hypabyssal kimberlite and volatile-undersaturated experiments suggest an origin in kimberlite magma that did not exsolve a fluid phase.

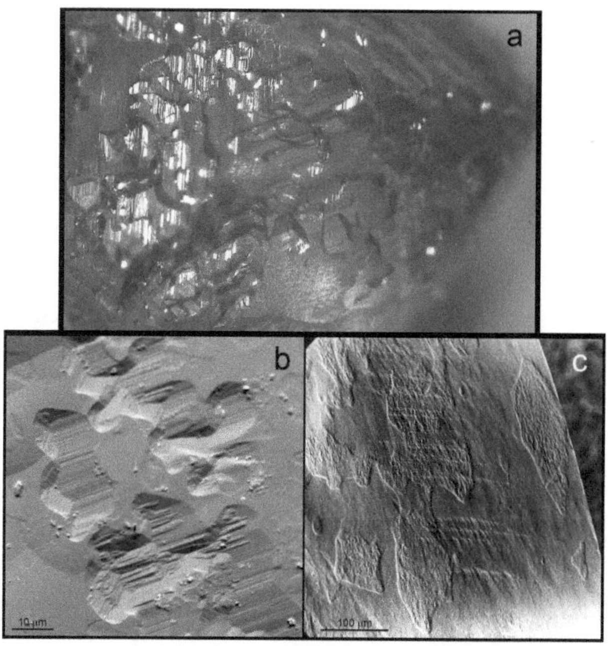

Figure 38. Examples of corrosive features on diamonds from hypabyssal kimberlites. (**a**) Corrosion sculpture on diamond from Gahcho Kué. (**b**) Corrosion sculpture on diamond from the Orapa kimberlite cluster. (**c**) Shallow depression on a South African diamond. [Image in (a) courtesy of Ingrid Chinn (De Beers Exploration). Image in (b) by Yana Fedortchouk. Image in (c) reprinted from Robinson (1979) with permission from the author.]

The occurrence of *frosting* on South African diamond productions was also investigated by Robinson et al. (1986) who divided this surface feature into coarse and fine frosting and showed the particular nature to be specific to the kimberlite source. Frosting on South African diamonds consisted of minute hexagonal to prismatic pits and occurred principally on dodecahedroids, but occasionally on octahedral diamonds. The frosting covered entire faces or only certain areas forming patches or islands. Examination of small to microdiamonds from Canadian productions (Fedortchouk et al. 2010) found the frosting to consist principally of micro-hillocks, which were either smooth drop-shaped (Fig. 39a) or sharp and pointy (Fig. 39b). Dodecahedroids with "smooth" *frosting* were common for all types of volcaniclastic kimberlite facies whereas those with "sharp" features occurred in hypabyssal kimberlites (Fedortchouk et al. 2010). Both authors considered this feature to be a very late-stage. Note that the general arrangement and the nature of *frosting* on diamonds in kimberlite is distinct from that caused by diamonds abraded in the alluvial environment (see below).

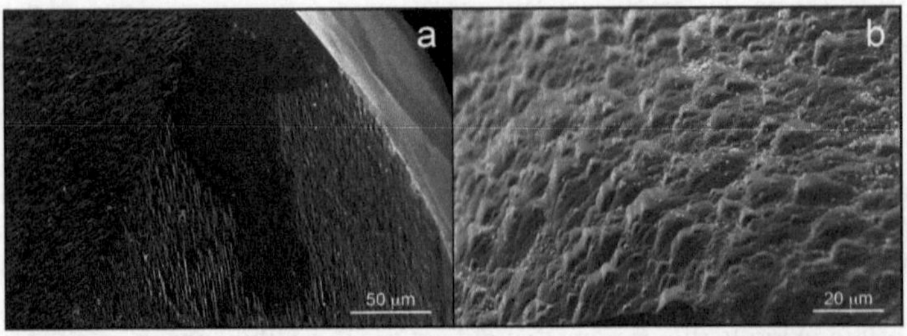

Figure 39. Secondary electron images of (**a**) frosting made of micro-hillocks with a smooth drop-like shape and (**b**) frosting made of sharp and pointy micro-hillocks. [Images by Yana Fedortchouk.]

Micro-disks are very minutely raised circular features on dodecahedroids and are described by Pandeya and Tolansky (1961), who proposed that the disks originated from an attachment of small fluid bubbles to the surface, momentarily protecting it from etching and leaving a raised disk behind. There is concentricity to some disks but others are more elliptical and curvilinear. The size of *micro-disks* is variable (Fig. 40). The disks overprint development of *terraces* and *hillocks,* and sometimes cover broken crystal surfaces indicating their origin during late-stage dissolution in kimberlite. The above authors also found an association with circular patterns known as *micro-pits*. Scanning electron microscope (SEM) images by Robinson (1979) showed these *micro-pits* to be nearly rectangular and about 10 μm in size. Microdiamond studies from the Ekati kimberlite mine identified two types of pit: one drop-shaped and the other with pyramidal walls (Zhang 2016) (Fig. 41). These *micro-pits* were found only on diamonds from volcaniclastic kimberlites where distribution varied from being present only on broken diamond surfaces in Panda and Koala pipes and occurring on dodecahedroid surfaces on diamonds from Misery and Fox pipes. In contrast, diamonds recovered from hypabyssal kimberlites (Leslie and Grizzly) did not display any *micro-pits*.

The limited experimental data on the formation of circular *micro-pits* reveal that they form in H_2O fluid at low pressure ($\leq$ 1 GPa), whereas addition of CO_2, dissolved SiO_2 and higher pressures suppress their development (Zhang et al. 2015). This could potentially make *micro-pits* (circular or other shapes) useful for deciphering conditions in kimberlite magma at the late stages of emplacement.

Ruts occur across some diamond surfaces and are of variable length, may be straight or sinuous, or may bifurcate (Fig. 42). According to Robinson (1979) the sides may be conspicuously etched and frosting sometimes extends into the rut. They are probably the result of weaknesses or pre-existing cracks being widened by an etchant passing between the diamond and its kimberlite matrix.

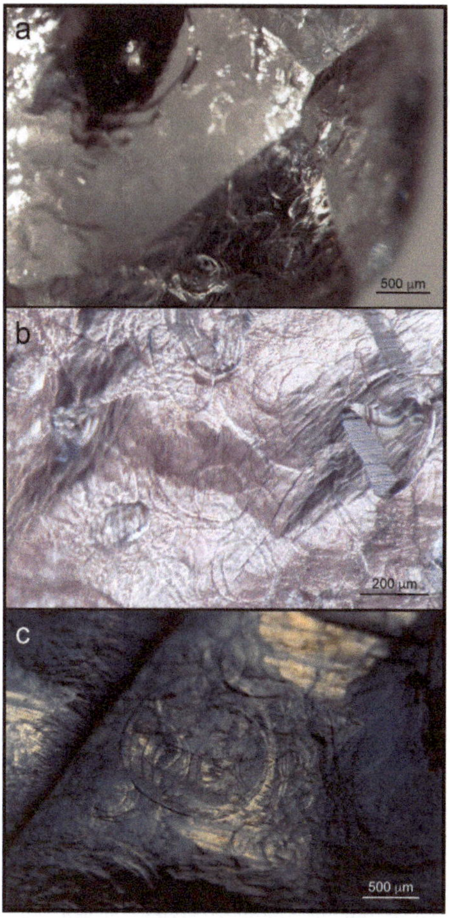

Figure 40. Examples of micro-disks on dodecahedroid and tetrahexahedroid surfaces. (**a**) Diamond from Koffiefontein (South Africa). (**b**) Diamond from Williamson (Tanzania). (**c**) Diamond from Letšeng (Lesotho). [Image in (a) by Nicole Meyer (University of Alberta). Images in (b,c) by Evan Smith (GIA).]

Rare surface features called *etch channels* or *spicules* are specifically considered by Lu et al. (2001) as hollow tubes open to the surface, ranging in form from straight to irregular and wormlike, some being interconnected and others showing crystallographic orientation. The openings have rhombic or modified rhombic shapes, widths of ~20–100 µm, lengths of ~200 µm–1 mm, and often terminate at solid inclusions. They occur in Type I and Type II diamonds and have a global distribution (Africa, Russia, South America, Australia and Canada) (Lu et al. 2001). *Etch channels* most likely originate from structural weaknesses on the diamond surface, such as dislocation bundles or where the edges of resorbed diamond intergrowths allow diamond dissolution to preferentially occur (Fig. 42a). These have then been aggressively etched and as dissolution proceeds, internal defects have either changed the direction or caused it to terminate, or the etching fluid has been depleted.

Surface features formed at the Earth's surface

Once diamond has been eroded out of its primary host rock, the new alluvial environment can cause additional surface features. These can be in the form of damage which overprints earlier characteristics, or surface color staining that is a consequence of radioactive bombardment. In the first case the damage can be very noticeable, affecting any diamond surface and can be found on all diamond shapes and most sizes, except the smaller stones (1–2 mm).

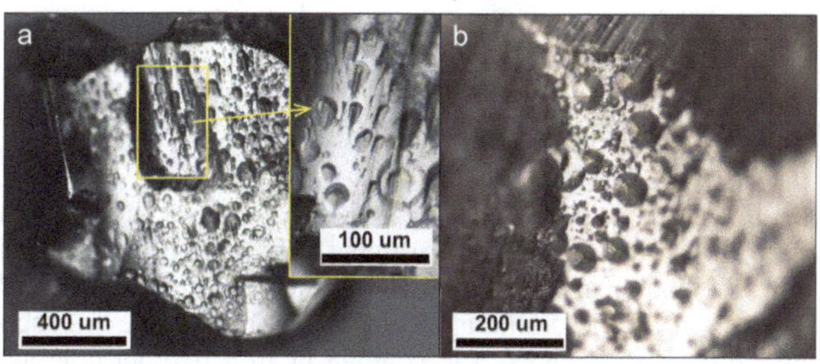

Figure 41. (**a**) Drop-shaped circular pits. (**b**) Pyramidal pits (transitional to tetragons). [Images by Zhihai Zhang (Dalhousie University).]

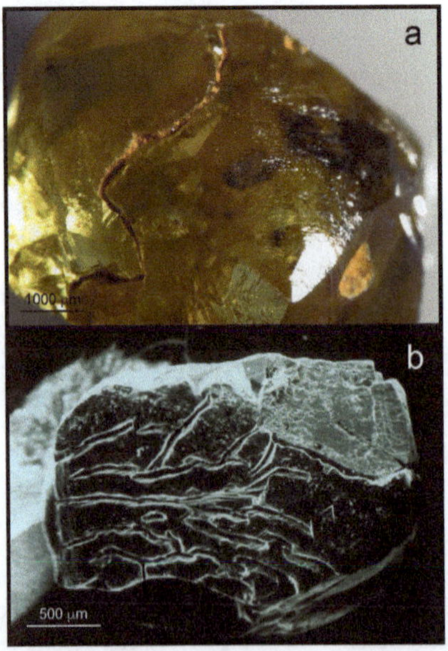

Figure 42. Ruts on diamond surfaces can be of variable length and shape, and result from fluid etching. (**a**) Diamond from Zimmi (Sierra Leone). (**b**) Secondary electron image of a Cullinan diamond exhibiting ruts across its octahedral surface. [Image in (a) by Karen Smit. Image in (b) by Jeff Harris.]

Radioactive damage can occur during diamond residence in primary kimberlite/lamproite after eruption, and/or more commonly in subsequent alluvial rocks. All shapes, surfaces and in this case, sizes can be affected. When irradiated in the primary kimberlite/lamproite, diamonds often have a totally transparent green coating. When irradiated in alluvial rocks, radiation damage is often observed as irregular and often localized green spotting. An exception is the diamonds in the Witwatersrand conglomerates that have a transparent green coating in addition to green radiation stains. Green radiation damage can turn brown if the diamond is heated above 600 °C (*discussed in more detail below*).

Alluvial surface damage. With diamond being the hardest known mineral, to be considering damage may seem superfluous, but diamond has a serious weakness in the form of a perfect cleavage. Such planes cause diamond to have a low brittle fracture. Thus, even if struck forcefully with an object less hard it may break along the {111} cleavage planes, whereas if progressively and slowly compressed the diamond will penetrate into the jaws of the press. In an alluvial environment, therefore, diamond can be damaged by local impacts or abraded during river transport. However, the probability of two diamonds colliding and causing damage to one another is extremely remote as the grade for alluvium is usually very low at about 1–2 carats per hundred tons. This diamond grade equates to one part diamond in 4.5×10^8 parts of rock.

The first sign of wear is abrasion damage to the *crystal vertices* and *crystal edges* particularly of octahedra, macles and dodecahedroids. Cubic diamonds would suffer the same fate except that many are either of poor quality and therefore do not survive transport, or if of gem quality are initially so exceptionally rare that damage has not been reported from an alluvial environment. The type of damage consists of very small breakages at the points and edges of the crystal (Fig. 43). If the *medial lines* of dodecahedroids are prominent or abrasion is relatively severe, then those edges may also be affected.

Impact damage takes the form of *percussion marks*, *chattermarks*, and *irregular surface scratches* (Fig. 44). In the case of *percussion marks*, the impact imparts an arcuate shape on the diamond surface, but directly within the diamond an internal fracture is created along the nearest cleavage plane to varying depths (micron scale). The size of this feature depends on the size and force of the impactor. *Chattermarks* are the result of a trail of closely spaced percussion marks whereas irregular surface scratches are aptly named random scratch-like marks on a diamond surface. The patterns of all these features can be quite varied.

A more structured alluvial abrasion feature which can be associated with those previously mentioned is a *network pattern*. The typical example shown in Figure 45 consists of two sets of {111} planes set at 109.5° apart. This configuration indicates they are probably the surface expression of two sets of plastic deformation lines/planes that were enhanced by abrasion. Finally, if a diamond becomes trapped, for example in an active pot-hole in a river course, mechanical wear may create a partial or all-over *frosting* giving the diamond a very dull appearance (Fig. 46).

Figure 43. An alluvial diamond from the Urals (Russia) showing abrasion of the edges of a tetrahexahedroid with lesser damage along the medial lines. [Image by Jeff Harris.]

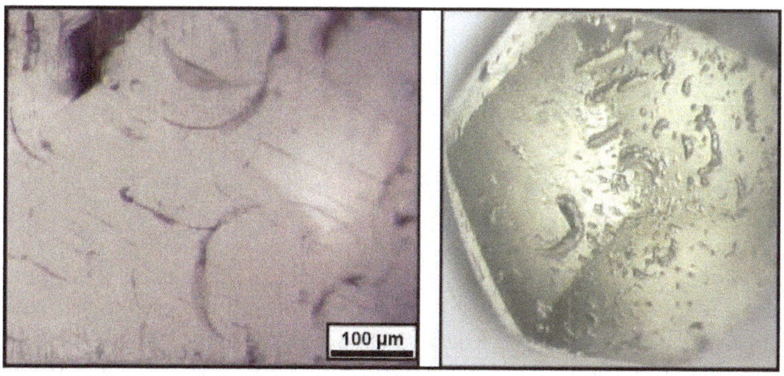

Figure 44. (**a**) Selection of different sized percussion marks on an alluvial diamond from the Urals (Russia). Note in the upper part of the picture, how the original percussion internally damages the diamond along a cleavage plane. (**b**) A 2–3 mm alluvial diamond from Namibia exhibits weak chattermarks with other irregular features of alluvial damage. [Images by Jeff Harris.]

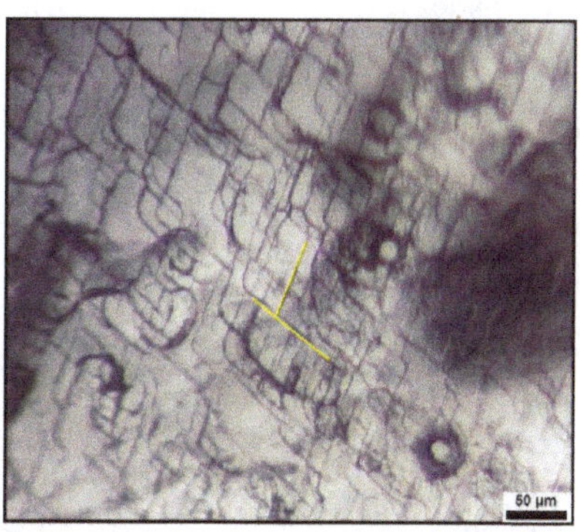

Figure 45. A diamond from the Urals (Russia) exhibiting a network pattern. [Image by Jeff Harris.]

Figure 46. A 2–3 mm alluvial diamond from Namaqualand (South Africa) exhibiting heavy frosting due to abrasion. [Image courtesy of De Beers Exploration.]

Green radiation stains. Radioactive damage and the resulting color—occurring either during diamond residence in post eruption primary rock (kimberlite/lamproite) or in subsequent alluvium—is the result of persistent surface bombardment of α-particles, which have a penetration depth of up to 30 μm (Nasdala et al. 2013). During penetration the α-particle has sufficient energy to displace a carbon atom in the diamond lattice causing an atomic vacancy, thereafter losing its energy and becoming a helium atom. Diamonds with lower nitrogen content may be more likely to be damaged by radiation, as vacancy concentration has been shown to correlate with nitrogen content (Kiflawi et al. 2007). The created vacancy (GR1) is a color center and if there is sufficient concentration of GR1, visible green coloration results. The time taken for the color to be seen depends on the intensity of the radiation and may vary between 10 million years for uraninite contact (Vance and Milledge 1972) to around one billion years for the radioactivity in zircons (Nasdala et al. 2013).

Kimberlites have low concentrations of radioactive uranium and thorium (ppm range; Mitchell 1986) and therefore are only able to impart limited radiation damage to diamonds while they are still *in situ*. Nevertheless, the majority of transparent green-coated diamonds (Fig. 47) are found only in kimberlites. They are confined to the upper weathered portion of the pipe where radioactive ground water exists, possibly sourced from potassic uranyl salts that formed from weathered phlogopite and scarce radioactive minerals in the kimberlite (Harris 1992). Much rarer examples of transparent green-coated diamonds recovered from alluvium are the Witwatersrand diamonds, where radioactivity in the groundwater probably originated from uraninite in the conglomerates.

In the alluvial environment a very low diamond concentration is matched by higher concentrations of radioactive minerals in the alluvium, which have been eroded from a wide range of surface crustal rocks over which a river system carrying diamonds has flowed. In this case radiation damage is the result of a particulate attack by α-particles from an adjacent radioactive mineral and the color intensity depends on the strength of the source and ranges in intensity from a pine green to a very dark green that almost appears black. The distribution of the radiation damage is controlled by the original distribution of the radioactive minerals in contact with the diamond (Fig. 47).

Brown radiation stains. If a green transparent coated or green stained diamond is heated in air to about 600 °C it very quickly turns irreversibly brown (Vance et al. 1973), perhaps indicating that brown naturally-irradiated diamonds have been heated to at least 600 °C post-eruption. Lawson et al. (1992) studied this transformation and showed that the reaction was kinetically controlled and heating experiments by Harris (unpublished) confirmed the time/temperature dependency. Attempts to use the kinetics of the reaction to determine the length of time to change green spots to brown spots has been thwarted because the error bar of the activation energy of the equation was relatively large. This resulted in a wide range of conversion periods for small adjustments of temperature (< 50 °C). For this reason, the equation was never used as a prospecting tool in diamond exploration.

The reason for the green-to-brown color change, is the temperature-driven local migration of vacancies (GR1) to one of two nitrogen aggregation sites (A and B centers), forming brown color centers known as H3 and H4 (Davies 1972; Mainwood 1994; Collins and Kiflawi 2009). The more this reaction proceeds, the darker the color. If the diamond with green stains is a Type II (containing virtually no nitrogen) then the second reaction does not occur. A detailed spectroscopic study of this color change is given by Nasdala et al. (2013). Figure 48 shows a typical array of brown radiation damage that can be associated on the same stone with green staining, signifying that radiation damage can occur at different stages in the life of an alluvial diamond.

Diamonds with green and brown radiation damage have been recovered from major placer deposits globally. Historically, irradiated diamonds were recovered from the Witwatersrand

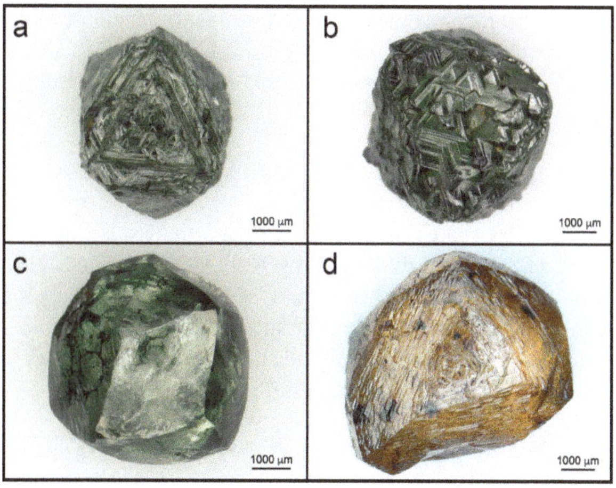

Figure 47. Alluvial diamonds from Guyana that have green and brown coloration from irradiation damage. [Images by Ellen Barrie (GIA).]

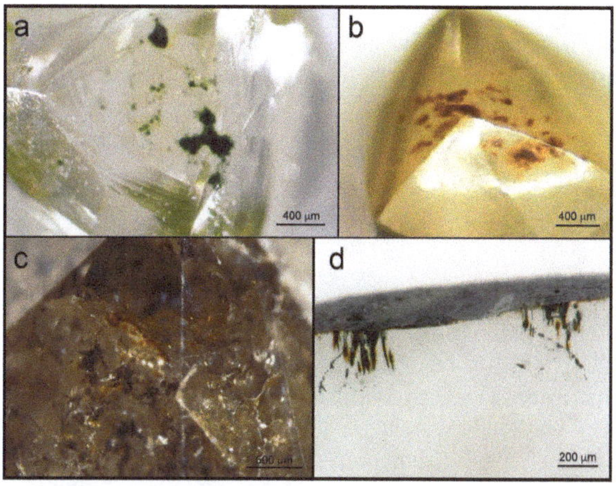

Figure 48. Alluvial diamonds with green and brown radiation stains **(a,b)** Diamonds from Namaqualand (South Africa) with staining on the surface. **(c,d)** Diamonds from Marange (Zimbabwe) with green and brown staining in fractures. [Images in (a,b) by Jeff Harris. Images in (c,d) by Karen Smit.]

basin in South Africa, where the presence of uranium-bearing minerals gave the diamonds significant color (Williams 1932; Gittus 1963; Raal and Robinson 1980). Such diamonds are now being recovered in the Diamantina district in Brazil (Karfunkel et al. 1995), the Quebrada Grande river basin in Venezuela (Kaminsky et al. 2000; Schulze et al. 2016) and most recently from the Marange deposit in Zimbabwe (Smit et al. 2018a).

An intriguing observation from color classification studies of diamonds from southern African mines (Harris et al. 1979) is that the high levels of transparent green coated diamonds recovered from the weathered upper portions of Finsch (South Africa) and Jwaneng (Botswana) (Harris et al. 1986) are not accompanied by a proportion of transparent brown diamonds. Eruption ages for Finsch and Jwaneng are 118 Ma and 240 Ma respectively (Gurney et al. 2005) and both diamond populations are not atypical. For the absence of a brown content the temperature in the primary environment had to be always below the temperature threshold for the second reaction to take place, or that radioactivity in the upper portion of the pipe must be a relatively recent event. The second possibility might explain the obvious presence of green and brown stains on coastal alluvial diamonds from Namaqualand and Namibia to which the above mines may have contributed (see Kjarsgaard et al. 2022, this volume, and references therein). For Namibia, the diamond influx from the interior was Oligocene (~30 Ma) (Jacob and Grobbelaar 2019).

Concluding remarks

We have not described every surface feature known for diamond, but the above descriptions represent all the significant features and some of the rarer ones. Descriptions of surface features from specific countries have been completed on Russian diamonds (the Urals and Siberia) by Orlov (1977) and southern African diamonds by Robinson (1979). A classification scheme for surface features is outlined in Robinson et al. (1989). Tappert and Tappert (2011) provide excellent images and descriptions of the main surface features. The most recent and very detailed Canadian and Botswanan microdiamond study is by Fedortchouk (2019).

MINERAL INCLUSION MORPHOLOGY

Mineral impurities in diamond can be conveniently divided into those that are associated with diamond formation (*synchronous*) and those that occurred subsequently (*epigenetic*). *Synchronous* minerals are crystalline, invariably totally enclosed by the diamond and normally range in size from < 50 µm up to rare examples > 500 µm. *Epigenetic* minerals may form by the alteration of synchronous minerals and can occur in wholly internal fracture systems or in fractures with access to the exterior of the diamond.

Until recently, synchronous minerals were considered to have grown solely syngenetically with the diamond. This view has been challenged due to crystallographic (Nestola et al. 2014; Agrosi et al. 2016) and isotopic evidence (Westerlund et al. 2004; Smit et al. 2016a, 2019b; Aulbach et al. 2018). These studies have demonstrated that even if inclusions are protogenetic and crystallized before the host diamond, they can display textures indicative of co-crystallization (or syngenesis) (Figs. 49 and 50). Smit et al. (2022, this volume) shows that a lack of syngenecity is not necessarily a hindrance for inclusion-based diamond dating.

Synchronous inclusions

Synchronous inclusions consist principally of oxides, carbonates, silicates, and sulfides. Rare examples where diamond crystals are seen as distinct inclusions are found in Yakutian (Bulanova 1995) and Koffiefontein diamonds (Fig. 51a). Although graphite is more commonly observed as epigenetic inclusions (see below), rare synchronous graphite inclusions have been identified in diamond. At Ekati (Fig. 51b), the book-like layering and the characteristic shape of graphite inclusions may indicate that it has not had diamond morphology imposed on it. As conditions changed, the graphite preferentially expanded and caused the (111) fracture plane around the inclusion (Glinnemann et al. 2003). On a micron-scale synchronous graphite has also been identified as evenly distributed inclusions in cubo-octahedral diamonds from Marange, Zimbabwe (Fig. 21). In this case, their small size prevented identification of their shape, but their association with methane suggests that both phases crystallized from the diamond source fluid (Smit et al. 2016b).

The mineralogy and chemistry of synchronous mineral inclusions indicate that diamonds form through a wide range of depths in the deep Earth. The majority of diamonds form in the sub-continental lithospheric mantle (SCLM), and inclusions mainly represent both peridotitic and eclogitic host rocks, but with a minor websteritic environment defined by mineral chemistry rather than mineralogy alone (see review by Stachel and Harris 2008). Diamonds also form in the sub-lithospheric mantle (see review by Stachel et al. 2005). Mineral inclusion chemistry indicates their formation in the upper mantle (above 440 km), transition zone (between 440 and 660 km) and the lower mantle (below 660 km) (Harte et al. 1999; Kaminsky 2012; Pearson

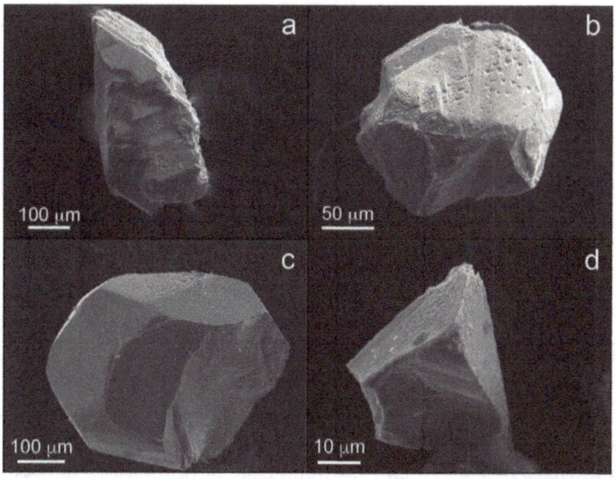

Figure 49. Secondary electron images showing cubo-octahedral morphologies in sulfide inclusions. These morphologies—including triangular step faces and trigons—are imposed by the diamond host, and are interpreted as evidence of co-crystallization of mineral inclusion and diamond host. (**a,b**) Eclogitic sulfides from Zimmi (Sierra Leone). (**c**) Eclogitic and (**d**) peridotitic sulfides from Ellendale (Australia). [Images by Karen Smit.]

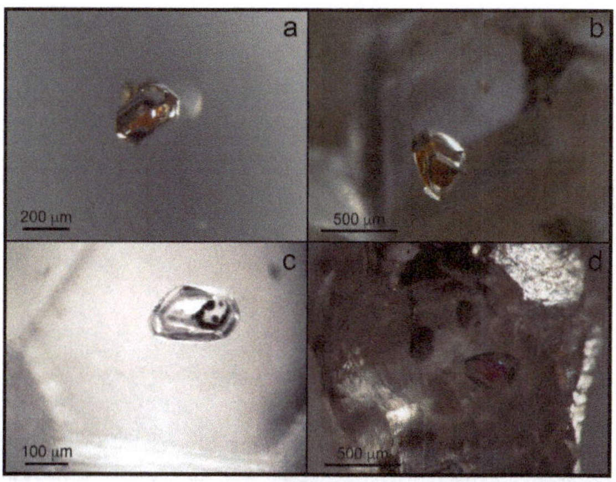

Figure 50. Cubo-octahedral morphologies in silicate inclusions. (**a,b**) Orange eclogitic garnet inclusions. (**c**) An elongate cubo-octahedral orthorhombic olivine inclusion in a matching cubo-octahedral Cullinan diamond (South Africa). (**d**) Purple-pink peridotitic garnet in a Koffiefontein diamond (South Africa). [Images in (a,b) by Karen Smit. Image in (c) by Jeff Harris. Image in (d) by Nicole Meyer (University of Alberta).]

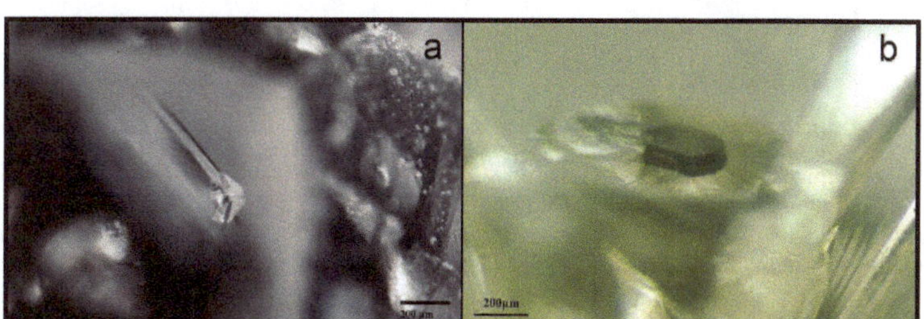

Figure 51. (**a**) Octahedral diamond crystal included in a diamond from Koffiefontein (South Africa). (**b**) Hexagonal graphite inclusion in an octahedral Ekati diamond (Canada). The graphite is stacked in 'book' form and is specifically oriented with respect to the diamond host. [Image in (a) by Nicole Meyer (University of Alberta). Image in (b) by Jeff Harris.]

et al. 2014; Burnham et al. 2016; Palot et al. 2017). The deepest diamonds that have been recovered formed in the lower mantle at depths of up to 800 km (Harte and Hudson 2013). Stachel et al. (2022, this volume) and Walter et al. (2022, this volume) give details on the mineral chemistry of lithospheric and sub-lithospheric inclusions.

Included minerals belong to all seven crystal systems and one might expect diverse inclusion morphology in diamond as a result. This is not the case, because all inclusion morphologies are influenced or imposed by the diamond host, leading to three shapes, common cubo-octahedra and rarer octahedra and cubes (Figs. 49, 50, 51, 52). Inclusions with cubo-octahedral, octahedral and cubic shapes may be equant or elongate, the latter invariably in diamond [110]. Morphologies may have recognizable faces over only part of the mineral, exhibit stepped surfaces parallel to those of the diamond, thicken or thin along particularly elongate inclusions (Fig. 52). Some inclusions may contain polymineralic intergrowths which can be clearly seen if the inclusion is elongate and there is a color distinction between two inclusions (Fig. 52d,f). More often such inclusions are identified after polishing for microprobe analyses. For examples of silicate and carbonate combinations see Phillips et al. (2004) and sulfide/garnet combinations are noted in Richardson et al. (2004) and Smit et al. (2019a). The crystallographic relationships between common peridotitic and eclogitic inclusions and their hosts are discussed in Angel et al. (2022, this volume).

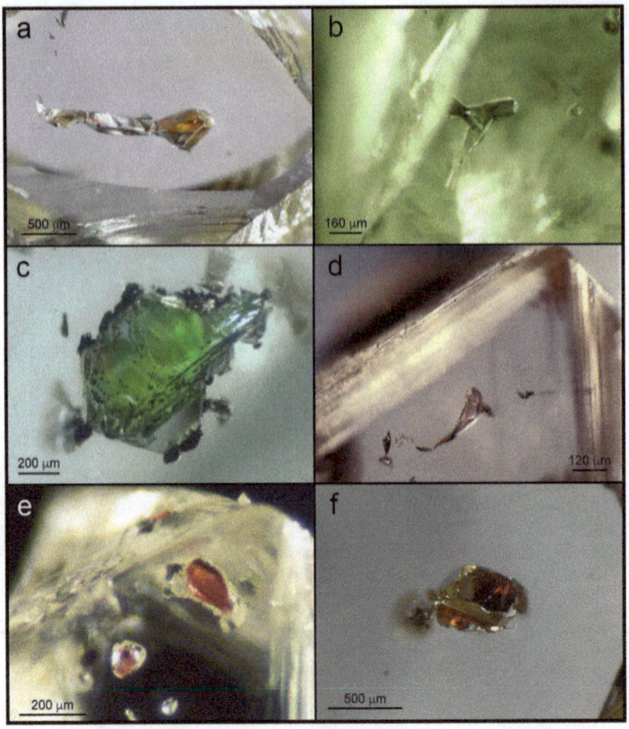

Figure 52. (**a**) Elongate orange garnet in diamond from the Urals (Russia), viewed through a (110) polished face. Well-defined cubo-octahedral faces give way to variable thickness and striations in reflectance along the length. (**b**) A bifurcation in a colorless inclusion in a Cullinan diamond (South Africa). (**c**) An elongate octahedral-shaped chrome diopside inclusion in a South African diamond. The dark material is graphite on the inclusion surface or in surrounding fractures. (**d**) A bi-mineralic peridotitic inclusion in a Cullinan diamond, comprising a purple garnet and a colourless mineral (either olivine or enstatite). (**e**) A flattish cubo-octahedral peridotitic garnet. (**f**) Bi-colored red and orange garnet inclusion—possibly eclogitic and pyroxenitic. For many inclusions, the color of the entire mineral is not visible due to the reflective interface between inclusion and diamond. [Images in (a–e) by Jeff Harris. Image in (f) by Karen Smit.]

Due to the reflective interface between mineral inclusion and diamond, inclusions can be partially obscured, and the color of the whole inclusion is not visible (Figs. 50 and 52). Along these interfaces between inclusions and diamonds, trapped fluids have recently been identified by Raman spectroscopy. In *lithospheric diamonds*, hydrous silicic fluid films of 1.5 μm maximum thickness were identified between diamond/inclusion interfaces (Nimis et al. 2016). All the principal peridotitic and eclogitic inclusions displayed fluid rims and the fluid composition was the same for both parageneses. A convenient explanation for the presence of this fluid is exsolution from the inclusions during eruptive decompression of the kimberlite or lamproite. Unfortunately, this is not the case, as chromite does not contain OH within its structure, yet still exhibited a fluid film. Thus the presence of these fluids reflects original incorporation of mantle fluids at the time of diamond growth (Nimis et al. 2016). In *sublithospheric diamonds*, fluid films comprised of reduced volatiles such as methane and hydrogen were found surrounding metallic inclusions in CLIPPIR diamonds and silicate inclusions in Type IIb diamonds (Fig. 53; Smith et al. 2016, 2018). In both these studies the fluid films were interpreted as a result of hydrogen escaping the inclusion and accumulating at the inclusion–host interface, with methane forming due to reaction of hydrogen with the surrounding carbon.

Metallic inclusions in newly-recognized CLIPPIR diamonds frequently occur along a <111> direction, as chains of inclusions that developed during brittle-plastic deformation in the mantle (Fig. 54; Smith et al. 2017).

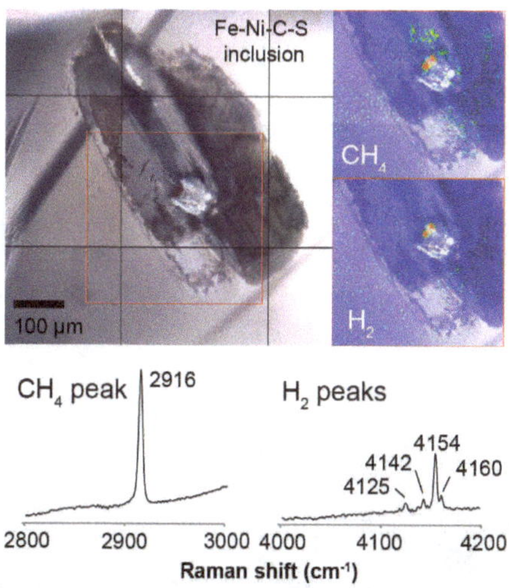

Figure 53. Minerals trapped in sub-lithospheric diamond may also develop internal fractures as a result of the extremely high residual pressure. Reduced fluid films around some sub-lithospheric inclusions have been identified by Raman spectroscopy. [Reproduced from Smith et al. (2016b) Large gem diamonds from metallic liquid in Earth's deep mantle. Science 354:1403–1405. Used with permission from the American Academy for the Advancement of Science.]

Epigenetic inclusions

Internal fractures in diamond arise either as a release of general stress within the stone or occur specifically around some synchronous inclusions, where differential thermal expansion is the cause (Figs. 53–55). Invariably in both cases fractures occur along diamond's {111} cleavage planes. Both of these occurrences often contain secondary dull black graphite of variable micron size. Stress release graphite takes the form of single or multiple disc-like clusters, probably occurring as a consequence of fracture formation, the {111} surfaces being the easiest way graphite can form from diamond (Fig. 56a). When analyzed by Raman spectroscopy, epigenetic graphite often shows highly amorphous/disordered structures (Khokhryakov and Nechaev 2015).

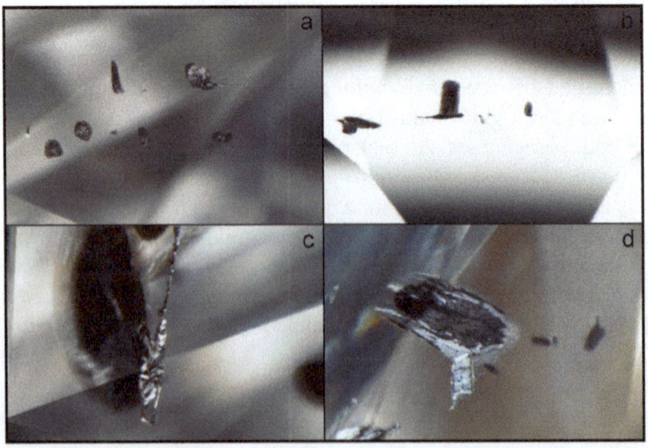

Figure 54. Examples of the typical appearance of metallic Fe-Ni-C-S inclusions in CLIPPIR diamonds. (**a**) Two parallel <111> oriented chains (field of view 2.34 mm). (**b**) Backlit inclusion chain (field of view 1.42 mm). (**c**) Elongate metallic inclusion with a round black fracture of {111} orientation that obliquely cuts the <111> elongation direction of the inclusion (field of view 1.42 mm). (**d**) A less elongate metallic inclusion with a small irregular tail on the lower left (field of view 1.42 mm). [Reproduced from Smith et al. (2017), Gems and Gemology. Used with permission.]

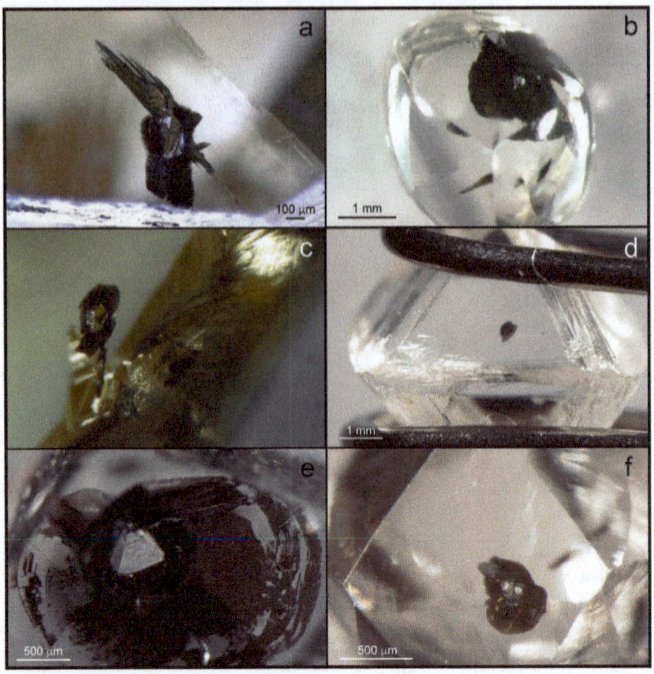

Figure 55. Fractures around sulfide inclusions, arising as a result of differential stress release during kimberlite eruption. Fractures typically develop along diamond's {111} cleavage planes. Size of the fractures is not always proportional to the size of the sulfide. [Images in (a-d) by Karen Smit. Images in (e-f) by Nicole Meyer (University of Alberta).]

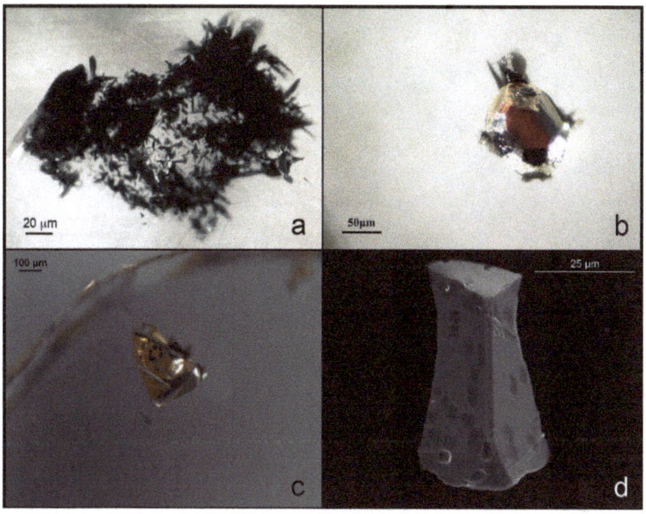

Figure 56. (**a**) A cluster of graphite flakes lying in the four equivalent cleavage planes of a South African diamond. (**b**) A peridotitic garnet inclusion in a diamond from Ghana showing graphite development principally at the inclusion coigns. (**c**) Hexagonal graphite flakes on an orange eclogitic garnet still encapsulated in a diamond. (**d**) Hexagonal graphite flakes on a sulfide inclusion broken out of a diamond. [Images in (a,b) by Jeff Harris. Images in (c,d) by Karen Smit.]

Where visible inclusions are involved, graphite may have an uneven distribution around the mineral itself with noticeable concentrations at inclusion coigns (Fig. 56b). Also if an inclusion face is parallel to an octahedral diamond plane some interfaces are decorated with graphite in the form of irregular flakes, or with hexagonal outlines (Fig. 56c–d).

Vacuum experiments by Harris and Vance (1972) created graphite around inclusions at temperature as low as 900 °C, the authors believing that the graphite reaction was caused by the release of carbon dioxide from the inclusion. With the later discovery of fluid films (Nimis et al. 2016; Smith et al. 2016b, 2018), a more adequate explanation is a reaction between diamond and the fluid. Graphite flakes on inclusion/diamond interfaces may relate to uneven distribution of the fluid film leading to only isolated reactions (Fig. 56b–d).

Sulfides have the greatest differential thermal expansion of all the inclusions and their internal fracture systems are relatively large and have a rosette appearance. The dark material in these fractures has a distinct metallic luster, which distinguishes it from graphite (Fig. 55). The fracture infilling is a consequence of a quantity of sulfide being extruded into the fractures from the central inclusion. This likely occurred at the time of fracturing and is due to sulfide plasticity at high temperatures.

Materials in external fractures have their origins either whilst diamond is still in the rocks that brought them to Earth's surface or within an alluvial environment once weathered out of those rocks. With alluvial diamonds, the more vigorous nature of that environment especially during transport certainly causes some diamonds to fracture and might allow ingress of secondary material. Because pseudomorphs are usually found with typical inclusion outlines (Fig. 57) their formation is most likely linked to an alteration of the inclusion through a surface-connected fracture, probably still in a kimberlitic/lamproitic environment where high temperature and volatiles aid the transformation.

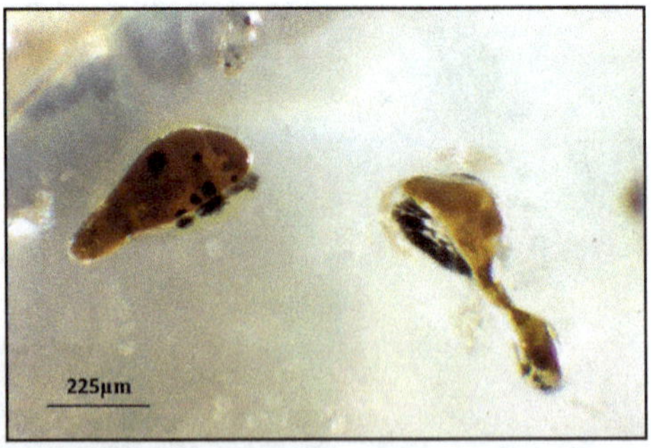

Figure 57. Two inclusions with the typical features of varying shape, as well as graphite on inclusion interfaces and in stress fractures. However, after X-ray analyses, both crystals were identified as powdered goethite, probably pseudomorphs after olivine or enstatite. Inclusions are in a diamond from Ghana. [Image by Jeff Harris.]

FINAL REMARKS

Much credit for our understanding of the morphology and surface features of diamond should go to those early notables, such as Fersmann and Goldschmidt (1911), Sutton (1928) and Williams (1932) who gathered together diamonds and laboriously drew beautifully detailed shape and surface feature characteristics of representative examples. Several early authors noted the presence of rounded surfaces on many diamonds, though it wasn't until the 1970s that it was accepted that these rounded surfaces were not growth features, but rather the result of dissolution (Moore and Lang 1972).

More recent works on diamond morphology have focused on understanding the formation conditions for the wide variety of surface features (e.g., Robinson 1979; Robinson et al. 1986). Experimental work covering a range of pressures, temperatures and fluid compositions has shown the specific formation conditions for features expected to form either in the diamond stability field, or during kimberlite and lamproite eruption (Khokhryakov and Pal'yanov 2007, 2010; Fedortchouk et al. 2019).

The literature on diamond morphology, internal growth, surface features and the shapes of inclusions is extensive. Tappert and Tappert (2011) have provided an excellent starting point, with an overview of the main morphological features of natural diamonds accompanied by superlative images. For those wishing to further consider the morphology of diamond and its mineral inclusions, the Proceedings Volumes and Extended Abstracts of the 11 International Kimberlite Conferences (1973–2018) will be very useful. The *Journal of the International Kimberlite Conferences* can be found at *https://journals.library.ualberta.ca/ikcabstracts.com.*

DIRECTIONS FOR FUTURE WORK

External and Internal Morphology. One area which might still be explored is the relationship between micro- and macro-diamonds to clarify whether the micros are simply smaller equivalents of the macro-populations, or whether they represent separate distinct growth events. This distinction may seem trivial but it has important implications during the economic evaluation of a kimberlite, and whether the characteristics of microdiamond populations can be extrapolated to larger sizes. This type of study likely requires a combined approach using CL to investigate internal growth features, inclusion studies, and geochemical studies of the diamond (carbon isotopes, nitrogen isotopes, trace elements).

We have shown the complexity of the internal growth of some diamonds. There are still many internal growth features that cannot be explained, and will only be resolved with further work. Up to now experimental work has largely concentrated on how an already formed diamond can change shape with respect to varying conditions. Perhaps further experiments can investigate how CL patterns develop in synthetic diamonds grown under a range of growth rates, pressure and temperature conditions and in different geological fluids.

Surface features. In almost any diamond parcel there are diamond crystals with unusual surface features, whose origins are still difficult to interpret. In the last decades, experimental studies were able to produce many dissolution features present on natural diamonds and to determine the controlling factors for their development. Some diamond surface features show a strong association with the host kimberlite lithology and with the behavior of volatiles during kimberlite ascent and emplacement. Using the information stored in the surface dissolution features of diamonds, further studies may help to constrain the formation of the different types of kimberlites and the mechanism of magma emplacement. Integrating studies of cathodoluminescence growth patterns of diamonds with dissolution surface features produced during mantle residence, should help to better understand the nature of diamond-growth and diamond-destructive metasomatic events in the sub-cratonic mantle. The study of fluids resulting in surface features on a wide variety of different diamonds may provide a broader understanding of carbon fluids in the deep Earth.

Inclusions. One of the main outstanding issues around mineral inclusion morphology is whether an inclusion was pre-existing in the lithosphere prior to diamond formation or whether it crystallized when the diamond source fluid was introduced into the lithosphere. Regardless, inclusions always have cubo-octahedral morphologies because they co-crystallize with the diamond. Though this may sound trivial, the distinction has implications for geothermobarometry (which depends on major and trace element equilibrium between different mineral inclusions) and geochronology (whether there was isotopic equilibrium between minerals). These issues are being addressed with the help of crystallographic work on mineral inclusions and their host diamonds (Angel et al. 2022, this volume). The implications for geothermobarometry and geochronology are discussed in Nimis (2022, this volume) and Smit et al. (2022, this volume).

Minerals found trapped in diamonds are the only pristine samples we have of the deep Earth, and there is considerable research emphasis on using these valuable mineral inclusions to understand mantle processes in the lithosphere and sub-lithospheric regions like the upper mantle, transition zone and lower mantle. Although the inclusion population in diamond is considerable and diverse, new minerals are continually discovered. All one needs is a supply of diamonds, a (Raman) microscope and a great deal of patience.

ACKNOWLEDGEMENTS

We thank Ingrid Chinn, Tom McCandless and the editor Graham Pearson, for their constructive reviews that improved the quality of our manuscript. Thank you to Wuyi Wang and Tom Moses for their permission to use unpublished GIA images.

REFERENCES

Agrosi G, Nestola F, Tempesta G, Bruno M, Scandale E, Harris J (2016) X-ray topographic study of a diamond from Udachnaya; implications for the genetic nature of inclusions. Lithos 248–251:153–159

Akagi T, Masuda A (1988) Isotopic and elemental evidence for a relationship between kimberlite and Zaire cubic diamonds. Nature 336:665–667

Angel RJ, Alvaro M, Nestola F (2022) Crystallographic methods for non-destructive characterization of mineral inclusions in diamonds. Rev Mineral Geochem 88:257–306

Arima M, Kozai Y (2008) Diamond dissolution rates in kimberlitic melts at 1300–1500 °C in the graphite stability field. Eur J Mineral 20:357–364

Aulbach S, Stachel T, Creaser RA, Heaman LM, Shirey SB, Muehlenbachs K, Eichenberg D, Harris JW (2009) Sulphide survival and diamond genesis during formation and evolution of Archaean subcontinental lithosphere; a comparison between the Slave and Kaapvaal Cratons. Lithos 112:747–757

Aulbach S, Creaser RA, Stachel T, Heaman LM, Chinn IL, Kong J (2018) Diamond ages from Victor (Superior Craton); intra-mantle cycling of volatiles (C, N, S) during supercontinent reorganisation. Earth Planet Sci Lett 490:77–87

Bovenkerk HP (1961) Some observations on the morphology and physical characteristics of synthetic diamond. Am Mineral 46:952–963

Boyd SR, Mattey DP, Pillinger CT, Milledge HJ, Mendelssohn M, Seal M (1987) Multiple growth events during diamond genesis; an integrated study of carbon and nitrogen isotopes and nitrogen aggregation state in coated stones. Earth Planet Sci Lett 86:341–353

Bragg WH, Bragg WL (1913) The structure of the diamond. Nature 91:557–557

Bragg WH, Bragg WL (1914) The structure of the diamond. Proc R Soc London A 89:277–291

Bulanova GP (1995) The formation of diamond. *In:* Diamond Exploration into the 21st Century. Vol 53; 1–3. Griffin WL, (ed) Elsevier, Amsterdam-New York, International, p 1–23

Bulanova GP, Griffin WL, Kaminsky FV, Davies RM, Spetsius ZV, Ryan CG, Andrew AS, Zahkarchenko OD (1999) Diamonds from Zarnitsa and Dalnaya kimberlites (Yakutia), their nature and lithospheric mantle source. *In:* Proceedings of the VIIth international kimberlite conference; Vol 1. Gurney JJ, Gurney JL, Pascoe MD, Richardson SH, (eds). Red Roof Design, Cape Town, South Africa, p 49–56

Bulanova GP, Pearson DG, Hauri EH, Griffin BJ (2002) Carbon and nitrogen isotope systematics within a sector-growth diamond from the Mir Kimberlite, Yakutia. Chem Geol 188:105–123

Bulanova GP, Wiggers de Vries DF, Pearson DG, Beard A, Mikhail S, Smelov AP, Davies GR (2014) An eclogitic diamond from Mir Pipe (Yakutia), recording two growth events from different isotopic sources. Chem Geol 381:40–54

Burnham AD, Bulanova GP, Smith CB, Whitehead SC, Kohn SC, Gobbo L, Walter MJ, Shirey SB (2016) Diamonds from the Machado River alluvial deposit, Rondonia, Brazil, derived from both lithospheric and sublithospheric mantle. Lithos 265:199–213

Chepurov AI, Khokhryakov AF, Sonin VM, Pal yYN (1985) Forms of dissolution of diamond crystals in high-temperature silicate solutions. Doklady Akademii Nauk SSSR 285:212–216 (in Russian)

Clement CR, Harris JW, Robinson DN, Hawthorne JB (1986) The De Beers kimberlite pipe; a historic South African diamond mine. *In:* Mineral deposits of Southern Africa. Anhaeusser CR, Maske S (eds). Geol Soc S Afr, Johannesburg, South Africa, p 2193–2214

Collins AT (1982) A spectroscopic survey of naturally-occurring vacancy-related colour centres in diamond. J Phys D: App Phys 15:1431–1438

Collins AT, Kiflawi I (2009) The annealing of radiation damage in type Ia diamond. J Phys Condens Matter 21:364209

Collins AT, Kanda H, Kitawaki H (2000) Colour changes produced in natural brown diamonds by high-pressure, high-temperature treatment. Diam Relat Mater 9:113–122

Davies G (1972) The effect of nitrogen impurity on the annealing of radiation damage in diamond. J Physics C: Solid State Phys 5:2534–2542

Davies RM, Griffin WL, O'Reilly SY, Doyle BJ, Mitchell RH, Gruetter HS, Heaman LM, Scott Smith BH, Stachel T (2004) Mineral inclusions and geochemical characteristics of microdiamonds from the DO27, A154, A21, A418, DO18, DD17 and Ranch Lake kimberlites at Lac de Gras, Slave Craton, Canada. Lithos 77:39–55

DeVries RC (1975) Plastic deformation and "work-hardening" of diamond. Mater Res Bull 10:1193–1199

Evans T, Sauter DH (1961) Etching of diamond surfaces with gases. Philos Mag 6:429–440

Evans T, Wild RK (1965) Plastic bending of diamond plates. Philos Mag 12:479–489

Fedortchouk Y (2015) Diamond resorption features as a new method for examining conditions of kimberlite emplacement. Contrib Mineral Petrol 170:36

Fedortchouk Y (2019) A new approach to understanding diamond surface features based on a review of experimental and natural diamond studies. Earth Sci Rev 193:45–65

Fedortchouk Y, Zhang Z (2011) Diamond resorption; link to metasomatic events in the mantle or record of magmatic fluid in kimberlitic magma? Can Mineral 49:707–719

Fedortchouk Y, Canil D, Semenets E (2007) Mechanisms of diamond oxidation and their bearing on the fluid composition in kimberlite magmas. Am Mineral 92:1200–1212

Fedortchouk Y, Canil D, Comodi P, Nestola F, Angel RJ (2009) Diamond oxidation at atmospheric pressure; development of surface features and the effect of oxygen fugacity. Eur J Mineral 21:623–635

Fedortchouk Y, Matveev S, Carlson JA (2010) H_2O and CO_2 in kimberlitic fluid as recorded by diamonds and olivines in several Ekati diamond mine kimberlites, Northwest Territories, Canada. Earth Planet Sci Lett 289:549–559

Fedortchouk Y, Chinn IL, Kopylova MG (2017) Three styles of diamond resorption in a single kimberlite; effects of volcanic degassing and assimilation. Geology (Boulder) 45:871–874

Fedortchouk Y, Liebske C, McCammon C (2019) Diamond destruction and growth during mantle metasomatism; an experimental study of diamond resorption features. Earth Planet Sci Lett 506:493–506

Fersman AE (1955) Crystallography of Diamond (Classics of Science). USSR Acad Sci, Moscow (in Russian)

Fersmann A, Goldschmidt V (1911) Der Diamant: Eine Studie. Carl Winter's Universitatbuchhandlung, Heidelberg

Fisher D, Sibley S, Kelly C (2009) Brown colour in natural diamond and interaction between the brown related and other colour-inducing defects. J Phys Condens Matter 21:364213

Frank FC, Lang AR (1965) X-ray topography of diamond. *In*: Physical Properties of Diamond. Berman R, (ed) Clarendon Press, Oxford, England, p 69–115; especially 86–102

Frank FC, Puttick KE, Wilks EM (1958) Etch pits and trigons on diamond: I. Philos Mag 3:1262–1272

Gaillou E, Post J, Bassim N, Zaitsev A, Rose T, Fries M, Stroud R, Steele A, Butler J (2010) Spectroscopic and microscopic characterizations of color lamellae in natural pink diamonds. Diam Relat Mater 19:1207–1220

Gaillou E, Post JE, Rose T, Butler JE (2012) Cathodoluminescence of natural, plastically deformed pink diamonds. Micros Microanal 18:1–11

Gittus JH (1963) Uranium. Metallurgy of the Rare Metals Vol 8. Butterworths, London

Glinnemann J, Kusaka K, Harris JW (2003) Oriented graphite single-crystal inclusions in diamond. Z Kristallog Cryst Mater 218:733–739

Graham RJ, Moustakas TD, Disko MM (1991) Cathodoluminescence imaging of defects and impurities in diamond films grown by chemical vapor deposition. J Appl Phys 69:3212–3218

Grantham DR, Allen JB (1960) Kimberlite in Sierra Leone. *In:* Overseas Geology and Mineral Resources. Quarterly Bulletin of the Overseas Geological Surveys. EH Beard (ed) Vol. 8, p 5–25

Green BL, Collins AT, Breeding CM (2022) Diamond spectroscopy, defect centers, color, and treatments. Rev Mineral Geochem 88:637–688

Gurney JJ, Hildebrand PR, Carlson JA, Fedortchouk Y, Dyck DR (2004) The morphological characteristics of diamonds from the Ekati property, Northwest Territories, Canada. Lithos 77:21–38

Gurney JJ, Helmstaedt HH, Le Roex AP, Nowicki TE, Richardson SH, Westerlund KJ, Hedenquist JW, Thompson JFH, Goldfarb RJ, Richards JP (2005) Diamonds; crustal distribution and formation processes in time and space and an integrated deposit model. Econ Geol 100:143–177

Hall AE, Smith CB (1984) Lamproite diamonds; are they different? *In:* Kimberlite occurrence and origin; a basis for conceptual models in exploration. Vol 8. Glover JE, Harris PG, (eds). University of Western Australia, Geology Department and Extension Service. Perth, Australia, p 167–212

Hanley PL, Kiflawi I, Lang AR (1977) On topographically identifiable sources of cathodoluminescence in natural diamonds. Phil Trans R Soc London Ser A 284:329–368

Harlow GE (1998) The Nature of Diamonds. Cambridge University Press, Cambridge, United Kingdom

Harris JW (1992) Diamond geology. *In:* The Properties of Natural and Synthetic Diamond. Field JE (ed) Academic Press, London, p 345–394

Harris JW, Vance ER (1972) Induced graphitisation around crystalline inclusions in diamond. Contrib Mineral Petrol 35:227–234

Harris JW, Hawthorne JB, Oosterveld M, Wehmeyer E (1975) A classification scheme for diamond and a comparative study of South African diamond characteristics. Phys Chem of the Earth 9:765–783

Harris JW, Hawthorne JB, Oosterveld MM (1979) Regional and local variations in the characteristics of diamonds from some southern African kimberlites. *In:* Kimberlites, diatremes, and diamonds; their geology, petrology, and geochemistry. Vol 1. Boyd FR, Meyer HOA, (eds). Am Geophys Union, Washington, United States, p 27–41

Harris JW, Hawthorne JB, Oosterveld MM (1983) A comparison of diamond characteristics from the De Beers Pool mines, Kimberley, South Africa. *In:* Kimberlites III; Documents. Vol 74. Kornprobst J, (ed) Ann Sci Univ Clermont II, Clermont-Ferrand, France, p 1–13

Harris JW, Hawthorne JB, Oosterveld MM (1986) A comparison of characteristics of diamonds from the Orapa and Jwaneng kimberlite pipes in Botswana. *In:* Vol 16 Extended Abstracts, IVth International Kimberlite Conference. Smith CB, (ed) Geol Soc Australia, Sydney, Australia, p 395–397

Harrison ER, Tolansky S (1964) Growth history of a natural octahedral diamond. Proc R Soc London A 279:490–496

Harte B, Hudson N (2013) Mineral associations in diamonds from the lowermost upper mantle and uppermost lower mantle. Proc 10th Int Kimberlite Confer, p. 235–253

Harte B, Harris JW, Hutchison MT, Watt GR, Wilding MC (1999) Lower mantle mineral associations in diamonds from Sao Luiz, Brazil. *In:* Mantle petrology; field observations and high-pressure experimentation; a tribute to Francis R (Joe) Boyd. Vol 6. Fei Y, Bertka Constance M, Mysen Bjorn O, (eds). Geochemical Society - University of Houston, Department of Chemistry, Houston, TX, United States, p 125–153

Howell D, Griffin WL, Piazolo S, Say JM, Stern RA, Stachel T, Nasdala L, Rabeau JR, Pearson NJ, O'Reilly SY (2013) A spectroscopic and carbon-isotope study of mixed-habit diamonds; impurity characteristics and growth environment. Am Mineral 98:66–77

Howell D, Stern RA, Griffin WL, Southworth R, Mikhail S, Stachel T (2015) Nitrogen isotope systematics and origins of mixed habit diamonds. Geochim Cosmochim Acta 157:1–12

Jacob J, Grobbelaar G (2019) Onshore and nearshore diamond mining on the south-western coast of Namibia: Recent activities and future exploration techniques. J Gemm 36:524–532
Jacob DE, Mikhail S (2022) Polycrystalline diamonds from kimberlites: Snapshots of rapid and episodic diamond formation in the lithospheric mantle. Rev Mineral Geochem 88:167–190
Kaminsky F (2012) Mineralogy of the lower mantle; a review of "super deep" mineral inclusions in diamond. Earth-Sci Rev 110:127–147
Kaminsky FV, Khachatryan GK (2004) The relationship between the distribution of nitrogen impurity centres in diamond crystals and their internal structure and mechanism of growth. Lithos 77:255–271
Kaminsky FV, Zakharchenko OD, Griffin WL, Channer DMD, Khachatryan BGK (2000) Diamond from the Guaniamo area, Venezuela. Can Mineral 38:1347–1370
Kamiya Y, Lang AR (1965) On the structure of coated diamonds. Philos Mag 11:347–356
Karfunkel J, Chaves MLSC, Svisero DP, Meyer HOA (1995) Diamonds from Minas Gerais, Brazil; an update on sources, origin, and production. Int Geol Rev 36:1019–1032
Khokhryakov AF, Nechaev DV (2015) Typomorphic features of graphite inclusions in diamond: experimental data. Russ Geol Geophys 56:232–238
Khokhryakov AF, Pal'yanov YN (2007) The evolution of diamond morphology in the process of dissolution; experimental data. Am Mineral 92:909–917
Khokhryakov AF, Pal'yanov YN (2009) Origin of drop-shaped hillocks on tetrahexahedroids of natural diamonds. Proceedings of the Annual Session of Russian Mineralogical Society "Ontogeny of minerals applied to scientific and industrial uses (to the 100 year anniversary of Prof. Dmitry Grigoriev)", St. Petersburg, 153–155 (in Russian)
Khokhryakov AF, Pal'yanov YN (2010) Influence of the fluid composition on diamond dissolution forms in carbonate melts. Am Mineral 95:1508–1514
Kiflawi I, Collins A, Lakoubovskii K, Fisher D (2007) Electron irradiation and the formation of vacancy–interstitial pairs in diamond. J Phys Condens Matter 19:046216
Kjarsgaard BA, de Wit M, Heaman LM, Pearson DG, Stiefenhofer J, Janusczcak N, Shirey SB (2022) A review of the geology of global diamond mines and deposits. Rev Mineral Geochem 88:1–118
Kozai Y, Arima M (2005) Experimental study on diamond dissolution in kimberlitic and lamproitic melts at 1300–1420 °C and 1 GPa with controlled oxygen partial pressure. Am Mineral 90:1759–1766
Lang AR (1964) Dislocations in diamond and the origin of trigons. Proc R Soc London A 278:234–242
Lang AR (1974a) On the growth-sectorial dependence of defects in natural diamonds. Proc R Soc London A 340:233–248
Lang AR (1974b) Glimpses into the growth history of natural diamonds. J Cryst Growth 24/25:108–115
Lang AR (1979) Internal structure. *In:*The properties of diamond. Field JE, (ed) Acad. Press, London, United Kingdom, p 425–469
Lang AR, Bulanova GP, Fisher D, Furkert S, Sarua A (2007) Defects in a mixed-habit Yakutian diamond: Studies by optical and cathodoluminescence microscopy, infrared absorption, Raman scattering and photoluminescence spectroscopy. J Cryst Growth 309:170–180
Lawson S, Davies G, Collins AT, Mainwood A (1992). Migration energy of the neutral vacancy in diamond. J Phys Condens Matter 4:L125–L131
Li ZY, Fedortchouk Y, Fulop A, Chinn IL, Forbes N (2018) Positively oriented trigons on diamonds from the Snap Lake kimberlite dike, Canada: Implications for fluids and kimberlite cooling rates. Am Mineral 103:1634–1648
Lu T, Shigley J, Koivula J, Reinitz I (2001) Observation of etch channels in several natural diamonds. Diam Relat Mater 10:68–75
Lüttge A (2006) Crystal dissolution kinetics and Gibbs free energy. J Electron Spectrosc 150:248–259
Machado WG, Moore M, Woods GS (1985) On the dodecahedral growth of coated diamonds. J Cryst Growth 71:718–727
Machado WG, Moore M, Yacoot A (1998) Twinning in natural diamond. II. interpenetrant cubes. J Appl Crystallogr 31:777–782
Mainwood A (1994) Nitrogen and nitrogen-vacancy complexes and their formation in diamond. Phys Rev B 49:7934–7940
Mineeva R, Speransky A, Titkov S, Zudin N (2007) The ordered creation of paramagnetic defects at plastic deformation of natural diamonds. Phys Chem Minerals 34:53–58
Mitchell RH (1986) Kimberlites: Mineralogy, Geochemistry, and Petrology. Plenum Press, New York
Moore M (1985) Diamond morphology. Industrial Diamond Rev 45:67–71
Moore M (2009) Imaging diamond with X-rays. J Phys Condens Matter 21:364217
Moore M, Lang AR (1972) On the internal structure of natural diamonds of cubic habit. Philos Mag 26:1313–1325
Moore M, Lang AR (1974) On the origin of the rounded dodecahedral habit of natural diamond. J Cryst Growth 26:133–139
Moses TM, Shigley JE, McClure S, Koivula J, Daele M (1999) Observations on GE-processed diamonds: A photographic record. Gem Gemol 35:14–22
Nailer SG, Moore M, Chapman J, Kowalski G (2007) On the role of nitrogen in stiffening the diamond structure. J Appl Crystallogr 40:1146–1152
Nasdala L, Grambole D, Wildner M, Gigler AM, Hainschwang T, Zaitsev AM, Harris JW, Milledge J, Schulze DJ, Hofmeister W, Balmer WA(2013) Radio colouration of diamond; a spectroscopic study. Contrib Mineral Petrol 165:843–861

Navon O, Hutcheon ID, Rossman GR, Wasserburg GJ (1988) Mantle-derived fluids in diamond micro-inclusions. Nature (London) 335:784–789

Nestola F, Nimis P, Angel RJ, Milani S, Bruno M, Prencipe M, Harris JW (2014) Olivine with diamond-imposed morphology included in diamonds; syngenesis or protogenesis? Int Geol Rev 56:1658–1667

Nimis P (2022) Pressure and temperature data for diamonds. Rev Mineral Geochem 88:533–566

Nimis P, Alvaro M, Nestola F, Angel RJ, Marquardt K, Rustioni G, Harris JW, Marone F (2016) First evidence of hydrous silicic fluid films around solid inclusions in gem quality diamonds. Lithos 260:384–389

Orlov YL (1977) The mineralogy of the diamond. John Wiley and Sons, New York N.Y., United States

Palot M, Pearson DG, Stachel T, Stern RA, Le Pioufle A, Gurney JJ, Harris JW (2017) The transition zone as a host for recycled volatiles: Evidence from nitrogen and carbon isotopes in ultra-deep diamonds from Monastery and Jagersfontein (South Africa). Chem Geol 466:733–749

Pandeya D, Tolansky S (1961) Micro-disk patterns on diamond dodecahedra. Proc Phys Soc 78:12–16

Pearson DG, Brenker FE, Nestola F, McNeill J, Nasdala L, Hutchison MT, Matveev S, Mather K, Silversmit G, Schmitz S, Vekemans B (2014) Hydrous mantle transition zone indicated by ringwoodite included within diamond. Nature 507:221–224

Petts DC, Stachel T, Stern RA, Hunt L, Fomradas G (2016) Multiple carbon and nitrogen sources associated with the parental mantle fluids of fibrous diamonds from Diavik, Canada, revealed by SIMS microanalysis. Contrib Mineral Petrol 171:17

Phillips D, Harris JW, Viljoen KS (2004) Mineral chemistry and thermobarometry of inclusions from De Beers Pool diamonds, Kimberley, South Africa. Lithos 77:155–179

Raal FA, Robinson DN (1980) Green for rarity. Nuclear Active 23:5–8

Rakovan J, Gaillou E, Post JE, Jaszczak JA, Betts JH (2014) Optically sector-zoned (star) diamonds from Zimbabwe, Rocks & Minerals 89:2:173–178

Richardson SH, Shirey SB, Harris JW (2004) Episodic diamond genesis at Jwaneng, Botswana, and implications for Kaapvaal Craton evolution. Lithos 77:143–154

Robinson DN (1979) Surface textures and other features of diamonds. PhD Dissertation, University of Cape Town

Robinson DN, Scott JA, Van NA, Anderson VG (1986) Events reflected in the diamonds of some Southern African kimberlites. *In:* Vol 16 Extended Abstracts, IVth International Kimberlite Conference. Smith CB, (ed) Geological Society of Australia, Sydney, Australia: 421–423

Robinson DN, Scott JA, Van NA, Anderson VG (1989) The sequence of events reflected in the diamonds of some Southern African kimberlites. *In:* Kimberlites and related rocks. Vol 14; 2. Ross J, Jaques AL, Ferguson J, Green DH, Y ORS, Danchin RV, Janse AJA (eds) Geological Society of Australia, Sydney, Australia, p 990–1000

Rondeau B, Fritsch E, Moore M, Thomassot E, Sirakian JF (2007) On the growth of natural octahedral diamond upon a fibrous core. J Cryst Growth 304:287–293

Schulze DJ, Nasdala L, Shirey SB (2016) Unusual paired pattern of radiohaloes on a diamond crystal from Guaniamo (Venezuela). Lithos 265:177–181

Seager AF (1979) The origin of a tetrahedral diamond. Mineral Mag 43:377–387

Seal M (1965) Structure in diamonds as revealed by etching. Am Mineral 50:105–123

Shirey SB, Richardson SH (2011) Start of the Wilson cycle at 3 Ga shown by diamonds from subcontinental mantle. Science 333:434–436

Shor R, Weldon R, Janse AJA, Breeding C, Shirey S (2015) Letšeng's unique diamond proposition. Gem Gemol 51:280–299

Smit KV, Shirey SB, Wang W (2016a) Type Ib diamond formation and preservation in the West African lithospheric mantle; Re/Os age constraints from sulphide inclusions in Zimmi diamonds. Precambrian Res 286:152–166

Smit KV, Shirey SB, Stern RA, Steele A, Wang W, Shirey SB (2016b) Diamond growth from C–H–N–O recycled fluids in the lithosphere; evidence from CH_4 micro-inclusions and $\delta^{13}C$–$\delta^{15}N$–N content in Marange mixed-habit diamonds. Lithos 265:68–81

Smit KV, Myagkaya E, Persaud S, Wang W (2018a) Black diamonds from Marange (Zimbabwe): a result of natural irradiation and graphite inclusions. Gem Gemol 11:177

Smit KV, D'Haenens-Johansson UFS, Howell D, Loudin LC, Wang W (2018b) Deformation-related spectroscopic features in natural Type Ib-IaA diamonds from Zimmi (West African craton). Miner Petrol 112:243–257

Smit KV, Stachel T, Luth RW, Stern RA (2019a) Evaluating mechanisms for eclogitic diamond growth; an example from Zimmi Neoproterozoic diamonds (West African Craton). Chem Geol 520:21–32

Smit KV, Shirey SB, Hauri EH, Stern RA (2019b) Sulfur isotopes in diamonds reveal differences in continent construction. Science 364:383

Smit KV, Timmerman S, Aulbach S, Shirey SB, Richardson SH, Phillips D, Pearson DG (2022) Geochronology of diamonds. Rev Mineral Geochem 88:567–636

Smith CB, Walter MJ, Bulanova GP, Mikhail S, Burnham AD, Gobbo L, Kohn C (2016a) Diamonds from Dachine, French Guiana: A unique record of early Proterozoic subduction. Lithos 265:82–95

Smith EM, Shirey SB, Nestola F, Bullock ES, Wang J, Richardson SH, Wang W (2016b) Large gem diamonds from metallic liquid in Earth's deep mantle. Science 354:1403–1405

Smith EM, Shirey SB, Wang W (2017) The very deep origin of the world's biggest diamonds. Gem Gemol 53:388–403

Smith EM, Shirey SB, Richardson SH, Nestola F, Bullock ES, Wang J, Wang W (2018) Blue boron-bearing diamonds from Earth's lower mantle. Nature 560:84–87
Spetsius Z (1998) Two generations of diamonds in the eclogite xenoliths. 7th Int Kimberlite Conf Abstr, p 844–846
Stachel T, Harris JW (2008) The origin of cratonic diamonds; constraints from mineral inclusions. Ore Geol Rev 34:5–32
Stachel T, Luth RW (2015) Diamond formation; where, when and how? Lithos 220–223:200–220
Stachel T, Brey GP, Harris JW (2005) Inclusions in sublithospheric diamonds; glimpses of deep earth. Elements 1:73–87
Stachel T, Banas A, Muehlenbachs K, Kurszlaukis S, Walker EC (2006) Archean diamonds from Wawa (Canada); samples from deep cratonic roots predating cratonization of the Superior Province. Contrib Mineral Petrol 151:737–750
Stachel T, Aulbach S, Harris JW (2022) Mineral inclusions in lithospheric diamonds. Rev Mineral Geochem 88:307–392
Sumida N, Lang AR (1981) Cathodoluminescence evidence of dislocation interactions in diamond. Philos Mag A 43:1277–1287
Sunagawa I (1995) The distinction of natural from synthetic diamonds. J Gemmol 24:485–499
Sutton JR (1928) Diamond: A Descriptive Treatise. T. Murby
Suzuki S, Lang AR (1976) Internal structures of natural diamonds revealing mixed-habit growth *In:* Diamond Research. Daniel P (ed) Industrial Diamond Information Bureau, Ascot, England, p 39–47
Tacy (2018) 2018 Tacy's Diamond Pipeline:
https://www.thediamondloupe.com/sites/awdcnewswall/files/attachments/TacyPipeline_TheDiamondLoupe_bis.pdf
Tappert R, Tappert MC (2011) Diamonds in Nature: A Guide to Rough Diamonds. Springer-Verlag, Berlin
Taylor WR, Canil D, Milledge HJ (1996) Kinetics of Ib to IaA nitrogen aggregation in diamond. Geochim Cosmochim Acta 60:4725–4733
Timmerman S, Chinn IL, Fisher D, Davies GR (2018) Formation of unusual yellow Orapa diamonds. Miner Petrol 112:209–218
Titkov SV, Krivovichev SV, Organova NI, Mills S (2012) Plastic deformation of natural diamonds by twinning; evidence from X-ray diffraction studies. Mineral Mag 76:143–149
Vance ER, Milledge HJ (1972) Natural and laboratory alpha-particle irradiation of diamond. Mineral Mag 38:878–881
Vance ER, Harris JW, Milledge HJ (1973) Possible origins of alpha-damage in diamonds from kimberlite and alluvial sources. Mineral Mag 39:349–360
Walter MJ, Thomson AR, Smith EM (2022) Geochemistry of silicate and oxide inclusions in sublithospheric diamonds. Rev Mineral Geochem 88:393–450
Welbourn CM, Rooney MLT, Evans DJF (1989) A study of diamonds of cube and cube-related shape from the Jwaneng Mine. J Cryst Growth 94:229–252
Welbourn CM, Cooper M, Spear PM (1996) De Beers natural versus synthetic diamond verification instruments. Gem Gemol 32:156–169
Westerlund KJ, Gurney JJ, Carlson RW, Shirey SB, Hauri EH, Richardson SH (2004) A metasomatic origin for late Archean eclogitic diamonds; implications from internal morphology of diamonds and Re–Os and S isotope characteristics of their sulfide inclusions from the Late Jurassic Klipspringer kimberlites. S Afr J Geol 107:119–130
Wiggers de Vries DF, Bulanova GP, de Corte K, Pearson DG, Craven JA, Davies GR (2013) Micron-scale coupled carbon isotope and nitrogen abundance variations in diamonds; evidence for episodic diamond formation beneath the Siberian Craton. Geochim Cosmochim Acta 100:176–199
Williams AF (1932) The Genesis of Diamond. Two Volumes. Ernest Benn Ltd., London
Wulff G (1901) On the question of speed of growth and dissolution of crystal surfaces. Z Kristallogr 34:449–530 (in German)
Yacoot A, Moore M (1993) X-ray topography of natural tetrahedral diamonds. Mineral Mag 57:223–230
Yacoot A, Moore M, Machado WG (1998) Twinning in natural diamond. I. Contact Twins. J Appl Crystallogr 31:767–776
Yamaoka S, Komatsu H, Kanda H, Setaka N (1977) Growth of diamond with rhombic dodecahedral faces. J Cryst Growth 37:349–352
Yamaoka S, Kanda H, Setaka N (1980) Etching of diamond octahedrons at high temperatures and pressure with controlled oxygen partial pressure. J Mater Sci 15:332–336
Yelisseyev AP, Pokhilenko NP, Steeds JW, Zedgenizov DA, Afanasiev VP (2004) Features of coated diamonds from the Snap Lake/King Lake kimberlite dyke, Slave Craton, Canada, as revealed by optical topography. Lithos 77:83–97
Zedgenizov DA, Harte B, Shatsky VS, Politov AA, Rylov GM, Sobolev NV (2006) Directional chemical variations in diamonds showing octahedral following cuboid growth. Contrib Mineral Petrol 151:45–57
Zedgenizov DA, Kalinina VV, Reutsky VN, Yuryeva OP, Rakhmanova MI, Shirey SB (2016) Regular cuboid diamonds from placers on the northeastern Siberian Platform. Lithos 265:125–137
Zhan Z (2020) Mechanisms and implications of deep earthquakes. Annu Rev Earth Planet Sci 48:147–174
Zhang Z (2016) Diamond resorption morphology as a fluid proxy in diamond-bearing environments: constraints from empirical and experimental studies. PhD Dissertation, Dalhousie University, Halifax, Canada
Zhang Z, Fedortchouk Y, Hanley JJ (2015) Evolution of diamond resorption in a silicic aqueous fluid at 1–3 GPa; application to kimberlite emplacement and mantle metasomatism. Lithos 227:179–193

Reviews in Mineralogy & Geochemistry
Vol. 88 pp. 167-190, 2022

Polycrystalline Diamonds from Kimberlites: Snapshots of Rapid and Episodic Diamond Formation in the Lithospheric Mantle

Dorrit E. Jacob*

Research School of Earth Sciences
The Australian National University
Canberra, Australia

dorrit.jacob@anu.edu.au

Sami Mikhail*

The School of Earth and Environmental Sciences
University of St. Andrews
United Kingdom

sm342@st-andrews.ac.uk

**both authors contributed equally to this chapter*

INTRODUCTION

Monocrystalline diamonds are the most valuable diamond type, economically. However, there are other varieties of diamond forged in Earth's lithospheric mantle, which, while not economically profitable, are of considerable value to the geosciences. Most prominent amongst these are fibrous diamonds (Weiss et al. 2022, this volume) and polycrystalline diamond aggregates (PDAs). Polycrystalline diamond aggregates are rocks in which the dominant mineral phase is diamond (Fig. 1), whereas fibrous diamonds are cuboid samples, sometimes with monocrystalline diamond cores and cloudy overgrowths ('coats'), or octahedral diamonds with fibrous cores (Weiss et al. 2022, this volume). The fibrous growth sectors are highly imperfect single crystals hosting millions of fluid and solid micro-inclusions (Navon et al. 1988). Polycrystalline diamond aggregates (PDAs) from kimberlites are the least well-studied of the diamond family. This chapter aims to showcase what we know of PDA-formation in the context of monocrystalline diamond formation, the origin of carbon enrichment in the cratonic lithosphere, and the identify the relationship(s) between polycrystalline diamond formation and plate tectonics.

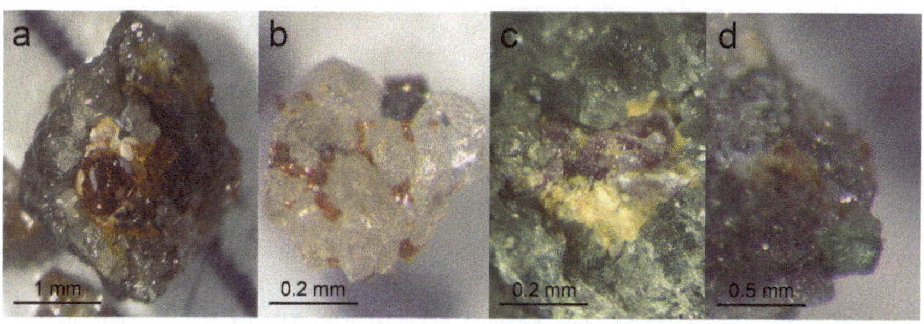

Figure 1. Images of silicate-bearing polycrystalline diamond aggregates from Orapa (**a–d**, from Mikhail et al. 2019b).

1529-6466/22/0088-0003$05.00 (print)
1943-2666/22/0088-0003$05.00 (online)
http://dx.doi.org/10.2138/rmg.2022.88.03

NOMENCLATURE

Nothing better serves to illuminate the blurred line between academia and industry in diamond science than the diversity of, and etymology behind, the nomenclature applied to polycrystalline diamond. The most commonly used names are framesite, stewartite, diamondite, boart, and carbonado (Table 1). The names used to sub-divide the group originate from the lexica of both industry and academia and are based on appearance (e.g., grain size), application, value, and in some cases, physical properties. For instance, if a polycrystalline diamond aggregate contains unidentified or identified magnetic phases (presumably magnetite) then it can be termed stewartite, but if the sample is non-magnetic then the name framesite can be applied (Heaney et al. 2005). These names derive from two persons named Stewart and Frames, former mine managers at the Cullinan Mine (Andy Moore, pers. comm. 2021). Both, stewartite and framesite can also be referred to as boart, which means the sample is not a single crystal and is not of gem quality, but that the sample is suitable for use as an abrasive. The etymology of the word diamondite follows a standard geological/petrographic classification scheme where rocks can be named according to their dominant mineral phase. For example, if a rock is dominantly comprised of clino- and/or ortho-pyroxene then it can be called a pyroxenite, and so the argument was put forward that mantle xenoliths dominated by diamond should be referred to as diamondite (Kurat and Dobosi 2000). Stewartite is the least commonly used name, either because most samples are non-magnetic or because most people do not check whether or not a framesite/diamondite is magnetic.

The carbonados are considered by most researchers to be a diamond sub-group of their own. Carbonados are dark-colored, equidimensional, microporphyritic and often have a glassy diamond patina that can show slickensides (Haggerty 2017). In short, carbonados do not look like the polycrystalline diamond aggregates found in kimberlites. Like PDAs associated with kimberlites, carbonados are porous, with about 10–15% pore space, but the pores are rounded rather than polygonal (Haggerty 2017). The primary host rock for carbonados is unknown. They are found exclusively in an alluvial setting in Mid-Proterozoic (1–1.5 Ga) metaconglomerates overlying the São Francisco and Congo-Kasai cratons in Brazil and in Mesozoic sandstones in the Central African Republic (Fettke and Sturges 1933; Leonardos 1937; Trueb and De Wys 1969; Haggerty 1999). The two localities are paleo-geographically connected and once formed part of the supercontinent Rodinia (De Waele et al. 2008). The intergrowths and inclusions described from carbonados feature phases not usually associated with diamond from the Earth's mantle. They comprise phases associated with crustal and sedimentary rocks, such as florencite-goyazite-gorceixite, xenotime, kaolinite, quartz, orthoclase, zircon, and also a suite of very reducing phases, such as SiC, TiN, and Fe, Fe–Ni, W–Fe–Cr–V, Ni–Cr, Si, Sn metallic

Table 1. Polycrystalline diamond aggregate terminology and characteristics.

Name	Defining characteristics
Framesite	Polycrystalline and non-magnetic
Stewartite	Polycrystalline and magnetic
Boart	Morphology is not economically viable and comprises >1 diamond crystal
Diamondite	Any polycrystalline aggregate where diamond is the most abundant mineral phase
Carbonado	Black, polycrystalline, and only found in the Mid-Proterozoic (1–1.5 Ga) metaconglomerate overlying the São Francisco and Congo-Kasai cratons

and alloyed phases (Heaney et al. 2005; Haggerty 2014). The mixture of uncommon syngenetic and epigenetic inclusions, including crustal inclusions, have been posing a challenge to develop models explaining their genesis which encompass a wide range of scenarios, including extraterrestrial origins (Heaney et al. 2005; Haggerty 2014). As carbonados are geologically much less constrained than polycrystalline diamond aggregates from kimberlites there is a risk that the new findings for the kimberlitic PDAs do not apply to carbonado. Consequently, a summary of the carbonado literature in this chapter would contribute little that has not been said before. For details on carbonado, we direct the reader to the extensive reviews existing in the literature for carbonados, such as Haggerty (2014, 2017) and Heaney et al. (2005).

Herein, we use the term "polycrystalline diamond aggregate" (PDA) to group all polycrystalline samples (Fig. 1). This review is concerned with the origin and geological significance of PDAs from kimberlites which host silicates and oxides. This is because their direct association with kimberlites and their intergrowths/inclusions provide *bona fide* geological context, without which all inferences are devoid of any meaningful relation with geological processes, or events.

POLYCRYSTALLINE DIAMOND AGGREGATES FROM KIMBERLITES

To our knowledge, polycrystalline diamond aggregates (Fig. 1) have, thus far, only been reported in the peer-reviewed literature from Group I kimberlites. PDAs amount to as much as 20% of the diamond production in Southern African and Yakutian kimberlites, including Orapa, Venetia, Cullinan (Premier), Jwaneng, Mirny, Aikhal, Yubileinaya and Sytykanskaya (Orlov 1977; Kaminsky et al. 1981; Gurney and Boyd 1982; McCandless et al. 1989; Smelova 1994; Kirkley et al. 1995; Jacob et al. 2000; Sobolev et al. 2016). PDAs can be up to several centimetres in size (several hundred carats) and comprise randomly oriented diamonds of variable size from ca. 5 μm to 5,000 μm (Sobolev 1977; Rubanova et al. 2012) intergrown with silicates, oxides and other phases (Jacob et al. 2011). An important and notable difference to inclusions in monocrystalline diamonds is that non-diamond phases in PDAs are *intergrown* with the diamond grains, confirming a petrogenetic relationship, rather than representing material trapped during diamond formation or random epigenetic phases. Individual diamonds in PDAs are also found included in the non-diamond phases, indicating the syngenetic formation of diamond, silicate, oxides, carbide, and fluid components (Kurat and Dobosi 2000; Jacob et al. 2004; Mikhail et al. 2019b).

The high porosity of some PDAs (up to 30%; Heaney et al. 2005; Jacob et al. 2011), the release of detectable ^{3}He when crushed under vacuum (Gautheron et al. 2005; Mikhail et al. 2019a), and the presence of trace amounts of hydrogen (e.g., 38 ± 5 ppm; Fourel et al. 2017) indicate the presence of fluids trapped during formation. In some cases, it appears that these intergrowths are not shielded from their surroundings and can thus be subject to metasomatic alteration. For example, a micro-computed tomography (μCT) study of a PDA sample from Orapa, Botswana (Jacob et al. 2011) presents spectacular hematite alteration rims around magnetite (Fig. 2), illustrating epigenetic metasomatic changes facilitated by the high porosity and permeability in some PDAs.

This chapter summarizes data on the diamond and non-diamond components in kimberlite-related PDAs. This information is synergized into a narrative of what we know and what we will still need to know as informed by these data. We show how the geochemical and petrological insights distinguish PDAs from the monocrystalline diamond types (gem and fibrous types) and explain what they reveal about our collective understanding of mantle metasomatism, tectonic volatile fluxing, and the deep carbon cycle.

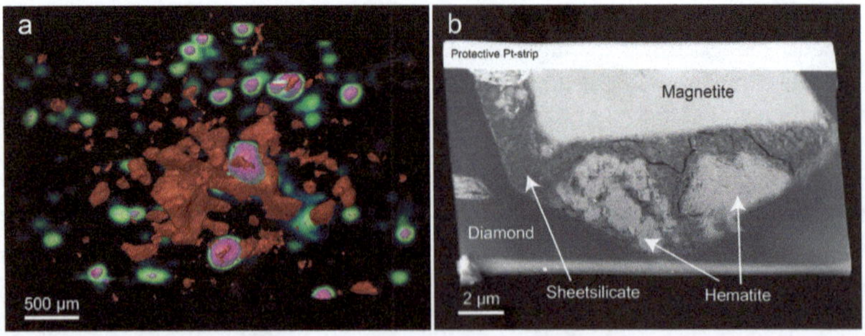

Figure 2. (**a**) µCT region of interest showing pore space (**red**) and magnetite grains (**purple**) covered by a replacement envelope (**green**) of a mixture of hematite and a sheet silicate. Diamonds are rendered transparent in this view. (**b**) TEM HAADF image of a FIB foil cut from a diamond grain showing a replacement rim around magnetite consisting of a mixture of hematite and a sheet silicate.

INSIGHTS FROM DIAMOND GEOCHEMISTRY

Polycrystalline diamond growth is the result of heterogeneous nucleation and rapid crystal growth driven mainly by carbon supersaturation (Sunagawa 2005). Hence, while monocrystalline diamonds can show extended histories of slow growth (Howell et al. 2012b; Timmerman et al. 2019a), PDA formation appears to be rapid (Orlov 1977). Therefore, because of their rapid formation, both fibrous diamonds and polycrystalline diamond aggregates provide important and complementary perspectives on diamond formation processes compared to the time-integrated story recorded by some monocrystalline diamonds.

Source(s) of diamond-forming metasomatic fluids

The origin of PDA-forming fluids has been evaluated using the $\delta^{13}C$ and $\delta^{15}N$ values for the diamond component of PDAs (Shelkov et al. 1997; Jacob et al. 2000; Maruoka et al. 2004; Mikhail et al. 2013, 2014c), and using helium isotope constraints from micro- to nano-inclusions in the rocks (Burgess et al. 1998; Gautheron et al. 2005; Mikhail et al. 2019a).

The stable isotope ratios of carbon and nitrogen for crustal and mantle derived material show distinctions between their average values (Hoefs 2009 and references therein) (Fig. 3). Hence, for diamonds, these stable isotope systems can be useful indicators for their sources and elucidate crust–mantle interaction (Stachel et al. 2022, this volume). This picture is complicated due to significant stable isotope fractionation effects, which can occur at high temperatures and pressures and in associated closed and open system diamond formation processes (Stachel et al. 2022, this volume). Owing to the similarity of the mass difference between the heavy and light isotopes for carbon and nitrogen, the equilibrium stable isotope fractionation factors for carbon and for nitrogen are similar in magnitude at mantle temperatures, and these are usually within 1–2‰ depending on the species involved (Richet et al. 1977; Deines 1980; Polyakov and Kharlashina 1995; Reutsky et al. 2008, 2015; Petts et al. 2016). However, the specific fractionation reactions in the Earth's mantle seem to play a bigger role for carbon (e.g., decarbonation) than for nitrogen (Cartigny et al. 2014; Petts et al. 2016). Progressive fractionation of an evolving fluid in a closed system (i.e., Rayleigh fractionation) has been suggested to explain monocrystalline diamonds that display continuous, coupled changes in carbon and nitrogen isotopic values (Petts et al. 2015). On the other hand, open system mixing of fluids (e.g., Petts et al. 2015) or separate fluid pulses with different isotopic compositions (e.g., Wiggers de Vries et al. 2013; Smit et al. 2016) have been suggested to explain the observed variability in the carbon and nitrogen isotope values of some other diamonds, including PDAs (Jacob et al. 2017 and Fig. 4). Thus, nitrogen concentrations of diamonds and their corresponding nitrogen isotope values ($\delta^{15}N$) complement the $\delta^{13}C$ values to trace the source of diamond-forming fluids in the deep Earth (Javoy et al. 1986; Boyd and Pillinger 1994; Cartigny et al. 1998a).

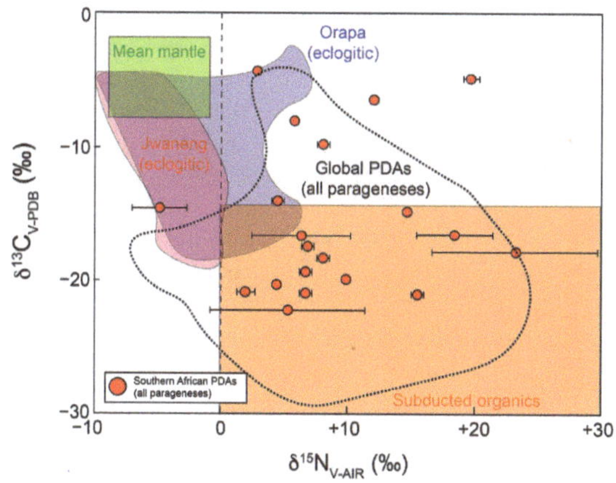

Figure 3. Plot of $\delta^{13}C$ vs $\delta^{15}N$ values for select diamond types. Shown here is the mean mantle C–N stable isotope field, within which most fibrous diamonds plot (compiled by Cartigny et al. 2014). Also shown are the fields for eclogitic monocrystalline diamonds from Jwaneng and Orapa (Cartigny et al. 1998b, 1999). The PDA data are from Gautheron et al. (2005), Maruoka et al. (2004), and Mikhail et al. (2013, 2014a, 2019b). The data for the $\delta^{13}C$ and $\delta^{15}N$ values for subducted organics are from Halama et al. (2010, 2014) Busigny et al. (2019), and the compilation in Cartigny et al. (2014).

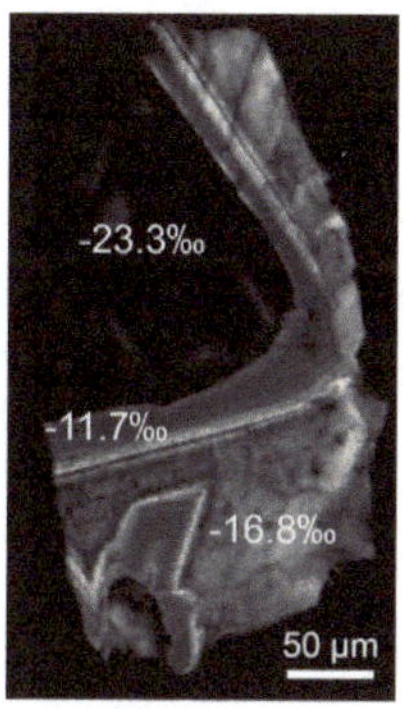

Figure 4. Cathodoluminescence image of a diamond fragment from a websteritic PDA from Venetia showing different growth zones with variable $\delta^{13}C$ values, indicating different growth pulses.

In general, PDA diamond geochemistry is characterized by ^{13}C-depletion and ^{15}N-enrichment, higher than average nitrogen abundances, a lower than average occurrence of nitrogen-free samples, and a range of nitrogen aggregation states (Mikhail et al. 2019b and references therein). The carbon isotope values for PDAs show a large range from $\delta^{13}C$ = −1‰ to −30‰ (Fig. 3) with modes at −5‰ and −18‰ (Deines 1980; Maruoka et al. 2004; Mikhail et al. 2013, 2019b; Jacob et al. 2014; Sobolev et al. 2016). However, when plotting only those data where combined C–N isotope systematics for garnet-bearing PDA samples are considered (to enable meaningful comparisons between the different isotopic and paragenetic systems), the mode at −5‰ is not present. It is noteworthy that, alongside peridotitic and websteritic PDAs only a small subset of eclogitic and websteritic monocrystalline diamonds show ^{13}C-depletion with mean values positioned lower than the −20‰ (e.g., eclogitic samples from Dachine (French Guiana; Smith et al. 2016) and Jericho (Northern Canada; De Stefano et al. 2009). However, what distinguishes the PDAs from monocrystalline diamonds is the fact that all parageneses (eclogitic, websteritic, and peridotitic) are found to show ^{13}C-depletion (Maruoka et al. 2004; Mikhail et al. 2014a, 2019b), with a mode at −18‰, while peridotitic monocrystalline diamonds plot with the mode at −5‰ (Table 2).

Table 2. Averages for samples where combined N and C data are available (no data considered when only one of the isotope systems is reported). Note the large standard deviations for the mean nitrogen concentrations for PDAs reflecting the large ranges of concentrations and the absence of a significant mean nitrogen abundance for PDAs.Data sources: monocrystalline diamond (Cartigny et al. 1997, 1998a, b, 1999, 2004, 2009; Palot et al. 2009; Thomassot et al. 2009; Cartigny 2010; Palot et al. 2012; Mikhail et al. 2014b; Smith et al. 2016), fibrous diamond (Javoy et al. 1986; Boyd et al. 1987, 1992; Klein-BenDavid et al. 2010), and PDAs (Javoy et al. 1986; Boyd et al. 1987, 1992; Shelkov 1997; Klein-BenDavid et al. 2010).

Diamond Type	*n*	$\delta^{13}C$ (‰)	StDev	$\delta^{15}N$ (‰)	StDev	N at.ppm	StDev
Monocrystalline (Peridotitic)	189	−4.6	1.9	−4.9	7.0	396	350
Monocrystalline (Eclogitic)	204	−9.5	8.2	−2.9	5.7	649	483
Fibrous	109	−6.5	3.1	−1.6	4.4	922	360
PDA	62	−18.1	5.4	+8.4	6.7	469	715

The $\delta^{15}N$ values range from −6.1‰ to +22.6‰ (Gautheron et al. 2005; Mikhail et al. 2014c; Jacob et al. 2017) with an average of +8.4‰ (Table 2)—an average value that is significantly more ^{15}N-enriched than for all other diamond types (Table 2). The ^{15}N-enrichment in PDAs is important because the interior and exterior of Earth show a pronounced and statistically significant isotopic disequilibrium for nitrogen, where the upper mantle is ^{15}N-depleted and the crustal reservoirs are ^{15}N-enriched (Boyd and Pillinger 1994). Thus, the ^{15}N-enriched nature of PDAs supports involvement of crustal material in their formation. In addition, oxygen isotopic values for garnets in PDAs from Venetia (Jacob et al. 2000) and Orapa (Mikhail et al. 2019b) show $\delta^{18}O$ values from +5.96 to +8.09‰, which are higher than $\delta^{18}O$ values for the ambient mantle ($\delta^{18}O$ = 5.5‰; Mattey and MacPherson 1993). The origin of ^{18}O-enriched or depleted oxygen in mantle samples is considered to be *bona fide* evidence for material altered in a thermal environment of too low energy to be assignable to the deep Earth, this is based on evidence that equilibrium stable isotope fractionation for $^{18}O/^{16}O$ at temperatures > 800 °C and at high pressures are negligible and cannot explain a shift of $\delta^{18}O$ value > ± 1‰ (Chacko et al. 2001).

The nitrogen concentrations in PDAs cover a large and unevenly distributed range from 4 to 3635 at.ppm N with an average nitrogen abundance of 496 ± 715 ppm (Table 2). These nitrogen concentration data are not resolvable from monocrystalline diamonds or fibrous diamond samples, but the latter show higher average nitrogen abundances of around 922 ± 360 ppm. The different average nitrogen concentrations unfortunately do not carry much significance due to the overlap of the datasets and considering the spread of data. In general, the spread of nitrogen concentrations in PDAs are not dissimilar from every other diamond type, where monocrystalline samples show lower average nitrogen abundances with 396 ± 350 and 649 ± 483 ppm for peridotitic and eclogitic monocrystalline samples, respectively (Cartigny 2005). These poorly defined averages for the nitrogen concentrations in mantle diamonds are sometimes used to argue for kinetic uptake of nitrogen in diamond during growth, where the slow diffusion of nitrogen in diamond (Koga et al. 2003) prohibits re-equilibration after formation (see Mikhail and Howell 2016, for a discussion). Compared to the concentrations of nitrogen, the stable isotope data are significantly distinct in their distribution and significance. Collectively, the role for a subducted crustal organic component for the significant ^{13}C-depletion is supported by the high nitrogen concentrations and very positive nitrogen isotope values for PDAs (Shelkov et al. 1997; Gautheron et al. 2005; Jacob et al. 2014, 2017; Mikhail et al. 2014c, 2019a) because the combined $\delta^{13}C$ vs $\delta^{15}N$ vs N ppm data are typical for material from Earth's surface (Boyd and Pillinger 1994).

When crushed under vacuum, PDAs release gas (Gautheron et al. 2005; Mikhail et al. 2019a), and this requires the presence of fluid micro-inclusions to explain the release of detectable ^{3}He after crushing (Gautheron et al. 2005; Mikhail et al. 2019a). Therefore, to shed more light on the origin of diamond-forming fluids one can apply noble gas isotope data to complement and test hypotheses developed using C–N stable isotope data (e.g., ^{3}He/^{4}He). This is useful because, while C and N isotope systems are powerful tracers of subducted material, ^{3}He is an equally powerful tracer for an otherwise hidden mantle signature within crustal-source dominated ^{13}C-depleted and ^{15}N-enriched systems. This is because ^{3}He is a primordial isotope and one of the most incompatible elements in silicate systems (Brooker et al. 2003; Jackson et al. 2013). Therefore, ^{3}He is not transferred into the solid Earth after degassing (instead, it is so light that it is lost to space). However, an unavoidable issue is ^{4}He ingrowth due to radioactive alpha decay. The degree of nitrogen aggregation in PDAs, fibrous and monocrystalline diamonds is consistent with millions to billions of years of mantle residence (Mikhail et al. 2019a and references therein) permitting ^{4}He growth due to alpha decay of U and Th in the sample, which decreases the ^{3}He/^{4}He ratios over time (Timmerman et al. 2019a, b). This results in a relationship between ^{3}He/^{4}He and ^{3}He concentration shown in Figure 5, reflecting radiogenic ^{4}He production since the fluids were trapped in diamond (see caption to Fig. 5).

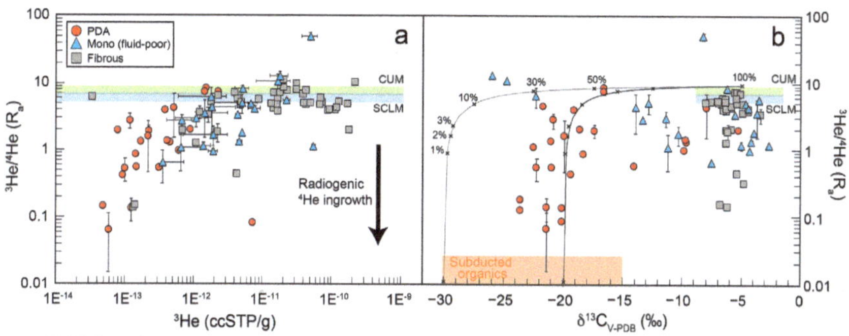

Figure 5. Helium isotope systematics of fluids released by in vacuo crushing and heating of PDAs, and monocrystalline and fibrous diamonds. The ^{3}He/^{4}He and δ^{13}C value of the modern convecting upper mantle (CUM) and the sub-continental lithospheric mantle (SCLM) are shown for reference (adapted from Mikhail et al. 2019a). The mixing lines shown in (**a**) are drawn between oceanic crustal-derived fluids with $\delta^{13}C = -20$ and -30‰, and $^3He/^4He = 0.01$ Ra, and mantle-derived fluids with $\delta^{13}C = -5$‰ and $^3He/^4He = 9$ Ra. The mixing lines plotted in (**b**) are hyperbolic as [^{4}He]mantle/[^{4}He]crust is assumed to be 10. Note, the mantle ^{3}He/^{4}He endmember is slightly higher than the present-day upper asthenosphere and lithosphere mantle value, reflecting the U–Th decay-driven temporal evolution of ^{3}He/^{4}He in the silicate Earth (Porcelli and Elliott 2008). Therefore, all ^{3}He/^{4}He ratios are minimum values, although it appears that diamonds with $>1\times10^{-12}$ ccSTP ^{3}He/g are largely immune to the effect of ingrowth because the ^{4}He production is too low to offset the R/Ra in this system (note the logarithmic scale of the *x*-axis). Comparative data are from Burgess et al. (1998), Gautheron et al. (2005), Broadley et al. (2018), Timmerman et al. (2018, 2019a, b), and Mikhail et al. (2019a).

After considering ingrowth of ^{4}He due to the decay of U and Th in the sample it is clear that the PDAs and some monocrystalline diamonds characterized by low δ^{13}C (commonly interpreted to indicate a crustal origin for the diamond-forming carbon), show high ^{3}He/^{4}He ratios typical for mantle fluids (Gautheron et al. 2005; Mikhail et al. 2019a). The spread of data can be explained by mixing between a He-rich, high ^{3}He/^{4}He mantle source with $\delta^{13}C = -5$‰ and a He-poor, low ^{3}He/^{4}He crustal component with δ^{13}C between −15 and −30‰ (akin to the crustal organic carbon field in Fig. 3). The involvement of a crustal component is strongly supported by the ^{15}N-enrichment in the ^{13}C-depleted PDA samples and best matches either altered oceanic crust (Cartigny et al. 2014) or organic material hosted in sedimentary rocks (Thomazo et al. 2009). Hence, the He–C–N isotope systematics in PDAs could be described by mixing between fluids released from subducted altered oceanic crust and mantle material, and mass balance skews the measured values towards the dominant component (Mikhail et al. 2019a). Therefore, although the ^{13}C-depleted carbon and ^{15}N-enriched nitrogen provide evidence for crustal sources for some of the PDA-forming fluids, the high ^{3}He/^{4}He ratios in PDAs from Orapa reveal an unambiguous mantle component for these strongly ^{13}C-depleted samples (Mikhail et al. 2019a).

In terms of PDA petrogenesis, these data characterize the PDA-forming metasomatic fluids as hybrids with subducted origins that have remobilized volatile elements in the subcratonic lithospheric mantle, initiated by subducted fluid percolating and interacting with potentially ancient material in the lithospheric mantle (first shown by Jacob et al. 2000).

Nitrogen aggregation and mantle residence times

Nitrogen is the most abundant lattice-bound impurity in diamond. Over time, nitrogen defects as a function of temperature from single nitrogen atoms (C centers, Type Ib), to pairs of atoms (A centres, Type IaA), to 4 nitrogen atoms tetrahedrally arranged about a vacancy (B centres, Type IaB; Evans and Qi 1982). This sequence is quantified and expressed as the degree of nitrogen aggregation in %. The first step in this process (C to A centre aggregation) occurs quite rapidly while the second step (A to B centre) occurs much more slowly, where the exact rate of conversion depends on the temperature and nitrogen concentration (Chrenko et al. 1977). The A to B centre aggregation follows a second-order kinetics law (Evans and Qi 1982), meaning nitrogen aggregation can be used to estimate either the average temperature of residence or the duration of diamond residence in the mantle (assuming the other is known). Furthermore, platelets—planar interstitial carbon aggregates found on the {100} crystal planes—are the by-products of B centre formation and are prone to degradation during residence in the mantle (Woods 1986). The nitrogen aggregation data are fitted to a second-order kinetic model (e.g., Howell et al. 2012a, b) to ascertain information about the number, or depth, of the diamond-forming events.

As revealed by infrared spectroscopy, nitrogen aggregation states for PDAs span the whole spectrum from pure IaA to pure IaB (Mikhail et al. 2014c). Platelet characteristics are found to be regular, meaning they preserve their time-dependent history and have not been degraded by deformation and/or heating. There is no correlation between paragenesis (as defined by garnet geochemistry) and any of the IR characteristics for PDAs from Orapa (Mikhail et al. 2019b).

The nitrogen aggregation state for several diamond grains from individual PDA samples are homogeneous, in contrast with monocrystalline diamonds which can show highly variable nitrogen aggregation states between different growth zones reflecting multiple growth events (Boyd et al. 1987; Palot et al. 2013; Bulanova et al. 2014; Timmerman et al. 2017). Orapa PDAs show varying degrees of aggregation (10–76% B) and plot along calculated isotherms as a function of nitrogen abundance which can be fit to residence temperatures between 1100 and 1175 °C. However, the calculations of average mantle residence temperatures for PDAs in Mikhail et al. (2019b) assume residence time of 1000 m.y. (Fig. 6). The aggregation process is controlled less by time than it is by temperature, and if we revere the calculation and vary temperature the reliability of these models is shown to be justifiably questionable. For example, if the model is computed with an elevated temperature of 1200 °C (increase of 25 °C from the maximum) or at 1400 °C (increase of 200 °C from the maximum), then the predicted mantle residence times for a sample with 76% B (and 600 ppm N) are 500 million years and 660,000 years, respectively.

Despite this, there is no evidence for higher-than-average residence temperatures (>1200 °C) for the Orapa PDAs, so the observation of regular and internally continuous nitrogen aggregation data (Mikhail et al. 2019b) means that the nitrogen aggregation data for each sample can be used to interpret their time–temperature history (with large uncertainties). Nevertheless, there are PDAs that show evidence for higher formation temperatures than 1200 °C, both at Orapa

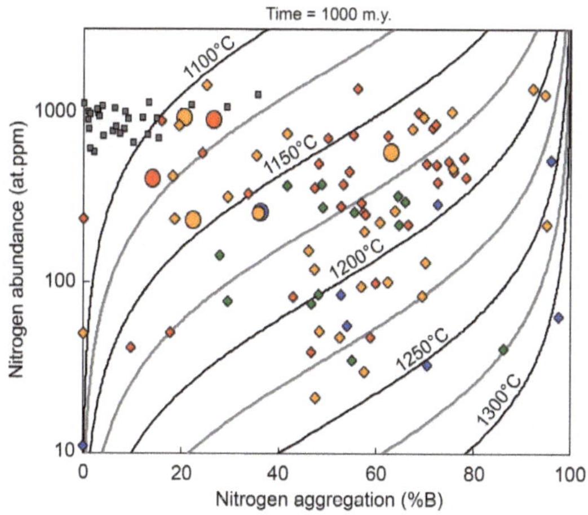

Figure 6. Variation diagrams for the average degree of nitrogen aggregation (%) vs. nitrogen abundance (at. ppm) for the diamonds from PDAs (**circles**) alongside the fibrous (**squares**) and monocrystalline diamonds (**diamonds**) from Orapa (adapted from Mikhail et al. 2019b). Isotherms chart the residence times of 1000 m.y. using the DiaMap software (Howell et al. 2012a, b). **Red** symbols–eclogitic paragenesis, **green**–peridotitic, **blue**–websteritic, **grey**–fibrous unknown, **orange**–unknown.

(Jacob et al. 2016) and also at Venetia, but nitrogen aggregation data unfortunately do not exist for these specific samples. One sample from the Venetia kimberlite, for example, was found to contain droplets of metallic iron and iron carbide and this assemblage requires temperatures above the liquidus of this system of 1370 °C at relevant pressures (Jacob et al. 2004).

All Orapa PDAs show B-centres of varying degrees, and this requires several hundred million years of residence in the mantle prior to emplacement in the crust at 91 Ma (Fig. 6). This is not the case for all PDAs, where data for PDAs of unknown provenance show nitrogen with very low and very high degrees of nitrogen aggregation (0–100% B; Mikhail et al. 2014c). These samples also show high temperature deformation and annealing structures in electron backscattered diffraction which perhaps could be argued to require significant mantle residence times (Rubanova et al. 2012). What can be stated, using these data, is that nitrogen aggregation data from PDA samples strongly oppose any notion that the carbon supersaturation event responsible for PDA-formation is temporally-associated with diamond-transporting kimberlite metasomatism (as has been proposed for fibrous samples; Boyd et al. 1994).

INSIGHTS FROM THE GEOCHEMISTRY OF NON-DIAMOND COMPONENTS IN POLYCRYSTALLINE DIAMOND AGGREGATES

Geochemical and petrological studies of the non-diamond constituents are beset with technical challenges, many of them caused by the hardness anisotropy of diamond combined with the random orientation of the grains in PDAs and the high refractive index of diamond. For example, preparing PDAs for *in situ* microanalysis is exceedingly difficult due to the almost impossible task of producing adequately well-polished surfaces (Rubanova et al. 2012; Jacob et al. 2017). Similarly, the *ex situ* optical identification of the non-diamond phases or textural relationships is commonly obscured by the opaque nature of most PDAs (see Fig. 1)—the result of the many opaque impurities combined with the high optical refractivity of diamond. These challenges are intensified by the fact that the non-diamond phases, such as silicates, which hold a wealth of geochemical information, are present only in minor or trace amounts and can be heterogeneously distributed within a single sample (Mikhail et al.

2019b). To date, effective examination of the non-diamond phases in PDAs, such as garnet, requires mechanical disaggregation using a crusher which destroys all in situ petrographic data. For example, on some occasions, only after the mechanical extraction of the garnet did it transpire that a single PDA was host to multiple distinct populations of websteritic garnets, and even revealing garnets of different paragenesis within a single PDA, such as sample ORF53 from Orapa which was host to one purple (peridotitic) and one orange (websteritic) garnet (Mikhail et al. 2019b). This 'lucky' find of two very different garnets in one sample, in fact highlights a notable characteristic of the non-diamond components in PDAs, which is that they contain disequilibrium assemblages at grain-scale. Importantly, touching pairs of chemically homogeneous silicates of a single paragenesis (e.g., peridotitic–websteritic–eclogitic) as they are encountered in inclusions in monocrystalline diamond, are yet to be identified in PDAs.

Modern high-resolution computed tomography (μCT) offers an alternative and non-destructive method to optical microscopy and has occasionally been applied in PDAs (Fig. 2; Jacob et al. 2011; Logvinova et al. 2015). Multi-scale petrographic study of PDAs by μCT is still expensive but will most likely gain popularity with increasing access (i.e., supply vs. demand).

Macro-inclusions and intergrowths

The most common non-diamond phases found in PDAs are garnets, followed by clinopyroxenes, but other phases can be present in significant abundances, including Mg-chromite, rutile, magnetite, sulphides and cohenite (iron carbide) (Gurney and Boyd 1982; Kirkley et al. 1991; Jacob et al. 2000, 2011, 2014, 2016; Dobosi and Kurat 2002, 2010; Mikhail et al. 2019b). Olivine and orthopyroxene are notably absent. Touching parageneses of silicate minerals are almost absent and often, PDAs host either only one mineral (most often garnet) or minerals of different parageneses which cannot have formed in equilibrium with each other. In addition, PDAs are known to display magnetism which requires the presence of magnetite and other magnetic phases (Jacob et al. 2011, 2016). It is worth pointing out that magnetite is a rare inclusion in PDAs, but its occurrence is more abundant in PDAs than in monocrystalline diamonds (Harris 1968; Sobolev et al. 1989) and may explain the magnetic properties of stewartite.

For diamond inclusions in general, garnet and clinopyroxene data are sub-divided into three groups, termed peridotitic, websteritic, and eclogitic, according to their silicate paragenesis. The assignment of inclusion paragenesis is empirical. For example, if assigned simply by color then websteritic garnets are sometimes wrongly classified as eclogitic, as both are orange, and indeed some workers have classified websteritic garnets as low-Ca eclogitic (e.g., Jacob et al. 2000; Kurat and Dobosi 2000). It is more accurate to use the major element geochemistry of garnet inclusions, such as the weight percent abundances of Cr_2O_3 and CaO (Fig. 7a), where peridotitic are categorized by high-Cr and eclogitic garnets by low Cr- and high-Ca (Sobolev et al. 1973), and websteritic garnets are intermediate in composition (Gurney and Boyd 1982). The paragenetic distinction(s) for pyroxenes is less clear (Fig. 7b), where a peridotitic paragenesis is assigned for clinopyroxenes with Mg-number [(Mg/Mg+Fe) × 100] >85, Cr-number [(Cr/Cr+Al) × 100] >10, and <2 wt.% Na_2O (Aulbach et al. 2002). Alternatively, all garnet and clinopyroxene inclusions can be classified according to their endmember compositions (Fig. 7c, d).

This dataset shows that garnets in PDAs are most commonly websteritic, with peridotitic and eclogitic compositions less common for PDAs than as inclusions in monocrystalline diamond (Fig. 8a). Considering the available data (from Orapa, Venetia, Mirny, and unknown) we find that 57% of the garnets from PDAs are websteritic (Fig. 8a), compared with only 3% of worldwide monocrystalline diamond garnet inclusions (Stachel and Harris 2008; Fig. 8b). In contrast to the websteritic character of garnets in PDAs, the clinopyroxenes found in PDAs are most often peridotitic (79%; Fig. 8c), which contrasts with clinopyroxenes in monocrystalline diamond inclusions (18%; Fig. 8d). Whether the relative abundances or formation mechanisms for websteritic diamond inclusions represent pre-metasomatic mineralogical heterogeneity in the upper mantle (Stachel and Harris 2008), eclogitic melt metasomatism (Aulbach et al. 2002; Kiseeva et al. 2016), or fluid metasomatism coeval with diamond formation (Mikhail et al. 2019b, 2021) is as yet unresolved. Characteristically, the most abundant mineral in the lithospheric mantle, olivine, is a notable absentee in PDAs. In addition, wehrlitic garnets and clinopyroxenes are absent, indicating that the striking absence of olivine is not a simple sampling bias, but is also reflected by the range of major element compositions for the silicates hosted in PDAs dominated by olivine-free websteritic/eclogitic parageneses (Figs. 7–8). The relatively high abundance of websteritic garnets in PDAs (Figs. 7–8) and the absence of olivine are a distinctive characteristic of PDAs that is not seen in the inclusion chemistry of monocrystalline diamonds from the same localities, such as Orapa (Mikhail et al. 2019b), or globally (Figs. 7–8), where peridotitic and eclogitic compositions represent the majority of the monocrystalline diamond yield (Figs. 7–8).

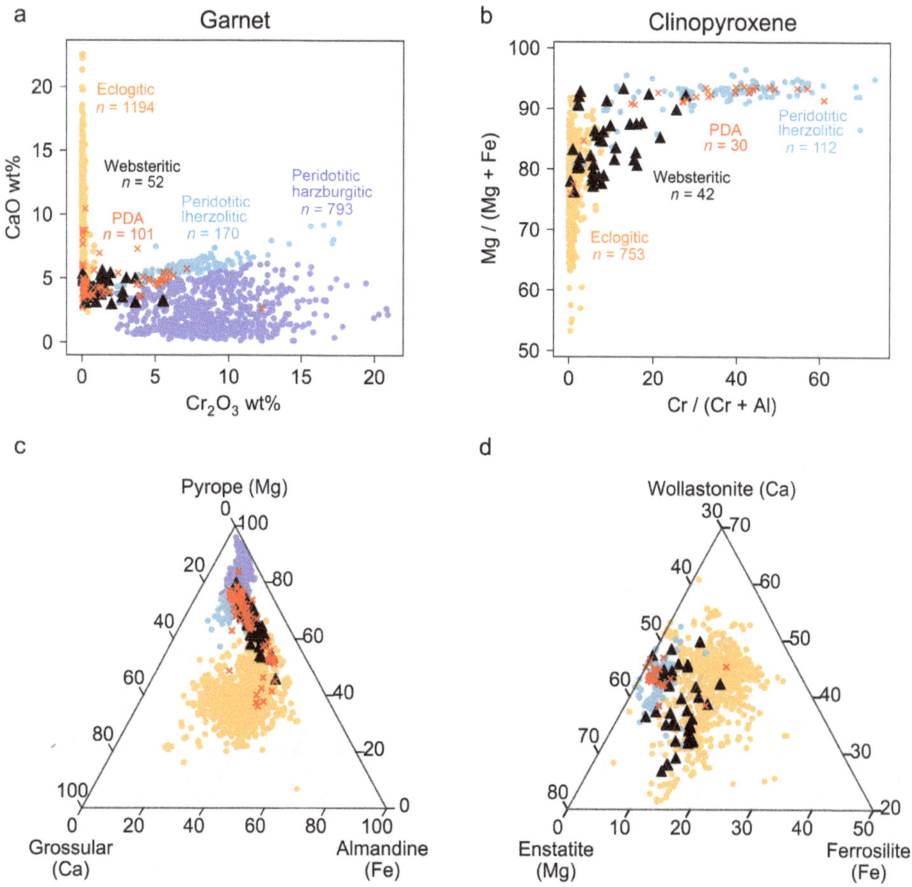

Figure 7. Inclusion parageneses and end-member compositions for garnets and clinopyroxenes from monocrystalline diamond and PDAs. The diamond inclusion database contains geochemical and metadata for > 9000 diamond inclusions. These plots utilize the clinopyroxene ($n = 926$), and garnet ($n = 2628$) data from lithospheric mantle diamonds. The data used here are sourced from Stachel and Harris (2008 and references therein) supplemented with more recent datasets (Viljoen et al. 1999; Cartigny et al. 2009; De Stefano et al. 2009; Hunt et al. 2009; Tappert et al. 2009; van Rythoven and Schulze 2009; Bulanova et al. 2010; Miller et al. 2014; Smith et al. 2016; Sobolev et al. 2016; Mikhail et al. 2019a; Jacob et al. unpublished).

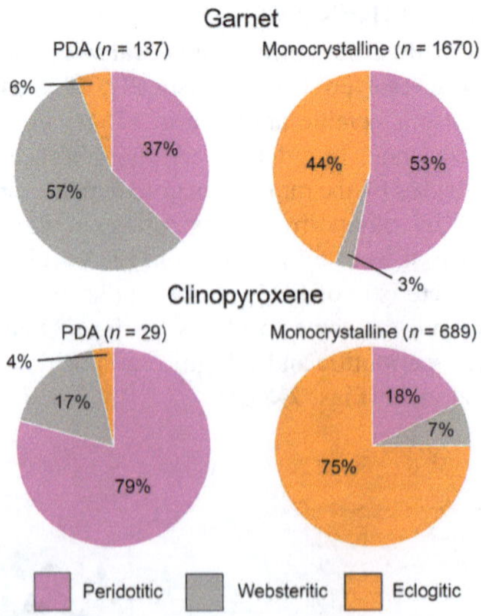

Figure 8. Pie charts reflecting the paragenesis for garnets (**left**) and clinopyroxenes (**right**) sourced from PDAs (**left**) and monocrystalline diamonds (**right**). The data for PDAs are sourced from Gurney and Boyd (1982), Gurney et al. (1984), Kirkley et al. (1995), Jacob et al. (2000, unpublished), Dobosi and Kurat (2010), Sobolev et al. (2016) and Mikhail et al. (2019a). The data for monocrystalline diamond inclusions are from Stachel and Harris (2008 and references therein).

Trace elements and radiogenic isotopes of silicates in PDAs

Garnets intergrown with diamond in PDAs show typical chondrite-normalized rare earth element (REE) patterns with depleted LREE and relatively flat to convex-upward MREE to HREE (Fig. 9a–c). Peridotitic garnets have mildly sinusoidal patterns (Fig. 9a), similar to those included in monocrystalline diamonds (Stachel et al. 2004). Most of the websteritic garnets show no LREE enrichment and only some have mildly enriched LREE patterns (Fig. 9b), while this group has the highest HREE abundances of up to 55 × chondrite. Europium anomalies are not pronounced and, if present, are mostly within 15%, with one exception of a markedly positive Eu-anomaly (50%) in an eclogitic garnet from Orapa (ORF78, Mikhail et al. 2019b). Out of the four published trace element datasets for clinopyroxenes (Fig. 9d), two peridotitic samples (solid lines in Fig. 9d) have the highest REE abundances followed by a websteritic clinopyroxene (dotted line) and an omphacite (dashed line). Neither clinopyroxenes nor garnets show a correlation for REE abundances with their chromium content.

The notion for the involvement of ancient lithospheric material is provided by the unradiogenic $^{87}Sr/^{86}Sr$ isotopic ratios (0.703189 to 0.703589) and unradiogenic Nd isotopic ratios of −15.9 to −21.7 (ε_{Ndi}) observed in four ^{18}O-enriched websteritic garnet samples from ^{13}C-depleted PDAs from Venetia (Jacob et al. 2000). These data fall towards the unradiogenic end of a large range of ε_{Ndi} values for individually analysed peridotitic diamond inclusions from Venetia of −62 to +157 (Koornneef et al. 2017), and are typical for ancient lithospheric material.

Micro-inclusions

While the release of detectable He, Ne, Ar and Xe after crushing of the PDAs under vacuum (Gautheron et al. 2005; Mikhail et al. 2019a) is circumstantial evidence for the presence of micro- to nano-fluid inclusions in PDAs, direct observation of small inclusions requires Focussed Ion Beam assisted Transmission Electron Microscopy (FIB-TEM) (Klein-BenDavid et al. 2006; Wirth 2009). For PDAs, these studies can relate minerals included in the diamond grains to those intergrown with them and thus can identify the type of metasomatic overprint in these rocks. A PDA from Orapa contained magnetite, pyrrhotite, rutile and omphacite as micro-inclusions in the diamond grains. However, magnetite grains that were not fully encapsulated by diamond showed hematite and silicate alteration rims (Fig. 2; Jacob et al. 2011), whereas pyrrhotite grains were present only as micro-inclusions in diamonds and not as intergrown phases. This indicated a primary or even protogenetic nature of pyrrhotite and an unambiguous metasomatic overprint for the magnetites, thus highlighting the importance of recognising epigenetic changes in PDAs before petrological models are developed.

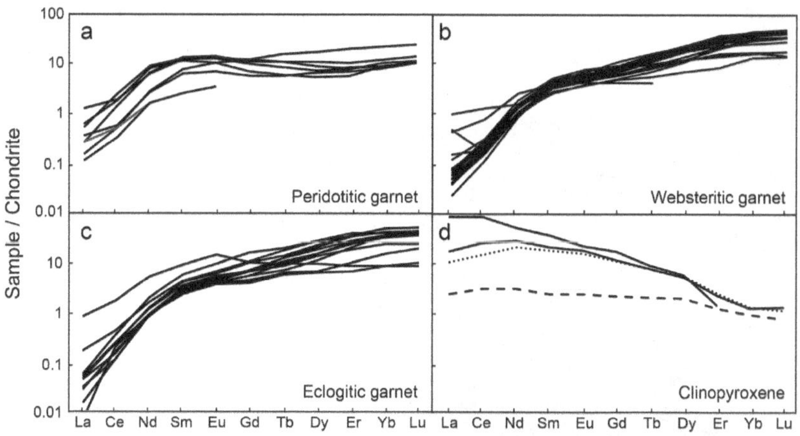

Figure 9. Chondrite-normalised rare earth element (REE) patterns for garnets (**a–c**) and clinopyroxenes (**d**) from PDAs. Different signatures in (**d**) denote peridotitic (**solid lines**), websteritic (**dotted line**) and eclogitic (**dashed line**) clinopyroxenes. Data from Jacob et al. (2000), Dobosi and Kurat (2010), Sobolev et al. (2016), and Mikhail et al. (2019b). Chondrite data from Sun and McDonough (1989).

Fluid inclusions in diamonds in this sample contained solid phases in the form of nanometre-sized grains of rutile, pyrrhotite and omphacites (but not magnetite), and silicic quench phases (i.e., non-stoichiometric Si-bearing phases detected in cavities that formerly contained fluid. (Jacob et al. 2011). Diamond-related fluids worldwide range compositionally between four end-members, namely a saline fluid rich in Cl, K, Na, H_2O and carbonate, a silicic end-member rich in Si, Al, K and H_2O, high Mg- and low Mg-carbonatitic endmembers rich in Mg, Ca, Fe, K and carbonate, where both show continuous compositional arrays with the silicic endmember on a ternary diagram (Weiss et al. 2009, 2011). The PDA sample from Orapa contained a 10 nm-sized carbonate inclusion in addition to the silicic phases; hence, the diamond-related fluid composition in this sample is silicic and water-bearing with rare carbonates, while halides are absent. In the context of the general composition of diamond fluids worldwide (Weiss et al. 2022, this volume), these observations place this specific sample close to the silicic fluid endmember (Weiss et al. 2022, this volume). Based on these data, it could be hypothesized that fluid inclusions in PDAs have a compositional range similar to that found in other varieties of diamond, consistent with REE modelling of garnets (Mikhail et al. 2019b), but testing this hypothesis will have to await further direct observations in the future.

Depth of origin

The scarcity of silicate inclusions (as opposed to silicate intergrowths) and the virtual absence of touching mineral phases in polycrystalline diamond aggregates has so far prevented attempts to derive a depth of formation for PDAs. Only one example is documented where a pressure estimate could be derived for the PDA from Orapa, described in *Micro-inclusions* above (Jacob et al. 2016). One of the pyrrhotite inclusions in diamond displayed a partially developed nanocrystalline reaction corona consisting of magnetite (Fig. 10).

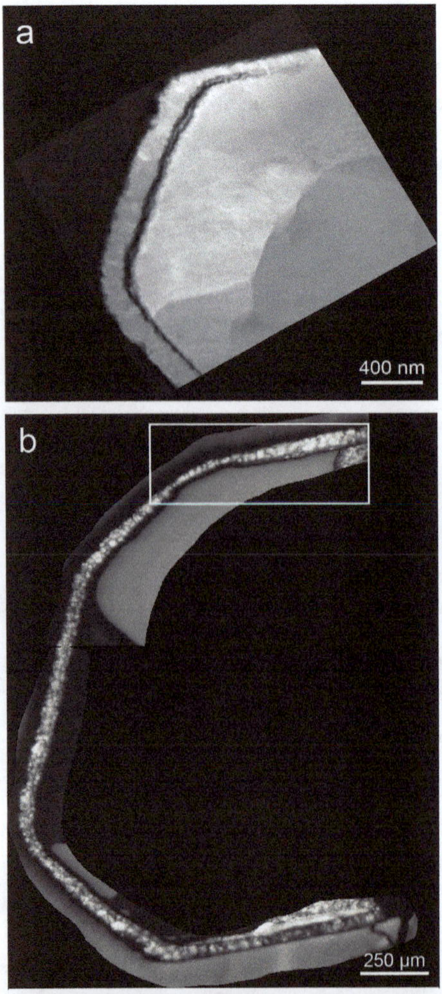

Figure 10. (**a**) TEM high angle annular dark field (HAADF) image of a diamond FIB foil showing a pyrrhotite inclusion with a magnetite rim. (**b**) Forescatter electron image of the magnetite corona. (**c**) Crystal preferred orientation of magnetite in the area indicated by a **white rectangle** in (**b**) color-coded according to the legend and the reference frame (**bottom right**). Grain boundaries are in **black**, **red lines** are twin boundaries. Modified after Jacob et al. (2016).

Transmission Kikuchi Diffraction (TKD) established epitaxy between pyrrhotite, magnetite corona and diamond host, indicating that the diamond nucleated on the magnetite corona which, in turn, formed at the expense of pyrrhotite via a redox reaction with the carbon-bearing fluid. Further TKD analysis of the nanocrystalline magnetite corona revealed textural evidence, namely twinning, which is typical for phase transitions and similar to twinning observed in back reactions of a high-pressure assemblage of Fe_4O_5 + Fe_2O_3 that replaces magnetite at pressures above 10–11 GPa at 1000–1600 °C (Schollenbruch et al. 2011; Woodland et al. 2012; Uenver-Thiele et al. 2017). The magnetite–(Fe_4O_5 + Fe_2O_3) phase boundary at ca. 10 GPa is thus a minimum pressure estimate for this sample, which must originate from ca. 320–330 km, close to the base of the Kaapvaal subcratonic lithosphere beneath the Orapa kimberlite cluster (Fouch et al. 2004).

Assuming no metasomatic overprinting for any of the PDA garnets, single garnet geobarometry performed using the machine learning-derived calibration of Thomson et al. (2021) yield average pressures of 6.0 (StDev ± 0.8) and 7.4 (StDev ± 0.5) GPa for peridotitic and websteritic garnets, respectively. The pressures for peridotitic and websteritic garnets were calculated using high and low Cr-calibration for P-type and E/W-type garnets from (Wijbrans et al. 2016). However, corresponding pressures calculated for samples containing multiple garnets yield pressure differences up to ±2 GPa, which could be explained by metasomatic overprinting for one of the garnets at an undefined time after diamond formation. These differences in pressure, however, also call into question the reliability of using single garnet geobarometry for PDAs without knowing if the garnets have been isolated or exposed to the mantle since PDA formation. In addition, the total fraction majorite content (i.e., Si-excess and considering Mj + NaMj) for garnets in PDAs shows a maximum of 0.06 with an average of only 0.01 (n = 104). Therefore, we find no evidence to support the notion that PDAs, as a group, are sourced from depths of below the subcratonic lithosphere.

Constraints on the formation age(s) of PDAs

There are no *bona fide* formation ages for PDAs, as yet. However, the preservation of significant trace element zonation in PDA-hosted garnets from Venetia (South Africa; Fig. 11) requires that these samples precipitated shortly before kimberlite eruption, based on modelling the trace element heterogeneity using known diffusion coefficients (Jacob et al. 2000). In addition, these same garnets show strongly unradiogenic ε_{Nd} with values from −15.9 to −21.7 ε_{Nd} (see above), typical for ancient lithospheric material, which was used to argue for recent remobilization of this material upon the formation of PDAs at Venetia (Jacob et al. 2000). Therefore, this means that the garnets contain older lithospheric material remobilized with the diamond-forming fluid at Venetia. In contrast, garnets from Orapa PDAs show very little, if any trace element variability within single garnets (Mikhail et al. 2019b). The geochemistry of garnets from Orapa PDAs therefore provides no evidence of a young formation age and no reliable evidence for a metasomatic event around the time of emplacement at 91 Ma (Mikhail et al. 2019b), in agreement with the nitrogen aggregation results for the diamonds hosting these garnets (discussed earlier).

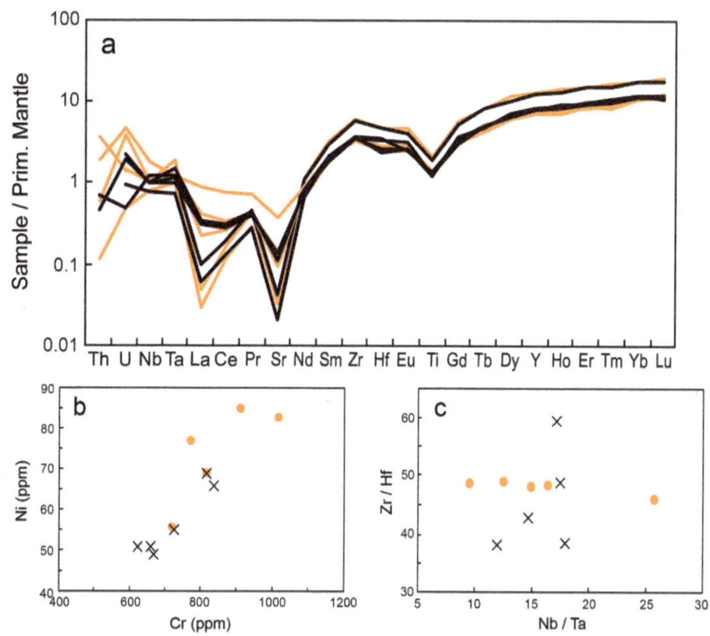

Figure 11. Trace element zonation in three different grains of websteritic garnet in PDA from Venetia (sample V948, Jacob et al. 2000). (**a**) Spidergrams (normalized to primitive mantle: Sun and McDonough 1989) show significant differences in most elements apart from titanium and the mid and heavy REE in the grains. (**b**) compatible element Ni and Cr variations (**crosses**: grain averages, **orange** are zones in two different grains), (**c**) variations in Nb/Ta and Zr/Hf ratios in the zoned grains.

Oxidation State

Polycrystalline diamond aggregates are broadly similar to subcratonic lithospheric monocrystalline diamonds with respect to the chemistry of the minerals included in and intergrown with the diamond grains. Like monocrystalline diamonds, PDAs span a large range of ca. 7 log units in oxygen fugacity (fO_2) as evidenced by their associated minerals: carbonates and magnetite in PDAs from Orapa indicate oxidising conditions of around $\log fO_2 = -6$ (Jacob et al. 2011), while cohenite (iron-carbide) as well as native iron in PDAs from Venetia record distinctly reducing conditions of about logfO_2 = −13 (Jacob et al. 2004, 2016); this characteristic reducing signature is shared with the monocystalline diamond suite at this locality (Deines et al. 2001).

Redox reactions, involving redox couples of carbon-bearing fluids and Fe–Ni-bearing phases, are proposed to be major diamond formation processes in the deep Earth's mantle (Rohrbach and Schmidt 2011). PDAs have delivered the first direct evidence for redox freezing of diamond by reduction of carbonatitic fluids with pyrrhotite to form magnetite and diamond (Jacob et al. 2016), while native iron and iron carbide (cohenite) included in a websteritic garnet (Jacob et al. 2004) are products of oxidation of a methane-bearing fluid. It should however be noted, that as of yet, the relationship between monocrystalline diamonds and polycrystalline diamond aggregates is unknown. Since the database for PDAs is much smaller than for monocrystalline diamonds, it is difficult to evaluate the abundance and potential prevalence of certain oxidation states, but the large range of fO_2 is significant and suggests a prominent role of redox gradients and transient, small-scale equilibria in their formation. Grain-scale equilibrium, instead, is scarce and most often absent in PDAs.

OVERVIEW OF CURRENT UNDERSTANDING FOR POLYCRYSTALLINE DIAMOND FORMATION

There can be no single model for polycrystalline diamond-formation, just like there is no single model for monocrystalline diamond-formation. For example, no single model can explain why some samples show high- and low- nitrogen aggregation (Mikhail et al. 2014c, 2019b) and no single model can reconcile the range of oxygen fugacities required to stabilize carbonates + magnetite at Orapa (Jacob et al. 2011, 2016) and cohenite + native iron at Venetia (Jacob et al. 2004).

The mineral inclusions and intergrowths show no indication for derivation from depths well below the subcratonic lithosphere, unlike some monocrystalline diamonds. Only one sample has yet allowed a reliable determination of a depth of formation, placing it close to the base of the subcratonic lithosphere of the Kaapvaal craton at the Orapa locality (Jacob et al. 2016), determined to be at 320–330 km by Fouch et al. (2004). This depth of origin, combined with the absence of significant Si-excess in the garnets, may mean that these diamond species primarily consist of remobilized material derived from the carbonate-enriched keels of the subcratonic lithosphere (Jacob et al. 2000; Foley 2009) and/or that carbon-rich fluids from depth reacted and formed diamond upon entering the ductile to brittle deformation transition zone at the base of the subcratonic lithospheric mantle as suggested by Jacob et al. (2004). The fact that PDAs have only been reported from plume-related Group I kimberlites and not from Group II kimberlites that originate from subcratonic lithospheric sources (Le Roux et al. 2003; Becker and Le Roux 2006) agrees well with the depth of derivation for the PDAs from the base of the lithosphere. What is most apparent is that some form of chemical interaction is involved between surface and mantle material, where mixing between subducted crustal material, namely altered oceanic crust, with ambient mantle carbon and the subcratonic lithospheric mantle is strongly supported by major, minor, and trace element systematics, as well as several isotopic systems (He–C–N–O–Sr–Nd) acquired carbon, oxides, and silicates in the PDAs.

Most garnets in PDAs are websteritic, but the clinopyroxenes are mostly peridotitic (Figs. 7–8). The predominance of websteritic garnets in PDAs (Figs. 7–8) and the absence of olivine (and of orthopyroxene) are a distinctive characteristic of PDAs that are not seen in the inclusion chemistry of monocrystalline diamonds (Figs. 7–8). Subcratonic garnet websterites are typically products of melt interaction with mantle rocks (e.g., Rehfeldt et al. 2008; Mallik and Dasgupta 2012). Hence, the websteritic silicates in the PDAs are most likely hybrid phases and are, like the magnetites, reaction products of the interactions of the carbon-bearing fluid with surrounding mantle material upon diamond formation. For example, a recent study used a predictive thermodynamic modelling approach and was able to fit the major element compositions websteritic garnet and clinopyroxenes for a large number of PDAs, and also those from monocrystalline diamonds (Mikhail et al. 2021). This predictive thermodynamic model was run using the extended Deep Earth Water model (Huang and Sverjensky 2019), which enables the assessment of a huge range of variables (*P–T–X*). The model produces a reaction pathway resulting from the reaction of an eclogitic fluid with a range of model peridotites (lherzolite, harzburgite, and dunite) at 5 GPa, 1000 °C, and across a range of redox conditions ($\log fO_2 = -1$ to -6). The outputs include the co-evolution for fluid composition (including fO_2 and pH) alongside carbon speciation under isothermal and isobaric conditions, and the major element composition and modal abundances for the precipitates (silicates and oxides in the case of Mikhail et al. 2021). The model does not encompass all conditions and compositions observed in nature—it is S- and Cr-free, limited to at 5 GPa and 1000 °C and adopts a smaller range of fO_2 than observed for PDAs—but it does satisfy a range of features common to all PDAs: Mikhail et al. (2021) show that fluid–rock interaction can result in the formation of (high-Mg) eclogitic, websteritic, and peridotitic garnets and clinopyroxenes from a single fluid during a diamond-forming metasomatic reaction (Fig. 12).

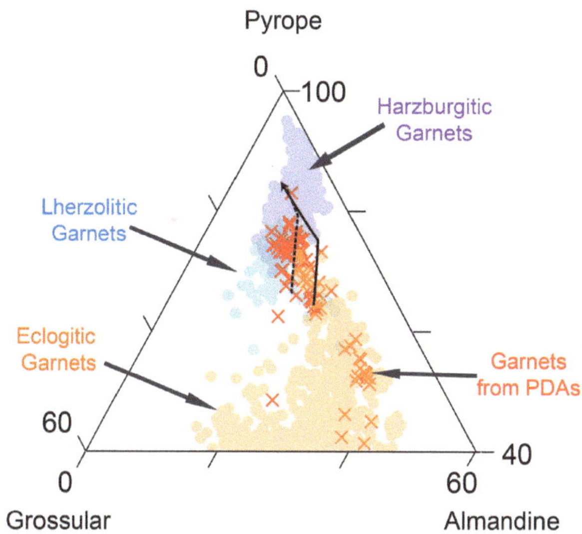

Figure 12. Selected model results for predicted garnet compositions during progressive metasomatism and diamond formation at 5 GPa and 1000 °C. The lines for the runs plotted are thick enough to represent the average trajectories for an eclogitic fluid reacting with model peridotites (lherzolite, harzburgite, and dunite) and model peridotites (lherzolite, harzburgite, and dunite) + aragonite (**dashed line**). Data are from Mikhail et al. (2021) and the input and output files are available at https://doi.org/10.17630/32ebd3c0-bba6-4aa1-9b6d-c53a0a3b61e0. All comparative data shown are taken from the same sources as listed in the caption of Figure 7c.

Interestingly, most models generate ca. 2% by volume magnetite, which could speak to the formation of stewartite, and because the model also generated websteritic garnets (predominantly), this might speak to the formation of framesite (Fig. 12). The model in Mikhail et al. (2021) would remobilize ancient material via dissolution in the fluid and re-precipitate diamond + silicates + oxides (Jacob et al. 2000). This model of diamond growth progressively trapping material during a reaction pathway could also satisfy the occurrence of chemical zonation in PDAs (Fig. 4) and the fact that some PDAs host disequilibrium assemblages (i.e., mixed paragenesis) because the diamond-, silicate-, and oxide-forming fluid follows an evolving pathway. For garnets this would be from low pyrope–eclogitic to high pyrope–peridotitic (Fig. 12). This would provide the mechanism to preserve snapshots of the metasomatic reaction by occasionally trapping garnets and clinopyroxenes at different stages of the reaction thus forming a diamond with inclusions of mixed paragenesis in the same place (no mechanically challenging spatial mobility is required to explain the occurrence) and during the same metasomatic event.

Noteworthy, PDAs also contain protogenetic and epigenetic phases in addition to syngenetic phases which highlights the need for petrographically contextualized data (e.g., μCT datasets). However, the timing of PDA formation is currently the least constrained aspect of their story, and one of the most integral to any realistic geological model. Several lines of evidence discussed in the sections above support the notion that some PDAs are geologically 'old' while others are 'young' in relation to kimberlite emplacement, hence pointing to episodic formation of PDAs likely linked to large-scale geodynamic events. Ergo, we are confident to say that PDA-formation is dynamic and usually involves the tectonically-induced chemical disequilibrium and fluid metasomatism at the base of the SCLM following the interaction of indigenous and subducted carbonaceous material. But we find no evidence for a single model for polycrystalline diamond-formation (note, the model of Mikhail et al. 2021 does not fit the low-Mg eclogitic PDAs). Excitingly, these data also suggest that there is no single polycrystalline diamond-forming event, globally.

PERSPECTIVES ON FUTURE RESEARCH

Placing diamond formation into the context of large-scale tectonothermal processes, such as subduction and plume–lithosphere interaction, is a fundamental requirement for understanding the deep carbon cycle. Therefore, placing PDA formation into the context of diamond formation is essential. This review has outlined what we know and exposed some glaring gaps in the knowledge base. Some of the most pressing questions focus on the what, when, and how of PDA formation.

When did PDAs form?

The temporal relationship between the PDAs and other diamond types, and their host kimberlites, requires immediate attention. While recent instrumental developments have brought radiogenic isotope analysis of very small samples within reach (Koornneef et al. 2014), we do not currently have any direct age information for PDAs. Circumstantial evidence from preserved trace element zonation (Fig. 11) and nitrogen aggregation point to young ages for some samples, whereas mantle residence times for other PDAs are required to be long to explain their nitrogen aggregation characteristics (Fig. 6). PDA formation could be as dynamic as monocrystalline diamond formation with episodic formation spanning billions of years (Gurney et al. 2010; Timmerman et al. 2017), or they could result from distinct metasomatic event(s) which may or may not be related to the formation of monocrystalline diamonds at specific localities (Mikhail et al. 2019b).

What formed PDAs?

The geochemistry of fluids for Orapa PDAs has been calculated using REE data from garnets and these data fit a model for a fluid geochemistry intermediate between the saline and carbonate endmembers (Mikhail et al. 2019b). Direct observations of the fluids by TEM in PDAs from Venetia instead showed carbonatitic-silicic chemistry (Jacob et al. 2014). More *in situ* analyses of the sub-micron fluid inclusions are required to quantify their major and minor element geochemistry to accurately chart their geochemistry within the established framework of the saline-silicic-carbonatitic ternary system (see Weiss et al. 2022, this volume).

How do PDAs form?

The source of PDA-forming fluids is variable and complex, of this there is no doubt. What is clear is that the source materials show mantle and crustal geochemical and isotopic signatures (Figs. 3, 4, 5), and the websteritic silicates are products of wallrock-fluid reactions (Mikhail et al. 2021). This requires some degree of mechanical exchange by subduction zone plate tectonics. However, the isotopic range of C–N data reveal that some PDAs form in distinct episodes with distinct isotopic sources. These data are recorded by spatially unconstrained stepped combustion (Mikhail et al. 2014a), but spatially resolved to be in the order of < 20 mm using SIMS data for some samples (Fig. 4). It is unclear, however, on which time scales these diamond-forming events were recorded in this particular sample suite. It is also unclear how widespread element and isotope zonation is within PDAs, both in the diamond grains (Fig. 4) as well as in the silicates (Fig. 11) because of the difficulty in polishing flat surfaces required to map out zonation using CL imaging or SIMS or LA-ICPMS profiling. Further elucidation of the processes and timescales that lead to the formation of this species of diamond carries the potential to provide an important different perspective into the deep carbon cycle and link global geotectonic processes with the redox-freezing of carbon in the deep lithosphere.

ACKNOWLEDGMENTS

The authors are grateful for reviews by Peter J. Heaney, Oded Navon and Tom McCandless. Karen Smit and Steve Shirey are thanked for their smooth editorial handling and support. DEJ acknowledges financial support by the German Sciences Foundation and the Australian Research Council (CE110001017). SM acknowledges financial support by a National Environmental Research Council standard grant (NE/PO12167/1) and a UK Space Agency Aurora grant (ST/T001763/1).

REFERENCES

Aulbach S, Stachel T, Viljoen KS, Brey GP, Harris JW (2002) Eclogitic and websteritic diamond sources beneath the Limpopo Belt—is slab-melting the link? Contrib Mineral Petrol 143:143–156

Becker M, Le Roux AP (2006) Geochemistry of South African on- and off-craton, Group I and Group II kimberlites: petrogenesis and source region evolution. J Petrol 47:673–703

Boyd SR, Pillinger CT (1994) A preliminary study of $^{15}N/^{14}N$ in octahedral growth form diamonds. Chem Geol 116:43–59

Boyd SR, Mattey DP, Pillinger CT, Milledge HJ, Mendelssohn M, Seal M (1987) Multiple growth events during diamond genesis: an integrated study of carbon and nitrogen isotopes and nitrogen aggregation state in coated stones. Earth Planet Sci Lett 86:341–353

Boyd SR, Pillinger CT, Milledge HJ, Seal MJ (1992) C and N isotopic composition and the infrared absorption spectra of coated diamonds: evidence for the regional uniformity of CO_2–H_2O rich fluids in lithospheric mantle. Earth Planet Sci Lett 108:139–150

Boyd SR, Pineau F, Javoy M (1994) Modelling the growth of natural diamonds. Chem Geol 116:29-42

Broadley MW, Kagi H, Burgess R, Zedgenizov D, Mikhail S, Imayrac M, Ragozin A, Pomazansky B, Sumino B (2018) Plume–lithosphere interaction, and the formation of fibrous diamonds. Geochem Perspect Lett 8:26–30

Brooker R, Du Z, Blundy JD, Keley SP, Allan NL, Wood BJ, Chamorro EM, Wartho JA, Purton JA (2003) The 'zero charge' partitioning behaviour of noble gases during mantle melting. Nature 432:738–741

Bulanova G, Walter M, Smith C, Kohn S, Armstrong L, Blundy J, Gobbo L (2010) Mineral inclusions in sublithospheric diamonds from Collier 4 kimberlite pipe, Juina, Brazil: subducted protoliths, carbonated melts and primary kimberlite magmatism. Contrib Mineral Petrol 160:489–510

Bulanova GP, Wiggers de Vries D, Mikhail S, Beard A, Davies GR, Pearson DG (2014) Carbon recycling and polygenetic origin of eclogitic diamonds recorded by a single crystal from the Mir pipe (Yakutia). Chem Geol 381:40–55

Burgess R, Johnson L, Mattey D, Harris J, Turner G (1998) He, Ar and C isotopes in coated and polycrystalline diamonds. Chem Geol 146:205–217

Busigny V, Cartigny P, Laverne C, Teagle D, Bonifacie M, Agrinier P (2019) A re-assessment of the nitrogen geochemical behavious in upper oceanic crust from Hole 504B: Implications for subduction budget in Central America. Earth Planet Sci Lett 525:115735

Cartigny P (2005) Stable isotopes and the origin of diamond. Elements 1:79–84

Cartigny P (2010) Mantle-related carbonados? Geochemical insights from diamonds from the Dachine komatiite (French Guiana). Earth Planet Sci Lett 296:329–339

Cartigny P, Boyd SR, Harris JW, Javoy M (1997) Nitrogen isotopes in peridotitic diamonds from Fuxian, China: the mantle signature. Terra Nova 9:175–179

Cartigny P, Harris JW, Javoy M (1998a) Eclogitic diamond formation at Jwaneng: no room for a recycled component. Science 280:1421–1424

Cartigny P, Harris JW, Phillips D, Girard M, Javoy M (1998b) Subduction-related diamonds? The evidence for a mantle-derived origin from coupled $\delta^{13}C$–$\delta^{15}N$ determinations. Chem Geol 147:147–159

Cartigny P, Harris JW, Javoy M (1999) Eclogitic, peridotitic and metamorphic diamonds and the problems of carbon recycling—the case of Orapa (Botswana). 7th International Kimberlite Conference Extended Abstracts, p 117–124

Cartigny P, Stachel T, Harris JW, Javoy M (2004) Constraining diamond metasomatic growth using C- and N-stable isotopes: examples from Namibia. Lithos 77:359–373

Cartigny P, Farquhar J, Thomassot E, Harris JW, Wing B, Masterson A, McKeegan K, Stachel T (2009) A mantle origin for Paleoarchean peridotitic diamonds from the Panda kimberlite, Slave Craton: evidence from ^{13}C-, ^{15}N- and $^{33,34}S$ stable isotope systematics. Lithos 112:852–864

Cartigny P, Palot M, Thomassot E, Harris JW (2014) Diamond formation: a stable isotope perspective. Annu Rev Earth Planet Sci 42:699–732

Chacko T, Cole DR, Horita J (2001) Equilibrium oxygen, hydrogen and carbon isotope fractionation factors applicable to geologic systems. Rev Mineral Geochem 43:1–81

Chrenko RM, Tuft RE, Strong HM (1977) Transformation of state of nitrogen in diamond. Nature 270:141–144

De Stefano A, Kopylova MG, Cartigny P, Afanasiev V (2009) Diamonds and eclogites of the Jericho kimberlite (Northern Canada). Contrib Mineral Petrol 158:295–315

De Waele B, Johnson SP, Pisarevsky SA (2008) Palaeoproterozoic to Neoproterozoic growth and evolution of the eastern Congo Craton: Its role in the Rodinia puzzle. Precambrian Res 160:127–141

Deines P (1980) The carbon isotopic composition of diamonds: relationship to diamond shape, color, occurrence and vapor composition. Geochim Cosmochim Acta 44:943–961

Deines P, Viljoen F, Harris JW (2001) Implications of the carbon isotope and mineral inclusion record for the formation of diamonds in the mantle underlying a mobile belt: Venetia, South Africa. Geochim Cosmochim Acta 65:813–838

Dobosi G, Kurat G (2002) Trace element abundances in garnets and clinopyroxenes from diamondites— a signature of carbonatitic fluid. Mineral Petrol 76:21–38

Dobosi G, Kurat G (2010) On the origin of silicate-bearing diamondites. Mineral Petrol 99:29–42

Evans T, Qi Z (1982) The kinetics of the aggregation of nitrogen atoms in diamond. Proc R Soc A 381:159–178

Fettke CR, Sturges FC (1933) Note on the structure of carbonado or black diamond. Am Mineral 18:172–174

Foley SF (2009) Rejuvenation and erosion of the cratonic lithosphere. Nat Geosci 1:503–510

Fouch MJ, James DE, VanDecar JC, van der Lee S, and Kaapvaal Seimsic Group (2004) Mantle seismic structure beneath the Kaapvaal and Zimbabwe cratons. S Afr J Geol 107:33–44

Fourel F, Lécuyer C, Demeny A, Boulvais P, Lange L, Jacob DE, Kovacs I (2017) $^2H/^1H$ measurements of amphiboles and nominally anhydrous minerals (clinopyroxene, garnet and diamond) using high-temperature CF-EA-PY-IRMS. Rapid Commun Mass Spectrom 31:2066–2072

Gautheron C, Cartigny P, Moreira M, Harris JW, Allegre CJ (2005) Evidence for a mantle component shown by rare gases, C and N isotopes in polycrystalline diamonds from Orapa (Botswana). Earth Planet Sci Lett 340:559–572

Gurney JJ, Boyd FR (1982) Mineral intergrowths with polycrystalline diamonds from the Orapa Mine, Botswana. Carnegie Institution Yearbook, p 267–273

Gurney JJ, Harris JW, Rickard RS (1984) Silicate and oxide inclusions in diamonds from the Orapa Mine, Botswana. *In:* Kimberlites II: The mantle and crust–mantle relationships. Vol. 2 Elsevier, p 3–9

Gurney JJ, Helmstaedt HH, Richardson SH, Shirey SB (2010) Diamonds through time. Econ Geol 105:689–712

Haggerty SE (1999) A diamond trilogy: Superplumes, suprecontinents, and supernovae. Science 285:851–860

Haggerty SE (2014) Carbonado: Physical and chemical properties, a critical evaluation of proposed origins, and a revised genetic model. Earth Sci Rev 130:49–72

Haggerty SE (2017) Carbonado diamond: A review of properties and origin. Gems Gemol 53:168–172

Halama R, Bebout GE, John T, Schenk V (2010) Nitrogen recycling in subducted oceanic lithosphere: the record in high- and ultrahigh-pressure metabasaltic rocks. Geochim Cosmochim Acta 74:1636–1652

Halama R, Bebout GE, John T, Scambelluri M (2014) Nitrogen recycling in subducted mantle rocks and implications for the global nitrogen cycle. Int J Earth Sci 103:2081–2099

Harris JW (1968) The recognition of diamond inclusions, Industrial Diamond Review. DeBeers Industrial Diamond Division

Heaney PJ, Vicenzi EP, De S (2005) Strange diamonds: the mysterious origins of carbonado and framesite. Elements 1:85–89

Hoefs J (2009) Stable Isotope Geochemistry, Springer, New York

Howell D, O'Neill CJ, Grant KJ, Griffin WL, O'Reilly SY, Pearson NJ, Stern RA, Stachel T (2012a) Platelet development in cuboid diamonds: insights from micro-FTIR mapping. Contrib Mineral Petrol 164:1011–1025

Howell D, O'Neill CJ, Grant KJ, Griffin WL, Pearson NJ, O'Reilly SY (2012b) μ-FTIR mapping: distribution of impurities in different types of diamond growth. Diamond Relat Mater 29:29–36

Huang F, Sverjensky DA (2019) Extended Deep Earth Water Model for predicting major element mantle metasomatism. Geochim Cosmochim Acta 254:192–230
Hunt L, Stachel T, Morton R, Grütter H, Creaser RA (2009) The Carolina kimberlite, Brazil—Insights into an unconventional diamond deposit. Lithos 112:843–851
Jackson CRM, Parman SW, Kelley SP, Cooper RF (2013) Constraints on light noble gas partitioning at the conditions of spinel-peridotite melting. Earth Planet Sci Lett 384:178–187
Jacob DE, Viljoen KS, Grassineau N, Jagoutz E (2000) Remobilization in the cratonic lithosphere recorded in polycrystalline diamond. Science 289:1182–1185
Jacob DE, Kronz A, Viljoen KS (2004) Cohenite and native iron inclusions in garnets from polycrystalline diamonds. Contrib Mineral Petrol 146:566–576
Jacob DE, Wirth R, Enzmann F, Kronz A, Schreiber A (2011) Nano-inclusion suite and High Resolution Micro-Computed-Tomography of polycrystalline diamond (framesite) from Orapa, Botswana. Earth Planet Sci Lett 308:307–316
Jacob DE, Dobrzhinetskaya L, Wirth R (2014) New insight into polycrystalline diamond genesis from modern nanoanalytical techniques. Earth Sci Rev 136:21–35
Jacob DE, Piazolo S, Schreiber A, Trimby P (2016) Redox-freezing and nucleation of diamond via magnetite formation in the Earth's mantle. Nat Commun 7:11891
Jacob DE, Stern RA, Chapman J, Piazolo S (2017) Insights into diamond formation from polycrystalline diamond aggregates. Goldschmidt Abstracts #1801
Javoy M, Pineau F, Delorme H (1986) Carbon and nitrogen isotopes in the mantle. Chem Geol 57:41–62
Kaminsky FV, Galimov EM, Genshaft YS, Ivanovskaya IN, Klyuev YA, Rovsha VS, Sandomirskaya SM, Smirnov VI (1981) Boart with a garnet from the Mir pipe (Yakutia). Doklady Akademii Nauk SSSR 256:674–677
Kirkley MB, Gurney JJ, Otter ML, Hill SJ, Daniels LR (1991) The application of C isotope measurements to the identification of the sources of C in diamonds: a review. Appl Geochem 6:477–494
Kirkley MB, Gurney JJ, Rickard RS (1995) Jwaneng Framesites: Carbon isotopes and intergrowth compositions. *In:* Diamonds: characterization, genesis and exploration. Meyer HOA, Leonardos OH (eds.) CPRM Spec Publ, Sao Paolo, p 127–135
Kiseeva ES, Wood BJ, Ghosh S, Stachel T (2016) The pyroxenite - diamond connection. Geochem Perspect Lett 2:1–9
Klein-BenDavid O, Wirth R, Navon O (2006) TEM imaging and analysis of microinclusions in diamonds: a close look at diamond-growing fluids. Am Mineral 91:353–365
Klein-BenDavid O, Pearson DG, Nowell GM, Ottley C, McNeill JCR, Cartigny P (2010) Mixed fluid sources involved in diamond growth constrained by Sr–Nd–Pb–C–N isotopes and trace elements. Earth Planet Sci Lett 289:123–133
Koga KT, van Orman JA, Walter MJ (2003) Diffusive relaxation of carbon and nitrogen isotope heterogeneity in diamond: a new thermochronometer. Phys Earth Planet Sci 139:35–43
Koornneef JM, Bouman C, Schwieters JB, Davies GR (2014) Measurement of small ion beams by thermal ionisation mass spectrometry using new 10^{13} Ohm resistors. Anal Chimica Acta 819:49–55
Koornneef JM, Gress MU, Chinn IL, Jelsma HA, Harris JW, Davies GR (2017) Archaean and Proterozoic diamond growth from contrasting styles of large-scale magmatism. Nat Commun 8:648
Kurat G, Dobosi G (2000) Garnet and diopside-bearing diamondites (framesites). Mineral Petrol 69:142–159
Le Roux AP, Bell DR, Davis P (2003) Petrogenesis of Group I kimberlites from Kimberley, South Africa: evidence from bulk-rock geochemistry. J Petrol 44:2261–2286
Leonardos OH (1937) Diamante e carbonado no estado da Bahia. Metallurgia servisio de fomento da produccao Mineral 19:1–23
Logvinova AM, Taylor LA, Fedorova EN. Yelisseyev AP, Wirth R, Howarth G, Reutsky VN, Sobolev NV (2015) A unique diamondiferous peridotite xenolith from Udachnaya kimberlite pipe, Yakutia: role of subduction in diamond formation. Russ Geol Geophys 56:306–320
Mallik A, Dasgupta R (2012) Reaction between MORB-eclogite derived melts and fertile peridotite and generation of ocean island basalts. Earth Planet Sci Lett 329–330:97–108
Maruoka T, Kurat G, Dobosi G, Koeberl C (2004) Isotopic composition of carbon in diamondites: record of mass fractionation in the upper mantle. Geochim Cosmochim Acta 68:1635–1644
Mattey DP, MacPherson CG (1993) High-precision oxygen isotope microanalysis of ferromagnesian minerals by laser-fluorination. Chem Geol (Isotope Geosci Sect) 105:305–318
McCandless TE, Kirkley MB, Robinson DN, Gurney JJ, Griffin WL, Cousens DR, Boyd FR (1989) Some initial observations on polycrystalline diamonds mainly from Orapa. Extended Abstracts 28th Int Geol Congr, p 47–51
Mikhail S, Howell D (2016) A petrological assessment of diamond as a recorder of the mantle nitrogen cycle. Am Mineral 101:780–787
Mikhail S, Kurat G, Verchovsky AB, Jones AP (2013) Peridotitic and websteritic diamondites provide new information regarding mantle melting and metasomatism induced through subduction of crustal volatiles. Geochim Cosmochim Acta 107:1–11
Mikhail S, Howell D, Hutchison MT, Verchovsky AB, Warburton P, Southworth R, Thompson AR, Jones AP, Milledge HJ (2014a) Constraining the internal variability of the stable isotopes of carbon and nitrogen within mantle diamonds. Chem Geol 366:14–23

Mikhail S, Guillermier C, Franchi IA, Beard AD, Verchovsky AB, Wood I, Jones AP, Milledge HJ (2014b) Empirical evidence for the fractionation of carbon isotopes between diamond and iron carbide from the Earth's mantle. Geochem Geophys Geosyst 15:855–866

Mikhail S, Howell D, McCubbin FM (2014c) Evidence for multiple diamondite forming events in the mantle. Am Mineral 99:1537–1543

Mikhail S, Crosby JC, Stuart FM, Di Nicola L, Abernethy FAJ (2019a) A secretive mechanical exchange between mantle and crustal volatiles revealed by helium isotopes in ^{13}C-depleted diamonds. Geochem Perspect Lett 11:39–43

Mikhail S, McCubbin FE, Jenner FE, Shirey SB, Rumble D, Bowden R (2019b) Diamondites: evidence for a distinct tectonothermal diamond forming event beneath the Kaapvaal craton. Contrib Mineral Petrol 174:71–86

Mikhail S, Rinaldi M, Mare ER, Sverjensky DA (2021) A genetic metasomatic link between eclogitic and peridotitic diamond inclusions. Geochem Perspect Lett 17:33–38

Miller CE, Kopylova M, Smith E (2014) Mineral inclusions in fibrous diamonds: constraints on cratonic mantle refertilization and diamond formation. Mineral Petrol 108:317–331

Navon O, Hutcheon ID, Rossman GR, Wasserburg GJ (1988) Mantle-derived fluids in diamond inclusions. Nature 335:784–789

Orlov JL (1977) The Mineralogy of Diamond. Wiley, New York

Palot M, Cartigny P, Viljoen F (2009) Diamond origin and genesis: A C and N stable isotope study on diamonds from a single eclogitic xenolith (Kaalvallei, South Africa). Lithos 112:758–766

Palot M, Cartigny P, Harris JW, Kaminsky FV, Stachel T (2012) Evidence for deep mantle convection and primordial heterogeneity from nitrogen and carbon stable isotopes in diamond. Earth Planet Sci Lett 357:179–193

Palot M, Pearson DG, Stachel T, Harris JW, Bulanova GP, Chinn I (2013) Multiple growth episodes or prolonged formation of diamonds? Inferences from infrared absorption data. *In:* Proceedings 10th International Kimberlite Conference. Vol 1. D Pearson DG, Grütter HS, Harris JW, Kjarsgaard BJ, O'Brien H, Chalpathi Rao NV, Sparks S (eds) Springer, p 281–296

Petts DC, Chacko T, Stachel T, Stern RA, Heaman LM (2015) A nitrogen isotope fractionation factor between diamond and its parental fluid derived from detailed SIMS analysis of a gem diamond and theoretical calculations. Chem Geol 410:188–200

Petts DC, Stachel T, Stern RA, Hunt L, Fomradas G (2016) Multiple carbon and nitrogen sources associated with the parental mantle fluids of fibrous diamonds from Diavik, Canada, revealed by SIMS microanalysis. Contrib Mineral Petrol 171:17–32

Polyakov VB, Kharlashina NN (1995) The use of heat capacity data to calculate carbon isotope fractionation between graphite, diamond, and carbon dioxide: a new approach. Geochim Cosmochim Acta 59:2561–2572

Porcelli D, Elliott, T (2008) The evolution of He isotopes in the convecting mantle and the preservation of high $^3He/^4He$ ratios. Earth Planet Sci Lett 269:175–185.

Rehfeldt T, Foley SF, Jacob DE, Carlson RW, Lowry D (2008) Contrasting types of metasomatism in dunite, wehrlite and websterite xenoliths from Kimberley, South Africa. Geochim Cosmochim Acta 72:5722–5756

Reutsky VN, Harte B, Borzdov YN, Palyanov YN (2008) Monitoring diamond crystal growth, a combined experimental and SIMS study. Eur J Mineral 20:365–374

Reutsky VN, Borzdov YN, Palyanov YN, Sokol A, Izokh O (2015) Carbon isotope fractionation during experimental crystallisation of diamond from carbonate fluid at mantle conditions. Contrib Mineral Petrol 170:41–50

Richet P, Bottinga Y, Javoy M (1977) A review of hydrogen, carbon, nitrogen, oxygen, sulphur, and chlorine stable isotope enrichment among gaseous molecules. Annu Rev Earth Planet Sci 5:65–110

Rohrbach A, Schmidt MW (2011) Redox freezing and melting in the Earth's deep mantle resulting from carbon-iron redox coupling. Nature 472:209–212

Rubanova EV, Piazolo S, Griffin WL, O'Reilly SY (2012) Deformation microstructures reveal a complex mantle history for polycrystalline diamond. Geochem Geophys Geosyst 13:Q10010

Schollenbruch K, Woodland AB, Frost DJ, Wang Y, Sanehira T, Langenhorst F (2011) In situ determination of the spinel–post-spinel transition in Fe_3O_4 at high pressure and temperature by synchrotron X ray diffraction. Am Mineral 96:820–827

Shelkov DA (1997) N and C isotopic composition of different varieties of terrestrial diamond and carbonado. PhD dissertation, The Open University, Milton Keynes, UK

Shelkov DA, Verchovsky AB, Milledge HJ, Pillinger CT (1997) Carbonado: a comparison between Brazillian and Ubangui sources with other forms of microcrystalline diamond based on carbon and nitrogen isotopes. Russ Geol Geophys 38:332–340

Smelova GB (1994) Origin of Diamond Aggregates from the Kimberlite Pipes of Yakutia (Yakutsk), Novosibirsk

Smit KV, Shirey SB, Stern RA, Steele A, Wang WY (2016) Diamond growth from C–H–N–O recycled fluids in the lithosphere: Evidence from CH_4 micro-inclusions and $\delta^{13}C$–$\delta^{15}N$–N content in Marange mixed-habit diamonds. Lithos 265:68–81

Smith CB, Walter MJ, Bulanova GP, Mikhail S, Burnham AD, Gobbo L, Kohn SC (2016) Diamonds from Dachine, French Guiana: a unique record of Early Proterozoic subduction. Lithos 265:82–95

Sobolev NV (1977) Deep-seated inclusions in kimberlites and the problem of the composition of the upper mantle. AGU, New York

Sobolev NV, Lavrentev YG, Pokhilenko NP, Usova LV (1973) Chrome-rich garnets from the kimberlites of Yakutia and their parageneses. Contrib Mineral Petrol 40:39–52

Sobolev NV, Yefimova ES, Pospelova LN (1989) Native iron in Yakutian diamonds and its paragenesis. Soviet Geol Geophys 22:18–21

Sobolev NV, Shatsky VS, Zedgenizov DA, Ragozin AL, Reutsky VN (2016) Polycrystalline diamond aggregates from the Mir kimberlite pipe, Yakutia: Evidence for mantle metasomatism. Lithos 265:257–266

Stachel T, Aulbach S, Brey GP, Harris JW, Leost I, Tappert R, Viljoen KS (2004) The trace element composition of silicate inclusions in diamonds: a review. Lithos 77:1–19

Stachel T, Cartigny P, Chacko T, Pearson DG (2022) Carbon and nitrogen in mantle-derived diamonds. Rev Mineral Geochem 88:809–876

Stachel T, Harris JW (2008) The origin of cratonic diamonds-constraints from mineral inclusions. Ore Geol Rev 34: 5–32

Sun S-S, McDonough W F (1989) Chemical and isotopic systematics of oceanic basalts: implications for mantle composition and processes. *In:* Magmatism in the Ocean Basins, Vol 42, Saunders AD, Norry MJ (eds.) Geol Soc Spec Publ, p 313–345

Sunagawa I (2005) Crystals. Growth, Morphology and Perfection. Cambridge University Press, Cambridge, New York

Tappert R, Foden J, Stachel T, Muehlenbachs K, Tappert M, Wills K (2009) The diamonds of South Australia. Lithos 112:806–821

Thomassot E, Cartigny P, Harris JW, Lorand JP, Rollion-Bard C, Chaussidon M (2009) Metasomatic diamond growth: A multi-isotope study (^{13}C, ^{15}N, ^{33}S, ^{34}S) of sulphide inclusions and their host diamonds from Jwaneng (Botswana). Earth Planet Sci Lett 282:79–90

Thomazo C, Pinti DL, Busigny V, Ader M, Hashizume K, Philippot P (2009) Biological activity and the Earth's surface evolution: Insights from carbon, sulfur, nitrogen and iron stable isotopes in the rock record. C R Palevol 8:665–678

Thomson AR, Kohn SC, Prabhu A, Walter MJ (2021) Evaluating the formation pressure of diamond-hosted majoritic garnets: A machine learning majorite barometer. J Geophys Res: Solid Earth 126:e2020JB020604

Timmerman S, Koornneef J, Chinn I, Davies GR (2017) Dated eclogitic diamond growth zones reveal variable recycling of crustal carbon through time. Earth Planet Sci Lett 463:178–188

Timmerman S, Honda M, Phillips D, Jaques AL, Harris JW (2018) Noble gas geochemistry of fluid inclusions in South African diamonds: implications for the origin of diamond-forming fluids. Mineral Petrol 112:181–195

Timmerman S, Krebs MY, Pearson DG, Honda M (2019a) Diamond-forming media through time—Trace element and noble gas systematics of diamonds formed over 3 billion years of Earth's history. Geochim Cosmochim Acta 257:266–283

Timmerman S, Krebs, MY, Burnham AD, Amelin, Y, Woodland S, Pearson DG, Jaques AL, Le Sosq C, Bennet VC, Bulanova GP, Smith CB, Harris JW, Tohver E (2019b) Primordial and recycled helium isotope signatures in the mantle transition zone. Science 365:692–694

Trueb LF, De Wys EC (1969) Carbonado: Natural polycrystalline diamond. Science 165:799–802

Uenver-Thiele L, Woodland AB, Boffa Ballaran T, Miyajima N, Frost DJ (2017) Phase relations of Fe-Mg spinels including new high-pressure post-spinel phases and implications for natural samples. Am Mineral 102:2054–2064

Viljoen KS, Phillips D, Harris JW, Robinson D (1999) Mineral inclusions in diamonds from the Venetia kimberlites, Northern Province, South Africa. 7th International Kimberlite Conferences, Extended Abstracts, p 943–945

Weiss Y, Kessel R, Griffin WL, Kiflawi I, Klein-BenDavid O, Bell DR, Harris JW, Navon O (2009) A new model for the evolution of diamond-forming fluids; evidence from microinclusion-bearing diamonds from Kankan, Guinea. Lithos 112:660–674

Weiss Y, Griffin WL, Bell DR, Navon O (2011) High-Mg carbonatitic melts in diamonds, kimberlites and the sub-continental lithosphere. Earth Planet Sci Lett 309:337–347

Weiss Y, Cazs J, Navon O (2022) Fluid inclusions in fibrous diamonds. Rev Mineral Geochem 88:475–532

Wiggers de Vries DF, Bulanova GP, De Corte K, Pearson DG, Craven JA, Davies GR (2013) Micron-scale coupled carbon isotope and nitrogen abundance variations in diamonds: evidence for episodic diamond formation beneath the Siberian Craton. Geochim Cosmochim Acta 100:176–199

Wijbrans CH, Rohrbach A, Klemme S (2016) An experimental investigation of the stability of majoritic garnet in the Earth's mantle and an improved majorite geobarometer. Contrib Mineral Petrol 171:50

Wirth R (2009) Focused Ion Beam (FIB) combined with SEM and TEM: advanced analytical tools for studies of chemical composition, microstructure and crystal structure in geomaterials on a nanometre scale. Chem Geol 261:217–229

Woodland AB, Frost DJ, Trots DM, Klimm K, Mezouar M (2012) In situ observation of the breakdown of magnetite (Fe_3O_4) to Fe_4O_5 and hematite at high pressures and temperatures. Am Mineral 97:1808–1811

Woods GS (1986) Platelets and the infrared-absorption of type-Ia diamonds. Proc R Soc Lond Ser A Math Phys Eng Sci 407:219–23

Reviews in Mineralogy & Geochemistry
Vol. 88 pp. 191-256, 2022

Non-cratonic Diamonds from UHP Metamorphic Terranes, Ophiolites and Volcanic Sources

Larissa F. Dobrzhinetskaya

Department of Earth and Planetary Sciences,
University of California at Riverside,
900 University Ave, Riverside, CA 92521, USA

larissa@ucr.edu

Earl F. O'Bannon III

Physical and Life Sciences, Physics Division,
Lawrence Livermore National Laboratory,
Livermore, CA 94550, USA

obannon2@llnl.gov

Hirochika Sumino

Department of General Systems Studies,
Graduate School of Arts and Science,
University of Tokyo, 3-8-1 Komaba, Tokyo, Japan

sumino@igcl.c.u-tokyo.ac.jp

INTRODUCTION

The discovery of diamonds in metamorphic rocks of continental affinities occurred shortly after the discovery of coesite in similar rocks. These important discoveries led to a revolution in our understanding of the subduction and exhumation of continental materials and the establishment of a new discipline, ultra-high-pressure metamorphism (UHPM). After these discoveries more ultra-high pressure (UHP) minerals were found and dozens of UHPM terranes were established. Diamond is a remarkable material which is chemically inert and stable over geological timescales making it the perfect "geological container" where gas, fluid, and solid inclusions can be preserved. Moreover, its presence is indicative of specific pressure and temperature conditions which implie subduction to a minimum depth of ~120 km. The inclusions trapped inside these diamonds can shed light on the composition, redox state, and evolution of the fluids related to UHPM diamond formation while the carbon isotope ratios of the diamond itself can inform us about the source of carbon. The purpose of this review is to provide a brief history on the discovery of microdiamonds starting with work that began in the mid-1960s in Kokchetav massif, Kazakhstan, through to the present day. Particular attention will be on more recent micro-diamond discoveries in the last decade. The current state of understating of the mechanisms of UPHM diamond formation, the misidentification of microdiamonds due to contamination from sample extraction/preparation, and the future of UHPM diamond research are discussed. The paper also considers the controversial topic of the occurrence of lonsdaleite within the population of microdiamonds from UHPM terranes as well as recently reported ophiolite- and volcanic-hosted diamonds.

1529-6466/22/0088-0004$10.00 (print)
1943-2666/22/0088-0004$10.00 (online)
http://dx.doi.org/10.2138/rmg.2022.88.04

A multitude of geological and geodynamic events are recorded in collisional orogens where large-scale processes such as deep subduction of lithospheric plates, mantle convection, and mountain building occur. Continental collision is accompanied by the subduction of continental materials into the Earth's deep interior. The exhumation of these materials back to the Earth's surface is one of the largest natural processes that occur on the Earth. These rocks contain minerals formed under pressures that are higher than the pressures that occur within the crust and they provide a considerable amount of information about processes that occur in the Earth's mantle. Evidence of mineral reactions and deformation, fluid and melt migrations, phase transformations, and geochemical recycling are recorded in these rocks. On a regional scale continental collision zones release "energy" through earthquakes and volcanic eruptions.

In the early stages of the theory of plate tectonics several concepts were established; one such concept was that the continental crust is buoyant and will always remain floating above the oceanic crust. Therefore, this early version of plate tectonic theory did not provide a mechanism for the subduction of continental lithosphere into the Earth's deep interior. This paradigm was significantly challenged in 1984 with two discoveries of coesite, first in the Dora Maira Massif, Alps (Chopin 1984) and the second, published only three months later, in the Western Gneiss Region, Norway (Smith 1984). Soon after these exciting discoveries, graphite pseudomorphs after diamond in the Beni Bousera alpine peridotite massif (Pearson et al. 1989) and then true diamonds, "microdiamonds" were discovered in the Kokchetav massif, Kazakhstan (Sobolev and Shatsky 1990). These startling revelations ushered in an entirely new discipline, ultra-high-pressure metamorphism (UHPM).

Crustal rocks of that have experienced recrystallization within or above the coesite and diamond stability fields (≥ 2.7–4.0 GPa, ~700–1000 °C) implying subduction to a depth of ≥ 90–120 km are considered as UHPM rocks. The presence of coesite and/or diamond in metamorphic rocks of continental affinities has become the standard for defining UHPM terranes. UHPM research flourished due to the detailed studies of outcrops and mineral/rock thin-sections in laboratories with the aid of advanced state-of-art analytical and synchrotron assisted instruments and technologies. This led to the establishment of more than 20 coesite- and nine diamond-bearing localities (Fig. 1) (e.g., Ogasawara 2005; Dobrzhinetskaya 2012; Schertl and Sobolev 2013; Liou et al. 2014 and references therein). By combining all the results of this research, an integrated picture of the processes operating in the deep Earth at converging plate boundaries is realized. Moreover, a greater understanding of how crustal materials are recycled into the deep mantle environment and incorporated into large-scale mantle dynamics is also gained from these studies.

Diamond is an important index-mineral of UHPM events. Since diamond is chemically inert and stable over geological timescales it is the perfect "geological container" where gas, fluid, and solid inclusions can be preserved. These inclusions can provide detailed information about the conditions and chemistry of diamond forming media as well as mantle mineralogy and geochemistry. Moreover, diamond, which is often included in zircon, can be a precise geochronological indicator of the peak of UHP metamorphism. Aside from the inclusions, $\delta^{13}C$ values of the diamond itself can provide information about the carbon reservoir from which it crystallized. The type of nitrogen aggregation in diamond can shed light on the residence time of diamond in the environment where the diamond crystallized.

There are many types of reviews published in international forums that discuss impact of UHP mineralogy and UHP metamorphism studies on refining thermobarometry and plate tectonic modelling (e.g., Ernst and Liou 1999, 2008; Chopin 2003; Liou et al. 2004, 2009, 2014; Green 2005; Gerya 2010; Beltrando et al. 2010; Green et al. 2010; Guillot et al. 2008; Schertl and Sobolev 2013; Shirey et al. 2013; Faryad and Cuthbert 2020). However, comprehensive reviews of the studies of UHPM diamonds as well as their detailed studies with modern state-of-the-art instruments and technologies are still scarce (e.g., Ogasawara 2005; Dobrzhinetskaya 2012).

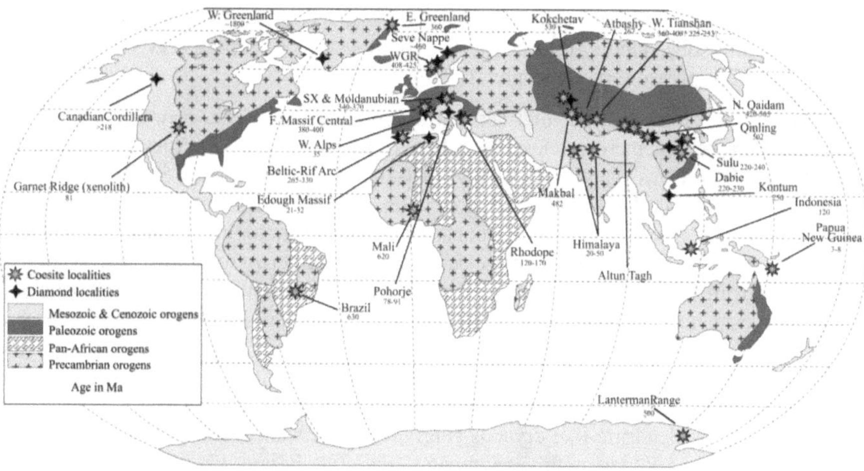

Figure 1. World map of coesite and non-cratonic microdiamond occurrences within Precambrian, Pan–African, Paleozoic and Mezozoic orogens. Numbers indicate age of UHP metamorphism in Ma. Sx–Saxoturingian area of Bohemian massif. Map is modified after Dobrzhinetskaya and Faryad (2011).

The goal of this review is to provide an update of new diamond discoveries that led to the establishment of new UHPM terranes, and to summarize and critically analyze existing concepts of diamond formation in orogenic belts, ophiolites and volcanic sources. An additional objective is to specifically emphasize certain aspects of knowledge from studies of diamonds which have been missed before and/or less highlighted in current research projects. The review outlines new constraints from many well-known collisional orogens which will lead to the improvement of geotectonic modelling that is currently being developed for very deep subduction of continental rocks, and their exhumation.

DIAMONDS FROM ULTRA-HIGH-PRESSURE METAMORPHIC TERRANES

The history of microdiamond discoveries in the Kokchetav massif, Kazakhstan

In the mid-1960s several detrital grains of diamond ranging in size from 7 to 200 mm were found in the Tertiary detrital deposits of titaniferous sands in the vicinity of the Kokchetav massif, northern Kazakhstan (Kashkarov and Polkanov 1964). The alleged source of the detrital material was determined to be the Precambrian metamorphic rocks of the Kokchetav massif (Essenov et al. 1968). By the 1970s extensive geological mapping and exploration for mineral resources began in Kazakhstan, led by A. A. Zaychkovsky from the Kokchetav Geological Survey. They collected nearly 10 m^3 of rocks from the eclogite lenses enclosed in garnet–biotite gneisses that were exposed at the shore of Kumdy Kol lake, Akmola Region, Kazakhstan. After crushing the rocks, the heavy minerals separation and flotation techniques were applied and, several microdiamonds were recovered from the residua (Rozen et al. 1972). It was concluded that these diamonds had formed under unusual crustal metamorphic conditions (Rozen et al. 1979).

The first detailed geological and petrographic descriptions of microdiamonds found *in situ* together with their host minerals was reported in Russian publications by Sobolev and Shatsky (1987, 1988). However, these unusual diamonds remained unknown to Western scientists until Sobolev and Shatsky (1990) published their results in *Nature*. They provided a detailed description of diamond morphologies and presented credible and convincing images of diamonds *in situ*. They were the first who postulated that these diamonds had formed in regional metamorphic rocks of continental affinities during deep subduction. They emphasized that new models are needed to explain how crustal rocks can be subducted to depths greater than ~100–120 km and subsequently returned to the Earth's surface with preserved mineralogical relicts of ultra-high-pressure metamorphic (UHPM) events.

Following the efforts of the international community, the Kokchetav diamond-bearing metamorphic terrane became the "classic locality" of ultra-high-pressure metamorphism (e.g., Ernst and Liou 1999; Ogasawara 2005; Dobrzhinetskaya 2012; Schertl and Sobolev 2013; Liou et al. 2014). The "classic UHP rocks" of Kokchetav massif include metamorphic rocks with sedimentary and volcanic protoliths. They consist of varying lithologies of felsic gneisses, calc-silicate rocks, schists, quartzites, marbles, eclogites and garnet pyroxenites. Microdiamonds ranging in size from ~10 to 100 μm (rare up to 150–300 μm) with an average size ~40–50 μm occur in all the above-mentioned lithologies (Fig. 2) in high concentrations (e.g., Dobrzhinetskaya 2012; Schertl and O'Brien 2013). According to Pechnikov and Kaminsky (2008) the Kokchetav microdiamond reserves are calculated to be as large as ~2.5 billion carats, with the average content of ~20 carats per metric tonne. The Kokchetav diamonds have varying morphologies such as rose-like and dendritic-like crystals (Fig. 3A), platy skeletal-like and shapeless single crystals (Fig. 3B,C), or polycrystalline aggregates (Fig. 2D), hopper-like cuboidal, or cuboidal and octahedral-like with truncated corners, and sometimes slightly rounded single crystals (Fig. 3E,F). They are often observed as inclusions in refractory minerals such as zircon and garnet where they remain well-preserved and protected from complete transformation to graphite during exhumation of the UHPM rocks accompanied by retrograde metamorphism. While zircon and garnet protect diamond inclusions from graphitization, the radiation damage (metamictization) of microdiamonds included in zircon is a common phenomenon. Several studies (e.g., Orwa et al. 2000; Shimizu and Ogasawara 2014) have reported that the radiation damage of the Kokchetav diamonds occurred due to α-particle emissions from radiactive decay of U and Th in host zircon. This damage can be qualitatively probed with Raman spectroscopy, the Raman mode of diamond in the core of zircon is broadened and shifted to lower wavenumbers relative to pristine diamond (Shimizu and Ogasawara 2014), additionally two Raman modes at 1490 cm^{-1} and 1630 cm^{-1} have been observed and attributed to the vacancy and the split interstital, respectively (e.g., Prawer et al. 2004).

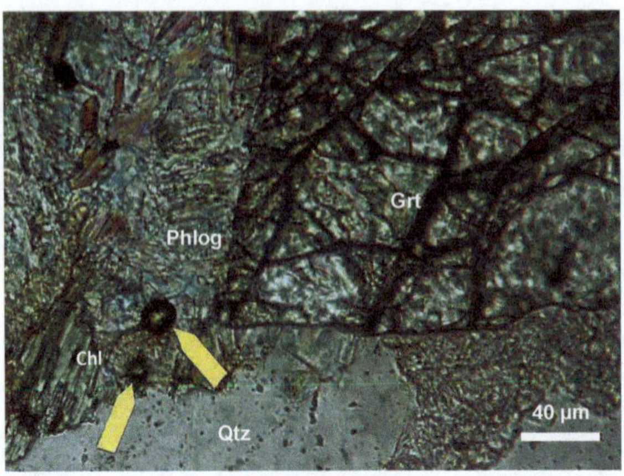

Figure 2. Photomicrograph of a standard petrographic thin section showing two diamond crystals (*in situ*) in quartz–calc–silicate rocks from Kumdy Kol lake locality, Kokchetav massif, Kazakhstan. Photomicrograph taken with a plane polarized transmitted light from an optical microscope equipped with a digital camera. **Yellow arrowheads** point to two diamond crystals situated at the phlogopite (Phlog)–chlorite (Chl) boundary; Grt–garnet; Qtz–quartz. Sample was collected by L. Dobrzhinetskaya from the abandoned underground mining gallery which crosses the diamond-bearing rocks of different lithologies; the petrographic thin section was prepared by L. Dobrzhinetskaya without any diamond abrasives.

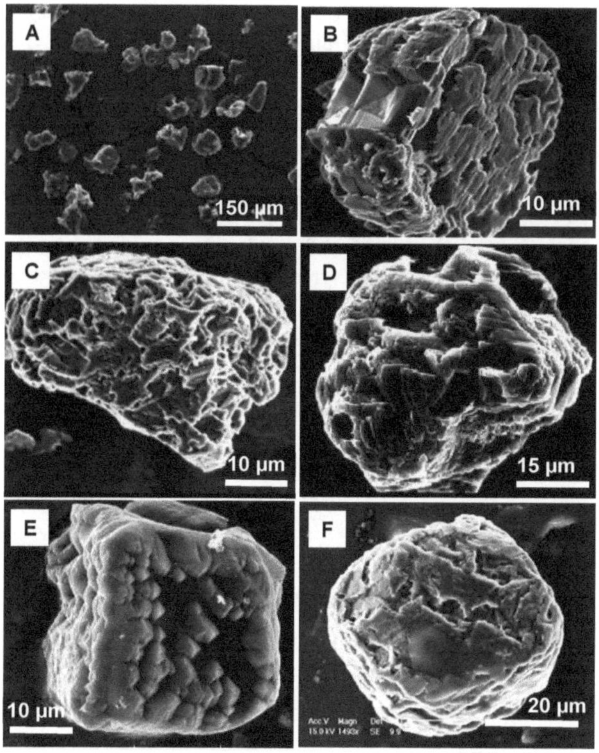

Figure 3. Secondary Electron Scanning Electron Microscope images of microdiamonds from Kokchetav massif, Kazakhstan (L. Dobrzhinetskaya's collection). **A**–diamonds separated from garnet–biotite gneiss; **B**–platy skeletal-like diamond single crystal from quartzite; **C**–shapeless diamond crystal from metacarbonate rocks; **D**–polycrystalline diamond from marble; **E**–truncated corners cuboid-like diamond with graphite (**black contrast**) from garnet–biotite gneiss; **F**–cuboidal diamond with truncated corners—from garnet–biotite gneiss.

The peak metamorphic conditions of the Kokchetav UHPM rocks are estimated by thermobarometry of the rock forming minerals to be P = ~4.5 GPa and T = 950–1000 °C(e.g., Ogasawara 2005). Studies of nanoinclusions in these diamonds suggest that the Kokchetav diamonds formed at a pressure range of 6–9 GPa, and a temperature range of 980–1200 °C (e.g., Dobrzhinetskaya et al. 2006a). The age of the UHPM event is Cambrian ~531 Ma whereas the protoliths of diamond-bearing rocks are as old as Precambrian ~2100–2700 Ma (Table 1).

Soon after Kazakhstan, similar microdiamonds (Fig. 4) were found in the Western Gneiss Region (WGR), Norway (Dobrzhinetskaya et al. 1995; van Roermund et al. 2002), the Erzgebirge Massif of Germany (Massonne 1999), in China in the metamorphic rocks of the Dabie mountains (Xu et al. 1992, 2003), and Qinling region (Yang et al. 2003), and the Kimi complex of the Greek Rhodope (Mposkos and Kostopoulos 2001, Perraki et al. 2006). Microdiamonds have also been discovered in UHP oceanic floor metasediments at Lago di Cignana, the western Alps, Italy (Frezzotti et al. 2011, 2014). The above-mentioned localities are unconditionally classified as UHPM terranes because aside from diamond other ultra-high-pressure minerals have been discovered there such as coesite, TiO_2-II (rutile with α-PbO_2 structure), majoritic garnet, Ca-Eskola-rich clinopyroxene (Ca-Eskola = $Ca\square Al_2Si_4O1_2$ where $\square$ represents a cation vacancy in the pyroxene structure, see Vogel 1966 and others). Since the detailed reviews on UHPM diamonds were published by Ogasawara (2005), Dobrzhinetskaya (2012) and Liou et al. (2014) several new diamond-bearing metamorphic rock localities have been reported.

Table 1 summarizes the occurrences of microdiamonds in metamorphic rocks starting from their earlier descriptions (e.g., Dobrzhinetskaya 2012) to present. Within them are new diamond localities reported from granulite and garnet peridotites from the northern part of the Bohemian massive (Kotková et al. 2011; Naemura et al. 2011; Perraki and Faryad 2014), detrital diamonds from Erzgebirge (Schönig et al. 2019), from felsic granulites of the Betic Rif Cordillera, NW Africa and SE Spain (Ruiz-Cruz and Sanz de Galdeano 2012, 2013, 2014), from Straumen area of Western Gneiss Region, Norway (Smith and Godard 2013), from gneisses and eclogites of the Tromsø Nappe of Caledonides (Janák et al. 2013), from gneisses of Seve Nappe of Caledonides, Sweden (Majka et al. 2014; Klonowska et al. 2017), from oceanic metasediments of Lago di Cignana, Western Alps, Italy (Frezzotti et al. 2011; 2014; Frezzotti 2019), from garnet–kyanite–quartz–feldspathic gneisses of Pohorje, Eastern Alps (Janák et al. 2015), from amphibolites of Quinling region, China (Wang et al. 2014), from West Greenland (Glassley et al. 2014), and from garnet megacryst in mélange of the Edough Massif, NE Algeria (Bruguier et al. 2017).

Table 1. Microdiamonds occurrences in worldwide ultrahigh-pressure metamorphic terranes. This table contains part of information adopted from Table 1 (Dobrzhinetskaya 2012).

UHPM terrane/ locality	Available data on diamond features	Age (Ma)	*P* (GPa), *T* (°C) estimate, and UHP index minerals associated with diamond	Data source
Kokchetav massif, Kazakhstan (Kumdy-Kol, Barchi-Kol)	size of crystals: 10–100 μm (average ~40 μm); skeletal, imperfect; cubes with truncated corners single crystals, and polycrystalline diamonds	531 (UHPM) 2100–2700 (protolith)	$P = 6–9$; $T = 980–1200$ coesite, titanatite with exsolution lamella of coesite; diamonds with inclusions of aragonite + $MgCO_3$	Sobolev and Shatsky 1990; Claoué-Long et al. 1991; Ogasawara et al. 2000, 2002; Okamoto et al. 2000; Parkinson 2000; Katayama et al. 2001a; Dobrzhinetskaya et al. 2001, 2003, 2006a
Dabie-Sulu, China	Diamonds (~60–150 μm size) included in garnet from eclogite; in situ in polished thick sections	220–240	$P > 2.7–5$; $T = 600–930$ coesite, diamond	Xu et al. 1992, 2003
North Qaidam, China	Diamond inclusion in Zr recovered from garnet peridotite	420–450	$P > 2.8–4$; $T = 620–740$ coesite, diamond, majoritic garnet, relicts of stishovite	Song et al. 2005; Mattisson et al. 2006; Zhang et al. 2006; Liu et al. 2007; Liou et al. 2009; Katsube et al. 2009
North Qinling, China	Diamond (≤1 μm size) inclusions in zircon extracted from eclogite body	507	$P > 2.6$; $T = 590–760$ diamond	Yang et al. 2003
	Diamond inclusion in zircon extracted from amphibolite body	490.4 ± 5.8	$P \sim 4.0 \pm 0.5$; $T = 670–750$	Wang et al. 2014

UHPM terrane/ locality	Available data on diamond features	Age (Ma)	P (GPa), T(°C) estimate, and UHP index minerals associated with diamond	Data source
Bohemian masif, Erzgebirge, Saidenbach, Germany	Diamonds (5–50 μm size) imperfect cubes, rose-like single crystals, and polycrystalline diamonds	360 (UHPM)	P>7; T = 900–1200 coesite, diamond, TiO_2 with αPbO_2 structure	Massonne 1999; Nasdala and Masssonne 2000; Hwang et al. 2000; Massonne and O'Brien 2003; Massonne et al. 2007
Bohemian massif, Moldanubian Zone, Ceské Stredohorí, Stráž nad Ohrí, Plesovice (Czech Republic)	Diamonds (~ 5–30 μm size) in situ graphitized diamond < 2 μm	340–380 (UHPM)	P > 3.5; T = 1100 diamond	Kotková et al. 2011; Naemura et al. 2011
Bohemian massif, Moldanobian Zone, Kutna Gora area, Czech Republic	Diamonds of ~7–8 μm Gföhl kyanite–bering granulites	~360 Ma (UHPM)	P > 4.5; T~680 Diamond, coesite, moissanite	Perraki and Faryad 2014
Western Gneiss Region, Norway (Fjortoft, Svartberget)	Diamons (1 nm–20 μm size) round-like crystals with "striation" and imperfect cubes with truncated edges	408–425 (UHPM)	P > 3.2–4; T = 800 coesite, majoritic garnet, diamond	Smith 1984; Dobrzhinetskaya et al. 1995; van Roermund et al. 2002; Spengler et al. 2006; Vrijmoed et al. 2006, 2008
Northern Norway, Tromso Nappe Caledonides	Diamond inclusions in garnet from kyanite–garnet–biotite–gneisses	452	P = 3.5 ± 0.5; T = 800	Janák et al. 2013
Western Gneiss Region, Norway, Straumen	Diamond inclusion in zircon from eclogite	408–425	P = 3.75 ± 0.75; T = 750 ± 150	Smith and Godard 2013
Rhodope, Greece (Kimi and Sideronero)	Diamonds of 3–9 mm size inclusions in garnet (Kimi and Xanthi area) and in garnet from garnet–kyanite–mica schist, Sideranero area	202 (UHPM) 39–186	P = 2.2; T = 750	Mposkos and Kostopoulos 2001; Perraki et al. 2006; Schmidt et al. 2010, 2011; Liati et al. 2011
Western Alps (Lago di Cignana)	Diamond (1–2 μm size) inclusions in Mn-rich garnet; associates with fluid inclusions of HCO_3	35	P ≥ 3.2; T ≤ 600 (cold subduction)	Frezzotti et al. 2011; 2014; Frezzotti 2019
Eastern Alps, Pohorje	Diamond + SiC inclusions in garnet	92–95	P ≥ 3.5 GPa, T = 800–850	Janák et al. 2015

UHPM terrane/ locality	Available data on diamond features	Age (Ma)	*P* (GPa), *T* (°C) estimate, and UHP index minerals associated with diamond	Data source
Betic Rif Cordilleras, Ceuta (NW Africa and SE Spain)	Diamond (5–15 μm size) intergrown with coesite; diamond inclusions in garnet, kyanite, apatite	330	$P > 4.3$–7; $T \rightarrow 1150$	Ruiz Cruz and Sanz de Galdeano 2012, 2013, 2014
Snasahögarna Mt., Seve Nappe, Sweden	Diamond (~1 μm size) diam + carbonate + graphite inclusions in garnet	441–445	$P = 4.1$–4.2; $T = 830$–840	Majka et al. 2014; Klonowska et al. 2017
Nagssugtoqidian Orogen, West Greenland	Diamond is partly replaced by graphite, majorite, reconstructed ringwoodite	1800	P ~7; T ~970	Glassley et al. 2014

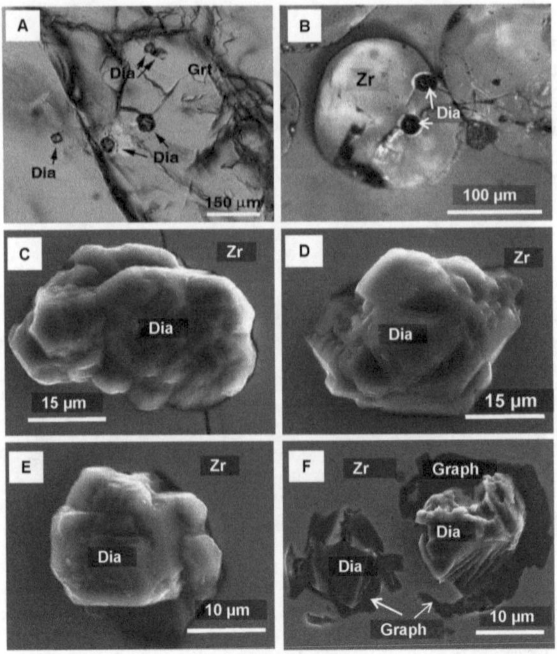

Figure 4. Diamonds from Erzgebirge UHPM terrane, Germany. Samples are collected by L. Dobrzhinetskaya from garnet–phengite–quatz–feldspathic gneisses occurred as small outcrops on the eastern shore of the Saidenbach Water Reservoir, Saxonian Erzgebirge, Germany. **A**–Photomicrograph of diamond inclusions in garnet from garnet–biotite–quartz–feldspathic gneiss (optical microscope equipped with a digital camera, polished thin section, reflected light). **B**–Photomicrograph of diamond inclusions in zircon separated from garnet–quartz–feldspathic gneiss (optical microscope equipped with a digital camera, zircons are glued on the petrographic glass slide and slightly polished, reflected light mode). **C**–**E**–Secondary electron SEM images of diamond inclusions in zircon (zircon was polished with a special technique using colloidal silica polishing compound). **F**–Secondary electron SEM images of diamond partly replaced by graphite included in zircon (sample preparation technique is like C–E).

New microdiamond localities discovered between 2011 and 2020

České Středohoří, Strážnad Ohří, Plešovice and Gföhl areas of Bohemian massif, Central Europe. The first *in situ* microdiamonds were reported as inclusions in garnet from felsic gneisses exposed in vicinities of Erzgebirge, northwestern part of the Bohemian Massif, Germany (Massonne 1999). The Erzgebirge locality became the second classical example of an UHPM terrane supported by extensive studies and publications (e.g., Nasdala and Massonne 2000; Stöckhert et al. 2001, 2009; Massonne 2003; Massonne and Tu 2007; Massonne et al. 2007b; Dobrzhinetskaya et al. 2003, 2006b, 2013). Since 2011 several new microdiamonds localities (Table 1) in the northern part of the Bohemian massif have also been reported (Kotková et al. 2011; Perraki and Faryad 2014).

Kotková et al. (2011) unambiguously documented microdiamonds ranging in size from 5–30 µm *in situ* in thin sections (Fig. 5) prepared from drill cores of felsic granulites extracted from two boreholes at the České Středohoří Mountains and from an outcrop in Stráž nad Ohří (Czech Republic).

Microdiamonds from the felsic granulites of these new localities occur as inclusions in grt, zr, and ky included in garnet. The correctly performed polishing technique provided indisputable evidence that the microdiamonds are *in situ* (Fig. 5). The microphotographs show that diamond inclusions "protrude" above the host minerals (zircon and/or garnet) well-polished surface. Coesite inclusions in kyanite included in garnet were also found in felsic granulite (Kotková et al. 2011). The authors hypothesized that during subduction (>120–150 km) the K-feldspar- and qtz-rich continental crust had "amalgamated" with small fragments of mantle materials and the latter are now presented as sparse lenses of the garnet peridotite containing relicts of Cr-spinel.

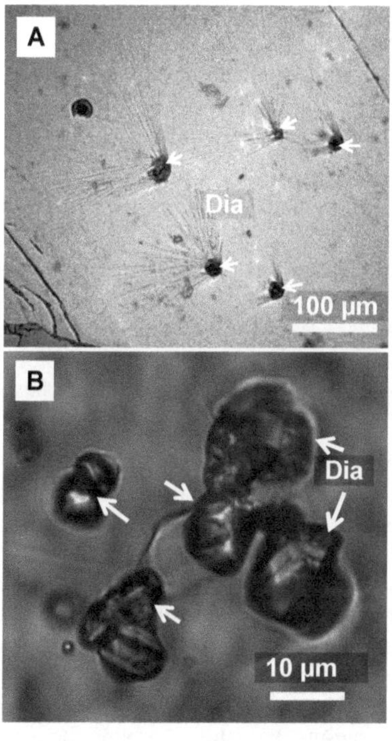

Figure 5. Diamonds from Saxony-type granulite of Bohemian massif (Kotková et al. 2011 with permission of Geology Research Czech Republic Moravia Periodicals). Photomicrographs from optical microscope: **A**–microdiamond (**white arrowheads**) inclusions in garnet (reflected light mode); polishing striation developed on the garnet surface around the microdiamonds. **B**–microdiamond (**white arrowheads**) inclusions in garnet (plain parallel light).

Kotková et al. (2011) emphasized that the protolith of these newly discovered diamond-bearing felsic granulites was "granitic crustal" rocks, which is different from the protolith of the Erzgebirge locality in Germany where the first diamond from this locality was reported by Massonne (1999). The Erzgebirge diamonds occur in garnet–biotite–phengite–quartz–feldspathic gneisses with a metasedimentary protolith. However, all these rocks belong to the same rock-unit of the Variscan orogeny, 340–380 Ma (e.g., Kröner and Willner 1998; Massonne and O'Brien 2003). Faryad and Cuthbert (2020) considered these diamond-bearing felsic granulites as former UHPM rocks that were overprinted by granulite facies metamorphism after their exhumation to the crust. Therefore, further studies of K-Fsp-rich felsic granulites with a primary granitic protolith may provide the opportunity to discover other UHPM minerals such as K-cymrite and/or K-wadeite which were first synthesized in laboratory (e.g., Fasshauer et al. 1997; Davies and Harlow 2002). Later K-cymrite was discovered in North Qaidam UHP eclogite, Western China (Zhang et al. 2009).

Perraki and Faryad (2014) discovered diamond, coesite, and moissanite inside of polished zircon grains. The zircons were extracted from the heavy minerals fraction that was separated from the crushed kyanite-bearing Gföhl granulite samples (the Moldanubian Zone of the Bohemian massif). The Gföhl kyanite-bearing granulites contain lenses and boudins of eclogites, garnet peridotites, and garnet pyroxenites. The thermobarometric calculations suggest that UHP metamorphism occurred at $P > 4$ GPa and $T = 680$ °C followed by granulite facies metamorphism and partial melting at $P < 2$ GPa and $T = 850–950$ °C. Perraki and Faryad (2014) agreed with Kotková et al. (2011) that the diamonds crystallized during UHP metamorphism of the Variscan Orogeny.

One novel aspect of the study of Perraki and Faryad (2014) is that they found SiO_2 (coesite) coexisting with diamond and moissanite (SiC). Moissanite formation requires extremely reducing conditions, e.g., 4.5–6 log units below the iron–wustite (IW) buffer, whereas diamond and SiO_2 (coesite) may coexist at an oxygen fugacity close to the diamond + CO fluid buffer (CCO) and the fayalite–magnetite–quartz (FMQ) oxygen buffers (e.g., Mathez et al. 1995; Frost and Frost 2014). Additional studies of the coexisting diamond, coesite, and moissanite inclusions in zircons from Gföhl granulites should be a high priority for future investigations, since such a study would shed light on the redox conditions of fluid–rock interactions and microdiamond formation in UHPM terranes.

Microdiamonds in detrital garnets from Erzgebirge, Germany. After several decades of the intensive studies motivated by the first discoveries of coesite and microdiamonds in rocks of continental affinities, it was shown that UHP metamorphism is not a rare phenomenon. However, despite the widespread exposure of presumably UHPM rocks in the continent–continent collisional orogens the mineralogical evidence of ultra-high-pressure metamorphism remains scarce, since during exhumation UHP minerals are typically overprinted with lower-grade metamorphism and partial melting. Schönig et al. (2019) proposed to study the distribution and characteristics of UHPM rocks by analyzing detrital garnets that have accumulated in surface sediments.

They collected detrital garnets from loose sediments deposited near the Erzgebirge massif where diamonds were described as inclusions in garnets and zircons (e.g., Nasdala and Massonne 2000; Dobrzhinetskaya et al. 2003). The sediments consist of sand from the creeks around the Saidenbach reservoir that have been draining the Erzgebirge UHPM rocks for a long time. After heavy mineral separation and examination of the residua with optical microscopy about 700 grains of garnets were chosen for further research. Out of the 700 inclusion bearing garnets they found 26 garnets containing 46 inclusions of monomineralic coesite, and 22 garnets containing 41 diamond inclusions, and they examined them with Raman spectroscopy (Schönig et al. 2019). These results provide undisputable evidence that the processes of erosion and weathering of UHPM rocks can relocate UHP minerals to sediments scattered around the hard rock exposures. Therefore, this method may be used as an additional technique for discovering new UHPM terranes. Similar methodology has been successfully applied to single zircon grains that were separated from UHPM rocks and has been successfully used for more than a decade to identify microdiamonds and other UHPM mineral inclusions.

The Betic and Rif Cordilleras, NW Africa, and SE Spain. A new microdiamond-bearing locality was discovered within the felsic gneisses of granulite facies in the Betic–Rif arc (NW Africa) and in the SE of Spain (Ruiz-Cruz and Sanz de Galdeano 2012). This tectonically complicated domain belongs to the pre-Mediterranean Alpine orogenic zone formed during the Africa–Eurasia collision (e.g., Platt et al. 2013). It should be noted that almost two decades ago cubic graphite pseudomorphos after diamonds were discovered in peridotites of Beni Bushera massif (Morocco), situated within the Rif orogenic belt (Pearson et al. 1989). This discovery was the first report that mantle rocks can be tectonically emplaced into the crustal environment from a depth corresponding to the diamond stability field. However, unambiguous proof of this concept was emphasized by finding similar cubic graphite in the Ronda peridotite massif, Spain (Davis et al. 1993). They showed that the basal planes {0001} of graphite are parallel to the {111} of the octahedra which demonstrated distinctly that graphite is pseudomorph after diamond. Studies of carbon isotopes in cubic graphite of both Beni Bushera and Ronda massifs show that the precursor diamonds crystallized from a ^{13}C-depleted reservoir most likely originating from subducted crust (Pearson et al. 1991). A decade later, El Attrasi et al. (2011) reported finding of the 0.5–2 μm size diamond crystals included in large graphite flakes extracted manually from a garnet pyroxenite layer of the Beni Bousera massif, thereby providing unequivocal confirmation that these rocks were at some point in the diamond stability field.

Ruiz-Cruz and Sanz de Galdeano (2012) reported findings of octahedral, cuboid and cubo-octahedral microdiamonds of <5 × 10–15 μm size (Table 1) as inclusions in garnet, kyanite, and quartz with relicts of coesite from felsic granulites exposed in the Ceuta area of the Northern Rif, NW Africa. The authors have estimated the metamorphic conditions of the Ceuta granulites as $P = 4.3$ GPa and $T = 1100$ °C (corresponding to a depth of >150 km) and suggested that the diamond-bearing crustal rocks and peridotites experienced a similar UHP metamorphic condition. They also suggested that the UHP metamorphism occurred ~330 Ma ago followed by partial melting and migmatization at ~265 Ma (Ruiz-Cruz and Sanz de Galdeano 2013). Given that the diamond-bearing garnet from felsic gneiss contains apatite' exsolution lamella and clusters, the authors calculated the phosphorus solubility in garnet and compared it with those obtained in high-pressure and high-temperature experiments (Konzett and Frost 2009; Konzett et al. 2012). Based on such a comparative analysis they concluded that the highest pressure reached during UHPM was 7 GPa, and temperature ~1150 °C (Ruiz-Cruz and Sanz de Galdeano 2013).

Straumen localiy, Western Gneiss Region, Norway. After the discovery of coesite inclusions in clinopyroxene occurring in the dolomite–eclogites from Grytting, Norway (Smith 1984) the Geological Survey of Norway (NGU) initiated a search in 1992 for possible microdiamonds in the WGR. Garnet–kyanite–phlogopite gneisses were collected at Fjørtoft island of WGR and crushed at the NGU laboratory. Several grains of microdiamonds (30–45 μm size) were separated from the heavy mineral concentrates extracted at the NGU chemical laboratory using a method of thermochemical decomposition (e.g., Dobrzhinetskaya et al. 1995). Later, microdiamonds were found *in situ* in thin sections prepared from hand-samples of garnet peridotite from Fjørtoft (van Roermund et al. 2002), and in Fe-Ti garnet peridotite from Svartberget (Vrijmoed et al. 2006, 2008).

In situ diamond has also been reported from two more localities in the central and northern parts of WGR, Norway (Smith and Godard 2013; Janák et al. 2013). Smith and Godard (2013) conducted Raman spectral point analysis and Raman mapping of the carbon-bearing inclusions in zircon from kyanite–phengite–coesite-bearing eclogites from the Straumen eclogite pod in the WGR, and they identified two graphitized diamonds. The authors noticed that the position of the

1st order Raman mode of the diamond inclusions varies from 1332 cm^{-1} (ambient value for single crystal diamond) down to as low as 1322 cm^{-1}. They observed additional features in the Raman spectra of these diamonds (doublets with peaks at 1328 cm^{-1} and 1322 cm^{-1}) which are like the Raman spectra of diamonds from the Kokchetav Massif. Given the Raman results and the *in situ* nature of the diamonds the geological origin of the Straumen diamonds is indisputable.

Tonsvika area, Tromsø nappe, the northern Norwegian Caledonides. Janák et al. (2013) documented microdiamonds (~10 μm) *in situ* in kyanite–garnet–biotite–gneisses exposed near Tonsvika, Tromsø nappe (452 Ma), in the northern part of the Norvegian Caledonides territory (Table 1). Microdiamonds were observed in thin sections with optical microscopy and confirmed with Raman spectroscopy. Microdiamonds occur as single crystals as well as multiphase inclusions of diamond + $MgCO_3$ and/or diamond + $CaCO_3$ in garnet. In addition to the diamond and diamond–carbonate inclusions, monocrystalline quartz with radial fractures was observed suggesting that this quartz is a product of decompression of coesite. Most of the Raman spectra collected from 21 microdiamonds in two petrographic thin sections have peaks that are observed between 1332 and 1330 cm^{-1} suggesting the presence of sp^3-hybridized carbon bonding (diamond). The presence of *G-bands* (1580 cm^{-1}) in Raman spectra indicate that sp^2-hybridized carbon bonding (graphite) is also present. The thermodynamic calculations reported by Janák et al. (2013) suggest that diamond crystallization took place at $P = 3.5 \pm 0.5$ GPa and $T = 770 \pm 50$ °C. The microdiamond discovery in Tromsø nappe combined with the previous discoveries of coesite, microdiamonds, majoritic garnets in the neighboring Western Gneiss Region extends this UHPM terrane by >100–120 km.

Jämtland and Årescutan areas, the Sweden Caledonides. Diamond inclusions are documented in 2014 for the first time in porphyroblastic garnets from garnet–sillimanite–biotite gneisses in the Seve Nappe Complex of northern Jämtland, Swedish Caledonides (Majka et al. 2014). Diamonds ranging in size from 4 to 7 μm with spheroidal and bleb-like forms occur inside of a multiphase assemblage of carbonate, quartz, and rutile (Fig. 6). The presence of sp^3 carbon bonding was confirmed with Raman spectroscopy, and scanning electron microscope (SEM) images acquired in secondary electron (SE) mode reveal negative relief of diamond inclusions with respect to the very flat, perfectly polished surface of the host garnet (Majka et al. 2014).

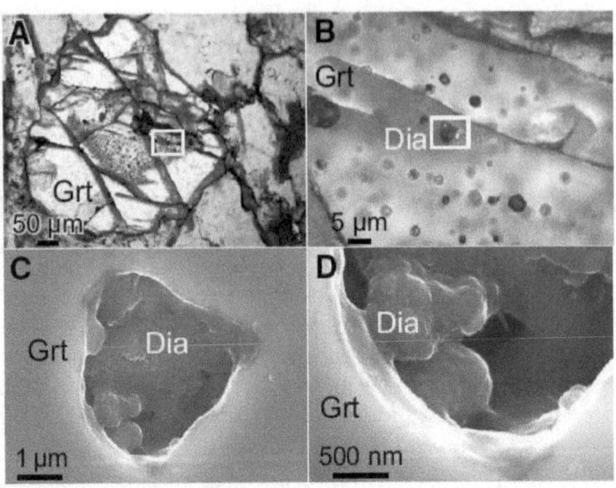

Figure 6. Microdiamond inclusions in garnet from garnet–sillimanite–biotite gneiss, Jämtland, Sweden Caledonides (Majka et al. 2014 with permission from the Geological Society of America). **A** and **B**–diamond (Dia) inclusion in garnet acquired at different magnifications (optical microscope: plane polarized transmitted light); **C** and **D**–Secondary electron SEM images of the same diamond shown in plates A and B (acquired at different magnifications).

Klonowska et al. (2017) discovered another microdiamond locality approximately 50 km from the locality described by Majka et al. (2014). Microdiamonds (~5 μm size) are found *in situ* as single and composite (diamond + carbonate) inclusions in garnet from garnet–phengite–kyanite–rutile gneisses in the Åreskutan area of the Seve Nappe Complex. Raman modes were observed at 1328 cm^{-1} and 1331 cm^{-1} confirming the presence of diamond (Solin and Ramdas 1970). Moreover, besides these diamonds, the gneisses contain textural relicts of decompressed coesite, observed as polycrystalline quartz surrounded by radial cracks. Thermodynamic calculations suggest that diamond crystallization took place at $P = 4.1–4.2$ GPa and $T = 830–840$ °C (Klonowska et al. 2017). The mineral assemblages of the UHP metamorphism were later overprinted under amphibolite and granulite facies conditions. The age of the UHP metamorphism recorded in the Åreskutan diamond-bearing gneisses occurred before ~445 Ma (Klonowska et al. 2017).

Diamonds from Lago di Cignana, Italy, Western Alps. Although UHP metamorphism in the Western Alps was proposed in the mid 1980s when coesite first was discovered by Chopin (1984) and later by Reinecke (1991), diamonds were not found in the Western Alps until almost two decades later (Frezzotti et al. 2011, 2014; Frezzotti 2019). Frezzotti et al. (2011) discovered microdiamond inclusions ranging from <2–30 μm in size in Mn-rich garnets from garnetite nodules and boudins enclosed within mica–schists and quartzites that are spatially associated with eclogite facies metabasites. The upper part of this metamorphosed ophiolitic section belongs to the Lago di Cignana tectonic unit of the Western Alps and appears to be the first recognized UHPM terrane which originated as a fragment of the Thetyan oceanic lithosphere. The tiny diamond inclusions are closely associated with COH fluid inclusions, and tiny inclusions of magnesite and SiO_2 (quartz). There is no evidence that the quartz is a product of decompression of coesite during exhumation. Instead, the SiO_2 clusters are products of lower temperature post–metamorphic reactions that occurred in adjacent layers of the mica-schists (Frezzotti et al. 2011, 2014).

Detailed studies of diamond and COH fluid enclosed in garnets using Raman spectroscopy and C and O isotope geochemistry revealed that organic carbon dissolved in hydrous fluids was transported to a depth ~110 km together with oceanic sediments during Alpine oceanic subduction about 35 Ma ago. Frezzotti et al. (2011, 2014) and Frezzotti (2019) concluded that during subduction the UHP metamorphism led to diamond crystallization from COH fluid at respectively high oxidation state [fO_2 = 0–1.5 log units higher than the fayalite–magnetite–quartz (FMQ) buffer]. This UHP metamorphism took place at $P \geq 3.2$ GPa and $T \leq 600$ °C and is classified as a cold subduction metamorphism (e.g., Frezzotti 2019).

Diamonds from Pohorje (Slovenia) in the Austro-Alpine UHPM terrane, Eastern Alps. The Pohorje Mountains have been considered as an UHPM terrane since Janák et al. (2004, 2006) constrained the pressure and temperature ($P \geq 3.5$ GPa and T =800–850 °C) for eclogite and peridotite enclosed in meta-carbonaceous rocks, although at that time no UHP index minerals had been known in this locality. Later, Janák et al. (2015) discovered ~10 μm microdiamond inclusions in garnets from metapelitic gneisses that consist of layers rich in biotite, white mica, garnet and kyanite alternating with felsic layers, veins and segregations of quartz and feldspar. Inside of the host-garnets diamond occurs as single cubes or cubo-octahedral crystals and as a part of composite inclusions where it is associated with moissanite (SiC) and fluid inclusions containing abundant CO_2 and traces of CH_4. Diamonds, associated minerals, and fluid inclusions were confirmed with Raman spectroscopy. The authors concluded that diamond formed at depth > 100 km during Late Cretaceous (ca. 92–95 Ma) subduction of the continental slab from reduced supercritical fluids that formed during UHP metamorphism of carbonaceous sediments. This diamond-forming media is similar to that reported by Perraki and Faryad (2014) in the Gföhl granulites of the Moldanubian Zone of the Bohemian massif, where they documented polyphase inclusions of diamond+moissanite+coesite in zircon.

More diamonds from Qinling region, China. Though the Central Orogenic belt (COB) of China is a well-established UHPM terrane where coesite (e.g., Okay et al. 1988) as well as quartz paramorphs possibly after stishovite (Liu et al. 2007, 2018) and majoritic garnet with lamella exsolution of ortho- and clinopyroxenes (Ye et al. 2000) are well-documented, diamonds in these rocks are scarce. There are two reports of diamond occurrences in the Dabie Mountains (Xu et al. 1992, 2003) and the Qinling region of China as inclusions in zircons (Yang et al. 2003). However, these reports were not confirmed by others for more than a decade.

Wang et al. (2014) documented more microdiamonds as inclusions in zircons that had been separated from amphibolite rocks from the North Qinling region. To date, all microdiamonds reported from the COB have only been confirmed with Raman spectroscopy. Advanced electron microscopy studies of these diamonds such as FIB and TEM have not yet been reported. Nevertheless, both FIB and TEM research on these unusual diamonds is very important since it would likely shed light onto the geochemical conditions under which these diamonds crystallized. Considering that very few microdiamonds have been documented in the UHPM rocks of the COB it is understandable that researchers may be skeptical of these reports. It is possible that the low concentration of diamonds in the COB may be explained by geochemical conditions that would impede diamond crystallization. Hypothetically, such conditions could be: (i) low concentration of carbon in the bulk rocks and/or in small amounts of fluid/hydrous melt that can penetrate subducting slabs during UHP metamorphism; (ii) perhaps the concentration of carbon was high enough to start diamond nucleation, but the oxygen fugacity was extremely high which would prevent diamond crystallization; (iii) or all the diamonds were replaced by graphite during slow exhumation of the UHPM rocks accompanied by low-pressure and high-temperature regional metamorphism.

Graphitized diamond in West Greenland UHPM terrane. A new UHPM terrane was documented by Glassley et al. (2014) within the Nagssugtoqidian Orogen (Proterozoic age, ca. 1.8 Ga) of West Greenland. The Orogen consists of undifferentiated metasedimentary/metavolcanic rocks and amphibolites. They include strongly deformed quartzo-feldspathic granitic and dioritic gneisses, garnet–sillimanite–graphite–schists, and amphibolites; all which were traditionally considered as upper amphibolite to granulite facies from regional metamorphism (e.g., Davidson 1979; Glassley and Sørensen 1980; Mengel 1983).

The mineralogical confirmation of different UHP events includes (i) the presence of relicts of orthopyroxene exsolved from a supersilicic garnet precursor, (ii) exsolution lamella of rutile from garnet and pyroxenes, (iii) exsolution lamella of magnetite from olivine, and (iv) unusual quartz needles exsolved from olivine—a hypothetical ringwoodite. Glassley et al. (2014) showed that the Raman spectra of three mm-sized carbon-bearing inclusions in garnet exhibit modes at 1335 cm^{-1}, which unambiguously suggests the presence of sp^3-carbon bonds, e.g., the diamond structure. Broad peaks at ~1420–1450 cm^{-1} which are consistent with hydrogenated carbon on the diamond surfaces and well-pronounced Raman bands of graphite at 1580 cm^{-1} suggest that diamond probably gradually transformed to graphite. The authors hypothesized that these diamonds were encapsulated in the garnet during their growth and were transformed to graphite during decompression accompanied by metamorphism to granulite facies. The UHP metamorphic conditions suggest a pressure of ~ 7 GPa and a temperature of ~975 °C.

The Western Mediterranean orogen—the Edough Massif, NE Algeria, new UHPM terrane. Several microdiamond crystals were discovered by Caby et al. (2014) as inclusions in garnet megacryst (≥ 5 cm) collected from a mélange zone exposed at the northern margin of Africa (the Edough Massif, NE Algeria). The garnet megacrysts were adjacent to actinolite-bearing and ultramafic boudins associated with tectonic fragments of marbles occurring within a major mylonite–ultramylonite band. Several cuboid-like diamond crystals ranging from 3–50 μm in size were identified with Raman spectroscopy (Caby et al. 2014).

Later, a single diamond crystal approximately 50 μm in size intergrown with rutile enclosed in an almandine-rich garnet was studied in detail with the aid of SEM techniques (Bruguier et al. 2017). The secondary electron SEM image presented in Figure 3a from Bruguier et al. (2017) is unequivocal evidence that the diamond–rutile inclusion is indigenous (e.g., not from sample preparation or contamination) to the Edough massif. The image shows that the diamond is clearly "protruding above" the perfectly polished surface of the host mineral, which is very typical for a diamond–garnet interface. The contact zone between the diamond–garnet interface, appears intact which strongly suggests that the diamond is naturally occurring and not contamination from a cutting with a diamond-saw and/or polishing with diamond grit. Other garnets from the same area contain rutile, zircon, apatite, and plagioclase inclusions.

Thermodynamic calculations suggest that an UHP metamorphic event took place at $P \geq 3.6$ GPa and $T = 750$ °C (Bruguier et al. 2017; Caby et al. 2014). Studies of the zircon inclusions suggest that the Alpine age of UHP metamorphism of the West Mediterranean Orogen is between 32.4 ± 3.3 Ma and 20.7 ± 2.3 Ma (Bruguier et al. 2017). This was followed by exhumation as slab rollback occurred in the Oligocene (Brun and Faccenna 2008; Tirel et al. 2013). Additional studies of this locality should find more evidence of UHPM, such as UHP minerals and/or traces of their decompression products and/or proof of retrograde metamorphism.

Diamonds from Nishisonogi unit, Nagasaki metamorphic complex, western Kyushu, Japan. Nishiyama et al. (2020) report the discovery of microdiamond aggregates in the matrix of a metapelite from the Nishisonogi unit, Nagasaki metamorphic complex, western Kyushu, Japan. The Nishisonogi unit represents a Cretaceous subduction complex which has been considered as an epidote–blueschist subfacies metamorphic unit, and the metapelite is a member of a serpentinite mélange in the Nishisonogi unit (Nishiyama et al. 2020). The glaucophane and phengite ages are reported to be 95–90 Ma using the $^{40}Ar/^{39}Ar$ method (Faure et al. 1988) and 85–60 Ma using the K–Ar method (Hattori and Shibata 1982). The aggregates of diamonds range in size from 10 to 50 μm and they are embedded in dolomite, phengite, and albite (see Fig. 4 in Nishiyama et al. 2020). The diamonds are not sticking out above the polished surface of the sample. However, it should be noted that they are surrounded by soft minerals. Using SiC and Al_2O_3 abrasives as the main polishing abrasives, one would expect to see the diamonds "standing above" the flat surface of the surrounding softer minerals (see Fig. 4 in Nishiyama et al. 2020). This is because of the hardness of diamond itself, and it would not be "eroded" or 'flattened" by either SiC or Al_2O_3 abrasives since these abrasives are not as hard as diamond. Unfortunately, images where the diamonds are completely enclosed by the host minerals are not presented in this publication.

The authors characterized the microdiamonds with electron diffraction and Raman spectroscopy and carefully discussed both possibilities of metastable and stable conditions of diamond crystallization in metapelites of the epidote–blueschist subfacies from a Cretaceous subduction complex in Western Kyushu, Japan. They considered several lines of evidence for HP to UHPM conditions such as the presence of quartz pseudomorphs after coesite and high Si content in phengite. They speculate that almost all other HP–UHPM mineral assamblages have beed oblitered due to the retrograde metamorphism. Ultimately the authors hypothesized that the diamond crystallized from a COH fluid at $P > 2.8$ GPa and $T = \sim 450$ °C.

The authors specify that no diamond abrasives were used for sample polishing, and that ony Al_2O_3 and SiC were used. However, they did not specify if any diamond-bearing tools were used to extract or cut the samples prior to their polishing. Though Nishiyama et al. (2020) used FIB-TEM techniques they did not carefully examine the diamond–host mineral interfaces. In Figure 5a,b from Nishiyama et al. (2020) they show STEM images at different magnifications aquired from a FIB milled foil from the sample area contaning microdiamond aggregate. In Figure 5 plate c, Nishiyama et al. (2020) show a *d*-spacing = 2.06 Å (direction [111]) for the inner spot. Importantly, these data neither prove nor disprove that the diamond is indigeneous.

To unambiguously classify the metapelites from a Cretaceous subduction complex in Western Kyushu, Japan as a new UHPM terrane more FIB-TEM research is required on both the existing samples and/or new samples from this locality, and studies of the diamond–host mineral interface are necessary to confirm that the diamond is indigenous.

Lonsdaleite from WGR, Norway and Kokchetav massif, Kazakhstan, and a problem with its identification

Since lonsdaleite, a hexagonal polymorph of diamond, has been reported to occur in UHPM terranes as well as in meteorites and impact craters, a short discussion is warranted. At General Electric Company Bundy and Kasper (1967) synthesized a wurtzite like polymorph of carbon at $P = 13$ GPa and $T = 1000$ °C, and they named it hexagonal diamond. In nature hexagonal diamond was first identified from the Canyon Diablo iron meteorite and Goalpara ureilite (e.g., Frondel and Marvin 1967; Hanneman et al. 1967). It was named lonsdaleite in honor of the distinguished crystallographer Professor Kathleen Lonsdale (Frondel and Marvin 1967). Following this initial report lonsdaleite has been found in many other meteorites and impact craters (e.g., Daulton et al. 1996; Clarke et al. 1981; Kvasnitsya et al. 2013; Nakmuta and Toh 2013; Nestola et al. 2020). It has been reported to occur in the following terrestrial impact craters: Ries Crater, Germany (e.g., Rost et al. 1978), Popigai impact structure (e.g., Masaitis et al. 1972; Koeberl et al. 1997; Murri et al. 2019), Sudbury impact structure, Ontario, Canada (e.g., Masaitis et al. 1997), and Lappajärvi structure, Finland (e.g., Masaitis et al. 1998). There is also one report of lonsdaleite occurrence in nodules from Pipe 50, Fuxian kimberlite field, China (Leung and Winston 2002). These discoveries led to the hypothesis that lonsdaleite forms during impact events and very high pressure and temperatures are needed. However, it has also been reported to occur in UHPM rocks which would require alternative hypothesis to explain their existence.

Lonsdaleite has been reported in WGR, Norway (Godard et al. 2003, 2004; Smith et al. 2004) and in Kumdy Kol, Kokchetav massif, Kazakhstan (Smith 2004; Smith et al. 2004, 2011; Dubinchuk et al. 2010; Shumilova et al. 2011). Since these localities have been proven to be UHPM terranes by numerous studies, which clearly showed that they did not form during impact events, alternative hypotheses are required to explain the existence of lonsdaleite in UHPM rocks. One hypothesis is that the formation of hexagonal polytypes such as lonsdaleite could be cumulative with the metamictization of diamond (Smith et al. 2011; Smith and Godard 2013). Another hypothesis, which will be discussed in greater detail below, is that they formed during an impact event with a comet (Tretiakova and Luykin 2016, 2017).

Raman spectroscopy has been one of the most common techniques for confirming the presence of natural lonsdaleite since it is non-destructive, and mm-sized samples can be studied *in situ* (Smith and Godard 2009; Shumilova et al. 2011; Goryainov et al. 2014). Lonsdaleite has a 2H structure (spacegroup $P6_3/mmc$) which has three theoretically predicted Raman active modes, the transverse optical oscillations of the E_{2g} mode, and the longitudinal A_{1g} and transverse E_{1g} optical modes. Wu and Xu (1998) report the following theoretically predicted active modes E_{2g} 1193 cm^{-1}, A_{1g} 1312 cm^{-1}, and the E_{1g} 1305 cm^{-1}, while Denisov et al. (2011) predict them to be at 1221, 1280, and 1338 cm^{-1} respectively. The E_{2g} mode is predicted to have very low intensity so the A_{1g} and E_{1g} modes are typically used to identify lonsdaleite. Since the predicted lonsdaleite modes occur near the cubic diamond Raman active T_{2g} mode it should be noted that the peak position and width of the cubic diamond mode can vary significantly. As discussed by Nasdala et al. (2016) the peak position of the Raman active diamond mode can be downshifted and broadened by local heating through absorption of the laser (Herchen and Cappelli 1991), nanometer sized crystals (Lipp et al. 1997), substitution of ^{13}C for ^{12}C (Anthony and Banholzer 1992), incorporation of B-impurities (Pedroza-Montero et al. 2005) and accumulation of structural damage from nearby radioactive phases (Nasdala et al. 2013). Hence, one needs to be cautious when interpreting Raman spectra of natural diamonds since shifts in the cubic diamond Raman peak position and peak width could be misinterpreted as lonsdaleite.

Notably, there appears to be no consensus on the existence of lonsdaleite as a discrete mineral (e.g., Németh et al. 2014). Németh et al. (2014) point out that despite extensive efforts lonsdaleite has never been produced as a pure material and they show that defects in cubic diamond provide an explanation for the characteristic *d*-spacings and reflections reported for lonsdaleite. Natural lonsdaleite has been described as stacking disordered diamond where cubic and hexagonal sequences are interlaced in a complex way (e.g., Salzmann et al. 2015; Murri et al. 2019). Thus, while numerous studies of lonsdaleite have been reported from UHPM terranes, more studies of natural lonsdaleite using high-resolution techniques are needed to resolve this controversy.

ISOTOPIC STUDIES OF UHPM DIAMONDS

Carbon and nitrogen isotopes, and nitrogen content in UHPM diamonds

Carbon isotopic composition (expressed as $\delta^{13}C$, see Stachel et al. 2022, this volume for definition) is a key parameter used to constrain the origin and formation environment of a diamond. Since UHPM diamonds formed *in situ* in subducted continental lithologies, their carbon isotope ratios should reveal the origin of carbon, and the isotope fractionation of the COH fluids from which UHPM diamond formed. These carbon isotope ratios record the formation conditions both before and during diamond crystallization.

Among the reports on carbon isotopic compositions, E-type (eclogitic) and P-type (peridotitic) diamonds (e.g., Harris et al. 1975; Cartigny 2005; Stachel et al. 2022, this volume) found in kimberlites clearly show sfome similarities; both types of diamonds have a peak at approximately −5‰, which is similar to mantle-derived carbon observed in mid-ocean ridge basalts (MORBs), carbonatites, and kimberlites (Table 2, Fig. 7). The E-type diamond carbon isotopes are also spread towards lighter carbon to −41‰ (De Stefano et al. 2009; Smart et al. 2011), suggesting their crustal origin (Fig. 7).

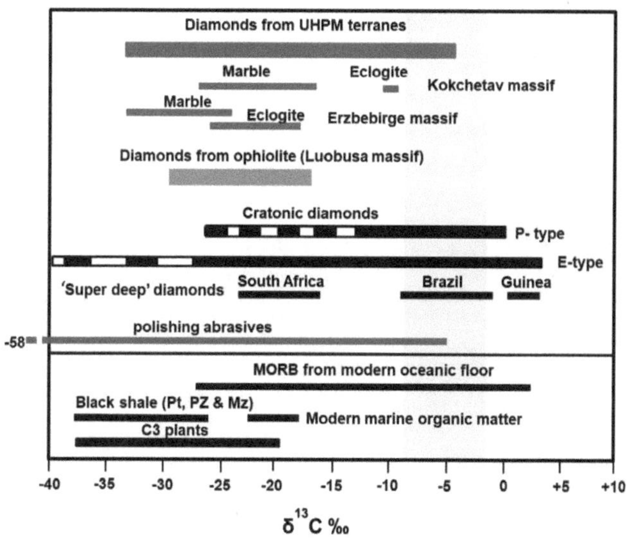

Figure 7. $\delta^{13}C$ characteristics of diamonds from UHPM terranes, diamonds from ophiolite (Luobasa chromitite massif), E-type and P-type of cratonic diamonds; modified after Liou et al. (2014), see Table 2 for references. $\delta^{13}C$ values of MORB, black shale of Proterozoic, Paleozoic, and Mesozoic ages and C3 plants adopted from Kohn (2010) and Meyers (2014).

Table 2. Carbon ($\delta^{13}C$) and nitrogen ($\delta^{15}N$) characteristics of diamonds, bulk rocks and carbon-bearing minerals from Kokchetav massif, Kazakhstan and Erzgebirge massif, Germany, kimberlitic xenoliths and ophiolitic chromitites.

Locality	Mineral/rock type	$\delta^{13}C$ value (‰)	$\delta^{15}N$ value (‰)	N content (ppm)	References
Kokchetav massif, Kazakhstan	diamonds-bulk measurements	−10.2 to −26.9			Chopin and Sobolev 1995; Lavrova et al. 1997
	diamond from garnet–clinopyroxene rocks;	−10.5			De Corte at al. 2000
	diamonds from marble	−10.2			
	diamond from garnet–clinopyroxene rock	−10 to −11	+5.9 (mean)	11,150	Cartigny et al. 2001
	diamond from marble	−8.5 to −10.19		2,650	
	S-diamond (outer rim) inclusion in garnet from dolomitic marble	−17 to −27			Imamura et al. 2013
	S-diamond (core) inclusion in garnet from dolomitic marble	−9 to −13			
	R-diamond inclusion in garnet from dolomitic marble	−8 to −15			
	Dolomite mineral from dolomitic marble	−4 to −7			Ohta 2003
	Dolomitic marble (bulk)	−2			
Erzgebirge Massif, Germany	Diamond inclusions in zircon (bulk analysis) from garnet–phengite gneiss	−24 to −33			Massonne and Tu 2007
	Diamond inclusions in zoned garnet (outer rim) from garnet–phengite–quartz–feldspathic gneiss	−22 to −26		740– 3,370	Dobrzhinetskaya et al. 2010
	Diamond inclusions in zoned garnet (core zone) from garnet–phengite–quartz–feldspathic gneiss	−17 to −19			Dobrzhinetskaya et al. 2010

Locality	Mineral/rock type	$\delta^{13}C$ value (‰)	$\delta^{15}N$ value (‰)	N content (ppm)	References
Worldwide UHPM terranes	UHPM metamorphic rocks	−3 to −30	−1.8 to +12.4		Cartigny 2005
Worldwide kimberlite-derived diamonds	Eclogitic diamonds	−40.7 to +2.5	−13.1 to +16.9	≤5 to 3833	Stachel et al. 2022
	Peridotitic diamonds	−34.5 to +2.3	−39.4 to +15	≤5 to 1923	Stachel et al. 2022
Luobusa ophiolitic massif (Tibet)	Chromitite	−18 to −28		20–670	Yang et al. 2013

A typical feature of UHPM diamonds is their yellow color (e.g., Schertl and Sobolev 2013) due to the presence of isolated nitrogen, which is the main substitutional impurity in diamond. Nitrogen has an extra electron compared to carbon, and the incorporation of a few ppm of isolated nitrogen atoms and their accompanying electrons changes the energy band gap in diamond to a level equivalent to absorption of blue and violet lights, resulting in a yellow color. The concentration of nitrogen, which is the main substitutional diamond impurity, is also an important parameter when it is combined with $\delta^{13}C$ to constrain the origin of metasomatic COH fluids and therefore diamond formation conditions (Table 2, Fig. 8). As noted by Cartigny et al. (2014), the behavior of nitrogen during diamond formation is unclear since nitrogen in diamond may be both compatible and incompatible (Smart et al. 2011; Palot 2013). On the contrary, if nitrogen compatibility is experimentally constrained, it would shed light on the composition, redox state, and evolution of the fluids related to UHPM diamond formation.

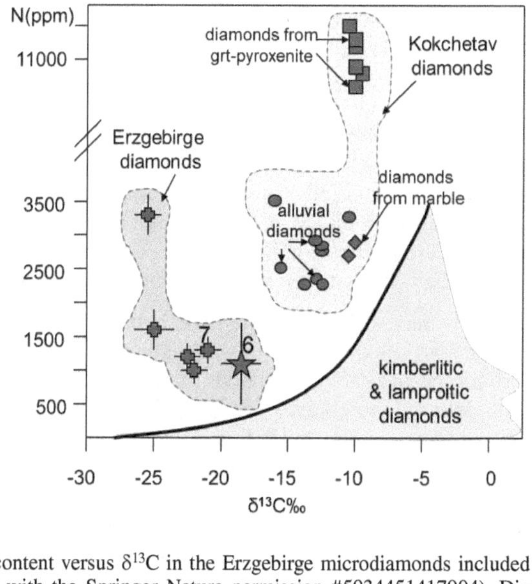

Figure 8. Nitrogen content versus $\delta^{13}C$ in the Erzgebirge microdiamonds included in garnets (Dobrzhinetskaya et al. 2010 with the Springer Nature permission #5034451417904). Diamonds #6 (**star**) and #7 (**cross**) are situated in the same garnet crystal in the core and the rim respectively. The fields of kimberlitic and lamproitic diamonds, and diamonds from garnet pyroxenite (**square**), marbles (**diamond**) and from alluvial deposits of the Kokchetav massif, Kazakhstan are adopted from Cartigny et al. (2001, 2003).

Nitrogen isotopic composition (expressed as $\delta^{15}N$, see Stachel et al. 2022, this volume for definition) also provide a clue to the origin of diamonds. Crustal nitrogen is enriched in ^{15}N, by approximately +5‰ compared with the atmosphere ($\delta^{15}N = 0$‰), whereas mantle nitrogen is depleted in ^{15}N by approximately −5‰. Being different to carbon, nitrogen isotopes are fractionated by only a few per mill during their incorporation into diamond, so nitrogen that is incorporated into diamond during crystallization is representative of the growth medium. Most E-type diamonds have a mantle-like $\delta^{15}N$, which is inconsistent with their origin from subducted organic carbon. On the other hand, UHPM diamonds systematically exhibit heavy $\delta^{15}N$ values ranging from +2 to +12 ‰ (Cartigny et al. 2001, 2014), suggesting their sedimentary source because devolatilization during sediment subduction would preferentially release ^{14}N, resulting in further enrichment of subducted material in ^{15}N (Bebout and Fogel 1992; Busigny et al. 2003). For thorough discussion on the origin of carbon and nitrogen isotopic compositions of diamond, the reader is refered to Stachel et al. (2022, this volume).

Carbon isotopes characteristics of the Kokchetav diamonds. Carbon isotopic composition studies of UHPM diamonds are limited due to their mm-size. One approach involves separating and grouping tens of thousands of microdiamonds and processing them together so that there is enough CO_2 to be analyzed with a conventional isotope-ratio mass spectrometer, this approach provides the average $\delta^{13}C$ ratio. For instance, Cartigny et al. (2001) reported $\delta^{13}C = -11$ ‰ to −10 ‰ in diamonds from garnet–clinopyroxene rocks and nitrogen contents of 11,150 ppm, while diamonds from marbles of the Kokchetav massif, Kazakhstan have a $\delta^{13}C = -10.2$‰ to −8.5‰ and 2650 ppm of nitrogen (Table 2, Fig. 8). Based on possible carbon isotope fractionation between coexisting carbonate and diamond, Cartigny et al. (2001) estimated diamond crystallization conditions as $T \leq 700$ °C and $P = 3$ GPa, e.g., slightly below the peak of UHPM, and concluded carbon and nitrogen stable isotopes strongly support their metasedimentrary origin. If Cartigny et al. (2001) data is combined with earlier $\delta^{13}C$ bulk measurements of the Kokchetav diamonds, they fall in range of −10.2 ‰ to −26.9 ‰ (e.g., Lavrova et al. 1997; Chopin and Sobolev 1995; Ogasawara 2005).

Later, Imamura et al. (2013) using Secondary Ion Mass Spectrometry (SIMS) technique analyzed carbon isotopes in diamonds (10–25 μm in size) included in garnet from dolomite marble of the Kokchetav Massif. These diamonds were subdivided on to morphological types: S-type consisting of single-crystal core and polycrystalline rim, and R-type presented by single crystal with rugged surfaces. In S-type diamonds the rim exhibits lighter $\delta^{13}C = -17$ to −27‰, whereas the core is much heavier, with $\delta^{13}C = -9$ to −13‰. The R-type diamonds fall into narrow ranges of $\delta^{13}C = -8$ to −15‰, which are like those in the core of the S-type diamond. This suggests that the R-type diamonds probably formed at the same stage as the core of the S-type, whereas rim growth during the second stage did not occur, or occurred, very weakly in R-type microdiamonds. Therefore, carbon isotopic data corroborate a two-stage growth of microdiamond in the Kokchetav dolomite marbles (Imamura et al. 2013).

Interestingly, the $\delta^{13}C$ value of dolomite from the host rocks ranges from −4 to −7‰, whereas the bulk dolomitic marble exhibits $\delta^{13}C = -2$‰ (Ohta et al. 2003). The marginal range of $\delta^{13}C = -8$ to −27‰ in S-type and R-type diamonds is different than that in mineral dolomite ($\delta^{13}C = -4$ to −7‰) and dolomitic marble host rock ($\delta^{13}C = -2$‰). It is difficult to attribute these differences to isotopic fractionation considering that the metamorphic temperature was very high (~1000 °C; e.g., Ogasawara et al. 2000). Under such conditions, the diamonds in dolomitic marble would not have been in carbon isotopic equilibrium with the dolomite. There appears to be no good candidate material or process that could occur in the dolomitic marble that would explain the origin of the very light $\delta^{13}C$ values between −17 and −27‰, measured in the rim of the S-type diamonds. Imamura et al. (2013) speculated that perhaps a fluid with a light carbon isotope ratio infiltrated into the dolomite marble from the surrounding gneisses, which might contain light carbon of organic origin. With regards to this explanation, it will be discussed below that the Erzgebirge microdiamonds from felsic gneiss have a similar light carbon, e.g., $\delta^{13}C = -17$ to $-27‰$ (Dobrzhinetskaya et al. 2010). However, Imamura et al. (2013) have not discussed these data because no carbon isotope measurements in diamonds from felsic gneisses of the Kokchetav massif have been reported.

Carbon isotope characteristics of the Erzgebirge diamonds. Massonne and Tu (2007b) used a bulk analysis approach to obtain the carbon isotope ratios of diamond inclusions in zircons from the Erzgebirge UHPM terrane, and they reported that lighter carbon was involved ($\delta^{13}C = -24‰$ to $-33‰$) in diamond formation where the precursor was graphite.

Dobrzhinetskaya et al. (2010) used SIMS to analyze the $\delta^{13}C$ and nitrogen contents in ~5 to 10 μm diameter diamonds included in garnet from the garnet–quartz–feldspathic gneiss of the Erzgebirge massif (Fig. 9). They showed that the diamonds situated within the garnets closer to the garnet rim zone showed some $\delta^{13}C$ variations up to 10‰, with average values ranging from −22 to −26‰ (Table 2, Fig. 9). Their nitrogen contents vary heterogeneously between each individual measurement and between all diamonds without any systematics, ranging from 240 to 4600 ppm. On the other hand, the diamonds situated closer to the central part of the zoned garnet exhibits $\delta^{13}C = -17$ to $-19‰$, with an average value of −17.8‰. The latter is heavier than another diamond in the intermediate zone of the same garnet ($\delta^{13}C = -23‰$) and the other diamonds situated closer to the outer zone of the garnet (Fig. 10). Nitrogen in the diamonds from intermediate zone varies from 100 to 1200 ppm, with an average value of 820 ppm; and average value of nitrogen in diamonds from outer garnet zone is 1,838 ppm. The nitrogen content data plotted versus $\delta^{13}C$ show a weakly pronounced tendency to increase with the increase of lighter carbon isotope content in the diamonds (Fig. 8).

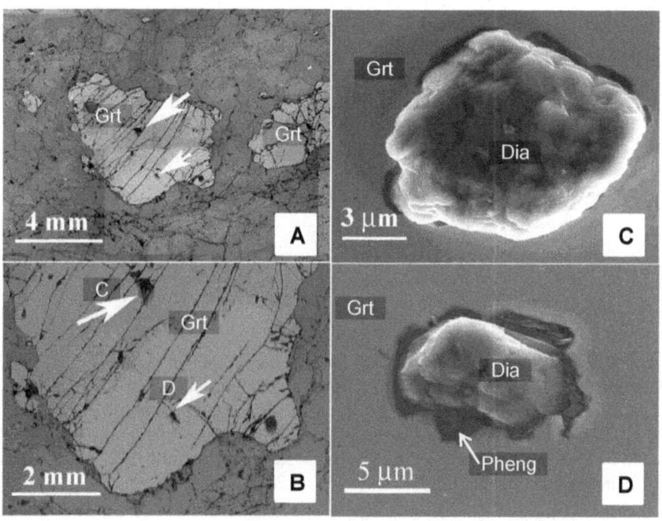

Figure 9. Carbon isotope characteristics in Erzgebirge diamonds (Dobrzhinetskaya et al. 2010, copyirght 2010 with the Springer Nature permission #5034451417904). **A** and **B**–optical microscope digital microphotograph shows (reflected light) the polished slide from Erzgebirge garnet–quartz–feldspathic gneiss. Garnet contains inclusions of microdiamonds (indicated by **white arrowheads**). Plates **C** and **D**–Secondary Electron SEM images of diamonds (shown as C and D in plate B) inclusions in garnet. Diamond in plate C which is situated in the central part of the garnet (see A) has $\delta^{13}C = -17$ to $-19‰$, whereas diamond D situated at the outer part of the zoned garnet is characterized by $\delta^{13}C = -22$ to -26 ‰.

Based on relationships between the diamond inclusion in the garnet (Fig. 10), diamond formation can be constrained by the chemical zoning of the host garnet, and $\delta^{13}C$ values of two diamonds. Dobrzhinetskaya et al. (2010) argued that there were two stages of diamond crystallization from a COH fluid rich in biogenic carbon and diverse in minor elements of crustal origin. The $\delta^{13}C$ of the first stage of crystallization falls within the range of −17 to −19‰, while the second stage of crystallization is characterized by $\delta^{13}C$ = −22 to −26‰. Both $\delta^{13}C$ ranges strongly suggest that the carbon from which they crystallizedbelongs to the same biogenic reservoir. Graphite from the 3.7 Ga-old turbidite greywacke and slate from West Greenland have primary $\delta^{13}C$ = −19‰, which is similar to the range of biologically reduced carbon, and it is within the range of reduced-carbon compositions found in most modern marine sediments (e.g., Rosing 1999). Dobrzhinetskaya et al. (2010) suggest that the 1st-stage of diamond crystallization was from a more reduced carbon than the second stage of crystallization. This work indicates that within Erzgebirge diamond populations there are two $\delta^{13}C$ signatures which are distinguished from each other not only by $\delta^{13}C$ values but also by the positions of the diamond within the garnet where different zones in the garnets are related to recrystallization events during UHP metamorphism. It also suggests that the carbon-rich fluid that deeply infiltrated the subducted sediments was probably reduced during an earlier stage, producing diamonds with slightly heavier isotopes than the diamonds that crystallized during the second stage.

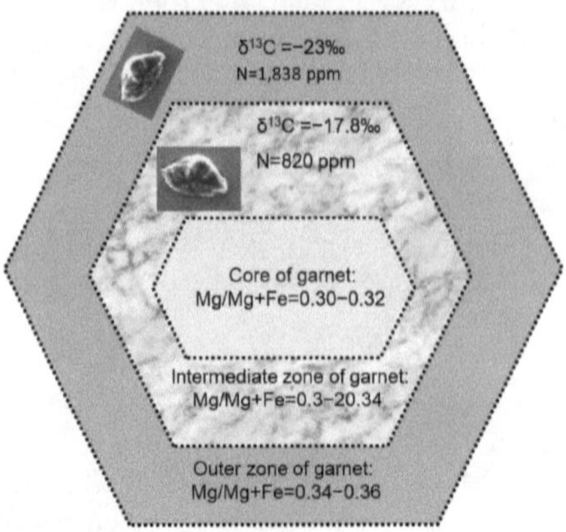

Figure 10. Sketch of the slightly zoned garnet with microdiamond inclusions situated within the intermediate zone between the core and the outer rim and close to the outer zone of garnet. Microdiamonds located in the intermediate zone of the garnet (Mg/Mg+Fe = 0.32–0.34) are characterized by "heavier" isotopes of carbon $\delta^{13}C$ = −17.8‰ and a N concentration of 820 ppm, whereas diamonds from outer garnet zone (Mg/Mg+Fe = 0.34–0.36) are characterized by "lighter" isotopes of carbon $\delta^{13}C$ = −23‰ and a N content of 1838 ppm (data are adopted from Dorzhinetskaya et al. 2010).

Additional carbon isotope analysis of UHPM diamonds from both Kokchetav and Erzgebirge UHPM terranes and any of the new UHPM localities at small scales (μm to nm) with SIMS are necessary to shed light on the evolution of carbon-bearing fluids in subducted continental lithologies before and during diamond formation.

Noble gas isotopes in diamonds from UHPM terranes

Noble gas isotopes in mantle-derived samples are key tracers of chemical heterogeneity of the Earth's mantle and the origin of the atmosphere (e.g., Ozima and Podosek 2002; Porcelli and Ballentine 2002). Samples of MORBs (quenched glass of mafic minerals) and ocean island basalts (OIB) have provided the most comprehensive understanding of mantle noble gases. MORBs show relatively uniform $^3He/^4He$ ratios around 1.1×10^{-5} and a trend in $^{22}Ne/^{22}Ne$–$^{20}Ne/^{22}Ne$ ratios toward a composition more enriched in nucleogenic ^{21}Ne. In contrast, OIB samples, which are derived from a deeper region of the mantle, exhibit higher $^3He/^4He$ ratios of up to 7×10^{-5} and less nucleogenic Ne isotope compositions (Honda et al. 1991; Trieloff et al. 2000; Graham 2002; Stuart et al. 2003). The latter characteristics are believed to be evidence for a mantle source in which primordial He, with $^3He/^4He$ of $(1.4–4.6) \times 10^{-4}$, and Ne ("solar" or "Ne–B" composition, Honda et al. 1991; Trieloff et al. 2000; Ballentine et al. 2005) has been less diluted by the addition of radiogenic ^4He and nucleogenic ^{21}Ne produced by decay of U- and Th-series elements. The primordial noble gas source is considered to have been isolated from whole mantle convection over geological timescales and sampled by upwelling mantle plumes (Jackson et al. 2010; Mukhopadhyay 2012). However, the reasons for the less radiogenic/nucleogenic character of the plume source are still under debate; it may be "undegassed" (e.g., Allègre et al. 1983; Kaneoka 1983; Kurz et al. 1982; Porcelli and Ballentine 2002; Porcelli and Elliott 2008; Mukhopadhyay 2012), or it may be depleted in U and Th by ancient melt extraction events (Parman 2007; Porcelli and Ballentine 2002; Jackson et al. 2010). The main advantage of analyzing noble gases contained in diamonds is their great potential to constrain the noble-gas state of the deep mantle because they are formed deeper than 150 km and act as chemically/mechanically stable "capsules". Most noble gas studies are focused on kimberlitic diamonds (e.g., Basu et al. 2013; Timmerman et al. 2018).

"Unprecedentedly" high Helium-3 content in the Kokchetav diamonds: myth or reality? Although currently there are dozens of UHPM diamond-bearing terranes that are known, only a few noble gas studies of microdiamonds from the Kokchetav massif have been reported (Begemann 1994; Pleshakov and Shukolyukov 1994; Shukolyukov et al. 1993; Verchovsky et al. 1993; Sumino et al. 2011). Due to the small (micrometer) size of UHPM diamonds, noble gas analyses were conducted using several tens-of-thousands (or even a greater number) of microdiamonds that were separated from their host rocks/minerals by treatments with HF and oxidizing acids. Pleshakov and Shukolyukov (1994) and Shukolyukov et al. (1993) reported "unprecedentedly" high $^3He/^4He$ ratio during in vacuo stepwise heating of Kokchetav microdiamonds with a maximum value of 0.47 in the low temperate release, which was the highest He content known among terrestrial diamonds (Pleshakov and Shukolyukov 1994; Shukolyukov et al. 1993). Moreover, this high $^3He/^4He$ value exceeds the maximum values reported for any terrestrial mantle-derived sample of $\leq 7 \times 10^{-5}$ (Stuart et al. 2003) and even the primordial $^3He/^4He$ value of $(1.7–4.6) \times 10^{-4}$ (Porcelli and Elliott 2008).

The most plausible source of the high $^3He/^4He$ ratio is nucleogenic ^3He production via the $^6Li(n, \alpha)^3H$ (β) ^3He reaction (Kurz et al. 1987). However, this requires an anomalously high Li content of ~100 ppm either in the diamonds, or within a ~20 μm vicinity in the host rocks (Begemann 1994; Verchovsky et al. 1993). Moreover, the host rock would need to contain more than 1% of U (Th/U = 3.3 is assumed) to provide a neutron source sufficient to produce the levels of ^3He observed (Begemann 1994; Verchovsky et al. 1993). An important feature of the Kokchetav diamonds was that high $^3He/^4He$ values were obtained from gas released over the relatively low temperature range of 200–1100 °C during in vacuo stepwise heating (Pleshakov and Shukolyukov 1994; Shukolyukov et al. 1993). These temperatures are significantly lower than the diamond graphitization temperature (over 1700 °C), and lower than the temperatures at which noble gases trapped in the diamond lattice are released. This suggests that the high $^3He/^4He$ component was not hosted by the diamond itself. Although Verchovsky et al. (1993) confirmed release of He with high $^3He/^4He$ (2.4×10^{-4}) during in vacuo heating of the Kokchetav diamonds at 820 °C. However, He released during diamond combustion under a controlled oxygen atmosphere showed a rather low $^3He/^4He$ ratio of $\leq 3 \times 10^{-6}$. Additionally, Verchovsky et al. (1993) showed that the amount of 'excess' ^3He (as well as $^3He/^4He$ ratio) strongly depends on the grain-size of diamond fractions: the coarser the fraction the less (by orders of magnitude) the

^{3}He concentration. These findings were interpreted to indicate that the high ^{3}He/^{4}He component may reside in Li-rich contaminants, which are quite resistant to treatments with HF and oxidizing acids during diamond separation from their host minerals/rocks, and therefore not trapped within the diamond itself (Begemann 1994; Verchovsky et al. 1993).

To clarify this "unprecedented" ^{3}He/^{4}He value Sumino et al. (2011) focused their attention on the presence of abundant nanometer-sized fluid inclusions in Kokchetav microdiamonds (Dobrzhinetskaya et al. 2001, 2005, 2006a), which could provide potential sites for preserving the noble gas signature of the diamond-forming COH fluid. Sumino et al. (2011) employed in vacuo sequential dynamic crushing extraction with which sample grains are mechanically crushed in vacuum in order to selectively extract noble gases from fluid/melt inclusions with significantly less noble gas release from the diamond lattice and any associated mineral inclusions (Kurz 1986; Kurz et al. 1987).

The crushing technique results were compared with those obtained from a conventional vacuum stepped heating method which releases noble gases from fluid and solid inclusions and the diamond lattice. According to Sumino et al. (2011) the diamond crushing process extracts most of the ^{3}He, indicating that ^{3}He occurs within inclusions trapped during the Kokchetav diamond formation. The inclusion-hosted ^{3}He/^{4}He of $(3.3–6.5) \times 10^{-5}$ is significantly higher than that of the MORB-source mantle (1.1×10^{-5}), but close to the maximum value observed in OIBs ($\sim 7 \times 10^{-5}$) containing noble gases enriched in a primordial component and delivered from the deep mantle by plumes. On the other hand, very low ^{3}He/^{4}He (5.0×10^{-8}) obtained by heating of the crushed powder supports the presence of radiogenic ^{4}He in the diamond lattice, resulting from α-particle implantation from U- and Th-series radioactive elements in the diamond-bearing minerals, which is consistent with the combustion results by Verchovsky et al. (1993). The results of Sumino et al. (2011) clearly indicate that the fluid inclusions in the Kokchetav diamonds preserve primordial He similar to those observed in OIBs and that abundant He with "unprecedentedly" high ^{3}He/^{4}He ratio reported by previous works (Shukolyukov et al. 1993; Pleshakov and Shukolyukov 1994) should be attributed to Li-rich contaminants resistant to separation treatments of the diamonds from their host rocks/minerals (Verchovsky et al. 1993; Begemann 1994).

Sumino et al. (2011) showed that isotope compositions of Ne (^{20}Ne/^{22}Ne and ^{21}Ne/^{22}Ne ratios) extracted from the Kokchetav diamonds during relatively low-temperature heating steps are on the trend defined by OIBs from deep-mantle-derived hotspots (e.g., Hawaii, Iceland, and Galapagos) which are enriched in primordial Ne (Honda et al. 1991; Trieloff et al. 2000; Kurz et al. 2009) but deviate from the trend defined by MORB data (Sarda et al. 1988) (Fig. 11). The Ne-isotope data indicate that deep-mantle-derived, primordial Ne is an intrinsic feature of the microdiamonds, but at higher temperatures during diamond graphitization, the Ne released shows a nucleogenic contribution. The Ne isotope data indicate that primordial Ne is an intrinsic feature of the microdiamonds, but at higher temperatures during diamond graphitization, the Ne released shows a nucleogenic contribution. The nucleogenic Ne resides in the diamond lattice as a result of implantation by recoiling Ne isotopes mainly produced during the reaction of α-particles (from U–Th decay) with oxygen and fluorine atoms in surrounding rocks (Kurz et al. 1987; Lal 1989; Verchovsky et al. 1993). The nucleogenic Ne composition released from the microdiamonds requires a very high F/O ratio of 0.02 (Fig. 11B) compared with the average crustal values of 0.0013 (Ballentine and Burnard 2002).

The observation that deep-mantle-derived, plume-like noble gases are contained in the Kokchetav microdiamonds paradoxically requires that deep-mantle-derived volatiles be present in the relatively shallow mantle wedge at 190–280 km depth. To solve this paradox, Sumino et al. (2011) suggest a few possible processes which are either capable of bringing deep-mantle-derived noble gases to the relatively shallow mantle: (i) a plume tapped from the lower mantle or core–mantle boundary; and/or (ii) a large-scale mantle convection cell which entrained a fragment of the deep mantle, and subsequently added it to the mantle wedge at depths of 190–280 km.

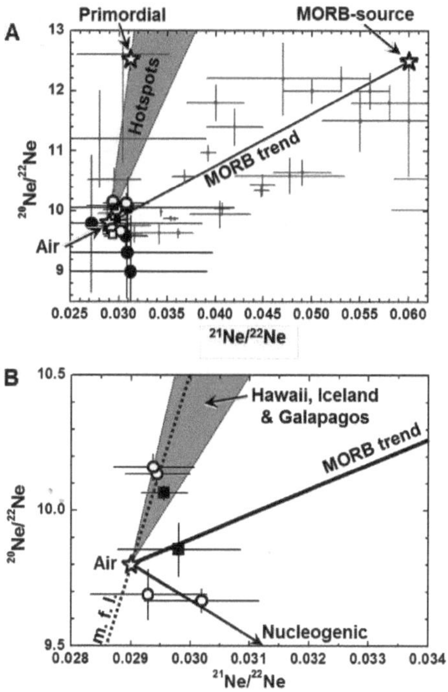

Figure 11. A–Neon three-isotope plot for the Kokchetav microdiamond, and plate **B** is a magnification of plate A. **Solid and open circles** were obtained by crushing and heating of the Kokchetav microdiamonds, respectively. Total isotope ratios of the heating steps (**solid squares**) and that of the crushed–powder heating (**open square**) are also shown. Reported data for kimberlitic diamonds (**solid small squares**, Gautheron et al. 2005; Honda et al. 1987, 2004; Ozima and Zashu 1988, 1991; Wada and Matsuda 1998; Broadley et al. 2018) are shown for comparison. End-member compositions (air, primordial, and MORB-source mantle) are shown as designated stars. Primordial and MORB-source mantle compositions and a trend which is formed by the data for hotspots (Hawaii, Iceland and Galapagos) enriched in the primordial component are taken from Trieloff et al. (2000), Sarda et al. (1988) and Kurz et al. (2009). Data point uncertainties are 1σ. Reference lines for nucleogenic production assuming the average crustal composition and F/O = 0.02 (Ballentine and Burnard 2002) and mass fractionation line (m.f.l.) of atmospheric Ne (Ozima and Podosek 2002) are shown as **solid and dotted lines**, respectively. Adapted from Sumino et al. (2011) with permission of Elsevier #5036211173353.

To date there is no geological evidence in the vicinity of the Kokchetav massif of mantle plume activity at the time of diamond formation (~530 Ma). A plume may have interacted with the subcontinental lithosphere which was beneath metasedimentary rocks before the Kokchetav slab was subducted and microdiamonds formed. To clarify the origin of the plume-like, deep-mantle-derived noble gases in the Kokchetav microdiamonds, more investigation is required particularly on UHPM diamonds from other localities. If metasomatism by plume-derived volatiles is essential for subduction and exhumation of continental crust as proposed by Seno and Rehman (2011), and the metasomatic volatiles play an important role in UHPM diamond formation, then plume-like noble gas is expected to be observed in UHPM diamonds from other localities.

MECHANISMS OF METAMORPHIC DIAMOND FORMATION

It is generally agreed upon that during a continent–continent collision crustal rocks can be subducted to depths of >150 km, which leads to the formation of UHPM terranes, however readers may be surprised to learn that there are countless hypotheses for the origin of microdiamonds found in these terranes. The debates are primarily focused around the crystallization conditions of diamond, some of the hypotheses include (i) from carbon-rich supercritical fluids vs. hydrous silicate and/or carbonatite melts, (ii) oxidation state of the media (e.g., see reviews: Ogasawara 2005 ; Frezzotti et al. 2011, 2014; Dobrzhinetskaya 2012; Schertl and Sobolev 2013; Liou et al. 2014; Zhang et al. 2017; Frezzotti 2019), (iii) stable vs. metastable conditions (e.g., Pechnikov and Kaminsky 2008, 2011; Simakov 2010), and/or (iv) from meteorite impact (Tretyakova and Luykin 2016, 2017). These hypotheses will be discussed individually in greater detail below.

During the last decade there have been attempts to resolve some of these disagreements which have resulted in considerable progress in our understanding of the formation conditions of microdiamonds. Particularly advancements in high resolution instruments such as STEM, FIB, a 'Structural & Chemical Analyser' (SCA) probe and 'inVia' Raman spectrometer (a Renishaw trademark) combined with SEM system (e.g., O'Bannon et al. 2020), μm and sub-μm monochromatic and white beam synchrotron X-ray diffraction (e.g., Stan et al. 2020) and synchrotron infrared (IR) micro-spectroscopy (e.g., Dobrzhinetskaya et al. 2006b) have all made it possible to perform analyses at the nano-scale of solid and gas/liquid inclusions in microdiamonds. Table 3 presents summarized information on nanoscale inclusions detected in diamonds from UHPM terranes by different techniques.

As new instruments and technology are adopted by researchers it has led to many other important discoveries and increased our knowledge about the origins of microdiamond. Moreover, the integration of high-pressure devices such as multianvil presses and diamond anvil cells into synchrotron X-ray facilities has enabled a wide range of experiments to study UHPM diamond crystallization at varying pressures, temperatures, fO_2, X_{H_2O}, organic/inorganic carbon and different bulk chemistry of the starting materials (e.g., Akaishi and Yamaoka 2000; Yamaoka et al. 2000; Kumar et al. 2001; Pal'yanov et al. 2002, 2013; Dobrzhinetskaya et al. 2004; Dobrzhinetskaya and Green 2007a; Zhang et al. 2011). These experiments have demonstrated that the kinetics of diamond crystallization varies depending on different experimental conditions such as hydrous or anhydrous, oxidation states, and if the pressure and temperature are high enough to stabilize sp^3 carbon bonding.

Table 3. Inclusions reported in microdiamonds from different UHPM terranes

Diamond locality	Rock type	Inclusions in diamonds	Identification method	References
Kokchetav massif, Kazakhstan	Garnet clinopyroxenite	CO_3^{2-}; H_2O; OH^-	IR spectroscopy	de Corte et al. 1998
	Garnet–biotite–gneiss	crystalline inclusions (40–80 nm)*: $Fe_2Si_2O_6$; TiO_2; Cr_2O_3; Fe_xO_y; SiO_2; $MgCO_3$; $ZrSiO_4$; $BaSO_4$; Th_xO_y ; almost all inclusions show trace components of: Si, Cr, Fe, Ti, Mg, Ca, Al, K, Na, S, P, Nb, Cl, Zn, Ni	FIB-TEM, SEM, EDX	Dobrzhinetskaya et al. 2001, 2003
	Garnet–biotite–gneiss	Si, K, P, Cl glass (1–5 nm)	AEM, EDX	Hwang et al. 2006
	Dolomite marble	Ultrapotassic fluid inclusions (<500 nm); low in SiO_2	AEM, EDX	Hwang et al. 2006
	Garnet–quartz–clinopyroxene	K rich Si poor fluid/melt in intergranular pockets within a microdiamond aggregate	AEM, EDX	Hwang et al. 2006
	Dolomite marble	$CaCO_3$ (Fe, Mg) aragonite (10–100 nm) along with COH fluid inclusions	FIB-TEM, EDS	Dobrzhinetskaya et al. 2005
	Dolomite marble	$CaCO_3$ (aragonite), $CaCO_3$, $MgCO_3$; N-bearing phase (5–250 nm)	FIB-TEM, EDS	Dobrzhinetskaya et al. 2006a

Diamond locality	Rock type	Inclusions in diamonds	Identification method	References
Erzgebirge, Germany	Gneiss lenses within migmatites of gneiss–eclogite	Silicates, apatite, rutile and diamond, altogether are included in a former fluid/melt pocket	SEM, EDS, EBSD	Stöckhert et al. 2001
	Quartz–feldspathic gneiss	Solid inclusions (20–60 nm)*: SiO_2, Al_2SiO_5, KH_2PO_4, Pb_xO_y or $PbCO_3$; unknown stoichiometry phases*: Si, K, P, Ti, O or Si, O; Si, Fe and O	FIB-TEM, SEM, EDX	Dobrzhinetskaya et al. 2003
	Quartz–feldspathic gneiss	P/K rich silica glass	AEM, EDS	Hwang et al. 2005
	Quartz–feldspathic gneiss	Molecular H_2O, OH^-, CO_3^{2-} and SiO_2; Pb_xO_y or $PbCO_3$; Al_2SiO_5	Synchrotron IR spectroscopy	Dobrzhinetskaya et al. 2006b
	Quartz–feldspathic gneiss	Silicates, apatite, rutile and diamond are included in former fluid/melt pockets located inside of garnet	SEM, EBSD, EDS	Stöckhert et al. 2009
	Garnet–phengite–quartz–feldspathic gneiss	Solid inclusions (5–20 nm)*: $KAlSi_3O_8$; Na_2SO_4; SiO_2; $KAlSi_3O_8$; $CaCO_3$; $ZrSiO_4$; $BaCO_3$ occurring inside of the diamond or at the diamond–diamond interface, almost all inclusions show trace amounts of some of these elements: Fe, Al, Cl, Mg, Ti, Pb, K, Ca, P, Cr	FIB-TEM, SEM, EDS, EELS	Dobrzhinetskaya et al. 2013
Lago di Cignana, western Alps, Italy	Oceanic metasedimentary rocks	Mg-calcite, rutile, graphite like amorphous carbon, SO_4^{2-}; HCO_3^-; CO_3^{2-}; $Si(OH)_4$ monomer; deprotonated monomers $SiO(OH)_3^-$ and $SiO_2(OH)_2^{2-}$	Raman and electron microprobe	Frezzotti et al. 2011

Note: *crystal structure was not determined due to the extremely small size of inclusions.

UHPM diamonds crystallized from a multicomponent supercritical COH fluid or a hydrous silicate melt. Studies of nanometric fluid and solid inclusions in microdiamonds (Table 3) have revealed important information on the crystallization conditions of microdiamonds. Conventional infrared (IR) spectroscopic studies revealed the presence of molecular H_2O and CO_3^{2-} radical from Kokchetav microdiamonds occurred in garnet clinopyroxenite, calc-silicate, garnet–pyroxene–quartz rocks and gneisses (e.g., de Corte et al. 1998, 2002; Sitnikova and Shatsky 2009). Dobrzhinetskaya et al. (2006b) carried out synchrotron IR spectroscopic studies of the Erzgebirge diamonds separated from garnet–phengite–quartz–feldspathic gneisses using a thermochemical microwave digesting method. They collected synchrotron IR spectra from two diamond crystals (Fig. 12).

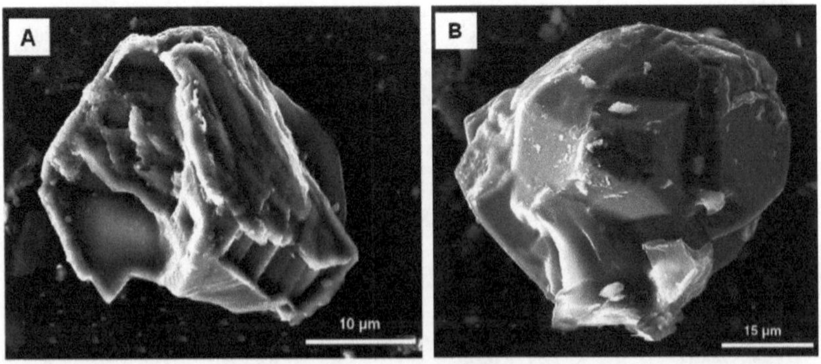

Figure 12. Secondary electron SEM images of diamonds extracted from garnet–phengite–quartz–feldspathic gneiss from the Erzgebirge massif, Germany: (**A**) diamond #1, (**B**) diamond #2 (Reprinted from Dobrzhinetskaya et al. 2006b, copyright 2006 with the permission of Elsevier # 5035711034058).

Both spectra (Fig. 13) show the presence of nitrogen impurities, molecular H_2O, OH^- hydroxyl radical, CO_3^{2-} radical, and silicate inclusions in addition the C–C diamond phonon absorption bonds. Dobrzhinetskaya et al. (2006b) intentionally studied the same samples with synchrotron IR that had been previously studied using FIB-TEM techniques (Dobrzhinetskaya et al. 2003). Abundant nanometric solid crystalline inclusions were found in the diamond FIB-foils; some of the identified phases are: (i) SiO_2-containing traces of K, P, Ti, Fe; (ii) inclusion of unknown stoichiometry which consists of three elements, K, P, and O, (iii) oxide Pb_xO_y, (iv) carbonate (Pb_2CO_3), and (v) Al_2SiO_5. Due to the extremely small size of these inclusions and their random orientations, their lattice parameters could not be determined by selected-area electron diffraction (SAED), however, all of them yield Bragg diffraction spots confirming their crystalline nature.

Besides the presence of crystalline nano-inclusions a considerable amount of nanometric-size euhedral cavities representing traces of decrepit fluid were documented. Many of these cavities were associated with dislocations related to diamond growth and are evidence of diamond nucleation and growth that took place in a COH fluid-rich media (Dobrzhinetskaya et al. 2003). The presence of larger scale diamond-bearing multiphase inclusions (or "pockets") inside of garnets and zircons (Fig. 14) observed in polished slides with optical microscopy and scanning electron microscopy corroborate perfectly with the nanoscale observations (Stöckhert et al. 2001; Dobrzhinetskaya et al. 2003, 2006a). Moreover, direct TEM observations of fluid-filled nanometric inclusion-bubbles preserved in FIB foil prepared from the Kokchetav diamond, and the serendipitous measurement of their contents immediately after bursting, showed that the fluid consists of C, H, O, Cl, S, Ca, Fe and K (Dobrzhinetskaya et al. 2005). This well-documented direct observation of residual fluid bubbles in microdiamonds unambiguously establishes their fluid growth medium.

Polycrystalline microdiamonds appear to be the best samples to study nano-fluid inclusions. There are several types of polycrystalline diamonds from the Kokchetav massif: (i) "spiral diamonds" which formed from a metal–sulfur–COH–silicate fluid (Hwang et al. 2003), (ii) S-type (star-shaped) and R-type (rugged face) diamonds that crystallized from a fluid during two stages (Ishida et al. 2003; Ogasawara 2005), and (iii) "overgrown" diamonds which crystallized from an aqueous-carbonate and aqueous-silicate fluid/melt (Sitnikova and Shatsky 2009). Polycrystalline diamond has also been described as daughter crystals associated with larger diamonds included in garnets from Erzgebirge felsic gneisses (Stöckhert et al. 2001, 2009). Stöckhert et al. (2009) concluded that the tiny daughter crystals precipitated from a supercritical COH + K-, Na-rich silicate fluid.

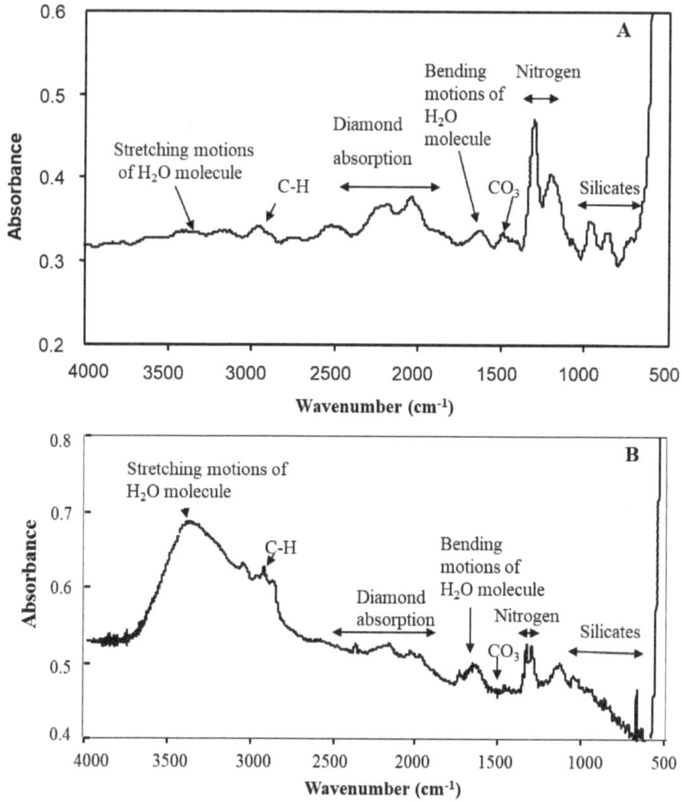

Figure 13. Synchrotron IR spectra (beam line U2A, Brookhaven National Laboratory, USA) obtained from Erzgebirge diamond #1 (**A**) and diamond #2 (**B**); diamonds are shown in plates A and B of Fig.12 (Reprinted from Dobrzhinetskaya et al. 2006b, copyright 2006 with the permission of Eslsevier # 5035711034058).

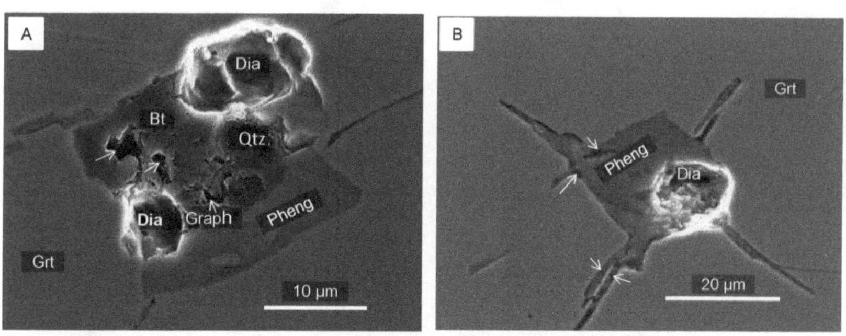

Figure 14. SEM secondary electron images of multiphase pockets in garnet from garnet–phengite–quartz–feldspathic gneiss, Erzgebirge, Germany (Reprinted from Dobrzhinetskaya et al. 2010, with the permission of Springer Nature #5034451417904). **A**—the multiphase pocket contains biotite (Bt), diamond (Dia), phengite (Pheng), quartz (Qtz), and graphite (Graph) shown by **arrowheads** (adapted from Dobrzhinetskaya et al. 2010); **B**—the phengite–diamond pocket contains abundant tiny cavities (**arrowheads**), e.g., traces of fluid inclusions decrepitated during sample preparation.

Dobrzhinetskaya et al. (2013) conducted SEM and FIB-assisted TEM studies of the Erzgebirge polycrystalline diamond inclusions in zircons (Fig.15). They studied five polycrystalline diamonds and found that they consist of 5–15 single crystals ranging in size from 0.3–5 μm with irregular sharp-angled grain boundaries (Fig. 16). Numerous triangular nanoscale voids situated along the diamond–diamond interfaces were filled with a COH fluid containing traces of Al, Co, F, V, Zn, Si, Cl, S, Ca, Mg, Fe, K in varying combinations and concentrations. The latter, if considered along with the "spider-like" dislocations, sharp-angled grain boundaries, and stacking faults that were observed in these diamonds, suggests that they rapidly nucleated and grew from a fluid media under an internal stress. These observations combined with previous results provide an additional constrain for the formation conditions of these UHPM microdiamonds. They likely crystallized from a supercritical fluid with a composition that was close to one of two end members, (1) a carbon-rich hydrous-silicic fluid and (2) a hydrous-saline fluid (Dobrzhinetskaya et al. 2006a, 2013; Jacob et al. 2014).

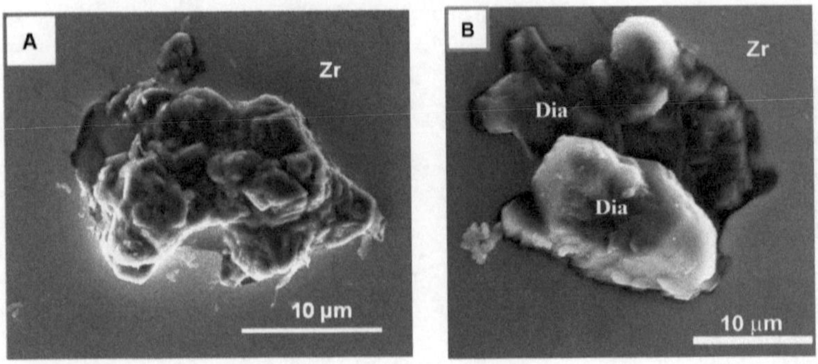

Figure 15. A and **B** – Polycrystalline diamond inclusions in zircons from garnet–phengite–quartz–feldspathic gneiss, Erzgebirge, Germany (Secondary Electron SEM images).

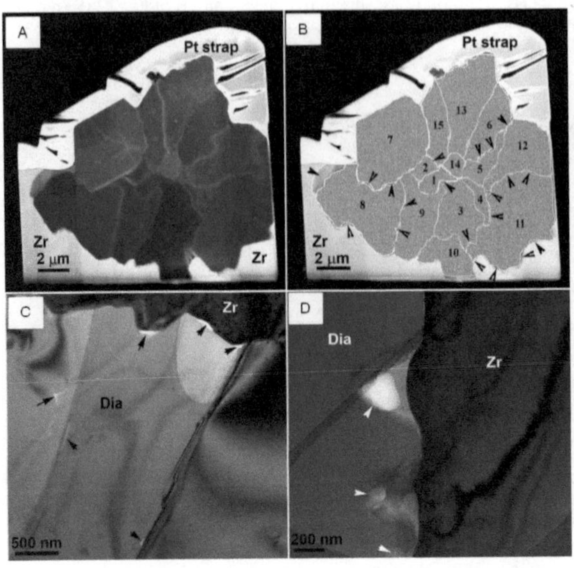

Figure 16. Bright Field TEM images of polycrystalline diamond inclusions in zircon (adapted from Dobrzhinetskaya et al. 2013). **A** – a FIB foil cut through the central part of a diamond shows that the polycrystalline diamond consists of at least 15 individual monocrystals situated closely together; **B** – a sketch of the individual monocrystals shows that the smaller crystals (#1, 2, and 14) are situated in the central part of the diamond whereas the large crystals have curved boundaries with each other and with the surrounding zircon. Plate **C** shows the "zig–zag" character of the diamond–zircon and diamond–diamond interfaces containing fluid pockets and a fluid film which are shown by the **black arrowheads**. Plate **D** shows triangle-like cavities at the diamond–zircon interface which are filled with fluid; these fluid pockets are marked by **white arrowheads**.

Zhang et al. (2017) also conducted studies of polycrystalline diamonds in zircons separated from quartz–feldspathic gneisses at Saidenbach water reservoir, Erzgebirge, Germany. Their conclusions confirm the studies by Dobrzhinetskaya et al. (2013), and unconditionally support the concept of UHPM diamond crystallization from a multicomponent COH fluid or COH fluid–hydrous silicate melt media.

Frezzotti et al. (2011) studied fluid inclusions associated with UHPM diamond formation in metamorphic schists from Lago di Cignana, Western Alps where the protolith of these rocks was likely oceanic sediments. Based on studies of the fluid geochemistry they concluded that the carbon was released from the subducting oceanic sediments at relatively shallow depths through the dissolution process (Frezzotti et al. 2011, 2014; Frezzotti 2019). Furthermore, their detailed investigation convincingly demonstrated that the deep subduction of oceanic sediments leads to a continuing saturation of the shallow fluid with carbon and precipitation of diamonds at UHPM conditions occurred. The Lago di Cignana diamond's nucleation from a COH fluid at UHPM conditions is unambiguously supported by the presence of coesite inclusions located in the same garnet growth zones where the primary fluid inclusions are located (Frezzotti et al. 2011, 2014; Frezzotti 2019). This is also in good agreement with thermodynamic calculations reported by Sverjensky et al. (2014) who investigated the stability of COH fluids as a function of pressure, temperature, and oxidation state.

A literature survey on UHPM diamond crystallization reveals that sometimes discussions are oriented around "fluid vs. melt" or "fluid vs. hydrous melt." Without clear a definition of how one should distinguish a silica-rich fluid from a hydrous silicate melt at high pressures and high temperatures, this controversy simply reflects a semantic issue. For example, experiments conducted by Bureau and Keppler (1999) suggest that there is a complete miscibility between silicate melt and water in most of the upper mantle, except at very shallow depths. In these experiments, the hydrous melt and silicate-bearing vapor coexisted separately at low T, whereas at higher T they mixed to become a single-phase supercritical fluid that was stable at $P = 1.5–2.5$ GPa. Experiments and direct studies of fluid inclusions in natural rocks conducted by others (e.g., Akaishi and Yamaoka 2000; Pal'yanov et al. 2002; Manning 2004; Ferrando et al. 2005; Frezzotti et al. 2007) also demonstrated that at $P \geq 4$ GPa and $T \geq 800$ °C there is no distinction between aqueous fluids and silicate melts because volatile-rich chemical systems are in a supercritical state.

A melt concept for Erzgebirge diamond crystallization was proposed by Massonne (2003) which was motivated by their studies of melting events in the bulk rocks of both Kokchetav and Erzgebirge diamond-bearing UHPM terranes which were constrained by zircon geochronology (Massonne et al. 2007a; Massonne and Fockenberg 2012). The cathodoluminescence images (Figs. 3 and 4 from Massonne et al. 2007a) clearly show that the cores of the zircons are typically eroded. The eroded zone of the zircon core is interpreted as being caused by a melting event. The second zone overgrew the eroded core and contains diamonds, while the third zone at the outer rim of the zircon grain does not contain any diamond inclusions. The authors report the following ages: metamorphic zircon core 337.0 ± 2.7 Ma (21 analyses); diamond-bearing zone of zircon 336.8 ± 2.8 Ma (23 analyses); and the outer rim zone of zircon 330.2 ± 5.8 Ma (12 analyses). Indeed, zircon ages combined with petrological observations could provide some constraints on the pressure–time history of these rocks, however there would be uncertaintnies in the pressure–time history that is derived from this approach. Moreover, the presence of an eroded core of zircon enveloped by diamond-bearing zircon zone does not provide any direct evidence that the diamonds originated from a melt; it can be interpreted as a metasomatic "erosion". We believe that information about the origin of the diamonds is still "hidden" inside these diamonds since Massonne et al. (2007a) did not study their inclusions. Thus, the media which these diamonds crystallized from, whether it be a carbon-rich silicate melt, and/or a COH fluid or a multicomponent fluid is not well constrained. Additional studies using high resolution techniques (e.g., synchrotron IR spectroscopy and FIB-assisted STEM or TEM coupled with nanoSIMS) would provide better constrains on the melt vs. fluid origin of these unique microdiamonds enclosed in zoned zircons.

Korsakov et al. (2004) suggested that diamonds from the Kokchetav UHPM terrane should be classified as "intragranular" type A (diamonds included in garnet, clinopyroxene and zircon; e.g., high-pressure minerals) and "intergranular" type B (diamonds occurring within pseudomorphs after high-pressure minerals), and "intergranular" type C which occur within the granitic-like material with a quartz+feldspar composition. They observed that type C "intergranular" diamonds occur in high concentrations within the quartz–feldspathic material (e.g., they could be called "granitic melt injections" or "migmatitic material") along the contact zone between the garnet–biotite–quartz and clinopyroxene–calcite layers. They proposed that these diamonds formed during the interaction of pelite–derived hydrous melt with a granitic composition with carbonate rocks during UHPM events. Korsakov et al. (2004) suggest that a granitic melt could act as a crystallization medium as well as a transport medium for type C "intergranular" diamonds. If one hypothesizes that three geologic events could have taken place, then three generations of diamonds could have formed, and they would have different ages, formation mechanisms, and media from which they crystallized.

To prove such a hypothesis, one should collect diamonds of each type along with: (i) geochronological data of diamond formation, (ii) type of inclusions preserved in diamonds (fluid, gas, solid matter) and their geochemical and crystal structure parameters, and (iii) morphology of diamonds. Until such data become available, the "three different type of diamonds" proposed by Korsakov et al. (2004) can more straightforwardly explained by one crystallization event that occurred during the same geological process (supercritical fluid/melt at the peak of the UHP metamorphism) which would imply that the diamonds all have the same age. Since diamond is chemically stable it may survive a retrograde metamorphic event, whereas other UHP minerals such as coesite, majorite, phengite, super Si-titanite, high Ca-Esk clinopyroxene and others can become re-equilibrated or replaced by their lower-pressure phases, e.g., pseudomorphs. Therefore, the same diamond (type A) which formed under peak UHP metamorphic conditions can be preserved within former UHPM minerals subjected to retrogressive metamorphism. However, in this case such a diamond would be classified type B according to Korsakov et al. (2004). If diamonds of type C are formed due to interaction between Si-rich and C-rich layers during their partial melting at UHPM conditions, such diamonds are expected to contain inclusions with chemistry that would support this origin hypothesis. Until such data are collected, the concepts of three types of diamonds (A, B and C) crystallizing during different UHPM events remain unproven.

Korsakov and Herman (2006) have also proposed that carbonate melt can be a medium where diamond crystallization occurred in the Kokchetav dolomite marbles. However, this hypothesis has been questioned by Imamura et al. (2013). They note that it is difficult to explain the extremely negative $\delta^{13}C$ values by diamond crystallization from a carbonate melt, because strong isotopic fractionation is not expected between diamond and carbonate melt at high pressures and temperatures (Bottinga 1968; Reutsky et al. 2015).

Oxidation state during UHPM diamond crystallization. Another important aspect of diamond crystallization is the oxidation state of the diamond-forming media. Association of Erzgebirge polycrystalline diamonds with carbonate ($CaCO_3$) inclusions suggests that they formed in an oxidizing environment close to the CCO buffer (Fig. 17) (e.g., Dobrzhinetskaya et al. 2013; Jacob et al. 2014). This is also supported by synchrotron infrared spectra of Erzgebirge diamond inclusions which show a well pronounced absorption band at 1430 cm^{-1}, which corresponds to the CO_3^{2-} radical (e.g., Dobrzhinetskaya et al. 2006b). Nanoinclusions of $MgCO_3$ and $CaCO_3$–aragonite were directly observed with TEM in diamonds from garnet–biotite gneiss and metacarbonate rocks of the Kokchetav massif (Dobrzhinetskaya et al. 2001, 2005, 2006a). This allows one to hypothesize that the Kokchetav diamonds also formed in an oxidizing environment close to the CCO buffer (Jacob et al. 2014).

Frezzotti et al. (2014) suggested that the Lago di Cignana (Western Alps) diamond nucleation began when the H_2O-rich fluid reached the excess concentration of carbon required for spontaneous nucleation of diamond. They showed that at high pressures the interaction of rock buffered fO_2 and the prograde *P–T* path may control the carbon saturation. The thermodynamic modeling (Frezzotti al. 2014) suggested that the diamond crystallized from COH fluids with a water concentration of about $0.992 < X_{H_2O} < 0.997$, and that the fO_2 was at the level required for EMOD (enstatite + magnesite = olivine + diamond) equilibrium (Fig. 17).

However, there are two cases where microdiamonds were found together with SiC as multiphase inclusions in zircon from Gföhl granulite samples of the Moldanubian Zone, Bohemian massif (Perraki and Faryad 2014) and in garnet from metapelitic gneisses of Pohorje, the Austro-Alpine UHPM terrane, Eastern Alps (Janák et al. 2015). Notably, Perraki and Faryad (2014) stated that they used SiC abrasives for one step of the polishing procedure, but to eliminate the possibility of contamination they removed the polish with an ultrasonic cleaner and polycrystalline-type diamond spray was used for the polishing of the thin section surface. Additionally, the occurrence of SiC and diamond grains beneath the polished surface was checked with both transmitted and reflected light microscopy.

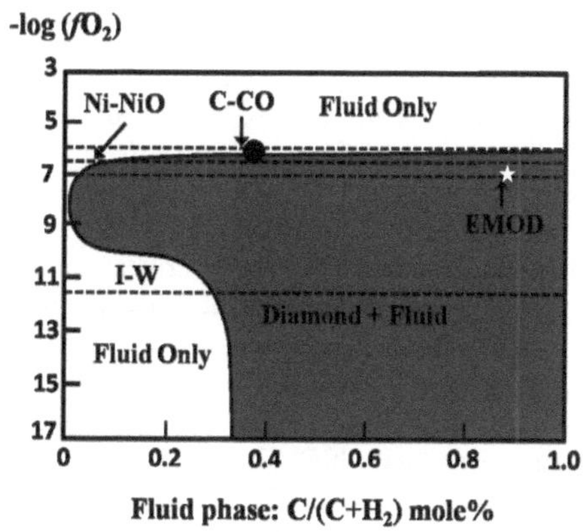

Figure 17. Diagram of –log fO_2 v. fraction of (C/C + H_2) mole % for COH fluids in equilibrium with diamond at 5 GPa and 1400 K (adopted from Jacob et al. 2011; Taylor 1990 and Dobrzhinetskaya et al. 2013). C–CO, Ni–NiO, EMOD (enstatite + magnesite + olivine + diamond), and IW (iron–wustite) are known buffer curves adopted from Eggler and Baker 1982, and Sokol et al. 2000. **Gray field** corresponds to conditions at which diamond coexists with a fluid (Diamond + Fluid); **blank fields** above and outside of the left part of the **black curve**, marked as Fluid Only, correspond to that oxidation state and concentrations of C and H_2 where diamond is not stable. The **black dot** corresponds to the position of the Erzgebirge diamonds within the C–CO buffer curve; **white star**—fluid oxidation state for diamonds from Lago di Cignana, Western Alps (Frezzotti et al. 2014).

Though Janák et al. (2015) hypothesized that diamond + moissanite inclusions in garnet from metasedimentary rocks of Pohorje were crystallized from a reduced COH fluid, they concluded that further detailed studies are needed to fully understand the oxidation state during their formation. Certainly, these unusual findings of diamond + SiC inclusions in garnets from UHPM terranes deserve additional close attention.

UHPM diamond formation under metastable conditions. Shortly after the discovery of diamonds in UHPM rocks, crystallization under metastable conditions was considered as an alternative to the classical concept of diamond formation in the Earth at $P \sim 4.0–6.0$ GPa and $T \sim 900–1400$ °C (Bundy 1980; Boyd et al. 1985).

Throughout the 1980s and into the mid of 1990s the concept of metastable diamond formation at low-pressure conditions of crustal metamorphism was proposed (e.g., Letnikov 1983; Lavrova 1991; Ekimova et al. 1992; Lavrova et al. 1995). Even at that time it was considered controversial to the new concept of diamond crystallization during UHP metamorphism of crustal rocks subducted to a depth ≥120 km (Sobolev and Shatsky 1988, 1990; Shatsky and Sobolev 1993; Shatsky et al. 1995). It has since been shown unambiguously that the Kokchetav massif is an UHPM terrane through the documentation of numerous UHPM mineral phases by hundreds of high-quality studies involving state-of-art and synchrotron assisted instruments (see reviews: Ernst and Liou 1999; Ogasawara 2005; Dobrzhinetskaya 2012; Schertl and Sobolev 2013; Liou et al. 2014). Despite all these convincing observations the concept of diamond crystallization outside of the diamond stability field has still occasionally appealed as a formation mechanism for diamonds found in UHPM terranes (e.g., Pechnikov and Kaminsky 2008, 2011; Simakov et al. 2008; Simakov 2010, 2015, 2018). Since this hypothesis has been invoked to describe the formation conditions of microdiamonds a short discussion about what is known from experiments on metastable diamond growth is warranted.

Pechnikov and Kaminsky (2008) focused their attention on the geometry of the "productive" diamond-bearing ore body using detailed mapping completed by the local Kokchetav Geological Survey, and they combined this information with their petrographic observations. They concluded that the diamonds were formed under crustal conditions (e.g., metastable) from crustal fluid–metasomatic processes that were active along the fault-zone superimposed on the Precambrian gneisses. They accepted the diamond bearing zircon age of 527–537 Ma (Claoué–Long et al. 1991; Jagoutz et al. 1990, 1991; Hermann et al. 2006) as the age of diamond formation and therefore the age of the fluid–metasomatic activities along the fault zone. The authors believe that the micrometer size of the diamonds, along with their isotopic signatures ($\delta^{13}C = -8.9$ ‰ through -27 ‰), support diamond formation under metastable crustal conditions. On the other hand, Pechnikov and Kaminsky (2008) note that their concept does not exclude the occurrence of episodes of UHP metamorphism and they accept the presence of coesite and other UHP minerals in the diamond-bearing rocks. However, their concept fails to explain when the UHP metamorphic episodes took place, and the authors did not discuss that fact that zircons containing diamonds and/or coesite, or pyroxene with high Ca-Eskola component and other UHP phases have the same age (e.g., Katayama et al. 2000, 2001a,b, 2002; Ogasawara 2005). Though the prevalent light carbon isotope characteristics of diamonds prove that such carbon has a crustal origin it does not reject the possibility of the subduction of crustal rocks to depths ≥ 120–150 km where the pressure and temperature are high enough to stabilize carbon in the form of diamond. Indeed, the arguments of Pechnikov and Kaminsky (2008) on the metastable origin of the Kokchetav diamonds are contrary to the conclusions reported by most other scientists who have collected and studied diamonds from this locality.

Simakov (2010) reports that they synthesized nano-diamonds from a water–alcohol solution which was the source of free carbon C, in a high-pressure reactor (volume ~500 cm^3) at $P = 0.1$ GPa and $T = 500$ °C. They directly applied the results of this experiment to explain the origin of mm-sized diamonds from UHPM terrane of the Kokchetav massif without taking into consideration all of the other UHP minerals associated with these diamonds (e.g., Katayama et al. 2000, 2001b, 2002; Ogasawara et al. 2002; Ogasawara 2005).

Furthermore, Simakov (2015, 2018) also discussed the metastable formation of diamond from a COHN fluid and applied this idea to explain the origin of UHPM diamonds. They discussed that nitrogen interaction decreases the total energy of diamond nuclei formation and the equilibrated methane pressure stabilizes the diamond, and they confirmed this by the hydrothermal synthesis of nano-diamond from nitrogen bearing organics at metastable *P–T* conditions (Simakov et al. 2008). Synthesis of nano diamonds at high temperatures and/or ambient–low pressure (e.g., metastable) are well-known in materials sciences processes such as laser ablation, chemical vapor deposition (CVD), autoclave synthesis from supercritical fluids, ion irradiation of graphite and ultrasound cavitation (e.g., Krueger 2008; Derjaguin and Fedoseev 1994; Fedoseev et al. 1971) as well detonation techniques (e.g., Danilenko 2004), which requires extremely high pressures. However, before invoking any of the processes mentioned above to explain how diamonds form in UHPM terranes, would require a critical evaluation of all the existing data that have been collected by many research groups. Thermodynamic calculations for *PT*-conditions of diamond-bearing rocks and the presence of UHP minerals associated with diamond are important petrological evidence indicating that the diamonds from Kokchetav formed under UHPM conditions and their formation does not easily fit to the models/experiments proposed by Simakov (2010, 2015, 2018). Additionally, experiments conducted by other groups using a starting material that is close to the bulk chemistry of the Kokchetav diamond-bearing rocks, under both hydrous and anhydrous, and at varying oxidation states have produced run products that are a good fit to the data collected from the rocks, diamond itself and other associated minerals (e.g., Hong et al. 1999; Pal'yanov et al. 1999, 2000, 2013; Akaishi and Yamaoka 2000; Yamaoka et al. 2000; Kumar et al. 2001; Hwang et al. 2003; Dobrzhinetskaya et al. 2004; Dobrzhinetskaya and Green 2007a,b). Therefore, while the models/experiments proposed by Simakov (2010, 2015, 2018) do, indeed, explain diamond formation outside the diamond stability field, they cannot be applied to diamonds from the Kokchetav massif or for diamonds from other well established UHPM terranes.

It should be emphasized that if diamond formation under metastable conditions is proposed the following data are expected to be presented for consideration: (i) thorough geologic mapping of the area where the sample was collected to find any other UHPM indicator minerals other than diamond; (ii) unambiguous proof of the indigenous origin of the diamond, the diamond should be *in situ* so that the interface between the diamond and its host mineral can be examined in detail with high resolution techniques, and (iii) all available thermodynamic data has been reviewed.

Comet impact hypothesis for the origin of diamonds from Kumdy Kol Lake, Kokchetav Massif. Recently a new impact–cosmic–metasomatic genesis of the Kokchetav diamond deposit was proposed (Tretiakova and Lyukhin 2016, 2017). It was provoked by an impact event followed by metamorphism and alteration of the rocks. The authors note that the ring structure of Kumdy Kol Lake, where most of the diamond-bearing metamorphic rocks of the Kokchetav massif are situated, has the form and size (~4 km diameter) resembling a small impact crater. They suggest that the diamond-bearing terrane formed by comet impact under an oblique angle which generated shock wave compression with peak pressures reaching > 50 GPa.

They argue that there is evidence of impact metamorphism, (1) presence of UHPM minerals diamond–lonsdaleite, coesite, omphacite, (2) the delivery by comet moissanite (SiC) and graphite spherules as well as "meteoritic matter" such as magnetite, hematite, iocite, troilite, ∝-Fe, Ni–Fe, (3) annealed metallic globules are observed having various fanciful forms (globules, small dump-bells, drops, spherules and so on) in host rock and rock-forming minerals, and (4) dislocation and birefringence in diamonds, planar structure in quartz, inclusions UHP minerals in rock forming minerals. The authors suggested that collision of huge velocity comet and the Earth generated rapid shock wave compression (pressure peak > 50 GPa) forming diamond, coesite and other UHPM minerals which later together with so-called "meteoritic matter' were subjected to regional metamorphism and metasomatism.

Additionally, they hypothesize that the "unprecedentedly" high $^3He/^4He$ isotopic ratios of the Kokchetav diamonds reported by Shukolyukov et al. (1993) are, indeed, due to the formation of these diamonds outside of the solar system. However, new measurements by Sumino et al. (2011) revealed that $^3He/^4He$ value of the Kokchetav diamonds are close to the values found for OIB, and that the extremely high $^3He/^4He$ ratio likely originated from Li-rich contaminants.

The "impact–cosmic–metasomatic" origin remains unproven based on the following evidence: (i) Sumino et al. (2011) have shown that the "unprecedentedly high" $^{3}He/^{4}He$ isotopic ratios are not likely a true feature of the Kokchetav diamonds, (ii) the authors have not presented unambiguous evidence of the claimed "oblique" impact (e.g., asymmetrical ejecta deposits), (iii) the authors do not propose that any of the other dozens of similar shaped lakes around Kumdy-Kol lake were caused by impact events, and (iv) and they do not report that the Kokchetav diamonds show unique moiré fringes which are well-known features observed in impact-formed diamonds (e.g., Langenhorst 2003; Ohfuji et al. 2015).

OPHIOLITE-HOSTED AND VOLCANIC DIAMONDS

Ophiolite-hosted diamonds

Liou et al. (2014) provided a comprehensive review of the history of the discovery of ultra-high-pressure minerals in ophiolitic chromitites and peridotites. To avoid unnecessary repetition here, we focus our reader's attention on the studies of diamonds and specifically *on the methods used for confirmation of whether the diamond is indigenous or not.* This topic is still highly controversial as is evidenced by several recent critical *comments* and *replies* on a few recent publications that have reported ophiolite-hosted diamonds. The crux of the matter is the indigenous nature of these reported diamonds. In this section we will discuss our vision and approach for how the community can reach a consensus about whether these diamonds are myth or reality.

To present ophiolite-hosted diamonds have been reported from more than a dozen geographic localities in the World (Fig. 18). Most of them have been found in the Tertiary Yarling-Zangbo suture zone of Tibet which formed during the India–Asia continental collision. The ophiolite complex situated along this suture zone is mainly composed of remnants of the Neo-Tethys ocean floor. Within them the Luobusa massif encompasses diamond-hosted chromitite and harzburgite (e.g., Fang and Bai 1981; Bai et al. 1993, 2001; Yang et al. 2007a), and the Purgan, Xigaze, Dongbo, Burang, Dangqiong, and Dongqiao massifs enclose diamond-hosted chromitites and peridotites (e.g., Yang et al. 2011, 2013, 2014; Lian and Yang 2019). Other worldwide ophiolite hosted diamond-bearing localities include podiform chromitites and/or peridotites in the Sartohai massif, NW China, Ray-Iz massif, Polar Ural, Russia (Yang et al. 2007b, 2015a); Hegenshan massif of Inner Mongolia, China (Huang et al. 2015); Mirdita massif, Albania (Wu et al. 2017, 2019; Xiong et al. 2017); Pasani-Karsanti massif, Turkey (Lian et al. 2017); Nidar ophiolite, India (Das et al. 2017); Myikyina massif, Myanmar (Chen et al. 2018), Tehuitzingo, Puebla State, southern Mexico where diamonds was documented *in situ* in chromitite-bearing serpentinites (Farré-de-Pablo et al. 2018), and diamond inclusions in magmatic olivine from ophiolite section of the Moa-Baracoa ophiolitic massif, eastern Cuba (Pujol-Solà et al. 2020).

Ophiolitic diamond formation hypotheses

Since the formation conditions for diamond and ophiolites are incompatible with each other several hypotheses have been put forward to explain their existence and some of them are very controversial (e.g., Arai 2013; Yang et al. 2014, 2015a,b; Zhou et al. 2014; McGowan et al. 2015; Robinson et al. 2015; Griffin et al. 2016; Rollinson 2016; Ballhaus et al. 2017; Dilek and Yang 2018; Lian and Yang 2019). We summarize them as three main hypotheses:

(i) Diamond is formed at depth in the mantle at high-pressure and high-temperature conditions with the source of carbon coming from the deep subduction of oceanic lithosphere fragments. These diamonds are then transported to the shallow mantle by mantle convection (e.g., Arai 2013; Mcgowan et al. 2015); and/or they are transported by mantle plumes to the shallow upper mantle (e.g., Xiong et al. 2015, 2017; Yang et al. 2015b; Griffin et al. 2016; Lian and Yang 2019).

(ii) Diamond crystallization at low T and low P (metastable conditions) from carbon-rich fluids that form during alteration of chromitites and other fragments of oceanic crust at shallow depths (Farré-de-Pablo et al. 2018, 2019, 2020; Pujol-Solà et al. 2020).

(iii) Diamond (and super-reduced alloy and minerals) formation from lightning strikes (Ballhaus et al. 2017, 2021).

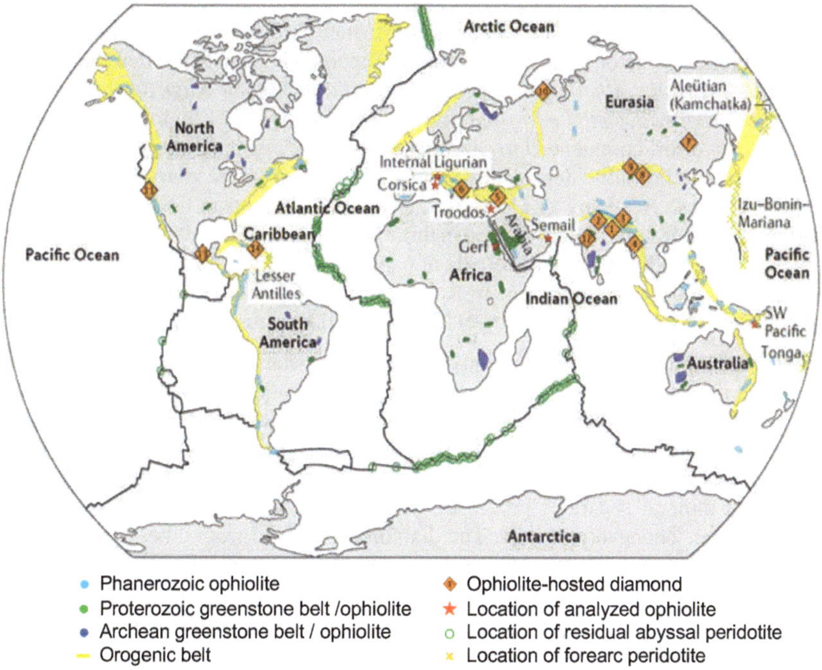

Figure 18. Global distribution of ophiolite belts and diamond-bearing ophiolites (Yang et al. 2021, with the Elsevier permission #5035651046919). Diamond-bearing ophiolites are typically located close to orogenic belts. The location of Archean (4–2.5 Ga) and Proterozoic (2.5–0.54 Ga) greenstone belts, Phanerozoic (<0.54 Ga) ophiolites and orogenic belts are from Dilek and Furnas 2011; Furnas et al. 2014 2015. Locations of known ophiolite-hosted diamonds: Liobasa, Zedang and Xigaze ophiolite (1) (Bai et al. 1993; Yang et al. 2014; Guo et al. 2015); Purang, Dongbo and Dangquong ophiolite (2) (Yang et al. 2011; Xu et al. 2015); Dingqing ophiolite (3) (Xiong et al. 2018); Myitkyina opiolite (4) (Chen et al. 2018); Pozanti–Karzanti ophiolite (5) (Lian et al. 2017); Mirdita ophilite (6) (Wu et al. 2017, 2019; Xiong et al. 2017); Hegenshan phiolite (7) (Huang et al. 2015); Qilian ophiolite (8), Sartohay ophiolite (9) (Tian et al. 2015); Ray-Iz ophiolite (10) (Yang et al. 2015); Josephine ophiolite (11), Nadar ophiolite (12) (Das et al. 2017); Tehuitzongo ophilite (13) (Farré-de-Pablo et al. 2018); Moa–Baracoa ophiolite (14) (Pujol-Solà et al. 2020). **Red stars** show locations discussed in Yang et al. 2021. **Open green circle** indicates locations of residual abyssal peridotites (Warren 2016). **Yellow crosses** show locations of the forearc peridotites (Dien et al. 2019).

Localities of ophiolites-hosted diamonds formed at mantle depth

Diamonds from ophiolites of the Yarling-Zangbo suture zone, Tibet. The first diamonds were found in Tibet during the examination of the heavy mineral separates retrieved from mechanically crashed chromitites using optical microscopy (Fang and Bai 1981). The heavy minerals separation techniques have different protocols and procedures in different countries; however, all are used for the exploration of industrial mineral deposits, which in the case of chromitites are platinum group elements (PGE) metals and alloys. Fang and Bai (1981) reported that 20 diamonds were recovered from the Luobusa chromitite and hargburgite, and six diamonds from similar rocks collected from Donqiao massif of the Bangong-Nujiang ophiolite belt situated 500 km northwest from the Luobusa massif (Bai et al. 1989, 1993). These diamonds were characterized as colorless and transparent euhedral/octahedral-like morphology; their sizes fall in range of 100–500 μm, though some broken crystals of 200–250 μm and diamonds as large as 900–1000 μm were also reported (Bai et al. 1989). Other than diamonds, additional minerals which are "unexpected" to occur in chromitites, such as SiC, graphite, Ni–Fe alloys, and Cr^{+2}-bearing chromium spinel were also documented in the heavy minerals separate.

These first reports of diamond findings in Tibetan chromite were eventually disproved by an international group of kimberlitic diamond experts who concluded that these diamonds "*...are consistent with synthetic contamination introduced at some stage during the sampling or concentration of heavy minerals from these ultramafic bodies*" (Taylor et al. 1995). After this report a new effort commenced to carefully process and separate heavy minerals from the Luobusa and other chromitite deposits of the Yarling-Zangbo suture zone, Tibet. The Institute of Geology, Chinese Academy of Geological Sciences, the Tibetan Geological Bureau, and international collaborators all participated in this effort. As a result, dozens of diamond crystals ranging from ~100–600 μm in size were documented again (Fig. 19) as well as more unusual new minerals including native Cr, Os–Ir–Ru alloys, Ir–Ni–Fe alloys, Pt–Fe alloys and others were recognized (e.g., see reviews Yang et al. 2007a 2014; Liou et al. 2014).

Additional explorations for ophiolitic diamonds in Tibetan and other worldwide ophiolitic terranes began; they used the same protocol, e.g., the collection of ~200 to 1000 kg of chromitites or peridotites, mechanical crushing and sieving of the rocks followed by heavy minerals separation including vibration, magnetic, heavy liquid floatation, and electrical conductivity techniques. Most Tibetan diamonds and diamonds from other ophiolitic belts were found in mineral separates processed in the Institute of Multipurpose Utilization of Mineral Resource, Zhengzhou, China. The diamonds retrieved from other Tibetan ophiolites (Purgan, Xigaze, Dongbo, Burang, Dangqiong, and Dongqiao massifs) share morphological similarities with those of Luobusa massif (Fig. 19), and they are always associated with of ultra-reduced minerals, metallic alloys and native elements (e.g., Liou et al. 2014; Yang et al. 2014; Xiong et al. 2017; Lian and Yang 2019).

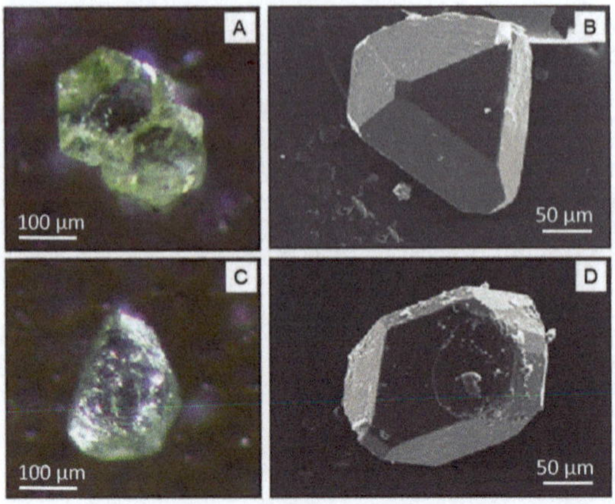

Figure 19. Topography of microdiamonds from chromitites and mantle peridotites (Xu et al. 2018, published under the terms of the CC–BY–NC license). **A** – Polycrystalline diamond from chromitite (Luobusa ophiolite, Tibet) exhibits cuboctahedral morphology and in light yellow color under a binocular microscope. **B** – Scanning electron microscope (SEM) image of diamond from chromitite (Luobusa ophiolite, Tibet) displaying octahedral-dominated cuboctahedron crystal habit. **C** – Colorless diamond from mantle peridotite (Luobusa ophiolite, Tibet) showing cuboctahedral morphology under a binocular microscope. **D** – SEM image of diamond from mantle peridotite (Luobusa ophiolite, Tibet) displaying hexoctahedron crystal morphology.

Tibetan diamond in situ, unusual UHP minerals and highly reduced minerals and alloys. A small diamond of ~6 μm size was found as an inclusion in an Os–Ir alloy that was recovered from heavy mineral separates from Luobusa chromitite. This diamond was confirmed using Raman spectroscopy and the first order Raman mode was reported to be centered at 1332.2 cm^{-1} (Yang et al. 2007a). The position of this diamond in the thin section (polished without diamond abrasives) was observed with both optical microscopy and with SEM in secondary electron mode. Although the authors claim this satisfies the minimum requirements for the diamond to be considered in situ, the diamond was not observed below the surface of the host mineral before it was exposed to the surface by polishing. Alsothe samplewas not studied with FIB-TEM technique to confirm the nature of the Os–Ir-alloy–diamond interface and to find/analyze any inclusions that may be encapsulated inside this diamond (Yang et al. 2007a). Observation of the diamond below the surface of the host mineral and detailed analysis of the diamond host mineral interface with FIB-TEM would have unambiguously proved this diamond is indigenous. It has been suggested that the *in situ* Tibetan diamonds were likely formed at depths of 150–300 km (Yang et al. 2007a, 2014) or even deeper than 380 km (Xiong et al. 2018).

Detailed studies were conducted on another unusual recovered mineral fragment, an Fe–Ti alloy which was retrieved from the same mineral separates as the Os–Ir alloy containing diamond (Yang et al. 2007a). The fragment of Fe–Ti alloy had a small portion of a silicate rock which contained coesite after stishovite and kyanite suggesting that coesite after stishovite and diamond, might be incorporated into the chromitites in the deep upper mantle, or that they have an impact origin with a prevalence of evidence for the deep mantle origin (Yang et al. 2007a).

Later Dobrzhinetskaya et al. (2009 2014b) studies of the same sample using the state-of-the-art FIB-TEM analytical techniques showed that the coesite and kyanite contained mm-sized inclusions of native iron, cubic–BN (qingsongite), TiC (osbornite) and TiO_2-II (rutile with α-PbS structure). The authors discussed two possible explanations that these data support: (i) the UHPM minerals have been encapsulated within chromitite at mantle depth, or (ii) they were formed by shock events due to a meteorite bombardment (Dobrzhinetskaya et al. 2009 2014b). Since the microstructures of UHPM minerals in the studied fragment are inconsistent with those formed due to shock metamorphism (e.g., Martini 1991; Stähle et al. 2008) the mantle origin of the fragment was deemed more plausible (Dobrzhinetskaya et al. 2009, 2014b). Dobrzhinetskaya et al. (2009, 2014b) have clearly informed the readers that the studied fragment of these metamorphic rocks, contained cubic-BN (quingsongite), TiC-osbornite, TiO_2-II, coesite after stishovite and some high Si-Al-rich amorphous material, all recovered from these Tibetan chromitites by "crushed rocks–minerals separation" technique.

Although Tibetan chromitites contain more than a dozen different native elements, alloys, and minerals only a few of them may be classified as UHP minerals (diamond, coesite, TiO_2 II). Most of the other reported minerals are highly reduced minerals (e.g., SiC-moissanite and FeO-wustite), native elements (e.g., Fe, Si, Ti, Cr), alloys (e.g., Os–Ir–Ru, Ni–Fe, Ir–Ni–Fe Pt–Fe), carbides [e.g., WC-qusongite, $(Cr,Fe,Ni)_9C_4$-yarlongite] and silicides (e.g., Fe_5Si_3-xifengite, $TiFeSi_2$-zanboite) to mention a few (see reviews: Robinson et al. 2004; Liou et al. 2014; Lian and Yang 2019; Yang et al. 2019a). Importantly verification of their stability field *at high pressures and high temperatures* requires further experimental studies.

Diamonds in situ from Rai-Iz ophiolite, Polar Ural, Russia. New reports of ophiolitic diamonds and ultra-reduced minerals in other countries of Euro-Asian continent (Russia, Albania, Turkey, India, and Myanmar) confirm that they are widespread and that new concepts are required to explain their occurrences in ophiolite formations. Some interesting results came from recent studies of Nano-inclusions in ophiolitic diamonds from the Ray-Iz massif, which is an early Paleozoic ophiolitic belt in the Polar Urals, Russia. Yang et al. (2015a) recovered 1000 grains of loose diamonds (~200–500 μm in size) from heavy mineral separates processed from 1500 kg of Ray-Iz chromitite samples. They were processed at the same facilities as the Tibetan diamonds.

To search for diamond *in situ* they cut 40 pieces of rock that were roughly 4 cm^2 in area and encased them in epoxy. The samples were grinded down in 100 μm increments using a diamond-coated disk with a maximum grain size of 37 μm. Optical microscope images show that two diamonds remained enclosed in chromite (Fig. 20, plates A and C). On plate B of Fig. 20 (sample Y–5B–17–2) one can see a sub-euhedral diamond crystal that is ~ 300 μm in diameter surrounded by thin patch of amorphous carbon. Sample Y5B-16-2 (plate D) exhibits a well-shaped hexagonal platy-like diamond crystal with dimensions of ~300 × 400 μm surrounded by an irregular-shaped mass of amorphous carbon—all are inside of a host chromite (Yang et al. 2015a).

Though the authors concluded that both crystals are diamonds, they provided only one spectrum (e.g., plate E shows only one Raman spectrum with band at 1330.8 cm^{-1}, see caption to Fig. 8, page 469, Yang et al. 2015a). Furthermore, the nature of the amorphous carbon remains unclear. The authors claim that it is amorphous based the absence of a Raman modes in the areas where elemental mapping showed that carbon was present (Yang et al. 2015a). However, even amorphous carbons with very little graphitic ordering show the presence of *G*-

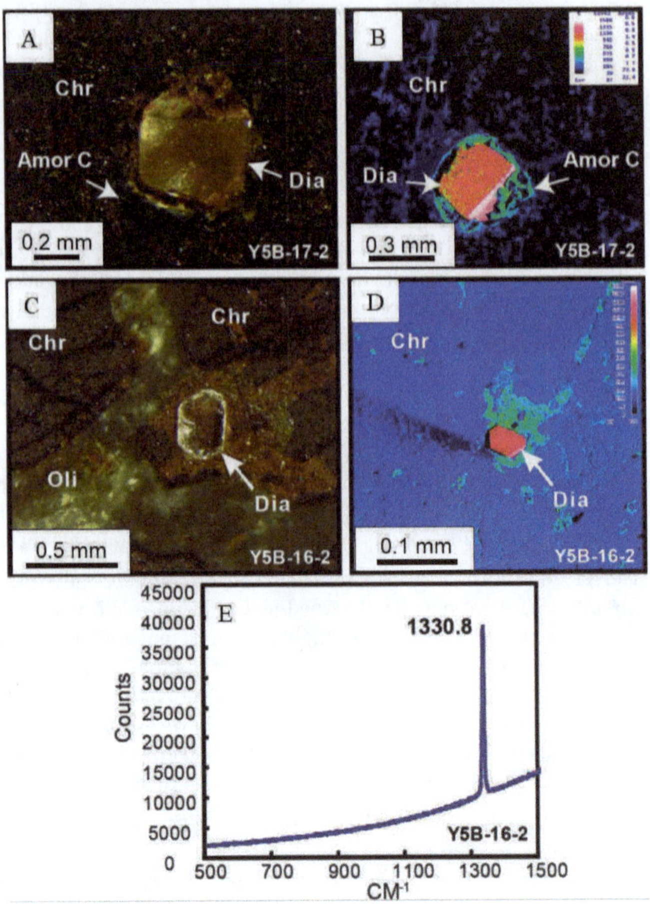

Figure 20. Photomicrographs (**A**, **C**) and, carbon element maps (**B**, **D**) of diamonds from the Ray-Iz ophiolite and (**E**) laser Raman pattern of diamond (reprinted from Yang et al. 2015a, copyright 2015 with the permission of Elsevier #5035691008088). The in-situ diamonds are hosted in spherical (sample 05B–17–2) to irregular (sample 05B–16–2) patches of amorphous carbon within chromitite of the Ray-Iz ophiolite. Chr–chromite, Oli–olivine, Dia–diamond. The **dark blue** color is chromite and olivine and the **light blue** color is amorphous carbon.

and *D*-like features in their Raman spectra (e.g., Ferrari and Robertson 2000). So, it remains a mystery as to why Yang et al. (2015a) did not observe any Raman peaks, they do not show these spectra and they do not report any instrument details from their Raman measurements (e.g., laser wavelength, spectrometer focal length, spectrometer resolution etc.). Moreover, presence of coesite inclusions in ophiolitic diamond is not an undisputable feature that can be used as a criterion to sort out if diamond is synthetic or indigenous. Milledge (1961) reported that they found coesite inclusions in one of the synthetic diamonds produced by the General Electric Company (G.E.C.) Laboratories in Schenectady county NY, USA. Therefore, careful characterization of sp^3-carbon (diamond), sp^2-carbon (graphite), and disordered or amorphous carbon should be provided in the future by Yang et al. (2015a) to prove unconditionally that the Ray-Iz diamonds are indigenous. They also need to show the differences between their so-called "amorphous carbon" and epoxy resin, which was used for this sample preparation.

Several loose diamonds were studied with FIB-TEM technique (Yang et al. 2015a,b). The results showed that the diamonds contain several ~2–10 μm inclusions of NiMnCo alloys, graphite and one inclusion of coesite. Further studies of inclusions in diamonds recovered from ophiolites are needed so that we can start to better understand the formation conditions of these diamonds.

Diamonds from the Skenderbeu chromitite massif of the Mirdita ophiolite, Albania. Wu et al. (2017) collected 600 kg of chromitite rocks from the Skenderbeu massif of the Mirdita ophiolite, Albania, the western part of Neo-Tethys ocean. After processing these samples at Institute of Multipurpose Utilization of Mineral Resources, Zhengzhou, China, the authors found 20 loose diamonds (100–200 μm in size) of octahedral, cubo-octahedral and irregular shapes (Fig. 21). Diamonds were confirmed with Raman spectroscopy and studied with TEM.

TEM studies of two FIB-prepared foils revealed that the Skenderbeu diamonds contain inclusions of NiMnCo alloy, traces of S and Cl, indicated escaped fluid, and $Ca_{0.81}Mn_{0.19}(SiO)_3$—which has a structure similar to wollastonite. They hypothesized that this phase may be a product of the decompression of a Ca-perovskite, although no undisputable data to support

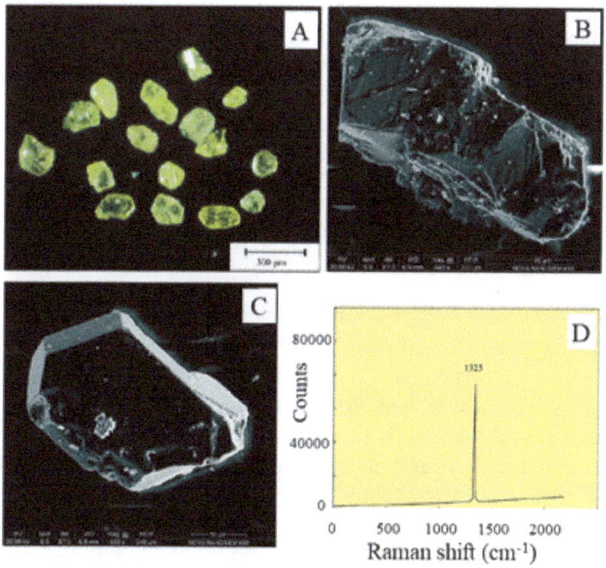

Figure 21. Phographs of diamonds recovered from Skenderbeu chromitites (Reprinted from Wu et al. 2017, copyright 2017, with the permission of John Wiley and Sons #5036200875863). **A** – a microphotograph showing yellow–light yellow diamonds. **B** – SEM image of irregular damond grain. **C** – SEM image of octahedral diamond. **D** – Raman spectrum showing typical diamond Raman shift around 1335 cm^{-1}.

this interpretation was provided. The authors proposed that diamonds were formed in the deep mantle from a carbon-saturated NiMnCo-rich melt acquired from a subducted slab of oceanic crust and lithosphere (Wu et al. 2017, 2020).

Diamonds from Pozanti-Karsanti podiform chromitite, Turkey. The Pozanti-Karsanti ophilite complex occurs within the eastern Tauride belt, in southern Turkey, and consists of well-preserved sections of mantle peridotite, ultramafic and mafic cumulates, podiform chromitites, gabbros, sheeted dikes, and basalts (Dilek et al. 1999). Lian et al. (2017) reported that they recovered more than 100 grains of euhedral and irregular shaped diamonds (50–250 μm in size) from heavy minerals separates from the Pozanti-Karsanti podiform chromitite. The rocks were processed at the same facilities in Zhengzhou, China, as all other ophiolitic diamonds described above. The authors also found SiC and many "crustal" minerals such as zircon, rutile, hematite, magnetite, quartz, sulfides, chlorite, monazite, and also olivine, pyroxene, and an unusual octahedral silicate pseudomorph with a complex chemistry (Mg, Si, Al, Cr, Fe, O and Mg,Si,O). Lian et al. (2018a) showed that the diamonds contain nanometric inclusions of Ni–Mn–Co alloys and $(Ca,Mn)SiO_3$ with triclinic symmetry similar to wollastonite. The authors have hypothesized that wollastonite may be a product of the decompression of $(Ca,Mn)SiO_3$-perovskite, and concluded that the diamonds may have crystallized from carbon-rich fluids in the mantle at a depth of more than 250 km and that they were rapidly transported to a shallower depth by an asthenospheric melt (Lian et al. 2017 2018a,b). Interestingly their interpretation is similar to what was proposed by Wu et al. (2017) for $Ca_{0.81}Mn_{0.19}(SiO)_3$ wollastonite inclusions in Skenderbeu diamonds from Albania.

Diamond in situ from Nadar ophiolites, India. The Nadar ophiolites are situated within the Indus suture zone, southeast Ladakh Himalaya, marking a collision zone between India and Asia. Das et al. (2017) discovered diamond in situ in a polished petrographic thin section of peridotite collected from the lower section of the Nadar ophiolites. Peridotite consists of olivine (50%), orthopyroxene (30%), clinopyroxene (10%), Cr-spinel (5%) and serpentine (5%). A diamond of ~50 μm in size is included in orthopyroxene; a Raman spectrum of the diamond exhibits a well-pronounced Raman band at 1326.7 cm^{-1} (characteristic of disordered sp^3-carbon bonding) and broad feature centered at 1583.2 cm^{-1} (sp^2 disordered graphite carbon bonding often referred to as the *G-band*). Some grains of olivine in this thin section contain numerous 2–5 μm size inclusions of graphite associated with fluid inclusions of C–H and H_2. For those grains of olivine which exhibit well-oriented exsolution lamellae of Cr-spinel and blebs of α-Fe_2O_3, the β-Mg_2SiO_4 was inferred to be a precursor. Das et al. (2017) suggested that diamond crystallized from C–H fluids that were derived from the bottom of the upper mantle—top of the mantle transition zone (depth ~ 410 km). Although 1, 3 and 6 μm diamond grits were used for sample preparation, Das et al. (2017) deduced that their diamond inclusion size (~50 μm) exceeds the size of diamonds from the polishing source, and therefore the diamond inclusion in orthopyroxene from Nadar peridotite is indigenous.

Carbon isotopes, nitrogen content and nitrogen aggregation in UHPM ophiolitic diamonds

According to Xu et al. (2018) carbon isotope characteristics of ophiolitic diamonds from chromitites are comparable with a light carbon reservoir: $\delta^{13}C = -29.0‰$ to $-16.9‰$ (Luobusa chromitite) and $\delta^{13}C = -28.7‰$ to $-18.3‰$ (Ray-Iz, Polar Urals, Russia). The carbon isotope values of ophiolitic diamonds bear similar features as metamorphic diamonds from UHPM terranes, and they are significantly different from cratonic diamonds (Fig. 7). Nitrogen content in Luobusa diamonds varies from 152 to 428 ppm, whereas in Ray-Iz diamonds it varies from 108 to 499 ppm, and both are significantly lower than the nitrogen content reported in diamonds from UHPM terranes and cratonic settings (see Table 2 for comparison). Nitrogen isotope characteristics for Tibetan ophiolite-hosted diamonds ranges from $\delta^{15}N = -28.4‰$ to $-18.8‰$ (e.g., Xu et al. 2018; Lian et al. (2018a) studied carbon and nitrogen isotope values in diamonds from Posanti-Karsanti chromitites, Turkey. The Pozanti-Karzanti diamonds are characterized by $\delta^{13}C = -28.4‰$ to $-18.8‰$ and $\delta^{15}N$ values ranging from $-19.1‰$ to $16.6‰$ with nitrogen contents ranging from 7 to 541 ppm. All these parameters are similar to those from diamonds of Loubasa and Ray-Iz ophiolites (Table 2) and their strongly ^{13}C-depleted more than likely originated from a subducted crustal carbon-reservoir.

Howell et al. (2015) studied nitrogen aggregation in Luobasa diamonds. They show that these diamonds contain nitrogen impurities in the form of singly substituted C centers (Type Ib diamond). According to Taylor et al. (1996) natural diamonds with similar nitrogen characteristics have a short residence time and they are rare in nature because isolated nitrogen aggregates to form nitrogen pairs (A centers–Type IaA diamond) and more aggregated nitrogen (B centers–Type IaB diamond). Type IaAB is most typical for cratonic diamonds which have long-term (~0.5 to 3 billion year) upper mantle residence, whereas Type Ib is a nitrogen aggregation characteristic, which is typical for synthetic diamond (Taylor et al. 1996). Thus, if ophiolitic diamonds are indigenous, their low aggregation state (Ib) would reflect a short residence time (millions of years' time scale) at temperatures < 900 °C (e.g., Xu et al. 2018). However, a literature survey shows that Ib diamonds were found within their cratonic populations (e.g., Hainschwang et al. 2013; Smit et al. 2016; Smith et al. 2016; Timmerman et al. 2018). Therefore, the nitrogen aggregation parameters do not allow us to unambiguously classify ophiolitic diamonds as either synthetic or indigenous.

Litasov et al. (2019a,b) have suggested that all ophiolitic diamonds are due to anthropogenic contamination. The main arguments are that the composition of the metallic inclusions in type Ib cuboctahedral diamonds in ophiolitic chromitites and peridotites is similar to the composition (Ni:Mn:Co = 70:25:5) of the catalyst used in industrial diamond synthesis. Yang et al. (2020) commented on Litasov et al. (2019b) stating that they have ignored the petrologic context related to microdiamond occurrences and inclusion assemblages discovered in ophiolite-hosted diamonds. Yang et al. (2020) also claim that they ignored the wide occurrence of ophiolite-hosted diamonds discovered in different locations by different research groups. Litasov et al. (2020) replied to this comment and made a simple thermodynamic and kinetic argument against the co-existence of diamond, moissanite, native elements, corundum, wadsleyite, coesite, and quisongite with major ophiolite minerals, they re-emphasized that the metallic inclusions have a similar composition to the catalysts used in industrial diamond synthesis, and they argue that the in situ findings discussed by Yang et al. (2019a) do not meet the basic requirements for in situ mineral grains (they were not observed below the surface before being exposed).

Yang et al. (2019a) also emphasized that the diamonds from ophiolites fall within the size ranging from 100–300 and ~500 μm, and they, therefore, are much larger than those which are used in typical laboratories as an abrasive material. Moreover, they asserted that the ophiolitic diamonds have well-developed crystal faces, whereas any other diamonds from drill bits, or stone-cutting saws would be presented by broken fragments of the crystals.

Shallow diamonds from ophiolites

Healed-crack diamond inclusions in chromite from altered ophiolitic chromitite pods in Tehuitzingo serpentinite (Southern Mexico). Farré-de-Pablo et al. (2019) report *in situ* diamond inclusions in chromite from ophiolitic chromitite pods hosted in the Tehuitzingo serpentinite, southern Mexico. They observed diamonds of 1 to 8 μm in size, quartz, clinochlore, serpentine, and amorphous carbon appeared as "healing-crack inclusions" in chromite. The authors concluded that diamond was formed at a shallow depth at T = 515–670 °C from reduced COH fluids, which infiltrated in during the shallow hydration of chromitite pods enveloped by serpentinite.

Yang et al. (2019b) and Massonne (2019) submitted comments on the Farré-de-Pablo et al. (2018) report raising different concerns. Yang et al. (2019b) specifically discuss that in suggesting a shallow origin for ophiolite-hosted diamonds Farré-de-Pablo et al. (2019) ignore the presence of many other UHPM phases occurring as both inclusions in diamonds and as associated grains. Yang et al. (2019b) propose an alternative scenario in which the diamonds, along with a carbon-rich fluid and some silicate grains were incorporated into the chromite grain at the time it crystallized in the diamond stability field thus sealing the fracture. Massonne (2019) raised concerns that these diamonds are not indigenous, and they could be considered contamination from abrasives or diamond-bearing tools. Massonne (2019) point out that only show diamonds that have been exposed to the surface, and that the five exposed grains are clearly xenomorphic.

In their reply Farré-de-Pablo et al. (2019) address the concerns of both Yang et al. (2019b) and Massonne (2019). They first address the concerns of Massonne (2019) and stress that they provided many lines of evidence supporting the presence of natural diamonds in the Tehuitzingo chromitites. They discuss that diamonds formed from COH fluids may also exhibit irregular morphologies and they cite Dobrzhinetskaya et al. (2004). In their reply to Yang et al. (2019b) they point out that UHPM minerals other than diamond were not found in the Tehuitzingo chromitites as evidence that the diamonds formed under low P and T conditions. The reply from Farré-de-Pablo et al. (2019) has not unambiguously proved the indigenous origin of these diamonds nor has it clarified the crystallization conditions of these diamonds.

Notably, the submission of comments on published papers and replies to the comments indicates that articles are being read by the community and that the topics are controversial. Moreover, they facilitate discussion not only between the authors but also the readers. These controversies should be considered positive since they are one way that science moves forward.

From our standpoint, finding diamonds as healing–crack inclusions in chromite (Farré-de-Pablo et al. 2018) is extremely interesting, and it may expand our understanding of microdiamond formation. However, the most important information that would unambiguously prove that the Tehuitzingo diamonds are indigenous is still missing. The authors presented FIB-TEM images accompanied with EELS spectroscopy data confirming both sp^3 carbon bonding (diamond) and amorphous carbon. Farré-de-Pablo et al. (2018, 2019) emphasized that the Tehuitzingo diamonds exhibit "cohesive" contacts with the host minerals and particularly with enveloping amorphous carbon which contains hollows interpreted as evidence of the escaped fluid. We scrutinized their FIB-TEM data (Fig. 2 in Farré-de-Pablo et al. 2019, and Supplemental Materials) and strongly believe that higher quality HRTEM images of diamond–amorphous carbon and diamond–chromite interfaces are necessary to prove or disprove the indigenous origin of this diamond. The authors should continue to do TEM studies (maybe with additional FIB milling) on the same foil to acquire HRTEM images accompanied by selected area electron diffraction (SAED) patterns of each phase. These data would provide a better understanding of the relationships between all solid phases (lattice fringes, d-spacings) and traces of fluid (e.g., as an example see Fig. 6 d in O'Bannon et al. 2020).

Diamonds inclusion in magmatic olivine from gabbro and chromitite of the Moa-Baracoa ophiolitic massif, eastern Cuba. Pujol-Solà et al. (2020) report that they found *in situ* diamond hosted by magmatic olivine occurring in gabbro and chromitite of an upper mantle section of ophiolite. The authors described 200–300 nm sized diamonds associated with low-temperature and low-pressure minerals such as lizardite, chrysotile and magnetite and they are hosted by trails of the healed cracks and CH_4-rich fluid inclusions inside of olivine. The authors concluded that the observed diamonds were formed at a shallow depth from reduced carbon-rich fluid at $P < 200$ MPa and $T < 350$ °C during serpentinization of olivine in gabbro and chromitite. They also hypothesized that the presence of super-reduced phases together with metastable diamond is clear evidence of a widespread infiltration process of CH_4-rich fluids in the oceanic lithosphere, and therefore the ultra-high-pressure origin for other ophiolitic diamonds can be dubious (Pujol-Solà et al. 2020).

Since these samples were prepared with abrasive diamond material, we believe that there would be fewer concerns of contamination if they provided more detailed data of the diamond–host minerals interface. Special attention should be focused on Fig. 2a,b,c of Pujol-Solà et al. (2020) where ~400 nm diamond inclusion is enveloped by "soft" serpentine material; the latter is situated between olivine and magnetite. The diamond–serpentine interface shown in Figure 2c of Pujol-Solà et al. (2020) does not provide clear information about the diamond–serpentine interface. HRTEM images are needed to observe lattice fringes of the diamond inclusion and its host serpentine. HRTEM images would be helpful to investigate the material between the diamond and serpentine which looks like a darker line surrounding the diamond (see their Fig. 2 c). It also would be helpful to inspect both the diamond and serpentine for any nanometric fluid or solid inclusions. The SAED analysis (Fig. 2c) and EELS spectrum (Fig. 2e) shown by Pujol-Solà et al. (2020) do confirm the presence of diamond, however, they do not prove or disprove that this diamond is indigenous.

Ophiolite-hosted diamonds and super-reduced minerals and alloys formed from natural lightning strikes

Ballhaus et al. (2017) proposed a new concept for ophiolitic diamond and associated ultra-reduced minerals formation due to natural lightning strikes. They conducted laboratory experiments in which basalt rock was subjected to electrical discharge (T~5700 °C) to simulate a natural cloud-to-ground lightning strike (Uman 1986). Ballhaus et al. (2017) report that the high-electrical energy discharge into basalt created a plasma followed by melting and rapid crystallization of Fe–Si alloys, W–Ti alloys, metallic Ti, SiC, amorphous carbon and spherulitic ejecta. Though no diamond was produced by these experiments, the authors believe that they produced potential precursors to diamond. Ballhaus et al. (2017) hypothesized that diamond could nucleate in the cores of fullerenes, when they contract during cooling similar to that described in earlier experiments reported by Banhart and Ayajan (1996). Furthermore, Ballhaus et al. (2017) explicated that coesite and stishovite could also be formed during discharging experiments, though these minerals were not found in the run products of their experiment. Ballhaus et al. (2017) concluded that all existing models of ophiolites formation should be scrutinized because the diamond (and some ultra-reduced phases) have dubious origin.

The lightning origin of diamonds and associated ultra-reduced minerals, alloys and metals hosted by ophiolites has been met with criticism. Griffin et al. (2018) disagreed with Ballhaus et al. (2017) hypothesis because diamonds from ophiolites exhibit characteristics which are not comparable with the environment created in experiments by the high-electrical energy discharge (equivalent of natural lightning).

Yang et al. (2018) argued that Ballhaus et al. (2017) experiments have never produced diamond with solid inclusions, whereas diamonds from ophiolites contain wide variety of UHPM and highly reduced mineral inclusions of SiC, NiMnCo and other alloys, Mn-rich olivine, garnet, Mn-oxides, chromite, coesite and fluids (Yang et al. 2015b; Wu et al. 2017; Moe et al. 2018). Additionally, Yang et al. (2018) asserted that the lightning process cannot produce so many diamonds in such geographically different places (China, Turkey, Russia, India, Albania to mention a few), and in the rocks which have been situated at the mantle depth during diamond crystallization.

Volcanic diamonds from Kamchatka, Russian Federation

A sensational discovery of microdiamonds occurring in lava and ash of the modern Tolbachik volcanic eruption of 2012–2013 (Kamchatka peninsula, Russian Federation) was reported by Karpov et al. (2014) and Gordeev et al. (2014). Later Galimov et al. (2016) proposed that carbon dispersed during the eruption could be the perfect source of carbon for these diamonds. They suggested "*...that Tolbachik microdiamonds were formed as a result of cavitation during the rapid movement of volcanic fluid*" (Galimov et al. 2016). According to this concept the ultra-high-pressure required for diamond formation is concentrated locally, e.g., inside of the carbon-rich fluid bubbles expelled by the volcano, while the external pressure remains not very high.

Further investigation of Tolbachik diamonds lead to conclusions that their geochemical and mineralogical data strongly support their origin from gases released during volcanic explosion (e.g., Galimov et al. 2020a; Galimov and Kaminsky 2021). They reported that Tolbachik diamonds are characterized by $\delta^{13}C = -22$ to -29‰, which are comparable with the $\delta^{13}C$ values of kimberlitic diamonds, and their $\delta^{15}N = -2.58$ and -2.32‰ are similar to $\delta^{15}N$ of volcanic gases, while atmospheric nitrogen has $\delta^{15}N = 0$‰ (Galimov et al. 2020a, b). The nanometric inclusions of Tolbachik diamonds are presented by Mn–Ni–Si alloys and silicides with compositions ranging from $(Mn,Ni)_4Si$ to $(Mn,Ni)_5Si_2$, Mn_5Si_2, and MnSi. These features resemble those in diamonds found *in situ* in ophiolitic chromitite and peridotite (e.g., Galimov et al. 2020a,b).

The Tolbachik diamonds existence as a product of a volcanic explosion was met with extreme criticism and suggestions that those diamonds and accompanied ultra-reduced minerals, alloys and metals are products of anthropogenic contamination (Litasov et al. 2019a,b, 2020; Pokhilenko et al. 2019). There is a series of Comments and Replies publications between those who considered Tolbachik diamonds (as well as their ophiolitic counterparts) as natural Earth materials (Galimov et al. 2020a,b; Galimov and Kaminsky 2021), and those who insisted that all of them are a result of anthropogenic contamination (e.g., Litasov et al. 2019a,b, 2020, 2021; Pokhilenko et al. 2019). It is clear from reading of these comments and replies that each group believes that their arguments are correct, and therefore they deny any disagreements.

Are diamonds from ophiolite and modern volcanic eruption myth or reality?

The narrative of discovery of both ophiolitic and volcanic diamonds to some degree reminds us of the story of ultra-high-pressure metamorphism. This was best described by Harry W. Green II as "a confused mixture of surprising, sometimes spectacular, discoveries and emotional reactions" (Green 2005, p.439). Green (2005) emphasized that science advances in a conservative fashion in a process where *accurate* ideas are added to the existing "library" of knowledge and *inaccurate* ideas are not. He also stressed that new ideas which attempt to shift the paradigm will be met with criticism proportional to how large a shift is needed (Green 2005). The example most relevant to the recent controversial reports of ophiolitic and volcanic diamonds, is the story of the discovery of microdiamonds in metamorphic rocks of continental affinities described at the beginning of this paper. First diamonds from Kokchetav massif, Kazakhstan became known to local geologists due to processing large volumes of the garnet–biotite gneisses using industrial scale methods for minerals separation (Rozen 1972; Rozen et al. 1979 and references therein). After Sobolev and Shatsky (1990) clearly demonstrated that diamonds from Kokchetav were indeed indigenous, Massonne (1999) reported the discovery of similar diamonds in metamorphic gneisses of Erzgebirge, Germany. This discovery was met with harsh criticism and denied at the 1999 Kimberlitic Conference (Green 2005), fortunately additional investigations were published soon after the conference supporting this new diamond find (Massonne 1999; Nasdala and Massonne 2000; Green 2005).

In general, the disbelief that diamond and other UHPM minerals could occur naturally in crustal metamorphic rocks was overcome by a step-by-step process which eventually disregarded the inaccurate observations and deductions. The process included major efforts of many scientific groups who organized an uncountable number of seminars and scientific schools associated with International meetings, where participants had the opportunity to see these "mysterious" diamonds *in situ* for themselves with the aid of an optical microscope. In parallel to these efforts more advanced analytical techniques were also being developed and presented to the scientific community. The resolution of SEM and FIB-TEM and the flux and brightness of synchrotron facilities continued to improve. As a result, there was a shift in the paradigm and diamond-bearing UHPM terrains are now well-documented and accepted by the community.

The main argument underlying the disbelief in ophiolitic and volcanic diamonds is that they have a lot of common characteristics with their synthetic counterparts which are widely used in industrial machines and tools. The increasing concerns arise because almost all worldwide ophiolite-hosted diamonds were separated at the same mineral processing facilities at the Institute of Multipurpose Utilization of Mineral Resources, Chinese Academy of Geological Sciences, Zhengzhou (e.g., Xu et al. 2009). There is, however, one encouraging report of an ophiolitic diamond that was extracted from the Luobasa chromitite by an individually developed method of minerals separation which excluded any possible anthropogenic/laboratory contamination (Howell et al. 2015; Griffin et al. 2016). This provides us evidence that some ophiolitic diamonds are indigenous.

We strongly believe that the indigenous origin of ophiolite-hosted diamonds can be unconditionally proven if diamond abrasives/tools are eliminated from every step of the samples preparation from extraction to polishing stages. The observations of the same diamond in a polished thin section surface should be conducted using a combination of techniques such as optical microscopy, Raman spectroscopy, and SEM, FIB-TEM confirmation of diamond–host mineral interface. The same approach should be applied to studies of volcanic diamonds. The critical samples should be shared between different research groups so that a wide range of analytical techniques can be utilized to study origin of these diamonds.

Once anthropogenic contamination can be ruled out, the community can move past this question and focus on more important aspects of these findings. This would include searching for natural processes that would be similar to the industrial diamond synthesis processes, as well as the formulation of new models that can adequately explain diamond growth in ophiolites and processes that can transport diamonds from elsewhere to the shallow mantle beneath spreading centers.

MISIDENTIFICATION OF MICRODIAMONDS DUE TO CONTAMINATION FROM SAMPLE PREPARATION

General precautions and examples

Since many diamonds from UHPM terranes as well as from ophiolites and volcanic sources are small ranging in size from ≤ 3–50 μm they can be misidentified with diamonds which were mechanically embedded into softer minerals during sample preparation (e.g., cutting, drilling, and/or polishing with diamond tools or abrasives). Recognition of possible contamination has been mentioned in the scientific literature as well as in numerous instruction manuals attached to polishing instruments. Microdiamond researchers are acutely aware that contamination from sample preparation can occur, and that distinguishing indigenous microdiamonds from diamonds embedded during sample preparation is not trivial.

Figure 22A–F contains a series of images showing the morphology of diamond particles found in a typical abrasive powder and those embedded in diamond saws. Figure 22G–H shows "loose" UHPM diamond crystals recovered from crushed rocks. If a diamond-bearing sample is properly polished (see detailed procedure in Dobrzhinetskaya et al. 2001), the diamonds will be elevated above the polished surface of host minerals (e.g., zircon and garnet) which is due to the difference in hardness between the diamond and host minerals (Fig. 23). Visual inspection of the images, especially those acquired with an optical microscope or SEM using low magnification, reveal that the morphology of diamonds from abrasive grits and natural diamonds are quite similar (Fig. 22). Therefore, the morphologies of the diamonds cannot always be used to distinguish between natural microdiamonds and diamonds from abrasive grits.

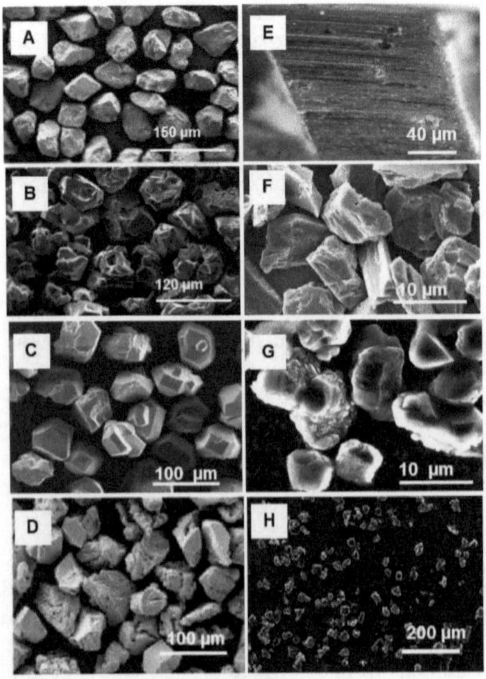

Figure 22. A–B: synthetic diamond abrasive powder, courtesy of Henan Baililai Superhard Material Co., Ltd; **C–D:** synthetic diamond grit, courtesy of Hiperion Materials & Technologies, France; **E**: a segment of a diamond saw blade, (Wiki Commons), **F:** synthetic diamond abrasive, TradeIndia Ltd (www.tradeindia.com); **G:** natural microdiamonds separated from Erzgebirge garnet–quartz–feldspathic gneiss (Dobrzhinetskaya's collection); **H:** natural microdiamonds separated from Kokchetav garnet–biotite gneisses (Dobrzhinetskaya's collection).

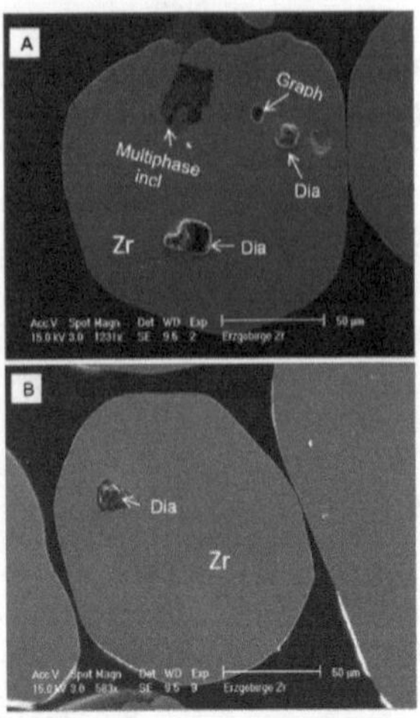

Figure 23. Secondary electron SEM images of zircon grains separated from the garnet–quartz–feldspathic gneiss from the Erzgebirge, Germany; zircons are mounted with Petropoxy on a standard petrographic glass slide and polished with SiC grit, Al_2O_3 and SiO_2–colloidal abrasive liquid (see polishing procedure protocol described in Dobrzhinetskaya et al. 2001). **A**: diamond inclusions are standing above the perfectly polished zircon surface, whereas graphite and multiphase inclusions (phengite + SiO_2 + $KAlSiO_3O_8$) remain flat; lattice parameters of SiO_2 and $KAlSi_3O_8$ are not measured due to the small size of inclusions. **B:** diamond inclusion in zircon; phengite (**black contrast**) occurs at the diamond–zircon interface.

Raman spectroscopy: indigenous diamond vs. its synthetic counterpart

One should be cautious when using Raman spectroscopy as the sole diagnostic for identifying indigenous diamonds. Diamond has a strong first order Raman mode at 1332 cm^{-1}, some abrasives show broadening of this mode as well as a strong shift of this mode to lower wavenumbers. However, Nasdala et al. (2016) show that the Raman parameters of diamond abrasives (collected from 11 different manufacturers) widely overlap with UHPM microdiamonds (Fig. 24A–C, E, F) making it nearly impossible to distinguish between a genuine UHPM relict or an introduced artefact based on Raman spectroscopy alone.

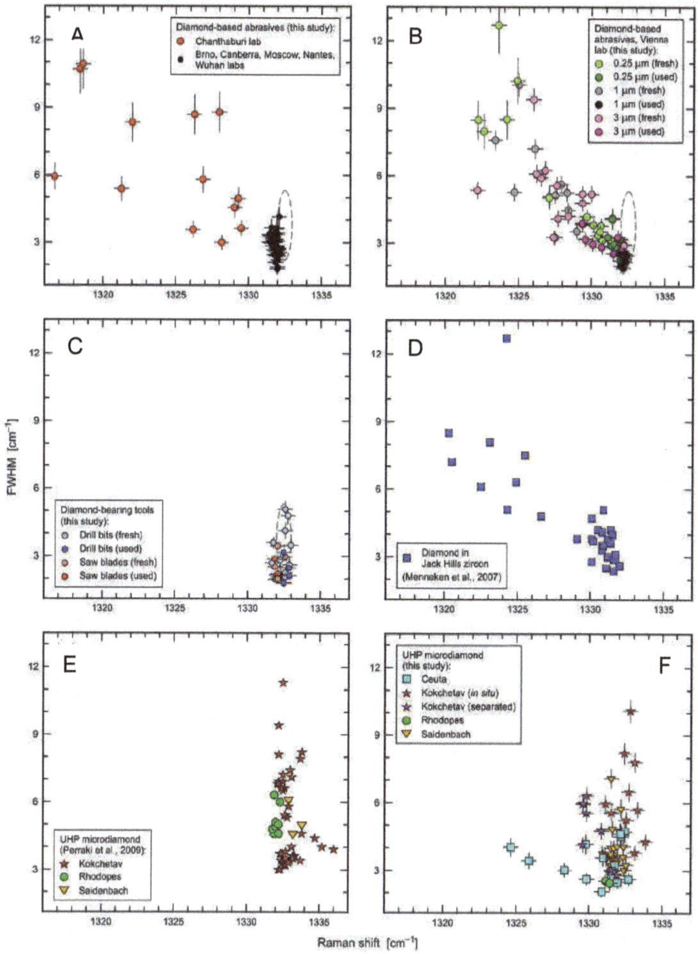

Figure 24. Plots of the FWHM as a function of peak position of the first–order Raman mode of diamond (LO = TO mode at 1332 cm^{-1} for ideal ambient single–crystal diamond), adopted from Nasdala et al. 2016—Fig. 5 (copyright 2016, reprinted with the permission of Elsevier #5032691073556). **A–C:** data from diamond-based abrasive and diamond tools (Nasdala et al. 2016's study), **dashed oval**—data from Perraki et al. 2009; **D:** Jack Hills diamonds in zircon (Menneken et al. 2007 Fig. 2); **E:** diamonds from UHPM terranes from Perraki et al. 2009; **F:** diamonds from UHPM terranes (Nasdala et al. 2016's study).

The study of Menneken et al. (2007) is a prominent example of diamond contamination during sample polishing. They reported the astonishing discovery of the oldest microdiamonds on Earth as inclusions in zircons from the Jack Hills conglomerate, Australia (~ 4.2–3.0 Ga, Nemchin et al. 2008). They used Raman spectroscopy as their primary diagnostic technique, particularly they relied on the FWHM of the diamond Raman mode to distinguish their diamonds from the diamond-based abrasive used for their sample preparation. Indeed, in Fig. 2 from Menneken et al. (2007) the plots of the FWHM vs. Raman shift show that the Jack Hills microdiamonds have larger FWHM's and are shifted to lower frequency than the abrasive diamonds used during their sample preparation. In addition to the Raman results Menneken et al. (2007) ruled out contamination from the diamond polishing powder based on (i) some of the diamonds are completely enclosed by zircon, and (ii) they argued that the size of the diamond inclusions are mostly larger than the size of the diamonds in the polishing materials.

However, several unreconciled discrepancies existed. One is that it does not matter if the zircons are considered as magmatic (Hopkins et al. 2008, 2010, 2012) or as metamorphic (Rasmussen et al. 2010, 2011, 2012) the oldest Jack Hills zircons have a shallow (<10 km) crustal origin. If diamond crystallized along with the zircon at these shallow depths, the diamonds would have formed outside the diamond stability field. Alternatively, the diamonds could have formed within the diamond stability field at a depth of ≥150 km and they were later transported to these shallow depths where they became incorporated into the zircons when they crystallized.

Another discrepancy is that other high-pressure minerals or evidence of their decompression have not been reported in the Jack Hills zircons, and that boundary between the diamond included in zircon is highly irregular and sharp, and even porous and filled with granular "debris" (Fig. 25A). Such "unusual" boundaries between two detached phases (diamond and zircon) look significantly different from the boundaries between an indigenous diamond and its host zircon. The contact zone in many samples generally shows that the diamond is intricately intergrown with its host zircon, which immediately makes it clear that such a diamond is indigenous (Fig. 25B and C).

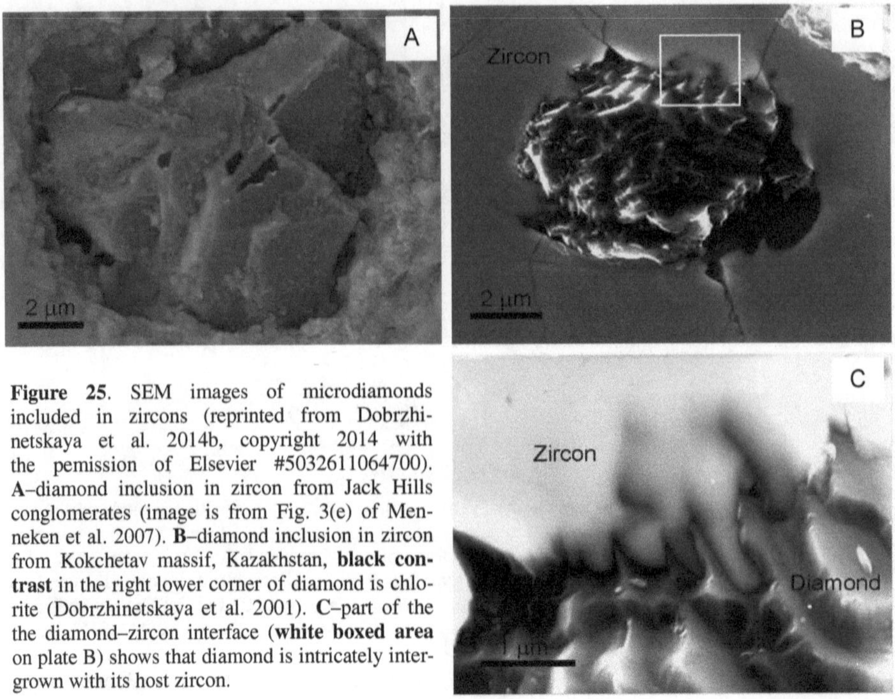

Figure 25. SEM images of microdiamonds included in zircons (reprinted from Dobrzhinetskaya et al. 2014b, copyright 2014 with the pemission of Elsevier #5032611064700). **A**–diamond inclusion in zircon from Jack Hills conglomerates (image is from Fig. 3(e) of Menneken et al. 2007). **B**–diamond inclusion in zircon from Kokchetav massif, Kazakhstan, **black contrast** in the right lower corner of diamond is chlorite (Dobrzhinetskaya et al. 2001). **C**–part of the the diamond–zircon interface (**white boxed area** on plate B) shows that diamond is intricately intergrown with its host zircon.

Considering these inconsistences, Dobrzhinetskaya et al. (2014a) examined the exact same diamond bearing zircon samples which were kindly provided by Menneken et al. (2007) and Nemchin et al. (2008). The SEM and TEM studies of FIB-prepared foils (Fig. 26 A,B,C) concluded that the "Jack Hills diamonds" are not indigenous, and that they are a product of contamination from the industrial diamond abrasive used for sample preparation (Dobrzhinetskaya et al. 2014a). Dobrzhinetskaya et al. (2014a) clearly demonstrated that fragments of broken diamond's ranging in size from ≤0.10–0.30 μm were mechanically embedded/inserted into surficial pores and cracks (Fig. 26 A), and they eroded softer minerals together with polishing dust and petropoxy resin all which occurred during sample preparation (Fig. 26 B,C).

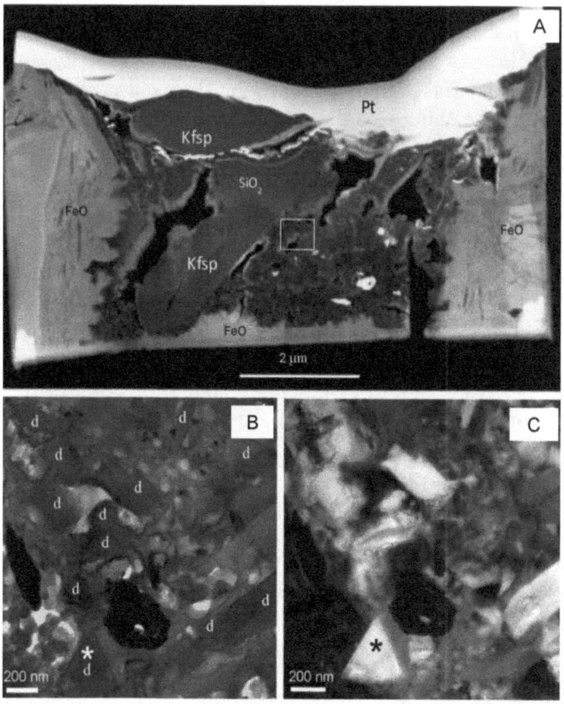

Figure 26. STEM–TEM images of FIB foils cut through inclusions identified by Raman spectroscopy as diamond (Dobrzhinetskaya et al. 2014b, copyright 2014 with the pemission of Elsevier #5032611064700). **A**–STEM image–overview of the place where diamond was identified with Raman spectroscopy (1st order diamond Raman mode observed at: 1332 cm^{-1}) shows the presence of Kfsp and amorphous SiO_2 inclusions. The platinum (Pt) film (**white contrast**) was deposited on the surface of the inclusion prior to the FIB foil preparation, and it has a concave–upward surface indicating that the material from the surface was preferentially eroded during the primary mechanical polishing, because the Kfsp inclusion is softer than the zircon. The **black contrast** areas between Kfsp and SiO_2 represent holes that were probably filled with volatile-rich solid materials or fluids that "evaporated" to the vacuum chamber during FIB milling. **B**–Bright Field TEM image of boxed area shown in plate (A) showing abundant diamond fragments (labeled as '**d**'); the dark contrasted grains with irregular shapes and surfaces are gold (from the previous gold sputtering). Note the unusual sharp–angle morphology of the diamond. The round bubble-like **bright–grey contrast** area is an epoxy resin mixed with the polishing debris. **C**–Image taken of the same area as (B) with characteristic X–rays of carbon (carbon–element map). Bright contrast indicates essentially pure carbon; * is labeled for an easy comparison with the diamond fragments shown on plates B and C.

Later, Menneken et al. (2014) used TEM and Raman spectroscopy on FIB cut sections of the embedded diamonds in their samples and they concluded that the embedded diamond particles originated from the diamond polishing powder. This study is evidence that diamonds that contaminated the sample during cutting and polishing can be misidentified as indigenous UHPM diamonds. In the case of the Jack Hills diamonds if this misidentification was not brought to the scientific forum, it could have had a profound impact on our understanding of the earliest phase of Earth's history. Since it is now known that the "Jack Hills diamonds" are not indigenous (Dobrzhinetskaya et al. 2014a; Mennekan et al. 2014, 2017) several new thorough systematic studies of synthetic diamonds from different manufacturers and microdiamonds from well-established UHPM terranes (Fig. 24) have been reported (Steger et al. 2013; Nasdala et al. 2016). Raman data from Jack Hills diamonds reported by Menneken et al. (2007) (Fig. 24 D) fit well with the Raman data from diamond-based abrasives and tools (Fig. 24A–C), and they partly overlap with Raman data collected from UHPM diamonds

(Fig. 24 E,F). As discussed above, Nasdala et al. (2016) demonstrated that Raman spectroscopy can be used as fast and nondestructive diagnostic method for diamond identification. However, they emphasize that Raman spectroscopy cannot unambiguously differentiate between natural diamonds and industrial diamond–abrasives. To avoid erroneous discoveries of new microdiamond localities one should avoid using Raman spectroscopic techniques as the *only* diagnostic for confirmation of the indigenous origin of the diamonds.

In an earlier chapter we discussed that microdiamond occurring together with SiC, moissanite (Perraki and Farayd 2014; Janák et al. 2015) would require ultra-reduced conditions. These samples are very important for future investigations since they imply a specific oxidation state of the diamond–moissanite inclusions in garnet. However, additional documentations are required that these SiC inclusions have indigenous origin. Nasdala et al. (2016) also note that the Raman spectra of geological moissanite and the Raman spectra of synthetic silicon carbide are not distinguishable (see their Fig. 4d). Consequently, the presence of moissanite in multiphase inclusions from UHPM assemblages should be confirmed in order to exclude any misinterpretation of contamination from SiC-based abrasives used for polishing materials.

How to prove that microdiamond is indigenous

In general, we believe that it should be a mandatory for acceptance of any new discovery of non-cratonic microdiamonds to demonstrate the nature of the diamond–host minerals interface to avoid misidentification with remnants of diamond-based polishing abrasives. In some cases, studies of the diamond–host mineral relationships with optical microscopy and their documentation with high resolution SE images from an SEM can be enough to confirm the indigenous nature of a microdiamond (e.g., Dobrzhinetskaya et al. 2001). For example, Figures 25 B and C present the secondary electron SEM images which demonstrate the intergrown features of zircon–diamond interface. Such relationships unconditionally confirm indigenous origin of the diamond inclusion in zircon. In other cases, it is necessary to use a TEM to investigate FIB-prepared foils that were cut through the diamond–host mineral interface (Fig. 26) (e.g., Dobrzhinetskaya et al. 2001, 2003, 2005, 2006a). As an example, see Figures 26 and 27 which demonstrate FIB-milled sample of the diamond–zircon interface. FIB-TEM techniques are the safest approach for avoiding the misidentification of industrial diamonds–abrasive as natural diamonds, especially if the diamond inclusions under question are smaller than 5 mm (e.g., Dobrzhinetskaya et al. 2003, 2006a).

Notably one robust approach for avoiding the misidentification of diamond and moissanite inclusions would be to not use diamond and SiC tools and abrasives during sample extraction and preparation, one could use corundum instead. Final finishing of the samples with colloidal SiO_2 liquid always provides a great polishing effect and makes diamond inclusions more visible on the flat polished surface of the host minerals.

FUTURE DIRECTIONS

Studies of UHPM microdiamonds have furthered our knowledge and understanding of geochemical and mineralogical processes during the deep subduction of crustal material and its interaction with the Earth's mantle. Despite their mm-size these diamonds have preserved solid, liquid, and gas inclusions over long time periods. These inclusions reflect pathways of the crustal rocks and organic carbon from the Earth's surface to its interior and back to the Earth's surface. The next decade of in situ microdiamond research may be focussed on the following:

(i) Confirmation of the indigenous nature of microdiamonds in newly discovered UHPM, ophiolites and volcanic localities requires the documentation of the diamond–host mineral interface with the aid of the high-resolution secondary electron SEM imaging, and/or TEM imaging and crystal structure data obtained from FIB-prepared samples.

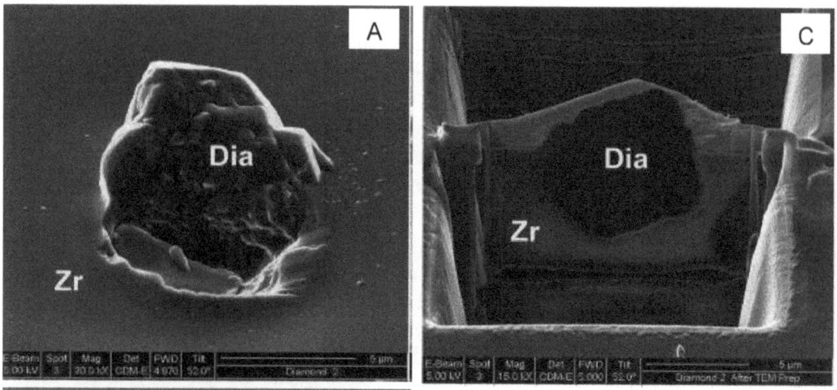

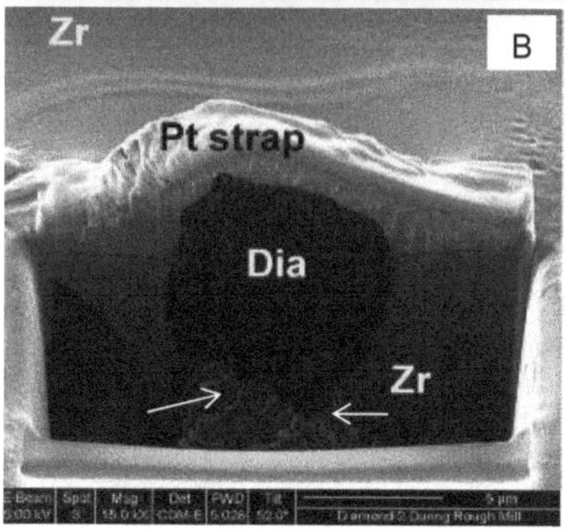

Figure 27. Secondary electron images (FIB-duobeam) of diamond included in zircon form Erzgebirge, Germany showing the FIB milling sequence (Dobrzhinetskaya et al. 2003, copyright 2013 reprinted with the pemission of Elsevier #5030491024109). **A:** polished cross–section of zircon host with a diamond inclusion. **B**: FIB section through the diamond crystal; vertical 'column-like' patterns (**white arrowheads**) are due to the initial 'rough' milling step. Pt strap covers top of the diamond crystal. **C**: FIB foil shows two–dimensional geometry of diamond inclusion and demonstrates the indigenous grain boundary between diamond and zircon host.

(ii) Studies of nanoscale inclusions in microdiamonds from different lithologies to improve our understanding of geochemical environments and PT conditions of diamond crystallization.

(iii) Raman studies of anomalies in microdiamonds which are interpreted as evidence of the "lonsdaleite structure" need to be investigated with FIB-TEM, and/or synchrotron assisted X-ray and spectroscopic techniques.

(iv) Studies of stable isotopes chemistry and noble gases from additional UHPM, ophiolite and volcanic diamond localities to increase our knowledge of processes that occur during continental crust–mantle interaction.

(v) Coordination of laboratory-assisted studies on microdiamonds with general questions of Earth's dynamics through studies of UHPM minerals assemblages associated with diamond formations.

(vi) Modeling and calculations of subduction and exhumation rates; and understanding of how much crustal materials have reached the "point-of-no-return" during deep subduction and remained in the Earth's deep interior being intermixed with the mantle.

ACKNOWLEDGEMENTS

LD research was performed as a part of the International Lithosphere Program, Task Force IV, project "Fate of the subducted continental lithosphere: insight through analytical mineralogy and microstructures", and were supported in different years by NSF grants EAR 0229666, EAR 0107118 and INT–EAR 0329596; Los Alamos National Laboratory through

Science and Technology Base Programs (grant 9949); the Consortium for Materials Properties Research in Earth Sciences (COMPRES); Pacific Rim Program of the University of California, and the University of California Lab Fee Research Grant. EFO acknowledges that a portion of this work was performed under the auspices of the US Department of Energy by Lawrence Livermore National Laboratory under Contract No. DE–AC52–07NA27344. HS research was supported by KAKENHI Grant Numbers JP20740314, JP23340169, JP24109702, JP26287139, and JP15KK0150 from MEXT/JSPS, Japan. Helpful discussions of several aspects of UHPM microdiamonds origin with C.V. Stan, M. Janák, J. Kotková and K. Naemura are acknowledged. Fruitful advices from the peer–reviewers H–P. Shertl, T. Tsujimori, S.W. Faryad and G. Pearson and editors K.V. Smit and S.B. Shirey of this special volume are greatly appreciated.

REFERENCES

Akaishi M, Yamaoka S (2000) Crystallization of diamond from C–O–H fluids under high-pressure and high-temperature conditions. J Cryst Growth 209:999–1003

Allègre CJ, Staudacher T, Sarda P, Kurz M (1983) Constraints on evolution of Earth's mantle from rare gas systematics. Nature 303:762–766

Anthony TR, Banholzer WF (1992) Properties of diamond with varying isotopic composition. Diam Relat Mater 1:717–26

Arai S (2013) Conversion of low–pressure chromitites to ultrahigh-pressure chromitites by deep recycling: a good inference. Earth Planet Sci Lett 379:81–87

Bai WJ, Zhou MF, Cai Y, Hu X (1989) The types andcharacteristics of basic and ultrabasic rocks from China. Bull Inst Geol Chin Acad Geol Sci (in Chinese) 20:51 –74

Bai WJ, Zhou M F, Robinson PT (1993) Possibly diamond-bearing mantle peridotites and podiform chromitites in the Luobusa and Dongqiao ophiolites, Tibet. Can J Earth Sci 30:1650–1659

Bai WJ, Yang JS, Fang QS, Yan BG, Zhang ZM (2001) Study on a storehouse of ultrahigh pressure mantle minerals—podiform chromite deposits. Earth Sci Front 8:111–121 (in Chinese, Abstr in English)

Ballentine CJ, Burnard PG, (2002) Production, release and transport of noble gases in the continental crust. Rev Mineral Geochem 47:481–538

Ballentine CJ, Marty B, Lollar BS, Cassidy M (2005) Neon isotopes constrain convection and volatile origin in the Earth's mantle. Nature 433:33–38

Ballhaus C, Helmy HM, Fonseca ROC, Wirth R, Schreiber A, Jöns W, Bragagni A, Nagel T, Dittrich S, Thome V, Hezel DC, Below R, Cieszynski H (2017) Ultra-high pressure and ultra-reduced minerals in ophiolites may form by lightning strikes. Geochem Perspect Lett 5:42–4

Ballhaus C, Helmy HM, Fonseca ROC, Wirth R, Schreiber A, Jöns N (2021) Ultra-reduced phases in ophiolites cannot come from Earth' mantle. Am Mineral 106:1053–1063

Banhart F, Ajayan PM (1996) Carbon onions as nanoscopic pressure cells for diamond formation. Nature 382:433–435

Basu S, Jones AP, Verchovsky AB, Kelley SP, Stuart FM (2013) An overview of noble gas (He, Ne, Ar, Xe) contents and isotope signals in terrestrial diamond. Earth Sci Rev 126:235–249

Bebout GE, Fogel ML (1992) Nitrogen-isotope compositions of metasedimentary rocks in the Catalina schist, California: implications for metamorphic devolatilization history. Geochim Cosmochim Acta 56:2839–2849

Begemann F (1994) Indigenous and extraneous noble gases in terrestrial diamonds. *In*: Noble Gas Geochemistry and Cosmochemistry. Matsuda J (ed.) Terra Scientific Publishing Company, Tokyo, p. 217–227

Beltrando M, Compagnony R, Lombardo B (2010) (Ultra)–high pressure metamorphism and orogenisis: an Alpine perspective. Gondwana Res 18:147–166

Bottinga Y (1968) Carbon isotope fractionation between graphite, diamond, and carbon dioxide. Earth Planet Sci Lett 5:301–307

Boyd FR, Gurney JJ, Richardson SH (1985) Evidence for a 150–200–km thick Archaean lithosphere from diamond inclusion thermobarometry. Nature 315:387–389

Broadley MB, Kagi H, Burgess R, Zedgenizov D, Mikhail S, Almayrac M, Ragozin A, Pomazansky B, Sumino H (2018) Plume–lithosphere interaction, and the formation of fibrous diamonds. Geochem Perspect Lett. 8:26–30

Bruguier O, Bosch D, Caby R, Vitale–Brovarone A, Fernandez L, Hammor D, Laouar R, Ouabadi A, Abdallah N, Mechati M (2017) Age of UHP metamorphism in the Western Mediterranean: Insight from rutile and minute zircon inclusions in a diamond-bearing garnet megacryst (Edough Massif, NE Algeria). Earth Planet Sci Lett 474:215–225

Brun J–P, Faccenna C (2008) Exhumation of high-pressure rocks driven by slab rollback. Earth Planet Sci Lett 272:1–7

Bundy FP (1980) The *P, T* phase and reaction diagram for elemental carbon. J Geophys Res Solid Earth 85:6930–6936

Bundy FP, Kasper JS (1967) Hexagonal diamond—A new form of carbon. J Chem Phys 46:3437–3446

Bureau H, Keppler H (1999) Complete miscibility between silicate melts and hydrous fluids in the upper mantle: experimental evidence and geochemical implications. Earth Planet Sci Lett 165:187–196

Busigny V, Cartigny P, Philippot P (2011) Nitrogen isotopes in ophiolitic metagabbros: a re–evaluation of modern nitrogen fluxes in subduction zones and implication for the early Earth atmosphere. Geochim Cosmochim Acta 75:7502–7521
Caby R, Bruguier O, Fernandez L, Hammor D, Boscha D, Mechati M, Laouar R, Ouabadi A, Abdallah N, Doucheta C (2014) Metamorphic diamonds in a garnet megacryst from the Edough Massif (northeastern Algeria). Recognition and geodynamic consequences. Tectonophysics 637:341–353
Cartigny P, de Corte K, Shatsky VS, Ader M, De Paepe P, Sobolev NV, Javoy M (2001) The origin and formation of metamorphic microdiamonds from the Kokchetav massif, Kazakhstan: a nitrogen and carbon isotopic study. Chem Geol 176:265–281
Cartigny P, Harris JW, Taylor A, Davies R, Javoy M (2003) On the possibility of a kinetic fractionation of nitrogen stable isotopes during natural diamond growth. Geochim Cosmochim Acta 67:1571–1576
Cartigny P (2005) Stable isotopes and the origin of diamond. Elements 1:79–84
Cartigny P, Palot M, Thomassot E, Harris JW (2014) Diamond formation: a stable isotope perspective. Annual Rev Earth Planet Sci 42:699–732
Chen Y, Yang JS, Xu Z, Tian Y, Lai S (2018) Diamonds and other unusual minerals from peridotites of the Myitkyina ophiolite, Myanmar. J Asian Earth Sci 164:179–193
Chopin C (1984) Coesite and pure pyrope in high-grade blueschists of the Western Alps: a first record and some consequences. Contrib Mineral Petrol 86:107–118
Chopin C (2003) Ultrahigh-pressure metamorphism: tracing continental crust into the mantle. Earth Planet Sci Lett 212:1–4
Chopin C, Sobolev NV (1995) Principal mineralogic indicators of UHP in crustal rocks. *In:* Ultrahigh-Pressure Metamorphism (Coleman RG, Wang X. eds) Cambridge University Press, p. 96–133
Claoué–Long JC, Sobolev NV, Shatsky VS, Sobolev AV (1991) Zircon response to diamond–pressure metamorphism in the Kokchetav massif, USSR. Geology 19:710–713
Clarke Jr RS, Appleman DE, Ross DR (1981) An Antarctic iron meteorite contains preterrestrial impact–produced diamond and lonsdaleite. Nature 291:396–398
Danilenko VV (2004) Shock-wave sintering of nanodiamonds. Phys Solid State 46:711–715
Das SA, Basu R, Mukherjee BK (2017) In situ peridotitic diamond in Indus ophiolite sourced from hydrocarbon fluids in the mantle transition zone. Geology 45:755–758
Daulton TL, Eisenhour DD, Bernatowicz TJ, Lewis RS, Buseck PR (1996) Genesis of presolar diamonds: Comparative high-resolution transmission electron microscopy study of meteoritic and terrestrial Nano-diamonds. Geochim Cosmochim Acta 60:4853–4872
Davidson LM (1979) Nagssugtoqidian granulite facies metamorphism in the Holsteinsborg region, West Greenland. Grønlands Geologiske Undersøgelse Rapport 89:97–108
Davies GR, Nixon PH, Pearson DG, Obata M (1993) Tectonic implications of graphitized diamonds from the Ronda, peridotite massif, southern Spain. Geology 21:471–474
Davies R, Harlow GE (2002) The high-pressure stability of K–cymrite and phases in the system Or–H_2O. Eos Trans AGU 83(47), Fall Meet Suppl Abstr, V72B–1308
de Corte K, Cartigny P, VShatsky VS, Sobolev NV, Javoy M (1998) First evidence of fluid inclusions in metamorphic microdiamonds from the Kokchetav massif, Northern Kazakhstan. Geochim Cosmochim Acta 62:3765–3777
de Corte K, Taylor WR, De Paepe P (2002) Inclusion contents of microdiamonds from UHP metamorphic rocks of the Kokchetav massif. *In:* The Diamond-bearing Kokchetav Massif. Parkinson CD, Katayma I, Liou JG, Maryama S (eds), Universal Academy Press Inc, Kazakhstan, p. 115–135
De Stefano A, Kopylova MG, Cartigny P, Afanasiev V (2009) Diamonds and eclogites of the Jericho kimberlite (Northern Canada). Contrib Mineral Petrol 158:295–315
Denisov VN, Mavrin BN, Serebryanaya NR, Dubitsky GA, Aksenenkov VV, Kirichenko AN, Kuzmin NV, Kulnitskiy BA, Perezhogin IA, Blank VD (2011) First–principles, UV Raman, X–ray diffraction and TEM study of the structure and lattice dynamics of the diamond–lonsdaleite system. Diam Relat Mater 20:951–953
Derjaguin BV, Fedoseev DV (1994) Physico–chemical synthesis of diamond in metastable range. Prog Surf Sci 45:71–80
Dien E, Gamal H, Arai S, Doucet L–S, Li ZX, Kil Y, Fougerouse D, Reddy SM, Saxey DW, Hamdy M (2019) Cr-spinel records metasomatism not petrogenesis of mantle rocks. Nat Commun 10:1–12
Dilek Y, Thy P, Hacker B, Grundvig S (1999) Structure and petrology of Tauride ophiolites and mafic dike intrusions (Turkey): Implications for the Neotethyan ocean. GSA Bull 111:1192–1216
Dilek Y, Furnes H (2011) Ophiolite genesis and global tectonics: geochemical and tectonic fingerprinting of ancient oceanic lithosphere. Geol Soc Am Bull 123:387–411
Dilek Y, Yang J (2018) Ophiolites, diamonds, and ultrahigh-pressure minerals: new discoveries and concepts on upper mantle petrogenesis. Lithosphere 10:3–13
Dobrzhinetskaya LF (2012) Microdiamonds—frontier of ultrahigh-pressure metamorphism: a review. Gondwana Res 21:207–223
Dobrzhinetskaya LF, Faryad SW (2011) Frontiers of ultrahigh-pressure metamorphism: view from field and laboratory. *In:* Ultrahigh-Pressure Metamorphism—25 Years After the Discovery of Coesite and Diamond. Dobrzhinetskaya LF, Faryad SW, Wallis S, Cuthbert S (eds), p 1–3

Dobrzhinetskaya LF, Green HW (2007a) Diamond synthesis from graphite in presence of water and SiO_2: implications for diamond formation in quartzites from Kazakhstan. Int Geol Rev 49:389–400

Dobrzhinetskaya LF, Green II HW (2007b) Experimental studies of mineralogical assemblages of metasedimentary rocks at Earth's mantle transition zone conditions. J Metamorph Geol 25:83–96

Dobrzhinetskaya LF, Eide EA, Larsen RB, Sturt BA, Trønnes RG, Smith DC, Taylor WR Posukhova TV (1995) Microdiamond in high-grade metamorphic rocks of the Western Gneiss region, Norway. Geology 23:597–600

Dobrzhinetskaya LF, Green II HW, Mitchell TE, Dickerson RM (2001) Metamorphic diamonds: mechanism of growth and inclusion of oxides. Geology 29:263–266

Dobrzhinetskaya LF, Green II HW, Weschler M, Darus M, Wang YC, Massonne HJ, Stöckhert B (2003) Focused ion beam technique and transmission electron microscope studies of microdiamonds from the Saxonian Erzgebirge, Germany Earth Planet Sci Lett 210:399–410

Dobrzhinetskaya LF, Renfro AP, Green II HW (2004) Synthesis of skeletal diamonds: Implications for microdiamond formation in orogenic belts. Geology 32:869–872

Dobrzhinetskaya LF, Wirth R, Green HW (2005) Direct observation and analysis of a trapped COH fluid growth medium in metamorphic diamond. Terra Nova 17:472–477

Dobrzhinetskaya LF, Wirth R, Green II HW (2006a) Nanometric inclusions of carbonates in Kokchetav diamonds from Kazakhstan: A new constraint for the depth of metamorphic diamond crystallization. Earth Planet Sci Lett 243:85–93

Dobrzhinetskaya LF, Liu Z, Cartigny P, Zhang J, Tchkhetia D, Hemley RJ, Green II HW (2006b) Synchrotron infrared and Raman spectroscopy of microdiamonds from Erzgebirge, Germany. Earth Planet Sci Lett 248:340–349

Dobrzhinetskaya LF, Wirth R, Yang J, Hatcheon ID, Weber P, Green HW (2009) Nitrides and oxides recording a highly reduced mantle environments form an ophiolite. PNAS 106:19233–19238

Dobrzhinetskaya LF, Green II HW, Takahata N, Sano Y, Shirai K (2010) Crustal signature of $\delta^{13}C$ and nitrogen content in microdiamonds from Erzgebirge, Germany: Ion microprobe studies. J Earth Sci 21:623–634

Dobrzhinetskaya LF, Wirth R, Green II HW, Schreiber A, O'Bannon III EF (2013) First find of polycrystalline diamond in ultrahigh-pressure metamorphic terrane of Erzgebirge, Germany. J Metamorph Geol 31:5–18

Dobrzhinetskaya LF, Wirth R, Green II HW (2014a) Diamonds in Earth's oldest zircons from Jack Hills conglomerate, Australia, are contamination. Earth Planet Sci Lett 387:212–218

Dobrzhinetskaya LF, Wirth R, Yang JS, Bai W, Green HW, Hutcheon ID, Weber PK, Grew E (2014b) Qingsongite, natural cubic boron nitride: The first boron mineral from the Earth's mantle. Am Mineral 99:764–772

Dubinchuk VT, Simakov SK, Pechnikov VA (2010) Lonsdaleite in diamond-bearing metamorphic rocks of the Kokchetav Massif. Dokl Earth Sci 430:40–42

Eggler DH, Baker DR (1982) Reduced volatiles in the system C–O–H: implications to mantle melting, fluid formation and diamond genesis. *In*: High Pressure Research in Geophysics. Akimoto S, Manghnani MH (eds), Center Academic Publisher, Tokyo, p. 237–250

Ekimova TE (1992) Diamond inclusions in rock–forming minerals of metamorphic rocks. Dokl Akad Nauk 322:366–368

El Atrassi F, Brunet F, Bouybaouene M, Chopin C, Chazot G (2011) Melting textures and microdiamonds preserved in graphite pseudomorphs from the Beni Bousera peridotite massif, Morocco. Eur J Mineral 23:157–168

Ernst WG, Liou JG (1999) Overview of UHP metamorphism and tectonics in well-studied collisional orogens. Int Geol Rev 41:477–493

Ernst WG, Liou JG (2008) High-and ultrahigh-pressure metamorphism: Past results and future prospects. Am Mineral 93:1771–86

Essenov CE, Efimov LA, Shiygin ED, Abduicabirova MA, Vedernicov NN, Nuriybaev AN (1968) To problem of the diamond deposits of the Northern Kazakhstan. Vestn Akad Nauk Kaz SSR 1:37–45

Fang Q, Bai WJ (1981) The discovery of Alpine-type diamond-bearing ultrabasic intrusions in Tibet. Geol Rev (Beijing) 27:455 –457

Farré-de-Pablo J, Proenza JA, González–Jiménez JM, Garcia–Casco A, Colás V, Roqué–Rossell J, Camprubí A, Sánchez–Navas A (2018) A shallow origin for diamonds in ophiolitic chromitites. Geology 47:75–78

Farré-de-Pablo J, Proenza JA, González–Jiménez JM, Garcia–Casco A, Colás V, Roqué–Rossell J, Camprubí A, Sánchez–Navas A (2019) A shallow origin for diamonds in ophiolitic chromitites: REPLY. Geology 47:e477–e478

Farré-de-Pablo J, Proenza JA, González–Jiménez JM, Aiglsperger T, Garcia–Casco A, Escuder–Viruete J, Colás V, Longo F (2020) Ophiolite hosted chromitite formed by supra–subduction zone peridotite –plume interaction. Geosci Front 11:2083–2102

Faryad SW, Cuthbert SJ (2020) High-temperature overprint in (U)HPM rocks exhumed from subduction zones; A product of isothermal decompression or a consequence of slab break–of (slab)? Earth Sci Rev 202:103108

Fasshauer DW, Chatterjee ND, Marler B (1997) Synthesis, structure, thermodynamic properties, and stability relations of K–cymrite, $K[AlSi_3O_8]\cdot H_2O$. Phys Chem Minerals 24:455–462

Faure M, Fabbri O, Monie P (1988) The Miocene bending of southwest Japan: new $^{39}Ar/^{40}Ar$ and microtectonic constraints from the Nagasaki schists (western Kyushu), an extension of the Sanbagawa high-pressure belt. Earth Planet Sci Lett 91:105–16

Fedoseev DV, Galimov EM, Varnin VP, Prokhorov VS, Deryagin BV (1971) Carbon isotopes fractionation at physical–chemical synthesis of diamond from gas. Doklady Akademii Nauk SSSR 201:1149–1150 (In Russian)

Ferrando S, Frezzotti ML, Dallai L, Compagnoni R (2005) Multiphase solid inclusions in UHP rocks (Su–Lu, China): Remnants of supercritical silicate-rich aqueous fluids released during continental subduction. Chem Geol 233:68–81

Ferrari AC, Robertson J (2000) Interpretation of Raman spectra of disordered and amorphous carbon. Phys Rev B 61:14095

Frezzotti ML (2019) Diamond growth from organic compounds in hydrous fluids deep within the Earth. Nat Comm 10:1–8

Frezzotti ML, Ferrando S, Dallai L, Compagnoni R (2007) Intermediate alkali–alumina–silicate aqueous solutions released by deeply subducted continental crust: fluid evolution in UHP OH-rich topaz–kyanite quartzites from Donghai (Sulu, China). J Petrol 48:1219–1241

Frezzotti ML, Selverstone J, Sharp ZD, Compagnoni R (2011) Carbonate dissolution during subduction revealed by diamond-bearing rocks from the Alps. Nat Geosci 4:703–706

Frezzotti ML, Huizenga JM, Compagnoni R, Selverstone J (2014) Diamond formation by carbon saturation in C–O–H fluids during cold subduction of oceanic lithosphere. Geochim Cosmochim Acta 143:68–86

Frondel C, Marvin UB (1967) Lonsdaleite, a hexagonal polymorph of diamond. Nature 214:587–589

Frost BR, Frost CD (2014) Essential of igneous and metamorphic petrology. Cambridge University Press, Cambridge UK

Galimov EM, Karpov GA, Sevast'yanov VS, Shilobreeva SN, Maksimov AP (2016) Diamonds in the products of the 2012–2013 Tolbachik eruption (Kamchatka) and mechanism of their formation. Geochem Int 54:829–833

Galimov EM, Kaminsky FV, Shilobreeva SN, Sevastyanov VS, Voropaev SA, Khachatryan GK, Wirth R, Schreiber A, Saraykin VV, Karpov GA, Anikin LP (2020a) Enigmatic diamonds from the Tolbachik volcano, Kamchatka Am Mineral 105:498–509

Galimov EM, Kaminsky FV, Karpov GA, Shilobreeva SN, Sevastyanova VS, Voropaev SA, Anikin LP, Wirth R, Khachatryan GK, Saraykin VV (2020b) The nature and compositional peculiarities of volcanogenic diamonds. Russ Geol Geophys 61:1065–1074

Galimov EM, Kaminsky FV (2021) Diamond in the oceanic lithosphere, volcanic diamonds and diamonds in ophiolites. Geochem Int 59:1–11

Gautheron C, Cartigny P, Moreira M, Harris JW, Allègre CJ (2005) Evidence for a mantle component shown by rare gases, C and N isotopes in polycrystalline diamonds from Orapa (Botswana). Earth Planet Sci Lett 240:559–572

Gerya TV (2010) Introduction to Numerical Geodynamic Modelling. Cambredge University Press, Cambridge UK

Glassley WE, Sørensen K (1980) Constant Ps–T amphibolite to granulite facies transition in Agto (West Greenland) metadolerites: implications and applications. J Petrol 21:69–105

Glassley WE, Korstgard J, Sorensen K, Platou SW (2014) A new UHP metamorphic complex in the ~1.8 Ga Nagssugtoqidian orogen of West Greenland. Am Mineral 99:1315–1334

Godard G, Smith DC, Thouvenin C (2003) Microdiamond within zircon in thin section in the Straumen coesite–kyanite–eclogite pod, Norway. *In*: The Alice Wain Memorial West Norway Eclogite Field Symposium, Abstr volume (Eide EA (ed). Norges Geologiske Undersøkelse, Rapport 2003.055

Godard G, Smith DC, Moreau M (2004) Raman mapping of microdiamonds and other carbon phases included in zircons in a Norwegian eclogite. 32nd Int Geol Cong, Florence 20–28 August 2004, Abstr 153–2

Gordeev EI, Karpov GA, Anikin LP, Krivovichev SV, Filatov SK, Ovsyannikov AA (2014) Diamonds in lavas of the Tolbachik fissure eruption in Kamchatka. Dokl Earth Sci 454:47–49

Goryainov SV, Likhacheva AY, Rashchenko SV, Shubin AS, Afanas'ev VP, Pokhilenko NP (2014) Raman identification of lonsdaleite in Popigai impactites. J Raman Spectrosc 45:305–13

Graham DW (2002) Noble gas isotope geochemistry of mid–ocean ridge and ocean island basalts: Characterization of mantle source reservoirs. Rev Mineral Geochem 47:247–317

Green HW (2005) Psychology of a changing paradigm: 40+ years of high-pressure metamorphism. Int Geol Rev 45:439–456

Green II HW, Dobrzhinetskaya LF, Bozhilov KN (2010) Alpe Arami story: triumph of the data over prejudice. J Earth Sci 21:731–743

Griffin WL, Afonso JC, Belousova EA, Gain SE, Gong XH, Gonzalez Jimenez JM, Howell D, Huang JX, McGowan N, Pearson NJ, Satsuawa T, Shi R, Williams P, Xiong Q, Yang JS, Zhang M, O'Reilly SY (2016) Mantle recycling: transition zone metamorphism of Tibetan ophiolitic peridotites and its tectonic implications. J Petrol 57:655–684

Griffin WL, Howell D, Gonzalez–Jimenez JM, Xiong Q, O'Reilly SY (2018) Comment on "Ultra-high pressure and ultra-reduced minerals in ophiolites may form by lightning strikes" Geochem Persp Let 7:1–2

Guillot S, Mahéo G, de Sigoyer J, Hattori KH, Pecher A (2008) Tethyan and Indian subduction viewed from the Himalayan high-to ultrahigh-pressure metamorphic rocks. Tectonophysics 451:225–241

Guo GL, Yang JS, Liu XD, Xu XZ, Zhang ZM, Tian YZ, Xiong FH, Wu Y (2015) Implications of unusual minerals in Zedang mantle peridotite. Tibet. Geol China 42:1483–1492

Hainschwang T, Fritsch E, Notari F, Rondeau B, Katrusha A (2013) The origin of color in natural C center bearing diamonds. Diam Rel Mater 39:27–40

Hanneman RE, Strong HM, Bundy FP (1967) Hexagonal diamonds in meteorites: Implications. Science 155:997–999

Harris JW, Hawthorne JB, Oosterveld MM, Wehmeyer E (1975) A classification scheme for diamonds and a comparative study of South African diamond characteristics. *In*: Physics and Chemistry of the Earth. Ahrens LH, Dawson JB, Duncan AR, Erlank AJ (eds) Pergamon Press p.765–783

Hattori H, Shibata K (1982) Radiometric dating of Pre–Neogene granitic and metamorphic rocks in northwestern Kyushu, Japan—with emphasis on geotectonics of the Nishisonogi zone. Bull Geol Surv Jpn 33:57–84
Herchen H, Cappelli MA (1991) First–order Raman spectrum of diamond at high temperatures. Phys Rev B 43:11740
Hermann J, Rubatto D, Korsakov A, Shatsky VS (2006) The age of metamorphism of diamondiferous rocks determined with SHRIMP dating of zircon. Rus Geol Geophys 47:513–520
Honda M, Reynolds JH, Roedder E, Epstein S (1987) Noble gases in diamonds: occurences of solarlike helium and neon. J Geophys Res 92:12521
Honda M, McDougall I, Patterson DB, Doulgeris A, Clague DA (1991) Possible solar noble–gas component in Hawaiian basalts. Nature 349:149–151
Honda M, Phillips D, Harris JW, Yatsevich I (2004) Unusual noble gas compositions in polycrystalline diamonds: preliminary results from the Jwaneng kimberlite, Botswana. Chem Geol 203:347–358
Hong SM, Akaishi M, Yamaoka S (1999) Nucleation of diamonds in the system of carbon and water under very high pressure and temperature. J Cryst Growth 200:326–329
Hopkins MD, Harrison TM, Manning CE (2008) Low heat flow inferred from > 4 Gyr zircons suggests Hadean plate boundary interactions. Nature 456:493–496
Hopkins MD, Harrison TM, Manning CE (2010) Constraints on Hadean geodynamics from mineral inclusions in > 4 Ga zircons. Earth Planet Sci Lett 298:367–376
Hopkins MD, Harrison TM, Manning CE (2012) Metamorphic replacement of mineral inclusions in detrital zircon from Jack Hills, Australia: Implications for the Hadean Earth: Comment. Geology 40:281–281
Howell D, Griffin WL, Yang J, Gain S, Stern RA, Huang JX, Jacob DE, Xu X, Stokes AJ, O'Reilly SY, Pearson, NJ (2015) Diamonds in ophiolites: Contamination or a new diamond growth environment? Earth Planet Sci Lett 430:284–295
Huang Z, Yang JS, Robinson PT, Wang Y, Xiong F, Zhang Z, Xu W (2015) The discovery of diamonds in chromitite of the Hegenshan ophiolite, Inner Mongolia. Acta Geol Sin–Eng 89:32–38
Hwang SL, Shen P, Chu HT, Yui TF (2000) Nanometer-size α-PbO_2-type TiO_2 in garnet: A thermobarometer for ultrahigh-pressure metamorphism. Science 288:321–324
Hwang SL, Shen P, Yui TF, Chu HT (2003) Metal–sulfur–COH–silicate fluid mediated diamond nucleation in Kokchetav ultrahigh-pressure gneiss. Eur J Mineral 15:503–511
Hwang SL, Shen P, Chu HT, Yui TF, Liou JG, Sobolev NV, Shatsky VS (2005) Crust-derived potassic fluid in terrestrial diamonds. Earth Planet Sci Lett 231:295–306
Hwang SL, Chu HT, Yui TF, Shen PY, Schertl H–P, Liou JG, Sobolev NV (2006) Nanometer–size P/K-rich silica glass (former melt) inclusions in microdiamond from gneisses of Kokchetav and Erzgebirge massifs: diversified characteristics of the formation media of metamorphic microdiamond in UHP rocks. Earth Planet Sci Lett 243:94–106
Imamura K, Ogasawara Y, Yurimoto H, Kusakabe M (2013) Carbon isotope heterogeneity in metamorphic diamond from the Kokchetav UHP dolomite marble, northern Kazakhstan. Int Geol Rev 55:453–467
Ishida H, Ogasawara Y, Ohsumi K, Saito A (2003) Two stage growth of microdiamond in UHP dolomite marble from Kokchetav Massif, Kazakhstan. J Metamorph Geol 21:515–522
Jackson MG, Carlson RW, Kurz MD, Kempton PD, Francis D, Blusztajn J (2010) Evidence for the survival of the oldest terrestrial mantle reservoir. Nature 466:853–856
Jacob DE, Dobrzhinetskaya LF, Wirth R (2014) New insight into polycrystalline diamond genesis from modern nanoanalytical techniques. Earth Sci Rev 136:21–35
Jacob DE, Wirth R, Enzmann F, Kronz A, Schreiber A (2011) Nano-inclusion suite and high resolution micro–computed–tomography of polycrystalline diamond (framesite) from Orapa, Botswana. Earth Planet Sci Lett 308:307–316
Jagoutz E, Shatsky VS, Sobolev NV (1990) Sr–Nd–Pb isotopic study of ultra-high *PT* rocks from Kokchetav massif. EOS 71:1707
Jagoutz E, Shatsky VS, Sobolev NV (1991) The origin and history of ultrahigh *PT* rocks from Kokchetav Massif. Terra Abstr 3:83
Janák M, Froitzheim N, Lupták B, Vrabec M, Ravna EJK (2004) First evidence for ultrahigh-pressure metamorphism of eclogites in Pohorje, Slovenia: Tracing deep continental subduction in the Eastern Alps. Tectonics 23:TC5014
Janák M, Froitzheim N, Vrabec M, Krogh Ravna EJ, De Hoog JCM (2006) Ultrahigh-pressure metamorphism and exhumation of garnet peridotite in Pohorje, Eastern Alps. J Metamorph Geol 24:19–31
Janák M, Krogh Ravna EJ, Kullerud K, Yoshida K, Milovský R, Hirajima T (2013) Discovery of diamond in the Tromsø Nappe, Scandinavian Caledonides (N. Norway). J Metamorph Geol 31:691–703
Janák M, Froitzheim N, Yoshida K, Sasinková V, Nosko M, Kobayashi T, Hirajima T, Vrabec M (2015) Diamond in metasedimentary crustal rocks from Pohorje, Eastern Alps: a window to deep continental subduction. J Metamorph Geol 33:495–512
Kaneoka I (1983) Noble gas constraints on the layered structure of the mantle. Nature 302:698–700
Karpov GA, Silaev VI, Anikin LP, Rakin VI, Vasil'ev EA, Filatov SK, Petrovskii VA, Flerov GB (2014) Diamonds and accessory minerals in products of the 2012–2013 Tolbachik fissure eruption. J Volcanol Seissmol 8:323–339
Kashkarov IF, Polkanov YA (1964) On discovery of diamonds in titanium–zirconium sands. Dokl Akad Nauk SSSR (Proc Acad Sci USSR) 157:1129 (in Russian)
Katayama I, Parkinson CD, Okamoto K, Nakajima Y, Maruyama S (2000) Supersilicic clinopyroxene and silica exsolution in UHPM eclogite and pelitic gneiss from the Kokchetav massif, Kazakhstan. Am Mineral 10:1368–1374

Katayama I, Maruyama S, Parkinson CD, Terada K, Sano Y (2001a) Ion micro–probe U–Pb zircon geochronology of peak and retrograde stages of ultrahigh-pressure metamorphic rocks from the Kokchetav massif, northern Kazakhstan. Earth Planet Sci Lett 188:185–198

Katayama I, Zayachkovsky AA, Maruyama S (2001b) Prograde pressure–temperature records from inclusions in zircons from ultrahigh-pressure–high-pressure rocks of the Kokchetav Massif, northern Kazakhstan. Island Arc 9:417–427

Katayama I, Ohta, M, Ogasawara Y (2002) Mineral inclusions in zircon from diamond-bearing marble in the Kokchetav massif, northern Kazakhstan. Eur J Mineral 14:1103–1108

Klonowska I, Janák M, Majka J, Petrik I, Froitzheim N, Gee DG, Sasinková V (2017) Microdiamond on Åreskutan confirms regional UHP metamorphism in the Seve Nappe Complex of the Scandinavian Caledonides. J Metamorph Geol 35:541–564

Koeberl C, Masaitis VL, Shafranovsky GI, Gilmour I, Langenhorst F, Schrauder M (1997) Diamonds from the Popigai impact structure, Russia. Geology 25:967–970

Kohn MJ (2010) Carbon isotope compositions of terrestrial C3 plants as indicators of (paleo) ecology and (paleo) climate. PNAS 107:19691–5

Konzett J, Frost DJ (2009) The high P–T stability of hydroxyl–apatite in natural and simplified MORB—an experimental study to 15 GPa with implications for transport and storage of phosphorus and halogens in subduction zones. J Petrol 50:2043–2062

Konzett J, Rhede D, Frost DJ (2012) The high *PT* stability of apatite and Cl partitioning between apatite and hydrous potassic phases in peridotite: an experimental study to 19 GPa with implications for the transport of P, Cl and K in the upper mantle. Contrib Mineral Petrol 163:277–296

Korsakov A, Herman J (2006) Silicate and carbonate melt inclusions associated with diamonds in deeply subducted carbonate rocks. Earth Planet Sci Lett 241:104–118

Korsakov A, Theunissen K, Smirnova LV (2004) Intergranular diamonds derived from partial melting of crustal rocks at ultrahigh-pressure metamorphic conditions. Terra Nova 16:146–151

Kotková J, O'Brien PJ, Ziemann MA (2011) Diamonds in the Bohemian Massif—evidence for ultrahigh-pressure metamorphism. Geologické výzkumy na Morave a ve Slezsku 18:35–38

Kröner A, Willner AP (1998) Time of formation and peak of Variscan HP–HT metamorphism of quartz–feldspar rocks in the central Erzgebirge, Saxony, Germany. Contrib Mineral Petrol 132:1–20

Krueger A (2008) Diamond nanoparticles: Jewels for chemistry and physics. Adv Mater 20:2445–2449

Kumar MS, Akaishi M, Yamaoka S (2001) Effect of fluid concentration on the formation of diamond in the CO_2–H_2O–graphite system under HP–HT conditions. J Cryst Growth 222:9–13

Kurz MD (1986) Cosmogenic helium in a terrestrial igneous rock. Nature 320:435–439

Kurz MD, Jenkins WJ, Hart SR (1982) Helium isotopic systematics of oceanic islands and mantle heterogeneity. Nature 297:43–47

Kurz MD, Gurney JJ, Jenkins WJ, Lott III DE (1987) Helium isotopic variability within single diamonds from the Orapa kimberlite pipe. Earth Planet Sci Lett 86:57–68

Kurz MD, Curtice J, Fornari D, Geist D, Moreira M (2009) Primitive neon from the center of the Galapagos hotspot. Earth Planet Sci Lett 286:23–34

Kvasnitsya V, Wirth R, Dobrzhinetskaya L, Matzel J, Jacobsen B, Hutcheon I, Tappero, R, Kovalyukh M (2013) New evidence of meteoritic origin of the Tunguska cosmic body. Planet Space Sci 84:131–140

Lal D (1989) An important source of ^{4}He (and ^{3}He) in diamonds. Earth Planet Sci Lett 96:1–7

Langenhorst F (2003) Nanostructures in ultrahigh-pressure metamorphic coesite and diamond: a genetic fingerprint. Mitt Österr Miner Ges 148:401–412

Lavrova LD (1991) New type of diamond deposits. Priroda (Nature) 12:62–68 (in Russian)

Lavrova LD, Petchnikov VA, Petrova MA, Ekimova TE, Nadezhdina ED (1995) New genetic type of diamond deposits–geological peculiarities and origin. *In*: International Kimberlite Conference: Extended Abstr 6:311–313

Lavrova LD, Pechnikov VA, Petrova MA, Ekimova TE, Karpenko SF, Lyalikov AV, Spiridonov VG, Bibikova EV, Fugzan MM, Shukolyukov YA (1997) Diamond formation in the age succession of geological events on the Kokchetav massif: evidence from isotopic geochronology. Geokhimiya (Geochemistry) 675–682 (in Russian)

Letnikov FA (1983) Diamond origin in deep–seated tectonic zones. Dokl Akad Nauk SSSR (Proc Acad Sci USSR) 271:433–435

Leung I, Winston R (2002) Microstructures and superelasticity in natural diamond. EOS Trans Am Geophys Union 83:MR61A–1028

Lian D, Yang J (2019) Ophiolite-hosted diamond: A new window for probing arbon cycling in the deep mantle. Engineering 5:406–420

Lian D, Yang J, Dilek Y, Wu W, Zhang Z, Xiong F, Liu F, Zhou W (2017) Deep mantle origin and ultra-reducing conditions in podiform chromitite: Diamond, moissanite, and other unusual minerals in podiform chromitites from the Pozanti–Karsanti ophiolite, southern Turkey. Am Mineral 102:1101–1113

Lian D, Yang JS, Wiedenbeck M, Dilek Y, Rocholl A, Wu W (2018a) Carbon and nitrogen isotope, and mineral inclusion studies on the diamonds from the Pozanti–Karsanti chromitite, Turkey. Contrib Mineral Petrol 173:72

Lian D, Yang JS, DilekY, Rocholl A (2018b) Mineralogy and geochemistry of peridotites and chromitites in the Aladag Ophiolite (southern Turkey): melt evolution of the Cretaceous Neotethyan mantle J Geol Soc 176:958–974

Liati A, Gebauer D, Fanning CM (2011) Geochronology of the Alpine UHP Rhodope zone: a review of isotopic ages and constraints on the geodynamic evolution. *In*: Ultrahigh-Pressure Metamorphism: 25 Years After The Discovery of Coesite and Diamond. Dobrzhinetskaya LF, Faryad SW, Wallis S, Cuthbert S (eds), Elsevier, London, p. 295–316

Liou JG, Tsujimori T, Zhang RY, Katayama I, Maruyama S (2004) Global UHP metamorphism and continent subduction/collision. The Himalayan Model. Int Geol Rev 46:1–27

Liou JG, Ernst WG, Zhang RY, Tsujimori T, Jahn BM (2009) Ultrahigh-pressure minerals and metamorphic terranes–the view from China. J Asian Earth Sci 35:199–231

Liou JG, Tsujimori T, Yang J, Zhang RY, Ernst WG (2014) Recycling of crustal materials through study of ultrahigh-pressure minerals in collisional orogens, ophiolites, and mantle xenoliths: A review. J Asian Earth Sci 96:386–420

Lipp MJ, Baonza VG, Evans WJ, Lorenzana HE (1997) Nanocrystalline diamond: Effect of confinement, pressure, and heating on phonon modes. Phys Rev B 56:5978

Litasov KD, Kagi H, Bekker TB, Hirata T, Makino Y (2019a) Cuboctahedral type Ib diamonds in ophiolitic chromitites and peridotites: the evidence for anthropogenic contamination. High Pres Res 39:480–488

Litasov KD, Kagi H, Voropaev SA, Hirata T, Ohfuji H, Ishibashi H, Makino Y, Bekker TB, Sevastyanov VS, Afanasiev VP, Pokhilenko NP (2019b) Comparison of enigmatic diamonds from the Tolbachik arc volcano (Kamchatka) and Tibetan ophiolites: Assessing the role of contamination by synthetic materials. Gondwana Res 75:16–27

Litasov KD, Bekker TB, Kagi H, Ohfuji H (2020) Reply to the comment on "Comparison of enigmatic diamonds from the Tolbachik arc volcano (Kamchatka) and Tibetan ophiolites: Assessing the role of contamination by synthetic materials" by Litasov et al. (2019) (Gondwana Research, 75, 16–27) by Yang et al. Gondwana Res 79:304–307

Litasov KD, Kagi H, Bekker TB, Makino Y, Hirata T, Brazhkin VV (2021) Why Tolbachik diamonds cannot be natural. Am Mineral 106:44–53

Liu L, Zhang J, Green, HW, Jin Z, Bozhilov KN (2007) Evidence of former stishovite in metamorphosed sediments, implying subduction to >350 km. Earth Planet Sci Lett 263:180–191

Liu L, Zhang J, Cao Y, Green, HW, Yang W, Xu H, Liao H, Kang L (2018) Evidence of former stishovite in UHP eclogite from the South Altyn Tagh, western China. Earth Planet Sci Lett 484:353–362

McGowan NM, Griffin WL, González–Jiménez JM, Belousova E, Afonso JC, Shi R, McCammon CA, Pearson NJ, O'Reilly SY (2015) Tibetan chromitites: Excavating the slab graveyard. Geology 43:179–182

Majka J, Rosén Å, Janák M, Froitzheim N, Klonowska I, Manecki M, Sasinková V, Yoshida K (2014) Microdiamond discovered in the Seve Nappe (Scandinavian Caledonides) and its exhumation by the "vacuum–cleaner" mechanism. Geology 42:1107–1110

Manning CE (2004) The chemistry of subduction–zone fluids. Earth Planet Sci Lett 223:1–16

Martini JEJ (1991) The nature, distribution and genesis of the coesite and stishovite associated with the pseudotachylite of the Vredefort Dome, South Africa. Earth Planet Sci Lett 101:285–300

Masaitis VL, Futergendler SI, Gnevushev MA (1972) Diamonds in impactites of the Popigai meteorite crater. All-Union Mineral Soc Proc 1:108–112

Masaitis VL, Shafranovsky GL, Grieve RA, Langenhorst F, Peredery W, Balmasov EL, Fedorova IG, Therriault A (1997) Discovery of diamonds at the Sudbury Structure. Lunar Planet Inst Contrib 922:33

Masaitis VL, Shafranovsky GI, Fedorova IG, Koivisto M, Korhonen JV (1998) Lappajarvi Astrobleme: The first find of impact diamonds on the Fennoscandian Shield. Lunar Planet Instit Contrib 1171

Massonne H–J (1999) A new occurrence of microdiamonds in quartzofeldspathic rocks of the Saxonian Erzgebirge, Germany, and their metamorphic evolution. Proc. 7th Int Kimberlite Confer, Cape Town 1998, P.H. Nixon Vol, p. 533–539

Massonne H–J (2003) A comparison of the evolution of diamondiferous quartz-rich rocks from the Saxonian Erzgebirge and the Kokchetav Massif: are so–called diamondiferous gneisses magmatic rocks? Earth Planet Sci Lett 216:347–364

Massonne H–J, O'Brien PJ (2003) The Bohemian massif and the NW Himalaya. *In*: Ultrahigh Pressure Metamorphism. Carswell DA, Compagnoni R, Rolfo F (eds) p 145–187

Massonne H–J, Kennedy A, Nasdala L, Theye T (2007a) Dating of zircon and monazite from diamondiferous quartzofeldspathic rocks of the Saxonian Erzgebirge—hints at burial and exhumation velocities. Mineral Mag 71:371–389

Massonne H–J, Tu W (2007b) $\delta^{13}C$ signature of early graphite and subsequently formed microdiamond from the Saxonian Erzgebirge, Germany. Terra Nova 19:476–480

Massonne H–J, Fockenberg T (2012) Melting of metasedimentary rocks at ultrahigh pressure—Insights from experiments and thermodynamic calculations. Lithosphere 4:269–285

Massonne H–J (2019) A shallow origin for diamonds in ophiolitic chromitites: COMMENT. Geology 47: e476

Mathez EA, Fogel RA, Hutcheon ID, Marshintsev VK (1995) Carbon isotopic composition and origin of SiC from kimberlites of Yakutia, Russia. Geochim Cosmochim Acta 59:781–791

McGowan NM, Griffin WL, González–Jiménez JM, Belousova E, Afonso JC, Shi R, McCammon CA, Pearson NJ, O'Reilly SY (2015) Tibetan chromitites: Excavating the slab graveyard. Geology 43:179–182

Mengel FC (1983) Chemistry of coexisting mafic minerals in granulite facies amphibolites from West Greenland: clues to conditions of metamorphism. Neues Jahrbuch für Mineralogie Abhandlungen 147:315–340

Menneken M, Nemchin AA, Geisler T, Pidgeon RT, Wilde SA (2007) Hadean diamonds in zircon from Jack Hills, Western Australia. Nature 448:917–920

Menneken M, Geisler T, Nemchin AA, Pollok K, Whitehouse M, Pidgeon RT, Wilde SA (2014) Is there really carbon in the detrital zircons from Jack Hills, Western Australia? Geophysical Research Abstr, EGU General Assembly 2014 16:EGU2014–13489

Menneken M, Geisler T, Nemchin AA, Whitehouse M, Wilde SA, Gasharova B, Pidgeon RT (2017) CO_2 fuid inclusions in Jack Hills zircons. Contrib Mineral Petrol 172:66–78

Meyers PA (2014) Why are the $\delta^{13}C_{org}$ values in Phanerozoic black shales more negative than in modern marine organic matter? Geochem Geophys 15:3085–106

Milledge HJ 1961 Coesite as an inclusion in G.E.C. synthetic diamonds. Nature 190:1181

Moe KS, Yang JS, Johnson P, Xu X, Wang W (2018) Spectroscopic analysis of microdiamonds in ophiolitic chromitite and peridotite. Lithosphere 10:133–141

Mposkos ED, Kostopoulos DK (2001) Diamond, former coesite and supersilicic garnet in metasedimentary rocks from the Greek Rhodope: a new ultrahigh-pressure metamorphic province established. Earth Planet Sci Lett 192:497–506

Mukhopadhyay S (2012) Early differentiation and volatile accretion recorded in deep–mantle neon and xenon. Nature 486:101–104

Murri M, Smith RL, McColl K, Hart M, Alvaro M, Jones AP, Németh P, Salzmann CG, Corà F, Domeneghetti MC, Nestola F. (2019) Quantifying hexagonal stacking in diamond. Sci Reports 9:1–8

Naemura K, Ikuta D, Kagi H, Odake S, Ueda T, Ohi S, Kobayashi T, Svojtka M, Hirajima T (2011) Diamond and other possible ultradeep evidence discovered in the orogenic spinel–garnet peridotite from the Moldanubian Zone of the Bohemian Massif, Czech Republic. *In:* Ultrahigh-Pressure Metamorphism. Elsevier, p. 77–111

Nakamuta Y, Toh S (2013) Transformation of graphite to lonsdaleite and diamond in the Goalpara ureilite directly observed by TEM. Am Mineral 98:574–581

Nasdala L, Massonne H–J (2000) Microdiamonds from the Saxonian Erzgebirge, Germany: in situ micro–Raman characterisation. Eur J Mineral 12:495–498

Nasdala L, Grambole D, Wildner M, Gigler AM, Hainschwang T, Zaitsev AM, Harris JW, Milledge J, Schulze DJ, Hofmeister W, Balmer WA (2013) Radio-colouration of diamond: a spectroscopic study. Contrib Mineral Petrol 165:843–61

Nasdala L, Steger S, Reissner C (2016) Raman study of diamond-based abrasives, and possible artefacts in detecting UHP microdiamond. Lithos 265:317–327

Nemchin AA, Whitehouse MJ, Menneken M, Geisler T, Pidgeon RT, Wilde SA (2008) A light carbon reservoir recorded in zircon-hosted diamond from the Jack Hills. Nature 454:92–95

Németh P, Garvie LA, Aoki T, Dubrovinskaia N, Dubrovinsky L, Buseck PR (2014) Lonsdaleite is faulted and twinned cubic diamond and does not exist as a discrete material. Nat Comm 5:1–5

Nestola F, Goodrich CA, Morana M, Barbaro A, Jakubek RS, Christ O (2020) Impact shock origin of diamonds in ureilite meteorites. PNAS 117:25310–25318

Nishiyama T, Ohfuji H, Fukuba K, Terauchi M, Nishi U, Harada K, Unoki K, Moribe Y, Yoshiasa A, Ishimaru S, Mori Y (2020) Microdiamond in a low–grade metapelite from a Cretaceous subduction complex, western Kyushu, Japan. Sci Rep 10:1–1

O'Bannon EF, Xia G, Wirth R, Shi F, King A, Dobrzhinetskaya L (2020) The transformation of diamond to graphite: Experiments reveal the presence of an intermediate linear carbon phase. Diam Relat Mater 108:107876

Ogasawara Y (2005) Microdiamonds in ultrahigh-pressure metamorphic rocks. Elements 1:91–96

Ogasawara Y, Ohta M, Fukusava K, Katayama I, Maruyama S (2000) Diamond bearing and diamond–free metacarbonate rocks from Kumdy–Kol from the Kokchetav massif, northern Kasakhstan. The Island Arc 9:400–416

Ogasawara Y, Fukasawa K, Maruyama S (2002) Coesite exsolution from supersilicic titanite in UHP marble from the Kokchetav Massif, northern Kazakhstan. Am Mineral 87:454–461

Ohfuji H, Irifune T, Litasov KD, Yamachita T, Isobe F, Afanasiev V, Pokhilenko NP (2015) Natural occurrence of pure Nano-polycrystalline diamond from impact crater. Sci Rep 5:14702

Ohta M, Mock T, Ogasawara Y, Rumble D (2003) Oxygen, carbon, and strontium isotope geochemistry of diamond-bearing carbonate rocks from Kumdy–Kol, Kokchetav Massif, Kazakhstan. Lithos 70:77–90

Okay A, Xu ST, Sengor AMC (1988) Coesite from the Dabie Shan ecologite, Central China. Eur J Mineral 1:595–598

Orwa JO, Nugent KW, Jamieson DN, Prawer S (2000) Raman investigation of damage caused by deep ion implantation in diamond. Phys Rev B 62:5461

Ozima M, Podosek FA (2002) Noble Gas Geochemistry. Cambridge Univ. Press, Cambridge

Ozima M, Zashu S (1988) Solar-type Ne in Zaire cubic diamonds. Geochim Cosmochim Acta 52:19–25

Ozima M, Zashu S (1991) Noble gas state of the ancient mantle as deduced from noble gases in coated diamonds. Earth Planet Sci Lett 105:13–27

Pal'yanov Y, Sokol A, Borzdov Y, Khokhryakov A, Sobolev NV (1999) Diamond formation from mantle carbonate fluids. Nature 400:417–418

Pal'yanov YuN, Sokol AG, Khokhryakov, AF, Pal'yanova GA, Borzdov YuM, Sobolev NV (2000) Diamond and graphite crystallization in COH fluid at PT parameters of the natural diamond formation. Doklady Nauk o Zemle (Proc Earth Sci) 375:1395–1399 (in Russian)

Pal'yanov N, Sokol AG, Borzdov M, Khokhryakov AF (2002) Fluid-bearing alkaline carbonate melts as the medium for the formation of diamonds in the Earth's mantle: an experimental study. Lithos 60:145–159

Pal'yanov YN, Bataleva YV, Sokol AG, Borzdov YM, Kupriyanov IN, Reutsky VN, Sobolev NV (2013) Mantle–slab interaction and redox mechanism of diamond formation. PNAS 110:20408–20413

Palot M, Pearson DG, Stern RA, Stachel T, Harris JW (2013) Multiple growth events, processes and fluid sources involved in diamond genesis: a micro–analytical study of sulphide-bearing diamonds from Finsch mine, RSA. Geochim. Cosmochim. Acta 106:51–70

Parman SW (2007) Helium isotopic evidence for episodic mantle melting and crustal growth. Nature 446:900–903

Pearson DG, Davies GR, Nixon PH, Milledge HJ (1989) Graphitized diamonds from a peridotite massif in Morocco and implications for anomalous diamond occurrences. Nature 338:60–62

Pearson DG, Davies GR, Nixon PH, Mattey DP (1991) A carbon isotope study of diamond facies pyroxenites from Beni Bousera, N. Morocco. Spec Ed J Petrol: Orogenic Lherzolites and Mantle Processes, 175–189

Pechnikov VA, Kaminsky FV (2008) Diamond potential of metamorphic rocks in the Kokchetav Massif, northern Kazakhstan. Eur J Mineral 20:395–413

Pechnikov VA, Kaminsky FV (2011) Structural and microstructural regularities of the distribution of diamond in metamorphic rocks of the Kumdy–Kol and Barchi–Kol deposits, Kokchetav Massif, Northern Kazakhstan. Can Mineral 49:673–690

Pedroza-Montero M, Chernov V, Castañeda B, Meléndrez R, Gonçalves JA, Sandonato GM, Bernal R, Cruz–Vázquez C, Brown F, Cruz–Zaragoza E, Barboza–Flores M (2005) TL, OSL, Raman spectroscopy and SEM characterization of boron doped diamond films. Phys Status Solidi B 202:2154–59

Perraki M, Proyer A, Mposkos E, Kaindl R, Hoinkes G (2006) Raman micro–spectroscopy on diamond, graphite and other carbon polymorphs from the ultrahigh-pressure metamorphic Kimi Complex of the Rhodope Metamorphic Province, NE Greece. Earth Planet Sci Lett 241:672–685

Perraki M, Korsakov AV, Smith DC, Mposkos E (2009) Raman spectroscopic and microscopic criteria for the distinction of microdiamonds in ultrahigh-pressure metamorphic rocks from diamonds in sample preparation materials. Am Mineral 94:546–56

Perraki M, Faryad SW (2014) First finding of microdiamond, coesite and other UHP phases in felsic granulites in the Moldanubian Zone: Implications for deep subduction and a revised geodynamic model for Variscan Orogeny in the Bohemian Massif. Lithos 202:157–166

Platt JP, Behr WM, Johanesen K, Williams JR (2013) The Betic–Rif arc and its orogenic hinterland: a review. Annu Rev Earth Planet Sci 41:313–357

Pleshakov AM, Shukolyukov YA (1994) Isotopic variations of helium in the diamonds of the Kokchetav massif's metamorphic rocks, Kazakhstan. in: Matsuda, J. (Ed.), Noble Gas Geochemistry and Cosmochemistry, Terra Scientific Publishing Company (TERRAPUB), Tokyo, p. 229–243

Pokhilenko NP, Shumilova TG, Afanas'ev VP, Litasov KD (2019) Diamonds in the Kamchatka peninsula (Tolbachik and Avacha volcanoes): Natural origin or contamination? Russian Geol Geoph 60:463–472

Porcelli D, Ballentine CJ (2002) Models for distribution of terrestrial noble gases and evolution of the atmosphere. Rev Mineral Geochem 47:411–48

Porcelli D, Elliott T (2008) The evolution of He isotopes in the convecting mantle and the preservation of high $^3He/^4He$ ratios. Earth Planet Sci Lett 269:175–185

Prawer S, Rosenblum I, Orwa JO, Adler J (2004) Identification of the point defects in diamond as measured by Raman spectroscopy: comparison between experiment and computation. Chem Phys Lett 390:458–461

Pujol–Sola N, Garcia–Casco A, Proenza JA, Gonzalez–Jimenez JM, del Campo A, Colas V, Canals À, Sánchez–Navas A, Roque–Rosell J (2020) Diamond forms during low pressure serpentinisation of oceanic lithosphere. Geochem Perspect Lett 15:19–24

Rasmussen B, Fletcher IR, Muhling JR, Wilde SA (2010) In situ U–Th–Pb geochronology of monazite and xenotime from the Jack Hills belt: Implications for the age of deposition and metamorphism of Hadean zircons. Precambrian Res 180:26–46

Rasmussen B, Fletcher IR, Muhling JR, Gregory CJ, Wilde SA (2011) Metamorphic replacement of mineral inclusions in detrital zircon from Jack Hills, Australia: Implications for the Hadean Earth. Geology 39:1143–1146

Rasmussen B, Fletcher IR, Muhling JR, Gregory CJ, Wilde SA (2012) Metamorphic replacement of mineral inclusions in detrital zircon from Jack Hills, Australia: Implications for the Hadean Earth: REPLY. Geology 40:282–283

Reinecke T (1991) Very–high-pressure metamorphism and uplift of coesite-bearing metasediments from the Zermatt–Saas zone, Western Alps. Eur J Mineral 3:7–17

Robinson PT, Bai WJ, Malpas J, Yang JS, Zhou MF, Fang QS, Hu XF, Cameron S, Staudigel H (2004) Ultra-high pressure minerals in the Luobusa Ophiolite, Tibet, and their tectonic implications. Geol Soc London Spec Publ 226:247–271

Robinson PT, Trumbull RB, Schmitt A, Yang JS, Li JW, Zhou MF, Erzinger J, Dare S, Xiong F (2015) The origin and significance of crustal minerals in ophiolitic chromitites and peridotites. Gondwana Res 27:486–506

Rollinson H (2016) Surprises from the top of the mantle transition zone. Geol Today 32:58–64

Rosing MT (1999) ^{13}C-depleted carbon microparticles in >3700–Ma sea–floor sedimentary rocks from West Greenland. Science 283:674–676

Rost R, Dolgov YA, Vishnevsky SA (1978) Gases in inclusions of impact glass in the Ries Crater, West Germany, and finds of high-pressure carbon polymorphs. Dokl Akad Nauk SSSR 241:695–698

Rozen OM (1972) A find of the diamonds linked with eclogites of the Precambrian Kokchetav massif. Dokl Akad Nauk SSSR 203:674–676

Rozen OM, Zayachkovsky AA, Kljuev YuA, Smirnov VL (1979) Peculiarities of trace–minerals composition and conditions of the formation of the North Kazakhstan eclogites. *In:* A.V. Sidorenko (Editor), Problems of Sedimentary Geology of Precambrian. Nauka Press, Moscow p. 170–186 (in Russian with English Abstract)

Reutsky V, Borzdov Y, Palyanov Y, Sokol A, Izokh O (2015) Carbon isotope fractionation during experimental crystallisation of diamond from carbonate fluid at mantle conditions. Cont Mineral Petrol 170:41

Ruiz-Cruz MD, de Galdeano CS (2012) Diamond and coesite in ultrahigh-pressure–ultrahigh-temperature granulites from Ceuta, Northern Rif, northwest Africa. Mineral Mag 76:683–705

Ruiz-Cruz MD, de Galdeano CS (2013) Coesite and diamond inclusions, exsolution microstructures and chemical patterns in ultrahigh pressure garnet from Ceuta (Northern Rif, Spain). Lithos 177:184–206

Ruiz-Cruz MD, de Galdeano CS (2014) Garnet variety and zircon ages in UHP meta-sedimentary rocks from the Jubrique zone (Alpujárride Complex, Betic Cordillera, Spain): evidence for a pre-Alpine emplacement of the Ronda peridotite. Int Geol Rev 56:845–868

Salzmann CG, Murray BJ, Shephard JJ (2015) Extent of stacking disorder in diamond. Diam Relat Mater 59:69–72

Sarda P, Staudacher T, Allègre CJ (1988) Neon isotopes in submarine basalts. Earth Planet Sci Lett 91:73–88

Schertl HP, O'Brien (2013) Continental crust at mantle depths: key minerals and microstructures. Elements 9:261–266

Schertl HP, Sobolev NV (2013) The Kokchetav massif, Kazakhstan: "Type locality" of diamond-bearing UHP metamorphic rocks. J Asian Earth Sci 63:5–38

Schönig J, von Eynatten H, Meinhold G, Lünsdorf NK (2019) Diamond and coesite inclusions in detrital garnet of the Saxonian Erzgebirge, Germany. Geology 47:715–718

Seno T, Rehman HU (2011) When and why the continental crust is subducted: Examples of Hindu Kush and Burma. Gondwana Res 19:327–333

Shatsky VS, Sobolev NV (1993) Some aspects of origin of diamonds in metamorphic rocks. Doklady Akademii Nauk (Proc Acad Sci) 331:1217–1219 (in Russian)

Shatsky VS, Sobolev NV, Vavilov MA (1995) Diamond–beraing metamorphic rocks of the Kokchetav massif (northern Kazakhstan). *In*: Ultrahigh Pressure Metamorphism, Coleman RG, Wang X (eds) Cambridge Univ Press, Cambridge, p 427–455

Shimizu R, Ogasawara Y (2014) Radiation damage to Kokchetav UHPM diamonds in zircon: varations in Raman, photoluminescence and cathodoluminiscence spestra. Lithos 206:201–213

Shirey S, Cartigny P, Frost D, Keshav S, Nestola F, Pearson G, Sobolev N, Walter MJ (2013) Diamonds and the geology of mantle carbon. Rev Mineral Geochem 75:355–421

Shukolyukov YuA, Pleshakov AM, Lavrova LD (1993) The unprecedentedly high $^{3}He/^{4}He$ ratio in diamonds from a metamorphic rock of the Kokchetav massif, Kazakhstan. Petrologia (Petrology) 1:110–119 (in Russian)

Shumilova TG, Mayer E, Isaenko SI (2011) Natural monocrystalline lonsdaleite. Dokl Earth Sci 441:1552–1554

Simakov SK (2010) Metastable nanosized diamond formation from a CHO fluid system. J Mat Res 25:2336–2340

Simakov SK (2015) Different sizes of diamond formation in natural processes. Dokl Earth Sci 461:419–421

Simakov SK (2018) Nano-and micron–sized diamond genesis in nature: an overview. Geosci Front 9:1849–1858

Simakov SK, Dubinchuk VT, Novikov MP, Drozdova IA (2008) Formation of diamond and diamond-type phases from the carbon-bearing fluid at *PT* parameters corresponding to processes in the Earth's crust. Dokl Earth Sci 421:835

Sitnikova ES, Shatsky VS (2009) New FTIR spectroscopy data on the composition of the medium of diamond crystallization in metamorphic rocks of the Kokchetav Massif. Russ Geol Geophys 50:842–849

Smart KA, Chacko T, Stachel T, Muehlenbachs K, Stern RA, Heaman LM (2011) Diamond growth from oxidized carbon sources beneath the Northern Slave Craton, Canada: a $\delta^{13}C$–N study of eclogite-hosted diamonds from the Jericho kimberlite. Geochim Cosmochim Acta 75:6027–47

Smit KV, Shirey SB, Wang W (2016) P Type Ib diamond formation and preservation in the West African lithospheric mantle: Re–Os age constraints from sulphide inclusions in Zimmi diamonds. Precambrian Res 286:152–156

Smith CB, Walter MJ, Bulanova GP, Mikhail S, Burhama AD, Gobbo L, Kohn SC (2016) Diamonds from Dachine, French Guiana: A unique record of early Proterozoic subduction. Lithos 265:82–95

Stähle V, Altherr R, Koch M, Nasdala L (2008) Shock–induced growth and metastability of stishovite and coesite in lithic clasts from suevite of the Ries impact crater (Germany). Contribut Mineral Petrol 155:457–472

Stachel T, Cartigny P, Chacko T, Pearson DG (2022) Carbon and nitrogen in mantle-derived diamonds. Rev Mineral Geochem 88:809–876

Stan CV, O'Bannon EF, Mukhin P, Tamura N, Dobrzhinetskaya LF (2020) X–ray Laue microdiffraction and Raman spectroscopic investigation of natural silicon and moissanite. Minerals 10:204–220

Steger S, Nasdala L, Wagner A (2013) Raman spectra of diamond abrasives and possible artefacts in detecting UHP microdiamond. CORALS Conf Raman Lumin Spectrosc Earth Sci, Vienna, Austria, p. 95–96

Stöckhert B, Duyster J, Trepmann C, Massonne H–J (2001) Microdiamond daughter crystals precipitated from supercritical COH + silicate fluids included in garnet, Erzgebirge, Germany. Geology 29:391–394

Stöckhert B, Trepmann CA, Massonne H–J (2009) Decrepitated UHP fluid inclusions: about diverse phase assemblages and extreme decompression rates (Erzgebirge, Germany). J Metamorph Geol 27:673–684
Stuart FM, Lass–Evans S, Fitton JG, Ellam RM (2003) High $^3He/^4He$ ratios in picritic basalts from Baffin Island and the role of a mixed reservoir in mantle plumes. Nature 424:57–59
Sumino H, Dobrzhinetskaya LF, Burgess R, Kagi H (2011) Deep-mantle-derived noble gases in metamorphic diamonds from the Kokchetav massif, Kazakhstan. Earth Planet Sci Lett 307:439–449
Sverjensky DA, Stagno V, Huang F (2014) Important role for organic carbon in subduction–zone fluids in the deep carbon cycle. Nat Geosci 7:909–913
Taylor WR (1990) A reappraisal of the nature of fluids included by diamond — a windowto deep–seated mantle fluids and redox conditions. *In*: Herbert HK, Ho SE (eds.) Stable Isotopes and Fluid Processes in Mineralization. University of Western Australia, Perth 23:333–349
Taylor WR, Milledge HJ, Griffin BJ, Nixon PH, Kamperman M, Mattey DP (1995) Caracteristics of microdiamonds from ultramafic massifs in Tibet: authentic ophiolitic diamonds or contamination? Int Kimberlite Conf: Extended Abstr 6:623–624
Taylor WR, Canil D, Milledge HJ (1996) Kinetics of Ib to IaA nitrogen aggregation in diamond. Geochim Cosmochim Acta 60:4725–4733
Tian Y, Yang JS, Zhang Z, Xiong F, Li Y, Liu Z, Liu F, Niu XL (2015) Discovery and implication of unusual mineral group from Sarthay high-Al chromitites, Xinjiang. Acta Petrol Sinica 31:3650–3662
Timmerman S, Chinn IL, Fisher D, Davies GR (2018) Formation of unusual yellow Orapa diamonds. Mineral Petrol 112:209–218
Timmerman S, Honda M, Phillips D, Jaques AL, Harris JW (2018) Noble gas geochemistry of fluid inclusions in South African diamonds: implication for the origin of diamond–forming fluids. Mineral Petrol 112:181–195
Tirel C, Brun J–P, Burov E, Wortel MJR, Lebedev S (2013) A plate tectonics oddity: Caterpillar–walk exhumation of subducted continental crust. Geology 41:555–558
Tretiakova LI, Lyukhin AM (2016) Impact–cosmic–metasomatic origin of microdiamonds from Kumdy–Kol deposit, northern Kazakhstan. National Geology (Otechestvennaya Geologiya) 2:69–77 (in Russian)
Tretiakova LI, Lyukhin AM (2017) Impact–cosmic–metasomatic origin of microdiamonds from Kumdy–Kol Deposit, Kokchetav Massiv, N. Kazakhstan. 11th Int Kimberlite Conf Extended Abstr 11IKC:4506
Trieloff M, Kunz J, Clague DA, Harrison D, Allègre CJ (2000) The nature of pristine noble gases in mantle plumes. Science 288:1036–1038
Uman MA (1986) All About Lightning. Dover Publications Inc New York 167
van Roermund HL, Carswell DA, Drury MR, Heijboer TC (2002) Microdiamonds in a megacrystic garnet websterite pod from Bardane on the island of Fjørtoft, western Norway: Evidence for diamond formation in mantle rocks during deep continental subduction. Geology 30:959–962
Verchovsky AB, Ott U, Begemann F (1993) Implanted radiogenic and other noble gases in crustal diamonds from Northern Kazakhstan. Earth Planet Sci Lett 120:87–102
Vogel DE (1966) Nature and chemistry of the formation of clinopyroxene–plagioclase symplectite from omphacite. Neues Jahrb Mineral Monatsh 6:185–189
Vrijmoed JC, Van Roermund HL, Davies GR (2006) Evidence for diamond–grade ultra-high pressure metamorphism and fluid interaction in the Svartberget Fe–Ti garnet peridotite–websterite body, Western Gneiss Region, Norway. Mineral Petrol 88:381–405
Vrijmoed JC, Smith DC, Van Roermund HLM (2008) Raman confirmation of microdiamond in the Svartberget Fe–Ti type garnet peridotite, Western Gneiss Region, Western Norway. Terra Nova 20:295–301
Wada N, Matsuda J (1998) A noble gas study of cubic diamonds from Zaire: constraints on their mantle source. Geochim Cosmochim Acta 62:2335–2345
Wang H, Wu Y–B, Gao S, Zheng J–P, Liu Q, Liu X–C, Qin Z–W, Yang S–H, Gong H–J (2014) Deep subduction of continental crust in accretionary orogen: Evidence from U–Pb dating on diamond-bearing zircons from the Qinling orogen, central China. Lithos 190:420–429
Warren JM (2016) Global variations in abyssal peridotite compositions. Lithos 248:193–219
Wu BR, Xu JA (1998) Total energy calculations of the lattice properties of cubic and hexagonal diamond. Phys Rev B 57:13355
Wu W, Yang J, Ma C, Milushi I, Lian D, Tian Y (2017) Discovery and significance of diamonds and moissanites in chromitite within the Skenderbeu massif of the Mirdita Zone Ophiolite, West Albania. Acta Geol Sin Eng 91:882–897
Wu W, Yang J, Wirth R, Zheng J, Lian D, Qiu T, Milushi I (2019) Carbon and nitrogen isotopes and mineral inclusions in diamonds from chromitites of the Mirdita ophiolite (Albania) demonstrate recycling of oceanic crust into the mantle. Am Mineral 104:485–500
Wu W, Yang J, Zheng J, Lian D, Qiu T, Rui H (2020) Origin of the Diamonds within Chromitite from the Mirdita Ophiolite (Albania) and its Geological Significance. Acta Geol Sin–Eng 94:64–64
Xiong F, Yang J, Robinson PT, Dilek Y, Milushi I, Xu X, Chen Y, Zhou W, Zhang Z, Lai S, Tian Y, Huang Z (2015) Petrology and geochemistry of high Cr-podiform chromitites of Bulqiza, Eastern Mirdita Ophiolite (EMO), Albania. Ore Geol Rev 70:188–207

Xiong F, Yang J, Robinson PT, Dilek Y, Milushi I, Xu X, Zhou W, Zhang Z, Rong H (2017) Diamonds discovered from high-Cr podiform chromitites of Bulqiza, eastern Mirdita ophiolite, Albania. Acta Geol Sin–Eng 9:455–468

Xiong F, Yang J, Robinson PT, Dilek Y, Xu X , Zhang Z (2018) Origin and significance of diamonds and other exotic minerals in the Dingqing ophiolite peridotites, eastern Bangong–Nujiang suture zone, Tibet. Lithosphere 10:142–155

Xu S, Okay AI, Ji S, Sengor AMC, Su W, Liu Y, Jiang L (1992) Diamond from the Dabie Shan metamorphic rocks and its implication for tectonic setting. Science 256:80–82

Xu S, Liu Y, Chen G, Compagnoni R, Rolfo F, He M, Liu H (2003) New finding of micro–diamonds in eclogites from Dabie–Sulu region in central–eastern China. Chinese Sci Bull 48:988–994

Xu X, Chen S, Yang J (2009) Unusual mantle mineral group from chromitite orebody Cr–11 in Luobusa ophiolite of Yarlung–Zangbo suture zone, Tibet. J Earth Sci 20:284–302

Xu XZ, Yang, JS, Ba DZ, Zhang ZM, Xiong FH, LiY (2015) Diamond discovered from the Dongbo mantle peridotite in the Yarlung Zangbo suture zone, Tibet. Geol China 42:1471–1482

Xu X, Cartigny P, Yang JS, Dilek Y, Xiong F, Guo G (2018) Fourier transform infrared spectroscopy data and carbon isotope characteristics of the ophiolite-hosted diamonds from the Luobusa ophiolite, Tibet, and Ray-Iz ophiolite, Polar Urals. Lithosphere 10:156–169

Yamaoka S, Kumar MS, Akaishi M, Kanda H (2000) Reaction between carbon and water under diamond–stable high pressure and high temperature conditions. Dia Relat Mater 9:1480–1486

Yang JS, Xu Z, Dobrzhinetskaya LF, Green II, HW, Pei X, Shi R, Wu C, Wooden JL, Zhang J, Wan Y, Li H (2003) Discovery of metamorphic diamonds in Central China: an indication of a >4000 km–long–zone of deep subduction resulting from multiple continental collisions. Terra Nova 15:370–379

Yang JS, Dobrzhinetskaya LF, Bai WJ, Fang QS, Robinson PT, Zhang JF, Green II HW (2007a) Diamond– and coesite-bearing chromitites from the Luobusa ophiolite, Tibet. Geology 35:875–878

Yang JS, Bai WJ, Fang Q S, Meng FC, Chen SY, Zhang ZM, Rong H (2007b) Discovery of diamond and an unusual mineral group from the podiform chromite, Polar Ural. Geol China 34:950–952

Yang JS, Xu XZ, Li Y, Li JY, Ba DZ, Rong H, Zhang ZM (2011) Diamonds recovered from peridotite of the Purang ophiolite in the Yarlung–Zangbo suture of Tibet: A proposal for a new type of diamond occurrence. Acta Petrol Sinica 27:3171–3178

Yang JS, Xu XZ, Zhang ZM, Rong H, LiY, Xiong FH, Liang FH, Liu Z, Liu F, Li JY, Li ZL, Chen SY, Guo GL, Robinson P (2013) Ophiolite-type diamonds and deep genesis of chromitite. Acta Geosci Sin 34:643–653 (in Chinese with English Abstract)

Yang JS, Robinson PT, Dilek Y (2014) Diamonds in ophiolites. Elements 10:127–130

Yang JS, Meng F, Xu X, Robinson PT, Dilek Y, Makeyev AB, Wirth R, Wiedenbeck M, Griffin WL, Cliff J (2015a) Diamonds, native elements and metal alloys from chromites of the Ray-Iz ophiolite of the Polar Urals. Gondwana Res 27:459–485

Yang JS, Robinson PT, Dilek Y (2015b) Diamond-bearing ophiolites and their geological occurrence. Episodes 38:344–364

Yang JS, Trumbull R, Robinson PT, Xiong FH, Lian DY (2018) Comment 2 on "Ultra-high pressure and ultra-reduced minerals in ophiolites may form by lightning strikes". Geochem Perspect Lett 8:6–7

Yang J, Robinson PT, Xu X, Xiong F, Lian D (2019a) Diamond in oceanic peridotites and chromitites: evidence for deep recycled mantle in the global ophiolite record. Acta Geol Sin–Eng 892:341–350

Yang J, Lian D, Robinson PT, Qiu T, Xiong F, Wu W (2019b) A shallow origin for diamonds in ophiolitic chromitites: COMMENT. Geology 47:e475

Yang J, Simakov SK, Moe K, Scribano V, Lian D, Wu W (2020) Comment on "Comparison of enigmatic diamonds from the Tolbachik arc volcano (Kamchatka) and Tibetan ophiolites: assessing the role of contamination by synthetic materials" by Litasov et al. 2019. Gondwana Res 79:301–303

Yang J, Lian D, Wu W, Rui H (2021) Peridotites, chromitites and diamonds in ophiolites. Nat Rev Earth Environ 2:198–212

Ye K, Cong BL, Ye DN (2000) The possible subduction of continental material to depths greater than 200 km. Nature 407:734–736

Zhang KJ, Cai JX, Zhang YX, Zhao TP (2006) Eclogites from central Qiangtang, northern Tibet (China) and tectonic implication. Earth Planet Sci Lett 245:722–729

Zhang RY, Liou JG, Iizuka Y, Yang JS (2009) First record of K–cymrite in North Qaidam UHP eclogite, Western China. Am Mineral 94:222–228

Zhang J, Prakapenka V, Kubo A, Kavner A, Green HW II, Dobrzhinetskaya LF (2011) Diamond formation from amorphous carbon and graphite in presence of COH fluids: an in situ high-pressure and–temperature laser–heated diamond anvil cell experiments. *In:* Ultrahigh Pressure Metamorphism: 25 Years After the Discovery of Coesite and Diamond, Dobrzhinetskaya LF, Faryad SW, Wallis S, Cuthbert S (eds) Elsevier, London, p.113–124

Zhang RY, Liou JG, Lo CH (2017) Raman spectra of polycrystalline microdiamond inclusions in zircons, and ultrahigh-pressure metamorphism of a quartzofeldspathic rock from the Erzgebirge terrane, Germany. Int Geol Rev 59:779–792

Zhou MF, Robinson PT, Su BX, Gao JF, Li JW, Yang JS, Malpasa J (2014) Compositions of chromite, associated minerals, and parental magmas of podiform chromite deposits: The role of slab contamination of asthenospheric melts in suprasubduction zone environments. Gondwana Res 26:262–283

Reviews in Mineralogy & Geochemistry
Vol. 88 pp. 257–306, 2022

Crystallographic Methods for Non-destructive Characterization of Mineral Inclusions in Diamonds

Ross J. Angel

IGG-CNR
Via G. Gradenigo 6
35131 Padova, Italy

rossjohnangel@gmail.com

Matteo Alvaro

Department of Earth and Environmental Sciences
University of Pavia
Via A. Ferrata 1, 27100 Pavia, Italy

matteo.alvaro@unipv.it

Fabrizio Nestola

Department of Geosciences
University of Padova
Via G. Gradenigo 6, 35131 Padova, Italy

fabrizio.nestola@unipd.it

INTRODUCTION

The mineralogy and chemical compositions of inclusions in diamonds are the primary source of information about the environment in which diamonds grow and help constrain the mechanisms of their growth. However, the vast majority of the information about inclusions has been gathered by extracting them from their diamonds, thus destroying all possibility of obtaining further information about the diamond–inclusion system as a whole with new experimental probes unavailable at the time of extraction. One such specific example is the recent discovery by X-ray tomography and *in situ* spectroscopy of the hydrous silicic fluid film that appears to be ubiquitous around silicate inclusions in lithospheric diamonds (Nimis et al. 2016); the films escaped detection in a multitude of analyses during more than 70 years of research involving the extraction of many thousands of such inclusions. Inclusions in diamond are under compressive stress as a result of their encapsulation at depth and ascent of the diamond to the Earth's surface; extraction also destroys this stress and thus prevents the depth of entrapment from being determined from the stress state by elastic geobarometry. The stress release on extraction can also lead to the phase changes and/or conversion of the inclusion to a powder (e.g., Joswig 2011). For inclusions from diamonds suspected as being from super-deep sources, extraction therefore risks the loss of rare or possibly unique samples. Non-destructive characterization of inclusions in diamonds should therefore be made *in situ* whenever possible.

Extraction of inclusions has often been motivated by a lack of experimental probes to measure the inclusions *in situ* in their host diamonds. However, developments in technology mean that a wide range of spectroscopies is now available that can be used on inclusions *in situ* to offer rapid initial phase identification (e.g., Raman), and to determine certain aspects

1529-6466/22/0088-0005$05.00 (print)
1943-2666/22/0088-0005$05.00 (online)
http://dx.doi.org/10.2138/rmg.2022.88.05

of chemical composition. Fourier transform infra-red spectrosocopy (FTIR) can be used to determine OH contents of minerals, synchrotron Mössbauer and μ-EXAFS to determine details about element oxidation states and structural environments in inclusions. None of these methods, however, provide a definitive identification of the mineral phase in the inclusion, nor the details of its crystallographic structure, nor its stress state. These require diffraction measurements. Single-crystal X-ray diffraction has been used since the work of Mitchell and Giardini (1953) to measure inclusions in diamonds. The presence of the diamond presents some challenges to obtaining the highest quality data and therefore the precision with which scientific conclusions can be derived from the data. But, by using methods developed over several decades to study crystals held in diamond-anvil pressure cells (DACs) and exploiting rapid developments in diffractometer technology and intensity of X-ray sources, diffraction measurements of inclusions can be of similar quality to those of crystals at ambient conditions, if the measurements are performed with care.

In this chapter we describe the methods that we have developed as best practice that, when used, will enable the reader to obtain the highest-quality data from most inclusions by using standard laboratory diffractometers. We focus on details of data collection and analysis methods and point out the pitfalls that can arise from poor set-up of the experiment. We show with examples how the crystallographic data measured from inclusions still buried deep within their diamonds can be used to provide simultaneously three key pieces of information to reconstruct the history of the diamond–inclusion pair. Intensity data collection and structure refinement confirms the identification of the mineral phase in the inclusion and allows the composition of the mineral to be determined, and the environment in which the host diamond grew to be characterized. Determination of the crystallographic orientations of inclusions with respect to their diamond hosts has recently provided new fundamental insights into how diamonds nucleate and grow in the mantle. The developments in methods to measure the stress state in inclusions and the theory of elastic geobarometry allows the depth of formation of diamonds to be inferred on the basis of a physical process that is independent from chemical equilibrium. The same principles and methods also apply to those inclusions that today are too small to be measured with laboratory X-ray sources and must still be measured at synchrotron beamlines. And they can be applied equally to all inclusions in diamond, whether the diamond hosts are believed to be lithospheric or sub-lithospheric (so-called super-deep diamonds). Indeed, the identification of the mineralogy of the inclusions and their stress state is the major determinant of the depth the formation of the diamond host.

DIAMOND GROWTH ENVIRONMENT: STRUCTURES OF INCLUSIONS

The mineralogy of inclusions is the primary source of information about the environment in which diamonds grow within the Earth. Structure determination and refinement from single-crystal X-ray diffraction data is a non-destructive method of determining the mineral phase or phases present in the inclusion and, for most minerals that are solid solutions, the composition of the phase. It is especially valuable for colorless minerals which are difficult to identify by optical methods and phases that have not had their Raman spectra recorded previously (e.g., Joswig et al. 1999). Our own experience is that up to 20% of olivine and orthopyroxene inclusions previously examined by optical methods alone (i.e., without Raman) were mis-identified. In addition, X-ray diffraction can also reveal the presence of hidden second phases in the inclusion, including inclusions within the main crystal (Joswig 2011; Howell et al. 2012b).

Further, in mineral structures such as olivines, garnets and pyroxenes, in which different cations are distributed over two or more cation sites, the structure refinement can provide information about the cation distribution and state of order which in turn may provide additional constraints on the conditions under which the host diamond grew. For example, cation order of Mg/Fe is believed to be relatively insensitive to pressure (e.g., Angel and Seifert 1999) and therefore could be used as a geothermometer, as has been done for crustal minerals. The specific cation content of a specific site in a mineral structure, such as the excess Si in the octahedral site in titanites (Knoche et al. 1998) or in garnet solid solutions (e.g., Wijbrans et al. 2016; Tao et al. 2018) can provide estimates of the pressure at the time of the growth of the inclusion. The key to the successful use of structural information from inclusions is a precise and accurate structure refinement, which in turn requires the measurement of high-quality X-ray intensity data from the inclusion embedded in the host diamond. This is not a trivial task.

When X-ray studies were first performed on inclusions within diamonds, data was collected on film (Mitchell and Giardini 1953; Futergendler and Frank-Kamenetsky 1961; Harris et al. 1970). The entire process of sample alignment and data collection was very time consuming, taking weeks to complete. Only a few inclusions could be studied which thereby limited the statistical significance of the results obtained. Further, the intensities of diffracted beams collected in this way were of low precision; the mineral species of an inclusion could be identified, but the composition could not be determined with any precision by structure refinement. The development of point detector diffractometers in the 1960s allowed more precise cell parameters to be measured through the use of peak profiling and diffracted-beam centering (Finger and King 1978), and quantitative measurements of diffracted beam intensities to be made. But point detectors are only able to collect one diffraction maxima at a time, meaning that data collections extended over days. And, because they only detect diffraction when the crystal is precisely oriented for a diffracted beam to enter the detector, finding the first few diffraction maxima from an inclusion crystal in a diamond was time consuming (e.g., Glinnemann et al. 2003; Joswig et al. 2003) because routine search procedures typically find the strong diffraction spots due to the diamond, and not the weaker ones of the inclusion (Fig. 1). For studies of inclusions on point detector instruments recourse was frequently made to using film methods to find the first diffraction spots, the same as was done for studies in diamond-anvil cells (Angel et al. 2000). This shows that an area detector such as film is far more efficient for finding diffraction maxima from a crystal of unknown structure and orientation. Therefore, the development of area detectors based on charge-coupled devices (CCDs) and image plates greatly simplified the process. These combined the intrinsic advantages of all area detectors, including film, with quantitative measurements of diffraction peak intensities. Some studies (e.g., Nestola et al. 2011; Howell et al. 2012a) have exploited the best aspects of both types of diffractometer in one study—using the area detector to find the peaks, the point detector to accurately center the inclusion and to determine its' precise cell parameters, and the area detector diffractometer a second time to perform the intensity data collection.

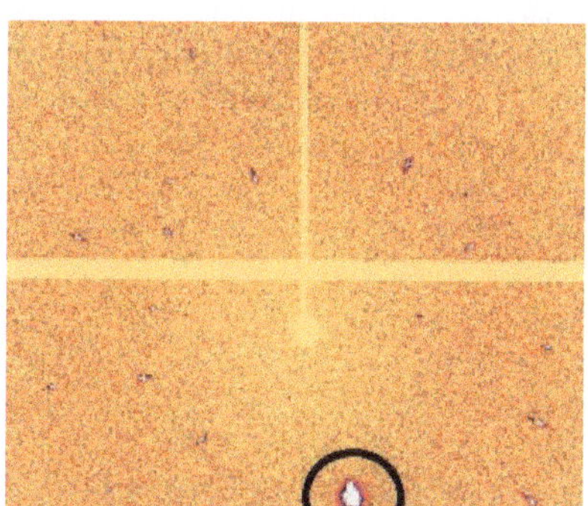

Figure 1. A typical frame image of the diffraction pattern from an olivine inclusion in a diamond. The circled spot is a diamond diffraction spot, while the other weaker spots are from the inclusion. Data collected on a laboratory diffractometer with MoKa radiation from a micro-source, and a Pilatus 200K detector. See Nestola (2015), Nimis et al. (2018) and Wenz et al. (2019) for further examples.

Image plates and CCDs have, however, three disadvantages compared to point detectors. First, they cannot be collimated, meaning that the diffraction maxima from the inclusion crystal may be often overlapped by diffraction from the diamond which often extends over large areas of the detector (Fig. 1). Second, the read-out time with these detector technologies is slow, often significantly greater than the exposure times for a single frame measurement; a 'frame' being a single exposure while the crystal is rotated by a small angle in the beam, the angular change during the exposure being called the 'frame width'. This prevents these detectors being used for data collections with small frame widths such as 0.2° to 0.5° that are necessary for peak profile measurements. Third, data collections with detectors of these two types are further slowed by the need to expose the detector and then to close the shutter while the data is read from the detector. The development of Hybrid Photon Counting (HPC) technology for area detectors (Forster et al. 2019) overcomes these last two problems of other area detectors by allowing the data signal to be read continuously from every pixel in the detector concurrently with the acquisition of diffraction data. This removes the need to stop the data collection to allow the data to be read from the detector, and so allows the data to be collected continuously during the rotation of the crystal, the same as was always done for diffractometers equipped with point detectors based on photo-multiplier tubes. HPC detectors also have effectively zero noise and very large dynamic range, making them ideal for collecting very weak diffraction intensities from small inclusions in the presence of strong diffraction spots from the diamond. These characteristics enables data to be collected with smaller frame widths (and thus shorter frame exposure times), the shortest time and smallest frame width only being limited by the frequency with which the detector is read, which is typically faster than 20 Hz.

These developments in technology, and the development of computing power and software algorithms to take advantage of the much-improved quality of data, have only become available within the last decade. In this section we describe the procedures for X-ray data collection and structure refinement that have been developed to exploit these advances and to obtain the highest-quality structure refinements of inclusion crystals trapped in diamonds. Many of these procedures have been modified from the techniques developed to measure high-quality structures from crystals under pressure in diamond-anvil pressure cells, and the reader is therefore encouraged to also study the methodological chapters in the Reviews in Mineralogy and Geochemistry volume 41 (Hazen and Downs 2000). The procedures described below for alignment of an inclusion on a diffractometer should also be followed even if only the orientation and cell parameters are required; failure to do so will often give results with significant systematic errors.

Single-crystal diffraction experiments

The most critical issue is that each inclusion must be accurately centered on the diffractometer prior to measurement. Optical centering alone is not sufficient because the high refractive index of diamond means that the image in the diffractometer telescope will be defocussed and also displaced if the face of the diamond is tilted even slightly with respect to the camera axis (Miletich et al. 2000). If an inclusion is not centered, the positions of the diffracted beams on the detector will include a contribution from the offset of the crystal. These will be interpreted as a wrong Bragg angle, giving rise to incorrect unit-cell parameters (Fig. 2). And the intensities of the diffracted beams will not be correct, at best leading to incorrect refined structures, at worst preventing structure refinement (e.g., Kunz et al. 2002). We therefore recommend the following procedures that have proved successful in a number of inclusion studies on laboratory sources (Joswig et al. 2003; Nestola et al. 2011; Kueter et al. 2016; Milani et al. 2016) and synchrotrons (Sobolev et al. 2000; Nestola et al. 2016a). A recent innovation at a synchrotron source is to use the much higher beam intensity to perform X-ray micro-tomography to map the inclusion locations in their host diamonds, and thereby center them in the beam for subsequent diffraction measurements (Wenz et al. 2019). In this section a working knowledge of single-crystal diffractometry and diffractometer terminology has been assumed.

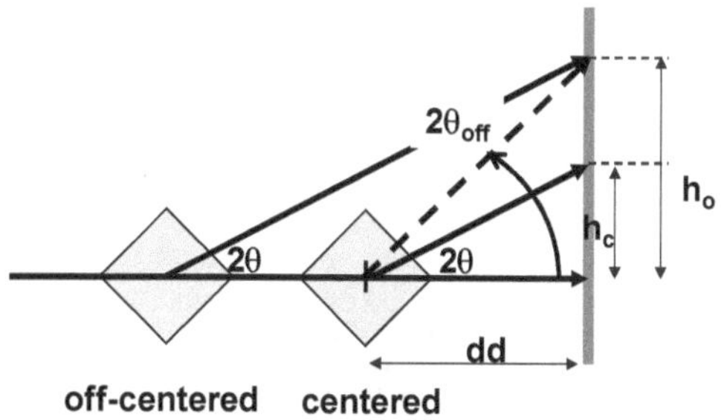

Figure 2. The effect of mis-centering a crystal on a diffractometer. Diffraction from the same plane in a crystal always generates a diffracted beam at twice the Bragg angle (i.e., 2θ) to the incident beam irrespective of the position of the sample crystal. If the crystal is centered the correct Bragg angle is determined as $\theta = \frac{1}{2}\arctan\left(h_c / dd\right)$, where h_c is the distance of the spot on the detector from the direct beam, and *dd* is the detector distance from the center of the diffractometer. The diffracted beam from a crystal displaced towards the X-ray source hits the detector at a greater distance h_o from the direct beam. This position is then interpreted as diffraction coming from the center of the diffractometer at the correct distance (*dd*) from the detector (dashed line) and therefore as a larger incorrect Bragg angle $\theta_{off} = \frac{1}{2}\arctan\left(h_o / dd\right)$.

Alignment. The exact orientation of the diamond on the goniometer head will be adjusted during the alignment procedure so the diamond must be mounted in a soft material like wax or plasticine, or on a goniometer head equipped with arcs. Mount the diamond so that a (111) face, or a polished window, through which you can best see the inclusion is vertical. In what follows, we call this the *viewing face*. If possible, mount the diamond in an orientation such that a ϕ rotation on the diffractometer will not bring a second inclusion into the X-ray beam. Then rotate the diamond to set the *viewing face* exactly perpendicular to the X-ray beam when the diffractometer is driven to zero.

Now drive the diffractometer to the position at which the viewing face is seen edge-on in the telescope. Adjust the physical orientation of the diamond until the *viewing face* is exactly vertical. Now drive the diffractometer to observe the inclusion through the *viewing face*, with this face exactly perpendicular to the telescope axis. Adjust the focus of the telescope to bring the inclusion into focus, and then center the inclusion in the telescope by using the translations on the goniometer head. If it is possible to see the inclusion through an opposite face on the diamond, rotate the diamond by 180°. The inclusion should still be centered, but it will probably be out of focus. If you believe that the inclusion is near the middle of the diamond, adjust both the video microscope focus and the goniometer head translation to bring it to focus. Repeat until the inclusion is in focus at both positions. If the inclusion is not in the middle of the diamond, you have to guess the correct position in depth for now.

This procedure should be sufficient to center the inclusion across the X-ray beam, but the centering along the X-ray beam at this point has relied on the optical focusing method, which in turn assumes that the inclusion lies in the middle of the diamond. That works for diamond-anvil cells in which the two anvils are of equal thickness. It does not work for inclusion crystals that lie away from the center of their host diamond. They will be off the center of rotation of the goniometer and thus as the ϕ or ω axes are rotated, the inclusion will be rotated off the centerline of the X-ray beam. On a conventional diffractometer with a sealed-tube source this is not a problem because the beam width is usually 300 μm or more, so the crystal stays in the beam during rotation and diffracted beam crystal centering (described below) can be used immediately to determine the crystal offsets which can then be corrected by moving the diamond with the translation slides on the goniometer head. Such a procedure works even if the inclusion cannot be seen optically (Joswig et al. 2003).

On a modern micro-focus laboratory diffractometer or a synchrotron beamline the full-width at half-maximum intensity (FWHM) of the beam at the crystal position may be 120 μm or significantly less. An inclusion that is not centered on the goniometer will be moved out of the beam during rotation, and it will either stop diffracting (if it is small compared to the beam diameter) or show reduced intensities if it is of similar size to the beam (e.g., see Angel et al. 2016). This phenomenon can be used to further improve the centering on diffractometers equipped with area detectors. Perform a rapid intensity data collection with a single ϕ-scan from 0 to 360°, with a step size of 3–6° per frame. This only takes a few minutes on a laboratory diffractometer equipped with a micro-source and a fast detector (e.g., HPC or CCD). The diffraction pattern should then be indexed and inspected for the disappearance of diffraction from the sample as it is rotated, which indicates that the inclusion has been moved out of the beam.

With the goniometer head translation slides, move the diamond across the beam by a reasonable amount (e.g., by the known FWHM of the beam) and repeat the fast data collection. Repeat this procedure until indexed diffraction spots from the inclusion are present throughout a 360° ϕ-scan. Final centering is then performed by integrating the intensities from the last of these ϕ-scans and observing the frame scale factors. The frame scale factors are determined by minimizing the internal agreement factor of the intensity data and they represent the relative deficit or surfeit in diffraction intensity on each frame (Angel et al. 2016). The measured intensities depend not only on how much of the inclusion crystal is in the beam, but also on the absorption path lengths in the diamond and the inclusion itself. For small silicate or graphite inclusions common in diamonds, the effects of self-absorption in the inclusion vary only slightly with orientation and can be ignored for this step. When an inclusion crystal is centered in the beam (Fig. 3a) the variation in scale factors as the crystal is rotated is only due to the changing absorption path lengths in the diamond and can be estimated quickly from the maximum and minimum dimensions of the diamond (Table 1). For extremely non-equant shapes like plates, the data collected in positions of high diamond absorption, and thus large-scale factors, should be ignored (Fig. 3b). When a mis-centered inclusion crystal moves out of the center of the beam during rotation on the goniometer, it diffracts less intensity so the frame scale factors are modified in combination with the effects of diamond absorption (compare Fig. 3c with 3a). The diamond should now be moved along the beam (when the diffractometer is at zero with the viewing face perpendicular to the beam) and the data collection repeated until the range in frame scale factors is reduced to that expected for the diamond shape (Table 1, and Fig. 3a,b). Normally, at most two iterations of this process of adjustment and 360° ϕ intensity data scan are required. We note at this point that it is critical to analyze the data with software that can immediately and automatically index the diffraction pattern of the inclusion even in the presence of the strong diffraction from the diamond. We use the program Crysalis Pro™ (Rigaku-Oxford-Diffraction 2016) for this purpose, which is available free on request from Rigaku Oxford Diffraction at their discretion. It can be used to analyze diffraction images from a wide variety of area detectors on both laboratory diffractometers and synchrotron beamlines.

Simple geometrical considerations indicate that the precision with which the inclusion can be centered by observing the frame scales depends on the FWHM of the beam and the size, shape and orientation of the crystal on the goniometer head. Typically, these procedures are normally sufficient to reduce the inclusion offset from the middle of the diffractometer to less than half the full width at half maximum of the X-ray beam, thus maybe an offset of < 40 μm on a laboratory diffractometer with a micro-focus source. Further precision in unit-cell parameters and the integrated intensities can be attained by eliminating the effects of these residual offsets by using the method of 'diffracted beam centering' (Hamilton 1974; King and Finger 1979; Angel et al. 2000). On a conventional point detector diffractometer this consists of measuring the same Bragg reflection from a crystal in eight different orientations on the diffractometer.

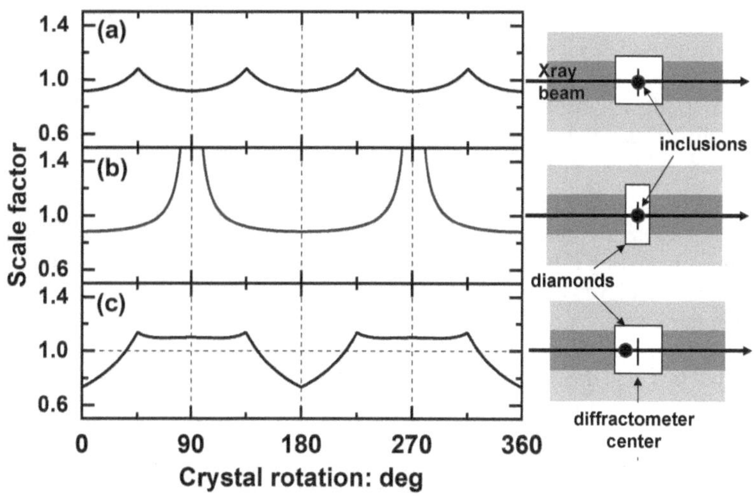

Figure 3. The effect of diamond shape and crystal offset on frame scale factors determined from diffracted intensities. Frame scale factors for a centered inclusion depend only upon the changing absorption by the diamond host, as the diamond is rotated. **(a)** Octahedral diamond, **(b)** diamond plate. Typical ranges of frame scale factors are listed in Table 1. **(c)** The frame scale factors for an off-centered inclusion show a more complex pattern because the crystal moves into areas of lower X-ray intensity as it is rotated.

Table 1. Absorption scale factor ratios for diamonds.

Shape	Dimensions (mm)	Scale factor ratio (max/min)
	Edge length	
Regular octahedron	5	2.10
	2	1.34
	1	1.16
	0.5	1.08
	Thickness × edge length	
Square plate	0.2 × 2	1.44
	0.5 × 2	1.35
	0.5 × 5*	2.5
	0.5 × 10*	6.8

Notes: Calculated for MoKα X-radiation (λ = 0.7093 Å), for which the linear absorption coefficient of diamond is 0.203 mm^{-1}. The scale factor ratios will be smaller for shorter wavelengths due to the reduced absorption coefficient. *Data from inclusions in diamonds of these shapes should be collected in fixed-ϕ mode (Angel et al. 2000).

From the differences in the peak positions (e.g., Fig. 2), the crystal offsets from the center of the diffractometer can be determined and these can be corrected by adjustment of the crystal prior to intensity data collection. These procedures have been standard on point detector diffractometers for many decades (e.g., Finger and King 1978; Angel and Finger 2011). On area detector diffractometers duplicate measurements of the same reflection appear on different image frames within the data collection. The corrections for small displacements of the crystal are therefore made during the analysis of the data after the full intensity data collection.

Intensity data collection. The general requirements for an intensity data collection on any diffractometer, whether equipped with a point detector or an area detector, are basically the same. Data should be collected, if possible, in orientations that minimize the absorption in the diamond host crystal. For approximately equant diamonds, such as octahedra, no useful optimization of diffractometer positions is possible. But data from inclusions within diamond plates (or polished mineral sections) should be collected in the same 'fixed-ϕ mode' that is used for DACs (Angel et al. 2000).

Each peak must be collected in a way that allows its profile shape to be determined, because this yields both more accurate unit-cell parameters, more accurate integrated intensities, and allows peaks affected by diffraction from the diamond to be identified. On a point detector instrument this is achieved by making the steps of the ω-scan through each peak significantly smaller than the intrinsic peak width (which depends on both the inclusion crystal and the divergence of the X-ray beam). Typically, a step size of 10–20% of the FWHM of the peaks is optimal, but this also depends on the methods used in the data reduction software to fit the data. With an area detector the minimum frame width is often restricted by motor speeds or detector read-out times, but it must be set so that each diffraction peak appears on several consecutive frames. For HPC detectors on laboratory micro-sources with significant beam divergence, we find that a frame width of 0.2–0.5° gives excellent results when processed in the Crysalis-Pro™ (Rigaku-Oxford-Diffraction 2016) software.

The coverage of the data collection should be as high as possible. The resolution (the highest Bragg angle to which data is collected) must be sufficiently high that a change in the resolution of the data does not significantly affect the structural parameters of interest. For critical cases this can be determined from test refinements to test datasets from crystals of the same mineral in air, as we describe below. But the basic guideline is that data to a resolution of ~0.7 Å is sufficient for routine structure identification and determination of bond lengths and angles. For the determination of reliable site occupancies in olivines a resolution of ~0.46 Å is required, corresponding to $2\theta_{max} = 100°$ for Mo Kα radiation. In order to obtain reliable frame scales, we collect data with an average redundancy of 6, with the frame exposure times set to obtain an average $I/\sigma(I) = 25$ for all reflections. With a small detector the optimal strategy is usually to collect at both positive and negative Bragg angles, and to set the exposure time at high angles to four times that at low angles. On a commercial diffractometer the software can be used to determine an optimal data collection strategy that meets these requirements. As a guide, on an early-generation Oxford Diffraction SuperNova diffractometer equipped with a Mo micro-source X-ray tube and a Dectris Pilatus-200 K detector a data collection on an olivine inclusion to these requirements typically produces more than 2,500 frames of data in 20 hours that yield in excess of 10,000 diffraction peak intensities (Angel et al. 2016). Improvements in source and detector technologies mean that this data collection time is significantly reduced on more modern diffractometer systems, and the same is true when data are collected at a synchrotron beamline because of the higher X-ray intensities.

Intensity data reduction. In order to obtain the highest-quality intensity data it is important to eliminate from the data both the effects of Bragg diffraction from the diamond, and the diffuse background (Fig. 1). This is achieved for both point detector and area detector data by fitting the peak profiles of every single diffraction peak from the inclusion crystal. The low background of diffuse scattering and the tails of the diffraction peaks of the diamonds is eliminated with a combination of local and global background refinement when the peak profiles are fit. Then the profiles are examined and those with unusual shapes are deleted on the assumption that they are affected by overlap with diffraction from either the diamond or another inclusion or they are affected by some other problem. The remaining profile parameters are then used to determine the 'normal' peak profile parameters for the sample crystal, and these are then applied to the fitting of the weaker peaks to obtain more precise intensities for them.

These procedures have been established for several decades as the 'learned-profile' method of intensity integration for point detector instruments (e.g., Diamond 1969; Angel 2003), and are incorporated into commercial software for area detectors. Experience shows that even with the advanced algorithms now available for eliminating peaks with poor profiles, it is necessary to check the results of the integration carefully. We find that the 'profilecorrint' plot provided in Crysalis-Pro™ (Rigaku-Oxford-Diffraction 2016) is a sensitive indicator of residual problems in the peak profiles. The frame scale factors should also be inspected to see that they lie within the range expected given the shape of the diamond host (Table 1). For olivine inclusions, a good quality dataset has an overall R_{int} less than 0.04, with $R_{int} < 0.02$ for low-angle data and $R_{int} < 0.10$ for high-angle data, with the average $F^2/\sigma(F^2) > 8$ in all angle ranges.

On an area-detector diffractometer the orientation of an inclusion can be determined within 1–2° by using peak positions found by a simple peak hunt in the dataset but the unit-cell parameters from such a process can be in error by several %. On the other hand, the integration provides peak positions from the fitting of the peak profiles of only those reflections unaffected by overlaps with diamond reflections. These peak positions should then be used to determine not only the crystal offsets and other diffractometer aberrations, but also the unit-cell parameters and orientation unaffected by the offsets and aberrations.

The data should then be averaged in a way to eliminate reflections with intensities that are inconsistent with their symmetry equivalents. These may be due to poor frame scaling or overlaps in the data frames from diamond reflections or reflections from other inclusions; rejection on the basis of peak profiles does not always catch all weak overlaps. Further, the diamond can also reduce the intensity of the inclusion peaks by diffracting away part of the incident beam, or by diffracting the diffracted beam from the sample (Loveday et al. 1990). These types of events do not change the diffraction profile, but only lower the intensity, and they can only be eliminated by averaging. We recommend detecting and eliminating affected reflections by applying the Blessing (1987) criteria for outlier rejection first to the low-intensity outliers, prior to averaging the remaining data in the normal way. The Average program (Angel 2004) is freely available for this task.

Structure refinements of inclusions

With the procedures described above for the collection and reduction of X-ray intensity data from inclusions, the data quality can be similar to that from a crystal in air and significantly better than for data collected from crystals at high pressures in DACs for which the reciprocal space access restrictions of the DAC mean that the datasets from one crystal are typically incomplete (Miletich et al. 2000). Therefore, it is possible to proceed with structure solution (if required) and routine refinement in exactly the same way as for data collected from crystals in air. Such procedures have yielded the structures of inclusions in lithospheric diamonds of olivine (Nestola et al. 2011; Kueter et al. 2016), garnets (Joswig 2011; Kueter et al. 2016), pyroxenes (Joswig 2011; Nestola et al. 2016a) and a high-pressure phase of $CaSiO_3$ (Joswig et al. 2003; Brenker et al. 2021). The quality of refined structures is sometimes improved by the use of robust-resistant weighting (Prince and Collins 1992) to identify and eliminate the influence of any remaining outliers in the intensity dataset. Secondary extinction effects in inclusion crystals seem to be surprisingly strong and should be corrected with a physically-realistic model such as that of Becker and Coppens (1974).

The determination of the composition of an inclusion is often the motivation for performing a structure refinement. For simple solid solutions such as (Mg,Fe)O the increase in cell edge and volume with increasing Fe content gives an initial indication of composition. But this is a minimum estimate of Fe content because the unit-cell parameter of the inclusion is also decreased from its room-pressure value because the inclusion is under pressure. The true composition of the inclusion can be determined *in situ* by structure refinement, in which the Fe/Mg ratio of the only cation site in the structure is refined (Nimis et al. 2018). For more complex structures in which the cations can be distributed over two or more sites, extra care has to be taken to use the appropriate refinement models established for the mineral species, and to verify that the model returns valid results for intensity datasets collected from inclusions.

Olivine. We use data collected from a suite of olivine inclusions in diamonds from the Udachnaya kimberlite to illustrate the further precautions that may be required to determine by structure refinement the composition of a mineral that is a solid solution. These inclusions are a subset of those whose orientations were determined and analyzed by Nestola et al. (2014). The composition of these olivines is required in order to estimate their residual inclusion pressures from their unit-cell volumes measured while still trapped in their diamond hosts, because the room-pressure volume of mantle olivines varies significantly with composition.

The most important point to remember while performing structure refinement is that an artificial model of the scattering power of the atoms in the crystal structure is being refined to the intensities of the Bragg diffraction peaks. The model is artificial because it considers the crystal to be perfect, without defects, and that all unit-cells within the crystal are identical in structure and composition. This is not true in a solid solution, such as olivine or garnet. Further, the independent atom model (IAM) most used for conventional structure refinement considers the atoms as independent scatterers of radiation, so there is no representation of the bonding electrons which also contribute to the scattering of the X-rays. While the IAM will certainly provide the correct structural architecture for the inclusion and quite precise and accurate bond lengths and angles it often results in incorrect site occupancies for cations (e.g., Angel and Nestola 2016). Thus, any refinement model must be first tested by refinement of the structure of a crystal of the same mineral of similar and known composition. We have found that the most reliable structure model is different for different mineral groups, but extensive reading of the crystallographic literature will provide good indications of the most appropriate model to use on most mineral groups, and its limitations.

Chemical analysis of olivine inclusions in lithospheric diamonds has shown their chemistry to be dominated by Mg and Fe, with trace amounts of Ni, Ca, and Mn (e.g., Stachel and Harris 2008). However, the total scattering power of each site in the structure is proportional to the number of electrons associated with the cation or mixture of cations occupying the site. Thus, it is impossible to determine how the several cation species are distributed across the M1 and M2 sites in the olivine structure. Cations therefore have to be put together in groups of similar number of electrons and, hence, scattering power. Trace amounts of Ni and Mn can be modelled by Fe, while the Ca content is effectively modelled by Mg and Fe equally. Structure refinements were then performed with just Mg and Fe allocated to the M1 and M2 sites and the tetrahedral site was assumed to be fully occupied by Si. The effective Mg number from the structure refinement is then simply the average of the refined Mg occupancies of the M1 and M2 sites, and it should correspond to the value of

$$Mg^* = (Mg + 0.5Ca) / (Mg + Fe + Mn + Ni + Ca) \quad (1)$$

determined from independent chemical analysis of the olivines.

Refinements of the structures of a crystal of known composition that is similar to those of inclusions and a crystal of pure Mg_2SiO_4, both measured in air, and DFT simulations of pure forsterite were combined to develop a robust refinement model. Conventional structure refinements using scattering factors for spherical atoms (the IAM) yield cation site occupancies for all three cases that are systematically in error (Angel et al. 2016; Angel and Nestola 2016). This can be attributed to the failure of the IAM to account for scattering by the bonding electrons (Kroll et al. 1997) and it also leads to the refined cation site occupancies being strongly correlated with both data resolution and the values of the atomic displacement parameters.The 'electron in bond' model (Kirfel and Will 1980; Heinemann et al. 1999) which models the bonding electrons by an additional 1 electron scatterer placed in line between the Si and O positions removes much of these correlations. In combination with robust-resistant weights in the least-squares, the data collection procedures and structure refinements of olivine inclusions yield values of Mg* in agreement with microprobe analyses of the same crystals (Angel and Nestola 2016).

These sorts of refinement tests also allow the optimal data collection parameters including resolution, coverage and redundancy, for a given type of inclusion to be determined prior to measurement. Single-crystal X-ray intensity datasets were then collected from 13 olivine inclusions in 5 diamonds with an Oxford Diffraction SuperNova diffractometer equipped with a Mo micro-source and a Dectris Pilatus-200K area detector, with the procedures that we described above. The number of averaged reflections used in the structure refinement ranged from ~1,000 to ~1,600, the number being strongly correlated with the average 'signal to noise' of the data $<F^2>/<\sigma(F^2)>$ which reflects differences in inclusion sizes and hence total scattering power. Longer data collections on the same instrument, or data collections with a more modern instrument or more intense synchrotron source would suppress these trends. Quality indices for the refinements ranged from $R_1 = 0.017$ and $wR_2 = 0.026$ for a dataset with average $<F^2>/<\sigma(F^2)> = 35$, to $R_1 = 0.071$ and $wR_2 = 0.095$ for a dataset with average $<F^2>/<\sigma(F^2)> = 11$. Uncertainties on bond lengths vary in the same way, for individual Si–O bonds from 0.0003 Å to 0.002 Å and those of M–O bonds are all better than 0.001 Å. The refined structures pass the most sensitive quality test for silicate structures; that the root-mean-square displacements of the Si and O atoms towards one another, that are calculated from refined anisotropic displacement parameters (Trueblood et al. 1996), are consistent with the strong bonding of the SiO_4 tetrahedra (Downs et al. 1990; Kunz and Armbruster 1990; Angel et al. 2016).

Figure 4 shows that the refined compositions of these olivine inclusions fall into the middle of the range of all chemically-analyzed olivine inclusions in cratonic diamonds (Stachel and Harris 2008). Further, while there are differences between the compositions of olivines in different diamonds, the compositions of all of the inclusions in any one diamond are the same within the estimated uncertainties (which come directly from the least-squares refinement of the structure). This result is consistent with the conclusion that some of these inclusions, such as all four in diamond 9 and inclusions 13B and 13C, are remnant pieces of the same original olivine crystals in the mantle at the time of diamond growth (Nestola et al. 2014). The geological reasonableness of the results summarized in Figure 4 demonstrates that the difficulties of data collection and structure refinement can be overcome to yield crystal structures of inclusions of a quality comparable to those obtained from samples measured in air.

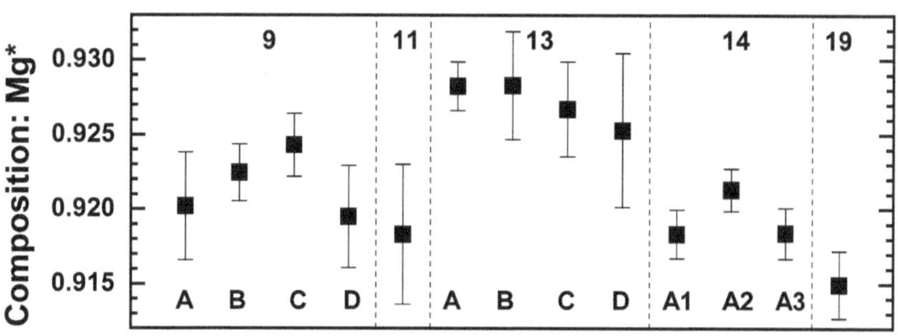

Figure 4. Compositions of 13 olivine inclusions from diamonds in the Udachnaya kimberlite as determined by *in situ* single-crystal X-ray diffraction. Mg* is the effective Mg number as defined in Equation (1). Numbers at the top designate the diamond hosts and the letters the individual inclusions as listed in Nestola et al. (2014). Vertical dashed lines divide inclusions from different host diamonds. The unit-cell volumes of the same inclusions are shown in Figure 18.

DIAMOND GROWTH MECHANISMS: ORIENTATIONS OF INCLUSION

The idea that the orientation of mineral inclusions within their host diamonds may imply something about whether the inclusion minerals grew prior to, or during, diamond growth dates back at least to the 1950s. The reasoning was that if an inclusion in diamond shows an epitaxial relationship with its diamond host, then it must have grown on a pre-existing free surface of the diamond. In this scenario, in order for the inclusion to be trapped, it is argued that diamond growth must have continued after growth of the inclusion crystal and therefore the diamond and inclusion growth occurred simultaneously. This would then constitute proof that the inclusions were syngenetic with the diamond and therefore geochemical analysis and dating of the inclusions would indicate the conditions at the time of the growth of the diamond and the age of the diamond. This logic is flawed; an epitaxial relationship could also result from a diamond nucleating and growing on a pre-existing mineral leading to a protogenetic inclusion.

A defining feature of 'epitaxy' is that the two minerals exhibit a very specific and crystallographically-simple orientation relationship that arises because of structural matching and bonding across the contact plane between them. The periodic nature of crystalline structures means that an orientation relationship that arises from epitaxy is exact. If epitaxy was an active process in the entrapment of a mineral phase in diamond, then it should result in a large number of inclusions of that mineral showing *exactly* the same crystallographic orientation with respect to their diamond hosts. Early studies reported the orientations of a small number of olivine, garnet and magnesiochromite inclusions (e.g., Mitchell and Giardini 1953; Futergendler and Frank-Kamenetsky 1961; Zyuzin 1967) relative to their host diamonds. Despite not meeting the strict requirements for epitaxy these orientations were then interpreted to reflect an epitaxial relationship (Hartman 1954; Futergendler and Frank-Kamenetsky 1961; Zyuzin 1967). However, further measurements showed that specific orientations between inclusions and diamond are not a common phenomenon (Harris et al. 1967; Harris 1968). And subsequent analysis showed that the claims made by Futergendler and Frank-Kamenetsky (1961) for structural matching between individual planes of the structures of olivine and diamond were crystallographically incorrect and physically impossible (Nestola et al. 2014). Nonetheless, despite the limited number of supporting data and the long-standing errors in crystallographic interpretation, the claims for epitaxy between diamond and its inclusions have long been used as evidence in favor of their syngenesis (e.g., Orlov 1977; Harris and Gurney 1979; Pearson and Shirey 1999). This was in large part due to the failure to clearly distinguish between the *measurements of orientations* and their interpretation in terms of the *genetic mechanisms* which may have created them, with the result that the interpretation in terms of epitaxy and syngenesis was promoted at the expense of clear and statistically-meaningful data evaluation.

In recent years, the advent of improved instrumentation has allowed many more inclusions to be measured *in situ* allowing statistically significant conclusions to be drawn about their crystallographic orientations relative to their host diamonds. We first describe the methods of measurement and data analysis, which yield the raw data on the orientation relationships exhibited by a given mineral when it occurs as an inclusion. Then we use several examples to show how different types of relative orientations can be used to suggest possible mechanisms for the entrapment of the inclusions by their host diamonds. These mechanisms often have implications for the environment in which the diamond grew. We also note that orientation information is also required if one wants to interpret the measured stress of inclusions of non-equant shapes in terms of entrapment conditions, as we discuss in the following section of this chapter.

Methods

Orientation measurements. The positions of the diffracted beams from a crystal (e.g., Fig. 1) on the detector of a diffractometer depend not only on the unit-cell parameters of the crystal and its symmetry, but also on the physical orientation of the crystal on the diffractometer; if the crystal is rotated the positions of the 'spots' on the detector will also be moved.

Therefore, the orientation of an inclusion crystal relative to the diffractometer can be determined by measuring its diffraction pattern and indexing it. This yields the 'orientation matrix' (Busing and Levy 1967) which describes the position of the diffraction spots in reciprocal space relative to the diffractometer. The orientation matrix defines both the unit-cell parameters of the crystal and the orientation of the unit-cell axes with respect to the external reference system of the diffractometer. The orientation of the inclusion relative to its host diamond can therefore be determined by measuring the diffraction patterns of both crystals simultaneously. Their respective orientation matrices will each define their orientations relative to the diffractometer, and appropriate mathematical manipulation of these two matrices allows the relative orientation of diamond host and inclusion to be determined (Angel et al. 2015b).

In order to determine the orientations of inclusions by single-crystal diffraction it is important that the inclusion is well-centered on the diffractometer by the methods described in the previous section '*Single-crystal diffraction experiments*'. Failure to accurately center the inclusion will result in incorrect peak positions (Fig. 2) and thus small errors in the orientation measurements and, perhaps more importantly, it will lead to difficulties in indexing the diffraction patterns as the inclusion is rotated out of the beam. Once the inclusion is centered, the host diamond adjacent to the inclusion will also contribute peaks to the diffraction pattern (Fig 1). In order to obtain the orientation data we recommend performing a relatively fast data collection scan with a small frame width which allows the peak positions to be determined by peak-profile fitting, as is done for integrating the intensities for a structure refinement (see previous section). If enough data are collected, then the offset of the inclusion from the center of the diffractometer can be calculated from the data and its effects on the orientation matrix eliminated (Hamilton 1974; King and Finger 1979).

Diamond spots appear together with the diffraction from the inclusion, but the orientations of both crystals can normally be determined unambiguously by indexing. For a small crystal in air, well-centered on the diffractometer, the uncertainty in orientation determined in this way is significantly less than 0.1°. The relatively large size of the diamonds results in large diffraction spots, whose size may result in an uncertainty in orientation of the diamonds alone of up to 1.5°. A practical test of the precision of orientation measurements is simply to remount the diamond in several different orientations on the diffractometer and remeasure the same inclusion each time. Nestola et al. (2014) showed that the uncertainties in the relative orientations of olivine inclusions relative to their diamond hosts is no more than 2°. The orientation of magnesiochromite inclusions, which have smaller unit cells than olivine and therefore fewer diffraction spots, are reproducible to better than 4° (Nimis et al. 2019). If the inclusions are oriented such that their reflections mostly overlap with those of the diamonds (e.g., Nimis et al. 2018), then the determination of the exact peak positions is difficult, but the nature of the orientation relationship will be obvious even if it cannot be measured precisely.

EBSD measurements can be made on partially exposed inclusions to determine their orientation with respect to their diamond hosts, because the polishing of the diamond and exposure (and consequent partial stress release) of the inclusion is not expected to rotate the inclusion by more than 1–2°. A single EBSD measurement measures the orientation of a small volume of the inclusion corresponding to the size of the electron beam. This contrasts with an X-ray diffraction measurement that provides the average orientation of the entire single-crystal inclusion. EBSD measurements should therefore be performed on many points in the inclusion. If the EBSD measurements show that the entire inclusion has the same orientation, then it should be used as a single datum in subsequent analysis. Orientations of crystals measured by EBSD are normally expressed in terms of Euler angles which characterize the rotations required to bring the laboratory or instrument axial reference system into coincidence with the crystallographic axes. Or vice-versa. Care therefore has to be taken to use the appropriate conventions for the definition of the Euler angles when interpreting data from any instrument. Once these conventions are known, the Euler angles can be used to calculate an orientation matrix, equivalent to that determined by single-crystal X-ray diffraction measurements, if that is required (Angel et al. 2015b).

Orientation analysis. Both EBSD and single-crystal diffraction yield the orientation of the diamond and the inclusion relative to an instrument coordinate system. These descriptions of the orientations of crystals relative to the laboratory reference axes are not unique, because of the symmetry of the crystal. For example, olivine has orthorhombic symmetry, which makes the direction of the positive *a*-axis symmetrically-equivalent to the opposite direction. Because these directions are symmetrically equivalent it is impossible by diffraction or any other physical measurement method to distinguish them. Thus, for example, an olivine described as having its '*a*-axis' vertically upwards on the instrument could be equally well described as having the same axis vertically downwards (Fig. 5). The orthorhombic symmetry of olivine therefore leads to 4 symmetrically equivalent and indistinguishable ways of indexing its diffraction pattern, and hence 4 symmetrically equivalent and indistinguishable descriptions of the orientation of one olivine crystal on the diffractometer or in an EBSD instrument. Figure 6a shows the directions of the *a*-, *b*- and *c*-axes of one olivine inclusion relative to its host diamond for all of these 4 symmetrically equivalent ways in which the olivine diffraction pattern can be indexed. It can be seen in Figure 6a that each crystallographic axis has two possible orientations, one in the upper hemisphere (closed symbol) closer to the **+c** direction than the **−c** direction of the host diamond, and one in the lower hemisphere (open symbol) closer to the **−c** direction of the diamond. Figure 6a therefore shows that even if all of the olivines in one diamond were measured with the diamond in the same orientation on the diffractometer, their orientations could appear different if this ambiguity caused by symmetry is not taken into account. A similar ambiguity occurs for indexing the diffraction patterns of the diamond host crystals; being cubic, the choice of how to label the three perpendicular crystallographic axes *a*, *b*, or *c* is arbitrary and this alone generates 24 equivalent relative orientations between the olivine inclusion and diamond (Fig. 6b). In combination the symmetry of the two crystals generates the 96 equivalent relative orientations shown in Figure 6c. Failure to account for this indexing ambiguity can result in olivine orientations relative to the diamond appearing more random than they really are. Data from multiple inclusions that resemble Figure 6c appear random, but in fact the inclusions have identical orientations when the symmetry is considered.

OrientXplot (Angel et al. 2015b) is a freely-available (*www.rossangel.net*) analysis program which addresses this problem by using the crystallographic symmetry of the inclusion and the diamond to generate all possible symmetry-equivalent relative orientations of the pair of crystals that are consistent with the diffraction patterns. All of the stereograms shown in this chapter were created directly with OrientXplot. The symmetry of both host and inclusion are entered into OrientXplot by the user, so that the effects of symmetry can be explored and accounted for. The program reads and correctly interprets the orientation matrices from all major commercial single-crystal diffractometers (Oxford Diffraction, Rigaku, Bruker, Stoe) so that data from different instruments can be analyzed together. User-defined conventions can also be used to analyze data collected on non-standard instruments such as often found on synchrotron beamlines. Software for EBSD measurements normally plots the data on a stereogram that is based on the instrument coordinate system (e.g., see Neuser et al. 2015) and we are not aware of any such software that accounts for the ambiguity due to symmetry. Therefore, OrientXplot accepts EBSD data as either Cartesian rotation matrices or as sets of Euler angles (with various conventions). Full details of input data formats, the mathematical methods used by the program, and worked examples are provided in Angel et al. (2015b) and the program documentation. Note that the common practice of plotting the poles of just the crystallographic axes of the diamond and the inclusion relative to the instrument coordinate system (e.g., Neuser et al. 2015; Davies et al. 2018) does not allow data from different diamond hosts to be quantitatively compared, and will often obscure the existence of anything but very specific orientational relationships. To clearly see the orientation relationships of a suite of inclusions to their hosts, OrientXplot can therefore plot the relative orientations of different inclusions in different diamond hosts simultaneously on the same stereogram.

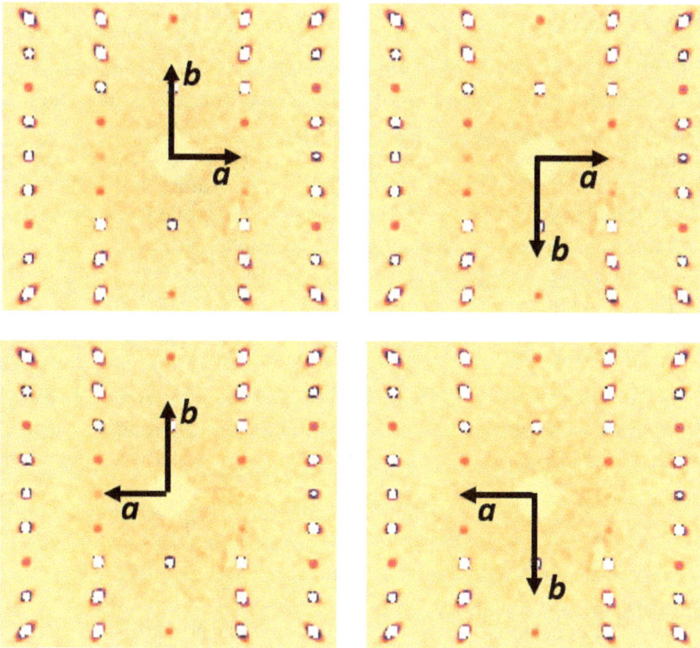

Figure 5. Four copies of the *hk0* section of the X-ray single-crystal diffraction pattern of an olivine. Because olivine has orthorhombic symmetry, the diffraction pattern can be indexed in four indistinguishable and entirely equivalent ways, as indicated by the axes on each diffraction pattern. The direction of the *c*-axis is fixed by the requirement that the crystallographic axes form a right-handed set. These four orientations are plotted in the stereogram of Figure 6a.

Figure 6. Stereograms created with OrientXplot (Angel et al. 2015b) illustrating symmetry-equivalent relative orientations of one single olivine inclusion in diamond. The orientations are illustrated by plotting the directions of the *a*-, *b*- and *c*-axes of olivine (red, green and blue respectively) relative to the crystallographic axes of the diamond host which form the basis of the stereograms. Poles in the upper hemisphere are solid symbols and open symbols indicate poles in the lower hemisphere. All stereograms are equal-angle projections. These conventions apply to all subsequent stereograms. Part (a) shows the 4 equivalent orientations of the olivine relative to the diamond arising from different indexings shown in Figure 5 of the olivine diffraction pattern alone. Part (b) shows the 24 equivalent relative orientations of the olivine arising from different indexings of the diamond diffraction pattern alone, and (c) both effects together; in this stereogram the solid poles overlie additional poles in the lower hemisphere.

For olivine inclusions, the ambiguity caused by symmetry means that the 96 symmetrically-equivalent relative orientations (Fig. 6c and Nestola et al. 2014) are all equally-valid descriptions of the orientation relationship between the diamond and the inclusion. There are several ways to choose which individual relative orientation to plot for analysis of the relative orientations of several inclusions of the same mineral. One is to choose the orientation which puts one particular axis of the inclusion within the asymmetric unit of the mutual orientation space of the two crystals. Another is to choose the relative orientation that puts one particular direction in the inclusion as close as possible to a direction in the diamond host. The criteria should be chosen carefully so as to test hypotheses about the possible relative orientations of inclusions and diamonds; the use of different choices is illustrated in the examples described below. Provided that the criteria always define one, no more and no less, of the possible symmetrically-equivalent orientations, the choice of which one to display does not affect the conclusions that are drawn from the resulting stereographic projections. If one choice indicates a strong orientation relationship, then all choices will do so. If there is no preferred orientation, then this will be evident in all plots of all valid choices of orientation pairs.

Crystallographic orientation relationships (CORs). The historical literature on the orientations of inclusions in diamonds has been confused for 70 years by the conflation of orientation measurements with their analysis and interpretation. For example, the word 'epitaxy' implies both a specific type of orientation relationship *and* the mechanism by which it was created. This has generated confusion over what the phrase 'orientation relationship' between inclusions and diamond means. Clarity has been brought to the subject by the concepts and terminology of Crystallographic Orientation Relationships (CORs) between a host and its inclusions (Griffiths et al. 2016, 2017; Habler and Griffiths 2017). A COR describes a type of orientation relationship between a suite of inclusions and their hosts. Different classes or types of COR are distinguished purely by geometrical criteria and, in particular, by their degrees of freedom. The COR concept is a data analysis tool and has no intrinsic genetic significance.

A *specific COR* describes pairs of crystals which have a specific and fixed orientational relationship to one another, for example as is found for true examples of epitaxial growth of one structure upon the substrate of another. A specific COR has no degrees of freedom, meaning that all inclusions belonging to the COR will have exactly the same orientation. Once allowance has been made for the ambiguity of indexing, poles of the crystallographic axes of the inclusions will all plot in exactly the same place in a stereogram based on the diamond host (Fig. 7a). Such CORs are well-known in minerals as the result of the solid-state exsolution of one mineral from another. One such example is the specific orientation of clinopyroxene grains within garnet inclusions in sub-lithospheric diamonds (Harte and Cayzer 2007) which are formed on decompression due to the decreasing capacity of the garnet to retain the Si associated with its majoritic component. Inclusions demonstrating a *rotational COR* have one specific direction of the inclusion parallel to one direction of the host diamond, but other directions will appear random as a result of rotation about the common axis. Crystallographic directions perpendicular to the common axis will plot on a single great circle in a stereogram (Fig. 7b). Rotational CORs therefore have one degree of freedom, the rotation around the common axis. *Random CORs* have two degrees of orientational variation, which is equivalent to saying that there is no specific orientational relationship between the inclusions and their host diamonds, which is immediately apparent in the scatter of the plots of crystallographic directions in a stereogram (Fig. 7c). Real measurements of inclusion orientations include experimental uncertainties, so the ideal cases of specific and rotational CORs are never obtained exactly, but only approximately. In addition, post-entrapment deformation of the diamond or modification of the inclusions may further scatter the orientation data.

The use of the COR concept does not imply anything about the mechanism by which the COR is developed for a particular type of mineral inclusion; it only describes the orientations of the inclusions. This approach, which we illustrate in the following examples, allows the analysis of inclusion orientation data to proceed in two well-defined and separated steps. First the type of COR or CORs is determined on the basis of statistical tests on the data, and these are generally not subject to significant uncertainty unless the number of data are not sufficient.

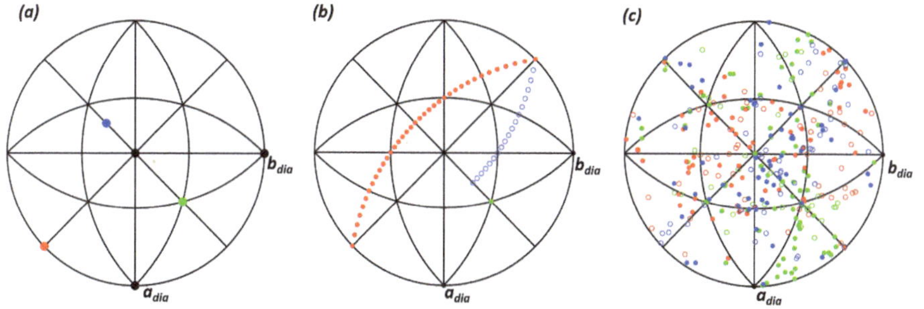

Figure 7. Stereograms illustrating different types of COR. **(a)** A specific COR with all of the inclusions in exactly the same orientation relative to the host. The poles of the crystallographic axes of all of the inclusions plot in the same place. The example has [010]inc // [111]diamond and [100]inc // [1$\bar{1}$0]diamond. **(b)** A rotational COR based on a rotation about the common axis [010]inc // [111]diamond. The poles of inclusion axes perpendicular to the rotation axis lie on a great circle. The poles of one-half of the *c*-axes are obscured by the poles of the *a*-axes in this plot. **(c)** A completely random COR illustrated by 113 published modern orientation measurements of olivine inclusions.

Then, once the statistical significance of the orientation measurements is clearly established, it can subsequently be interpreted in terms of the growth mechanisms of the diamonds and inclusions. If such interpretations subsequently prove to be incorrect, the underlying information about the COR remains valid and open to further alternative genetic interpretations. The COR concept also avoids the bias implicit in many earlier studies which often looked for evidence of specific CORs rather than describing the full statistical distribution of inclusion orientations present.

Examples

Olivine, clinopyroxene and garnet inclusions. Mitchell and Giardini (1953) in perhaps the first measurement of the orientations of inclusions in diamonds by diffraction showed that two elongated olivine inclusions in two diamonds had an orientation (010)olivine // (111)diamond and [101]olivine // [$\bar{1}$01]diamond. If generally true, this would constitute a specific COR and would probably imply some kind of epitaxial control on the orientations, as indeed claimed by Futergendler and Frank-Kamenetsky (1961) even though they also found three other olivines with significantly different orientations. Frank-Kamenetsky (1964) reported three more flattened olivines with the same orientation, but also found two other types of orientation. In recent years, individual measurements of at least 113 individual olivine inclusions *in situ* in diamond have been published from both EBSD (Neuser et al. 2015) and X-ray diffraction measurements (Nestola et al. 2014; Milani et al. 2016; Seryotkin et al. 2017; Sobolev et al. 2020). The as-published orientations appear random (Fig. 7c). When the randomization due to the indexing ambiguity is removed by plotting the *b*-axis of the olivine within the joint asymmetric unit of the diamond and olivine, they effectively fill the entire orientational space (Fig. 8a). This confirms that olivine inclusions in diamonds exhibit a *random* COR. It is also obvious from Figure 8a that a small proportion of olivines that are part of the random COR have orientations that are approximately in a specific alignment with the diamond, including the orientation reported by Mitchell and Giardini (1953). Such specific orientations will always be found, of course, as a subset of inclusions exhibiting a COR of more degrees of freedom.

The presence of a proposed specific COR can be easily tested with this data, by selecting the one symmetry-equivalent orientation of each measurement that is closest to the proposed COR. To test for the specific COR of Mitchell and Giardini (1953), the orientation with (010)olivine closest to (111)diamond is plotted on a stereogram. Figure 8b shows that the *b*-axes, equivalent to the (010) plane normals, do not plot systematically close to the [111] axis of the diamond; in fact, the angles between the two span the whole possible range of values from 0 to 54.7°.

And the [101] directions of the olivine are also widely scattered and not coincident with [$\bar{1}$01] of the diamond. This excludes the proposed specific COR. Comparison of this plot with that expected for a rotational COR around [111]diamond (Fig. 7b) clearly also excludes the possibility that olivines in diamonds exhibit a rotational COR. This example demonstrates that distinguishing specific CORs from rotational CORs and rotational CORs from random CORs requires large numbers of inclusions to be measured, and conclusions must not be based on a handful of measurements.

The reason for olivines exhibiting a random COR, irrespective of the morphology of the inclusion, appears to be the recently-discovered ubiquitous presence of aqueous fluid films coating the inclusions, which must have been trapped along with the olivine at the time of diamond growth (Nimis et al. 2016). This fluid would prevent any direct interaction or bonding between the surfaces of the olivine and the diamond and thus prevent the development of any preferred orientation through a process such as epitaxy. The presence of the film also explains the absence of dislocations in the diamond adjacent to the inclusions (Agrosi et al. 2016). The incorporation of the fluid into the inclusion at the time of entrapment is due to the solid surfaces of olivine and diamond being mutually repulsive, as shown by DFT simulations (Bruno et al. 2016). In the presence of fluid, the olivine and diamond surfaces would therefore separate and fluid would be drawn in by capillary action to separate the olivine and diamond surfaces at the time of inclusion entrapment.

However, there are also many examples of two or more olivine inclusions in a single diamond that exhibit the same orientation as one another (Nestola et al. 2014; Milani et al. 2016; Seryotkin et al. 2017), and these inclusions also share the same composition as determined by *in situ* single-crystal diffraction (Fig. 4). Recent measurements of the orientations of garnet inclusions within lithospheric diamonds (Davies et al. 2018; Nestola et al. 2019b; Sobolev et al. 2020) show exactly the same characteristics as olivine inclusions; the overall COR between the garnets and diamond is also random (Fig. 8c), while iso-oriented groups of garnets occur in the same host diamond (Nestola et al. 2019b). Given that both olivine and garnet inclusions in diamond exhibit a random COR (Fig. 8), this grouping of orientations in one diamond requires another explanation that does not invoke interaction between the inclusion and the diamond or any influence of surface energies. Post-entrapment dissolution and re-precipitation could not lead to different isolated inclusions developing identical orientations. The only rational explanation proposed so far is that these iso-oriented inclusions were originally parts of one single crystal within the host rock prior to diamond growth (Nestola et al. 2014). Such single crystals would have been partially dissolved by the fluids from which the diamond grew in order to create space for the growth of the diamond, which then entrapped the remnants of the original crystals (Fig. 9). The mutual slight mis-orientations of iso-oriented inclusions, in the case of olivine, is consistent with the common observation of sub-grain boundaries in olivine within peridotite diamond host rocks (e.g., Baptiste et al. 2012).

Groups of two or three iso-oriented sets of inclusions suggest that diamonds nucleated on grain boundaries, or triple junctions, respectively, and then grew while dissolving the surrounding minerals (Fig. 9). Four differently-oriented groups of iso-oriented inclusions have not yet been found in a single diamond. The spacing and spatial arrangement of all such groups of iso-oriented inclusions is consistent also with the estimated grain size of minerals in mantle rocks, while the preferred orientation exhibited by olivines in a single diamond (but not with respect to the diamond) is consistent with the preferred orientation found in olivines in mantle peridotites (Nestola et al. 2014). There is insufficient published data to determine the nature of the COR of clinopyroxene inclusions with their host diamonds. But the observation that a single clinopyroxene inclusion with no specific orientation with respect to its diamond host did however have an identical orientation to an adjacent clinopyroxene in the garnet peridotite xenolith (Nestola et al. 2017), supports the model for diamond growth shown in Figure 9.

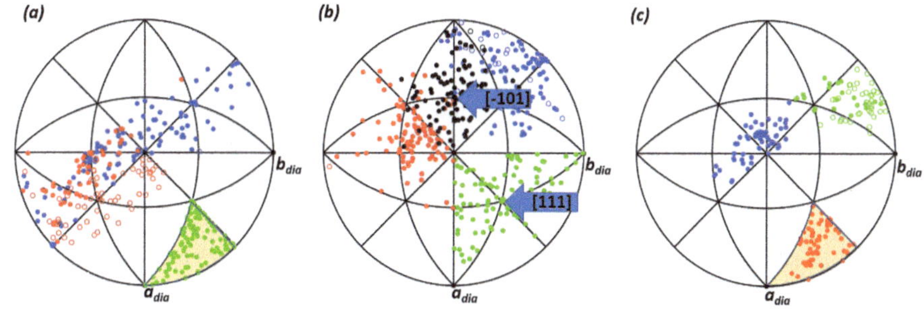

Figure 8. **(a)** Crystallographic orientations of 113 modern published measurements of olivine inclusions plotted with their *b*-axis in the joint asymmetric unit of olivine plus diamond (**yellow shaded region**). **(b)** The same data plotted for the orientation with the *b*-axis closest to [111]diamond, so as to test for the presence of the Mitchell and Giardini (1953) specific COR. Note that the *b*-axes are spread widely from [111] diamond and that the [101] directions of the olivine (**black symbols**) do not coincide with [$\bar{1}$01]diamond. **(c)** Orientations of 61 modern published measurements of garnet inclusions in diamonds with their *a*-axes in the joint asymmetric unit of garnet plus diamond (**yellow shaded region**). While the *a*-axes show a random distribution similar to that of olivine, the scatter of the *b*- and *c*-axes is less because of the higher symmetry of garnet.

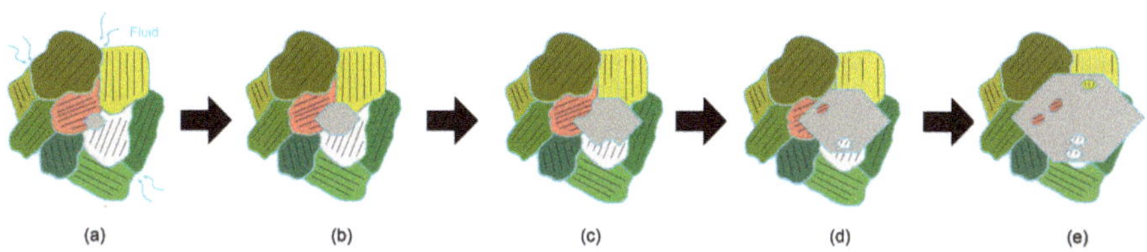

Figure 9. Cartoon showing how after **(a)** infiltration of fluid (**blue**) and initial nucleation and then **(b)** growth of a diamond on a grain boundary of the host rock **(c)** it starts to incorporate fragments of the host minerals as a consequence of unequal growth rates on the faces of the diamond. **(d,e)** These fragments become trapped as inclusions after further growth of the diamond, but retain their orientations inherited from their parent mineral grain, leading to groups of iso-oriented inclusions being found in a single diamond. At the same time, fluid is also trapped as a thin film around each inclusion. Modified and extended from Nestola et al. (2014).

Such a scenario of dissolution of pre-existing mantle minerals to form the inclusions found in lithospheric diamonds means that, in strict application of the terminology, these inclusions are protogenetic with respect to their diamonds. Given that the datasets (Fig. 8) are from both inclusions with faces parallel to (111)-type planes of diamond and many other shapes, these data show that inclusion shape is not relevant for determining the genetic relationship of these inclusions to their host diamonds. The common occurrence of pseudo-octahedral shapes of inclusions is simply a reflection of the much lower surface energies of (111) faces of diamond compared to any other face of diamond. Thus, modern orientation data on olivine and garnet inclusions has shown that the morphology of the inclusions in diamond has no bearing on whether they are syngenetic or protogenetic. These conclusions open once again the question of whether ages and geochemistry of inclusions record the conditions and time of diamond growth. If the inclusion sizes are small enough then solid-state diffusion of elements could be sufficiently rapid to ensure re-equilibration of the inclusion with the diamond growth environment, including the resetting of isotopic systems used for dating (Davies et al. 2018; Nestola et al. 2019b). Whether this is also true for all inclusions is a matter for further research.

Ferropericlase inclusions. The limited published data on the orientations of these inclusions in sub-lithospheric diamonds suggest that they approximate a specific COR (Fig. 10a) in which the crystallographic axes of the cubic inclusions are very close to those of the diamonds (Nimis et al. 2018). If the COR was exact and the small scatter in orientations was just a consequence of measurement uncertainties, then all of the crystallographic axes would be scattered in a similar way when plotted on a stereogram. However, the [112] directions (and those close to it, such as [111]) of the ferropericlase inclusions and diamonds all coincide to within the 2° uncertainty in the measurement. This suggests that the ferropericlase inclusions exhibited a specific COR when they grew as a result of interfacial energy minimisation, in turn implying that no fluid film separated the inclusions and diamond during growth and that these inclusions are a product of the diamond-forming reaction (Nimis et al. 2018). This would mean that the compositions of such inclusions are not representative of bulk mantle ferropericlase, but only of the reaction itself. The interpretation is that a specific COR was subsequently transformed into a rotational COR as the result of post-entrapment plastic deformation of the diamonds (Nimis et al. 2018). These measurements are from just two diamonds, so many more measurements are required to determine whether all ferropericlase inclusions show this COR, or whether a range of CORs for ferropericlase are present in diamonds which would indicate a range of growth mechanisms and environments.

Graphite inclusions. Graphite is a common inclusion in diamond often associated with other minerals. This graphite is believed to be produced by internal graphitization (e.g., Harris and Vance 1972), a process perhaps enhanced by the presence of fluid films around silicate inclusions (Nimis et al. 2016). Graphite is more rarely found as isolated single-phase euhedral inclusions where evidence suggests that the diamonds may have nucleated on the graphite (Nasdala et al. 2005). Glinnemann et al. (2003) showed that out of 13 measured isolated graphite inclusions of the latter type, 12 have their *c*-axes within 4° of a <111> direction of the diamond. Insufficient data are given in the paper to completely define the relative orientations, so in drawing Figure 10b we have assumed that the reported misalignments are due to measurement uncertainties, and that the true relationship is exact. Graphite has hexagonal symmetry, so the *a*-axes of the graphite inclusions are perpendicular to the *c*-axes and therefore plot on a great circle around the [111]diamond, and appear to define a rotational COR. However, the 12 measured inclusions fall into two main groups, one with the <100> directions of the graphite within a few degrees of the <110> directions of the host diamond, and the other with the angle between these two directions being 34–38° (Fig. 10b). The fact that only one inclusion from 12 has an intermediate rotational angle suggests that these data might represent two distinct specific CORs, a conclusion supported by analysis of the structures of the interface planes of graphite and diamond which shows that these two orientations correspond to the best two structural matches (Glinnemann et al. 2003). Further careful measurements of graphite inclusions are needed to determine whether these conclusions are statistically significant, or instead whether graphite displays a full rotational COR. If only the specific CORs are found this would support the interpretation that the diamond nucleated on the graphite and subsequently entrapped it (Nasdala et al. 2005).

Magnesiochromite inclusions. Of the types of inclusions for which there are sufficient data to attempt a meaningful analysis, these show the most complex combination of CORs. The modern dataset consists of 58 inclusions in lithospheric diamonds from various Siberian kimberlites (Nimis et al. 2019; Sobolev et al. 2020) plus 7 from Damtshaa, Botswana and 6 from Panda, Canada (Nimis et al. 2019). A stereographic plot of the data (Fig. 11a) with the *a*-axis plotted in the common asymmetric unit appears, at first sight, very similar to the random CORs exhibited by garnets (Fig. 8c). But closer inspection shows that there is a clustering of inclusions with their *a*-axes close to the *a*-axis of the diamond (compare Fig. 11a carefully with Fig. 8a and 8c) and there appears also to be a clustering of the [111]mgchr directions towards [111]diamond. This is confirmed by replotting the data with the orientation that has [111]mgchr closest to [111]diamond; then 33 out of the 71 inclusions (46%) have the angle between these two directions less than the estimated measurement uncertainties of 4° (Fig. 11b). This means that their octahedral faces are parallel within error to an octahedral face of their host diamond. These inclusions form a rotational COR which is especially evident when the <110> directions which lie in the (111) plane of the inclusions are plotted on the stereogram (Fig. 11b). With the limited numbers of measurements, the group whose [111] mgchr are more than 4° from [111]diamond (Fig. 11c) appear to represent a random COR. The interpretation of these data is that magnesiochromite inclusions exhibit two CORs; one random and one rotational. The random COR is entirely consistent with the observation of fluid rims around these inclusions (Nimis et al. 2016; Nimis et al. 2019) that would, as for olivine and garnet inclusions, prevent formation of an exact COR through epitaxial growth. The only mechanism so far proposed for the creation of the rotational COR is that it is the result of mechanical impingement. Although spinels in mantle rocks do not normally display octahedral shapes, dissolution by the fluids present at the time of diamond growth will tend to dissolve the higher-energy faces of the magnesiochromites which will then tend to develop a habit that is octahedral. In the presence of fluid and under lithostatic load the tendency will

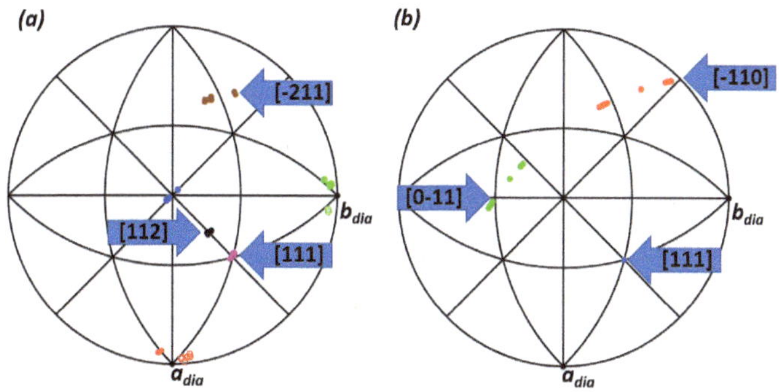

Figure 10. **(a)** Crystallographic orientations of ferropericlase inclusions which approximate a specific COR in which their crystallographic axes are parallel to those of the diamond. Note that the [112] axes of all of the inclusions plot together, indicating that the small rotational component of the COR has been created by post-entrapment plastic deformation of the diamond. **(b)** The orientations of graphite inclusions exhibit a rotational COR with [111]diamond // [001]graphite that may in fact consist of two exact CORs represented by the two clusters of *a*-axes (**red**), with one outlier between them.

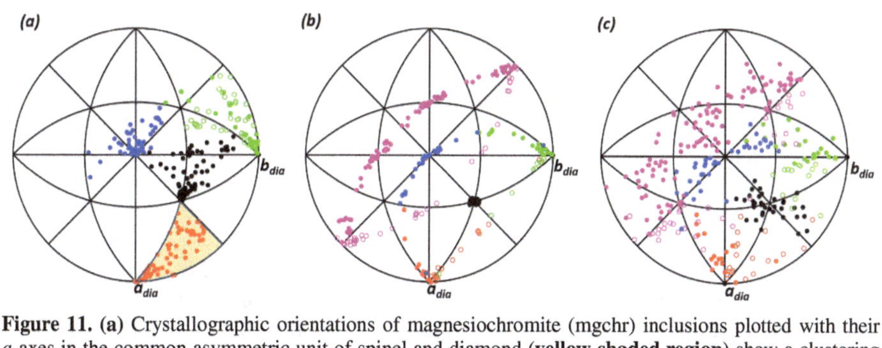

Figure 11. **(a)** Crystallographic orientations of magnesiochromite (mgchr) inclusions plotted with their *a*-axes in the common asymmetric unit of spinel and diamond (**yellow shaded region**) show a clustering of *a*-axes near to the *a*-axis of their diamond hosts. **(b)** The orientations of inclusions with their [111] axes (**black poles**) within 4° of [111]diamond form an approximate rotational COR, outlined by the poles of the <110>mgchr directions lying in the [111] zone, shown as magenta poles. The crystallographic axes of these inclusions fall on a small circle around the [111] direction. **(c)** The remaining inclusions with their [111] axes greater than 4° from [111]diamond form a random COR.

be for the dominant octahedral faces of adjacent spinels and diamonds to become parallel to one another, simply by being pushed together mechanically. The apparent tendency for some of these inclusions to exhibit an approximate specific COR can then be explained as the mechanical interaction of a spinel remnant not with a single plane of the diamond, but with etch pits in, or growth steps on, the (111) surface of the diamond (Nimis et al. 2019).

DEPTH OF DIAMOND GROWTH: ELASTIC THERMOBAROMETRY

Evidence for inclusion over-pressures in the form of induced birefringence haloes in diamond adjacent to inclusions has been known for many years (e.g., Sorby and Butler 1869; Meyer quoted in Cohen and Rosenfeld 1979). Inclusions are over-pressured because diamond is so much stiffer than the silicate minerals, or fluids, found as inclusions, so the diamond expands less on exhumation than would a free crystal of the inclusion mineral. Therefore, when the diamond is in the laboratory the inclusion crystal is constrained to have a smaller volume than a free crystal of the same mineral would have at room conditions. The inclusion will therefore exhibit an over-pressure. Elastic thermobarometry (or piezobarometry) is the process of determining the P and T of entrapment of inclusions from the over-pressure measured in them when their host mineral is at room conditions. It has recently provided new constraints on the P–T conditions attained during metamorphism (e.g., Zhong et al. 2018; Gonzalez et al. 2019; Korsakov et al. 2020; Kosminska et al. 2020). When applied to inclusions in diamond it can provide estimates of their depth of formation and growth beyond that available from identification of the mineral phases present as inclusions. Elastic thermobarometry has been applied to various types of inclusions in diamonds with variable degrees of apparent success (e.g., Harris et al. 1970; Cohen and Rosenfeld 1979; Izraeli et al. 1999; Sobolev et al. 2000; Howell and Nasdala 2008; Howell et al. 2010, 2012a; Nestola et al. 2011; Anzolini et al. 2016; Kueter et al. 2016; Anzolini et al. 2018) that reflect the concurrent development of the theory and techniques for characterizing the stress and strain states of inclusions in diamond.

The application of elastic thermobarometry relies on several fundamental assumptions. The first is that when the inclusions are trapped by the host diamond, both experience the same temperature and the same hydrostatic pressure. This is not unreasonable, since the evidence of the 'fluid' films around many inclusions (Nimis et al. 2016, 2019; Smith et al. 2016, 2018), the presence of fluid inclusions (e.g., Smith et al. 2015; Gubanov et al. 2019) and the resorption of inclusion minerals (Nimis et al. 2019) indicate that diamonds large enough to host macroscopic inclusions probably grow in fluid-saturated environments. The established theory also assumes that the inclusion space in the diamond is filled completely and exactly by the inclusion crystal without any void space, but the theory can be adapted to account for the presence of fluid as we describe below. The basic analysis further assumes that all volume changes in the system following entrapment are elastic. If there has been plastic deformation of the diamond after diamond growth and inclusion entrapment, some or all of the inclusion stress may be released. Similarly, the amount of inclusion stress released by cracking of the diamond around the inclusion depends on whether the inclusion also contains fluid in addition to the mineral. Cracks and plastic deformation can be detected by a variety of techniques, including X-ray tomography (e.g., Rustioni et al. 2015; Mazzucchelli et al. 2016; Nimis et al. 2016; Wenz et al. 2019) and topography (e.g., Agrosi et al. 2017). In all of these cases, the measured inclusion pressure will be reduced but this does not invalidate the use of elastic thermobarometry. Proper characterization of the inclusion often allows either the minimum entrapment pressure to be estimated, or the conditions of the resetting of the inclusion stress to be calculated. Phase changes in the inclusions after entrapment to lower-pressure polymorphs may result in the inclusion pressure being buffered (e.g., Gillet et al. 1984; Perrillat et al. 2003; Anzolini et al. 2016) to a higher value than expected for a single-phase inclusion of the higher-pressure polymorph present at the time of entrapment, and also requires more elaborate analysis.

The over-pressure in the inclusion leads to elastic deformation of the diamond host around the inclusion and this is what gives rise in otherwise optically-isotropic diamond to the strain-induced birefringence which has been used to characterize the inclusion stress state (Howell and Nasdala 2008; Campomenosi et al. 2020a). The deformation of the host diamond leads to elastic stresses within it which balance the over-pressure in the inclusion. The retention of stress or over-pressure inside inclusions is thus the result of a mechanical equilibrium between the host diamond around the inclusion and the inclusion itself. Removal of this confining stress by removal of the diamond by polishing it away, partially or completely, or by breaking out the inclusion, will lead to partial or complete loss of inclusion stress (e.g., Campomenosi et al. 2018; Mazzucchelli et al. 2018; Zhong et al. 2019). For example, Glinnemann et al. (2003) showed that graphite inclusions near the surface of their host diamonds exhibit less compression than those deep within their hosts, while garnet inclusions close to the outside surfaces of the diamond show no significant difference in Raman shifts compared to garnets measured at room pressure (Liu et al. 1990). Therefore, unlike the determinations of the orientation of inclusions or their crystal structure that we have discussed in the preceding sections, meaningful measurements of inclusion stress can only be made on fully-buried inclusions at least three inclusion radii from the diamond surface, other inclusions and any other defects (e.g., Zhang 1998; Tajčmanová et al. 2014). It is therefore not possible for EBSD measurements on partially exposed inclusions to return any meaningful data with respect to the true inclusion stress or strain state (Campomenosi et al. 2018; Zhong et al. 2019).

Theory

Forward problem. The development of stress and strain in an inclusion trapped within diamond is an elastic effect and therefore, by definition, reversible. We first describe the forward problem, i.e., what happens to the inclusion from encapsulation at depth until its diamond host is transported to the Earth's surface. Diamond is cubic and extremely stiff compared to the minerals found within it as inclusions, so we can use the theory developed for elastically isotropic systems as an initial basis for most analyses of inclusions in diamonds. The various extensions and adaptations of the isotropic theory are described in the 'examples' below.

We start our analysis by considering an isotropic mineral inclusion in diamond, and we assume that the inclusion is a single mineral phase which does not undergo reactions or phase transformations. At the time of diamond growth and entrapment of the inclusion, both the diamond and the inclusion are at the same *P* and *T*. The inclusion fills the void in the diamond completely, so the diamond void and inclusion also share the same volume, *V*. If, after entrapment, the *P* and *T* changes in such a way that the natural expansion and contraction of a free crystal of the inclusion exactly matches that of the host diamond, the inclusion will continue to exactly fit the void space in the diamond without the application of any additional stress. This line in *P–T* space is called an isomeke (Rosenfeld and Chase 1961; Adams et al. 1975), and is defined by the difference in the volume thermal expansion coefficients (α_{inc} and α_{dia}) and compressibilities (β_{inc} and β_{dia}) of the inclusion phase and the diamond. The volume compressibility of a phase is the inverse of its bulk modulus, thus; $\beta_{inc} = \frac{1}{K_{inc}} = \frac{-1}{V}\left(\frac{dV}{dP}\right)_T$. The slope of an isomeke through any *P–T* point is:

$$\left(\frac{\partial P}{\partial T}\right)_{isomeke} = \frac{\alpha_{inc} - \alpha_{dia}}{\beta_{inc} - \beta_{dia}} \tag{2}$$

Isomekes can therefore be calculated from the EoS of the diamond and of the inclusion phase. Isomekes have also been called 'zero volume difference' (ZVD) curves (Izraeli et al. 1999) and 'isovolume loci' (Barron 2003). An example of an isomeke of mantle-composition olivine with diamond is given in Figure 12. It has a positive slope because, as for most silicates, olivines have

Table 2. Elastic properties and equations of state.

	K_{0T} : GPa	β_{0T} : TPa^{-1}	$\alpha_0 \times 10^5$: K^{-1}	EoS Type
Diamond	444(2)	2.25(1)	0.2672(3)	BM2 + HP thermal-pressure Angel et al. (2015a)
Pyrope	169.8(4)	5.89(2)	2.465(5)	BM3 + MGD (q = 0) this work
Grossular	166.9(3)	5.99(1)	2.041(3)	BM3 + MGD, this work
Almandine	175.0(5)	5.71(2)	1.974(5)	BM3 + MGD (q-compromise) this work
Mantle olivine	126.0(2)	7.94(2)	2.667(9)	BM4 + isothermal-type EoS Angel et al. (2018)
Coesite	101.9(3)	9.81(3)	0.646(2)	BM4 + MGD (q-compromise) this work
Graphite	32.0	31.3	1.65	Tait + HP thermal-pressure Holland and Powell (2011)

Note: K_{0T} is the isothermal Reuss bulk modulus, β_{0T} is the corresponding compressibility, and α_0 is the volume thermal expansion coefficient of each phase at room conditions.

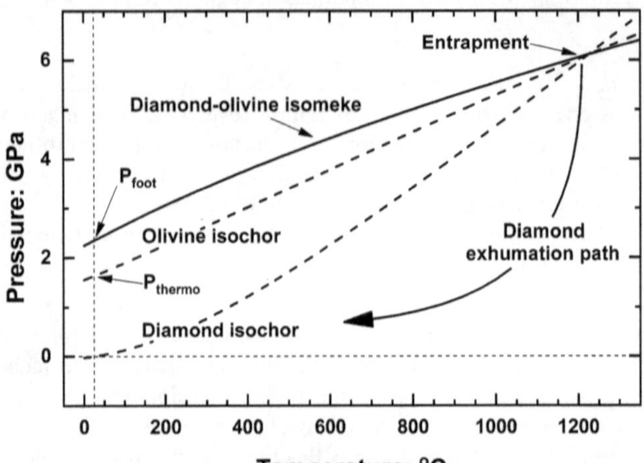

Figure 12. Isochors of diamond and olivine, and the diamond–olivine isomeke, passing through the point P = 6.2 GPa and T = 1230 °C. Note that the isomeke, which is the line along which an olivine inclusion in diamond would have the same pressure as the external pressure, diverges significantly from the isochors. After entrapment, the olivine inclusion pressure will be higher than the external pressure at all P, T conditions below the entrapment isomeke. Because the diamond isochor through these particular entrapment conditions also passes through room conditions, the P_{thermo} of the olivine inclusion is the pressure of the olivine isochor at room T. This is *not* the case for other entrapment conditions that do not lie on the diamond isochor passing through room conditions.

volume thermal expansion and compressibility coefficients larger than diamond ($\alpha_{inc} > \alpha_{dia}$ and $\beta_{inc} > \beta_{dia}$), Table 2. Although diamond is very stiff, neither α_{dia} nor β_{dia} are zero, and the isomeke of the olivine with diamond is *not* the same as an isochor of either phase (Fig. 12).

The importance of the isomeke concept is that for P and T conditions on the entrapment isomeke, the inclusion will have the same pressure as the external pressure on the diamond host crystal. However, the typical eruption path of diamonds in kimberlites involves rapid pressure decrease and is obviously not along an isomeke (Fig. 12). When the external P and

T drop below that of the diamond–inclusion isomeke through entrapment conditions, the inclusion crystal wants to expand more than it is allowed by the diamond because $\beta_{inc} > \beta_{dia}$, but it is restricted to the volume change of the stiffer diamond. Therefore, when the diamond is erupted the inclusion develops an over-pressure with respect to the external pressure applied to the diamond (Fig. 12) at all external pressures and temperatures that lie below the entrapment isomeke (e.g., Angel et al. 2014b; Ferrero and Angel 2018). Room conditions lie below the entrapment isomekes of silicate inclusions in diamonds (e.g., Fig. 12), which is why they exhibit positive remnant pressures.

The first step in the calculation of the over-pressure in the inclusion is to use the volume change of the host diamond from entrapment to room conditions (calculated from its *P–V–T* Equation of State or EoS) in the EoS of the inclusion mineral. This is just a thermodynamic calculation, so the resulting pressure has become designated as P_{thermo} (previously PI* in Angel et al. 2014b). It only depends on the EoS of the two phases, and does not depend on the size, shape or any other geometrical aspect of the system. As a simple example, consider entrapment of a mantle-composition olivine at 1230 °C and 6.2 GPa. At these conditions the molar volume of diamond calculated from its EoS (Table 2) is 3.420 cm^3/mol, the same as its molar volume at room conditions; these entrapment conditions lie on the isochor of diamond that passes through room conditions (Fig. 12). Therefore, for this particular example, on recovery to room conditions the void in the diamond that contains the inclusion has the same volume as it had at entrapment. However, the olivine at entrapment is calculated to have V = 43.34 cm^3/mol, 1.26% smaller than the volume of a free olivine at room conditions of 43.89 cm^3/mol. Thus, because the olivine inclusion is restricted by the diamond to maintain the same volume as it had at entrapment, it is under 1.26% volume compression at room temperature, corresponding to a remnant pressure in the inclusion of ~1.64 GPa (Fig. 12).

This is a fictive state that is never attained in real systems because there is a pressure difference between the host, at ambient pressure, and the inclusion at P_{thermo}. The system is not in mechanical equilibrium. The pressure P_{thermo} is greater than ambient pressure for nearly all silicate inclusions in diamonds, so it drives expansion of the inclusion and reduces the inclusion pressure by an amount ΔP_{relax} (Fig. 13). By convention (Angel et al. 2014b) a relaxation that decreases the inclusion pressure has $\Delta P_{relax} < 0$. The final inclusion pressure, P_{inc}, which is what is always measured, is therefore related to the thermodynamic pressure by:

$$P_{inc} = P_{thermo} + \Delta P_{relax} \quad (3)$$

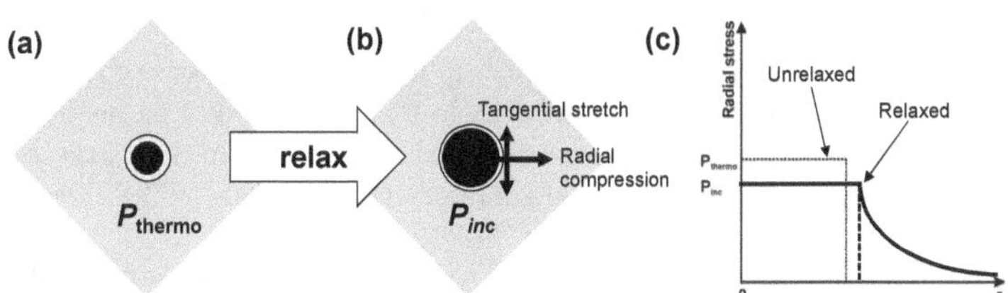

Figure 13. Cartoon of the inclusion relaxation process. **(a)** An unrelaxed inclusion at a pressure P_{thermo}. Consider the shell of diamond (**circle**) immediately surrounding the inclusion. **(b)** Inclusion expansion during relaxation compresses this shell in a radial direction and stretches it in the directions tangential to the inclusion wall. **(c)** The radial expansion of the inclusion decreases the inclusion pressure from P_{thermo} to P_{inc}, and the radial compression of the diamond creates a stress gradient within it. See also Angel et al. (2014b).

The expansion of the inclusion crystal compresses the host in a radial direction, thereby increasing the radial stress in the host. Mechanical equilibrium is attained when the pressure in the inclusion becomes equal to the radial stress in the host at the wall of the inclusion (Fig. 13c).

The radial expansion of the inclusion also causes expansion in the diamond in directions parallel to the inclusion surface. Thus, it is this elastic relaxation that creates the stress and strain fields around the inclusion which can be detected as the strain-induced birefringence (e.g., Sorby and Butler 1869; Howell and Nasdala 2008; Howell et al. 2010), or as changes in the Raman spectra of the host diamond (Nasdala et al. 2003; Campomenosi et al. 2020a). For most inclusions, the stresses drop off in proportion to the distance from the center of the inclusion cubed (Fig. 13c). At distances of three or more radii from the inclusion, the induced stresses in the diamond become insignificant. At shorter distances, these stresses can interact with any inhomogeneity in the host diamond, such as another inclusion, cracks, external surfaces etc. These interactions can lead to a modification of the elastic relaxation that therefore changes the final inclusion pressure away from that expected for an isolated inclusion.

Diamond is cubic and therefore expands and contracts isotropically in response to changes in P and T. But, it is elastically anisotropic because elasticity is a 4th-rank tensor property (Nye 1957). This means that the elastic properties of diamond change with direction in the crystal. Therefore, the amount of elastic relaxation (Fig. 13) around an inclusion will be different in different directions, leading to small deviations in the inclusion stress and strain from isotropic. However, because diamond is so stiff the effect of its elastic anisotropy on P_{inc} is so small that it can be ignored and the diamond host can be treated as being elastically isotropic. The amount of relaxation of an isolated isotropic inclusion then depends on the contrast between the elastic properties of the inclusion and the diamond, and the shape of the inclusion. For a spherical inclusion in diamond, the fractional volume change of the inclusion due to relaxation (Zhang 1998; Angel et al. 2017) is:

$$\frac{\Delta V_{relax}}{V} = \frac{3P_{inc}}{4G_{dia}} \tag{4}$$

when the diamond is at room pressure (considered to be effectively zero). The shear modulus of diamond at room conditions is G_{dia}=535 GPa. The pressure change in the inclusion, due to the relaxation, ΔP_{relax}, can be calculated from this volume change and the EoS of the inclusion (Angel et al. 2017). For the example of olivine shown in Figure 12, ΔP_{relax} = −0.26 GPa for a spherical inclusion, and the final measurable inclusion pressure is P_{inc}=1.38 GPa. For small inclusion pressures a rough estimate can be made by using the definition of the bulk modulus to obtain:

$$\Delta P_{relax} \sim \frac{3K_{inc}P_{inc}}{4G_{dia}} \tag{5}$$

We have described the theory of piezobarometry in terms of recovery of the host diamond to room conditions. But the theory applies equally to the calculation of inclusion pressures at any other conditions, for example while the diamond is still deep in the Earth, or while the diamond and its inclusion are heated in the laboratory (Harris et al. 1970). For non-ambient conditions the volume relaxation is given by $\frac{3(P_{inc} - P_{ext})}{4G_{dia}}$ and the value used for G_{dia} in Equation (4) must be changed to that for the external P and T. Equation (5) does not apply, so the ΔP_{relax} must be calculated from this volume change and the EoS of the inclusion at the temperature of measurement (Angel et al. 2017).

Inverse problem. For an isotropic spherical inclusion the determination of possible entrapment conditions consists of reversing the steps described in the preceding section. The inclusion pressure P_{inc} is first measured and the ΔP_{relax} is calculated through Equation (4) and the EoS of the inclusion, to obtain P_{thermo}. This is the one and only constraint on the entrapment conditions from the measurement of the inclusion. It is therefore not possible to determine a unique P and T of entrapment (two variables) from this single observation. But it is possible to determine the entrapment isomeke, the line in P–T space from which the inclusion would exhibit the same P_{thermo} and P_{inc}.

The point on the entrapment isomeke at room temperature can be calculated by considering the isothermal compression of the diamond host and the inclusion, starting from a common volume (Fig. 14) at which the diamond is at room pressure, and the inclusion is at P_{thermo}. Because the diamond is stiffer than the inclusion, the PV curve of the diamond is shallower than that of the inclusion and the two EoS curves converge with increasing pressure. At the point at which they cross both the diamond and the inclusion have the same P and V (and the same T). This is a point on the entrapment isomeke. This particular pressure on the entrapment isomeke at the final temperature T_{end} of the inclusion measurement has been called P_{foot} (Angel et al. 2014b). Once P_{foot} is determined, the full entrapment isomeke corresponding to the measured P_{inc} can be calculated from the EoS of the two phases, either step-wise by using Equation (2), or more directly by finding the P at each T at which the fractional volume change of both the inclusion and the diamond from P_{foot}, T_{end} are exactly the same (e.g., Angel et al. 2020).

Whether an inclusion phase makes a good geobarometer depends on the slopes of its isomekes with diamond. These are determined by the contrast between the elastic properties of the inclusion and those of diamond (Eqn. 2). Diamond is significantly stiffer than all reported inclusion phases, so the thermal expansion coefficient of the inclusion phase is the major determinant of the isomeke slope through its contrast to that of diamond, ($\alpha_{inc} - \alpha_{dia}$). Thus, phases with a small α_{inc} similar to that of diamond, such as coesite and graphite, have isomekes that are relatively flat in P–T space (Fig 15) that make them good geobarometers. Inclusions of these phases also therefore have P_{foot} similar to the pressures of entrapment (Fig. 15), although the measured P_{inc} is significantly lower because the elastic relaxation term is relatively large. Thus, a coesite inclusion trapped at the same time and conditions as an olivine or a garnet should exhibit a much higher residual inclusion pressure when measured in the laboratory. In principle, the best geothermometers would be stiff inclusion phases with a β_{inc} similar to that of diamond, with a large α_{inc} so as to make the isomekes with diamond as steep as possible (Eqn. 2). Unfortunately, stiff minerals tend to display small thermal expansion.

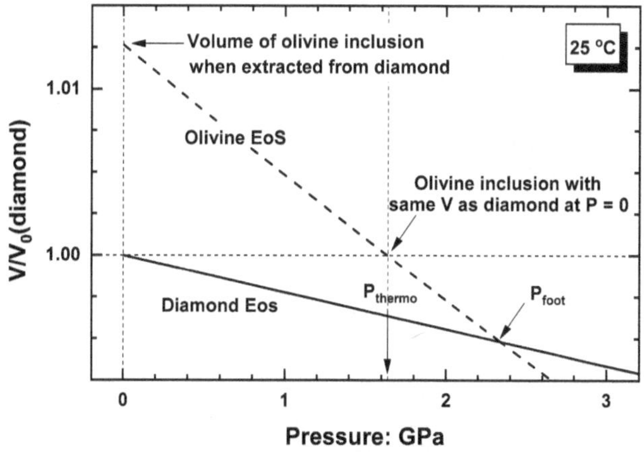

Figure 14. The determination of P_{foot} in graphical form with the pressure–volume curves of olivine and diamond. The olivine volume curve has been scaled so that at P_{thermo} it is the same volume as the diamond. This is equivalent to the inclusion fitting exactly the inclusion cavity in the diamond. The two EoS curves cross in this example at 2.36 GPa, where the olivine and diamond would have the same P, V and T. This pressure is therefore P_{foot}, a point on the entrapment isomeke of this inclusion, as drawn in Figure 12.

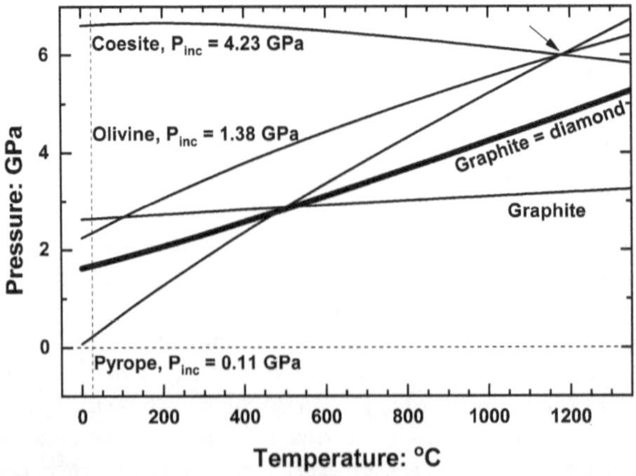

Figure 15. The isomekes of coesite, olivine and pyrope garnet with diamond have very different slopes, mainly because of their different thermal expansion coefficients (Table 2), so that the values of P_{foot} (the isomeke P at room T) and P_{inc} (**labelled**) are very different even when inclusions are trapped at the same P and T (**arrowed**). An isomeke for graphite trapped within the stability field of graphite is also shown—it also is very shallow because of the low thermal expansion coefficient of graphite. The diamond = graphite equilibrium phase boundary (Day 2012) is shown as a **grey line**.

Methods

Calculation of entrapment conditions of an inclusion requires three steps. The first step is the determination of the stress or the strain state of the inclusion crystal when the host crystal is at known conditions of P and T, normally room conditions. This can be achieved by *in situ* diffraction or Raman spectroscopy. The second step is to calculate the amount of stress and strain relaxation that the inclusion has undergone as a consequence of the contrast between the stress in the inclusion and the external pressure on the host, and thereby calculate the strain or stress that the inclusion would have in the fictive unrelaxed state. The third step is to use this information about the unrelaxed state to calculate possible entrapment conditions of the inclusion by using the thermodynamic equations of state of the two minerals.

Measuring inclusion pressures. The fundamental requirement for success in thermobarometry is an accurate and precise measurement of the stress in the inclusion. It is important to understand that even in an isotropic inclusion its pressure is *never* measured directly as a pressure. Instead, the pressure is inferred from an *in situ* measurement of some physical property of the inclusion. Most commonly, the Raman spectrum of the inclusion is measured, and the shift in the frequencies of the Raman lines from their values at room pressure is used to calculate the inclusion pressure from their known rate of shift with pressure. This is strictly incorrect as we describe in detail below. More rarely, the unit-cell parameters and volume of an inclusion are determined with an X-ray diffraction measurement, and the inclusion pressure inferred from their known variation with pressure (e.g., Nestola et al. 2011). While the determination of orientation of the inclusion crystal by diffraction is not sensitive to small offsets of the crystal from the center of the diffractometer, the unit-cell parameters and volume values can be significantly changed by failure to center the inclusion exactly. It is therefore crucial to first follow the entire procedure for centering the inclusion, that is outlined in the sections above, before making the measurements to determine its unit-cell parameters and volume.

It is clear that diffraction measurements of the unit-cell parameters of an inclusion do not directly measure its stress state. But the difference from the values of the unit-cell parameters measured on a free crystal at room conditions can be turned into the *average* strain experienced by the inclusion trapped in the host relative to a free crystal of the same material at room conditions. For example, the strain along the crystallographic *a*-axis of an inclusion crystal measured as a_{inc} can be defined as:

$$\varepsilon_1 = \frac{a_{inc} - a_0}{a_0} \tag{6}$$

where a_0 is the unit-cell parameter the inclusion crystal would have when it is free of the host and at room pressure and temperature. In general, strain is a second rank tensor with six components (e.g., Nye 1957). The three normal components (ε_1, ε_2 and ε_3 in Voigt notation) describe the change in dimensions in three mutually perpendicular directions in the inclusion, and for crystals of orthorhombic symmetry and higher can be defined purely in terms of the cell parameters *a*, *b*, and *c*, as in Equation (6). Three shear components (ε_4, ε_5 and ε_6) characterize the strains in three mutually-perpendicular planes. The definition of the individual strain components in monoclinic and triclinic crystals in terms of the unit-cell parameters depends on the choice of axial system for the tensor (e.g., Schlenker et al. 1978; Knight 2010). Equation (6) defines the Lagrangian infinitesimal strains, which are usually suitable for the analysis of the remnant strains in inclusions when the host is at room conditions because the strains are small.

The volume strain $\varepsilon_V = \varepsilon_1 + \varepsilon_2 + \varepsilon_3$ is defined as the fractional volume difference:

$$\varepsilon_V = \frac{V_{inc} - V_0}{V_0} \tag{7}$$

What is rarely made clear in the geological literature is that Raman shifts also do not measure stress or pressure. The change in the peak position or wavenumber of a Raman peak is determined *not* directly by the pressure or the temperature but is primarily a function of the strains applied to a crystal. A change in *P* or *T* changes the unit-cell parameters and this change (or strain, Eqn. 6) determines the wavenumber of the Raman mode. This concept was first established by Grüneisen (1926) for isotropic solids, who defined what is now known as the thermal or thermodynamic Grüneisen parameter which relates the change in internal energy of an isotropic solid to the change in pressure along an isochor. The quasi-harmonic approximation (QHA) for the thermodynamics of solids amounts to assuming that the thermal Grüneisen parameter is a constant which does not change with either *P* or *T*. For anisotropic crystals one can define a phonon or mode Grüneisen tensor γ^m (Ziman 1960; Key 1967; Cantrell 1980; Angel et al. 2019) for each Raman-active mode. This is a second rank tensor which defines the fractional change in the phonon wavenumber, $\Delta\omega/\omega$, arising from the strain applied to the crystal. Written out in full in Voigt (1910) notation:

$$\frac{-d\omega}{\omega} = \gamma_1^m \varepsilon_1 + \gamma_2^m \varepsilon_2 + \gamma_3^m \varepsilon_3 + \gamma_4^m \varepsilon_4 + \gamma_5^m \varepsilon_5 + \gamma_6^m \varepsilon_6 \tag{8}$$

This equation means that the changes in the Raman peak positions depend on all of the strains in three dimensions experienced by the crystal, not just the volume. The values of γ_i^m are different for different modes. If the crystal follows the QHA then the values of the six components, γ_i^m, of a mode Grüneisen tensor will be constants and will not change with pressure and temperature, or with the magnitude of the strains. The use of the mode Grüneisen tensors has several advantages over using polynomials in *P* and *T* to describe the variation in peak shifts (e.g., Schmidt and Ziemann 2000; Ashley et al. 2017; Morana et al. 2020). First, the Grüneisen tensor has a sound physical basis, and reflects the underlying physics that the shifts due to *P* and *T* are not independent in materials that exhibit quasi-harmonic behavior. Second, they account for deviatoric strains, as required for the analysis of host–inclusion systems, with the same approach as used for hydrostatic conditions. And, third, less coefficients are needed to describe the peak shifts than are often required for polynomials in *P* and *T*.

For spherical inclusions the strain is homogeneous (Eshelby 1957) irrespective of the symmetry of the host or inclusion crystals. In a faceted inclusion the strains vary across the inclusion volume, reflecting stress concentration at edges and corners. As a consequence, the Raman shift of a single band varies across a shaped inclusion (e.g., Zhukov and Korsakov 2015; Campomenosi et al. 2018; Murri et al. 2018), and a single measurement at one random point in an inclusion cannot therefore be interpreted as being representative of the inclusion stress and strain state. But the shear strains at the center of a faceted inclusion are normally orders of magnitude less than the normal strains (e.g., Mazzucchelli et al. 2018). And, for spherical inclusion crystals in diamond with symmetries that are not triclinic or monoclinic, $\gamma_4^m = \gamma_5^m = \gamma_6^m = 0$ by symmetry. In all of these cases the shifts of the Raman lines *measured at the center of the inclusion*, compared to a free crystal at room conditions, become governed by:

$$\frac{-\Delta\omega}{\omega} = \gamma_1^m \varepsilon_1 + \gamma_2^m \varepsilon_2 + \gamma_3^m \varepsilon_3 \tag{9}$$

with the strains of the inclusion also being defined relative to a free crystal of the same mineral at room conditions. Equation (9) applies to triclinic, and monoclinic inclusions when they experience no shear strains, and to orthorhombic inclusion crystals. For higher symmetries, discussed below and in detail in Angel et al. (2019), some or all of the components of the mode Grüneisen tensor are equal.

Therefore, the strains in an inclusion can be determined if the values of the components of its mode Grüneisen tensors are known, and sufficient Raman lines are measured to enable the strains ε_i to be determined through Equation (9). The values of the tensor components are not easy to obtain experimentally, because this requires different strain states to be imposed on a crystal. In general, it is not sufficient to measure the Raman shifts under hydrostatic pressure, because this only generates a single strain state and not the independent ranges of strains required. A practical alternative is to use computer simulations of the mineral based on density-functional theory (DFT) to calculate the Raman spectra of a crystal under a range of deviatoric strains from which the tensor components are determined by a least-squares fit of Equation (9). Tensor components have been calculated in this way for quartz (Murri et al. 2018; Murri et al. 2019), zircon (Stangarone et al. 2019) and rutile (Musiyachenko et al. 2021). They can be used in the Strainman program (Angel et al. 2019) to determine the strains in the inclusions from the difference in the Raman line positions measured at the center of the inclusion from the positions of the same lines measured on a free reference crystal of the same composition in air (Murri et al. 2018; Alvaro et al. 2020; Campomenosi et al. 2020b; Musiyachenko et al. 2021). In particular, care must be taken to only measure isolated inclusions as the presence of surfaces, cracks or other inclusions within 3 inclusion radii will modify the elastic relaxation (Fig. 13) and thus the final inclusion pressure. Entrapment conditions calculated from non-isolated inclusions will therefore be incorrect.

Calculating relaxation and entrapment. The volume relaxation of a spherical isotropic inclusion in an isotropic host crystal at room pressure is given by Equation (4) and can be converted into a ΔP_{relax} via the EoS of the inclusion phase, from which P_{thermo} can be calculated. However, inclusions in diamonds are never spherical, and the shape of the inclusion changes the amount of relaxation. Finite element model (FEM) calculations have been used to model the shape effects. Mazzucchelli et al. (2018) defined a shape factor Γ that can be used to 'correct' the pressure measured by Raman spectroscopy *at the center of the inclusion* to the one expected for the same phase trapped at the same conditions but as a spherical inclusion:

$$P_{inc}^{spherical} = \frac{P_{inc}^{measured}}{1+\Gamma} \tag{10}$$

If the $P_{inc}^{measured}$ is determined from X-ray diffraction measurements, this is an average over the entire inclusion volume, and a similar factor can be calculated from a FEM of the inclusion (Anzolini et al. 2019).

A shape factor of $\Gamma = 0$ represents an inclusion whose relaxation is the same as that of an isotropic sphere. The biggest influence on the shape factor Γ is the aspect ratio of the inclusion; platy inclusions have larger shape factors than more equant ones (Mazzucchelli et al. 2018). The shape factor for an octahedral inclusion of pyrope garnet in diamond (Mazzucchelli, pers. comm.) is −0.026, which gives a 2.6% correction to P_{inc} which is insignificant compared to the uncertainties associated with the measurement of inclusion pressures. However, for plate and needle-shaped inclusions the values of Γ become significant (Table 3). Values of the shape factor for other inclusions in diamond have not been calculated, but octahedrally-shaped inclusions of olivine, coesite and walstromite-structured $CaSiO_3$ (now known as breyite, Brenker et al. 2021) which are significantly softer than garnets will have smaller correction factors as indicated by the trends found by Mazzucchelli et al. (2018).

Table 3. Shape correction factors for pyrope inclusions in diamond.

Shape	**Aspect ratio**	Γ
Regular octahedron*	1:1:1	−0.026
Ellipsoids	5:1:1	−0.034
	2:1:1	−0.013
	1:1:1	−0.000
	1:2:2	−0.027
	1:5:5	−0.154
Prisms	5:1:1	−0.065
	2:1:1	−0.045
Cube	1:1:1	−0.037
Plates	1:2:2	−0.055
	1:5:5	−0.188

Note: Shape factors from Mazzucchelli et al. (2018) except for *pers. comm.

When the inclusion is a non-cubic mineral, or a cubic mineral with significant elastic anisotropy, then the relaxation of the stress itself is anisotropic and depends not only upon the contrast of the elastic properties but also on the orientation of the inclusion crystal with respect to both the host and the shape of the inclusion. Thus, the relaxation of a prismatic inclusion of olivine in diamond (e.g., Mitchell and Giardini 1953) will depend on which crystallographic axis olivine is parallel to the prism axis, and the amount of strain and stress relaxed in each crystallographic direction will be different. The characterization and quantification of this relaxation requires the numerical calculation of the relaxation tensor for each combination of orientation and shape for each host and inclusion pair of interest (Mazzucchelli et al. 2019). This tensor can be used to correct the measured inclusion stress or strain to that expected from a spherical inclusion and subsequent calculations can be performed based on the isotropic analysis for spherical inclusions (Bonazzi et al. 2019).

Therefore, after measurement of the inclusion pressure it should be corrected for shape effects, and the $P_{inc}^{spherical}$ can then be used in the EosFit-Pinc program (Angel et al. 2017) to calculate the relaxation, the P_{thermo}, the P_{foot} and the entrapment isomeke. The EosFit-Pinc program, available for free download at www.rossangel.net, has a simple graphical user interface and is not restricted to specific inclusion phases because the EoS can be chosen and loaded by the user. Most common forms of *P–V–T* EoS are supported by the EosFit program suite (Angel et al. 2014a), which gives the user the opportunity to explore the effect on the predicted entrapment conditions of issues such as uncertainties and differences between various published EoS.

Simple linear EoS models have been used in the past for inclusion calculations (e.g., Zhang 1998; Barron 2003) in which the properties such as thermal expansion and compressibility are assumed not to change significantly with temperature and pressure. This can be a good approximation for the conversion of the ΔV_{relax} of the inclusion into ΔP_{relax}, because the pressures involved are small compared to the bulk moduli. However, it results in calculated isomekes that are linear in P–T space, whereas they are actually curved (e.g., Fig. 12) as a result of the stiffening of solids and decrease in thermal expansion of solids with increasing pressure (and the counter effects with increasing temperature). This is especially important for inclusions in diamond, as its room- temperature thermal expansion coefficient is less than 20% of its value at lithospheric depths. This is one of the main factors leading to curved isomekes between diamonds and their inclusions. If the variation in properties was ignored and the isomeke shown in Figure 12 was extrapolated from P_{foot} as a straight line with a slope equal to its initial slope (i.e., with constant compressibilities and thermal expansivities for both phases), then it is clear that the entrapment pressure at 1230 °C would be overestimated by more than 1 GPa.

The calculation of an isomeke from a single inclusion with the isotropic model only provides a line of possible entrapment conditions in P–T space (e.g., Fig. 12). We emphasize again that the entrapment isomeke *does not* represent the path of exhumation of the diamond (Fig. 12), but it is only a set of P–T conditions at which the inclusion may have been entrapped; if it has not been modified by other processes following entrapment, the inclusion pressure P_{inc} holds no information about the exhumation path. Further information is required to obtain a pressure of entrapment and thus depth of diamond growth. One possibility is to use the intersection of the isomeke with a mantle geotherm (e.g., Cohen and Rosenfeld 1979; Anzolini et al. 2019) and a minimum estimate of entrapment pressure can be obtained from the point where the entrapment isomeke enters the stability field of diamond (Graham and Cybriwsky 1981). If two different phases are present as separate inclusions in a diamond and belong to the same growth stage, then the intersection of their isomekes provides an estimate of the P and T of entrapment, a calculation that can be readily performed in the EntraPT software (Mazzucchelli et al. 2021).

The EntraPT program (Mazzucchelli et al. 2021), freely available at *www.mineralogylab.com*, also provides further shape corrections based on finite element modeling (Mazzuchelli et al. 2018), and the calculation of the relaxation tensor for non-ellipsoidal elastically anisotropic inclusions in an isotropic host (Mazzucchelli et al. 2019; Morganti et al. 2020). EntraPT implements the methods applied by Bonazzi et al. (2019) to correctly calculate the mean stress in inclusions from the measured strains, and then to calculate entrapment isomekes with the isotropic model including the propagation of uncertainties, providing a clear workflow through the complex issues surrounding the handling of data from shaped anisotropic inclusions.

Examples

The key factor in measuring and interpreting inclusion stress states is the symmetry of the stress field in the inclusion when it is measured in the laboratory. It has important and different consequences for all three steps required to calculate the entrapment conditions of any inclusion crystal. Given that diamond is so much stiffer than the inclusions, the symmetry of the stress field depends only on the symmetry and the shape of the inclusion crystal. The following examples of common inclusions in diamonds have therefore been divided by the symmetry of the inclusion to illustrate how to use measurements of their remnant pressures to determine the depths of their entrapment by the growing diamond. In particular, we emphasize the analytical and computational procedures necessary to overcome, or work around, the fact that we are trying to use a model for inclusion pressures that is based on assumptions of isotropic properties and shape for inclusions that are neither spherical nor elastically isotropic.

Uncertainties in the EoS parameters of diamond and the inclusion phase contribute to uncertainties in isomeke, and thus entrapment, pressures. However, because diamond is so stiff, and its EoS parameters are relatively well determined (Table 2) the uncertainties in them contribute far less to uncertainties in entrapment pressures than do those arising from the uncertainties in measuring P_{inc} (Angel et al. 2015a). Uncertainties in published EoS of silicates vary considerably, as does their sensitivity to compositional variation in solid solutions, so the consequences for calculations of entrapment conditions must be evaluated carefully for each mineral system.

Cubic inclusions. When a cubic crystal is subject to hydrostatic stress or a change in temperature its size changes equally in all directions; the strain (Eqn. 6) is isotropic. Consequently, the application of an isotropic strain by a diamond on a spherical inclusion with cubic symmetry creates a hydrostatic stress in the inclusion, and the symmetry of the inclusion remains cubic. As explained above, the influence of the elastic anisotropy of diamond on the relaxation is extremely small and well below the resolution of measurements (Mazzucchelli et al. 2019), so the final stress state can be treated as hydrostatic. The pressure in the inclusion, P_{inc}, can therefore be determined by measuring its unit-cell volume by diffraction and using the known EoS for the mineral. Or the pressure can be determined by measuring the shift of Raman lines of an inclusion.

For cubic crystals the components of the mode Grüneisen tensors are $\gamma_1^m = \gamma_2^m = \gamma_3^m$, and the other components are zero, so Equation (9) reduces to:

$$\frac{-d\omega}{\omega} = \gamma_1^m\left(\varepsilon_1 + \varepsilon_2 + \varepsilon_3\right) = \gamma_1^m \frac{dV}{V} \tag{11}$$

This says that the Raman shift depends only upon the total volume change. Therefore P_{inc} can be determined from the measured shifts of its Raman bands from those of a free crystal of the same composition at room pressure, by using calibrations from high-pressure hydrostatic experiments. Note that the positions of Raman bands of free crystals that are solid solutions, such as garnets, can depend significantly on composition, as do their dependence on pressure and this must be allowed for in the analysis as well as the effects of peak broadening on apparent peak positions (Gillet et al. 2002; Kalugina and Zedgenizov 2020). Hydrostatic conditions in a cubic inclusion can be verified by comparing the pressures indicated by different Raman bands; if they differ significantly this is an indication that non-hydrostatic stresses are significant, and the inclusion should be treated as one of lower symmetry.

The P_{inc} should then be corrected by the appropriate shape factor (e.g., Table 3) to obtain $P_{inc}^{spherical}$, from which the entrapment isomekes can be calculated. Entrapment isomekes for several garnet endmembers are shown in Figure 16 which illustrates several issues critical for the correct interpretation of inclusion pressures as entrapment conditions. Most importantly, the P_{inc} expected for different garnet compositions trapped at the same pressure and temperature differs by up to 1 GPa. For example, a pure pyrope trapped at 6.1 GPa and 1200 °C will exhibit a P_{inc} = 0.1 GPa when measured at room conditions, whereas a pure almandine will have P_{inc} = 1.1 GPa. A measured P_{inc} = 0.5 GPa corresponds to entrapment pressures at 1200 °C of 6.9 GPa for pyrope, 5.6 GPa for grossular, and 4.8 GPa for almandine, a difference of more than 60 km in inferred depth of entrapment. Second, the thermodynamic calculations suggest that any inclusion trapped at temperatures higher (and pressures lower) than the zero-pressure isomeke should exhibit a negative P_{inc}. This would require bonding between the internal diamond surface and the inclusion to stretch the inclusion to a volume greater than that of a free crystal at room conditions. However, no negative P_{inc} values have ever been reported for inclusions in diamond. Therefore, such inclusions must expand within the diamond to their normal volume at room conditions, but no more, and they exhibit no pressure when measured at room conditions. Figure 16 therefore shows that pyrope, and pyrope-rich inclusions, cannot be used to determine entrapment conditions along geotherms corresponding to surface heat flows of more than ~37 mW/m^2. By contrast, grossular and almandine garnets will still exhibit $P_{inc} > 0$ for these entrapment conditions (Fig. 16).

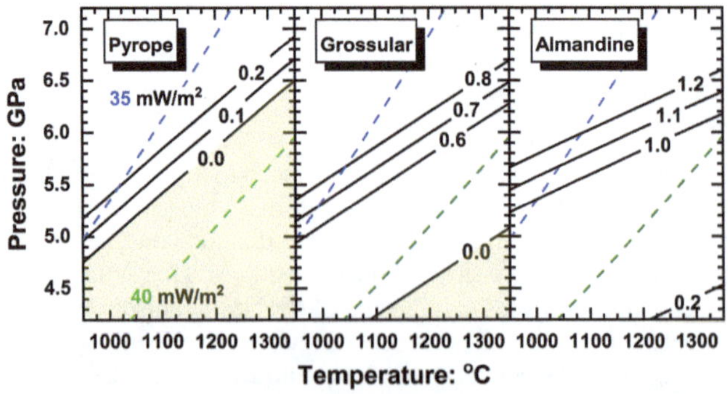

Figure 16. Entrapment isomekes for garnet endmembers in diamond, labelled with the corresponding P_{inc} calculated with the EoS listed in Table 2. Note that the isomeke slopes become shallower as the thermal expansion coefficients decrease from pyrope to grossular to almandine. Inclusions trapped in the **yellow areas** will exhibit zero pressure when measured at room conditions (see text for explanation). The **colored dashed lines** are mantle isotherms for different surface heat flows (Hasterok and Chapman 2011).

The reason for these significant changes in P_{inc} with composition can be understood in terms of the slopes of the isomekes (Eqn. 2) shown in Figure 16. The total range of bulk moduli of garnets is < 20 GPa (Table 2), which is < 10% of the difference from the bulk modulus of diamond (444 GPa). Therefore, the difference in compressibilities $\beta_{inc} - \beta_{dia}$ between different garnet inclusions and diamond varies by less than 10%. This contrast also controls the pressure spacing of isomeke lines, which are therefore very similar for all of the garnets with diamond (Fig. 16). In contrast, thermal expansion coefficients, α, of garnets vary between ~2.0 and 2.5×10^{-5} K^{-1}. Given that $\alpha_{dia} = 0.27 \times 10^{-5}$ K^{-1}, this means that the slopes of the isomekes vary because of a 25% change in $(\alpha_{inc} - \alpha_{dia})$. Garnets with lower thermal expansion coefficients, such as almandine, therefore have shallower isomekes that result in higher P_{foot} values and thus P_{inc} pressures, than pyrope that has higher thermal expansion. In fact, the influence of bulk modulus on P_{inc} for pyrope–almandine–grossular garnets is insignificant compared to the influence of thermal expansion.

It is therefore critical to determine the composition of garnet inclusions in order to obtain correct estimates of their entrapment depths, for example by *in situ* X-ray diffraction as described above. The composition also allows the EoS to be determined from the known variation of EoS parameters of garnets with composition; for example, within the pyrope–almandine–grossular garnets, the bulk modulus can be considered constant for these purposes, and the thermal expansion coefficient appears to vary approximately linearly with composition (e.g., Bosenick and Geiger 1997). Once the appropriate EoS is established from the composition, the P_{inc} can be determined from the measured unit-cell volume, and the volume of a free crystal of the same composition determined from the systematics of how garnet cell volumes change with composition. In the calculation of P_{inc} the biggest effect of composition is on the determination of V_0. After correction for the shape effects the entrapment isomeke can be calculated with the EoS for that particular composition.

The effects of the uncertainties in the EoS parameters of end-member garnets and the differences between various EoS formulations for the thermal effects on P_{trap} and therefore estimated depths of entrapment are small for garnets, amounting to no more than 0.2–0.3 GPa in the worst cases, corresponding to less than 10 km uncertainty in depth. On the other hand, if the composition is unknown, then Figure 16 shows that depths cannot be determined from P_{inc}. Figure 16 also suggests that an alternative approach of calculating the P_{trap} for each of the endmembers and interpolating between these in proportion to the composition of the garnet might give reasonable estimates of entrapment pressures for eclogitic garnets in the absence of detailed EoS data for the composition of interest. Another alternative is to calculate the entrapment conditions for a mixture of the end-member components, as if they were acting as separate phases. The last caution to note is that at least some eclogitic garnets in diamond have a fluid rim (Nimis et al. 2016), whose influence on the inclusion pressure should be included in calculations, as we discuss below.

If the change in elasticity, the Raman bands, or even the unit-cell volume with composition is not known for the inclusion mineral, an alternative approach is to extract the inclusion after measurement of P_{inc} and use the free inclusion crystal itself to determine its EoS (Nestola et al. 2019a). This procedure is necessary for inclusions such as spinels whose elastic properties and unit-cell volumes are a complex function of not just composition but also state of cation order.

Non-cubic inclusions. These minerals have elastic properties that vary with direction. When subject to the isotropic strain imposed by the diamond as a result of recovery to room conditions, they should therefore develop anisotropic stresses. The mutual elastic relaxation of the host and the inclusion reduces the anisotropy of the stress, but not to the hydrostatic state (Mazzucchelli et al. 2019); if the diamond and inclusion have behaved perfectly elastically from entrapment, the inclusion should not be under hydrostatic stress when measured in the laboratory. Whether this is reduced to insignificant levels or not depends on the mineral properties and on their crystallographic orientation with respect to their morphology. This can be calculated by FEM calculations for each specific inclusion (Mazzucchelli et al. 2019) but is time consuming.

Therefore, a first step in analyzing non-cubic inclusions is to determine their stress state experimentally. This can be achieved by plotting the measured inclusion unit cell parameters, normalized to the room-pressure values for the mineral, against the fractional volume change of the inclusion normalized in the same way (e.g., Fig. 17). In the case of coesite, a monoclinic mineral, one of the principal axes of compression is the (100) plane normal, so this plane spacing is calculated from the unit-cell parameters and plotted along with b/b_0 and c/c_0. The measured inclusion data for both coesite (Sobolev et al. 2000) and graphite (Glinnemann et al. 2003) fall near to the lines calculated from hydrostatic compression measurements, indicating that these inclusions have a stress field that is close to hydrostatic. The one exception appears to be the lowest-pressure graphite inclusion measured by Glinnemann et al. (2003) which plots on the line of isotropic strain.

As noted above, the positions of Raman bands are dependent upon the strain applied to the crystal (Eqns. 8,9), and not to the mean stress or pressure. Therefore, another way to determine whether the stress state of an inclusion is hydrostatic is to convert the measured Raman shifts into pressures by using the appropriate hydrostatic calibration for the mineral. If different Raman

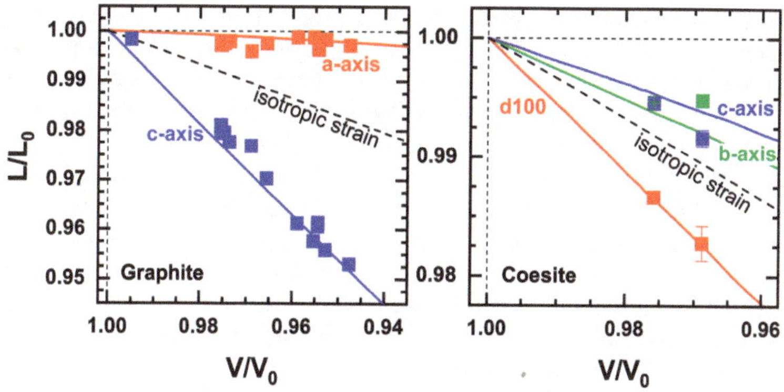

Figure 17. Measured strain states of graphite (Glinnemann et al. 2003) and coesite (Sobolev et al. 2000) inclusions in diamonds plotted as symbols. Colored lines are the cell parameters of crystals under hydrostatic stress measured by Hanfland et al. (1989) and Angel et al. (2001).

bands give the same pressure, then the stress state in the inclusion is approximately hydrostatic and can be determined from the Raman shifts, and this pressure should agree with that determined by diffraction, as is the case for the coesite inclusions shown in Figure 17 (Sobolev et al. 2000).

When the stress state of a significantly anisotropic inclusion such as coesite or graphite is approximately hydrostatic, this indicates that some other physical process beyond mutual elastic relaxation has modified the inclusion stress state. Such processes may include the presence of fluid in the inclusion or plastic flow in the host diamond to reduce the inclusion pressure. Given that the graphite inclusions measured by Glinnemann et al. (2003) appear to retain a P_{inc} of 0.1 to 2.6 GPa, consistent with 'entrapment' in the thermodynamic stability field of graphite (Fig. 15), the observed hydrostatic stress may be the result of internal redistribution of stress due to internal plastic flow in the inclusion. One might then interpret the as indicating that the graphite formed internally by inversion from diamond, perhaps due to the presence of trapped fluids. However, Glinnemann et al. (2003) also report that the majority of these inclusions show a series of specific CORs with respect to the diamond (Fig. 10b) which they interpret in terms of a coherent interface; such coherency could result in strains that are significantly different from that expected for a spherical inclusion with no bonding to the diamond, in which case the estimates of entrapment conditions with the current simple model of inclusion pressures would be incorrect.

If the stress state of the inclusion is shown to be significantly non-hydrostatic, use of the Strainman program (Angel et al. 2019) allows the strain state of the inclusion to be calculated from the measured positions of the Raman lines. The strains measured in this way, or directly by diffraction, can be converted into stresses by use of the elastic modulus tensor of the inclusion crystal (Nye 1957); see Mazzucchelli et al. (2021) for applications of this approach to inclusions. Of course, minerals become stiffer with increasing pressure, so use of a room-pressure elastic tensor will lead to an under-estimate of the true stress values of the inclusion by a few %. This is insignificant for most inclusions, except those that retain pressures in excess of 1 GPa. For quartz inclusions in garnets Bonazzi et al. (2019) and Gonzalez et al. (2019) have demonstrated that the stress and strain state is non-hydrostatic, but that a good estimate of entrapment pressures can be obtained by using the mean normal stress as P_{inc} in subsequent calculations. A similar approach should work for non-cubic inclusions in diamonds.

For a number of low-symmetry inclusion phases neither the elastic tensors are known, nor have the mode Grüneisen tensors been determined, and therefore the detailed stress state in the inclusion cannot be determined. An alternative approach in these cases is to use DFT simulations to determine which Raman line or lines is least affected by non-hydrostatic stresses (Anzolini et al. 2018; Nestola et al. 2018a). The position of this line in the Raman spectrum of the inclusion is then measured, and the shift from the position of the same line in a free crystal is then converted to a pressure using the measured hydrostatic calibration. For a kyanite inclusion in a diamond from the Voorspoed kimberlite, without any detectable fluid present in the inclusion, this method gave a calculated entrapment pressure for the measured FTIR N-aggregation residence temperature that overlaps with *P–T* estimates obtained by traditional chemical geobarometry for diamonds from other kimberlites from the Kaapvaal craton (Nestola et al. 2018a). The calculated entrapment conditions for a $CaSiO_3$ inclusion (breyite) determined in this way are in the thermodynamic stability field of the phase (Anzolini et al. 2018).

The elastic anisotropy of olivine is significantly less than coesite or graphite and available measurements (e.g., Nestola et al. 2011) do not clearly distinguish whether the stress state of olivine inclusions is hydrostatic or not. Given the low elastic anisotropy and that the mean stress in olivine inclusions is typically less than 1 GPa, no significant error is introduced by determining the P_{inc} from the unit-cell volume of the inclusion as measured by diffraction. Figure 18 shows the measured volumes of 13 olivine inclusions from the Udachnaya kimberlite, plotted against their composition as determined by *in situ* X-ray data collection and structure refinement, as described in a previous section of this chapter. In order to determine the P_{inc} it is also necessary

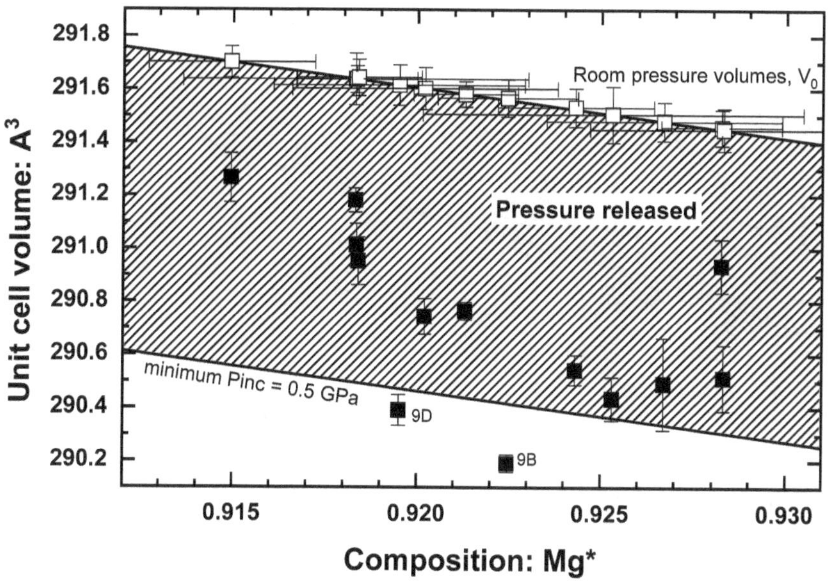

Figure 18. The measured unit-cell volumes (filled symbols) of 13 olivine inclusions in 4 different diamonds from the Udachnaya kimberlite, plotted against their composition determined by *in situ* structure refinement (Rustioni et al. 2015; Mazzucchelli et al. 2016). Mg* is the effective Mg number, defined in Equation 1, and plotted in Figure 4. The open symbols are the V_0 values for each inclusion, calculated from the composition and the known variation of unit-cell volume with composition (Schwab and Küstner 1977). A P_{inc} of approximately 0.5 GPa is the minimum pressure that should be found in any olivine inclusion trapped in the stability field of diamond (Fig. 19).

to know the unit-cell volume that the inclusion would have at room pressure, V_0. The latter varies with composition. If the inclusion cannot be freed from the diamond the V_0 of the inclusion can be calculated from the composition determined by *in situ* structure refinement, and the known variation of V_0 of olivines with Mg/Fe ratio (Schwab and Küstner 1977). Great care has to be taken in normalizing the unit-cell measurements because the exact values depend on diffractometer alignment, calibration and the peak-centering algorithms used (Angel et al. 2000); the cell parameters of the same crystal measured on different diffractometers will often differ by many times the estimated standard uncertainties. The simple solution is to measure in air a crystal or series of crystals of known composition with exactly the same protocols on the same diffractometer on which the inclusions are measured, at a similar time. Tests with data sets from other mantle olivine crystals collected in air show that this procedure yields the correct V_0 with uncertainties dominated by the uncertainty in composition from structure refinement (Fig. 18).

The equilibrium phase boundary between graphite and diamond (Day 2012) at mantle temperatures lies between the entrapment isomekes for olivine in diamond that would give rise to $P_{inc} = 0.5$ and 0.6 GPa (Fig. 19). Allowing for the small effects of shape on relaxation, a P_{inc} of approximately 0.5 GPa is the minimum pressure that should be found in any olivine inclusion. This corresponds to a volume compression of $V/V_0 = 0.996$, a boundary that is also shown in Figure 18. This type of graph then becomes a useful diagnostic tool because it immediately shows which inclusions have volumes, and hence P_{inc} values, consistent with entrapment during growth of the diamond within its thermodynamic stability field. At these minimum depths and pressures, uncertainties in the EoS of olivine (Angel et al. 2018) result in an uncertainty in P_{inc} of less than 0.03 GPa, corresponding to an uncertainty in unit-cell volume of 0.06 $Å^3$, about the size of the error bars in volume in Figure 18. It is therefore clear from this figure that these olivine

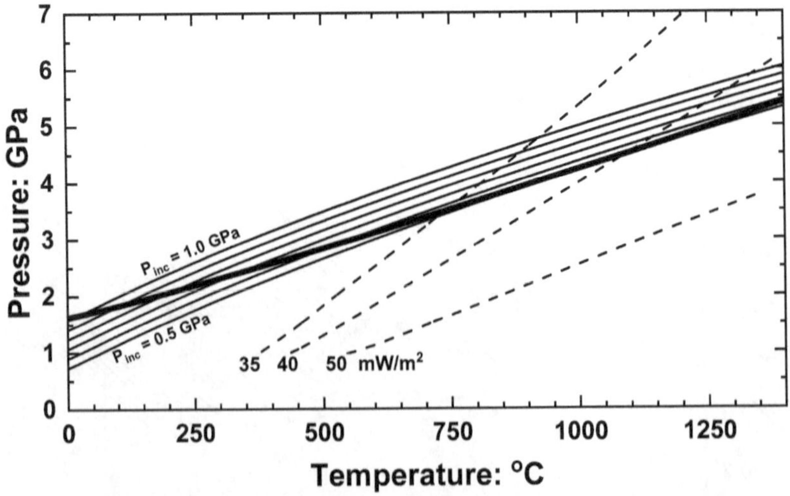

Figure 19. Entrapment isomekes of mantle composition olivine in diamond at 0.1 GPa intervals of P_{inc}. The **colored lines** are mantle isotherms for different surface heat flows (Hasterok and Chapman 2011). Note that in the range of mantle temperatures, the diamond = graphite equilibrium phase boundary (**grey line**, Day 2012) lies between the isomekes with P_{inc} = 0.5 and 0.6 GPa.

inclusions have had their inclusion stress state modified by some other process that is not considered in the simple elastic model. The fact that the measured volumes are larger than the maximum expected from the elastic analysis is consistent with partial, but not complete, pressure release which could occur either by brittle fracture of the diamond at low temperatures or by plastic flow at higher temperatures. Certainly, a number of these inclusions have very fine fractures of a few μm width leading from them. If a crack of thickness t were planar and extended around the entire equatorial plane of a spherical inclusion of radius r it would create a fractional volume increase in the inclusion of $3t/4r$. A planar fracture 1 μm thick around an inclusion of radius 50 μm would increase the volume by 1.5%, which is more than enough to account for all of the inclusion volumes shown in Figure 18. However, microscopy and X-ray absorption tomography (e.g., Rustioni et al. 2015; Mazzucchelli et al. 2016; Nimis et al. 2016) show that cracks do not surround the inclusions in this way and therefore they can only contribute a much smaller increase in inclusion volume that is insufficient by orders of magnitude to account for the observed pressure release indicated by Figure 18.

The influence of fluids. It was recently discovered that olivine inclusions in gem-quality diamonds are surrounded by a thin film of hydrous silicic fluid that is no more than 1.5 μm thick (Nimis et al. 2016). Such films seem to be almost ubiquitous around silicate and oxide inclusions in lithospheric diamonds (e.g., Nestola et al. 2018a; Nimis et al. 2019) and can modify the stress state of the inclusion through several processes. First, because they are a fluid, they cannot support shear stresses which should ensure that the stress state of the solid part of the inclusion is hydrostatic. This may explain the hydrostatic stress state of some of the inclusions shown in Figure 17, and it allows the model of isotropic elasticity to be used to calculate entrapment conditions, provided that allowance is made for the presence and equation of state of the fluid. The latter is unknown because the exact composition of the fluid is not known. If the properties of the fluid can be modelled with those of water (Zhang and Duan 2005), then the final pressure of a coesite inclusion is decreased compared to a dry inclusion; a dry inclusion would exhibit P_{inc} ~3 GPa if it had been trapped at 4.2 GPa and 1000 °C, and 10 vol% water would decrease P_{inc} to ~2.1 GPa. This means that if the presence of water in a coesite inclusion is ignored the entrapment pressures will be under-estimated.

By contrast, in olivine the expected final inclusion pressures are *increased* by the presence of water for entrapment near to the lower stability limit of diamond, and the effect of water decreases with increasing depth of entrapment (Fig. 20). This sounds like a surprising and counter-intuitive result, because fluids are much more compressible than solid minerals, the bulk modulus of liquid water being less than 2 GPa at room conditions (Zhang and Duan 2005). However, their thermal expansion coefficients are also an order of magnitude higher than solids. As a consequence, the isomekes of water with diamond have similar slopes to those of olivine with diamond (but not coesite and diamond; Fig. 15) and hence the presence of pure water in an inclusion in olivine has a small effect on the inclusion pressures (Fig. 20).

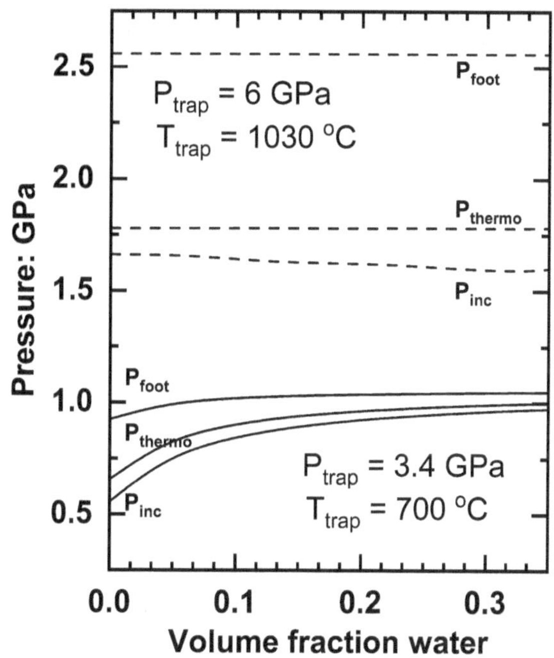

Figure 20. The variation of the pressures of a composite olivine plus water inclusion as a function of water content for two different entrapment conditions. Calculated with the BM4 + isothermal-type EoS for olivine (Angel et al. 2018) and the water EoS of Zhang and Duan (2005). Calculations performed with the mphase utility of EosFit7c (Musiyachenko et al. 2021).

Thus, the presence of water *alone* therefore cannot explain the low inclusion pressures that are uniformly observed in olivine inclusions, which would be expected to exceed 1 GPa if they were entrapped significantly deeper than the depths of the graphite–diamond phase boundary (Figs. 15, 19, 20). But, the presence of apparently amorphous solid on the diamond surface and in contact with the fluid rim in these inclusions (Nimis et al. 2016) suggests that the fluid may have precipitated solids that were originally dissolved in the fluid at the time of entrapment, just as brucite is found as a precipitate in some super-deep diamonds (Palot et al. 2016). If the precipitation is associated with an overall reduction in the volume of the fluid plus precipitate it will contribute to a further decrease in inclusion pressure. The same reasoning applies to the precipitation of additional olivine onto the olivine, or carbon onto the diamond during ascent of the diamond. The evaluation of the magnitude of these effects awaits the full characterization of the chemical composition of the various components of the inclusion rims, and thermodynamic modelling.

The presence of fluid also greatly magnifies the effects of cracks around the inclusion compared to cracks around purely solid inclusions. The formation of the crack allows the fluid to flow out from the rim and into the crack, effectively increasing the volume of the fluid rim by the volume of the crack. The bulk modulus of fluid water at shallow lithospheric depths is of the order of 10 GPa (Zhang and Duan 2005). Therefore, the drop in pressure of an inclusion fluid as the result of the opening of a crack of just 1/3 of the volume of the fluid rim would be enough to drop the inclusion pressure by ~3 GPa. This is enough to reduce to zero the P_{inc} of any olivine trapped at any depth in the lithosphere. Smaller crack volumes are therefore sufficient to explain the range of pressures and unit-cell volumes measured for olivine inclusions (Fig. 18 and Nestola et al. 2011) and why olivine P_{inc} values are often much less than expected. The conclusion is that without the characterization of the fluid in detail and the volumes of the cracks and fluid rim, olivine inclusion pressures should not be used to infer growth conditions of the diamonds. Both the sulfide-carbide and silicate inclusions in CLIPPIR and blue diamonds (Smith et al. 2016; Smith et al. 2018) also contain both CH_4 and H_2 as fluid phases, and calculations similar to those that we have outlined here will have to be made in order to calculate entrapment conditions from the observed inclusion stress states.

Phase transformations. Many mineral phases that are stable in the deep Earth are thermodynamically unstable at lower pressures and, in the absence of constraints, transform to lower density polymorphs or unmix into a less dense assemblage of phases. For example, the density changes associated with the transformations of ringwoodite to wadsleyite and to olivine are responsible for the well-known seismic discontinuities that define the Earth's transition zone. In the mantle, such phase changes can occur because the materials are not constrained. By contrast, an inclusion phase is constrained by the surrounding diamond and is not free to expand. As a consequence, the state of thermodynamic equilibrium is changed to one in which the boundary conditions are no longer P and T, but the T and the volume imposed by the diamond (Ferrero and Angel 2018). This concept is also the origin of inclusion pressures that we measure as P_{inc}. If the inclusion pressure drops into the stability field of a lower pressure polymorph and the temperature is sufficiently high, whether or not the transformation takes place depends on whether the volume change due to the transformation can be accommodated by the diamond expansion on exhumation, unless the necessary volume is provided by processes such as plastic flow or brittle fracturing of the diamond host.

If a mineral transformation is accompanied by a large density change, then the volume is not available to accommodate the volume change due to the transition, and the high-pressure phase can be preserved. This is why it is possible to recover single-phase inclusions of phases such as jeffbenite, a high-pressure polymorph of garnet (Armstrong and Walter 2012; Nestola et al. 2016b) and ringwoodite (Pearson et al. 2014), as well as calcium silicate perovskite with minor exsolution (Nestola et al. 2018b). The reverse argument has been made to conclude that $CaSiO_3$ inclusions with the walstromite structure (the mineral breyite) cannot be inverted from $CaSiO_3$ perovskite, because the 30% volume change could not be accommodated by the diamond host (Anzolini et al. 2016). Indeed, the measured P_{inc} of 4.3 GPa is consistent with entrapment of the $CaSiO_3$ inclusion within the stability field of breyite (Anzolini et al. 2018; Woodland et al. 2020). The P_{inc} is also higher than the lower-pressure limit to breyite stability, which explains why it did not invert to its lower-pressure wollastonite polymorph. Fractured inclusions of breyite exhibit much lower pressures (Anzolini et al. 2016) which indicates that the fracturing occurred at temperatures that were too low to allow breyite to invert to wollastonite.

Transformations that involve smaller volume changes can be accommodated by the expansion of the diamond as the pressure of the inclusion drops below the equilibrium phase boundary. Then, the high-pressure phase will start to transform to the low-pressure polymorph. However, because the low-pressure polymorph is less dense, the pressure in the inclusion will increase as the transformation proceeds, until the pressure returns to that of the phase boundary. At this point the transformation will stop. Further exhumation may lead to further expansion of the inclusion volume, allowing further transformation to occur, so the pressure will be maintained at approximately the pressure of the phase boundary until the transformation of the inclusion is complete. This is the concept of pressure buffering. The various possibilities for the subsequent evolution of the inclusion pressure, which depend on whether the transformation proceeds to completion, are discussed for the inversion of coesite to quartz in garnet host crystals by Ferrero and Angel (2018). These principles will also apply to inclusions in diamonds. For coesite inclusions in diamonds, the evidence of the measurements of Sobolev et al. (2000) and Smith et al. (2015), which are in agreement with EoS calculations (Fig. 15) is that sealed and isolated coesite inclusions can remain at pressures in the stability field of coesite and no transformation to quartz occurs.

Mineral inclusions that are single-phase solid solutions at the time of entrapment often exsolve other phases due to the reduced mutual solubility of chemical components at lower temperatures and pressures. Examples include clinopyroxene exsolution from majoritic garnets (Harte and Cayzer 2007), spinel exsolution from magnesiowustite (Wirth et al. 2014; Palot et al. 2016) and the exsolution of titanite perovskite from silicate perovskite (Nestola et al. 2018b). As for simple isochemical phase transformations, the density changes associated with these reactions mean that it is possible for the pressure in the inclusion to be buffered to a higher value than expected for a single phase. The interpretation of the residual pressures of such multiphase inclusions is further complicated by the need to know the volume fractions of all of the phases present, their individual equations of state and, in the case of coherent exsolution, the requirement to consider the effects of the coherency between the phases on the stress and strain evolution of the composite inclusion. While calculations of coherency strains have been made for crustal minerals at ambient pressure (e.g., Robin 1974; Williame and Brown 1974), we are not aware of any being applied to mantle phases trapped in diamonds.

Plastic deformation. Diamond can accommodate stresses by plastic deformation generated by the motion of dislocations. This is a thermally activated process whose rate also depends on the applied stress. Plastic deformation is not significant in diamond below 1000 °C (Weidner et al. 1994) but at 1700 °C it is quite rapid even when the applied deviatoric stresses are low (Howell et al. 2012b). Therefore, inclusion pressures can be reduced by plastic flow driven by the pressure difference between P_{inc} and the external pressure during diamond ascent at high temperatures; the driving force is zero at the time of entrapment because the inclusion is trapped at the external pressure but increases while the diamond travels towards the crust. In general, lower temperatures and rapid exhumation rates will allow less plastic deformation to occur. As a consequence, there is little evidence for significant plastic flow around inclusions in lithospheric diamonds (Agrosi et al. 2016), and consideration of plastic flow is not needed to interpret the inclusion pressures measured in these diamonds. However, pervasive dislocation networks are present in sub-lithospheric diamonds (Cayzer et al. 2008; Agrosi et al. 2017; Smith et al. 2018). It is clear that this plastic deformation is the mechanism by which these diamonds can accommodate the volume changes associated with the inversion of high-pressure phases to the lower-pressure polymorphs sometimes found in these diamonds (Cayzer et al. 2008). The effect on inclusion pressures can, in principle, be calculated either by explicit finite element modelling of the inclusion (Anzolini et al. 2019) or by a 1-D visco-elasto-plastic mechanical model which enables the effects of different exhumation rates on the inclusion pressure to be explored (Zhong et al. 2018, 2020) and, in principle, constraints to be placed on the speed of ascent of the kimberlite hosting the diamond.

FUTURE PERSPECTIVES

The history of the study of inclusions in diamonds shows that it is extremely important to characterize mineral inclusions *in situ* so as to preserve them for measurement by techniques that are either not yet capable of the sensitivity or resolution required, or perhaps not yet even developed. An obvious example is the discovery of new mineral phases as inclusions in diamond which has been made possible by the far greater X-ray intensities available at the synchrotron compared to the laboratory (e.g., Meyer et al. 2019). The scarcity of examples discussed in this chapter shows how few data have been collected *in situ* from inclusions in diamonds, compared to the thousands of *ex situ* measurements. Yet these examples have shown the new discoveries (e.g., the fluids found by Nimis et al. 2016) that can be made, and the far-reaching questions that can be answered, by *in situ* measurements with equipment available in any crystallography laboratory. There are thus many opportunities to use modern analytical methods and computational power to answer very specific questions about inclusions in diamonds and thus how they were formed and encapsulated, and at what conditions in the Earth.

The techniques that we have described here in some detail for certain mineral inclusions in diamond can be readily applied to other mineral inclusions in diamonds. In particular, a large number of orientation measurements will enable the CORs of inclusions with respect to diamonds to be unambiguously identified and tests performed to determine whether there is a correlation between COR types and the type of host rock or other aspects of the diamond growth environment. Such measurements should continue to be applied to inclusions such as olivine, garnet or graphite, for which there is insufficient data to determine with statistical significance whether multiple CORs exist as sub-sets of the data; for example, it is not clear whether the graphite inclusions shown in Figure 10b represent two specific CORs, or a single rotational COR. Many more determinations of graphite orientations in diamond would answer this simple question and help determine how such inclusions, and their host diamond, were formed. Similarly, there is only published data for ferropericlase inclusions in two diamonds; many more measurements are required to determine whether the specific COR observed is a general phenomenon and whether it has genetic significance. Orientation measurements of inclusion clusters can be combined with imaging techniques such as X-ray tomography to provide clear insights into the mechanism by which the inclusions were trapped and, therefore, how the diamond grew (Nimis et al. 2019). Similar insights may well be obtained by further measurements of inclusions not only *in situ* in the diamond, but while the diamond is still in contact with its current host rock (Nestola et al. 2017).

The major element chemistry of an inclusion can be determined by *in situ* X-ray diffraction and spectroscopy. Doing so preserves the inclusion for future research while providing useful data. For inclusions in sub-lithospheric diamonds that may represent the only known terrestrial example of a mineral, *in situ* measurements avoid the risk of the destruction of unique samples either through phase transformation or amorphization, or by simple loss of a sample much smaller than its host diamond. Thus the discovery of the ringwoodite as a single inclusion in diamond (Pearson et al. 2014) was proven by a combination of *in situ* single-crystal diffraction and Raman spectroscopy. The *in situ* measurement of its infra-red absorption spectrum was the first ever proof that the mantle transition zone is significantly hydrated, all obtained without risk of sample loss.

Despite studies published over the last half-century (e.g., Harris et al. 1970) the use of inclusion stress states to determine depths of diamond growth is only now becoming feasible through concurrent development of theory with the improvement in the precision of both the *in situ* measurements that we have described and the elastic properties of minerals. The presence of fluid around many lithospheric inclusions (Nimis et al. 2016) both complicates and simplifies the analysis of their stress state. It complicates analysis because it is now necessary to determine the phase equilibria of the trapped fluid and the associated amorphous rims in order to know how the total volume of the rim varies with P and T, not only from their respective EoS but also as a result of reactions within the fluid and the precipitation of the amorphous material (Palot et al. 2016). If that is done, then the problem of inclusion stress is simplified in two ways. The mineral component of the inclusion will be under hydrostatic stress and thus the shifts of its Raman lines will directly yield the inclusion pressure via calibrations of Raman shifts with pressure. The elasticity of the composite fluid + mineral inclusion can be treated as isotropic and the entrapment conditions calculated from P_{inc} using the Reuss bound on the elastic properties of the mineral plus those of the rim (Musiyachenko et al. 2021).

The current models for determining inclusion entrapment conditions from the inclusion stress assume free-slip boundary conditions between the host and the inclusion. Therefore, if fluid is absent then orientation measurements must first be performed to determine whether or not the inclusions have an exact COR to the diamond. If they do, this may indicate epitaxial bonding between the diamond host and the inclusion which may modify the stress state significantly through coherency, as is probably the case for graphite (Glinnemann et al. 2003) and ferropericlase (Nimis et al. 2018) inclusions. Interpretation of the inclusion stress state then requires explicit calculations by FEM of the contribution from the coherency strains and how they vary as the external P and T evolve during diamond ascent. In the absence of coherency the mean stress in the inclusion can be treated as a pressure in the isotropic model to obtain a good estimate of the entrapment isomeke (Bonazzi et al. 2019; Gonzalez et al. 2019). Beyond that approach, the anisotropy of the elastic properties of an inclusion should allow the isomekes of different directions in a single inclusion crystal to be used to determine the entrapment conditions uniquely, as recently demonstrated for quartz inclusions in garnets (Alvaro et al. 2020).

We have not discussed the characterization of the strain in the diamond created by the pressure difference between the inclusion and the external ambient pressure on the diamond because its quantitative analysis is far more complicated and at the limits of current capabilities in mineral physics. The difficulty arises from the fact that the symmetry of the diamond in the vicinity of the inclusion is broken by the combination of radial compressional strains and tensional tangential strains arising from elastic relaxation (Fig. 13) and it is therefore no longer cubic. The reduction in the symmetry of the diamond means that its physical tensor properties change, including the optical properties which lead to the optical birefringence haloes around inclusions. Similarly, the values of the mode Grüneisen tensor components may be expected to change from their values in cubic diamond. Once these have been determined by DFT simulation and confirmed to agree with experimental measurements (Grimsditch et al. 1978), the shifts of the diamond Raman modes in the host diamond have the potential to be used to infer the inclusion pressure (Nasdala et al. 2005). Similarly, it is possible that the intensities of the Raman line of diamond could be used to map the strain field around inclusions, similar to the method applied to garnets that host inclusions (Campomenosi et al. 2020a).

The general conclusion to be drawn from this review is that rapidly-developing instrumental technologies along with new concepts from mineral and crystal physics and the application of advanced computational methods such as DFT and FEM, can be used in combination to provide new ways to characterize inclusions in diamonds and yield new insights into how and where they nucleate and grow within the Earth.

ACKNOWLEDGMENTS

Our own research in this field has been supported by funding from the European Research Council under the European Union's Horizon 2020 research and innovation program through grant agreements 307322 to Fabrizio Nestola and 714936 to Matteo Alvaro, and by the SIR-MIUR Grant MILE DEEp (no. RBSI140351) to Matteo Alvaro. We would like to thank

Mattia Mazzucchelli for providing calculations on shape correction factors, Paolo Nimis for recalculating geotherms, and both for discussions and comments. For more than a decade Mathias Meyer (Rigaku Oxford Diffraction) has given us much valuable advice about collecting and processing X-ray data and has willingly developed new features in the Crysalis^{TM} software for our research programs. We thank Michelle Wenz, Martin Kunz, Daria Pasqual, and the editors Karen Smit and Graham Pearson for their very helpful and thorough reviews and comments.

REFERENCES

Adams HG, Cohen LH, Rosenfeld JL (1975) Solid inclusion piezothermometry I: comparison dilatometry. Am Mineral 60:574–583

Agrosi G, Nestola F, Tempesta G, Bruno M, Scandale E, Harris J (2016) X-ray topographic study of a diamond from Udachnaya: Implications for the genetic nature of inclusions. Lithos 248:153–159

Agrosi G, Tempesta G, Della Ventura G, Guidi MC, Hutchison M, Nimis P, Nestola F (2017) Non-destructive in-situ study of plastic deformations in diamonds: X-ray diffraction topography and μ–FTIR mapping of two super deep diamond crystals from Sao Luiz (Juina, Brazil). Crystals 7

Alvaro M, Mazzucchelli ML, Angel RJ, Murri M, Campomenosi N, Scambelluri M, Nestola F, Korsakov A, Tomilenko AA, Marone F, Morana M (2020) Fossil subduction recorded by quartz from the coesite stability field. Geology 48:24–28

Angel R, Seifert F (1999) The effect of pressure on cation ordering in minerals: Problems and perspectives. Phase Transitions 69:1–16

Angel RJ (2003) Automated profile analysis for single-crystal diffraction data. J Appl Crystallogr 36:295–300

Angel RJ (2004) Average. www.rossangel.net

Angel RJ, Finger LW (2011) SINGLE: a program to control single-crystal diffractometers. J Appl Crystallogr 44:247–251

Angel RJ, Nestola F (2016) A century of mineral structures: how well do we know them? Am Mineral 101:1036–1045

Angel RJ, Downs RT, Finger LW (2000) High-temperature and high-pressure diffractometry. *In*: High-pressure and high-temperature crystal chemistry. Vol 41. Hazen RM, Downs RT, (eds). MSA, p 559–596

Angel RJ, Mosenfelder JL, Shaw CSJ (2001) Anomalous compression and equation of state of coesite. Phys Earth Planet Int 124:71–79

Angel RJ, Gonzalez-Platas J, Alvaro M (2014a) EosFit7c and a Fortran module (library) for equation of state calculations. Z Kristallogr 229:405–419

Angel RJ, Alvaro M, Nestola F (2018) 40 years of mineral elasticity: a critical review and a new parameterisation of equations of state for mantle olivines and diamond inclusions. Phys Chem Mineral 45:95–113

Angel RJ, Alvaro M, Nestola F, Mazzucchelli ML (2015a) Diamond thermoelastic properties and implications for determining the pressure of formation of diamond–inclusion systems. Russian Geol Geophys 56:211–220

Angel RJ, Milani S, Alvaro M, Nestola F (2015b) OrientXplot—a program to process and display relative crystal orientations. J Appl Crystallogr 48:1330–1334

Angel RJ, Milani S, Alvaro M, Nestola F (2016) High quality structures at high pressure? Insights from inclusions in diamonds. Z Kristallogr 231:467–473

Angel RJ, Mazzucchelli ML, Alvaro M, Nestola F (2017) EosFit-Pinc: A simple GUI for host–inclusion elastic thermobarometry. Am Mineral 102:1957–1960

Angel RJ, Murri M, Mihailova B, Alvaro M (2019) Stress, strain and Raman shifts. Z Kristallogr 234:129–140

Angel RJ, Mazzucchelli ML, Alvaro M, Nestola F (2020) Response to the discussion by Zhong et al. of 'EosFit-Pinc: a simple GUI for host–inclusion elastic thermobarometry'. Am Mineral 105:1587–1588

Angel RJ, Mazzucchelli ML, Alvaro M, Nimis P, Nestola F (2014b) Geobarometry from host–inclusion systems: the role of elastic relaxation. Am Mineral 99:2146–2149

Anzolini C, Prencipe M, Alvaro M, Romano C, Vona A, Lorenzon S, Smith EM, Brenker FE, Nestola F (2018) Depth of formation of super-deep diamonds: Raman barometry of $CaSiO_3$–walstromite inclusions. Am Mineral 103:69–74

Anzolini C, Angel RJ, Merlini M, Derzsi M, Tokár K, Milani S, Krebs MY, Brenker FE, Nestola F, Harris JW (2016) Depth of formation of $CaSiO_3$-walstromite included in super-deep diamonds. Lithos 265:138–147

Anzolini C, Nestola F, Mazzucchelli ML, Alvaro M, Nimis P, Gianese A, Morganti S, Marone F, Campione M, Hutchison MT, Harris JW (2019) Depth of diamond formation obtained from single periclase inclusions. Geology 47:219–222

Armstrong LS, Walter MJ (2012) Tetragonal almandine pyrope phase (TAPP): retrograde Mg-perovskite from subducted oceanic crust? Euro J Mineral 24:587–597

Ashley K, Barkoff DW, Steele-MacInnis M (2017) Barometric constraints based on apatite inclusions in garnet. Am Mineral 102:743–749

Baptiste V, Tommasi A, Demouchy D (2012) Deformation and hydration of the lithospheric mantle beneath the Kaapvaal craton, South Africa. Lithos 149:31–50

Barron LM (2003) A simple model for the pressure preservation index of inclusions in diamond. Am Mineral 88:1615–1619

Becker PJ, Coppens P (1974) Extinction within the limit of validity of the Darwin transfer equations. I General formalisms for primary and secondary extinction and their application to spherical crystals. Acta Crystallogr A30:129–147

Blessing R (1987) Data reduction and error analysis for accurate single crystal diffraction intensities. Crystallogr Rev 1:3–58

Bonazzi M, Tumiati S, Thomas J, Angel RJ, Alvaro M (2019) Assessment of the reliability of elastic geobarometry with quartz inclusions. Lithos 350–351:105201

Bosenick A, Geiger C (1997) Powder X ray diffraction study of synthetic pyrope-grossular garnets between 20 and 295 K. J Geophys Res B 102:22649–22657

Brenker FE, Nestola F, Brenker L, Peruzzo L, Harris JW (2021) Origin, properties and structure of breyite: the second most abundant mineral inclusion in super-deep diamonds. Am Mineral 106:38–43

Bruno M, Rubbo M, Aquilano D, Massaro FR, Nestola F (2016) Diamond and its olivine inclusions: A strange relation revealed by ab initio simulations. Earth Planet Sci Lett 435:31–35

Busing W, Levy H (1967) Angle calculations for 3- and 4- circle X-ray and neutron diffractometers. Acta Crystallogr 22:457–464

Campomenosi N, Mazzucchelli ML, Mihailova B, Angel RJ, Alvaro M (2020a) Using polarized Raman spectroscopy to study the stress gradient in mineral systems with anomalous birefringence. Contrib Mineral Petrol 175:16

Campomenosi N, Rubatto D, Hermann J, Mihailova B, Scambelluri M, Alvaro M (2020b) Establishing a protocol for the selection of zircon inclusions in garnet for Raman thermobarometry. Am Mineral 105:992–1001

Campomenosi N, Mazzucchelli ML, Mihailova B, Scambelluri M, Angel RJ, Nestola F, Reali A, Alvaro M (2018) How geometry and anisotropy affect residual strain in host inclusion systems: coupling experimental and numerical approaches. Am Mineral 103:2032–2035

Cantrell JH (1980) Generalized Grüneisen tensor from solid nonlinearity parameters. Phys Rev B 21:4191–4195

Cayzer NJ, Odake S, Harte B, Kagi H (2008) Plastic deformation of lower mantle diamonds by inclusion phase transformations. Euro J Mineral 20:333–339

Cohen LH, Rosenfeld JL (1979) Diamond: depth of crystallisation inferred from compressed included garnet. J Geol 87:333–340

Davies GR, van den Heuval Q, Matveev S, Drury MR, Chinn IL, Gress MU (2018) A combined cathodoluminescence and electron backscatter diffraction examination of the growth relationships between Jwaneng diamonds and their eclogitic inclusions. Mineral Petrol 112:S231–S242

Day HW (2012) A revised diamond-graphite transition curve. Am Mineral 97:52–62

Diamond R (1969) Profile analysis in single-crystal diffractometry. Acta Crystallogr Sect A 25:43–55

Downs RT, Gibbs GV, Boisen MB (1990) A study of the mean-square displacement amplitudes of Si, Al and O atoms in framework structures: evidence for rigid bonds, order, twinning and stacking faults. Am Mineral 75:1253–1267

Eshelby JD (1957) The determination of the elastic field of an ellipsoidal inclusion, and related problems. Proc R Soc London Ser A 241:376–396

Ferrero S, Angel RJ (2018) Micropetrology: are inclusions grains of truth? J Petrol 59:1671–1700

Finger LW, King H (1978) A revised method of operation of the single-crystal diamond cell and refinement of the structure of NaCl at 32 kbar. Am Mineral 63:337–342

Forster A, Brandstetter S, Schulze-Briese C (2019) Transforming X-ray detection with hybrid photon counting detectors. Philos Trans R Soc A 377:15

Frank-Kamenetsky VA (1964) The nature of structural impurities and inclusions in minerals. LGU Press, Leningrad

Futergendler SI, Frank-Kamenetsky V (1961) Oriented inclusions of olivine, garnet and chrome-spinel in diamonds (in Russian). Zapiski Vsesoyuznogo Mineralogicheskogo Obshchestva 90:230–236

Gillet P, Ingrin J, Chopin C (1984) Coesite in subducted continental crust: P–T history deduced from an elastic model. Earth Planet Sci Lett 70:426–436

Gillet P, Sautter V, Harris JW, Reynard B, Harte B, Kunz M (2002) Raman spectroscopic study of garnet inclusions in diamonds from the mantle transition zone. Am Mineral 87:312–317

Glinnemann J, Kusaka K, Harris JW (2003) Oriented graphite single-crystal inclusions in diamond. Z Kristallogr 218:733–739

Gonzalez JP, Thomas JB, Baldwin SL, Alvaro M (2019) Quartz-in-garnet and Ti-in-quartz thermobarometry: Methodology and first application to a quartzofeldspathic gneiss from eastern Papua New Guinea. J Metamorph Geol 37:1193–1208

Graham E, Cybriwsky Z (1981) Garnet inclusions in diamond and the state of the upper mantle. Phys Chem Mineral 7:216–222

Griffiths TA, Habler G, Abart R (2016) Crystallographic orientation relationships in host–inclusion systems: New insights from large EBSD data sets. Am Mineral 101:690–705

Griffiths TA, Habler G, Schantl P, Abart R (2017) Crystallographic Orientation Relationships (CORs) between rutile inclusions and garnet hosts: towards using COR frequencies as a petrogenetic indicator. EGU General Assembly Conference Abstracts 19:7523

Grimsditch MH, Anastassakis E, Cardona M (1978) Effect of uniaxial stress on the one-center optical phonon of diamond. Phys Rev B 18:901–904

Grüneisen E (1926) Zustand des festen Körpers. Handbuch der Physik 1:1–52

Gubanov N, Zedgenizov D, Sharygin I, Ragozin A (2019) Origin and evolution of high-Mg carbonatitic and low-Mg carbonatitic to silicic high-density fluids in coated diamonds from Udachnaya kimberlite pipe. Minerals 9:734

Habler G, Griffiths TA (2017) Crystallographic orientation relationships. *In*: Mineral Reaction Kinetics: Microstructures, textures, chemical and isotopic signatures. Heinrich W, Abart R, (eds). European Mineralogical Union and the Mineralogical Society of Great Britain & Ireland, London, p 541–585

Hamilton WC (1974) Angle settings for four-circle diffractometers. *In*: Vol IV Kynoch Press, Birmingham, International Tables for X-ray Crystallography, p 273–284

Hanfland M, Beister H, Syassen K (1989) Graphite under pressure: Equation of state and first-order Raman modes. Phys Rev B 39:12598–12603

Harris JW (1968) The recognition of diamond inclusions. Pt. 1: Syngenetic inclusions. Ind Diamond Rev 28:402–410

Harris JW, Vance ER (1972) Induced graphitisation around crystalline inclusions in diamond. Contrib Mineral Petrol 35:227–234

Harris JW, Gurney J (1979) Inclusions in diamond. *In*: Properties of Diamond. Field JE, (ed) Academic Press, London, p 555–591

Harris JW, Henrique R, Meyer HOA (1967) Orientation of silicate mineral inclusions in diamonds. Crystal Growth 7:118–123

Harris JW, Milledge HJ, Barron THK, Munn RW (1970) Thermal expansion of garnets included in diamond. J Geophys Res 75:5775–5792

Harte B, Cayzer N (2007) Decompression and unmixing of crystals included in diamonds from the mantle transition zone. Phys Chem Mineral 34:647–656

Hartman H (1954) A discussion on 'Oriented olivine inclusions in diamond'. Am Mineral 39:674–675

Hasterok D, Chapman D (2011) Heat production and geotherms for the continental lithosphere. Earth Planet Sci Lett 307:59–70

Hazen RM, Downs RT (eds) (2000) High-Pressure and High-Temperature Crystal Chemistry. Reviews in Mineralogy Vol. 41, Mineral Soc Am

Heinemann R, Staack V, Fischer A, Kroll H, Vad T, Kirfel A (1999) Temperature dependence of Fe,Mg partitioning in Acapulco olivine. Am Mineral 84:1400–1405

Holland TJB, Powell R (2011) An improved and extended internally consistent thermodynamic dataset for phases of petrological interest, involving a new equation of state for solids. J Metamorph Geol 29:333–383

Howell A, Nasdala L (2008) Using strain birefringence in diamond to estimate the remnant pressure on an inclusion. Aust J Earth Sci 55:1175–1178

Howell D, Wood IG, Nestola F, Nimis P, Nasdala L (2012a) Inclusions under remnant pressure in diamond: a multi-technique approach. Euro J Mineral 24:563–573

Howell D, Wood IG, Dobson DP, Jones AP, Nasdala L, Harris JW (2010) Quantifying strain birefringence halos around inclusions in diamond. Contrib Mineral Petrol 160:705–717

Howell D, Piazolo S, Dobson DP, Wood IG, Jones AP, Walte N, Frost DJ, Fisher D, Griffin WL (2012b) Quantitative characterization of plastic deformation of single diamond crystals: A high pressure high temperature (HPHT) experimental deformation study combined with electron backscatter diffraction (EBSD). Diamond Relat Mater 30:20–30

Izraeli E, Harris J, Navon O (1999) Raman barometry of diamond formation. Earth Planet Sci Lett 173:351–360

Joswig W (2011) In situ X-ray diffraction investigations of a multiphase inclusion in a diamond from the Juina district, Brazil. Z Kristallogr 226:226–228

Joswig W, Paulus E, Winkler B, Milman V (2003) The crystal structure of $CaSiO_3$–walstromite, a special isomorph of wollastonite-II. Z Kristallogr 218:811–818

Joswig W, Stachel T, Harris JW, Baur W, Brey GP (1999) New Ca-silicate inclusions in diamonds —tracers from the lower mantle. Earth Planet Sci Lett 173:1–6

Kalugina AD, Zedgenizov DA (2019) Raman discrimination of garnet inclusions in Siberian diamonds. J Raman Spectr 51:1438–1444

Key SW (1967) Grüneisen tensor for anisotropic materials. J Appl Phys 38:2923–2928

King H, Finger LW (1979) Diffracted beam crystal centering and its application to high-pressure crystallography. J Appl Crystallogr 12:374–378

Kirfel A, Will G (1980) Bonding in $[S_2O_6]^{2-}$: Refinement and pictorial representation from an X-ray diffraction study of $Na_2S_2O_6{\cdot}2H_2O_6$ and $Na_2S_2O_6{\cdot}2D_2O$ Acta Crystallogr B 36:512–523

Knight K (2010) Analytical expressions to determine the isothermal compressibility tensor and the isobaric thermal expansion tensor for monoclinic crystals: application to determine the direction of maximum compressibility in jadeite. Phys Chem Mineral 37:529–533

Knoche R, Angel RJ, Seifert F, Fliervoet TF (1998) Complete substitution of Si for Ti in titanite $Ca(Ti_{1-x}Si_x)^{VI}Si^{IV}O_5$. Am Mineral 83:1168–1175

Korsakov AV, Kohn MJ, Perraki M (2020) Applications of Raman spectroscopy in metamorphic petrology and tectonics. Elements 16:105–110

Kosminska K, Spear FS, Majka J, Faehnrich K, Manecki M, Piepjohn K, Dallmann WK (2020) Deciphering late Devonian-early Carboniferous *P–T–t* path of mylonitized garnet-mica schists from Prins Karls Forland, Svalbard. J Metamorph Geol 38:471–493

Kroll H, Lueder T, Schlenz H, Kirfel A, Vad T (1997) The Fe^{2+},Mg distribution in orthopyroxenes: a critical assessment of its potential as a geospeedometer. Euro J Mineral 9:705–733

Kueter N, Soesilo J, Fedortchouk Y, Nestola F, Belluco L, Troch J, Wälle M, Guillong M, Von Quadt A, Driesner T (2016) Tracing the depositional history of Kalimantan diamonds by zircon provenance and diamond morphology studies. Lithos 265:159–176

Kunz M, Armbruster T (1990) Difference displacement parameters in alkali feldspars: effects of (Si,Al) order–disorder. Am Mineral 75:141–149

Kunz M, Gillet P, Fiquet G, Sautter V, Graafsma H, Conrad P, Harris J (2002) Combined in situ X-ray diffraction and Raman spectroscopy on majoritic garnet inclusions in diamonds. Earth Planet Sci Lett 198:485–493

Liu L-G, Mernagh TP, Jaques AL (1990) A mineralogical Raman spectroscopy study on eclogitic garnet inclusions in diamonds from Argyle. Contrib Mineral Petrol 105:156–161

Loveday JS, McMahon MI, Nelmes RJ (1990) The effect of diffraction by the diamonds of a diamond-anvil cell on single-crystal sample intensities. J Appl Crystallogr 23:392–396

Mazzucchelli M, Angel RJ, Rustioni G, Milani S, Nimis P, Domeneghetti MC, Marone F, Harris JW, Nestola F, Alvaro M (2016) Elastic geobarometry and the role of brittle failure on pressure release. Geophys Res Abstr 18:13569

Mazzucchelli ML, Angel RJ, Alvaro M (2021) EntraPT: An online platform for elastic geothermobarometry. Am Mineral 106:830–837

Mazzucchelli ML, Reali A, Morganti S, Angel RJ, Alvaro M (2019) Elastic geobarometry: relaxation of elastically anisotropic inclusions. Lithos 350–351:105218

Mazzucchelli ML, Burnley P, Angel RJ, Morganti S, Domeneghetti MC, Nestola F, Alvaro M (2018) Elastic geothermobarometry: corrections for the geometry of the host–inclusion system. Geology 46:231–234

Meyer NA, Wenz MD, Walsh JPS, Jacobsen SD, Locock AJ, Harris JW (2019) Goldschmidtite, (K,REE,Sr)(Nb,Cr)O_3: A new perovskite supergroup mineral found in diamond from Koffiefontein, South Africa. Am Mineral 104:1345–1350

Milani S, Nestola F, Angel RJ, Nimis P, Harris J (2016) Crystallographic orientations of olivine inclusions in diamonds. Lithos 265:312–316

Miletich R, Allan DR, Kuhs WF (2000) High-pressure single-crystal techniques. *In*: High-Pressure and High-Temperature Crystal Chemistry. Vol 41. Hazen RM, Downs RT, (eds). MSA, p 445–519

Mitchell RS, Giardini AA (1953) Oriented olivine inclusions in diamond. Am Mineral 38:136–138

Morana M, Mihailova B, Angel RJ, Alvaro M (2020) Quartz metastability at high pressure: what new can we learn from polarized Raman spectroscopy? Phys Chem Mineral 47:34

Morganti S, Mazzucchelli M, Alvaro M, Reali A (2020) A numerical application of the Eshelby theory for geobarometry of non-ideal host–inclusion systems. Meccanica 55:751–764

Murri M, Alvaro M, Angel RJ, Prencipe M, Mihailova BD (2019) The effects of non-hydrostatic stress on the structure and properties of alpha-quartz. Phys Chem Mineral 46:487–499

Murri M, Mazzucchelli ML, Campomenosi N, Korsakov AV, Prencipe M, Mihailova B, Scambelluri M, Angel RJ, Alvaro M (2018) Raman elastic geobarometry for anisotropic mineral inclusions. Am Mineral 103:1869–1872

Musiyachenko KA, Murri M, Prencipe M, Angel RJ, Alvaro M (2021) A Grüneisen tensor for rutile and its application to host-inclusion systems. Am Mineral 106:1586–1595

Nasdala L, Hofmeister W, Harris JW, Glinnemann J (2005) Growth zoning and strain patterns inside diamond crystals as revealed by Raman maps. Am Mineral 90:745–748

Nasdala L, Brenker FE, Glinnemann J, Hofmeister W, Gasparik T, Harris JW, Stachel T, Reese I (2003) Spectroscopic 2D-tomography: residual pressure and strain around mineral inclusions in diamonds. Euro J Mineral 15:931–935

Nestola F (2015) The crucial role of crystallography in diamond research. Rendiconti Lincei Scienze Fisiche e Naturali 26:225–233

Nestola F, Jung H, Taylor LA (2017) Mineral inclusions in diamonds may be synchronous but not syngenetic. Nat Commun 8:14168

Nestola F, Nimis P, Angel RJ, Milani S, Bruno M, Prencipe M, Harris JW (2014) Olivine with diamond-imposed morphology included in diamond. Syngenesis or protogenesis? International Geology Review 56:1658–1667

Nestola F, Prencipe M, Nimis P, Sgreva N, Peritt SH, Chinn IL, Zaffirio G (2018a) Toward a robust elastic geobarometry of kyanite inclusions in eclogitic diamonds. J Geophys Res: Solid Earth 123:6411–6423

Nestola F, Nimis P, Ziberna L, Longo M, Marzoli A, Harris JW, Manghnani MH, Fedortchouk Y (2011) First crystal-structure determination of olivine in diamond: composition and implications for provenance in the Earth's mantle. Earth Planet Sci Lett 305:249–255

Nestola F, Alvaro M, Casati MN, Wilhelm H, Kleppe A, Jephcoat A, Domeneghetti M, Harris J (2016a) Source assemblage types for cratonic diamonds from X-ray synchrotron diffraction. Lithos 265:334–338

Nestola F, Burnham AD, Peruzzo L, Tauro L, Alvaro M, Walter MJ, Gunter M, Anzolini C, Kohn SC (2016b) Tetragonal Almandine-Pyrope Phase, TAPP: finally a name for it, the new mineral jeffbenite. Mineral Mag 80:1219–1232

Nestola F, Korolev N, Kopylova M, Rotiroti N, Pearson DG, Pamato MG, Alvaro M, Peruzzo L, Gurney JJ, Moore AE, Davidson J (2018b) $CaSiO_3$ perovskite in diamond indicates the recycling of oceanic crust into the lower mantle. Nature 555:237–241

Nestola F, Zaffiro G, Mazzucchelli ML, Nimis P, Andreozzi GB, Periotto B, Princivalle F, Lenaz D, Secco L, Pasqualetto L, Logvinova AM (2019a) Diamond–inclusion system recording old deep lithosphere conditions at Udachnaya (Siberia). Sci Rep 9:12586

Nestola F, Jacob DE, Pamato MG, Pasqualetto L, Oliveira B, Greene S, Perritt S, Chinn I, Milani S, Kueter N, Sgreva N (2019b) Protogenetic garnet inclusions and the age of diamonds. Geology 47:431–434

Neuser RD, Schertl H-P, Logvinova AM, Sobolev NV (2015) An EBSD study of olivine inclusions in Siberian diamonds: evidence for syngenetic growth? Russ Geol Geophys 56:321–329

Nimis P, Angel RJ, Alvaro M, Nestola F, Harris JW, Casati N, Marone F (2019) Crystallographic orientations of magnesiochromite inclusions in diamonds: what do they tell us? Contrib Mineral Petrol 174:29

Nimis P, Alvaro M, Nestola F, Angel RJ, Marquardt K, Rustioni G, Harris JW, Marone F (2016) First evidence of hydrous silicic fluid films around solid inclusions in gem-quality diamonds. Lithos 260:384–389

Nimis P, Nestola F, Schiazza M, Reali R, Agrosi G, Mele D, Tempesta G, Howell D, Hutchison MT, Spiess R (2018) Fe-rich ferropericlase and magnesiowüstite inclusions reflecting diamond formation rather than ambient mantle. Geology 47:27–30

Nye JF (1957) Physical Properties of Crystals. Oxford University Press, Oxford

Orlov YL (1977) The Mineralogy of the Diamond. John Wiley & Sons, New York

Palot M, Jacobsen SD, Townsend JP, Nestola F, Marquardt K, Miyajima N, Harris JW, Stachel T, McCammon CA, Pearson DG (2016) Evidence for H_2O-bearing fluids in the lower mantle from diamond inclusion. Lithos 265:237–243

Pearson DG, Shirey SB (1999) Isotopic dating of diamonds. *In*: Application of Radiogenic Isotopes to Ore Deposit Research and Exploration. Lambert DD, Ruiz J, (eds). Society of Economic Geologists, p 143–171

Pearson DG, Brenker FE, Nestola F, McNeill J, Nasdala L, Hutchison MT, Matveev S, Mather K, Silversmit G, Schmitz S, Vekemans B (2014) Hydrous mantle transition zone indicated by ringwoodite included within diamond. Nature 507:221–223

Perrillat JP, Daniel I, Lardeaux JM, Cardon H (2003) Kinetics of the quartz–coesite transition: application to the exhumation of ultrahigh-pressure rocks. J Petrol 44:773–788

Prince E, Collins DM (1992) Refinement of structural parameters, 8.2: Other refinement methods. *In*: International Tables for X-ray Crystallography. Vol C Wilson A, (ed) International Union of Crystallography, Kluwer Academic Publishers, Dordrecht

Rigaku-Oxford-Diffraction (2016) Crysalis-Pro Software system, Rigaku Corporation, Oxford, UK

Robin PF (1974) Stress and strain in cryptoperthite lamellae and the coherent solvus of alkali feldspars. Am Mineral 59:1299–1318

Rosenfeld JL, Chase AB (1961) Pressure and temperature of crystallization from elastic effects around solid inclusion minerals? Am J Sci 259:519–541

Rustioni G, Angel RJ, Milani S, Mazzucchelli ML, Nimis P, Domeneghetti MC, Marone F, Alvaro M, Harris JW, Nestola F (2015) Elastic geobarometry for host–inclusion systems: Pressure release and the role of brittle failure. Rendiconti Online della Società Geologica Italiana 35:137

Schlenker JL, Gibbs GV, Boison Jr. MB (1978) Strain-tensor components expressed in terms of lattice parameters. Acta Crystallogr A 34:52–54

Schmidt C, Ziemann MA (2000) In-situ Raman spectroscopy of quartz: A pressure sensor for hydrothermal diamond-anvil cell experiments at elevated temperatures. Am Mineral 85:1725–1734

Schwab R, Küstner D (1977) Präzisionsgitter konstantenbestimmung zur festlegung röntgenographisher Bestimmungskurven für synthetische Olivine der Mischkristallreihe Forsterit–Fayalit. Neues Jahrb Mineral Monatsh:205–215

Seryotkin YV, Skvortsova VL, Logvinova AM, Sobolev NV (2017) Results of study of crystallographic orientation of olivine and diamond from Udachnaya kimberlite pipe, Yakutia. Dokl Earth Sci 476:1155–1158

Smith E, Kopylova M, Frezzotti M, Afanasiev V (2015) Fluid inclusions in Ebelyakh diamonds: Evidence of CO_2 liberation in eclogite and the effect of H_2O on diamond habit. Lithos 216–217:106–117

Smith E, Shirey BS, Nestola F, Bullock E, Wang J, Richardson SH, Wang W (2016) Large gem diamonds from metallic liquid in Earth's deep mantle. Science 354:1403–1405

Smith EM, Shirey SB, Richardson SH, Nestola F, Bullock ES, Wang JH, Wang WY (2018) Blue boron-bearing diamonds from Earth's lower mantle. Nature 560:84–87

Sobolev NV, Seryotkin YV, Logvinova AM, Pavlushin AD, Ugap'eva SS (2020) Crystallographic orientation and geochemical features of mineral inclusions in diamonds. Russ Geol Geophys 61:634–649

Sobolev NV, Fursenko BA, Goryainov SV, Shu JF, Hemley RJ, Mao HK, Boyd FR (2000) Fossilized high pressure from the Earth's deep interior: The coesite-in-diamond barometer. PNAS 97:11875–11879

Sorby HC, Butler PJ (1869) On the structure of rubies, sapphire, diamonds and some other minerals. Proc R Soc London 17:291–302

Stachel T, Harris JW (2008) The origin of cratonic diamonds–constraints from mineral inclusions. Ore Geol Rev 34:5–32

Stangarone C, Angel R, Prencipe M, Campomenosi N, Mihailova BD, Alvaro M (2019) Measurement of strains in zircon inclusions by Raman spectroscopy. Euro J Mineral 31:685–694

Tajčmanová L, Podladchikov Y, Powell R, Moulas E, Vrijmoed J, Connolly JAD (2014) Grain-scale pressure variations and chemical equilibrium in high-grade metamorphic rocks. J Metamorph Geol 32:195–207

Tao RBA, Fei YW, Bullock ES, Xu C, Zhang LF (2018) Experimental investigation of Fe^{3+}-rich majoritic garnet and its effect on majorite geobarometer. Geochim Cosmochim Acta 225:1–16

Trueblood KN, Bürgi H-B, Burzlaff H, Dunitz JD, Gramaccioli CM, Schulz HH, Schuemli U, Abrahams SC (1996) Atomic displacement parameter nomenclature. Acta Crystallogr A 52:770–781

Voigt W (1910) Lehrbuch der Kristallphysik. Teubner, Leipzig

Weidner D, Wang Y, Vaughan M (1994) Strength of diamond. Science 266:419–422

Wenz MD, Jacobsen SD, Zhang DZ, Regier M, Bausch HJ, Dera PK, Rivers M, Eng P, Shirey SB, Pearson DG (2019) Fast identification of mineral inclusions in diamond at GSECARS using synchrotron X-ray microtomography, radiography and diffraction. J Synchrot Radiat 26:1763–1768

Wijbrans CH, Rohrbach A, Klemme S (2016) An experimental investigation of the stability of majoritic garnet in the Earth's mantle and an improved majorite geobarometer. Contrib Mineral Petrol 171:20

Williame C, Brown WL (1974) A coherent elastic model for the determination of the orientation of exsolution boundaries: application to feldspars. Acta Crystallogr A 30:316–331

Wirth R, Dobrzhinetskaya LF, Harte B, Schreiber A, Green HW (2014) High-Fe (Mg,Fe)O inclusion in diamond apparently from the lowermost mantle. Earth Planet Sci Lett 404:365–375

Woodland AB, Girnis AV, Bulatov VK, Brey GP, Hofer HE (2020) Breyite inclusions in diamond: experimental evidence for possible dual origin. Euro J Mineral 32:171–185

Zhang Y (1998) Mechanical and phase equilibria in inclusion–host systems. Earth Planet Sci Lett 157:209–222

Zhang Z, Duan Z (2005) Prediction of the PVT properties of water over wide range of temperatures and pressures from molecular dynamics simulation. Phys Earth Planet Int 149:335–354

Zhong X, Moulas E, Tajčmanová L (2018) Tiny timekeepers witnessing highrate exhumation processes. Sci Rep 8:2234

Zhong X, Dabrowski M, Jamtveit B (2019) Analytical solution for the stress field in elastic half-space with a spherical pressurized cavity or inclusion containing eigenstrain. Geophys J Int 216:1100–1115

Zhong X, Moulas E, Tajčmanová L (2020) Post-entrapment modification of residual inclusion pressure and its implications for Raman elastic thermobarometry. Solid Earth 11:223–240

Zhukov VP, Korsakov AV (2015) Evolution of host–inclusion systems: a visco-elastic model. J Metamorph Geol 33:815–828

Ziman JM (1960) Electrons and phonons: the theory of transport phenomena in solids. Oxford University Press, Oxford

Zyuzin N (1967) On nature of orientation of garnet inclusion from Yakutian diamond. Geologiya i Geofizika 8:126–128

Reviews in Mineralogy & Geochemistry
Vol. 88 pp. 307-392, 2022

Mineral Inclusions in Lithospheric Diamonds

Thomas Stachel

Earth and Atmospheric Sciences
University of Alberta
Edmonton, AB, Canada

tstachel@ualberta.ca

Sonja Aulbach

Institut für Geowissenschaften
Goethe Universität Frankfurt
Frankfurt am Main, Germany

s.aulbach@em.uni-frankfurt.de

Jeffrey W. Harris

School of Geographical and Earth Sciences
University of Glasgow
Glasgow, UK

Jeff.Harris@glasgow.ac.uk

INTRODUCTION

Initially, it may seem extremely daunting to be sitting in front of several thousand carats of small diamonds, knowing that the likelihood of finding a diamond with a meaningful inclusion is about 1% of that parcel. Like other events in life, however, the joy of finding such material soon outweighs the perceived hardship. Mineral inclusions in diamond provide unique and pristine information about inner Earth, from depths of about 120 to just over 800 km and over 3.5 to 0.1 billion years of Earth history. Such is the scope and importance of inclusion research.

Once the inclusion-bearing diamonds have been collected, hopefully from a known source, an intricate set of procedures falls into place. A microscopic description of the physical features of the diamond and its valuable cargo are needed, a preliminary paragenesis is assigned based on inclusion color, and both diamond and inclusion are photographed in detail. Then comes the release of the μm-sized inclusion by crushing of the diamond host, the hope being that the mineral is recovered in one piece, an infrequent event. The next stage is mounting in epoxy and polishing, minimizing rounding of the inclusion surface *en route* to quarter μm perfection. When this stage is complete, the researcher can afford to relax as modern-day analytical techniques are non-destructive (electron microprobe) or dig only small holes in the perfect inclusion surface (ion microprobe and to a lesser degree, laser ablation-based micro-sampling).

In this chapter, we present results of the above procedures for minerals included in diamonds that originate in the lithospheric mantle. A comprehensive review of the mineral chemistry of inclusions, based on an already largely saturated database of almost 5000 inclusion analyses, was previously presented by Stachel and Harris (2008) and is not repeated here. Instead, the two main sections of the review focus on detailed comparisons of the major and trace element compositions of peridotitic and eclogitic suite inclusions with the equivalent minerals in cratonic xenoliths.

1529-6466/22/0088-0006$10.00 (print)
1943-2666/22/0088-0006$10.00 (online)
http://dx.doi.org/10.2138/rmg.2022.88.06

A history of inclusion discovery in lithospheric diamonds

When John Evelyn was advised by his parents in 1642 that an extended European tour would be better than fighting on the Royalist's side in the English Civil War (1642–1651), a connection between this sojourn and the probable first description of a mineral inclusion in diamond might be considered distinctly tenuous. Nevertheless, on Friday, September 29th, 1645 his diary records that he was in Venice and he wrote, "I went to see the collection of a noble Venetian Signor Rugini: he has a stately Palace richly furnished, …. but above all was a diamond which had growing in it a faire ruby" (de Beer 1955). It is not surprising that he identified a ruby, a mineral known from ancient times. However, since ruby is exceptionally rare as an inclusion in diamond, the inclusion was most likely a peridotitic garnet.

Another 17th century traveler, Jean Baptist Tavernier, noted black material in some Indian diamonds he bought, referring to them as "l'herbe pourrie", translated as either "rotten weed" or "decomposed vegetable matter" (Tavernier 1676). The burning of a diamond by Lavoisier in 1772 and similar quantitative experiments with charcoal convinced the French scientist that these two substances were distinct forms of carbon. When graphite was further shown to be a third form of carbon, the origin and formation of diamond was linked to processes involving biogenic matter and gases, a concept that lasted for most of the 19th century. Göppert (1864) outlines the beliefs of over twenty scientists published between 1816 and 1860. The majority view was that diamond formation was linked to biogenic matter, although water, topaz and gold were surmised as other inclusions. One author had a different view, suggesting that diamond formation was the result of compression of carbonic acid in Earth, which caused carbon to crystallize as diamond. This rather perceptive view of diamond formation was not appreciated by the monograph's author, a botanist and paleontologist. Brewster (1861) examined fifty Indian diamonds and remarked that some had small cavities having "shapes that resembled those of insects and lobsters" (see below).

If one steps back in time for a moment and assumes a scientific knowledge base of the mid-19th century, Figures 1 to 4 illustrate what the early scientists may have been looking at. Figure 1 shows graphite clusters in a diamond from Sierra Leone exhibiting a 'plant-like' impression. The realistic, encapsulated flying insect inclusion in Figure 2 is from the Karowe Mine in Botswana, the head and feelers being the result of some minute inclusion, which also caused two disproportionately large graphite-filled cleavage fractures to create the wings. The very rounded outline of the clear inclusion in Figure 3 might easily confirm the presence of water or gas within that diamond. Finally, Figure 4 would reinforce some of these initial observations as vegetable matter (graphite flakes) occurring on the interfacial surfaces between the water or gas inclusions and the diamond.

With the discovery of the first primary diamond deposits in South Africa in 1870, inclusion identification was still by observation only, and the new igneous environment did not have an immediate impact on research thinking. Boutan (1886) divided inclusions into crystalline and non-crystalline, and post-1870 research proposed hematite (iron glance), ilmenite and pyrite as inclusions. Lewis and Bonney (1897) considered that compressed gas was responsible for the common yellow color of diamonds from Kimberley. Bauer (1904) believed he had recognized fluid enclosures in diamond and, from their behavior when heated, considered the liquid to consist of either carbon dioxide and water or a saline solution. They also mention diamond as an inclusion.

With a new century came the understanding that the minerals found in diamond were linked to the mineralogy of the kimberlite rock and its xenoliths. In perhaps the shortest letter ever to be published in the periodical Nature (14 column lines long), Sutton (1907) reasonably argued the first identification of garnet (color unspecified) inside a South African diamond. Additionally, refractive index work on inclusions by Wagner (1914) identified pyrope and

Figure 1. An octahedral diamond from Sierra Leone showing a graphite cluster in the central portion of the diamond.

Figure 2. An 'insect'-shaped inclusion in a Botswana diamond. See text for details.

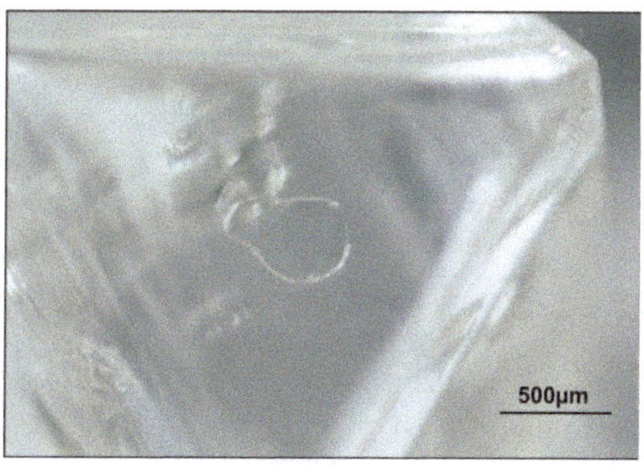

Figure 3. The rounded nature of this colorless inclusion might easily be considered as a liquid by a 19th century observer.

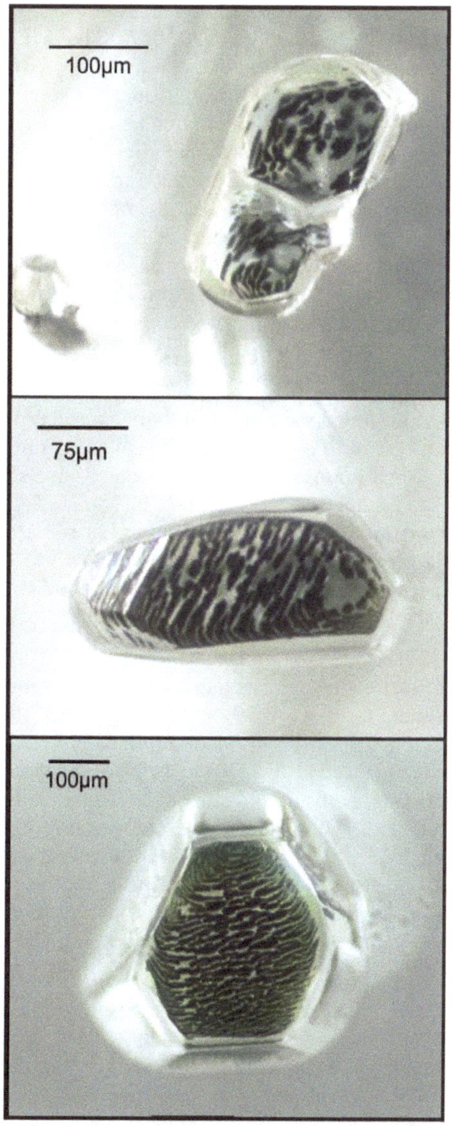

Figure 4. The presence of graphite crystals on {111} diamond/inclusion interfaces may have helped reinforce the presence of 'vegetable matter' in diamond to early inclusion researchers.

recorded the presence of (chrome) diopside in a diamond. Sutton (1921) listed garnet, ilmenite, pyrite, and mentions olivine for the first time as definite inclusions and notes zircon, mica and cognate graphite as probable inclusions. Sutton also studied epigenetic material in diamond fractures proposing calcite, an apophyllite mineral and pectolite. Magnetite was identified as an inclusion by Spencer (1924). The book on diamonds by Williams (1932) listed the known inclusion minerals up to that time and provided clear evidence of diamond as an inclusion: a mauve stone embedded in a colorless one. Williams concludes, as might be expected from the evidence at hand, that diamond was a phenocryst in kimberlite.

The first color micro-photographs of a suite of inclusions, mostly in polished diamonds, were described by Gübelin (1952). The inclusion in the sixth diamond in the color set likely resolves why zircons were commonly noted by previous authors as an inclusion in diamond, as the color of that mineral looks very much like the typical pale orange of an eclogitic garnet. The connection to zircon may also have been influenced by the typical honey color of zircons recovered from kimberlite heavy media concentrate. Two colorless inclusions were visually identified as quartz, a rare mineral in diamond, though they were likely the much more commonly found colorless mineral, olivine. Apart from crystalline inclusions, this work also captured photographic detail of fractures within diamonds containing dark/black material. For example, what the author called hematite 'iron-roses' look very much like typical sulfide inclusions within their rosette fractures. Graphite was identified by experiment, but much of the other dark material was considered as original magma. There were also observations and descriptions about diamond and cloud-like masses as inclusions.

Diamonds containing clouds were specifically examined by Giardini and Melton (1975). Similar to Gübelin, these authors considered such inclusions as gaseous. This is because the four, six or twelve sided cavities of 1 to 5 μm size contained no solid material when broken or polished through. Subsequent studies have shown this not to be the case (Weiss et al. 2022). Clouds range in color from white to mid-grey with an overall shape that can be irregular but is more often distinctly cuboid or star-shaped and visually aligned with the shape of the diamond. In the first case, the sides of the cube are aligned at right-angles to the fourfold direction of the diamond and in the second orientation, the six arms of the star cloud end at the apices of the octahedron, if that is the enclosing diamond shape.

The identification of an olivine inclusion by Mitchell and Giardini (1953) using X-ray diffraction techniques quickly developed into the principal way of defining inclusions in diamond. Orlov (1959) and Futergendler (1960) used this procedure to document the newly discovered (in 1954) Siberian diamonds, and the latter author reviewed her inclusion work over the second half of the 1950s. Futergendler first identified enstatite and chrome spinel, and from both authors other documented inclusions were olivine, chrome diopside, garnet and diamond. With Orlov, some cell parameters and refractive indices are given and inclusions are divided into syngenetic and epigenetic, the latter being serpentine, as a pseudomorph after olivine, red quartz and red iron hydroxide both filling diamond fractures. Much of Orlov's inclusion work can be found in his translated book (Orlov 1977). Also, orientation studies between diamond and some of the principal inclusions were developed by Futergendler and Frank-Kamenetsky (1961), a subject fully considered in Angel et al. (2022).

Eppler (1961) studied the specific gravity and refractive indexes of some inclusions. From the associated description to his Figure 4, the previously common descriptions of ilmenite as an inclusion are most likely chrome spinel. Also identified were a high Mg# (molar 100Mg/(Mg+Fe)) olivine, a garnet of a composition approaching pyrope, the aforementioned ilmenite (likely chrome spinel) and apatite. He also remarked upon gaseous cloud-like inclusions, black expansion fractures around minerals, and considered that graphite and quartz were not likely inclusions.

From large batches of diamond samples from all the South African mines then operated by De Beers (and with one stone from the DRC), Sharp (1966) selected 89 diamonds containing black inclusions. After diamond crushing, brassy-colored non-diamond material was collected and individually examined by X-ray powder diffraction. Apart from a single cohenite (from a Jagersfontein diamond), pyrrhotite was identified as the dominant phase. Cell parameters indicated that two inclusions were nearly stoichiometric (i.e., showed no Fe deficiency) and therefore troilite. Most of the identified specimens were intergrown with pentlandite and to a lesser extent pyrite. Although there was some uncertainty about whether pyrite was primary, Sharp (1966) had found most of the minerals in the typical sub-solidus re-equilibration sequence of sulfide inclusions in diamond.

Harris (1968a,b) outlined criteria for recognizing whether inclusions in diamond were syngenetic or epigenetic. These works were based principally on the nature of the inclusions and their response to X-rays. Of the syngenetic inclusions, rutile and coesite were new identifications along with a dispersed dark cloud of microcrystalline olivine. Epigenetic inclusions were grouped into pseudomorphs after primary inclusions (goethite, quartz and possibly mica), fractures surrounding inclusions (graphite, pyrrhotite and pentlandite), and material in surface fractures (calcite, hematite, kaolinite, sellaite and xenotime).

Although commercial electron microprobe analyzers (EMPA) were available from the late 1950s, their use in determining the chemistry of inclusions in diamond only began in the late 1960s. This development essentially eliminated the need for X-ray techniques in inclusion recognition and came at a time when there was a growing awareness of the mineralogical and chemical links between inclusions in diamond and the equivalent minerals in peridotitic and eclogitic xenoliths found in kimberlites. The earliest EMPA study was completed by Meyer (1968), on a suite of inclusion garnets. Subsequent work by Meyer and Boyd (1972) separated garnets into peridotitic and eclogitic types, with inclusion assemblages further emphasizing the two principal growth environments for diamond. At this time, EMPA did not find many new inclusion minerals, although kyanite, ilmenite and zircon are examples, but instead rather precisely defined the chemistry of the existing mineralogies.

During the 1970s two major inclusion research centres emerged, one in South Africa through John Gurney at the University of Cape Town, the other through Nick Sobolev at the Siberian Branch of the Russian Academy of Sciences in Novosibirsk. Part of the inclusion research at Cape Town was linked to a prospecting program established by De Beers Consolidated Mines (organized by Barry Hawthorne and managed by Jeff Harris), who, over a ten-year period starting in 1970/71, provided representative inclusions for analysis from each of the De Beers mines in South Africa and Botswana. Similar independent inclusion studies were conducted in Novosibirsk on diamonds from the developing Siberian mines and linked to xenolith and xenocryst studies, providing important new insights into the crystal chemistry of inclusions. Much of the work on southern African localities, but also studies from other localities around the world, can be found in the 1st to 4th IKC Proceedings volumes (1975–1989 and references therein), where, for example, such rare inclusions as kyanite, sanidine, albite, phlogopite, zircon and moissanite are identified by EMPA and linked to a diamond paragenesis. Much of the early results of the Novosibirsk group were reviewed in a seminal book by Sobolev (1977; translated from the Russian edition of 1974).

With the southern African work, additional projects involving inclusions were set in place. For example, one was the determination of inclusion abundance tables and the relationship between inclusions and specific diamond shapes for some of the mines (Harris and Gurney 1979). A second project led to the first successful determination of genesis ages for diamond. Although by the mid-to-late 70s the eruption ages of several diamondiferous kimberlites were known (Davis 1977), the question still remained as to whether diamond was a phenocryst or xenocryst in the rock that brought it to the surface. The purity of diamond excluded any kind of dating analysis but the visual characteristics of the inclusions indicated a probably syngenetic growth with the diamond, and some of those minerals contained radioactive isotopes. Whilst the first studies used U–Pb in sulfides (Welke et al. 1974; Kramers 1979), the most successful work was started by Richardson et al. (1984), by determining the Sm–Nd isotopes in aliquoted harzburgitic garnet inclusions from the De Beers Pool and Finsch mines in South Africa. This work showed beyond doubt that diamonds were xenocrysts in kimberlite and easily justified the collection of 200 inclusion-bearing diamonds from both sources for the analyses. The Sm–Nd chronometer was then applied to date separate aliquots of eclogitic garnet and clinopyroxene inclusions (Richardson 1986) and, over the subsequent 20 years, to determine diamond ages in southern Africa, Russia and Australia (reviewed in Smit et al. 2022). With the considerable improvement in analytical precision, the Sm–Nd dating technique can now obtain an age from an array formed by single garnet or clinopyroxene inclusions (Koornneef et al. 2017; Gress et al. 2021).

Other dating techniques followed; the principal one being Re–Os isotopes from sulfide inclusions, first developed by Pearson et al. (1998) in diamonds from the Koffiefontein mine in South Africa specifically to date single diamonds. Results confirmed a xenocrystal origin for diamond but also, where comparison was possible, gave similar ages to those obtained from the silicates, indicating that the common dates were not coincidences (see Table 9 in Stachel and Harris 2008). The use of Ar–Ar isotopes from peridotitic or eclogitic clinopyroxene inclusions, although initially considered to be a genesis dating technique (see Burgess et al. 1989), became more useful in determining diamond eruption ages, with particular importance in diamond prospecting (see, for example, Phillips et al. 2018).

From the mid-1980s, therefore, not only had many different inclusion minerals been assigned to the peridotitic and eclogitic growth environments, or to the minor websteritic suite (Gurney et al. 1984), but also, inclusions were providing important information about mantle evolution. A significant consequence of the first two diamond dating programs, for example, was that the results generally correlated with tectonothermal events which had affected the lithospheric mantle where the diamonds occurred; a subject more fully considered by Smit et al. (2022). Also, the discovery and subsequent study of eclogitic inclusions in lamproite-hosted diamonds at Argyle, in Australia (Sobolev 1985, Jaques et al. 1989) led to the first proposal that the mantle source for a predominantly eclogitic diamond deposit was generated during subduction (in this case along the Kimberley Craton margin). Additionally, inclusion chemistry linked to HPHT experimental work enabled the temperature and pressure of inclusion formation to be determined (see Boyd and Gurney 1986; Nimis 2022), and broader inclusion studies began to explore the relationship between mineralogy and particularly the chemistry of inclusions and the carbon and nitrogen isotopes and nitrogen abundances of the host diamonds (see Stachel et al. 2022).

Returning specifically to the inclusions, a significant review of the two principal parageneses was provided by Meyer (1987) and for all three parageneses by Stachel and Harris (2008). Examples of further studies that have added to our inclusion knowledge are the finding of rare single crystals of primary graphite in diamond from the Panda kimberlite in Canada (Glinnemann et al. 2003) and a thorough study of rare phlogophite inclusions, identified in peridotitic and eclogitic diamonds from Siberia, Arkhangelsk and Venezuela (Sobolev et al. 2009). Thus, inclusions in diamonds seemed to firmly place diamond genesis in the subcratonic lithospheric mantle.

However, after four crystalline isotropic reddish-brown inclusions of magnesio-wüstite (now more commonly called ferropericlase) were recovered from four diamonds from the Koffiefontein mine in South Africa in the early 1970s (later studied by Cardoso 1980), this belief began to be queried. It was the recovery of a lower mantle inclusion assemblage of low-nickel enstatite (retrogressed bridgmanite) and a ferropericlase in a single diamond from the Orroroo kimberlite dykes in South Australia (Scott Smith et al. 1984) that established a whole new dimension to studies of inclusions in diamond. Thus were born sublithospheric and lower mantle diamond studies, a subject covered by Walter et al. (2022).

In conclusion to this historical overview of cratonic diamonds the reader is referred to Table 1, which updates the inclusion compilation of Stachel (2014) with subsequent discoveries.

The inclusion–diamond relationship

The research on the geochemistry of inclusions in diamond captured in the database underpinning this review generally relied on sampling of run-of-mine diamonds between 1.8 and 3.5 mm in size (corresponding to round aperture Diamond Trading Company (DTC) sieve sizes +5 to −11), although in some rare instances inclusions were recovered from diamonds measuring less than 1.1 mm (DTC sieve size −1; e.g., Stachel et al. 2006). Based on the examination of over 1 million diamonds ~2 mm in size mined from nine kimberlites in South Africa and Botswana, the abundance of diamonds with visually identifiable mineral inclusions is 1.0% (see Table 1 of Stachel and Harris 2008). The size of studied mineral inclusions commonly is about 100 μm in longest dimension, although there are exceptions both to smaller and larger sizes. Our observations suggest an only minor increase in inclusion size towards larger diamonds and for the same inclusion mineral, the chemical composition does not significantly change with diamond size.

Disequilibrium among non-touching inclusions is occasionally observed, reflecting incorporation during diamond growth in chemically evolving environments (e.g., Griffin et al. 1988; Bulanova 1995; Wang 1998). Based on multiple lines of evidence (reviewed, e.g., in Gurney 1989 and Stachel and Luth 2015), disequilibrium between non-touching inclusions is nevertheless considered the exception rather than the rule.

Inclusions exhibit an often cubo-octahedral shape, irrespective of mineral symmetry or paragenesis (Harris et al. 2022). For many years, this characteristic was considered to reflect syngenetic growth with that of the host diamond (e.g., Harris 1968a; Bulanova 1995). This hypothesis was recently queried by Nestola et al. (2014), who observed an absence of an overall preferred orientation of inclusions in diamond indicating a protogenetic origin. Occurrence of both syngenetic and protogenetic relationships was subsequently documented through a combination of cathodoluminescence imaging and crystallographic orientation (EBSD) data (Davies et al. 2018). Based on small inclusion sizes (see above) and fast diffusion speeds in hot mantle environments, pre-existing host-rock minerals are nevertheless expected to generally equilibrate during fluid-mediated diamond growth events and thus the term "synchronous" inclusion was introduced (Nestola et al. 2017, 2019a; Pamato et al. 2021). That a lack of syngeneity does not necessarily preclude inclusion-based diamond dating is established in detail in Smit et al. (2022).

Table 1. Mineral inclusions in cratonic diamond.

Peridotitic
Common: Cr-pyrope, Forsteritic olivine, Enstatite, Cr-diopside, Mg-chromite, Fe–Ni sulphides.
Rare: Coesite*, Mg-ilmenite, Magnesite, Calcite, Native Fe, Zircon, Phlogopite, Yimengite.
Eclogitic
Common: Grossular–Almandine–Pyrope, Omphacitic Clinopyroxene, Fe-sulphides.
Occasional: Rutile, Coesite*, Kyanite.
Rare: Corundum, Ilmenite, Magnetite, Fe–Mg-chromite, Phlogopite, K-feldspar, Titanite, Staurolite, Zircon, Moissanite, Calcite, Dolomite.
Websteritic
Common: Almandine–Pyrope, Diopside–Augite, Enstatite.
Occasional: Coesite*, Olivine.
Rare: Phlogopite, Goldschmidtite
Origin Uncertain
Common: Graphite (mostly epigenetic to all parageneses)
Rare: Primary graphite, Diamond, Perovskite, Amphibole, Apatite, Eskolaite, Sr-titanate, Monazite.

Notes: * Refers to a mineral inclusion where an original structure is inferred.
See text for most of the epigenetic mineralogy.

Inclusion suites and parageneses

A first order division of mantle-derived diamonds is based on their origin in either the lithospheric mantle (our topic) or the sublithospheric mantle, the latter including the convecting upper mantle, the transition zone and the lower mantle (reviewed in Walter et al. 2022). Correcting for an extreme overrepresentation of sublithospheric diamonds in the literature, Stachel and Harris (2008) estimated that about 99% of diamonds produced worldwide ultimately derive from the lithospheric mantle and only about 1% from sublithospheric sources. The subsequent recognition that the CLIPPIR subpopulation of large, irregular, nitrogen-free (Type II) diamonds largely or entirely derives from transition zone to lower mantle environments (Smith et al. 2016) suggests that the estimate for sublithospheric diamonds needs to be increased to about 2% and the proportion of lithospheric diamonds decreased to 98%. Based on their mineral inclusions, lithospheric diamonds are conventionally subdivided into two main suites, peridotitic and eclogitic, plus a minor websteritic or pyroxenitic (websterites are a subclass of pyroxenites) suite (Meyer 1987; Gurney 1989; Stachel and Harris 2008). Based on 2844 diamonds with analyzed inclusions, Stachel and Harris (2008) derived relative abundances of 65% for the peridotitic, 33% for the eclogitic and 2% for the websteritic suite.

Peridotitic suite. Based on a primitive mantle Mg# of 89–90 (see McDonough and Sun 1995; their Table 4), olivine, orthopyroxene and clinopyroxene inclusions from depleted cratonic peridotite substrates[1] have Mg# ≥ 90, with infrequent Mg# values between 90–85 indicating increasingly melt-modified peridotitic substrates, which become transitional to websterites (Viljoen et al. 2018). In addition to their more magnesian character, peridotitic inclusions are separated from the eclogitic inclusion suite by Cr_2O_3 in garnet > 1 wt% (Schulze 2003) and molar Cr# (100Cr/(Cr+Al)) in clinopyroxene ≳7–10 (Stachel and Harris 2008). The peridotitic inclusion suite may be further subdivided into harzburgitic (to dunitic), lherzolitic, and wehrlitic parageneses, based principally on the groundbreaking work on garnet classification using Cr and Ca abundances by Sobolev et al. (1969, 1973). In this review, we employ the updated garnet classification scheme of Grütter et al. (2004) for this purpose. It should be noted that classification of peridotites based on garnet compositions is distinct from classification based on petrographic modes (Streckeisen 1976): garnet chemistry tracks equilibration with (1) orthopyroxene only (harzburgitic paragenesis), (2) clinopyroxene only (wehrlitic paragenesis), or (3) both pyroxenes (lherzolitic paragenesis), whilst petrographically defined boundaries are based on 5 vol% of either clinopyroxene (for the harzburgite/lherzolite distinction) or orthopyroxene (for the lherzolite/wehrlite distinction). Consequently, using garnet chemistry, highly depleted rocks may classify as lherzolites in spite of very minor (< 1 vol%) clinopyroxene contents. However, in contrast to xenolith studies, classification based on mineral modes is not feasible in inclusion research.

The minerals encountered as peridotitic suite inclusions in diamond are given in Table 1 and here we will exclusively focus on the common inclusion types (garnet, olivine, orthopyroxene, clinopyroxene and spinel), with the exception of sulfides that are discussed elsewhere in this volume (Smit et al. 2022). A table with inclusion abundance counts for lithospheric diamonds is given by Stachel and Harris (2008; their Table 2).

Eclogitic and websteritic/pyroxenitic suites. Eclogites are metamorphic rocks with a broadly basaltic composition, composed predominantly of jadeite-rich clinopyroxene and grossular-almandine-rich garnet (cf. Figs. 31, 32 in section *Eclogitic–Pyroxenitic Substrates*). In one more formal definition of eclogite, the clinopyroxene has $Al/(Al + Fe^{3+}) > 0.5$ on the M1 site and $0.2 \leq Na/(Na + Ca) < 0.8$ on the M2 site (Clark and Papike 1968). Conversely,

[1] In this review, we use the term "substrate" to refer to the rock in which diamond crystallised and which may have been the source of protogenetic and synchronous inclusions. The chemical composition and/or mineralogy of this rock may evolve subsequent to diamond formation, but this later modified diamond source did not form the substrate for diamond precipitation.

garnet–clinopyroxene rocks with jadeite-poor clinopyroxene are classified as pyroxenite, which, *sensu* Streckeisen (1976), encompasses all ultramafic rocks with <40 vol% olivine. There are gradations between garnet clinopyroxenites, which can be mineralogically and texturally similar to mantle eclogite, and websterites containing additional orthopyroxene ± olivine.

The assignment of silicate inclusions in diamond to the eclogitic suite is relatively straightforward and based on low Cr_2O_3 (< 1 wt%; see above) at elevated CaO contents in garnet, and low Cr_2O_3 and elevated Al_2O_3 contents in clinopyroxene (Gurney 1984; Grütter et al. 2004; Stachel and Harris 2008). In contrast, the websteritic inclusion suite, with compositions intermediate between the peridotitic and eclogitic suites, is less well defined, and relies on elevated Cr# and Mg#, and low Ca# in garnet, and/or the presence of orthopyroxene (Deines et al. 1993; Stachel and Harris 2008). Eclogitic and pyroxenitic garnet occurring as inclusions, as xenocrysts or in xenoliths can be grouped, based on Cr and Ca contents and Mg#, into classes G3 (eclogitic), G4 (eclogitic and pyroxenitic) and G5 (pyroxenitic) (Grütter et al. 2004). More recent work aimed at distinguishing Cr-poor garnet sourced from granulite and orogenic (crustal) eclogite vs. mantle eclogite recognized Ti/Si and Mg/Fe as useful discriminants (Hardman et al. 2018). The classification used here follows Aulbach and Jacob (2016) by employing a simplified version of the definition of Clark and Papike (1968), whereby cpx with $0.2 \leq Na/(Na+Ca) < 0.8$ classifies as eclogitic and that with lower ratios as pyroxenitic. This translates into distinctive bulk-rock compositions and is therefore not controlled by a pressure-dependent change in mineral chemical composition.

In addition, using a combination of major and trace elements, gabbroic eclogites with positive Eu anomalies indicative of high normative plagioclase modes (Eu/Eu* = chondrite-normalized $Eu/(Sm \times Gd)^{0.5} > 1.05$; Rudnick and Fountain 1995) are distinguished from basaltic eclogites with lower Eu/Eu*. The latter are classified into high-Ca, high-Mg and low-Mg eclogites. High-Ca eclogites have garnet with Ca# (molar 100Ca / (Ca + Fe^{total} + Mg + Mn)) > 20, high-Mg eclogites garnet with Ca# ≤ 20 and Mg# > 60, and low-Mg eclogites garnet with Ca# ≤ 20 and Mg# ≤ 60 (cf. Fig. 32a in section *Eclogitic–Pyroxenitic Substrates*). This classification contains petrogenetic information, reviewed below, but relies on the availability of both major and trace element compositions for both garnet and clinopyroxene. Also, these groupings do not typically have sharp distinctions, especially when samples from various localities with their own endemic compositional ranges are considered.

Inclusion minerals associated with the eclogitic and websteritic suites are given in Table 1 and again, our focus is on the common minerals (garnet and clinopyroxene) with the exception of sulfides, which are discussed by Smit et al. (2022).

Origin and evolution of subcratonic lithospheric mantle

As previously discussed, diamonds are principally associated with peridotitic and eclogitic substrates occurring in subcratonic lithospheric mantle. Through dating of inclusions following pioneering developments by Richardson et al. (1984) and Pearson et al. (1998) (for a review and full references see Smit et al. 2022) it emerged that formation of diamonds in harzburgitic substrates principally occurred in the Paleoarchean (Gurney et al. 2010; Shirey and Richardson 2011). Formation of lherzolitic diamonds preferentially occurred in the Paleoproterozoic but extends through the Mesoproterozoic and, with diamonds from the Victor Mine (Aulbach et al. 2018), even into the Neoproterozoic. For individual diamonds with presumably lherzolitic sulfides from Koffiefontein and Jagersfontein, even Mesozoic formation ages are observed (Pearson et al. 1998; Aulbach et al. 2009). Probably as a consequence of younger formation ages, lherzolitic diamonds are not exclusively restricted to undisturbed Archean lithospheric mantle but also occur in areas impacted by younger tectonothermal events (Shirey et al. 2002; Richardson and Shirey 2008; Smit et al. 2014a; Stachel et al. 2018b) or underpinned by entirely post-Archean lithospheric mantle (e.g., Banas et al. 2007; Hunt et al. 2009; Czas et al. 2020;

Liu et al. 2021). Episodic addition of eclogitic diamonds to the subcratonic lithospheric mantle was associated with subduction along craton margins immediately following the presumed onset of modern style Wilson cycles at about 3 Ga (Helmstaedt et al. 2010; Shirey and Richardson 2011) and likely continues to the present, with the youngest eclogitic diamonds dated at 0.14 (± 0.09) Ga (Timmerman et al. 2017).

The formation of the most important diamond substrate, cratonic peridotites, is associated with intense primary melt depletion (Jordan 1978; Boyd 1989; Herzberg 2004) that reached up to the point of orthopyroxene exhaustion from the melting residues, i.e., requiring up to about 40% melt extraction (Walter 1999; Bernstein et al. 2007). These large degrees of melt extraction are presumed to relate to 100–300 °C higher mantle potential temperatures in the Archean (Labrosse and Jaupart 2007; Korenaga 2008; Aulbach and Arndt 2019a). The very low FeO of some cratonic peridotites (filtered for orthopyroxene addition and refertilization), which requires melt extraction at pressures > 5 GPa, has been used among other arguments to infer melting associated with ascending mantle plumes (e.g., Boyd 1989; Griffin et al. 1999b; Aulbach 2012). Based, for example, on Cr/Al ratios (Stachel et al. 1998) and concentrations of mildly incompatible trace element contents including HREE (Canil 2004; Pearson and Wittig 2008), a principally low-pressure origin in the Archean equivalent of oceanic ridges and/or arc mantle is favoured by others (reviewed in Brey and Shu 2018 and Pearson et al. 2021), which implies a metamorphic origin for cratonic garnet peridotites from Cr-spinel peridotite precursors. Suggested modes of emplacement into the subcratonic lithosphere then involve imbrication associated with subduction (Helmstaedt and Schulze 1989; Kesson and Ringwood 1989), or lateral compression and thickening (Jordan 1978; Wang et al. 2018; Pearson et al. 2021) or vertical attaching (Herzberg and Rudnick 2012; Perchuk et al. 2020).

Irrespective of the exact mode of origin, in several cratons (e.g., Kaapvaal, Yakutia, Tanzania) the composition of cratonic peridotites points to enrichment in silica (i.e., enrichment in orthopyroxene; Boyd 1989; Kesson and Ringwood 1989; Kelemen et al. 1998). Major element relationships (Al/Si, Mg/Si) of cratonic peridotites indicate involvement of high-temperature aluminous pyroxene in this process, inconsistent with Si enrichment through infiltration of siliceous melts in cold supra-subduction zones (Aulbach 2012). Also, the absence of detectable perturbations in $\delta^{18}O$ values of orthopyroxene-enriched cratonic peridotites clearly precludes slab-derived melts as the principal agent of Si enrichment (Regier et al. 2018) and points to melt–rock reactions involving mantle-derived melts instead (e.g., Kelemen et al. 1992; Koornneef et al. 2009; Tomlinson and Kamber 2021). For the Kimberley area of the Kaapvaal Craton, orthopyroxene enrichment in the lithospheric mantle has been linked to the 2.7 Ga Ventersdorp large igneous province (Branchetti et al. 2021), which agrees well with the model of Si enrichment with komatiite infiltration (Tomlinson and Kamber 2021). More subtle metasomatic enrichment of principally trace elements (cryptic metasomatism; Dawson 1984) in Kaapvaal cratonic peridotites was attributed to percolation of small-volume melts of the carbonatite-kimberlite spectrum, peaking between 3.3 and 2.7 Ga and then tapering off in the Mesoproterozoic (Shu and Brey 2015). As a final important stage of metasomatic overprint of the Kaapvaal Craton lithospheric mantle, Simon et al. (2007) identified localized clinopyroxene ± garnet re-enrichment, likely associated with kimberlite magmatism < 200 Ma.

Mantle eclogites and associated inclusions in diamond are widely interpreted as representing various parts of subducted oceanic crust, which inherited major element, trace element and isotopic signatures from low-pressure accumulation-differentiation processes of igneous protoliths, seawater alteration and subsequent dehydration and partial melting (MacGregor and Manton 1986; Beard et al. 1996; Jacob 2004; Stachel et al. 2004). Strong indices for a subduction origin of inclusions and xenoliths comprise non-mantle oxygen isotope compositions, implying low temperature crustal fractionation, and Eu anomalies, requiring the participation of plagioclase during protolith formation (Schulze et al. 2003; Jacob 2004;

Stachel et al. 2018a). Moreover, sulfide inclusions have yielded a range of mass-dependently fractionated S isotope compositions, with $\delta^{34}S$ from −13 to +14‰, similar to sediments and altered oceanic crust (Chaussidon et al. 1987; Eldridge et al. 1991). Finally, mass-independently fractionated S isotope compositions reflect processes unique to the O_2-poor Archean atmosphere, which were incorporated first into sediments and, after subduction, into sulfides trapped by diamond (Farquhar et al. 2002; Thomassot et al. 2009; Smit et al. 2019a). Apart from such signatures of a low-pressure origin and the stable isotope evidence, direct formation from mafic melts in the mantle is precluded because major and trace element relationships are inconsistent with the participation of garnet, which is on the liquidus of mafic melts at depths from which eclogites are entrained (Aulbach and Arndt 2019b). The occurrence of eclogite with low-pressure signatures in the cratonic lithosphere, combined with age constraints (Smit et al. 2022), indicates emplacement during two principal processes: (1) Archean amalgamation of cratonic nuclei into larger cratonic masses, with closure of intervening ocean basins and (2) subsequent accretion/collision at craton margins, best exemplified by the Slave craton.

Databases

For the present review, we compiled two databases for inclusions in diamond of all suites from the literature, one focusing on major element compositions (8412 inclusion analyses), the other on trace element compositions (949 inclusion analyses). For the presented data, a blanket detection limit of 0.02 wt% was assumed for major element oxides. For trace elements, collated analytical tables in the literature typically only report data above the specific limits of detection. These databases are available for download at https://doi.org/10.7939/DVN/EJUE1G (major element database) and https://doi.org/10.7939/DVN/581P99 (trace element database).

Peridotitic inclusions and xenolith minerals. For comparison of inclusion compositions with the same mineral phases in cratonic peridotites, we assembled a further database of minerals in well-equilibrated mantle xenoliths, including 967 samples from on-craton kimberlite localities (used in plots presented here), 909 being garnet peridotites. This database is not yet in a publication ready state but can be obtained from us on request. Mineral analyses of olivine, orthopyroxene, clinopyroxene, garnet or spinel were filtered for analytical totals between 98.5 and 101.5 (98.5–102.0 for garnet and 97.5–102.0 for spinel) and for cation per formula unit ranges that match accepted bounds for these mineral species. To assess differences and similarities in the trace element composition of inclusion and xenolith/xenocryst garnets, we compiled an additional database with trace element compositions of 152 harzburgitic and 281 lherzolitic garnets from xenoliths (or in a few instances occurring as xenocrysts) in on-craton kimberlites. For both xenolith databases, a paragenesis was assigned based on garnet major element chemistry (Grütter et al. 2004), which for the major element database in a few instances was revised based on assignments in the original publications (where garnets with "lherzolitic" compositions sometimes come from pyroxenite xenoliths). Of the 843 peridotitic garnets that pass the analytical filters described above, 679 (80.5%) derive from lherzolites (classified as G1 = megacrysts ($n = 64$), G9 = lherzolitic ($n = 460$), and G11 = high-TiO_2 peridotitic ($n = 155$)), 154 (18.0%) derive from harzburgites (with 105 of the G10 garnets classifying as "diamond facies" (G10D)), and 12 (1.4%) derive from wehrlites (G12). Only samples with an assigned harzburgitic or lherzolitic paragenesis were used in this review. Textural information (allowing to separate, e.g., strongly deformed varieties) is generally not available for these samples.

For the inclusion database, orthopyroxenes and clinopyroxenes with Mg# < 90 recovered from diamonds from the Voorspoed Mine were excluded from the peridotitic data set because of their compositionally unique character related to intense melt metasomatism (Viljoen et al. 2018).

Eclogitic inclusions and xenolith minerals. For comparison with their xenolithic counterparts, only garnet and/or clinopyroxene-bearing diamonds are considered. The database comprises reported eclogitic and websteritic diamonds with 23 touching clinopyroxene–garnet pairs and 249 non-touching pairs, 567 unpaired clinopyroxenes, and 906 unpaired garnet inclusions. Multiple non-touching inclusions per mineral and diamond are not counted individually but were averaged instead. However, touching inclusions were not averaged with non-touching ones as they may have re-equilibrated in response to changing pressure–temperature conditions, contrary to non-touching inclusions.

The xenolith database encompasses on-craton samples for which both major and trace element compositions are reported for both garnet and clinopyroxene, and excluding those where trace element compositions were affected by kimberlite contamination evident in anomalous Ba and Nb concentrations (cf., Aulbach and Jacob 2016). Cut-off values for contamination are estimated using mineral–melt distribution coefficients and concentrations in the contaminant, typically kimberlite, which determine how much Ba and Nb can be incorporated in garnet and clinopyroxene under equilibrium conditions. Garnet/kimberlite distribution coefficients for Ba of 0.0002 and for Nb of 0.0008, and clinopyroxene/kimberlite distribution coefficients for Ba of 0.001 and for Nb of 0.0002 are taken from Girnis et al. (2013). Combined with Ba and Nb concentrations in primitive kimberlites of 1250 and 260 ppm, respectively (Becker and Le Roex 2006), this yields maximum values of 0.25 ppm for Ba and 0.052 ppm for Nb in garnet, and of 1.25 ppm for Ba and 0.208 ppm for Nb in clinopyroxene at equilibrium. Allowing for metasomatism by more incompatible trace element enriched melts, we apply cut-offs for Ba in garnet and clinopyroxene of 0.5 and 1.5 ppm, respectively, and for Nb of 0.5 and 1 ppm, respectively. For a comparison of mineralogy, only xenoliths for which at least clinopyroxene and garnet major element data are reported are considered, which is the case for 1907 eclogite and 544 pyroxenite xenoliths. For a comparison of compositions, diamond-absent xenoliths are only considered when REE data are available, allowing full classification (695 samples), whereas diamond-bearing xenoliths (n = 251) are retained regardless of trace-element data availability (cf. Table 3 in section *Eclogitic–Pyroxenitic Substrates*). Cratons with corresponding data coverage encompass Kaapvaal, Zimbabwe, West Africa, India, Siberia, Slave, Sask, Superior, the North Atlantic craton in West Greenland, and the East European Platform (references in Appendix Table 1).

Values below the detection limit are often not given and can therefore not be counted for calculation of average or median compositions, which implies that these are maxima. Since this applies only to a small proportion of the data for elements considered here, the resultant bias is considered minor.

MAJOR ELEMENT COMPOSITION OF PERIDOTITIC SUITE INCLUSIONS AND COMPARISON WITH CRATONIC PERIDOTITE XENOLITH MINERALS

Olivine

Of the 1468 analysed peridotitic olivine inclusions in diamond, the majority (79%) cannot be assigned to a specific paragenesis due to an absence of coexisting garnet or clinopyroxene, 16% are harzburgitic, 5% are lherzolitic and a single inclusion (equivalent to 0.07%) is of wehrlitic paragenesis. The olivine inclusions are high in forsterite content, with a median Mg# of 92.8 (mean is 92.7 ± 1.3; 1 sigma). Olivine from cratonic garnet peridotites is lower in Mg# (median value of 91.8, mean of 91.7 ± 1.3). Comparing the Mg# of harzburgitic (median and mean of 93.1) and lherzolitic (median and average of 91.8) olivine inclusions, harzburgitic inclusions are more magnesian (Mg# difference is 1.3; Fig. 5). The same distinction is seen for harzburgitic (median is 92.8, average is 92.9) and lherzolitic (median is 91.6, average is 91.5) olivines in cratonic garnet peridotites (average difference is 1.3; Fig. 5). For both parageneses, equality of means in Mg# of olivines included in diamond and contained in xenoliths cannot be rejected, based on a two tailed Student's t-test for two independent samples, assuming

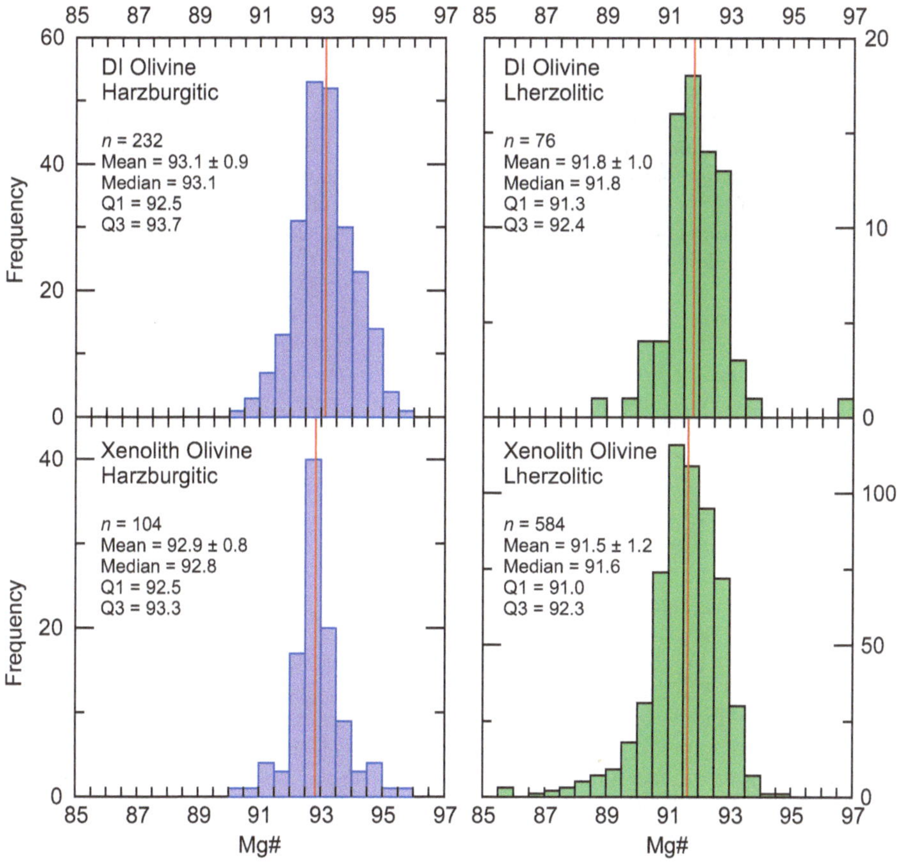

Figure 5. Histograms of molar Mg# (100Mg / (Mg + Fe)) of olivine included in diamond (DI, **top**) and from cratonic garnet peridotite xenoliths (**bottom**). The paragenesis of olivine inclusions is derived from coexisting garnet or the presence of clinopyroxene (lherzolitic). The harzburgitic paragenesis is shown on the **left**, the lherzolitic paragenesis on the **right**. The majority (n = 1157) of olivine inclusions are of unspecified paragenesis (not shown), due to an absence of coexisting garnet or clinopyroxene. Their mean Mg# (92.8 ± 1.2) is the same as that of the combined data set for harzburgitic and lherzolitic inclusions (92.8 ± 1.1), indicating that peridotitic olivines of unspecified paragenesis represent a mixture of harzburgitic and lherzolitic olivines, in proportions similar to the 3:1 ratio for olivines of known paragenesis. For each set, the number of analyzed olivines (n), the mean (± 1 standard deviation), the median value and the interquartile range (as middle values of the lower half (Q1) and the upper half (Q3) of the data set) are given as well. The median value is also indicated as a **red line** in each histogram.

unequal variances and an alpha value of 5%. Olivine Mg# is widely interpreted to gauge the degree of depletion of the protoliths of subcratonic lithospheric mantle (Boyd 1989; Herzberg 2004; Bernstein et al. 2007; Pearson and Wittig 2014) with typically only minor subsequent modification. The observation of equal means then suggests that, on the paragenesis level, diamond substrates and diamond-free garnet peridotites experienced effectively the same depletion history. Equality of means on the paragenesis level is reconciled with a higher Mg# of inclusion olivines relative to xenolith olivines based on very different relative proportions of the harzburgitic and lherzolitic parageneses, with 75% of inclusions with an assigned paragenesis being harzburgitic, as opposed to 15% of olivines from garnet harzburgite xenoliths.

A long-recognized difference between olivine included in diamond and from garnet peridotite xenoliths is a higher Cr content of the former (Meyer and Boyd 1972; Meyer 1987). Peridotitic olivine inclusions have a median Cr_2O_3 content of 0.05 wt% (mean is 0.06 ± 0.08 wt%), with an interquartile range of 0.03–0.14 wt% and a maximum value of 1.36 wt%. Olivine in cratonic garnet peridotites has a median Cr_2O_3 content of 0.03 wt% (mean is 0.03 ± 0.02 wt%), with an interquartile range from below the limit of detection (< 0.02 wt%) to 0.05 wt%, with a maximum value of 0.12 wt%. Higher Cr in inclusion olivines has been related to incorporation of Cr^{2+} (Meyer and Boyd 1972) during crystallization at reducing conditions and at high temperature (Li et al. 1995). High Cr, however, is not a universal characteristic of olivine inclusions and only 10% have Cr_2O_3 > 0.10 wt%.

Orthopyroxene

As with olivine inclusions, the majority of the 417 analysed peridotitic orthopyroxene inclusions (57%) are of unspecified paragenesis, 32% are harzburgitic and 11% are of lherzolitic paragenesis. Inclusion orthopyroxene is almost pure endmember enstatite, with a median Mg# of 94.1 (mean is 94.0 ± 1.9)[2]. These Mg# values are 1.3 higher than the median and mean for olivine inclusions in diamond (see above). Orthopyroxene in cratonic garnet peridotites is less magnesian than inclusions (median Mg# of 92.9, mean of 92.8 ± 1.1). On the paragenesis level, harzburgitic orthopyroxene inclusions (median value of 94.9, mean of 94.8 ± 1.2) are more magnesian than lherzolitic inclusions (median Mg# of 92.8, mean of 92.4 ± 1.9; Fig. 6). The harzburgitic inclusion dataset is strongly biased by a large number (n = 59) of highly magnesian orthopyroxenes from De Beers Pool (the combined production of the Kimberley Mines), and after removing this sample set, the median Mg# drops to 94.1 (mean of 94.2 ± 1.1; Fig. 6). These paragenesis level values are very similar to harzburgitic (median Mg# of 93.8, mean of 93.9 ± 0.7) and lherzolitic (median Mg# of 92.7, mean of 92.6 ± 1.0; Fig. 6) orthopyroxenes from cratonic garnet peridotites. A two tailed Student's t-test for two independent samples, assuming unequal variances and an alpha value of 5%, however, rejects the hypothesis of equal means for both the harzburgitic and lherzolitic inclusion and xenolith orthopyroxenes. Based on the nevertheless high similarity of inclusion and xenolith orthopyroxene on the paragenesis level, the lower Mg# of xenolith orthopyroxenes for the combined peridotitic data sets largely reflects a much lower proportion of the harzburgitic paragenesis, which makes up only 16% (as opposed to 75% of inclusions). This matches the observation made for olivine inclusions.

Among minor elements, NiO in orthopyroxene is higher in inclusions in diamond (median value is 0.13 wt%, mean is 0.13 ± 0.03 wt%) compared to cratonic garnet peridotites (median value is 0.09 wt%, mean is 0.07 ± 0.05 wt%). There is no variation in NiO with paragenesis: for inclusions of known lherzolitic and harzburgitic paragenesis the median NiO content is 0.13 wt% and the mean is 0.14 ± 0.03 wt%; for xenoliths the two parageneses have the same median and mean as the combined peridotitic orthopyroxene dataset. Given the similarity in Mg# of orthopyroxene inclusions and in xenoliths on the paragenesis level and the similarity of NiO in olivine inclusions (median NiO is 0.36 wt%, mean is 0.36 ± 0.05 wt%) and in xenoliths (median NiO is 0.38 wt%, mean is 0.36 ± 0.08 wt%), this observation cannot be interpreted as a bulk rock signal but suggests a temperature control on Ni in orthopyroxene instead, similar to Ni in garnet (Griffin et al. 1989a). Indeed, a blurred positive correlation (r^2 = 0.2, not illustrated) between temperature of last equilibration (enstatite in clinopyroxene-based estimates; Brey and Köhler 1990) and NiO content is observed for our database of 189 orthopyroxene analyses from cratonic garnet lherzolite xenoliths.

[2] As previously noted, excluded from the peridotitic data set is a suite of "lherzolitic" orthopyroxenes (and clinopyroxenes) from Voorspoed (Kaapvaal Craton) derived from strongly melt-metasomatised substrates, which shows characteristics, such as a mean Mg# of 88.9 and mean MnO content of 0.28 wt%, that are clearly transitional to the websteritic suite (see Viljoen et al. 2018 for a detailed discussion).

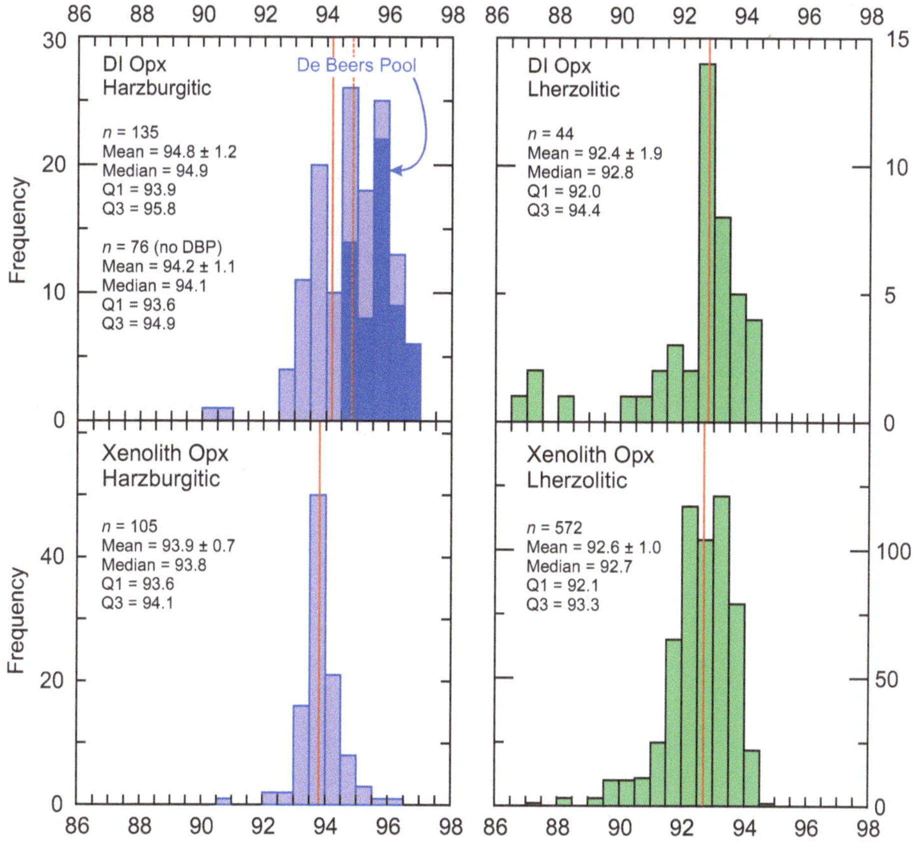

Figure 6. Histograms of molar Mg# (100Mg / (Mg + Fe)) of orthopyroxene included in diamond (DI, **top**) and from cratonic garnet peridotite xenoliths (**bottom**). The harzburgitic paragenesis is shown on the **left**, the lherzolitic paragenesis on the **right**. As with olivine inclusions, the majority (n = 235) of orthopyroxene inclusions are of unspecified paragenesis (not shown), due to an absence of coexisting garnet or clinopyroxene, but again their mean Mg# (94.1 ± 1.3) is very similar to that of the combined data set for harzburgitic and lherzolitic inclusions (93.8 ± 2.2). For each set, the number of analyzed olivines (n), the mean (±1 standard deviation), the median value and the interquartile range (as middle values of the lower half (Q1) and the upper half (Q3) of the data set) are given as well. For harzburgitic orthopyroxene inclusions, samples from De Beers Pool (shown in **dark blue**) are highly overrepresented and, due to their highly magnesian character, bias the dataset. Therefore, a second set of statistical parameters is given that excludes the De Beers Pool orthopyroxenes (labelled "no DBP"). The median value is also indicated as a red line in each histogram. For harzburgitic inclusions, the higher median value including the De Beers Pool data is shown as a **dashed line**. 'De Beers Pool' is the pooled diamond production from the four mines in Kimberley: Bultfontein, De Beers, Dutoitspan, Wesselton.

Due to their incompatible behavior during mantle melting, Na and Ti were efficiently removed from the Archean melting residues that now form subcratonic lithospheric mantle (O'Hara et al. 1975; McDonough and Rudnick 1998). This is well reflected in harzburgitic orthopyroxene inclusions, where over 50% of analyses for Na_2O and TiO_2 are below the limit of detection (≤ 0.02 wt%). For inclusions of the lherzolitic paragenesis (and presumably lherzolitic inclusions contained in the "unspecified peridotitic" group), metasomatic re-enrichment of the diamond substrates is documented in maximum Na_2O contents of

Table 2. Comparison of selected major elements in lherzolitic clinopyroxenes included in diamond (DI) and from cratonic garnet peridotite xenoliths.

	Mg#		TiO_2		Cr_2O_3		Al_2O_3		Na_2O		K_2O	
	DI	Xenolith	DI	Xenolith	DI	Xenolith	DI	Xenolith	DI	Xenolith	DI	Xenolith
Min	88.4	86.9	≤ 0.02	≤ 0.02	≤ 0.02	0.17	0.05	0.36	0.18	0.24	≤ 0.02	≤ 0.02
Q1	91.9	91.3	≤ 0.02	0.05	0.81	1.02	0.89	1.61	0.67	1.38	0.05	≤ 0.02
Median	92.8	92.3	0.05	0.13	1.08	1.37	1.42	1.96	1.17	1.63	0.10	0.03
Q3	93.5	93.4	0.10	0.26	1.64	1.73	2.12	2.28	1.90	1.92	0.18	0.05
Max	96.4	96.4	1.85	0.65	15.6	3.96	5.83	5.65	10.10	3.76	1.68	0.26

Notes: Interquartile ranges given as middle values of the lower half (Q1) and the upper half (Q3) of the data sets. Because of strongly non-normal distributions and common outliers, means and standard deviations are not quoted. A blanket detection limit of ≤ 0.02 wt% is assumed for all oxides analysed and to avoid biasing the data set, values below the limit of detection were not omitted but treated as ½ the detection limit instead. Mg# is given as Mg / (Mg+Fe) × 100

0.57 wt% and maximum TiO_2 contents of 0.29 wt%. For orthopyroxene from cratonic garnet peridotites, the metasomatic re-enrichment of Na and Ti can be much more intense (Fig. 7). From the xenolith data set, a positive correlation between Na and Ti is apparent, which on a cation basis has a slope of about 2, consistent with a coupled substitution of the type $2Na^{1+}$-$^{VI}Ti^{4+}$ (Cameron and Papike 1981). Using the 95th percentile values for Na_2O (≤ 0.16 wt%) and TiO_2 (≤ 0.06 wt%) of orthopyroxene inclusions as upper cutoffs, which in combination encompass 93% of the inclusion data set, permits (empirical) definition of a diamond inclusion field for orthopyroxenes. Forty two percent of orthopyroxenes from cratonic garnet peridotites fall outside this inclusion field, indicating that substantive metasomatic re-enrichment of Na and Ti is much more widespread in xenolith samples.

Clinopyroxene

Lherzolitic clinopyroxene inclusions (n = 148) are Cr_2O_3-rich, with contents typically near 1 wt%. Their median Mg# (92.8) is the same as that of lherzolitic orthopyroxene inclusions and mildly elevated relative to clinopyroxene from cratonic garnet lherzolites (92.3; Table 2). Compositional differences between inclusion and xenolith clinopyroxenes are evident for Al, Ti, Na, but largely overlapping compositional ranges preclude the definition of compositional cut-offs between the two groups (Table 2). Al_2O_3 in inclusions shows a strongly right (towards higher Al) skewed distribution with a mode at 0.8 wt% (median value 1.42 wt%) whereas xenolith-derived clinopyroxene displays a near-normal distribution with a mode at 1.8 wt% (median value 1.96 wt%; see Fig. 8). Inclusion clinopyroxenes also have lower median and 25th percentile (Q1) values in TiO_2 and Na_2O contents (Table 2). The strongest compositional difference exists for K_2O, where the inclusion distribution shows a tail up to 1.68 wt%, whereas xenolith clinopyroxenes do not exceed 0.26 wt% (Fig. 9). The occurrence of strong to extreme metasomatic enrichment in potassium is a unique characteristic of inclusions, which also have a higher median value and interquartile range (Q1 and Q3 values Table 2), but nevertheless, the bulk of clinopyroxenes from diamonds and cratonic garnet lherzolites overlap in K_2O content (Fig. 9).

Garnet

The "classical" Cr_2O_3 versus CaO diagram (Sobolev et al. 1973; Gurney and Switzer 1973; Gurney 1984; Grütter et al. 2004; Fig. 10), with fields separating the three principal peridotitic parageneses, harzburgitic

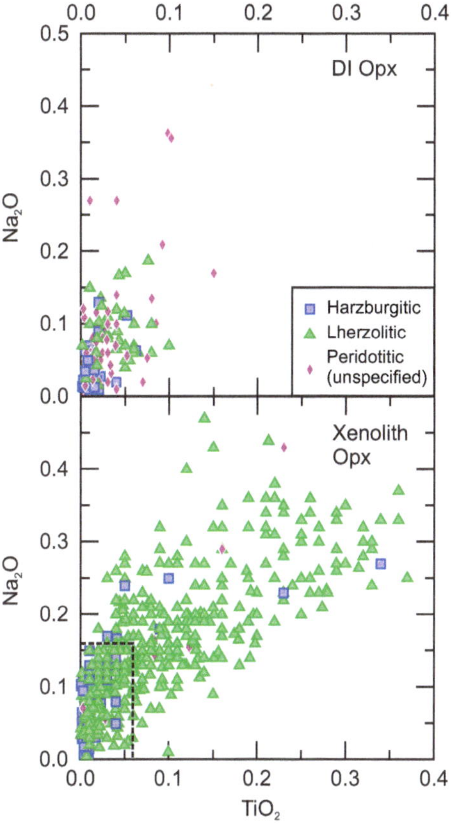

Figure 7. Co-variations of Na_2O and TiO_2 (wt%) in orthopyroxene included in diamond (DI, **top**) and from cratonic garnet peridotite xenoliths (**bottom**). The **dashed box** in the bottom diagram is based on the 95th percentile in Na_2O and TiO_2 contents of inclusion orthopyroxenes.

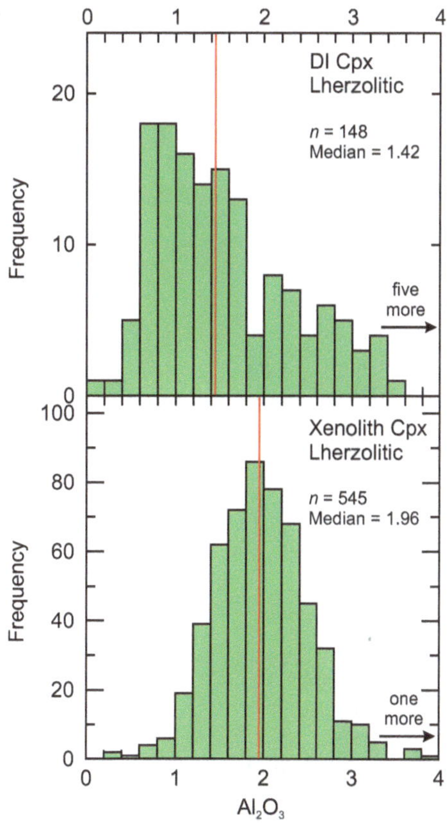

Figure 8. Histogram of Al_2O_3 content (wt%) in lherzolitic clinopyroxene included in diamond (DI, **top**) and from cratonic garnet peridotite xenoliths (**bottom**). A few outliers, five inclusions with Al_2O_3 up to 5.8 wt% and one xenolith clinopyroxene with Al_2O_3 of 5.7 wt%, exceed the shown range. Median values are indicated as **red lines**.

(-dunitic), lherzolitic and wehrlitic, is widely used in assessing the compositions of mantle-derived garnets. For our database (n = 1175), 81% of inclusion garnets are of harzburgitic paragenesis, 18% are lherzolitic and 1% are wehrlitic. Ultra-depleted harzburgitic and dunitic diamond substrates are represented by garnets with CaO contents < 1.8 wt% (Grütter et al. 1999; Fig. 10), which applies to 38% of peridotitic inclusions (or 47% of garnets classified as harzburgitic). In evaluating these relative proportions, it should, however, be noted that the published garnet inclusion data set is strongly biased towards a few well studied locations, with about 40% of analyses in our database being derived from only four occurrences (De Beers Pool, Finsch and Premier (Cullinan Mine) on the Kaapvaal Craton and Victor on the Superior Craton), with a clear under-representation of inclusion garnets from Yakutian occurrences (9%). For that reason, we consider the original proportion of 85% harzburgitic (-dunitic) or "G10" garnet inclusions established by Gurney (1984), using a much smaller data set, to be still equally valid. Garnets from cratonic peridotite xenoliths (n = 843) show an exactly inverse relationship in the proportions of the harzburgitic (18% in our database) and lherzolitic (81%) parageneses, with again 1% of garnets with wehrlitic compositions (Fig. 10). Ultra-depleted (< 1.8 wt% CaO) xenolith garnets are rare (2% of all peridotitic and 13% of harzburgitic garnets).

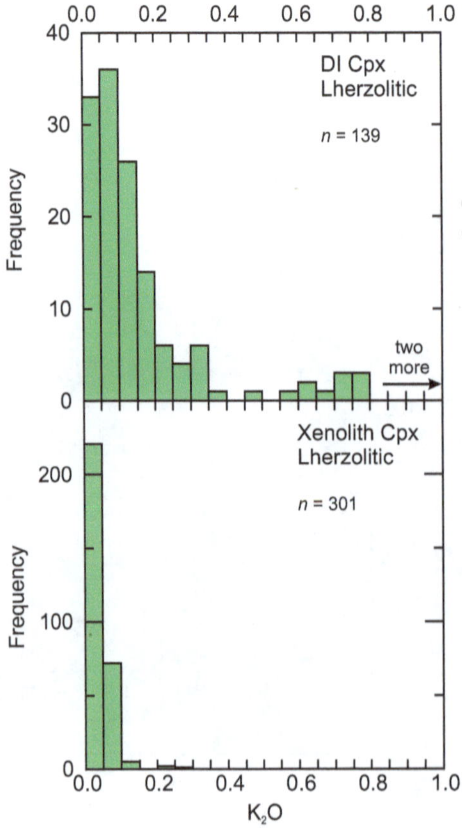

Figure 9. Histogram of K_2O content (wt%) in lherzolitic clinopyroxene included in diamond (DI, **top**) and from cratonic garnet peridotite xenoliths (**bottom**). Two outliers, inclusion clinopyroxenes with K_2O of 1.57 and 1.68 wt%, exceed the shown range.

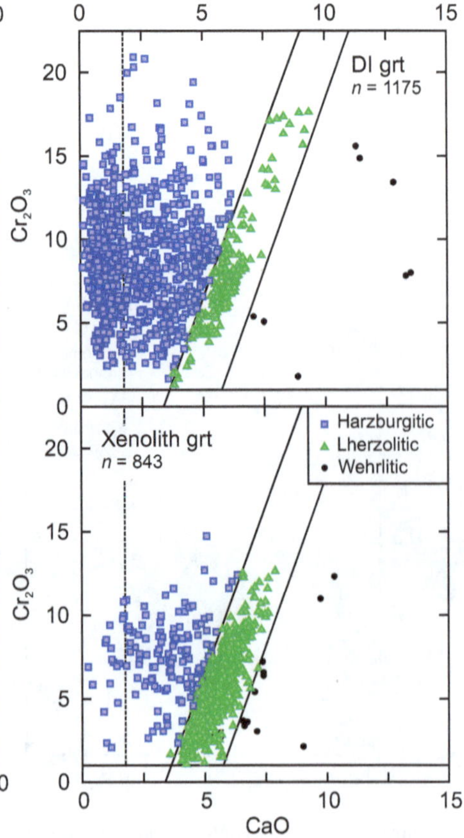

Figure 10. Cr_2O_3 versus CaO (wt%) in peridotitic garnet included in diamond (DI, **top**) and from cratonic xenoliths (**bottom**). Fields for the different peridotitic parageneses (harzburgitic, lherzolitic and wehrlitic) and cut-off towards eclogitic garnets at 1 wt% Cr_2O_3 are from Grütter et al. (2004). Ultra-depleted harzburgitic and dunitic garnets fall to the left of the **dashed line** at 1.8 wt% CaO (Grütter et al. 1999). The total number of analyzed peridotitic garnets is given for both inclusions in diamond and xenoliths.

Calcium intercept values (calculated after Grütter et al. 2004) are a one-dimensional measure of how far a garnet is removed in Ca content from the harzburgite–lherzolite division, which intersects the *x*-axis in a Cr_2O_3–CaO plot (Fig. 10) at 3.375 wt% CaO. In garnet compositional space, the transition to Ca intercept values below 3.375 is considered petrologically equivalent to the exhaustion of clinopyroxene from the residues during primary melt depletion and corresponds to about 20% melt removal (Jaques and Green 1980; Kushiro 1994). Further decreasing Ca intercept values of garnet document increasing degrees of primary melt depletion, progressing to the exhaustion of orthopyroxene in the residues at about 40% melt extraction (Jaques and Green 1980; Bernstein et al. 2007 and references therein). For harzburgitic garnets, Ca intercept values, therefore, are a measure of the primary chemical depletion of their substrates (see Grütter et al. 1999 for a detailed discussion).

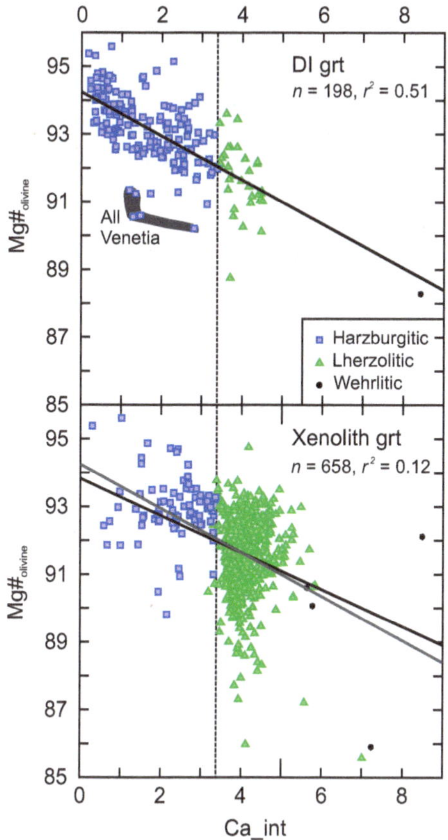

Figure 11. Olivine Mg# versus Ca intercept value (Ca_int) of coexisting garnets included in diamond (DI, **top**) and from cratonic peridotite xenoliths (bottom). Ca intercept values measure how far a garnet is removed from the harzburgite–lherzolite division (see Fig. 10) shown as a **dashed line** with a Ca intercept value of 3.375 (for details of calculation, see Grütter et al. 2004). Linear regressions (**black lines**) are given for both the inclusion- and xenolith-based datasets. Five harzburgitic garnet–olivine inclusion pairs that fall distinctly below the regression line are derived from Venetia Mine diamonds, for which crystallization following modification of their mantle substrates through the 2.05 Ga Bushveld large igneous province was inferred (Richardson et al. 2009). For comparison purposes, the inclusion-based regression is shown as a **dark grey line** in the xenolith garnet plot (**bottom**).

Correspondingly, Mg# and Ca intercept values are negatively correlated for both inclusion and xenolith garnets (not shown). Garnet Mg# is, however, controlled by temperature-dependent Mg–Fe partitioning with olivine, orthopyroxene and clinopyroxene and hence only provides a distorted image of bulk rock Mg#. Pairing garnet Ca intercept values with the Mg# of coexisting olivines (Fig. 11; see also Eaton et al. 2009, their Fig. 4) sidesteps this distortion problem, with the drawback of greatly reducing the data set for inclusions (from 1175 individual garnets to 198 garnet–olivine pairs). Despite some scatter, regression of the inclusion data set yields a clear linear trend (r^2 = 0.51; Fig. 11), with olivine Mg# increasing from 92.0 at the transition from lherzolitic to harzburgitic garnets (Ca intercept value = 3.375) to 94.2 for Ca-free garnets. Note that for the lherzolitic paragenesis some deviation from this linear trend is expected, as (1) varying degrees of primary melt depletion prior to the point of exhaustion of clinopyroxene in the protoliths would not be reflected in the Ca intercept values of garnet and (2) metasomatic addition of clinopyroxene to original harzburgitic melting residues may have occurred. In spite of that, including only harzburgitic garnet–olivine pairs results in an indistinguishable regression line ($y = 94.2 - 0.62x$ versus $y = 94.2 - 0.65x$ for the entire data set) but with a lower r^2 (0.31). For the xenolith data set, the same trend of decreasing Mg# with increasing Ca intercept values is visible but with much more scatter (Fig. 11). For the lherzolitic paragenesis, the variation in olivine Mg# at a given garnet Ca intercept value visually appears more pronounced than for the inclusion data set. Given the equality of mean Mg# of lherzolitic olivines inside and outside of diamond (see above), this apparently increased variability likely only reflects a much larger number of analyzed lherzolitic xenoliths compared to inclusion pairs. Highly similar median and mean Mg# values of lherzolitic garnets inside and outside diamond (inclusions: median = 83.0 and mean = 83.3 ± 2.8; xenolith garnets: median = 83.4, mean = 83.1 ± 2.2) support this interpretation. The overlapping trends in olivine Mg# versus garnet Ca intercept value document that peridotitic diamond substrates and cratonic garnet peridotites followed the same melt depletion path, with the key difference that carbon-bearing fluids preferentially react with the cpx-absent, depleted peridotitic substrates during mineralization events that produce diamond.

As a consequence of intense primary melt depletion, Na and Ti contents in peridotitic diamond substrates and cratonic peridotites are expected to be very low (see above). For harzburgitic garnets included in diamond and from cratonic peridotite xenoliths, median Na_2O and TiO_2 contents are indeed at or below the limit of detection (≤ 0.02 wt%; averages are not reported because of strongly non-normal distributions), with concentrations above the limit of detection being considered to indicate secondary metasomatic re-enrichment (Figs. 12 and 13). For the lherzolitic paragenesis, median contents are above the limit of detection for TiO_2

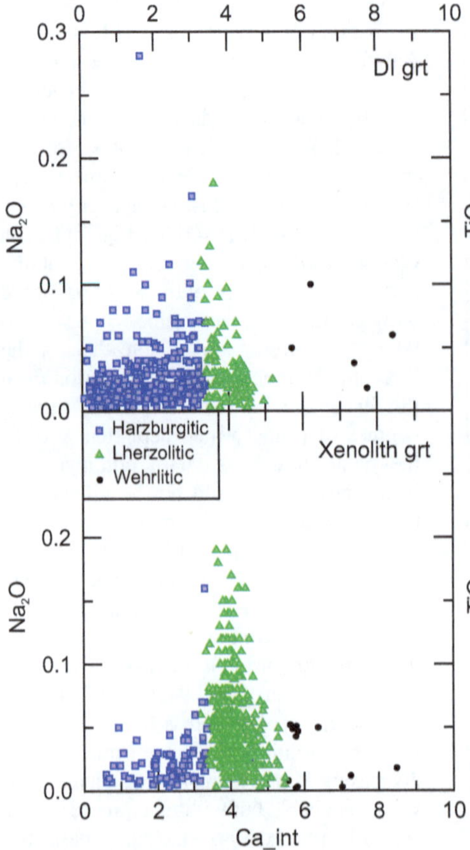

Figure 12. Na_2O content (wt%) versus Ca intercept value (Ca_int; see Fig. 11 for explanation) for peridotitic garnets included in diamond (**top**) and from cratonic xenoliths (**bottom**).

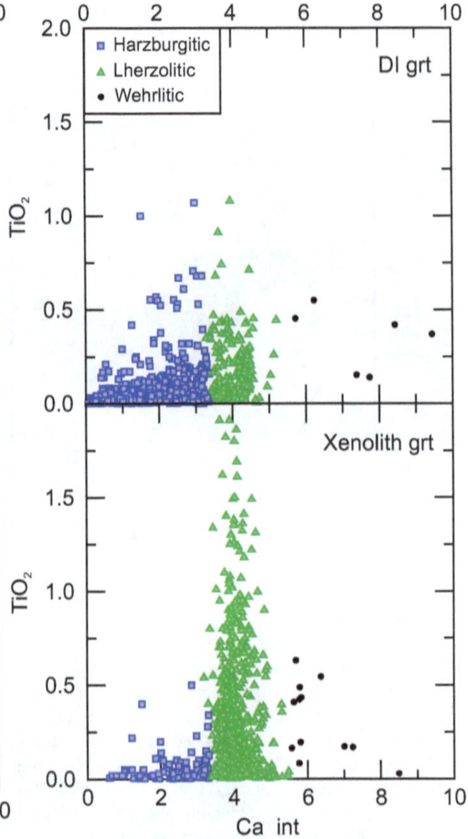

Figure 13. TiO_2 content (wt%) versus Ca intercept value (Ca_int; see Fig. 11 for explanation) for peridotitic garnets included in diamond (DI, **top**) and from cratonic xenoliths (**bottom**). For the harzburgitic paragenesis, comparison of the upper and lower diagrams gives the impression of higher TiO_2 contents in inclusions compared to xenolith garnets. This impression, however, relates to the much higher number of harzburgitic inclusion garnets analysed for TiO_2 (n = 864 versus n = 134 for xenolith garnets), with median values for both inclusions and xenolith garnets being ≤ 0.02 wt% (i.e., below the limit of detection) and Q3 values (middle value of the upper half of the data set) being nearly identical (0.06 wt% for inclusions and 0.05 wt% for xenolith garnets).

(0.13 wt% for inclusions and 0.24 wt% for xenolith garnets) and for xenolith garnets also for Na_2O (median of 0.04 wt%). In this work, we repurpose $Na_2O \leq 0.02$ wt% (below the limit of detection) and $TiO_2 \leq 0.04$ wt% from our previous work (Stachel and Harris 2008) to represent upper-limit thresholds for garnets included in diamond and from xenoliths, above which cratonic lherzolites are considered enriched in Na and Ti by secondary metasomatism.

On that basis, Figures 12 and 13 document widespread metasomatic modification of the harzburgitic and lherzolitic substrates of both inclusion and xenolith garnets. Maximum Na_2O concentrations in garnet are 0.31 wt% (a harzburgitic inclusion outside the range of Figure 12) and 0.19 wt% (xenoliths) and extreme TiO_2 values reach 1.08 wt% (inclusions) and 1.91 wt% (xenoliths). For xenolith garnets there is a clear positive correlation of Na with Ti ($r^2 = 0.67$; Figure not shown), which is barely discernible for inclusions ($r^2 = 0.15$). A slope of Na versus Ti of about 0.2 (on a cation basis) indicates that Na likely is fully accommodated by the substitution mechanism $^{VIII}M^{2+} + {}^{VI}Al^{3+} = {}^{VIII}Na^{1+} + {}^{VI}Ti^{4+}$ (Bishop et al. 1976). The strong excess of Ti over Na reflects additional substitution mechanisms of the type $2^{VI}Al^{3+} = {}^{VI}Ti^{4+} + {}^{VI}M^{2+}$ or $^{VI}Al^{3+} + {}^{IV}Si^{4+} = {}^{VI}Ti^{4+} + {}^{IV}Al^{3+}$ (Ackerson et al. 2017).

Intense metasomatic addition of Ti is particularly common for lherzolitic xenolith garnets, with 25% of samples containing > 0.5 wt% TiO_2 (Q3 value), as opposed to only 2.4% of inclusions. A large set of pressure–temperature estimates for well-equilibrated cratonic garnet lherzolites (n = 372) allows the relationship of Ti-metasomatism and temperature of last equilibration to be assessed. This relationship indicates that TiO_2 contents > 0.5 wt% (the above quoted Q3 value for lherzolitic xenolith garnets), with one exception, are restricted to temperatures > 1130 °C (Fig. 14). A comparable data set does not currently exist for inclusions in diamond but the observed paucity of inclusion garnets with TiO_2 > 0.5 wt% documents that the style of high-temperature, Ti-rich ("melt") metasomatism (Griffin and Ryan 1995) documented in one quarter of lherzolite xenoliths either did not significantly impact diamond

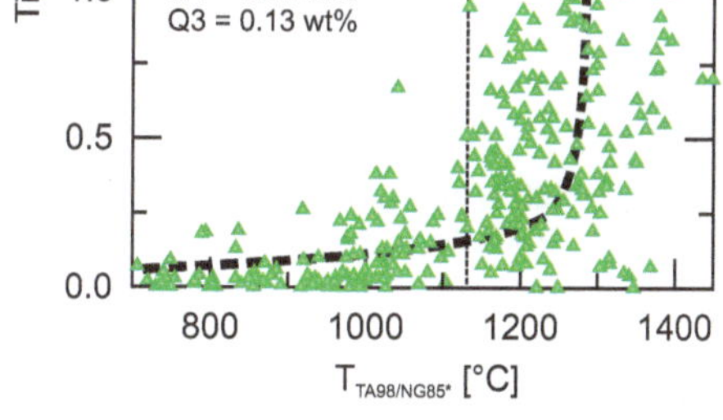

Figure 14. TiO_2 content (wt%) in xenolith garnets versus temperature of last equilibration (°C) of their cratonic lherzolite hosts. Temperature is calculated using the enstatite in clinopyroxene thermometer of Taylor (1998) in combination with the garnet–orthopyroxene Al exchange barometer of Nickel and Green (1985) (with Al in the M1 site of orthopyroxene calculated after Carswell and Gibb 1987). With a single exception, TiO_2 contents in garnet remain relatively low (≤ 0.02 to 0.40 wt%) at equilibration temperatures below 1130 °C. At higher temperatures, a large range in TiO_2 contents is observed (from 0.04 to 1.91 wt%), consistent with the melt metasomatic trend of Griffin and Ryan (1995). For the low- and high-temperature groups, median values and interquartile ranges (as middle values of the lower half (Q1) and the upper half (Q3) of the data sets) are given as well.

substrates or only occurred in a different, much milder form (note footnote 1 in section *Introduction* here). Given that typical temperatures of diamond formation in peridotitic substrates are estimated at 1130 ± 120 °C (Stachel and Luth 2015; see also Nimis 2022), this near absence of Ti metasomatism of diamond substrates is not a temperature effect but either indicates an increasing abundance of Ti-rich melt metasomatism in the more recent geological past (Mesoproterozoic to Phanerozoic) and/or a diamond unfriendly effect of such metasomatism that renders affected potential substrates infertile.

Mg-chromite

Low Al contents and consequent high bulk rock Cr/Al ratios cause spinel to persist to diamond-stable pressures in cratonic peridotites and, in particular, in peridotitic diamond substrates (Doroshev et al. 1997; Klemme 2004; Ziberna et al. 2013). Of the 1001 peridotitic spinel inclusions in diamond in our database, most (97%) are associated with neither garnet nor clinopyroxene and hence are of unspecified peridotitic paragenesis. Based on the observation that only 9% of the Mg-chromite inclusions with an assigned paragenesis are lherzolitic and 91% are harzburgitic, a mostly harzburgitic paragenesis may also be assumed for the inclusions with unspecified paragenesis. Based on high pressure experiments, spinel, in equilibrium with garnet and orthopyroxene, at the pressure–temperature conditions of the graphite–diamond transition along a typical cratonic geotherm, should have a molar Cr# (100Cr/(Cr+Al)) of ≥ 80 (Doroshev et al. 1997; Girnis and Brey 1999; Klemme 2004). In good agreement with these experimental outcomes, the median Cr# of spinel inclusions in diamond is 86.8 (mean = 86.6 ± 3.5) and the few samples of unspecified peridotitic paragenesis falling towards lower Cr# (Fig. 15) likely represent orthopyroxene-free (i.e., dunitic) assemblages, for which spinel + olivine remains stable for all lithospheric pressures.

For comparison of spinels from cratonic garnet peridotite xenoliths with inclusions in diamond, we excluded all xenolith spinels with Cr# < 60 because of a likely shallow garnet facies (< 100 km, see Klemme 2004) or retrograde origin. Despite this cut-off, the Cr# of spinel from garnet peridotites (median 80.5, mean 78.8 ± 6.8) is lower than that of inclusions. The Mg# of all but two (0.02%) spinel inclusions is > 50 (median = 69.1, mean = 68.9 ± 5.2) and compared to xenolith spinels (median Mg# = 62.6, mean = 61.6 ± 6.0) higher by about 6–7. Besides bulk rock composition, the Mg# of spinel depends on the temperature dependent exchange of Mg–Fe with olivine (O'Neill and Wall 1987); elevated Mg# for inclusion spinels may thus, at least partially, represent a temperature (depth) effect. Combined with high Cr#, spinel Mg# generally >50 allows classification as Mg-chromite. The ferric iron number (Fe^{3+}# = 100Fe^{3+} / (Fe^{2+} + Fe^{3+}); calculated from stoichiometry after Droop 1987) of spinel inclusions (median Fe^{3+}# = 15.9, mean = 15.7 ± 7.4; Fig. 16) is lower than that of xenolith spinels (median = 18.8, mean = 19.6 ± 7.6).

Overall low Fe^{3+}# in both inclusions and xenolith spinels is consistent with the moderately reducing conditions observed in the deep lithospheric mantle (Frost and McCammon 2008; Stagno et al. 2013; Luth et al. 2022). In view of the almost exclusively harzburgitic association of inclusion spinel, focusing on garnet harzburgites only, the Cr# (median = 83.1, mean = 82.8 ± 4.2) and Mg# (median = 63.9, mean = 64.2 ± 4.3) in xenolith spinel increases whilst the Fe^{3+}# (median = 18.1, mean = 17.5 ± 6.9) decreases slightly, but still not to the levels observed for inclusions (Figs. 15 and 16). TiO_2 contents are generally similarly low in both inclusions (median = 0.09 wt%; no mean reported for strongly non-normal distribution) and xenolith spinels (median = 0.10 wt%) but elevated 75th percentile values (Q3 = 0.22 wt% for inclusions and 0.33 wt% for xenolith spinels) indicate that moderate metasomatic re-enrichment of Ti is relatively widespread.

Key observations and conclusions

The principal distinction between diamond substrates and the cratonic mantle roots as sampled by garnet peridotite xenoliths is the well-established, much higher proportion of harzburgite (-dunite) to lherzolite (-wehrlite) in the former (~85:15%; Gurney 1984) compared to the latter (18:82% in our xenolith database). Potentially dunitic mineral assemblages, indicated by CaO contents in garnet < 1.8 wt%, are common diamond substrates (~38%) but rarely documented in xenoliths (~2%). This latter discrepancy may be somewhat enhanced by sampling bias during xenolith collection, as garnet- and/or clinopyroxene-rich peridotites are more likely to be picked. Using mineral Mg# as an indicator of source depletion through melt extraction again documents the overall more depleted character of diamond substrates relative to the deep (garnet-facies) cratonic lithospheric mantle represented in the xenolith record. Olivine, because of its very high modal abundance in cratonic peridotites (72 vol% as mean and median value; Pearson et al. 2004), is the most robust proxy for this purpose, as its Mg# effectively becomes independent of temperature and instead equivalent to bulk rock Mg#. The observation that on the paragenesis level olivine inside and outside of diamond has statistically indistinguishable means in Mg#, therefore, has fundamental implications: (1) that the major element composition of inclusions is imposed by the substrate and not by the diamond forming medium and (2) that widespread Fe-rich metasomatism of the lithospheric mantle did not occur subsequent to diamond formation, which is focused in the Paleoarchean for harzburgitic substrates and the Paleoproterozoic–Mesoproterozoic for lherzolitic substrates. The latter conclusion precludes neither metasomatism by small melt fractions/fluids subsequent to diamond formation, as such events have limited impact on bulk rock Mg#, nor localized metasomatic shifts in Mg#.

A distinguishing feature between inclusions and xenolith minerals lies in the higher Cr content and Cr/Al ratio of garnet and Mg-chromite in diamond. The extensive melt depletion events that characterize the protoliths of subcratonic lithospheric mantle did not affect the Cr content of the melting residues but efficiently removed Al (from primitive mantle to average cratonic harzburgite, Cr contents remain constant but about 90% of Al is removed; e.g., McDonough and Rudnick 1998, their Table 1). Higher Cr contents for inclusions are not limited to garnets with very low Ca intercept values and high Mg# (in fact, the highest Cr contents are associated with harzburgitic garnets with relatively low Mg# of 84–86) but occur equally for mildly subcalcic and even for lherzolitic garnets. This suggests that the almost exclusive restriction of Cr_2O_3 contents greater than 13 wt% to inclusion garnets is not a consequence of higher degrees of primary melt depletion being restricted to diamond substrates.

Nevertheless, the systematic offset to higher Cr contents, as best seen for spinel inclusions, may, at least partially, represent a source signal. Cr is not usually considered as a mobile element during mantle metasomatism but high pressure (4–6 GPa) experiments indicate an order of magnitude increase in the solubility of Cr in hydrous fluids associated with increased salinity (addition of 1.3–3 wt% of KCl; Klein-BenDavid et al. 2011). These authors consider that during percolation of such saline high-pressure fluids, Cr could dissolve from the peridotitic wall rocks without significantly affecting their high Cr concentrations. The ability of such saline fluids to impose their high-Cr characteristics on garnet inclusions in diamond, however, is likely limited, given that their mean Cr# (100Cr/(Cr+Al)) is only 7.5 (high-density fluid compilation of Weiss et al. 2022) compared to Cr# of 38–65 for the unusually Cr_2O_3-rich (> 13 wt%) garnet inclusions discussed here. In contrast, enrichment of Al in cratonic peridotites is frequently observed and possibly relates to fluids derived from recycled sediments (e.g., Rapp et al. 2017). Consequently, widespread but relatively mild re-enrichment in Al may have decreased Cr/Al ratios and increased modal garnet in cratonic peridotites sampled as xenoliths, with this effect being strongest for rocks with extreme primary Al depletion (very high Cr/Al ratios). This type of metasomatism may be responsible

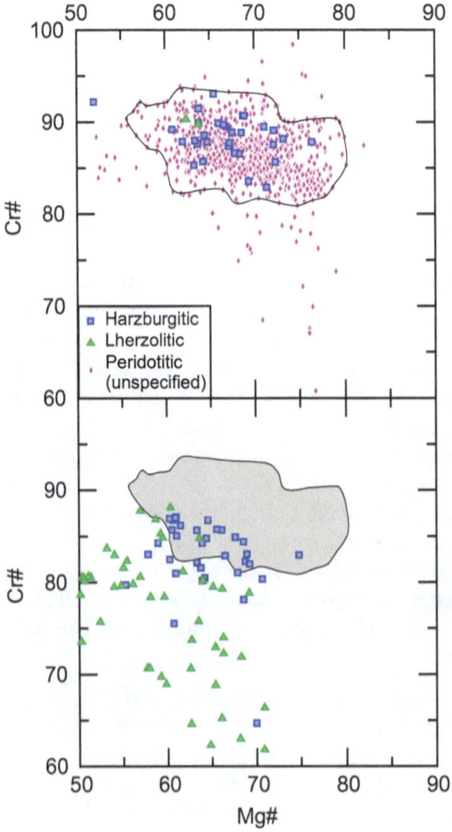

Figure 15. Molar Cr# (100Cr / (Cr + Al)) versus molar Mg# (100Mg / (Mg + Fe^{2+}), with Fe^{2+} calculated from total Fe based on stoichiometry; Droop 1987) for peridotitic Mg-chromites included in diamond (DI, **top**) and from cratonic garnet (-spinel) peridotite xenoliths (**bottom**). The **grey area** in the bottom diagram corresponds to the outline of the area of common spinel inclusions in diamond in the upper diagram (cf., Fipke et al. 1995). The paragenesis of Mg-chromite inclusions is derived from coexisting garnet or the presence of clinopyroxene (lherzolitic), in their absence the paragenesis is unspecified.

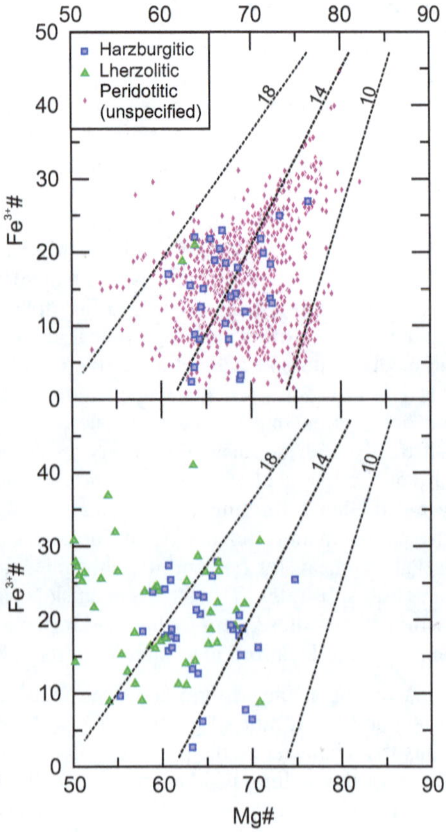

Figure 16. Molar Fe^{3+}# (100Fe^{3+}/ (Fe^{2+} + Fe^{3+})) versus molar Mg# (100Mg / (Mg + Fe^{2+})) for peridotitic Mg-chromites included in diamond (DI, **top**) and from cratonic garnet (-spinel) peridotite xenoliths (**bottom**). Fe^{2+} and Fe^{3+} were calculated from total Fe content based on stoichiometry (Droop 1987). The **dashed lines** and associated numbers indicate approximately constant total iron content (FeO^{total} in wt%).

for both the lower average and maximum Cr contents in xenolith-hosted garnet and spinel; it may have affected diamond substrates as well, just to a lesser degree.

An alternate explanation relates to the pressure and temperature dependence of the distribution of Cr between garnet and spinel (Doroshev et al. 1997; Grütter et al. 2006). Experiments demonstrated inclusion-like high Cr/Al ratios for coexisting Cr-pyrope and Cr-spinel in harzburgite at high pressures and temperatures (> 5 GPa and > 1200 °C; Malinovsky and Doroshev 1975; Girnis and Brey 1999). High Cr/Al inclusion compositions are accordingly interpreted as indicative of some diamond growth occurring over a wide range of temperatures, elevated above a cratonic geotherm during high-temperature thermal perturbations (Girnis et al. 1999; Grütter 2006).

The most sensitive indicators of metasomatism on the major-minor element level are Na, Ti and, for clinopyroxene, K. In orthopyroxene, clinopyroxene and garnet, metasomatic enrichment of Na and Ti in both inclusions and xenoliths is most prominent in the lherzolitic paragenesis and very intense Ti-rich metasomatism is almost entirely restricted to lherzolite xenoliths. Intense Ti metasomatism (leading to > 0.5 wt% TiO_2 in garnet) is restricted to temperatures of last equilibration in excess of 1130 °C, i.e., occurs above the hydrous solidus of lherzolite along typical cratonic geotherms (e.g., Wyllie and Ryabchikov 2000). Since equilibration temperatures of > 1130 °C are frequently observed for inclusions, the almost complete absence of evidence of intense Ti-metasomatism in inclusions (mild Ti-metasomatism is more frequent) likely relates to either a diamond unfriendly character of such metasomatism or an increase in Ti-metasomatic intensity or frequency after the formation of most peridotitic diamonds in the Archean-Mesoproterozoic. The opposite is true for potassic metasomatism documented in clinopyroxene, where only fairly moderate levels of enrichment are seen in xenoliths ($K_2O \leq 0.26$ wt%) whilst clinopyroxene inclusions show a tail in their frequency distribution of up to 1.68 wt% K_2O. This type of metasomatism may relate to potassic brines that are frequently observed as high-density fluid inclusions in fibrous diamonds (Izraeli et al. 2001; Weiss et al. 2022). That evidence for such metasomatism is best preserved inside diamond may relate to reactivation of potassic mantle metasomes and migration to shallower mantle in the course of transient heating events (e.g., Waters and Erlank 1988; Tappe et al. 2008). The overall low abundance of strongly potassic clinopyroxene inclusions suggests that potassic brines may be one medium of diamond formation in peridotitic substrates, but not the dominant one.

TRACE ELEMENT COMPOSITION OF PERIDOTITIC SUITE INCLUSIONS AND COMPARISON WITH CRATONIC PERIDOTITE XENOLITH MINERALS

Garnet

To our knowledge, the first REE pattern for peridotitic garnet was obtained by Philpotts et al. (1972), via isotope dilution thermal ionization mass spectrometry on separated minerals from a phlogopite garnet lherzolite xenolith (GSFC #187) from Roberts Victor (Kaapvaal Craton). From today's perspective, the reported humped REE_N pattern with a prominent peak at Dy would be considered anomalous.

The two common types of REE_N patterns of garnet from cratonic peridotites, *normal* and *sinuous* or *sinusoidal*, were first reported by Shimizu (1975): (1) garnets from three sheared (mosaic-porphyroclastic texture) lherzolite xenoliths from kimberlites in Lesotho showed steep positive slopes in the $LREE_N$ and fairly flat MREE–HREE at ≥10 times chondritic abundance, similar to garnets from eclogite xenoliths and megacrysts (hence the term "*normal*" patterns), (2) garnets from three granular (coarse texture) lherzolite xenoliths from Lesotho and the Kimberley area had sinusoidal patterns, with a prominent peak at Sm_N–Eu_N and an upward kink in slope near Er_N. For inclusions in diamond, similar sinusoidal REE_N pattern were subsequently reported by Shimizu and Richardson (1987) for harzburgitic garnets in diamonds from Kimberley and Finsch, while normal REE_N patterns were first observed for lherzolitic garnet inclusions from Akwatia, Ghana (Stachel and Harris 1997).

REE_N patterns of garnet inclusions of harzburgitic and lherzolitic paragenesis. Since the pioneering work of Shimizu and Richardson (1987), the number of REE analyses of garnet inclusions has risen considerably as ion microprobes enjoyed wider use, with 223 harzburgitic and 72 lherzolitic garnets in our database (see Fig. 17). These analyses clearly establish sinusoidal REE_N patterns as the hallmark of both harzburgitic and lherzolitic inclusion garnets. On average, the degree of sinuosity, however, is more pronounced for harzburgitic garnets (see median, 10th and 90th quartile patterns in Fig. 17).

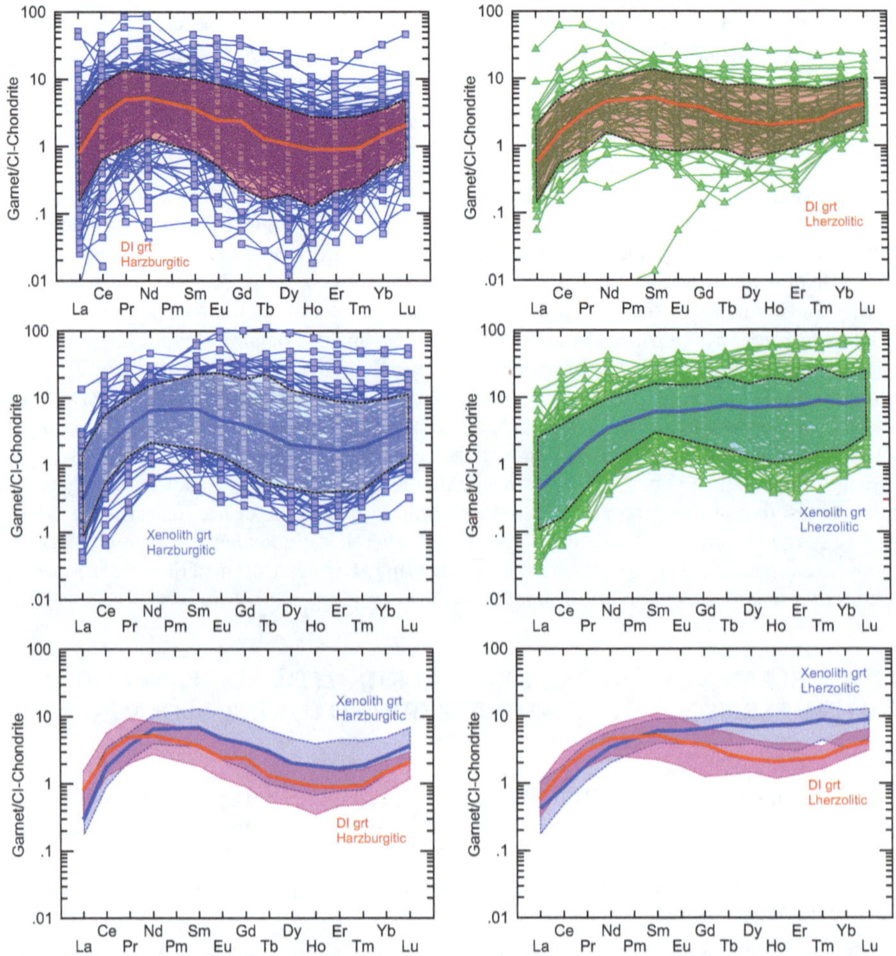

Figure 17. Chondrite-normalized REE patterns for harzburgitic (n = 223) and lherzolitic (n = 72) garnet inclusions in diamond (DI, **top**) and harzburgitic (n = 152) and lherzolitic (n = 280) garnets from xenoliths (**middle**). **Thick red or blue lines** are median values, transparent fields are ranges from the 10th to 90th percentile. The **two bottom plots** compare the median patterns and interquartile ranges (i.e., 25th to 75th percentile) of inclusion (**red**) and xenolith (**blue**) garnets of harzburgitic and lherzolitic paragenesis. CI-chondrite composition is from McDonough and Sun (1995).

In detail, the range of REE_N patterns displayed by peridotitic garnet inclusions is much more complex than apparent from Figure 17. Based on overall shape and the presence and position of a peak in the $LREE_N$–$MREE_N$, we identified 10 types of REE_N patterns, which due to uncertainties in the classification of garnet analyses that do not include the LREE praseodymium (Pr) were assigned into 12 classes. The REE_N patterns for these classes are shown in Appendix Figure 1 and the frequency of their occurrence for garnets of the harzburgitic and lherzolitic paragenesis is shown in Figure 18. Patterns with peaks at Ce, Pr, Nd, Sm and Eu are all sinusoidal. Non-sinusoidal patterns are either V-shaped with a peak at La, show steep *positive* or nearly *flat* slopes from $LREE_N$ to $HREE_N$, or classify as *normal* (rise in $LREE_N$, flat and superchondritic MREE–HREE). *Abnormal* patterns fit none of these categories and typically are "noisy" with multiple peaks. For harzburgitic garnets, sinusoidal patterns peaking at Nd are clearly dominant (Fig. 18). Including patterns with a peak at Nd

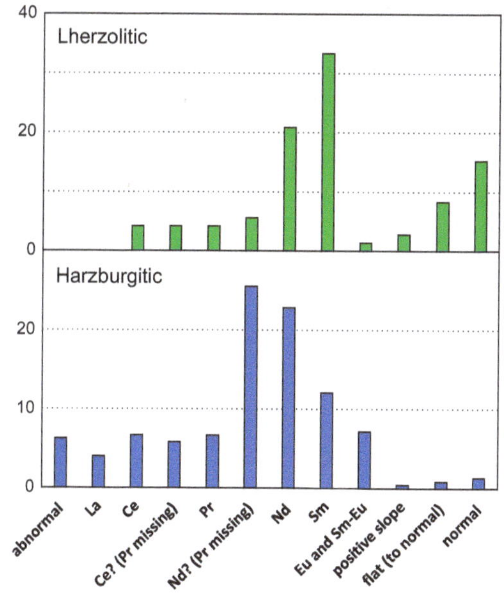

Figure 18. The relative abundance (in %) of the various inclusion garnet REE_N patterns, calculated separately for the harzburgitic and lherzolitic parageneses (based on 295 peridotitic garnets: 223 harzburgitic and 72 lherzolitic). The classes from Ce to Eu represent the different peak positions for sinusoidal REE_N patterns. The groups Ce? and Nd? show peaks at these elements but since the neighboring LREE Pr was not analyzed, a true peak at Pr cannot be excluded – hence the question marks. REE_N patterns peaking at La are V-shaped. Other types (1) show positive slopes from $LREE_N$ to $HREE_N$, (2) are nearly flat, or (3) show a positive slope within the $LREE_N$ but nearly flat $MREE_N$–$HREE_N$ (normal patterns). Patterns that fit none of these categories are classified as abnormal. REE_N patterns for all groups are shown in Appendix Figure 1.

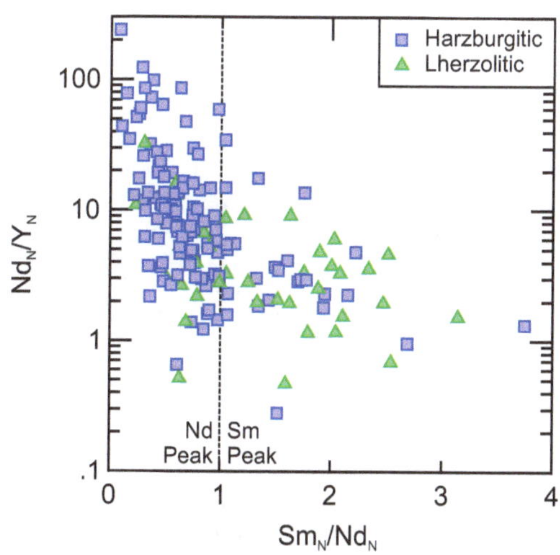

Figure 19. Chondrite-normalized ratios of Nd/Y and Sm/Nd for inclusion garnets with REE_N patterns peaking (1) at Nd (77% harzburgitic), (2) at Nd? (Pr missing, 93% harzburgitic) and (3) at Sm (with about equal abundance of harzburgitic and lherzolitic garnets). Garnets with REE_N patterns peaking at Nd ($Sm_N/Nd_N < 1$) have a much steeper slope in the MREE (high Nd_N/Y_N) than samples with a peak at Sm ($Sm_N/Nd_N > 1$).

with Pr not analyzed, 49% of harzburgitic garnets peak at Nd. By contrast, 33% of lherzolitic garnets peak at Sm and only 21% at Nd (27% including the peaks at Nd with Pr missing) and 15% of lherzolitic garnets show normal patterns, as opposed to 1% of harzburgitic garnets (Fig. 18). The transition between the two most important REE_N patterns, peak at Nd (including samples with Pr missing) to peak at Sm, is associated with a decrease in the degree of sinuosity, measured as the Nd_N/Y_N ratio (MREE$_N$ slope) in Figure 19.

Other trace elements. Other commonly and reliably analyzed trace elements in garnet inclusions are Sr, Nb, Ti, Hf, Zr and Y (in sequence of increasing compatibility; Fig. 20). Whilst Sr concentrations are higher in harzburgitic garnets (median value of 2.4 ppm, compared to 0.67 ppm for lherzolitic garnets; likely controlled by the absence/presence of clinopyroxene, see Stachel et al. 2004), the patterns for the remaining elements are nearly parallel but

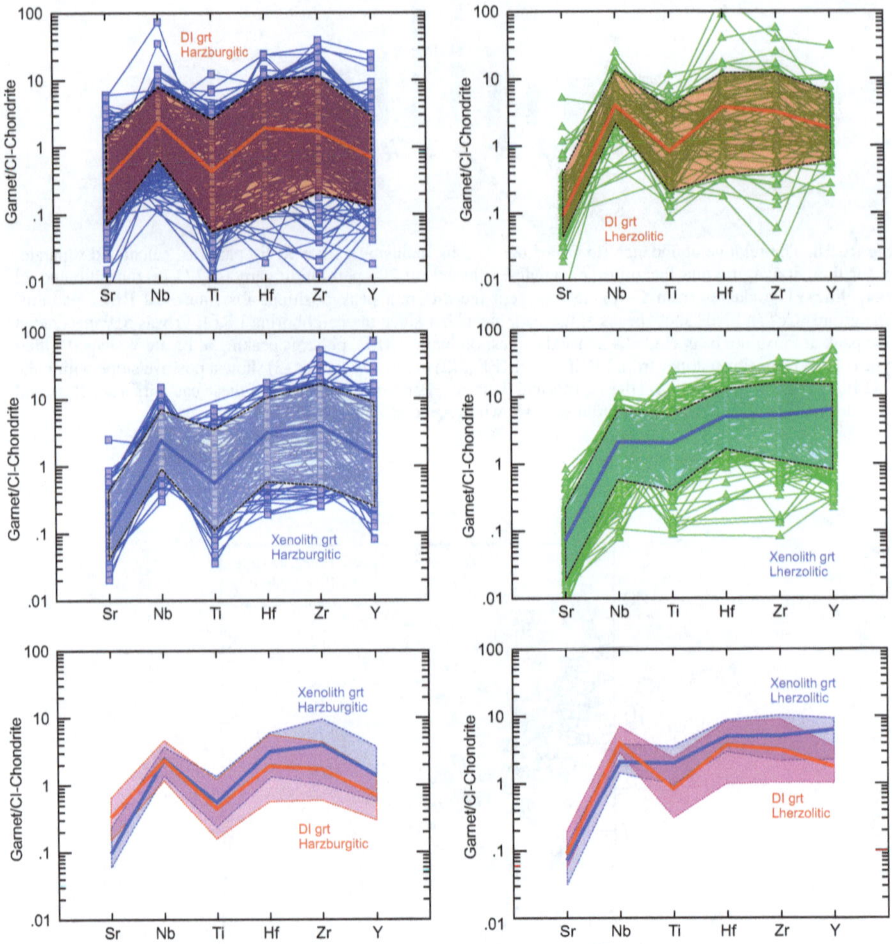

Figure 20. Chondrite-normalized patterns for large ion lithophile (Sr) and high field strength elements (Nb, Ti, Hf, Zr and Y) in harzburgitic and lherzolitic garnet inclusions in diamond (**top**) and from cratonic garnet peridotite xenoliths (**middle**). Compatibility in garnet (based on garnet–melt partition coefficients; e.g., Green 1994) increases from left to right. **Thick red or blue lines** are median values, transparent fields are ranges from the 10th to 90th percentile. The **two bottom diagrams** compare the median patterns and interquartile ranges (i.e., 25th to 75th percentile) of inclusion (**red**) and xenolith (**blue**) garnets of harzburgitic and lherzolitic paragenesis.

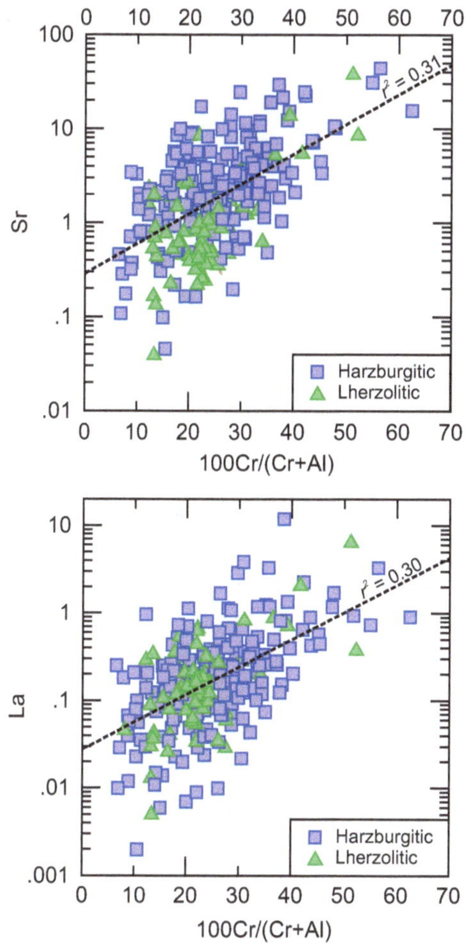

Figure 21. Sr (**top**) and La (**bottom**) in ppm versus molar Cr# (100Cr / (Cr + Al)) for peridotitic garnet inclusions in diamond. The **dashed lines** are linear regressions with the coefficient of determination (r^2) being indicated.

with higher concentrations in lherzolitic garnets (Fig. 20). Both parageneses show distinct negative anomalies for Sr and Ti (subchondritic median values), flat and typically superchondritic Nb, Hf and Zr, and slightly depressed Y_N (with a subchondritic median for harzburgitic garnets).

Correlations with major elements. Griffin et al. (1999a) highlighted the usefulness of garnet Cr# as a measure for the degree of chemical depletion of cratonic peridotites. In addition to possible metasomatic modification (e.g., Burgess and Harte 1999; Grütter and Menzies 2017), the interpretation of garnet Cr# is, however, complicated by the pressure-controlled exchange of Cr and Al between spinel and garnet (in spinel–garnet peridotites, garnet Cr# is a geobarometer; Grütter et al. 2006). In garnet, increasing Cr# is accompanied by broad trends of enrichment in strictly incompatible trace elements such as Sr and the LREE (Fig. 21).

A more unambiguous indicator in garnet of the degree of primary melt extraction from its source protolith is its position relative to the lherzolitic trend in a Cr/Ca plot, which can be expressed as the Ca intercept value (Grütter et al. 2004; see above) or more simply but less accurately, as the Ca# (molar 100Ca/(Ca+Mg+Fe)). In addition to primary melt depletion, garnet Ca intercept values (or Ca#) also track secondary metasomatic modification of the diamond substrates. Figure 22 shows a positive correlation between the Yb content and Ca# of garnet. To interpret this correlation as a straightforward geochemical trend may, however, be misleading as crystal chemical controls have to be considered as well. For eclogites, increasing partitioning of REE into garnet with increasing Ca content was recognized by Harte and Kirkley (1997). In a study of trace element partitioning between mantle minerals and silico-carbonate melts, Girnis et al. (2013) confirmed these findings and observed a strong increase in garnet–melt partition coefficients for REE with Ca#. This effect renders HREE like Yb and Lu incompatible in strongly subcalcic garnets (Girnis et al. 2013). Figure 22 shows a predicted trend for Yb in garnet, based purely on the crystal chemical effect imposed by varying the Ca# number from 15 (approximately equivalent to the lherzolite/harzburgite boundary) to 0. The predicted effect is over an order of magnitude stronger than the observed correlation in the inclusion data set, suggesting that either the crystal chemical effect was overestimated by Girnis et al. (2013) or that other effects (e.g., bulk rock chemistry) partially compensate the crystal chemically predicted strong HREE depletion in low-Ca garnets.

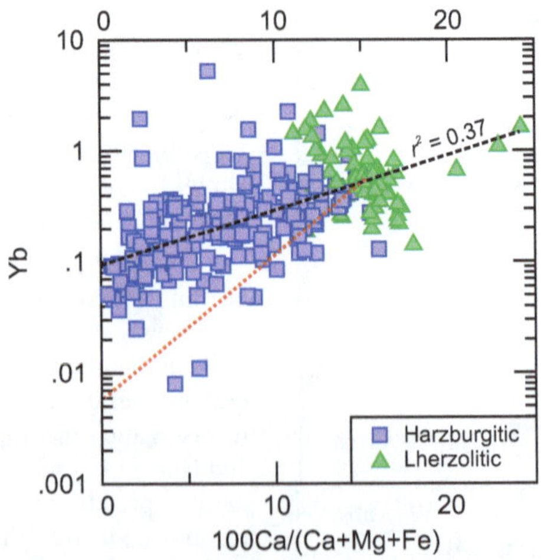

Figure 22. Yb (ppm) versus molar Ca# (100Ca / (Ca + Mg + Fe)) for peridotitic garnet inclusions in diamond (DI). The **black dashed line** is a linear regression with the coefficient of determination (r^2) being indicated. The **red stippled line** is a hypothetical trend, showing how garnet should evolve away from compositions at the lherzolite/harzburgite boundary (Ca# of ~15) towards highly subcalcic, harzburgitic–dunitic compositions, based on trace element partitioning with a silico-carbonatitic melt of constant composition (using the X_{Ca} dependent *D*-value for Yb of Girnis et al. 2013).

Comparison with garnet from cratonic peridotites. In cratonic peridotites, harzburgitic garnets inside and outside of diamond share overall similar, sinusoidal REE_N patterns (Fig. 17). On average, xenolith garnets, however, have higher MREE and HREE, resulting in a shift of the peak from Nd (for inclusion garnets) to Sm for the median REE_N pattern. Sr contents in xenolith garnets tend to be lower and Zr and Y contents higher than in inclusions garnets (Fig. 20). Distinct Nd and Sr isotope composition of harzburgitic garnets included in diamond and from concentrate were first reported by Richardson et al. (1984) and are discussed in detail by Smit et al. (2022).

For lherzolitic garnets, differences are even more pronounced: xenolith garnets have significantly higher contents in the MREE–HREE from Gd onwards, resulting in normal REE_N patterns for median and 25th and 75th quartile compositions, as opposed to sinusoidal patterns for the median and interquartile range of lherzolitic garnet inclusions in diamond (Fig. 17). Lherzolitic xenolith garnets generally also lack the pronounced negative Ti_N anomalies characteristic for inclusion garnets and on average have higher Y contents (Fig. 20).

Clinopyroxene

Shimizu (1975) not only defined the two principal types of REE_N patterns of peridotitic garnet but also characterized the REE_N patterns of associated clinopyroxene: (1) clinopyroxene occurring together with garnet with normal REE_N in the three studied sheared lherzolite xenoliths showed mildly fractionated REE_N patterns with about 10× chondritic LREE and chondritic to mildly subchondritic HREE, similar to clinopyroxene in eclogite xenoliths and megacrysts. (2) Clinopyroxene in equilibrium with garnet with sinusoidal REE_N in three coarse lherzolites was found to have much more fractionated LREE–HREE, falling from ~20–30 chondritic LREE to about 0.4× chondritic HREE.

Types of REE_N patterns. The two groupings introduced by Shimizu (1975) largely hold for peridotitic clinopyroxene inclusions in diamond. For inclusions, the mildly fractionated group is characterized by *humped* patterns peaking at Nd_N (Fig. 23), with La_N/Sm_N between 0.3 and 2.5 with a median value of 0.6. Two clinopyroxenes associated with the humped

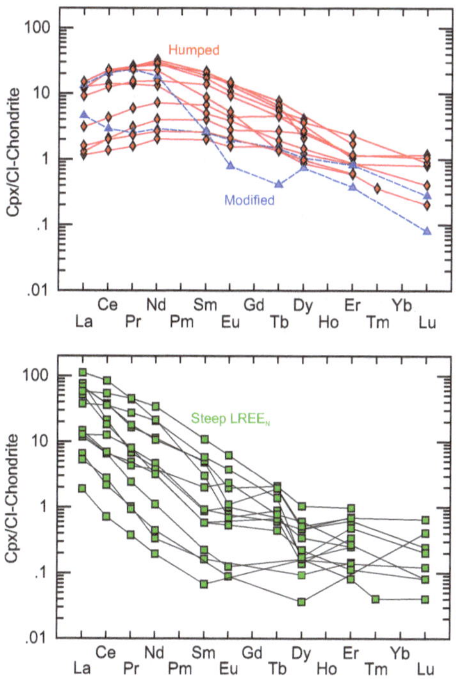

Figure 23. Chondrite-normalized REE patterns for peridotitic clinopyroxene inclusions in diamond. **Top**: mildly fractionated humped patterns (peaking at Nd_N) and similar modified patterns, distinct through either additional enrichment in La and Ce or a trough in the $MREE_N$ (Sm–Tb). **Bottom**: clinopyroxenes with steep $LREE_N$ slopes and fairly unfractionated Dy_N/Lu_N.

group show *modified* patterns, which by comparison with the humped patterns show enrichment of La and Ce in one case and enriched LREE and a trough in $MREE_N$ in the other. In slight contrast to the xenolith data of Shimizu (1975), the key distinguishing criterion of the second group are steep negative $LREE_N$ slopes (*steep-$LREE_N$*) and not necessarily a high LREE–HREE ratio (Fig. 23). La_N/Sm_N ranges between 4.6 and 87 with a median value of 22.

Clinopyroxene inclusions with humped REE_N patterns coexist with garnets with normal patterns (n = 2 for inclusion database), characterized by steep positive slopes in the $LREE_N$. This coexistence matches observations made for peridotite xenoliths beyond the limited sample set of Shimizu (1975), with, e.g., the occurrence of clinopyroxene with humped patterns together with garnet with normal patterns being characteristic for the Type 1 garnet lherzolites of Grégoire et al. (2003). Clinopyroxene inclusions with steep $LREE_N$ coexist with garnets with sinusoidal or positively sloped REE_N (n = 4), the latter characterized by a less positive slope in the LREE compared to garnets with normal REE_N patterns. Clinopyroxene with steep $LREE_N$ (but with a peak at Ce) and garnets with sinusoidal or positively sloped REE_N patterns also characterize the Type 2 garnet lherzolites of Grégoire et al. (2003).

In major/minor element composition, clinopyroxenes with humped REE_N patterns have lower Mg# (87.9–93.1, median of 90.6) and higher TiO_2 contents (0.04–0.23 wt%, median of 0.10 wt%) than clinopyroxenes with steep-$LREE_N$ (Mg# 91.5–94.8, median of 92.8; TiO_2 < 0.02–0.05 wt%, median of 0.03).

Olivine

Olivine is the last of the three peridotitic inclusion minerals that have received attention for trace element studies. Following up on an initial study (De Hoog et al. 2010) on trace elements in olivine from peridotite xenoliths and their application to thermobarometry, De Hoog et al. (2019) analyzed 10 trace elements in 36 olivine inclusions in diamonds from Akwatia, Ghana with an ion microprobe. They found that the contents of Ca and Na and their ratios with Al can be used in combination to constrain the paragenesis (harzburgitic versus lherzolitic) of peridotitic olivine inclusions, while Ti in olivine may be employed as a tracer of metasomatic re-enrichment of diamond substrates. Based on new reference materials and recent advances in the analysis of trace elements in olivine via LA-ICPMS (Bussweiler et al. 2019), we expect a large increase in trace element data for olivine inclusions over the coming years. Only such an expanded data set will allow us to conclusively assess whether the paragenetic fields defined by De Hoog et al. (2019) on the basis of Akwatia inclusions can be applied universally.

Discussion of trace elements in peridotitic inclusions

Normalization to primitive (J4) garnet and cpx. To derive the trace element characteristics of the rock a mineral derives from, normalization to CI-chondrite or primitive mantle is of limited use, as the derived patterns are strongly determined by the degree of compatibility of the studied trace elements in the available lattice sites. To overcome this problem, Stachel et al. (1998) instead suggested normalization to minerals that are well equilibrated in a bulk rock with a primitive mantle-like composition and chose high-temperature sheared garnet-lherzolite xenolith J4 from Jagersfontein (Jagoutz and Spettel, unpublished data; Wolff-Boenisch 1994). Since then, other samples with near-primitive compositions have been recognized (e.g., Vitim xenolith 313-105 of Ionov et al. 2005) but despite offering a more complete range of analyzed trace elements, their bulk rock REE patterns are neither a perfect match for primitive nor for depleted MORB-source mantle. For that reason, we carefully re-analyzed mineral separates from J4 using laser ablation sector field ICPMS (Appendix Table 2) and continue the use of J4 garnet and clinopyroxene for normalization purposes.

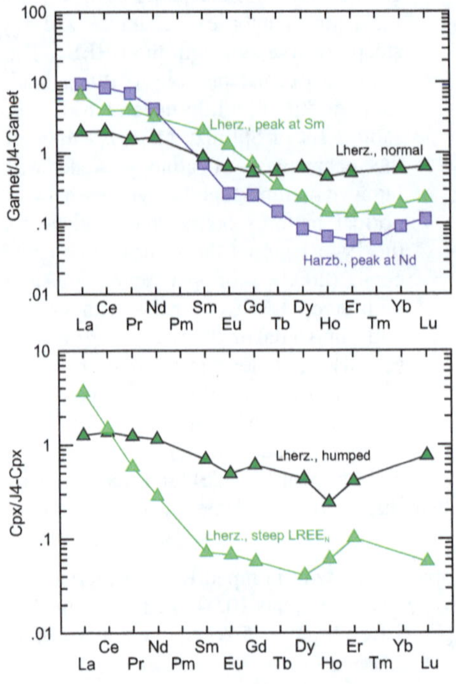

Figure 24. Median REE contents of the principal garnet (**top**) and clinopyroxene (**bottom**) groups, normalized to REE concentrations of the same minerals in a well equilibrated garnet lherzolite xenolith with a primitive mantle-like bulk composition (J4 from Jagersfontein; major and trace element mineral compositions in Appendix Table 2). Based on their chondrite normalized REE$_N$ patterns (Fig. 18 and Appendix Figure 1), the three principal garnet groups shown are: (1) harzburgitic garnets peaking at Nd, (2) lherzolitic garnets peaking at Sm and (3) lherzolitic garnets with normal REE$_N$. The two principal types of chondrite-normalized clinopyroxene REE patterns are (1) humped and (2) steep LREE$_N$.

For simplicity, for J4-normalization we only use the median patterns of the principal REE$_N$ pattern types (see Fig. 18 and Appendix Fig. 1): harzburgitic garnet inclusions peaking at Nd, lherzolitic garnet inclusions peaking at Sm, and lherzolitic garnet inclusions with normal REE$_N$ patterns. Normalization to primitive garnet reveals that harzburgitic garnets with strongly sinusoidal patterns reflect sources with REE contents steadily decreasing from mildly enriched LREE through the MREE to a trough in the HREE at Er, with increasingly less-depleted HREE from Er to Lu, overall resulting in a V-shaped REE$_{J4}$ pattern (Fig. 24). On the other end of the spectrum, normal REE$_N$ patterns of lherzolitic garnet inclusions, which are dominant among garnets from lherzolite xenoliths (Fig. 17), represent sources with near-primitive trace element compositions similar to J4 (Fig. 24). Lherzolitic garnet inclusions with REE$_N$ patterns peaking at Sm fall in between these two endmembers, with near primitive LREE concentrations, a less prominent drop towards Er$_{J4}$ than for harzburgitic garnets, and a shallower increase towards Lu$_{J4}$. Given the higher distribution coefficients for LREE in clinopyroxene relative to garnet (e.g., Johnson 1998), minor relative depletion of LREE in lherzolitic relative to harzburgitic garnets may partially relate to metamorphic equilibration with clinopyroxene for the lherzolitic paragenesis.

For lherzolitic substrates, the median values of the two principal pattern types in clinopyroxenes, humped and steep-$LREE_N$, may be used to assess the validity of the garnet-based findings. HREE concentrations in clinopyroxene inclusions in diamond generally are low (in part below detection), resulting in noisy median HREE patterns. In addition, Yb analyses in clinopyroxene via SIMS appear to frequently suffer from an analytical problem (possibly an unrecognized interference) and hence are not reported in a number of studies. This drawback relative to garnet inclusions is compensated by higher and therefore more accurately determined LREE concentrations (e.g., La is approximately 1–100 × chondritic in abundance, compared to 0.03–15 × chondritic for garnet inclusions). Normalized to J4-clinopyroxene, inclusions with humped REE_N patterns also reveal a near-primitive composition (Fig. 24). This agrees well with the observation that such clinopyroxenes occur together with garnets with normal REE_N patterns. Clinopyroxene with steep $LREE_N$ patterns shows enrichment in LREE and steep negative slopes towards the MREE, when normalized to J4 (Fig. 24). Overall, this fits with the observation made for lherzolitic garnets with sinusoidal patterns peaking at Sm but suggests a more continuous negative slope from La–Sm in the diamond substrates than indicated by these garnets. Due to the problems outlined above, the HREE patterns of clinopyroxenes are considered less informative than those of garnet and not further discussed here.

Considering the well-established origin of cratonic peridotites as residues of intense melt extraction events in the Archean (*see above*), the LREE-enriched character of peridotitic diamond substrates appears counterintuitive. Based on REE patterns of unmodified MORB residues (e.g., Sharma and Wasserburg 1996), the positive slope within the $HREE_{J4}$ from Tm to Lu, observed for garnets with chondrite-normalized patterns peaking at Nd and Sm (Fig. 24), fits the steep positive $LREE_N$–$HREE_N$ slopes in such residues. The common negative slopes from La_{J4}–Er_{J4} then are a consequence of metasomatic re-enrichment in incompatible trace elements, first recognized as a universal process in lithospheric peridotites by Frey and Green (1974). From harzburgitic garnets, through lherzolitic garnets with sinusoidal REE_N and associated clinopyroxenes with steep-$LREE_N$ patterns, to lherzolitic garnets with normal REE_N and associated clinopyroxenes with humped REE_N patterns, the signature of primary depletion becomes increasingly obliterated and eventually completely erased in the J4-normalized mineral trace elements patterns (and hence the bulk rocks). The decrease in Mg# and increase in TiO_2 for the transition from clinopyroxenes with steep-$LREE_N$ to humped patterns implies that this metasomatic progression is not limited to trace elements but falls in the categories of "enrichment in major–minor–trace elements without modal change" or "modal metasomatism" of Harte (1983, 1987).

Fluid versus melt metasomatism. Having established (1) that garnets with sinusoidal REE_N patterns, when normalized to a primitive garnet composition, document preferential re-enrichment in LREE and (2) that the progression from sinusoidal towards normal REE_N patterns (or from V-shaped to primitive J4-normalized patterns) increasingly also involves addition of moderately incompatible trace elements and major/minor elements, raises the question what metasomatic agent(s) drive this re-enrichment process. Trace element signatures, in particular ratios of strongly to mildly incompatible elements (LREE/HREE, Zr/Y, Sm/Hf), have long been used to make a first order division between fluid- and melt-driven metasomatism (Stosch and Lugmair 1986; Shimizu and Richardson 1987; Griffin and Ryan 1995; Stachel et al. 1998; Viljoen et al. 2014). Griffin and Ryan (1995) empirically developed a widely used diagram of Y versus Zr (Fig. 25), with vectors for low-temperature phlogopite (fluid) metasomatism (high Zr/Y) and high-temperature melt metasomatism (intermediate Zr/Y). These two trends emerge from a field for "depleted" garnets near the origin, where a distinction is no longer possible (approximately at $Y < 5$ ppm and $Zr < 20$ ppm).

For inclusion garnets exceeding this threshold, harzburgitic garnets mostly fall onto or close to the fluid metasomatic trend whilst lherzolitic garnets generally have lower Zr/Y, compatible with melt metasomatic re-enrichment of their substrates. A notable exception is the large population of lherzolitic garnet inclusions from the Victor Mine in Canada, which mainly displays Zr/Y ratios indicative of fluid metasomatism or intermediate between fluid and melt metasomatism (Fig. 25). Excluding Victor garnets, this overall distribution was linked (Stachel and Harris 1997; Stachel and Luth 2015) to harzburgitic diamond substrates mostly residing below their hydrous solidus (thus permitting percolation of fluids only), while lherzolitic diamonds form generally above the hydrous (± carbonated) solidus (allowing for the percolation of melt). Geothermobarometric data for lherzolitic inclusions in diamonds from Victor indicate their derivation from just below the hydrous lherzolite solidus, consistent with their comparatively high Zr/Y.

For the study of garnets included in diamond, however, Y and Zr generally are not sensitive enough as tracers of metasomatism and 70% of garnets fall below the threshold of Y < 5 and Zr < 20 ppm. This is in stark contrast to cratonic peridotite xenoliths, where only 29% of garnets fall below the Y and Zr threshold. This much higher proportion among inclusions of garnets below the threshold again highlights the overall much more depleted nature of peridotitic diamond substrates compared to the xenolith and xenocryst sample. Despite their low Y and Zr, garnets below the threshold nevertheless are almost invariably affected by metasomatic re-enrichment. Garnets reflecting the intense primary melt depletion characteristic for cratonic peridotites should have trace element patterns with $LREE_N/MREE_N \ll 1$ (see above), a signature that is virtually absent among inclusion garnets even below the Y and Zr threshold. Of the inclusion garnets below the threshold, only 7% have $Nd_N/Y_N < 1$, and one sample (0.5%) has $Nd_N/Y_N < 0.1$. Xenolith garnets below the threshold show a very similar picture, with only 11% having $Nd_N/Y_N < 1$ and 0.7% < 0.1. With over 2/3 of garnet inclusions being depleted below the threshold in Y–Zr space, it is evident that diamond formation usually is associated with mild metasomatic events, reflecting low fluid–rock or melt–rock ratios. Note that low

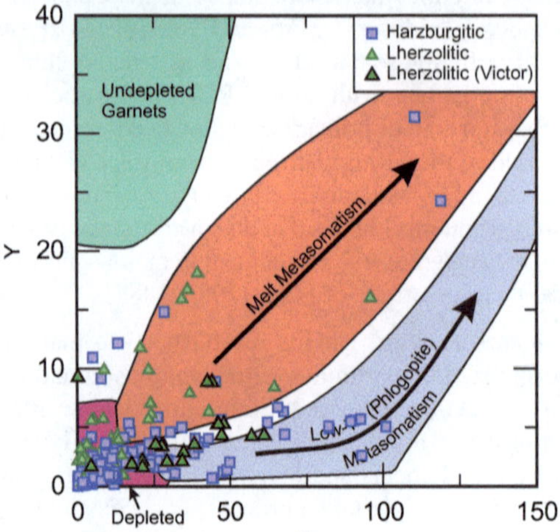

Figure 25. Y versus Zr (ppm) in peridotitic garnet inclusions in diamond. Compositional fields and trends are from Griffin and Ryan (1995). The major melt extraction event that preceded the formation of sub-cratonic lithospheric mantle in the Archean is reflected in Y and Zr contents of garnets falling into the depleted field (shown in **purple**). Secondary metasomatic re-enrichment may occur (1) through fluids at low (subsolidus) temperatures, associated with high Zr/Y (**blue field**) or (2) through melts (supersolidus conditions), associated with lower Zr/Y (**red field**). The bulk of inclusion garnets falls into the depleted field, but for garnets where Zr and Y contents are sufficiently high (Y ≥ 5 ppm and Zr ≥ 20 ppm) to assign a trend, fluid metasomatism dominates for harzburgitic garnet inclusions and melt metasomatism for lherzolitic inclusions. Assuming an average cratonic model geotherm, this observation is consistent with harzburgitic substrates in the lithospheric mantle generally residing at pressure–temperature conditions below their solidus and lherzolitic substrates residing mostly at supersolidus conditions (see main text for a more detailed discussion). An exception is formed by lherzolitic garnets in diamonds from the Victor Mine, for which geothermobarometric results indicate formation below the wet lherzolite solidus.

fluid/melt–rock ratios do not imply diamond formation under fluid-limited conditions: protracted diamond growth involving multiple metasomatic pulses evidently involves sufficiently high ratios of fluid/melt–diamond to preclude widespread Rayleigh fractionation of stable isotopes and nitrogen in diamond (Howell et al. 2020; Stachel et al. 2022).

On the level of mineral inclusions, the effects of multiple metasomatic events cannot be separated using only trace elements but require radiogenic isotope approaches. For example, based on unsupported (low Rb) highly radiogenic Sr in harzburgitic garnet inclusions, Richardson et al. (1984) proposed that a melt metasomatic event affected the harzburgitic diamond substrates beneath the Kimberley area of the Western Kaapvaal Craton within 300 m.y. before diamond formation (at about 3.3–3.2 Ga).

Fluids as metasomatic agents. The concept of fluid-driven metasomatism of lithospheric mantle was widely popularized by Bailey (1982) and two key arguments support the involvement of supercritical CHO fluids in mantle metasomatism, particularly of harzburgitic substrates: (1) As discussed above, extremely high ratios of strongly to moderately incompatible trace elements and a negligible metasomatic effect on both moderately incompatible trace elements (Ti, Y, HREE) and major elements (e.g., Mg#, Ca, Al and Cr) clearly are not compatible with melt involvement. (2) The average cratonic model geotherm (40 mW/m^2; Hasterok and Chapman 2011) intersects the carbonated, hydrous solidus of harzburgite (Wyllie 1987) at ~1300 °C. This implies that with the exception of the deepest regions ($\gtrsim$180 km) located near the lithosphere–asthenosphere boundary, harzburgite in the lithospheric mantle is impermeable to melt percolation, as migration along grain boundaries causes both thermal and compositional equilibration, which inevitably leads to freezing of melts in subsolidus substrates.

The range in composition of possible metasomatic fluids can be considerably narrowed by examining a few basic constraints:

1. *The fluid must be stable in equilibrium with peridotite.* This constraint is not met for CO_2-rich fluids, which may occur only on the oxidizing side of the EMOD/G (enstatite + magnesite = olivine + diamond/graphite; Eggler and Baker 1982) buffer. Along cratonic geotherms throughout the entire peridotitic lithospheric mantle, CO_2 is buffered through carbonation reactions involving olivine ± pyroxene (Wyllie and Huang 1976; Brey et al. 1983). Depending on the redox conditions, H_2O-rich and CH_4-rich fluids can both be stable in equilibrium with peridotite. Excluding a subgroup of unusually oxidized samples from the Slave craton, the xenolith record for the depth range 140–180 km has a main range in $\Delta \log fO_2$ (FMQ), i.e., the oxygen fugacity in log unit difference relative to the FMQ buffer, of ~ −3 to −1.5 (see, e.g., Luth and Stachel 2014). For the average conditions of diamond formation (~1140 °C/5 GPa, based on garnet–olivine thermometry; Stachel and Luth 2015), this fO_2 range implies that CHO fluids are at the water maximum, with water contents ranging between 85 and 98 mol% (calculated using GFluids of Zhang and Duan 2010; see Fig. 1 of Stachel et al. 2017). Matjuschkin et al. (2019) showed that on the reducing side of the water maximum (lower end of the main range considered here), GFluids may significantly underestimate CH_4 contents but this does not affect the principal conclusion that the xenolith record over the 140–180 km depth range is centered at fO_2 conditions corresponding approximately to the water maximum.
2. *The fluid must be able to percolate through peridotite at low fluid:rock ratios.* High dihedral angles for both CO_2–olivine and CH_4–olivine (Watson et al. 1990; Wyllie 1992; Ferrando et al. 2017) preclude efficient percolation of such fluids through the lithospheric mantle. The dihedral angle for aqueous fluid–olivine decreases with increasing pressure (Watson et al. 1990), permitting efficient percolation at pressures ≥ 3 GPa (Mibe et al. 1998). Added salinity decreases dihedral angles of aqueous fluids with olivine even further (Huang et al. 2019).

3. *The fluid must dissolve and transport the chemical components associated with mantle metasomatism.* Low solubility of trace elements such as REE is well documented for CO_2-rich fluids under mantle conditions (e.g., Meen et al. 1989). An experimental study of trace element solubility in CH_4-rich fluids is still outstanding but given the non-polar character of the symmetrical CH_4 tetrahedron, CH_4 fluids are expected to be poor solvents. In contrast, the solubility of solids in aqueous fluids increases strongly with pressure and is on the wt% level at mantle conditions (e.g., Ayers et al. 1997). At diamond stable conditions (≥ 4 GPa), trace element partitioning between peridotite minerals and aqueous fluids is similar to partitioning with silicate melts, with strong LREE/HREE fractionation (Fig. 26), and about 36 wt% (15 mol%) dissolved solids in the fluid at subsolidus conditions (1000 °C, 5–6 GPa; Kessel et al. 2015).

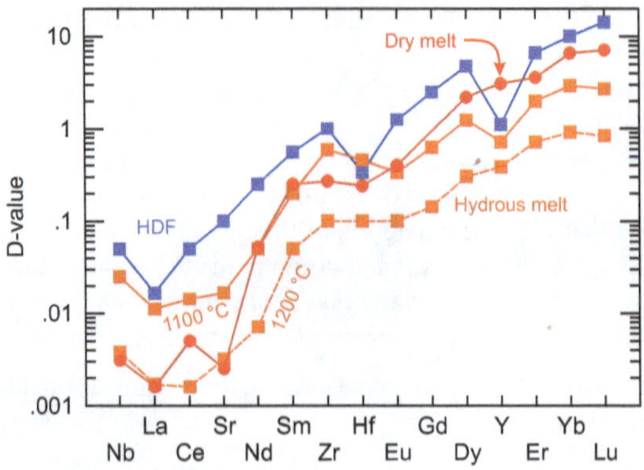

Figure 26. Trace element distribution coefficients (D-values) between garnet and possible metasomatic agents ranging from hydrous high-density fluids (HDF) through hydrous melts to "dry" melts. The 6 GPa experiments for HDF (1000 °C; 64 wt% H_2O) and hydrous melts (1100 and 1200 °C; 39 and 24 wt% H_2O, respectively) are from Kessel et al. (2015), the experiments for garnet in equilibrium with dry Hawaiian olivine tholeiite are from Johnson (1998). The experiments of Kessel et al. (2015) suggest that trace element partitioning into garnet decreases by about an order of magnitude from HDF to hydrous melt significantly above the solidus (1200 °C). Overall slopes and predicted LREE/HREE fractionations are, however, quite similar for all four sets of partition coefficients. Element order reflects increasing compatibility in the garnet structure (after Green 1994).

In conclusion, (1) the redox state recorded in mantle xenoliths from the main portion of the lithospheric diamond stability field falls at or near the water maximum, (2) H_2O is the only CHO fluid species with a sufficiently low dihedral angle with olivine to readily percolate along grain boundaries and (3) aqueous fluids also are the only CHO species with the ability to dissolve incompatible trace elements in the required quantities. The high load of dissolved solids carried by aqueous fluids at the pressure–temperature conditions inside the diamond stability field justifies using the term "high-density fluid" (HDF; Klein-BenDavid et al. 2007; Weiss et al. 2022) instead. Research on primary fluid inclusions in fibrous, clouded and coated diamonds indicates that for peridotitic substrates, hydrous HDFs generally range from potassium-rich brines to Mg-carbonatitic, while in eclogitic substrates an evolution from saline through low-Mg carbonatitic to silicic HDFs is observed (Weiss et al. 2015, 2022). The carbonatitic and silicic inclusions are transitional between hydrous HDFs and hydrous melts. For example, inclusions representing an original carbonate melt with molar $100H_2O$ / $(H_2O + CO_2)$ of only 5–20, were recognized in cuboid diamonds of unknown paragenesis from Udachnaya (Zedgenizov et al. 2004). Fluid inclusion-rich diamonds are often interpreted as having grown at high levels of carbon supersaturation in the HDF, to be genetically linked to kimberlite or proto-kimberlite magmatism, and to be typically of similar age as the transporting host kimberlite (e.g., Boyd et al. 1987; Gurney et al. 2010; Weiss et al. 2011). On that basis, the HDFs observed in clouded and fibrous diamonds may represent a growth medium that is distinct from that for generally much older (Gurney et al. 2010; Smit et al. 2022) smooth-faced monocrystalline diamonds. The discovery of rare but compositionally highly similar, saline to carbonatitic hydrous HDFs as inclusions in gem-quality monocrystalline diamonds from Kankan, Finsch, Voorspoed and Venetia (Weiss et al. 2014; Jablon and Navon 2016), however, clearly established a much wider role of HDFs in diamond formation, both with respect to diamond types and time. We emphasize, however, that the presence of HDFs as inclusions in some smooth-faced monocrystalline diamonds does not automatically imply that all diamonds formed from HDFs or that the types of HDFs observed in diamond are the principal agents of mantle metasomatism. These conjectures will have to be validated or disproven through future research and will be discussed further below.

Melts as metasomatic agents. Hydrous silicate, silico-carbonate and carbonate melt compositions have all been suggested as agents of mantle metasomatism, based on studies of very distinct aspects of subcratonic lithospheric mantle:

1. A group of peridotite xenoliths with fertile compositions—often displaying "sheared" textures, equilibration temperatures that exceed the local conductive geotherm, and evidence for rim-ward enrichment in Fe, Ti, Zr, Y and Ga in constituent minerals—document re-fertilization in the course of (likely repeated) infiltration of melts. Implicated melt compositions range from proto-kimberlitic (megacryst magma) through primitive OIB-like asthenospheric melts and alkali picrites to komatiitic (Gurney and Harte 1980; Nixon et al. 1981; Hervig et al. 1986; Smith 1988; Griffin et al. 1989b; Harte et al. 1993; Mahotkin et al. 2000).

2. A subset of peridotite xenoliths that are enriched in orthopyroxene from, e.g., the Kaapvaal and Siberian cratons (Boyd 1989; Boyd et al. 1997) is interpreted as evidence for Si-enrichment through melts (Rudnick et al. 1994; Kelemen et al. 1998). Based on the narrow, mantle-like $\delta^{18}O$ record of olivine from cratonic peridotites, including those with an opx-enriched signature, Si-enrichment through dacitic (trondhjemitic–tonalitic–granodioritic) slab melts (Rapp and Watson 1995) can be excluded, as it would inevitably lead to ^{18}O enriched signatures (Regier et al. 2018). Instead, Si-enrichment in cratonic peridotites may have occurred by melts generated through hydrous melting of peridotite (Mitchell and Grove 2015); reaction with basic melts, which initially dissolve clinopyroxene and, migrating further into the wall rock, precipitate orthopyroxene (Kelemen et al. 1992); or reaction with komatiite at high pressure (> 4 GPa; Tomlinson and Kamber 2021).

3. A comparative study of diamond suites from areas with normal (Finsch Mine) and anomalously low (Premier kimberlite at Cullinan Mine) seismic P-wave velocities in the diamond-stable lithospheric mantle beneath the Kaapvaal Craton allowed the identification of contrasting styles of metasomatic re-enrichment (Viljoen et al. 2014; see also Shirey et al. 2002). While harzburgitic diamond substrates at Finsch document low-temperature fluid style metasomatism, the peridotitic diamond sample at Premier shows a high proportion of the lherzolitic paragenesis among peridotitic diamonds, with evidence of high temperature melt metasomatism introducing Ca, Fe, Ti, Zr, Y and REE (Viljoen et al. 2014). The refertilization in major and trace elements of the peridotitic diamond substrates beneath Premier and the P-wave anomaly linked to the associated changes in mineralogy (conversion of harzburgite to lherzolite) is attributed to the formation of the 2 Ga Bushveld large igneous province (Shirey et al. 2002; Viljoen et al. 2014) and thus an example of extensive silicate melt metasomatism associated with a mantle plume. Another example of

intense modification of diamond substrates associated with plume magmatism is the 2.7 Ga Ventersdorp large igneous province and the inclusions in diamonds from the proximal Voorspoed Mine, which document the partial to complete conversion of depleted peridotite to websterite (Viljoen et al. 2018).

4. Over a wide range of pressures in the lithospheric and sublithospheric mantle, incipient melting of carbonated peridotites yields magnesio-carbonatites (e.g., Dalton and Presnall 1998). Their low density, high mobility, and high contents in incompatible trace elements make carbonate melts highly efficient agents of mantle metasomatism. An important feature of carbonate metasomatism is a minimal impact on the major element composition of affected substrates, typically with the exception of Ca only (Rudnick et al. 1993). Characteristic geochemical signatures of carbonatite metasomatism are high LREE/HREE, Nb/La, Ca/Al and Zr/Hf ratios and low Ti/Eu ratios (Rudnick et al. 1993; Yaxley 1993). Metasomatism through sodic magnesiocarbonatite, derived from melting of carbonated pargasite peridotite, was identified, e.g., in spinel peridotites from Mt Leura, Victoria, and may lead to conversion of spinel lherzolite to wehrlite at pressure < 2.1 GPa (Green and Wallace 1988). Sodic magnesio-carbonate metasomatism of peridotites is, however, limited to a narrow pressure range (2.1–3.1 GPa) bracketed by the breakdown of carbonate at lower pressures and the breakdown of pargasite at higher pressures.

In diamond stable lithospheric mantle, fO_2 conditions generally are on the reducing side of the EMOD buffer (see Stagno et al. 2013; Luth and Stachel 2014), limiting the *in situ* generation of carbonated melts to localized domains with elevated redox conditions, as observed for some samples from the Central Slave Craton (Creighton et al. 2010). In the absence of such carbonated domains, carbonate metasomatism of the deep lithospheric mantle must be caused by melts derived from below, as inferred, e.g., for the metasomatic modification of highly depleted peridotitic lithosphere beneath the Gregory Rift Valley in northern Tanzania (Rudnick et al. 1993). Asthenosphere-derived carbonatites may originate from upwelling metasomatized mantle that experiences redox melting via oxidation of diamond and reduction of Fe^{3+} when reaching depths < 250 km (Rohrbach and Schmidt 2011). In present day Earth, subducting oceanic crust only exceeds the anhydrous solidus of carbonated eclogite at depth > 400 km (i.e., ~ transition zone depth; Dasgupta et al. 2004; Thomson et al. 2016) but would have melted at increasingly shallower depth with the mantle potential temperature increasing towards the Archean. At some intermediate time, this may have allowed for carbonatite metasomatism of the lithospheric mantle directly from melting of carbonated eclogites and possibly sediments. Depending on the exact estimate for the temporal evolution of mantle potential temperature, Archean and Paleoproterozoic subduction zones may have been sufficiently hot to cause shallow carbonate melting and direct recycling back to the surface (Dasgupta and Hirschmann 2010), precluding subduction-driven mantle metasomatism through carbonated melts. Observational evidence to the contrary was, however, produced by Shu and Brey (2015), who employed subcalcic garnets in peridotite xenoliths and included in diamonds to study Archean metasomatism in the Kaapvaal lithospheric mantle. Based principally on Ti/Eu and Zr/Hf characteristics, they identified carbonate and carbonate–silicate (kimberlite) melts as the principal agents of metasomatism and associated diamond growth in harzburgitic substrates.

Ti/Eu and Zr/Hf systematics of metasomatic agents. The ratios of Ti/Eu and Zr/Hf of whole rocks, garnet and clinopyroxene are used to trace metasomatic interactions and to constrain the nature of metasomatic agents (Rudnick et al. 1993; Yaxley 1993; Shu and Brey 2015; Uenver-Thiele et al. 2017). Directly comparing the Ti/Eu and Zr/Hf systematics of garnets with those of potential metasomatic agents requires that garnet does not strongly fractionate these ratios during metasomatic equilibration. Experimentally determined partitioning data between garnet and dry basaltic melt (Johnson 1998), strongly hydrous basanite (18–27 wt% H_2O; Green et al. 2000) and hydrous fluid (Stalder et al. 1998) show that garnet faithfully reflects the Zr/Hf ratio of the melt/fluid. The same applies to strongly subcalcic garnet (Ca# = 100Ca / (Ca + Mg + Fe) ≤ 5, approximately equivalent to CaO ≤ 2 wt%) in equilibrium with silico-carbonate ("kimberlite") melts, while more calcic lherzolitic garnets (average Ca# ~15) have D_{Zr}/D_{Hf} of ~ 0.6, i.e., Zr/Hf ratios 40% lower than the coexisting melt (Girnis et al. 2013). Garnet in equilibrium with carbonate melt at high pressure (6.6–8.6 GPa; Dasgupta et al. 2009) exhibits an offset to 25% lower ratios (D_{Zr}/D_{Hf} of ~ 0.75). With the exception of lherzolitic garnets in equilibrium with kimberlite melt, garnet Zr/Hf ratios thus are a good match for the fluid/melt they last equilibrated with. More limited partitioning data for Ti and Eu indicate that garnet in equilibrium with dry and hydrous (4.8 wt% H_2O) basalt will have approximately 30–40% lower Ti/Eu (Johnson 1998; Gaetani et al. 2003), while garnet in equilibrium with carbonatite has a 7 times higher ratio than the melt (Dasgupta et al. 2009). However, the extent of the offset in Ti/Eu of garnet relative to basaltic melts is not relevant on the very large scale of observed variations in Ti/Eu (Fig. 27). Also, considering the extremely low Ti/Eu of magnesio-carbonatites (average of 52; Chakhmouradian 2006; compared to, e.g., 7451 for N-MORB; Sun and McDonough 1989), the actual effect of preferential partitioning of Ti over Eu into garnet in equilibrium with carbonatites is very limited.

Strongly subchondritic Ti/Eu and superchondritic Zr/Hf (Rudnick et al. 1993; Yaxley 1993) of most of the studied Kaapvaal subcalcic garnets was used by Shu and Brey (2015) to establish mantle metasomatism through carbonate and silico-carbonate melts. Our database of xenolith-derived peridotitic garnets overall agrees with this finding, but suggests that HDFs should also be considered as a potential metasomatic agent for garnets with near-chondritic Zr/Hf (Fig. 27). For garnet inclusions in diamond with super-chondritic Zr/Hf (> 37), a number of samples have very low Ti/Eu (< 500) and closely match average compositions of magnesio-carbonatite while samples with more elevated Ti/Eu fall close to the primitive Group I and II kimberlite compositions (Fig. 27). Nearly half of the inclusions (55% of harzburgitic and 47% of lherzolitic garnets), however, have subchondritic Zr/Hf. Overall, HDFs in fibrous diamonds and the melts in equilibrium with MARID (mica–amphibole–rutile–ilmenite–diopside) and PIC (phlogopite–ilmenite–diopside) xenoliths from the Kaapvaal Craton (see caption of Fig. 27 for references), provide a better fit for the metasomatic agent required for the majority of inclusions, with some harzburgitic inclusions indicating even lower Zr/Hf in the metasomatic agent. Considering the principal types of REE_N patterns observed for peridotitic garnet inclusions, the median composition for harzburgitic garnets peaking at Nd (Zr/Hf = 30; Ti/Eu = 1297) closely matches the field for MARID and PIC melts, lherzolitic garnets peaking at Sm_N (median Zr/Hf = 53; Ti/Eu = 292) nearly overlap average magnesio-carbonatites, and lherzolitic garnets with normal REE_N (median Zr/Hf = 40; Ti/Eu = 5734) evolve towards primitive mantle (Zr/Hf = 37; Ti/Eu = 7825). These observations indicate that the carbonatitic metasomatism invoked for xenolith garnets (Shu and Brey 2015) likely affected some inclusion garnets, but is not the dominant metasomatic process reflected by inclusions. This interpretation is consistent with the observation that carbonate inclusions in smooth-faced monocrystalline diamonds are very rare; for the peridotitic suite, only one well established example is documented, consisting of a magnesite inclusion in a lherzolitic diamond from Namibia. As an alternative, equilibration with MARID-PIC melts (for inclusions with low Zr/Hf), HDFs or kimberlites (carbonate–silicate melts) provides better matches for a large number of inclusion garnets.

Considering the evolution of the entire garnet data set shown in Figure 27, the starting point are the harzburgite residues left behind by major Archean melt extraction events (see above). After exhaustion of clinopyroxene (D_{Zr}/D_{Hf} of clinopyroxene–melt is ~0.5–0.6; Johnson 1998; Norman et al. 2005), orthopyroxene controlled the continued decrease in the Zr/Hf ratio of the melting residues (D_{Zr}/D_{Hf} of orthopyroxene–melt is ~0.2–0.5; Ulmer 1989; Green et al. 2000). After elimination of cpx from the residues (D_{Ti}/D_{Eu} of clinopyroxene–melt is ~ 1; Johnson 1998; Norman et al. 2005), an increase in Ti/Eu ratio of the residues is

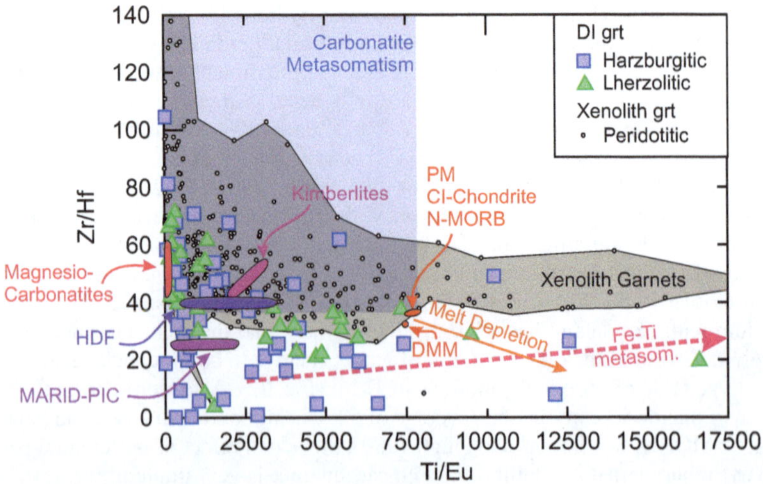

Figure 27. Zr/Hf versus Ti/Eu for harzburgitic and lherzolitic garnet inclusions in diamond. The **grey field** (with small dots indicating individual samples) outlines our database of garnets from cratonic garnet peridotite xenoliths. Two average compositions for magnesio-carbonatites worldwide (Chakhmouradian 2006; Chakhmouradian et al. 2009) are shown as a **red field**. Following Rudnick et al. (1993) and Yaxley (1993), carbonatite metasomatism is characterized by superchondritic Zr/Hf and subchondritic Ti/Eu, shown as a **pale blue field**. The **pink field** connects the primitive Group I (high Zr/Hf and Ti/Eu end) and Group II kimberlite compositions of Becker and Le Roex (2006). The **dark purple field** for high density fluids (HDFs) connects the two endmember trace element patterns ("planed" at the high Ti/Eu end and "ribbed" at the low Ti/Eu end) defined by Weiss et al. (2013). These two endmembers are not specific to particular major element compositions of HDFs, but the particular examples chosen by Weiss et al. (2013) are both hydrous silicic HDFs. The **light purple field** is for average melts in equilibrium with clinopyroxene from MARID (low Ti/Eu end) and PIC (high Ti/Eu end) xenoliths, calculated from clinopyroxene analyses of Fitzpayne et al. (2019) using $D_{\text{cpx-melt}}$ of Johnson (1998). The **small red oval** encompasses the compositions of primitive mantle, normal mid-ocean ridge basalt (both Sun and McDonough 1989) and CI-chondrite (McDonough and Sun 1995), the **small red dot** right below is the depleted MORB mantle (DMM) composition of Workman and Hart (2005). The **orange arrow** is an approximate trend for intense melt depletion leaving a spinel-harzburgite residue behind (after Shu and Brey 2015; see text for details). The **dashed pink arrow** points towards the composition of megacryst garnets (average composition for the Grib kimberlite, Arkhangelsk, with Zr/Hf of 39 and Ti/Eu of 28,060; Kargin et al. 2016), which likely precipitated from the metasomatic agent associated with Fe–Ti metasomatism (see text for details).

controlled by orthopyroxene (D_{Ti}/D_{Eu} of orthopyroxene–melt is ~3; Ulmer 1989; Green et al. 2000) and spinel (compatibility of Ti, at strict incompatibility of Eu, in Cr- and Fe^{3+}-rich spinels; Wijbrans et al. 2015). The approximate evolution of residues of large-scale Archean melt depletion is indicated as a solid arrow in Figure 27. The resulting high Ti/Eu and low Zr/Hf of the depleted protoliths of subcratonic lithospheric mantle were associated with very low concentrations of Ti, Eu, Zr and Hf, making these rocks highly susceptible to metasomatic overprint. This metasomatism involved a range of fluids/melts (see above) with variable Zr/Hf. High Eu (reflecting high LREE/HREE) and low Ti in the metasomatic agent(s) moved the affected diamond substrates to very low Ti/Eu.

Focusing only on inclusion garnets with Ti/Eu ≳2500 and subchondritic Zr/Hf: if the observed spread from high Ti/Eu, akin to melt-depleted compositions, to ~2500 was due to metasomatic enrichment of Eu (trend of decreasing Ti/Eu), then a negative correlation of Ti/Eu with Eu content would be expected, but this is not the case (Fig. 28). Instead, there is a crude positive correlation of Ti/Eu with Ti content (Fig. 28), suggesting that the spread in Ti/Eu is largely a consequence of preferential re-enrichment in Ti that occurred as a second metasomatic event (dashed trend of increasing Ti/Eu in Fig. 27). Compared to the original

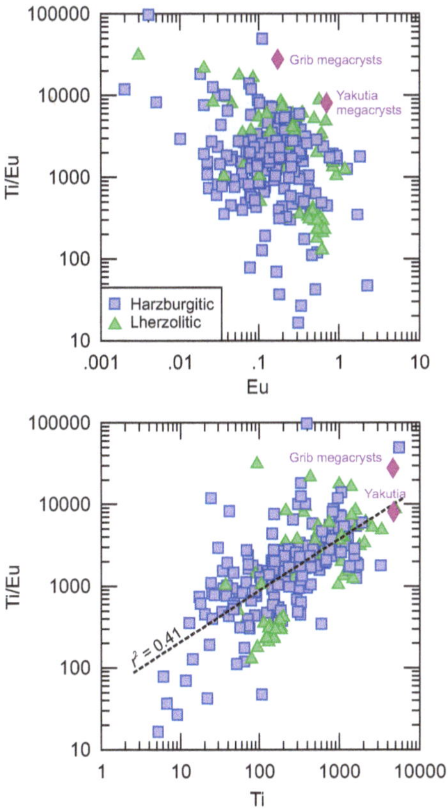

Figure 28. Ti/Eu versus Eu (**top**) and Ti content (**bottom**) in ppm in harzburgitic and lherzolitic garnet inclusions in diamond. The absence of a statistically significant negative correlation between Ti/Eu and Eu suggests that the spread in Ti/Eu is not principally due to variable metasomatic (LREE and) MREE enrichment. The presence of a crude positive correlation between Ti/Eu and Ti suggests that metasomatic enrichment in Ti (with little effect on Eu contents) is the principal factor in determining Ti/Eu ratios. Average garnet megacryst compositions for the Grib kimberlite, Arkhangelsk ($n = 20$; Kargin et al. 2016) and Yakutian kimberlites ($n = 3$; Kostrovitsky et al. 2013) are indicated as pink diamonds.

melt-depleted harzburgites, which also have high Ti/Eu (melt depletion trend in Fig. 27), the inclusion garnets with Ti/Eu > 2500 have higher Eu (metasomatic event 1, affecting all garnets inclusions) and higher Ti contents (metasomatic event 2, affecting only some garnets). Very high Ti/Eu ratios are documented as a consequence of melt metasomatism in garnet rim compositions from hot sheared peridotites, with Ti/Eu ratios of, e.g., 9800 from Lesotho (Smith et al. 1993) and 23,000 from Arkahangelsk (Kargin et al. 2017). They also characterize garnets precipitated from the megacrysts magma, with Ti/Eu of 4600–44,000 (Yakutia: Kostrovitsky et al. 2013; Arkahangelsk: Kargin et al. 2016, 2017). Thus, a possibly modified, generally mild form (cf. Figs. 7 and 13) of the Fe–Ti metasomatism documented in xenoliths (see above) and tentatively linked to the proto-kimberlite/megacryst magma (e.g., Kargin et al. 2016; Ashchepkov et al. 2017a; Shchukina et al. 2017) may have affected some diamond substrates. The inferred second stage re-enrichment in Ti appears to have had little effect on Zr/Hf ratios, implying moderate to low Zr/Hf and/or low Zr and Hf concentrations in the metasomatic melt (consistent with an absence of positive correlations of either Zr or Hf with Ti/Eu). With D_{Zr}/D_{Hf} of garnet–melt being in the range 0.6–1 (*see above*), the near-chondritic Zr/Hf ratio of megacrysts (34–44; Kargin et al. 2017) is approximately representative of their parental melt. Combined with its very high Ti/Eu ratio, the megacryst magma would consequently be consistent with the evolution seen in Figure 27 (dashed pink arrow labelled Fe-Ti metasomatism, pointing towards megacryst garnets).

Calculated melts in equilibrium with garnet inclusions. A different approach to constrain the nature of metasomatic agents is to calculate the composition of melts or fluids in equilibrium with garnet or clinopyroxene inclusions and to compare these calculated compositions to observed mantle melts or fluids (e.g., Shimizu 1975). For this approach to be valid in the context of diamond studies, the inclusion mineral and the metasomatic agent must have reached equilibrium. For syngenetic inclusions, which precipitate together with their diamond hosts from a metasomatic agent, this precondition is implicitly met. For preexisting minerals, equilibration is a diffusion-controlled process and for centimeter-sized crystals in coarse peridotites may take over a billion years at the temperatures of diamond stable lithospheric mantle. For protogenetic inclusions in diamond (see above), which are typically recovered in the 0.1–0.2 mm range, diffusive equilibration during a fluid or melt infiltration event will, however, be rapid, e.g., ~20,000 years for REE in a 0.1 mm defect-free garnet residing at 1150 °C (Nestola et al. 2019b). Faster diffusion along dislocations could decrease the required time significantly. On that basis, we consider equilibrium between inclusion minerals and diamond-forming metasomatic fluids/melts highly likely even in the case of protogenetic inclusions.

A significant recent breakthrough for modelling the composition of metasomatic agents came from the experimental work of Kessel et al. (2015), who provided a large set (including the REE) of partition coefficients for both garnet and clinopyroxene with subsolidus HDFs (aqueous fluids with up to 36 wt% dissolved solids). For the calculation of model melts, we use the three principal garnet groups derived in Figure 18: harzburgitic garnets with REE_N peaking at Nd, lherzolitic garnets with sinusoidal patterns peaking at Sm, and lherzolitic garnets with normal REE_N patterns. For each of these garnet groups, we chose the 25th percentile, the median and the 75th percentile patterns to calculate the equilibrium compositions of HDF (6 GPa, 1000 °C; Kessel et al. 2015), hydrous melt (6 GPa, 1100 °C, melt with 39 wt% H_2O; Kessel et al. 2015) and "dry" melt (Johnson 1998). The suffix "dry" relates to the nominally dry Kilauea olivine tholeiite used in the experiments by Johnson (1998), but does not imply that the calculated melt compositions are only applicable to volatile-free systems. Because of the similarity in Zr/Hf and Ti/Eu of the median composition of lherzolitic garnets peaking at Sm and average magnesio-carbonatite (Fig. 27), for this garnet group we also use the partition coefficients for garnet-carbonatite of Dasgupta et al. (2009). The calculated equilibrium melts (Fig. 29) are then compared to (1) the spectrum of HDFs observed in diamonds, consisting of the typical "planed" and "ribbed" patterns of Weiss et al. (2013) and a potentially parental, saline HDF observed in a diamond from the Fox kimberlite (Weiss et al. 2015), (2) the primitive Group I and II kimberlite compositions of Becker and Le Roex (2006) and (3) in the case of the lherzolitic garnets peaking at Sm, the average magnesio-carbonatite composition of Chakhmouradian et al. (2009). The REE_N patterns for average MARID melt overlap the primary kimberlite compositions of Becker and Le Roex (2006), while the pattern for PIC melt shows the same slope but at about 1/3 of the REE concentrations (see caption of Fig. 27 for details on the calculation of MARID and PIC melt compositions). On the scale of Figure 29, the resulting offset, however, is so small that the kimberlite patterns may be viewed as representative for MARID and PIC melts as well.

For harzburgitic garnets peaking at Nd, all calculated HDF and melt REE_N patterns have very steep negative slopes (Fig. 29, top). The upward kink in the $HREE_N$ is interpreted to be imposed by the substrates rather than as a characteristic of the primary metasomatic medium (see above). As harzburgitic substrates generally reside at subsolidus conditions, only the calculated HDF pattern is relevant for comparison in this case (Fig. 29, top, right panel). The absolute trace element concentrations in HDFs observed in fibrous diamonds are mainly a reflection of inclusion density and hence were normalized to match the Dy content of the calculated HDF. The comparison reveals a modest fit, which is closest for the primitive saline pattern ("saline Fox"; Fig. 29, top). The $LREE_N$–$HREE_N$ fractionation in the calculated HDF covers almost three orders of magnitude, "ribbed" patterns span two orders of magnitude and "planed" patterns one order. Extrapolating the HREE slope (below the limit of quantification; Weiss et al. 2015) for the primitive saline pattern results in more similar overall $LREE_N$–$HREE_N$ fractionation, but differences to the calculated HDF are still prominent in the much steeper slope from Nd_N to Eu_N for the latter. This misfit in REE patterns between calculated HDF and primitive saline fluid is exacerbated when considering the experimental study of Rustioni et al. (2021), which showed that with increasing salinity the LREE/HREE ratio of garnet in equilibrium with fluid of a given REE content decreases strongly. For saline fluids, HDF compositions (Fig. 29) calculated using the partition coefficients of Kessel et al. (2015), therefore, underestimate the $LREE_N$–$HREE_N$ fractionation.

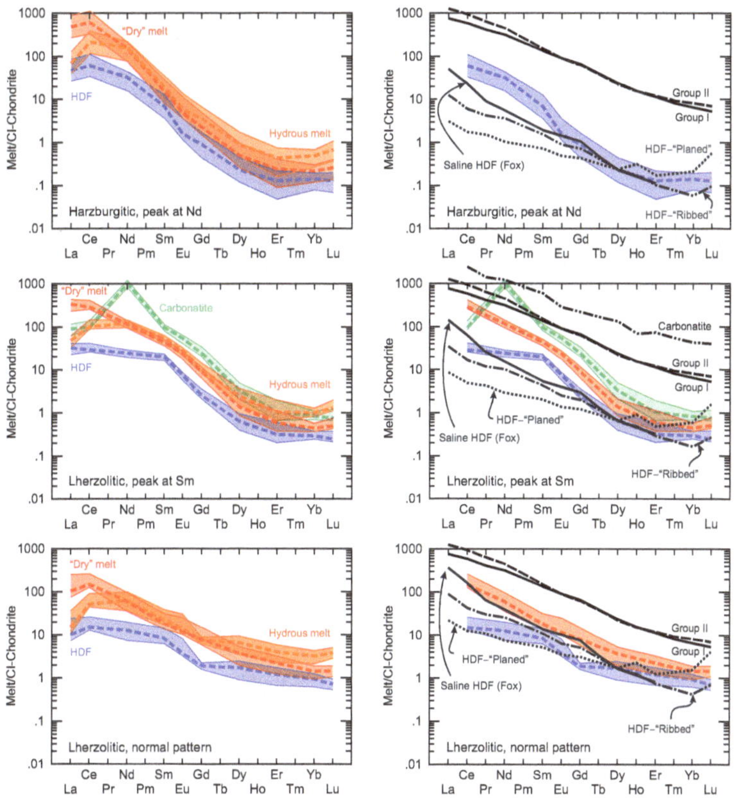

Figure 29. The **left panels** shows REE_N patterns of calculated HDFs and melts in equilibrium with peridotitic garnet inclusions in diamond that fall into the three principal groups of observed patterns: harzburgitic garnets peaking at Nd (**top**), lherzolitic garnets peaking at Sm (**middle**) and lherzolitic garnets with normal REE_N patterns (**bottom**). For all three garnet groups, the median compositions (**dashed lines**) and the interquartile range (25th and 75th percentile; **colored bands**) are used for the calculations. Hypothetical HDF and melt compositions were calculated from garnet compositions using three different sets of experimentally determined partition coefficients: for (1.) aqueous high-density-fluid (1000 °C, 6 GPa; **blue**) and (2.) hydrous melt (1100 °C, 6 GPa; **orange**) in equilibrium with metasomatized garnet lherzolite (Kessel et al. 2015) and (3) Kilauea olivine tholeiite in equilibrium with near-liquidus garnet and clinopyroxene ("dry" melt; 1430 °C and 3 GPa; Johnson 1998; **red**). In addition, for the group of lherzolitic garnets showing sinusoidal patterns peaking at Sm (**middle**), partition coefficients for carbonate melt in equilibrium with garnet lherzolite were used (1265–1470 °C and 6.6–8.6 GPa; Dasgupta et al. 2009; **green**). Kinks at Ce for the calculated melt/fluid compositions are interpreted as artifacts, related to generally low concentrations of La in inclusion garnets (affecting analytical precision) and difficulties in determining extremely small (10^{-2} to 10^{-3} range) partition coefficients from experimental charges with high accuracy. In the right panels, the calculated HDF and melt compositions are compared to primitive Group I and Group II kimberlite compositions (Becker and Le Roex 2006; **solid and dashed black lines**, respectively) and HDFs in fibrous diamonds with "planed" and "ribbed" traced element patterns (Weiss et al. 2013; **dotted and dotted-dashed grey lines**, respectively) and their potentially parental, saline fluid composition (Fox diamond E217; Weiss et al. 2015; **solid grey line**). Absolute trace element concentrations of fibrous diamonds are not only a function of the HDF inclusion composition but are strongly controlled by inclusion density. Consequently, for the comparison with calculated garnet fluids/melts, the "ribbed", "planed" and saline-Fox patterns were normalized to the Dy content of the median calculated HDF pattern. The average composition of magnesio-carbonatites worldwide is from Chakhmouradian et al. (2009). Since harzburgitic diamonds generally form at subsolidus temperatures, only the calculated HDF composition is used for the comparison in the top row. For lherzolitic garnets, both the calculated HDF and "dry" melt compositions are shown. The hydrous melt composition (see left panel) was only left out to avoid overcrowding. Because of likely poor precision, La in calculated HDF/melt is omitted for the comparisons with observed fluid/melt compositions.

For lherzolitic garnets peaking at Sm, the calculated HDF pattern shows a kink at Sm but otherwise is again most similar to the primitive saline HDF pattern (Fig. 29, middle), with the above caveat that the $LREE_N$–$HREE_N$ fractionation of the calculated HDF would increase if the effect of salinity was considered. The patterns for calculated silicate and carbonate melt compositions are even steeper than for the calculated HDF and hence do not match observed kimberlite and magnesio–carbonatite compositions. In comparison, the calculated HDF and melt compositions for lherzolitic garnets with normal REE_N patterns show much lower levels of $LREE_N$–$HREE_N$ fractionation. The calculated HDF and hydrous melt compositions (the latter shown only in the left panel of Fig. 29, bottom) are intermediate in slope between the observed "planed" and "ribbed" patterns, while the "dry" melt composition parallels primitive Group I and II kimberlites. The formation of lherzolitic diamonds generally above the wet lherzolite solidus makes the match with kimberlites particularly relevant.

An inherent problem of visual comparisons of calculated and observed fluid/melt trace element patterns is that REE concentrations in the various media cover almost five orders of magnitude. Consequently, as long as the sign of the slope is the same, very distinct patterns appear visually similar. To examine the differences in slope in the REE_N patterns of calculated and observed fluid and melt compositions in more detail, we use the chondrite-normalized ratio of Nd/Dy (Fig. 30). In addition, focusing exclusively on REE blurs the distinction between HDFs with "planed" and "ribbed" trace element patterns, for which a key distinguishing criterion is the presence ("ribbed") or absence ("planed") of a strong negative Nb anomaly (Weiss et al. 2013), expressed as a strongly elevated chondrite-normalized Ce/Nb ratio in Figure 30. Calculated HDF, hydrous melt and "dry" melt compositions all show similar Nd_N/Dy_N, decreasing strongly from equilibrium with garnets with sinusoidal REE_N peaking and Nd to those with normal patterns. Only for the melts in equilibrium with garnets with normal patterns are Nd_N/Dy_N ratios observed similar to HDFs and low volume mantle melts. Using partition coefficients for garnet-carbonatite (Dasgupta et al. 2009) leads to even more extreme LREE–MREE fractionation, with Nd_N/Dy_N ratios much higher than natural magnesio-carbonatites for all three garnet groups utilized for melt calculation. In terms of Ce_N/Nb_N ratio, only calculated carbonate melts in equilibrium with garnets with sinusoidal REE_N patterns peaking at Nd would show negative Nb anomalies (Ce_N/Nb_N of about 5). This excludes HDFs with "ribbed" trace element patterns as a principal diamond-forming metasomatic agent and instead highlights a similarity in the occurrence of mildly positive Nb anomalies ($Ce_N/Nb_N < 1$) for calculated HDF/melts with Group I kimberlites and with natural HDFs with "planed" trace element patterns.

Key observations and conclusions

Garnet, and for the lherzolitic paragenesis also clinopyroxene, are the principal carriers of incompatible trace elements among peridotitic inclusions in diamond. The REE_N patterns of garnet inclusions are typically sinusoidal, with harzburgitic garnet most commonly peaking at Nd and lherzolitic garnets at Sm. As another end-member type of REE_N patterns, some lherzolitic and very few harzburgitic garnets have normal patterns, characterized by flat $MREE_N$–$HREE_N$. This evolution from sinusoidal to normal garnet REE_N patterns is mirrored for clinopyroxene inclusions by a transition from patterns with a characteristic steep negative slope in the $LREE_N$ to humped patterns peaking at Nd_N.

Normalization to "primitive mantle" garnet/clinopyroxene documents a two-stage evolution of the protoliths (cf. Frey and Green 1974), starting with intense melt depletion causing overall REE depletion and steep positive $LREE_N/HREE_N$ slopes in the bulk rocks (preserved only in the HREE from Er to Lu). This depletion is typically followed by metasomatic re-enrichment through a medium with very high LREE and LREE/HREE and distinctly subchondritic HREE. Garnet with normal REE_N patterns and coexisting clinopyroxene with humped patterns indicate primitive mantle-like bulk rock REE concentrations and are a consequence of metasomatism through media with only moderately fractionated trace element contents. Whilst such near-

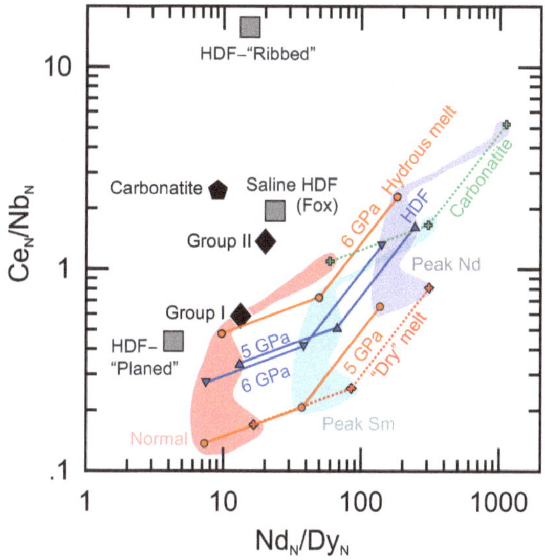

Figure 30. The chondrite normalized ratios of Ce/Nb and Nd/Dy are used to assess how well calculated melts for inclusion garnets representing the three principal groups of REE_N patterns match HDFs in fibrous diamonds and kimberlite and carbonatite melts. The three principal garnet patterns are (1) harzburgitic garnets peaking at Nd_N, (2) lherzolitic garnets peaking at Sm_N and (3) lherzolitic garnets with normal REE_N patterns. In all three cases the median compositions are used. Hypothetical melt compositions in equilibrium with the median garnet compositions were calculated using six different sets of experimentally determined partition coefficients: aqueous HDFs (1000 °C, pressures of 5 and 6 GPa; **blue**) and hydrous melts (1100 °C, pressures of 5 and 6 GPa; **orange**) in equilibrium with metasomatized garnet lherzolite (Kessel et al. 2015), Kilauea olivine tholeiite in equilibrium with near-liquidus garnet and clinopyroxene ("dry" melt; 1430 °C and 3 GPa; Johnson 1998; **red**) and carbonate melt in equilibrium with garnet lherzolite (1265–1470 °C and 6.6–8.6 GPa; Dasgupta et al. 2009; **green**). The calculated melt compositions are compared to primitive Group I and Group II kimberlite compositions (Becker and Le Roex 2006) and high-density-fluids in fibrous diamonds with "planed" and "ribbed" traced element patterns (Weiss et al. 2013) and their potentially parental saline fluid composition (Fox diamond E217; Weiss et al. 2015). The presence of a strong negative Nb anomaly for "ribbed" HDFs and the absence of such anomalies in "planed" HDFs is a key criterion for their distinction (Weiss et al. 2013). We use the Ce_N/Nb_N ratio as a measure for the presence and extent of negative ($Ce_N/Nb_N > 1$) and positive (< 1) Nb anomalies. The Nd_N/Dy_N ratio is a measure for the slope of REE_N patterns. Nd and Dy were chosen because of their comparatively high absolute abundances, with especially the HREE abundances in the analyzed fibrous diamonds being extremely low.

primitive REE_N patterns are rare among inclusions – about 5% of all garnet inclusions have normal REE_N—they strongly dominate among peridotite xenolith and xenocryst garnets. This is not only a function of reversed proportions of the harzburgitic and lherzolitic parageneses for inclusions (high ratio) and xenoliths (low ratio) but is also observed when looking at lherzolitic garnets only. Whether this observation reflects the well-established affinity of diamond to the most depleted peridotitic substrates (e.g., Gurney 1984), or an increasing metasomatic modification through time that inclusions were shielded from by their diamond hosts, or probably a combination of both, is not well constrained by currently existing data.

The nature of the metasomatic agents that re-enriched the diamond substrates and presumably also caused repeated episodes of diamond formation cannot be fully resolved. The HDF inclusions observed in coated and fibrous diamonds and sometimes detected in clear, smooth-faced monocrystalline diamonds are too "melt-like" in their trace element patterns to create the extreme LREE/HREE fractionation seen in the vast majority of garnet and clinopyroxene inclusions. Whether that reflects a temporal evolution, with Archean diamond-forming HDFs showing overall stronger trace element fractionation, or if the HDFs precipitating either fibrous diamonds or diamond coats are more primitive members of a HDF fractionation series (e.g., through chromatographic column type fractionation; Navon and Stolper 1987; Bodinier et al. 1990) will have to be established through future research. Near-primitive trace element patterns are consistent with melt-like metasomatic agents (including the HDFs seen in fibrous diamonds) and, given the relationship between cratonic geotherms and the hydrous solidus of lherzolite, likely reflect infiltration of low volume hydrous melts, similar to primitive kimberlites. Conversely, the high temperatures required for hydrous melting of strongly depleted peridotites (harzburgites-dunites) shield such rocks against pervasive melt metasomatism, as such melts must freeze upon equilibration with their wall rocks.

ECLOGITIC–PYROXENITIC SUBSTRATES

Occurrence, mineralogy and classification of eclogite/pyroxenite inclusions and xenoliths

Occurrence. As broadly metabasaltic rocks sometimes containing coesite or diamond, eclogite occurs in dynamic settings where (ultra-)high-pressure metamorphism takes place. This encompasses deeply subducted and exhumed portions of oceanic crust consisting of pillow lavas, sheeted dikes and gabbros (e.g., White and Klein 2014), which are now sampled as lenses in orogens (e.g., Liou et al. 2004) and as xenoliths in kimberlites and related rocks (e.g., Jacob 2004). Diamond populations dominated by eclogitic inclusions are often associated with craton margins or cratonic lithosphere affected by later tectono-magmatism, suggestive of a subduction origin of the source rock (Sobolev 1985; Shirey et al. 2002; Stachel and Harris 2008). Some mantle xenolith suites are dominated by eclogite (e.g., at the Roberts Victor and Orapa mines in the Kaapvaal craton, and at Koidu in Sierra Leone in the West African craton; MacGregor and Carter 1970; Hills and Haggerty 1989). As the lithospheric mantle contains only a few percent (Schulze 1989), at most several 10s of percent (Garber et al. 2018) eclogite, this local dominance highlights preferential preservation of eclogite versus peridotite and/or non-representative sampling by kimberlite and lamproite magmas (e.g., Moss et al. 2018).

A proportion of garnet–clinopyroxene rocks in xenolith suites with jadeite-poor clinopyroxene classify as pyroxenite, although these are sometimes referred to as high-Mg or Group A eclogite (Coleman et al. 1965; Taylor and Neal 1989). While bimineralic pyroxenites grading into high-Mg eclogites form part of most cratonic xenolith suites (Aulbach et al. 2020a), orthopyroxene-bearing pyroxenites are more common in craton-margin and off-craton localities (Jacob 2004), where they record generally lower equilibration pressures than, and have an origin different from, eclogite (Schmädicke et al. 2011).

Mineralogy. Stachel and Harris (2008) found that garnet and clinopyroxene, the main rock-forming minerals in eclogite, comprised 56% and 39% of the total of inclusions in 1311 eclogitic diamonds; other common silicate and oxide inclusions being rutile (2.7%) and coesite/quartz (1.8%). In contrast, garnet, clinopyroxene, orthopyroxene, coesite and olivine constitute 38%, 37%, 21%, 3% and 1%, respectively of inclusions in 136 websteritic diamonds (Stachel and Harris 2008). Despite their accessory mineral status in common lithospheric mantle sources, sulfides are the most common inclusion in diamond (Harris and Gurney 1979; Haggerty 1986). However, they will not be further considered here, as their occurrence as an inclusion in diamond is considered as part of the diamond age-dating work and is reviewed in Smit et al. (2022).

Of 2451 xenoliths considered for comparison of mineralogy, commonly reported primary accessory and minor silicate and oxide minerals include diamond (n = 251), rutile (n = 83), kyanite (n = 67), graphite (n = 59; of these 10 also contain diamond), corundum (n = 25) and coesite/quartz (n = 11). Sulfides are common, though not necessarily reported as part of the assemblage (pers. observation). Garnet clinopyroxenites (n = 334), which grade into websterite and olivine-poor lherzolite, more commonly contain orthopyroxene (n = 30 vs. 8 in eclogite xenoliths), but only 15 of 544 pyroxenite xenoliths (3%) are diamondiferous, compared to 12% of eclogite xenoliths. As the description of websteritic inclusion mineralogy provided in the previous paragraph is based on a different classification scheme than that applied to pyroxenite xenoliths (see Inclusion suites and parageneses), these findings are not entirely comparable.

The proportion of diamondiferous eclogite (12%) and pyroxenite samples (3%) may be high because such rocks are frequently specifically targeted for study. On the other hand, accessory minerals may be more common than they appear, because they need not be exposed at the section scale or may not have been reported, depending on the aims of the study. Whilst other minerals in the xenoliths, such as amphibole, plagioclase and phlogopite are likely to be of secondary, metasomatic origin, zircon, apatite and ilmenite could be primary, although more likely to be metasomatically introduced (Aulbach et al. 2020a, and references therein). In addition, Jacob (2004) notes the typical purity of minerals in mantle eclogite, which rarely contain inclusions, as opposed to their orogenic counterparts. Eclogites rich in CaO and Na_2O are particularly affected by texturally late alteration, resulting in the partial or complete replacement of primary jadeite-rich clinopyroxene by "spongy" diopside-rich clinopyroxene plus K-rich glass and other minerals (e.g., Snyder et al. 1997; Spetsius and Taylor 2002). This texturally late assemblage reflects the greater instability of jadeite-rich clinopyroxene in contact with an ultramafic carbonated agent such as kimberlite (e.g., Misra et al. 2004). Also, the presence of amphibole and phlogopite in the kelyphite rims of garnet and of K-rich glass in the spongy clinopyroxene suggests partial melting and kelyphite formation catalyzed by kimberlite-related fluids (Snyder et al. 1997; Spetsius and Taylor 2002).

Classification. Of the aforementioned 2451 xenoliths considered for mineralogy, only 946 have reported major element abundances for both garnet and clinopyroxene, allowing distinction of eclogites and pyroxenites. This yields 776 mantle eclogite samples (236 diamondiferous) and 170 pyroxenite samples (15 diamondiferous). Only a small proportion of diamondiferous samples have reported REE concentrations that further allow assignment to gabbroic, with positive Eu anomalies, vs. non-gabbroic sources (Table 3).

When inclusions are recast in the scheme applied to mantle xenoliths, clinopyroxene–garnet inclusion pairs comprise 235 eclogitic and 37 pyroxenitic diamonds. Of unpaired clinopyroxene, 481 are eclogitic (15 of 26 with REE data have a gabbroic signature) and 86 pyroxenitic (2 of 3 with REE data are gabbroic). Unpaired garnet can be assigned to high-Ca, high-Mg, low-Mg and gabbroic groups, bearing in mind that some high-Mg samples may actually derive from pyroxenite sources (~70% of pyroxenite xenoliths have high-Mg garnet) and some high-Ca samples without REE data may actually derive from gabbroic sources (67% of gabbroic xenoliths contain high-Ca garnet) (Table 3).

Ages of eclogite/pyroxenite xenoliths and inclusions

Obtaining accurate and precise formation ages from mantle eclogite and pyroxenite is challenging because bulk rocks are affected by kimberlite contamination, and because two-mineral isochrons yield kimberlite eruption or cooling ages. Also, zircon is rare in these rocks and can be of metasomatic origin (Jacob 2004; Aulbach et al. 2019a). Nevertheless, ages and the methods by which they were derived are summarized in Appendix Table 1. Worldwide, the oldest age arrays obtained from Pb–Pb, Sm–Nd and Re–Os geochronology on bulk rocks or mineral separates are often Mesoarchean to Neoarchean, (3.2 to 2.5 Ga) reflecting the suturing of continental nuclei and haphazard incorporation of intervening oceanic lithosphere to form larger cratonic entities. Exclusively Paleoproterozoic ages (2.5 to 1.6 Ga) have been obtained from eclogite and pyroxenite xenoliths in the northern and central Slave craton, which coincide with collision/accretion at the western craton margin, preceded by subduction of an intervening ocean floor.

Table 3. Number of eclogitic and pyroxenitic inclusions in diamond and of diamondiferous and barren xenoliths used in this study, and their classification.

Sample type/ classification	Total	High-Ca eclogite	High-Mg eclogite	Low-Mg eclogite	Gabbroic eclogite	Unclassified eclogite	Pyroxenite	Gabbroic pyroxenite	Unclassified pyroxenite
Inclusions in diamond									
Touching pairs*	23	0	0	0	3	12	1	0	7
% of classified		0	0	0	75		25	0	
Non-touching pairs*	249	4	5	4	10	197	3	1	25
% of classified		15	19	15	37		11	4	
Unpaired clinopyroxene**	567		11 non-gabbroic		15	466	1	2	83
% of classified			38		52		3	7	
Unpaired garnet***	906	20	10	7	71	798	?	?	?
% of classified		19	9	6	66				
Xenoliths									
Barren	695	86	110	50	294	0	102	53	0
% of classified		12	16	7	42	0	15	8	0
Diamondiferous*	251	1	7	6	56	166	1	0	14
% of classified		1	10	8	79		1	0	

Notes: Eclogite = Na/(Na+Ca) in clinopyroxene ≥ 0.2, pyroxenite < 0.2; gabbroic = whole-rock Eu/Eu* > 1.05; full classification is only possible when major and REE element data are available for both clinopyroxene and garnet, which is the case for all barren xenoliths chosen for this study. * A proportion of unclassified samples (no REE data) may be gabbroic. ** 11 eclogitic clinopyroxene inclusions with REE data are non-gabbroic, but cannot be classified as high-Ca, high-Mg or low-Mg in the absence of garnet. *** A proportion of garnet (in particular high-Mg) may be pyroxenitic, but cannot be attributed in the absence of clinopyroxene; a proportion of unclassified samples may be gabbroic (in particular high-Ca), but cannot be attributed without REE data.

Comparison between eclogitic/pyroxenitic inclusions in diamond and their barren and diamondiferous xenolithic counterparts

In the following, inclusions in diamond are compared with similar minerals in xenoliths, the latter distinguished according to whether they are reported as diamondiferous or barren. Median and average concentrations are typically very similar, especially for large numbers (100s) of samples in a given group. Differences are more apparent for pyroxenites, for which smaller numbers are available and which are a heterogeneous group with respect to their compositions and origins (Stachel and Harris 2008). Median compositions are used for comparison to safeguard against outliers, and median absolute deviations are reported as the corresponding uncertainty. Nevertheless, the relatively small number of inclusions (especially pyroxenitic) with trace-element or $\delta^{18}O$ data make comparisons and unambiguous interpretation tenuous. Also, minor elements, such Cr, Ni and K, are not consistently analysed and reported. Appendix Table 3 shows the statistical analysis for four eclogite and two pyroxenite classes. Minerals in diamond and similar ones in diamondiferous xenoliths are distinguished according to whether they are eclogitic or pyroxenitic and whether they are touching or non-touching, as the latter relationship affects calculated temperatures, and also some distribution coefficients. To simplify the following compositional comparison, touching and non-touching inclusions are considered together, and unpaired clinopyroxene, which allows classification into eclogite and pyroxenite, is subsumed with paired clinopyroxene. Finally, the gabbroic pyroxenite class shares the cumulate signature with gabbroic eclogites and the enriched signature with non-gabbroic pyroxenite xenoliths. For the sake of clarity, the gabbroic pyroxenite class is not described below or shown in most figures.

Mineral major and minor element compositions. Clinopyroxene in eclogitic diamond has a median Mg# of 76.5, a diopside content of 0.46 and a jadeite (Jd) component of 0.32 (Appendix Table 3). For clinopyroxene in diamondiferous eclogite these values are a higher Mg# (82.4) and a diopside content of (0.41) but similar Jd (0.35). In contrast, clinopyroxene in all classes of barren eclogite (gabbroic, high-Ca, high-Mg, low-Mg) have Mg# of 78.3–85.3, consistently higher diopside content (0.54–0.65) and mostly lower Jd, varying between 0.33 in gabbroic and 0.20 in high-Mg eclogites. Median K_2O contents of clinopyroxene in diamond (0.20 wt%) are more than twice as high as in diamondiferous eclogite (0.09 wt%) and 3–4 times as high as in barren eclogites ($\leq$ 0.06 wt%). In a diagram of Jd vs. Mg#, (Fig. 31), clinopyroxene inclusions in eclogitic diamond span a large range, with diamondiferous eclogites shifted to higher Mg#, and barren eclogites to both higher Mg# and lower Jd. In pyroxenitic diamond, clinopyroxene has a median Mg# of 83.5 and a diopside content of 0.72, which is much lower than the Mg# ($\geq$ 90.1) or diopside content ($\geq$0.77) in clinopyroxene from diamondiferous or barren pyroxenite, but still higher than in eclogitic clinopyroxene. By definition, jadeite contents are lower in pyroxenitic than in eclogitic clinopyroxene, but Cr_2O_3 contents are higher ($\geq$ 0.13 wt%), though overlapped by clinopyroxene in high-Mg eclogite with 0.14 wt%.

Garnet in eclogitic diamond has distinctly higher TiO_2 (0.61 wt%) and Na_2O contents (0.20 wt%) than garnet in diamondiferous eclogite (0.36 and 0.13 wt%, respectively) and various barren eclogite classes ($\leq$ 0.27 and 0.09 wt%, respectively) (Appendix Table 3). Median Mg# of included garnet is 52.7. Garnet in diamondiferous eclogite has Mg# 61.7, which is very similar to that in gabbroic barren eclogite (62.2). All the three garnet groups referred to above have a Ca# (100Ca / (Ca + Mg + Fe^{total} + Mn)) of 24. Garnet in high-Mg barren eclogite has the highest Cr_2O_3 contents (0.13 wt%) and Mg# (68.6) and the lowest Ca# (11.2) of all eclogite classes, the nearest comparison being the unpaired high-Mg garnet in diamond (Cr_2O_3 0.17 wt%, Mg# 66.1, Ca# 11.1). Compared to paired inclusions, unpaired garnet specifically extends into the field of diamondiferous eclogites with high Mg#–Ca# typical of some gabbroic eclogites (Fig. 32a). Garnet inclusions in pyroxenitic diamond have again lower Mg# (63.2) and higher Ca# (14.3) than in diamondiferous pyroxenites (77.7 and

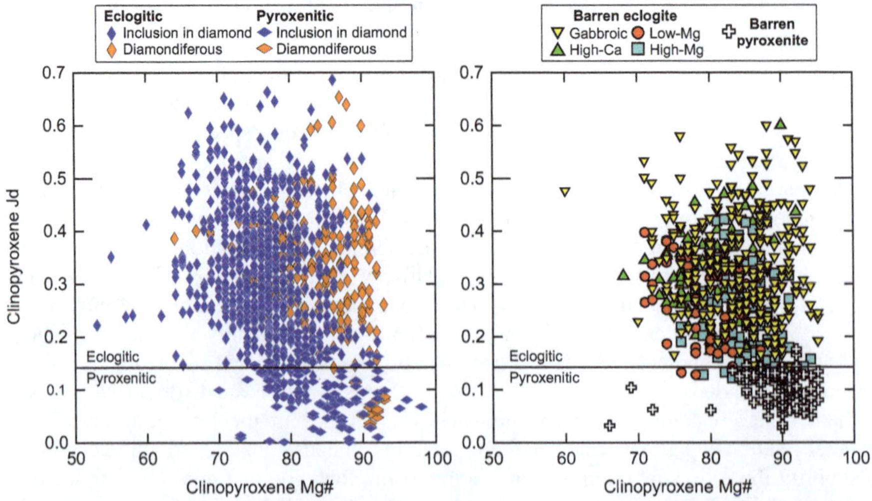

Figure 31. Jadeite content of clinopyroxene (Jd) as a function of Mg# (100Mg / (Mg + Fe^{total})). Eclogite and pyroxenite are roughly separated according to Jd (formally Na / (Na + Ca)), as indicated on **right**. Inclusions (paired and unpaired) and diamondiferous xenoliths on **left**, barren xenoliths on **right**.

10.2, respectively). Median Cr_2O_3 contents in pyroxenitic garnet (≥ 0.29 wt% for all classes) are higher than in eclogitic garnet (≤ 0.13 wt% for all classes), and many fall on a "lherzolite" trend of CaO vs. Cr_2O_3 that indicates buffering of CaO contents by orthopyroxene (Fig. 32b).

Ca# in garnet shows a broad positive correlation with Jd, where inclusions range to higher Jd at a given Ca# and are only weakly represented at Ca# ≳37 compared to diamondiferous eclogites (Fig. 33a). Combined, Na_2O in garnet and K_2O in clinopyroxene show a strongly diminished range both in diamondiferous and in barren eclogites/pyroxenites compared to inclusions (Fig. 33b).

Mineral trace element compositions. Trace elements for which at least tens of analyses for inclusions in diamond are available include Ni, Sr, Y, Zr, Nb, the REE and Hf (Appendix Table 3). For clinopyroxene, median Ni contents are lowest in inclusions (160 ppm, vs. 251 ppm in clinopyroxene from diamondiferous eclogite and 196–312 ppm from barren eclogite). Median Sr contents in inclusions (136 ppm) and in clinopyroxene from diamondiferous eclogite (142 ppm) are similar and intermediate between those from high-Ca eclogite (118 ppm) and gabbroic eclogite (162 ppm), whereas that in high-Mg and low-Mg eclogite is higher (186 and 193 ppm, respectively). Clinopyroxene in diamondiferous eclogite has low Y content (0.9 ppm), between that in gabbroic eclogite (0.6 ppm) and high-Ca eclogite (0.9 ppm), but inclusions have higher abundances (2.4 ppm). Median Zr contents of clinopyroxene in diamond (14.1 ppm) and diamondiferous xenoliths (10.5 ppm) are again similar to those in gabbroic and high-Ca eclogite (10.3 ppm and 16.4 ppm, respectively), being higher still in high-Mg and low-Mg eclogites. Conversely, by far the highest median Nb contents (0.25 ppm) are recorded for clinopyroxene in diamondiferous eclogites, followed by inclusions (0.10 ppm), whereas those in barren eclogites are lowest (≤ 0.08 ppm). The comparison for pyroxenite classes is hampered by the low number of trace-element analyses both for inclusions and for diamondiferous xenoliths (Appendix Table 3), and by the heterogeneous nature of their petrogenesis (Stachel and Harris 2008; Aulbach and Jacob 2016). Median concentrations of trace elements in clinopyroxene from pyroxenitic diamond are very different from those of their xenolithic counterpart, which are, in turn, more similar to those of high-Mg eclogite xenoliths (Appendix Table 3).

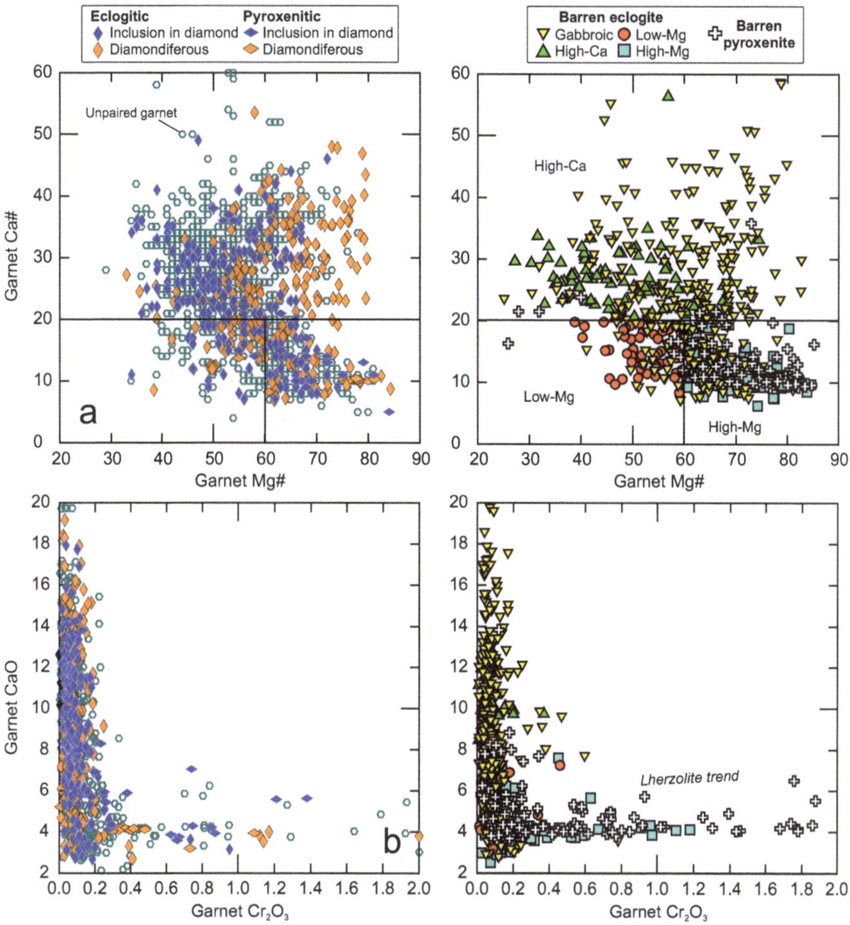

Figure 32. a. Ca# (100Ca / (Ca + Mg + Fe^{total} + Mn) as a function of Mg# in garnet. Garnet from non-gabbroic eclogite (whole-rock Eu/Eu* < 1.05; Eu/Eu* = chondrite-normalized Eu / $(Sm \times Gd)^{1/2}$; Rudnick and Fountain 1995) is classified into high-Mg, low-Mg and high-Ca varieties, as shown on the right. Garnet from gabbroic eclogite is typically high-Ca, but also occurs at lower Ca#. **b.** CaO (wt%) as a function of Cr_2O_3 (wt%). At Cr_2O_3 ≳0.3 wt%, garnet falls on a CaO–Cr_2O_3 trend typical of websterite and lherzolite, and indicative of buffering by orthopyroxene. Inclusions (paired and unpaired) and diamondiferous xenoliths are on the on **left**, barren xenoliths on the **right**.

Chondrite-normalized median REE patterns of clinopyroxene (Fig. 34a) show systematically decreasing abundances between Gd and Lu for all classes, except clinopyroxene in diamondiferous eclogite, which shows an upturn in Yb-Lu. The pattern of clinopyroxene in eclogitic diamond shows a hump in Nd_N, most similar to that in high-Ca eclogite, but differs from xenolithic clinopyroxene in having a less steep slope in the $MREE_N$–$HREE_N$ and higher concentrations of these elements. REE abundances in clinopyroxene from high-Mg and low-Mg eclogites are higher than for inclusions. Clinopyroxene in gabbroic eclogite shows a positive Eu anomaly (Eu/Eu* = 1.22), shared to a lesser extent by that in diamondiferous eclogite (1.19) and that in diamond (1.07). Individual REE_N patterns of clinopyroxene in pyroxenitic diamond are extremely heterogeneous, not only with respect to abundances, which vary by several orders of magnitude, but also to shape (not shown), which precludes calculation of a meaningful median composition. Individual inclusions have Eu/Eu* from 0.61 to 2.4, with a median of 1.07,

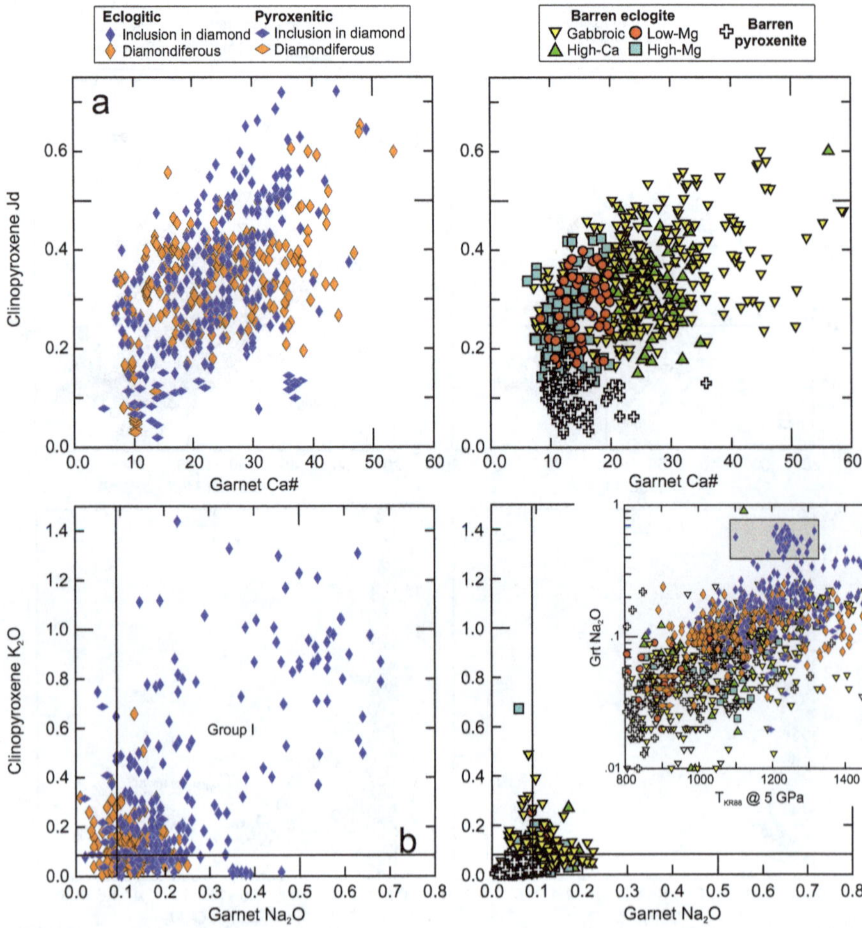

Figure 33. a. Jadeite content in clinopyroxene (Jd) as a function of Ca# (100Ca / (Ca + Mg + Fe^{total} + Mn)) in garnet. **b.** K_2O content (wt%) in clinopyroxene as a function of Na_2O content (wt%) in garnet. Samples with K_2O in clinopyroxene ≥ 0.08 wt% and Na_2O in garnet ≥ 0.09 wt%, with a higher likelihood of being diamondiferous, have been classified as Group I eclogites by McCandless and Gurney (1989); those in the lower left quadrant are Group II eclogites. **Inset in right panel** shows that Na_2O content in garnet increases with temperature, with a group offset to higher Na_2O. Inclusions and diamondiferous xenoliths on **left**, barren xenoliths on **right**.

i.e., slightly gabbroic. Clinopyroxene in the single diamondiferous pyroxenite with REE data, together with its barren counterpart, has the highest LREE abundances of all classes, but shows a stronger negative slope in the $MREE_N$–$HREE_N$, similar to that in gabbroic pyroxenite (Fig. 34a).

For garnet, median Ni and Sr contents are highest for inclusions in diamond (62 and 1.76 ppm, respectively), followed by that in diamondiferous eclogite (49 and 0.92 ppm, respectively) and barren eclogite classes (41–14 ppm and 0.62–0.26 ppm, respectively). Median Y contents of garnet in diamond (30 ppm), high-Ca eclogite (34 ppm) and low-Mg eclogite (36 ppm) are highest, whereas those of garnet in diamondiferous eclogite (19 ppm) are more similar to those of garnet in gabbroic eclogite (17 ppm). Median Zr contents are identical for garnet in diamond and in diamondiferous xenoliths (19 ppm), whereas those for garnet in high-Ca and low-Mg eclogite (11 ppm) and in gabbroic eclogite (9 ppm) are much lower. Median Nb contents vary widely, and are lowest for garnet in gabbroic eclogite (0.03 ppm), followed by

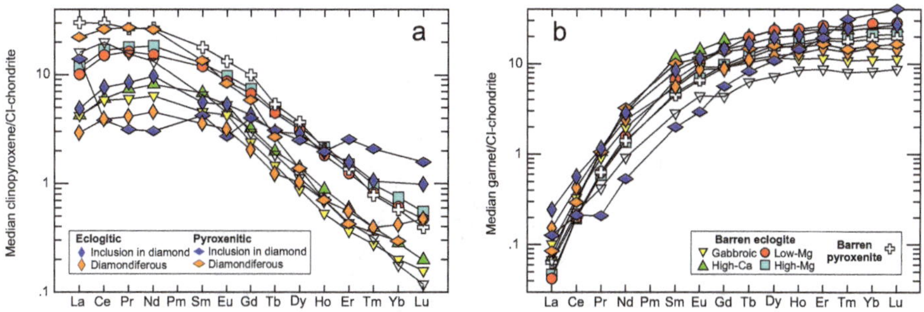

Figure 34. Chondrite-normalized REE patterns of **a.** clinopyroxene and **b.** garnet. Only median values per class are shown for clarity. CI-chondrite of Sun and McDonough (1989).

inclusions (0.06 ppm) and highest for garnet in diamondiferous eclogite (0.20 ppm), as applies to coexisting clinopyroxene.

Chondrite-normalized median REE patterns for garnet in the various eclogite classes are very similar at first glance, with low LREE, more or less strongly increasing abundances to Dy_N and then small or no increases to Lu_N (Fig. 34b). Garnet in diamond and in diamondiferous eclogite have the highest La and Ce abundances, but, in contrast to inclusions, garnet in diamondiferous eclogite shows a flatter $HREE_N$ pattern, similar to garnet in gabbroic eclogite. Contrary to clinopyroxene, REE_N patterns of individual garnet inclusions in pyroxenitic diamond are relatively coherent (not shown). The median REE_N pattern for garnet in pyroxenitic diamond shows a strong increase between Pr and Lu, and is very different from that of garnet in diamondiferous pyroxenite, which increases strongly between La and Sm and then flattens rather abruptly. This in turn differs from the REE_N pattern of garnet in barren pyroxenites, which does not attain as high LREE abundances.

Garnet oxygen isotopes For six paired eclogitic garnet inclusions the median $\delta^{18}O$ is +7.4‰, whereas in diamondiferous eclogite it is +6.1‰ and in barren xenoliths it is lower still, varying between +6.0‰ for garnet in low-Mg eclogite (n = 9) and +5.4‰ for that in high-Ca and gabbroic eclogite (both +5.4‰, n = 23 and 81, respectively). Unpaired high-Ca and low-Mg garnet inclusions, which are likely eclogitic, both have median $\delta^{18}O$ of +6.8‰ (n = 32 and 3, respectively), whereas for high-Mg garnet $\delta^{18}O$ is +6.6‰ (n = 9). The variability of $\delta^{18}O$ is high (> 6‰ for garnets in eclogite xenoliths with Mg# < 50 but diminishes to < 3‰ at Mg# > 75, especially in barren pyroxenite xenoliths (Fig. 35). Inclusions show a crude negative correlation of $\delta^{18}O$ vs. Mg# (r^2 = 0.2, n = 50) and are barely represented in the field of sub-mantle $\delta^{18}O$ at low Mg#, where a group of garnets from barren high-Ca and gabbroic eclogite xenoliths plots (Fig. 35). The latter are exclusively from the Roberts Victor eclogite. No $\delta^{18}O$ data are available for clinopyroxene inclusions in the database, or for garnet in pyroxenitic diamond or diamondiferous pyroxenite. Garnet in barren pyroxenite has $\delta^{18}O$ of +5.5‰.

Temperature estimates. For all samples, temperatures are calculated with the Mg–Fe exchange thermometer of Krogh (1988; T_{KR88}), either at a pre-set pressure of 5 GPa using total Fe contents, in keeping with Stachel and Luth (2015), or solved iteratively with regional model conductive geotherms from Hasterok and Chapman (2011; hereafter HC11). The geotherm for non-touching inclusions corresponds to a surface heat flow of 38 mW/m^2, chosen to yield median pressures identical to those obtained for touching inclusions (see below). Samples yielding temperatures < 800 °C are not displayed or counted because temperatures may be underestimated at T < 900 °C (Krogh 1988) and because there is overlap with lower crustal xenoliths, e.g., in the Kaapvaal craton (Pearson et al. 1995). Temperatures > 1500 °C are inferred to reflect disequilibrium, although the cut-off may actually be lower depending on lithosphere thickness.

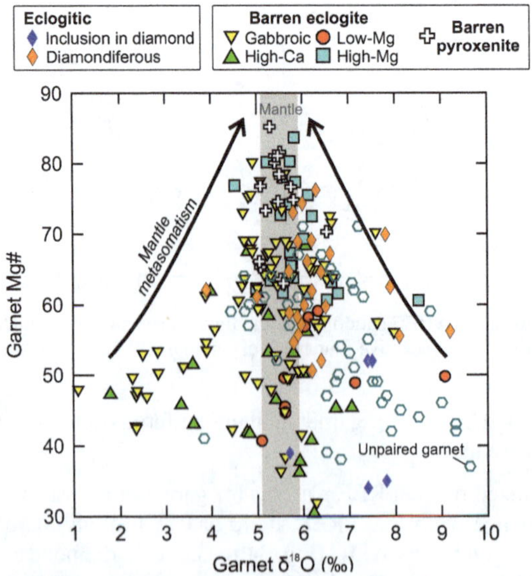

Figure 35. Mg# (100Mg / (Mg + Fe^{total})) in garnet as a function of $\delta^{18}O$ (‰ deviation of $^{18}O/^{16}O$ from the VSMOW standard). **Arrows** qualitatively indicate how the range of $\delta^{18}O$ at the highest Mg# in barren eclogites narrows towards typical mantle values (5.1–5.9‰; Mattey et al. 1994) in response to mantle metasomatism (Aulbach et al. 2020a).

Non-touching and touching eclogitic inclusions record median temperatures of 1225 and 1132 °C, respectively, reduced to 1207 and 1099 °C, respectively when the inclusions in Argyle diamonds, which are derived from an atypical high-temperature environment, are not considered (cf. Stachel and Harris 2008). These values compare with 1134 °C for diamondiferous eclogites (Appendix Table 3). Barren eclogites yield lower median temperatures, between 1087 °C for high-Mg eclogites and 962 °C for low-Mg eclogites. Non-touching and touching pyroxenitic inclusions give 1198 and 1065 °C, respectively, whereas temperature estimates for diamondiferous and barren pyroxenites are much lower, 920 and 974 °C, respectively. Temperature distributions are displayed in Figure 36a.

Trace element distribution. The distribution of trace elements between garnet and clinopyroxene in mantle eclogite and pyroxenite xenoliths has been shown to variably depend on garnet Ca#, jadeite (Jd) in clinopyroxene, and temperature and pressure (O'Reilly and Griffin 1995; Harte and Kirkley 1997). Mantle metasomatism of eclogite tends to lower Ca# and jadeite content, with preferred trace element partitioning into clinopyroxene, which can lead to the paradoxical observation that garnet from metasomatized eclogite has abundances of some incompatible elements similar to or even lower than garnet in unmetasomatized specimens (Aulbach et al. 2020a).

Garnet Ca# and jadeite content in clinopyroxene from diamond and diamondiferous eclogite show a broad positive covariation (r^2 = 0.25, n = 515). This is also recognizable in barren eclogite (r^2 = 0.34, n = 694), albeit at a lower jadeite intercept value, reflecting a higher proportion of pyroxenite (Fig. 33a). Thus, Ca# in garnet and Jd in clinopyroxene, which do not co-vary significantly with temperature, are coupled. With the exception of one outlier, inclusions, diamondiferous xenoliths and barren xenoliths show a negative correlation of log D_{Sr} (the distribution coefficient D of Sr between clinopyroxene and garnet) with garnet Ca# (r^2 = 0.64, n = 24; r^2 = 0.29, n = 78 and r^2 = 0.38, n = 635, respectively). These trends are

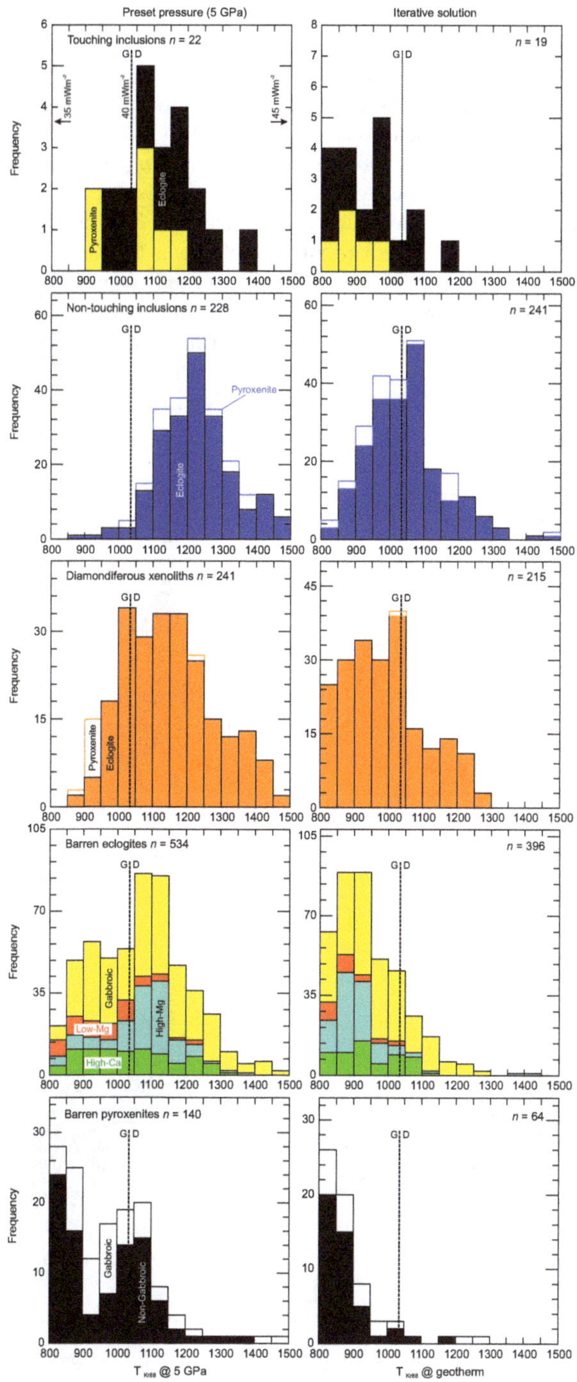

Figure 36. Distribution of estimated temperatures, based on **a.** T_{KR88} at 5 GPa, and **b.** T_{KR88} solved iteratively with peridotite-derived regional conductive geotherms parameterized from Hasterok and Chapman (2011), for touching and non-touching inclusions and diamondiferous and barren xenoliths. The temperature where diamond becomes stable along conductive geotherms corresponding to surface heat flows of 35, 40 and 45 mW/m^2 (Hasterok and Chapman 2011) is indicated in the **top left panel**, and for the 40 mW/m^2 instance, as a **dashed line** in both panels.

offset to increasingly higher log D_{Sr} at a given garnet Ca# for diamondiferous and barren xenoliths (Fig. 37a). This variation is also true for the more incompatible trace element Ce (excluding four outliers), where inclusions have $r^2 = 0.35$ ($n = 31$), diamondiferous xenoliths have $r^2 = 0.44$ ($n = 71$) and barren xenoliths have $r^2 = 0.59$ ($n = 674$) (Fig. 37b). This trace-element relationship similarly applies to Zr and Nb (not shown), except that there is greater overlap between inclusions in diamond and barren eclogites. The distribution of Y, which is compatible in bulk eclogite (see, e.g., distribution coefficients for garnet and clinopyroxene in equilibrium with tonalite melt in Barth et al. 2002), is strongly negatively correlated with Ca# ($r^2 = 0.64$, $n = 783$), but there are no distinct trends for diamondiferous and barren samples (not shown). Both log D_{Sr} and log D_{Ce} show marked negative correlation with $T_{KR88(5\ GPa)}$ ($r^2 = 0.43$, $n = 747$ and $r^2 = 0.31$, $n = 752$, respectively), whereas log D_Y shows no correlation. As trace element data are available for only one touching inclusion, the effect of cooling and re-equilibration on trace element distribution cannot be assessed.

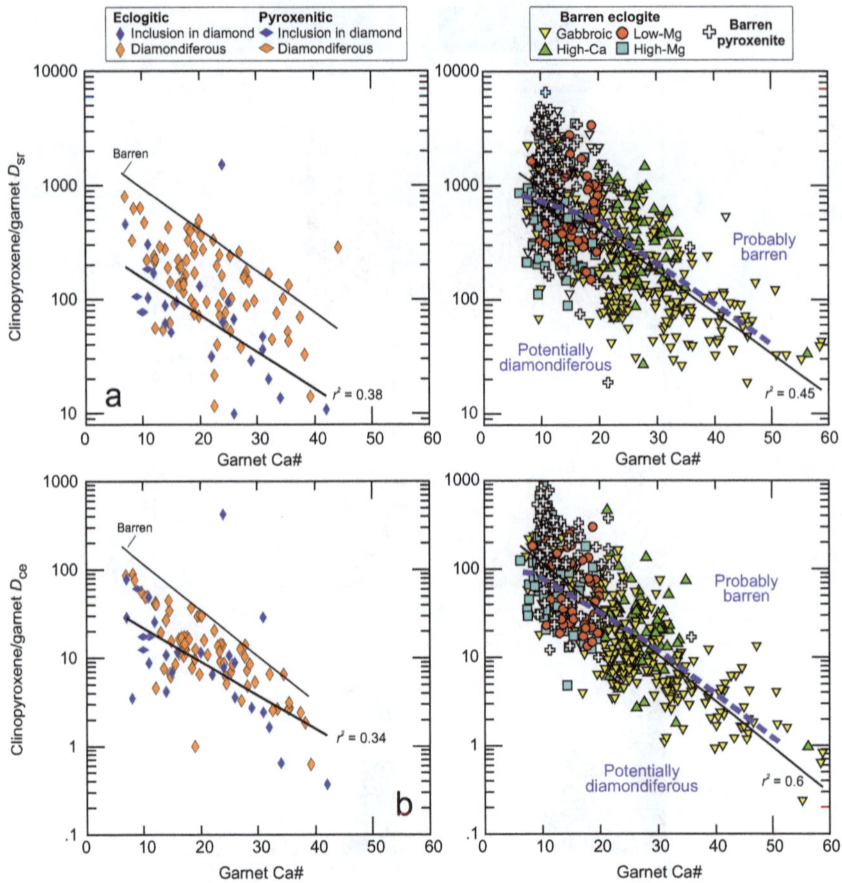

Figure 37. Log distribution of **a.** D_{Sr} and **b.** D_{Ce} between clinopyroxene and garnet as a function of garnet Ca# (100Ca / (Ca + Mg + Fe^{total} + Mn)). Separate regressions for inclusions and barren xenoliths are indicated. **Thick stippled line** separates the field encompassing inclusions and diamondiferous xenoliths from the field where almost exclusively barren xenoliths occur. Inclusions and diamondiferous xenoliths on **left**, barren xenoliths on **right**.

Reconstructed whole rocks. Following Aulbach and Jacob (2016), whole rocks are reconstructed from 55 wt% garnet and 45 wt% clinopyroxene, bearing in mind that HFSE abundances are minima when rutile is not considered. The presence of minor kyanite and/or orthopyroxene can further adversely affect accuracy of whole rock estimates, which is the case for only < 3% kyanite-bearing and < 2% orthopyroxene-bearing xenoliths, coesite being reported for even fewer xenoliths in this study. The above mineral modes correspond to the median modes determined for large eclogite xenoliths reported in Agashev et al. (2018), with a standard deviation of 5%. While bulk major-element compositional variations are largely accommodated by both garnet and clinopyroxene compositions (Aulbach and Jacob 2016), a single mode is unlikely to be representative of the true variation, and, in particular, pyroxenites may have higher clinopyroxene/garnet ratios. Typical uncertainties in major-element contents of reconstructed bulk rocks for a 10% modal uncertainty are displayed in Figure 38.

From the compilations in Figure 38, median whole-rock FeO contents for inclusions (11.8 wt%) are higher than for diamondiferous eclogites (9.8 wt%), and most similar to barren high-Ca eclogites (12.4 wt%), whereas those in high-Mg and gabbroic eclogites are much lower (9.7 and 9.4 wt%) (Appendix Table 3). Conversely, whole-rock MgO contents for inclusions (10.1 wt%) are lower than those for diamondiferous eclogites (11.5 wt%), and are lowest for barren high-Ca eclogites (9.6 wt%) and, by definition, highest for high-Mg eclogites (14.8 wt%). Also by definition, CaO contents are highest in high-Ca eclogites (12.5 wt%), whereas diamondiferous eclogites have 11.6 wt% and inclusions have 10.8 wt%. For other oxides not shown in Figure 38, the whole-rock Al_2O_3 contents of inclusions (16.3 wt%) and diamondiferous eclogites (16.5 wt%) are most similar to gabbroic eclogites (16.3 wt%), and higher than in other barren eclogites, the lowest values being observed in high-Mg eclogites (14.7 wt%). Similar variations are mirrored by Na_2O contents: 2.5 wt% for inclusions and diamondiferous eclogites, 2.3 wt% for gabbroic and high-Ca eclogites, and only 1.6 wt% for high-Mg eclogites. Whole-rock Cr_2O_3 contents are lowest for high-Ca eclogites (0.05 wt%) and inclusions (0.06 wt%), and highest for high-Mg eclogites (0.14 wt%).

Whole rock reconstruction for pyroxenitic inclusions shows FeO contents (11.1 wt%) that are higher than for diamondiferous pyroxenites (7.0 wt%) or barren pyroxenites (7.3 wt%), as applies to eclogites. Conversely, MgO contents are lowest for inclusions (15.9 wt%) and highest for diamondiferous pyroxenites (18.6 wt%), whereas CaO (11.6–11.7 wt%) and Cr_2O_3 contents (0.28–0.30 wt%) are indistinguishable and Al_2O_3 contents similar (13.4 wt% for inclusions, 13.9 wt% for diamondiferous and barren pyroxenites). Moreover, Na_2O contents are higher for inclusions (0.90 wt%) than for diamondiferous pyroxenites (0.65 wt%), but similar to barren pyroxenites (0.92 wt%). The major-element relationships of Figure 38 illustrate the offset of xenolith chemistry towards overall lower FeO and higher MgO, Cr_2O_3 and CaO contents compared to inclusions.

Chondrite-normalized median REE patterns for bulk rocks reconstructed from inclusions (Fig. 39) show strong LREE depletion, similar to reconstructed diamondiferous, high-Ca and gabbroic eclogite. The REE pattern flattens out for gabbroic eclogite and, at twice higher abundances, high-Ca eclogite. In contrast, both diamond and diamondiferous eclogite show a continuous increase towards the HREE. Low-Mg and high-Mg eclogites are characterized by much higher LREE abundances and slightly increasing abundances in the HREE. Positive Eu anomalies are common to whole rocks reconstructed from inclusions (1.05), for diamondiferous eclogites (1.16) and, by definition, for gabbroic eclogites (1.21). The other eclogite classes have lower median Eu/Eu* (0.96–0.98). Reconstruction of pyroxenites from inclusions is hampered by the small number of samples and their wide compositional variation, as described for individual minerals. The single diamondiferous pyroxenite has a REE_N pattern similar to its barren non-gabbroic counterpart (Fig. 39). Barren gabbroic pyroxenite shares positive Eu/Eu* and low HREE with barren eclogite, but is LREE-enriched.

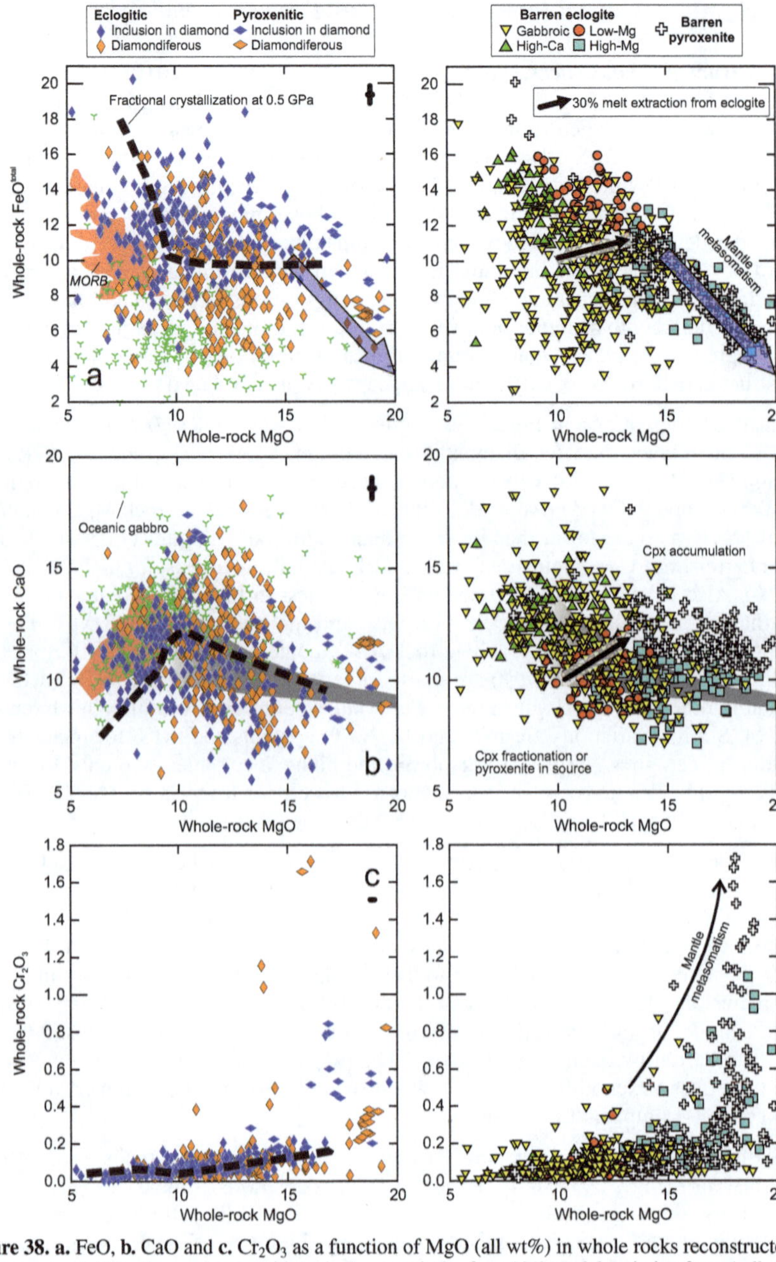

Figure 38. a. FeO, **b.** CaO and **c.** Cr_2O_3 as a function of MgO (all wt%) in whole rocks reconstructed from 55% garnet and 45% clinopyroxene, with typical uncertainties for a 10% modal variation from Aulbach et al. (2020a; indicated in top right of each graph in the left panel). "Mantle metasomatism" trend in a. reflects addition of high-temperature pyroxene from carbonated, ultramafic melt (Aulbach et al. 2020a). **Orange field in the left panels** encompasses MORBs (crystallising at ~0.05 GPa) from Jenner and O'Neill (2012), **green symbols** show gabbros (Eu/Eu* > 1.05) from PetDB (www.earthchem.org/petdb), and the **thick stippled line** is for fractional crystallization of picrite at a pressure of 0.5 GPa from Aulbach and Jacob (2016). The **grey field in b.** shows expected compositions for differentiating primitive mantle-derived melts, with higher and lower CaO suggestive of clinopyroxene accumulation and fractionation, respectively (Herzberg and Asimow 2008). **Right panels for a. and b.** show **arrows** corresponding to the quantitative compositional change after 30% melt extraction from eclogite, as calculated from experimental melting residues (see Aulbach and Jacob 2016). Inclusions and diamondiferous xenoliths on left, barren xenoliths on right. High Cr_2O_3. **c.** is associated with intense mantle metasomatism, where almost exclusively barren xenoliths occur.

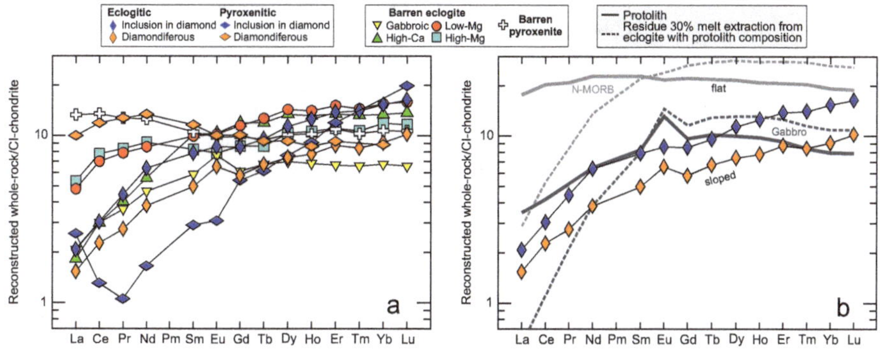

Figure 39. a. Chondrite-normalized REE patterns of whole rocks reconstructed from inclusions and xenoliths. Median values per class are shown for clarity. Only a single pyroxenite inclusion pair has reported Tm and Yb, which causes a spurious spike in the median pattern and is therefore not considered. **b.** Patterns of inclusions and diamondiferous eclogites, with positive slopes throughout the REE. Shown for comparison, NMORB (Gale et al. 2013) and average gabbro (Benoit et al. 1996), and hypothetical eclogites with NMORB and gabbro composition that lost 30% partial melt (from Aulbach et al. 2020b; **dashed lines**), with flat MREE–HREE patterns. CI-Chondrite of Sun and McDonough (1989).

Thermobarometry for eclogitic/pyroxenitic inclusions, and kinetically inhibited element redistribution

Some thoughts on inclusion thermobarometry. Accurate barometry for eclogitic and pyroxenitic garnet–clinopyroxene pairs remains elusive (cf., Nimis 2022), although some formulations have been proposed (e.g., Beyer et al. 2015; Ashchepkov et al. 2017b). In most instances, the thermal gradient of the regional lithospheric mantle has been independently determined from peridotitic minerals or assemblages, so that pressures can be estimated by solving T_{KR88} iteratively with the appropriate gradient, assuming equilibration to the conductive geotherm (e.g., Aulbach et al. 2020a). Because in xenoliths, minerals used for geothermobarometry remain in contact and equilibrate to ambient mantle temperatures until entrainment, they reflect the thermal state of the lithosphere at the time of kimberlite emplacement. The same should apply to touching garnet–clinopyroxene pairs included in diamond. Thus, for xenoliths and touching inclusions, T_{KR88} is solved iteratively with geothermal gradients parameterized from Hasterok and Chapman (2011; HC11), corresponding to surface heat flows of 38–35 mW/m^2. These model geotherms either correspond to those directly determined based on peridotitic assemblages or clinopyroxene relative to HC11 geotherms, as is the case for Koidu, West Africa. In other instances, they were roughly translated from the older Pollack and Chapman (1977) family of geotherms as quoted in older literature (Appendix Table 1). Application of iterative solutions moves a fraction of xenoliths to temperatures < 800 °C, and results in pressures > 8 GPa (> 250 km) for high-temperature samples equilibrated to cool geotherms (35 mW/m^2). Keeping all solutions, this results in average pressures of 4.9 ± 1.5 GPa and temperatures of 950 ± 110 °C for the small number of touching eclogitic inclusions (n = 15). These values compare relatively well with 5.2 ± 1.2 and 970 ± 130 °C for 228 diamondiferous eclogites (37 mW/m^2). With touching pyroxenite inclusions (n = 8) average pressures and temperatures vary between 3.8 ± 0.5 GPa and 860 ± 80 °C (38 mW/m^2), whilst 14 diamondiferous pyroxenites yield 4.1 ± 0.7 and 740 ± 100 °C (36 mW/m^2). Of the latter, ten are from the northern Slave craton where the lithosphere is characterized by a cold geotherm (Grütter 2009).

The appropriate choice of a geotherm for non-touching inclusions, which last equilibrated at the time of encapsulation, is less certain. As outlined above, eclogitic diamond formation commenced around 3 Ga, when lithospheric conductive gradients would have been higher. One could assume that the lithosphere experienced only cooling, and that touching and non-touching inclusions formed over the same pressure range. The conductive geotherm for iterative calculations is then chosen such that the median pressure for non-touching inclusion pairs equals that of touching inclusions. According to Hasterok and Chapman (2011), this situation occurs when the geotherm corresponds to a surface heat flow of 38 mW/m^2, and 220 non-touching eclogitic inclusions give values of 4.9 ± 0.8 GPa and 1040 ± 130 °C. The resultant cooling by only some 100 °C is consistent with stabilization of cool continental lithosphere by the Mesoarchean, before substantial eclogitic diamond formation. In addition, pressure–temperature conditions for non-touching inclusions can be underestimated if they experienced decompression due to extension, as suggested for the Kimberley area in the Kaapvaal craton (Phillips et al. 2004; Stachel and Luth 2015) and the western Zimbabwe craton margin (Aulbach et al. 2017). A comparison of Figure 36a and 36b shows a marked difference in the determined temperatures using pre-set and iteratively determined estimates.

Age constraints indicate that at least some eclogitic diamond suites (Slave, Kaapvaal) formed in close temporal relationship to major subduction events (for details on suggested diamond forming processes during subduction, see further below). If so, they could have followed ancient subduction geotherms, different from those in the coeval continental lithosphere or modern subduction zones. These thermobaric gradients must have been lower than those obtained for exhumed ≥ 2 Ga metamorphic rocks (> 375 °C/GPa; Brown and Johnson 2018), in order to allow a pressure–temperature path consistent with diamond stability. A possible Archean slab gradient is roughly reconstructed in Figure 40a, based on following observations: Orogenic eclogite with an oceanic crustal protolith from the Archean Belomorian Belt indicates peak conditions of 1.85 GPa and 710 °C (Li et al. 2015). Elastic geobarometry on quartz inclusions in garnet from mantle eclogite at Mir (Siberian craton) yielded a pressure of 3 GPa at 850 °C, and was interpreted as reflecting entrapment during prograde metamorphism corresponding to a surface heat flow of 40–50 mW/m^2 (Alvaro et al. 2020). This may be linked to Archean subduction, as eclogite xenoliths from the Siberian craton have yielded Mesoarchean ages (Appendix Table 1). Combined with the pressure–temperature estimates for non-touching inclusions, these provide a predictive pressure–temperature path for subducting Archean oceanic crust (Fig. 40a).

In contrast to the low temperatures obtained from iterative solutions on non-touching inclusions (average 1040 ± 120 °C), constraints from the aggregation of N impurities in eclogitic diamonds, which is a function of N concentration, temperature and age, indicate average mantle residence temperatures of 1140 ± 50 °C for a 3 Ga eclogitic diamond age (Stachel and Harris 2008). As N-based temperatures reflect time-integrated thermal states, they may include transient heating events associated with melt migration. Moreover, not all eclogitic diamonds formed during relatively cold subduction, but may instead be associated with (renewed) growth during later melt metasomatism (Stachel and Luth 2015), as supported by multiple Proterozoic eclogitic diamond formation events based on inclusion dating (Wiggers de Vries et al. 2013; Gress et al. 2021), and by ca. 1.6 Ga eclogitic diamonds from Argyle recording atypically high temperatures (Stachel and Harris 2008).

Inhibited minor element redistribution in touching inclusions. As described above, the distribution of some minor and trace elements between clinopyroxene and garnet is temperature-dependent. Thus, touching inclusions and their xenolithic counterparts should show similar systematics, whereas non-touching inclusions might differ due to higher equilibration temperatures. Specifically, the distribution coefficient of both Na_2O and TiO_2 between clinopyroxene and garnet decreases with increasing temperature, as higher concentrations of these elements are accommodated in garnet (Aulbach 2020). If this should be the case, then D_{Na_2O,TiO_2} for clinopyroxene–garnet would be expected to be higher for touching inclusions than for non-touching inclusions. However, when D_{Na_2O} is considered as a function of T_{KR88} calculated at 5GPa, it becomes evident that several touching inclusions

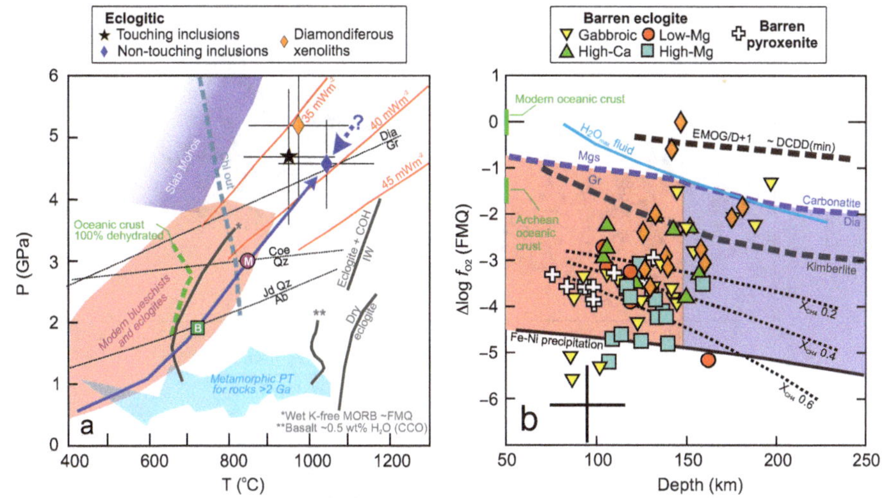

Figure 40. a. Pressure (GPa)–temperature (°C) diagram showing average values and 1σ error bars for iterative solutions for touching and non–touching inclusions and diamondiferous xenoliths. The geotherm for non–touching inclusions was chosen such that they give an average pressure similar to touching inclusions although this may yield underestimated pressure–temperature conditions if the latter experienced decompression in addition to cooling (indicated by **dashed arrow with "?"**). Also shown are: solidi (solid grey lines) for dry, nearly–dry or wet eclogite at various oxygen buffering equilibria (CCO carbon–carbon monoxide, FMQ fayalite–magnetite–quartz, IW iron–wüstite) from Schmidt and Poli (2014), Litasov et al. (2014) and Ziaja et al. (2014); a **curve** depicting 100% dehydration of oceanic crust (Laurie et al. 2013); oceanic and continental high–pressure fields for rocks in modern subduction zones (Penniston–Dorland et al. 2015); and the peak pressure–temperature conditions of metamorphic rocks older than 2 Ga from Brown and Johnson (2018). Shown in addition is a peak pressure–temperature for Neoarchaean orogenic eclogites ("B" – Belomorian Belt, Russia) from Li et al. (2015) and the pressure–temperature of entrapment of quartz inclusions in garnet from a presumably Archaean eclogite xenolith from Mir ("M" – Siberian craton) from Alvaro et al. (2020); these points define a suggested subduction path for Archaean oceanic crust, shown by the **thick dark blue arrow**. Model conductive geotherms corresponding to surface heat flows of 35, 40 and 45 mW/m^2 from Hasterok and Chapman (2011), the diamond–graphite boundary from Day (2012), and the coesite–quartz boundary and albite–jadeite–quartz reaction after Brown and Johnson (2018). The chlorite–out curve and pressure–temperature conditions for warm slab Mohos (**purple field**) allowing deserpentinization within the diamond stability field are from Syracuse et al. (2010). **b.** Oxygen fugacity estimates (fO_2) based on the eclogite oxybarometer of Stagno et al. (2015) for mantle eclogite and pyroxenite xenoliths relative to the fayalite–magnetite–quartz buffer (expressed as $\Delta\log fO_2$ (FMQ)) as a function of depth (samples in Aulbach et al. 2019b and references therein, plus those from Burness et al. 2020 and Mikhailenko et al. 2020). Data for diamondiferous eclogites are from a single study (Mikhailenko et al. 2020) and may not be representative. Estimates for modern and Archaean oceanic crust are from Aulbach et al. (2019b). Shown for comparison are the fO_2-dependent fraction of CH_4 in CHO fluids, and conditions with maximum H2O in the fluid (from Luth and Stachel 2014), the stability of carbonatite, carbonated silicate melt with 10 mol% CO_2 (~kimberlite) vs. diamond and graphite (from Stagno et al. 2013), and the minimum fO_2 for the reaction dolomite + coesite = diopside + diamond (DCDD), which lies at least one log unit above the reaction enstatite + magnesite = olivine + graphite/diamond (EMOG/D; Luth 1993; Stachel and Luth 2015), all applicable for a 40 mW/m^2 geotherm as calculated by the cited authors.

in eclogitic and pyroxenitic diamonds have low *D* values at a given temperature (Fig. 41a). These low *D* values are more similar to those of (higher-temperature) non-touching inclusions than to (similar-temperature) diamondiferous xenoliths, which suggests that at least minor element redistribution for inclusions is inefficient. Figure 41b shows that a similar situation applies for TiO_2. The decreasing TiO_2 solubility in both garnet and clinopyroxene with decreasing temperature leads to exsolution of rutile, with the implication that this mineral is more likely to be observed in lower-temperature specimens, and that TiO_2 concentrations

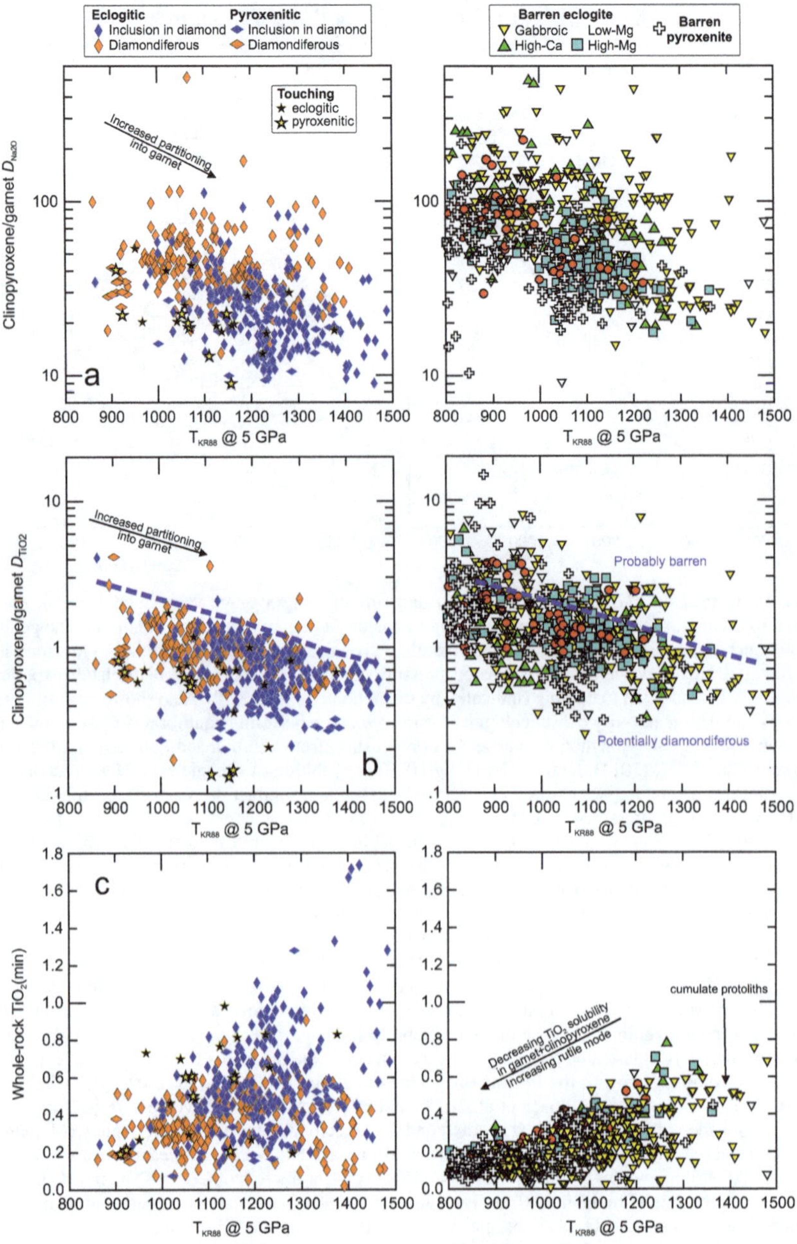

Figure 41. Log distribution of **a.** D_{Na_2O} and **b.** D_{TiO_2} between clinopyroxene and garnet as a function of T_{KR88} at 5 GPa. High temperatures (pressures along conductive geotherms) lead to increased partitioning of Na_2O and TiO_2 into garnet. High D values at a given temperature may be associated with barren xenoliths, whereas those with low D values may be diamondiferous (indicated by **blue stippled line**). **c.** Minimum TiO_2 content (wt%; reconstructed without rutile) as a function of T_{KR88} at 5 GPa, illustrating that inclusions have higher average TiO_2 contents consistent with higher temperatures compared to xenoliths, which favors TiO_2 uptake in both garnet and clinopyroxene, whereas free rutile is required to accommodate bulk rock TiO_2 contents at low temperatures (Aulbach 2020). Many touching inclusions have high TiO_2 at their estimated temperatures, presumably because they have not purged excess TiO_2 from their lattice, whereas continued Mg–Fe exchange allowed them to equilibrate to lower average temperatures than isolated inclusions. The **vertical arrow in right panel** illustrates the effects of accumulation in low-pressure protoliths, leading to low bulk TiO_2 contents in particular in gabbroic eclogites. Inclusions and diamondiferous xenoliths on **left**, barren xenoliths on **right**.

in bulk rocks are increasingly underestimated when only garnet and clinopyroxene are considered (Aulbach 2020). Minimum (rutile-free) bulk-rock TiO_2 contents reconstructed from non-touching inclusions as a function of T_{KR88} loosely follow the xenolith trend, but with higher average contents consistent with higher entrapment temperatures and rutile solubility (Fig. 41c). However, there is also an off-set to higher TiO_2 at a given temperature compared to xenoliths, which may reflect the inability of included garnet to purge the excess TiO_2 from its lattice, first via nucleation of lamellae and then by recrystallization and formation of discrete rutile grains. One may speculate that this is due to the dryness of the inclusions where structural hydrogen, once lost to the inclusion–diamond interface (Nimis et al. 2016), is not replenished in contrast to matrix minerals. Conversely, the similar temperature estimates for touching inclusions and diamondiferous eclogites suggest that the redistribution of MgO and FeO, on which T_{KR88} is based, is more efficient.

Source rocks of eclogitic and pyroxenitic diamond

Source rocks of eclogitic diamond. Inclusions in diamonds are expected to reflect the mineralogy and composition of their various substrates (Boyd and Gurney 1986; Meyer 1987), in this case mantle eclogite. With respect to major elements, eclogitic inclusions cover nearly the entire range displayed by xenoliths, but are more common at bulk-rock MgO < 13 wt%. They seemingly overlap mostly with high-Ca and gabbroic eclogites (Fig. 38). Mantle eclogites with positive Eu anomalies (gabbroic class) and low total REE abundances (ΣREE) derive from sources that experienced plagioclase–olivine accumulation, partially followed by reaction with melt, which explains the wide range of compositions in this group (Aulbach and Jacob 2016). The fractionated melts develop negative Eu anomalies. Broadly, high-Ca and some high-Mg eclogites have more- and less-differentiated basaltic protoliths, respectively, whereas low-Mg eclogites may require Fe-rich protoliths. Inclusions follow the expected systematics of low-pressure cumulates vs. differentiated melts with respect to Eu/Eu* and ΣHREE (Fig. 42a; only HREE are summed to exclude effects of LREE loss via melt depletion and gain via metasomatism). Also, 65% of unpaired garnet and 58% of unpaired clinopyroxene inclusions have Eu/Eu* > 1.05. The proportion of gabbroic xenoliths (i.e., showing positive Eu anomalies) amongst diamondiferous xenoliths with reported REE contents is much higher at 80%. These numbers agree well with estimates that some 70% of modern oceanic crust is constituted by a gabbroic layer, the remainder being pillows, flows and sheeted dike complexes where magma freezes in layers and conduits (White and Klein 2014). Conversely, as noted for cratonic eclogite (Aulbach and Arndt 2019a), a mantle cumulate origin is precluded because eclogitic inclusions do not show the predicted positive correlation of Y with MgO, but weak negative correlation instead ($r^2 = 0.25$, $n = 21$; Fig. 42b).

Following the low-pressure igneous stage at spreading ridges, low- and high-temperature seawater alteration produces minerals with higher and lower $\delta^{18}O$, respectively, compared to the mantle (Muehlenbachs and Clayton 1972a,b). These signatures are still recognizable in many mantle eclogites (Jacob 2004), and in garnet inclusions (Fig. 35). Due to the cross-over of alteration signatures and the oxygen isotope mantle range at intermediate temperature conditions of crustal alteration, or just a lack of alteration, mantle-like $\delta^{18}O$ is not proof against low-pressure protoliths (Smart et al. 2012). Subsequent subduction and metamorphism of oceanic crust causes dehydration and loss of a LREE-enriched silicic partial melt in some, though not all eclogites (Schulze et al. 2003), especially in warm slabs. Eclogitic diamond sources now residing at temperatures $\gtrsim$ 800 °C would have crossed the solidus of wet metabasalt (Schmidt and Poli 2014), which is supported by the pressure–temperature paths of exhumed ancient orogenic eclogites (Fig. 40a). Experimental phase relationships further indicate that this entails loss of SiO_2 and Na_2O, accompanied by an increase in MgO, FeO and CaO, while the presence of substantial amounts of garnet in the residue leads to compatible behaviour of

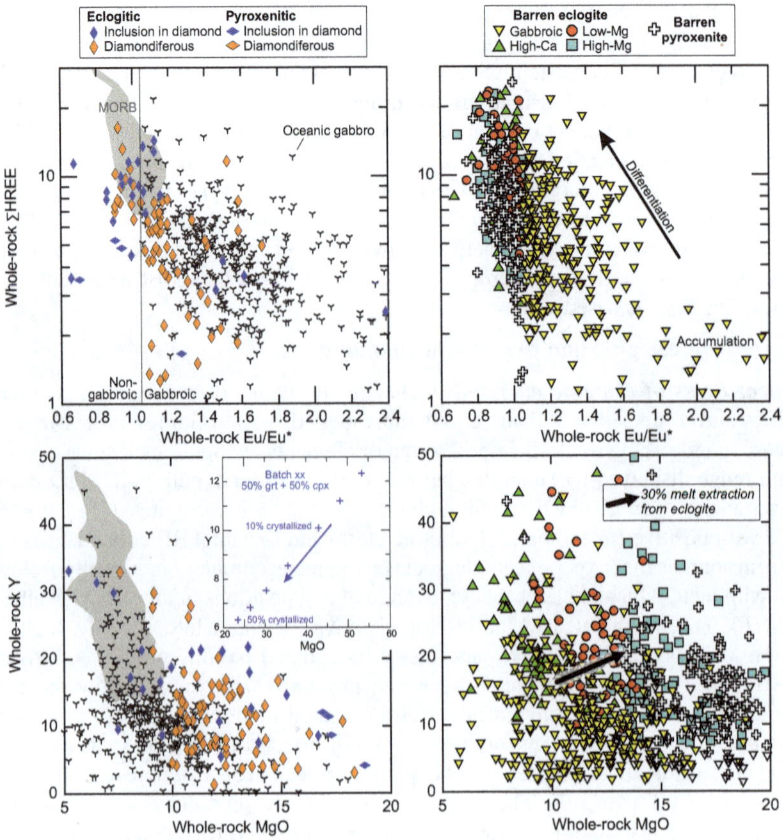

Figure 42. a. ΣHREE (ppm, summed from Tb to Lu) as a function of Eu/Eu* (chondrite-normalized Eu / (Sm × Gd)$^{1/2}$; CI-chondrite of Sun and McDonough 1989) in reconstructed bulk rocks, illustrating the effects of low-pressure olivine + plagioclase accumulation and evolution of residual melts. Shown for comparison is the field for MORB (Jenner and O'Neill 2012) and gabbros (PetDB database). **b.** Y (ppm) as a function of MgO (wt%) in reconstructed bulk rocks, illustrating that mantle eclogites and inclusions do not follow the trend (**inset in left panel**) expected for high-pressure crystallization ("xx") of picritic melts (calculated trend from Aulbach and Arndt 2019b). **Arrow in right panel** shows quantitatively the compositional change associated with extraction of 30% melt from eclogite. Inclusions and diamondiferous xenoliths on **left**, barren xenoliths on **right**.

the HREE and Y, and incompatible behaviour of the LREE during melting (Aulbach and Jacob 2016). Indeed, high MgO, low SiO_2 and Na_2O as well as low LREE/HREE in both inclusions and mantle eclogites have been interpreted as signatures of melt loss from eclogite (e.g., Ireland et al. 1994; Jacob 2004; Stachel et al. 2004), although low SiO_2 and LREE/HREE can also be a primary signature of oceanic crustal cumulates (Aulbach and Jacob 2016). The strong LREE-depletion in high-Ca and gabbroic eclogite xenoliths (Fig. 39) may causally reflect their higher original Na_2O content, because the jadeite component in clinopyroxene, together with quartz/coesite and H_2O, participate in melt production (Laurie and Stevens 2012). A higher initial Na level will ultimately cause higher levels of depletion after melt loss. Again, the majority of bulk rocks reconstructed from inclusions have LREE-depleted REE_N patterns most similar to gabbroic and high-Ca eclogites, but with some important differences (Fig. 39) that provide insights into diamond formation processes discussed below.

Depending on the fertility after initial dehydration melting at low pressure and on the redox state of the eclogite and infiltrating fluid, oceanic crust may experience second-stage partial melting during fluid ingress (Fig. 40a). Diamondiferous eclogites and inclusions do not extend to the extremely melt-depleted compositions (very low Ce/Yb_{NMORB}) observed in some barren eclogites, perhaps due to the failure of diamond to crystallize from fluids in such depleted rocks. Otherwise, as noted above, the proportion of inclusions (~60%) and diamondiferous eclogites (80%) with gabbroic protoliths is similar to that in modern oceanic crust, indicating that diamond formation occurs in all portions of the oceanic crust, with a preference for intrusive sections indicated by diamondiferous eclogites. Given mechanisms for eclogitic diamond formation discussed below, this need not, however, reflect the extent to which C is originally distributed in these rocks. Curiously, few inclusions have strongly gabbroic signatures (low whole-rock Y or HREE abundances, strong positive Eu anomalies, Fig. 42), possibly indicating that deep crustal cumulate sections are less propitious substrates for the formation of inclusion-bearing diamonds.

Source rocks of pyroxenitic/websteritic diamond. For inclusions in diamond, the websteritic paragenesis is compositionally less well defined than the eclogitic one, which probably reflects multiple formation mechanisms (Stachel and Harris 2008). Similarly, pyroxenite xenoliths, which include rocks referred to as "high-Mg eclogites" in the literature, have multiple suggested origins (Aulbach and Jacob 2016). Bimineralic pyroxenite xenoliths grading into orthopyroxene-bearing websterites and, for inclusions in diamond, the websteritic inclusion suite from Venetia have been interpreted as the products of the interaction of slab-derived silicic melt with peridotite (Aulbach et al. 2002, 2007; Smit et al. 2014b), based on the similarity to experiments where peridotite was hybridized with eclogite-derived melts (Yaxley and Green 1998). On the other hand, a large proportion of bimineralic pyroxenite grading into high-Mg eclogites (20–40% of most eclogite-bearing xenolith suites) shows evidence for an origin via metasomatic interaction of mantle eclogite with a small-volume carbonated ultramafic melt similar to kimberlite (Aulbach et al. 2020a). Some pyroxenitic inclusions share major element features with the metasomatized group. This includes high MgO and low FeO (Fig. 38a), high CaO (Fig. 38b) and high Cr_2O_3 at high MgO (Fig. 38c), all of which have been interpreted as reflecting addition of a high-temperature pyroxene from the metasomatic melt (Aulbach et al. 2020a). A single diamondiferous pyroxenite with trace element data (from the Siberian craton) has a pattern similar to median barren pyroxenite (Fig. 39). Diamond in pyroxenite xenoliths from the northern Slave craton, where kimberlite-like metasomatism has been recognized, may have a similar origin (Smart et al. 2009). However, the few inclusions in pyroxenitic diamonds with REE data have spoon-shaped patterns with steep positive slopes in the $MREE_N$–$HREE_N$ that cross-cut those of pyroxenite xenoliths (Fig. 39). Similarly high resultant La_N/Pr_N and Lu_N/Gd_N are seen in only a few pyroxenite xenoliths. Finally, some pyroxenitic inclusions have lower MgO contents combined with relatively high FeO (Fig. 38a), and low ΣHREE contents (Fig. 42b), perhaps reflecting a deep crustal cumulate protolith, with insufficient Na_2O to metamorphose into eclogite with Jd-rich clinopyroxene.

The dearth of trace-element data makes it difficult to confidently distinguish between different scenarios for the origin of pyroxenitic diamond source rocks. As applies to high-Mg eclogite, an origin of pyroxenitic inclusions as direct mantle cumulates is precluded by the lack of the predicted element relationships proposed for precipitation of substantial amounts of garnet (e.g., correlated Y-MgO; Fig. 42b). The small proportion of pyroxenitic diamond (15%, of all pyroxenitic–eclogitic inclusions, based on paired and isolated clinopyroxene) suggests that this rock type is a subordinate diamond source. Note that Stachel and Harris (2008), based on different classification criteria (elevated Mg# and Cr_2O_3, low Ca content in garnet, presence of orthopyroxene), obtain a relative abundance for websteritic inclusions (amongst lithospheric diamonds) of only 2.3%, as opposed to 5.3% in this study.

Diamond formation mechanisms in eclogitic/pyroxenitic source rocks

Diamond formation from saline fluids. A group of eclogitic inclusions has very high Na_2O in garnet and K_2O in clinopyroxene, occupying a field where diamondiferous or barren eclogite does not occur (Fig. 33b). Garnets with Na_2O content > 0.4 wt% are off-set from trends of increasing Na concentrations with increasing temperature (inset in Fig. 33b), and therefore reflect a bulk compositional effect, as previously noted by Grütter and Quadling (1999). The same interpretation applies for K-rich clinopyroxenes, and therefore, these inclusion compositions reflect Na–K-rich bulk rocks. They are not associated with any major or trace element signature except low Cr_2O_3 and high Sr contents, although not enough trace-element data are available to further establish the nature of the metasomatic agent. This may point to diamond precipitation from saline high-density fluids, which are observed in (though not limited to) eclogitic diamond (Izraeli et al. 2001, 2004), and may be ultimately linked to subducted slabs based on observational data (Weiss et al. 2014, 2015, 2022), modelling (Sverjensky and Huang 2015) and experiments (Shatskiy et al. 2019). Omphacite occurring with saline high-density fluid in the same eclogitic diamond has elevated Na_2O (6.7 wt%) and K_2O (0.6 wt%; Weiss and Goldstein 2018). Any medium involved in the formation of diamond cannot be dissociated from the substrate in which the diamond forms (i.e., mineral inclusions and eclogite have the same composition at the locus of encapsulation in diamond). Therefore, the exclusive occurrence of high Na_2O in garnet and high K_2O in clinopyroxene inclusions in diamonds (Fig. 33b), whereas eclogite xenoliths lack these signatures in their garnet or clinopyroxene, indicates that such compositions do not survive during subsequent mantle eclogite evolution. This suggests efficient removal of this signature from the substrate during subsequent interaction with Na_2O- and K_2O undersaturated fluids or melts. If this interpretation is correct, 14% of eclogitic diamond (containing garnets with Na_2O content > 0.4 wt%) may have formed from saline high-density fluids.

Diamond formation from small-volume melts or COH fluids. Mantle eclogites are dominantly LREE-depleted, indicative of partial melt loss without subsequent re-enrichment. This is gauged by normalization to a putative low-pressure protolith with flat REE_N patterns, such as NMORB, and equally applicable to oceanic crust formed from a higher melt fraction, such as picrite. A sizable proportion of bulk high-Mg eclogites and pyroxenites (37%) show LREE-enrichment (Nd/Yb_{NMORB} or $La/Sm_{NMORB} > 1$), accompanied by a lowering of Jd (Fig. 43) and also of Ca# in garnet, which are both hallmarks of metasomatism by kimberlite-like melt (Aulbach et al. 2020a). As the lowering of garnet Ca# is accompanied by decreased partitioning of many incompatible elements into garnet, this leads to weak signatures of metasomatism based on trace elements in garnet alone (Aulbach et al. 2020a). From an fO_2 perspective, carbonated silicate melts, such as kimberlite, are stable to lower fO_2 than pure carbonatite (Fig. 40b), allowing for coexistence of melts with dilute carbonate content and diamond over a large fO_2 range (Stagno et al. 2015; Mikhailenko et al. 2020). The ability of kimberlite-like melt to be stable with eclogite to lower fO_2 than with peridotite implies carbon mobility over a wider range of conditions, and may thus help explain the overabundance of eclogitic relative to peridotitic diamonds (Stagno et al. 2015). In addition, small-volume melts (like kimberlite) or fluids imparting LREE-enrichment are important drivers of diamond formation in peridotite source rocks (Stachel et al. 2004; Stachel and Luth 2015; this work). However, less than a quarter of eclogitic-pyroxenitic bulk rocks reconstructed from inclusions (23%) show LREE-enrichment and associated high $LREE_N/HREE_N$ (Fig. 43), the hallmark of mantle metasomatism by diamond-forming CHO-fluids and small-volume melts in peridotite. Thus, although the trace-element database for eclogitic/pyroxenitic inclusions in diamond is limited, it appears that diamond formation in the eclogite system is not dominantly linked to a typical mantle metasomatic signature.

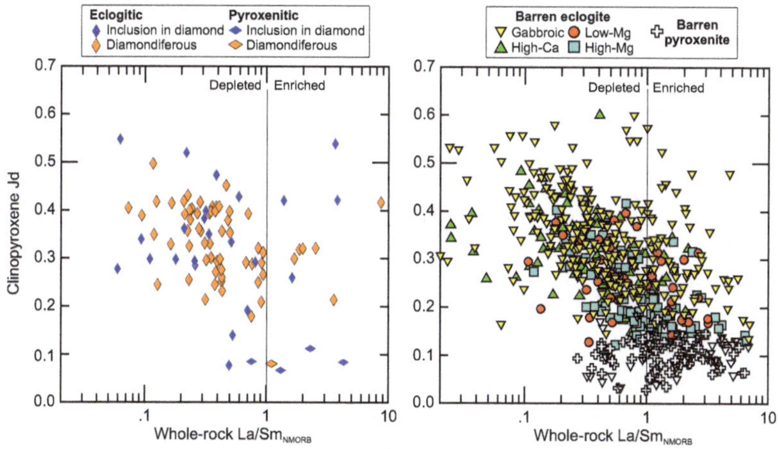

Figure 43. Jadeite content in clinopyroxene as a function of NMORB-normalized La/Sm of reconstructed whole rocks, illustrating enrichment and depletion. NMORB of Gale et al. (2013). Inclusions and diamondiferous xenoliths on **left**, barren xenoliths on **right.**

Diamond formation during subduction. Age distributions (Smit et al. 2022) suggest formation of the first eclogitic diamond populations in or associated with Archean (Kalahari Craton) and Paleoproterozoic (Slave craton) subduction zones, perhaps linked to the onset of plate tectonics at ~ 3 Ga (Shirey and Richardson 2011). This initial formation was followed by renewed growth or formation of eclogitic diamond during Proterozoic tectonothermal events, up to at least the Mesozoic (Timmerman et al. 2017). The coincidence of eclogite xenolith ages with eclogitic diamond ages and with craton amalgamation in the Kaapvaal craton (ca. 3.0–2.9 Ga) and with collision at the craton margin in the Slave craton (ca. 1.9 Ga) is most readily explained by interaction of oceanic crust with C-bearing fluids liberated in subduction zones. The median composition of diamondiferous eclogites is most similar to that of barren gabbroic eclogites with respect to strong positive Eu anomalies and LREE depletion (Fig. 39). The median composition of whole rocks reconstructed from eclogitic inclusions does not have as strong a gabbroic character and is more similar to barren high-Ca eclogite. The $MREE_N$–$HREE_N$ patterns of bulk rocks reconstructed from both eclogitic and pyroxenitic inclusions show marked positive slopes that cut across the flatter pattern of high-Ca eclogite (Fig. 39), resulting in high NMORB-normalized Yb/Gd. This slope is unlike that of potential protoliths and can also not be achieved by extraction of partial melt (modelled compositions in Fig. 39b). Similar patterns have been ascribed to disequilibrium processes involving large fluid volumes, such as those encountered in subduction zones (Aulbach et al. 2020b). In particular, serpentinite has been implicated as a major fluid source in subducting slabs (Rüpke et al. 2004), and disequilibrium REE_N patterns with LREE–MREE depletion and HREE enrichment have been observed in orogenic eclogites affected by fluids derived from deserpentization (Beinlich et al. 2010). In this scenario, the low LREE contents of diamondiferous eclogites and inclusions need not be (solely) due to melt extraction, but can result from auto-metasomatism (i.e., fluid indigenous to the slab), whereby the compatible HREE partition into garnet, and the incompatible LREE are removed with the fluid (Aulbach et al. 2020b).

The viability of a scenario of diamond formation via interaction of slab-derived fluids with oceanic crust is supported by thermal modelling, which indicates that carbon and volatiles in subducting slabs can escape shallow decarbonation and dehydration (Shirey et al. 2021). It can be assessed by considering phase relations of oceanic slabs and the stability of diamond vs. other carbon minerals as a function of pressure, temperature and oxygen fugacity. Oceanic crust dehydrates, accompanied by loss of a small amount of melt, at forearc depth (Spandler and Pirard 2013), where diamond is not stable (Fig. 40a). The small melt fraction explains the persistence of coesite (which participates in the melting reaction) in some mantle eclogites (Schulze et al. 2003). In contrast, pulsed dehydration of deep crust and oceanic mantle, which follows cooler subduction geotherms than slab tops (John et al. 2012; Laurie and Stevens 2012; Spandler and Pirard 2013), could flush oceanic crust within the diamond stability field. If ancient oceanic crust followed a cool subduction path, aided by the proximity of early-cooled cratonic lithosphere (Eaton and Perry 2013), it could enter the diamond stability field at pressures < 4 GPa (Fig. 40a). Whether the high-pressure interaction of eclogite with fluids causes partial melting, as suggested by Stachel and Luth (2015), depends on the refractoriness of the oceanic crust after initial dehydration melting and the redox state of the fluid, which determine the eclogite solidus.

Carbonate, disordered carbonaceous matter, and CaO, are efficiently mobilized in crustal COH fluids (Frezzotti et al. 2011; Menzel et al. 2020; Vitale Brovarone et al. 2020). These materials may be one factor contributing to a preponderance of Ca-rich compositions amongst inclusions and diamondiferous eclogites. As graphite and dolomite are observed in eclogite-hosted veins (Li et al. 2020; Zhu et al. 2020), and microdiamond occurs in fluid inclusions in oceanic subduction zones (Frezzotti et al. 2014), clearly subduction-related fluids can mobilize and supersaturate in carbon or carbonate. Almost all mantle eclogites have fO_2 where simple CHO fluids are near the H_2O-maximum or on the reduced side, increasingly rich in CH_4 (Fig. 40b). However, fluids at high pressures are more complex, containing dissolved solids (Kessel et al. 2005) and possibly a multitude of additional hydrocarbon species (Huang et al. 2017). Since massive fluid flow is invoked in the subduction model, diamondiferous eclogites may represent a fluid-buffered system, in which reducing fluids would impose their fO_2 on eclogite. This lends support to models of diamond formation via oxygen-conserving reactions in CHO fluids (Stachel and Luth 2015), occurring during interaction of deep deserpentinization fluids with overlying oceanic crust. At the relatively low, subsolidus temperatures in a subducting slab, oxygen-conserving diamond formation from fluids may also be achieved due to a drop in pH (Sverjensky and Huang 2015).

The small number of eclogite xenoliths reflecting fO_2 conditions above the H_2O-maximum (Fig. 40b) suggests that the stability of diamond in eclogite to 1 log unit higher fO_2 values compared to peridotite (Luth 1993) plays a subordinate role in explaining the relative overabundance of eclogitic diamonds relative to the small volume of eclogite in the lithospheric mantle. Deserpentization fluids have been inferred to be both oxidising (Debret and Sverjensky 2017) and reducing (Piccoli et al. 2019). Thus, percolation of more oxidized, CO_2-rich fluid into originally reduced eclogite would allow for diamond formation via reduction of CO_2. But the mixing of such fluids with different redox character alone would be a highly effective process. At the water maximum, only about 2% of other species (CO_2, CH_4) are permitted, and if, as an example, fluid 1 contains 4% CO_2 and fluid 2 contains 4% CH_4, they will precipitate 50% of their carbon as diamond upon mixing in an oxygen preserving reaction (Deines 1980; Luth and Stachel 2014). A CH_4-dominated fluid at lower fO_2 interacting with initially more oxidized eclogite at mantle depth might precipitate diamond coupled to reduction of Fe^{3+} in eclogite. Both interpretations are consistent with an evaluation of diamond formation mechanisms in eclogite calling for isochemical diamond precipitation from CH_4-dominated fluids or oxidation of methane on infiltration of eclogite (Smit et al. 2019b). In summary, the preponderance of diamonds forming in eclogitic over peridotitic substrates may reflect that serpentinized harzburgite can carry far more water than the crust, and, as it dewaters at greater depth than the oceanic crust, those fluids leave to form diamond in the overlying crustal portion of the slab now sampled as eclogites.

Evolution of the mantle eclogite reservoir and consequences for diamond preservation

A composite present-day mantle eclogite/pyroxenite reservoir. Inclusions in eclogitic/pyroxenitic diamonds represent the most pristine record of the mantle eclogite reservoir at the time of encapsulation, with the caveat that the diamond-forming process itself may have introduced some modification (e.g., Ca–HREE enrichment during disequilibrium interaction with dehydration fluids in subduction zones). Inclusions do not reflect the kimberlite-related late (pre-eruption) metasomatism typically affecting the mantle lithosphere (and xenoliths during transport), unless their diamond host only formed during this process, as discussed for fibrous diamonds (Boyd et al. 1987; Gurney et al. 2010). Conversely, mantle xenoliths reflect the state of the eclogite/pyroxenite source at the time of typically much later kimberlite magmatism. In order to facilitate a comparison between inclusions, diamondiferous eclogites/pyroxenites and their barren counterparts, we calculate the composition of a composite present-day mantle eclogite/pyroxenite source, taking median compositions for each of the six classes weighted by their proportion. Pyroxenite xenoliths are included in the composite as many have geochemical signatures indicative of interaction of carbonated ultramafic melt with mantle eclogite (Aulbach et al. 2020a), and they therefore constitute an integral part of the evolution of the mantle eclogite source. Median compositions of the composite eclogite, diamondiferous eclogite and bulk rocks reconstructed from eclogitic inclusions are given in Table 4.

Mantle eclogite evolution. The mantle eclogite reservoir is affected by mantle metasomatism, as also recognized in its peridotitic counterpart (e.g., Dawson 1984). It encompasses stealth effects (addition of minerals forming part of the primary mineral assemblage), modal effects (addition of clearly metasomatic minerals such as phlogopite) and cryptic effects (changes to the trace element composition such as LREE-enrichment) (Harte 1983; O'Reilly and Griffin 2013). Metasomatic effects on mantle eclogite include a reduction of the jadeite component in clinopyroxene and grossular component in garnet, a decrease in bulk-rock FeO and Al_2O_3 contents and increase in MgO, Cr_2O_3, ± SiO_2, as well as LREE-enrichment accompanied by higher Sr, Pb, Th, U, ± Zr, ± Nb, and lower Li, Cu, ± Zn (Aulbach et al. 2020a). Strong metasomatism was suggested to be accompanied by a muting of typically crustal signatures, such as non-mantle $\delta^{18}O$ and the size of Eu anomalies. Rare studies on eclogite xenoliths with inclusion-bearing diamonds have shown that the inclusions retain evidence for partial melting, while the host eclogite has been subsequently enriched in MgO, Cr_2O_3 and incompatible elements, and depleted in Na_2O and CaO by a kimberlite-like metasomatic agent (Ireland et al. 1994; Smart et al. 2012). A comparison of composite mantle eclogite with bulk rocks reconstructed from eclogitic inclusions shows that the mantle eclogite reservoir has gained 0.52 wt% CaO (+5% relative), 3.0 wt% MgO (+29%) and 0.06 wt% Cr_2O_3 (+87%), and lost 0.91 wt% Al_2O_3 (–6%), 2.2 wt% FeO (–18%), 0.58 wt% Na_2O (–24%), and 0.04 wt% K_2O (–56%). These changes are ascribed to the effects of mantle metasomatism over time. Lower minimum TiO_2 content (–0.35 wt%, –63%) in the mantle eclogite reservoir is linked to lower average temperatures (1050 °C at 5 GPa vs. 1220 °C for inclusions), and corresponding lower TiO_2 solubility in garnet, clinopyroxene and consequently reconstructed bulk rock, and higher inferred rutile modes (Aulbach 2020). With respect to trace elements, evolution of the mantle eclogite reservoir comprises an increase in Sr (+43%) and LREE abundances (e.g., +140% for La), and a dilution of Y (–22%) and HREE abundances (e.g., –42% for Lu), whereby Sm is least affected (–2%), and Zr and Hf also change little (–2%) (Table 4). These systematics are consistent with interaction with a small-volume melt that is LREE-enriched and HREE-depleted relative to bulk eclogite, such as kimberlite. The similar relative abundance of garnet with positive Eu anomalies (Eu/Eu* > 1.05) amongst inclusions (65%) and barren eclogites and pyroxenites (66%) suggests that the metasomatic evolution does not lead to significant muting of this crustal signature in garnet, whereas it is muted in clinopyroxene. Like TiO_2, lower Nb abundances in the xenoliths (–46%) are not (solely) related to metasomatism, but reflect the increased presence of accessory rutile, which strongly partitions Nb and Ta, at lower average temperatures (Aulbach 2020).

Diamond-destructive effect of mantle metasomatism. Diamondiferous xenoliths are expected to have experienced the metasomatic evolution recorded by the barren eclogite reservoir, with the important exception that they do not capture mantle processes that were diamond-destructive. A direct comparison with inclusions must consider that diamondiferous eclogites have a much stronger gabbroic signature (bulk-rock Eu/Eu*=1.16) than eclogitic inclusions (1.05). This difference may at least in part explain 1.7% higher bulk Al_2O_3, 7% higher CaO contents, nearly 20% lower FeO, and lower REE abundances (Table 4; Figs. 38a,b, 39b) in diamondiferous xenoliths. The increase in bulk MgO and Cr_2O_3 is not as marked as for barren xenoliths (+14% vs. +29% for Mg, and +10% vs. +87% for Cr), and there is no dilution of bulk Na_2O content in contrast to barren xenoliths (Table 4). Similarly, in diamondiferous xenoliths, Sr is less enriched (+12%) than in barren xenoliths (+43%), and the strong positive slope in the REE is retained (i.e., high Yb/Gd_{NMORB}). These systematics are taken to indicate that strong mantle metasomatism, with greater enrichment in MgO, Cr_2O_3 and some incompatible elements leads to diamond destruction, via dissolution into C-undersaturated melts. The intermediate TiO_2 content in diamondiferous eclogite (0.36 wt%; a minimum estimate as rutile is not considered) is consistent with its intermediate median temperature (1130 °C at 5 GPa). In contrast, for unclear reasons, median Nb abundances are highest in diamondiferous eclogites (0.24 ppm).

The higher median temperatures (pressures along conductive geotherms) recorded in diamondiferous compared to barren eclogites suggest that the low-pressure population was thinned out by diamond-destructive mantle metasomatism that particularly affects the mid-lithosphere, close to the lower stability limit of diamond (Aulbach et al. 2020a). The low Na_2O contents in garnet from barren eclogite and pyroxenite xenoliths therefore reflect the superposition of two effects: Firstly, the effect of lower average temperatures, and pressures along conductive geotherms, where diamond is not stable, and secondly, the effect of diamond-destructive mantle metasomatism, during which Na_2O is diluted (Aulbach et al. 2020a). Diamond destruction during pyroxenitization, with a marked lowering of bulk Na_2O and Jd component associated with strong metasomatism by kimberlite-like melt, may also explain why diamondiferous pyroxenite xenoliths make up only 6% of all diamondiferous xenoliths. Indeed, the diamond-destructive effect of kimberlite melt at mantle depth has been experimentally demonstrated (Fedortchouk et al. 2017; 2019). Thus, the enrichment in LREE over MREE (La/Sm) in almost a quarter of bulk eclogites and pyroxenites reconstructed from inclusions is probably not linked to such mantle metasomatism, except in a few localities (e.g., northern Slave craton; Smart et al. 2009), suggesting a delicate balance between diamond saturation vs. resorption in kimberlite-like melts, which is also documented by the inferred precipitation of fibrous diamond from kimberlites (Boyd et al. 1987; Gurney et al. 2010).

Compositional indicators for diamondiferous vs. barren sources. When both clinopyroxene and garnet are available for study, log D_{Ce} for clinopyroxene–garnet at a given garnet Ca# may be useful to predict the presence of diamond in eclogite, whereby the combination of low garnet Ca# and high log D_{Ce} reflects the small-volume melt metasomatism dominating in barren eclogites and pyroxenites (Fig. 37b).

The following relationship can be formulated for the recognition of diamondiferous eclogite substrates:

$$\log D_{Ce} < -5.06 \times Ca\#_{garnet}/100 + 2.54$$

This equation predicts that 50% of diamond-free eclogite and pyroxenite may actually contain diamond. It accurately determines the presence of diamond in 97% of diamondiferous xenoliths and in 94% of the eclogitic inclusions. Based on garnet only, the density of inclusions and diamondiferous xenoliths is smaller at garnet MgO content > 16 wt%, as only 11% of diamond-associated garnets have this feature, compared to 31% of barren xenoliths,

Table 4. Temperatures and salient compositional characteristics of calculated bulk eclogite compositions, reconstructed from inclusions (DI) and diamondiferous and barren xenoliths.

Sample type	**TKR885 °C**	**$TiO_2$1 wt%**	**Al_2O_3 wt%**	**Cr_2O_3 wt%**	**FeO^{total} wt%**	**MgO wt%**	**CaO wt%**	**Na_2O wt%**	**K_2O wt%**	**Sr ppm**	**Y ppm**
Eclogitic DI median	1221	0.55	16.3	0.07	11.8	10.1	10.8	2.5	0.07	53	17
Diamondiferous eclogite median	1134	0.36	16.5	0.07	9.8	11.5	11.6	2.5	0.09	59	11
% change vs. DI		–34	2	10	–17	14	7	1	33	12	–36
Composite eclogite median[6]	1046	0.20	15.3	0.12	9.7	13.1	11.4	1.9	0.03	76	13
% change vs. DI		–63	–6	87	–18	29	5	–24	–56	43	–22
Sample type	**Zr_{min} ppm[1]**	**Nb_{min}[1] ppm**	**La ppm**	**Sm ppm**	**Lu ppm**	**Hf_{min}[1] ppm**	**Nd/Yb[2] NMORB**	**Yb/Gd[2] NMORB**	**Eu/Eu*[3]**	**ΣHREE[4]**	
Eclogitic DI median	16	0.08	0.49	1.2	0.41	0.55	0.29	1.69	1.05	8.6	
Diamondiferous eclogite median	19	0.24	0.36	0.8	0.26	0.50	0.36	1.51	1.16	5.7	
% change vs. DI	17	192	–26	–37	–38	–8	24	–11	10	–34	
Composite eclogite median[6]	16	0.06	1.17	1.2	0.24	0.54	0.70	1.25	1.09	6.5	
% change vs. DI	–2	–32	139	–2	–42	–2	139	–26	4	–25	

Notes:
[1]Concentrations in the bulk rock are minima because rutile, which dominates the Nb-Ta budget and concentrates Zr, was not considered in bulk rock reconstruction. The deficit is higher for lower-temperature samples than for higher-temperature samples.
[2]Normalized to NMORB of Gale et al. (2013).
[3]Defined as chondrite-normalized Eu / $(Sm \times Gd)^{1/2} > 1.05$ (Rudnick and Fountain 1995); chondrite of Sun and McDonough (1989).
[4]Sum of HREE abundances from Tb to Lu.
[5]Temperature using the thermometer of Krogh (1988) and calculated at a pre-set pressure of 5 GPa.
[6]Reconstructed from median compositions of barren eclogite and pyroxenite classes reported in Appendix 2, weighted by their relative abundances as per Table 3.

typically high-Mg eclogites and pyroxenites (Fig. 44a). Finally, not only elevated Na_2O (here > 0.04 wt%) in garnet can be indicative of a diamondiferous eclogite source (Robinson et al. 1984; McCandless and Gurney 1989), but also elevated TiO_2 contents (> 0.1 wt%). Despite unquestionable bulk compositional effects (Grütter and Quadling 1999), both of these minor elements increase in concentration in garnet with increasing temperature (pressure along conductive geotherms), therefore representing graphite- vs. diamond-stable conditions. More than 66% of diamond-free eclogite and pyroxenite xenoliths and 98% of inclusions plus diamondiferous xenoliths are correctly assigned by this combined criterion (Fig. 44b). The number of inclusions and diamondiferous xenoliths increases at values of $Na_2O \geq 0.07$ wt% in garnet (Gurney 1984) and $TiO_2 \geq 0.2$ wt% in garnet, matched by 37% of barren xenoliths, and 89% of inclusions plus diamondiferous xenoliths.

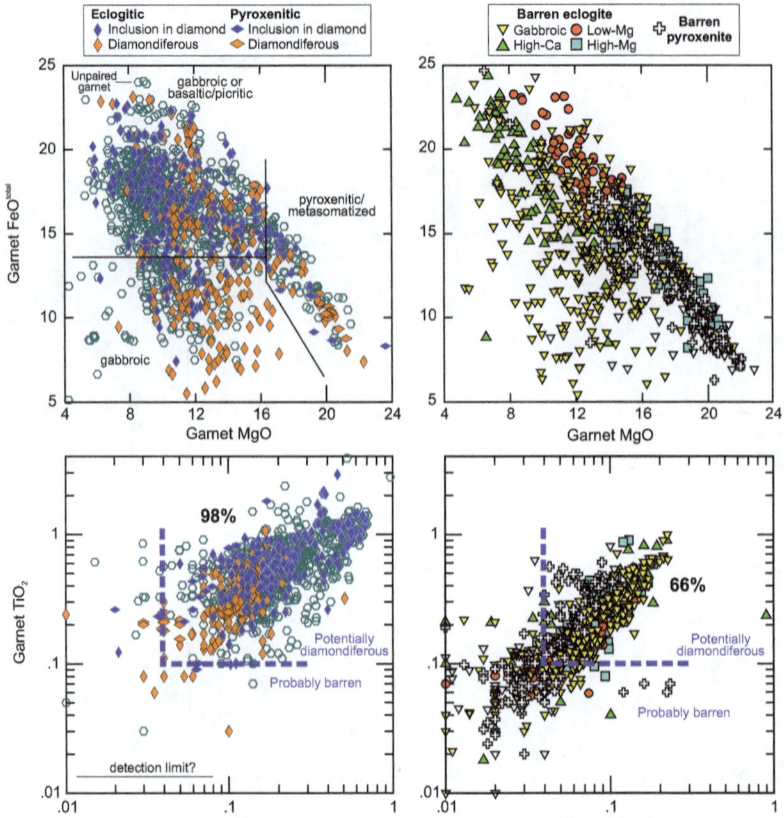

Figure 44. a. FeO (wt%) as a function of MgO (wt%) in garnet. Fields on left roughly separate garnet derived from pyroxenites or metasomatized eclogites from garnet derived from gabbroic eclogite and that derived from either gabbroic or non-gabbroic eclogite. **b.** TiO_2 (wt%) as a function of Na_2O (wt%) in garnet. 98% of inclusions and diamondiferous xenoliths occur at high abundances of these minor elements in a field occupied by 66% of barren xenoliths. Inclusions and diamondiferous xenoliths on **left**, barren xenoliths on **right**.

Key observations and conclusions

Eclogitic inclusions in lithospheric diamond have major and trace element systematics broadly similar to those of mantle eclogites, which originated as variably differentiated oceanic crust. Direct formation of eclogite from magma at mantle pressures is not supported.

Non-mantle $\delta^{18}O$ compositions in some inclusions and xenoliths point to a variably seawater-altered oceanic crust (AOC) protolith, which lost a small volume partial melt during subsequent dehydration melting in the course of subduction, causing some LREE-depletion. Known oxygen fugacity values of mantle eclogite are mostly below the FMQ buffer. Thus, the stability of diamond in eclogite to higher fO_2 than in peridotite plays a subordinate role in explaining the overabundance of eclogitic diamonds relative to the abundance of eclogite in the mantle lithosphere. Less than a quarter of inclusions (23%) and even fewer diamondiferous xenoliths (11%) show the LREE-enrichment that ensues from kimberlite–eclogite or CHO–fluid–eclogite interaction. Thus, eclogitic diamond formation is not dominantly linked to a typical mantle metasomatic signature, which in eclogite appears to be responsible for diamond destruction rather than growth. Therefore, the stability of diamond together with carbonated silicate melt at fO_2 conditions much lower than is the case for pure carbonatite, also is not important in explaining eclogitic diamond formation. Inclusions and diamondiferous xenoliths are depleted in $MREE_N$ relative to $HREE_N$, which is not seen in barren xenoliths and possible crustal protoliths. This depletion is interpreted as a disequilibrium feature caused by interaction with large fluid volumes during subduction-related dehydration of underlying oceanic mantle. If so, oxygen fugacity values indicate that diamond may have formed either by reduction of CO_2 in relatively oxidized fluids, or during mixing of CHO fluids falling on either side of the water maximum, a feature observed during deserpentinization. The subduction environment, with the availability of large redox contrasts and large volumes of mobile fluids, may be favourable for the disproportionately higher amount of eclogitic than peridotitic diamond growth. Whether the carbon involved in this growth is derived directly from the dehydration of serpentinized peridotite or picked up by the dehydration fluid as it migrates through the oceanic crust may be resolved using multiple stable isotopes (e.g., C–N isotopes of diamond and O–S isotopes of inclusions compared to various portions of oceanic lithosphere). While elevated MgO and Cr_2O_3 contents in garnet and high D_{Ce} between clinopyroxene–garnet at a given Ca# are associated with barren eclogites, high combined TiO_2 and Na_2O in garnet are indicative of high-temperature (pressure along conductive geotherm) sources that were little affected by mantle metasomatism and have high diamond potential.

ACKNOWLEDGEMENTS

Kelsey Graversen (University of Alberta) is thanked for her tremendous help in adding full references (DOIs) and in error checking the inclusion-in-diamond database. Amber Yu (University of Alberta) upgraded the xenolith database into a truly useful tool and assembled an additional database for trace elements in xenolith minerals. Fanus Viljoen (University of Johannesburg), Steve Richardson (University of Cape Town), as formal reviewers, and Steve Shirey (Carnegie Institution), as our editor, significantly improved the manuscript. Herman Grütter (SRK Consulting) is thanked for valuable comments on an early version of the manuscript. Saebyul Choe of The American Museum of Natural History kindly shipped crush and a fragment of the primitive J4 xenolith to us. Andrew Locock and Yan Luo (both University of Alberta) are thanked for their help in analysing J4 mineral separates with EPMA and sector field ICPMS, respectively. T.S. acknowledges long-term support for his diamond research program through the Discovery and the Collaborative Research Development (CRD) grant programs of the Natural Sciences and Engineering Research Council of Canada (NSERC), and additional grants from the Canada Research Chairs program and the Canada Foundation for Innovation (CFI). S.A. thanks the German Research Foundation for their continued support, currently via grant DFG AU356/11. Diamond research would not happen without the continued generous support of major (principally De Beers and Rio Tinto) and many junior companies active in diamond exploration and mining. We sincerely thank all the enthusiastic geologists, engineers and diamond sorters in these companies who helped us with samples, gave of their expertise and occasionally organized some financial support.

REFERENCES

Ackerson MR, Watson EB, Tailby ND, Spear FS (2017) Experimental investigation into the substitution mechanisms and solubility of Ti in garnet. Am Mineral 102:158–172

Agashev AM, Pokhilenko LN, Pokhilenko NP, Shchukina EV (2018) Geochemistry of eclogite xenoliths from the Udachnaya kimberlite pipe: Section of ancient oceanic crust sampled. Lithos 314:187–200

Alvaro M, Mazzucchelli ML, Angel RJ, Murri M, Campomenosi N, Scambelluri M, Nestola F, Korsakov A, Tomilenko AA, Marone F, Morana M (2020) Fossil subduction recorded by quartz from the coesite stability field. Geology 48:24–28

Angel RJ, Alvaro M, Nestola F (2022) Crystallographic methods for non-destructive characterization of mineral inclusions in diamonds. Rev Mineral Geochem 88:257–306

Ashchepkov IV, Ntaflos T, Spetsius ZV, Salikhov RF, Downes H (2017a) Interaction between protokimberlite melts and mantle lithosphere: Evidence from mantle xenoliths from the Dalnyaya kimberlite pipe, Yakutia (Russia). Geosci Front 8:693–710

Ashchepkov IV, Ntaflos T, Logvinova AM, Spetsius ZV, Downes H, Vladykin NV (2017b) Monomineral universal clinopyroxene and garnet barometers for peridotitic, eclogitic and basaltic systems. Geosci Front 8:775–795

Aulbach S (2012) Craton nucleation and formation of thick lithospheric roots. Lithos 149:16–30

Aulbach S (2020) Temperature-dependent rutile solubility in garnet and clinopyroxene from mantle eclogite: Implications for continental crust formation and V-based oxybarometry. J Petrol 61:egaa065

Aulbach S, Jacob DE (2016) Major- and trace-elements in cratonic mantle eclogites and pyroxenites reveal heterogeneous sources and metamorphic processing of low-pressure protoliths. Lithos 262:586–605

Aulbach S, Arndt NT (2019a) Eclogites as palaeodynamic archives: Evidence for warm (not hot) and depleted (but heterogeneous) Archaean ambient mantle. Earth Planet Sci Lett 505:162–172

Aulbach S, Arndt NT (2019b) Origin of high-Mg bimineralic eclogite xenoliths in kimberlite—reply to comment from Claude Herzberg. Earth Planet Sci Lett 510:234–237

Aulbach S, Stachel T, Viljoen KS, Brey GP, Harris JW (2002) Eclogitic and websteritic diamond sources beneath the Limpopo Belt - is slab-melting the link? Contrib Mineral Petrol 143:56–70

Aulbach S, Pearson NJ, O'Reilly SY, Doyle BJ (2007) Origins of xenolithic eclogites and pyroxenites from the central Slave craton, Canada. J Petrol 48:1843–1873

Aulbach S, Shirey SB, Stachel T, Creighton S, Muehlenbachs K, Harris JW (2009) Diamond formation episodes at the southern margin of the Kaapvaal Craton: Re-Os systematics of sulfide inclusions from the Jagersfontein Mine. Contrib Mineral Petrol 157:525–540

Aulbach S, Jacob DE, Cartigny P, Stern RA, Simonetti SS, Wörner G, Viljoen KS (2017) Eclogite xenoliths from Orapa: Ocean crust recycling, mantle metasomatism and carbon cycling at the western Zimbabwe craton margin. Geochim Cosmochim Acta 213:574–592

Aulbach S, Creaser RA, Stachel T, Heaman LM, Chinn IL, Kong J (2018) Diamond ages from Victor (Superior Craton): Intra-mantle cycling of volatiles (C, N, S) during supercontinent reorganisation. Earth Planet Sci Lett 490:77–87

Aulbach S, Heaman LM, Jacob DE, Viljoen KS (2019a) Ages and sources of mantle eclogites: ID-TIMS and in situ MC-ICPMS Pb-Sr isotope systematics of clinopyroxene. Chem Geol 503:15–28

Aulbach S, Woodland AB, Stern RA, Vasilyev P, Heaman LM, Viljoen KS (2019b) Evidence for a dominantly reducing Archaean ambient mantle from two redox proxies, and low oxygen fugacity of deeply subducted oceanic crust. Sci Rep 9:20190

Aulbach S, Viljoen KS, Gerdes A (2020a) Diamondiferous and barren eclogites and pyroxenites from the western Kaapvaal craton record subduction processes and mantle metasomatism, respectively. Lithos 368:105588

Aulbach S, Massuyeau M, Garber JM, Gerdes A, Heaman LM, Viljoen KS (2020b) Ultramafic carbonated melt- and auto-metasomatism in mantle eclogites: Compositional effects and geophysical consequences. Geochem Geophys Geosyst 21:e2019GC008774

Ayers JC, Dittmer SK, Layne GD (1997) Partitioning of elements between peridotite and H_2O at 2.0–3.0 GPa and 900–1100°C, and application to models of subduction zone processes. Earth Planet Sci Lett 150:381–398

Bailey DK (1982) Mantle metasomatism—continuing chemical change within the Earth. Nature 296:525–530

Banas A, Stachel T, Muehlenbachs K, McCandless TE (2007) Diamonds from the Buffalo Head Hills, Alberta: Formation in a non-conventional setting. Lithos 93:199–213

Barth MG, Foley SF, Horn I (2002) Partial melting in Archean subduction zones: constraints from experimentally determined trace element partition coefficients between eclogitic minerals and tonalitic melts under upper mantle conditions. Precambrian Res 113:323–340

Bauer M (1904) Precious Stones. Translated from German with Additions by L. J. Spencer. Charles Griffin & Company, London

Beard BL, Fraracci KN, Taylor LA, Snyder GA, Clayton RA, Mayeda TK, Sobolev NV (1996) Petrography and geochemistry of eclogites from the Mir kimberlite, Yakutia, Russia. Contrib Mineral Petrol 125:293–310

Becker M, Le Roex AP (2006) Geochemistry of South African on- and off-craton, Group I and Group II kimberlites: Petrogenesis and source region evolution. J Petrol 47:673–703

Beinlich A, Klemd R, John T, Gao J (2010) Trace-element mobilization during Ca-metasomatism along a major fluid conduit: Eclogitization of blueschist as a consequence of fluid–rock interaction. Geochim Cosmochim Acta 74:1892–1922

Benoit M, Polvé M, Ceuleneer G (1996) Trace element and isotopic characterization of mafic cumulates in a fossil mantle diapir (Oman ophiolite). Chem Geol 134:199–214

Bernstein S, Kelemen PB, Hanghoj K (2007) Consistent olivine Mg# in cratonic mantle reflects Archean mantle melting to the exhaustion of orthopyroxene. Geology 35:459–462

Beyer C, Frost DJ, Miyajima N (2015) Experimental calibration of a garnet–clinopyroxene geobarometer for mantle eclogites. Contrib Mineral Petrol 169:18

Bishop FC, Smith JV, Dawson JB (1976) Na, P, Ti and coordination of Si in garnet from peridotite and eclogite xenoliths. Nature 260:696–697

Bodinier JL, Vasseur G, Vernières J, Dupuy C, Fabriès J (1990) Mechanisms of mantle metasomatism: Geochemical evidence from the Lherz orogenic peridotite. J Petrol 31:597–628

Boutan ME (1886) Le diamant. *In*: Frémy's Encyclopédie Chimique. Paris

Boyd FR (1989) Compositional distinction between oceanic and cratonic lithosphere. Earth Planet Sci Lett 96:15–26

Boyd FR, Gurney JJ (1986) Diamonds and the African lithosphere. Science 232:472–477

Boyd SR, Mattey DP, Pillinger CT, Milledge HJ, Mendelssohn M, Seal M (1987) Multiple growth events during diamond genesis: an integrated study of carbon and nitrogen isotopes and nitrogen aggregation state in coated stones. Earth Planet Sci Lett 86:341–353

Boyd FR, Pokhilenko NP, Pearson DG, Mertzman SA, Sobolev NV, Finger LW (1997) Composition of the Siberian cratonic mantle: evidence from Udachnaya peridotite xenoliths. Contrib Mineral Petrol 128:228–246

Branchetti M, Zepper JCO, Peters STJ, Koornneef JM, Davies GR (2021) Multi-stage garnet formation and destruction in Kimberley harzburgitic xenoliths, South Africa. Lithos 390–391:106119

Brewster D (1861) On the pressure cavities in topaz, beryl and diamond and their bearing on geological theories. Trans R Soc Edinburgh 23:39–44

Brey GP, Köhler T (1990) Geothermobarometry in four-phase lherzolites II. New thermobarometers, and practical assessment of existing thermobarometers. J Petrol 31:1353–1378

Brey GP, Shu Q (2018) The birth, growth and ageing of the Kaapvaal subcratonic mantle. Mineral Petrol 112:23–41

Brey G, Brice WR, Ellis DJ, Green DH, Harris KL, Ryabchikov ID (1983) Pyroxene-carbonate reactions in the upper mantle. Earth Planet Sci Lett 62:63–74

Brown M, Johnson T (2018) Secular change in metamorphism and the onset of global plate tectonics. Am Mineral 103:181–196

Bulanova GP (1995) The formation of diamond. J Geochem Explor 53:1–23

Burgess R, Turner G, Laurenzi M, Harris JW (1989) $^{40}Ar/^{39}Ar$ laser probe dating of individual clinopyroxene inclusions in Premier eclogitic diamonds. Earth Planet Sci Lett 94:22–28

Burgess SR, Harte B (1999) Tracing lithospheric evolution trough the analysis of heterogeneous G9/G10 garnets in peridotite xenoliths, I: Major element chemistry. *In*: The JB Dawson Volume, Proceedings of the VIIth International Kimberlite Conference. Vol 45. Gurney JJ, Gurney JL, Pascoe MD, Richardson SH (eds) Red Roof Design, Cape Town, p 66–80

Burness S, Smart KA, Tappe S, Stevens G, Woodland AB, Cano E (2020) Sulphur-rich mantle metasomatism of Kaapvaal craton eclogites and its role in redox-controlled platinum group element mobility. Chem Geol 542:119476

Bussweiler Y, Pearson DG, Stachel T, Kjarsgaard BA (2018) Cr-rich megacrysts of clinopyroxene and garnet from Lac de Gras kimberlites, Slave Craton, Canada—implications for the origin of clinopyroxene and garnet in cratonic lherzolites. Mineral Petrol 112:583–596

Bussweiler Y, Giuliani A, Greig A, Kjarsgaard BA, Petts D, Jackson SE, Barrett N, Luo Y, Pearson DG (2019) Trace element analysis of high-Mg olivine by LA-ICP-MS – Characterization of natural olivine standards for matrix-matched calibration and application to mantle peridotites. Chem Geol 524:136–157

Cameron M, Papike JJ (1981) Structural and chemical variations in pyroxenes. Am Mineral 66:1–50

Canil D (2004) Mildly incompatible elements in peridotites and the origins of mantle lithosphere. Lithos 77:375–393

Cardoso P (1980) A study of mantle inclusions in the Koffiefontein Kimberlite pipe, South Africa. MSc thesis University of Cape Town

Carswell DA, Gibb FGF (1987) Evaluation of mineral thermometers and barometers applicable to garnet herzolite assemblages. Contrib Mineral Petrol 95:499–511

Chakhmouradian AR (2006) High-field-strength elements in carbonatitic rocks: Geochemistry, crystal chemistry and significance for constraining the sources of carbonatites. Chem Geol 235:138–160

Chakhmouradian AR, Boehm CO, Demeny A, Reguir EP, Hegner E, Creaser RA, Halden NM, Yang P (2009) "Kimberlite" from Wekusko Lake, Manitoba: Actually a diamond-indicator-bearing dolomite carbonatite. Lithos 112:347–357

Chaussidon M, Albarede F, Sheppard SMF (1987) Sulfur isotope heterogeneity in the mantle from ion microprobe measurements of sulfide inclusions in diamonds. Nature 330:242–244

Clark JR, Papike JJ (1968) Crystal-chemical characterization of omphacites. Am Mineral 53:840–868

Coleman RG, Lee DE, Beatty LB, Brannock WW (1965) Eclogites and eclogites: their differences and similarities. Geol Soc Amer Bull 76:483–508

Creighton S, Stachel T, Eichenberg D, Luth RW (2010) Oxidation state of the lithospheric mantle beneath Diavik diamond mine, central Slave craton, NWT, Canada. Contrib Mineral Petrol 159:645–657

Czas J, Pearson DG, Stachel T, Kjarsgaard BA, Read GH (2020) A Palaeoproterozoic diamond-bearing lithospheric mantle root beneath the Archean Sask Craton, Canada. Lithos 356–357:105301

Dalton JA, Presnall DC (1998) Carbonatitic melts along the solidus of model lherzolite in the system CaO–MgO–Al_2O_3–SiO_2–CO_2 from 3 to 7 GPa. Contrib Mineral Petrol 131:123–135

Dasgupta R, Hirschmann MM (2010) The deep carbon cycle and melting in Earth's interior. Earth Planet Sci Lett 298:1–13

Dasgupta R, Hirschmann MM, Withers AC (2004) Deep global cycling of carbon constrained by the solidus of anhydrous, carbonated eclogite under upper mantle conditions. Earth Planet Sci Lett 227:73–85

Dasgupta R, Hirschmann MM, McDonough WF, Spiegelman M, Withers AC (2009) Trace element partitioning between garnet lherzolite and carbonatite at 6.6 and 8.6 GPa with applications to the geochemistry of the mantle and of mantle-derived melts. Chem Geol 262:57–77

Davies GR, van den Heuvel Q, Matveev S, Drury MR, Chinn IL, Gress MU (2018) A combined cathodoluminescence and electron backscatter diffraction examination of the growth relationships between Jwaneng diamonds and their eclogitic inclusions. Mineral Petrol 112:231–242

Davis GL (1977) The ages and uranium contents of zircons from kimberlites and associated rocks. Carnegie Inst Wash Yearb 76:631–654

Dawson JB (1984) Contrasting types of upper mantle metasomatism. *In*: Kimberlites II. The Mantle and Crust–Mantle Relationships. Kornprobst J (ed) Elsevier, Amsterdam, p 289–294

Day HW (2012) A revised diamond-graphite transition curve. Am Mineral 97:52–62

de Beer ES (ed) (1955) The diary of John Evelyn, Vol. 2: Kalendarium, 1620–1649. Oxford Universirty Press, Oxford

De Hoog JCM, Gall L, Cornell DH (2010) Trace-element geochemistry of mantle olivine and application to mantle petrogenesis and geothermobarometry. Chem Geol 270:196–215

De Hoog JCM, Stachel T, Harris JW (2019) Trace-element geochemistry of diamond-hosted olivine inclusions from the Akwatia Mine, West African Craton: implications for diamond paragenesis and geothermobarometry. Contrib Mineral Petrol 174:100

Debret B, Sverjensky DA (2017) Highly oxidising fluids generated during serpentinite breakdown in subduction zones. Sci Rep 7:10351

Deines P (1980) The carbon isotopic composition of diamonds: relationship to diamond shape, color, occurrence and vapor composition. Geochim Cosmochim Acta 44:943–961

Deines P, Harris JW, Gurney JJ (1993) Depth-related carbon isotope and nitrogen concentration variability in the mantle below the Orapa kimberlite, Botswana, Africa. Geochim Cosmochim Acta 57:2781–2796

Doroshev AM, Brey GP, Girnis AV, Turkin AI, Kogarko LN (1997) Pyrope–knorringite garnets in the Earth's mantle: Experiments in the MgO–Al_2O_3–SiO_2–Cr_2O_3 system. Russ Geol Geophys 38:559–586

Droop GTR (1987) A general equation for estimating Fe^{3+} concentrations in ferromagnesian silicates and oxides from microprobe analyses, using stoichiometric criteria. Geol Mag 51:431–435

Eaton DW, Perry HKC (2013) Ephemeral isopycnicity of cratonic mantle keels. Nat Geosci 6:967–970

Eaton DW, Darbyshire F, Evans RL, Grütter H, Jones AG, Yuan X (2009) The elusive lithosphere–asthenosphere boundary (LAB) beneath cratons. Lithos 109:1–22

Eggler DH, Baker DR (1982) Reduced volatiles in the system C–O–H: Implications for mantle melting, fluid formation and diamond genesis. *In*: High pressure research in geophysics. Akimoto S, Manghnani MH, (eds). Center for Acad Publ, Tokyo, p 237–250

Eldridge CS, Compston W, Williams IS, Harris JW, Bristow JW (1991) Isotope evidence for the involvement of recycled sediments in diamond formation. Nature 353:649–653

Eppler WF (1961) Inclusions in diamond. J Gemmol 8:1–13

Farquhar J, Wing BA, McKeegan KD, Harris JW, Cartigny P, Thiemens MH (2002) Mass-independent sulfur of inclusions in diamond and sulfur recycling on early Earth. Science 298:2369–2372

Ferrando S, Castelli D, Frezzotti ML (2017) Fluid properties control degassing or storage of abiogenic CH_4 during slab exhumation: the fluid inclusion record from the Western Alps. American Geophysical Union, Fall Meeting 2017, abstract #V41A-04

Fedortchouk Y, Chinn IL, Kopylova MG (2017) Three styles of diamond resorption in a single kimberlite: Effects of volcanic degassing and assimilation. Geology 45:871–874

Fedortchouk Y, Liebske C, McCammon C (2019) Diamond destruction and growth during mantle metasomatism: An experimental study of diamond resorption features. Earth Planet Sci Lett 506:493–506

Frezzotti ML, Selverstone J, Sharp ZD, Compagnoni R (2011) Carbonate dissolution during subduction revealed by diamond-bearing rocks from the Alps. Nat Geosci 4:703–706

Frezzotti M-L, Huizenga J-M, Compagnoni R, Selverstone J (2014) Diamond formation by carbon saturation in C-O-H fluids during cold subduction of oceanic lithosphere. Geochim Cosmochim Acta 143:68–86

Fipke CE, Gurney JJ, Moore RO (1995) Diamond exploration techniques emphasising indicator mineral geochemistry and Canadian examples. Geol Surv Canada Bull 423:24–36

Fitzpayne A, Giuliani A, Maas R, Hergt J, Janney P, Phillips D (2019) Progressive metasomatism of the mantle by kimberlite melts: Sr–Nd–Hf–Pb isotope compositions of MARID and PIC minerals. Earth Planet Sci Lett 509:15–26

Frey FA, Green DH (1974) The mineralogy, geochemistry and origin of lherzolite inclusions in Victorian basanites. Geochim Cosmochim Acta 38:1023–1059

Frost DJ, McCammon CA (2008) The redox state of Earth's mantle. Annu Rev Earth Planet Sci 36:389–420

Futergendler SI (1960) X-ray study of solid inclusions in Ural and Yakutia diamonds. Mat Vses Nauchi Issled Geol Inst 40:83–87 (in Russian)

Futergendler SI, Frank-Kamenetsky VA (1961) Oriented inclusions of olivine, garnet and chrome-spinel in diamonds. Zapisky Vses Mineral Obsh 90:230–236 (in Russian)

Gaetani GA, Kent AJR, Grove TL, Hutcheon ID, Stolper EM (2003) Mineral/melt partitioning of trace elements during hydrous peridotite partial melting. Contrib Mineral Petrol 145:391–405

Gale A, Dalton CA, Langmuir CH, Su Y, Schilling J-G (2013) The mean composition of ocean ridge basalts. Geochem Geophys Geosyst 14:489–518

Garber JM, Maurya S, Hernandez JA, Duncan MS, Zeng L, Zhang HL, Faul U, McCammon C, Montagner JP, Moresi L, Romanowicz BA (2018) Multidisciplinary constraints on the abundance of diamond and eclogite in the cratonic lithosphere. Geochem Geophys Geosyst 19:2062–2086

Giardini AA, Melton CE (1975) Nature of cloud-like inclusions in two Arkansas diamonds. Am Mineral 60:931–933

Girnis AV, Brey GP (1999) Garnet–spinel–olivine–orthopyroxene equilibria in the FeO–MgO–Al_2O_3–SiO_2–Cr_2O_3 systems; II, Thermodynamic analysis. Eur J Mineral 11:619–636

Girnis AV, Stachel T, Brey GP, Harris JW, Phillips D (1999) Internally consistent geothermobarometers for garnet harzburgites: model refinement and application. *In*: The J.B. Dawson Volume, Proceedings of the VIIth International Kimberlite Conference. Gurney JJ, Gurney JL, Pascoe MD, Richardson SH (eds) Red Roof Design, Cape Town, 247–254

Girnis AV, Bulatov VK, Brey GP, Gerdes A, Höfer HE (2013) Trace element partitioning between mantle minerals and silico-carbonate melts at 6–12 GPa and applications to mantle metasomatism and kimberlite genesis. Lithos 160–161:183–200

Glinnemann J, Kusaka K, Harris JW (2003) Oriented graphite single-crystal inclusions in diamond. Z Kristallogr 218:733–739

Göppert HR (1864) Ueber Einschlüsse im Diamant. Die Erben Loosjes, Haarlem

Green TH (1994) Experimental studies of trace-element partitioning applicable to igneous petrogenesis—Sedona 16 years later. Chem Geol 117:1–36

Green DH, Wallace ME (1988) Mantle metasomatism by ephemeral carbonatite melts. Nature 336:459–462

Green TH, Blundy JD, Adam J, Yaxley GM (2000) SIMS determination of trace element partition coefficients between garnet, clinopyroxene and hydrous basaltic liquids at 2–7.5 GPa and 1080–1200°C. Lithos 53:165–187

Grégoire M, Bell DR, Le Roex AP (2003) Garnet lherzolites from the Kaapvaal Craton (South Africa): Trace element evidence for a metasomatic history. J Petrol 44:629–657

Gress MU, Pearson DG, Chinn I, Thomassot E, Davies GR (2021) Mesozoic to Paleoproterozoic diamond growth beneath Botswana recorded by Re Os ages from individual eclogitic and websteritic inclusions. Lithos 388–389:106058

Griffin WL, Ryan CG (1995) Trace-elements in indicator minerals - Area selection and target evaluation in diamond exploration. J Geochem Explor 53:311–337

Griffin WL, Jaques AL, Sie SH, Ryan CG, Cousens DR, Suter GF (1988) Conditions of diamond growth—A proton microprobe study of inclusions in west Australian diamonds. Contrib Mineral Petrol 99:143–158

Griffin WL, Cousens DR, Ryan CG, Sie SH, Suter GF (1989a) Ni in chrome pyrope garnets: a new geothermometer. Contrib Mineral Petrol 103:199–202

Griffin WL, Smith D, Boyd FR, Cousens DR, Ryan CG, Sie SH, Suter GF (1989b) Trace-element zoning in garnets from sheared mantle xenoliths. Geochim Cosmochim Acta 53:561–567

Griffin WL, O'Reilly SY, Ryan CG (1999a) The composition and origin of subcontinental lithospheric mantle. *In*: Mantle Petrology: Field Observations and High Pressure Experimentation: A tribute to Francis R (Joe) Boyd. Vol Special Publication No. 6. Fei Y, Bertka CM, Mysen BO, (eds). The Geochemical Society, Houston, p 13–45

Griffin WL, Doyle BJ, Ryan CG, Pearson NJ, O'Reilly SY, Davies R, Kivi K, VanAchterbergh E, Natapov LM (1999b) Layered mantle lithosphere in the Lac de Gras area, Slave Craton: Composition, structure and origin. J Petrol 40:705–727

Grütter HS (2006) Cr-pyrope and chromite inside and outside diamond: - a tribute to the work of Jeff Harris. Abstract, Goldschmdt Conference 2006, Geochim Cosmochim Acta 70:A219-A219

Grütter HS (2009) Pyroxene xenocryst geotherms: Techniques and application. Lithos 112:1167–1178

Grütter HS, Quadling KE (1999) Can sodium in garnet be used to monitor eclogitic diamond potential? *In*: The JB Dawson Volume, Proceedings of the VIIth International Kimberlite Conference. Gurney JJ, Gurney JL, Pascoe MD, Richardson SH, (eds). Reed Roof Design, Cape Town, p 314–319

Grütter HS, Menzies AH (2017) Discrete Ti-Al±Ca metasomatism at ~53 kbar in chromite–garnet peridotites from Newlands kimberlite, South Africa. 11th International Kimberlite Conference: Extended Abstracts

Grütter HS, Apter DB, Kong J (1999) Crust-mantle coupling: evidence from mantle-derived xenocrystic garnets. *In*: The JB Dawson Volume, Proceedings of the VIIth International Kimberlite Conference. Gurney JJ, Gurney JL, Pascoe MD, Richardson SH (eds) Red Roof Design, Cape Town, p 307–313

Grütter HS, Gurney JJ, Menzies AH, Winter F (2004) An updated classification scheme for mantle-derived garnet, for use by diamond explorers. Lithos 77:841–857

Grütter HS, Latti D, Menzies A (2006) Cr-saturation arrays in concentrate garnet compositions from kimberlite and their use in mantle barometry. J Petrol 47:801–820

Gübelin E (1952) Inclusions in diamonds. J Gemmol 3:174–187

Gurney JJ (1984) A correlation between garnets and diamonds in kimberlites. *In*: Kimberlite Occurrence and Origin: A Basis for Conceptual Models in Exploration. Vol 8. Glover JE, Harris PG, (eds). Publs Geol. Dept. & Univ. Extension, Univ. West. Aust., Perth, p 143–166

Gurney JJ (1989) Diamonds. *In*: Kimberlites and Related Rocks GSA Spec Publ 14. Vol Vol 2. Ross J, *et al.* (eds). Blackwell, Carlton, p 935–965

Gurney JJ, Harte B (1980) Chemical variations in upper mantle nodules from Southern African kimberlites. Philos Trans R Soc London Ser A 297:273–293

Gurney JJ, Switzer GS (1973) The discovery of garnets closely related to diamonds in the Finsch pipe, South Africa. Contrib Mineral Petrol 39:103–116

Gurney JJ, Harris JW, Rickard RS (1984) Minerals associated with diamonds from the Roberts Victor Mine. *In*: Kimberlites II: the mantle and crust-mantle relationships. Kornprobst J (ed) Elsevier, Amsterdam, p 25–32

Gurney JJ, Helmstaedt HH, Richardson SH, Shirey SB (2010) Diamonds through time. Econ Geol 105:689–712

Haggerty SE (1986) Diamond genesis in a multiply-constrained model. Nature 320:34–37

Hardman MF, Pearson DG, Stachel T, Sweeney RJ (2018) Statistical approaches to the discrimination of crust- and mantle-derived low-Cr garnet—Major-element-based methods and their application in diamond exploration. J Geochem Explor 186:24–35

Harris JW (1968a) The recognition of diamond inclusions. Part 1: Syngenetic mineral inclusions. Ind Diam Rev 28:402–410

Harris JW (1968b) The recognition of diamond inclusions. Part 2: Epigenetic mineral inclusions. Ind Diam Rev 28:558–561

Harris JW, Gurney JJ (1979) Inclusions in diamond. *In*: The Properties of Diamond. Field JE (ed) Academic Press, London, p 555–591

Harris JW, Smit KV, Fedortchouk Y, Moore M (2022) Morphology of monocrystalline diamond and its inclusions. Rev Mineral Geochem 88:119–166

Harte B (1983) Mantle peridotites and processes - the kimberlite sample. *In*: Continental basalts and mantle xenoliths. Hawkesworth CJ, Norry MJ (eds) Shiva Publishing Ltd, Nantwich, p 46–91

Harte B (1987) Metasomatic events recorded in mantle xenoliths: an overview. *In*: Mantle Xenoliths. Nixon PH (ed) John Wiley & Sons Ltd., Chichester, p 625–640

Harte B, Kirkley MB (1997) Partitioning of trace elements between clinopyroxene and garnet: Data from mantle eclogites. Chem Geol 136:1–24

Harte B, Hunter RH, Kinny PD (1993) Melt geometry, movement and crystallization, in relation to mantle dykes, veins and metasomatism. Philos Trans R Soc London Ser A342:1–21

Hasterok D, Chapman DS (2011) Heat production and geotherms for the continental lithosphere. Earth Planet Sci Lett 307:59–70

Helmstaedt HH, Schulze DJ (1989) Southern African kimberlites and their mantle sample: implications for Archean tectonics and lithosphere evolution. *In*: Kimberlites and Related Rocks GSA Spec Publ 14. Vol. 1. Ross J, *et al.* (eds). Blackwell, Carlton, p 358–368

Helmstaedt HH, Gurney JJ, Richardson SH (2010) Ages of cratonic diamond and lithosphere evolution: constraints on Precambrian tectonics and diamond exploration. Can Mineral 48:1385–1408

Hervig RL, Smith JV, Dawson JB (1986) Lherzolite xenoliths in kimberlites and basalts: petrogenetic and crystallochemical significance of some minor and trace elements in olivine, pyroxenes, garnet and spinel. Trans R Soc Edinburgh: Earth Sci 77:181–201

Herzberg C (2004) Geodynamic information in peridotite petrology. J Petrol 45:2507–2530

Herzberg C, Asimow PD (2008) Petrology of some oceanic island basalts: PRIMELT2.XLS software for primary magma calculation. Geochem Geophys Geosyst 9:Q09001

Herzberg C, Rudnick R (2012) Formation of cratonic lithosphere: An integrated thermal and petrological model. Lithos 149:4–15

Hills DV, Haggerty SE (1989) Petrochemistry of eclogites from the Koidu kimberlite complex, Sierra-Leone. Contrib Mineral Petrol 103:397–422

Howell D, Stachel T, Stern RA, Pearson DG, Nestola F, Hardman MF, Harris JW, Jaques AL, Shirey SB, Cartigny P, Smit KV (2020) Deep carbon through time: Earth's diamond record and its implications for carbon cycling and fluid speciation in the mantle. Geochim Cosmochim Acta 275:99–122

Huang F, Daniel I, Cardon H, Montagnac G, Sverjensky DA (2017) Immiscible hydrocarbon fluids in the deep carbon cycle. Nat Commun 8:15798

Huang Y, Nakatani T, Nakamura M, McCammon C (2019) Saline aqueous fluid circulation in mantle wedge inferred from olivine wetting properties. Nat Commun 10:5557

Hunt L, Stachel T, Morton R, Grütter H, Creaser RA (2009) The Carolina kimberlite, Brazil—Insights into an unconventional diamond deposit. Lithos 112:843–851
Ionov DA, Blichert-Toft J, Weis D (2005) Hf isotope compositions and HREE variations in off-craton garnet and spinel peridotite xenoliths from central Asia. Geochim Cosmochim Acta 69:2399–2418
Ireland TR, Rudnick RL, Spetsius Z (1994) Trace-elements in diamond inclusions from eclogites reveal link to Archean granites. Earth Planet Sci Lett 128:199–213
Izraeli ES, Harris JW, Navon O (2001) Brine inclusions in diamonds: a new upper mantle fluid. Earth Planet Sci Lett 187:323–332
Izraeli ES, Harris JW, Navon O (2004) Fluid and mineral inclusions in cloudy diamonds from Koffiefontein, South Africa. Geochim Cosmochim Acta 68:2561–2575
Jablon BM, Navon O (2016) Most diamonds were created equal. Earth Planet Sci Lett 443:41–47
Jacob DE (2004) Nature and origin of eclogite xenoliths from kimberlites. Lithos 77:295–316
Jaques AL, Green DH (1980) Anhydrous melting of peridotite at 0–15 kb pressure and the genesis of tholeiitic basalts. Contrib Mineral Petrol 73:287–310
Jaques AL, Hall AE, Sheraton JW, Smith CB, Sun S-S, Drew RM, Foudoulis C, Ellingsen K (1989) Composition of crystalline inclusions and C-isotopic composition of Argyle and Ellendale diamonds. *In*: Kimberlites and related rocks. GSA Spec Publ 14. Vol 2. Ross J, *et al.* (eds). Blackwell, Carlton, p 966–989
Jenner FE, O'Neill HS (2012) Analysis of 60 elements in 616 ocean floor basaltic glasses. Geochem Geophys Geosyst 13:Q02005
John T, Gussone N, Podladchikov YY, Bebout GE, Dohmen R, Halama R, Klemd R, Magna T, Seitz HM (2012) Volcanic arcs fed by rapid pulsed fluid flow through subducting slabs. Nat Geosci 5:489–492
Johnson KTM (1998) Experimental determination of partition coefficients for rare earth and high-field-strength elements between clinopyroxene, garnet, and basaltic melt at high pressures. Contrib Mineral Petrol 133:60–68
Jordan TH (1978) Composition and development of the continental tectosphere. Nature 274:544–548
Kargin AV, Sazonova LV, Nosova AA, Tretyachenko VV (2016) Composition of garnet and clinopyroxene in peridotite xenoliths from the Grib kimberlite pipe, Arkhangelsk diamond province, Russia: Evidence for mantle metasomatism associated with kimberlite melts. Lithos 262:442–455
Kargin AV, Sazonova LV, Nosova AA, Pervov VA, Minevrina EV, Khvostikov VA, Burmii ZP (2017) Sheared peridotite xenolith from the V. Grib kimberlite pipe, Arkhangelsk Diamond Province, Russia: Texture, composition, and origin. Geosci Front 8:653–669
Kelemen PB, Dick HJB, Quick JE (1992) Formation of harzburgite by pervasive melt/rock reaction in the upper mantle. Nature 358:635–641
Kelemen PB, Hart SR, Bernstein S (1998) Silica enrichment in the continental upper mantle via melt/rock reaction. Earth Planet Sci Lett 164:387–406
Kessel R, Ulmer P, Pettke T, Schmidt MW, Thompson AB (2005) The water-basalt system at 4 to 6 GPa: Phase relations and second critical endpoint in a K-free eclogite at 700 to 1400 °C. Earth Planet Sci Lett 237:873–892
Kessel R, Pettke T, Fumagalli P (2015) Melting of metasomatized peridotite at 4–6 GPa and up to 1200 °C: an experimental approach. Contrib Mineral Petrol 169:1–19
Kesson SE, Ringwood AE (1989) Slab-mantle interactions. 2. The formation of diamonds. Chem Geol 78:97–118
Klein-BenDavid O, Izraeli ES, Hauri E, Navon O (2007) Fluid inclusions in diamonds from the Diavik mine, Canada and the evolution of diamond-forming fluids. Geochim Cosmochim Acta 71:723–744
Klein-BenDavid O, Pettke T, Kessel R (2011) Chromium mobility in hydrous fluids at upper mantle conditions. Lithos 125:122–130
Klemme S (2004) The influence of Cr on the garnet–spinel transition in the Earth's mantle: experiments in the system $MgO–Cr_2O_3–SiO_2$ and thermodynamic modelling. Lithos 77:639–646
Koornneef JM, Davies GR, Döpp SP, Vukmanovic Z, Nikogosian IK, Mason PRD (2009) Nature and timing of multiple metasomatic events in the sub-cratonic lithosphere beneath Labait, Tanzania. Lithos 112:896–912
Koornneef JM, Gress MU, Chinn IL, Jelsma HA, Harris JW, Davies GR (2017) Archaean and Proterozoic diamond growth from contrasting styles of large-scale magmatism. Nat Commun 8:648
Korenaga J (2008) Urey ratio and the structure and evolution of Earth's mantle. Rev Geophys 46:2007RG000241
Kostrovitsky SI, Solov'eva LV, Yakovlev DA, Suvorova LF, Sandimirova GP, Travin AV, Yudin DS (2013) Kimberlites and megacrystic suite: Isotope-geochemical studies. Petrology 21:127–144
Kramers JD (1979) Lead, uranium, strontium, potassium and rubidium in inclusion-bearing diamonds and mantle-derived xenoliths from southern Africa. Earth Planet Sci Lett 42:58–70
Krogh EJ (1988) The garnet–clinopyroxene Fe–Mg geothermometer—A reinterpretation of existing experimental data. Contrib Mineral Petrol 99:44–48
Kushiro I (1994) Recent experimental studies on partial melting of mantle peridotites at high pressures using diamond aggregates. J Geol Soc Jpn 100:103–110
Labrosse S, Jaupart C (2007) Thermal evolution of the Earth: Secular changes and fluctuations of plate characteristics. Earth Planet Sci Lett 260:465–481
Laurie A, Stevens G (2012) Water-present eclogite melting to produce Earth's early felsic crust. Chem Geol 314:83–95
Laurie A, Stevens G, van Hunen J (2013) The end of continental growth by TTG magmatism. Terra Nova 25:130–136

Lewis HC, Bonney TG (1897) Papers and notes on the genesis and matrix of the diamond. Longmans, Green and Co., London
Li JP, O'Neill HSC, Seifert F (1995) Subsolidus phase-relations in the system MgO–SiO_2–Cr–O in equilibrium with metallic Cr, and their significance for the petrochemistry of chromium. J Petrol 36:107-132
Li XL, Zhang LF, Wei CJ, Slabunov AI (2015) Metamorphic *PT* path and zircon U–Pb dating of Archean eclogite association in Gridino complex, Belomorian province, Russia. Precambrian Res 268:74–96
Li J-L, Schwarzenbach EM, John T, Ague JJ, Huang F, Gao J, Klemd R, Whitehouse MJ, Wang X-S (2020) Uncovering and quantifying the subduction zone sulfur cycle from the slab perspective. Nature Commun 11:514
Liou JG, Tsujimori T, Zhang RY, Katayama I, Maruyama S (2004) Global UHP metamorphism and continental subduction/collision: The Himalayan model. Int Geol Rev 46:1–27
Litasov KD, Shatskiy A, Ohtani E (2014) Melting and subsolidus phase relations in peridotite and eclogite systems with reduced C–O–H fluid at 3–16 GPa. Earth Planet Sci Lett 391:87–99
Liu J, Pearson DG, Wang LH, Mather KA, Kjarsgaard BA, Schaeffer AJ, Irvine GJ, Kopylova MG, Armstrong JP (2021) Plume-driven recratonization of deep continental lithospheric mantle. Nature 592:732–736
Luth RW (1993) Diamonds, eclogites, and the oxidation-state of the Earth's mantle. Science 261:66–68
Luth RW, Stachel T (2014) The buffering capacity of lithospheric mantle: implications for diamond formation. Contrib Mineral Petrol 168:1–12
Luth RW, Palyanov YN, Bureau H (2022) Experimental petrology applied to natural diamond growth. Rev Mineral Geochem 88:755–808
MacGregor ID, Carter JL (1970) The chemistry of clinopyroxenes and garnets of eclogite and peridotite xenoliths from the Roberts Victor mine, South Africa. Phys Earth Planet Sci 3:391–397
MacGregor ID, Manton WI (1986) Roberts Victor eclogites—Ancient oceanic crust. J Geophys Res 91:14063–14079
Mahotkin IL, Gibson SA, Thompson RN, Zhuravlev DZ, Zherdev PU (2000) Late Devonian diamondiferous kimberlite and alkaline picrite (proto-kimberlite?) magmatism in the Arkhangelsk region, NW Russia. J Petrol 41:201–227
Malinovsky IY, Doroshev AM (1975) Effect of temperature and pressure on the composition of Cr-bearing phases of the garnet + enstatite + spinel + forsterite associations. Experimental research on mineralogy. Bull Inst Geol Geophys, Novosibirsk, p 121–125 (in Russian)
Matjuschkin V, Woodland AB, Yaxley GM (2019) Methane-bearing fluids in the upper mantle: an experimental approach. Contrib Mineral Petrol 174:1
Mattey D, Lowry D, Macpherson C (1994) Oxygen-isotope composition of mantle peridotite. Earth Planet Sci Lett 128:231–241
McCandless TE, Gurney JJ (1989) Sodium in garnet and potassium in clinopyroxene: Criteria for classifying mantle eclogites. *In*: Kimberlites and Related Rocks, Vol 2 Their Mantle/Crust Setting, Diamonds and Diamond Exploration. Vol 14. Ross J, Jacques AL, Ferguson J, Green DH, O'Reilly SY, Danchin RV, Janse AJA (eds) Geological Society of Australia Special Publication No. 14. Blackwell Scientific, Carlton, p 827–832
McDonough WF, Rudnick RL (1998) Mineralogy and composition of the upper mantle. Rev Mineral 37:139–164
McDonough WF, Sun S-S (1995) The composition of the Earth. Chem Geol 120:223–253
Meen JK, Eggler DH, Ayers JC (1989) Experimental evidence for very low solubility of rare-earth elements in CO_2-rich fluids at mantle conditions. Nature 340:301–303
Menzel MD, Garrido CJ, López Sánchez-Vizcaíno V (2020) Fluid-mediated carbon release from serpentinite-hosted carbonates during dehydration of antigorite–serpentinite in subduction zones. Earth Planet Sci Lett 531:115964
Meyer HOA (1968) Chrome pyrope: an inclusion in natural diamond. Science 160:1446–1447
Meyer HOA (1987) Inclusions in diamond. *In*: Mantle xenoliths. Nixon PH, (ed) John Wiley & Sons Ltd., Chichester, p 501–522
Meyer HOA, Boyd FR (1972) Composition and origin of crystalline inclusions in natural diamonds. Geochim Cosmochim Acta 36:1255–1273
Mibe K, Fujii T, Yasuda A (1998) Connectivity of aqueous fluid in the Earth's upper mantle. Geophys Res Lett 25:1233–1236
Mikhailenko D, Stagno V, Korsakov AV, Andreozzi GB, Marras G, Cerantola V, Malygina EV (2020) Redox state determination of eclogite xenoliths from Udachnaya kimberlite pipe (Siberian craton), with some implications for the graphite/diamond formation. Contrib Mineral Petrol 175:107
Misra KC, Anand M, Taylor LA, Sobolev NV (2004) Multi-stage metasomatism of diamondiferous eclogite xenoliths from the Udachnaya kimberlite pipe, Yakutia, Siberia. Contrib Mineral Petrol 146:696–714
Mitchell AL, Grove TL (2015) Melting the hydrous, subarc mantle: the origin of primitive andesites. Contrib Mineral Petrol 170:13
Mitchell RS, Giardini AA (1953) Oriented olivine inclusions in diamond. Am Mineral 38:136–138
Moss SW, Kobussen A, Powell W, Pollock K (2018) Kimberlite emplacement and mantle sampling through time at A154N kimberlite volcano, Diavik Diamond Mine: lessons from the deep. Mineral Petrol 112:397–410
Muehlenbachs K, Clayton RN (1972a) Oxygen isotope studies of fresh and weathered submarine basalts. Can J Earth Sci 9:172–184
Muehlenbachs K, Clayton RN (1972b) Oxygen isotope geochemistry of submarine greenstones. Can J Earth Sci 9:471–478

Navon O, Stolper E (1987) Geochemical consequences of melt percolation: The upper mantle as a chromatographic column. J Geol 95:285–307

Nestola F, Nimis P, Angel RJ, Milani S, Bruno M, Prencipe M, Harris JW (2014) Olivine with diamond-imposed morphology included in diamonds. Syngenesis or protogenesis? Int Geol Rev 56:1658–1667

Nestola F, Jung H, Taylor LA (2017) Mineral inclusions in diamonds may be synchronous but not syngenetic. Nat Commun 8:14168–14168

Nestola F, Jacob DE, Pamato MG, Pasqualetto L, Oliveira B, Greene S, Perritt S, Chinn I, Milani S, Kueter N, Sgreva N (2019a) Protogenetic garnet inclusions and the age of diamonds. Geology 47:431–434

Nestola F, Zaffiro G, Mazzucchelli ML, Nimis P, Andreozzi GB, Periotto B, Princivalle F, Lenaz D, Secco L, Pasqualetto L, Logvinova AM (2019b) Diamond-inclusion system recording old deep lithosphere conditions at Udachnaya (Siberia). Sci Rep 9:12586

Nickel KG, Green DH (1985) Empirical geothermobarometry for garnet peridotites and implications for the nature of the lithosphere, kimberlites and diamonds. Earth Planet Sci Lett 73:158–170

Nimis P (2022) Pressure and temperature data for diamonds. Rev Mineral Geochem 88:533–566

Nimis P, Alvaro M, Nestola F, Angel RJ, Marquardt K, Rustioni G, Harris JW, Marone F (2016) First evidence of hydrous silicic fluid films around solid inclusions in gem-quality diamonds. Lithos 260:384–389

Nixon PH, Rogers NW, Gibson IL, Grey A (1981) Depleted and fertile mantle xenoliths from southern African kimberlites. Annu Rev Earth Planet Sci 9:285–309

Norman M, Garcia MO, Pietruszka AJ (2005) Trace-element distribution coefficients for pyroxenes, plagioclase, and olivine in evolved tholeiites from the 1955 eruption of Kilauea Volcano, Hawai'i, and petrogenesis of differentiated rift-zone lavas. Am Mineral 90:888–899

O'Hara MJ, Saunders M, Mercy ELP (1975) Garnet peridotite, possible ultrabasic magmas and eclogite: interpretation of upper mantle processes in kimberlite. Phys Chem Earth 9:681–713

O'Neill HSC, Wall VJ (1987) The olivine-orthopyroxene-spinel oxygen-geobarometer, the nickel precipitation curve and the oxygen fugacity of the earth's upper mantle. J Petrol 28:1169–1191

O'Reilly SY, Griffin WL (1995) Trace-element partitioning between garnet and clinopyroxene in mantle-derived pyroxenites and eclogites—*P–T–X* controls. Chem Geol 121:105–130

O'Reilly SY, Griffin WL (2013) Mantle metasomatism. *In*: Metasomatism and the Chemical Transformation of Rock. Harlov DE, Austrheim H (eds) Springer Verlag, Berlin Heidelberg, p 471–533

Orlov YL (1959) Oriented olivine inclusions in diamond. Tr Mineral Muz, Akad Nauk SSSR 10:103–120 (in Russian)

Orlov YL (1977) The mineralogy of the diamond. John Wiley & Sons, New York

Pamato MG, Novella D, Jacob DE, Oliveira B, Pearson DG, Greene S, Afonso JC, Favero M, Stachel T, Alvaro M, Nestola F (2021) Protogenetic sulfide inclusions in diamonds date the diamond formation event using Re–Os isotopes. Geology 49:941–945

Pearson DG, Wittig N (2008) Formation of Archaean continental lithosphere and its diamonds: the root of the problem. J Geol Soc 165:895–914

Pearson DG, Wittig N (2014) The formation and evolution of cratonic mantle lithosphere—Evidence from mantle xenoliths. *In*: Treatise on Geochemistry (Second Edition). Turekian HDHK, (ed) Elsevier, Oxford, p 255–292

Pearson NJ, O'Reilly SY, Griffin WL (1995) The crust–mantle boundary beneath cratons and craton margins: A transect across the south-west margin of the Kaapvaal craton. Lithos 36:257–287

Pearson DG, Shirey SB, Harris JW, Carlson RW (1998) Sulphide inclusions in diamonds from the Koffiefontein kimberlite, S Africa: constraints on diamond ages and mantle Re–Os systematics. Earth Planet Sci Lett 160:311–326

Pearson DG, Canil D, Shirey SB (2004) Mantle samples included in volcanic rocks: xenoliths and diamonds. *In*: Treatise on Geochemistry Volume 2: The Mantle and Core. Carlson RW, (ed) Elsevier-Pergamon, Oxford, p 171–275

Pearson DG, Scott JM, Liu J, Schaeffer A, Wang LH, van Hunen J, Szilas K, Chacko T, Kelemen PB (2021) Deep continental roots and cratons. Nature 596:199–210

Penniston-Dorland SC, Kohn MJ, Manning CE (2015) The global range of subduction zone thermal structures from exhumed blueschists and eclogites: Rocks are hotter than models. Earth Planet Sci Lett 428:243–254

Perchuk AL, Gerya TV, Zakharov VS, Griffin WL (2020) Building cratonic keels in Precambrian plate tectonics. Nature 586:395–401

Phillips D, Harris JW, Viljoen KS (2004) Mineral chemistry and thermobarometry of inclusions from De Beers Pool diamonds, Kimberley, South Africa. Lithos 77:155–179

Phillips D, Harris JW, de Wit MCJ, Matchan EL (2018) Provenance history of detrital diamond deposits, west coast of Namaqualand, South Africa. Mineral Petrol 112:259–273

Philpotts JA, Schnetzler CC, Thomas HH (1972) Petrogenetic implications of some new geochemical data on eclogitic and ultrabasic inclusions. Geochim Cosmochim Acta 36:1131–1166

Piccoli F, Hermann J, Pettke T, Connolly JAD, Kempf ED, Duarte JFV (2019) Subducting serpentinites release reduced, not oxidized, aqueous fluids. Sci Rep 9:19573

Pollack HN, Chapman DS (1977) On the regional variation of heat-flow, geotherms, and lithospheric thickness. Tectonophysics 38:279–296

Rapp RP, Watson EB (1995) Dehydration melting of metabasalt at 8–32 kbar - Implications for continental growth and crust-mantle recycling. J Petrol 36:891–931

Rapp RP, Timmerman S, Lowczak J, Jaques AL (2017) Metasomatism of cratonic lithosphere by hydrous, silica-rich, fluids derived from recycled sediment: experimental insights at 5–7 GPa. International Kimberlite Conference: Extended Abstracts #11
Regier ME, Mišković A, Ickert RB, Pearson DG, Stachel T, Stern RA, Kopylova M (2018) An oxygen isotope test for the origin of Archean mantle roots. Geochem Perspect Lett 9:6–10
Richardson SH (1986) Latter-day origin of diamonds of eclogitic paragenesis. Nature 322:623–626
Richardson SH, Shirey SB (2008) Continental mantle signature of Bushveld magmas and coeval diamonds. Nature 453:910–913
Richardson SH, Gurney JJ, Erlank AJ, Harris JW (1984) Origin of diamonds in old enriched mantle. Nature 310:198–202
Richardson SH, Shirey SB, Harris JW, Carlson RW (2001) Archean subduction recorded by Re–Os isotopes in eclogitic sulfide inclusions in Kimberley diamonds. Earth Planet Sci Lett 191:257–266
Richardson SH, Pöml PF, Shirey SB, Harris JW (2009) Age and origin of peridotitic diamonds from Venetia, Limpopo Belt, Kaapvaal–Zimbabwe craton. Lithos 112:785–792
Robinson DN, Gurney JJ, Shee SR (1984) Diamond eclogite and graphite eclogite xenoliths from Orapa, Botswana. *In*: Kimberlites, II The Mantle and Crust-Mantle Relationships. Kornprobst J, (ed) Elsevier, Amsterdam, p 11–24
Rohrbach A, Schmidt MW (2011) Redox freezing and melting in the Earth's deep mantle resulting from carbon-iron redox coupling. Nature 472:209–212
Rudnick RL, Fountain DM (1995) Nature and composition of the continental crust: A lower crustal perspective. Rev Geophys 33:267–309
Rudnick RL, McDonough WF, Chappell BW (1993) Carbonatite metasomatism in the northern Tanzanian mantle: Petrographic and geochemical characteristics. Earth Planet Sci Lett 114:463 - 475
Rudnick RL, Donough WL, Orpin A (1994) Northern Tanzanian peridotite xenoliths: a comparison with Kaapvaal peridotites and inferences on metasomatic interactions. *In*: Kimberlites, Related Rocks and Mantle Xenoliths. Vol 1/A. Meyer HOA, Leonardos OH, (eds). CPRM Spec Publ Jan/94, Brasilia, p 336–353.
Rüpke LH, Morgan JP, Hort M, Connolly JAD (2004) Serpentine and the subduction zone water cycle. Earth Planet Sci Lett 223:17–34
Rustioni G, Audetat A, Keppler H (2021) The composition of subduction zone fluids and the origin of the trace element enrichment in arc magmas. Contrib Mineral Petrol 176:51
Schmädicke E, Okrusch M, Rupprecht-Gutowski P, Will TM (2011) Garnet pyroxenite, eclogite and alkremite xenoliths from the off-craton Gibeon Kimberlite Field, Namibia: A window into the upper mantle of the Rehoboth Terrane. Precambrian Res 191:1–17
Schmidt MW, Poli S (2014) Devolatilization during subduction. *In*: Treatise on Geochemistry 2nd Edition—The Crust. Holland HD, Turekian KK, Rudnick RL, (eds). Elsevier/Pergamon, Amsterdam, p 669-701
Schulze DJ (1989) Constraints on the abundance of eclogite in the upper mantle. J Geophys Res-Solid Earth Planets 94:4205–4212
Schulze DJ (2003) A classification scheme for mantle-derived garnets in kimberlite: a tool for investigating the mantle and exploring for diamonds. Lithos 71:195–213
Schulze DJ, Harte B, Valley JW, Brenan JM, Channer DMD (2003) Extreme crustal oxygen isotope signatures preserved in coesite in diamond. Nature 423:68–70
Scott Smith BH, Danchin RV, Harris JW, Stracke KJ (1984) Kimberlites near Orroroo, South Australia. *In*: Kimberlites I: Kimberlites and related rocks. Kornprobst J, (ed) Elsevier, Amsterdam, p 121–142
Sharma M, Wasserburg GJ (1996) The neodymium isotopic compositions and rare earth patterns in highly depleted ultramafic rocks. Geochim Cosmochim Acta 60:4537–4550
Sharp WE (1966) Pyrrhotite: a common inclusion in South African diamonds. Nature 211:402–403
Shatskiy A, Arefiev AV, Podborodnikov IV, Litasov KD (2019) Origin of K-rich diamond-forming immiscible melts and CO_2 fluid via partial melting of carbonated pelites at a depth of 180–200 km. Gondwana Res 75:154–171
Shchukina EV, Agashev AM, Pokhilenko NP (2017) Metasomatic origin of garnet xenocrysts from the V. Grib kimberlite pipe, Arkhangelsk region, NW Russia. Geosci Front 8:641–651
Shimizu N (1975) Rare earth elements in garnets and clinopyroxenes from garnet lherzolite nodules in kimberlites. Earth Planet Sci Lett 25:26–32
Shimizu N, Richardson SH (1987) Trace element abundance patterns of garnet inclusions in peridotite-suite diamonds. Geochim Cosmochim Acta 51:755–758
Shirey SB, Richardson SH (2011) Start of the Wilson Cycle at 3 Ga shown by diamonds from subcontinental mantle. Science 333:434–436
Shirey SB, Harris JW, Richardson SH, Fouch MJ, James DE, Cartigny P, Deines P, Viljoen KS (2002) Diamond genesis, seismic structure, and evolution of the Kaapvaal-Zimbabwe craton. Science 297:1683–1686
Shirey SB, Wagner LS, Walter MJ, Pearson DG, van Keken PE (2021) Slab transport of fluids to deep focus earthquake depths - thermal modeling constraints and evidence from diamonds. AGU Adv 2:e2020AV000304
Shu Q, Brey GP (2015) Ancient mantle metasomatism recorded in subcalcic garnet xenocrysts: Temporal links between mantle metasomatism, diamond growth and crustal tectonomagmatism. Earth Planet Sci Lett 418:27–39

Simon NSC, Carlson RW, Pearson DG, Davies GR (2007) The origin and evolution of the Kaapvaal cratonic lithospheric mantle. J Petrol 48:589–625

Smart KA, Heaman LM, Chacko T, Simonetti A, Kopylova M, Mah D, Daniels D (2009) The origin of high-MgO diamond eclogites from the Jericho Kimberlite, Canada. Earth Planet Sci Lett 284:527–537

Smart KA, Chacko T, Stachel T, Tappe S, Stern RA, Ickert RB, EIMF (2012) Eclogite formation beneath the northern Slave craton constrained by diamond inclusions: Oceanic lithosphere origin without a crustal signature. Earth Planet Sci Lett 319:165–177

Smit KV, Stachel T, Stern RA (2014a) Diamonds in the Attawapiskat area of the Superior craton (Canada): evidence for a major diamond-forming event younger than 1.1 Ga. Contrib Mineral Petrol 167:1–16

Smit KV, Stachel T, Creaser RA, Ickert RB, DuFrane SA, Stern RA, Seller M (2014b) Origin of eclogite and pyroxenite xenoliths from the Victor kimberlite, Canada, and implications for Superior craton formation. Geochim Cosmochim Acta 125:308–337

Smit KV, Shirey SB, Hauri EH, Stern RA (2019a) Sulfur isotopes in diamonds reveal differences in continent construction. Science 364:383–385

Smit KV, Stachel T, Luth RW, Stern RA (2019b) Evaluating mechanisms for eclogitic diamond growth: An example from Zimmi Neoproterozoic diamonds (West African craton). Chem Geol 520:21–32

Smit KV, Timmerman S, Aulbach S, Shirey SB, Richardson SH, Phillips D, Pearson DG (2022) Geochronology of diamonds. Rev Mineral Geochem 88:567–636

Smith D (1988) Implications of zoned garnets for the evolution of sheared lherzolites: Examples from northern Lesotho and the Colorado Plateau. J Geophys Res: Solid Earth 93:4895–4905

Smith D, Griffin WL, Ryan CG (1993) Compositional evolution of high-temperature sheared lherzolite PHN 1611. Geochim Cosmochim Acta 57:605–613

Smith EM, Shirey SB, Nestola F, Bullock ES, Wang JH, Richardson SH, Wang WY (2016) Large gem diamonds from metallic liquid in Earth's deep mantle. Science 354:1403–1405

Snyder GA, Taylor LA, Crozaz G, Halliday AN, Beard BL, Sobolev VN, Sobolev NV (1997) The origins of Yakutian eclogite xenoliths. J Petrol 38:85–113

Sobolev NV (1977) Deep-seated inclusions in kimberlites and the problem of the composition of the upper mantle. (Translated from the Russian edition, 1974). AGU, Washington

Sobolev NV (1985) Ultramafic and eclogitic types of paragenesis of diamonds. Conference on Native Elements Formation in Endogenic Processes, Extended Abstracts. Acad Sci USSR, Siberian Division, Yakutsk, (not paginated)

Sobolev NV, Lavrent'ev YG, Pospelova LN, Sobolev VC (1969) Chrome pyrope from Yakutian diamonds. Doklady Akademii Nauk SSSR 189:133–136

Sobolev NV, Lavrent'ev YG, Pokhilenko NP, Usova LV (1973) Chrome-rich garnets from the kimberlites of Yakutia and their paragenesis. Contrib Mineral Petrol 40:39–52

Sobolev NV, Logvinova AM, Efimova ES (2009) Syngenetic phlogopite inclusions in kimberlite-hosted diamonds: implications for role of volatiles in diamond formation. Russ Geol Geophys 50:1234–1248

Spandler C, Pirard C (2013) Element recycling from subducting slabs to arc crust: A review. Lithos 170:208–223

Spencer LJ (1924) An inclusion of magnetite in diamond. Mineral Mag 20:245–247

Spetsius ZV, Taylor LA (2002) Partial melting in mantle eclogite xenoliths: Connections with diamond paragenesis. Int Geol Rev 44:973–987

Stachel T (2014) Diamond. Mineralogical Association of Canada Short Course Series 44:1–28

Stachel T, Harris JW (1997) Diamond precipitation and mantle metasomatism - evidence from the trace element chemistry of silicate inclusions in diamonds from Akwatia, Ghana. Contrib Mineral Petrol 129:143–154

Stachel T, Harris JW (2008) The origin of cratonic diamonds—Constraints from mineral inclusions. Ore Geol Rev 34:5–32

Stachel T, Luth RW (2015) Diamond formation — Where, when and how? Lithos 220–223:200–220

Stachel T, Viljoen KS, Brey G, Harris JW (1998) Metasomatic processes in lherzolitic and harzburgitic domains of diamondiferous lithospheric mantle: REE in garnets from xenoliths and inclusions in diamonds. Earth Planet Sci Lett 159:1–12

Stachel T, Aulbach S, Brey GP, Harris JW, Leost I, Tappert R, Viljoen KS (2004) The trace element composition of silicate inclusions in diamonds: a review. Lithos 77:1–19

Stachel T, Banas A, Muehlenbachs K, Kurszlaukis S, Walker EC (2006) Archean diamonds from Wawa (Canada): samples from deep cratonic roots predating cratonization of the Superior Province. Contrib Mineral Petrol 151:737–750

Stachel T, Chacko T, Luth RW (2017) Carbon isotope fractionation during diamond growth in depleted peridotite: Counterintuitive insights from modelling water-maximum CHO fluids as multi-component systems. Earth Planet Sci Lett 473:44–51

Stachel T, Harris JW, Hunt L, Muehlenbachs K, Kobussen AF, EIMF (2018a) Argyle diamonds: How subduction along the Kimberley Craton edge generated the world's biggest diamond deposit. *In*: Geoscience and Exploration of the Argyle, Bunder, Diavik, and Murowa Diamond Deposits. Vol 20. Davy AT, Smith CB, Helmstaedt H, Lynton Jaques A, Gurney JJ, (eds). Society of Economic Geologists, p 145–167

Stachel T, Banas A, Aulbach S, Smit KV, Wescott P, Chinn IL, Kong J (2018b) The Victor Mine (Superior Craton, Canada): Neoproterozoic lherzolitic diamonds from a thermally-modified cratonic root. Mineral Petrol 112:325–336

Stachel T, Cartigny P, Chacko T, Pearson DG (2022) Carbon and nitrogen in mantle-derived diamonds. Rev Mineral Geochem 88:809–876

Stagno V, Ojwang DO, McCammon CA, Frost DJ (2013) The oxidation state of the mantle and the extraction of carbon from Earth's interior. Nature 493:84–88

Stagno V, Frost DJ, McCammon CA, Mohseni H, Fei Y (2015) The oxygen fugacity at which graphite or diamond forms from carbonate-bearing melts in eclogitic rocks. Contrib Mineral Petrol 169:16

Stalder R, Foley SF, Brey GP, Horn I (1998) Mineral-aqueous fluid partitioning of trace elements at 900–1200°C and 3.0–5.7 GPa: new experimental data for garnet, clinopyroxene, and rutile, and implications for mantle metasomatism. Geochim Cosmochim Acta 62:1781–1801

Stosch HG, Lugmair GW (1986) Trace element and Sr and Nd isotope geochemistry of peridotite xenoliths from the Eifel (West Germany) and their bearing on the evolution of the subcontinental lithosphere. Earth Planet Sci Lett 80:281–298

Streckeisen A (1976) To each plutonic rock its proper name. Earth Sci Rev 12:1–33

Sun S-S, McDonough WF (1989) Chemical and isotopic systematics of oceanic basalts: implications for mantle compositions and processes. *In*: Magmatism in the ocean basins. Vol 42. Saunders AD, Norry MJ (eds) Geol Soc Spec Publ, p 313–345

Sutton JR (1907) The relationship between diamonds and garnets. Nature 75:488–488

Sutton JR (1921) Inclusions in diamond from South Africa. Mineral Mag 19:208–210

Sverjensky DA, Huang F (2015) Diamond formation due to a pH drop during fluid–rock interactions. Nature Communications 6:8702

Syracuse EM, van Keken PE, Abers GA (2010) The global range of subduction zone thermal models. Phys Earth Planet Sci 183:73–90

Tappe S, Foley SF, Kjarsgaard BA, Romer RL, Heaman LM, Stracke A, Jenner GA (2008) Between carbonatite and lamproite - Diamondiferous Torngat ultramafic lamprophyres formed by carbonate-fluxed melting of cratonic MARID-type metasomes. Geochim Cosmochim Acta 72:3258–3286

Tavernier JB (1676) Travels in India. Translated from French original by V. Ball, 1889. Macmillan and Co., London

Taylor WR (1998) An experimental test of some geothermometer and geobarometer formulations for upper mantle peridotites with application to the thermobarometry of fertile Iherzolite and garnet websterite. N Jb Miner Abh 172:381–408

Taylor LA, Neal CR (1989) Eclogites with oceanic crustal and mantle signatures from the Bellsbank kimberlite, South Africa. 1. Mineralogy, petrography and whole rock chemistry. J Geol 97:551–567

Thomassot E, Cartigny P, Harris JW, Lorand JP, Rollion-Bard C, Chaussidon M (2009) Metasomatic diamond growth: A multi-isotope study (^{13}C, ^{15}N, ^{33}S, ^{34}S) of sulphide inclusions and their host diamonds from Jwaneng (Botswana). Earth Planet Sci Lett 282:79–90

Thomson AR, Walter MJ, Kohn SC, Brooker RA (2016) Slab melting as a barrier to deep carbon subduction. Nature 529:76–79

Timmerman S, Koornneef JM, Chinn IL, Davies GR (2017) Dated eclogitic diamond growth zones reveal variable recycling of crustal carbon through time. Earth Planet Sci Lett 463:178–188

Tomlinson EL, Kamber BS (2021) Depth-dependent peridotite–melt interaction and the origin of variable silica in the cratonic mantle. Nat Commun 12:1082

Uenver-Thiele L, Woodland AB, Seitz H-M, Downes H, Altherr R (2017) Metasomatic processes revealed by trace element and redox signatures of the lithospheric mantle beneath the Massif Central, France. J Petrol 58:395–422

Ulmer P (1989) Partitioning of high field strength elements among olivine, pyroxenes, garnet and calc-alkaline picrobasalt: Experimental results and application. Inter J Mass Spectrom Ion Phys 1989:42–47

Viljoen KS, Harris JW, Ivanic T, Richardson SH, Gray K (2014) Trace element chemistry of peridotitic garnets in diamonds from the Premier (Cullinan) and Finsch kimberlites, South Africa: Contrasting styles of mantle metasomatism. Lithos 208–209:1–15

Viljoen KS, Perritt SH, Chinn IL (2018) An unusual suite of eclogitic, websteritic and transitional websteritic-lherzolitic diamonds from the Voorspoed kimberlite in South Africa: Mineral inclusions and infrared characteristics. Lithos 320–321:416–434

Vitale Brovarone A, Tumiati S, Piccoli F, Ague JJ, Connolly JAD, Beyssac O (2020) Fluid-mediated selective dissolution of subducting carbonaceous material: Implications for carbon recycling and fluid fluxes at forearc depths. Chem Geol 549:119682

Wagner PA (1914) The diamond fields of Southern Africa. Transvaal Leader, Johannesburg

Walter MJ (1999) Melting residues of fertile peridotite and the origin of cratonic lithosphere. *In*: Mantle Petrology: Field Observations and High Pressure Experimentation: A tribute to Francis R (Joe) Boyd. Vol 6. Fei Y, Bertka CM, Mysen BO, (eds). The Geochemical Society, Houston, p 225–239

Walter MJ, Thomson AR, Smith EM (2022) Geochemistry of silicate and oxide inclusions in sublithospheric diamonds. Rev Mineral Geochem 88:393–450

Wang H, van Hunen J, Pearson DG (2018) Making Archean cratonic roots by lateral compression: A two-stage thickening and stabilization model. Tectonophysics 746:562–571

Wang W (1998) Formation of diamond with mineral inclusions of ''mixed'' eclogite and peridotite paragenesis. Earth Planet Sci Lett 160:831–843

Waters FG, Erlank AJ (1988) Assessment of the vertical extent and distribution of mantle metasomatism below Kimberley, South Africa. J Petrol Spec Vol:185–204

Watson EB, Brenan JM, Baker DR (1990) Distribution of fluids in the continental mantle. *In*: Continental Mantle. Vol 16. Menzies MA (ed) Oxford Monographs on Geology and Geophysics, p 111–125

Weiss Y, Goldstein SL (2018) The involvement of diamond-forming fluids in the metasomatic "cocktail' of kimberlite sources. Mineral Petrol 112:149–167

Weiss Y, Griffin WL, Bell DR, Navon O (2011) High-Mg carbonatitic melts in diamonds, kimberlites and the sub-continental lithosphere. Earth Planet Sci Lett 309:337–347

Weiss Y, Griffin WL, Navon O (2013) Diamond-forming fluids in fibrous diamonds: The trace-element perspective. Earth Planet Sci Lett 376:110–125

Weiss Y, Kiflawi I, Davies N, Navon O (2014) High-density fluids and the growth of monocrystalline diamonds. Geochim Cosmochim Acta 141:145–159

Weiss Y, McNeill J, Pearson DG, Nowell GM, Ottley CJ (2015) Highly saline fluids from a subducting slab as the source for fluid-rich diamonds. Nature 524:339–342

Weiss Y, Cazs J, Navon O (2022) Fluid inclusions in fibrous diamonds. Rev Mineral Geochem 88:475–532

Welke HJ, Allsopp HL, Harris JW (1974) Measurements of K, Rb, U, Sr, and Pb in diamonds containing inclusions. Nature 252:35–36

White WM, Klein EM (2014) Composition of the oceanic crust. *In*: Treatise on Geochemistry 2nd Edition. Vol 2. Holland HD, Turekian KK, Carlson RW (eds) Elsevier Pergamon, Amsterdam, p 457–496

Wiggers de Vries DF, Pearson DG, Bulanova GP, Smelov AP, Pavlushin AD, Davies GR (2013) Re–Os dating of sulphide inclusions zonally distributed in single Yakutian diamonds: Evidence for multiple episodes of Proterozoic formation and protracted timescales of diamond growth. Geochim Cosmochim Acta 120:363–394

Wijbrans CH, Klemme S, Berndt J, Vollmer C (2015) Experimental determination of trace element partition coefficients between spinel and silicate melt: the influence of chemical composition and oxygen fugacity. Contrib Mineral Petrol 169:45

Williams AF (1932) The genesis of the diamond. Ernest Benn Limited, London

Wolff-Boenisch B (1994) Aufbau der Analytik zur Messung der Osmium-Isotopie sowie die Strontium-, Neodym- und Osmium-Systematik eines Hoch-Temperatur-Granat-Lherzoliths (J4) von Jagersfontein (Südafrika). Dipl Thesis, University of Mainz (Germany)

Workman RK, Hart SR (2005) Major and trace element composition of the depleted MORB mantle (DMM). Earth Planet Sci Lett 231:53–72

Wyllie PJ (1987) Metasomatism and fluid generation in mantle xenoliths. *In*: Mantle Xenoliths. Nixon PH (ed) John Wiley & Sons Ltd., Chichester, p 609–621

Wyllie PJ (1992) Experimental petrology: Earth materials science. *In*: Understanding Earth. Brown GC, Hawkesworth CJ, Wilson RCL, (eds). Cambridge University Press, Cambridge, p 67–87

Wyllie PJ, Huang W-L (1976) Carbonation and melting reactions in the system CaO–MgO–SiO_2–CO_2 at mantle pressures with geophysical and petrological applications. Contrib Mineral Petrol 54:79–107

Wyllie PJ, Ryabchikov ID (2000) Volatile components, magmas, and critical fluids in upwelling mantle. J Petrol 41:1195–1206

Yaxley GM (1993) Carbonatite metasomatism in the mantle: sources and roles of carbonate in metasomatic enrichment processes in the lithosphere. PhD thesis University of Tasmania

Yaxley GM, Green DH (1998) Reactions between eclogite and peridotite: Mantle refertilisation by subduction of oceanic crust. Schweizerische Mineralogische und Petrographische Mitteilungen 78:243–255

Zedgenizov DA, Kagi H, Shatsky VS, Sobolev NV (2004) Carbonatitic melts in cuboid diamonds from Udachnaya kimberlite pipe (Yakutia): evidence from vibrational spectroscopy. Mineral Mag 68:61–73

Zhang C, Duan ZH (2010) GFluid: An Excel spreadsheet for investigating C–O–H fluid composition under high temperatures and pressures. Computers Geosci 36:569–572

Zhu J, Zhang L, Tao R, Fei Y (2020) The formation of graphite-rich eclogite vein in S.W. Tianshan (China) and its implication for deep carbon cycling in subduction zone. Chem Geol 533

Ziaja K, Foley SF, White RW, Buhre S (2014) Metamorphism and melting of picritic crust in the early Earth. Lithos 189:173–184

Ziberna L, Klemme S, Nimis P (2013) Garnet and spinel in fertile and depleted mantle: insights from thermodynamic modelling. Contrib Mineral Petrol 166:411–421

Reviews in Mineralogy & Geochemistry
Vol. 88 pp. 393-450, 2022

7

Geochemistry of Silicate and Oxide Inclusions in Sublithospheric Diamonds

Michael J. Walter

Earth and Planets Laboratory
Carnegie Institution for Science
5241 Broad Branch Road NW
Washington D.C. 20015
USA

mwalter@carnegiescience.edu

Andrew R. Thomson

Department of Earth Sciences
University College London
Gower Street
London WC1E 6BT
UK

a.r.thomson@ucl.ac.uk

Evan M. Smith

Gemological Institute of America
50 west 47th Street
New York, NY 10036
USA

evan.smith@gia.edu

INTRODUCTION

Minerals included in diamonds provide direct information about the petrologic and chemical environment of diamond crystallization. They record information relating to local and regional mantle processes and provide important contextual information for global-scale tectonic interpretations (Stachel et al. 2005; Stachel and Harris 2008; Harte 2010; Shirey et al. 2013, 2019). Most mined inclusion-bearing diamonds originate in sub-continental, cratonic mantle lithosphere but a small percentage host mineral inclusions consistent with an origin beneath the lithosphere (~1%, Stachel and Harris 2008). Key among these inclusions are silicate and oxide minerals that provide either direct (e.g., majoritic garnet, ringwoodite) or circumstantial (e.g., $CaSiO_3$-rich and $MgSiO_3$-rich phases; ferropericlase) evidence for a high-pressure origin deep in the convecting mantle; we refer to these diamonds as "sublithospheric" although they are also commonly called "superdeep". Studies over the past four decades have provided a wealth of information to draw upon to interrogate the origins of sublithospheric diamonds and their inclusions and to speculate on broader geologic and geodynamic implications.

In the 1980s researchers began to recognize that some diamonds carry inclusions indicative of an origin beneath continental lithosphere, extending to depths even into the lower mantle (Scott-Smith et al. 1984; Moore et al. 1986; Wilding et al. 1991; Harte and Harris 1994;

1529-6466/22/0088-0007$10.00 (print)
1943-2666/22/0088-0007$10.00 (online)
http://dx.doi.org/10.2138/rmg.2022.88.07

Harris et al. 1997; Stachel et al. 1998a; Harte et al. 1999). Paramount among these are inclusions with (Mg,Fe)O and (Mg,Fe)SiO_3 stoichiometry, and on the basis of co-occurrence in the same diamond they were interpreted as ferropericlase and retrograde Mg-silicate perovskite (bridgmanite) from the shallow lower mantle. Discoveries of inclusions with $CaSiO_3$ stoichiometry, sometimes also co-occurring with $MgSiO_3$-rich phases and/or ferropericlase and interpreted as retrograde Ca-silicate perovskite, supported the view of a lower mantle genesis related to mantle peridotite (Harte et al. 1999; Joswig et al. 1999; Stachel et al. 2000b; Kaminsky et al. 2001; Hayman et al. 2005). Garnet inclusions with excess octahedrally coordinated silicon per formula unit (Moore and Gurney 1985, 1989; Moore et al. 1991; Stachel and Harris 1997; Stachel et al. 1998a) provided further evidence for a sublithospheric origin on the basis of experiments that revealed the pressure dependence of elemental substitutions (Akaogi and Akimoto 1977).

Over several decades numerous studies have uncovered many new examples of sublithospheric diamonds hosting these key indicator phases while also identifying a wide variety of other mineral inclusions interpreted to have an origin in the deep upper mantle to lower mantle, including but not limited to ringwoodite, stishovite, CF-phase, NAL-phase, K-hollandite, CAS phase, and phase Egg (Wirth et al. 2007; Bulanova et al. 2010; Walter et al. 2011; Thomson et al. 2014; Zedgenizov et al. 2015). The reader is referred to several recent review papers that provide an inventory of inclusion types in sublithospheric diamonds (Stachel and Harris 2008; Harte 2010; Kaminsky 2012; Shirey et al. 2013, 2019).

On the basis of mineralogical, petrological and geochemical data it has become increasingly apparent that many sublithospheric diamonds record processes that are related to subduction of lithospheric plates (Stachel et al. 2000a,b; Stachel 2001; Walter et al. 2008; Tappert et al. 2009b; Bulanova et al. 2010; Kiseeva et al. 2013b; Thomson et al. 2014; Burnham et al. 2015; Ickert et al. 2015; Shirey et al. 2019). The major and trace element geochemistry of majoritic garnet and Ti-rich $CaSiO_3$-rich phases in particular point to an origin involving subducted basaltic oceanic crust, as does the presence of rare inclusions interpreted as retrograde CF-phase and NAL-phase. The prevalence of light carbon isotopic compositions in diamonds and heavy oxygen isotopes in hosted inclusions provide additional supporting evidence for this hypothesis (Burnham et al. 2015; Ickert et al. 2015).

Sublithospheric diamonds have distinctly low N with ~70% considered Type II (e.g., <~20 at.ppm N) and with > 90% having <100 at.ppm N. When measurable the N is highly aggregated and dominated by B centers (~87% have >50 %B), consistent with storage in the mantle at high temperature. In comparison lithospheric diamonds have higher N, are typically classified as Type I, averaging ~250 at.ppm N but extending to >1000 at.ppm N, and with <20% low N Type II. Lithospheric diamonds also commonly exhibit poorly aggregated N (e.g., <50 %B) indicative of storage at cooler cratonic temperatures (Harlow 1998; Stachel et al. 2002; Pearson et al. 2003; Shirey et al. 2013; Smith and Kopylova 2014). Sublithospheric diamonds tend to be irregular in shape, show weak cathodoluminescence, have textures indicating residence in a high-strain environment and sometimes exhibit multiple nucleation sites, resorption and regrowth. Like their lithospheric counterparts, precipitation of sublithospheric diamonds is thought to occur primarily from C-saturated fluids or melts, with carbonatitic, hydrous, methane-rich and metallic liquids all implicated on the basis of the mineralogy and geochemistry of the inclusions (Walter et al. 2008; Harte 2010; Harte and Richardson 2012; Shirey et al. 2013, 2019; Smith et al. 2016b, 2018).

Here we review the mineralogy, major and trace element geochemistry of key silicate and oxide mineral inclusions in sublithospheric diamonds from global data sets assembled from the literature. The purpose of this synthesis is to focus on inclusions that have compositions of major mineral phases in upper mantle, transition zone and lower mantle assemblages in both meta-peridotite (e.g., peridotite, harzburgite, dunite) and meta-basalt (e.g., basalt, pyroxenite).

A further requirement is that inclusions occur commonly enough for substantial geochemical data to be available from locations spanning multiple continents and cratons. Accordingly, we focus on inclusions of majoritic garnet, $MgSiO_3$- and $CaSiO_3$-rich phases, ferropericlase, olivine and clinopyroxene, assembling datasets comprising 659 inclusions. We will not ignore other inclusion types entirely but will rather discuss them in relation to these more abundant inclusions.

Depending on when they formed or equilibrated relative to their diamond hosts, inclusions in diamonds are classified as protogenetic (preceding diamond formation), syngenetic (co-crystallizing with diamond) or epigenetic (crystallizing after diamond formation). Typically, a syngenetic origin for inclusions has been inferred if, regardless of their crystal system, the inclusions show a cubo-octahedral morphology that is imposed by their diamond hosts, which is most commonly the case (Harris and Gurney 1979; Meyer 1987; Stachel and Harris 2008; Bulanova et al. 2010). On the basis of the commonly observed diamond-imposed morphology the sublithospheric inclusions described herein are all assumed to be syngenetic, or at least have equilibrated synchronous with diamond crystallization and were trapped at the time of diamond growth, recording a geochemical snapshot of this process. These issues are discussed more extensively, from a geochronology perspective, in Smit et al. (2022, this volume).

Throughout this review the geochemistry and mineralogy of sublithospheric inclusions will be discussed relative to observations from petrological experiments performed at mantle conditions. Figure 1 shows experimentally derived estimates of the modal mineralogy expected in primitive mantle peridotite (e.g., pyrolite), harzburgite and mid-ocean ridge basalt (MORB) compositions (Ishii et al. 2018, 2019), illustrating how majoritic garnet, bridgmanite, Ca-perovskite, ferropericlase, olivine polymorphs and clinopyroxene dominate mineral assemblages at the depths of the deep upper mantle, transition zone and shallow lower mantle. Of the inclusions in our global dataset, ~42% are ferropericlase inclusions, 32% are majoritic garnet, $MgSiO_3$-rich and $CaSiO_3$-rich inclusions comprise about 8% each, 6% are clinopyroxene and 4% are olivine. We also discuss the occurrence of SiO_2 and retrograde CF and NAL phases that have been reported in sublithospheric diamonds, but these make up only a small fraction (< 1%) of the silicate inclusion population with reported chemistry.

Thus, the most common phases in sublithospheric diamonds mirror those in meta-basaltic and meta-peridotitic lithologies at high pressure (Harte 2010) but have been reported in proportions inconsistent with expected modal abundances in natural mantle assemblages (Fig. 1). This observation has been used to suggest the mantle might not accurately reflect a model primitive peridotite composition (Kaminsky 2012, 2017). However, diamond growth and trapping processes combined with sampling and preservation biases make observed inclusion proportions unreliable for assessing the modal mineralogy of the ambient mantle at depth (Liu 2002; Nimis et al. 2019). In this review we provide an analysis of the geochemistry of the common sublithospheric diamond inclusion types and use experimentally and theoretically derived phase relations and element partitioning data to constrain their depth of origin, plausible lithological associations and formation processes.

MAJORITIC GARNET

Garnet is the dominant aluminous mineral in mantle assemblages at depths greater than ~30–70 km, eventually becoming the second most abundant mineral in mantle peridotite and the dominant mineral in basaltic compositions throughout the deeper upper mantle and transition zone (Irifune 1987; Irifune and Ringwood 1993; Ishii et al. 2019) (Fig. 1). Garnet is chemically diverse, following the ideal formula $A_3B_2Si_3O_{12}$, where A cations occupy dodecahedral sites and B cations occupy octahedral sites. It is generally the case that divalent cations occupy the A-site in garnet, while the octahedral B-site is normally filled with trivalent cations. There are many exceptions to this simplified scheme (Grew et al. 2013)

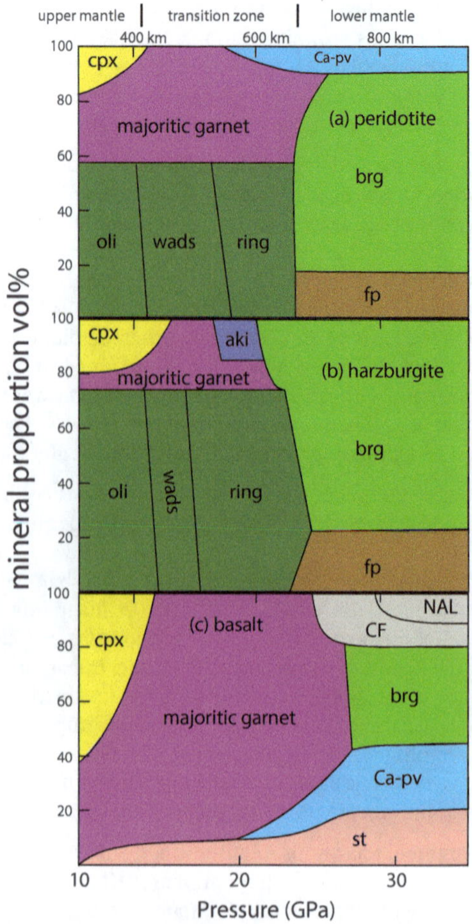

Figure 1. Mineral proportions (vol%) at 1600 °C based on the experiments of Ishii et al. (2018, 2019) in **(a)** primitive mantle peridotite, **(b)** harzburgite and **(c)** basalt bulk compositions as a function of pressure (GPa). Phase labels: cpx = clinopyroxene; oli = olivine; wads = wadsleyite; ring = ringwoodite; brg = bridgmanite; Ca-pv = Ca-silicate perovskite; st = stishovite; fp = ferropericlase; NAL = NAL phase; CF = CF phase.

but the most significant for understanding mantle garnets is that of titanium and phosphorus cations. Titanium occurs almost exclusively as Ti^{4+} in natural garnets (Locock 2008; Grew et al. 2013), and at lithospheric conditions can occupy either the tetrahedral Si site or the B-site (Waychunas 1987; Proyer et al. 2013; Ackerson et al. 2017a,b). However, at conditions relevant to sublithospheric inclusions, it is assumed that all Ti occupies the octahedral site in garnet (Ackerson et al. 2017b), charge balanced by monovalent Na^+ (or K^+) on the A-site (Ono and Yasuda 1996; Locock 2008; Proyer et al. 2013). Phosphorus, which is assumed to be exclusively pentavalent, is believed to occupy the tetrahedral silicon site, predominantly charge balanced by monovalent cations on the A-site (Haggerty et al. 1994).

Lithospheric garnets have a maximum of 3 silicon atoms per formula unit (pfu), whereas those that formed at pressures greater than about 6 to 8 GPa (equivalent to ~180–240 km depth) contain additional silicon, which is commonly referred to as the "majorite" component. The additional silicon is a consequence of the increasing solubility of pyroxene into garnet as equilibration pressure increases, and can be described by two principal substitution mechanisms:

$$2B^{3+} = {}^{VI}M^{2+} + {}^{VI}Si^{4+} \text{ [maj]} \quad (1)$$

$$A^{2+} + B^{3+} = {}^{VIII}X^{+} + {}^{VI}Si^{4+} \text{ [Na-maj]} \quad (2)$$

where ${}^{VI}M^{2+}$ is a divalent octahedral cation and ${}^{VIII}X^{+}$ is a monovalent dodecahedral cation. Both substitutions have been shown to increase with pressure, indicating that the octahedral silicon content (majorite component) in garnet is pressure dependent.

In the $MgO–Al_2O_3–SiO_2$ system, completion of substitution (1) leads to the formation of the majorite (Mj) garnet endmember ${}^{VIII}Mg_3{}^{VI}[MgSi]^{IV}Si_3O_{12}$ (Akaogi and Akimoto 1977; Irifune 1987), whereas substitution (2) produces the Na-majorite (NaMj) component ${}^{VIII}[Na_2Mg]^{VI}Si_2{}^{IV}Si_3O_{12}$ in the $Na_2O–MgO–Al_2O_3–SiO_2$ system (Irifune et al. 1989; Dymshits et al. 2013). The extent to which substitutions (1) and (2) occur is a complex function of pressure, temperature and composition (Collerson et al. 2010; Wijbrans et al. 2016; Beyer and Frost 2017; Tao et al. 2018), but substitution (1) tends to dominate in meta-peridotitic assemblages and substitution (2) in meta-basaltic assemblages (Kiseeva et al. 2013b; Thomson et al. 2021).

Importantly, any measurable majorite component can be used as a single mineral barometer via several available calibrations (Collerson et al. 2010; Wijbrans et al. 2016; Beyer and Frost 2017; Tao et al. 2018; Thomson et al. 2021). Majoritic garnets are classified as those that have a discernible majorite component (e.g., > 3 Si pfu) in their reported chemical analysis. In contrast with many previous studies, we follow the approach of Thomson et al. (2021) and explicitly account for tetrahedral phosphorus and monovalent charge balanced titanium, with the majorite component defined as:

$$\text{majorite component [maj]} = \text{Si} + \text{P} - 3 \text{ (pfu)} \quad (3)$$

Majoritic garnet inclusions provide direct evidence of an origin at depths greater than ~ 200 km on the basis of their pressure sensitive substitutions (Eqns. 1, 2) and are the only inclusions that provide a quantifiable, chemistry-based barometer. Inclusions of majoritic garnet are widespread and have been observed in diamonds collected from a wide range of localities, including cratons in South Africa, Brazil, Western Africa, Canada, Russia and China. We have compiled major element chemical analyses from 214 garnet inclusions that have a majorite component ≥ 0.005. Data and references are provided in Table 1 (Available at: https://doi.org/10.5683/SP3/LIVK1K). Most of the majoritic garnet inclusions are reported as single or multiple occurrences in a single diamond (>60%) or co-occurrences with clinopyroxene (~20%), and there are seven co-occurrences with a $CaSiO_3$-rich phase and seven with SiO_2. Thus, the co-occurring mineralogy indicates crystallization of majoritic garnet inclusions throughout the deep upper mantle and transition zone (Fig. 1).

Many, if not most, majoritic garnet inclusions have undergone retrograde re-equilibration and exsolution (Fig. 2). Exsolution of omphacitic clinopyroxene sometimes with other minor phases has frequently been reported (Harte and Cayzer 2007; Thomson et al. 2014; Zedgenizov et al. 2014a, 2016; Burnham et al. 2016; Sobolev et al. 2016), often as volumetrically small rinds at the extremities of individual inclusions (Fig. 2). In many studies where majoritic garnet inclusions are reported, exsolution features are not described or analysed. Our experience suggests that exsolution features may have commonly gone unreported, possibly as a consequence of poorer imaging capabilities in early generations of electron microprobes. This may be especially prevalent in studies where inclusions were broken out of diamonds rather than exposed by polishing. Omission of exsolved clinopyroxene affects the bulk inclusion chemistry such that it will always lead to underestimation of pressure using empirical, chemistry-based barometers. Analysis of entire majoritic garnet inclusions is also critical for future attempts to accurately date the inclusions.

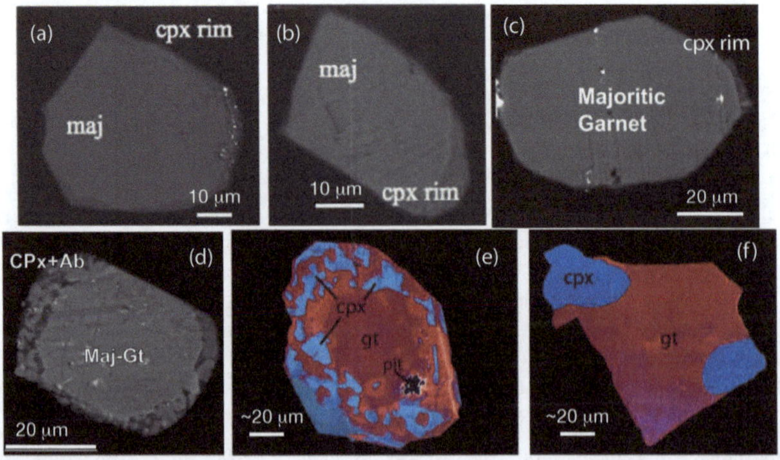

Figure 2. Backscattered electron images of majoritic garnet inclusions from the literature: **(a, b)** CC BY from Thomson et al. (2014); **(c)** Reprinted by permission from Springer Nature Customer Service Centre GmbH: Springer Nature. Bulanova et al.(2010), Copyright 2010; **(d)** Reprinted from Zedgenizov et al. (2014a) Copyright (2014), with permission from Elsevier; **(e, f)** false color: Reprinted by permission from Springer Nature Customer Service Centre GmbH: Springer Nature. Harte and Cayzer (2007), Copyright 2007. Majoritic garnet inclusions commonly exhibit exsolution of clinopyroxene, often as rinds around the margin of the inclusions (a–d), although more pervasive unmixing features also occur (e, f).

Major and minor element compositions

The major and minor element compositions of 215 majoritic garnet inclusions as determined by electron probe microanalysis are provided in Table 1 (Available at: https://doi.org/10.5683/SP3/LIVK1K) along with references describing analytical protocols. We compare these inclusion chemistries to a dataset of synthetic experimental majoritic garnets crystallized in peridotitic, harzburgitic, basaltic, sedimentary and hybrid bulk compositions across a wide range of upper mantle and transition zone pressure and temperature conditions. To our knowledge, the compositional data presented in Table 1 are based on microprobe analysis of the garnet portion of exposed/extracted inclusions without reincorporation of any exsolved material, making their derived pressures minimum estimates.

Figure 3a is a plot of CaO vs Cr_2O_3, a scheme originally constructed for classifying lithospheric garnets, which effectively delineates the majoritic garnet inclusions into two types: ***Low-Cr*** garnet inclusions exhibit low Cr_2O_3 (< ~1 wt.%), a wide range of Ca content (~ 2–18 wt.% CaO) and have Mg#s (Mg/(Mg+Fe)) predominantly lower than 0.7; ***High-Cr*** garnet inclusions exhibit high Cr_2O_3 (1–20 wt.%), low CaO (< 6 wt.%) and have Mg#s > 0.7 and typically > 0.8. In comparison with majoritic garnets from experimental studies, low-Cr inclusions are unlike garnets that crystallize in meta-peridotitic assemblages but overlap extensively with those produced in meta-basaltic assemblages that generally have extremely low Cr_2O_3 (< 0.1 wt.%) contents (Fig. 3, dark field) and a similarly wide range of CaO contents. We note that Cr_2O_3 contents in majoritic garnets produced in meta-sediment experiments invariably are not reported but are also expected to yield low-Cr garnet. Several low-Cr inclusions extend towards higher Cr_2O_3 and somewhat lower CaO contents and appear intermediate between majoritic garnet produced in meta-peridotitic and meta-basaltic assemblages (Thomson et al. 2016a).

High-Cr inclusions, with the exception of just a few, have much higher Cr than reported experimental garnets in fertile meta-peridotitic assemblages (Fig. 3, light-grey field) but overlap considerably with garnets produced in meta-harzburgitic assemblages (Fig. 3, red hex-stars), suggesting the high Cr_2O_3 may originate in highly depleted mantle compositions (Moore and Gurney 1985; Stachel et al. 2000a; Wang et al. 2000b; Schulze et al. 2008; Motsamai et al. 2018).

Figure 3b presents an alternative chemography based on CaO and TiO_2 contents. Experimental garnets generally occupy separate regions of this diagram depending on bulk composition, with meta-peridotitic garnets having low CaO and TiO_2, and meta-basaltic and meta-sedimentary garnets having higher CaO and TiO_2. Low-Cr inclusions, as distinguished primarily by lower Mg# on Figure 3b, are best represented by meta-basaltic and meta-sedimentary experimental garnets. In contrast the high-Cr inclusions have TiO_2 contents on the low-side of those observed in meta-peridotitic assemblages but are akin to some experimental meta-harzburgitic garnets. We note also that experimental meta-sediment garnets typically have Mg#s $\ll 0.4$, generally inconsistent with the observed compositional range of majoritic garnet inclusions. Overall, low-Cr majorite inclusions are similar to lithospheric garnet inclusions that have been classified as eclogitic (Stachel et al. 2000a) and which we refer to as meta-basaltic, whereas high-Cr inclusions are similar to garnets with depleted, meta-harzburgitic affinity.

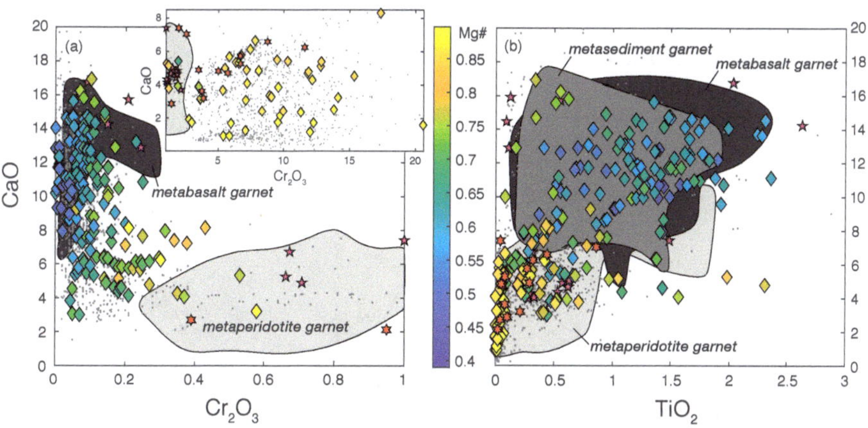

Figure 3. (a) CaO vs Cr_2O_3 (wt. %) and **(b)** CaO vs TiO_2 in the global majoritic garnet database (Table 1). Shaded fields depict the range of majoritic garnet compositions formed in meta-peridotitic (**light grey**), meta-basaltic (**dark grey**), and meta-sediment (**medium grey**) assemblages in high pressure-temperature experiments. Also shown are compositions for garnets formed in meta-harzburgitic assemblages (**red hex-stars**) and those produced by reaction of carbonated melt with meta-peridotite (**pink stars**). **Diamond symbols** show majoritic garnet inclusions in diamonds from the global sublithospheric database and are colored for Mg# ($Mg/[Mg+Fe_{tot}]$). **Small grey dots** are lithospheric garnet inclusions (see Stachel et al. 2022, this volume). The inset in (a) shows an expanded range for Cr_2O_3. Experimental data are from: ***meta-peridotite:*** Akaogi and Akimoto 1979; Irifune 1987; Irifune and Ringwood 1987; Herzberg and Zhang 1996; Walter 1998; Wang and Takahashi 2000; Wood 2000; Nishihara and Takahashi 2001; Hirose 2002; Ulmer and Sweeney 2002; Dasgupta et al. 2004; Kawamoto 2004; Walter et al. 2004; Brey et al. 2008; Dalou et al. 2009; Ghosh et al. 2009, 2014; Rohrbach and Schmidt 2011; Corgne et al. 2012; Tao et al. 2018; Rzehak et al. 2020; ***meta-harzburgite***: Irifune and Ringwood 1987; Ishii et al. 2019; ***meta-basalt***: Irifune et al. 1986; Yasuda et al. 1994; Ono and Yasuda 1996; Ono 1998; Okamoto and Maruyama 1999, 2004; Hirose and Fei 2002; Hammouda 2003; Litasov and Ohtani 2005; Kiseeva et al. 2013a; Thomson et al. 2016a; Beyer and Frost 2017; Zhang et al. 2020; ***meta-sediment***: Irifune et al. 1994; Ono 1998; Grassi and Schmidt 2011; Ishii et al. 2012; Bulatov et al. 2014; Brey et al. 2015; ***reaction***: Thomson et al. 2016a; Sun and Dasgupta 2019; Zhang et al. 2020.

Substitution mechanisms. Figure 4 is a plot showing majoritic garnet substitutional components (Eqns. 1–3) for inclusion and experimental garnets expressed as variation of monovalent and divalent cations. Majoritic garnets stable in meta-peridotitic assemblages predominantly follow the [maj] substitution vector. These meta-peridotitic garnets possess few, but often not zero, monovalent cations that are not charge balancing titanium (Fig. 4a), and they have an increasing proportion of divalent cations with increasing majorite component (Fig. 4b). The scatter around the ideal substitution vectors in Figure 4b may be partially explained by unidentified Fe^{3+}, but the incorporation of "extra" monovalent cations, especially at higher values of (Si + P – 3), demonstrates the occurrence of both substitution mechanisms in the experimental garnet compositions but with the [maj] substitution predominant. In contrast, experimental majoritic garnets from meta-basaltic and meta-sedimentary assemblages predominantly follow the ideal [Na-maj] substitution vector. This behavior reflects the higher alkali and lower magnesium contents in these bulk compositions, resulting in increasing monovalent cations with increasing majorite component (Fig. 4a), whereas divalent cations decrease (Fig. 4b).

Also shown on Figure 4 (as diamonds) are global majoritic garnet inclusion compositions from Table 1. High-Cr inclusions (green diamonds) cluster around the origin and extend solely along the [maj] vector to approximately 0.3. Low-Cr inclusions (blue diamonds) do not exclusively follow either substitution but rather span the range of compositions between both the [maj] and [Na-maj] substitutions. This could indicate lower Na basaltic protoliths, perhaps due to more Mg-rich basalts produced earlier in Earth's history (Pearson et al. 2003), but has previously been interpreted to indicate an association with hybrid or pyroxenitic compositions (Kiseeva et al. 2013b; Thomson et al. 2016a, 2021). Low-Cr inclusions with larger [maj] components possess higher but variable magnesium contents, 0.6 < Mg# < 0.85, and generally follow the [maj] substitution. However, these inclusion compositions are clearly not tied exclusively to the [maj] vector, and some skew significantly towards the [Na-maj] substitution.

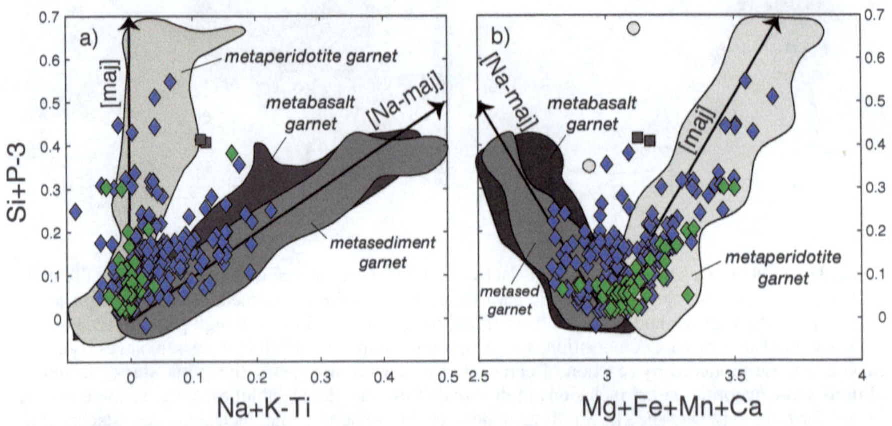

Figure 4. Majoritic garnet substitution vectors expressed in terms of Si+P–3 versus **(a)** variation in Na+K–Ti, which specifically accounts for monovalent cations charge balancing octahedral titanium, thus isolating only the components of pressure dependent substitutions, and **(b)** Mg+Fe+Mn+Ca. Since very few experimental samples have been explicitly analysed for ferric/ferrous iron content (Kiseeva et al. 2018), total iron content is calculated as Fe^{2+} together with Mg^{2+}, Ca^{2+} and Mn^{2+}. [maj] = majorite substitution; [Na-maj] = Na-majorite substitution. **Blue diamonds** are low-Cr inclusions and **green diamonds** are high-Cr inclusions. Fields showing experimental garnets are shaded as in Figure 3.

Majoritic garnet barometry

The chemical variation expressed by elemental substitutions in experimental majoritic garnet data sets has been used to calibrate empirical barometers for quantifying crystallization pressures, providing important constraints on the depths of diamond formation. The reader is referred to Nimis 2022 (this volume) for additional coverage of mineral barometry.

All published majoritic garnet barometers exclude effects of temperature from their experimental calibrations, relying solely on a parametrization of the major element chemistry to predict inclusion formation pressures (Collerson et al. 2010; Wijbrans et al. 2016; Beyer and Frost 2017; Tao et al. 2018; Thomson et al. 2021). The accuracy of calculated pressures in each study is based on the ability to reproduce their respective calibration datasets and is estimated to be ± 1–2 GPa. However, uncertainties can be much larger when these barometers are applied to experimental majoritic compositions outside the range of the calibration data. For example, when applied to majoritic garnets in the entire experimental database, Thomson et al. (2021) demonstrated much larger uncertainties among extant barometers, and a tendency for pressure underestimation, sometimes by as much as 10 GPa, when applied to the highest-pressure experimental garnets. In contrast to previous studies that used experimental subsets with limited compositional range, Thomson et al. (2021) trained a machine learning algorithm with all available experimental data to produce a barometer calibrated across the entire experimental pressure, temperature and composition space, with a significantly improved accuracy in reproducing the complete experimental dataset, especially at the highest pressures.

Shown on Figure 5 are histograms of garnet inclusion pressures calculated using majoritic garnet barometers. Despite differences in absolute pressures, all barometers exhibit a bimodal pressure distribution with distinct pressure modes at ~ 7–10 and ~ 12–15 GPa. The barometer of Thomson et al. (2021) predicts the highest-pressure modes with some inclusions indicating pressures as high as ~ 22 GPa (~ 600 km depth). However, we emphasize that many, if not all, majoritic garnet inclusions contain small amounts of exsolved omphacitic pyroxene, whose omission leads to pressure underestimation. This exsolution is presumably the effect of partial re-equilibration at lower pressures, post-entrapment, during diamond exhumation. Based on eight inclusions available from the entire global dataset where adequate data is available to estimate bulk inclusion compositions and correct for exsolution, Thomson et al. (2021) demonstrate that inclusion pressures may be underestimated by ~4 ± 2.5 GPa if exsolution features are ignored.

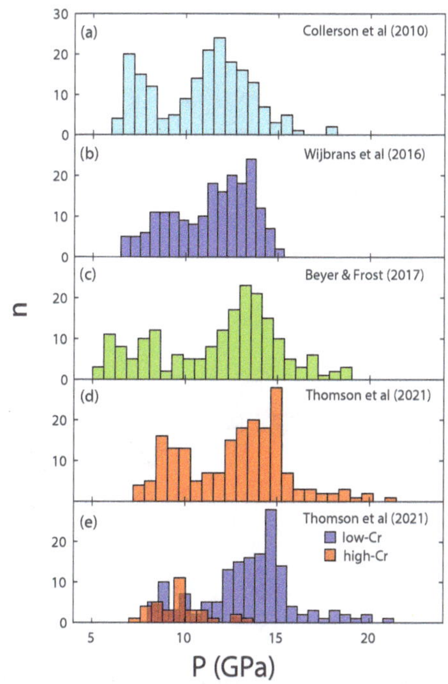

Figure 5. Pressure histograms of the global majoritic garnet inclusion dataset based on four recent experimentally calibrated empirical barometers: **(a)** Collerson et al. 2010; **(b)** Beyer and Frost 2017; **(c)** Wijbrans et al. 2016; **(d)** Thomson et al. 2021. The barometer of Tao et al. (2018) yields a distribution almost identical to Collerson et al. (2010). Note that there is no correction applied for pyroxene exsolution. **(e)** Pressure histogram applying the barometer of Thomson et al. (2021) with garnets separated on the basis of Cr-content.

Figure 5e shows histograms of majoritic garnet inclusion pressures calculated using the barometer of Thomson et al. (2021) separated according to low-Cr and high-Cr varieties. The distribution demonstrates that nearly all of the high-pressure mode is occupied by low-Cr garnets, which are meta-basaltic in composition. The concentration of the high-Cr majoritic garnets at lower pressures (e.g., < ~10 GPa) suggests that the diamonds hosting these depleted meta-peridotitic inclusions formed in a unique environment compared to high-Cr inclusions, possibly even in deep cratonic lithosphere rather than in the convecting mantle.

Trace element compositions

Measurements of trace element concentrations, mostly made using SIMS analyses at the Edinburgh Ion Probe Facility (EIMF), are reported for fifty-eight majoritic garnet inclusions in Table 2 (Available at: https://doi.org/10.5683/SP3/LIVK1K). Figure 6 shows trace element 'spidergrams' for majoritic garnet inclusions with element abundances normalized to the silicate Earth model of McDonough and Sun (1995), which we refer to as BSE (bulk silicate Earth). We have opted to maintain the low-Cr (Fig. 6a,c) and high-Cr (Fig. 6b,d) divisions and find that trace element abundance patterns are also generally distinct between these two groups.

Low-Cr majorite inclusions are systematically more enriched in trace elements than high-Cr inclusions, generally by about one order of magnitude. Virtually all majoritic garnet inclusions possess a negative Sr anomaly and many have positive Zr and Hf anomalies. Rare earth element (REE) patterns generally show depletions in the light rare earths (LREE), with Lu/La ratios ranging from ~ 0.15–1800 with >80% of inclusions greater than unity. Low-Cr inclusions generally have higher Lu/La than high-Cr inclusions. Where measured,

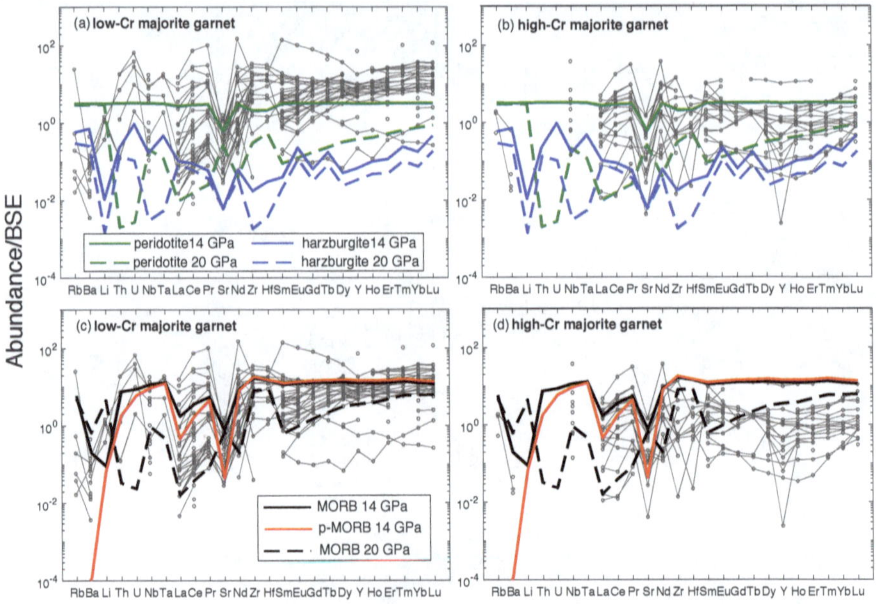

Figure 6. Trace element abundances in majoritic garnet inclusions from the global dataset. **(a)** and **(c)** show low-Cr garnets and **(b)** and **(d)** high-Cr garnets. Shown in panels **(a)** and **(b)** are calculated abundance patterns for majoritic garnet expected in meta-peridotitic and meta-harzburgitic mineral assemblages at 14 GPa (garnet + wadsleyite + clinopyroxene) and 20 GPa (garnet+ ringwoodite + Ca-silicate perovskite). Shown in panels **(c)** and **(d)** are calculated abundance patterns for majoritic garnet expected in meta-basaltic (MORB) mineral assemblages at 14 GPa (garnet + clinopyroxene + stishovite) and 20 GPa (garnet + Ca-silicate perovskite + stishovite), and in 'processed' MORB at 14 GPa (garnet + clinopyroxene + stishovite).

Th, U and Nb are relatively enriched relative to BSE whereas Ba, Li and Rb, with a few exceptions, are variably depleted. Small negative Eu and Y anomalies are present in some low-Cr majoritic garnets, whereas a number of the high-Cr inclusions exhibit large Y anomalies.

Also shown on Figure 6 for comparison with observed inclusions are calculated trace element concentrations estimated for majoritic garnet in subsolidus meta-peridotitic, meta-harzburgitic and meta-basaltic (MORB) mineral assemblages. Subsolidus majoritic garnet compositions were calculated using published mineral/melt partition coefficients for experimentally constrained phase assemblages following the approach of Thomson et al. (2016b).

For example, the following mass balance defines the trace element contents of any single phase in a multi-phase assemblage:

$$X_i^{\text{total}} = \alpha X_i^{\text{A}} + \beta X_i^{\text{B}} + \gamma X_i^{\text{C}} + \ldots \quad (4)$$

$$X_i^{\text{A}} = X_i^{\text{total}} \Big/ \left(\alpha + \beta \frac{D_i^{\text{B/melt}}}{D_i^{\text{A/melt}}} + \gamma \frac{D_i^{\text{C/melt}}}{D_i^{\text{A/melt}}} + \ldots \right) \quad (5)$$

where X_i^A, X_i^B, X_i^C are the concentrations of trace element i in phase A, B, and C, D_i are mineral/melt partition coefficients, and α, β and γ are the proportions of phases A, B and C in the phase assemblage. Bulk trace element contents (X_i^{total}) for peridotite are taken as BSE (McDonough and Sun 1995), for harzburgite are taken from the average of nine samples formed by melt depletion in a subduction zone environment as reported in Secchiari et al (2020), and mean mid-ocean ridge basalt (ALL-MORB) is used to represent basaltic compositions (Gale et al. 2013). 'Processed' MORB is calculated as described in Thomson et al. (2016b) and is used as an estimate of subducted MORB crust post sub-arc dehydration. Table 3 (Available at: https://doi.org/10.5683/SP3/LIVK1K) provides the source of partition coefficients and the phase proportions used in each phase assemblage to calculate trace element abundances in coexisting phases in mineral assemblages at pressures from the transition zone to the lower mantle.

Figure 6a shows that low-Cr inclusions are unlike those expected in meta-peridotitic assemblages at conditions of the transition zone; majoritic garnet in equilibrium with Ca-silicate perovskite in meta-peridotite or meta-harzburgite are significantly more depleted than the low-Cr inclusions. Peridotitic majoritic garnets at shallow transition zone conditions in equilibrium with wadsleyite and clinopyroxene have similar overall levels of enrichment relative to BSE as some low-Cr inclusions but the overall pattern and especially the abundances and slope of the REE and mild Sr anomaly are unlike the majoritic garnet inclusions. Consistent with their Ca and Cr contents, trace elements show that low-Cr majoritic garnets do not have meta-peridotitic affinity.

In contrast, Figure 6c demonstrates that low-Cr majoritic inclusions share characteristics of majoritic garnet compositions expected in meta-basaltic assemblages. The calculated trace element abundances of garnet at 14 GPa are generally within the range observed in low-Cr inclusions, while at 20 GPa calculated abundances are at the lower range of the inclusions due to coexistence with Ca-silicate perovskite. We note that the depletions in some large ion lithophile elements (LILE) and Sr are best reproduced in the 'processed' MORB composition, consistent with loss during sub-arc slab devolatilization. While some low-Cr inclusions have relatively flat middle to heavy REE similar to meta-basaltic garnet at 14 GPa (e.g., Lu/Sm near unity), many show depletions in LREE and MREE similar to MORB at 20 GPa, although with higher overall abundances by up to an order of magnitude. Thus, while low-Cr inclusions are more like meta-basaltic garnet compositions they generally have trace elements abundances that are elevated relative to expectations for trapped portions of subsolidus materials in processed MORB.

Overall abundance levels in high-Cr inclusions are generally similar to those expected in meta-peridotitic lithologies. They tend to possess negative Sr anomalies and generally have relatively flat REE abundance patterns (Fig. 6b), most akin to meta-peridotite at lower pressures where Ca-silicate perovskite is not stable, which is consistent with their lower calculated pressures (Fig. 5e). There are very few measurements of LILE, Th, U, Nb, Ta for high-Cr inclusions; while this may suggest they were very depleted in these components, it is important to note that these elements were not analyzed in all studies. Two additional features of the high-Cr inclusions are that many have negative Y anomalies, whose origin is unclear but are suggested to be associated with Earth surface processes (Thomson et al. 2016b). Additionally, several of the high chromium samples possess sinusoidal REE patterns, features that are common amongst lithospheric xenoliths and are thought to record the influence of metasomatic fluids (Stachel et al. 1998b; Wang et al. 2000b; Stachel et al. 2004), potentially consistent with their origin in the deep lithospheric mantle.

$CaSiO_3$-RICH AND $MgSiO_3$-RICH INCLUSIONS

Inclusions in sublithospheric diamonds with ABO_3 stoichiometry occur in both calcium-rich and magnesium-rich varieties. On the basis of their mineralogy and chemistry these inclusions are commonly interpreted to represent high-pressure phases with a former 'perovskite' structure that have retrogressed to lower-pressure polymorphs or phase assemblages (Harte and Harris 1994; Stachel et al. 1998a, 2000b, 2005; Harte et al. 1999; Joswig et al. 1999; Hutchison et al. 2001; Kaminsky et al. 2001; Brenker et al. 2002, 2005, 2021; Davies et al. 2004b; Hayman et al. 2005; Walter et al. 2008, 2011; Tappert et al. 2009b; Harte 2010; Thomson et al. 2014; Zedgenizov et al. 2015, 2016, 2020; Burnham et al. 2016; Nestola et al. 2018). As such, these inclusions are some of the few known samples thought to originate from the deep transition zone and shallow lower mantle and, therefore, can provide insight into the lithologies and processes occurring at these depths.

The mineral perovskite *sensu stricto* has a $CaTiO_3$ composition, is orthorhombic, and crystallizes in the *Pnma* space group. Perovskite-structured phases with both $MgSiO_3$ (bridgmanite) and $CaSiO_3$ compositions crystallize in high-pressure and temperature experiments in meta-basaltic and meta-peridotitic assemblages (Liu and Ringwood 1975; Yagi et al. 1978; Ito et al. 1984; Irifune 1987; Kesson et al. 1994; Kesson et al. 1995). $MgSiO_3$-rich inclusions interpreted as former bridgmanite occur as retrograde enstatite. $CaSiO_3$ and $Ca(Si,Ti)O_3$ inclusions in diamond are typically interpreted as products of originally perovskite-structured phases that have retrogressed to lower-pressure polymorphs, with $CaSiO_3$ most often occurring as breyite (formerly known as calcium walstromite) but wollastonite has also been observed (Nestola et al. 2018; Smith et al. 2018). $CaTiO_3$ perovskite coexisting with $CaSiO_3$ is also observed as part of composite inclusion assemblages.

Here we review the chemistry of $MgSiO_3$-rich and $CaSiO_3$-rich phases that occur as inclusions in sublithospheric diamonds. We recognize that the inclusions do not occur as high-pressure perovskite-structured polymorphs but rather as retrograde minerals, and we will review the evidence for their identification as former bridgmanite and Ca-silicate perovskite minerals, respectively.

$CaSiO_3$-RICH INCLUSIONS

The compilation of $CaSiO_3$-rich inclusion compositions includes fifty-three samples in diamonds from four cratons, forty-one of which are from South America. Mineralogical identification of the observed inclusions is often assumed on the basis of major element stoichiometry, although in some studies crystal structures have been determined by Raman spectroscopy or X-ray diffraction (Joswig et al. 1999; Brenker et al. 2005; Walter et al. 2011; Thomson et al. 2014; Burnham et al. 2016; Korolev et al. 2018; Nestola et al. 2018; Smith et al. 2018). Inclusions are either single phase $CaSiO_3$ (breyite or wollastonite) or composite mixtures of $CaSiO_3$ and other phases including $CaTiO_3$ (perovskite), $CaSi_2O_5$ (titanite-structured), Ca_2SiO_4

(larnite) and ZrO_2 (Fig. 7). Composite inclusions with overall $Ca(Si,Ti)O_3$ stoichiometry are typically interpreted to represent unmixing of an originally homogeneous phase; the alternative to the unmixing interpretation is that a portion of a 'rock' or melt was trapped encapsulating exactly a composition with ABO_3 stoichiometry, which is exceedingly improbable.

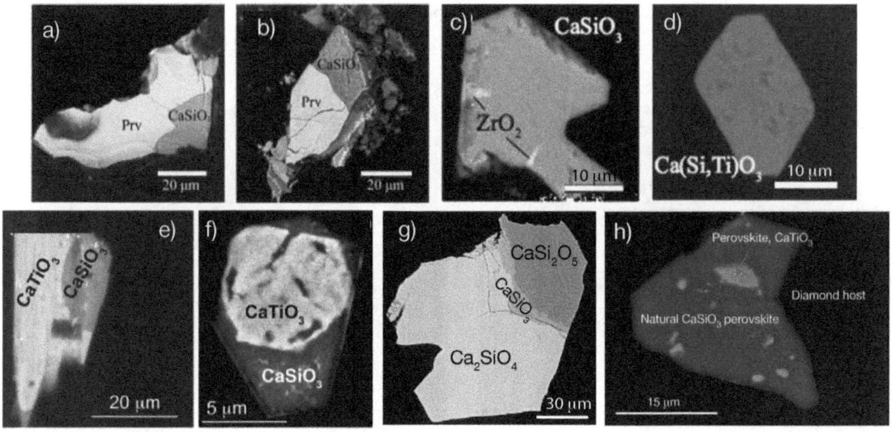

Figure 7. Backscattered electron images of composite $CaSiO_3$-rich inclusions from previous studies: **(a–b)** Prv = $CaTiO_3$ perovskite. Reprinted by permission from Springer Nature Customer Service Centre GmbH: Springer Nature. Hayman et al. (2005), Copyright 2005; **(c–d)** CC BY from Thomson et al. (2014); **(e, f)** Walter et al. 2008, reprinted by permission from Springer Nature Customer Service Centre GmbH: Springer Nature. Bulanova et al. (2010), Copyright 2010; **(g)** Reprinted from Joswig et al. (1999) Copyright (1999), with permission from Elsevier; **(h)** Reprinted by permission from Springer Nature Customer Service Centre GmbH: Springer Nature. Nestola et al. (2018b), Copyright 2018.

In five instances $CaSiO_3$-rich phases co-occur in the same diamond with $MgSiO_3$-rich phases, in nine cases they co-occur with ferropericlase, and in only three diamonds with both $MgSiO_3$ and ferropericlase. These non-touching, co-occurring assemblages are generally attributed to a lower mantle association (Stachel et al. 2000b; Kaminsky et al. 2001; Davies et al. 2004a; Hayman et al. 2005; Zedgenizov et al. 2014a) (Fig. 1). In five cases a $CaSiO_3$-rich phase occurs in the same diamond with majoritic garnet (Kaminsky et al. 2001; Hayman et al. 2005; Bulanova et al. 2010; Zedgenizov et al. 2014a). Other phases co-occurring with $CaSiO_3$-rich phases include SiO_2 (likely former stishovite), merwinite ($Ca_3Mg(SiO_4)_2$), $CaSi_2O_5$-titanite, chromite, Fe–Ni-metal and sulphide (Stachel et al. 2000b; Kaminsky et al. 2001; Brenker et al. 2005, 2021; Hayman et al. 2005; Walter et al. 2008; Bulanova et al. 2010; Thomson et al. 2014; Zedgenizov et al. 2014a,b; Burnham et al. 2016; Smith et al. 2016b).

Major and minor element compositions

Reported major and minor element compositions of $CaSiO_3$-rich inclusions are provided in Table 4 (Available at: https://doi.org/10.5683/SP3/LIVK1K). Many workers have noted that $CaSiO_3$-rich inclusions tend to be nearly phase pure, with analyzed compositions showing only minor amounts of MgO, Al_2O_3 and FeO in nearly all occurrences (Harte et al. 1999; Joswig et al. 1999; Stachel et al. 2000b; Kaminsky et al. 2001; Davies et al. 2004b; Hayman et al. 2005; Walter et al. 2008; Bulanova et al. 2010; Thomson et al. 2014; Zedgenizov et al. 2014a; Nestola et al. 2018). Most notably, MgO abundances are < 0.1 wt% in 29 of 40 inclusions where MgO was measured, and in those cases where MgO is not reported inclusions presumably also have exceptionally low abundances of this routinely measured oxide. Such low MgO contents are inconsistent with expectations for Ca-silicate perovskite in a meta-peridotitic assemblage at mantle temperatures (Wang et al. 2000a; Walter et al. 2008; Bulanova et al. 2010; Armstrong et al. 2012; Thomson et al. 2014).

We divide the $CaSiO_3$-rich inclusions into two groups based on a distinct compositional gap in TiO_2 contents, resulting in forty ***low-Ti*** inclusions with $TiO_2 < 0.7$ wt% and thirteen ***high-Ti*** inclusions with $TiO_2 > 2$ wt%. Figure 8 is a plot of Ti/(Ti + Si) versus (a) Mg/(Mg + Ca) and (b) Al (per formula unit), illustrating the unusual bulk compositions of many of the inclusion relative to the compositions of Ca-silicate perovskite in meta-peridotitic and meta-basaltic mineral assemblages synthesized in experiments.

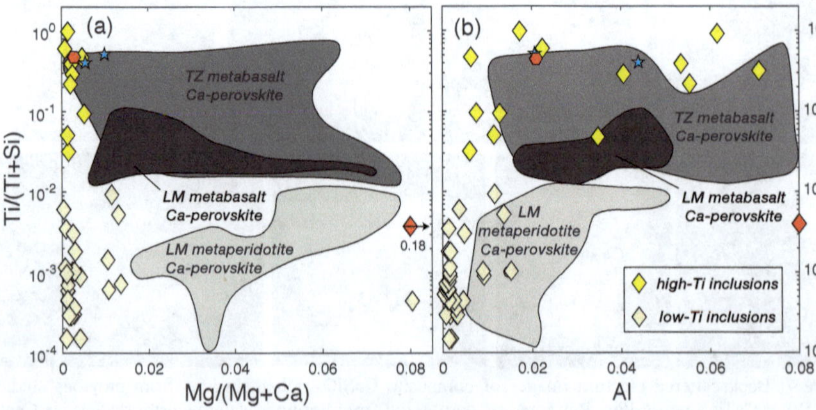

Figure 8. Molar Ti/(Ti+Si) versus **(a)** Mg/(Mg+Ca) and versus **(b)** Al (per formula unit) for $CaSiO_3$-rich inclusions (low-Ti = **light yellow diamonds**; high-Ti = **dark yellow diamonds**) in the global database. The **red diamond** is the calculated composition of an alkali and Cr-rich inclusion reported in Tschauner et al. (2021). Shown for comparison are fields for Ca-silicate perovskite synthesized in experimental studies: **Light grey field** shows Ca-silicate perovskite equilibrated with a lower mantle meta-peridotitic assemblage (LM = ± bridgmanite / ferropericlase / majorite); **Medium grey field** shows Ca-silicate perovskite equilibrated with a transition zone meta-basaltic assemblage (TZ = ± majorite / st / CAS phase / K-hollandite / aragonite / melt) and the **dark grey field** Ca-silicate perovskite equilibrated with a lower mantle meta-basaltic assemblage (LM = ± Al-bridgmanite /majorite / st /NAL / CF). Experimental data are from (Irifune 1994; Wood 2000; Ono et al. 2001; Hirose 2002; Hirose and Fei 2002; Litasov and Ohtani 2005; Ishii et al. 2011, 2018; Kiseeva et al. 2013a; Thomson et al. 2016a; Kuwahara et al. 2018). **Blue stars** are Ca-silicate perovskite produced by reaction between carbonated melt derived from basaltic crust and reducing primitive mantle (+majorite / ringwoodite / fp) (Thomson et al. 2016a). The **red hexagon** shows Ca-silicate perovskite equilibrated with majoritic garnet, magnesite and carbonated melt at 20 GPa, 1475 °C, in the system $CaO–MgO–Al_2O_3–TiO_2–SiO_2– CO_2$ (Walter et al. 2008).

Low-Ti $CaSiO_3$ inclusions. Most of the low-Ti $CaSiO_3$-rich inclusions are unlike those synthesized in experiments, having both exceptionally low MgO and TiO_2 contents (Fig. 8a), and also typically very low Al_2O_3 (Fig. 8b) and FeO contents. Three of the low-Ti inclusions co-occur in diamonds together with $MgSiO_3$-rich inclusions that have been interpreted to be of lower mantle origin, yet in each case the MgO contents are < 0.1 wt% (Stachel et al. 2000b; Davies et al. 2004b; Hayman et al. 2005; Zedgenizov et al. 2014a). Such low MgO contents are inconsistent with Ca-silicate perovskite in equilibrium with bridgmanite ($MgSiO_3$) in a lower mantle assemblage at mantle temperatures (Fig. 8), which have much higher MgO contents (Irifune et al. 2000; Walter et al. 2008). Only one inclusion, from Machado River in Brazil, has an MgO content high enough to potentially be consistent with an origin as part of a meta-peridotitic assemblage at lower mantle pressures and temperatures, albeit with much lower Ti (Burnham et al. 2016). Also shown on Figure 8 are experiments where Ca-silicate perovskite is in equilibrium with transition zone phases like majoritic garnet and ringwoodite in metabasaltic compositions. These experiments produce some Ca-silicate perovskites with lower Mg contents but with much higher Ti contents. In any case, the compositions of nearly all low-Ti $CaSiO_3$ inclusions are very unlike Ca-silicate perovskites produced in equilibrium with bridgmanite in primitive mantle peridotite at temperatures appropriate for ambient lower mantle (e.g., > 1700 °C).

A potential explanation for the low MgO content of the low-Ti $CaSiO_3$ inclusions is that they formed initially as Ca-silicate perovskite in equilibrium with bridgmanite but at low temperatures, considerably lower than in the experiments plotted on Figure 8 (Irifune et al. 2000; Armstrong et al. 2012). Irifune et al. (2000) demonstrated that at temperatures of 1500 °C and above, substantial (~10× higher than inclusions) MgO dissolves into Ca-silicate perovskite and suggested that the low-MgO content in $CaSiO_3$-rich inclusions reported in Harte et al. (1999) might reflect equilibration and inclusion entrapment at <1200 °C where the solvus widens, possibly in cool subducted lithosphere. Currently the solvus at temperatures below ~1400 °C is poorly constrained experimentally but could potentially be used as a thermometer for low-Ti $CaSiO_3$ inclusions.

High-Ti $CaSiO_3$ inclusions. Of the thirteen inclusions identified as having high titanium, ten are described as composite inclusions of $CaSiO_3$ + $CaTiO_3$ (e.g., Fig. 7). Reconstructing the bulk composition of composite inclusions, when attempted, has been done either through broad beam analysis of entire inclusions, or by analyzing phases separately and recombining them based on estimates of their modal abundance; both of these approaches can have considerable uncertainties (Walter et al. 2008, 2011; Bulanova et al. 2010; Thomson et al. 2014). Like low-Ti inclusions, high-Ti inclusions also have very low MgO contents but have higher Al_2O_3 and FeO (Fig. 8). Armstrong et al. (2012) showed that as the Ti-content in Ca-silicate perovskite in equilibrium with bridgmanite increases so does the solubility of MgO, such that the high-Ti inclusions should have levels of MgO at the several weight percent level if they formed in the lower mantle. On this basis, equilibrium of the high-Ti inclusions with bridgmanite at lower mantle pressures is excluded for all high-Ti inclusions.

As shown on Figure 8, some experimental Ca-silicate perovskites produced in meta-basaltic assemblages at transition zone pressures in equilibrium with majoritic garnet have low-Mg contents and high-Ti contents consistent with the high-Ti inclusions. We note that the experiments that best reproduce the inclusions are at relatively low temperatures and were produced in equilibrium with hydrous fluids or carbonatitic melts. Especially noteworthy are two experiments at 1000 °C where Ca-silicate perovskite is equilibrated with majoritic garnet, stishovite and a hydrous fluid, and has very low MgO contents but relatively high Al_2O_3 contents (Litasov and Ohtani 2005). Walter et al. (2008) reported on experiments in a simplified carbonated basalt system that showed $Ca(Si,Ti)O_3$-perovskite with very low MgO contents in equilibrium with majoritic garnet (red hexagon, Fig. 8). Similarly, Ca-silicate perovskite compositions in equilibrium with majoritic garnet and carbonatitic melt in experiments with basaltic starting compositions, or where carbonated melts were reacted with peridotite, also have high-TiO_2, low-MgO and high Al_2O_3 (Fig. 8b) and FeO similar to the inclusions (Walter et al. 2008; Thomson et al. 2016a).

Phase Relations

$CaSiO_3$-rich inclusions are often cited as evidence for diamond formation in the lower mantle (Harte et al. 1999; Joswig et al. 1999; Stachel et al. 2000b; Hayman et al. 2005; Harte 2010; Walter et al. 2011; Burnham et al. 2016; Smith et al. 2016b, 2018; Nestola et al. 2018). However, phase relations do not require a lower mantle or even transition zone origin for perovskite-structured $CaSiO_3$-rich phases to occur as inclusions in diamond (Kubo et al. 1997; Walter et al. 2008; Bulanova et al. 2010; Woodland et al. 2020; Brenker et al. 2021). Because the inclusions typically have only minor amounts of MgO, Al_2O_3 and FeO, phase relations for $CaSiO_3$-rich inclusions are well represented in the system $CaO–SiO_2–TiO_2$.

Low-Ti $CaSiO_3$ inclusions. Figure 9a shows phase relations in the $CaSiO_3$ system. Ca-silicate perovskite is stable at pressures above about 13 to 14 GPa at temperatures expected in the mantle, in contrast to the higher pressures at which Ca-silicate perovskite stabilizes in meta-basaltic or meta-peridotitic mantle assemblages (~20 GPa, Fig. 1). $CaSiO_3$ decomposes

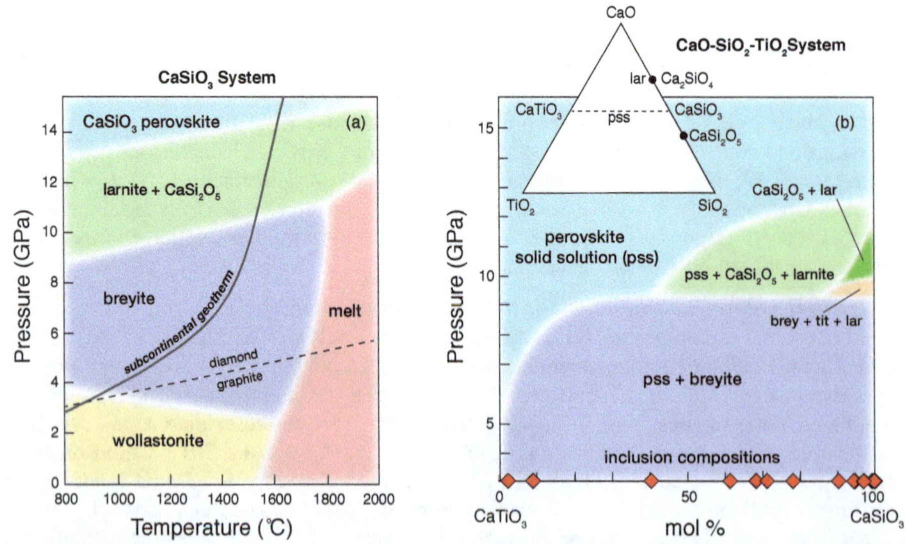

Figure 9. **(a)** Phase relations in the system $CaSiO_3$, after Figure 1 in Woodland et al. (2020). **(b)** Phase relations in the system $CaTiO_3$-$CaSiO_3$ after Kubo et al. (1997) in mol%. Phase relations are binary at compositions < ~40 mol% $CaSiO_3$ above ~9 GPa, with the ternary phase chemography depicted in the inset figure. **Red diamonds** are $CaSiO_3$-rich inclusions from the global dataset. Phase labels: pss = perovskite solid solution; lar = larnite; brey = breyite.

to a mixture of larnite (Ca_2SiO_4) plus titanite-structured $CaSi_2O_5$ between about 10 and 13 GPa, transforming to breyite at pressures below about 9 to 10 GPa and to wollastonite below ~3 GPa. Of the low-Ti inclusions where crystal structure was determined by Raman or X-ray diffraction, the majority indicate breyite as the $CaSiO_3$ phase (Table 4—Available at: https://doi.org/10.5683/SP3/LIVK1K). If the inclusions were originally formed as Ca-silicate perovskite, phase relations suggest the diamonds were transported from depths of at least ~ 400 km (~13 GPa) to depths of < 300 km where breyite is stable (~10 GPa) prior to exhumation by kimberlite. Wollastonite has also been observed in one inclusion (Nestola et al. 2018; Smith et al. 2018), which has a stability at less than about 3 GPa (Chatterjee et al. 1984; Sokolova and Dorogokupets 2021), requiring temperatures below ~1000 °C for both wollastonite and diamond to co-exist in equilibrium (Fig. 9a).

Four of the low-Ti $CaSiO_3$ inclusions show evidence of retrograde phase unmixing consistent with decompression. Joswig et al. (1999) reported on a low-Ti $CaSiO_3$ inclusion from Kankan (Guinea) with the composite assemblage breyite + larnite + titanite (Fig. 7g), which would ostensibly place its last equilibration directly on the phase boundary at ~10 GPa (Fig. 9a). Burnham et al. (2016) report clinopyroxene exsolution in two inclusions from Machado River (Brazil), one with a reconstructed bulk composition that could be in equilibrium with bridgmanite as described below. Two of the low-Ti $CaSiO_3$ inclusions from Juina (Brazil) co-occur with majoritic garnet (Tables 1 and 4—Available at: https://doi.org/10.5683/SP3/LIVK1K) and barometry yields pressures of ~13 and ~8 GPa but no information about possible clinopyroxene exsolution is provided so these are minimum pressures. Nine of the low-Ti inclusions co-occur with ferropericlase with Mg#s ranging from 0.75 to 0.9, with three of these co-occurring with an $MgSiO_3$-rich phase, indicating a deep transition zone or lower mantle origin related to a meta-peridotitic assemblage. However, we reiterate, the low MgO contents preclude equilibration with bridgmanite along a mantle geotherm and indicate either that the inclusions did not equilibrate with bridgmanite or did so at a significantly lower temperature, possibly in cold subducted lithosphere.

High-Ti $CaSiO_3$ inclusions. Figure 9b shows phase relations along the $CaTiO_3$-$CaSiO_3$ join. Of the ten composite inclusions exhibiting unmixing of $CaSiO_3$ and $CaTiO_3$ phases (Fig. 7), in five cases $CaSiO_3$ breyite and $CaTiO_3$ perovskite were confirmed through Raman or X-ray diffraction (Table 4—Available at: https://doi.org/10.5683/SP3/LIVK1K). Figure 9b shows that the pressure at which a single-phase $Ca(Si,Ti)O_3$ perovskite solid solution stabilizes depends on Ti-content, ranging from ~13 GPa at the $CaSiO_3$-rich end to <5 GPa at the $CaTiO_3$-rich end. If the high-Ti inclusions originated as perovskite solid solutions then crystallization pressures of greater than ~6 to 13 GPa are indicated, with most > 9 GPa. Five of the high-Ti $CaSiO_3$ inclusions co-occur with majoritic garnet, and barometry yields pressures of ~11, 12, 13, 14 and 22 GPa with no correction for clinopyroxene exsolution (Thomson et al. 2021). These pressures are consistent with formation of the composite $CaSiO_3$ inclusions originally as perovskite solid solutions but at depths spanning the deep upper mantle and transition zone rather than the lower mantle, consistent with their low MgO contents.

Pressures based on co-occurring majoritic garnet for the entire $CaSiO_3$-rich inclusion suite are generally consistent with minimum estimates from elastic barometry (Anzolini et al. 2018) and suggest that pressures of Ca-silicate perovskite entrapment in diamond may be considerably lower than expectations based on phase relations of mantle lithologies (Fig. 1), possibly due to crystallization from Ca-rich fluids or melts (Brenker et al. 2005, 2021; Walter et al. 2008; Bulanova et al. 2010). The unmixing exhibited in many of the $CaSiO_3$-rich inclusions to breyite-bearing assemblages requires transport of the diamond from the perovskite stability field to shallower depths in the mantle, with suggested mechanisms including mantle convection (e.g., transport in a plume) or with a percolating melt (Davies et al. 2004b; Harte and Cayzer 2007; Walter et al. 2008; Bulanova et al. 2010; Sun et al. 2020).

While exsolution as well as co-occurring minerals indicate an origin as Ca-silicate perovskite for many of the $CaSiO_3$-rich inclusions, it is possible that in some cases inclusions crystallized directly as single phase breyite at upper mantle pressures (e.g., ~ 3 to 9 GPa, Fig. 9a) rather than as Ca-silicate perovskite. It has been suggested that this might occur in Ca-rich lithologies like subducted meta-sediment or through reactions of melts derived from such Ca-rich sediments and peridotitic mantle (Brenker et al. 2005, 2021; Woodland et al. 2020). It is noteworthy that merwinite ($Ca_3MgSi_2O_8$) has been recognized as an inclusion co-occurring with low-Ti $CaSiO_3$ in two cases, suggestive of a Ca-rich association. We also note a few Ca-rich inclusions have a Ca/Si ratio greater than unity in composite assemblages of breyite + larnite (Brenker et al. 2005, 2021; Smith et al. 2018), although no bulk compositions have been reported and these are not part of our data compilation. It is currently unclear whether these composite inclusions represent a Ca-rich silicate phase formed in a unique lithology, or they contain a mass-balancing $CaSi_2O_5$-titanite phase that has gone undetected, or whether such inclusions might represent a trapped melt phase.

Perovskite-structured $CaSiO_3$ inclusions. Two studies present evidence for $CaSiO_3$-rich inclusions retaining a perovskite structure, both concluding petrogenesis within the lower mantle and preservation to the surface. Nestola et al. (2018) combined Raman, X-ray diffraction and EBSD on a composite $CaSiO_3$ + $CaTiO_3$ inclusion from South Africa (Cullinan) and interpreted the $CaSiO_3$ portion of the inclusion to be in an orthorhombic perovskite structure (Fig. 7h). However, this interpretation cannot be reconciled with the phase relations in Figure 9b, as there is no stability field where $CaTiO_3$-perovskite and $CaSiO_3$-perovskite coexist; experiments demonstrate a complete solid solution between these phases and unmixing should yield $CaTiO_3$ perovskite + breyite or wollastonite. Given the proximity of both $CaSiO_3$ and $CaTiO_3$ regions in the inclusions (Fig. 7h) and their very similar Raman spectra, as well as the large uncertainty in unit cell volume from the X-ray diffraction data, this interpretation requires further evaluation.

Tschauner et al. (2021) presented evidence from X-ray diffraction coupled with compositional information derived from a bulk LA-ICP-MS analysis (diamond plus inclusions) to argue that the core of a coated diamond from Botswana (GRR-1507) contains inclusions of an unusual alkali and chrome-rich variety of $CaSiO_3$ in the cubic perovskite structure. The data presented for both the diamond and the inclusion are more easily reconciled with an origin in cratonic lithospheric mantle. The high N content and poorly aggregated N of the diamond core are inconsistent with an origin at temperatures of the convecting mantle but are consistent with storage at lithospheric temperatures. X-ray diffraction data are not unique and can be well matched to phases common in micro-inclusion-bearing lithospheric diamonds. The calculated (not directly measured) bulk inclusion composition is too imprecise to confirm a phase with $CaSiO_3$ stoichiometry. Most notably, the remarkably high K, Na and Cr contents of the calculated inclusion are unlike any known $CaSiO_3$-rich inclusions in our data set, greater by factors of the order 10× (Fig. 8, Table 4—Available at: https://doi.org/10.5683/SP3/LIVK1K), but are similar to micro-inclusions found in the same suite of samples and other coated lithospheric diamonds (Navon et al. 1988; Weiss et al. 2014).

Both of these results require further evaluation, but we suggest the geological implausibility of recovering a sample of perovskite-structured $CaSiO_3$ at Earth's surface that originated in the transition zone or lower mantle. Experiments indicate that lower pressure polymorphs of $CaSiO_3$ (e.g., breyite or wollastonite) equilibrate in experiments in a matter of minutes to hours at 1200 °C (Kubo et al. 1997; Sueda et al. 2006), and these minerals are commonly observed in our $CaSiO_3$-rich inclusion database both as mono-crystalline phases and as unmixed components of composite inclusions (Fig. 7; Table 4—Available at: https://doi.org/10.5683/SP3/LIVK1K). Further, no experimentally synthesized perovskite-structured $CaSiO_3$ phase has ever been recovered from high pressure to 1 atmosphere conditions to our knowledge, converting instead to an amorphous phase upon decompression (Mao et al. 1989; Wang and Weidner 1994; Thomson et al. 2019). More work is needed to better evaluate potential *P–T* paths that may permit a stable perovskite-structured phase to be retained to the surface, but currently the data presented in these studies do not, in our view, support the interpretation of stable perovskite-structured $CaSiO_3$-rich phases as inclusions in diamonds.

Trace element compositions

Trace elements have been analyzed in twenty of the $CaSiO_3$-rich inclusions, eleven from the low-Ti group and nine from the high-Ti group. These data are provided in Table 5 (Available at: https://doi.org/10.5683/SP3/LIVK1K) and plotted as spidergrams on Figure 10, normalized to BSE. Eighteen of the twenty inclusions were measured by SIMS at the Edinburgh Ion Probe Facility (EIMF) and two by LA-ICP-MS. Also shown on these diagrams are the calculated abundance patterns for Ca-silicate perovskite in equilibrium with assemblages predicted for meta-peridotite and meta-basalt (MORB) at transition zone and lower mantle conditions.

Low-Ti $CaSiO_3$ inclusions. Figure 10a shows trace element abundances for the low-Ti inclusions. The inclusions are enriched relative to BSE in most trace elements by up to more than two orders of magnitude. Most trace elements are also enriched by up to an order of magnitude relative to models for the trace element abundances expected for sub-solidus Ca-silicate perovskite in meta-peridotitic or meta-harzburgitic assemblages in either the lower mantle or deep transition zone. They are most akin to Ca-silicate perovskite in either MORB or processed MORB, but this is not consistent with their low Ti contents (Fig. 8). Many of the inclusions have relative depletions in Rb, Ba, Sr, Zr, Hf, Nb, Ta and Y and enrichments in light REE relative to heavy REE. It is noteworthy that the least enriched inclusion could be consistent with an origin in a meta-peridotitic lower mantle assemblage, and this inclusion, from Machado River in Brazil (Burnham et al. 2016), is also the only inclusion in the entire dataset with an MgO content plausibly consistent with equilibration with bridgmanite at lower mantle temperatures. Thus, the low-Ti inclusions have a meta-peridotitic major element affinity but are generally enriched in incompatible trace elements.

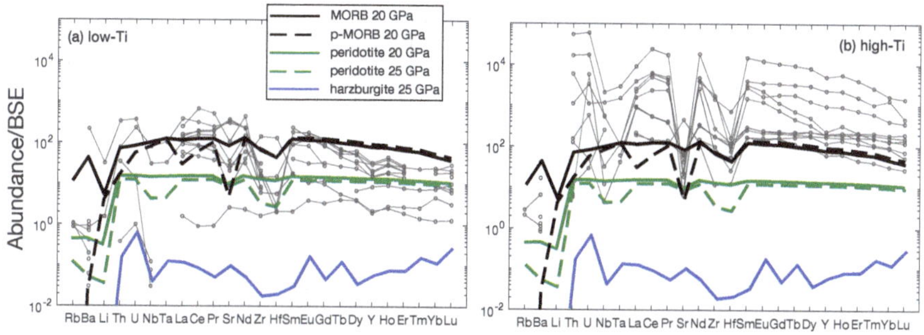

Figure 10. Trace element abundances in $CaSiO_3$-rich inclusions from the global dataset normalized to BSE: **(a)** low-Ti inclusions, and **(b)** high-Ti inclusions. Also shown are calculated abundance patterns for Ca-silicate perovskite expected in meta-peridotitic, meta-harzburgitic, and meta-basaltic (MORB and processed MORB) mineral assemblages in the transition zone and lower mantle; meta-basalt (MORB/p-MORB) at 20 GPa = Ca-silicate perovskite + majoritic garnet + ringwoodite; meta-peridotite at 20 GPa = Ca-silicate perovskite + majoritic garnet + ringwoodite; meta-peridotite/harzburgite at 25 GPa = Ca-silicate perovskite + bridgmanite + ferropericlase. Mineral modes and partition coefficients are provided in Table 3.

High-Ti $CaSiO_3$ inclusions. Figure 10b shows trace element abundances for the high-Ti inclusions. A distinguishing feature of these inclusions is their extremely elevated trace element abundances relative to BSE. For example, the most enriched inclusions have 1000 to 70,000 × BSE in incompatible elements like Th, U and the light REE. The inclusions also exhibit a large negative Sr anomaly in all but one inclusion and are characterized by relative depletions in Hf, Zr, Nb and Ta, and also Rb and Ba when measured (these elements were below detection levels in several inclusions).

The trace element abundances in the inclusions are not consistent with subsolidus Ca-silicate perovskite in either meta-peridotitic or meta-basaltic mineral assemblages in the upper to lower mantle. Abundances in the least enriched inclusions overlap with modeled Ca-silicate perovskite in MORB or processed MORB but have much more pronounced Sr and HFSE anomalies. Most inclusions are significantly enriched relative to expectations for MORB, by up to more than two orders of magnitude for many elements (Wang et al. 2000a; Walter et al. 2008; Bulanova et al. 2010; Thomson et al. 2016b). Models for the overall enrichments in trace elements and the characteristics of the abundance patterns in $CaSiO_3$ inclusions are generally consistent with equilibration involving low-degree melts, likely derived from meta-basaltic assemblages, as described further in the '*Discussion*' section.

$MgSiO_3$-RICH INCLUSIONS

The compilation of $MgSiO_3$-rich phases with ABO_3 stoichiometry includes fifty-five inclusions in diamonds from six cratons, forty-two of which are from South America (Table 6—Available at: https://doi.org/10.5683/SP3/LIVK1K). Both single phase and composite (Fig. 11) inclusions have been reported. Thirty-five of the inclusions occur in diamonds with assemblages that include ferropericlase and four co-occur with $CaSiO_3$-rich phases. There are three co-occurrences with both ferropericlase and $CaSiO_3$ in the same diamond. The co-occurrence of $MgSiO_3$-rich phases with ferropericlase and/or $CaSiO_3$-rich phases has provided the basis for the interpretation of a meta-peridotitic, lower mantle association for these inclusion assemblages (Harte et al. 1999; Stachel et al. 2000b; Hutchison et al. 2001; Kaminsky et al. 2001; Davies et al. 2004b; Hayman et al. 2005; Harte 2010; Burnham et al. 2016). No $MgSiO_3$-rich inclusions have been found with a bridgmanite crystal structure, and it has been commonly assumed that the observed enstatite structured $MgSiO_3$ inclusions, identified on the basis of X-ray diffraction and Raman spectroscopy (Hutchison et al. 2001; Walter et al. 2011; Thomson et al. 2014; Burnham et al. 2016), represent retrogression from bridgmanite.

Minerals observed together with $MgSiO_3$ in composite inclusions and interpreted as exsolved phases during retrogression include olivine, ferropericlase, jeffbenite (formerly known as TAPP, ideally $Mg_3Al_2Si_3O_{12}$), ilmenite, magnetite, spinel and magnesian ulvöspinel (Fig. 11) (Harte et al. 1999; Stachel et al. 2000b; Hutchison et al. 2001; Kaminsky et al. 2001; Hayman et al. 2005; Walter et al. 2011; Thomson et al. 2014; Zedgenizov et al. 2014a). Other phases co-occurring with $MgSiO_3$-rich phases in the same diamond include clinopyroxene, olivine, carbonate and Ni-rich metal. We note that eight jeffbenite inclusions have been included in our $MgSiO_3$-rich inclusion compilation.

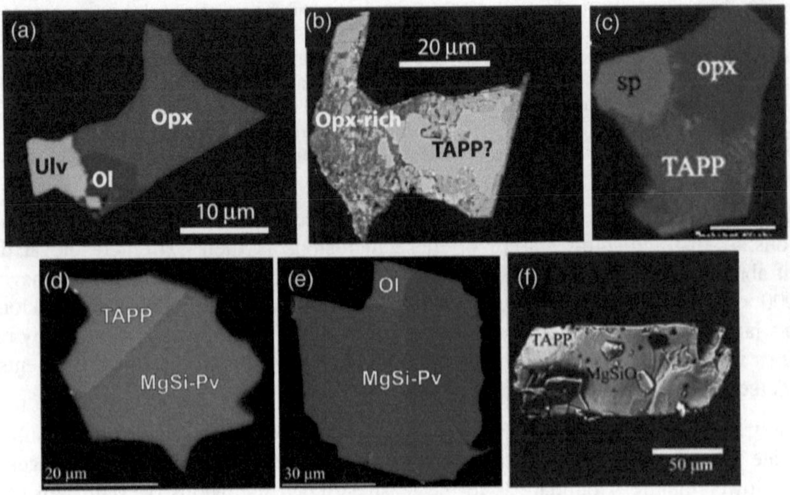

Figure 11. Backscattered electron images of composite $MgSiO_3$-rich inclusions from previous studies: (**a–b**) From Walter et al. (2011). Reprinted with permission from AAAS; (**c**) CC BY Thomson et al. (2014a); (**d–e**) Reprinted from Zedgenizov et al. (2014a) Copyright (2014), with permission from Elsevier; (**f**) Reprinted by permission from Springer Nature Customer Service Centre GmbH: Springer Nature. Hayman et al. (2005), Copyright 2005. Opx = orthopyroxene; Ol = olivine; Ulv = ulvöspinel; TAPP = jeffbenite; sp = spinel; MgSi-Pv = $MgSiO_3$-phase.

Major element compositions

Published major element compositions of the $MgSiO_3$-rich inclusions as determined by electron microprobe are provided in Table 6 (Available at: https://doi.org/10.5683/SP3/LIVK1K). The inclusions are separated into two distinctive groups on the basis of a gap in their alumina contents: a ***low-Al*** group with $Al_2O_3 < 3.5$ wt% and a ***high-Al*** group with $Al_2O_3 > 7$ wt%.

The *low-Al* group comprises thirty-four $MgSiO_3$-rich inclusions, twenty-four of which are reported as single-phase inclusions and a further ten are composite and contain minor exsolved phases that include olivine, ferropericlase and jeffbenite. The EPMA analyses of the composite inclusions, to the best of our knowledge, do not include exsolved phases but represent the $MgSiO_3$-rich portion of the inclusion. The exceptions are two composite inclusions reported by Burnham et al. (2016) where bulk inclusion compositions are reported by recombination of observed phases. Twenty-seven of the low-Al inclusions co-occur with ferropericlase and four with ferropericlase and $CaSiO_3$-rich phases.

The *high-Al* group comprises twenty-one inclusions, all from South America. Seven of these are single-phase inclusions and six are composite inclusions (Fig. 11) containing enstatite ± jeffbenite/olivine/spinel-ulvöspinel/magnetite/SiO_2. The bulk compositions of four of the six composite inclusions were estimated either through broad beam or multiple analyses of entire inclusions or by combining spot analyses of individual phases on the basis of their mode estimated from image analysis (Walter et al. 2011; Thomson et al. 2014). The remaining eight inclusions in this group are jeffbenite.

Armstrong et al (2012) noted the similarity of jeffbenite inclusions to high-Al bridgmanite produced in experiments on basaltic starting compositions and speculated on this basis that jeffbenite could represent retrograde aluminous bridgmanite. These authors also located a low-pressure stability field for jeffbenite at < 10 GPa that is consistent with its formation as a lower-pressure phase. Three high-alumina, single phase inclusions co-occur with ferropericlase inclusions, as does one of the Na-rich inclusions and five of the jeffbenite inclusions.

Figure 12 shows $MgSiO_3$-rich inclusions plotted in the compositional ternary diagram $(Mg+Fe^{2+})-(Si+Ti)-(Al+Cr+Fe^{3+})$. The *low-Al* inclusions plot in a well-defined region that partially overlaps the field of experimental bridgmanites synthesized in meta-peridotitic assemblages. However, most experimental bridgmanites made using primitive mantle compositions have higher trivalent cations than the inclusions. Many *low-Al* inclusions have similarity with bridgmanite produced in experiments using harzburgitic bulk compositions, and also overlap with akimotoite synthesized in peridotitic bulk compositions; akimotoite is an ilmenite-structured $MgSiO_3$-rich phase that is stable over a limited pressure-temperature range in meta-peridotite near the base of the transition zone (Fig. 16; Stixrude and Lithgow-Bertelloni 2011).

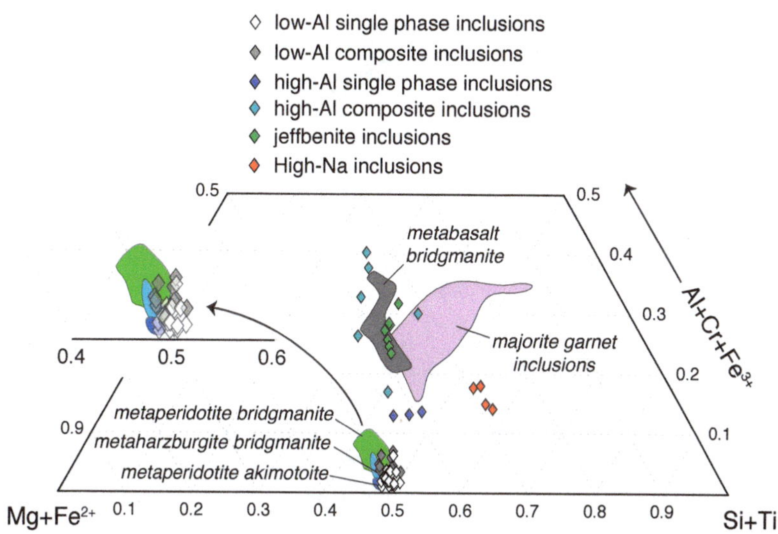

Figure 12. Projection of $MgSiO_3$-rich inclusion compositions (Table 6) into the pseudo-ternary $(Mg+Fe^{2+})$-$(Si+Ti)$-$(Al+Cr+Fe^{3+})$. Iron is treated as a 50:50 mixture of Fe^{2+} and Fe^{3+} for plotting purposes based on the established proclivity for bridgmanite to accommodate ferric iron (McCammon et al. 1997; McCammon et al. 2004). Inclusions types are separated as follows: **white diamonds**, low-Al single phase inclusions; **grey diamonds**, low-Al composite inclusions; **blue diamonds**, high-Al single phase inclusions (Type-II of Hutchison et al. 2001); **cyan diamonds**, high-Al composite inclusions; **green diamonds**, jeffbenite inclusions; **red diamonds**, high-Na (Type III of Hutchison et al. 2001). Fields are also shown for experimental bridgmanite and akimotoite produced in meta-basaltic, meta-peridotitic and meta-harzburgitic assemblages at pressures from 23 to 47 GPa (Irifune and Ringwood 1987; Irifune et al. 1994, 2010; Wood 2000; Ono et al. 2001; Hirose 2002; Hirose and Fei 2002; Tronnes and Frost 2002; Nishiyama and Yagi 2003; Nishiyama et al. 2004; Walter et al. 2004; Litasov and Ohtani 2005; Ishii et al. 2011, 2018, 2019). Also shown is the field of majoritic garnet inclusions, this study.

In contrast, the *high-Al* inclusions show considerable compositional variation, with Figure 12 showing that the six composite inclusions (cyan diamonds) and the jeffbenite inclusions (green diamonds) are generally similar to bridgmanite produced in meta-basalt on this projection, whereas the three single phase inclusions (blue diamonds; Type II $MgSiO_3$ inclusions of Hutchison et al. 2021) plot between experimental meta-peridotitic and meta-basaltic bridgmanites. The four Na-rich (~4–6 wt% Na_2O) inclusions (red diamonds; Type III $MgSiO_3$ inclusions of Hutchison et al. 2021) are unlike any experimental bridgmanite or other $MgSiO_3$-rich inclusions.

Figure 13 shows the Mg# of $MgSiO_3$-rich inclusions plotted against the Mg# of ferropericlase inclusions that co-occur in the same diamond. Also shown are fields for bridgmanite and ferropericlase equilibrated together in experiments on fertile peridotite bulk compositions and in a harzburgite composition. Ferropericlase with Mg#s less than ~0.8 are inconsistent with equilibration with co-occurring bridgmanite in meta-peridotitic or meta-harzburgitic assemblages. In diamonds hosting ferropericlase with Mg#s greater than 0.8, it is striking that very few of the inclusion pairs plot within the field of fertile meta-peridotite. Many of the *low-Al* bridgmanite-ferropericlase inclusion pairs overlap with or plot close to the field of meta-harzburgite, with $MgSiO_3$-rich inclusions tending to have very high Mg#s. Three single-phase high-Al $MgSiO_3$-ferropericalse pairs and two jeffbenite-ferropericlase pairs plot just within or close to the experimental meta-peridotite field.

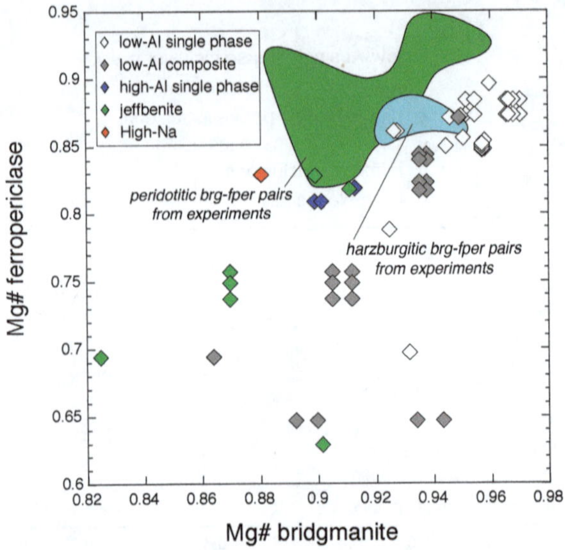

Figure 13. Mg# of $MgSiO_3$-rich inclusions versus the Mg# of co-occurring (same diamond) ferropericlase inclusions (all iron calculated as divalent). The fields show bridgmanite and ferropericlase compositions from experiments on fertile primitive peridotite compositions (**green field**) (Irifune 1994; Wood 2000; Hirose 2002; Tronnes and Frost 2002; Nishiyama and Yagi 2003; Walter et al. 2004; Ishii et al. 2011, 2018; Kuwahara et al. 2018) and harzburgite composition (**cyan field**) (Ishii et al. 2019), where bridgmanite coexists with ferropericlase ± Ca-silicate perovskite.

Low-Al inclusions. Figure 14 shows NiO, Al_2O_3 and CaO versus Mg# for $MgSiO_3$-rich inclusions (diamonds) compared with bridgmanite synthesized in experiments on peridotitic bulk compositions. The low-Al inclusions (white diamonds) occur over a range of Mg# from ~ 0.86 to 0.97, most concentrated between 0.92 and 0.97. Comparatively, bridgmanites observed in primitive meta-peridotitic assemblages have bulk compositions with lower Mg#s

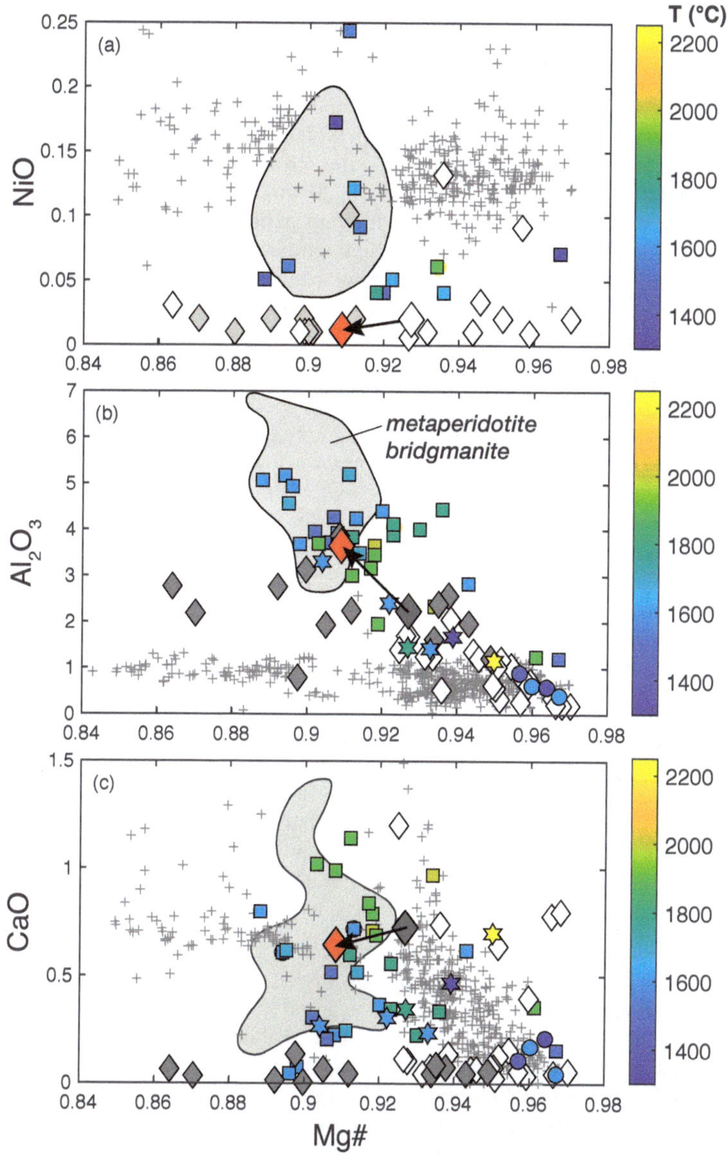

Figure 14. NiO **(a)**, Al_2O_3 **(b)** and CaO **(c)** versus Mg# showing $MgSiO_3$-rich inclusions: **white diamonds**, low-Al single phase inclusions; **light grey diamonds** in (a) are high-Al inclusions. In (b) and (c) **dark grey diamonds are** low-Al composite inclusions; **red diamond**, re-constructed low-Al, high Ca inclusion from Burnham et al. (2016), connected to the measured $MgSiO_3$ portion of the inclusion by an arrow. Experimental bridgmanites coexisting with ferropericlase ± Ca-silicate perovskite in meta-peridotitic assemblages are shown as a **light grey field** (Ito and Takahashi 1987; Irifune 1994; Wood 2000; Hirose 2002; Tronnes and Frost 2002; Nishiyama and Yagi 2003; Nishiyama et al. 2004; Walter et al. 2004; Irifune et al. 2010; Ishii et al. 2011, 2018, 2019; Kuwahara et al. 2018); **squares**, bridgmanite in experiments on meta-peridotite coexisting with majorite ± ferropericlase / Ca-silicate perovskite / ringwoodite; **circles**, akimotoite in experiments on meta-peridotitic compositions coexisting with majorite ± Ca-silicate perovskite/ringwoodite/stishovite; **six pointed stars**, bridgmanite in experiments on harzburgite composition; **crosses**, orthopyroxene inclusions in lithospheric diamonds (see Stachel et al. 2022 for references). Experiments denoted by **symbols** are shaded for experimental temperature.

that concentrate between 0.88 and 0.92, with some extending as high as 0.97. The NiO contents of the $MgSiO_3$-rich inclusions are low, generally less than 0.05 wt%; only in a few exceptions do inclusions possess values exceeding 0.1 wt%.

NiO contents of the $MgSiO_3$-rich inclusions are low relative to enstatite inclusions in lithospheric diamonds (gray crosses in Fig. 14a), which has been used as evidence to support an origin in the lower mantle because Ni is expected to partition strongly into coexisting ferropericlase (Harte et al. 1999; Stachel et al. 2000b). However, Figure 14a shows that experimental bridgmanite in equilibrium with ferropericlase in experiments at pressures between 23 and 43 GPa and over a range of high temperatures have NiO contents distinctly higher than the majority of the inclusions, ranging from about 0.05 to 0.25 wt%. This implies that equilibration with ferropericlase in a primitive mantle composition does not account for the low Ni contents of most inclusions.

Figures 14b and 14c show that like NiO, the Al_2O_3 and CaO contents of $MgSiO_3$-rich inclusions are generally lower than in experimental peridotitic bridgmanite compositions. The Al_2O_3 contents of the inclusions show a negative correlation with Mg#, ranging from about 3 to 0.2 wt% at Mg#s between 0.92 and 0.97. In composite low-Al inclusions the $MgSiO_3$ portions with Mg#s less than 0.92 are higher in Al_2O_3 (~ 3 wt%) than in single phase low-Al inclusions. Experimental bridgmanites in equilibrium with a lower mantle assemblage of ferropericlase ± Ca-silicate perovskite range from ~3 to 7 wt% Al_2O_3 and are unlike the inclusion compositions. Experimental bridgmanites in equilibrium with a deep transition zone assemblage of majoritic garnet ± Ca-silicate perovskite/ringwoodite/ferropericlase are shown as squares on Figure 14. A few of these experimental bridgmanites trend to very low Al_2O_3 and high Mg#, and we note that two experimental bridgmanites with ~1 wt% Al_2O_3 occur in majoritic garnet + ringwoodite-bearing (±Ca-silicate perovskite/ferropericlase) assemblages at ~ 23 GPa.

Also intriguing are the compositions of akimotoite in four experiments coexisting with majoritic garnet ± ringwoodite/Ca-silicate perovskite/stishovite (circles on Fig. 14), providing an alternative interpretation for original polymorph of some $MgSiO_3$-rich inclusions. Notably, bridgmanite formed in experiments on depleted harzburgite (hex-stars) can also have similarly low Al_2O_3 contents and high Mg#s that overlap with many of the low-Al inclusions. We note that while there is overlap with the Al_2O_3 contents of enstatite inclusions from lithospheric diamonds at high Mg#s, overall, the $MgSiO_3$-rich inclusions are distinct from lithospheric enstatites.

CaO contents in experimental bridgmanites in equilibrium with ferropericlase ±Ca-silicate perovskite assemblages (Fig. 14c) are also notably higher than most of the observed inclusions. Bridgmanites in several majorite-bearing experiments at lower temperatures have similarly low CaO contents, and as observed with Al_2O_3, experimental akimotoites have CaO contents similar to the inclusions as do bridgmanites in meta-harzburgite assemblages. Like Al_2O_3, there is overlap with the CaO contents of enstatite inclusions from lithospheric diamonds at high Mg#s but, overall, the $MgSiO_3$-rich inclusions have lower CaO and are distinct from lithospheric enstatite inclusions. An exception is a low-Al inclusion with high NiO that is akin to lithospheric inclusions from Eurelia (Australia) but co-occurs with ferropericlase (Tappert et al. 2009b).

Four of the low-Al inclusions that have high CaO (> 0.5 wt%) are from the Machado River deposit in Brazil as reported in the study of Burnham et al. (2016). Three of these inclusions have low NiO (<0.05 wt%) and low Al_2O_3 (<0.5 wt%), while the fourth composite inclusion has ~3.8 wt% Al_2O_3. All four of these inclusions co-occur with high Mg# ferropericlase that could be in equilibrium with bridgmanite. It is noteworthy that a $CaSiO_3$-inclusion from the Machado locality is also the only such inclusion with an MgO content consistent with an origin in primitive mantle peridotite. Burnham also reports a reconstructed $MgSiO_3$-rich inclusion that is shown as a red diamond on Figure 14. This inclusion, like other low-Al composite inclusions, has an exsolved aluminous phase with a jeffbenite composition. Reconstruction of the bulk composition results in slightly lower NiO and CaO but higher Al_2O_3. We expect that other low-Al composite inclusions may also require such corrections. This would imply that alumina contents in some cases may be underestimated, possibly reflected in the nearly constant alumina content of the low-Al composite inclusions irrespective of Mg#.

High-Al inclusions. Figure 15 shows TiO_2, Na_2O, CaO and Al_2O_3 versus Mg# for *high-Al* $MgSiO_3$-rich inclusions compared with bridgmanites synthesized in experiments on basaltic bulk compositions. The six composite inclusions and one of the jeffbenite inclusions, all from the Juina region of Brazil, have low Mg#s (~0.43–0.65) consistent with experimental bridgmanites formed in meta-basaltic assemblages (Walter et al. 2011; Armstrong and Walter 2012; Pla Cid et al. 2014; Thomson et al. 2014). The composite inclusions include high-Ti and high-Al contents, although the two inclusions reported by Pla Cid et al. (2014) are notable in their low TiO_2. The Ca-contents of all the composite inclusions are lower than observed in experimental high-Al bridgmanite phases.

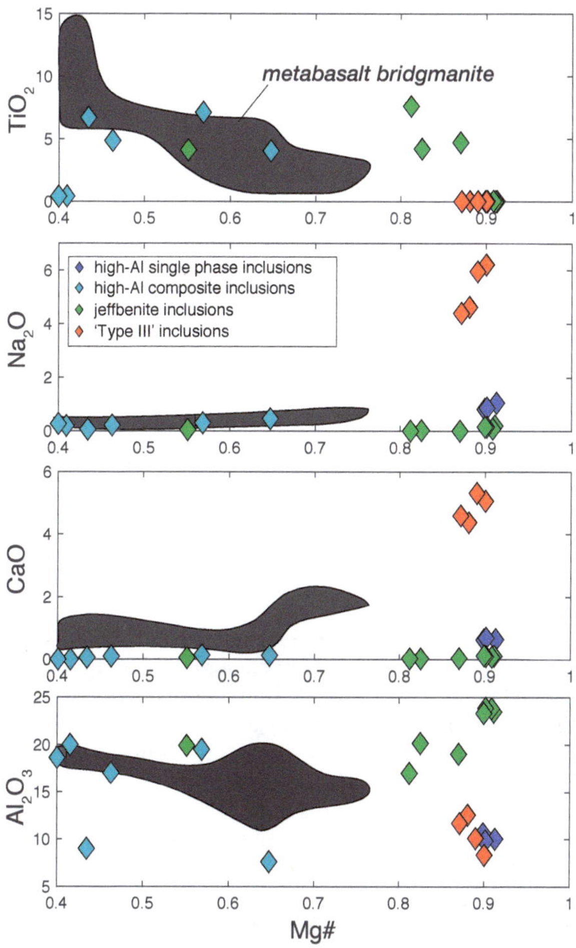

Figure 15. TiO_2 **(a)**, Na_2O **(b)**, CaO **(c)** and Al_2O_3 **(d)** versus Mg# showing high-Al $MgSiO_3$-rich inclusions: **blue diamonds**, low-Al single phase inclusions; **cyan diamonds** high-Al composite inclusions; **green diamonds**, jeffbenite inclusions; **red** diamonds, Type III inclusions of Hutchison et al. (2001); **dark grey field**, experimental high-Al bridgmanite in meta-basaltic assemblages ± majorite / Ca-silicate perovskite / CF phase / NAL phase / stishovite (Ono et al. 2001; Hirose and Fei 2002; Litasov and Ohtani 2005; Kuwahara et al. 2018).

The remaining fourteen high-alumina inclusions have Mg#s that are much higher than experimental bridgmanites in meta-basaltic assemblages and are more akin to those in meta-peridotitic assemblages. The seven high-Mg# jeffbenite inclusions have very high Al_2O_3, coupled with very low CaO and Na_2O, and several of these coexist with iron-rich ferropericlase (Mg#<0.8, Fig. 13). The TiO_2 contents of these inclusions vary, with TiO_2 present either as a minor component (<0.1 wt%) or a major element (~4–8 wt%). Thus, the high Mg# jeffbenite inclusions, while having some features consistent with bridgmanite, appear to be unique relative to bridgmanite formed in either meta-peridotitic or meta-basaltic assemblages.

The three high-Al single phase inclusions ("Type II" inclusions of Hutchison et al, 2001) have alumina contents that are distinctly higher (~10 wt% Al_2O_3) than produced in experiments on primitive peridotite (<7 wt% Al_2O_3), but also have low CaO. These inclusions co-occur with ferropericlase with Mg#s of 0.81 to 0.82, nominally consistent with expectations from experiments on peridotite compositions (Fig. 13). The four "Type III" inclusions of Hutchison et al (2001) are unique in their very high Na_2O and CaO contents. None of these high Mg#, high-Al, high-Na inclusions are consistent with any experimental bridgmanites in the literature and may represent a unique association. On the basis of experiments on the inclusion bulk compositions, Hutchison et al. (2001) interpreted these to have a unique origin at pressures corresponding to the lower transition zone, albeit at temperatures several hundred degrees higher than the mantle geotherm.

Phase relations. Figure 16 shows calculated phase relations in the system $MgSiO_3$–Al_2O_3 with pressure at 1600 °C (Fig. 16a), and for a pyrolite (fertile peridotite) composition (Fig. 16b) at pressures and temperatures of the deep upper mantle and shallow lower mantle (Stixrude and Lithgow-Bertelloni 2011).

The alumina content in bridgmanite is a potential barometer, and Figure 16a shows the Al_2O_3 contents (mole fraction) in the low-Al and high-Al inclusions for comparison with phase relations in this simplified system. If the low-Al $MgSiO_3$-inclusions are former bridgmanite then phase relations either indicate a pressure of origin of ~ 22.5 to 26 GPa if the inclusions equilibrated with majoritic garnet, or formation at unconstrained higher pressures if they did not because alumina becomes increasingly soluble in bridgmanite at higher pressures. Because none of the low-Al inclusions are reported to co-occur with majoritic garnet, their low-Al contents likely indicate formation involving a low-alumina protolith (e.g., harzburgite) as suggested by their high Mg#s and depletion in CaO. We note that low-Al $MgSiO_3$-rich inclusions have Al_2O_3-contents that are also generally consistent with that expected for akimotoite at ~20 to 22.5 GPa.

The high-Al single phase inclusions have alumina contents consistent with a pressure of about 27 GPa if they co-existed with corundum, and we note that two of the three inclusions co-occur with ruby (corundum) in addition to relatively Mg-rich ferropericlase (Harte et al. 1999), suggesting that these inclusions may have originated in an alumina-rich, peridotitic protolith at the top of the lower mantle. The high-Al composite inclusions indicate pressures of ~27 to 34 GPa if they formed in equilibrium with corundum (none co-occur with ruby) but these are minimum pressure if they did not. If the jeffbenite inclusions were former bridgmanite then their alumina contents indicate minimum pressures of ~32 to 34 GPa but, again, none of these inclusions co-occur with ruby. It is noteworthy that the few low-Al $MgSiO_3$ inclusions and jeffbenite inclusions that have been analyzed for ferric iron have elevated $Fe^{3+}/\sum Fe$ that is generally compatible with expectations for bridgmanite in the shallow lower mantle (McCammon et al. 1997, 2004).

Phase relations for a pyrolytic composition are shown relative to ambient mantle and slab geotherms in Figure 16b. Along an ambient mantle geotherm, bridgmanite forms at about 24 GPa and coexists with majoritic garnet, Ca-silicate perovskite and ferropericlase, and there is only a very small akimotoite stability field. The akimotoite field expands at lower

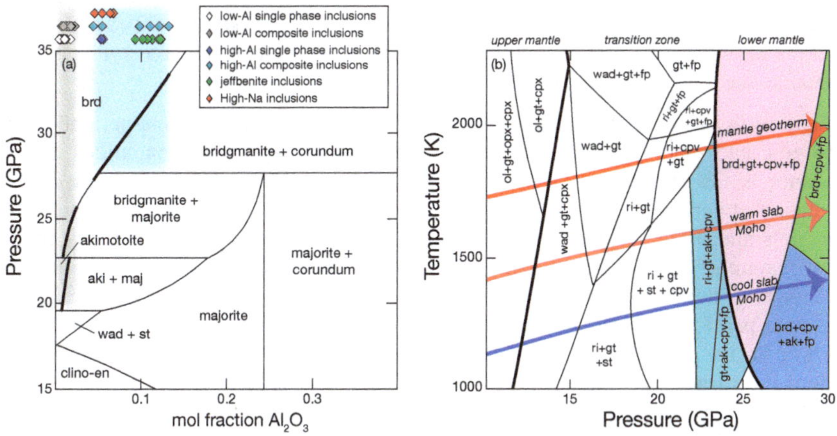

Figure 16. (a) Phase relations in the system $MgSiO_3$-Al_2O_3 (mole fraction Al/(Al+Mg)) as a function of pressure at 1600 °C (Stixrude and Lithgow-Bertelloni 2011). Also shown are the alumina contents of the $MgSiO_3$-rich inclusions (see symbol key). The **grey bar** shows the range of Al-content in low-Al inclusions and the **blue bar** the range in high-Al inclusions. **(b)** Phase relations for 'pyrolitic' mantle (Stixrude and Lithgow-Bertelloni 2011). The **cyan region** shows akimotoite + majoritic garnet stability, the **purple shaded region** shows bridgmanite + majoritic garnet stability, the **dark blue region** shows bridgmanite + akimotoite stability, and the **green region** bridgmanite stability. Also shown are model geotherms for the mantle (Katsura et al. 2010) and warm and cool slabs at the Moho (Shirey et al. 2021) (deflections at phase boundaries omitted). Phase labels are: ak = akimotoite; brd = bridgmanite; cpx = clinopyroxene; cpv = Ca-silicate perovskite; fp = ferropericlase; gt = garnet; ol = olivine; opx = orthopyroxene; ri = ringwoodite; st = stishovite; wad = wadsleyite.

temperatures such that along a warm or cold slab Moho geotherm akimotoite is stabilized over an ~2 GPa pressure interval at the base of the transition zone (Ishii et al. 2011). Whether or not the low-Al $MgSiO_3$-rich inclusions represent bridgmanite or in some cases akimotoite remains an open question. However, we re-iterate that the low CaO contents are not consistent with bridgmanite in fertile mantle peridotite at temperatures of the mantle geotherm but could be produced either at lower temperatures (Irifune et al. 2000) or in a depleted harzburgitic lithology, or both, which is also consistent with their low Al_2O_3 and high Mg#s and plausibly places their origin in subducted depleted lithospheric mantle along a cool mantle geotherm.

Trace element compositions

Trace elements have been analyzed in twenty-two of the $MgSiO_3$-rich inclusions; ten low-Al inclusions, six high-Al inclusions and six jeffbenite inclusions. Data are provided in Table 7 (Available at: https://doi.org/10.5683/SP3/LIVK1K) and plotted as spidergrams on Figure 17, normalized to BSE. All inclusions were measured by SIMS at EIMF (Harte et al. 1999; Stachel et al. 2000b; Bulanova et al. 2010; Burnham et al. 2016; Thomson et al. 2016b). Also shown on Figure 17b are the calculated abundance patterns for bridgmanite in equilibrium with assemblages predicted for meta-peridotite and meta-basalt (MORB) at transition zone and lower mantle conditions.

The low-Al inclusions have lithophile trace element abundances that, overall, are depleted or similar to BSE. Patterns exhibit notable depletions in Ba, Sr and Y relative to the generally flat REE (Fig. 17a). The two high-Al, single-phase inclusions have similar patterns to the low-Al inclusions and are among the most depleted of the inclusions. The high-Al composite inclusions are more enriched overall and exhibit conspicuous enrichments, up to three orders of magnitude relative to BSE, in Nb, Ta, Zr and Hf, while also having depletions in Ba, Sr and Y. The jeffbenite inclusions have generally similar abundance patterns to other $MgSiO_3$-rich inclusions, with some resembling closely the low-Al inclusions and others exhibiting similar enrichments in Nb, Ta, Zr and Hf as the composite inclusions. These similarities are suggestive that all the $MgSiO_3$-rich inclusions share a common mineralogical pedigree, with an origin as bridgmanite the commonly held interpretation.

Figure 17b shows modeled abundance patterns for bridgmanite in meta-peridotitic, meta-harzburgitic and meta-basaltic assemblages at shallow lower mantle conditions (25 GPa). Abundance levels of the low-Al and high-Al inclusions are similar to those predicted for bridgmanite in peridotitic mantle. However, depletions in Ba, Sr and Y are not predicted in any models and, if these inclusions are former bridgmanite minerals, this likely reflects a distinct feature of the source lithology or the melts and fluids they may have equilibrated with. However, the relative enrichments in Nb, Ta, Zr and Hf are predicted for bridgmanite in the meta-peridotitic and meta-basaltic models, which reflects the presence of Ca-silicate perovskite in the coexisting assemblage. These patterns emerge as Ca-silicate perovskite partitions most trace elements strongly with the exception of Nb, Ta, Zr and Hf. In contrast, bridgmanite has a predilection for Nb, Ta, Zr and Hf, such that strong relative enrichments in these elements occur in assemblages where both phases coexist. Note that bridgmanite in the meta-harzburgitic assemblage has a relatively flat and depleted pattern. Overall, the abundance patterns of $MgSiO_3$-rich inclusions are consistent with expectations for bridgmanite, and many indicate the control of coexisting Ca-silicate perovskite on trace element abundances.

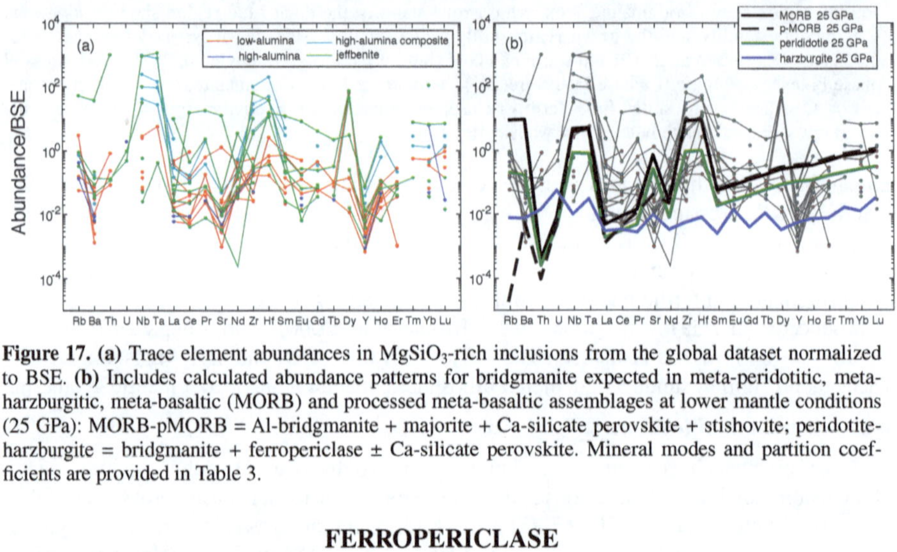

Figure 17. **(a)** Trace element abundances in $MgSiO_3$-rich inclusions from the global dataset normalized to BSE. **(b)** Includes calculated abundance patterns for bridgmanite expected in meta-peridotitic, meta-harzburgitic, meta-basaltic (MORB) and processed meta-basaltic assemblages at lower mantle conditions (25 GPa): MORB-pMORB = Al-bridgmanite + majorite + Ca-silicate perovskite + stishovite; peridotite-harzburgite = bridgmanite + ferropericlase ± Ca-silicate perovskite. Mineral modes and partition coefficients are provided in Table 3.

FERROPERICLASE

Ferropericlase is an oxide mineral with the general formula (Mg,Fe)O, representing a complete solid solution between periclase (MgO) and wüstite (FeO). The term is often used synonymously with magnesiowüstite, but for simplicity we will use ferropericlase in reference to the entire range of solid solution. Ferropericlase has cubic symmetry, crystallizing in the $Fm\bar{3}m$ space group in the B1, or rock salt structure, and is stable throughout the Earth's entire pressure range from crust to the core (Duffy et al. 2015). Ferropericlase is the most common inclusion type in sublithospheric diamonds, and here a literature dataset of 269 inclusions in diamonds collected from five continents has been compiled. More than 60% of the ferropericlase inclusions occur as the only mineral observed in their diamond hosts, often occurring in multiples in a single diamond (we note this observation could be biased by unreported co-occurring phases, especially colorless phases that are difficult to detect).

About a quarter of the reported inclusions co-occur with silicate minerals and about 15% co-occur with $MgSiO_3$-rich and/or $CaSiO_3$-rich phases.

Ferropericlase comprises about 15% by volume of a primitive, peridotitic lower mantle assemblage coexisting with bridgmanite and Ca-silicate perovskite (Fig. 1). In the numerous cases where it occurs as the only inclusion type in a diamond it is commonly used to infer a sublithospheric origin, as ferropericlase is rare as a co-occurring mineral in diamonds that are demonstrably lithospheric (Harte et al. 1999; Stachel et al. 2000b, 2005; Kaminsky et al. 2001; Davies et al. 2004a; Hayman et al. 2005; Zedgenizov et al. 2014a). In the absence of other phases that can potentially provide barometric constraints, the composition of ferropericlase provides no direct information about the depth of diamond and ferropericlase crystallization, which may occur at upper mantle or transition zone conditions and be directly related to diamond forming redox reactions (Stachel and Harris 1997; Brey et al. 2004; Thomson et al. 2016a; Seitz et al. 2018; Bulatov et al. 2019; Nimis et al. 2019). Barometric estimates based on elasticity and elastoplasticity theory can help constrain the depth of origin, for example a recent estimate for two ferropericlase inclusions from a single diamond from Brazil indicate minimum depths of entrapment of ~16 GPa (Anzolini et al. 2019), leaving open the possibility of a transition zone or lower mantle origin.

Major element compositions

Major element compositions based on a literature compilation of electron microprobe analyses of 269 ferropericlase inclusions are provided in Table 8 (Available at: https://doi.org/10.5683/SP3/LIVK1K), with Figure 18 plotting NiO, Al_2O_3, Cr_2O_3 and Na_2O versus Mg#. Also shown are compositions of ferropericlase coexisting with bridgmanite ±Ca-silicate perovskite/garnet/ringwoodite/melt in experiments on primitive peridotitic and harzburgitic bulk compositions at pressures of the deep transition zone and shallow lower mantle.

Ferropericlase inclusions span a wide range of Mg# from about 0.15 to 0.95 (Fig. 18). In comparison, ferropericlase coexisting with lower mantle phases in experiments show a limited range of Mg# from about 0.83 to 0.95, and with NiO contents between about 0.25 and 1.5 wt%. The NiO contents of the ferropericlase inclusions (Fig. 18a) range from near zero at the lowest Mg#s to about 2 wt%, with an apparent positive correlation between NiO and Mg# (Kaminsky et al. 2001; Davies et al. 2004a; Kaminsky 2012; Thomson et al. 2016a). Ferropericlase inclusions that co-occur together with both low-Al and high-Al $MgSiO_3$-rich inclusions are highlighted on Figure 18. These and other ferropericlase inclusions with high Mg#s (> 0.8) have NiO contents that are generally higher or at the extreme high end of the experimental distribution. In contrast, co-occurring $MgSiO_3$-rich inclusions generally have low NiO contents relative to experiments (Fig. 14). The average measured NiO partition coefficient between bridgmanite and ferropericlase ($D^{fp/brg} = X^{Ni,fp}/X^{Ni,brg}$) in experiments where NiO is reported is 16 ± 9, whereas in the co-occurring inclusions it is 86 ± 56. Thus, as with Mg# (Fig. 13), co-occurring bridgmanite-ferropericlase pairs are not consistent with those produced in experiments on primitive mantle peridotite. As discussed above, the Mg#s of these bridgmanite–ferropericlase pairs do not match those from primitive mantle peridotite but are more akin to those expected in depleted meta-harzburgite assemblages.

Figure 18b–d show that the Cr_2O_3, Al_2O_3 and Na_2O contents of most ferropericlase inclusions are on the low side or lower than ferropericlase compositions produced in experiments on primitive mantle peridotite and these elements exhibit no apparent correlation with Mg#. The ferropericlase inclusions that co-occur with $MgSiO_3$-rich inclusions are also depleted in these elements. However, we note that the Cr_2O_3, Al_2O_3 and Na_2O contents in high Mg# ferropericlase inclusions overlap with ferropericlase from experiments on harzburgite composition. This depletion in high Mg# ferropericlase is consistent with the $MgSiO_3$-rich inclusions they co-occur with, which also have low Al_2O_3 (and CaO) contents relative to

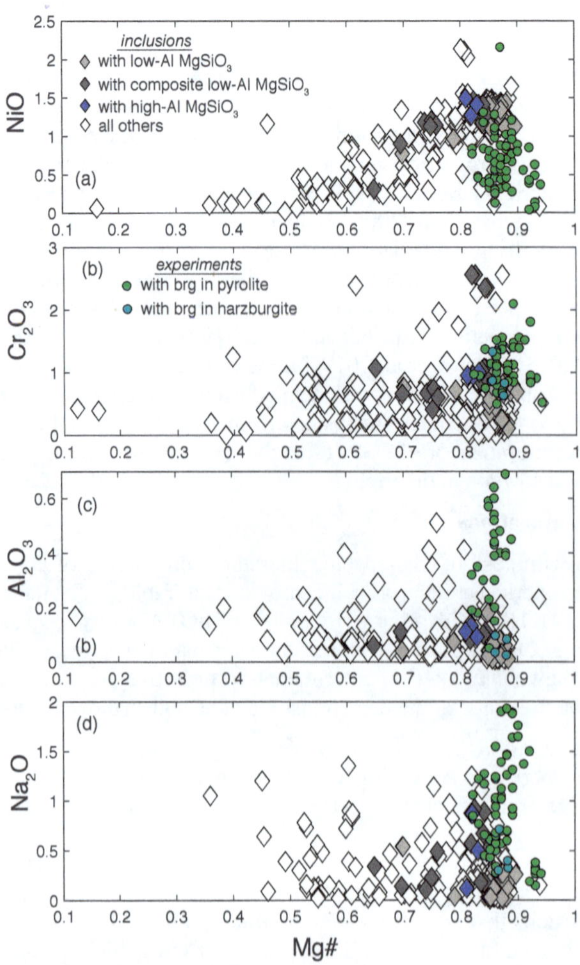

Figure 18. **(a)** NiO, **(b)** Al_2O_3, **(c)** Cr_2O_3 and **(d)** Na_2O versus Mg# showing ferropericlase inclusions from the global dataset: **white diamonds**, no co-occurring phases reported; **light grey diamonds**, co-occurring with single-phase, low-Al $MgSiO_3$-rich inclusions; **dark grey diamonds**, co-occurring with composite, low-Al $MgSiO_3$-rich inclusions; **blue diamonds**, co-occurring with high-Al $MgSiO_3$-rich inclusions; **green circles,** ferropericlase in experiments on primitive mantle peridotite in equilibrium with bridgmanite ± Ca-silicate perovskite/majorite/ringwoodite (Irifune 1994; Wood 2000; Hirose 2002; Tronnes and Frost 2002; Nishiyama and Yagi 2003; Nishiyama et al. 2004; Walter et al. 2004; Irifune et al. 2010; Ishii et al. 2011, 2018; Kuwahara et al. 2018); **cyan circles,** ferropericlase in experiments on harzburgite in equilibrium with bridgmanite ± majorite/ringwoodite (Ishii et al. 2019).

bridgmanite in meta-peridotite assemblages (Fig. 14). Thus, while more than half of the population of ferropericlase inclusions have Mg#s generally consistent with an origin related to meta-peridotite at lower mantle conditions, most of these have minor element abundances suggesting a relationship to a depleted composition such as harzburgite rather than primitive mantle. The low MgO contents of $CaSiO_3$-rich inclusions and low CaO and Al_2O_3 of $MgSiO_3$-rich inclusions that co-occur with ferropericlase together indicate a low temperature equilibration in depleted peridotite, implicating an association with the harzburgitic portion of cold subducted slab lithosphere.

The ferropericlase inclusions that have Mg#s below ~0.85 trend to low-NiO and Cr_2O_3 contents and have uniformly low Al_2O_3 but highly variable Na_2O contents. These low Mg# ferropericlase inclusions are too iron-rich to have equilibrated as part of an assemblage associated with primitive mantle peridotite or harzburgite. Several possible modes of origin have been postulated for these ferropericlase inclusions with lower Mg#s, including:

1. *The composition of the lower mantle is vastly different than primitive upper mantle* (Kaminsky et al. 2001; Kaminsky 2012). We consider this explanation improbable because the proportion and compositional range of syngenetic ferropericlase inclusions are expected to record diamond forming reactions (syngenesis) rather than entrapment of ambient mantle phases (protogenesis) in proportions or with compositions reflecting its bulk composition. For example, in the study by Nimis et al. (2019) nine iron-rich ferropericlase inclusions in two diamonds from Juina (Brazil) displayed a clear crystallographic orientation relationship between the diamond host and the inclusions indicative of co-crystallization during the diamond forming process.

2. *Ferropericlase crystallized in the deep lower mantle* where a spin-transition in iron (> ~1700 km) results in more iron-rich ferropericlase, or in the D" layer at base of the lower mantle due to an iron-rich composition or preferential partitioning of iron into ferropericlase relative to post-perovskite (Harte et al. 1999; Hayman et al. 2005; Wirth et al. 2014; Palot et al. 2016). Magnesioferrite $(Mg,Fe^{3+})Fe_2O_4$ exsolution blebs observed in ferropericlase, sometimes accompanied by sub-micron blebs of Fe-Ni metal, have been used as evidence in support of a deep lower mantle origin (Wirth et al. 2014; Palot et al. 2016). However, recent experimentally determined phase relations show that a stability field of magnesioferrite occurs at ~8 to 10 GPa (1000–1600 °C) and there is no indication of a high-pressure magnesioferrite stability field up to ~ 20 GPa (Uenver-Thiele et al. 2017), although one may exist at higher pressures (Andrault and Bolfan-Casanova 2001). We suggest that the simplest explanation for magnesioferrite blebs observed in ferropericlase inclusions is that they represent either exsolution from original ferropericlase with excess Fe_2O_3 under upper mantle conditions or oxidation by coexisting, carbonated fluids as described below (Thomson 2017; Uenver-Thiele et al. 2017). This exsolution is consistent with the low-pressure unmixing exhibited in both $CaSiO_3$-rich and $MgSiO_3$-rich inclusions described above. While the inclusions may have originated at higher pressures, neither the presence of magnesioferrite in ferropericlase nor their iron-rich compositions necessitate a lower mantle origin.

3. *Ferropericlase crystallized as a product of redox reactions* involving oxidized carbonate or carbonated melt and reducing peridotite at upper mantle, transition zone or lower mantle conditions (McCammon et al. 1997, 2004; Liu 2002; Bulatov et al. 2014, 2019; Thomson et al. 2016a; Seitz et al. 2018). Liu (2002) proposed that the range of iron-rich ferropericlase compositions may reflect the continuous subsolidus decarbonation of ferromagnesite according to the reaction:

$$\begin{aligned}2\,Mg_xFe_{1-x}CO_3\ \text{(ferromagnesite)} &= Mg_yFe_{1-y}CO_3\ \text{(ferromagnesite)} \\ &+ Mg_zFe_{1-z}O\ \text{(ferropericlase)} + C\ \text{(diamond)} + O_2\end{aligned} \quad (6)$$

In this reaction, the product ferropericlase solid solution becomes progressively iron enriched at the expense of ferromagnesite solid solution. As discussed below, there is ample evidence for the role of fluids or melts in sublithospheric diamond formation rather than subsolidus decarbonation, however, the essence of the decarbonation reaction suggested by Liu (2002) may equally apply to decarbonation in the liquid phase. Thomson et al. (2016b) performed experiments at 20 GPa in which a model carbonated melt of basaltic oceanic crust was reacted with reducing peridotite and both diamond and iron-rich ferropericlase were observed among reaction products. Similarly, Bulatov et al. (2019) showed experimentally that iron-rich ferropericlase and diamond can crystallize simultaneously during the reduction of carbonate-silicate melt in equilibrium with olivine at upper mantle conditions. Seitz et al. (2018) measured Li isotopes in iron-rich ferropericlase inclusions from Juina and observed a range that encompasses that of serpentinized ocean floor peridotites, fresh and altered MORB, seafloor sediments and of eclogites. They suggest that dehydration and redox reactions in altered portions of slabs subducted into the transition zone and shallow lower mantle led to the formation of diamond and iron-rich ferropericlase inclusions.

Iron redox state in ferropericlase. The redox state of iron in a small population of ferropericlase inclusions from both Kankan (high Mg#) and Juina (low Mg#) has been measured by Mossbauer spectroscopy (McCammon et al. 1997, 2004) and compared to experimentally determined Fe^{3+} solubility in ferropericlase applicable to depths at the top of the lower mantle (Otsuka et al. 2013). The oxygen fugacities calculated from measured $Fe^{3+}/\sum Fe$ are close to the upper stability limit of diamond and higher than expected in ambient mantle peridotite in the shallow lower mantle (Frost and McCammon 2008; Otsuka et al. 2013). Thus, if these ferropericlase inclusions are formed in the shallow lower mantle, especially those co-occurring with $MgSiO_3$-rich phases that also have high ferric iron content, their high Fe^{3+} concentrations may record diamond formation in a region of redox gradients. Such regions in the upper mantle or shallow lower mantle may arise from subduction of oxidized material into reducing mantle, and the inclusions may have precipitated from oxidized, carbonate-bearing melts or fluids (McCammon et al. 2004; Rohrbach and Schmidt 2011; Otsuka et al. 2013; Thomson et al. 2016a).

Trace element compositions

Trace elements have been analyzed in thirty-eight of the ferropericlase inclusions and data are provided in Table 9 (Available at: https://doi.org/10.5683/SP3/LIVK1K) and plotted as spidergrams on Figure 19, normalized to BSE. Twenty-one of the inclusions were measured by SIMS at the EIMF (Hutchison 1997; Harte et al. 1999; Burnham et al. 2016) and seventeen by LA-ICP-MS (Kaminsky et al. 2001). Also shown on this diagram are the calculated abundance patterns for ferropericlase in meta-peridotitic and meta-harzburgitic assemblages at lower mantle conditions.

Ferropericlase inclusions are generally depleted in lithophile trace elements relative to BSE but show a wide variation, with many elements spanning several orders of magnitude.

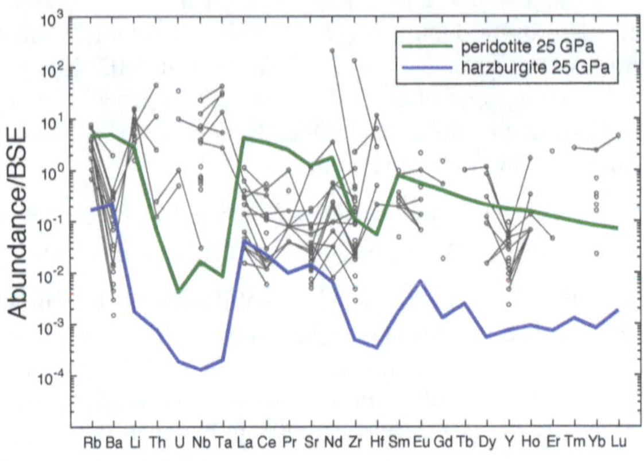

Figure 19. Trace element abundances in ferropericlase inclusions from the global dataset normalized to BSE. Also shown are calculated abundance patterns for ferropericlase expected in meta-peridotite (bridgmanite + Ca-silicate perovskite + ferropericlase) and meta-harzburgite (bridgmanite perovskite + ferropericlase) in the lower mantle at 25 GPa. Mineral modes and partition coefficients are provided in Table 3.

Relative depletions are apparent in Ba and Y, as are enrichments in Rb and Li, and in some samples, Th, U, Nb and Ta. Overall, the REE appear to be relatively unfractionated, although data are sparse for many elements. Abundances are generally in the range predicted in models for peridotitic ferropericlase in the lower mantle but with notable differences. The REE generally fall between ferropericlase in primitive meta-peridotite and meta-harzburgite assemblages, whereas Th, U, Nb and Ta are notably enriched relative to the predicted depletions for these elements in lower mantle ferropericlase. The depletions in Ba and Y are also not predicted in these lithologies and likely are inherited from a distinct source.

OLIVINE

Olivine (orthorhombic, *Pbnm*) and its higher-pressure polymorphs, wadsleyite (orthorhombic, *I2/m*) and ringwoodite (cubic, $Fd\bar{3}m$), comprise approximately 60 vol% of primitive mantle lithologies in the upper mantle and transition zone as shown on Figure 1. Inclusions with $(Mg,Fe)_2SiO_4$ stoichiometry are one of the most common inclusions in lithospheric diamonds and are typically interpreted as representative of a peridotitic association (Stachel et al. 2022, this volume) but appear to be notably rare in diamonds that are demonstrably sublithospheric in origin. We compiled a global dataset consisting of twenty eight $(Mg,Fe)_2SiO_4$ inclusions observed in diamond suites that have been identified as sublithospheric (Table 10). Fifteen $(Mg,Fe)_2SiO_4$ inclusions co-occur with ferropericlase of which six also co-occur with an $MgSiO_3$-rich phase, and one co-occurs with both $MgSiO_3$-rich and $CaSiO_3$-rich phases. Two $(Mg,Fe)_2SiO_4$ inclusions occur with an $MgSiO_3$-rich phase and one with a $CaSiO_3$-rich phase. The remainder are reported to occur either in isolation or with other rare inclusion phases.

To our knowledge, all of the inclusions included in Table 10 (Available at: https://doi.org/10.5683/SP3/LIVK1K) occur in the olivine structure. There is a single occurrence of a reported inclusion with the ringwoodite structure on the basis of Raman and X-ray diffraction measurements taken while the inclusion remained within the diamond (i.e., unexposed at the surface) with the data indicating an Mg# of ~0.75 ± 0.2 and containing about 1.5 wt% water (Pearson et al. 2014); geochemical data is not available from this inclusion so it is not part of our dataset.

The major element compositions of the $(Mg,Fe)_2SiO_4$ inclusions, as determined by electron microprobe analyses, are provided in Table 10 (Available at: https://doi.org/10.5683/SP3/LIVK1K) and plotted in Figure 20 along with 1478 olivine inclusion from lithospheric diamonds for comparison. Also shown are olivine, wadsleyite and ringwoodite compositions produced in experiments using primitive peridotite compositions at pressures >10 GPa.

Figure 20 shows that, overall, $(Mg,Fe)_2SiO_4$ inclusions are unlike the bulk of olivine inclusions in lithospheric diamonds. Lithospheric inclusions commonly have Mg#s between ~0.91–0.95, whereas the majority of sublithospheric $(Mg,Fe)_2SiO_4$ inclusions have Mg#s <0.91, with the remainder greater than 0.94. Many sublithospheric $(Mg,Fe)_2SiO_4$ inclusions have low NiO, high CaO, high Al_2O_3 and high Cr_2O_3. The low NiO contents of the $(Mg,Fe)_2SiO_4$ inclusions coexisting with an $MgSiO_3$-rich phase is consistent with the low-NiO contents of those inclusions as well. In comparison to experimentally produced $(Mg,Fe)_2SiO_4$ phases, many of the inclusions are most akin to ringwoodite and least like olivine. Like many of the inclusions, ringwoodite in experiments are notably lower in Mg# than wadsleyite and olivine. While sharing some features with higher pressure $(Mg,Fe)_2SiO_4$ polymorphs, many of the inclusions have unique compositions making it difficult to assign them to a certain polymorph, yet it is clear that they are different from the bulk of lithospheric olivine inclusions.

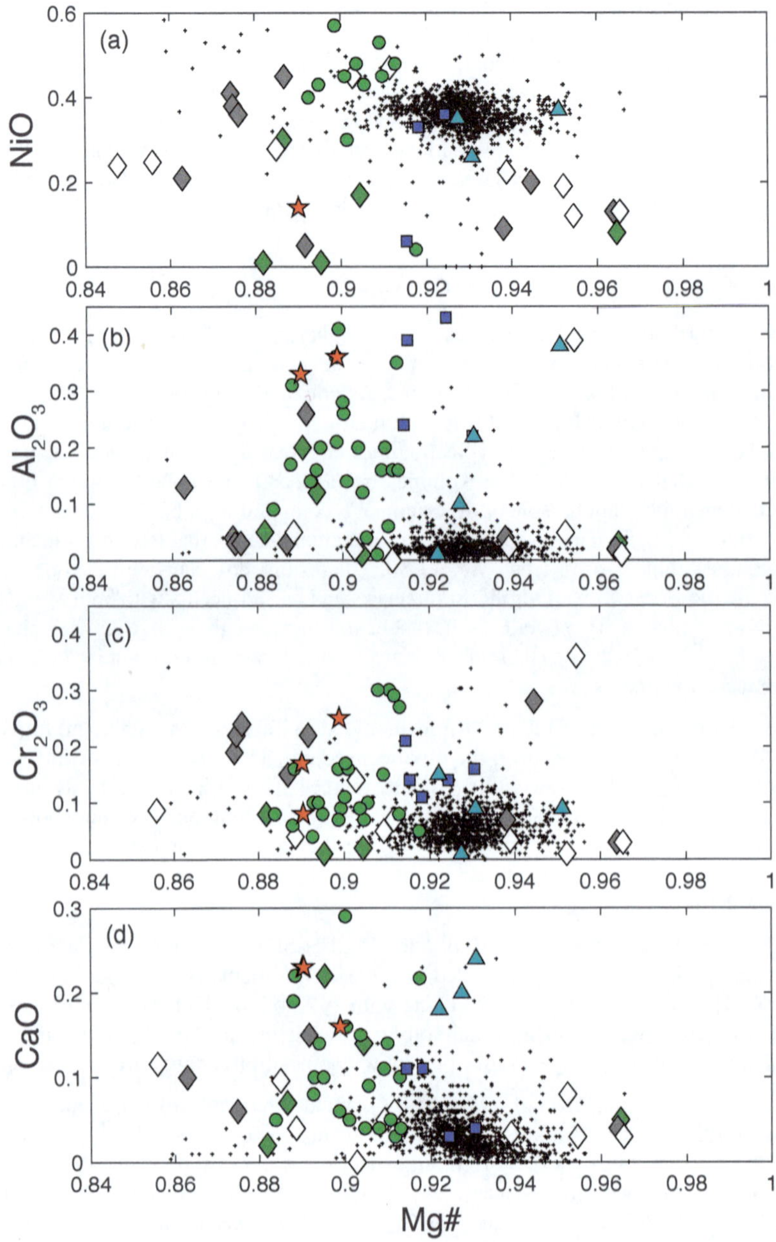

Figure 20. **(a)** NiO, **(b)** Al_2O_3, **(c)** Cr_2O_3 and **(d)** CaO versus Mg# showing olivine inclusions from the global dataset: **green diamonds**, co-occurring with ferropericlase, $MgSiO_3$, $\pm CaSiO_3$; **grey diamonds**, co-occurring with ferropericlase ±cpx ±jeffbenite; **white diamonds**, other; **crosses**, olivine inclusions in lithospheric diamonds (see Stachel et al. 2022, this volume for references); **green circles,** experimental ringwoodite in meta-peridotite; **red stars**, experimental ringwoodite coexisting with akimotoite; **blue squares**, experimental wadsleyite in meta-peridotite; **cyan triangles**, experimental olivine in meta-peridotite at 13–14 GPa in equilibrium with majoritic garnet ± cpx (Irifune and Ringwood 1987; Irifune 1994; Wood 2000; Hirose 2002; Ishii et al. 2011, 2018, 2019).

CLINOPYROXENE

Clinopyroxene (monoclinic) with $XY(Si,Al)_2O_6$ stoichiometry is a major constituent of meta-peridotitic (~20 vol%) and meta-basaltic assemblages (~60 vol%) at depths of ~300 km (Fig. 1) but disappears in these assemblages by ~400 to 500 km as it dissolves into majoritic garnet. Clinopyroxene inclusions, ranging from diopside to jadeite, are common in lithospheric diamonds but are much less common in sublithospheric diamonds. We compiled a global dataset consisting of forty clinopyroxene inclusions from four cratons in diamonds that have been identified as sublithospheric (Table 11—Available at: https://doi.org/10.5683/SP3/LIVK1K).

Clinopyroxene inclusions can be separated into two groups based on their Na_2O contents. ***High-Na*** clinopyroxene includes thirty inclusions with Na_2O ranging from 1.4 to 13.1 wt%, with bulk compositions that are generally augitic to omphacitic, but with one jadeite. The high-Na clinopyroxenes co-occur with garnet in twenty-six of the inclusions with compositions reported for twenty of these, all of which are majoritic and yield pressures ranging from ~9 to 18 GPa, providing direct evidence for their sublithospheric origin. ***Low-Na*** clinopyroxene includes ten inclusions with extremely low Na_2O of < 0.13 wt% and have compositions that are augitic to diopsidic. Five of these inclusions co-occur with ferropericlase and two of these with both ferropericlase and $MgSiO_3$-rich inclusions. Three clinopyroxene inclusions occur with a $CaSiO_3$-rich phase, and two of these also contain merwinite.

The major element compositions of clinopyroxene inclusions, as determined by electron microprobe analyses, are provided in Table 11 (Available at: https://doi.org/10.5683/SP3/LIVK1K) and plotted in Figure 21 along with 1321 clinopyroxene inclusions from lithospheric diamonds for comparison. Figure 21a shows a clear distinction between the low-Na and high-Na groups. High-Na pyroxene are generally much more aluminous and overlap extensively with 'eclogitic' lithospheric garnets in terms of Na_2O, CaO and Mg#. In contrast, the low-Na clinopyroxenes are relatively distinct, with extremely low Na_2O which does not increase with Al_2O_3 content, high CaO contents that are not seen in lithospheric inclusions and high Mg#s.

Also shown on Figure 21 are experimental clinopyroxene compositions from meta-peridotitic and meta-basaltic assemblages at pressures of 8 to 19 GPa. The high-Na inclusions are generally consistent with clinopyroxenes expected in meta-basaltic or meta-pyroxenitic assemblages, although we note that they generally do not overlap with clinopyroxenes produced in experiments on hydrous or carbonated eclogitic that are typically more alumina and sodium-rich and calcium poor relative to the inclusions. The low-Na group are generally unlike compositions produced in experiments on peridotitic compositions, especially in their very high CaO contents and low-Na_2O for a given alumina content.

SiO_2

An SiO_2 phase is reported to co-occur in fifteen diamonds hosting sublithospheric inclusions in our data sets. While coesite (monoclinic, C2/c) has been identified, the assumption is that the original inclusions were formed in the stishovite structure (tetragonal rutile-type, $P4_2/mnm$), which is the stable SiO_2 phase from ~ 9 to 75 GPa (Zhang et al. 1996; Fischer et al. 2018). SiO_2 occurs with majoritic garnet in seven diamonds in our dataset and the co-occurring majoritic garnets yield pressures of ~ 10 to 22 GPa (Table 1—Available at: https://doi.org/10.5683/SP3/LIVK1K). Two of the SiO_2 inclusions also exhibit exsolved kyanite indicating unmixing of alumina during retrogression. Two diamonds containing $CaSiO_3$-rich inclusions, both low-Ti $CaSiO_3$ from Brazil, also contain SiO_2 inclusions (Table 4—Available at: https://doi.org/10.5683/SP3/LIVK1K), whereas there are no reported co-occurrences of SiO_2 with $MgSiO_3$-rich, olivine or clinopyroxene inclusions in our data sets. SiO_2 inclusions

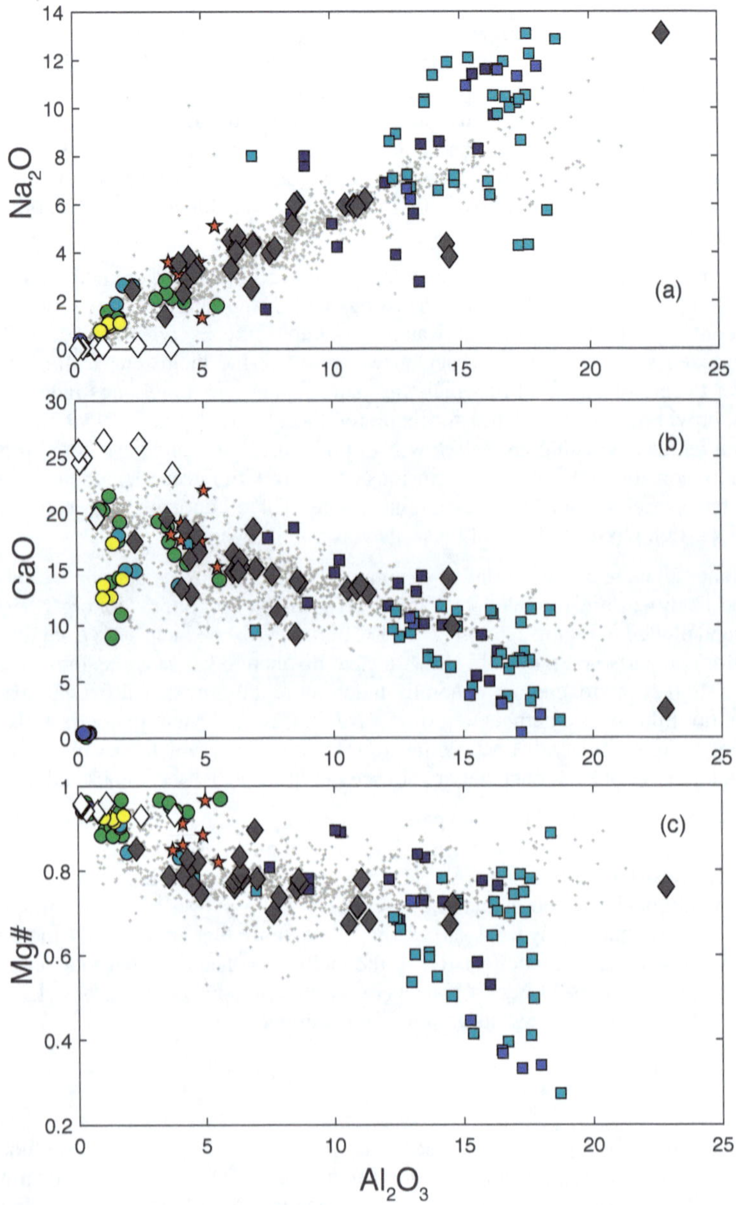

Figure 21. (a) Na_2O, **(b)** CaO and **(c)** Mg# versus Al_2O_3 showing clinopyroxene inclusions from the global dataset: **white diamonds**, low-Na inclusions; **dark gray diamonds, high-Na inclusions;** grey **crosses**, clinopyroxene inclusions in lithospheric diamonds (see Stachel et al. 2022, this volume, for references). Experimental clinopyroxene from the literature include: **green circles,** anhydrous fertile peridotite (Akaogi and Akimoto 1979; Irifune 1987; Ishii et al. 2018; Rzehak et al. 2020); **cyan circles,** carbonated peridotite (Ghosh et al. 2009); **yellow circles,** carbonated peridotite melt reaction experiments (Sun and Dasgupta 2019); **blue circles,** anhydrous harzburgite (Irifune and Ringwood 1987; Ishii et al. 2019); **dark blue squares,** anhydrous basalt (Irifune et al. 1986; Beyer and Frost 2017; Ishii et al. 2019); **light blue squares,** hydrous basalt (Okamoto and Maruyama 2004); **cyan squares,** carbonated basalt (Kiseeva et al. 2013a; Thomson et al. 2016a); **red stars,** anhydrous pyroxenite (Beyer and Frost 2017).

co-occur with ferropericlase in five diamonds, with examples from three cratons (Table 8—Available at: https://doi.org/10.5683/SP3/LIVK1K).

In addition to diamonds included in our data sets, nine coesite inclusions were reported in diamonds from the Juina-5 and Collier-4 kimberlites, Brazil (Burnham et al. 2015), and as isolated inclusions in diamonds from Sao Luiz, Brazil (Zedgenizov et al. 2014a). With the exception of inclusions containing evidence for exsolved kyanite, SiO_2 inclusions are reported to be nearly phase pure with only trace amounts of other elements including TiO_2 and Al_2O_3 (Bulanova et al. 2010; Kaminsky 2012; Thomson et al. 2014; Zedgenizov et al. 2014a; Burnham et al. 2015).

The occurrence of stishovite associated in diamonds with other inclusions of meta-basaltic affinity (e.g., Ti-rich $CaSiO_3$, low-Cr majoritic garnet) is expected based on phase relations (Fig. 1). Burnham et al. (2015) measured the carbon isotopic compositions of host diamonds and the oxygen isotopic composition of coesite inclusions from the Collier-4 and Juina-5 kimberlites, Brazil, two localities that have produced a variety of inclusions of meta-basaltic affinity and found a range of negative carbon isotopic compositions and positive oxygen isotopic compositions consistent with an origin related to subducted oceanic crust.

However, the co-occurrence of SiO_2 with ferropericlase, which is generally attributed to a peridotitic association, requires a different explanation. At lower mantle conditions in the MgO–FeO–SiO_2 system, ferropericlase and stishovite occur together once the FeO solubility in bridgmanite is exceeded, and at ambient lower mantle temperatures (e.g., ~1600 °C) this occurs at ~12 mol% FeO (Fei et al. 1996). However, at < ~1100 °C, bridgmanite breaks down to ferropericlase plus stishovite at < ~5 mol% FeO, and in a fertile mantle composition a field of ferropericlase coexisting with stishovite and Ca-silicate perovskite occurs at ~25 GPa at temperatures < ~900 °C (Stixrude and Lithgow-Bertelloni 2011). Thus, the association of stishovite and ferropericlase may represent bridgmanite breakdown associated with either iron-rich lithologies or low temperatures. The Mg#s of ferropericlase co-occurring with stishovite in our data set range from 0.78 to 0.86, with most 0.84 and above. For a fertile mantle these are far too magnesian to be in equilibrium with stishovite at mantle temperatures, and thus are either not in equilibrium with co-occurring stishovite (Stachel et al. 2000b), or were equilibrated at much lower temperatures.

CF AND NAL PHASES

Composite inclusions with bulk stoichiometries consistent with the calcium ferrite (CF) structured phase and new aluminous (NAL) phase that occur in meta-basaltic assemblages at conditions of the lower mantle (Fig. 1) have been described in diamonds from Brazil (Walter et al. 2011; Thomson et al. 2014; Zedgenizov et al. 2014a). Within their stability fields the CF phase is orthorhombic (*Pbnm*) and has the general formula XY_2O_4 ($X = K^+, Na^+, Ca^{2+}, Mg^{2+}$; $Y = Al^{3+}$ and Si^{4+}), whereas the NAL phase is hexagonal ($P6_3/m$) and has the general formula $AX_2Y_6O_{12}$ ($A = Na^+, K^+, Ca^{2+}$; $X = Mg^{2+}, Fe^{2+}$; $Y = Al^{3+}, Si^{4+}$ (Miyajima et al. 2001; Wicks and Duffy 2016). Inclusions interpreted as retrograde CF phase are found as composite mixtures of spinel $(Mg,Fe)Al_2O_4$ and nepheline $NaAlSiO_4$, whereas NAL phases as composite mixtures of spinel and a nepheline–kalsilite phase, $(Na,K)AlSiO_4$. Bulk inclusion compositions as determined by wide beam EPMA analysis or reconstruction from phase modes (Walter et al. 2011; Thomson et al. 2014) yield stoichiometries close to the ideal CF and NAL phases produced in experiments on basaltic compositions (Ono et al. 2001; Ricolleau et al. 2010; Ishii et al. 2019), providing strong evidence for the role of subducted oceanic crust in their origin. Trace elements have been reported for six NAL phases and two CF phases (Thomson et al. 2016b) and abundance patterns generally display depletion in REE and large negative Y anomalies and relative enrichments in Th, U, Nb, Ta and Rb relative to BSE.

DISCUSSION

We assembled comprehensive datasets of silicate and oxide inclusions in sublithospheric diamonds that ostensibly represent major rock forming minerals in the mantle (majoritic garnet, Ca-silicate perovskite, bridgmanite, ferropericlase, olivine and clinpyroxene). The major and trace element compositions of the inclusions combined with experimental phase equilibrium and element partitioning data provide a basis for conceptual models of their origin and reveal information about mantle geodynamic processes leading to diamond formation.

The geochemical features of the sublithospheric inclusions generally permit a distinction between a meta-basaltic association (low-Cr majoritic garnet; high-Ti $CaSiO_3$; low Mg#, high-Al $MgSiO_3$; CF and NAL phases) and a meta-peridotitic association (high-Cr majoritic garnet; low-Ti $CaSiO_3$; low-Al, high Mg# $MgSiO_3$; ferropericlase) (Stachel et al. 2005; Harte 2010). However, it is also apparent from experimental major element systematics and trace element modeling that inclusion compositions generally do not conform, with very few exceptions, to expectations for primary subsolidus minerals in primitive meta-peridotitic assemblages (e.g., pyrolite) or meta-basaltic (e.g., MORB) assemblages at upper mantle, transition zone or lower mantle conditions.

The geochemistry of syngenetic sublithospheric inclusions cannot be separated from models for how the host diamonds form, and like their lithospheric counterparts, sublithospheric diamonds exhibit abundant evidence for crystallization from fluids or melts. Therefore, to provide context for general models of inclusion genesis we first discuss observations regarding diamond crystallization.

Diamond precipitation from fluids and melts

It is well-established that diamonds originating in cratonic lithospheric mantle precipitate primarily from carbon-bearing fluids (Deines 1980; Sunagawa 1984; Haggerty 1986; Bulanova 1995; Shirey et al. 2013) and they provide a baseline for comparison with sublithospheric diamonds. On the basis of fluid inclusions trapped in fibrous diamonds from the lithosphere the parental fluids exhibit a range in composition, including high- and low-Mg carbonatitic, chlorine-rich and silica-rich aqueous fluids (Klein-BenDavid et al. 2007, 2009; Weiss et al. 2013, 2014). Cathodoluminescence imaging of lithospheric diamonds reveals internal growth textures with intricate, concentric zoning, as well as evidence of resorption and recrystallization, textures indicative of precipitation from carbon-saturated fluids as opposed to solid-state transformation from graphite (Bulanova 1995; Shirey et al. 2013). Further evidence for fluid-mediated diamond precipitation includes fracture infillings (Czas et al. 2018) and systematic changes in carbon and nitrogen abundance and isotopic composition, features that are consistent with crystallization from a fractionating fluid phase (Boyd et al. 1987; Smart et al. 2011).

Sublithospheric diamonds can potentially form through subsolidus decarbonation reactions in the mantle, for example, through reaction with silica in oceanic crust (Maeda et al. 2017; Li et al. 2018; Drewitt et al. 2019) or through reaction of carbonate with reduced phases such as iron or iron carbide (Liu 2002; Zhu et al. 2019). However, sublithospheric diamonds typically have internal textural features that are similar to lithospheric diamonds, displaying intricate, complex growth layering, and in some cases multiple nucleation centers, indicative of crystallization from fluids or melts, examples of which are provided in Figure 22 (Hayman et al. 2005; Bulanova et al. 2010; Thomson et al. 2014; Zedgenizov et al. 2014a; Palot et al. 2017). An exception may be CLIPPIR and Type IIb diamonds, which typically have no discernible growth structure but show abundant dislocation networks indicative of plastic deformation and annealing (Smith et al. 2016b, 2018). Sublithospheric diamonds commonly exhibit deformation textures indicative of residence in a high-strain environment at high temperature (Bulanova et al. 2010; Thomson et al. 2014; Smith et al. 2016a, 2018; Shirey et al. 2019). Sublithospheric diamonds also preserve both small-scale and large-scale intra-diamond carbon isotope variations among growth zones (Fig. 22), consistent with growth from fractionating fluids/melts or multiple episodes of growth from fluids/melts of variable composition (Stachel et al. 2002; Bulanova et al. 2010; Shirey et al. 2013, 2019; Thomson et al. 2014; Zedgenizov et al. 2014a). In contrast to lithospheric diamonds, which are often regular crystal forms and can exhibit fluid-inclusion-rich fibrous diamond growth, sublithospheric diamonds tend to have more irregular morphologies and fibrous diamond growth has not been observed.

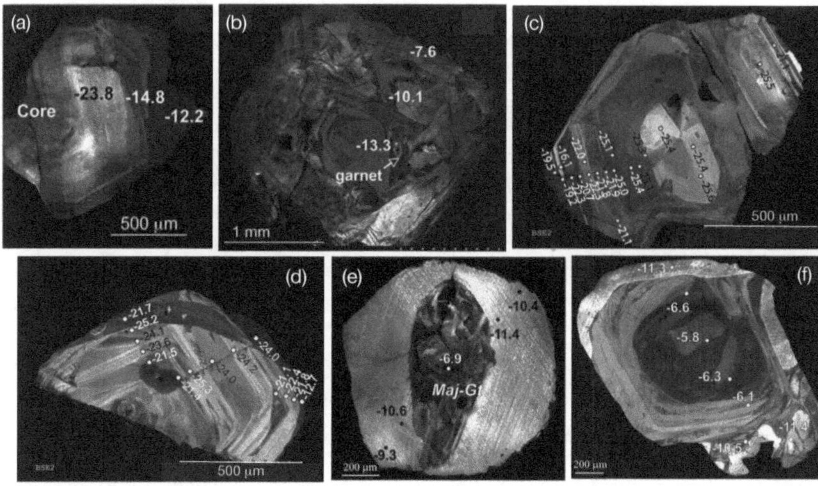

Figure 22. Cathodoluminescence images of sublithospheric diamonds from the Juina region of Brazil. Carbon isotopic compositions $\delta^{13}C$ (‰ relative to PDB standard) are shown at the positions they were analyzed by SIMS. **(a)** diamond with weak octahedral zonation and a large change in $\delta^{13}C$ from core to rim (J2, Collier 4); **(b)** diamond with a core zone with a rounded octahedral/rounded shape and resorption surrounded by an intermediate region showing irregular-aggregate growth with features indicative of deformation and inclusions of diamond-in-diamond in this zone indicating nucleation of multiple diamond seeds, and a narrow rim zone with fine octahedral zoning and plastic deformation (J9, Collier 4); **(c)** octahedral crystal with sector zoned cubo-octahedral core and octahedral zonation (Ju5-115, Juina 5); **(d)** blocky octahedral core and regular octahedral zonation with rounded resorbed corners (Ju5-71, Juina 5); **(e)**; rounded octahedral core with complex zonation with a thick rim showing evidence of plastic deformation present as a 'tatami' pattern with lines that do not correspond to the core zoning structure (36, Sao Luiz); **(f)** Rounded octahedral core showing evidence of intermittent growth and resorption with irregular outer zonation and a rim showing evidence of plastic deformation (34, Sao Luiz). ***Credits***: **a,b** Reprinted by permission from Springer Nature Customer Service Centre GmbH: Springer Nature. Bulanova et al. (2010), Copyright 2010); **c,d** CC BY Thomson et al. (2014); **e,f** Reprinted from Zedgenizov et al. (2014a), Copyright (2014), with permission from Elsevier.

Diamonds precipitate from fluids or melts when carbon species, such as CO_2, CH_4, CO_3^{2-}, or HCO_3^-, become reduced or oxidized as a consequence of changes in temperature (Luth and Stachel 2014; Stachel and Luth 2015), pH (Sverjensky and Huang 2015) and redox conditions (Haggerty 1986; Frost and McCammon 2008; Bulanova et al. 2010; Rohrbach and Schmidt 2011; Shirey et al. 2013; Stagno et al. 2013; Thomson et al. 2016a) (see also Luth et al. 2022, this volume). Gradients in these variables exist, for example, when carbon-bearing fluids and melts permeate and migrate through rocks. On the basis of tomographic imaging of eclogitic mantle xenoliths from cratonic lithosphere, diamonds precipitate at silicate grain boundaries, exhibiting growth along intergranular planes that likely served as pathways for fluid flow (Anand et al. 2004; Liu et al. 2009b; Czas et al. 2018). There are no such examples of diamondiferous sublithospheric mantle xenoliths, but the diamond textural similarities described above are suggestive of a similar fluid-mediated crystallization. By inference, crystalline silicate and oxide inclusions in sublithospheric diamonds likely equilibrated with or crystallized directly from dissolved components in fluids or melts migrating through rocks in subducted lithosphere or the mantle.

Because sublithospheric diamonds are believed to grow at much greater depths than lithospheric diamonds, the fluids and melts will also possess a different chemical character. For example, H_2O-rich fluids are expected to be well beyond their second critical endpoints in both peridotitic and basaltic systems, transitioning into hydrous melts with a large fraction of dissolved silicate component (Kessel et al. 2005; Mibe et al. 2007; Liu et al. 2009a; Mibe et al. 2011; Kawamoto et al. 2012; Wang et al. 2020). Before considering the types of fluids and melts that sublithospheric inclusions may crystallize from, we consider evidence from stable isotopes, particularly carbon, that provide further context for potential source lithologies and for understating different populations of sublithospheric diamonds and their inclusions.

Stable isotope compositions of diamonds and inclusions

The stable isotope compositions of sublithospheric diamonds and their inclusions have been used extensively to inform interpretations of their origin (Deines 1980; Deines et al. 1991; Hutchison et al. 1999; Cartigny 2005; Bulanova et al. 2010; Palot et al. 2012; Shirey et al. 2013, 2019; Thomson et al. 2014; Zedgenizov et al. 2014a; Burnham et al. 2015; Ickert et al. 2015) (see also Stachel et al. 2022b, this volume). Carbon and nitrogen isotopes of the diamonds and oxygen isotopes of inclusions can provide evidence for potential source lithologies of carbon-bearing fluids and melts.

Carbon isotopes have been measured most extensively using SIMS, often at a spatial scale of individual growth layers, revealing distinct populations among sublithospheric diamonds. Figure 23 shows frequency histograms of the carbon isotopic composition for diamonds hosting inclusions in our database, separated by inclusion type. The primitive mantle is assumed to have a carbon isotopic composition of about $\delta^{13}C = -5 \pm 2‰$ (relative to a PDB carbonate standard), whereas carbon in subducted lithosphere lithologies varies from about $\delta^{13}C = 0‰$ (e.g., seawater carbonate) to values lower than $\delta^{13}C = -25‰$ that plausibly represent a source of either biogenic or abiogenic organic carbon (Cartigny 2005). A remarkable distinction is apparent between diamonds hosting inclusions with compositions consistent with a meta-peridotitic association relative to those with a meta-basaltic association.

Both major element and trace element characteristics of low-Cr majoritic garnet inclusions indicate a meta-basaltic or mixed basaltic-peridotitic (meta-pyroxenitic) association. The carbon isotopic composition of diamonds hosting low-Cr garnet inclusions (Fig. 23a) show a wide range of values between about $\delta^{13}C = -3$ to $-25‰$, with the majority being substantially isotopically lighter than mantle carbon, suggestive of a source of carbon in subducted slab lithologies, in particular basaltic oceanic crust (Kaminsky et al. 2001; Stachel et al. 2002; Cartigny 2005; Bulanova et al. 2010; Palot et al. 2012, 2017; Cartigny et al. 2014; Thomson et al. 2014; Zedgenizov et al. 2014a). Supporting this interpretation are measurements of isotopically heavy oxygen exhibited by garnet and SiO_2 inclusions hosted by diamonds with isotopically light carbon, with heavy oxygen attributed to interaction of oceanic crust with seawater (Burnham et al. 2015; Ickert et al. 2015). Further support for this interpretation comes from nitrogen isotopes. Although sublithospheric diamonds generally have very low N contents (~70% are Type II and >90% have N < 100 at.ppm), where it has been analyzed in diamonds that contain low-Cr majoritic garnet, N isotopes are heavy relative to primitive mantle, consistent with an oceanic crustal source (Palot et al. 2012; Regier et al. 2020). Very few of the diamonds hosting high-Cr majoritic garnets have been measured for C isotopes. Four of the five have generally low mantle-like values with one much lighter, but the data is too sparse to draw any firm conclusions.

Diamonds hosting $CaSiO_3$-rich inclusions (Fig. 23b) show a range of $\delta^{13}C$ from about 0 to $-25‰$. Although the measurements are relatively few, there is an apparent distinction between diamonds hosting low-Ti and high-Ti inclusions. Diamonds containing low-Ti inclusions,

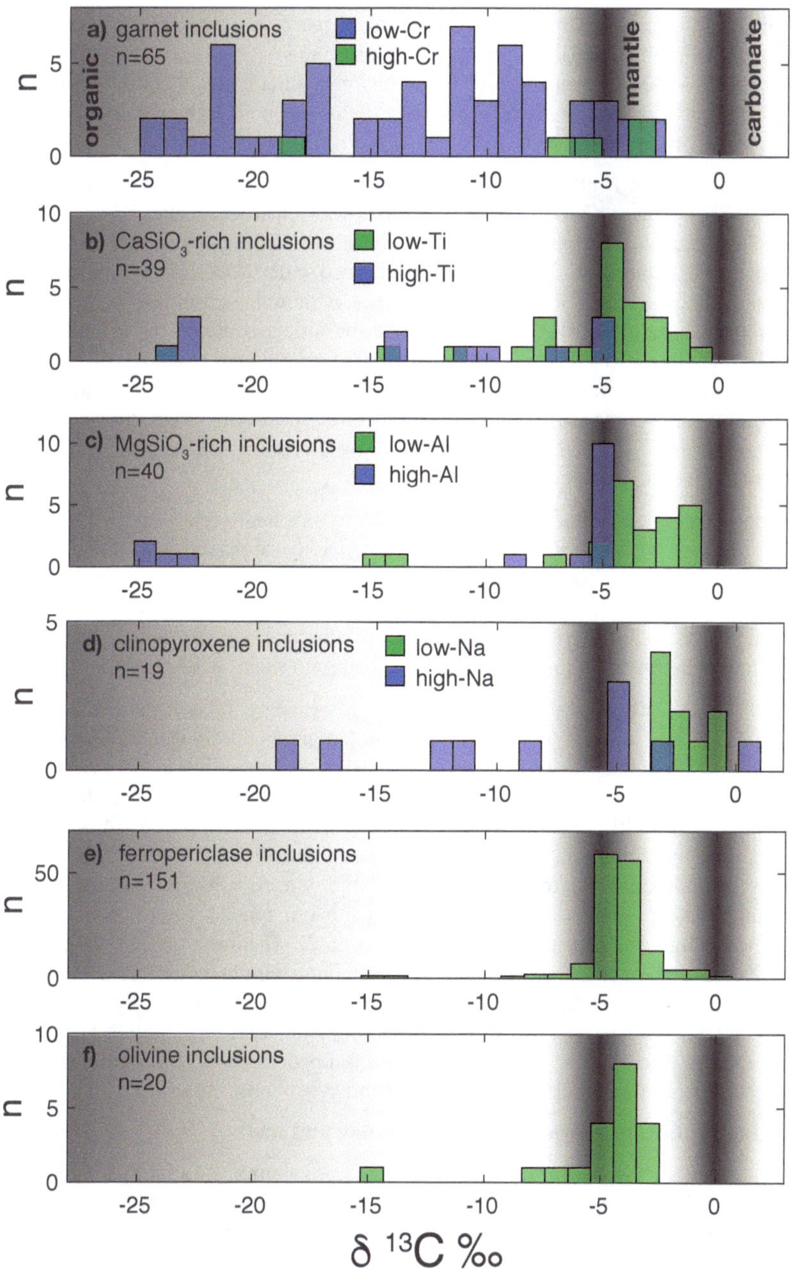

Figure 23. Histograms of carbon isotope ($\delta^{13}C$ ‰) compositions of diamonds hosting inclusions in our data compilation (data in Tables 1, 4, 6, 8, 10, 11). Values of mantle carbon, and recycled carbon (seawater carbonate and 'organic' carbon) are based on ranges given in Cartigny (2005). **(a)** Garnet inclusions separated into low-Cr and high-Cr groups; **(b)** $CaSiO_3$-rich inclusions separated into low-Ti and high-Ti groups; **(c)** $MgSiO_3$ inclusions separated into low-Al and high-Al groups; **(d)** clinopyroxene inclusions separated into low-Na and high-Na groups; **(e)** ferropericlase inclusions; **(f)** olivine inclusions.

with chemical features most consistent with a meta-peridotitic association, generally have heavier C, with isotope compositions exhibiting a peak overlapping with mantle carbon; exceptions include one diamond with very light carbon (~ –23‰) and one with anomalously heavy carbon (~ –1‰). In contrast, high-Ti inclusions, with chemical features consistent with a meta-basaltic association, generally occur in diamonds with lighter carbon, with a range similar to low-Cr majoritic garnet.

Diamonds hosting $MgSiO_3$-rich inclusions also show a range of $\delta^{13}C$ from about 0 to –25‰ (Fig. 23c). Low-Al $MgSiO_3$-rich inclusions with chemical features indicating a meta-peridotitic association predominantly occur in diamonds with heavy C, $\delta^{13}C$ from about 0 to –4‰, with two exceptions at $\delta^{13}C$ of ~ –15‰. Indeed, carbon isotope signatures among this group tend to show heavier C than in normal mantle. Although measurements are sparse, high-Al inclusions that are more consistent with a meta-basaltic association occur in diamonds that have $d^{13}C$ from about –5 to –25‰, similar to low-Cr majoritic garnet and high-Ti $CaSiO_3$-rich inclusion-bearing diamonds. Similarly, diamonds with clinopyroxene inclusions (Fig. 23d) with high Na that are compositionally akin to a meta-basaltic association show a wide range of $\delta^{13}C$ from ~ 0 to –18 ‰, whereas diamonds with low Na clinopyroxene are isotopically heavy, generally intermediate between mantle and carbonate.

Ferropericlase inclusions (Fig. 23e) occur in diamonds with a relatively narrow distribution of carbon isotopes with a peak at $\delta^{13}C$ ~ –4‰, generally consistent with mantle carbon but with a distribution toward both heavier and lighter carbon. Diamonds hosting olivine (Fig. 23f) inclusions exhibit a similar distribution to ferropericlase but with a few notable outliers to much lighter C. Both ferropericlase and olivine are minerals typically associated with a meta-peridotitic association, which is generally consistent with their carbon isotope compositions.

Variations in carbon isotopes within diamonds hosting inclusions with compositions indicating a meta-basaltic association (e.g., low-Cr garnet, high-Ti $CaSiO_3$) can be large across growth zones (Fig. 22) (Bulanova et al. 2010; Palot et al. 2012; Thomson et al. 2014; Zedgenizov et al. 2014a). Such large variations can be attributed to changes in the source of carbon in fluids or melts contributing to diamond growth rather than fractionation of carbon during diamond precipitation. In cases where measurements have been made in diamonds from core to rim, examples are found for both large core to rim increases and decreases in carbon isotopic compositions, although there is an apparent tendency that when diamond cores are isotopically light the rims tend to be heavier (Fig. 22) (Bulanova et al. 2010; Thomson et al. 2014; Zedgenizov et al. 2014a; Palot et al. 2017). For example, in ten cases where cores had isotopic compositions lower than $\delta^{13}C$ = –20‰, the compositions of rims were notably heavier extending in some cases to near mantle values (Thomson et al. 2014). Such variations might represent mixing of carbon sources as melts and fluids derived from oceanic crust mix with carbon derived from the mantle as they migrate and evolve (Burnham et al. 2015).

The origin of sublithospheric inclusions in fluids and melts

On the basis of the major, trace and isotopic composition of sublithospheric diamonds and their mineral inclusions both carbonated melts and carbon-bearing hydrous fluids or melts have been postulated as potential diamond-forming media (Walter et al. 2008; Bulanova et al. 2010; Harte 2010; Stachel and Luth 2015; Palot et al. 2016; Thomson et al. 2016a,b; Smith et al. 2018; Timmerman et al. 2019; Zhu et al. 2019) (see also Luth et al. 2022, this volume). We note these kinds of fluids/melts are not mutually exclusive and any melt or fluid in the mantle will likely contain carbon, hydrogen and other volatile incompatible elements. Also implicated as a potential fluid growth medium are Fe–Ni–S–C metallic alloy or sulfide melts (Smith et al. 2016b). Subducting lithospheric slabs are most commonly suggested as the source of volatile-rich fluid components, and here we assess these model diamond-forming fluids.

Low-degree carbon-rich melts and the origin of meta-basaltic inclusions. Inclusions indicating a genetic relationship with meta-basaltic assemblages have been most closely linked to an origin involving low-degree melts of subducted oceanic crust. Lithophile trace element abundances observed in low-Cr majorite and high-Ti $CaSiO_3$-rich phases (Fig. 6 and 9), coupled with considerations of melting phase relations, have been interpreted to reflect equilibration with a low-degree melt of subducted oceanic crust, generally in the range of 300 to 600 km (Walter et al. 2008; Bulanova et al. 2010; Zedgenizov et al. 2014a; Thomson et al. 2016a,b). In this model, diamond co-crystallizes from carbon-bearing low-degree melts through redox reactions as melts migrate away from the oceanic crust and into peridotite, either within the slab or outside the slab in ambient mantle (Bulanova et al. 2010; Rohrbach and Schmidt 2011; Walter et al. 2011; Sun et al. 2020). Modeling indicates that melts can migrate through channelized porous flow and depending on the flux from the slab, remain molten for millions of years, reacting and differentiating as they move (Sun and Dasgupta 2019; Sun et al. 2020, 2021).

Figure 24 shows calculated trace element abundances (normalized to MORB) in melts that could coexist with low-Cr majoritic garnet, high-Ti Ca-silicate perovskite and low Mg#/high-Al bridgmanite on the basis of experimental mineral–melt partition coefficients (Table 3—Available at: https://doi.org/10.5683/SP3/LIVK1K). Also shown for comparison are calculated low-degree melts ($F = 0.01$) of model MORB and processed MORB at 20 GPa as well as the compositions of a suite of oceanic carbonatites (Hoernle et al. 2002). Overall, there is a good correspondence among the calculated coexisting melts from the three meta-basaltic inclusion types, suggesting a similar parental melt compositions, including enrichments in highly incompatible elements (e.g., Th, U, Nb, Ta, La, Ce) and common relative depletions in Sr, Zr, Hf and Pb (not shown, typically unmeasured or below detection limit). Bulanova et al. (2010) and Thomson et al. (2016a) presented trace element models that included Ca-silicate perovskite and majoritic garnet fractionation that can plausibly explain much of the observed trace element abundance variations in inclusion sample suites from individual pipes in the Juina region (e.g., Collier-4, Juina-5). In particular the trace element abundances of the $CaSiO_3$-rich inclusions are strikingly similar to low-degree melts of oceanic crust and with carbonatites from oceanic settings. The expected high U/Pb and relatively unfractionated Rb/Sr and Lu/Hf makes these melts putative candidates for imposing a chemical character to the mantle they interact with that could grow a HIMU like isotopic character over time (Sun et al. 2021).

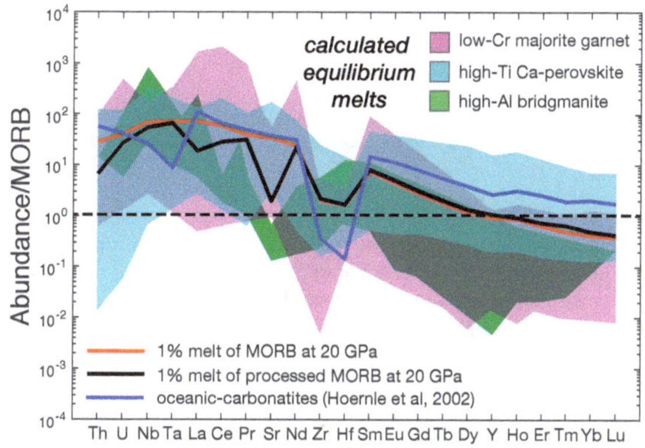

Figure 24. Trace element spider diagram showing MORB normalized abundances of calculated melts that could be in equilibrium with low-Cr majorite (**pink field**), high-Ti Ca perovskite (**cyan field**) and high-Al bridgmanite (**green field**) and model melts for 1% partial melting of MORB at 20 GPa (**red line**) and 1% melting of 'processed' MORB (**black line**) at 20 GPa, using partition coefficients from Table 3. Also shown is an average composition of oceanic carbonatite (Hoernle et al. 2002).

Figure 25 shows the solidus of carbonated oceanic crust (Kiseeva et al. 2013a; Thomson et al. 2016a; Zhang et al. 2020) relative to calculated pressure-temperature profiles for subducting lithosphere at the slab top and Moho in modern subduction zone settings (Shirey et al. 2021). Differences in the solidi among the experimental studies can generally be attributed to variations in the Ca/Mg of the source basalt and should reflect expected variations in subducted, altered MORB. The "slab-therms" indicate that carbonated oceanic crust will melt at pressures of about 13 GPa or greater, with only the coldest slabs potentially avoiding melting. Addition of water to the carbon-bearing system will reduce the solidus further, making melting of oceanic crust an inevitable consequence of subduction to deep upper mantle and transition zone depths.

Subducting slabs are believed to often stagnate in the transition zone due to a combination of buoyancy forces related to phase transitions (Bina 1997; Billen 2008), such that most or all slab crust may be expected to undergo carbonated melting in the transition zone at ~1100–1200 °C. We note that majoritic garnet inclusions indicate two pressure modes at about 9 and 14 GPa (Fig. 4), or with a plausible exsolution correction, at about 13 and 18 GPa (Thomson et al. 2021). The high pressure mode in particular, which is dominated by low-Cr garnets of the meta-basaltic association (Fig. 5b), occurs at pressures generally consistent with a model for melting of carbonated oceanic crust. The experiments of Thomson et al. (2016b) also showed that the products of reaction between carbonated melt from oceanic crust and mantle peridotite include low-Cr, high Ca majorite garnet with compositions intermediate to meta-peridotitic and meta-basaltic majoritic garnet, high-Ti Ca-perovskite with low MgO, and ferropericlase with variable but low Mg#s, all features that are consistent with observations from meta-basaltic inclusions in sublithospheric diamonds.

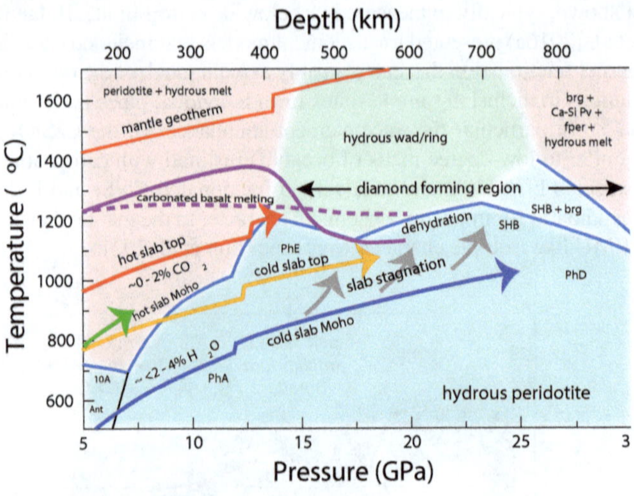

Figure 25. Pressure–temperature diagram showing generalized phase relations for water-saturated peridotite (Poli and Schmidt 2002; Iwamori 2004; Kawamoto 2004; Komabayashi and Omori 2006; Shirey et al. 2021). Phase labels designate generalized regions of stable hydrous magnesian silicates: ant = antigorite; PhA = phase A; PhE = phase E; PhD = phase D; SHB = superhydrous phase B; wad = wadsleyite; ring = ringwoodite; brg = bridgmanite. Average geotherms for cold and hot slabs are shown for peridotite at the Moho (**blue and green arrows**) and for the top of the oceanic crust (**orange and red arrows**) and are from Shirey et al. (2021). Reactions where hydrous phases dehydrate are shown as **solid blue lines**. Melting curves for carbonated basalt are shown as **purple curves** (solid: Thomson et al. 2016b; dashed: Kiseeva et al, 2013). The mantle geotherm is from Katsura et al. (2010). Cold slab mantle can transport H_2O beyond antigorite breakdown and into the transition zone. Slab stagnation and heating may cause dehydration throughout the transition zone and into the shallow lower mantle at temperatures of ~1100 to 1300 °C. Carbonated oceanic crust melts at approximately the same temperature as slab mantle dehydrates, potentially creating a diamond forming region throughout the transition zone and shallow lower mantle related to decarbonation and dehydration of slab crust and mantle.

Hydrous fluids/melts and the origin of meta-peridotitic inclusions. The discovery of a hydrous ringwoodite (~1.5 wt% H_2O) inclusion in a diamond from Juina, Brazil, provides primary evidence for the involvement of H_2O-rich fluids in sublithospheric diamond genesis (Pearson et al. 2014), although only one such inclusion has been identified to date. Crystallographic data indicate an Mg# of 0.75 ± 0.2, and the inclusion co-occurs with a $CaSiO_3$-rich inclusion with a breyite structure, presumably of the low-Ti variety due to a lack of evidence for an exsolved $CaTiO_3$ phase. Thus, the association appears nominally meta-peridotitic. Further evidence for the role of an H_2O-rich peridotitic association comes from exsolved brucite and magnesioferrite in a composite ferropericlase (Mg# = 0.84) inclusion from Juina (Brazil).

Relatively high levels of boron (reaching up to a few ppm) in blue (Type IIb) diamonds that host $MgSiO_3$-rich phases, $CaSiO_3$-rich phases, SiO_2 and ferropericlase inclusions, are postulated to result from dehydration of post-serpentine minerals on the basis of the high-partition coefficient for boron in serpentine during seawater alteration (Smith et al. 2018). Inclusions of ferropericlase, $MgSiO_3$-rich phases, $CaSiO_3$-rich phases, SiO_2, and CF- or NAL-phase in Type IIb diamonds have been found to have a thin layer of methane ± hydrogen fluid coexisting alongside the solid phases within the inclusion cavity, tentatively suggesting the involvement of a hydrous component in diamond growth (Smith et al. 2018). The co-occurring phases reported in Type IIb diamonds also appear to be part of a meta-peridotitic association although chemical analyses were not reported. Evidence for the role of an H_2O-rich basaltic association comes from reports of micro- or nano-inclusions of phase Egg and δ-AlOOH (Wirth et al. 2007; Kaminsky 2017).

Figure 25 shows a representative phase diagram for hydrous peridotite that, together with modeled pressure–temperature paths at the slab Moho for modern subduction zones, can be used to assess the fate of water in the oceanic lithospheric mantle as it subducts into the transition zone. As pointed out by previous workers, retaining water in subducted lithosphere requires that temperatures in the slab mantle remain below the temperature minimum (~ 7 GPa and 700°C) along the dehydration curve where antigorite and 10 Å phase (±Mg-sursassite) remain stable (Iwamori 2004; Komabayashi et al. 2004; Shirey et al. 2021), which has been referred to as a "choke point" (Iwamori 2004). Estimates for the water storage capacity in antigorite-bearing slab mantle is about 4–5 wt% (Iwamori 2004; Komabayashi and Omori 2006), whereas 10 Å phase-bearing (±Mg-sursassite) mantle assemblages have storage capacity of ~1–2 wt% H_2O (Iwamori 2004; Fumagalli and Poli 2005). Warm slabs have temperatures higher than the choke point leading to dehydration of slab mantle at depths shallower than ~250 km, and such slabs are unlikely to transport significant amounts of water deeper into the mantle either in basaltic crust or peridotitic lithosphere (Okamoto and Maruyama 2004; van Keken et al. 2011; Shirey et al. 2021). However, cooler slabs have *P–T* paths at the Moho and in the even cooler interior portions of the lithosphere that can remain well below the choke point. Depending on the efficiency of hydration of mantle lithosphere near the surface, cooler slabs can potentially transport as much 4–5 wt% water deeper into the mantle, at least locally, in serpentinized regions formed in deep fractures related to slab bending near the Earth's surface (Faccenda 2014). In colder slabs antigorite transforms at ~250 km and deeper to a series of dense hydrous magnesium silicate phases (DHMS) that in mantle peridotite compositions have the capacity to store at least 5 wt% water and potentially more than 10 wt% (Iwamori 2004).

Water retention in DHMS phases in subducting mantle lithosphere has been postulated in a number of previous studies (Thompson 1992; Frost 1999; Poli and Schmidt 2002; Iwamori 2004; Ohtani et al. 2004; Omori et al. 2004; Harte 2010; van Keken et al. 2011; Ohtani 2015; Maurice et al. 2018). As described above, most slabs are expected to slow down and deform in the transition zone (Bina 1997; Billen 2008). During this stagnation H_2O-bearing slabs will heat by conduction before descending into the lower mantle. Figure 25 illustrates that heating by only a few hundred degrees in the transition zone, which can occur in an ~10 m.y. timeframe (Shirey et al. 2021), would result in breakdown of DHMS phases in the slab mantle at ~1200–1300 °C to wadsleyite or ringwoodite-bearing assemblages and a hydrous fluid.

Wadselyite- and ringwoodite-bearing mantle can accommodate ~1–2 wt% water but a free fluid phase is expected for more water rich-regions of slab mantle. If slabs do not heat sufficiently in the transition zone to dehydrate DHMS phases, dehydration is unavoidable at ~ 700–800 km due to another deep trough, or second 'choke point', as they penetrate into the lower mantle. Depending on temperature, phase D, superhydrous phase B or ringwoodite will transform into a nominally anhydrous assemblage of bridgmanite, Ca-silicate perovskite and ferropericlase with a much lower bulk water storage capacity (< ~0.1 wt%) (Fu et al. 2019), possibly resulting in a hydrous melt at the top of the lower mantle (Schmandt et al. 2014; Ohtani 2015, 2020; Walter et al. 2015).

The composition of H_2O-rich, supercritical fluids or melts released at transition zone and lower mantle depths are poorly known but should have a considerable dissolved silicate component derived from either meta-peridotitic or meta-basaltic mantle sources, being well beyond their second critical endpoints (Mibe et al. 2007, 2011; Wang et al. 2020). Such fluids may also have a dissolved carbon component acquired from mantle peridotite or oceanic crust. Precipitation of diamond and inclusions may occur through subsequent reaction as fluids migrate within the slab and potentially out of the slab (Harte 2010). The meta-peridotitic association of low-Al $MgSiO_3$-rich phases, low-Ti $CaSiO_3$-rich phases and ferropericlase may possibly be best explained by crystallization involving such H_2O-rich fluids/melts at low-temperature in subducted lithospheric mantle and especially in a depleted peridotite composition (e.g., harzburgite) (Harte 2010; Smith et al. 2018; Shirey et al. 2021).

Figure 16 illustrates that at ~1200 °C, where dehydration and fluid release is expected to occur, assemblages are dominated by ringwoodite, akimotoite, bridgmanite, Ca-perovskite, garnet and ferropericlase. The generally low Mg content of low-Ti $CaSiO_3$-rich phases and the generally low-Al and low-Ca contents of $MgSiO_3$ inclusions may indicate equilibrium and crystallization from cool, hydrous fluids or melts carrying dissolved peridotitic components. This may either reflect equilibrium assemblages of Ca-silicate perovskite and akimotoite at the base of the transition zone or relatively low-temperature equilibration between Ca-perovskite and bridgmanite where the solvus results in low levels of solid solution (Irifune et al. 2000). Such low temperatures may also explain the co-occurrence of ferropericlase and stishovite.

The role of Fe–Ni–S–C metallic melts. Iron-rich metal and sulfide inclusions have been reported in diamonds containing sublithospheric inclusions (Bulanova et al. 2010; Kaminsky and Wirth 2011; Smith and Kopylova 2014; Smith et al. 2016b, 2018). Bulanova et al. (2010) reported a metallic iron phase co-occurring in a diamond containing a majoritic garnet, and composite inclusions interpreted as retrograde CAS-phase and K-hollandite, and also observed low-Ni (<3 wt%) pyrrhotite inclusions co-occurring with Ti-rich $CaSiO_3$, majoritic garnet and SiO_2 inclusions in seven diamonds from the Collier-4 kimberlite, Juina (Brazil). Kaminsky and Wirth (2011) reported iron carbides (Fe_3C, Fe_2C and $Fe_{23}C_6$) associated with native iron, also in a diamond from the Juina area, and although not occurring with other sublithospheric inclusions, they concluded a sublithospheric origin based on phase relations in the Fe–C system.

Smith et al. (2016b) reported that many high-quality, Type IIa, gem diamonds, exemplified by the Cullinan, Constellation, and Koh-i-Noor, commonly contain iron-nickel-carbon-sulfur inclusions. Metallic inclusions have been found in 67 out of 83 diamonds in the so-called CLIPPIR suite (Cullinan-like, inclusion-poor, relatively pure, irregularly shaped, and resorbed), and some of these diamonds also contain majoritic garnet or $CaSiO_3$-rich (including Ti-rich $CaSiO_3$) phases which, based on the arguments discussed above, indicate a sublithospheric origin (Smith et al. 2016b, 2017, 2021). A thin fluid layer of methane ± hydrogen was also found associated with the metallic inclusions likely dissolved in the original iron melt under reducing conditions. Smith et al. (2018) also report on rare metallic phases similar to those common in the CLIPPIR suite in the boron-bearing (Type IIb) suite of sublithospheric diamonds.

These diamonds contain a wide range of inclusion types with $CaSiO_3$-phases being the most common, but also including majoritic garnet, SiO_2, and possibly CF or NAL phase in a diamond containing a metallic phase. Smith et al. (2016b) interpret the metallic inclusions as the solidified products of a metallic liquid phase in the deep mantle and suggested strong N partitioning into the melt phase, which may explain the characteristically low nitrogen content in CLIPPIR diamonds and perhaps in sublithospheric diamonds more generally (Smith and Kopylova 2014).

Sublithospheric diamonds containing Fe-rich metal alloy and sulfide inclusions are also generally characterized by isotopically variable and often extremely light carbon ($\delta^{13}C$ from −26.9 to −3.8 ‰ in CLIPPIR diamonds; $\delta^{13}C$ −26 to −8‰ in Collier-4 diamonds). Although the number of metal/sulfide-bearing diamonds co-occurring with sublithospheric silicate inclusions is small, the suite observed to date contains phases suggestive of basaltic. A recent study of the Fe isotopic composition of the metallic inclusions reveals a strikingly heavy isotopic signature ($\delta^{56}Fe = 0.79$ to 0.90‰) thought to be inherited from magnetite and/or Fe–Ni alloys precipitated during serpentinization of oceanic peridotite (Smith et al. 2021). This heavy iron signature is suggestive that the metallic liquid trapped in CLIPPIR diamonds has a lithological connection to subducted, serpentinized peridotite.

SUMMARY AND PERSPECTIVES

On the basis of an analysis of published geochemical data of predominant silicate and oxide inclusions in sublithospheric diamonds and building upon observations and ideas that have been discussed in the literature for over three decades, a framework emerges for understanding processes occurring in the deep mantle related to the subduction of lithospheric plates. Few inclusions have major and trace element chemistry consistent with expectations for primitive meta-peridotitic mantle assemblages at upper mantle, transition zone or lower mantle pressures and temperatures, and we conclude that sublithospheric diamonds do not typically incorporate minerals representative of ambient mantle.

Generally speaking, the inclusions define two populations that appear to reflect different source lithologies, or more specifically, the fluids and melts derived from and/or that interact with those lithologies. Consistent with previous interpretations we assign the two populations of inclusions to meta-peridotitic and meta-basaltic varieties, with the former broadly representative of peridotitic, harzburgitic and dunitic lithologies and the latter including basaltic or pyroxenitic ones. The host diamonds themselves have carbon isotopic compositions that also reveal a distinction among these groups, with meta-peridotitic inclusions hosted by diamonds with a restricted range of predominantly isotopically heavier, mantle-like carbon, whereas the meta-basaltic group exhibits a wider range that extends to isotopically light carbon associated with subducted oceanic crust.

The meta-peridotite group includes low-Al $MgSiO_3$, low-Ti $CaSiO_3$, ferropericlase, high-Cr majoritic garnet, olivine and low-Na clinopyroxene inclusions. Low-Al $MgSiO_3$ (former bridgmanite and/or akimotoite), low-Ti $CaSiO_3$ (former Ca-silicate perovskite) and high Mg# ferropericlase are generally consistent with diamond formation in the shallow lower mantle or deep transition zone, and the observed inclusion suite provides evidence for equilibration at low temperatures (e.g., low Mg in Ca-perovskite; low Ca in $MgSiO_3$; the ferropericlase plus stishovite association) perhaps in the range of 1000–1200 °C. Many inclusions also indicate a depleted source lithology (e.g., low Al, low Ca, high Mg# in some $MgSiO_3$; low Cr, Al and Na in ferropericlase), which when combined with low equilibration temperatures suggests an origin related to depleted slab mantle lithosphere.

Harte (2010) summarized many of these features and suggested an origin related to dehydration of subducted lithospheric mantle near the base of the transition zone, and phase relations for dehydration in subducted mantle support this hypothesis. The heavy iron isotopes in iron-alloy inclusions that may be related to magnetite formation during serpentinization and the generally mantle-like carbon isotopic compositions of the diamonds hosting the meta-peridotitic inclusions all support this hypothesis. This may indicate that slab-derived hydrous fluids acquire their carbon from the mantle portion of the down-going slab or from the ambient mantle, with little or no interaction with subducted oceanic crust.

The presence of methane in some inclusions is suggestive of reduced fluids possibly equilibrated with metal alloy phases. The high-Cr majoritic garnets also have a meta-peridotite chemistry but their apparent formation pressures indicate a predominantly upper mantle origin unrelated to the deeper inclusion assemblage, suggesting these garnet inclusions may have formed in the deepest portions of the lithospheric mantle. Low-Na clinopyroxene, with their unusual co-occurrences with ferropericlase and Ca-rich phases, also remain enigmatic but seemingly reflect an upper mantle and transition zone origin.

The meta-basaltic group is characterized by low-Cr garnet, high-Ti $CaSiO_3$, high-Al $MgSiO_3$, high-Na clinopyroxene as well as rare CF and NAL inclusions (Fig. 1). These inclusions have major element compositions akin to high pressure phases in meta-basaltic to meta-pyroxenitic lithologies. Majoritic garnet barometry places their origin throughout the deep upper mantle to the transition zone, which is consistent with phase equilibrium constraints on the origin of high-Ti $CaSiO_3$ inclusions. In contrast, high-Al $MgSiO_3$, CF and NAL inclusions provide evidence of a shallow lower mantle assemblage. Thus, the meta-basaltic group apparently forms over a wide depth range.

Meta-basaltic inclusions have trace element concentrations that are generally enriched over levels expected in MORB, sometimes by many orders of magnitude, with elemental abundance patterns linking them to deeply subducted oceanic crust. Low-degree melts from oceanic crust, possibly carbonatitic, are implicated in their origin and diamond formation may occur through reduction within slab mantle or ambient mantle through redox freezing. Low temperatures of equilibration are indicated by the high-Ti $CaSiO_3$ inclusions (low-Mg content) consistent with melting of carbonated/hydrated oceanic crust at ~1200 °C as constrained by melting phase relations. Many majoritic garnet and clinopyroxene compositions indicate an origin related to meta-pyroxenite and these may represent hybrid reaction products when melts from subducted oceanic crust infiltrate and react with the slab mantle or ambient mantle. The carbon isotopic compositions of the diamond hosts are consistent with this scenario and may reflect mixing of carbon sourced from subducted oceanic crust and peridotitic mantle sources.

The common theme among models for sublithospheric diamond and inclusion formation is the key role of subducted lithosphere, and particularly the fluids and melts that are derived from lithologies in the slab that infiltrate and react with their surroundings both within and external to the slab, precipitating diamonds and their inclusions. The overall narrative presented here for the generation of sublithospheric diamonds related to subducted slab lithosphere stagnating in the transition zone and shallow lower mantle is reminiscent of the megalith model of Ringwood (1991), an observation echoed in many studies over the intervening years. Yet many questions remain as we continue to develop our understanding of sublithospheric diamond and inclusion formation and what these samples reveal in detail about deep mantle processes.

In this review we have taken a global perspective, highlighting the clear similarities within inclusion groups that span across all sampled cratons. However, processes related to diamond and inclusion formation will also likely be dependent on the specific tectonic and geodynamic setting of subduction, and regional differences can be expected that can only become readily apparent with larger, more diverse data sets. Some outstanding questions that can be addressed in future studies might include:

1. What is the actual distribution of co-occurring phases in sublithospheric diamonds? In many cases, especially in pioneering studies where cracking diamonds to retrieve inclusions was a common practice, the co-occurring phases within individual diamonds are unknown or unreported creating data biases. Future studies should concentrate on identifying all major inclusions co-occurring in single diamonds. This requires a combination of techniques including X-ray diffraction and tomography, especially at synchrotron facilities (Wenz et al. 2019). Diamonds should be fully characterized by cutting and polishing to reveal inclusions, ideally with full major and trace element chemical analysis of inclusions and isotopic analysis of the host diamonds.

2. How are iron-rich ferropericlase inclusions formed and at what depth? Are these lower mantle or upper mantle phases? Do they represent remnants of the redox process during interaction of melts/fluids with the mantle during diamond crystallization? What is the source of the pronounced Ba and Y anomalies? More data on the redox state of the iron-rich inclusions, trace element data and iron isotopic compositions of inclusions are needed to understand these important and abundant inclusions.

3. What is the true majoritic garnet pressure distribution? Although reliable majorite barometers are available and a large dataset reveals a range of pressures from the asthenosphere to the transition zone, majoritic garnet inclusions can only provide accurate depths of origin if their bulk chemistry is accurately known. The common unmixing of clinopyroxene in sublithospheric garnet inclusions currently masks this information. Future studies of majoritic garnet inclusions should concentrate on reconstructing bulk compositions through approaches such as X-ray tomography, sequential sectioning and analysis of all unmixed components.

4. What is the upwelling history of composite inclusions? Unmixing of phases is common in sublithospheric inclusions but the uplift history remains a mystery in most cases. What is their final residence depth and what is the uplift mechanism (i.e., solid-state upwelling in a mantle plume; transport in a percolating melt) remain important unresolved questions.

5. What are the ages of diamond and inclusion formation and can they be linked in space and time to paleo-subduction? How do they relate to the timing of ascent and volcanic emplacement at surface? There is scarce information about the age of sublithospheric diamond formation, yet this is critical for placing models within a geodynamic and tectonic context. Dating of small sulfide grains using Re–Os and $CaSiO_3$-rich phases using U–Pb, although tremendously challenging, likely represent the best opportunity to provide key age constraints.

6. Fluids and melts generated by subducting slab lithosphere are the likely media of diamond formation and inclusion equilibration, yet little is known about the phase equilibria or composition of such melts. What are the compositions of the deep melts and fluids from which the diamonds precipitate and the inclusions likely equilibrate? What reactions occur to form diamonds and their inclusions? Experiments on a range of compositions with mixed volatile phases are required to develop a firmer understanding of the fluids, melts and reactions that form sublithospheric diamonds and their inclusions.

ACKNOWLEDGEMENTS

The tables associated with this chapter can be found at https://doi.org/10.5683/SP3/LIVK1K (Walter et al. 2022). We would like to thank Galina Bulanova, Antony Burnham, Rick Carlson, Ben Harte, Dan Howell, Simon Kohn, Nico Kueter, Sami Mikhail, Fabrizio Nestola, Peng Ni, Graham Pearson, Anat Shahar, Steve Shirey, Chris Smith, Lara Speich, Thomas Stachel, Peter van Keken and Lara Wagner for the many conversations and insights that helped shaped this review. Special thanks to Thomas Stachel for access to his incredible inclusion database. We thank Ben Harte, Thomas Stachel, Karen Smit and Graham Pearson for constructive reviews that improved the clarity of ideas and presentation in this chapter.

REFERENCES

Ackerson MR, Tailby ND, Watson EB (2017a) XAFS spectroscopic study of Ti coordination in garnet. Am Mineral 102:173–183

Ackerson MR, Watson EB, Tailby ND, Spear FS (2017b) Experimental investigation into the substitution mechanisms and solubility of Ti in garnet. Am Mineral 102:158–172

Akaogi M, Akimoto S (1977) Pyroxene–garnet solid-solution equilibria in systems $Mg_4Si_4O_{12}$–$Mg_3Al_2Si_3O_{12}$ and $Fe_4Si_4O_{12}$–$Fe_3Al_2Si_3O_{12}$ at high-pressures and temperatures. Phys Earth Planet Sci 15:90–106

Akaogi M, Akimoto S (1979) High-pressure phase-equilibria in a garnet lherzolite, with special reference to Mg^{2+}–Fe^{2+} partitioning among constituent minerals. Phys Earth Planet Sci 19:31–51

Anand M, Taylor LA, Misra KC, Carlson WD, Sobolev NV (2004) Nature of diamonds in Yakutian eclogites: views from eclogite tomography and mineral inclusions in diamonds. Lithos 77:333–348

Andrault D, Bolfan-Casanova N (2001) High-pressure phase transformations in the $MgFe_2O_4$ and Fe_2O_3–$MgSiO_3$ systems. Phys Chem Mineral 28:211–217

Anzolini C, Prencipe M, Alvaro M, Romano C, Vona A, Lorenzon S, Smith EM, Brenker FE, Nestola F (2018) Depth of formation of super-deep diamonds: Raman barometry of $CaSiO_3$-walstromite inclusions. Am Mineral 103:69–74

Anzolini C, Nestola F, Mazzucchelli ML, Alvaro M, Nimis P, Gianese A, Morganti S, Marone F, Campione M, Hutchison MT, Harris JW (2019) Depth of diamond formation obtained from single periclase inclusions. Geology 47:219–222

Armstrong LS, Walter MJ (2012) Tetragonal almandine pyrope phase (TAPP): retrograde Mg-perovskite from subducted oceanic crust? Eur J Mineral 24:587–597

Armstrong LS, Walter MJ, Tuff JR, Lord OT, Lennie AR, Kleppe AK, Clark SM (2012) Perovskite phase relations in the system CaO-MgO-TiO_2-SiO_2 and implications for deep mantle lithologies. J Petrol 53:611–635

Beyer C, Frost DJ (2017) The depth of sub-lithospheric diamond formation and the redistribution of carbon in the deep mantle. Earth Planet Sci Lett 461:30–39

Billen MI (2008) Modeling the dynamics of subducting slabs. Annu Rev Earth Planet Sci 36:325–356

Bina CR (1997) Patterns of deep seismicity reflect buoyancy stresses due to phase transitions. Geophys Res Lett 24:3301–3304

Boyd SR, Mattey DP, Pillinger CT, Milledge HJ, Mendelssohn M, Seal M (1987) Multiple growth events during diamond genesis - an integrated study of carbon and nitrogen isotopes and nitrogen aggregation state in coated stones. Earth Planet Sci Lett 86:341–353

Brenker FE, Stachel T, Harris JW (2002) Exhumation of lower mantle inclusions in diamond: ATEM investigation of retrograde phase transitions, reactions and exsolution. Earth Planet Sci Lett 198:1–9

Brenker FE, Vincze L, Vekemans B, Nasdala L, Stachel T, Vollmer C, Kersten M, Somogyi A, Adams F, Joswig W, Harris JW (2005) Detection of a Ca-rich lithology in the Earth's deep (> 300 km) convecting mantle. Earth Planet Sci Lett 236:579–587

Brenker FE, Nestola F, Brenker L, Peruzzo L, Harris JW (2021) Origin, properties, and structure of breyite: The second most abundant mineral inclusion in super-deep diamonds. Am Mineral 106:38–43

Brey GP, Bulatov V, Girnis A, Harris JW, Stachel T (2004) Ferropericlase—a lower mantle phase in the upper mantle. Lithos 77:655–663

Brey GP, Bulatov VK, Girnis AV (2008) Geobarometry for peridotites: Experiments in simple and natural systems from 6 to 10 GPa. J Petrol 49:3–24

Brey GP, Girnis AV, Bulatov VK, Hofer HE, Gerdes A, Woodland AB (2015) Reduced sediment melting at 7.5–12 GPa: phase relations, geochemical signals and diamond nucleation. Contrib Mineral Petrol 170:18

Bulanova GP (1995) The formation of diamond. J Geochem Explor 53:1–23

Bulanova GP, Walter MJ, Smith CB, Kohn SC, Armstrong LS, Blundy J, Gobbo L (2010) Mineral inclusions in sublithospheric diamonds from Collier 4 kimberlite pipe, Juina, Brazil: subducted protoliths, carbonated melts and primary kimberlite magmatism. Contrib Mineral Petrol 160:489–510

Bulatov VK, Brey GP, Girnis AV, Gerdes A, Hofer HE (2014) Carbonated sediment–peridotite interaction and melting at 7.5–12 GPa. Lithos 200:368–385

Bulatov VK, Girnis AV, Brey GP, Woodland AB, Hofer HE (2019) Ferropericlase crystallization under upper mantle conditions. Contrib Mineral Petrol 174:45

Burnham AD, Thomson AR, Bulanova GP, Kohn SC, Smith CB, Walter MJ (2015) Stable isotope evidence for crustal recycling as recorded by superdeep diamonds. Earth Planet Sci Lett 432:374–380

Burnham AD, Bulanova GP, Smith CB, Whitehead SC, Kohn SC, Gobbo L, Walter MJ (2016) Diamonds from the Machado River alluvial deposit, Rondonia, Brazil, derived from both lithospheric and sublithospheric mantle. Lithos 265:199–213

Cartigny P (2005) Stable isotopes and the origin of diamond. Elements 1:79–84

Cartigny P, Palot M, Thomassot E, Harris JW (2014) Diamond formation: A stable isotope perspective. Annu Rev Earth Planet Sci 42:699–732

Chatterjee ND, Johannes W, Leistner H (1984) The system $CaO–Al_2O_3–SiO_2–H_2O$—New phase-equilibria data, some calculated phase-relations, and their petrological applications. Contrib Mineral Petrol 88:1–13

Collerson KD, Williams Q, Kamber BS, Omori S, Arai H, Ohtani E (2010) Majoritic garnet: A new approach to pressure estimation of shock events in meteorites and the encapsulation of sub-lithospheric inclusions in diamond. Geochim Cosmochim Acta 74:5939–5957

Corgne A, Armstrong LS, Keshav S, Fei YW, McDonough WF, Minarik WG, Moreno K (2012) Trace element partitioning between majoritic garnet and silicate melt at 10–17 GPa: Implications for deep mantle processes. Lithos 148:128–141

Czas J, Stachel T, Pearson DG, Stern RA, Read GH (2018) Diamond brecciation and annealing accompanying major metasomatism in eclogite xenoliths from the Sask Craton, Canada. Mineral Petrol 112:311–323

Dalou C, Koga KT, Hammouda T, Poitrasson F (2009) Trace element partitioning between carbonatitic melts and mantle transition zone minerals: Implications for the source of carbonatites. Geochim Cosmochim Acta 73:239–255

Dasgupta R, Hirschmann MM, Withers AC (2004) Deep global cycling of carbon constrained by the solidus of anhydrous, carbonated eclogite under upper mantle conditions. Earth Planet Sci Lett 227:73–85

Davies RM, Griffin WL, O'Reilly SY, Andrew AS (2003) Unusual mineral inclusions and carbon isotopes of alluvial diamonds from Bingara, eastern Australia. Lithos 69:51–66

Deines P (1980) The carbon isotopic composition of diamonds—Relationship to diamond shape, color, occurrence and vapor composition. Geochim Cosmochim Acta 44:943–961

Deines P, Harris JW, Gurney JJ (1991) The carbon isotopic composition and nitrogen-content of lithospheric and asthenospheric diamonds from the Jagersfontein and Koffiefontein kimberlite, South Africa. Geochim Cosmochim Acta 55:2615–2625

Drewitt JWE, Walter MJ, Zhang HL, McMahon SC, Edwards D, Heinen BJ, Lord OT, Anzellini S, Kleppe AK (2019) The fate of carbonate in oceanic crust subducted into earth's lower mantle. Earth Planet Sci Lett 511:213–222

Duffy T, Madhusudhan N, Lee KKM (2015) Mineralogy of super-earth planets. *In*: Treatise on Geophysics. Elsevier, p 149–178

Dymshits AM, Bobrov AV, Bindi L, Litvin YA, Litasov KD, Shatskiy AF, Ohtani E (2013) Na-bearing majoritic garnet in the $Na_2MgSi_5O_{12}–Mg_3Al_2Si_3O_{12}$ join at 11–20 GPa: Phase relations, structural peculiarities and solid solutions. Geochim Cosmochim Acta 105:1–13

Faccenda M (2014) Water in the slab: A trilogy. Tectonophysics 614:1–30

Fei Y, Wang Y, Finger LW (1996) Maximum solubility of FeO in $(Mg,Fe)SiO_3$-perovskite as a function of temperature at 26 GPa: Implication for FeO content in the lower mantle. J Geophys Res-Solid Earth 101:11525–11530

Fischer RA, Campbell AJ, Chidester BA, Reaman DM, Thompson EC, Pigott JS, Prakapenka VB, Smith JS (2018) Equations of state and phase boundary for stishovite and $CaCl_2$-type SiO_2. Am Mineral 103:792–802

Frost DJ (1999) The stability of dense hydrous magnesium silicates in Earth's transition zone and lower mantle. *In*: Mantle Petrology: Field Observations and High-Pressure Experimentation A Tribute to Francis R (Joe) Boyd. Vol 6. Fei Y, Bertka CM, Mysen BO (eds). The Geochemical Society, Houston, TX, p 283–296

Frost DJ, McCammon CA (2008) The redox state of Earth's mantle. Annu Rev Earth Planet Sci 36:389–420

Fu S, Yang J, Karato SI, Vasiliev A, Presniakov MY, Gavriliuk AG, Ivanova AG, Hauri EH, Okuchi T, Purevjav N, Lin JF (2019) Water concentration in single-crystal (Al,Fe)-bearing bridgmanite grown from the hydrous melt: Implications for dehydration melting at the topmost lower mantle. Geophys Res Lett 46:10346–10357

Fumagalli P, Poli S (2005) Experimentally determined phase relations in hydrous peridotites to 6.5 GPa and their consequences on the dynamics of subduction zones. J Petrol 46:555–578

Gale A, Dalton CA, Langmuir CH, Su YJ, Schilling JG (2013) The mean composition of ocean ridge basalts. Geochem Geophys Geosyst 14:489–518

Ghosh S, Ohtani E, Litasov KD, Terasaki H (2009) Solidus of carbonated peridotite from 10 to 20 GPa and origin of magnesiocarbonatite melt in the Earth's deep mantle. Chem Geol 262:17–28

Ghosh S, Litasov K, Ohtani E (2014) Phase relations and melting of carbonated peridotite between 10 and 20 GPa: a proxy for alkali- and CO_2-rich silicate melts in the deep mantle. Contrib Mineral Petrol 167:964

Grassi D, Schmidt MW (2011) Melting of carbonated pelites at 8–13 GPa: generating K-rich carbonatites for mantle metasomatism. Contrib Mineral Petrol 162:169–191

Grew ES, Locock AJ, Mills SJ, Galuskina IO, Galuskin EV, Halenius U (2013) Nomenclature of the garnet supergroup. Am Mineral 98:785–810

Haggerty SE (1986) Diamond genesis in a multiply-constrained model. Nature 320:34–37

Haggerty SE, Fung AT, Burt DM (1994) Apatite, phosphorus and titanium in eclogitic garnet from the upper-mantle. Geophys Res Lett 21:1699–1702

Hammouda T (2003) High-pressure melting of carbonated eclogite and experimental constraints on carbon recycling and storage in the mantle. Earth Planet Sci Lett 214:357–368

Harlow GE (1998) The Nature of Diamonds. Cambridge University Press, Cambridge, UK

Harris J, Gurney J (1979) Inclusions in diamonds. *In*: Properties of Diamond. Field JE (ed) Academic Press, London
Harris J, Hutchison MT, Hursthouse M, Light M, Harte B (1997) A new tetragonal silicate mineral occurring as inclusions in lower-mantle diamonds. Nature 387:486–488
Harte B (2010) Diamond formation in the deep mantle: the record of mineral inclusions and their distribution in relation to mantle dehydration zones. Mineral Mag 74:189–215
Harte B, Cayzer N (2007) Decompression and unmixing of crystals included in diamonds from the mantle transition zone. Phys Chem Mineral 34:647–656
Harte B, Harris JW (1994) Lower mantle mineral associations preserved in diamonds. Mineral Mag A58:384–385
Harte B, Richardson S (2012) Mineral inclusions in diamonds track the evolution of a Mesozoic subducted slab beneath West Gondwanaland. Gondwana Res 21:236–245
Harte B, Harris JW, Hutchison MT, Watt GR, Wilding MC (1999) Lower mantle mineral associations in diamonds from Sao Luiz, Brazil. *In*: Mantle Petrology: Field Observations and High Pressure Experimentation, A Tribute to Francis R (Joe) Boyd. Vol 6. Fei Y, Mysen BO, Bertka CM, (eds). The Geochemical Society, Houston, p 125–153
Hayman PC, Kopylova MG, Kaminsky FV (2005) Lower mantle diamonds from Rio Soriso (Juina area, Mato Grosso, Brazil). Contrib Mineral Petrol 149:430–445
Herzberg C, Zhang JZ (1996) Melting experiments on anhydrous peridotite KLB-1: Compositions of magmas in the upper mantle and transition zone. J Geophys Res-Solid Earth 101:8271–8295
Hirose K (2002) Phase transitions in pyrolitic mantle around 670-km depth: Implications for upwelling of plumes from the lower mantle. J Geophys Res-Solid Earth 107(B4):ECV-3
Hirose K, Fei YW (2002) Subsolidus and melting phase relations of basaltic composition in the uppermost lower mantle. Geochim Cosmochim Acta 66:2099–2108
Hoernle K, Tilton G, Le Bas MJ, Duggen S, Garbe-Schonberg D (2002) Geochemistry of oceanic carbonatites compared with continental carbonatites: mantle recycling of oceanic crustal carbonate. Contrib Mineral Petrol 142:520–542
Hutchison M (1997) Constitution of the Deep Transition Zone and Lower Mantle shown by Diamonds and their Inclusions. PhD, University of Edinburgh
Hutchison M, Cartigny P, Harris J (1999) Carbon and nitrogen compositions and physical characteristics of transition zone and lower mantle diamonds from São Luiz, Brazil. *In*: Proc VII Int Kimberlite Conf. Vol 2. Gurney JJ, Gurney JL, Pascoe MD, Richardson SH, (eds). Red Roof Designs, Cape Town, p 372–382
Hutchison MT, Hursthouse MB, Light ME (2001) Mineral inclusions in diamonds: associations and chemical distinctions around the 670-km discontinuity. Contrib Mineral Petrol 142:119–126
Ickert RB, Stachel T, Stern RA, Harris JW (2015) Extreme O-18-enrichment in majorite constrains a crustal origin of transition zone diamonds. Geochem Perspect Lett 1:65–73
Irifune T (1987) An experimental investigation of the pyroxene garnet transformation in a pyrolite composition and its bearing on the constitution of the mantle. Phys Earth Planet Sci 45:324–336
Irifune T (1994) Absence of an aluminous phase in the upper part of the earths lower mantle. Nature 370:131–133
Irifune T, Ringwood AE (1987) Phase-transformations in a harzburgite composition to 26 GPa—Implications for dynamical behavior of the subducting slab. Earth Planet Sci Lett 86:365–376
Irifune T, Ringwood AE (1993) Phase-transformations in subducted oceanic-crust and buoyancy relationships at depths of 600–800 km in the mantle. Earth Planet Sci Lett 117:101–110
Irifune T, Sekine T, Ringwood AE, Hibberson WO (1986) The eclogite–garnetite transformation at high-pressure and some geophysical implications. Earth Planet Sci Lett 77:245–256
Irifune T, Hibberson WO, Ringwood AE (1989) Eclogite–garnetite transformation at high pressure and its bearing on the occurrence of garnet inclusions in diamond. *In*: Kimberlites and Related Rocks. Vol GSA Spec Publ 14. Ross J (ed) Blackwell, Carlton, p 877–882
Irifune T, Ringwood AE, Hibberson WO (1994) Subduction of continental-crust and terrigenous and pelagic sediments—An experimental-study. Earth Planet Sci Lett 126:351–368
Irifune T, Miyashita M, Inoue T, Ando J, Funakoshi K, Utsumi W (2000) High-pressure phase transformation in $CaMgSi_2O_6$ and implications for origin of ultra-deep diamond inclusions. Geophys Res Lett 27:3541–3544
Irifune T, Shinmei T, McCammon CA, Miyajima N, Rubie DC, Frost DJ (2010) Iron partitioning and density changes of pyrolite in earth's lower mantle. Science 327:193–195
Ishii T, Kojitani H, Akaogi M (2011) Post-spinel transitions in pyrolite and Mg_2SiO_4 and akimotoite–perovskite transition in $MgSiO_3$: Precise comparison by high-pressure high-temperature experiments with multi-sample cell technique. Earth Planet Sci Lett 309:185–197
Ishii T, Kojitani H, Akaogi M (2012) High-pressure phase transitions and subduction behavior of continental crust at pressure-temperature conditions up to the upper part of the lower mantle. Earth Planet Sci Lett 357:31–41
Ishii T, Kojitani H, Akaogi M (2018) Phase relations and mineral chemistry in pyrolitic mantle at 1600–2200 degrees C under pressures up to the uppermost lower mantle: Phase transitions around the 660-km discontinuity and dynamics of upwelling hot plumes. Phys Earth Planet Sci 274:127–137

Ishii T, Kojitani H, Akaogi M (2019) Phase relations of harzburgite and MORB up to the uppermost lower mantle conditions: precise comparison with pyrolite by multisample cell high-pressure experiments with implication to dynamics of subducted slabs. J Geophys Res-Solid Earth 124:3491–3507

Ito E, Takahashi E (1987) Melting of peridotite at uppermost lower-mantle conditions. Nature 328:514–517

Ito E, Takahashi E, Matsui Y (1984) The mineralogy and chemistry of the lower mantle—An implication of the ultrahigh-pressure phase-relations in the system MgO–FeO–SiO_2. Earth Planet Sci Lett 67:238–248

Iwamori H (2004) Phase relations of peridotites under H_2O-saturated conditions and ability of subducting plates for transportation of H_2O. Earth Planet Sci Lett 227:57–71

Joswig W, Stachel T, Harris JW, Baur WH, Brey GP (1999) New Ca-silicate inclusions in diamonds—Tracers from the lower mantle. Earth Planet Sci Lett 173:1–6

Kaminsky F (2012) Mineralogy of the lower mantle: A review of 'super-deep' mineral inclusions in diamond. Earth Sci Rev 110:127–147

Kaminsky FV (2017) Lower-mantle mineral associations. *In*: Earth's Lower Mantle: Composition and Structure. p 23–46

Kaminsky FV, Wirth R (2011) Iron carbide inclusions in lower-mantle diamond from Juina, Brazil. Can Mineral 49:555–572

Kaminsky FV, Zakharchenko OD, Davies R, Griffin WL, Khachatryan-Blinova GK, Shiryaev AA (2001) Superdeep diamonds from the Juina area, Mato Grosso State, Brazil. Contrib Mineral Petrol 140:734–753

Katsura T, Yoneda A, Yamazaki D, Yoshino T, Ito E (2010) Adiabatic temperature profile in the mantle. Phys Earth Planet Sci 183:212–218

Kawamoto T (2004) Hydrous phase stability and partial melt chemistry in H_2O-saturated KLB-1 peridotite up to the uppermost lower mantle conditions. Phys Earth Planet Sci 143:387–395

Kawamoto T, Kanzaki M, Mibe K, Matsukage KN, Ono S (2012) Separation of supercritical slab-fluids to form aqueous fluid and melt components in subduction zone magmatism. PNAS 109:18695–18700

Kessel R, Ulmer P, Pettke T, Schmidt MW, Thompson AB (2005) The water–basalt system at 4 to 6 GPa: Phase relations and second critical endpoint in a K-free eclogite at 700 to 1400 degrees C. Earth Planet Sci Lett 237:873–892

Kesson SE, Fitz Gerald JD, Shelley JMG (1994) Mineral chemistry and density subducted basaltic crust at lower-mantle pressures. Nature 372:767–769

Kesson SE, Fitz Gerald JD, Shelley JMG, Withers RL (1995) Phase-relations, structure and crystal-chemistry of some aluminous silicate perovskites. Earth Planet Sci Lett 134:187–201

Kiseeva ES, Litasov KD, Yaxley GM, Ohtani E, Kamenetsky VS (2013a) Melting and phase relations of carbonated eclogite at 9–21 GPa and the petrogenesis of alkali-rich melts in the deep mantle. J Petrol 54:1555–1583

Kiseeva ES, Yaxley GM, Stepanov AS, Tkalcic H, Litasov KD, Kamenetsky VS (2013b) Metapyroxenite in the mantle transition zone revealed from majorite inclusions in diamonds. Geology 41:883–886

Kiseeva ES, Vasiukov DM, Wood BJ, McCammon C, Stachel T, Bykov M, Bykova E, Chumakov A, Cerantola V, Harris JW, Dubrovinsky L (2018) Oxidized iron in garnets from the mantle transition zone. Nat Geosci 11:144–147

Klein-BenDavid O, Izraeli ES, Hauri E, Navon O (2007) Fluid inclusions in diamonds from the Diavik mine, Canada and the evolution of diamond-forming fluids. Geochim Cosmochim Acta 71:723–744

Klein-BenDavid O, Logvinova AM, Schrauder M, Spetius ZV, Weiss Y, Hauri EH, Kaminsky FV, Sobolev NV, Navon O (2009) High-Mg carbonatitic microinclusions in some Yakutian diamonds-a new type of diamond-forming fluid. Lithos 112:648–659

Komabayashi T, Omori S (2006) Internally consistent thermodynamic data set for dense hydrous magnesium silicates up to 35 GPa, 1600 degrees C: Implications for water circulation in the Earth's deep mantle. Phys Earth Planet Sci 156:89–107

Komabayashi T, Omori S, Maruyama S (2004) Petrogenetic grid in the system MgO-SiO_2-H_2O up to 30 GPa, 1600 degrees C: Applications to hydrous peridotite subducting into the Earth's deep interior. J Geophys Res-Solid Earth 109:B03206

Korolev NM, Kopylova M, Bussweiler Y, Pearson DG, Gurney J, Davidson J (2018) The uniquely high-temperature character of Cullinan diamonds: A signature of the Bushveld mantle plume? Lithos 304:362–373

Kubo A, Suzuki T, Akaogi M (1997) High pressure phase equilibria in the system $CaTiO_3$–$CaSiO_3$: stability of perovskite solid solutions. Phys Chem Mineral 24:488–494

Kuwahara H, Nomura R, Nakada R, Irifune T (2018) Simultaneous determination of melting phase relations of mantle peridotite and mid-ocean ridge basalt at the uppermost lower mantle conditions. Phys Earth Planet Sci 284:36–50

Li XY, Zhang ZG, Lin JF, Ni HW, Prakapenka VB, Mao Z (2018) New High-pressure phase of $CaCO_3$ at the topmost lower mantle: Implication for the deep-mantle carbon transportation. Geophys Res Lett 45:1355–1360

Litasov KD, Ohtani E (2005) Phase relations in hydrous MORB at 18–28 GPa: implications for heterogeneity of the lower mantle. Phys Earth Planet Sci 150:239–263

Liu LG (2002) An alternative interpretation of lower mantle mineral associations in diamonds. Contrib Mineral Petrol 144:16–21

Liu LG, Ringwood AE (1975) Synthesis of a perovskite-type polymorph of $CaSiO_3$. Earth Planet Sci Lett 28:209–211

Liu X, Zhang LF, Hack CA, Zheng HF, Hu XM, Chang LL, He Q (2009a) Effect of water on the partial melting process of some silicate systems: Important implication of the second critical endpoint. Acta Petrologica Sinica 25:3407–3421

Liu Y, Taylor LA, Sarbadhikari AB, Valley JW, Ushikubo T, Spicuzza MJ, Kita N, Ketcham RA, Carlson W, Shatsky V, Sobolev NV (2009b) Metasomatic origin of diamonds in the world's largest diamondiferous eclogite. Lithos 112:1014–1024

Locock AJ (2008) An Excel spreadsheet to recast analyses of garnet into end-member components, and a synopsis of the crystal chemistry of natural silicate garnets. Computers Geosci 34:1769–1780

Luth RW, Stachel T (2014) The buffering capacity of lithospheric mantle: implications for diamond formation. Contrib Mineral Petrol 168:1083

Luth RW, Palyanov YN, Bureau H (2022) Experimental petrology applied to natural diamond growth. Rev Mineral Geochem 88:755–808

Maeda F, Ohtani E, Kamada S, Sakamaki T, Hirao N, Ohishi Y (2017) Diamond formation in the deep lower mantle: a high-pressure reaction of $MgCO_3$ and SiO_2. Sci Rep 7:40602

Mao HK, Chen LC, Hemley RJ, Jephcoat AP, Wu Y, Bassett WA (1989) Stability and equation of state of $CaSiO_3$-perovskite to 134 GPa. J Geophys Res-Solid Earth Planets 94:17889–17894

Maurice J, Bolfan-Casanova N, Padron-Navarta JA, Manthilake G, Hammouda T, Henot JM, Andrault D (2018) The stability of hydrous phases beyond antigorite breakdown for a magnetite-bearing natural serpentinite between 6.5 and 11 GPa. Contrib Mineral Petrol 173:86

McCammon C, Hutchison M, Harris J (1997) Ferric iron content of mineral inclusions in diamonds from Sao Luiz: A view into the lower mantle. Science 278:434–436

McCammon CA, Stachel T, Harris JW (2004) Iron oxidation state in lower mantle mineral assemblages—II. Inclusions in diamonds from Kankan, Guinea. Earth Planet Sci Lett 222:423–434

McDonough WF, Sun SS (1995) The composition of the Earth. Chem Geol 120:223–253

Meyer HOA (1987) Inclusions in diamond. *In*: Mantle xenoliths. Nixon PH (ed) John Wiley & Sons Ltd, Chichester, p 501–522

Mibe K, Kanzaki M, Kawamoto T, Matsukage KN, Fei YW, Ono S (2007) Second critical endpoint in the peridotite–H_2O system. J Geophys Res-Solid Earth 112:B03201

Mibe K, Kawamoto T, Matsukage KN, Fei YW, Ono S (2011) Slab melting versus slab dehydration in subduction-zone magmatism. PNAS 108:8177–8182

Miyajima N, Yagi T, Hirose K, Kondo T, Fujino K, Miura H (2001) Potential host phase of aluminum and potassium in the Earth's lower mantle. Am Mineral 86:740–746

Moore RO, Gurney JJ (1985) Pyroxene solid-solution in garnets included in diamond. Nature 318:553–555

Moore AE, Gurney JJ (1989) Mineral inclusions in diamond from Monastery kimberlite, South Africa. *In*: Kimberlites and Related Rocks. GSA Spec Publ Vol. 14. J Ross et al. (ed) Blackwell, Carlton, p 1029–1041

Moore RO, Otter ML, Rickard RS, Harris JW, Gurney JJ (1986) The occurance of moissanite and ferr-periclase as inclusions in diamond. Geol Soc Aust Abstracts 16:409–411

Moore RO, Gurney JJ, Griffin WL, Shimizu N (1991) Ultra-high pressure garnet inclusions in Monastery diamonds—Trace-element abundance patterns and conditions of origin. Euro J Mineral 3:213–230

Motsamai T, Harris JW, Stachel T, Pearson DG, Armstrong J (2018) Mineral inclusions in diamonds from Karowe Mine, Botswana: super-deep sources for super-sized diamonds? Mineral Petrol 112:169–180

Navon O, Hutcheon ID, Rossman GR, Wasserburg GJ (1988) Mantle-derived fluids in diamond micro-inclusions. Nature 335:784–789

Nestola F, Korolev N, Kopylova M, Rotiroti N, Pearson DG, Pamato MG, Alvaro M, Peruzzo L, Gurney JJ, Moore AE, Davidson J (2018) $CaSiO_3$ perovskite in diamond indicates the recycling of oceanic crust into the lower mantle. Nature 555:237–241

Nimis P (2022) Pressure and temperature data for diamonds. Rev Mineral Geochem 88:533–566

Nimis P, Nestola F, Schiazza M, Reali R, Agrosi G, Mele D, Tempesta G, Howell D, Hutchison MT, Spiess R (2019) Fe-rich ferropericlase and magnesiowustite inclusions reflecting diamond formation rather than ambient mantle. Geology 47:27–30

Nishihara Y, Takahashi E (2001) Phase relation and physical properties of an Al-depleted komatiite to 23 GPa. Earth Planet Sci Lett 190:65–77

Nishiyama N, Yagi T (2003) Phase relation and mineral chemistry in pyrolite to 2200 degrees C under the lower mantle pressures and implications for dynamics of mantle plumes. J Geophys Res-Solid Earth 108(B5):2255

Nishiyama N, Irifune T, Inoue T, Ando J, Funakoshi K (2004) Precise determination of phase relations in pyrolite across the 660 km seismic discontinuity by in situ X-ray diffraction and quench experiments. Phys Earth Planet Sci 143:185–199

Ohtani E (2015) Hydrous minerals and the storage of water in the deep mantle. Chem Geol 418:6–15

Ohtani E (2020) The role of water in Earth's mantle. Natl Sci Rev 7:224–232

Ohtani E, Litasov K, Hosoya T, Kubo T, Kondo T (2004) Water transport into the deep mantle and formation of a hydrous transition zone. Phys Earth Planet Sci 143:255–269

Okamoto K, Maruyama S (1999) The high-pressure synthesis of lawsonite in the MORB+H_2O system. Am Mineral 84:362–373

Okamoto K, Maruyama S (2004) The eclogite-garnetite transformation in the MORB+H_2O system. Phys Earth Planet Sci 146:283–296

Omori S, Komabayashi T, Maruyama S (2004) Dehydration and earthquakes in the subducting slab: empirical link in intermediate and deep seismic zones. Phys Earth Planet Sci 146:297–311
Ono S (1998) Stability limits of hydrous minerals in sediment and mid-ocean ridge basalt compositions: Implications for water transport in subduction zones. J Geophys Res-Solid Earth 103:18253–18267
Ono S, Yasuda A (1996) Compositional change of majoritic garnet in a MORB composition from 7 to 17 GPa and 1400 to 1600 degrees C. Phys Earth Planet Sci 96:171–179
Ono S, Ito E, Katsura T (2001) Mineralogy of subducted basaltic crust (MORB) from 25 to 37 GPa, and chemical heterogeneity of the lower mantle. Earth Planet Sci Lett 190:57–63
Otsuka K, Longo M, McCammon CA, Karato SI (2013) Ferric iron content of ferropericlase as a function of composition, oxygen fugacity, temperature and pressure: Implications for redox conditions during diamond formation in the lower mantle. Earth Planet Sci Lett 365:7–16
Palot M, Cartigny P, Harris JW, Kaminsky FV, Stachel T (2012) Evidence for deep mantle convection and primordial heterogeneity from nitrogen and carbon stable isotopes in diamond. Earth Planet Sci Lett 357:179–193
Palot M, Jacobsen SD, Townsend JP, Nestola F, Marquardt K, Miyajima N, Harris JW, Stachel T, McCammon CA, Pearson DG (2016) Evidence for H_2O-bearing fluids in the lower mantle from diamond inclusion. Lithos 265:237–243
Palot M, Pearson DG, Stachel T, Stern RA, Le Pioufle A, Gurney JJ, Harris JW (2017) The transition zone as a host for recycled volatiles: Evidence from nitrogen and carbon isotopes in ultra-deep diamonds from Monastery and Jagersfontein (South Africa). Chem Geol 466:733–749
Pearson DG, Canil D, Shirey SB (2003) Mantle samples included in volcanic rocks: xenoliths and diamonds. *In*: Treatise on Geochemistry. Vol 2. RW C, (ed), p 171–277
Pearson DG, Brenker FE, Nestola F, McNeill J, Nasdala L, Hutchison MT, Matveev S, Mather K, Silversmit G, Schmitz S, Vekemans B (2014) Hydrous mantle transition zone indicated by ringwoodite included within diamond. Nature 507:221–224
Petts DC, Chacko T, Stachel T, Stern RA, Heaman LM (2015) A nitrogen isotope fractionation factor between diamond and its parental fluid derived from detailed SIMS analysis of a gem diamond and theoretical calculations. Chem Geol 410:188–200
Pla Cid J, Nardi LVS, Pla Cid C, Gisbert PE, Balzaretti NM (2014) Acid compositions in a veined-lower mantle, as indicated by inclusions of (K,Na)-hollandite + SiO_2 in diamonds. Lithos 196:42–53
Poli S, Schmidt MW (2002) Petrology of subducted slabs. Annu Rev Earth Planet Sci 30:207–235
Proyer A, Habler G, Abart R, Wirth R, Krenn K, Hoinkes G (2013) TiO_2 exsolution from garnet by open-system precipitation: evidence from crystallographic and shape preferred orientation of rutile inclusions. Contrib Mineral Petrol 166:211–234
Regier ME, Pearson DG, Stachel T, Luth RW, Stern RA, Harris JW (2020) The lithospheric-to-lower-mantle carbon cycle recorded in superdeep diamonds. Nature 585:234–238
Ricolleau A, Perrillat JP, Fiquet G, Daniel I, Matas J, Addad A, Menguy N, Cardon H, Mezouar M, Guignot N (2010) Phase relations and equation of state of a natural MORB: Implications for the density profile of subducted oceanic crust in the Earth's lower mantle. J Geophys Res-Solid Earth 115:B08202
Ringwood AE (1991) Phase-transformations and their bearing on the constitution and dynamics of the mantle. Geochim Cosmochim Acta 55:2083–2110
Rohrbach A, Schmidt MW (2011) Redox freezing and melting in the Earth's deep mantle resulting from carbon–iron redox coupling. Nature 472:209–212
Rzehak LJA, Rohrbach A, Vollmer C, Hofer HE, Berndt J, Klemme S (2020) Ferric–ferrous iron ratios of experimental majoritic garnet and clinopyroxene as a function of oxygen fugacity. Am Mineral 105:1866–1874
Schmandt B, Jacobsen SD, Becker TW, Liu ZX, Dueker KG (2014) Dehydration melting at the top of the lower mantle. Science 344:1265–1268
Schulze DJ, Coopersmith HG, Harte B, Pizzolato LA (2008) Mineral inclusions in diamonds from the Kelsey Lake Mine, Colorado, USA: Depleted Archean mantle beneath the Proterozoic Yavapai Province. Geochim Cosmochim Acta 72:1685–1695
Scott-Smith BH, Danchnin RV, Harris JW, Stracke KJ (1984) Kimberlites near Orrorroo, South Australia. *In*: Kimberlites I: Kimberlites and Related Rocks. Kornprobst J (ed) Elsevier, Amsterdam, p 121–242
Secchiari A, Montanini A, Bosch D, Macera P, Cluzel D (2020) Sr, Nd, Pb and trace element systematics of the New Caledonia harzburgites: Tracking source depletion and contamination processes in a SSZ setting. Geosci Front 11:37–55
Seitz HM, Brey GP, Harris JW, Durali-Muller S, Ludwig T, Hofer HE (2018) Ferropericlase inclusions in ultradeep diamonds from Sao Luiz (Brazil): high Li abundances and diverse Li-isotope and trace element compositions suggest an origin from a subduction melange. Mineral Petrol 112:291–300
Shirey SB, Cartigny P, Frost DJ, Keshav S, Nestola F, Nimis P, Pearson DG, Sobolev NV, Walter MJ (2013) Diamonds and the geology of mantle carbon. *In*: Carbon in Earth. Vol 75. Hazen RM, Jones AP, Baross JA (eds) p 355–421
Shirey S, Smit K, Pearson D, Walter M, Aulbach S, Brenker F, Bureau H, Burnham A, Cartigny P, Chacko T, Frost D (2019) Diamonds and the mantle geodynamics of carbon: deep mantle carbon evolution from the diamond record. *In*: Deep Carbon: Past to Present. Orcutt BN, Daniel I, Dasgupta R, (eds). Cambridge University Press, Cambridge, U.K., p 89–128
Shirey SB, Wagner LS, Walter MJ, Pearson DG, van Keken PE (2021) Slab transport of fluids to deep focus earthquake depths – thermal modeling constraints and evidence from diamonds. AGU Adv 2:e2020AV000304

Smart KA, Chacko T, Stachel T, Muehlenbachs K, Stern RA, Heaman LM (2011) Diamond growth from oxidized carbon sources beneath the Northern Slave Craton, Canada: A delta C-13–N study of eclogite-hosted diamonds from the Jericho kimberlite. Geochim Cosmochim Acta 75:6027–6047

Smit KV, Shirey SB, Stern RA, Steele A, Wang WY (2016) Diamond growth from C–H–N–O recycled fluids in the lithosphere: Evidence from CH micro-inclusions and delta C-13–delta N-15–N content in Marange mixed-habit diamonds. Lithos 265:68–81

Smit KV, Stachel T, Luth RW, Stern RA (2019) Evaluating mechanisms for eclogitic diamond growth: An example from Zimmi Neoproterozoic diamonds (West African craton). Chem Geol 520:21–32

Smit KV, Timmerman S, Aulbach S, Shirey SB, Richardson SH, Phillips D, Pearson DG (2022) Geochronology of diamonds. Rev Mineral Geochem 88:567–636

Smith EM, Kopylova MG (2014) Implications of metallic iron for diamonds and nitrogen in the sublithospheric mantle. Can J Earth Sci 51:510–516

Smith CB, Walter MJ, Bulanova GP, Mikhail S, Burnham AD, Gobbo L, Kohn SC (2016a) Diamonds from Dachine, French Guiana: A unique record of early Proterozoic subduction. Lithos 265:82–95

Smith EM, Shirey SB, Nestola F, Bullock ES, Wang JH, Richardson SH, Wang WY (2016b) Large gem diamonds from metallic liquid in Earth's deep mantle. Science 354:1403–1405

Smith EM, Shirey SB, Wang WY (2017) The very deep origin of the world's biggest diamonds. Gems Gemol 53:388–403

Smith EM, Shirey SB, Richardson SH, Nestola F, Bullock ES, Wang JH, Wang WY (2018) Blue boron-bearing diamonds from Earth's lower mantle. Nature 560:84–87

Smith EM, Ni P, Shirey SB, Richardson SH, Wang WY, Shahar A (2021) Heavy iron in large gem diamonds traces deep subduction of serpentinized ocean floor. Sci Adv 7: eabe9773

Sobolev NV, Wirth R, Logvinova AM, Yelisseyev AP, Kuzmin DV (2016) Retrograde isochemical phase transformations of majoritic garnets included in diamonds: A case study of subcalcic Cr-rich majoritic pyrope from a Snap Lake diamond, Canada. Lithos 265:267–277

Sokolova TS, Dorogokupets PI (2021) Equations of state of Ca-silicates and phase diagram of the $CaSiO_3$ system under upper mantle conditions. Minerals 11:322

Stachel T (2001) Diamonds from the asthenosphere and the transition zone. Euro J Mineral 13:883–892

Stachel T, Harris JW (1997) Syngenetic inclusions in diamond from the Birim field (Ghana)—A deep peridotitic profile with a history of depletion and re-enrichment. Contrib Mineral Petrol 127:336–352

Stachel T, Harris JW (2008) The origin of cratonic diamonds—Constraints from mineral inclusions. Ore Geol Rev 34:5–32

Stachel T, Luth RW (2015) Diamond formation—Where, when and how? Lithos 220:200–220

Stachel T, Harris JW, Brey GP (1998a) Rare and unusual mineral inclusions in diamonds from Mwadui, Tanzania. Contrib Mineral Petrol 132:34–47

Stachel T, Viljoen KS, Brey G, Harris JW (1998b) Metasomatic processes in lherzolitic and harzburgitic domains of diamondiferous lithospheric mantle: REE in garnets from xenoliths and inclusions in diamonds. Earth Planet Sci Lett 159:1–12

Stachel T, Brey GP, Harris JW (2000a) Kankan diamonds (Guinea) I: from the lithosphere down to the transition zone. Contrib Mineral Petrol 140:1–15

Stachel T, Harris JW, Brey GP, Joswig W (2000b) Kankan diamonds (Guinea) II: lower mantle inclusion parageneses. Contrib Mineral Petrol 140:16–27

Stachel T, Harris JW, Aulbach S, Deines P (2002) Kankan diamonds (Guinea) III: delta C-13 and nitrogen characteristics of deep diamonds. Contrib Mineral Petrol 142:465–475

Stachel T, Aulbach S, Brey GP, Harris JW, Leost I, Tappert R, Viljoen KS (2004) The trace element composition of silicate inclusions in diamonds: a review. Lithos 77:1–19

Stachel T, Brey GP, Harris JW (2005) Inclusions in sublithospheric diamonds: Glimpses of deep Earth. Elements 1:73–78

Stachel T, Aulbach S, Harris JW (2022a) Mineral inclusions in lithospheric diamonds. Rev Mineral Geochem 88:307–392

Stachel T, Cartigny P, Chacko T, Pearson DG (2022b) Carbon and nitrogen in mantle-derived diamonds. Rev Mineral Geochem 88:809–876

Stagno V, Ojwang DO, McCammon CA, Frost DJ (2013) The oxidation state of the mantle and the extraction of carbon from Earth's interior. Nature 493:84–88

Stixrude L, Lithgow-Bertelloni C (2011) Thermodynamics of mantle minerals-II. Phase equilibria. Geophys J Int 184:1180–1213

Sueda Y, Irifune T, Yamada A, Inoue T, Liu X, Funakoshi KI (2006) The phase boundary between $CaSiO_3$ perovskite and Ca_2SiO_4+$CaSi_2O_5$ determined by in situ X-ray observations. Geophys Res Lett 33:L10307

Sun CG, Dasgupta R (2019) Slab-mantle interaction, carbon transport, and kimberlite generation in the deep upper mantle. Earth Planet Sci Lett 506:38–52

Sun YZ, Hier-Majumder S, Xu YG, Walter M (2020) Stability and migration of slab-derived carbonate-rich melts above the transition zone. Earth Planet Sci Lett 531:1160000

Sun YZ, Hier-Majumder S, Tauzin B, Walter MJ, Ballmer M, Xu YG, Kim S (2021) Evidence of volatile-induced melting in the Northeast Asian Upper Mantle. J Geophys Res-Solid Earth 126:e2021JB022167

Sunagawa I (1984) Morphology of natural and synthetic diamond crystals. *In*: Materials Science of the Earth's Interior. Sunagawa I, (ed) Terra Science, Tokyo, p 303–330

Sverjensky DA, Huang F (2015) Diamond formation due to a pH drop during fluid–rock interactions. Nat Commun 6:8702

Tao RBA, Fei YW, Bullock ES, Xu C, Zhang LF (2018) Experimental investigation of Fe^{3+}-rich majoritic garnet and its effect on majorite geobarometer. Geochim Cosmochim Acta 225:1–16

Tappert R, Stachel T, Harris JW, Muehlenbachs K, Ludwig T, Brey GP (2005) Diamonds from Jagersfontein (South Africa): messengers from the sublithospheric mantle. Contrib Mineral Petrol 150:505–522

Tappert R, Foden J, Stachel T, Muehlenbachs K, Tappert M, Wills K (2009a) The diamonds of South Australia. Lithos 112:806–821

Tappert R, Foden J, Stachel T, Muehlenbachs K, Tappert M, Wills K (2009b) Deep mantle diamonds from South Australia: A record of Pacific subduction at the Gondwanan margin. Geology 37:43–46

Thompson AB (1992) Water in the Earth's upper mantle. Nature 358:295–302

Thomson AR (2017) Diamonds from the lower mantle? Am Mineral 102:929–930

Thomson AR, Kohn SC, Bulanova GP, Smith CB, Araujo D, Walter MJ, Eimf (2014) Origin of sub-lithospheric diamonds from the Juina-5 kimberlite (Brazil): constraints from carbon isotopes and inclusion compositions. Contrib Mineral Petrol 168:1081

Thomson AR, Walter MJ, Kohn SC, Brooker RA (2016a) Slab melting as a barrier to deep carbon subduction. Nature 529:76–79

Thomson AR, Kohn SC, Bulanova GP, Smith CB, Araujo D, Walter MJ (2016b) Trace element composition of silicate inclusions in sub-lithospheric diamonds from the Juina-5 kimberlite: Evidence for diamond growth from slab melts. Lithos 265:108–124

Thomson AR, Crichton WA, Brodholt JP, Wood IG, Siersch NC, Muir JMR, Dobson DP, Hunt SA (2019) Seismic velocities of $CaSiO_3$ perovskite can explain LLSVPs in Earth's lower mantle. Nature 572:643–647

Thomson AR, Kohn SC, Prabhu A, Walter MJ (2021) Evaluating the formation pressure of diamond-hosted majoritic garnets; A machine learning majorite barometer. J Geophys Res-Solid Earth 126:e2020JB020604

Timmerman S, Honda M, Burnham AD, Amelin Y, Woodland S, Pearson DG, Jaques AL, Le Losq C, Bennett VC, Bulanova GP, Smith CB (2019) Primordial and recycled helium isotope signatures in the mantle transition zone. Science 365:692–694

Tronnes RG, Frost DJ (2002) Peridotite melting and mineral–melt partitioning of major and minor elements at 22–24.5 GPa. Earth Planet Sci Lett 197:117–131

Tschauner O, Huang SC, Yang SY, Humayun M, Liu WJ, Corder SNG, Bechtel HA, Tischler J, Rossman GR (2021) Discovery of davemaoite, $CaSiO_3$-perovskite, as a mineral from the lower mantle. Science 374:891–894

Uenver-Thiele L, Woodland AB, Ballaran TB, Miyajima N, Frost DJ (2017) Phase relations of $MgFe_2O_4$ at conditions of the deep upper mantle and transition zone. Am Mineral 102:632–642

Ulmer P, Sweeney RJ (2002) Generation and differentiation of group II kimberlites: Constraints from a high-pressure experimental study to 10 GPa. Geochim Cosmochim Acta 66:2139–2153

van Keken PE, Hacker BR, Syracuse EM, Abers GA (2011) Subduction factory: 4. Depth-dependent flux of H_2O from subducting slabs worldwide. J Geophys Res-Solid Earth 116:B01401

Walter MJ (1998) Melting of garnet peridotite and the origin of komatiite and depleted lithosphere. J Petrol 39:29–60

Walter MJ, Nakamura E, Tronnes RG, Frost DJ (2004) Experimental constraints on crystallization differentiation in a deep magma ocean. Geochim Cosmochim Acta 68:4267–4284

Walter MJ, Bulanova GP, Armstrong LS, Keshav S, Blundy JD, Gudfinnsson G, Lord OT, Lennie AR, Clark SM, Smith CB, Gobbo L (2008) Primary carbonatite melt from deeply subducted oceanic crust. Nature 454:622–625

Walter MJ, Kohn SC, Araujo D, Bulanova GP, Smith CB, Gaillou E, Wang J, Steele A, Shirey SB (2011) Deep mantle cycling of oceanic crust: Evidence from diamonds and their mineral inclusions. Science 334:54–57

Walter MJ, Thomson AR, Wang W, Lord OT, Ross J, McMahon SC, Baron MA, Melekhova E, Kleppe AK, Kohn SC (2015) The stability of hydrous silicates in Earth's lower mantle: Experimental constraints from the systems $MgO–SiO_2–H_2O$ and $MgO–Al_2O_3–SiO_2–H_2O$. Chem Geol 418:16–29

Walter MJ, Thomson AR, Smith EM (2022) Major and trace element composition of sublithospheric inclusions in diamond. https://doi.org/10.5683/SP3/LIVK1K, Scholars Portal Dataverse, V1.

Wang YB, Weidner DJ (1994) Thermoelasticity of $CaSiO_3$ perovskite and implications for the lower mantle. Geophys Res Lett 21:895–898

Wang WY, Takahashi E (2000) Subsolidus and melting experiments of K-doped peridotite KLB-1 to 27 GPa: Its geophysical and geochemical implications. J Geophys Res-Solid Earth 105:2855–2868

Wang WY, Gasparik T, Rapp RP (2000a) Partitioning of rare earth elements between CaSiO3 perovskite and coexisting phases: constraints on the formation of $CaSiO_3$ inclusions in diamonds. Earth Planet Sci Lett 181:291–300

Wang WY, Sueno S, Takahashi E, Yurimoto H, Gasparik T (2000b) Enrichment processes at the base of the Archean lithospheric mantle: observations from trace element characteristics of pyropic garnet inclusions in diamonds. Contrib Mineral Petrol 139:720–733

Wang JT, Takahashi E, Xiong XL, Chen LL, Li L, Suzuki T, Walter MJ (2020) The water-saturated solidus and second critical endpoint of peridotite: Implications for magma genesis within the mantle wedge. J Geophys Res-Solid Earth 125:e2020JB019452

Waychunas GA (1987) Synchrotron radiation XANES spectroscopy of Ti in minerals—Effects of Ti bonding distances, Ti valence, and site geometry on absorption-edge structure. Am Mineral 72:89–101

Weiss Y, Griffin WL, Navon O (2013) Diamond-forming fluids in fibrous diamonds: The trace-element perspective. Earth Planet Sci Lett 376:110–125

Weiss Y, Kiflawi I, Davies N, Navon O (2014) High-density fluids and the growth of monocrystalline diamonds. Geochim Cosmochim 141:145–159

Wenz MD, Jacobsen SD, Zhang DZ, Regier M, Bausch HJ, Dera PK, Rivers M, Eng P, Shirey SB, Pearson DG (2019) Fast identification of mineral inclusions in diamondat GSECARS using synchrotron X-ray microtomography, radiography and diffraction. J Synchrotron Radiat 26:1763–1768

Wicks JK, Duffy T (2016) Crystal structures of minerals in the lower mantle. *In*: Deep Earth: Physics and Chemistry of the Lower Mantle and Core. Terasaki H, Fischer RA, (eds). Americam Geophysical Union, Washington, D.C., p 69–87

Wijbrans CH, Rohrbach A, Klemme S (2016) An experimental investigation of the stability of majoritic garnet in the Earth's mantle and an improved majorite geobarometer. Contrib Mineral Petrol 171

Wilding MC, Harte B, Harris JW (1991) Evidence for a deep origin for Sao Luiz diamonds. Extd Abstrs 5th Int Kimb Conf:456–458

Wirth R, Vollmer C, Brenker F, Matsyuk S, Kaminsky F (2007) Inclusions of nanocrystalline hydrous aluminium silicate "Phase Egg" in superdeep diamonds from Juina (Mato Grosso State, Brazil). Earth Planet Sci Lett 259:384–399

Wirth R, Dobrzhinetskaya L, Harte B, Schreiber A, Green HW (2014) High-Fe (Mg, Fe)O inclusion in diamond apparently from the lowermost mantle. Earth Planet Sci Lett 404:365–375

Wood BJ (2000) Phase transformations and partitioning relations in peridotite under lower mantle conditions. Earth Planet Sci Lett 174:341–354

Woodland AB, Girnis AV, Bulatov VK, Brey GP, Hofer HE (2020) Breyite inclusions in diamond: Experimental evidence for possible dual origin. Euro J Mineral 32:171–185

Yagi T, Mao HK, Bell PM (1978) Structure and crystal-chemistry of perovskite-type $MgSiO_3$. Phys Chem Mineral 3:97–110

Yasuda A, Fujii T, Kurita K (1994) Melting phase-relations of an anhydrous midocean ridge basalt from 3 to 20 GPa—Implications for the behavior of subducted oceanic–crust in the mantle. J Geophys Res-Solid Earth 99:9401–9414

Zedgenizov DA, Kagi H, Shatsky VS, Ragozin AL (2014a) Local variations of carbon isotope composition in diamonds from Sao-Luis (Brazil): Evidence for heterogenous carbon reservoir in sublithospheric mantle. Chem Geol 363:114–124

Zedgenizov DA, Shatskiy A, Ragozin AL, Kagi H, Shatsky VS (2014b) Merwinite in diamond from Sao Luiz, Brazil: A new mineral of the Ca-rich mantle environment. Am Mineral 99:547–550

Zedgenizov DA, Shatsky VS, Panin AV, Evtushenko OV, Ragozin AL, Kagi H (2015) Evidence for phase transitions in mineral inclusions in superdeep diamonds of the Sao Luiz deposit (Brazil). Russian Geol Geophys 56:296–305

Zedgenizov DA, Ragozin AL, Kalinina VV, Kagi H (2016) The mineralogy of Ca-rich inclusions in sublithospheric diamonds. Geochem Int 54:890–900

Zedgenizov D, Kagi H, Ohtani E, Tsujimori T, Komatsu K (2020) Retrograde phases of former bridgmanite inclusions in superdeep diamonds. Lithos 370: 105659

Zhang JZ, Li B, Utsumi W, Liebermann RC (1996) In situ X-ray observations of the coesite–stishovite transition: Reversed phase boundary and kinetics. Phys Chem Mineral 23:1–10

Zhang YF, Wang C, Jin ZM (2020) Decarbonation of stagnant slab in the mantle transition zone. J Geophys Res-Solid Earth 125:e2020JB019533

Zhu F, Li J, Liu JC, Lai XJ, Chen B, Meng Y (2019) Kinetic control on the depth distribution of superdeep diamonds. Geophys Res Lett 46:1984–1992

Reviews in Mineralogy & Geochemistry
Vol. 88 pp. 451-474, 2022

Raman Identification of Inclusions in Diamond

Evan M. Smith

Gemological Institute of America
New York, NY 10036
USA

evan.smith@gia.edu

Mandy Y. Krebs

Department of Renewable Resources
University of Alberta
Edmonton, AB T6G 2H1
Canada

krebs@ualberta.ca

Philomena-Theresa Genzel, Frank E. Brenker

Department of Geosciences
Goethe University Frankfurt
60438 Frankfurt
Germany

INTRODUCTION

Diamonds and their inclusions are some of the most scientifically valuable samples of the Earth (Haggerty 1999; Shirey et al. 2019). Among the analytical techniques used to study diamonds, Raman spectroscopy offers several advantages that make it an appealing tool for characterizing inclusions. It is a relatively low-cost, rapid, and non-destructive option, requiring minimal sample preparation, if any. Inclusions can often be characterized in-situ, while still fully enclosed in their diamond host, which ensures that no material is inadvertently lost (e.g., fluid) and the remnant pressure of the inclusion–host system is preserved. The pressure within inclusions can be on the order of several gigapascals (e.g., Nasdala et al. 2003) and is especially important for stabilizing the crystal structure of certain high-pressure minerals, such as ringwoodite, in sublithospheric diamonds (Pearson et al. 2014). Ideally, Raman spectroscopy can be complemented by other in-situ methods, such as infrared spectroscopy (FTIR), micro-beam X-ray diffraction (XRD), X-ray computed tomography (CT), and synchrotron X-ray fluorescence (XRF). Raman spectroscopy can serve as a first step to help characterize inclusions before employing more time-consuming or destructive analytical techniques, but it can also serve as a powerful tool in its own right for diamond research (e.g., Liu et al. 1990; Gillet et al. 2002; Nasdala et al. 2003, 2005; Brenker et al. 2005; Walter et al. 2011; Howell et al. 2012; Pearson et al. 2014; Nimis et al. 2016; Smit et al. 2016; Smith et al. 2016b, 2018; Anzolini et al. 2018; Kemppinen et al. 2018). This chapter is the first Raman spectroscopy review specifically applied to inclusions in diamond.

Collecting Raman spectra can be straightforward, often with little training for users. However, obtaining good quality data and making reasonable interpretations can be challenging

1529-6466/22/0088-0008$05.00 (print)
1943-2666/22/0088-0008$05.00 (online)
http://dx.doi.org/10.2138/rmg.2022.88.08

and varies with each application of Raman spectroscopy. This chapter provides an overview of the application of Raman spectroscopy to inclusions in diamond, with a focus on highlighting relevant phases likely to be encountered in this context. A collection of Raman spectra is available as supplementary material (Smith 2021), intended to provide a diamond-specific resource for identifying included phases. Many but not all of these spectra come from inclusions in diamond. For some minerals, spectra from inclusions are not available or available spectra collected from non-inclusion samples are higher quality and more representative, making them better suited for use as a reference. Additional spectral databases, from the RRUFF Project (Lafuente et al. 2015) for example, can be found online, although users should be mindful of occasional erroneous data. With the correct mineralogical parameters, Raman spectra can also be predicted from first principles, such as in the WURM database of computed spectra, which sometimes can be useful for identifying Raman peak distributions of an unknown mineral phase (Caracas and Bobocioiu 2011). Helpful background information on Raman spectroscopy and its geological applications can be found in reviews elsewhere (e.g., Burke 2001; Fries and Steele 2010; Frezzotti et al. 2012; Neuville et al. 2014; Elements 2020).

PRINCIPLES AND METHODS

Raman spectroscopy

When visible light intercepts matter, some of the light is scattered. Although most of the scatter occurs elastically, meaning the photon energy (or wavelength) remains unchanged, a small proportion of photons (approximately 1 in 10^7) scatter inelastically and their energy is measurably changed (Pasteris and Beyssac 2020). This is the Raman effect. The energy change, or Raman shift, is a consequence of some energy being transferred into (Stokes) or being released from (anti-Stokes) the vibrations of bonded atoms in the light-scattering material. This means that the spectrum of all the inelastically scattered photon energies is not random, but has bands (also called peaks) that correspond to the discrete vibrational frequencies in the material (i.e., the sample).

Raman spectroscopy exploits this phenomenon, whereby the spectrum of inelastically scattered light contains structural and, to some extent, chemical information about the sample. Typically, a laser source with photons of uniform energy is focused onto the sample and the scattered light is collected. The intense elastic scattering of the incident laser wavelength is excluded or removed from the signal using a Rayleigh rejection filter, triple monochromator or, in the case of a Fourier Transform system, an interferometer. It is important to note that the Raman shift is relative to the incident photon energy, meaning the energy or wavelength of a Raman band depends on the laser wavelength. For this reason, Raman spectra are plotted on a wavenumber-shift or Raman shift scale, which shows the energy change of scattered photons *relative* to the incident laser photons (in units of cm^{-1}). This is not to be confused with an *absolute* wavenumber scale of energy (conventionally used for infrared absorption spectroscopy).

Instrumentation and sample considerations

The basic Raman spectrometer (or Raman microprobe) consists of a laser light source, a microscope, and a spectrometer. In the context of examining inclusions in diamonds, an important variable to consider is the use of a confocal arrangement of the optical pathway (e.g., Everall 2009). A two-dimensional confocal aperture blocks out-of-focus light, constraining the effectively analyzed sample volume and improving spatial resolution. A quasi-confocal arrangement can be achieved using a slit rather than an aperture. The spatial resolution also depends on how tightly the objective lens focuses the laser, which is a function of the lens's numerical aperture (NA) and the laser excitation wavelength (λ), and is proportional to λ/NA. For example, with a confocal setup, a 0.9 NA, 100× magnification objective lens and 514.5 nm laser can achieve a spatial resolution of approximately 1 × 1 × 5 μm^3, elongate in the vertical axis (i.e., the axis of the incident laser beam). Note, however, that there are many additional factors affecting the spatial resolution (e.g., Kim et al. 2020) and that focusing the laser inside the diamond will further distort the spatial resolution due to refraction, surface roughness, and other effects. Ideally, inclusions should be as close to the surface of the diamond as possible, but realistically, they are often hundreds of micrometers deep or more and the working distance becomes an issue. For this reason, long working distance lenses are practical, at 100×, 50× or 20× magnification, for example, and can permit analysis of inclusions as deep as a few millimeters. Long working distance lenses have the tradeoff of a lower NA (e.g., a 50× long working distance lens might have a NA of 0.5), which increases the analyzed volume, chiefly in the vertical dimension because the height of the analyzed volume is inversely proportional to the square of the NA. This lowers the spatial resolution and can result in a diminished Raman signal intensity from the inclusion of interest.

The choice of laser is also important. Blue, green, and red lasers can all be used with varying success depending on the sample. Shorter wavelengths lead to smaller spot sizes and more intense Raman scattering, but the drawback is an increased likelihood of fluorescence from the diamond (or inclusion) that may drown out the Raman signal and decrease spectral resolution. Heating of the inclusion by the laser spot is not normally a concern, but it is possible and measures such as reducing the laser power may be prudent for sensitive hydrous or opaque phases. Smaller grain sizes and coexistence with phases that act as insulators are also factors that can facilitate heating. In this sense, Raman spectroscopy is not always non-destructive (e.g., de Faria et al. 1997). When striving to avoid damage, a good approach is to start at the lowest laser power and iteratively work up to a point where the Raman signal strength is acceptable. If possible, measuring the laser at the sample using a power meter is preferable. Additional factors such as the depth of the inclusion will affect the laser power reaching the inclusion.

It is possible to analyze through rough diamond surfaces, especially smooth and flat crystal or cleavage surfaces. Generally, the better an inclusion can be resolved visually, the greater the chances of being able to analyze it. A well-polished diamond surface (facet or window) is therefore ideal and becomes increasingly important for inclusions that are smaller or located deeper inside the diamond. If polishing is to be pursued, it may be helpful to collect X-ray CT images to locate all inclusions to aid planning, although this has the potential for a high radiation dose that could affect the sample. During polishing, good technique is important to avoid excessive sample heating, which could hypothetically modify or destabilize sensitive inclusions, such as hydrous ringwoodite.

When using a Raman microscope, good control over the lighting conditions is critical for navigating the subsurface space inside a diamond to find specific inclusions or parts of inclusions and bring them into focus for analysis. Transmitted light, with a diffuser plate or frosted microscope slide, or even an external light source, such as an intense fiber optic light, can be especially helpful. Depending on the sample, it can be possible to analyze inclusions even smaller than 1 μm, although successfully targeting the laser on the inclusion can present a challenge and the Raman signal may be weak. If a sufficiently dense cloud of inclusions is present, it can be possible to train the spot on the cloud and have a good chance of intersecting inclusions (e.g., Navon et al. 2017). Alternatively, when small inclusions are more dispersed, an automated mapping approach with a small step size can sequentially analyze a grid of many points and maximize the chances of hitting inclusions (e.g., Smit et al. 2016). Mapping capabilities are also helpful when analyzing multi-phase inclusions, where a traditional point analysis approach could allow some phases to go unnoticed, such as the presence of small amounts of molybdenite in sulfide inclusions (Kemppinen et al. 2018). Newer instruments capable of faster automated Raman mapping, in combination with the higher spatial resolution afforded by confocal optics, are likely to bolster the use of 2-D and 3-D mapping for inclusion analysis.

Common challenges

Diamond has a very strong first-order Raman band, located at 1332.5 cm^{-1} in relatively strain- and impurity-free crystals (Schiferl et al. 1997) (larger grains than the excitation beam diameter), which is conveniently isolated from the 100–1300 cm^{-1} region where the key spectral features of most minerals lie. However, the diamond host can still present a challenge to inclusion analysis when it luminesces, imparting a strong broad background that can drown out the Raman spectrum, or when it introduces more discrete photoluminescence bands into the spectrum. Laser-induced photoluminescence can sometimes be overcome by controlling the laser spot size to excite less of the diamond (Kemppinen et al. 2018), or attempting analysis again with a different laser wavelength, if possible. Photoluminescence from defects in the diamond lattice, which may sometimes occur only locally around an inclusion (e.g., Fig. 9 in Gu and Wang 2018), can produce bands at discrete wavelengths that overlap the Raman spectral features. It can be helpful to analyze the host diamond itself, adjacent to an inclusion of interest, for comparison. Also, photoluminescence features can be diagnosed by changing the laser excitation wavelength. These features depend on electronic transitions and will be detected at the same *absolute* energy or wavenumber, and therefore a different apparent Raman shift (wavenumber-shift), upon changing the laser wavelength. True Raman bands will plot at constant wavenumber-shift values, meaning their position in the Raman shift spectrum will not change upon changing the laser wavelength. Having the ability to choose between multiple laser wavelengths thus presents an advantage.

It is not uncommon for inclusions to contain multiple phases, which can be overlapped in Raman spectra, making phase identification more difficult. For example, this could occur as the result of co-trapping of multiple mineral grains together or as a result of retrogression of a single high-pressure mineral into multiple lower-pressure phases. Retrogression is a common phenomenon observed in sublithospheric diamonds. If an inclusion is not homogeneous, the Raman spectrum can exhibit substantial spatial variability and recognizing all the phases present can be difficult. Some phases may be very weakly active or even Raman inactive. Analyzing multiple points on the inclusion, potentially by automated mapping (e.g., Walter et al. 2011), is crucial to being able to recognize and deconvolve the spectra of mixed phases.

Another challenging aspect of inclusions in diamond is the presence of remnant pressure. As a diamond makes its way to Earth's surface, the changes in pressure and temperature allow the diamond to relax, but its inclusions are constrained by the host diamond. As a result, many inclusions are under residual stress, which can often be seen visually with crossed polarizing filters as a halo of anomalous birefringence in the diamond surrounding an inclusion (Howell et al. 2010). The remnant pressure can be on the order of a few gigapascals (Nasdala et al. 2003) and, rarely, might be greater than 10 GPa for some exotic micrometer- to nanometer-sized inclusions (Navon et al. 2017; Tschauner et al. 2018). The pressure can make it difficult to identify phases because residual pressure affects the bonding geometry and shifts the position of some Raman bands. For example, the main band of coesite in an unstrained state resides at 520.6 cm^{-1} but under high remnant pressure as an inclusion this band has been recorded at 538 cm^{-1} (Smith et al. 2018). The host diamond Raman band also shifts under stress (Grimsditch et al. 1978; Sharma et al. 1985), which can be exploited to map out the stress distribution around inclusions (Nasdala et al. 2003; Howell et al. 2012). As a final note of caution, mentioned by Nasdala and Schmidt (2020), Raman spectroscopy can be deceptively simple and does not always receive sufficient oversight by trained users, so it is important to keep in mind that not all published Raman results are reliable.

INCLUSION IDENTIFICATION

For most users, the principal goal is to identify the phase or phases that make up a visible inclusion in a diamond. Mineral inclusions are categorized below in terms of depth of origin (lithospheric and sublithospheric) and major host rock types. Lithospheric and sublithospheric diamonds are presented separately because the latter often change from their original high-pressure mineralogy to lower-pressure phases during exhumation from the mantle, adding a degree of complexity to their interpretation.

Phase identification is accomplished by carefully comparing band positions, intensities, and shapes against those of known spectra, a process that can be aided by software that can automatically select possible matches from a database. It is important to keep in mind that band positions may be shifted due to remnant pressure in inclusions and that relative band intensities of a crystal can change as a function of its orientation with respect to the incident laser and its polarization. For identifying inclusions in diamond, it is helpful to consider the most plausible candidates as those minerals and mineral assemblages that have been encountered previously (Tables 1 and 2). Our list is not exhaustive and does not include the various phases that have been observed in micro- or nano-scale inclusions (e.g., Logvinova et al. 2008; Kaminsky 2012) that are less conducive to Raman analysis. A collection of relevant Raman spectra, including examples collected in-situ from inclusions, is available as supplementary material (Smith 2021). As visual inspection of inclusions under a microscope is often integral to Raman analysis, users may wish to refer to published images to see some good examples of different inclusion types (e.g., Koivula 2000; Tappert and Tappert 2011). In addition to minerals, fluids or amorphous solids are also possible inclusions in diamond, and these too can often be identified from their Raman spectra. However, scrutiny is required and phase identification is not always possible based on Raman spectra alone.

Inclusions in lithospheric diamonds

Inclusions in diamonds from the continental lithospheric mantle are divided into three paragenetic associations, reflecting their host rocks: peridotitic (comprising lherzolitic, harzburgitic and wehrlitic parageneses), eclogitic, and websteritic (Stachel and Harris 2008). Diamond from websteritic sources is relatively uncommon, accounting for only about 2% of inclusion-bearing lithospheric diamonds, whereas peridotitic and eclogitic parageneses account for 65% and 33%, respectively (Stachel 2014). Mineral phases encountered in lithospheric diamonds are laid out in Table 1 (also see Table 1 in Shirey et al. 2013) and some examples of corresponding Raman spectra from inclusions are shown in Figure 1. These inclusions are often single mineral phases and correspond to familiar minerals encountered in crustal rocks. As long as a reasonable Raman spectrum is collected, the identification of these phases is usually straightforward. In contrast, inclusions in sublithospheric diamonds often contain coexisting minerals representing a retrogression assemblage and may contain obscure mineral phases, adding a greater challenge to identification and interpretation.

Inclusions in sublithospheric diamonds

Inclusions derived from the sublithospheric mantle broadly mirror the peridotitic and eclogitic host rock associations of lithospheric diamond suites, but with mineralogy dictated by higher pressures and temperatures (see reviews by Stachel et al. 2005; Harte 2010; Kaminsky 2012; Harte and Hudson 2013; Shirey et al. 2013; Nestola 2017). The mineralogy of peridotitic and eclogitic rocks as a function of depth, which has been delineated experimentally, provide a practical guide for interpreting inclusions in sublithospheric diamonds (e.g., Walter et al. 2011). Although the high-pressure mineralogy of peridotitic or eclogitic bulk compositions does explain many inclusion assemblages in sublithospheric diamonds, there are also inclusions that deviate from this framework, such as the occurrence of carbonates, or of Ca-silicate inclusions

Table 1. Mineral inclusions in lithospheric diamond, modified from Stachel (2014).

Peridotitic	**Eclogitic**	**Websteritic**	**Uncertain**
	Common		
Cr-pyrope	Grossular–almandine–pyrope	Almandine-pyrope	Graphite
Olivine	Omphacitic clinopyroxene	Diopside-augite	
Enstatite	Fe sulfides*	Enstatite	
Cr-diopside			
Mg-chromite			
Fe–Ni sulfides*			
	Occasional		
	Rutile	Coesite	
	Coesite	Olivine	
	Rare		
Coesite	Kyanite	Phlogopite	Diamond
Mg-ilmenite	Corundum		Calcite
Magnesite	Ilmenite		Dolomite
Calcite	Magnetite		Perovskite
Native Fe	Fe–Mg-chromite		Amphibole
Zircon	Phlogopite		Moissanite
Phlogopite	K-feldspar		Apatite
Yimengite	Titanite		Eskolaite
	Staurolite		Sr-titanate
	Zircon		Monazite
	Moissanite		
	Calcite		
	Dolomite		

Notes: *pyrrhotite ± pentlandite ± chalcopyrite ± molybdenite (Kemppinen et al. 2018), with bulk Ni content being >17 wt% for peridotitic and <10 wt% for eclogitic sulfides

with Ca:Si ratios above 1:1 that cannot be interpreted simply as a back-transformation product of $CaSiO_3$-perovskite (Brenker et al. 2005, 2007). Other host rock types, such as subducted serpentinites or oceanic sediments, may also be relevant for sublithospheric diamond formation. In detail, many sublithospheric inclusions do not appear to be pristine grains of ambient mantle rocks, but instead bear evidence of variable chemical modification by, or crystallization from, low-degree melts or fluids prior to their entrapment in diamond, possibly related to diamond formation (Stachel et al. 2000; Walter et al. 2008; Thomson et al. 2016).

Table 2 lists mineral phases encountered in sublithospheric diamonds and some examples of their Raman spectra are shown Figure 2. It should be noted that few inclusions are found as well-preserved high-pressure minerals. The large pressure-temperature change during exhumation from the sublithospheric mantle destabilizes some mineral inclusions, causing inversion (change in structure) or retrogression (breakdown into multiple phases) to lower-pressure polymorphs or assemblages. Observed inclusion phases must be carefully interpreted to judge the identity of the original inclusion phase that was trapped during diamond growth (also see Table 2 in Shirey et al. 2013). For example, inclusions of bridgmanite ($MgSiO_3$-perovskite) invert to orthopyroxene, but the original inclusion identity can be inferred based on

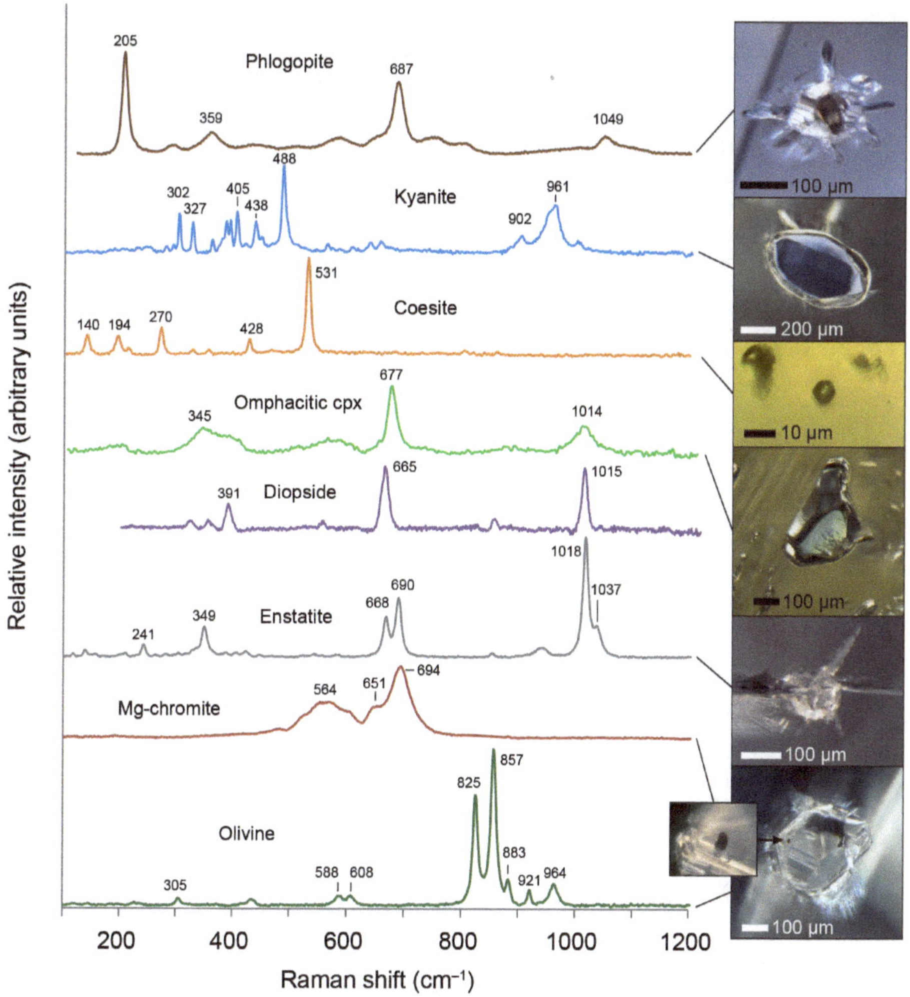

Figure 1. Raman spectra of a variety of mineral inclusions in lithospheric diamonds, along with corresponding inclusion images. The phlogopite inclusion, which also has an O–H stretching band at 3732 cm^{-1}, is in a facetted diamond (sample 110206866925) that also contains olivine inclusions. The kyanite inclusion is in a faceted diamond (sample 110209083424) that also contains orange eclogitic garnet inclusions.The coesite inclusion is in a yellow facetted diamond (sample 110208419590) and exhibits pressure-induced band shifts (Smith et al. 2016a). The omphacitic clinopyroxene (cpx) inclusion is in a yellow-green facetted diamond (sample 890000082714) that also contains orange eclogitic garnet (Fig. 3) inclusions (Smith and Wang 2017). The diopside inclusion (no image) is in a diamond from Sao Luiz (Juina area), Brazil. The enstatite is in a facetted diamond (sample 110208779832). The Mg-chromite is a small (~10 μm) inclusion trapped within a larger forsteritic olivine inclusion in a facetted diamond (sample 100722703657).

low NiO contents (300 ppm or less) compared to typical upper mantle orthopyroxene (1000 ppm or more) (Stachel et al. 2005). Sometimes bridgmanite contains Al, and the inclusion undergoes retrogression to an assemblage of orthopyroxene plus aluminous phases, such as jeffbenite and spinel (Walter et al. 2011; Harte and Hudson 2013). Jeffbenite [$(Mg,Fe)_3Al_2Si_3O_{12}$], formerly known as TAPP (tetragonal almandine-pyrope phase), may appear by retrogression of aluminous bridgmanite or majoritic garnet, though its possible occurrence as an original/

Table 2. Mineral inclusions observed in sublithospheric diamonds, adapted from Harte and Hudson (2013) and Stachel (2014).

Single or multi-phase mineral inclusion	Possible original inclusion phase
Common	
*Ferropericlase–magnesiowüstite	
Ca-rich inclusions: Breyite ($CaSiO_3$-walstromite), larnite (β-Ca_2SiO_4), $CaSi_2O_5$-titanite, wollastonite, pseudowollastonite, perovskite ($CaTiO_3$), titanite ($CaTiSiO_5$), ZrO_2, (additional Ca-silicates?)	Various: $CaSiO_3$-perovskite (or $Ca(Si,Ti)O_3$), breyite, larnite, $CaSi_2O_5$-titanite
Majoritic garnet ± cpx, olivine, plagioclase	Majoritic garnet
Jeffbenite	Majoritic garnet? Bridgmanite?
Jeffbenite + NaAl-pyroxene	Majoritic garnet
Olivine	Olivine, wadsleyite, ringwoodite
Quartz or coesite ± kyanite	Stishovite
Nepheline, spinel ± olivine	CF phase
Kalsilite, nepheline, spinel ± cpx	NAL phase
Opx ± jeffbenite, spinel, ilmenite, olivine, cpx	Bridgmanite (may be Al-bearing)
*Omphacitic cpx	
Rarer occurrences	
$CaSiO_3$-perovskite ± perovskite ($CaTiO_3$)	$Ca(Si,Ti)O_3$-perovskite
K-feldspar	Liebermannite
Ca-feldspar	Stöfflerite
Calcite	Aragonite
*Ringwoodite	
*Phase Egg	
*Sulfides	
*Merwinite	
*Magnesite	
*Ilmenite	
*Corundum	
*Chromite	
*Spinel	
*Native Fe–Ni	
*Siderite	
*Titanite	

Note: *Inferred as the original inclusion phase, without inversion/retrogression. Abbreviations: opx = orthopyroxene; cpx = clinopyroxene; CF = Calcium-ferrite-structured phase; NAL = New aluminous phase.

primary inclusion is not ruled out (Nestola et al. 2016). Retrogression reactions can be complex, especially if multiple high-pressure phases have been trapped together (e.g., Brenker et al. 2002). Consequently, it is not always possible to accurately interpret the original inclusion mineralogy without additional tools beyond Raman spectroscopy.

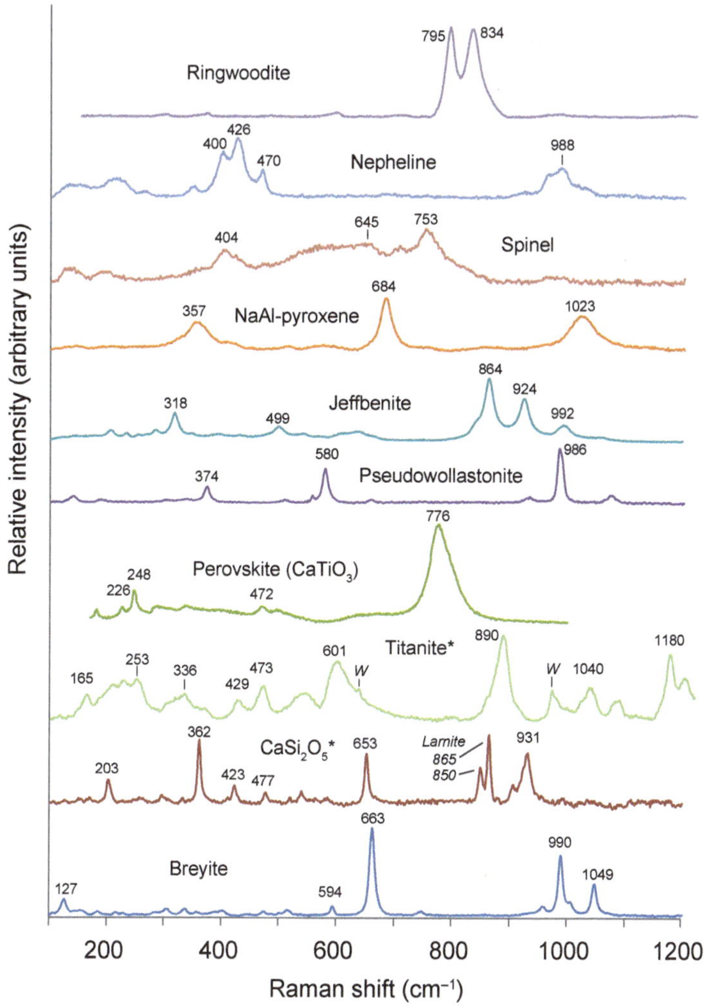

Figure 2. Raman spectra of a variety of phases included in sublithospheric diamonds. The ringwoodite shown is a synthetic crystal, R090003 from RRUFF (Lafuente et al. 2015). Nepheline and spinel were found coexisting in a multiphase inclusion in a boron-bearing (Type IIb) diamond (sample 110208245246 in Smith et al. 2018). NaAl-pyroxene was found as part of a composite inclusion with jeffbenite in a Type IIb diamond (sample 880000037816 in Smith et al. 2018). The jeffbenite shown is in a diamond from Sao Luiz (Juina area), Brazil (Nestola et al. 2016). Phases related to Ca- and Ca,Ti-silicate inclusions are pseudowollastonite (composite with breyite in sample 110208423120), perovskite (composite with $CaSiO_3$-perovskite) (Nestola et al. 2018), titanite (composition uncertain) with wollastonite (bands labeled W) in the same analyzed volume (coexisting with other Ca-silicates in sample 100918637486), $CaSi_2O_5$ (structure uncertain) with larnite in the same analyzed volume (sample 110208104790), and breyite (sample 880000037816). (*There may be two different $CaSi_2O_5$ polymorphs, one with monoclinic (titanite-structured) and one with triclinic structure.) Note that olivine, kyanite, coesite, diopside, and enstatite (orthopyroxene) from Figure 1 are also important phases for sublithospheric diamonds.

Interpretation of the depth of origin and the likely host rock association can be made on the basis of assemblages of coexisting inclusions (Stachel et al. 2005; Harte 2010; Kaminsky 2012; Harte and Hudson 2013; Shirey et al. 2013; Nestola 2017). Caution is required, as some inclusions that are common in sublithospheric diamonds provide ambiguous depth information when found as lone occurrences. Ferropericlase, for example, is found among inclusion assemblages representing host rocks with peridotitic composition in the lower mantle (deeper than 660 km), but ferropericlase inclusions on their own do not indicate a lower mantle origin (Stachel et al. 2005). Regions of low silica activity in the upper mantle can stabilize ferropericlase (Brey et al. 2004). Similar care is required for lone inclusions of breyite, which cannot be automatically assumed to originate from $CaSiO_3$-perovskite in the transition zone to lower mantle (Woodland et al. 2020; Brenker et al. 2021).

Remarks on some specific inclusion types

Graphite. Graphite is a common occurrence in diamond, sometimes as protogenetic, euhedral crystals (Nasdala et al. 2005) and possibly as a metastable syngenetic phase (Smit et al. 2016), but more often as a result of decompression in fractures surrounding inclusions, such as around sulfide inclusions. In addition to the most intense G band (~1580 cm^{-1}), as well as the D and 2D bands of graphite, a weak but sharp band at 867 cm^{-1} that corresponds to "forbidden" out-of-plane vibrations (Kawashima and Katagiri 1999) is occasionally observed in graphitic material in inclusions.

In Raman spectroscopy, the presence of graphite can be helpful when analyzing inclusions that happen to be weak or Raman inactive, because detecting graphite from around the inclusion confirms that spectral information is, in fact, being collected from the inclusion. In this case, finding no discernible Raman spectrum from the major volume of the inclusion itself can be a useful observation.

Distinguishing among garnets. Garnet is widespread in the lithosphere, asthenosphere, and mantle transition zone and it is regularly found as an inclusion in diamond. In general, most garnet inclusions have similar features (Fig. 3) but their spectra change as a function of composition (Kolesov and Geiger 1998) and structural disorder. It is possible to determine garnet compositions by considering their spectral features as linear combinations between end-members (Bersani et al. 2009). For diamond, Kalugina and Zedgenizov (2020) have evaluated suites of eclogitic and peridotitic garnet inclusions and found reasonable agreement between compositions measured by EPMA analysis and compositions estimated from Raman spectra using the approach of Bersani et al. (2009). Kalugina and Zedgenizov (2020) used the rotation ($R[SiO_4]$) and stretching (Si–O) vibrations of SiO_4-tetrahedra near 360 and 910 cm^{-1}, respectively, to estimate relative molar proportions of pyrope, almandine, and grossular. For their samples, $R[SiO_4]$ was the most sensitive band position, with eclogitic garnet inclusions falling within 355.9–361.2 cm^{-1} and peridotitic garnet inclusions being within 361.2–365.2 cm^{-1} (Kalugina and Zedgenizov 2020). This difference in $R[SiO_4]$ band position mainly reflects the difference in contribution from almandine and pyrope end-members at 342 cm^{-1} and 364 cm^{-1}, respectively (Kolesov and Geiger 1998).

It can also be possible to recognize majoritic garnet (i.e., garnet from the sublithospheric mantle, with more than 3 Si per formula unit), from the width of the strongest Raman mode near 910 cm^{-1} (McMillan et al. 1989). Gillet et al. (2002) proposed that widened bands (FWHM greater than about 25 cm^{-1}) are due to disorder of Al, Mg, and Si at the octahedral sites in majoritic garnet and can be considered diagnostic. If a garnet were to exhibit grain sizes smaller than the excitation laser beam diameter, this could also cause broadening. Narrower bands, however, do not necessarily rule out a majoritic component.

Sulfides. Sulfide inclusions can often be recognized visually, by their metallic grey or yellowish appearance and black graphitic fracture rosettes. The major sulfide phase, pyrrhotite, is only weakly Raman active and is usually not observable by Raman in-situ. Bands at about 376 cm^{-1} and 341 cm^{-1} are associated with monoclinic pyrrhotite (Fig. 4), but can vary with Fe/S ratio and structure (Hope et al. 2001), whereas hexagonal pyrrhotite may be inactive (Mernagh and Trudu 1993). Pentlandite (main band at 370 cm^{-1}) and chalcopyrite in sulfide

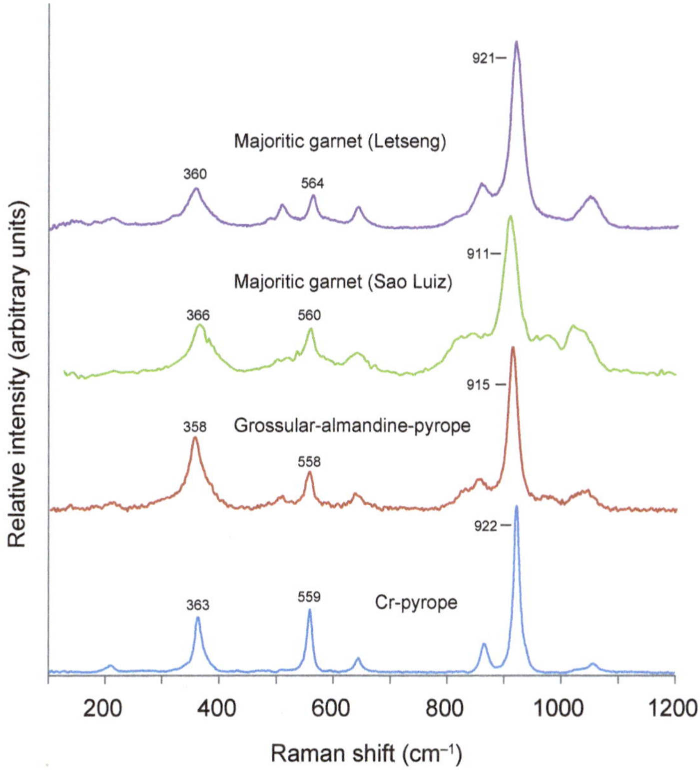

Figure 3. Examples of Raman spectra from garnet inclusions in diamond. Majoritic garnets in diamonds from Letseng (sample 110207974892 in Smith et al. 2016b) and Sao Luiz (sample BZ43 in Gillet et al. 2002) have broadening of the strongest band and some additional weak and broad underlying features between 800 and 1000 cm^{-1}. The lithospheric garnet inclusions grossular–almandine–pyrope (sample 110209083424) and Cr-pyrope (Siberian diamond sample in Kalugina and Zedgenizov 2020) may potentially be distinguished from one another based on the position of the ~360 cm^{-1} band (Kalugina and Zedgenizov 2020).

inclusions can also be difficult to detect. Studies on sulfide inclusions have more often relied on exposing the inclusion to characterize it. However, a recent Raman-based identification of small amounts of molybdenite in a significant number of sulfide inclusions in diamonds from several localities has led to the suggestion that Raman should be used as part of the initial characterization of sulfides prior to Re–Os dating studies (Kemppinen et al. 2018). Molybdenite and chalcopyrite (Fig. 4) are more Raman active and are detected more easily than pyrrhotite and pentlandite.

Calcium silicates. Calcium silicate inclusions, most notably breyite (Brenker et al. 2021; $CaSiO_3$-walstromite), have popularly been interpreted as a breakdown product from more or less pure $CaSiO_3$-perovskite, which is part of the experimentally-predicted mineralogy of the lower part of the transition zone and lower mantle for almost all rock types, including peridotitic, eclogitic and most sedimentary bulk compositions (e.g., Irifune and Ringwood 1987, 1993). A single example of non-inverted $CaSiO_3$-perovskite in diamond shows that the entrapment of $CaSiO_3$-perovskite indeed occurs, although, unfortunately, $CaTiO_3$- and $CaSiO_3$-perovskite seem to show almost indistinguishable Raman spectra (Nestola et al. 2018). In detail, however, most calcium silicate inclusions are more complicated (Brenker et al. 2021) and exhibit multiple phases (Brenker et al. 2005). Their interpretation as a breakdown product from $CaSiO_3$-perovskite should not be automatic.

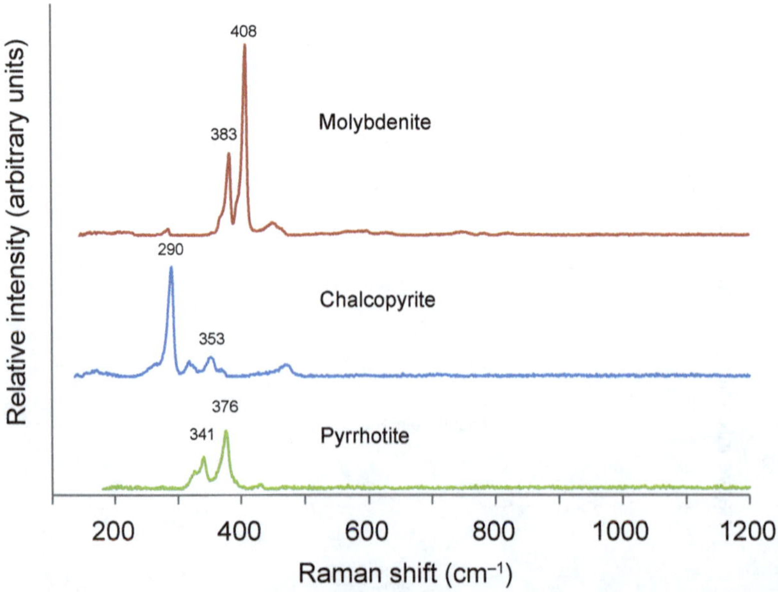

Figure 4. Raman spectra of phases found in sulfide inclusions in diamond. Sulfides can be challenging to characterize by Raman spectroscopy, but it has been suggested as a useful way to check for the presence of molybdenite, which may be important for Re–Os geochronology studies (Kemppinen et al. 2018). Pyrrhotite (R060440), chalcopyrite (R050018), and molybdenite (R060124) are from RRUFF (Lafuente et al. 2015).

For a long time, it was presumed that a pure $CaSiO_3$ phase would not be stable in any likely mantle rock composition at depths shallower than about 520 km. However, on the basis of inclusions in diamond, Brenker et al. (2005) have demonstrated that Ca,Si-rich environments exist in the sublithospheric mantle and that these regions can enable the formation and entrapment of breyite in diamond at depths of less than 300 km.

In some instances, a simple $CaSiO_3$ precursor is precluded by bulk inclusion Ca:Si ratios that are markedly different from 1:1 (Brenker et al. 2005). If diamond is formed in the lower part of the transition zone or lower mantle, $CaSiO_3$-perovskite might be captured together with stishovite (SiO_2) if a basaltic or sedimentary precursor is assumed. In this case the Ca:Si ratio of the whole inclusion will be lower than 1:1. However, it was experimentally demonstrated that breyite could crystallize together with diamond in the upper mantle (at 6–10 GPa) in SiO_2-rich compositions (Woodland et al. 2020) resulting in the same phase assemblage.

For inclusions with a bulk Ca:Si ratio above 1:1, which usually consist of coexisting breyite and larnite (β-Ca_2SiO_4), the interpretation is less straightforward. In this case, diamond crystallization in the lower mantle is unlikely for peridotitic or basaltic bulk compositions. In order to account for such an inclusion, the diamond could have formed at conditions corresponding to the two-phase stability field of larnite and $CaSi_2O_5$-titanite. Here, a high Ca:Si ratio could be achieved by capturing a higher amount of larnite relative to $CaSi_2O_5$-titanite, breyite or $CaSiO_3$-perovskite. The formation depth in this case will be between 300 and 400 km, that is, in the upper mantle above the transition zone. Furthermore, Fedoraeva et al. (2019) have experimentally shown that with an aragonite-breyite starting assemblage, an assemblage of breyite-larnite can be formed at 6 GPa and about 1700 °C due to incongruent melting, which again is at upper mantle conditions. Considering these possibilities, it is clear that the finding of breyite (with or without other Ca-silicates) inclusions alone as an inclusion in diamond does not necessarily indicate a transition zone or lower mantle depth (Brenker et al. 2021).

Besides breyite, a surprising number of different calcium silicates can sometimes be found in inclusions, even co-existing metastably (Table 2). Some uncertainty surrounds the Raman identification of $CaSi_2O_5$-titanite. Two different $CaSi_2O_5$ polymorphs have been observed in high pressure experimental products, corresponding to monoclinic (titanite-structure) and triclinic structures (Angel 1997; Kubo et al. 1997; Kudoh and Kanzaki 1998; Akaogi et al. 2004). In some experiments, monoclinic $CaSi_2O_5$-titanite produced at high pressure has inverted to the triclinic structure upon decompression (Akaogi et al. 2004). It might therefore be expected that both $CaSi_2O_5$ polymorphs could potentially be found in calcium silicate inclusions in diamond (compare $CaSi_2O_5$* and titanite* in Fig. 2).

The Raman spectrum corresponding to a synthetic $CaSi_2O_5$ sample produced by Gasparik et al. (1994) has been reported as $CaSi_2O_5$-titanite ($CaSi_2O_5$* in Fig. 2) in multi-phase calcium silicate inclusions (Nasdala et al. 2003; Anzolini et al. 2016; Smith et al. 2016b). However, it is not clear if the crystal structure of this sample, after decompression, has been verified as monoclinic $CaSi_2O_5$-titanite or if it might potentially be the triclinic polymorph.

In some multi-phase calcium silicate inclusions, a Raman spectrum resembling true monoclinic titanite ($CaTiSiO_5$) with a prominent band at ~600 cm^{-1} (titanite* in Fig. 2) has been recorded (Smith et al. 2017). The composition of this phase was not determined, but based on the observation that the Raman spectra of $CaSiO_3$-perovskite and $CaTiO_3$-perovskite might be virtually indistinguishable (Nestola et al. 2018), it is considered possible that the spectra of $CaSi_2O_5$-titanite and true titanite ($CaTiSiO_5$) are similar. If so, the $CaSi_2O_5$ sample of Gasparik et al. is not actually in the titanite structure but in the triclinic structure.

With the variety of calcium silicates described in diamonds to date, it may be reasonable to expect some additional phases to be encountered in future studies. Wollastonite-II, for instance, has been found in multiphase carbonatitic microinclusions (Kaminsky et al. 2009), and could potentially be part of larger calcium silicate inclusions. However, it may not be readily distinguishable from breyite by Raman spectroscopy, because their crystal structures are nearly identical (Joswig et al. 2003; Dörsam et al. 2009; Barkley et al. 2011). Some other phases that may be important for Ca-rich inclusion investigations are rankinite ($Ca_3Si_2O_7$), kilchoanite ($Ca_3Si_2O_7$), hartrurite/post-hartrurite (Ca_3SiO_5), tilleyite/post-tilleyite ($Ca_5Si_2O_7(CO_3)_2$), trabzonite ($Ca_4(Si_3O_9OH)OH$), and chegemite ($Ca_7(SiO_4)_3(OH)_2$), as well as hydrous post-hartrurite, hydrous larnite, and hydrous $CaSiO_3$-perovskite (e.g., Németh et al. 2017).

Ferropericlase–magnesiowüstite. Inclusions of ferropericlase–magnesiowüstite are one of the most commonly reported phases in sublithospheric diamonds. This phase is often transparent and brown, but can possess a colorful iridescence at the inclusion surface that is a good indicator for identification (Fig. 5). Inclusions can exhibit a wide range of Mg/(Mg+Fe) ratios, but most are within 0.60–0.82 (Stachel et al. 2005). For simplicity, the term ferropericlase is sometimes applied to the whole ferropericlase–magnesiowüstite series in inclusions.

Pure periclase (MgO) is Raman inactive, whereas wüstite (FeO) is weakly active with one somewhat broad band near 652 cm^{-1} (de Faria et al. 1997). The Raman spectra of periclase–wüstite solid solutions have not been studied. As there is no accepted Raman spectrum for ferropericlase, a Raman-based identification is strictly not possible. However, ferropericlase inclusions whose identities have been confirmed using other techniques do exhibit some consistent Raman features. Figure 5 shows Raman spectra associated with confirmed (Fig. 5a–f) and inferred (Fig. 5g) ferropericlase inclusions in diamond, though the exact assignment of these features is uncertain. The ferropericlase inclusions in one of these diamonds (Fig. 5d–f) have also been identified by both EPMA analysis (Mg/(Mg+Fe) = 0.91) and X-ray diffraction (Smith et al. 2018). The possibility of exsolved magnesioferrite within ferropericlase inclusions (e.g., Palot et al. 2016) may be important, although the spectra in Figure 5 do not resemble magnesioferrite (e.g., D'Ippolito et al. 2015). One possibility is that the features are related

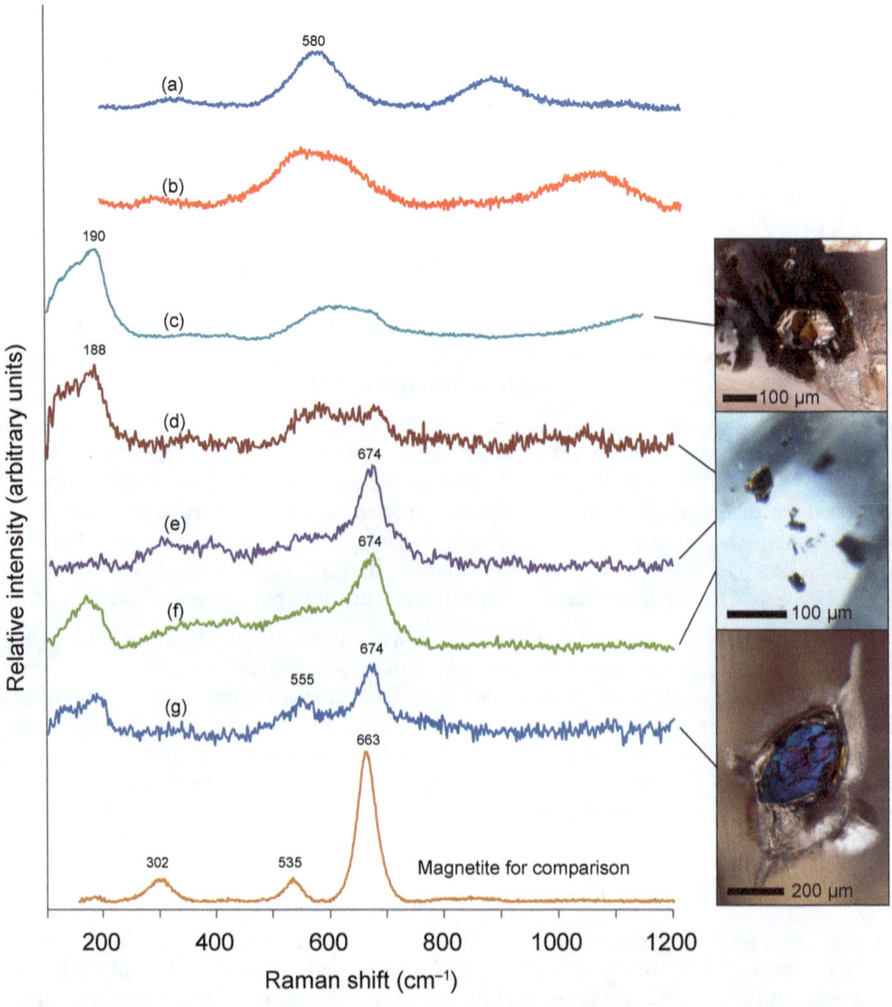

Figure 5. Raman spectra tentatively associated with ferropericlase inclusions. The spectral features themselves may not necessarily belong to ferropericlase. **(a)** Ferropericlase inclusion in a diamond from Juina, Brazil. **(b)** Ferropericlase in a diamond from Sao Luiz (Juina area), Brazil. **(c)** Ferropericlase in a pink-brown diamond (sample CBP-0937) also containing two-phase composite inclusions of breyite and $CaTiO_3$-perovskite. This spectrum was collected from an exposed surface after breaking open the inclusion. **(d, e, f)** Three spectra collected from different inclusions in a boron-bearing (Type IIb) diamond (sample 110208425476 in Smith et al. 2018). **(g)** Large and iridescent inferred ferropericlase in a facetted pink diamond (sample 110206861072). Magnetite reference spectrum R080025 from RRUFF (Lafuente et al. 2015).

to magnetite (shown in Fig. 5). Laser-induced decomposition during Raman analysis could potentially cause ferropericlase to break down at the inclusion surface to produce magnetite, similar to the destabilization of wüstite noted by de Faria et al. (1997) that was found to produce α-Fe and magnetite (Fe_3O_4) under intense laser excitation. It may also be possible to partially destabilize ferropericlase by overheating the diamond during polishing. Alternatively, ferropericlase inclusions can potentially contain naturally-occurring micro-exsolutions of metal and magnetite (Anzolini et al. 2020). Further study is needed to characterize the Raman spectrum of ferropericlase inclusions.

Metallic melt inclusions. A recently recognized variety of sublithospheric diamonds, called CLIPPIR diamonds, contain metallic Fe–Ni–C–S melt as their most abundant inclusion (Smith et al. 2016b, 2017). These inclusions were trapped as liquid, so are technically not mineral inclusions and are not listed in Table 2. However, they deserve mention in the discussion of sublithospheric diamonds. Although not well-represented in the geological literature from inclusion studies, many inclusion-free, nitrogen-deficient (Type IIa) diamonds can tentatively be assigned to the CLIPPIR category, in which case they may account for approximately 1% of the gem marketplace (Smith et al. 2017). The physical characteristics of CLIPPIR diamonds, often being colorless, large, and inclusion-poor, place them among the most valuable gem diamonds, seldom accessible for research.

The Fe–Ni–C–S metallic melt inclusions solidify to an assemblage of cohenite, pyrrhotite, and Fe–Ni alloy (± minor Fe–Cr-oxide, Fe-phosphate) (Smith et al. 2016b, 2017). These phases are virtually undetectable by Raman spectroscopy in-situ. However, Raman spectroscopy can still be helpful to increase the confidence in identifying them, beyond visual recognition. It is often possible to detect methane (as sharp band around 2915–2918 cm^{-1}), and sometimes molecular hydrogen, at the inclusion-diamond interface (Smith et al. 2016b, 2017). However, methane has also been recorded in other kinds of inclusions in diamond (Smit et al. 2016) and is not diagnostic of Fe–Ni–C–S metallic inclusions. Similar metallic melt inclusions, potentially of variable composition, have been observed more rarely in sublithospheric Type IIb (boron-bearing) diamonds (Smith et al. 2018). Various inclusions with native Fe, Ni, Fe–Ni alloys, and Fe-carbides have also been described in other kinds of sublithospheric diamonds, although it is unclear if these are related to the recurring Fe–Ni–C–S melt found in CLIPPIR diamonds (Hayman et al. 2005; Bulanova et al. 2010; Gurney et al. 2010; Kaminsky and Wirth 2011).

Additional high-pressure phases

As mentioned above, high-pressure phases are commonly found as retrograde transformation products and only a handful of mineral inclusions with their original mineralogy intact have been reported. One such mineral is ringwoodite, the most dominant mineral at the lower part of the mantle transition zone (~520–660 km depth) (e.g., Irifune and Ringwood 1987), which was first synthesized in Fe_2SiO_4 composition in 1958 (Ringwood 1958) and not long after discovered in the Tenham L6-chondrite (Binns et al. 1969). The first terrestrial sample, however, has only recently been found, as an inclusion in diamond, and was first identified using Raman spectroscopy (Pearson et al. 2014). Thus, while finding pristine high-pressure phases as inclusions in diamond is rare, it is possible, and Raman spectroscopy is an ideal tool to identify them. Unfortunately, not all experimentally derived phases have published Raman spectra. An example is the Al-rich phases expected to form in basaltic/eclogitic bulk compositions at lower mantle conditions, called NAL (new aluminous) phase (e.g., Akaogi et al. 1999; Gasparik et al. 2000) and CF (calcium-ferrite type structure) phase (e.g., Irifune and Ringwood 1993; Hirose et al. 2005; Ono et al. 2005), that are listed as possible original inclusions in Table 2. However, our spectral database (Smith 2021) includes a number of other possible deep mantle phases, from experiments and meteorites studies, that could plausibly occur in diamonds and the most important ones are introduced in this section.

Wadsleyite. While ringwoodite has been found in diamond, wadsleyite, the high-pressure equivalent of olivine replacing it as the principal rock-forming mineral below the 410 km discontinuity (e.g., Irifune and Ringwood 1987) in the upper portion of the mantle transition zone, has not yet been found. Because the entrapment conditions of high-pressure inclusions observed in terrestrial diamonds extend well within the stability of field of wadsleyite (Navon et al. 2017; Tschauner et al. 2018), researchers are actively searching for a pristine specimen.

Dense hydrous magnesium silicates. Experimental studies have shown that a number of dense hydrous magnesium silicates (DHMS) are stable at mantle pressures, corresponding to depths greater than ~200 km (e.g., Ohtani et al. 2005; Komabayashi and Omori 2006). These phases are denoted by alphabetical names (e.g., A, B, etc.) and are expected to form either within deeply subducted cold slabs by metamorphism of serpentinites or within the convecting mantle by metasomatism from slab-derived hydrous fluid (Tschauner 2019). The DHMS phases proposed to be relevant to sublithospheric diamond formation (Harte 2010) are phase A, phase E, superhydrous phase B, and phase D (Litasov and Ohtani 2007 and references therein). Serpentinites and DHMS phases formed from them have further been suggested as potential carriers of water and boron for the formation of Type IIb diamonds in the transition zone and lower mantle (Smith et al. 2018).

Sediment assemblages. Geochemical, geophysical and experimental studies have shown evidence suggesting the subduction of continental crust and terrigenous and pelagic sediments to the mantle transition zone and possibly even to the lower mantle, despite their relative buoyancy compared to the surrounding mantle (e.g., von Huene and Scholl 1991; Irifune et al. 1994; Workman et al. 2004; Afonso and Zlotnik 2011). Once this material reaches depths greater than 250–300 km, its density increases significantly due to the formation of dense silicates, marking a "point of no return" for subducted continental crust. The volume of continental material that can be subducted is mainly dependent on the rheology of the continental crust, with strong crusts favoring deep subduction.

Based on experimental work on the phase relations in dry and hydrous sediment systems at transition zone and lower mantle pressures and temperatures, summarized by Litasov and Ohtani (2007), as yet undiscovered high-pressure minerals stable at such depths are liebermannite ($KAlSi_3O_8$-hollandite) and zagamiite (CAS phase). Liebermannite (Urakawa et al. 1994; Ma et al. 2018) is considered to be the most abundant phase in a continental crust bulk composition at pressure and temperature conditions corresponding to the mantle transition zone (Irifune et al. 1994). While both liebermannite and zagamiite (Gautron et al. 1996; Beck et al. 2004; Ma et al. 2018) have been found in shocked meteorites but not as terrestrial samples, one inclusion in diamond (Kankan, Guinea) has been reported where liebermannite might have been the precursor (Stachel et al. 2000). In hydrous sediment systems, two additional experimentally predicted phases are topaz-OH and δ-AlOOH (Wunder et al. 1993, 1999; Suzuki et al. 2000; Ohtani et al. 2001; Xue et al. 2006, 2010). Topaz-OH is thought to be stable above the mantle transition zone, between 250–350 km, while δ-AlOOH is expected to be a breakdown product of *phase egg*, along with stishovite, at depths below 450 km.

Tuite. Tuite (γ-$Ca_3(PO_4)_2$) is the main product formed from of apatite, a common accessory mineral occurring in a variety of terrestrial rocks, at high pressures and has been shown to be stable at upper and lower mantle conditions (Murayama et al. 1986; Zhai et al. 2013). The natural occurrence of tuite was first found in the Suizhou L6 chondrite (Xie et al. 2003). The apatite-tuite transformation has been shown experimentally in both apatite–MORB (Konzett and Frost 2009) and apatite–peridotite (Konzett et al. 2011) systems, but so far no terrestrial samples have been discovered.

VOLATILE COMPONENTS

In addition to detecting characteristic vibrations from mineral phases, Raman spectroscopy is well-suited for detecting liquids or gases trapped as inclusions in diamond. Mineral inclusions in diamond can sometimes have small amounts of fluid coexisting with the main solid portion of the inclusion. For example, based on Raman analyses, it has been found that some common inclusions in lithospheric diamonds are surrounded by a thin layer (< 1.5 μm) of hydrous silicic fluid that contains $Si_2O(OH)_6$, $Si(OH)_4$, and molecular H_2O (Nimis et al. 2016). Inclusions in some sublithospheric diamonds also possess an invisible fluid layer, made up of CH_4, with or without H_2 (Smith et al. 2016b, 2018) and possibly other species.

The ferropericlase inclusion shown in Figure 5g even has a layer of CH_4. Methane has also been found associated with graphite in mixed-habit diamonds from Marange (Smit et al. 2016).

Fluid and melt inclusions have been reported along healed cracks in diamond, in which case Raman spectroscopy has revealed N_2 and CO_2 in the fluids (Tomilenko et al. 2001; Smith et al. 2014, 2015). Nitrogen has also been recorded in solid form, as nano-inclusions in some sublithospheric diamonds, where it is proposed to have exsolved from the substitutional nitrogen in the diamond lattice (Rudloff-Grund et al. 2016; Navon et al. 2017). Reviews on Raman analysis of fluid inclusions are a useful resource for characterizing volatile species (e.g., Burke 2001; Frezzotti et al. 2012; Bodnar and Frezzotti 2020).

RAMAN BAROMETRY

At the moment of entrapment, an inclusion is assumed to be enclosed by the host without any anomalous stresses at the interface (i.e., a uniform uninterrupted stress state with no squeezing or stretching and a perfect geometrical fit between the inclusion and host). When a diamond is carried to Earth's surface, it can relax freely in response to the change in pressure and temperature, but its inclusions are nearly fixed in volume and are unable to expand freely. As a result, many inclusions are under considerable remnant pressures, reaching a few gigapascals (e.g., Nasdala et al. 2003).

Raman spectra can provide barometric information of the host-inclusion system because crystal lattice strains distort bonds, thereby changing vibrational frequencies and the position of Raman bands. It is possible to quantify the remnant pressure in an inclusion if a calibration is available for pressure-induced band shifts for the included phase, having the same chemical composition and structure (e.g., Sobolev et al. 2000; Gillet et al. 2002; Nasdala et al. 2003). By combining this remnant pressure with thermoelastic parameters of the mineral inclusion and the diamond host, it is possible to constrain the pressure–temperature conditions of diamond formation (Izraeli et al. 1999; Angel et al. 2014, 2015; Anzolini et al. 2018). Software such as EoSFit-Pinc (Angel et al. 2017) is useful for this purpose. However, non-elastic behavior of the diamond host around the inclusion, due to brittle or plastic deformation, can effectively alleviate some inclusion pressure, leading to underestimation of the depth of diamond growth (Howell et al. 2012).

Some additional obstacles limit this technique. The simplest inclusion would be a sphere, with isotropic properties, trapped in a host with isotropic properties. A somewhat equant, round shaped inclusion of a cubic mineral trapped in diamond (also cubic) is a reasonable approximation of the ideal scenario (Howell et al. 2010). Care must be taken when examining pressures in anisotropic mineral inclusions in diamond because of the possibility for deviatoric stresses that will lead to Raman band shifts that do not follow congruently with hydrostatic pressure calibrations (Anzolini et al. 2018). There is currently no reliable method among existing analytical techniques and hydrostatic calibrations to determine strains in minerals subject to deviatoric stresses and interpret them in terms of a single "inclusion pressure." Another issue is that strains and Raman shifts can vary spatially within an inclusion, which is exacerbated by anisotropy and the effects of inclusion geometry (Mazzucchelli et al. 2018). In these cases, the center of the inclusion, away from edges, provides the best measurement for calculating entrapment conditions (Murri et al. 2018).

Host diamond barometry

Distortion of the host diamond around inclusions can be easily observed using crossed polarizers. This anomalous birefringence can be used to gage residual pressure and the distribution of strains in a diamond (Howell et al. 2010). The distortions also affect both the position and width of its main Raman band, normally located at 1332.5 cm^{-1} (also often stated as 1332 cm^{-1}) in relatively strain- and impurity-free diamond (Schiferl et al. 1997).

Growth patterns and bulk deformation features of the whole diamond can also be manifested in spatial variations of the diamond Raman features. The distribution of strain around an inclusion is highly variable, but Raman mapping has been successfully applied to reveal its maxima/minima and geometry (Nasdala et al. 2003; Howell et al. 2012). Pressure exerted on the diamond in the <111> axis (perpendicular to one of its octahedral planes) causes a band shift of + 0.7 cm^{-1}/GPa and along the <001> axis (perpendicular to one of its cube planes) a band shift of + 2.2 cm^{-1}/GPa (Grimsditch et al. 1978; Sharma et al. 1985). As the crystallographic direction is relevant for the band shift calibration, it is necessary to understand the geometry of stresses and Raman analysis in order to quantify pressure adjacent to an inclusion (Nasdala et al. 2003). The band shift of the diamond adjacent to an inclusion can nevertheless provide some indication of remnant pressure inside the inclusion.

Inclusion barometry

Using Raman spectroscopy, the pressure has been quantified in several kinds of inclusions in diamond, such as coesite, olivine, clinopyroxene, garnet, breyite, and solid nitrogen (Liu et al. 1990; Izraeli et al. 1999; Sobolev et al. 2000; Gillet et al. 2002; Nasdala et al. 2003; Barron et al. 2008; Howell et al. 2012; Smith et al. 2015; Navon et al. 2017; Anzolini et al. 2018).

Coesite. The strongest band of coesite at 520.6 cm^{-1} in an unstressed crystal shifts by +2.9 (±0.1) cm^{-1}/GPa under hydrostatic compression (Hemley 1987). Beginning with Sobolev et al. (2000), this calibrated pressure-induced band shift has been used for inclusion barometry in several studies. It is good practice to check multiple coesite bands to ensure they exhibit similar pressure-induced shifts in their position (e.g., Table 3 in Hemley 1987). Otherwise, if the band shifts are incoherent, it suggests anisotropic stresses and the shift of the 520.6 cm^{-1} band alone cannot be simply translated into an inclusion pressure. Calculation of trapping conditions of coesite inclusions based on remnant pressure often yields depths shallower than the diamond stability field, possibly due to anisotropic effects or plastic deformation around inclusions (Howell et al. 2010).

Breyite. An experimental calibration of pressure-induced Raman shifts up to 7.5 GPa under hydrostatic conditions was accompanied by ab initio calculations to assess the influence of non-hydrostatic conditions (Anzolini et al. 2018). The latter considerations revealed that some bands in the spectrum are more strongly influenced by non-hydrostatic conditions than others. The 977 cm^{-1} band of breyite was proposed to be the least sensitive to deviatoric stresses and most appropriate for determining the residual pressure, using its calibrated shift of +5.16(±0.09) cm^{-1} / GPa (Anzolini et al. 2018). It is notable that although many breyite inclusions are interpreted to be of sublithospheric origin in the transition zone or lower mantle, the remnant pressures in most breyite inclusions are relatively modest. The greatest band shifts yet observed in breyite inclusions are only sufficient to constrain the minimum inclusion entrapment depth to about 240–280 km (Anzolini et al. 2018; Smith et al. 2018). Brittle and plastic deformation of the surrounding diamond host could be responsible for some of this discrepancy. Some inclusions also have textural evidence of expansion of inclusion material out into decompression cracks (e.g., Smith et al. 2017, 2018), which could violate the assumption of a constant inclusion volume.

CONCLUDING REMARKS

Raman spectroscopy is well-suited as a reconnaissance tool for inclusion characterization and can be complementary to other in-situ methods such as FTIR, micro-beam XRD, X-ray CT, and synchrotron-XRF. Basic identification of mineral inclusions is often straightforward, especially by referring to the spectra of those minerals that have been found previously in diamond, or even those that are anticipated to occur in the deep mantle based on high-pressure experiments (Smith 2021). When used in conjunction with techniques to measure inclusion composition, such as EPMA analysis, or *in situ* with synchrotron-XRF (Laforce et al. 2014; De Pauw et al. 2020), Raman spectroscopy provides a simple way to confirm the crystal structure of mineral phases. The technique can also reveal hidden phases that might otherwise go unnoticed, such as molybdenite in sulfide inclusions (Kemppinen et al. 2018), or invisible layers of fluid trapped at the interface between solid inclusions and the host diamond (e.g., Nimis et al. 2016; Smith et al. 2018). In addition to identifying trapped phases, Raman data can reveal information about stresses in inclusions and permit calculation of minimum trapping conditions. The speed, low cost, minimal sample preparation, and capability to analyze inclusions in-situ make Raman spectroscopy a valuable method in the analytical toolkit for diamond research.

ACKNOWLEDGEMENTS

The authors wish to thank L. Nasdala and A. Steele for their careful reviews, as well as S. Shirey and K. Smit for their editorial handling. Thanks to F. Nestola for discussion and providing the spectrum for jeffbenite, and to P. Diggle for discussion. F.E.B. would like to thank the DFG (project BR2015/36-1) and the Dr. Rolf M. Schwiete Stiftung Foundation for financial support.

REFERENCES

Afonso JC, Zlotnik S (2011) The subductability of continental lithosphere: the before and after story. *In*: Arc-Continent Collision, Frontiers in Earth Sciences. Vol 4. Brown D, Ryan PD (eds). Springer, Berlin, Heidelberg, p 53–86

Akaogi M, Hamada Y, Suzuki T, Kobayashi M, Okada M (1999) High pressure transitions in the system $MgAl_2O_4$–$CaAl_2O_4$: a new hexagonal aluminous phase with implication for the lower mantle. Phys Earth Planet Inter 115:67–77

Akaogi M, Yano M, Tejima Y, Iijima M, Kojitani H (2004) High-pressure transitions of diopside and wollastonite: phase equilibria and thermochemistry of $CaMgSi_2O_6$, $CaSiO_3$ and $CaSi_2O_5$–$CaTiSiO_5$ system. Phys Earth Planet Inter 143–144:145–156

Angel R (1997) Transformation of fivefold-coordinated silicon to octahedral silicon in calcium silicate, $CaSi_2O_5$. Am Mineral 82:836–839

Angel RJ, Mazzucchelli ML, Alvaro M, Nimis P, Nestola F (2014) Geobarometry from host-inclusion systems: The role of elastic relaxation. Am Mineral 99:2146–2149

Angel R, Alvaro M, Nestola F, Mazzucchelli M (2015) Diamond thermoelastic properties and implications for determining the pressure of formation of diamond-inclusion systems. Russ Geol Geophys 56:211–220

Angel RJ, Mazzucchelli ML, Alvaro M, Nestola F (2017) EosFit-Pinc: A simple GUI for host-inclusion elastic thermobarometry. Am Mineral 102:1957–1960

Anzolini C, Angel R, Merlini M, Derzsi M, Tokár K, Milani S, Krebs M, Brenker F, Nestola F, Harris J (2016) Depth of formation of $CaSiO_3$-walstromite included in super-deep diamonds. Lithos 265:138–147

Anzolini C, Prencipe M, Alvaro M, Romano C, Vona A, Lorenzon S, Smith EM, Brenker FE, Nestola F (2018) Depth of formation of super-deep diamonds: Raman barometry of $CaSiO_3$-walstromite inclusions. Am Mineral 103:69–74

Anzolini C, Marquardt K, Stagno V, Bindi L, Frost DJ, Pearson DG, Harris JW, Hemley RJ, Nestola F (2020) Evidence for complex iron oxides in the deep mantle from FeNi(Cu) inclusions in superdeep diamond. PNAS:202004269

Barkley MC, Downs RT, Yang H (2011) Structure of walstromite, $BaCa_2Si_3O_9$, and its relationship to $CaSiO_3$-walstromite and wollastonite-II. Am Mineral 96:797

Barron LM, Barron BJ, Mernagh TP, Birch WD (2008) Ultrahigh pressure macro diamonds from Copeton (New South Wales, Australia), based on Raman spectroscopy of inclusions. Ore Geol Rev 34:76–86

Beck P, Gillet P, Gautron L, Daniel I, El Goresy A (2004) A new natural high-pressure (Na,Ca)-hexaluminosilicate $[(Ca_xNa_{1-x})Al_{3+x}Si_{3-x}O_{11}]$ in shocked Martian meteorites. Earth Planet Sci Lett 219:1–12

Bersani D, Andò S, Vignola P, Moltifiori G, Marino I-G, Lottici PP, Diella V (2009) Micro-Raman spectroscopy as a routine tool for garnet analysis. Spectrochim Acta Part A 73:484–491

Binns RA, Davis RJ, Reed SJB (1969) Ringwoodite, natural $(Mg,Fe)_2SiO_4$ spinel in the Tenham meteorite. Nature 221:943

Bodnar RJ, Frezzotti ML (2020) Microscale chemistry: raman analysis of fluid and melt inclusions. Elements 16:93–98

Brenker FE, Stachel T, Harris JW (2002) Exhumation of lower mantle inclusions in diamond: ATEM investigation of retrograde phase transitions, reactions and exsolution. Earth Planet Sci Lett 198:1–9

Brenker FE, Vincze L, Vekemans B, Nasdala L, Stachel T, Vollmer C, Kersten M, Somogyi A, Adams F, Joswig W (2005) Detection of a Ca-rich lithology in the Earth's deep (> 300 km) convecting mantle. Earth Planet Sci Lett 236:579–587

Brenker FE, Vollmer C, Vincze L, Vekemans B, Szymanski A, Janssens K, Szaloki I, Nasdala L, Joswig W, Kaminsky F (2007) Carbonates from the lower part of transition zone or even the lower mantle. Earth Planet Sci Lett 260:1–9

Brenker FE, Nestola F, Brenker L, Peruzzo L, Harris JW (2021) Origin, properties and structure of breyite: The second most abundant mineral inclusion in super-deep diamonds. Am Mineral 106:38–43

Brey GP, Bulatov V, Girnis A, Harris JW, Stachel T (2004) Ferropericlase—a lower mantle phase in the upper mantle. Lithos 77:655–663

Bulanova GP, Walter MJ, Smith CB, Kohn SC, Armstrong LS, Blundy J, Gobbo L (2010) Mineral inclusions in sublithospheric diamonds from Collier 4 kimberlite pipe, Juina, Brazil: Subducted protoliths, carbonated melts and primary kimberlite magmatism. Contrib Mineral Petrol 160:489–510

Burke EAJ (2001) Raman microspectrometry of fluid inclusions. Lithos 55:139–158

Caracas R, Bobocioiu E (2011) The WURM project—a freely available web-based repository of computed physical data for minerals. Am Mineral 96:437–443

D'Ippolito V, Andreozzi GB, Bersani D, Lottici PP (2015) Raman fingerprint of chromate, aluminate and ferrite spinels. J Raman Spectrosc 46:1255–1264

de Faria DLA, Venâncio Silva S, de Oliveira MT (1997) Raman microspectroscopy of some iron oxides and oxyhydroxides. J Raman Spectrosc 28:873–878

De Pauw E, Tack P, Lindner M, Ashauer A, Garrevoet J, Vekemans B, Falkenberg G, Brenker FE, Vincze L (2020) Highly sensitive nondestructive rare earth element detection by means of wavelength-dispersive X-ray fluorescence spectroscopy enabled by an energy dispersive pn-charge-coupled-device detector. Anal Chem 92:1106–1113

Dörsam G, Liebscher A, Wunder B, Franz G, Gottschalk M (2009) Crystal structure refinement of synthetic $Ca_{0.43}Sr_{0.57}[SiO_3]$-walstromite and walstromite–fluid Ca–Sr distribution at upper-mantle conditions. Eur J Mineral 21:705–714

Elements (2020) Raman spectroscopy in earth and planetary sciences. Pasteris JD, Beyssac O (eds) Elements 16:77–148

Everall NJ (2009) Confocal Raman microscopy: performance, pitfalls, and best practice. Appl Spectrosc 63:245a–262a

Fedoraeva AS, Shatskiy A, Litasov KD (2019) The join $CaCO_3$–$CaSiO_3$ at 6 GPa with implication to Ca-rich lithologies trapped by kimberlitic diamonds. High Press Res 39:547–560

Frezzotti ML, Tecce F, Casagli A (2012) Raman spectroscopy for fluid inclusion analysis. J Geochem Explor 112:1–20

Fries M, Steele A (2010) Raman spectroscopy and confocal Raman imaging in mineralogy and petrography. *In:* Confocal Raman Microscopy. Springer, p 111–135

Gasparik T, Wolf K, Smith CM (1994) Experimental determination of phase relations in the $CaSiO_3$ system from 8 to 15 GPa. Am Mineral 79:1219–1222

Gasparik T, Tripathi A, Parise JB (2000) Structure of a new Al-rich phase, $[K, Na]_{0.9}[Mg, Fe]_2[Mg, Fe, Al, Si]_6O_{12}$, synthesized at 24 GPa. Am Mineral 85:613–618

Gautron L, Kesson SE, Hibberson WO (1996) Phase relations for $CaAl_2Si_2O_8$ (anorthite composition) in the system CaO–Al_2O_3–SiO_2 at 14 GPa. Phys Earth Planet Inter 97:71–81

Gillet P, Sautter V, Harris J, Reynard B, Harte B, Kunz M (2002) Raman spectroscopic study of garnet inclusions in diamonds from the mantle transition zone. Am Mineral 87:312–317

Grimsditch MH, Anastassakis E, Cardona M (1978) Effect of uniaxial stress on the zone-center optical phonon of diamond. Phys Rev B: Condens Matter 18:901–904

Gu T, Wang W (2018) Optical defects in milky type IaB diamonds. Diamond Relat Mater 89:322–329

Gurney JJ, Helmstaedt HH, Richardson SH, Shirey SB (2010) Diamonds through time. Econ Geol 105:689–712

Haggerty SE (1999) A diamond trilogy: superplumes, supercontinents, and supernovae. Science 285:851–860

Harte B (2010) Diamond formation in the deep mantle: the record of mineral inclusions and their distribution in relation to mantle dehydration zones. Mineral Mag 74:189–215

Harte B, Hudson NFC (2013) Mineral associations in diamonds from the lowermost upper mantle and uppermost lower mantle. Proceedings of 10th International Kimberlite Conference Vol 1, p 235–253

Hayman P, Kopylova M, Kaminsky F (2005) Lower mantle diamonds from Rio Soriso (Juina area, Mato Grosso, Brazil). Contrib Mineral Petrol 149:430–445

Hemley RJ (1987) Pressure dependence of Raman spectra of SiO_2 polymorphs: α-quartz, coesite, and stishovite. *In*: High-Pressure Research in Mineral Physics. Vol 39. Manghnani MH, Syono Y, (eds). AGU, Washington, D. C., p 347–359

Hirose K, Takafuji N, Sata N, Ohishi Y (2005) Phase transition and density of subducted MORB crust in the lower mantle. Earth Planet Sci Lett 237:239–251

Hope GA, Woods R, Munce CG (2001) Raman microprobe mineral identification. Miner Eng 14:1565–1577

Howell D, Wood IG, Dobson DP, Jones AP, Nasdala L, Harris JW (2010) Quantifying strain birefringence halos around inclusions in diamond. Contrib Mineral Petrol 160:705–717

Howell D, Wood IG, Nestola F, Nimis P, Nasdala L (2012) Inclusions under remnant pressure in diamond: a multi-technique approach. Eur J Mineral 24:563–573

Irifune T, Ringwood AE (1987) Phase transformations in primitive morb and pyrolite compositions to 25 GPa and some geophysical implications. *In*: High-Pressure Research in Mineral Physics. Manghnani MH, Syono Y, (eds). Tokyo/American Geophysical Union, Terra Scientific Publishing Company, Washington DC, p 231–242

Irifune T, Ringwood AE (1993) Phase transformations in subducted oceanic crust and buoyancy relationships at depths of 600–800 km in the mantle. Earth Planet Sci Lett 117:101–110

Irifune T, Ringwood AE, Hibberson WO (1994) Subduction of continental crust and terrigenous and pelagic sediments: an experimental study. Earth Planet Sci Lett 126:351–368

Izraeli ES, Harris JW, Navon O (1999) Raman barometry of diamond formation. Earth Planet Sci Lett 173:351–360

Joswig W, Paulus EF, Winkler B, Milman V (2003) The crystal structure of $CaSiO_3$-walstromite, a special isomorph of wollastonite-II. Z Kristallogr Cryst Mater 218:811

Kalugina AD, Zedgenizov DA (2020) Raman discrimination of garnet inclusions in Siberian diamonds. J Raman Spectrosc 51:1438–1444

Kaminsky F (2012) Mineralogy of the lower mantle: A review of 'super-deep' mineral inclusions in diamond. Earth Sci Rev 110:127–147

Kaminsky FV, Wirth R (2011) Iron carbide inclusions in lower-mantle diamond from Juina, Brazil. Can Mineral 49:555–572

Kaminsky F, Wirth R, Matsyuk S, Schreiber A, Thomas R (2009) Nyerereite and nahcolite inclusions in diamond: evidence for lower-mantle carbonatitic magmas. Mineral Mag 73:797–816

Kawashima Y, Katagiri G (1999) Observation of the out-of-plane mode in the Raman scattering from the graphite edge plane. Phys Rev B Condens Matter 59:62–64

Kemppinen LI, Kohn SC, Parkinson IJ, Bulanova GP, Howell D, Smith CB (2018) Identification of molybdenite in diamond-hosted sulphide inclusions: Implications for Re–Os radiometric dating. Earth Planet Sci Lett 495:101–111

Kim Y, Lee EJ, Roy S, Sharbirin AS, Ranz L-G, Dieing T, Kim J (2020) Measurement of lateral and axial resolution of confocal Raman microscope using dispersed carbon nanotubes and suspended graphene. Curr Appl Phys 20:71–77

Koivula JI (2000) The MicroWorld of Diamonds. Gemworld International, Inc., Northbrook, IL

Kolesov BA, Geiger CA (1998) Raman spectra of silicate garnets. Phys Chem Miner 25:142–151

Komabayashi T, Omori S (2006) Internally consistent thermodynamic data set for dense hydrous magnesium silicates up to 35 GPa, 1600 °C: Implications for water circulation in the Earth's deep mantle. Phys Earth Planet Inter 156:89–107

Konzett J, Frost DJ (2009) The high *P–T* stability of hydroxyl-apatite in natural and simplified MORB—an experimental study to 15 GPa with implications for transport and storage of phosphorus and halogens in subduction zones. J Petrol 50:2043–2062

Konzett J, Rhede D, Frost DJ (2011) The high PT stability of apatite and Cl partitioning between apatite and hydrous potassic phases in peridotite: an experimental study to 19 GPa with implications for the transport of P, Cl and K in the upper mantle. Contrib Mineral Petrol 163:277–296

Kubo A, Suzuki T, Akaogi M (1997) High pressure phase equilibria in the system $CaTiO_3$–$CaSiO_3$: stability of perovskite solid solutions. Phys Chem Miner 24:488–494

Kudoh Y, Kanzaki M (1998) Crystal chemical characteristics of α-$CaSi_2O_5$, a new high pressure calcium silicate with five-coordinated silicon synthesized at 1500°C and 10 GPa. Phys Chem Miner 25:429–433

Laforce B, Schmitz S, Vekemans B, Rudloff J, Garrevoet J, Tucoulou R, Brenker FE, Martinez-Criado G, Vincze L (2014) Nanoscopic X-ray fluorescence imaging of meteoritic particles and diamond inclusions. Anal Chem 86:12369–12374

Lafuente B, Downs RT, Yang H, Stone N (2015) The power of databases: The RRUFF project. *In*: Highlights in Mineralogical Crystallography. Armbruster T, Danisi RM (eds). W. De Gruyter, Berlin, p 1–30

Litasov KD, Ohtani E (2007) Effect of water on the phase relations in Earth's mantle and deep water cycle. *In*: Geological Society of America Special Papers. Vol 421. Ohtani E (ed), p 115–156

Liu L-g, Mernagh TP, Jaques AL (1990) A mineralogical Raman spectroscopy study on eclogitic garnet inclusions in diamonds from Argyle. Contrib Mineral Petrol 105:156–161

Logvinova A, Wirth R, Fedorova E, Sobolev N (2008) Nanometre-sized mineral and fluid inclusions in cloudy Siberian diamonds: new insights on diamond formation. Eur J Mineral 20:317–331

Ma C, Tschauner O, Beckett JR, Rossman GR, Prescher C, Prakapenka VB, Bechtel HA, MacDowell A (2018) Liebermannite, $KAlSi_3O_8$, a new shock-metamorphic, high-pressure mineral from the Zagami Martian meteorite. Meteorit Planet Sci 53:50–61

Mazzucchelli ML, Burnley P, Angel RJ, Morganti S, Domeneghetti MC, Nestola F, Alvaro M (2018) Elastic geothermobarometry: corrections for the geometry of the host–inclusion system. Geology 46:231–234

McMillan P, Akaogi M, Ohtani E, Williams Q, Nieman R, Sato R (1989) Cation disorder in garnets along the $Mg_3Al_2Si_3O_{12}$–$Mg_4Si_4O_{12}$ join: an infrared, Raman and NMR study. Phys Chem Miner 16:428–435

Mernagh TP, Trudu AG (1993) A laser Raman microprobe study of some geologically important sulphide minerals. Chem Geol 103:113–127

Murayama JK, Nakai S, Kato M, Kumazawa M (1986) A dense polymorph of $Ca_3(PO_4)_2$: a high pressure phase of apatite decomposition and its geochemical significance. Phys Earth Planet Inter 44:293–303

Murri M, Mazzucchelli ML, Campomenosi N, Korsakov AV, Prencipe M, Mihailova BD, Scambelluri M, Angel RJ, Alvaro M (2018) Raman elastic geobarometry for anisotropic mineral inclusions. Am Mineral 103:1869–1872

Nasdala L, Schmidt C (2020) Applications of Raman spectroscopy in mineralogy and geochemistry. Elements 16:99–104

Nasdala L, Brenker FE, Glinnemann J, Hofmeister W, Gasparik T, Harris JW, Stachel T, Reese I (2003) Spectroscopic 2D-tomography: Residual pressure and strain around mineral inclusions in diamonds. Eur J Mineral 15:931–935

Nasdala L, Hofmeister W, Harris JW, Glinnemann J (2005) Growth zoning and strain patterns inside diamond crystals as revealed by Raman maps. Am Mineral 90:745–748

Navon O, Wirth R, Schmidt C, Jablon BM, Schreiber A, Emmanuel S (2017) Solid molecular nitrogen (δ-N_2) inclusions in Juina diamonds: Exsolution at the base of the transition zone. Earth Planet Sci Lett 464:237–247

Németh P, Leinenweber K, Ohfuji H, Groy T, Domanik KJ, Kovács IJ, Kovács JS, Buseck PR (2017) Water-bearing, high-pressure Ca-silicates. Earth Planet Sci Lett 469:148–155

Nestola F (2017) Inclusions in super-deep diamonds: windows on the very deep Earth. Rend Lincei-Mat Appl 28:595–604

Nestola F, Burnham AD, Peruzzo L, Tauro L, Alvaro M, Walter MJ, Gunter M, Anzolini C, Kohn SC (2016) Tetragonal Almandine-Pyrope Phase, TAPP: finally a name for it, the new mineral jeffbenite. Mineral Mag 80:1219–1232

Nestola F, Korolev N, Kopylova M, Rotiroti N, Pearson DG, Pamato MG, Alvaro M, Peruzzo L, Gurney JJ, Moore AE, Davidson J (2018) $CaSiO_3$ perovskite in diamond indicates the recycling of oceanic crust into the lower mantle. Nature 555:237–241

Neuville DR, de Ligny D, Henderson GS (2014) Advances in Raman spectroscopy applied to earth and material sciences. Rev Mineral Geochem 78:509–541

Nimis P, Alvaro M, Nestola F, Angel RJ, Marquardt K, Rustioni G, Harris JW, Marone F (2016) First evidence of hydrous silicic fluid films around solid inclusions in gem-quality diamonds. Lithos 260:384–389

Ohtani E, Litasov K, Suzuki A, Kondo T (2001) Stability field of new hydrous phase, δ-AlOOH, with implications for water transport into the deep mantle. Geophys Res Lett 28:3991–3993

Ohtani E, Hirao N, Kondo T, Ito M, Kikegawa T (2005) Iron-water reaction at high pressure and temperature, and hydrogen transport into the core. Phys Chem Miner 32:77–82

Ono S, Ohishi Y, Isshiki M, Watanuki T (2005) In situ X-ray observations of phase assemblages in peridotite and basalt compositions at lower mantle conditions: Implications for density of subducted oceanic plate. J Geophys Res Solid Earth 110:B02208

Palot M, Jacobsen SD, Townsend J, Nestola F, Marquardt K, Miyajima N, Harris J, Stachel T, McCammon C, Pearson D (2016) Evidence for H_2O-bearing fluids in the lower mantle from diamond inclusion. Lithos 265:237–243

Pasteris JD, Beyssac O (2020) Welcome to Raman spectroscopy: Successes, challenges, and pitfalls. Elements 16:87–92

Pearson DG, Brenker FE, Nestola F, McNeill J, Nasdala L, Hutchison MT, Matveev S, Mather K, Silversmit G, Schmitz S, Vekemans B (2014) Hydrous mantle transition zone indicated by ringwoodite included within diamond. Nature 507:221–224

Ringwood AE (1958) Olivine–spinel transition in fayalite. GSA Bull 69:129–130

Rudloff-Grund J, Brenker F, Marquardt K, Howell D, Schreiber A, O'Reilly S, Griffin W, Kaminsky F (2016) Nitrogen nanoinclusions in milky diamonds from Juina area, Mato Grosso State, Brazil. Lithos 265:57–67

Schiferl D, Nicol M, Zaug JM, Sharma SK, Cooney TF, Wang S-Y, Anthony TR, Fleischer JF (1997) The diamond $^{13}C/^{12}C$ isotope Raman pressure sensor system for high-temperature/pressure diamond-anvil cells with reactive samples. J Appl Phys 82:3256–3265

Sharma SK, Mao HK, Bell PM, Xu JA (1985) Measurement of stress in diamond anvils with micro-Raman spectroscopy. J Raman Spectrosc 16:350–352

Shirey SB, Cartigny P, Frost DJ, Keshav S, Nestola F, Nimis P, Pearson DG, Sobolev NV, Walter MJ (2013) Diamonds and the geology of mantle carbon. Rev Mineral Geochem 75:355–421

Shirey S, Smit K, Pearson D, Walter M, Aulbach S, Brenker F, Bureau H, Burnham A, Cartigny P, Chacko T, Frost D (2019) Diamonds and the mantle geodynamics of carbon: Deep mantle carbon evolution from the diamond record. *In*: Deep Carbon: Past to Present. Orcutt BN, Daniel I, Dasgupta R, (eds). Cambridge University Press, Cambridge, p 89–128

Smit KV, Shirey SB, Stern RA, Steele A, Wang W (2016) Diamond growth from C–H–N–O recycled fluids in the lithosphere: Evidence from CH_4 micro-inclusions and $\delta^{13}C$–$\delta^{15}N$–N content in Marange mixed-habit diamonds. Lithos 265:68–81

Smith EM (2021) Raman spectra catalogue for inclusions in diamond. https://doi.org/10.7939/DVN/JEHGBW

Smith EM, Wang W (2017) Diamond with concentric inclusions. Gems Gemol 53:228

Smith EM, Kopylova MG, Frezzotti ML, Afanasiev VP (2014) N-rich fluid inclusions in octahedrally-grown diamond. Earth Planet Sci Lett 393:39–48

Smith EM, Kopylova MG, Frezzotti ML, Afanasiev VP (2015) Fluid inclusions in Ebelyakh diamonds: Evidence of CO_2 liberation in eclogite and the effect of H_2O on diamond habit. Lithos 216–217:106–117

Smith EM, Vendrell C, Johnson P (2016a) Coesite inclusions with filaments in diamond. Gems Gemol 52:410–412

Smith EM, Shirey SB, Nestola F, Bullock ES, Wang J, Richardson SH, Wang W (2016b) Large gem diamonds from metallic liquid in Earth's deep mantle. Science 354:1403–1405

Smith EM, Shirey SB, Wang W (2017) The very deep origin of the world's biggest diamonds. Gems Gemol 53:388–403

Smith EM, Shirey SB, Richardson SH, Nestola F, Bullock ES, Wang J, Wang W (2018) Blue boron-bearing diamonds from Earth's lower mantle. Nature 560:84–87

Sobolev NV, Fursenko BA, Goryainov SV, Shu J, Hemley RJ, Mao A, Boyd FR (2000) Fossilized high pressure from the Earth's deep interior: the coesite-in-diamond barometer. PNAS 97:11875–11879

Stachel T (2014) Diamond. *In*: Geology of Gem Deposits, Second Edition. Groat LA (ed) Mineralogical Association of Canada, p 1–28

Stachel T, Harris JW (2008) The origin of cratonic diamonds – Constraints from mineral inclusions. Ore Geol Rev 34:5–32
Stachel T, Harris JW, Brey GP, Joswig W (2000) Kankan diamonds (Guinea) II: lower mantle inclusion parageneses. Contrib Mineral Petrol 140:16–27
Stachel T, Brey G, Harris JW (2005) Inlcusions in sublithospheric diamonds: glimpses of deep earth. Elements 1:73–78
Suzuki A, Ohtani E, Kamada T (2000) A new hydrous phase δ-AlOOH synthesized at 21 GPa and 1000 °C. Phys Chem Miner 27:689–693
Tappert R, Tappert MC (2011) Diamonds in nature: a guide to rough diamonds. Springer Science & Business Media
Thomson AR, Kohn SC, Bulanova GP, Smith CB, Araujo D, Walter MJ (2016) Trace element composition of silicate inclusions in sub-lithospheric diamonds from the Juina-5 kimberlite: Evidence for diamond growth from slab melts. Lithos 265:108–124
Tomilenko AA, Ragozin AL, Shatskii VS, Shebanin AP (2001) Variation in the fluid phase composition in the process of natural diamond crystallization. Dokl Earth Sci 379:571–574
Tschauner O (2019) High-pressure minerals. Am Mineral 104:1701–1731
Tschauner O, Huang S, Greenberg E, Prakapenka VB, Ma C, Rossman GR, Shen AH, Zhang D, Newville M, Lanzirotti A, Tait K (2018) Ice-VII inclusions in diamonds: Evidence for aqueous fluid in Earth's deep mantle. Science 359:1136–1139
Urakawa S, Kondo T, Igawa N, Shimomura O, Ohno H (1994) Synchrotron radiation study on the high-pressure and high-temperature phase relations of $KAlSi_3O_8$. Phys Chem Miner 21:387–391
von Huene R, Scholl DW (1991) Observations at convergent margins concerning sediment subduction, subduction erosion, and the growth of continental crust. Rev Geophys 29:279–316
Walter MJ, Bulanova GP, Armstrong LS, Keshav S, Blundy JD, Gudfinnsson G, Lord OT, Lennie AR, Clark SM, Smith CB, Gobbo L (2008) Primary carbonatite melt from deeply subducted oceanic crust. Nature 454:622–625
Walter MJ, Kohn SC, Araujo D, Bulanova GP, Smith CB, Gaillou E, Wang J, Steele A, Shirey SB (2011) Deep mantle cycling of oceanic crust: Evidence from diamonds and their mineral inclusions. Science 334:54–57
Woodland AB, Girnis AV, Bulatov VK, Brey GP, Höfer HE (2020) Breyite inclusions in diamond: experimental evidence for possible dual origin. Eur J Mineral 32:171–185
Workman RK, Hart SR, Jackson M, Regelous M, Farley KA, Blusztajn J, Kurz M, Staudigel H (2004) Recycled metasomatized lithosphere as the origin of the Enriched Mantle II (EM2) end-member: Evidence from the Samoan volcanic chain. Geochem Geophys Geosyst 5:Q04008
Wunder B, Rubie DC, Ross CR, Medenbach O, Seifert F, Schreyer W (1993) Synthesis, stability, and properties of $Al_2SiO_4(OH)_2$: A fully hydrated analogue of topaz. Am Mineral 78:285–297
Wunder B, Andrut M, Wirth R (1999) High-pressure synthesis and properties of OH-rich topaz. Eur J Mineral 11:803–813
Xie X, Minitti ME, Chen M, Mao H-K, Wang D, Shu J, Fei Y (2003) Tuite, γ-$Ca_3(PO_4)_2$: a new mineral from the Suizhou L6 chondrite. Eur J Mineral 15:1001–1005
Xue X, Kanzaki M, Fukui H, Ito E, Hashimoto T (2006) Cation order and hydrogen bonding of high-pressure phases in the Al_2O_3–SiO_2–H_2O system: An NMR and Raman study. Am Mineral 91:850–861
Xue X, Kanzaki M, Fukui H (2010) Unique crystal chemistry of two polymorphs of topaz-OH: A multi-nuclear NMR and Raman study. Am Mineral 95:1276
Zhai S, Yamazaki D, Xue W, Ye L, Xu C, Shan S, Ito E, Yoneda A, Yoshino T, Guo X, Shimojuku A (2013) P–V–T relations of γ-$Ca_3(PO_4)_2$ tuite determined by in situ X-ray diffraction in a large-volume high-pressure apparatus. Am Mineral 98:1811

Reviews in Mineralogy & Geochemistry
Vol. 88 pp. 475–532, 2022

Fluid Inclusions in Fibrous Diamonds

Yaakov Weiss, Janina Czas, Oded Navon

Institute of Earth Sciences
The Hebrew University
Jerusalem, 9190401
Israel
yakov.weiss@mail.huji.ac.il

**Now at Earth and Planets Lab, Carnegie Institution for Science jczas@carnegiescience.edu*

INTRODUCTION

Most diamonds are metasomatic minerals, which means they grew from fluids or melts that moved through solid mantle rocks and chemically interacted with them. The involvement of fluids was recognized in the early days of diamond research (e.g., Sobolev 1960; Kennedy and Nordie 1968; Meyer and Boyd 1972; Harte et al. 1980) and was mostly attributed to kimberlite melts (either the erupting kimberlite, or a deeper proto-kimberlite composition). Based on Rb–Sr and Sm–Nd systematics of garnet inclusions in diamonds, Richardson et al. (1984) suggested that it involves "introduction and entrapment of asthenosphere-derived alkali, LREE and CO_2 enriched interstitial melt, which remained liquid until the time of diamond crystallization". However, only sporadic reports on the occurrence of alkali-rich mineral inclusions or fluid- or melt-inclusions in diamonds were reported at that time. In later years, evidence for such fluids and the metasomatic origin of diamonds accumulated, including, the association of diamonds with other metasomatic minerals and with alteration zones in xenoliths (e.g., Schulze et al. 1996; Anand et al. 2004; Zedgenizov et al. 2018); trace element abundance patterns of garnet and clinopyroxene inclusions in diamonds that point to interaction with, or growth from melts that are extremely enriched in incompatible trace elements (Shimizu and Richardson 1987; Stachel and Harris 1997; Stachel et al. 2004 and references therein) and observations of fluids or melts in microinclusions in diamonds, which are the subject of this review.

Before proceeding to the observations of fluids in diamonds, it is necessary to define the terms 'fluid' and 'melt'. The definition of 'fluid' in dictionaries is commonly "capable of flowing" (Merriam-Webster Dictionary), or "a substance that flows" (Cambridge and Oxford Learner's dictionaries). Physics textbooks use a similar definition "a substance (or a phase) that flows under the application of a shearing stress". This definition includes both liquids and gases, as well as supercritical fluids (Landau and Lifshitz 1987). In inclusion studies, a similar broad definition was suggested by Roedder (1984) and Bodnar (2003). The latter explained that some researchers relate to the state of the inclusion at room temperature and exclude glassy inclusions. However, he suggests including all inclusions that were fluid at the time of entrapment. This definition covers both high- and low-density fluids (Bodnar 2003). High-density fluids (HDFs) include silicate, sulfide, metallic, carbonatitic or saline supersolidus melts and dense supercritical fluids that are above the second critical endpoint (the termination of the solidus in a P–T space; Wyllie and Ryabchikov 2000; Kessel et al. 2005). At lithospheric pressures and temperatures, low-density fluids would have densities of ~1000 kg/m^3. The most common are C–O–H-bearing supercritical fluids like H_2O, CO_2, CH_4, H_2, and their mixtures, together with the solute they carry (which can reach up to a few tens of percent). The values for the high-density ones are ~2000 kg/m^3 or more, but past the second critical endpoint, the transition between low-and high-density fluids is continuous.

1529-6466/22/0088-0009$10.00 (print)
1943-2666/22/0088-0009$10.00 (online)
http://dx.doi.org/10.2138/rmg.2022.88.09

This review concentrates on HDFs that are encapsulated in diamonds. It begins with the historical perspective of the research of fluids in diamonds. We then review the shape and texture of fibrous diamonds which commonly carry HDFs, including the description of the three main forms of such diamonds: 'fully fibrous' where the whole diamond consists of fibers (~10 μm in diameter) and grows to develop cuboid {100} faces; 'cloudy' where a cuboid fibrous core is surrounded by a non-fibrous, octahedral zone; and 'coated' where a clear core is surrounded by fibrous overgrowth. This section also includes a short review of the chemistry of the diamonds themselves, focusing on nitrogen speciation as well as carbon and nitrogen isotope composition. The next section describes the various analytical techniques employed to explore the nature of the trapped fluids. Here the attitude is historic, and in the case of analytical techniques that are not common, or that do not provide compositional information (e.g., X-ray diffraction or transmission electron microscopy), we also present a review of the results. The following section focuses on the geochemistry of the HDFs trapped in fibrous diamonds and their four compositional types: **silicic**, which are rich in SiO_2, K_2O, Al_2O_3 and water, and form a continuous array with **low-Mg carbonatitic** compositions, rich in CaO, FeO, MgO and CO_2; **High-Mg carbonatitic** HDFs are distinguished by their very high MgO (>15 wt%) and high CaO, FeO, K_2O and CO_2; **saline** HDFs have high K, Na and Cl values, and contain varying proportions of FeO, BaO, CO_2 and water. This section ends by reviewing the direct and indirect evidence for the occurrence of such HDFs and other fluids in non-fibrous diamonds.

The discussion begins with a section on the temporal and geographical distribution of HDFs and a possible connection between HDF type and lithospheric lithology. The following section deals with estimations of pressure and temperature of diamond growth and HDF entrapment, and the current hypotheses on the origin of the four compositional types. We then review the role of HDFs in the formation of diamonds, where the collection of observations to date suggest they are the media from which the majority of lithospheric diamonds grew. The final sections highlight that all HDF types are excellent metasomatic agents, which impact the chemical composition of the continental lithosphere and possibly other mantle environments, emphasize the close relation between high-Mg carbonatitic HDF type and kimberlites, and discuss the possible role of HDFs in alkaline basaltic magmatism.

FLUIDS IN DIAMONDS—EARLY OBSERVATIONS

The earliest mention of fluids in diamonds was, most probably, by Bauer in his book "*Edelsteinkunde*" (1896, translated into *Precious Stones*, 1904). He described observations of liquid and gas-bearing microscopic inclusions within diamonds (as quoted by Melton et al. 1972 and Fesq et al. 1975).

Early studies of the three types of microinclusion-bearing diamonds were conducted by Shah and Lang (1963), Kamiya and Lang (1965) and Moore and Lang (1972), who applied X-ray topography to study their internal structure: a) cloudy diamonds with an internal cloud of microinclusions, b) coated diamonds in which a clear octahedral core is surrounded by fibrous growth, and c) cuboid diamonds that consist mostly of fibrous growth, sometimes with a small core (Fig. 1). Orlov (1977) summarized optical studies of diamonds and designated the cuboid diamonds as Variety III and the coated diamonds as Variety IV.

Angress and Smith (1965) used infrared spectroscopy to investigate the coat of a coated diamond but reported mainly on the nitrogen concentration. In a subsequent study, Chrenko et al. (1967) detected characteristic bands of molecular water (~3400 and ~1640 cm^{-1}), carbonate (1430 and 880 cm^{-1}) and nitrate (1360 and 835 cm^{-1}) in the IR spectra of the opaque coat of a diamond, but not in its transparent core (the presence of nitrate was not confirmed by later studies).

Melton and Giardini (1974, 1975) reported the composition of gases released by crushing diamonds under vacuum. Gas signals of transparent or light-yellow diamonds were similar to the system blank levels ($0.2–2\times10^{-5}$ cm^3 STP, for an empty crushing chamber at 0 °C and 1 bar). However, the crushing of cubic and translucent diamonds at 200 °C yielded gas volumes that were an order of magnitude larger. The main gases released were H_2O and CO_2. Gas released by graphitization of diamonds contained mostly CO with minor H_2 and the quantities were $>10^{-2}$ cm^3 STP (Kaiser and Bond 1959; Melton and Giardini 1976), but the H_2 may be attributed to a reaction of water during graphitization (Fesq et al. 1975).

Early attempts to determine the nature of the material trapped in the microinclusions by X-ray diffraction and X-ray fluorescence were not conclusive (Lonsdale and Milledge 1965; Seal 1966, 1970; Harris 1968). Prinz et al. (1975) found "fluffy, filamentous material" in two diamonds from the Democratic Republic of the Congo (DRC), but did not describe the diamonds. The composition obtained by electron probe microanalysis (EPMA) of this material was SiO_2 67 wt%; TiO_2 2.5 wt%; Al_2O_3 14–17 wt%; FeO 2–3 wt%; MgO 1 wt%; Na_2O 0.2–2.5 wt% and K_2O 8.5–10 wt%. As mentioned below, this composition agrees roughly with that of later studies of microinclusion-bearing fibrous diamonds.

Fesq et al. (1975) used instrumental neutron activation analysis (INAA) to investigate the nature of impurities in diamonds. They examined groups of diamonds (~10–20 crystals in each sample) with mineral inclusions as well as diamonds with no visible inclusions. They concluded that in addition to mineral inclusions, the diamonds also carried picritic (high-Mg basaltic) melt associated with an H_2O+CO_2 rich component or phase that carried incompatible trace elements. The presence of a sulfide-rich phase was inferred from the correlation between Fe, Ni, Cu and Co. These impurities were found in diamonds with and without visible inclusions.

INAA of coated diamonds was attempted by Kodochigov and co-workers (according to Orlov 1977) and Glazunov et al. (1967). Bibby (1979) used INAA to determine the trace-element composition of a coated diamond. He found low levels in the core, but the coat carried up to 50 ppm Fe and a few ppm of Na, K, Ba and Ce. Based on the good intercorrelation of K, Na, Ba, and the REE and their poor correlation with Sc, Cr, and Mn, he suggested the presence of carbonate microinclusions.

FIBROUS DIAMONDS

Morphology and texture

Fibrous diamonds are imperfect single crystals, consisting of parallel microscopic fibers that are slightly misaligned from the diamond crystallographic axes. They have one set of crystal faces and they produce a single X-ray diffraction pattern (Orlov et al. 1980; Smith et al. 2011). However, they are distinct from the non-fibrous monocrystalline diamonds (i.e., octahedral, cubic and macles) that grew layer by layer with a higher degree of perfection. Although fibrous diamonds are monocrystalline, in view of these differences and the absence of a better distinction, in the present paper we will use the term 'non-fibrous monocrystalline' (NFMC) when addressing other monocrystalline varieties as opposed to fibrous ones throughout the paper.

Optical microscopy is a straightforward method for identifying fibrous growth. Microinclusions trapped in the fibrous zones result in a translucent or opaque appearance of the diamond. Under high magnification, the tiny microinclusions (<1 μm) can be observed (but not their internal structure; Fig. 1d). Crossed nicols reveal high birefringence by the diamond due to stresses that may be the result of the fibrous growth or the presence of microinclusions.

Fibrous growth leads to the development of cuboid faces (Lang 1974a), thus 'fully fibrous' diamonds have a cubic shape (Fig. 1a). If octahedral, layer by layer growth develops

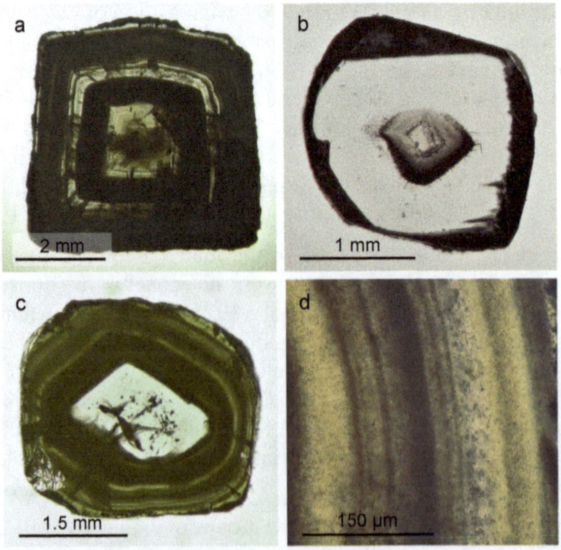

Figure 1. Photomicrographs of polished slabs of microinclusion-bearing fibrous diamonds. (**a**) 'fully fibrous' cuboid diamond from Canada (exact mine is unknown). The concentric layers define variation in microinclusion-density from center to edge. In this specific diamond (and in most fibrous diamonds) the composition of the microinclusions is homogeneous and does not show chemical radial variation, indicating it formed in pulses, but during a single growth event. (**b**) 'cloudy' diamond from Finsch, South Africa (Weiss and Goldstein 2018); with a cuboid internal zone of fibrous growth that is full of microinclusions, overgrown by a clear octahedral outer zone. (**c**) 'coated' diamond from Kankan, Guinea (Weiss et al. 2009); consists of an octahedral non-fibrous monocrystalline (NFMC) core and a fibrous coat. In this diamond, the concentric layers in the coat are characterized by different microinclusion compositions (i.e., a zoned diamond). (**d**) high magnification image of the fibrous coat in (c), showing thousands of microinclusions (black dots). See section '*Morphology and Texture' and 'Homogeneity and zoning in individual diamonds*' for additional details.

on a fibrous diamond, a 'cloudy' diamond forms with a cuboid, fibrous, inclusion-rich, translucent or opaque inner zone and an outer transparent NFMC zone (Fig. 1b). External fibrous overgrowth on an existing NFMC diamond produces a 'coated diamond' where the triangular, octahedral {111} faces conserve their original size, while the cubic {100} faces expand with the growing thickness of the coat (Fig. 1c).

X-ray topography can be used to reveal the details of the fibrous texture (e.g., Lang 1974a,b; Yelisseyev et al. 2004; Zedgenizov et al. 2006; Rondeau et al. 2007; Skuzovatov et al. 2011). The fiber diameter varies from 10 μm down to the resolution of the technique (<1 μm). The fibers grow with an average orientation parallel to the <111> direction, but individual fibers may be misoriented by up to 0.5° (Lang 1974b). Branching allows the fibers to fill the full volume of the growing diamond and leads to the formation of cuboid surfaces (parallel to {100}; Moore and Lang 1972; Orlov et al. 1982; Rondeau et al. 2007). The microinclusions (<1 μm) are scattered between the fibers (Moore and Lang 1972).

The fibrous zones have weak cathodoluminescence (CL), but it is still useful to emphasize different stages of growth. Most studies record the total intensity of the luminescence, which is sufficient to produce a map of the diamond growth layers (similar to the growth rings of a tree). An example is Fig. 1c of Klein-BenDavid et al. (2004) that reveals a transparent core and 5 concentric growth layers. In coated diamonds, the NFMC core is usually brighter than the fibrous coat, whereas, in cloudy diamonds, the fibrous inner zone is darker than the NFMC outer one (e.g., Fig. 3 of Zedgenizov et al. 2006, but see Rondeau et al. 2007 for a counterexample).

In summary, fibrous diamonds or growth zones are defined by the presence of fine fibers, as revealed by X-ray topography. They commonly carry microinclusions in variable numbers, but may also be clear of inclusions. Branching of the fibers leads to space-filling and the development of cuboid external faces.

Nitrogen concentration and aggregation

Nitrogen is the main element that substitutes for carbon in the diamond lattice (see Green et al. 2022). Its concentration can reach levels of a few thousand ppm. Fibrous diamonds typically carry a high concentration of N (~60% with more than 500 ppm and 20% with more than 1000 ppm) that resides mostly in A-centers (two nitrogen atoms replacing two carbon atoms, see Green et al. 2022), but some fibrous diamonds show the yellow tint caused by C-centers (a single nitrogen atom substituting for a single carbon atom) and others carry both A- and B-centers (4 nitrogen atoms around a vacancy) and platelets (interstitial planes carrying carbon and nitrogen atoms; Speich et al. 2018). The concentration is determined from the IR spectrum of the diamond based on the unique absorption of each of the three centers (e.g., the CAXBD97.xls spreadsheet produced by Dr. David Fisher, Diamap of Howell et al. 2012 or Quiddit by Speich and Kohn 2020). When analyzing the spectrum of microinclusion-rich diamonds, care must be taken for the overlap between nitrogen absorption at 900–1400 cm^{-1} and the major peaks of the silicates and phosphates. Most fibrous diamonds carry only A-centers and allow an easy escape from the problem. However, when B- or C-centers are present or suspected, one cannot simply use the standard fits and more elaborate treatment of the baseline and the deconvolution of the various peaks is needed (e.g., Kopylova et al. 2010; Weiss et al. 2018).

Taylor et al. (1990, 1996) compiled experimental data on the systematics of nitrogen aggregation (Evans and Qi 1982; Clark and Davey 1984) with ages and thermometric data of natural diamonds, and quantified the aggregation kinetics. They showed that it is controlled by the residence temperature and the initial amount of nitrogen (assumed to be equal to the total nitrogen measured). Most fibrous diamonds, with only A-centers, had to reside in the mantle long enough to allow the merging of C-centers into A-centers, but not too long to avoid conversion of the A- into B-centers (Fig. 2). For example, according to Taylor's calibrations, a diamond that carries 1000 ppm nitrogen with 10% in C-centers could not have spent more than 10 Myr at 950 °C or 0.1 Myr at 1050 °C. Formation of a few percent of B-centers calls for longer times, ~100 Myr at 1150 °C and billions of years at ~1000 °C (Fig. 2). Figure 2 demonstrates that the estimation of absolute times is very imprecise (as the time axis is logarithmic), but the aggregation data allow for the separation of distinct growth events. In Fig. 2, an individual diamond would appear as a diagonal line representing the interdependence of time and temperature for a given nitrogen concentration and B/(A+B) ratio. A population of diamonds formed in a single event would appear as a group of closely spaced parallel lines (e.g., Kempe et al. 2021a).

Carbon and nitrogen isotopes

NFMC diamonds cover a wide range of carbon isotope composition ($\delta^{13}C$) with a distinct peak at −5‰ (relative to the PDB standard). Diamonds carrying eclogitic mineral inclusions peak at −5±1.2‰ (at half maximum), but about one-third of them have lower isotopic compositions (−10 to −42‰) and a few have more positive $\delta^{13}C$ values (0 to +3‰; Cartigny et al. 2014). Those with peridotitic inclusions have a somewhat wider peak of $\delta^{13}C$ at −4.5 ± 1.5‰ (Cartigny et al. 2014), but very few diamonds fall outside this range (most of the light grey area in Fig. 3c consists of eclogitic diamonds). Fibrous diamonds have more negative $\delta^{13}C$ values with a mode at −6.5 ± 1.2‰ (Fig. 3a,c). Early studies of fibrous diamonds found a narrow range between −6 and −8‰ (Galimov 1984; Javoy et al. 1984; Boyd et al. 1987; Cartigny et al. 2001), but later studies (Klein-BenDavid et al. 2004, 2007; Shiryaev et al. 2005; Zedgenizov et al. 2006, 2009, 2011, 2016, 2017; Skuzovatov et al. 2016; Timmerman et al. 2018a; Gubanov et al. 2019) widened the range, mainly towards heavier values (Fig. 3a).

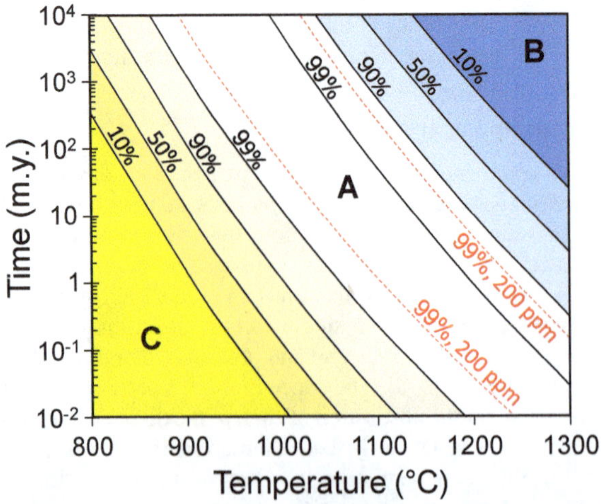

Figure 2. Mantle residence time–temperature (*t–T*) relations implied by the total nitrogen concentration and level of aggregation in diamonds. The **yellow region** at the lower left side shows the aggregation of nitrogen from C-centers (single nitrogen atom) to A-centers (paired nitrogen atoms), and the blue region in the upper right side indicates the combination of A-centers to form B-centers (aggregates of four nitrogen atoms surrounding a vacancy), as a function of *t–T*. The **black solid lines** are calculated for a total nitrogen concentration of 1000 atomic ppm and differ by the level of aggregation expressed as the percentage of nitrogen in A-centers. Diamonds with lower concentrations take longer or need higher temperatures to aggregate, as shown by the shift between the 99% lines for 1000 (**black**) and 200 atomic ppm (**red**). The concentration and aggregation state of the nitrogen in a specific diamond allow the calculation of *t* as a function of *T*. The resulting diagonal line for that diamond is similar and parallel to the lines that appear on the diagram. Most fibrous diamonds show only A-centers in their FTIR spectra and plot in the **white central region** (> 99% A-centers with a small, unresolved fraction of either C- or B-centers). Note that they are not necessarily young and can reside at low lithospheric temperatures for a considerable time. The method is very sensitive to residence temperature but provides only rough constraints on time. The parameters used for the calculations are an activation energy of $E_a = 6.0$ eV and a rate constant of $\ln(A) = 18.81$ (Taylor et al. 1996) for the C to A transition and $E_a = 7.0$ eV ($E_a/R = 81160$) and $\ln(A) = 12.59$ for the A to B transition (Taylor et al. 1996; Leahy and Taylor 1997). These are our preferred values for fibrous diamonds. Other estimations are slightly different and variations between octahedral and cubic growth zones were also reported for the A to B reaction (Taylor et al. 1990, 1996; Zedgenizov et al. 2017). We prefer the values for octahedral diamonds because the diamond fibers grow in the <111> octahedral direction.

Although more data is needed, the figure also shows that there is no correlation between the type of HDF and a specific range of carbon isotope compositions of its host fibrous diamonds.

The range of nitrogen isotope compositions ($\delta^{15}N$) of fibrous diamonds is also narrow (–8 to –2‰, relative to air standard) compared with the full range of NFMC diamonds (–24 to +12‰; see Figs. 7 and 8 of Cartigny et al. 2014). Only a handful of fibrous diamonds were analyzed for both the isotopic composition of their nitrogen and the composition of their HDF microinclusions (Fig. 3b), not enough to search for regularities and correlations.

Most fibrous diamonds exhibit uniform carbon isotope compositions, or small variations (<2‰), but some are zoned. Some present a continuous evolution during growth and appear to evolve in different directions (e.g., the arrows in Fig. 3c). A few others exhibit sharp changes that are accompanied by a change in the chemistry of the trapped HDF (Klein-BenDavid et al. 2004; Shiryaev et al. 2005; Zedgenizov et al. 2011). As discussed in *'Homogeneity and zoning'* below, this coupled variation is a strong indication that the diamond carbon is supplied by the HDF.

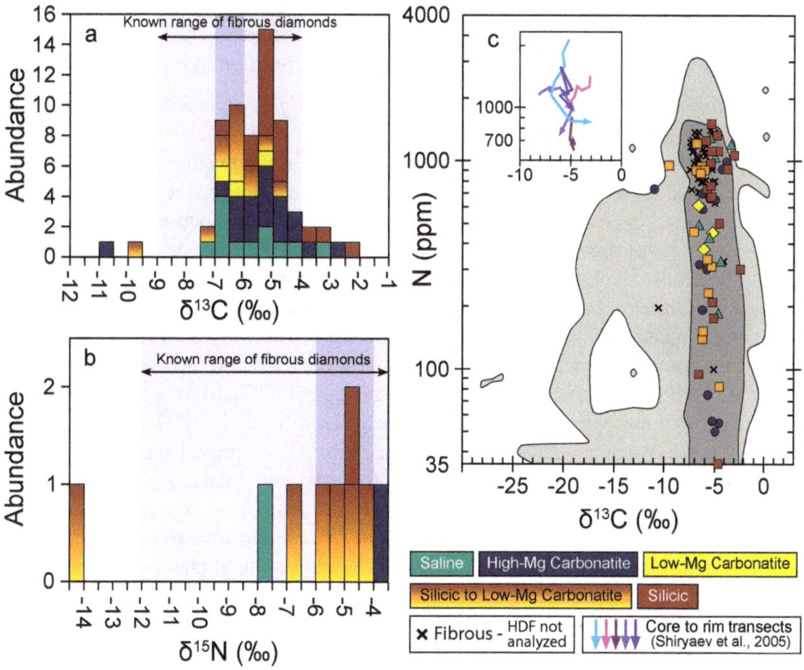

Figure 3. Stable isotope compositions of HDF-bearing fibrous diamonds. (**a**) and (**b**) histograms of carbon and nitrogen isotope composition of fibrous diamonds with known HDF type. The previously reported $\delta^{13}C$ and $\delta^{15}N$ ranges for fibrous diamonds (Cartigny et al. 2014) are shown in **light purple**. They include diamonds where the HDF composition was not analyzed. The main modes of the data are highlighted in **dark purple**. (**c**) $\delta^{13}C$ versus N content of fibrous diamonds. The **light grey outline** indicates the known range of all diamond compositions (Cartigny et al. 2014), with **dark grey** indicating the highest density of the dataset. The arrows in the inset indicate the core to rim evolution in isotopically zoned diamonds analyzed by Shiryaev et al. (2005). The Xs mark the isotopic composition and nitrogen content of diamonds that were not analyzed for their HDF composition.

THE ANALYSIS OF FLUID INCLUSIONS IN FIBROUS DIAMONDS:HISTORICAL PERSPECTIVE AND METHODS

The microinclusions and the mineralogy of their secondary assemblage

Optical microscopy of the microinclusions. High magnification optical microscopy allows observations of the individual microinclusions but is insufficient to resolve their internal structure (Fig. 1d). In some cases, it is possible to see the microinclusions arranged in linear arrays (e.g., Fig. 1 of Navon et al. 1988), but in most cases, their distribution appears to be random, except for radial variation in the number-density of the microinclusions (Fig. 1d). The number-density varies widely, from no microinclusions at all (commonly close to the rim) to more than 10^8 inclusions/mm^3 in densely populated bands, corresponding to distances of ~2 μm between neighboring microinclusions (Klein-BenDavid et al. 2006). Light reflected from polished surfaces also reveals the microinclusion-bearing zones. The microinclusions appear as tiny holes that are the origin of fine polishing grooves that extend from each hole.

Except for very rare cases, the fluid inclusions trapped in fibrous diamonds are smaller than 1 μm in size and are designated as microinclusions throughout the paper (where it is clear that we refer to microinclusions we sometimes use 'inclusions' for short). Most mineral inclusions in NFMC diamonds are larger than 1 μm and are referred to as mineral inclusions.

However, some fibrous diamonds carry mineral inclusions that are smaller than 1 μm. We refer to these as micro-mineral inclusions. Larger mineral inclusions are extremely rare in the fibrous zones, while micro-mineral inclusions are not common in NFMC diamonds.

Transmission electron microscopy (TEM) and high-resolution scanning electron microscopy (SEM). The first glimpse into the microinclusions was achieved using TEM (Lang and Walmsley 1983). In these early days, a 0.5 mm thick diamond wafer was sawn and thinned first by conventional polishing (down to ~50 μm) and then by ion beam down to electron transparency. More recent works by Wirth and coworkers used a focused ion beam (FIB) of gallium ions to cut ultra-thin foils (15 × 10 × 0.1–0.2 μm) that can be mounted directly for examination (Wirth 2004). Lang and Walmsley (1983) found numerous, mostly empty holes, 100–1000 nm in diameter, but in a few cases some material remained, and they succeeded to identify one of the phases as apatite. Energy dispersive spectroscopy (EDS) revealed clear peaks of P and Ca and convergent beam electron diffraction (CBED) revealed hexagonal symmetry with $a = 9.55$ Å. Using similar methods, Walmsley and Lang (1992a,b) found biotite (with excess silica) and ferroan dolomite in other fibrous diamonds. Guthrie et al. (1991), Klein-BenDavid et al. (2006) and Logvinova et al. (2008a, 2011, 2015, 2018, 2019a) provided more comprehensive descriptions of the microinclusions, studying both open and intact inclusions. Bright-field images reveal microinclusions, a few tens to a few hundred nanometers in size, that were sliced open during the preparation of the 10 nm thick foils. Microinclusions in the fibrous diamonds are dominantly multiphase assemblages that consist of a few mineral phases and a hole that held a low-density fluid before the inclusion was opened (Fig. 4). In some cases, amorphous material is also present. The mineral assemblage is secondary and grew from the trapped HDF during cooling, leaving a residual, low-density hydrous phase.

TEM allows the use of various techniques to identify the phases and determine their composition and structure. As CBED leads, in many cases, to fast destruction of the phase under the focused beam, high-resolution electron microscopy (HREM) is used for lattice fringe images

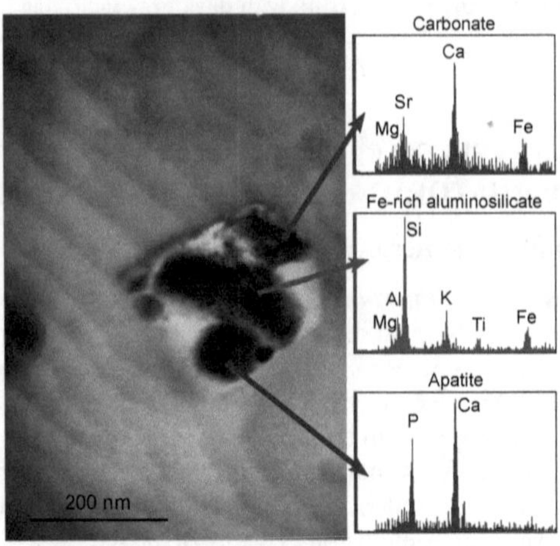

Figure 4. Transmission electron microscope (TEM) image of a microinclusion in a fibrous diamond from the DRC (courtesy of George Guthrie, Los Alamos National Laboratory). The microinclusion shows a fairly well-defined shape imposed by the diamond preferred growth habit. The inclusion contains a multiphase assemblage of daughter minerals (**dark phases**) and a hydrous low-density fluid (**light areas**). In this specific inclusion, energy-dispersive spectra (EDS) revealed carbonates, silicates and phosphates. The solid phases take most of the volume, but water and its solute occupy ~20% of the area. The diamond trapped a uniform HDF, and the secondary assemblage crystallized upon cooling during or after the eruption. For additional details see section '*Transmission electron microscopy (TEM) and high-resolution scanning electron microscopy (SEM)*'.

of the various phases, and a Fast Fourier Transform algorithm is used to compute diffraction patterns from these images. Cell symmetry and parameters derived from the patterns are compared with libraries and literature data and allow identification of the phases. EDS spectra can be obtained from very small volumes and yield the composition of phases, down to a few nanometer scales (Fig. 4), but care should be taken as the beam can pass through two, and even more, phases that overlie each other. Electron energy loss spectroscopy (EELS) is used to detect the presence of various elements and to distinguish the chemical environment of various elements, e.g., to tell carbon in carbonates from the diamond carbon. Further details on the FIB and TEM-related techniques are provided in Wirth (2004, 2009) and Jacob et al. (2014).

The inclusions and their internal structure can also be imaged using high-resolution scanning electron microscopy (SEM) and the various crystals can be analyzed. However, the need to coat the surface disturbs either the imaging (carbon coating) or the analyses (gold coating).

Microinclusions with **high-Mg carbonatitic** compositions carry dolomite and phlogopite as the major phases with magnesite, Ba- and Sr-bearing carbonates, apatite, sylvite, rutile, ferropericlase, magnetite, clinohumite and sulfide, and a fluid phase that was lost during sample preparation (Klein-BenDavid et al. 2006; Logvinova et al. 2011, 2019a). Pyroxene was found in a single microinclusion (Klein-BenDavid et al. 2006), but it may have been trapped as a solid mineral together with the fluid (as described by Izraeli et al. 2001). Inclusions carrying **saline** HDFs crystallized KCl, ferroan dolomite, Ca–Sr–Ba-carbonate, mica, and apatite and single occurrences of halite and Ba–P phosphate (Klein-BenDavid et al. 2006; Logvinova et al. 2011). In inclusions with **low-Mg carbonatitic** HDFs the secondary assemblage consists of dolomite and Ca-carbonate (sometimes with Ba and Sr), sheet-silicates, apatite, KCl, ilmenite, magnetite, ferropericlase and sulfides. The sheet silicates were identified as biotite (Walmsley and Lang 1992b) or phlogopite (Logvinova et al. 2008a) and the sulfides include pyrrhotite, Ni-sulfide and K–Fe-sulfide (Logvinova et al. 2008a). When compositions are more **silicic**, quartz crystallizes and the sheet silicates are phengite and in some cases paragonite (Guthrie et al. 1991; Kvasnytsya and Wirth 2009; Logvinova et al. 2019a). Both are rich in silica and alumina. Klein-BenDavid et al. (2006) obtained compositions that fit high-silica mica with up to 7.6 Si atoms per formula unit (22 oxygen atoms). The above phases are joined by Ca ± Sr ± Ba-carbonate, a single occurrence of sulfide and some unidentified Al-Si rich phases (Logvinova et al. 2019a).

Fourier-transform infrared spectroscopy (FTIR). Infra-red spectroscopy was the analytical technique that first detected the presence of water and carbonates in the microinclusions (Chrenko et al. 1967). Navon et al. (1988) used FTIR spectroscopy to detect water, carbonate, apatite, a sheet silicate and an additional silicate phase (which was later identified as quartz, Navon 1991) in coated and cuboid diamonds from the DRC. They also estimated the carbonate and water concentrations based on absorption coefficients they measured or adapted from the literature. Fig. 5 presents a typical spectrum of a diamond carrying silicic to low-Mg carbonatitic HDF.

In 1991 Navon recognized that the observed absorption peaks at 782–786 and 809–813 cm^{-1} were caused by secondary quartz in microinclusions. These peaks were shifted from their 1 bar position (779 and 798 cm^{-1}) because of elevated internal pressures of (1.5–2.1 GPa) in the inclusions. He used this pressure at room temperature to show that the diamonds originated from the shallow part of the diamond stability field in the lithospheric mantle. A somewhat lower and uniform pressure of 1.5 ± 0.1 GPa was reported by Schrauder and Navon (1994) for diamonds from Botswana. Kagi et al. (2000) used near-IR to provide another indication for high internal pressure within the microinclusions. In addition to liquid water absorption at 5180 cm^{-1}, they identified a shoulder at 5000 cm^{-1}, which they attribute to ice-VI at 1.9 GPa, similar to the pressure indicated by the shift of the quartz absorption.

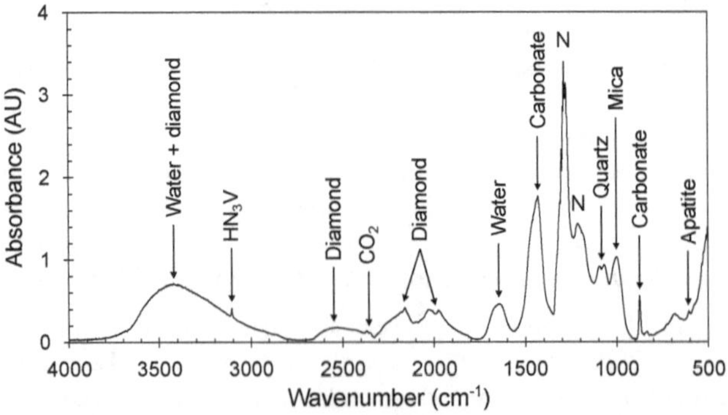

Figure 5. FTIR spectrum of a fibrous diamond from Diavik, Canada. The intrinsic diamond absorbance is at 1800–2700 cm^{-1}. The two absorption bands of nitrogen (N) are typical of aggregation in A-centers; no B- or C-centers are detected in this spectrum. A small peak at 3107 cm^{-1} is due to C–H vibration in VN_3H centers in the diamond matrix. Absorbance due to phases in the microinclusions is also observed. The peak at ~1650 cm^{-1} and the broad band at ~3440 cm^{-1} are caused by bending and stretching modes, respectively, of liquid water. Hydrogen bonding leads to lower frequencies, down to 2800 cm^{-1} (Weiss et al. 2013a). Carbonates (calcite, dolomite and magnesite) absorb at ~1430–1450 cm^{-1} and can be distinguished by their absorption peak at ~880–870 cm^{-1}. Quartz, apatite and mica in the microinclusions absorb at ~1100, 1066 and 1004 cm^{-1}, respectively. The presence of apatite is also indicated by the small peak at 605 cm^{-1}.

Zedgenizov et al. (2004) pointed out that Udachnaya diamonds are richer in the carbonatitic component relative to diamonds from DRC and Botswana. They also noted that water absorption in the Udachnaya diamonds extends to lower wavenumbers and, as a result the intensity ratio of the stretching and bending modes of water I_{3420}/I_{1650} is lower than that of the African diamonds. Weiss et al. (2010) calibrated the absorption of calcite, dolomite, magnesite, apatite, quartz and phlogopite, and combined them with that of water (Thompson. 1965; Venyaminov and Prendergast 1997) to measure the concentration of the various secondary phases in the diamonds. The IR data agree well with modal calculations based on EPMA data (see '*H_2O and CO_2 in the HDFs: Connecting EPMA and FTIR*' below). Weiss et al. (2013a) found that the width of the water stretching mode and the position of the bending mode vary between HDF types and used it to identify the nature of the trapped HDF (saline, high-Mg carbonatitic, or silicic to low-Mg carbonatitic; see '*Major element compositions*' below) based on its IR spectrum. This explains the observed differences in I_{3420}/I_{1650} mentioned above between Siberian and African diamonds (Zedgenizov et al. 2004).

Together all these studies turn FTIR analyses into a powerful tool that yields important information on the type of the trapped HDF, the identity and abundance of most of the major phases: carbonates, silicates, phosphates, and water (but not halides), and in the case of silicic HDFs, even an estimation of the internal pressure within the inclusions. Additional absorption bands at 2340 and 669 cm^{-1} appear in many spectra but in most cases, they are caused by atmospheric CO_2 in the spectrometer. A similar case can be made for the absorption at 2700–3000 cm^{-1} that may be the result of surface hydrocarbons, but it is also possible that these bands are contributed by CO_2 and/or organic compounds in the inclusions (Kopylova et al. 2010).

Micro-Raman spectroscopy. Raman spectroscopy was used by Zedgenizov et al. (2004) and allowed the identification of carbonates, silica glass and apatite in a few microinclusions in fibrous diamonds from Udachnaya which were large enough for analysis. They also found an olivine inclusion and graphite. The analyses did not capture the multiphase nature of the inclusions (aside from the detection of apatite and carbonate in a single inclusion). Most of the inclusions are too small and do not yield useful spectra (Kopylova et al. 2010).

X-ray diffraction (XRD). Another method used to identify the secondary assemblage in the microinclusions is X-ray diffraction. Harris (1968) pioneered the application of this technique to diamond studies and it is now used extensively to characterize mineral inclusions in diamonds and to determine their internal pressure. However, its application to fibrous diamonds was always problematic and did not yield useful results. Kopylova et al. (2010) reported lines corresponding to micas and clay minerals, but noted that the technique failed to identify carbonates or apatite, two main phases in the microinclusions. Smith et al. (2011) conducted the most comprehensive XRD study of fibrous diamonds. They also concluded that the technique fails to identify some important phases. Nevertheless, they were able to confirm the presence of sylvite, halite and Ba–Cl hydrate; dolomite, ikaite ($CaCO_3 \cdot 6H_2O$), Na–Mg carbonate, two types of Ba–Mg carbonates and an unidentified phase with a 2.63 Å band. The powder-like patterns confirmed the random orientation of the secondary phases within the microinclusions.

Major elements analysis

Secondary ion mass-spectrometry (SIMS). This technique was the first to yield a complete chemical analysis of the major elements in the microinclusions (Navon et al. 1988). They used the O^- beam of the Cameca IMS-3F at Caltech to drill into fibrous diamonds and read the intensities of the secondary positive ions formed by the diamond carbon and the various major elements in the microinclusions. Ion yields were calibrated via EPMA analysis of the volume that was later drilled by SIMS, and by analysis of mixed graphite and glass pellets. The yields were high and the background, measured on gem-quality diamonds, was very low. However, the analysis of a volume with a few hundred microinclusions, some of them larger than others, produced large fluctuations in the intensity of the various ions. As the ions were counted sequentially, this was problematic and very long counting times were needed to reach a stable average counting rate (Navon 1989).

Navon et al. (1988) compared the SIMS results to those of EDS analyses of individual, subsurface microinclusions and instrumental neutron activation analysis (INAA) and reported good agreement between element ratios determined by the three methods. However, the complex sample preparation and the long duration of the SIMS analysis needed to obtain precise counting statistics, compared with the ease and accuracy of EPMA-EDS analysis of the major elements in individual microinclusions (see next section) led to the prevalence of the latter in following studies.

Electron-probe microanalysis (EPMA). Electron-probe microanalysis of individual, subsurface microinclusions is the main technique used to analyze the major element composition of the trapped HDFs (e.g., Schrauder and Navon 1994; Izraeli et al. 2001; Tomlinson et al. 2006; Zedgenizov et al. 2009). Navon (1989) showed that it is more accurate than broad beam analysis of many inclusions. The inclusions are identified by comparing back-scattered electron (BSE) and secondary-electron (SE) images of the same area. In the BSE images, the microinclusions appear as bright fuzzy dots due to their high electron density relative to the diamond. However, they can be mistaken for dust particles. This is solved by examination of the SE image, which presents surface topography. The dust is clear and sharp, but the subsurface microinclusions are not visible. An electron beam with a typical current of 10–15 nA is accelerated to 15 KeV and focused to a ~0.5 μm diameter or less at the center of the inclusion. The X-rays generated from the microinclusion and the surrounding carbon matrix are analyzed by an energy dispersive spectrometer (EDS) and wavelength dispersive spectrometers (WDS). Using a standard correction procedure and mineral standards for each of the elements of interest, the number of counts for each element is translated into element concentrations.

Monte Carlo simulations of electron scattering indicate a penetration depth of ~2 μm into the diamond and a similar range of scattering to the sides (at 15 kV). As a result, most X-rays are emitted by the diamond carbon atoms and only a few percent of the calculated analysis are due to elements in the microinclusions. The correction procedure calculates the sum of chlorine and the major oxides and assumes that the difference to 100% (commonly >95%) is all carbon. Finally, to get the composition of the HDF itself, the total of chlorine and oxides is normalized to 100%, plus accounting for excess calculated oxygen due to the presence of Cl (see '*Introduction*' in the supplementary compilation of HDF data, Weiss et al. 2022). In reality, the concentration of the oxides in the HDF is lower, as water and CO_2 (in carbonates) must be included as well. However, because they are measured by FTIR in a different volume, we must first normalize each EPMA analysis in order to combine them and calculate the average composition for the diamond (or a zone in the diamond). Only then, the average EPMA composition and the FTIR dataset can be merged to provide the full major element composition including the volatiles (see next section).

The very low background produced by the carbon matrix allows observations of very small peaks, but the low concentrations of the various elements still lead to large relative errors in the composition of individual inclusions. Jablon and Navon (2016, their Fig. A3) estimated the precision and concluded that the error for major elements of the inclusion is a few percent (*relative*), but it increases to tens of percent for the minor elements. However, except for a few zoned diamonds, the composition of the HDFs in any one diamond is uniform. Thus, the standard error on the average composition of the HDF in each diamond is reduced by a factor of $\sqrt{n-1}$, relative to the standard deviation of the individual analyses. Commonly, 30–50 inclusions along a few profiles are analyzed for each diamond. Plotting the profiles allows verification of homogeneity. Most diamonds are homogeneous, and the standard errors are smaller by factors of 5–7 relative to the standard deviation of the individual analyses. When calculating the average composition, it is important to remove analyses of microinclusions that trapped discrete minerals or mixtures of mineral+fluid (Izraeli et al. 2001; Tomlinson et al. 2006; Weiss et al. 2018; Weiss and Goldstein 2018). As the diamonds are cleaned with HF, fluorine concentration is monitored, and F-rich analyses, which are indicative of exposed inclusions, are also removed.

Other methods. Information on the concentration of some major elements can be obtained using INAA and LA-ICP-MS, along with data on many trace elements. The two techniques are discussed in section '*Trace element analyses*' below.

H_2O and CO_2 in the HDFs: Connecting EPMA and FTIR

To fully characterize the HDF major element composition, it is necessary to integrate both EPMA and FTIR data. Izraeli et al. (2001) provided a rough estimate of carbonate ions as the sum of the molar proportions of Mg + Fe + Ca and calculated the water content using the relative proportions of carbonates and water determined from the IR data. Klein-BenDavid et al. (2007) accounted, in a simplified manner, for cations taken by chlorides, silicates and phosphates and Weiss et al. (2010) provided a better framework. Below, we present their method and examine one more case.

We begin with translating the EPMA data into modal proportions, using the various minerals of the secondary assemblage as a guide. Silicates, phosphates and halides are determined, and the excess cations are assumed to be compensated by carbonate ions. First, the molar proportion of halides and apatite is calculated based on the Cl and P_2O_5 concentrations determined by EPMA. If Cl < K + Na, the excess Na is added to the carbonate. If Cl > K + Na, the excess Cl bonds with Ca. The case of the silicates is more complex. IR and TEM data indicate that the silicate phase is mostly phlogopite-biotite solid solution, $K(Mg,Fe)_3AlSi_3O_{10}(OH)_2$, but that silica-rich mica is found in the silicic HDF microinclusions (Klein-BenDavid et al. 2006). In most cases, the Si/Al ratio of HDFs is larger than 3 (Table 1), and both micas are necessary.

In pure phlogopite-biotite, Fe + Mg = Si = 3Al and K/2 = Si/6. We assume that Ti^{4+} resides in the phlogopite, and obtain:

$$\text{Carbonate} = (\text{Mg}+\text{Fe}+\text{Ca}+\text{Ba}+(\text{Na}+\text{K})/2)-\text{Cl}/2-5\text{P}/3-\text{Si}-\text{Si}/6+2\text{Ti} \quad (1)$$

If Al < Si/3, we assume that the mica is: $K(Mg,Fe)_{3-x}Al_{1-x}Si_{3+x}O_{10}(OH)_2$, which leads to: $Mg_{(mica)} + Fe_{(mica)} = (5Si + 9Al)/8$, and $K_{(mica)}/2 = (Al+Si)/8$ in the mica.

The carbonate is then (Weiss et al. 2010):

$$\text{Carbonate} = (\text{Mg}+\text{Fe}+\text{Ca}+\text{Ba}+(\text{Na}+\text{K})/2)-\text{Cl}/2-5\text{P}/3-(6\text{Si}+10\text{Al})/8+2\text{Ti} \quad (2)$$

In the case of silicic HDFs, where Fe + Mg < 3Al < Si, we assume that all the iron and magnesium, as well as the Al, Ti and Si are in the silicate phases (mica and quartz) and possibly in titanates, and estimate the K bound in the silicates as $K_{(silicates)} = Al$. In this case, the carbonate is (Weiss et al. 2010):

$$\text{Carbonate} = \text{Ca}+\text{Ba}+(\text{Na}+\text{K}-\text{Cl}-\text{Al})/2-5\text{P}/3 \quad (3)$$

If Al > Si/3, the excess Al should enter the octahedral sites (muscovite solid solution) and with $Mg_{(mica)} + Fe_{(mica)} = 3(Si-Al)/2$ and $K_{(mica)}/2 = Si/6$, the proper equation is:

$$\text{Carbonate} = (\text{Mg}+\text{Fe}+\text{Ca}+\text{Ba}+(\text{Na}+\text{K})/2)-\text{Cl}/2-5\text{P}/3-(10\text{Si}-9\text{Al})/6+2\text{Ti} \quad (4)$$

Once the weight fraction of the CO_2 component is determined from the molar proportion of carbonates, the water is estimated based on the $CO_2/(CO_2 + H_2O)$ ratio determined from the FTIR absorption bands of water and carbonates. The main obstacle here is that there is no available absorption coefficient for the residual hydrous fluid in the microinclusions, and even for pure water, the two values in the literature differ by ~25%. Venyaminov and Prendergast (1997) determined $\varepsilon_{water} = 109$ (AU/cm)/(mole/liter), compared with the old value of 86 (AU/cm)/(mole/liter) of Thompson (1965). We use the newer value. Using the older, the water proportion will be higher by 15–20%. Now it is possible to determine the composition of the HDF as wt% of all oxides + Cl + CO_2 + H_2O (Table 2); in molal proportions (moles per 1 kg or 55.5 moles of water, Table 3) or as wt% of the major components: silicates (+titanates), carbonates, halides, phosphates and water (Table 4).

As can be seen in Fig. 6, this procedure works well. The comparison is close for high-Mg carbonatitic HDFs, which have relatively simple secondary mineral assemblage—mostly dolomite, calcite or magnesite, and less (K,Na)Cl (not active in IR), phlogopite, apatite and a hydrous fluid with dissolved halides and carbonates. The fit is reasonable in the case of the low-Mg carbonatitic compositions, but poor for the silicic HDFs. The likely reasons are that the mica is no longer phlogopite, but rather, a Si-rich mica (Klein-BenDavid et al. 2006; Logvinova et al. 2019a) with different absorption coefficient, the presence of an unidentified Al-Si-rich phase (Logvinova et al. 2019) or silicic glass (Zedgenizov et al. 2004). Still, it shows that FTIR spectroscopy alone can be used for identifying the type of HDF present and to provide a semiquantitative estimate of the mineralogy and hence the chemistry of all the major constituents of an HDF.

Trace element analysis

Instrumental neutron activation analyses (INAA). This technique involves irradiation of the sample with neutrons in a nuclear reactor. Nuclear reactions transform (activate) the elements in the sample into short-lived isotopes that emit γ-rays upon their decay. The γ-rays are counted to determine the concentrations of the new isotopes and consequently, the abundance of their parents in the diamond.

Table 1. Average composition* of the four endmembers on H_2O- and CO_2-free basis (wt%).

	Silicic	Low-Mg carbonatitic	High-Mg carbonatitic	Saline
*n**	37	11	32	24
SiO_2	56.5	13.2	9.0	4.2
TiO_2	2.9	2.7	1.2	0.8
Al_2O_3	9.1	3.8	2.8	0.8
FeO	6.0	23.6	13.9	8.0
MgO	3.9	11.7	23.7	3.7
CaO	4.7	20.7	22.8	6.5
BaO	0.5	0.4	2.0	8.4
Na_2O	1.9	6.1	6.3	11.8
K_2O	12.4	13.0	12.9	30.0
P_2O_5	1.5	3.0	2.8	0.6
Cl	0.9	2.3	3.4	32.6
Total	100.2*	100.5	100.8	107.4

Note: **n* - Number of HDF compositions averaged in calculating the composition of each type (ignoring isolated extreme analyses). Totals are higher than 100 wt% because of excess oxygen due to the presence of chlorine.

Table 2. Average composition of the four endmembers, including H_2O and CO_2 (wt%).

	Silicic	Low-Mg carbonatitic	High-Mg carbonatitic	Saline
SiO_2	48.6	8.3	5.5	2.8
TiO_2	2.5	1.7	0.7	0.5
Al_2O_3	7.8	2.4	1.7	0.6
FeO	5.1	14.9	8.4	5.5
MgO	3.3	7.4	14.4	2.5
CaO	4.0	13.1	13.8	4.4
BaO	0.4	0.3	1.2	5.7
Na_2O	1.6	3.9	3.8	8.1
K_2O	10.7	8.2	7.8	20.4
P_2O_5	1.3	1.9	1.7	0.4
Cl	0.8	1.4	2.1	22.1
CO_2	4.2	26.1	31.7	10.7
H_2O	9.8	10.7	7.6	21.3
Total	100.2	100.3	100.5	105.0

Bibby (1979) reported the first INAA study of a coated diamond, but did not have all the major elements needed to obtain the composition of the HDFs. Fragments of coated and fully fibrous diamonds from DRC and Botswana were also analyzed by Palme and Spettel (unpublished, reported in Navon 1989). Schrauder et al. (1996) used INAA to determine a large number of elements in a suite of diamonds from Botswana. Diamonds were fragmented and hand-picked before cleaning in acids in order to expose and avoid cracks. They were then irradiated, cleaned once more, counted 5, ~10 and ~40 days after the irradiation, and analyzed for the concentrations of Na, K, Fe and 28 trace elements in the diamonds. Na/K and Fe/K ratios obtained via INAA compared well with EPMA measurements (Schrauder and Navon 1994) and allowed re-normalization and determination of the trace element concentrations of the HDFs themselves. Br, I and U (along with a few major elements) were also measured following their transformation to noble gases via neutron activation and noble gas mass spectrometry (see '*Noble gas and halogen analysis*' below).

Table 3. Average composition of the four endmembers (molal units).

	Silicic	Low-Mg carbonatitic	High-Mg carbonatitic	Saline
Si	82.7	13.0	12.0	2.2
Ti	3.2	2.0	1.2	0.3
Al	15.7	4.4	4.4	0.5
Fe	7.3	19.4	15.4	3.6
Mg	8.4	17.1	46.9	2.9
Ca	7.3	21.9	32.4	3.7
Ba	0.3	0.2	1.0	1.7
Na	5.3	11.7	16.2	12.2
K	23.2	16.3	21.9	20.3
P	1.8	2.5	3.1	0.3
Cl	2.3	3.8	7.7	29.3
CO_2	9.8	55.5	94.5	11.4
H_2O	55.5*	55.5	55.5	55.5
Total	100.2	100.3	100.5	105.0

Note: *55.5 moles of water weigh one kilogram and hence, the concentrations of all other components is in molal units (moles/1 kg water).

Table 4. Melt components* in the average composition of the endmembers (wt%).

	Silicic	Low-Mg carbonatitic	High-Mg carbonatitic	Saline
Silicates	75	22	14	7
Carbonates	11	60	71	27
Phosphates	3	4	4	1
Halides	2	3	4	43
Water	10	11	8	21
Total	100	100	100	100

Note: *See section '*H_2O and CO_2 in the HDFs: Connecting EPMA and FTIR*' for the calculations of the various components.

Proton induced X-ray emission (PIXE). Rege et al. (2005) presented results obtained by Griffin et al. (1993) using PIXE (also known as proton probe, Ryan et al. 1990). Two diamonds previously studied by Schrauder et al. (1996) were re-analyzed and the duplicate analysis of Ca, Ti, Cr, Fe, Ni, Cu, Zn, Sr, Zr and Ba were used to calibrate a standard for LA-ICP-MS.

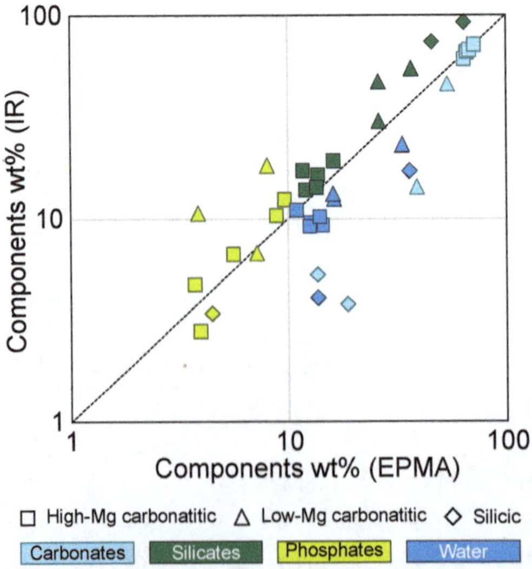

Figure 6. Daughter phase concentrations (wt%) in microinclusions as determined by FTIR and calculated based on EPMA analyses for carbonatitic and silicic HDFs. **Colors** denote the phases and the **symbols** stand for the HDF type. Selected analyses for 10 fibrous diamonds are from Weiss et al. (2009) and Klein-BenDavid et al. (2009). The proportions determined using FTIR are based on the absorbance (AU/cm) at the main peak of each phase and its conversion factor (ppm·cm/AU; Weiss et al. 2010). Oxide concentrations of HDFs as determined by EPMA are translated to modal proportions according to the scheme outlined in section '*H_2O and CO_2 in the HDFs: Connecting EPMA and FTIR*'.

Early mass-spectrometry—combustion and crushing. Early analyses of trace elements in diamonds used emission spectroscopy to analyze impurities in the ash left after burning the diamond, but no specific analysis of fibrous diamond was reported (see Bibby 1982 for review). Akagi and Masuda (1988) burnt cuboid fibrous diamonds in a platinum foil within a quartz tube under oxygen flow. They collected the residue from the Pt foil and the gas-trap and analyzed the solution using isotope dilution mass spectrometry. Akagi (1999) crushed the diamonds, leached soluble elements first in water and then with hydrochloric acid, and determined the concentration of Rb, Sr, Ba, REE and $^{87}Sr/^{86}Sr$ ratios.

Laser-ablation inductively coupled plasma mass-spectrometry (LA-ICP-MS). The development of LA-ICP-MS allows for a much faster and easier measurement of some major and many trace elements in fibrous diamonds. The laser beam is focused into a 20–100 μm spot on the diamond surface ablating the diamond and its microinclusions. Part of the ablated material accumulates around the ablation pit, but most of it is carried into the plasma torch, ionized and analyzed using a quadrupole mass-spectrometer.

The technique was first applied to NFMC diamonds by Watling et al. (1995), who defined their investigation as preliminary. It took 8 more years for Resano et al. (2003) to try again, using an improved laser and mass spectrometer set-up, but here too, the purpose was looking for chemical fingerprint of diamond provenance for identification of conflict diamonds, and no fibrous diamonds were analyzed. Wang et al. (2003) also reported concentrations of 28 elements determined using LA-ICP-MS, but provided no details.

To date, most LA-ICP-MS analyses of fibrous diamonds have been carried out at the GEMOC lab at Macquarie University, Australia. The method was developed by Rege et al. (2005) and utilizes a 266 nm Nd:YAG laser focused into a custom-made ablation cell. Helium is used as the carrier gas and is mixed with argon before its introduction into the plasma torch of an Agilent 7500 ICP-MS. They introduced a cellulose standard that is doped with ~1000 ppm of a variety of elements. Carbon is also measured on the sample and the standard, allowing the translation of count-rate ratios into the concentration of the various elements in the diamond. A comparison of the concentrations determined by LA-ICP-MS with those determined by INAA (Schrauder et al. 1996) and by PIXE (Griffin et al. 1993) indicated good accuracy with most elements falling close to the 1:1 line (Rege et al. 2005). Rege et al. (2010), Araújo et al. (2009), Zedgenizov et al. (2007, 2009, 2011, 2018), Weiss et al. (2009, 2011, 2013b, 2018) and Weiss and Goldstein (2018) all used the Macquarie system and procedures for determining trace element concentration in the diamonds and their HDFs.

Tomlinson et al. (2005) used an ArF excimer laser at 193 nm, a GeoLes illumination system and an Elan 6100 ICP-MS to analyze 5 coated diamonds from DRC. They used a NIST 612 glass standard, but no carbon standard and thus could report only element ratios. In a later paper, Tomlinson et al. (2009) assumed that matrix effects are not too important and used the NIST 612 to derive concentration in the diamond. They used the carbonate content as measured by FTIR and EPMA in the diamond and the HDF, respectively, to get the trace element content of the HDF. Dalpé et al. (2010) used a CRM-S2150 doped oil standard.

As diamonds have a simple matrix, consisting mostly of carbon and minor nitrogen, the number of significant interferences is limited to ion pairs of C, N, O and Ar (see Supplementary Table 2 in Rege et al. 2010). These interferences affect many of the major elements, up to $^{58}Fe = {}^{40}Ar^{18}O$. However, the high abundance of the major elements of the HDFs still allows for reasonable precision in their determination. Weiss et al. (2008) compared element/Fe ratios for the major elements determined by LA-ICP-MS with those measured using EPMA and found good agreement between the two methods. They followed Schrauder et al. (1996) in using potassium that was measured by both EPMA and LA-ICP-MS to translate the concentrations in the diamond to concentrations in the HDF.

We note that, in principle, a carbon standard is not needed. A preliminary study in our lab at the Hebrew University found similar counting-rate ratios for various element pairs comparing analyses of the pure NIST 610 with analyses of pressed pellets made of graphite powder to which 20% of ground NIST 610 were added (Elazar and Weiss 2021). The similarity suggests that it is sufficient to measure the NIST standard. That means that if the major elements are determined by EPMA, and potassium, and/or other elements, are measured by both techniques, the element/K concentration-ratios determined using the NIST standard can be converted to the trace element concentrations in the HDF.

For example, for the case of La:

$$La_{ppm}^{HDF} = \frac{La_{cps}^{d}}{K_{cps}^{d}} \frac{K_{cps}^{st}}{K_{ppm}^{st}} \frac{La_{ppm}^{st}}{La_{cps}^{st}} [K_2O]_{wt\%}^{HDF} \frac{10^4 \cdot 78.2}{94.2}$$

where d = diamond,
st = standard,
ppm = parts per million,
cps = count per second,
$[K_2O]_{wt\%}^{HDF}$ = the concentration of K_2O in the HDF as obtained by normalizing the concentrations of all major oxides and Cl measured by EPMA to 100 wt% (or, alternatively, allowing for water and carbonate as well) and the factor at the end translates $[K_2O]_{wt\%}^{HDF}$ into $[K]_{ppm}^{HDF}$.

Noting the similar concentration of K_2O in silicic and carbonatitic HDFs, we suggest that in the lack of major element data, the trace element concentrations in the HDFs can be estimated using K_2O abundance of 9 wt% for silicic and carbonatitic HDFs and 20 wt% for saline compositions (Table 2). The nature of the fluid can be determined using IR spectroscopy.

Although this is an approximation, the added inaccuracy will only lead to a minor uniform vertical shift of all elements on the trace element abundance diagram.

Off-line laser ablation of fibrous diamonds. In an effort to lower the limit of quantification and to open the way for trace element analysis of NFMC diamonds and Sr–Nd–Pb isotopic analysis of fibrous diamonds, Pearson and co-workers developed the off-line approach (McNeill et al. 2009; Klein-BenDavid et al. 2010, 2014). Following cleaning in ultra-pure HF and HNO_3 and weighing, the diamond was placed in a sealed PTFE beaker with an optical window and ablated using a UP-213 New wave laser system. The sample was moved under the laser so that a large area (a few hundred μm in size) was ablated. The ablated material was collected in HCl that was dried and taken in HNO_3 before splitting into two aliquots: 10% for trace element and 90% for isotopic analysis. The carefully pre-cleaned beakers and windows as well as the use of ultrapure acids in an ultraclean lab led to blanks at the picogram levels and allowed trace element and isotopic analyses of ~500 μg of ablated diamond.

The technique yielded very accurate results for fibrous diamonds, confirming the on-line analyses (see Fig. 2 and 4 of Klein-BenDavid et al. 2014). It was used successfully in trace element analyses of a few NFMC gemmy diamonds, where concentrations were very low, but the patterns resembled those of fibrous diamonds (Melton et al. 2012; Krebs et al. 2019), and for radiogenic isotopic analysis of fibrous diamonds (following section). Recently, a new custom-made off-line ablation system equipped with a Spectra-Physics Talon 532-20 laser was installed in Yaakov Weiss' lab in Jerusalem.

Radiogenic isotope analysis

Radiogenic isotopes are an important tool in dating geologic events and as tracers of melt and fluid sources. Their use as tracers is based on the preservation of isotopic ratios of heavy elements during melting, fractionation, and separation of immiscible liquids. Akagi and Masuda (1988) were the first to determine the Sr isotopic composition of HDFs after burning a diamond in a platinum crucible under flowing oxygen. They dissolved the residue in HCl and used half of the material for the analysis of Sr isotopes, using standard wet chemical methods. The other half was used for analyzing the trace element concentrations. Klein-BenDavid et al. (2010, 2014) replaced the burning of the diamond by laser ablation and the quartz tubes by Teflon beakers fitted with an optical window. The details of the ablation and dissolution procedures were described above, in the off-line laser ablation of fibrous diamonds ('*Trace element analysis*' section). Separation chemistry and isotopic analysis of the derived solution were performed using standard techniques. To date, only 42 fibrous diamonds from Canada, Botswana, DRC, Guinea, and Siberia were analyzed for their Sr isotopic compositions and a handful of those for their Nd and Pb isotopic compositions (Klein-BenDavid et al. 2010, 2014; Smith et al. 2012; Weiss et al. 2015).

Noble gas and halogen analysis

Fibrous diamonds were first analyzed for their Ar content and isotopic composition using noble gas mass spectrometry by Zashu et al. (1986), without noticing their unique character and the presence of microinclusions. They determined the K, ^{36}Ar, and ^{40}Ar content of a group of diamonds from DRC and obtained linear correlations between K and ^{40}Ar concentrations and between $^{40}K/^{36}Ar$ and $^{40}Ar/^{36}Ar$ ratios. Interpreting the slope of the correlation between the two ratios as age (an isochron diagram) they obtained 6 ± 0.3 Ga, older than the age of the Earth. It took three more studies to reach the correct explanation - the correlations reflect the presence of microinclusions full of HDFs that contain not only K_2O and other oxides, but also excess argon that came with the fluid and is not the result of potassium decay in the diamond. The very old age reflects the stronger enrichment of the original trapped fluid in Ar relative to K. Measurement of Cl, after its activation into ^{38}Ar, revealed an even tighter $^{40}Ar/^{36}Ar$–$Cl/^{36}Ar$ correlation, demonstrating that the Ar–K correlation has nothing to do with the diamond age (Turner et al. 1990).

Following studies yielded data on the concentrations and the isotopic composition of He, Ne, Ar, Kr and Xe in fibrous diamonds from DRC, South Africa, Botswana, Guinea, Yakutia and Canada (Ozima and Zashu 1991; Burgess et al. 1998, 2009; Wada and Matsuda 1998; Weiss et al. 2017, 2021 Broadley et al. 2018; Timmerman et al. 2018b, 2019; Johnson et al. 2000). Noble gas analyses of fibrous diamonds were performed by diamond graphitization and gas released at $T > 1800$ °C (commonly at ~2000 °C) using a low blank, high-temperature resistance furnace (e.g., Wada and Matsuda 1998), a laser heating system (Broadley et al. 2018), or by crushing the diamond in a high-vacuum crushing vessel (e.g., Burgess et al. 1998). While the heating techniques extract all the gas from the diamond, including microinclusions and diamond lattice, the crushing technique preferentially extracts the gases trapped in the microinclusions (Burgess et al. 1998; Sumino et al. 2011). This issue is important because nuclear production of noble gases can add ^3He, ^4He and ^{21}Ne in diamonds (Basu et al. 2013 and references therein), almost entirely to the diamond lattice and not in the microinclusions (which constitute only ~0.1% of the diamond by mass, Navon et al. 1988). Crushing the diamonds rather than burning them, reveals the signature of the microinclusions, and allows the distinction, for example, between the He compositions of the trapped HDFs (mantle source) and He confined to the diamond lattice (from secondary sources; Burgess et al. 1998).

Burgess et al. (1998), Weiss et al. (2017, 2021), Timmerman et al. (2018b, 2019) and Broadley et al. (2018) reported results on He, Ne, Ar, Kr, and Xe concentrations and isotopic composition obtained by crushing fibrous diamonds, rather than graphitizing them. If the extraction efficiency by crushing is not high enough, only part of the gas trapped in the microinclusions is released. This may affect the absolute gas concentrations but should not influence their isotopic ratios. Timmerman et al. (2019) graphitized the crushed diamond material and estimated that only 4–20% of the gas in the microinclusions is released by crushing. However, these measurements yielded a significant variation in $^3He/^4He$ ratios, of up to 10 Ra values, between crushing and heating techniques (where Ra is the atmospheric ratio of 1.39×10^{-6}). In contrast, Broadley et al. (2018) reported ^4He concentrations during crushing that are higher by 2–3 orders of magnitude compared with final heating results, and similar $^3He/^4He$ ratios for the two modes of extraction. Thus, gas extraction efficiency by crushing may be much higher (as a result of different crushing methods and initial sample size), and the measured concentrations and isotopic ratios represent the trapped HDF compositions.

Following the activation of diamonds by neutrons, they provide information on additional elements, whose isotopes react to form synthetic noble gas isotopes that can be measured as well. The best-known example is the 'Ar–Ar dating' method where ^{39}K reacts to ^{39}Ar. Turner et al. (1990), Wada and Masuda (1998) and Burgess and co-workers (Burgess et al. 1998, 2002, 2004, 2009; Johnson et al. 2000) have used neutron activation to determine K (via ^{39}Ar), Cl (^{38}Ar), Ca (^{37}Ar), Br (^{80}Kr), I (^{128}Xe) and U (^{134}Xe) concentrations and study noble gases and halogens in fibrous diamonds from Africa, Siberia and Canada.

THE COMPOSITION OF HIGH-DENSITY FLUIDS

The discovery of the various types of high-density fluids

More than 20 years of research lead to the discovery of the four types of HDFs. Navon et al. (1988) first published compositions that span a range between carbonatitic and silicic HDFs. Thirteen years later Izraeli et al. (2001) described a new saline HDF. Last, Klein-BenDavid et al. (2009) distinguished between low-Mg carbonatitic endmember that forms a continuous array towards the silicic one, and high-Mg carbonatitic HDFs that were defined as a new, fourth type. Many other studies confirmed the global distribution of all four compositional types (Figs. 7 and 8, Tables 1–4), but no new major element composition has been discovered since. Trace element analyses revealed two repeating patterns of enrichment in incompatible elements (Weiss et al. 2008b; Rege et al. 2010): highly fractionated patterns with negative anomalies of alkalis, HFSE and in most cases Sr, and less fractionated, smoother patterns. These two types of patterns have been identified in all four types of HDFs (Fig. 9). Radiogenic isotope ratios correlate with the trace elements indicating a relationship between metasomatic enrichment of the source rocks and HDFs signatures (Klein-BenDavid et al. 2014).

Homogeneity and zoning in individual diamonds

Many fibrous diamonds exhibit radial variation in the number density of microinclusions (Fig. 1). However, out of ~300 fibrous diamonds analyzed for the major element composition of their microinclusion, only a handful are chemically zoned and show significant radial (core-to-rim) variation of HDF composition. Even in these diamonds, there is no consistent chemical evolution of fluid composition (e.g., from silicic to carbonatitic) or systematic isotopic trend in the diamond carbon (Klein-BenDavid et al. 2004; Shiryaev et al. 2005; Weiss et al. 2009; Zedgenizov et al. 2011; see arrows in the inset in Fig. 3c). Thus >95% of known HDF-bearing fibrous diamonds recorded growth from a homogeneous medium. Oxide concentration in individual microinclusions varied randomly with no clear core-to-rim evolution and by less than 20% relative. Hence, a single, major element composition achieved by averaging a few tens of HDF inclusions is representative of the diamond to better than ±15% (Weiss et al. 2008a), reflecting analytical errors and a limited real variation between the individual microinclusions.

LA-ICP-MS analyses determine the average trace element composition of thousands of microinclusions in the ablated volume of the diamond. Multiple ablation analyses of different parts of a single diamond agree to better than 15% (Tomlinson et al. 2009; Weiss et al. 2009, 2013b, 2018; Rege et al. 2010; Klein-BenDavid et al. 2014), thus showing that HDFs in an individual diamond are also homogenous for trace elements.

We note that the available isotopic data on HDFs in fibrous diamonds provide contrasting evidence. In some cases, repeated analyses of the same diamond reveal homogeneity (Weiss unpublished data), in other cases significant variation is documented (Klein-BenDavid et al. 2010). To clarify this issue of uniform vs. variable HDF isotopic compositions in a single diamond, additional data are necessary.

Most zoned diamonds reflect two or more distinct growth events, for example, two Kankan (Guinea) fibrous diamonds exhibit inner zones with silicic HDFs and outer ones with low-Mg carbonatitic compositions (Weiss et al. 2009). Similar variations from silicic to carbonatite HDFs were reported by Shiryaev et al. (2005) and the reversed direction by Zedgenizov et al. (2011). Weiss et al. (2014) reported two distinct microinclusion-rich layers in a clear octahedral NFMC diamond from Finsch (South Africa); the inner one carries a low-Mg carbonatitic HDF and the outer one a saline composition. The contrasting radial change of the composition and the sharp boundaries that are also reflected by CL imaging suggest that zoning is the result of growth from different HDFs, rather than *in situ* evolution of a single composition.

Such zoning allows constraining whether the HDFs carry the carbon from which their host diamond grows. Klein-BenDavid et al. (2004) reported saline and high-Mg carbonatitic HDFs in the inner and outer zones, respectively, of a diamond from Diavik (Canada). The carbon isotope compositions of the two zones show a clear difference: −8.1 to −8.9‰ in the inner vs. −11.4 to −12.8‰ in the outer. Shiryaev et al. (2005) reported similar relationships in a cuboid fibrous diamonds from Brazil. The inner part was silicic, while the outer part carried low-Mg carbonatitic HDFs and the carbon isotopic composition changed sharply at the boundary. The simultaneous changes of CL, major element compositions of the HDF and the carbon isotope compositions of the host diamonds thus indicate that the carbon and nitrogen for diamond growth are carried and supplied by the HDFs.

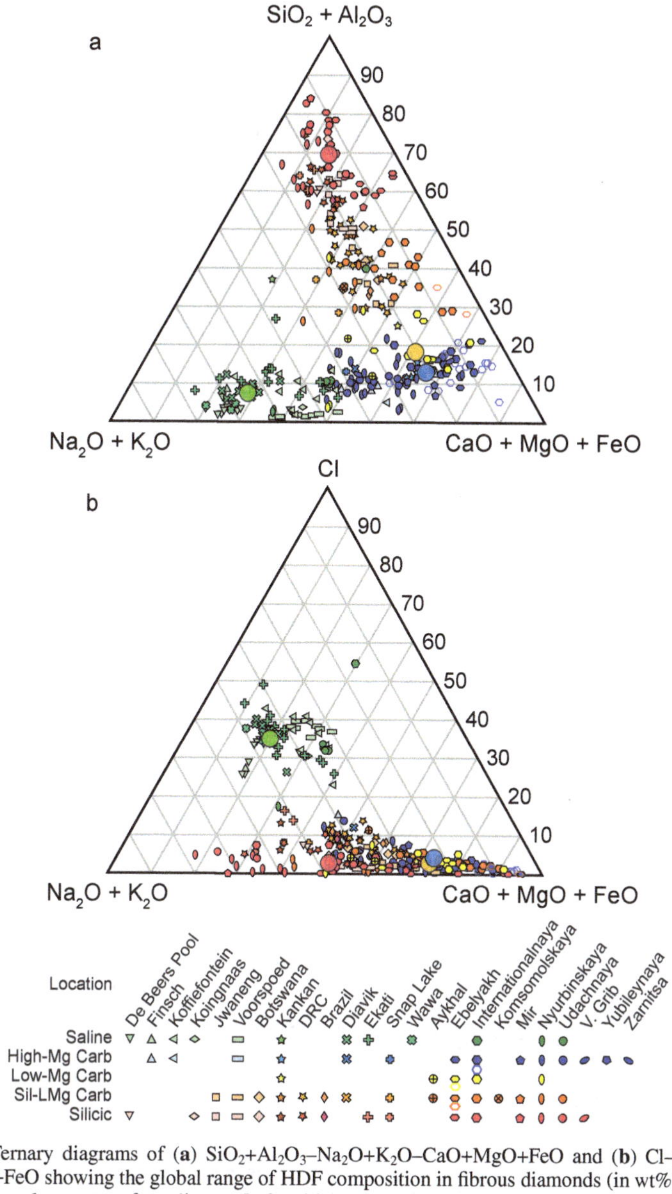

Figure 7. Ternary diagrams of (**a**) SiO_2+Al_2O_3–Na_2O+K_2O–CaO+MgO+FeO and (**b**) Cl–Na_2O+K_2O–CaO+MgO+FeO showing the global range of HDF composition in fibrous diamonds (in wt%). HDF types are coded by color: **greens** for saline, **reds** for silicic, **orange** shading for silicic to low-Mg carbonatitic, **yellows** for low-Mg carbonatitic, and **blues** for high-Mg carbonatitic HDFs. The different mine localities are indicated by shape. Each data point represents an average of >10 microinclusions analyzed in an individual diamond. The compositions of the four HDF types from Table 1 are shown as large **semitransparent circles with black rims**. HDFs with MgO > 40 and CaO > 35 wt% are presented by open symbols (see also Fig 8a,b); all are from Siberia (Internationalnaya, Ebelyakh), and likely represent a mixture between HDF and micro-mineral inclusions (carbonates/silicates that were not identified in the original studies). See the supplementary compilation of HDF data (Weiss et al. 2022) for classification of the HDF compositional types based on oxide wt% parameters, and for the full database and references.

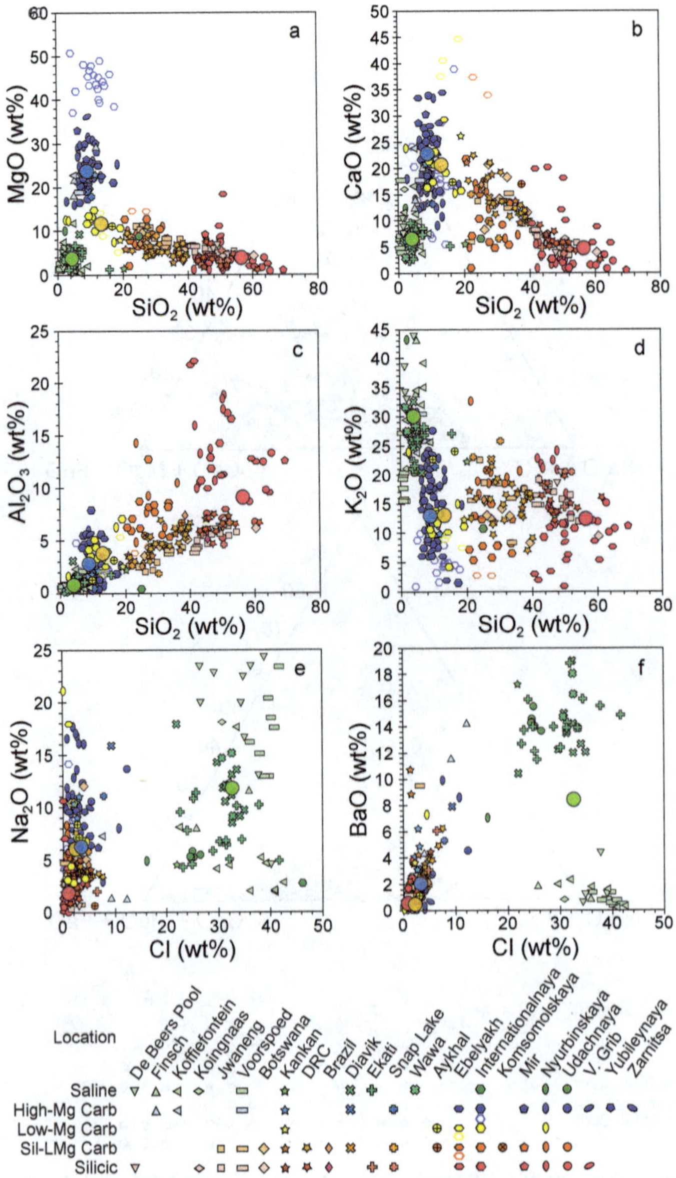

Figure 8. Major oxide variation diagrams of HDF compositions in fibrous diamonds (in wt%, normalized on water- and CO_2-free basis). The compositions of the four HDF types (Table 1) are shown as large **semi-transparent circles with black rims**. (**a**) MgO vs. SiO_2 best expresses the four compositional HDF types: high-Mg carbonatitic HDFs with very high MgO, low silica and alumina, silicic to low-Mg carbonatitic omposition with varying amount of carbonates and saline HDFs that carry mostly K, Na, Cl and water and are depleted in carbonates and silicates. **Open symbols** represent HDFs from Siberia with MgO >40 and CaO > 35 wt% (See Fig. 7 for explanation). (**b–d**) CaO, Al_2O, and K_2O vs. SiO_2; note the high-Al trend and low K_2O values of some HDFs from Siberia compared to similar HDF compositions from all other localities. (**e, f**) Na_2O and BaO vs. Cl show large variations in Na_2O of saline HDFs, and the distinction between high-Ba saline compositions of Canadian (and Siberian) and low-Ba ones of South African HDFs. Full data and references are provided as a supplementary compilation of HDF data (Weiss et al. 2022).

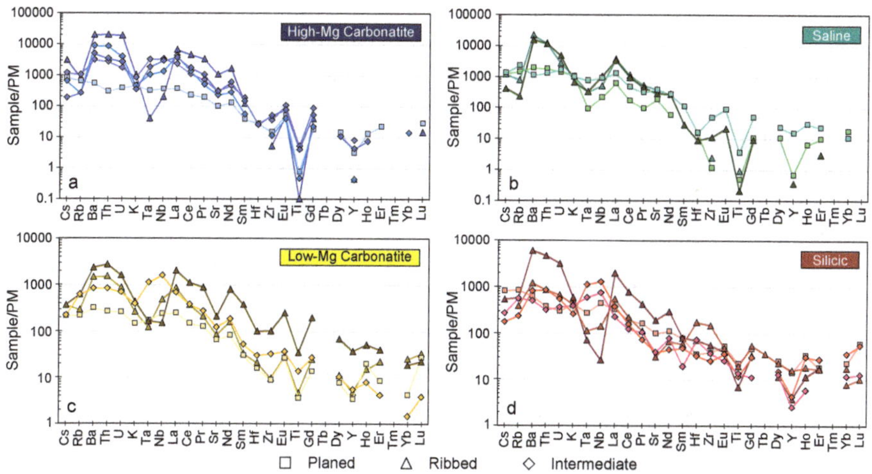

Figure 9. Primitive-mantle (PM)-normalized trace-element patterns of representative saline, silicic, low-Mg carbonatitic and high-Mg carbonatitic HDFs. Two repeating patterns of enrichment in the most incompatible elements are observed in all HDF compositional types: 'Ribbed' with LILE and LREE enrichments and HFSE and alkali depletions similar to calcalkaline magmas and continental crustal rocks, and 'Planed' which has lower LILE and LREE abundances and 'smoother' overall patterns similar to oceanic basalts. Intermediate patterns with a positive Ta–Nb anomaly are also observed in all HDF types. See sections '*Trace element compositions*' for additional details and '*Trace element analysis*' for the conversion of concentration in the diamond to concentration in the HDF. Primitive-mantle (PM) values are from McDonough and Sun (1995).

Major element composition

The compositional space spanned by HDFs in fibrous diamonds is huge, from highly silicic melts to carbonatitic and saline compositions. The high water and carbonate content distinguishes them from any of the melts we know at the surface such as basalts or granites and even from erupted kimberlites, lamproites, or carbonatites. Yet, fibrous diamonds from mines all over the world carry the same four types of HDFs (see '*Geographical distribution of HDFs*' below). The major element composition of all diamonds is presented in Figs. 7 and 8, and in the Supplementary compilation of HDF data Excel file that accompanies this paper (Weiss et al. 2022). Tables 1–4 present the composition of the four HDF types. These compositions (shown in Figs. 7 and 8) were compiled by averaging 11–37 diamonds for the various HDF types, ignoring isolated extreme analyses.

The tables present the compositions in four ways. Table 1 provides the major elements that were measured by EPMA normalized to 100 wt% and allowing for excess oxygen due to chlorine (Cl*15.999/2/35.45). This is the standard way to present the composition, avoiding the uncertainty associated with the IR determination of water and CO_2. The **silicic** HDFs are highly siliceous (SiO_2 = 57 wt%, with some HDFs reaching up to 65 wt%), and rich in K_2O (12.4 wt%, with more scatter and values up to 22 wt%) and Al_2O_3 (10 wt%). Many diamonds form a continuous array between this component and **low-Mg carbonatitic** HDFs that are richer in CaO, FeO and MgO (up to 15 wt% MgO). **High-Mg carbonatitic** HDFs form a distinct group, especially through their very high MgO (>15 wt%). CaO (~20 wt%) and FeO (~14 wt%) are also high, while SiO_2 and Al_2O_3 are low (9 and 3 wt% respectively). The K_2O content varies between localities (see '*Geographical distribution of HDFs*' below; Fig. 8d) with some carrying more than 20 wt%, while many others only contain 5–12 wt% K_2O. The **saline** HDFs are mostly (K,Na)Cl with a few percent of Fe-, Ca-, Ba- and Mg-carbonates and little SiO_2, Al_2O_3, TiO_2 and P_2O_5. The BaO concentration in Table 1 averages two distinct groups, saline HDFs with high BaO (~13 wt%) in Canadian and Russian diamonds and those with low BaO content (~1 wt%) in South-African diamonds.

In Table 2 we added the calculated CO_2 (based on the EPMA analysis, see section '*H_2O and CO_2 in the HDFs: Connecting EPMA and FTIR*' above) and the H_2O components (based on the IR data) and renormalized to 100 wt% (again accounting for excess oxygen due to chlorine). These two values are less accurate but are useful for understanding the nature of the HDFs. CO_2 makes up 25 and 32 wt% of the low- and high-MgO carbonatitic HDFs, respectively, but only 11 wt% of the saline HDFs and <5 wt% of the silicic ones. The water content of the HDFs varies between 8 wt% in the high-Mg carbonatitic HDFs to 21 wt% in the saline ones. We note that the water content we calculate depends on the choice of the absorption coefficient, which is not well constrained. The results presented here were calculated using the high value of ε_{water} = 109 (AU/cm)/(mole/liter). If the lower value is used the water content increases by a few percent, and up to ~26 wt% in the saline endmember.

Table 3 presents the data in atomic proportions. The water content was normalized to 55.5 moles or 1 kg of water, so all elements are in molal concentration (moles/1 kg water). With 1.6 to ~4.6 ions per water molecule, it is clear that the HDFs are not hydrous solutions, but rather melts or supercritical fluids with a high content of carbonate and water. To get a feeling for the nature of these melts, Table 4 groups the various components by their anions: silicates, carbonates (including both metal and carbonate ions), phosphates, halides and water. Together, the HDFs show a wide range of miscibility among the various components. Carbonate and water are always present. Phosphate is always a minor constituent. Silicates and halides are variable but also appear in all HDFs. During analysis, the best way to distinguish an HDF from a micromineral inclusion is that the HDF analysis carries all the major elements.

The sole indication for the presence of sulfide HDFs in diamonds is provided by three fibrous diamonds from Yubileinaya, Siberia (Klein-BenDavid et al. 2003). The diamonds contain carbonatitic HDF with $CO_2/(CO_2+H_2O)$ ratio between 0.26–0.65, but also carry other microinclusions rich in S, Ni and Fe with Ni/(Ni+Fe) that ranges between 0.3–0.7. The S/(S+Ni+Fe) ratio of individual inclusions varies between 0.35–0.64, suggesting that this is a melt rather than a mineral composition. The presence of small amounts of alkali elements and Mg are indicative of a melt as well. The sulfide melts coexist with carbonatitic HDFs, but the data is insufficient for defining the relation between the two fluids. TEM studies of microinclusions in other diamonds (e.g., Logvinova et al. 2019a) showed that sulfides, if present, make up only a trivial fraction of most inclusions.

Trace element composition

Trace element abundance patterns of HDFs (Fig. 9) reveal very high concentrations of incompatible elements relative to primitive mantle (PM) values. The REE are fractionated and variable, mostly negative anomalies exist for the alkali elements (K, Rb, and Cs), high field strength elements (HFSE: Ti, Zr, Nb, Hf, Ta), Sr and Y relative to elements of similar compatibility (e.g., Schrauder et al. 1996; Zedgenizov et al. 2007, 2009, 2011; Tomlinson et al. 2009; Weiss et al. 2009, 2013b; Skuzovatov et al. 2012 ; Smith et al. 2012; Klein-BenDavid et al. 2014; Gubanov et al. 2019; Timmerman et al. 2019). A closer look reveals two main patterns (see below), which are present in all the four HDF types.

Correlations between the trace element composition and HDF types were only identified within a few studies, and mainly within diamonds of the same locality. Namely, the concentrations of Rb, Cs, Ba, Th, U, LREE and Sr decrease linearly from silicic to low-Mg carbonatitic HDFs in diamonds from Jwaneng, Botswana (Schrauder et al. 1996). Similarly, La/Dy ratios decreasing from silicic to low-Mg carbonatitic compositions were observed in HDFs in diamonds from Yakutia, Siberia, (Zedgenizov et al. 2007, 2009, 2011). These observations suggest that carbonate-rich HDFs are more enriched in incompatible elements than silicic ones.

Conversely, Ba, U and Th concentrations and La/Dy ratios all increase from silicic to low-Mg carbonatitic HDFs in diamonds from Kankan, Guinea (Weiss et al. 2009). Compared with the silicic to low-Mg carbonatitic array, the high-Mg carbonatitic compositions exhibit the highest (Nb,Ta,Ba,Th,U)/alkalis, LREE/MREE and LREE/HREE ratios, both in Yakutia and in Kankan (e.g., Zedgenizov et al. 2007; Weiss et al. 2009, 2011; Skuzovatov et al. 2012; Gubanov et al. 2019). In general, high-Mg carbonatitic HDFs in Siberian diamonds are characterized by high levels of Nb and Ta compared to Ba, Th, U, alkalis and LREE contents (i.e., PM normalized (La,Th,Ba,U)/Nb_{PM} ratios <1; Figs. 10 and 11). But lower Nb and Ta levels were recorded in some other Siberian diamonds (Zedgenizov et al. 2009, 2011), in diamonds from Kankan and Diavik (Weiss et al. 2011, 2015), and are even lower in high-Mg carbonatitic HDFs from Finsch (i.e., (La,Th,Ba,U)/$Nb_{PM} \gg 1$; Fig. 9; Weiss and Goldstein 2018).

The saline HDFs are distinct from the other types. This is seen in their elevated La/Pr and La/Dy ratios compared to all other HDFs (Fig. 11); Th/U and Ba/U ratios are also higher in the saline HDFs. Moreover, a few studies pointed out that the Sr* $(Sr/\sqrt{(Pr \times Nd)})_{PM}$ anomaly is positive for most saline HDFs and negative for all other types (Fig. 11d; Klein-BenDavid et al. 2014; Weiss et al. 2015). Similarly, the Eu* anomaly (expressed here as Eu/Sm_{PM}, because of the lack of sufficient Gd data) in the saline HDFs is > 1, compared to the majority of all other HDFs, which have either negative anomalies ($Eu/Sm_{PM} < 1$) or no anomaly (Fig. 11c; Smith et al. 2012; Weiss et al. 2013b, 2015; Klein-BenDavid et al. 2014).

When all the available HDF trace element data are examined (Supplementary compilation of HDF data, Weiss et al. 2022), the observed relationship between Ba, Th, U and LREE enrichment and Nb, Ta and alkalis depletion is the most persistent. This is expressed as positive correlations between ratios such as La/Nb, Th/Nb, Ba/Rb and U/K, regardless of the HDF major-element compositions or their host diamond provenance (Fig. 10a,b).

Based on these characteristics, Weiss at al. (2008b, 2013b) identified two main trace element patterns in all four HDF types (Fig. 9). One with large ion lithophile element (LILE) and LREE enrichment, and HFSE and alkali depletion, similar to calcalkaline magmas and continental crustal rocks. The other has lower LILE and LREE abundances and 'smoother' overall trace-element patterns similar to oceanic basalts. These patterns were designated first as 'Tables' and 'Benches' (Weiss et al. 2008b), and later as 'fibrous-low' and 'fibrous-high' (Rege et al. 2010) or 'Ribbed' and 'Planed' patterns (Weiss et al. 2013b), respectively. The two are best distinguished using co-variation diagrams of (Th,U)/(Nb,Rb) vs. (La,Ce)/(Nb,Rb) (Fig. 10), where (La,Th)/$Nb_{PM} \approx 100$, La/$Rb_{PM} \approx 25$ and Th/$Rb_{PM} \approx 50$ for the fractionated 'Ribbed' pattern and (La,Th)/$Nb_{PM} \approx 0.5$, La/$Rb_{PM} \approx 0.2$ and Th/$Rb_{PM} \approx 0.5$ for the unfractionated 'Planed' pattern (excluding a few extreme compositions). The two trace element patterns are observed in all four major element HDF types (Fig. 9). Intermediate patterns that vary through the complete range between the two also exist. The consistency of the two patterns in HDFs of differing compositions, from different mantle localities, well separated in space and time, overcomes the small trace element variations between HDF types observed in single localities. This emphasizes the main feature of major-trace element systematics—the trace elements are decoupled from the major element composition of the HDF (Weiss et al. 2013b).

Sr, Nd and Pb isotopic compositions

The pioneering work of Akagi and Masuda (1988) provided the first Sr isotope data of HDFs in five diamonds from DRC. The $^{87}Sr/^{86}Sr$ ratios ranged between 0.7038 to 0.7052 and showed no correlation with $^{87}Rb/^{86}Sr$ values indicative of an isochronous relationship. Similarly, all later studies of HDFs from Botswana, Canada, Siberia, DRC, and Guinea (42 diamonds in total), yielded no Rb–Sr isochron systematics (Klein-BenDavid et al. 2010, 2014; Smith et al. 2012; Weiss et al. 2015).

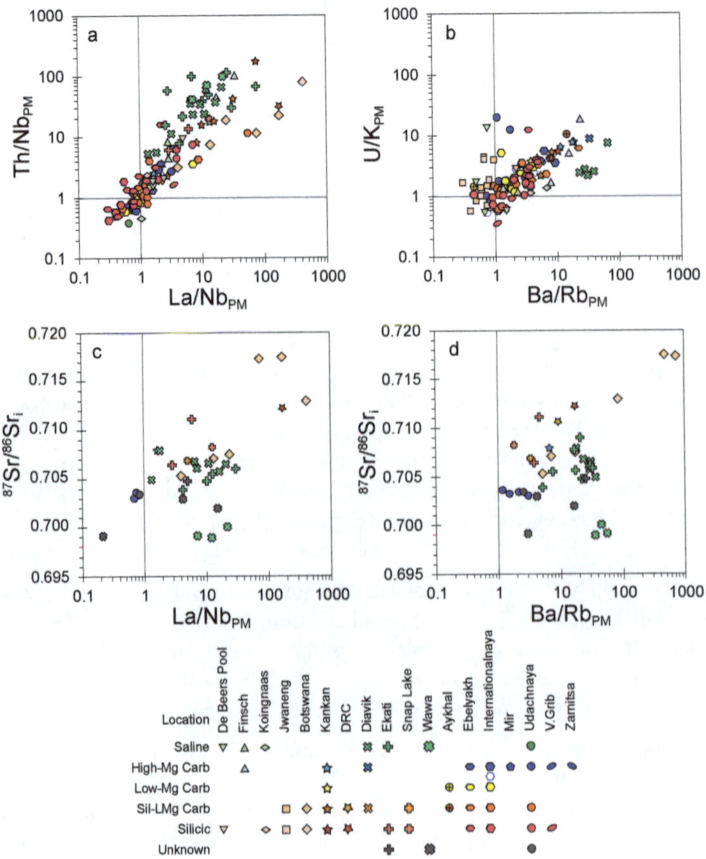

Figure 10. Trace element and Sr isotope ratios of HDFs in fibrous diamonds. The observed relationships between La/Nb, Th/Nb, Ba/Rb and U/K (**a, b**) reflect the variation between the two incompatible element patterns, 'Ribbed' and 'Planed' (Fig. 9), that characterize HDFs, regardless of their major-element compositional type or their host diamond provenance. It emphasizes the decoupling of trace elements from the major element composition of the HDF. Despite the relatively small number of available radiogenic isotope analyses of HDFs, silicic to low-Mg carbonatitic HDFs (as well as high-Mg carbonatite compositions) show a broad correspondence between initial $^{87}Sr/^{86}Sr_i$ isotope values and incompatible element ratios (**c, d**), which suggests a relationship between metasomatic enrichment of the source rocks and the elevated $^{87}Sr/^{86}Sr$ signature. PM - primitive mantle values of McDonough and Sun (1995).

Saline HDFs in diamonds from the ~2.7 Ga Wawa deposit (Smith et al. 2012) show very low initial $^{87}Sr/^{86}Sr$ ratios ($^{87}Sr/^{86}Sr_i$ = 0.699–0.703, Fig. 10c). In contrast, all four types of HDFs from Phanerozoic kimberlites span a wide range of $^{87}Sr/^{86}Sr_i$ ratios, from MORB-like depleted values (i.e., 0.703–0.704) to more radiogenic compositions. Highly radiogenic compositions of $^{87}Sr/^{86}Sr_i$ >0.710 and up to 0.723 are observed solely in silicic to low-Mg carbonatitic HDFs (Fig. 10c,d; Klein-BenDavid et al. 2010, 2014). These HDFs, mainly in DRC and Botswanan diamonds, also exhibit positive correlations between Sr isotopic composition and trace element concentrations. Klein-BenDavid et al. (2010, 2014) suggested that the correlation indicates a relationship between metasomatic enrichment of the source rocks and the elevated $^{87}Sr/^{86}Sr$ signature of HDFs derived from these sources. For silicic to low-Mg carbonatitic HDFs, a general and robust relationship is apparent between higher $^{87}Sr/^{86}Sr$ isotopic values and elevated La/Nb, Th/Nb and Ba/Rb ratios (Fig. 10c,d). Similar systematics exist for high-Mg carbonatitic compositions. These relationships suggest that a

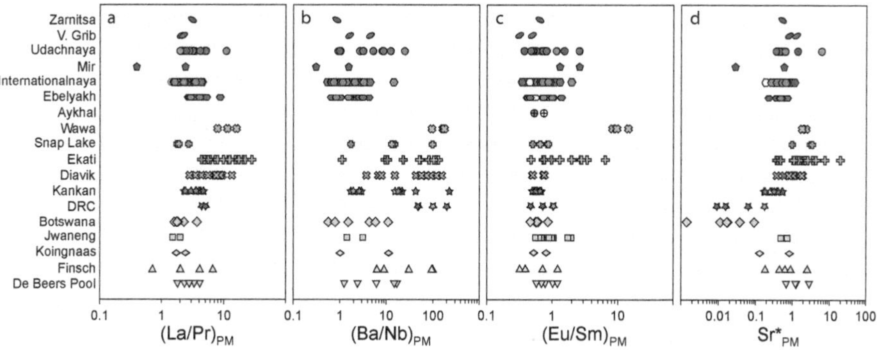

Figure 11. Comparison of La/Pr, Ba/Nb, Eu* (expressed here as Eu/Sm) and Sr* ($Sr/\sqrt{(Pr \times Nd)}$) anomalies of HDFs from different mines and lithosphere provinces. Symbol color relates to HDF composition: saline HDFs are shown in **green** and high-Mg carbonatitic HDFs in **blue**. Low-Mg carbonatitic HDFs are shown in **yellow**, silicic HDFs in **red** and intermediate compositions between them in **orange**. For additional information refer to the legend of Fig. 8. PM – concentrations are normalized to the primitive mantle values of McDonough and Sun (1995).

radiogenic Sr isotopic signature is related to a more strongly fractionated ('Ribbed') trace element pattern within HDFs of the same type. Saline HDFs from the Northwest Territories (Ekati and Diavik) do not show these relationships; rather they exhibit positive correlations between $^{87}Sr/^{86}Sr_i$ and Th/U, La/Pr, and La/Sm, as well as Sr* and Eu* anomalies (see Fig. 3 of Weiss et al. 2015).

Radiogenic isotopic data for Nd and Pb from HDFs are very scarce, only five diamonds (all silicic to low-Mg carbonatitic) were analyzed for their Nd and only three for their Pb isotopic composition (Klein-BenDavid et al. 2010, 2014). The $^{143}Nd/^{144}Nd$ ratios range between 0.5118 to 0.5105 and correlate negatively with the Sr isotopic composition. The few diamonds analyzed for Pb isotopic compositions display high $^{207}Pb/^{204}Pb$ ratios (15.62–15.67) compared to depleted MORB mantle (DMM) or kimberlites and modest $^{206}Pb/^{204}Pb$ ratios (17.87–18.12), placing them at a slightly more radiogenic position than calculated primitive mantle sources (Kamber and Collerson 1999; Murphy et al. 2003; Klein-BenDavid et al. 2010, 2014).

Noble gas compositions

Fibrous diamonds are highly enriched in noble gases. Crushing yields ^{4}He concentrations of 10^{-6}–10^{-5} cm^3 STP/g and ^{40}Ar between 10^{-7}–10^{-6} cm^3 STP/g (Burgess et al. 1998; Broadley et al. 2018; Timmerman et al. 2018b, 2019; Weiss et al. 2021). In most cases, similar concentrations were measured by fibrous diamond graphitization (Ozima and Zashu 1991; Wada and Matsuda 1998; Burgess et al. 2009), which suggests that most of the gas is trapped in the microinclusions and released by crushing alone. Equivalent high He concentrations were only found in "popping rock" mid-ocean ridge basalts (Moreira et al. 1998). Burgess et al. (1998) combined these data with $^{40}Ar/K$ and $^{4}He/^{40}Ar^{*}$ (*corrected for atmospheric ^{40}Ar) and determined that He and Ar are present in the trapped HDFs (rather than HDF+diamond) at concentrations of 10^{-3} cm^3 STP/g, corresponding to enrichment by 3 orders of magnitude over present-day upper mantle values.

The $^{3}He/^{4}He$ isotopic ratios of HDFs in most analyzed diamonds from Africa, Siberia, and Canada vary between 2–8 Ra (the atmospheric ratio of 1.39×10^{-6}; Burgess et al. 1998, 2009; Weiss et al. 2017, 2021; Timmerman et al. 2018b), similar to the range of MORB ($^{3}He/^{4}He$ = 8 ± 1 Ra; Graham 2002), recycled subducted surface material (Kurz et al. 1982) and subcontinental lithospheric mantle (SCLM, 6.1 ± 2.1 Ra; Day et al. 2015). Only a few HDFs have higher ratios (10–11.2 Ra) or lower, radiogenic values of <1 Ra (Burgess et al. 1998;

Broadley et al. 2018; Timmerman et al. 2019; Weiss et al. 2021). Broadley et al. (2018) suggested that the variation between SCLM and the upper range of $^3He/^4He$ ratios of the HDFs reflects a mixture between lithospheric and deep volatile noble gas components originating from a mantle plume, whereas $^3He/^4He<1$ was postulated to result from the accumulation of radiogenic ^4He atoms produced by α-decay of U and Th (Burgess et al. 1998). Weiss et al. (2017) showed that $^3He/^4He$ ratios of HDFs correlate negatively with the amount of K_2O and positively with the carbonate component in fibrous diamonds from Africa, which reflects increasing $^3He/^4He$ from saline to low-Mg carbonatitic to high-Mg carbonatitic HDFs. Exceptional are the South African silicic and carbonatitic HDFs which have much lower $^3He/^4He$ compared to silicic and carbonatitic compositions from Botswana and Guinea, to the degree that their $^3He/^4He$ ratios are lower (more radiogenic) compared to diamonds with saline HDFs from the same locations. This could suggest a complex evolution of the source of the fluids or represent different metasomatic events at different times over the history of the local continental mantle. The latter alternative is consistent with the observed negative correlation between ^4He contents and Ra values and positive correlation with increasing U–Th concentrations in HDFs from South Africa and DRC (Timmerman et al. 2018b, 2019).

It was suggested that by combining He isotope analyses with U–Th-He abundance measurements, the geochronology of HDF-bearing diamonds and C–O–H metasomatism can be constrained (Weiss et al. 2017; Timmerman et al. 2019). Indeed, the slope of ^4He vs. U–Th variation of HDFs in the few DRC diamonds deviates from the ~70 Ma eruption age of their possibly related kimberlite, indicating that these diamonds formed 10 to several hundred Myr before their host kimberlites (Timmerman et al. 2019). However, that study did not take into account the possibility of diffusive loss of He, which is critical for a meaningful interpretation of (U–Th)/He data and for constraining ages (Farley 2002). The issue was recently addressed by Weiss et al. (2021), who showed that the budget of He in a diamond is affected by diffusion over geological time scales. Taking into account the impact of possible He diffusion on the $^3He/^4He$ ratios of South African diamonds and considering the thermal and tectonic history of the Kaapvaal craton, they suggested an upper limit of $\sim 1.8 \pm 0.2 \times 10^{-19}$ cm^2/s for the diffusivity of He in fibrous diamonds. This diffusion limit provides a means to directly place an upper and lower limit on the timing of diamond crystallization and C–O–H metasomatism of the SCLM. For the analyzed diamonds from South Africa, it constrained the timing of diamond formation during three metasomatic events in the late-Mesozoic, Paleozoic and Proterozoic, each by a different HDF type (saline, silicic and carbonatitic, respectively; Weiss et al. 2021).

The $^{40}Ar/^{36}Ar$ values of fibrous diamonds are high, typically >3000, and reach values above 40,000 (Turner et al. 1990). Wada and Matsuda (1998) showed that in fibrous diamonds from DRC, such isotopic variation form a trend with ^{36}Ar concentrations, between air-like Ar ($^{36}Ar > 10^{-9}$ cm^3 STP/g and $^{40}Ar/^{36}Ar \approx 3$ 00; Lee et al. 2006) and a mantle component similar to MORB 'popping rock' source ($^{36}Ar < 10^{-10}$ and $^{40}Ar/^{36}Ar$ >30,000; Moreira et al. 1998). However not all fibrous diamonds fall on such a trend, as evident by those from Canada which exhibit high ^{36}Ar and high $^{40}Ar/^{36}Ar$ (Johnson et al. 2000; Burgess et al. 2009). These isotopic values indicate that the high content of ^{36}Ar in the HDF microinclusions is derived from their mantle source. $^{20}Ne/^{22}Ne$ and $^{21}Ne/^{22}Ne$ ratios reach values up to ~11 and ~0.06, respectively (Wada and Matsuda 1998; Broadley et al. 2018). Within the $^{20}Ne/^{22}Ne$–$^{21}Ne/^{22}Ne$ space, the HDFs Ne isotopic compositions are distinct from those of air and fall between the air–MORB mixing line and the air–solar mixing line. For Xe, the $^{129}Xe/^{130}Xe$ vs. $^{136}Xe/^{130}Xe$ isotopic ratios of a few fibrous diamonds from DRC correlate and lie on a mixing line defined by air and MORB (Wada and Matsuda 1998). Timmerman et al. (2018b, 2019) are the only available studies thus far that provide information about noble gases heavier than He for fibrous diamonds with known HDF compositions. Using the crushing technique, these authors measured $^{40}Ar/^{36}Ar$

ratios between 480–3600 for fibrous diamonds containing silicic to low-Mg carbonatitic HDFs, $^{40}Ar/^{36}Ar$ ratios between 390–1940 for high-Mg carbonatitic HDFs, and values in the range of 380–30,000 for saline compositions. The $^{40}Ar/^{36}Ar$ ratios obtained by step-heating diamond graphitization of a few of the silicic to low-Mg carbonatitic HDFs yielded values that varied between 670 and 15,400 (Timmerman et al. 2019). $^{129}Xe/^{130}Xe$ vs. $^{136}Xe/^{130}Xe$ isotopic compositions of all HDF types in 7 diamonds that they analyzed lay on mixing lines between air and MORB or air and crust (Staudacher and Allègre 1982). Timmerman et al. (2018b) noted that the $^{36}Ar/^{130}Xe$ and $^{84}Kr/^{130}Xe$ ratios for the analyzed saline, silicic to low-Mg carbonatitic, and high-Mg carbonatitic HDFs all fall in the compositional field for sea-water, sediments, and crustal endmembers and are distinct from atmospheric Ar/Xe and Kr/Xe compositions.

Halogen compositions

Unfortunately, only four diamonds were measured for both their major element and halogen compositions (Jwaneng diamonds analyzed by Schrauder and Navon 1994 and Johnson et al. 2000). However, K is commonly analyzed together with the halogens and its concentration in the HDFs is overall uniform, allowing for the use of K/Cl to identify the type of the HDF and normalize the results. The average K/Cl of diamonds from the Ekati mine, Canada (Johnson et al. 2000; Burgess et al. 2009) is close to one, which is suggestive of saline HDFs, similar to most major element analyses of diamonds from that mine (Fig. 12).

Cl concentrations in the Ekati HDFs vary between 20–35 wt%, the Br content of most samples is 1000–2000 ppm, with Br/Cl ratios of 0.001–0.003. Some diamonds reach Br levels of up to 3 wt% and extreme molar Br/Cl ratios (up to 0.063, Johnson et al. 2000). Iodine is also enriched: 10–30 ppm and I/Cl of $1–3 \times 10^{-5}$ in most samples, and up to 1200 ppm and I/Cl = 1.7×10^{-3} in the most fractionated sample. The ratios of most samples fall close to the MORB field and extend towards values typical of altered oceanic crust. However, Br concentrations are highly enriched in comparison to PM ($\sim 10^4 \times$ PM) and even more in

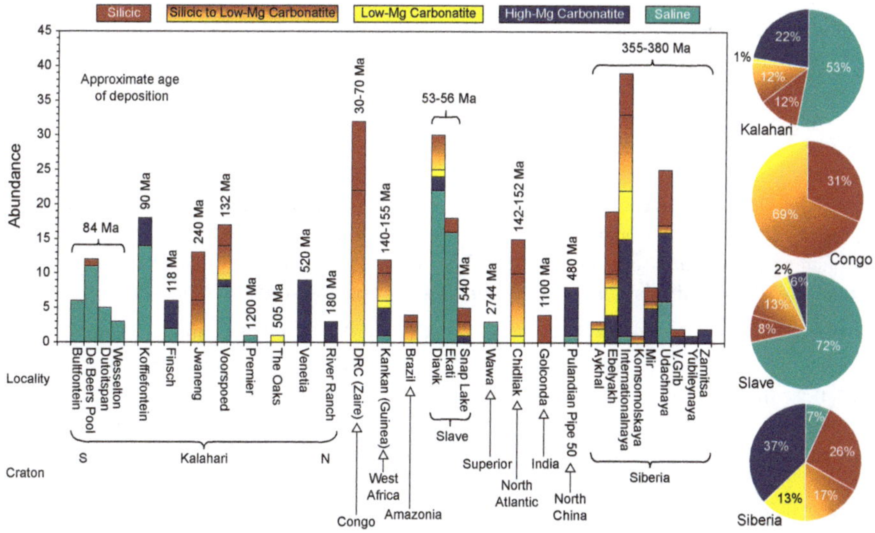

Figure 12. Worldwide abundance and distribution of HDF compositions, separated by locations and cratons. Relative abundances are presented for selected cratons. The database for this figure includes available published sources (Weiss et al. 2022) and data for Bultfontein, Dutoitspan, Wesselton, Premier, The Oaks, River Ranch (Weiss, unpublished data); Venetia (Kempe et al. 2021b); Chidliak (Elazar and Weiss 2021); Golconda (Schrauder 1997) and Pulandian (Mershon et al. 2021). The Kalahari Craton includes the Kaapvaal and Zimbabwe Cratons with the Limpopo Belt between them.

comparison to the depleted MORB reservoir. Iodine is enriched to levels of 1500–2500 × PM. Separation of a saline fluid from a silicate melt was proposed as a possible reason for the extreme values (Burgess et al. 2009). Fractionation of halides, which was observed experimentally (Förster et al. 2019) may also play a role in removing Cl, K and Na and enrich the residual fluid in Br and I (Johnson et al. 2000).

Diamonds from DRC, Jwaneng (Botswana) and Aikhal (Siberia) have K/Cl of 2–10 and are silicic to low-Mg carbonatitic. Assuming K_2O content of 13 wt%, (Table 2), we obtain Cl concentration of 1–3 wt%, similar to the EPMA measurements, 30–60 ppm Br, and 1–3 ppm I (with some diamonds reaching up to 17 ppm). Br/Cl ratios vary between $1–2 \times 10^{-3}$, close to the seawater value, somewhat higher than MORB and similar to altered oceanic crust (Burgess et al. 2009; Broadley et al. 2018). I/Cl ratios of $0.5–3 \times 10^{-5}$ are higher than seawater and span a broad area with MORB values at the center (see Fig. 5 of Burgess et al. 2009). These ratios represent enrichment of 300–800 over the PM value for Cl, 250–700 × PM for Br and 80–300 for I, similar to levels of other incompatible elements. Halogen concentrations are extremely high in saline HDFs, though they are highly enriched in the other types as well. Such enrichment in halogen concentrations and elevated Br/Cl and I/Cl ratios further strengthen the argument for the involvement of a subducted component in the generation of HDFs.

High-density fluid microinclusions in non-fibrous diamonds

The formation mechanism for fibrous diamonds from the HDFs they trapped is broadly accepted, whereas the mode of formation of NFMC diamonds is less clear. The differences in texture, nitrogen aggregation and age, as well as the tight carbon isotopic composition of fibrous diamonds relative to the wide range of NFMC and polycrystalline diamonds, left the door open for other mechanisms of growth. Thus, it is interesting to examine whether HDFs, their fingerprints, or other fluids are found in non-fibrous diamonds.

Bulanova et al. (1988) and Novgorodov et al. (1990) reported 10–30 μm inclusions of potassium aluminosilicate glasses associated with inclusions of K-feldspar and other eclogitic minerals in colorless NFMC diamonds from Mir. Shatsky et al. (2019) reported two melt+clinopyroxene inclusions, with a SiO_2- and Al_2O_3-rich bulk composition, in two alluvial NFMC diamonds from Ebelyakh, Siberia. A single microinclusion composed of Fe, Si, P, Mg, K, Ca, Al and carbonate was identified by Jacob et al. (2011) in a FIB foil cut from a polycrystalline framesite from Orapa, but no additional information was provided. Using TEM on foils cut from Juina (Brazil) NFMC diamonds, Kaminsky et al. (2009, 2013) described a suite of HDF-related multiphase inclusions, mostly smaller than 1 μm in size. The suite included carbonates (magnesite, $Na_2Mg(CO_3)_2$ and Ba- and Sr-carbonates), phosphates, phlogopite, chlorides, fluorides, sulfides and pores (that probably included residual low-density fluids), and was designated: carbonate-halide association.

Weiss et al. (2014) were the first to identify HDFs, similar to those found in fibrous diamonds, within NFMC diamonds. They found and analyzed two microinclusion-rich layers near the edge of an NFMC Finsch diamond and used X-ray topography to show that the diamond is monocrystalline throughout. The inner layer carried carbonatitic HDFs, while the outer layer was saline. They also characterized six saline inclusions in the NFMC core of a coated diamond from Kankan. Recently, Kempe et al. (2021a) identified 3 silicic microinclusions in the NFMC core of a coated diamond from Voorspoed. A systematic search for inclusions in NFMC diamonds was conducted by Jablon and Navon (2016), who examined twinning planes of 30 twinned diamonds (macles) from Venetia and Voorspoed, as well as the diamond adjacent to the planes. Inclusions were absent in the majority of diamonds however, 32 shallow, subsurface microinclusions were identified in 8 of the macles. All were found where the twinning plane formed corners. Twelve inclusions in two diamonds carried submicrometer orthopyroxene and a K–Mg carbonate. The other 20 carried HDFs. Twelve inclusions in four Venetia diamonds carried high-Mg carbonatitic HDF and a silicic inclusion was found in another. In Voorspoed, three saline microinclusions were found in one diamond, and four silicic to carbonatitic ones in another. The level of nitrogen aggregation of the inclusion-bearing diamonds is the same as that of other 40 macles and 23 octahedral diamonds from the same batch. This is true for both localities (Jablon and Navon 2016).

Larger carbonate inclusions were found in a few Siberian diamonds. Sobolev et al. (1997) reported phlogopite associated with calcite (and dolomite, see Sobolev et al. 2009) in three NFMC diamonds from the Sputnik kimberlite. They did not provide the proportions of the two phases, but the combined composition of $CaCO_3$ and phlogopite brings the analysis close to that of silicic–carbonatitic HDFs. On a TiO_2 vs. Mg# diagram, the phlogopites fall in the eclogitic field that is associated with the silicic–carbonatitic HDFs (Sobolev et al. 2009). Raman spectroscopy of a particularly large (43×30×23 μm) carbonate inclusion in a NFMC diamond (Logvinova et al. 2019b) revealed that it consists of three distinct carbonates, which according to TEM-EDS analyses have a combined bulk composition of $0.1(Na_{0.5}K_{1.5})CO_3{\cdot}0.9(Ca_{0.57}Mg_{0.43})CO_3$. This is very close to the composition of high-Mg carbonatitic HDFs, but with slightly elevated CaO content and no measurable SiO_2, Al_2O_3, TiO_2, BaO or Cl, which are normally present at the few wt% level in such HDFs.

Aiming their micro-Raman at the interface between mineral inclusions and the host NFMC diamond, Nimis et al. (2016) characterized bands of $Si_2O(OH)_6$ and $Si(OH)_4$ at 650 and 800 cm^{-1} as well as water bands at 1650 and >3300 cm^{-1}. These bands are caused by a thin film of a hydrous residual fluid with dissolved silicates. The presence of the thin layer (<1.5 μm) of low-density material at the diamond-inclusion interface was confirmed by X-ray tomography. CO_2 could not be identified as its Raman bands are too close to the intense diamond line. Neither CH_4 nor N_2 were detected.

Another line of evidence for the presence of HDFs in NFMC diamonds derives from trace element analysis. Early efforts were reviewed in section '*Trace elements analyses*' above. A major improvement came with the introduction of the off-line laser ablation method (McNeill et al. 2009; Klein BenDavid et al. 2010). McNeill et al. (2009) studied gem-quality NFMC diamonds from South Africa, Siberia and Venezuela, distinguished between a group enriched in Ba, U and La over Nb and another with smoother, less fractionated patterns, and concluded that the two groups bear resemblance to HDFs in fibrous diamonds (Weiss et al. 2008a,b). Melton et al. (2012) examined fragments of 10 clear NFMC diamonds that carried garnet inclusions. The trace element abundance pattern of the diamond richest in trace elements was similar to that of HDFs in fibrous diamonds. Krebs et al. (2019) conducted a more comprehensive study of gem and near-gem NFMC diamonds from Finsch and Newlands mines, South Africa. They reported that they "see the same spectrum of fluids in both high-purity gem and near-gem diamonds that was previously documented in fibrous diamonds. 'Planed' and 'Ribbed' trace element patterns characterize not only the HDF inclusions in fibrous diamonds but also in gem diamonds". Another significant result of this study was the identification of similar patterns in silicate-bearing and sulfide-bearing NFMC diamonds from the Victor mine, Canada. Timmerman et al. (2019b) reached a similar conclusion for silicate- and sulfide-bearing NFMC diamonds from the Koffiefontein, Letlhakane and Orapa mines, South Africa; their trace element signatures are similar to those of fibrous diamonds and they record both 'Planed' (unfractionated) and 'Ribbed' (significant inter-element fractionation) patterns.

There is little evidence for other fluids trapped in NFMC diamonds. Chepurov et al. (1994), Tomilenko et al. (1997) and Smith et al. (2014, 2015) reported the presence of CO_2, N_2 and hydrocarbons in secondary fluid inclusions, larger than 1 μm, along healed cracks in diamonds from North-East Siberia and Africa. Nanoinclusions carrying solid molecular nitrogen were described in sublithospheric diamonds from Juina, the Urals and South Africa (Kagi et al. 2016, Rudloff-Grund et al. 2016, Navon et al. 2017, Sobolev et al. 2019). Navon et al. (2017) suggested that they formed by exsolution of nitrogen from the diamond matrix and not by direct trapping of nitrogen-rich fluid. Solid CO_2 was reported in diamonds from Colorado (Chinn et al. 1995a,b) and an unknown source (Schrauder and Navon 1993). Mixed-habit diamonds from Zimbabwe contain measurable CH_4 (Smit et al. 2016) and Smith et al. (2016) found thin films of CH_4+H_2 around metallic inclusions in clear sublithospheric diamonds.

In summary, the present dataset of direct measurements of fluids and trace elements in NFMC diamonds clearly shows the presence of HDFs similar to those found in fibrous diamonds. Evidence for other fluids are sporadic and their source is less clear.

TEMPORAL AND GEOGRAPHICAL DISTRIBUTION OF HIGH-DENSITY FLUIDS

Temporal distribution of high-density fluids

A decade ago, the prevailing concept on the formation of fibrous diamonds and their HDFs asserted that they formed episodically in the mantle, close in time to the eruption of their host kimberlite over the last ~500 Ma. This assumption, that fibrous diamonds are young, was based on the short mantle residence time of fibrous diamonds at mantle temperatures, as evident by their low nitrogen aggregation (~100% A-centers, no B-centers and no platelets; Fig. 2), and the Phanerozoic emplacement ages of most kimberlites that carried these diamonds to the surface (e.g., Boyd et al. 1987; Navon 1999; Faure 2010; Gurney et al. 2010). However, in recent years, new data changed this idea: a) even diamonds with only A-centers may survive for many millions of years at 900–1000 °C (Fig. 2); b) HDFs were discovered in fibrous diamonds from Proterozoic and Archean deposits; c) fibrous diamonds with both A- and B-centers and platelets were identified at various mines d) in a few mines, nitrogen aggregation indicates several events of fibrous diamond formation, each with its own characteristic HDF type and e) U–Th/He dating of fibrous diamonds yielded ages older than 750 Myr.

The 2.7 Ga metaconglomerates near Wawa (Smith et al. 2012), as well as the 1.2 Ga Premier kimberlite (Weiss unpublished data), host fibrous diamonds that contain HDFs (Fig. 12). This is direct evidence for the antiquity of HDFs and their presence in the mantle at least since late Archean times. The fibrous diamonds, in both localities, have nitrogen in A-centers only, indicative of a short mantle residence time. The HDFs in both cases are solely saline, and more specifically, of the "high-Ba" group (Fig. 8f).

Reports of fibrous diamonds that carry nitrogen in both A- and B-centers came first from the Siberian craton. Nitrogen aggregation ratios varying between 10–70 %B (where $\%B = 100N_B/(N_A+N_B)$, and N is the nitrogen content in A- and B-centers) were found in fibrous diamonds from the Aikhal, Internationalnaya, Mir, Nyurbinskaya and Udachnaya kimberlites (Zedgenizov et al. 2006, 2009, 2020; Skuzovatov et al. 2011). These diamonds carried silicic to low-Mg carbonatitic and high-Mg carbonatitic HDFs. Later, fibrous diamonds with B-centers were documented at Finsch, De Beers Pool and Voorspoed on the Kaapvaal craton (Weiss et al. 2018; Weiss and Goldstein 2018; Kempe et al. 2021a). The aggregation ratios varied between 10–35 %B, and the diamonds carried all four HDF types (saline, silicic, low- and high-Mg carbonatitic compositions). In these localities, a few generations of fibrous diamonds were identified, each with its distinct HDF composition. Lastly, ratios of 10–40 %B were recently detected in fibrous diamonds from Pulandian, North China craton (a dozen of those carried high-Mg carbonatitic HDFs and one contained saline compositions; Mershon et al. 2021). HDFs were also found in a NFMC diamond from Finsch with 17 %B (low-Mg carbonatitic HDFs; Weiss et al. 2014), and in clear NFMC diamonds from Voorspoed and Venetia with 40–65 %B (saline, silicic to low-Mg carbonatitic and high-Mg carbonatitic HDFs; Jablon and Navon 2016).

Together, all these diamonds indicate much longer mantle residence times compared with the majority of fibrous diamonds that carry only A-centers at similar nitrogen concentrations. They show that fibrous diamond and the HDF they carry are not necessarily connected to kimberlite eruptions and formed episodically over extended geological periods. This is true in various lithospheric provinces and for all four HDF types. The above ideas are supported by the recent U–Th/He geochronology of HDF-bearing fibrous diamonds from the Congo and Kaapvaal cratons, indicating penetration of HDFs and formation and survival of fibrous diamonds in the SCLM over a time span of >1 billion years (Timmerman et al. 2019b; Weiss et al. 2021).

Geographical distribution of high-density fluids

Fibrous diamonds are documented in many localities from all Archean cratons (Fig. 12). This indicates the global distribution, the long history, and the ubiquity of the four HDF types in the deep continental lithosphere. However, some differences exist when comparing the compositions of similar HDF types from various localities.

Silicic to low-Mg carbonatitic HDFs. Silicic to low-Mg carbonatitic compositions were reported from the Kaapvaal craton (Schrauder and Navon 1994; Weiss et al. 2018; Kempe et al. 2021a; Weiss unpublished data), the Congo craton (Navon et al. 1988; Kopylova et al. 2010; Kosman et al. 2016; Timmerman et al. 2019), the West African craton (Weiss et al. 2009), the Amazonian craton (Shiryaev et al. 2005), the Slave craton (Klein-BenDavid et al. 2007; Logvinova et al. 2008b; Weiss et al. 2015), the North Atlantic craton (Elazar and Weiss 2021), India (Schrauder et al. 1994), and the Siberian craton (Klein-BenDavid et al. 2009; Zedgenizov et al. 2009, 2011; Skuzovatov et al. 2016; Gubanov et al. 2019). The Siberian HDFs have the most extreme compositions on both ends of the silicic to low-Mg carbonatitic array, with SiO_2 both >65 and <15 wt% (on a carbonate- and water-free basis; Figs. 7a and 8a). As noted by Elazar et al. (2021) the HDFs from Siberia also form a distinct high-Al trend compared to similar HDF compositions from all other localities/cratons (Fig. 8c). CaO enrichment, while more restricted, is observed for HDFs in Ebelyakh fibrous diamonds (Fig. 8b) and elevated FeO values were recorded for Internationalnaya (not shown). These enrichments are compensated by lower K_2O contents (typically <10 wt%, Fig. 8d), as well as lower P_2O_5 (not shown) and Cl values (Fig. 7b). Another distinction of the Siberian silicic to low-Mg carbonatitic HDFs is their lower (Th,La,Ba)/Nb ratios (Figs. 10a and 11).

High-Mg carbonatitic HDFs. The high-Mg carbonatitic HDFs in Siberian fibrous diamonds (e.g., Klein-BenDavid et al. 2009; Zedgenizov et al. 2009, 2011; Skuzovatov et al. 2016; Gubanov et al. 2019) also show some unique compositional characteristics compared with those from the Slave craton (Klein-BenDavid et al. 2007; Logvinova et al. 2008b), the Kalahari craton (Timmerman et al. 2018a; Weiss and Goldstein 2018; Kempe et al. 2021a,b; Weiss unpublished data), the West African craton (Weiss et al. 2009) and the North China craton (Mershon et al. 2021). At Internationalnaya, Al_2O_3 and FeO contents are higher than at any other locality worldwide (Fig. 8c). All Siberian high-Mg carbonatitic HDFs have lower BaO contents resulting in lower Ba/Rb and Ba/K ratios compared to high-Mg carbonatitic HDFs from other localities/cratons (Figs. 10b and 11). In addition, the general enrichment of incompatible elements in such HDFs from Siberia is moderate, characterized by $La/Sm_{PM} < 20$, compared with values between 20–80 in other localities; (La,Th)/(Nb,Rb) and Sr/Rb show similar behavior, i.e., lower ratios in Siberia compared to other provinces.

Saline HDFs. Saline HDFs were identified in fibrous diamonds from the Kaapvaal craton (Izraeli et al. 2001; Timmerman et al. 2018a,b; Weiss and Goldstein 2018; Weiss et al. 2018; Kempe et al. 2021a; Weiss unpublished data), the West African craton (Weiss et al. 2009), the Slave craton (Klein-BenDavid et al. 2007; Tomlinson et al. 2006; Weiss et al. 2015), the Superior craton (Smith et al. 2012), the North China craton (Mershon et al. 2021), and the Siberian craton (Zedgenizov et al. 2018). No diamonds with saline HDFs were found in the

Congo craton. Saline HDFs from Udachnaya have higher TiO_2 concentrations (~5 wt%), the ones from Koffiefontein are higher in FeO (10–20 wt%), and those from De Beers Pool have elevated Na_2O (mostly >15 wt%), compared to all other localities (Fig. 8e).

A striking dichotomy is observed for BaO (Fig. 8f), which is very high in saline HDFs from the Slave craton (10–18 wt%) and Udachnaya (13–15 wt%), but much lower in those from the Kaapvaal craton (BaO < 8 wt%). HDFs from the 1.2 Ga Premier kimberlite, on the Kaapvaal craton, have the highest BaO concentration (8 wt%, Weiss unpublished data) and those from the >2.7 Ga Wawa deposit (14–17 wt%) are as enriched as other Canadian HDFs, indicating that high Ba levels in saline HDFs existed for billions of years. The Ba enrichment in the Slave craton HDFs is reflected in higher Ba/Rb and Ba/Nb ratios, which coincide with well-developed fractionated 'Ribbed' trace element patterns of the Slave saline HDFs (i.e., high (La,Th)/(Nb,Rb) and LREE/(MREE,HREE) ratios; Figs. 10a,b and 11).

High-density fluids and lithospheric lithology

The petrology of the source rocks of the HDFs, as well as the wall-rocks they interact with during migration, must be important in determining their major and trace element composition (e.g., Tomlinson et al. 2009; Zedgenizov et al. 2009; Weiss et al. 2011, 2013b; Gubanov et al. 2019). Figure 12 presents the abundances of the various HDF types globally and emphasizes that it varies between localities and cratons. For example, only silicic to low-Mg carbonatitic HDFs were reported in fibrous diamonds from Jwaneng, Botswana (Schrauder and Navon 1994), DRC (mostly from the Mbuji-Mayi area, Navon et al. 1988; Kopylova et al. 2010) and the CH-7 kimberlite at Chidliak (Baffin Island, Canada, Elazar and Weiss 2021). In Jwaneng and Chidliak, an eclogite-dominated source for diamonds is indicated by mineral inclusions in diamonds and kimberlite indicator minerals (Gurney et al. 1995; Pell et al. 2013; Gress et al. 2018; Xia 2018). Similarly, eclogitic minerals dominate the assemblage in Mbuji Mayi diamond inclusions (Mvuemba-Ntanda et al. 1982) and ~90% of the xenoliths are eclogitic (El-Fadili et al. 1999). The association between HDF type and the dominant diamond-host lithology in these three deposits agrees with the strong connection that has been established between silicic to low-Mg carbonatitic HDF compositions and hydrous/carbonated eclogites (see '*Formation of silicic to low-Mg carbonatitic HDFs*' below for details). Similar relations can be identified at other localities, although the correlations are not as defined as in the examples above.

In Siberia, most of the 100 analyzed fibrous diamonds are from the Udachnaya and Internationalnaya kimberlites (25 and 39 diamonds, respectively). In both kimberlites, fibrous diamonds carry all the HDF types (Figs. 7 and 12), but the representation of the various types is not uniform. The proportions of {saline}:{silicic to low-Mg carbonatitic}:{high-Mg carbonatitic} HDFs is 25:36:39 in Udachnaya, and 3:62:35 in Internationalnaya. Peridotite is the predominant diamond host lithology in Yakutian kimberlites, as evident by diamond inclusions and xenoliths (Sobolev 1977; Kharkiv et al. 1997, as cited by Ionov et al. 2010; Sobolev et al. 2004). This explains the predominance of high-Mg carbonatitic HDFs in the two localities, as this HDF type has been linked to a carbonated peridotite source (see '*Formation of high-Mg carbonatitic HDFs*' below for details). However, it cannot account for the significant abundance of silicic to low-Mg carbonatitic HDFs. We also note that saline HDFs are more abundant in fibrous diamonds from the Udachnaya kimberlite, where Cl was detected in melt inclusions in groundmass olivine (Kamenetsky et al. 2009). In comparison, at the Ebelyakh placers the majority of the 19 analyzed fibrous diamonds carry silicic to low-Mg carbonatitic HDFs (90%), with a minor representation of high-Mg carbonatitic (10%) and no saline compositions, quite distinct from the observed proportions in Udachnaya and Internationalnaya (Fig. 12). Within the framework of the apparent relationship between HDF type and diamond-host lithology, these differences follow the known distribution of mineral inclusions in diamonds from the Siberian placers, which are dominated by an eclogitic suite (>70%; Shatsky et al. 2015), compared to largely peridotitic inclusions in diamond at Udachnaya and Internationalnaya (Sobolev et al. 2004).

High-density fluid types in the southwestern Kaapvaal and the central Slave craton are dominated by saline compositions (>80%; Fig. 12). Many of the fibrous diamonds with saline HDFs carry micro-mineral inclusions or mixed mineral-HDF microinclusions and can therefore be assigned to either an eclogitic or a peridotitic paragenesis. Most of those samples are associated with peridotitic hosts (Izraeli et al. 2001; Tomlinson et al. 2006; Klein-BenDavid et al. 2007; Weiss et al. 2015, 2018). This is consistent with the predominance of peridotite in both lithosphere provinces based on mineral inclusions in diamonds and xenoliths abundances (e.g., Gurney et al. 1979; Lawless et al. 1979; Skinner 1989; Stachel et al. 2003; Menzies et al. 2004; Tappert et al. 2005). In the southwestern Kaapvaal, the minority (25%) of saline HDFs associated with micro-mineral inclusions are of eclogitic affinity (most of which are from Koffiefontein; Izraeli et al. 2001; Timmerman et al. 2018a,b; Weiss and Goldstein 2018; Weiss et al. 2018; Weiss unpublished data), while the remaining 75% are associated with peridotitic inclusions. In contrast, in the central Slave eclogitic minerals are only associated with silicic HDFs and only peridotitic minerals are found in diamonds with saline HDFs (Tomlinson et al. 2006; Weiss et al. 2015). High-Mg carbonatitic HDFs are solely linked to peridotitic micro-mineral inclusions in both cratons (including the Witwatersrand East Block of the Kaapvaal; Klein-BenDavid et al. 2004; Weiss et al. 2018; Kempe et al. 2021a; Weiss unpublished data).

In addition, a temporal gap was identified between saline HDFs and other compositional types in the Kaapvaal (Weiss and Goldstein 2018; Weiss et al. 2018, 2021; Kempe et al. 2021a). Such a gap, based on nitrogen aggregation of the host diamonds, was not observed in the central Slave (Tomlinson et al. 2006; Klein-BenDavid et al. 2007; Weiss et al. 2015). In both cases, saline HDFs invaded the lithosphere and interacted with different lithologies, as evident by major and trace elements and isotopic variations of the HDFs, and the HDF-associated micro-mineral inclusions (Izraeli et al. 2001, 2004; Tomlinson et al. 2009; Weiss et al. 2015, Weiss and Goldstein 2018). However, while in the Kaapvaal this fluid-rock interaction did not lead to melting of silicates and formation of non-saline HDFs (Weiss et al. 2018), in the Slave both silicic and carbonatitic HDFs formed by *in situ* melting as the saline HDFs traversed mixed peridotite-eclogite lithosphere (Weiss et al. 2015). This difference in the impact of saline HDF percolation through both lithospheres is attributed to the extremely refractory nature of diamond-bearing harzburgite/dunite in the Kaapvaal, compared to the significantly less depleted Slave lithosphere (as well as the Siberian one; Stachel et al. 2003; Stachel and Harris 2008).

THE FORMATION OF HIGH-DENSITY FLUIDS

Pressure and temperature of fibrous diamond formation

The lithospheric origin of fibrous diamonds is manifested in a few independent lines of evidence:

a) The majority of fibrous diamonds carry high nitrogen concentrations that reside only in A-centers, and most of those with both A- and B-centers have low B/(B+A) ratios (see '*Temporal distribution of HDFs*' above). At the top of the asthenosphere, at ~1400 °C, B-centers would develop in a very short time, a few thousand years or less (Fig. 2, Taylor et al. 1996; Leahy and Taylor 1997), indicating that the fibrous diamonds grew in the lithosphere at cooler temperatures.

b) Based on the shift of the IR peaks of secondary quartz in silicic HDF-bearing diamonds, Navon (1991) estimated that fibrous diamond originated at lithospheric pressures of 4–7 GPa.

c) Geothermobarometry on micro-mineral inclusions that are found together with the HDF microinclusions yielded low temperatures of 1000–1185 ± 200 °C for Koffiefontein fibrous diamonds (Izraeli et al. 2001) and 1000 ± 100 °C for De Beers Pool ones (Weiss et al. 2018). Tomlinson et al. (2006) obtained a range of 930–1010 °C from larger mineral inclusions they found in fibrous diamonds from Panda. These are all lithospheric temperatures that correspond to pressures of 4–6 GPa along cratonic geotherms (e.g., Hasterok and Chapman 2011), and fall in the lower part of the temperature range obtained from mineral inclusions in NFMC lithospheric diamonds (e.g., Stachel and Harris 2008; Nimis et al. 2020; Nimis 2022).

High-density fluids as low-volume melts

The major and trace element chemistry of the HDFs indicates that they are low-volume melts, e.g., low-degree partial melts, percolating melts that interacted with a large volume of rock, or the very last residue following crystallization of a volumetrically more abundant parental melt. The high levels of enrichment in many incompatible elements can only be reached if the HDFs represent small fractions. Water comprises ~8–21 wt% of the HDFs (Table 2) compared with 100–500 ppm in the primitive mantle (McDonough and Sun 1995), corresponding to enrichment by a factor of 300–1500. Carbon, K and Cl yield similar factors of ~450, 500 and 1,000–10,000, respectively. Even if the source was enriched relative to primitive mantle, the melt/rock ratio must be small. Trace elements reveal similar enrichment in incompatible elements (Fig. 9) as do estimates based on the enrichment in He and Ar (~10^3–10^4, Turner et al. 1990, Burgess et al. 1998). Also, experimental studies of melting of peridotite or eclogite in the presence of CO_2 and/or H_2O have shown that melts of similar major element composition only form close to the solidus (see below).

Trace element abundance patterns of HDFs are best explained by equilibration with mantle peridotite without accessory phases for the 'Planed', and with the presence of mica and titanates for 'Ribbed' patterns (Weiss et al. 2013b). Such equilibration may happen during melting, but also during percolation or crystallization. The higher abundances of Ba, Th, U and LREE in the 'Ribbed' patterns, suggest an even lower fraction of partial melting for such HDFs. This allows for the survival of mica, titanates, and zircon in the source (or their growth during percolation) and leads to the formation of the typical negative anomalies in alkalies, Ti, Nb, Ta, Zr and Hf.

The positive correlation of Sr isotopic ratios with La/Nb ratios (low in 'Planed' and high in 'Ribbed' HDFs, Klein-BenDavid et al. 2014) suggests that the different patterns originate at the source or during percolation, as it is difficult to explain the isotope-trace-element correlation by fractionation. The origin of the 'Planed' patterns with low $^{87}Sr/^{86}Sr$ ratios can be linked to a source in the convective asthenosphere (similar to that of Group I kimberlites and ocean-island basalts, OIBs, e.g., Hofmann 1997; Paton et al. 2007), while the high $^{87}Sr/^{86}Sr$ ratios and the 'Ribbed' patterns reflect old enrichment of the source region that lies, most probably, in the lithosphere (e.g., Smith 1983; Giuliani et al. 2015).

Immiscible separation between the various high-density fluids

Schrauder and Navon (1994) considered three processes for the formation of the silicic to low-Mg carbonatitic HDFs: melting, crystallization, and immiscibility. The immiscibility scenario gained popularity with the discovery of the saline HDFs (Izraeli et al. 2001; Klein-BenDavid et al. 2004, 2007). Klein-BenDavid et al. (2007) pointed out the absence of intermediate compositions between the silicic and the saline endmembers (Fig. 7) and suggested that crystallization of carbonates drives carbonatitic melts into an immiscibility gap where silicic and saline melts segregate. Safonov et al. (2007, 2009) conducted experiments in the $CaMgSi_2O_6$–($CaCO_3$, Na_2CO_3)–KCl system and found such an immiscibility gap. However, in their water-free system, the evolution was from the immiscible saline and silicic melts at high temperatures towards one carbonatitic composition at lower temperatures. Contrasting fluid evolution recorded in zoned diamonds (Shiryaev et al. 2005; Weiss et al. 2009) and the variability of $^{87}Sr/^{86}Sr$ led to the rejection of the immiscibility model (Weiss et al. 2009).

Formation of high-Mg carbonatitic high-density fluids

Zedgenizov et al. (2007, 2009, 2011) reviewed a few possible models for the formation of high- and low-Mg carbonatitic HDFs. They recognized the similarity of the high-Mg compositions to kimberlites, but also pointed out differences between these HDFs and kimberlitic magmas. Following the distinction between low-Mg and high-Mg carbonatitic HDFs (Klein-BenDavid et al. 2009), Weiss et al. (2009, 2011) established a strong connection between high-Mg carbonatitic HDFs and a carbonated peridotite source, based on the results of near-solidus melting experiments (Dalton and Presnall 1998; Dasgupta and Hirschmann 2007; Brey et al. 2008). This is supported by the association of high-Mg carbonatitic HDFs solely with peridotitic micro-mineral inclusions (e.g., Klein-Ben David et al. 2004; Weiss et al. 2018; Gubanov et al. 2019).

In view of the possible connection between high-Mg carbonatitic and saline HDFs (Klein-BenDavid et al. 2004), Weiss et al. (2009) examined potential options for the formation of such an array. Their preferred scenario was the penetration of saline HDF into a carbonated peridotite. The saline component provides a source for K, Cl and Ba, and may provide many of the trace elements of the carbonatitic HDF. Later, Weiss et al. (2011) distinguished between Udachnaya HDFs with high-K_2O and 'Planed' patterns and those from Kankan with low-K_2O and more 'Ribbed' pattern. They emphasized the close similarity of the trace element abundance patterns to those of Group I and Group II kimberlites and suggested that both high-Mg carbonatitic HDFs and kimberlites form by melting of carbonated peridotite, with or without phlogopite. The HDFs represent a lower degree of melting (~0.5% for Udachnaya and ~0.2% for Kankan) compared with Group I and II kimberlites (2% and 0.8%, respectively; see Fig. 6 and 7 of Weiss et al. 2011). Klein-BenDavid et al. (2014) reported the Sr isotope compositions of Udachnaya and Kankan HDFs. The values measured for Udachnaya (0.7043–0.7053) fall in the range of Group I kimberlites, and those measured in a Kankan diamond (0.7079) overlap with Group II kimberlites.

More evidence for the similarities between kimberlites and high-Mg carbonatitic HDFs comes from melting experiments of carbonated peridotite, where the first melt at the solidus is similar in composition to the HDFs (e.g., Dasgupta and Hirschmann 2007) and approaches that of kimberlite at higher melt fractions (e.g., Dalton and Presnall 1998; Brey et al. 2009). In summary, melting of a carbonated peridotite, either due to heating or due to penetration of saline fluids is the preferred scenario for the formation of high-Mg carbonatitic HDFs.

Formation of silicic to low-Mg carbonatitic high-density fluids

The high silica content of the silicic HDFs is not at equilibrium with peridotite and calls for another source rock. Experiments in the eclogite-H_2O system at 4–6 GPa produce highly silicic, water-rich fluids that resemble HDFs (Kessel et al. 2005). Kiseeva et al. (2012) demonstrated that even in the eclogite–CO_2 system, the first melt is silicic, rich in SiO_2, Al_2O_3 and K_2O, which is joined by an immiscible carbonatitic melt at higher temperatures. The carbonatitic melt they produced is low in MgO and its composition resembles that of the low-Mg carbonatitic HDFs, but with low K_2O, reflecting the high CO_2/K_2O ratio of 11 of the starting material (much higher than the mantle value of ~1.5; McDonough and Sun 1995). Other experimental melts in the eclogite–CO_2 system are also close to the low-Mg carbonatitic HDFs (e.g., Hammouda 2003; Yaxley and Brey 2004). The connection between silicic to low-Mg carbonatitic HDFs and eclogites is also supported by the association of such compositions with omphacite and eclogitic garnet micromineral inclusions (e.g., Klein-Ben David et al. 2009; Weiss et al. 2009, 2015; Gubanov et al. 2019). Subducted sediments are another possible source for producing some silicic to low-Mg carbonatitic HDFs (e.g., Bulatov et al. 2014).

Weiss et al. (2009) suggested that the continuous array from the hydrous-silicic to low-Mg carbonatitic HDFs is the result of dissolution of carbonates in H_2O- and K_2O-rich silicic melts with increasing temperature. Litasov et al. (2010) favored evolution in the opposite direction and argue that in an eclogite–CO_2–Cl system, more silicates are dissolved into an initial carbonatitic melt with increasing temperature and degree of melting.

Recently Elazar et al. (2019) have shown that for eclogites containing both water and CO_2 the initial melt is not silicic or carbonatitic, but rather of intermediate composition. They suggested that it is the CO_2/H_2O ratio of the source rather than temperature that controls the position of the near-solidus melt along the silicic to low-Mg carbonatitic array. Dvir et al. (2017) provided evidence that near solidus fluids and melts in the lherzolite–H_2O–CO_2 system also fall close to the central part of the silicic to low-Mg carbonatitic array. In their experiments with CO_2/H_2O ratios of ~0.1 and 0.5 (by weight), the melt of the CO_2-rich composition has higher MgO and lower SiO_2 content compared with the melt of the H_2O-rich one, supporting the role of the source CO_2/H_2O ratio in controlling the silicic–carbonatitic character of the HDF. Further evidence for the control exerted by the source CO_2/H_2O ratio rather than by temperature and the degree of melting is provided by the composition of the HDFs. Silicic and low-Mg carbonatitic compositions have similar levels of K_2O, TiO_2 and Cl (Table 2). Temperature control should lead to increasing melt fraction and dilution of these incompatible elements, but such a trend is not observed in the analytical data. The similar (U,Th,La)/K of the silicic and low-Mg carbonatitic HDF types (Fig. 10b) further strengthens this suggestion.

In summary, the array represents the melting of eclogites that contain water and carbonates in variable amounts. Figures 7 and 8 show that most of the silicic to low-Mg carbonatitic HDFs fall in the middle of the array, indicating that the majority originate from rocks that carry both water and carbonates. Sediments are less probable, but possible sources of some of the HDFs.

Formation of saline high-density fluids

The saline HDFs are the most intriguing as they do not resemble any mantle melts that erupt on Earth's surface. Melting of peridotite or eclogite source-rocks is unlikely to form K-rich saline HDFs. K and Cl have similar incompatibility, but while the average molar K/Cl ratio of the HDFs is 0.68 ± 0.14 1σ (Supplementary compilation of HDF data, Weiss et al. 2022), the mantle ratio, as defined by MORB and OIB glasses, falls in the range of 6–45 (e.g., Kendrick et al. 2013, 2017). Fractionation of mantle melts is also unlikely. Most carry low concentrations of both K and Cl and require extensive fractionation that is not expected to produce a uniform residual K- and Cl-rich saline HDF.

Klein-BenDavid et al. (2007) noted a consistent negative correlation of MgO and CaO with Cl content in saline HDFs. They examined fractionation of Mg-rich carbonates from a carbonatitic HDF, concluded that extreme fractionation is needed, and noted the difficulty to remove potassium (relative to Cl). Carbonates, phlogopite and high-Si mica micro-mineral inclusions were found in association with saline HDFs (Izraeli et al. 2004, Weiss et al. 2018), but mica removal is limited by the low abundance of silica and alumina, and crystallization of K-bearing carbonates is required. This is not impossible (e.g., Shatskiy et al. 2016, Arefiev et al. 2019), but such phases are rare (Logvinova et al. 2019a). Last, HDFs of intermediate compositions between carbonatitic and saline compositions are not common in fibrous diamonds (Figs. 7 and 8, but see Klein-BenDavid et al. 2004). Similar difficulties exist for a possible fractionation from a silicic composition to form saline HDFs.

The high Cl content of the saline HDFs and the association with water and carbonates strongly suggests a subducted source (Izraeli et al. 2001; Tomlinson et al. 2006; Weiss et al. 2015, 2018; Timmerman et al. 2018a). Indeed, the K/Cl ratio of the saline HDFs is within the range of altered oceanic crust (K/Cl = 0.3–2; Ito et al. 1983; Sano et al. 2008).

This alternative is also supported by the pronounced positive Eu and Sr anomalies of saline HDFs in fibrous diamonds from the Slave craton (Weiss et al. 2015), which reflect the involvement of plagioclase during low-pressure crustal processes of protolith formation. Low $^3He/^4He$ isotopic ratios of South African saline HDF (2.7–4.4 Ra), further strengthen a connection with recycled subducted surface material (Weiss et al. 2017, 2021a; Timmerman et al. 2018b).

The chemical evolution of saline HDFs is not yet clear. Weiss et al. (2015) suggested that they originated as pore fluids trapped in the crust during low-pressure hydrothermal alteration. The high K/Na ratios evolved through the consumption of H_2O and Na during spilitization (hydration of basalt) and/or due to preferential incorporation of sodium into eclogitic minerals (principally omphacite) at high pressure (Izraeli et al. 2001). Dehydration at 150–200 km depth can generate highly saline fluids. Marine sediments are another possible source for the saline HDFs, based on high-pressure–temperature reaction experiments with peridotite (Förster et al. 2019). The equilibrium fluid in these experiments was not analyzed, but the crystallization of euhedral alkali chloride with K/Na = 0.3–6 and Cl/(Na+K) = 1.03 ± 0.44 molar ratios suggest growth from a fluid that is comparable to the saline HDFs.

Once formed, saline HDFs flow through different mantle lithologies. This is indicated by their association with micro-mineral inclusions. The most common are olivine, pyroxene and garnet with compositions typical for 'normal' mantle peridotites, websterites or eclogites (Izraeli et al. 2001, 2004; Klein-BenDavid et al. 2004; Tomlinson et al. 2006; Weiss et al. 2015, 2018; Timmerman et al. 2018a,b). These compositions and the association of similar HDFs with different mantle lithologies suggest that the minerals represent a preexisting mantle through which the saline HDF percolates rather than the lithology of their source. Fluid–rock interaction is also reflected in the trace element and isotopic composition of the saline HDFs.

Weiss et al. (2009, 2015) further suggested that the introduction of saline HDFs may induce *in situ* melting of carbonated-peridotite or eclogite, leading to the formation of high-Mg carbonatitic or silicic to low-Mg carbonatitic HDFs, respectively. This model connects all the HDF types to a common parental saline composition.

HIGH-DENSITY FLUIDS AND DIAMOND FORMATION

High-density fluids are excellent media for the nucleation and growth of diamonds. The lowest conditions for diamond synthesis, 1150 °C, 5.7 GPa, were achieved in the system $(K,Na)_2CO_3 + C_2H_2O_4$ (oxalic acid) + C (graphite) (Pal'yanov et al. 1999), which is a simple analog of carbonatitic HDFs. Similar temperatures were calculated for minerals included in fibrous diamonds (Izraeli et al. 2001; Weiss et al. 2018). Still, while the growth of fibrous diamonds from the HDFs they trap is widely accepted, the formation mode of NFMC diamonds is debated (e.g., Cartigny et al. 2014; Stachel et al. 2017).

It is agreed that mantle diamonds are metasomatic minerals and that their carbon is generally supplied by the metasomatic fluids (Haggerty 1986; Stachel and Luth 2015). However, two issues are still debated: a) the state of the fluids—subsolidus low-density fluids such as water or methane, or high-density fluids such as the diamond HDFs, kimberlites and similar melts. b) the oxidation state of the fluid—evidence was collected supporting the full wide range of carbon redox states, from carbonates and CO_2 (with C^{+4}) to CH_4 (C^{-4}).

Stachel and Luth (2015) used a solidus curve sketched by Wyllie (1987) to argue that the solidus of harzburgites+H_2O+CO_2 lies at ~1270 °C (at P >6 GPa), some 170 °C above that for lherzolite. However, Brey et al. (2011) have shown that at 1100 °C and 6 GPa K-rich carbonatitic melts coexist with harzburgites. The solidus of lherzolite + K_2O + CO_2 overlaps that of the harzburgite system. Adding water may lower the solidi to even lower temperatures.

Low-density hydrous fluids with 60–80 wt% water also exist with peridotite under such conditions (5–6 GPa, 1000 °C, Kessel et al. 2015). But if carbonate is also present, the water fraction decreases to 35–45 wt% (Dvir et al. 2017), not much higher than in the HDFs. At 1100 °C, above the solidus it falls to 28 wt%. It may be even lower if the K_2O/H_2O in the experiment (0.03–0.08) would be closer to the mantle ratio of ~1 (Kessel et al. 2015). We conclude that hydrous fluids at 5–6 GPa and 1100 °C would dissolve incompatible components such as carbonates, K_2O, Cl and incompatible trace elements and would resemble the HDFs documented in diamonds. Presumably, the water fraction in equilibrium with harzburgites is somewhat higher, but we expect that the difference between such assumed hydrous fluid and HDFs is rather small.

The carbonate-bearing HDFs present the best choice for the oxidation state of carbon in diamond-forming fluids, not only for fibrous diamonds, but for most NFMC diamonds as well. HDFs were identified in NFMC diamonds (Weiss et al. 2014; Jablon and Navon 2016; Kempe et al. 2021) and their trace element signature was also found in gem and semi-gem diamonds (Krebs et al. 2019). We note that they may also carry a small proportion of reduced carbon compounds (Kopylova et al. 2010).

A common argument against an extensive role for HDFs as the source of most NFMC diamonds is the oxidation state of the lithosphere that is below the EMOD buffer ($MgSiO_3+MgCO_3 = Mg_2SiO_4+C+O_2$). However, as shown by Stagno et al. (2013), pure carbonate melt is stable at more reducing conditions than EMOD, and if the carbonate fraction is diluted, its stability in the melt extends to cover even more reducing conditions (by up to ~2 log units). As pointed out by Stachel and Luth (2015) the buffering capacity of the lithosphere is small, and an oxidized melt can migrate through a reduced lithosphere, react with it and deposit diamonds over a long distance.

Can HDFs carry other species of carbon besides carbonates? This interesting possibility was presented by Stachel et al. (2017) for a C–O–H low-density fluid. Based on thermodynamic modeling, they argued that CO_2 (or carbonates) and CH_4 can coexist and react with each other to form diamonds. The idea is also supported by thermodynamic modeling of water-peridotite and water-eclogite systems that predict the coexistence of diverse oxidized and reduced carbon species (Sverjensky and Huang 2015; Huang and Sverjensky 2020). Can this be applied to HDFs? Kopylova et al. (2010) reported absorption at 2800–3000 cm^{-1} in IR spectra of fibrous diamonds from DRC and related it to CH_3 and CH_2 groups. We note that this region is prone to contamination on the diamond surface, but future studies should examine carefully the IR spectra of fibrous diamonds and look for the existence of organic reduced and oxidized species predicted by the thermodynamic models. If found, this will increase the diamond forming capacity of HDFs.

The collection of observations to date identifies HDFs as the best candidates for the formation of most lithospheric diamonds, however they are not the only ones. Observations of CH_4 in diamonds are rare (Smit et al. 2016) and growth from CH_4, or CH_4-H_2O low-density fluids was suggested based on modeling of isotopic fractionation (e.g., Thomassot et al. 2007; Smit et al. 2019). Experimental studies found that "in oxidized fluids (CO_2; CO_2–H_2O; H_2O), the intensity of spontaneous nucleation, the rate of diamond growth on seed crystals, and the degree of graphite-diamond transformation are considerably higher than those in reduced fluids (H_2O–CH_4; CH_4–H_2)" (Palyanov et al. 2015), but a recent study does suggest nucleation and growth of diamond in the harzburgite+CH_4 system (Matjuschkin et al. 2020). Smith et al. (2016) reported the finding of iron–nickel–carbon–sulfur inclusion in lower mantle diamonds, but metallic iron inclusions are known from shallower diamonds as well (Bulanova et al. 1998), and this mode of growth may be important and efficient where metallic melts exist (e.g., below 250 km, Frost and McCammon 2008).

HIGH-DENSITY FLUIDS, DIAMONDS, AND METASOMATISM

Experimental studies showed that hydrous, saline, carbonate and carbonate–silicate melts with compositions that are similar to all HDF types can efficiently crystallize diamonds (e.g., Akaishi et al. 1990; Pal'yanov et al. 1999, 2002; Litvin 2009; Bureau et al. 2016). Carbon isotopes confirm the growth of diamonds from carbon supplied by the HDFs (Klein-BenDavid et al. 2004: Shiryaev et al. 2005). Growth from a fluid is also consistent with the morphology and symmetric growth patterns of diamonds (Sunagawa 1984; Bulanova 1995; Tappert and Tappert 2005). Their close association with veins and alteration zones in xenoliths (e.g., Schulze et al. 1996; Howarth et al. 2014; Logvinova et al. 2015), and metasomatic minerals (e.g., Meyer 1987; Leost et al. 2003; Bulanova et al. 2004; Zedgenizov et al. 2018), indicate that diamond formation is a by-product of HDF-rock metasomatic interaction.

Elevated concentrations of incompatible elements in garnet and clinopyroxene, which would otherwise be refractory and depleted in composition, are evidence for metasomatism (Shimizu 1975; Hoal et al. 1994; Shimizu and Richardson 1987). Among these, sub-calcic garnets globally exhibit sinusoidal chondrite-normalized (CN) REE pattern, which is, perhaps, the most identified geochemical feature attributed to the formation of diamonds (Stachel et al. 1998, 2004; Grütter et al. 2004). The varying REE enrichment patterns in metasomatized garnets, both in xenoliths and as mineral inclusions in diamonds, are commonly explained as the result of a two-stage evolution of the continental lithospheric mantle, i.e., intensive chemical depletion by melting, followed by re-fertilization via enriched fluids or melts (e.g., Stachel and Harris 2008; Klein-BenDavid and Pearson 2009; Lazarov et al. 2012; Aulbach et al. 2013; Shu and Brey 2015; Shchukina et al. 2019). Stachel et al. (2004) attributed the sinusoidal REE patterns to the interaction of a C–O–H fluid highly enriched in incompatible elements with a depleted 'pre-metasomatic' garnet. Alternatively, Pearson et al. (1995) and Malkovets et al. (2007) suggested that the sinusoidal pattern is a primary feature and was formed upon direct garnet crystallization during a reaction between a silica-bearing CH_4-rich fluid and chromite-bearing harzburgite, which represent a pre-metasomatized Archean lithospheric mantle (Griffin et al. 2008). In either case, the fluids responsible for the sinusoidal REE pattern in mantle garnets are also associated with diamond formation.

Methane-rich fluid inclusions were identified in olivine phenocrysts from orogenic harzburgite (Song et al. 2009) and ophiolitic dunite (Liu and Fei 2006), but trace element data of such fluids are not available. Thus, the formation of sinusoidal patterns in lherzolite and harzburgite garnets and garnet inclusions in diamonds by interaction with methane-rich fluids cannot be assessed.

Ziberna et al. (2013) and Shu and Brey (2015) modeled the sinusoidal patterns in peridotitic garnets and showed that they can result through interaction between a representative kimberlite or a hypothetical enriched carbonatitic melt with a depleted garnet. The choice of kimberlites (or any other highly alkaline melt) as a metasomatic agent is reasonable, as they satisfy the need for an LREE/HREE enriched fluid to interact with the 'pre-metasomatic' garnet (Stachel et al. 2004). However, their composition at depth is still debatable. HDFs have equal, and some even higher, LREE/HREE ratios and are well characterized, pristine samples of deep fluids capable of metasomatizing the continental lithosphere.

Weiss et al. (2009, 2014) used a simple mass balance expression, $\phi C_{m,i}+(1-\phi)C_{g,i}=\phi C_{m,f}+(1-\phi)C_{g,f}$, to model the evolution of REE patterns in a garnet during interaction with a HDF. Where ϕ is the melt mass fraction, which in an open system metasomatic process may be larger than the actual porosity, and C_i and C_f are the initial and final concentrations of an element in the melt (m) or the garnet (g). This simplified expression ignores the other phases in the system, but provides a good approximation for harzburgite where garnet is the main depository of many trace elements. Using $C_{m,i}$ = HDF, $C_{g,i}$ = 'pre-metasomatic' garnet of Stachel et al.

(2004), and assuming equilibrium at the final stage, $C_{m,f} = C_{g,f}/D_{(garnet/HDF)}$, Weiss et al. (2009, 2014) showed that the reacted garnet ($C_{g,f}$) can develop sinusoidal and 'normal' REE patterns that are close to those of harzburgite and lherzolite garnets (Fig. 13). Similarly, 'U' or 'V' shaped REE patterns in ultradepleted harzburgite garnets can be formed due to interaction with a very small amount of HDFs ($\phi \ll 1\%$; Gibson et al. 2013). The important terms controlling the amplitude and shape of the garnet pattern are the REE pattern of the initial HDF and ϕ. For example, a more pronounced sinusoidal pattern is developed in garnet interacting with HDFs that have $La/Dy_{CN} \approx 100$ compared to those having $La/Dy_{CN} \approx 10$ (Fig. 13a), and the 'Dip' around Dy gets shallower (Lu/Dy decreases) with increasing ϕ (Fig. 13b). Based on this, Weiss et al. (2014) suggested that the progression of peridotite garnets from strongly sinusoidal (mostly in harzburgite) to normal REE patterns (common in lherzolite) can be explained by interaction with increasing amounts of the same HDF, rather than attributing the harzburgite patterns to interaction with low-density fluids and the lherzolite patterns to interaction with melts (Griffin et al. 1999; Stachel and Harris 2008; Banas et al. 2009). Here, we demonstrate that introducing HDFs to an eclogitic 'depleted' garnet produces an REE pattern with flat and elevated HREE–MREE and decreasing LREE values, which closely resemble the characteristic 'normal' REE pattern of eclogite garnets in mantle xenoliths and diamond inclusions (Fig. 13c; e.g., Stachel et al. 2004; Satchel and Harris 2008; Jacob 2004). Thus, HDFs can account for metasomatism in both peridotite and eclogite, two lithologies that are in close association with them, as evident from the micro-mineral inclusions recorded in fibrous diamonds (e.g., Tomlinson et al. 2006; Weiss et al. 2018).

High-density fluids provide a unique record of the compositions and nature of deep mantle fluids. Their high volatile content (Table 2) and the strong enrichment in incompatible elements of all HDF types (Fig. 9), make them effective metasomatic agents. The presence of HDFs in eclogitic garnets (Elazar et al. 2021) demonstrates their ability to migrate and react with lithospheric rocks over a range of depths (McKenzie 1989). Indeed, the mineralogy and chemical composition of xenoliths/xenocrysts and some alkaline magmas suggest a prevalent involvement of HDFs, in addition to other melts, in mantle metasomatism through time (e.g., Richardson et al. 1984; Erlank et al. 1987; Menzies 1990; O'Reilly and Griffin 2013).

HIGH-DENSITY FLUIDS, KIMBERLITES, AND ALKALI MANTLE MELTS

Diamonds are xenocrysts in kimberlites, as indicated by their occurrence in mantle xenoliths (Sobolev and Kuznetsova 1966; Viljoen et al. 1992; Smart et al. 2009; Aulbach et al. 2011), by the high pressures of inclusions in diamond (e.g., Meyer 1987; Liu et al. 1990; Izraeli et al. 2001; Stachel and Harris 2008; Nimis 2020), and by the old ages of most diamonds, much older than that of their host kimberlite (e.g., Richardson et al. 1984; Gurney et al. 2010; Koornneef et al. 2017, and references therein). Still, Akagi and Masuda (1988) and Schrauder et al. (1996) proposed a genetic relation between diamond-forming HDFs and kimberlites at depth, based on their similar trace element patterns. This similarity has been broadly confirmed for HDFs and kimberlites from different lithospheric provinces (e.g., Tomlinson et al. 2009; Zedgenizov et al. 2007, 2017; Rege et al. 2010; Weiss et al. 2011; Skuzovatov et al. 2012). Among the HDFs, the closest resemblance to kimberlites is demonstrated by the high-Mg carbonatitic type. This is based on their common high Nb,Ta/La ratio and their depletion in K, Rb and Cs compared with other highly incompatible elements like Ba, U, Th and La (Fig. 14a; Weiss et al. 2011). Notably, this trace element pattern characterizes the unique HIMU alkali ocean island basalts (OIB; having a high U/Pb ratio, μ) from the Cook-Austral islands (Chauvel et al. 1997; Willbold and Stracke 2006), St. Helena (Kawabata et al. 2011), and Bermuda (Mazza et al. 2019), as well as lavas from the Grande Comore island (Class et al. 1998) and oceanic and continental basanites (e.g., Weinstein et al. 2006; Pilet et al. 2008, 2011).

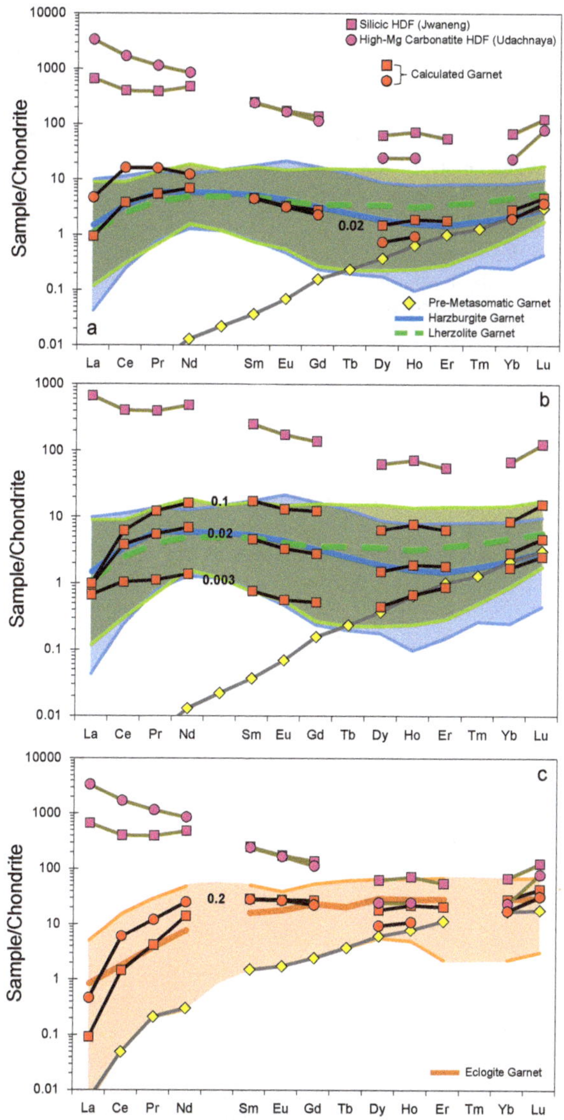

Figure 13. Rare earth element (REE) patterns of HDFs and of garnet inclusions in diamonds. (**a**) Calculated equilibrium patterns for garnet (**black line with red symbols**), formed by the interaction of HDFs (**gray line with pink symbol**) with a pre-metasomatic garnet (**gray line with yellow diamond symbols**; Stachel et al. 2004). The sinusoidal pattern amplitude is controlled by the HDF LREE/HREE enrichment. The average patterns of harzburgitic and lherzolitic garnets (**thick blue solid and green dashed lines**) and the corresponding ranges (**shaded blue and green areas**, Stachel et al. 2004) are shown for comparison. (**b**) The effect of the melt mass fraction (**black numbers**) on the amplitude and shape of the calculated sinusoidal patterns of the garnet. (**c**) Introducing HDFs into a 'depleted' eclogitic garnet produces REE patterns that closely resemble the characteristic 'normal' REE pattern of eclogitic garnets in mantle xenoliths and diamond inclusions (**thick orange solid line and shaded area**; Stachel et al. 2004). The calculations for peridotitic garnet (a and b) use the partition coefficients of Brey et al. (2008), though other sets of partition coefficients lead to similar results (e.g., Green 1994; Salters and Stracke 2004; Dasgupta et al. 2009), except for D_{La}, which varies significantly between different studies and is set here to 0.0015. For eclogitic garnet (c), the partition coefficients of Green et al. (2000) are used. Chondrite REE values are from McDonough and Sun (1995). See section '*HDFs, diamonds and metasomatism*' and Weiss et al. (2014) for additional details, and section '*Trace element analysis*' for the conversion of concentration in the diamond to concentration in the HDF.

The trace element patterns of HDFs, kimberlites and HIMU-basalts are subparallel (Fig. 14a), with their main difference being the abundances of highly incompatible elements and varying depletion in Ti, Zr and Hf (and Sr in the case of some basanites) compared to rare earth elements of similar incompatibility. These differences may reflect source heterogeneity, the degree of partial melting, the retention of accessory phases during melting, or different element partitioning in carbonatitic versus silicate melts. An important issue is the close link between the trace-element composition of HDFs trapped at depth and surface kimberlites and basalts, which indicates that the incompatible element content of the latter does not change much on their way to the surface. Moreover, it suggests similar enrichment and depletion processes experienced by their mantle source over geologic time, because batch melting of either a depleted source or a primitive mantle composition cannot reproduce the characteristic negative anomalies of the alkalis together with high Nb and Ta levels of these enriched melts.

Both carbonates (as minerals or melts) and peridotite are required for the generation of high-Mg carbonatitic HDFs and kimberlites (see section '*Formation of high-Mg carbonatitic HDFs*' above). The same constituents were suggested for the source of HIMU OIBs, based on their high CaO and Ca/Al ratios (Dasgupta et al. 2007; Jackson and Dasgupta 2008; Castillo 2015) and the high Mn/Fe and Ca/Al ratios of their olivine phenocrysts compared to other oceanic basalts (Weiss et al. 2016; Mazza et al. 2019). Similarly, a metasomatized peridotite source was proposed for basanites based on experimentally produced key major- and trace-element features (Pilet et al. 2008, 2011). Within this framework, carbonate can be introduced into peridotite through metasomatic interaction with HDFs and/or kimberlite melts, which form almost identical trace elements enriched sources at different mixing ratios (Weiss et al. 2011). Such interaction, for example, the addition of 0.5% of high-Mg carbonatitic HDF or 2.5% of kimberlitic melt into a depleted peridotite, mimics the trace-element patterns of some Archean metasomatized mantle xenoliths (e.g., Schmidberger and Francis 2001; Grégoire et al. 2003; Simon et al. 2007; Wittig et al. 2008) and the average post-Archean SCLM (McDonough 1990). These mantle-derived samples are a natural manifestation that illustrates the impact of carbonatite/kimberlite on depleted peridotite within the lithosphere. A source with similar trace element compositions is expected to form in the mantle transition-zone and deep upper mantle, as a result of similar interaction between carbonatitic melts which originate from subducting carbonated oceanic crust and the ambient mantle (Walter et al. 2008; Litasov et al. 2013; Regier et al. 2020).

To test the possible petrogenetic link between high-Mg carbonatitic HDFs and kimberlites, Weiss et al. (2011) used the trace element composition of available Archean xenoliths and the average post-Archean SCLM as natural representatives for a metasomatized carbonated peridotite and examined the products of melting. Figure 14b shows that simple batch melting of such a source closely reproduces the pattern of kimberlites. The same source can account for the pattern of HIMU basalts and basanites at higher melt fraction; however, the more compatible elements are too low in the produced melt. Weiss et al. (2016) found that batch melting of a mixture of 'normal' (trace-element-depleted) and metasomatized peridotite closely reproduces the complete trace element pattern of such basalts (Fig. 14c). They assumed a simple mixing of 80:20, respectively, between the two lithologies to produce a mixed peridotite source. The more enriched Bermuda lavas lie above the calculated melt, but with parallel pattern (Fig. 14c, Mazza et al. 2019). Such a pattern is reproduced at a lower melt fraction of a source with higher proportions of metasomatized peridotite.

The unique trace element patterns of kimberlites and HIMU-flavored alkali basalts require a distinctive enriched source that carries the signature of carbonatitic metasomatism (Hauri et al. 1993; Dalton and Presnall 1998; Saal et al. 1998; Greenwood et al. 1999; Le Roex et al. 2003; Dasgupta et al. 2007; Brey et al. 2008; Jackson and Dasgupta 2008). The impact of high-Mg carbonatitic HDFs on mantle peridotite can produce such a source. Other HDF types have different compositions and trace element patterns, but similar metasomatic

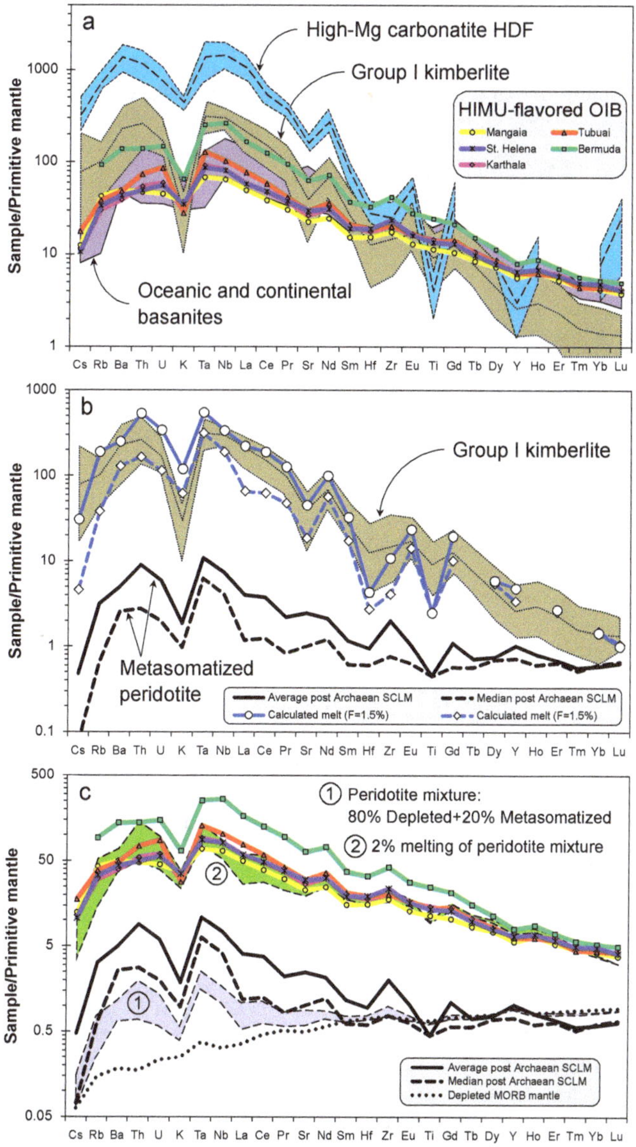

Figure 14. HDFs, kimberlites and alkali mantle melts. (**a**) Normalized trace element patterns comparing high-Mg carbonatitic HDFs from Udachnaya, global Group I kimberlites, HIMU-flavored alkali OIB, and oceanic and continental basanites. (**b**) Batch melting of a metasomatized peridotite (represented by average and median post Archean SCLM; McDonough 1990), with mantle mineralogy of Ol =66%, Opx =26%, Cpx = 3%, Grt = 5% and Carb < 1%. Using partition coefficients for mineral/carbonatitic-melt (Dasgupta et al. 2009) closely reproduces the pattern of kimberlites at partial melt fraction (F) of 1.5–2%. (**c**) Batch melting of a mixture of 80% 'normal' depleted MORB mantle (Salters and Stracke 2004) and 20% of the metasomatized peridotite of (b) result in a trace element pattern comparable to many HIMU-flavored alkali OIBs and oceanic and continental basanites. The calculations assume equilibrium with a residual modal mineralogy of Ol = 72%, Opx = 20%, Cpx = 2%, Gr = 5% and Carb = 1%; partition coefficients for silicate/basaltic melt from alphaMELTS (Smith and Asimow 2005), and F = 2%. The pattern of the metasomatized SCLM can be formed by interaction with HDFs. See section '*HDFs, kimberlites and alkali mantle melts*' and Weiss et al. (2011, 2016) for additional details.

potential upon interaction with mantle lithologies. The passage of such HDFs through and the interaction with different mantle lithologies impart chemical 'stains' that result in enriched mantle sources. In turn, the melting of such sources conveys these unique chemical and isotope compositional signatures to mantle-derived magmas that erupt at the Earth's surface. However, this is not a "chicken & egg" situation. HDFs with 'Planed' pattern (Fig. 9) may be derived directly from melting of depleted or primitive mantle and evolve into the other patterns as they percolate through the mantle (Weiss et al. 2013b).

SUMMARY AND CONCLUSIONS

Lithospheric diamonds form at depths >140 km beneath the Earth's surface from circulating carbon- and water-rich (C–O–H) fluids. High-density fluids (HDFs) encapsulated in microinclusions in fibrous diamonds are the best-preserved and most common samples of such fluids. Fibrous diamonds and the HDFs they carry are documented in localities of diverse ages and geographical provenance, mostly from Africa, Canada and Siberia, but also from Brazil, India and China, indicating their global distribution and long history in the subcontinental lithospheric mantle.

About 300 fibrous diamonds have been analyzed to date for their HDF major element compositions, roughly 200 of those have also been analyzed for trace element content, and some for their radiogenic isotope composition (mostly Sr isotope ratios). Four types of major element compositions, i.e., saline, silicic to low-Mg carbonatitic and high-Mg carbonatitic HDFs, are ubiquitous worldwide. Trace element compositions of HDFs also have consistent patterns globally, regardless of their host diamond provenance. Two repeating patterns are revealed: a highly fractionated one and a smoother one. The first is characterized by LILE and LREE enrichment, and HFSE and alkali depletion, similar to calcalkaline magmas and continental crustal rocks. The second has lower LILE and LREE abundances and 'smoother' overall trace-element pattern similar to oceanic basalts. These two patterns, as well as intermediate compositions, have been identified in all four types of HDFs, indicating the decoupling between major and trace elements of HDFs. The available Sr isotopic signatures suggest a relationship to the characteristics of their mantle sources, as also reflected in the trace element patterns.

A strong connection has been established between silicic to low-Mg carbonatitic HDF compositions and hydrous/carbonated eclogites, based on the association of these HDFs with eclogitic mineral inclusions, as well as their similarity to experimental near-solidus melts in the H_2O-CO_2-eclogite system. Similar arguments have been used to establish a connection between high-Mg carbonatitic HDFs and a carbonated peridotite source. Saline HDFs are more enigmatic and their chemical evolution and source rock are not as clear. At present, the most robust hypothesis is that saline HDFs originate from a subducted source, which is supported by their high Cl content, low K/Cl ratio, positive Eu and Sr anomalies and low $^3He/^4He$ isotopic ratios. Once formed, saline HDFs can percolate through different mantle lithologies, and induce *in situ* melting of carbonated-peridotite or eclogite, leading to compositional transition and the formation of silicic and carbonatitic HDF types.

The varying abundance of HDF types in different cratons is fascinating, and likely related to the petrological and physical characteristics of the local lithospheres. For example, the abundance of silicic to low-Mg carbonatitic HDFs in DRC and Jwaneng, two lithospheres with dominance of eclogitic diamonds, is prominent. Other observations, like the high abundance of saline HDFs in the Slave and southwestern Kaapvaal, are clear, but not yet reconciled. In addition, the distinct high-Al trend of silicic to low-Mg carbonatitic HDFs in Siberia compared to other lithosphere provinces, or the striking dichotomy between high-Ba saline HDFs from the Slave and low-Ba ones in the Kaapvaal, are unresolved. Integrated studies on HDF compositions and the local host lithosphere, as well as additional HDF isotopic analyses which are still scarce, are likely to shed light on such issues.

High-density fluids are excellent media for the nucleation and growth of diamonds. Here we summarized recent observations that indicate the involvement of HDFs in the formation of most (but not all) diamonds, fibrous and non-fibrous alike. The most widely recognized geochemical feature attributed to the formation of lithospheric diamonds are metasomatized garnets, which globally exhibit enriched REE patterns. Such patterns can form by the interaction of any HDF type with a depleted garnet in both peridotite and eclogite lithologies, thus strengthening the connection between diamond formation and HDFs. The prevalence of similar metasomatized garnets in mantle-derived xenoliths, together with the similarity in trace elements between HDFs (in particular high-Mg carbonatitic ones), kimberlites, HIMU-flavored alkali OIB and oceanic and continental basanites, reflect the impact of HDFs on the deep lithospheric mantle and their role in deep Earth processes.

High-density fluids provide a unique record of the compositions and nature of deep mantle C–O–H-bearing fluids. Their high volatile content, strong enrichment in incompatible elements and possible relation to subduction processes, make HDFs a key player in the global circulation of volatiles. However, to date, there are no robust age constraints for HDFs, nor for their host fibrous diamonds. Future progress in analytical techniques and/or sampling methods for analyzing HDFs, especially for determining their radiogenic isotope composition, are likely to offer a remedy and provide comprehensive chemical characteristics and meaningful geological ages of HDFs in diamonds. It will potentially elucidate the origin of all HDF types, the relationship between them, and the ages of their host diamonds and the metasomatic processes that led to their formation. Such new data will provide the opportunity to understand the complex history of alteration events in different lithospheric provinces, the type of fluids involved, and relate these deep mantle events to regional tectonics and magmatic history. Since HDFs trapped in diamonds are the most pristine samples of an abundant C–O–H-bearing fluids in Earth's mantle, they are likely to provide important contributions to our understanding of the deep carbon and water cycle throughout Earth's history.

ACKNOWLEDGMENTS

We thank Ofra Klein-BenDavid, Oded Elazar, Yael Kempe, and Reed Mershon for sharing their unpublished data and Dan Howell for a modified version of the DiaMap software. We thank Thomas Stachel and Karen Smit, the editors, for handling the paper and Dimitry Zedgenizov, Thomas Stachel, and an anonymous reviewer for constructive reviews that greatly improved the paper. This research was supported by the Israel Science Foundation grant No. 2015/18 to YW and grant No. 734/17 to ON.

REFERENCES

Akagi T (1999) Relationship of strontium isotope ratio and chemical composition of micro inclusions in fibrous diamonds: Implication for isotopically different components of the upper mantle. Geochem J 33:101–107

Akagi T, Masuda A (1988) Isotopic and elemental evidence for a relationship between kimberlite and Zaire cubic diamonds. Nature 336:665–667

Akaishi M, Kanda H, Yamaoka S (1990) Synthesis of diamond from graphite-carbonate system under very high temperature and pressure. J Cryst Growth 104:578–581

Anand M, Taylor LA, Misra KC, Carlson WD, Sobolev NV (2004) Nature of diamonds in Yakutian eclogites: Views from eclogite tomography and mineral inclusions in diamonds. Lithos 77:333–348

Angress JF, Smith, SD (1965) The observation of defect-activated one-phonon infra-red absorption in diamond coat. Philos Mag 12:415–417

Araújo DP, Griffin WL, O'Reilly SY, Grant KJ, Ireland T, Holden P, van Achterbergh E (2009) Microinclusions in monocrystalline octahedral diamonds and coated diamonds from Diavik, Slave Craton: Clues to diamond genesis. Lithos 112:724–735

Arefiev AV, Shatskiy A, Podborodnikov IV, Litasov KD (2019) The K_2CO_3–$CaCO_3$–$MgCO_3$ system at 6 GPa: Implications for diamond forming carbonatitic melts. Minerals 9:558

Aulbach S, Stachel T, Heaman LM, Carlson JA (2011) Microxenoliths from the Slave craton: Archives of diamond formation along fluid conduits. Lithos 126:419–434
Aulbach S, Griffin WL, Pearson NJ, O'Reilly SY (2013) Nature and timing of metasomatism in the stratified mantle lithosphere beneath the central Slave craton (Canada). Chem Geol 352:153–169
Banas A, Stachel T, Phillips D, Shimizu N, Viljoen KS, Harris JW (2009) Ancient metasomatism recorded by ultra-depleted garnet inclusions in diamonds from DeBeers Pool, South Africa. Lithos 12:736–746
Basu S, Jones AP, Verchovsky AB, Kelley SP, Stuart FM (2013) An overview of noble gas (He, Ne, Ar, Xe) contents and isotope signals in terrestrial diamond. Earth Sci Rev 126:235–249
Bauer MH (1896) Edelsteinkunde (English translation, 1904, by LJ Spencer) Reprinted 1968. Dover Publications, New York
Bibby DM (1979) Zonal distribution of impurities in diamond. Geochim Cosmochim Acta 43:415–423
Bibby DM (1982) Impurities in natural diamond. *In:* Chemistry and Physics of Carbon. Vol 18. Thrower PA (ed) Marcel Dekker CRC Press, New York, p 1–91
Bodnar RJ (2003) Introduction to fluid inclusions. *In:* Fluid inclusions: Analysis and interpretation. Vol 32. Samson I, Anderson A, Marshall D (eds) Mineralogical Association of Canada, Short Course, p 1–8
Boyd SR, Mattey DP, Pillinger CT, Milledge HJ, Mendelssohn M, Seal M (1987) Multiple growth events during diamond genesis: an integrated study of carbon and nitrogen isotopes and nitrogen aggregation state in coated stones. Earth Planet Sci Lett 86:341–353
Brey GP, Bulatov VK, Girnis AV, Lahaye Y (2008) Experimental melting of carbonated peridotite at 6–10 GPa. J Petrol 49:797–821
Brey GP, Bulatov VK, Girnis AV (2009) Influence of water and fluorine on melting of carbonated peridotite at 6 and 10 GPa. Lithos 112:249–259
Brey GP, Bulatov VK, Girnis AV (2011) Melting of K-rich carbonated peridotite at 6–10 GPa and the stability of K-phases in the upper mantle. Chem Geol 281:333–342
Broadley MB, Kagi H, Burgess R, Zedgenizov D, Mikhail S, Almayrac M, Ragozin A, Pomazansky B, Sumino H (2018) Plume-lithosphere interaction, and the formation of fibrous diamonds. Geochem Perspect Lett 8:26–30
Bulanova GP (1995) The formation of diamond. J Geochem Explor 53:1–23
Bulanova GP, Novgorodov PG, Pavlova LA (1988) The first find of a melt inclusion in diamond from the Mir pipe. Geokhimiya 5:756–765
Bulanova GP, Griffin WL, Ryan CG (1998) Nucleation environment of diamonds from Yakutian kimberlites. Mineral Mag 62:409–419
Bulanova GP, Muchemwa E, Pearson DG, Griffin BJ, Kelley SP, Klemme S, Smith CB (2004) Syngenetic inclusions of yimengite in diamond from Sese kimberlite (Zimbabwe)—evidence for metasomatic conditions of growth. Lithos 77:181–192
Bulatov VK, Brey GP, Girnis AV, Gerdes A, Höfer HE (2014) Carbonated sediment–peridotite interaction and melting at 7.5–12 GPa. Lithos 200:368–385
Bureau H, Frost DJ, Bolfan-Casanova N, Leroy C, Esteve I, Cordier P (2016) Diamond growth in mantle fluids. Lithos 265:4–15
Burgess R, Johnson LH, Mattey DP, Harris JW, Turner G (1998) He, Ar and C isotopes in coated and polycrystalline diamonds. Chem Geol 146:205–217
Burgess R, Layzelle E, Turner G, Harris JW (2002) Constraints on the age and halogen composition of mantle fluids in Siberian coated diamonds. Earth Planet Sci Lett 197:193–203
Burgess R, Kiviets GB, Harris JW (2004) Ar–Ar age determinations of eclogitic clinopyroxene and garnet inclusions in diamonds from the Venetia and Orapa kimberlites. Lithos 77:113–124
Burgess R, Cartigny P, Harrison D, Hobson E, Harris J (2009) Volatile composition of microinclusions in diamonds from the Panda kimberlite, Canada: Implications for chemical and isotopic heterogeneity in the mantle. Geochim Cosmochim Acta 73:1779–1794
Cartigny P, Harris JW, Javoy M (2001) Diamond genesis, mantle fractionations and mantle nitrogen content: a study of δ^{13}C–N concentrations in diamonds. Earth Planet Sci Lett 185:85–98
Cartigny P, Palot M, Thomassot E, Harris JW (2014) Diamond formation: a stable isotope perspective. Annu Rev Earth Planet Sci 42:699–732
Castillo PR (2015) The recycling of marine carbonates and sources of HIMU and FOZO ocean island basalts. Lithos 216:254–263
Chauvel C, McDonough W, Guille G, Maury R, Duncan R (1997) Contrasting old and young volcanism in Rurutu Island, Austral chain. Chem Geol 139:125–143
Chepurov AI, Tomilenko AA, Shebanin AP, Sobolev NV (1994) Fluid inclusions in natural diamonds from placers of Yakutia. Dokl Akad Nauk 336:662–665
Chinn IL, Gurney JJ, Milledge JH, Taylor WR, Woods PA (1995a) Cathodoluminescence properties of CO_2-bearing and CO_2-free diamonds from the George Creek K1 kimberlite dike. Int Geol Rev 37:254–258
Chinn IL, Gurney JJ, Milledge HJ, Taylor WR, McCallum ME (1995b) CO_2-bearing diamonds from the George Creek K1 kimberlite dyke of the Colorado-Wyoming State Line District. International Kimberlite Conference: Extended Abstracts 6:113–115

Chrenko RM, McDonald RS, Darrow KA (1967) Infra-red spectra of diamond coat. Nature 213:474–476

Clark CD, Davey ST (1984) One-phonon infrared absorption in diamond. J Phys C: Solid State 17:1127–1140

Class C, Goldstein SL, Altherr R, Bachèlery P (1998) The process of plume–lithosphere interactions in the ocean basins—the case of Grande Comore. J Petrol 39:881–903

Dalpé C, Hudon P, Ballantyne DJ, Williams D, Marcotte D (2010) Trace element analysis of rough diamond by LA-ICP-MS: A case of source discrimination? J Forensic Sci 55:1443–1456

Dalton JA, Presnall DC (1998) The continuum of primary carbonatitic–kimberlitic melt compositions in equilibrium with lherzolite: data from the system $CaO–MgO–Al_2O_3–SiO_2–CO_2$ at 6 GPa. J Petrol 39:1953–1964

Dasgupta R, Hirschmann MM (2007) A modified iterative sandwich method for determination of near-solidus partial melt compositions. II. Application to determination of near-solidus melt compositions of carbonated peridotite. Contrib Mineral Petrol 154:647–661

Dasgupta R, Hirschmann MM, Smith ND (2007) Partial melting experiments of peridotite + CO_2 at 3 GPa and genesis of alkalic ocean island basalts. J Petrol 48:2093–2124

Dasgupta R, Hirschmann MM, McDonough WF, Spiegelman M, Withers AC (2009) Trace element partitioning between garnet lherzolite and carbonatite at 6.6 and 8.6 GPa with applications to the geochemistry of the mantle and of mantle-derived melts. Chem Geol 262:57–77

Day JM, Barry PH, Hilton DR, Burgess R, Pearson DG, Taylor LA (2015) The helium flux from the continents and ubiquity of low-$^3He/^4He$ recycled crust and lithosphere. Geochim Cosmochim Acta 153:116–133

Dvir O, Pettke T, Fumagalli P, Kessel R (2011) Fluids in the peridotite–water system up to 6 GPa and 800 C: new experimental constrains on dehydration reactions. Contrib Mineral Petrol 161:829–844

Elazar O, Weiss Y (2021) Extension tectonics and mantle metasomatism in the North Atlantic Craton mirrored in diamond microinclusions. Israel Geological Society Annual meeting, Yeruham 2021, Abstract Book, p 33–34, https://en.igs.org.il/annual-conference

Elazar O, Frost D, Navon O, Kessel R (2019) Melting of H_2O and CO_2-bearing eclogite at 4–6 GPa and 900–1200 °C: Implications for the generation of diamond-forming fluids. Geochim Cosmochim Acta 255:69–87

Elazar O, Kessel R, Huang J-X, Marquardt K, Navon O (2021) Silicic microinclusions in a metasomatized eclogite from Roberts Victor mine, South Africa. Lithos 388:106057

El Fadili S, Demaiffe D, Gurney JJ (1999) Petrology of eclogite and granulite nodules from the Mbuji Mayi kimberlites (Kasai, Congo): the significance of kyanite–omphacite intergrowths. *In:* Proceedings of the 7th International Kimberlite Conference. Vol 1. Gurney JJ, Gurney JL, Pascoe MF, Richardson SH (eds) Red Roof Designs, Cape Town, p 205–213

Erlank AJ, Waters FG, Hawkesworth CJ, Haggerty SE, Allsopp HL, Rickard RS, Menzies MA (1987) Evidence for mantle metasomatism in peridotite nodules from the Kimberley pipes, South Africa. *In:* Mantle Metasomatism. Menzies MA, Hawkesworth J (eds) Academic, London, p 221–311

Evans T, Qi Z (1982) Aggregation of nitrogen in diamond. *In:* The properties of natural and synthetic diamond. Field JE (ed) Academic, London, p 259–290

Evans T, Qi Z, Maguire J (1981) The stages of nitrogen aggregation in diamond. J Phys C: Solid State 14:L379

Farley KA (2002) (U–Th)/He dating: Techniques, calibrations and applications. Rev Mineral Geochem 47:819–844

Faure S (2010) World Kimberlites CONSOREM Database (Version 3)

Fesq HW, Bibby DM, Erasmus CS, Kable EJD, Sellschop JPF (1975) A comparative trace element study of diamonds from Premier, Finsch and Jagersfontein mines, South Africa. Phys Chem Earth 817–836

Förster MW, Foley SF, Marschall HR, Alard O, Buhre S (2019) Melting of sediments in the deep mantle produces saline fluid inclusions in diamonds. Sci Adv 5:eaau2620

Frost DJ, McCammon CA (2008) The redox state of Earth's mantle. Annu Rev Earth Planet Sci 36: 389–420

Galimov EM (1984) Variations in isotopic composition of diamonds and inferences for diamond genesis conditions. Geokhimiya 8:1091–1118

Gibson SA, McMahon SC, Day JA, Dawson JB (2013) Highly refractory lithospheric mantle beneath the Tanzanian craton: evidence from Lashaine pre-metasomatic garnet-bearing peridotites. J Petrol 54:1503–1546

Giuliani A, Phillips D, Woodhead JD, Kamenetsky VS, Fiorentini ML, Maas R, Soltys A, Armstrong RA (2015) Did diamond-bearing orangeites originate from MARID-veined peridotites in the lithospheric mantle? Nat Commun 6:6837

Glazunov MP, Kobochnikov PN, Orlov YL (1967) Determination of impurities and investigation of their distribution in the natural diamonds from various places of origin with the neutron activation technique. Isotopenpraxis 3:231–235

Graham DW (2002) Noble gas isotope geochemistry of mid-ocean ridge and ocean island basalts: Characterization of mantle source reservoirs. Rev Mineral Geochem 47:247–317

Green TH (1994) Experimental studies of trace-element partitioning applicable to igneous petrogenesis—Sedona 16 years later. Chem Geol 117:1–36

Green TH, Blundy J D, Adam J, Yaxley GM (2000) SIMS determination of trace element partition coefficients between garnet, clinopyroxene and hydrous basaltic liquids at 2–7.5 GPa and 1080–1200 C. Lithos 53:165–187

Green BL, Collins AT, Breeding CM (2022) Diamond spectroscopy, defect centers, color, and treatments. Rev Mineral Geochem 88:637–688

Greenwood JC, Gibson SA, Thompson RN, Weska RK, Dickin AP (1999) Cretaceous kimberlites from the Paranatinga-Batovi region, central Brazil: geochemical evidence for subcratonic lithospheric mantle heterogeneity. *In:* Proceedings of the 7th International Kimberlite Conference. Vol 1. Gurney JJ, Gurney JL, Pascoe MF, Richardson SH (eds) Red Roof Designs, Cape Town, p 291–298

Grégoire M, Bell DR, Le Roex AP (2003) Garnet lherzolites from the Kaapvaal Craton (South Africa): trace element evidence for a metasomatic history. J Petrol 44:629–657

Gress MU, Howell D, Chinn IL, Speich L, Kohn SC, van den Heuvel Q, Schulten E, Pals ASM, Davies GR (2018) Episodic diamond growth beneath the Kaapvaal craton at Jwaneng mine, Botswana. Mineral Petrol 112:219–229

Griffin WL, Ryan CG and Win TT (1993) Trace elements in diamond: a proton microprobe study. CSIRO Restricted Report, 363R

Griffin WL, Shee SR, Ryan CG, Win TT, Wyatt BA (1999) Harzburgite to lherzolite and back again: metasomatic processes in ultramafic xenoliths from the Wesselton kimberlite, Kimberley, South Africa. Contrib Mineral Petrol 134:232–250

Griffin WL, O'Reilly SY, Afonso JC, Begg G (2008) The composition and evolution of lithospheric mantle: a re-evaluation and its tectonic implications. J Petrol 50:1185–1204

Grütter HS, Gurney JJ, Menzies AH, Winter F (2004) An updated classification scheme for mantle-derived garnet, for use by diamond explorers. Lithos 77 841–857

Gubanov N, Zedgenizov D, Sharygin I, Ragozin A (2019) Origin and evolution of high-Mg carbonatitic and low-Mg carbonatitic to silicic high-density fluids in coated diamonds from Udachnaya kimberlite pipe. Minerals 9:734

Gurney JJ, Zweistra P (1995) The interpretation of the major element compositions of mantle minerals in diamond exploration. J Geochem Explor 53:293–309

Gurney JJ, Jacob WRO, Dawson JB (1979) Megacrysts from the Monastery kimberlite pipe, South Africa. *In:* The Mantle Sample: Inclusions in Kimberlites and Other Volcanics. Boyd FR, Meyer HOA (eds), AGU, Washington DC, p 227–243

Gurney JJ, Helmstaedt HH, Richardson SH, Shirey SB (2010) Diamonds through time. Econ Geol 105:689–712

Guthrie Jr GD, Veblen DR, Navon O, Rossman GR (1991) Submicrometer fluid inclusions in turbid-diamond coats. Earth Planet Sci Lett 105:1–12

Haggerty SE (1986) Diamond genesis in a multiply-constrained model. Nature 320:34–38

Hammouda T (2003) High-pressure melting of carbonated eclogite and experimental constraints on carbon recycling and storage in the mantle. Earth Planet Sci Lett 214:357–368

Harris JW (1968) Recognition of diamond inclusions. Part 1. Syngenetic mineral inclusions. Part 2. Epigenetic Mineral Inclusions. Ind Diamond Rev 28:402–410:458–461

Harte B, Gurney JJ, Harris JW (1980) The formation of peridotitic suite inclusions in diamonds. Contrib Mineral Petrol 72:181–190

Hasterok D, Chapman DS (2011) Heat production and geotherms for the continental lithosphere. Earth Planet Sci Lett 307:59–70

Hauri EH, Hart SR (1993) Re–Os isotope systematics of HIMU and EMII oceanic island basalts from the south Pacific Ocean. Earth Planet Sci Lett 114353–371

Hoal KEO, Hoal BG, Erlank AJ, Shimizu N (1994) Metasomatism of the mantle lithosphere recorded by rare earth elements in garnets. Earth Planet Sci Lett 126:303–313

Hofmann AW (1997) Mantle geochemistry: the message from oceanic volcanism. Nature 385:219–229

Howarth GH, Sobolev NV, Pernet-Fisher JF, Barry PH, Penumadu D, Puplampu S, Ketcham RA, Maisano JA, Taylor D, Taylor LA (2014) The secondary origin of diamonds: multi-modal radiation tomography of diamondiferous mantle eclogites. Int Geol Rev 56:1172–1180

Howell D, O'Neill CJ, Grant KJ, Griffin WL, Pearson NJ, O'Reilly SY (2012) μ-FTIR mapping: Distribution of impurities in different types of diamond growth. Diamond Relat Mater 29:29–36

Huang F, Sverjensky DA (2020) Mixing of carbonatitic into saline fluid during panda diamond formation. Geochim Cosmochim Acta 284:1–20

Ionov DA, Doucet LS, Ashchepkov IV (2010) Composition of the lithospheric mantle in the Siberian craton: new constraints from fresh peridotites in the Udachnaya-East kimberlite. J Petrol 51:2177–2210

Ito E, Harris DM, Anderson Jr AT (1983) Alteration of oceanic crust and geologic cycling of chlorine and water. Geochim Cosmochim Acta 47:1613–1624

Izraeli ES, Harris JW, Navon O (2001) Brine inclusions in diamonds: a new upper mantle fluid. Earth Planet Sci Lett 187:323–332

Izraeli ES, Harris JW, Navon O (2004) Fluid and mineral inclusions in cloudy diamonds from Koffiefontein, South Africa. Geochim Cosmochim Acta 68:2561–2575

Jablon BM, Navon O (2016) Most diamonds were created equal. Earth Planet Sci Lett 443:41–47

Jackson MG, Dasgupta R (2008) Compositions of HIMU, EM1, and EM2 from global trends between radiogenic isotopes and major elements in ocean island basalts. Earth Planet Sci Lett 276:175–186

Jacob DE (2004) Nature and origin of eclogite xenoliths from kimberlites. Lithos 77:295–316

Jacob DE, Wirth R, Enzmann F, Kronz A, Schreiber A (2011) Nano-inclusion suite and high resolution micro-computed-tomography of polycrystalline diamond (framesite) from Orapa, Botswana. Earth Planet Sci Lett 308:307–316

Jacob DE, Dobrzhinetskaya L, Wirth R (2014) New insight into polycrystalline diamond genesis from modern nanoanalytical techniques. Earth Sci Rev 136:21–35

Javoy M, Pineau F, Demaiffe D (1984) Nitrogen and carbon isotopic composition in the diamonds of Mbuji Mayi (Zaire). Earth Planet Sci Lett 68:399–412

Johnson LH, Burgess R, Turner G, Milledge HJ, Harris JW (2000) Noble gas and halogen geochemistry of mantle fluids: comparison of African and Canadian diamonds. Geochim Cosmochim Acta 64:717–732

Kagi H, Lu R, Davidson P, Goncharov AF, Mao HK, Hemley RJ (2000) Evidence for ice VI as an inclusion in cuboid diamonds from high *PT* near infrared spectroscopy. Mineral Mag 64:1089–1097

Kagi H, Zedgenizov DA, Ohfuji H, Ishibashi H (2016) Micro-and nano-inclusions in a superdeep diamond from São Luiz, Brazil. Geochem Int 54:834–838

Kaiser W, Bond WL (1959) Nitrogen, a major impurity in common type I diamond. Phys Rev 115:857

Kamber BS, Collerson KD (1999) Origin of ocean island basalts: a new model based on lead and helium isotope systematics. J Geophys Res: Solid Earth 104:25479–25491

Kamenetsky VS, Maas R, Kamenetsky MB, Paton C, Phillips D, Golovin AV, Gornova MA (2009) Chlorine from the mantle: magmatic halides in the Udachnaya-East kimberlite, Siberia. Earth Planet Sci Lett 285:96–104

Kaminsky F, Wirth R, Matsyuk S, Schreiber A, Thomas R (2009) Nyerereite and nahcolite inclusions in diamond: evidence for lower-mantle carbonatitic magmas. Mineral Mag 73:797–816

Kaminsky FV, Wirth R, Schreiber A (2013) Carbonatitic inclusions in deep mantle diamond from Juina, Brazil: new minerals in the carbonate-halide association. Can Mineral 51:669–688

Kamiya Y, Lang AR (1965) On the structure of coated diamonds. Philos Mag 11:347–356

Kawabata H, Hanyu T, Chang Q, Kimura JI, Nichols AR, Tatsumi Y (2011) The petrology and geochemistry of St. Helena alkali basalts: evaluation of the oceanic crust-recycling model for HIMU OIB. J Petrol 52:791–838

Kempe Y, Navon O, Chinn I, Weiss Y (2021a) Multiple metasomatic diamond-forming events in a cooling lithosphere beneath Voorspoed, South Africa. Lithos 106285

Kempe Y, Weiss Y, Navon O (2021b) Time-temperature relations and formation history of diamonds from the Venetia and Voorspoed Mines, South Africa. Israel Geological Society Annual meeting, Yeruham 2021, Abstract Book, p 66–67, https://en.igs.org.il/annual-conference

Kendrick MA, Arculus R, Burnard P, Honda M (2013) Quantifying brine assimilation by submarine magmas: Examples from the Galápagos Spreading Centre and Lau Basin. Geochim Cosmochim Acta 123:150–165

Kendrick MA, Hémond C, Kamenetsky VS, Danyushevsky L, Devey CW, Rodemann T, Jackson MG, Perfit MR (2017) Seawater cycled throughout Earth's mantle in partially serpentinized lithosphere. Nat Geosci 10:222–228

Kennedy GC, Nordlie BE (1968) The genesis of diamond deposits. Econ Geol 63:495–503

Kessel R, Ulmer P, Pettke T, Schmidt MW, Thompson AB (2005) The water–basalt system at 4 to 6 GPa: Phase relations and second critical endpoint in a K-free eclogite at 700 to 1400 °C. Earth Planet Sci Lett 237:873–892

Kessel R, Pettke T, Fumagalli P (2015) Melting of metasomatized peridotite at 4–6 GPa and up to 1200 °C: an experimental approach. Contrib Mineral Petrol 169:37

Kharkiv AD, Zinchuk NN, Zuev VM (1997) History of Diamond. Nedra, Moscow

Kiseeva ES, Yaxley GM, Hermann J, Litasov KD, Rosenthal A, Kamenetsky VS (2012) An experimental study of carbonated eclogite at 3.5–5.5 GPa—implications for silicate and carbonate metasomatism in the cratonic mantle. J Petrol 53:727–759

Klein-BenDavid O, Pearson DG (2009) Origins of subcalcic garnets and their relation to diamond forming fluids—case studies from Ekati (NWT-Canada) and Murowa (Zimbabwe). Geochim Cosmochim Acta 73:837–855

Klein-BenDavid O, Logvinova AM, Izraeli ES, Sobolev NV, Navon O (2003) Sulfide melt inclusions in Yubileinaya (Yakutia) diamonds. International Kimberlite Conference: Extended Abstracts 8

Klein-BenDavid O, Izraeli ES, Hauri E, Navon O (2004) Mantle fluid evolution—a tale of one diamond. Lithos 77:243–253

Klein-BenDavid O, Wirth R, Navon O (2006) TEM imaging and analysis of microinclusions in diamonds: A close look at diamond-growing fluids. Am Mineral 91:353–365

Klein-BenDavid O, Izraeli ES, Hauri E, Navon O (2007) Fluid inclusions in diamonds from the Diavik mine, Canada and the evolution of diamond-forming fluids. Geochim Cosmochim Acta 71:723–744

Klein-BenDavid O, Logvinova AM, Schrauder M, Spetius ZV, Weiss Y, Hauri EH, Kaminsky FV, Sobolev NV, Navon O (2009) High-Mg carbonatitic microinclusions in some Yakutian diamonds—a new type of diamond-forming fluid. Lithos 112:648–659

Klein-BenDavid O, Pearson DG, Nowell GM, Ottley C, McNeill JC, Cartigny P (2010) Mixed fluid sources involved in diamond growth constrained by Sr–Nd–Pb–C–N isotopes and trace elements. Earth Planet Sci Lett 289:123–133

Klein-BenDavid O, Pearson DG, Nowell GM, Ottley C, McNeill JC, Logvinova A, Sobolev NV (2014) The sources and time-integrated evolution of diamond-forming fluids–Trace elements and isotopic evidence. Geochim Cosmochim Acta 125:146–169

Koornneef JM, Gress MU, Chinn IL, Jelsma HA, Harris JW, Davies GR (2017) Archaean and Proterozoic diamond growth from contrasting styles of large-scale magmatism. Nat Commun 8:648

Kopylova M, Navon O, Dubrovinsky L, Khachatryan G (2010) Carbonatitic mineralogy of natural diamond-forming fluids. Earth Planet Sci Lett 291:126–137

Kosman CW, Kopylova MG, Stern RA, Hagadorn JW, Hurlbut JF (2016) Cretaceous mantle of the Congo craton: Evidence from mineral and fluid inclusions in Kasai alluvial diamonds. Lithos 265:42–56

Krebs MY, Pearson DG, Stachel T, Laiginhas F, Woodland S, Chinn I, Kong J (2019) A common parentage-low abundance trace element data of gem diamonds reveals similar fluids to fibrous diamonds. Lithos 324:356–370

Kurz MD, Jenkins WJ, Hart SR (1982) Helium isotopic systematics of oceanic islands and mantle heterogeneity. Nature 297:43–47

Kvasnytsya VM, Wirth R (2009) Nanoinclusions in microdiamonds from Neogenic sands of the Ukraine (Samotkan'placer): A TEM study. Lithos 113:454–464

Landau LD, Lifshitz EM (1987) Fluid Mechanics: Course of Theoretical Physics. Vol. 6 Pergamon

Lang AR (1974a) Glimpses into the growth history of natural diamonds. J Cryst Growth 24:108–115

Lang AR (1974b) Space-filling by branching columnar single-crystal growth: An example from crystallisation of diamond. J Cryst Growth 23:151–153

Lang AR, Walmsley JC (1983) Apatite inclusions in natural diamond coat. Phys Chem Mineral 9:6–8

Lawless PJ, Gurney JJ, Dawson JB (1979) Polymict peridotites from the Bultfontein and De Beers mines, Kimberley, South Africa. *In:* The Mantle Sample: Inclusions in Kimberlites and Other Volcanics. Boyd FR, Meyer HOA (eds), AGU, Washington DC, p 145–155

Lazarov M, Brey GP, Weyer S (2012) Evolution of the South African mantle—a case study of garnet peridotites from the Finsch diamond mine (Kaapvaal craton); Part 2: Multiple depletion and re-enrichment processes. Lithos 154:210–223

Le Roex AP, Bell DR, Davis P (2003) Petrogenesis of Group I Kimberlites from Kimberley, South Africa: Evidence from Bulk-rock Geochemistry. J Petrol 44:2261–2286

Leahy K, Taylor WR (1997) The influence of the Glennie domain deep structure on the diamonds in Saskatchewan kimberlites. Russ Geol Geophys 38:481–490

Lee J-Y, Marti K, Severinghaus JP, Kawamura K, Yoo H-S, Lee JB, Kim JS (2006) A redetermination of the isotopic abundances of atmospheric Ar. Geochim Cosmochim Acta 70:4507–4512

Leost I, Stachel T, Brey GP, Harris JW, Ryabchikov ID (2003) Diamond formation and source carbonation: mineral associations in diamonds from Namibia. Contrib Mineral Petrol 145:15–24

Litasov KD, Safonov OG, Ohtani E (2010) Origin of Cl-bearing silica-rich melt inclusions in diamonds: Experimental evidence for an eclogite connection. Geology 38:1131–1134

Litasov KD, Shatskiy A, Ohtani E, Yaxley GM (2013) Solidus of alkaline carbonatite in the deep mantle. Geology 41:79–82

Litvin YA (2009) The physicochemical conditions of diamond formation in the mantle matter: experimental studies. Russ Geol Geophys 50:1188–1200

Liu W, Fei PX (2006) Methane-rich fluid inclusions from ophiolitic dunite and post-collisional mafic–ultramafic intrusion: the mantle dynamics underneath the Palaeo-Asian Ocean through to the post-collisional period. Earth Planet Sci Lett 242:286–301

Liu LG, Mernagh T, Jaques A (1990) A mineralogical Raman spectroscopy study on eclogitic garnet inclusions in diamonds from Argyle. Contrib Mineral Petrol 105:156–161

Logvinova AM, Wirth R, Fedorova EN, Sobolev NV (2008a) Nanometre-sized mineral and fluid inclusions in cloudy Siberian diamonds: new insights on diamond formation. Eur J Mineral 20:317–331

Logvinova AM, Klein-BenDavid O, Wirth R, Pearson DG, Navon O, Weiss Y, Pokhilenko NP (2008b) Diamond forming fluids from Snap Lake–a comprehensive study. International Kimberlite Conference: Extended Abstracts 9

Logvinova AM, Wirth R, Tomilenko AA, Afanas'ev VP, & Sobolev NV (2011) The phase composition of crystal-fluid nanoinclusions in alluvial diamonds in the northeastern Siberian Platform. Russ Geol Geophys 52:1286–1297

Logvinova AM, Taylor LA, Fedorova EN, Yelisseyev AP, Wirth R, Howarth G, Reutsky VN, Sobolev NV (2015) A unique diamondiferous peridotite xenolith from the Udachnaya kimberlite pipe, Yakutia: role of subduction in diamond formation. Russ Geol Geophys 56:306–320

Logvinova AM, Wirth R, Zedgenizov DA, Taylor LA (2018) Carbonate–silicate–sulfide polyphase inclusion in diamond from the Komsomolskaya Kimberlite Pipe, Yakutia. Geochem Int 56:283–291

Logvinova A, Zedgenizov D, Wirth R (2019a) Specific multiphase assemblages of carbonatitic and Al-rich silicic diamond-forming fluids/melts: TEM observation of microinclusions in cuboid diamonds from the placers of Northeastern Siberian Craton. Minerals 9:50

Logvinova AM, Shatskiy A, Wirth R, Tomilenko AA, Ugap'eva SS, Sobolev NV (2019b) Carbonatite melt in type Ia gem diamond. Lithos 342:463–467

Lonsdale K, Milledge HJ (1965) X-ray diffraction studies on diamond and some related materials. *In:* Physical Properties of Diamond. Berman R (ed) Clarendon Press, Oxford, p 12–68

Malkovets VG, Griffin WL, O'Reilly SY, Wood BJ (2007) Diamond, subcalcic garnet, and mantle metasomatism: Kimberlite sampling patterns define the link. Geology 35:339–342

Matjuschkin V, Woodland AB, Frost DJ, Yaxley GM (2020). Reduced methane-bearing fluids as a source for diamond. Sci Rep 10:6961

Mazza SE, Gazel E, Bizimis M, Moucha R, Béguelin P, Johnson EA, McAleer RJ, Sobolev AV (2019) Sampling the volatile-rich transition zone beneath Bermuda. Nature 569:398–403
McDonough WF (1990) Constraints on the composition of the continental lithospheric mantle. Earth Planet Sci Lett 101:1–18
McDonough WF, Sun SS (1995) The composition of the Earth. Chem Geol 120:223–253
McKenzie D (1989) Some remarks on the movement of small melt fractions in the mantle. Earth Planet Sci Lett 95:53–72
McNeill J, Pearson DG, Klein-BenDavid O, Nowell GM, Ottley CJ, Chinn I (2009) Quantitative analysis of trace element concentrations in some gem-quality diamonds. J Phys: Condens Matter 21:364207
Melton CE, Giardini AA (1974) The composition and significance of gas released from natural diamonds from Africa and Brazil. Am Mineral 59:775–782
Melton, CE, Giardini AA (1975) Experimental results and a theoretical interpretation of gaseous inclusions found in Arkansas natural diamonds. Am Mineral 60:413–417
Melton CE, Giardini AA (1976) Experimental evidence that oxygen is the principal impurity in natural diamonds. Nature 263:309–310
Melton CE, Salotti CA, Giardini AA (1972) The observation of nitrogen, water, carbon dioxide, methane and argon as impurities in natural diamonds. Am Mineral 57:1518–1523
Melton GL, McNeill J, Stachel T, Pearson DG, Harris JW (2012) Trace elements in gem diamond from Akwatia, Ghana and DeBeers Pool, South Africa. Chem Geol 314:1–8
Menzies A, Westerlund K, Grütter H, Gurney J, Carlson J, Fung A, Nowicki T (2004) Peridotitic mantle xenoliths from kimberlites on the Ekati Diamond Mine property, NWT, Canada: Major element compositions and implications for the lithosphere beneath the central Slave craton. Lithos 77:395–412
Menzies MA (1990) Continental mantle. Oxford University Press, New York
Mershon RB, Navon O, Weiss Y (2021) Microinclusions and Nitrogen Aggregation in Diamonds from Pulandian, China. Israel Geological Society Annual meeting, Yeruham 2021, Abstract Book, p 83–84, https://en.igs.org.il/annual-conference
Meyer HOA (1987) Inclusions in diamond. *In:* Mantle xenoliths. Nixon P (ed) John Wiley & Sons Ltd, Chichester, p 501–522
Meyer HOA, Boyd FR (1972) Composition and origin of crystalline inclusions in natural diamonds. Geochim Cosmochim Acta 36:1255–1273
Moore M, Lang AR (1972) On the internal structure of natural diamonds of cubic habit. Philos Mag 26:1313–1325
Moreira M, Kunz J, Allègre C (1998) Rare gas systematics in popping rock: Isotopic and elemental compositions in the upper mantle. Science 279:1178–1181
Murphy DT, Kamber BS, Collerson KD (2003) A refined solution to the first terrestrial Pb-isotope paradox. J Petrol 44:39–53
Mvuemba-Ntanda FM, Moreau J, Meyer HOA (1982) Particularites des inclusions cristallines primaires des diamants du Kasai, Zaire. Can Mineral 20:217–230
Navon O (1989) Chemical and mineralogical characterization of micro-inclusions in diamonds. PhD Dissertation, California Institute of Technology, Pasadena, California
Navon O (1991) High internal pressures in diamond fluid inclusions determined by infrared absorption. Nature 353:746–748
Navon O (1999) Formation of diamonds in the Earth's mantle. *In*: Proceedings of the 7th International Kimberlite Conference. Vol 2. Gurney JJ, Gurney JL, Pascoe MF, Richardson SH (eds), Red Roof Designs, Cape Town, p 584–604
Navon O, Hutcheon ID, Rossman GR, Wasserburg GJ (1988) Mantle-derived fluids in diamond micro-inclusions. Nature 335:784–789
Navon O, Wirth R, Schmidt C, Jablon BM, Schreiber A, Emmanuel S (2017) Solid molecular nitrogen (δ-N2) inclusions in Juina diamonds: Exsolution at the base of the transition zone. Earth Planetary Sci Lett 464:237–247
Nimis P (2022) Pressure and temperature data for diamonds. Rev Mineral Geochem 88:533–566
Nimis P, Alvaro M, Nestola F, Angel RJ, Marquardt K, Rustioni G, Harris JW, Marone F (2016) First evidence of hydrous silicic fluid films around solid inclusions in gem-quality diamonds. Lithos 260:384–389
Nimis P, Preston R, Perritt SH, Chinn IL (2020) Diamond's depth distribution systematics. Lithos 376:105729
Novgorodov PG, Bulanova GP, Pavlova LA, Mikhailov VN, Ugarov VV, Shebanin AP, Argunov KP (1990) Inclusions of potassic phases, coesite and omphacite in the coated diamond crystal from the "Mir" pipe. Dokl Akad Nauk 310:439–443
O'Reilly SY, Griffin WL (2013) Mantle Metasomatism. *In:* Metasomatism and the Chemical Transformation of Rock: The Role of Fluids in Terrestrial and Extraterrestrial Processes. Harlov DE, Austrheim H (eds) Springer Berlin, Heidelberg, p 471–533
Orlov IL (1977) The Mineralogy of the Diamond. John Wiley & Sons, New York
Orlov YL, Bulienkov NA, Martovitsky VP (1980) The spherocrystals of diamond-new type of natural single crystals having a fibrous structure. Dokl Akad Nauk 252:703–707
Orlov YL, Bulienkov NA, Martovitsky VP (1982) A study of the internal structure of variety III diamonds by X-ray section topography. Phys Chem Mineral 8:105–111
Ozima M, Zashu S (1991) Noble gas state of the ancient mantle as deduced from noble gases in coated diamonds. Earth Planet Sci Lett 105:13–27

Pal'yanov YN, Sokol AG, Borzdov YM, Khokhryakov AF, Sobolev NV (1999) Diamond formation from mantle carbonate fluids. Nature 400:417–418

Pal'yanov N, Sokol AG, Borzdov M, Khokhryakov AF (2002) Fluid-bearing alkaline carbonate melts as the medium for the formation of diamonds in the Earth's mantle: an experimental study. Lithos 60:145–159

Palyanov YN, Kupriyanov IN, Khokhryakov AF, Ralchenko VG (2015) Crystal growth of diamond. *In*: Handbook of Crystal Growth. Vol 2. Rudolph P (ed) Elsevier, Amsterdam, p 671–713

Paton C, Hergt JM, Phillips D, Woodhead JD, Shee SR (2007) New insights into the genesis of Indian kimberlites from the Dharwar Craton via in situ Sr isotope analysis of groundmass perovskite. Geology 35:1011–1014

Pearson DG, Shirey SB, Carlson RW, Boyd FR, Pokhilenko NP, Shimizu N (1995) Re-Os, Sm-Nd, and Rb-Sr isotope evidence for thick Archaean lithospheric mantle beneath the Siberian craton modified by multistage metasomatism. Geochim Cosmochim Acta 59:959–977

Pell J, Grütter H, Neilson S, Lockhart G, Dempsey S, Grenon H (2013). Exploration and discovery of the Chidliak kimberlite province, Baffin Island, Nunavut: Canada's newest diamond district. *In:* Proceedings of 10th International Kimberlite Conference. Vol 2. Pearson DG, Grütter HS, Harris JW, Kjarsgaard BA, O'Brien H, Rao NVC, Sparks S (eds) Springer, New Delhi, p 209–227

Pilet S, Baker MB, Stolper EM (2008) Metasomatized lithosphere and the origin of alkaline lavas. Science 320:916–919

Pilet S, Baker MB, Müntener O, Stolper EM (2011) Monte Carlo simulations of metasomatic enrichment in the lithosphere and implications for the source of alkaline basalts. J Petrol 52:1415–1442

Prinz M, Mansoni DV, Hlava PF, Keil K (1975) Inclusions in diamonds: Garnet lherzolite and eclogite assemblages. Phys Chem Earth 797–815

Rege S, Jackson S, Griffin WL, Davies RM, Pearson NJ, O'Reilly SY (2005) Quantitative trace-element analysis of diamond by laser ablation inductively coupled plasma mass spectrometry. J Anal At Spectrom 20:601–611

Rege S, Griffin WL, Pearson NJ, Araújo D, Zedgenizov D, O'Reilly SY (2010) Trace-element patterns of fibrous and monocrystalline diamonds: Insights into mantle fluids. Lithos 118:313–337

Regier ME, Pearson DG, Stachel T, Luth RW, Stern RA, Harris JW (2020) The lithospheric-to-lower-mantle carbon cycle recorded in superdeep diamonds. Nature 585:234–238

Resano M, Vanhaecke F, Hutsebaut D, De Corte K, Moens L (2003) Possibilities of laser ablation-inductively coupled plasma-mass spectrometry for diamond fingerprinting. J Anal At Spectrom 18:1238–1242

Richardson SH, Gurney JJ, Erlank AJ, Harris J (1984) Origin of diamonds in old enriched mantle. Nature 310:198–202

Roedder E (1984) Fluid inclusions. Reviews in Mineralogy. Vol 12. Mineralogical Society of America, Washington DC

Rondeau B, Fritsch E, Moore M, Thomassot E, Sirakian JF (2007) On the growth of natural octahedral diamond upon a fibrous core. J Cryst Growth 304:287–293

Rudloff-Grund J, Brenker FE, Marquardt K, Howell D, Schreiber A, O'Reilly SY, Griffin WL, Kaminsky FV (2016) Nitrogen nanoinclusions in milky diamonds from Juina area, Mato Grosso State, Brazil. Lithos 265:57–67

Ryan CG, Cousens DR, Sie SH, Griffin WL, Suter GF, Clayton E (1990) Quantitative PIXE microanalysis of geological material using the CSIRO proton microprobe. Nucl Instrum Methods Phys Res, Sect B 47:55–71

Saal AE, Hart SR, Shimizu N, Hauri EH, Layne GD (1998) Pb isotopic variability in melt inclusions from oceanic island basalts, Polynesia. Science 282:1481–1484

Safonov OG, Perchuk LL, Litvin YA (2007) Melting relations in the chloride–carbonate–silicate systems at high-pressure and the model for formation of alkalic diamond–forming liquids in the upper mantle. Earth Planet Sci Lett 253:112–128

Safonov OG, Chertkova NV, Perchuk LL, Litvin YA (2009) Experimental model for alkalic chloride-rich liquids in the upper mantle. Lithos 112:260–273

Salters VJ, Stracke A (2004) Composition of the depleted mantle. Geochem Geophys Geosyst 5:Q05B07

Sano T, Miyoshi M, Ingle S, Banerjee NR, Ishimoto M, Fukuoka T (2008) Boron and chlorine contents of upper oceanic crust: basement samples from IODP Hole 1256D. Geochem Geophys Geosyst 9:Q12O15

Schmidberger SS, Francis D (2001) Constraints on the trace element composition of the Archean mantle root beneath Somerset Island, Arctic Canada. J Petrol 42:1095–1117

Schrauder M (1997) Fluids in Diamonds from Africa, Russia and India. PhD Dissertation. University of Vienna, Vienna, Austria

Schrauder M, Navon O (1993) Solid carbon dioxide in a natural diamond. Nature 365:42–44

Schrauder M, Navon O (1994) Hydrous and carbonatitic mantle fluids in fibrous diamonds from Jwaneng, Botswana. Geochim Cosmochim Acta 58:761–771

Schrauder M, Navon O, Szafranek D, Kaminsky FV, Galimov EM (1994) Fluids in Yakutian and Indian diamonds. Goldschmidt Conference Edinburgh, Mineral Mag A 58:813–814

Schrauder M, Koeberl C, Navon O (1996) Trace element analyses of fluid-bearing diamonds from Jwaneng, Botswana. Geochim Cosmochim Acta 60:4711–4724

Schulze DJ, Wiese D, Steude J (1996) Abundance and distribution of diamonds in eclogite revealed by volume visualization of CT X-ray scans. J Geol 104:109–114

Seal M (1966) Nature of diamond coat. Philos Mag 13:645–648

Seal M (1970) Some observations on inclusions and localized type II structures in diamond. *In:* Proceedings of Diamond Conference, Oxford, Chapter 16

Shah CJ, Lang AR (1963) An unusual distribution of precipitates in a diamond. Mineral Mag J M Soc 33:594–599

Shatskiy A, Litasov KD, Palyanov YN, Ohtani E (2016) Phase relations on the K_2CO_3–$CaCO_3$–$MgCO_3$ join at 6 GPa and 900–1400 °C: Implications for incipient melting in carbonated mantle domains. Am Mineral 101:437–447
Shatsky VS, Zedgenizov DA, Ragozin AL, Kalinina VV (2015) Diamondiferous subcontinental lithospheric mantle of the northeastern Siberian Craton: Evidence from mineral inclusions in alluvial diamonds. Gondwana Res 28:106–120
Shchukina EV, Agashev AM, Pokhilenko NP (2017) Metasomatic origin of garnet xenocrysts from the V. Grib kimberlite pipe, Arkhangelsk region, NW Russia. Geosci Front 8:641–651
Shimizu N (1975) Rare earth elements in garnets and clinopyroxenes from garnet lherzolite nodules in kimberlites. Earth and Planet Sc Lett 25:26–32
Shimizu N, Richardson SH (1987) Trace element abundance patterns of garnet inclusions in peridotite-suite diamonds. Geochim Cosmochim Acta 51:755–758
Shiryaev, AA, Izraeli ES, Hauri EH, Zakharchenko OD, Navon O (2005) Chemical, optical and isotopic investigation of fibrous diamonds from Brazil. Russ Geol Geophys 46:1185–1201
Shu Q, Brey GP (2015) Ancient mantle metasomatism recorded in subcalcic garnet xenocrysts: Temporal links between mantle metasomatism, diamond growth and crustal tectonomagmatism. Earth Planet Sci Lett 418:27–39
Simon NSC, Carlson RW, Pearson DG, Davies GR (2007) The origin and evolution of the Kaapvaal cratonic lithospheric mantle. J Petrol 48:589–625
Skinner CP (1989) The petrology of peridotite xenoliths from the Finsch kimberlite, South Africa. S Afr J Geol 92:197–206
Skuzovatov SY, Zedgenizov DA, Shatsky VS, Ragozin AL, Kuper KE (2011) Composition of cloudy microinclusions in octahedral diamonds from the Internatsional'naya kimberlite pipe (Yakutia). Russ Geol Geophys 52:85–96
Skuzovatov SY, Zedgenizov DA, Ragozin AL, Shatsky VS (2012) Growth medium composition of coated diamonds from the Sytykanskaya kimberlite pipe (Yakutia). Russ Geol Geophys 53:1197–1208
Skuzovatov S, Zedgenizov D, Howell D, Griffin WL (2016) Various growth environments of cloudy diamonds from the Malobotuobia kimberlite field (Siberian craton). Lithos 265:96–107
Smart KA, Heaman LM, Chacko T, Simonetti A, Kopylova M, Mah D, Daniels D (2009) The origin of high-MgO diamond eclogites from the Jericho Kimberlite, Canada. Earth Planet Sci Lett 284:527–537
Smit KV, Shirey SB, Stern RA, Steele A, Wang W (2016) Diamond growth from C–H–N–O recycled fluids in the lithosphere: Evidence from CH4 micro-inclusions and $\delta^{13}C$–$\delta^{15}N$–N content in Marange mixed-habit diamonds. Lithos 265:68–81
Smit KV, Stachel T, Luth RW, Stern RA (2019) Evaluating mechanisms for eclogitic diamond growth: An example from Zimmi Neoproterozoic diamonds (West African craton). Chem Geol 520:21–32
Smith CB (1983) Pb, Sr and Nd isotopic evidence for sources of southern African Cretaceous kimberlites. Nature 304:51–54
Smith PM, Asimow PD (2005) Adiabat_1ph: A new public front-end to the MELTS, pMELTS, and pHMELTS models. Geochem Geophys Geosyst 6:Q02004
Smith EM, Kopylova MG, Dubrovinsky L, Navon O, Ryder J, Tomlinson EL (2011) Transmission X-ray diffraction as a new tool for diamond fluid inclusion studies. Mineral Mag 75:2657–2675
Smith EM, Kopylova MG, Nowell GM, Pearson DG, Ryder J (2012) Archean mantle fluids preserved in fibrous diamonds from Wawa, Superior craton. Geology 40:1071–1074
Smith EM, Kopylova MG, Frezzotti ML, Afanasiev VP (2014) N-rich fluid inclusions in octahedrally-grown diamond. Earth Planet Sci Lett 393:39–48
Smith EM, Kopylova MG, Frezzotti ML, Afanasiev VP (2015) Fluid inclusions in Ebelyakh diamonds: Evidence of CO2 liberation in eclogite and the effect of H_2O on diamond habit. Lithos 216:106–117
Smith EM, Shirey SB, Nestola F, Bullock ES, Wang J, Richardson SH, Wang W (2016) Large gem diamonds from metallic liquid in Earth's deep mantle. Science 354:1403–1405
Sobolev VS (1960) Conditions of formation of diamond deposits. Geol Geofiz 1:3–20
Sobolev N (1977) Deep-seated inclusions in kimberlites and the problem of the composition of the upper mantle. (Translated from the Russian edition, 1974). AGU, Washington DC
Sobolev NV, Kuznetsova IK (1966) Mineralogy of diamond-bearing eclogites. Dokl Akad Nauk 167:1365
Sobolev NV, Kaminsky FV, Griffin WL, Yefimova ES, Win TT, Ryan CG, Botkunov AI (1997) Mineral inclusions in diamonds from the Sputnik kimberlite pipe, Yakutia. Lithos 39:135–157
Sobolev NV, Logvinova AM, Zedgenizov DA, Seryotkin YV, Yefimova E, Floss C, Taylor LA (2004) Mineral inclusions in microdiamonds and macrodiamonds from kimberlites of Yakutia: a comparative study. Lithos 77:225–242
Sobolev NV, Logvinova AM, Efimova ES (2009) Syngenetic phlogopite inclusions in kimberlite-hosted diamonds: implications for role of volatiles in diamond formation. Russ Geol Geophys 50:1234–1248
Sobolev NV, Logvinova AM, Tomilenko AA, Wirth R, Bul'bak TA, Luk'yanova LI, Fedorova EN, Reutsky VN, Efimova ES (2019) Mineral and fluid inclusions in diamonds from the Urals placers, Russia: Evidence for solid molecular N2 and hydrocarbons in fluid inclusions. Geochim Cosmochim Acta 266:197–219
Song S, Su L, Niu Y, Lai Y, Zhang L (2009) CH_4 inclusions in orogenic harzburgite: evidence for reduced slab fluids and implication for redox melting in mantle wedge. Geochim Cosmochim Acta 73:1737–1754
Speich L, Kohn SC (2020) QUIDDIT-QUantification of infrared active defects in diamond and inferred temperatures. Comput Geosci 144:104558
Speich L, Kohn SC, Bulanova GP, Smith CB (2018) The behaviour of platelets in natural diamonds and the development of a new mantle thermometer. Contrib Mineral Petrol 173:39

Stachel T, Harris JW (1997) Diamond precipitation and mantle metasomatism–evidence from the trace element chemistry of silicate inclusions in diamonds from Akwatia, Ghana. Contrib Mineral Petrol 129:143–154
Stachel T, Harris JW (2008) The origin of cratonic diamonds — Constraints from mineral inclusions. Ore Geol Rev 34:5–32
Stachel T, Luth RW (2015) Diamond formation—Where, when and how? Lithos 220:200–220
Stachel T, Viljoen KS, Brey G, Harris JW (1998) Metasomatic processes in lherzolitic and harzburgitic domains of diamondiferous lithospheric mantle: REE in garnets from xenoliths and inclusions in diamonds. Earth Planet Sci Lett 159:1–12
Stachel T, Harris JW, Tappert R, Brey GP (2003) Peridotitic diamonds from the Slave and the Kaapvaal cratons—similarities and differences based on a preliminary data set. Lithos 71:489–503
Stachel T, Aulbach S, Brey GP, Harris JW, Leost I, Tappert R, Viljoen KF (2004) The trace element composition of silicate inclusions in diamonds: A review. Lithos 77:1–19
Stachel T, Chacko T, Luth RW (2017) Carbon isotope fractionation during diamond growth in depleted peridotite: Counterintuitive insights from modelling water-maximum CHO fluids as multi-component systems. Earth Planet Sci Lett 473:44–51
Stagno V, Ojwang DO, McCammon CA, Frost DJ (2013) The oxidation state of the mantle and the extraction of carbon from Earth's interior. Nature 493:84–88
Staudacher T, Allègre CJ (1982) Terrestrial xenology. Earth Planet Sci Lett 60:389–406
Sumino H, Dobrzhinetskaya LF, Burgess R, Kagi H (2011) Deep-mantle-derived noble gases in metamorphic diamonds from the Kokchetav massif, Kazakhstan. Earth Planet Sci Lett 307:439–449
Sunagawa T (1984) Morphology of natural and synthetic diamond crystals. *In:* Materials Science of the Earth's Interior. Terra Scientific Publishing, Tokyo, p 303–330
Sverjensky DA, Huang F (2015) Diamond formation due to a pH drop during fluid–rock interactions. Nat Commun 6:8702
Tappert R, Tappert MC (2011) Diamonds in nature: a guide to rough diamonds. Springer Science & Business Media, Berlin, Heidelberg
Tappert R, Stachel T, Harris JW, Shimizu N (2005) Mineral inclusions in diamonds from the Panda kimberlite, Slave Province, Canada. Eur J Mineral 17:423–440
Taylor WR, Jaques AL, Ridd M (1990) Nitrogen-defect aggregation characteristics of some Australasian diamonds; time-temperature constraints on the source regions of pipe and alluvial diamonds. Am Mineral 75:1290–1310
Taylor WR, Canil D, Milledge HJ (1996) Kinetics of Ib to IaA nitrogen aggregation in diamond. Geochim Cosmochim Acta 60:4725–4733
Thomassot E, Cartigny P, Harris JW, Viljoen KF (2007) Methane-related diamond crystallization in the Earth's mantle: stable isotope evidences from a single diamond-bearing xenolith. Earth Planet Sci Lett 257:362–371
Thompson WK (1965) Infra-red spectroscopic studies of aqueous systems. Part 1. Molar extinction coefficients of water, deuterium oxide, deuterium hydrogen oxide, aqueous sodium chloride and carbon disulphide. Trans Faraday Soc 61:2635–2640
Timmerman S, Jaques AL, Weiss Y, Harris JW (2018a) N-δ13C-inclusion profiles of cloudy diamonds from Koffiefontein: Evidence for formation by continuous Rayleigh fractionation and multiple fluids. Chem Geol 483:31–46
Timmerman S, Honda M, Phillips D, Jaques AL, Harris JW (2018b) Noble gas geochemistry of fluid inclusions in South African diamonds: Implications for the origin of diamond-forming fluids. Mineral Petrol 112:181–195
Timmerman S, Yeow H, Honda M, Howell D, Jaques AL, Krebs MY, Woodland S, Pearson DG, Ávilaa JN, Ireland TR (2019) U–Th/He systematics of fluid-rich 'fibrous' diamonds–Evidence for pre-and syn-kimberlite eruption ages. Chem Geol 515:22–36
Tomilenko AA, Chepurov AI, Palyanov YN, Pokhilenko LN, Shebanin AP (1997) Volatile components in the upper mantle (from data on fluid inclusions). Geol Geofiz 38:294–303
Tomlinson EL, Müller W (2009) A snapshot of mantle metasomatism: Trace element analysis of coexisting fluid (LA-ICP-MS) and silicate (SIMS) inclusions in fibrous diamonds. Earth Planet Sci Lett 279:362–372
Tomlinson E, De Schrijver I, De Corte K, Jones AP, Moens L, Vanhaecke F (2005) Trace element compositions of submicroscopic inclusions in coated diamond: A tool for understanding diamond petrogenesis. Geochim Cosmochim Acta 69:4719–4732
Tomlinson EL, Jones AP, Harris JW (2006) Co-existing fluid and silicate inclusions in mantle diamond. Earth Planet Sci Lett 250 581–595
Turner G, Burgess R, Bannon M (1990) Volatile-rich mantle fluids inferred from inclusions in diamond and mantle xenoliths. Nature 344:653–655
Venyaminov SY, Prendergast FG (1997) Water (H_2O and D_2O) molar absorptivity in the 1000–4000 cm^{-1} range and quantitative infrared spectroscopy of aqueous solutions. Anal Biochem 248:234–245
Viljoen KS, Swash PM, Otter ML, Schulze DJ, Lawless PJ (1992) Diamondiferous garnet harzburgites from the Finsch kimberlite, Northern Cape, South Africa. Contrib Mineral Petrol 110:133–8
Wada N, Matsuda JI (1998) A noble gas study of cubic diamonds from Zaire: Constraints on their mantle source. Geochim Cosmochim Acta 62:2335–2345
Walmsley JC, Lang AR (1992a) On sub-micrometre inclusions in diamond coat: crystallography and composition of ankerites and related rhombohedral carbonates. Mineral Mag 56:533–543

Walmsley JC, Lang AR (1992b) Oriented biotite inclusions in diamond coat. Mineral Mag 56:108–111
Walter MJ, Bulanova GP, Armstrong LS, Keshav S, Blundy JD, Gudfinnsson G, Lord OT, Lennie AR, Clark SM, Smith CB, Gobbo L (2008) Primary carbonatite melt from deeply subducted oceanic crust. Nature 454:622–625
Wang W, Moses TM, Shigley JE (2003) Physical and chemical features of a large coated natural diamond crystal. Diamond Relat Mater 12:330–335
Watling, JR, Herbert HK, Barrow IS, Thomas AG (1995) Analysis of diamonds and indicator minerals for diamond exploration by laser ablation–inductively coupled plasma mass spectrometry. Analyst 120:1357–1364
Weinstein Y, Navon O, Altherr R, Stein M (2006) The role of lithospheric mantle heterogeneity in the generation of Plio-Pleistocene alkali basaltic suites from NW Harrat Ash Shaam (Israel). J Petrol 47:1017–1050
Weiss Y, Goldstein SL (2018) The involvement of diamond-forming fluids in the metasomatic 'cocktail' of kimberlite sources. Mineral Petrol 112:149–167
Weiss Y, Griffin WL, Elhlou S, Navon O (2008a) Comparison between LA-ICP-MS and EPMA analysis of trace elements in diamonds. Chem Geol 252:158–168
Weiss Y, Griffin WL, Harris JW, Navon O (2008b) Diamond-forming fluids in fibrous diamonds: The trace-element perspective. International Kimberlite Conference: Extended Abstracts 9
Weiss Y, Kessel R, Griffin WL, Kiflawi I, Klein-BenDavid O, Bell DR, Harris JW, Navon O (2009) A new model for the evolution of diamond-forming fluids: Evidence from microinclusion-bearing diamonds from Kankan, Guinea. Lithos 112:660–674
Weiss Y, Kiflawi I, Navon O (2010) IR spectroscopy: quantitative determination of the mineralogy and bulk composition of fluid microinclusions in diamonds. Chem Geol 275:26–34
Weiss Y, Griffin WL, Bell DR, Navon O (2011) High-Mg carbonatitic melts in diamonds, kimberlites and the sub-continental lithosphere. Earth Planet Sci Lett 309:337–347
Weiss Y, Kiflawi I, Navon O (2013a) The IR absorption spectrum of water in microinclusion-bearing diamonds. *In:* Proceedings of 10th International Kimberlite Conference. Vol 1. Pearson DG, Grütter HS, Harris JW, Kjarsgaard BA, O'Brien H, Rao NVC, Sparks S (eds) Springer, New Delhi, p 271–280
Weiss Y, Griffin WL, Navon O (2013b) Diamond-forming fluids in fibrous diamonds: The trace-element perspective. Earth Planet Sci Lett 376:110–125
Weiss Y, Kiflawi I, Davies N, Navon O (2014) High-density fluids and the growth of monocrystalline diamonds. Geochim Cosmochim Acta 141:145–159
Weiss Y, McNeill J, Pearson DG, Nowell, GM, Ottley CJ (2015) Highly saline fluids from a subducting slab as the source for fluid-rich diamonds. Nature 524:339–342
Weiss Y, Class C, Goldstein SL, Hanyu T (2016) Key new pieces of the HIMU puzzle from olivines and diamond inclusions. Nature 537:666–670
Weiss Y, Class C, Goldstein SL, Winckler G (2017) Ages of mantle metasomatism from U–Th–He systematics of diamond-forming COH fluids. International Kimberlite Conference: Extended Abstracts 11
Weiss Y, Navon O, Goldstein SL, Harris JW (2018) Inclusions in diamonds constrain thermo-chemical conditions during Mesozoic metasomatism of the Kaapvaal cratonic mantle. Earth Planet Sci Lett 491:134–147
Weiss Y, Kiro Y, Class C, Winckler G, Harris JW, Goldstein SL (2021) Helium in diamonds unravels over a billion years of craton metasomatism. Nat Commun 12:2667
Weiss Y, Czas J, Navon O (2022) Global published compositions of high-density fluids (HDFs) in fibrous diamonds, Version 1.0. Interdisciplinary Earth Data Alliance (IEDA)
Willbold M, Stracke A (2006) Trace element composition of mantle end-members: Implications for recycling of oceanic and upper and lower continental crust. Geochem Geophys Geosyst 7:Q04004
Wirth R (2004) Focused Ion Beam (FIB) A novel technology for advanced application of micro-and nanoanalysis in geosciences and applied mineralogy. Eur J Mineral 16:863–876
Wirth R (2009) Focused Ion Beam (FIB) combined with SEM and TEM: Advanced analytical tools for studies of chemical composition, microstructure and crystal structure in geomaterials on a nanometre scale. Chem Geol 261:217–229
Wittig N, Pearson DG, Webb M, Ottley CJ, Irvine GJ, Kopylova M, Jensen SM, Nowell GM (2008) Origin of cratonic lithospheric mantle roots: A geochemical study of peridotites from the North Atlantic Craton, West Greenland. Earth Planet Sci Lett 274:24–33
Wyllie PJ (1987) Metasomatism and fluid generation in mantle xenoliths. *In:* Mantle xenoliths. Nixon P (ed) John Wiley & Sons Ltd, Chichester, p 609–621
Wyllie PJ, Ryabchikov ID (2000) Volatile components, magmas, and critical fluids in upwelling mantle. J Petrol 41:1195–1206
Xia X (2018) Mineral inclusions in diamonds from Chidliak (Nunavut, Canada): constraining the diamond substrates. MSc Dissertation, University of Alberta, Edmonton, Alberta
Yaxley GM, Brey GP (2004) Phase relations of carbonate-bearing eclogite assemblages from 2.5 to 5.5 GPa: implications for petrogenesis of carbonatites. Contrib Mineral Petrol 146:606–619
Yelisseyev AP, Pokhilenko NP, Steeds JW, Zedgenizov DA, Afanasiev VP (2004) Features of coated diamonds from the Snap Lake/King Lake kimberlite dyke, Slave craton, Canada, as revealed by optical topography. Lithos 77:83–97
Zashu S, Ozima M, Nitoh O (1986) K–Ar isochron dating of Zaire cubic diamonds. Nature 323:710–712

Zedgenizov DA, Kagi H, Shatsky VS, Sobolev NV (2004) Carbonatitic melts in cuboid diamonds from Udachnaya kimberlite pipe (Yakutia): Evidence from vibrational spectroscopy. Mineral Mag 68:61–73
Zedgenizov DA, Harte B, Edinburgh Ion Microprobe Facility (EIMF), Shatsky VS, Politov AA, Rylov GM, Sobolev NV (2006) Directional chemical variations in diamonds showing octahedral following cuboid growth. Contrib Mineral Petrol 151:45–57
Zedgenizov DA, Rege S, Griffin WL, Kagi H, Shatsky VS (2007) Composition of trapped fluids in cuboid fibrous diamonds from the Udachnaya kimberlite: LAM-ICPMS analysis. Chem Geol 240:151–162
Zedgenizov DA, Ragozin AL, Shatsky VS, Araújo D, Griffin WL, Kagi H (2009) Mg and Fe-rich carbonate–silicate high-density fluids in cuboid diamonds from the Internationalnaya kimberlite pipe (Yakutia). Lithos 112:638–647
Zedgenizov DA, Ragozin AL, Shatsky VS, Araújo D, Griffin WL (2011) Fibrous diamonds from the placers of the northeastern Siberian Platform: Carbonate and silicate crystallization media. Russ Geol Geophys 52:1298–1309
Zedgenizov DA, Kalinina VV, Reutsky VN, Yuryeva OP, Rakhmanova MJ (2016) Regular cuboid diamonds from placers on the northeastern Siberian platform. Lithos 265:125–137
Zedgenizov DA, Malkovets VG, Griffin WL (2017a) Composition of diamond-forming media in cuboid diamonds from the V. Grib kimberlite pipe (Arkhangelsk province, Russia). Geochem J 51:205–213
Zedgenizov, D, Reutsky V, Wiedenbeck M (2017b) The Carbon and nitrogen isotope characteristics of type Ib-IaA Cuboid diamonds from alluvial placers in the northeastern Siberian platform. Minerals 7:178
Zedgenizov DA, Ragozin AL, Shatsky VS, Griffin WL (2018) Diamond formation during metasomatism of mantle eclogite by chloride-carbonate melt. Contrib Mineral Petrol 173:84
Zedgenizov DA, Skuzovatov SY, Griffin WL, Pomazansky BS, Ragozin AL, Kalinina VV (2020) Diamond-forming HDFs tracking episodic mantle metasomatism beneath Nyurbinskaya kimberlite pipe (Siberian craton). Contrib Mineral Petrol 175:106
Ziberna L, Nimis P, Zanetti A, Marzoli A, Sobolev NV (2013) Metasomatic processes in the central Siberian cratonic mantle: Evidence from garnet xenocrysts from the Zagadochnaya kimberlite. J Petrol 54:2379–2409

Reviews in Mineralogy & Geochemistry
Vol. 88 pp. 533–566, 2022

Pressure and Temperature Data for Diamonds

Paolo Nimis

Dipartimento di Geoscienze
Università di Padova
via G. Gradenigo 6, 35131 Padova, Italy

paolo.nimis@unipd.it

INTRODUCTION

One of the key scientific questions about diamonds is "how are they formed?" To answer this question, we need to know the diamond-forming reactions and the physicochemical conditions under which these reactions take place. The pressure (*P*) and temperature (*T*) of diamond formation are an essential part of this knowledge and their assessment is pivotal to develop predictive scenarios of diamond distribution in the Earth interior. These scenarios may contribute to our understanding of global Earth processes, such as the long-term carbon cycle, and might also eventually improve our capability to select potential targets for diamond exploration (Shirey et al. 2013; Nimis et al. 2020).

The evaluation of the *P* and *T* of diamond formation can be carried out at two levels of investigation. The first is concerned with formation conditions for individual diamonds or small populations of diamonds from specific sources. This approach has been so far the most widely practiced. The second level considers the statistical distribution of *P–T* conditions for diamond formation at either local or global scale. This type of investigation is hampered by the difficulty of obtaining large sets of suitable samples from a specific locality or for a statistically significant number of localities, and is therefore unavoidably affected to some degree by sampling bias. Despite inherent limitations, the latter approach is the most appropriate to reveal systematics in diamond *P–T* distributions and, ultimately, in diamond depth distribution within the Earth.

Early reviews of *P–T* distributions for lithospheric diamonds were made by Nimis (2002), based on thermobarometry of chromian diopside inclusions, and by Stachel and Harris (2008) and Stachel (2014), using a more comprehensive set of thermobarometers. More recently, Nimis et al. (2020) investigated diamond depth distributions for a set of South African kimberlites and provided evidence for systematic trends of likely global significance. The depth distribution for sublithospheric diamonds worldwide was reviewed by Harte (2010). In this contribution, I first describe the methods that can be used to estimate the *P–T* conditions of diamond formation, highlighting their respective strengths and weaknesses. I then review existing diamond *P–T* data and their implications for diamond distribution with depth from both a local and a global perspective.

METHODS FOR DIAMOND THERMOBAROMETRY

Thermobarometry of diamonds can be carried out by estimating *P–T* conditions of chemical or elastic equilibrium of mineral inclusions contained within them. With some assumptions, the aggregation state of nitrogen substituting for carbon in the diamond lattice can also be used as a thermometer. In some cases, it is possible to derive both *P* and *T* estimates for a diamond by combining independent thermobarometric methods. In most instances, however, either *P*

1529-6466/22/0088-0010$05.00 (print)
1943-2666/22/0088-0010$05.00 (online)
http://dx.doi.org/10.2138/rmg.2022.88.10

or T estimates can be directly retrieved with sufficient confidence. Below is a list of currently available methods for diamond thermobarometry. Those based on chemical equilibria are specific for different minerals or mineral pairs. Since many inclusions in diamonds occur as isolated grains, detached from their original mineral assemblage, particular emphasis is given in this review to methods that use single minerals to retrieve P and/or T. The following descriptions emphasize thermobarometry applications rather than principles, and the reader is referred to the original publications for details of the specific thermobarometers.

Thermometry

Pyroxenes. Many ultramafic mantle rocks contain both clinopyroxene and orthopyroxene. This mineral pair forms one of the most widely used geological thermometers, which is based on the net-transfer reaction of the enstatite component between the two pyroxenes $MgSiO_3(Opx) \leftrightarrow MgSiO_3(Cpx)$. The very small P-dependency of this reaction limits propagation of errors when T is calculated by iteration in combination with an independent barometer. Several versions of this thermometer have been proposed in the literature. Tests based on experimental results and pyroxene compositions in natural xenoliths suggest that the calibration of Taylor (1998) is the most reliable for mantle peridotites and pyroxenites in a T range of 700 to at least 1400 °C (Nimis and Grütter 2010), which covers the entire range of the lithospheric diamond window. The much more popular Brey and Köhler (1990) calibration gives similar results for clinopyroxenes with relatively low Na contents around 0.05 atoms per 6-oxygen formula unit (apfu), but was proven to overestimate with increasing Na_{Cpx} (e.g., +150 °C at Na_{Cpx} = 0.25 apfu; Nimis and Grütter 2010). For inclusions in diamonds and diamond-bearing peridotitic xenoliths (average Na_{Cpx} = 0.12 apfu, median = 0.09 apfu, standard deviation = 0.08 apfu), the Brey and Köhler (1990) calibration will yield estimates that are on average ~50 °C higher that those using the Taylor (1998) calibration. Larger discrepancies may be expected for unusually Na-rich clinopyroxenes.

Since clinopyroxene–orthopyroxene inclusion pairs are very rare in diamonds, it is generally more practical to use the alternative single-clinopyroxene version of Nimis and Taylor (2000). The two-pyroxene Taylor (1998) and single-clinopyroxene Nimis and Taylor (2000) methods provide almost indistinguishable results when applied to mantle ultramafic rocks that contain both pyroxenes (Nimis and Grütter 2010). Although the Nimis and Taylor (2000) thermometer uses only one pyroxene to calculate T, its application expressly requires that both pyroxenes were part of the mineral assemblage and were in chemical equilibrium. Therefore, this method is only suitable for diamonds belonging to the lherzolitic or websteritic suites and containing clinopyroxene inclusions, either isolated or associated with other minerals. If orthopyroxene was not part of the original mineral assemblage, as is the case for wehrlitic inclusions, single-clinopyroxene thermometry would only provide a minimum T estimate. An anomalously low T estimate (e.g., one that lies much below the local geotherm) may be a sign that the clinopyroxene is wehrlitic. Ziberna et al. (2016) suggested that a Ca/(Ca + Mg) molar ratio of > 0.5 should also be considered as suspicious, as only ~1% of orthopyroxene-saturated mantle clinopyroxenes lie above this value. Nonetheless, discrimination of wehrlitic clinopyroxenes is not generally possible based merely on compositional criteria.

Simakov (2008) proposed a different, more complex calibration of the single-clinopyroxene thermometer, which definitely improves performance above 1500 °C, though not over the $T <$ 1400 °C range typical of 'lithospheric' temperatures (note that the $T_{\text{Nimis and Taylor 2000}}$ estimates reported in Fig. 10 of Simakov 2008 are incorrect and unduly suggest overestimation below 1300 °C).

The Ca-in-Opx thermometer of Brey and Köhler (1990) is the single-orthopyroxene version of the pyroxene thermometer and can provide estimates that are complementary to, and independent from those obtained with the single-clinopyroxene thermometer. The two single-pyroxene methods were shown to be mutually consistent, provided a correction is applied below 900 °C (Nimis and Grütter 2010). However, since inclusions of likely harzburgitic (i.e., clinopyroxene-free) affinity are relatively common in diamonds (Stachel and Harris 2008), the possibility that $T_{Ca\text{-}in\text{-}Opx}$ estimates for orthopyroxene inclusions are only minimum T values is much higher than for the single-clinopyroxene thermometer. Therefore, its usefulness is rather limited in diamond studies.

Garnet. The Ni content of garnet in equilibrium with olivine is very sensitive to T and apparently independent of P (Ryan et al. 1996; Canil 1999). Since the Ni content of olivine shows small variations in both mantle xenoliths (O'Reilly et al. 1997) and inclusions in diamonds (Griffin et al. 1992; Sobolev et al. 2008), T can be retrieved from Ni_{garnet} alone, by assuming an appropriate Ni content for coexisting forsteritic olivine. A useful reference values for $Ni_{olivine}$ is the mean value for mantle olivines (mean ± standard deviation = 2900 ± 360 ppm; Ryan et al. 1996), which is also close to the average value for olivine inclusions in diamonds worldwide (Stachel and Harris 2008; Sobolev et al. 2009). If available, the mean $Ni_{olivine}$ value for olivine inclusions in diamonds from the same locality may be used (e.g., 3150 ± 200 ppm for 51 inclusions from the Kalahari craton, Griffin et al. 1992; 2700 ppm for 88 inclusions from Arkhangel'sk province, Malkovets et al. 2011). Choosing one or the other value changes the final T estimate by at most a few tens of degrees (Fig. 1). The concentrations of Ni_{garnet} can be measured with a laser-ablation inductively-coupled plasma mass spectrometer (LA-ICP-MS), an ion microprobe (SIMS) or a proton microprobe (PIXE), which typically ensure a precision within a few ppm. The assumption of equilibrium with olivine may require compositional filtering (e.g., Grütter et al. 2004) to exclude any non-peridotitic garnet.

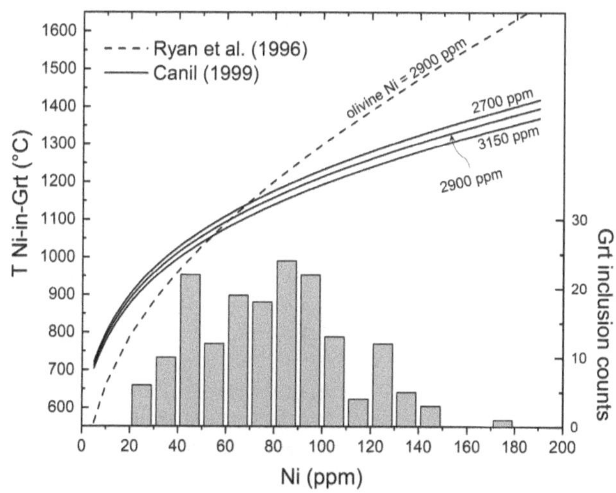

Figure 1. Relationships between Ni contents in garnet and calculated temperatures using the Ni-in-Grt thermometers of Ryan et al. (1996) and Canil (1999). The Ryan et al. (1996) version assumes a Ni content in olivine of 2900 ppm. Temperatures for the Canil (1999) version were calculated assuming a Ni content in coexisting olivine equal to the average for mantle olivines (2900 ppm) and to the average for olivine inclusions in Kalahari diamonds (3150 ppm) and in Arkhangel'sk diamonds (2700 ppm), respectively. The frequency distribution for 171 Cr-pyrope inclusions in diamonds from various localities (Griffin et al. 1992, 1993; Davies et al. 2004a; Viljoen et al. 2014; De Hoog et al. 2019) is shown for comparison, excluding two outliers at 236 and 431 ppm Ni.

The Ni-in-garnet thermometer has seen wide application in studies of Cr-pyrope garnets included in diamonds (e.g., Griffin et al. 1992, 1993; Davies et al. 2004a; Viljoen et al. 2014; De Hoog et al. 2019). Its calibration, however, is somewhat controversial. Ryan et al. (1996)

calibrated it against mantle xenoliths, using T values derived from a combination of the olivine–garnet Fe–Mg-exchange thermometer (O'Neill and Wood 1979) and the MacGregor (1974) and Brey and Köhler (1990) orthopyroxene–garnet barometers. Canil (1999) calibrated it against experiments at $T \geq 1200$ °C. The two calibrations give identical results at ~1100 °C, but progressively diverge at lower and higher T (Fig. 1). As a result, estimates obtained through the Canil (1999) formulation will typically span over narrower T intervals. Considering the most typical range for Ni in garnet inclusions, the maximum difference T_{Canil}–T_{Ryan} is ca. +100 °C at 20 ppm Ni and ca. –250 °C at 180 ppm Ni. It has been variously claimed that either calibration, or even their 'average', agree best with other independent thermometers when applied to mantle xenoliths (e.g., Ryan et al. 1996; Canil 1999; De Hoog et al. 2019; Czas et al. 2020; Nimis et al. 2020), but a definitive assessment using internally consistent thermobarometers as reference is still lacking. The latest attempt to refine the Ni-in-garnet thermometer was made by Sudholz et al. (2021a), who recalibrated it against new experiments in a relatively narrow T range (1100–1325 °C) and introduced correction terms for Ca and Cr contents in the garnet. When tested against independent estimates for xenoliths using the Nimis and Taylor (2000) enstatite-in-clinopyroxene thermometer, the Sudholz et al. (2021a) calibration shows improved overall accuracy above 1100 °C relative to the Canil (1999) experimental calibration, but poorer overall precision and slightly stronger progressive overestimation at lower T (see Fig. 7 in Sudholz et al. 2021a).

The Mn content of garnet in equilibrium with mantle olivine is sensitive to T (Smith et al. 1991) and can be used as a single-mineral thermometer, assuming a constant Mn content in olivine. This thermometer relies on electron microprobe data and was proposed as a substitute for the Ni-in-garnet thermometer when trace element data are not available (Grütter et al. 1999; Creighton 2009). The declared precision is rather poor (mostly ±150 °C) and, in the absence of independent olivine data, severe underestimation may occur at high T. Therefore, this method may have some use in surveys of large populations of garnet xenocrysts recovered during diamond exploration (e.g., Grütter and Tuer 2009), but its utility for diamond thermobarometry is limited.

Ashchepkov et al. (2010) calibrated two monomineral thermometers for garnets in equilibrium with clinopyroxene or olivine, respectively. These methods are simplified versions of Fe–Mg-exchange thermometers for clinopyroxene–garnet or olivine-garnet pairs, in which the clinopyroxene and olivine compositions are modeled from the composition of the garnet. Their accuracy cannot be better than that of the original two-mineral formulations, which are problematic themselves (see below). Also, P must be independently known to calculate T. In a test on mantle xenoliths from the Udachnaya kimberlite, these thermometers reproduced temperatures calculated with the pyroxene thermometer of Brey and Köhler (1990), at pressures calculated with the orthopyroxene–garnet barometer, to within ca. ±100 °C. Additional uncertainties due to problematic estimation of P in the absence of compositional data for coexisting orthopyroxene (see below) probably make these thermometers of little interest for diamond thermobarometry, especially when investigating small populations of diamonds.

Olivine. The Al content in peridotitic olivine is sensitive to T and can be used as a thermometer. The first Al-in-olivine thermometer for peridotitic olivine at conditions relevant to diamond was empirically calibrated on mantle xenoliths by De Hoog et al. (2010), using P–T estimates derived by a combination of the two-pyroxene thermometer and orthopyroxene–garnet barometer of Brey and Köhler (1990). The method was recalibrated by Bussweiler et al. (2017) by analysis of olivine from high-P–T experimental charges. The Al content in cratonic mantle olivine ranges from ~1 ppm to ~250 ppm and is conveniently analyzed by techniques such as LA-ICP-MS or SIMS, which ensure a precision within a few ppm. Electron microprobe analysis (EMPA) using very high probe currents (e.g., ≥ 200 nA) and long count times (≥ 100 s) may be a valid alternative (Batanova et al. 2018; D'Souza et al. 2020).

Al-in-olivine temperatures for spinel peridotites or strongly metasomatized garnet peridotites may be significantly underestimated or, respectively, overestimated. Compositional screening using an Al vs V diagram may help to discriminate these 'unsafe' olivines (Bussweiler et al. 2017). Screening is especially important if olivine is not physically associated with garnet, as occurs with many inclusions in diamonds. The Al-in-olivine thermometer gives systematically higher estimates (~50 °C, slightly increasing below 900 °C) than the single-clinopyroxene thermometer (Nimis and Taylor 2000) for natural mantle peridotites, probably reflecting slight suppression of Al incorporation in the experimental olivines due to Na loss (Bussweiler et al. 2017). The discrepancy is sufficiently small to ensure successful application of the method, but should be considered when comparing estimates obtained using different thermometers. The Al-in-olivine thermometer has not yet been widely used in diamond studies (see Korolev et al. 2018 for an example of application), owing to its recent development and the necessity of non-routine analysis of Al. Considering the great abundance of olivine among inclusions in diamonds (Stachel and Harris 2008), its application is likely to increase significantly in the future.

Spinel. The Zn content of spinel in equilibrium with mantle olivine is sensitive to T. Since the Zn content of mantle olivine is nearly constant (mean ± standard deviation = 52 ± 14 ppm), the Zn content of spinel can directly be used as a thermometer (Ryan et al. 1996). The concentrations of Zn in spinel can be measured by SIMS or PIXE, which ensure a precision within a few ppm. The current version of the Zn-in-spinel thermometer was calibrated over the range 680–1180 °C against T for coexisting garnets using the Ryan et al. (1996) version of the Ni-in-garnet thermometer. Therefore, it is subject to at least the same uncertainties as the Ni-in-garnet thermometer (see above).

Fe–Mg exchange thermometers. These popular thermometers are based on the distribution of Fe^{2+} and Mg between two mineral phases and can be used to derive T for inclusions in which garnet is associated with olivine, orthopyroxene or clinopyroxene. Tests using mineral compositions from mantle xenoliths and experiments showed that these thermometers suffer from either low precision or systematic errors or both (see review of Nimis and Grütter 2010). In peridotitic systems, the olivine-garnet and the clinopyroxene–garnet thermometers show the lowest precision, with possible discrepancies exceeding 200 °C. The Harley (1984) orthopyroxene–garnet formulation shows the highest precision, but systematically overestimates at $T < 1100$ °C and underestimates at $T > 1100$ °C (by up to ~150 °C, on average, for mantle xenoliths). All these discrepancies may reflect a modest sensitivity to T of the exchange reaction (particularly for olivine-garnet), oversimplification in the solid solution models used (especially for garnet and clinopyroxene), and $Fe^{3+}/\Sigma Fe$ ratios in the minerals different from those in the calibration samples. Matjuschkin et al. (2014) experimentally demonstrated the large effect of neglected Fe^{3+} on orthopyroxene–garnet and olivine-garnet thermometers. Their experiments in the Na-free $CaO–FeO–Fe_2O_3–MgO–Al_2O_3–SiO_2$ system at 5 GPa and 1000–1400 °C suggest that temperature estimates may be significantly improved if calculations are made using Fe^{2+} instead of total Fe in the garnet. Mössbauer data for orthopyroxene and garnet in peridotitic xenoliths, however, suggest that the partitioning of Fe^{3+} between these two minerals is affected by pressure and Na content in the orthopyroxene (Nimis et al. 2015). Therefore, a test over a range of relevant pressures in a Na-bearing system would be desirable. Nimis and Grütter (2010) empirically corrected the Harley (1984) thermometer using two-pyroxene thermometry of well equilibrated mantle xenoliths as calibration. The corrected version generally yields lower T estimates at low T and higher T estimates at high T relative to the uncorrected thermometer and may be more accurate for orthopyroxene–garnet pairs equilibrated under 'average' mantle redox conditions. On the other hand, it may produce large errors (> to $\gg$100 °C) if redox conditions are far from 'average', especially for strongly oxidized conditions at very high pressure and when both P and T are calculated by iteration in combination with an orthopyroxene–garnet barometer (Nimis et al. 2015). Considering that at deep lithospheric levels diamond is stable over a relatively wide range of redox conditions ($\Delta\log fO_2$ FMQ ≈ –5 to –2), this problem can be serious. Also, application to some xenolith suites tends to produce frequent high-*P–T* outliers (Grütter, pers. comm.). In spite of its recognized systematic discrepancies, the original Harley (1984) version may still be preferable for diamond thermobarometry due to its lower *P* dependence, which reduces error propagation during iterative *P–T* calculations. However, the systematic deviations of the Harley (1984) thermometer at *T* far from 1100 °C, especially if all Fe is treated as Fe^{2+}, should be considered in data evaluation.

The clinopyroxene–garnet thermometer represents the only viable method to determine *T* from major element compositions of eclogitic inclusions in diamonds. There are many versions of this thermometer in the literature. The most recent calibration (Nakamura 2009) is based on an expanded experimental database covering mafic and ultramafic compositions across the *P* and *T* ranges relevant to diamond (1.5–7.5 GPa, 800–1820 °C) and incorporates updated solid solution models for both clinopyroxene and garnet. Older, more simplified, but in some cases still very popular versions of this thermometer do not reproduce the same experiments equally well and often show systematic deviations with changing *T* and composition. The nominal uncertainty of the Nakamura (2009) thermometer (±74 °C at the 1-sigma level) is still about double that of the best performing thermometers for peridotitic inclusions. None of the available versions of the clinopyroxene–garnet thermometer include corrections for the content of jadeite in clinopyroxene, therefore application to clinopyroxenes with Na > 0.5 apfu is not recommended. This excludes 15% of reported compositions for inclusions in eclogitic diamonds. As regards the Fe^{3+} problem, Purwin et al. (2013) and Matjushkin et al. (2014) measured $Fe^{3+}/\Sigma Fe$ ratios of coexisting clinopyroxenes and garnets in a $CaO–FeO–Fe_2O_3–MgO–Al_2O_3–SiO_2$ system at 2.5–5.0 GPa and 800–1400 °C and found no systematic deviations in the results of clinopyroxene–garnet Fe–Mg-exchange thermometry using the Krogh (1988) thermometer when all Fe was treated as Fe^{2+}. However, since the system investigated was Na-free, it did not allow for the incorporation of Fe^{3+} in clinopyroxene as aegirine component. This mechanism was shown to be relevant in clinopyroxenes from garnet peridotites (Woodland 2009) and may be even more so in eclogitic clinopyroxenes. The effect of Na on Fe^{3+} uptake in clinopyroxene probably contributes to the lower precision of the clinopyroxene–garnet thermometer in applications to natural samples.

Trace element-based clinopyroxene–garnet thermometers. The partitioning of REE between clinopyroxene and garnet is sensitive to *T* and has been used as a base for a clinopyroxene–garnet thermometer (Witt-Eickschen and O'Neill 2005; Sun and Liang 2015; Pickles et al. 2016). A possible advantage of REE thermometry over classical Fe–Mg-exchange thermometry is that REE diffusivity is small compared to that of Fe and Mg. Therefore, REE thermometry is predictably more robust against reequilibration of touching inclusions in diamonds during kimberlite transport or subsequent cooling. The version by Pickles et al. (2016) is particularly interesting, as it is independent of other thermometers and reproduces experimental temperatures moderately well (mean deviation of ±76 °C) with a small *P* dependence (~50 °C/1 GPa). This method is also independent of variations in $Fe^{3+}/\Sigma Fe$ ratio and may thus pose a promising alternative to the traditional clinopyroxene–garnet Fe–Mg exchange thermometer. A practical disadvantage is the requirement of high-quality REE data, which cannot be obtained with routine electron microprobe analysis. To the author's knowledge, there are no *T* data for diamonds using this method.

Nitrogen-aggregation thermochronometry. Nitrogen is the most common impurity in natural diamonds. With time, nitrogen atoms in the diamond crystal lattice tend to aggregate, proceeding from initial C-centers (single nitrogen) in Type Ib diamond, to A-centers (nitrogen pairs) in Type IaA diamond and, finally, to B-centers (nitrogen in tetrahedral arrangement) in Type IaB diamond. The first transformation is completed in relatively short geological times (for a N content of 1000 ppm: ~100 yr at 1350 °C, and ~16 m.y at 950 °C; Taylor et al. 1996). Therefore, most natural diamonds are of the intermediate IaAB variety, the proportion of B-centers being a function of nitrogen content, temperature and time (Taylor et al. 1990; Mendelssohn and Milledge 1995). If the ages of the diamond and of its host kimberlite are known, one can calculate the integrated average temperature under which the diamond has resided in the mantle since its formation, i.e., the mantle residence temperature, T_{res}. Radiometric dating of inclusions within a particular diamond is rarely available. More commonly, the age of the diamond is assumed on the basis of radiometric dating of other diamonds from the same source. However, nitrogen aggregation is much more sensitive to temperature than to time and can be used as an effective thermometer, assuming the diamond has resided in the mantle at constant T since formation (Taylor et al. 1990, 1996). A diagram of nitrogen content vs the relative percentage of B-centers permits appreciation of age uncertainties on thermometry, since isotherms can be drawn for different mantle residence times (Fig. 2). Chronometry errors of one b.y. propagate thermometry errors of only a few tens of degrees on calculated T_{res}, unless the diamond is unusually young.

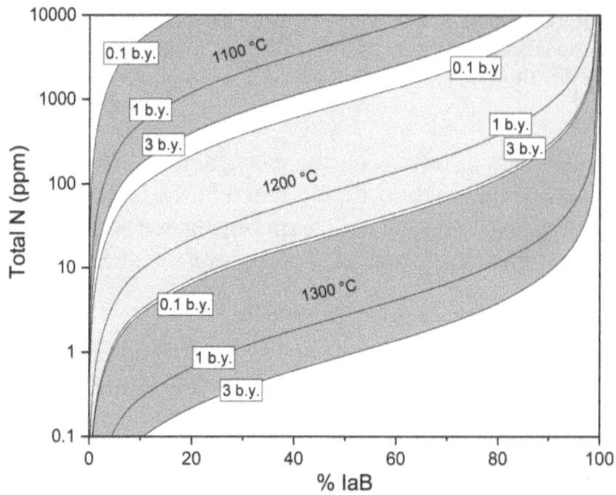

Figure 2. Plot of total nitrogen content in diamond vs percentage of nitrogen as B-centers. Mantle residence isotherms (1100, 1200 and 1300 °C) are calculated for residence times of 0.1, 1 and 3 b.y. (Taylor et al. 1990), using $E_a = 7.0$ eV and $\ln(A) = 12.59$ (Taylor et al. 1996).

Nitrogen content and aggregation can be measured by Fourier-Transform Infrared (FTIR) spectroscopy, using an infrared microscope in transmission mode. This method logically provides nitrogen content and aggregation data averaged over the volume of diamond analyzed. Working on thinned (ideally, a few hundred μm) samples at high spatial resolution may help to detect zoning and to obtain more robust estimates from distinct diamond growth zones. The reader is referred to Kohn et al. (2016) for practical guidelines on spatially resolved FTIR thermochronometry and for interpretation of data for zoned diamonds. Detection limits for nitrogen content in diamonds are typically on the order of 5–10 ppm for high-quality spectra. Nitrogen contents below detection limit (Type-II diamonds) are relatively rare in lithospheric diamonds, but are more common in sublithospheric diamonds and are clearly unsuitable for nitrogen-aggregation thermochronometry. It must be clear that, contrary to thermodynamic equilibria used in inclusion thermobarometry, nitrogen aggregation is a kinetic process and so does not provide a snapshot of temperature conditions at a certain time. For instance, a short period of elevated temperature, e.g., due to circulation of hot fluids in the mantle during or after diamond growth, may have a large effect on nitrogen aggregation and could lead to ambiguous interpretations of calculated T_{res}. In general, unless the diamond is very young relative to its kimberlite host, heating on a time

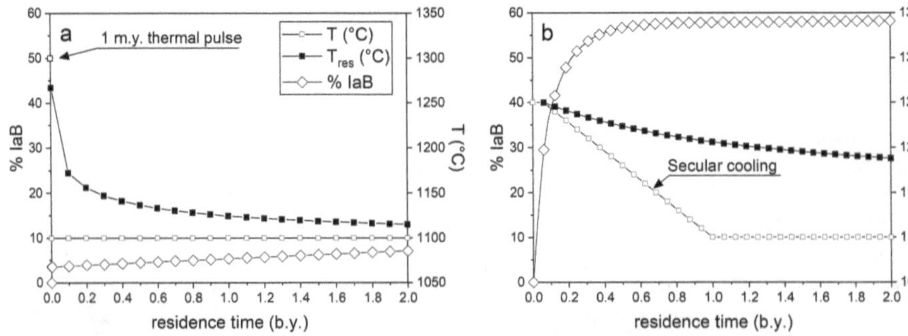

Figure 3. Variation of nitrogen aggregation and calculated T_{res} vs. mantle residence time for two diamonds with different thermal histories. (**a**) Diamond is formed at 1300 °C during a short-term (1 m.y) thermal pulse, followed by rapid return to ambient conditions of 1100 °C. Unless the residence time is very short, T_{res} rapidly diverges from the formation *T*. After a few 100 m.y., T_{res} approaches final ambient conditions. Note that the timing of the short thermal pulse does not influence the values of T_{res} that are calculated after the pulse at any given residence time. (**b**) Diamond formation at 1250 °C is followed by secular cooling to 1100 °C over a 1 b.y. period and then by constant *T* conditions for a further 1 b.y. The apparent T_{res} remains close to the formation *T* for short residence times, but progressive underestimation of formation *T* occurs with increasing time.

scale of ~1 m.y. will have small effects on the calculated T_{res}, which will approach ambient mantle conditions rather than conditions during diamond formation (Fig. 3a). If the diamond experienced secular cooling over a b.y. time scale subsequent to formation, the calculated T_{res} may both significantly underestimate the formation *T* and overestimate the final ambient *T* (Fig. 3b).

Barometry

Orthopyroxene–garnet. The Al content of orthopyroxene in equilibrium with garnet is strongly sensitive to *P* and forms the basis of the most widely used barometer for garnet-bearing peridotites. In this barometer, *P* mostly depends on the Al content in the orthopyroxene, but the compositions of both minerals must be considered to obtain robust estimates. Of the many calibrations of this barometer, that of Brey and Köhler (1990) is by far the most popular, but the older Nickel and Green (1985) calibration reproduces experimental pressures for most peridotitic systems to somewhat better precision (Nimis and Grütter 2010). Still, the Nickel and Green (1985) version was formulated under simple model-system assumptions and requires correction of Al-component activity to avoid *P* underestimation for orthopyroxenes with non-negligible contents of jadeite component. Carswell and Gibb (1987) and Carswell (1991) proposed two simple correction schemes, based on different assumptions on how Ti is coupled with Na and Al in orthopyroxene. The two corrections only apply to orthopyroxenes with $Na > (Cr + Fe^{3+} + 2{\cdot}Ti)$ or, respectively, $Na > (Cr + Fe^{3+} + Ti)$ atoms per formula unit, Fe^{3+} usually being neglected in practical applications. That of Carswell and Gibb (1987), which was also adopted by Brey and Köhler (1990) for their barometer, produces slightly smaller *P* variations, but there is as yet no solid experimental basis to prefer one correction over the other. The *T*-dependence of the orthopyroxene–garnet barometer is reasonably small (~0.2–0.3 GPa / 50 °C), but still sufficiently large to require an accurate independent estimate of *T*. This fact should be considered when this barometer is used in combination with the Harley (1984) orthopyroxene–garnet thermometer, which is known to overestimate at low *T* and underestimate at high *T* (see above): the artificial compression of *T* estimates produced by the Harley (1984) thermometer will necessarily result in non-natural compression also of *P* estimates. In general, errors in input *T* will displace the calculated *P*–*T* estimates roughly along a typical cratonic geotherm, thus limiting the 'apparent' scatter in *P*–*T* plots for samples equilibrated on the geotherm (Fig. 4). This may be an advantage if the aim is to define the mantle geotherm, but may give the misperception of a high overall precision by masking the true dispersion of the data.

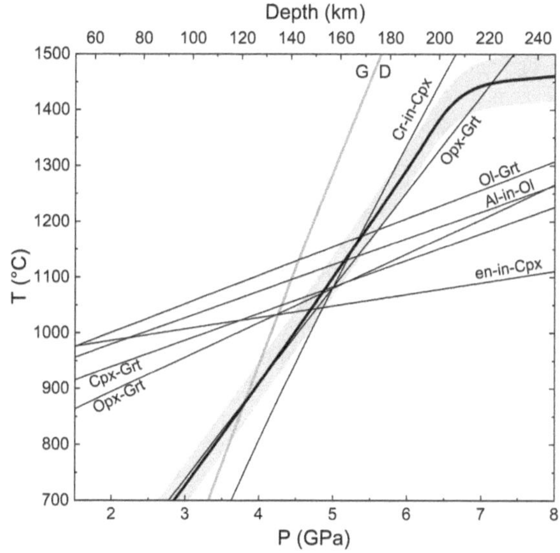

Figure 4. *P* and *T* dependence for some of the most widely used thermometers and barometers suitable for inclusions in diamonds, using typical inclusion compositions. Thermometers: Cpx–Grt – Nakamura (2009); Opx–Grt – Harley (1984), en-in-Cpx– Nimis and Taylor (2000); Al-in-Ol – Bussweiler et al. (2017); Ol–Grt – Wu and Zhao (2007). Barometers: Opx–Grt – Nickel and Green (1985); Cr-in-Cpx – Nimis and Taylor (2000). The thick solid line with 95% confidence bands is an example of a typical model cratonic geotherm calculated with FITPLOT from xenolith *P–T* data (Mather et al. 2011). GD: the graphite–diamond boundary after Day (2012).

Clinopyroxene–garnet. Barometry of bimineralic eclogites and of eclogitic clinopyroxene–garnet inclusion pairs in diamonds has long been a problem for mantle scientists. There have been several attempts to develop a suitable barometer based on the *P*-sensitive equilibrium grossular + pyrope ↔ diopside + Ca-Tschermak. According to a recent test by Beyer et al. (2018), the Beyer et al. (2015) formulation reproduces pressures for experiments in natural systems significantly better than the earlier formulations of Simakov and Taylor (2000) and Simakov (2008). Nonetheless, in the Beyer et al. (2018) test on natural xenoliths from the Jericho kimberlite, Slave craton, *P* estimates for eclogites were systematically lower (by ~1 GPa) than orthopyroxene–garnet *P* estimates for peridotites recording similar *T*. As a result, the Jericho eclogites oddly appeared to fall on a hotter geotherm. The *T*-dependence of the clinopyroxene–garnet barometer is ca. 0.25 GPa / 50 °C, i.e., similar to that of the orthopyroxene–garnet barometer. Therefore, also in this case, errors in input *T* would displace *P–T* estimates roughly along the geotherm and cannot explain the observed discrepancy between peridotite-system and eclogite-system thermobarometry results.

A further test of clinopyroxene–garnet thermobarometry is illustrated in Figure 5, for touching or non-touching inclusions in diamonds and diamond- or graphite-bearing xenoliths. The *P–T* estimates reflect iterations of the Beyer et al. (2015) barometer and the Nakamura (2009) Fe–Mg exchange thermometer. Since the uncertainty of the barometer increases rapidly with decreasing concentrations of $^{[4]}Al$ in clinopyroxene, application to clinopyroxenes with Si contents >1.985 atoms per 6-oxygen formula is not recommended by Beyer et al. (2015). These compositions are relatively common for inclusions in diamonds (35% of 224 inclusions) and diamondiferous xenoliths (42% of 233 xenoliths), thus numerous samples had to be excluded. Even if further severe filters were applied to discard potentially low-quality chemical analyses, the results are disconcerting: although graphite-bearing xenoliths plot in the graphite stability field, most diamond-bearing xenoliths and inclusions in diamonds yield too low pressures well outside the diamond stability field (Fig. 5). Iteration of earlier formulations of the same barometer and thermometer does not improve the results (cf. Fig. 10 in Shirey et al. 2013). The reason for this discrepancy is unclear. Simakov and Taylor (2000) cautioned that a barometer based on the Ca-Tschermak content in clinopyroxene may lead to incorrect estimates when applied to kyanite- or SiO_2-oversaturated assemblages, but the Beyer et al. (2015) barometer shows no systematic deviations in calculated pressures for experiments in these systems. The large chemical complexity of eclogitic garnets and, especially, eclogitic clinopyroxenes might in part explain the poor performance of the barometer in applications to natural systems, in which variable amounts of Na, Fe^{3+}, Cr, K and vacancy-bearing (i.e., Ca-Eskola) components occur. A further possible explanation is the high sensitivity of the barometer to even small errors in the chemical analyses of clinopyroxene. Beyer et al. (2015) provided a formula to predict relative errors on calculated pressures due to propagation of uncertainties on SiO_2 contents, assuming an uncertainty of 1% relative in electron microprobe analysis: $\%err = 1.94 \times 10^{-8} e^{10.18398[Si\ apfu]}$. These model errors do not consider the additional errors that derive from iterative calculation of P and T (Fig. 5). A 1% relative increase in SiO_2 contents may move many inaccurate P–T estimates into the diamond field, suggesting that application of the clinopyroxene–garnet barometer may require very high-quality, well-standardized electron microprobe analyses to yield meaningful results. Many of the analyses reported in the literature are possibly of routine quality and do not meet the necessary standards for robust clinopyroxene–garnet barometry. Still, analytical quality alone is unlikely to account for all of the observed discrepancies (Fig. 5).

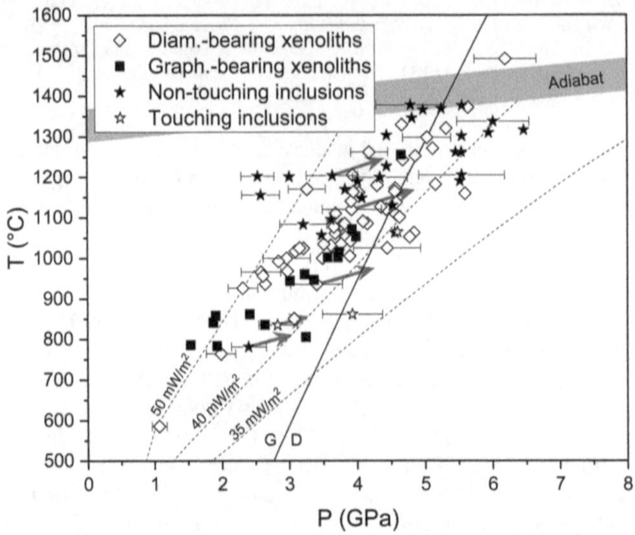

Figure 5. P–T estimates for diamond- or graphite-bearing eclogitic xenoliths and eclogitic clinopyroxene–garnet inclusion pairs in diamonds from worldwide sources, based on iteration of the Beyer et al. (2015) clinopyroxene–garnet barometer and Nakamura (2009) clinopyroxene–garnet thermometer. The following quality filters were adopted for clinopyroxenes: oxide totals = 99–101 wt%, cation totals per 6 oxygens = 3.99–4.02 apfu, Si < 1.985 apfu, X_{jd} < 0.5. The less restrictive upper limit for cation totals relative to the ideal value of 4 apfu allows for the presence of some Fe^{3+} in the clinopyroxene. These filters excluded 85% of 224 inclusions in diamonds and 73% of 233 diamond–eclogite xenoliths! Estimated P error bars based on Si contents (Beyer et al. 2015) are only shown for some representative samples. None of the data actually had errors smaller than the symbols. The grey arrows indicate the change in P–T estimates that is obtained by increasing the SiO_2 content in selected clinopyroxenes by 1% relative. Model-conductive geotherms for 35- and 40- and 50-mW/m² surface heat flow are after Hasterok and Chapman (2011). Mantle adiabat was calculated assuming a mantle potential temperature of 1327 ± 40 °C (Katsura et al. 2010). GD: graphite–diamond boundary after Day (2012).

Sun and Liang (2015) proposed that REE partitioning between clinopyroxene and garnet could be used not only as a thermometer (see above) but also as a barometer. Unfortunately, the discrepancies between calculated P_{REE} and run pressures for a set of independent validation experiments appear to be very large (up to ~3 GPa). A test conducted on twelve quartz eclogites, two graphite eclogites and nine diamond eclogites passed the appropriate stability-field constraints for all but two of the nine diamond eclogites; these fell at 0.2 and 0.5 GPa lower *P* than the diamond-graphite phase boundary. The proposed REE thermobarometry approach is promising, but further tests are necessary before this method can be used with confidence in diamond studies.

If trace element data are available, the partitioning of Li between clinopyroxene and garnet can also be used as a barometer (Hanrahan et al. 2009). Pressures for fifteen calibration experiments on a Mg-rich eclogitic composition at 4–13 GPa and 1100–1500 °C are reproduced to a very reasonable ±0.2 GPa (1 sigma) and the *T* dependence of the barometer is small (~0.2 GPa / 50 °C). Preliminary tests yielded conditions compatible with diamond for three diamond-bearing xenoliths and for non-touching inclusion pairs in four diamonds, whereas two other diamonds yielded conditions up to 2 GPa lower than the diamond stability field. Hanrahan et al. (2009) suggested that the two spurious *P* estimates could reflect disequilibrium between the non-touching inclusions. However, the use of non-touching inclusions is essential for this barometer, as the high diffusivity of Li may lead to reequilibration of small touching inclusions during and after transport to surface. Similar to the REE barometer, further tests are needed before this method can be recommended for diamond studies.

Monomineral barometers. Nimis and Taylor (2000) developed an empirical Cr-in-clinopyroxene barometer for chromian diopsides in equilibrium with garnet. In combination with the Nimis and Taylor (2000) single-clinopyroxene thermometer, this barometer uniquely permits derivation of both *P* and *T* conditions based on the composition of a single mineral inclusion. The *T*-dependence of the Cr-in-clinopyroxene barometer (0.15–0.25 GPa / 50 °C for diamond inclusion compositions) is generally lower than that of the orthopyroxene–garnet barometer (~0.2–0.3 GPa / 50 °C) and the *P*-dependence of the complementary single-clinopyroxene thermometer is very small (Fig. 4). Therefore, if *P* and *T* are calculated by iteration, errors in input *T* or *P* will displace a clinopyroxene *P–T* estimate at an angle relative to a cratonic geotherm and will contribute to 'apparent' scatter in *P–T* results for samples equilibrated on a steady-state geotherm. The Nimis and Taylor (2000) Cr-in-clinopyroxene barometer tends to progressively underestimate above 4.5 GPa (by up to ~1 GPa at 7 GPa; Nimis 2002; Ziberna et al. 2016). This is a known artifact and the discrepancy should be considered when comparing Cr-in-clinopyroxene estimates with those obtained using other methods. Nimis et al. (2020) proposed an empirical correction to minimize the differences with *P* estimates obtained using the orthopyroxene–garnet barometer at high *P*, at the cost of somewhat lower precision. These authors cautioned that their correction represented a temporary measure to reduce inconsistencies between independent barometric estimates. More recently, Sudholz et al. (2021b) provided an experimental recalibration of the Cr-in-clinopyroxene barometer that yields results very similar to the empirically corrected calibration for clinopyroxenes associated with diamonds or hosted in well equilibrated peridotitic xenoliths at $P > 5$ GPa (Fig. 6a). At lower *P* the two revised calibrations diverge slightly, but the Nimis et al. (2020) calibration maintains somewhat better overall consistency with orthopyroxene–garnet barometry (Fig. 6b,c). For proper application of the Cr-in-clinopyroxene barometer, the analyses must undergo compositional filtering in order to select clinopyroxenes in equilibrium with garnet and eliminate certain compositions for which analytical uncertainties propagate excessive uncertainties on the calculated *P*. Ziberna et al. (2016) provided a useful cookbook to perform the necessary compositional screening, which slightly improved an earlier one proposed by Grütter (2009). According to this cookbook, some compositions require that electron microprobe analyses are carried out using beam

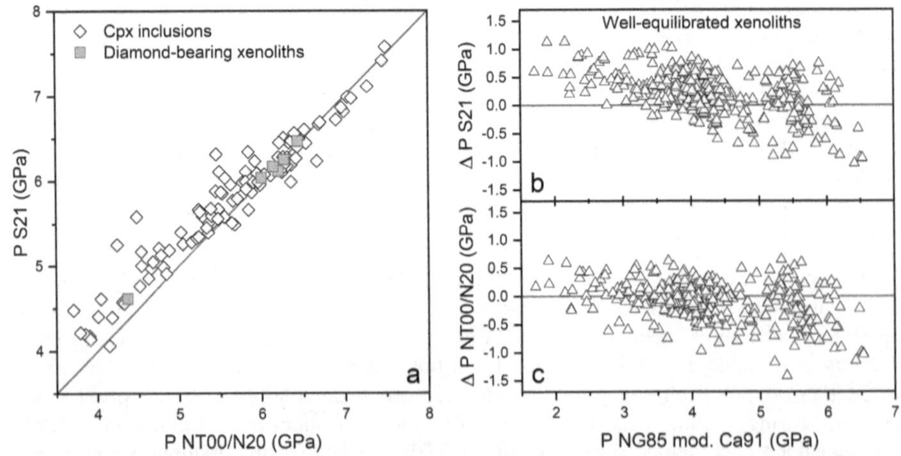

Figure 6. (a) Comparison between *P* estimates obtained with the Cr-in-clinopyroxene barometer, as recalibrated by Sudholz et al. (2021b) and Nimis et al. (2020), respectively, for clinopyroxenes included in diamonds and in diamond-bearing peridotitic xenoliths. (b-c) Deviations of *P* calculated with the Sudholz et al. (2021b) and Nimis et al. (2020) recalibrations of the Cr-in-clinopyroxene barometer from *P* estimated with the Nickel and Green (1985) orthopyroxene–garnet barometer for 349 well-equilibrated xenoliths (xenolith compilation after Nimis and Grütter 2010, with clinopyroxene compositions filtered according to Ziberna et al. 2016). For 87 xenoliths containing orthopyroxenes with Na > Cr + Ti (molar fractions), the Nickel and Green (1985) barometer was modified as suggested by Carswell (1991). The alternative modification proposed by Carswell and Gibb (1987) would produce somewhat lower *P* estimates for these samples and increased scatter at high *P* in both (b) and (c). All clinopyroxene pressures in (a), (b) and (c) were calculated by iteration with the Nimis and Taylor (2000) single-clinopyroxene thermometer. Orthopyroxene–garnet pressures in (b) and (c) were calculated by iteration with the Taylor (1998) two-pyroxene thermometer.

currents and counting times higher than routine to minimize analytical uncertainties, whereas some unfavorable compositions should simply be discarded. In particular, the recommended limitation to $Cr/(Cr+Al)_{mol} > 0.1$ excludes about two thirds of reported inclusions classified as websteritic. When applied to properly filtered graphite- or diamond-bearing xenoliths and inclusions in peridotitic diamonds, the Cr-in-clinopyroxene barometers yield results consistent with or within ~0.3 GPa of the stability field of the respective carbon phase (Nimis and Taylor 2000; see also Fig. 8a). The lack of adequate filtering in previously published thermobarometric results greatly contributed to the larger overall scatter in clinopyroxene *P*–*T* plots relative to those for orthopyroxene–garnet pairs (cf. Stachel and Harris 2008).

Griffin and Ryan (1995) devised an empirical barometer based on the Cr content of garnet in equilibrium with orthopyroxene and spinel. This Cr-in-garnet barometer was later refined by Ryan et al. (1996) and further revised, using a different Cr/Ca-in-garnet approach, by Grütter et al. (2006). The last two versions are the most widely used, typically in combination with a Ni-in-garnet thermometer. That of Grütter et al. (2006) gives up to ~10% lower estimates and a better agreement with petrological constraints imposed by the graphite–diamond transition, and is therefore preferable. If the coexistence of garnet with spinel is unknown, as is the case for many garnet inclusions in diamonds, the Cr-in-garnet methods only provide an estimate of minimum pressure.

Simakov (2008) proposed a simplified version of the clinopyroxene–garnet barometer for eclogites that solely relies on the Ca-Tschermak content in the clinopyroxene. The accuracy of this monomineral barometer cannot be better than that of the native two-phase barometer and is thus prone to the same problematic issues (see above). Ashchepkov et al. (2017) calibrated or recalibrated a series of empirical monomineral thermobarometers, which should be suitable for peridotitic and eclogitic garnets and clinopyroxenes. When applied to mantle xenoliths and diamond inclusions from individual localities, these methods produce strongly scattered *P–T* estimates (e.g., Figs. 9–12 in Ashchepkov et al. 2017), which are difficult to compare with the more regular *P–T* distributions that are typically obtained with other, more widely tested methods. The usefulness of the Ashchepkov et al. (2017) monomineral formulations for diamond thermobarometry is therefore questionable.

With increasing pressure beyond ~7 GPa, mantle garnets incorporate increasing proportions of majoritic (i.e., pyroxene-like) component by substitution of Si for octahedrally coordinated Al and Cr. The Na and Ti contents also tend to increase. Collerson et al. (2000) utilized the relationship observed between majorite fractions and pressure in experiments on mafic and ultramafic systems to devise an empirical single-mineral barometer. The method was later revised by Collerson et al. (2010), Wijbrans et al. (2016) and Beyer and Frost (2017). Only Wijbrans et al. (2016) proposed two distinct calibrations, one for peridotitic and another for eclogitic systems. Experimental tests by Beyer and Frost (2017) and Beyer et al. (2018) showed systematic deviations or decreased precision for all but the Beyer and Frost (2017) formulation. This was particularly relevant for eclogitic compositions, which were underrepresented in previous calibration datasets. The Beyer and Frost (2017) barometer was calibrated at pressures of 6 to 20 GPa and temperatures of 900 to 2100 °C. These conditions cover sublithospheric mantle conditions up to ~575 km depth, corresponding to the lower transition zone. Experimental pressures are reproduced with a standard error of estimate of 0.86 GPa. In relative terms, considering the enormous range of pressures covered by the calibration, this uncertainty is not much larger than that of the barometers used for lithospheric diamonds. The effect of temperature on pressure estimates is apparently negligible. Note that application of the majorite barometer to lithospheric garnets equilibrated at $P < 6$ GPa is not recommended, because under such conditions the barometer becomes extremely sensitive to analytical errors (Beyer and Frost 2017). Tao et al. (2018) showed that high Fe^{3+} contents in the garnet tend to produce underestimated pressures with previous majorite barometers and provided a further revised calibration of the Collerson et al. (2010) barometer that explicitly takes into account Fe^{3+}. The Tao et al. (2018) calibration may be preferable when Fe^{3+} data are available and measured $Fe^{3+}/\Sigma Fe$ ratios are greater than 0.2. All these majorite barometers require that garnet occurs in equilibrium with a clinopyroxene phase. This may be an issue, because with increasing pressure pyroxene progressively dissolves into garnet and at pressures above ~15–19 GPa, depending on bulk composition and temperature, mantle rocks no longer contain pyroxene (Irifune et al. 1986; Irifune 1987; Okamoto and Maruyama 2004). Pyroxene-undersaturated condition can be suggested by the presence of $CaSiO_3$ phases along with majoritic garnet as inclusions in diamonds or by high Ca contents in majoritic garnet. Under such conditions, the barometer would give erroneously low pressures (Harte and Cayzer 2007). If there is no independent evidence that the garnet was pyroxene-saturated, any pressure estimates close to that of the clinopyroxene-out reaction should be considered as estimates of the minimum *P* of entrapment. Thomson et al. (2021) devised a novel type of majorite barometer using machine-learning algorithms. Compared to traditional majorite barometers, this new barometer is apparently insensitive to petrological limitations (most notably, the absence of clinopyroxene) and reproduces experimental majoritic garnet compositions over a wider range of bulk compositions and experimental pressures (6–25 GPa) with much better overall precision (mostly within ±2 GPa). Thomson et al. (2021) caution that the compositions of majoritic garnet inclusions in diamonds lie in a region with relatively fewer experiments and machine-learning regressions may not be reliable in extrapolation. Thus, pressure predictions for inclusions in diamonds may have somewhat larger uncertainties. Still, this is the only available barometric method that allows estimating pressures for majoritic garnet inclusions beyond the clinopyroxene-out reaction. When applied to inclusions in diamonds, differences between pressure estimates obtained with the Thomson et al. (2021) and Beyer and Frost (2017) barometers lie in the range –2.5 to +6 GPa. Whatever the majorite barometric method used, if the garnet reequilibrated with clinopyroxene after entrapment, as is typically the case in unmixed inclusions (see below), underestimation of P will occur.

Elastic methods. When an inclusion is entrapped in a diamond, both the inclusion and the diamond are initially at the same pressure. On eruption, the pressure acting on the diamond drops to 1 atm, but a residual pressure, P_{inc}, of up to several GPa may develop on the inclusion as a consequence of the inclusion and host having different elastic properties. If we know the P_{inc} and the elastic properties of the inclusion and host, we can back-calculate the conditions under which the two minerals would be at the same pressure. These conditions describe a line in P–T space, known as the entrapment isomeke, and the entrapment conditions (i.e., the diamond formation P–T) will be one of the points along the isomeke. This forms the basis for elastic barometry of diamonds. The principles, problems and on-going developments of this method are described in detail in Angel et al. (2022, this volume).

Projection onto a geotherm. When only thermometric estimates are available due to a lack of suitable barometers, a tentative estimate of pressure can be obtained by projecting temperature estimates on a reference geotherm. This will most conveniently be a geotherm derived by fitting P–T data for mantle xenoliths or xenocrysts from the same kimberlitic source. The FITPLOT computational software by Mather et al. (2011) provides a useful aid to model geotherms from xenolith P–T data. There may be significant uncertainties in some of the input parameters that are used for the modeling, specifically the thickness and heat production of the upper and lower crust and the mantle potential temperature, but in most cases these uncertainties will not significantly affect the shape of the geotherm within the lithospheric diamond window, which will be chiefly constrained by the xenolith P–T data. Nonetheless, assuming different mantle potential temperatures will affect the estimated thickness of the lithosphere and the depth extent of the conductive portion of the geotherm. This may have some relevance for some old, high-T diamonds, as mantle potential temperatures may have varied over geological time. If this variation is not considered, thermometric estimates may unduly be projected onto the adiabatic portion of the geotherm and lead to severe overestimation of P. Values for the present-day mantle potential temperature and its change with time can be found in Katsura et al. (2010) and Ganne and Feng (2017). The output data that describe the relevant section of the model geotherm can conveniently be fitted through a polynomial expression to derive P as a function of T. The final P–T estimate will be given by the intersection of this polynomial with the isotherm described by the chosen thermometer. Note that this isotherm will not generally be a flat line in a P–T plot, due to the generally significant P-dependence of the thermometers (Fig. 4).

The basic assumption underlying this approach is that the inclusions last equilibrated at P–T conditions lying on the mantle geotherm at the time of eruption. This is generally acceptable for touching inclusion pairs, which can reequilibrate in mostly the same way as touching minerals in xenoliths (but see caveat in section below dedicated to touching vs. non-touching inclusions), whereas it is much less obviously so for non-touching inclusions. Comparisons between P–T estimates for inclusions and xenoliths from the same sources showed that the assumption often holds also for non-touching inclusions (Nimis 2002). Nonetheless, there is also evidence of some diamond inclusions recording conditions significantly hotter or colder than the xenolith geotherm (Griffin et al. 1993; Sobolev and Yefimova 1998; Nimis 2002; Stachel et al. 2003, 2004; Weiss et al. 2018; Nimis et al. 2020). Also, multiple non-touching inclusions within individual diamonds may record a range of conditions, indicating thermal fluctuations during the growth history of the diamond (Griffin et al. 1993). Projection onto a geotherm should therefore be used with caution and relatively large uncertainties should be allowed. Assuming a maximum possible T difference from a xenolith geotherm of ~250 °C, consistent with available estimates for diamonds using single-clinopyroxene (Nimis and Taylor 2000), single-garnet (Canil 1999) and orthopyroxene–garnet (Harley 1984) thermometers, projection on a typical cratonic geotherm may result in a P uncertainty of up to ~1.5 GPa (Fig. 4).

Projection onto the adiabatic portion of the mantle geotherm may provide a necessary independent constraint for the application of elastic barometric methods to sublithospheric diamonds (e.g., Anzolini et al. 2019).

Mineral stability. Lithospheric mantle rocks provide little opportunity to estimate conditions of diamond formation based only on the stability of mineral assemblages. For example, across the entire diamond window, the only significant mineralogical change is the spinel-to-garnet transition in peridotite. The spinel-to-garnet reaction is not univariant and in depleted cratonic peridotites with strongly elevated Cr/Al, spinel + garnet assemblages may persist over a large range of *P–T* conditions (e.g., MacGregor 1970; Webb and Wood 1986; Klemme 2004; Ziberna et al. 2013). In general, spinel is stabilized to higher pressures in more refractory, Cr-rich and Al-poor bulk compositions. This allows magnesiochromite to be incorporated as inclusions in diamonds that were formed in very deep, refractory environments (e.g., 6.5 GPa, corresponding to a depth of over 200 km, for a diamond from the Udachnaya kimberlite, Siberia; Nestola et al. 2019a).

Mineral stabilities become more interesting in the sublithospheric mantle, where a number of characteristic mineralogical changes occur (Harte 2010; Harte and Hudson 2013). Although these changes do not permit bracketing of *P–T* conditions with a resolution comparable to that of conventional thermobarometers, in some cases they constrain the formation of a diamond to within specific mantle depth regions. For instance, the coexistence of Fe-bearing periclase with enstatite, interpreted to be inverted bridgmanite (previously known as MgSi-perovskite), has long provided evidence that some diamonds were formed at lower-mantle depths (Scott Smith et al. 1984; Moore et al. 1986). Even in the absence of periclase, inverted bridgmanite from the lower mantle can readily be distinguished from upper-mantle enstatite by its very low Ni contents (Stachel et al. 2000). The finding of a rare inclusion of ringwoodite in a diamond from Juina, Brazil, unequivocally demonstrated its origin from the mantle transition zone (Pearson et al. 2014). Also, the reconstructed compositions of a suite of exsolved inclusions in Juina diamonds nicely matched those of minerals expected to form in a basaltic system in the lower mantle (Walter et al. 2011).

However, using incomplete mineral assemblages found as inclusions in diamonds as depth markers is not always free from ambiguity. For instance, inclusions of breyite (previously known as CaSi-walstromite) have long been considered to represent retrogressed $CaSiO_3$-perovskite from depths greater than ~600 km (e.g., Harte et al. 1999; Joswig et al. 1999), but there is compelling evidence that at least some breyite inclusions originated at much shallower depths within the upper mantle (Brenker et al. 2005; Anzolini et al. 2016). Woodland et al. (2020) recently provided experimental support for a possible upper-mantle origin of breyite. Also, Fe-bearing periclase is the most common inclusion of interpreted lower-mantle origin (Kaminsky 2012), but there is evidence that Fe-bearing periclase participates in mineral parageneses straddling the upper mantle-lower mantle boundary (Hutchison et al. 2001) and that it may be a co-product of diamond-forming reactions that may occur in the transition zone and even in the overlying upper mantle (Thomson et al. 2016; Bulatov et al. 2019).

Best practices in diamond thermobarometry

Data quality. A sometimes underrated prerequisite for robust thermobarometry is precise and accurate chemical analyses. Inclusions in diamonds are often small and their chemical analysis can be challenging. Moreover, some thermobarometers may be particularly sensitive to analytical uncertainties on elements that occur at critical concentration levels. For instance, Ziberna et al. (2016) explored propagation of analytical errors on *P* estimates using the single-clinopyroxene barometer of Nimis and Taylor (2000) and found that routine electron microprobe analytical conditions may be insufficient for meaningful barometry of many inclusions in diamonds. Clearly, not only poor counting statistics but also bad sample preparation and improper standardization can be an issue and may lead to spurious results. A survey of published chemical data for over 600 clinopyroxene inclusions in lithospheric diamonds reveals that ~30% of them have oxide total weight percentages <98.5% or >101%, or cation sums < 3.98 or > 4.03 atoms per formula unit and, thus, are definitely not of good quality! Note that the above cut-offs generously take into account the possible effects of considering all Fe as Fe^{2+} in electron microprobe analyses.

The next critical step when studying inclusions that were separated from coexisting mineral phases is the definition of their original paragenesis. Even if single-mineral thermobarometers may often come in handy, they invariably assume that the investigated mineral last equilibrated with specific other minerals. As for clinopyroxene and garnet, which are major constituents in several types of mantle rocks, simple compositional screening based on either major or trace element concentrations may help to discriminate between the most typical paragenetic varieties (e.g., Griffin et al. 2002; Grütter et al. 2004; Nimis et al. 2009; Ziberna et al. 2016). Unusual compositions that are not sensitively discriminated by available classification schemes or fall outside the compositional range over which the thermobarometers were calibrated, as well as false positives that unduly survive compositional filtering, may undermine the reliability of thermobarometric data. This problem may be minimized by investigating large datasets, when this is permitted by the number and nature of the available samples.

Even when all cautionary steps have been made to select suitable compositional data, thermobarometric estimates are subject to errors. As discussed in the previous section, some of these errors are systematic and lead to predictable inconsistencies between estimates obtained using different methods. This fact should always be considered when comparing data obtained using different thermobarometers or even different formulations of the same thermobarometer. Thermobarometric methods with demonstrated internal consistency must be prioritized if the purpose is to explore the distribution of data for heterogeneous inclusion populations.

In this respect, thermobarometry of eclogitic diamonds remains problematic. Considering the compositional limitations required by the 'best' available barometric methods (i.e., $Si_{Cpx} < 1.985$ apfu and $Na_{Cpx} < 0.5$ apfu), ~45% of over 200 reported eclogitic clinopyroxene–garnet inclusion pairs are automatically excluded. In addition, the reliability of barometric methods for eclogitic inclusions is questionable (see above).

When *P* estimates are too uncertain or not available, for example, due to a lack of suitable barometers for the specific inclusions investigated, projection of temperatures onto the local xenolith geotherm can provide a last-resort solution to estimate *P*. The resulting uncertainties can be too large for an accurate characterization of genesis conditions for individual diamonds, but the procedure can have some utility for exploring general trends in large datasets (e.g., Korolev et al. 2018; Nimis et al. 2020). It probably remains the safest available method for barometry of eclogitic inclusions, notwithstanding potential over- or underestimation for diamonds formed at conditions hotter or, respectively, colder than ambient mantle at the time of eruption. A fairly popular alternate has been to calculate *T* for these diamonds at a fixed *P* of 5 GPa, approximately corresponding to the average *P* for diamonds worldwide. This method is not as effective and may overcompress the range of *T* estimates, due to the generally significant *P*-dependence of the thermometers (Fig. 4).

Touching vs. non-touching. When a pair of touching mineral grains included in a diamond reequilibrate to changing external conditions, their total encapsulated volume remains constrained by the surrounding diamond. This forces the included minerals to follow a distinct pressure path and to achieve final compositions that are different from those of the same minerals outside the diamond (see Angel et al. 2015 and Ferrero and Angel 2018 for more general discussion of this phenomenon). The most common changes in ambient conditions are probably isobaric cooling or heating after diamond growth, in response to secular variations in mantle thermal state or following short-term thermal pulses related to diamond-forming processes. In most cases of diamond-hosted inclusions, cooling will cause a decrease in the inclusion *P* even if the external *P* remains constant, leading to underestimation of the actual diamond *P* using chemical thermobarometers. The extent of this underestimation can be calculated if the thermoelastic properties of the inclusion and host minerals are known (Angel et al. 2015). For various combinations of garnet, pyroxene and olivine inclusions in diamond, the error would amount to ~0.3 GPa for a temperature drop of 100 °C. Heating will have the opposite effect. These errors are of a similar magnitude as those of the orthopyroxene–garnet and single-clinopyroxene barometers and may thus contribute to some of the scatter in *P–T* plots for touching inclusions. The resulting errors may be lower, however, if plastic deformation in the diamond is able to reduce the under- or overpressures developed on the inclusions.

If some thermal reequilibration occurs, the use of single-mineral thermobarometers on touching inclusions may generate further errors when some key minerals are not part of the inclusion assemblage. For instance, touching inclusions of clinopyroxene-orthopyroxene devoid of garnet will reliably record the ambient thermal state at the time of eruption, but will be poor candidates for single-clinopyroxene barometry. In response to a thermal change, the clinopyroxene will easily reequilibrate with orthopyroxene, but not with garnet, and so a false apparent encapsulation *P* will be calculated at the final equilibrium *T* (Nimis 2002). Touching inclusions of clinopyroxene ± garnet ± olivine would instead produce estimates close to the diamond formation conditions, because the composition of the clinopyroxene in orthopyroxene-free assemblages would be little affected by a variation in temperature. Polymineralic inclusions consisting of touching clinopyroxene-orthopyroxene–garnet ± olivine will fully reequilibrate and will provide *P and T* conditions close to those at the time of eruption (except for the small systematic errors described in the previous paragraph).

Non-touching inclusions are immune to the above problems, because the inert diamond protects them from post-entrapment reequilibration, and are unequivocally the best candidates to determine the diamond formation *P–T*, provided reliable monomineral thermobarometers can be used. In all other cases, non-touching inclusion pairs are prone to potential errors, because the individual inclusions may have been incorporated at different times and under different conditions (see above) and may thus not be in mutual equilibrium. The largest *T* difference between multiple inclusions in individual diamonds was reported for garnets in a diamond from the Mir kimberlite, Siberia (Griffin et al. 1993). The calculated difference varies greatly depending on which calibration of the Ni-in-garnet thermometer is used, i.e., ca. 400 °C with the Ryan et al. (1996) calibration or ca. 230 °C with the Canil (1999) calibration. Disequilibrium amongst non-touching inclusions in diamonds, however, appears to be the exception rather than the rule, as multiple inclusions in individual diamonds most often yield similar *P–T* estimates once the systematic deviations between different thermobarometers are taken into account (Stachel and Harris 2008; Stachel and Luth 2015).

Syngenesis vs. protogenesis. For several decades, thermobarometry of inclusions in diamonds has relied on the assumption of syngenesis, i.e., the inclusions formed or completely recrystallized during the growth of diamond. A corollary of this assumption was that the inclusions record the conditions of diamond formation, excepting for potential post-entrapment modifications in touching inclusions. The main proof for syngenesis was that diamond typically imposed its shape on the inclusions. This widely accepted paradigm was challenged by Nestola et al. (2014), who found evidence of protogenesis (i.e., formation before the diamond) for some olivine inclusions that would have been classified as syngenetic based on morphological criteria. Similar evidence was then found for other olivine, clinopyroxene, garnet and magnesiochromite inclusions (Milani et al. 2016; Nestola et al. 2017, 2019b; Nimis et al. 2019). The possibility that inclusions in diamonds are protogenetic suggests that they might record conditions predating diamond formation. This might be particularly relevant for isolated inclusions, if their host diamonds were formed from hot fluids or melts intruding through relatively cool mantle rocks. Did the preexisting mineral grains have enough time to fully reequilibrate to the changing thermal conditions during diamond growth and before their incorporation was complete?

The time required by a mineral to adjust its composition to new physicochemical conditions strongly depends on temperature and grain size, and can be estimated from diffusion kinetics. For instance, a clinopyroxene grain with a diameter of 200 μm would fully reset its Ca/Mg ratio (the main temperature sensor in pyroxene thermometry) in less than 100,000 yr above 1200 °C and in a few m.y. below 900 °C (Fig. 7). Whether these time spans are short or long enough for complete reequilibration depends on the diamond growth rate, which is not precisely known. It is tempting to infer that incomplete reequilibration of protogenetic inclusions, especially the largest and shallowest (i.e., coldest) ones, may account for some of the scatter in *P–T* plots for non-touching inclusions (Fig. 8). Also, the similarity of *P–T* conditions occasionally observed for inclusions in diamonds and xenoliths from the same sources may alternatively be ascribed to diamond formation from thermally-equilibrated media (Nimis 2002) or to lack of reequilibration of protogenetic inclusions during short-lived diamond-forming processes (Nestola et al. 2014).

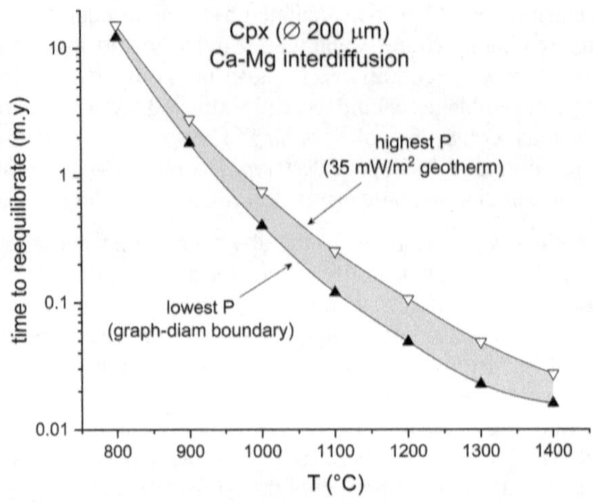

Figure 7. Time required to reequilibrate the Ca/Mg ratio in a spherical clinopyroxene grain with diameter of 200 μm as a function of temperature. The two curves are calculated from Ca-Mg inter-diffusion data (Zhang et al. 2010) for compositions on the pyroxene solvus at pressures corresponding to a 35 mW/m^2 geotherm (≈ maximum *P* for lithospheric diamonds) and to the graphite–diamond boundary (minimum *P* for diamonds), respectively. Increasing or decreasing the grain diameter by a factor of 2 would increase or, respectively, decrease the reequilibration time by a factor of ~4.

Exsolved inclusions. Some inclusions in diamonds show evidence of post-entrapment unmixing. Reported examples are clinopyroxenes with orthopyroxene exsolution lamellae (± coesite), representing unmixed high-*T* clinopyroxenes, and intergrowths of more or less majoritic garnet and clinopyroxene, representing unmixed high-*P* majoritic garnets.

Exsolved clinopyroxene inclusions are uncommon and have only been reported in diamonds from Namibian placers and from the Voorspoed kimberlite, South Africa (Leost et al. 2003; Viljoen et al. 2018). The coexistence of clinopyroxene and orthopyroxene makes these inclusions ideal for pyroxene thermometry, but the exsolved assemblage does not provide barometric estimates. Since pyroxene thermometers have a small *P*-dependence, projection onto the local xenolith geotherm provides in these cases reasonable estimates of the depth of

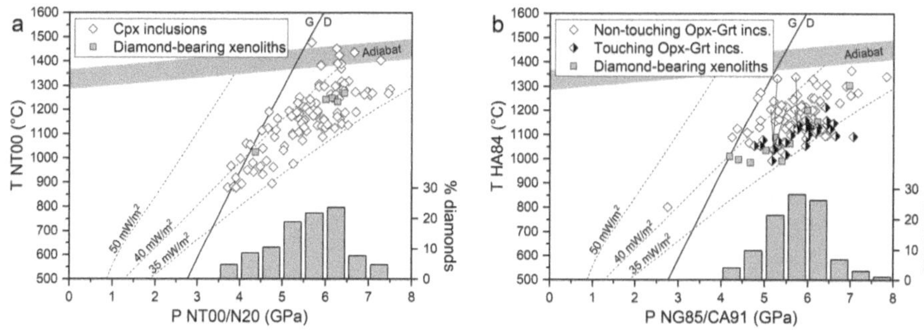

Figure 8. (**a**) *P–T* estimates for chromian clinopyroxene inclusions in worldwide diamonds ($n = 100$) and diamond-bearing lherzolitic xenoliths ($n = 6$) calculated with the Nimis and Taylor (2000) single-clinopyroxene barometer (as corrected by Nimis et al. 2020) and thermometer. Using the Sudholz et al. (2021b) recalibration of the clinopyroxene barometer would have produced broadly similar results (cf. Fig. 6). All compositions were filtered as recommended by Ziberna et al. (2016). Twenty-six (retained) compositions would have required better than routine EMPA counting statistics to minimize *P* uncertainties, but insufficient analytical details were provided. These records did not produce any particular deviations from the overall or local trends and therefore were not discarded. The great majority of the inclusions were isolated within the diamonds. Inclusions of clinopyroxene in contact with orthopyroxene but not with garnet were not considered, because post-entrapment reequilibration in these inclusions may produce spurious barometric results (Nimis 2002). (**b**) *P–T* estimates for orthopyroxene–garnet inclusion pairs in worldwide diamonds ($n = 92$) and diamond-bearing xenoliths ($n = 11$) calculated with a combination of the Harley (1984) thermometer and Nickel and Green (1985) barometer, with the modification to the barometer introduced by Carswell (1991) for orthopyroxenes with Na > Cr + Ti (molar fractions). Six orthopyroxene analyses with cation sums <3.98 atoms per formula unit were not considered. The compositions of multiple inclusions in individual diamonds were averaged, so that any single *P-T* point corresponds to one diamond. Tie lines in (b) connect touching inclusions in diamonds from Mwadui, Jwaneng and Jagersfontein with non-touching inclusions in other diamonds from the same localities and similar depths. Some of the apparent discrepancies between the two plots can be ascribed to known systematic inconsistencies between the thermobarometers used (see text for further explanation). Conductive geotherms for 35- and 40- and 50-mW/m^2 surface heat flow after Hasterok and Chapman (2011). The mantle adiabat was calculated assuming a mantle potential temperature of 1327 ± 40 °C (Katsura et al. 2010). GD: graphite–diamond boundary after Day (2012). Sources of data for diamond inclusions and diamondiferous xenoliths used in the plots: Aulbach (1999), Boyd (1984), Boyd and Finnerty (1980), Creighton et al. (2008), Daniels and Gurney (1989), Davies et al. (2004a,b), Dawson and Smith (1975), De Stefano et al. (2009), Deines and Harris (2004), Geological Survey of Canada (1989), Gurney et al. (1984), Harris et al. (1994, 2004, and unpubl.), Hervig et al. (1986), Hutchison (1997), Jaques et al. (1989, 1990), Kopylova et al. (1997), Korolev et al. (2018), Logvinova et al. (2015), McDade and Harris (1999), Moore and Gurney (1989), Nimis et al. (2020), Otter and Gurney (1989), Phillips et al. (2004), Pokhilenko et al. (1977, 2004), Ponomarenko et al. (1980), Rickard et al. (1989), Shee et al. (1982), Sobolev (1977), Sobolev et al. (1976, 1997a,b, 1999, 2009), Stachel and Harris (1997a,b), Stachel et al. (1998, 2003, 2004, 2018), Tappert et al. (2005, 2006), Tsai et al. (1979), University of Cape Town (unpubl. reports), Van Rythoven and Schulze (2009), Van Rythoven et al. (2011), Viljoen et al. (1992), Wang et al. (2000), Wilding et al. (1994).

origin of the diamonds (Nimis et al. 2020). If the compositions and relative proportions of the exsolved phases are known (e.g., by combining electron microprobe data with 2D or, better, 3D image analyses), the integrated pre-exsolution composition of the clinopyroxene can be reconstructed. The integrated composition can provide a *T* estimate at the time of diamond growth, assuming equilibrium with a separate orthopyroxene during entrapment, or a minimum *T* estimate of diamond formation, in the absence of orthopyroxene. In the reported cases, thermometry of the reconstructed clinopyroxenes provided evidence for diamond formation at temperatures near the mantle adiabat and ~200 °C hotter than the ambient mantle at the time of eruption. Attempts to apply the single-clinopyroxene barometer to the reconstructed compositions yielded unreasonably low pressure estimates, suggesting formation in strongly modified mantle environments (Nimis et al. 2020).

Exsolution textures are common in majoritic garnet inclusions in sublithospheric diamonds (Moore and Gurney 1989; Wilding 1990; Harte and Cayzer 2007). These textures are interpreted to reflect decompression during slow ascent of the diamonds in convecting mantle or mantle plumes. The presence of exsolutions implies that pressure estimates based on majoritic garnet compositions are minima. Using reconstructed pre-exsolution garnet compositions will yield conditions closer to those of diamond formation. Even so, the resulting estimates may be minima, because the garnet may have originated at depths beyond the clinopyroxene-out reaction (see previous Barometry section).

P–T DATA FOR DIAMONDS

Lithospheric diamonds

Figure 8 shows a compilation of *P–T* data for lithospheric diamonds from worldwide sources, obtained using the most robust available thermobarometer combinations. The compilation comprises clinopyroxene inclusions in 100 lherzolitic diamonds and orthopyroxene–garnet inclusion pairs in 77 harzburgitic, 12 lherzolitic and 10 websteritic diamonds. The great majority of the clinopyroxene inclusions were isolated within their host diamonds and should generally record conditions close to those of diamond formation. The orthopyroxene–garnet pairs consist of both touching and non-touching assemblages. The touching pairs may have reequilibrated after their incorporation and should provide conditions close to the final conditions of residence in the mantle. In addition, *P–T* data were calculated, using the same methods, for diamondiferous xenoliths for which suitable mineral analyses were available.

In comparing the two plots for single-clinopyroxene and orthopyroxene–garnet thermobarometry (Fig. 8a,b), it is important to consider the known systematic inconsistencies between the thermobarometric methods used (see above). The orthopyroxene–garnet thermometer tends to overestimate temperature below 1100 °C and to underestimate above 1100 °C. This accounts at least in part for the more restricted temperature range obtained for the orthopyroxene–garnet inclusions. Nonetheless, disregarding a single obvious outlier, all *P–T* values fall within 0.3 GPa of the diamond stability field, confirming the general reliability of the thermobarometers. Most of the *P–T* results lie between the 35- and 40-mW/m^2 conductive geotherms of Hasterok and Chapman (2011), i.e., at conditions typical for cratonic lithospheres. Most exceptions record higher temperature conditions, which in the case of clinopyroxene may reach the mantle adiabat. All of these exceptions consist of non-touching inclusions. A possible explanation is that several diamonds were formed from thermally non-equilibrated media derived from the sublithospheric convective mantle. Another potential source of mantle temperature inhomogeneity during diamond formation can be incomplete cratonization at the time of diamond growth. For instance, in the Kaapvaal craton a stable geothermal gradient was established by about 2.5 Ga and followed by secular cooling (e.g., Michaut and Jaupart 2007; Brey and Shu 2018) and re-heating (Griffin et al. 2003), but many Kaapvaal diamonds are significantly older than 2.5 Ga (see review in Shirey and Richardson 2011) or may have formed during important perturbations of the local lithosphere (Griffin et al. 2003; Korolev et al. 2018). Therefore, it is not surprising that many of these diamonds may record conditions hotter or colder than those at the time of kimberlite eruption (e.g., Nimis et al. 2020).

The touching orthopyroxene–garnet inclusion pairs deserve further discussion. In the available dataset, the majority of them (23 out of 31) refer to diamonds from the De Beers Pool (Kimberley, South Africa). These inclusions were studied by Phillips et al. (2004), who carefully documented differences between *P–T* estimates recorded by touching vs. non-touching inclusions. These workers found that the non-touching inclusions yielded higher *P–T* values than the touching inclusions and ascribed this observation to post-entrapment reequilibration of the touching inclusions under decreasing *T* conditions, possibly accompanied by mantle uplift and decompression. They also correctly noted that post-entrapment elastic modifications were not of sufficient magnitude to reconcile the observed *P–T* differences between touching and non-touching inclusions. However, Nimis et al. (2020) observed that both touching and non-touching inclusions in the Phillips et al. (2004) dataset record a colder geothermal regime than mantle xenoliths and xenocrysts in the same kimberlites, casting doubts on the secular cooling hypothesis. Whereas the relatively cold signature of the non-touching inclusions might reflect old conditions pre-dating the final thermal equilibration of the lithosphere recorded by the xenoliths, the data for the touching inclusions are difficult to interpret. Weiss et al. (2018) speculated that the low temperatures recorded by the touching inclusions could be related to late infiltration of cold, slab-derived fluids, remnants of which were preserved in some cloudy and coated diamonds from the same locality. Yet it remains unclear why infiltration of these fluids should leave virtually no traces in the sampled mantle xenoliths.

Touching orthopyroxene–garnet inclusions from other localities are probably too few to be statistically significant. Nonetheless, in all these cases the touching inclusions yield temperatures significantly lower than non-touching inclusions in other diamonds from about the same depth (Fig. 8b). Additional thermobarometric evidence that diamond formation may be followed by cooling was provided by Stachel and Luth (2015) and Viljoen et al. (2018). However, Stachel and Luth (2015) also described touching inclusions yielding temperatures similar to or even slightly higher (up to 40 °C) than non-touching inclusions in the same diamonds.

Diamond's depth systematics. Irrespective of the nature of the inclusions and the thermobarometric methods used, the frequency distribution of pressures for lithospheric diamond is essentially unimodal, with a mode at ~6 GPa, corresponding to depths around ~190 km (Fig. 8; see also Stachel and Harris 2008 and Stachel 2014 for broadly similar distributions obtained using different thermobarometers and partly different datasets). The slight differences between the pressure distributions obtained for clinopyroxene and orthopyroxene–garnet inclusions may largely be ascribed to recognized inconsistencies between the different thermobarometer combinations used and, perhaps, to sampling bias. Therefore, these differences are probably not significant. The hard constraints imposed by the graphite–diamond and lithosphere-asthenosphere boundaries cannot fully explain the observed depth distributions, which show decreased diamond frequencies approaching these boundaries (Fig. 8). Nimis et al. (2020) obtained very similar depth distributions with depth modes at 180 ± 10 km for diamonds from three individual South African kimberlite sources by using various combinations of thermobarometric methods on different inclusion types. They also found no clear correlation with the depth distributions of mantle xenocrysts from the same sources calculated using the same thermobarometric methods. They concluded that the observed diamond depth distributions are unrelated to the kimberlite sampling efficiency and, thus, have genetic meaning and likely global significance.

The global significance of the observed depth distributions is also supported by the abundant $T_{\text{Ni-in-garnet}}$ data for garnet inclusions (Fig. 9). These data show a mode at ~1200 °C, but no pressure constraints are generally available. Assuming that the garnets last equilibrated at 'average' conditions between the 35- and 40- mW/m^2 model geotherms, similar to the clinopyroxene and orthopyroxene–garnet inclusions (Fig. 8), the Ni-in-garnet temperature mode would correspond, again, to a pressure mode of ~6 GPa and a depth mode of ~190 km. The gentler increase of frequencies to ~1200 °C and their steeper decrease at higher temperatures (Fig. 9) also mirror the frequency variations with increasing pressure to ~6 GPa and beyond exhibited by the clinopyroxene and orthopyroxene–garnet inclusions (Fig. 8).

The diamond depth distributions in the lithosphere may be the result of both constructive and destructive processes (Nimis et al. 2020). In particular, the progressive decrease in diamond concentrations at depths shallower than ~170–190 km may reflect (i) decreasing precipitation rates for diamond from ascending parental media or (ii) progressive decrease

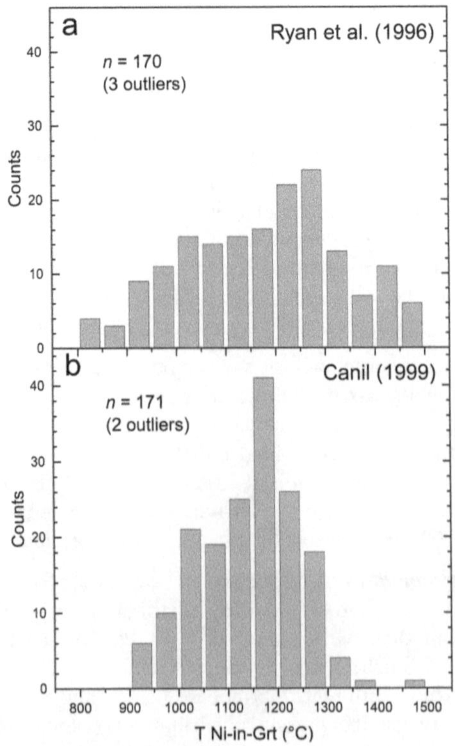

Figure 9. Frequency distribution of Ni-in-Grt temperatures for garnet inclusions in diamonds, assuming 2900 ppm Ni in olivine and using the Ryan et al. (1996) and Canil (1999) calibrations. The distribution shows a distinct mode at ~1175 °C for the Canil (1999) calibration and a less distinct mode at ~1250 °C for the Ryan et al. (1996) calibration. Sources of inclusion data as in Fig. 1.

in the amount of these deep-sourced media at shallower levels. Probably, both mechanisms concur to limit the diamond endowment in the shallow portion of the diamond window. The rapid decrease of diamonds near the base of the lithosphere may reflect (i) a positive balance between the opposed effects that an increase of P and an increase of T may have on carbon solubility in some mantle fluids/melts along specific P–T paths or (ii) a progressive decrease in diamond endowment due to reactions with carbon-undersaturated asthenospheric fluids/melts. These two alternative hypotheses are difficult to prove or disprove, because our knowledge of the behavior of natural carbon-bearing melts and fluids at high P–T is still limited. The prevalence of melt-metasomatized lithologies near the base of typical lithospheres coincides with diminished diamond frequency and suggests that melt-driven metasomatism may be an important destructive agent for diamond (e.g., McCammon et al. 2001). Remobilization of carbon from these deep lithospheric roots may eventually contribute to build up the diamond concentration at shallower depths (Nimis et al. 2020).

The recognition that depths around ~190 km are particularly favorable for diamonds may have interesting implications for their mineralogical exploration. Areas characterized by the occurrence of xenocrystic mantle minerals (e.g., clinopyroxenes or garnets) from these particular depths may be considered as high-priority targets. Nonetheless, since the xenocrysts and diamond depth distributions may be uncorrelated, even limited xenocryst records from the diamond's most favorable depth range may be indicative of significant diamond potential (Nimis et al. 2020).

Sublithospheric diamonds

Figure 10 shows the global distribution of pressure estimates for diamonds containing majoritic garnets. At present, these are the only types of inclusions that provide abundant, quantitative barometric data for sublithospheric diamonds. A difficulty in investigating these data lies in the fact that they may represent minimum *P* estimates. As explained in the previous section, this limitation is due to the frequent occurrence of exsolutions and, for all but the Thomson et al. (2021) barometer, to the possible absence of accompanying clinopyroxene in the original mineral assemblage from which the majoritic garnet inclusion was separated on entrapment. If one considers only inclusions that are reportedly free of clinopyroxene exsolutions, two distinct unimodal distributions are obtained for peridotitic and eclogitic/pyroxenitic garnets, respectively (Fig. 10). Using the Beyer and Frost (2017) barometer, the peridotitic inclusions are mostly clustered in the 175–225 and 225–275 km depth intervals, near the lithosphere-asthenosphere boundary. Most of these inclusions would thus more appropriately be classified as deep lithospheric. The eclogitic/pyroxenitic inclusions are, on average, much deeper and mostly fall in the depth interval between 300 and 500 km. If one considers also inclusions that do contain exsolved pyroxenes, the peak of eclogitic/pyroxenitic inclusions at ~400 km becomes more prominent. Using the Thomson et al. (2021) barometer, the distribution for the eclogitic/pyroxenitic inclusions is stretched to slightly higher pressures, with a maximum estimated depth of 613 ± 44 km, whereas many of the peridotitic inclusions are shifted to significantly greater depths, falling well into the sublithospheric region (Fig. 10). Some of the deepest (>450 km) majoritic garnets fall in the region where clinopyroxene disappears and thus the Beyer and Frost (2017) barometer may underestimate pressure for these samples. However, the Thomson et al. (2021) machine-learning barometer, which is immune to this effect, does not suggest a much deeper origin for these garnets (Fig. 10). Differences between depth estimates for these very deep samples using the two alternative barometers are between –65 and +81 km (average +17 km).

It should be considered that, in some cases, in the absence of accurate imaging data, the occurrence of fine clinopyroxene exsolutions may have been overlooked. The use of improved BSE and EBSD imaging might allow recognition of exsolved clinopyroxene in nominally 'clinopyroxene-free' majoritic inclusions (cf. Harte and Cayzer 2007; Zedgenizov et al. 2014). The thirteen Juina majoritic inclusions for which no coexisting clinopyroxene has been reported to date give depth estimates in a restricted range between 353 and 455 km (using the Beyer and Frost 2017 barometer) or 368 to 462 km (using the Thomson et al. 2021 barometer) and are in part responsible for the prominent peak at ~400 km in Figure 10. If these inclusions also contained exsolutions and the available electron microprobe data do not faithfully reflect their original compositions, then the overall peak could be shifted to somewhat greater depths or, at least, the overall depth distribution would be smoother than shown in Figure 10. Thomson et al. (2021) suggested that all majoritic garnet inclusions in South American diamonds probably contain exsolutions. They also noted that in the eight inclusions for which both exsolved and reconstructed bulk compositions were available the amount of exsolution is a simple function of the calculated pressure. Accordingly, they devised an empirical correction that can be applied to all inclusions for which reconstructed bulk compositions are not available. When this correction is applied to South American diamonds, the peak for eclogitic majoritic garnet inclusions is smoothed and shifted to ca. 100 km greater depth. It remains unclear if this correction, which is based on a restricted number of inclusions, has any general validity at either global or local scale. In any case, the high abundance of depth estimates between ~300 and ~500 km, even for inclusions that are certainly clinopyroxene-free or whose original composition was reconstructed, suggests that most of the eclogitic/pyroxenitic majoritic garnets indeed originated from this depth range (Harte 2010), or perhaps from a slightly more extended range of ~300 to ~600 km if the Thomson et al. (2021) correction for South

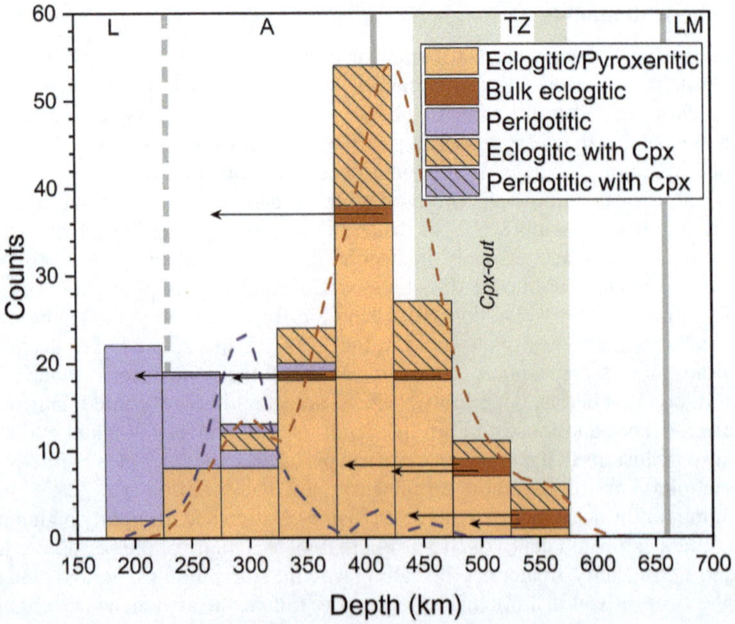

Figure 10. Distribution of estimated depths for inclusions of majoritic garnet (Si > 3.05 atoms per formula unit) in worldwide diamonds, based on barometric estimates (Beyer and Frost 2017) and using the Preliminary Reference Earth Model (PREM; Dziewonski and Anderson 1981) for pressure-depth conversion. 'Bulk' refers to exsolved eclogitic inclusions for which the original pre-exsolution composition was reconstructed; **arrows** connect their depth estimates with those obtained using the compositions of the respective exsolved garnets, where available. Other inclusions in which garnet was reported to be in contact with clinopyroxene or probably contained fine clinopyroxene exsolutions are shown separately, as these garnets certainly reequilibrated to lower pressure conditions during their long journey to the surface. The **dashed curves** show the kernel density distribution of depths for the same eclogitic (**brown**) and peridotitic (**purple**) samples obtained using the machine-learning barometer of Thomson et al. (2021). The **shaded green** area corresponds to the depth range over which clinopyroxene disappears in different bulk compositions (Irifune et al. 1986; Irifune 1987; Okamoto and Maruyama 2004); depth estimates falling within this area using the Beyer and Frost (2017) barometer may be underestimated (see text for further explanation). L: lithosphere. A: asthenosphere. TZ: transition zone. LM: lower mantle. Sources of data: Bulanova et al. (2010), Burnham et al. (2015, 2016), Davies et al. (1999, 2004b), Deines et al. (1991), Harte and Caizer (2007), Hutchison (1997), Kaminsky et al. (2001), Korolev et al. (2018), Meyer and Mahin (1986), Meyer et al. (1994), Moore and Gurney (1985, 1989), Moore et al. (1991), Motsamai et al. (2018), Pokhilenko et al. (2004), Schulze et al. (2008), Shatskii et al. (2010), Shatsky et al. (2015), Smith et al. (2009, 2016), Sobolev et al. (1997b, 2004), Stachel and Harris (1997b), Stachel et al. (2000, 2006), Tappert et al. (2005), Thomson et al. (2014, 2021), Tsai et al. (1979), University of Cape Town (unpubl. reports), Wilding (1990), Zedgenizov et al. (2014).

American diamonds is valid. This range overlaps with independent depth estimates obtained from phase stability constraints for other sublithospheric diamonds containing Ca-silicate inclusions (300–360 km; Brenker et al. 2005; Anzolini et al. 2016) and with a minimum depth estimate obtained from elastic barometry for a single ferropericlase inclusion (~450 km; Anzolini et al. 2019). Of further interest, the 300–600 km range corresponds to the depths at which carbonated MORB slabs are most likely to intersect their solidus, providing an ideal environment for focused melting and potential diamond growth (Thomson et al. 2016, 2021). Considering that many periclase-magnesiowüstite inclusions, for which barometric data are not available, may also have formed from subduction-related melts (Thomson et al. 2016; Nimis et al. 2018) the actual proportion of diamonds from the 300–600 km interval may be even larger than shown in Figure 10. The reader is referred to Walter et al. (2022, this volume) for a thorough discussion on these and related aspects.

A second favorable region for diamond formation is probably located near the upper mantle–lower mantle boundary and uppermost lower mantle (Harte 2010). Evidence for diamond formation in this region is provided by a number of key mineral associations within individual diamonds, mostly consisting of various combinations of periclase, (Mg,Fe)SiO_3, $(Mg,Fe)_2SiO_4$, jeffbenite (formerly known as TAPP), breyite, corundum and SiO_2, and of a NaAl-pyroxene phase (believed to have formed as highly majoritic garnet) alongside with periclase or jeffbenite (Hutchison et al. 2001; Harte 2010). Although in several diamonds only part of the key mineral assemblages is actually represented, the general consistency between the observed mineral associations led Harte (2010) to suppose that a large number of diamonds could come from the relatively narrow depth range of 600 to 800 km. Most inclusions in diamonds from this depth range are ultramafic, as opposed to the mafic inclusions that dominate in the 300–600 km interval.

CONCLUSIONS AND FUTURE WORK

Mineral inclusions are very rare in diamonds and occur in only about 1% of examined stones, with significant differences between individual localities (Stachel and Harris 2008). Moreover, only a few mantle minerals or mineral assemblages that may occur as inclusions in diamonds are suitable for thermobarometry. Therefore, only a small proportion of natural diamonds may yield robust information on the *P–T* conditions of their formation, and these diamonds are mostly peridotitic. This proportion may increase in the future, in parallel with the development of new thermobarometric methods or refinement of existing ones. A significant field of potential and desirable improvement concerns the eclogitic inclusions, which represent a major fraction of inclusions in lithospheric diamonds, but for which existing barometric methods still provide inconsistent results. A dedicated test on clinopyroxene–garnet pairs similar to that performed on chromian diopsides by Ziberna et al. (2016) might shed more light on the minimum analytical quality that is required for reliable barometry of these inclusions. Extending the applicability of non-traditional methods, such as elastic barometry, to non-ideal configurations and to a more diverse list of inclusion minerals might also have a dramatic effect on the number of diamonds suitable for thermobarometry, although the need for elastically preserved inclusions generally restricts applications to relatively small inclusions not surrounded by fractures. An interesting advantage of the elastic method is that it can rely on non-destructive techniques and, thus, can leave the samples intact and ready for further investigation with other methods. Such investigations require the use of undamaged samples in which internal strains are well preserved. Legacy destructive techniques used to expose minerals included in diamond, unfortunately, prohibit future examination of these unique samples for purposes of elastic barometry.

Reanalysis of some previously investigated inclusions or diamondiferous xenoliths might provide a relatively effortless way to increase the number of diamond *P–T* data. Four of 195 published chromian clinopyroxene analyses and 6 of 156 published orthopyroxene analyses had to be discarded while preparing this review only because of sub-standard analytical quality. Moreover, twenty-five clinopyroxenes were rejected because their compositions fell just off the major-element compositional fields that are used to classify garnet-associated chromian diopsides (Ziberna et al. 2016). Trace element analysis of these diopsides might have allowed more robust discrimination (e.g., based on Sc/V relationships; Nimis et al. 2009) of the garnet-associated samples and many of them might have turned out to be suitable for thermobarometry. Reintegrating all these samples would increase the number of *P–T* data for chromian diopside-bearing diamonds by a remarkable 27% and for diamonds overall by 17%.

An enlargement of the number of *P–T* data for diamonds would have significant positive outcomes in terms of sample bias reduction. Sample bias is presently a major problem, since the number of localities for which a statistically significant number of sufficiently reliable *P–T* estimates are currently available is limited at best to three and all are from the Kaapvaal craton (Nimis et al. 2020). Consequent benefits of sample bias reduction would be a better assessment of diamond depth distributions at individual localities, and a more robust evaluation of diamond *P–T* systematics and of their significance in terms of diamond forming processes and preservation.

If the number of *P–T* data for diamonds is important, their internal consistency may be even more so. In fact, most of the combinations of thermobarometric methods that may be applied to diamond inclusions do show some degree of inconsistency with one another. These inconsistencies mostly stem from simplified thermodynamic treatment of solid solutions in minerals and from uncertainties in the data used for their calibration and derived either from experiments or from independent xenolith thermobarometry. The internal consistency of thermobarometric methods is particularly critical when comparing data for different inclusion types, as commonly is necessary in diamond studies. This is another field of potential scientific advance that may eventually contribute to an improved understanding of diamond forming processes.

ACKNOWLEDGMENTS

This review is the result of the knowledge, expertise and generosity of many people. Fruitful collaboration through the years with many of them, and particularly with W.R. Taylor, H. Grütter, L. Ziberna, F. Nestola and R.J. Angel, has been of decisive importance in producing my personal contributions to diamond thermobarometry. Yet many more people have in fact contributed to this review through their enormous scientific work and by providing the community with invaluable data over about five decades. Most of them appear in the reference list below. I express my sincere gratefulness to all of these people and, in particular, to T. Stachel, who generously shared a diamond inclusion database, which served as a useful basis to build the updated database used in this work. I am also grateful to A. Thomson, for help in majorite barometry calculations, and to H. Grütter, G. Brey and editor T. Stachel for their very helpful and thorough reviews and comments.

REFERENCES

Angel RJ, Nimis P, Mazzucchelli ML, Alvaro M, Nestola F (2015) How large are departures from lithostatic pressure? Constraints from host-inclusion elasticity. J Metamorph Geol 33:801–813

Angel RJ, Alvaro M, Nestola F (2022) Crystallographic methods for non-destructive characterization of mineral inclusions in diamonds. Rev Mineral Geochem 88:257–306

Anzolini C, Angel RJ, Merlini M, Derzsi M, Tokár K, Milani S, Krebs MY, Brenker FE, Nestola F, Harris JW (2016) Depth of formation of $CaSiO_3$-walstromite included in super-deep diamonds. Lithos, 265:138–147

Anzolini C, Nestola F, Mazzucchelli M, Alvaro M, Nimis P, Gianese A, Morganti S, Marone F, Campione M, Hutchison MT, Harris J (2019) Depth of diamond formation obtained from single periclase inclusions. Geology 47:219–222

Ashchepkov IV, Pokhilenko NP, Vladykin NV, Logvinova AM, Kostrovitsky SI, Afanasiev VP, Pokhilenko LN, Kuligin SS, Malygina LV, Alymova NV, Khmelnikova OS, Palessky SV, Nikolaeva IV, Karpenko MA, Stagnitsky YB (2010) Structure and evolution of the lithospheric mantle beneath Siberian craton, thermobarometric study. Tectonophysics 485:17–41

Ashchepkov IV, Ntaflos T, Logvinova AM, Spetsius ZV, Downes H, Vladykin NV (2017) Monomineral universal clinopyroxene and garnet barometers for peridotitic, eclogitic and basaltic systems. Geosci Frontiers 8:775–795

Aulbach S (1999) The chemistry of syngenetic mineral inclusions in diamonds from Venetia and the stable isotope composition of diamonds from Mwadui and the Kankan district. MSc Thesis, JW Goethe University, Frankfurt, Germany

Batanova VG, Sobolev AV, Magnin V (2018) Trace element analysis by EPMA in geosciences: detection limit, precision and accuracy. IOP Conf Series: Mater Sci Eng 304:012001

Beyer C, Frost DJ (2017) The depth of sub-lithospheric diamond formation and the redistribution of carbon in the deep mantle. Earth Planet Sci Lett 461:30–39

Beyer C, Frost DJ, Miyajima N (2015) Experimental calibration of a garnet–clinopyroxene geobarometer for mantle eclogites. Contrib Mineral Petrol 169:1–21

Beyer C, Rosenthal A, Myhill R, Crichton WA, Yu T, Wang Y, Frost DJ (2018) An internally consistent pressure calibration of geobarometers applicable to the Earth's upper mantle using in situ XRD. Geochim Cosmochim Acta 222:421–435

Boyd FR (1984) Siberian geotherm based on lherzolite xenoliths from the Udachnaya kimberlite, U.S.S.R. Geol 12:528–530

Boyd FR, Finnerty AA (1980) Conditions of origin of natural diamonds of peridotite affinity. J Geophys Res 85:6911–6918

Brenker FE, Vincze L, Vekemans B, Nasdala L, Stachel T, Vollmer C, Kersten M, Somogyi A, Adams F, Joswig W (2005) Detection of a Ca-rich lithology in the Earth's deep (>300 km) convecting mantle. Earth Planet Sci Lett 236:579–587

Brey GP, Köhler T (1990) Geothermobarometry in four-phase lherzolites. II. New thermobarometers, and practical assessment of existing thermobarometers. J Petrol 31:1353–1378

Brey GP, Shu Q (2018) The birth, growth and ageing of the Kaapvaal subcratonic mantle. Mineral Petrol 112:23–41

Bulanova G, Walter M, Smith C, Kohn S, Armstrong L, Blundy J, Gobbo L (2010) Mineral inclusions in sublithospheric diamonds from Collier 4 kimberlite pipe, Juina, Brazil: subducted protoliths, carbonated melts and primary kimberlite magmatism: Contrib Mineral Petrol 160:489–510

Bulatov VK, Girnis AV, Brey GP, Woodland AB, Höfer HE (2019) Ferropericlase crystallization under upper mantle conditions, Contrib Mineral Petrol. 174:45

Burnham AD, Thomson AR, Bulanova GP, Kohn SC, Smith CB, Walter MJ (2015) Stable isotope evidence for crustal recycling as recorded by superdeep diamonds. Earth Planet Sci Lett 432:374–380

Burnham AD, Bulanova GP, Smith CB, Whitehead SC, Kohn SC, Gobbo L, Walter MJ (2016) Diamonds from the Machado River alluvial deposit, Rondônia, Brazil, derived from both lithospheric and sublithospheric mantle. Lithos 265:199–213

Bussweiler Y, Brey GP, Pearson DG, Stachel T, Stern RA, Hardman MF, Kjarsgaard BA, Jackson SE (2017) The aluminum-in-olivine thermometer for mantle peridotites experimental versus empirical calibration and potential applications. Lithos 272–273:301–314

Canil D (1999) The Ni-in-garnet geothermometer: calibration at natural abundances. Contrib Mineral Petrol 136:240–246

Carswell DA (1991) The garnet–orthopyroxene Al barometer: problematic application to natural garnet lherzolite assemblages. Mineral Mag 55:19–31

Carswell DA, Gibb FG (1987) Evaluation of mineral thermometers and barometers applicable to garnet lherzolite assemblages. Contrib Mineral Petrol 95:499–511

Collerson KD, Hapugoda S, Kamber BS, Williams Q (2000) Rocks from the mantle transition zone: majorite-bearing xenoliths from Malaita, southwest Pacific. Science 288:1215–1223

Collerson KD, Williams Q, Kamber BS, Omori S, Arai H, Ohtani E (2010) Majoritic garnet: a new approach to pressure estimation of shock events in meteorites and the encapsulation of sub-lithospheric inclusions in diamond. Geochim Cosmochim Acta 74:5939–5957

Creighton S (2009) A semi-empirical manganese-in-garnet single crystal thermometer. Lithos 112S:177–182

Creighton S, Stachel T, McLean H, Muehlenbachs K, Simonetti A, Eichenberg D, Luth R (2008) Diamondiferous peridotitic microxenoliths from the Diavik Diamond Mine, NT. Contrib Mineral Petrol 155:541–554

Czas J, Pearson DG, Stachel T, Kjarsgaard BA, Read GH (2020) A Palaeoproterozoic diamond-bearing lithospheric mantle root beneath the Archean Sask Craton, Canada. Lithos 356–357:105301

Daniels LRM, Gurney JJ (1989) The chemistry of garnets, chromites and diamond inclusions from the Dokolwayo kimberlite, Kingdom of Swaziland. *In*: Kimberlites and Related Rocks. Ross J et al. (eds) GSA Spec Publ 14. Blackwell, Carlton, p 1012–1021

Davies RM, Griffin WL, Pearson NJ, Andrew AS, Doyle BJ, O'Reilly SY (1999) Diamonds from the deep: pipe DO-27, Slave craton, Canada. *In*: The J.B. Dawson Volume, Proc VIIth Int Kimberlite Conf. Gurney JJ, Gurney JL, Pascoe MD, Richardson SH (eds) Red Roof Design, Cape Town, p 148–155

Davies RM, Griffin WL, O'Reilly SY, Doyle BJ (2004a) Mineral inclusions and geochemical characteristics of microdiamonds from the DO27, A154, A21, A418, DO18, DD17 and Ranch Lake kimberlites at Lac de Gras, Slave Craton, Canada. Lithos 77:39–55

Davies RM, Griffin WL, O'Reilly SY, McCandless TE (2004b) Inclusions in diamonds from the K14 and K10 kimberlites, Buffalo Hills, Alberta, Canada: diamond growth in a plume? Lithos 77:99–111

Dawson JBM, Smith JV (1975) Occurrence of diamond in a mica–garnet lherzolite xenolith from kimberlite. Nature 254:580–581

Day HW (2012) A revised diamond–graphite transition curve. Am Mineral 97:52–62

De Hoog JCM, Gall L, Cornell DH (2010) Trace-element geochemistry of mantle olivine and application to mantle petrogenesis and geothermobarometry. Chem Geol 270:196–215

De Hoog JCM, Stachel T, Harris JW (2019) Trace-element geochemistry of diamond-hosted olivine inclusions from the Akwatia Mine, West African Craton: implications for diamond paragenesis and geothermobarometry. Contrib Mineral Petrol 174:100

Deines P, Harris JW, Gurney JJ (1991) The carbon isotopic composition and nitrogen content of lithospheric and asthenospheric diamonds from the Jagersfontein and Koffiefontein kimberlite, South Africa. Geochim Cosmochim Acta 55:2615–2625

Deines P, Harris JW (2004) New insights into the occurrence of ^{13}C-depleted carbon in the mantle from two closely associated kimberlites: Letlhakane and Orapa, Botswana. Lithos 77:125–142
De Stefano A, Kopylova MG, Cartigny P, Afanasiev V (2009) Diamonds and eclogites of the Jericho kimberlite (Northern Canada). Contrib Mineral Petrol 158:295–315
D'Souza RJ, Canil D, Coogan LA (2020) Geobarometry for spinel peridotites using Ca and Al in olivine. Contrib Mineral Petrol 175:5
Dziewonski AM, Anderson DL (1981) Preliminary reference Earth model. Phys Earth Planet Inter 25:297–356
Ferrero S, Angel RJ (2018) Micropetrology: are inclusions grains of truth? J Petrol 59:1671–1700
Ganne J, Feng X (2017) Primary magmas and mantle temperatures through time. Geochem Geophys Geosyst 18:872–888
Geological Survey of Canada (1989) The development of advanced technology to distinguish between diamondiferous and barren diatremes. Open File 2124
Griffin WL, Ryan CG (1995) Trace elements in indicator minerals: area selection and target evaluation in diamond exploration. J Geochem Expl 53:311–337
Griffin WL, Gurney JJ, Ryan CG (1992) Variations in trapping temperatures and trace elements in peridotite-suite inclusions from African diamonds: evidence for two inclusion suites, and implications for lithosphere stratigraphy. Contrib Mineral Petrol 110:1–15
Griffin WL, Sobolev NV, Ryan CG, Pokhilenko NP, Win TT, Yefimova ES (1993) Trace elements in garnets and chromites: diamond formation in the Siberian lithosphere. Lithos 29:235–256
Griffin WL, Fisher NI, Friedman JH, O'Reilly SY, Ryan CG (2002) Cr-pyrope garnets in the lithospheric mantle: 2. Compositional populations and their distribution in time and space. Geochem Geophys Geosyst 3:1073
Griffin WL, O'Reilly SY, Natapov LM, Ryan CG (2003) The evolution of lithospheric mantle beneath the Kalahari Craton and its margins. Lithos 71:215–241
Grütter HS (2009) Pyroxene xenocryst geotherms: Techniques and application. Lithos 112S:1167–1178
Grütter H, Apter DB, Kong J (1999) Crust–mantle coupling: evidence from mantle-derived xenocrystic garnets. *In*: The J.B. Dawson Volume, Proc VIIth Int Kimberlite Conf. Gurney JJ, Gurney JL, Pascoe MD, Richardson SH (eds), Red Roof Design, Cape Town, p. 307–313
Grütter HS, Gurney JJ, Menzies AH, Winter F (2004) An updated classification scheme for mantle-derived garnet, for use by diamond explorers. Lithos 77:841–857
Grütter H, Latti D, Menzies A (2006) Cr-saturation arrays in concentrate garnet compositions from kimberlite and their use in mantle barometry. J Petrol 47:801–820
Grütter HS, Tuer J (2009) Constraints on deep mantle tenor of Sarfartoq-area kimberlites (Greenland), based on modern thermobarometry of mantle-derived xenocrysts. Lithos 112S, 124–129
Gurney JJ, Harris JW, Rickard RS (1984) Minerals associated with diamonds from the Roberts Victor Mine. *In*: Kimberlites II: The Mantle and Crust-Mantle Relationships. Kornprobst J (ed) Elsevier, Amsterdam, p 25–32
Hanrahan H, Brey G, Woodland A, Seitz H-M, Ludwig T (2009) Li as a barometer for bimineralic eclogites: Experiments in natural systems. Lithos 112S:992–1001
Harley SL (1984) An experimental study of the partitioning of Fe and Mg between garnet and orthopyroxene. Contrib Mineral Petrol 86:359–373
Harris JW, Duncan DJ, Zhang F, Mia Q, Zhu Y (1994) The physical characteristics and syngenetic inclusion geochemistry of diamonds from pipe 50, Liaoning Province, People's Republic of China. *In*: Diamonds: Characterization, Genesis and Exploration. Meyer HOA, Leonardos OH (eds) CPRM Spec Publ Jan/94, Brasilia, p 106–115
Harris JW, Stachel T, Leost I, Brey GP (2004) Peridotitic diamonds from Namibia: constraints on the composition and evolution of their mantle source. Lithos 77:209–223
Harte B (2010) Diamond formation in the deep mantle: the record of mineral inclusions and their distribution in relation to mantle dehydration zones. Mineral Mag 74:189–215
Harte B, Cayzer N (2007) Decompression and unmixing of crystals included in diamonds from the mantle transition zone. Phys Chem Minerals 34:647–656
Harte B, Hudson NFC (2013) Mineral associations in diamonds from the lowermost upper mantle and uppermost lower mantle. *In*: Proc 10th Int Kimb Conf. Vol 1. Pearson DG et al. (eds) Spec Issue J Geol Soc India, p 235–253
Harte B, Harris J, Hutchison M, Watt G, Wilding M (1999) Lower mantle mineral associations in diamonds from São Luiz, Brazil. *In*: Mantle Petrology: Field Observations and High-Pressure Experimentation: A tribute to Francis R. (Joe) Boyd. Fei Y, Bertka CM, Mysen BO (eds) Geochem Soc Spec Publ No 6, p 125–153
Hasterok D, Chapman DS (2011) Heat production and geotherms for the continental lithosphere. Earth Planet Sci Lett 307:59–70
Hervig RL Smith JV and Dawson JB (1986) Lherzolite xenoliths in kimberlites and basalts: petrogenetic and crystallographic significance of some minor and trace elements in olivine, pyroxenes, garnet and spinel. Trans R Soc Edin Earth Sci 77:181–201
Hutchison MT (1997) Constitution of the deep transition zone and lower mantle shown by diamonds and their inclusions. PhD Thesis, University of Edinburgh, Scotland, UK
Hutchison MT, Hursthouse MB, Light ME (2001) Mineral inclusions in diamonds: Associations and chemical distinctions around the 670-km discontinuity. Contrib Mineral Petrol 142:119–126
Irifune T (1987) An experimental investigation of the pyroxene-garnet transformation in a pyrolite composition and its bearing on the constitution of the mantle. Phys Earth Planet Inter 45:324–336

Irifune T, Ringwood AE, Hibberson WO (1986) The eclogite–garnetite transformation at high pressure and some geophysical implications. Earth Planet Sci Lett 77:245–256

Jaques AL, Hall AE, Sheraton JW, Smith CB, Sun S-S, Drew RM, Foudoulis C, Ellingsen K (1989) Composition of crystalline inclusions and C-isotopic composition of Argyle and Ellendale diamonds. *In*: Kimberlites and Related Rocks. Ross J et al. (eds) Geol Soc Australia Spec Publ Spec Publ 14. Blackwell, Carlton, p 966–989

Jaques AL, O'Neill HStC, Smith CB, Moon J, Chappell BW (1990) Diamondiferous peridotite xenoliths from the Argyle (AK1) lamproite pipe, Western Australia. Contrib Mineral Petrol 104:255–276

Joswig W, Stachel T, Harris JW, Baur WH, Brey GP (1999) New Ca-silicate inclusions in diamonds – tracers from the lower mantle. Earth Planet Sci Lett 173:1–6

Kaminsky FV (2012) Mineralogy of the lower mantle: A review of 'super-deep' mineral inclusions in diamond. Earth-Sci Rev 110:127–147

Kaminsky FV, Zakharchenko OD, Davies R, Griffin WL, Khachatryan-Blinova GK, Shiryaev AA (2001) Superdeep diamonds from the Juina area, Mato Grosso State, Brazil. Contrib Mineral Petrol 140:734–753

Katsura T, Yoneda A, Yamazaki D, Yoshino T, Ito E (2010) Adiabatic temperature profile in the mantle. Phys Earth Planet Int 183:212–218

Klemme S (2004) The influence of Cr on the garnet–spinel transition in the Earth's mantle: experiments in the system $MgO–Cr_2O_3–SiO_2$ and thermodynamic modelling. Lithos 77:639–646

Kohn SC, Speich L, Smith CB, Bulanova GP (2016) FTIR thermochronometry of natural diamonds: a closer look. Lithos 265:148–158

Kopylova MG, Gurney JJ, Daniels LRM (1997) Mineral inclusions in diamonds from the River Ranch kimberlite, Zimbabwe. Contrib Mineral Petrol 129:366–384

Korolev NM, Kopylova M, Bussweiler Y, Pearson DG, Gurney J, Davidson J (2018) The uniquely high-temperature character of Cullinan diamonds: A signature of the Bushveld mantle plume? Lithos 304–307:362–373

Krogh EJ (1988) The garnet–clinopyroxene Fe–Mg-geothermometer—a reinterpretation of existing experimental data. Contrib Miner Pet 99:44–48

Leost I, Stachel T, Brey GP, Harris JW, Ryabchikov ID (2003) Diamond formation and source carbonation: mineral associations in diamonds from Namibia. Contrib Mineral Petrol 145:15–24

Logvinova AM, Taylor LA, Fedorova EN, Yelisseyev AP, Wirth R, Howarth G, Reutsky VN, Sobolev NV (2015) A unique diamondiferous peridotite xenolith from the Udachnaya kimberlite pipe, Yakutia: role of subduction in diamond formation. Russ Geol Geophys 56:306–320

MacGregor ID (1970) The effect of CaO, Cr_2O_3, Fe_2O_3 and Al_2O_3 on the stability of spinel and garnet peridotites. Phys Earth Planet Inter 3:372–377

MacGregor ID (1974) The system $MgO-A1_2O_3-SiO_2$: solubility of $A1_2O_3$ in enstatite for spinel and garnet peridotite compositions. Am Mineral 59:110–119

Malkovets VG, Zedgenizov DA, Sobolev NV, Kuzmin DV, Gibsher AA, Shchukina EV, Golovin NN, Verichev EM, Pokhilenko NP (2011) Contents of trace elements in olivines from diamonds and peridotite xenoliths of the V. Grib kimberlite pipe (Arkhangel'sk Diamondiferous Province, Russia). Dokl Earth Sci 436:219–223

Mather KA, Pearson DG, McKenzie D, Kjarsgaard BA, Priestley K (2011) Constraints on the depth and thermal history of cratonic lithosphere from peridotite xenoliths, xenocrysts and seismology. Lithos 125:729–742

Matjuschkin V, Brey GP, Höfer HE, Woodland AB (2014) The influence of Fe^{3+} on garnet–orthopyroxene and garnet–olivine geothermometers. Contrib Mineral Petrol 167:972

McCammon CA, Griffin WL, Shee SR, O'Neill HSC (2001) Oxidation during metasomatism in ultramafic xenoliths from the Wesselton kimberlite, South Africa: implications for the survival of diamond. Contrib Mineral Petrol 141:287– 296

McDade P, Harris JW (1999) Syngenetic inclusion bearing diamonds from Letseng-la-Terai, Lesotho. *In*: The P.H. Nixon Volume, Proc VIIth Int Kimberlite Conf. Gurney JJ, Gurney JL, Pascoe MD, Richardson SH (eds) Red Roof Design, Cape Town, p 557–565

Mendelssohn MJ, Milledge HJ (1995) Geologically significant information from routine analysis of the mid-infrared spectra of diamonds. Int Geol Rev 37:95–110

Meyer HOA, Mahin RA (1986) The kimberlites of Guinea, West Africa. 4th Int Kimberlite Conf Abstr, Geol Soc Aus Abstr Series 16:66

Meyer HOA, Zhang A, Milledge HJ, Mendelssohn MJ (1994) Diamonds and inclusions in diamonds from Chinese kimberlites. *In*: Proc 5th Int Kimberlite Conf, vol 1, Diamonds: characterization, genesis and exploration. Meyer HOA, Leonardos OH (eds) CPRM, Brasilia, p 98–105

Michaut C, Jaupart C (2007) Secular cooling and thermal structure of continental lithosphere. Earth Planet Sci Lett 257:83–96

Milani S, Nestola F, Angel RJ, Nimis P, Harris JW (2016) Crystallographic orientations of olivine inclusions in diamonds. Lithos 265:312–316

Moore RO, Gurney JJ (1985) Pyroxene solid solution in garnets included in diamond. Nature 318:553–555

Moore RO, Gurney JJ (1989) Mineral inclusions in diamond from the Monastery kimberlite, South Africa. *In*: Kimberlites and Related Rocks. Vol 2. Their Mantle/Crust Setting, Diamonds and Diamond Exploration, Proc 4th Int Kimberlite Conf. Ross J et al. (eds) Geol Soc Aust Spec Publ 14:1029–1041

Moore RO, Otter ML, Rickard RS, Harris JW, Gurney JJ (1986) The occurrence of moissanite and ferro-periclase as inclusions in diamond. 4th Int Kimberlite Conf Ext Abstr, Perth. Geol Soc Aust Abstr, 16:409–411

Moore RO, Gurney JJ, Griffin WL, Shimizu N (1991) Ultra-high pressure garnet inclusions in Monastery diamonds: trace element abundance patterns and conditions of origin. Eur J Mineral 3:213–230

Motsamai T, Harris JW, Stachel T, Pearson DG, Armstrong J (2018) Mineral inclusions in diamonds from Karowe Mine, Botswana: Super-deep sources for super-sized diamonds? Miner Petrol 112:169–180

Nakamura D (2009) A new formulation of garnet–clinopyroxene geothermometer based on accumulation and statistical analysis of a large experimental data set. J Metamorph Geol 27:495–508

Nestola F, Nimis P, Angel RJ, Milani S, Bruno M, Prencipe M, Harris JW (2014) Olivine with diamond-imposed morphology included in diamonds. Syngenesis or protogenesis? Int Geol Rev 56:1658-1667

Nestola F, Haemyeong J, Taylor LA (2017) Mineral inclusions in diamonds may be synchronous but not syngenetic. Nat Commun 8:14168

Nestola F, Zaffiro G, Mazzucchelli ML, Nimis P, Andreozzi GB, Periotto B, Princivalle F, Lenaz D, Secco L, Pasqualetto L, Logvinova AM, Sobolev NV, Lorenzetti A, Harris JW (2019a) Diamond-inclusion system recording old deep lithosphere conditions at Udachnaya (Siberia). Sci Rep 9:12586

Nestola F, Jacob DE, Pamato MG, Pasqualetto L, Oliveira B, Greene S, Perritt S, Chinn I, Milani S, Kueter N, Sgreva N, Nimis P, Secco L, Harris JW (2019b) Protogenetic garnet inclusions and the age of diamonds. Geology 47:431–434

Nickel KG, Green DH (1985) Empirical geothermobarometry for garnet peridotites and implications for the nature of the lithosphere, kimberlites and diamonds. Earth Planet Sci Lett 73:158–170

Nimis P (2002) The pressures and temperatures of formation of diamond based on thermobarometry of chromian diopside inclusions. Can Mineral 40:871–884

Nimis P, Grütter H (2010) Internally consistent geothermometers for garnet peridotites and pyroxenites. Contrib Mineral Petrol 159:411–427

Nimis P, Taylor WR (2000) Single clinopyroxene thermobarometry for garnet peridotites; Part I, Calibration and testing of a Cr-in-Cpx barometer an enstatite-in-Cpx thermometer. Contrib Mineral Petrol 139:541–554

Nimis P, Zanetti A, Dencker I, Sobolev NV (2009) Major and trace element composition of chromian diopsides from the Zagadochnaya kimberlite (Yakutia, Russia): Metasomatic processes, thermobarometry and diamond potential. Lithos 112:397–412

Nimis P, Goncharov A, Ionov DA, McCammon C (2015) Fe^{3+} partitioning systematics between orthopyroxene and garnet in mantle peridotite xenoliths and implications for thermobarometry of oxidized and reduced mantle rocks. Contrib Mineral Petrol 169:6

Nimis P, Nestola F, Schiazza M, Reali R, Agrosì G, Mele D, Tempesta G, Howell D, Hutchison M, Spiess R (2018) Fe-rich ferropericlase and magnesiowüstite inclusions reflecting diamond formation rather than ambient mantle. Geology 47:27–30

Nimis P, Angel RJ, Alvaro M, Nestola F, Harris JW, Casati N, Marone F (2019) Crystallographic orientations of magnesiochromite inclusions in diamonds: What do they tell us? Contrib Mineral Petrol 174:29

Nimis P, Preston R, Perritt SH, Chinn IL (2020) Diamond's depth distribution systematics. Lithos 376–377:105729

Okamoto K, Maruyama S (2004) The eclogite–garnetite transformation in the MORB + H_2O system. Phys Earth Planet Inter 146:283–296

O'Neill HStC, Wood BJ (1979) An experimental study of Fe–Mg partitioning between garnet and olivine and its calibration as a geothermometer. Contrib Mineral Petrol 70:59–70

O'Reilly SY, Chen D, Griffin WL, Ryan CG (1997) Minor elements in olivine from spinel lherzolite xenoliths: implications for thermobarometry. Mineral Mag 61:257–269

Otter ML, Gurney JJ (1989) Mineral inclusions in diamond from the Sloan diatremes, Colorado-Wyoming State Line kimberlite district, North America. *In*: Kimberlites and Related Rocks. Ross J et al. (eds) Geol Soc Australia Spec Publ Spec Publ 14. Blackwell, Carlton, p 1042–1053

Pearson D, Brenker F, Nestola F, McNeill J, Nasdala L, Hutchison M, Matveev S, Mather K, Silversmit G, Schmitz S, (2014) Hydrous mantle transition zone indicated by ringwoodite included within diamond. Nature 507:221–224

Phillips D, Harris JW, Viljoen KS (2004) Mineral chemistry and thermobarometry of inclusions from De Beers Pool diamonds, Kimberley, South Africa. Lithos 77:155–179

Pickles JR, Blundy JD, Brooker RA (2016) Trace element thermometry of garnet-clinopyroxene pairs. Am Mineral 101:1438–1450

Pokhilenko NP, Sobolev NV, Lavrent'ev YuG (1977) Xenoliths of diamondiferous ultramafic rocks from Yakutian kimberlites. 2nd Int Kimberlite Conf Ext Abs: pages not numbered

Pokhilenko NP, Sobolev NV, Reutsky VN, Hall AE, Taylor LA (2004) Crystalline inclusions and C isotope ratios in diamonds from the Snap Lake/King Lake kimberlite dyke system: evidence of ultradeep and enriched lithospheric mantle. Lithos 77:57–67

Ponomarenko AI, Spetsius ZV, Sobolev NV (1980) New type of diamond-bearing rock - pyroxenite. Dokl Akad Nauk SSSR 251:438–441

Purwin H, Lauterbach S, Brey GP, Woodland AB, Kleebe H-J (2013) An experimental study of the Fe oxidation states in garnet and clinopyroxene as a function of temperature in the system $CaO–FeO–Fe_2O_3–MgO–Al_2O_3–SiO_2$: implications for garnet–clinopyroxene geothermometry. Contrib Mineral Petrol 165:623–639

Rickard RS, Harris JW, Gurney JJ, Cardoso P (1989) Mineral inclusions from Koffiefontein mine. *In*: Kimberlites and Related Rocks. Ross J et al. (eds) Geol Soc Australia Spec Publ Spec Publ 14. Blackwell, Carlton, p 1054–1062

Ryan CG, Griffin WL, Pearson NJ (1996) Garnet geotherms: Pressure–temperature data from Cr-pyrope garnet xenocrysts in volcanic rocks. J Geophys Res 101:5611–5625

Schulze DJ, Coopersmith HG, Harte B, Pizzolato L-A (2008) Mineral inclusions in diamonds from the Kelsey Lake Mine, Colorado, USA: Depleted Archean mantle beneath the Proterozoic Yavapai province. Geochim Cosmochim Acta 72:1685–1695

Scott Smith BH, Danchin RV, Harris JW, Stracke KJ (1984) Kimberlites near Orroroo, South Australia. *In*: Kimberlites I: Kimberlites and Related Rocks. Kornprobst, J (ed) Elsevier, Amsterdam, p 121–142

Shatskii VS, Zedgenizov DA, Ragozin AL (2010) Majoritic garnets in diamonds from placers of the northeastern Siberian Platform. Doklady Earth Sci 432:835–838

Shatsky VS, Zedgenizov DA, Ragozin AL, Kalinina VV (2015) Diamondiferous subcontinental lithospheric mantle of the northeastern Siberian Craton: Evidence from mineral inclusions in alluvial diamonds. Gondwana Res 28:106–120

Shee SR, Gurney JJ, Robinson DN (1982) Two diamond-bearing peridotite xenoliths from the Finsch kimberlite, South Africa. Contrib Mineral Petrol 81:79–87

Shirey SB, Richardson SH (2011) Start of the Wilson cycle at 3 Ga shown by diamonds from subcontinental mantle. Science 333:434–436

Shirey SB, Cartigny P, Frost DJ, Keshav S, Nestola F, Nimis P, Pearson DG, Sobolev NV, Walter MJ (2013) Diamonds and the Geology of Mantle Carbon. *In*: Carbon in Earth. Hazen RM, Jones AP, Baross JA (eds) Rev Mineral Geochem 75:355–421

Simakov SK (2008) Garnet–clinopyroxene and clinopyroxene geothermobarometry of deep mantle and crust eclogites and peridotites. Lithos 106:125–136

Simakov SK, Taylor LA (2000) Geobarometry for mantle eclogites: solubility of Ca-Tschermaks in clinopyroxene. Int Geol Rev 42:534–544

Smith D, Griffin W, Ryan C, Sie S (1991) Trace-element zonation in garnets from The Thumb: heating and melt infiltration below the Colorado Plateau. Contrib Mineral Petrol 107:60-79

Smith CB, Pearson DG, Bulanova GP, Beard AD, Carlson RW, Wittig N, Sims K, Chimuka L, Muchemwa E (2009) Extremely depleted lithospheric mantle and diamonds beneath the southern Zimbabwe Craton. Lithos 112:1120–1132

Smith CB, Walter MJ, Bulanova GP, Mikhail S, Burnham AD, Gobbo L, Kohn SC (2016) Diamonds from Dachine, French Guiana: A unique record of early Proterozoic subduction. Lithos 265:82–95

Sobolev NV (1977) Deep-Seated Inclusions in Kimberlites and the Problem of the Composition of the Upper Mantle. Am Geophys Un, Washington, DC

Sobolev NV, Yefimova ES (1998) Compositional variations of chromite inclusions as an indicator of the zonation of diamond crystals. Dokl Earth Sci 359:163–166

Sobolev NV, Botkunov AI, Lavrent'ev YG, Usova LV (1976) New data on the minerals associated with the diamonds in the "Mir" kimberlite pipe in Yakutia. Geologiya I Geofizika, 17:1–10

Sobolev NV, Kaminsky FV, Griffin WL, Yefimova ES, Win TT, Ryan CG, Botkunov AI (1997a) Mineral inclusions in diamonds from the Sputnik kimberlite pipe, Yakutia. Lithos 39:135–157

Sobolev NV, Yefimova ES, Reimers LF, Zakharchenko OD, Makhin AI, Usova LV (1997b) Mineral inclusions in diamonds of the Arkhangelsk kimberlite province. Russ Geol Geophys 38:379–393

Sobolev NV, Yefimova ES, Koptil VL (1999) Mineral inclusions in diamonds in the Northeast of the Yakutian diamondiferous province. *In*: The P.H. Nixon Volume, Proc VIIth Int Kimberlite Conf. Gurney JJ, Gurney JL, Pascoe MD, Richardson SH (eds) Red Roof Design, Cape Town, p 816–822

Sobolev NV, Logvinova AM, Zedgenizov DA, Seryotkin YV, Yefimova ES, Floss C, Taylor LA (2004) Mineral inclusions in microdiamonds and macrodiamonds from kimberlites of Yakutia: a comparative study. Lithos 77:225–242

Sobolev NV, Logvinova AM, Zedgenizov DA, Pokhilenko NP, Kuzmin DV, Sobolev AV (2008) Olivine inclusions in diamonds: high-precision approach to minor elements. Eur J Mineral 20:305–315

Sobolev NV, Logvinova AM, Efimova ES (2009) Syngenetic phlogopite inclusions in kimberlite-hosted diamonds: implications for role of volatiles in diamond formation. Russ Geol Geophys 50:1234–1248

Stachel T (2014) Diamonds. Mineral Ass Can Short Course 44:1–28

Stachel T, Harris JW (1997a) Syngenetic inclusions in diamond from the Birim field (Ghana) - A deep peridotitic profile with a history of depletion and re-enrichment. Contrib Mineral Petrol 127:336–352

Stachel T, Harris JW (1997b) Diamond precipitation and mantle metasomatism – evidence from the trace element chemistry of silicate inclusions in diamonds from Akwatia, Ghana. Contrib Mineral Petrol 129:143–154

Stachel T, Harris JW (2008) The origin of cratonic diamonds - constraints from mineral inclusions. Ore Geol Rev 34:5–32

Stachel T, Luth RW (2015) Diamond formation—where, when and how? Lithos 220–223:200–220

Stachel T, Harris JW, Brey GP (1998) Rare and unusual mineral inclusions in diamonds from Mwadui, Tanzania. Contrib Mineral Petrol 132:34–47

Stachel T, Harris JW, Brey GB, Joswig W (2000) Kanakan diamonds (Guinea) II: lower mantle inclusion parageneses. Contrib Mineral Petrol 140:16–27

Stachel T, Harris JW, Tappert R, Brey GP (2003) Peridotitic diamonds from the Slave and the Kaapvaal cratons — similarities and differences based on a preliminary data set. Lithos 71:489–503

Stachel T, Viljoen KS, McDade P, Harris JW (2004) Diamondiferous lithospheric roots along the western margin of the Kalahari Craton—the peridotitic inclusion suite in diamonds from Orapa and Jwaneng. Contrib Mineral Petrol 147:32–47
Stachel T, Banas A, Muelenbachs K, Kurslaukis S, Walker EC (2006) Archean diamonds from Wawa (Canada): samples from deep cratonic roots predating cratonization of the Superior Province. Contrib Mineral Petrol 151:737–750
Stachel T, Banas A, Aulbach S, Smit KV, Wescott P, Chinn IL, Kong J (2018) The Victor Mine (Superior Craton, Canada): Neoproterozoic lherzolitic diamonds from a thermally modified cratonic root. Mineral Petrol 112 (Suppl 1) S325–S336
Sudholz ZJ, Yaxley GM, Jaques AL, Chen J (2021a) Ni-in-garnet geothermometry in mantle rocks: a high pressure experimental recalibration between 1100 and 1325 °C. Contrib Mineral Petrol 176:32
Sudholz ZJ, Yaxley GM, Jaques AL, Brey GP (2021b) Experimental recalibration of the Cr-in-clinopyroxene geobarometer: improved precision and reliability above 4.5 GPa. Contrib Mineral Petrol 176:11
Sun C, Liang Y (2015) A REE-in garnet-clinopyroxene thermobarometer for eclogites, granulites and garnet peridotites. Chem Geol 393–394:79–92
Tao R, Fei Y, Bullock ES, Xu C, Zhang L (2018) Experimental investigation of Fe^{3+}-rich majoritic garnet and its effect on majorite geobarometer. Geochim Cosmochim Acta 225:1–16
Tappert R, Stachel T, Harris JW, Muelenbachs K, Ludwig T, Brey GP (2005) Diamonds from Jagersfontein (South Africa): messengers from the sublithospheric mantle. Contrib Mineral Petrol 150:505–522
Tappert R, Stachel T, Harris JW, Muehlenbachs K, Brey GF (2006) Placer diamonds from Brazil: Indicators of the composition of the Earth's mantle and the distance to their kimberlitic sources. Econ Geol 101:453–470
Taylor WR (1998) An experimental test of some geothermometer and geobarometer formulations for upper mantle peridotites with application to the thermobarometry of fertile lherzolite and garnet websterite. N Jb Miner Abh 172:381–408
Taylor WR, Jacques AL, Ridd M (1990) Nitrogen-defect aggregation characteristics of some Australasian diamonds, Time-temperature constraints on the source regions of the pipe and alluvial diamonds. Am Mineral 75:1290–1310
Taylor WR, Canil D, Milledge HJ (1996) Kinetics of Ib to IAa nitrogen aggregation in diamond. Geochim Cosmochim Acta 60:4725–4733
Thomson AR, Kohn SC, Bulanova GP, Smith CB, Araujo D, EIMF, Walter MJ (2014) Origin of sub-lithospheric diamonds from the Juina-5 kimberlite (Brazil): constraints from carbon isotopes and inclusion compositions. Contrib Mineral Petrol 168:1081
Thomson AR, Walter MJ, Kohn SC, Brooker RA (2016) Slab melting as a barrier to deep carbon subduction. Nature 529:76–79
Thomson AR, Kohn SC, Prabhu A, Walter MJ (2021) Evaluating the formation pressure of diamond-hosted majoritic garnets: A machine learning majorite barometer. J Geophys Res: Solid Earth 126:e2020JB020604
Tsai H, Meyer HOA, Moreau J, Milledge HJ (1979) Mineral inclusions in diamond: Premier, Jagersfontein and Finsch kimberlites, South Africa, and Williamson Mine, Tanzania. *In*: Kimberlites, Diatremes, and Diamonds: their Geology, Petrology, and Geochemistry. Proc 2nd Int Kimberlite Conf. Boyd FR, Meyer HOA (eds) Am Geophys Un, Washington, DC, p 16–26
Van Rythoven AD, Schulze DJ (2009) In-situ analysis of diamonds and their inclusions from the Diavik mine, Northwest Territories, Canada: mapping diamond growth. Lithos 112:870–879
Van Rythoven AD, McCandless TE, Schulze DJ, Bellis A, Taylor LA, Liu Y (2011) Diamond crystals and their mineral inclusions from the Lynx kimberlite dyke complex, central Quebec. Can Mineral 49:691–706
Viljoen KS, Swash PM, Otter ML, Schulze DJ, Lawless PJ (1992) Diamondiferous garnet harzburgites from the Finsch kimberlite, Northern Cape, South Africa. Contrib Mineral Petrol 110:133–138
Viljoen KS, Harris JW, Ivanic T, Richardson SH, Gray K (2014) Trace element chemistry of peridotitic garnets in diamonds from the Premier (Cullinan) and Finsch kimberlites, South Africa: contrasting styles of mantle metasomatism. Lithos 208:1–15
Viljoen KS, Perritt SH, Chinn IL (2018) An unusual suite of eclogitic, websteritic and transitional websteritic-lherzolitic diamonds from the Voorspoed kimberlite in South Africa: mineral inclusions and infrared characteristics. Lithos 320–321:416–434
Walter MJ, Kohn SC, Araujo D, Bulanova GP, Smith CB, Gaillou E, Wang J, Steele A, Shirey SB (2011) Deep mantle cycling of oceanic crust: evidence from diamonds and their mineral inclusions. Science 334:54–57
Walter MJ, Thomson AR, Smith EM (2022) Geochemistry of silicate and oxide inclusions in sublithospheric diamonds. Rev Mineral Geochem 88:393–450
Wang WY, Sueno S, Takahashi E, Yurimoto H, Gasparik T (2000) Enrichment processes at the base of the Archean lithospheric mantle: observations from trace element characteristics of pyropic garnet inclusions in diamonds. Contrib Mineral Petrol 139:720–733
Webb SAC, Wood BJ (1986) Spinel-pyroxene-garnet relationships and their dependence on Cr/Al ratio. Contrib Mineral Petrol 92:471–480
Weiss Y, Navon O, Goldstein SL, Harris JW (2018) Inclusions in diamonds constrain thermo-chemical conditions during Mesozoic metasomatism of the Kaapvaal cratonic mantle. Earth Planet Sci Lett 491:134–147
Wijbrans CH, Rohrbach A, Klemme S (2016) An experimental investigation of the stability of majoritic garnet in the Earth's mantle and an improved majorite geobarometer. Contrib Mineral Petrol 171:50

Wilding MC (1990) A study of diamonds with syngenetic inclusions. PhD thesis, University of Edinburgh, Scotland, UK
Wilding MC, Harte B, Fallick AE, Harris JW (1994) Inclusion chemistry, carbon isotopes and nitrogen distribution in diamonds from the Bultfontein Mine, South Africa. *In*: Diamonds: Characterization, Genesis and Exploration. Meyer HOA, Leonardos OH (eds) CPRM Spec Publ Jan/94, Brasilia, p 116–126
Witt-Eickschen G, O'Neill H (2005) The effect of temperature on the equilibrium distribution of trace elements between clinopyroxene, orthopyroxene, olivine and spinel in upper mantle peridotite. Chem Geol 221:65–101
Woodland AB (2009) Ferric iron contents of clinopyroxene from cratonic mantle and partitioning behaviour with garnet. Lithos 112S:1143–1149
Woodland AB, Girnis AV, Bulatov VK, Brey GP, Höfer HE (2020) Breyite inclusions in diamond: experimental evidence for possible dual origin. Eur J Mineral 32:171–185
Wu CM, Zhao GC (2007) A recalibration of the garnet–olivine geothermometer and a new geobarometer for garnet peridotites and garnet–olivine–plagioclase-bearing granulites. J Metamorph Geol 25:497–505
Zedgenizov DA, Kagi H, Shatsky VS, Ragozin AL (2014) Local variations of carbon isotope composition in diamonds from São-Luis (Brazil): Evidence for heterogenous carbon reservoir in sublithospheric mantle. Chem Geol 363:114–124
Zhang Z, Ganguly J, Ito M (2010) Ca–Mg diffusion in diopside: tracer and chemical inter-diffusion coefficients. Contrib Mineral Petrol 159:175–186
Ziberna L, Klemme S, Nimis P (2013) Garnet and spinel in fertile and depleted mantle: insights from thermodynamic modelling. Contrib Mineral Petrol 166:411–421
Ziberna L, Nimis P, Kuzmin D, Malkovets VG (2016) Error sources in single-clinopyroxene thermobarometry and a mantle geotherm for the Novinka kimberlite, Yakutia. Am Mineral 101:2222–2232

Reviews in Mineralogy & Geochemistry
Vol. 88 pp. 567-636, 2022

11

Geochronology of Diamonds

Karen V. Smit*, Suzette Timmerman

Department of Earth and Atmospheric Sciences
University of Alberta
Edmonton T6G 2E3 Alberta, Canada

*Now at the University of the Witwatersrand: *karen.smit@wits.ac.za*,
suzette@ualberta.ca

Sonja Aulbach

Institute of Geosciences
Goethe-Universität
Frankfurt am Main, Germany

s.aulbach@em.uni-frankfurt.de

Steven B. Shirey

Earth and Planets Laboratory
Carnegie Institution for Science
5241 Broad Branch Road NW, Washington DC 20015, U.S.A.

sshirey@carnegiescience.edu

Stephen H. Richardson

Department of Geological Sciences
University of Cape Town
Rondebosch 7701, South Africa

steve.richardson@uct.ac.za

David Phillips

School of Geography, Earth and Atmospheric Sciences
The University of Melbourne
Parkville, VIC 3010, Australia

dphillip@unimelb.edu.au

D. Graham Pearson

Department of Earth and Atmospheric Sciences
University of Alberta
Edmonton T6G 2E3 Alberta, Canada

gdpearso@ualberta.ca

INTRODUCTION

Diamond crystallization ages are essential information for science and industry, leading to an understanding of how, why, and where diamonds form in Earth and how best to find them. Such knowledge is of prime importance for constraining the source and mobility of volatile-rich diamond-forming fluids in the mantle and the nature of the geodynamic processes that release these fluids.

1529-6466/22/0088-0011$10.00 (print)
1943-2666/22/0088-0011$10.00 (online)
http://dx.doi.org/10.2138/rmg.2022.88.11

Geochronology using radiogenic isotopes relies on the differential partitioning of parent and daughter elements into different phases, including minerals and melts, and determining the time elapsed after parent–daughter fractionation and closure of the system. However, the purity of monocrystalline diamond presents a fundamental problem: the diamonds themselves have such low abundances of the elements of interest that precise and accurate age determinations are precluded. Therefore, dating the formation of diamond has to rely on dating its inclusions, an effort which started nearly 50 years ago and since then has undergone a profound evolution. The reliance on inclusions to obtain diamond ages is predicated on the hypothesis that the encapsulating diamond prevents further isotopic exchange between the mineral inclusion and the surrounding mantle.

In the 1970s, sulfides were the first inclusions for which age determinations were attempted using Pb isotopes (Welke et al. 1974; Kramers 1979). Due to low U–Th content, only Pb–Pb model ages are obtainable for sulfides. Diamonds from the Finsch and Kimberley kimberlites (Kaapvaal craton, South Africa) gave Pb–Pb model ages in excess of 2 Ga, much older than their Cretaceous host kimberlites. Even though the analytical data were subject to large blank corrections, as well as uncertainties in mantle Pb growth curves, these results provided some of the first evidence that diamonds are xenocrysts in the host kimberlite (Kramers 1979).

Richardson et al. (1984) produced ~3.3 Ga Sm–Nd and Rb–Sr model ages on composites of harzburgitic garnet inclusions from Kimberley and Finsch diamonds. In addition to proving that diamonds occur as xenocrysts in kimberlite, these results also clearly indicated that diamonds are billions of years old and established a Paleoarchean age for the continental lithospheric mantle sources for the diamonds.

Richardson and co-authors subsequently provided age constraints for many diamond suites from South Africa, Botswana, Australia, and Russia (e.g., Richardson 1986; Richardson et al. 1990, 1993; Richardson and Harris 1997). At the time of Richardson's Sm–Nd and Rb–Sr work, single grains of garnet or clinopyroxene from gem-quality monocrystalline diamonds needed to be grouped together with other similar grains—each broken from its own diamond host—in order to obtain sufficient Sm, Nd, Rb, and Sr to generate a signal on the mass spectrometer. This problem appeared to have been overcome in the late 1980s, when Burgess et al. (1989) and Phillips et al. (1989) reported $^{40}Ar/^{39}Ar$ ages for single clinopyroxene inclusions in diamonds from the Premier kimberlite (*Cullinan Mine*) in South Africa. However, subsequent work revealed unforeseen Ar redistribution problems with the approach used at the time, which has limited wider application of the technique (Burgess et al. 2004). Nevertheless, the $^{40}Ar/^{39}Ar$ method has proved useful for detrital diamond provenance studies (Phillips et al. 2004, 2018; Laiginhas et al. 2009).

In the mid-1990s, the analytical methods developed for Re–Os on whole-rock samples (e.g., Shirey and Walker 1998) were miniaturized and modified so that single sulfide inclusions (typically between 2 and 10 μg but also larger) could be analyzed, obviating the need to use composites (Pearson et al. 1998b; Pearson and Shirey 1999). Since the initial Re–Os study on Koffiefontein (Kaapvaal craton) sulfide-bearing diamonds (Pearson et al. 1998b), Re–Os has become the most widely used chronometer for determining diamond ages world-wide (e.g., Richardson et al. 2001; Westerlund et al. 2006; Richardson and Shirey 2008; Aulbach et al. 2009a,c, 2018; Smit et al. 2010, 2016; Wiggers de Vries et al. 2013; Gress et al. 2021a,b).

The recent advent in mass spectrometry of 10^{13} Ohm feedback resistors in Faraday cup current amplifiers, resulting in more sensitive detectors (Koornneef et al. 2014), has made it possible to measure the isotopes of Sm, Nd, Rb, and Sr at the extremely low amounts in which they occur in single garnet and clinopyroxene inclusions. This technique has been applied to inclusions in diamond from Botswana and South Africa (Koornneef et al. 2017; Timmerman et al. 2017; Gress et al. 2021a, b, c), affording the clearer detection of multiple diamond-forming events within a given mantle region, in particular when used in conjunction with Re–Os dating of sulfide inclusions in the same diamond suite (e.g., Gress et al. 2021b).

A new method that shows promise in this evolving field is the dating of fluid-rich fibrous diamonds using the U–Th/He isotope system (Timmerman et al. 2019c; Weiss et al. 2021). Fluid-rich mircoinclusions contain sufficient $^{235,238}U$ and ^{232}Th that decay over time to ^{4}He. Where mineral inclusions are used to date monocrystalline gem-quality diamonds, fluid inclusions can be used to date fibrous diamonds and give important constraints on whether they formed contemporaneously or in separate events.

In the 35 ensuing years since the Richardson et al. (1984) study, the ages of diamonds from almost every significant diamond-producing locality have been determined from their mineral inclusions. This allows viewing diamond formation less as an isolated event in the mantle under each mine, and more in the geological context of global tectonothermal events, such as the supercontinent cycle.

In this chapter we describe the radioisotope systems used for diamond geochronology and the principles on which they are based. Different approaches to using mineral inclusions for age constraints have different implications regarding accuracy and equilibration between diamond-forming fluids, the inclusions themselves and their host lithologies. Practical considerations of determining and interpreting diamond ages, including sources of uncertainty, the relationship of the dated inclusions to their host diamond, other (e.g., non-geochronological) evidence for ancient diamond ages, and a practical formalism for diamond ages are reviewed. A final discussion of terranes where diamond ages have been applied to constraining the geological evolution of continents illustrates one of the reasons why diamond ages are important.

DATING PRINCIPLES AND GLOSSARY

Terminology

Since the production of the first radiometric ages (Rutherford 1905) geologists have debated the significance of the event being *'dated'* by geochronology. All geochronology comes down to accurately determining the time of parent–daughter fractionation. The geological meaning of this fractionation often leads to multiple interpretations and disagreements.

Diamonds are typically dated to understand the time of crystallization: when carbon atoms in a fluid or melt took the form of a diamond crystal lattice. Dating the crystallization age of diamond is one of the most significant challenges in geochronology, and a practice that has generated much debate in the geological literature (e.g., Pidgeon 1989; Richardson 1989; Navon 1999; Pearson and Shirey 1999).

Dating diamond is difficult because the diamond lattice itself cannot be directly dated. There are several reasons for this. Despite being composed of nearly pure carbon, the ages of the vast majority of diamonds are older than the time-span accessible to ^{14}C dating (~50 k.y.). Radioactive trace elements with much slower decaying radionuclides such as K, Rb, Sm, Re, and U do not substitute directly into the diamond lattice itself, but may instead be hosted in fluid and mineral inclusions (e.g., Weiss et al. 2022, this volume; Stachel et al. 2022a, this volume).

Even though fluid inclusions hold promise for radiometric dating, the radioisotopes of interest are often present in such low abundances in gem diamonds that precise isotopic measurement is precluded (e.g., McNeill et al. 2009; Melton et al. 2012; Krebs et al. 2019; Timmerman et al. 2019a). When present in higher concentrations in fibrous or fluid-rich diamonds, these radioisotopes may show complex isotopic systematics that are not always conducive to absolute age determination (Klein-BenDavid et al. 2010, 2014). Recently, however, fibrous fluid-rich diamonds have been successfully dated with U–Th/He systematics (Timmerman et al. 2019c; Weiss et al. 2021).

Mineral inclusions trapped inside diamond during growth are most frequently used for radiogenic isotope analyses. These inclusions are rare in gem-quality diamonds and often tiny—less than 100 μm. The geological events that give the least ambiguous dates involve the crystallization of a mineral at high enough temperatures to reset any pre-existing chronological information and chemically homogenize the isotopic system being considered. This is followed by very rapid cooling, to '*freeze in*' the element distribution created at the time of crystallization, allowing the geochronometer to be '*set*'. The challenge of determining the time of diamond crystallization can be well illustrated by considering some of the assumptions underlying the various types of radiometric age calculations.

To assist the non-specialist reader in understanding the differences, advantages and disadvantages in these calculations, we provide some definitions of the various terms used, prior to the main discussion. These ideas will be amplified with examples given through consideration of the different dating techniques.

Decay constant and half-life. The decay constant (λ) can be thought of as the probability that a given nuclide will decay over some specified period of time, usually expressed in units of time^{-1}. The decay constant (λ) for ^{187}Re is given as:

$$\lambda = 1.6689 \pm 0.0031 \times 10^{-11}\ a^{-1} \text{ (Selby et al. 2007)}$$

where a = year

The term '*half-life*' ($t_{1/2}$) is slightly more tangible because it is given in units of years (for geological systems) and is expressed as the time taken for the original number of parent atoms to decrease by half. Re-arrangement of the standard equation for radioactive decay leads to the relationship between the half-life and the decay constant being given as:

$$t_{1/2} = 0.693/\lambda$$

Parent–daughter ratio. The ratio of the unstable radioactive '*parent*' nuclide to the stable '*daughter*' nuclide produced by radioactive decay, e.g., ^{187}Re to ^{187}Os, ^{147}Sm to ^{143}Nd.

Common. All the stable isotopes of an element, including the pre-existing daughter isotope, that are inherited at the time the isotopic system is closed. Radioactive decay adds new atoms of the daughter isotope to the '*common*' or extant isotopes of the element. In K–Ar geochronology this is equivalent to '*extraneous*' (includes excess and inherited) Ar.

Isotopic system closure. The time at which atomic diffusion of parent or daughter isotope effectively ceases, at the sample-scale being considered, e.g., at the micron-scale of the lattice of a given mineral, or at the mineral-to-mineral scale of a rock. In the case of a single mineral trapped in a diamond it can be argued that '*closure*' of the system occurs when diamond encloses an inclusion and isolates it from isotopic exchange with its surroundings, because the diamond lattice prevents any measurable diffusional exchange with the components of the entrapped mineral. In such a case, isotopic closure can be argued to reflect the '*date of diamond crystallization*'.

If, however, the daughter nuclide is a gas (^{40}K–^{40}Ar or U–Th–^{4}He decay) or a highly incompatible element in the mineral lattice (e.g., Pb produced from the decay of U in zircon or rutile), the daughter nuclide may diffuse to the interface between the diamond and the inclusion over million-to-billion-year residency at mantle temperatures. In these cases, failure to analyze the whole system—the gas liberated on releasing the inclusion from the diamond or the fluid film at the inclusion-diamond interface—will yield an erroneous time since diamond crystallization.

Initial isotopic ratio. The isotopic ratio of the daughter element, e.g., $^{187}Os/^{188}Os_i$ existing at the time of system equilibration and closure by igneous or metamorphic mineral crystallization. Subsequent accumulation of the daughter isotope by radioactive decay produces a new, more radiogenic isotopic ratio compared to the initial isotopic ratio. In the case of mineral inclusions in diamond, this is the isotopic ratio of Sr, Nd or Os, that the inclusion possessed at the time of diamond encapsulation. This ratio can be calculated from the age of diamond formation, the measured isotopic composition of the daughter element and the parent/daughter isotopic ratio.

Types of ages—what is being dated and what is in a name?

Prior to an assessment of accuracy and precision, the most important consideration about any geological age is an understanding of *what event is being dated.* In the context of diamond geochronology, the resulting age may reflect any, or a combination of:

i. the instant of diamond crystallization, or the time of percolation of diamond-forming fluids into the host mantle,

ii. cooling of the diamond and its dated inclusion to a certain temperature after diamond formation,

iii. crystallization of a *'protogenetic'* mineral now included in diamond, prior to its encapsulation in the diamond,

iv. mixing of geochemical signatures of different fluid and solid components that are mobilized during the events that cause diamond formation.

Below we will consider the main types of geological ages employed in diamond dating studies. For a more complete explanation of the principles involved, the reader is referred to Reiners et al. (2018).

Model ages. A model age traces the time of divergence of the isotopic evolution of a system (*rock or mineral*) from an assumed reference reservoir—e.g., the primitive mantle or mantle with a chondritic composition—due to a change or fractionation of the parent–daughter isotope ratio in response to a geological event such as partial melting. For example, a Re–Os model age will give the time of separation from a primitive or depleted mid-ocean ridge basalt (MORB)-like mantle source, in other words, the time of initial melt formation. This is because Re is 'mildly' incompatible and partitioned into the basaltic melt during mantle melting, whereas Os is 'strongly' compatible and remains in the residue.

Model ages often *do not* reflect the time of diamond formation; and *in the simplest case* reflect the time that the isotopic evolution of the mineral diverged from a given reservoir. For example, model ages obtained for lithospheric mineral inclusions would reflect separation of the source mantle from the convecting asthenospheric mantle. In many cases that is likely to be the time of cratonic lithospheric mantle formation because of the accompanying, extensive parent/daughter fractionation. In the case of a peridotitic garnet inclusion in a diamond, its Nd isotope model age is a calculation of the length of time required for the garnet to develop its measured Nd isotopic composition (at its measured Sm/Nd ratio) with two main assumptions:

1. that it was derived from a stated reference mantle source, and

2. there was a single stage evolution. More complex two-stage model ages are possible but are rarely used in diamond dating studies.

The two options for a reference mantle reservoir are usually a mantle with time-averaged Sm–Nd isotopic characteristics similar to the parent–daughter ratio of chondritic meteorites—a T_{CHUR} model age—or a mantle source slightly depleted in melt components—a T_{DM} model age. In effect, the calculation estimates the intersection of the time-integrated Nd isotope evolution of the sample, with that of the chosen reference reservoir. The resulting '*model age*', in the case of peridotitic garnet, may estimate the time of melt depletion or metasomatic enrichment of the garnet-bearing source , if it occurred close to the time of melt depletion. A third assumption is involved if we relate either of these events to the simultaneous crystallization of the host diamond.

In reality, many model ages integrate the *time-averaged effects* of multiple events, including any significant pre-history of an inclusion, where it inherits some isotopic evolution different from the assumed host reservoir before the diamond-entrapment event. Model ages must be treated with caution due to these underlying assumptions and the problems caused by possible protogenetic inclusions with significant prior histories, though the reader is referred to the recent discussion of these issues by Pamato et al. (2021) and the sections below.

Most sulfides take in very large amounts of *common* Os from their environment that becomes integrated with the *radiogenic* Os. Most mantle sulfides (and inclusions in diamonds) have experienced significant pre-history, and in the case of eclogitic source rocks, acquired radiogenic Os between formation of the oceanic crustal protolith and diamond formation during or after subduction. Even if such a pre-existing sulfide is later isotopically '*reset*' during diamond crystallization, retention of a radiogenic initial isotopic ratio and possible loss of Re from the sulfide at the time of resetting will lead to a model age calculation that will *not reflect* the diamond crystallization age when calculated with reference to a model evolution curve for the primitive or depleted mantle.

The most robust model or absolute ages are given by systems where a particular mineral may have little or almost no pre-existing radiogenic or *common* daughter isotope. This is generally the case with U–Pb ages acquired from zircon. It is also the case where the mineral is old enough with such an elevated parent–daughter isotopic ratio, such as some sulfides with extremely high Re/Os, that its evolution relative to the assumed reference reservoir has a very steep slope. In such cases, very little uncertainty results from the nature of the assumed parent reservoir. Conversely, great caution must be given to model ages in systems where the isotopic evolution of a sample diverges only slightly from its assumed parent reservoir (i.e., where the parent–daughter fractionation is subtle), because the very similar parent–daughter isotope ratios lead to little divergence of the sample from the parent reservoir and large uncertainties depending on the choice of reference reservoir (e.g., mantle that is chondritic, depleted, or enriched).

Isochron ages. Isochrons are graphical constructs that rely on the cogenetic relationship between a whole rock and its mineral constituents, a suite of different rocks or, in the case of diamond geochronology, mineral inclusions. The conventional diagrams plot the parent/daughter isotope ratio of the system on the *x*-axis and the daughter isotope ratio on the *y*-axis (Fig. 1). A suite of minerals all derived from the same rock—or geologic event generated in a larger mantle region with the same initial isotope ratio—will evolve to produce linear relationships whose slope is proportional to the time since the system was homogenized. Isotopic homogenization is attained or approached, typically above the closure temperature of

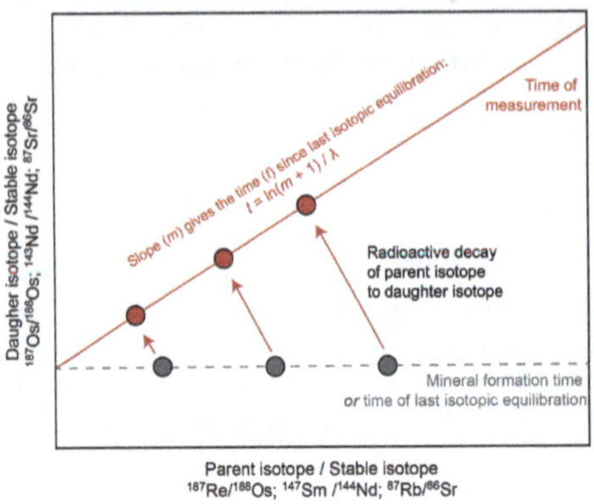

Figure 1. Principles of a conventional isochron diagram. The *x*-axis displays the ratio of the radioactively decaying isotope to a stable isotope of the daughter. The minerals with higher amounts of parent and/or lower amounts of daughter plot further to the right. The higher the content of parent isotope, the more daughter isotope is produced. These minerals plot at a higher position on the *y*-axis, which is the ratio of the radiogenic daughter to the same stable daughter isotope as on the *x*-axis. Radioactive decay preserves the slope that is generated by different amounts of parent relative to daughter in the different minerals. The slope of a line fitted through the present-day measured isotopic ratios, and the independently determined decay rate (accepted by the scientific community as a constant), yield the time since last isotopic equilibration.

the isotopic system of interest, especially if aided by a fluid or melt. The isotopic evolution of these mineral systems is then isolated by cooling of a hot rock to below the closure temperature, where inter-mineral diffusive equilibrium effectively ceases. The displacement of the different samples along the line is a positive correlation with their parent–daughter isotope ratio.

Isochrons require analyses of at least two samples to define a straight line, and more than two samples are highly preferred. Isochron ages are preferred over model ages, though they too require certain assumptions to be met for ages to be valid. Isochron ages also require careful interpretation, especially when the scatter in the data exceeds that expected from analytical uncertainties alone. For instance, a key requirement for the isochron approach is that all samples should start out with the same initial isotopic ratio so that they develop statistically significant linearity required of a *'true'* isochron relationship. This is rare enough in diamond dating that the term *'linear data array'* is used, which more accurately reflects the evident geological scatter.

For diamond inclusion dating, the attainment of initial isotopic *'equilibrium'* or uniformity is probably the single most difficult assumption to satisfy. Unlike a rock outcrop, we do not know the spatial relationship of the diamonds and their inclusions to each other in the mantle. We have to use additional data such as diamond internal structure (growth zones), inclusion chemistry and nitrogen aggregation systematics in the diamonds to relate different diamonds to each other. Even with these criteria, it is likely that one of the largest sources of uncertainty in diamond inclusion isochron dating is within *'over-dispersed'* datasets. This is where the uncertainties are dominated by the fact that inclusions from different diamonds may not have identical initial ratios during diamond formation, and only partial isotopic equilibration was achieved (Fig. 2).

Multiple inclusions within the same diamond growth zone are more likely to have the same initial ratio than even multiple inclusions from different growth zones within the same diamond (e.g., Wiggers de Vries et al. 2013). In the case of protogenetic inclusions, high melt/fluid–rock ratios during diamond formation should help to homogenize the isotopic ratio of the system, but any remaining variation will make a contribution to the statistical scatter and resulting uncertainty on the isochron regression.

Despite these caveats and challenges, many of which reflect natural mineralogical variability of samples beyond the control of the analyst, it is still possible to recover valuable age information from the isochron arrays defined by diamond inclusion suites. Geologically significant ages can be obtained as long as inclusions from the same diamond formation event are regressed together, as illustrated by the various examples dealt with in the remainder of this chapter.

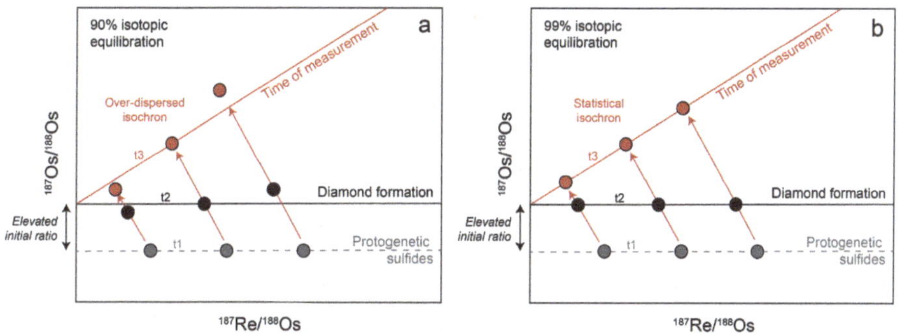

Figure 2. Illustrative isochron diagrams at 90 and 99% chemical diffusive equilibration. With increasing isotopic equilibration, the statistical quality of the isochron increases (see better alignment of datapoints with the isochron in **b)** because of closer adherence to a *'uniform initial ratio'* required to produce a statistically perfect isochron. Modified from Pamato et al. (2021).

RADIOGENIC ISOTOPIC DATING METHODS AND APPLICABLE INCLUSION TYPES

$^{40}Ar/^{39}Ar$

The possibility of dating diamonds using the $^{40}Ar/^{39}Ar$ laser probe technique was first investigated by Burgess et al. (1989) and Phillips et al. (1989). The ability to analyze single inclusions represented a major advance at the time compared to Rb–Sr and Sm–Nd analyses (e.g., Richardson et al. 1984; Richardson 1986), which required composites of several tens to hundreds of inclusions to achieve reasonable age precision.

Principles. The $^{40}Ar/^{39}Ar$ dating method is a variant of the K-Ar technique and is based on the natural decay of ^{40}K to ^{40}Ar, with a half-life of 1250 m.y. (McDougall and Harrison 1999 and references therein). In the case of the $^{40}Ar/^{39}Ar$ method, the parent ^{40}K isotopic content is determined indirectly with the stable isotope ^{39}K converted to ^{39}Ar by neutron irradiation in a nuclear reactor. The $^{40}Ar^{*}/^{39}Ar$ ratio is then proportional to age, when calculated relative to one or more co-irradiated neutron fluence monitors (reference minerals) of known age. Here, $^{40}Ar^{*}$ refers to radiogenic ^{40}Ar produced from ^{40}K decay. As the atmosphere contains 0.93% Ar (by volume), most rock/mineral samples include some proportion of atmospheric Ar adsorbed on mineral surfaces or contained within the crystal lattice. Atmospheric Ar has a measured $^{40}Ar/^{36}Ar$ ratio of 298.56 ± 0.31 (Lee et al. 2006), and measurement of ^{36}Ar can be used to correct $^{40}Ar_{total}$ for (assumed) atmospheric Ar contamination. The assumption of an initial atmospheric $^{40}Ar/^{36}Ar$ ratio can be tested using standard isochron plots, or, more typically, $^{36}Ar/^{40}Ar$ vs. $^{39}Ar/^{40}Ar$ correlation diagrams. If a mineral contains common argon with a composition above the atmospheric ratio, the additional ^{40}Ar is termed extraneous (excess or inherited) argon. The age of a sample is then given by the age equation:

$$t_{\text{unk}} = \frac{1}{\lambda}\ln\left(F \times J_{\text{ref}} + 1\right)$$

where unk = unknown;
$\lambda = 5.543 \times 10^{-10}\ \text{yr}^{-1}$ (Steiger and Jäger 1977);
$F = {}^{40}Ar^{*}/{}^{39}Ar$;
ref = reference mineral (or fluence monitor);
J = J-value:

$$J = \frac{(e^{\lambda t_{\text{ref}}} - 1)}{F_{\text{ref}}}$$

Commonly used reference minerals include the Fish Canyon Tuff sanidine (FCTs = 28.201 Ma; Kuiper et al. 2008), Alder Creek Rhyolite sanidine (ACRs = 1.816 Ma; Niespolo et al. 2017) and Mt Dromedary biotite (GA1550 = 99.125 Ma; Phillips et al. 2017). The irradiation process also produces so-called interfering isotopes such as ^{40}Ar from K and $^{36}Ar + {}^{39}Ar$ from Ca. Correction for these interfering isotopes is accomplished by analyzing co-irradiated, zero-age K-glasses and/or salts (e.g., CaF_2, K_2SO_4, KCl).

Argon is a noble gas, and its retention in a mineral lattice is a function of temperature, grain size (or more accurately diffusion dimension) and cooling rate. Typical closure (or blocking) temperatures for key K-bearing minerals are ~300 °C in biotite, ~400 °C in muscovite and ~550 °C in hornblende (McDougall and Harrison 1999 and references therein). As the $^{40}Ar/^{39}Ar$ method has a lower closure temperature than most of the other methods described herein, it is used for both geochronology and thermochronology (McDougall and Harrison 1999 and references therein).

Inclusion types. The $^{40}Ar/^{39}Ar$ method is applicable to K-bearing minerals. Common K-bearing phases such as micas and feldspars are rare or absent as inclusions in diamonds. However, many clinopyroxene inclusions of both eclogitic (omphacite) and peridotitic (Cr-diopside) parageneses, contain small quantities of K, typically ~1000 ppm, with percent levels of potassium (~1.5 wt%) reported for some inclusions (Harlow and Veblen 1991; Harlow 1997). Elevated potassium levels have also been reported for eclogitic garnet inclusions from Orapa diamonds (Burgess et al. 1989, 2004). Clinopyroxene inclusions of eclogitic paragenesis are generally more abundant than the peridotitic Cr-diopside inclusions (Stachel and Harris 2008), but both types have been the target of $^{40}Ar/^{39}Ar$ dating studies. Bulanova et al. (2004) identified inclusions of yimengite [$K(Cr,Ti,Mg,Fe,Al)_{12}O_{19}$], a rare potassium magnetoplumbite mineral, in a diamond from the Sese kimberlite (Zimbabwe). Thus, other relatively K-rich mineral inclusions may be suitable candidates for $^{40}Ar/^{39}Ar$ dating.

Practical applications. Initial attempts to determine diamond ages using $^{40}Ar/^{39}Ar$ laser probe methods involved analyses of eclogitic clinopyroxene inclusions from Cullinan (formerly Premier) Mine diamonds, which had been cleaved to expose the inclusions (Burgess et al. 1989; Phillips et al. 1989). The two studies yielded analogous average ages of 1198 ± 28 Ma (2σ) (Phillips et al. 1989) and 1185 ± 95 Ma (2σ) (Burgess et al. 1989), respectively—indistinguishable from the eruption age of the Premier kimberlite (~1.2 Ga; Kramers and Smith 1983). Prior to this, Richardson (1986) determined a Sm–Nd age of 1150 ± 60 Ma, for pooled eclogitic inclusions from Cullinan diamonds and the $^{40}Ar/^{39}Ar$ ages were therefore interpreted as the time of diamond formation.

However, subsequent $^{40}Ar/^{39}Ar$ analyses of similarly (partially) exposed clinopyroxene and garnet inclusions in diamonds from Orapa, Jwaneng (Botswana), Udachnaya (Russia) and Argyle (Australia), revealed unexpected complexities with this experimental approach, with most ages intermediate between the kimberlite eruption event and Sm–Nd (and later Re–Os) diamond formation ages (Richardson et al. 1990; Burgess et al. 1992). The favored explanation for this behavior was diffusion of pre-eruption radiogenic Ar ($^{40}Ar^*$) to the diamond/inclusion interface region during mantle residence, with partial to complete loss of this component when the diamond was cleaved to expose the inclusion (Burgess et al. 1992). Step-heating $^{40}Ar/^{39}Ar$ analyses of clinopyroxene inclusions extracted from their host diamonds produced similar age results, but with lower temperature steps yielding older apparent ages than the fusion steps (with minor exceptions) (Phillips et al. 2004; Phillips and Harris 2008). The 'intermediate' age results were attributed to partial Ar diffusion to the diamond/inclusion interface and defect structures in the clinopyroxene inclusions; this process is discussed in more detail in the section on '*Diffusion of elements from inclusion to inclusion-diamond interface*'.

The $^{40}Ar/^{39}Ar$ dating of exposed or partially exposed inclusions produces only minimum estimates of diamond formation ages. To determine diamond genesis ages, Burgess et al. (1992) suggested laser drilling to encapsulated inclusions within diamonds, to recover radiogenic Ar located in both the diamond/inclusion interface and the clinopyroxene inclusion. The difficulty with this approach is that diamond becomes opaque (black) in response to neutron irradiation, which makes locating buried inclusions a non-trivial issue. To address this issue, Burgess et al. (1994) ground and polished selected diamond faces to within ~0.5 mm of the inclusion and marked inclusion positions using laser spots. Using this method, Burgess et al. (1994) and Bulanova et al. (2004) reported $^{40}Ar/^{39}Ar$ ages for clinopyroxene-bearing diamonds from the Venetia kimberlite and yimengite-bearing diamond from the Sese kimberlite, respectively. Burgess et al. (2004) also attempted to laser-drill garnet- and clinopyroxene-bearing diamonds from the Orapa kimberlite but were unsuccessful as the inclusions were too deeply buried. Instead, these authors carried out furnace step-heating analyses of the Orapa diamonds, which released Ar in two pulses, one corresponding to Ar release from melting of the inclusion and decrepitation of the diamond (1200–1800 °C) and the second during diamond graphitization

(>2000 °C). Some of the Venetia and Sese diamonds produced ages approaching the time of host kimberlite eruption (Bulanova et al. 2004; Burgess et al. 2004), possibly due to decrepitation of the polished surface prior to analysis (Bulanova et al. 2004). However, three Orapa diamonds yielded ages of ~1.0 Ga, similar to previous Sm–Nd diamond formation ages (Richardson et al. 1990), with other Orapa ages possibly supporting Re–Os ages of ~2.9 Ga (Shirey et al. 1999). One Venetia inclusion-bearing diamond gave an unrealistically old age of ~4.6 Ga, suggesting that at least some diamonds may contain excess Ar. Clearly, further studies involving laser drilling and/or furnace step-heating are required to investigate the utility of the $^{40}Ar/^{39}Ar$ method for dating diamond formation events.

Despite the challenges in dating diamond formation events, the observation that exposed inclusions often give ages approaching the times of host kimberlite eruption highlighted the possibility of constraining the provenance of detrital diamond deposits (Burgess et al. 1997; Phillips et al. 2004, 2017; Phillips and Harris 2008, 2009; Laiginhas et al. 2009). This rather unique application of the $^{40}Ar/^{39}Ar$ method is documented in more detail in the section below on provenance studies of detrital diamond deposits.

U–Th/He

The potential of the U–Th/He method for dating minerals was first recognized by Strutt (1905), and only much later applied to dating diamonds (Kurz et al. 1987). Through *in vacuo* crushing and step-heating experiments, Kurz et al. (1987) showed that monocrystalline diamonds had limited variation in ^{3}He but large variation in ^{4}He, even within individual diamonds from Orapa. This variability could not be explained by alpha implantation alone, but was probably caused by differences in U–Th and/or ^{4}He concentrations in different diamond growth zones. With assumptions about the U–Th content, Kurz et al. (1987) estimated that monocrystalline Orapa diamonds were at least 500 Ma old, thereby showing that U–Th/He isotopes have potential for dating diamonds, and confirming earlier studies that diamonds are much older than the kimberlite (Kramers 1979; Richardson et al. 1984).

Principles. The U–Th/He dating system is based on the decay of radioactive isotopes where ^{4}He atoms are created by capturing two electrons by α-particles. Most of the ^{4}He on Earth is formed by decay of U and Th (^{4}He from longer-lived decays such as $^{147}Sm \rightarrow {}^{143}Nd$, is of lesser importance in diamonds, because only single α-particles are produced per decay):

$$^{238}U \rightarrow {}^{206}Pb + 8\alpha$$

$$^{235}U \rightarrow {}^{207}Pb + 7\alpha$$

$$^{232}Th \rightarrow {}^{208}Pb + 6\alpha$$

Alpha-particles from U and Th decay are formed in a decay chain via several intermediate radionuclides. The intermediate radionuclides with relatively long half-lives (^{231}Pa, ^{234}U, ^{230}Th) can be fractionated from ^{235}U and ^{238}U, caused by differences in chemical properties of the various elements (e.g., Th, U, Pa), resulting in an imbalance called secular disequilibrium. This can occur during crystallization/partial melting and is important for many minerals when the crystallization period of a mineral is shorter than ~5× the half-life of the intermediate daughter nuclide (e.g., Farley 2002).

Secular disequilibrium is negligible for dating fluid inclusions in fibrous diamonds for two main reasons:

1. trace elements (U, Th) are primarily located in fluid inclusions in diamond (Timmerman et al. 2019; Krebs et al. 2019) and there is likely no fractionation between the diamond-forming fluid and encapsulated fluid inclusions;
2. after 5× the longest half-life (^{234}U; 5× = 1.23 million years) any initial disequilibrium returns to a secular equilibrium (Bourdon et al. 2003) and all studied fibrous diamonds and their fluid inclusions are significantly older than 1.23 million years.

Therefore, the total radiogenic ^{4}He (including the usually negligible ^{147}Sm decay for completeness) is defined as:

$$^{4}\text{He} = 8\,^{238}\text{U}\left(e^{\lambda 238t} - 1\right) + 7\,^{235}\text{U}\left(e^{\lambda 235t} - 1\right) + 6\,^{232}\text{Th}\left(e^{\lambda 232t} - 1\right) + 1\,^{147}\text{Sm}\left(e^{\lambda 147t} - 1\right)$$

where:

$\lambda_{238} = 1.55125 \times 10^{-10}\ \text{yr}^{-1}$
$\lambda_{235} = 9.8485 \times 10^{-10}\ \text{yr}^{-1}$
$\lambda_{232} = 4.9475 \times 10^{-11}\ \text{yr}^{-1}$
$\lambda_{147} = 6.54 \times 10^{-12}\ \text{yr}^{-1}$

This equation can be iteratively solved for time using the software package HelioPlot (Vermeesch 2010), which is also incorporated into IsoPlotR (Vermeesch 2018).

Formation of ^{4}He by radioactive decay of Sm is negligible and ^{235}U can be expressed as ^{238}U/137.88 at present-day. Therefore, ^{4}He production can be approximated by a simplified formula (see supplementary information in Timmerman et al. 2019c for derivation). Because diamond-forming fluids often contain '*common*' ^{4}He, U–Th/He ages are considered to be maximum ages unless a 'common' ^{4}He correction is made. This issue can potentially be solved when cogenetic samples form an isochron array, or by using the measured stable ^{3}He concentration and assuming an initial ^{4}He/^{3}He ratio.

$$^{4}\text{He} = t\left(1.29\text{e}^{-9}\,^{238}\text{U} + 2.97\text{e}^{-10}\,^{232}\text{Th}\right)$$

$$\frac{^{4}\text{He}}{^{3}\text{He}}\text{measured} = \frac{^{4}\text{He}}{^{3}\text{He}}\text{initial} + t\frac{\left(1.29e^{-9}\,^{238}\text{U} + 2.97e^{-10}\,^{232}\text{Th}\right)}{^{3}\text{He}}$$

The closure temperature of the U–Th/He system varies for different minerals, it is one of the lowest closure temperatures of all radiogenic systems, with temperatures of ~100 °C for apatite (Lippolt et al. 1994; Wolf et al. 1996; Warnock et al. 1997) and ~600 °C for garnet (Dunai and Roselieb 1996). Although diamond is formed above the U–Th/He closure temperature, diffusion rates for He (He > U–Th) in diamond are extremely slow: ±4 × 10^{-21} cm^2/s (Shelkov et al. 1998) to 1.8 × 10^{-19} cm^2/s (Weiss et al. 2021). Due to these slow diffusion rates in diamond, the U–Th/He system is effectively closed in diamond, and makes it an ideal mineral for U–Th/He dating.

One of the limiting factors in U–Th/He dating applications of diamonds is the ability to precisely and accurately measure the low U, and Th abundances in diamonds. The lack of diamond reference materials hampers the ability to assess the accuracy of direct laser ablation measurements of diamonds. Induced neutron activation techniques have been used (e.g., Schrauder et al. 1996) but this approach renders the diamond unusable for noble gas measurements. The application of the method has been considerably helped by the advent of the '*off-line*' laser ablation ICP-MS measurement approach that allows pre-concentration of analytes to orders of magnitude higher than possible using conventional laser ablation ICP-MS approaches (McNeill et al. 2009; Klein-BenDavid et al. 2010).

Inclusion types. Polycrystalline and fibrous diamonds have significantly higher concentrations of trace elements and noble gases compared to monocrystalline gem-quality diamonds. This difference is thought to be due to the fact that fibrous diamonds contain abundant fluid inclusions (Navon et al. 1988; Weiss et al. 2022, this volume), suggesting that trace elements and noble gases are primarily located in these fluid inclusions rather than in the diamond lattice itself. Because the abundances of U, Th are in the ppt range for monocrystalline gem-quality diamonds (McNeill et al. 2009; Krebs et al. 2019; Timmerman et al. 2019a), the accuracy and precision of these measurements is likely to remain one of the largest sources

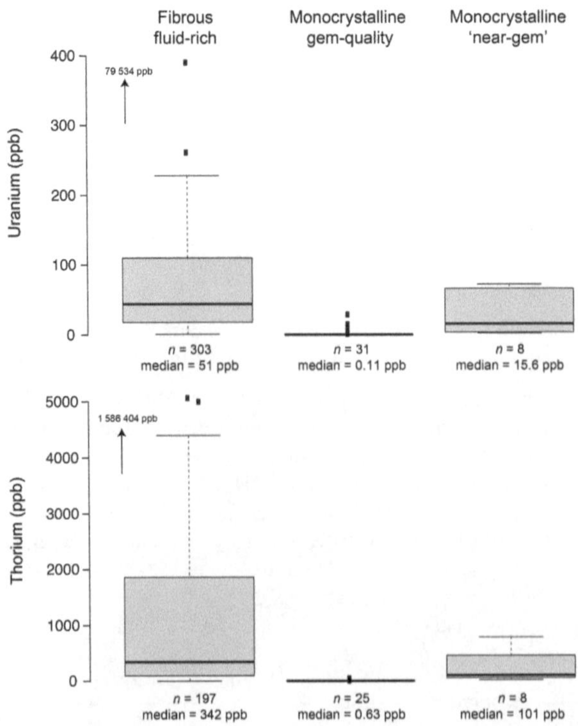

Figure 3. 'Box and whisker' plots of the U and Th contents in fibrous fluid-rich diamonds compared to monocrystalline gem-quality and near-gem diamonds. Because the abundances of U and Th are prohibitively low in monocrystalline gem-quality diamonds, the U–Th/He method is limited to fibrous diamonds that have higher U–Th concentrations. Plotted using the data compilation in Weiss et al. (2022b).

of uncertainty in U–Th/He age determination of gem diamonds. This currently limits the U–Th/He method to 'fluid-rich' fibrous diamonds with higher U–Th concentrations (Fig. 3).

Polycrystalline diamonds—such as carbonado and framesite—have much smaller crystal sizes (generally smaller than 100 μm) and in such small diamonds, diffusion of He and implantation of ^{4}He from the host rock needs to be taken into consideration (Ozima et al. 1991). Furthermore, polycrystalline diamonds have abundant mineral inclusions and He is unlikely to be released from mineral inclusions during crushing, resulting in lower apparent ages (Burgess et al. 1998).

Practical applications. Despite the recognition in the 1980s that the U–Th/He method may be useful for dating diamonds (Kurz et al. 1987), it was not until 1998 that the first diamonds were analyzed for both U and He (Burgess et al. 1998). In these early studies He was released by crushing diamond in a vacuum (Burgess et al. 1998, 2009). Larger fragments that remained after crushing were irradiated and analyzed for U ($^{134}Xe_U$) using an extended Ar(–Kr–Xe) step-heating method (Burgess et al. 1998) or analyzed for U and Th by neutron activation (Schrauder et al. 1996).

Early He isotope studies on diamonds showed the importance of cosmogenic ^{3}He accumulation in alluvial diamonds residing near Earth's surface for long periods. Cosmogenic ^{3}He near the outside of the rough diamond (Honda et al. 1987; Lal 1989) adds to nucleogenic ^{3}He that accumulates through the neutron bombardment of Li (^{6}Li(n,αβ)^{3}He) (Honda et al. 1987; Kurz et al. 1987). It was also recognized that the outer ~30 μm of a diamond crystal are affected by implanted ^{4}He from U–Th decay of minerals in host rock for the diamond (Shelkov et al. 1998; Fig. 4).

Based on fragment sizes, modeling, vacuum crushing and step heating of fibrous diamonds, it was recognized that crushing is not an efficient way to extract all He from the diamond (releasing 4–20% of the total He), resulting in an underestimation of the age (Timmerman et al. 2019c). This is because α particles ejected from fluid/mineral inclusions can diffuse over a distance of a few μm and may accumulate within the diamond lattice, depending on how long the diamond resides in the mantle. The effects of α-ejection distance from fluid inclusions (12 μm) and He release from mineral inclusions (14–15 μm) were modeled and, in the case of two diamonds from the Democratic Republic of the Congo (DRC), α-ejection and/or fluid inclusion abundance were shown to be more important factors for He accumulation than α-implantation near the diamond surface (Fig. 4; Timmerman et al. 2019c).

The first U–Th/He dating study on fibrous diamonds (Timmerman et al. 2019c) assumed a closed system with negligible He diffusion ($D = 4 \times 10^{-21}$ cm^2/s; Shelkov et al. 1994), which provided a minimum age. Recently, the upper limit of He diffusion was estimated based on U, Th, and He contents of measured diamonds, an assumed initial R/Ra and a major craton stabilization age of 2.6 Ga and gives an upper diffusion rate of 1.8×10^{-19} cm^2/s (Weiss et al. 2021). Taking He diffusion into account gave an upper age limit, and showed saline high density fluid (HDF)-bearing diamonds from Finsch and the Kimberley mines (also known as '*De Beers Pool*') are within error of kimberlite eruption ages (Weiss et al. 2021). On the other hand, silicic and carbonatitic HDF-bearing diamonds from Congo (Timmerman et al. 2019c), Finsch and Kimberley (Weiss et al. 2021) are significantly older than kimberlite eruption, with the oldest carbonatitic metasomatic event occurring between 0.75 and 2.6 Ga (Weiss et al. 2021). These age results show that not all fibrous diamonds formed shortly before kimberlite eruption, as was previously suggested from consideration of their N aggregation systematics and the similarity of their Sr isotope and trace element signatures to kimberlitic melts (e.g., Smith et al. 2012 and references therein).

Major improvements in our understanding of He evolution in the continental lithospheric mantle are still required to provide more accurate initial ratio constraints and model ages, though data are becoming available (e.g., Timmerman 2018; Timmerman et al. 2019b; Pernet-Fisher et al. 2019; Broadley et al. 2021). Corrections for initial trapped ^{4}He using the measured amount of nucleogenic ^{21}Ne and the mantle $^{21}Ne^*/^{4}He^*$ production ratio are possible and require evaluation. The influence of micro mineral inclusions and the relative release of mantle and radiogenic He during crushing *versus* heating also need further investigation.

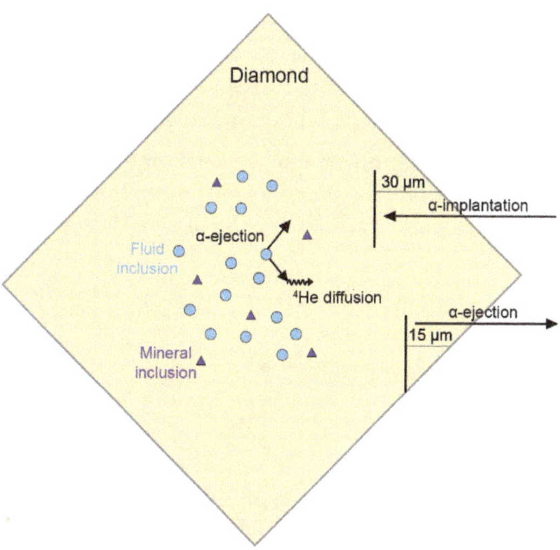

Figure 4. Processes influencing the ^{4}He budget in diamond. This includes α-implantation into diamond from the host rock and α-ejection from diamonds and its fluid inclusions. Fluid and mineral inclusions are not to scale. Implantation and ejection distances are average values. Modified from Timmerman et al. (2019c).

U–Pb and Pb–Pb

The earliest studies aimed at dating inclusion-bearing diamonds began in the 1970s and focused on the U–Th–Pb isotope system in sulfides. Early analyses needed large numbers of inclusions due to the high procedural blanks at the time, and also suffered from large uncertainties in the mantle growth curve for Pb. It was another two decades before *in situ* techniques enabled further work on the Pb–Pb model ages.

Principles. The U–Th–Pb system encompasses three separate chronometers based on the decay chains of the long-lived parent isotopes ^{238}U, ^{235}U and ^{232}Th, which terminate in daughter isotopes ^{206}Pb, ^{207}Pb and ^{208}Pb respectively. Natural Pb has three stable radiogenically derived isotopes and only one stable non-radiogenic isotope ^{204}Pb. The latter is by far the least abundant Pb isotope, typically ≤1.4% of common Pb. The U and Th decays (see equations above) produce three equations of the same form that can be used to determine an age:

$$^{206}\text{Pb} = {}^{238}\text{U}\,(e^{\lambda_{238}t} - 1) \text{ where } \lambda_{238} = 1.55125 \times 10^{-10}\text{ yr}^{-1}$$

$$^{207}\text{Pb} = {}^{235}\text{U}\,(e^{\lambda_{235}t} - 1) \text{ where } \lambda_{235} = 9.8485 \times 10^{-10}\text{ yr}^{-1}$$

$$^{208}\text{Pb} = {}^{232}\text{Th}\,(e^{\lambda_{232}t} - 1) \text{ where } \lambda_{232} = 4.9475 \times 10^{-11}\text{ yr}^{-1}$$

One of the key advantages of this system is that the two U–Pb age equations can be combined into a Pb–Pb equation whereby U is eliminated as a variable. In this equation, the (present-day) ^{235}U/^{238}U term is assumed to be constant, although we now know it is not exactly constant. In principle only the Pb isotopic composition of a sample needs to be measured to obtain a ^{207}Pb/^{206}Pb age:

$$\left(\frac{^{207}\text{Pb}}{^{206}\text{Pb}}\right) = 1/137.88\left[\frac{(e^{\lambda_{235}t} - 1)}{(e^{\lambda_{238}t} - 1)}\right]$$

In practice, both U and Pb concentrations are normally measured in order to compare U–Pb and Pb–Pb ages as a check on closed system behavior. This can be illustrated on a U–Pb Concordia diagram (Wetherill 1956) where the Concordia curve represents the locus of points of equal ^{206}Pb/^{238}U and ^{207}Pb/^{235}U ages. For a concordant mineral sample, the ^{207}Pb/^{206}Pb age is the most precise and accurate of such absolute age determinations because any analytical errors in measurement of the U/Pb ratio have been eliminated.

This approach is ideal for mineral inclusions in diamond such as zircon (Kinny and Meyer 1994) and perovskite (Hamilton et al. 2003) that carry significant amounts of U and Th but largely exclude Pb at the time of crystallization. $CaSiO_3/CaTiO_3$ polymorphs (Bulanova et al. 2010), and rutile inclusions in diamond (Schmitt et al. 2019) have also been analyzed, however, the occurrence of these phases as inclusions in diamond is extremely rare.

Sulfides and clinopyroxene are the most commonly encountered inclusion in diamonds that contain measurable Pb. Because sulfides do not incorporate U and Th, only Pb–Pb ages are obtainable, based on an expansion of the radiogenic ^{207}Pb/^{206}Pb term in the Pb–Pb equation to include initial or common Pb:

$$\frac{\left[\left(\frac{^{207}\text{Pb}}{^{204}\text{Pb}}\right)_p - \left(\frac{^{207}\text{Pb}}{^{204}\text{Pb}}\right)_i\right]}{\left[\left(\frac{^{206}\text{Pb}}{^{204}\text{Pb}}\right)_p - \left(\frac{^{206}\text{Pb}}{^{204}\text{Pb}}\right)_i\right]} = 1/137.88\left[\frac{(e^{\lambda_{235}t} - 1)}{(e^{\lambda_{238}t} - 1)}\right]$$

where subscripts p and i refer to present and initial isotope ratios respectively.

There is no certain way of establishing the prevailing (initial) Pb isotope ratios for mantle sources from which the sulfide inclusions are derived. Theoretical mantle growth curves therefore need to be constructed for plausible values of $^{238}U/^{204}Pb$ (μ) and Th/U (κ) from which Pb–Pb model ages can then be determined. Such growth curves can be single-stage or multi-stage and lead to model age uncertainty just from the choice of growth curve alone. Alternatively, where an array of data can be regressed, for multiple inclusions, secondary Pb–Pb isochrons could be used to establish ages. No such datasets are currently available as sulfides are typically analyzed for Re–Os isotopes and the Re–Os separation chemistry (Pearson et al. 1998b) is not easily adapted for low-blank Pb isotope measurement, though in theory it could be achieved.

A variety of instrumental techniques are available for making U–Th–Pb isotope measurements. In order of development, they are isotope dilution thermal ionization mass spectrometry (ID-TIMS), secondary ion mass spectrometry (SIMS) [or Sensitive High Resolution Ion MicroProbe (SHRIMP)] and laser ablation inductively coupled plasma mass spectrometry (LA-ICP-MS). Of these, ID-TIMS is the most precise and accurate absolute dating technique but requires inclusion dissolution, chemical separation of U (Th) and Pb, and different instrumental protocols for each element. Conversely, SIMS/SHRIMP and LA-ICP-MS are *in situ* techniques requiring matrix-matched standards to be run in parallel with samples. The latter techniques are therefore less precise and accurate but do offer greater spatial resolution within individual sectioned inclusions.

Inclusion types. The U–Th–Pb dating technique is applicable to two main categories of inclusions. The more useful but least abundant type are silicate and oxide minerals that incorporate lithophile U and Th and exclude (initial or common) Pb. These are minerals such as zircon ($ZrSiO_4$), perovskite ($CaTiO_3$), breyite ($Ca_3Si_3O_9$), the Ca-$TiSiO_3$ phases, and rutile (TiO_2), in which U^{4+} and Th^{4+} substitute for Zr^{4+} and Ti^{4+}. The less useful but most abundant type are sulfide minerals that incorporate substantial chalcophile Pb but exclude (parent) U and Th. These are various forms of monosulfide solid solution and its exsolution products, principally pyrrhotite ($Fe_{1-x}S$), pentlandite ($[Fe,Ni]_9S_8$) and chalcopyrite ($CuFeS_2$) in which Pb^{2+} readily substitutes for Fe^{2+} and Cu^{2+}. Sulfide inclusions of both peridotitic (high Ni, Os) and eclogitic (low Ni, Os) parageneses have been recognized. However, a robust division of peridotitic sulfides into harzburgitic versus lherzolitic types remains elusive. Lead contents of sulfide inclusions from global localities measured by SHRIMP range from <1 ppm to 1,500 ppm (Eldridge et al. 1991; Rudnick et al. 1993). Eclogitic sulfides were found to have the lowest and peridotitic sulfides the highest Pb concentrations but with considerable overlap at the low abundance end (Eldridge et al. 1991; Rudnick et al. 1993). The latter observation is borne out by other Re–Os studies of sulfide inclusions. For example, sulfide inclusion Pb concentrations measured by ID-TIMS for three eclogitic diamonds from Cullinan (Richardson and Shirey 2008) and two diamonds, one eclogitic and one peridotitic, from Letseng (Shirey and Richardson 2011) were indistinguishable from total procedural blank (10 pg).

Practical applications. In the earliest work, Welke et al. (1974) investigated a single composite of 'black' sulfide inclusions from the ~1153 Ma old Premier kimberlite (Cullinan Mine, South Africa). The mixed diamond composite containing inclusions of unknown paragenesis was burnt to produce an inclusion residue for U–Pb analysis by ID-TIMS. Unfortunately, the analytical data were subject to large blank corrections in addition to uncertainties in model mantle growth curves but the data suggested an age for the sulfides of around 1.5 Ga. Subsequently, Kramers (1979) extended this approach to similar 'black' sulfide inclusion composites from Cullinan, Finsch and the four mines in Kimberley—Bultfontein, De Beers, Dutoitspan, Wesselton—sometimes collectively referred to as '*De Beers Pool*'. The data for Cullinan yielded a Pb–Pb model age closer to the 1153 Ma kimberlite emplacement age based on mantle Pb growth curves in favor at the time (Cumming and Richards 1975;

Stacey and Kramers 1975). In contrast, the Finsch and Kimberley data gave significantly older Pb–Pb model ages in excess of 2 Ga for diamonds from Cretaceous kimberlites. These results provided some of the first solid evidence that the diamonds were ancient and thus xenocrysts in the host kimberlite (Kramers 1979).

It took another two decades before the advent of *in situ* techniques enabled further work on the Pb–Pb model ages of sulfide inclusions in diamonds. In the early 1990s, the then novel SHRIMP at Australian National University (ANU) was used to investigate individual sulfide inclusions from a range of African and Siberian kimberlite localities (Eldridge et al. 1991; Rudnick et al. 1993; Hamilton et al. 1997). While the sulfide inclusions in these studies could be assigned to eclogitic and peridotitic parageneses based on Ni and Pb contents and in some cases coexisting minerals, the limitations of Pb–Pb model age interpretation became increasingly apparent. This problem was particularly clear for eclogitic sulfides with more radiogenic Pb isotopic compositions that gave anomalously young (relative to kimberlite emplacement) or future model ages. On the other hand, peridotitic sulfides varied in age between young and much older ages up to a maximum of 2.5 Ga.

Unfortunately, mantle Pb growth curves suffer from complex assumptions that limit their geochronological utility. Despite these limitations, the global set of Pb–Pb model ages for sulfide inclusions reported by Eldridge et al. (1991) and Rudnick et al. (1993) indicated that eclogitic diamonds are generally younger than peridotitic diamonds (for a more detailed review, see Pearson and Shirey 1999). Note, however, that there is considerable overlap between eclogitic and peridotitic sulfide-inclusion bearing diamond ages at the young end of the published Pb–Pb model age spectrum (Eldridge et al. 1991; Rudnick et al. 1993; Hamilton et al. 1997; Pearson and Shirey 1999) suggesting that *'young' (i.e., non-Archean)* peridotitic sulfide inclusion-bearing diamonds from Cretaceous kimberlites such as Koffiefontein and Finsch may be lherzolitic rather than harzburgitic (Shirey and Richardson 2011; Harvey et al. 2016). For example, the Finsch kimberlite carries diamondiferous lherzolite xenoliths in addition to (disaggregated) harzburgitic diamond host rocks (Shee et al. 1982).

In situ U–Pb dating using ion microprobe techniques (SIMS/SHRIMP) has been applied to silicate and oxide inclusion minerals such as zircon, perovskite and rutile with some success. However, the abundance of these accessory minerals as inclusions in diamonds is extremely low. So far only one zircon inclusion in a small diamond from the Mbuji Mayi kimberlite in the DRC has been successfully dated using the SHRIMP instrument (Kinny and Meyer 1994). The zircon yielded a $^{206}Pb/^{238}U$ age of 628 ± 12 Ma, based on the average of three spot analyses, as compared to a late Cretaceous kimberlite emplacement age of 70 ± 1 Ma (Schärer et al. 1997). Surprisingly, U, Th and Pb concentrations in the center of the crystal were found to vary by a factor of two (U = 360–660 ppm; Th = 250–560 ppm; Pb = 40–80 ppm) suggesting little internal redistribution of U and Th or loss of Pb (Kinny and Meyer 1994). However, the possibility that radiogenic Pb could have been partially lost by diffusion to the inclusion–diamond interface could not be entirely discounted and remains an issue to be addressed in future studies (Kinny and Meyer 1994; Pearson and Shirey 1999). Furthermore, the absence of any other inclusion phases in this diamond meant that little could be said about its paragenetic association and conditions of formation beyond speculating that the inclusion was eclogitic (Kinny and Meyer 1994) and the host diamond was most likely of continental lithospheric origin.

In a similar one-off study using the SHRIMP II instrument at the Geological Survey of Canada, a perovskite inclusion in a diamond from the Sytykanskaya kimberlite (Siberia) was investigated by Hamilton et al. (2003). Five spot analyses showed a twofold range of U (70–130 ppm) and Th (1,200–2,900 ppm) concentrations. Mean $^{206}Pb/^{238}U$ and $^{208}Pb/^{232}Th$ ages of around 350 ± 20 Ma were obtained, well within error of the 344 Ma host kimberlite age (Davis et al. 1980). The same small diamond was reported to carry olivine and chromite inclusions consistent with a peridotitic paragenesis but further work is needed to verify the syngenetic versus epigenetic nature of this perovskite inclusion (Hamilton et al. 2003). However, the perovskite-structured phase goldschmidtite ($KNbO_3$) has been reported as an unaltered inclusion in a diamond from Koffiefontein, South Africa (Meyer et al. 2019) indicating that such minerals can be syngenetic.

Within a broader study of some 20 sublithospheric diamonds from the Collier-4 kimberlite in the Juina district, Brazil, one CaTi-perovskite + breyite inclusion was dated using the Cameca IMS1270 ion microprobe at Edinburgh University (Bulanova et al. 2010). The inclusion yielded a relatively well-constrained $^{206}Pb/^{238}U$ age of 101 ± 7 Ma (Bulanova et al. 2010) within error of the kimberlite emplacement age of 93 ± 2 Ma (Heaman et al. 1998). This is the first absolute age for a sublithospheric diamond with an inferred oceanic crustal protolith (Bulanova et al. 2010) subducted beneath the Amazonian craton in Cretaceous time (Harte and Richardson 2012).

A suite of rare rutile inclusions in eclogitic diamonds from the Mir kimberlite, Siberia, were dated using the Cameca IMS1270 ion microprobe at the University of California, Los Angeles (Schmitt et al. 2019). Rutile inclusions with the highest U concentrations (up to 23 ppm) yielded a mean U–Pb concordia age of 375 ± 7 Ma for 19 spot analyses. This age is slightly older than U–Pb ages of 362–357 Ma for Mir zircon megacrysts (Davis et al. 1980; Spetsius et al. 2002) but within the published range (376–346 Ma) for regional kimberlite magmatism (Kjarsgaard et al. 2022, this volume). It also differs substantially from Re–Os isochron ages of 0.9 and 2.1 Ga for eclogitic and lherzolitic sulfide inclusion-bearing diamonds from Mir (Wiggers de Vries et al. 2013). Schmitt et al. (2019) offer four possible scenarios to explain this difference. While their preferred interpretation is that the rutile U–Pb age is a valid diamond crystallization age, they do allow the possibility of diffusive Pb loss whereby incompatible radiogenic Pb could accumulate at the inclusion-diamond interface (see Fig. 9C in Schmitt et al. 2019). If correct, the latter interpretation would be consistent with the recent identification of hydrous-silicic fluid films (which could harbor radiogenic Pb and other cations) around silicate and oxide mineral inclusions in gem diamonds from global sources (Nimis et al. 2016).

Sm–Nd and Rb–Sr

Dating inclusion-bearing diamonds using the combined Sm–Nd and Rb–Sr isotope systems first became feasible in the early 1980s. It was around this time that elucidation of the oceanic mantle array and bulk earth evolution curve (Allègre et al. 1979; DePaolo and Wasserburg 1979; Jacobsen and Wasserburg 1980) allowed for sensible interpretation of the correlated Nd and Sr isotopic signatures of kimberlite-borne mantle xenoliths and xenocrysts. Both Nd and Sr tend to reside in the same mantle minerals (garnet and clinopyroxene) and can be analyzed in the same samples using sequential ion exchange column chemical separation and mass spectrometric techniques. Furthermore, Sm, Nd, Rb and Sr are all variably incompatible elements and their geochemical behavior and relative partitioning between different silicate phases is well understood.

Principles. The Rb–Sr dating technique is based on β decay of ^{87}Rb to ^{87}Sr with a long half-life (50 b.y). Given strong fractionation of Rb from Sr during various igneous differentiation processes, Sr isotope ratios can increase quite rapidly in phases with high Rb/Sr:

$$\left(\frac{^{87}\mathrm{Sr}}{^{86}\mathrm{Sr}}\right)_\mathrm{p} = \left(\frac{^{87}\mathrm{Sr}}{^{86}\mathrm{Sr}}\right)_\mathrm{i} + \left(\frac{^{87}\mathrm{Rb}}{^{86}\mathrm{Sr}}\right)(e^{\lambda_{87}t} - 1)$$

where

$\lambda_{87} = 1.42 \times 10^{-11}$ yr^{-1} (conventional value; Steiger and Jäger 1977),
1.397×10^{-11} yr^{-1} (refined experimental value; Rotenberg et al. 2012),
and subscripts p and i refer to present and initial isotope ratios, respectively.

Rubidium is an alkali element, highly mobile in aqueous fluids, and has only two isotopes (^{85}Rb, ^{87}Rb), making it difficult to correct for instrumental mass fractionation. The high abundances of Rb that are typical of crustal environments relative to the mantle mean that great care must be taken to guard against secondary, laboratory-origin contaminants or natural Rb enhancement due to secondary alteration (e.g., via cracks in the diamond). Conversely, Sr is an alkaline earth element that is less mobile in aqueous fluids. Strontium has four isotopes (^{84}Sr, ^{86}Sr, ^{87}Sr, ^{88}Sr), providing separate reference isotope ratios for mass fractionation correction ($^{86}Sr/^{88}Sr$) and isotope dilution ($^{84}Sr/^{88}Sr$). The precision of Rb–Sr age determinations is thus ultimately constrained by how well intrinsic Rb concentrations can be measured.

The Sm–Nd dating technique is based on the α-decay of ^{147}Sm to ^{143}Nd, which has a very long half-life of 106 billion years. Combined with minimal fractionation of Sm from the geochemically similar Nd, neodymium isotope ratios tend to increase very slowly over time:

$$\left(\frac{^{143}\mathrm{Nd}}{^{144}\mathrm{Nd}}\right)_\mathrm{p} = \left(\frac{^{143}\mathrm{Nd}}{^{144}\mathrm{Nd}}\right)_\mathrm{i} + \left(\frac{^{147}\mathrm{Sm}}{^{144}\mathrm{Nd}}\right)(e^{\lambda_{147}t} - 1)$$

where
$\lambda_{147} = 6.54 \times 10^{-12}\ \mathrm{yr}^{-1}$
and subscripts p and i refer to present and initial isotope ratios, respectively.

Samarium and Nd are both rare earth elements (REE) that are much less mobile in aqueous fluids than Rb or Sr. Both Sm and Nd have seven isotopes that are more than enough to correct for isobaric interferences and provide separate reference isotope ratios for mass fractionation correction and isotope dilution. However, the precision of Sm–Nd age determinations remains limited by the small spread of parent/daughter ratio observed for REE that are only two atomic numbers apart. This problem is exacerbated by the longer half-life of ^{147}Sm relative to ^{87}Rb, ^{187}Re or ^{238}U.

The most precise and accurate instrumental technique for Rb–Sr and Sm–Nd isotope measurements is isotope dilution for accurate measurement of parent/daughter abundances by thermal ionization mass spectrometry (ID-TIMS) to obtain isotopic compositions. *In situ* techniques using these isotope systems are generally not favored for diamond dating studies, on account of the very small quantities of Sr and Nd present in silicate inclusions and the five- to six-figure precision required for meaningful measurements of Sr and Nd isotope ratios.

Given the very low Rb contents of mantle minerals such as garnet and clinopyroxene, $^{87}Rb/^{86}Sr$ ratios measured for these inclusion phases are insignificant and age corrections are negligible. Present-day measured $^{87}Sr/^{86}Sr$ ratios are effectively equivalent to initial $^{87}Sr/^{86}Sr$ ratios at the time of diamond encapsulation and independent Rb–Sr ages are therefore unobtainable. However, $^{87}Sr/^{86}Sr$ ratios can serve as a proxy for initial $^{143}Nd/^{144}Nd$ ratios to check whether inclusions from different diamonds dated using the Sm–Nd isochron approach are indeed cogenetic.

Inclusion types. The use of the Rb–Sr and Sm–Nd isotopic systems is generally applicable to two important silicate inclusion minerals, namely, garnet and clinopyroxene. In garnet, Sm^{3+} and Nd^{3+} replace Mg^{2+} and Fe^{2+} (and Ca^{2+}) on the X site via coupled substitution. At the same time, Sr^{2+} readily substitutes for Ca^{2+} whereas the much larger Rb^{+} cation (like K^{+}) is crystallographically excluded at the time of crystallization. In diopside and omphacite, Sm^{3+} and Nd^{3+} replace Mg^{2+} and Fe^{2+} on M1 via coupled substitution whereas Sr^{2+} substitutes for Ca^{2+} on M2 while a very small amount of Rb^{+} can be accommodated alongside K^{+} in place of Na^{+} on M2.

Subcalcic Cr-pyrope garnet is the most important carrier phase of both Nd (2–10 ppm) and Sr (2–25 ppm) in harzburgitic diamonds from type localities such as Finsch and Kimberley

(Richardson et al. 1984; Viljoen et al. 2014)), but Sr content extends to almost 50 ppm in similar garnets from xenoliths and diamonds from Siberia (Shimizu and Sobolev 1995; Pearson et al. 1995; Shimizu et al. 1997). In lherzolitic diamonds, calcium-saturated Cr-pyrope garnet and Cr-diopside are both carrier phases for Nd (1–2 ppm and 2–4 ppm, respectively) whereas diopside is the dominant carrier phase for Sr (≥ 100 ppm) as observed in some Cullinan diamonds (e.g., Richardson et al. 1993). In eclogitic diamonds, pyrope-almandine garnet and omphacite are significant carrier phases for both Nd (1–8 ppm and 1–20 ppm, respectively) and Sr (1–30 ppm and 50–1800 ppm, respectively) as observed in studies of diamonds from the Argyle, Cullinan, Finsch, Jwaneng, Orapa and Letlhakane localities (Richardson 1986; Richardson et al. 1990, 1999; Smith et al. 1991; Timmerman et al. 2017). The Nd and Sr contents of other common silicate (olivine, orthopyroxene), oxide (chromite) and sulfide (pyrrhotite) inclusion phases are negligible. Kyanite has also been measured (Mei Yan Lai and Suzette Timmerman, unpublished data) and contains negligible amounts of either element. In sublithospheric inclusions, Sr and Nd abundances are high enough to be analyzed, but the potential for geochronology is still being evaluated (Pearson et al. 2007; Thomson et al. 2016). Rare mineral inclusions of, e.g., breyite, rutile, and loparite could potentially contain high enough Sr and Nd contents for isotope analysis. A compilation of Rb–Sr and Sm–Nd contents for worldwide lithospheric diamond inclusions is given in Stachel et al. 2022a, this volume)

Practical applications. The use of Sm–Nd and Rb–Sr isotopic systems differs substantially according to the paragenesis of the diamond suite to be dated. In the case of harzburgitic diamonds, garnet is the only significant carrier phase for Sm, Nd and Sr while Rb is crystallographically excluded. Furthermore, such garnets show little spread in low (i.e., strongly sub-chondritic) Sm/Nd ratios for a given locality (Richardson et al. 1984; Shimizu and Richardson 1987). This means that Sm–Nd isochron ages are unobtainable for harzburgitic diamond suites from type localities like Kimberley and Finsch. However, such low Sm/Nd ratios lend themselves to the Sm–Nd model age approach by virtue of their shallow evolution curves relative to the reference reservoir, as seen on a standard Nd isotope evolution diagram. This is the approach adopted by Richardson et al. (1984) in obtaining Paleoarchean Nd model ages of 3.4–3.3 Ga (or 3.5–3.4 Ga for a depleted mantle source) for suites of several hundred subcalcic garnet inclusion-bearing diamonds from the Cretaceous Kimberley and Finsch kimberlites.

While the reliability of such Paleoarchean model ages for Finsch and Kimberley harzburgitic diamonds has frequently been questioned (Pidgeon 1989; Pearson et al. 1995; Shimizu and Sobolev 1995; Navon 1999; Pearson and Shirey 1999; Spetsius et al. 2002), these challenges have for the most part been shown to be unsustainable based on the following considerations. First, the use of composites of subcalcic garnet inclusions to obtain enough material for reasonably precise Nd and Sr isotope analysis (based on then available instrumental techniques) has been criticized as producing averages with no direct age significance. However, the extremely unradiogenic Nd isotopic character of these averages, as well as independent replication of results, precludes a wide distribution of individual inclusion values (Richardson et al. 1984; Caro et al. 2008), though this apparent uniformity is not present in every suite (Koornneef et al. 2017). Second, the highly radiogenic Sr isotope signature of unencapsulated subcalcic garnet macrocrysts (from disaggregated garnet harzburgite host rocks) contrasts with the markedly less radiogenic Sr in inclusions and strongly supports the long-term isolation of the inclusions by encapsulation in diamond and storage in the old-thick-cold continental lithospheric mantle beneath the Kaapvaal craton since the Paleoarchean (Fig. 5; Richardson et al. 1984; Boyd et al. 1985; Shirey and Richardson 2011).

In the case of lherzolitic diamonds, Sm and Nd (and Sr) are partitioned between calcium-saturated Cr-pyrope garnet and (modally rare) Cr-diopside with garnet acquiring the highest and diopside the lowest Sm/Nd ratios. Lherzolitic diamond suites are therefore suitable for obtaining Sm–Nd isochron ages while Sr isotopes provide a means for verifying whether inclusions from different diamonds are cogenetic. This is the approach adopted in the early 1990s, again using composites of similar inclusions, for dating lherzolitic diamonds from the Cullinan Mine (Richardson et al. 1993) and harzburgitic to lherzolitic diamonds from Udachnaya (Richardson and Harris 1997). In both studies, Sm–Nd isochron ages of approximately 2.0 Ga were obtained notwithstanding some variation in garnet initial $^{87}Sr/^{86}Sr$ (discussed more in Pearson and Shirey 1999) attributed to a contribution of radiogenic Sr from one or more extraneous grains in some composites. The ages thus obtained for lherzolitic diamond formation would be consistent with major refertilization of the continental lithospheric mantle via melt metasomatism associated (for example) with widespread Bushveld magmatism at that time (Richardson et al. 1993; Shirey et al. 2002; Richardson and Shirey 2008).

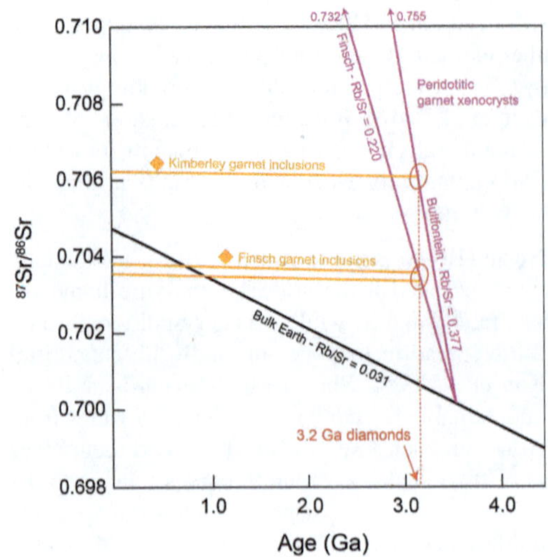

Figure 5. Strontium isotope evolution diagram for composite garnet inclusions from Kimberley and Finsch. Garnet inclusions in Finsch and Bultfontein (Kimberley) diamonds have $^{87}Sr/^{86}Sr$ ratios that intersect with the Sr isotope evolution curve for peridotitic garnet xenocrysts at around 3.2 Ga, here termed '*Sr model ages*'. These same composite garnet inclusions yielded 3.4–3.3 Ga Sm–Nd model ages, that indicate the age of the earliest Kaapvaal lithospheric mantle. Modified from Richardson et al. (1984).

In the case of eclogitic diamonds, Sm, Nd and Sr are invariably partitioned between pyrope–almandine garnet and omphacite, with other accessory phases such as kyanite, rutile and corundum being of little importance. Again, garnet exhibits the highest Sm/Nd ratios, and omphacite the lowest Sm/Nd ratios, whereas Sr isotopes provide the means for verifying whether inclusions from the same diamond or different diamonds are cogenetic. This approach was first applied in the 1980s to suites of eclogitic diamonds from Argyle, Cullinan, Orapa and Finsch at a time when composites of similar inclusions were still required to obtain enough analyte for TIMS analysis of small mineral inclusions (Richardson 1986; Richardson et al. 1990). Even so, the Sm–Nd isochron relationships obtained were sufficiently robust to indicate that eclogitic diamond formation only became prevalent in the Proterozoic. Further corroboration of these results was provided by work on a set of big individual inclusions from Finsch diamonds, which yielded Sm–Nd model ages of 2.4–1.4 Ga albeit with large errors (Smith et al. 1991). The last study of composite eclogitic inclusions was on diamonds from Jwaneng (Richardson et al. 1999), and this approach essentially came to an end with the advent of single sulfide inclusion Re–Os dating in the late 1990s (Pearson et al. 1998b, 1999a,b; Pearson and Shirey 1999).

Two decades later, technological advances in mass spectrometry, in particular the use of 10^{13} Ω resistors in ion current amplifiers (Koornneef et al. 2014), have rejuvenated the Sm–Nd isochron approach to dating eclogitic and peridotitic diamond formation. Using single garnet and clinopyroxene inclusions in Letlhakane and Orapa diamonds, Timmerman et al. (2017) have documented multiple eclogitic diamond growth events from the Proterozoic to the Cretaceous that can be related to regional tectono-magmatic events. The key to this work has been to group individual inclusions on separate Sm–Nd isochrons according to their trace element chemistry, Sr isotope ratio, and nitrogen aggregation state of the host diamond (Timmerman et al. 2017). The grouping of inclusions for Sm–Nd isochrons based on similar chemical characteristics was subsequently applied to single harzburgitic and lherzolitic garnet inclusions from Venetia diamonds. Two separate isochrons—2.95 and 1.15 Ga—were defined based on differences in model ages, CaO contents and REE patterns, and were linked to fluid-dominated metasomatism and basaltic melt-dominated metasomatism, respectively (Koornneef et al. 2017). The Archean isochron age array consisting of low Sm/Nd garnets was not detected in an earlier study of composite harzburgitic-lherzolitic garnets from Venetia (Richardson et al. 2009), though the presence of an older Archean component was inferred from the initial $^{143}Nd/^{144}Nd$ ratio. In addition, over a quarter of the individual inclusions showed higher Sm/Nd and $^{143}Nd/^{144}Nd$ ratios than the earlier composite study, highlighting the need to re-evaluate pooled isochron ages and analyze individual inclusions (Koornneef et al. 2017).

The single inclusion approach for silicate inclusions in diamonds is clearly the way of the future (Gress et al. 2021a,b,c). The most compelling results will ultimately come from the analysis of rare coexisting (but non-touching) garnet and clinopyroxene inclusions from the same growth zone within a diamond. In rare instances where sulfide inclusions exist in the same growth zone as either garnet or clinopyroxene, the Sm–Nd and Re–Os systems yield the same ages within analytical uncertainty. This provides confirmation that the isochron ages reflect the diamond growth event (Gress et al. 2021b).

Re–Os

With the advent of new analytical tools in the 1990s, in particular the development of high-precision negative thermal ionization mass-spectrometry (Creaser et al. 1991; Völkening et al. 1991), the Re–Os isotope system became a commonly used dating tool and isotopic tracer. Several reviews through the years discussed the application of the Re–Os isotope system in cosmochemistry and high-temperature geochemistry and the dating of mantle peridotites and eclogites (Shirey and Walker 1998; Pearson 1999; Carlson et al. 2008; Rudnick and Walker 2009; Aulbach et al. 2016; Becker and Dale 2016; Harvey et al. 2016; Luguet and Reisberg 2016; Luguet and Pearson 2019; Reisberg 2021).

Because Re is mildly incompatible during partial melting of peridotite—whereas Os is compatible—this system is particularly suitable for dating of melt depletion events leading to the formation of lithospheric mantle. This leads to large differences in the Re/Os of melt and residue, which then evolve to highly radiogenic and unradiogenic $^{187}Os/^{188}Os$, respectively (Shirey and Walker 1998). Thus, this system was quickly adopted to date mantle samples, such as mantle xenoliths (e.g., Walker et al. 1989; Pearson et al. 1995; Pearson 1999) and exhumed lithospheric mantle sections (e.g., Reisberg and Lorand 1995; Saal et al. 2001). The high partition coefficients for Re and Os into sulfide and the possibility of their strong fractionation made the Re–Os system a likely candidate for the dating of single sulfide inclusions in diamonds. The first attempt at dating sulfides on Koffiefontein diamonds (Pearson et al. 1998b) was immediately successful—producing Archean and Proterozoic ages that eventually proved to be major eclogitic diamond forming events in the Kaapvaal craton. This study also yielded one of the first convincing young peridotitic diamond ages, matching the mantle residence suggested for this diamond from nitrogen aggregation systematics. A very useful aspect of the Re–Os system is being able to relate the sulfide inclusions in diamonds to the sulfides in their eclogitic and peridotitic hosts, i.e., a paragenetic distinction, because of clear differences in the concentrations of Re and Os in these two lithologies (Pearson et al. 1998), and because of the difference in their major element compositions (e.g., Deines and Harris 1995) that can be analyzed from the same digestion that is used for Re–Os isotopes.

Principles. Use of the Re–Os isotope system as a geochronometer is based on the β-decay of ^{187}Re:

$$^{187}\text{Re} \rightarrow {}^{187}\text{Os} + e^- + 1 \text{ electron antineutrino}$$

As such, this chronometer is suitable for dating via the isochron approach, with the usual premise that the samples used in the regression formed, or isotopically re-equilibrated, at the same time, in the same source, and that the system remained closed afterwards.

$$\left(\frac{^{187}\text{Os}}{^{188}\text{Os}}\right)_p = \left(\frac{^{187}\text{Os}}{^{188}\text{Os}}\right)_i + \left(\frac{^{187}\text{Re}}{^{188}\text{Os}}\right)\left(e^{\lambda_{187}t} - 1\right)$$

where
$\lambda_{187} = 41.5 \times 10^{-9}$ yr^{-1}
and subscripts p and i refer to present and initial isotope ratios, respectively

System closure is often not met by mantle-derived samples such as peridotite xenoliths (e.g., Walker et al. 1989; Pearson et al. 1995), which have scattered Re–Os isotope systematics due to their open system behavior up to and even beyond eruption to Earth's surface. This is because subsequent to melt depletion, the continental lithospheric mantle remained open to interaction with fluids or melts carrying Re, making Re more susceptible to secondary overprinting than Os. There is also a lack of isotopic equilibration of isolated accessory phases controlling Os (see Luguet and Pearson 2019 for a detailed review). It is, however, this peculiarity of Os ± Re strongly partitioning into sulfides and alloys (Lorand and Luguet 2016) that has given rise to the application of Re–Os geochronology to individual sulfide inclusions in diamond. Although many diamond suites have been shown to be of Archean or Proterozoic age (see section '*Categorization of ages*'), and diamond is considered the ultimate chemically inert container, inclusions in diamond do not always yield statistically precise isochrons. This is because diamond is precipitated during secondary processes involving fluid or melt ingress, which requires homogenization of the Os reservoir in the pre-existing mantle and that of the incoming fluid or melt, a process that produces cogenetic diamonds with some scatter in the initial Os isotopic composition (see section '*Protogenetic versus syngenetic debate*').

As an alternative to *'linear data arrays'*, Re–Os model ages can be calculated, preferably for ancient diamonds that host sulfides with extremely high Re/Os (e.g., Richardson et al. 2001), although model ages by themselves can be subject to large uncertainties. This is because the model age approach requires the source composition (often some primitive or chondritic upper mantle composition) to be known rather well back in time. Projection to the mantle evolution curve can have large age uncertainty for young samples due to mantle heterogeneity (e.g., Luguet and Pearson 2019) and large age uncertainty for old samples—especially those with low Re/Os ratios—due to projection errors intersecting the mantle composition. The latter problem does not hold for sulfides with very high Re/Os (eclogitic), whose Os isotope evolution intersects mantle evolution curves at a high angle, producing minimal differences in calculated model ages. Two types of model ages—T_{MA} and T_{RD}—may be calculated (Shirey and Walker 1998; Walker et al. 1989) (Fig. 6). A T_{MA} age dates the time of separation from a reference reservoir such as an upper mantle with a primitive or chondritic (chon) composition using measured $^{187}Re/^{188}Os$ and $^{187}Os/^{188}Os$:

$$T_{MA} = 1/\lambda \times \ln\left[\left(\frac{^{187}\text{Os}/^{188}\text{Os}_{\text{chon}} - {}^{187}\text{Os}/^{188}\text{Os}_{\text{sample}}}{^{187}\text{Re}/^{188}\text{Os}_{\text{chon}} - {}^{187}\text{Re}/^{188}\text{Os}_{\text{sample}}}\right)\right] + 1$$

A T_{RD} age also dates the time of separation from a reference reservoir, but assumes that Re is perfectly incompatible and is completely extracted during partial melting (i.e., $^{187}Re/^{188}Os = 0$).

Using the measured $^{187}Os/^{188}Os$ yields:

$$T_{RD} = 1/\lambda \times \ln\left[\left(\frac{^{187}Os/^{188}Os_{chon} - {}^{187}Os/^{188}Os_{sample}}{^{187}Re/^{188}Os_{chon}}\right)\right] + 1$$

An inherent assumption in the application of T_{RD} ages is that Os in the sample is unradiogenic. The resultant age is a minimum age that deviates increasingly from T_{MA} the smaller the degree of melt extraction (and by definition, the smaller the resulting Re/Os fractionation), and with elapsed time. In mantle peridotite residues produced from very large degrees of melting, where Re/Os has been reduced to nearly zero, T_{RD} should equal the T_{MA} model age. In mantle xenoliths that have been overprinted by interaction with their host magma, typically involving Re addition, a third model age (T_{RD}, erupt) is calculated, by subtracting the amount of ^{187}Os added by the decay of ^{187}Re since emplacement of the host magma, and using this initial $^{187}Os/^{188}Os$ to calculate T_{RD} (Pearson et al. 1995; Fig. 6). This model age is not applicable to sulfide inclusions in diamond that have remained closed systems since encapsulation. While the underlying principles have not changed, potential pitfalls and best practices have been finessed through the years, along with further developments in sample preparation and analytical protocols. The reader is referred to Rudnick and Walker (2009), Meisel and Horan (2016), Luguet and Pearson (2019), Pearson et al. (2021) and Reisberg (2021) for recent updates.

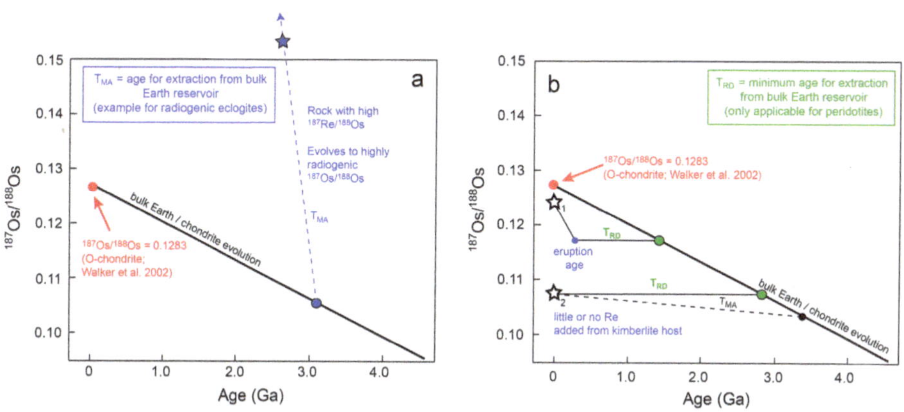

Figure 6. Graphical illustration of Re–Os model ages, that are calculated by projection back to a primitive mantle or bulk Earth (chondrite) evolution curve (Shirey and Walker 1998; Pearson and Shirey 1999). Both types of model ages give the time of extraction from a bulk Earth reservoir, and may not necessarily give the time of diamond formation. For eclogitic rocks and sulfides that have high Re/Os, T_{MA} ages are calculated. For peridotite rocks and sulfides, T_{RD} ages are more appropriate since the sample Re/Os isotopic ratio is significantly reduced by melt extraction. Because the mantle has become more isotopically heterogeneous with time, projection back to a bulk Earth evolution curve can have large age uncertainties for young samples.

Inclusion types. Rhenium and Os are highly siderophile elements, which, in the absence of metal phases, behave in a strongly chalcophile manner, such that they are concentrated in—and their behavior is controlled by—sulfide minerals (Mitchell and Keays 1981; Lorand 1989; Fleet et al. 1996; Lorand and Luguet 2016). Combined with the fact that sulfide minerals are the most common inclusions in diamond (Harris and Gurney 1979), the potential of the Re–Os isotope system as a diamond dating tool was recognized early on (Pearson et al. 1998b, 1999a,b; Pearson and Shirey 1999). Because melts have high Re/Os and low Os concentrations, whereas residues have low Re/Os and high Os concentrations, it follows that peridotitic diamonds contain sulfides with higher Os contents (median 42 ppm, $n = 52$) than sulfides in eclogitic diamonds (median 0.034 ppm, $n = 82$; Aulbach et al. 2016), which formed in a host rock that is the metamorphic equivalent of a mantle-derived melt, i.e., basalt. Sulfide-bearing eclogitic diamonds are more difficult to date for two reasons: (1) Os concentrations may be too low to obtain precisions sufficient for meaningful age calculation, depending on the size of the inclusion and total Os available for analysis, and (2) they invariably form in a high-Re/Os rock that has evolved to radiogenic $^{187}Os/^{188}Os$ by the time of eclogitization and diamond formation (Richardson et al. 2001; Aulbach et al. 2009c). This implies that meaningful model ages, requiring assumptions on the isotopic evolution (Re/Os) of the source, cannot be calculated with any confidence, and age constraints must rely on the isochron approach. An exception are sulfide grains with low common Os contents, where a large proportion of the measured Os is ^{187}Os from the *in situ* decay of ^{187}Re and from which absolute ages may be obtained (Richardson et al. 2001), as described in more detail below. The low amount of Os available for measurement in that study (20–120 fg) means that these are analytically highly challenging measurements to make. Additional challenges common to both peridotitic and eclogitic sulfide inclusions are discussed in section '*Sources of uncertainty*'.

Practical applications. The first constraints on diamond formation ages via Re–Os dating of sulfide inclusions were presented in Pearson et al. (1998b) who obtained an age for mineral inclusions in a single peridotitic Koffiefontein diamond that yielded a date similar to the kimberlite emplacement age. In this same study, five eclogitic inclusions yielded ages of ~1.05 ± 0.12 Ga and 2.9 Ga, depending on the interpretations of this small data set. This was followed by a steady succession of papers reporting Re–Os isotope systematics of sulfide inclusions in diamond that recorded ages spanning the range from close to craton formation to eruption of the host kimberlite (see worldwide data compilation: Smit et al. 2021: https://doi.org/10.7939/DVN/DRAJGT). These studies are reviewed in more detail in a later section. To date, Re–Os dating of whole sulfide inclusions liberated by breaking the host diamond is performed in only few laboratories world-wide because the small amount of analyte requires ultra-low blanks and therefore ultraclean reagents and laboratories, coupled to sensitive, state-of-the-art mass spectrometers that are dedicated to the analysis of small samples. Initiated at the Carnegie Institution for Science in Washington DC (Pearson et al. 1998b), the technique has since only been adopted at The University of Durham (UK) and the University of Alberta (Canada).

SOURCES OF UNCERTAINTY WHEN DETERMINING DIAMOND FORMATION AGES

While diamonds, as chemically inert containers, offer ideal conditions for closed-system evolution of their inclusions, it can be difficult to ascertain whether the inclusions formed at the same time in the same mantle source, or alternatively were isotopically re-equilibrated in response to the same event. That is, whether the inclusions are cogenetic with each other and with the diamond. Given the protogenetic origin of many inclusions (e.g., Nestola et al. 2014, 2019; Smit et al. 2016, 2019b; Aulbach et al. 2018), isotopic re-equilibration between the various inclusions comprising an isochron is typically required to obtain a meaningful age (see section *'Protogenetic versus syngenetic debate'* below). This can only be achieved, imperfectly, if the mantle volume is affected by events with high fluid–rock or melt–rock ratios (Aulbach et al. 2018). Because of the requirement of isotopic homogenization of a pre-existing reservoir, there is often significant scatter in the initial ratios defined by either mantle rocks or inclusions from diamonds, which inevitably deteriorates the quality of the isochron. Below we discuss some of the factors that may contribute to uncertainty in isochronous relationships, including partial isotopic re-equilibration, open system behavior, and analytical considerations such as over/under spiking.

Protogenetic versus syngenetic debate

A long-standing premise for diamond dating has been that the inclusions and the diamond co-crystallized from the same diamond-forming fluid. In other words, that there is a *syngenetic* relationship between diamond and inclusion, as opposed to the diamond incorporating a pre-existing (*protogenetic*) mineral grain. The assumption of a *syngenetic* relationship has traditionally been based on textures observed in diamond inclusions that show that the diamond's cubo-octahedral morphology has been imposed onto the inclusion (Harris 1968; Harris and Gurney 1979; Harris et al. 2022, this volume). Epitaxial crystallographic relationships between inclusions and diamond (e.g., Mitchell and Giardini 1953) can also be a key indicator of this *syngenetic* link. For example, most sulfide inclusions show cubo-octahedral morphologies (Harris et al. 2022, this volume) that are not expected for sulfide minerals, but rather are imposed by the host diamond, either onto (*partially?*) molten sulfides, or where sulfide compositions dictate that trapping maybe sub-solidus, by simultaneous growth with the diamond dominating the surface energy. These textures were traditionally interpreted as syngenetic, meaning that the sulfide and diamond precipitated from the same fluid/melt.

Later work showed that some inclusions could have a *syngenetic* interface with diamond even if crystallographic evidence suggests the inclusion was *protogenetic* (Agrosi et al. 2016). Similarly, Archean ages for sulfides enclosed in Paleozoic mantle zircon were interpreted as evidence that diamond sulfide inclusion-based ages are unrelated to diamond growth (Spetsius et al. 2002). Thus there is increasing evidence that mineral inclusions in diamonds can be pre-existing rather than *syngenetic*. Observations of compositional features of inclusions requiring a multi-stage evolution of their mantle source prior to their incorporation into diamond have led to concerns about the meaning of inclusion ages with respect to the diamond formation event (Nestola et al. 2017; Davies et al. 2018).

In the previous sections, we discussed the ample compositional evidence for pre-history of both silicate and sulfide inclusions prior to their incorporation in diamond. The faceted nature and small size—relative to the grain size of mantle xenoliths—of most inclusions in diamonds may belie an origin as dissolution-remnants of larger "protogenetic" crystals. However, this lack of syngeneity does not preclude a diamond formation age being obtained from inclusion-based dating studies. Because diamond formation is associated with fluids and melts, minerals from the surrounding rocks are either incorporated during partial melting, or fluid-enhanced diffusion takes place so that isotopes are re-equilibrated (Nestola et al. 2019; Pamato et al. 2021). Isotopic equilibration can also be achieved if the inclusion recrystallizes as the diamond is precipitated at high fluid– or melt–rock ratios during major diamond formation events (Aulbach et al. 2018). Such mineral inclusions are better referred to as *synchronous* inclusions (Nestola et al. 2017), and due to re-equilibration at the time of diamond formation will still yield geologically relevant diamond ages.

Implicit in this discussion, is the recognition that model ages obtained for a mineral inclusion in diamond are insufficient to define a diamond formation age. As discussed above (section '*Dating Principles*'), model ages provide the minimum age of the continental lithospheric mantle rather than the diamond age. In systems where a high degree of radiogenic common Os is likely to be inherited by the inclusion, even if growing synchronously with the diamond, model ages are potentially less reliable. Conversely, an isochron or age array can be interpreted to date a diamond formation event, so long as the assumptions behind the approach are broadly met. The robustness of sulfide Re–Os isochron ages has been recently demonstrated by age-equivalent Sm–Nd isochrons from the same diamonds (Gress et al. 2021b). The issues relating to dating diamonds using the isochron approach are similar to those posed by dating any metamorphic rock where new minerals may have grown perhaps by metasomatism of some components. A large-scale diamond-formation event with high fluid–rock ratios can explain why inclusions in diamonds are sometimes considered *cogenetic*—they retain similar initial radiogenic isotope compositions—even if they are sampled as spatially unrelated xenocrysts (e.g., Allègre 2008). In smaller-scale events with lower fluid–rock ratios, isotopic compositions are expected to show scatter due to the open-system environment in the mantle, with repeated metasomatic events causing elemental fractionation and isotopic overprint. The observation in most isochron diagrams is that inclusions tend to show some scatter around the linear regression, suggesting that inclusions are from sources with varying initial radiogenic isotope ratios that were imperfectly re-equilibrated and mixed during diamond formation, all leading to higher uncertainties in the determined age. Even so, this scattered age is likely to be a useful approximation of the age of the diamond-forming event.

Sources of uncertainty in silicate studies

Diffusion of elements from inclusion to inclusion-diamond interface. Diffusion is faster at higher temperatures. For this reason, diffusion of elements between mineral inclusions and the inclusion–diamond interface or fluid film may occur during mantle residence, possibly causing elemental fractionation and changed parent/daughter isotopic ratios over time. In this case, the mineral inclusions themselves will no longer represent 'closed' systems and may not yield diamond formation ages.

One example of diffusion is the loss of argon from clinopyroxene inclusions during mantle residence. Analyses of exposed or extracted clinopyroxene inclusions yield $^{40}Ar/^{39}Ar$ ages close to kimberlite eruption, with most ages between the times of kimberlite eruption and diamond formation (e.g., Burgess et al. 1994, 2004; Phillips et al. 2004, 2018; Phillips and Harris 2008, 2009). The favored explanation for this behavior was diffusion of pre-eruption radiogenic Ar ($^{40}Ar^*$) to the diamond/inclusion interface region during mantle residence, with partial to complete loss of this component when the diamond was cleaved to expose the inclusion (Burgess et al. 1992). However, step-heating experiments have shown that lower temperature steps yield older apparent ages than the fusion steps (Phillips et al. 2004; Phillips and Harris 2008). These steps indicate that the diffusional behavior is more complex. Partial accumulation of pre-eruption Ar in defect structures within the clinopyroxene lattice is suggested with concomitant or subsequent partial Ar diffusion to the diamond/inclusion interface. Fracturing around the inclusion in response to changing pressure and temperature during the eruption process (Bulanova et al. 2004; Burgess et al. 2004; Phillips et al. 2004; Phillips and Harris 2008) is likely the cause. Therefore, laser drilling to enclosed inclusions or furnace graphitization of the diamond plus inclusion is required to sample the radiogenic Ar residing at both the interface and within the inclusion. As there are technical challenges with both approaches, the $^{40}Ar/^{39}Ar$ method has not been widely used to determine diamond formation ages.

Lead loss is common in U-bearing minerals, due to the different chemical behavior of U compared to its daughter Pb isotopes, as well as alpha-recoil damage in the mineral. A recent U–Pb study on rutile inclusions from Mir eclogitic diamonds yielded ages consistent with kimberlite eruption (Schmitt et al. 2019) and are much younger than Re–Os isochron ages for sulfide inclusions from eclogitic diamonds of the same pipe (Wiggers de Vries et al. 2013). Pb loss from rutile into cracks, a Pb-rich phase on inclusion margins, or Pb loss into a fluid film surrounding inclusions (Nimis et al. 2016) cannot be excluded. Currently, the (trace element) composition of the fluid film is unknown and therefore it is not known whether there is a differential Pb isotopic composition. Thus, the rutile inclusions were suggested to record diamond formation shortly before kimberlite eruption (Schmitt et al. 2019). However, as U partitions into rutile 250× more strongly than Pb (Foley et al. 2000), if Pb is lost to a fluid film then U–Pb dating of rutile likely does not record diamond formation. Other studies of high U–Pb inclusion phases in diamond have given similarly young ages (Hamilton et al. 1997) similar to kimberlite emplacement, indicating that this may be a common phenomenon, at least for rutile and $CaTiO_3$ perovskite inclusions. In contrast to U–Pb, diffusion rates and partition coefficients of Sm and Nd are similar to each other, in the same order of magnitude, suggesting limited parent–daughter fractionation between inclusion and fluid film. Further *in situ* studies are required, but currently there is no evidence for significantly modified Nd isotopic compositions of silicate inclusions through loss of elements to a fluid film or into cracks.

Extraneous argon. *'Excess Ar'* is defined as ^{40}Ar, apart from atmospheric ^{40}Ar, that is incorporated into rocks and minerals by processes other than *in situ* radioactive decay of ^{40}K (McDougall and Harrison 1999). Excess Ar results in $^{40}Ar/^{36}Ar$ values above the atmospheric ratio of >298.56 (Lee et al. 2006), and may be identified from isochron plots. Inherited Ar is radiogenic Ar ($^{40}Ar^*$) that is included in rocks and minerals through contamination by older material or involves the retention of $^{40}Ar^*$ from a previous event (McDougall and Harrison 1999). Extraneous Ar is a term that includes both excess and inherited Ar.

If it is assumed that clinopyroxene inclusions extracted from their host diamond have lost all pre-eruption Ar, $^{40}Ar/^{39}Ar$ dating of the inclusions should yield the age of the host kimberlite (e.g., Phillips et al. 2004). If a clinopyroxene inclusion retains a proportion of the pre-eruption Ar, this Ar component is considered '*inherited*' Ar. In cases where $^{40}Ar/^{39}Ar$ ages are anomalously old compared to other dating methods or the age of the Earth, then the inclusions would be deemed to contain excess Ar. For example, Burgess et al. (2004) reported $^{40}Ar/^{39}Ar$ ages of >4.5 Ga for enclosed inclusions from an Orapa diamond and concluded that the inclusions (or diamond) must contain excess Ar.

Inherited compositional characteristics and isotopic dis/equilibrium. There are many compositional features observed in silicate inclusions that require them to have resided in the continental lithospheric mantle prior to diamond formation. For example, common sinusoidal rare earth element (REE) patterns in harzburgitic garnet inclusions suggest a complex history, and are caused by LREE metasomatism in a melt-depleted peridotite prior to diamond formation (Stachel et al. 1998; Taylor et al. 2003) or interaction of the diamond-forming fluid with melt-depleted peridotite (Koornneef et al. 2017).

Large variations are observed in Sr isotope ratios between silicate inclusions in monocrystalline diamonds (e.g., Richardson et al. 1990), but also even within single fibrous diamonds (0.709–0.717; Klein-BenDavid et al. 2010). In particular, radiogenic $^{87}Sr/^{86}Sr$ ratios in garnet inclusions in diamond (Richardson et al. 1984, 1993; Timmerman et al. 2017), fluid inclusions in diamond (Klein-BenDavid et al. 2010,2014), and garnets recovered from the kimberlite host (Richardson et al. 1984) are unsupported by their low $^{87}Rb/^{86}Sr$ ratios. Initial ratios are well above ratios of the bulk Earth at that time (Richardson et al. 1990). This implies a multi-stage isotopic evolution of the diamond-forming fluid source and complicates using Rb–Sr for age constraints, resulting in Rb–Sr model ages that are unlikely to have geological significance (Timmerman et al. 2017).

The isotopic composition of relatively high atomic number elements like Sr do not change during high-temperature geological processes such as immiscibility or fluid–mineral fractionation during diamond growth. Based on trace elements and Sr–Nd–Pb isotopic compositions, an Archean metasomatized lithospheric mantle with phlogopite (high Rb/Sr) or subducted sediment (Mazzone and Haggerty 1986; Viljoen et al. 1996) can be the source of highly radiogenic Sr (Klein-BenDavid et al. 2010, 2014). More recent element fractionation in the metasomatized ancient source or mixing with depleted fluid/s can lower Rb/Sr ratios, providing a large range in $^{87}Sr/^{86}Sr$ ratios and radiogenic Sr that is unsupported by the present-day low Rb/Sr ratios. Rb–Sr isotopic compositions need to be treated with caution before being applied to diamond dating. However, they can still be useful since similar initial Sr isotope ratios are a good indicator that the inclusions are cogenetic, or protogenetic inclusions whose isotopic compositions have been equilibrated during the diamond-forming event and hence can be used to determine grouping for Sm–Nd isochron regressions.

As Sm and Nd show similar chemical behaviour, the radiogenic ingrowth of ^{143}Nd is less variable due to smaller relative differences in Sm/Nd ratios (Richardson et al. 1984). Since diffusion of Sm and Nd in garnet is about 75,000× slower than Os diffusion in sulfide at mantle temperatures of 1100 °C (Tirone et al. 2005), variability in initial $^{143}Nd/^{144}Nd$ ratios is more likely to be preserved in different silicate inclusions during diamond formation.

Samarium diffusion modeling by Nestola et al. (2019) indicates that a 0.1 mm garnet grain will reach equilibrium within 200,000 years at 1150 °C (which is considered to be the average temperature for diamond formation within the continental lithospheric mantle; Stachel and Luth 2015). At the same temperature, ~1 mm garnets, would take around 20 million years to equilibrate (Fig. 7). These diffusion calculations use the slowest diffusion coefficients in the literature (Van Orman et al. 2002), and equilibration times are two orders of magnitude faster if diffusion coefficients from Tirone et al. (2005) are used. Regardless, these equilibration times are significantly less than the uncertainty on most Sm–Nd diamond ages. Under hydrous mantle conditions during diamond growth, garnet would have even faster diffusion rates (Katayama and Karato 2008). Based on (anhydrous) garnet diffusion modeling and assuming protogenesis, Nestola et al. (2019) recommended that silicate inclusions in diamonds that are >100 μm in size and formed at temperatures <1000 °C should not be used for age determinations, though more diffusion data, relevant to the diamond growth environment, is needed to evaluate these criteria (Koornneef et al. 2017).

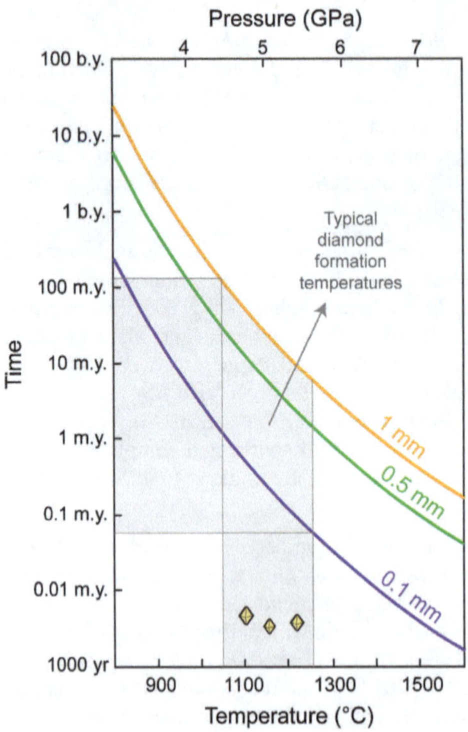

Figure 7. Samarium diffusion times for garnets of different sizes (0.1 mm, 0.5 mm, 1 mm), along a geotherm corresponding to 40 mW/m^2 surface heat flow. The **grey box** represents typical diamond formation conditions in the sub-continental lithospheric mantle ($T = 1150 \pm 100$ °C) (Stachel and Luth 2015). Equilibration times at these temperatures are roughly between 0.1 m.y and 100 m.y, which is within the typical uncertainties associated with diamond age determinations. For example, a 0.1 mm garnet grain will reach equilibrium within 200,000 years at 1150 °C, calculated using the slowest diffusion coefficients in the literature (Van Orman et al. 2002). Equilibration times are two orders of magnitude faster if diffusion coefficients from Tirone et al. (2005) are used. Diffusion coefficients have so far only been determined under anhydrous conditions, and diffusion may be even faster under hydrous conditions that may be more typical of diamond formation. Modified from Nestola et al. (2019).

Sources of uncertainty in sulfide studies

Pre-encapsulation history of sulfide inclusions. There are several eclogitic sulfide-derived isochrons with strongly radiogenic initial $^{187}Os/^{188}Os$, that exceed estimates for a co-eval mantle with chondritic composition (Walker et al. 2002) at the time of diamond formation. These supra-chondritic initial $^{187}Os/^{188}Os$ ratios require ingrowth of ^{187}Os in a high Re/Os environment prior to diamond formation and sulfide entrapment. Given the eclogitic nature of these inclusions, this evolution plausibly occurred in the oceanic crustal protolith, and radiogenic Os was subsequently inherited when sulfide and diamond grew from subduction-related fluids or melts. Alternatively, since MORBs are sulfide-saturated (e.g., Patten et al. 2013), the sulfides may have existed all through the evolution of the oceanic crust, and recrystallized upon metamorphism. Many eclogitic sulfides have Re/Os that is too low to account for their radiogenic $^{187}Os/^{188}Os$, suggesting that Re was lost during partial melting as the eclogite was invaded by diamond forming fluid (e.g., Westerlund et al. 2004; Smit et al. 2016).

Figure 8 illustrates how highly radiogenic $^{187}Os/^{188}Os$ recorded in eclogitic sulfide inclusions can be inherited from high Re/Os protoliths. Sulfide inclusions in the 650 Ma Zimmi diamonds, West African craton have highly radiogenic $^{187}Os/^{188}Os$ (1.5–2.2) and mass-independently fractionated sulfur isotopes that were generated in Archean protoliths subducted into the continental lithospheric mantle (Smit et al. 2016, 2019b). Neoproterozoic (~700–900 Ma) oceanic crust would require $^{187}Re/^{188}Os \geq 2000$ to evolve to these radiogenic $^{187}Os/^{188}Os$ within 100–200 million years, i.e., the assumed time between formation of the igneous precursors at an oceanic spreading ridge and their metamorphism during subsequent subduction. Various components of modern oceanic crust have highly variable $^{187}Re/^{188}Os$ ranging from 100s to 1000s (Peucker-Ehrenbrink et al. 2003). Eclogite xenoliths from the nearby Koidu kimberlite, which most likely formed in the Neoarchean (Barth et al. 2002; Aulbach et al. 2019), have $^{187}Re/^{188}Os$ from 2.4 to 159 (Barth et al. 2002). Such Archean oceanic crust could produce the $^{187}Os/^{188}Os$ measured in Zimmi sulfides in the two billion years after eclogite emplacement. Archean sulfur derived from associated subducted Archean sediments could explain the mass-independently fractionated sulfur isotopes. Fluid/melt fluxing of the mantle during Neoproterozoic subduction then facilitated partial melting, Re–Os isotopic re-homogenization, and Zimmi diamond formation.

In another example, eclogitic sulfide inclusions from both the 2.89 ± 0.06 Ga Kimberley and 1.89 ± 0.19 Ga Diavik diamonds (Richardson et al. 2001; Aulbach et al. 2009c) have radiogenic $^{187}Os/^{188}Os$ requiring Re/Os ~50–100. These Re/Os are lower than those typical of young oceanic crust (Re/Os up to ~1,000; Shirey and Walker 1998) but similar to oceanic crust which loses Re during the subduction, dehydration, and metamorphism (Becker 2000; Dale et al. 2007) that takes place within 10–200 million years after the oceanic crust is formed at a mid-ocean ridge (Fig. 8). This scenario can generate the initial $^{187}Os/^{188}Os$ ratios recorded in Kimberley and Diavik sulfide inclusions—0.156 ± 0.011 and 0.13 ± 0.10, respectively—and suggests eclogitic diamond formation shortly after or during subduction, where sufficient fluids are liberated to isotopically homogenize the Os isotopes in portions of the oceanic crust. Both these localities record coeval eclogitization, with eclogite xenoliths recording the same ages as craton amalgamation (Kaapvaal: Shirey et al. 2001; Menzies et al. 2003) and craton–margin collisional events (Slave: Schmidberger et al. 2007; Aulbach et al. 2009b).

Evidence for isotopic equilibration of pre-existing sulfide inclusions. Osmium diffuses readily in sulfides and has a low closure temperature in pyrrhotite (Brenan et al. 2000) which favors the establishment of isotopic equilibrium with the diamond-forming fluid. Several diamond localities clearly show that pre-existing sulfides were isotopically re-equilibrated at the time of diamond formation, including Klipspringer eclogitic sulfides, Victor lherzolitic sulfides, and Zimmi eclogitic sulfides (Westerlund et al. 2004 ; Smit et al. 2016, 2019; Aulbach et al. 2018).

Based on 14 sulfide inclusions, a 718 ± 49 Ma age array (MSWD = 6.1) was obtained for Victor lherzolitic diamonds, with initial $^{187}Os/^{188}Os$ of 0.1177 ± 0.0016 (Aulbach et al. 2018). This initial ratio is less radiogenic than the convecting mantle at 720 Ma, indicating that at least a portion of the Os was inherited from older sources in the continental lithospheric mantle and that these sulfide inclusions precipitated or re-precipitated along with the Victor diamonds. The observation that an age array was obtained despite contributions from isotopically heterogeneous sources in the continental lithospheric mantle—a region known to experience repeated and ongoing metasomatism—suggests the participation of sufficient amounts of melt during sulfide-bearing diamond formation to allow for near-re-homogenization of Os isotopic compositions from various sources (Aulbach et al. 2018).

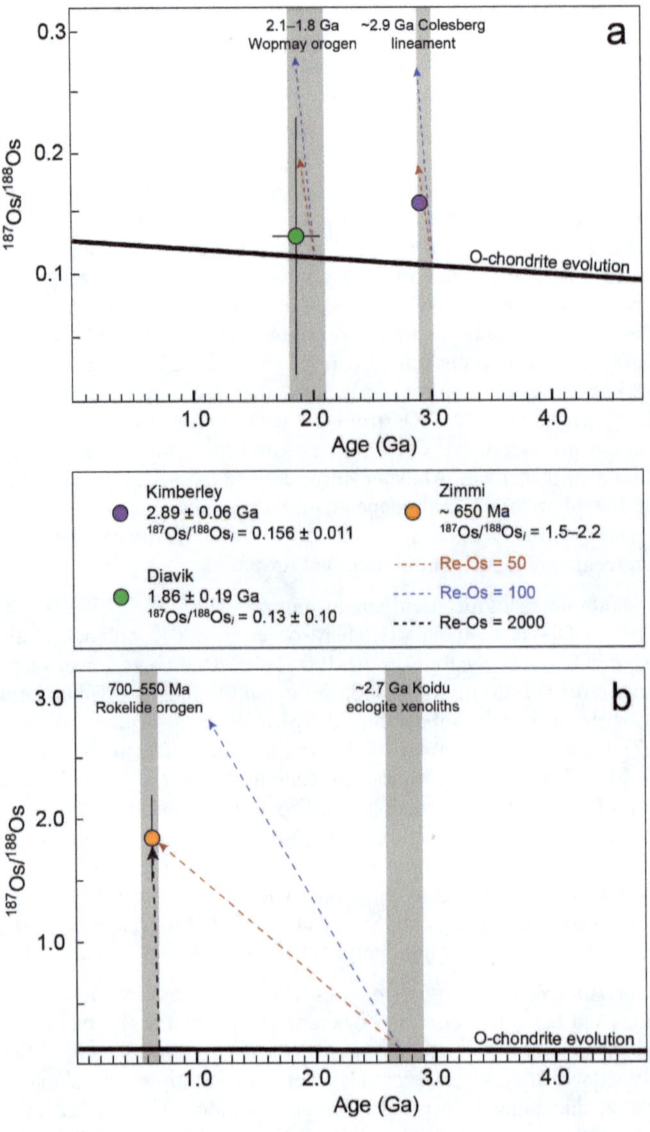

Figure 8. Os isotope evolution diagram for dated eclogitic diamonds from three localities: Kimberley, Kaapvaal craton (Richardson et al. 2001), Diavik, Slave craton (Aulbach et al. 2009c), Zimmi and Koidu, West African craton (Smit et al. 2016); illustrating ingrowth of radiogenic Os in (meta)basaltic protoliths with variably high Re/Os. For Kimberley and Diavik, subduction-related eclogitization of oceanic crust around 10–100 million years after its formation can explain the moderately radiogenic initial $^{187}Os/^{188}Os$ obtained from Re–Os isochrons. Eclogite xenoliths from Diavik have measured $^{187}Re/^{188}Os$ of 0.4 to 52 (Aulbach et al. 2009b), although the Re/Os of the igneous precursor prior to dehydration ± partial melting would have been higher (Becker 2000). At Zimmi, strongly radiogenic initial $^{187}Os/^{188}Os$ could be explained by remobilization of Os from an eclogite reservoir emplaced in Neoarchean time, as inferred for eclogite xenoliths from nearby Koidu, which have measured $^{187}Re/^{188}Os$ from 2.4 to 159 (Barth et al. 2002). Note trade-off between the age of the oceanic crust at the time of diamond formation and its Re/Os.

Similarly, five eclogitic sulfide inclusions from Klipspringer, South Africa, record a 2.55 ± 0.15 Ga isochron (MSWD = 0.87), with an elevated initial $^{187}Os/^{188}Os$ of 0.187 ± 0.046 (Westerlund et al. 2004). This initial Os isotope ratio requires pre-evolution in a high Re/Os environment, such as Archean eclogite in the continental lithospheric mantle. Ventersdorp magmatism at 2.7 Ga heated the lithosphere, causing partial melting of sulfide-bearing eclogite and homogenization of the Re–Os system, and resulting in isochronous relationships between these pre-existing sulfides (Westerlund and Gurney 2004; Westerlund et al. 2004).

Exsolution of monosulfide solid solution to a low-temperature assemblage. At mantle conditions, sulfide exists as monosulfide solid solution (*mss*), that is Ni-rich in peridotitic diamonds and Ni-poor grading into Ni-rich pyrrhotite in eclogitic diamonds. Upon low-temperature re-equilibration, for example during transport to Earth's surface, *mss* exsolves into a low-temperature assemblage, typically consisting of pentlandite ($(Fe,Ni)_9S_8$), pyrrhotite ($Fe_{(1-x)}S$) and chalcopyrite ($CuFeS_2$) in varying proportions (e.g., Lorand and Grégoire 2006; Fig. 9). In addition, molybdenite (MoS_2) has recently been described in the Raman spectra of 73 out of 80 Mir eclogitic sulfides (Kemppinen et al. 2018; Smith et al. 2022, this volume). Molybdenum though, is not an uncommon trace element in eclogitic sulfides. For example, Diavik eclogitic sulfide inclusions have Mo contents between 44 and 160 ppm (Aulbach et al. 2012). During exsolution, both molybdenite and chalcopyrite concentrate towards the exterior surfaces of sulfide inclusions, as observed in backscattered electron images and EDS maps of sulfides (Fig. 10).

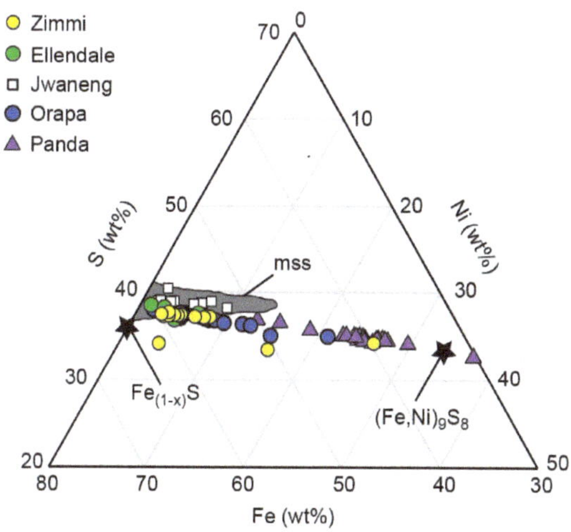

Figure 9. Compositions of sulfide inclusions in diamonds from different localities. Field for high temperature *mss*—monosulfide solid solution—from Ebel and Naldrett (1997). Pyrrhotite ($Fe_{(1-x)}S$) and pentlandite ($(Fe,Ni)_9S_8$) are low-temperature sulfides that are produced during exsolution. Data for Zimmi from Smit et al. (2016); Ellendale, Jwaneng and Orapa from Karen Smit (unpublished data) and Richardson et al. (2004); and Panda (Ekati Mine) from Westerlund et al. (2006).

It has been shown that during low-temperature re-equilibration, trace elements—including Re and Os—will partition differently between the different sulfide minerals, as observed in both orogenic peridotites and mantle xenoliths (van Acken et al. 2010; Holwell et al. 2011) and in sulfide inclusions in diamond (Richardson et al. 2001). As Re is preferentially partitioned into the combination of chalcopyrite + molybdenite relative to pyrrhotite + pentlandite, low-temperature re-equilibration has the potential to cause problems for Re–Os age dating if the whole inclusion is not recovered. This is a particular challenge when sulfides, due to their softness, break into multiple pieces during extraction from the diamond, as small fragments including Re-rich rims may be lost. Practitioners of the Re–Os sulfide method go to extraordinary lengths to ensure that they recover all the material liberated by cracking a diamond, because they have long been aware of these issues. This also means that great caution needs to be applied to Os isotope measurements made on all sulfides because of exsolution, especially by laser-ablation approaches where polishing of thin sections or grain mounts is a pre-requisite for analysis. Indeed, some scatter observed in Re–Os data sets may be related to incomplete inclusion recovery and may be dominated by variation in the Re/Os ratio.

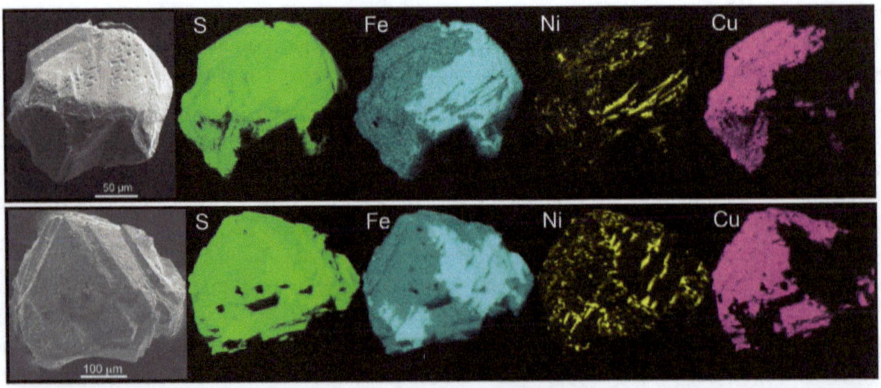

Figure 10. Energy-dispersive X-ray spectroscopy maps showing the spatial distribution of major elements in sulfide inclusions. During kimberlite eruption and cooling of monosulfide solid solution (*mss*), pentlandite- and chalcopyrite-rich sulfides are exsolved. Reproduced from Smit et al. (2016).

Increased scatter in Re–Os isotopic data naturally results in higher uncertainties on the determined age, but need not significantly affect the slope of the regression in a Re–Os isochron diagram. This is illustrated in the Re–Os isotope study of peridotitic sulfides from Ellendale (NW Australia) where one peridotitic sulfide falls off the main isochron formed by six other sulfides because it broke during extraction and was only partially recovered (Smit et al. 2010). Its higher $^{187}Re/^{188}Os$ is explained by sampling more Cu-rich portions of the sulfide, and incorporation of this sulfide in the isochron regression yields an identical 1426 Ma age, though the uncertainty on the age increases from 130 Ma to 390 Ma. In contrast, any effect on $^{187}Os/^{188}Os$ is typically minor because Re–Os fractionation during low-temperature re-equilibration occurs during eruption, and most kimberlites and lamproites are very young relative to the age of the diamond (Gurney et al. 2010). This is best demonstrated in sulfide DP9 from Kimberley (also known as '*De Beers Pool*'), where one fragment of this sulfide had double the Cu concentration and 30% higher $^{187}Re/^{188}Os$, while maintaining similar $^{187}Os/^{188}Os$ to the other fragments (Richardson et al. 2001). Regression of both fragments yields an isochron age equivalent to that of the kimberlite eruption.

Considering the potential for increased scatter in the Re–Os isotopic compositions of heterogeneous sulfides and the blocking temperatures well below the storage temperature in the lithospheric mantle, it is remarkable that the overwhelming majority of ages obtained on sulfide inclusions in diamonds appear to have geological significance. In rare cases where silicate and sulfide inclusions co-exist in the same diamond growth zone, similar ages have been obtained (Gress et al. 2021b), providing increased confidence in the robustness of inclusion-based diamond ages.

Homogenization of sulfide inclusions back to their high temperature *mss* mantle composition can be achieved by heating the diamond (and its inclusions) in a furnace (McDonald et al. 2017). By using this homogenization approach, Re–Os isotopes are more evenly distributed through the sulfide, and scatter resulting from incomplete recovery of the sulfide inclusion may be significantly reduced. Another benefit is that Re–Os isotope measurements can be done utilizing only part of the sulfide, so that the remainder of the inclusion can be used for other measurements. Sulfur isotope measurements by ion microprobe are however excluded, since there is currently no standard available for *mss*.

Low Os content. Due to the mildly incompatible nature of Re, and the compatible nature of Os during partial melting, eclogitic sulfides formed in a metabasaltic source typically have higher Re/Os than peridotitic sulfides due to both higher Re and lower Os contents (see section on '*Re–Os*' above). Because eclogitic sulfides typically have lower Os than peridotitic sulfides, larger eclogitic sulfide grains need to be dissolved in order to measure enough Os to obtain acceptable measurement statistics and reduce blank corrections. Eclogitic inclusions are often barely large enough to achieve these measurement goals. Thus, during mass spectrometry, the total amount of ^{187}Os available for ionization is extremely low, and when combined with the very low ^{188}Os result in high in-run uncertainties on $^{187}Os/^{188}Os$. In particular, this is true for tiny (< 7 μg) eclogitic inclusions. Data compiled for several eclogitic sulfide suites show that eclogitic sulfides that weigh less than 7 μg (less than ~130 μm diameter), yield excessively high uncertainties in $^{187}Os/^{188}Os$ (Fig. 11). Judging the size of sulfide inclusions can be difficult in some diamonds due to extensive "rosette" fractures that may surround the inclusion, leading to substantial over-estimates of inclusion size in some cases. The low total Os available for analysis in relatively small sulfide inclusions leads to more substantial blank corrections on both Os content and $^{187}Os/^{188}Os$ (Fig. 11). This propagates through to higher uncertainties in final measured values compared to sulfides with higher Os content (either larger eclogitic inclusions, or peridotitic sulfides with higher Os concentration).

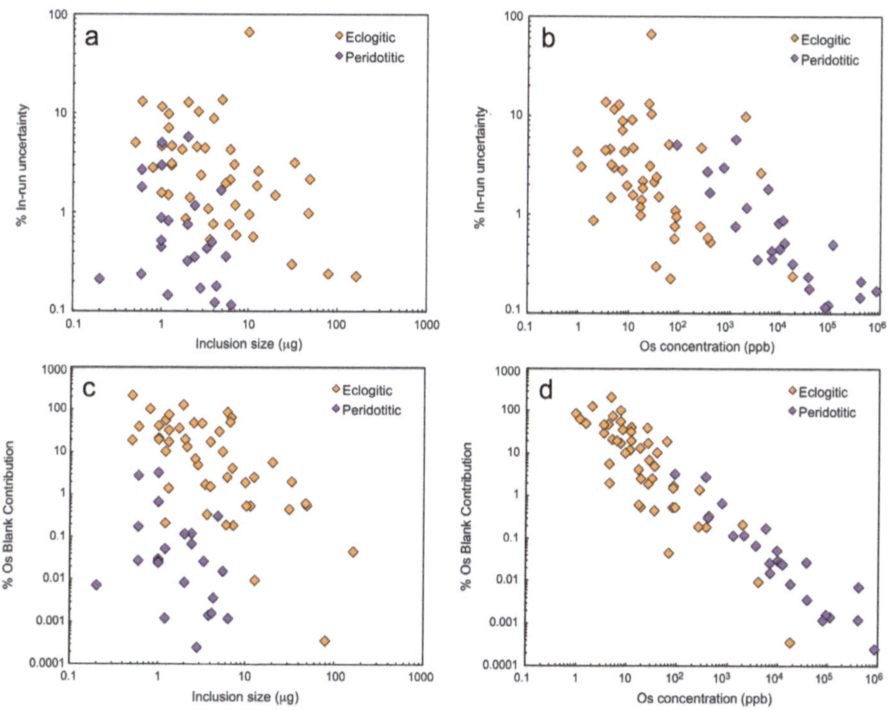

Figure 11. The influence of inclusion size and Os concentration on the **a,b)** in-run analytical uncertainty, and **c,d)** estimated blank contribution on the sulfide Os content. Smaller inclusions with lower Os concentration have the highest contribution of blank (*background*) Os. Smaller inclusions that give low Os ion yields during mass spectrometry have the highest in-run uncertainties. For small eclogitic sulfide inclusions from Sewa River and Victor, the in-run uncertainties and blank contributions were prohibitively high and no age determinations could be made. Peridotitic sulfides from Sewa River (Karen Smit, unpublished data), Victor (Aulbach et al. 2018) and Ellendale (Smit et al. 2010). Eclogitic sulfides from Orapa, Juina, Sewa River (Karen Smit, unpublished data), Zimmi (Smit et al. 2016), Victor (Aulbach et al. 2018) and Ellendale (Smit et al. 2010).

Sources of uncertainty in both silicate and sulfide studies

Under/Over spiking. While isotope dilution is generally regarded as the '*gold standard*' of elemental abundance determinations, a weakness it has is the need for prior information on likely Re, Os, Sm, Nd, Rb, Sr concentrations and elemental ratios in the inclusions. An informed estimate needs to be made in order to calculate the appropriate amount of isotopic spike to be added prior to dissolution. For Re and Os in sulfides, such an estimate is typically based on whether it is eclogitic (Ni-poor) or peridotitic (Ni-rich). Sulfide paragenesis is determined prior to dissolution by EDS analyses of the sulfide, to get an approximation of the bulk Ni content. If the inclusion breaks during extraction, it is better to get a qualitative composition from an interior part of the grain to avoid the chalcopyrite-rich portions along the exterior. For Sm, Nd and Rb, Sr in silicates, the identity and paragenesis of the mineral inclusion can usually be recognized based on colour (and/or EMP analysis on a flat portion of the unpolished grain) and concentrations estimated based on the range observed in corresponding xenolith and xenocryst minerals.

For sublithospheric diamonds, where paragenetic distinction is not as simple, estimating the Re and Os concentrations in the inclusion can be more complicated. For example, Re and Os concentrations for two metallic inclusions in a CLIPPIR diamond (Smith et al. 2017, 2021) could not be accurately determined for practical reasons. To facilitate handling of these 50–100 μm inclusions, which are highly susceptible to electrostatic/magnetic forces, Os was distilled directly from the inclusions while they were still embedded in the cleaved diamond surface. Their weights were thus difficult to estimate and their Os contents much higher than expected. The resulting sample solutions were critically under-spiked. The very high isotope dilution error magnification (EM) applicable to these sample-spike mixtures meant that measured isotope dilution ratios were within EM-propagated error of the normal (unspiked) isotope composition. Under such circumstances the ability to determine precise Os isotopic compositions is unaffected because all that is being measured is the normal Os in the sample (i.e., negligible spike Os correction). In addition, the chromic–sulfuric reagent used for sulfide dissolution preferentially attacks the pyrrhotite and its dissolution efficacy on Fe–Ni metal, cohenite and other accessory phases such as Cr–Fe-oxide is unknown. Since the exsolution of the homogeneous Fe–Ni–C–S phase on exhumation to pyrrhotite, cohenite and Fe–Ni metal will fractionate Re from Os differently in each phase, actual concentrations and Re/Os ratio would have been impossible to interpret. Again, this effect does not hinder the determination of accurate Os isotopic compositions (Smith et al. 2017, 2021) but serves to illustrate the complexities that can be encountered.

Both over- and under-spiking can introduce substantial propagated uncertainties in concentration and isotopic ratios measurements. If the isotope dilution ratio (spike isotope/reference isotope; e.g., $^{190}Os/^{188}Os$, $^{150}Nd/^{146}Nd$, $^{84}Sr/^{86}Sr$) of the sample–spike mixture (R_M) is too high, it is very difficult to determine the natural isotopic composition of the sample. Conversely, if a sample is underspiked, error magnification on the abundance estimate increases more rapidly, though at minimal effect on the radiogenic isotope ratio (Fig. 12).

Open system. Cracks that connect an inclusion to the diamond surface and permit exchange with fluids in the mantle and/or kimberlite, can compromise the goal of determining the diamond formation age. Although plastic deformation of diamond is more common at high temperatures in the mantle, brittle deformation is known to occur. Routine practice during age determinations is to exclude mineral inclusions with visible cracks to the diamond surface, however the presence of annealed cracks cannot always be ruled out despite cathodoluminescence investigation of the diamond surrounding the inclusion. Older cracks sealed with younger diamond growth have been documented from a few localities, and may certainly be more common (e.g., Taylor et al. 1995; Smith et al. 2014).

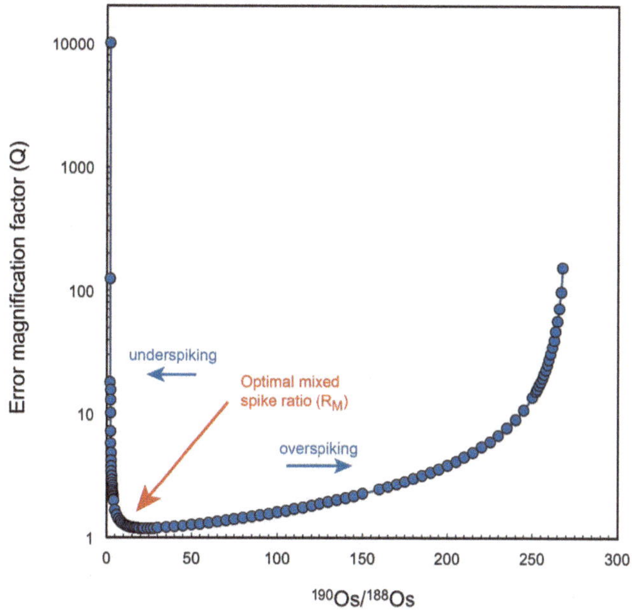

Figure 12. An informed estimate of Re, Os, Sm, Nd, Rb and Sr concentrations and elemental ratios in inclusions needs to be made in order to calculate the amount of isotopic spike to be added prior to dissolution. Both over- and under-spiking can introduce uncertainties in concentrations and isotopic ratios. For Re–Os isotope analyses, isotopic spikes typically contain ^{190}Os as the tracer isotope. If the ^{190}Os/^{188}Os in the sample-spike mixture (R_M) is too low, error magnification (Q) of the Os abundance estimate increases rapidly. If overspiking occurs (i.e., ^{190}Os/^{188}Os is too high) then it becomes difficult to determine the isotopic composition of the sample.

The only isotopic dataset that indicates connection to the diamond surface by obvious cracks is for six sulfide inclusions in Jagersfontein diamonds (Sonja Aulbach, unpublished data). Interestingly, the Re–Os isotopic compositions of these sulfides have a linear relationship and yield an imprecise age of 252.5 ± 41.9 Ma (Fig. 13). This age is significantly younger than the other diamond ages determined for Jagersfontein (1.7 Ga, 1.1 Ga; Aulbach et al. 2009a), but older than kimberlite eruption (86 Ma; Smith 1983). Thus, while inclusions in such diamonds can no longer inform us on diamond formation ages, they may still be interpretable. In this case, the age overlaps with those reported for strongly metasomatized peridotite xenoliths from the 84 Ma Bultfontein kimberlite (Giuliani et al. 2014), and may reflect isotopic resetting due to mantle metasomatism known to affect all continental lithospheric mantle sections prior to kimberlite eruption (Wass and Rogers 1980).

CLASSIFICATION OF AGES BY TYPE OF DIAMOND SUITE

Diamond ages can be classified into types by the number of stones involved, the relationship of inclusions to growth zones in the diamond, the mineralogy of the inclusion(s), and way the ages are determined (e.g., model age vs. isochron age). Such classification can provide a useful framework for interpreting the ages geologically—especially because diamonds are xenocrysts physically removed from the geologic context of their mantle hosts by the nature of kimberlitic sampling, except for rare diamondiferous xenoliths.

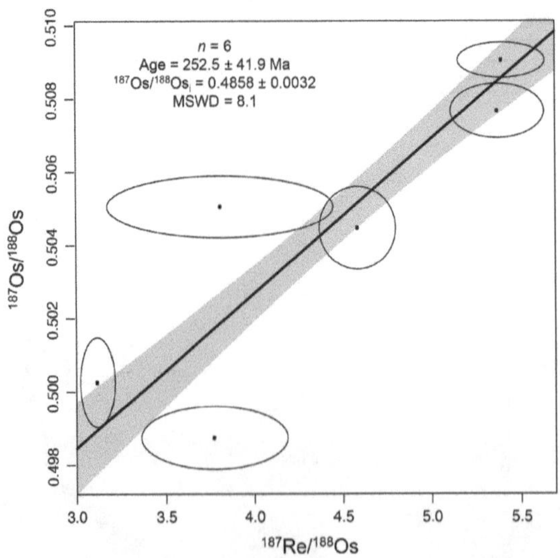

Figure 13. Jagersfontein isochron obtained from sulfides with visible cracks to the surface of the diamond. Six inclusions yield an imprecise date of 252 ± 42 Ma (Sonja Aulbach, unpublished data), which is younger than other diamond ages determined for Jagersfontein and significantly older than kimberlite eruption (86 Ma; Smith 1983). See text for further details.

Single diamond ages

Although it would be ideal for each diamond dating study to have many diamonds with inclusions, the researcher is often confronted with few or maybe even only one propitious stone because diamonds with big enough inclusions for analysis are quite rare. Surprisingly, useful ages from sulfides using the Re–Os system can be obtained from just one diamond in four different ways. These ages are possible with the Re–Os system because of the huge range in the abundance of parent ^{187}Re and daughter ^{187}Os, whereas they are not always practical with the Sm–Nd or Rb–Sr systems because of the very limited range of parent to daughter abundance in these systems.

Model ages. Model ages (see '*Principles*' section above for a general outline) are calculated from the measured isotopic composition of a sulfide inclusion and the Os isotopic evolution of some reference reservoir (typically the Earth's mantle, but also chondritic reservoirs). This calculation is essentially a correction for radiogenic growth of daughter ^{187}Os using the time-averaged ^{187}Re/^{187}Os ratio from the point at which the inclusion had the same isotopic composition as the reference reservoir. Uncertainty comes from a combination of factors: a) the uncertainty in the isotopic growth of the reference reservoir, b) analytical uncertainty in the Re/Os of the inclusion, and c) the assumption that the inclusion measured experienced a single-stage evolution on separating from the reference reservoir. While the insights that can be gained from just one determination can be quite remarkable (e.g., the ability to determine whether a sulfide originated in Mesoarchean lithosphere versus Paleoproterozoic lithosphere) the precision and accuracy are often only known to within 200–300 million years, and, thus, such model ages are not suited to unraveling fine-scale age differences. While model ages provide invaluable constraints on the stabilization age and evolution of diamond source rocks, one model age from a diamond inclusion does not necessarily date a diamond formation event.

Peridotitic sulfides nearly always have very high Os content and very low Re content. The low resultant ^{187}Os/^{188}Os of peridotitic sulfides often gives *minimum ages* for the depletion or '*freezing-in*' of an initial Os isotopic composition. A good example is the one sulfide inclusion in a diamond analyzed from Murowa (Zimbabwe craton). This sulfide produced a T_{RD} of 3.25 Ga, within uncertainty of the oldest model age on a mantle xenolith from the same locality (T_{RD} of 3.31 Ga). Although the model age obtained from the diamond may not necessarily reflect the time of diamond formation, these ages confirm the existence of protoliths older than the Mesoarchean in the Zimbabwean continental lithospheric mantle (Smith et al. 2009; Pearson et al. 2018), which had long been suspected due to Paleoarchean granite-greenstones, detrital zircons, and chromite deposits in the Zimbabwe craton (Dodson et al. 1988; Wilson 1990; Nägler et al. 1997; Zeh et al. 2014)

Peridotitic sulfides with very low Re/Os are often more reflective of the inclusion protolith, whereas isochrons (and model ages that are in agreement with isochrons) most likely reflect diamond formation (e.g., Pearson et al. 1999a,b). This could be the case for 3.5–3.1 Ga model ages from Udachnaya peridotitic sulfide inclusions that likely reflect the minimum age of the Siberian continental lithosphere rather than the age of diamond formation. Other age determinations for Udachnaya peridotitic diamonds give 2 Ga isochron ages (e.g., Richardson and Harris 1997), similar to the ages for eclogitic diamonds (Wiggers de Vries et al. 2013), granulite xenoliths (Koreshkova et al. 2008) and peridotite xenoliths (Doucet et al. 2015; Ionov et al. 2015).

Single eclogitic sulfides can produce model ages by extrapolation from their extremely high $^{187}Os/^{188}Os$, back to a bulk Earth or mantle growth curve. For eclogitic sulfides, the accuracy of the model age depends on the closed system behavior of Re and Os and thus the fidelity of the Re/Os ratio since entrapment in the diamond, as well as how much radiogenic Os has accumulated. Thus, while a model age can, in principle, be obtained on a single grain, prudence dictates use of multiple eclogitic sulfides on an age array (see below) to provide some indication that the Re/Os has not varied and to obtain more accuracy on the projection.

Model ages for the Sm–Nd system in silicate minerals also have the potential to reflect diamond growth if the Sm/Nd fractionation is substantial enough. Occasionally, single diamond model ages have successfully been obtained using the Sm–Nd system for sufficiently large silicate inclusions. For example, Smith et al. (1991) obtained a Nd model age of 2.11 ± 0.12 Ga for a websteritic clinopyroxene inclusion in a Kimberley diamond based on this approach. Model ages are less definitive where Sm/Nd fractionation is subtle relative to the reference reservoir, for example, the Cullinan lherzolitic garnets (Richardson et al. 1993).

Radiogenic daughter (absolute) ages. Occasionally, sulfides are uncovered that contain only a radiogenically-derived isotope of Os—^{187}Os. These minerals contain no '*common*' or unradiogenic Os, which indicates that at the time of incorporation into the diamond the sulfide contained only Re and an unmeasurably small amount of Os. In a study of sulfides from the De Beers Pool diamonds in Kimberley (South Africa), Richardson et al. (2001) discovered five sulfide inclusions with these characteristics. These sulfides were so overspiked with the typical ^{190}Os spike that error magnification leads to large uncertainties in the exceptionally low Os concentrations. In this situation, $^{187}Re/^{188}Os$ values become extraordinarily high ($\gg 10{,}000$) and fail to plot on conventional isochron diagrams with normal samples (cf. Richardson et al. 2001). Sulfides like these appear to be rare but they present an interesting case study in dating methodology.

Core–rim ages. A common question that arises is how long it takes for a diamond to crystallize. When a diamond has grown with the fortuitous entrapment of several inclusions in different locations relative to the core and rim of the diamond, the minimum time for a diamond's growth can be estimated from the core–rim ages. The most detailed examples of this approach are on diamonds from the Siberian and Kalahari cratons. Wiggers de Vries et al. (2013) documented two main episodes of diamond crystallization (~2.1–1.8 Ga and ~1.0–0.9 Ga) in the Siberian craton (23rd Party Congress, Mir, and Udachnaya kimberlites). One diamond, sample 4239, had sulfides defining an isochron at the older, Paleoproterozoic age in the core as well as sulfides defining the younger Mesoproterozoic isochron age in the rim. These age differences confirm that this diamond took around 1 billion years to form, probably in two distinct growth pulses of much shorter duration. Similarly, multiple episodes of diamond formation are present at Letlhakane and Orapa, with two diamonds containing inclusions in different growth zones showing age differences up to 2 b.y. (Timmerman et al. 2017). Sample LK75 from Letlhakane had a clinopyroxene in an intermediate growth zone that falls along a 2.3 Ga isochron and a garnet in the rim that falls along a 0.25 Ga isochron. Sample OR02 from Orapa had a garnet in an inner zone that falls along a 1.1 Ga isochron, and a garnet in the rim that is within error of the kimberlite eruption age.

These results demonstrate that these diamonds are aggregated products of at least two episodes of diamond growth. Unfortunately, these ages do not reveal the crystallographic growth rate of the diamond over a billion years. Such composite diamond crystals seem to be the norm—as can be seen from CL textural studies of many diamonds from kimberlites worldwide, where the most striking examples are monocrystalline diamonds with younger fibrous overgrowths (e.g., Petts et al. 2016; see also Jacob and Mikhail 2022, this volume; Weiss et al. 2022, this volume). Even if one were to obtain a perfect diamond where the intricate growth zones contained inclusions that were spaced at just the right interval to determine a zone-to-zone age, it is likely that the uncertainties for each zone would overlap enough to preclude the determination of a growth rate.

Mineral isochrons. The gold standard for dating a diamond is the 'mineral isochron' approach, explained in more detail in the '*Principles*' section above. The value of mineral isochrons from single diamonds comes from the proximity of the mineral grains, which enhances the likelihood that they were in isotopic equilibrium at the time of closure of the isotopic system. In diamonds containing inclusions that define mineral isochrons, the date obtained is the time of system closure and the age of formation of the diamond because it records the blocking of interdiffusion between the mineral grains by the diamond.

Yellow, Type Ib alluvial diamonds from Zimmi (Sierra Leone) provide one of the clearest mineral isochron examples (Fig. 14). Smit et al. (2016) analyzed three different diamonds each with 3–4 separate sulfide inclusions, and each diamond yielded a 650 Ma age regression from its own '*linear data array*' whose individual uncertainties overlapped. Each diamond has a high but not identical initial $^{187}Os/^{188}Os$ (between 1.5 and 2.2) so it is not sensible to include all the data in one regression. Even with these differences in the initial $^{187}Os/^{188}Os$, this high or so-called 'radiogenic' $^{187}Os/^{188}Os$ requires long-term evolution of the sulfides in a high Re/Os eclogitic host and confirms that the Zimmi diamonds all formed during one diamond-forming episode in the continental lithospheric mantle.

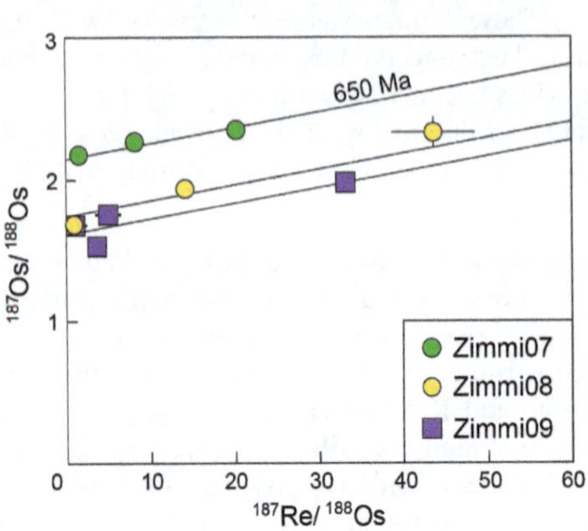

Figure 14. Three Zimmi diamonds—each with three to four sulfide inclusions—yielding identical 650 Ma ages. These data have too much scatter to meet the requirements for true isochrons, but are termed '*linear data arrays*'. Scatter is likely due to partial isotopic re-equilibration of pre-existing sulfides at the time of encapsulation in the diamond. Where not visible, uncertainties are smaller than the size of the symbol. Modified from Smit et al. (2016).

In peridotitic diamonds from the Ekati mine in the Slave craton, Westerlund et al. (2006) obtained sulfide Re–Os isochrons on five separate diamonds with 2–3 inclusions per diamond—all of which can be combined to give an aggregate 3523 ± 170 Ma isochron with 11 inclusions (Fig. 15). Because the sulfide isochrons can be regressed together with the inclusions from other diamonds, this multi-phase isochron firmly establishes the age of the oldest diamonds ever dated. Equally important, the elevated initial $^{187}Os/^{188}Os$ of the isochron is based on inclusions that crystallized in the outer growth zones of the diamonds—not their cores. This textural evidence supports the interpretation that the last stage of diamond growth was from fluids that had incorporated some radiogenic crustal Os derived from the recycling of oceanic lithosphere. The local vs global scale implications of this recycling are a source of debate.

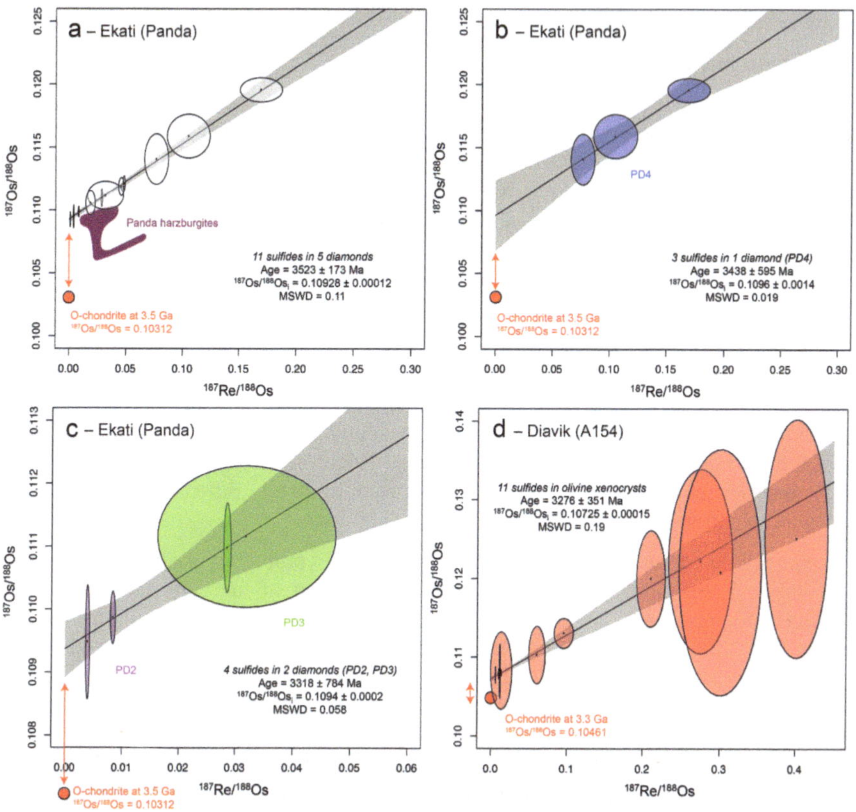

Figure 15. Re–Os isochron plots from sulfide inclusions in diamonds and harzburgites from the Ekati mine (Westerlund et al. 2006) and olivine xenocrysts from the Diavik mine (Aulbach et al. 2004). The Ekati and Diavik mines are both in the Lac de Gras area of the Slave craton and appear to contain diamonds of similar age range hosted in Paleoarchean lithospheric mantle. **a)** Eleven sulfides from five diamonds chosen for their co-linearity on the isochron plot. Three of these diamonds each have multiple inclusions which support the eleven-inclusion array as indicating that these diamonds formed at the same time from a fluid that equilibrated isotopically at the time of diamond crystallization. Note that this fluid has an initial $^{187}Os/^{188}Os$ that is significantly enriched relative to a depleted mantle that had evolved with a chondritic $^{187}Re/^{188}Os$. **b, c)** Detail of the three diamonds with sulfide mineral isochrons are shown in these two Re–Os isochron panels. **d)** Sulfides in individual olivine grains from Diavik have a similar age to the diamonds from Ekati, and a similar $^{187}Os/^{188}Os$ to the more scattered harzburgite xenoliths from Ekati (Westerlund et al. 2006). This corroborates the age of the lithospheric mantle in the Lac de Gras region and the enriched initial $^{187}Os/^{188}Os$ of the diamond-forming fluid that percolated through the lithospheric mantle beneath Ekati. These datasets show the importance of being able to obtain mineral isochrons or linear data arrays to bolster the accuracy of initial isotopic compositions.

Occasionally during sample preparation, a sulfide inclusion will break into several pieces leading to a serendipitous variant of the mineral isochron. Richardson et al. (2001) showed that two pieces from a single sulfide plotted on a single mineral isochron giving an age within uncertainty of the 85 Ma kimberlite eruption age. The weighted average of the two grains plotted on the 2.89 ± 0.26 Ga aggregate isochron for the single sulfides from 11 individual diamonds from Kimberley. This kind of mineral isochron is strong evidence that the inclusions remain totally isolated from alteration (and potential Re addition) during kimberlite eruption and throughout the 85 million years since. The exsolution of a monosulfide to an assemblage of sulfide minerals, each of which partitions the parent and daughter isotopes differently, will produce data that fall off isochrons unless the entire assemblage is extracted and analyzed as discussed above. Striving to recover as much of the sulfide inclusion as possible is critical to obtaining accurate ages.

Multiple diamond ages

When the supply of diamonds with inclusions is plentiful, a good test of a meaningful age is the adherence of the data to linearity on an isochron plot. This test is widely used in geochronology because it signifies that the individual minerals in igneous rocks or different rocks in an igneous complex were in isotopic equilibrium at the time of closure of the isotopic system. In other words, the diffusion length scale through the melt for equilibration before closure was much larger than the distance between the individual minerals or rocks. When these conditions are met or at least can be evaluated rigorously by the fit to linearity, a true '*isochron*' is obtained. In the case of inclusions, system closure is obtained by encapsulation within diamonds.

For suites of monocrystalline diamonds, the very low occurrence of sulfide and silicate inclusions within run-of-mine production implies that the majority of diamonds were growing apart in the mantle such that the inclusions might not have been in isotopic equilibrium, except through contact with diamond-forming fluids. Thus, data on an isochron diagram are unlikely to meet the true geochronological definition of an isochron or pass tests of rigorous fit to linearity. Instead, the term '*linear data array*' is used, to clearly differentiate data sets with more scatter than true isochrons.

Composite isochron (linear data array) ages. The composite isochron age approach using combined Sm–Nd and Rb–Sr systems was adopted in the early 1990s for dating lherzolitic diamonds from Cullinan (Richardson et al. 1993) and harzburgitic-to-lherzolitic diamonds from Udachnaya (Richardson and Harris 1997). At that time, composites of tens of mostly small inclusions (average weight = 20 μg) were required to obtain enough material for sufficiently precise Nd and Sr isotope analysis by ID-TIMS. In principle, compositing or grouping of cogenetic inclusions with a range in major and trace element concentrations can reduce the spread in Sm/Nd ratio but will not affect the overall isochron relationship, so long as the inclusions are cogenetic and had the same initial isotope ratio. In practice, if non-cogenetic inclusions are composited, important information about multiple diamond-forming events at a single locality may be lost.

The most important feature of lherzolitic diamond suites (i.e., containing calcic Cr-pyrope and Cr-diopside) for dating purposes is the observed spread in Sm/Nd ratio of garnet inclusions which is positively correlated with calcium content and differs from pyroxene; Cr-pyrope acquires the highest and Cr-diopside the lowest Sm/Nd ratios. In the Cullinan study (Richardson et al. 1993), individual inclusions were grouped in different categories based on Ca, Cr, Ti and Na contents measured by electron microprobe on tiny fragments of each inclusion (Richardson et al. 1993), a more precise refinement of the previous selection method by optical properties. The best-defined lherzolitic garnet category had significant Ti and Na contents (up to 0.9% TiO_2, 0.2% Na_2O) and a distinctive reddish tinge. In rare bimineralic inclusion-bearing diamonds, this variety of garnet can be seen to coexist with bright emerald-green diopside with correspondingly higher Ti, Al, Cr and Na contents than pale emerald-green diopside that can be paired with more calcic, pale-purple lherzolitic garnets. The best-defined lherzolitic garnet–diopside pair of composites gave a two-point Sm–Nd isochron age of 1.93 ± 0.04 Ga. The other three lherzolitic garnet composites with intermediate Sm–Nd ratios also lie on this isochron within larger analytical errors. However, the garnet and clinopyroxene composites show a significantly large range in initial $^{87}Sr/^{86}Sr$ ratio (0.7037–0.7058). In particular, the garnet composites with lower Sr contents extend to the higher end of this range, suggesting the presence of one or more grains with radiogenic Sr.

This compositional variability is not unexpected given the formation of lherzolitic diamonds in pre-existing continental lithospheric mantle via melt metasomatism of refractory garnet harzburgite laced with radiogenic Sr (Richardson et al. 1984), resulting in small-scale variability in initial isotope ratio for inclusions from different diamonds. The more conservative five-point Sm–Nd isochron age of 1.94 ± 0.10 Ga (Richardson et al. 1993) lies close to a 2.054 Ga U–Pb zircon age for the Merensky Reef of the Bushveld Complex (Scoates and Friedman 2008), suggesting that the Archean harzburgitic continental lithosphere was modified by Bushveld-type magmas (Richardson et al. 1993; Shirey et al. 2002; Richardson and Shirey 2008). Further Sm–Nd and Rb–Sr work on individual inclusions from global lherzolitic diamond localities is clearly warranted.

Age arrays from single inclusions in multiple diamonds. It is extremely rare to find multiple large inclusions within a single growth zone in one diamond. For this reason, it is the more usual case to analyze individual inclusions from different diamonds to obtain age constraints. The main requirements for a true isochron in all radiogenic isotope systems (see above), are that the inclusions are cogenetic, that they had identical initial radiogenic isotope ratios, and that the system remained closed and has not been disturbed since isotopic closure.

However, for inclusions in diamonds these conditions are often not precisely met. Observed variability is often greater than can be explained from experimental (analytical) uncertainty alone, especially due to the high analytical precision and sensitivity of modern mass spectrometry compared to older studies. This is usually attributed to the '*geological uncertainty*' that encompasses the natural variability of non-touching mineral grains that actually could have been tens of meters apart in the mantle. The uncertainties associated with these assumptions are discussed above in Section '*Sources of uncertainty when determining diamond ages*'. The resulting isochrons are referred to as '*over-dispersed*', with the variability likely coming from a variation in the initial isotopic ratio, e.g., due to incomplete isotopic re-equilibration (Fig. 2).

A cogenetic relationship for individual inclusions can be evaluated based on the characteristics of the host diamond growth zone—C isotopic compositions, N aggregation characteristics and subtle N abundance variations in growth zones as imaged by cathodoluminescence—or the compositions of the inclusions to be dated. Dating sulfide versus silicate minerals requires different approaches to evaluating that they belong to the same diamond generation event. The studies on Slave craton diamonds by Westerlund et al. (2006) and Siberian diamonds by Wiggers de Vries et al. (2013) are good examples of the latter feature; multiple inclusions in diamonds were extracted and correlated directly with separate episodes of fluid influx and hence diamond growth that could be seen as isotopically distinct or revealing significant differences in age for different zones (see section on 'core to rim analyses' above). '*Linear data arrays*' of $Re/^{188}Os$ versus $^{187}Os/^{188}Os$ have provided valid ages from the world's most significant diamond localities (see worldwide data compilation in Smit et al. 2021). Diamond ages ranging from the Paleoarchean through to the Neoproterozoic and even Mesozoic have been obtained, which mostly correlate to known geological processes (see section '*Geological applications: What diamonds and their ages have revealed*').

For sulfides, it is critical before isotopic analysis to ascertain the peridotitic ([Ni] typically > ~6 wt%) vs. eclogitic ([Ni] typically < ~6 wt%) paragenesis by a non-destructive analytical technique, such as energy dispersive X-ray microanalysis on polished diamond plates or isolated sulfide grains released from the diamond. Besides being a first-order screen for potential co-genesis, the paragenesis of the sulfide is critical to estimating the Os content of the sulfide so that an isotopic spike of the right concentration can be used in isotope dilution analyses.

For silicates, $^{147}Sm/^{144}Nd$ versus $^{143}Nd/^{144}Nd$ arrays are made from single garnet and clinopyroxene inclusions that are grouped at least by similar chemical characteristics in the same manner as composite isochron ages. In the study by Timmerman et al. (2017), eclogitic inclusions were additionally grouped by similar diamond host growth structure, nitrogen aggregation state, trace element abundances, and strontium isotope compositions. In the study by Koornneef et al. (2017) different peridotitic diamond populations from Venetia were distinguished by Sm–Nd depleted mantle model ages and subsequently by Ca contents of garnets. The group with model ages of 3.0 ± 0.1 Ga and CaO contents < 2.5 wt% had trace element patterns with higher LREE and lower HREE relative to the group with 1.1 ± 0.2 Ga model ages and CaO > 2.5 wt%. This resulted in two isochron ages (2.95 and 1.15 Ga) similar to the model ages. While some other Venetia garnet inclusions fall onto these isochrons, their model ages and/or different REE patterns do not justify including them. The difference between the mixed Sm–Nd isochron array at 2.30 ± 0.04 Ga of pooled inclusions from Venetia (Richardson et al. 2009) and isochron arrays of 2.95 ± 0.07 and 1.15 ± 0.11 Ga from individual inclusions (Koornneef et al. 2017) showcase the differences that can exist between composited grains and single grains on an isochron or '*linear data array*' (Fig. 16). In the former study, inclusions from diamonds of different ages might get combined; in the latter study, an atypical inclusion in one diamond might control the age obtained. In both cases, best practice dictates the use of other compositional information (e.g., mineral chemistry, Sr isotopic composition, diamond type and C isotopic composition) to regress inclusions from diamonds that have grown in the same episode of fluid infiltration. Of course, the main goal is to be able to resolve these different episodes as distinct outside the age uncertainties of each isochron.

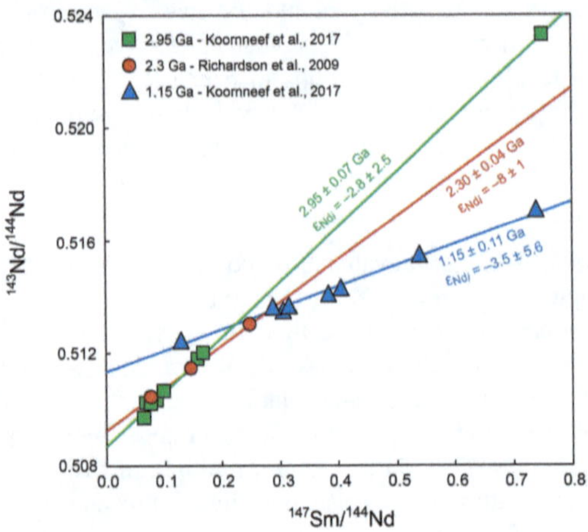

Figure 16. Samarium–Nd isochrons for peridotitic diamonds from Venetia (Kaapvaal craton). Garnet inclusions with similar Ca content and REE characteristics, from diamonds with similar $\delta^{13}C$-values, form two distinct '*linear data arrays*' at ~2.95 and 1.15 Ga (Koornneef et al. 2017). Three groups of composite garnet inclusions plot along a ~2.3 Ga '*linear data array*' that is intermediate between arrays obtained from individual inclusions (Richardson et al. 2009). There are additional inclusion data from Venetia that do not fit these two events. See text for further discussion.

Practical definition for geologically meaningful diamond ages

Our discussion of diamond geochronology so far raises the question of what geological events are being dated with any ages, whether they are '*model ages*', '*single diamond ages*', or '*linear data arrays*'. Diamonds crystallize from carbon- and nitrogen-bearing fluids that pass though peridotites and eclogites. Isotopic closure is obtained when diamond grows and encapsulates the inclusion (see Stachel et al. 2022b, this volume for various diamond formation mechanisms), and essentially stops diffusive exchange between the inclusion and other minerals in the host rock or the fluid. For a '*model age*' from an inclusion in a diamond to represent the time of diamond formation, it should be considered as the time that the inclusion was in equilibrium with a defined reservoir (such as an estimated mantle composition). It is then assumed that this time is similar to the time of encapsulation in the diamond. For any '*single diamond age*' obtained from multiple inclusions (core-to-rim, mineral isochron) or '*linear data array*' age, we must be dating individual sulfides or silicates that have shared a common geological history.

Since the overall goal of these different types of ages is to determine the time of diamond formation, the geological prehistory of any protogenetic (pre-existing) inclusion is secondary compared to the time of encapsulation by diamond as given by the model age, the core-to-rim age, the mineral isochron age, or the age of the linear data array. Considering the haphazard nature of mantle sampling during a kimberlitic eruption, the chaotic nature of diamond transport in the rising kimberlitic magma, and the distances over which an eruption occurs, it is remarkable that linear age arrays in suites of xenocrystic diamonds exist at all. The final important test is to see if the determined radiometric ages fit into the larger geodynamic history that might indicate when metasomatism and accompanying melts and fluids invaded the continental lithospheric mantle.

Diamond encapsulation shuts down parent–daughter fractionation and diffusive re-equilibration on the mineral inclusion scale—an event that can be dated geochronologically. Geological and experimental studies show that the dynamic nature of the mantle produces the melts/fluids that make diamonds in two basic ways: the dehydration of altered and hydrated slabs when they subduct into the mantle or the generation of melts/fluids during the adiabatic ascent of mantle plumes/upwellings. Often the age of the events that are capable of producing diamond-forming fluids can be determined independently through paleomagnetism studies, plate reconstructions, timing of volcanism, crustal evolution histories, or continental collision events. The alignment of the timing of such events with the ages obtained from diamond geochronology is tacit supporting evidence for the veracity of the diamond ages. Good examples from the Kaapvaal, Siberian, Slave, and West African cratons are given below.

Indirect evidence for the antiquity of diamonds

Besides the practical geological tests described above—meaning that the determined ages fit within the larger regional geological framework—there is abundant indirect and non-inclusion evidence that diamonds are indeed as old as the ages given by radiometric age determinations. Consideration of this evidence is useful in light of the generally accepted protogenecity of many inclusions, and previous arguments that diamonds are much younger than the inclusions that they host (e.g., Navon 1999; Spetsius et al. 2002).

Diamond deposit age. The first line of evidence is the occurrence of diamonds in volcaniclastic and sedimentary rocks that are known to be Archean and Proterozoic. Note though, that diamonds purportedly found as micro-inclusions in 4.252–3.058 Ga detrital zircon from the Jack Hills metasedimentary belt in Australia (Menneken et al. 2007) have since been identified as laboratory contamination (Dobrzhinetskaya et al. 2014).

The oldest known diamond-bearing sedimentary rocks are the 2.89–2.82 Ga gold-bearing Central Rand sequence of the Witwatersrand basin in South Africa (Wagner 1914; Williams 1932; Raal 1969). Diamond association with typical diamond indicator minerals was suggested to point to a kimberlitic source rock (Helmstaedt et al. 2010 and references therein). A similar association of diamonds and indicator minerals has recently been reported from 2.85 Ga conglomerates from the northern Slave craton (Timmerman et al. 2020). Diamonds in the 2.75 Ga Fortescue Group conglomerates in the Hamersley basin of the Pilbara craton in western Australia were linked to the recycling of diamond detritus into a once continuous basin connecting the Pilbara and Kaapvaal cratons (Helmstaedt et al. 2010).

Archean diamonds also occur in the Michipicoten greenstone belt in the Wawa-Abitibi Terrane of the Superior craton (Canada). Diamonds occur in 2.74–2.68 Ga lamprophyre dikes and metaconglomerates, and carry inclusions compositionally similar to diamonds from the worldwide cratonic lithospheric mantle (Stachel et al. 2006; Helmstaedt et al. 2010; Kopylova et al. 2011; Miller et al. 2012). The continental lithospheric mantle at 2.7 Ga could be shown to be exceptionally thick (300 km) and refractory based on majorite components in harzburgitic garnet inclusions (Stachel et al. 2006), suggesting formation of a diamondiferous cratonic nucleus prior to 2.7 Ga amalgamation of the Superior craton (Stachel et al. 2006; Kopylova et al. 2011; Miller et al. 2012). Other work on Wawa diamonds showed a high proportion of mixed peridotitic and eclogitic inclusions in a single diamond, tentatively ascribed to mixing of subducted oceanic crust and mantle in a subduction environment (De Stefano et al. 2006). This is in accord with a potential subduction imprint revealed by slightly heavy C-isotopes of octahedral diamonds (Stachel et al. 2006).

In northern Brazil, Venezuela and Guyana, diamonds and gold are mined from the Roraima supergroup 1.95–1.78 Ga, although the diamonds appear to derive from > 1.98 Ga conglomerates in the Arai Formation (Meyer and McCallum 1993; Reis et al. 2017; Bassoo et al. 2021). These diamonds have predominantly peridotitic inclusions, with a minor eclogitic population (Bassoo et al. 2021).

The Marange diamonds in eastern Zimbabwe occur in >1.1 Ga conglomerates of the Umkondo Group, and have also been locally deposited into softer Quaternary sediments (Hanson et al. 2004; Moore et al. 2009). The kimberlite source(s) for the diamonds have not been found and the age of the diamonds is unknown. However, from the age of the Umkondo conglomerates the diamonds are at least Proterozoic.

Although the vast majority of kimberlites have Phanerozoic ages, some diamondiferous kimberlites (e.g., 1.2 Ga Premier in South Africa; 1.1 Ga Wajrakarur in India and 1.1 Ga Kyle Lake in Canada; compilation in Kjarsgaard et al. (2022, this volume), and the richly diamondiferous Argyle lamproite in western Australia (Jaques et al. 1990) were emplaced in the Proterozoic, providing incontrovertible evidence for the antiquity of the diamonds contained therein.

Nitrogen aggregation state. The level of nitrogen aggregation in the diamond lattice provides an independent indication of the billion-year geological history of most diamonds. Substitution of nitrogen atoms into the diamond lattice, and later diffusion into aggregated nitrogen is described in detail in Green et al. (2022, this volume) to which the reader is directed for background. Diffusion of isolated substitutional nitrogen (N_S^0– C centers) and aggregation to 2N (A centers) and N_4V (B centers with a vacancy – V) is dependent on three variables: temperature, time and nitrogen concentration. Concentration of A and B centers can be relatively well constrained using Fourier Transform Infrared (FTIR) spectroscopy. However, without information about the average temperature a diamond experienced over its geological history, nitrogen aggregation state can only be a blunt and imprecise chronometer at best.

For example, constructed isotherms for varying mantle residence times show that 1-billion-year time variation gives less than 50 °C temperature variation, which is smaller than the overall uncertainties for most geothermometers (Nimis 2022, this volume). Similarly, the presence of relatively unaggregated A centers in a diamond cannot be used as evidence for '*young*' diamond growth, since aggregation is much more dependent on temperature. The oldest dated diamonds from Ekati retain relatively unaggregated nitrogen (<30% B centers) over their 3.5 billion-year mantle history, likely due to their residence in a shallower (cooler) part of the Slave continental lithospheric mantle (Westerlund et al. 2006).

Despite these caveats, using the average temperatures for diamond growth obtained from mineral inclusions along with nitrogen aggregation characteristics in the host diamond (Stachel 2014; Nimis 2022, this volume), the billion-year mantle residence history of the majority of lithospheric Type IaAB diamonds is fully supported.

Retarded isotopic growth of inclusions versus mantle minerals. Another independent measure of diamond antiquity comes from the general isotopic uniqueness of mineral grains included in diamonds observed by Richardson et al. (1984) who compared harzburgitic garnet inclusions from diamonds from Finsch ($^{87}Sr/^{86}Sr_i = 0.7037$) and Kimberley ($^{87}Sr/^{86}Sr_i = 0.7062$) with similar concentrate garnets separated from the Finsch ($^{87}Sr/^{86}Sr_i = 0.7073$ to 0.7321) and Bultfontein ($^{87}Sr/^{86}Sr_i = 0.7102$ to 0.7551) kimberlites (Fig. 5). The large difference in initial Sr isotopic composition indicated billions of years of encapsulation of the inclusion minerals in their diamond host compared to unencapsulated garnets of similar composition, especially at the very low measured $^{87}Rb/^{86}Sr$. This difference supported the more precise 3.4–3.3 Ga Sm–Nd model ages obtained on the harzburgitic garnet inclusions (that indicate the minimum age for the western Kaapvaal lithosphere). Pearson and Shirey (1999) summarized the Sr isotopic data for peridotitic and eclogitic garnet inclusion suites from Premier, Udachnaya, Orapa, and Argyle (Richardson 1986; Richardson et al. 1990, 1993, 1998; Richardson and Harris 1997), and noted that long lithospheric mantle residence time of billions of years is implicit in the Sr isotopic data and supports the Sm–Nd isochrons of these studies.

GEOLOGICAL APPLICATIONS: WHAT DIAMONDS AND THEIR AGES HAVE REVEALED

Detrital diamond deposits: provenance, glaciation, geomorphology, and plate reconstruction

Prior to the discovery of kimberlites in the Kimberley area of South Africa in 1871, all historical diamonds were recovered from alluvial deposits, most notably occurrences in India (e.g., Krishna River) and Borneo, Kalimantan Province (e.g., Spencer et al. 1988 and references therein). Alluvial diamond deposits are now known from many cratonic regions worldwide (Kjarsgaard et al. 2022, this volume). In some cases, the alluvial diamond deposits can be traced back to primary kimberlite or related rock sources (e.g., Kwango River – Lunda Norte kimberlites, Angola). However, many detrital diamond deposits have no clear relationships to known primary sources. Well documented examples include the deposits along the West Coast of southern Africa, alluvial occurrences across southeast Asia, southeast Australia, the Ural Mountains in Russia, the Krishna River in India, the Yuan region of China, and the Paraná Basin in Brazil (Kjarsgaard et al. 2022, this volume). These so-called '*headless*' placers have generated significant controversy and have spawned a number of '*non-traditional*' explanations for the source(s) of these diamonds. These include primary rock types other than kimberlites/lamproites and alternative diamond formation theories, for example, subduction in eastern Australia and Kalimantan, obduction of ophiolite complexes and meteorite impact (White et al. 2016 and references therein).

Determining the source(s) of detrital diamond deposits is important to exploration for primary diamond occurrences. Detrital sources also provide information on past kimberlite and related rock eruption episodes, erosional histories across source continental terranes, paleo-drainage patterns, glacial movements and the tectonic evolution of affected terranes. Conversely, knowledge of the age of the source kimberlites and/or related rocks provides an important constraint on the provenance of detrital diamonds. The actual host rocks of detrital diamonds provide minimum age constraints on the primary source(s) of contained diamonds. This information is particularly important for older sediments, given the limited number of available primary sources. In some cases, maximum age constraints have been provided by $^{40}Ar/^{39}Ar$ analyses of clinopyroxene inclusions extracted from detrital diamonds.

Application of the $^{40}Ar/^{39}Ar$ method to provenance studies. The $^{40}Ar/^{39}Ar$ method presents an important tool for constraining the provenance of detrital diamond populations because it provides ages on single diamonds with silicate inclusions. While these genesis ages can be complicated by the loss of Ar to the diamond–inclusion interface during diamond residence in the mantle, or accumulation of pre-eruption Ar in lattice defects in the diamond, these ages can be useful when there is only partial Ar loss (Bulanova et al. 2004; Burgess et al. 2004; Phillips et al. 2004; Phillips and Harris 2008).

The basis for interpretation of $^{40}Ar/^{39}Ar$ ages obtained from detrital diamonds stems from a study of Orapa mine diamonds (Phillips and Harris 2008) where it was shown that many $^{40}Ar/^{39}Ar$ ages on clinopyroxene approached the age of the host kimberlite. In this study, eclogitic clinopyroxene inclusions were extracted from 26 diamonds. Of these, 35% (9 of 26) yielded $^{40}Ar/^{39}Ar$ ages within error of the age of the Orapa kimberlite (~90 Ma) (Phillips and Harris 2008). The remaining inclusions gave older apparent ages, up to 436 ± 84 Ma (2σ). Overall, 77% (20 of 36) of inclusion ages were within 50 Ma, and 92% (33 of 36) within 100 million years, of the age of the Orapa eruption event. Therefore, $^{40}Ar/^{39}Ar$ ages obtained from clinopyroxene inclusions represent maximum estimates for the time of source kimberlite eruption, although the youngest ages may approximate the time of kimberlite eruption. In the case of multiple diamond populations, only the youngest detrital population, or populations separated by ~100 million years, may be resolvable. This novel $^{40}Ar/^{39}Ar$ approach has been used to constrain the provenance of detrital diamonds from the West Coast of southern Africa (Namibia; Namaqualand, South Africa), the Urals in Russia, the NW Province alluvial deposits in South Africa and the Copeton alluvial deposit in Australia. Each of these deposits is discussed in turn.

Revealing southern African paleo-drainage systems. Extensive diamond placer deposits are located along the west coasts of Namibia and Namaqualand (South Africa), with >120 million carats of gem-quality diamonds recovered (Oosterveld 2003). Considerable controversy exists as to the provenance of these deposits. The interior of Namibia hosts several kimberlite localities, but none are diamondiferous. The nearest primary diamond-bearing occurrences are located more than 700 km inland, on the Kalahari craton. This cratonic area hosts numerous kimberlites, with emplacement ages ranging from ~1350 Ma to ~80 Ma (Kjarsgaard et al. 2022, this volume). The two main competing theories for the origin of the West coast diamonds involve either: 1) erosion of distal Cretaceous/Jurassic kimberlites and orangeites, and transport of the diamonds to the West Coast via the paleo-Orange River (e.g., de Wit 1993, 1999; Bluck et al. 2005); or 2) recent erosion of Permo-Carboniferous (~300 Ma) glacial deposits (Dwyka Group), with the primary source of the diamonds being pre-Dwyka kimberlites on the Kalahari craton (e.g., Van Wyk and Pienaar 1986; Maree 1987, 1988; Moore and Moore 2004).

Phillips and Harris (2009) conducted $^{40}Ar/^{39}Ar$ step-heating experiments on 50 peridotitic and eclogitic clinopyroxene inclusions from Namibian diamond production. Ages ranged from 62 ± 30 Ma to 1441 ± 700 Ma, with the majority of ages <300 Ma. These data indicate that the majority of West Coast diamonds could not have been sourced from the Dwyka glacial deposits but were sourced from post-Dwyka kimberlites/lamproites. In addition, six inclusions yielded ages unique to Cretaceous (90–80 Ma) kimberlites on the Kalahari Craton. The authors concluded that the Namibian diamonds likely originated from erosion of Cretaceous/Jurassic kimberlites and/or lamproites (previously termed 'orangeites' or 'Group II kimberlites') located on the western part of the Kalahari Craton, with only minor contributions possible from syn-Karoo (e.g., Jwaneng, Botswana) and/or pre-Karoo (e.g., Venetia, National) kimberlites.

$^{40}Ar/^{39}Ar$ analyses of clinopyroxene inclusions extracted from Namaqualand diamonds (West Coast, South Africa) were conducted in two batches, the first using an older generation single-collector mass spectrometer and the latter using a state-of-the-art multi-collector mass spectrometer (Phillips et al. 2018). The two batches yielded similar results, with apparent ages ranging from 118 ± 44 to 3684 ± 191 Ma and 121 ± 15 Ma to 689 ± 5 Ma, respectively. Similar to the Namibia study, the majority of Namaqualand inclusions (88%) gave ages consistent with an origin from post-Dwyka kimberlites/orangeites, with few (if any) diamonds originating from >500 Ma kimberlites. However, in this case, no inclusion ages could be uniquely assigned to Cretaceous age Kalahari Craton kimberlites. Therefore, it was concluded that the Namaqualand diamonds were mainly sourced from older Cretaceous/Jurassic (~200–115 Ma) orangeites (also known as Group II kimberlites) and transported to the West Coast via the paleo-Karoo River (de Wit et al. 1992). In contrast, the Namibian diamonds appear to have been sourced from both Cretaceous/Jurassic kimberlites (~90–80 Ma) and orangeites (~200–115 Ma) and were transported to the Namibian coast by the paleo-Kalahari River, which captured the paleo-Karoo River and formed a pre-cursor to the modern Orange River.

New diamond sources in the East European craton. Alluvial diamond deposits occur along the western margin of the Middle and North Ural Mountains in Russia and are located in Devonian (~400 Ma) sediments of the Takaty Formation (Kukharenko 1955; Bekker et al. 1970). Re–Os analyses of eclogitic sulfide inclusions from several diamonds produced an isochron age of 1280 ± 310 Ma (2σ), interpreted as the age of diamond formation (Laiginhas et al. 2009). $^{40}Ar/^{39}Ar$ analyses of eclogitic clinopyroxene inclusions from five Urals alluvial diamonds produced similar ages averaging 472 ± 28 Ma (2σ). These data constrain the ages of source kimberlites/lamproites to ~500–400 Ma. As there are no known kimberlites/lamproites of this age in the region, Laiginhas et al. (2009) suggested the existence of undiscovered primary sources in the Volga-Uralia terrane of the East European Craton.

Glacial transport of diamonds in South Africa and Australia. Several alluvial diamond deposits occur in the North West Province of South Africa. These include unusual diamond-bearing gravels in the Lichtenburg-Ventersdorp area, which occur within sinuous ridges and in sinkholes above an erosional surface of the ~2.7 Ga Malmani dolomites (de Wit 2016 and references therein). Based on agate populations as well as Cambrian 'mantle' zircon ages of 664–489 Ma, and a $^{40}Ar/^{39}Ar$ clinopyroxene inclusion age of ~ 600 Ma, de Wit (2016) concluded that the diamonds were transported from Cambrian age kimberlites by Dwyka glaciers and deposited in paleo-eskers, to form the sinuous ridges in the area.

The Copeton alluvial diamond occurrence in New South Wales (Australia) is associated with Tertiary gravels underlying Tertiary basalts. The gravels are devoid of kimberlitic indicator minerals and some of the diamonds contain unusual calc-silicate mineral inclusions (Burgess et al. 1998; Davies et al. 2002). This has led to suggestions that the diamonds might be derived from local Tertiary leucitite and/or basaltic volcanism, with the diamonds originally formed in subducted oceanic slabs (Taylor 1990; Sutherland et al. 1994; Barron et al. 1998, 2011). Pearson et al. (1998a) obtained 3.4 and 2.1 Ga model ages from two peridotitic sulfides from Copeton diamonds, suggesting that they were of cratonic origin. Burgess et al. (1998) obtained $^{40}Ar/^{39}Ar$ ages from three clinopyroxene inclusions from two diamonds, all of which yielded similar ages averaging 340 ± 28 Ma. This result was considered to represent the age of the source kimberlite/lamproite. Given the lack of kimberlites/lamproites in the region, the absence of kimberlitic indicator minerals and the well-sorted nature of the diamond population, Burgess et al. (1998) raised the possibility that the Copeton diamonds could have originated from Antarctica, with subsequent northward transportation by Permian age glaciers.

Diamond ages and craton evolution

A worldwide association between ancient cratons and diamond occurrences has long been known (e.g., Kennedy 1964; Clifford 1966; Gurney and Switzer 1973; Boyd and Gurney 1986). Typical usage of the term '*craton*' refers to ancient portions of the Earth's crust that have been stable since 2.5 Ga, even though early work recognized the association of diamonds with ancient cratons of at least 1.5 Ga age (Kennedy 1964; Clifford 1966). This strict usage based on age is problematic since many '*cratons*' worldwide actually show significant evidence for later reworking during the Proterozoic and Phanerozoic. Instead of crustal age, an updated definition of 'craton' uses thickness of the stabilizing lithospheric mantle below continental crust, visualized through fast seismic shear wave speeds in global seismic velocity models. In the updated definition, cratons are regions of the Earth's continental crust that are underlain by 150–200 km thick continental lithospheric mantle providing long-term stability since at least 1 Ga (Pearson et al. 2021). Using this updated definition, around 63% of exposed continental crust and 18% of Earth's surface can be considered to be cratonic.

The association of diamonds with cratonic regions, and the intersection of peridotite-derived conductive geotherms with the diamond stability field, led to the suggestion that diamonds reside in the lithospheric mantle beneath cratons (e.g., Sobolev 1977; Boyd et al. 1985; Boyd and Gurney 1986; Haggerty 1986). The first sulfide Pb–Pb ages (Kramers 1979) and the first harzburgitic-garnet-inclusion-based diamond ages proved that diamonds originate in old lithospheric mantle (Richardson et al. 1984).

Diamonds have formed through nearly all of Earth's history (Figs. 17, 18), in distinct episodes that can often be linked to larger-scale tectonic processes (Richardson et al. 2004; Pearson and Wittig 2008; Shirey and Richardson 2011; Howell et al. 2020). Diamonds, and their formation ages, are the ideal time-resolved mantle samples that can provide an overview of continent formation and evolution. A classic example is how the creation, assembly, and modification of the Kaapvaal-Zimbabwe craton is reflected in the age and chemistry of multiple generations of diamonds formed and stored in its lithospheric mantle (Shirey et al. 2002, 2004).

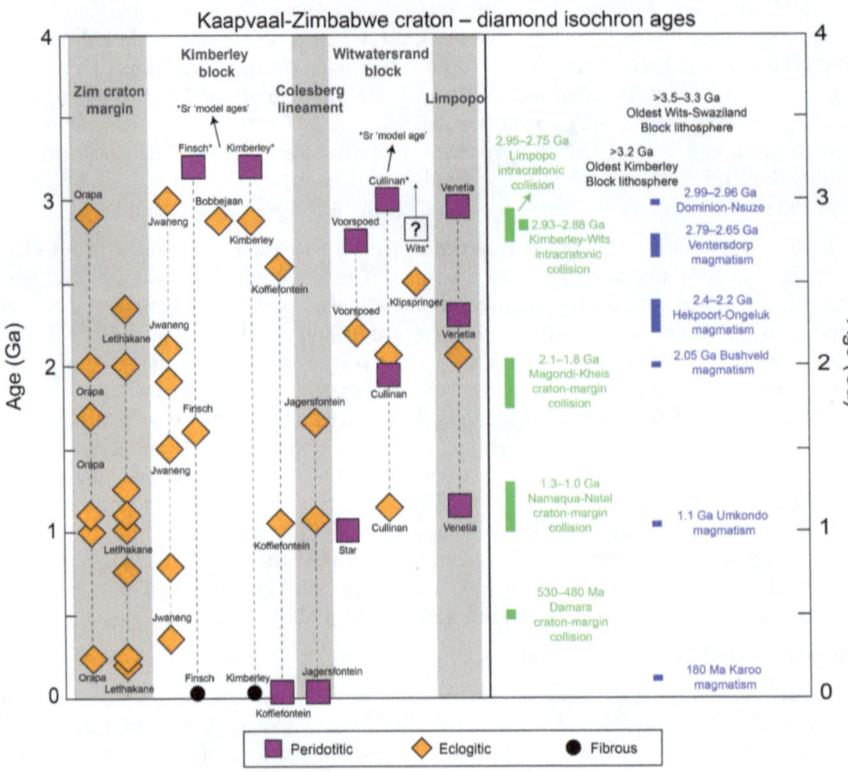

Figure 17. Isochron or *'linear data array'* ages for eclogitic and peridotitic diamonds, from the Kaapvaal–Zimbabwe cratons in southern Africa. Many localities have multiple growth events with diamonds forming in both eclogite and peridotite. Archean ages for Finsch, Kimberley and Cullinan diamonds are Sr 'model ages' (see Fig. 5). Diamonds occurring in Archean rocks in the Witwatersrand *'Wits'* Basin are shown as their minimum age (age of their sedimentary host rocks). Diagram updated and modified from Howell et al. (2020). See worldwide data compilation (Smit et al. 2021; https://doi.org/10.7939/DVN/DRAJGT) for an overview of all isochron ages, and their references.

Formation of the first cratonic blocks, craton assembly and the onset of plate tectonics

Models for the formation of the thick buoyant lithospheric mantle beneath stable cratons range from *horizontal tectonic processes*, such as compressional thickening (Jordan 1978;

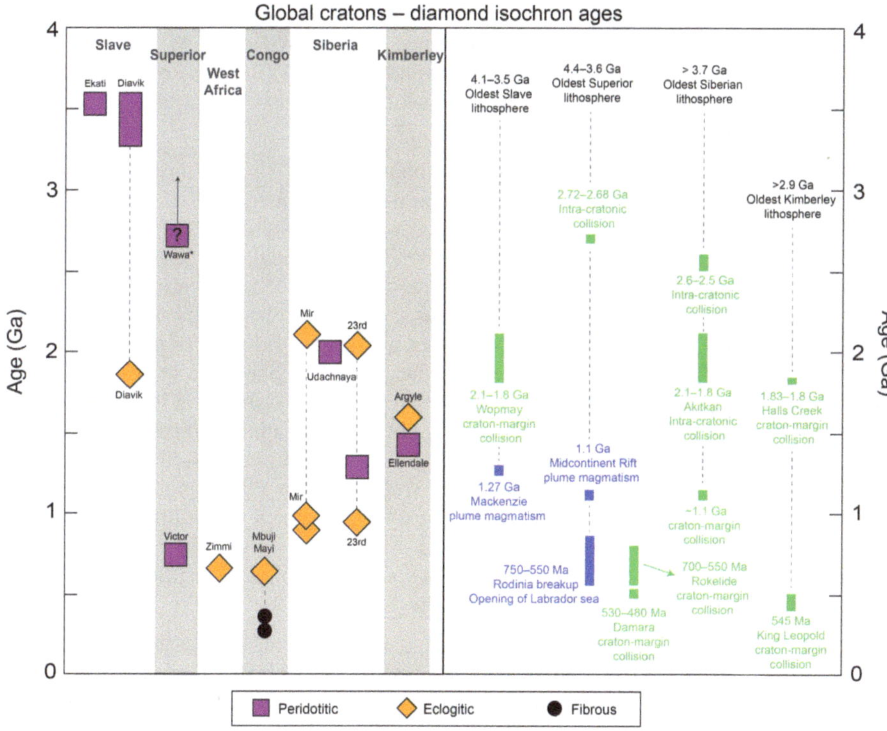

Figure 18. All isochron or *'linear data array'* ages for eclogitic and peridotitic diamonds, from global cratons. Diamonds occurring in Archean rocks at Wawa (Superior craton) are shown as the age of their sedimentary host rocks. Updated and modified from Howell et al. (2020). See worldwide data compilation (Smit et al. 2021; https://doi.org/10.7939/DVN/DRAJGT) for an overview of all isochron ages, and their references.

Wang et al. 2018; Pearson et al. 2021) accompanying Wilson cycle continental collision (Shirey and Richardson 2011) or stacking of subducted oceanic lithosphere (Helmstaedt and Schulze 1989; Pearson and Wittig 2008) to *vertical tectonic processes*, such as formation in mantle plumes (e.g., Aulbach 2012), melting at the base of oceanic plateaus (Nair and Chacko 2008; Reimink et al. 2014), or ambient mantle melting below hot ridges (Herzberg and Rudnick 2012). These divergent models have led to debate between proponents of horizontal versus vertical processes in the formation of the earliest continental lithosphere. As the only stabilized relicts of mantle left from the Archean, continental lithospheric mantle and the key minerals it contains (e.g., diamonds) provide important constraints on the geodynamics of the early Earth and the onset of plate tectonics—Earth's present, global, mantle-convection-driven process (Pearson et al. 2021).

The global record for diamond ages shows that prior to 3.0 Ga, only peridotitic diamonds formed, whereas after 3.0 Ga, eclogitic diamonds became prevalent (Figs. 17, 18). From this observation and supercontinent-style rifting of the Kaapvaal and Pilbara cratons at 3.2 Ga (Zegers et al. 1998; see summary in Van Kranendonk and Kirkland 2016), Shirey and Richardson (2011) suggested that the prevalence of eclogitic diamonds after 3.0 Ga resulted from the capture of eclogite and diamond-forming fluids in continental lithospheric mantle via subduction and continental collision, marking the onset of the Wilson cycle of plate tectonics. The absence of >3.2 Ga eclogite xenoliths and eclogitic diamonds from the lithospheric mantle in any craton may suggest that the earliest cratonic nuclei formed by non-Wilson cycle processes. Similarly, the presence of mass-independently fractionated sulfur in sulfide inclusions from various cratons can be used to assess the role of plate tectonics in craton formation (Smit et al. 2019b). However, the absence of a true Wilson cycle does not rule out the presence of some form of more localized recycling or even shallow plate subduction, and the formation processes behind cratonic nuclei may differ from that responsible for the larger cratonic units extant today (Pearson et al. 2021).

The oldest known diamonds that inform us about the earliest craton formation processes are the 3.5–3.3 Ga Ekati and Diavik diamonds on the Slave craton (Westerlund et al. 2006; Aulbach et al. 2009c, 2011). Sulfide inclusions in diamonds from the Panda kimberlite (Ekati Mine) and their associated harzburgites have elevated initial Os isotopic compositions that are best explained by some form of crustal recycling (Fig. 15; Westerlund et al. 2006). However, the lack of both atmospherically-modified sulfur isotopes in the sulfide inclusions (Cartigny et al. 2009), and non-mantle O-isotopes in garnet peridotites (Regier et al. 2018) suggest that depleted peridotitic lithologies became part of the Slave continental lithosphere in a way that did not involve substantial surface material. This requirement is consistent with models proposed for formation of the earliest shallow Slave continental lithospheric mantle, by crust formation in an oceanic plateau setting (Reimink et al. 2014), or by collisional compression (Regier et al. 2018; Wang et al. 2018), followed by thickening via underplating of plume mantle residues (Griffin et al. 1999), dated to 3.27 ± 0.34 Ga (Aulbach et al. 2004). This plume model and the suggested timing are in excellent agreement with the recently reported 3.19 ± 0.12 Ga age obtained from detrital chromite linked to komatiite magmatism in the Slave craton (Haugaard et al. 2021).

Preservation of ultra-depleted residues juxtaposed with enriched signatures in diamond forming fluids coupled with the absence of trapped and recognizable oceanic lithospheric components (e.g., eclogite, MIF-sulfur) highlights the special mix of geologic conditions that converged to form one of Earth's earliest cratons. It is important to consider, however, that isolated evidence for Paleoarchean recycling processes in the Slave craton does not imply that Earth had already transitioned to sustained global plate tectonics.

Amalgamation of different cratonic nuclei, each with their own lithospheric mantle, into a larger stable cratonic region, a process that may well reflect the inception of the supercontinent cycle (e.g., van Kranendonk and Kirkland 2016) is reflected by diamond ages in the Kaapvaal–Zimbabwe craton. Kimberley, Bobbejaan, Jwaneng and Orapa eclogitic diamonds all record 2.9 Ga ages (Fig. 19) (Pearson et al. 1998b; Shirey et al. unpublished, 2013; Richardson et al. 2001; Richardson et al. 2004). These diamond ages overlap with the timing of subduction-collision along the Colesberg lineament that is recorded in crustal xenoliths (Schmitz et al. 2004), and the occurrence of 2.9 Ga diamondiferous eclogite xenoliths with basaltic to komatiitic compositions (Shirey et al. 2001; Menzies et al. 2003). Along with their enriched initial $^{187}Os/^{188}Os$ and atmospherically-modified sulfur isotopes (Orapa and Jwaneng; Farquhar et al. 2002; Thomassot et al. 2009) these 2.9 Ga eclogitic diamond ages suggest the involvement of enriched recycled components during diamond formation, likely derived from subducted oceanic lithosphere involved in amalgamation of the Kimberley and Witwatersrand blocks of the Kaapvaal craton.

Craton modification associated with diamond growth

The continental lithospheric mantle has been recognized as an important carbon reservoir, which, after initial melt extraction imposing highly refractory and reducing conditions, was gradually re-enriched in volatiles and sulfur and re-oxidized through infiltration of fluids and melts from passive upwelling, episodically impinging plumes and subducting slabs (e.g., Foley and Fischer 2017). We now know that lithospheric diamonds form during episodes of fluid infiltration into the lithospheric mantle keel in response to large-scale geotectonic processes (Figs. 17, 18), and studies of lithospheric diamonds are the key means to look at carbon cycling between the crust and mantle over the past 3.5 billion years.

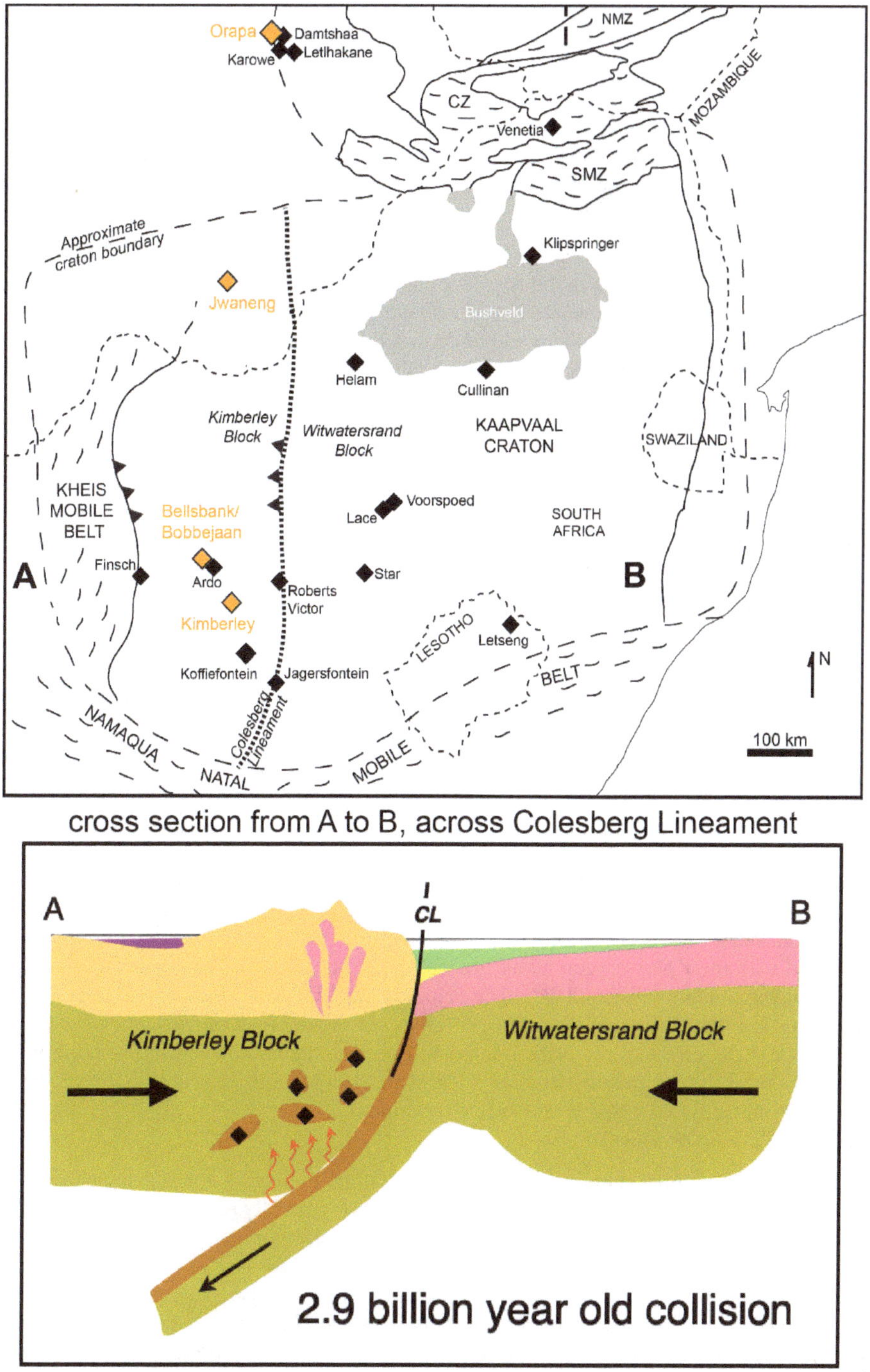

Figure 19. Eclogitic diamond formation at Orapa, Jwaneng, Bobbejaan and Kimberley (**orange diamond symbols**) associated with 2.9 Ga subduction and amalgamation of the Kaapvaal craton. Collision schematic from Shirey et al. (2013). See worldwide data compilation (Smit et al. 2021; https://doi.org/10.7939/DVN/DRAJGT) for an overview of all isochron ages, and their references.

Subduction along craton margins. Diamond ages have changed our thinking about the association of diamonds with ancient continental mantle keels. Contrary to traditional thinking (e.g., Clifford 1966), diamond age determinations have shown that diamonds also form during active tectonism in the orogenic belts around cratons. Incorporation of such areas into the spectrum of Earth's lithosphere that has diamond potential has expanded diamond exploration targets and will continue to contribute to finding new economic diamond deposits (Sobolev 1985; Jaques et al. 1990). Eclogitic diamonds have become powerful tracers of supercontinent assembly, and how subduction introduces carbon-bearing fluids to the continental lithospheric mantle, thereby affecting its composition and stability (e.g., Shirey and Richardson 2011).

Perhaps the best example is the Argyle mine in Australia, where '*subduction along the Kimberley craton edge generated the world's biggest diamond deposit*' (Stachel et al. 2018a). Argyle is a famous supplier of pink and red diamonds (Shigley et al. 2001) and known for its exceptionally high diamond grades (Jaques et al. 1990; Rayner et al. 2018). Argyle occurs within the Proterozoic Halls Creek orogen (1.83-1.92 Ga; Hancock and Rutland 1984). The Argyle eclogitic diamonds formed at 1580 ± 30 Ma (Richardson 1986) in "close" temporal association with Halls Creek collisional processes, likely in Archean mantle (Luguet et al. 2009). On the other side of the Kimberley craton, ~20 Ma Ellendale lamproites occur in the King Leopold orogen. The lherzolitic Ellendale diamonds formed at 1426 ± 130 Ma, also in >2.9 Ga Archean mantle (Smit et al. 2010). The presence of Archean mantle below Argyle and likely Ellendale shows that early Proterozoic convergent orogens (Tyler et al. 1999; Griffin et al. 2000) play a key role in diamond formation within the cratonic mantle of the Kimberley craton, which extended beneath these areas.

Even if diamond-bearing kimberlites and lamproites do not erupt through the surface expression of an orogenic belt, there are several worldwide diamond localities that have allowed us to link diamond formation to known subduction processes along craton margins. The Diavik eclogitic diamonds formed at 1.86 ± 0.19 Ga (Aulbach et al. 2009c) coincident with collision and accretion of the Wopmay orogen along the western margin of the Slave craton (Hoffman et al. 1989), an event that is also recorded in eclogite xenoliths (Schmidberger et al. 2007; Aulbach et al. 2009b).

In southern Africa, there are several localities where diamonds formed in temporal or spatial association with the Magondi-Kheis and Namaqua-Natal orogens along the Kaapvaal craton margin. At Jagersfontein, five eclogitic diamonds record formation at 1.65 ± 0.18 Ga associated with the Kheis orogen along the western Kaapvaal craton (Aulbach et al. 2009a). Also at Jagersfontein, a second younger eclogitic population formed at 1070 ± 110 Ma, coincident with the Namaqua-Natal orogen along the southern Kaapvaal margin (Aulbach et al. 2009a). The effect of the Namaqua-Natal event on the continental lithospheric mantle is also evident in the diamonds from Koffiefontein (1048 ± 120 Ma; Pearson et al. 1998b) and Star (1000 ± 33 Ma; see worldwide data compilation in Smit et al. 2021; https://doi.org/10.7939/DVN/DRAJGT).

In West Africa, three eclogitic diamonds from Zimmi in Sierra Leone each yield identical ~650 Ma ages (Smit et al. 2016). Diamond formation is temporally related to subduction along the SW margin of the Man craton, recorded in the Rokelide orogen (700–550 Ma; Lytwyn et al. 2006). In Siberia, lherzolitic and eclogitic diamonds formed in the Paleoproterozoic (Richardson and Harris 1997; Wiggers de Vries et al. 2013) in association with the orogenies that created the 2.1–1.8 Ga Angara and Akitkan orogens (Cherepanova and Artemieva 2015 and references therein). In diamonds sampled by the 23rd Party Congress kimberlite, there are both 1.30 ± 0.29 Ga lherzolitic diamonds and 0.97 ± 0.31 Ga eclogitic diamonds (Wiggers de Vries et al. 2013). These Mesoproterozoic ages may be linked to subduction processes related to the assembly of the Rodinia supercontinent (Gladkochub et al. 2006).

Examples of diamond localities that occur in off-craton (or craton-margin) settings are George Creek, Sloan and other kimberlites in the Colorado--Wyoming province (Otter 1989; McCallum et al. 1991; Chinn 1995), the Prairie Creek lamproite province in Arkansas (Dunn 2003), and Buffalo Head Hills in Alberta (Banas et al. 2007). However, there are no age constraints on any of these diamonds to determine whether diamond formation occurred due to influx of subduction fluids and melts, or whether they are older diamonds that survived in the mantle through tectonic reworking. Clearly these areas should be the targets of future dating projects.

'Non-subduction' refertilization. Lherzolitic diamond formation is a widespread phenomenon (e.g., Richardson and Harris 1997; Smit et al. 2010; Aulbach et al. 2018) (see also worldwide data compilation in Smit et al. 2021), corresponding to ~12% of peridotitic-inclusion-bearing lithospheric diamonds and dominating the inclusion population at some locations (Stachel et al. 2018b, 2022a, this volume). Refertilization (or lherzolitization) of the initially depleted lithosphere (van Achterbergh et al. 2001; Pearson et al. 2002; Simon et al. 2007) and minor associated diamond growth likely occurred in the presence of small-volume melts, producing garnet inclusions with characteristic, generally mildly sinusoidal REE patterns (Stachel and Luth 2015). Lherzolitic diamond formation has been suggested to represent the deep and early component of the refertilization and re-oxidation of the lithosphere required for successful kimberlite eruption (Aulbach et al. 2018; Shirey et al. 2019). So far, all known lherzolitic diamonds formed during the Proterozoic, though additional dating studies may extend this age range.

Diamonds from the Venetia mine in the Limpopo belt (between the Kaapvaal and Zimbabwe cratons) have a range of ages from Neoarchean to Neoproterozoic. A Sm–Nd isochron array from pooled inclusions yielded a 2.30 ± 0.04 Ga age (Richardson et al. 2009). Isochrons constructed from individual inclusions from different diamonds gave 2.95 ± 0.07 and 1.15 ± 0.11 Ga (Koornneef et al. 2017) (Fig. 16). Richardson et al. (2009) interpreted the 2.3 Ga isochron age as mixing between 2 Ga Bushveld basaltic magmatism with older (>3 Ga) harzburgite. Similarly, diamond formation at 2.95 and 1.15 Ga were also ascribed to metasomatism in the continental lithospheric mantle, where Archean diamonds formed in association with rifting in the southern Zimbabwe craton, and Proterozoic diamonds formed during magmatism associated with the Umkondo large igneous province (Koornneef et al. 2017).

Udachnaya (*Siberia*), Cullinan (*Kaapvaal*), Ellendale (*Kimberley*) and Victor (*Superior*) all have Proterozoic lherzolitic diamonds. Ages for Udachnaya (2.01 ± 0.6 Ga; Richardson and Harris 1997) and Cullinan (1.93 ± 0.04 Ga; Richardson et al. 1993) diamonds were determined from composite silicates, and may represent an average diamond age. In the latter case, however, there is a strong association with Bushveld magmatism at 2.05 Ga (Richardson et al. 1993; Shirey et al. 2002; Richardson and Shirey 2008). Regardless, these localities show that peridotitic diamond formation is not restricted to Archean depleted harzburgite, but that re-enriched lherzolite is also an important economic diamond host, challenging existing paradigms (Stachel et al. 2018b). A Re–Os sulfide isochron shows that Ellendale lherzolitic diamonds formed at 1.43 ± 0.03 Ga, within Archean lithospheric mantle (Smit et al. 2010), however influx of carbon-bearing fluids and/or melts has not yet been correlated with known tectonic or magmatic events in the Kimberley continental lithosphere. The Victor lherzolitic diamonds formed at 719 ± 49 Ma (Aulbach et al. 2018), during rifting and breakup of the Rodinia supercontinent which promoted mobilization of fluids/melts into the Superior continental lithosphere. Trace elements in the Victor diamonds suggest a melt-like character to the diamond-forming fluid (Krebs et al. 2019). The one Koffiefontein diamond that formed at 69 ± 30 Ma in association with proto-kimberlite magmatism is also likely to be lherzolitic (Pearson et al. 1998b).

The Klipspringer diamonds (South Africa) are also a good example of diamond formation due to metasomatic reworking of the continental lithospheric mantle, although they are not lherzolitic. Here, eclogitic diamonds formed at 2.55 ± 0.15 Ga, likely due to partial melting of eclogite in response to Ventersdorp magmatism and heating at 2.7 Ga (Westerlund et al. 2004).

Sublithospheric diamond ages

Despite the extensive dating studies of mineral inclusions in lithospheric diamonds, the dating of sublithospheric diamonds is still in its infancy because mineral inclusions with sufficient Sm, Nd, U, Pb, Re, and Os are extremely rare and often small in size. Extant studies are limited as only four Re–Os and Sm–Nd model ages and two U–Pb ages have been obtained. However, as might be expected from geodynamic considerations, ages on sublithospheric diamonds display none of the Archean ages so prevalent in the lithospheric diamond populations—they are all Mesoproterozoic or younger.

A complication is that while lithospheric diamonds have inclusions that can be clearly defined by compositional similarity to major mantle minerals of peridotitic, websteritic, and eclogitic parageneses (Stachel et al. 2022a, this volume), sublithospheric diamonds do not (Walter et al. 2022, this volume). Ultrahigh PT phases such as Ca-silicate perovskite, majorite, bridgmanite, breyite, and ferropericlase-magnesiowustite are stable in the deeper mantle below the continental lithospheric mantle. These phases often revert to lower-pressure assemblages, display intermediate compositions between expected endmembers (e.g., Thomson et al. 2016), and have non-stochiometric solid solutions. Thus, while some inclusions can show a loose affinity to peridotitic or eclogitic precursors, other inclusions cannot be easily classified into the same simple parageneses as lithospheric diamonds (Stachel et al. 2022a, this volume). Sublithospheric phases that are suitable for U–Pb or Sm–Nd dating include breyite ($Ca_3Si_3O_9$ in walstromite structure), $CaTiO_3$, and majoritic garnet. Rare sulfides (Hutchison 1997; Hayman et al. 2005; Kaminsky et al. 2009; Smit and Shirey, unpublished; Bulanova et al. 2010) are also suitable for Re–Os dating. Ferropericlase is not datable with commonly-used radioactive decay systems.

The isotopic dataset for sublithospheric inclusions is currently biased by: 1) the type of inclusions, and 2) the location of the studied inclusions. Sampling biases aside, there is an increasing amount of evidence that the majority of recovered sublithospheric diamonds are associated with subducted material (Shirey et al. 2021; Walter et al. 2022, this volume). This is based on the carbon isotopic compositions of the diamonds themselves and abundant evidence from inclusions, such as mineral assemblage bulk compositions that have slab associations, trace element abundances, Sr, Pb and O isotopic compositions (e.g., Stachel et al. 2005; Pearson et al. 2007; Walter et al. 2008, 2011; Shirey et al. 2013, 2019; Ickert et al. 2015; Burnham et al. 2015; Thomson et al. 2016; Timmerman et al. 2019d).

While no *bona fide* single crystals of lower mantle minerals have been recovered, retrograde assemblages interpreted as lower mantle equivalents coming from both the crust and the mantle portion of the slab do exist (Harte et al. 1999; Stachel et al. 2000; Walter et al. 2011; Smith et al. 2018). There has been only one attempt at dating of inclusions from Letseng (Lesotho—Smith et al. 2017, 2021), and all the other dated inclusions are from the Juina area in Brazil, where the younger diamond ages may be associated with Mesoproterozoic to Phanerozoic subduction. Dating sublithospheric inclusions from other locations, such as Kankan, Monastery, Williamson, Jagersfontein, and DO27, is needed to determine if *'young'* sublithospheric diamond formation is a global phenomenon. Although the lack of older (>3 Ga) sublithospheric diamonds can easily be explained by the current limited dataset, potential geological reasons for the difference between sublithospheric and lithospheric diamond ages are discussed in the section below.

Currently, the most precise age for a sublithospheric inclusion in diamond is a 101 ± 7 Ma U–Pb age from a breyite / $CaTiO_3$-perovskite in the outer portion of a Collier-4 diamond (Bulanova et al. 2010). Although no matrix-matched standards were used during the SIMS analyses, the age of 101 ± 7 Ma seems accurate as it is close to the age of kimberlite eruption. Subsequent analysis on an eclogitic sulfide in the interior of this same Collier-4 diamond by Smit and Shirey (unpublished data) yielded a Neoproterozoic (650 ± 150 Ma) depleted mantle model age. This core–rim age difference leads to contrasting interpretations: either this is a Neoproterozoic diamond in which the U–Pb system in breyite remained open until kimberlite eruption or it is a Cretaceous diamond that incorporated a Neoproterozoic sulfide in the deep convecting mantle in which it grew. The former scenario is roughly consistent with Re–Os data (Hutchison et al. 2012) on a sulfide inclusion in an alluvial diamond also recovered from the Juina area. This eclogitic (based on Re, Os, and Ni content) sulfide produced a Re–Os mantle model age of ~1271 Ma. These two Proterozoic ages are significantly older than the kimberlitic magmatism in the Juina area (94–79 Ma; Davis 1977; Heaman et al. 1998; Kaminsky et al. 2010).

Uranium–Pb ratios from ICP-MS analyses in combination with Pb isotopic compositions give another young age estimate for a Collier-4 sublithospheric diamond (Timmerman et al. 2019d). Using a Stacey and Kramers (1975) Pb growth curve yields an average age of 42 Ma with a maximum of 162 Ma. Using a high common $^{207}Pb/^{206}Pb$ value of 1.17, as common Pb correction, results in a maximum U–Pb age of 660 Ma. While the diamond does contain substantial common Pb, these data provide further evidence for *'young'* sublithospheric diamond formation.

Another age estimate for sublithospheric diamonds that is younger than those from the continental lithospheric mantle comes from the pooled analysis of Sr and Nd isotopic compositions of around 40 majoritic garnet inclusions from Sao Luiz, Brazil (Harte and Richardson 2012). Two groups of pooled inclusions have Sr and Nd isotopic compositions that fall on the oceanic mantle Sr–Nd array between depleted MORB and enriched OIB endmembers. These isotopic compositions clearly retain younger oceanic mantle isotopic signatures—not those of the ancient continental lithospheric mantle. From these isotopic compositions, Harte and Richardson (2012) suggest diamond formation in subducted oceanic crust of Mesozoic age.

Sublithospheric diamonds that retain distinctive metallic inclusions—known as CLIPPIR diamonds (Smith et al. 2016)—present a yet unknown age story, since we only have one age determination to rely on so far (Smith et al. 2017, 2021). CLIPPIR diamonds have been consistently recovered from the Cullinan, Karowe and Letseng mines in southern Africa, where kimberlite eruption occurred at 1153 Ma, 93 Ma and 90 Ma, respectively. Thus, purely from their known occurrences, CLIPPIR diamonds show that diamond formation has been occurring in metallic domains of the sub-lithosphere/asthenosphere since at least the Mesoproterozoic (Smith et al. 2016). The sulfide portion of two metallic inclusions in a Letseng CLIPPIR diamond yielded an unradiogenic $^{187}Os/^{188}Os$ isotope composition of 0.111 that is equivalent to a circa 2.5 Ga old model age. While it is tempting to conclude that this diamond is Archean, Fe isotopic analyses on the metallic iron show that it is derived from the magnetite–awaruite assemblages created by low-temperature serpentinization of ocean-floor peridotite. As the current oceanic lithosphere contains common Proterozoic to even Neoarchean Os isotope signatures that result from prior depletion (e.g., Liu et al. 2008; Harvey et al. 2006; Scott et al. 2019) this old age is interpreted as inherited from the oceanic lithosphere (Smith et al. 2016, 2021).

Diamond ages and implications for recycling of volatiles into the mantle

As detailed above, diamond ages have placed important time constraints on the recycling of fluid mobile compounds and elements such as water, carbon, nitrogen, boron, and sulfur into the continental lithospheric mantle especially in relation to the geologic evolution of the continents. For sublithospheric diamonds, the lack of measured ages makes time constraints on recycling somewhat speculative. But as there is no other way to trace the recycling of these volatiles far back in geologic time without the framework provided by diamond ages, some basic observations can be made.

For lithospheric diamonds, the comparison of their ages to the crustal rock geochronologic record provides evidence of a major change in the style of mantle geodynamics to produce differentiated, stable crust and attendant mantle keel from a vertical mode (e.g., Van Kranendonk 2010; Sizova et al. 2015; Fischer and Gerya 2016; Bédard 2020) to a horizontal mode (e.g., Jordan 1978; Shirey and Richardson 2011; Scott et al. 2019). The *vertical mode* encompasses mantle plume upwelling, thickening of the lithospheric mantle by underplating, vertical crustal tectonics, and internal (autochthonous) igneous crustal differentiation whereas the *horizontal mode* encompasses lateral oceanic lithosphere accretion, horizontal crustal tectonics, advective thickening, and external (allochthonous) igneous crustal differentiation. Looking at all cratons, examples of both modes exist from the Eoarchean to the Mesoarchean (e.g., Van Kranendonk et al. 2007, 2015; Reimink et al. 2019; Bauer et al. 2020; O'Neill et al. 2020). But since the latter mode looks more like modern plate tectonics, these essential questions are currently being vigorously debated: a) whether subduction/horizontal geodynamics became dominant in this time period (e.g., Stern 2018; Brown et al. 2020), b) if so when in this billion-year interval it became dominant (e.g., Van Kranendonk et al. 2007; Bauer et al. 2019; Capitanio et al. 2019; Bauer et al. 2020), and c) did it look like modern plate tectonics?

Using seismicity and specifically seismic tomography to image subduction (Wadati-Benioff) zones into the mantle transition zone and occasionally into the upper part of the lower mantle (e.g., Li et al. 2008), we can relate this style of plate subduction to modern plate tectonics and to the formation of the wide variety of sublithospheric diamonds (Shirey et al. 2021). Thus good, extensive sublithospheric diamond ages would provide one very important constraint on the onset of modern plate tectonics on Earth. As the 1153 Ma Premier kimberlite (Wu et al. 2013) contains CLIPPIR diamonds where metallic inclusions have heavy Fe isotopic compositions linked to serpentinite (Smith et al. 2021), we know that modern-style plate subduction extends at least to the Mesoproterozoic.

Most diamondiferous kimberlites are Mesozoic to Cenozoic, and though some are Paleozoic to even Proterozoic, it is highly unlikely that very old diamondiferous kimberlites are yet to be discovered. Only when we have a significant dataset of sublithospheric diamond ages, whose average age proves to be Mesoarchean to Paleoproterozoic will we be able to confidently extend deep slab subduction back this far. This circumstance rests entirely on the ability of sublithospheric diamonds to survive for more than 2 billion years in the mobile, convecting mantle and to fortuitously find their way to being sampled by erupting kimberlitic magma. These prospects seem geologically unlikely. More likely is the expectation that sublithospheric diamonds are geologically much younger than most lithospheric diamonds and that they and the kimberlites that bring them to the surface are both products of modern-style deep plate tectonics.

HOW FAR HAVE WE COME AND WHERE SHOULD WE GO?

The last review on diamond dating with radiogenic isotopes by Pearson and Shirey (1999) outlined six main future directions for isotopic dating of diamonds. Below we briefly outline how far these suggestions have been met, along with some recommendations for future work.

1) Relationship between silicate and sulfide ages

A major analytical development since the last review is the ability to measure the low Rb–Sr and Sm–Nd content in single silicate inclusions (Timmerman et al. 2017). This development makes it possible to investigate the ages of a wider range of inclusion-bearing diamonds rather than the biased sample set of sulfide-bearing diamonds. Recently,

Gress et al. (2021b) showed that silicate (853 ± 55 Ma) and sulfide (786 ± 250 Ma) isochron ages from a single Letlhakane diamond agree within uncertainty. Further work from worldwide localities is needed to establish whether silicate and sulfide diamond formation always occurs contemporaneously, but unfortunately not all diamond suites contain both sets of inclusions. However, these results clearly validate the approach of using isochrons defined by mineral inclusions to date diamond formation.

2) Isotopic dating of diamonds and their host rock

Given that most diamonds are products of metasomatic fluids, ongoing evaluation of any age differences between the host rock and the diamonds is required. While diamond-dating research has strengthened the idea that diamonds form much younger than their host rocks, some studies have shown that older components from the host rock are likely included into the diamond fluid (e.g., Smit et al. 2016, 2019b; Aulbach et al. 2018). However there has not yet been a benchmark study of the age of diamonds and host minerals in the same mantle xenolith. Such a study requires an exceptional specimen: containing diamonds with sufficient inclusions to date, and with host rock minerals that are sufficiently pristine to yield geochronological information.

3) Ability of the diamond-forming event to affect different isotopic systems

A key issue remaining in diamond geochronology at the turn of the millennium was whether it was possible that diamond-formation ages could be reflected by protogenetic inclusions, allied with the question of how common protogenetic inclusions were. This concept has been addressed by isotopic studies (e.g., Westerlund et al. 2004; Smit et al. 2016, 2019b; Aulbach et al. 2018), crystallographic studies and Nd diffusion calculations (Nestola et al. 2019) and by Os diffusion calculations (Pamato et al. 2021). Crystallographic studies (Nestola et al. 2017) showed that the lattice orientations of many silicate inclusions were not parallel to the orientation of the diamond, raising the prospect that many inclusions, especially silicates, could be protogenetic. New modeling of diffusion and careful consideration of parent–daughter element isotope fractionation (e.g., Pamato et al. 2021) make convincing arguments that in most cases, the diamond formation event is sufficient to re-set isotope systematics in protogenetic inclusions. While the re-setting is not, in most cases, extensive enough to yield isochrons that pass rigorous statistical thresholds, it is sufficient for the arrays defined on isochron constructs to reflect the timing of diamond formation, for both Sm–Nd and Re–Os systems. More work is required to understand why the Rb–Sr system is generally not amenable to yielding isochron ages. In addition, the powerful ^{176}Lu–^{176}Hf decay system has been under-exploited for constraining diamond formation ages. Increased instrumental sensitivity ought to lead to the exploitation of this decay system to date diamond formation as the parent–daughter element fractionations it displays in garnets are often significantly larger than the Sm–Nd system (e.g., Baxter and Scherer 2013)

4) Experiments on Ar release from inclusions, meaning of Ar ages, and comparison of Ar ages with other isotopic systems such as U–Th/He

Work by Phillips and Harris (2008; 2009) and others has demonstrated that argon contained in inclusions within diamonds is mobile under mantle temperature and pressure conditions and diffuses to the diamond/inclusion interface and/or inclusion defect sites over time. Definitive measurements of argon diffusivity parameters in inclusion minerals such as Cr-diopside and pyrope garnet remain outstanding, and therefore the argon closure temperature for these minerals is unknown. The consequence of argon mobility is that diamond formation ages are only attainable by analyzing totally encapsulated inclusions, either by laser drilling/fusion analyses or furnace step-heating/fusion analysis. Extraction or exposure of the inclusions results in partial or complete loss of pre-emplacement radiogenic argon. However, this behavior can be exploited to constrain the provenance of detrital diamonds (e.g., Phillips et al. 2018).

Improvement in trace element measurements recently resulted in U–Th/He isochron and model ages for fibrous diamonds (Timmerman et al. 2019c; Weiss et al. 2021). Based in part on their nitrogen-aggregation characteristics, these fluid-rich diamonds were previously assumed to form in association with proto-kimberlitic fluids/melts. These new U–Th/He ages have shown that this is not always the case, with fibrous diamonds forming up to several 100 m.y. (Timmerman et al. 2019c; Weiss et al. 2021) and possibly 1000 m.y. (2.5 Ga; Weiss et al. 2021) prior to kimberlite eruption. Future studies using U–Th/He isotopes may yield a much wider range of formation ages for fibrous diamonds and allow for improved understanding of any association of diamond-forming fluids with (proto-) kimberlitic melts in the lithosphere. Similarly, by the time of the next diamond-dating review, U–Th/He dating may have allowed us to demonstrate the variable formation ages of fibrous vs. monocrystalline diamonds.

5) Future diamond inclusion dating studies need to be combined with detailed CL and FTIR studies of the diamonds prior to destruction

This recommendation has now become routine procedure for diamond dating studies since it is essential to understand the relationship between inclusions and growth zones in each diamond. Similar to completed studies (Wiggers de Vries et al. 2013; Timmerman et al. 2017, 2019a; Smit et al. 2016, 2019a; Gress et al. 2021a,b; Howell et al. 2020), future studies should continue to combine age dating of inclusions with an attempt to determine the *composition of the diamond-forming fluid* to obtain a complete picture of diamond composition through time. This can be achieved through redox modelling of the diamond parental fluid (core-to-rim $\delta^{13}C$–N content profiles), trace element analyses of the diamond, as well as Raman and/or TEM identification of trapped fluid inclusions.

6) Global studies of diamond ages to determine whether diamond formation can be linked to specific tectonic events and processes

The advent of both single sulfide Re–Os dating (Pearson et al. 1998b) and single silicate Nd isotope dating (Timmerman et al. 2017) has led to step-wise increases in the number of diamond dating studies; allowing for meaningful reviews of global data, both in Howell et al. (2020) and this review. There are now numerous examples where diamond formation is linked to plate tectonic processes like subduction, collision and rifting.

The antiquity of diamonds and their formation in the continental lithospheric mantle has been established since the first diamond-dating study (Richardson et al. 1984). However, more robust constraints on the age of cool, deep lithospheric keels capable of hosting diamonds—as opposed to young, hot early cratonic nuclei—can be made if sufficient diamonds with suitable inclusions are found from Archean sediments (e.g., Timmerman et al. 2020). A focus of on-going studies is also on locations where continental lithospheric mantle is much younger—such as the predominantly Proterozoic Sask craton (Czas et al. 2020)—to further explore the apparent pulses of Proterozoic peridotitic diamond growth.

From a global perspective, there are currently no diamond ages available from the known deposits in the North Atlantic craton, the Wyoming craton, the Kola-Karelian cratons in the Baltic shield, or the cratons of India and China. Although diamonds from the Juina region are the focus of sublithospheric diamond dating, diamonds from other deposits in Brazil, and other areas in South America (São Francisco craton and Guyana shield) still need diamond-dating studies. These are obvious gaps in the database.

Lastly, new on-going studies are also focusing on establishing the age of sublithospheric diamonds to add geochronological data to the plethora of new information coming from studies of the transition zone and lower mantle (Shirey et al. 2021; Walter et al. 2022, this volume). There is much to look forward to in the next few years of diamond dating!

ACKNOWLEDGEMENTS

The supplementary tables are available at *https://doi.org/10.7939/DVN/DRAJGT*. Janne Koornneef, Roberta Rudnick, and the editor Thomas Stachel provided constructive comments that helped us clarify the ideas presented in this chapter. S.T. acknowledges funding from the Natural Sciences and Engineering Research Council (NSERC) and the Government of Canada through a Banting post-doctoral fellowship. S.A. thanks the German Research Foundation for their continued support, currently via grant DFG AU356/11.

REFERENCES

Agrosi G, Nestola F, Tempesta G, Bruno M, Scandale E, Harris J (2016) X-ray topographic study of a diamond from Udachnaya; implications for the genetic nature of inclusions. Lithos 248–251:153–159

Allègre CJ (2008) Isotope Geology. Cambridge University Press

Allègre CJ, Othman DB, Polve M, Richard P (1979) The Nd–Sr isotopic correlation in mantle materials and geodynamic consequences. Phys Earth Planet Sci 19:293–306

Aulbach S (2012) Craton nucleation and formation of thick lithospheric roots. Lithos 149:16–30

Aulbach S, Griffin WL, Pearson NJ, O'Reilly SY, Kivi K, Doyle BJ (2004) Mantle formation and evolution, Slave Craton: constraints from HSE abundances and Re–Os isotope systematics of sulfide inclusions in mantle xenocrysts. Chem Geol 208:61–88

Aulbach S, Shirey SB, Stachel T, Creighton S, Muehlenbachs K, Harris JW (2009a) Diamond formation episodes at the southern margin of the Kaapvaal Craton; Re–Os systematics of sulfide inclusions from the Jagersfontein Mine. Contrib Mineral Petrol 157:525–540

Aulbach S, Creaser RA, Pearson NJ, Simonetti SS, Heaman LM, Griffin WL, Stachel T (2009b) Sulfide and whole rock Re–Os systematics of eclogite and pyroxenite xenoliths from the Slave Craton, Canada. Earth Planet Sci Lett 283:48–58

Aulbach S, Stachel T, Creaser RA, Heaman LM, Shirey SB, Muehlenbachs K, Eichenberg D, Harris JW (2009c) Sulphide survival and diamond genesis during formation and evolution of Archaean subcontinental lithosphere; a comparison between the Slave and Kaapvaal Cratons. Lithos 112:747–757

Aulbach S, Stachel T, Heaman LM, Carlson JA (2011) Microxenoliths from the Slave Craton; archives of diamond formation along fluid conduits. Lithos 126:419–434

Aulbach S, Stachel T, Seitz H-M, Brey GP (2012) Chalcophile and siderophile elements in sulphide inclusions in eclogitic diamonds and metal cycling in a Paleoproterozoic subduction zone. Geochim Cosmochim Acta 93:278–299

Aulbach S, Mungall JE, Pearson DG (2016) Distribution and processing of highly siderophile elements in cratonic mantle lithosphere. Rev Mineral Geochem 81:239–304

Aulbach S, Creaser RA, Stachel T, Heaman LM, Chinn IL, Kong J (2018) Diamond ages from Victor (Superior Craton): Intra-mantle cycling of volatiles (C, N, S) during supercontinent reorganisation. Earth Planet Sci Lett 490:77–87

Aulbach S, Höfer HE, Gerdes A (2019) High-Mg and low-Mg mantle eclogites from Koidu (West African craton) linked by Neoproterozoic ultramafic melt metasomatism of subducted Archaean plateau-like oceanic crust. J Petrol 60:723–754

Banas A, Stachel T, Muehlenbachs K, McCandless TE (2007) Diamonds from the Buffalo Head Hills, Alberta: Formation in a non-conventional setting. Lithos 93:199–213

Barron LM, Lishmund S, Oakes G, Barron B (1998) Primary and related diamond occurrences within a Phanerozoic subduction regime in eastern New South Wales, Australia. International Kimberlite Conference: Extended Abstracts 7:46–48

Barron LM, Mernagh T, Barron B, Pogson R (2011) Spectroscopic research on ultrahigh pressure (UHP) macrodiamond at Copeton and Bingara NSW, Eastern Australia. Spectrochim Acta A 80:112–118

Barth MG, Rudnick RL, Carlson RW, Horn I, McDonough WF (2002) Re–Os and U–Pb geochronological constraints on the eclogite–tonalite connection in the Archean Man Shield, West Africa. Precambrian Res 118:267–283

Bassoo R, Befus KS, Liang P, Forman SL, Sharman G (2021) Deciphering the enigmatic origin of guyana's diamonds. Am Mineral 106:54–68

Bauer A, Reimink JR, Chacko T, Foley BJ, Shirey SB, Pearson G (2019) Zircon Hf isotope evidence for a global transition between stagnant-and mobile-lid tectonics. AGUFM 2019:V31G-0135

Bauer A, Reimink J, Chacko T, Foley B, Shirey S, Pearson D (2020) Hafnium isotopes in zircons document the gradual onset of mobile-lid tectonics. Geochem Perspect Lett 14:1–6, doi: 10.7185/geochemlet.2015

Baxter EF, Scherer EE (2013) Garnet geochronology: timekeeper of tectonometamorphic processes. Elements 9:433–438

Becker H (2000) Re–Os fractionation in eclogites and blueschists and the implications for recycling of oceanic crust into the mantle. Earth Planet Sci Lett 177:287–300

Becker H, Dale CW (2016) Re–Pt–Os isotopic and highly siderophile element behavior in oceanic and continental mantle tectonites. Rev Mineral Geochem 81:369–440

Bédard JH (2020) From the LIPS of a serial killer: Endogenic retardation of biological evolution on unstable stagnant-lid planets. Planet Space Sci 192:105068

Bekker YR, Bekasova NB, Ishkov AD (1970) Diamond-bearing placers in the Devonian rocks of the Northern Urals. Lithol Mineral Resour 4:421–430
Bluck BJ, Ward JD, de Wit MCJ (2005) Diamond mega-placers; Southern Africa and the Kaapvaal Craton in a global context. Geol Soc Spec Publ 248:213–245
Boyd FR, Gurney JJ (1986) Diamonds and the African lithosphere. Science 232:472–477
Boyd FR, Gurney JJ, Richardson SH (1985) Evidence for a 150–200-km thick Archaean lithosphere from diamond inclusion thermobarometry. Nature 315:387–389
Brenan JM, Cherniak DJ, Rose LA (2000) Diffusion of osmium in pyrrhotite and pyrite: implications for closure of the Re–Os isotopic system. Earth Planet Sci Lett 180:399–413
Broadley M, Byrne D, Kopylova MG, Thomassot E, Almayrac M, Marty B (2021) Determining the volatile composition of the ancient mantle through the analysis of Archean aged fibrous diamonds. Goldschmidt abstract, https://doi.org/10.7185/gold2021.7301
Brown M, Johnson T, Gardiner NJ (2020) Plate tectonics and the Archean Earth. Annu Rev Earth Planet Sci 48:291–320
Bulanova GP, Muchemwa E, Pearson DG, Griffin BJ, Kelley SP, Klemme S, Smith CB (2004) Syngenetic inclusions of yimengite in diamond from Sese Kimberlite (Zimbabwe); evidence for metasomatic conditions of growth. Lithos 77:181–192
Bulanova GP, Walter MJ, Smith CB, Kohn SC, Armstrong LS, Blundy J, Gobbo L (2010) Mineral inclusions in sublithospheric diamonds from Collier 4 kimberlite pipe, Juina, Brazil; subducted protoliths, carbonated melts and primary kimberlite magmatism. Contrib Mineral Petrol 160:489–510
Burgess R, Turner G, Harris JW (1992) ^{40}Ar–^{39}Ar laser probe studies of clinopyroxene inclusions in eclogitic diamonds. Geochim Cosmochim Acta 56:389–402
Burgess R, Turner G, Harris JW (1997) ^{40}Ar–^{39}Ar laser probe dating of diamonds. *In*: Seventh annual V M Goldschmidt conference. Vol 921. Lunar and Planetary Institute, Houston, TX, United States, p 37–38
Burgess R, Turner G, Laurenzi M, Harris JW (1989) ^{40}Ar–^{39}Ar laser probe dating of individual clinopyroxene inclusions in Premier eclogitic diamonds. Earth Planet Sci Lett 94:22–28
Burgess R, Kelley SP, Turner G, Harris JW (1994) Diamond formation and eruption ages by laser ^{40}Ar– ^{39}Ar dating of syngenetic inclusions. *In*: Abstracts of the eighth international conference on Geochronology, cosmochronology, and isotope geology. Lanphere MA, Dalrymple GB, Turrin BD, (eds) U S Geological Survey, Reston, VA, USA, p 43
Burgess R, Johnson LH, Mattey DP, Harris JW, Turner G (1998) He, Ar and C isotopes in coated and polycrystalline diamonds. Chem Geol 146:205–217
Burgess R, Kiviets GB, Harris JW (2004) Ar–Ar age determinations of eclogitic clinopyroxene and garnet inclusions in diamonds from the Venetia and Orapa kimberlites. Lithos 77:113–124
Burgess R, Cartigny P, Harrison D, Hobson E, Harris J (2009) Volatile composition of microinclusions in diamonds from the Panda Kimberlite, Canada; implications for chemical and isotopic heterogeneity in the mantle. Geochim Cosmochim Acta 73:1779–1794
Burnham AD, Thomson AR, Bulanova GP, Kohn SC, Smith CB, Walter MJ (2015) Stable isotope evidence for crustal recycling as recorded by superdeep diamonds. Earth Planet Sci Lett 432:374–380
Capitanio F, Nebel O, Cawood P, Weinberg R, Clos F (2019) Lithosphere differentiation in the early Earth controls Archean tectonics. Earth Planet Sci Lett 525:115755
Carlson RW, Shirey SB, Schonbachler M (2008) Applications of PGE radioisotope systems in geo-and cosmochemistry. Elements 4:239–245
Caro G, Bennett VC, Bourdon B, Harrison TM, von Quadt A, Mojzsis SJ, Harris JW (2008) Application of precise ^{142}Nd/^{144}Nd analysis of small samples to inclusions in diamonds (Finsch, South Africa) and Hadean zircons (Jack Hills, Western Australia). Chem Geol 247:253–265
Cartigny P, Farquhar J, Thomassot E, Harris JW, Wing B, Masterson A, McKeegan K, Stachel T (2009) A mantle origin for Paleoarchean peridotitic diamonds from the Panda kimberlite, Slave Craton; evidence from ^{13}C, ^{15}N and ^{33,34}S stable isotope systematics. Lithos 112:852–864
Cherepanova Y, Artemieva I (2015) Density heterogeneity of the cratonic lithosphere: a case study of the Siberian craton. Gondwana Res 28:1344–1360
Chinn IL (1995) A study of unusual diamonds from the George Creek K1 kimberlite dyke, Colorado. PhD thesis, University of Cape Town, Cape Town, South Africa
Clifford TN (1966) Tectono-metallogenic units and metallogenic provinces of Africa. Earth Planet Sci Lett 1:421–434
Creaser R, Papanastassiou D, Wasserburg G (1991) Negative thermal ion mass spectrometry of osmium, rhenium and iridium. Geochim Cosmochim Acta 55:397–401
Cumming G, Richards J (1975) Ore lead isotope ratios in a continuously changing Earth. Earth Planet Sci Lett 28:155–171
Czas J, Pearson DG, Stachel T, Kjarsgaard BA, Read GH (2020) A Palaeoproterozoic diamond-bearing lithospheric mantle root beneath the Archean Sask Craton, Canada. Lithos 356:105301
Davies RM, O'Reilly SY, Griffin WL (2002) Multiple origins of alluvial diamonds from New South Wales, Australia. Econ Geol Bull Soc Econ Geol 97:109–123
Davies GR, van den Heuvel Q, Matveev S, Drury MR, Chinn IL, Gress MU (2018) A combined cathodoluminescence and electron backscatter diffraction examination of the growth relationships between Jwaneng diamonds and their eclogitic inclusions. Mineral Petrol 112:231–242

Davis GL (1977) The ages and uranium contents of zircons from kimberlites and associated rocks. International Kimberlite Conference: Extended Abstracts 2:67–69

Davis G, Sobolev N, Kharkiv A (1980) New data on the age of Yakutian kimberlites (U–Pb zircon method). Doklady Akademii Nauk SSSR 254:175–179

De Stefano A, Lefebvre N, Kopylova M (2006) Enigmatic diamonds in Archean calc-alkaline lamprophyres of Wawa, southern Ontario, Canada. Contrib Mineral Petrol 151:158–173

de Wit M (1999) Post-Gondwana drainage and the development of diamond placers in western South Africa. Econ Geol 94:721–740

de Wit MCJ (1993) Cainozoic evolution of the drainage systems in the north-western Cape. PhD thesis, Univeristy of Cape Town, Cape Town, South Africa

de Wit MC (2016) Early Permian diamond-bearing proximal eskers in the Lichtenburg/Ventersdorp area of the North West Province, South Africa. S Afr J Geol 119:585–606

de Wit MJ, de Ronde CE, Tredoux M, Roering C, Hart RJ, Armstrong RA, Green RW, Peberdy E, Hart RA (1992) Formation of an Archaean continent. Nature 357:553–562

DePaolo D, Wasserburg G (1979) Sm–Nd age of the Stillwater Complex and the mantle evolution curve for neodymium. Geochim Cosmochim Acta 43:999–1008

Deines P, Harris JW (1995) Sulfide inclusion chemistry and carbon isotopes of African diamonds. Geochim Cosmochim Acta 59:3173–3188

Dobrzhinetskaya L, Wirth R, Green H (2014) Diamonds in Earth's oldest zircons from Jack Hills Conglomerate, Australia, are contamination. Earth Planet Sci Lett 387:212–218

Dodson MH, Compston W, Williams IS, Wilson JF (1988) A search for ancient detrital zircons in Zimbabwean sediments. J Geol Soc London 145:977–983

Doucet LS, Ionov DA, Golovin AV (2015) Paleoproterozoic formation age for the Siberian cratonic mantle: Hf and Nd isotope data on refractory peridotite xenoliths from the Udachnaya kimberlite. Chem Geol 391:42–55

Dunai T, Roselieb K (1996) Sorption and diffusion of helium in garnet: implications for volatile tracing and dating. Earth Planet Sci Lett 139:411–421

Dunn DP (2003) Diamond economics of the Prairie Creek Lamproite, Murfreesboro, AR, USA. Ore Geol Rev 22:251–262

Ebel D, Naldrett A (1997) Crystallization of sulfide liquids and the interpretation of ore composition. Can J Earth Sci 34:352–365

Eldridge CS, Compston W, Williams IS, Harris JW, Bristow JW (1991) Isotope evidence for the involvement of recycled sediments in diamond formation. Nature 353:649–653

Farley KA (2002) (U–Th)/He dating: Techniques, calibrations, and applications. Rev Mineral Geochem 47:819–844

Farquhar J, Wing BA, McKeegan KD, Harris JW, Cartigny P, Thiemens MH (2002) Mass-independent sulfur of inclusions in diamond and sulfur recycling on early Earth. Science 298:2369–2372

Fischer R, Gerya T (2016) Early Earth plume-lid tectonics: A high-resolution 3D numerical modelling approach. J Geodyn 100:198–214

Fleet M, Crocket J, Stone W (1996) Partitioning of platinum-group elements (Os, Ir, Ru, Pt, Pd) and gold between sulfide liquid and basalt melt. Geochim Cosmochim Acta 60:2397–2412

Foley SF, Fischer TP (2017) An essential role for continental rifts and lithosphere in the deep carbon cycle. Nat Geosci 10:897–902

Foley SF, Barth MG, Jenner GA (2000) Rutile/melt partition coefficients for trace elements and an assessment of the influence of rutile on the trace element characteristics of subduction zone magmas. Geochim Cosmochim Acta 64:933–938

Giuliani A, Phillips D, Maas R, Woodhead JD, Kendrick MA, Greig A, Armstrong RA, Chew D, Kamenetsky VS, Fiorentini ML (2014) LIMA U–Pb ages link lithospheric mantle metasomatism to Karoo magmatism beneath the Kimberley region, South Africa. Earth Planet Sci Lett 401:132–147

Gladkochub D, Pisarevsky S, Donskaya T, Natapov L, Mazukabzov A, Stanevich A, Sklyarov E (2006) The Siberian Craton and its evolution in terms of the Rodinia hypothesis. Episodes 29:169–174

Green BL, Collins AT, Breeding CM (2022) Diamond spectroscopy, defect centers, color, and treatments. Rev Mineral Geochem 88:637–688

Gress MU, Koornneef JM, Thomassot E, Chinn I, van Zuilen K, Davies GR (2021a) Sm–Nd isochron ages coupled with CN isotope data of eclogitic diamonds from Jwaneng, Botswana. Geochim Cosmochim Acta 293:1–17

Gress MU, Pearson DG, Chinn IL, Thomassot E, Davies GR (2021b) Mesozoic to Paleoproterozoic diamond growth beneath Botswana recorded by Re–Os ages from individual eclogitic and websteritic inclusions. Lithos 388–389:106058

Gress MU, Timmerman S, Chinn I, Koornneef JM, Thomassot E, van der Valk EAS, van Zuilen K, Bouden N, Davies GR (2021c) Two billion years of episodic and simultaneous websteritic and eclogitic diamond formation beneath the Orapa kimberlite cluster, Botswana. Contrib Mineral Petrol 176:54

Griffin WL, Doyle BJ, Ryan CG, Pearson NJ, O'Reilly SY, Davies R, Kivi K, Van Achterbergh E, Natapov LM (1999) Layered mantle lithosphere in the Lac de Gras area, Slave Craton: Composition, structure and origin. J Petrol 40:705–727

Griffin T, Page R, Sheppard S, Tyler I (2000) Tectonic implications of Palaeoproterozoic post-collisional, high-K felsic igneous rocks from the Kimberley region of northwestern Australia. Precambrian Res 101:1–23

Gurney JJ, Switzer GS (1973) The discovery of garnets closely related to diamonds in the Finsch Pipe, South Africa. Contrib Mineral Petrol 39:103–116

Haggerty SE (1986) Diamond genesis in a multiply-constrained model. Nature 320:34–38

Hamilton MA, Stern R, Pearson DG, Harris JW, Shirey SB (1997) Dating diamond growth: SHRIMP II ion microprobe Pb isotopic analyses of syngenetic sulphide inclusions from Koffiefontein diamonds. GAC-MAC:A62

Hamilton MA, Sobolev NV, Stern RA, Pearson DG (2003) SHRIMP U–Pb dating of a perovskite inclusion in diamond: evidence for a syneruption age for diamond formation, Sytykanskaya kimberlite pipe, Yakutia region, Siberia. International Kimberlite Conference: Extended Abstracts, 8. https://doi.org/10.29173/ikc3245

Hancock S, Rutland R (1984) Tectonics of an early Proterozoic geosuture: the Halls Creek orogenic sub-province, northern Australia. J Geodynamics 1:387–432

Hanson R, Crowley J, Bowring S, Ramezani J, Gose W, Dalziel I, Pancake J, Seidel E, Blenkinsop T, Mukwakwami J (2004) The age of the Umkondo Group, eastern Zimbabwe, and implications for palaeomagnetic correlations. Science 304:1126–1129

Harlow GE (1997) K in clinopyroxene at high pressure and temperature; an experimental study. Am Mineral 82:259–269

Harlow GE, Veblen DR (1991) Potassium in clinopyroxene inclusions from diamonds. Science 251:652–655

Harris JW (1968) The relative orientation of solid mineral inclusions in diamond

Harris JW, Gurney JJ (1979) Inclusions in diamond. *In*: The Properties of Diamond. Field JE (ed) Acad Press, London, UK, p 555–591

Harris JW, Smit KV, Fedortchouk Y, Moore M (2022) Morphology of monocrystalline diamond and its inclusions. Rev Mineral Geochem 88:119–166

Harte B, Richardson S (2012) Mineral inclusions in diamonds track the evolution of a Mesozoic subducted slab beneath west Gondwanaland. Gondwana Res 21:236–245

Harte B, Harris JW, Hutchison MT, Watt GR, Wilding MC (1999) Lower mantle mineral associations in diamonds from Sao Luiz, Brazil. *In*: Mantle Petrology; Field Observations and High-pressure Experimentation; a Tribute to Francis R (Joe) Boyd. Vol 6. Fei Y, Bertka Constance M, Mysen Bjorn O, (eds). Geochemical Society - University of Houston, Department of Chemistry, Houston, TX, United States, p 125–153

Harvey J, Gannoun A, Burton KW, Rogers NW, Alard O, Parkinson IJ (2006) Ancient melt extraction from the oceanic upper mantle revealed by Re Os isotopes in abyssal peridotites from the Mid-Atlantic ridge. Earth Planet Sci Lett 244:606–621

Harvey J, Warren JM, Shirey SB (2016) Mantle sulfides and their role in Re–Os and Pb isotope geochronology. Rev Mineral Geochem 81:579–649

Haugaard R, Wateron P, Ootes L, Pearson DG, Luo Y, Konhauser K (2021) Detrital chromites reveal Slave craton's missing komatiite. Geology 49:1079–1083

Hayman PC, Kopylova MG, Kaminsky FV (2005) Lower mantle diamonds from Rio Soriso (Juina area, Mato Grosso, Brazil). Contrib Mineral Petrol 149:430–445

Heaman L, Teixeira N, Gobbo L, Gaspar J (1998) U–Pb mantle zircon ages for kimberlites from the Juina Paranatinga Provinces, Brazil. International Kimberlite Conference: Extended Abstracts 7:322–324

Helmstaedt HH, Schulze DJ (1989) Evidence for subduction and spreading in the Archean rock record; implications for Archean tectonic style and the evolution of the subcontinental lithosphere. *In*: Workshop on the Archean Mantle. Vol 89–05. Ashwal Lewis D, (ed) Lunar and Planetary Institute, Houston, TX, USA, p 42–44

Helmstaedt HH, Gurney JJ, Richardson SH (2010) Ages of cratonic diamond and lithosphere evolution; constraints on Precambrian tectonics and diamond exploration. Can Mineral 48:1385–1408

Herzberg C, Rudnick R (2012) Formation of cratonic lithosphere: an integrated thermal and petrological model. Lithos 149:4–15

Hoffman PF, Bally A, Palmer A (1989) Precambrian geology and tectonic history of North America. Chapter 16 *In:* The Geology of North America—An Overview,. Geol Soc Am p 447–512

Holwell D, McDonald I, Butler I (2011) Precious metal enrichment in the Platreef, Bushveld Complex, South Africa: evidence from homogenized magmatic sulfide melt inclusions. Contrib Mineral Petrol 161:1011–1026

Honda M, Reynolds JH, Roedder E, Epstein S (1987) Noble gases in diamonds; occurrences of solarlike helium and neon. J Geophys Res B 92; 12:12507–12521

Howell D, Stachel T, Stern R, Pearson D, Nestola F, Hardman M, Harris JW, Jaques A, Shirey S, Cartigny P (2020) Deep carbon through time: Earth's diamond record and its implications for carbon cycling and fluid speciation in the mantle. Geochim Cosmochim Acta 275:99–122

Hutchison M (1997) Constitution of the deep transition zone and lower mantle shown by diamonds and their inclusions. PhD thesis, University of Edinburgh, Edinburgh, United Kingdom

Hutchison MT, Dale CW, Nowell GM, Laiginhas FA, Pearson DG (2012) Age constraints on ultra-deep mantle petrology shown by Juina diamonds. International Kimberlite Conference: Extended Abstracts, 10. https://doi.org/10.29173/ikc3733

Ickert RB, Stachel T, Stern RA, Harris JW (2015) Extreme 18O-enrichment in majorite constrains a crustal origin of transition zone diamonds. Geochem Perspect Lett 1:65–74

Ionov DA, Doucet LS, Carlson RW, Golovin AV, Korsakov RV (2015) Post-Archean formation of the lithospheric mantle in the central Siberian craton: Re–Os and PGE study of peridotite xenoliths from the Udachnaya kimberlite. Geochim Cosmochim Acta 165:466–483

Jacob DE, Mikhail S (2022) Polycrystalline diamonds from kimberlites: Snapshots of rapid and episodic diamond formation in the lithospheric mantle. Rev Mineral Geochem 88:167–190

Jacobsen SB, Wasserburg G (1980) Sm–Nd isotopic evolution of chondrites. Earth Planet Sci Lett 50:139–155

Jaques AL, O'Neill HSC, Smith CB, Moon J, Chappell BW (1990) Diamondiferous peridotite xenoliths from the Argyle (AK1) lamproite pipe, Western Australia. Contrib Mineral Petrol 104:255–276

Jordan TH (1978) Composition and development of the continental tectosphere. Nature 274:544–548

Kaminsky FV, Khachatryan GK, Andreazza P, Araujo DP, Griffin WL (2009) Super-deep diamonds from kimberlites in the Juina area, Mato Grosso State, Brazil. Lithos 112:833–842

Kaminsky FV, Sablukov SM, Belousova EA, Andreazza P, Tremblay M, Griffin WL (2010) Kimberlitic sources of super-deep diamonds in the Juina area, Mato Grosso State, Brazil. Lithos 114:16–29

Katayama I, Karato S-I (2008) Effects of water and iron content on the rheological contrast between garnet and olivine. Phys Earth Planet Inter 166:57–66

Kemppinen LI, Kohn SC, Parkinson IJ, Bulanova GP, Howell D, Smith CB (2018) Identification of molybdenite in diamond-hosted sulphide inclusions: Implications for Re–Os radiometric dating. Earth Planet Sci Lett 495:101–111

Kennedy W (1964) The structural differentiation of Africa in the Pan-African (+/-500 m.y.) tectonic episode. *In:* University of Leeds, Research Institute of African Geology, Annual Report on Scientific Results, p 48–49

Kinny PD, Meyer HOA (1994) Zircon from the mantle; a new way to date old diamonds. J Geol 102:475–481

Kjarsgaard BA, de Wit M, Heaman LM, Pearson DG, Stiefenhofer J, Janusczcak N, Shirey SB (2022) A review of the geology of global diamond mines and deposits. Rev Mineral Geochem 88:1–118

Klein-BenDavid O, Pearson DG, Nowell GM, Ottley C, McNeill JCR, Cartigny P (2010) Mixed fluid sources involved in diamond growth constrained by Sr–Nd–Pb–C–N isotopes and trace elements. Earth Planet Sci Lett 289:123–133

Klein-BenDavid O, Pearson DG, Nowell GM, Ottley C, McNeill JC, Logvinova A, Sobolev NV (2014) The sources and time-integrated evolution of diamond-forming fluids–Trace elements and isotopic evidence. Geochim Cosmochim Acta 125:146–169

Koornneef J, Bouman C, Schwieters J, Davies G (2014) Measurement of small ion beams by thermal ionisation mass spectrometry using new 10^{13} Ohm resistors. Analytica Chimica Acta 819:49–55

Koornneef JM, Gress MU, Chinn IL, Jelsma HA, Harris JW, Davies GR (2017) Archaean and Proterozoic diamond growth from contrasting styles of large-scale magmatism. Nat Commun 8:1–8

Kopylova MG, Afanasiev VP, Bruce LF, Thurston PC, Ryder J (2011) Metaconglomerate preserves evidence for kimberlite, diamondiferous root and medium grade terrane of a pre-2.7 Ga Southern Superior protocraton. Earth Planet Sci Lett 312:213–225

Kramers JD (1979) Lead, uranium, strontium, potassium and rubidium in inclusion-bearing diamonds and mantle-derived xenoliths from southern Africa. Earth Planet Sci Lett 42:58–70

Kramers J, Smith C (1983) A feasibility study of U–Pb and Pb–Pb dating of kimberlites using groundmass mineral fractions and whole-rock samples. Chem Geol 41:23–38

Krebs M, Pearson D, Stachel T, Laiginhas F, Woodland S, Chinn I, Kong J (2019) A common parentage-low abundance trace element data of gem diamonds reveals similar fluids to fibrous diamonds. Lithos 324:356–370

Kuiper K, Deino A, Hilgen F, Krijgsman W, Renne P, Wijbrans JR (2008) Synchronizing rock clocks of Earth history. Science 320:500–504

Kukharenko A (1955) Diamonds of the Urals. PhD thesis, Gosgeolizdat, Moscow, Russia

Kurz MD, Gurney JJ, Jenkins WJ, Lott III DE (1987) Helium isotopic variability within single diamonds from the Orapa kimberlite pipe. Earth Planet Sci Lett 86:57–68

Laiginhas FA, Pearson DG, Phillips D, Burgess R, Harris JW (2009) Re–Os and $^{40}Ar/^{39}Ar$ isotope measurements of inclusions in alluvial diamonds from the Ural Mountains; constraints on diamond genesis and eruption ages. Lithos 112:714–723

Lal D (1989) An important source of ^{4}He (and ^{3}He) in diamonds. Earth Planet Sci Lett 96:1–7

Lee J-Y, Marti K, Severinghaus JP, Kawamura K, Yoo H-S, Lee JB, Kim JS (2006) A redetermination of the isotopic abundances of atmospheric Ar. Geochim Cosmochim Acta 70:4507–4512

Lippolt HJ, Leitz M, Wernicke RS, Hagedorn B (1994) (Uranium+ thorium)/helium dating of apatite: experience with samples from different geochemical environments. Chem Geol 112:179–191

Liu C-Z, Snow JE, Hellebrand E, Brügmann G, von der Handt A, Büchl A, Hofmann AW (2008) Ancient, highly heterogeneous mantle beneath Gakkel ridge, Arctic Ocean. Nature 452:311–316

Lorand J (1989) Abundance and distribution of CuFeNi sulfides, sulfur, copper and platinum-group elements in orogenic-type spinel lherzolite massifs of Ariege (Northeastern Pyrenees, France). Earth Planet Sci Lett 93:50–64

Lorand J-P, Grégoire M (2006) Petrogenesis of base metal sulphide assemblages of some peridotites from the Kaapvaal craton (South Africa). Contrib Mineral Petrol 151:521

Lorand J-P, Luguet A (2016) Chalcophile and siderophile elements in mantle rocks: Trace elements controlled by trace minerals. Rev Mineral Geochem 81:441–488

Luguet A, Pearson G (2019) Dating mantle peridotites using Re–Os isotopes: The complex message from whole rocks, base metal sulfides, and platinum group minerals. Am Mineral 104:165–189

Luguet A, Reisberg L (2016) Highly siderophile element and ^{187}Os signatures in non-cratonic basalt-hosted peridotite xenoliths: Unravelling the origin and evolution of the post-Archean lithospheric mantle. Rev Mineral Geochem 81:305–367

Luguet A, Jaques AL, Pearson DG, Smith CB, Bulanova GP, Roffey SL, Rayner MJ, Lorand JP (2009) An integrated petrological, geochemical and Re/Os isotope study of peridotite xenoliths from the Argyle lamproite, Western Australia and implications for cratonic diamond occurrences. Lithos 112:1096–1108

Lytwyn J, Burke K, Culver S (2006) The nature and location of the suture zone in the Rokelide orogen, Sierra Leone: Geochemical evidence. J Afr Earth Sci 46:439–454

Maree B (1988) Outeur se antwoord op bespreking. S Afr J Geol 91:421–422

Maree BD (1987) Die afsetting en verspreiding van spoeldiamante in Suid-Afrika Translated Title: The deposition and distribution of alluvial diamonds in South Africa. Verhandelinge van die Geologiese Vereniging van Suid Afrika [Trans Geol Soc S Afr] 90:428–447

Mazzone P, Haggerty S (1986) Corganites and corgaspinites. Two new types of aluminous assemblages from the Jagersfontein kimberlite pipe. International Kimberlite Conference: Extended Abstracts 4:279–281

McCallum ME, Huntley PM, Falk RW, Otter ML (1991) Morphological, resorption and etch feature trends of diamonds from kimberlites within the Colorado-Wyoming state line district, USA International Kimberlite Conference: Extended Abstracts 5:261–263

McDonald I, Hughes HSR, Butler IB, Harris JW, Muir D (2017) Homogenisation of sulphide inclusions within diamonds; a new approach to diamond inclusion geochemistry. Geochim Cosmochim Acta 216:335–357

McDougall I, Harrison TM (1999) Geochronology and Thermochronology by the $^{40}Ar/^{39}Ar$ Method. Oxford University Press

McNeill J, Pearson D, Klein-BenDavid O, Nowell G, Ottley C, Chinn I (2009) Quantitative analysis of trace element concentrations in some gem-quality diamonds. J Phys Condens Matter 21:364207

Meisel T, Horan MF (2016) Analytical methods for the highly siderophile elements. Rev Mineral Geochem 81:89–106

Melton GL, McNeill J, Stachel T, Pearson DG, Harris JW (2012) Trace elements in gem diamond from Akwatia, Ghana and DeBeers Pool, South Africa. Chem Geol 314–317:1–8

Menneken M, Nemchin AA, Geisler T, Pidgeon RT, Wilde SA (2007) Hadean diamonds in zircon from Jack Hills, Western Australia. Nature 448:917–920

Menzies AH, Carlson RW, Shirey SB, Gurney JJ (2003) Re–Os systematics of diamond-bearing eclogites from the Newlands Kimberlite. Lithos 71:323–336

Meyer HOA, McCallum ME (1993) Diamonds and their sources in the Venezuelan portion of the Guyana Shield. Econ Geol Bull Soc Econ Geol 88:989–998

Meyer NA, Wenz MD, Walsh JP, Jacobsen SD, Locock AJ, Harris JW (2019) Goldschmidtite,(K, REE, Sr)(Nb, Cr) O_3: A new perovskite supergroup mineral found in diamond from Koffiefontein, South Africa. Am Mineral 104:1345–1350

Miller CE, Kopylova MG, Ryder J (2012) Vanished diamondiferous cratonic root beneath the Southern Superior province: evidence from diamond inclusions in the Wawa metaconglomerate. Contrib Mineral Petrol 164:697–714

Mitchell RS, Giardini AA (1953) Oriented olivine inclusions in diamond. Am Mineral 38:136–138

Mitchell RH, Keays RR (1981) Abundance and distribution of gold, palladium and iridium in some spinel and garnet lherzolites: implications for the nature and origin of precious metal-rich intergranular components in the upper mantle. Geochim Cosmochim Acta 45:2425–2442

Moore A, Cotterill F, Broderick T, Plowes D (2009) Landscape evolution in Zimbabwe from the Permian to present, with implications for kimberlite prospecting. S Afr J Geol 112:65–88

Moore JM, Moore AE (2004) The roles of primary kimberlitic and secondary Dwyka glacial sources in the development of alluvial and marine diamond deposits in Southern Africa. J Afr Earth Sci 38:115–134

Nägler TF, Kramers JD, Kamber BS, Frei R, Prendergast MDA (1997) Growth of sub- continental lithospheric mantle beneath Zimbabwe started at or before 3.8 Ga: Re–Os study on chromites. Geology 25:983–986

Nair R, Chacko T (2008) Role of oceanic plateaus in the initiation of subduction and origin of continental crust. Geology 36:583–586

Navon O (1999) Diamond formation in the Earth's mantle. *In*: Proceedings of the 7th International Kimberlite conference. Vol 2. Gurney John J, Gurney James L, Pascoe Michelle D, Richardson Stephen H, (eds). Red Roof Publishing, Cape Town, p 584–604

Navon O, Hutcheon ID, Rossman GR, Wasserburg GJ (1988) Mantle-derived fluids in diamond micro-inclusions. Nature 335:784–789

Nestola F, Nimis P, Angel RJ, Milani S, Bruno M, Prencipe M, Harris JW (2014) Olivine with diamond-imposed morphology included in diamonds; syngenesis or protogenesis? Int Geol Rev 56:1658–1667

Nestola F, Jung H, Taylor LA (2017) Mineral inclusions in diamonds may be synchronous but not syngenetic. Nat Commun 8:14168

Nestola F, Jacob DE, Pamato MG, Pasqualetto L, Oliveira B, Greene S, Perritt S, Chinn I, Milani S, Kueter N (2019) Protogenetic garnet inclusions and the age of diamonds. Geology 47:431–434

Niespolo EM, Rutte D, Deino AL, Renne PR (2017) Intercalibration and age of the Alder Creek sanidine $^{40}Ar/^{39}Ar$ standard. Quat Geochron 39:205–213

Nimis P (2022) Pressure and temperature data for diamonds. Rev Mineral Geochem 88:533–566

Nimis P, Alvaro M, Nestola F, Angel RJ, Marquardt K, Rustioni G, Harris JW, Marone F (2016) First evidence of hydrous silicic fluid films around solid inclusions in gem quality diamonds. Lithos 260:384–389

O'Neill C, Marchi S, Bottke W, Fu R (2020) The role of impacts on Archaean tectonics. Geology 48:174–178

Oosterveld M (2003) Evaluation of alluvial diamond deposits. *In*: Workshop Alluvial Diamonds in South Africa. Vol 2, p 1–7

Otter ML (1989) Diamonds and their mineral inclusions from the Sloan diatremes of the Colorado-Wyoming State Line kimberlite district, North America. PhD thesis, University of Cape Town, Cape Town, South Africa

Ozima M, Zashu S, Tomura K, Matsuhisa Y (1991) Constraints from noble-gas contents on the origin of carbonado diamonds. Nature 351:472–474

Pamato MG, Novella D, Jacob DE, Oliveira B, Pearson DG, Greene S, Afonso JC, Favero M, Stachel T, Alvaro M, Nestola F (2021) Protogenetic sulfide inclusions in diamonds date the diamond formation event using Re–Os isotopes. Geology 49:941–945

Patten C, Barnes S-J, Mathez EA, Jenner FE (2013) Partition coefficients of chalcophile elements between sulfide and silicate melts and the early crystallization history of sulfide liquid: LA-ICP-MS analysis of MORB sulfide droplets. Chem Geol 358:170–188

Pearson DG (1999) The age of continental roots. Lithos 48:171–194

Pearson DG, Shirey SB (1999) Isotopic dating of diamonds. *In*: Application of radiogenic isotopes to ore deposit research and exploration. Vol 12. Lambert David D, Ruiz J (eds) Soc Econ Geol, Boulder, CO, USA, p 143–171

Pearson DG, Wittig N (2008) Formation of Archaean continental lithosphere and its diamonds; the root of the problem. J Geol Soc London 165:895–914

Pearson DG, Carlson RW, Shirey SB, Boyd FR, Nixon PH (1995) Stabilisation of Archaean lithospheric mantle; a Re–Os isotope study of peridotite xenoliths from the Kaapvaal Craton. Earth Planet Sci Lett 134:341–357

Pearson DG, Davies R, Shirey SB, Carlson RW, Griffin W (1998a) The age and origin of eastern Australian diamonds: Re–Os isotope evidence from sulfide inclusions in two diamonds from Wellington, New South Wales. International Kimberlite Conference: Extended Abstracts 7:664–666

Pearson DG, Shirey SB, Harris JW, Carlson RW (1998b) Sulphide inclusions in diamonds from the Koffiefontein kimberlite, S Africa; constraints on diamond ages and mantle Re–Os systematics. Earth Planet Sci Lett 160:311–326

Pearson DG, Shirey SB, Bulanova GP, Carlson RW, Milledge HJ (1999a) Re–Os isotope measurements of single sulfide inclusions in a Siberian diamond and its nitrogen aggregation systematics. Geochim Cosmochim Acta 63:703–711

Pearson DG, Shirey SB, Bulanova GP, Carlson RW, Milledge HJ (1999b) Dating and paragenetic distinction of diamonds using the Re–Os isotope system; application to some Siberian diamonds. *In*: Proceedings of the 7th International Kimberlite Conference. Vol 2. Gurney John J, Gurney James L, Pascoe Michelle D, Richardson Stephen H, (eds). Red Roof Publishing, Cape Town, p 637–643

Pearson DG, Irvine G, Carlson RW, Kopylova M, Ionov D (2002) The development of lithospheric keels beneath the earliest continents: time constraints using PGE and Re–Os isotope systematics. Geol Soc London Spec Publ 199:65–90

Pearson DG, Harlou R, Hayman P, Cartigny P, Kopylova M (2007) Sr isotopic compositions of ultra-deep inclusions in diamonds; implications for mantle chemical structure and evolution. Geochim Cosmochim Acta 71:A769

Pearson DG, Smith C, Liu J, Mather K, Krebs M, Bulanova G, Kobussen A (2018) Characteristics and origin of the mantle root beneath the Murowa diamond mine: Implications for craton and diamond formation. Chapter 19 *In:* Geoscience and Exploration of the Argyle, Bunder, Diavik and Murowa Diamond Deposit. Davy AT, Smith CB, Helmstaedt H, Jaques AL, Gurney JJ (eds) Spec Publ Soc Econ Geol Vol 20, p 403–424

Pearson DG, Scott JM, Liu J, Schaeffer A, Wang LH, Van Hunen J, Szilas K, Chacko T, Kelemen PB (2021) Deep continental roots and cratons. Nature 592:199–210

Pernet-Fisher JF, Barry PH, Day JM, Pearson DG, Woodland S, Agashev AM, Pokhilenko LN, Pokhilenko NP (2019) Heterogeneous kimberlite metasomatism revealed from a combined Re–Os isotope study of Siberian megacrystalline dunite xenoliths. Geochim Cosmochim Acta 266:220–236

Petts DC, Stachel T, Stern RA, Hunt L, Fomradas G (2016) Multiple carbon and nitrogen sources associated with the parental mantle fluids of fibrous diamonds from Diavik, Canada, revealed by SIMS microanalysis. Contrib Mineral Petrol 171:17

Peucker-Ehrenbrink B, Bach W, Hart SR, Blusztajn JS, Abbruzzese T (2003) Rhenium-osmium isotope systematics and platinum group element concentrations in oceanic crust from DSDP/ODP Sites 504 and 417/418. Geochem Geophys Geosystems 4:8911

Phillips D, Harris JW (2008) Provenance studies from $^{40}Ar/^{39}Ar$ dating of mineral inclusions in diamonds; methodological tests on the Orapa Kimberlite, Botswana. Earth Planet Sci Lett 274:169–178

Phillips D, Harris JW (2009) Diamond provenance studies from $^{40}Ar/^{39}Ar$ dating of clinopyroxene inclusions; an example from the west coast of Namibia. Lithos 112:793–805

Phillips D, Onstott TC, Harris JW (1989) $^{40}Ar/^{39}Ar$ laser-probe dating of diamond inclusions from Premier kimberlite. Nature 340:460–462

Phillips D, Harris JW, Kiviets GB (2004) $^{40}Ar/^{39}Ar$ analyses of clinopyroxene inclusions in African diamonds; implications for source ages of detrital diamonds. Geochim Cosmochim Acta 68:151–165

Phillips D, Matchan E, Honda M, Kuiper K (2017) Astronomical calibration of $^{40}Ar/^{39}Ar$ reference minerals using high-precision, multi-collector (ARGUSVI) mass spectrometry. Geochim Cosmochim Acta 196:351–369

Phillips D, Harris JW, de Wit MC, Matchan EL (2018) Provenance history of detrital diamond deposits, West Coast of Namaqualand, South Africa. Mineral Petrol 112:259–273

Pidgeon RT (1989) Archaean diamond xenocrysts in kimberlites; how definitive is the evidence? *In*: Kimberlites and Related Rocks. Vol 14; 2. Ross J, Jaques AL, Ferguson J, Green DH, Y ORS, Danchin RV, Janse AJA, (eds). Geological Society of Australia, Sydney NSW, Australia, p 1006–1011

Raal FA (1969) A study of some gold mine diamonds. Am Mineral 54:292–296

Rayner MJ, Jaques A, Boxer G, Smith C, Lorenz V, Moss SW, Webb K, Ford D (2018) The geology of the Argyle (AK1) diamond deposit, Western Australia. Chapter 4 *In:* Geoscience and Exploration of the Argyle, Bunder, Diavik and Murowa Diamond Deposit. Davy AT, Smith CB, Helmstaedt H, Jaques AL, Gurney JJ (eds).Spec Publ Soc Econ Geol Vol 20, p 89–117

Regier M, Miškovic A, Ickert RB, Pearson DG, Stachel T, Stern RA, Kopylova M (2018) An oxygen isotope test for the origin of Archean mantle roots. Geochem Perspect Lett 9:6–10

Reimink JR, Chacko T, Stern RA, Heaman LM (2014) Earth's earliest evolved crust generated in an Iceland-like setting. Nat Geosci 7:529–533

Reimink J, Pearson D, Shirey S, Carlson R, Ketchum J (2019) Onset of new, progressive crustal growth in the central Slave craton at 3.55 Ga. Geochem Perspect Lett 10:8–13

Reiners PW, Carlson RW, Renne PR, Cooper KM, Granger DE, McLean NM, Schoene B (2018) Geochronology and Thermochronology. John Wiley and Sons

Reis NJ, Nadeau S, Fraga LM, Betiollo LM, Faraco MTL, Reece J, Lachhman D, Ault R (2017) Stratigraphy of the Roraima Supergroup along the Brazil–Guyana border in the Guiana shield, Northern Amazonian Craton-results of the Brazil-Guyana geology and geodiversity mapping project. Brazilian J Geol 47:43–57

Reisberg L (2021) Osmium isotope constraints on formation and refertilization of the non-cratonic continental mantle lithosphere. Chem Geol 574:120245

Reisberg L, Lorand J-P (1995) Longevity of sub-continental mantle lithosphere from osmium isotope systematics in orogenic peridotite massifs. Nature 376:159–162

Richardson SH (1986) Latter-day origin of diamonds of eclogitic paragenesis. Nature 322:623–626

Richardson SH (1989) As definitive as ever: A Reply to Archean diamond xenocrysts in kimberlites; how definitive is the evidence? by R T Pidgeon. *In*: Kimberlites and Related Rocks. Vol 14; 2. Ross J, Jaques AL, Ferguson J, Green DH, O'Reilly SY, Danchin RV, Janse AJA, (eds). Geol Soc Aust, Sydney NSW, Australia, p 1070–1072

Richardson SH, Shirey SB (2008) Continental mantle signature of Bushveld magmas and coeval diamonds. Nature 453:910–913

Richardson SH, Gurney JJ, Erlank AJ, Harris JW (1984) Origin of diamonds in old enriched mantle. Nature 310:198–202

Richardson SH, Erlank AJ, Harris JW, Hart SR (1990) Eclogitic diamonds of Proterozoic age from Cretaceous kimberlites. Nature 346:54–56

Richardson SH, Harris JW, Gurney JJ (1993) Three generations of diamonds from old continental mantle. Nature 366:256–258

Richardson SH, Harris JW (1997) Antiquity of peridotitic diamonds from the Siberian Craton. Earth Planet Sci Lett 151:271–277

Richardson SH, Chinn IL, Harris JW (1998) Age and origin of eclogitic diamonds from the Jwaneng kimberlite, Botswana. International Kimberlite Conference: Extended Abstracts 7:734–736

Richardson SH, Chinn IL, Harris JW (1999) Age and origin of eclogitic diamonds from the Jwaneng Kimberlite, Botswana. *In*: Proceedings of the 7th International Kimberlite Conference. Vol 2. Gurney John J, Gurney James L, Pascoe Michelle D, Richardson Stephen H, (eds). Red Roof Publishing, Cape Town, p 709–713

Richardson SH, Shirey SB, Harris JW, Carlson RW (2001) Archean subduction recorded by Re–Os isotopes in eclogitic sulfide inclusions in Kimberley diamonds. Earth Planet Sci Lett 191:257–266

Richardson SH, Shirey SB, Harris JW (2004) Episodic diamond genesis at Jwaneng, Botswana, and implications for Kaapvaal Craton evolution. Lithos 77:143–154

Richardson SH, Poeml PF, Shirey SB, Harris JW (2009) Age and origin of peridotitic diamonds from Venetia, Limpopo Belt, Kaapvaal-Zimbabwe Craton. Lithos 112:785–792

Rotenberg E, Davis DW, Amelin Y, Ghosh S, Bergquist BA (2012) Determination of the decay-constant of ^{87}Rb by laboratory accumulation of ^{87}Sr. Geochim Cosmochim Acta 85:41–57

Rudnick RL, Walker RJ (2009) Interpreting ages from Re–Os isotopes in peridotites. Lithos 112:1083–1095

Rudnick RL, Eldridge CS, Bulanova GP (1993) Diamond growth history from in situ measurement of Pb and S isotopic compositions of sulfide inclusions; with Suppl. Data 9305. Geology 21:13–16

Rutherford E (1905) Radio-activity. Cambridge University Press

Saal A, Takazawa E, Frey F, Shimizu N, Hart S (2001) Re–Os isotopes in the Horoman peridotite: evidence for refertilization? J Petrol 42:25–37

Schärer U, Corfu F, Demaiffe D (1997) U–Pb and Lu–Hf isotopes in baddeleyite and zircon megacrysts from the Mbuji-Mayi kimberlite: constraints on the subcontinental mantle. Chem Geol 143:1–16
Schmidberger SS, Simonetti A, Heaman LM, Creaser RA, Whiteford S (2007) Lu–Hf, in-situ Sr and Pb isotope and trace element systematics for mantle eclogites from the Diavik diamond mine; evidence for Paleoproterozoic subduction beneath the Slave Craton, Canada. Earth Planet Sci Lett 254:55–68
Schmitt AK, Zack T, Kooijman E, Logvinova AM, Sobolev NV (2019) U–Pb ages of rare rutile inclusions in diamond indicate entrapment synchronous with kimberlite formation. Lithos 350:105251
Schmitz MD, Bowring SA, de Wit MJ, Gartz V (2004) Subduction and terrane collision stabilize the western Kaapvaal craton tectosphere 2.9 billion years ago. Earth Planet Sci Lett 222:363–376
Schrauder M, Koeberl C, Navon O (1996) Trace element analyses of fluid-bearing diamonds from Jwaneng, Botswana. Geochim Cosmochim Acta 60:4711–4724
Scoates JS, Friedman RM (2008) Precise age of the platiniferous Merensky Reef, Bushveld Complex, South Africa, by the U–Pb zircon chemical abrasion ID-TIMS technique. Econ Geol 103:465–471
Scott J, Liu J, Pearson D, Harris G, Czertowicz T, Woodland S, Riches A, Luth R (2019) Continent stabilisation by lateral accretion of subduction zone-processed depleted mantle residues; insights from Zealandia. Earth Planet Sci Lett 507:175–186
Selby D, Creaser RA, Stein HJ, Markey RJ, Hannah JL (2007) Assessment of the ^{187}Re decay constant by cross calibration of Re–Os molybdenite and U–Pb zircon chronometers in magmatic ore systems. Geochim Cosmochim Acta 71:1999–2013
Shee SR, Gurney JJ, Robinson DN (1982) Two diamond-bearing peridotite xenoliths from the Finsch Kimberlite, South Africa. Contrib Mineral Petrol 81:79–87
Shelkov DA, Verchovsky AB, Milledge HJ, Pillinger CT (1998) The radial distribution of implanted and trapped ^{4}He in single diamond crystals and implications for the origin of carbonado. Chem Geol 149:109–116
Shigley JE, Chapman J, Ellison RK (2001) Discovery and mining of the Argyle diamond deposit, Australia. Gems Gemol 37:26–41
Shimizu N, Richardson SH (1987) Trace element abundance patterns of garnet inclusions in peridotite-suite diamonds. Geochim Cosmochim Acta 51:755–758
Shimizu N, Sobolev NV (1995) Young peridotitic diamonds from the Mir kimberlite pipe. Nature 375:394–397
Shimizu N, Pokhilenko NP, Boyd FR, Pearson DG (1997) Geochemical characteristics of mantle xenoliths from the Udachnaya kimberlite pipe. Russ Geol Geophys 38:205–217
Shirey SB, Richardson SH (2011) Start of the Wilson cycle at 3 Ga shown by diamonds from subcontinental mantle. Science 333:434–436
Shirey SB, Walker RJ (1998) The Re–Os isotope system in cosmochemistry and high-temperature geochemistry. Annu Rev Earth Planet Sci 26:423–500
Shirey SB, Harris JW, Carlson RW (1999) Re–Os systematics of sulfide inclusions in diamonds from the Orapa Kimberlite, Botswana; implications for multiple generations of diamond growth. Eos Trans Am Geophys Union 80 suppl 46:1191
Shirey SB, Carlson RW, Richardson SH, Menzies A, Gurney JJ, Pearson DG, Harris JW, Wiechert U (2001) Archean emplacement of eclogitic components into the lithospheric mantle during formation of the Kaapvaal Craton. Geophys Res Lett 28:2509–2512
Shirey SB, Harris JW, Richardson SH, Fouch MJ, James DE, Cartigny P, Deines P, Viljoen F (2002) Diamond genesis, seismic structure, and evolution of the Kaapvaal-Zimbabwe Craton. Science 297:1683–1686
Shirey SB, Richardson SH, Harris JW (2004) Integrated models of diamond formation and craton evolution. Lithos 77:923–944
Shirey SB, Cartigny P, Frost DJ, Keshav S, Nestola F, Nimis P, Pearson DG, Sobolev NV, Walter MJ (2013) Diamonds and the geology of mantle carbon. Rev Mineral Geochem 75:355–421
Shirey SB, Smit KV, Pearson DG, Walter MJ, Aulbach S, Brenker FE, Bureau H, Burnham AD, Cartigny P, Chacko T, Frost DJ, Hauri EH, Jacob DE, Jacobsen SD, Kohn SC, Luth RW, Mikhail S, Navon O, Nestola F, Nimis P, Palot M, Smith EM, Stachel T, Stagno V, Steele A, Stern RA, Thomassot E, Thomson AR, Weiss Y (2019) Diamonds and the mantle geodynamics of carbon: deep mantle carbon evolution from the diamond record. *In*: Deep Carbon: Past to Present. Orcutt BN, Daniel I, Dasgupta R, (eds). Cambridge University Press, Cambridge, p 89–128
Shirey SB, Wagner LS, Walter MJ, Pearson DG, van Keken PE (2021) Slab transport of fluids to deep focus earthquake depths—thermal modeling constraints and evidence from diamonds. AGU Adv 2: e2020AV000304
Simon NSC, Carlson RW, Pearson DG, Davies GR (2007) The origin and evolution of the Kaapvaal cratonic lithospheric mantle. J Petrol 48:589–625
Sizova E, Gerya T, Stüwe K, Brown M (2015) Generation of felsic crust in the Archean: a geodynamic modeling perspective. Precambrian Res 271:198–224
Smit KV, Milne SEM, Richardson SH, Timmerman S, Pearson DG (2021) Radiogenic isotope compositions of inclusions in diamond. https://doi.org/10.7939/DVN/DRAJGT, Scholars Portal Dataverse, V1
Smit KV, Shirey SB, Richardson SH, le Roex AP, Gurney JJ (2010) Re/Os isotopic composition of peridotitic sulphide inclusions in diamonds from Ellendale, Australia; age constraints on Kimberley cratonic lithosphere. Geochim Cosmochim Acta 74:3292–3306

Smit KV, Shirey SB, Wang W (2016) Type Ib diamond formation and preservation in the West African lithospheric mantle; Re/Os age constraints from sulphide inclusions in Zimmi diamonds. Precambrian Res 286:152–166
Smit KV, Stachel T, Luth RW, Stern RA (2019a) Evaluating mechanisms for eclogitic diamond growth: An example from Zimmi Neoproterozoic diamonds (West African craton). Chem Geol 520:21–32
Smit KV, Shirey SB, Hauri EH, Stern RA (2019b) Sulfur isotopes in diamonds reveal differences in continent construction. Science 364:383–385
Smith CB (1983) Pb, Sr and Nd isotopic evidence for sources of southern African Cretaceous kimberlites. Nature 304:51–54
Smith CB, Gurney JJ, Harris JW, Otter ML, Kirkley MB, Jagoutz E (1991) Neodymium and strontium isotope systematics of eclogite and websterite paragenesis inclusions from single diamonds, Finsch and Kimberley Pool, RSA. Geochim Cosmochim Acta 55:2579–2590
Smith CB, Pearson DG, Bulanova GP, Beard AD, Carlson RW, Wittig N, Sims K, Chimuka L, Muchemwa E (2009) Extremely depleted lithospheric mantle and diamonds beneath the southern Zimbabwe Craton. Lithos 112:1120–1132
Smith EM, Kopylova MG, Nowell GM, Pearson DG, Ryder J (2012) Archean mantle fluids preserved in fibrous diamonds from Wawa, Superior craton. Geology 40:1071–1074
Smith EM, Kopylova MG, Frezzotti ML, Afanasiev VP (2014) N-rich fluid inclusions in octahedrally-grown diamond. Earth Planet Sci Lett 393:39–48
Smith EM, Shirey SB, Nestola F, Bullock ES, Wang J, Richardson SH, Wang W (2016) Large gem diamonds from metallic liquid in Earth's deep mantle. Science 354:1403–1405
Smith EM, Richardson SH, Shirey SB, Wang W (2017) Old unradiogenic Os in deep mantle metallic liquid from large gem IIa diamonds. International Kimberlite Conference: Extended Abstracts, 11. https://doi.org/10.29173/ikc3990
Smith EM, Shirey SB, Richardson SH, Nestola F, Bullock ES, Wang J, Wang W (2018) Blue boron-bearing diamonds from Earth's lower mantle. Nature 560:84–87
Smith EM, Ni P, Shirey SB, Richardson S, Wang W, Shahar A (2021) Heavy iron in large gem diamonds traces deep subduction of serpentinized ocean floor. Sci Adv 7:eabe9773
Smith EM, Krebs MY, Genzel P-T, Brenker FE (2022) Raman identification of inclusions in diamond. Rev Mineral Geochem 88:451–474
Sobolev NV (1977). Deep-seated Inclusions in Kimberlites and the Problem of the Composition of the Upper Mantle. (Translated from the Russian edition 1974). AGU, Washington, 279 pp
Sobolev NV (1985) Ultramafic and eclogitic types of paragenesis of diamonds. Conference extended abstract: Native elements formation in endogenic processes, Acad Sci USSR, Siberian Division, Yakutsk (not paginated)
Spencer LK, Dikinis SD, Keller PC, Kane RE (1988) The diamond deposits of Kalimantan, Borneo. Gems Gemol 24:67–80
Spetsius ZV, Belousova E, Griffin WL, O'Reilly SY, Pearson N (2002) Archean sulfide inclusions in Paleozoic zircon megacrysts from the Mir kimberlite, Yakutia: implications for the dating of diamonds. Earth Planet Sci Lett 199:111–126
Stacey Jt, Kramers J (1975) Approximation of terrestrial lead isotope evolution by a two-stage model. Earth Planet Sci Lett 26:207–221
Stachel T (2014) Diamond. Geology of Gem Deposits. Groat LA (ed) Mineral Assoc Canada, Short Course Ser 44:1–28
Stachel T, Harris JW (2008) The origin of cratonic diamonds; constraints from mineral inclusions. Ore Geol Rev 34:5–32
Stachel T, Luth RW (2015) Diamond formation; where, when and how? Lithos 220–223:200–220
Stachel T, Viljoen KS, Brey G, Harris JW (1998) Metasomatic processes in lherzolitic and harzburgitic domains of diamondiferous lithospheric mantle; REE in garnets from xenoliths and inclusions in diamonds. Earth Planet Sci Lett 159:1–12
Stachel T, Harris JW, Brey GP, Joswig W (2000) Kankan diamonds (Guinea) II: lower mantle inclusion parageneses. Contrib Mineral Petrol 140 :16–27
Stachel T, Brey GP, Harris JW (2005) Inclusions in sublithospheric diamonds; glimpses of deep earth. Elements 1:73–87
Stachel T, Harris J, Hunt L, Muehlenbachs K, Kobussen A, EIMF (2018a) Argyle diamonds-how subduction along the Kimberley craton edge generated the world's biggest diamond deposit. Chapter 6 *In:* Geoscience and Exploration of the Argyle, Bunder, Diavik and Murowa Diamond Deposit. Davy AT, Smith CB, Helmstaedt H, Jaques AL, Gurney JJ (eds).Spec Publ Soc Econ Geol Vol 20, p 145–167
Stachel T, Banas A, Aulbach S, Smit KV, Wescott P, Chinn IL, Kong J (2018b) The Victor Mine (Superior Craton, Canada): Neoproterozoic lherzolitic diamonds from a thermally-modified cratonic root. Mineral Petrol 112:325–336
Stachel T, Aulbach S, Harris JW (2022a) Mineral inclusions in lithospheric diamonds. Rev Mineral Geochem 88:307–392
Stachel T, Cartigny P, Chacko T, Pearson DG (2022b) Carbon and nitrogen in mantle-derived diamonds. Rev Mineral Geochem 88:809–876
Stachel T, Banas A, Muehlenbachs K, Kurszlaukis S, Walker EC (2006) Archean diamonds from Wawa (Canada); samples from deep cratonic roots predating cratonization of the Superior Province. Contrib Mineral Petrol 151:737–750
Steiger RH, Jäger E (1977) Subcommission on geochronology: convention on the use of decay constants in geo-and cosmochronology. Earth Planet Sci Lett 36:359–362
Stern RJ (2018) The evolution of plate tectonics. Philos Trans R Soc A 376:20170406
Strutt RJ (1905) On the radio-active minerals. Proc R Soc London Ser A 76:88–101

Sutherland FL, Raynor LR, Pogson RE (1994) Spinel to garnet lherzolite transition in relation to high temperature palaeogeotherms, eastern Australia. Aust J Earth Sci 41:205–220

Taylor WR (1990) A reappraisal of the nature of fluids included by diamond; a window to deep-seated mantle fluids and redox conditions. *In*: Proceedings of the Conference on Stable Isotopes and Fluid Processes in Mineralization. Vol 23. Herbert Hugh K, Ho Susan E, (eds). University of Western Australia, Geology Department and Extension Service. Perth, West, Aust., Australia, p 333–349

Taylor WR, Bulanova G, Milledge HJ (1995) Quantitative nitrogen aggregation study of some Yakutian diamonds; constraints on the growth, thermal, and deformation history of peridotitic and eclogitic diamonds. International Kimberlite Conference: Extended Abstracts 6:608–610

Taylor LA, Anand M, Promprated P, Floss C, Sobolev NV (2003) The significance of mineral inclusions in large diamonds from Yakutia, Russia. Am Mineral 88:912–920

Thomassot E, Cartigny P, Harris JW, Lorand JP, Rollion-Bard C, Chaussidon M (2009) Metasomatic diamond growth; a multi-isotope study (^{13}C, ^{15}N, ^{33}S, ^{34}S) of sulphide inclusions and their host diamonds from Jwaneng (Botswana). Earth Planet Sci Lett 282:79–90

Thomson AR, Kohn SC, Bulanova GP, Smith CB, Araujo DP, Walter MJ, Shirey SB (2016) Trace element composition of silicate inclusions in sub-lithospheric diamonds from the Juina-5 kimberlite; evidence for diamond growth from slab melts. Lithos 265:108–124

Timmerman S (2018) Diamonds-Time capsules of volatiles and the key to dynamic Earth evolution. PhD thesis, Australian National University, Canberra, Australia

Timmerman S, Koornneef JM, Chinn IL, Davies GR (2017) Dated eclogitic diamond growth zones reveal variable recycling of crustal carbon through time. Earth Planet Sci Lett 463:178–188

Timmerman S, Krebs MY, Pearson DG, Honda M (2019a) Diamond-forming media through time–Trace element and noble gas systematics of diamonds formed over 3 billion years of Earth's history. Geochim Cosmochim Acta 257:266–283

Timmerman S, Honda M, Zhang X, Jaques A, Bulanova G, Smith C, Burnham A (2019b) Contrasting noble gas compositions of peridotitic and eclogitic monocrystalline diamonds from the Argyle lamproite, Western Australia. Lithos 344:193–206

Timmerman S, Yeow H, Honda M, Howell D, Jaques AL, Krebs MY, Woodland S, Pearson DG, Ávila JN, Ireland TR (2019c) U–Th/He systematics of fluid-rich 'fibrous' diamonds–Evidence for pre-and syn-kimberlite eruption ages. Chem Geol 515:22–36

Timmerman S, Honda M, Burnham A, Amelin Y, Woodland S, Pearson DG, Jaques A, Le Losq C, Bennett V, Bulanova G (2019d) Primordial and recycled helium isotope signatures in the mantle transition zone. Science 365:692–694

Timmerman S, Pearson G, Nestola F, Banas A, Stachel T, Stern RA, Reimink JR, Vezinet A, Ielpi A, Mircea C (2020) Diamond-bearing metasediments point to thick, cool lithospheric root established by the Mesoarchean beneath parts of the Slave Craton (Canada). AGUFM:DI020-0002

Tirone M, Ganguly J, Dohmen R, Langenhorst F, Hervig R, Becker H-W (2005) Rare earth diffusion kinetics in garnet: experimental studies and applications. Geochim Cosmochim Acta 69:2385–2398

Tyler I, Page R, Griffin T (1999) Depositional age and provenance of the Marboo Formation from SHRIMP U–Pb zircon geochronology: implications for the early Palaeoproterozoic tectonic evolution of the Kimberley region, Western Australia. Precambrian Res 95:225–243

van Achterbergh E, Griffin WL, Stiefenhofer J (2001) Metasomatism in mantle xenoliths from the Letlhakane kimberlites: estimation of element fluxes. Contrib Mineral Petrol 141:397–414

van Acken D, Becker H, Hammerschmidt K, Walker RJ, Wombacher F (2010) Highly siderophile elements and Sr–Nd isotopes in refertilized mantle peridotites—a case study from the Totalp ultramafic body, Swiss Alps. Chem Geol 276:257–268

Van Kranendonk MJ (2010) Two types of Archean continental crust: Plume and plate tectonics on early Earth. Am J Sci 310:1187–1209

Van Kranendonk MJ, Kirkland CL (2016) Conditioned duality of the Earth system: Geochemical tracing of the supercontinent cycle through Earth's history. Earth Sci Rev 160:171–187

Van Kranendonk MJ, Smithies RH, Hickman AH, Champion D (2007) secular tectonic evolution of Archean continental crust: interplay between horizontal and vertical processes in the formation of the Pilbara Craton, Australia. Terra Nova 19:1–38

Van Kranendonk MJ, Smithies RH, Griffin WL, Huston DL, Hickman AH, Champion DC, Anhaeusser CR, Pirajno F (2015) Making it thick: a volcanic plateau origin of Palaeoarchean continental lithosphere of the Pilbara and Kaapvaal cratons. Geol Soc London Spec Publ 389:83–111

Van Orman JA, Grove TL, Shimizu N, Layne GD (2002) Rare earth element diffusion in a natural pyrope single crystal at 2.8 GPa. Contrib Mineral Petrol 142:416–424

Van Wyk JP, Pienaar LF (1986) Diamondiferous gravels of the Lower Orange River, Namaqualand. *In*: Mineral deposits of Southern Africa. Anhaeusser CR, Maske S, (eds). Geol Soc S Afr, Johannesburg, South Africa, p 2309–2321

Vermeesch P (2010) HelioPlot, and the treatment of overdispersed (U–Th–Sm)/He data. Chem Geol 271:108–111

Vermeesch P (2018) IsoplotR: A free and open toolbox for geochronology. Geosci Front 9:1479–1493

Viljoen KS, Smith CB, Sharp ZD (1996) Stable and radiogenic isotope study of eclogite xenoliths from the Orapa kimberlite, Botswana. Chem Geol 131:235–255

Viljoen KS, Harris JW, Ivanic T, Richardson SH, Gray K (2014) Trace element chemistry of peridotitic garnets in diamonds from the Premier (Cullinan) and Finsch kimberlites, South Africa; contrasting styles of mantle metasomatism. Lithos 208–209:1–15

Völkening J, Walczyk T, Heumann KG (1991) Osmium isotope ratio determinations by negative thermal ionization mass spectrometry. Int J Mass Spectrom Ion Processes 105:147–159

Wagner PA (1914) The Diamond Fields of Southern Africa. Johannesburg: The Transvaal Leader; London: The Technical Book-Shop

Walker R, Carlson R, Shirey S, Boyd F (1989) Os, Sr, Nd, and Pb isotope systematics of southern African peridotite xenoliths: implications for the chemical evolution of subcontinental mantle. Geochim Cosmochim Acta 53:1583–1595

Walker R, Horan M, Morgan J, Becker H, Grossman J, Rubin A (2002) Comparative ^{187}Re–^{187}Os systematics of chondrites: Implications regarding early solar system processes. Geochim Cosmochim Acta 66:4187–4201

Walter M, Bulanova G, Armstrong L, Keshav S, Blundy J, Gudfinnsson G, Lord O, Lennie A, Clark S, Smith C (2008) Primary carbonatite melt from deeply subducted oceanic crust. Nature 454:622–625

Walter MJ, Kohn SC, Araujo D, Bulanova GP, Smith CB, Gaillou E, Wang J, Steele A, Shirey SB (2011) Deep mantle cycling of oceanic crust; evidence from diamonds and their mineral inclusions. Science 334:54–57

Walter MJ, Thomson AR, Smith EM (2022) Geochemistry of silicate and oxide inclusions in sublithospheric diamonds. Rev Mineral Geochem 88:393–450

Wang H, van Hunen J, Pearson DG (2018) Making Archean cratonic roots by lateral compression: a two-stage thickening and stabilization model. Tectonophysics 746:562–571

Warnock A, Zeitler P, Wolf R, Bergman S (1997) An evaluation of low-temperature apatite UTh/He thermochronometry. Geochim Cosmochim Acta 61:5371–5377

Wass SY, Rogers N (1980) Mantle metasomatism—precursor to continental alkaline volcanism. Geochim Cosmochim Acta 44:1811–1823

Weiss Y, Cazs J, Navon O (2022a) Fluid inclusions in fibrous diamonds. Rev Mineral Geochem 88:475–532

Weiss Y, Czas J, Navon O (2022b) Global published compositions of high-density fluids (HDFs) in fibrous diamonds, Version 1.0. Interdisciplinary Earth Data Alliance (IEDA), doi: 10.26022/IEDA/112016

Weiss Y, Kiro Y, Class C, Winckler G, Harris JW, Goldstein SL (2021) Helium in diamonds unravels over a billion years of craton metasomatism. Nat Commun 12:2667

Welke HJ, Allsopp HL, Harris JW (1974) Measurements of K, Rb, U, Sr and Pb in diamonds containing inclusions. Nature 252:35–37

Westerlund KJ, Gurney JJ (2004) Silicate and oxide inclusion characteristics and infra-red absorption analysis of diamonds from the Klipspringer kimberlites, South Africa. S Afr J Geol 107:131–146

Westerlund KJ, Gurney JJ, Carlson RW, Shirey SB, Hauri EH, Richardson SH (2004) A metasomatic origin for late Archean eclogitic diamonds; implications from internal morphology of diamonds and Re–Os and S isotope characteristics of their sulfide inclusions from the Late Jurassic Klipspringer kimberlites. S Afr J Geol 107:119–130

Westerlund KJ, Shirey SB, Richardson SH, Carlson RW, Gurney JJ, Harris JW (2006) A subduction wedge origin for Paleoarchean peridotitic diamonds and harzburgites from the Panda Kimberlite, Slave Craton; evidence from Re–Os isotope systematics. Contrib Mineral Petrol 152:275–294

Wetherill GW (1956) Discordant uranium–lead ages I. Eos, Trans Am Geophys Union 37:320–326

White LT, Graham I, Tanner D, Hall R, Armstrong RA, Yaxley G, Barron L, Spencer L, van Leeuwen TM (2016) The provenance of Borneo's enigmatic alluvial diamonds; a case study from Cempaka, SE Kalimantan. Gondwana Res 38:251–272

Wiggers de Vries DF, Pearson DG, Bulanova GP, Smelov AP, Pavlushin AD, Davies GR (2013) Re–Os dating of sulphide inclusions zonally distributed in single Yakutian diamonds; evidence for multiple episodes of Proterozoic formation and protracted time scales of diamond growth. Geochim Cosmochim Acta 120:363–394

Williams AF (1932) The genesis of the diamond. London Ernest Benn

Wilson JF (1990) A craton and its cracks: some of the behaviour of the Zimbabwe block from the Late Archaean to the Mesozoic in response to horizontal movements, and the significance of some of its mafic dyke fracture patterns. J Afr Earth Sci 10:483–501

Wolf R, Farley K, Silver L (1996) Helium diffusion and low-temperature thermochronometry of apatite. Geochim Cosmochim Acta 60:4231–4240

Wu F, Mitchell RH, Li Q, Sun J, Liu C, Yang Y (2013) In situ U–Pb age determination and Sr–Nd isotopic analysis of perovskite from the Premier (Cullinan) kimberlite, South Africa. Chem Geol 353:83–95

Zegers TE, de Keijzer M, Passchier CAW, White SH (1998) The Mulgandinnah shear zone, and Archean crustal-scale strike-slip zone, eastern Pilbara, Western Australia. Precambrian Res 88:233–248

Zeh A, Stern RA, Gerdes A (2014) The oldest zircons of Africa—their U–Pb–Hf–O isotope and trace element systematics, and implications for Hadean to Archean crust–mantle evolution. Precambrian Res 241:203–230

Reviews in Mineralogy & Geochemistry
Vol. 88 pp. 637–688, 2022

Diamond Spectroscopy, Defect Centers, Color, and Treatments

Ben L. Green

Department of Physics
University of Warwick
Coventry CV4 7AL, UK

Alan T. Collins

Department of Physics
King's College London
Strand
London WC2R 2LS, UK

Christopher M. Breeding

Gemological Institute of America
The Robert Mouawad Campus
5345 Armada Drive
Carlsbad, CA92008, USA

b.green@warwick.ac.uk; alan.collins@kcl.ac.uk; mbreedin@gia.edu

OPTICAL SPECTROSCOPY OF DEFECT-FREE DIAMOND

Optical spectroscopy is a valuable non-destructive technique for the study of many materials. For diamond it is used for fundamental research to enhance our knowledge about this remarkable material; it is used in gem-testing laboratories to determine whether a diamond is natural or synthetic and whether its color is natural or has been enhanced by treatment; and it is used in the characterization and selection of diamonds for certain high-technology applications—laser windows, for example.

The nature of light

The study of a material's spectral interaction with light is known as optical spectroscopy, and we can begin by asking the simple question, "What is light?". The light that we observe with our eyes (and also the light that we can't see) is due to the propagation of energy in space as a combination of electric and magnetic fields, known as an *electromagnetic wave*. This wave can be characterized by its wavelength, and for light in the visible region the wavelengths span from 400 nm in the violet region to 700 nm in the red region. We are all familiar with the sight of a rainbow, and if the colors of the rainbow are shown on a diagram where the wavelengths increase vertically downwards, then below the red is the infrared, and above the violet we have the ultraviolet. These regions cannot be detected by the human eye, but they can be investigated using suitable instruments which are sensitive to these wavelengths. For diamond, we will find that measurements in all three regions are needed.

Light can also be regarded as a stream of *photons*. For light that has a single wavelength (and therefore a single color, i.e., *monochromatic* light), each photon has a *quantum* of energy (measured in Joules) given by

$$E = \mathrm{h}\nu = \frac{\mathrm{h}c}{\lambda}$$

1529-6466/22/0088-0012$10.00 (print)
1943-2666/22/0088-0012$10.00 (online)
http://dx.doi.org/10.2138/rmg.2022.88.12

where h is Planck's constant, ν is the frequency, c is the speed of light and λ is the wavelength. The Joule is not a very convenient unit, and in spectroscopy E is usually expressed in electron volts (eV) as

$$E = \frac{hc}{e\lambda} = \frac{1239.84}{\lambda(\text{nm})}$$

where e is the electronic charge.

For high-precision measurements it is important to realize that the wavelength in this formula is the wavelength in a vacuum. Most measurements are made in air, and the refractive index of air must be taken into account. A more appropriate numerical factor is then

$$E \approx \frac{1239.50}{\lambda(\text{nm})}$$

and therefore, an infrared electromagnetic wave of energy 1 eV has a wavelength of 1239.50 nm in air. The approximate sign is because the refractive index of air changes slightly with temperature, pressure and wavelength.

In the infrared spectral region, and in Raman spectroscopy, the preferred unit is wavenumbers. The wavenumber $\bar{\nu}$ is given by the reciprocal of the vacuum wavelength in cm and is therefore expressed in cm^{-1}. This is equivalent to

$$\bar{\nu} = \frac{10^7}{\lambda(\text{nm})}\,\text{cm}^{-1}$$

The following spectral positions (measured *in vacuo*) are therefore identical: 500.00 nm, 2.4797 eV, 20,000 cm^{-1}. We explicitly note that wavenumbers are proportional to energy, whereas wavelength is proportional to its inverse.

Phonons and Raman scattering

To interpret some of the optical spectra of diamond we need to consider the vibrations of the carbon bonds. Consider a one-dimensional crystal with the ends fixed; Figure 1 shows such a "crystal" with seven atoms. The possible frequencies of vibration are represented by one half-wavelength, two half-wavelengths, three half-wavelengths etc., up to the point where adjacent atoms are vibrating in antiphase. The last corresponds to the maximum possible frequency of vibration. The same considerations apply in a three-dimensional crystal, although the full details are more complicated than suggested by this simple model.

By analogy with light waves, we can represent a vibrational wave by a particle called a *phonon*. It is possible for photons and electrons to interact with the vibrational waves in a crystal, giving *photon–phonon* and *electron–phonon* scattering. The energy of a phonon is usually written as $\hbar\Omega$, where $\hbar$ is $h/2\pi$ and $\Omega = 2\pi\nu$ (c.f. $E = h\nu$ for a photon).

Raman scattering is an important example of the photon–phonon interaction. Incident photons with energy $h\nu$ are weakly scattered by the phonons in a crystal to produce Raman-scattered photons at energies:

$h\nu_S = h\nu - \hbar\Omega$ (phonon creation; Stokes scattering), and
$h\nu_{AS} = h\nu + \hbar\Omega$ (phonon destruction; anti-Stokes scattering).

In the first process the incoming photon loses energy and generates a vibrational wave in the crystal; in the second case the photon gains energy from vibrational waves that are already present in the material. At room temperature, and below, the anti-Stokes scattering in diamond is very weak compared with the Stokes scattering. The Raman energy of diamond is 1332.5 cm^{-1}, and this corresponds to the maximum vibrational frequency of the diamond lattice at room temperature and lower. A Raman spectrum may be produced by using an intense monochromatic excitation source (typically a laser) and plotting the number of scattered photons versus the photon energy separation (in wavenumbers) from the photon energy of the source. Figure 2 shows the first-order Raman line at 1332 cm^{-1} and the weak, characteristic, second-order spectrum (which is not simply a single peak) with a sharp spike at 2667 cm^{-1}. As Raman scattering is intimately connected with vibrations within the crystal, a Raman microscope can be used to study compressive and tensile strain inside a crystal via shifts in the local Raman frequency (Crisci et al. 2011), and even give information on the environment surrounding a non-diamond inclusion (Howell et al. 2012).

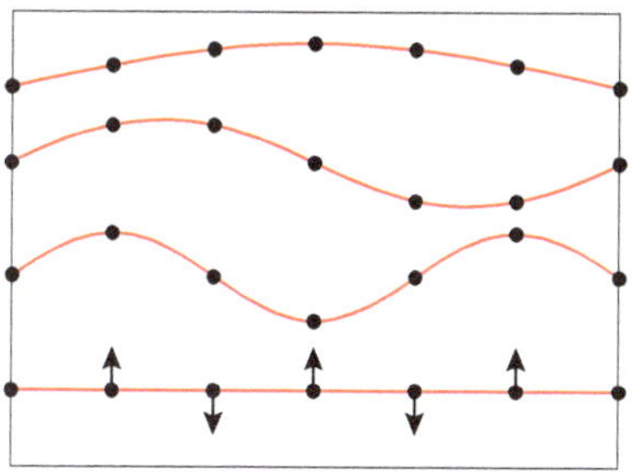

Figure 1. Illustration of the possible vibrational modes of a finite one-dimensional chain of atoms whose ends are fixed. The energy of the vibrations increases from a single half-wavelength (**top**) to nearest-neighbor atoms vibrating in antiphase (**bottom**). In a real crystal, the number of atoms, and hence vibrational modes, is of the order of 10^{14} per cubic millimeter, but we may still gain a basic but intuitive picture of crystal vibrations from this model.

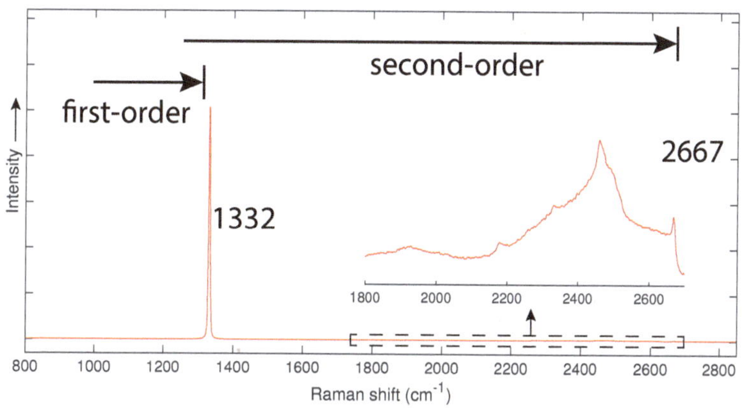

Figure 2. First- and second-order Raman spectra collected from a high-purity diamond using excitation at 514.5 nm. These spectral shapes are intrinsic to high-quality, high-purity diamond and can be used to distinguish diamond from analogues. In high-resolution spectroscopy, the width and absolute position of the Raman line can be used to analyze strain within a sample.

Optical absorption

When light is incident on one side of a transparent material, some light is reflected, and some is transmitted. If the intensities of the incident light $I_0(\lambda)$ and the transmitted light $I_t(\lambda)$ at each wavelength λ are measured, a quantity called the *absorption coefficient,* α, may be calculated at each wavelength using the expression

$$\alpha = \frac{1}{t}\log_e\left(\frac{I_0(\lambda)}{I_t(\lambda)}\right)$$

This is an approximate expression which ignores reflection at the crystal surfaces; the major effect of the approximation is to offset the zero of the absorption coefficient. The sample thickness, t, is usually measured in cm, and so the units of α are cm^{-1}.

Commercial spectrophotometers and Fourier Transform Infrared (FTIR) spectrometers usually display the output as *absorbance, A*. This is defined by

$$A = \log_{10}\left(\frac{I_0(\lambda)}{I_t(\lambda)}\right)$$

Absorbance may be converted to absorption coefficient using $\alpha = \log_e(10)A/t$. The importance of using absorption coefficient it is that is proportional to the concentration of the absorbing species and is not impacted by the thickness of the material or pathlength.

Absorption processes in diamond

In the diamond crystal structure, each carbon atom is neighbored by four other carbon atoms in a tetrahedral arrangement, as shown in Figure 3: other arrangements of carbon are possible and result in different materials such as amorphous carbon, graphite, etc. Each carbon atom has four outer electrons (the so-called *valence electrons*), and when the carbon atoms bond to each other in the diamond crystal there is an electron pair bond between each carbon atom pair: this sharing of electrons between nearest neighbors forms the archetypal *covalent bond*.

There are two ways in which this structure can absorb light:

- if the energy of the light is high enough, electrons can be knocked out of the bonds; this requires short wavelength light (ultraviolet)
- at long wavelengths (infrared), the absorbed light generates phonons, setting the carbon atoms into vibration

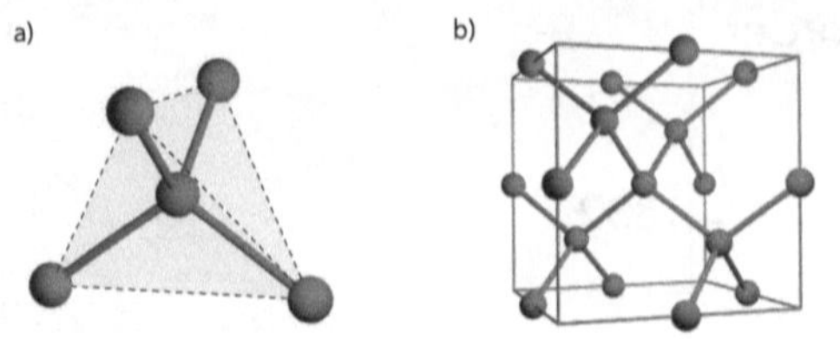

Figure 3. a) In diamond, each carbon atom is covalently bonded to its four nearest neighbors in a tetrahedral configuration: each bond is formed by the sharing of one of four valence electrons with a nearest neighbor. **b)** The diamond structure is frequently illustrated using the conventional unit cell, which provides greater context and displays a section of the lattice which is repeated throughout the entire crystal.

The absorption spectrum of a defect-free diamond for wavelengths between 0.2 and 20 μm (the visible region is between 0.4 and 0.7 μm) is given in Figure 4. At short wavelengths (around 0.23 μm) the ultraviolet light has sufficient energy to remove electrons from the electron-pair bonds, and at wavelengths shorter than this the diamond absorbs strongly, giving rise to the so-called *absorption edge*. In the infrared part of the spectrum, between 2.5 and 6.5 μm, there are the characteristic two- and three-phonon absorption bands at wavelengths where the infrared radiation can set the carbon atoms into vibration. For a defect-free diamond, as a result of its homonuclear tetrahedral symmetry, there is no one-phonon absorption (i.e., absorption of light which results in the creation of a single phonon, corresponding to the region with wavenumbers less than 1332 cm^{-1}, or equivalently wavelengths of greater than 7.5 μm).

The absorption edge, produced when a photon knocks out a valence electron from a bond, occurs at ~0.23 μm, corresponding to an energy of 5.49 eV. For a brief time that electron becomes available to conduct electricity, as it is no longer tightly bound to its original atom. It is useful to summarize this situation on an *energy level diagram* such as that shown in Figure 5. For a defect-free diamond in the dark, the valence band (comprised of the bonding, valence electrons, which are not free to move through the crystal) is completely filled with

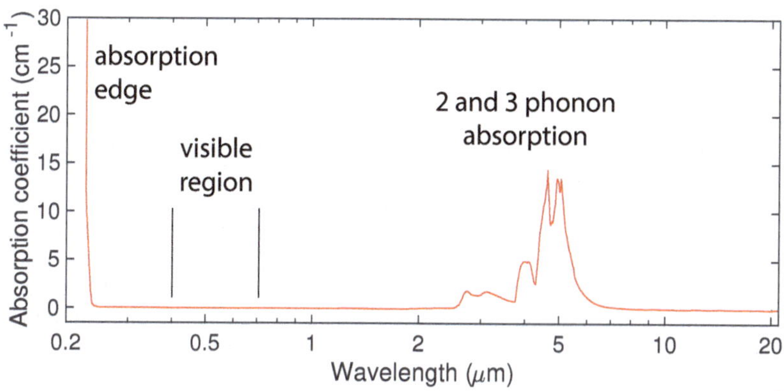

Figure 4. Aside from characteristic two- and three-phonon absorption in the mid-infrared, defect-free diamond is transparent from >20 μm to the band gap threshold at 225 nm (5.49 eV).

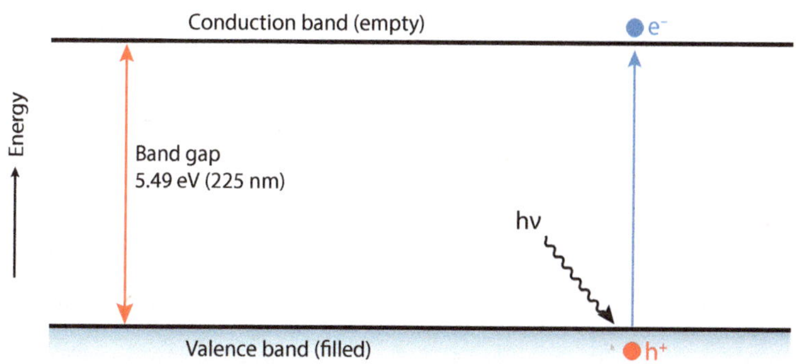

Figure 5. Energy level diagram for defect-free diamond. At room temperature and in the absence of light, all electrons reside in the valence band and participate in the covalent bonding of diamond. UV light of energy hν >5.49 eV (<225 nm) can excite valence band electrons (e^-) into the conduction band, allowing them briefly to contribute to electrical conductivity within the diamond, and leaving behind a net positive charge (hole, h^+) in the valence band.

electrons; consequently, the conduction band (which corresponds to high-energy states that do not participate in bonding and in which the electrons are instead free to move through the crystal) contains no electrons. The energy separation of the bands is 5.49 eV and is known as the *band gap*. Illuminating the diamond with ultraviolet light of wavelength shorter than 0.23 μm excites an electron from the valence band to the conduction band and leaves a *hole* in the valence band. This is a greatly simplified picture; nevertheless, it is useful in illustrating the processes behind some optical transitions. Note that it is also possible to *thermally* excite electrons from the valence to the conduction band; however, for this to occur, the available thermal energy ~k_BT—where k_B is the Boltzmann constant and T is the temperature—must be at least comparable to the band gap, which only occurs in diamond for extreme temperatures well in excess of 1000 °C but is of crucial importance in other semiconductors such as silicon.

For a defect-free diamond there is no absorption in the visible region, and it is therefore completely colorless. In reality, most diamonds are colored, and so must absorb light in the visible region. This absorption is due to the presence of defects within the colored diamond: these defects and their associated effects are discussed from the next section onward.

Infrared absorption

Figure 6 shows the infrared spectrum of a defect-free (intrinsic) diamond recorded using a FTIR spectrometer. To retain a similar appearance to historical spectra (which generally used μm for the horizontal axis), most FTIR instruments plot the data with wavenumbers increasing from right to left, as shown here. As discussed above, defect-free diamond possesses no one-phonon absorption; however, there are processes which allow absorption involving combinations of two or more phonons. This characteristic absorption spectrum unmistakably identifies diamond. As the spectrum is intrinsic to diamond, it is also invaluable in the analysis of infrared spectra of diamond. The absorption coefficient of a 1 cm thick diamond is 12.3 cm^{-1} at 2000 cm^{-1} (Tang et al. 2005): this enables the conversion from absorbance to absorption coefficient to be made without measuring the path length through the diamond, which can be difficult in the case of gem-cut samples that are designed to produce multiple internal reflections.

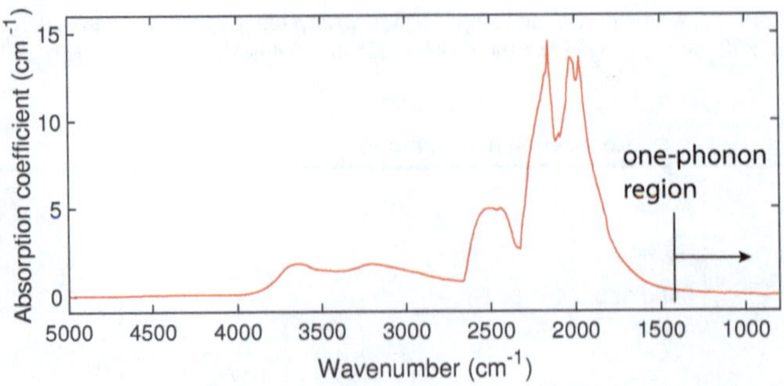

Figure 6. Infrared spectrum of intrinsic diamond, displaying the characteristic multi-phonon absorption at approximately 1500 cm^{-1} to 4000 cm^{-1}. The homonuclear tetrahedral symmetry of defect-free diamond means that bulk absorption processes which create only one phonon are not possible, yielding no absorption in the one-phonon region of 1332 cm^{-1} and below.

SPECTROSCOPY AND PROPERTIES OF IMPERFECT DIAMOND

Introduction

Defect-free diamond is an important physics playground and technological tool due to its well-known list of superlative properties, such as its hardness, thermal conductivity, transparency window and so on. However, for geological, gemological and quantum technology applications, it is the *flaws* within a diamond which are most important.

Defects in crystals can be classified into two broad categories: point defects and extended defects. Point defects encompass those defects which are "point-like", and which typically can be described within only one or two diamond unit cells, with fewer than ten atoms deviating significantly from the perfect, pure-carbon diamond structure. Extended defects include crystallographic faults such as dislocations and very large defect aggregates such as interstitial or vacancy clusters, and they may measure upwards of hundreds of microns in at least one dimension (a vacancy is simply a lattice site from which a carbon atom has been removed, leaving a vacant carbon site). It is essential to take both categories into account when determining the growth and subsequent history of a given diamond, and both are routinely employed in the discrimination between natural, synthetic, and treated diamond (see the final section below, and D'Haenens-Johansson et al. 2022, this volume).

This chapter is concerned primarily with the properties, treatment, and interaction of point defects in diamond. Most point defects in diamond are identified and studied through their interaction with light of various wavelengths. These interactions are defined by the geometrical, elemental and electronic structure of the defect in question and so are unique to a given defect, yielding a powerful spectroscopic "fingerprinting" tool.

Ultraviolet and visible absorption and luminescence

An absorption peak at 415 nm in diamond was first reported by Walter (1891). A more recent measurement of the absorption spectrum, which has a sharp line at 415 nm (violet) and a structured band that extends into the ultraviolet part of the spectrum and is commonly referred to as N3, is given in Figure 7. The line at 415 nm is much sharper at 77 K than it is at room temperature (293 K), and the structure in the band is also better defined at low temperature. When this absorption band is strong it gives the diamond a yellow color. Most of the perceived color is produced by a related absorption band known as N2 with its major peak at 478 nm (shown later in Fig. 32a).

Optical absorption in the visible range is the result of an electronic excitation, whereby an electron is "excited" to a higher energy state—analogous to a more energetic orbit in a hydrogen atom. These higher-energy electrons will attempt to lower their energy by a number of potential processes. For defects which absorb in the visible range, this frequently occurs by the subsequent emission of light, known generally as luminescence. Therefore, most colored diamonds will also luminesce under appropriate conditions.

Figure 8 shows a selection of uncut natural diamonds illuminated at 365 nm, many of which display a blue emission. An emission spectrum from one of the blue-luminescing diamonds is also given, and we see that there is a sharp line at 415 nm and a structured band at *longer* wavelengths (i.e., lower energy). It is clear that most of the band lies in the range 420 to 480 nm, which is why the luminescence appears blue.

Overlaying the absorption and luminescence spectra of N3 on a single diagram demonstrates that the spectra are approximately mirror images (Fig. 9). In order to understand this, the absorption transitions can be represented on an energy level diagram (Fig. 10). The sharp spectral line corresponds to an *electronic* transition from the ground state of the optical center to the excited state. If more energy is provided, the optical center and the surrounding lattice can also be set into vibration at certain characteristic frequencies. In luminescence the excited center can either relax directly to the ground state or, via vibrational levels, produce luminescence at lower energies. The ground and excited energy levels are located somewhere between the conduction band and the valence band (within the band gap). The energy separation of the ground and excited states can be measured with high precision, but in the majority of cases it is not possible (or very difficult) to determine their absolute positions in the band gap.

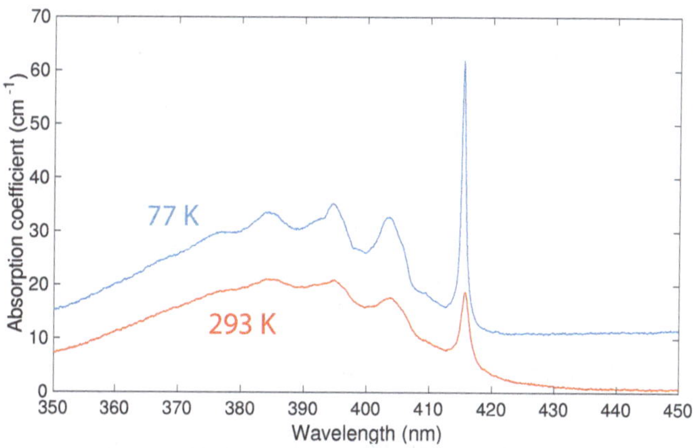

Figure 7. Optical absorption spectra of a diamond containing the 415 nm defect. The characteristic 415 nm peak and spectral sideband are better defined when the spectrum is collected with the sample at liquid nitrogen temperature (77 K) versus room temperature: this effect is typical of optically active point defects in diamond (and other materials). The spectra have been offset for clarity, and the total integrated absorption is the same at both temperatures.

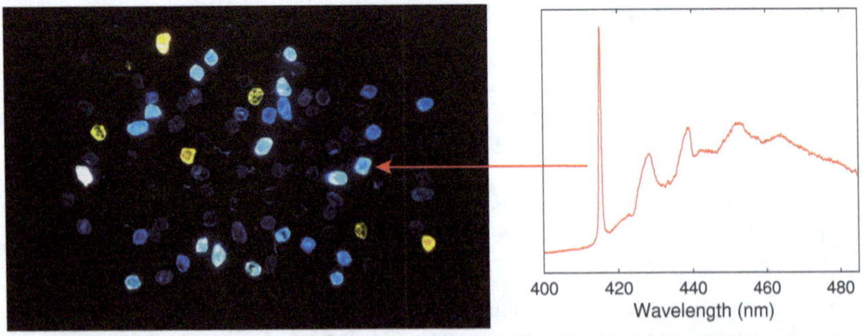

Figure 8. Visible luminescence of natural diamonds excited with 365 nm light. The light blue emission is due to the presence of the 415 nm (N3) defect, an exemplar liquid nitrogen temperature (~80 K) luminescence spectrum of which is given to the right.

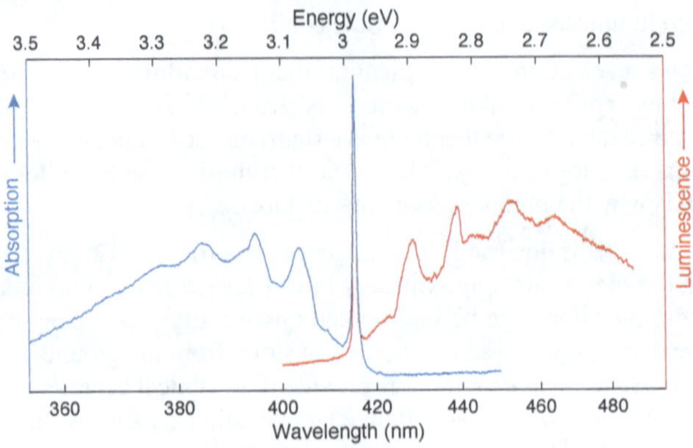

Figure 9. The characteristic luminescence and absorption spectra of the "415 nm (N3)" center at 80 K. When plotted on an energy scale (top), they are almost mirror images of one another.

The spectral structure of a point defect is a result of the electronic structure of the point defect and its interaction with the surrounding crystal. In an isolated atom or monatomic gas, optical transitions are very narrow in wavelength (energy) corresponding to well-defined electronic transitions between discrete electronic orbital states. Point defects in diamond are sometimes considered as trapped atoms or molecules: nevertheless, there is inevitably some interaction between the electronic structure of the point defect and the vibration continuum of the diamond lattice. This combination of interactions gives rise to so-called *vibronic* transitions, which involve both vibrational and electronic interactions.

Purely electronic transitions between the ground vibrational states of the ground and excited electronic states give rise to a spectral feature known as the zero-phonon line (ZPL). Optical spectra of point defects in diamond are often characterized by their ZPL wavelength, which is (typically) identical in absorption and emission. Vibronic transitions between the ground and excited electronic states involve the creation of lattice vibrations with characteristic energy $\hbar\Omega$ yielding a continuum of transitions which manifests as a broad, vibronic sideband. This sideband appears at higher energies than the ZPL in absorption, and lower energies than the ZPL in emission (Fig. 10).

Many defects in diamond have absorption and luminescence spectra similar to those shown in Figure 9 for the N3 center, though the particulars of a point defect's vibronic spectrum are a result of its electronic structure and coupling to vibrational modes. When identifying a particular optical vibronic spectrum, care must be exercised to compare both the ZPL wavelength and the

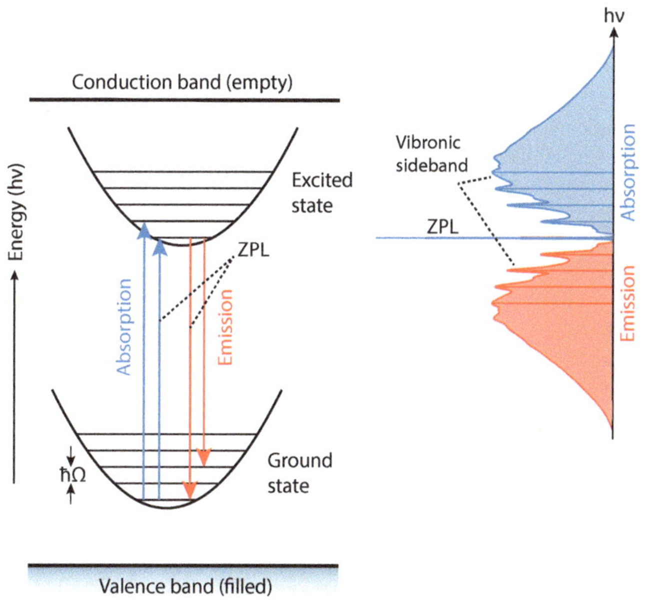

Figure 10. Illustration of a vibronic absorption and emission process at a point defect. Some point defects introduce additional electronic states into the band gap, giving rise to optical absorption in the UV, visible, or near-IR regions. If the energy (hν) of the incoming light is precisely the separation of the ground- and excited-states, then a sharp zero-phonon line (ZPL) is observed; if light of higher energy is incident on the defect then phonons are created in addition to the electronic excitation, causing the lattice to vibrate. When the defect subsequently relaxes, it emits light which is at the ZPL energy or lower. In both cases, the structure of the absorption / emission band is characteristic of the defect and related to the phonon energies $\hbar\Omega$ with which the defect interacts.

corresponding sideband to the reference spectrum, as multiple defects may emit at very similar wavelengths. For example, the unfortunately named H3 and 3H centers, with very similar ZPL wavelengths of 503.2 and 503.4 nm at 80 K, respectively, have different compositions but can be identified by their drastically different vibronic sidebands (Walker 1979).

Following excitation via the absorption of light, the time for which an optical center stays in the excited state covers a huge range, depending on the defect. In simple cases the time is determined by a statistical process rather similar to radioactivity. After the excitation is stopped, the luminescence decays exponentially with a characteristic decay time τ.

$$I(t) = I(0)e^{-\frac{t}{\tau}}$$

where $I(t)$ is the intensity at time t and $I(0)$ is the initial intensity. Luminescence decay times typically range from a few nanoseconds to 100s of microseconds. If there are no competing processes the measured decay time is equal to the true luminescence decay time of the optical center. However, there are often non-radiative transitions which make the measured decay time shorter and also decrease the luminescence efficiency; the luminescence is then said to have been *quenched* (see *Generation of luminescence*).

With a few exceptions, almost all the luminescence observed from diamond originates from vibronic bands, and there are quite literally several hundred ZPLs documented in the literature (Zaitsev 2001; Dischler 2012): several of the most important are detailed later in this chapter. In early studies it was fashionable to give a name (such as GR1, H3 or N3, where GR, H, and N indicated general radiation, heated, and natural, respectively) to an optical center, depending on its origin (Clark et al. 1956a,b). With optical centers studied more recently it is conventional to refer either to the atomic structure, or the wavelength or energy of the ZPL. (However, atomic structures have been established for a relatively small number of optical centers in diamond, compared to the plethora of identified optical signatures.) Most optical centers in diamond can occur in both the neutral and negative charge states—for example, the nitrogen-vacancy center NV^0 and NV^-, where V indicates vacancy henceforth. To understand this, we consider a model that is used for *donors* and *acceptors* in semiconductors.

Donors, acceptors, and charge transfer

The best-studied donor in diamond is substitutional nitrogen (Fig. 11a), where substitutional indicates that the nitrogen atom remains approximately on the original carbon lattice site. Nitrogen is from group 5 of the periodic table and is trivalent in diamond: in its neutral charge state, three of its valence electrons are used for bonding as with the group 4 carbon atoms. The remaining two valence electrons form a lone pair that does not form a bond with the remaining carbon, leading to a distortion where this unique carbon and the nitrogen relax away from one another (Fig. 11a). It is much easier to remove one electron from this lone pair than it is to remove an electron from one of the bonds (i.e., it requires less energy), and if this happens, the nitrogen relaxes back to undistorted lattice site and the nitrogen is said to have *donated* an electron to the conduction band, leaving behind a positively charged substitutional nitrogen atom. On an energy-level diagram (Fig. 12) we show the nitrogen donor at 1.7 eV below the conduction band. Its position in the band gap has been determined by measuring, as a function of temperature (available thermal energy), the electrical conductivity of a diamond rich in substitutional nitrogen (Farrer 1969). When another defect is also present, the negative charge on the nitrogen atom may be transferred to that defect (see below).

Optical transitions are allowed from the donor level to the conduction band, but, because of additional restrictions, the threshold energy of 2.2 eV for optical absorption is higher than that determined by temperature-dependent electrical measurements. The 2.2 eV threshold is in the middle of the visible region at approximately 560 nm, and substitutional nitrogen's optical absorption ramps from this wavelength into the blue (Fig. 11b), producing a yellow color. This absorption structure is typical of electronic transitions between a defect and the conduction or valence band, where a continuum of band states leads to a broad, relatively featureless band. Contrast this spectrum to that of the 415 nm vibronic band which can also produce a yellow color (Fig. 7). Natural diamonds containing isolated substitutional N are rare (~0.1%; Dyer et al. 1965), but most diamond produced by high pressure high temperature (HPHT) synthesis contains nitrogen in this form.

A similar picture is used for substitutional boron in diamond (Fig. 11c). Boron is in group 3 of the periodic table and has only 3 valence electrons—one too few to complete diamond's 4-fold covalent bonding. It is conventional to refer to that missing electron as a *hole*, and under the influence of an electric field this hole moves as though it has a positive charge. Removing the hole is equivalent to the boron atom *accepting* an electron, and on an energy level diagram (Fig. 12) the boron *acceptor* lies 0.37 eV above the valence band, as determined by temperature-dependent electrical conductivity measurements (Collins and Williams 1971).

For boron, the ionization energy (the energy required to remove the positive hole) measured optically is the same as that determined from electrical conductivity measurements (Collins et al. 1969). In the absorption spectrum (Fig. 11d), at energies below the ionization energy, we see absorption due to transitions to internal excited states, followed by a continuum absorption at higher energy corresponding to photoionization of the hole. The spectrum shown is in the infrared region, but the continuum absorption extends into the visible region, giving the diamond a blue color. Natural diamonds containing boron as the major impurity are also rare, but it is not difficult to grow blue synthetic HPHT diamonds by removing nitrogen from the growth capsule and introducing ("doping with") boron.

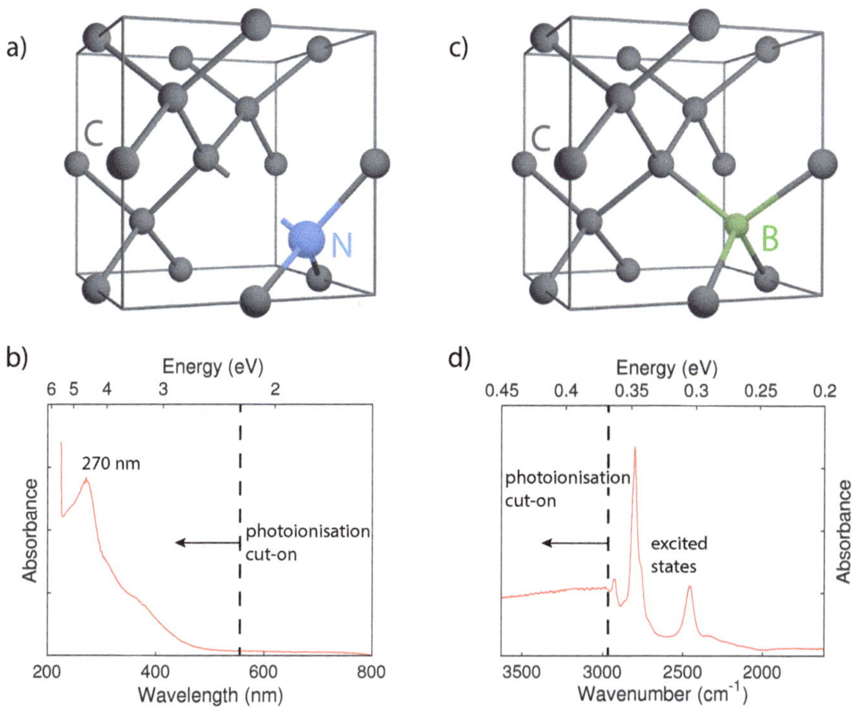

Figure 11. **a)** The structure of neutral substitutional nitrogen in diamond. Whereas carbon possesses four valence electrons (one for each nearest neighbor), the group-5 nitrogen atom possesses five, leaving an additional electron to be accommodated into one of its bonds, raising the bond energy and lengthening it to approximately 130% of a normal C–C bond (Etmimi et al. 2009). Incomplete bonds between atoms indicate that the atoms do not form a bond, and are included simply to provide a guide to the eye. **b)** It is possible to optically ionize neutral substitutional nitrogen with light of $\lambda < 560$ nm, leading to its characteristic optical absorption spectrum and associated 270 nm absorption peak (Jones et al. 2009). **c)** Boron, belonging to periodic group 3, possesses one fewer valence electron than carbon and therefore it is energetically favorable for it to accept one: this results in one bond contracting slightly (to approximately 98% of a normal C–C bond; Goss and Briddon 2006). **d)** Approximately 0.37 eV is required to thermally or optically excite an electron from the valence band to the acceptor state of substitutional boron. Boron's photoionization continuum extends into the visible region, lending a blue color to the diamond. Note that the intrinsic diamond absorption has been subtracted from this spectrum.

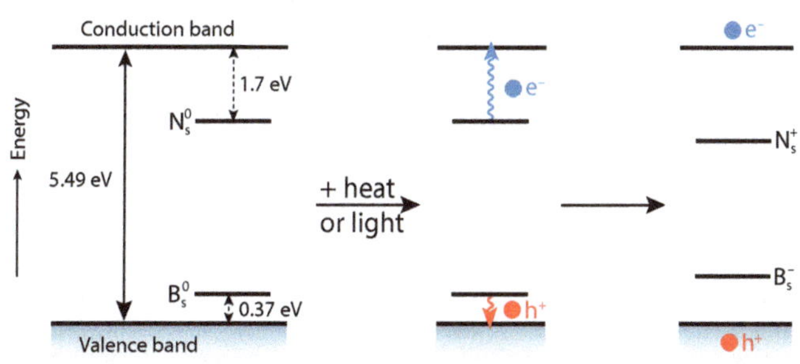

Figure 12. Illustration of the donor and acceptor processes for substitutional nitrogen and boron, the neutral charge states of which lie relatively close to the conduction and valence bands, respectively. The addition of sufficient heat or light can excite an electron from the nitrogen to the conduction band, resulting in an electron in the conduction band and a positively charged defect: the same is true for the boron, with hole, boron, valence, and negatively in place of electron, nitrogen, conduction, and positively, respectively. The charge carrier is then free to migrate through the band, potentially to be trapped by another point defect.

In diamonds with a high single-nitrogen (N_s) concentration, the negative charge state of a given defect—such as NV—is typically dominant, whereas the neutral charge state is more prominent in diamonds with a low single-nitrogen concentration (Collins 2002). The interpretation here is that when a neutral defect center (X^0) and a neutral substitutional nitrogen defect (N_s^0) are in close proximity, the nitrogen can donate an electron, converting the defect to its negatively charged version and leaving the nitrogen in a positive charge state:

$$X^0 + N_s^0 \leftrightarrow X^- + N_s^+$$

This effect is known as *charge transfer*, and the defect X can be considered as a *charge trap*. In many cases it is possible to drive the reverse process, ionizing X^- and reverting back to the neutral charge states of the relevant defects: however, the particulars of the charge transfer process, and the extent to which it can be driven in either direction, are sample- and defect-specific. In extreme cases this can lead to significant visible color changes which can be driven by light (photochromism) or heat (thermochromism) (Fig. 13): natural diamonds which exhibit photochromism are sometimes referred to as chameleon (Hainschwang et al. 2005). These changes are typically reversible as they do not involve the creation or destruction of defect centers, but only the migration of charge between charge traps: the processes are therefore distinct from high-temperature annealing, which involves the migration, creation, and destruction of defect centers. The only defect center which has been definitively identified as positively charged in diamonds containing B as the major impurity is the positively charged substitutional nitrogen center, N_s^+ (Lawson et al. 1998).

Generation of luminescence

There are several distinct ways in which luminescence can be generated in diamond. All of the following excitation mechanisms have been reported—we will focus our discussion on the first two as the most common and of practical significance in diamond:

- Photoluminescence (photon beam)
- Cathodoluminescence (electron beam)
- Electroluminescence (current injection)
- Ionoluminescence (ion beam)
- Thermoluminescence (heat)
- Triboluminescence (friction / polishing)

Some diamonds may also emit light for milliseconds to tens of seconds after the excitation is shut off—a phenomenon known as phosphorescence.

Below-band gap photoluminescence. In photoluminescence (PL) with below-band gap excitation (i.e., using light of <5.49 eV), a vibronic optical center is excited using a wavelength lying within its absorption band, directly exciting an internal transition within the defect; after a short time, the center relaxes to the ground state, emitting light in the luminescence sideband (see Fig. 10). For this process to occur, the incident light (for spectroscopy, typically a laser beam) must lie within the absorption band of the defect center (Fig. 14). Excitation using wavelengths outside the defect's absorption band will not induce photoluminescence from that center, enabling selective excitation of defects. This can be a powerful tool in characterizing samples which contain multiple defects whose spectra overlap.

The physical mechanisms behind the PL and Raman processes are distinct; however, they are experimentally measured in the same way. Therefore, care should be taken to ensure that spectral features are correctly ascribed to luminescence or Raman. Raman shift is relative to the

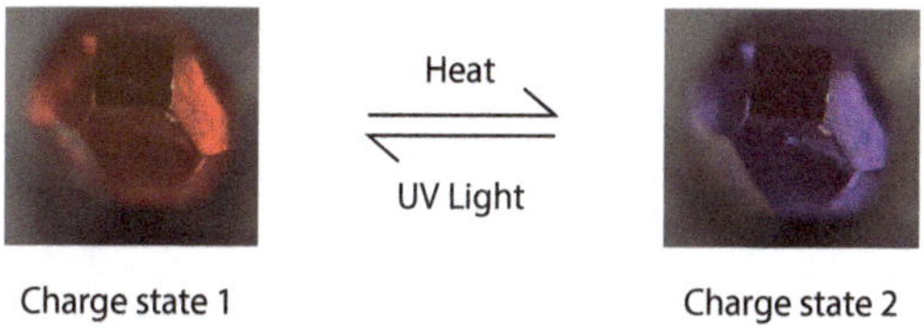

Figure 13. An extreme example of photo- and thermo-chromism. The HPHT-grown sample contains N_2V and NV defects in abundance. By driving charge between substitutional nitrogen and these defects, it is possible to bias towards N_2V^0 and NV^0 giving a red hue (**left**); or N_2V^- and NV^- yielding a purple color. Adapted from Dale (2015).

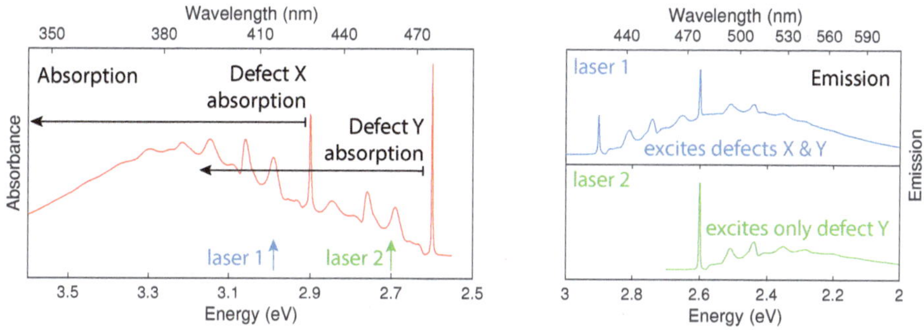

Figure 14. An absorption measurement (**left**) measures the total absorption of the sample at a given wavelength. Where the absorption profiles of multiple defects overlap, the absorbance at that wavelength is the sum of the absorption due to all defects simultaneously. In this example, the absorbance at 3.0 eV is a result of absorption by both X & Y defects, while at 2.7 eV only defect Y contributes to the spectrum. Using sub-band gap excitation, photoluminescence can be excited from multiple defect centers selectively. A sub-band gap laser incident on a crystal containing multiple defects will only induce photoluminescence from those centers which absorb it. Using laser 1 (3.0 eV), both defects X & Y will absorb the laser light and subsequently emit (**top right**); using laser 2 (2.7 eV), only defect Y will absorb and emit (**bottom right**).

excitation wavelength, and spectral features arising from a Raman process will always manifest at the same energy shift from the excitation energy. Photoluminescence ZPL energies are a property of the emitting defect and are fixed in absolute energy. We therefore distinguish between Raman and PL by measuring the same area of the crystal using different energy excitations and comparing spectral features plotted against energy shift from the excitation energy.

Photoluminescence can be quenched in diamonds with high concentrations of defects, because of non-radiative energy transfer between defect centers following optical excitation (Davies and Crossfield 1973; Crossfield et al. 1974). The quenching defects do not themselves necessarily luminesce: for example, substitutional nitrogen (PL-inactive) quenches PL from NV^- centers (Manson et al. 2018). Photoluminescence is therefore NOT a quantitative method for measuring defect concentrations. However, in diamonds with low concentrations of defects, a semi-quantitative estimate of defect concentrations can be made by comparing the intensity of the PL with the intensity of the Raman spectrum (or, in the case of a high-sensitivity instrument measuring a diamond which contains very low defect densities, by directly counting the defects—see *Quantification of defect concentrations*). Significant care must be taken when using a polarized light source such as a laser, as the intensity of the Raman line can change by over 100× depending on the crystallographic face being probed (Solin and Ramdas 1970; Grimsditch et al. 1978; Steele et al. 2016).

We note that the charge state in which a defect is measured can be sensitive to the laser wavelength used to excite it, in addition to the absorption-based considerations described above. For example, 532 nm is efficient at photoionizing NV^0 to produce NV^-, while 560 nm is not, and therefore the measured charge ratio NV^0:NV^- is highly sensitive to the laser wavelength used to perform the measurement (Aslam et al. 2013).

The strengths of PL as a diagnostic and spectroscopic tool include its speed, relative insensitivity to surface contaminants, and production of spectra with sharp peaks which enable precise and potentially automatic analysis. By moving the sample under the laser beam (or rastering the laser beam itself), it is possible to map the relative concentrations of defects (Fig. 15), which can be important in understanding the incorporation mechanisms of defects or even aid in origin identification.

Cathodoluminescence. Electron-beam excitation, using electrons of energy typically 5 to 50 keV, excites electrons from the valence band to the conduction band, leaving holes in the valence band and producing electron–hole pairs. Penetration depths of the electron beam are ~0.2 μm at 5 keV and ~15 μm at 50 keV (Davies 1979). After excitation the electron and hole capture each other and form a short-lived center called an exciton, with a hydrogenic-like energy structure. In very pure diamond the electron and hole then annihilate, producing free-exciton recombination, which will not concern us here. In defect-containing diamond the excitons may be captured by optical centers which themselves become excited and then decay, producing characteristic luminescence—see the simplified schematic in Figure 16.

Cathodoluminescence (CL) images can be produced using a scanning electron microscope (SEM) or simply with a suitable electron gun. In the SEM an image can be produced by measuring the light generated by the electron beam via a detector which produces a signal proportional to the incident intensity. By synchronizing the data collection with the raster scan a digital map can be generated in which each pixel represents the intensity at that point. This information can be processed to produce a gray-scale image or a false color image. If the CL is passed through filters or a monochromator before reaching the detector, images can be produced which show the intensity distributions in different wavelength regions (a hyperspectral image).

An optical image can be obtained directly by photographing the specimen (using film or a digital camera) as it is scanned by the electron beam. Alternatively, a stationary defocused electron beam can be used to illuminate the full area of the specimen, from which a photographic image can be produced. This technique was used by Sumida and Lang (1981) in their seminal study of dislocation networks in natural Type IIb diamond. For less demanding applications the whole area of the specimen can be excited with a simple electron gun. The specimen may then be examined and photographed using an optical microscope placed above the viewing window (Henry 2020). An example of this technique is given in Figure 17, which is a photograph of a slice of natural diamond being irradiated with a defocused electron beam produced by a cold-cathode electron gun of the type described by Henry (2020). The mainly blue luminescence shows evidence of a complex growth history, and we can also see yellow/green CL originating from lines of plastic deformation. High magnification reveals patches of yellow luminescence from extended defects known as *giant platelets*. Using suitable collection optics, the spectra of these defects can be determined (Wight et al. 1971; Collins and Woods 1982). Many diamonds exhibit "Band A" CL, but there are, in fact, several different bands, and the CL may be blue or green. Much of the luminescence in Figure 17 originated from the blue band A.

Typical voltages encountered in an SEM are not high enough to directly induce lattice damage in diamond. For an electron beam focused to a small spot the excitation densities can be very high, ~10^{23} electron–hole pairs per cubic centimeter per second. The luminescence decay time of many optical centers in diamond is very short (typically 10 to 50 ns), and so even very small concentrations of such defects can produce bright cathodoluminescence, because each optical center can be excited and decay 10^7 to 10^8 times per second. However, some defect centers are not efficiently excited by an electron beam, even though they may have a short lifetime. CL from the negative NV center, for example, is exceedingly weak as a result of its charge dynamics, with emission instead via the neutral NV center as a result of electron-beam-induced charge conversion (Fedyanin and Agio 2016). Just like photoluminescence, CL is also quenched by high concentrations of defects (typically nitrogen-related).

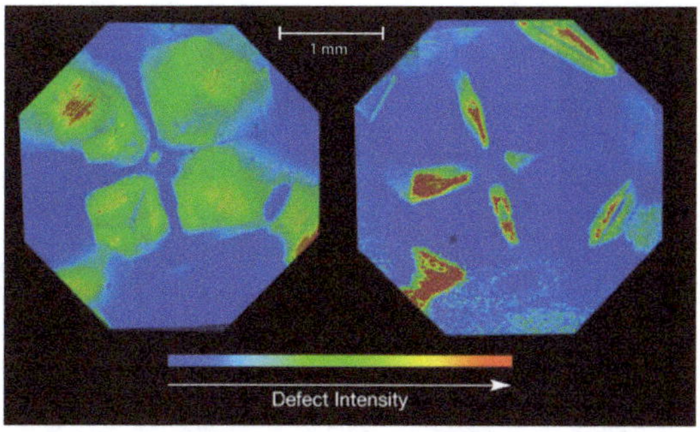

Figure 15. Mapping the intensity of spectral luminescence peaks by PL reveals the relative incorporation of point defects (637 nm, NV^- [**left**]; and 815 nm [**right**]) into different growth sectors of an HPHT-grown sample which was subsequently irradiated. Map collected from the table of a facetted gem. Modified from Loudin (2017).

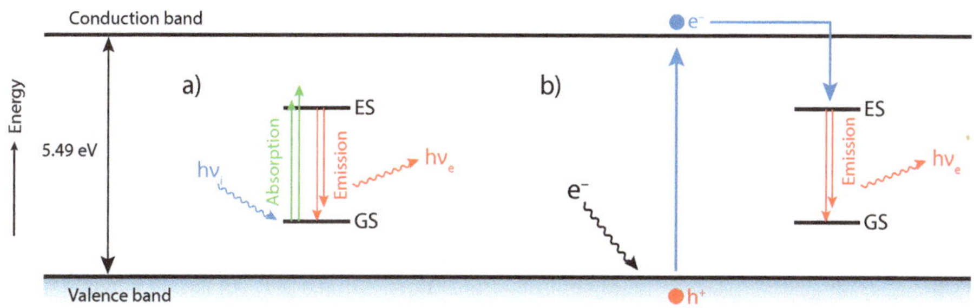

Figure 16. a) For below-band gap photoluminescence, the energy of the incoming photon ($h\nu_i$) is enough to excite an electron from a defect's ground state (GS) to its excited state (ES): the defect subsequently emits a lower-energy photon ($h\nu_e$). **b)** For cathodoluminescence, a high-energy electron excites an electron to the conduction band. The electron is then trapped by the excited state of a defect which subsequently emits.

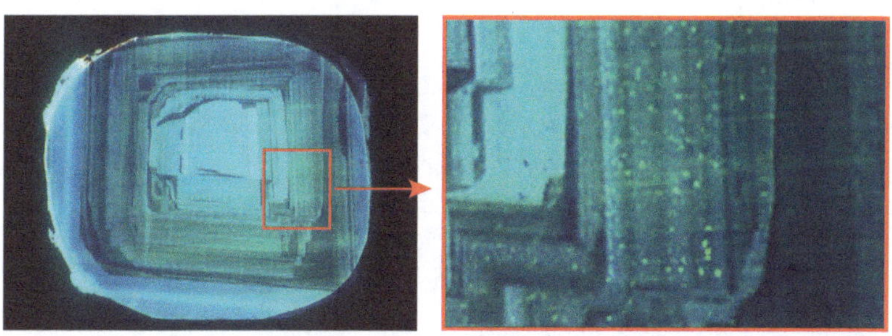

Figure 17. Growth horizons of a natural diamond as imaged by cathodoluminescence: the growth habit of the crystal at different times can be determined by the structure within the horizons. The blue emission is largely due to a broad band centered at approximately 435 nm known as "band A". High magnification (right) reveals the presence of yellow emission from giant platelets. Modified from Collins and Woods (1982).

As diamond is an effective electrical insulator, the incident electron beam in CL can induce local surface charging: to mitigate this charging, the diamond is typically coated in a thin conductive layer (frequently gold) before measurement. Due to this additional experimental complexity over PL, CL is less routinely employed in gemological and geological surveys.

Above-band gap photoluminescence. Electron–hole pairs may also be produced in diamond by the absorption of electromagnetic radiation with energy greater than the band gap. This is schematically similar to the process given in Figure 16b, with the incident electron being replaced by an ultraviolet photon with wavelength 225 nm or lower. The resulting luminescence is similar to that produced by electron excitation (Fig. 16b). The DiamondView instrument developed by the Diamond Trading Company (now De Beers Group Technologies) uses this technique. Figure 18 shows an image of a faceted CVD diamond excited in the DiamondView. The luminescence is produced by small concentrations of nitrogen-vacancy (NV) centers, and features striations which indicate the differential production of NV centers on different crystallographic surfaces during growth.

Finally, in this section we note that caution is required in comparing luminescence spectra from different publications. The "true" spectrum $L(\lambda)$ is measured with a spectrometer which has a response function $R(\lambda)$. The measured signal is therefore $L(\lambda) \times R(\lambda)$. The $R(\lambda)$ varies from spectrometer to spectrometer (for example, due to the use of different detection cameras). In principle, $R(\lambda)$ can be determined using a calibrated source (such as a tungsten lamp) but most published spectra are uncorrected.

Phosphorescence

Fluorescence and phosphorescence are two names given to luminescence, based on their decay times relative to typical human perception. Fluorescence is the name given to luminescence which decays effectively instantly on human timescales once the excitation source is removed, while phosphorescence is the name given to luminescence which persists for human-perceivable timescales after the excitation source is removed. As such, phosphorescence describes an empirical phenomenon rather than an underlying mechanism. Phosphorescence may arise from several distinct mechanisms, including internal transitions within a particular defect; and thermal trapping or release of charge carriers and their subsequent recombination (Dean 1965).

The most common phosphorescence spectra observed in diamond are related to donor-acceptor pair recombination, typically in conjunction with the presence of boron as a shallow acceptor. As a result, phosphorescence is routinely observed from both natural and synthetic semiconducting diamond with bands centered at approximately 500 and 660 nm (Walsh et al. 1971; Klein et al. 1995; Watanabe et al. 1997; Eaton-Magaña and Lu 2011). HPHT-grown near-colorless diamond, which often contains donors (nitrogen) and acceptors (boron) at the level of approximately 0.001–0.1 ppm, may also phosphoresce via this mechanism (D'Haenens-Johansson et al. 2015; Eaton-Magaña et al. 2017). Some features in synthetic diamond, such as the 499 nm system, may have internal electronic structure which results in intrinsic delayed luminescence (Khong et al. 1994; Wassell et al. 2018).

Phosphorescence spectra (example given in Fig. 19) are typically excited using near- or above-band gap excitation to generate large, transient concentrations of electron–hole pairs; however, it is possible to generate phosphorescence in some samples even with mid-gap energies (Freitas et al. 1994). Phosphorescence is routinely employed in the discrimination between natural and synthetic diamond (D'Haenens-Johansson et al. 2022, this volume), and several instruments have been built which specifically exploit this property.

Infrared absorption

As stated previously, defect-free diamond exhibits no absorption in the infrared one-phonon region. The addition of defects destroys the perfect symmetry of the diamond lattice, and some absorption is then created in the one-phonon region. As an example, Figure 20 shows the IR spectrum of a Type IaA diamond—one which contains most of the nitrogen in the form

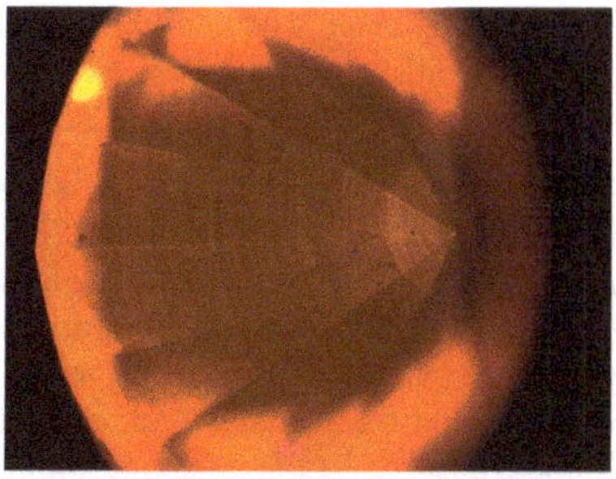

Figure 18. Typical DiamondView image of a synthetic diamond grown by chemical vapor deposition. The general orange color is the result of emission from nitrogen-vacancy (NV) defects. The striations are a result of differential production of NV defects on different crystallographic surfaces during growth.

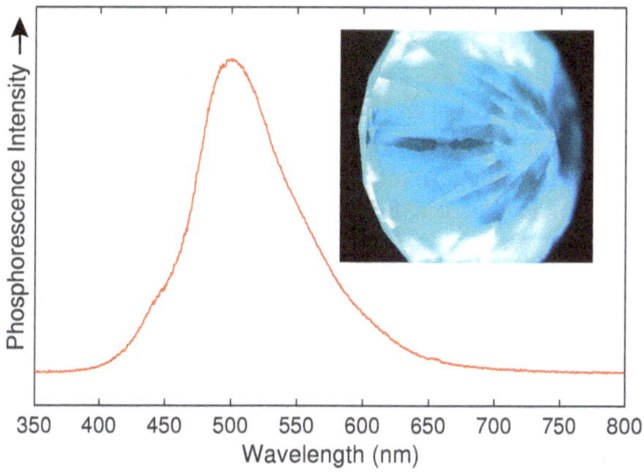

Figure 19. Typical phosphorescence spectrum from a near-colorless HPHT diamond. Inset: image of typical phosphorescence from a near-colorless HPHT-grown diamond. The difference in intensity is a combination of differential impurity uptake in different growth sectors (e.g., the dark line across the center) and the cut of the sample itself. Modified from D'Haenens-Johansson et al. (2015).

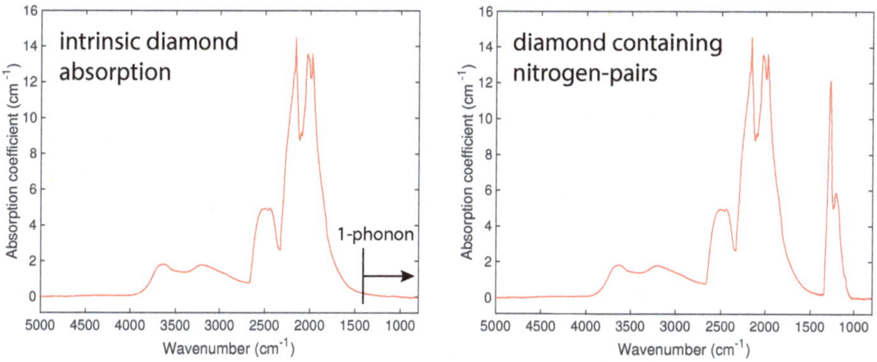

Figure 20. Infrared spectra of intrinsic diamond (**left**) and a diamond which contains a substantial quantity of defects in the form of nearest-neighbor nitrogen pairs. The nitrogen-pair defect is one of the most common found in natural diamond and is known as the A center.

of nearest-neighbor nitrogen pairs, known as *A centers* and with the atomic structure N_2. Note that the two neighboring nitrogen atoms in an A center do not chemically bond (and therefore do not represent a trapped diatomic nitrogen molecule): N_2 is employed henceforth on this understanding.

It has been shown by several researchers that there is a good correlation between the absorption coefficient at the peak of the one-phonon band (at 1282 cm^{-1}) and the concentration of A centers. Similar correlations have been established for the B aggregate of nitrogen (known as B centers, atomic structure N_4V, see *Aggregated nitrogen* later in this chapter) and for isolated substitutional nitrogen (often called the C center). These calibration factors enable the determination of the concentration of nitrogen in diamond, and the form in which it is present. This illustrates the power of optical absorption spectroscopy applied to diamond. In addition to quantifying nitrogen aggregates, IR absorption provides indispensable insight into many other defects and impurities, most notably hydrogen.

Electron paramagnetic resonance

It is important to note that while visible or infrared luminescence is an indication of the presence of structural or chemical defects, their absence does not imply a defect-free crystal. A particular defect may possess characteristic optical absorption or luminescence, or it may be entirely optically inactive, in which case we must turn to alternative techniques to assess a defect's presence and concentration. Optical phenomena such as absorption and emission of visible light in diamond are associated with the interaction between the *electric* field of an electromagnetic wave and electronic dipoles within the material. It is also possible for the material to interact with the *magnetic* component of an electromagnetic wave via magnetic dipoles.

Pure, crystallographically perfect diamond essentially does not interact with an external magnetic field, making it *non-magnetic*. However, defects within diamond are often *paramagnetic*, which means that they have intrinsic magnetic fields which align with an external magnetic field, enhancing it. This effect is due to an interaction between the external magnetic field and an electron's intrinsic magnetic moment, known as spin, within the defect. Spin can be visualized as a miniature bar magnet or compass needle, centered on the electron in question and possessing both a north and south pole.

Whether or not a particular defect is paramagnetic is dependent on its electronic configuration. Each defect electron is considered to contribute a spin which can either be "up" (+½) or "down" (-½): the overall spin of a defect, S, is the sum of these contributions over all local electrons. Defects with $S \geq ½$ are paramagnetic, and defects with $S = 0$ are non-magnetic. Therefore, defects with an odd number of electrons are always paramagnetic; defects with an even number of unpaired electrons are paramagnetic if the unpaired electrons' spin axes align, and non-magnetic if the unpaired electrons' spin axes are anti-aligned (Fig. 21).

The spin states of a paramagnetic defect will split into a total of $2S+1$ energy levels in the presence of an external magnetic field (the Zeeman effect). Magnetic dipole transitions can be driven between these states using electromagnetic radiation of the correct energy—typically in the microwave region for moderate magnetic fields. By monitoring the absorption of a monochromatic microwave field and sweeping an external magnetic field, a spectrum of absorption against magnetic field strength can be generated. This form of spectroscopy is known as electron paramagnetic resonance (EPR) or electron spin resonance (ESR).

Unpaired electrons are incredibly sensitive to their local environment, and it is this property which gives EPR its spectroscopic power. For example, many chemical isotopes possess an intrinsic magnetic moment (nuclear spin) which interacts with the unpaired electron, giving an unambiguous fingerprint of that element's presence within the defect structure (Fig. 22) when combined with contextual information such as known impurities within the sample and their

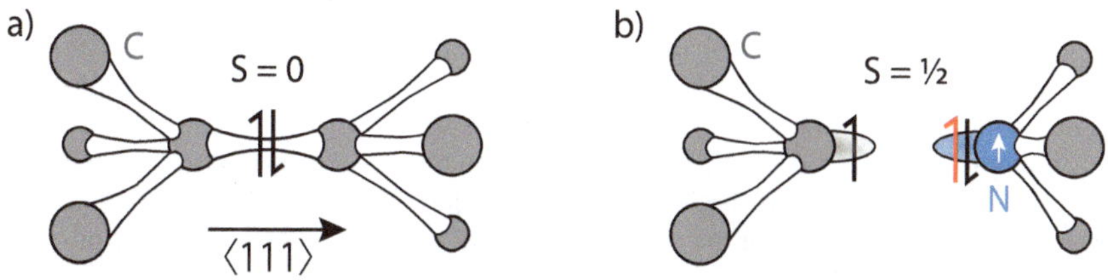

Figure 21. a) In defect-free diamond, each carbon atom possesses four valence electrons, and pairs each electron with one from each of its nearest neighbors, forming diamond's strong covalent bonds each of which has total spin $S = 0$ and therefore is EPR-inactive. **b)** In neutrally charged substitutional nitrogen, the nitrogen and one of its nearest-neighbor carbon atoms do not bond, and the additional valence electron of the nitrogen has no pair from its nearest neighbors, resulting in overall spin $S = ½$ and a measurable EPR signal. The nitrogen atom itself also has a small intrinsic magnetic moment (known as nuclear spin), as indicated by the white arrow.

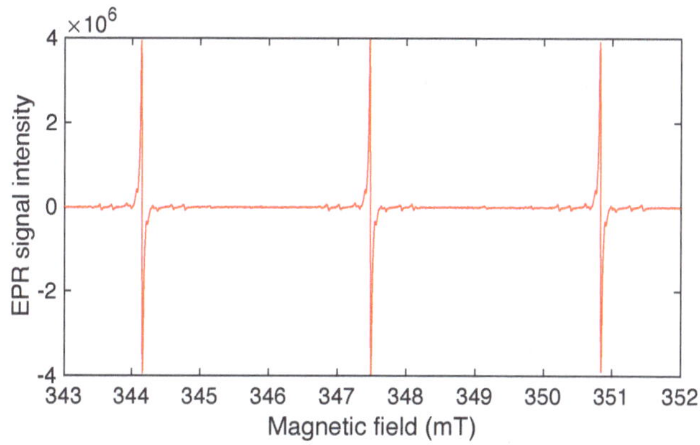

Figure 22. Typical EPR spectrum of a diamond containing neutrally charged substitutional nitrogen, N_s^0, with the magnetic field parallel to the <001> crystal axis. The three most intense lines are a result of magnetic interactions between the defect's unpaired electron spin and the nuclear spin of ^{14}N. The other, smaller "satellite" lines surrounding each primary transition are a result of further interactions with nearby ^{13}C nuclei (within a few angstroms). The magnetic field position and intensity of each satellite identifies the atomic location occupied by each corresponding ^{13}C, demonstrating the power of EPR spectroscopy (Cox et al. 1994; Peaker et al. 2016).

corresponding relative isotopic abundances. Thus, EPR allows the identification of previously unidentified magnetic spectra directly with their corresponding defect's chemical constituents and geometric configuration (and even neighboring defects): this approach can be particularly powerful for defect identification when combined with modern modeling techniques such as density functional theory (DFT) (Breeze et al. 2020). Additionally, the measured EPR spectrum depends on the relative orientation of the defect's symmetry axis and the external field. By measuring the EPR spectrum as a function of sample angle, an EPR spectrum can therefore yield information on a defect's symmetry, chemical constituents and electronic structure.

In addition to EPR's role in defect structure determination, it is also highly sensitive, non-destructive, and absolutely quantitative. Substitutional nitrogen concentrations down to 1 ppb can be routinely measured even in small (<0.5 ct) samples. The primary drawback of EPR as a spectroscopic tool is the requirement for the presence of unpaired spin, and as such it is most useful when employed in conjunction with traditional forms of optical spectroscopy for a holistic view of a given sample: this is particularly true in cases where the EPR spectrum of a defect has been reported but not its optical analogue, and hence EPR is required to identify the presence of the defect. In some particular cases it is possible to access an EPR-active ($S > 0$) state of an otherwise EPR-inactive ($S = 0$) defect via e.g., the application of appropriate optical illumination during the measurement (Felton et al. 2008).

EPR of diamond has a rich history within physics and has been instrumental in the identification and quantification of diamond point defects (Smith et al. 1959; Davies et al. 1978; Loubser and van Wyk 1978; van Wyk and Loubser 1993; Baker and Newton 1994; Newton 1994; van Wyk and Woods 1995; Nadolinny et al. 2017; Breeze et al. 2020). It has not been employed as a routine discrimination technique in gem labs due to the significant measurement timescales and the high operator skill required. Nevertheless, recent gemological publications have begun to adopt it to better understand both synthetic and natural diamond gemstones and their provenance (Wang et al. 2012).

Quantification of defect concentrations

Quantification of defect concentration is possible by several different techniques. As discussed previously, routine absolute quantification of defect concentrations is difficult to impossible in bulk using luminescence techniques. Unexpectedly, it is relatively simple to employ PL to quantify particular defects once their concentration is below approximately 1 μm^{-3}. With very high-sensitivity microscopes and single-photon-counting detectors it is possible to spectrally identify, and map, each emitting defect individually. This can be employed to study growth processes, defect diffusion and aggregation kinetics on a truly microscopic level (Chakravarthi et al. 2020), with resolutions of several hundred nanometers typically achieved (Fig. 23). However, for routine quantification absorption techniques must be employed, namely UV–Visible absorption (typically at liquid nitrogen temperature), IR absorption, and EPR spectroscopies.

For optical techniques, the measurement of absorption coefficient is not sufficient to absolutely quantify the defect concentration: identical concentrations of different defects will yield different absorption coefficients in a measurement, i.e., some defects absorb more of the incoming light *per defect* than others—a quantity known as the *absorption cross-section*. Accurately computing the absorption cross-section of a defect from first principles is nontrivial and would nevertheless require experimental verification.

To relate the concentration of a defect to its integrated optical absorption, we require a *calibration constant*. This constant enables us to measure the optical spectrum of a defect in a standard measurement and compute a defect density (in number per normalized volume). Here, we turn to other techniques for aid. For example, EPR is quantitative because one electron in defect A will absorb the same amount of energy as one electron in defect B. Therefore, by having a suitably well-characterized reference sample of known concentration, it is possible to quantify the concentration of other EPR-active defects. In cases where EPR-active defects have a known optical analogue, the concentration as measured by EPR can be related to the defect's integrated optical absorption, yielding a linear plot of concentration vs. integrated absorption, the gradient of which is the calibration constant for that defect. This approach using EPR is just one example of the ways in which optical absorption may be correlated with the corresponding defect concentration. In general, the integrated intensity of a sharp line, or the peak intensity of a broad band, is compared with a defect concentration measured by EPR, or nuclear activation analysis, or burning followed by residual gas analysis, or another suitable technique. For a vibronic band it is better to use the ZPL, rather than the peak intensity of the band, because the former is less susceptible to errors in determining the zero of absorption. A number of calibration constants for common defects are given in Davies (1999) and Dale (2015).

A research grade UV–Vis spectrophotometer will achieve a noise floor of approximately 0.001 A, which corresponds to approximately 0.2% transmission noise. Using NV^- as an exemplar defect in diamond which absorbs in the visible region: this corresponds to detection limits of approximately 0.1 and 1 ppb for diamonds which are 1 cm and 0.1 cm thick, respectively. EPR detection limits are on the order of 10^{12} spins, or approximately 0.1 ppb for a 1 ct diamond.

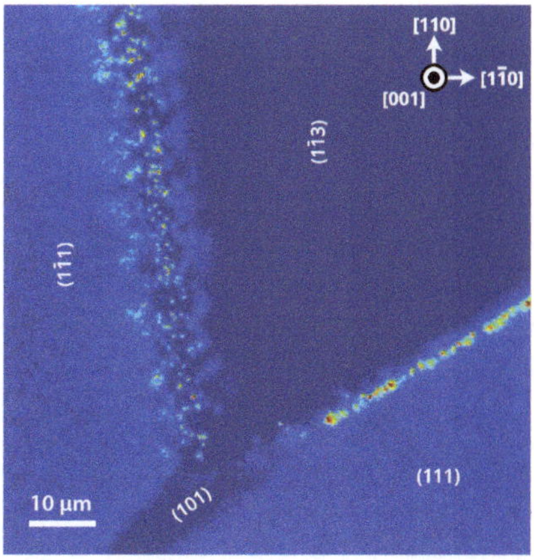

Figure 23. Decoration of growth sector boundaries in an as-grown HPHT sample, as imaged by a single-photon counting fluorescence microscope. The interfaces between each of the labelled growth sectors are decorated with isolated nitrogen-vacancy centers. Each of the bright spots on the $(1\bar{1}1) \rightarrow (1\bar{1}3)$ transition corresponds to a single atomic nitrogen-vacancy pair, which can be detected with high efficiency using an advanced fluorescence microscope. In this case it is possible to routinely reach ultimate sensitivity i.e., the detection of a single defect. The less bright fluorescence from the {111} planes arises from nickel uptake during growth in the {111} sectors, but not {110} or {113}. Modified from Diggle et al. (2020).

THE CLASSIFICATION OF DIAMOND

The classification scheme of diamond was largely developed between 1934 and 1986 and is based on the behavior in diamond of the two elements on either side of carbon in the periodic table—namely nitrogen and boron. The scheme is convenient for allocating a given diamond to a particular group with certain gross attributes, but there will be properties of that diamond that are not described by the classification scheme. In addition, there are some diamonds which cannot adequately be categorized by this scheme. A number of these exceptions will be described after the basic scheme has been outlined.

Nitrogen

Nitrogen is the dominant impurity in the majority of natural diamonds. Diamonds which contain sufficient nitrogen to produce easily measurable nitrogen-related absorption in the defect-induced one-phonon IR region are generically known as *Type I*. Other diamonds are generically known as *Type II*. It is important to recognize that a diamond classified as Type II may, nevertheless, contain small concentrations of nitrogen—"electronic grade" synthetic Type IIa CVD-grown diamond is typically specified as $[N_s^0]$ <5 ppb, which is undetectable by standard IR spectrometers, but routinely measurable using EPR (Cann 2009).

Nitrogen may be present as isolated substitutional atoms or in aggregated forms. Experiments carried out in the laboratory on the aggregation of nitrogen have played a crucial part in appreciating how these processes occur in nature. Nitrogen is believed to be incorporated in natural diamond initially on isolated substitutional sites. Natural diamonds which contain nitrogen of this type are rare (<0.1% of natural gem diamond; Smit et al. 2016), but nitrogen in high-pressure, high-temperature (HPHT) synthetic diamond is almost exclusively in this form. If present in sufficient concentration the substitutional nitrogen produces a characteristic absorption in the one-phonon region, shown in Figure 24. The major features are a broad band with a maximum at about 1130 cm^{-1}, and a sharp line at 1344 cm^{-1} associated with a *localized vibrational mode*. This description is used because the frequency is above the bulk lattice phonon maximum frequency of 1332 cm^{-1} and the vibration is therefore physically localized at the defect. This effect also accounts for the peak's narrow width. By measuring the absorption coefficient at either of these locations the concentration of isolated substitutional nitrogen can be determined (Woods et al. 1990b; Lawson et al. 1998). The single nitrogen also produces absorption in the visible and ultraviolet spectral regions which gives the diamond a yellow color. Diamonds that contain isolated nitrogen in their absorption spectra are known as *Type Ib*.

Aggregated nitrogen

After residing at geological temperatures over a billion-year timescale, most of the nitrogen in diamonds is present in aggregated forms. During the first stage of aggregation the A *center* is formed. Otherwise defect-free diamonds in which all the nitrogen is present as A centers have no absorption in the visible region (and so are colorless) but do exhibit characteristic absorption in the ultraviolet region (Fig. 24). A centers consist of two nearest-neighbor substitutional nitrogen atoms (Davies 1976). Diamonds containing aggregated nitrogen are known collectively as *Type Ia* and when most of the nitrogen is present as A centers the description *Type IaA* is used.

The one-phonon absorption for a Type IaA diamond is given in Figure 24. Nitrogen concentrations can be determined from the absorption coefficient in this region (usually measured at the most intense peak at 1282 cm^{-1}—see Woods et al. 1990a) and are typically 1000 ppm or higher. We see that nature has been very kind to the gem trade; if this concentration of nitrogen were present in the isolated substitutional form the diamonds would have a very unattractive deep yellow/brown color. However, as noted above, many Type IaA diamonds are colorless.

During the second stage of aggregation the B *center* is formed, and samples which have most of the nitrogen in this form are called *Type IaB* diamonds. The most intense peak in the one-phonon region is at approximately 1175 cm^{-1}. Studies of the B center indicate that it is comprised of 4 nitrogen atoms surrounding a vacancy (Loubser and van Wyk 1981). Near-surface carbon atoms may diffuse to the surface; in the bulk, they may diffuse to internal surfaces (voids) or become interstitial atoms. In the majority of Type IaB diamonds these carbon interstitials come together (perhaps with some nitrogen atoms) to form extended planar defects called *platelets* (Goss et al. 2003). Absorption at these platelets produces a moderately sharp peak in the region of 1365 cm^{-1}, frequently known as the *B′ peak*, or simply the *platelet peak* (Woods 1986) (Fig. 25). In what are termed *regular diamonds* the intensity of the B′ peak is proportional to the intensity of the absorption produced by the B centers. In the remaining (*irregular*) Type IaB diamonds the intensity of the B' peak is less than expected from this proportionality or may be completely absent (Woods 1986) as in Figure 24. Analysis of irregular diamonds demonstrates a lower concentration of platelets, which themselves possess typically smaller dimensions than those found in regular diamonds. This disparity is interpreted as platelet degradation and is understood to be indicative of higher mantle storage temperatures allowing its use as a mantle thermometer (Speich et al. 2018).

There is some weak absorption in the one-phonon region, called the *D band*, which is associated with the platelet peak, and this must be taken into account when the concentrations of the A and B centers are being calculated from the infrared spectra. Nitrogen concentrations in Type IaB diamond can be determined from the absorption coefficient due to B centers in the one-phonon region (usually measured at the plateau at 1282 cm^{-1}—see Woods et al. 1990a). Woods (1986) shows data for Type IaB diamonds with nitrogen concentrations up to 2000 ppm, but in most Type IaB diamonds the nitrogen concentration is lower.

The B center does not cause any known absorption in the ultraviolet or visible regions. However, diamonds containing B centers generally have absorption due to the N3 center. The N3 defect has three nitrogen atoms on a {111} plane, surrounding a common vacancy, N_3V^0 (i.e., a B center with one fewer nitrogen atoms). Woods (1986) showed that in regular Type IaB diamonds the intensity of the N3 absorption is proportional to the concentration of platelets. In such diamonds the platelet concentration is also proportional to the concentration of B centers,

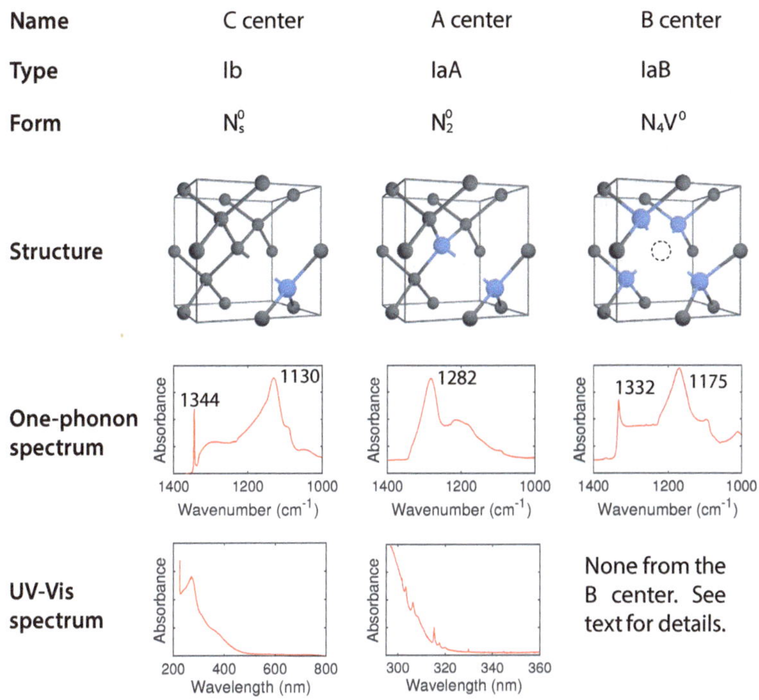

Figure 24. Summary of the nitrogen-related defects which form the basis for the Type I diamond classification. Diamonds which contain a similar concentration of both A and B centers are referred to as Type IaAB or simply Type Ia. Incomplete bonds between atoms indicate that the atoms do not form a bond and are included simply to provide a guide to the eye.

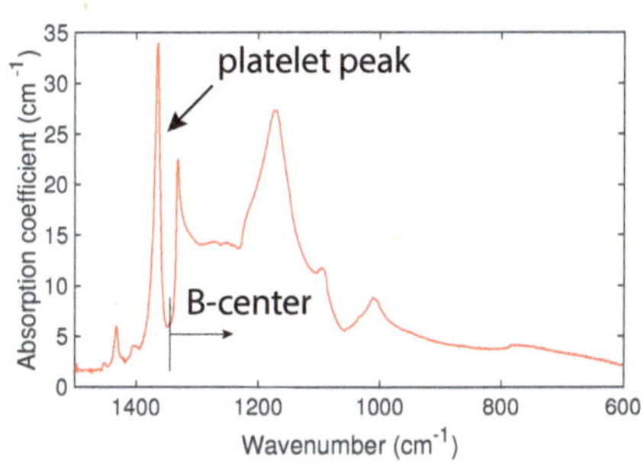

Figure 25. Infrared absorption spectrum of a regular Type IaB diamond, displaying a prominent B′ platelet peak at 1365 cm^{-1}.

and therefore Woods concluded that N3 centers are correlated with the absorption produced by B centers and that they are formed in a minor side reaction during the production of these B centers. That implies that the intensity of the N3 absorption is proportional to the intensity of the absorption produced by B centers in both regular and irregular Type IaB diamonds, though no systematic study has been carried out.

The nitrogen defects have been discussed separately, but most natural Type Ia diamonds contain a mixture of the A and B forms of nitrogen. These are normally described as *Type IaAB,* or simply *Type Ia*. Additionally, many natural diamonds containing C centers are better classified as Type Ib-IaA (Smit et al. 2016; Lai et al. 2019).

Boron

In a very small fraction of natural diamonds (approximately 0.1% (King et al. 1998) and less than 0.02% of gem quality diamonds (Smith et al. 2018)), substitutional boron is the major detectable impurity. These are *Type IIb* diamonds, and, as illustrated in Figure 11, such diamonds have an optical absorption with a continuum absorption starting at an energy of 0.37 eV in the infrared and extending into the red visible region: this is an electronic absorption corresponding to excitation of electrons from the valence band which are captured by the boron defect. In favorable cases this absorption gives the diamond an attractive light- to dark-blue color. Boron also produces some vibrational absorption in the defect-induced one-phonon region.

The ionization energy of boron is sufficiently low that about 1% can be thermally excited at room temperature, producing positive holes in the valence band (see Fig. 12). Type IIb diamond is therefore a *p-type semiconductor*—a semiconductor in which the majority charge carrier is holes (if electrons are the dominant carrier then the material is known as an *n-type semiconductor*). Wedepohl (1957) carried out Hall effect studies on Type IIb diamond and showed that, in order to explain the results satisfactorily, the diamonds must also contain donors at a smaller concentration than the concentration of boron acceptors. It is probable that this donor is nitrogen in the majority of natural diamonds.

The concentration of neutral boron acceptors $[B^0]$ is equal to the total boron concentration minus the donor concentration and can be determined (Collins and Williams 1971) from the area of the excited-state absorption peak at 0.348 eV (2802 cm^{-1}, Fig. 11d). Other regions of the absorption spectrum can also be used if the absorption at 0.348 eV is too strong (Howell et al. 2019). Boron is a widespread element in the Earth's crust with relatively low abundance, and it is probable that many diamonds contain boron. However, it is only those diamonds for which boron is the major impurity which are semiconductors and referred to as Type IIb. Those Type II diamonds which are not semiconductors are known as *Type IIa*. Recent evidence suggests that Type IIb diamonds originate deep within the Earth's mantle, beneath the transition zone (Smith et al. 2018).

Summary

The information given in this section is summarized in Figure 26. It is important to note that the type system does not capture many important properties: the color of a sample may change between many hues without changing type, as the type system does not capture information on visible properties. This means that many nitrogen-containing defects such as 3107 cm^{-1} (see later section on *Hydrogen*), and defects which do not present recognized IR absorption spectra, such as NV and N_3V, are not considered by the type system, and a diamond which presents no nitrogen-related features in its one-phonon absorption spectrum may nevertheless contain easily detectable concentrations of nitrogen in other forms. Additionally, many diamonds have absorption bands in the one-phonon region that are not included in this scheme. For example, a broad absorption and corresponding sharp peak at 1332 cm^{-1} has been shown by Lawson et al. (1998) to arise at positively charged single nitrogen, N_s^+. Hainschwang et al. (2012) found that many natural diamonds, classified as Type Ib because of the presence of absorption at 1344 cm^{-1}, have an additional "Y component" in the one-phonon absorption, with a maximum at 1145 cm^{-1}. Collins and Mohammed (1982) showed that, for all the samples in their study of natural brown diamonds which have an absorption band at 480 nm, the one-phonon region also has an absorption peak at 1240 cm^{-1} that is not seen in other diamonds. Finally, there is a whole range of absorption spectra associated with other impurities (such as hydrogen, nickel etc.) which are not captured by the type system.

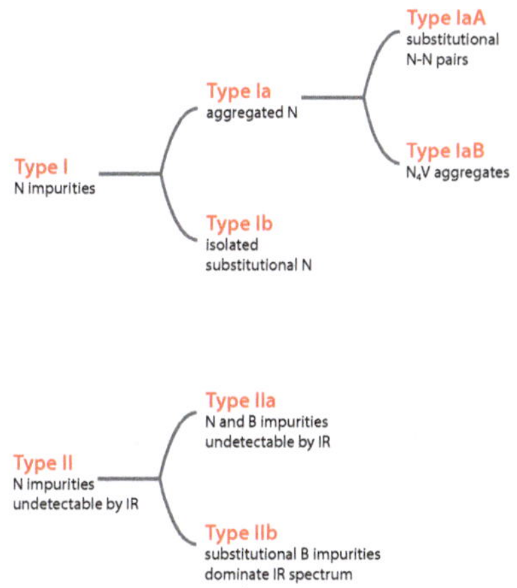

Figure 26. Summary of diamond type classification system, which is based on the absence or presence of infrared absorption due to specific forms of boron and nitrogen incorporated into the diamond. The type system can provide a gross idea of sample properties in the majority of natural samples, but by no means does it fully describe a given sample's attributes. As an example, it is possible to modify a sample's color from brown to green to pink without changing its type.

POINT DEFECTS IN DIAMOND

Defects are ubiquitous in both natural and synthetic diamond and give rise to many of the properties that make diamond such an extraordinary material technologically, gemologically and geologically. Even the most extraordinary Type IIa colorless natural diamond contains a significant concentration of defects, though not in a configuration that produces color. Having established the methods by which point defects in diamond are typically analyzed, and discussed the type system, which allows gross classification of diamonds into groups, we now discuss several important point defects and their properties directly. Defect interactions and their macroscopic effects on a crystal are summarized in the subsequent sections.

Intrinsic

Intrinsic point defects are those defects which involve carbon as the only element (i.e., no impurities). The most important of these are the vacancy and the interstitial—a vacant lattice site and a carbon atom off-site, respectively. These defects are the primary result of irradiation of diamond, whether by electrons, gamma rays or other particle sources: incoming radiation imparts enough energy locally to break the carbon–carbon bonds, displacing the atom to an off-site position and leaving behind a vacant lattice site (Fig. 27).

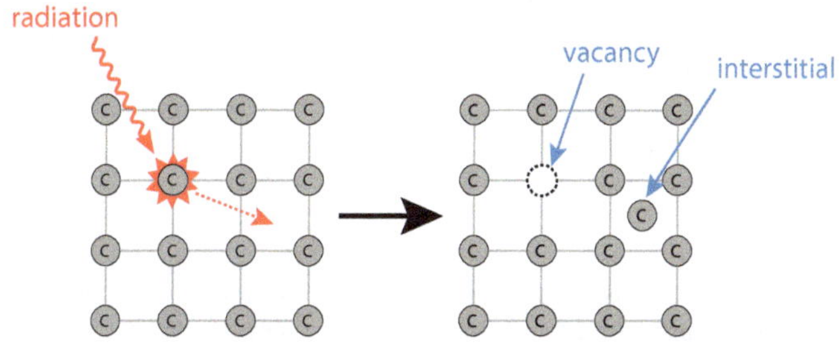

Figure 27. Intrinsic damage created by high energy radiation (electrons, gamma rays, ions etc.). The vacancy-interstitial pair may immediately self-annihilate or may freeze in, whereupon subsequent heating will allow them to diffuse separately through the lattice.

If the vacancy and interstitial do not self-annihilate during the irradiation, they are "frozen" into the lattice. The concentrations of vacancies and interstitials may not be equal, and their ratio depends on the sample temperature during irradiation, with lower concentrations of interstitials being detected for irradiation performed at higher temperature (Newton et al. 2002). This apparent discrepancy has been attributed to charge-related effects during the irradiation itself, which produces transient, high-mobility, interstitials which can subsequently combine with other defects or impurities to form defect complexes, or annihilate at internal or external surfaces (Newton et al. 2002). In the absence of irradiation, interstitials are mobile at relatively low temperatures of ≥400 °C (Iakoubovskii et al. 2003), at which point a small proportion (~30%) of vacancies are also lost, presumably to vacancy–interstitial recombination (Allers et al. 1998). Vacancies themselves begin to diffuse at approximately 700 °C (Davies et al. 1992; Newton et al. 2002). These (lab-timescale) threshold temperatures are frequently encountered in the defect investigations and in color treatments, as the migration and subsequent combination of vacancies and interstitials with other defects can drastically modify a diamond's properties (see *Treatments to change the color of diamond*).

Spectroscopically, the neutral and negatively charged vacancy possess ZPLs at 741 and 394 nm and are referred to as GR1 and ND1, respectively: the ratio between them is typically controlled by the substitutional nitrogen concentration (see *Donors, acceptors, and charge transfer*). There are several configurations of intrinsic interstitial, but the most common gives rise to a weak absorption at approximately 667 nm (Smith et al. 2004). Other optically-active configurations such as TR12 and 3H are less-well characterized and typically estimated to be significantly lower in concentration (Walker 1977; Davies 1981). Both the common interstitial (Twitchen et al. 1996) and the negatively-charged vacancy (Twitchen et al. 1999) can also be studied by EPR.

Macroscopic stress within a diamond, caused by e.g., metallic inclusions or crystallographic faults, can be visualized using a microscope equipped with two linear polarizers, one either side of the sample and with their polarization axes mutually perpendicular (colloquially known as "crossed polarizers"). Under this configuration, stress-induced birefringence causes those areas of the crystal which are under stress to become bright, while unstressed regions remain dark. The vast majority of natural diamonds contain fibrous networks of dislocations (extended defects that can span macroscopic dimensions of a crystal) (Fig. 28). In natural Type II diamond, or very low nitrogen concentration Type I, the birefringence pattern often appears "cross-hatched" when viewed through crossed polarizers—a result of the random strains to which the crystal has been exposed during residence time in the mantle. For natural Type Ia diamonds with significant nitrogen impurities, mottled or linear patterns dominate the birefringence seen under crossed polarizers. Dislocation networks in Type II diamonds can often be visualized directly (rather than indirectly via their effect on crystal stress, as we observe in crossed polarizers) using DiamondView or cathodoluminescence (see Sumida and Lang 1981, Harris et al. 2022, this volume). In natural Type IIa samples, plastic deformations can yield an overall brown color due to absorption at vacancy clusters (Mäki et al. 2009, 2011), which can be subsequently removed by HPHT treatment (see *Brown diamonds*). In contrast, synthetic crystals frequently display dislocations which follow the growth direction (in the case of CVD) or the interfaces between growth sectors (in the case of HPHT), yielding characteristic patterns which are indicative of the crystal habit during growth

Nitrogen

As discussed in the previous section, nitrogen is present in significant quantities in the vast majority of natural diamond. For synthetics, atmospheric nitrogen is incorporated into HPHT-grown diamond in significant quantities (>100 ppm) unless specific chemical traps ("getters") are added to the synthesis capsule, which are often employed to target colorless Type IIa material with <0.1 ppm nitrogen. In CVD growth, the addition of even small amounts of nitrogen

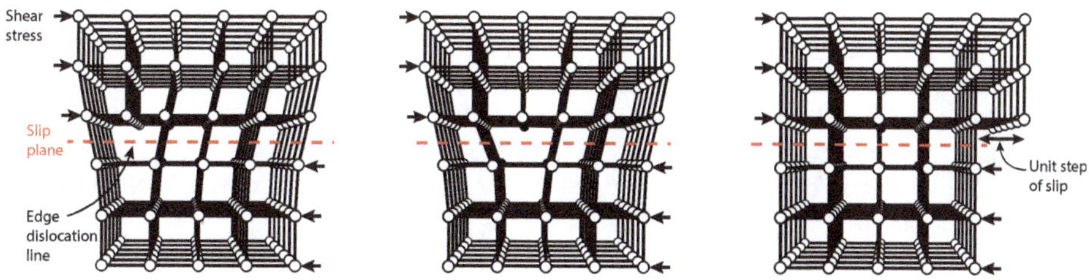

Figure 28. Example of an edge dislocation propagating through a crystal as a result of differential (shear) stresses. In diamond these dislocations do not anneal out at mantle temperatures, even over geological timescales. Adapted from Callister (2006).

(10–200 ppm) to the gas phase can increase the growth rate by 2–10× (Tallaire et al. 2006, 2013; Achard et al. 2007), with incorporation into the diamond typically <100× lower than the gas phase concentration. Therefore, the vast majority of colorless CVD gems contain <1 ppm nitrogen to achieve commercially viable growth rates while retaining relatively high color grades.

The aggregation of nitrogen from substitutional towards higher-order aggregates (N_2 and N_4V) forms the basis of the Type I classification system. The additional electron present in substitutional nitrogen (vs. carbon) introduces strain into the lattice, extending one of the N–C bonds by up to 30% (Breuer and Briddon 1996). The nitrogen aggregation process therefore achieves a reduction in lattice strain via the clustering of nitrogen atoms around a vacancy (or one another). The energy required to directly swap a nitrogen and carbon atom is extremely high, and thus nitrogen migration through the lattice is understood to be mediated by vacancies and interstitials (Collins 1980; Pinto et al. 2012; Jones et al. 2015) (Fig. 29).

At its simplest level, the aggregation of nitrogen from substitutional to the N_4V (B center) form can be expressed as

$$4N_s \rightarrow N_4V + C_I$$

where C_I is an interstitial carbon atom: large aggregates of interstitials form the platelets associated with high levels of nitrogen aggregation. The balance between aggregation termination and substitutional nitrogen (or other aggregates) is determined by temperature: lower temperatures drive towards higher aggregation while extreme temperatures are capable of destroying aggregates, re-producing substitutional nitrogen (Brozel et al. 1978; Evans and Qi 1982).

Chemical kinetics. The aggregation of nitrogen is typically described in terms of chemical reaction kinetics. Early, seminal work on nitrogen aggregation laid the foundations for the application of quantitative chemical kinetics to the study of defect production, migration and diffusion in diamond (Chrenko et al. 1977; Collins 1980; Allen and Evans 1981; Evans and Qi 1982). Almost universally, energy to drive an aggregation process in diamond is provided as heat by raising the temperature of the sample (though it is possible to perform "local" annealing via non-thermal-equilibrium methods, Chen et al. 2019).

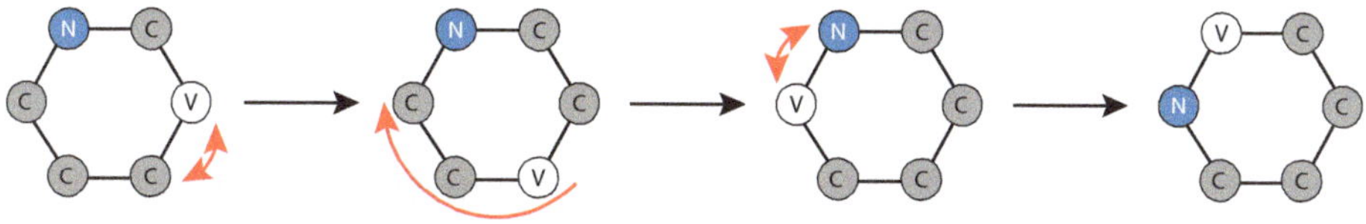

Figure 29. Vacancy-assisted migration of nitrogen in diamond. The direct carbon-nitrogen exchange process is extremely high in energy and is not expected to occur at any practicable temperature. Instead, the nitrogen moves by exchange with a vacancy, which becomes mobile at approximately 700 °C. It is possible for the same vacancy to be trapped and re-trapped, allowing the pseudo-NV to migrate through the lattice as a unit. Adapted from Pinto et al. (2012).

Chemical kinetics can occur with different qualitative behavior, with the relevant processes being described as "first-order", "second-order", or "mixed second-order". A first-order process simply requires enough energy for the chemical reaction to occur—sources and sinks are assumed to be infinite and the rate depends only on the remaining concentration. An example would be the destruction of a point defect (say, NV→ N + V). Within a first-order process, the change in concentration of the defect in question ([*X*]) is simply proportional to its concentration

$$\frac{d[X]}{dt} = -k[X]$$

The rate constant *k* depends exponentially on the ratio of the energy required to drive the process (its activation energy, E_A) to the available thermal energy ($k_B T$, where k_B is the Boltzmann constant and *T* the temperature in thermal equilibrium). The attempt frequency, *f*, describes the rate at which the constituents involved in the process encounter one another during random diffusion through the lattice:

$$k = f \exp\left(-\frac{E_A}{k_B T}\right) \quad (1)$$

A second-order process depends on the energy required for the process to occur, the energy available to drive the process, and the concentration of the reactants. Taking for example the aggregation of 2 $N_S \rightarrow N_2$, the process will run quickly at first but slow down as the available reactants (N_S) decrease appreciably in concentration throughout the process. Mathematically, the change in concentration of the defect is now non-linear in concentration

$$\frac{d[X]}{dt} = -k[X]^2$$

A mixed second-order process is one in which the reactants are different—for instance, N + V → NV—in which case the process depends on the changing concentrations of both precursors,

$$\frac{d[X]}{dt} = \frac{d[Y]}{dt} = -k[X][Y]$$

Practically, the relevant chemical kinetics and their associated activation energies are determined in experiments where the same sample is heated multiple times at the same temperature and the concentrations of defects are measured at each interval (isothermal annealing), or by heating the same sample for the same time at multiple temperatures (isochronal annealing). A particular defect concentration may not straightforwardly fit any of the above descriptions if multiple, otherwise independent, processes are occurring simultaneously. The determination of the attempt frequency *f* in Equation (1) can be particularly difficult and often an attempt frequency matching the diamond lattice vibration frequency is assumed (though not necessarily always justified).

To illustrate the impact of different kinetic orders, we consider their behavior in different concentration regimes. A first-order process rate is independent of concentration, and as long as samples are held at the same temperature, the same percentage change in concentration will be observed for any non-zero starting concentration and a fixed annealing time. For second-order (and mixed second-order) processes, the percentage change will depend strongly on the starting concentrations, with higher concentrations yielding higher percentage changes. Therefore, the temperature at which a measurable change in concentration is observed will differ between different samples. This insight, combined with careful measurement and reasoning, gives us a route to employ nitrogen aggregation to study a diamond's residence conditions over geological timescales (Allen and Evans 1981; Taylor et al. 1990, 1996; Bulanova 1995; Mendelssohn and Milledge 1995; Kohn et al. 2016; Speich et al. 2018)—see Nimis (2022, this volume) for further information.

Production of A and B centers. The aggregation of C centers to A centers in synthetic Type Ib material is found to obey second-order kinetics, with the production of N_2 (A centers) being second-order in the substitutional nitrogen (C center) concentration (Chrenko et al. 1977). Aggregation occurs on lab timescales at approximately ≥1700 °C; at 2400 °C the rate of production and destruction of A centers reaches equilibrium at approximately 95% in the A center form (Evans and Qi 1982; Collins et al. 2000). The rate of aggregation can be dramatically enhanced at low temperatures by first irradiating the material before thermal annealing, introducing vacancies and interstitials into the lattice and hence greatly increasing the mobility of nitrogen (as discussed previously) (Collins 1980; Allen and Evans 1981). Additionally, the aggregation rate is sensitive to the pressure during annealing, with higher pressure associated with decreased aggregation rates (Kiflawi et al. 1997).

Aggregation to N_4V (B centers) occurs at approximately ≥2400 °C on lab timescales. Precise evaluations of experiments become difficult in these conditions due to the extreme temperatures and pressures (typically >8 GPa). However, upon heating Type IaA diamond, the resulting B center concentration was found to be second-order in A centers, suggesting that it is A centers themselves which aggregate to form B centers. During this process little change is observed in the N_3V (N3) concentration, suggesting that it is a byproduct of aggregation, rather than necessarily the primary aggregation route (Brozel et al. 1978). Alternatively, it may be that the capture cross-section of N_3V for diffusing nitrogen (presumably in the NV or N_I form) is very high, making it difficult to generate high concentrations in the presence of other decomposing or diffusing nitrogen species (Kiflawi and Bruley 2000). The production of B centers is associated with the production of platelets, which initially were interpreted as large nitrogen-containing aggregates but which are now recognized as interstitial carbon aggregates (see previous subsection) (Woods 1986; Kiflawi et al. 1998; Goss et al. 2003).

Degradation of platelets in Type IaB diamond is associated directly with the formation of dislocation loops and the production of <111>-oriented crystalline structures called voidites, with diameters of approximately 2–200 nm (Hirsch et al. 1986; Evans et al. 1995; Navon et al. 2017), in addition to changes in size distribution and concentration of platelets (Kiflawi and Bruley 2000). The precise constituents of voidites are unclear, though there is evidence that they contain solid NH_3 (Rudloff-Grund et al. 2016) or crystalline molecular nitrogen under high pressure (Bruley and Brown 1989; Kiflawi and Bruley 2000; Navon et al. 2017), and they may therefore arise due to the release of nitrogen impurities trapped at platelets or due to decomposition of B centers themselves.

Diffusion, aggregation, and formation of nitrogen-related defects. While significant study has been dedicated to the aggregation and behavior of nitrogen in diamond, there are still some aspects which remain unclear. In material containing mainly single nitrogen and which is subsequently irradiated, the vacancy-assisted aggregation of N to N_2V is relatively well-understood: vacancies produced during irradiation diffuse to substitutional nitrogen at 700–800 °C, forming NV; the NV unit becomes mobile around ≥ 1400 °C, and diffuses to remaining substitutional nitrogen to form N_2V. This may capture an interstitial or emit a vacancy, forming A centers typically at ≥1800 °C. However, the subsequent formation routes for N_3V and N_4V remain under debate. As discussed above, the concentration of N_3V is typically low even in samples which eventually produce significant B centers (N_4V) at ≥2400 °C.

Irradiation of material which already contains aggregated nitrogen produces different defects. In irradiated Type IaA material, N_2V production occurs at much lower temperatures (~800 °C) via vacancy capture at the A centers themselves (Collins and Kiflawi 2009). Irradiation of Type IaB diamond, followed by annealing at ~ 800 °C, leads to the formation of $N_4V_2^0$, a color center known as H4, via vacancy capture at B centers (Loubser and van Wyk 1981).

Nitrogen itself is an efficient trap for interstitials as well as vacancies. Carbon interstitials which diffuse to substitutional nitrogen are understood to become nitrogen interstitials (Kiflawi and Mainwood 1996). These nitrogen interstitials are mobile at relatively low temperatures (~600 °C, Liggins et al. 2010) and may themselves capture another nitrogen to become the di-nitrogen interstitial, which is responsible for the infrared H1a absorption at 1450 cm^{-1} (Liggins et al. 2010). The temperature at which carbon interstitial capture occurs is type-dependent (Woods 1984; Fisher and Lawson 1998), but H1a typically anneals out at 1400–1500 °C (Liggins et al. 2010).

In natural diamond it is assumed that nitrogen is incorporated substitutionally. It is unclear which mechanism mediates nitrogen aggregation from substitutional to A centers in the majority of natural diamond, but it is likely that both interstitial and vacancy-driven processes are involved. The spontaneous thermal formation energies for interstitials and vacancies are both much higher than their migration energies (Jones et al. 2015), so that if they are introduced to the diamond by means other than direct thermal formation, they can assist with diffusion of atoms through the material at more moderate temperatures. Natural diamonds frequently include flaws (e.g., dislocations, damage from low-flux natural irradiation, material around inclusions) which can act as sources of vacancies and interstitials and it is likely that they provide the mechanism for early-stage aggregation in these diamonds.

Within the nitrogen-vacancy family, the most important color centers and their corresponding ZPLs are: $NV^{0/-}$ (575 / 637 nm); $N_2V^{0/-}$ (H3: 503 nm / H2: 986 nm); N_3V^0 (N3: 415 nm); $N_4V_2^0$ (H4: 496 nm). With the exception of N_3V^0, which is almost ubiquitous among Type Ia diamonds, these nitrogen-vacancy related centers are typically introduced during irradiation and annealing, though they can also be present in natural diamonds typically at low concentrations. An in-depth summary of nitrogen-related point defects and their interactions in diamond can be found in Ashfold et al. (2020), and their effect on diamond color is discussed in the following sections. Characteristic spectral features of dominant nitrogen-related defects are summarized in Table 1.

Hydrogen

Hydrogen is a common impurity in diamond, with some natural diamonds reported to contain in excess of 1000 ppm (Hudson and Tsong 1977). Hydrogen content in gem-quality single crystal synthetic diamond is typically much lower (<50 ppm). It is unclear where the majority of hydrogen resides: CVD diamond which is irradiated and subsequently annealed can produce high concentrations of NVH (Hartland 2014), suggesting a latent source of hydrogen which is as-yet unidentified. Hydrogen can be diffused into the diamond lattice under elevated temperature and pressure, resulting in the passivation of boron acceptors (Chevallier et al. 1998) and the destruction of NV centers by hydrogen capture (which become NVH) (Stacey et al. 2012), further providing evidence for the solubility of hydrogen without revealing its location.

The presence of (currently detectable) hydrogen in diamond is primarily monitored through the vibration of carbon–hydrogen (C–H) stretch modes, which occur at characteristically high frequencies (~2800–3300 cm^{-1}) in IR absorption spectra. Of these C–H modes, the most important is the 3107 cm^{-1} mode, which in so-called hydrogen-rich diamond can be stronger in peak intensity than the intrinsic multi-phonon absorption (Fritsch et al. 2007). This mode, and its corresponding 1405 cm^{-1} bend mode, were recently recognized as belonging to N_3VH^0— a defect which is identical to the N3 (N_3V^0) defect but with a hydrogen bonded to the single

Table 1. Summary of some common spectroscopic optical features encountered in diamond (where NIR and MIR denote near- and mid-infrared, respectively). For a more comprehensive review of spectroscopic features, their properties and origins, see the reviews by Zaitsev (2001) and Dischler (2012). LVM is Local Vibrational Mode (see *Ultraviolet and visible absorption and luminescence*).

Name	Structure	ZPL wavelength (nm)	Mid-IR features (cm^{-1})
		UV–Vis–NIR vibronic bands listed by ZPL wavelength	
ND1	V^-	394	
N3	N_3V_0	415	
H4	N_4V_{20}	496	
H3	N_2V_0	503	
	NV_0	575	
594 nm		594	
	NV^-	637	
	C_I	666 broad (absorption only)	
	SiV^-	737	
GR1	V_0	741	
	SiV_0	946	
H2	N_2V^-	986	
		Infrared LVMs and vibronic transitions	
H1c			5171 peak
H1b			4941 peak
H1a	N_{2I}		1450 peak
	N_3VH^0		1405, 3107 peaks
	B_s^0	Decreasing ramp from red to blue	2464, 2802; ramp. Weak one-phonon
		Infrared vibrational one-phonon absorption	
C center	N_s^0	Ramp from green to blue	1344 LVM; 1130 peak
	N_s^+		1332, 1046 peaks
B center	N_4V^0		1332, 1175 peaks
A center	N_2^0	UV ramp	1282 peak

carbon neighbor (Goss et al. 2014) (Fig. 30a,b). Studies of high-hydrogen, high-nitrogen CVD diamond have identified hydrogen-containing analogues of other nitrogen aggregates, including NVH^0 (3123 cm^{-1}, Cruddace 2007; Khan et al. 2013, 520 nm band, Khan et al. 2009) and NVH^- (EPR, Glover et al. 2003) (Fig. 30c), N_2VH^0 (EPR, Hartland 2014), and N_3VH^0 (1405 and 3107 cm^{-1}, Goss et al. 2014; Coxon et al. 2020). Of these, only N_3VH^0 has been definitively reported in natural diamond, though diamonds high in hydrogen often exhibit absorption at 3123 cm^{-1}, which is likely to correspond to low concentrations of NVH^0 (Fig. 30a,c). The N_4VH configuration is not expected to be energetically favorable, and therefore N_3VH^0 presumably represents the termination of this aggregation cycle over geological timescales. N_3VH^0 is stable to at least 2300 °C (Zaitsev et al. 2018).

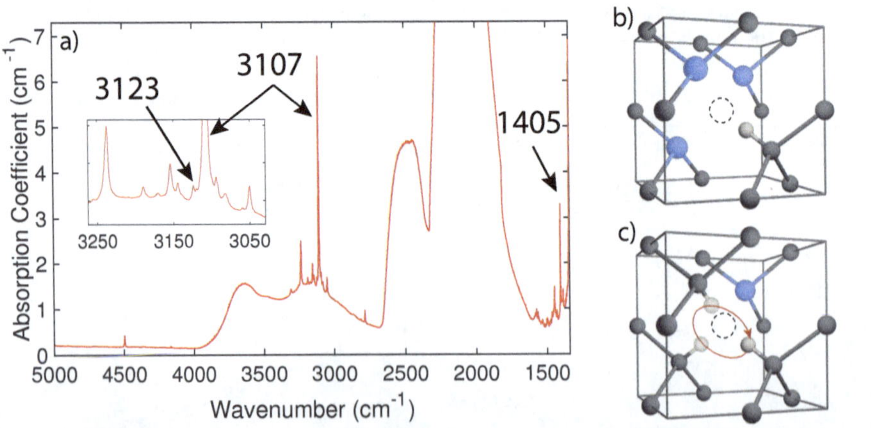

Figure 30. a) Infrared spectrum of a high-hydrogen natural sample. The one phonon region has been omitted due to spectrometer saturation. A plethora of C–H modes at 3000–3300 cm^{-1} is evident, with the most intense being the 3107 cm^{-1} stretch mode associated with N_3VH^0: its corresponding bend mode at 1405 cm^{-1} and bend-stretch overtone at 4496 cm^{-1} are also typically observed in this material. Atomic structure of **b)** N_3VH^0 and **c)** $NVH^{0/-}$. The former is identical to the defect which produces N3 (415 nm) absorption, but with a hydrogen bonded to the lone carbon neighbor. In NVH, the hydrogen quantum-mechanically tunnels between the three equivalent carbon neighbors.

The presence of significant hydrogen in CVD-grown high-nitrogen material means that the aggregation of nitrogen in this material does not necessarily follow the route described in the previous subsection. Typically, both the standard nitrogen aggregates (N_nV, with $n = 1$–4) and their hydrogen-containing analogues (N_nVH, with $n = 1$–3) are produced simultaneously with the latter typically the dominant route (Zaitsev et al. 2018). As the hydrogen-containing analogues are not associated with strong color (except a weak pink induced by NVH), then similar colors to those obtained with HPHT-grown diamond may often be produced in CVD material, but with a lower efficiency in terms of starting nitrogen to target color. A complete understanding of annealing behavior and dominant aggregation mechanisms in this material is yet to be reached (Ashfold et al. 2020).

Many other C–H IR modes have been reported but in the vast majority of cases their atomic origin is unknown—summaries are given in Fritsch et al. (2007) and Meng et al. (2008). Analysis of these modes is complicated by the fact that they occur within a relatively narrow frequency window—an indication that the C–H frequency is only weakly perturbed by the atomic configuration of the corresponding defect. Therefore, direct assignment via theoretical modelling methods is difficult without supporting experimental evidence (e.g., isotopic shift or annealing behavior). Unlike nitrogen and many other elements, there are very few sharp absorption or emission lines which are associated with hydrogen-related defects.

Boron

Boron forms part of the diamond classification system as a result of its dramatic effect on diamond color. Despite its prominence within the diamond community, very few boron-related centers have been definitively identified. This is especially surprising given its high solubility in diamond, which allows boron incorporation up to 10%. At these concentrations, diamond becomes metallic (rather than semiconducting) and is even capable of superconductivity at low temperature (Ekimov et al. 2004). Its metallic behavior can be exploited in many electrochemical applications in conjunction with diamond's supreme chemical resistance (Macpherson 2015).

Several luminescence centers are associated with Type IIb diamond (e.g., 648 and 776 nm, Green 2013; Eaton-Magaña and Ardon 2016) but they are universally weak, having no effect on perceived color: corresponding absorption spectra of these samples typically show only broad absorption with no sharp features (King et al. 1998). Even intrinsic features such as neutral vacancies can be suppressed in Type IIb diamond following irradiation (Collins 1977; Green 2013). It seems likely that this is a result of the boron acting as an acceptor and biasing defects towards otherwise unattainable charge states; however, there is as-yet no convincing evidence for V^+, and the suppression of V^0 may instead be a result of boron-vacancy (or boron-interstitial) interactions.

Other elements

A huge number of defects have been identified in diamond, primarily through visible or infrared optical measurements and covering much of the periodic table (Zaitsev 2001; Dischler 2012). The intentional introduction of other elements into diamond has accelerated due to its potential in optically-active quantum computation, communication, and sensing technologies (Aharonovich and Neu 2014; Atatüre et al. 2018; Awschalom et al. 2018; Barry et al. 2020). This is typically performed either by doping during synthesis (Tallaire et al. 2015; Ekimov and Kondrin 2017; Achard et al. 2020) or by post-growth ion implantation (Pezzagna et al. 2011; Lühmann et al. 2018). As a result of the plethora of possible defects we make no attempt to be comprehensive and simply mention those elements which are most prominent.

Silicon is incredibly rare in natural diamond but common in CVD-grown synthetics, typically arising due to contaminants in the growth reactor. This typically results in ZPLs at 737 nm (which splits into a doublet at liquid nitrogen temperature) and 946 nm corresponding to SiV^- and SiV^0, respectively, and which can be visible in both luminescence and absorption (D'Haenens-Johansson et al. 2011). Unlike NV, where the nitrogen resides effectively in a carbon lattice site, in $SiV^{0/-}$ the silicon relaxes back towards the vacancy, adopting a so-called semi-vacancy position (Fig. 31a,b). At high SiV^0 concentrations, samples can become moderately gray–blue colored and be heavily photochromic (Breeze et al. 2020). Recent measurements have also identified substitutional silicon, which is detectable only as a broad peak around 1338 cm^{-1} in the infrared (Breeze et al. 2020).

A large number of transition-metal-related defects (predominantly nickel and cobalt) have also been identified, primarily as a result of their use as a solvent-catalyst during HPHT synthesis. Nickel-related defects are relatively frequently observed in natural diamond (though typically not in concentrations high enough to affect color; Breeding et al. 2018). Much of the work on transition metal defects has been completed using EPR (Baker and Newton 1994; Nadolinny and

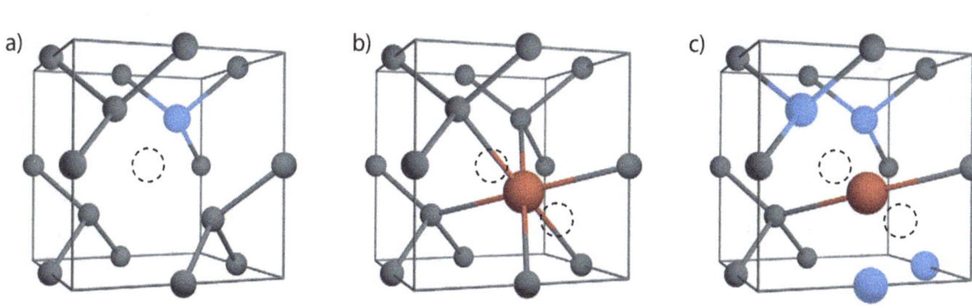

Figure 31. Comparison of the **a)** NV and **b)** SiV / NiV / CoV structures. In both cases, only one atom (by number) is missing from the lattice, but in NV the nitrogen remains substitutional, whereas for the larger atoms the impurity relaxes back to a bond-centered location between two lattice sites; the so-called semi-vacancy position. **c)** The NE8 defect, which emits at 794 nm, is comprised of a nickel atom in the semi-vacancy position with four of its nearest-neighbors replaced by nitrogen (Kupriyanov et al. 1999; Nadolinny et al. 2005).

Yelisseyev 1994; Nadolinny et al. 1999, 2005). Several well-known optical centers are related to nickel (e.g., the 494, 658, and 883/885 nm centers, Lawson and Kanda 1993; Yelisseyev and Kanda 2007; Breeding et al. 2018) and cobalt (Lawson et al. 1996; Kupriyanov et al. 2001) though in most cases their atomic configuration is not definitively known.

The presence of nickel or cobalt in Type Ib diamond (as would be expected for an HPHT crystal grown with a Ni or Co solvent) can significantly enhance nitrogen aggregation processes at lower temperatures via the efficient generation of highly-mobile nitrogen interstitials (Fisher and Lawson 1998; Nadolinny et al. 2000). It is also possible for the nickel itself to trap nitrogen atoms, producing defects such as NE8 (Fig. 31c). The uptake of nickel and cobalt during HPHT synthesis is sector-dependent (Fisher and Lawson 1998), with nickel-related optical activity detected only in the {111} sectors (Collins et al. 1990)—this can be revealed by luminescence mapping (Fig. 15, 23).

As of writing, color centers in diamond are attracting significant attention for quantum sensing, quantum computation, and quantum communication. The most-studied defect in this field is NV^- (see reviews at (Doherty et al. 2013; Rondin et al. 2014)), though other vacancy-containing defects (such as the group-IV-vacancies, $SiV^{0/-}$, GeV^-, SnV^- and PbV^-) are also attractive as a result of their much higher ZPL to sideband ratio, which is critical in many photonic applications (Aharonovich and Neu 2014; Atatüre et al. 2018; Bassett et al. 2019). Fundamentally, diamond is an ideal host for many of these defects as the vibrational and electronic coupling between the defect and the bulk lattice is often weak, allowing the defects themselves to be treated and manipulated as "trapped atoms": this is highly unusual among solid state defects, as in most materials the coupling between the defect and the material is large, leading to broad emission bands with weak ZPLs and poor electron spin lifetimes—critical metrics for their intended applications.

NATURALLY COLORED DIAMONDS

The variety of point defects and extended defects that can occur in diamond is vast, but only a few of those defects (the so-called color centers) give rise to the beautiful colors that make many gem diamonds so sought-after. Although most people think of colorless diamonds as the purest and most valuable gems, the occurrence of certain defects in diamond in just the right concentrations is so rare that stones with saturated red, orange, blue, and green hues have commanded sums of up to US $4M per carat (Associated Press 2015).

The geological history of a diamond also has an important impact on the color centers that are present. Some color-producing point defects like isolated substitutional nitrogen and boron are incorporated during crystallization and growth deep in the Earth's mantle. More complex color centers involving nitrogen are created during the aggregation process when the diamond resides for long time periods in the mantle under elevated temperatures and pressures. Many of the more common extended defects in diamond that produce color are thought to be created by differential stresses within the mantle, resulting in plastic deformation of the diamond lattice (Fig. 28). Finally, radiation-related color centers involving vacancies are often created in the diamond lattice in alluvial settings in the shallow crust where diamonds may soak in radioactive fluids for thousands of years (Breeding et al. 2018). A closer look at the causes of the most common colors in gem diamonds gives a better understanding of how these stones can inform a complex world of diamond geology.

Yellow diamonds

Yellow is one of the most common colors seen in gem diamond and can occur with a huge range in saturations from slight yellow to deeply saturated yellow, depending on the particular nitrogen color center present. In most pale to moderately colored yellow diamonds, high concentrations of N3 defects, created by the nitrogen aggregation process during mantle residence, produce a series of absorptions in the visible range, ~415 nm (N3) as well as 451 nm, and 478 nm (N2), that give rise to a "straw yellow" color (Fig. 32a). These diamonds are collectively referred to in the gem industry as "Cape" diamonds due to their original association with South Africa, but they occur in nearly all diamond deposits (King et al. 2005). Much of the D to Z color grading system developed by GIA—where D is colorless and Z is light yellow—is a gradation in the concentration of Cape defects, and thus yellow color, in gem diamonds (King et al. 2008). With significant concentrations or in larger stones with longer optical pathlengths, Cape defects can produce strongly saturated yellow hues as well.

Some of the more intensely colored yellow diamonds, often known in the gem trade as "canary" yellows, owe their color to the presence of C centers that were incorporated at crystallization, though most canary yellow diamonds have undergone significant aggregation and occur as Type IaA diamonds with traces of isolated substitutional nitrogen. These isolated substitutional nitrogen atoms strongly absorb light in the ultraviolet to blue parts of the spectrum (Fig. 32b), to create deeply saturated yellow hues with relatively low concentrations of C centers present (a few tens of ppm). Due to the aggregation pathway that most nitrogen impurities in diamond follow, these strongly colored yellow Type Ib diamonds are very rare and are thought to be preserved by either very short residence times in the mantle or crystallization and residence in regions of unusually cold mantle (Smit et al. 2016). Those rare examples of nearly pure Type Ib diamond typically have very low total nitrogen concentrations. Nearly all yellow color produced in laboratory grown diamonds is also a result of C center absorption.

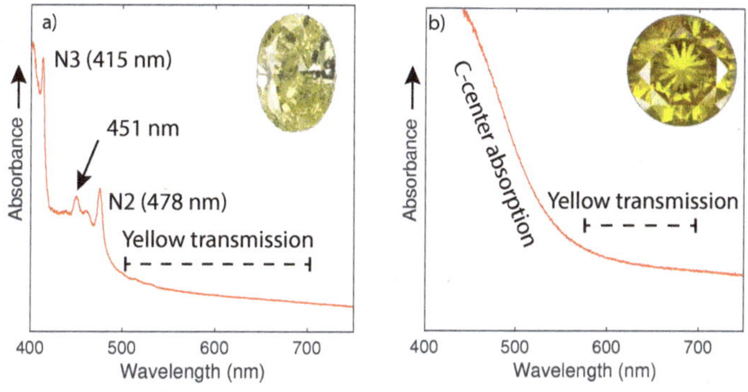

Figure 32. "Cape" diamonds show absorption features from N3 defects **(a)** while "canary" diamonds rely on substitutional nitrogen **(b)** to produce their characteristic yellow colors. Spectra collected with the samples at 80 K.

Other notable, but rare, causes of yellow color include the previously-discussed H3 defect, and 480 nm visible absorption band. The 480 nm visible absorption band is a poorly understood defect that is intimately associated with the presence of C centers, with theoretical modelling suggesting it could be oxygen-related (Gali et al. 2001). Orange colored diamonds are closely related to yellows and are primarily produced by 480 nm band absorption or by a combination of C centers and "Y components" (Hainschwang et al. 2012).

Brown diamonds

Brown is also one of the most common hues encountered in natural diamonds and, in conjunction with yellow, forms the basis for the D to Z color grading range (King et al. 2008). Unlike many other color centers involving complex point defects, brown color is thought to be a product of the extended defects produced by plastic deformation of the diamond lattice.

As planes of atoms are offset during deformation, clusters of up to 70 vacancies form (Fujita et al. 2009). Vacancy clusters absorb light across the visible range, increasing toward the ultraviolet, and produce varying shades of brown depending on their abundance (Fig. 33). Brown diamonds occur in all diamond deposits but are particularly common in the Argyle mine in Australia.

These vacancy clusters must form at temperatures where diamond is still capable of being deformed plastically (>900 °C, DeVries 1975), meaning that these defects are created at some point after diamond has crystallized in the mantle but before it is transported to the crust. Presumably some form of differential stress produced by mantle flow or the initial stages of incorporation and transport by the kimberlite magma may be responsible.

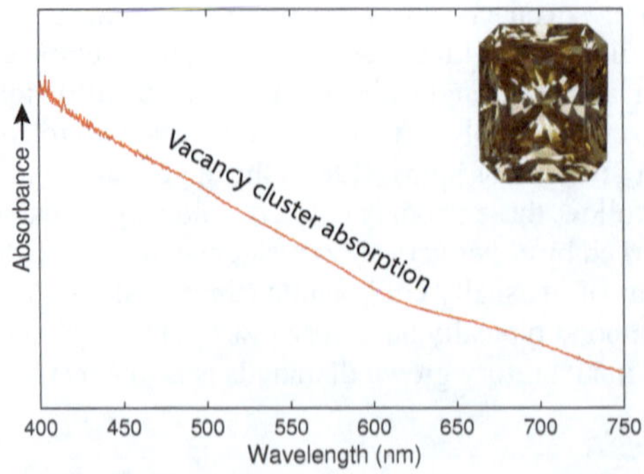

Figure 33. Brown color in most natural diamonds is a result of absorption by large vacancy clusters which are produced by plastic deformation. The absorption itself is characteristically broad and devoid of sharp optical transitions.

Pink to red diamonds

Some of the most valuable gem diamonds occur with pink to red hues that are also thought to be partially produced by plastic deformation of the diamond lattice. Red diamonds were among the first colored diamonds to approach US $1M per carat at auction (the Hancock Red; Christies 2020). Unlike their brown counterparts, pink and red diamonds are extremely rare and apparently represent a unique circumstance of plastic deformation that produces a broad absorption band centered around 550 nm (Fig. 34a). The 550 nm band occurs both in diamonds with and without nitrogen impurities (Type Ia and IIa), but the pink to red color distribution is quite different depending on the aggregation state of the impurities (Gaillou et al. 2010; Eaton-Magaña et al. 2018b, 2020). Pink Type IIa diamonds show even color distribution, while the pink color in Type Ia diamonds is concentrated in colored lamellae that indicate discrete zones of plastic deformation. These lamellae even take on a different appearance depending on the aggregation state and concentration of nitrogen in the diamonds. Dominantly Type IaA diamonds which mostly hail from Russia (the Lomonosov mine and others) show well developed pink lamellae separated by zones with no color at all (Howell et al. 2015; Eaton-Magaña et al. 2018b). Lower nitrogen, dominantly Type IaAB to IaB, pink to red diamonds from the Argyle mine in Australia tend to have much more closely spaced pink lamellae where the color blends between them (Eaton-Magaña et al. 2018b). Little is known about the atomic structure of the defect which produces the 550 nm absorption band, but it has been shown to be temporarily reduced in intensity by exposure to strong UV light, suggesting some charged analogue also exists to facilitate a charge transfer bleaching effect (Khan et al. 2009, 2010; Eaton-Magaña et al. 2018b, 2020). Recent work has linked the 550 nm absorption band to a broad photoluminescence band at ~660 nm (Gaillou et al. 2010; Eaton-Magaña et al. 2018b, 2020).

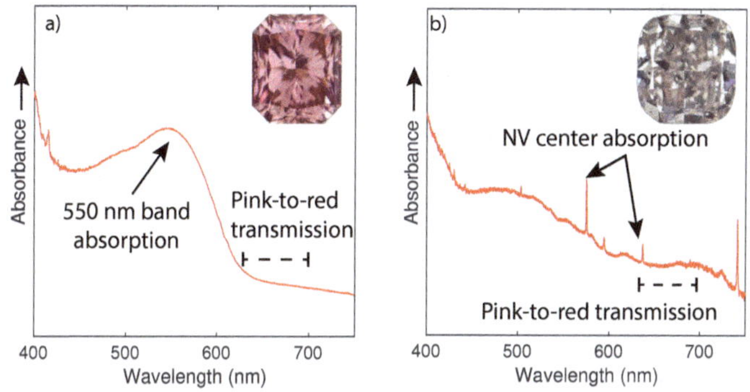

Figure 34. Natural pink to red diamonds are usually colored by a 550 nm absorption band caused by plastic deformation **(a)**, but on extremely rare occasions can be colored by nitrogen-vacancy center absorption **(b)**. In contrast, nitrogen-vacancy absorption and emission is extremely common in synthetic diamond grown by chemical vapor deposition. Spectra collected with samples at 80 K.

Pink to red color can also be generated in diamond by relatively high concentrations of nitrogen-vacancy (NV) centers (Fig. 34b). In natural diamonds, this occurrence is extremely rare and typically produces pale pink colors, many of which are thought to originate from the historically important Golconda region in India (Fritsch 1998) so that diamonds of this hue are known as Golconda pink regardless of origin. However, through irradiation and annealing treatment of diamonds containing C centers, high concentrations of NV can be created to produce a saturated pink to red color. Additionally, many synthetic diamonds also contain C centers and can be treated to produce pink to red color via absorption and emission at the NV defect.

Blue diamonds

Blue color in diamond is most commonly produced by the presence of uncompensated boron in the lattice, also making the stones semiconducting. Boron absorbs light strongly in the infrared region extending as a continuum into the visible range to produce a blue hue (Fig. 35). Boron absorbs so strongly that only a few hundred parts per billion are adequate to produce blue color in most diamonds (Eaton-Magaña et al. 2018a). Many famous blue gem diamonds like the Hope diamond owe their color to boron. In 2015, US \$4M/carat was paid for a 12.03 carat blue diamond known as the Blue Moon of Josephine (Associated Press 2015), then the highest per-carat price ever achieved for a diamond.

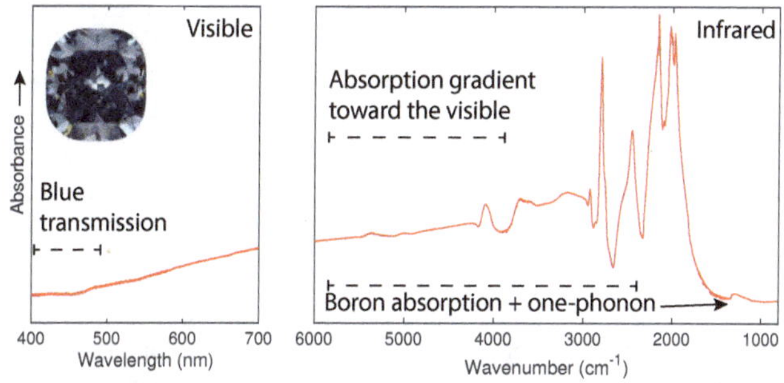

Figure 35. The most common cause of blue color in diamond is boron impurities that absorb strongly in the infrared region (**right**) with an absorption continuum that extends into the visible range (**left**) to produce the blue color.

Type IIb diamonds like these are among the rarest of gem diamonds. Recent work has shown that many Type IIb diamonds (as well as large, pure Type IIa stones) contain mineral inclusions originating from as deep as 660 km in the Earth (Smith et al. 2018). This model suggests boron introduction from seawater to an altered slab that subducted to those depths where the diamond was formed. While Type IIb diamonds occur in deposits around the world, a few mines, like the Cullinan mine (formerly Premier) in South Africa are known to produce higher numbers of these stones. Blue Type IIb diamonds can also be grown in a laboratory through introduction of boron into the growth chamber (for CVD growth) or growth capsule (for HPHT growth). Typically, the more boron added, the deeper the color saturation. At higher boron concentration, above a few ppm, the diamond starts appearing black.

Blue color can also be produced by relatively high amounts of radiation damage to the diamond lattice, either naturally or during irradiation treatment. Neutral vacancy defects (GR1) absorb very strongly in the red end of the visible spectrum, and when they occur in high enough concentrations, that absorption overwhelms any other absorptions due to nitrogen impurities and produces a blue hued diamond (Fig. 36a). In nature, blue color only occurs very rarely as a product of strong localized beta or gamma irradiation that penetrates to great depths in the diamond. A few such natural diamonds have been recovered in South America and central Africa (Draper 1951; Themelis 1987). Surficial alpha damage produces green color, as discussed below. Blue color can readily be added to natural or synthetic diamond using electron or gamma-ray irradiation to produce large quantities of vacancies.

Green diamonds

Most green diamonds derive their color from vacancy defects (GR1) that absorb light in the red region of the visible spectrum. These individual vacancies are created, along with interstitial defects, when diamonds are exposed to a source of radiation, either naturally in the Earth or in a laboratory. As discussed in previous sections, vacancies become mobile at relatively low temperatures (~700 °C) and can dominate the dynamics of defect and color center production (Collins 2001).

The appearance of radiation damage in diamond varies depending on the source of radiation. Alpha particles only penetrate to depths of 20–30 μm and usually produce green colored spots (called radiation stains by gemmologists) at or just below the diamond natural surface. These spots change almost instantly to a brown color at around 550 °C or possibly lower temperatures with prolonged residence in the Earth (Nasdala et al. 2013; Eaton-Magaña and Moe 2016). Beta radiation is more penetrative and causes internal zones of color. Gamma radiation usually affects the entire diamond. In nature, most radiation damage is thought to occur through prolonged interaction with fluids in shallow crustal environments, either in the kimberlite host or, more commonly, in an alluvial setting. However, the occurrence of radioactive mineral grains adjacent to diamond in its host rock cannot be ruled out as a source of damage in many cases. Regardless of the source, due to the relatively low thermal stability of vacancy defects in diamond, the radiation damage must occur at shallow levels in the Earth's crust. For natural diamonds, the uranium-thorium decay sequence is the most likely radiation source as it is the most abundant in the Earth. Laboratory electron irradiation (or occasionally gamma irradiation) treatments are commonly used to change the color of both natural and synthetic diamonds to create green diamonds.

Generally, as vacancy concentration increases, the saturation of green color also increases until the color of the diamond starts to take on a blue color component at higher amounts of radiation damage (Fig. 36a). The presence of other defects in combination with vacancies may significantly impact the diamond color. Most natural diamonds contain nitrogen impurities that absorb strongly in the blue region of the visible spectrum. The combination of nitrogen absorption with GR1 absorption usually results in a green to yellow-green diamond color (Fig. 36b). Another important consideration for green color in diamond is distribution of the radiation damage.

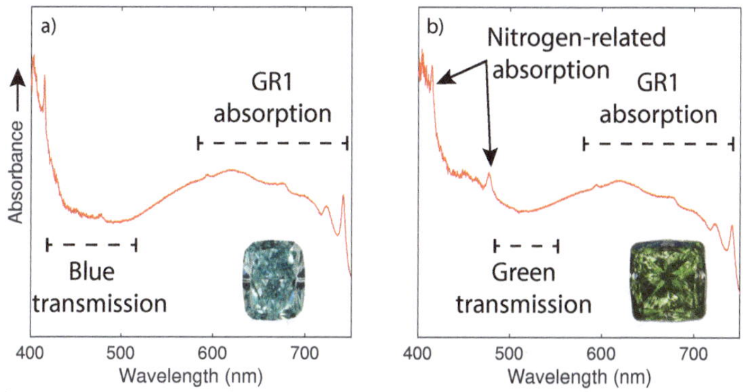

Figure 36. Isolated neutral vacancies, produced by radiation damage, produce an absorption known as GR1 (with its ZPL at 741 nm and a broad vibronic band which reaches approximately 575 nm) and can result in a blue color in diamond when their concentrations are very high and other defects are absent (**a**). When nitrogen impurities are present, GR1 defects are the most common cause of green color (**b**). Spectra collected with samples at 80 K.

Most natural diamonds only see significant surficial damage with very shallow color zones near radiation stains or thin "skins" of green color on the outer rim of the crystals produced by alpha irradiation (Vance and Milledge 1972; Vance et al. 1973; Breeding et al. 2018) Natural diamonds with green body color (internal color zones) are extremely rare and represent unique conditions in the Earth where significant beta or gamma exposure has occurred over geologic time. One of the most famous green diamonds is the Dresden Green, a 41-carat stone thought to originate from India (Hough 1913; Rosch 1957; Bosshart 1989, 1994). While green diamonds can occur in many deposits worldwide, they are most commonly found in South America, Zimbabwe, and the Central African Republic (Draper 1951; Themelis 1987; Breeding et al. 2018).

Violet diamonds

Amongst the rarest of gem diamonds are those that exhibit a deep violet color. The violet color is a result of a combination of defects including nitrogen defects such as N3 that absorb <450 nm, a broad band centered at ~530–540 nm that may be due to plastic deformation, and a broad absorption band centered at ~720 nm that is only observed in diamonds with high concentrations of hydrogen-related defects (van der Bogert et al. 2009; Eaton-Magaña et al. 2018a) (Fig. 37). These stones are only sourced from the Argyle mine in Australia, suggesting that they may very well be an analogue of the pink diamonds discovered there, but

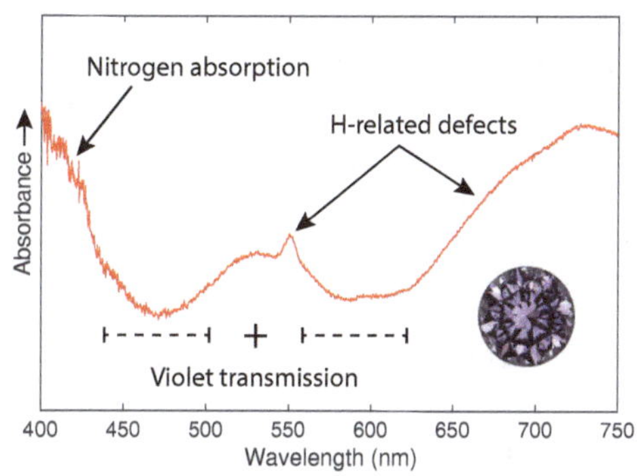

Figure 37. Violet color in diamond results from a combination of hydrogen-related absorptions, nitrogen absorptions and a broad band at 530–540 nm. Those absorptions labelled "H-related" have only been observed in high hydrogen-diamonds, but their origin is unknown (van der Bogert et al. 2009; Eaton-Magaña et al. 2018a). Spectrum collected with sample at 80 K.

with additional hydrogen impurities. Stones with similar hydrogen and nitrogen defects but missing the ~530 nm band tend to be grey to brown to green in color.

TREATMENTS TO CHANGE THE COLOR OF DIAMOND

Many Type IaA diamonds and some Type IIa diamonds have very little absorption in the visible region and so are colorless or near-colorless. Most diamonds, however, have some tint of color to them (either yellowish from nitrogen or brownish from plastic deformation), and the origins of these colors have been discussed in the preceding section. Occasionally some of these colors are very attractive, and diamonds which have a "fancy" color can be command a very high selling price. Many diamonds, however, have an insipid color (usually pale yellow, pale brown, or grey) which makes them less desirable than near-colorless or fancy-colored diamonds of the same size. In certain cases, the color of a diamond can be improved by some form of treatment. In this section we briefly consider various methods by which the color of a diamond can be changed. The defects involved have been considered in the preceding two sections.

In the gemological world, treatment is considered anything other than cutting and polishing that is done to a gemstone to change its natural appearance and thus improve its desirability. In the case of diamond, treatments generally cause enhancements to either the color or clarity of the stone and may include coatings, laser drilling, filling of fractures with glass, or rearrangement of atoms in the lattice. Our discussion will focus on ways in which the atomic structure of diamond is re-engineered to produce a particular color or attempt to remove color completely. The primary lattice-based treatments commonly applied to diamond are irradiation (with or without subsequent annealing), high pressure, high temperature annealing (usually performed at ~5–6 GPa and ~1800–2100 °C), or a combination thereof.

Irradiation

Irradiation of Type Ia or Type IIa diamond with high-energy electrons creates vacancies and interstitials, leading to a blue, blue-green, or green color, depending on the energy of the electrons used for the irradiation and absorptions from pre-existing nitrogen-related defects (Fig. 38). Irradiation with neutrons produces vacancies and interstitials, as well as more extensive radiation-damage products which absorb more strongly in the blue region of the visible spectrum. Neutron-irradiated diamonds therefore have a green color, and at high doses will simply appear black. Most commercial diamond treaters perform a "stabilization" annealing at ~500 °C following any irradiation treatment to remove many of the numerous low-binding-energy defects which are created by irradiation, in addition to the small population of close-proximity and highly strained vacancy-interstitial defects (Iakoubovskii et al. 2003). These defects have a minor effect on the color overall, but when removed, the color of the diamond shifts slightly to a more stable hue, suggesting that the interstitial defects may undergo some minor photo- or thermo-chromism, presumably due to changes in charge state.

When an irradiated Type Ia diamond is annealed at approximately 800 °C, the vacancies become mobile and wander through the diamond until they are trapped at one of the forms of nitrogen. A vacancy trapped at an A center produces an H3 center (N_2V^0), and a vacancy trapped at a B center results in an H4 center ($N_4V_2^0$; Fig. 38). The absorption bands associated with these centers occur at the blue end of the visible region and the diamonds consequently have a yellow color—similar to that of some fancy yellow diamonds in which the color is due to single nitrogen (Figs. 11a, 32b) or the N2 and N3 systems (Figs. 7, 32a).

The heating of an electron-irradiated Type Ia diamond to approximately 1400 °C, may generate absorption by the H2 (N_2V^-) center as a result of the production of H3 (N_2V^0) and subsequent charge transfer between H3 and isolated substitutional nitrogen (C centers). In favorable cases this treatment can give the diamond an attractive green color (Collins 2001).

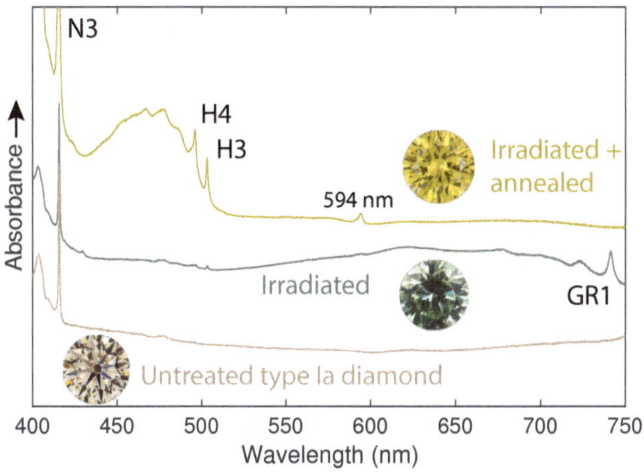

Figure 38. Irradiation treatment of a Type Ia diamond creates GR1 (V0) defects that produce green color. Annealing of the diamond at 800 °C produces H3 and H4 defects due to migration and combination of vacancy defects with A and B centers and creates yellow color. Spectra collected with samples at 80 K.

Irradiation of Type Ib diamond generates vacancies principally in the negative charge state (because of the large concentration of nitrogen donors). The absorption associated with V^- is referred to spectroscopically as the ND1 center, and is in the UV region with a ZPL at 394 nm. The absorption by the isolated substitutional nitrogen, in combination with weak GR1 absorption produced by the irradiation, gives the diamond an unattractive drab green color (Fig. 39). After annealing at 800 °C, vacancies become trapped at the single-nitrogen centers

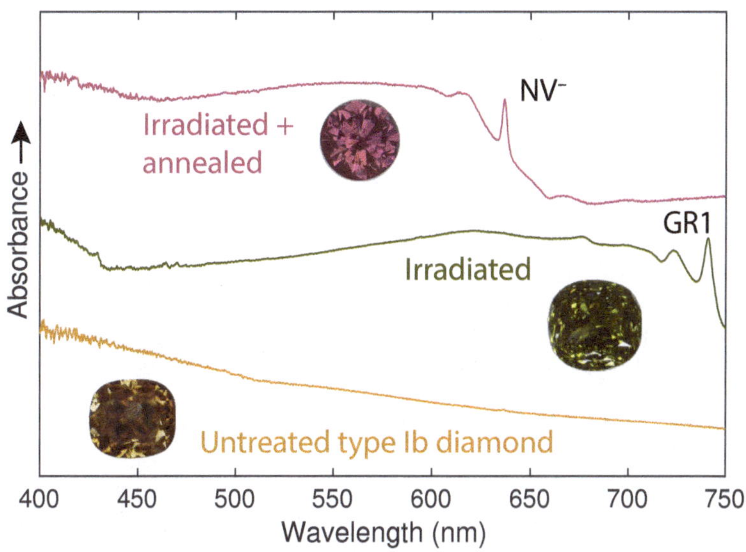

Figure 39. Irradiation treatment of a brown Type Ib diamond creates V^- (not shown) as well as the V^0 that produces a dull green color. Annealing of the diamond at 800 °C generates a combination of vacancy-related defects (largely negatively charged due to C centers being donors) and creates pink to red color. Spectra collected with samples at 80 K.

to produce the negative nitrogen-vacancy center NV^-. The absorption produced by this center gives the diamond a pink to red color which, to the eye, is similar to that of natural pink diamonds described in the previous section.

HPHT treatment

Many natural diamonds have a brown color produced by the presence of vacancy clusters (see *Brown diamonds*). These vacancy clusters can be partially dissociated by annealing under HPHT conditions. In Type IIa diamond the released vacancies are trapped at dislocations or diffuse to the surface, and the net effect is to reduce the brown color. Occasionally HPHT treatment of Type IIa diamonds yields samples with a slight pink hue if a 550 nm absorption band originally coexisted with the vacancy clusters causing the brown color, or blue color if compensated boron is freed by the treatment to become uncompensated (Hall and Moses 2000).

When Type Ia diamonds are HPHT treated, the vacancies become trapped at A centers, producing H3 centers. The net effect is to reduce the brown component and increase the yellow component of the color, resulting in an overall improvement in appearance. H3 centers also produce green luminescence to visible light which sometimes appears as a greenish tint on the yellow body color. The results described are achieved with annealing temperatures around 1800 °C. If the treatment temperature is increased to 2000 to 2100 °C then, in addition, some of the A nitrogen dissociates to form single nitrogen atoms. As with low-temperature annealing described above, charge transfer between H3 and substitutional nitrogen (N_2V^0 and N_s^0, yielding N_2V^- and N_s^+) can add a green component to the color due to the H2 (N_2V^-) absorption.

Natural Type Ia diamonds containing a modest concentration of N3 centers may have a "tinted" yellow color which is neither near-colorless nor fancy yellow. HPHT annealing at sufficiently high temperatures to dissociate some of the A nitrogen produces absorption at the blue end of the visible spectrum (Figs. 11a, 32b), due to the production of single nitrogen, thereby enhancing the yellow color or adding a significant orange color component if enough single nitrogen atoms are created.

Combination treatments

Starting with an off-color Type Ia diamond it is also possible to produce a pink diamond by a three-stage process: HPHT annealing to produce single nitrogen; irradiation to produce vacancies: annealing at 800 °C to produce NV^- centers. Similarly, a two-stage process is now commonly used to create vivid blue color. The off-color diamond is HPHT annealed to remove brown color (but not at high enough temperature to create isolated nitrogen) and subsequently irradiated with high energy electrons to impart the saturated blue hue. This treatment allows for a much more attractive product than simply irradiating an off-color diamond.

In the last decade, treaters have also begun to employ a very low dose electron irradiation treatment and sometimes low temperature annealing following HPHT processing of Type II diamonds. This is an attempt to add back certain defects that are known to be removed by HPHT and are used for identification of natural and treated diamonds by gem laboratories. These post-HPHT treatments have no effect on the color of the diamond and are performed with the sole purpose of attempting to fraudulently pass treated diamonds as naturally colored stones (Eaton-Magana and Breeding 2016).

Although the colors produced by various treatments are similar to colors that occur naturally, examination of the diamonds using optical spectroscopy allows most naturally colored and treated diamonds to be differentiated.

USING OPTICAL SPECTROSCOPY TO DETECT TREATED DIAMOND

Proper disclosure of gemstone treatments is explicitly required by law in nearly every country, but many treated diamonds are bought, sold, or submitted to gem laboratories undisclosed every day. Treated-color gem diamonds are worth anywhere from 50% less to several orders of magnitude less than their natural color equivalents, simply due to the rarity of the finest natural color stones. This huge discrepancy in pricing generally requires that colored diamonds be sent to gemological laboratories for certification before sale. While traditional gemological tools like the microscope, handheld spectroscope, and fluorescence lamp are still useful for some identification needs, modern atomic lattice-based treatments of diamond require the use of spectroscopic analytical instruments for proper detection (Overton and Shigley 2008; Breeding et al. 2010). The value of diamond dictates that analysis be done in a non-destructive manner, making optical spectroscopy the primary tool used for identification. Specifically, infrared (IR) and ultraviolet–visible (UV–Vis) absorption spectroscopy and photoluminescence (PL) spectroscopy are the most commonly employed methods of detecting treatments in gem diamond, and are employed even in the analysis of low-value gemstones such as melee. Certain analytical techniques have proven more effective for identification of each of the major diamond color treatments discussed in the previous section—HPHT treatment and irradiation treatment—and combinations of these treatments usually show some characteristics of both.

Detection of HPHT treatment

As mentioned in the preceding section, HPHT treatment has drastically different goals and results depending on the diamond type of the sample being treated. In general, most treated Type II diamonds have undergone HPHT treatment to remove brown color and produce a near colorless stone, or in some cases a pink or blue diamond color (Hall and Moses 2000; Fisher 2009). Most natural diamonds are Type Ia, with approximately 1.3% being Type II (Smith et al. 2016) and having the potential to be decolorized. For near colorless diamonds, then, the basic screening method is to evaluate diamond type using IR spectroscopy to separate Type II samples. With only a few rare exceptions, Type Ia near colorless diamonds can safely be concluded to be natural color. Among those stones determined to be Type II, PL spectroscopy is employed to search for defects either added or removed by the treatment process (Fisher and Spits 2000; Eaton-Magaña and Breeding 2016). Fortunately, in order to dramatically increase the speed at which the atoms rearrange (depending exponentially on temperature), temperatures employed during HPHT treatment are significantly higher than those experienced by most natural diamonds of approximately 1050–1200 °C (Stachel and Harris 2008; Stachel and Luth 2015). Several natural centers are unstable at HPHT conditions and are annealed from the stone or reconfigured into other defects. Other defects change their charge state due to the production of new electron donors by dissociation of minute amounts of aggregated nitrogen (A centers primarily) to form C centers (Brozel et al. 1978). Despite being classified as Type IIa, most near-colorless diamonds of this type contain some nitrogen impurities in the form of NV or H3 defects that can be detected down to several parts-per-billion by PL spectroscopy.

The creation of C center donors during HPHT treatment causes some of the NV centers to convert from the neutral state (NV^0) to the negative charge state (NV^-) and H3 centers (N_2V^0) to convert to their negatively charged equivalents, H2 centers (N_2V^-) (Chalain et al. 2000; Collins et al. 2000; Fisher and Spits 2000; Lim et al. 2010) (Fig. 40). Thus, a detected ZPL height ratio of $NV^-/NV^0 > 1$ (at 80 K using 514 nm laser excitation: the laser wavelength will affect this ratio) and significant H2 emission in near-colorless diamonds strongly suggests that HPHT treatment has occurred.

For brownish Type Ia diamonds, HPHT treatment at approximately 1800 °C typically adds color by creation of more color centers like H3 and C centers. As vacancy clusters are

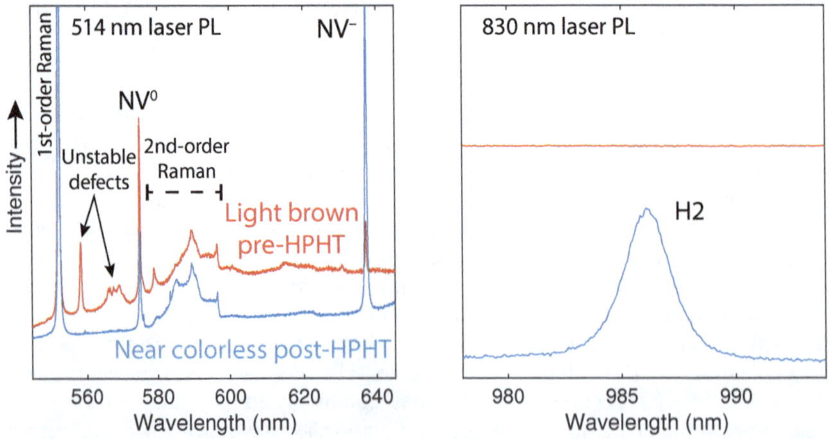

Figure 40. Photoluminescence spectroscopy reveals that unstable defects are removed by HPHT treatment of Type IIa brown diamonds to decolorize them. In addition, the NV^-/NV^0 ratio increases to >1 in treated diamonds (as measured with PL using a 514 nm laser). The H2 defect is rare in untreated natural samples and can be used in combination with other spectroscopic markers to identify treated diamonds. Spectra collected with the samples at 80 K.

destroyed by HPHT treatment (Mäki et al. 2009), the individual vacancies migrate through the lattice and become trapped at abundant A centers to generate H3 defects that produce yellow color (and sometimes a green luminescence overtone as well) (Collins et al. 2000). H4 centers—produced at lower temperatures (around 800 °C) when vacancies are trapped at B centers—are not stable at HPHT conditions. H4 centers do not form during HPHT treatment even in diamonds containing high concentrations of B centers. At higher treatment temperatures (>2000 °C), aggregated nitrogen dissociates to create significant numbers of C centers which also add yellow color (Collins et al. 2000). Absorption (UV–Vis and IR) spectroscopies are more useful for identification in these cases. Liquid nitrogen temperature UV–Vis shows that H3, C centers, and sometimes H2 are responsible for the color, a relatively uncommon occurrence for natural yellow or green color as discussed in *Green diamonds*. Infrared spectroscopy often shows the occurrence of C centers (1344 cm^{-1}) along with both A and B centers, another condition that is extremely rare in untreated natural diamonds (Fig. 41). In addition, platelets are usually significantly reduced by the treatment, a feature that can easily be recognized in IR spectroscopy in combination with the occurrence of C centers (Reinitz et al. 2000) (Fig. 41).

Detection of irradiation treatment (including subsequent annealing)

Unlike HPHT treatment where the temperatures are higher than natural diamond settings, irradiation treatment involves nearly identical conditions and results to those that occur naturally in the Earth (Fig. 42). The only major difference is the timescale—a few minutes in a lab compared to hundreds to millions to billions of years in the Earth—but, unfortunately, few measurable spectroscopic characteristics seem to be tied to the timescale. Consequently, whether irradiation occurred in the Earth or a laboratory is sometimes difficult to determine, leading to a higher percentage of "undetermined" color origin assessments on laboratory reports of green diamonds.

Photoluminescence spectroscopy has proven useful in detecting, in treated diamonds, some interstitial-related defects that are typically annealed out naturally by long residence times at elevated temperatures experienced by natural color diamonds in the Earth (Breeding et al. 2018). Observations of natural radiation damage on the surface of these stones provides

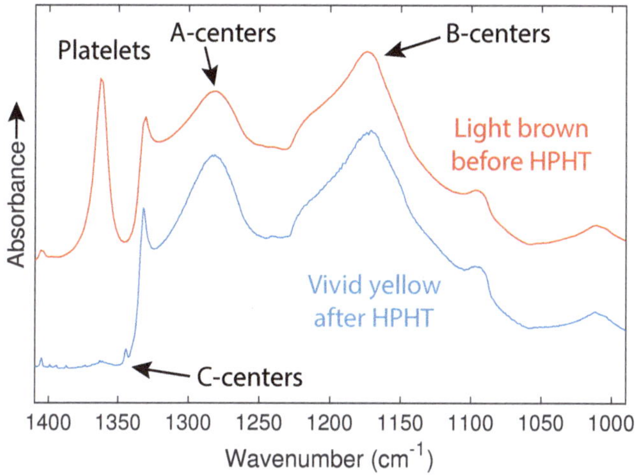

Figure 41. HPHT treatment of Type Ia diamonds usually causes a noticeable decrease in platelet concentration and, if the treatment temperature is >2000 °C, the presence of C centers can usually be detected along with A and B centers. Baseline offset for clarity.

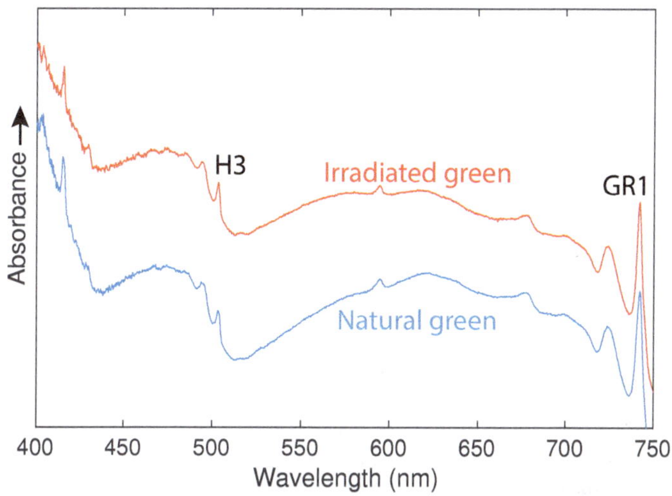

Figure 42. Natural irradiaton in the Earth and laboratory irradiation using electrons often results in very similar defect structures and green colors in gem diamond, making the separation difficult in some cases, especially if the pre-treatment material is natural (as is the case here). Spectra collected with the samples at 80 K.

some evidence that a natural radiation source was present at some time in the diamond's history. More useful, however is the post-irradiation annealing treatment process that is often done at different temperatures to produce yellow, pink to red, or orange diamonds. Heating irradiated Type Ia diamonds at 800 to 1400 °C causes vacancies to become trapped at nitrogen impurities to create H3, H4, or NV defects that give rise to the newly treated diamond colors. A by-product of engineering these color centers is the creation of several other defects that do not typically occur with high absorption intensity in equivalent natural color stones, including the 594 nm center, H1a (1450 cm^{-1}), H1b (4935 cm^{-1}), and H1c (5165 cm^{-1}) (Zaitsev 2001;

Collins 2003, Fig. 43). The 594 nm absorption center can be detected by UV–Vis absorption spectroscopy (at liquid nitrogen temperature), while the others absorb distinctly in IR spectra (Fig. 43). These centers are believed to be nitrogen-containing interstitial centers which have been shown to anneal out between 1100 and 1400 °C (Collins et al. 2005; Liggins et al. 2010). Therefore, the presence of these absorption peaks with significant intensity in Type Ia diamonds strongly suggests the possibility of irradiation and annealing treatments.

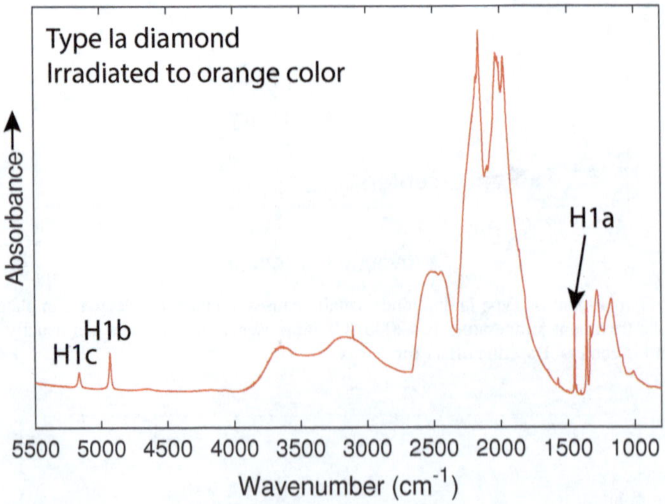

Figure 43. After irradiation and annealing treatments, Type Ia diamonds often develop H1a, H1b, and H1c defects in significant concentrations. These defects are extremely rare in natural color diamonds.

SUMMARY

Diamond is an extreme material with a rich history at the forefront of physical and spectroscopic understanding, driven by its economic value in both cosmetic and technological applications. The relatively recent ability to synthesise high-quality diamond with superb control over impurities and growth morphology has been instrumental in the development of our understanding of defect and impurity processes within natural diamond. Combined with recent spectroscopic advances such as the ability to optically interrogate and monitor a single atomic defect, diamond continues to drive spectroscopy theory and practice. The authors are confident that these updated techniques will continue to shed further light on diamond growth and evolution, and ultimately the geological processes which drive them in nature.

ACKNOWLEDGEMENTS

Thank you to Jon Goss, an anonymous reviewer, and the editor, Karen Smit, for their constructive reviews. Ben Green gratefully acknowledges support from the Royal Academy of Engineering in the form of a Research Fellowship, and thanks Matthew Dale and Ben Breeze for useful discussions.

REFERENCES

Achard J, Silva F, Brinza O, Tallaire A, Gicquel A (2007) Coupled effect of nitrogen addition and surface temperature on the morphoFy and the kinetics of thick CVD diamond single crystals. Diamond Relat Mater 16:685–689

Achard J, Jacques V, Tallaire A (2020) Chemical vapour deposition diamond single crystals with nitrogen-vacancy centres: A review of material synthesis and technology for quantum sensing applications. J Phys D: Appl Phys 53:313001

Aharonovich I, Neu E (2014) Diamond nanophotonics. Adv Opt Mater 2:911–928

Allen BP, Evans T (1981) Aggregation of nitrogen in diamond, including platelet formation. Proc R Soc London, Ser A 375:93–104

Allers L, Collins AT, Hiscock J (1998) The annealing of interstitial-related optical centres in Type II natural and CVD diamond. Diamond Relat Mater 7:228–232

Ashfold MNR, Goss JP, Green BL, May PW, Newton ME, Peaker CV (2020) Nitrogen in Diamond. Chem Rev 120:5745–5794

Aslam N, Waldherr G, Neumann P, Jelezko F, Wrachtrup J (2013) Photo-induced ionization dynamics of the nitrogen vacancy defect in diamond investigated by single-shot charge state detection. New J Phys 15:013064

Associated Press (2015) Rare blue diamond sells for record $48.5 million at auction. NBC News, www.nbcnews.com/business/markets/rare-blue-diamond

Atatüre M, Englund D, Vamivakas N, Lee S-Y, Wrachtrup J (2018) Material platforms for spin-based photonic quantum technologies. Nat Rev Mater 3:38–51

Awschalom DD, Hanson R, Wrachtrup J, Zhou BB (2018) Quantum technologies with optically interfaced solid-state spins. Nat Photonics 12:516–527

Baker JM, Newton ME (1994) The use of ENDOR to identify the atomic structure of defects in diamond. Appl Magn Reson 7:209–235

Barry JF, Schloss JM, Bauch E, Turner MJ, Hart CA, Pham LM, Walsworth RL (2020) Sensitivity optimization for NV-diamond magnetometry. Rev Mod Phys 92:015004

Bassett LC, Alkauskas A, Exarhos AL, Fu KC (2019) Quantum defects by design. Nanophotonics

Bosshart G (1989) The Dresden Green. J Gemmol 21:351

Bosshart G (1994) Investigations mordernes du diamant vert de Dresden et interpretation des resultats. Revue de Gemmologie 121:5

Breeding CM, Shen AH, Eaton-Magaña S, Rossman GR, Shigley JE, Gilbertson AM (2010) Developments in gemstone analysis techniques and instrumentation during the 2000s. Gems Gemol 46:241–257

Breeding CM, Eaton-Magaña S, Shigley JE (2018) Natural-color green diamonds: a beautiful conundrum. Gems Gemol 54:2–27

Breeze BG, Meara CJ, Wu XX, Michaels CP, Gupta R, Diggle PL, Dale MW, Cann BL, Ardon T, D'Haenens-Johansson UFS, Friel I (2020) Doubly charged silicon vacancy center, Si-N complexes, and photochromism in N and Si codoped diamond. Phys Rev B 101:184115

Breuer SJ, Briddon PR (1996) Energy barrier to reorientation of the substitutional nitrogen in diamond. Phys Rev B 53:7819

Brozel MR, Evans T, Stephenson RF (1978) Partial dissociation of nitrogen aggregates in diamond by high temperature-high pressure treatments. Proc R Soc London, Ser A 361:109–127

Bruley J, Brown LM (1989) Quantitative electron energy-loss spectroscopy microanalysis of platelet and voidite defects in natural diamond. Philos Mag A 59:247–261

Bulanova GP (1995) The formation of diamond. J Geochem Explor 53:1–23

Callister WD (2006) Materials Science and Engineering: An Introduction, 7th ed. Wiley, Hoboken

Cann BL (2009) Thesis: Magnetic resonance studies of point defects in diamond. University of Warwick

Chakravarthi S, Moore C, Opsvig A, Pederson C, Hunt E, Ivanov A, Christen I, Dunham S, Fu KMC (2020) Window into NV center kinetics via repeated annealing and spatial tracking of thousands of individual NV centers. Phys Rev Materials 4:023402

Chalain J-P, Fritsch E, Hänni HA (2000) Identification of GE POL diamonds: A second step. J Gemmol 27:73–78

Chen Y-C, Griffiths B, Weng L, Nicley S, Ishmael SN, Lekhai Y, Johnson S, Stephen CJ, Green BL, Morley GW, Newton ME (2019) Laser writing of individual atomic defects in a crystal with near-unity yield. Optica 6:662–666

Chevallier J, Theys B, Lusson A, Grattepain C, Deneuville A, Gheeraert E (1998) Hydrogen-boron interactions in p-type diamond. Phys Rev B 58:7966–7969

Chrenko RM, Tuft RE, Strong HM (1977) Transformation of the state of nitrogen in diamond. Nature 270:141–144

Christies (2020). https://www.christies.com/features/10-jewels-that-created-auction-history-and-changed-the-market-9549-3.aspx. Retrieved on 2021/02/05

Clark CD, Ditchburn RW, Dyer HB (1956a) The absorption spectra of irradiated diamonds after heat treatment. Proc R Soc London, Ser A 237:75–89

Clark CD, Ditchburn RW, Dyer HB (1956b) The absorption spectra of natural and irradiated diamonds. Proc R Soc London, Ser A 234:363–381

Collins AT (1977) Radiation damage in natural semiconducting diamond. In NB Urli and JW Corbett, Eds., Radiation Effects in Semiconductors 1976 (Institute of Physics Conference Series 31), p 346–353

Collins AT(1980) Vacancy enhanced aggregation of nitrogen in diamond. J Phys C: Solid State Phys 13:2641–2650

Collins AT (2001) The colour of diamond and how it may be changed. J Gemmol 27:341–359

Collins AT (2002) The Fermi level in diamond. J Phys: Condens Matter 14:3743–3750
Collins AT (2003) The detection of colour-enhanced and synthetic gem diamonds by optical spectroscopy. Diamond Relat Mater 12:1976–1983
Collins AT, Kiflawi I (2009) The annealing of radiation damage in Type Ia diamond. J Phys: Condens Matter 21:364209
Collins AT, Mohammed K (1982) Optical studies of vibronic bands in yellow luminescing natural diamonds. J Phys C: Solid State Phys 15:147–158
Collins AT, Williams AWS (1971) The nature of the acceptor centre in semiconducting diamond. J Phys C: Solid State Phys 4:1789–1800
Collins AT, Woods GS (1982) Cathodoluminescence from "giant" platelets, and of the 2.526 eV vibronic system, in Type Ia diamonds. Philos Mag Part B 45:385–397
Collins AT, Lightowlers EC, Dean PJ (1969) Role of phonons in the oscillatory photoconductivity spectrum of semiconducting diamond. Phys Rev 183:725
Collins AT, Kanda H, Burns RC (1990) The segregation of nickel-related optical centres in the octahedral growth sectors of synthetic diamond. Philos Mag B: 61:797–810
Collins AT, Kanda H, Kitawaki H (2000) Colour changes produced in natural brown diamonds by high-pressure, high-temperature treatment. Diamond Relat Mater 9:113–122
Collins AT, Connor A, Ly C-H, Shareef A, Spear PM (2005) High-temperature annealing of optical centers in Type-I diamond. J Appl Phys 97:083510–083517
Cox A, Newton ME, Baker JM (1994) 13 C, 14 N, 15 N ENDOR measurements on the single substitutional nitrogen centre (P1) in diamond. J Phys: Condens Matter 6:551–563
Coxon DJL, Staniforth M, Breeze BG, Greenough SE, Goss JP, Monti M, Lloyd-Hughes J, Stavros VG, Newton ME (2020) An ultrafast shakedown reveals the energy landscape, relaxation dynamics, and concentration of the N_3VH^0 defect in diamond. J Phys Chem Lett 11:6677–6683
Crisci A, Baillet F, Mermoux M, Bogdan G, Nesladek M, Haenen K (2011) Residual strain around grown-in defets in CVD diamond single crystals: A 2D and 3D Raman imaging study. Physica status solidi A 208:2038–2044
Crossfield MD, Davies G, Collins AT, Lightowlers EC (1974) The role of defect interactions in reducing the decay time of H3 luminescence in diamond. J Phys C: Solid State Phys 7:1909–1917
Cruddace RJ (2007) Thesis: Magnetic resonance and optical studies of point defects in single crystal CVD diamond. University of Warwick
D'Haenens-Johansson UFS, Edmonds AM, Green BL, Newton ME, Davies G, Martineau PM, Khan RUA, Twitchen DJ (2011) Optical properties of the neutral silicon split-vacancy center in diamond. Phys Rev B 84:245208
D'Haenens-Johansson UFS, Katrusha AN, Moe KS, Johnson P, Wang W (2015) Large colorless HPHT-grown synthetic gem diamonds from new diamond technology, Russia. Gems Gemol 51:260–279
D'Haenens-Johansson UFS, Butler JE, Katrusha AN (2022) Synthesis of diamonds and their identification.Rev Mineral Geochem 88:689–754
Dale MW (2015) Colour Centres on Demand in Diamond. Thesis, University of Warwick
Davies G (1976) The A nitrogen aggregate in diamond-its symmetry and possible structure. J Phys C: Solid State Phys, 9, L537–L542
Davies G (1979) Cathodoluminescence. *In:* The properties of diamond. JE Field (Ed.) Academic Press, London, p. 165–181
Davies G (1981) The Jahn–Teller effect and vibronic coupling at deep levels in diamond. Rep Prog Phys 44:787–830
Davies G (1999) Current problems in diamond: towards a quantitative understanding. Physica B 273–274:15–23
Davies G, Crossfield M (1973) Luminescence quenching and zero-phonon line broadening associated with defect interactions in diamond. J Phys C: Solid State Phys 6:L104–L108
Davies G, Welbourn CM, Loubser JHN (1978) Optical and electron paramagnetic effects in yellow Type Ia diamonds. Diamond Research, p. 23–30
Davies G, Lawson SC, Collins AT, Mainwood A, Sharp SJ (1992) Vacancy-related centers in diamond. Phys Rev B, 46:13157–13170
Dean PJ (1965) Bound excitons and donor-acceptor pairs in natural and synthetic diamond. Phys Rev 139:A588–A602
DeVries RC (1975) Plastic deformation and "work-hardening" of diamond. Mater Res Bull 10:1193–1199
Diggle PL, D'Haenens-Johansson UFS, Green BL, Welbourn CM, Tran Thi TN, Katrusha A, Wang W, Newton ME (2020) Decoration of growth sector boundaries with nitrogen vacancy centers in as-grown single crystal high-pressure high-temperature synthetic diamond. Phys Rev Materials 4:093402
Dischler B (2012) Handbook of Spectral Lines in Diamond. Springer-Verlag, Berlin
Doherty MW, Manson NB, Delaney P, Jelezko F, Wrachtrup J, Hollenberg LCLL (2013) The nitrogen-vacancy colour centre in diamond. Phys Rep 528:1–45
Draper T (1951) The diamond mines of Diamantina—Past and present. Gems Gemol 7:49–57
Dyer HB, Raal FA, Du Preez L, Loubser JHN (1965) Optical absorption features associated with paramagnetic nitrogen in diamond. Philos Mag 11:763–774
Eaton-Magaña S, Ardon T (2016) Temperature effects on luminescence centers in natural Type IIb diamonds. Diamond Relat Mater 69:86–95
Eaton-Magaña S, Breeding CM (2016) An introduction to photoluminescence spectroscopy for diamond and its applications in gemology. Gems Gemol 52:2–17

Eaton-Magaña S, Lu R (2011) Phosphorescence in Type IIb diamonds. Diamond Relat Mater 20:983–989

Eaton-Magaña SC, Moe KS (2016) Temperature effects on radiation stains in natural diamonds. Diamond Relat Mater 64:130–142

Eaton-Magaña S, Shigley JE, Breeding CM (2017) Observations on HPHT-grown synthetic diamonds: A review. Gems Gemol 53:262–285

Eaton-Magaña S, Breeding CM, Shigley JE (2018a) Natural-color blue, gray, and violet diamonds: Allure of the deep. Gems Gemol 54:112–131

Eaton-Magaña S, Ardon T, Smit KV, Breeding CM, Shigley JE (2018b) Natural-color pink, purple, red, and brown diamonds: Band of many colors. Gems Gemol 54:352–377

Eaton-Magaña S, McElhenny G, Breeding CM, Ardon T (2020) Comparison of gemological and spectroscopic features in Type IIa and Ia natural pink diamonds. Diamond Relat Mater 105:107784

Ekimov EA, Kondrin MV (2017) Vacancy–impurity centers in diamond: prospects for synthesis and applications. Physics-Uspekhi 60:539–558

Ekimov EA, Sidorov VA, Bauer ED, Mel'nik NN, Curro NJ, Thompson JD, Stishov SM (2004) Superconductivity in diamond. Nature 428:542–5

Etmimi K, Ahmed ME, Briddon PR, Goss JP, Gsiea AM (2009) Nitrogen-pair paramagnetic defects in diamond: A density functional study. Phys Rev B 79:205207

Evans T, Qi Z (1982) The kinetics of the aggregation of nitrogen atoms in diamond. Proc R Soc London, Ser A 381:159–178

Evans T, Kiflawi I, Luyten W, Tendeloo G Van, Woods GS (1995) Conversion of platelets into dislocation loops and voidite formation in Type IaB diamonds. Proc R Soc London, Ser A 449:295–313

Farrer RG (1969) On the substitutional nitrogen donor in diamond. Solid State Commun 7:685–688

Fedyanin DY, Agio M (2016) Ultrabright single-photon source on diamond with electrical pumping at room and high temperatures. New J Phys 18:073012

Felton S, Edmonds AM, Newton ME, Martineau PM, Fisher D, Twitchen DJ (2008) Electron paramagnetic resonance studies of the neutral nitrogen vacancy in diamond. Phys Rev B 77:081201

Fisher D (2009) Brown diamonds and high pressure high temperature treatment. Lithos 112:619–624

Fisher D, Lawson SC (1998) The effect of nickel and cobalt on the aggregation of nitrogen in diamond. Diamond Relat Mater 7:299–304

Fisher D, Spits RA (2000) Spectroscopic Evidence of GE POL HPHT-Treated Natural Type IIa Diamonds. Gems Gemol 36:42–49

Freitas JA, Klein PB, Collins AT (1994) Evidence of donor-acceptor pair recombination from a new emission band in semiconducting diamond. Appl Phys Lett 64:2136–2138

Fritsch E (1998) The nature of color in diamonds. *In:* The Nature of Diamonds. GE Harlow (ed) Cambridge University Press, p. 23–47

Fritsch E, Hainschwang T, Massi L, Rondeau B (2007) Hydrogen-related optical centres in natural diamond: An update. New Diamond Front Carbon Technol 17:63–89

Fujita N, Jones R, Öberg S, Briddon PR (2009) Large spherical vacancy clusters in diamond—Origin of the brown colouration? Diamond Relat Mater 18:843–845

Gaillou E, Post J., Bassim ND, Zaitsev AM, Rose T, Fries MD, Stroud RM, Steele A, Butler JE (2010) Spectroscopic and microscopic characterizations of color lamellae in natural pink diamonds. Diamond Relat Mater 19:1207–1220

Gali A, Lowther JE, Deák P (2001) Defect states of substitutional oxygen in diamond. J Phys: Condens Matter 13:11607–11613

Glover C, Newton M, Martineau P, Twitchen D, Baker J (2003) Hydrogen incorporation in diamond: The nitrogen-vacancy–hydrogen complex. Phys Rev Letters 90:185507

Goss JP, Briddon PR (2006) Theory of boron aggregates in diamond: First-principles calculations. Phys Rev B 73:085204

Goss JP, Coomer BJ, Jones R, Fall CJ, Briddon PR, Öberg S (2003) Extended defects in diamond: The interstitial platelet. Phys Rev B 67:165208

Goss JP, Briddon PR, Hill V, Jones R, Rayson MJ (2014) Identification of the structure of the 3107 cm^{-1} H-related defect in diamond. J Physics: Condens Matter 26:145801

Green BL (2013) Optical and Magnetic Resonance Studies of Point Defects in Single Crystal Diamond. Thesis, University of Warwick

Grimsditch MH, Anastassakis E, Cardona M (1978) Effect of uniaxial stress on the zone-center optical phonon of diamond. Phys Rev B 18:901–904

Hainschwang T, Simic D, Fritsch E, Deljanin B, Woodring S, DelRe N (2005) A gemological study of a collection of chameleon diamonds. Gems Gemol 41:20–35

Hainschwang T, Fritsch E, Notari F, Rondeau B (2012) A new defect center in Type Ib diamond inducing one phonon infrared absorption: The y center. Diamond Relat Mater 21:120–126

Hall M, Moses TM (2000) Gem Trade Lab Notes: Diamond—blue and pink HPHT annealed. Gems Gemol 36:254–255

Harris JW, Smit KV, Fedortchouk Y, Moore M (2022) Morphology of monocrystalline diamond and its inclusions. Rev Mineral Geochem 88:119–166

Hartland CB (2014) Thesis: A Study of Point Defects in CVD Diamond Using Electron Paramagnetic Resonance and Optical Spectroscopy. University of Warwick

Henry D (2020) Optical cathodoluminescence (Optical-CL). Retrieved 2021/01/20 from https://serc.carleton.edu/research_education/geochemsheets/opticalcl.html

Hirsch PB, Pirouz P, Barry JC (1986) Platelets, dislocation loops and voidites in diamond. Proc R Soc London, SerA 407:239–258

Hough E (1913) The great green diamond. Metropolitan Mag 38:9

Howell DW, Wood IG, Nestola F, Nimis P, Nasdala L (2012) Inclusions under remnant pressure in diamond: a multi-technique approach. Eur J Mineral 24:563–573

Howell DW, Fisher D, Piazolo S, Griffin WL, Sibley SJ (2015) Pink color in type I diamonds: Is deformation twinning the cause? Am Mineral 100:1518–1527

Howell DW, Collins AT, Loudin LC, Diggle PL, D'Haenens-Johansson UFS, Smit KV, Katrusha AN, Butler JE, Nestola F (2019) Automated FTIR mapping of boron distribution in diamond. Diamond Relat Mater 96:207–215

Hudson PRW, Tsong IST (1977) Hydrogen impurity in natural gem diamond. J Mater Sci 12:2389–2395

Iakoubovskii K, Kiflawi I, Johnston K, Collins A, Davies G, Stesmans A (2003) Annealing of vacancies and interstitials in diamond. Physica B, 340–342:67–75

Jones R, Goss JP, Briddon PR (2009) Acceptor level of nitrogen in diamond and the 270-nm absorption band. Phys Rev B 80:033205

Jones R, Goss JP, Pinto H, Palmer DW (2015) Diffusion of nitrogen in diamond and the formation of A-centres. Diamond Relat Mater 53:35–39

Khan RUA, Martineau PM, Cann BL, Newton ME, Twitchen DJ (2009) Charge transfer effects, thermo and photochromism in single crystal CVD synthetic diamond. J Phys: Condens Matter 21:364214

Khan RUA, Martineau PM, Cann BL, Newton ME, Dhillon HK, Twitchen DJ (2010) Color alterations in CVD synthetic diamond with heat and UV exposure: implications for color grading and identification. Gems Gemol 46:18–27

Khan RUA, Cann BL, Martineau PM, Samartseva J, Freeth JJP, Sibley SJ, Hartland CB, Newton ME, Dhillon HK, Twitchen DJ (2013) Colour-causing defects and their related optoelectronic transitions in single crystal CVD diamond. J Phys: Condens Matter 25:275801

Khong YL, Collins AT, Allers L (1994) Luminescence decay time studies and time-resolved cathodoluminescence spectroscopy of CVD diamond. Diamond Relat Mater 3:1023–1027

Kiflawi I, Bruley J (2000) The nitrogen aggregation sequence and the formation of voidites in diamond. Diamond Relat Mater 9:87–93

Kiflawi I, Mainwood A (1996) Nitrogen interstitials in diamond. Phys Rev B 54:16719

Kiflawi I, Sittas G, Kanda H, Fisher D (1997) The irradiation and annealing of Si-doped diamond single crystals. Diamond Relat Mater 6:146–148

Kiflawi I, Bruley J, Luyten W, Van Tendeloo G (1998) 'Natural' and 'man-made' platelets in Type-Ia diamonds. Philos Mag B 78:299–314

King JM, Moses TM, Shigley JE, Welbourn CM, Lawson SC, Cooper M (1998) Characterizing natural-color Type IIb blue diamonds. Gems Gemol 34:246–268

King JM, Shigley JE, Gelb TH, Guhin SS, Hall M, Wang W (2005) Characterization and grading of natural-color yellow diamonds. Gems Gemol 41:88–115

King JM, Geurts RH, Gilbertson AM, Shigley JE (2008) Color grading "D-to-Z" diamonds at the GIA Laboratory. Gems Gemol 44:296–321

Klein PB, Crossfield MD, Freitas JA, Collins AT (1995) Donor-acceptor pair recombination in synthetic Type-IIb semiconducting diamond. Phys Rev B 51:9634–9642

Kohn SC, Speich L, Smith CB, Bulanova GP (2016) FTIR thermochronometry of natural diamonds: A closer look. Lithos 265:148–158

Kupriyanov IN, Gusev VA, Borzdov YM, Kalinin AA, Palyanov YN (1999) Photoluminescence study of annealed nickel- and nitrogen-containing synthetic diamond. Diamond Relat Mater 8:1301–1309

Kupriyanov IN, Gusev VA, Pal'yanov YN, Borzdov YM, Sokol AG (2001) Photoluminescence excitation study of cobalt-related optical centers in high-pressure high-temperature diamond. Diamond Relat Mater 10:59–62

Lai MY, Stachel T, Breeding CM, Stern R (2019) Yellow diamonds with colourless cores—evidence for episodic diamond growth beneath Chidliak and the Ekati mine, Canada. Mineral Petrol 114:91–103

Lawson SC, Kanda H (1993) An annealing study of nickel point defects in high-pressure synthetic diamond. J Appl Phys 73:3967–3973

Lawson SC, Kanda H, Watanabe K, Kiflawi I, Sato Y, Collins AT (1996) Spectroscopic study of cobalt-related optical centers in synthetic diamond. J Appl Phys 79:4348–4357

Lawson SC, Fisher D, Hunt DC, Newton ME (1998) On the existence of positively charged single-substitutional nitrogen in diamond. J Phys: Condens Matter 10:6171–6180

Liggins S, Newton ME, Goss JP, Briddon PR, Fisher D (2010) Identification of the dinitrogen <001>-split interstitial H1a in diamond. Phys Rev B 81:085214

Lim H, Park S, Cheong H, Choi HM, Kim YC (2010) Discrimination between natural and HPHT-treated Type IIa diamonds using photoluminescence spectroscopy. Diamond Relat Mater 19:1254–1258

Loubser JHN, van Wyk JA (1978) Electron spin resonance in the study of diamond. Rep Prog Phys 41:1201–1248

Loubser JHN, van Wyk JA (1981) Models for the H3 and H4 centres based on ESR measurements. *In:* 32nd Diamond Conference. Diamond Conference, Reading

Loudin LC (2017) Photoluminescence mapping of optical defects in HPHT synthetic diamond. Gems Gemol 53:180

Lühmann T, Raatz N, John R, Lesik M, Rödiger J, Portail M, Wildanger D, Kleißler F, Nordlund K, Zaitsev A, Roch JF (2018) Screening and engineering of colour centres in diamond. J Phys D: Appl Phys 51:483002

Macpherson JV (2015) A practical guide to using boron doped diamond in electrochemical research. Phys Chem Chem Phys 17:2935–2949

Mäki J-M, Tuomisto F, Kelly CJ, Fisher D, Martineau PM (2009) Properties of optically active vacancy clusters in Type IIa diamond. J Phys: Condens Matter 21:364216

Mäki J-M, Tuomisto F, Varpula A, Fisher D, Khan RUA, Martineau PM (2011) Time dependence of charge transfer processes in diamond studied with positrons. Phys Rev Letters 107:217403

Manson NB, Hedges M, Barson MSJ, Ahlefeldt R, Doherty MW, Abe H, Ohshima T, Sellars MJ (2018) NV–N^+ pair centre in 1b diamond. New J Phys 20:113037

Mendelssohn MJ, Milledge HJ (1995) Geologically significant information from routine analysis of the mid-infrared spectra of diamonds. Int Geol Rev 37:95–110

Meng YF, Yan CS, Lai J, Krasnicki S, Shu H, Yu T, Liang Q, Mao HK, Hemley RJ (2008) Enhanced optical properties of chemical vapor deposited single crystal diamond by low-pressure/high-temperature annealing. PNAS 105:17620–17625

Nadolinny V, Komarovskikh A, Palyanov Y (2017) Incorporation of large impurity atoms into the diamond crystal lattice: EPR of split-vacancy defects in diamond. Crystals 7:237

Nadolinny VA, Yelisseyev AP (1994) New paramagnetic centres containing nickel ions in diamond. Diamond Relat Mater 3:17–21

Nadolinny VA, Yelisseyev AP, Baker JM, Newton ME, Twitchen DJ, Lawson SC, Yuryeva OP, Feigelson BN (1999) A study of ^{13}C hyperfine structure in the EPR of nickel-nitrogen-containing centres in diamond and correlation with their optical properties. J Phys: Condens Matter 11:7357–7376

Nadolinny VA, Yelisseyev AP, Baker JM, Twitchen DJ, Newton ME, Feigelson BN, Yuryeva OP (2000) Mechanisms of nitrogen aggregation in nickel- and cobalt-containing synthetic diamonds. Diamond Relat Mater 9:883–886

Nadolinny VA, Baker JM, Yuryeva OP, Newton ME, Twitchen DJ, Palyanov YN (2005) EPR study of the peculiarities of incorporating transition metal ions into the diamond structure. Appl Magn Reson 28:365–381

Nasdala L, Grambole D, Wildner M, Gigler AM, Hainschwang T, Zaitsev AM, Harris JW, Milledge J, Schulze DJ, Hofmeister W, Balmer WA (2013) Radio-colouration of diamond: A spectroscopic study. Contrib Mineral Petrol 165:843–861

Navon O, Wirth R, Schmidt C, Jablon BM, Schreiber A, Emmanuel S (2017) Solid molecular nitrogen (δ-N2) inclusions in Juina diamonds: Exsolution at the base of the transition zone. Earth Planet Sci Lett 464:237–247

Newton ME (1994) Electron paramagnetic resonance of radiation damage centres in diamond. *In:* Properties and Growth of Diamond. G Davies (Ed.), IEE/INSPEC, p. 153–158

Newton ME, Campbell BA, Twitchen DJ, Baker JM, Anthony TR (2002) Recombination-enhanced diffusion of self-interstitial atoms and vacancy–interstitial recombination in diamond. Diamond Relat Mater 11:618–622

Nimis P (2022) Pressure and temperature data for diamonds. Rev Mineral Geochem 88:533–566

Overton TW, Shigley JE (2008) A history of diamond treatments. Gems Gemol 44:32–55

Peaker CV, Atumi MK, Goss JP, Briddon PR, Horsfall AB, Rayson MJ, Jones R (2016) Assignment of ^{13}C hyperfine interactions in the P1-center in diamond. Diamond Relat Mater 70:118–123

Pezzagna S, Rogalla D, Wildanger D, Meijer J, Zaitsev A (2011) Creation and nature of optical centres in diamond for single-photon emission-overview and critical remarks. New J Phys, 13: 035024

Pinto H, Jones R, Palmer DW, Goss JP, Briddon PR, Öberg S (2012) On the diffusion of NV defects in diamond. Phys status solidi (a) 209:1765–1768

Reinitz IM, Buerki PR, Shigley JE, McClure SF, Moses TM (2000) Identification of HPHT-treated yellow to green diamonds. Gems Gemol 36:128–137

Rondin L, Tetienne J-P, Hingant T, Roch J-F, Maletinsky P, Jacques V (2014) Magnetometry with nitrogen-vacancy defects in diamond. Rep Prog Phys 77:056503

Rosch S (1957) Das Grune Gewölbe in Dresden. Z Dtsch Ges Edelsteinkunde, Sonderheft, p. 79–84

Rudloff-Grund J, Brenker FE, Marquardt K, Howell D, Schreiber A, O'Reilly SY, Griffin WL, Kaminsky FV (2016) Nitrogen nanoinclusions in milky diamonds from Juina area, Mato Grosso State, Brazil. Lithos 265:57–67

Smit KV, Shirey SB, Wang W (2016) Type Ib diamond formation and preservation in the West African lithospheric mantle: Re–Os age constraints from sulphide inclusions in Zimmi diamonds. Precambrian Res 286:152–166

Smith W, Sorokin P, Gelles I, Lasher G (1959) Electron-spin resonance of nitrogen donors in diamond. Phys Rev 115:1546–1552

Smith HE, Davies G, Newton ME, Kanda H (2004) Structure of the self-interstitial in diamond. Phys Rev B 69:045203

Smith EM, Shirey SB, Nestola F, Bullock ES, Wang J, Richardson SH, Wang W (2016) Large gem diamonds from metallic liquid in Earth's deep mantle. Science 354:1403–1405

Smith EM, Shirey SB, Richardson SH, Nestola F, Bullock ES, Wang J, Wang W (2018) Blue boron-bearing diamonds from Earth's lower mantle. Nature 560:84–87

Solin SA, Ramdas AK (1970) Raman spectrum of diamond. Phys Rev B 1:1687–1698

Speich L, Kohn SC, Bulanova GP, Smith CB (2018) The behaviour of platelets in natural diamonds and the development of a new mantle thermometer. Contrib Mineral Petrol 173:39

Stacey A, Karle TJ, McGuinness LP, Gibson BC, Ganesan K, Tomljenovic-Hanic S, Greentree AD, Hoffman A, Beausoleil RG, Prawer S (2012) Depletion of nitrogen-vacancy color centers in diamond via hydrogen passivation. Appl Phys Lett 100:071902

Stachel T, Harris JW (2008) The origin of cratonic diamonds—Constraints from mineral inclusions. Ore Geol Rev 34:5–32
Stachel T, Luth RW (2015) Diamond formation —Where, when and how? Lithos, 220–223:200–220
Steele JA, Puech P, Lewis RA (2016) Polarized Raman backscattering selection rules for (hhl)-oriented diamond- and zincblende-type crystals. J Appl Phys 120:055701
Sumida N, Lang AR (1981) Cathodoluminescence evidence of dislocation interactions in diamond. Philos Mag A 43:1277–1287
Tallaire A, Collins AT, Charles D, Achard J, Sussmann R, Gicquel A, Newton ME, Edmonds AM, Cruddace RJ (2006) Characterisation of high-quality thick single-crystal diamond grown by CVD with a low nitrogen addition. Diamond Relat Mater 15:1700–1707
Tallaire A, Achard J, Silva F, Brinza O, Gicquel A (2013) Growth of large size diamond single crystals by plasma assisted chemical vapour deposition: Recent achievements and remaining challenges. C R Phys 14:169–184
Tallaire A, Lesik M, Jacques V, Pezzagna S, Mille V, Brinza O, Meijer J, Abel B, Roch JF, Gicquel A, Achard J (2015) Temperature dependent creation of nitrogen-vacancy centers in single crystal CVD diamond layers. Diamond Relat Mater 51:55–60
Tang CJ, Neves AJ, Carmo MC (2005) On the two-phonon absorption of CVD diamond films. Diamond Relat Mater 14:1943–1949
Taylor WR, Jaques AL, Ridd M (1990) Nitrogen-defect aggregation characteristics of some Australasian diamonds: time-temperature constraints on the source regions of pipe and alluvial diamonds. Am Mineral 75:1290–1310
Taylor WR, Canil D, Milledge HJ (1996) Kinetics of Ib to IaA nitrogen aggregation in diamond. Geochimica et Cosmochimica Acta 60:4725–4733
Themelis T (1987) Diamonds from Venezuela. Lapidary Journal 41:59–64
Twitchen D, Newton M, Baker J, Tucker O, Anthony T, Banholzer W (1996) Electron-paramagnetic-resonance measurements on the di-<001>-split interstitial center (R1) in diamond. Phys Rev B 54:6988–6998
Twitchen DJ, Hunt DC, Smart V, Newton ME, Baker JM (1999) Correlation between ND1 optical absorption, the concentration of negative vacancies determined by electron paramagnetic resonance (EPR). Diamond Relat Mater 8:1572–1575
van der Bogert C, Smith C, Hainschwang T, Mcclure S (2009) Gray-to-blue-to-violet hydrogen-rich diamonds from the Argyle Mine, Australia. Gems Gemol 45:20–37
van Wyk JA, Loubser JHN (1993) ENDOR of the P2 centre in type-Ia diamonds. J Phys: Condens Matter 5:3019–3026
van Wyk JA, Woods GS (1995) Electron spin resonance of excited states of the H3 and H4 centres in irradiated type Ia diamonds. J Phys: Condens Matter 7:5901–5911
Vance ER, Milledge HJ (1972) Natural and laboratory α -particle irradiation of diamond. Minerall Mag 38:878–881
Vance ER, Harris JW, Milledge HJ (1973) Possible origins of α-damage in diamonds from kimberlite and alluvial sources. Mineral Mag 39:349–360
Walker J (1977) An optical study of the TR12 and 3H defects in irradiated diamond. J Phys C: Solid State Phys 10:3031–3037
Walker J (1979) Optical absorption and luminescence in diamond. Rep Prog Phys 42:1605–1659
Walsh PS, Lightowlers EC, Collins AT (1971) Thermoluminescence and phosphorescence in natural and synthetic semiconducting diamond. J Lumin 4:369–392
Walter B (1891) Eine charakteristiche Absorptionserscheinung des Diamanten. Ann Physik Chem 3:42, 504–510.[1]
Wang W, D'Haenens-Johansson UFS, Johnson P, Moe KS, Emerson E, Newton ME, Moses TM (2012) CVD synthetic diamonds from Gemesis Corp. Gems Gemol 48:80–97
Wassell AM, McGuinness CD, Hodges C, Lanigan PMP, Fisher D, Martineau PM, Newton ME, Lynch SA (2018) Anomalous green luminescent properties in CVD synthetic diamonds. Phys Status Solidi A 215:1800292
Watanabe K, Lawson SC, Isoya J, Kanda H, Sato Y (1997) Phosphorescence in high-pressure synthetic diamond. Diamond Relat Mater 6:99–106
Wedepohl PT (1957) Electrical and optical properties of Type IIb diamonds. Proc Phys Soc Sect B 70:177–185
Wight DR, Dean PJ, Lightowlers EC, Mobsby CD (1971) Luminescence from natural and man-made diamond in the near infrared. J Lumin 4:169–193
Woods GS (1984) Infrared absorption studies of the annealing of irradiated diamonds. Philos Mag B 50:673–688
Woods GS (1986) Platelets and the infrared Absorption of Type Ia Diamonds. Proc R Soc London, Ser A 407:219–238
Woods GS, Purser GC, Mtimkulu ASS, Collins AT (1990a) The nitrogen content of Type Ia natural diamonds. J Phys Chem Solids 51:1191–1197
Woods GS, Van Wyk JA, Collins AT (1990b) The nitrogen content of Type Ib synthetic diamond. Philos Mag Part B 62:589–595
Yelisseyev AP, Kanda H (2007) Optical centers related to 3d transition metals in diamond. New Diamond Front Carbon Technol 17:127
Zaitsev AM (2001) Optical Properties of Diamond. Springer
Zaitsev AM, Moe KS, Wang W (2018) Defect transformations in nitrogen-doped CVD diamond during irradiation and annealing. Diamond Relat Mater 88:237–255

[1] Series 3, volume 42, edited by G. Wiedemann, and also known as Wiedemann's Annalen, is volume 278 (issue 3) in the Wiley online archive.

Reviews in Mineralogy & Geochemistry
Vol. 88 pp. 689–754, 2022

Synthesis of Diamonds and Their Identification

Ulrika F. S. D'Haenens-Johansson

Gemological Institute of America (GIA)
New York City, New York 10036, USA

ujohansson@gia.edu

James E. Butler

Cubic Carbon Ceramics
Huntingtown, Maryland 20639, USA

jimbutler29@comcast.net

Andrey N. Katrusha

Alkor-D Research
Kiev 03134, Ukraine

akatdps@gmail.com

INTRODUCTION

Since 1797, when Tennant demonstrated that diamond consists solely of elemental carbon by comparing the volume of carbon dioxide formed by burning identical weights of charcoal and diamond (Tennant 1797), scientists have been trying to synthesize diamond by converting various carbon-containing substances. Several attempts followed, but it would take over 150 years of continued research before the first successful report of diamond growth was published by scientists at General Electric (Bundy et al. 1955). Diamond synthesis by Union Carbide in 1952 and Allmänna Svenska Elektriska Aktiebolaget (ASEA) in 1953 predate the work at General Electric, but were not reported until later (Liander and Lundblad 1960; Eversole 1962; Angus 2014). The most familiar carbon allotropes are the crystalline phases of graphite and diamond. The hexagonal form of graphite consists of sheets of sp^2 hybridized carbon atoms in a hexagonal array, with each atom bonded to three equidistant nearest neighbor atoms. The layers are attracted to each other by weak van der Waals forces and may be arranged in a hexagonal, or rarely rhombohedral, stacking sequence. Meanwhile, the carbon orbitals in diamond are sp^3 hybridized, with each atom covalently bonded to four nearest neighbors in a tetrahedral arrangement. The prevalent diamond structure is cubic, though the hexagonal lonsdaleite form also exists. Figure 1 illustrates the pressure and temperature (P, T) phase and transition diagram for pure carbon. Theoretically and experimentally determined conditional phase boundary lines separate the diamond and graphite stability fields. The high cohesive and activation energies associated with the different carbon phases mean that other metastable forms can occur under conditions at which they are not thermodynamically stable. For instance, diamond exists at room temperatures and pressures, whereas graphite can survive pressures well into the diamond stability field (Bundy 1980; Bundy et al. 1996).

High quality single crystal diamonds can be synthesized using either the high pressure, high temperature (HPHT) or chemical vapor deposition (CVD) techniques. In the former, growers use an HPHT press to crystallize diamond from a carbon source such as graphite either by direct or indirect (catalytic) conversion methods (Fig. 2), applying temperatures and

1529-6466/22/0088-0013$10.00 (print)
1943-2666/22/0088-0013$10.00 (online)
http://dx.doi.org/10.2138/rmg.2022.88.13

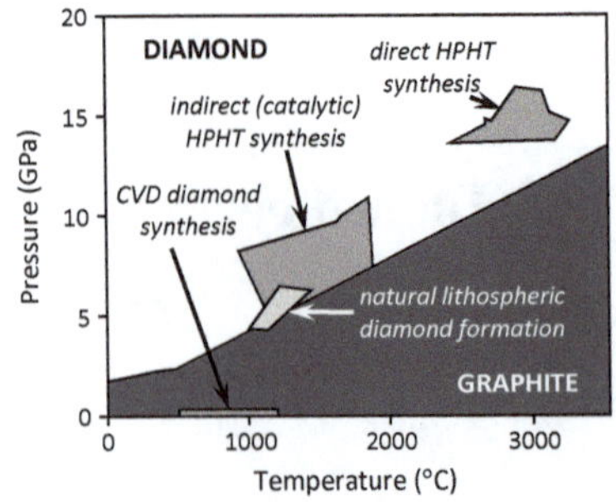

Figure 1. Carbon phase diagram as a function of pressure and temperature, illustrating the conditions for natural diamond formation in the lithosphere as well as those for laboratory-grown diamonds synthesized by chemical vapor deposition (CVD) and high pressure, high temperature (HPHT) methods. [Based on phase diagrams compiled by Bundy (1980) and Bundy et al. (1996), including natural diamond data presented by Stachel and Harris (2008).]

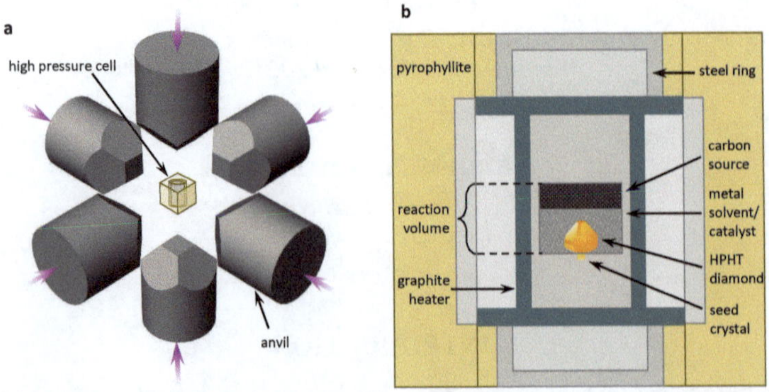

Figure 2. Illustrations of (**a**) the pressure transmitting anvil arrangement for a standard cubic press and (**b**) the high pressure cell (HPC). The internal reaction volume in the HPC contains the key ingredients for HPHT diamond synthesis: a carbon source (such as graphite), metal solvent/catalyst (typically Fe or its alloys), and one or more diamond seed crystals. For the temperature gradient method, the temperature is controlled such that the region with the carbon source is hotter than the seed crystal. The carbon dissolves into the metal solvent/catalyst melt, reaching supersaturation and recrystallizing as HPHT diamond material at the seed.

pressures at which diamond is the dominant phase. The latter technique is instead based on gas-phase chemical reactions involving hydrocarbon and hydrogen gases to deposit metastable diamond at low pressure in a CVD reaction chamber (Fig. 3). Though thermodynamically unstable, under the selected conditions diamond growth is kinetically favorable. This chapter reviews the principles, technology and current capabilities of these two different techniques, as well as the properties of the diamonds that they yield.

Although diamond's fame originates from its use as an attractive gemstone, research on the synthesis and characterization of their high laboratory-grown counterparts has historically been driven by a scientific community eager to manufacture materials specifically tailored to harness some of the remarkable combination of properties that this material or its lattice defects boast. Industrial and technological applications include diamond abrasives and cutting tools, quantum sensing, high power electronics, thermal management, optical windows, x-ray optics, and electrochemical devices for water treatment (Wort and Balmer 2008; Balmer et al. 2009; Field 2012; Shvyd'ko et al. 2017; Awschalom et al. 2018; Cobb et al. 2018; Koizumi et al. 2018; Donato et al. 2019; Yang et al. 2019; Achard et al. 2020). In the last couple of decades, laboratory-grown diamond quality improved and started to obtain gemological properties akin to natural diamonds, sparking interest in the gem and jewelry industry. Though long considered a niche product with limited distribution, in 2021 laboratory-grown diamonds were reported to account for 5 to 8% of the total global diamond (natural + laboratory-grown)

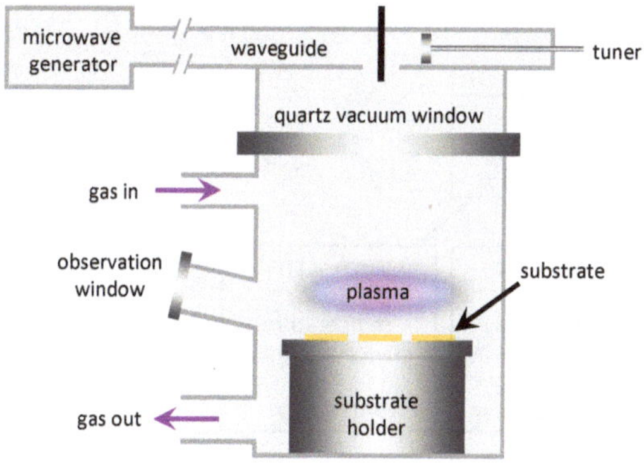

Figure 3. Schematic of a microwave CVD reactor. A hydrocarbon gas in an excess of hydrogen is introduced into the growth chamber. In this design, microwaves are used to activate a plasma, dissociating the gas components into species which can react with diamond substrates heated to 600–1200 °C, depositing new diamond material. [Modified after Martineau et al. (2004) *Gems and Gemology*, Vol. 40, Fig. 2, p. 4.]

jewelry market; this value was under 1.5% as recently as 2018 (Zimnisky 2021; Branstrator 2022). Market analysis estimated the annual production for laboratory-grown diamonds created for gem applications to be 6–7 million carats (1 carat = 1 ct = 200 mg) in 2020 (Linde et al. 2021). For reference, approximately 40% of the 140 million carats of natural rough diamonds mined annually are gem-quality (Zimnisky 2020). Competition between growers in this sector has led to dramatic rise in diamond sizes, producing faceted gems up to ~15 ct (e.g., Poon and Wang 2016; Wang et al. 2022). Further size increases may be tempered by consumer demand, though growers have been developing methods to synthesize 100 ct crystals. Despite global focus being placed on >1 ct faceted laboratory-grown diamond manufacturing, China has become a significant mass producer of melee-sized goods. Melee diamonds are loosely defined as those weighing 0.001–0.2 ct, but this may vary by country.

The introduction of laboratory-grown diamonds in the gem and jewelry industry has also led to concern that they may infiltrate the natural diamond stream and be sold without full disclosure of their man-made origin. In order to maintain transparency and protect gem and jewelry buyers, scientists and gemologists at grading laboratories have worked in tandem with the diamond physics and geology communities, as well as diamond growers from across the globe, to develop identification methods. This chapter presents a comprehensive account of the key gemological and spectroscopic characteristics that are used to confidently separate natural and laboratory-grown diamonds.

HIGH PRESSURE, HIGH TEMPERATURE GROWTH OF DIAMOND

Principles of growth

This section discusses the growth of high pressure, high temperature (HPHT) diamonds, synthesized at pressure and temperature conditions where diamond is the thermodynamically stable form of carbon. Commercial HPHT synthesis is generally conducted at higher temperatures and pressure with respect to the graphite–diamond boundary line (see Fig. 1), accounting for kinetic effects related to uncertainty in the boundary line position which may negatively affect the crystallization of diamond from the graphite phase, leading to growth plane incompleteness or inclusion entrapment.

Research teams at General Electric investigated two primary routes for HPHT diamond growth in the 1950s through 1970s: the direct and indirect (stimulated) conversion of graphite to diamond as presented in Figure 4 (Bundy et al. 1955; Bovenkerk et al. 1959; Bundy 1962, 1963; Wentorf 1971; Strong 1989). These approaches differ in technological complexity. Although the former growth method is faster, it requires specialized high-pressure control equipment

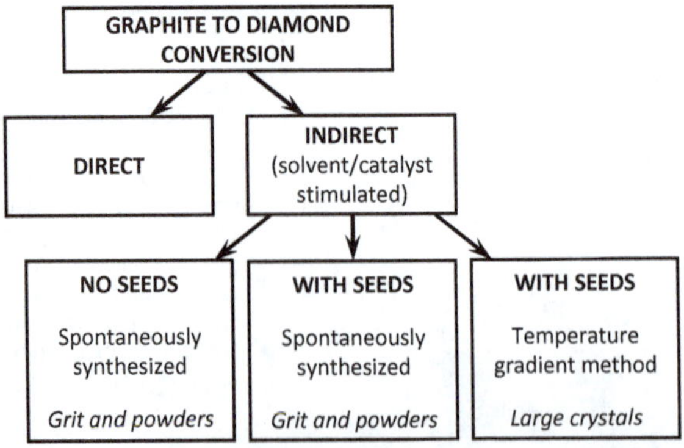

Figure 4. Schematic of the different graphite to diamond conversion processes.

that can generate and maintain pressures higher than 13 GPa at temperatures above 2000 °C, as well as technical staff capable of manipulating these parameters simultaneously. Thus direct graphite to diamond conversion tends to yield small diamond crystals ranging from a few nanometers to millimeters in size, which may be polycrystalline in nature (Bovenkerk et al. 1959; Bundy 1962, 1963; Irifune et al. 2003, 2004, 2014). Commercial diamond production is based on the indirect graphite to diamond conversion approach, where a carbon source material, usually graphite powder, is mixed with a solvent/catalyst and placed in a growth cell (the "reaction volume"). The solvent/catalyst is used to stimulate growth, transport carbon to the growing surface, and decrease the pressure and temperature requirements for diamond nucleation, with typical conditions of 5–6 GPa and 1300–1600 °C. The reaction volume may also contain diamond seed crystals to aid in nucleation. Diamond grit and powder production can occur with or without diamond seeds, whereas large diamond crystal production requires them. Generally, diamonds grown by the spontaneous nucleation method (no seeds) readily incorporate nitrogen impurities, resulting in yellow color (Chrenko et al. 1971), but specialized approaches can result in colorless diamonds with good morphology, as shown in Figure 5. Color-producing impurities in HPHT diamonds will be discussed in the section *Identification of HPHT diamonds: Absorption spectroscopy and color.* The temperature gradient catalytic growth method has been successfully used to synthesize multi-carat sized HPHT diamonds (Burns et al. 1999; Kanda 2000; Sumiya and Tamasaku 2012; Sumiya 2014; Sumiya et al. 2015; D'Haenens-Johansson et al. 2015a). Operating under pressures within the diamond stability field, the temperature of the reaction volume is increased so that the solvent/catalyst melts. A temperature gradient of ~20–50 °C is produced, resulting in a hotter environment for the carbon source than the seed region. The carbon source material dissolves in the melt, with

Figure 5. Colorless Type IIa HPHT diamond crystals with predominantly octahedral {111} morphology obtained by the spontaneous nucleation process (~50–150 μm in size). The gray shiny pieces are remains of the reaction medium after initial acid cleaning. [Diamonds synthesized by ANK with Alkor-D Ltd. Photograph by ANK.]

the carbon atoms being transported toward the cooler seeds. Under the selected conditions the carbon becomes supersaturated in the melt, crystallizing as HPHT diamond on the seeds. Figure 2 illustrates the key components of this method. HPHT growth occurs along different crystal directions with differing growth rates, typically resulting in crystals with well-developed cube {100} and octahedral {111} faces (cuboctahedral morphology) as shown in Figure 6 (Strong and Chrenko 1971; Kanda et al. 1980, 1989; Sunagawa 1984, 1995; Shigley et al. 1986, 1987, 1992, 1993a,b; Burns et al. 1999). Presently metal solvent/catalysts consisting of pure or alloy forms of Fe, Ni, Co, Mn, Cr, Ta, or Nb are used for industrial and gem diamond synthesis (see the upcoming section *Selection of metal solvent/catalyst*). Diamond crystallization using alternative non-metallic growth media such as carbonate-silicate, sulfide-carbon, sulfur-carbon, carbon-oxygen-hydrogen, and phosphorus-carbon systems have been reported, with experiments primarily geared toward gaining a better understanding of fundamental diamond formation processes in the mantle or creating unusually doped specimens. Such synthesis falls outside the scope of this chapter, as they are not used for large crystal diamond growth, however Palyanov et al. (2015a) present a detailed review.

Figure 6. HPHT diamond crystals (0.5–0.69 ct) grown by Jinan Zhongwu New Materials Company Ltd. of Shandong, China, showing characteristic cuboctahedral growth morphology. [Used by permission of the Gemological Institute of America, from Eaton-Magaña et al. (2017) *Gems and Gemology*, Vol. 53, Fig. 2, p. 266.]

Technology of growth

High pressure equipment capable of reaching pressures greater than 5 GPa—the minimum necessary for HPHT diamond nucleation—have been developed by several research groups across the world using different approaches for pressure generation and transmission. Excluding rare multi-stage high pressure machines with small (<1 mm^3) reaction volumes, the designs can be grouped into four main types: the belt, toroid, BARS, and cubic presses. Figure 7 illustrates the relationship between investment costs, reaction volumes, maintenance costs, and difficulty of usage for these presses. The belt and toroid presses are based on axial compression of the reaction cell, differing in side support design. With the former, a pair of pistons push into the chamber and are supported by binding rings forming a "belt" around the structure (Hall 1960a,b). In the latter, toroidal sub-cavities hold the pressure inside the main semi-spherical cavity (Kolchin et al. 1981; Khvostantsev et al. 2004). In contrast, the BARS and cubic systems apply pressure along multiple axes. BARS presses (referred to by the Russian abbreviation for their description) have a split-sphere multi-anvil block design, using hydraulic oil to generate pressure through an elastic membrane usually made from rubber or similar material (Shigley et al. 1993a; Palyanov et al. 2017). Cubic presses (e.g., Fig. 2) are based on six perpendicular anvils coupled to independent hydraulic cylinders compressing a cubic chamber (Zeitlin et al. 1961). The compression mechanism largely defines the weight and size of the presses. To generate the equivalent load on a similarly sized sample chamber, the mechanical piston hydraulics used for belt and toroid systems are approximately 2–3 times larger than those for BARS systems.

HPHT presses are manufactured globally, with certain regions specializing in a particular type. Factories in former USSR countries produce the most popular series of BARS presses, with the model names (300, 350, and 400) describing the diameter of the sphere in millimeters.

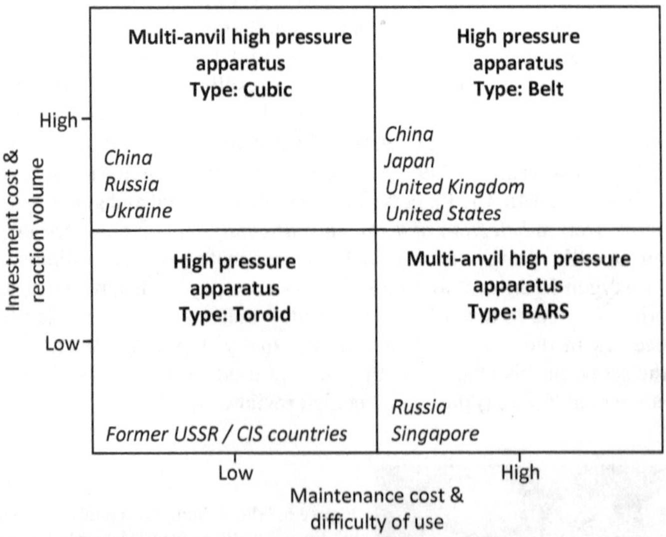

Figure 7. Comparison of the costs, reaction volumes, and difficulty of usage associated with high pressure, high temperature (HPHT) equipment presently existing in the market. Considering 2020 prices, the "low" regions indicate average investment costs of about \$20,000–\$30,000, while the "high" regions represent those greater than \$100,000. For reaction volumes, the regions indicate either 1–5 cm^3 or 10–300 cm^3, respectively.

BARS presses use an oxide/halide ceramic cube (or parallelepiped) high pressure cell (HPC), where the edge defines the reaction volume, e.g., a 32 mm cube contains a 2 cm^3 reaction volume. These presses can produce loads of 6–12 MN (Feigelson 1983). Toroid systems also originate from this region, with central recess (lune) diameters of 20, 30 and 40 mm in combination with hydraulic press models DO043 or DO044. These presses are capable of generating 20 and 25 MN loads, respectively. The diameter of the lune defines the reaction volume, e.g., a 40 mm diameter provides an approximately 2 cm^3 reaction volume. Reaction volumes up to 10 cm^3 have been reported (Palyanov et al. 2015a). Belt systems from the United States are used for diamond grit production, with reaction volume sizes up to 1000 cm^3. The Chinese development of cubic systems has benefited from government support, leading to a continuous improvement in reaction volume sizes, ranging from the series 460 to the series 1000 (as of June 2020). The series number represent the diameter of the hydraulic cylinder in millimeters. The series 560, 650 and 750 cubic presses are the most widely used models for both grit and large diamond growth, attaining loads of 30–50 MN. As for BARS presses, the cubic HPC edge defines the reaction volume, e.g., 74 mm cubes have reaction volumes of about 40 cm^3.

All four press types consist of three main parts: the pressure generating machine, the pressure transmitting unit (referred to as the high pressure apparatus, HPA), and the sample containing HPC. In order to generate and hold the pressure on the sample, the force is transmitted from the press machine through the respective HPAs to the HPCs (Figs. 2 and 8). The HPA is made from cemented carbides composed of tungsten carbide and cobalt (WC-Co alloys, such as WC-6% Co, WC-8% Co and WC-10% Co). The geometry of the HPC is dictated by the shape of the compressive and tensile components of the HPA. Presently, the hardness and toughness of the HPA hard alloy set the upper limit for the achievable pressure in systems used for large volume diamond mass production. For further details regarding HPA and HPC design see Bundy (1988). Future improvements in large volume superhard HPAs would propel the diamond industry to increase production volumes, improve sample quality, and further explore promising unconventional growth media for diamond synthesis.

(a) (b)

Figure 8. (**a**) High-pressure cells (vessels) used for different high-pressure apparatus (HPA): toroid (**left**), belt (**middle**), and cubic (**right**). Belt and cubic systems benefit from higher reaction volumes. (**b**) High-pressure cells used in cubic high-pressure presses (series 650/750/850/950) from the 1990s till present. The largest version of the cell, with dimensions 88 × 88 × 88 mm (shown on the **right**) is currently the largest commercially available cell, and widely used. Larger cells exist, but are still in the experimental development phase. Cell dimensions from left to right: 58 × 58 × 58, 74 × 74 × 74 mm, and 88 × 88 × 88 mm. [Photographs by ANK.]

Heavy investment in diamond synthesis using cubic presses have led to design upgrades and regular breakthroughs in diamond volumes and qualities, warranting further detailed description of this press type. Cubic high pressure cells (Fig. 2b and 8b) are made of powder pressed pyrophyllite with a central cavity containing tens of concentric and axially assembled components designed for three main functions:

1. Pressure transmission to the sample reaction chamber;
2. Resistive heat generation to melt the metal solvent/catalyst;
3. Providing thermal and chemical insulation to prevent damage to other press components during high pressure and temperature experiments over long periods of time.

These functions are generally applicable to all HPHT systems. Components are made of advanced materials in order to fulfill these functions and withstand the combination of extreme pressures (5–7 GPa) and temperatures (1300–1800 °C) produced in the sample reaction chamber. Certain salts and refractory ceramics possess the required chemical inertness, high melting temperatures, and phase stability at high pressures. Boride and chloride salts, as well as ceramics in the form of oxides (Mg, Al, Zr, or Y) are commonly used by the diamond synthesis industry (Strong 1977; Zhu et al. 2019). The component materials may differ for systems that require higher temperatures, such as those used for HPHT annealing treatments to change diamond color.

The temperature in the cell is generated by means of graphite or composite material resistive heaters. A precise temperature field distribution is formed in the diamond synthesis region by careful selection of the geometry and physical properties of the electric system, considering how these may be affected when subjected to simultaneous temperature and pressure changes. Temperature gradients across the cell are undesirable for diamond grit production, with efforts made to minimize fluctuations close to the cell sides. In contrast, controlled axial and radial temperature gradients are the basis of the temperature gradient growth method used for large high quality diamond synthesis (e.g., D'Haenens-Johansson et al. 2015a). Careful component designs which consider detailed electric and heat conductance properties in the cell enable control of the heat generation and dissipation distributions, offering the flexibility to grow samples using either scheme. Finite element method (FEM) simulations are widely used to characterize the temperature field in the growth cell under different conditions (Li et al. 2013). The main drawback with this approach is that it cannot consider the component material properties at high temperatures and pressures. Results from complementary experiments analyzing real temperature distributions using thermocouples, as well as observations of phase transitions, provide the necessary feedback to improve the simulated material properties. Temperature gradients as fine as 1 °C/mm can be predicted and realized within the cells, facilitating the set-up of the optimal carbon supersaturation conditions necessary for growth of high quality diamond crystals with flat faces and good morphology. In the case of suboptimal carbon supersaturation, growth is unstable, resulting in the formation of multi-crystal aggregates known as druzes (Fig. 9).

Figure 9. The simultaneous presence of optimal and over-saturated carbon flow during the HPHT diamond growth process resulted in the formation of high quality flat-surfaced single crystals (**upper right**) and multi-crystal aggregates (druzes, **lower left**) due to growth instability, respectively. The larger crystals weigh approximately 4 ct. The diamond materials are here imbedded in the remains of the growth medium after a short acid clean. [Diamonds grown by ANK with Alkor-D Ltd. Photograph by ANK.]

Selection of metal solvent/catalyst

Historically, HPHT diamond growers focused on selecting metal solvent/catalysts with the lowest possible diamond nucleation temperatures due to practical and economic reasons. Lower growth temperatures in turn also meant that lower pressures could be used, easing the demands on the press and increasing the lifespan of the HPA components. The costs of the metals used were also considered. Initially, the solvent/catalysts were Mn and Ni alloys (Bundy 1963). However, there was a great disadvantage to their use: the growing crystals readily trapped inclusions of carbides associated with these metals. Thus diamond crystal powders produced at the time were characterized by poor toughness. Inclusion uptake could be decreased, and the material properties improved, by operating at higher temperatures and pressures using Fe, Ni and Co alloys. Growth using a variety of solvent/catalysts composed of combinations of invar (Fe–Ni) and kovar (Fe–Ni–Co) alloys have been carefully analyzed (Strong and Chrenko 1971; Kocherzhinski and Kulik 1996; Sugano et al. 1996). Eutectic compositions are used as they lower the melting temperature.

Diamond growers devise proprietary melts with additional elements to decrease the pressure and temperature parameters or improve the crystal characteristics. Such additives may either be elements which are fairly chemically inert with respect to carbon (Cu, Sn, In, As, Zn, or Au), or they may be known to interact, forming carbides (Cr, Si, V, Mg, Nb, or Zr). These latter carbide formers are also strong nitride formers, and are usually added specifically to getter nitrogen that would otherwise readily incorporate into the growing diamond. As the molten solvent/catalyst should contain more than 10 at.% carbon, additives may be used to increase carbon solubility (Strong and Chrenko 1971; Kanda 2000). The resulting complex metallic mixtures have been tested for diamond production at a range of pressure and temperature conditions to optimize crystal quality. The conclusions are generally that growth benefits from higher temperature and pressure parameters. Intentional or inadvertent doping of the diamond crystals due to impurities in the melt ingredients, carbon source, or cell components further complicate growth and expand the variables that growers can customize. These dopants, further discussed in the section *Control of color center content and extended defects*, affect the HPHT diamond properties, most notably color. Large, high quality HPHT diamonds that are as-grown colorless, blue, green, and yellow can be produced.

The metal solvent/catalyst is one of the factors that affect the morphology of the growing crystal. For instance, Kanda et al. (1989) demonstrated that pure Ni melts resulted in cubic {100} and octahedral {111} faces, whereas Ni-alloys which included Co and Fe promoted the development of dodecahedral {110} and trapezohedral {113} faces. The solvent/catalyst composition also influences surface structures on the crystal faces, which may form dendritic, vein-like, lamellar or etch pit patterns (Kanda et al. 1980). The patterns correspond to the solvent/metal textures that are created during the quenching process that occur at the end of growth runs.

Diamond seed considerations

Large HPHT diamond growth using the temperature gradient method is initiated homoepitaxially on a diamond seed. Growth on the seed can follow two different approaches. Either all sides of the seed can be overgrown with HPHT diamond (like natural diamond growth), or growth can be restricted to one face, with greater control over the orientation. Figure 10 shows diamonds grown using the first method, with a yellow diamond seed enveloped by new colorless diamond material. For the second method the selected growth face of the seed is placed in contact with the metal solvent/catalyst, while the other faces are shielded by some barrier material (Wentorf 1971). The latter is the most commonly used approach for HPHT diamond growth.

Figure 10. Experimental colorless Type IIa HPHT diamond crystals with cubic {100} faces obtained by the overgrowth method on yellow Type Ib laboratory-grown diamond seeds from usual diamond powder production. The size of the colorless crystals range between 1.0 and 1.5 mm. The cubic external morphologies seen here are extraordinarily rare for diamonds grown using high growth temperatures (>1850 °C), and were possible due to the use of non-traditional growth media based on magnesium. [Diamonds grown and photographed by ANK (Alkor-D Ltd.).]

The maximum size of each HPHT diamond is defined by the number of seeds simultaneously used as the growing diamonds compete for a limited amount of carbon source material and space within the reaction cell. It may be possible to increase the crystal output if the reaction cell is large enough to be segregated into a multi-layered growth structure. An example of multiple seeds for a two-layered growth cycle is shown in Figure 11. During growth the layers were vertically adjacent to each other, and advanced techniques were used to generate the necessary temperature gradient across the system to initiate diamond nucleation and growth under the appropriate pressure and temperature conditions.

The orientation of the diamond nucleation face on the seed defines the crystallographic orientation of the first atomic-scale HPHT overgrowth layer. The growth directions of subsequent atomic layers are influenced by the pressure and temperature parameters as well as the chemical composition of the growth environment, usually resulting in the characteristic cuboctahedral morphology (Strong and Chrenko 1971; Kanda et al. 1980, 1989; Burns et al. 1999). Typically, the upper growth face of the HPHT diamond has the same orientation as the nucleation face on the seed, unless the sample has experienced significant variations in growth conditions. Consequently, growers are able to exercise some control over the growing crystal morphology. Figures 6, 9, 11, and 12 present examples of HPHT diamonds grown on (100)-oriented (cubic) seeds, while Figure 13 shows samples grown on (111)-oriented (octahedral) seeds. Although intentional growth on alternative seed facet orientations such as (113), (115), or (110) are also possible, in practice it is mainly restricted to fundamental growth research as seed preparation is challenging and availability is limited.

Seed size selection has to balance material costs (larger seeds are more expensive) and the grower's ability to initiate and control growth without multiple nucleation sites. Seeds range from tens of micrometers to a few millimeters in size. There is a plentiful supply from the industrial diamond grit and powder manufacturing sector.

Figure 11. Typical example of HPHT diamonds resulting from multiple seed and multiple layer growth methods. The view is from the seed side (the bottom of the reaction volume illustrated in Fig. 2b), where the growth medium has been partially dissolved by acid cleaning. Average crystal size is ~4 mm each. [Diamonds grown and photographed by ANK.]

Figure 12. Nitrogen doping influences the color of HPHT diamonds, with yellow color being observed with increasing isolated substitutional nitrogen concentrations ($[N_S^0]$). These concentrations were controlled by including nitrogen getters in the growth capsule. Each diamond crystal is ~7 ct in size. [HPHT diamonds synthesized and photographed by ANK.]

Figure 13. Deep blue to black Type IIb HPHT diamonds grown on {111}-oriented diamond seeds. Sizes range from approximately 6 to 15 ct. The upper face has a {111} orientation, reflecting that of the seed. The sizes and orientations of the other faces depend on the pressure and temperature conditions in the growth cell. [HPHT diamonds synthesized and photographed by ANK.]

Effect of growth temperature and pressures

Growth temperature affects the crystal morphology, with the primary faces changing from {100} to {100} + {111} and then {111} as temperatures are increased when using standard ferrous solvent/catalysts (Satoh et al. 1990; Sumiya et al. 2002). Temperature ranges for good crystal quality are narrow, with reported windows spanning only about 10–40 °C, depending on the applied pressure and the type of diamond being synthesized. At lower temperatures skeletal crystals start to form, where growth primarily proceeds along corners and edges rather than at the crystal faces. At higher temperatures inclusion uptake increases. However, there are exceptions to this behavior. Rare pure rhombidodecohedral {110} (12-faced crystal form of {110} with cubic crystal symmetry) diamonds can be grown under certain pressure and temperature conditions for a given metal solvent/catalysts, as shown in Figure 14. The relationship between the pressure, temperature, solvent/catalyst, and resulting grown diamond morphology is complex. For example, the HPHT diamonds shown in Figure 10 were grown at high temperatures (>1850 °C), yet are dominated by {100} faces. These diamonds were grown using an atypical Mg-based solvent/catalyst.

The solubility of impurities such as nitrogen into the solvent/catalyst rises with temperature, potentially leading to lower impurity concentrations in the growing diamond (Burns et al. 1999). Furthermore, increasing temperatures leads to a decrease in the efficiency of impurity getters (Kanda and Lawson 1995). Hence drifting or fluctuating local temperatures at the growth fronts can modify the distribution of impurities within growth sectors, defined as volumes within a diamond that were grown with common growth planes (Kanda and Yamaoka 1993; Burns et al. 1999). Further discussion of growth sectors is presented in the section *Identification of HPHT diamonds: Growth sectors*. Higher temperatures in nitrogen-containing growth can also lead to the formation of aggregated nitrogen defects (nearest neighbor nitrogen-pairs known as A-centers) (Davies 1976; Kanda et al. 1990; Vins et al. 1991; Babich et al. 2000, 2004).

Pressure influences significantly the nucleation rate in the indirect graphite to diamond conversion growth method (Bundy et al. 1955), whereas for the temperature gradient method its influence is relatively small and restricted to the effect of "widening" the effective technological growth zone. It has been observed that pressure has some influence on the solubility of carbon in melted solvent/catalysts, so that fluctuations also affect the crystal growth.

While diamond nucleation and/or seeded growth have been achieved using many pressure and temperature combinations, temperatures exceeding 1600 °C and pressures above 7 GPa have been identified as suitable for stable mass production of diamonds. However presently such parameters are not economically viable to generate in larger volume high pressure systems since production costs would be much higher than the sale price of the product, without much improvement in the diamond properties.

Figure 14. Example of extremely rare HPHT diamond single crystals with pure rhombic dodecahedral {110} external morphologies. The upper and lower rows show crystals with 3-point and 4-point joining tips, respectively. The light yellow hue of the crystals is a result of nitrogen doping occurring due to loss of nitrogen-getter efficiency at high growth temperatures (>1850 °C). [HPHT diamonds grown by ANK with Alkor-D Ltd. Photograph by ANK.]

Doping and control of color center content

Diamond is a wide-bandgap semiconductor with a bandgap energy of ~5.48 eV, making it transparent from the ultraviolet, through the visible, and into the infrared range (~225 nm–2.6 μm) (Clark 1959; Collins 1982). Consequently, perfect, impurity-free diamond consisting solely of carbon atoms is a colorless material. The growth environment, as well as post-growth conditions or treatments may however introduce color-producing impurities and defect complexes. Diamond has one of the highest atomic densities (atoms/cm^3) of any terrestrial material. This is due to the small size of carbon atoms and the four strong and short carbon to carbon sp^3 bonds per each carbon atom in the lattice. As a result, it is difficult to fit larger atoms into the lattice. Nitrogen, boron, and possibly phosphorus atoms are known to occupy lattice sites, while larger atoms such as silicon and nickel usually are accompanied by vacancies for them to fit in the lattice. Defects related to carbon interstitials are observed and there is evidence for atomic hydrogen and protons associated with elemental impurities. Defects can change the electrical, optical, thermal, and structural properties of diamonds. Thus the ability to manipulate the defect content in laboratory-grown diamonds through doping or growth control measures are important. Many defect centers absorb light, fluoresce, luminesce, and/or phosphoresce, enabling their identification, as discussed in detail in the section *Separation of natural and laboratory-grown diamonds*. Collins (1982, 1992, 1993a), Ashfold et al. (2020), and Green et al. (2022, this volume) present reviews on diamond defects.

Nitrogen and boron are the most widespread impurities in both natural and laboratory-grown diamonds (Collins 1993a,b; Ashfold et al. 2020). Unintentional contamination with them during HPHT synthesis is common due to their availability in the growth environment. Air, which has 78.09% nitrogen by volume, is easily trapped in pores and cavities within the reaction cell or its components. Trace amounts of both nitrogen and boron can be found in the carbon source, metal solvent/catalyst, and cell materials. Whereas the purity of the solvents and graphite can be controlled with high accuracy (up to 0.001 wt%), the purification process of the total assembled growth capsule is highly complicated and not reproducible due to constraints in press operation. For these reasons, precise control of nitrogen and boron content in the grown diamonds primarily relies on a chemical approach—the use of impurity getters (Mg, Al, Zr, Ti, V, Cr) or compensators (B, S, N, P). Getters act as chemical absorbers of the undesirable elemental impurity, for instance by forming nitrides, minimizing the element's availability for dissolution into the diamond lattice (Strong and Chrenko 1971; Wentorf 1971; Sumiya and Satoh 1996; Burns et al. 1999; Wakatsuki et al. 1999). Rather than preventing impurity incorporation into the growing crystal, compensators form lattice defects that interact with those formed by impurities, donating or accepting electrons and reducing the concentration of a specific defect charge state which produces unwanted properties (e.g., color) (Burns et al. 1990).

Figure 12 shows the effect of careful addition of getters to control the concentration of nitrogen in HPHT diamonds. As the getter content is decreased, there is an increase in the concentration of isolated nitrogen defects and a transition from Type IIa, to IIa + Ib (where the crystal contains both Type IIa and Ib material), to Ib (see *Identification tools* section below for an explanation of diamond types). There is a stark change in the appearance of the diamonds as the isolated nitrogen defect is associated with yellow color (Chrenko et al. 1971), as discussed in *Identification of HPHT diamonds: Absorption spectroscopy and color*. This control opens up the possibility of defect engineering, where specific concentrations of desirable color centers are produced, either in "as-grown" (untreated) material, or through treatment. Nitrogen-vacancy (NV) centers, which can produce an attractive pink color in gem quality diamonds (e.g., Fig. 15), are also of great interest to the scientific community as they develop methods to harness properties of the negatively charged state of the center (NV^-) for uses in quantum technology and other fields. NV centers can be created by irradiating and annealing HPHT diamonds with suitable isolated nitrogen concentrations (Moses et al. 1993; Shigley et al. 2004a). The addition of nitrogen getters narrows the range of temperatures at which diamonds with good crystallinity can be grown, and leading to higher uptake of melt inclusions and associated carbides (Sumiya and Satoh 1996; Burns et al. 1999, 2009; Sumiya et al. 2002). It is thought that nitrogen getters decrease the solubility of carbon in the metal solvent/catalyst, or prevent the diffusion of carbon. Group Ib metals (Cu, Ag, Au, etc.) can be used to decompose the carbides (Wakatsuki 1966). Nitrogen getters can also affect the crystal morphology (D'Haenens-Johansson et al. 2015a).

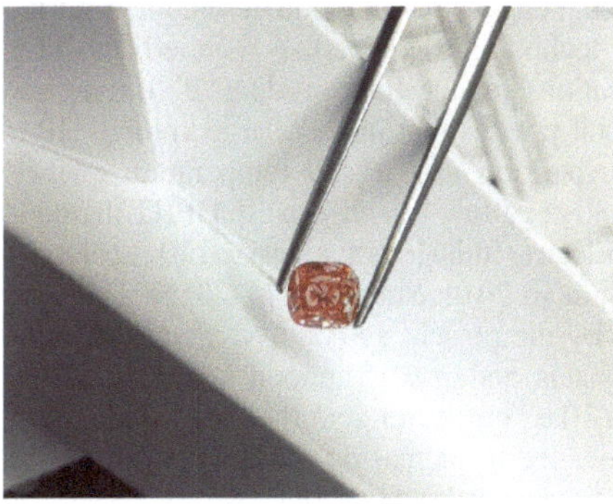

Figure 15. 2.55 ct HPHT-grown gem diamond with a saturated pink color due to high concentrations of nitrogen-vacancy centers. These color centers were obtained by a multi-step process: the original Type IIa + Ib HPHT diamond, containing a low concentration of substitutional nitrogen, was first electron irradiated and finally annealed at a moderate temperature. [HPHT diamond produced and photographed by ANK.]

Uncompensated boron gives rise to blue color in HPHT diamond (Collins and Williams 1971) (see also *Identification of HPHT diamonds: Absorption spectroscopy and color*). Impurity uptake in HPHT diamonds is sensitive to the orientation of the growing face, growth rates and temperature (Strong and Chrenko 1971; Burns et al. 1990, 1999; Satoh et al. 1990). Boron concentrations are typically higher in {111} growth sectors, followed by {110} sectors, whereas nitrogen concentrations follow {111} > {100} > {113} > {110} (Burns et al. 1990). Even blue color can only be achieved by reducing nitrogen concentrations in the diamond to less than 10^{18} cm^{-3}. Otherwise, the resulting material will contain both yellow Type Ib and blue Type IIb growth sectors, which may produce an overall green hue (Shigley et al. 1992, 2004a,b; Fritsch and Shigley 1993; Rooney et al. 1993; Hainschwang and Katrusha 2003). Both Type IIb or IIa + IIb HPHT diamonds are thus grown by adding a nitrogen getter and a boron dopant (e.g., amorphous boron). An example of sector-dependent boron uptake is shown in Figure 16. Further discussion on this phenomenon is presented in *Identification of HPHT diamonds: Growth sectors.*

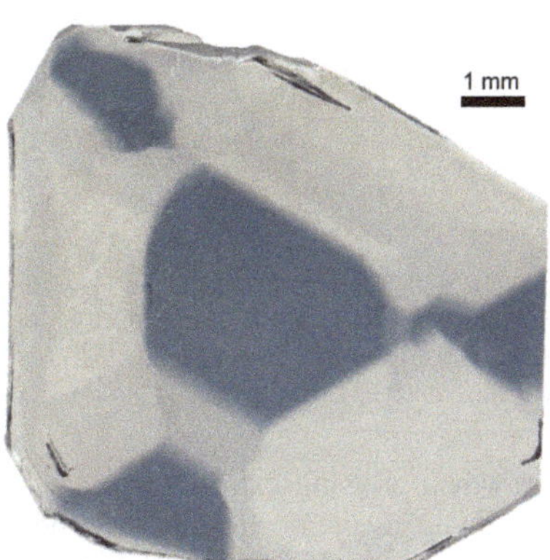

Figure 16. Example of the growth sector dependent incorporation of impurities in large laboratory-grown diamonds. The blue zones indicate the higher concentrations of uncompensated boron present in {111} sectors. [Modified after Howell et al. (2019) *Diamond Related Materials*, Vol. 96, Fig. 1, p. 209.]

Although nitrogen and boron are the most studied impurities in HPHT diamonds, several other elements such as nickel, cobalt, silicon, and germanium can be introduced during synthesis, forming vacancy-containing defect complexes (Collins and Spear 1982; Collins et al. 1990; Lawson and Kanda 1993a,b; Lawson et al. 1996; Sittas et al. 1996; Palyanov et al. 2015b, 2017). These impurities often originate from the solvent/catalyst and its additives.

Large HPHT diamond growth

The growth of extra-large diamond single crystals with high crystalline quality has been simultaneously a dream and one of the biggest challenges for diamond researchers during the last 60 years. Progress was tempered by the technological expertise required, the high associated costs, as well as underdeveloped applications for the final product. Before around 2010 large diamonds were grown using modified belt presses that include large volume HPAs with precise cooling water, hydraulic pressure suspension and in-cell temperature control. De Beers Diamond Research Laboratory grew a 34.80 ct yellow-brown Type Ib HPHT diamond in 1992 (pictured in Kanda (2005), while Sumitomo Electric Industries synthesized 10 ct colorless Type IIa diamonds (Sumiya and Tamasaku 2012; Sumiya 2014; Sumiya et al. 2015). The team at New Diamond Technology (NDT) Ltd. (headed by the present author ANK) adapted cubic series 750 and 850 presses to grow large diamond crystals, and in 2015 announcing the growth of a 32.26 ct colorless diamond that was later cut into a 10.02 ct E color gem (Deljanin et al. 2015). Continuous improvements in their growth capabilities led to increasing rough HPHT diamond sizes from 1 to 60 ct for a reasonable cost and in short time periods up to about 2 weeks. Previously the generally accepted averaged mass growth rate for good single crystals was 1–2 mg/hour, whereas the NDT team could achieve rates as high as 45 mg/hour in those record stones (Deljanin et al. 2015; D'Haenens-Johansson et al. 2015a). To date NDT has produced the largest faceted gem laboratory-grown diamonds reported, including 20.23 ct Fancy Vivid yellowish orange, 15.32 ct H color, and 10.08 ct Fancy deep blue diamonds (Poon and Wang 2016; Ardon and Eaton-Magaña 2018; Wang et al. 2022). As of the year 2020 at the facilities of Alkor-D Ltd. the author ANK has broken the 100 ct limit of single crystal rough HPHT diamonds, as shown in Figure 17. They have also produced a Guinness World Record-breaking 150.42 ct blue HPHT diamond (28.55 × 28.25 × 22.53 mm), not pictured (IGI 2022). Notably these Type IIa and IIb diamonds were grown using extremely fast average growth rates (>64 mg/hour). The limited reaction volume resulted in some deterioration of the exterior of the samples.

Figure 17. Examples of super large HPHT diamonds grown by ANK with Alkor-D Ltd. (**a**) This Type IIa "colorless" crystal weighed 94.30 ct (22.16 × 23.70 × 18.88 mm), while the (**b**) Type IIb nearly-black crystals shown here left and right weigh 109.81 ct (25.72 × 24.74 × 19.84 mm) and 115.65 ct (28.42 × 24.71 × 23.36 mm), respectively. [Photographs taken by (a) ANK and (b) Jia Xin (Jae) Liao for GIA.]

There are three factors that are extremely important when growing extra-large diamond single crystals: temperature control and reproducibility within the given in-cell isotherm pattern, high chemical stability of the hardware materials used to withstand prolonged exposure times (up to 400 hours), and the availability of a large growth volume for the solvent bath and the necessary quantity of carbon source. It is expected that increasing the maximum achievable cubic press load up to 80–90 MN and increasing the high pressure cubic cells to span 90–100 mm will make it possible to grow diamonds weighing as much as 500 ct (100 g). In addition to attracting interest from the gem industry, future applications include manufacturing single crystal diamond plates with pre-determined characteristics (e.g., high crystalline purity, charge carrier mobility, thermal conductivity or electrical conductivity), with uses as varied as substrates for CVD growth or diamond x-ray optics components (Stoupin et al. 2016; Shvyd'ko et al. 2017).

CHEMICAL VAPOR DEPOSITION OF DIAMOND

Principles of growth

This section discusses the growth of diamond by chemical vapor deposition (CVD) under conditions where diamond is not the thermodynamically stable form of carbon, as highlighted in Figure 1. The earliest reports of verifiable diamond growth under metastable conditions (low pressure) date to the 1950s to 1970s, initially in the United States, Russia, and then Japan (Angus 1994; Fedoseev 1994; Setaka 1994). A detailed summary of the history of chemical vapor deposition of diamond is provided by Angus (2014). The inclusion of hydrogen as a key reactant/solvent initiated the use of hot filaments and plasmas for creating atomic hydrogen and activating the reactants which demonstrated growth of mm-sized crystals. The mid 1980s then saw an explosion of interest in CVD diamond research forming the basis of our current understanding and diverse applications of diamond materials by groups primarily in Japan, US, and Europe (Angus 2014).

Crystalline diamond can be grown by chemical vapor deposition with a multitude of different gaseous reactant activation techniques: hot surfaces (hot filaments), plasmas (DC, RF, microwave, e.g., Fig. 3), and combustion flames. As will be presented later, there are many common chemical processes to these apparently dissimilar techniques. In the hot filament and plasma-based techniques, the reactant gases contain primarily hydrogen and a dilute amount of some hydrocarbon (generally less than 10%). The reactants are activated at pressures between ca. 1 mbar to several bar (typically between 20 and 500 mbar) to create many molecular, radical, and ionic species. Figure 18 presents a schematic of the diamond CVD process. In this schematic, the reactants are introduced into a pressure containment vessel (reactor) and the pressure is maintained with a throttle valve to the exhausting system (pump). The hot filaments (typically tungsten at 2000–2500 °C) or the plasma (gas temperature in the range of 2000–5000 °C) dissociate the reactants and initiate complex gaseous chemical reactions while the gas is transported by forced flow, convection, and/or diffusion toward the growing diamond crystals whose surface temperature is maintained at between 600 and 1200 °C. "Boundary" layers are formed just above the growth surfaces with strong temperature gradients, reactant gradients, and gas stagnation. Growth then proceeds on the surfaces of the diamond crystals. Generally, diamond seed crystals of sizes between several nm to several cm are provided to start the process, commonly referred to as "substrates." However, nucleation techniques for forming the diamond seeds from stable carbon/carbide clusters on the surface are also used (Liu and Dandy 1995).

CVD diamond materials can be polycrystalline agglomerates with a continuum of grain sizes from nm to cm, or single crystal blocks or plates. Various properties of these materials can also be tuned by controlling the grain size, defects, and impurities (Butler and Windischmann 1998; Butler and Sumant 2008).

The CVD growth of crystalline diamond is one of the most complex of any chemical vapor deposition or physical vapor deposition process (PVD) (Butler and Woodin 1993; Goodwin and Butler 1997; Butler et al. 2009a,b). Its study involves many scientific disciplines: chemical kinetics, plasma chemistry, fluid dynamics, thermal transport, surface chemistry, and possibly, diffusion processes in solids.

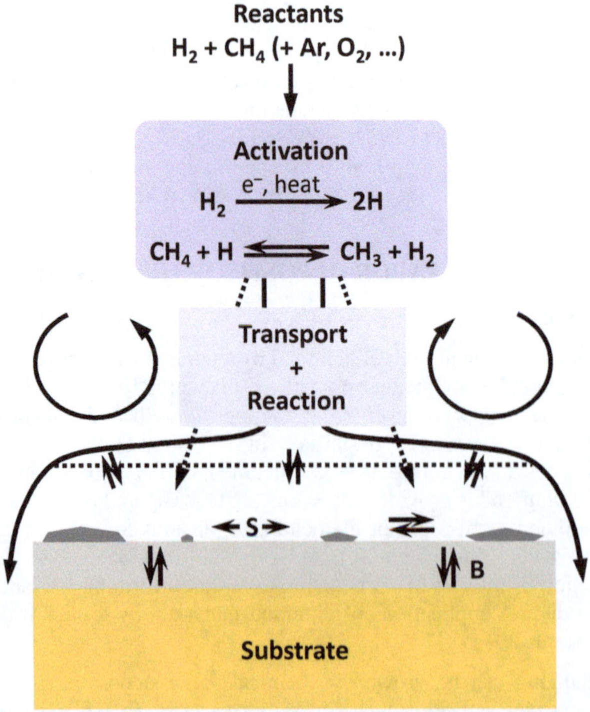

Figure 18. A schematic of the diamond CVD process. Various reactive and transport processes are shown in the gas-phase and surface (**S**). Diffusion through a boundary layer (**dashed line**), and into the crystal bulk (**B**) are also depicted. [Modified after Celii and Butler (1991), *Annual Review of Physical Chemistry*, Vol. 42, Fig. 2, p. 647.]

Surface vs. bulk temperatures

The temperature of a gas, surface, or solid is an extremely important parameter for controlling the kinetics of reactions and transport of species. While there are reports of very slow, poor-quality diamond growth by CVD at temperatures as low as ~300 °C, the optimal surface temperature is generally between 800 and 1200 °C for excellent crystallinity and reasonable growth rates. Surfaces at over 800 °C are sufficiently hot to break even the strongest of single chemical bonds that an adsorbate might make with the surface. Reactants can only bond to the surface with very strong bonds, and even those that can bond are likely to desorb back to the gas after a short time. Hence, the incorporation of species into the growing bulk will likely involve the formation of multiple chemical bonds to the surface, while desorption of reactants to the gas will dominate over the mobility or diffusion of reactants across the surface.

Bulk processes like structural rearrangements or diffusion of atoms are very unlikely at the bulk temperatures of diamond CVD (less than 1200 °C), because of the extremely high "Debye" temperature of bulk diamond, ~1800 °C. The Debye temperature is a parameter measuring the transition of the heat capacity of a solid from quantum phonon statistics to classical thermodynamics, and as such is a measure of the thermal motion of atoms in the bulk. Heating diamond to approximately its Debye temperature initiates the transition to the more thermodynamically stable form of carbon—graphite—catalyzed at defects and surfaces (Evans 1979). A consequence of this is that there is very little thermal motion in the diamond lattice during growth, with diffusion possible only for lattice vacancies and interstitials, and with little repair of structural defects.

Nucleation vs. growth

The nucleation of diamond on non-diamond substrates can happen, but is a poorly understood process. Important factors include a source of carbon to form clusters of carbon on or within the surface, bonding of the carbon cluster to the substrate to prevent desorption at the surface temperatures, and etching/rearrangement reactions within the cluster to convert to local sp^3 diamond structure. See Pehrsson et al. (1993) and Liu and Dandy (1995) for more detailed discussions. For the most part, current researchers and growers of CVD diamond who grow on non-diamond substrates will prepare the substrate for growth by treating or seeding the surface with crystalline diamond particles (Clausing et al. 1989; Inspektor et al. 1989; Philip et al. 2003; Buijnsters et al. 2005). Depending on the application, these particles can range in size from nm to mm. Then subsequent growth is homoepitaxial growth of diamond on diamond.

Diagnostics of growth environment

The important insight enabling the growth of diamond at low pressure was the recognition of the role of hydrogen in stabilizing the surfaces of diamond and in etching non-diamond carbon. This is discussed at length in the review of the early history by Angus (2014). One major advantage of CVD diamond growth over HPHT techniques, or diamond growth in the mantle of the earth, is that scientists can employ a range of in situ analytical techniques to study the process. The first in situ examination of the gas phase composition during diamond CVD with infrared absorption spectroscopy detected the conversion of the reactants (methane 0.5% in hydrogen) to C_2H_2 and CH_3 in a hot filament activated deposition chamber (Celii et al. 1988). Direct detection of ground state atomic H in the gas phase of hot filament and plasma based reactors followed (Celii and Butler 1989; Celii et al. 1990a,b). Optical emission from excited states of atomic hydrogen was also observed in plasma activated diamond CVD (Stalder and Sharpless 1990; Chu et al. 1991; Muranaka et al. 1991). Mass spectroscopy sampled in situ from a hot filament reactor detected CH_4, C_2H_2, and C_2H_4 inferred the presence of methyl (CH_3) radical and H atoms in the complex gaseous chemical processes (Harris et al. 1988; Harris and Weiner 1990; Hsu 1991). Sampling of the exhaust gases from a microwave plasma detected CH_4, C_2H_2, and C_2H_4; and CO and H_2O when O_2 was added to the reactant gases (Johnson et al. 1990; Hsu et al. 1994).

Which chemical species are important for the CVD growth of diamond? The conclusion reached by many experiments is atomic H and the CH_3 radical. Kinetic and isotopic experiments support that the bulk of the carbon incorporated into the diamond crystal is from CH_3. For an extensive review, see Goodwin and Butler (1997). The typical growth environment involves a gas pressure above 30 mbar and a relatively long gas residence time in the reactor, allowing for the complex gaseous kinetics to approach a stable equilibrium. An approximate model of the observed gas phase composition was presented and is shown in Figure 19 (Butler et al. 1989). In this model, the detailed composition of the gas phase can be approximated by the thermodynamic composition of the input reactants placed in a container at an equilibrium temperature of the plasma or hot filament source. An important consequence of the model is the implication that diamond growth is not sensitive to the specific nature of the hydrocarbon reactant, but just to the ratio of carbon to hydrogen in the reactants. While this "zero order" approximation is useful in a general sense, the gaseous chemistry is far more complex than this due to the spatial and temperature gradients in the growth environment.

The interconversion of the dilute hydrocarbon species in atomic and molecular hydrogen is shown schematically in Figure 20. In the hotter portions of the gas (in the plasma), the chemistry is driven toward the lower right of the figure, by the conversion of CH_n to C_2H_n, while in the colder portions of the reactor, near the reactor walls and the growing surface, the chemistry is driven to the upper left by atomic hydrogen reactions (C_2H_n to CH_n) (Ma et al. 2009).

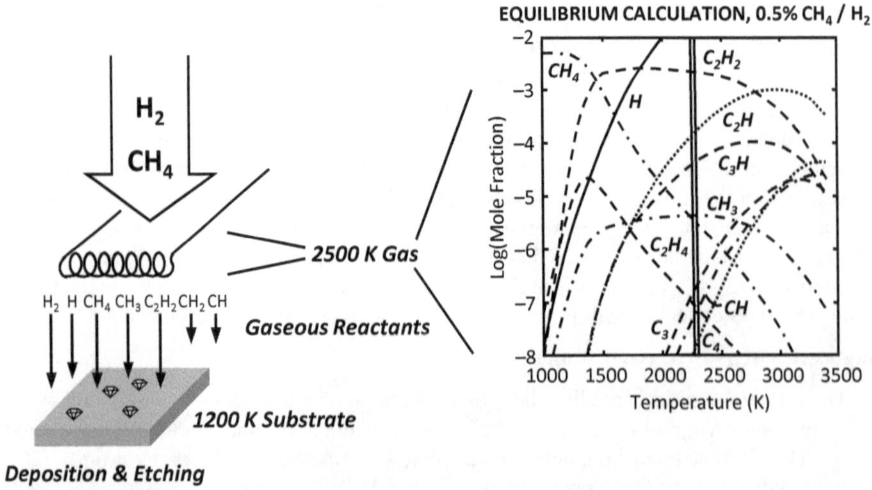

Figure 19. A simple model of the gaseous processes occurring in filament-assisted diamond CVD. On the right are the results of a thermodynamic equilibrium calculation of the mole fraction of various species present versus the equilibrium temperature. The schematic on the left summarizes the gaseous transport of species. [Modified after Butler et al. (1989), *High Temperature Science*, Vol. 27, Fig. 8, p. 193.]

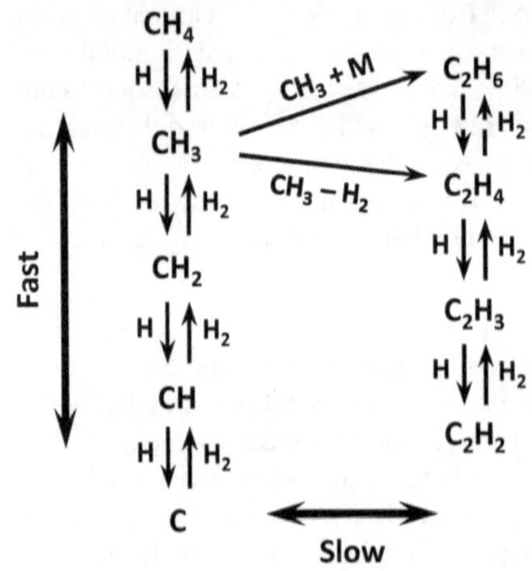

Figure 20. The principal gas phase reactions involve the rapid hydrogen transfer reactions amongst the C_1 and C_2 species, and to a lesser degree, the bimolecular hydrocarbon reactions forming C_2 and higher species. [Modified after Butler and Woodin (1993), *Philosophical Transactions: Physical Sciences and Engineering*, Vol. 342, Fig. 2, p. 211.]

The composition of the gas near the hot filaments or in the plasma (>2400 °C) is very different than the flux of gaseous species colliding with the growing diamond surface. This is because the stagnant gas at or near the growing surface has (nearly) the same temperature as the growing surface (800–1200 °C). Thermal and compositional boundary layers form with sharp gradients between the bulk gas and the surface.

Since diamond growth occurs on the surface of the diamond substrates, the critical chemical species causing growth must collide with the surface at a rate at least as fast as carbon is incorporated in diamond. Extensive experimental measurement and kinetic modeling have determined that the only species meeting this criterion in the CVD environment are H_2, H, CH_4, CH_3, and C_2H_2, and that the dominant carbon contributor is the CH_3 radical. Goodwin and Butler (1997) provide a detailed review and discussion.

The surface of the growing diamond is terminated by hydrogen, forming C-H bonds, very strong single bonds with strengths ~ 96 to 102 kcal/mole (Calvert and Pitts 1966). Temperatures above 850 °C are sufficient to slowly break these strong bonds and desorb hydrogen (Koleske et al. 1994, 1995). In ultra-high vacuum, "clean" non-hydrogenated surfaces can be produced by annealing to over 850 °C (Pate 1986; Thomas et al. 1992), but in a gas containing atomic and molecular hydrogen and hydrocarbon species, the surfaces are partially hydrogenated by the balance of desorption and reactive adsorption. The limitations on the surface temperatures for CVD diamond growth are set by the hydrogen termination of the surface: for complete hydrogen coverage there are no reactive sites, and when there are no terminating hydrogen atoms, the surface reverts to sp^2 bonding and converts to graphite.

Given the shortness of the C-C bond in diamond, the surfaces have an extremely high areal density of surface carbon atoms (1 to $3 \times 10^{15}\,cm^{-2}$). Termination of all the carbon surface bonds with any non-carbon atom except H(D) is impossible. Even H is too large to terminate all the bonds on the (100) surface. This surface undergoes a (2×1) surface reconstruction to form carbon dimer chains on the surface (Hamza et al. 1990). Growth surfaces are not necessarily flat low index planes, but contain many steps and local geometric environments. Figure 21 displays ball and stick models of steps on the (111), (110), and (100) 2×1 surfaces.

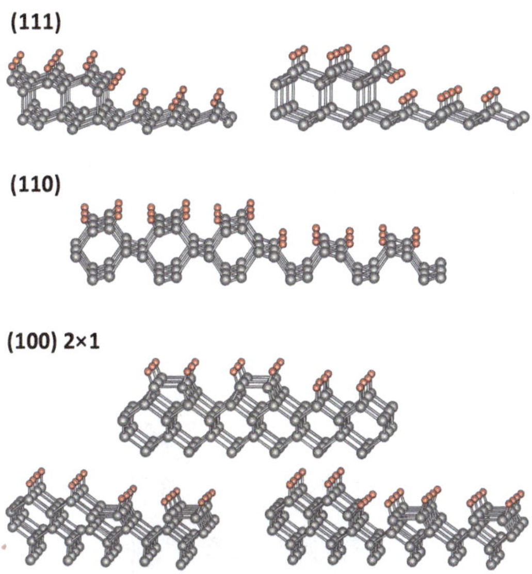

Figure 21. Idealized structures for steps on the indicated low index hydrogen-terminated diamond surfaces. The **large gray balls** represent carbon atoms and the **small red balls** represent the surface bound hydrogen atoms. [Modified after Butler et al. (2009b), *Journal of Physics: Condensed Matter*, Vol. 21, Fig. 4, p. 364201-9.]

A "standard mode" of diamond CVD has evolved over the years with many contributors and is summarized in (Butler and Woodin 1993; Goodwin and Butler 1997; Butler et al. 2009a,b). The key elements of the model are:

The diamond lattice is stabilized and prevented from rearrangement to graphitic carbon by termination with hydrogen atoms (or a similar chemical species), while the temperature is too low for spontaneous bulk rearrangement to occur (i.e., below the Debye temperature of diamond)

The gaseous activation process dissociates molecular hydrogen into atomic hydrogen which reacts with the source hydrocarbon to create a complex mixture of hydrocarbon species including reactive carbon-containing radicals. The H atoms created by the gaseous activation process also abstract hydrogen from the surface C-H bonds, thereby creating surface radical sites. These radical sites will occasionally react with gas phase carbon-containing radicals such as methyl radicals (as discussed above), resulting in a strongly bound adsorbed carbon species. Much more frequently, however, the radical sites are simply refilled by recombining with gaseous H atoms. This constant turnover of the surface-terminating species (hydrogen) drives the surface chemistry to dehydrogenate the adsorbed carbon species and to incorporate carbon into the lattice by creating multiple bonds to the lattice

The atomic hydrogen, and, to a lesser extent, other gaseous species, react with any *sp* or sp^2 carbon sites on the surface, converting them into sp^3 bonded carbon

Single crystal diamond growth

CVD epitaxial growth of single crystal begins with a single crystal diamond substrate, usually in the form of a thin plate (ca. 200 to 500 μm thick). The surfaces of the plate may be polished, laser cut, or cleaved. The most popular orientation of the upper surface is near or on-axis (100). Other orientations like (110), (111), and (113) are less frequently used. The quality and orientation of the seed crystal may be important for some applications. For example, dislocations in the seed generally propagate in the epitaxial layer, and impurities or dopants tend to incorporate preferentially on certain crystal faces and local structures (Martineau et al. 2004; Gaukroger et al. 2008; Friel et al. 2009; Tallaire et al. 2017a). An example of CVD diamond growth on a yellow Type Ib seed is shown in Figure 22, where the epitaxial growth was performed in two sequential growth runs. When viewed from the growth side to the seed, the color is dominated by that of the seed. Dark polycrystalline diamond rims readily grow on the sample edges, potentially encroaching on the growing diamond and shrinking the top surface, unless precautions are taken to minimize their formation (Mokuno et al. 2005; Nad et al. 2015, 2016). Such blocks are often cored with a laser machining system to remove the outer rim and the initial seed. Figure 23 shows the result after coring. Figures 22 and 23 display a common result, however with the proper recipe and conditions, it is possible to expand the sides of the seed crystal without the polycrystalline rim (e.g., Tallaire et al. 2017b).

Figure 22. As-grown CVD diamonds are typically thick films with dark polycrystalline diamond rims along the edges. During sample processing, the substrate is removed and the central gem-quality portion is laser cut into a cylinder or cuboid shape which can then be faceted into a gemstone. Treatments may be used to change CVD diamond color. In this example, a colorless CVD layer was grown in a two-step process to maximize thickness. The yellow color observed is from the underlying Type Ib HPHT substrate. [Modified after Nad et al. (2015) *Diamond Related Materials*, Vol 60, Fig. 16, p. 32.]

Since sizable natural and HPHT-grown crystals are rare or expensive, larger areas of quasi single crystal material are generated either by tiling multiple plates of well oriented single crystals and growing them together (Yamada et al. 2011, 2013, 2014; Tallaire et al. 2013), or by nucleating highly oriented diamonds seeds on uniquely prepared non-diamond substrates and growing them to a thickness where they appear as single crystal material with a high dislocation density (Gallheber et al. 2018).

Figure 23. The polycrystalline rim of a CVD-grown single crystal after laser coring. [Photograph by JEB.]

Twinning

Crystal twinning occurs in all forms of crystalline diamond (Kohn 1958; Varma 1970; Walmsley and Lang 1983; Angus et al. 1992). It is particularly common to CVD-grown material and material grown at very high growth rates, e.g., HPHT diamonds. It may occur at the pressures and temperatures of natural diamond or HPHT diamond synthesis by atomic rearrangement of the lattice (annealing) to relieve stress due to plastic deformation of the crystal after growth, or as part of the growth of the crystal (Lang 1979; Palyanov et al. 2015a). During growth by CVD where there is little or no motion of the atoms in the lattice, only one model has been proposed for how a twin is formed during growth (Butler and Oleynik 2008). In this model a twin can only form on a local part of the diamond surface with an exposed {111} plane by the building of a three-atom bridge between adjacent surface carbon atoms to form a four-atom island nucleating the next layer of growth. Due to rapid isomerization of the three-atom bridge at the growth temperature, the result is a nearly 50:50 chance that the four-atom island will be twinned relative to the underlying crystal structure.

Doping

In the CVD growth of diamond, intentional doping is accomplished by adding volatile compounds of the intended dopant to the gaseous reactants, for example N_2, B_2H_6, SiH_4 or PH_3, or by placing a solid compound containing the intended dopant in the reactor and exposing it to the plasma, such as solid Si or B. Both "p-type" and "n-type" electrical conduction is observed with B and P doping, respectively (Lightowlers and Collins 1976; Koizumi et al. 1998). A dopant of particular note is nitrogen, which is known to have a catalytic effect on the growth rate (Chayahara et al. 2004; Tallaire et al. 2006; Achard et al. 2007; Butler and Oleynik 2008). Nitrogen contamination can occur from gases, leaks or desorption from reactor components, but careful growth can produce high purity CVD diamonds with nitrogen concentrations below 1 ppb (parts per billion carbon atoms, 10^{14} cm^{-3}), which are attractive for electronic applications (Isberg et al. 2002; Tallaire et al. 2006; Edmonds et al. 2012). Intentional doping using N_2 can result in the incorporation of up to tens of ppm (parts per million) of nitrogen (Edmonds et al. 2012). For the growth of optically transparent material or larger samples, nitrogen-doping can be beneficial, with even low concentrations of nitrogen in the gas phase (~10 ppm) resulting in nearly a tripling of the growth rate (Tallaire et al. 2006). However, a balance must be struck as high concentrations of nitrogen introduce brown color in the material, attributed to the formation of vacancy clusters and/or nitrogen-related defects (Barnes et al. 2006; Jones et al. 2007; Mäki et al. 2007; Khan et al. 2013; Zaitsev et al. 2020). The presence of nitrogen in the growth gases also has a pronounced effect on the growth mode, transitioning from layer-by-layer growth to step-flow with macroscopic step-bunching (Yamada et al. 2016). Unlike natural diamonds, nitrogen in CVD diamonds is found primarily as isolated substitutional nitrogen (N_S), nitrogen-vacancy (NV), or nitrogen-vacancy-hydrogen (NVH) forms (Edmonds et al. 2012). Since CVD diamond is grown at low temperatures (well below its Debye temperature), is produced in relatively short time scales (~ days to months, not years), and has a comparatively low nitrogen concentration, there is little opportunity to form the more complicated nitrogen-related defects common to natural diamonds, such as the A- (nitrogen pairs), B- (N_4V^0) aggregates, or N3 (N_3V^0) centers, unless subjected to post-growth annealing at high temperatures.

Technology of growth

A source of atomic hydrogen is a critical component in the CVD growth of diamond materials. In the commercial production of CVD diamond, two general sources of atomic hydrogen are employed: plasmas and hot wires or surfaces (at over 1800 °C). Each approach has technical advantages and disadvantages, as well as economic and intellectual property concerns. Almost all growth of single crystal and high purity material is done with plasma assisted CVD, generally driven by microwave sources (primarily at 2.45 GHz or 915 MHz), although some manufacturers use lower frequencies and even DC driven plasmas. Figure 3 shows a schematic of a microwave CVD reactor, while Figure 24 displays a picture of a single seed diamond growth in a typical commercial reactor operating at 2.45 GHz. Figure 25 displays a multiple seed diamond growth run in a different CVD reactor. Hot wire (tungsten, tantalum, etc.) methods of dissociating hydrogen are used mostly for growth covering larger areas (up to 1 m^2), thin film, and polycrystalline coatings. CVD diamond growth has also been demonstrated in combustion flames and molecular beams (Hansen et al. 1988; Lee et al. 1994), but neither of these is currently used in commercial production.

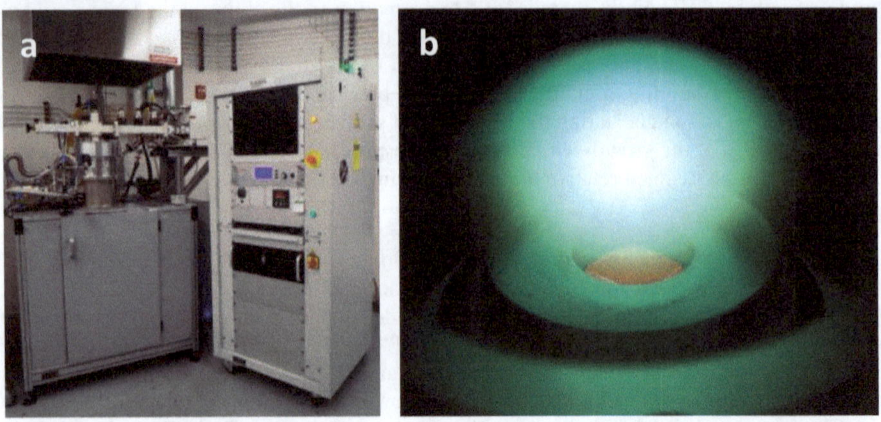

Figure 24. (**a**) A commercially available plasma tool (PLASSYS) for the growth of CVD diamond operating at 2.45 GHz. (**b**) CVD diamond synthesis showing the growing diamond in a pocket substrate holder beneath the plasma. [Photographs by Adrian Chan for GIA.]

Figure 25. Multiple single crystal seeds growing in a CVD diamond reactor. [Used by permission of Element Six (UK), Ltd. and Lightbox Jewelry, Inc.]

Applications and commercial production of CVD diamond

Many companies are involved in the commercial production of CVD diamond products. Due to the competitive nature of the business, it is difficult to determine the exact size and nature of the production factories and few detailed images are available. Figure 26 displays two examples of parts of CVD diamond materials growth facilities.

Figure 26. (**a**) A CVD diamond growth facility based on commercially available tools; and (**b**) A CVD diamond growth facility based on proprietary designed tools. [Used by permission of (**a**) ABD Diamonds Pvt. Ltd. and (**b**) Diamond Foundry, Inc.]

There are a wide range of applications for diamond materials in technology due to its many extreme properties (thermal, mechanical, optical, etc.). Figure 27 displays images of many of the applications of CVD diamond materials. A unique advantage of CVD diamond is the ability to control and monitor the growth process, to "engineer" the material to the application. This can be done by controlling the purity, grain size, thermal diffusivity, optical transparency, toughness, where and what dopants are introduced, surface termination, isotopic purity, etc. The diverse array of applications of CVD diamond span the fields of mechanical, optical, electronic, medical, quantum, electrochemical, communications, and high-power electronics, to name a few. These have been reviewed in many places such as (Prelas et al. 1997; Butler and Windischmann 1998; Wort and Balmer 2008; Balmer et al. 2009; Auciello and Sumant 2010; Nemanich et al. 2014; Awschalom et al. 2018; Yang et al. 2019; Achard et al. 2020).

Figure 27. Composite image demonstrating innovative applications derived from CVD diamonds, including ultra-precision machining, drilling, thermal management, optics, wastewater management and quantum-enabled sensing. [Used by permission of Element Six (UK) Ltd.]

The largest widely publicized initial commercial application for blocks of single crystal diamond grown by CVD methods is for gems. Figure 28 displays a 9.04 ct round brilliant cut CVD diamond, while Figure 29 shows a unique all-diamond ring fabricated from a single block of CVD diamond. As of early 2022, a 16.41 ct (G color, VVS_2 clarity) princess-cut specimen is the largest reported gem CVD diamond (Wang et al. 2022). Major players in CVD diamond gem production can be found in the United States, United Kingdom, India, Singapore and China (Linde et al. 2021).

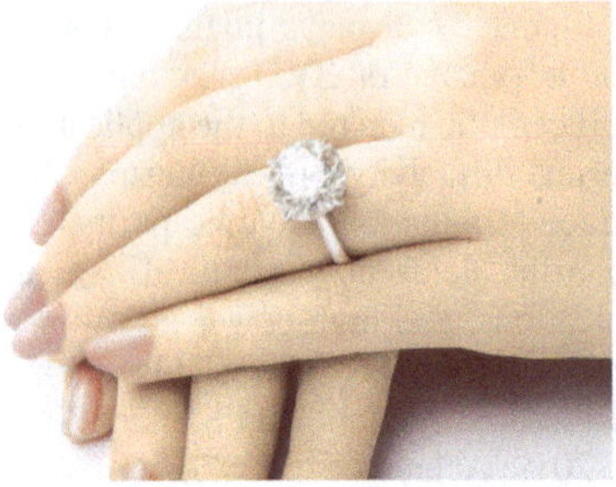

Figure 28. A ring containing a 9.04 ct round brilliant CVD gem diamond. [Used by permission of WD Lab Grown Diamonds.]

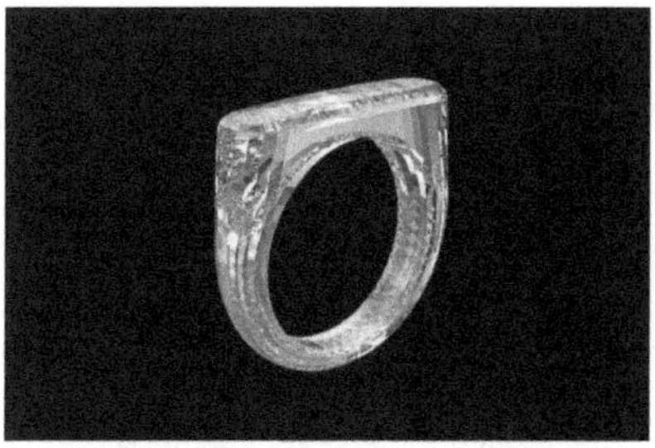

Figure 29. A unique single crystal CVD diamond ring created for a charity auction which raised $461,250 for the fight against AIDS. [Used by permission of Diamond Foundry, Inc.]

SEPARATION OF NATURAL AND LABORATORY-GROWN DIAMONDS

Background

As the quality and purity of laboratory-grown diamonds have improved over the years, visual inspection of polished diamonds—in particular if colorless or near-colorless—is rarely sufficient for their identification as to origin. Instead, conclusive identification is largely based on the presence, absence, relative concentrations and distributions of point defects in the diamond lattice, detected by advanced microscopic and spectroscopic means. The formation of these defects are a result of the growth environment and associated processes as well as any subsequent residence at elevated temperatures and/or treatment conditions, consequently providing an insight into the diamond in question's history. It is important to note that most point defects are not unique to either natural or laboratory-grown diamonds. Laboratory-grown diamonds may also be treated to further expand the types of possible defects. Thus care needs to be taken when combining several pieces of evidence to reach a conclusion on a diamond's natural or laboratory-grown origin.

Certain defects, termed color centers, absorb electromagnetic radiation in the visible range, influencing the color of the diamond material if present in sufficiently high concentrations. Both natural and laboratory-grown diamonds exist in the full rainbow spectrum of colors. Collins (1982, 1992, 1993a, 2001), Ashfold et al. (2020), and Green et al. (2022, this volume) review some of the main color-producing defects in diamonds. Defects in natural and laboratory-grown diamonds may be as-grown or produced through treatments such as HPHT or low pressure high temperature (LPHT) annealing, irradiation, or combinations thereof. In the gem trade natural diamonds are most commonly found in hues ranging from colorless to light yellow, and are typically characterized by letter color grades ranging from D-to-Z, respectively, based on the diamond color grading system invented by the GIA (King et al. 2008). Brown and gray diamonds that are more colorless than a K color reference master stone are also assigned letter grades. For color saturations equivalent to the grades K-to-Z, brown-hued diamonds are given a letter grade as well as a color word description (K–M = "Faint," N–R = "Very Light," and S–Z = "Light"), whereas gray diamonds are only given word descriptions. Diamonds with stronger yellow color saturation than the letter grade Z, or other color hues, are known as "fancy color" diamonds and are described using the fancy color grade terminology as explained by King et al. (1994). Certain gemological laboratories offer laboratory-grown diamond reports, characterizing the materials using conventional diamond grading boundaries and terminology.

In the following sections, the identification tools used by gemologist and scientists for diamond analysis will be summarized. Subsequently, the signature characteristics of both HPHT- and CVD-grown diamonds will be reviewed, highlighting the features that can be used for their identification.

Identification tools

The identification of natural and laboratory-grown diamonds relies on a range of different observations, with gemological laboratories developing confidential testing protocols to protect the gem and jewelry buying public from deceptive or unintentional misrepresentation of laboratory-grown diamonds as natural. The tools used for identification can be separated into observational and spectroscopic tools, with many of the former being the building blocks of standard gemological studies. Foremost amongst them is a microscope allowing visual inspection of samples under magnification (10–64× for most gemological applications) with reflected, bright field and dark field illumination capabilities (Renfro 2015). Coupled with crossed-polarized filter plates, birefringence patterns related to crystallographic strain in the diamond can be revealed, which may be indicative of certain growth or post-growth processes. These patterns are also sometimes referred to as anomalous double refraction patterns (Kane 1980). Howell (2012) presents a comprehensive review on strain birefringence in natural diamonds.

Defects in diamond can interact with incident light through a variety of processes which can allow their detection (Collins 1982, 1992, 1993a; Ashfold et al. 2020; Green et al. 2022, this volume). Zaitsev (2001) and the references therein catalog several hundred features associated with defects in diamond materials, with numerous defect structures identified. Table 1 provides a brief summary of some of the key features discussed in this chapter. Many defects absorb light at characteristic wavelengths in the ultraviolet (UV) to visible and infrared (IR) range, and can be detected using absorption spectroscopy. Once appropriately normalized, the integrated intensity of the absorption spectrum for a specific defect is proportional to its concentration, with defect-dependent detection limits that may be as low as a few parts per billion carbon atoms (Collins and Williams 1971; Davies 1981; Kiflawi et al. 1994; Boyd et al. 1994, 1995; Lawson et al. 1998; Howell et al. 2019). Typically, fast Fourier transform infrared (FTIR) absorption spectra are collected at room temperature. The sensitivity of UV-visible absorption is greatly improved by cooling the samples to liquid nitrogen temperature (77 K).

Defect absorptions from nitrogen- and boron-related defects in the IR range are the basis for a diamond classification system (Robertson et al. 1934), where Type I diamonds contain detectable nitrogen, in either the aggregated (Type Ia) or isolated (Type Ib) forms. The aggregated nitrogen in Type Ia diamonds may be in the A- or B-center forms (nearest neighbor nitrogen pairs or N_4V^0, respectively) with the type refined into Type IaA, IaB or IaAB (Davies 1976, 1981; Jones et al. 1992; Boyd et al. 1994, 1995). Isolated substitutional nitrogen is also referred to as the C-center (Type Ib) (Dyer et al. 1965). If nitrogen is below the detection limit, it is a Type II diamond, further divided into those that do (Type IIb) or do not (Type IIa) contain detectable boron. The IR features associated with these defects are listed in Table 1. For more on the diamond classification system and its use in diamond gemology, refer to Collins (1993a), Breeding and Shigley (2009), Ashfold et al. (2020), and Green et al. (2022, this volume).

The absorption of light may also stimulate luminescence, the emission of light from defects in the material (Collins 1993a; Eaton-Magaña and Breeding 2016; Green et al. 2022, this volume). Conventionally, diamond luminescence is evaluated by observing the sample using a shortwave (254 nm) and longwave (365 nm) UV combination gem lamp, preferably in a dark room (Luo and Breeding 2013). The luminescence color (wavelength and energy) is related to the defect species that are effectively stimulated, and their distribution through the sample may also be observed. Luminescence detected while the sample is under illumination is known as fluorescence or photoluminescence, whereas phosphorescence is the emission of light after the excitation source is turned off. The fluorescence and phosphorescence response of diamonds depend upon the defects present in the material, and as such vary for the different synthesis chemistries, potential treatments, diamond types, and colors. Further detail can be recognized using an instrument such as the DiamondView, where deep-UV illumination with energies greater than the diamond bandgap (i.e., wavelengths <225 nm, energies >5.5 eV) is coupled with video imaging under magnification. Nearly all diamonds will fluoresce under these conditions, and the resulting patterns will be sharp and surface-specific (Welbourn et al. 1996). These patterns are primarily associated with changes in the defect concentrations as the diamonds grow, whether naturally or in a laboratory, and as such are excellent features for their distinction. Importantly, although treatments may modify the fluorescent defect concentrations, and thus potentially the colors and distributions observed, the fundamental growth pattern will remain. Several viewing angles should be used when attempting to observe fluorescence patterns.

Additionally, gemological laboratories often use photoluminescence (PL) spectroscopy at 77 K to confirm the presence of specific luminescent defects in diamond, allowing sensitivities which may be of the order of parts per trillion. This is particularly useful for defect analysis of colorless diamonds, which have low concentrations of color centers. Unlike absorption spectroscopy, the relationship between the intensity of luminescent features and their concentration is complex, with the presence of certain defects (e.g., A-centers) potentially quenching the luminescence of others (e.g., Vasil'ev et al. 2004). A color center's excitation efficiency also depends upon the illuminating laser wavelength, so spectra are typically collected using a suite of different lasers with wavelengths such as 325, 488, 514, 532, 633, 830 nm. More rarely, cathodoluminescence imaging and spectroscopy, where incident electrons are used to excite luminescence, can also be used for defect analysis.

Identification of HPHT diamonds

External morphology. Gem quality HPHT diamonds are grown using the temperature gradient method, where the morphology is sensitive to growth conditions such as the temperature, pressure, and metal solvent/catalyst, dictating which crystallographic faces will dominate (Strong and Chrenko 1971; Kanda et al. 1980, 1989; Burns et al. 1999). HPHT diamond growth originates at a diamond seed crystal, simultaneously growing upwards and outwards along several crystallographic directions, dominated by growth along the cube ⟨100⟩ and octahedral ⟨111⟩ directions. Consequently HPHT diamonds typically exhibit well-developed octahedral {111} and cube {100} growth faces, along with minor dodecahedral {110} and trapezohedral {113} faces, as shown in Figures 6 (Strong and Chrenko 1971; Kanda et al. 1980, 1989; Sunagawa 1984, 1995; Shigley et al. 1986, 1987, 1992, 1993a,b; Burns et al. 1999). {115} and {117} faces have also been reported (Strong and Chrenko 1971; Rooney 1992; Shigley et al. 1992). The distinctive cuboctahedral morphology can be used to distinguish HPHT crystals from natural diamonds, which generally have octahedral, resorbed, or irregular shapes (Sunagawa 1995; Harris et al. 2022, this volume). However, occasionally HPHT diamonds can also be irregularly shaped.

Dendritic, vein-like, spiral, and lamellar or striated surface patterns can be observed on the external faces (Tolansky and Sunagawa 1959; Kanda et al. 1980; Koivula and Fryer 1984; Shigley et al. 1986, 1987, 1992, 1997; Sunagawa 1995). In particular, the dendritic surface patterns are considered to be indicative of HPHT diamonds. Unlike many natural rough diamonds these crystals do not show significant mechanical abrasion or chemical etching. Though rarely seen, triangular depressions (commonly known as "trigons") or triangular pyramids may be present on octahedral {111} faces and are thus not exclusive to natural diamonds (Kanda et al. 1980; Koivula and Fryer 1984; Shigley et al. 1987, 1993a, 1997; Rooney 1992). The surface patterns on HPHT diamonds have been attributed to phase transformations and the microstructure of the metal solvent/catalyst during the final quenching stages of synthesis (Kanda et al. 1980; Burns et al. 1999). The bottom growth surface of the crystal may bear the imprint of the seed.

Table 1. Summary of key defects observed in natural and laboratory-grown diamonds. Listed references pertain to descriptions of spectral features and associated defect structures. See main text for details about their observation in laboratory-grown diamonds. Collins (1982, 1992, 1993a), Zaitsev (2001), Ashfold et al. (2020) and Green et al. (2022, this volume) present reviews on defects in diamonds.

Feature	Spectroscopic technique	Details	References
λ^{-3} ramp	UV-visible-NIR absorption	Absorption ramp rising from ~1200 nm to 225 nm with a profile ~λ^{-3} seen in as-grown CVD diamonds. Structure still under debate. Correlates with the 360 nm band. Can be removed by high temperature annealing.	Barnes et al. 2006; Mäki et al. 2007; Jones 2009; Khan et al. 2013
270 nm 4.6 eV	UV-visible-NIR absorption	Neutral substitutional nitrogen, N_S^0. Also known as the C-center. Broad peak at 270 nm overlapping a rise in absorption from ~560 nm to the intrinsic diamond absorption edge. Observed in both natural and laboratory-grown diamonds.	Dyer et al. 1965; Chrenko et al. 1971
360 nm 3.49 eV	UV-visible-NIR absorption	Broad band seen in as-grown CVD diamonds. Can be removed by high temperature annealing. Correlates with the λ^{-3} ramp.	Martineau et al. 2004; Khan et al. 2009, 2010, 2013
389 nm 3.188 eV	PL	Possibly nitrogen-related. Seen in as-grown CVD diamonds, but can be removed by high temperature annealing. Rarely observed in HPHT and irradiated natural diamonds.	Collins and Woods 1987; Collins and Lawson 1989; Robins et al. 1989; Zaitsev 2000; Martineau et al. 2004
390 nm 3.18 eV	UV-visible-NIR absorption	Broad band associated with pink color in natural diamonds. Seen in conjunction with 550 nm band. Defect responsible is unidentified, but linked to plastic deformation.	Collins 1982; King et al. 2002; Fisher et al. 2009; Byrne et al. 2012, 2014a,b; Howell et al. 2015; Eaton-Magaña et al. 2018a
393.3 nm 3.15 eV	UV-visible-NIR absorption	Negatively charged vacancy defect, V^-, also known as ND1. Observed in some natural diamonds. Can be produced in laboratory-grown diamonds through treatment.	Dyer and Du Preez 1965; Davies 1977; Twitchen et al. 1999
415 nm 2.985 eV	UV-visible-NIR absorption, PL	N_3V^0 defect, known as N3. In absorption it is also referred to as the "Cape series." Common in natural diamonds. Can be produced in HPHT diamonds synthesized or annealed at high temperatures, and in CVD diamonds by high temperature annealing.	Davies et al. 1978; Collins 1982

Feature	Spectroscopic technique	Details	References
451–459 nm 2.70–2.75 eV	PL	Series of unidentified peaks introduced by high temperature annealing in CVD diamonds.	Martineau et al. 2004; Wang et al. 2012
468 nm 2.65 eV	PL	Also reported at 467 nm. Sometimes referred to as TR12'. Possibly related to nitrogen, interstitials, or vacancy-clusters. Observed in as-grown CVD diamonds, but can be removed by high temperature annealing.	Iakoubovskii and Adriaenssens 2000a,b; Iakoubovskii et al. 2000; Rzepka et al. 2001; Iakoubovskii and Stesmans 2002; Zaitsev et al. 2021
478 nm 2.596 eV	UV-visible-NIR absorption	A transition at the N3 (N_3V^0) defect, known as the N2 band. Part of the "Cape series" of peaks.	Davies et al. 1978; Collins 1982
484 nm 2.56 eV	UV-visible-NIR absorption, PL	Complex multiplet with four thermalized components (483.67/483.97/484.28/484.56 nm), each with a fine doublet structure. Nickel-related defect, possibly including a negatively charged substitutional ion. Seen in nickel-containing HPHT diamonds. Rare in natural diamonds.	Collins and Spear 1983; Collins et al. 1990; Lawson and Kanda 1993a,b; Pereira et al. 1994, 1995; Nazaré et al. 1995; Collins 2000
489.2 nm 2.534 eV	UV-visible-NIR absorption, PL	Known as the S2 or NE2 center. Has additional peaks at 477.8, 472.6, and 470.6 nm. Nickel ion in a divacancy position surrounded by three nitrogen atoms $(C_2–N–Ni–N_2–C)^0$ with a triclinic symmetry. Can be observed in HPHT and natural diamonds.	Nadolinny and Yelisseyev 1993; Yelisseyev and Nadolinny 1995; Yelisseyev et al. 1996, 2002; Nadolinny et al. 1997, 1998, 1999; Kanda and Watanabe 1999; Yelisseyev and Kanda 2007
494 nm 2.51 eV	UV-visible-NIR absorption	Possibly the negatively charged substitutional nickel defect, Ni_S^-. Seen in some nickel- and nitrogen-containing HPHT diamonds. Reported in some natural diamonds.	Collins and Spear 1982; Collins et al. 1990; Lawson and Kanda 1993a,b; Nazaré et al. 2001; Yelisseyev and Kanda 2007
495.9 nm 2.498 eV	UV-visible-NIR absorption, PL	N_4V^0 defect, known as H4. Seen in natural diamonds. Can be introduced by treatments in laboratory-grown diamonds.	Clark et al. 1956; de Sa and Davies 1977; Jones et al. 1994; van Wyk and Woods 1995

Feature	Spectroscopic technique	Details	References
496.7 nm 2.495 eV	UV-visible-NIR absorption, PL	Known as the S3 or NE1 defect. Nickel ion in a divacancy position surrounded by two nitrogen atoms (C_2–N–Ni–N_2–C)$^-$. Can be observed in HPHT-grown and natural diamonds.	Nadolinny and Yelisseyev 1993; Yelisseyev et al. 1996, 2006; Nadolinny et al. 1997, 1998, 1999; Yelisseyev and Kanda 2007
503.2 nm 2.463 eV	UV-visible-NIR absorption, PL	N_2V^0 defect, known as H3. Common in natural diamonds. Occasionally observed in as-grown CVD diamonds and Type Ib HPHT diamonds. Can be produced by treatment in CVD and HPHT diamonds.	Clark et al. 1956; Davies 1976; Davies et al. 1976, 1992; Mita et al. 1990; Jones et al. 1994; van Wyk and Woods 1995
503.5 nm 2.462 eV	UV-visible-NIR absorption, PL	Known as the 3H defect. Possibly the ⟨100⟩ split self-interstitial (also called the di-interstitial) in diamond. Rarely observed in natural diamonds. Can be introduced in natural and laboratory-grown diamonds by treatments.	Walker 1977; Steeds et al. 1999a,b; Iakoubovskii et al. 2001, 2003
520 nm 2.39 eV	UV-visible-NIR absorption	Broad band, sometime referred to as the 515 nm band. Possibly related to NVH^0. Seen in CVD diamonds, but be removed by high temperature annealing.	Martineau et al. 2004; Khan et al. 2009, 2010, 2013
520–580 nm 2.14–2.38 eV	PL	Unidentified series of peaks introduced by high temperature annealing in CVD diamonds.	Wang et al. 2012
523.3 nm 2.326 eV	UV-visible-NIR absorption, PL	Known as the NE3 defect. Nickel ion in a di-vacancy position surrounded by three nitrogen atoms (C_2–N–Ni–N_2–C)^0 with a monoclinic symmetry. Observed in natural and HPHT diamonds.	Nadolinny and Yelisseyev 1993; Yelisseyev et al. 1996; Nadolinny et al. 1997, 1998, 1999; Yelisseyev and Kanda 2007
533 nm 2.326 eV	PL	Also reported at 534 nm. Stress can split line into a doublet. Possibly a nitrogen-vacancy type complex. Observed in CVD diamonds, but can be removed by high temperature annealing.	Ruan et al. 1991; Burton et al. 1995; Charles et al. 2004; Martineau et al. 2004
550 nm 2.25 eV	UV-visible-NIR absorption	Broad band associated with pink color in natural diamonds. Often seen in conjunction with 390 nm band. Unidentified defect linked to plastic deformation.	Collins 1982; King et al. 2002; Fisher et al. 2009; Gaillou et al. 2010; Byrne et al. 2012, 2014a,b; Eaton-Magaña et al. 2018a, 2020

Feature	Spectroscopic technique	Details	References
575 nm 2.156 eV	UV-visible-NIR absorption, PL	The neutral nitrogen-vacancy defect, NV^0. Observed in natural and laboratory-grown diamonds.	Davies 1979; Collins and Lawson 1989; Mita 1996, Doherty et al. 2013
594.3 nm 2.086 eV	UV-visible-NIR absorption	Sometimes referred to as the 594 or 595 nm defect. Unidentified defect, possibly nitrogen-related. Can be introduced by treatment in natural and laboratory-grown diamonds.	Collins 1978, 1982; Davies and Nazaré 1980
596 nm 2.08 eV	UV-visible-NIR absorption	Unidentified defect seen in CVD diamonds. Not the same defect as that responsible for the 596/597 nm doublet in PL. Can be removed by high temperature annealing.	Wang et al. 2003; Martineau et al. 2004
596/597 nm 2.077/2.080 eV	PL	Unidentified defect seen in CVD diamonds. Not the same defect as that observed at 596 nm in absorption. Can be removed by high temperature annealing.	Wang et al. 2003; Martineau et al. 2004
637 nm 1.945 eV	UV-visible-NIR absorption, PL	Negatively charged nitrogen-vacancy defect, NV^-. Can be observed in natural and laboratory-grown diamonds.	Davies and Hamer 1976; Loubser and van Wky 1977; Mita 1996; Doherty et al. 2013
658 nm 1.883 eV	UV-visible-NIR absorption, PL	System includes peaks at 658.4 (dominant), 650.8 and 648.1 nm, as well as a vibronic band at ~638 nm in absorption (~682 nm in PL). Sometimes reported at 659 nm. Detected in nickel- and nitrogen-containing HPHT diamonds and thought to involve both elements. Has been observed in natural diamonds.	Collins and Spear 1982; Collins et al. 1990; Lawson and Kanda 1993a,b; Nazaré and Rino 1993; Nazaré et al. 1996; Yelisseyev and Kanda 2007
694 nm 1.787 eV	PL	Doublet (~692.9/691.5 nm) that has been attributed to nickel-, iron-, nitrogen- and hydrogen-related defects. Can be observed in natural and HPHT diamonds.	Nadolinny et al. 1999; Iakoubovskii and Adriaenssens 2002; Yelisseyev and Kanda 2007; Kupriyanov et al. 2016
736.7/736.9 nm 1.682 eV	UV-visible-NIR absorption, PL	The negatively charged silicon-vacancy defect, SiV^-. Common in CVD diamonds, and rare in natural and HPHT diamonds.	Clark and Dickerson 1991; Clark et al. 1995; Goss et al. 1996; Hepp et al. 2014; Rogers et al 2014; Breeze et al. 2020

Feature	Spectroscopic technique	Details	References
741 nm 2.673 eV	UV-visible-NIR absorption, PL	The neutral vacancy defect, V^0, also known as GR1. In absorption has associated transitions labeled GR2–GR8. Present in some natural diamonds. Can be introduced in laboratory-grown diamonds by treatment.	Coulson and Larkins 1971; Clark and Walker 1973; Davies 1977; Collins 1982; Lowther 1993; Twitchen et al. 1999
793.5 nm 1.563 eV	UV-visible-NIR absorption, PL	Known as the NE8 defect. A nickel atom surrounded by four nitrogen atoms. In absorption shows vibronic structure at ~700 nm. Can be observed in HPHT and natural diamonds.	Lawson and Kanda 1993b; Nadolinny et al. 1999; Yelisseyev and Kanda 2007
796.9 nm 1.56 eV	PL	Unidentified defect. Observed in some treated CVD diamonds.	Wang et al. 2010
806.4 nm 1.54 eV	PL	Unidentified defect. Observed in some treated CVD diamonds.	Wang et al. 2010
850 nm 1.46 eV	PL	Unidentified defect. Observed in some high temperature annealed CVD diamonds	Wang et al. 2012
875.5 nm 1.42 eV	PL	Unidentified defect. Observed in some high temperature annealed CVD diamonds	Wang et al. 2012
884 nm 1.40 eV	UV-visible-NIR absorption, PL	Doublet split at ~885.1/883.4 nm with an associated broad absorption band at ~685 nm. Nickel-related, possibly described by a trigonally distorted Ni^+ ion or a Ni–B complex. The dominant feature seen in nickel-containing HPHT diamonds. Occasionally observed in natural and CVD diamonds.	Collins and Spear 1983; Collins 1989; Davies et al. 1989; Collins et al. 1990; Isoya et al. 1990; Nazaré et al. 1991; Lawson and Kanda 1993a,b; Goss et al. 2004; Yelisseyev and Kanda 2007
946 nm 1.31 eV	UV-visible-NIR absorption, PL	The neutral silicon-vacancy defect, SiV^0. Only reported in CVD diamonds, but may be present in rare HPHT and naturals as the negative charge state of the defect has been observed.	D'Haenens-Johansson et al. 2010, 2011; Green et al. 2017, 2019; Rose et al. 2018; Breeze et al. 2020; Zhang et al. 2020
949 nm 1.31 eV	PL	Unidentified defect. Observed in some treated CVD diamonds.	Wang et al. 2010
986 nm 1.257 eV	UV-visible-NIR absorption, PL	N_2V^- defect, known as H2. Observed in some natural and HPHT diamonds. Can be produced in CVD diamonds by HPHT annealing.	Mita et al. 1990; Lawson et al. 1992

Feature	Spectroscopic technique	Details	References
7354 cm^{-1}	FTIR absorption	Hydrogen-related defect. Associated peaks reported at 6428, 6524, 6525, and 8753 cm^{-1}. Observed in CVD diamonds, but can be removed by high temperature annealing.	Fuchs et al. 1995a,b; Glover 2003; Charles et al. 2004; Martineau et al. 2004; Cruddace 2007
6856 cm^{-1}	FTIR absorption	Hydrogen-related defect. Observed in CVD diamonds, but can be removed by high temperature annealing.	Fuchs et al. 1995a,b; Martineau et al. 2004
5564 cm^{-1}	FTIR absorption	Hydrogen-related defect. Observed in CVD diamonds, but can be removed by high temperature annealing.	Fuchs et al. 1995a,b; Martineau et al. 2004
3323 cm^{-1}	FTIR absorption	Also reported at 3324 cm^{-1}. Attributed to a N–H–C defect. Observed in CVD diamonds, but can be removed by high temperature annealing.	Fuchs et al. 1995a,b; Charles et al. 2004; Martineau et al. 2004; Cruddace 2007; Liggins 2010; Goss et al. 2014
3123 cm^{-1}	FTIR absorption	The neutral nitrogen-vacancy-hydrogen defect, NVH^0. Common in as-grown CVD diamonds. Removed by high temperature annealing. Rare in natural diamonds, but a peak at this position has been reported in Type Ia natural diamonds with strong N_3VH^0 (3107 cm^{-1}) absorption.	Fuchs et al. 1995a,b; Cruddace 2007; Khan et al. 2013; Lai et al. 2020
3107 cm^{-1}	FTIR absorption	The N_3VH^0. Associated peak at 1405 cm^{-1}. Common in natural diamonds. Present in some HPHT diamonds that have been grown or annealed at high temperatures. Can be produced in CVD diamonds by high temperature annealing.	Woods and Collins 1983; Goss et al. 2014
~2803 cm^{-1}	FTIR absorption	The uncompensated substitutional boron defect, B_S^0. The most easily identifiable feature is a broad peak at 2803 cm^{-1} peak. Spectrum also includes bands at 2458 and 2930 cm^{-1}, as well as a rise in absorption from the 2930 cm^{-1} feature through the infrared into the red region of the visible spectrum. Can be observed in natural and laboratory-grown diamonds.	Collins and Williams 1971; Collins 1993b; Howell et al. 2019; Green et al. 2022, this volume

Feature	Spectroscopic technique	Details	References
1502 cm^{-1}	FTIR absorption	Possibly a stretching mode vibration of a C–N bond. Can be introduced by treatment in natural and laboratory-grown diamond.	Woods 1984; Collins et al. 1988
1450 cm^{-1}	FTIR absorption	The dinitrogen ⟨001⟩ split-interstitial defect, known as H1a. Can be introduced by treatment in natural and laboratory-grown diamonds	Clark et al. 1956; Collins 1980; Wood and Collins 1982; Woods 1984; Collins et al. 1988; Liggins et al. 2010
1344 cm^{-1}	FTIR absorption	The neutral substitutional nitrogen defect, N_S^0, also known as the C-center. Spectrum includes a broad band at 1130 cm^{-1}. Also produces absorption in the visible range (270 nm peak). Observed in natural and laboratory-grown diamonds, though more common in the latter.	Dyer et al. 1965; Chrenko et al. 1971; Collins 1980; Collins and Woods 1982; Kiflawi et al. 1994; Lawson et al. 1998
1332 cm^{-1}	FTIR absorption	The positively-charged substitutional nitrogen defect, N_S^+. Other defects can produce absorption at 1332 cm^{-1}, but the N_S^+ spectrum also shows bands at 950 and 1046 cm^{-1}. Can be seen in natural and laboratory-grown diamonds.	Lawson et al. 1998
~1282 cm^{-1}	FTIR absorption	Neutral nearest-neighbor substitutional nitrogen pairs, N_2^0, known as A-centers. Common in natural diamonds. Present in some as-grown yellow- or orange-hued HPHT diamonds that have been grown or annealed at high temperatures.	Davies 1976, 1981; Jones et al. 1992; Boyd et al. 1994
~1175 cm^{-1}	FTIR absorption	The N_4V^0 defect, known as the B-center. Common in natural diamonds. Present in some as-grown yellow- or orange-hued HPHT diamonds that have been grown or annealed at high temperatures.	Davies 1981; Jones et al. 1992; Boyd et al. 1994, 1995

Note: "Annealing" refers to high temperature treatment, typically done at atmospheric pressure. "Treatment" refers to some combination of annealing at low (LPHT) and/or high (HPHT) pressures, and irradiation. Some treatment sequences are disclosed in the literature, whereas others are not. Diamonds (either natural or laboratory-grown) treated at HPHT conditions should not be confused for those that are grown by HPHT methods.

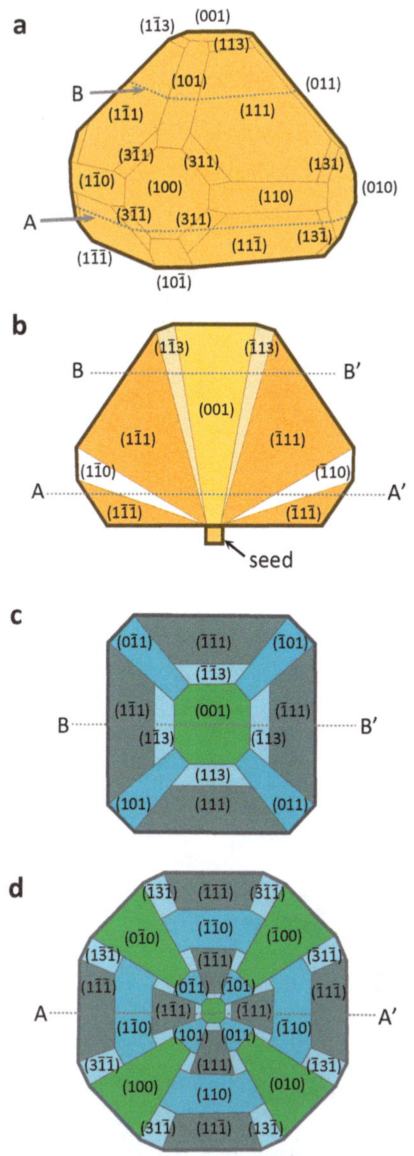

Figure 30 (**a**) A classic HPHT diamond showing the dominant octahedral {111} and cube {100} growth faces, and the minor dodecahedral {110} and trapezohedral {113} faces. (**b**) A cross-sectional view of the crystal parallel to the (110) plane, revealing the growth sectors and seed. The saturation of the yellow color correlates with the nitrogen concentration. (**c**) and (**d**) illustrate the growth patterns that could be observed by fluorescence techniques if the sample was sectioned in the (001) cube planes labelled **A** or **B**. The cross-like pattern is often referred to as a "Maltese cross." [Modified after Welbourn et al. (1996) *Gems and Gemology*, Vol. 32, Fig. 5, p. 162.]

Growth sectors. A growth sector is defined as the region of the sample that was grown with a common crystallographic growth plane (Burns et al. 1990), as illustrated in Figure 30. The incorporation efficiency for impurities differs for the growth sectors, leading to sector-dependent distributions throughout the crystal (Strong and Chrenko 1971; Burns et al. 1990, 1999; Collins et al. 1990; Satoh et al. 1990; Lawson et al. 1996). For instance, for most as-grown HPHT diamonds isolated nitrogen defect concentrations in the sectors can be arranged from highest to lowest as {111} > {100} > {113} > {110} (Burns et al. 1990). The behavior may be reversed for {100} and {111} sectors when grown at low temperatures (Satoh et al. 1990). Meanwhile for boron, another common contaminant or dopant of HPHT diamonds, the highest concentrations are found in the {111} sectors, followed by the {110} sectors (Burns et al. 1990). The distribution of nickel- and cobalt-related defects have also been found to be sector-dependent (Collins et al. 1990; Lawson et al. 1996). Importantly, as this distribution is the result of the intrinsic growth structure, it can still be perceived if the sample has been cut and polished to remove the distinctive external morphology. Furthermore, although defect species may be modified through treatment, the growth pattern will remain (Shigley et al. 1993a, 2004a; Moses et al. 1993). Consequently, the positive identification of HPHT origin is often based on techniques that can reveal the distinguishing growth pattern.

HPHT diamonds with sufficiently high concentrations of color-producing defects can show distinct color zoning (Figs. 16 and 31). For cut samples, the appearance will depend not only on the defect concentrations of the growth sectors, but also on their distribution relative to the faceting arrangements. The zoning may be more obvious by viewing samples under magnification and diffuse lighting while immersed in methylene iodide, water, or colorless alcohol to minimize reflections. There are many examples of brown-to-yellow-to-orange color zoning due to single substitutional nitrogen, and blue due to boron, documented in the literature (Crowningshield 1971; Strong and Chrenko 1971; Koivula and Fryer 1984; Shigley et al. 1986, 1987, 1992, 1993a, 2002, 2004a,b; Burns et al. 1990; Fritsch and Shigley 1993; Rooney et al. 1993; Howell et al. 2019). Mixed Type Ib + IIb HPHT diamonds may show both yellow and blue color zoning, and may have a green face-up appearance if faceted (Shigley et al. 1992, 2004a; Fritsch and Shigley 1993; Rooney et al. 1993; Hainschwang and Katrusha 2003). Rarely, the concentration of nickel-related defects in HPHT diamonds can be sufficiently high to produce green color, preferentially incorporating into the {111} growth sectors (Collins and Spear 1982; Collins et al. 1990; Johnson and Myagkaya 2017; Eaton-Magaña 2019). Red or pink hued HPHT diamonds can be created by irradiation followed by annealing of Type Ib or Ib + IaA HPHT material, producing high concentrations of nitrogen-vacancy (NV) centers responsible for the color (Moses et al. 1993; Shigley et al. 2004a). Color zoning is not exclusive to HPHT diamonds. For instance, ~8% of natural yellow diamonds have a patchy "scotch and water" color distribution (King et al. 2005), while pink diamonds often show pink lamellae (Hofer 1985; King et al. 2002; Gaillou et al. 2010; Eaton-Magaña et al. 2018a), and subtle blue parallel bands may be observed in blue diamonds (King et al. 1998). Treatments that include irradiation can also produce regional color concentrations in both natural and laboratory-grown samples (Fritsch and Shigley 1989; Wang et al. 2005a). However, sharp geometric patterns—in particular squares, octagonal "stop signs," and the "Maltese Cross" (Figs. 30c,d)—are rarely observed in natural diamonds and should be treated as suspicious.

Figure 31. Distinctive color zoning in faceted HPHT diamonds stemming from growth sector dependent impurity distributions. [Used by permission of the Gemological Institute of America, from Shigley et al. (2004b) *Gems and Gemology*, Vol. 40, Fig. 5, p. 307.]

In addition to color zoning, "color graining" may be observed at the growth sector boundaries in the interior of the diamond or on its surface (Fig. 32). As with color zoning, graining may reveal a stop-sign, Maltese Cross, funnel, or hourglass patterns (Shigley et al. 1986, 1987, 1992, 1993a, 2004a; Fritsch and Shigley 1993). Grain lines may also be found in the {100} sectors in planes aligned parallel to the cube crystal faces (Shigley et al. 1987, 1992). In contrast, grain lines in natural diamonds are usually parallel to octagonal planes (Kane 1980). For brown-to-yellow-to-orange HPHT diamonds, this distinctive graining is attributed to variations in the index of refraction from the changing nitrogen content (Burns et al. 1990; Frank et al. 1990). For a review on graining in natural diamonds see Kane (1980).

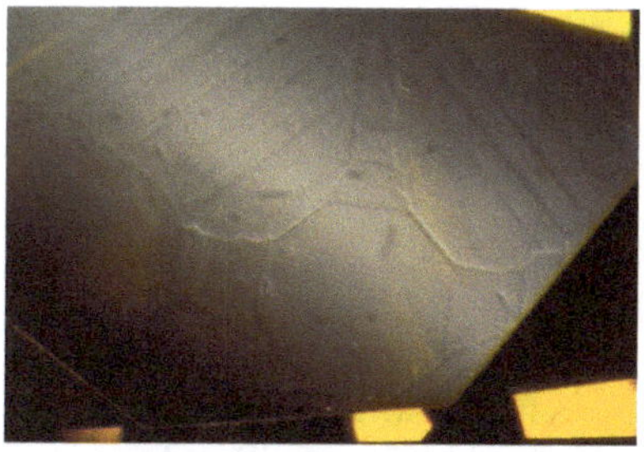

Figure 32. Octagonal stop-sign shaped graining on the polished surface of an HPHT diamond (field of view 3.5 mm). [Photomicrograph by John I. Koivula for GIA.]

By definition, near-colorless, and colorless HPHT diamonds cannot show color zoning. However, the underlying growth sectors are still present, and can be detected using techniques that are more sensitive to minute defect concentrations, such as deep UV fluorescence imaging. The application of fluorescence methods to reveal growth zoning in both colored and colorless HPHT diamonds will be detailed below in *Luminescence behavior*.

Inclusions. Like natural diamonds, HPHT diamonds are susceptible to trapping materials from their environment during growth. Traces of the metal solvent/catalyst can be encapsulated in the crystal, often having elongated, triangular tabular, or irregular shapes, as shown in Figures 33a,b (Crowningshield 1971; Strong and Chrenko 1971; Koivula and Fryer 1984; Shigley et al. 1986, 1987, 1992, 1993a,b, 1997, 2002, 2004a,b; Burns et al. 1990, 1999; Fritsch and Shigley 1993; Moses et al. 1993; Lang et al. 1995; Sumiya and Satoh 1996; Sumiya et al. 2002; D'Haenens-Johansson et al. 2014, 2015a). Using transmitted light, they are black and opaque, while reflected light reveals a gray and metallic appearance. The metal solvent/catalyst composition, growth rate, temperature, and pressure can affect the number, size, and shapes of

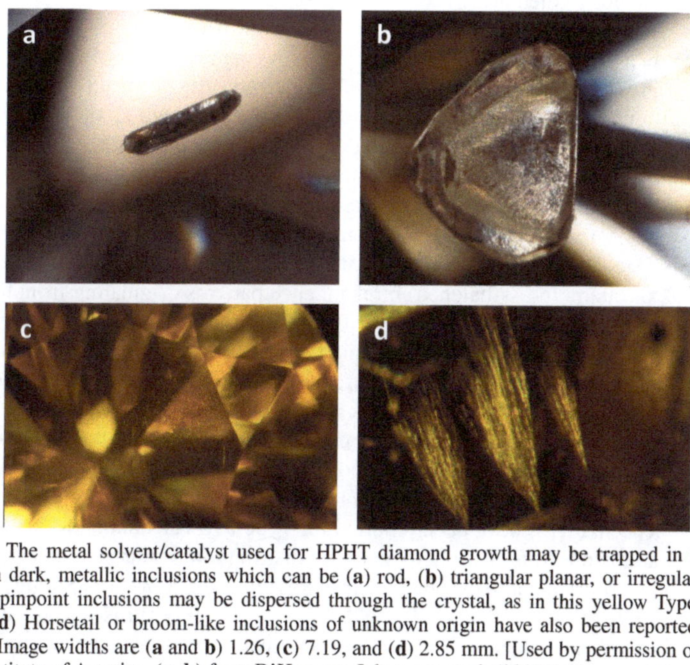

Figure 33. The metal solvent/catalyst used for HPHT diamond growth may be trapped in the crystal, resulting in dark, metallic inclusions which can be (**a**) rod, (**b**) triangular planar, or irregularly shaped. (**c**) Sparse pinpoint inclusions may be dispersed through the crystal, as in this yellow Type Ib HPHT diamond. (**d**) Horsetail or broom-like inclusions of unknown origin have also been reported in HPHT diamonds. Image widths are (**a** and **b**) 1.26, (**c**) 7.19, and (**d**) 2.85 mm. [Used by permission of the Gemological Institute of America, (**a, b**) from D'Haenens-Johansson et al. (2015a) *Gems and Gemology*, Vol. 51, Fig. 6, p. 267, (**c**) from Shigley et al. (2002) *Gems and Gemology*, Vol. 38, Fig. 6, p. 305, and (**d**) from Koivula and Fryer (1984) *Gems and Gemology*, Vol. 20, Fig. 12, p. 153.]

the inclusions. The addition of nitrogen getters to exclude nitrogen from the growing diamond by forming stable nitrides can further increase the uptake of inclusions (Sumiya and Satoh 1996; Burns et al. 1999, 2009; Sumiya et al. 2002). Due to their ferrous origin, an inclusion-containing HPHT diamond may be attracted to a strong magnet (Koivula and Fryer 1984; Rossman and Kirschvink 1984; Shigley et al. 1986, 1987, 1993a,b, 1997, 2004a,b; D'Haenens-Johansson et al. 2014, 2015a; Eaton-Magaña et al. 2017). For example, D'Haenens-Johansson et al. (2014, 2015a) found that faceted HPHT diamonds with clarities of VS_2 or poorer were weakly magnetic. Rare metallic inclusions have been observed in Type IIa, IaB, and IIb natural diamonds, but differ in appearance, with natural inclusions found adjacent to graphitic decompression cracks unlike those in the laboratory-grown samples (Sharp 1966; Rossman and Kirschvink 1984; Jacob et al. 2004; Kaminsky and Wirth 2011; Mikhail et al. 2014; Smith et al. 2016, 2018a,b).

Dilutely dispersed whitish pinpoint inclusions, as shown in Figure 33c, are frequently seen in fancy colored HPHT diamonds (Crowningshield 1971; Koivula and Fryer 1984; Shigley et al. 1986, 1987, 1993a,b, 1997, 2002, 2004a). Their sparse, even distribution is distinct from that of pinpoints in natural diamonds, which are typically concentrated in wispy clouds or associated with growth zones. Pinpoints in HPHT diamonds may consist of metallic and silicate phases (Lang et al. 1995). In diamonds grown using nitrogen getters, pinpoints have been attributed to the formation of stable carbides associated with the nitrogen getter element (Sumiya and Satoh 1996; Burns et al. 1999). The addition of copper to the growth capsule can suppress carbide formation.

Figure 33d presents another unusual inclusion seen in HPHT diamonds, described as a "horsetail" (Crowningshield 1971; Koivula and Fryer 1984). They form near the crystal surfaces and are typically lost during faceting.

Strain birefringence patterns. Examination of untreated near-colorless, colorless, and blue HPHT diamonds (which are always Type II, see the *Absorption spectroscopy and color* section) between crossed-polarized filters has been shown to be a simple method to alert the viewer of their laboratory-grown nature. Their low concentration of impurities, coupled with the even pressure that they are subjected to during growth, result in high crystalline perfection and few extended defects such as dislocations (Sumiya et al. 1997; Burns et al. 2009; Sumiya and Tamasaku 2012; Tran Thi et al. 2017). Consequently, they have low internal strain levels and only show extremely weak (black, gray or blue interference colors)—or more typically no—strain birefringence patterns, except adjacent to inclusions and cracks (Fig. 34) (Crowningshield 1971; Lang et al. 1991; Fritsch and Shigley 1993; Rooney et al. 1993; Shigley et al. 1993b; D'Haenens-Johansson 2014, 2015a). Natural diamonds commonly show mottled, banded, radial, or tatami (cross-hatched) birefringence patterns originating from the presence of defects, inclusions, and deformation events from their growth and mantle residency history (Tolansky 1966; Lang 1967; Kane 1980; Sumiya et al. 1997; Howell 2012). Natural Type II diamonds are characterized by high levels of strain and dislocations, forming distinctive tatami patterns. Conversely, natural Type I diamonds occasionally do not show clear strain patterns, so the absence of a pattern in an unknown diamond is not conclusive proof that it is HPHT-grown, and should thus be analyzed using another complementary method for confirmation (e.g., FTIR for diamond type) (Koivula and Fryer 1984). Radial, cross-hatched, and banded strain birefringence patterns have been reported for Type I HPHT diamonds due to variations in the lattice parameters from the inhomogeneous distribution of nitrogen defects, as well as surrounding both pinpoint and larger inclusions (Koivula and Fryer 1984; Shigley et al. 1986, 1987, 1992, 1993a, 2004b; Frank et al. 1990; Lang et al. 1991; Fritsch and Shigley 1993; Sumiya et al. 1997; Palyanov et al. 2010). Many note that the patterns are nevertheless weaker compared to those for natural diamonds.

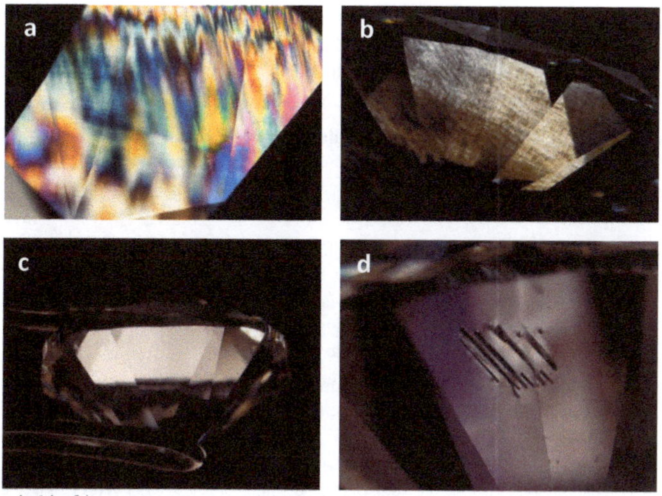

Figure 34. Strain birefringence patterns produced when viewing a diamond sample between crossed polarizing filters can help alert the operator of potential HPHT diamonds. Natural diamonds [(**a**) and (**b**)], as well as those grown by CVD methods (see Fig. 38), show a range of patterns including patchy/mottled, tatami-like/cross-hatched, or banded. On the other hand, HPHT diamonds that fall in the D–Z color range [(**c**) and (**d**)] are characterized by extremely low strain levels and the absence of patterns, except adjacent to clarity features such as inclusions. Sample details: (**a**) 4.59 ct, Fancy Intense pink, VS_2 clarity (field of view 4.77 mm), (**b**) 6.92 ct, D color, IF clarity (field of view 9.61 mm), (**c**) 2.30 ct, E color, SI_1 clarity (field of view 14.48 mm), and (**d**) 0.54 ct, E color, SI_1 clarity (field of view 3.57 mm). [Photomicrographs by UFSDJ.]

Luminescence behavior. Table 2 summarizes common fluorescence and phosphorescence colors and intensities for as-grown and treated HPHT diamonds analyzed using shortwave and longwave UV (Shigley et al. 2004b). Representative data for individual samples can be found in Crowningshield (1971), Koivula and Fryer (1984), Shigley et al. (1986, 1987, 1992, 1993a,b, 1997, 2002, 2004a), Moses et al. (1993), Rooney et al. (1993), and D'Haenens-Johansson et al. (2014, 2015a), while Eaton-Magaña et al. (2017) present statistical analysis of several thousand HPHT diamonds submitted to GIA. Most HPHT diamonds show a stronger fluorescence response to the shortwave rather than the longwave UV light produced by a gem lamp, a useful quality as it is opposite to that typical to natural diamonds (e.g., Moses et al. 1997; King et al. 1998, 2002, 2005). It is important to note that there are exceptions to this behavior for both HPHT and natural diamonds. The sector-dependent impurity incorporation in HPHT diamonds can lead to zonation in the fluorescence patterns, often taking distinctive square or cross shapes which may be perceived in fancy colored specimens (Fig. 35a). Cuboctahedral patterns are extremely rare for natural diamonds.

Colorless to near-colorless HPHT diamond are typically inert to longwave UV light, with Eaton-Magaña et al. (2017) reporting only ~2% of colorless HPHT diamonds showing fluorescence, which was orange or yellow in color. Yet among natural diamonds, 35% show longwave UV fluorescence, and of those 99% fluoresce blue due to the presence of N3 (N_3V^0) centers (Moses et al. 1997). The strong, persistent phosphorescence observed for colorless to near-colorless HPHT Type IIa and IIb diamonds is also particularly useful for their detection due to the rarity of phosphorescence in similarly colored natural diamonds (Fig. 35b) (Crowningshield 1971; Koivula and Fryer 1984; Rooney et al. 1993; Shigley et al. 1993b, 1997; D'Haenens-Johansson et al. 2014, 2015a). The phosphorescence, associated with boron (Watanabe et al. 1997), indicates the presence of this element even in nominally Type IIa HPHT diamonds. This behavior is the basis for certain diamond screening equipment used by gemologists. However, this useful property can be removed by irradiation treatments (Robinson 2018).

Table 2. Fluorescence and phosphorescence behavior of as-grown and treated HPHT diamonds stimulated by longwave (LWUV, 365 nm) and shortwave (SWUV, 254 nm) ultraviolet light from a combination gem lamp. [Adapted from the fold-out chart accompanying Shigley et al. (2004b) *Gems and Gemology*, Vol. 40, p. 303–313.]

HPHT diamond	LWUV fluorescence	SWUV fluorescence	Fluorescence intensities	Phosphorescence to SWUV
yellow/brown Ib (+ IIa), as-grown	usually inert	weak to moderate green-yellow, yellow-green, green, or inert	typically SWUV > LWUV	occasionally weak yellow or greenish yellow
green Ib + IIb, as-grown	usually inert or weak orange	weak to moderate yellow, green-yellow, or orange	typically SWUV > LWUV	weak to moderate yellow, often persistent
blue IIb (+ IIa), as-grown	usually inert or weak orange	weak to moderate green, yellow, or orange	typically SWUV > LWUV	moderate to strong yellow or blue, often persistent
colorless IIa (+ IIb*), as-grown	usually inert	weak to moderate green, yellow, blue, or orange	typically SWUV > LWUV	weak to moderate blue, yellow, or green-yellow, often persistent
yellow/brown IaA +Ib, HPHT annealed	moderate to strong green, yellow-green, green-yellow, yellow, or orange	moderate to strong green, yellow-green, green-yellow, yellow, or orange	SWUV > LWUV, SWUV = LWUV, LWUV > SWUV	occasionally weak to moderate yellow or yellow-green, often persistent
pink Ib, irradiated + annealed	moderate to strong orange or orange-red	weak to moderate orange	often LWUV > SWUV	occasionally weak orange or orange-red
red/purple Ib + IaA, irradiated + annealed	inert or weak to moderate orange or orange-red	weak to moderate orange or orange-red	SWUV > LWUV, SWUV = LWUV, LWUV > SWUV	weak green
green IIa or Ib, irradiated	very weak to weak red-orange or orange-red	weak to moderate green-yellow, yellow-green, green, or inert	SWUV > LWUV, SWUV = LWUV, LWUV > SWUV	rarely observed

Note: * Colorless HPHT diamonds can contain low concentrations of boron (D'Haenens-Johansson et al. 2014, 2015a).

Inspection under deep-UV (wavelength <225 nm) can reveal further detail in the growth patterns, stimulating sufficient contrast even in near-colorless to colorless Type IIa or weak Type IIb HPHT diamonds such as those shown in Figure 36 (Welbourn et al. 1996; Shigley et al. 1997, 2002, 2004b; D'Haenens-Johansson et al. 2014, 2015a; Eaton-Magaña et al. 2017). The majority of natural diamonds show octahedral, or occasionally cubic, growth patterns which may also reveal a complex sequence of dissolution and growth events (Welbourn et al. 1996; Harris et al. 2022, this volume). In contrast, cuboctahedral patterns such as those seen for HPHT diamonds are unusual. Growth zones are rarely detected for natural Type II diamonds, whose fluorescence patterns are instead dominated by polygonized dislocation networks (Fig. 36d) and can thus be clearly separated from HPHT diamonds (Sumida and Lang 1981; Martineau et al. 2004). The fluorescence colors depend upon the defect compositions of individual samples. When considering colorless and near-colorless natural and HPHT-grown Type II diamonds it is common for naturals to show clear blue "Band A" fluorescence associated with dislocations or N3 centers (Dean 1965; Ruan et al. 1992; Takeuchi et al. 2001; Martineau et al. 2004), whereas HPHT diamonds show blue-green or green-blue fluorescence colors (D'Haenens-Johansson et al. 2014, 2015a). Furthermore, deep-UV light stimulates strong, long-lived green-blue phosphorescence centered about 500 nm and may also show yellow

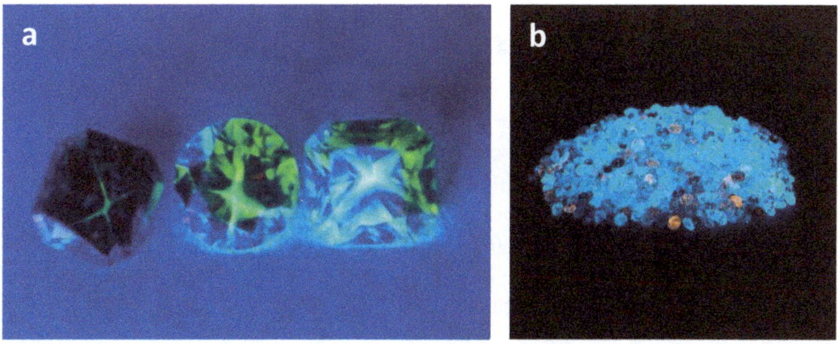

Figure 35. Analysis of HPHT diamonds under UV illumination may reveal their laboratory-grown origin. (**a**) These yellow diamonds display characteristic cross- and square-shaped green fluorescence patterns under longwave UV, indicative of HPHT growth. The fluorescence colors in HPHT diamonds may include green, yellow-green, green-yellow, yellow, orange, or orange-red, depending on the diamond type and treatment history (Table 2). It is typical for them to show a stronger response to shortwave (254 nm) rather than longwave (365 nm) excitation. For comparison, 99% of natural fluorescing diamonds emit blue light and most have a stronger response to longwave UV. (b) Colorless and blue HPHT diamonds display long-lived phosphorescence to shortwave UV due to the presence of boron impurities, a feature that is seldom seen for natural diamonds. [Used by permission of the Gemological Institute of America. (**a**) Kammerling et al. (1995) *Gems and Gemology*, Vol. 31, Fig. 7, p. 123. (**b**) Persaud et al. (2018) *Gems and Gemology*, Vol. 54, Fig. 10, p. 209.]

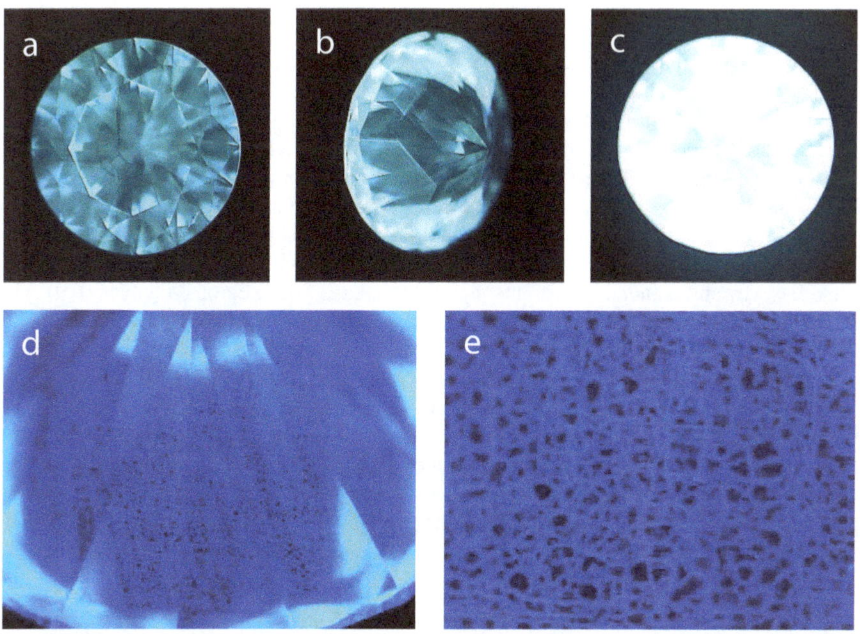

Figure 36. Cuboctahedral growth patterns revealed using a DiamondView deep-UV imaging system for a 0.41 ct G color HPHT diamond. (**a**) and (**b**) are fluorescence images collected under illumination, whereas (**c**) demonstrates the strong, long-lived phosphorescence that is characteristic of untreated Type IIa (generally contaminated with traces of boron) or IIb HPHT diamonds. In contrast, natural Type IIa or IIb diamonds rarely show growth zones, with fluorescence patterns dominated by polygonized dislocation networks as shown in (**d**) and (**e**) for a 2.09 ct D color diamond. Phosphorescence observed for natural Type IIb diamonds typically decay more rapidly than for HPHT diamonds. [Images (**a–c**) used by permission of the Gemological Institute of America, D'Haenens-Johansson et al. (2014) *Gems and Gemology*, Vol. 50, Fig. 4, p. 36. Images (**d**) and (**e**) by UFSDJ.]

emission at 575 nm (Watanabe et al. 1997; Eaton-Magaña et al. 2008; Eaton-Magaña and Lu 2011; D'Haenens-Johansson et al. 2014, 2015a). Under deep-UV, the overall phosphorescence of natural Type IIb diamonds may appear blue, gray, violet, or red due to emission bands centered about 500 nm and/or 660 nm, decaying more rapidly than that of laboratory-grown diamonds (Eaton-Magaña et al. 2008; Eaton-Magaña and Lu 2011).

Largely superseded by the comparatively user-friendly DiamondView deep-UV imaging system, cathodoluminescence has also been shown to be a powerful technique for detailed analysis of HPHT diamond growth patterns and their luminescent defects (Woods and Lang 1975; Shigley et al. 1987, 1992, 1993a,b; Collins et al. 1990; Ponahlo 1992; Fritsch and Shigley 1993; Rooney et al. 1993; Burns et al. 2009).

Absorption spectroscopy and color. HPHT diamonds belonging to several classification types can be produced (Table 2). The different impurity uptakes of growth sectors mean that oftentimes a single crystal will contain minor sectors belonging to different diamond types. Unless steps are taken to minimize nitrogen incorporation standard HPHT synthesis results in predominantly Type Ib material, similar to that mass-produced for diamond grit (Dyer et al. 1965; Chrenko et al. 1971; Crowningshield 1971; Strong and Chrenko 1971; Koivula and Fryer 1984; Shigley et al. 1986, 1987, 1992, 1993a, 2002, 2004a,b; Burns et al. 1990; Fritsch and Shigley 1993; Rooney et al. 1993). They may have yellow, to greenish or orangy or brownish yellow, to orange-yellow color. Type Ib diamonds contain single substitutional nitrogen dispersed through the crystal lattice. This color center produces absorption both in the IR and visible regions. IR absorption spectra show a peak at 1344 cm^{-1} and a broad band at 1130 cm^{-1} (Dyer et al. 1965; Chrenko et al. 1971; Collins 1980; Collins and Woods 1982; Kiflawi et al. 1994; Lawson et al. 1998), while UV-visible spectra show a rise in absorption from ~560 nm toward the intrinsic diamond absorption edge at 225 nm, with a broad peak at 270 nm, responsible for the yellow-to-orange-to-brown color (e.g., Fig. 12) (Dyer et al. 1965; Chrenko et al. 1971; Collins 1982, 1993a, 2001). Nickel-related defects introduced by the metal solvent/catalyst may produce additional absorption features at 494, 658, 793.5 (with vibronic structure at ~700 nm), and a doublet about 884 nm (broad band at ~685 nm), imparting a green hue (Collins and Spear 1982; Shigley et al. 1987, 1993a, 2004a; Collins et al. 1990; Lawson and Kanda 1993a,b; Yelisseyev and Nadolinny 1995; Yelisseyev et al. 1996, 2002; Yelisseyev and Kanda 2007; Johnson and Myagkaya 2017; Eaton-Magaña 2019). These peaks are not exclusive to HPHT diamonds, but may indicate a laboratory-grown diamond that needs further inspection (e.g., Chalain 2003; Lang et al. 2004; Wang and Moses 2007; Wang et al. 2007a; Breeding et al. 2018). Figure 37a illustrates a representative UV-visible absorption spectrum for a Type Ib yellow HPHT diamond. In nature, Type Ib diamonds are very rare, accounting for ~0.1% of diamonds, and often contain natural needle-like inclusions to help in their identification (Dyer et al. 1965; Shigley et al. 1986; Breeding and Shigley 2009). Most yellow-hued natural diamonds are Type Ia and are instead colored by the "Cape series" of peaks that include the N3 (N_3V^0) zero phonon line at 415 nm, vibronic structure to shorter wavelengths, as well as the N2 transition at the same defect at 478 nm (Davies et al. 1978; Collins 1982, 2001; King et al. 2005; Green et al. 2022, this volume).

Isolated nitrogen atoms may aggregate to A-centers (nitrogen pairs) if synthesis times and temperatures are elevated, resulting in Type IaA material, and may also progress further to form B-centers (N_4V^0) (Kanda et al. 1990; Vins et al. 1991; Kanda and Yamaoka 1993; Shigley et al. 1993a, 2002; Babich et al. 2000, 2004; Yelisseyev et al. 2002). HPHT or LPHT annealing of as-grown Type Ib diamond is an alternative route for nitrogen aggregation (Chrenko et al. 1977; Evans et al. 1981; Evans and Qi 1982; Shigley et al. 1993a; Kiflawi et al. 1994, 1998; Kim et al. 2011; Kazuchits et al. 2016; Chen et al. 2018). For both the as-grown and treated cases, a portion of the isolated nitrogen centers will remain (e.g., they are mixed Type Ib + IaA) as the nitrogen aggregates cycle through formation and dissociation (Brozel et al. 1978).

A- and B-centers show characteristic IR spectra with absorption features at 1282 and 1175 cm^{-1}, respectively (Davies 1981; Boyd et al. 1994, 1995). They do not produce diamond color as they do not absorb light in the visible range. However, the isolated nitrogen present will contribute, resulting in similarly hued material as the Type Ib HPHT diamonds (Fig. 37a) (Shigley et al. 1993a, 2002; Yelisseyev et al. 2002; Kim et al. 2011; Kazuchits et al. 2016; Chen et al. 2018). Unlike natural diamonds, there are no Type Ia colorless or near-colorless HPHT diamonds. This is significant as ~98% of natural diamonds are Type Ia (Dyer et al. 1965), making diamond typing a useful tool for screening out colorless to near-colorless HPHT diamonds. Nitrogen-containing HPHT diamonds that have been annealed or grown at temperatures sufficiently high to induce nitrogen aggregation can also form the N3 and N_3VH^0 (3107 cm^{-1}) centers (Chrenko et al. 1977; Evans and Qi, 1982; Kiflawi et al. 1996; Goss et al. 2014). These centers are common in natural Type Ia diamonds, and occasionally observed in Type IIa diamonds.

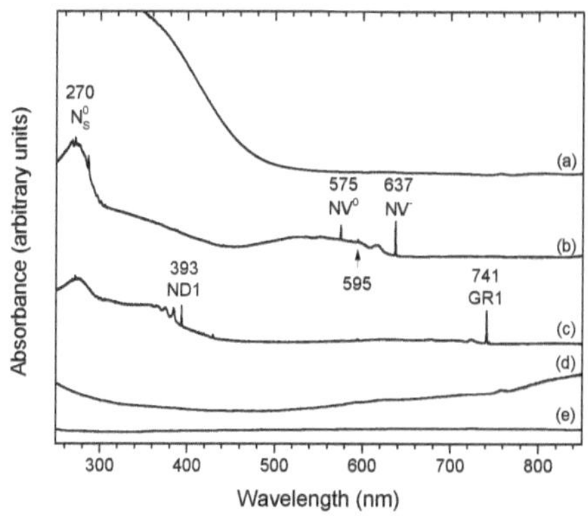

Figure 37. UV-visible-NIR absorption spectra collected at 77 K for HPHT diamonds reveal the defects responsible for their color. (**a**) This 1.25 ct Fancy Vivid orange yellow diamond is colored by isolated nitrogen, producing strong absorption at wavelengths below 350 nm. (**b**) This 0.30 ct Fancy Vivid purplish pink diamond owes its color from nitrogen-vacancy centers that were produced by irradiation and annealing of Type Ib diamond. The spectrum includes isolated nitrogen absorption at 270 nm, as well as the treatment-related 595 nm center and neutral vacancy (GR1, too weak to observed at this scale). (**c**) HPHT diamonds that are irradiated to a blue-to-green color, as this 0.41 ct Fancy bluish green sample, contain high concentrations of vacancies in the neutral (GR1) and negatively charged (ND1) states. This sample also has detectable isolated nitrogen. (**d**) Boron in this 0.55 ct Fancy Intense blue HPHT diamond results in featureless absorption from the infrared to red region of the spectrum. (**e**) Colorless HPHT diamonds like this D color 0.71 ct have defect concentrations that are too low to produce detectable absorption in the visible range.

Type Ib and Ib + IaA HPHT diamonds are candidates for irradiation and annealing to form pink/red to orangy or brownish pink/red to purplish pink or purple material, as show in Figure 15 (Moses et al. 1993; Shigley et al. 2004a,b). This treatment sequence combines vacancies produced by the irradiation with the resident isolated nitrogen centers to form nitrogen-vacancy (NV) centers. NV centers in the neutral and negative charge states absorb in the visible, with zero phonon lines at 575 and 637 nm, respectively (Collins 1982, 2001). Due to the high isolated nitrogen content in Type Ib and Ib + IaA, which acts as an electron donor, a majority of the NV centers will be negatively charged, resulting in absorption from its associated vibronic structure between 500 and 600 nm, producing the pink-to-red hued color component (Fig. 37b). This, combined with the yellow-to-brown component of the isolated nitrogen absorption produces the characteristic color range of these treated HPHT diamonds. The treatment may also introduce features including GR1 (neutral vacancy, V^0, 741 nm), H3 (N_2V^0, 503.2 nm), and 595 nm peaks as tabulated by Moses et al. (1993) and Shigley et al. (2004a). In the infrared range, absorption from the dinitrogen ⟨001⟩ split-interstitial H1a (1450 cm^{-1}) and a peak at 1502 cm^{-1}, both produced through irradiation and annealing treatment, may be detected (Clark et al. 1956; Collins 1980; Woods and Collins 1982; Woods 1984; Collins et al. 1988; Liggins 2010; Liggins et al. 2010). Increased control during growth and treatment can also be used produce weak Type Ib (nominally Type IIa) pink HPHT diamonds (Eaton-Magaña et al. 2017). The vast majority of similarly hued natural diamonds are colored by unidentified broad absorption bands at 390 and 550 nm, an underlying brown absorption continuum attributed to vacancy clusters, as well as possibly N3 (Collins 1982, 2001; King et al. 2002; Fisher et al. 2009; Gaillou et al. 2010; Byrne et al. 2012, 2014a,b; Howell et al. 2015; Eaton-Magaña et al. 2018a, 2020; Green et al. 2022, this volume), making the detection of NV centers suspicious. Untreated natural Type IIa pink diamonds colored by NV centers, known as "Golconda" diamonds, only account for ~0.6% of natural pink diamonds (Fritsch 1998; Eaton- Magaña et al. 2018a). However, note that NV centers can also be responsible for pink color in treated natural and CVD diamonds (Collins 1982, 2001; Wang et al. 2005a, 2010; Eaton-Magaña and Shigley 2016).

Blue or grayish blue HPHT diamonds containing sufficient uncompensated boron (i.e., an excess of boron compared to isolated nitrogen) to be classified as Type IIb show absorption features at ~2458, 2803, and 2930 cm^{-1}, with the latter feature having a tail which extends through the infrared into the red region of the visible spectrum (Collins and Williams 1971; Strong and Chrenko 1971; Koivula and Fryer 1984; Collins 1993b; Rooney et al. 1993; Shigley et al. 2004a,b; Howell et al. 2019; Green et al. 2022, this volume). Boron readily incorporates into the growing diamond lattice, and may be added intentionally to the growth capsule. Boron contamination may originate from the carbon source material or the metal solvent/catalyst (Sumiya and Satoh 1996). Colorless and near-colorless HPHT diamonds can be Type IIa or IIa + IIb, where the uncompensated boron concentration is insufficient to produce appreciable color (Strong and Chrenko 1971; Koivula and Fryer 1984; Rooney et al. 1993; Shigley et al. 1993b, 1997, 2004b; Sumiya and Satoh 1996; D'Haenens-Johansson et al. 2014, 2015a). Pure Type IIa diamonds only show intrinsic diamond absorption in the infrared region, and are transparent through the visible to UV range down to the diamond absorption edge at 225 nm. Figure 37 shows a representative UV-visible absorption spectrum for both blue (d) and colorless (e) HPHT diamond, demonstrating an absence of sharp features. Their absorption spectra are often indistinguishable to those of natural Type IIa and IIb diamonds, highly valuable groups that account for <1.3% and <0.02% of natural gem diamonds, respectively (King et al. 1998; Smith et al. 2016, 2018a; Eaton-Magaña et al. 2018a).

Samples have also been grown containing both Type IIb and Ib sectors (Shigley et al. 1992, 2004a,b; Fritsch and Shigley 1993; Rooney et al. 1993; Hainschwang and Katrusha 2003). When faceted they may have a yellowish green to green to bluish green appearance due to the absorption contributions from boron and isolated nitrogen (Fig. 37). The authors are not aware of any reports of mixed Type Ib + IIb natural diamonds, in agreement with Shigley et al. (1992).

Type IIa and weak Ib HPHT diamonds may be irradiated to produce vacancies, introducing GR1 (V^0) absorption with a zero phonon line at 741 nm and vibronic structure in the red, and may also include sharp irradiation-related features such as ND1 (V^-, 393.5 nm), GR2–GR8 peaks (412–430 nm), and the 595 nm peak (Fig. 37c) (Davies 1977; Shigley et al. 2004a,b). The treated diamond will have a yellowish green to green to bluish green to blue color, depending on the strength of nitrogen-related absorption present in the as-grown material and the ratio of GR1 to ND1 defects. The GR1 center is also the primary defect responsible for green color in both natural and treated diamonds, complicating their distinction (Collins 1982, 2001; Shigley and Fritsch 1990; Breeding et al. 2018; Green et al. 2022, this volume). Natural diamonds colored by GR1 may be Type IIa or Ia. For a review on green color in natural diamonds, including spectroscopic and gemological features such as irradiation spots or coatings, see Breeding et al. (2018).

Photoluminescence spectroscopy. The increased detection sensitivity provided by PL is particularly useful for probing defects in colorless to near-colorless Type IIa or IIb diamonds. Nevertheless, PL spectra for such HPHT diamonds typically show very few, and oftentimes weak, emission peaks (D'Haenens-Johansson et al. 2014, 2015a). The most prominent features are nickel-related, such as the 884 nm doublet and a 4-peak multiplet at about 484 nm (Collins and Spear 1982, 1983; Davies et al. 1989; Collins et al. 1990; Nazaré et al. 1991, 1995; Lawson and Kanda 1993a,b; Collins 2000; Yelisseyev and Kanda 2007; D'Haenens-Johansson et al. 2014, 2015a), likely due to nickel contamination from the metal solvent/catalyst. Occasionally doublet emission at 736.7/736.9 nm from the negatively charged silicon-vacancy center (SiV^-) is detected (Clark et al. 1995, D'Haenens-Johansson et al. 2014, 2015a), a defect more commonly associated with CVD diamonds (Wang et al. 2003, 2005b, 2007b, 2012; Martineau et al. 2004). These centers are seldom observed for natural diamonds, and are thus useful indicators of HPHT synthesis (Chalain 2003; Noble et al. 1998; Wang and Moses 2007; Wang et al. 2007a; Breeding and Wang 2008).

Absorption spectroscopy, coupled with gemological observations and fluorescence behavior provide sufficient information to support the identification of fancy colored HPHT diamonds. Many of the features that would be observed using PL can be detected in absorption. Consequently, there are not many detailed reports presenting their PL spectra for this purpose (Shigley et al. 2002, 2004a; Kennedy and Johnson 2016; Eaton-Magaña et al. 2017). Most notable are peaks associated with nickel-related defects (484 nm; S2/NE2, 489.2 nm; S3/NE1, 496.7 nm; NE3, 523.3 nm; 658 nm; 694 nm; NE8, 793.5 nm; 884 nm), nitrogen-related defects ($H3/N_2V^0$, 503.2 nm; NV^0, 575 nm; NV^-, 637 nm; $H2/N_2V^-$, 986 nm), or the neutral vacancy defect GR1 (741 nm) if the sample had undergone irradiation treatment (Yelisseyev et al. 1996, 2002, 2003, 2006; Kanda and Watanabe 1999; Nadolinny et al. 1999; Yelisseyev and Kanda 2007). Peak details are summarized in Table 1.

Identification of CVD diamonds

External morphology. As-grown uncut CVD diamonds can be readily identified due to their unique appearance, as shown in Figure 22. CVD diamond morphology depends upon the orientation of the underlying seed and growth conditions (Silva et al. 2006, 2008, 2009; Brinza et al. 2008; Tallaire et al. 2013). Gem-quality material is usually synthesized using (100)-oriented substrates, consequently growing as cubic (001)-oriented thick films with (100) and (010) edges. Minor dodecahedral {110} faces may form along the edges, as well as octahedral {111} and trapezohedral {113} faces on the crystal corners. The {111} faces are undesirable as they are susceptible to twinning, resulting in stress buildup and crack development. The {110} faces have also been linked to higher internal stress levels and sample breakage (Tallaire et al. 2013). Uncut CVD diamonds may also show a dark polycrystalline diamond rim along the edges (Mokuno et al. 2005; Nad et al. 2015, 2016). The top surface topography of a CVD crystal is affected by many factors including the substrate orientation, misalignment angle, and crystalline quality, gas composition, and temperature, and may show macro-bunching steps that flow across the surface and/or hillocks (Janssen et al. 1992; de Theije et al. 2000; Takeuchi et al. 2000; Wang et al. 2005b; Achard et al. 2007; Martineau et al. 2009).

Visual features. CVD diamonds generally obtain high clarities. They may contain small, black, irregularly shaped amorphous carbon or disordered graphitic inclusions or pinpoints (Wang et al. 2003, 2005b, 2007b, 2012; Martineau et al. 2004; Crisci et al. 2011). Sometimes these inclusions are cloud-like or lie in planes oriented parallel to the substrate. In samples that have been HPHT treated for decolorization, inclusions may be surrounded by radial fractures stemming from high levels of localized strain (Wang et al. 2012). Inclusions in CVD diamonds are unfortunately not diagnostic since they have a similar appearance to those in some natural diamonds.

Color concentration is generally even in fancy colored pink or brown CVD diamonds, though fluctuations in growth parameters or interruptions may lead to colored bands aligned parallel to the substrate plane (Wang et al. 2003, 2005b, 2010; Martineau et al. 2004). Graining associated with threading dislocations may be apparent (Wang et al. 2007b, 2010).

Strain birefringence patterns. Threading dislocations aligned approximately perpendicular to the growing surface are known to nucleate at or near the interface between the substrate and the CVD diamonds, often forming bundles (Fujita et al. 2006; Gaukroger et al. 2008; Martineau et al. 2009; Pinto and Jones 2009). They may lead to an anisotropic strain birefringence, most easily observed parallel to the growth direction in unfaceted samples (Martineau et al. 2004). Irregular, banded, and cross-hatched (tatami-like) birefringence patterns have been seen for CVD diamonds (Fig. 38), with interference colors ranging from subdued grays and blues to vivid reds and greens (Wang et al. 2003, 2005b, 2007b, 2010, 2012). Localized strain fields may be found surrounding inclusions (Wang et al. 2003, 2005b, 2010, 2012; Pinto and Jones 2009; Crisci et al. 2011). Unlike many HPHT diamonds, strain birefringence patterns in CVD diamonds are generally difficult to distinguish from those in natural samples.

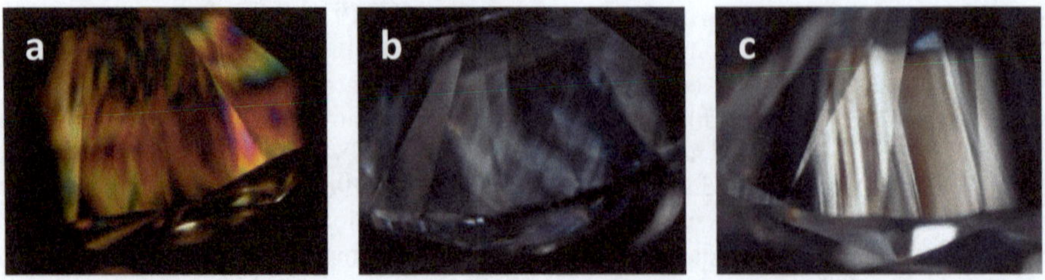

Figure 38. Strain birefringence patterns for faceted CVD diamonds can be (**a**) patchy, (**b**) tatami-like or (**c**) banded, making them difficult to separate from natural diamonds (e.g., Fig. 34a,b). These samples are characterized as follows: (**a**) 0.28 ct, H color, VS_2 clarity (field of view 3.10 mm); (**b**) 0.47 ct, G color, VVS_2 clarity (field of view 4.33 mm); and (**c**) 0.70 ct, I color, VVS_2 clarity (field of view 5.11 mm). [Photomicrographs by Wuyi Wang for GIA.]

Luminescence behavior. Most colorless and near-colorless CVD diamonds are inert to longwave UV light. Analysis of CVD diamonds submitted to GIA between 2003 and 2016 suggests only approximately 7% of samples show very faint to faint fluorescence (Eaton-Magaña and Shigley 2016). However, it was noted that the incidence of longwave UV fluorescence decreased over the years as CVD products evolved, with all samples tested since 2013 being inert. When the color of such fluorescence was noted, as-grown CVD diamonds emitted orange, orange-yellow, or yellow light originating from nitrogen-vacancy centers (Martineau et al. 2004; Wang et al. 2005b, 2007b). The brown hue common in CVD diamonds grown with intentional or unintentional nitrogen doping can be removed by high temperature annealing (usually using an HPHT press, but LPHT annealing has also been reported), resulting in a more valuable gem product (Wang et al. 2003, 2012; Charles et al. 2004; Martineau et al. 2004; Mäki et al. 2007; Meng et al. 2008; Liang et al. 2009; Eaton-Magaña and Shigley 2016; Eaton-Magaña et al. 2021). This treatment will reduce NV content, while introducing H3 (N_2V^0) centers, producing green fluorescence (Wang et al. 2012). These orange and green fluorescence colors can alert viewers of their laboratory-grown origin as 99% of natural similarly colored fluorescing diamonds emit blue light due to N3 (N_3V^0) (Moses et al. 1997). Furthermore, CVD diamonds show a stronger fluorescence response to shortwave UV, in contrast to most naturals.

The second most common color of CVD diamonds are pink-hued (Eaton-Magaña and Shigley 2016; Eaton-Magaña et al. 2021), with the vast majority colored by high concentrations of NV centers produced through treatment (see *Absorption spectroscopy and color* section below). The dominance of NV centers in pink CVD diamonds gives rise to moderate to strong red or orange fluorescence to both longwave and shortwave UV light (Wang et al. 2010; Eaton-Magaña and Shigley 2016). Extremely rare Golconda pink natural diamonds (~0.6% of natural pink diamonds) as well as most treated pink natural-grown and HPHT diamonds (Table 2) will also fluoresce in similar hues due to their elevated NV concentrations (Eaton-Magaña and Shigley 2016; Eaton-Magaña et al. 2018a). Meanwhile, 58% and 9% of natural pink diamonds colored by the 550 nm band show blue (N3) or yellow fluorescence, respectively, with the remainder being inert (Eaton-Magaña et al. 2018a).

Inspection of CVD diamonds under magnification and deep-UV excitation (<225 nm) can reveal characteristic growth patterns and help with conclusive identification. CVD synthesis of gem-quality diamond usually proceeds through step-flow growth on (001)-oriented substrates, with riser and terrace segments oriented along [001] and [101] directions. As they present differently oriented planes of growth, the distinct impurity uptake of terraces and risers can result in the formation of a fluorescence pattern of tightly spaced lines known as "striations," illustrated in Figures 39 and 40 (Martineau et al. 2004, 2009). Additionally, deliberate or accidental growth events such as variations in deposition conditions or interruptions can

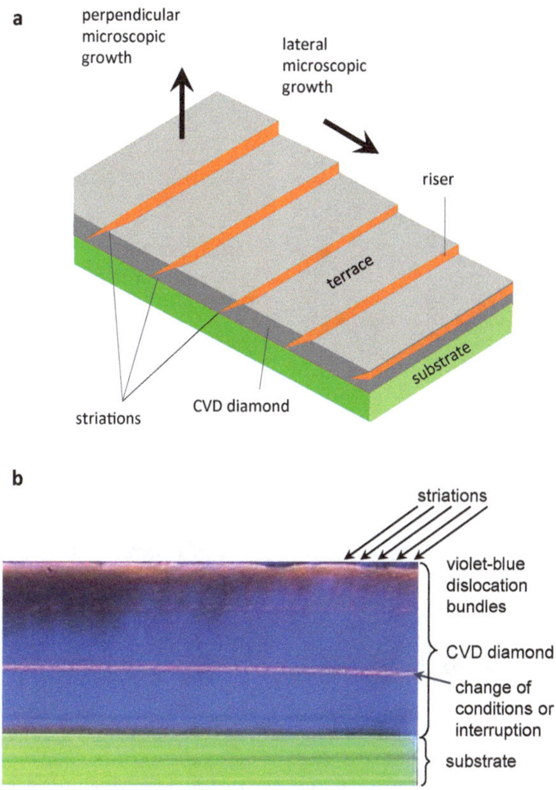

Figure 39. (**a**) The different incorporation efficiency of defects on risers and terraces during step-flow growth of CVD diamond can lead to formation of striations. (**b**) This DiamondView deep-UV fluorescence image of a cross-sectional slice of a CVD diamond layer grown on an HPHT diamond substrate displays several characteristic growth patterns. Orange striations from the differential incorporation of NV centers are inclined diagonally relative to the substrate plane. The horizontal line traversing the CVD layer indicates a change in growth conditions or potential interruption. Violet blue dislocation bundles appear to originate close to the substrate–CVD layer interface, threading upwards through the CVD material. [Image (**a**) modified after Martineau et al. (2004) *Gems and Gemology*, Vol. 40, Fig. 15, p. 13. Image (**b**) modified after Martineau et al. (2009) *Physics Status Solidi C*, Vol 6, Fig. 4, p. 1955.]

dramatically modify the impurity content, resulting in an abrupt change in fluorescence color and/or brightness (Figs. 39b and 40a,d). In a cross-sectional view this will appear as a line lying parallel to the growth plane. Evidence of these growth events, in particular in larger CVD diamonds, suggests that they are being grown in stages as a series of layers (Mokuno et al. 2005; Asmussen et al. 2008; Dieck et al. 2015; Nad et al. 2015; Law and Wang 2016). Figure 22 displays an as-grown CVD diamond crystal grown in two stages. Blue fluorescing threading dislocations, typically nucleating at or close to the boundary between the substrate and CVD diamond layer, may propagate through the CVD material aligned approximately parallel to the bulk growth direction (Martineau et al. 2004). Step-flow growth may cause the dislocations to deviate, following a zigzag path (Martineau et al. 2009). Depending on the viewing direction, dislocations may appear as threads or bundles with a clear linear habit, or may be patchy (Martineau et al. 2004, 2009; Wang et al. 2005b; Dieck et al. 2015). The anisotropy of the patterns emphasizes the importance of inspecting diamonds from multiple viewing angles.

Deep-UV fluorescence images for faceted CVD diamonds are presented in Figure 40. Variations in growth and treatment histories mean that CVD diamond fluorescence can span the full rainbow spectrum of colors (Eaton-Magaña and Shigley 2016). Importantly, treatments cannot remove the intrinsic growth patterns of diamonds, but will affect the relative concentrations of defects and thus potentially fluorescence colors. Here we review the most commonly observed

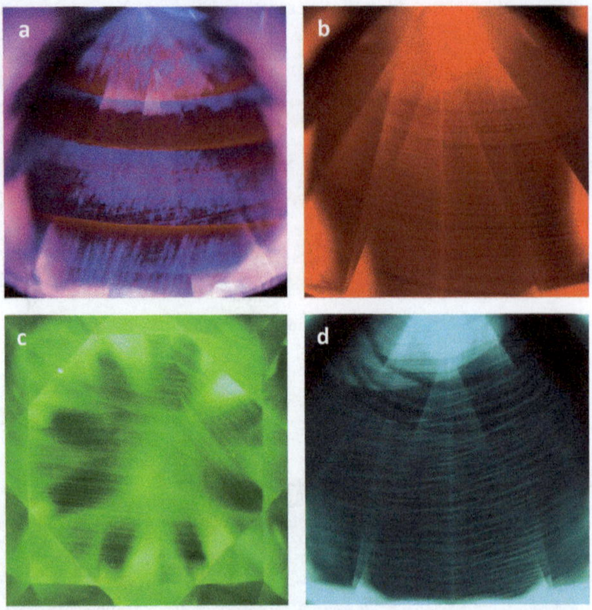

Figure 40. DiamondView deep-UV fluorescence images of CVD diamonds can help unravel their growth and treatment histories. (**a**) This 3.23 ct I color untreated CVD diamond was grown as a series of layers. Violet blue dislocation bundles are seen to propagate through the layers, with striations oriented diagonally with respect to the boundaries. The sample's overall pinkish red fluorescence originates from NV centers. Color treatments of CVD diamonds change defect concentrations and hence can modify their fluorescence colors, yet characteristic patterns such as striations and layer boundaries will remain unchanged. (**b**) This 1.23 ct Fancy Intense pink CVD diamond underwent multi-treatment to produce high concentrations of NV centers, responsible for the red fluorescence observed. HPHT decolorization treatment of these (**c**) 2.55 ct J color and (**d**) 0.41 ct G color CVD diamonds created H3 (N_2V^0) and N3 (N_3V^0) defects, producing the green to greenish blue fluorescence colors. Additionally, these two samples show green-blue phosphorescence othat may be related to N–B donor–acceptor pairs (not shown). [Image (**a**) used by permission of the Gemological Institute of America from Dieck et al. (2015) *Gems and Gemology*, Vol. 51, Fig. 23, p. 438. Images (**b**), (**c**), and (**d**) taken for GIA by Kyaw Soe Moe, Pui Lai Law, and Ellen Barrie, respectively.]

fluorescence reactions. High purity CVD diamonds with low nitrogen concentrations are characterized by low fluorescent defect content, typically only showing patchy or thread-like blue fluorescence from dislocation bundles, yet can be separated from natural Type IIa polygonized dislocation networks by careful examination (Martineau et al. 2004; Wang et al. 2005b). Such CVD diamonds are rarely grown for gem purposes due to cost. Instead, deep-UV fluorescence images of as-grown colorless, near-colorless or brown CVD diamonds usually show orange, red or pink bulk fluorescence due to NV center emission (Wang et al. 2003, 2005b, 2007b; Martineau et al. 2004; Wassell et al. 2018; McGuinness et al. 2020). They may also weakly phosphoresce orange or blue (Wang et al. 2007b; McGuinness et al. 2020). HPHT annealing for decolorization introduces H3 (N_2V^0), N3 (N_3V^0), and a series of unidentified emission peaks and bands between 430 and 500 nm (Wang et al. 2012 ; Wassell et al. 2018; McGuinness et al. 2020). Consequently, the overall fluorescence excited by deep-UV changes from the as-grown orange, red, or pink response colors to green or green-blue. Additionally, the HPHT annealed diamonds show strong green-blue phosphorescence that may be associated with N–B donor–acceptor pairs (Watanabe et al. 1997; Wang et al. 2007b, 2012; Wassell et al. 2018). Multi-treatment of CVD diamonds to produce pink body color yield high NV concentrations, resulting in intense yellowish orange, orange, reddish orange, or red fluorescence (Wang et al. 2007b, 2010).

Absorption spectroscopy and color. The vast majority of D–Z color CVD diamonds can be classified as Type IIa, although trace amounts of isolated nitrogen may be detected for poorer colors (N_S^0 peak at 1344 cm^{-1}). The rarity of this diamond type in nature, and its frequency for both HPHT- and CVD-grown diamonds emphasize that such diamonds should be treated as suspicious and undergo careful testing. FTIR spectra may aid in determining whether a CVD diamond is as-grown or treated. Hydrogen-related peaks at 8753, 7354, 6856, 6425 (also reported at 6428 cm^{-1}), 5564, and 3323 cm^{-1} (N–C–H, also reported at 3324 cm^{-1}) are exclusive to CVD diamonds, but can be reduced or removed by LPHT or HPHT decolorizing treatments (Fuchs et al. 1995a,b; Glover 2003; Charles et al. 2004; Martineau et al. 2004; Cruddace 2007; Meng et al. 2008; Liang et al. 2009; Liggins 2010). The 3123 cm^{-1} peak, identified as the neutral nitrogen-vacancy-hydrogen center (NVH^0), is also common in CVD diamonds (Fuchs et al. 1995a,b; Cruddace 2007; Khan et al. 2013). It may be useful for identification as it has not been observed for Type IIa natural diamonds, yet may be detected in some Type Ia diamonds with intense N_3VH^0 center absorption (3107 cm^{-1}) (Lai et al. 2020). NVH^0 also anneals out during HPHT treatment, but may survive LPHT treatment (Martineau et al. 2004; Cruddace 2007; Meng et al. 2008; Liang et al. 2009). The N_3VH^0 center, a common defect in natural diamonds, can be introduced by annealing at temperatures >1700 °C (Charles et al. 2004; Martineau et al. 2004; Liggins 2010; Wang et al. 2012). For more detail regarding FTIR absorption features present in as-grown and HPHT treated D–Z color CVD diamonds see the work by Martineau et al. (2004) and Wang et al. (2003, 2005b, 2007b, 2012).

UV-visible-NIR absorption spectra of CVD diamonds may have a combination of broad bands and sharper peaks contributing to their color (Fig. 41). At room temperature, colorless and near-colorless CVD diamonds are often featureless, possibly showing an absorption ramp rising from ~1200 nm to the diamond absorption edge at 225 nm. The underlying "profile" is proportional to λ^{-3} (Khan et al. 2013). The defect responsible for this gradual increase in absorption is still unclear, though it may be caused by vacancy clusters or defects with internal surfaces with sp^2-like bonding (Barnes et al. 2006; Mäki et al. 2007; Jones 2009). Thus the absorption spectra may appear similar to those of rare natural Type IIa diamonds, yet differ from those of the much more common Type Ia diamonds with Cape features. Poorer color as-grown CVD diamonds tend to have yellow, brown, or pinkish undertones, with the hue determined by the relative intensities of broad absorption features at ~270 (N_S^0), 360, 520 nm (related to NVH^0, and sometimes referred to as the 515 nm band), and the underlying profile (Martineau et al. 2004; Khan et al. 2009, 2010, 2013). Khan et al. (2009, 2010, 2013) demonstrated that these three bands are thermo- and photochromic, weakening (becoming more colorless) after heating

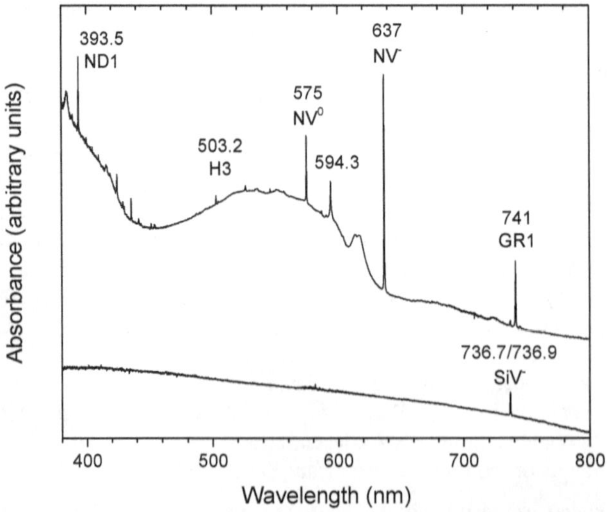

Figure 41. Representative visible-NIR absorption spectra collected at 77 K for (**bottom**) colorless and (**top**) pink CVD diamonds. The spectrum for the colorless sample (0.41 ct, F color) only shows weak absorption from the SiV^- center, transmitting most light across the visible range. A 0.48 ct Fancy Vivid purplish pink treated CVD was used for the pink example. Strongly absorbing NV centers create transmission windows around 450 nm and above 637 nm, contributing red and blue color components. Peaks at 393.5 (ND1, V^-), 594.3 (the "595 nm" center), 741 (GR1, V^0) indicate treatment. [The data for the pink CVD diamond was previously reported as sample APD-8094 in Wang et al. (2010) *Gems and Gemology*, Vol. 46, Fig. 10, p. 10.]

at 525–550 °C due to charge transfer involving $N_S^{0/+}$, $NVH^{0/-}$, and other currently unidentified defects. This effect is reversible, with UV light stimulating their increase. Annealing at temperatures greater than 1400 °C causes permanent changes to the defect structures, reducing the intensities of the featureless absorption profile, as well as the 360 and 520 nm bands, and decreasing the brown color saturation. Differences in their thermal stabilities mean that annealing at temperatures greater than 1800 °C—at which high pressures are typically used to minimize sample graphitization—results in more efficient color removal (Khan et al. 2013).

Absorption spectroscopy with cooling to liquid nitrogen temperatures may reveal the presence of additional defects, including H3 (N_2V^0, 503.2 nm), NV^0 (575 nm), NV^- (637 nm), SiV^- (736.7/736.9 nm), and an unknown CVD-specific center absorbing at 596 nm, depending on the growth and treatment history (Wang et al. 2003, 2005b, 2007b, 2012; Martineau et al. 2004). SiV^- can be detected using the more sensitive photoluminescence technique in a few natural (<1%) and HPHT diamonds (Breeding and Wang 2008; D'Haenens-Johansson et al. 2014, 2015a), but has not been reported using absorption methods, rendering it a good indicator of CVD diamond material. A peak at 946 nm has been identified as the neutral charge state of the silicon-vacancy (SiV^0), but is seldom detected (D'Haenens-Johansson et al. 2010, 2011; Breeze et al. 2020). Silicon is often unintentionally introduced into the growing CVD material by plasma etching of silicon-containing reactor components such as silica windows (Robins et al. 1989; Barjon et al. 2005). Unlike many other indicator features in CVD diamond, silicon-vacancies have excellent thermal stability and only start to anneal out at 2200 °C (Clark and Dickerson 1991). Thus SiV^- may be detected in CVD diamonds of many colors.

CVD diamonds that have been selected for irradiation and annealing treatments to produce pink color from NV centers are very weak Type Ib (sometimes referred to as "nominally Type IIa" by the gemological community), showing a sharp peak at 1344 cm^{-1} and broad band at 1130 cm^{-1} (both from N_S^0) in the IR. Their absorption features are similar to those observed for pink treated HPHT diamonds. The UV-visible-NIR absorption spectra (Fig. 41) are dominated by NV centers, but also show other nitrogen-related features such as the 270 nm broad band from N_S^0 and the 503.2 nm peak from H3 (N_2V^0). Additionally, they include features associated with irradiation such as ND1 (V^-, 393.5 nm), GR1 (V^0, 741 nm), or irradiation and annealing such as the "595 nm" (594.3 nm) defect (Wang et al. 2010; Zaitsev et al. 2018; Song et al. 2021). FTIR spectra further indicate treatment, containing the dinitrogen ⟨001⟩ split-interstitial H1a (1450 cm^{-1}) and the 1502 cm^{-1} peak (Wang et al. 2010; Song et al. 2021). As "NV-pink" CVD samples are nitrogen-containing, the CVD-specific 3123 cm^{-1} (NVH^0) peak may be detected (Wang et al. 2010; Zaitsev et al. 2018). However, it is worth noting that this feature will be absent or weakened in samples that underwent HPHT treatment for brown color removal prior to irradiation and annealing (Wang et al. 2010; Song et al. 2021). Diamonds colored pink by NV centers are likely to be treated natural or laboratory-grown diamonds and thus should be thoroughly tested (Moses et al. 1993; Shigley et al. 2004a,b; Wang et al. 2005a,b, Eaton-Magaña et al. 2018a). Eaton-Magaña and Shigley (2016) present a succinct table comparing characteristics for such samples.

Unlike natural and HPHT diamonds, blue gem-quality CVD diamonds are rarely Type IIb crystals colored by boron. Controlled boron doping can be achieved by the addition of boron containing chemicals such as diborane (B_2H_6) to the growth gases, but synthesis is challenged by poor crystal morphology, low growth rates, and soot formation at high doping levels (Achard et al. 2012). Martineau et al. (2004) reported growing experimental boron-doped CVD diamonds with an inhomogeneous layered appearance. These samples were Fancy Light grayish blue to Fancy Deep blue in color, showed absorption spectra similar to boron-containing natural diamonds (e.g., King et al. 1998), and did not show nitrogen, hydrogen or silicon-related features to help identify their laboratory-grown origin. Boron-doped CVD overgrowth layers on natural diamonds have been observed at GIA, where 80–200 μm CVD diamond were deposited onto the stones to impart a blue color when viewed face-up. In absorption, these hybrid diamonds share properties of both the underlying natural diamond and the CVD layer, allowing identification, especially when combined with fluorescence imaging (Fritsch and Phelps 1993; Moe et al. 2017; Ardon and McElhenny 2019).

Instead of boron-doping, the most popular route to create blue colored CVD diamonds is by post-growth irradiation treatment of Type IIa samples (Ardon and Wang 2014; Moe et al. 2015; Eaton-Magaña 2018; Song et al. 2020). This introduces the GR1 (V^0) defect with a peak at 741 nm and vibronic structure absorbing in the red, as well as irradiation features such as TR12 (469.9 nm) and the 595 nm center. Irradiated CVD diamonds have a green modifier, with the color described as greenish blue or green blue, hues which account for less than 9% of blue natural diamonds (Eaton-Magaña et al. 2018b).

Unusual gray to blue color due to strong absorption from the SiV^0 center and its associated photoconductivity spectrum has also been reported for highly silicon-doped CVD diamonds (Peretti et al. 2014; D'Haenens-Johansson et al. 2015b; Eaton-Magaña and Shigley 2016; Breeze et al. 2020; Eaton-Magaña et al. 2021). SiV^0 and SiV^- are photo- and thermochromic, where the blue color can be temporarily deepened by UV illumination. Exposure to strong white light returns such diamonds to their base color, which may be gray, brownish pink or less intense blue.

Photoluminescence spectroscopy. Photoluminescence spectra collected at liquid nitrogen temperature are particularly useful for CVD diamond identification (e.g., Fig. 42), though multiple laser excitations are necessary to detect defects spanning the UV to NIR region. As-grown CVD spectra typically show emission from NV^0 (575 nm), NV^- (637 nm), often SiV^- (736.7/736.9 nm), and occasionally SiV^0 (946 nm), H3 (N_2V^0, 503.2 nm), and a peak at 389 nm (Iakoubovskii et al. 2000; Wang et al. 2003, 2005b, 2007b; Charles et al. 2004; Martineau et al. 2004; Eaton-Magaña and Shigley 2016). The latter peak, thought to be nitrogen-related, has been reported in some HPHT diamonds and in irradiated nitrogen-containing diamonds

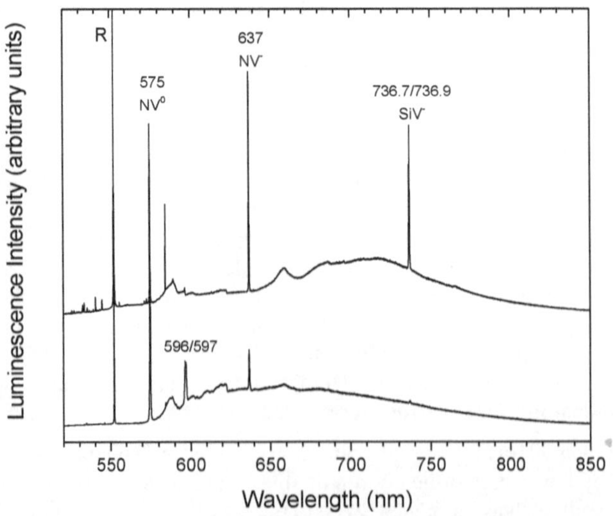

Figure 42. Photoluminescence spectra for as-grown (**bottom**) and HPHT annealed (**top**) CVD diamonds collected using 514 nm laser excitation at 77 K. Nitrogen-vacancy and silicon-vacancy centers are observed for both. SiV^- is often detected for CVD diamonds, yet extremely rare for natural diamonds (<1%), making it a good indicator for synthesis. The as-grown CVD diamond also shows a CVD-specific unassigned doublet emission at 596/597 nm. HPHT treatment removes this feature, but introduces a series of sharp peaks in the 520–580 nm region. The peak labelled R is the intrinsic diamond Raman peak.

(Collins and Woods 1987; Collins and Lawson 1989; Robins et al. 1989; Zaitsev 2000; Martineau et al. 2004). CVD-specific PL features include peaks at 468 nm (also reported at 467 nm) and 533 nm (or 534 nm), and a doublet at 596/597 nm. The center responsible for the 468 nm peak (also referred to as TR12') may involve nitrogen as the peak's intensity increases with nitrogen concentration, but has also been attributed to interstitial- or vacancy-cluster-related defects (Iakoubovskii and Adriaenssens 2000a,b; Iakoubovskii et al. 2000; Rzepka et al. 2001; Zaitsev 2001; Iakoubovskii and Stesmans 2002; Zaitsev et al. 2021). The 533 nm peak may originate from a nitrogen-vacancy type complex (Ruan et al. 1991; Zaitsev 2001). The defect emitting at 596/597 nm is unassigned, but is thought to differ from that associated with the 596 nm absorption peak in brown CVD diamonds (Martineau et al. 2004).

HPHT annealing destroys the defects responsible for the 389, 468, 533, and 596/597 nm features, while also reducing the intensity of NV centers (Charles et al. 2004; Martineau et al. 2004; Wang et al. 2012; Zaitsev et al. 2021). Yet these conditions result in the formation of more complex nitrogen-related defects such as N3 (N_3V^0, 415 nm), H3 (N_2V^0, 503.2 nm), and its negatively charged counterpart, H2 (986 nm). Additionally, the treatment introduces unidentified peaks at 850 and 875.5 nm, and a series of lines in the 451–459 and 520–580 nm regions.

The PL spectra for CVD diamonds that have been treated pink by irradiation and annealing for nitrogen-vacancy formation are unsurprisingly dominated by emission from both NV^0 and NV^- centers, but also include peaks at 389, 495.9 (H4, N_4V^0), 503.5 (3H, interstitial-related), 796.9, 806.4, and 949.0 nm (Wang et al. 2010; Eaton-Magaña and Shigley 2016; Song et al. 2021). Strong absorption from the "595 nm" (594.3 nm) and GR1 (V^0, 741 nm) centers, which overlap with the vibronic bands associated with NV^0 and NV^- luminescence, are detected as negative peaks. Further peaks will be present as outlined above, depending on whether the material was as-grown or underwent HPHT treatment prior to irradiation and annealing.

CONCLUDING REMARKS

The development and global expansion of diamond synthesis using HPHT and CVD technologies has led to rapid improvements in material quality and sizes. These dynamic and competitive fields continue to be actively researched by academic and industrial scientists alike, who characterize diamond and its crystal defects to exploit many of its unrivalled properties and find new avenues of use. The nascent laboratory-grown gem diamond sector is showing signs of maturity, benefiting from decades of synthesis research for industrial and technological applications. Despite the similarities between natural and laboratory-grown diamonds, the fundamental differences in their formation provide evidence of their growth origin, enabling their conclusive identification.

ACKNOWLEDGEMENTS

UFSDJ would like to thank the diamond research and gemology team members at GIA, whose comprehensive data collection and examination of both natural and laboratory-grown diamonds are the basis of the identification section. Also, she would like to acknowledge the work by Alex Balter (GIA) who conducted a lot of the referenced statistical analysis of diamonds submitted to our laboratories. JEB thanks the CVD diamond research community for many years of open, supportive, and collaborative discussions which rapidly advanced our understanding prior to the current commercialization efforts. ANK is deeply grateful to all the specialists who contributed to the fascinating research process of growing new and ultra large diamond crystals. Special recognition go to Mr. Roman Isakov, the Chief Technology Officer of New Diamond Technology Ltd. (NDT), for his exceptional engineering talent in the diamond production field, and to Dr. Alexander Koliadin, the former CTO of NDT, whose innovative ideas made it possible to start the growth project. Thank you also to all the teams at NDT and Alkor-D Ltd. who daily provide the technological support necessary to create world record-breaking laboratory-grown diamonds. The authors are also thankful to Dr. Sally Eaton-Magaña and Dr. Christopher M. Breeding for providing helpful comments on an early draft of the manuscript. Dr. Evan M. Smith is thanked for help with Figure 1. We are also grateful for reviews by Prof. Michael Ashfold, Dr. Boris N. Feigelson, and Dr. Daniel J. Twitchen.

REFERENCES

Achard J, Silva F, Brinza O, Tallaire A, Gicquel A (2007) Coupled effect of nitrogen addition and surface temperature on the morphology and the kinetics of thick CVD diamond single crystals. Diamond Relat Mater 16:685–689

Achard J, Issaoui R, Tallaire A, Silva F, Barjon J, Jomard F, Gicquel A (2012) Freestanding CVD boron doped diamond single crystals: A substrate for vertical power electronic devices? Phys Stat Sol (a) 209:1651–1658

Achard J, Jacques V, Tallaire A (2020) CVD diamond single crystals with NV centres: A review of material synthesis and technology for quantum sensing applications. J Phys D: Appl Phys, 53:313001

Angus JC (1994) Development of CVD diamond growth in the United States. *In*: Synthetic Diamond: Emerging CVD Science and Technology, The Electrochemical Society Series. Spear KE, Dismukes JP (eds) John Wiley, Interscience, New York, p. 21–39

Angus JC (2014) Diamond synthesis by chemical vapor deposition: The early years. Diamond Relat Mater 49:77–86

Angus JC, Sunkara M, Sahaida SR, Glass JT (1992) Twinning and faceting in early stages of diamond growth by chemical vapor deposition. J Mater Res 7:3001–3009

Ardon T, Eaton-Magaña S (2018) Lab Notes: 15 carat HPHT synthetic diamond. Gems Gemol 54:217–218

Ardon T, McElhenny G (2019) Lab Notes: CVD layer grown on natural diamond. Gems Gemol 55:97–99

Ardon T, Wang W (2014) Lab Notes: Heavily irradiated CVD synthetic diamond. Gems Gemol 50:240–241

Ashfold MNR, Goss JP, Green BL, May PW, Newton ME, Peaker CV (2020) Nitrogen in diamond. Chem Rev 120:5745–5794

Asmussen J, Grotjohn TA, Schuelke T, Becker MF, Yaran MK, King DJ, Wicklein S, Reinhard DK (2008) Multiple substrate microwave plasma-assisted chemical vapor deposition single crystal diamond synthesis. Appl Phys Lett 93:130–132

Auciello O, Sumant AV (2010) Status review of the science and technology of ultrananocrystalline diamond (UNCD ™) films and application to multifunctional devices. Diamond Relat Mater 19:699–718
Awschalom DD, Hanson R, Wrachtrup J, Zhou BB (2018) Quantum technologies with optically interfaced solid-state spins. Nat Photonics 12:516–527
Babich YV, Feigelson BN, Fisher D, Yelisseyev AP, Nadolinny VA, Baker JM (2000) The growth rate effect on the nitrogen aggregation in HPHT grown synthetic diamonds. Diamond Relat Mater 9:893–896
Babich YV, Feigelson BN, Yelisseyev AP (2004) Nitrogen aggregation and linear growth rate in HPHT synthetic diamonds. Diamond Relat Mater 13:1802–1806
Balmer RS, Brandon JR, Clewes SL, Dhillon HK, Dodson JM, Friel I, Inglis PN, Madgwick, Markham ML, Mollart TP, Perkins N, Scarsbrook GA, Twitchen DJ, Whitehead AJ, Wilman JJ, Woollard SM. (2009) Chemical vapour deposition synthetic diamond: Materials, technology and applications. J Phys: Condens Matter 21:364221
Barjon J, Rzepka E, Jomard F, Laroche, JM, Ballutaud D, Kociniewski T, Chevallier J (2005) Silicon incorporation in CVD diamond layers. Phys Stat Sol (a) 202:2177–2181
Barnes R, Bangert U, Martineau P (2006) HR-TEM imaging and image simulation of vacancy clusters in brown diamond. Phys Stat Sol (a) 203:3081–3087
Bovenkerk HP, Bundy FP, Hall HT, Strong HM, Wentorf RH (1959) Preparation of diamond. Nature 194:1094–1098
Boyd SR, Kiflawi I, Woods GS (1994) The relationship between infrared absorption and the A defect concentration in diamond. Philos Mag B 69:1149–1153
Boyd SR, Kiflawi I, Woods GS (1995) Infrared absorption by the B nitrogen aggregate in diamond. Philos Mag B 72:351–361
Branstrator B (2022) Measuring the lab-grown diamond market: Size, growth, and future opportunities. National Jeweler, *https://www.nationaljeweler.com/articles/10624-measuring-the-lab-grown-diamond-market-size-growth-and-future-opportunities*
Breeding CM, Shigley JE (2009) The "Type" classification system of diamonds and its importance in gemology. Gems Gemol 45:96–111
Breeding CM, Wang W (2008) Occurrence of the Si-V defect center in natural colorless gem diamonds. Diamond Relat Mater 17:1335–1344
Breeding CM, Eaton-Magaña S, Shigley JE (2018) Natural-color green diamonds: A beautiful conundrum. Gems Gemol 54:2–27
Breeze BG, Meara CJ, Wu XX, Michaels CP, Gupta R, Diggle PL, Dale MW, Cann BL, Ardon T, D'Haenens-Johansson UFS, Friel I, Rayson MJ, Briddon PR, Goss JP, Newton ME, Green BL (2020) Doubly charged silicon vacancy center, Si–N complexes, and photochromism in N and Si codoped diamond. Phys Rev B 101:184115
Brinza O, Achard J, Silva F, Bonnin X, Barroy P, De Corte K, Gicquel A (2008) Dependence of CVD diamond growth rate on substrate orientation as a function of process parameters in the high microwave power density regime. Phys Stat Sol (a) 205:2114–2120
Brozel MR, Evans T, Stephenson RF (1978) Partial dissociation of nitrogen aggregates in diamond by high temperature-high pressure treatments. Proc R Soc A 361:109–127
Buijnsters JG, Shankar P, van Enckevort WJP, Schermer JJ, ter Meulen JJ (2005) Adhesion analysis of polycrystalline diamond films on molybdenum by means of scratch, indentation and sand abrasion testing. Thin Solid Films 474:186–196
Bundy FP (1962) Direct conversion of graphite to diamond in static pressure apparatus. Science 137:1057–1058
Bundy FP (1963) Direct conversion of graphite to diamond in static pressure apparatus. J Chem Phys 38:631–643
Bundy FP (1980) The P, T phase and reaction diagram for elemental carbon, 1979. J Geophys Res 85:6930–6936
Bundy FP (1988) Ultra-high pressure apparatus. Phys Rep 167:133–176
Bundy FP, Hall HT, Strong HM, Wentorf RH (1955) Man-made diamonds. Nature 176:51–55
Bundy FP, Bassett WA, Wathers MS, Hemley RJ, Mao HK, Goncharov AF (1996) The pressure-temperature phase and transformation diagram for carbon; updated through 1994. Carbon 34:141–153
Burns RC, Cvetkovic V, Dodge CN, Evans DJF, Rooney MLT, Spear PM, Welbourn CM (1990) Growth-sector dependence of optical features in large synthetic diamonds. J Cryst Growth 104:257–279
Burns RC, Hansen JO, Spits RA, Sibanda M, Welbourn CM, Welch DL (1999) Growth of high purity large synthetic diamond crystals. Diamond Relat Mater 8:1433–1437
Burns RC, Chumakov AI, Connell SH, Dube D, Godfried HP, Hansen JO, Härtwig J, Hoszowska J, Masiello F, Mkhonza L, Rebak M, Rommevaux A, Setshedi R, Van Vaerenbergh P (2009) HPHT growth and x-ray characterization of high-quality Type IIa diamond. J Phys: Condens Matter 21:364224
Burton NC, Steeds JW, Meaden GM, Shreter YG, Butler JE (1995) Strain and microstructure variation in grains of CVD diamond film. Diamond Relat Mater 4:1222–1234
Butler JE, Oleynik I (2008) A mechanism for crystal twinning in the growth of diamond by chemical vapour deposition. Philos Trans R Soc, A 366:295–310
Butler JE, Sumant AV (2008) The CVD of nanodiamond materials. Chem Vap Deposition 14:145–160
Butler JE, Windischmann H (1998) Developments in CVD-diamond synthesis during the past decade. MRS Bull 23:22–27
Butler JE, Woodin RL (1993) Thin-film diamond growth mechanisms. Philos Trans R Soc, A 342:209–224

Butler JE, Celii FG, Oakes DB, Hanssen LM, Carrington WA, Snail KA (1989) Studies of diamond chemical vapor-deposition. High Temp Sci 27:183–197
Butler JE, Cheesman A, Ashfold MNR (2009a) Recent progress in the understanding of CVD growth of diamond. *In*: CVD Diamond for Electronic Devices and Sensors. Sussmann RS (ed) John Wiley & Sons, Chichester, p 103–124
Butler JE, Mankelevich YA, Cheesman A, Ma J, Ashfold MNR (2009b) Understanding the chemical vapor deposition of diamond: Recent progress. J Phys: Condens Matter 21:364201
Byrne KS, Anstie JD, Chapman JG, Luiten AN (2012) Optically reversible photochromism in natural pink diamond. Diamond Relat Mater 30:31–36
Byrne KS, Chapman JG, Luiten AN (2014a) Photochromic charge transfer processes in natural pink and brown diamonds. J Phys: Condens Matter 26:035501
Byrne KS, Chapman JG, Luiten AN (2014b) Corrigendum: Photochromic charge transfer processes in natural pink and brown diamonds (2014 J. Phys.: Condens. Matter 26 035501). J Phys: Condens Matter 26:239502
Calvert JG, Pitts JN (1966) Photochemistry. John Wiley & Sons, New York
Celii FG, Butler JE (1989) Hydrogen atom detection in the filament-assisted diamond deposition environment. Appl Phys Lett 54:1031–1033
Celii FG, Butler JE (1991) Diamond chemical vapor deposition. Annu Rev Phys Chem 42:643–684
Celii FG, Pehrsson PE, Wang HT, Butler JE (1988) Infrared detection of gaseous species during the filament-assisted growth of diamond. Appl Phys Lett 52:2043–2045
Celii FG, Thorsheim HR, Butler JE (1990a) In-situ THG detection of gas phase atomic hydrogen in diamond CVD processes. Carbon 28:808
Celii FG, Thorsheim HR, Butler JE, Plano LS, Pinneo JM (1990b) Detection of ground-state atomic-hydrogen in a dc plasma using third-harmonic generation. J Appl Phys 68:3814–3817
Chalain J-P (2003) Gem News International: A natural yellow diamond with nickel-related optical centers. Gems Gemol 39:325–326
Charles SJ, Butler JE, Feygelson BN, Newton ME, Carroll DL, Steeds JW, Darwish H, Yan CS, Mao HK, Hemley RJ (2004) Characterization of nitrogen doped chemical vapor deposited single crystal diamond before and after high pressure, high temperature annealing. Phys Stat Sol (a) 201:2473–2485
Chayahara A, Mokuno Y, Horino Y, Takasu Y, Kato H, Yoshikawa H, Fujimori N (2004) The effect of nitrogen addition during high-rate homoepitaxial growth of diamond by microwave plasma CVD. Diamond Relat Mater 13:1954–1958
Chen N, Ma H, Yan B, Chen L, Chen L, Guo L (2018) Characterization of various centers in synthetic Type Ib diamond under HPHT annealing. Cryst Growth Des 18:3870–3876
Chrenko RM, Strong HM, Tuft RE (1971) Dispersed paramagnetic nitrogen content of large laboratory diamonds. Philos Mag 23:313–318
Chrenko RM, Tuft RE, Strong HM (1977) Transformation of the state of nitrogen in diamond. Nature 270:141–144
Chu HN, Denhartog EA, Lefkow AR, Jacobs J, Anderson LW, Lagally MG, Lawler JE (1991) Measurements of the gas kinetic temperature in a CH_4-H_2 discharge during the growth of diamond. Phys Rev A 44:3796–3803
Clark CD (1959) The absorption edge spectrum of diamond. J Phys Chem Solids 8:481–484
Clark CD, Dickerson CB (1991) The 1.681 eV centre in polycrystalline diamond. Surf Coat Tech 47:336–343
Clark CD, Walker J (1973) The neutral vacancy in diamond. Proc R Soc Lond A 334:241–257
Clark CD, Ditchburn RW, Dyer HB (1956) The absorption spectra of irradiated diamonds after heat treatment. Proc R Soc Lond A 237:75–89
Clark CD, Kanda H, Kiflawi I, Sittas G (1995) Silicon defects in diamond. Phys Rev B 51:16681–16688
Clausing RE, Heatherly L, More KL, Begun GM (1989) Electron-microscopy of the growth features and crystal-structures of filament assisted CVD diamond films. Surf Coat Technol 39:199–210
Cobb SJ, Ayres ZJ, Macpherson JV (2018) Boron doped diamond: A designer electrode material for the twenty-first century. Ann Rev Anal Chem 11:463–484
Collins AT (1978) Investigating artificially coloured diamonds. Nature 273:654–655
Collins AT (1980) Vacancy enhanced aggregation of nitrogen in diamond. J Phys C: Solid State Phys 13:2641–2650
Collins AT (1982) Colour centres in diamond. J Gemmol 18:37–75
Collins AT (1989) The polarised absorption and cathodoluminescence associated with the 1.40 eV centre in synthetic diamond. J Phys: Condens Matter 1:439–450
Collins AT (1992) The characterisation of point defects in diamond by luminescence spectroscopy. Diamond Relat Mater 1:457–469
Collins AT (1993a) Intrinsic and extrinsic absorption and luminescence in diamond. Phys B 185:284–296
Collins AT (1993b) The optical and electronic properties of semiconducting diamond. Phil Trans R Soc Lond A 342:233–244
Collins AT (2000) Spectroscopy of defects and transition metals in diamond. Diamond Relat Mater 9:417–423
Collins AT (2001) The colour of diamond and how it may be changed. J Gemmol 27:341–359
Collins AT, Lawson SC (1989) Cathodoluminescence studies of isotope shifts associated with localised vibrational modes in synthetic diamond. J Phys: Condens Matter 39:6929–6937
Collins AT, Spear PM (1982) Optically active nickel in synthetic diamond. J Phys D: Appl Phys 15:83–87
Collins AT, Spear PM (1983) The 1.40 eV and 2.56 eV centres in synthetic diamond. J Phys C: Solid State Phys 16:963–973

Collins AT, Williams AWS (1971) The nature of the acceptor centre in semiconducting diamond. J Phys C: Solid State Phys 4:1789–1800
Collins AT, Woods GS (1982) An anomaly in the infrared absorption spectrum of synthetic diamond. Philos Mag B 46:77–83
Collins AT, Woods GS (1987) Isotope shifts of nitrogen-related localised mode vibrations in diamond. J Phys C: Solid State Phys 20:L797–L801
Collins AT, Davies G, Kanda H, Woods GS (1988) Spectroscopic studies of carbon-13 synthetic diamond. J Phys C: Solid State Phys 21:1363–1376
Collins AT, Kanda H, Burns RC (1990) The segregation of nickel-related optical centres in the octahedral growth sectors of synthetic diamond. Philos Mag B 61:797–810
Coulson CA, Larkins FP (1971) Isolated single vacancy in diamonds - I. Electronic structure. J Phys Chem Solids 32:2245–2257
Crisci A, Baillet F, Mermoux M, Bogdan G, Nesládek M, Haenen K (2011) Residual strain around grown-in defects in CVD diamond single crystals: A 2D and 3D Raman imaging study. Phys Stat Sol (a) 208:2038–2044
Crowningshield R (1971) General Electric's cuttable synthetic diamonds. Gems Gemol 2:302–314
Cruddace RC (2007) Magnetic Resonance and Optical Studies of Point Defects in Single Crystal CVD Diamond. PhD Dissertation, University of Warwick, Coventry, United Kingdom
Davies G (1976) The A nitrogen aggregate in diamond—its symmetry and possible structure. J Phys C: Solid State Phys 9: L537–L542
Davies G (1977) Charge state of the vacancy in diamond. Nature 269:498–500
Davies G (1981) Decomposing the IR absorption spectra of diamonds. Nature 290:40–41
Davies G, Hamer MF (1976) Optical studies of the 1.945 eV vibronic band in diamond. Proc R Soc Lond A 348:285–298
Davies G, Nazaré MH (1980) Uniaxial stress splitting of E to E transitions at trigonal centres in cubic crystals: The 594 nm band in diamond. J Phys C: Solid State Phys 13:4127–4136
Davies G, Nazaré MH, Hamer MF (1976) H3 (2.463 eV) vibronic band in diamond: Uniaxial stress effects and the breakdown of mirror symmetry. Proc R Soc Lond A 351:245–265
Davies G, Welbourn CM, Loubser JHN (1978) Optical and electron paramagnetic effects in yellow Type Ia diamonds. Diamond Res 23–30
Davies G, Neves AJ, Nazaré MH (1989) Nickel isotope effects in the 1.4 eV centre in synthetic diamond. Europhys Lett 9:47–52
Davies G, Lawson SC, Collins AT, Mainwood A, Sharp SJ (1992) Vacancy-related centers in diamond. Phys Rev B 46:13157–13170
Dean PJ (1965) Bound exciton and donor-acceptor pairs in natural and synthetic diamond. Phys Rev 139:A588-A602
Deljanin B, Alessandri M, Peretti A, Åström (2015) NDT breaking the 10 carat barrier: World record faceted and gem-quality synthetic diamonds investigated. Contrib Gemol 15:1–7, *http://gemresearch.ch/ndt-breaking-the-10-carat-barrier-world-record-faceted-and-synthetic-diamonds-investigated/*
de Sa ES, Davies G (1977) Uniaxial stress studies of the 2.498 eV (H4), 2.417 eV and 2.536 eV vibronic bands in diamond. Proc R Soc Lond A 357:231–251
de Theije FK, Schermer JJ, van Enckevort WJP (2000) Effects of nitrogen impurities on the CVD growth of diamond: Step bunching in theory and experiment. Diamond Relat Mater 9:1439–1449
D'Haenens-Johansson UFS, Edmonds AM, Newton ME, Goss JP, Briddon PR, Baker JM, Martineau PM, Khan RUA, Twitchen DJ, Williams SD (2010) EPR of a defect in CVD diamond involving both silicon and hydrogen that shows preferential alignment. Phys Rev B 82:155205
D'Haenens-Johansson UFS, Edmonds AM, Green BL, Newton ME, Davies G, Martineau PM, Khan RUA, Twitchen DJ (2011) Optical properties of the neutral silicon split-vacancy center in diamond. Phys Rev B 84:245208
D'Haenens-Johansson UFS, Moe KS, Johnson P, Wong SY, Lu R, Wang W (2014) Near-colorless HPHT synthetic diamonds from AOTC Group. Gems Gemol 50:30–45
D'Haenens-Johansson UFS, Katrusha AN, Moe KS, Johnson P, Wang W (2015a) Large colorless HPHT-grown synthetic gem diamonds from New Diamond Technology, Russia. Gems Gemol 51:260–279
D'Haenens-Johansson U.F.S., Ardon T., Wang W. (2015b) CVD synthetic gem diamonds with high silicon-vacancy concentrations. Conference on New Diamond and Nano Carbons, Shizuoka, Japan, May 2015
Dieck C, Loudin L, D'Haenens-Johansson UFS (2015) Lab Notes: Two large CVD-grown synthetic diamonds tested by GIA. Gems Gemol 51:437–439
Doherty MW, Manson NB, Delaney P, Jelezko F, Wrachtrup J, Hollenberg LCL (2013) The nitrogen-vacancy colour centre in diamond. Phys Rep 528:1–45
Donato N, Rouger N, Pernot J, Longobardi G, Udrea F (2019) Diamond power devices: State of the art, modelling, figures of merit and future perspective. J Phys D: Appl Phys 53:093001
Dyer HB, Du Preez L (1965) Irradiation damage in Type I diamond. J Chem Phys 42:1898–1906
Dyer HB, Raal FA, Du Preez L, Loubser JHN (1965) Optical absorption features associated with paramagnetic nitrogen in diamond. Philos Mag 11:763–774
Eaton-Magaña S (2018) Lab Notes: Three irradiated CVD synthetic diamonds. Gems Gemol 54:215–216
Eaton-Magaña S (2019) Lab Notes: Faint green HPHT synthetic diamonds. Gems Gemol 55:99–101

Eaton-Magaña S, Breeding CM (2016) An introduction to photoluminescence spectroscopy for diamond and its application in gemology. Gems Gemol 52:2–17

Eaton-Magaña S, Lu R (2011) Phosphorescence in Type IIb diamonds. Diamond Relat Mater 20:983–989

Eaton-Magaña S, Shigley JE (2016) Observations on CVD-grown synthetic diamonds: A review. Gems Gemol 52:222–245

Eaton-Magaña S, Post JE, Heaney PJ, Freitas J, Klein P, Walters R, Butler JE (2008) Using phosphorescence as a fingerprint for the Hope and other blue diamonds. Geology 36:83–86

Eaton-Magaña S, Shigley JE, Breeding CM (2017) Observations on HPHT-grown synthetic diamonds: a review. Gems Gemol 53:262–284

Eaton-Magaña S, Ardon T, Smit KV, Breeding CM, Shigley JE (2018a) Natural-color pink, purple, red, and brown diamonds: Band of many colors. Gems Gemol 54:352–377

Eaton-Magaña S, Breeding CM, Shigley JE (2018b) Natural-color blue, gray, and violet diamonds: Allure of the deep. Gems Gemol 54:112–131

Eaton-Magaña S, McElhenny G, Breeding CM, Ardon T (2020) Comparison of gemological and spectroscopic features in Type IIa and Ia natural pink diamonds. Diamond Relat Mater 105:107784

Eaton-Magaña S, Ardon T, Breeding CM (2021) Laboratory-grown diamond: A gemological laboratory perspective. J Gems Gemmol 23:25–39

Edmonds AM, D'Haenens-Johansson UFS, Cruddace RJ, Newton ME, Fu K-MC, Santori C, Beausoleil RG, Twitchen DJ, Markham ML (2012) Production of oriented nitrogen-vacancy color centers in synthetic diamond. Phys Rev B 86:035201

Evans T (1979) Changes produced by high temperature treatment of diamond. *In*: The Properties of Diamond. Field JE, (ed) Academic Press, London, p 403–424

Evans T, Qi Z (1982) The kinetics of the aggregation of nitrogen atoms in diamond. Proc R Soc Lond A 381:159–178

Evans T, Qi Z, Maguire J (1981) The stages of nitrogen aggregation in diamond. J Phys C 14:379–384

Eversole WG (1962) Synthesis of diamond, US3030187 and US3003188

Fedoseev DV (1994) Summary of research on diamond growth from the gas phase in Russia. *In*: Synthetic Diamond: Emerging CVD Science and Technology, The Electrochemical Society Series. Spear KE, Dismukes JP (eds) John Wiley, Interscience, New York, p 41–56

Feigelson BN (1983) Some questions of pressure generation in the multianvil apparatus. *In:* Silicate Systems at High Pressures, Proceedings of the Institute of Geology and Geophysics. Siberian Branch of the Academy of Sciences of the USSR, Novosibirsk, p 62–76 (in Russian)

Field JE (2012) The mechanical and strength properties of diamond, Rep Prog Phys 72:126505

Fisher D, Sibley SJ, Kelly CJ (2009) Brown colour in natural diamond and interaction between the brown related and other colour-inducing defects. J Phys: Condens Matter 21:364213

Frank FC, Lang AR, Evans DJF, Rooney MLT, Spear PM, Welbourn CM (1990) Orientation-dependent nitrogen incorporation on vicinals on synthetic diamond cube growth surfaces. J Cryst Growth 100:354–376

Friel I, Clewes SL, Dhillon HK, Perkins N, Twitchen DJ, Scarsbrook GA (2009) Control of surface and bulk crystalline quality in single crystal diamond grown by chemical vapour deposition. Diamond Relat Mater 18:808–815

Fritsch E (1998) The nature of color in diamonds. *In*: The Nature of Diamonds. Harlow G (ed) Cambridge University Press, Cambridge, p 23–47

Fritsch E, Phelps AW (1993) Type IIb diamond thin films deposited onto near-colorless natural gem diamonds. Diamond Relat Mater 2:70–74

Fritsch E, Shigley J (1989) Contribution to the identification of treated colored diamonds: diamonds with peculiar color-zoned pavilions. Gems Gemol 25:95–101

Fritsch E, Shigley JE (1993) The separation of natural from synthetic gem-quality diamonds on the basis of crystal growth criteria. J Cryst Growth 128:425–428

Fuchs F, Wild C, Schwarz K, Koidl P (1995a) Hydrogen-related IR absorption in chemical vapour deposited diamond. Diamond Relat Mater 4:652–656

Fuchs F, Wild C, Schwarz K, Müller-Sebert W, Koidl P (1995b) Hydrogen induced vibrational and electronic transitions in chemical vapor deposited diamond, identified by isotopic substitution. Appl Phys Lett 66:177–179

Fujita N, Blumenau AT, Jones R, Öberg S, Briddon PR (2006) Theoretical studies on ⟨100⟩ dislocations in single crystal CVD diamond. Phys Stat Sol (a) 203:3070–3075

Gaillou E, Post JE, Bassim ND, Zaitsev AM, Rose T, Fries MD, Stroud RM, Steele A, Butler JE (2010) Spectroscopic and microscopic characterizations of color lamellae in natural pink diamonds. Diamond Relat Mater 19:1207–1220

Gallheber BC, Fischer M, Mayr M, Straub J, Schreck M (2018) Growth, stress, and defects of heteroepitaxial diamond on Ir/YSZ/Si(111). J Appl Phys 123:225302

Gaukroger MP, Martineau PM, Crowder MJ, Friel I, Williams SD, Twitchen DJ (2008) X-ray topography studies of dislocations in single crystal CVD diamond. Diamond Relat Mater 17:262–269

Glover C (2003) A Study of Defects in Single Crystal CVD Diamond. PhD Dissertation, University of Warwick, Coventry, United Kingdom

Goodwin DG, Butler JE (1997) Theory of diamond chemical vapor deposition. *In*: Handbook of Industrial Diamonds. Prelas MA, Popovici G, Bigelow LK (eds) Marcel Dekker, New York, p 527–581

Goss JP, Jones R, Breuer SJ, Briddon PR, Öberg S (1996) The twelve-line 1.682 eV luminescence center in diamond and the vacancy-silicon complex. Phys Rev Lett 77:3041–3044

Goss JP, Briddon PR, Jones R, Öberg S (2004) The lattice location of Ni in diamond: A theoretical study. J Phys C: Condens Matter 16:4567–4578

Goss JP, Briddon PR, Hill V, Jones R, Rayson MJ (2014) Identification of the structure of the 3107 cm^{-1} H-related defect in diamond. J Phys: Condens Matter 14:145801

Green BL, Mottishaw S, Breeze BG, Edmonds AM, D'Haenens-Johansson UFS, Doherty MW, Williams SD, Twitchen DJ, Newton ME (2017) Neutral silicon-vacancy center in diamond: Spin polarization and lifetimes. Phys Rev Lett 119:096402

Green BL, Doherty MW, Nako E, Manson NB, D'Haenens-Johansson UFS, Williams SD, Twitchen DJ, Newton ME (2019) Electronic structure of the neutral silicon-vacancy center in diamond. Phys Rev B 99:161112(R)

Green BL, Collins AL, Breeding CM (2022) Diamond spectroscopy, defect centers, diamond color and treatments. Rev Mineral Geochem XX:xxx-xxx

Hainschwang T, Katrusha A (2003) Gem News International: A bicolored synthetic diamond. Gems Gemol 39:163–164

Hall HT (1960a) High temperature high pressure apparatus, US2941248

Hall HT (1960b) Ultra-high-pressure, high-temperature apparatus: The "belt." Rev Sci Instrum 31:125–131

Hamza AV, Kubiak GD, Stulen RH (1990) Hydrogen chemisorption and the structure of the diamond C(100)-(2×1) surface. Surf Sci 237:35–52

Hanssen LM, Carrington WA, Butler JE, Snail KA (1988) Diamond synthesis using an oxygen-acetylene torch. Mater Lett 7:289–292

Harris JW, Smit KV, Fedortchouck Y, Moore M (2022) Morphology of monocrystalline diamond and its inclusions. Rev Mineral Geochem XX:xxx-xxx

Harris SJ, Weiner AM (1990) Methyl radical and H-atom concentrations during diamond growth. J Appl Phys 67:6520–6526

Harris SJ, Weiner AM, Perry TA (1988) Measurement of stable species present during filament-assisted diamond growth. Appl Phys Lett 53:1605–1607

Hepp C, Müller T, Waselowski V, Becker JN, Pingault B, Sternschulte H, Steinmüller-Nethl D, Gali A, Maze JR, Atatüre M, Becher C (2014) Electronic structure of the silicon vacancy color center in diamond. Phys Rev Lett 112:036405

Hofer S (1985) Pink diamonds from Australia. Gems Gemol 21:147–155

Howell D (2012) Strain-induced birefringence in natural diamond: A review. Eur J Mineral 24:575–585

Howell D, Fisher D, Piazolo S, Griffin WL, Sibley SJ (2015) Pink color in Type I diamonds: Is deformation twinning the cause? Am Mineral 100:1518–1527

Howell D, Collins AT, Loudin LC, Diggle PL, D'Haenens-Johansson UFS, Smit KV, Katrusha AN, Butler JE, Nestola F (2019) Automated FTIR mapping of boron distribution in diamond. Diamond Relat Mater 96:207–215

Hsu WL (1991) Mole fractions of H, CH_3, and other species during filament-assisted diamond growth. Appl Phys Lett 59:1427–1429

Hsu WL, McMaster MC, Coltrin ME, Dandy DS (1994) Molecular-beam mass-spectrometry studies of chemical-vapor-deposition of diamond. Jpn J Appl Phys 33:2231–2239

Iakoubovskii K, Adriaenssens GJ (2000a) Luminescence excitation spectra in diamond. Phys Rev B 61:174–182

Iakoubovskii K, Adriaenssens GJ (2000b) Optical study of some interstitial-related centres in CVD diamond. Phys Stat Sol (a) 181:59–64

Iakoubovskii K, Adriaenssens GJ (2002) Evidence for a Fe-related defect centre in diamond. J Phys C: Condens Matter 14:L95–L98

Iakoubovskii K, Stesmans A (2002) Chemical vapour deposition diamond studied by optical and electron spin resonance techniques. J Phys: Condens Matter 14:R467–R499

Iakoubovskii K, Adriaenssens GJ, Vohra YK (2000) Nitrogen incorporation in diamond films homoepitaxially grown by chemical vapour deposition. J Phys C: Condens Matter 12:L519–L524

Iakoubovskii K, Adriaenssens GJ, Dogadkin NN, Shiryaev AA (2001) Optical characterization of some irradiation-induced centers in diamond. Diamond Relat Mater 10:18–26

Iakoubovskii K, Kiflawi I, Johnston K, Collins A, Davies G, Stesmans A (2003) Annealing of vacancies and interstitials in diamond. Phys B 340–342:67–75

IGI (2022) IGI analyzes record 150 carat rough lab grown diamond. IGI Press Room, *https://www.igi.org/press/igi-analyzes-record-150-carat-rough-lab-grown-diamond*

Inspektor A, Liou Y, McKenna T, Messier R (1989) Plasma CVD diamond deposition in C–H–O systems. Surf Coat Technol 39:211–221

Irifune T, Kurio A, Sakamoto S, Inoue T, Sumiya H (2003) Ultrahard polycrystalline diamond from graphite. Nature 421:599–560

Irifune T, Kurio A, Sakamoto S, Inoue T, Sumiya H, Funakoshi K-I (2004) Formation of pure polycrystalline diamond by direct conversion of graphite at high pressure and high temperature. Phys Earth Planet Inter 143:593–600

Irifune T, Isobe F, Shinmei T (2014) A novel large-volume Kawai-type apparatus and its application to the synthesis of sintered bodies of nano-polycrystalline diamond. Phys Earth Planet Inter 228:255–261

Isberg J, Hammersberg J, Johansson E, Wikström T, Twitchen DJ, Whitehead AJ, Coe SE, Scarsbrook GA (2002) High carrier mobility in single-crystal plasma-deposited diamond. Science 297:1670–1672

Isoya J, Kanda H, Uchida Y (1990) EPR studies of interstitial Ni centers in synthetic diamond crystals. Phys Rev B 42:9843–9852

Jacob DE, Kronz A, Viljoen KS (2004) Cohenite, native iron and troilite inclusions in garnets from polycrystalline diamond ag- gregates. Contrib Mineral Petrol 146:566–576

Janssen G, van Enckevort WJ, Vollenberg W, Giling LJ (1992) Characterization of single-crystal diamond grown by chemical vapour deposition processes. Diamond Relat Mater 1:789–800

Johnson CE, Weimer WA, Hasting MAS (1990) Mass-spectroscopy of deposition exhaust-gases and oxidation of diamond films. Carbon 28:791–791

Johnson P, Myagkaya E (2017) Lab Notes: HPHT synthetic diamond with intense green color. Gems Gemol 53:96–98

Jones R (2009) Dislocations, vacancies and the brown colour of CVD and natural diamond. Diamond Relat Mater 18:820–826

Jones R, Briddon PR, Öberg S (1992) First-principles theory of nitrogen aggregates in diamond. Philos Mag Lett 66:67–74

Jones R, Torres VJB, Briddon PR, Öberg S (1994) Theory of nitrogen aggregates in diamond: The H3 and H4 defects. Mater Sci Forum 143–147:45–50

Jones R, Hounsome LS, Fujita N, Öberg S, Briddon PR (2007) Electrical and optical properties of multivacancy centres in diamond. Phys Stat Sol (a) 204:3059–3064

Kaminsky FV, Wirth R (2011) Iron carbide inclusions in lower- mantle diamond from Juina, Brazil. Can Mineral 49:555–572

Kammerling RC, Reinitz I, Fritsch E (1995) Lab Notes: Synthetic diamond suite. Gems Gemol 31:122–123

Kanda H (2000) Large diamonds grown at high pressure conditions. Braz J Phys 30:482–489

Kanda H (2005) Crystal growth of diamond. *In:* Bulk Crystal Growth of Electronic, Optical and Optoelectronic Materials. Capper P (ed) John Wiley & Sons, Ltd., Chichester, p 407–432

Kanda H, Lawson SC (1995) Growth temperature effects of impurities in HP/HT diamonds. Industrial Diamond Rev. 55:56–61

Kanda H, Watanabe K (1999) Distribution of nickel related luminescence centers in HPHT diamond. Diamond Relat Mater 8:1463–1469

Kanda H, Yamaoka S (1993) Inhomogeneous distribution of nitrogen impurities in {111} growth sectors of high pressure synthetic diamond. Diamond Relat Mater 2:1420–1423

Kanda H, Akaishi M, Setaka N, Yamaoka S, Fukunaga O (1980) Surface structures of synthetic diamonds. J Mater Sci 15:2743–2748

Kanda H, Ohsawa T, Fukunaga O, Sunagawa I (1989) Effect of solvent metals upon the morphology of synthetic diamonds. J Cryst Growth 94:115–124

Kanda H, Ohsawa T, Yamaoka S (1990) Formation of nitrogen pairs in synthetic diamond during growth. *In*: Science and Technology of New Diamond. Saito S, Fukunaga O, Yoshikawa M (eds) KTK Scientific Publishers/Terra Scientific Publishing Co., Tokyo, p 1420–1423

Kane RE (1980) The elusive nature of graining in gem quality diamonds. Gems Gemol 16:294–314

Kazuchits NM, Rusetsky MS, Kazuchits, VN, Zaitsev AM (2016) Aggregation of nitrogen in synthetic diamonds annealed at high temperature without stabilizing pressure. Diamond Relat Mater 64:202–207

Kennedy L, Johnson P (2016) Lab Notes: Yellow synthetic diamond with nickel-related green fluorescence. Gems Gemol 52:196–197

Khan RUA, Martineau PM, Cann BL, Newton ME, Twitchen DJ (2009) Charge transfer effects, thermo and photochromism in single crystal CVD synthetic diamond. J Phys: Condens Matter 21:364214

Khan RUA, Martineau PM, Cann BL, Newton ME, Dhillon HK, Twitchen DJ (2010) Color alterations in CVD synthetic diamond with heat and UV exposure: Implications for color grading and identification. Gems Gemol 46:18–26

Khan RUA, Cann BL, Martineau PM, Samartseva J, Freeth JJP, Sibley SJ, Hartland CB, Newton ME, Dhillon HK, Twitchen DJ (2013) Colour-causing defects and their related optoelectronic transitions in single crystal CVD diamond. J Phys: Condens Matter 25:275801

Khvostantsev LG, Slesarev VN, Brazhkin VV (2004) Toroid type high-pressure device: History and prospects. High Pressure Res 24:317–383

Kiflawi I, Mayer AE, Spear PM, Van Wyk JA, Woods GS (1994) Infrared absorption by the single nitrogen and A defect centres in diamond. Philos Mag B 69:1141–1147

Kiflawi I, Fisher D, Kanda H, Sittas G (1996) The creation of the 3107 cm^{-1} hydrogen absorption peak in synthetic diamond single crystals. Diamond Relat Mater 5:1516–1518

Kiflawi I, Kanda H, Mainwood A (1998) The effect of nickel and the kinetics of the aggregation of nitrogen in diamond. Diamond Relat Mater 7:327–332

Kim JR, Kim D-K, Zhu H, Abbaschian R (2011) High pressure and high temperature annealing on nitrogen aggregation in lab-grown diamonds. J Matter Sci 46:6264–6272

King JM, Moses TM, Shigley JE, Liu Y (1994) Color grading of colored diamonds in the GIA Gem Trade Laboratory. Gems Gemol 30:220–242

King JM, Moses TM, Shigley JE, Welbourn CM, Lawson SC, Cooper M (1998) Characterizing natural-color Type IIb blue diamonds. Gems Gemol 34:246–268

King JM, Shigley JE, Guhin SS, Gelb TH, Hall M (2002) Characterization and grading of natural-color pink diamonds. Gems Gemol 38:128–147

King JM, Shigley JE, Gelb TH, Guhin SS, Hall M, Wang W (2005) Characterization and grading of natural-color yellow diamonds. Gems Gemol 41:88–115

King JM, Geurts RH, Gilbertson AM, Shigley JE (2008) Color grading "D-to-Z" diamonds at the GIA Laboratory. Gems Gemol 44:296–321

Kocherzhinski YA, Kulik OG (1996) Equilibrium phase diagrams and manufacture of synthetic diamonds. Powder Metall Met Ceram 35:470–483

Kohn JA (1958) Twinning in diamond-type structures: a proposed boundary-structure model. Am Mineral 43:263–284

Koivula JI, Fryer CW (1984) Identifying gem-quality synthetic diamonds: An update. Gems Gemol 20:146–158

Koizumi S, Kamo M, Sato Y, Mita S, Sawabe A, Reznik A, Uzan-Saguy C, Kalish R (1998) Growth and characterization of phosphorus doped n-type diamond thin films. Diamond Relat Mater 7:540–544

Koizumi S, Umezawa H, Pernot J, Suzuki M (eds) (2018) Power Electronics Device Applications of Diamond Semiconductors. Woodhead Publishing, Duxford

Kolchin AV, Veprintsev VI, Klyachko LI, Novgorodov AS, Zubkov VM, Vereschagin LF, Khvostantsev LG, Novikov AP, Yanshin SI, Yanshina KM, Yanshina ES, Yanshin IS, Yanshina ZP (1981) Device for building-up high pressure. UA4290741

Koleske DD, Gates SM, Thoms BD, Russell JN, Butler JE (1994) Isothermal desorption of hydrogen from polycrystalline diamond films. Surf Sci 320:L105–L111

Koleske DD, Gates SM, Thoms BD, Russell JN, Butler JE (1995) Hydrogen on polycrystalline diamond films: Studies of isothermal desorption and atomic deuterium abstraction. J Chem Phys 102:992–1002

Kupriyanov IN, Khokhryakov AF, Borzdov YM, Palyanov YN (2016) HPHT growth and characterization of diamond from a copper-carbon system. Diamond Relat Mater 69:198–206

Lai MY, Breeding CM, Stachel T, Stern RA (2020) Spectroscopic features of natural and HPHT-treated yellow diamonds. Diamond Relat Mater 101:107642

Lang AR (1967) Causes of birefringence in diamond. Nature 213:248–251

Lang A (1979) Internal structure. *In*: The Properties of Diamond. Field JE, (ed) Academic Press, London, p 425–469

Lang AR, Moore M, Makepeace APW, Wierzchowski W, Welbourn CM (1991) On the dilatation of synthetic Type Ib diamond by substitutional nitrogen impurity. Phil Trans R Soc Lond A 337:497–520

Lang AR, Vincent R, Burton NC, Makepeace APW (1995) Studies of small inclusions in synthetic diamonds by optical microscopy, microradiography and transmission electron microscopy. J Appl Cryst 28:690–699

Lang AR, Yelisseyev AP, Pokhilenko NP, Steeds JW, Wotherspoon A (2004) Is dispersed nickel in natural diamonds associated with cuboid growth sectors in diamonds that exhibit a history of mixed-habit growth? J Cryst Growth 263:575–589

Law PL, Wang W (2016) Lab Notes: CVD synthetic diamond over 5 carats identified. Gems Gemol 52:414–416

Lawson SC, Kanda H (1993a) An annealing study of nickel point defects in high-pressure synthetic diamond. J Appl Phys 73:3967–3973

Lawson SC, Kanda H (1993b) Nickel in diamond: an annealing study. Diamond Relat Mater 2:130–135

Lawson SC, Davies G, Collins AT, Mainwood A (1992) The 'H2' optical transition in diamond: The effects of uniaxial stress perturbations, temperature and isotopic substitution. J phys: Condens Matter 4:3439–3452

Lawson SC, Kanda H, Watanabe K, Kiflawi I, Sato Y, Collins AT (1996) Spectroscopic study of cobalt-related optical centers in synthetic diamond. J Appl Phys 79:4348–4357

Lawson SC, Fisher D, Hunt DC, Newton ME (1998) On the existence of positively charged single-substitutional nitrogen in diamond. J Phys: Condens Matter 10:6171–6180

Lee SS, Minsek DW, Vestyck DJ, Chen P (1994) Growth of diamond from diamond from atomic-hydrogen and a supersonic free jet of methyl radicals. Science 263:1596–1598

Li Z-C, Jia X-P, Huang G-F, Hu M-H, Li Y, Yan B-M, Ma H-A (2013) FEM simulations and experimental studies of the temperature field in a large diamond crystal growth cell. Chin Phys B 22:014701

Liander H, Lundblad E (1960) Some observations on the synthesis of diamonds. Ark Kemi 16:139–149

Liang Q, Yan C-S, Meng Y, Lai J, Krasnicki S, Mao H-K, Hemley RJ (2009) Recent advances in high-growth rate single-crystal CVD diamond. Diamond Relat Mater 18:698–703

Liggins S (2010) Identification of Point Defects in Treated Single Crystal Diamond. PhD Dissertation, University of Warwick, Coventry, United Kingdom

Liggins S, Newton ME, Goss JP, Briddon PR, Fisher D (2010) Identification of the dinitrogen ⟨001⟩ split interstitial H1a in diamond. Phys Rev B 81:085214

Lightowlers EC, Collins AT (1976) Determination of boron in natural semiconducting diamond by prompt particle nuclear microanalysis and Schottky barrier differential-capacitance measurements. J Phys D: Appl Phys 9:951–963

Linde O, Kravchenko S, Epstein A, Rentmeesters (2021) The Global Diamond Industry 2020–21, Bain & Company, *https://www.bain.com/insights/global-diamond-industry-2020-21/*
Liu H, Dandy DS (1995) Diamond Chemical Vapor Deposition—Nucleation and Early Growth Stages. Noyes Publications, Park Ridge
Loubser JHN, van Wyk JA (1977) A triplet state in irradiated and annealed Type Ib diamonds. Diamond Res 11:11–14
Lowther JE (1993) Excited states of the vacancy in diamond. Phys Rev B 48:11592–11601
Luo Y, Breeding CM (2013) Fluorescence produced by optical defects in diamond: Measurement, characterization, and challenges. Gems Gemol 40:82–97
Ma J, Cheesman A, Ashfold MNR, Hay KG, Wright S, Langford N, Duxbury G, Mankelevich YA (2009) Quantum cascade laser investigations of CH_4 and C_2H_2 interconversion in hydrocarbon/H_2 gas mixtures during microwave plasma enhanced chemical vapor deposition of diamond. J Appl Phys 106:033305
Mäki JM, Tuomisto F, Kelly C, Fisher D, Martineau P (2007) Effects of thermal treatment on optically active vacancy defects in CVD diamonds. Phys B 401–402:613–616
Martineau PM, Lawson SC, Taylor AJ, Quinn SJ, Evans DJF, Crowder MJ (2004) Identification of synthetic diamond grow using Chemical Vapor Deposition (CVD). Gems Gemol 40:2–25
Martineau P, Gaukroger M, Khan R, Evans D (2009) Effect of steps on dislocations in CVD diamond grown on {001} substrates. Phys Status Solidi C 6:1953–1957
McGuinness CD, Wassell AM, Lanigan MP, Lynch SA (2020) Separation of natural from laboratory-grown diamond using time-gated luminescence imaging. Gems Gemol 56:220–229
Meng YF, Yan CS, Lai J, Krasnicki S, Shu H, Yu T, Liang Q, Mao HK, Hemley RJ (2008) Enhanced optical properties of chemical vapor deposited single crystal diamond by low-pressure/high-temperature annealing. PNAS 105:17620–17625
Mikhail S, Guillermier C, Franchi IA, Beard AD, Crispin K, Verchovsky AB, Jones AP, Milledge HJ (2014) Empirical evidence for the fractionation of carbon isotopes between diamond and iron carbide from the Earth's mantle. Geochem Geophys Geosyst 15:855–866
Mita, Y (1996) Change of absorption spectra in Type Ib diamond with heavy neutron irradiation. Phys Rev B 53:11360–11364
Mita Y, Nisida Y, Suito K, Onodera A, Yazu S (1990) Photochromism of H2 and H3 centres in synthetic Type Ib diamonds. J Phys: Condens Matter 2:8567–8574
Moe KS, D'Haenens-Johansson U, Wang W (2015) Lab Notes: Irradiated green-blue CVD synthetic diamonds. Gems Gemol 51:320–321
Moe KS, Johnson P, D'Haenens-Johansson U, Wang W (2017) Lab Notes: A synthetic diamond overgrowth on a natural diamond. Gems Gemol 53:237–239
Mokuno Y, Chayahara A,Soda Y, Horino Y, Fujimori N (2005) Synthesizing single-crystal diamond by repetition of high rate homoepitaxial growth by microwave plasma CVD. Diamond Relat Mater 14:1743–1746
Moses TM, Reinitz I, Fritsch E, Shigley JE (1993) Two treated-color synthetic red diamonds seen in the trade. Gems Gemol 29:182–190
Moses TM, Reinitz IM, Johnson ML, King JM, Shigley JE (1997) A contribution to understanding the effect of blue fluorescence on the appearance of diamonds. Gems Gemol 22:244–259
Muranaka Y, Yamashita H, Miyadera H (1991) Synthesis and purification of diamond films using the microwave plasma of a CO-H_2 system. J Mater Sci 26:3235–3243
Nad S, Gu Y, Asmussen J (2015) Growth strategies for large and high quality single crystal diamond substrates. Diamond Relat Mater 60:26–34
Nad S, Charris A, Asmussen, J (2016) MPACVD growth of single crystalline diamond substrates with PCD rimless and expanding surfaces. Appl Phys Lett 109:162103
Nadolinny VA, Yelisseyev AP (1993) New paramagnetic centres containing nickel ions in diamond. Diamond Relat Mater 3:17–21
Nadolinny VA, Yelisseyev AP, Yuryeva OP, Feygelson BN (1997) EPR study of the transformations in nickel containing centres at heated synthetic diamonds. Appl Magn Resol 12:543–554
Nadolinny V, Yelisseyev, A, Yurjeva, O, Hofstaetter A, Meyer B, Feigelson B (1998) Relationship between electronic states of nickel-containing centres and donor nitrogen in synthetic and natural diamonds. Diamond Relat Mater 7:1558–1561
Nadolinny VA, Yelisseyev AP, Baker JM, Newton ME, Twitchen DJ, Lawson SC, Yuryeva OP, Feigelson BN (1999) A study of ^{13}C hyperfine structure in the EPR of nickel-nitrogen-containing centres in diamond and correlation with their optical properties. J Phys: Condens Matter 11:7357–7376
Nazaré MH, Rino LM (1993) Synthetic diamond: the optical band at 1.883 eV. J Mater Sci Eng B 21:329–332
Nazaré MH, Neves AJ, Davies G (1991) Optical studies of the 1.40-eV Ni center in diamond. Phys Rev B 43:14196
Nazaré MH, Mason PW, Watkins GD, Kanda H (1995) Optical detection of magnetic resonance of nitrogen and nickel in high-pressure synthetic diamond. Phys Rev B 51:16741–16745
Nazaré MH, Rino LM, Kanda H (1996) High pressure synthetic diamond: optical studies of nickel related defects. Mater Sci Eng A 290:302–305
Nazaré MH, Lopes JC, Neves AJ (2001) Nickel related defects in diamond: The 2.51eV band. Phys B 308–310:616–620

Nemanich RJ, Carlisle JA, Hirata A, Haenen K (2014) CVD diamond—Research, applications, and challenges. MRS Bull 39:490–498

Noble CJ, Pawlik Th, Spaeth J-M (1998) Electron paramagnetic resonance investigations of nickel defects in natural diamonds. J Phys: Condens Matter 50:11781–11793

Palyanov YN, Borzdov YM, Khokhryakov AF, Kupriyanov IN, Sokol AG (2010) Effect of nitrogen impurity on diamond crystal growth processes. Cryst Growth Des 10:3169–3175

Palyanov YN, Kupriyanov IN, Khokhryakov AF, Ralchenko VG (2015a) Crystal growth of diamond. *In:* Handbook of Crystal Growth. Rudolph P (ed) Elsevier, Boston, p 671–713

Palyanov YN, Kupriyanov IN, Borzdov YM, Surovtsev NV (2015b) Germanium: A new catalyst for diamond synthesis and a new optically active impurity in diamond. Sci Rep 5:1–8. doi: 10.1038/srep14789

Palyanov YN, Kupriyanov IN, Khokhryakov AF, Borzdov YM (2017) High-pressure crystallization and properties of diamond from magnesium-based catalysts. CrystEngComm 19:4459–4475

Pate BB (1986) The diamond surface: Atomic and electronic structure. Surf Sci 165:83–142

Pehrsson PE, Celii FG, Butler JE (1993) Chemical Mechanisms of Diamond CVD. *In*: Diamond Films and Coatings. Davis R, (ed) Noyes Publications, Park Ridge, p 68–146

Pereira E, Santos L, Pereira L, Hofmann DM, Christmann P, Stadler W, Meyer BK (1994) Slow emission of the 2.56 eV centre in synthetic diamond. Diamond Relat Mater 4:53–58

Pereira E, Pereira L, Hofmann DM, Stadler W, Meyer BK (1995) Excited levels of the 2.56 eV emission in synthetic diamond. Radiat Eff Defects Solids 135:1–4

Peretti A, Herzog F, Bieri W, Alessandri M, Günther D, Frick DA, Cleveland E, Zaitsev AM, Deljanin B (2014) New generation of synthetic diamonds reches the market. Part A: Identification of CVD-grown blue diamonds. Contrib Gemol 14:3–20

Persaud S, Wang W, Johnson P, Balov IP (2018) Lab Notes: One natural melee diamond found in large batch of HPHT synthetic melee. Gems Gemol 54:208–209

Philip J, Hess P, Feygelson T, Butler JE, Chattopadhyay S, Chen KH, Chen LC (2003) Elastic, mechanical, and thermal properties of nanocrystalline diamond films. J Appl Phys 93:2164–2171

Pinto H, Jones R (2009) Theory of the birefringence due to dislocations in single crystal CVD diamond. J Phys Condens Matter 21:364220

Ponahlo J (1992) Cathodoluminescence (CL) and CL spectra of De Beers' experimental synthetic diamonds. J Gemmol 23:3–17

Poon TPY, Wang W (2016) Lab Notes: Blue HPHT synthetic diamond over 10 carats. Gems Gemol 52:416

Prelas MA, Popovici G, Biglow LK (eds) (1997) Handbook of Industrial Diamonds and Diamond Films. Marcel Dekker, Inc., New York

Renfro N (2015) Digital photomicrography for gemologists. Gems Gemol 51:144–159

Robertson R, Fox, JJ, Martin AE (1934) Two types of diamond. Philos Trans R Soc, A 232:463–535

Robins LH, Cook LP, Farabaugh EN, Feldman A (1989) Cathodoluminescence of defects in diamond films and particles grown by hot-filament chemical-vapor deposition. Phys Rev B 39:13367–13377

Robinson A (2018) Irradiated HPHT-grown diamonds might escape detection as synthetics, says lab. IDEX Online, *http://www.idexonline.com/FullArticle?Id=43871*

Rogers LJ, Jahnke KD, Doherty MW, Dietrich A, McGuinness LP, Müller C, Teraji T, Sumiya H, Isoya J, Manson NB, Jelezko F (2014) Electronic structure of the negatively charged silicon-vacancy center in diamond. Phys Rev B 89:235101

Rooney MLT (1992) {115} Growth in boron-doped synthetic diamonds. J Cryst Growth 116:15–21

Rooney MLT, Welbourn CM, Shigley JE, Fritsch E, Reinitz I (1993) De Beers near colorless-to-blue experimental gem-quality synthetic diamonds. Gems Gemol 29:38–45

Rose BC, Huang D, Zhang ZH, Stevenson P, Tyryshkin AM, Sangtawesin S, Srinivasan S, Loudin L, Markham ML, Edmonds AM, Twitchen DJ, Lyon SA, De Leon NP (2018) Observation of an environmentally insensitive solid-state spin defect in diamond. Science 361:60–63

Rossman G, Kirschvink J (1984) Magnetic properties of gem-quality synthetic diamonds. Gems Gemol 20:163–166

Ruan J, Choyke WJ, Partlow WD (1991) Cathodoluminescence and annealing study of plasma-deposited polycrystalline diamond films. J Appl Phys 69:6632–6636

Ruan J, Kobashi K, Choyke WJ (1992) On the "band-A" emission and boron related luminescence in diamond. Appl Phys Lett 60:3138–3140

Rzepka E, Silva F, Lusson A, Rivière A, Gicquel A (2001) Luminescence study of polycrystalline CVD diamond films containing a small amount of nitrogen. Diamond Relat Mater 10:542–545

Satoh S, Sumiya H, Tsuji K, Yazu S (1990) Differences in nitrogen concentration and aggregation among (111) and (100) growth sectors of large synthetic diamonds. *In*: Science and Technology of New Diamond. Saito S, Fukunaga O, Yoshikawa M (eds) KTK Scientific Publishers/Terra Scientific Publishing Co., Tokyo, p 351–355

Setaka N (1994) History of diamond synthesis in Japan. *In*: Synthetic Diamond: Emerging CVD Science and Technology, The Electrochemical Society Series. Spear KE, Dismukes JP (eds) John Wiley, Interscience, New York, p 57–90

Sharp WE (1966) Phyrrhotite: A common inclusion in South African diamonds. Nature 211:402–403

Shigley JE, Fritsch E (1990) Optical properties of some greenish blue-to-green diamonds. SPIE Proc 1325:315–324

Shigley JE, Fritsch E, Stockton CM, Koivula JI, Fryer CW, Kane RE (1986) The gemological properties of the Sumitomo gem-quality synthetic yellow diamonds. Gems Gemol 22:192–208

Shigley JE, Fritsch E, Stockton CM, Koivula JI, Fryer CW, Kane RE, Hargett DR, Welch CW (1987) The gemological properties of the De Beers gem-quality synthetic diamonds. Gems Gemol 23:187–206

Shigley JE, Fritsch E, Reinitz I, Moon M (1992) An update on Sumitomo gem-quality synthetic diamonds. Gems Gemol 28:116–122

Shigley JE, Fritsch E, Koivula JI, Sobolev NV, Malinovsky IY, Pal'yanov YN (1993a) The gemological properties of Russian gem-quality synthetic yellow diamonds. Gems Gemol 29:228–248

Shigley JE, Fritsch E, Reinitz I (1993b) Two near-colorless General Electric Type-IIa synthetic diamond crystals. Gems Gemol 29:191–197

Shigley JE, Moses TM, Reinitz I, Elen S, McClure SF, Fritsch E (1997) Gemological properties of near-colorless synthetic diamonds. Gems Gemol 33:42–53

Shigley JE, Abbaschian R, Clarke C (2002) Gemesis laboratory-created diamonds. Gems Gemol 38:301–309

Shigley JE, McClure SF, Breeding CM, Shen AHT, Muhlmeister SM (2004a) Lab-grown colored diamonds from Chatham created gems. Gems Gemol 40:128–145

Shigley JE, Breeding CM, Shen AHT (2004b) An updated chart on the characteristics of HPHT-grown synthetic diamonds. Gems Gemol 40:303–313

Shvyd'ko Y, Blank V, Terentyev S (2017) Diamond x-ray optics: Transparent, resilient, high-resolution, and wavefront preserving. MRS Bull 42:437–444

Silva F, Achard J, Bonnin X, Michau A, Tallaire A, Brinza O, Gicquel A (2006) 3D crystal growth model for understanding the role of plasma pre-treatment on CVD diamond crystal shape. Phys Stat Sol (a) 203:3049–3055

Silva F, Bonnin X, Achard J, Brinza O, Michau A, Gicquel A (2008) Geometric modeling of homoepitaxial CVD diamond growth: I. The {100}{111}{113} system. J Cryst Growth 310:187–203

Silva F, Achard J, Brinza O, Bonnin X, Hassouni K, Anthonis A, De Corte K, Barjon J (2009) High quality, large surface area, homoepitaxial MPACVD diamond growth. Diamond Relat Mater 18:683–697

Sittas G, Kanda H, Kiflawi I, Spear PM (1996) Growth and characterization of Si-doped diamond single crystals grown by the HPHT method. Diamond Relat Mater 5:866–869

Smith EM, Shirey SB, Nestola F, Bullock ES, Wang J, Richardson SH, Wang W (2016) Large gem diamonds from metallic liquid in Earth's deep mantle. Science 354:1403–1405

Smith EM, Shirey SB, Richardson SH, Nestola F, Bullock ES, Wang J, Wang W (2018a) Blue boron-bearing diamonds from Earth's lower mantle. Nature 560:84–87

Smith EM, Shirey SB, Wang W (2018b) The very deep origin of the world's biggest diamonds. Gems Gemol 53:388–403

Song Z, Lu T, Liu H, Dai H, Ke J, Zhu W, Zhang J (2020) Identification of Type IIa blue CVD synthetic diamonds from Huzhou SinoC Semiconductor Co. in China. J Gemmol 37:306–313

Song Z, Dai H, Gao B, Zhu W (2021) Identification of pink-coloured CVD synthetic diamonds from Huzhou Sino-C Semiconductor Co. in China. Crystals 11:872

Stalder KR, Sharpless RL (1990) Plasma properties of a hydrocarbon arcjet used in th eplasma deposition of diamond thin-films. J Appl Phys 68:6187–6190

Stachel T, Harris JW (2008) The origin of cratonic diamonds—Constraints from mineral inclusions. Ore Geol Rev 34:5–32

Steeds JW, Charles S, Davis TJ, Gilmore A, Hayes J, Pickard D, Butler JE (1999a) Creation and mobility of self-interstitials in diamond by use of a transmission electron microscope and their subsequent study by photoluminescence microscopy. Diamond Relat Mater 8:94–100

Steeds JW, Davis TJ, Charles SJ, Hayes JM, Butler JE (1999b) 3H luminescence in electron-irradiated diamond samples and its relationship to self-interstitials. Diamond Relat Mater 8:1847–1852

Stoupin S, Antipov S, Butler JE, Kolyadin AV, Katrusha A (2016) Large-surface-area diamond (111) crystal plates for applications in high-heat-load wavefront-preserving X-ray crystal optics. J Synchrotron Radiat 23:1118–1123

Strong HM (1977) Novel diamond products and the manufacture thereof. US4042673

Strong HM (1989) Early diamond making at General Electric. Am J Phys 57:794–801

Strong HM, Chrenko RM (1971) Further studies on diamond growth rates and physical properties of laboratory-made diamond. J Phys Chem 75:1838–1843

Sugano T, Ohashi N, Tsurumi T, Fukunaga O (1996) Pressure and temperature region of diamond formation in systems graphite and Fe containing alloy. Diamond Relat Mater 5:29–33

Sumida N, Lang AR (1981) Cathodoluminescence evidence of dislocation interactions in diamond. Philos Mag A 43:1277–1287

Sumiya H (2014) HPHT synthesis of large, high-quality, single crystal diamonds. *In*: Comprehensive Hard Materials. Sarin V (ed) Elsevier, Oxford, p 195–215

Sumiya H, Satoh S (1996) High-pressure synthesis of high-purity diamond crystal. Diamond Relat Mater 5:1359–1365

Sumiya H, Tamasaku K (2012) Large defect-free synthetic Type IIa diamond crystals synthesized via high pressure and high temperature. Jpn J Appl Phys 51:090102

Sumiya, H, Toda N, Nishibayashi Y, Satoh S (1997) Crystalline perfection of high purity synthetic diamond crystal. J Cryst Growth 178:485–494
Sumiya H, Toda N, Satoh S (2002) Growth rate of high-quality large diamond crystals. J Cryst Growth 237–239:1281–1285
Sumiya H, Harano K, Tamasaku K (2015) HPHT synthesis and crystalline quality of large high-quality (001) and (111) diamond crystals. Diamond Relat Mater 58:221–225
Sunagawa I (1984) Morphology of natural and synthetic diamond crystals. *In*: Materials Science of the Earth's Interior. Sunagawa I (ed) Terra Scientific Publishing Co., Tokyo, p 303–330
Sunagawa I (1995) The distinction of natural from synthetic diamond. J Gemmol 24:485–499
Takeuchi D, Watanabe H, Yamanaka S, Okushi H, Kajimura K. (2000) Homoepitaxial diamond films grown by step-flow mode in various misorientation angles of diamond substrates. Diamond Relat Mater 9:231–235
Takeuchi D, Watanabe H, Yamanaka S, Okushi H, Sawada H, Ichinose H, Sekiguchi T, Kajimura K (2001) Origin of band-A emission in diamond thin films. Phys Rev B 63:245328
Tallaire A, Collins AT, Charles D, Achard J, Sussmann R, Gicquel A, Newton ME, Edmonds AM, Cruddace RJ (2006) Characterisation of high-quality thick single-crystal diamond grown by CVD with a low nitrogen addition. Diamond Relat Mater 15:1700–1707
Tallaire A, Achard J, Silva F, Brinza O, Gicquel A (2013) Growth of large size diamond single crystals by plasma assisted chemical vapour deposition: Recent achievements and remaining challenges. C R Phys 14:169–184
Tallaire A, Mille V, Brinza O, Thi TNT, Brom JM, Loguinov Y, Katrusha A, Koliadin A, Achard J (2017a) Thick CVD diamond films grown on high-quality Type IIa HPHT diamond substrates from New Diamond Technology. Diamond Relat Mater 77:146–152
Tallaire A, Brinza O, Mille V, William L, Achard J (2017b) Reduction of dislocations in single crystal diamond by lateral growth over a macroscopic hole. Adv Mater 29:1604823
Tennant S (1797) On the nature of the diamond. Phil Trans R Soc Lond 87:123–127
Thomas RE, Rudder RA, Markunas RJ (1992) Thermal desorption from hydrogenated and oxygenated diamond (100) surfaces. J Vac Sci Technol A 10:2451–2457
Tolansky S (1966) Birefringence of diamond. Nature 211:158–160
Tolansky S, Sunagawa I (1959) Spiral and other growth forms of synthetic diamonds: A distinction between natural and synthetic diamond. Nature 184:1523–1527
Tran Thi TN, Morse J, Caliste D, Fernandez B, Eon D, Hartwig J, Barbay C, Mer-Calfati C, Tranchant N, Arnault JC, Lafford TA, Baruchel J (2017) Synchrotron Bragg diffraction imaging characterization of synthetic diamond crystals for optical and electronic power device applications. J Appl Cryst 50:561–569
Twitchen DJ, Hunt DC, Smart V, Newton ME, Baker JM (1999) Correlation between ND1 optical absorption and the concentration of negative vacancies determined by electron paramagnetic resonance (EPR). Diamond Relat Mater 8:1572–1575
van Wyk JA, Woods GS (1995) Electron spin resonance of excited states of the H3 and H4 centres in irradiated Type Ia diamonds. J Phys: Condens Matter 7:5901–5911
Varma CKR (1970) Evidence of deformation twinning in natural diamond crystals. J Phys Chem Solids 31:890
Vasil'ev EA, Ivanov-Omskii VI, Pomazanskii BS, Bogush IN (2004) The N3 center luminescence quenched by nitrogen impurity in natural diamond. Tech Phys Lett 30:802–803
Vins VG, Feigelson BN, Eliseev AP, Patrin NS, Patrenin YV, Nehaev PY (1991) Optically active defects in diamonds grown at 1350–1750 °C. Sverkhtverd. Mater. 3:21–26 (in Russian)
Wakatsuki M (1966) New catalysts for synthesis of diamond. Jap J Appl Phys 5:337
Wakatsuki M, Ohnishi A, Jia X, Kurosawa M, Sueno S, Hayakawa S, Gohshi Y (1999) Growth of diamond with Zr-containing molten metal solvents and metal elements as incorporated impurities. Diamond Relat Mater 8:1438–1440
Walker J (1977) An optical study of the TR12 and 3H defects in irradiated diamond. J Phys C: Solid State Phys 10:3031–3037
Walmsley JC, Lang AR (1983) Transmission electron-microscopic observations of deformation and microtwinning in a synthetic diamond compact. J Mater Sci Lett 2:785–788
Wang W, Moses T (2007) Lab Notes: Type IIa diamond with intense green color introduced by Ni-related defects. Gems Gemol 43:156–158
Wang W, Moses TM, Linares RC, Shigley JE, Hall MS, Butler JE (2003) Gem-quality synthetic diamonds grown by a chemical vapor deposition (CVD) method. Gems Gemol 39:268–283
Wang W, Smith CP, Hall MS, Breeding CM, Moses TM (2005a) Treated-color pink-to-red diamonds from Lucent Diamonds Inc. Gems Gemol 41:6–19
Wang W, Tallaire A, Hall MS, Moses TM, Achard J, Sussmann RS, Gicquel A (2005b) Experimental CVD synthetic diamonds from LIMHP-CNRS, France. Gems Gemol 41:234–244
Wang W, Hall M, Breeding CM (2007a) Natural Type Ia diamond with green-yellow color due to Ni-related defects. Gems Gemol 43:240–243
Wang W, Hall MS, Moe KS, Tower J, Moses TM (2007b) Latest-generation CVD-grown synthetic diamonds from Apollo Diamond Inc. Gems Gemol 43:294–312
Wang W, Doering P, Tower J, Lu R, Eaton-Magaña S, Johnson P, Emerson E, Moses TM (2010) Strongly colored pink CVD lab-grown diamonds. Gems Gemol 46:4–17

Wang W, D'Haenens-Johansson UFS, Johnson P, Moe KS, Emerson E, Newton ME, Moses TM (2012) CVD synthetic diamond from Gemesis Corp. Gems Gemol 48:80–97

Wang W, Persaud S, Myagkaya E (2022) New record size for CVD laboratory-grown diamond. GIA Research News, *https://www.gia.edu/gia-news-research/january-2022-largest-cvd*

Wassell AM, McGuiness CD, Hodges C, Lanigan PMP, Fisher D, Martineau PM, Newton ME, Lynch SA (2018) Anomalous green luminescent properties in CVD synthetic diamonds. Phys Stat Sol (a) 215:1800292

Watanabe K, Lawson SC, Isoya J, Kanda H, Sato Y (1997) Phosphorescence in high-pressure synthetic diamond. Diamond Relat Mater 6:99–106

Welbourn CM, Cooper M, Spear PM (1996) De Beers natural versus synthetic diamond verification instruments. Gems Gemol 32:156–169

Wentorf RH (1971) Some studies of diamond growth rates. J Phys Chem 75:1833–1837

Woods GS (1984) Infrared absorption studies of the annealing of irradiated diamonds. Philos Mag B 50:673–688

Woods GS, Collins AT (1982) The 1450 cm^{-1} infrared absorption in annealed, electron-irradiated Type I diamonds. J Phys C: Solid State Phys 15:L949–L952

Woods GS, Collins AT (1983) Infrared absorption spectra of hydrogen complexes in type I diamonds. J Phys Chem Solids 44:471–475

Woods G, Lang A (1975) Cathodoluminescence, optical absorption and x-ray topographic studies of synthetic diamonds. J Cryst Growth 28:215–226

Wort CJH, Balmer RS (2008) Diamond as an electronic material. Mater Today 11:22–28

Yamada H, Chayahara A, Mokuno Y, Umezawa H, Shikata SI, Fujimori N (2011) Fabrication of 1 inch mosaic crystal diamond wafers. Appl Phys Express 3:051301

Yamada H, Chayahara A, Mokuna Y, Tsubouchi N, Shikata SI (2013) Uniform growth and repeatable fabrication of inch-sized wafers of a single-crystal diamond. Diamond Relat Mater 33:27–31

Yamada H, Chayahara A, Mokuno Y, Kato Y, Shikata S (2014) A 2-in. mosaic wafer made of a single-crystal diamond. Appl Phys Lett 104:102110

Yamada H, Chayahara A, Mokuno Y (2016) Effects of intentionally introduced nitrogen and substrate temperature on growth of diamond bulk single crystals. Jpn J Appl Phys 55:01AC07

Yang NJ, Yu SY, Macpherson JV, Einaga Y, Zhao HY, Zhao GH, Swain GM, Jiang X (2019) Conductive diamond: synthesis, properties, and electrochemical applications. Chem Soc Rev 48:157–204

Yelisseyev A, Kanda H (2007) Optical centers related to 3d transition metals in diamond. New Diamond Front Carbon Technol 17:127–178

Yelisseyev AP, Nadolinny VA (1995) Photoinduced absorption lines related to nickel impurity in annealed synthetic diamonds. Diamond Relat Mater 4:177–185

Yelisseyev A, Nadolinny V, Feigelson B, Terentyev S, Nosukhin S (1996) Spatial distribution of impurity defects in synthetic diamonds obtained by the BARS technology. Diamond Relat Mater 5:1113–1117

Yelisseyev A, Babich Yu, Nadolinny V, Fisher D, Feigelson B (2002) Spectrospic study of HPHT synthetic diamonds, as grown at 1500 °C. Diamond Relat Mater 11:22–37

Yelisseyev A, Lawson S, Sildos I, Osvet A, Nadolinny V, Feigelson B, Baker JM, Newton M, Yuryeva O (2003) Effect of HPHT annealing on the photoluminescence of synthetic diamonds grown in the Fe–Ni–C system. Diamond Relat Mater 12:2147–2168

Yelisseyev AP, Steeds JW, Babich YV, Feigelson BN (2006) A new approach to investigation of nickel defect transformation in the HPHT synthetic diamonds using local optical spectroscopy. Diamond Relat Mater 15:1886–1890

Zaitsev AM (2000) Vibronic spectra of impurity-related optical centers in diamond. Phys Rev B 61:909–922

Zaitsev AM (2001) Optical Properties of Diamond: A Data Handbook. Springer, Berlin

Zaitsev AM, Moe KS, Wang W (2018) Defect transformations in nitrogen-doped CVD diamond during irradiation and annealing. Diamond Relat Mater 88:237–255

Zaitsev AM, Kazuchits NM, Kazuchits VN, Moe KS, Rusetsky MS, Korolik OV, Kitajima K, Butler JE, Wang W (2020) Nitrogen-doped CVD diamond: Nitrogen concentration, color and internal stress. Diamond Relat Mater 105:107794

Zaitsev AM, Kazuchits NM, Moe KS, Butler JE, Korolik OV, Rusetsky MS, Kazuchits VN (2021) Luminescence of brown CVD diamond: 468 nm luminescence center. Diamond Relat Mater 113:108255

Zeitlin A, Herman H, Brayman J (1961) Super-high pressure apparatus. US 2968837

Zhang Z-H, Stevenson P, Thiering G, Rose BC, Huang D, Edmonds AM, Markham ML, Lyon SA, Gali A, De Leon NP (2020) Optically detected magnetic resonance in neutral silicon vacancy centers in diamond via bound exciton states. Phys Rev Lett 125:237402

Zhu H, Pearson K, Sunset Peak International Limited (2019) Multi-heater method for growing high quality diamond. US10166518 B2

Zimnisky P (2020) State of the Diamond Market—Independent Monthly Data Analysis, August 2020, Volume 3, Issue 12

Zimnisky P (2021) Lab-grown diamond jewelry market forecast to almost double in size by 2025. Paul Zimnisky Diamond Analytics, *https://www.paulzimnisky.com/Lab-Grown-Diamond-Jewelry-Market-Forecast-to-Almost-Double-in-Size-by-2025*

Reviews in Mineralogy & Geochemistry
Vol. 88 pp. 755–808, 2022

14

Experimental Petrology Applied to Natural Diamond Growth

Robert W. Luth

Department of Earth and Atmospheric Sciences
University of Alberta
1-26 Earth Sciences Building
Edmonton, Alberta T6G 2E3
Canada

robert.luth@ualberta.ca

Yuri N. Palyanov

V.S. Sobolev Institute of Geology and Minerology
Siberian Branch of the Russian Academy of Sciences
3, Koptyug ave., Novosibirsk 630090
Russia
and
Novosibirsk State University
2, Pirogov str., Novosibirsk 630090
Russia

palyanov@igm.nsc.ru

Hélène Bureau

Institut de Minéralogie de Physique de la Matière et de Cosmochimie
Sorbonne Université, UMR CNRS 7590, Muséum National d'Histoire Naturelle
Campus Jussieu boite 115
4, place Jussieu, 75005 Paris
France

helene.bureau@upmc.fr

INTRODUCTION

How can experiments help?

Perhaps the first point to address in this chapter is what role, or roles, experimental petrology can play in understanding the formation of natural diamond. As can be seen from the other chapters in this volume, there is a rich diversity of diamonds—even confining ourselves to terrestrial examples, there are diamonds that form in impacts, those that are found in ophiolites and in UHP metamorphic rocks, as well as those that are found in mantle-derived volcanics such as kimberlites and lamproites. Some of these—such as those found in ophiolites (e.g., Farré-de-Pablo et al. 2018 and references therein; Litasov et al. 2019a,b)—may form under *P*, *T* conditions at which diamond is metastable rather than stable, although such occurrences are controversial (Farré-de-Pablo et al. 2019; Massonne 2019; Yang et al. 2019). Others, such as mantle-derived monocrystalline, fibrous, and coated diamonds, presumably form at *P*, *T* conditions at which diamond is the stable polymorph of carbon. This diversity presents a rich *P*, *T* composition space for experimentalists to explore to provide some insights into diamond formation.

1529-6466/22/0088-0014$10.00 (print)
1943-2666/22/0088-0014$10.00 (online)
http://dx.doi.org/10.2138/rmg.2022.88.14

Experiments allow researchers to explore first-order questions such as

- Can diamond form from this particular composition of fluid or melt?
- Can diamond form from a specific redox reaction, such as carbonate reacting with a reduced fluid?

Experiments can also address questions such as how the diamond-forming melts or fluids originally form, and how they evolve as they interact with different lithologies in the Earth.

The vast majority of natural diamonds form in the Earth's lithospheric mantle (See Nimis 2022, this volume; Stachel et al. 2022a, this volume), and so unsurprisingly most of the experimental work to date has focused on this pressure range. There has been some work at much higher pressure and temperature conditions extending to those of the deep Earth, however, and we shall touch on those as well in this review.

CONCEPTUAL BACKGROUND

Stability of diamond

In the one-component carbon system, diamond is the high-pressure polymorph stable at pressures above the graphite—diamond transition (Kennedy and Kennedy 1976; Day 2012). Adding more components makes the situation more complex: in C–O, diamond would be stable in a region of P, T, fO_2 space bounded by the reactions

$$\text{C (graphite)} = \text{C (diamond)} \tag{1}$$

and

$$\text{C (diamond)} + O_2 = CO_2 \text{ (fluid)} \tag{2}$$

If the fluid phase contains H as well, then the relevant system is C–O–H (Fig. 1), in which the diamond saturation curve separates a two-phase field of diamond + fluid from a field of C-undersaturated fluid. It should be noted that the fO_2 of the fluid increases from left to right along the diamond saturation curve, and for context the value of fO_2 for the iron-wüstite buffer (IW), and the locations of fluids with values of $\log(fO_2)$ three and four units below the calculated (metastable) location of the fayalite-magnetite-quartz buffer (FMQ) are shown.

The situation in the Earth's mantle is more complex, given that CO_2 becomes unstable in the presence of olivine (Newton and Sharp 1975) because of the carbonation reaction

$$Mg_2Si_2O_6 \text{ (enstatite)} + 2MgCO_3 \text{ (magnesite)} = 2Mg_2SiO_4 \text{ (forsterite)} + 2CO_2. \tag{3}$$

A consequence of the stability of this reaction is the well-known EMOD reaction

$$\begin{gathered} Mg_2Si_2O_6 \text{ (enstatite)} + 2MgCO_3 \text{ (magnesite)} = \\ 2Mg_2SiO_4 \text{ (forsterite)} + 2C \text{ (diamond)} + O_2 \end{gathered} \tag{4}$$

(Eggler and Baker 1982), which further constrains the stability of diamond. Schematically, Figure 2 shows the fields of stability for carbonate, diamond, and fluid in T–$\log(fO_2)$ space at constant pressure within the diamond P, T stability field. Under conditions where a carbonate melt rather than crystalline magnesite is stable, an analogous reaction to (3) where $MgCO_3$ is a component in the melt would apply, as discussed by Stagno and Frost (2010).

Under more reducing conditions such as those proposed to occur with increasing depth in the Earth's mantle (see overview by Stagno 2019), carbon can dissolve in the alloy or form carbides such as cohenite, $(Fe, Ni)_3C$, or Fe_7C_3. Another complication is the presence of sulfur—with a bulk silicate Earth estimated concentration on the same order as that for carbon

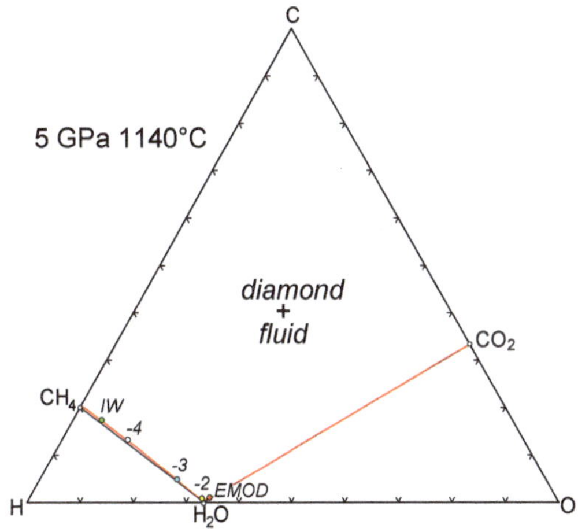

Figure 1. The system C–O–H at 5 GPa, 1140 °C. The red line denotes the composition of fluid in equilibrium with diamond as a function of composition. Numbers along the line are values of log fO_2 relative to that of the FMQ reference buffer. Also for reference are shown the locations of the fluids in equilibrium with the iron–wüstite reference buffer (IW) and the EMOD reaction (see text). Calculated with GFluid (Zhang and Duan 2010).

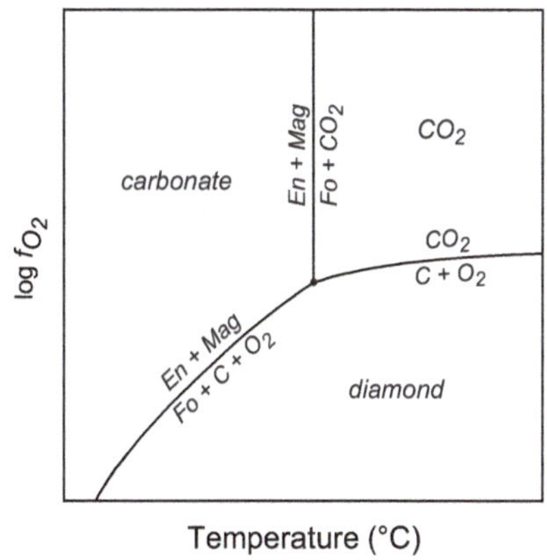

Figure 2. Schematic log fO_2—temperature section at constant pressure, showing the relationship between reactions (2), (3), and (4) in the text. The fields are labeled with the stable carbon-bearing form that would coexist with forsterite and enstatite. See text for discussion.

(~250 ppmw, McDonough and Sun 1995). Although there is ongoing debate about the relative importance of alloys, carbides, and sulfide melts as hosts of carbon in the Earth's mantle, the key point for the experimentalist is that all three of these are at least potential sources of carbon that could precipitate diamond upon oxidation—a process tractable to experimental study.

Source of carbon

Considerations of the large-scale carbon cycle (e.g., Lee et al. 2019) reveal two basic categories of carbon in the Earth's mantle: primordial carbon surviving from early Earth accretion and differentiation, and carbon re-introduced into the Earth's interior via subduction. The present carbon content for the primitive mantle is ~90–130 ppm, although as outlined in the recent review by Dasgupta and Grewal (2019) estimates range from ~100 to >500 ppm. Subsequent to accretion and large-scale initial differentiation leading to core formation, it is not clear how carbon is hosted in the mantle, but it seems reasonable that this carbon would be present in its thermodynamically stable form(s) at the *P*, *T* and oxidation state of the ambient mantle. This presupposes that ferric–ferrous equilibria will serve as a sink or source for oxygen, which is reasonable at these carbon concentrations.

Subduction transports carbon down into the mantle, but the amount of carbon that survives sub-arc processing to be subducted into the mantle remains an active area of research (e.g., Kelemen and Manning 2015; Galvez and Pubellier 2019; Lee et al. 2019). Carbon would be present in both subducted sediment and altered oceanic crust (AOC) as carbonate, both biogenic and abiogenic, as well as some fraction as organic matter. The relative contribution of sediment and AOC likely depends on the specific thermal regime of a subducting slab, but a recent study by Li et al. (2019) implicates the latter as the dominant carrier of carbon as both abiogenic and biogenic carbonate into the mantle. Whether all this carbonate is reduced to graphite during subduction (Galvez et al. 2013) or survives as carbonate until partial melting ensues is unclear, so the experimentalist is faced with addressing both possibilities as carbon sources for diamond formation.

Fluids, melts, and supercritical fluids proposed to form diamond

Diamond can be formed directly from graphite in the absence of a solvent or catalyst, but this transformation requires pressures above ~10 GPa and temperatures in excess of 1800 °C (e.g., Irifune et al. 2004 and references therein). Diamond growth at lower pressures and temperatures requires some sort of solvent or growth media, and thus the first question is what that medium might be. From the examination of natural diamonds, it is well accepted that diamonds grow through metasomatic processes in C–O–H–N–S–Cl bearing mobile fluids, melts or supercritical fluids, and occurs over times and depths.

As outlined in other chapters in this volume, study of the inclusions in natural diamonds led to the identification of syngenetic inclusions, those that form as the diamond was growing (Harris 1968). The mineral inclusions enabled researchers to distinguish between diamonds that were growing in peridotitic and eclogitic lithologies, for example. Other inclusions were thought to sample the growth medium from which the diamond is precipitating. Fibrous diamonds in particular were fruitful to study in this context because they contain the so-called "HDF" (high-density fluid) inclusions. Among fibrous diamonds, the coated ones, with a monocrystalline pure core surrounded by a fibrous rim, inspired the idea that both kind of diamonds may share the same parent fluids and that a single diamond may grow in a few or several events. Chrenko et al. (1967) reported the first indication for the presence of carbonates and water in inclusions in diamond, but Navon et al. (1988) were the first to propose that coated diamonds sample mantle melts or volatile-rich fluids (enriched in H_2O, CO_3^{2-}, K_2O, Cl, depleted in MgO and including several other elements) from which they grow. Numerous publications have followed (see Weiss et al. 2022, this volume), including some significant studies such as: the finding of solid carbon dioxide inclusions in fibrous diamonds (Schrauder and Navon 1993), the discovery of brine inclusions in cloudy diamonds (Izraeli et al. 2001), the chemical and isotopic connection of fluid compositions trapped in coated diamond's inclusions from various cratons showing that fluids derived from mantle source would not be affected by local heterogeneities (Klein-BenDavid et al. 2004), the idea of miscibility/immiscibility processes linking mantle fluids of different compositions or "endmembers" during diamond growth (Klein-BenDavid et al. 2007), and many others.

In brief, the bulk compositions of these HDFs can be described in terms of four endmembers: (i) silicic HDFs that are rich in Si, Al, K, and water with some carbonate; (ii) saline HDFs that are rich in Cl, K, Na, water, and carbonate; (iii) and (iv) hydrous carbonatitic HDFs that are Mg-rich and Mg-poor, respectively.

Along with the relationship between these endmembers and their origin, a long-standing question has been whether the more common monocrystalline diamonds form from the same type of fluids. Strong evidence for this idea has been established from a variety of studies. For example: the common sinusoidal REE patterns between garnet inclusions trapped in the monocrystalline area of coated diamonds and HDFs trapped in the fibrous rims (Weiss et al. 2009); the similarity of hydrosilicic HDFs that precipitated the rims of coated diamonds with the fluids that precipitated most monocrystalline diamonds (Rege et al. 2010); the first finding of HDF micro-inclusions in a monocrystalline diamond, similar in major and trace elements compositions to those of fibrous diamonds (Weiss et al. 2014), suggesting that these HDF are involved during the growth of many monocrystalline diamonds. As a last example, Nimis et al. (2016) discovered thin "fluid" hydrous silicic films around solid mineral inclusions in gem-quality monocrystalline diamonds from peridotitic and eclogitic suites that were interpreted as potential vestiges of the original fluid present during diamond growth.

Are fibrous, coated, and monocrystalline diamonds growing by the same process? In our view, the debate was definitively closed in the affirmative—at least for lithospheric diamonds, by the study of Jablon and Navon (2016), who found micro-inclusions of HDFs trapped in the twinning planes of twin gem diamonds (macles) together with silicate minerals. They concluded that the mechanism of diamond formation is similar for most diamonds. It is worth noting that this process also includes the growth of micro diamonds in ultra-high-pressure metamorphic rocks (e.g., Stöckhert et al. 2001; Dobrzhinetskaya et al. 2007; Frezzotti et al. 2014).

Most of the studies based on fluid inclusions document the important role played by water during diamond growth, a role possibly not only limited to subduction zones where water is likely recycled from slabs, but also in the cratonic mantle. Although most of the HDFs observed in lithospheric diamonds are oxidized, some diamonds such as mixed-habit ones containing octahedral and cuboid sectors also host inclusions with reduced fluids such as CH_4 (Smit et al. 2016). Diamond growth in fluids evolving in redox state is also able to reconcile the apparent contradictory messages delivered from diamond stable isotope studies, because depending on the carbon source, CO_2 (Boyd et al. 1994) or CH_4 (Thomassot et al. 2007), the core to rim evolution of the diamond's $\delta^{13}C$ signature will be reversed. Furthermore, both eclogitic and peridotitic diamonds can be derived from the same isotopically homogeneous carbon source with metasomatic growth (Cartigny 2005).

These HDFs may not be limited to the lithosphere; some evidences of volatile-bearing inclusions are found in sublithospheric diamonds, such as ice VII inclusions (Tschauner et al. 2018); CH_4 and H_2 associated with inclusions of solidified iron–nickel–carbon–sulfur melts in large gem diamonds (Smith et al. 2016); or unknown fluids in association with inclusions of iron carbides (Kaminsky and Wirth 2011). These metallic melts may take part in diamond growth in the lower mantle and transition zone, but possibly also in the lithosphere, in both cases in very reduced environments.

In this review, we will try to guide the reader into the voluminous literature on experimental studies of diamond formation. To provide some structure to this, we will look at different growth media in turn, moving from "simple" to more complex systems. In each section, we will tabulate studies chronologically, and the reader will recognize how the experimental studies have evolved as studies of natural diamonds have continued to provide insights and constraints to help ground the experiments.

C–O–H FLUIDS

Overview

There have been extensive studies over the last 30 years of diamond nucleation and growth from initially graphite-saturated C–O–H fluids. Diamond grows from a variety of these fluids, from CH_4- to CO_2-rich (Table 1). The studies tabulated here are those in which graphite was present in the starting material (or generated upon breakdown of the organic fluid source); studies examining the ability of fluids to participate in oxidation-reduction reactions, such as those in which carbonate is reduced to diamond, are discussed in a subsequent section.

In general, these studies demonstrate that diamond growth is higher in H_2O-rich systems, lowest in CH_4-rich systems, reflecting a dependence on the activity of H_2O in the fluid (which is clearly fO_2 dependent). Diamond growth is characterised by temperature-dependent induction times, certainly for nucleation of new diamond crystals, and in some cases even for growth on pre-existing seed diamonds. More recent studies such as Matjuschkin et al. (2020) have demonstrated the ability of these fluids to grow diamonds at lower temperatures (e.g., that are more realistic for lithospheric mantle geotherms). In this study, extensive efforts were devoted to improving experimental design to minimize changes in fluid composition over the course of the experiments.

Table 1. Studies of diamond nucleation and growth in C–O–H fluids.

Study	System	*P* (GPa)	*T* (°C)	Duration	Seeds	Capsule/ buffering	Results
Akaishi et al. (1990b)	H_2O from $Mg(OH)_2$ or $Ca(OH)_2$	7.7	2150	20 min	N	Mo capsule	SN. CM: {111}
Yamaoka et al. (1992)	H_2O	7.7	1800–2200	2 h	Y	Ta capsule	GS at 2000°, 2200°. CM: {111}
Onodera et al. (1992)	Hydrocarbons from camphene, adamantane (both $C_{10}H_{16}$), and fluorene ($C_{13}H_{10}$)	7–9	710– 1325	30 min	N	Ta or Mo capsule	SN from camphene reported as low as 820°, 8 GPa or 710°, 6 GPa (in different parts of text). SN from adamantane at 8 GPa and 1325° but not at lower *T*. No SN from fluorene (7–9 GPa, 800–975°). These temperatures may be underestimated based on the location of their thermocouple relative to the sample (their Fig. 1)
Latourrette and Holloway (1994)	CO_2	8	950–1150	1–17 h	N	Pt	SN reported in all expts. Starting material was Fe_3O_4 + Gr, $NiCO_3$ + Gr, or Ni + PdO + Gr
Hong et al. (1999)	H_2O	7.7	1600– 2200	0.5–4 h	In some	Graphite-lined Mo	SN at 2000°, 2200°, GS at 1700°, 1800°. CM: {111}
Yamaoka et al. (2000)	H_2O	5.5, 7.7	1200–1500	24 h	In some	Pt	At 5.5 GPa, No SN in unseeded expts; GS at >1300°. At 7.7 GPa, in unseeded experiments SN at ≥1400°, not below. 100% conversion at 1500°. In seeded experiments, GS at >1300°. CM: {111}
Sun et al. (2000)	CO_2 from Ag_2CO_3 or Ag_2O + Gr	7.7	1500–2000	0.5–27 h	N	Graphite	Complete conversion at ≥1800° at ≥0.5 h. Some SN at 1700°/5 h but not 1700°/2 h. Some SN at 1600°/12 h and 1500°/27 h. CM: {111}
Akaishi and Yamaoka (2000)	CO_2–H_2O from OAD	7.7	1400–2000	0.5–360 h	N	Graphite-lined Mo	SN at 2000°/0.5 h; SN at 1800°/2 h but not at 0.5 h; SN at 1700°/5 h but not at 2 h; SN at 1600°/24 h but not at 9 h; SN at 1500°/88 h but not at 48 h; SN at 1400°/360 h. CM: {111}
Shaji Kumar et al. (2000)	CO_2–H_2O from OA	7.7	1300–1500	24–240 h	N	Pt	SN at 1500°, 10% at 24 h, 50% at 36 h, ~complete at 42 h; SN at 1400°, ~10% at 96 h; SN at 1300°, <10% at 240 h. CM: {111}
Akaishi et al. (2000)	CO_2–H_2O from OAD	7.7	1600	6–17 h	N	Pt	SN starting at 8 h, complete at 17 h. CM: {111}

Study	System	*P* (GPa)	*T* (°C)	Duration	Seeds	Capsule/ buffering	Results
Pal'yanov et al. (2000)	CO_2 from AgOx, H_2O, CO_2–H_2O from OAD, CH_4–H_2 from Anth"	5.7	1300	84 h	Y	Pt or Au	H_2O: SN in Pt; CO_2: SN in Pt; CO_2–H_2O: SN in Au and Pt; CH_4–H_2: negligible (< 1 μm) GS in Au. No fluid: No SN, No GS. CM: {111}
Akaishi et al. (2001)	CH_4–H_2O from StA and OAD	7.7	1500	1–48 h	N	Pt	No SN at 24 h, small amount of SN at 48 h. CM: {111}
Shaji Kumar et al. (2001)	CO_2–H_2O from OA	7.7	1500	24 h	N	Pt	Examined effect of mixing OA and graphite and differing OA:graphite ratios. CM: {111}
Sokol et al. (2001a)	CO_2 from AgOx, H_2O, CO_2–H_2O from OAD, CH_4–H_2 from Anth	5.7	1200–1420	42–136 h	Y	Pt or Au	CO_2: SN at 1300°/84 h, 1420°/42 h. H_2O: SN at 1300°/84 h, 1420°/42 h, GS at 1200°/135 h. CO_2–H_2O: SN at 1200°/136 h, 1300°/84 h, 1420°/42 h. CH_4–H_2: No SN at 1200°/136 h, 1300°/84 h, 1420°/42 h. Possible GS at 1200°, negligible (< 1 μm) GS at 1300°, 1420°. CM: {111}
Sun et al. (2001)	H_2O	5.5, 7.7	1300–2200	0.25–24 h	N	Ta, Mo, or Pt	H_2O: SN at 5.5, 1400°/24 h and 7.7, 1500°/24 h in Pt; SN at higher *T* in Mo, Ta—capsule leakage issue.
	CO_2 from AgOx	7.7	1500–2000	0.5–27 h	N	Graphite-lined Mo	CO_2: SN in all expts. CO_2–H_2O OA): SN in Mo at 2000°/0.5, 1800°/2, 1600°/20. No SN at 1800/°0.5, 1600°/12. SN in Pt down to 1300°/240.
	CO_2–H_2O from OA, CO_2–H_2O from MA	7.7	1300–2000	0.5–240 h		Graphite-lined Mo, Pt	CO_2–H_2O (MA): SN at 1500°/48, not 1400°/86. CM: {111}
Okada et al. (2002a,b)	H_2O from $Mg(OH)_2$	7.7	1835	see Results		Mo	In situ study; SN upon heating to 1835°. CM: {111}
Yamaoka et al. (2002c)	CO_2 from C+PtO_2	7.7	1500	8–24 h	N	Pt	SN at 24h, not 8 h. CM: {111}
Yamaoka et al. (2002b)	H_2O from glucose	7.7	1500	1–33 h 10 min–40 h	Y N	Pt	SN starting at 10 h, 100% at 33 h GS at ≥1 hr. CM: {111}
Dobrzhinetskaya et al. (2004)	H_2O from $Mg(OH)_2$	7.5–8	1200–1500	see results	N	Pt	1200°: graphite only to 24 h, SN after 122 h; 1500°: SN at 5 h, 100% at 138 h
	CO_2–H_2O from OAD	8.5	1500				SN at 32 h (~20–25%) and ~100% at 120 h. CM: skeletal diamonds

Study	System	*P* (GPa)	*T* (°C)	Duration	Seeds	Capsule/ buffering	Results
Okada et al. (2004)	H_2O from $Mg(OH)_2$	6.6–8.9	1400–1835	See Results		Mo	In-situ study of reaction kinetics; SN at $T > 1400°$. CM: {111}
Palyanov et al. (2007a)	H_2O	7.5	1500–1600	15 h	Y	Pt	SN at both 1500° and 1600°. CM: {111}
Sokol et al. (2009)	Various reduced fluids from H_2O- to CH_4-rich	6.3	1400–1600	15–48 h		Pt capsule w/ Mo–MoO_2 or Fe-FeO buffer	Focus on analysis of quench fluid phase; diamond SN in unbuffered H_2O expt at 1400°/42 h, GS in unbuffered H_2O–CH_4 expts at 1400°/42 h. GS only in 1 buffered expt (H_2O-rich, 1600°/24 h). CM: {111}
Palyanov et al. (2010a)	CO_2 from AgOx	7.5	1300–1600	2–80 h		Pt capsule, Fe_3O_4–Fe_2O_3 buffer	SN at 1600°/2 h, 1600°/15 h; GS from lowest *T* expts (1400°/40 h at 6.3 GPa, 1300°/80 h at 7.5 GPa). CM: {111}
Zhang et al. (2011)	H_2O from glucose. H_2O–CO_2 from OAD	9–11	1000–1400 K	15–50 min	DAC		Diamond-anvil cell expts. Growth reported in both expts w/ glucose, none in OAD expt
Matjuschkin et al. (2020)	CH_4–H_2 from StA	5–7	1100–1300	4–23 hr	N	Au and olivine caps, Mo–MoO_2 buffer	Expts contain olivine and orthopyroxene. SN at 5 GPa, 1250°/15 h, 6 GPa 1100–1300°, 7 GPa 1200–1300°

Notes: All studies contained graphite in the starting composition. Gr – graphite. OAD – oxalic acid dihydrate. OA – anhydrous oxalic acid ($H_2C_2O_4$). AgOx – silver oxalate ($Ag_2C_2O_4$). StA – stearic acid ($C_{18}H_{36}O_2$). Anth – anthracene ($C_{14}H_{10}$). MA – malonic acid ($C_3H_4O_4$). glucose – $C_6H_{12}O_6$. SN – spontaneous nucleation and growth of diamond. GS – diamond growth on seeds. CM – diamond crystal morphology (where given in article). Temperatures in *Results* column in °C.

Experimental issues

In experiments at diamond-stable conditions, C–O–H volatiles are added to the capsule as either a fluid by micro-syringe, or as solid materials that break down to fluids at the conditions of the experiment. C–O bearing materials such as oxalic acid dihydrate ($H_2C_2O_4{\cdot}2H_2O$) and silver oxalate ($Ag_2C_2O_4$) have a long history of use in experimental petrology (e.g., Holloway et al. 1968; Boettcher et al. 1973). These materials, as well as others such as anthracene ($C_{14}H_{10}$) and glucose ($C_6H_{12}O_6$), have been used to generate C–O–H fluids for diamond-synthesis experiments. A key assumption is that the solid starting material break downs to the expected fluid composition at experimental conditions; progress is being made on both *in situ* studies of these fluids (McCubbin et al. 2014) and careful characterisation of fluids following quenching of the experiment (e.g., Tiraboschi et al. 2016; Sokol et al. 2017). Other issues arise for specific compounds; for example, the elemental silver liberated by decomposition of silver oxalate can alloy with the sample capsule, which can lead to melting, capsule rupture, and loss of fluid (e.g., Brey et al. 1991).

Because of the potential for hydrogen diffusion across the capsule wall during the experiment, evolution of the composition of the fluid phase during the experiment is an issue relevant to these experiments (see discussions in Sokol et al. 2004, for example; Palyanov et al. 2010a). This long-standing experimental issue has led to various approaches to buffer the fluid composition (see overviews by Rubie 1999; Stagno 2019), and Table 1 notes studies in which buffering of the fluid phase by an external hydrogen buffer was employed.

Possible future directions

Continued focus on experiments with well-constrained redox conditions to maximize experimental run durations and to maintain constant fluid compositions will undoubtably improve our understanding of the dependence of diamond growth rate on fluid composition. Given the solubility of silicates in C–O–H fluids at high P and T, continued study of diamond growth in fluids saturated with peridotitic or eclogitic mineral assemblages would seem to be the most fruitful in terms of direct relevance to diamond growth in the Earth's mantle. Finally, studies of diamond growth in C–O–H fluids will benefit from continued efforts to improve thermodynamic models of fluids at high P,T. As an example, the solubility of carbon in CO_2 fluid at 6.3 GPa and 1250–1400 °C measured by Palyanov et al. (2010a) is ~1 order of magnitude larger than that predicted by the GFluid thermodynamic model (Zhang and Duan 2010). Furthermore, fluid models such as GFluid are restricted to fluid constituents in the C–O–H system. Extending these models to quantitatively model the effects of the solubility of silicates and oxides is sorely needed—and will require an enormous amount of experimental data to calibrate.

CARBONATES

Overview

Like C–O–H fluids, carbonate melts have attracted significant attention of experimentalists as potential growth media for diamond over the past three decades. A rich variety of alkali- and alkaline earth carbonates have been explored in these studies (Table 2), including those most likely to be relevant to diamond growth in the Earth's mantle, such as those in the Na_2CO_3–K_2CO_3–$CaCO_3$–$MgCO_3$–$FeCO_3$ system. All studies in Table 2 have graphite present in the starting material (or likely formed by breakdown of oxalate or other solid sources in the initial stages of the experiment) and use carbonate melts as media for graphite transformation into diamond. Experiments that explore diamond nucleation and growth as a result of redox reactions involving carbonates are described in another section.

Table 2. Studies of diamond nucleation and growth in carbonate systems.

Study	System	P (GPa)	T (C)	Duration	Seeds	Capsule/ buffering	Results
Akaishi et al. (1990a)	Li_2CO_3–Gr, Na_2CO_3–Gr, $SrCO_3$–Gr, $CaCO_3$–Gr, $MgCO_3$–Gr	7.7	2150	20 min	N	Mo	All had SN at 2150°. Expts with $CaCO_3$–Gr: No SN at 1800°, ~50% conversion at 2000°. CM: rounded shape
Akaishi et al. (1996)	$MgCO_3$–Gr	7.7	1800–2450	5, 30 min	Y	Ta	Sintering of diamond powder at >2000°
Taniguchi et al. (1996)	$MgCO_3$–Gr, $K_2Mg(CO_3)_2$–Gr	9–10	1300–1700	20 min	N	Gr	$MgCO_3$–Gr: SN at 9.5 and 10 GPa at ≥1600°; none observed at 9 GPa 1400–1750°. $K_2Mg(CO_3)_2$–Gr: SN above 1650° at 9–10 GPa. CM: {111}>{100}
Litvin et al. (1997)	$K_2Mg(CO_3)_2$–Gr	8–11	1700	5–10 min	N	MgO–BN	SN at 9–11 GPa, 1700°, 10 min. 8–9 GPa, 1700°, 5 min recrystallized Gr only. CM: {111}, {100}
Srikanth et al. (1997)	$MnCO_3$–Gr	6.5–7.7	1700–2100	20 min	N	Mo	No SN at 6.5 GPa 1700° or 1800°, or at 7.7 GPa below 2000°. SN at 7.7 GPa, 2000° and 2100°. CM: {111}>{100}
Litvin et al. (1998a,b,c)	$Na_2Mg(CO_3)_2$– $K_2Mg(CO_3)_2$–$NaKMg(CO_3)_2$–Gr	8–10	1700–1800	?	Y	Gr	SN and GS. CM: {111}, {100}
Sokol et al. (1998)	Na_2CO_3–Gr	6.8	1700	10 min, 16 h	Y	Gr	SN. CM: {111}
Pal'yanov et al. (1998a)	Li_2CO_3–Gr, Na_2CO_3–Gr, K_2CO_3–Gr, Cs_2CO_3–Gr, $CaCO_3$–Gr, $MgCO_3$–Gr, $SrCO_3$–Gr, $CaMg(CO_3)_2$–Gr	7	1700–1750	18.5 h			Ranked carbonates in terms of diamond growth efficacy as $Li_2CO_3 > Na_2CO_3 > K_2CO_3 > Cs_2CO_3$ and $CaMg(CO_3)_2 > CaCO_3 > MgCO_3 > SrCO_3$. CM: Na_2CO_3 {111}>{100}; K_2CO_3 {111}, {111}>{100}; Li_2CO_3 {111}>{100}, {*hhk*}; Cs_2CO_3 {111}; alkaline-earth carbonates {111}, {*hhk*}, {*hll*}
Borzdov et al. (1999)	K_2CO_3–Gr	7	1700–1750	2–11.5 h	Y	Pt	SN at 2 h. 100% diamond at 11.5 h.CM: {111}, {111}≫{100}
Litvin et al. (1999a)	$K_2Ca(CO_3)_2$–Gr, $Na_2Ca(CO_3)_2$–Gr	8.5, 9.5	1680–1800	40–65 min	Y	Gr	$K_2Ca(CO_3)_2$–Gr: 100% diamond at 9.5 GPa, 1750°, 40 min. $Na_2Ca(CO_3)_2$–Gr: GS at 8.5 GPa, 1750° and 9.5 GPa, 1800°. CM: {111}
Litvin et al. (1999b)	$Na_2Mg(CO_3)_2$–Gr, $NaKMg(CO_3)_2$–Gr	8–10	1700–1800	~30 min?	Y	Gr or MgO	GS reported. Figure shows SN down to ~1550° at 8.5 and 9 GPa. CM: {111}
Litvin and Zharikov (1999)	$K_2Fe(CO_3)_2$–Gr, Mixed K–Na–Ca–Mg–Fe carbonate + Gr	7–9	1650–1800	20–60 min	Y	Gr	$K_2Fe(CO_3)_2$–G: SN at 9 GPa, 1800°. Mixed carbonate: GS reported at 7–9 GPa, 1650–1800°. CM: {111}
Pal'yanov et al. (1999b)	Li_2CO_3–Gr, Na_2CO_3–Gr, K_2CO_3–Gr, Cs_2CO_3–Gr	7	1700–1750	10 min–18.5 h	Y	Pt	SN in all systems at 2 h; not at 20 min for Na_2CO_3 and K_2CO_3 or at 30 min for Cs_2CO_3. Catalytic activity trend seen: $Li_2CO_3 \gg Na_2CO_3 > K_2CO_3 > Cs_2CO_3$. CM: Li_2CO_3 {111}>{100}, {*hkk*}; Na_2CO_3 {111}>{100}; K_2CO_3 {111}, {111}≫{100}; Cs_2CO_3 {111}

Study	System	P (GPa)	T (C)	Duration	Seeds	Capsule/ buffering	Results
Pal'yanov et al. (1999a)	Na_2CO_3–Gr, Na_2CO_3–CO_2–H_2O (OAD)–Gr, K_2CO_3–Gr, K_2CO_3–CO_2–H_2O (OAD)–Gr	5.7	1150–1420	20–120 h	Y	Pt	Na_2CO_3–Gr: No SN but GS at 1360°/40 h; SN at 30 h at 1420°, GS but no SN at 20 h. K_2CO_3–Gr: No GS at 1250°/40 h, GS but no SN at 1300–1420°. Na_2CO_3–CO_2–H_2O (OAD)–Gr: SN at 1150/120 h, GS but no SN at 1250/40 h; SN at 1360/40 h and 1420°/20 h. K_2CO_3–CO_2–H_2O (OAD)–Gr: SN at 1150°/120 h, GS but no SN at 1250°/40 h and 1420°/20 h. CM: Na_2CO_3 {111}, {100}; K_2CO_3 and Na_2CO_3–CO_2–H_2O {111}
Sato et al. (1999)	$CaCO_3$–Gr, $MgCO_3$–Gr, $Ca_{0.4}Mg_{0.6}CO_3$–Gr	7.7	1600–2000	1–12.5 h	N	Mo with Gr liner, two in Pt	$CaCO_3$–Gr: SN at 1800°/6 h, not at 1 h; SN 1 h at 2000° but not 1900°. $MgCO_3$–Gr: SN at 2000°/1 h, not at 1900°/1 h or 1800°/6 h. $Ca_{0.4}Mg_{0.6}CO_3$–Gr: No SN at 1600°/12.5 h or 1600°/1 h; SN at 1700° at 11 h, not at 9 or 6 h; SN at 1800° at 6 h, not at 1 h; SN at 1900°/1 h and 2000°/1 h. CM: {111}
Sumiya and Satoh (1999)	$CaCO_3$–Gr, $MgCO_3$–Gr, $CoCO_3$–Gr	7.7	2000–2100	15 min	N	Mo	SN at 2000°, complete at 2100° in $CaCO_3$ and $MgCO_3$, SN at 1850° in $CoCO_3$. CM: {111}
Sokol et al. (2000)	Li_2CO_3–Gr, Na_2CO_3–Gr, K_2CO_3–Gr, Cs_2CO_3–Gr, $CaCO_3$–Gr, $CaMg(CO_3)_2$–Gr	7	1700–1750	4–19 h	Y	Pt	SN reported for all expts; focus of study was analysis of quenched fluid phase.
Liu et al. (2001)	$MnCO_3$	6–12	>2000	?	DAC		Laser–heated DAC; reported formation of diamond by breakdown of carbonate at P ~12 GPa
Sokol et al. (2001b)	$CaMg(CO_3)_2$–Gr, $CaMg(CO_3)_2$–H_2O–Gr, $CaMg(CO_3)_2$–CO_2+H_2O (OAD)–Gr, $CaMg(CO_3)_2$– $Na_2C_2O_4$–Gr	5.7, 7	1300–1700	2–42 h	Y	Pt	$CaMg(CO_3)_2$–Gr: SN at 7 GPa, 1700° at ≥4 h; G at 2 h. No SN at 5.7 GPa, 1420/°42 h. $CaMg(CO_3)_2$–H_2O–Gr: SN at 5.7 GPa, 1420°/42 h. $CaMg(CO_3)_2$– CO_2–+H_2O–Gr: SN at 5.7 GPa, 1420°/42 h; GS only at 1300°/42 h. $CaMg(CO_3)_2$– Na_2C_2O4–Gr: GS only at 5.7 GPa, 1300°/42 h
Pal'yanov et al. (2002a)	Na_2CO_3–Gr, K_2CO_3–Gr, Na_2CO_3–H_2O–Gr, K_2CO_3–H_2O–Gr, Na_2CO_3–CO_2–Gr, K_2CO_3– CO_2–Gr, Na_2CO_3–H_2O+ CO_2 (OAD)–Gr, K_2CO_3–H_2O + CO_2 (OAD)–Gr	5.7	1150–1420	5–136 h	Y	Mostly Pt, 1 in Gr, 2 in Au	Study focused on how the induction period changes with P, T, and presence of fluids. Found reactivity much higher in fluid–bearing expts. Na_2CO_3–Gr: At 7 GPa and 1700°, SN at 2 h, not 0.3 h; at 5.7 GPa 1420°, SN at 30 h, not 20 h; No SN at lower T (1250–1360°) at 40 h. K_2CO_3–G: SN at 7 GPa, 1700/2 h. No SN at 5.7 1250°/84 h, 1300°/40 h 1420°/40 h. For both with CO_2 or H_2O, have SN at 5.7 GPa, 1420°/40 h. With H_2O–CO_2, had SN at 1150°/120 h, and with decreasing time with increasing T to 20 h at 1420° (for Na_2CO_3). CM: Na_2CO_3 {111}>{100}; K_2CO_3 {111}; Na_2CO_3–CO_2–H_2O {111}; K_2CO_3–CO_2–H_2O {111}
Shatsky et al. (2002)	K_2CO_3–Gr	6.3	1650	40 h	Y	Gr	SN (~25–29% transformation). CM: {111}

Study	System	*P* (GPa)	*T* (C)	Duration	Seeds	Capsule/ buffering	Results
Litvin and Spivak (2003)	Mixed K–Na–Ca–Mg–Fe carbonate + Gr, Natural limestone + Gr	7.5–8.5	1500–2000	< 1 h	N	Gr	SN at >7.5 GPa, > 1500° in 45–50 min expts reported for both starting materials. CM: {111}
Sokol and Pal'yanov (2004)	Review of fluid, carbonate+fluid results	5.7–7.7	1150–2000				Review of previous work; discusses induction period preceding spontaneous SN of diamond, which increases with decreasing *T*. Report increasing intensity of diamond formation in the order: $K_2CO_3(Na_2CO_3)–H_2O–CO_2–C > CO_2–C \approx H_2O–CO_2–C \approx H_2O–C \approx CaMg(CO_3)_2–H_2O–C \gg CH_4–H_2O–C \gg CH_4–H_2–C$
Spivak and Litvin (2004)	Mixed K–Na–Ca–Mg–Fe carbonate + Gr	5.5–8.5	~1200–2250	?	?	?	Updated and expanded on Litvin and Zharikov (1999). Few expt'l details. Photos of diamonds formed at 7.8 GPa 1990°, 8 GPa 1500°. Figure has diamond growth down to ~1200° at ~6 GPa. CM: {111}
Palyanov et al. (2007a)	K_2CO_3 + Gr, H_2O–Gr, $K_2CO_3 + H_2O$ + Gr	7.5	1400–1600	15 h	Y	Pt	K_2CO_3–Gr: SN at ≥1500°, none at 1400°. H_2O–Gr: SN at 1500° and 1600° $K_2CO_3 + H_2O$ + Gr: SN at 1500° and 1600°, increasing amt of diamond formation as H_2O content increases. CM: K_2CO_3 {111}, {100}; $K_2CO_3 + H_2O$ {111}
Spivak et al. (2008)	Mixed K–Na–Ca–Mg–Fe carbonate + Gr	7–8.1	1500–1600	1–24 min	N	Gr	SN reported in 3 expts (7 GPa 1500°/8 min, 7 GPa 1600°/1 min, 8.1 GPa 1530°/24 min). CM: {111}
Tomlinson et al. (2011)	$MgCO_3$ + Gr	10–20	1900–2100	15 min	N	Re	Diamond formation reported at 1900° at 10 GPa and 2000° at 15 and 20 GPa. CM: {111} + {100}
Spivak et al. (2012)	$CaCO_3$	11–43	1600–3900 K	5 min	N	DAC (Re gasket + Ne) and MAP	Laser–heated DAC. Diamond formation from breakdown of $CaCO_3$ melt at ~3500 K, ~16 and ~43 GPa. CM: {111}
Solopova et al. (2013)	Mixed K–Na–Ca–Mg carbonate + Gr	7–8.5	1500–1800	5–60 min	N	?	SN at 7.5 GPa 1600° to 8.5 GPa 1800°. CM: {111}
Palyanov et al. (2016)	Na_2CO_3–Gr, Na_2C_2O4	6.3, 7.5	1250–1700	10–66 h	Some expts	Pt	$Na_2C_2O_4$ w/o graphite added: Breaks down via $2Na_2C_2O_4 = 2Na_2CO_3 + C + CO_2$. Formed diamond at 6.3 GPa, ≥1300°, 7.5 GPa, ≥1350° from $Na_2CO_3–CO_2$ melt. Control expts with Na_2CO_3–C grew diamond at 6.3 GPa, 1400–1570°. CM: $Na_2CO_3 + CO_2$ {111}, {100} with convex faces and fibrous structure
Khokhryakov et al. (2016)	$Na_2C_2O4–CaN_2$	6.3	1500	2, 30 h	Y	Pt	Some GS at 2 h, no SN. SN at 30 h. Observed nitrogen conc up to 1100 ppm, depending on growth sector. CM: Hexaoctahedron with {100}, fibrous structure

Notes: DAC – diamond-anvil cell. MAP– multi-anvil press. Gr – Graphite. OAD – oxalic acid dihydrate. SN – spontaneous nucleation and growth of diamond. GS – diamond growth on seeds. CM – diamond crystal morphology (where given in article). Temperatures in *Results* column in °C.

Review of Table 2 reveals that the early experiments were conducted at temperatures well above those reasonable for diamond growth in the mantle for two reasons: (1) diamond nucleation and growth require temperatures significantly overstepping of the graphite–diamond transition and (2) diamond growth requires the carbonate to be partially molten in order to serve as a viable growth medium. The solidus temperatures of alkaline-earth carbonates are in excess of ~1400–1600 °C at ≥6 GPa for successful growth of diamond. Alkali carbonates, with their lower melting temperatures, allow melt-present experiments at lower temperatures, which facilitates crystallization of diamond at lower temperatures. The addition of volatiles such as water and CO_2 to the systems further encourages diamond crystallization to proceed at lower temperatures.

Experimental issues

Carbonate melts are characterised by low viscosity at least to pressures of ~6 GPa (c.f. Jones et al. 2013 and references therein) although their viscosity may increase at higher pressures (Wilding et al. 2019). In the experiments, possible escape of the liquid through a graphite or MgO capsules must be considered—most of the experiments use Pt or Au in part for this reason.

Oxygen (hydrogen) fugacity control is also a consideration; hydrogen diffusion through the capsule wall can drive carbon precipitation via $2H_2 + CO_2$ (melt) $= C + 2H_2O$, which also introduces H_2O in the system (e.g., see discussion in Palyanov et al. 2016). However, it is worth emphasizing that use of an external oxygen buffer in a double-capsule configuration with hydrogen diffusion across the capsule wall equalizes the chemical potential of hydrogen, not the fO_2 (e.g., Whitney 1972), and these buffers are best used as hydrogen sinks unless there is a fluid coexisting with the carbonate melt in the experimental charge.

Alkali carbonates, particularly K_2CO_3, are notoriously hygroscopic and special care must be taken in experiments with these carbonates. Figure 7 of Shatskiy et al. (2015a) illustrates the effect of adsorbed H_2O on phase relationships: the lowest-T eutectic in the K_2CO_3–$CaCO_3$(–H_2O) drops from ~1200 °C to ~1000 °C at 6 GPa when the samples were dried at 100 °C rather than 300 °C. Some of the differences in the results of the studies involving alkali carbonates may result from this issue. Parenthetically, the temperature effect Shatskiy et al. (2015a) observed also implies high solubility of H_2O in carbonate melts at high pressures—a field relatively unexplored except at much lower pressures (0.025–0.225 GPa; Keppler 2003).

Possible future directions

Further studies of diamond growth in carbonate melts could focus on exploring more of the composition space relevant to natural melts. These studies will benefit from insights provided by recent experimental studies of phase relationships in carbonate systems at mantle pressures (see review by Shatskiy et al. 2015b, for example). The study of systems such as K_2CO_3-$CaCO_3$-$MgCO_3$ (Arefiev et al. 2019) and Na_2CO_3–$CaCO_3$–$MgCO_3$ (Podborodnikov et al. 2019) hold particular promise in this regard.

Given that it has been firmly established that diamond can indeed grow in carbonate melts, another fruitful line of research is to take advantage of these melts to grow diamond to continue to address issues such as nitrogen incorporation into diamond (e.g., Khokhryakov et al. 2016) and isotopic fractionation between diamond and coexisting melts or fluids (Reutsky et al. 2015a,c, 2018; Bureau et al. 2018) (see later sections).

On another front, there remains another unresolved question: How does carbon dissolve in carbonate melts? Can it dissolve as a neutral carbon species, or does it dissolve as carbonate? A key challenge in resolving this question is the fact that almost all carbonate liquids do not quench to glass, with the intriguing exceptions of some K–Mg carbonates and some more complex melts in the $BaSO_4$–$La(OH)_3$–$Ca(OH)_2$–CaF_2–$CaCO_3$ system (see *Carbonate glasses* in Jones et al. 2013 and references therein). This issue naturally complicates both measurements of solubility and solution mechanisms. The available data on C solubility in carbonate melts show low solubility: <0.3 wt.% at 6.8 GPa and 1700 °C in Na_2CO_3 melt (Sokol et al. 1998). On the other hand, CO_2 appears to dissolve readily in carbonate melts: Palyanov et al. (2016) observed a solidus depression of ~100 °C at 6.3 and 7.5 GPa in the system Na_2CO_3-CO_2–C compared to Na_2CO_3, and argued that the 16.4 wt.% of CO_2 produced upon decomposition of sodium oxalate ($Na_2C_2O_4$) dissolved completely in the melt at these conditions. This solubility is higher than the 11.5 wt.% and 6.5 wt.% CO_2 dissolved in $CaCO_3$ and $MgCO_3$ melts, respectively, at liquidus conditions at 2.7 GPa (Huang and Wyllie 1976). To our knowledge, there are no spectroscopic data on solution mechanisms of either C or CO_2 in carbonate melts at diamond-stable *P, T*. This topic would of course be an ideal *in situ* study, such as those done at lower pressures (~700 MPa e.g., Mysen 2018), when technology has evolved sufficiently to make measurements at these conditions possible.

SILICATE MELTS

Overview

The existing experimental data indicate that volatile-free silicate melts cannot provide either diamond nucleation or growth on seeds in the range of *P, T*-parameters that are of interest for natural diamond formation. Silicate and oxide systems become diamond-forming only when H_2O is added (Table 3). Clearly, part of this effect is a result of water lowering the solidus and liquidus temperatures to stabilize the melt to lower temperatures more representative of conditions of natural diamond formation. In most systems, however, we lack basic information about the solubility of H_2O in the melt as a function of *P* and *T*—or indeed the basic phase equilibria—needed to differentiate the effects of melt fraction and melt composition on diamond growth. Nevertheless, the studies in Table 3 show that the nucleation and growth of diamond on seeds was established at temperatures of 1500–1600 °C in water-bearing silicate melts and silicate-bearing aqueous fluids. An increase in the H_2O content in such systems results in a significant increase in the degree of graphite into diamond transformation and is accompanied by a reduction in the induction period preceding diamond nucleation and an increase in carbon mass transfer. As a result, the most favorable conditions for nucleation and growth of diamond occur in the water-rich fluid phase, containing small amounts of silicate or oxide solute. A decrease in temperature leads to a decrease in the diamond-forming ability of silicate–aqueous media, which is usually attributed to more sluggish kinetics at lower temperature. The minimum nucleation temperature for diamond was 1500 °C at 6.3 and 7.5 GPa. The minimum temperature for diamond growth on seeds is 1400 °C at 6.3 GPa (Table 3).

Experimental issues

The overwhelming majority of experiments in the silicate-water systems were carried out in Pt ampoules without special buffering. The presence of graphite and water among the initial reagents due to H_2 diffusion through the walls of Pt ampoules, in the absence of buffering, can lead to the formation of an uncontrolled amount of CO_2 in a predominantly aqueous fluid.

Future directions

One of the unsolved problems is the determination of the solubility of carbon depending on temperature and on the content of H_2O. Experimental studies in this direction will make it possible to evaluate the real scale of diamond formation in the processes of evolution of the system composition and with a decrease in the *P, T*-parameters.

As seen in Table 3, most experiments in the silicate–water systems were carried out without the participation of alkalis. The addition of alkaline components can change the pH of the medium and bring the model system closer to the eclogitic fluid and, possibly, allow to reduce the minimum *P, T*-parameters.

Table 3. Studies of diamond nucleation and growth in silicate melts.

Study	System	*P* (GPa)	*T*(°C)	Duration	Seeds	Capsule/ buffering	Results
Borzdov et al. (1999)	$KAlSi_2O_8$–Gr	7.0	1700–1750	2–18 h	Y	Gr	No SN
Sokol et al. (1999)	$CaMgSi_2O_6$–OAD (20 wt.%)–Gr, $NaAlSi_2O_6$–OAD (20 wt.%)–Gr	7.0	1700–1750	4 h	Y	Pt	SN in silicate–H_2O–CO_2–C systems. CM: {111}
Pal'yanov et al. (2005a)	SiO_2–H_2O (4–74 wt.%)–Gr, Mg_2SiO_4–H_2O (4–74 wt.%)-Gr	7.5	1600	40 h	Y	Pt	SN of diamond. With an increase in the H_2O content, the degree of graphite into diamond transformation significantly increases. CM: {111}
Dobrzhinetskaya and Green (2007)	Gr–SiO_2 ± muscovite ± albite–H_2O (6–15 wt.%)	7.0–8.5	1500	1–43 h	Y/N	Pt/Ni–NiO	No diamonds found
	Gr–SiO_2–H_2SiO_3(10 wt.%) ± Al_2O_3	7.0–8.5	1500	2–62 h	Y/N	Pt/Fe–FeO	SN of diamond occurred in all experiments with a duration more than 2 hours, mainly in assemblage with SiC and coesite
Sokol and Pal'yanov (2008)	SiO_2–H_2O–Gr w/ $H_2O/(H_2O + SiO_2) = 1 - 0.05$, Mg_2SiO_4–Gr w/$H_2O/(H_2O + Mg_2SiO_4) = 1 - 0.08$	7.5	1600	15, 40 h	Y	Pt	SN of diamond occurred in the entire composition range. The degree of graphite to diamond transformation is a function of the H_2O content, which controls both the kinetics of diamond nucleation and the intensity of carbon mass transfer. Decreasing H_2O content decreases degree of transformation. CM: {111}
Palyanov and Sokol (2009)	Mg_2SiO_4–H_2O (13–86 wt.%)–Gr	7.5	1500–1600	15 h	Y	Pt	SN of diamond occurred in all expts. A decrease in temperature results in a decrease in the intensity of diamond forming processes and the formation of metastable graphite. CM: {111}
Sokol et al. (2010)	Mg_2SiO_4 (72 wt.%)–H_2O (8 wt.%)–Gr (20 wt.%)	6.3	1200–1600	16–40 h	Y	Pt	No SN of diamond at 1200°/40h. Only GS at 1400°/40h. SN of diamond at 1600°/16h, 40h. CM: {111}
Fagan and Luth (2011)	$SiO_2 \cdot 0.36 H_2O + 1.5 Mg(OH)_2$ (16.7 wt.% H_2O)	5.5–7.0	1300–1600	4–24 h	Y	Pt	No SN at 1300°/24h ($P = 6$ and 7 GPa) and at 1500°/4h ($P = 6$ GPa). GS at 1500–1600°/4h ($P = 7$ GPa). CM: {111}

Notes: Gr – Graphite. OAD – oxalic acid dihydrate. SN – spontaneous nucleation and growth of diamond. GS – diamond growth on seeds. CM – diamond crystal morphology (where given in article). Temperatures in *Results* column in °C.

BRINES

Early studies showed that diamond can grow in anhydrous alkali halides (Wang and Kanda 1998; Litvin 2003), but such studies are unlikely to be directly relevant to diamond growth in the Earth because of the absence of H_2O. Brines, or hydrous solutions of alkali halides, on the other hand, do form one of the "endmembers" of HDF inclusions in natural diamonds. Remarkably few studies address diamond growth in such systems. The two studies in Table 4 demonstrate that both KCl + H_2O and NaCl + H_2O fluids can nucleate and grow diamonds at 7.5 GPa and high temperatures (1500–1600 °C). The more extensive work in the former system documents a composition-dependence on the degree of transformation of graphite to diamond (see Table 4). Subsequent work in brine-containing systems has shifted to systems coexisting with silicates or carbonates (see subsequent sections).

CARBONATE–CHLORIDE SYSTEMS

There has been limited study of chloride-carbonate systems (Table 5). From these results, it is clear that diamond nucleation and growth can proceed in either anhydrous or hydrous carbonate-chloride melts. An unresolved issue is the minimum temperature at which this occurs: Tomlinson et al. (2004) observed growth as low as 1050 °C at 7 GPa, whereas Palyanov et al. (2007a) did not see diamond growth at 1400 °C at 7.5 GPa. Given the limited data, it is impossible to say whether the presence of H_2O in the system allows diamond nucleation and/or growth at lower temperatures.

MODEL CARBONATE–SILICATE SYSTEMS

Overview

These studies explore the efficacy of carbonate–silicate melts as a medium for diamond nucleation and growth. In effect, these studies can be considered extensions of carbonate-melt studies that explore the effect of dissolution of silicate constituents into the melt. At the same time, these studies mark a significant step towards simulating systems representative of those potentially present in the Earth's mantle. Examination of Table 6 shows that these melts are effective growth media for diamond, particularly at higher temperatures, which in part reflects the temperatures required for melt to be present, but also the temperature-dependence of diamond nucleation and growth as seen in the previous systems.

Future directions

Unresolved questions in these systems include whether the effect of adding carbonate is strictly a result of stabilising a melt to lower temperatures, a compositional effect of the presence of carbonate, or a result of decreasing viscosity of the melt with increasing carbonate content. Systematic studies to tease out these possibilities may be warranted if such systems are viewed to be useful models for natural diamond growth.

Table 4. Diamond growth in brines.

Study	System	*P* (GPa)	*T* (°C)	Duration	Seeds	Capsule	Results
Palyanov et al. (2007a)	KCl + H_2O + Gr	7.5	1500, 1600	15 h	Y	Pt	1500°: No SN at 19 wt.% H_2O, SN from 35–91 wt.% H_2O with increasing % diam, slight decrease in total % diam in 100% H_2O. 1600°: SN from 17–100 wt. % H_2O, maximum transformation at 66–69 wt.% H_2O. CM: {111}
Khokhryakov et al. (2009)	NaCl + H_2O + Gr	7.5	1600	40 h	Y	Pt	Growth and entrapment of graphite inclusions. CM: {111}

Notes: Gr – graphite. SN – spontaneous nucleation and growth of diamond. GS – diamond growth on seeds. CM – diamond crystal morphology. Temperatures in *Results* column in °C

Table 5. Studies of diamond nucleation and growth in carbonate–chloride (±H_2O) systems.

Study	System	*P* (GPa)	*T* (°C)	Duration	Seeds	Capsule	Results
Tomlinson et al. (2004)	KCl + K_2CO_3 + Gr	7–7.7	1050–1420	5–60 min	Y	Pt	GS at 1050°/60 min, 1260°/60 min, 1420°/<5 min. CM: {111}
Palyanov et al. (2007a)	KCl + K_2CO_3 + Gr	7.5	1400–1600	15 h	Y	Pt	No SN at 1400°, SN at 1500° (more in K_2CO_3-rich) and 100% diam at 1600° w/KCl varying from 25–80 wt.%. CM: {111}, {100}
Palyanov et al. (2007a)	KCl + K_2CO_3 + H_2O + Gr	7.5	1500	15 h	Y	Pt	SN in three compositions with variable KCl:K_2CO_3:H_2O. CM: {111}

Notes: Gr – graphite. SN – spontaneous nucleation and growth of diamond. GS – diamond growth on seeds. CM – diamond crystal morphology. Temperatures in *Results* column in °C.

Table 6. Studies of diamond nucleation and growth in model carbonate–silicate systems.

Study	System	*P* (GPa)	*T* (°C)	Duration	Seeds	Capsule	Results
Borzdov et al. (1999)	66.6 wt.% K_2CO_3 + 33.4 wt.% SiO_2 + Gr, 80 wt.% K_2CO_3 + 11.4 wt.% MgO + 8.6 wt. % SiO_2 + Gr, 80 wt.% K_2CO_3 + 5.96 wt.% MgO + 5.08 wt. % Al_2O_3 + 8.96 wt. %SiO_2 + Gr	7	1700–1750	1–17.5 h	Y	Pt	K_2CO_3–SiO_2 + Gr: SN in 1 h K_2CO_3–MgO–SiO_2 + Gr: SN in 17.5 h K_2CO_3–MgO–Al_2O_3–SiO_2 + Gr: SN in 17 h. Contrasts w/ no SN or GS in 2–18 h expts with $KAlSi_3O_8$ + Gr. Carbonate– or carbonate ± silicate ± oxide melts fluxed diamond growth. CM: {111}, {111}≫{100}
Shatsky et al. (2002)	K_2CO_3 + SiO_2 (5–75 wt.% SiO_2) + Gr, K_2CO_3 + Mg_2SiO_4 (Fo) (5–95 wt.% Fo) + Gr	6.3	1650	40 h	Y	Gr	K_2CO_3 + SiO_2 + Gr: SN up to 25 wt.% SiO_2; GS only at higher SiO_2; K_2CO_3 + Fo + Gr: SN up to 50 wt.% Fo; GS only in 50–90 wt.% Fo; no GS at 95 wt.% Fo. In both series, maximum intensity of diamond growth at ~10 wt.% silicate. All expts contained K–bearing carbonate–silicate melt. CM: {111}, {100}
Litvin et al. (2008a)	$NaAlSi_3O_8$ + K_2CO_3 (50–90 wt.% Ab) + Gr, Mg_2SiO_4 + K_2CO_3 (60–80 wt.% Fo) + Gr, SiO_2 + K_2CO_3 (30–50 wt.% SiO_2) + Gr	7.6–8.5	1620–1800	4–40 min	Y	Gr	GS in all expts in presence of carbonate–silicate melt. SN reported in expts except for Mg_2SiO_4–K_2CO_3 at <40 wt.% carbonate and SiO_2–K_2CO_3 at < 60 wt.% carbonate. CM: {111}
Spivak et al. (2008)	$NaAlSi_3O_8$ + K_2CO_3 (60–80 wt.% Ab) + Gr	8.5	1730, 1800	6 min 30 min	N	Gr	SN reported in presence of melt in both experiments. CM: {111}
Bataleva et al. (2012)	(Ca,Mg)CO_3–SiO_2–Al_2O_3–(Mg,Fe)(Cr,Fe,Ti)O_3 (16.1 wt.% CO_2)	6.3	1350–1650	20 h	Y	Pt, Pt + Gr	GS at 1550° and 1650° in Pt capsules (not at 1350° or 1450°), at 1350° and 1450° in Pt–Gr capsules. Melt present in all expts, amount increased from <10 wt.% at 1350–1450° to >20 wt.% at 1550° and 1650°. Observed dissolution of diamond seeds in expts w/ Pt capsules but not in Pt–Gr capsules (fO_2 dependent). CM: growth {111}, regeneration of {100}→{111}

Notes: Gr – graphite. SN – spontaneous nucleation and growth of diamond. GS – diamond growth on seeds. CM – diamond crystal morphology. Temperatures in *Results* column in °C.

MULTICOMPONENT CARBONATE–SILICATE MEDIA AND "DIAMOND-BEARING" ROCKS

Overview

The main regularities in diamond crystallization, as revealed in simple model systems (above), are also valid for multicomponent media and are as follows:

- Silicate melts do not facilitate nucleation and diamond growth; it is the addition of water makes the silicate + H_2O systems diamond-forming. The silicate component acts as a diamond formation inhibitor (e.g., Sokol and Pal'yanov 2008), with all other parameters being constant.
- In carbonate–silicate environments, the silicate component also inhibits diamond formation.
- The most effective diamond-forming media are alkaline carbonate melts as well as H_2O- and CO_2-bearing fluids; nucleation and growth of diamond in these media are realized at the lowest *P*, *T*, probably reflecting their stability as melts or fluids at these conditions.
- The induction period preceding the nucleation and growth of diamond depends on *P*, *T* and the composition of the crystallization medium.
- With an increase in *P* and *T*, the diamond-forming ability of all media increases. A decrease in *P* and *T* in all media leads to crystallization of metastable graphite.
- An increase in pressure from 5.7 to 12 GPa and above allows the reduction of the minimum diamond nucleation temperature in all systems, including multicomponent carbonate–silicate media.
- All other parameters being equal, the nucleation of a diamond requires higher temperatures than the growth of a diamond on seeds.
- Crystallization of diamond or graphite in a carbonated eclogite at 9–23 GPa was suggested to result from oxidation of Fe^{2+} to Fe^{3+} in coexisting garnet (Kiseeva et al. 2013).

Experimental issues

Experiments in multicomponent carbonate–silicate systems and using the "diamond-bearing" rocks were carried out mainly in Pt ampoules (Table 7). The effect of the ampoule material (for example, Pt and Au) on the diamond nucleation process in non-metallic solvents has been studied insufficiently. For example, in long-term experiments (tens of hours), the dissolution of Au and its crystallization on diamond in the C–O–H fluid were shown (Sokol et al. 2001a). The effect of Pt on diamond nucleation was established in long-term experiments in the kimberlite–graphite system (Palyanov et al. 2015); however, the real scale of the effect of Pt is not clear. It can be most significant at high *P* and *T*, especially when using small diameter Pt ampoules.

Future directions

New experimental studies need to be performed in these fields:

- Despite a large number of experiments carried out in a wide range of *P*, *T*-parameters and in various media, there is still no clarity and experimental confirmation of how diamond is formed in eclogites and peridotites in the pressure range of 4–7 GPa and 900–1400 °C, characteristic of formation of most natural diamonds (Stachel and Harris 2008).

- There is still insufficient experimental data to assess the effect of various components of the C–O–H–N–S system fluid on the crystallization of diamond in multicomponent media. The data on the defect-impurity composition of diamonds synthesized in various model media are still scarce, which does not allow us to fully assess the overall picture of the effect of the composition of diamond-forming media on the real structure of diamond.

SULFIDE MELTS

The studies in this section (Table 8) document that diamond can nucleate and grow in iron-sulfide melts, but typically at temperatures higher than those appropriate for corresponding pressures in the mantle. From these studies, sulfide melts appear to be less effective at mediating diamond growth than either H_2O-rich fluids or carbonate melts.

CARBONATE REDUCTION

To this point, all of the studies had graphite in the starting materials, and the primary focus was on whether a particular melt or fluid would mediate the transformation of graphite to diamond at conditions approaching those appropriate for the Earth. In contrast, the studies in this section all contain carbonate as a carbon source and use different reducing agents to produce diamond by reduction (Table 9). The early studies demonstrated the viability of this mechanism, using metal, metal carbides, or metal hydrides as reducing agents. Subsequent studies looked at the possible role of C–H fluids, sulfides and oxides. Overall, reduction of carbonate has been shown to be a viable mechanism for formation of diamond; the key question in terms of applicability to the Earth is what the appropriate reducing agent might be in different environments, particularly in the lithospheric mantle at depths shallower than those at which iron metal or carbide are stabilized.

DIAMOND-ANVIL CELL EXPERIMENTS

Overview

Whereas they are not designed for that purpose for most of the cases, laser heated diamond anvil cells (LHDAC) experiments may give significant information about the formation of superdeep diamonds, and demonstrate progress towards multiphase mantle petrology at conditions of the lowermost mantle.

Most of these studies (Table 10) are looking at high *T*- and *P*-dependent breakdown of carbonates like $CaCO_3 = CaO + C + O_2$ in the context of subduction recycling, or to characterize the formation of new mineralogical species of carbonates in the deep Earth. For example, Dorfman et al. (2018) show that $CaCO_3$ is preserved at pressure and temperature conditions reaching those of the deep lower mantle, supporting its relative stability in carbonate-rich lithologies under reducing lower mantle conditions. Boulard et al. (2011) provide evidence for the presence of both diamonds and an oxidized C-bearing phase, suggesting that oxidized and reduced forms of carbon might coexist in the deep mantle.

Litvin et al. (2014) took the benefit of these HPHT tools to study the diamond-forming lower mantle systems, by investigating melting phase relations of simple carbonates of Ca, Mg, Na and multicomponent Mg–Fe–Na–carbonate up to 60 GPa and 3500–4000 K (LHDAC and multi-anvil presses). They grew diamonds that showed that 'Super-deep' diamonds can crystallize in the system carbonate-magnesiowüstite-Mg-perovskite-carbon.

Table 7. Studies of diamond nucleation and growth in multicomponent carbonate–silicate media and diamond-forming rocks.

Study	System	*P* (GPa)	*T* (C)	Duration	Seeds	Capsule/ buffering	Results
Arima et al. (1993)	Kimberlite–Gr	7–7.7	1800–2200	20–900 min	Y/N	Gr (Mo, Ta)	SN
Palyanov et al. (2001b)	Natural pyroxene–carbonate (~22 wt.% SiO_2)–Gr	5.7 7	1420 1700	41 h 3 h	Y	Pt	SN of diamond in both expts w/ this composition. Contrasts with garnet–pyroxene rock at 5.7, 1420° that did not produce diamond. Analysis of gas phase after expt showed $H_2O \gg CO_2$
Litvin et al. (2001)	Natural calcio–carbonatites (~12–~22 wt.% SiO_2) + Gr	7	1450–1550	10–40 m	Y	Gr	SN reported for all four compositions. Melt present in all expts
Litvin et al. (2003)	Natural garnet–cpx–carb rock	5.5–7.5	1420–1720	1–31 min	Y (in some)	Gr	Reported GS at *T* as low as 1420° (6.5 GPa) in presence of melt. SN in expts from 1420–1720°; dissolution of seeds in other expts from 1450° at 6.1 GPa to 1660° at 5.5 GPa
Dobrzhinets-kaya et al. (2004)	Carbonate–talc–Gr	7.5–8.5	1200–1500	1–138 h	N	Pt	SN at 1500° and 8.1 GPa/34 h
Bobrov et al. (2004)	Natural carbonate–silicate rock	4-7	1200–1700	20–125 min	Y	$Pt_{60}Rh_{40}$	Reported diamond formation from carbonatitic melt
Litvin et al. (2005a)	Natural carbonatite	7–8.5	1500–2000	40 min	N	Gr	Liquidus reported to be at ~1700–1730°. Diamondite formed at 7–8.5 GPa, 1800° in carbonate–silicate melt
Litvin et al. (2008b)	Eclogite + dolomite, eclogite + K_2CO_3, eclogite + K–Na–Mg–Ca–Fe carbonate	8.5	1570–1800	4–42 min	Y	Gr	GS in all expts. SN reported in carbonate–silicate melt at low % (~30–40 wt.%) eclogite
Litvin and Bobrov (2008)	Peridotite + K–Na–Mg–Ca–Fe carbonate	8.5	1750–1780	12–15 min	Y	Gr	SN in carbonate–silicate melt up to ~ 30 wt.% peridotite in starting material
Litvin et al. (2008a)	Eclogite + dolomite, eclogite + K_2CO_3, eclogite + K–Na–Mg–Ca–Fe carbonate	7–8.5	1000–1800	2–45 min	Y	Gr	GS except one expt at 1000° (rest ≥1500°). Reported barrier to nucleation of diamond in the melts that varied depending on the carbonate composition
Spivak et al. (2008)	cpx–gt + K_2CO_3 + Gr, Eclogite + dolomite	7.3–8.5	1650–1800	6–40 min	N	Gr	SN reported in presence of melt in all 5 expts (7.3 GPa 1650°/40 min, 8.5 GPa 1720–1800°)
Bobrov and Litvin (2009, 2011)	Peridotite + carbonate, eclogite + carbonate. Carbonates: K_2CO_3, $CaMg(CO_3)_2$, K–Na–Mg–Ca–Fe carbonate	7–8.5	1750–1820	10–15 min	Y	Gr	GS in all expts at 8.5 GPa, 1750–1820°. Reported barrier to nucleation of diamond in the melts that varied depending on the carbonate and silicate composition of the melt

Study	System	*P* (GPa)	*T* (C)	Duration	Seeds	Capsule/ buffering	Results
Bureau et al. (2012)	MELD–SiO_2–Al_2O_3–TiO_2 –(Ca,Mg,Na_2,K_2)CO_3–H_2O–Gr ± D	7–9	1200–1700	10 min–144 h	In most	Pt	MELD = average inclusion composition from Navon et al. (1988). Observed SN in unseeded expts at 1500°/~5 h but not 1450°/10 min at 7 GPa. In seeded expts, observed GS at 7 GPa as low as 1200° (144 h expt), SN in most other expts at higher *P* and *T*
Spivak and Litvin (2012)	Calcio–carbonatite w/~28 wt.% SiO_2 + Gr	8.5	1320–1750	30 min	Y	Gr	Sample previously studied by Litvin et al. (2001) at 7 GPa. Varied C content in SM. Melt present in expts at >1325°. Diamond reported in expts at 20 and 40 wt.% C, not in 0 or 10 wt.% C expts. Diamond reported in subsolidus expt at 1320°, 40 wt.% C
Kiseeva et al. (2013)	Natural eclogites	9–21	1100–1800	12–80 h	N	$Au_{75}Pd_{25}$ capsules	Report diamond SN as accessory minerals in experiments on one of their eclogite compositions at 9–23 GPa; attributed to reduction of carbonate concomitant with Fe^{2+} oxidation in coexisting garnet
Litasov et al. (2014)	Peridotite + Fe (3 wt.%) + $C_{18}H_{36}O_2$ Eclogite + Fe (3 wt.%) + $C_{18}H_{36}O_2$	3–16	1200–1600	1–24 h	N	AuPd or Pt/Mo/ MoO_2 or Fe/FeO	SN at $T \geq 1400°$ and 12 GPa/12 h. SN at $T \geq 1200°$ and 16 GPa/24 h
Litvin et al. (2014)	$CaCO_3$ Na_2CO_3 $MgCO_3$ $FeCO_3$ Gr, $MgCO_3$–Gr,$CaCO_3$–Gr, Na_2CO_3–Gr and Mg–Fe–Na–carbonate–magnesiowüstite–Mg–perovskite–Gr	10–60	1100–4000 K	About 5 min	N	DAC (in Ne) and MAP	SN of nano diamonds
Palyanov et al. (2015)	Kimberlite–Gr	6.3	1300–1570	40 h	Y	Pt	GS of diamond at $T \geq 1520°$. SN at $T \geq 1520°$ (Pt). No SN in volume
		7.5	1450–1570	40 h	Y	Pt	GS at $T \geq 1400°$. SN at $T \geq 1450°$ (Pt). SN at $T \geq 1570°$ (in volume)
Brey et al. (2015)	Na–gloss + Gr with 6 wt.% CO_2 and 7 wt.% H_2O	7.5	900–1300	18–73 h	?	Pt with Re	No SN of diamond
		9.5	1000,1300	26–72 h			No SN at 1000°/72 h. SN at 1300°/26 h
		12	1200,1400,1600	2–48 h			No SN at 1200°/48 h. SN at 1400°/27 h and 1600°/2 h
	K–gloss + Gr	7.5	800–1300	24–100 h		Pt with Re	No SN at 800–1300°/48 h. SN at 1300°/25 h
		8.5–9.5	900–1300	25–101 h			No SN at 900–1200°. SN at 1300°/25 h
		10.5	900–1300	4–98 h			No SN at 900°, 1200°/4–15 h. SN at 1200°/27 h, 1300°/25 h
		12	800–1400	24–100 h			No SN at 800°/100 h. SN at 1200°/24 h at 1400°/27.5 h

Study	System	*P* (GPa)	*T* (C)	Duration	Seeds	Capsule/ buffering	Results
Bureau et al. (2016)	MORB + Pelagic sediments + H_2O + NaCl + Gr	6–7	1300–1400	6–10 h	Y	Pt	No SN, GS only at 1350° and 7 GPa
Girnis et al. (2018)	H_2O– and CO_2–bearing model sediment + garnet harzburgite +Gr (+ *T* gradient)	7.5–10.5	1300–1500	6–119 h	N	Pt(Re)	SN above 1300° at 7.5 GPa and 1200° at 10.5 GPa
	H_2O– and CO_2–bearing model sediment + garnet harzburgite + Dm + Gr (No T gradient)	7.5	1100–1400	2–210 h	Y	Pt(Re)	SN observed above 1200° at 7.5 GPa
Palyanov et al. (2021b)	$Mg_3Si_4O_{10}(OH)_2$ (Talc) + magnesite or dolomite or dolomite + Al_2O_3 + SiO_2	6.3–7.5	1300–1600	6–40 h	N	Pt, Pt loop (cathode)	SN (in the cathode zone) from carbon of carbonates under electric field of 0.4–1 v

Notes: Gr – graphite. DAC – Diamond-anvil cell. MAP – Multi-anvil press. SN – spontaneous nucleation and growth of diamond. GS – diamond growth on seeds. Temperatures in *Results* column in °C.

Table 8. Studies of diamond nucleation and growth in sulfide melts.

Study	System	*P* (GPa)	*T* (C)	Duration	Seeds	Capsule/ buffering	Results
Litvin et al. (2002)	$CuFeS_2$–$(Fe,Ni)_9S_8$ + Gr, $CuFeS_2$–$Fe_{1-x}S$ + Gr, CuS + Gr, Ag_2S + Gr	7–7.5	1450–1500	40–80 min	Y	Gr	SN reported in all compositions
Pal'yanov et al. (2003)	$(Fe,Ni)_9S_8$ + Gr	6.3, 7, 7.5	1450–1800	8.5–65 h	Y	Gr	SN at ≥ 1600° at 7.5 GPa, GS at 1450° at 6.3 GPa, ≥ 1550° at 7 GPa. CM: {111}
Litvin and Butvina (2004)	$CuFeS_2$–$(Fe,Ni)_9S_8$ + Gr, $CuFeS_2$–$Fe_{1-x}S$ + Gr, Ag_2S + Gr	7–8	1650–1800	10–180 min	Y	Gr	SN reported in all sulfide + Gr expts; unclear which SM mix used in each expt
Palyanov et al. (2006)	$(Fe,Ni)_9S_8$ + Gr, FeS + Gr	6.3–7.5	1450–2200	2–40 h	Y	Gr	$(Fe,Ni)_9S_8$ + Gr: SN at ≥1900° at 7 GPa, ≥1600° at 7.5 GPa; GS at 1450° at 6.3 GPa, ≥1550° at 7 and 7.5 GPa. FeS + Gr: No SN at 6.3 GPa, 1500°, 7 GPa 1550–1800°. SN at ≥ 1600° at 7.5 GPa; GS in all expts (6.3 GPa, 1500°; 7 GPa 1550–1800°; 7.5 GPa 1550–2200°)
Litvin et al. (2005b)	$Fe_{1-x}S$ + Gr	7.5–8.9	1800–2100	5–60 min	N	Gr	SN reported at these conditions; photo of diam formed at 8.2 GPa, 1910°/30 min
Shushkanova and Litvin (2006)	$Fe_{1-x}S$ + Gr	6.7	1660	4 min	N	Gr	SN of polycrystalline diamond

Study	System	*P* (GPa)	*T* (C)	Duration	Seeds	Capsule/ buffering	Results
Shushkanova and Litvin (2008a,b)	$Fe_{1-x}S$ + Gr	6–7.1	1409–1695	?	Y	Gr	Determined *P*, *T* range of SN and GS in sulfide melt
Spivak et al. (2008)	FeS	6.7	1660	4 min	N	Gr	SN observed
Chepurov et al. (2009a)	Fe–Co–S + Gr	5.5	1300	5–48 h	Y	?	Varied Fe:Co:S ratios; SN reported in most expts. Used Ti in sample as nitrogen sink
Zhimulev et al. (2012)	Fe–Co–S + Gr, Fe–Ni–S + Gr	5.5	1300	5–51 h	Y	MgO	Observed SN in metal–rich, S–poor melts: up to 14 wt.% S in Fe–Co–S; up to 9.9 wt.% S in Fe–Ni–S
Zhimulev et al. (2013)	Fe–S + Gr	5.5	1350	21–25 h	Y	MgO	Observed SN in 4/5 expts with ~5 wt.% S
Zhimulev et al. (2016a)	Fe–S + Gr	5.3–5.5	1300–1370	1–25 h	Y	MgO	SN in 4 wt.% S at 5.5 GPa, ≥ 1350° but no SN at 5.3 GPa, 1300° in 4 wt.% S. GS observed at 5.3 GPa, 1300° at 0.8–3.2 wt.% S.
Palyanov et al. (2020a)	Fe_9Ni_1 + Gr + FeS_2, S from 0 to 20 wt %	6	1400	20 h	Y	MgO	With an increase in the S content from 0 to 20 wt%, it was established a decrease in the degree of transformation Gr to Diam from 100 to 0, the formation of metastable graphite, a decrease in the solubility of carbon from 6.6 to 0.8 wt% and a decrease in the nitrogen content in diamonds from 50–100 ppm. CM: {111}, specific antisketal crystals

***Notes*:** Pn – pentlandite; Gr – graphite. SN – spontaneous nucleation and growth of diamond. GS – diamond growth on seeds. CM – diamond crystal morphology (where given in article). Temperatures in *Results* column in °C.

Table 9. Studies of diamond nucleation and growth by reduction of carbonate.

Study	System	*P* (GPa)	*T* (C)	Duration	Seeds	Capsule/ buffering	Results
Arima et al. (2002)	$CaMg(CO_3)_2$–Si, $CaMg(CO_3)_2$–SiC	7.7	1500–1800	1–24 h	In some	Pt	$CaMg(CO_3)_2$: No SN at 1800°. $CaMg(CO_3)_2$–Si: SN at 1600°/4 h, 1800°/1 h. GS at 1800° except in expt w/ lowest Si content. $CaMg(CO_3)_2$–SiC: SN at 1500°/24 h, 1600°/8 h but not at 4 h, 1800°/1 h. GS at 1800°/1 h, 1600°/4.2 h but not at 2 h. CM: {111}, specific triangular pyramids on {111} faces of seed crystals
Pal'yanov et al. (2002b)	$MgCO_3 + SiO_2$, $MgCO_3 + Na_2CO_3 + SiO_2$. No Gr. High fH_2 from $TiH_{1.9}$ in sample assembly.	6–7	1350–1800	10–43 h	Y (in most)	Pt, high fH_2 from $TiH_{1.9}$ in sample assembly.	$MgCO_3$+SiO_2: GS at 1400°, SN at 1450° at 6 GPa; At 7 GPa, no SN or GS at 1500°, 1600°; SN at 1750, 1800°. $MgCO_3$+Na_2CO_3+SiO_2: SN at 1400°, 1500° at 6 GPa and 1800° at 7 GPa. $MgCO_3$ present in all expts. Proposed diamond–forming reactions: $MgCO_3 + SiO_2 = MgSiO_3 + CO_2$ and $MgCO_3 + MgSiO_3 = Mg_2SiO_4 + CO_2$, then $CO_2 + 2H_2 = C + 2H_2O$. CM: {111} and fibrous patterns on the {100} face of seed crystals
Yamaoka et al. (2002a)	$CaCO_3$ + stearic acid ($C_{18}H_{36}O_2$)	7.7	1500	0.5–48 h	N	Pt	Two capsule configurations. Single capsule w/ $CaCO_3$ + StA produced Gr in short runs, diamond at ≥24 h. In double capsule expts ($CaCO_3$ in IC, StA in OC), found Gr in IC in 2 h expt, Gr + Diam in 12 h expt, Diam in 24 and 48 h expts. CM: {111}
Pal'yanov et al. (2005b)	$MgCO_3 + SiO_2$, $CaMg(CO_3)_2 + SiO_2$, $MgCO_3 + Al_2O_3 + SiO_2$	5.2–7	1200–1800	10–91 h	Y	Pt, most expts had high fH_2 from $TiH_{1.9}$ in sample assembly	More diamond growth in $MgCO_3$–SiO_2–Al_2O_3 than in $MgCO_3$–SiO_2. Diamond growth from either subsolidus fluid or carbonate–silicate melt, depending on conditions. Analysis of quench fluid phase documented high $X(H_2O)$ in $TiH_{1.9}$–present expts. CM: {111}
Siebert et al. (2005)	$FeCO_3$ + Si, $FeCO_3$ + FeSi	10–25	1700–1800	1–7 min	N	Gr or MgO	$FeCO_3$ + Si: SN at 10–25 GPa at 1700°, 25 GPa at 1800°. $FeCO_3$ + FeSi: SN in sole expt at 20 GPa, 1800°. CM: {111}
Gunn and Luth (2006)	$MgCO_3$ + (Mg,Fe)SiO_3 + Fe–S–O mix	6–7.5	1300	6—8 h	N	MgO	Gr produced in 6, 24, 48 h expts. No diamond observed
Palyanov et al. (2007b)	$MgCO_3$–SiO_2–Al_2O_3–FeS	6.3	1250–1800	8–44 h	Y	MgO+Pt, MgO+BN, MgO+ Gr	Gr produced in all expts. SN in expts at 1650°, 1700°. GS in all expts except at 1800°/8 h. CM: {111} and fibrous patterns on the {100} face of seed crystals
Chepurov et al. (2011)	$CaCO_3$ + $Ca(OH)_2$ + Fe	4	1350	5 h	Y	Pt	Gr reported in expt
Martin and Hammouda (2011)	$CaMg(CO_3)_2 + SiO_2$	4.25–6	800–1300	~2–94 h	N	AuPd or Mo-lined Pt	AuPd expts: Gr produced at 800° at 4.25, 5.5, 6 GPa and at 1000° at 6 GPa. Mo–Pt expts: Gr produced at 950–1150° at 4.25 GPa, 1150° and 1300° at 5.5, and 900°, 1050°, 1150°, 1250° at 6 GPa

Study	System	*P* (GPa)	*T* (C)	Duration	Seeds	Capsule/ buffering	Results
Bataleva et al. (2012)	$(Ca,Mg)CO_3$–SiO_2–Al_2O_3–$(Mg,Fe)(Cr,Fe,Ti)O_3$	6.3	1350–1650	20 h	Y	Pt, Pt + Gr	GS at 1550° and 1650° in Pt capsules (not at 1350° or 1450°), at 1350° and 1450° in Pt–Gr capsules. Observed dissolution of diamond seeds in expts w/ Pt capsules, not in those with Pt–Gr capsules. Diamond formation in Pt capsules attributed to decarbonation of carbonate; in Pt–Gr capsules, recrystallization of Gr from capsule. CM: Growth form {111}, regeneration of {100} → {111}
Palyanov et al. (2013b)	Fe–$(Mg,Ca)CO_3$	6.5, 7.5	1000–1650	8–60 h	Y	Pt	Graphite produced in all expts. Diam SN at 1350–1600° at 6.5 GPa, at ≥1300° at 7.5 GPa. GS at 1200–1400° at 7.5 GPa. CM: {111} in metal melt and in carbonate–oxide melt
Bataleva et al. (2016b)	Fe–$(Mg,Ca)CO_3$–S	6.3	900–1400	18–20 h	N	Gr, talc	Graphite produced in all expts (900–1400°). Carbonate seen only at 900°.
Bataleva et al. (2016d)	$(Mg,Ca)CO_3$–Al_2O_3–SiO_2–FeO	6.3	1150–1650	20 h	Y	Pt	Diamond growth at 1150° and 1250°, diamond dissolution at 1350–1650°. Gr reported at 1150–1450°. CM: Growth form {111}, regeneration of {100} → {111}
Bataleva et al. (2019)	Fe–$(Mg,Ca)CO_3$–S	6.3	1500, 1600	18 h	Y	Gr	Gr present in both expts; Diam SN in both. CM: {111}
Zdrokov et al. (2019)	olivine–ankerite–S, olivine–ankerite–FeS_2	6.3	1050–1550	20–60 h	Y	Gr	Formation of graphite (1350–1550° in S system; 1450–1550° in FeS_2 system) and diamond (1550°) by interaction of coexisting Fe–Ni–S and carbonate melts. CM: Growth form {111}, regeneration of {100} → {111}

***Notes*:** SN – spontaneous nucleation and growth of diamond; GS – diamond growth on seeds. CM – diamond crystal morphology. Temperatures in *Results* column in °C.

Table 10. Studies of diamond formation at very high pressures–diamond anvil cell experiments.

Study	System	*P* (GPa)	*T*	Duration	Seeds	Capsule/ buffering	Results
Liu et al. (2001)	$MnCO_3$	6–12	>2000 °C	?	DAC		Laser–heated DAC; reported formation of diamond by break-down of carbonate at *P* ~12 GPa
Tschauner et al. (2001)	CO_2	30–80	1500–3000 K	?	DAC	DAC	Laser–heated DAC; reported formation of diamond by breakdown of CO_2. Raman peaks attributed to new diamond observed between 36 and 72 GPa
Seto et al. (2008)	MORB + $CaCO_3$, SiO_2 + $MgCO_3$, $MgCO_3$	31–80	1700–3300 K	1–2 h	DAC	DAC	Laser–heated DAC expts. Reported diamond formation at >3000 K at 33–80 GPa in MORB + $CaCO_3$, at 3300 K at 31, 44 GPa and at 2500 K at 73 GPa in SiO_2 + $MgCO_3$

Study	System	*P* (GPa)	*T*	Duration	Seeds	Capsule/ buffering	Results
Bayarjargal et al. (2010)	$CaCO_3$ in NaCl pressure medium	9–21	1000–4000 K	?	DAC	DAC	Formed graphite by decomposition of $CaCO_3$. Annealing of sample produced diamond
Boulard et al. (2011)	$MgCO_3$, $(Fe_{0.75}Mg_{0.25})CO_3$, $MgO + CO_2$, $(Mg_{0.6}Fe_{0.4})O + CO_2$	80–105	2400–2805 K RT	?	DAC	DAC	Reported diamond formation (nanodiamonds)
Litasov et al. (2011)	CO_2	40–70	1800–2000 K		DAC	DAC	Diamond formation from CO_2 at 40–70 GPa
Zhang et al. (2011)	H_2O from glucose, $H_2O–CO_2$ from OAD	9–11	1000–1400 K	15–50 min	DAC	DAC	Growth reported in both expts w/ glucose, none in OAD expt
Boulard et al. (2012)	$FeO + CO_2$, $FeO(OH) + CO_2$, $Fe_2O_3 + CO_2(Mg_{0.6}Fe_{0.4})O + CO_2$	40–105	1460–3650 K	About 2 h	DAC	DAC	Reported diamond formation (nanodiamonds) in all exp. except with Fe_2O_3
Spivak et al. (2012)	$CaCO_3$	11–43	1600–3900 K	5 min	N	DAC (Re gasket + Ne) and MAP	Laser–heated DAC. Diamond formation from breakdown of $CaCO_3$ melt at ~3500 K, ~16 and ~43 GPa
Litvin et al. (2014)	$CaCO_3$ Na_2CO_3 $MgCO_3$ $FeCO_3$ Gr, $MgCO_3$–Gr, $CaCO_3$–Gr, Na_2CO_3–Gr and Mg–Fe–Na–carbonate–magnesio-wüstite– Mg–perovskite–Gr	10–60	1100–4000 K	About 5 min	N	DAC (in Ne) and MAP	SN of nano diamonds
Solopova et al. (2015)	$MgCO_3$	12–84	1600–3300 K	1 min	DAC	DAC (and MAP at 23 GPa)	Reported MgO + diamond in MAP at 23 GPa, 2700 K, inferred diamond formation in DAC at >2700 K by presence of MgO
Thomson et al. (2016)	Peridotite + Fe + carbonate	20	1590 °C	124 min	DAC	DAC	Reduction of carbonate melt by Fe metal to diamond reported
Maeda et al. (2017)	$MgCO_3 + SiO_2$ + Pt (in NaCl or SiO_2 pressure media)	~30–152	300–3100 K	10–120 min	N	DAC	Laser-heated diamond anvil cell. Diamond nucleation at 70 GPa
Litvin et al. (2017)	$MgO–FeO–SiO_2$ + Mg–Fe–Na–K carbonate + Gr	26	?	?	DAC	DAC	Reported diamond formation at 26 GPa coexisting with oxide + silicates
Dorfman et al. (2018)	$CaMg(CO_3)_2$ + Fe	51–113	1800–2500 K	10–20 min	DAC	DAC	Reported diamond formation in all 4 expts (51, 66, 77, 113 GPa)
Drewitt et al. (2019)	$FeO–MgO–SiO_2–CO_2$ and $CaO–MgO–SiO_2–CO_2$ systems	48–89	1580–2160 K	24–61 min	DAC	DAC	Report formation of diamond in higher–T expts (1745–2160 K at 54–88 GPa in FMSC, 1905–2160 K at 40–84 GPa in CMSC) by decarbonation of carbonate
Martirosyan et al. (2019)	Natural $MgCO_3$–Fe	70–150	800–2600 K	10–20 min	DAC	DAC	Reported diamond formation in part of the experiments

Notes: DAC – Diamond anvil cell. MAP – Multi-anvil press. SN – Spontaneous nucleation and growth of diamond.

Recently, Martirosyan et al. (2019) performed LHDAC experiments in the $MgCO_3$–Fe system combined with *in situ* synchrotron X-ray diffraction and *ex situ* transmission electron microscopy. Based on the results they suggest that the interaction of carbonates with Fe^0 or Fe^0-bearing rocks can produce Fe-carbide and diamond, which can accumulate in the D" region (the lowermost portion of the mantle that sits just above the molten iron-rich outer core).

Future Directions

Because of the pressure limitation of large volume presses, DAC experiments may be precious tools to understand the growth of superdeep diamonds in the lower mantle. Another potentially interesting approach is the use of dynamic compression experiments (i.e., laser-compressed materials), to generate diamonds at ultrahigh pressures and temperatures in order to study impact diamonds or the formation of diamonds in planetary interiors conditions (Kraus et al. 2017).

CARBIDES: SOURCES OF CARBON AND REDUCING AGENTS

There has been significant interest in exploring the possibility for iron carbides to act as both carbon sources and reducing agents since the general acceptance of the idea that depth-driven reduction in the Earth's mantle stabilizes Fe–Ni metal and carbide in C-bearing mantle regimes. Studies have examined interaction of carbides, particularly cohenite (Fe_3C), with oxides, silicates, carbonates, and sulfides, as can be seen in Table 11 below. Cohenite has proved to be an effective source of carbon upon reaction, usually forming metastable graphite. Growth of diamond is not typically observed in the absence of a melt. In many of the studies in Table 11, it is notable that the presence of melt in itself is insufficient, and diamond growth is inhibited until a sufficiently high temperature is achieved. In carbonate-bearing experiments, elemental carbon is produced by both breakdown of cohenite as well as reduction of CO_2 produced by decarbonation reactions involving the carbonate.

INCLUSIONS IN DIAMOND: EXPERIMENTAL CONSTRAINTS

Inclusions trapped during growth of synthetic diamonds at high pressure and temperature have been observed since the first syntheses in metallic melt solvents. In one of the early reports from the G.E. group, Bovenkerk (1961) stated "It has not been possible so far to grow diamond without trace inclusions." For industrial purposes, of course, inclusions are something to avoid or minimize (e.g., discussion in Sumiya et al. 2002). On the other hand, given that inclusions in natural diamonds provide valuable clues to the nature of the growth medium, it is of interest to deliberately try to grow diamonds with inclusions to see, for example, if the inclusions provide an accurate picture of the fluid or melt from which the diamond is growing. Early studies of synthetic diamond documented metal inclusions inferred to be from the growth medium (Lonsdale et al. 1959), and, intriguingly, coesite (Milledge 1961). The possibility of formation of single fluid inclusions during diamond crystallization in metal-carbon system was shown by Osorgin et al. (1987). Later work documented the presence of a rich diversity of inclusions (taenite, wüstite, spinel, silicate, diamond, and fluid) in synthetic diamond grown in the Fe–Ni–C system (e.g., Pal'yanov et al. 1997). In this study, the oxide and silicate inclusions resulted from diffusion of elements from the container materials and from impurities in the initial reagents. These examples suggested that detailed study of inclusions in diamond growing in other systems would be fruitful in exploring how representative these inclusions are of the growth medium.

As shown in Table 12, many studies report the presence of inclusions in passing, such as graphite (e.g., Pal'yanov et al. 1999b) and quenched melt, either reduced or oxidized—sometimes in the same experiment (Palyanov et al. 2013b), but there have been also some studies designed specifically to synthesize inclusion-bearing diamonds (e.g., Khokhryakov et al. 2009; Bureau et al. 2012, 2016; Bataleva et al. 2016c). The first of these documented the rich variety of ways in which graphite inclusions in diamond can be formed; importantly it showed that graphite inclusions could form during growth of the diamond crystal from a carbonate melt—in the stability field of diamond. The others showed a good fidelity between the materials trapped as inclusions and those present in the environment in which the diamonds were growing. These studies provide a foundation for further studies on inclusion formation and trapping in a variety of media. For example, Bataleva et al. (2016c) suggested that slower growth rates would decrease the relative proportion of fluid or melt inclusions in the growing diamond, as had been previously documented for inclusions in magmatic minerals (Roedder 1984). Applying these methodologies to other growth media—such as the saline-rich endmember seen in natural fibrous diamonds—would be of considerable interest, as another example.

NITROGEN AND BORON STUDIES

Overview

Natural diamonds are known to contain impurities, the most common being nitrogen. The concentrations of nitrogen depend on the paragenesis and vary from below 10 ppm to > 3500 ppm (see Stachel et al. 2022b, this volume). Boron and hydrogen are also naturally present in the diamond lattice (Green et al. 2022, this volume).

Synthesis of commercial HPHT diamonds in, for example Fe–Co melts under standard synthesis conditions, contain ~100–300 ppm N in {111} growth sectors, and usually about half that in {100} growth sectors when grown under "standard conditions" (i.e., 1300–1400 °C, 5.5 GPa, Burns et al. 1999); lower growth temperatures can reverse this difference, and increase N contents up ~1000 ppm because of the decreased solubility of N in the melt at lower temperatures. In these syntheses, nitrogen is introduced as a contaminant in the starting materials, and its concentration in the diamond can be reduced by addition of so-called nitrogen "getters" such as Al, Ti, or Zr, which form nitrides at the experimental conditions.

In general, non-metallic solvents—carbonate, carbonate–silicate, sulfide, and sulfur—tend to crystallize diamond with higher (500–1500 ppm) N contents. Use of N-bearing compounds such as BN, NaN_3, $Ba(N_3)_2$, $CaCN_2$, Fe_3N, P_3N_5, or $C_3H_6N_6$ can produce diamond with >3000 ppm N (Table 13).

With the methodology for growing diamond with specific amounts of N, the next frontier is to use modern ion probe methods to examine N concentrations and isotopic compositions across diamond crystals, as pioneered by Reutsky et al. (2008b, 2017), for example. An attempt to find the interconnection between morphology and defect-impurity composition of diamond with its formation conditions was performed by Palyanov et al. (2021a).

Boron is another significant impurity found in diamonds, which gives them a blue color. Defined as Type IIb diamonds, they are very rare <0.1% of the worldwide extracted diamonds and naturally contain almost no N and up to 10 ppm B (Gaillou et al. 2012). Blue diamonds are synthesized for industry because B gives diamonds semi-conductive properties, given that significant amounts of boron (> 1000 at.ppm) can be incorporated in synthetic CVD diamonds (Polyakov et al. 2001). Synthetic boron-doped diamonds are also used as heaters either in large volume presses or in diamond anvil cells for HPHT experimentation (Shatskiy et al. 2009).

Table 11. Studies of diamond nucleation and growth in carbide-bearing systems.

Study	System	*P* (GPa)	*T* (C)	Duration	Seeds	Capsule/ buffering	Results
Palyanov et al. (2013b)	$(Mg,Ca)CO_3$–Fe_3C	7.5	1000–1400	60 h	Y	Pt	Graphite formed in all expts. GS at ≥1200°, SN at 1400°. Carb + oxide melt present in all diamond–forming expts. CM: {111} in carbonate–oxide melt
Bataleva et al. (2015b)	Fe_3C–S	6.3	900–1600	18–20 h	Y	Gr (mostly), Talc, MgO	Graphite formed in all expts. At $T < 1200°$, reaction $2Fe_3C + 3S_2 \rightarrow 6FeS + 2C$. At ≥1200°, formed melt + C. GS at 1400–1600°. Expts in talc and MgO capsules confirmed C production by breakdown of Fe_3C. CM: Growth form {111}, regeneration of {100} → {111} on seed crystals
Bataleva et al. (2015a)	$(Mg,Ca)CO_3$– Al_2O_3–SiO_2–Fe	6.3, 7.5	1150–1650	8–20 h	N	Pt	Formed Fe_3C and graphite in all expts. Subsolidus at 6.3 GPa 1150–1450°. At 7.5 GPa formed carb–silicate melt + graphite at 1450°–1650°. No diamond SN reported, but no diamond seeds in expts. See paper for full phase assemblages
Bataleva et al. (2016a)	$Fe_3C + Fe_{0.7}Ni_{0.3}S$	6.3	1100–1500	20 h	Y	Gr	Graphite formed in all experiments. Coh + 2 liquids at 1100–1200°. One liquid + Gr at 1300–1500° with GS of diamond. CM: Growth form {111}, regeneration of {100} → {111} on seed crystals
Bataleva et al. (2016e)	Fe_3C–Fe_2O_3, Fe_3C–Fe_2O_3–MgO–SiO_2	6.3	900–1600	18–20 h	Y	Gr (mostly), Talc, MgO	Graphite formed in all expts in both systems, including those in talc or MgO capsules. Both systems subsolidus at 900–1300°. GS in presence of Fe–C liquid at 1400–1600° in Fe_3C–Fe_2O_3. GS in presence of a Fe–C liquid at 1400° and two liquids (Fe–C and silicate–oxide) at 1500° and 1600° in Fe_3C–Fe_2O_3–MgO–SiO_2. CM: Growth form {111}, regeneration of {100} → {111} on seed crystals
Bataleva et al. (2017)	Fe_3C–S, Fe_3C–FeS_2, Fe_3C–S–MgO–SiO_2	6.3	900–1600	18–40 h	Y	Gr (mostly), Talc, MgO	All systems: Graphite formed in all experiments, including those in talc and MgO capsules (Fe_3C–S, Fe_3C–S–MgO–SiO_2 systems). GS at ≥1300° in all three systems in presence of liquid(s), but no GS at 1200° despite presence of liquid(s). CM: Growth form {111}, regeneration of {100} → {111} on seed crystals
Bataleva et al. (2018a)	Fe_3C–SiO_2–Al_2O_3–$(Mg,Ca)CO_3$	6.3, 7.5	1100–1650	8–40 h	Y	Pt	Graphite formed in all expts. No Coh or carb at 6.3 GPa 1100°–1500°, and diamond GS only at 1500° at 6.3 GPa where melt was seen in inclusions in garnet. At 7.5 GPa, GS seen in all expts (1250–1650°). Liquid present at ≥1450°. C formed from both carbide and decarbonation–produced CO_2. CM: {111} and {100}
Bataleva et al. (2018b)	Fe_3C–SiO_2–$(Mg,Ca)CO_3$	6.3	1100–1500	20–40 h	Y	Gr	Graphite formed in all expts. Diamond GS only at 1500°, not at 1100°–1400°. Fe–C melt present at ≥1300°. C formed from both carbide and decarbonation–produced CO_2. CM: {111} and {100}
Bataleva et al. (2018c)	Fe_3C–SiO_2–$(Mg,Ca)CO_3$, Fe_3C–SiO_2–Al_2O_3–$(Mg,Ca)CO_3$	6.3	1300–1500	20 h	Y	Gr	Graphite formed in all expts. Diamond GS reported at 1400° (both systems) and at 1500° in the Al_2O_3–bearing system. Fe–C melt in all expts at least as inclusions in silicates. CM: {111} and {100}
Bataleva et al. (2018d)	Fe_3C–SiO_2–MgO	6.3	1100–1500	20–40 h	Y	Gr	Graphite formed in all expts; Fe–C melt present at ≥1300°. Diamond GS at 1500°. CM: {111} and {100}
Bataleva et al. (2019)	Fe_3C– $(Mg,Ca)CO_3$–S	6.3	900–1400	18–20 h	Y	Gr (mostly), Talc, MgO	Graphite formed in all expts; two immiscible liquids at ≥1100°. Diamond GS at ≥1400° (including in MgO capsule), SN at ≥1400° in Gr capsules. CM: {111} in both melts

Notes: SN – spontaneous nucleation and growth of diamond. GS – diamond growth on seeds. CM – diamond crystal morphology. Temperatures in *Results* column in °C.

Table 12. Studies of inclusion formation.

Study	System	P (GPa)	*T* (C)	Duration	Seeds	Capsule/ buffering	Results
Lonsdale et al. (1959)	G.E. synthetic diamonds						XRD identification of Ni inclusions in synthetic diamonds; observation of other inclusions but were not able to identify
Milledge (1961)	G.E. synthetic						XRD identification of coesite inclusions in synthetic diamonds from GE
Osorgin et al. (1987)	Ni–Mn–C	5–6	1300–1400	2–8 h	No	talc	Fluid and metal–fluid inclusions (CO, H_2, N_2)
Pal'yanov et al. (1994, 1997)	Ni–Mn–C, Fe–Ni–C	5–6	1300–1600	1–150 h	Y		Inclusions of taenite, wüstite, spinel, garnet, diamond, and hydrocarbon fluid
Tomilenko et al. (1998)	Ni–Mn–C and Ni–Fe–C	5–6	1350–1550				Hydrocarbon fluid inclusions
Pal'yanov et al. (1999b)	Li_2CO_3–Gr, Na_2CO_3–Gr, K_2CO_3–Gr, Cs_2CO_3–Gr	7	1700–1750	10 min–18.5 h	Y	Pt	Graphite inclusions observed
Burns et al. (1999)							Summary of DeBeers syntheses. Discusses inclusion formation as function of solvent/catalyst composition, growth rates, synthesis *P*, *T*. Notes use of Ti as N "getter" produces abundant TiC inclusions
Kanda (2000)	Metal solvent (Ni, Co, Fe)–catalyst (Cu, Mg) + Gr + NaCl	5	1500–1800	1–3 mg/h	Y		Inclusions of metallic solvents trapped in large diamonds, possibly also oxides and halides
Pal'yanov et al. (2002a)	Na_2CO_3–Gr, K_2CO_3–Gr, Na_2CO_3–H_2O–Gr, K_2CO_3– H_2O–Gr, Na_2CO_3–CO_2–Gr, K_2CO_3– CO_2–Gr, Na_2CO_3–H_2O+ CO_2 (OAD)–Gr, K_2CO_3– H_2O+ CO_2 (OAD)–Gr	5.7	1150–1420	5–136 h	Y	Mostly Pt, 1 in Gr, 2 in Au	Observed "minute black inclusions" in all expts where diamond grew on seed crystals and "transparent inclusions of carbonate melt" in some expts
Sumiya et al. (2002)							Review paper, discusses synthesis conditions to minimize trapping of metal inclusions
Yin et al. (2002)	Fe–Ni + Gr	5.5	1300	12 min	?	?	TEM study of nanometer inclusions identified graphite, $(FeNi)_{23}C_6$ and γ–FeNi
Chepurov et al. (2007)	$Ni_{0.7}Fe_{0.3}$+ Gr + Ti + B_2O_3	5.5–6	1350–1450	80–120 h	N		Diamond nucleation and growth with inclusions of metallic solvent
Khokhryakov et al. (2009)	NaCl + H_2O + Gr, KCl + K_2CO_3 + H_2O + Gr, K_2CO_3 + H_2O + Gr, SiO_2 + H_2O + Gr, Mg_2SiO_4 + H_2O + Gr, $(Ca,Mg)CO_3$ + SiO_2 + H_2O + CO_2 + Gr, kimberlite + H_2O + CO_2 + Gr	7.5	1500–1600	15–40 h			Examine formation of graphite inclusions. Found they formed by three mechanisms:1) protogenetic (pre–existing graphite incorporated into growing diamond) in NaCl + C and Na_2CO_3 + C systems. 2) Syngenetic, forming during joint crystallization of diamond + graphite. Seen only in H_2O–bearing systems. 3) Post–grown inclusions resulting from graphitization

Study	System	P (GPa)	*T* (C)	Duration	Seeds	Capsule/ buffering	Results
Palyanov et al. (2009)	S + Gr	6.3–7.5	1550–2000				Observed red and black inclusions, suggested to form from trapping of CS_2 melt during growth
Palyanov et al. (2010b)	$Ni_{0.7}Fe_{0.3}$ + Gr + Fe_3N and $CaCN_2$	5.5	1400	65 h	N	MgO	Diamond growth with metallic inclusions
Tomlinson et al. (2011)	$Fe_{0.9}Ni_{0.1}$ + Gr	15 20	2000 2100	15 min 15 min	N	Re–Gr	Noted common occurrence of metal inclusions
Bureau et al. (2012)	MELD (SiO_2–Al_2O_3–TiO_2–(Ca,Mg,Na_2,K_2) CO_3) + H_2O + Gr	7–9	1250–1700	10 min–144 h	In most	Pt	Observed inclusions containing phengite or glass, sometimes with empty void in inclusion, and discrete empty inclusions. Voids and empty inclusions interpreted to have been fluid-filled before breach of inclusion wall during sample prep
Palyanov et al. (2012)	$Ni_{0.7}Fe_{0.3}$ + $Mg(OH)_2$ + SiO_2 + Gr	6	1370	15 h	Y	Pt	Nucleation and growth on seeds, Observed inclusions of metallic solvent
Palyanov et al. (2013a)	$Ni_{0.7}Fe_{0.3}$ + Gr +$Mg(OH)_2$ + SiO_2	6	1300–1370	15 h	N	Pt?	Diamond nucleation and growth inclusions of metal–carbon melt and metal-free fluids of hydrocarbon composition
Palyanov et al. (2013b)	Fe–(Mg, Ca)CO_3	6.5, 7.5	1000–1400	8–60 h	Y	Pt	Diamonds in reduced part of sample contained inclusions of quenched Fe–C melt, those that grew in the carbonate melt contained carbonate inclusions
Zhimulev et al. (2013)	Fe–S + Gr	5.5	1350	21–25 h	Y	MgO	Metal–sulfide inclusions in diamonds led to fracture of the crystals
Palyanov et al. (2015)	Gp I kimberlite (Udachnaya–East), synthetic Gp. II kimberlite	6.3, 7.5	1300–1570	40 h	Y	Pt w/Gr liner	Alkaline carbonate–silicate melt and silicate (olivine and pyroxene) inclusions observed in polycrystalline diamond aggregates
Bataleva et al. (2016c)	SiO_2–(Ca,Mg)CO_3–(Fe,Ni)S	6.3	1650–1750	2–24 h	Y	Gr	Study designed to trap inclusions during diamond growth. Inclusions of gr, carb–silicate liquid and sulfide liquid, opx, diamond, and CO_2-rich fluid. Inclusions characterised by Raman spectroscopy. Abundant inclusions of melt and fluid attributed to high growth rate of diamond
Bureau et al. (2016)	MELD + H_2O + NaCl + Gr, MELD + $FeCO_3$ + H_2O ± NaCl + Gr	7	1300–1400	6–30 h	Y	Pt or AuPd	Observed both monocrystalline, pure fluid, and multi-phase inclusions of same mineralogy as found in matrix: diam, carbonate, phengite, coesite, rutile in MELD expts, olivine + carbonate in MELD + $FeCO_3$ expts
Khokhryakov et al. (2016)	$Na_2C_2O_4$–$CaCN_2$	6.3	1500	2, 30 h	Y	Pt	Carbonate inclusions inferred from IR spectra
Palyanov et al. (2016)	$Na_2C_2O_4$ ± Gr	6.3, 7.5	1250–1700	10–66 h	Y	Pt	Inclusions of Na-carbonate melt identified by Raman. Peaks for CO_2 interpreted as having exsolved from melt on cooling
Smith and Wang (2016)	HPHT diamonds						Found CH_4 and H_2 coexisting with metallic inclusions in commercially grown HPHT diamonds
Bataleva et al. (2019)	Fe–(Mg,Ca)CO_3–S	6.3	1500, 1600	18 h	Y	Gr	Reported inclusions of magnesiowüstite, graphite, sulfide, and carbonate

***Notes*:** MELD – average inclusion composition from Navon et al. (1988).

Table 13. Studies of nitrogen and boron incorporation into diamond.

Study	System	*P* (GPa)	*T* (C)	Duration	Seeds	Capsule/ buffering	Results
Kanda and Yamaoka (1993)	Co + Gr	6	~1500	12–18 h	?	?	Grew diamond w/ ~200 ppm N, found heterogeneous distribution of A and C centres
Sumiya and Satoh (1996)							Review paper of diamond growth in metallic melts. Discusses nitrogen getters to reduce N to <0.1 ppm and techniques for minimizing B
Fisher and Lawson (1998)	Ni + Gr, Ni–Fe + Gr, Co–Fe + Gr						N concentrations from ~170–455 ppm. Examined effect of Ni concentration on N aggregation in synthetic diamonds, mostly in {111} growth sectors
Burns et al. (1999)	Co–Ti, Fe–Al, Fe–Co, Fe–Al–B						Review of DeBeers research; N contents (ppm) reported as <0.01–2 for Co–Ti, 0.4–1 for Fe–Al, 100–1000 for Fe–Co, below detection for Fe–Al–B solvent/catalysts. Note that addition of B in addition to N getter to solvent/ catalyst allowed growth of blue Type IIb diamonds; higher concentrations in {111} than in {100}. Colour depends on B/N
Borzdov et al. (2002)	Fe_3N + Gr	7	1550–1850	20 h	Y	Gr	GS at ≥ 1600°, SN at ≥ 1700°. 100% conversion at 1850°, decreasing to <1% at 1700°. N up to ~3300 ppm
Pal'yanov et al. (1998b)	Ni–Fe + Gr	5.5–6.5	1500–1600	100–150 h	Y	?	~130–200 ppm N
Kanda et al. (1999)	Na_2SO_4 + Gr	7.7	2000	30 min	N	Gr, surrounded by NaCl–ZrO_2 or BN	Grew diamond with 1200–1900 ppm N with BN container, ~200 ppm N with NaCl–ZrO_2
Kanda (2000)							Review paper; discusses B and N in HPHT synthetic diamonds grown in metallic systems
Pal'yanov et al. (2001a)	S + Gr	7	1750–1850	3–7 h	Y	Gr	Up to 700 ppm N
Pal'yanov et al. (2002b)	$MgCO_3$ + SiO_2, $MgCO_3$ + Na_2CO_3 + SiO_2. No Gr. High fH_2 from $TiH_{1.9}$ in sample assembly	6–7	1350–1800	10–43 h	Y (in most)	Pt, high fH_2 from $TiH_{1.9}$ in sample assembly	300–800 ppm N in diamonds from 1750–1800° expts at 7 GPa, both starting materials
Liang et al. (2005)	Fe–Ni + NaN_3 + Gr	5–5.8	1230–1480	15 min	N	ceramic	N conc varied from 500 to ~2000–2500 ppm with increasing NaN_3 from 0–0.5 wt.%
Liang et al. (2006)	Fe–Ni + NaN_3 + Gr	5–5.8	1230–1480	15 min	N	ceramic	N conc varied from 350–500 to 1700–2000 ppm w/ increasing NaN_3 from 0–0.5 wt.%

Study	System	*P* (GPa)	*T* (C)	Duration	Seeds	Capsule/ buffering	Results
Palyanov et al. (2006)	$(Fe,Ni)_9S_8$ + Gr	7	1900	16, 40 h	Y	Gr	~1000 ppm N
Chepurov et al. (2007)	Fe–Ni–C, Fe–Co–C + Ti + B_2O_3	5.5–6	1350–1450	80–120 h	N	Pt?	Growth of Type IIb boron-bearing monocrystalline diamonds. Almost no N
Palyanov et al. (2007b)	$MgCO_3$–SiO_2–Al_2O_3–FeS	6.3	1650, 1700	23.5 h 8 h	Y	MgO+ Gr	Measured diamonds formed in two expts, both with ~1500 ppm N
Zhang et al. (2007)	Gr + $Fe_{0.9}Ni_{0.1}$+ boron	5.4–5.6	1550–1700 K		N		Growth of large monocrystals of Type IIb diamonds
Reutsky et al. (2008b)	Fe–Ni + Gr	5.5	1450	90 h			SIMS study. Observed zoning in N across diamond. Also measured N isotopes
Spivak et al. (2008)							Reported up to 850 ppm N in diamonds from carbonate-bearing expts
Yu et al. (2008)	Fe–Ni + $Ba(N_3)_2$ + Gr	5–5.7	1230–1430	13 min	?	?	Report N conc varied from 200–400 to 1600–2400 ppm as $Ba(N_3)_2$ increased from 0 to 0.5 wt.%
Zhang et al. (2008)	Fe–Co–Ni + NaN_3 + Gr	5.4	1200	2–20 h	N	pyrophyllite	Reported up to 1520 ppm N
Chepurov et al. (2009b)	Fe–Co + Gr, Fe–Co–TiO_2 +Gr, Fe–Co–Ti + Gr	5.5	1400	46–69 h	Y	?	N contents: 133 ppm in Fe–Co, 121 ppm in Fe–Co–TiO_2, 9 ppm in Fe–Co–Ti
Palyanov et al. (2009)	S + Gr	6.3–7.5	1550–2000	3–40 h	Y	Gr	≤ 10–1500 ppm N found
Shatskiy et al. (2009)	B + graphite mixture	20	1600			MAP	Growth of B-doped diamond to use them as heaters at HT (>3000°)
Palyanov et al. (2010b)	Fe–Ni + Gr, Fe_3N and $CaCN_2$ N sources	5.5	1400	65 h	Y	MgO	127–1077 ppm N with varying amounts of Fe_3N; 124–850 ppm N with varying $CaCN_2$
Huang et al. (2010a)	Ni–Mn–Co + NaN_3 + Gr	~5.5	1240–1300	?	Y	pyrophyllite	N conc ~200 to ~1250 ppm w/ increasing NaN_3 from 0 to 0.8 wt.%
Huang et al. (2010b)	Fe–Ni–Co + NaN_3 + Gr	~5.5	~1230–1280	11 h	Y	ceramic material	Report range of N conc from 700–750 ppm to 1671–1742 ppm N w/ increasing NaN_3 from 0.3 to 0.6 wt.%
Liu et al. (2011)	Carbonyl iron powder + Gr	5.2–6.8	1200–1800	10–20 min	?	?	Report 1100–1500 ppm N in diamond grown in carbonyl iron + Gr system
Palyanov et al. (2012)	Ni_7Fe_3 + $Mg(OH)_2$ + SiO_2 + Gr	6	1370	15 h	Y	MgO	Observed decrease in N from 220–230 ppm to 40–50 ppm with increasing H_2O in melt
Zhang et al. (2012)	Fe–Ni + P_3N_5 + Gr	5–6.3	1250–1550	15 min	N	?	Report N conc varied from 200–400 ppm to 1300–1600 ppm as P_3N_5 increased from 0 to 0.4 wt.%

Study	System	*P* (GPa)	*T* (C)	Duration	Seeds	Capsule/ buffering	Results
Palyanov et al. (2013b)	Fe–(Mg,Ca)CO_3	6.5, 7.5	1000–1400	8–60 h	Y	Pt	Diamonds grown in metal-carbon melt contained 100–200 ppm N, diamonds formed in carbonate melt had 1000–1500 ppm N
Liu et al. (2016)	Fe–Ni + Gr, Fe–Ni–Co + Gr. P_3N_5 or $C_3H_6N_6$ as N source	5–6.3	1300–1650	15 min–30 h	Y	MgO	N conc up to ~2300 ppm in FeNi-C + P_3N_5, up to ~3400 pm in FeNiCo–C + $C_3H_6N_6$
Khokhryakov et al. (2016)	$Na_2C_2O_4$ + $CaCN_2$	6.3	1500	2, 30 h	Y	Pt	No nucleation in 2 h expt, SN in 30 h expt. Found variable N content (100–1100 ppm) depending on growth direction
Palyanov et al. (2016)	$Na_2C_2O_4$	6.3, 7.5	1300–1700	10–66 h	Y	Pt	50–150 ppm N in $Na_2C_2O_4$ expts compared to 100–200 ppm N in Na_2CO_3 + Gr expts
Reutsky et al. (2017)	$Na_2C_2O_4$	6	1400			Pt	SIMS study of carbonate-grown diamonds observed difference in conc of N between {111} and {100} of ~20×
Chen et al. (2018)	Fe–Ni + Al + NaN_3 + Gr	5.8	1380–1400	?	Y	ZrO_2–MgO	Found N decreased with Al content: 80 ppm w/o Al, 28 ppm w/10% Al, <1 ppm N w/ 20% Al. Found N linearly increases from 372–1573 ppm w/ increasing NaN_3 from 0.1–0.5%
Sokol et al. (2019)	Fe_3N + FeS	7.8	1600–1800	30–60 min	?	BN + Gr	Report 2100–2600 ppm N and 130–150 ppm B in diamond in equilibrium w/ BN
Palyanov et al. (2020b)	Ni_7Fe_3 + Fe_2O_3 + Gr	6	1400	40 h	Y	MgO	Found N increases from 200–250 ppm w/ no O additive to 1100–1200 ppm w/ 10 wt.% O

Notes: GS – diamond growth on seeds. SN – spontaneous nucleation and growth of diamond. MAP – Multi-anvil press. Temperatures in *Results* column in °C.

Future directions

The growth of diamonds with or without impurities is of great importance for understanding the mechanisms at depth in the Earth. For example, experiments performed by Palyanov et al. (2013b) show that simultaneous diamond growth in metal-rich and in carbonate-rich parts of the same capsule would form diamonds containing only 100–200 ppm N when in contact with metallic Fe, whereas those formed in the carbonate melt zone contain 1000–1500 ppm N. This difference is explained by the siderophilic affinity of N to metallic liquids at high *P*, *T* conditions. The occurrence of Type II diamonds may be explained by growth in the presence of metallic melts (Smith and Kopylova 2013). Further experimental studies are necessary to confirm such a hypothesis.

CARBON ISOTOPE STUDIES

In general, the bulk carbon isotope composition of HPHT diamonds solely depends upon the initial carbon isotope composition of the diamond-forming system. All the diamonds from HPHT experiments have $\delta^{13}C$ in a range from –20 to –30‰ PDB, which is inherited from the initial graphite sources used for synthesis (Hoering 1961; Laptev et al. 1978; Ivanovskaya et al. 1981; Taniguchi et al. 1996; Reutsky et al. 2008a, 2015a) (Table 14). In the film-growth (FG) experiments, no significant isotope inhomogeneity is reported for produced diamond material (Hoering 1961; Laptev et al. 1978; Ivanovskaya et al. 1981). When the temperature gradient growth (TGG) method is used, carbon isotope fractionation, which accompanied the diamond crystallization from carbon solution in metal melt, produces a certain isotope profile in the direction of crystal growth (Reutsky et al. 2008a,b). In Fe-Co+Gr, Fe-Ni+Gr and Fe+Gr systems, the carbon isotope fractionation coefficients at 5.5 GPa and 1400–1500 °C are the same and vary from about 2.0 to 4.5‰ (Reutsky et al. 2008b). Therefore, within the distance of 1.5 mm from the seed these diamond crystals show gradual decrease of $\delta^{13}C$ on a scale of 2–4‰, followed by slight fluctuation around carbon isotope composition of initial graphite (Reutsky et al. 2008a). In the metal–carbon systems the $\delta^{13}C$ of diamond always equal or higher than $\delta^{13}C$ of the initial graphite. The carbon isotope fractionation coefficients associated with diamond growth are strongly depends on the linear growth rate and may vary from 0 to 4.5‰ (Reutsky et al. 2012).

Carbon isotope fractionation was also studied in the $NaCO_3$+CO_2+C model system with a single carbon source also studied (Reutsky et al. 2015a). The carbon isotope fractionation in those experiments show opposite direction, compared to metal–carbon systems, and the diamond here is always depleted in heavy (^{13}C) isotope in comparison with carbonate–CO_2 fluid, exactly as expected from calculations of isotope equilibrium (e.g., Richet et al. 1977). At 7.5 GPa temperature-dependence of carbon isotope fractionation coefficient was determined as the function: $\Delta_{\text{Carbonate fluid-Diamond}} = 7.38 \times 10^6 \times T^{-2}$ and established to be dependent on carbonate:CO_2 ratio in the fluid (Reutsky et al. 2015a). The scale of this depletion is comparable to that documented for metal-carbon systems and reaches 4‰. Later work on diamond experimental crystallization from carbonate–silicate–water fluid reveal a similar estimate of fluid/diamond carbon isotope fractionation at 7 GPa and 1400 °C (Bureau et al. 2018).

Detailed investigation of different crystallographic sectors of HPHT diamonds reveal surface-controlled carbon isotope distribution within diamond crystals showing higher isotope fractionation on faces {100} in comparison with face {111} (Reutsky et al. 2017). This is supported by ab-initio calculations and provide explanation of systematic isotope lightening of growth sectors {100} relative to {111}.

Useful information on diamond growth can be obtained from investigation of paired carbon and nitrogen isotope fractionation during diamond growth in certain crystallographic directions using high spatial resolution technique such as secondary ion mass spectrometry.

Table 14. Studies of carbon isotope fractionation at HPHT diamond crystallization.

Study	System	*P* (GPa)	*T* (C)	Duration	Seeds	IRMS technique	Results
Hoering (1961)	Metal melt + Gr	7	1700		N	Bulk combustion	Presumably FG. No difference of carbon isotope composition between starting graphite and bulk diamond is observed
Laptev et al. (1978)	Fe + Gr	4, 5	1100–1200		N	Bulk combustion	Same result
Ivanovskaya et al. (1981)	Fe–Mn + Gr	10 6.5	1500 1800			Bulk combustion	Same result
Boyd et al. (1988)	Metal melt + Gr	?	?		G	Bulk combustion	Slight but analytically significant 0.6‰ heterogeneity across the crystal
Taniguchi et al. (1996)	$K_2Mg(CO_3)_2$ +Gr	9–10				Bulk combustion	Bulk diamond inherits $\delta^{13}C$ of initial graphite. No influence of carbonate carbon was recognized
Arima et al. (2002)	$CaMg(CO_3)_2$ + Si	7.7	1800	60 m	N	Bulk combustion	Broad correspondence of the resulting diamond composition to initial carbonate $\delta^{13}C$. Used for carbon source verification
Reutsky et al. (2008a)	Fe–Ni + Gr, Fe–Co + Gr	5.5	1450	17.5 h	N	Bulk combustion	TGG. 3.2‰ difference between very last portion of diamond and residual carbon in quenched metal melt was documented
Reutsky et al. (2008b)	Fe–Ni + Gr	5.5	1450	90 h	G	SIMS	Detailed carbon isotope profiles along growth sectors of different crystal faces documented for the first time. Partition coefficients for carbon isotopes were obtained for {100} and {111} sectors of growth. Influence of diamond growth rate for particular crystal faces was recognized
Satish-Kumar et al. (2011)	Fe + Gr	5.0	1200–2100	0.5–15 h	N	Bulk combustion	C isotope fractionation between diamond and carbon solution in metal melt is determined
Reutsky et al. (2012)	Fe–Ni + Gr	5.5	1450	17.5–160 h	N	Bulk combustion	Influence of linear growth rate to the carbon isotope partition coefficient in the metal-carbon system at HPHT was documented. Correspondence of observed relation to Burton–Prim–Slichter model was established
Palyanov et al. (2013b)	$Mg_{0.9}Ca_{0.1}CO_3$ + Fe^0	6.5	1350–1550	8–60 h	N	Bulk combustion	Up to 6.5‰ carbon isotope fractionation at diamond-producing redox interaction of carbonate and metal iron
Reutsky et al. (2015b)	Fe–Ni + Gr	6.3	1600–1400	5 h	N	Bulk combustion	Distribution of carbon isotopes between metal melt, crystalline Fe_3C and diamond near the peritectic point was established
Reutsky et al. (2015a)	$Na_2C_2O_4$	6.3, 7.5	1300–1500 1400–1700	10–40 h	N	Bulk combustion	C isotope effect for carbonate–CO_2 interaction in a sodium carbonate fluid at experimental *P, T* is close to 3‰. The carbon isotope fractionation at diamond crystallization from a carbonate fluid at 7.5 GPa decreases with increasing *T* based on the function: $\Delta_{\text{Carbonate fluid-Diamond}} = 7.38 \times 10^6 \times T^{-2}$
Reutsky et al. (2017)	$Na_2C_2O_4$	6	1400		G	SIMS	Surface induced C isotope fractionation during diamond crystallization was established from observed difference between {111} and {100} supported by ab-initio calculations
Reutsky et al. (2018)	(Mg,Ca)CO_3 + Fe	6.5	1550	10 h	N	SIMS	A set of diamond crystals ranges in $\delta^{13}C$ from –0.5 to –17.1‰ was collected from a single experiment of carbonate-metal iron interaction with the single carbon source (initial carbonate $\delta^{13}C$ = +0.2‰), establishing huge kinetic carbon isotope fractionation at redox interaction at *P,T* conditions of diamond stability

DIAMOND DISSOLUTION EXPERIMENTS

Overview

Natural diamonds exhibit various surface textures, acquired during their history and corresponding to growth and dissolution events. For a long time, the dissolution textures were attributed to the effect of the kimberlite melts while diamonds were brought up to the surface (Robinson 1979), and eventually to later secondary processes mostly due to transport and erosion.

Experimental studies evidenced that resorption features are also acquired from mantle-stable corrosive fluids when diamonds are residing at depth (Table 15). In the studies of Chepurov et al. (1985) and Khokhryakov and Pal'yanov (1990), it was experimentally demonstrated for the first time that dissolution of flat-faced diamonds in water-bearing melts results in the formation of rounded crystals, with morphology identical to natural dodecahedroids and octahedroids. The reader is invited to read the detailed review of Fedortchouk (2019), which combines the observation of the morphology of natural diamonds with the description of dissolution and etching experiments, and demonstrates the usefulness of the experimental approach. It is shown that the shape and size of the etch pits on diamond surface depends on the temperature and H_2O/CO_2 ratio in the fluid, whereas pressure affects the efficiency of diamond crystal shape transformation from octahedral into rounded resorbed forms. Thanks to experiments, the identification of resorption produced in the mantle source from that in the kimberlite magma is possible.

Resorption processes possibly occur in between multiple growth events, such as observed in fibrous diamonds (Klein-BenDavid et al. 2007), recently it was proposed that such corrosive mantle fluids would likely be of "melt" nature (Fedortchouk et al. 2019). Surprisingly, it is also observed that both dissolution and growth may occur in very similar fluids/melts, such as in the system carbonate–silicate enriched in $H_2O–CO_2$ (Khokhryakov and Pal'yanov 2007a). Growth and dissolution events may alternate during diamond formation, possibly depending on carbon saturation, water versus CO_2 proportions, or redox conditions.

Future directions

More experiments are necessary to decrypt the complexity of diamond's histories and to understand the sequence of growth and dissolution events that may occur to diamonds from their birth in the mantle to the exposure to the near surface.

CONCLUSIONS

As can be seen from the preceding sections, there has been an impressive body of experimental work related to diamond formation over the past decades. The widespread adoption of experimental apparatus such as the belt, the multi-anvil, and the laser-heated diamond-anvil systems have allowed researchers to access the necessary *P*, *T* space to address a variety of questions with respect to diamonds. It should be noted that the experimental data reviewed here are not always in unambiguous agreement with each other. This is largely resulting from the use in different scientific groups of different equipment and methods for preparing and conducting experiments.

Building on the work done in industrial laboratories growing diamonds in a variety of metal melts, the ability of a rich variety of fluids and melts, including those possibly stable in the Earth's mantle, to mediate crystallization of diamond has been firmly established. Actually, growing diamonds in these melts or fluids allows researchers to address questions such as isotopic fractionation during diamond growth and the incorporation of impurities into a growing diamond, which will provide valuable constraints to understanding the formation of diamonds in nature.

Table 15. Studies of diamond resorption.

Study	System	*P* (GPa)	*T* (C)	Duration	Seeds	Capsule/ buffering	Results
Evans and Sauter (1961)	Air and gas mixtures (O_2, N or Ar, H_2O)		800–1400	5 min–16 h	Y	Tubes	Diamond etching and graphitization
Evans and Phaal (1962)	O_2	RP	650–1350	?	Y	Silica and alumina	Diamond oxidation (dissolution) and etching
Davies and Evans (1972)	Air	RP and 4.8	1850–2000 and 1950–2200	Up to 200 min	Y	Pt	Diamond graphitization
Harris and Vance (1974)	Natural kimberlites	RP and 0.1	900–1600	30 min–24 h	Y	G, Pt and AgPd	Vacuum graphitization and etching by degassing volatiles from natural kimberlite powder
Kanda et al. (1977)	H_2O	5	1100–1500	5–30 min	Y	Pt	Etching of natural diamond octahedron
Yamaoka et al. (1980)	Fe–O, Mn–O+ SiO_2	1.5–4	800–1400	5 min–5 h	Y	Pt. Mn and Fe oxide-based oxygen buffers	Shape of etching pits from redox conditions
Chepurov et al. (1985)	Alkaline basalt melt ± H_2O	2.5	1300–1500	0.5–4 h	Y	Pt	Influence of H_2O on the shape of diamond dissolution
Cull and Meyer (1986)	CO–CO_2	RP	900–1000	12–48 h	Y	Pt. fO_2 QFM–NNO	Diamond oxidation
Khokhryakov and Palyanov (1990)	Alkaline basalt melt + H_2O ± CO_2, lamproite melt	2.5–5.5	1100–1450	5–180 min	Y	Pt	Different resorption morphologies
Arima (1998)	Kimberlite melts	2.5	1300–1500	10–240 min	Y	Pt	Diamond resorption
Khokhryakov and Pal'yanov (2000)	$CaCO_3$	7.0	1700–1750	5 h	Y	Pt	Diamond dissolution
Sonin et al. (2000)	Air	RP	700–1200	5–120 min	Y	Pt crucible HM, NNO, CCO buffers	Diamond oxidation and etching at temperature and for various oxygen fugacities
Khokhryakov et al. (2001)	Na_2CO_3, $CaCO_3$, $MgCO_3$, $CaMgSi_2O_6 + H_2O$	5.7	1400	10 min–40 h	Y	Pt	Diamond dissolution, evolution of crystal morphology

Study	System	*P* (GPa)	*T* (C)	Duration	Seeds	Capsule/ buffering	Results
Sonin et al. (2001)	Alkaline basalt melt	3	1300	30–90 min	Y	Pt	Diamond etching
Khokhryakov et al. (2002)	Na_2CO_3 +Fe_2O_3, Ag_2CO_3, H_2O, Ti, MgO + Ti	5.7	1400	1–20 h	Y	Pt, various fO_2 HM buffer	Diamond dissolution, effect of redox
Sonin et al. (2003)	Silicate melt–C–O–H–S	3	1300	15–30 min	Y	Pt (MAP)	Etching
Zhimulev et al. (2004)	Mantle xenoliths (dunite, eclogite, lherzolith) + C–O–H fluid	5.5–6	1450–1500	1 h	Y	Pt (MAP)	Diamond etching
Kozai and Arima (2005)	Kimberlite, lamproite melts+ $CaMg(CO_3)_2$	1	1300–1420	20–600 min	Y	Pt,IW, MW, HM buffers	Dissolution
Khokhryakov and Palyanov (2006)	KNO_3 and $NaNO_3$ melts	RP	600–900	5–60 min	Y	alundum crucible	Dislocation etching
Sonin et al. (2006)	Basaltic melt	RP	1130	30–60 min	Y	HM NNO buffers	Etching at various oxygen fugacities
Fedortchouk et al. (2007)	Kimberlite melt, Carbonate melt, Alkaline basalt+brucite and $CaCO_3$ and SiO_2	1	1150–1500	6 min	Y	Pt	Graphitization or resorption, depending on the fluid phase: H_2O and CO_2 (no reaction with the melt)
Khokhryakov and Pal'yanov (2007a)	$CaMg(CO_3)_2$, $CaCO_3$, $CaMgSi_2O_6$, kimberlite + H_2O	5.7	1400	0.17–37 h	Y	Pt	Diamond dissolution, evolution of crystal morphology
Khokhryakov and Palyanov (2007b)	$NaNO_3$ and KNO_3 melts	RP	750–800	5–60 min	Y	alundum crucible	Etching to detect planar defects
Arima and Kozai (2008)	Natural and synthetic kimberlite and $MgCO_3$	1–2.5	1300–1500	3–600 min	Y	Pt + H_2O IW buffer	Dissolution
Sonin et al. (2008)	NaCl and NaF melts	RP	1300–1350	30–120 min	Y	Pt	Etching
Sonin et al. (2009)	Alkali basalt + NaCl, NaF	3	1350	30 min	Y	Pt	Etching
Fedortchouk and Canil (2009)	Gas mixtures: CO–CO_2	RP	1000–1100	10 min–17 h	Y	Pt	Variable fO_2 from –9.5 to –16.1 Diamond oxidation, morphology depending on the fluid

Study	System	*P* (GPa)	*T* (C)	Duration	Seeds	Capsule/ buffering	Results
Khokhryakov and Pal'yanov (2010)	Na_2CO_3, $CaCO_3$, $CaMg(CO_3) \pm H_2O \pm CO_2$	5.7–7.5	1400–1750	1–37 h	Y	Pt	Influence of fluid composition on the form of diamond dissolution
Sonin et al. (2010)	NaCl, $Ca(OH)_2$	3	1350	30–120 min	Y	Pt	Diamond etching
Khokhryakov et al. (2014)	KNO_3	RP	700	5–60 min	Y	alundum crucible	Effect of N impurity on etching
Khokhryakov and Palyanov (2015b)	KNO_3	RP	700	15–60 min	Y	alundum crucible	Effect of N impurity on etching
Fedortchouk (2015)	H_2O–rich CO_2–rich	1	1150–1350	2240–2880 min	Y	Pt	Diamond resorption
Khokhryakov and Palyanov (2015a)	H_2O–carbonate melts ($CaCO_3$)	5.7	1300	30-45 h	Y	Pt	Dissolution of block diamond crystals
Sokol et al. (2015)	Kimberlite and carbonatite melts $\pm H_2O$	6.3	1400	0.5–10 h	Y	Pt ± Re foil, Re–ReO_2 and HM buffers	Dissolution and resorption of diamond as indicators for redox of resorption
Zhang et al. (2015)	$H_2O \pm MgO \pm SiO_2$	1–3	1150–1400	60–4320 min	Y	Pt	Shape and size of the etching pits depends on temperature and fluid composition
Zhimulev et al. (2016b)	$Fe_{0.7}S_{0\text{-}3}$ melt	4	1400	1 h	Y	ZrO_2, CaO, and MgO	Diamond resorption
Sonin et al. (2018b)	$Fe_{0.7}S_{0\text{-}3}$ melt	4	1400	1 h	Y	ZrO_2, CaO, and MgO	Diamond resorption
Sonin et al. (2018a)	Fe melt + S in various amounts	3.5	1400	1 h	Y	MgO	Diamond dissolution
Chepurov et al. (2018)	Fe–S melt + kimberlite	4	1400	1 h	Y	MgO	Resorption in metal-sulfide-silicate melt
Khokhryakov et al. (2018)	Mg–Si–C	7–7.5 RP	1800, 700	30 min 5 min	Y	alundum crucible	Growth and selective dislocation etching
Khokhryakov and Palyanov (2018)	Carbonate melts H_2O–carbonate–silicate melts Sulfide melts	5.7–7.5	1100–2000	0.5–5 h	Y	Pt and G	Dissolution and graphitization as indicators of the sectorial structure of crystals

Study	System	*P* (GPa)	*T* (C)	Duration	Seeds	Capsule/ buffering	Results
Fedortchouk et al. (2019)	$MgO–H_2O(–CO_2)$ $MgO–SiO_2–H_2O(–CO_2)$ $CaO–MgO–SiO_2–CO_2(–H_2O)$ $CaO–MgO–CO_2(–H_2O)$	6	1200–1500	0.5–6 h	Y	Pt	Different resorption morphologies Only metasomatism by melts is destructive
Gryaznov et al. (2019)	Fe–Ni–S melts	3.5	1400		Y	MgO	Dissolution features due to metallic melts in the mantle
Khokhryakov et al. (2020)	Kimberlite melt + H_2O	6.3	1400	10 h	Y	Pt + Re foil, $Re–ReO_2$ and HM buffers	Etching and dissolution, used as indicators for redox of resorption

Notes: RP – room pressure. MAP – Multi-anvil press. min – minutes. h – hours. G – graphite.

Where to from here?

Despite all our collective efforts, some fundamental questions remain. For example, what triggers diamond formation? What is the carbon source, what is the nature of the diamond's parents: fluids, melts, supercritical fluids? Is it depending on tectonic setting, or only on depth? How can we use the incorporation of impurities in diamonds to identify their growth mechanisms?

The experimental studies modeling slab-mantle reactions, whereby carbonate-bearing lithologies—representing subducted slab material—interact with reduced, metal-bearing mantle collectively are an excellent example of addressing a process question. But is this how all diamonds form? Classic models of diamond formation by, for example, oxidized fluids or melts interacting with reduced (but metal-free) mantle, or vice-versa, remain elusive to test experimentally. Other models, such as diamond formation by decompression or cooling of fluids (Stachel and Luth 2015), by partial melting in the presence of a hydrous, carbon-bearing fluid (Luth 2017; Smit et al. 2019), by pH changes in a fluid interacting with different mantle lithologies (Sverjensky and Huang 2015), under the influence of an electric field (Palyanov et al. 2021b), or by mixing of different fluids or melts (e.g., Huang and Sverjensky 2020) await experimental testing as well. There are still many experiments to perform to understand the growth of super deep diamonds in the transition zone and lower mantle. Another challenge for the future is the use of diamond growth to understand the mechanism of inclusions trapping: protogenetic versus syngenetic, or synchronous (Nestola et al. 2017) and thereby to provide information to properly decipher the complex messages delivered by inclusions about the deep Earth mineralogy and chemistry.

To conclude, there is still work to do! We need the experimental approach, in constant interactions with the study of natural diamonds and with models, to map the cartography of diamond growth at any depth into the Earth.

ACKNOWLEDGMENTS

The authors gratefully acknowledge support for their research from their respective funding agencies: NSERC Discovery Grants (RWL), Russian Science Foundation under grant no. 19-17-00075 and state assignment of IGM SB RAS (YNP), and the program Emergences 2019 Sorbonne-Université, CNRS, contract 193256 Hydrodiams (HB). Thorough and constructive reviews by Yana Fedortchouk and an anonymous reviewer significantly improved the manuscript. Finally, the patience of the editors was much appreciated.

REFERENCES

Akaishi M, Yamaoka S (2000) Crystallization of diamond from C–O–H fluids under high-pressure and high-temperature conditions. J Cryst Growth 209:999–1003

Akaishi M, Kanda H, Yamaoka S (1990a) Synthesis of diamond from graphite–carbonate system under very high temperature and pressure. J Cryst Growth 104:578–581

Akaishi M, Kanda H, Yamaoka S (1990b) High pressure synthesis of diamond in the systems of graphite-sulfate and graphite-hydroxide. Jpn J Appl Phys 2 29:L1172–L1174

Akaishi M, Yamaoka S, Ueda F, Ohashi T (1996) Synthesis of polycrystalline diamond compact with magnesium carbonate and its physical properties. Diamond Relat Mater 5:2–7

Akaishi M, Shaji Kumar MD, Kanda H, Yamaoka S (2000) Formation process of diamond from supercritical H_2O–CO_2 fluid under high pressure and high temperature conditions. Diamond Relat Mater 9:1945–1950

Akaishi M, Shaji Kumar MD, Kanda H, Yamaoka S (2001) Reactions between carbon and a reduced C–O–H fluid under diamond-stable HP-HT condition. Diamond Relat Mater 10:2125–2130

Arefiev AV, Shatskiy A, Podborodnikov IV, Litasov KD (2019) The K_2CO_3–$CaCO_3$–$MgCO_3$ system at 6 GPa: Implications for diamond forming carbonatitic melts. Minerals 9:558

Arima M (1998) Experimental study of growth and resorption of diamond in kimberlitic melts at high pressures and temperatures. Int Kimberlite Confer: Extended Abstr 7:32–34

Arima M, Kozai Y (2008) Diamond dissolution rates in kimberlitic melts at 1300–1500 °C in the graphite stability field. Eur J Mineral 20:357–364

Arima M, Kozai Y, Akaishi M (2002) Diamond nucleation and growth by reduction of carbonate melts under high-pressure and high-temperature conditions. Geology 30:691–694

Arima M, Nakayama K, Akaishi M, Yamaoka S, Kanda H (1993) Crystallization of diamond from a silicate melt of kimberlite composition in high-pressure and high-temperature experiments. Geology 21:968–970

Bataleva YV, Palyanov YN, Sokol AG, Borzdov YM, Palyanova GA (2012) Conditions for the origin of oxidized carbonate–silicate melts: Implications for mantle metasomatism and diamond formation. Lithos 128–131:113–125

Bataleva YV, Palyanov YN, Sokol AG, Borzdov YM, Bayukov OA (2015a) The role of rocks saturated with metallic iron in the formation of ferric carbonate–silicate melts: Experimental modeling under *PT*-conditions of lithospheric mantle. Russ Geol Geophys 56:143–154

Bataleva YV, Palyanov YN, Borzdov YM, Bayukov OA, Sobolev NV (2015b) Interaction of iron carbide and sulfur under *P–T* conditions of the lithospheric mantle. Dokl Earth Sci 463:707–711

Bataleva YV, Palyanov YN, Borzdov YM, Sobolev NV (2016a) Graphite and diamond formation via the interaction of iron carbide and Fe,Ni–sulfide under mantle *P–T* parameters. Dokl Earth Sci 471:1144–1148

Bataleva YV, Palyanov YN, Borzdov YM, Bayukov OA, Sobolev NV (2016b) The formation of graphite upon the interaction of subducted carbonates and sulfur with metal-bearing rocks of the lithospheric mantle. Dokl Earth Sci 466:88–91

Bataleva YV, Palyanov YN, Borzdov YM, Kupriyanov IN, Sokol AG (2016c) Synthesis of diamonds with mineral, fluid and melt inclusions. Lithos 265:292–303

Bataleva YV, Palyanov YN, Sokol AG, Borzdov YM, Bayukov OA (2016d) Wüstite stability in the presence of a CO_2 –fluid and a carbonate–silicate melt: Implications for the graphite/diamond formation and generation of Fe-rich mantle metasomatic agents. Lithos 244:20–29

Bataleva YV, Palyanov YN, Borzdov YM, Bayukov OA, Sobolev NV (2016e) Conditions for diamond and graphite formation from iron carbide at the *P–T* parameters of lithospheric mantle. Russian Geol Geophys 57:176–189

Bataleva YV, Palyanov YN, Borzdov YM, Bayukov OA, Zdrokov EV (2017) Iron carbide as a source of carbon for graphite and diamond formation under lithospheric mantle *P–T* parameters. Lithos 286–287:151–161

Bataleva Y, Palyanov Y, Borzdov Y, Novoselov I, Bayukov O (2018a) Graphite and diamond formation in the carbide–oxide–carbonate interactions (Experimental modeling under mantle *P*, *T*-conditions). Minerals 8:522

Bataleva YV, Palyanov YN, Borzdov YM, Bayukov OA, Sobolev NV (2018b) Experimental modeling of C^0-forming processes involving cohenite and CO_2–fluid in a silicate mantle. Dokl Earth Sci 483:1427–1430

Bataleva YV, Palyanov YN, Borzdov YM, Novoselov ID, Bayukov OA, Sobolev NV (2018c) Conditions of formation of iron–carbon melt inclusions in garnet and orthopyroxene under *P–T* conditions of lithospheric mantle. Petrology 26:565–574

Bataleva YV, Palyanov YN, Borzdov YM, Zdrokov EV, Novoselov ID, Sobolev NV (2018d) Formation of the Fe,Mg-silicates, Fe^0, and graphite (diamond) assemblage as a result of cohenite oxidation under lithospheric mantle conditions. Dokl Earth Sci 479:335–338

Bataleva YV, Palyanov YN, Borzdov YM, Novoselov ID, Bayukov OA (2019) An effect of reduced S-rich fluids on diamond formation under mantle-slab interaction. Lithos 336–337:27–39

Bayarjargal L, Shumilova TG, Friedrich A, Winkler B (2010) Diamond formation from $CaCO_3$ at high pressure and temperature. Eur J Mineral 22:29–34

Bobrov AV, Litvin YA (2009) Peridotite-eclogite-carbonatite systems at 7.0–8.5 GPa: Concentration barrier of diamond nucleation and syngenesis of its silicate and carbonate inclusions. Russian Geol Geophys 50:1221–1233

Bobrov AV, Litvin YA (2011) Mineral equilibria of diamond-forming carbonate–silicate systems. Geochem Int 49:1267–1363

Bobrov AV, Litvin YA, Divaev FK (2004) Phase relations and diamond synthesis in the carbonate–silicate rocks of the Chagatai Complex, Western Uzbekistan: Results of experiments at P=4–7 GPa and T=1200–1700°C. Geochem Int 42:39–48

Boettcher AL, Mysen BO, Allen JC (1973) Techniques for the control of water fugacity and oxygen fugacity for experimentation in solid-media high-pressure apparatus. J Geophys Res 78:5898–5901

Borzdov YM, Sokol AG, Pal'yanov YN, Kalinin AA, Sobolev NV (1999) Studies of diamond crystallization in alkaline silicate, carbonate and carbonate–silicate melts. Doklady Akademii Nauk 366:530–533

Borzdov Y, Pal'yanov Y, Kupriyanov I, Gusev V, Khokhryakov A, Sokol A, Efremov A (2002) HPHT synthesis of diamond with high nitrogen content from an Fe_3N–C system. Diamond Relat Mater 11:1863–1870

Boulard E, Gloter A, Corgne A, Antonangeli D, Auzende AL, Perrillat JP, Guyot F, Fiquet G (2011) New host for carbon in the deep Earth. PNAS 108:5184–5187

Boulard E, Menguy N, Auzende AL, Benzerara K, Bureau H, Antonangeli D, Corgne A, Morard G, Siebert J, Perrillat JP, Guyot F (2012) Experimental investigation of the stability of Fe-rich carbonates in the lower mantle. J Geophys Res: Solid Earth 117

Bovenkerk HP (1961) Some observations on the morphology and crystal characteristics of synthetic diamonds. Am Mineral 46:952–963

Boyd SR, Pillinger CT, Milledge HJ, Mendelssohn MJ, Seal M (1988) Fractionation of nitrogen isotopes in a synthetic diamond of mixed crystal habit. Nature 331:604–607

Boyd SR, Pineau F, Javoy M (1994) Modeling the growth of natural diamonds. Chem Geol 116:29–42
Brey GP, Kogarko LN, Ryabchikov ID (1991) Carbon dioxide in kimberlitic melts. Neues Jahrb Mineral Monatsh 4:159–168
Brey GP, Girnis AV, Bulatov VK, Höfer HE, Gerdes A, Woodland AB (2015) Reduced sediment melting at 7.5–12 GPa: phase relations, geochemical signals and diamond nucleation. Contrib Mineral Petrol 170:1–25
Bureau H, Langenhorst F, Auzende AL, Frost DJ, Esteve I, Siebert J (2012) The growth of fibrous, cloudy and polycrystalline diamonds. Geochim Cosmochim Acta 77:202–214
Bureau H, Frost DJ, Bolfan-Casanova N, Leroy C, Esteve I, Cordier P (2016) Diamond growth in mantle fluids. Lithos 265:4–15
Bureau H, Remusat L, Esteve M, Pinti DL, Cartigny P (2018) The growth of lithospheric diamonds. Sci Adv 4: eaat1602
Burns RC, Hansen JO, Spits RA, Sibanda M, Welbourn CM, Welch DL (1999) Growth of high purity large synthetic diamond crystals. Diamond Relat Mater 8:1433–1437
Cartigny P (2005) Stable isotopes and the origin of diamond. Elements 1:79–84
Chen L, Miao X, Ma H, Guo L, Wang Z, Yang Z, Fang C, Jia X (2018) Synthesis and characterization of diamonds with different nitrogen concentrations under high pressure and high temperature conditions. CrystEngComm 20:7164–7169
Chepurov AI, Khokhriakov AF, Sonin VM, Palianov IN, Sobolev NV (1985) The shapes of diamond crystal dissolution in silicate melts under high-pressure. Doklady Akademii Nauk 285:212–216
Chepurov AI, Zhimulev EI, Fedorov, II, Sonin VM (2007) Inclusions of metal solvent and color of boron-bearing monocrystals of synthetic diamond. Geol Ore Deposits 49:648–651
Chepurov AI, Zhimulev EI, Sonin VM, Chepurov AA, Pokhilenko NP (2009a) Crystallization of diamond in metal-sulfide melts. Dokl Earth Sci 428:1139–1141
Chepurov AI, Zhimulev EI, Eliseev AP, Sonin VM, Fedorov, II (2009b) The genesis of low-N diamonds. Geochem Int 47:522–525
Chepurov AI, Sonin VM, Zhimulev EI, Chepurov AA, Tomilenko AA (2011) On the formation of element carbon during decomposition of $CaCO_3$ at high *P–T* parameters under reducing conditions. Dokl Earth Sci 441:1738–1741
Chepurov AI, Sonin VM, Zhimulev EI, Chepurov AA, Pomazansky BS, Zemnukhov AL (2018) Dissolution of diamond crystals in a heterogeneous (metal–sulfide–silicate) medium at 4 GPa and 1400 degrees C. J Mineral Petrol Sci 113:59–67
Chrenko RM, McDonald RS, Darrow KA (1967) Infra-red spectrum of diamond coat. Nature 213:474–476
Cull FA, Meyer HOA (1986) Oxidation of diamond at high temperature and 1 atm total pressure with controlled oxygen fugacity. Int Kimberlite Confer: Extended Abstracts 4:377–379
Dasgupta R, Grewal DS (2019) Origin and early differentiation of carbon and associated life-essential volatile elements on Earth. *In*: Deep Carbon: Past to Present. Orcutt BN, Daniel I, Dasgupta R, (eds). Cambridge University Press, Cambridge, p 4–39
Davies G, Evans T (1972) Graphitization of diamond at zero pressure and at a high pressure. Proc R Soc London Ser A 328:413–427
Day HW (2012) A revised diamond-graphite transition curve. Am Mineral 97:52–62
Dobrzhinetskaya LF, Green HW (2007) Diamond synthesis from graphite in the presence of water and SiO_2: Implications for diamond formation in quartzites from Kazakhstan. International Geology Review 49:389–400
Dobrzhinetskaya LF, Renfro AP, Green HW (2004) Synthesis of skeletal diamonds: Implications for microdiamond formation in orogenic belts. Geology 32:869–872
Dobrzhinetskaya LF, Wirth R, Green HW (2007) A look inside of diamond-forming media in deep subduction zones. PNAS 104:9128–9132
Dorfman SM, Badro J, Nabiei F, Prakapenka VB, Cantoni M, Gillet P (2018) Carbonate stability in the reduced lower mantle. Earth Planet Sci Lett 489:84–91
Drewitt JWE, Walter MJ, Zhang H, McMahon SC, Edwards D, Heinen BJ, Lord OT, Anzellini S, Kleppe AK (2019) The fate of carbonate in oceanic crust subducted into earth's lower mantle. Earth Planet Sci Lett 511:213–222
Eggler DH, Baker DR (1982) Reduced volatiles in the system C–O–H: Implications to mantle melting, fluid formation, and diamond genesis. *In*: High-pressure research in geophysics. Vol 12. Akimoto S, Manghnani MH, (eds). Center for Academic Publications, Tokyo, p 237–250
Evans T, Phaal C (1962) The kinetics of the diamond–oxygen reaction. *In*: Proceedings of the Fifth Conference on Carbon. Pergamon, p 147–153
Evans T, Sauter DH (1961) Etching of diamond surfaces with gases. Philos Mag 6:429–440
Fagan AJ, Luth RW (2011) Growth of diamond in hydrous silicate melts. Contrib Mineral Petrol 161:229–236
Farré-de-Pablo J, Proenza JA, González-Jiménez JM, Garcia-Casco A, Colás V, Roqué-Rossell J, Camprubí A, Sánchez-Navas A (2018) A shallow origin for diamonds in ophiolitic chromitites. Geology 47:75–78
Farré-de-Pablo J, Proenza JA, González-Jiménez JM, Garcia-Casco A, Colás V, Roqué-Rosell J, Camprubi A, Sánchez-Navas A (2019) A shallow origin for diamonds in ophiolitic chromitites Reply. Geology 47:E477-E478
Fedortchouk Y (2015) Diamond resorption features as a new method for examining conditions of kimberlite emplacement. Contrib Mineral Petrol 170:1–19
Fedortchouk Y (2019) A new approach to understanding diamond surface features based on a review of experimental and natural diamond studies. Earth Sci Rev 193:45–65

Fedortchouk Y, Canil D (2009) Diamond oxidation at atmospheric pressure: development of surface features and the effect of oxygen fugacity. Eur J Mineral 21:623–635

Fedortchouk Y, Canil D, Semenets E (2007) Mechanisms of diamond oxidation and their bearing on the fluid composition in kimberlite magmas. Am Mineral 92:1200–1212

Fedortchouk Y, Liebske C, McCammon C (2019) Diamond destruction and growth during mantle metasomatism: An experimental study of diamond resorption features. Earth Planet Sci Lett 506:493–506

Fisher D, Lawson SC (1998) The effect of nickel and cobalt on the aggregation of nitrogen in diamond. Diamond Relat Mater 7:299–304

Frezzotti M-L, Huizenga J-M, Compagnoni R, Selverstone J (2014) Diamond formation by carbon saturation in C–O–H fluids during cold subduction of oceanic lithosphere. Geochim Cosmochim Acta 143:68–86

Gaillou E, Post JE, Rost D, Butler JE (2012) Boron in natural type IIb blue diamonds: Chemical and spectroscopic measurements. Am Mineral 97:1–18

Galvez ME, Beyssac O, Martinez I, Benzerara K, Chaduteau C, Malvoisin B, Malavieille J (2013) Graphite formation by carbonate reduction during subduction. Nat Geosci 6:473–477

Galvez ME, Pubellier M (2019) How do subduction zones regulate the carbon cycle? *In*: Deep Carbon: Past to Present. Orcutt BN, Daniel I, Dasgupta R, (eds). Cambridge University Press, Cambridge, p 276–312

Girnis AV, Brey GP, Bulatov VK, Höfer HE, Woodland AB (2018) Graphite to diamond transformation during sediment–peridotite interaction at 7.5 and 10.5 GPa. Lithos 310–311:302–313

Green BL, Collins AT, Breeding CM (2022) Diamond spectroscopy, defect centers, color, and treatments. Rev Mineral Geochem 88:637–688

Gryaznov IA, Zhimulev EI, Sonin VM, Lindenblot ES, Chepurov AA, Sobolev NV (2019) Morphological features of diamond crystals resulting from dissolution in a Fe–Ni–S melt under high pressure. Dokl Earth Sci 489:1449–1452

Gunn SC, Luth RW (2006) Carbonate reduction by Fe-S-O melts at high pressure and high temperature. Am Mineral 91:1110–1116

Harris JW (1968) Recognition of diamond inclusions. 1. Syngenetic mineral inclusions. Ind Diam Rev 28:402–410

Harris JW, Vance ER (1974) Studies of reaction between diamond and heated kimberlite. Contrib Mineral Petrol 47:237–244

Hoering TC (1961) The carbon isotope effect in the synthesis of diamond. Year Book–Carnegie Institution of Washington 60:204

Holloway JR, Burnham CW, Millhollen GL (1968) Generation of H_2O–CO_2 mixtures for use in hydrothermal experimentation. J Geophys Res 73:6598–6600

Hong SM, Akaishi M, Yamaoka S (1999) Nucleation of diamond in the system of carbon and water under very hight pressure and temperature. J Cryst Growth 200:326–328

Huang F, Sverjensky DA (2020) Mixing of carbonatitic into saline fluid during panda diamond formation. Geochim Cosmochim Acta 284:1–20

Huang WL, Wyllie PJ (1976) Melting relationships in the systems CaO–CO_2 and MgO-CO_2 to 33 kilobars. Geochim Cosmochim Acta 40:129–132

Huang GF, Jia XP, Li SS, Zhang YF, Li Y, Zhao M, Ma HA (2010a) Synthesis of large diamond crystals containing high-nitrogen concentration at high pressure and high temperature using Ni-based solvent by temperature gradient method. Chin Phys B 19

Huang GF, Jia XP, Li SS, Hu MH, Li Y, Zhao M, Yan BM, Ma HA (2010b) Effects of additive NaN_3 on the HPHT synthesis of large single crystal diamond grown by TGM. Sci China Phys Mech Astronom 53:1831–1835

Irifune T, Kurio A, Sakamoto S, Inoue T, Sumiya H, Funakoshi K (2004) Formation of pure polycrystalline diamond by direct conversion of graphite at high pressure and high temperature. Phys Earth Planet Inter 143–44:593–600

Ivanovskaya IN, Shterenberg LY, Makhov SF, Musina AR, Filonenko VP (1981) On carbon isotope fractionation in solid-phase synthesis of diamond. Geokhimiya 9:1415–1417

Izraeli ES, Harris JW, Navon O (2001) Brine inclusions in diamonds: a new upper mantle fluid. Earth Planet Sci Lett 187:323–332

Jablon BM, Navon O (2016) Most diamonds were created equal. Earth Planet Sci Lett 443:41–47

Jones AP, Genge M, Carmody L (2013) Carbonate melts and carbonatites. *In*: Carbon in Earth. Vol 75. Hazen RM, Jones AP, Baross JA, (eds). p 289–322

Kaminsky FV, Wirth R (2011) Iron carbide inclusions in lower-mantle diamond from Juina, Brazil. Can Mineral 49:555–572

Kanda H (2000) Large diamonds grown at high pressure conditions. Braz J Phys 30:482–489

Kanda H, Yamaoka S (1993) Inhomogeneous distribution of nitrogen impurities in {111} growth sectors of high-pressure synthetic diamond. Diamond Relat Mater 2:1420–1423

Kanda H, Yamaoka S, Setaka N, Komatsu H (1977) Etching of diamond octahedrons by high-pressure water. J Cryst Growth 38:1–7

Kanda H, Akaishi M, Yamaoka S (1999) Synthesis of diamond with the highest nitrogen concentration. Diamond Relat Mater 8:1441–1443

Kelemen PB, Manning CE (2015) Reevaluating carbon fluxes in subduction zones, what goes down, mostly comes up. PNAS 112:E3997–E4006

Kennedy CS, Kennedy GC (1976) The equilibrium boundary between graphite and diamond. J Geophys Res 81:2467–2470

Keppler H (2003) Water solubility in carbonatite melts. Am Mineral 88:1822–1824

Khokhryakov AF, Pal'yanov YN (1990) Morphology of diamond crystals dissolved in hydrous silicate melts. Mineralogischeskiy Zhurnal 12:14–23

Khokhryakov AF, Pal'yanov YN (2000) The dissolution forms of diamond crystals in $CaCO_3$ melt at 7 GPa. Russian Geol Geophys 41:682–687

Khokhryakov AF, Palyanov YN (2006) Revealing of dislocations in diamond crystals by the selective etching method. J Cryst Growth 293:469–474

Khokhryakov AF, Pal'yanov YN (2007a) The evolution of diamond morphology in the process of dissolution: Experimental data. Am Mineral 92:909–917

Khokhryakov AF, Palyanov YN (2007b) Revealing of planar defects and partial dislocations in large synthetic diamond crystals by the selective etching. J Cryst Growth 306:458–464

Khokhryakov AF, Pal'yanov YN (2010) Influence of the fluid composition on diamond dissolution forms in carbonate melts. Am Mineral 95:1508–1514

Khokhryakov AF, Palyanov YN (2015a) Effect of crystal defects on diamond morphology during dissolution in the mantle. Am Mineral 100:1528–1532

Khokhryakov AF, Palyanov YN (2015b) Effect of nitrogen impurity on etching of synthetic diamond crystals. J Cryst Growth 430:71–74

Khokhryakov AF, Palyanov YN (2018) Manifestation of diamond sectoriality during dissolution and graphitization. J Cryst Growth 502:1–6

Khokhryakov AF, Pal'yanov YN, Sobolev NV (2001) Evolution of crystal morphology of natural diamond in dissolution processes: Experimental data. Dokl Earth Sci 381:884–888

Khokhryakov AF, Pal'yanov YN, Sobolev NV (2002) Crystal morphology as an indicator of redox conditions of natural diamond dissolution at the mantle PT parameters. Dokl Earth Sci 385:534–537

Khokhryakov AF, Nechaev DV, Sokol AG, Palyanov YN (2009) Formation of various types of graphite inclusions in diamond: Experimental data. Lithos 112:683–689

Khokhryakov AF, Palyanov YN, Kupriyanov IN, Borzdov YM, Sokol AG (2014) Effect of nitrogen impurity on the dislocation structure of large HPHT synthetic diamond crystals. J Cryst Growth 386:162–167

Khokhryakov AF, Palyanov YN, Kupriyanov IN, Nechaev DV (2016) Diamond crystallization in a CO_2-rich alkaline carbonate melt with a nitrogen additive. J Cryst Growth 449:119–128

Khokhryakov AF, Palyanov YN, Borzdov YM, Kozhukhov AS, Sheglov DV (2018) Dislocation etching of diamond crystals grown in Mg-C system with the addition of silicon. Diamond Relat Mater 88:67–73

Khokhryakov AF, Nechaev DV, Sokol AG (2020) Microrelief of rounded diamond crystals as an indicator of the redox conditions of their resorption in a kimberlite melt. Crystals 10

Kiseeva ES, Litasov KD, Yaxley GM, Ohtani E, Kamenetsky VS (2013) Melting and phase relations of carbonated eclogite at 9–21 GPa and the petrogenesis of alkali-rich melts in the deep mantle. J Petrol 54:1555–1583

Klein-BenDavid O, Izraeli ES, Hauri E, Navon O (2004) Mantle fluid evolution–A tale of one diamond. Lithos 77:243–253

Klein-BenDavid O, Izraeli ES, Hauri E, Navon O (2007) Fluid inclusions in diamonds from the Diavik mine, Canada and the evolution of diamond-forming fluids. Geochim Cosmochim Acta 71:723–744

Kozai Y, Arima M (2005) Experimental study on diamond dissolution in kimberlitic and lamproitic melts at 1300–1420° C and 1 GPa with controlled oxygen partial pressure. Am Mineral 90:1759–1766

Kraus D, Vorberger J, Pak A, Hartley NJ, Fletcher LB, Frydrych S, Galtier E, Gamboa EJ, Gericke DO, Glenzer SH, Granados E (2017) Formation of diamonds in laser-compressed hydrocarbons at planetary interior conditions. Nat Astron 1:606–611

Laptev VA, Ivanovskaya IN, Galimov EM (1978) A study of carbon isotope fractionation in the process of diamond synthesis. VII All-union symposium on stable isotopes in geochemistry:95

Latourrette T, Holloway JR (1994) Oxygen fugacity of the diamond plus C–O fluid assemblage and CO_2 fugacity at 8-GPa. Earth Planet Sci Lett 128:439–451

Lee C-TA, Jiang H, Dasgupta R, Torres M (2019) A framework for understanding whole-Earth carbon cycling. *In*: Deep carbon: Past to present. Orcutt BN, Daniel I, Dasgupta R, (eds). Cambridge University Press, Cambridge, p 313–357

Li K, Li L, Pearson DG, Stachel T (2019) Diamond isotope compositions indicate altered igneous oceanic crust dominates deep carbon recycling. Earth Planet Sci Lett 516:190–201

Liang ZZ, Jia X, Ma HA, Zang CY, Zhu PW, Guan QF, Kanda H (2005) Synthesis of HPHT diamond containing high concentrations of nitrogen impurities using NaN_3 as dopant in metal–carbon system. Diamond Relat Mater 14:1932–1935

Liang ZZ, Kanda H, Jia X, Ma HA, Zhu PW, Guan QF, Zang CY (2006) Synthesis of diamond with high nitrogen concentration from powder catalyst-C-additive NaN3 by HPHT. Carbon 44:913–917

Litasov KD, Goncharov AF, Hemley RJ (2011) Crossover from melting to dissociation of CO_2 under pressure: Implications for the lower mantle. Earth Planet Sci Lett 309:318–323

Litasov KD, Shatskiy A, Ohtani E (2014) Melting and subsolidus phase relations in peridotite and eclogite systems with reduced COH fluid at 3–16 GPa. Earth Planet Sci Lett 391:87–99

Litasov KD, Kagi H, Bekker TB, Hirata T, Makino Y (2019a) Cuboctahedral type Ib diamonds in ophiolitic chromitites and peridotites: the evidence for anthropogenic contamination. High Press Res 39:480–488

Litasov KD, Kagi H, Voropaev SA, Hirata T, Ohfuji H, Ishibashi H, Makino Y, Bekker TB, Sevastyanov VS, Afanasiev VP, Pokhilenko NP (2019b) Comparison of enigmatic diamonds from the Tolbachik arc volcano (Kamchatka) and Tibetan ophiolites: Assessing the role of contamination by synthetic materials. Gondwana Res 75:16–27

Litvin YA (2003) Alkaline-chloride components in processes of diamond growth in the mantle and high-pressure experimental conditions. Dokl Earth Sci 389:388–391

Litvin YA, Butvina VG (2004) Diamond-forming media in the system eclogite-carbonatite-sulfide-carbon: Experiments at 6.0–8.5 GPa. Petrology 12:377–387

Litvin YA, Bobrov A (2008) Experimental study of diamond crystallization in carbonate-peridotite melts at 8.5 GPa. Dokl Earth Sci 422:1167–1171

Litvin YA, Spivak AV (2003) Rapid growth of diamondite at the contact between graphite and carbonate melt: Experiments at 7.5–8.5 GPa. Dokl Earth Sci 391A:888–891

Litvin YA, Zharikov VA (1999) Primary fluid-carbonatitic inclusions in diamond: Experimental modeling in the system K_2O–Na_2O–CaO–MgO–FeO–CO_2 as a diamond formation medium in experiment at 7–9 GPa. Dokl Earth Sci 367A:801–805

Litvin YA, Chudinovskikh LT, Zharikov VA (1997) Crystallization of diamond and graphite from alkaline-carbonate melts at 7–11 GPa. Doklady Akademii Nauk 355:669–672

Litvin YA, Chudinovskikh LT, Zharikov VA (1998a) Seed growth of diamond in the system $Na_2Mg(CO_3)_2$-$K_2Mg(CO_3)_2$-C at 8–10 GPa. Doklady Akademii Nauk 359:818–820

Litvin YA, Chudinovskikh LT, Zharikov VA (1998b) Crystallization of diamond in the system $Na_2Mg(CO_3)_2$-$K_2Mg(CO_3)_2$-C at 8–10 GPa. Doklady Akademii Nauk 359:668–670

Litvin YA, Chudinovskikh LT, Saparin GV, Obyden SK, Chukichev MV, Vavilov VS (1998c) Peculiarities of diamonds formed in alkaline carbonate-carbon melts at pressures of 8–10 GPa: Scanning electron microscopy and cathodoluminescence data. Scanning 20:380–388

Litvin YA, Aldushin KA, Zharikov VA (1999a) Synthesis of diamond at 8.5–9.5 GPa in the system $K_2Ca(CO_3)_2$-$Na_2Ca(CO_3)_2$-C modeling compositions of fluid-carbonatitic inclusions in kimberlitic diamonds. Dokl Earth Sci 367A:813–816

Litvin YA, Chudinovskikh LT, Saparin GV, Obyden SK, Chukichev MV, Vavilov VS (1999b) Diamonds of new alkaline carbonate-graphite HP syntheses: SEM morphology, CCL-SEM and CL spectroscopy studies. Diamond Relat Mater 8:267–272

Litvin YA, Spivak AV, Matveev YA (2003) Experimental study of diamond formation in the molten carbonate–silicate rocks of the Kokchetav metamorphic complex at 5.5–7.5 GPa. Geochem Int 41:1090–1098

Litvin YA, Jones AP, Beard AD, Divaev FK, Zharikov VA (2001) Crystallization of diamond and syngenetic minerals in melts of diamondiferous carbonatites of the Chagatai Massif, Uzbekistan: Experiment at 7.0 GPa. Dokl Earth Sci 381:1066–1069

Litvin YA, Butvina VG, Bobrov AV, Zharikov VA (2002) The first synthesis of diamond in sulfide-carbon systems: The role of sulfides in diamond genesis. Dokl Earth Sci 382:40–43

Litvin YA, Kurat G, Dobosi G (2005a) Experimental study of diamondite formation in carbonate–silicate melts; a model approach to natural processes. Russ Geol Geophys 46:1285–1299

Litvin YA, Shushkanova AV, Zharikov VA (2005b) Immiscibility of sulfide-silicate melts in the mantle: Role in the syngenesis of diamond and inclusions (based on experiments at 7.0 GPa). Dokl Earth Sci 403:719–722

Litvin YA, Litvin VY, Kadik AA (2008a) Experimental characterization of diamond crystallization in melts of mantle silicate-carbonate-carbon systems at 7.0–8.5 GPa. Geochem Int 46:531–553

Litvin YA, Litvin V, Kadik A (2008b) Study of diamond and graphite crystallization from eclogite-carbonatite melts at 8.5 GPa: The role of silicates in diamond genesis. Dokl Earth Sci 419:486–491

Litvin Y, Spivak A, Solopova N, Dubrovinsky L (2014) On origin of lower-mantle diamonds and their primary inclusions. Phys Earth Planet Inter 228:176–185

Litvin YA, Spivak AV, Simonova DA, Dubrovinsky LS (2017) On origin and evolution of diamond-forming lower-mantle systems: physicochemical studies in experiments at 24 and 26 GPa. J Phys Conf Ser 950:042045

Liu LG, Lin CC, Yang YJ (2001) Formation of diamond by decarbonation of $MnCO_3$. Solid State Commun 118:195–198

Liu X, Jia X, Zhang Z, Li Y, Hu M, Zhou Z, Ma H-a (2011) Crystal growth and characterization of diamond from carbonyl iron catalyst under high pressure and high temperature conditions. Cryst Growth Des 11:3844–3849

Liu X, Jia X, Fang C, Ma H-A (2016) Diamond crystallization and growth in N–H enriched environment under HPHT conditions. CrystEngComm 18:8506–8515

Lonsdale K, Milledge HJ, Nave E (1959) X-ray studies of synthetic diamonds. Mineral Mag 32:185–201

Luth R (2017) Diamond formation during partial melting in the Earth's mantle. GSA Annual Meeting in Seattle, Washington, USA-2017. Geol Soc Am 49:Abstr #20–26

Maeda F, Ohtani E, Kamada S, Sakamaki T, Hirao N, Ohishi Y (2017) Diamond formation in the deep lower mantle: a high-pressure reaction of $MgCO_3$ and SiO_2. Sci Rep 7:40602

Martin AM, Hammouda T (2011) Role of iron and reducing conditions on the stability of dolomite + coesite between 4.25 and 6 GPa—a potential mechanism for diamond formation during subduction. Eur J Mineral 23:5–16

Martirosyan NS, Litasov KD, Lobanov SS, Goncharov AF, Shatskiy A, Ohfuji H, Prakapenka V (2019) The Mg–carbonate–Fe interaction: Implication for the fate of subducted carbonates and formation of diamond in the lower mantle. Geosci Front 10:1449–1458

Massonne H-J (2019) A shallow origin for diamonds in ophiolitic chromitites: Comment. Geology 47:e476-e476

Matjuschkin V, Woodland AB, Frost DJ, Yaxley GM (2020) Reduced methane-bearing fluids as a source for diamond. Sci Rep 10:6961

McCubbin FM, Sverjensky DA, Steele A, Mysen BO (2014) In-situ characterization of oxalic acid breakdown at elevated P and T: Implications for organic C–O–H fluid sources in petrologic experiments. Am Mineral 99:2258–2271

McDonough WF, Sun SS (1995) The composition of the Earth. Chem Geol 120:223–253

Milledge HJ (1961) Coesite as an inclusion in G.E.C. synthetic diamonds. Nature 190:1181–1181

Mysen B (2018) Mass transfer in the Earth's interior: fluid–melt interaction in aluminosilicate–C–O–H–N systems at high pressure and temperature under oxidizing conditions. Prog Earth Planet Science 5:6

Navon O, Hutcheon ID, Rossman GR, Wasserburg GJ (1988) Mantle-derived fluids in diamond micro-inclusions. Nature 335:784–789

Nestola F, Jung H, Taylor LA (2017) Mineral inclusions in diamonds may be synchronous but not syngenetic. Nat Commun 8:14168

Newton RC, Sharp WE (1975) Stabilty of forsterite + CO_2 and its bearing on the role of CO_2 in the mantle Earth Planet Sci Lett 26:239–244

Nimis P (2022) Pressure and temperature data for diamonds. Rev Mineral Geochem 88:533–566

Nimis P, Alvaro M, Nestola F, Angel RJ, Marquardt K, Rustioni G, Harris JW, Marone F (2016) First evidence of hydrous silicic fluid films around solid inclusions in gem-quality diamonds. Lithos 260:384–389

Okada T, Utsumi W, Shimomura O (2002a) In situ x-ray observations of the diamond formation process in the C–H_2O–MgO system. J Phys Condens Matter 14:11331–11335

Okada T, Utsumi W, Kaneko H, Yamakata M, Shimomura O (2002b) In situ X-ray observations of the decomposition of brucite and the graphite–diamond conversion in aqueous fluid at high pressure and temperature. Physi Chem Minerals 29:439–445

Okada T, Utsumi W, Kaneko H, Turkevich V, Hamaya N, Shimomura O (2004) Kinetics of the graphite-diamond transformation in aqueous fluid determined by in-situ X-ray diffractions at high pressures and temperatures. Phys Chem Minerals 31:261–268

Onodera A, Suito K, Morigami Y (1992) High-pressure synthesis of diamond from organic compounds. Proceedings of the Japan Academy Series B-Physical and Biological Sciences 68:167–171

Osorgin NI, Palianov IN, Sobolev NV, Khokhriakova IP, Chepurov AI, Shugurova NA (1987) Inclusions of liquefied gases in diamond crystals. Doklady Akademii Nauk Sssr 293:1214–1217

Palyanov YN, Sokol AG (2009) The effect of composition of mantle fluids/melts on diamond formation processes. Lithos 112:690–700

Pal'yanov YN, Khokhryakov AF, Borzdov YM, Doroshev AM, Tomilenko AA, Sobolev NV (1994) Inclusions in synthetic diamonds. Doklady Akademii Nauk 338:78–80

Pal'yanov YN, Khokhryakov AF, Borzdov YM, Sokol AG, Gusev VA, Rylov GM, Sobolev NV (1997) Growth conditions and real structure of synthetic diamond crystals. Russian Geol Geophys 38:920–945

Pal'yanov YN, Sokol AG, Borzdov YM, Sobolev NV (1998a) Experimental study of diamond crystallization in carbonate–carbon systems in connection with the problem of diamond genesis in magmatic and metamorphic rocks. Geologiya I Geofizika 39:1780–1792

Pal'yanov YN, Borzdov YM, Sokol AG, Khokhriakov AF, Gusev VA, Rylov GM, Sobolev NV (1998b) High-pressure synthesis of high-quality diamond single crystals. Diamond Relat Mater 7:916–918

Pal'yanov YN, Sokol AG, Borzdov YM, Khokhryakov AF, Sobolev NV (1999a) Diamond formation from mantle carbonate fluids. Nature 400:417–418

Pal'yanov YN, Sokol AG, Borzdov YM, Khokhryakov AF, Shatsky AF, Sobolev NV (1999b) The diamond growth from Li_2CO_3, Na_2CO_3, K_2CO_3 and Cs_2CO_3 solvent-catalysts at $P = 7$ GPa and $T = 1700–1750\,°C$. Diamond Relat Mater 8:1118–1124

Pal'yanov YN, Sokol AG, Khokhryakov AF, Pal'yanova GA, Borzdov YM, Sobolev NV (2000) Diamond and graphite crystallization in COH fluid at PT parameters of the natural diamond formation. Dokl Earth Sci 375:1395–1398

Pal'yanov Y, Borzdov Y, Kupriyanov I, Gusev V, Khokhryakov A, Sokol A (2001a) High-pressure synthesis and characterization of diamond from a sulfur-carbon system. Diamond Relat Mater 10:2145–2152

Palyanov YN, Shatsky VS, Sokol AG, Tomilenko AA, Sobolev NV (2001b) Crystallization of metamorphic diamond: An experimental modeling. Dokl Earth Sci 381:935–938

Pal'yanov YN, Sokol AG, Borzdov YM, Khokhryakov AF (2002a) Fluid-bearing alkaline carbonate melts as the medium for the formation of diamonds in the Earth's mantle: an experimental study. Lithos 60:145–159
Pal'yanov YN, Sokol AG, Borzdov YM, Khokhryakov AF, Sobolev NV (2002b) Diamond formation through carbonate–silicate interaction. Am Mineral 87:1009–1013
Pal'yanov YN, Borzdov YM, Ovchinnikov IY (2003) Experimental study of the interaction between pentlandite melt and carbon at mantle *PT* parameters: Condition of diamond and graphite crystallization. Dokl Earth Sci 392:1026–1029
Pal'yanov YN, Sokol AG, Sobolev NV (2005a) Experimental modeling of mantle diamond-forming processes. Russian Geol Geophys 46:1271–1284
Pal'yanov YN, Sokol AG, Tomilenko AA, Sobolev NV (2005b) Conditions of diamond formation through carbonate–silicate interaction. Eur J Mineral 17:207–214
Palyanov YN, Borzdov YM, Khokhryakov AF, Kupriyanov IN, Sobolev NV (2006) Sulfide melts–graphite interaction at HPHT conditions: Implications for diamond genesis. Earth Planet Sci Lett 250:269–280
Palyanov YN, Shatsky VS, Sobolev NV, Sokol AG (2007a) The role of mantle ultrapotassic fluids in diamond formation. PNAS 104:9122–9127
Palyanov YN, Borzdov YM, Bataleva YV, Sokol AG, Palyanova GA, Kupriyanov IN (2007b) Reducing role of sulfides and diamond formation in the Earth's mantle. Earth Planet Sci Lett 260:242–256
Palyanov YN, Kupriyanov IN, Borzdov YM, Sokol AG, Khokhryakov AF (2009) Diamond crystallization from a sulfur–carbon system at HPHT conditions. Cryst Growth Des 9:2922–2926
Palyanov YN, Sokol AG, Khokhryakov AF, Sobolev NV (2010a) Experimental study of interaction in the CO_2-C system at mantle *PT* parameters. Dokl Earth Sci 435:1492–1495
Palyanov YN, Borzdov YM, Khokhryakov AF, Kupriyanov IN, Sokol AG (2010b) Effect of nitrogen impurity on diamond crystal growth processes. Cryst Growth Des 10:3169–3175
Palyanov YN, Borzdov YM, Kupriyanov IN, Khokhryakov AF (2012) Effect of H_2O on diamond crystal growth in metal-carbon systems. Cryst Growth Des 12:5571–5578
Palyanov YN, Khokhryakov AF, Borzdov YM, Kupriyanov IN (2013a) Diamond growth and morphology under the influence of impurity adsorption. Cryst Growth Des 13:5411–5419
Palyanov YN, Bataleva YV, Sokol AG, Borzdov YM, Kupriyanov IN, Reutsky VN, Sobolev NV (2013b) Mantle–slab interaction and redox mechanism of diamond formation. PNAS 110:20408–20413
Palyanov YN, Sokol AG, Khokhryakov AF, Kruk AN (2015) Conditions of diamond crystallization in kimberlite melt: experimental data. Russian Geol Geophys 56:196–210
Palyanov YN, Kupriyanov IN, Sokol AG, Borzdov YM, Khokhryakov AF (2016) Effect of CO_2 on crystallization and properties of diamond from ultra-alkaline carbonate melt. Lithos 265:339–350
Palyanov YN, Borzdov YM, Khokhryakov AF, Bataleva YV, Kupriyanov IN (2020a) Effect of sulfur on diamond growth and morphology in metal-carbon systems. Crystengcomm 22:5497–5508
Palyanov YN, Borzdov YM, Kupriyanov IN, Bataleva YV, Nechaev DV (2020b) Effect of oxygen on diamond crystallization in metal–carbon systems. ACS Omega
Palyanov YN, Khokhryakov AF, Kupriyanov IN (2021a) Crystallomorphological and crystallochemical indicators of diamond formation conditions. Crystal Rep 66:142–155
Palyanov YN, Borzdov YM, Sokol AG, Bataleva YV, Kupriyanov IN, Reutsky VN, Wiedenbeck M, Sobolev NV (2021b) Diamond formation in an electric field under deep Earth conditions. Sci Adv 7:eabb4644
Podborodnikov IV, Shatskiy A, Arefiev AV, Bekhtenova A, Litasov KD (2019) New data on the system Na_2CO_3–$CaCO_3$–$MgCO_3$ at 6 GPa with implications to the composition and stability of carbonatite melts at the base of continental lithosphere. Chem Geol 515:50–60
Polyakov VI, Rukovishnikov AI, Rossukanyi NM, Ralchenko VG (2001) Electrical properties of thick boron and nitrogen contained CVD diamond films. Diamond Relat Mater 10:593–600
Rege S, Griffin WL, Pearson NJ, Araujo D, Zedgenizov D, O'Reilly SY (2010) Trace-element patterns of fibrous and monocrystalline diamonds: Insights into mantle fluids. Lithos 118:313–337
Reutsky VN, Borzdov YM, Palyanov YN (2008a) Carbon isotope fractionation associated with HPHT crystallization of diamond. Diamond Relat Mater 17:1986–1989
Reutsky VN, Harte B, Eimf, Borzdov YM, Palyanov YN (2008b) Monitoring diamond crystal growth, a combined experimental and SIMS study. Eur J Mineral 20:365–374
Reutsky VN, Borzdov YM, Palyanov YN (2012) Effect of diamond growth rate on carbon isotope fractionation in Fe–Ni–C system. Diamond Relat Mater 21:7–10
Reutsky V, Borzdov Y, Palyanov Y, Sokol A, Izokh O (2015a) Carbon isotope fractionation during experimental crystallisation of diamond from carbonate fluid at mantle conditions. Contrib Mineral Petrol 170
Reutsky VN, Borzdov YM, Palyanov YN (2015b) Carbon isotope fractionation during high pressure and high temperature crystallization of Fe-C melt. Chem Geol 406:18–24
Reutsky VN, Palyanov YN, Borzdov YM, Sokol AG (2015c) Isotope fractionation of carbon during diamond crystallization in model systems. Russian Geol Geophys 56:239–244

Reutsky VN, Kowalski PM, Palyanov YN, Wiedenbeck M, Eimf (2017) Experimental and theoretical evidence for surface-induced carbon and nitrogen fractionation during diamond crystallization at high temperatures and high pressures. Crystals 7:190

Reutsky VN, Palyanov YN, Wiedenbeck M (2018) Carbon isotope composition of diamond crystals grown via redox mechanism. Geochem Int 56:1398–1404

Richet P, Bottinga Y, Javoy M (1977) A review of hydrogen, carbon, nitrogen, oxygen, sulphur, and chlorine stable isotope fractionation among gaseous molecules. Annu Rev Earth Planet Sci 5:65–110

Robinson DN (1979) Surface textures and other features of diamonds. PhD University of Cape Town, Cape Town

Roedder E (ed) (1984) Fluid Inclusions. Reviews in Mineralogy, Vol. 12, Mineralogical Society of America, Chantilly, Virginia

Rubie DC (1999) Characterising the sample environment in multianvil high-pressure experiments. Phase Transitions 68:431–451

Satish-Kumar M, So H, Yoshino T, Kato M, Hiroi Y (2011) Experimental determination of carbon isotope fractionation between iron carbide melt and carbon: ^{12}C-enriched carbon in the Earth's core? Earth Planet Sci Lett 310:340–348

Sato K, Akaishi M, Yamaoka S (1999) Spontaneous nucleation of diamond in the system $MgCO_3$–$CaCO_3$–C at 7.7 GPa. Diamond Relat Mater 8:1900–1905

Schrauder M, Navon O (1993) Solid carbon dioxide in a natural diamond. Nature 365:42–44

Seto Y, Hamane D, Nagai T, Fujino K (2008) Fate of carbonates within oceanic plates subducted to the lower mantle, and a possible mechanism of diamond formation. Physi Chem Minerals 35:223–229

Shaji Kumar MD, Akaishi M, Yamaoka S (2000) Formation of diamond from supercritical H_2O–CO_2 fluid at high pressure and high temperature. J Cryst Growth 213:203–206

Shaji Kumar MD, Akaishi M, Yamaoka S (2001) Effect of fluid concentration on the formation of diamond in the CO_2–H_2O–graphite system under HP–HT conditions. J Cryst Growth 222:9–13

Shatskiy A, Borzdov YM, Litasov KD, Sharygin IS, Palyanov YN, Ohtani E (2015a) Phase relationships in the system K_2CO_3-$CaCO_3$ at 6 GPa and 900–1450 °C. Am Mineral 100:223–232

Shatskiy A, Yamazaki D, Morard G, Cooray T, Matsuzaki T, Higo Y, Funakoshi K, Sumiya H, Ito E, Katsura T (2009) Boron-doped diamond heater and its application to large-volume, high-pressure, and high-temperature experiments. Rev Sci Instrum 80:023907

Shatskiy AF, Litasov KD, Palyanov YN (2015b) Phase relations in carbonate systems at pressures and temperatures of lithospheric mantle: review of experimental data. Russian Geol Geophys 56:113–142

Shatsky AF, Borzdov YM, Sokol AG, Pal'yanov YN (2002) Phase formation and diamond crystallization in carbon-bearing ultrapotassic carbonate–silicate systems. Geologiya I Geofizika 43:940–950

Shushkanova AV, Litvin YA (2006) Formation of diamond polycrystals in pyrrhotite-carbonic melt: Experiments at 6.7 GPa. Dokl Earth Sci 409:916–920

Shushkanova AV, Litvin YA (2008a) Diamond nucleation and growth in sulfide-carbon melts: An experimental study at 6.0–7.1 GPa. Eur J Mineral 20:349–355

Shushkanova AV, Litvin YA (2008b) Diamond formation in sulfide pyrrhotite-carbon melts: Experiments at 6.0–7.1 GPa and application to natural conditions. Geochem Int 46:37–47

Siebert J, Guyot F, Malavergne V (2005) Diamond formation in metal–carbonate interactions. Earth Planet Sci Lett 229:205–216

Smit KV, Stachel T, Luth RW, Stern RA (2019) Evaluating mechanisms for eclogitic diamond growth: An example from Zimmi Neoproterozoic diamonds (West African craton). Chem Geol 520:21–32

Smit KV, Shirey SB, Stern RA, Steele A, Wang W (2016) Diamond growth from C–H–N–O recycled fluids in the lithosphere: Evidence from CH_4 micro-inclusions and $\delta^{13}C$–$\delta^{15}N$–N content in Marange mixed-habit diamonds. Lithos 265:68–81

Smith EM, Kopylova MG (2013) Implications of metallic iron for diamonds and nitrogen in the sublithospheric mantle. Can J Earth Sci 51:510–516

Smith EM, Wang W (2016) Fluid CH_4 and H_2 trapped around metallic inclusions in HPHT synthetic diamond. Diamond Relat Mater 68:10–12

Smith EM, Shirey SB, Nestola F, Bullock ES, Wang J, Richardson SH, Wang W (2016) Large gem diamonds from metallic liquid in Earth's deep mantle. Science 354:1403–1405

Sokol AG, Pal'yanov YN (2004) Diamond crystallization in fluid and carbonate-fluid systems under mantle P-T conditions: 2. An analytical review of experimental data. Geochem Int 42:1018–1032

Sokol AG, Pal'yanov YN (2008) Diamond formation in the system MgO–SiO_2–H_2O–C at 7.5 GPa and 1,600 °C. Contrib Mineral Petrol 155:33–43

Sokol AG, Pal'yanov YN, Borzdov YM, Khokhryakov AF, Sobolev NV (1998) Crystallization of diamond from Na_2CO_3 melt. Dokl Earth Sci 361:388–391

Sokol AG, Borzdov YM, Khokhriakov AF, Pal'yanov YN, Sobolev NV (1999) Diamond crystallization in silicate-fluid systems at P = 7.0 GPa and T = 1700–1750 °C. Doklady Akademii Nauk 368:99–102

Sokol AG, Tomilenko AA, Pal'yanov YN, Borzdov YM, Pal'yanova GA, Khokhryakov AF (2000) Fluid regime of diamond crystallisation in carbonate-carbon systems. Eur J Mineral 12:367–375

Sokol AG, Pal'yanov YN, Pal'yanova GA, Khokhryakov AF, Borzdov YM (2001a) Diamond and graphite crystallization from C–O–H fluids under high pressure and high temperature conditions. Diamond Relat Mater 10:2131–2136
Sokol AG, Borzdov YM, Pal'yanov YN, Khokhryakov AF, Sobolev NV (2001b) An experimental demonstration of diamond formation in the dolomite-carbon and dolomite-fluid-carbon systems. Eur J Mineral 13:893–900
Sokol AG, Pal'yanov YN, Pal'yanova GA, Tomilenko AA (2004) Diamond crystallization in fluid and carbonate-fluid systems under mantle P-T conditions: 1. Fluid composition. Geochem Int 42:830–838
Sokol AG, Palyanova GA, Palyanov YN, Tomilenko AA, Melenevsky VN (2009) Fluid regime and diamond formation in the reduced mantle: Experimental constraints. Geochim Cosmochim Acta 73:5820–5834
Sokol AG, Palyanov YN, Litasov KD (2010) Effect of oxygen fugacity on the H_2O storage capacity of forsterite in the carbon-saturated systems. Geochim Cosmochim Acta 74:4793–4806
Sokol AG, Khokhryakov AF, Palyanov YN (2015) Composition of primary kimberlite magma: constraints from melting and diamond dissolution experiments. Contrib Mineral Petrol 170:1–18
Sokol AG, Palyanov YN, Tomilenko AA, Bul'bak TA, Palyanova GA (2017) Carbon and nitrogen speciation in nitrogen-rich C–O–H–N fluids at 5.5–7.8 GPa. Earth Planet Sci Lett 460:234–243
Sokol AG, Khokhryakov AF, Borzdov YM, Kupriyanov IN, Palyanov YN (2019) Solubility of carbon and nitrogen in a sulfur-bearing iron melt: Constraints for siderophile behavior at upper mantle conditions. Am Mineral 104:1857–1865
Solopova NA, Spivak AV, Litvin YA, Shiryaev AA, Tsel'movich VA, Nekrasov AN (2013) Kinetic peculiarities of diamond crystallization in K–Na–Mg–Ca–carbonate–carbon melt–solution. Phys Solid State 55:373–376
Solopova NA, Dubrovinsky L, Spivak AV, Litvin YA, Dubrovinskaia N (2015) Melting and decomposition of $MgCO_3$ at pressures up to 84 GPa. Physi Chem Minerals 42:73–81
Sonin VM, Fedorov, II, Pokhilenko LN, Pokhilenko NP (2000) Diamond oxidation rate as related to oxygen fugacity. Geol Ore Deposits 42:496–502
Sonin VM, Zhimulev EI, Fedorov, II, Tomilenko AA, Chepurov AI (2001) Etching of diamond crystals in a dry silicate melt at high P-T parameters. Geochem Int 39:268–274
Sonin VM, Zhimulev EI, Chepurov AI, Afanas'ev VP, Tomilenko AA (2003) Etching of diamond crystals in the system silicate melt–C–O–H–S fluid under a high pressure. Geochem Int 41:688–693
Sonin VM, Zhimulev EI, Fedorov, II, Chepurov AI (2006) Effect of oxygen fugacity on the etching rate of diamond crystals in silicate melt. Geol Ore Deposits 48:499–501
Sonin VM, Zhimulev EI, Chepurov AI, Fedorov II (2008) Diamond stability in NaCl and NaF melts at high pressure. Dokl Earth Sci 420:641–643
Sonin VM, Zhimulev EI, Chepurov AI, Pokhilenko NP (2009) Diamond stability in silicate–halogenide melts at high pressure. Dokl Earth Sci 425:441–443
Sonin VM, Zhimulev EI, Chepurov AI, Afanas'ev VP, Pokhilenko NP (2010) High-pressure etching of diamond in chloride melt in the presence of aqueous fluid. Dokl Earth Sci 434:1359–1361
Sonin VM, Zhimulev EI, Chepurov AA, Chepurov AI, Pokhilenko NP (2018a) Influence of the sulfur concentration in a Fe–S melt on diamond preservation under *P–T* conditions of the Earth's mantle. Dokl Earth Sci 481:922–924
Sonin VM, Zhimulev EI, Pomazanskiy BS, Zemnuhov AL, Chepurov AA, Afanasiev VP, Chepurov AI (2018b) Morphological features of diamond crystals dissolved in $Fe_{0.7}S_{0.3}$ melt at 4 GPa and 1400 °C. Geol Ore Deposits 60:82–92
Spivak AV, Litvin YA (2004) Diamond syntheses in multicomponent carbonate-carbon melts of natural chemistry: Elementary processes and properties. Diamond Relat Mater 13:482–487
Spivak AV, Litvin YA (2012) Paragenetic relations of diamond with silicate and carbonate minerals in the carbonatite-diamond system: Experiments at 8.5 GPa. Geochem Int 50:217–226
Spivak AV, Litvin YA, Shushkanova AV, Litvin VY, Shiryaev AA (2008) Diamond formation in carbonate–silicate-sulfide-carbon melts: Raman- and IR-microspectroscopy. Eur J Mineral 20:341–347
Spivak AV, Litvin YA, Ovsyannikov SV, Dubrovinskaia NA, Dubrovinsky LS (2012) Stability and breakdown of $Ca^{13}CO_3$ melt associated with formation of ^{13}C-diamond in static high pressure experiments up to 43 GPa and 3900 K. J Solid State Chem 191:102–106
Srikanth V, Akaishi M, Yamaoka S, Yamada H, Taniguchi T (1997) Diamond synthesis from graphite in the presence of $MnCO_3$. Journal of the American Ceramic Society 80:786–790
Stachel T, Harris JW (2008) The origin of cratonic diamonds–Constraints from mineral inclusions. Ore Geol Rev 34:5–32
Stachel T, Luth RW (2015) Diamond formation — Where, when and how? Lithos 220–223:200–220
Stachel T, Aulbach S, Harris JW (2022a) Mineral inclusions in lithospheric diamonds. Rev Mineral Geochem 88:307–392
Stachel T, Cartigny P, Chacko T, Pearson DG (2022) Carbon and nitrogen in mantle-derived diamonds. Rev Mineral Geochem 88:809–876
Stagno V (2019) Carbon, carbides, carbonates and carbonatitic melts in the Earth's interior. J Geol Soc 176:375–387
Stagno V, Frost DJ (2010) Carbon speciation in the asthenosphere: Experimental measurements of the redox conditions at which carbonate-bearing melts coexist with graphite or diamond in peridotite assemblages. Earth Planet Sci Lett 300:72–84
Stöckhert B, Duyster J, Trepmann C, Massonne H-J (2001) Microdiamond daughter crystals precipitated from supercritical COH + silicate fluids included in garnet, Erzgebirge, Germany. Geology 29:391–394

Sumiya H, Satoh S (1996) High-pressure synthesis of high-purity diamond crystal. Diamond Relat Mater 5:1359–1365

Sumiya H, Satoh S (1999) Synthesis of polycrystalline diamond with new non-metallic catalyst under high pressure and high temperature. Inter J Refractory Metals Hard Mater17:345–350

Sumiya H, Toda N, Satoh S (2002) Growth rate of high-quality large diamond crystals. J Cryst Growth 237:1281–1285

Sun LL, Akaishi M, Yamaoka S (2000) Formation of diamond in the system of Ag_2CO_3 and graphite at high pressure and high temperatures. J Cryst Growth 213:411–414

Sun LL, Wu Q, Wang WK (2001) Bulk diamond formation from graphite in the presence of C–O–H fluid under high pressure. High Press Res 21:159–173

Sverjensky DA, Huang F (2015) Diamond formation due to a pH drop during fluid–rock interactions. Nat Commun 6:8702

Taniguchi T, Dobson D, Jones AP, Rabe R, Milledge HJ (1996) Synthesis of cubic diamond in the graphite-magnesium carbonate and graphite–$K_2Mg(CO_3)_2$ systems at high pressure of 9–10 GPa region. J Mater Res 11:2622–2632

Thomassot E, Cartigny P, Harris JW, Viljoen KS (2007) Methane-related diamond crystallization in the Earth's mantle: Stable isotope evidences from a single diamond-bearing xenolith. Earth Planet Sci Lett 257:362–371

Thomson AR, Walter MJ, Kohn SC, Brooker RA (2016) Slab melting as a barrier to deep carbon subduction. Nature 529:76–79

Tiraboschi C, Tumiati S, Recchia S, Miozzi F, Poli S (2016) Quantitative analysis of COH fluids synthesized at HP-HT conditions: an optimized methodology to measure volatiles in experimental capsules. Geofluids 16:841–855

Tomilenko AA, Chepurov AI, Pal'yanov YN, Shebanin AP, Sobolev NV (1998) Hydrocarbon inclusions in synthetic diamonds. Eur J Mineral 10:1135–1141

Tomlinson E, Jones A, Milledge J (2004) High-pressure experimental growth of diamond using C–K_2CO_3–KCl as an analogue for Cl-bearing carbonate fluid. Lithos 77:287–294

Tomlinson EL, Howell D, Jones AP, Frost DJ (2011) Characteristics of HPHT diamond grown at sub-lithosphere conditions (10–20 GPa). Diamond Relat Mater 20:11–17

Tschauner O, Mao HK, Hemley RJ (2001) New transformations of CO_2 at high pressures and temperatures. Physical Review Letters 87:075701

Tschauner O, Huang S, Greenberg E, Prakapenka VB, Ma C, Rossman GR, Shen AH, Zhang D, Newville M, Lanzirotti A, Tait K (2018) Ice-VII inclusions in diamonds: Evidence for aqueous fluid in Earth's deep mantle. Science 359:1136

Wang Y, Kanda H (1998) Growth of HPHT diamonds in alkali halides: possible effects of oxygen contamination. Diamond Relat Mater 7:57–63

Weiss Y, Cazs J, Navon O (2022) Fluid inclusions in fibrous diamonds. Rev Mineral Geochem 88:475–532

Weiss Y, Kessel R, Griffin WL, Kiflawi I, Klein-BenDavid O, Bell DR, Harris JW, Navon O (2009) A new model for the evolution of diamond-forming fluids: Evidence from microinclusion-bearing diamonds from Kankan, Guinea. Lithos 112:660–674

Weiss Y, Kiflawi I, Davies N, Navon O (2014) High-density fluids and the growth of monocrystalline diamonds. Geochim Cosmochim Acta 141:145–159

Whitney JA (1972) The effect of reduced H_2O fugacity on the buffering of oxygen fugacity in hydrothermal experiments. Am Mineral 57:1902–1908

Wilding M, Bingham PA, Wilson M, Kono Y, Drewitt JWE, Brooker RA, Parise JB (2019) CO_{3+1} network formation in ultra-high pressure carbonate liquids. Sci Rep 9:15416

Yamaoka S, Kanda H, Setaka N (1980) Etching of diamond octahedrons at high-temperatures and pressure with controlled oxygen partial-pressure. J Mater Sci 15:332–336

Yamaoka S, Akaishi M, Kanda H, Osawa T (1992) Crystal growth of diamond in the system of carbon and water under very high pressure and temperature. J Cryst Growth 125:375–377

Yamaoka S, Shaji Kumar MD, Akaishi M, Kanda H (2000) Reaction between carbon and water under diamond-stable high pressure and high temperature conditions. Diamond Relat Mater 9:1480–1486

Yamaoka S, Shaji Kumar MD, Kanda H, Akaishi M (2002a) Formation of diamond from $CaCO_3$ in a reduced C–O–H fluid at HP–HT. Diamond Relat Mater 11:1496–1504

Yamaoka S, Shaji Kumar MD, Kanda H, Akaishi M (2002b) Thermal decomposition of glucose and diamond formation under diamond-stable high pressure-high temperature conditions. Diamond Relat Mater 11:118–124

Yamaoka S, Shaji Kumar MD, Kanda H, Akaishi M (2002c) Crystallization of diamond from CO_2 fluid at high pressure and high temperature. J Cryst Growth 234:5–8

Yang JS, Lian DY, Robinson PT, Qiu T, Xiong FH, Wu WW (2019) A shallow origin for diamonds in ophiolitic chromitites. Geology 47:E475-E475

Yin LW, Li MS, Gong ZG, Xu B, Song YJ, Hao ZY (2002) Analysis of nanometer inclusions in high pressure synthesized diamond single crystals. Chem Phys Lett 355:490–496

Yu RZ, Ma HA, Liang ZZ, Liu WQ, Zheng YJ, Jia X (2008) HPHT synthesis of diamond with high concentration nitrogen using powder catalyst with additive $Ba(N_3)_2$. Diamond Relat Mater 17:180–184

Zdrokov E, Novoselov I, Bataleva Y, Borzdov Y, Palyanov Y (2019) Experimental modeling of silicate and carbonate sulfidation under lithospheric mantle P, T-parameters. Minerals 9

Zhang C, Duan Z (2010) GFluid: An Excel spreadsheet for investigating C–O–H fluid composition under high temperatures and pressures. Computers and Geosciences 36:569–572

Zhang J, Prakapenka V, Kubo A, Kavner A, Green HW, Dobrzhinetskaya LF (2011) Diamond formation from amorphous carbon and graphite in the presence of COH fluids: An *in situ* high-pressure and -temperature laser-heated diamond anvil cell experimental study. *In*: Ultrahigh-pressure metamorphism: 25 years after the discovery of coesite and diamond. Dobrzhinetskaya LF, Faryad SW, Wallis S, Cuthbert S (eds). Elsevier, London, p 113–124

Zhang JQ, Ma HA, Jiang YP, Liang ZZ, Tian Y, Jia X (2007) Effects of the additive boron on diamond crystals synthesized in the system of Fe-based alloy and carbon at HPHT. Diamond Relat Mater 16:283–287

Zhang YF, Zang CY, Ma HG, Zhongzhu LZ, Lin Z, Li SS, Jia XP (2008) HPHT synthesis of large single crystal diamond doped with high nitrogen concentration. Diamond Relat Mater 17:209–211

Zhang Z, Fedortchouk Y, Hanley JJ (2015) Evolution of diamond resorption in a silicic aqueous fluid at 1–3 GPa: Application to kimberlite emplacement and mantle metasomatism. Lithos 227:179–193

Zhang ZF, Jia XP, Liu XB, Hu MH, Li Y, Yan BM, Ma HA (2012) Synthesis and characterization of a single diamond crystal with a high nitrogen concentration. Chin Phys B 21

Zhimulev EI, Shein MA, Pokhilenko NP (2013) Diamond crystallization in the Fe–S–C system. Dokl Earth Sci 451:729–731

Zhimulev EI, Sonin VM, Mironov AM, Chepurov AI (2016a) Effect of sulfur concentration on diamond crystallization in the Fe–C–S system at 5.3–5.5 GPa and 1300–1370 °C. Geochem Int 54:415–422

Zhimulev EI, Sonin VM, Afanasiev VP, Chepurov AI, Pokhilenko NP (2016b) Fe–S melt as a likely solvent of diamond under mantle conditions. Dokl Earth Sci 471:1277–1279

Zhimulev EI, Sonin VM, Fedorov, II, Tomilenko AA, Pokhilenko LN, Chepurov AI (2004) Diamond stability with respect to oxidation in experiments with minerals from mantle xenoliths at high *P–T* parameters. Geochem Int 42:520–525

Zhimulev EI, Chepurov AI, Sinyakova EF, Sonin VM, Chepurov AA, Pokhilenko NP (2012) Diamond crystallization in the Fe–Co–S–C and Fe–Ni–S–C systems and the role of sulfide–metal melts in the genesis of diamond. Geochem Int 50:205–216

Reviews in Mineralogy & Geochemistry
Vol. 88 pp. 809–876, 2022

Carbon and Nitrogen in Mantle-Derived Diamonds

Thomas Stachel

Earth and Atmospheric Sciences
University of Alberta
Edmonton, AB T6G 2E3, Canada

tstachel@ualberta.ca

Pierre Cartigny

Institut de physique du globe de Paris
Université de Paris
CNRS, F-75005 Paris, France

cartigny@ipgp.fr

Thomas Chacko, D. Graham Pearson

Earth and Atmospheric Sciences
University of Alberta
Edmonton, AB T6G 2E3, Canada

tchacko@ualberta.ca; gdpearso@ualberta.ca

INTRODUCTION

This chapter is devoted to the carbon and nitrogen stable isotope compositions of terrestrial diamonds with a strong focus on monocrystalline diamonds formed in Earth's mantle. The wealth of C- and N-stable isotope studies forces us to make some choices to keep this chapter within an acceptable length. Here, we focus on both the ground-breaking early diamond stable isotope studies as well as the latest developments in the field. Using a comprehensive database of diamond stable isotope data compiled from literature, we examine key constraints on diamond-forming processes, the origins of diamond growth media and the cycling of C and N through shallow and deep Earth reservoirs.

Diamond as a unique probe to study mantle carbon and nitrogen cycles

Given its inert and resistant nature, diamond can be preserved during and following ascent from its primary stability field, located at depths > 110 km in ancient cratonic roots or the sublithospheric mantle beneath. As discussed elsewhere in this volume, through micro-scale inclusions of minerals, primarily silicates, oxides and sulfides (Stachel et al. 2022, this volume; Walter et al. 2022, this volume), and fluids (Weiss et al. 2022, this volume) entrapped during diamond growth, diamond provides unique information on the physical and chemical processes responsible for its genesis, the mineralogy of Earth's mantle, and the evolution of the lithospheric and the convecting sublithospheric mantle. Mineral- and fluid-inclusion-bearing diamonds are, however, rare, constituting only a few percent of typical mine productions.

This chapter focusses instead on the information that is carried by every diamond, i.e., its carbon (Lavoisier 1776, p. 615) and, with the exception of rare Type II diamonds, its main substitutional impurity nitrogen (Kaiser and Bond 1959). The definition of Type II diamond as containing N below the limit of detection is method-dependent but for the commonly

1529-6466/22/0088-0015$10.00 (print)
1943-2666/22/0088-0015$10.00 (online)
http://dx.doi.org/10.2138/rmg.2022.88.15

applied technique of micro-infrared spectroscopy implies N-contents below about 10 atomic ppm (at.ppm). Other substitutional or interstitial impurities, including H, B, and O, have not been applied in a consistent way to understand diamond formation and consequently are not considered here.

A diamond's journey to the surface is *unpleasant* and *unsafe* and, in some instances, only a small percentage of the original diamond load in the transporting magma may be preserved. Transport of diamond in hot, potentially oxidizing and fluid-rich kimberlite magma at depths outside its primary stability field may lead to graphitization, resorption to rounded dodecahedra, or etching (Robinson et al. 1989; Harris et al. 2022, this volume). Xenoliths and xenocrysts from the deep lithospheric mantle are often affected by alteration, re-equilibration or kimberlite metasomatism (e.g., Richardson et al. 1985; Boyd et al. 1997) and any carbonate minerals present would typically break down during final ascent. Other carbon hosts, such as graphite or carbide, cannot be unambiguously ascribed to a particular depth (Shiryaev and Gaillard 2014). Consequently, when dealing with the deep carbon cycle, diamond represents a unique probe into Earth's mantle with respect to both depth (down to ~ 800 km) and time (formation ages up to 3.5 Ga; see Smit et al. 2022, this volume). Carbon-bearing magmas, such as kimberlites, carbonatites, mid-ocean ridge basalts and ocean island basalts represent complementary sources of information but they sample/average much larger mantle volumes. Degassing and low-temperature alteration of samples are also major concerns when aiming to address, e.g., the C-isotope variability and the origin of carbon in such magmas. This explains why diamond, which retains its original isotopic composition, is a particularly valuable source of information in the study of the deep carbon cycle.

On Earth, diamond has been reported from a wide variety of geological contexts, some not relevant to this review. Meteorites can contain substantial amounts (~ 1500 ppm) of nanometer-sized diamonds: some of these nano-diamonds formed from material originally present in the solar nebula and some are of pre-solar origin (Huss 2005). The occurrence of diamond in meteorites does not necessarily require the high pressure and temperature conditions under which mantle diamonds typically grow. Considering surface energy minimization, even at low P and T conditions it is energetically more favorable to grow nanometer-sized diamond instead of graphite or other polycyclic aromatics (Badziag et al. 1990). This metastable formation mechanism allows for the precipitation of diamond from a gas phase under low-vacuum conditions (0.01–0.27 atmospheric pressure; diamond synthesis through chemical vapor deposition (CVD)). A metastable formation mechanism may also apply to nano-diamonds in melt inclusions in xenoliths from Hawaii (Wirth and Rocholl 2003) and, possibly some metamorphic diamonds, found in subducted and exhumed metamorphic rocks (Simakov 2011). At least some metamorphic diamonds, including those from the Kokchetav massif in Kazakhstan (Sobolev and Shatsky 1990), however, grew in their primary stability field, as supported by several independent geobarometers (see Chopin 2003 for review). Impact diamonds, occasionally up to 1 cm in size, occur, for example, in the Popigai and Ries impact craters (Hough et al. 1995; Koeberl et al. 1997) and are also commonly considered as being formed in their primary stability field. Impact diamonds, however, differ from mantle and metamorphic diamonds by being formed through solid state conversion of a graphite-rich target whereas mantle and metamorphic diamonds likely grow from C-rich fluids/melts (see below and Luth et al. 2022, this volume). Nano-diamonds found in uranium-rich carbonaceous sediments probably also form in the solid state but under crustal P and T conditions, through alpha particle-induced transitioning of graphite sp^2-bonds into diamond sp^3-bonds (Daulton and Ozima 1996). Among the many other forms of natural diamond, carbonado, a sintered type of polycrystalline diamond, remains the most enigmatic. Carbonados may form during meteorite impact, in Earth's mantle or in extraterrestrial environments (see Garai et al. 2006; Kagi and Fukura 2008; Cartigny 2010; Haggerty 2014 and reference therein).

Diamond types

Here we focus on the C- and N-stable isotope geochemistry of mantle-derived diamonds, which include a number of subdivisions. Not considering orogenic peridotite massifs, mantle-derived diamonds are transported to Earth's surface by kimberlites or, less commonly, by lamproites, ultramafic lamprophyres and, in one instance, komatiite (see Kjarsgaard et al. 2022, this volume).

Under the generic term '*smooth-surfaced monocrystalline diamonds*' we include smooth-faced monocrystalline diamonds that crystallized as primary octahedra, macles or cuboids and their resorption forms. All these diamonds formed prior to kimberlite activity and were therefore xenocrysts in their host kimberlite, with mantle residence times varying from several b.y. to a few hundred m.y. (Smit et al. 2022; Green et al. 2022, both this volume). The xenocrystic nature of smooth-surfaced monocrystalline diamonds is supported by advanced nitrogen aggregation states (typically Type IaAB, see Green et al. 2022, this volume) but the reverse is not true (i.e., a diamond with poorly aggregated nitrogen is not necessarily young). Based on the study of their mineral inclusions (typically about 1% of diamonds contain inclusions ≥ 100 μm in size, Stachel and Harris 2008), smooth-surfaced monocrystalline diamonds can be assigned to three particular mantle depth intervals:

1. *Lithospheric mantle (110–250 km)*. Based on their mineral inclusions, (smooth-surfaced monocrystalline) lithospheric diamonds grow in peridotitic, eclogitic or more rarely webstertic (pyroxenitic) substrates. On the basis of garnet composition and the presence of clinopyroxene, the peridotitic diamond substrates can be further subdivided into dunites-harzburgites, lherzolites, and wehrlites (Sobolev 1977; Meyer 1987; Stachel et al. 2021). Being both the most abundant and most studied type of diamond, the vast majority (> 90%) of C- and N-isotope data originate from lithospheric diamonds. Sulfide inclusion-bearing diamonds, assigned to either the peridotitic or eclogitic suite based on sulfide Ni-contents, may show geochemical characteristics, such as higher N-contents and lower N-aggregation states, that are distinct from other peridotitic or eclogitic diamonds from the same kimberlite (Cartigny et al. 2009; Thomassot et al. 2009). In such cases, a distinct diamond-forming event, possibly from a different growth medium, for sulfide-included diamonds may be invoked.
2. *Asthenosphere and transition zone (250–660 km)*. Such diamonds can be identified from the mineralogy and chemical composition of their inclusions, in particular the occurrence of majoritic garnet inclusions (Moore and Gurney 1985; Walter et al. 2021). Most diamonds of this type have silicate inclusions indicative of metabasaltic to metapyroxenitic substrates. Diamonds containing inclusions (e.g., ringwoodite; Pearson et al. 2014) indicative of a meta-peridotitic paragenesis are extremely rare. There are no mantle xenoliths that are directly derived from sublithospheric depth (see Sautter et al. 1991; Deines and Haggerty 2000 for descriptions of retrogressed, originally sublithospheric xenolith samples) and therefore, transition-zone diamonds represent a unique source of information. Although rare, nearly every available *asthenospheric and transition-zone* diamond has been studied for its stable isotopic composition.
3. *Lower mantle (> 660 km)*. Diamonds from the lower mantle can be identified from their co-existing silicate (bridgmanite and CaSi-perovskite) and oxide (ferropericlase) inclusions (see Walter et al. 2021). Although diamond sustains high-internal pressures, these are generally not sufficient to preserve the original high-pressure crystallographic structure of the inclusion. A lower mantle origin is thus usually inferred from the chemical composition of the inclusions and associations that cannot occur under equilibrium conditions at lower pressures (e.g., combinations of ferropericlase with bridgmanite or CaSi-perovskite). Most lower mantle diamonds derive from meta-peridotitic substrates, with meta-basaltic associations being rare

(see Stachel et al. 2005; Thomson et al. 2014; Walter et al. 2021). Of the three principal minerals constituting lower mantle associations, only bridgmanite is exclusively restricted to depth exceeding the 660 km seismic discontinuity. This creates some uncertainty when assigning diamonds containing bridgmanite-free associations (e.g., CaSi-perovskite ± ferropericlase) to the lower mantle, as a transition-zone origin may also be possible. The recently recognized superdeep origin of an inclusion-bearing subset of so-called CLIPPIR (nitrogen-free, large, resorbed, inclusion-poor, and irregularly shaped) diamonds is characterized by phase assemblages that may occur in both the transition zone and the lower mantle (Smith et al. 2016).

Four additional groups of terrestrial diamonds occur that are not included in the data set reviewed here but are listed for completeness:

1. *Polycrystalline diamond aggregates (PDA).* Polycrystalline diamonds are a common component of kimberlite-hosted diamond populations, with typical abundances ranging from several percent to > 10% (e.g., at the Orapa and Jwaneng mines in Botswana; Harris et al. 1986). Their polycrystalline nature requires high supersaturation conditions leading to high nucleation rates (Sunagawa 1990). This type of diamond is addressed in detail by Jacob and Mikhail (2022, this volume).

2. *Fibrous diamonds, cloudy diamonds and diamond coats.* These diamond varieties commonly are associated with rough fibrous diamond growth occurring under extreme levels of supersaturation (Sunagawa 1990). Their typically opaque nature relates to abundant micrometer- to nanometer-sized inclusions of 'mantle fluids', trapped during diamond growth (Kamiya and Lang 1964; Navon et al. 1988). The chapter by Weiss et al. (2022, this volume) addresses these diamonds and the nature and composition of the '*fluids*' included in them.

3. *Metamorphic diamonds.* Such diamonds have been reported from various localities and are found in subducted rocks that subsequently experienced rapid exhumation to the surface (see Dobrzhinetskaya et al. 2022, this volume). Although there is an increasing number of ultra-high-pressure (UHP) metamorphic terranes being recognized (through the presence of coesite, diamond or other high-pressure indicators), some described localities remain contentious and may relate to sample contamination (see Howell et al. 2015a), misidentification (e.g., Beyssac and Chopin 2003) or metastable diamond formation (e.g., Simakov 2010, 2011). Due to the typically very small diamond size (often ≤ 10 μm) in these occurrences, C- and N-isotope data remain scarce, with the Kokchetav massif in Kazakhstan representing the best studied locality (Sobolev and Shatsky 1990; De Corte et al. 1998, 1999).

4. *Graphitized octahedra after diamond in high-temperature peridotite massifs.* Graphite crystals up to 2 cm have been identified in pyroxenite dikes within the high-temperature peridotite massifs of Beni-Boussera (Morroco) and Ronda (Spain). The pyroxenite veins are typically ≤ 30 cm in width, though the C-rich varieties (up to 25 wt% locally) are commonly > 1 m thick. Graphite occurs as aggregates of distorted octahedra and/or in irregular habit. The occurrence of sharp-edged octahedra, a crystal form that is in the cubic system rather than the hexagonal system of graphite, is striking (see Fig. 1 in Pearson et al. 1989) and, along with the orientation of the graphite relative to the octahedral form, constitute the primary basis of their identification as graphitized diamonds (Slodkevich 1980, 1983; Pearson et al. 1989; Davies et al. 1993). On the basis of paleogeothermal gradients, the presence of diamond pseudomorphs requires exhumation of these mantle sections from depths in excess of 200 km (e.g., Davies et al. 1993). Little work has subsequently been devoted to this type of occurrence, but available studies still support the concept that graphitized diamonds do occur and, accordingly, that sections of both the crust and the mantle can be exhumed from the diamond stability field in orogenic cycles (e.g., Leech and Ernst 1998; Korsakov et al. 2019).

We will not venture into a detailed discussion of two complex topics related to diamond formation: the composition of diamond-forming fluids/melts and the role of oxygen fugacity in diamond stability. The following statements summarize our view on these two topics:

1. Terrestrial diamonds, with the exception of impact diamonds, form from a 'mobile phase' commonly refered to as 'diamond growth medium', 'carbon-bearing fluid/melt' or 'COH-fluid'. This deliberately vague wording illustrates that little is known about the exact chemical composition or redox state (i.e., whether it is oxidized or reduced) of this mobile phase. Other wording includes 'CO_2-bearing fluids' or 'CH_4-bearing fluids', in which cases the redox state of the fluid and the diamond-forming meachanism via either either reduction or oxidation are specified, but again the fluids overall chemical composition is not specified. High-density-fluids (HDFs) of 'saline', 'hydrous–silicic' and 'carbonatitic' character are directly observed in fluid-rich *fibrous* diamonds (see Weiss et al. 2021) and clearly form an important part of the spectrum of diamond-forming fluids/melts.

2. Diamond is an allotrope of elemental carbon (C^0) with a stability field that is not only restricted by pressure and temperature but also limited to a narrow range of redox conditions (oxygen fugacities or fO_2). At fO_2 conditions more oxidizing than the enstatite–magnesite–olivine–diamond (EMOD) buffer, carbon is stored in the form of carbonate rather than as diamond (Eggler and Baker 1982). For the strongly reducing conditions prevailing in Earth's mantle beyond about 300 km depth, where fO_2 is buffered by the Fe–FeO (IW) equilibrium, carbon either dissolves in an Fe–Ni metal phase or occurs as carbides or methane rather than as diamond (Frost and McCammon 2008; Rohrbach et al. 2014). Thus when a diamond originates from these great depths, e.g., from the transition zone (410–660 km depth), then it must have either formed in substrates more oxidized than the ambient mantle at those depths or been infiltrated by an oxidizing agent such as a carbonated melt. During the infiltration of metasomatic fluids in the lithospheric mantle, the fO_2 of the diamond substrate must remain in the sweet spot between the IW and EMOD buffers or diamond destruction will occur. For further reading into this aspect of diamond formation/survival and oxygen fugacity, we refer the reader to Luth (1993), O'Neill et al. (1993), Frost and McCammon (2008), Rohrbach and Schmidt (2011), Stagno et al. (2013), Stachel and Luth (2015), and Luth et al. (2022, this volume).

A brief history of diamond stable isotope studies (1940–2000)

By convention, the stable isotope composition of carbon is expressed using the delta-notation, where the isotope ratio is stated as relative deviation to an international standard, which is generally taken to be a belemnite from the Peedee Formation of South Carolina, USA[1].

$$\delta^{13}C = \left(\frac{\left(\frac{^{13}C}{^{12}C}\right)\text{sample}}{\left(\frac{^{13}C}{^{12}C}\right)\text{VPDB}} - 1 \right) \times 1000 [‰]$$

By definition, the Vienna-Pee Dee Belemnite has a $\delta^{13}C_{VPDB}$ = 0‰. Expressed differently, a diamond with $\delta^{13}C$ = −40‰ has a $^{13}C/^{12}C$ ratio (0.01078771) that is 40 parts per thousand (i.e., 4%) lower than the $^{13}C/^{12}C$ ratio of the VPDB standard (0.01123720).

[1] Note that this is the 'old' way of expressing the δ-notation, the 'new' and actually more rigorous way of expressing δ is omitting the '× 1000' which is already included when using the '‰'. We retained the old notation to avoid any misunderstanding with the literature that has and mostly still uses the 'old' notation. Importantly, this has no influence on the C- and N-isotope data reported here or in any previous or future papers.

Consisting of pure carbon, diamond was among the first materials to be analyzed for $^{13}C/^{12}C$ (Nier and Gulbransen 1939) but the measurements were still imprecise until the advent of dual-inlet mass spectrometers (Nier 1947). The first systematic C-isotope study of diamonds was conducted by Craig (1953) on six octahedral diamonds from Kimberley, South Africa (*'De Beers Pool'*—Bultfontein, De Beers, Dutoitspan, Wesselton). He reported $\delta^{13}C$-values ranging from −4.7 to −2.7‰. Subsequently, Wickman (1956) analyzed 37 diamonds from kimberlites from central and southern Africa with $\delta^{13}C$-values ranging from −9.6 to −3.2‰ and concluded that diamonds have a globally homogenous C-isotope composition. Actually, Wickman discarded a value of −13.9‰ measured on one of the diamonds, the sample being, according to him, contaminated with organic matter, while another sample from New South Wales (Australia) had a $\delta^{13}C$-value of +2.7‰. Vinogradov et al. (1965) reported $\delta^{13}C$-values ranging from −8.8 to −5.6‰ on diamonds from Yakutia (Siberia) and confirmed the previous conclusions. These early studies used diamonds to characterize the isotopic composition and variability of mantle carbon rather than aiming to understand the origin of diamond itself. The first three studies all suggested that mantle carbon has a $\delta^{13}C$-value of ~ −6 to −5‰. This was an important outcome that allowed the average C-isotope composition of surface carbon to be established. Preceding the theory of plate tectonics, the $\delta^{13}C$-value of bulk surface carbon was presumed to be the same as that of the mantle. With sedimentary carbonate and organic matter having $\delta^{13}C$-values of ~ 0‰ and ~ −25‰, respectively, a mantle value of −6 to −5‰ implies that these two surface carbon reservoirs are in about 80:20 proportions (e.g., Javoy et al. 1982). This estimate is still valid globally today, although it is now well documented that large heterogeneities exist among different subduction zones (e.g., Plank and Manning 2019). From a diamond perspective, during the 1950s and 1960s, Earth's mantle was seen as being isotopically homogeneous.

The first strongly negative $\delta^{13}C$-values for diamond were obtained by Vinogradov et al. (1966) but determined on carbonados, whose origin (Earth's mantle or extraterrestrial) is still debated. Subsequently, Koval'skiy et al. (1972) measured $\delta^{13}C$-values between −22.2 and −21.4‰ on diamonds from Yakutian placer deposits and Smirnov et al. (1979) obtained values down to −20.5‰ in diamonds from kimberlites in Lesotho, demonstrating that the assumption of C-isotope homogeneity among diamonds was invalid. By the end of the 1970s, a considerable amount of additional data were obtained, primarily by Russian groups (mostly published in Russian, with some papers translated in Geochemistry International). These data are not all easily accessible but it is worth noting that the histogram for the C-isotope composition of diamonds compiled by Deines (1980; his Fig. 4) was primarily built on data produced by Russian laboratories. By the end of the 1970s it had become clear that diamond can be strongly depleted in ^{13}C, down to −35‰ (today, the lowest known value in a terrestrial natural diamond is −41.4‰; De Stefano et al. 2009). The study of Sobolev et al. (1979) was seminal in demonstrating that diamonds belonging to the peridotitic suite have a narrower range in $\delta^{13}C$-values than those belonging to the eclogitic suite. This finding paved the way for a series of studies examining the C-isotope composition of diamond in relation to source paragenesis, inclusion mineralogy and chemistry, and other physical characteristics (e.g., shape or color) of diamond, most notably through the systematic studies of Peter Deines and coworkers (e.g., Deines et al. 2009 and references therein). The original finding of Sobolev et al. (1979) was generally confirmed, although not every diamond mine studied contained eclogitic diamonds with $\delta^{13}C$-values < −10‰ (e.g., the Finsch Mine; Deines et al. 1989) and the abundance of eclogitic diamonds with low $\delta^{13}C$-values was found to be variable (ranging from 0 to 100% of the eclogitic diamond population). Sobolev et al. (1979) suggested that the non-mantle like C-isotope compositions of eclogitic diamonds derive from surface 'sedimentary' signatures subducted into the mantle, making diamond a tracer of plate tectonics. Milledge et al. (1983) observed that Type II diamonds from the Cullinan Mine (formerly known as Premier) cover a large range of $\delta^{13}C$-values (from ~ −32 to ~ 0‰, i.e., almost the entire $\delta^{13}C$-range known at that time) and attributed this finding also to variations in the *nature of the carbon reservoir source*. Nisbet et al. (1994) expressed this differently as

'*Can* [eclogitic] *diamond be dead bacteria*?'. This model has profound implications: suggesting the possibility that sediments may be subducted without drastic (isotope) re-equilibration with their surrounding environment so as to preserve low $\delta^{13}C$-values.

With the improvement of analytical techniques, the study of diamond internal variability became viable. This occurred first by analyzing laser-cut sections of diamonds (cubes < 0.25 mm in each dimension, weighting as little as 0.05 mg; e.g., Swart et al. 1983; Boyd et al. 1987) and subsequently using secondary ion mass spectrometry (SIMS) with a spatial resolution of 10–20 µm in diameter and 1–5 µm in depth (Harte et al. 1992, 1999; Hauri et al. 1999; Fitzsimons et al. 2000). These studies typically showed little diamond-internal variability (generally < 3‰ and often < 1‰; see Fig. 7 in Galimov 1991; Fig. 4 in Cartigny et al. 2004; and Harte et al. 1999 for the first SIMS-based assessment) compared to the worldwide diamond $\delta^{13}C$-range. These intra-grain variations in isotopic composition are discussed further below with respect to the negligible diffusion of carbon and nitrogen in diamond and models of diamond formation.

Although the content and speciation of nitrogen in diamond is not the focus of this review, nitrogen forms the basis of diamond classification and thus a brief summary is given here (for details and appropriate references, see Green et al. 2022, this volume). Based on FTIR (Fourier transform infrared) analyses, Robertson et al. (1934) recognized two types of diamond (Type I and II), the former displaying extra-absorption in the 1400–900 cm^{-1} region. Only in 1959 was it understood that this extra absorption relates to nitrogen defects (Kaiser and Bond 1959). Nitrogen and carbon have similar ionic radii and charge, enabling nitrogen to substitute as an impurity into the diamond lattice. Nitrogen concentrations reach up to 3800 at.ppm in mantle-derived diamonds (Donnelly et al. 2007) but typically are much lower (median value of 160 at.ppm; see Stachel 2014 and Cartigny et al. 2014 for reviews). Based on the infrared characteristics of diamond and without necessarily understanding the nature of the underlying defect(s), further work (for details see Green et al. 2022) led to a subdivision into Type Ib (single N-impurities, typical of synthetic diamonds and occurring in < 0.1% of natural diamonds), Type Ia (aggregated nitrogen impurities), Type IIa (no nitrogen, no boron) and Type IIb (no nitrogen, boron-bearing). Type Ia diamonds were subsequently subdivided in IaA (N-pairs) and IaB (4 N-atoms around a vacancy). Nitrogen is a substitutional defect (Kaiser and Bond 1959) being strongly bonded to C-atoms. The occurrence of the various N-defects in diamond is not arbitrary and follows a well-accepted sequence of aggregation, starting from single N-atoms (Ib diamond) to N-pairs (IaA diamonds) to 4 N-atoms around a vacancy (IaB diamonds). The aggregation sequence follows a second-order kinetic diffusion process (Chrenko et al. 1977; Evans and Qi 1982) and allows constraints to be placed on either the residence time or the residence temperature of diamond in the mantle. Because of the high activation energy of the IaA to IaB conversion, N-aggregation is, however, primarily used as a geothermometer (Evans and Harris 1989; Taylor et al. 1990). Since the early 1980s, infrared spectroscopy has been the main method to determine diamond N-content and aggregation state (e.g., Milledge et al. 1983; Deines et al. 1984). Given typical infrared beam sizes of ~ 100 µm and thickness of analyzed diamond fragments up to 1 mm, these analyses correspond to bulk measurements. SIMS analyses now determine nitrogen contents (but not aggregation state) in diamond to sub-ppm levels and with high spatial resolution. Both methods show that the abundance of N in superdeep (asthenosphere to lower mantle) diamonds is low and, from FTIR data, associated with advanced aggregation (typically < 100 ppm in IaB state; e.g., Tappert et al. 2005; Palot et al. 2012 and reference therein). Fibrous diamonds are N-rich with little N-aggregation (~ 1000 ppm and IaA, e.g., Cartigny et al. 2003 and reference therein). The abundance of N in eclogitic and peridotitic diamonds, for which the largest number of data are available, can be variable across locations and across diamond size classes (with diamonds < 1 mm often showing a high abundance of Type II diamonds; Tolansky and Komatsu 1967). On average, peridotitic diamonds have lower N-contents (mean: 219 at.ppm; median: 109 at.ppm) than eclogitic diamonds (mean: 472 at.ppm; median: 454 at.ppm; see supplementary data set).

In some locations, sulfide-bearing diamonds have higher N-content and lower N-aggregation states than silicate-included diamonds belonging to the same paragenesis. Metamorphic diamonds can have much higher N-contents (up to 10,000 at.ppm) and are invariably low in aggregation state (Ib-IaA) (Xu et al. 2018 and references therein).

The first systematic N-isotope[2] study was undertaken on fibrous diamonds from Mbuji Mayi (Democratic Republic of the Congo) using conventional dual-inlet techniques and thus requiring large (0.87 to 1.55 g) amounts of diamond (Javoy et al. 1984). Although fibrous diamonds constitute only a few percent of the worldwide production, these diamonds have the advantage of being N-rich (average ~ 1000 at.ppm N) and, at least at this locality, are in part quite large (up to a few cm). This paper was seminal for several reasons: first, it demonstrated that fibrous diamonds record ^{15}N-depleted isotope compositions (average $\delta^{15}N$ of ~ −5‰ relative to atmospheric nitrogen taken as the reference, i.e., standard value), which was unexpected, and secondly, it demonstrated the viability of using N-isotopes as a tracer of crustal recycling (see below). Javoy et al. (1984) explained the observed variations in C- and N-isotope compositions by open-system fractionation of kimberlite volatiles (through Rayleigh distillation, a model that was imposed by the most extreme sample (D3), which was actually not a fibrous diamond). Importantly, their model seemed to imply that fibrous diamonds are 'young', having roughly the age of the kimberlite. In the same year, Richardson et al. (1984) reported the first Sm–Nd isotope model ages of ~ 3.2 Ga for non-fibrous, garnet-included diamonds from the Finsch mine and the pooled production of the four mines in Kimberley ('*De Beers Pool*'), establishing that monocrystalline diamond formation predated kimberlite magmatism by several billion years. These Archean diamond ages for non-fibrous diamonds are probably the main reason why the model of Javoy et al. (1984), linking fibrous diamond formation to (proto)-kimberlite derived volatiles, received little attention although it has been supported through a number of subsequent studies (for early studies see Boyd et al. 1987, 1994; Navon et al. 1988; Akagi and Masuda 1998; for a different viewpoint, see Klein-BenDavid et al. 2010; Timmerman et al. 2019). These early C- and N-isotope data on fibrous diamonds from Mbuji Mayi were subsequently confirmed and extended worldwide (Boyd et al. 1987, 1992; Burgess et al. 2009; Klein-BenDavid et al. 2010; Timmerman et al. 2019). What remains striking in all these studies is the small C- and N-isotope variability of fibrous diamonds, with values nearly identical to mid-ocean ridge basalts (see Fig. 1 in Cartigny and Marty 2013); this observation is commonly used to argue that fibrous diamonds, and by inference kimberlites, form from volatiles derived from the convecting mantle. The studies of Boyd et al. (1987, 1992) were also important because they introduced the use of static vacuum mass spectrometry (the same type of instrument used for the analysis of noble gases), allowing the analysis of three orders of magnitude smaller (sub-mg, a 1 mm^3 diamond is 3.5 mg) or much more N-poor samples. Based on this technical break-through, the diamond internal variability of N-isotope compositions could be studied together with N-content and C-isotope compositions. The first systematic study of smooth-surfaced monocrystalline diamonds was undertaken by Boyd and Pillinger (1994) on samples of unknown paragenesis, followed by a series of studies (starting with Cartigny et al. 1997) on diamonds containing mineral inclusions of known paragenesis (peridotitic or eclogitic) and derived from known localities (see reference list associated with the supplementary data set).

The fine-scale C- and N-isotope variability of diamonds became truly open to study with use of ion microprobes (Fitzsimons et al. 1999; Harte et al. 1999; Bulanova et al. 2002; Zedgenizov and Harte 2004; Craven et al. 2009; Smart et al. 2011; Peats et al. 2012; Palot et al. 2014, 2017; Howell et al. 2015b; Petts et al. 2015; Smit et al. 2016, 2019a), although not every study using MC-SIMS (multi-collector secondary ion mass spectrometer) actually focused on detailed C- and N-mapping (e.g., Johnson et al. 2012; Krebs et al. 2016; Howell et al. 2020). It should be noted that the later (2009 onwards) studies in this list achieved a factor of 2 to 5 improvement

[2] $\delta^{15}N_{diamond} = [(^{15}N/^{14}N)_{diamond}/(^{15}N/^{14}N)_{AIR}-1] \times 1000$ (in ‰)

in analytical precision of the isotopic analyses relative to the earlier studies, which permitted more subtle internal variations to be detected. Although limited to a few multi-collector SIMS instruments worldwide configured to do these isotopic analyses and requiring very good sample preparation, SIMS analysis nowadays represents a powerful way to investigate the internal C- and N-isotope variability of diamonds. Below ~ 100 ppm (using Faraday cups), the precision on $\delta^{15}N$-measurements on SIMS instruments is still limited by counting statistics and bulk techniques—despite being destructive—remain central; e.g., Palot et al. (2012) reported the nitrogen isotopic composition of a lower mantle diamond containing only 6 ppm nitrogen.

Early models: some still in consideration

Primordial heterogeneity. Until the acquisition of larger datasets, no one had made a serious attempt to differentiate diamond sub-populations based on paragenesis, C- and N-isotope composition or aggregation state. Primarily relying on statistical analysis, Deines et al. (1984) used a data set of 66 diamonds from Cullinan and 91 diamonds from Finsch to determine that many sub-populations of diamond exist (for example, McDade and Harris 1999 identified 10 sub-populations among 48 studied diamonds from the Letseng Mine in Lesotho). From this original and subsequent studies, Deines et al. (1984) favored a model where the C-isotope variability observed in diamond is primordial, i.e., inherited from the heterogeneity of Earth's accretionary building blocks - represented by chondritic meteorites. This primordial heterogeneity would have needed to survive the formation of magma oceans early in Earth's history and the subsequent differentiation, stirring and homogenization of bulk silicate Earth that is documented in many geochemical proxies, among them the observation of consistent mass-dependent $^{17}O/^{16}O$–$^{18}O/^{16}O$ signatures (e.g., Robert et al. 1992; Rumble et al. 2007). From a diamond perspective, the primordial heterogeneity model also fails to demonstrate why strongly negative $\delta^{13}C$-values mostly occur among eclogitic diamonds; the model of Deines and coworkers actually predicts that peridotitic rather than eclogitic diamonds should display the largest $\delta^{13}C$-variability. Today, this model is generally no longer considered a valid hypothesis and evidence for primordial stable isotopic heterogeneity was subsequently only proposed for very few diamonds (Cartigny et al. 1997; Palot et al. 2012), based on extremely low $\delta^{15}N$-values (down to –40‰) that were actually accompanied by mantle-like $\delta^{13}C$-values (around –4‰).

In retrospect, Deines et al. (1984) probably considered too many parameters in distinguishing diamond sub-populations. For example, nitrogen aggregation is very sensitive to mantle residence temperature (besides N-content and residence time) and consequently, in hindsight, dividing diamond populations based on N-aggregation state was not the best choice. Like N-aggregation, plastic deformation and associated brown coloration postdate diamond formation; in addition, the shear stresses leading to plastic deformation may be highly localized. Distinguishing diamond sub-populations based on evidence for plastic deformation, therefore, was not pertinent. Finally, the assumption that diamond with a low nitrogen content formed in a N-poor mantle environment was subsequently contradicted by observations that diamonds from a single mantle xenolith (Thomassot et al. 2007) and growth zones within single diamonds (e.g., Fitzsimons et al. 1999; Smart et al. 2011) can vary in N-content by several orders of magnitude, illustrating that partitioning of nitrogen between diamond and its growth medium is an additional, very important parameter to consider.

High temperature stable isotope fractionation. As discussed in detail below (see section *Origin of covariations among $\delta^{13}C$, $\delta^{15}N$ and nitrogen content*), the high temperatures of mantle processes generally result in small equilibrium isotope fractionation factors. As such, the generation of large stable isotope variations in mantle samples either requires very extensive Rayleigh fractionation or derivation of the element in question from multiple sources with widely different isotopic compositions. The models described in this section call for Rayleigh fractionation to produce the observed isotopic variations.

Early contributions (e.g., Javoy et al. 1986) proposed a possible role for within-mantle isotope fractionation processes in producing large ranges of δ^{13}C-values from a single, homogenous, starting composition with a mantle-like δ^{13}C-value of ~ –5‰. However, these studies did not point to a specific process that would account for the distinct δ^{13}C-distributions of eclogitic versus peridotitic diamonds. Instead, they considered the frequency distribution of diamonds worldwide without paying attention to source paragenesis, despite the earlier evidence of the difference in diamond paragenesis (Sobolev et al. 1979). Javoy et al. (1986) proposed that large carbon isotope fractionation between CO_2 and magma could lead to the observed large variability in diamond δ^{13}C-values. This model is analogous to CO_2 degassing during volcanic eruptions, which has been shown to produce up to 12‰ range in the δ^{13}C-values of the evolved CO_2 (e.g., Aubaud et al. 2006). Criticism of this model focusses on the large (> 4‰) isotope fractionation factors used by Javoy et al. (1986) in their calculations, which rely on a *preliminary* value determined by Javoy et al. (1978). More recent experiments indicate somewhat smaller CO_2–melt C-isotope fractionation factors (at 1200 °C) of ~ 2.0–2.4‰ for basalt and soda-melilite melts (Mattey et al. 1990; Mattey 1991) and 3.0‰ for alkali (Li–Na–K) carbonate melts (Appora-Gnekindy 1998). Smaller fractionation factors would lead to less dispersion of δ^{13}C-values during Rayleigh fractionation. Some additional empirical evidence from kimberlites and carbonatites can also be considered. Carbonatites and kimberlites show restricted variation in δ^{13}C, i.e., 95% of values being between –8 and –2‰ (e.g., Wilson et al. 2007; Giuliani et al. 2012), yet in some rare cases, δ^{13}C is reported to spread over –35 to +35‰ (Galimov 1991).

Galimov (1991) was the first to address differences in δ^{13}C-distribution among eclogitic and peridotitic diamonds, which he related again to Rayleigh fractionation in the mantle but without identifying a specific reaction. Importantly, Javoy et al. (1986) and Galimov (1991) both realized that the carbon stable isotope variability of diamond cannot relate to the crystallization process itself but instead must be produced prior to diamond formation, with diamond largely behaving as a *passive recorder*, which faithfully logs but does not significantly contribute to C-isotope variations in the mantle. Otherwise, peridotitic and eclogitic diamonds would display the same δ^{13}C-distributions. Building on these two previous models, the new observation of negative δ^{15}N-values for most eclogitic diamonds (see also below), and the experimental work of Luth (1993) and Knoche et al. (1999) on the poor CO_2-buffering capacity of olivine-free mantle lithologies, Cartigny et al. (1998) proposed that the distinct δ^{13}C-distributions of eclogitic and peridotitic diamonds record characteristic pathways in the chemical and isotopic evolution of initially isotopically indistinguishable, oxidized melts/fluids. In this model, the differences in evolution ultimately reflect the role of olivine in peridotites in buffering CO_2 through carbonation reactions, whereas in olivine-free eclogites decarbonation and CO_2-escape from carbonated fluids/melts is permissible. Such decarbonation of fluids/melts in eclogite would allow the escape of ^{13}C-enriched CO_2, leaving a ^{13}C-depleted residue from which eclogitic diamonds could crystallize.

Rayleigh fractionation predicts δ^{13}C frequency distributions varying with the logarithm of the residual carbon fraction. Consequently, even for large fractionation factors (e.g., 5‰) only a very small fraction of diamond (1%) will reach isotopic compositions < –27‰, based on an initial δ^{13}C of –5‰ for the diamond-forming medium. In other words, to derive δ^{13}C-distributions as observed in specific cases such as at Argyle (Australia) or Dachine (French Guiana) from a starting composition at δ^{13}C ~ –5‰, over 99% of the fractionated carbon reservoir must remain 'unsampled'; either the diamonds were not sampled/preserved or the bulk of the carbon did not crystallize as diamond. Such a possibility is, however, unlikely and for these and similar cases (e.g., diamonds from Jagersfontein, New South Wales and Jericho as extreme examples) an initial δ^{13}C-value less than –5‰ is required.

Forming diamond from subducted carbon (isotope fingerprinting). In isotope forensics, the isotopic compositions of two *compounds* (a bulk sample, or specific molecules, or even the intra-molecular composition of a single molecule) are compared to constrain whether these are the same or not, in order to understand, for example, whether this is *real tequila*, or whether an *orange juice contains added sugar*. The principle of relating eclogitic diamond formation to recycled sedimentary carbon follows the same approach, best illustrated in Nisbet et al. (1994) and McCandless and Gurney (1997). In order to demonstrate diamond formation from recycled organic matter and sedimentary carbonate carbon with typical $\delta^{13}C$-values of −25 and 0‰, respectively (e.g., Plank and Manning 2019; their Fig. 2a and references therein), Nisbet et al. pointed to the observation that the $\delta^{13}C$ ranges covered by diamond and these sedimentary carbon sources are very similar and predicted that even lower $\delta^{13}C$-values would be found in diamond to achieve a perfect match (i.e., the authors addressed the range of values, not the resulting distributions). This agreement between $\delta^{13}C$ ranges has been a central argument for the formation of eclogitic diamond from recycled carbon (see section *Origin of large ranges in* $\delta^{13}C$ *and* $\delta^{15}N$) and is supported by the indisputably recycled nature of most eclogite xenoliths.

The key difference between isotope forensics and diamond geology, however, lies in the metamorphic history of the carbon and nitrogen source rocks preceding diamond formation, which may include processes (e.g., re-equilibration with other carbon phases in the rock, devolatilization, melting) that shift the isotope compositions of the C- and N-bearing phases relative to their original values. A straight comparison between subducted carbon, hosted in sediments and in altered oceanic crust (organic carbon with $\delta^{13}C \sim -25‰$, with or without associated carbonates), and diamond assumes that the original isotopic compositions are preserved during subduction and associated metamorphism. Yet, when in direct contact, carbonate and organic carbon (graphite) undergo recrystallization and associated C-isotope exchange at a temperature of $\geq \sim 600$ °C (e.g., Valley and O'Neil 1981), i.e., at less than 100 km depth along a typical subduction geotherm. This is supported by the observation that the C-isotope composition of CO_2 at arcs covers only a restricted range of $\delta^{13}C$-values (from −8 to 0‰; Fig. 2 of Plank and Manning 2019). Furthermore, to preserve the extreme $\delta^{13}C$-values present in subducted organic carbon and biogenic carbonate, an additional pre-requisite is that no significant mixing with mantle carbon must occur. The suggestion of eclogitic diamond formation *exclusively* from recycled organic carbon thus relies on several assumptions that often are not explicitly considered (e.g., Nisbet et al. 1994; Tappert et al. 2005).

The present-day budget of subducted sedimentary carbon illustrates a large variability in the ratio of carbonate/organic carbon (with a relative proportion of organic carbon from 0 to 80%, averaging close to 20%; see Plank and Manning 2019); therefore, the opportunity to subduct sediment with a high ratio of organic carbon to carbonate (implying low bulk $\delta^{13}C$) exists, but is compensated by higher average $\delta^{13}C$-values in other subduction settings. The most recent development (Li et al. 2019) emphasizes that the altered igneous portion of oceanic crust (AOC) has a total carbon content of similar magnitude to that of subducted sediments and the combined $\delta^{13}C$ range (−24 to +11‰) of normal and biogenic carbonate and organic matter in AOC is of similar extent to diamond (−41 to +3‰). An AOC origin of carbon in eclogitic diamonds (inferred from $\delta^{13}C$ outside the mantle range) is consistent with the observation that eclogite xenoliths and eclogitic inclusions in diamonds differ significantly from sedimentary protolith compositions. An important role of mantle-derived carbon-bearing fluids during diamond formation in AOC-derived substrates was highlighted by Ickert et al. (2013), with the observed "mixing" relationship implying low carbon contents in most of the AOC-derived substrates, with the exception of the originally shallowest and most altered sea floor basalts. With a mean $\delta^{13}C$ of about −5‰, AOC and in particular its organic matter and biogenic carbonate components must still escape metamorphic homogenization, at least locally, to impart the prominent ^{13}C-depleted tail to the eclogitic diamond population. Melting of subducted oceanic crust in the Archean is often invoked to account for the occurrence of TTG magmas and the magnesian character of cratonic eclogite xenoliths. The behavior of carbon during melting is however difficult to predict, as carbon may behave both compatibly or incompatibly in the melting residue, depending on various parameters—in particular speciation, chemical composition and CO_2/SiO_2 ratio (see Yaxley et al. 2019 for a review); we will return to this aspect further below, but it is clear that models relating diamond formation to subducted carbon require that carbon must survive devolatilization, decarbonation and melting.

Negligible diffusion and homogenization of C and N in diamond

The fairly homogenous C-isotope composition of individual diamond crystals does not reflect diffusion and homogenization over time. This is implicit from fine scale (down to at least 1 μm) zoning in cathodoluminescence (CL) images and sharp contrasts in N-content and stable isotopic composition (down to the analytical resolution limit of ~ 10 μm), and was ultimately experimentally demonstrated for C and N (Koga et al. 2003, 2005; Harte et al. 2009). Put simply, diffusion of C and N in diamond on the millimeter scale may only be possible for mantle residence times corresponding to the age of the Earth at *very high* temperatures (> 2000 °C), such as those expected in the lower mantle. In other words, the limited C-isotope variability of most diamonds—which form in the lithospheric upper mantle at mean temperatures of ~ 1150–1200 °C (Stachel and Luth 2015)—is a primary feature that directly relates to diamond formation, not its post-crystallization history. Nitrogen diffuses in the diamond lattice, allowing it to aggregate over time, but only on the nanometer level. For diamonds from the lower mantle, significant diffusion is possible but again not supported by the observation of fine-scale features in CL images and abrupt changes in carbon isotope composition and nitrogen concentration along SIMS transects (e.g., Palot et al. 2014).

Database and methods

As the basis for this review, we compiled an analytical database for 5115 diamonds from literature. Of these samples, 4307 were analyzed for their carbon isotope composition, 840 for their nitrogen isotope composition and all 5115 for their nitrogen content. The emphasis in compiling the database was on inclusion-bearing diamonds of known paragenesis (3474 diamonds), but an additional 1541 diamonds of unknown paragenesis are included (also in the statistical analysis of "all diamonds" in Figs. 1 and 6). For the vast majority of diamonds, analyses either are single bulk combustion analyses or represent averages of multiple bulk combustion or MC-SIMS analyses. For a small number of samples, where SIMS profiles revealed distinct growth events and, consequently, average values are not representative of real compositions, individual growth zones where averaged as "core" and "rim" (76 averages for 38 diamonds) and sometimes additionally as "middle" (for 23 diamonds). Note that "core" and "rim" of diamond fragments may not correspond to true centers and rims of the original crystals. The full database, a list of references and a "dictionary" explaining data columns and abbreviations are available online at https://doi.org/10.7939/DVN/B8VYHV. For the sake of readability of this chapter, the data sources referenced online are not duplicated here.

Raleigh isotope fractionation during diamond precipitation can be modeled assuming either a single carbon species or two carbon species in the fluid. The modeling spreadsheets (DiaRIF.xlsx and RIFMS.xlsx, respectively) used by us to calculate fractionation trends are provided online at https://doi.org/10.7939/DVN/JAEQKZ. Temperature dependent carbon isotope fractionation factors for diamond–$CaCO_3$, diamond–CO_2 and diamond–CH_4 are based on calculations or experimental data reported in Richet et al. (1977), Chacko et al. (1991), Polyakov and Kharlashina (1995), and Horita (2001). See also Table 1 in Stachel et al. (2017). As detailed in a later section, C-isotope fractionation factors are relatively well established for many systems relevant to diamond formation but there is considerably more uncertainty associated with N-isotope fractionation factors for diamond, which, for the most part, have only been investigated empirically based on fractionation trends observed in natural diamonds.

For data presentation we chose to employ kernel density estimation, a non-parametric approach, where kernel density plots (KDPs) visualize the density over which a large population is distributed, based on the assumption that the data represent a random sample of the overall population. For the C-isotope database investigated here, at the individual deposit scale, this underlying assumption might be debatable, but it is likely true when considering the large overall population of diamonds analyzed globally. KDPs use a uniform bandwidth that should be calculated objectively from the data, by the adopted program, rather than by the user. Unlike histograms, KDPs do not require the user to choose end-point bins which can create spurious features to the plots. When used in an objective way, KDPs are thus less subject to the creation of artificial peaks in the visualized distribution. In a kernel density estimate, each data point in the sample population is featured as a Gaussian curve centered on the data point value, whose standard deviation is given by the bandwidth. The smooth curve of the KDP is then the sum of the individual Gaussian distributions.

To assess the proportion of data that may be contained within the various peaks in the data distributions, we use a mixture modeling approach. Mixture modeling (e.g., Fraley and Raftery 2002), is a parametric approach that typically assumes that the population density is the sum of a small number of Gaussian distributions. The key goal of the approach, for a given sample, is to identify the number of components, their means, standard deviations and proportions of the components making up the overall populations. Where large distributions approach a Gaussian form, this approach is usually more successful than other approaches, such as feature recognition, in identifying the main features of the population (see Rudge 2008 for more detail).

Here, we compute Kernel density estimations and mixture models using the program DensityPlotter (Vermeesch 2012). The kernel width (bandwidth) is calculated in an adaptive approach based on local density, rather than determined by analytical uncertainty (Vermeesch 2012). Mixture modeling of stable isotope data with DensityPlotter uses the algorithm of Galbraith (2005) to assess the number of components that may constitute a given distribution of data. The program calculates peak locations for each "component" and their proportions.

Statistical tests were calculated using the Real Statistics Resource Pack (Excel Add-in) written by Charles Zaiontz (*http://www.real-statistics.com*). Two-tailed Student's *t-tests* for equality of means of two independent samples were performed assuming unequal variances and an alpha-value of 5% (i.e., a decision not to reject the null hypothesis of equal means carries 95% confidence). Normality of sample distributions was assessed using both the Shapiro–Wilk and d'Agostino-Pearson (Omnibus version) tests, again using an alpha-value of 5%. Both tests are robust with symmetric and long-tailed distributions that feature in several of our datasets. The Shapiro-Wilk test is vulnerable to losing power in data sets with a significant number of identical values with small standard deviations, which are present in some C-isotope datasets. In these cases, the d'Agostino–Pearson test is favoured as it is less susceptible to identical values. This test first calculates skewness and kurtosis to examine how far the distribution is from Gaussian, then it calculates how far each of these parameters differs from the expected value for a Gaussian distribution, computing a P value from the sum of these differences.

CARBON ISOTOPE COMPOSITION OF DIAMOND AND ITS RELATIONSHIP TO INCLUSION PARAGENESIS

On the $\delta^{13}C$ scale, the carbon isotope composition of Earth's mantle, based on analyses of carbonatites, kimberlite carbonates, mantle xenoliths and volcanic CO_2, is about −5‰ (Deines 2002). The median value (−5.2‰) and the main mode in the kernel density estimation of mantle-derived diamonds (−5.1‰; Fig. 1; Table 1) coincide with this mantle value (c.f., Deines 1980; Galimov 1991). Defining the $\delta^{13}C$ mantle array as -5 ± 2‰ (Cartigny et al. 2014) includes the

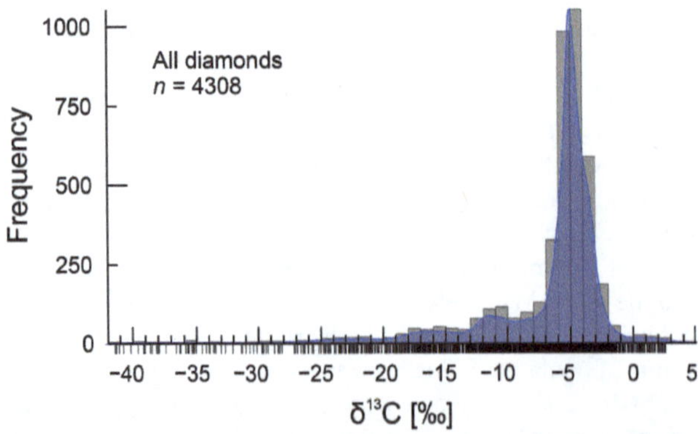

Figure 1. Distribution in carbon isotope composition ($\delta^{13}C$) for (mostly) inclusion-bearing diamonds worldwide. The blue curve is a Kernel density estimation, shown in the background is a histogram with a 1‰ binning interval. **Black dashes** below the *x*-axis indicate individual data points.

Table 1. The carbon isotope compositions ($\delta^{13}C$) of inclusion-bearing diamonds from worldwide sources.

Suite	All diamonds	Peridotitic			Eclogitic	Websteritic	Asthenospheric-TZ "Eclogitic"	Lower Mantle
Paragenesis		All	Lherzolitic	Harzburgitic				
n	4308	1906	216	503	1204	58	55	154
Min (‰)	–41.4	–34.5	–11.8	–26.4	–40.7	–41.3	–24.8	–28.3
Max (‰)	2.5	2.3	0.3	2.3	2.5	–3.7	0.9	0.7
Average (‰)	–6.9	–4.9	–4.5	–5.1	–9.4	–11.5	–14.5	–6.1
SD (‰)	5.5	1.9	1.5	1.7	6.7	9.1	7.2	5.6
Median (‰)	–5.2	–4.9	–4.8	–5.1	–6.5	–6.0	–15.3	–4.6
MAD (‰)	1.1	0.8	0.7	0.7	2.3	2.2	6.5	0.7
Q1 (‰)	–6.8	–5.5	–5.3	–5.7	–11.9	–17.9	–21.2	–5.2
Q3 (‰)	–4.2	–4.0	–3.7	–4.2	–5.0	–4.8	–8.5	–3.8
IQR (‰)	2.6	1.5	1.6	1.5	6.9	13.1	12.7	1.4
Mode (‰)	–5.1	–5.2	–5.2	–5.3	–5.0	–4.9		–4.9

Note: Statistical parameters quoted are: *n*—number of samples; Min—lowest values; Max—highest values; Average—arithmetic mean; SD—1 standard deviation calculated based on a sample; Median—median value; MAD—median absolute deviation; Q1—25th percentile; Q3—75th percentile; IQR—interquartile range (IQR = Q3–Q1); Mode—maximum value of the kernel density estimation.

majority of diamonds (69%) but the total distribution extends far beyond this range, from –41.4 to +2.5‰. To further evaluate the origin of this variability, we split mantle-derived diamonds into suites and further into parageneses based on their mineral inclusion content.

Peridotitic suite

The carbon isotope composition of lithospheric diamonds with peridotitic inclusions forms a narrow distribution about the mantle value (median $\delta^{13}C$-value of –4.9‰; mode at –5.2‰; Fig. 2), with 85% of samples falling between –7 and –3‰ and 95% of samples between –8 and –2‰. Only 1% of samples fall below –10‰ (extending down to –34.5‰) and only four samples (0.2%) exceed 0‰ (Table 1). The distribution is non-normal (Shapiro-Wilk and d'Agostino-Pearson tests) though the strong negative skewness value (–4.6) is due to a significant number of outliers biasing the value: 18 samples or 1.1% of the entire population have $\delta^{13}C$-values three

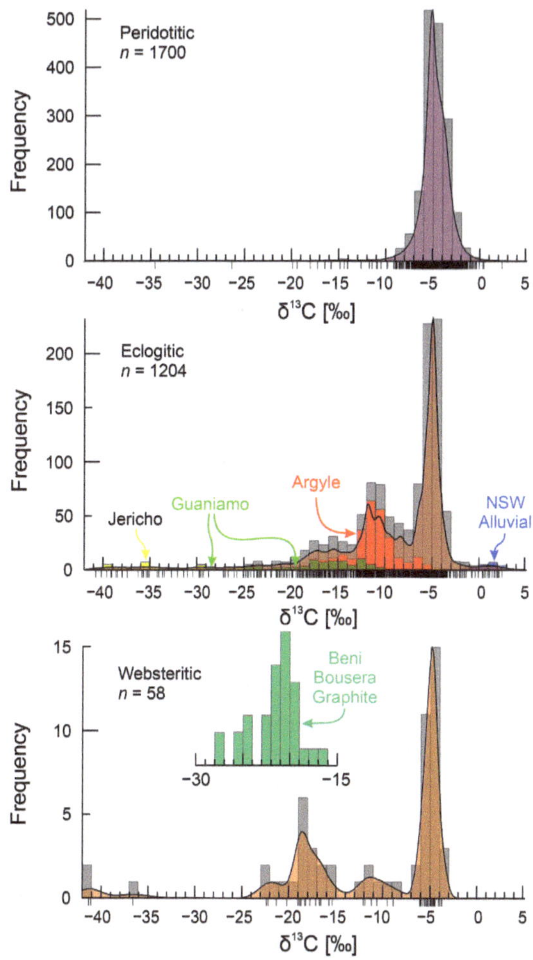

Figure 2. Distribution in carbon isotope composition for diamonds with peridotitic **(top)**, eclogitic (**middle**) and websteritic **(bottom)** suite inclusions. The **colored curves** are Kernel density estimations, shown in the background are histograms with a 1‰ binning interval. **Black dashes** below the *x*-axis indicate individual data points. For the distribution of eclogitic diamonds, four locations with distinctly non-mantle-like distributions (Jericho in the Northern Slave Craton, Guaniamo on the Amazon Shield, Argyle on the Kimberley Craton and the New South Wales alluvials) are highlighted. The insert above the websteritic diamond distribution shows the carbon isotope composition of graphite pseudomorphs after diamond in pyroxenite layers (dikes) in the Beni Bousera peridotite complex (Pearson et al. 1991) for comparison.

standard deviations below the mean (i.e., < −10.7‰; see Table 1), versus the 5 expected at the 99.7 probability level. These outliers thus likely represent a separate population of peridotitic diamonds. Through the presence of clinopyroxene and garnet inclusions, for close to half of the peridotitic samples a specific paragenesis, harzburgitic (–dunitic) or lherzolitic, can be assigned. Diamonds containing wehrlitic inclusions are rare (0.6% of inclusion-bearing diamonds; Stachel and Harris 2008) and consequently are ignored here. For the harzburgitic paragenesis, the median value of –5.1‰ is well within the Median Absolute Deviation (MAD[3]) about the median of the peridotitic distribution as a whole. The kernel density plot, however, reveals a

[3] In statistics MAD is considered a more robust measure of variability than standard deviation because the former does not rely on the assumption of a normal distribution.

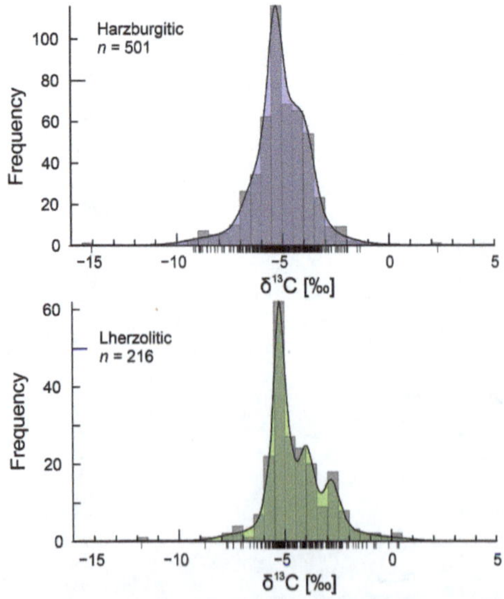

Figure 3. Distribution in carbon isotope composition for diamonds with peridotitic inclusions of harzburgitic **(top)** and lherzolitic **(bottom)** paragenesis. The **colored curves** are Kernel density estimations, shown in the background are histograms with a 1‰ binning interval. **Black dashes** below the *x*-axis indicate individual data points.

minor peak on the ^{13}C-enriched side of the distribution (Fig. 3) reflecting the contribution of diamond populations from Akwatia in Ghana and Boa Vista in Brazil (alluvial deposits), both with median values of −3.9‰, Williamson (formerly Mwadui) in Tanzania, with a median of −3.8‰, and Pipe 50 in China, at −3.5‰ (c.f., Stachel et al. 2009, their Fig. 2). This skewness is even more pronounced in the distribution for lherzolitic diamonds (Fig. 3), with the appearance of distinct secondary modes at −4.0 and −2.8‰. The two secondary modes principally reflect a bimodal distribution of lherzolitic diamonds at Cullinan (formerly Premier) Mine in South Africa, with a subordinate contribution of overall ^{13}C-enriched lherzolitic diamonds from Kankan and Akwatia in West Africa and Williamson in Tanzania. Although it is important to note the mildly ^{13}C-enriched nature of peridotitic diamonds from the West African and Tanzania cratons and of the potentially Bushveld event-associated lherzolitic diamonds from Cullinan Mine (Richardson et al. 1993) on the Kalahari Craton, the principal observation for peridotitic diamonds is their tight distribution about the mantle value of −5‰.

Eclogitic suite

Most of the variability in carbon isotope composition seen in the worldwide diamond distribution derives from samples with eclogitic (meta-basaltic) inclusions. The eclogitic suite of lithospheric diamonds shows a prominent mode at −5.0‰ but is distinctly skewed towards more negative values extending to −40.7‰, with a second mode at −11.8‰ (Fig. 2). The record for eclogitic diamonds is dominated by samples from the Kalahari Craton of Southern Africa (603 of 1204 analyses), which together with diamonds from the Central Slave Craton, the Siberian Platform and the Ural alluvials account for the prominent mode at -5‰. A dominance of mantle-like carbon is, however, not a universal feature of eclogitic diamonds, with datasets, e.g., for the deposits at Argyle (Kimberley Craton), Guaniamo (Amazon Craton) and Jericho (Northern Slave Craton) having overall ^{13}C-depleted distributions with an almost complete absence of mantle-like carbon (see Cartigny et al. 2014, their Fig. 5). The large analytical data set for Argyle (221 analyses) is responsible for the second mode at −11.8‰ in the eclogitic distribution (Fig. 2). To limit the influence of localities that are overrepresented in the worldwide database (Argyle and Premier/Cullinan), Cartigny et al. (2014) adjusted each locality to have the same statistical weight but still obtained the same second mode at ~ −11‰. $\delta^{13}C$-values above 0‰ are restricted to diamonds from the New South Wales alluvial deposits (Fig. 2), with an unusually Ca-rich inclusion suite that may relate to meta-rodingites as diamond substrates (Davies et al. 2003). The most ^{13}C-depleted values among eclogitic diamonds, down to −40.7‰, are reported from the former Jericho mine (Northern Slave Craton).

Websteritic (pyroxenitic) suite

Based on the chemistry of websteritic inclusions in lithospheric diamonds, the websteritic (or more general, pyroxenitic) suite forms in substrates that are intermediate between those of the eclogitic and peridotitic suites (Gurney et al. 1984, Stachel and Harris 2008). Despite a much smaller sample number, websteritic diamonds have a very similar $\delta^{13}C$-distribution to eclogitic diamonds, with a prominent mode at −4.9‰ and a tail towards ^{13}C-depleted compositions extending to −41.3‰ (again a diamond from Jericho), with secondary modes peaking at −11.2 and −18.5‰ (Fig. 2). The key difference to the eclogitic diamond distribution is an absence of ^{13}C-enriched websteritic diamonds, with the maximum $\delta^{13}C$-value being −3.7‰. A further distinguishing criterion is that websteritic diamonds from individual deposits generally either all fall into the mantle range (e.g., Venetia on the Kalahari Craton, the Namibian coast, A154 at Diavik on the Central Slave Craton) or are all distinctly depleted in ^{13}C ($\delta^{13}C < -10$‰; e.g., Jagersfontein and the Orapa Cluster on the Kalahari Craton and Jericho).

Graphite pseudomorphs after diamond in pyroxenites (garnet–clinopyroxenites) from Beni Bousera (Morocco) provide an additional data set for the carbon isotope composition of "websteritic diamonds". The Beni Bousera pseudomorphs range in $\delta^{13}C$ from −27.6 to −16.4‰ (Fig. 2), with a median at −21.0‰, and fall into the ^{13}C-depleted group of websteritic diamonds.

Asthenospheric and transition zone suite

Only very few diamonds have been described that contain majoritic garnet inclusions of harzburgitic, lherzolitic or wehrlitic paragenesis. In all cases, these peridotitic majoritic garnet inclusions carry a signature of depleted lithospheric substrates, by either being subcalcic (having CaO contents below the lherzolitic Ca-Cr array, indicative of clinopyroxene-free substrates; see Sobolev et al. 1973), or through high Cr_2O_3 contents (≥ 4 wt%), i.e., their compositions are not consistent with an origin in fertile convecting mantle but instead point to their derivation from either over-thickened subcratonic lithospheric mantle (Pokhilenko et al. 2004) or deeply subducted oceanic lithospheric mantle. In total, carbon isotope analyses are available for eight such diamonds and range from −12.2 to −2.7‰, with a median of −5.2‰. Due to their rarity, uncertain origin and low number of analyses, these samples are not further considered here.

Far more common in asthenosphere- and transition zone-derived diamonds are inclusions with "eclogitic" (meta-basaltic) compositions, mostly low-Cr (below detection to 0.37 wt% Cr_2O_3) majoritic garnets. The carbon isotope compositions of these sublithospheric diamonds cover a wide range from −24.8 and +0.9‰ (Fig. 4; Table 1), similar to lithospheric eclogitic and websteritic diamonds. Individual localities seem to carry specific carbon isotope signatures. Diamonds from the Juina area in Brazil fall into two clusters, one with strongly negative values (−24.8 to −23.4‰) and the other with moderate to no ^{13}C depletion (−15.3 to −4.6‰) relative to the canonical mantle value. Other localities have much narrower ranges, with, e.g., Jagersfontein diamonds being strongly ^{13}C-depleted ($\delta^{13}C$ between −23.0 and −17.2‰), Koffiefontein (Kalahari Craton) diamonds showing less negative $\delta^{13}C$ (−18.6 to −15.4‰) and Kankan (W. African Craton) diamonds being ^{13}C-enriched (−3.1 to +0.9‰). Overall, these "eclogitic" sublithospheric diamonds are unlike any of the other suites discussed here, showing a relatively flat $\delta^{13}C$ frequency distribution, with no mode near the mantle value, and high variance. This characteristic is, however, shared with a small number of individual deposits (Argyle, Guaniamo, Jericho, NSW alluvials) contributing to the eclogitic diamond population (see Fig. 2).

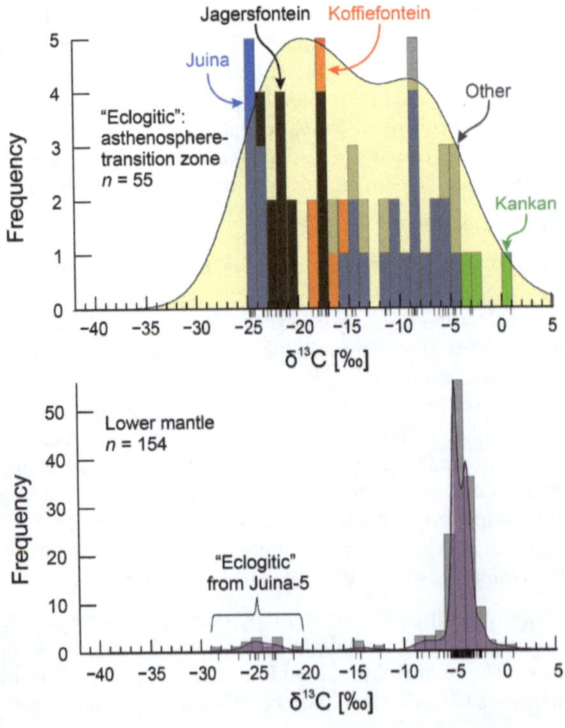

Figure 4. Distribution in carbon isotope composition for diamonds with superdeep inclusions. The distribution for diamonds from the asthenosphere and transition zone **(top)** is entirely based on samples containing inclusions with "eclogitic" (meta-basaltic) mineral compositions, mostly low-Cr majoritic garnets. Most of these inclusions are derived from four deposits, Juina (Amazon Craton), Jagersfontein and Koffiefontein (Kalahari Craton) and Kankan (West African Craton). Lower mantle diamonds **(bottom)** generally derive from meta-peridotitic substrates, with the exception of a small number of samples from the Juina-5 kimberlite containing inclusions with re-constructed bulk compositions that document metabasaltic ("eclogitic") diamond substrates (Thomson et al. 2014). The **colored curves** are Kernel density estimations, shown in the background are histograms with a 1‰ binning interval. **Black dashes** below the *x*-axis indicate individual data points.

Lower mantle suite

The bulk of diamonds with an assigned lower mantle origin carries associations including the mineral ferropericlase. These diamonds are generally assigned to "meta-peridotitic" substrates and predominantly come from the Juina area, Kankan and the DO27 kimberlite from the Central Slave Craton. A definitive lower-mantle paragenesis of ferropericlase-bearing diamonds is only substantiated by co-existence with $MgSiO_3$, with the appropriate Ni partitioning for equilibration at lower mantle pressures (Stachel et al. 2000b), or $CaSiO_3$. For diamonds carrying ferropericlase inclusions only, a shallower origin (Brey et al. 2004) is possible but, for the purpose of this review, such an origin is considered unlikely as long as other diamonds from the same locality contain associations unique to the lower mantle. For a small subgroup of Juina diamonds, exclusively recovered from the Juina 5 kimberlite, Thomson et al. (2014) reconstructed inclusion compositions indicative of a meta-basaltic ("eclogitic") association. The meta-peridotitic diamonds have a $\delta^{13}C$-distribution focused on a prominent mode at −4.9‰, with an associated secondary mode at −3.9‰, and a range from −14.8 to +0.7‰ (Fig. 4). The distribution is non-normal (Shapiro-Wilk and d'Agostino-Pearson tests) and has a negative skewness (−2.5). The meta-basaltic diamonds are all strongly depleted in ^{13}C ($\delta^{13}C$ between −28.3 to −20.9‰; Fig. 4).

The $\delta^{13}C$-distributions of meta-peridotitic lower mantle diamonds and peridotitic lithospheric diamonds are a close match. Using a Student's *t-test*, equality of means between metaperidotitic lower mantle diamonds and peridotitic lithospheric diamonds cannot be rejected; on the paragenesis level the same is true for lherzolitic but not for harzburgitic lithospheric diamonds. With the lherzolitic paragenesis (Fig. 3) the lower mantle diamonds share a rapid drop-off in abundance at $\delta^{13}C$ below −6‰ and a secondary mode at about −4‰. The secondary mode for meta-peridotitic lower mantle diamonds is caused by samples from Kankan (West African Craton), which also contribute to the secondary mode at −4‰ for lithospheric lherzolitic diamonds. This documents that lower mantle and lithospheric peridotitic (lherzolitic) diamonds not only share the same principal mode but also subtle isotopic shifts typical for a particular cratonic region.

Commonalities and differences of the isotopic composition of diamond carbon across mantle reservoirs

The carbon isotope compositions of the four principal diamonds suites (peridotitic, eclogitic, asthenosphere-transition zone, and lower mantle) are compared in Figure 5. Peridotitic lithospheric and meta-peridotitic lower mantle diamonds share a narrow principal mode and median value close to or at the mantle value, as defined by other mantle-derived samples such as carbonatites, kimberlites, mid-ocean ridge and ocean-island basalts. This strong similarity is particularly true for the lherzolitic paragenesis of lithospheric diamonds, whilst harzburgitic diamonds have a statistically different mean than meta-peridotitic lower mantle diamonds. The peridotitic lithospheric and metaperidotitic lower mantle diamond suites evidently tap the same mantle carbon reservoir and the observation of common minor isotopic shifts for lower mantle and lithospheric diamonds from the West African Craton documents a vertical flux of carbon between superdeep and shallow diamond substrates.

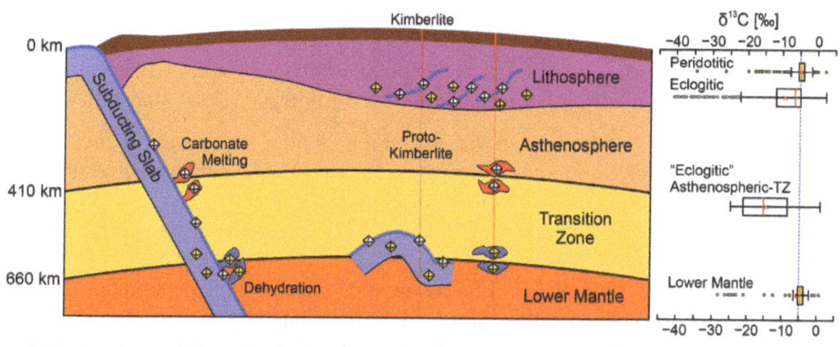

Figure 5. Environment of formation and carbon isotope composition for the four principal diamond suites. Lithospheric diamonds of the peridotitic (**yellow**) and eclogitic (**white**) suites grow in depleted peridotitic (purple) and meta-basaltic (eclogitic; **dark blue**) substrates (e.g., Meyer and Boyd 1972; Sobolev 1977). In the asthenosphere and transition zone, diamonds form in subducting oceanic crust (**dark blue**; Stachel et al. 2000a) or in reaction zones between slab-derived carbonatitic melts and pyrolytic wall rock (**red**; Walter et al. 2008; Thomson et al. 2016). Inclusions in these diamonds (**white**) have meta-basaltic ("eclogitic") affinity. In the lowermost transition zone and the lower mantle, diamonds may form in the meta-peridotitic (**light blue with yellow diamonds**) or meta-basaltic (**dark blue with white diamonds**) portions of subducting slabs, or in adjacent pyrolytic wall rock (**blue jets with yellow diamonds**), driven by dehydration reactions (Harte 2010) and the associated expulsion of oxidized fluids into ambient mantle (Stachel et al. 2005; Regier et al. 2020). Sampling of superdeep diamonds likely occurs through ascending proto-kimberlite magma (**red dotted lines**), either from stagnant slabs (megaliths; **light and dark blue**) or from remnant reaction zones between slab-derived hydrous fluids (**blue**) or carbonatitic melts (**red**) and pyrolytic wall rock. Lithospheric diamonds are collected during rapid final ascent of kimberlite magma (**red line**) through subcratonic lithospheric mantle. The carbon isotopic composition ($\delta^{13}C$ in ‰) of the four principal diamond suites is given in the right-hand column as median values (**red lines**), averages (**red crosses**), 25th to 75th percentiles (interquartile range; **boxes**) and ranges (**whiskers**). Outliers are indicated as **small circles** (**white** for meta-basaltic and **yellow** for meta-peridotitic affinity). The **blue dashed line** indicates the assumed mantle value of −5‰ (Deines 2002).

Diamonds formed in meta-basaltic substrates in the lithosphere and in the asthenosphere-transition zone have similar ranges (Fig. 5) but clearly differ through the strong dominance of mantle-like carbon in lithospheric eclogitic diamonds that is not as evident in the sub-lithospheric diamonds (Fig. 2). This not only documents that mantle-derived carbon is negligible for diamond formation associated with subducting slabs in the asthenosphere and transition zone but also suggests a common origin of both ^{13}C-depleted and ^{13}C-enriched diamond carbon for meta-basaltic environments in lithospheric and sublithospheric substrates. This aspect will be discussed further below.

NITROGEN ISOTOPE COMPOSITION OF DIAMOND AND ITS RELATIONSHIP TO INCLUSION PARAGENESIS

The nitrogen isotope composition of Earth's mantle, based on analyses of mid-ocean ridge basalts and fibrous diamonds, is −5 ± 2‰ (Cartigny and Marty 2013). The nitrogen isotope composition of diamond has a unimodal distribution with minor skewness to positive values and a range over 56‰ (from −39.4 to +16.9‰, Fig. 6). Compared to the mantle value, the mode (−3.5‰) and median (−2.1‰) are shifted to slightly higher values (Table 2). The range from −8.5 to +1.5‰ (i.e., the mode of −3.5 ± 5‰) contains about 2/3 of the data. The specific nitrogen isotope distributions of the various suites and parageneses of mantle-derived inclusion-bearing diamonds are discussed in detail below.

Peridotitic suite

The $\delta^{15}N$-distribution of diamonds with peridotitic inclusions has a principal mode at +2.0‰ with a secondary mode (or shoulder; Fig. 7) at −4.1‰. This Kernel density estimation can be approximated as a mixture of two Gaussian distributions with modes at −7.5‰ (44%) and +3.4‰ (56%; see section Database and Methods), which may or may not reflect geological reality. The peridotitic $\delta^{15}N$-distribution principally results from different modes in $\delta^{15}N$ for different localities, with ^{15}N-depleted diamond populations, e.g., at Pipe 50 (Liaoning, China) and the Victor Mine (Superior Craton), and ^{15}N-enriched populations, e.g., at the Kimberley mines ('*De Beers Pool*'), Cullinan Mine (both Kalahari Craton) and Ellendale (Kimberley Craton).

Broken down further by paragenesis, diamonds with harzburgitic inclusions have a mode at +2.1‰, coinciding with the upper mode for the peridotitic suite, and a distribution that is skewed towards more negative values (Fig. 8; Table 2). Lherzolitic diamonds show a distribution with two pronounced modes (−6.3 and +2.9‰), which can be largely resolved into two Gaussian distributions with modes at −5.9‰ (42%) and +3.9‰ (58%)—suggesting two overlapping distributions rather than a continuous but under-sampled population. Similar to the entire peridotitic suite, the two lherzolitic sub-populations largely reflect isotopically light nitrogen in diamonds from Victor and isotopically heavy nitrogen in diamonds from Cullinan Mine. The distribution of diamonds with lherzolitic inclusions (−11.7 to +9.6‰, Fig. 8) lacks the tail to strongly ^{15}N-depleted samples seen for harzburgitic diamonds (−24.1 to +12.4‰). The peridotitic diamond with the lowest $\delta^{15}N$-value (−39.4‰; diamond KK-21 from Kankan) cannot be further classified, containing olivine inclusions only (see Stachel et al. 2022, this volume).

Eclogitic suite

Diamonds with eclogitic inclusions show a nitrogen isotope distribution with a mode at −6.2‰, a shoulder at about −2.0‰, and a skewness towards high values (Fig. 7). Breaking the $\delta^{15}N$-distributions down by carbon isotope composition reveals a hitherto unrecognized systematic shift (Fig. 9): diamonds with a carbon isotope composition close to the mantle value ($\delta^{13}C$ between −7 and −3‰) are very similar to the overall eclogitic distribution in terms of their $\delta^{15}N$-values (same mode at −6.2‰ and a similar shoulder at about -2.0‰), with the exception of an absence of values greater than +11.1‰. Diamonds with distinctly ^{13}C-depleted isotope compositions ($\delta^{13}C \leq -15$‰) are ^{15}N-enriched, with a near normal distribution about

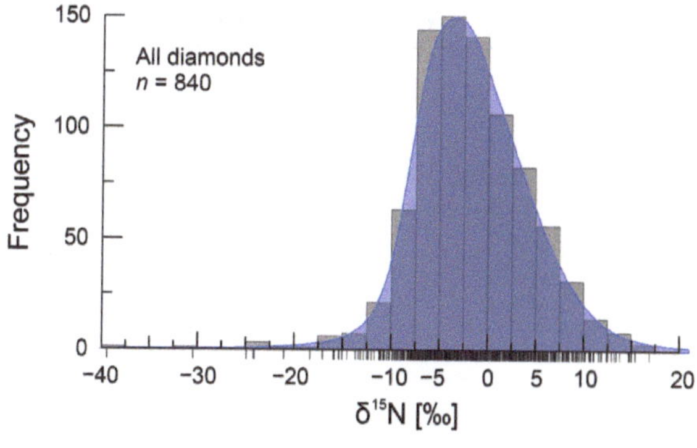

Figure 6. Distribution in nitrogen isotope composition ($\delta^{15}N$) for (mostly) inclusion-bearing diamonds worldwide. The **blue curve** is a Kernel density estimation, shown in the background is a histogram with a 1‰ binning interval. **Black dashes** below the x-axis indicate individual data points.

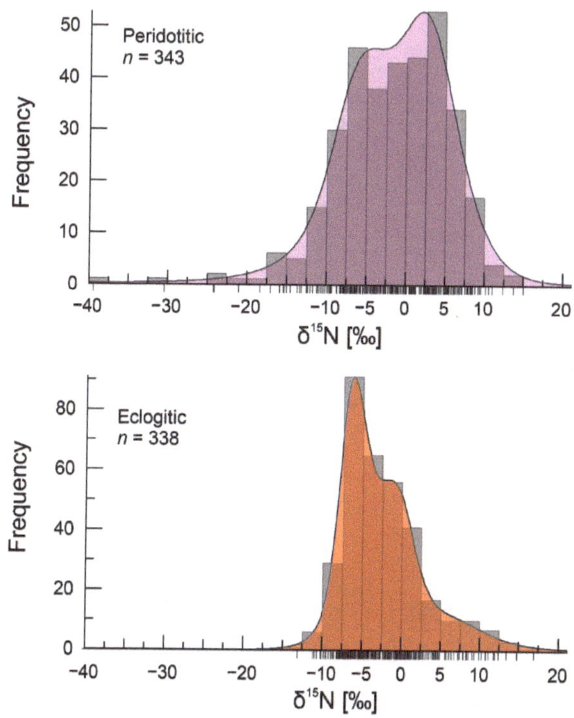

Figure 7. Distribution in nitrogen isotope composition for diamonds with peridotitic **(top)** and eclogitic suite **(bottom)** inclusions. The **colored curves** are Kernel density estimations, shown in the background are histograms with a 1‰ binning interval. **Black dashes** below the x-axis indicate individual data points.

a mode at +1.3‰ and a range from −5.3 to +14.8‰ (Fig. 9; Table 2). The data set for the ^{13}C-depleted diamonds is small (n = 22) but derives from seven different locations (Jwaneng, Orapa, Namibian southern coastal mines, Kimberley [all from the Kalahari Craton], Ural alluvials, Panda [Ekati Mine, Central Slave Craton] and Chidliak [likely North Atlantic Craton]) and, therefore, represents more than just local variability.

Table 2. Statistical assessment of the nitrogen isotope composition ($\delta^{15}N$) of inclusion-bearing diamonds from worldwide sources.

Suite	All diamonds	Peridotitic			Eclogitic			Lower Mantle
Paragenesis		All	Lherzolitic	Harzburgitic	All	$\delta^{13}C$ −7 to −3‰	$\delta^{13}C$ ≤ −15‰	
n	840	343	76	102	338	239	22	29
Min (‰)	−39.4	−39.4	−11.7	−24.1	−13.1	−13.1	−5.3	−24.9
Max (‰)	16.9	15.0	9.6	12.4	16.9	11.1	14.8	2.6
Average (‰)	−1.7	−1.7	−0.3	−0.1	−2.4	−3.6	2.6	−3.6
SD (‰)	6.0	7.0	5.5	6.8	5.2	4.1	5	4.9
Median (‰)	−2.1	−1.1	1.2	1.2	−3.4	−4.6	1.5	−2.8
MAD (‰)	3.9	4.8	4.5	4.1	3.2	2.7	3.2	1.7
Q1 (‰)	−5.8	−6.2	−5.7	−3.6	−6.4	−6.7	−1.7	−4.6
Q3 (‰)	2.1	3.2	3.7	4.9	0.1	0.6	6.1	−1.2
IQR (‰)	7.9	9.4	9.4	8.5	6.5	7.3	7.8	3.4
Mode (‰)	−3.5	2.0	2.9	2.1	−6.2	−6.2	1.3	−2.7

Note: Statistical parameters quoted are: *n*—number of samples; Min—lowest values; Max—highest values; Average—arithmetic mean; SD—1 standard deviation calculated based on a sample; Median—median value; MAD—median absolute deviation; Q1—25th percentile; Q3—75th percentile; IQR—interquartile range (IQR = Q3–Q1); Mode—maximum value of the kernel density estimation.

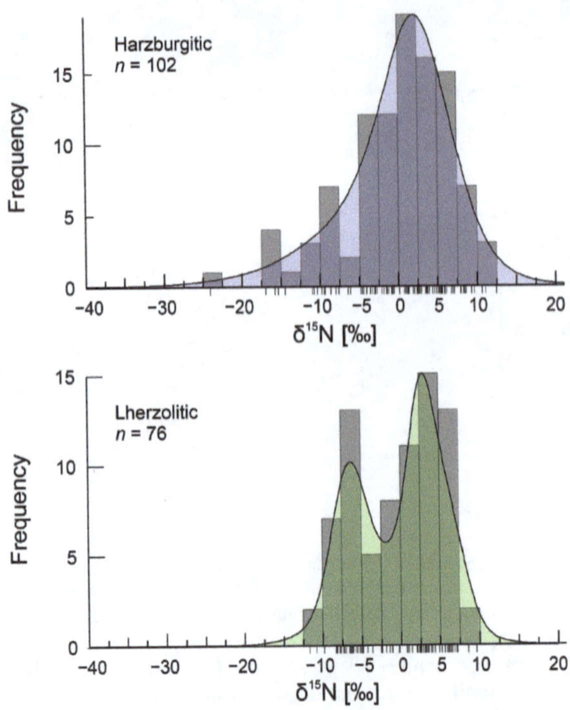

Figure 8. Distribution in nitrogen isotope composition for diamonds with peridotitic inclusions of harzburgitic **(top)** and lherzolitic **(bottom)** paragenesis. The **colored curves** are Kernel density estimations, shown in the background are histograms with a 1‰ binning interval. **Black dashes** below the *x*-axis indicate individual data points.

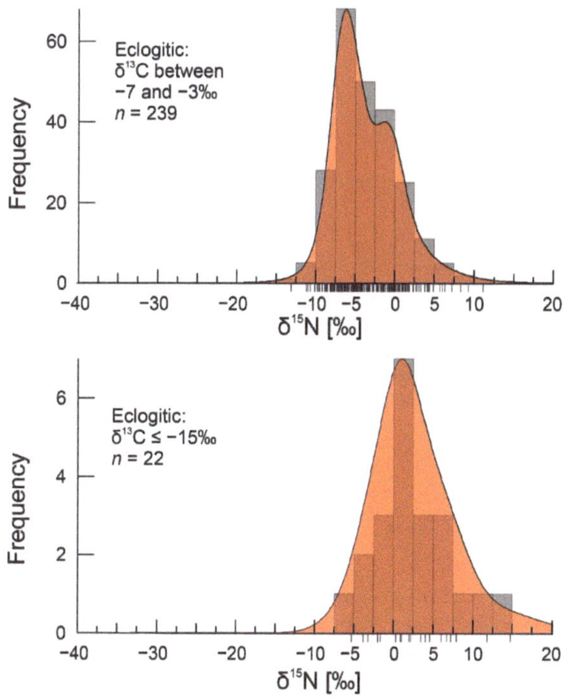

Figure 9. Distribution in nitrogen isotope composition for diamonds with eclogitic inclusions, split into two groups based on their carbon isotope composition: diamonds with $\delta^{13}C$-values very close to the accepted mantle value (−5 ± 2‰; **top**) and with strongly ^{13}C-depleted compositions (≤ −15‰; **bottom**). The **colored curves** are Kernel density estimations, shown in the background are histograms with a 1‰ binning interval. **Black dashes** below the *x*-axis indicate individual data points.

Lower mantle suite

For superdeep diamonds, nitrogen contents that are typically either low or below the limit of detection lead to a small number of samples with associated nitrogen isotope analyses, which is only large enough for lower mantle diamonds (n = 29) to be included here. For a limited $\delta^{15}N$ dataset on asthenospheric and transition zone diamonds with "eclogitic" majorite inclusions, the reader is referred to Palot et al. (2012, 2017).

The analyzed diamonds with lower mantle inclusions (13 from Juina and 16 from Kankan) show a very tight normal distribution about a $\delta^{15}N$ mode of −2.7‰ (median −2.8‰; Fig. 10; Table 2). The diamonds have a $\delta^{13}C$ range from −7.6 to −1.3‰ and contain inclusions considered to reflect meta-peridotitic substrates. If we assume that nitrogen in the meta-peridotitic substrates for the analyzed lower mantle diamonds is not subduction-related, the narrow $\delta^{15}N$-distribution of these diamonds may be used to bracket the nitrogen isotope composition of the lower mantle. For diamond formation at expected lower mantle temperatures in excess of 1600 °C (e.g., Stixrude and Lithgow-Bertelloni 2007), nitrogen fractionation factors between diamond and neutral to variably reduced nitrogen species (from N_2 through NH_3 and NH_4^+ to nitrides) will be fairly small (see Fig. 19), placing the nitrogen isotope composition of the lower mantle between about −4 and 0‰. This clearly supports the view of a convectively (largely) homogenized isotopic composition of nitrogen across the transition zone—lower mantle boundary (Palot et al. 2012).

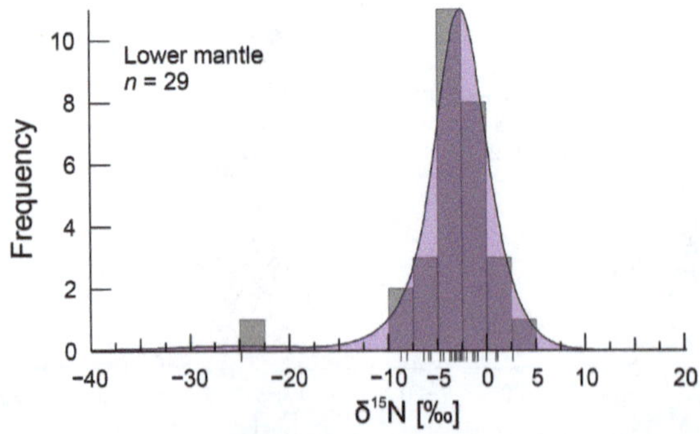

Figure 10. Distribution in nitrogen isotope composition for diamonds with lower mantle inclusions. Based on their inclusion assemblages (including the mineral ferropericlase in 22 of the 29 analyzed samples), derivation from principally meta-peridotitic substrates is inferred. The one outlier with a δ^{15}N-value of −24.9‰ also represents a diamond containing a ferropericlase inclusion. The **colored curve** is a Kernel density estimation, shown in the background is a histogram with a 1‰ binning interval. **Black dashes** below the *x*-axis indicate individual data points.

Commonalities and differences of the nitrogen isotope composition of diamond across mantle reservoirs

Comparing the δ^{15}N-distributions presented in Figures 6 to 10 reveals commonalities across diamond suites: (1) The ^{15}N-depleted peridotitic (mainly lherzolitic) diamond productions from Victor and Pipe 50 have a combined mode in δ^{15}N at −6.6‰, coinciding with the main mode for eclogitic diamonds (−6.2‰). This suggests derivation from a common nitrogen reservoir. (2) The main modes in nitrogen isotope composition for peridotitic diamonds, their harzburgitic and lherzolitic sub-divisions, and strongly ^{13}C-depleted ($\delta^{13}C \leq -15$‰) eclogitic diamonds fall in the δ^{15}N range +1 to +3‰, again suggesting tapping of a common nitrogen reservoir. (3) Lower mantle diamonds have a narrow distribution with a mode (−2.7‰) that falls between the two common modes for lithospheric diamonds, but coincides with the mode for all analyzed diamonds (−3.5‰). Given that the latter is a composite distribution of mostly peridotitic (41%) and eclogitic (40%) diamonds with distinct modes, this agreement may be purely coincidental.

The overall similarity in the nitrogen isotope compositions of peridotitic and eclogitic diamonds is further highlighted in Figure 11. The interquartile ranges and full ranges of eclogitic and lower mantle diamonds are contained within the peridotitic ranges and a

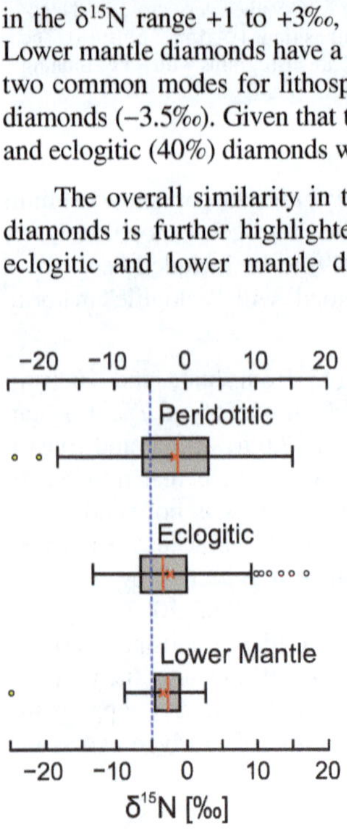

Figure 11. Nitrogen isotope composition (δ^{15}N) of peridotitic, eclogitic and lower mantle suite diamonds. Median values (**red lines**), averages (**red crosses**), 25th to 75th percentiles (interquartile range; **boxes**) and ranges (**whiskers**) are shown. Outliers are indicated as small circles (**white** for meta-basaltic and **yellow** for (meta-)peridotitic affinity). Two outliers for the peridotitic suite (at −39.4 and −30.4‰) are outside the δ^{15}N range shown. The **blue dashed line** indicates the assumed mantle value of −5 (± 2)‰ (Cartigny and Marty 2013).

single ^{15}N-enriched eclogitic diamond exceeds the peridotitic range. Based on the greatly expanded dataset presented here, the notion that positive $\delta^{15}N$-values are a clear indication of subduction-related diamond associations cannot be maintained. In fact, $\delta^{15}N$-values greater than 0‰ are more commonly observed among peridotitic (45% > 0‰) than among eclogitic (25%) diamonds. The main mode of peridotitic diamonds ($\delta^{15}N$ of +2‰) may instead indicate involvement of plume-derived nitrogen ($\delta^{15}N$ of +3 ±2‰; Dauphas and Marty 1999; Marty and Dauphas 2003). The fact that most eclogitic and peridotitic diamonds share similar $\delta^{15}N$-values strongly suggests similar origins of their nitrogen (see also Cartigny 2005; Cartigny and Marty 2013; Smart et al. 2016 for further reviews and a discussions).

RELATIONSHIPS AMONG $\delta^{13}C$, $\delta^{15}N$ AND NITROGEN CONTENT

$\delta^{13}C$ and nitrogen content

Nitrogen is the most abundant impurity in diamond (see *Introduction*), with average and median N-contents of 314 and 166 at.ppm (or 366 and 194 ppm on a weight basis), respectively, based on our worldwide database ($n = 3899$ for N_{FTIR}). Estimates of the nitrogen content of depleted (MORB source) mantle, based on ratios of nitrogen with noble gases and with carbon, range between 0.09 and 0.27 ppm (see review in Marty 2012). Accordingly, the average and median nitrogen content of diamond is 2100–4100 and 700–1400 times higher, respectively, than that of depleted mantle, which makes diamond an excellent target mineral for investigating the behavior of nitrogen in mantle processes.

Whether nitrogen preferentially partitions into diamond or the fluid/melt from which diamond precipitates is a subject of longstanding debate. One end-member view is that nitrogen is strictly incompatible in the diamond structure (i.e., $K_N = (C_{N\text{-diamond}}/ C_{N\text{-fluid/melt}}) < 1$) but that kinetic effects play an important role. Specifically, rapidly-grown diamonds approach compatibility for nitrogen ($K_N \sim 1$) whereas slowly-grown diamonds largely exclude nitrogen (Boyd et al. 1994; Cartigny et al. 2001). The observation of sector zoning of nitrogen in diamond (Boyd et al. 1988)—during mixed habit growth octahedral sectors of natural diamonds incorporate about 12% more nitrogen than cuboid sectors (Howell et al. 2012 and reference therein)—clearly documents that kinetic effects play a role in controlling the partitioning behavior of nitrogen between diamond–fluid/melt. The opposite view, a compatible behavior of nitrogen in diamond ($K_N > 1$), is based on the observation of fractionation trends involving progressively decreasing N-contents in multiple diamonds from a single xenolith (Thomassot et al. 2007) and in individual diamonds from both natural occurrences (Smart et al. 2011; Wiggers de Vries et al. 2013; Petts et al. 2015; Smit et al. 2016, 2019a) and high-pressure/high-temperature diamond synthesis (Reutsky et al. 2008; Stachel et al. 2009). Mikhail and Howell (2016) suggested that nitrogen can partition either compatibly or incompatibly during diamond growth, depending on nitrogen speciation (N_2, NH_3 or NH_4^+) and how the different nitrogen species dissociate to create the transient monatomic nitrogen necessary for incorporation into the diamond lattice. Nitrogen speciation, in turn, depends on oxygen fugacity (Li and Keppler 2014) and, for aqueous diamond-forming fluids, on pH (Mikhail and Sverjensky 2014). Empirical evidence from fractionation trends, however, suggests that nitrogen is compatible under both strongly reducing (CH_4 as the principal carbon species: $K_N = 2.0$–3.5; Thomassot et al. 2007; Wiggers de Vries et al. 2013; Smit et al. 2019a) and more oxidizing conditions (carbonate or CO_2 as the principal carbon species: $K_N = 2.0$–4.4; Wiggers de Vries et al. 2013; Petts et al. 2015). The K_N-values in these studies were all estimated assuming diamond formation from fluids composed entirely of a single carbon species. Based on the speciation of fluids in the system COH (Zhang and Duan 2009) and the redox stability of pure carbonate melts versus dilute carbonate components (dilution extends stability to lower oxygen fugacities; Stagno and Frost 2010), diamond formation from a single pure carbon species is considered highly unlikely. As a consequence, the above values for K_N represent minimum estimates (Petts et al. 2015). For example, the water-rich (up to ~ 98% H_2O) COH fluids predicted to exist in diamond-stable subcratonic mantle near ~ 150 km depth require K_N-values > 70 to account for the observed N-content fractionation trends (Stachel et al. 2017). The extreme average enrichment of nitrogen in diamond relative to depleted mantle is also circumstantial evidence for the compatible nature of nitrogen in diamond.

Early SIMS-based studies of internal co-variations of $\delta^{13}C$ and N-content within individual diamonds (e.g., Fitzsimons et al. 1999; Harte et al. 1999; Zedgenizov and Harte 2004) principally focused on differences between successive growth zones and were limited by the comparatively low analytical precision (~0.6–1.4‰, 2 sigma) of $\delta^{13}C$ measurements using single-collector instruments. These studies indicated large variability in nitrogen content among different growth zones, accompanied by (within uncertainty) near constant or uncorrelated and abruptly changing carbon isotope compositions. With the advent of multi-collector SIMS (e.g., Craven et al. 2009; Smart et al. 2011) analytical uncertainty improved significantly (< 0.2‰, 2 sigma), allowing for subtle variations in carbon isotope composition to be detected, but the focus remained on tracing variations in $\delta^{13}C$-values and N-content from core to rim across multiple growth zones (e.g., Peats et al. 2012; Palot et al. 2013). Growth zone boundaries visible in cathodoluminescence (CL) images, however, represent discontinuities that may relate to abrupt fluctuations in fluid composition (fluid re-charge) and/or physical conditions during continued diamond growth. Alternatively, there may have been temporal gaps in diamond growth of unknown duration with re-initiation of diamond growth from new, potentially completely unrelated fluid pulses. Consequently, smooth continuous trends characterized by correlated, outward evolving carbon isotope compositions and nitrogen concentrations can typically only be detected on the level of individual growth zones with internally fairly homogenous appearance in CL images (first documented by Smart et al. 2011). Despite a series of subsequent studies, using multi-collector SIMS and also a step-wise oxidation approach (Mikhail et al. 2014), the number of samples with convincing continuous co-variations in $\delta^{13}C$ and N-content has remained extremely small, indicating that Rayleigh fractionation during the growth of individual diamonds is relatively rare. Nevertheless, smooth trends of both outward increasing (Smart et al. 2011; Wiggers de Vries et al. 2013; Smit et al. 2016) and decreasing (Wiggers de Vries et al. 2013; Reutsky et al. 2017; Smit et al. 2019a) $\delta^{13}C$-values associated with decreasing nitrogen contents have been documented. The interpretation of such trends is discussed further below (see section *Origin of covariations among* $\delta^{13}C$, $\delta^{15}N$ *and nitrogen content*).

On the level of our worldwide diamond database, $\delta^{13}C$ and nitrogen content are uncorrelated (Fig. 12). For the peridotitic suite, a comparatively small variance in carbon isotope composition relative to eclogitic diamonds creates a tight cluster with highly variable N-contents (below detection to 2390 at.ppm) mainly falling between −7 and −3‰ (Fig. 12). Over this $\delta^{13}C$-range, the moving median and average nitrogen contents drop by 80–90 at.ppm towards both sides of their respective maxima at ~ 5‰ (170 and 260 at. ppm nitrogen, respectively; see Fig. 13). If this drop was related to Rayleigh fractionation during diamond precipitation (Stachel et al. 2009), then based on the K_N-values discussed above, N-contents should fall more sharply (between 50 and 99%, depending on carbon speciation and exact choice of K_N; see fractionation trends in Fig. 12). If unrelated to fractionation, the maximum in median and average nitrogen content at $\delta^{13}C$ ~ 5‰ implies that fluids carrying the putative mantle carbon component are also the most enriched in nitrogen. For diamonds of the eclogitic suite, Stachel and Harris (1997) observed a decrease in maximum nitrogen content as $\delta^{13}C$ decreases away from the mantle value (−5‰). This relationship was subsequently formalized by Cartigny et al. (2001) as the "limit sector" and interpreted as resulting from Rayleigh fractionation during outgassing of CO_2 from carbonated melts (causing depletion of ^{13}C in the residual melt) and associated loss of nitrogen from the melt. Subsequent studies, e.g., on eclogitic suite diamonds from the Olenek River (Yakutia; Shatsky et al. 2014) and micro-diamonds from the Orapa Mine

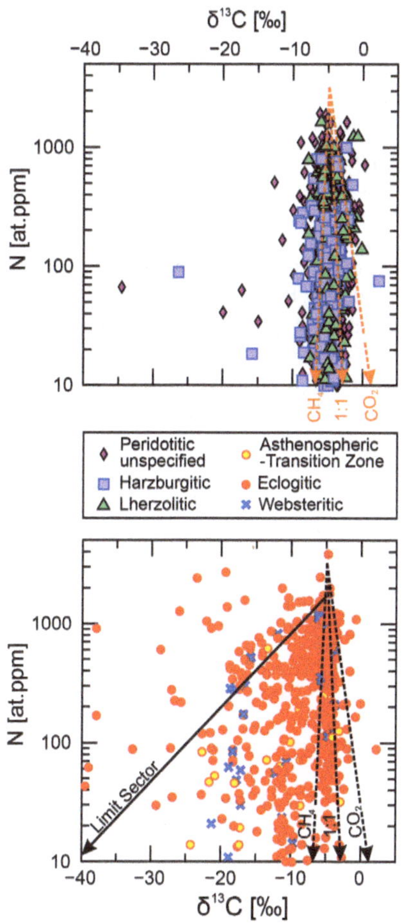

Figure 12. Nitrogen content (at.ppm) versus carbon isotope composition for inclusion-bearing diamonds **(top)** of the peridotitic suite (harzburgitic, lherzolitic and unspecified paragenesis shown separately) and **(bottom)** of the eclogitic, websteritic, and ("eclogitic") asthenospheric-transition zone suites. The **dashed lines** indicate Rayleigh fractionation trends calculated for fluids made up entirely of diamond-forming carbon species (pure CH_4, a 1:1 mix of CH_4 and CO_2, and pure CO_2), using diamond–fluid isotope fractionation factors appropriate for a temperature of 1150 °C (approximate average temperature of diamond formation) and a diamond–fluid nitrogen partition coefficient (K_N) of 4.4 (taken from Petts et al. 2015). The **solid black line** is the "limit sector" of Cartigny et al. (2001) and corresponds to a fractionation trend for a carbonated diamond-forming fluid/melt caused by fluid loss.

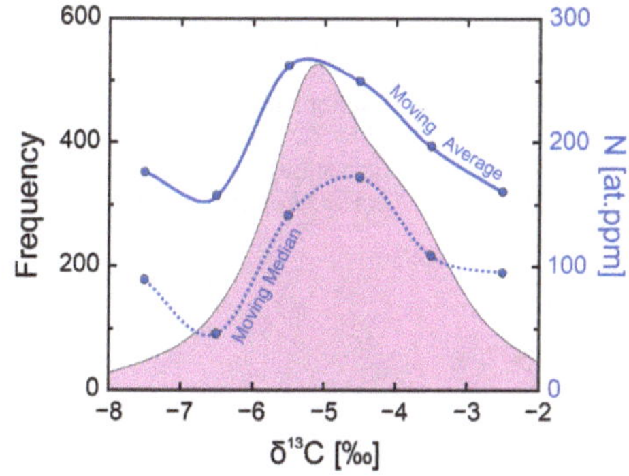

Figure 13. Kernel density estimation of the distribution in carbon isotope composition ($\delta^{13}C$) for diamonds worldwide containing inclusions of the peridotitic suite. The **blue curves** represent variations in nitrogen concentration (at.ppm, axis on right) with $\delta^{13}C$. The **solid curve** and **filled blue circles** are average values, the **dashed line** and **filled blue circles** are median values, calculated in one permille increments.

(Chinn et al. 2018; not included in Fig. 12), however, contradicted the limit sector relationship by showing high nitrogen contents (up to 3600 at.ppm) at $\delta^{13}C$-values below −20‰. Overall, the lack of correlation between carbon isotope composition and nitrogen content for both the worldwide peridotitic and eclogitic suites suggests that either (1) carbon and nitrogen are at best weakly coupled in diamond-forming fluids, (2) variably evolved fluids carry distinct nitrogen contents, (3) a range of K_N-values exists, or (4) that precipitation/break-down of other N-bearing mantle phases (e.g., clinopyroxene or phlogopite) accompanies diamond growth.

$\delta^{15}N$ and nitrogen content

For the worldwide database, nitrogen in diamond and its isotopic composition (Fig. 14) show a surprising lack of coupling. Under the precondition that natural diamond-forming processes do not involve a highly variable elemental or isotopic partitioning behavior of nitrogen (i.e., large variations in K_N-value or N-isotope fractionation factor), this decoupling suggests that derivation of diamond-forming fluids from a single source (mantle or subducted crust), with constant nitrogen concentration and $\delta^{15}N$-values, may be excluded. Additionally, if the large variability of nitrogen concentration and isotopic composition was due a single process, e.g., Rayleigh fractionation during diamond growth (see below), then a highly correlated behavior would be expected. A Rayleigh fractionation trend, induced by either diamond crystallization (Fig. 14) or nitrogen loss from a melt, clearly cannot explain the variability of diamonds of either the peridotitic or eclogitic suites.

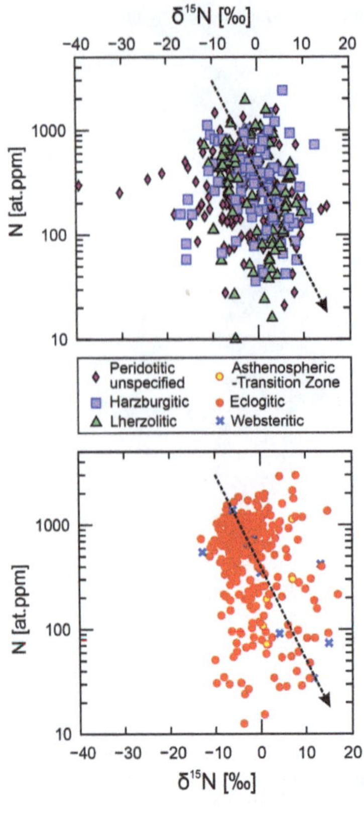

Figure 14. Nitrogen content (at.ppm) versus nitrogen isotope composition for inclusion-bearing diamonds **(top)** of the peridotitic suite (harzburgitic, lherzolitic and unspecified paragenesis shown separately) and **(bottom)** of the eclogitic, websteritic, and ("eclogitic") asthenospheric-transition zone suites. The **dashed line** indicates a Rayleigh fractionation trend using a diamond–fluid N-isotope fractionation factor appropriate for a temperature of 1150 °C (approximate average temperature of diamond formation) and a diamond–fluid nitrogen partition coefficient (K_N) of 4.4 (Petts et al. 2015).

For the peridotitic suite, the data form a cloud from low to high $\delta^{15}N$-values and nitrogen contents (Fig. 14). From this cloud extends an array of peridotitic diamonds of unspecified paragenesis with intermediate nitrogen contents (251 to 385 at.ppm) towards extremely ^{15}N-depleted compositions ($\delta^{15}N$ < 20‰). If these isotopic compositions indeed represented mixing with a primordial component similar to enstatite chondrite (Cartigny et al. 1997; Palot et al. 2012), then this primordial component would be characterized by N-contents similar to the growth medium of "regular" diamonds.

For diamonds of the eclogitic suite, samples with broadly mantle-like $\delta^{15}N$ (−10 to 0‰) cluster at fairly high nitrogen concentrations, from which the data spread to higher $\delta^{15}N$ and/or lower N-contents. Two outliers, representing the highest nitrogen contents in the data set (~ 2900 at.ppm), are observed at $\delta^{15}N$ of +4.3‰ and +7.7‰ and relate to diamonds from Wellington (SE Australia) and Panda.

Considering diamonds for individual suites/parageneses on the level of individual mines does not significantly change this general picture, although, on the local level some eclogitic diamonds show a more consistent behavior in $\delta^{15}N$–N space with clear differences between certain occurrences (e.g., Jwaneng vs. Orapa mines). Hogberg et al. (2016) undertook a comprehensive SIMS-based study of 94 small diamonds from the Chidliak CH-6 and CH-7 kimberlites, producing 251 paired nitrogen content and $\delta^{15}N$ spot analysis. Although no mineral inclusions were recovered during this study, Chidliak diamonds are presumed to predominantly derive from eclogitic substrates. In diamonds with significant internal variations in nitrogen content and $\delta^{15}N$, subparallel trends of decreasing nitrogen content and increasing $\delta^{15}N$ were observed (Hogberg et al. 2016). Still, most of the population forms a cluster of mildly negative $\delta^{15}N$ at variable nitrogen contents and only a subset of diamonds fall on a positively correlated array in log(N)–$\delta^{15}N$ space. This latter group was interpreted to reflect limited degrees of Rayleigh fractionation during the waning stages of mantle fluid influx (Hogberg et al. 2016).

On the level of individual diamond plates, only two SIMS-based studies (Petts et al. 2015; Smit et al. 2016) observed smooth rim-ward trends of increasing $\delta^{15}N$ and decreasing nitrogen contents. The opposite observation, a trend of jointly decreasing $\delta^{15}N$ and nitrogen contents, was made for a set of cogenetic diamonds recovered from a single lherzolite xenolith from the Cullinan mine (Thomassot et al. 2007).

Based on the assumption that diamond nitrogen content decreases in the course of Rayleigh fractionation processes, associated with either melt degassing prior to diamond formation (Cartigny et al. 2001) or diamond formation in fluid-limited systems, high nitrogen concentrations in diamond should indicate primitive fluid compositions. The opposite, however, is not true and nitrogen-poor primitive components may exist but will exert only weak leverage on composite $\delta^{15}N$-values during fluid mixing. The variance in the nitrogen content of diamond at any given $\delta^{15}N$-value is large and, consequently, to reveal underlying trends we use both moving (1‰ steps) median N-contents calculated over 2‰ intervals in $\delta^{15}N$ and 5th- or 6th-order polynomials fitted through the nitrogen content versus $\delta^{15}N$ data (Fig. 15 and 16). The nitrogen isotope compositions of identified primitive (high-N) components as recorded by diamond are offset from the true value by the appropriate nitrogen isotope fractionation factors (depending on nitrogen speciation and temperature) and true values likely will be several permille higher (see section *Equilibrium nitrogen isotope fractionation factors related to diamond*).

For the peridotitic suite, a nitrogen-rich component with $\delta^{15}N$ ~ −3‰ is apparent, falling between the principal peak at +2.0‰ and the shoulder at −4.1‰ in the $\delta^{15}N$ kernel density distribution (Fig. 15). A second primitive component, with a $\delta^{15}N$ value of about −10 to −12‰, falls outside the principal $\delta^{15}N$-distribution and, consequently, is poorly represented in the data set. Analyzing the peridotitic data set on the paragenesis level, the nitrogen-rich component at about –3‰ (to –2‰) is also present in the lherzolitic diamonds while the ^{15}N-depleted component at about –10 to –12‰ is clearly seen in harzburgitic diamonds (Fig. 15). The polynomial fit of the harzburgitic N-content and $\delta^{15}N$ data suggests a second nitrogen-rich component at about +3‰, overlapping with the mode in the Kernel density distribution for harzburgitic diamonds at +2.1‰, but this component is not clearly evident in the moving medians. The lherzolitic diamond distribution suggests a further nitrogen-rich component at the ^{15}N-enriched end of the frequency distribution (> +7‰; Fig. 15). Combining the information on nitrogen-rich primitive components for the various groups of peridotitic diamonds, a low frequency ^{15}N-depleted component (–10 to –12‰), a higher frequency component with a $\delta^{15}N$-value of about –3‰ and a possible strongly ^{15}N-enriched component (> +7‰) are indicated. The negatively skewed unimodal distribution in $\delta^{15}N$ of harzburgitic diamonds could be explained by a dominant nitrogen component at ~ +3‰ (coinciding with the $\delta^{15}N$-value of plume derived nitrogen; Dauphas and Marty 1999), with the negative skew of the distribution resulting from mixing with a potentially primordial nitrogen-rich component with $\delta^{15}N < -10‰$ (Cartigny et al. 1997). For the lherzolitic suite the nitrogen-rich component at about –3 to –2‰ falls between the two $\delta^{15}N$ modes, suggesting that the principal mode at +2.9‰ represents a nitrogen-poor component (in the +2 to +6‰ range, Fig. 15) with the ^{15}N-depleted smaller mode representing either a less N-rich component, defined by the moving medians to lie at –7‰, or fairly constant mixing between the nitrogen rich component at –3 to –2‰ with a dominant nitrogen poor component with low $\delta^{15}N$ (< –10‰).

For the eclogitic suite, a ^{15}N-depleted, nitrogen-rich component (localized at ~ –6‰ by the 5th order polynomial) and a much less abundant, even more nitrogen-rich component at ~ +7‰ are clearly present (Fig. 16).

$\delta^{13}C$ and $\delta^{15}N$

For peridotitic suite diamonds worldwide, $\delta^{13}C$-values fall mostly in the mantle range whilst $\delta^{15}N$-values vary by almost 55‰, creating a near-horizontal array in Figure 17. The scatter in $\delta^{13}C$ at any given $\delta^{15}N$ is too large, even on the level of individual parageneses, to allow for statistically significant covariations between the two parameters to be determined; separate regressions for lherzolitic and harzburgitic paragenesis diamonds indicate shallow positive and negative (apparent) slopes, respectively, but with a low $r^2 = 0.14$ in both cases. The peridotitic data set can be interpreted in terms of two end-member models (or a combination thereof): (1) Rayleigh isotope fractionation of fluids carrying mantle-derived carbon and nitrogen creates the main portion of the array (see section *Origin of covariations among $\delta^{13}C$, $\delta^{15}N$ and nitrogen content*). Especially for fluids carrying a CO_2–CH_4 mix or CH_4 as the diamond-forming carbon species, the effects of carbon isotope fractionation would disappear in the noise of the data set. The outliers with $\delta^{13}C$-values below the mantle array and $\delta^{15}N$-values below about –10‰, however, cannot be explained by this model (Fig. 17). (2) The $\delta^{13}C$–$\delta^{15}N$ array represents mixing between several reservoirs (see section *Origin of large ranges in $\delta^{13}C$ and $\delta^{15}N$*; see also Mikhail et al. 2014), involving a mantle component, several distinct subducted components (with low and high $\delta^{13}C$ and low and high $\delta^{15}N$) and a potentially primordial component with extreme ^{15}N depletion (explaining values down to –39.4‰). To create large variations in $\delta^{15}N$ without much effect on $\delta^{13}C$ requires that the subducted components are nitrogen-rich, consistent with indications for nitrogen-rich components with positive $\delta^{15}N$ discussed in the preceding section. Considering individual mines would not change the above pictures. A SIMS-based study (Smit et al. 2016) of diamond plates of inferred peridotitic origin, however, revealed rare smooth internal trends of rim-ward increasing $\delta^{13}C$ and $\delta^{15}N$. A fractionation trend of increasing $\delta^{15}N$ but decreasing $\delta^{13}C$ was observed for multiple lherzolitic diamonds from a single xenolith (Thomassot et al. 2007). In both cases, diamond precipitation is interpreted as driving the recorded N-content and isotope variability.

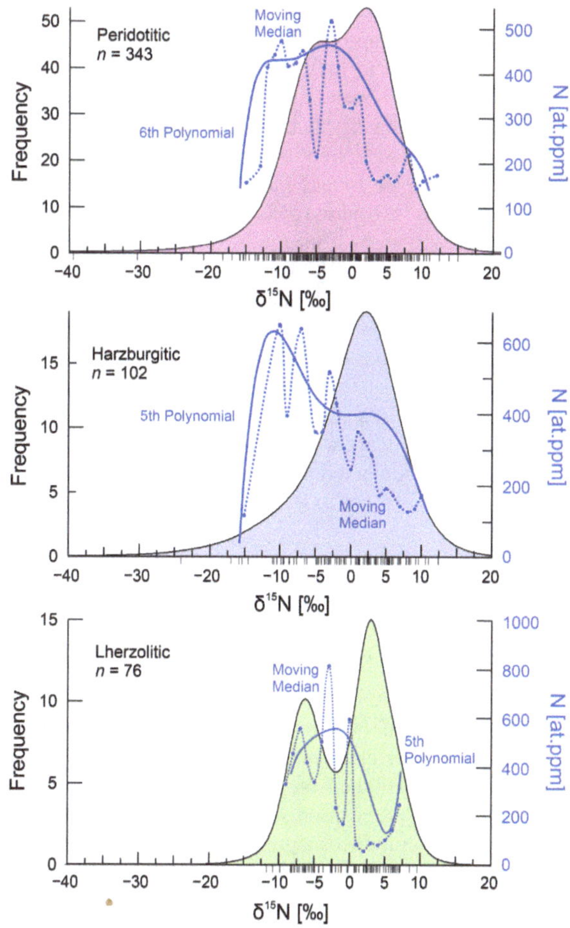

Figure 15. Kernel density estimation of the distribution in nitrogen isotope composition ($\delta^{15}N$) for diamonds worldwide containing inclusions of the peridotitic suite **(top)** and the harzburgitic (**middle**) and lherzolitic **(bottom)** parageneses (which both are also included in the peridotitic suite data set). The **blue curves** represent variations in nitrogen concentration (at.ppm, axis on right) with $\delta^{15}N$. The **solid curve** is a higher order polynomial fit; the **filled blue circles** connected by a **dashed line** are median values, typically calculated in one permille steps for increments that are two permille wide. Only at the two ends of the distributions, where data density strongly decreases, is a wider step and increment size used (typically steps of two and increments of four permille). **Black dashes** below the *x*-axis indicate individual data points and, together with the Kernel density estimations, indicate how robust median values and polynomial fits are in particular $\delta^{15}N$ ranges.

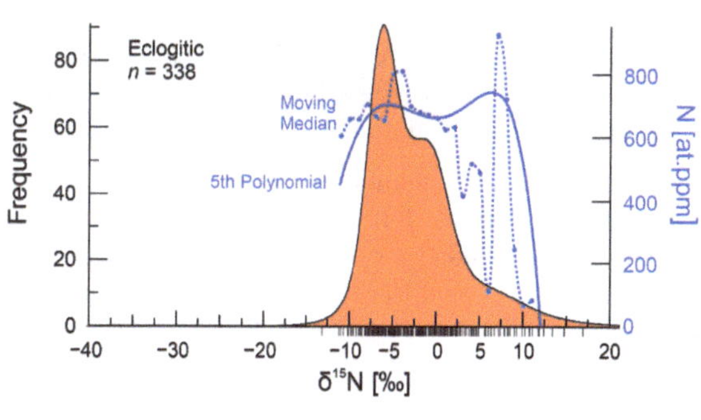

Figure 16. Kernel density estimation of the distribution in nitrogen isotope composition ($\delta^{15}N$) for diamonds worldwide containing inclusions of the eclogitic suite. The **blue curves** represent variations in nitrogen concentration (at.ppm, axis on right) with $\delta^{15}N$. The **solid curve** is a higher order polynomial fit; the **filled blue circles** connected by a **dashed line** are median values, calculated in one permille steps for bins that are two permille wide. **Black dashes** below the *x*-axis indicate individual data points and, together with the Kernel density estimations, indicate how robust median values and polynomial fits are in particular $\delta^{15}N$ ranges.

For diamonds of the eclogitic suite, similar to their peridotitic counter-parts, a near-horizontal array with mantle-like carbon isotope compositions and $\delta^{15}N$ ranging between about −10 and +5‰ is apparent (Fig. 17). In contrast to the peridotitic suite, however, there are a number of samples (including all "eclogitic" diamonds of sublithospheric origin) that are moderately to strongly ^{13}C-depleted. To explain this data distribution through mixing, requires a mantle-like component and a strongly ^{13}C-depleted and ^{15}N-enriched component with variable nitrogen contents (Fig. 17; see section *Origin of large ranges in* $\delta^{13}C$ *and* $\delta^{15}N$ for details). On the level of individual mines, the large sample sets with paired $\delta^{13}C$-$\delta^{15}N$ analyses for Jwaneng (n = 129) and Orapa (n = 49) both display a continuum from low (~ −20‰) to mantle-like (−5 ± 2‰; Orapa) or slightly elevated (−2.0‰; Jwaneng) $\delta^{13}C$-values. For Jwaneng, however, there is only a mild overall increase in $\delta^{15}N$ from samples with mantle-like $\delta^{13}C$ (median $\delta^{15}N$-value −5.4‰, average −4.5‰) to strongly ^{13}C-depleted ($\delta^{13}C < -15$‰; median $\delta^{15}N$-value −1.8‰, average −0.9‰) compared to Orapa (mantle-like samples: median $\delta^{15}N$-value −2.7‰, average 0.9‰; strongly ^{13}C-enriched samples: median +6.0‰, average +6.2‰). A SIMS-based study on a diamond plate of the eclogitic suite observed a smooth trend of rim-ward increasing $\delta^{13}C$ and $\delta^{15}N$ (combined with decreasing nitrogen content), which was interpreted as a Rayleigh fractionation trend (Petts et al. 2015). The study documented a one order of magnitude higher sensitivity of nitrogen isotopes compared to carbon isotopes to Rayleigh fractionation.

ORIGIN OF COVARIATIONS AMONG $\delta^{13}C$, $\delta^{15}N$ AND NITROGEN CONTENT

Correlated variations in C- and N-isotope composition and N-content observed within individual diamonds or within a suite of diamonds have been attributed to several different models:

1. An *equilibrium fractional crystallization model* whereby the precipitation of diamond and the attendant partitioning of isotopes or elements between the diamond and its parental fluid/melt causes progressive shifts in the isotopic and chemical composition of this fluid and, in turn, shifts in the composition of the later-formed diamond (i.e., a model where diamond precipitation induces significant isotope and chemical variability in the residual fluid, an active recorder). Notably, in this model, each growth zone of the diamond forms in isotopic and chemical equilibrium with the fluid present at the time of its growth.
2. A *kinetic fractional crystallization model*, which is analogous to the equilibrium model described above except that the isotopic and chemical partitioning between the diamond rim and growth medium reflects a kinetic fractionation rather than thermodynamic equilibrium.
3. A *mixing model* in which variations in the diamond's isotopic and elemental composition is attributed to the mixing of fluids derived from two or more sources with different isotopic compositions and N concentrations.

Which of these three models applies to a particular diamond or suite of diamonds affects interpretations regarding the nature of the diamond-forming fluid (e.g., was the fluid oxidizing or reducing?) and also affects inferences about the source(s) of these fluids. In this section, we begin with a detailed discussion of equilibrium C- and N-isotope partitioning between diamond and the likely diamond growth media and also summarize what we consider to be the best currently available diamond-related equilibrium fractionation factors for these two isotope systems. We then discuss kinetic isotope fractionation and whether isotope fractionation that occurs during the growth of natural diamonds more closely reflects an equilibrium or a kinetic process. Finally, we consider the role of fluid mixing in producing the isotopic and elemental variations observed in natural diamonds.

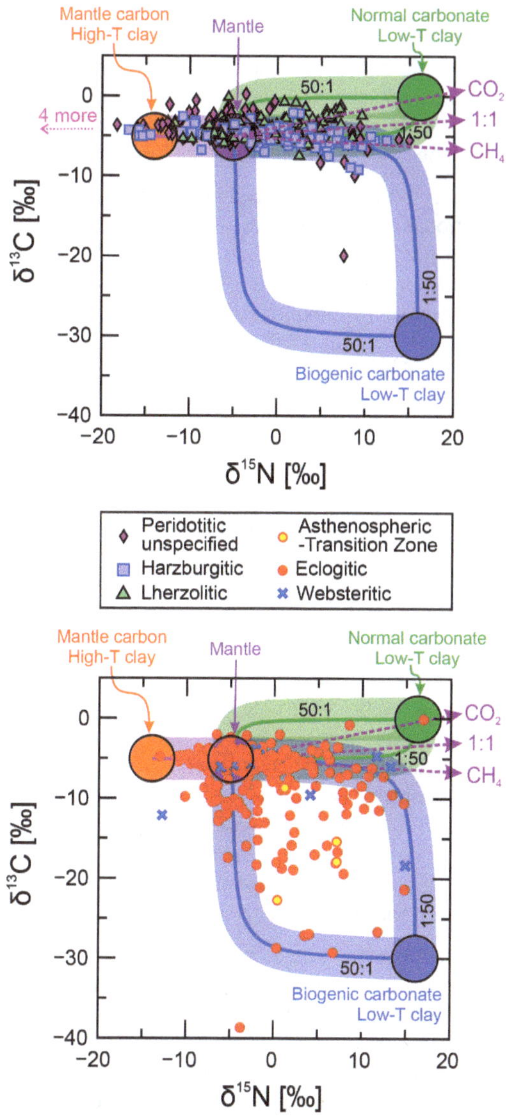

Figure 17. Carbon versus nitrogen isotope composition for inclusion-bearing diamonds **(top)** of the peridotitic suite (harzburgitic, lherzolitic and unspecified paragenesis shown separately) and **(bottom)** of the eclogitic, websteritic, and ("eclogitic") asthenospheric-transition zone suites. The four end-member components shown are taken from Li et al. (2019) and correspond to a mantle component and three components hosted in subducted former altered oceanic crust: (1) in normal marine carbonate, which precipitated in isotopic equilibrium with dissolved inorganic carbon (DIC) in seawater, and nitrogen in low-temperature clay; (2) in biogenic carbonate, again together with nitrogen in low-temperature clay; (3) in AOC that was altered at relatively high temperatures, containing negligible C but ^{15}N-depleted clay. Mixing lines between the mantle component and the AOC components correspond to high nitrogen content ($[N/C]_{Mantle}/[N/C]_{AOC}$ of 1:50) and low nitrogen content (50:1) in the subducted components (Li et al. 2019). The **purple dashed lines** indicate Rayleigh fractionation trends calculated for fluids made up entirely of diamond-forming carbon species (pure CH_4, a 1:1 mix of CH_4 and CO_2, and pure CO_2), using diamond–fluid isotope fractionation factors appropriate for a temperature of 1150 °C (approximate average temperature of diamond formation) and a diamond–fluid nitrogen partition coefficient (K_N) of 4.4 (taken from Petts et al. 2015).

High-temperature stable isotope fractionation

Only a brief introduction to equilibrium stable isotope fractionation theory is given here. The reader is referred to the seminal paper of Urey (1947) and the review papers by Richet et al. (1977) and Chacko et al. (2001) for a more detailed description of theory and the calculation of isotope reduced partition function ratios from which isotope fractionation factors can be calculated.

Chemical bonds involving the heavier and lighter stable isotopes of a particular element have slightly different vibrational energies, which cause co-existing C- or N-bearing phases or species (e.g., diamond and a fluid) to have distinct isotopic compositions. These differences in isotopic composition are expressed in terms of a fractionation factor, α. For example, the C-isotope fractionation between diamond and CO_2 can be expressed as $\alpha_{diamond\text{-}CO_2} = (^{13}C/^{12}C)_{diamond}/(^{13}C/^{12}C)_{CO_2} = (\delta^{13}C_{diamond} + 1000)/(\delta^{13}C_{CO_2} + 1000)$. Because α-values are numbers close to 1, fractionation factors are commonly reported in terms of $1000 \ln \alpha_{diamond\text{-}CO_2} = \Delta^{13}C_{diamond\text{-}CO_2}$. In this chapter, we will refer to fractionation factors in one of the latter two forms.

At high temperatures, where thermodynamic equilibrium is more likely to prevail, the fractionation factor, Δ, scales approximately linearly with $1/T^2$; the higher the temperature, the smaller the isotope fractionation, which eventually becomes zero at infinite temperature. Stable isotope fractionation is commonly assumed to be largely independent of pressure over the pressure ranges of the crust and upper mantle because the volume change associated with isotope substitution is small (Clayton et al. 1975). Fractionations at the extreme pressures of the lower mantle may in principle be somewhat different than those in the crust or upper mantle but the magnitude of any potential pressure effect decreases markedly with increasing temperature (Polyakov and Kharlashina 1994; Horita and Polyakov 2015). As such, extrapolation of fractionations determined for low-pressure conditions to deep mantle conditions is unlikely to lead to large errors because of the very high temperatures associated with the deeper parts of the mantle. Finally, chemical bonds in oxidized phases are generally stronger than those in reduced phases and the heavy isotope favours substances with stronger bonds. Consequently, at equilibrium, the most reduced species (carbides, nitrides) are more depleted in the heavy isotope of an element (^{13}C relative to ^{12}C, ^{15}N relative to ^{14}N) than more oxidized species (elemental forms such as diamond/graphite, N_2-gas), which in turn are more depleted than the most oxidized forms (carbonates, CO_2, nitrates).

Carbon and nitrogen isotope fractionation factors relevant to diamonds have been estimated using several different methods including laboratory experiments, theoretical calculations, and the analysis of natural samples. Each of these methods has advantages and disadvantages. Laboratory experiments are the most direct and in principle most reliable method for determining fractionation factors but are limited in a number of ways. Experiments can typically only be conducted over a restricted temperature range so as not to exceed the thermal stability limit of the phases of interest. At the same time, the experiments need to be carried out at temperatures high enough to permit significant isotopic exchange to occur between those phases. In many cases, these experimental fractionation data, obtained over a limited temperature range, must then be extrapolated outside that range for their application to natural samples. Another challenge for isotope exchange experiments, especially those involving the isotopically refractory mineral diamond, is the demonstration of the attainment of isotopic equilibrium during the course of the experiment.

Two criteria must be met to unequivocally prove that equilibrium was established:

Firstly, the same phases should be present at the start and end of the experiment, i.e., isotopic exchange should be accomplished either by diffusional processes or recrystallization of pre-existing phases in the experimental charges rather than by the growth of new phases. Although the synthesis of new phases during the experiment greatly facilitates isotopic exchange, the free energy changes associated with these synthesis reactions are orders of magnitude greater than those associated with isotope exchange reactions, which can result in unidirectional, kinetic isotope effects and produce spurious, non-equilibrium fractionations (Matsuhisa et al. 1978; O'Neil 1986; Chacko et al. 2001).

Secondly, experiments should be conducted in such a way that the equilibrium fractionation factor can be 'reversed'. That is, at any given temperature, two or more companion experiments should be carried out where the initial isotopic fractionations between the phases of interest lie on opposite sides of the equilibrium fractionation factor (i.e., initial fractionations both smaller and larger than the equilibrium value). Measuring the same final fractionation factor in all the companion experiments is the clearest demonstration possible that true isotopic equilibrium has been achieved.

It is important to note that fractionation factors obtained in experiments that do not meet these two criteria should be regarded with caution, as they may not record equilibrium values. To date, no carbon or nitrogen isotope exchange experiments directly involving diamond have met these criteria, which reflects the difficulty in conducting laboratory experiments with a mineral with such slow diffusion rates for C and N (Koga et al. 2003; Harte et al. 2009).

Theoretical methods, following the protocols originally developed by Urey (1947) for ideal gases and then later extended to solids (e.g., Bottinga 1968; Kieffer 1982), have also been used to determine diamond-related fractionation factors. Unlike experiments, the calculations are not restricted to a relatively narrow temperature range. However, the calculations, particularly for solid phases, require a number of simplifying approximations, the accuracy of which must be verified independently (e.g., through well-constrained experimental data). Despite this limitation, an especially useful aspect of theoretical calculations is their ability to accurately predict the temperature dependence of fractionation factors. Thus, although theory may or may not accurately capture the absolute magnitude of fractionation factors, it does provide robust constraints on the basic shape of fractionation factor versus temperature curves. As such, theoretical constraints are essential for reliable interpolation of fractionations within the temperature range of experimental studies and particularly the extrapolation of experimental data outside that temperature range.

A third method of obtaining fractionation factors related to diamonds is through the assessment of isotopic evolution trends measured in natural diamonds (e.g., Javoy et al. 1986; Thomassot et al. 2007; Petts et al. 2015). The distinct advantage of this approach is that nature has had much longer timescales to establish equilibrium during diamond growth than is possible in the laboratory. Unfortunately, the natural sample approach often lacks key information, including the temperature at which the diamonds used for calibration formed and the chemical composition of the diamond-growth medium. Such information is required for a full understanding of the fractionation factor that is being estimated. In some cases, especially in connection with nitrogen isotope fractionations, natural sample approaches take on greater importance as no reliable experimental studies or rigorous theoretical calculations have yet been done.

Equilibrium carbon isotope fractionation factors related to diamond

The C-isotope fractionation factors most relevant to this chapter are those between diamond and possible diamond-growth media, including CO_2-, CH_4- or carbonate-bearing fluids/melts and, at more reducing deeper mantle conditions, carbide-bearing melts/fluids and minerals.

Experiments. To our knowledge, there are only two published experimental studies directly involving diamond that attempted to calibrate equilibrium C-isotope fractionation factors for that mineral (Reutsky et al. 2015a; Bureau et al. 2018). In the Reutsky et al. (2015a) experiments, the breakdown of a sodium oxalate starting material at high *P*–*T* (6.3–7.5 GPa, 1300–1700 °C) in the diamond stability field produced a mixture of diamond, sodium carbonate melt and CO_2. The authors inferred that CO_2, which was present as a separate fluid phase after the experiment, was entirely dissolved in the carbonate melt at experimental conditions. Accordingly, 1/3 of the total carbon in the carbonate melt/fluid phase was dissolved CO_2. Reutsky et al. (2015a) interpreted the isotopic data from four of their experiments to record equilibrium C-isotope fractionations between carbonate fluid/melt and diamond, which could be represented by the equation: $\Delta^{13}C_{\text{carbonate melt/fluid–diamond}} = 7.38 \times 10^6/T^2$ (T in Kelvin).

The starting materials for the Bureau et al. (2018) experiments consisted of varying mixtures of oxides, carbonates and seed diamonds along with H_2O and dissolved salts. These were subjected to P–T conditions in the diamond stability field of 7.5 GPa and 1400 °C. The experimental run products, which were complex and included newly-grown diamond, were analyzed isotopically by NanoSIMS. It should be noted that this technique allows very small sample volumes to be analyzed but has large analytical uncertainties of > 1‰. Bureau et al. reported a $\Delta^{13}C_{\text{carbonate fluid–diamond}} = +2.8‰$ at 1400 °C. Given that the Reutsky et al. (2015a) and Bureau et al. (2018) experiments did not meet either of the two criteria required for demonstrating the establishment of isotopic equilibrium (new phases must not form; experiments should be reversed), the accuracy of the resulting fractionation factors must be evaluated in light of theoretical calculations and data derived from better-constrained experimental studies on related systems.

Well-constrained C-isotope exchange experiments have been conducted in three systems indirectly related to diamond formation, CO_2–calcite (Chacko et al. 1991; Scheele and Hoefs 1992; Rosenbaum 1994), calcite–graphite (Deines and Eggler 2009; Kueter et al. 2019a) and CO_2–CH_4 (Horita 2001; Kueter et al. 2019b). In terms of the CO_2–calcite system, the experimental study that most closely met the criteria for the demonstration of equilibrium is that of Rosenbaum (1994). Similarly, the high-temperature (1200–1400 °C) calcite–graphite experiments of Deines and Eggler (2009) used those two minerals as starting materials (i.e., no growth of new minerals during the experiments) and reversed the equilibrium factor with companion experiments at each temperature. The experiments of Kueter et al. (2019a) in the carbonate–graphite system were of a different style than those of Deines and Eggler in that they used starting materials comprising organic matter (tartaric acid and sucrose) and a mixture of $CaCO_3$ and Na_2CO_3, which decomposed under experimental conditions to form graphite and carbonate melt. As such, these are synthesis experiments that cannot in and of themselves unambiguously demonstrate that true isotopic equilibrium has been achieved. The two experimental studies in the CO_2–CH_4 system collectively investigated a broad range of temperatures (200–1000 °C), using a Ni catalyst to facilitate isotopic exchange between the gas species (Horita 2001; Kueter et al. 2019b).

C-isotope experiments were also carried out between graphite and Fe_3C (Satish-Kumar et al. 2011) and between Fe_3C and Fe-C melt (Reutsky et al. 2015b), with some of the latter experiments crystallizing diamond in addition to Fe_3C. Both of these experimental studies involved the crystallization of new phases during the course of the experiments, which leads to some ambiguity as to whether the measured isotopic fractionations represent equilibrium values. A refitting of the Satish-Kumar et al.'s (2011) data points to a straight line in which the fractionation factor is forced to go to a value of 0 at infinite temperature results in the following equation: $\Delta^{13}C_{\text{graphite-}Fe_3C} = 12.2 \times 10^6/T^2$. Reutsky et al. (2015b) measured a $\Delta^{13}C_{Fe_3C\text{-FeC melt}} \sim +2‰$ at 6.3 GPa and 1400 °C. They also measured a $\Delta^{13}C_{\text{diamond-}Fe_3C} = \sim +5‰$ but, because their experiments were intentionally polythermal (1600–1400 °C) and diamond formed earlier in the crystallization sequence than Fe_3C, the authors did not interpret this +5‰ value to reflect an equilibrium fractionation factor.

Theoretical calculations. Theory enables the calculation of isotope reduced partition function ratios, commonly referred to as β-factors, for individual fluid or mineral species. The β-factor calculations either use measured spectroscopic data for species as input for statistical mechanical computations or are based on a first-principles, lattice dynamics approach. These β-factors can then be used to compute isotopic fractionation factors between species. More specifically, the fractionation between species *a* and *b* at a given temperature can be represented as $\alpha_{a\text{-}b} = \beta_a / \beta_b$ or as $\Delta_{a\text{-}b} = 1000 \ln \beta_a - 1000 \ln \beta_b$.

Many calculations of C-isotope β-factors have been done for CO_2 (e.g., Bottinga 1968; Richet et al. 1977; Chacko et al. 1991; Polyakov and Kharlashina 1995) and calcite (Bottinga 1968; Chacko et al. 1991; Deines 2004; Schauble et al. 2006) and have yielded quite comparable results (mostly within ~ 0.2‰ at high temperature), despite differences in input data and calculation methodology[4]. Deines (2004) and Schauble et al. (2006) also did β-factor calculations for two other carbonate minerals relevant to the mantle, magnesite and dolomite. Fewer high-temperature β-factor calculations have been done for CH_4 than for CO_2 but the two available studies yielded very comparable results (Bottinga 1969a; Richet et al. 1977). Here, we have chosen to use the calculations of Chacko et al. (1991) for CO_2 and calcite, Schauble et al. (2006) for magnesite and dolomite and Richet et al. (1977) for CH_4.

Two sets of C-isotope β-factor calculations have been done for graphite and diamond, a pioneering study by Bottinga (1969b) and a later study by Polyakov and Kharlashina (1995). Significantly different calculation methodologies and types of input data were used in the two studies but they produced very similar β-factors that are within 0.3 and 0.1‰ of each other for graphite and diamond, respectively, at $T \geq 800$ °C. Polyakov and Kharlashina (1994) investigated the pressure dependence of β-factors (the β-factors for all phases increase with pressure) and noted that β-factors for graphite are more strongly affected by pressure than those of either diamond or calcite. As a consequence, although diamond is slightly enriched in ^{13}C relative to graphite at ambient pressure, that small degree of ^{13}C partitioning is predicted to reverse at higher pressure (i.e., graphite becomes the more ^{13}C-enriched phase).

Recently, β-factor calculations have also been done for carbide minerals (Fe_3C and SiC–Horita and Polyakov 2015; Fe_3C–Liu et al. 2019). The two sets of calculations disagree somewhat in their β-factors for Fe_3C but both indicate that carbides should be strongly depleted in ^{13}C relative to all the other carbon species described above. For instance, Liu et al.'s (2019) calculations indicate $\Delta^{13}C_{MgCO_3\text{-}Fe_3C} = +6.1‰$ at 15 GPa and 1500 °C compared to $\Delta^{13}C_{CaCO_3\text{-}Fe_3C} = +6.7‰$ ($\Delta^{13}C_{MgCO_3\text{-}CaCO_3} = +0.3‰$ at 1500 °C) at that *P–T* condition according to Horita and Polyakov (2015). It is not clear which of these two studies provides the more accurate β-factors for Fe_3C but we have chosen here to use the values of Horita and Polyakov as only they provide equations for reproducing their β-factor calculations as a function of temperature.

Comparison of experiment and theory. The data from the experimental studies, particularly the well-constrained fractionation data in the CO_2–calcite, calcite–graphite and CO_2-CH_4 systems, can be used to assess the validity of the theoretical calculations of β-factors for CO_2, CH_4, calcite and graphite, and indirectly, also the reliability of the calculations for diamond. The theoretical calculations are in reasonable agreement with all the experimental data in the CO_2–calcite system but the best constrained experimental datum in that system, that at 900 °C by Rosenbaum (1994) (2.70 ± 0.18‰ (1σ)), is in near perfect agreement with the fractionation at that temperature (2.68‰) derived from β-factor calculations for CO_2 and calcite by Chacko et al. (1991). The excellent agreement between experiment and theory indicates that the approximations inherent to the calculations, especially that for the solid phase, calcite, are valid.

The correspondence between theory and experiment is not quite as good for the CO_2–CH_4 system. As noted by both Horita (2001) and Kueter et al. (2019b), a theoretically-derived fractionation curve based on the calculations of Richet et al. (1977) is similar to but displaced to slightly smaller fractionations than their experimental data. Here, we have left the β-factor calculations for CO_2 as is and applied a multiplicative factor (0.9846) to the β-factors of

[4] The β-factor calculations of Bottinga (1968) contained an error, which was corrected by Chacko et al. (1991). The corrected Bottinga (1968) calculations, those of Chacko et al. (1991) and Schauble et al. (2006) give virtually identical high-temperature C-isotope β-factors for calcite whereas those of Deines (2004) give values about 3% lower than the other studies cited here.

CH_4 so as to minimize the weighted sum of squares deviation of the calculations from the experimental data points. This small (~ 1.5%) correction to the β-factors of CH_4 improves the fit of the calculations to the experimental data but maintains the basic shape of the fractionation curve prescribed by theory.

Theory and experiment can also be compared in the calcite–graphite system. The reversed experiments of Deines and Eggler (2009), conducted at 2 GPa, indicate $\Delta^{13}C_{calcite\text{-}graphite}$ of +1.34 ± 0.17‰ and +1.09 ± 0.03‰ at 1200 and 1400 °C, respectively. These data correspond almost exactly with the fractionations of 1.35 and 1.05‰ obtained by combining the β-factors for calcite (Chacko et al. 1991) and graphite (Polyakov and Kharlashina 1995) calculated for a pressure of 2 GPa. The theoretical calculations also agree with the results of the 1 GPa synthesis experiments of Kueter et al. (2019a), which indicate fractionations of 1.4 ± 0.2‰ and 1.1 ± 0.2‰ at 1300 and 1500 °C, respectively, compared to calculated fractionations of 1.32 and 1.04‰ at 1 GPa and those temperatures. This excellent agreement between theory and experiment strongly suggests that theoretical calculations for graphite are accurate. Moreover, given that the calculation methodology and input data for the diamond calculations are completely analogous to those for graphite (Polyakov and Kharlashina 1995), it can reasonably be inferred that the β-factors for diamond are also likely to be accurate.

Given the demonstrated ability of the theoretical calculations to predict fractionation factors involving CO_2, calcite and graphite (and by analogy diamond), it is instructive to critically evaluate the diamond-related fractionation factors reported in the experimental studies of Reutsky et al. (2015a) and Bureau et al. (2018). The former study proposed the fractionation equation: $\Delta^{13}C_{carbonate\ melt/fluid\text{–}diamond} = 7.38 \times 10^6/T^2$ based on their experiments in which diamond crystallized from a melt/fluid comprising 2/3 sodium carbonate and 1/3 dissolved CO_2. The theoretical calculations indicate that the C-isotope fractionation between a melt/fluid containing calcium carbonate[5] and dissolved CO_2 in 2:1 proportions and diamond can be described by: $\Delta^{13}C_{Ca\text{-}carbonate\ melt/fluid\text{–}diamond} = 4.0 \times 10^6/T^2$. The temperature coefficient for this theoretically-based fractionation equation is nearly a factor of 2 smaller than that given by the equation of Reutsky et al. (2015a). Similarly, the 1400 °C carbonate-diamond fractionation factor indicated by the theoretical calculations ($\Delta^{13}C_{CaCO_3\text{–}diamond} = +1.2‰$) is more than a factor of 2 smaller than that reported by Bureau et al. (2018) ($\Delta^{13}C_{carbonate\ fluid\text{–}diamond} = +2.8‰$). For the reasons discussed above, it is highly unlikely that the theoretical calculations are wrong to this degree. Rather, we conclude that the two experimental studies, both of which involved isotopic exchange associated with new mineral growth, record kinetic rather than equilibrium C-isotope fractionations. As discussed further in the section *Kinetics* below, chemical reactions in which isotopic fractionations are kinetically controlled generally favor concentration of the light isotope (^{12}C) in the reaction products. As such, rapid synthesis of diamond from carbonate fluids as occurred in both sets of experiments will lead to crystallization of diamond that is anomalously ^{12}C-enriched and, in turn, carbonate–diamond fractionation factors that are spuriously large relative to equilibrium values.

The accuracy of the β-factor calculations for iron carbide (Fe_3C) cannot be rigorously assessed on the basis of the available experimental data. The synthesis experiments of Satish-Kumar et al. (2011) in the graphite-Fe_3C system recorded a $\Delta^{13}C_{graphite\text{–}Fe_3C} = +3.64‰$ at 5 GPa and 1500 °C, which can be compared to calculated $\Delta^{13}C_{diamond\text{–}Fe_3C}$ at that *P–T* condition of ~ +5.5‰ (Horita and Polyakov 2015) and $\Delta^{13}C_{diamond\text{–}Fe_3C} = +4.6‰$ at 15 GPa and 1500 °C (Liu et al. 2019). Because increasing pressure produces a larger increase in the β-factors of graphite than diamond (see above), the fractionation between graphite and Fe_3C is predicted

[5] The calculations of Deines (2004) indicate that sodium carbonate (Na_2CO_3) has a slightly lower affinity for ^{13}C than calcite. The fractionation between a sodium carbonate melt (with 33% dissolved CO_2) and diamond can be described by: $\Delta_{Na\text{-}carbonate\ melt/fluid\text{–}diamond} = 3.7 \times 10^6/T^2$.

to be larger than that between diamond and Fe_3C at high pressure (Polyakov and Kharlashina 1994) and so the discrepancy between experiment and calculation is quite large. Given that synthesis-style experiments cannot demonstrate the attainment of true isotopic equilibrium, it is unclear whether it is the experiments or the calculations that are in error.

Recommended C-isotope fractionation factors. Following the reasoning and criteria discussed above, Table 3 presents equations for calculating what we believe to be the best equilibrium C-isotope fractionation factors currently available for diamond-related systems. This compilation is provided for convenience but the reader is urged to cite the original authors if these fractionation factors are used. A spreadsheet (Calc_C_N-isotope-fractionation-factors.xlsm) for calculating these fractionation factors as a function of temperature is provided online at https://doi.org/10.7939/DVN/XZNF6V.

Figure 18 plots these fractionation factors as a function of $1/T^2$. Relative to diamond, CO_2 and carbonates are enriched in ^{13}C (positive $\Delta^{13}C_{phase\text{-}diamond}$) whereas CH_4 and carbides are depleted in ^{13}C (negative $\Delta^{13}C_{phase\text{-}diamond}$). Also shown in the figure is the fractionation between diamond and the so-called water-maximum fluid, which comprises dominantly H_2O along with equal parts CO_2 and CH_4. This fluid favors ^{13}C relative to diamond but slightly less so than carbonates. The curve describing the fractionation between diamond and Fe_3C must be regarded as tentative as it is the only one not corroborated by robust experimental fractionation data. One potential test of these diamond–Fe_3C fractionations is based on the natural sample data of Mikhail et al. (2014), who reported a $\Delta^{13}C_{diamond\text{–}Fe_3C} = +7.2 \pm 1.3‰$, derived from what they interpreted to be syngenetic iron carbide inclusions in two Jagersfontein diamonds. Assuming the two phases formed in C-isotope equilibrium, this diamond-Fe_3C fractionation corresponds to a temperature of 1256 ± 137 °C according to the equations of Table 3. This is a plausible temperature for diamonds originating in the deep lithosphere or formed during subduction of an oceanic slab into the asthenosphere and transition zone. However, other calibrations of this fractionation (e.g., Satish-Kumar et al. 2011; Liu et al. 2019) would also yield plausible (but lower) temperatures.

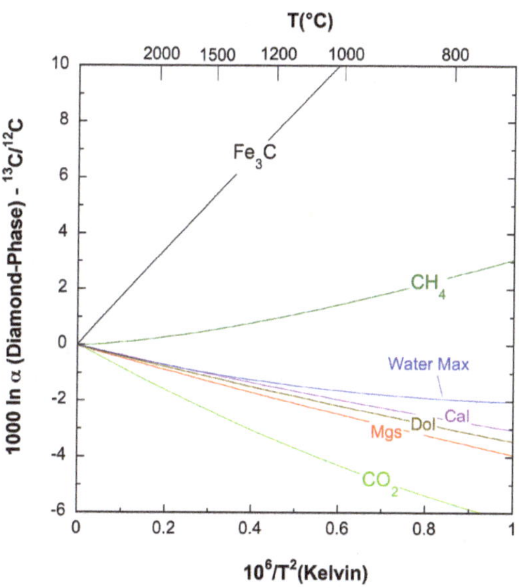

Figure 18. Summary of carbon isotope fractionation factors related to diamond plotted versus $10^6/T^2$ (*T* in Kelvin). The fractionation curves are based on calculations of β factors for the various phases (sources of β-factor calculations given in Table 3). Mgs = magnesite; Dol = dolomite; Cal = calcite. 'Water max' fluid is a water-rich fluid comprising equal amounts of CO_2 and CH_4 (see text).

As emphasized above, fractionation factors vary little with pressure, and the β-factors of Table 3 are those based on calculations done for ambient pressure conditions. The β-factors of

Table 3. Carbon isotope β-factors for species relevant to diamond.

Species	C_0	C_1	C_2	C_3	C_4	C_5	C_6
CO_2	−0.079	30.551	40.120	59.435	57.452	30.238	65.049
CH_4	0.086	20.234	31.262	48.914	47.887	25.030	53.071
calcite	0.001	25.180	15.383	13.297	9.5195	4.2857	8.5026
magnesite	0.020	25.948	14.000	7.6798	2.3338	0.23526	−0.09332
dolomite	0.020	25.436	13.525	7.3632	2.2372	0.23045	−0.07713
Fe_3C	0.000	4.5140	2.7791	0.01299	0.0000	0.0000	0.0000
diamond	0.000	21.649	9.790	5.2834	1.8790	0.30747	0.0000

Note: C-isotope β-factors reported here as 1000 ln β, which can be computed for each phase/species using the following expression: $1000 \ln \beta = C_0 + C_1x - C_2{\times}10^{-1} x^2 + C_3{\times}10^{-2} x^3 - C_4{\times}10^{-3} x^4 + C_5{\times}10^{-4} x^5 - C_6{\times}10^{-6} x^6$, where $x = 10^6/T^2$ and T is temperature in Kelvin. The equations are applicable at temperatures between 25 and 10000 °C, except the equation for Fe_3C, which is only valid above 450 °C. These equations are polynomial fits to β-factor calculations reported in the following sources: CO_2 and calcite (Chacko et al. 1991); magnesite and dolomite (Schauble et al. 2006); diamond (Polyakov and Kharlashina 1995); Fe_3C (Horita and Polyakov 2015); CH_4 – the β-factors of Richet et al. (1977) multiplied by a factor of 0.9846 (see text). The C-isotope fractionation factor between any two carbon-bearing species a and b at a particular temperature is given by the arithmetic difference in their 1000 ln β values ($\Delta^{13}C_{a\text{-}b} = 1000 \ln \alpha_{a\text{-}b} = 1000 \ln \beta_a - 1000 \ln \beta_b$) at that temperature.

all phases increase with pressure (Polyakov and Kharlashina 1994) but at somewhat different rates depending on the properties of the phase. As a consequence, there is a small change in the magnitude of fractionation factors with increasing pressure, which only becomes detectable at extreme pressure. For example, Horita and Polyakov (2015) estimate that at 1500 °C, $\Delta^{13}C_{diamond\text{–}Fe_3C}$ increases from +5.38 to +5.85‰ as pressure increases from 0.001 GPa (surface pressure) to 30 GPa and $\Delta^{13}C_{diamond\text{-}CaCO_3}$ increases from −1.07 to −1.16‰ over that same pressure interval. These small pressure effects on the magnitude of fractionation factors (typically < 0.3 percent change per GPa) should not be significant for most purposes.

We would add one cautionary note regarding the preferred fractionation factors listed in Table 3. These fractionation factors are based on theoretical calculations or experiments in relatively simple chemical systems (e.g., CO_2–calcite) and thus do not consider the complex nature of diamond-forming mantle fluids/melts. Although fractionations are dominantly controlled by the vibrational characteristics of the simple fluid species carrying the element of interest (e.g., CO_2, CH_4, CO_3^{2-}), the presence of other dissolved constituents in the fluid phase can have second-order effects on the fractionation behavior of the simple fluid species. The exact magnitude of these second-order effects is difficult to ascertain but a crude maximum estimate can be obtained by considering the impact of different cations on C-isotope fractionations in carbonate minerals, whose fractionation behavior is largely determined by the vibrational characteristics of the CO_3^{2-} anion. Theoretical calculations indicate that changing the cation bonded to the carbonate anion from Mg to alkalis (strongest and weakest affinities for ^{13}C, respectively) has < ~ 0.5‰ effect on fractionations at a typical diamond-forming temperature of 1150 °C (Deines 2002; Schauble et al. 2006). Therefore, we anticipate that the presence of other dissolved species in natural diamond-forming fluids will only have a relatively small impact on the fractionation equations of Table 3.

Equilibrium nitrogen isotope fractionation factors related to diamond

As was the case with C-isotopes, the N-isotope fractionation factors most pertinent to this chapter are those between diamond and likely N-bearing species in diamond-growth media. Many workers have proposed that N_2 and NH_3 are the two dominant N-fluid species in the upper mantle under more oxidizing and reducing conditions, respectively (e.g., Deines et al. 1989; Li and Keppler 2014; Sokol et al. 2017). However, on the basis of laboratory experiments or theoretical calculations, other N-bearing species have also been suggested, including NH_2^+ and NH_2^- (Mysen 2018, 2019) and NH_4^+ (Mikhail and Sverjensky 2014). In addition, at the more reducing conditions of the deep mantle, N may exist in the form of a nitride-bearing melt or mineral.

To our knowledge, no experiments aimed specifically at calibrating equilibrium N-isotope fractionation factors between diamond and potential N-bearing species in diamond-growth media have been carried out. The available fractionation factors are derived either from theoretical calculations or extracted from natural sample datasets.

Theoretical calculations. The first attempt at constraining N-isotope fractionations for diamond was that of Deines et al. (1989), who based their estimate on Richet et al.'s (1977) N-isotope β-factor calculations for N_2, NH_3 and HCN. In particular, the β-factors of HCN were taken as an analogue for the fractionation behavior of the C-N bond in diamond. Petts et al. (2015) followed the same general theoretical approach as Deines et al. (1989) but used β-factor calculations for the CN^- molecule (rather than HCN) as the analogue for N in diamond. Both studies concluded that diamond would be depleted in ^{15}N relative to the likely upper mantle N fluid species. Petts et al. (2015) reported the following sequence in terms of affinity for ^{15}N relative to ^{14}N: $NH_4^+ > N_2 > NH_3 > CN^-$ (Fig. 19). They also calculated N-isotope fractionation factors between the CN^- diamond analogue and these N-bearing fluid species of −1 to −4‰ at temperatures between 1000–1300 °C. No theoretical calculations have been done for the specific NH_2 or nitride species that may exist in the mantle. However, Hanschmann (1981) did report β-factor calculations for NH_2CH_3 and BN at a selected number of temperatures below 727 °C. These calculations indicate that NH_2^- fractionates N-isotopes broadly like N_2 and that boron nitride is depleted in ^{15}N relative to the CN^- diamond analogue.

It should be noted that the β-factor calculations for the HCN and CN^- gas molecules are, at best, crude analogues for the N-isotope fractionation behavior of diamond. For instance, carbon and nitrogen in these gas molecules are connected by a strong triple bond compared to a weaker single bond in diamond. Other things being equal, bond strength generally correlates with a substance's relative affinity for the heavy isotope, which predicts that the weaker C–N bond in diamond would have a lower affinity for ^{15}N than either HCN or CN^- (Petts et al. 2015). On the other hand, condensed phases can partition isotopes differently than gases of the same composition (e.g., Horita and Wesolowski 1994), which may cause an isotopic effect in the opposite direction. That is, N in the solid phase diamond may have a stronger affinity for the heavy isotope than predicted by calculations on the HCN or CN^- gas molecules. Despite these uncertainties, the general prediction of the calculations is that diamond would favor the light nitrogen isotope relative to most N fluid species in the mantle. It appears that nitrides are the only currently known mantle N species that is likely to have a lower affinity for ^{15}N than diamond.

Natural sample calibrations. Several studies have attempted to extract N-isotope fractionation factors from data acquired on natural diamonds. This approach, which was pioneered by Thomassot et al. (2007) and also used by Palot et al. (2014) and Petts et al. (2015), is a bootstrap method based on co-variations in C- and N-isotope compositions and N-contents, either within a single diamond or within a suite of diamonds that are inferred to be co-genetic. The method implicitly assumes that the observed isotopic and elemental variations are the product of a closed-system (i.e., fluid-limited) Rayleigh fractionation process and that isotopic and chemical equilibrium is always maintained between the rim of the growing diamond and the growth medium. It also assumes that diamond is the only C- and N-bearing phase crystallizing from the growth medium (i.e., no other C- or N-bearing carbonate, silicate or nitride phases are co-crystallizing with diamond). Finally, the method assumes knowledge of the C-isotope fractionation factor between diamond and that growth medium. The papers by Cartigny et al. (2001), Thomassot et al. (2007), Smart et al. (2011) and Petts et al. (2015) provide details of the procedure. In essence, the slope of data arrays in $\delta^{13}C$ versus N abundance plots, combined with the known C-isotope fractionation factor between the growth medium and diamond ($\Delta^{13}C_{diamond-fluid}$) can be used to derive a diamond–fluid N partition coefficient (K_N). Armed with K_N and $\Delta^{13}C_{diamond-fluid}$ values, the slope of co-variations between $\delta^{13}C$ and $\delta^{15}N$ can be used to deduce $\Delta^{15}N_{diamond-fluid}$, the N-isotope fractionation factor between the growth medium and diamond.

To implement this approach, Thomassot et al. (2007) carried out high-precision, bulk isotopic and elemental analysis of 59 diamonds extracted from a single, highly diamondiferous peridotite xenolith recovered from the Premier kimberlite (*Cullinan mine*) of South Africa. They interpreted the trends in their data to reflect Rayleigh-style fractionation of diamond from a single batch of fluid rather than two-component mixing of fluids with different isotopic compositions and N abundances. They also inferred that their data trends were best explained if the early-formed diamond had relatively high $\delta^{13}C$- and $\delta^{15}N$-values and N-contents, which progressively evolved to lower values during fractional crystallization of diamond from the growth medium. In an equilibrium process, evolution of diamond to lower $\delta^{13}C$-values requires crystallization from a reducing (e.g., CH_4-rich) fluid. Moreover, the trend towards decreasing $\delta^{15}N$-values implies that $\Delta^{15}N_{diamond-fluid}$ is positive (Thomassot et al. computed $K_N \sim 2$ and $\Delta^{15}N_{diamond-fluid} \sim +1.2‰$). In turn, a positive value for $\Delta^{15}N_{diamond-fluid}$ (see above) requires that these diamonds crystallized from a nitride-bearing growth medium (c.f., Fig. 19) or a growth medium containing some other, heretofore unknown N species with a lower affinity for ^{15}N than diamond.

Smart et al. (2011) and Petts et al. (2015) conducted detailed ion microprobe (SIMS) analyses of isotopic and N abundance zoning profiles in a gem-quality diamond recovered from an eclogite xenolith from the Jericho kimberlite of northern Canada. The core zone of this diamond shows smooth rimward increases in both $\delta^{13}C$- and $\delta^{15}N$-values coupled to a decrease in N concentrations. Broadly similar zoning patterns are seen in diamonds from Marange (Smit et al. 2016). In the case of the Jericho diamond, the increase in $\delta^{13}C$-values was interpreted to reflect crystallization of the diamond from an oxidizing, carbonate-bearing fluid/melt ($\Delta^{13}C_{diamond-fluid} = -1.7‰$ at 1100 °C) (Smart et al. 2011; Petts et al. 2015). Notably, the shift in $\delta^{15}N$-values observed in the core zone of this diamond is ~ 8.5 times larger than the shift in $\delta^{13}C$-values over the same interval. From these data trends, Petts et al. (2015) extracted a $K_N \geq 4.4$ (i.e., N is highly compatible in diamond relative to this growth medium) and a $\Delta^{15}N_{diamond-fluid} = -4.0 \pm 1.2‰$ (2σ). The numerical accuracy of this fractionation factor determination depends on the validity of the various assumptions noted above, namely that isotopic and chemical equilibrium was maintained between the growth medium and the diamond rim and that the

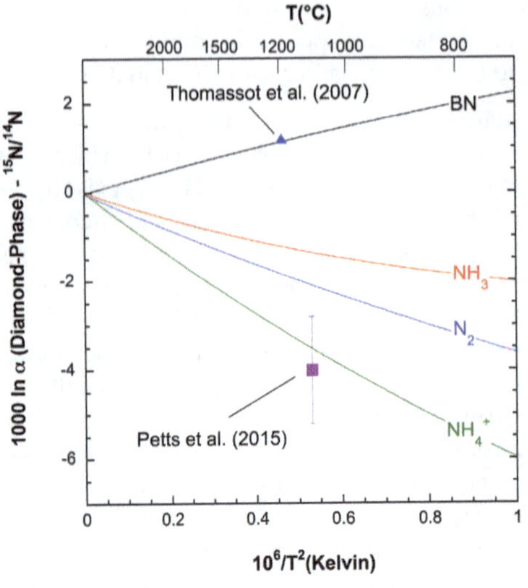

Figure 19. Summary of nitrogen isotope fractionation factors related to diamond plotted versus $10^6/T^2$ (T in Kelvin). The fractionation curves are based on calculations of β-factors for NH_4^+, N_2, NH_3 and CN^- by Petts et al. (2015) and BN by Hanschmann (1981). In these calculations, CN^- is taken to be an analogue for the N-isotope fractionation behavior of diamond. Also shown are two estimates of the N-isotope fractionation factor between the growth medium and diamond by Thomassot et al. (2007) and Petts et al. (2015) derived from analyses of natural diamonds. See text for discussion.

C-isotope fractionation factor between the diamond and the growth medium was −1.7‰. It is safe to conclude from this analysis, however, that the sign of $\Delta^{15}N_{diamond-fluid}$ must be negative for this diamond and also that the magnitude of $\Delta^{15}N_{diamond-fluid}$ must be a relatively large in order to account for the much bigger shift in $\delta^{15}N$ compared to $\delta^{13}C$-values.

Recommended N-isotope fractionation factors. There is much more uncertainty regarding equilibrium N-isotope than C-isotope fractionation factors for diamond. The two detailed natural sample studies that have been carried out have proposed opposite signs for $\Delta^{15}N_{diamond-fluid}$. More systematic studies on natural diamonds should be carried out to see if a dominant fractionation pattern and direction emerges. It is recommended, however, that these studies be based on detailed documentation of zoning profiles in individual diamonds rather than the bulk analysis of multiple diamonds as it is much easier to unambiguously determine the direction of isotopic and elemental fractionation with the former approach. In terms of theoretical estimates, a first-principles, lattice dynamics determination of β-factors for N-bearing diamond is clearly preferable to calculations based on gas molecule diamond proxies such as HCN or CN^-. Such first-principles calculations, although challenging, should be attempted.

In the interim, Table 4 and the on-line spreadsheet (Calc_C_N-isotope-fractionation-factors.xlsm at https://doi.org/10.7939/DVN/XZNF6V) presents equations for reproducing the N-isotope calculations reported by Petts et al. (2015) for N_2, NH_3, NH_4^+ and the diamond proxy, CN^-. The table also presents a polynomial fit to the β-factor calculations for BN by Hanschmann (1981), which should provide a general indication of how nitride-bearing fluids or melts may fractionate N-isotopes relative to diamond.

Table 4. Nitrogen isotope β-factors for species relevant to diamond.

Species	C_0	C_1	C_2	C_3	C_4	C_5	C_6
NH^{4+}	−0.044	188.20	352.73	610.15	632.58	342.59	74.349
N_2	0.014	153.77	227.38	332.47	311.22	158.20	32.941
NH_3	0.054	140.35	274.93	490.58	518.25	284.11	62.186
CN^-	0.009	110.10	135.48	175.35	151.81	73.265	14.706
BN	0.000	83.403	83.659	70.669	11.188	−24.428	−11.308

Note: N-isotope β-factors reported here as 1000 ln β, which can be computed for each gas molecule s using the following expression: $1000 \ln \beta = C_0 + C_1x - C_2{\times}10^{-1}x^2 + C_3{\times}10^{-2}x^3 - C_4{\times}10^{-3}x^4 + C_5{\times}10^{-4}x^5 - C_6{\times}10^{-6}x^6$, where $x = 10^6/T^2$ and T is temperature in Kelvin. The CN^- gas molecule is taken to be a diamond analogue. The equations are applicable at temperatures between 0 and 5000 °C. These equations are polynomial fits to β-factor calculations for NH_4^+, N_2, NH_3 and CN^- by Petts et al. (2015) and for boron nitride (BN) by Hanschmann (1981).

Kinetic effects

As is the case with equilibrium fractionation, kinetic fractionation processes can also generate systematic and continuous changes in the isotopic and elemental compositions of diamonds. Kinetic (i.e., non-equilibrium) isotopic fractionation corresponds to a unidirectional process, which, in contrast to equilibrium fractionation, is not reversible. For chemical reactions in which isotopic fractionation is kinetically controlled, the products of the reaction are typically depleted in the heavy isotope relative to the reactants because bonds in the reactants made with the light isotope are weaker and therefore more reactive than those made with the heavy isotope. However, cases of inverse kinetic isotope fractionation also exist in which the products are enriched in the heavy isotope (see Casciotti 2009).

That kinetic rather than equilibrium isotope fractionation can occur during diamond growth is well known from laboratory experiments, including some of the C-isotope experimental studies discussed above. In addition, strong sector zoning in $\delta^{15}N$-values was observed in a synthetic diamond grown at elevated pressure-temperature conditions (Boyd et al. 1988). Sector zoning refers to crystallographically-controlled differences in chemical or isotopic composition within different parts (sectors) of the same crystal. Diamond is in fact the first mineral for which isotope-sector zoning was reported. The finding that cubic and octahedral growth sectors of a synthetic diamond could differ by ~ 45‰ (Boyd et al. 1988) calls into question the use of diamond to infer the $\delta^{15}N$-values of mantle fluids and to assess the origin of nitrogen and, by implication, carbon in diamonds. Two subsequent studies on natural mixed-habit (with cuboid and octahedral growth sectors) diamonds, however, found $\delta^{15}N$ values within error along the [100] and [111] growth directions (Bulanova et al. 2002; Cartigny et al. 2003) and concluded that kinetic isotope fractionation during sector growth is not relevant to natural diamond formation. The very detailed study of Howell et al. (2015b) generally confirmed the above conclusion by noting differences in N-isotope composition between cuboid and octahedral growth sectors of only 0.4 to 1.0‰, i.e., within or just slightly exceeding the reported analytical uncertainty. These data were therefore taken as evidence that nitrogen uptake in natural diamond is not associated with significant kinetic isotope effects and that isotopic variations can generally be interpreted assuming the establishment of isotopic equilibrium between the diamond-forming medium and diamond.

Contrary to this conclusion, two recent studies argue that the C-isotope and by implication the N-isotope compositions of natural diamonds are largely controlled by kinetic isotope fractionation processes (Reutsky et al. 2017; Kueter et al. 2020). The conclusions of these important studies, if correct and widely applicable, would fundamentally alter our interpretation of isotopic and elemental co-variations in diamonds. As such, the two studies are discussed in detail below.

Reutsky et al. (2017) model. Reutsky et al. based their interpretations on zoning profiles observed in two gem-quality cuboid placer diamonds from the Yakutia region of Russia combined with kinetic modeling of these zoning profile data. The core zones of these diamonds are characterized by rimward decreases in $\delta^{13}C$-values coupled to decreasing N abundances; $\delta^{15}N$-values show no obvious zoning trend over the same core-zone interval. The rims of these diamonds have low N-contents, which also correspond to a precipitous drop in $\delta^{15}N$ to values as low as −63‰. If correct, this would be the lowest $\delta^{15}N$-value ever reported for diamonds. The present authors are not convinced of the accuracy of these extreme $\delta^{15}N$-values, which were obtained on low N-content diamond using the SIMS analytical technique. The ^{14}N–^{12}C peak, which is used to monitor the abundance of ^{14}N in SIMS diamond analyses, is isobaric with the ^{13}C–^{13}C peak. Although the two peaks are nominally resolved at the instrument mass resolution setting used by Reutsky et al. (2017), the tail of the much larger ^{13}C–^{13}C peak still contributes appreciably to the ^{14}N–^{12}C peak in the case of very low N-content (< 50 ppm) diamond (Stern et al. 2014). Analyses of such low N diamond by SIMS can therefore result in the measurement of spuriously low $\delta^{15}N$-values unless a non-routine analytical protocol is followed with the instrument operated at a much higher mass resolution (R.A. Stern, personal communication 2020).

Setting aside potential difficulties with the N-isotope analyses at the rims of these Yakutian placer diamonds, Reutsky et al. (2017) propose an innovative kinetic model for the origin of the correlated C-isotope and N-content zoning profiles observed in the core zones of their diamonds. The impetus for suggesting a kinetic rather than an equilibrium model for these diamonds stems from the observation of rimward decreases in $\delta^{13}C$-values. In an equilibrium scenario, a trend towards lower $\delta^{13}C$-values requires diamond precipitation from a reduced (e.g., CH_4) fluid as these are the only known fluid species that partition ^{12}C relative to diamond. However, previous studies had reported the presence of micro-carbonate inclusions in the Yakutian placer diamonds examined by Reutsky et al. (2017), which would imply that these diamonds crystallized from an oxidized fluid.

To resolve this paradox, Reutsky et al. (2017) proposed a decreasing linear growth rate (DLGR) model, originally developed in the material sciences for trace-element uptake during crystal growth (Burton et al. 1953). The model assumes that the diamond-forming fluid maintains a constant C-isotope composition throughout diamond growth and attributes the development of compositional zoning in diamond to a decrease in its linear growth rate as crystal size increases. That is, the model assumes that the volume growth rate of diamond remains constant, which in turn requires that the linear growth rate (the rate at which individual crystal faces move outward from their starting point) continuously decreases because the surface areas of crystal faces are getting larger during diamond growth. According to the DLGR model, the effective diamond–fluid C-isotope fractionation factor is zero when the crystal first forms and the linear growth rate is fast (i.e., initially no C-isotope fractionation occurs between diamond and fluid) but becomes larger as this growth rate decreases. More specifically, the effective fractionation factor progressively approaches the equilibrium fractionation factor as the linear growth rate decreases and should theoretically equal the equilibrium fractionation factor when the linear growth rate goes to zero. Because the model assumes that the fluid's isotopic composition remains constant throughout diamond growth, the increase in the magnitude of the effective fractionation factor with decreasing linear growth rate results in a progressively changing C-isotope composition of the newly-grown diamond rim. Counterintuitively, the model predicts rimward decreases in the $\delta^{13}C$-values of diamond crystallizing from an oxidized fluid and rimward increases in the $\delta^{13}C$-values of diamond crystallizing from a reduced fluid, which is opposite to what is predicted by the equilibrium model. In the case of the studied Yakutian placer diamonds, which are inferred to have precipitated from an oxidized (carbonate) fluid, crystallization led to progressively decreasing $\delta^{13}C$-values in the diamond.

Reutsky et al. (2017) explain the observed rimward decrease in the N-content of their diamonds in an analogous way. There is no preferential partitioning of N between diamond and its parental fluid during the early stages of crystal growth when the linear growth rate is fast (i.e., $K_N = 1$) but the N partition coefficient progressively approaches the equilibrium value as the linear growth rate slows. Such a scenario would result in a trend towards decreasing N-content during diamond growth provided that the equilibrium K_N-value is < 1 (i.e., N behaves incompatibly in diamond).

The following lines of evidence suggest that, even if the DLGR model is valid for the particular Yakutia diamonds that Reutsky et al. (2017) studied, it is unlikely to be broadly applicable to natural diamonds. If correct, the constraints of the DLGR model should apply to all growing diamonds. That is, all diamonds should experience a progressive decrease in their linear growth rates as they increase in size because that phenomenon is an inevitable consequence of the constant volume growth rate assumption inherent to the model. Accordingly, if the model is generally applicable, C-isotope zoning profiles should be very common if not ubiquitous in natural diamonds. That prediction, however, is at odds with the observation that most diamonds that have been investigated to date do not show appreciable C-isotope zoning (see sections *Introduction* and *Diamond through time*). Although problematic for the DLGR model, the lack of isotopic zoning in many diamonds can readily be accommodated by the equilibrium fractional crystallization model by invoking a fluid-dominated (i.e., the fluid is effectively an infinite reservoir that does not change its isotopic composition as diamond precipitates) rather than a fluid-limited (i.e., where diamond crystallization induces isotopic shifts in the residual fluid) process. Only in the latter scenario will diamond crystallization lead to significant changes in the fluid isotopic composition and in turn the development of isotopic zoning profiles in the diamond. With the DLGR model, a fluid-limited system would produce reversals in C-isotope zoning profiles. In other words, for diamond crystallization from an oxidized fluid, $\delta^{13}C$-values would first decrease and then increase (or visa-versa) as crystallization proceeds because of the combined effects of both changing fractionation factor and changing fluid isotopic composition during diamond crystallization. Such smooth (rather that abrupt reversals across growth zone boundaries) and systematic reversals in zoning profiles have not yet been documented in natural diamonds.

Although Reutsky et al. (2017) applied the DLGR model to gem-quality placer diamonds, their model and indeed kinetic models in general are more likely applicable to rapidly-grown fibrous diamonds than to slowly-grown gem diamonds (Sunagawa 1990). However, the available data suggest that C-isotope zoning is even less common in fibrous diamonds than in gem diamonds (namely $\delta^{13}C$-values down to a minimum of ~ −8‰ at the core of fibrous diamonds increasing up to ~ −5‰ towards the outer rim; e.g. Boyd et al. 1992). We conclude therefore that while the DLGR model may be applicable, in some instances, to diamond growth on laboratory timescales, it does not provide a general explanation for the origin of zoning profiles in either fibrous or gem-quality natural diamonds.

Kueter et al. (2020) model. Like Reutsky et al. (2017), Kueter et al. (2020) also suggest that kinetic rather than equilibrium processes may control C-isotope compositions during diamond growth but propose a different kinetic model to account for their observations. The model is based on experiments in which stearic acid ($C_{18}H_{36}O_2$), sealed in a gold capsule, was rapidly decomposed at 0.2 GPa and 800 °C to form a mixture comprising primarily solid elemental carbon and CH_4 gas, along with lesser amounts of H_2O, H_2, and various other trace gases. Diffusion of H_2 gas out of the gold capsule over time promoted oxidation of the CH_4 to elemental carbon globules by the reaction: $CH_4 = C + H_2$. Under equilibrium conditions, elemental carbon (graphite) partitions ^{13}C relative to methane and, as a result, graphite precipitation should drive residual CH_4 to progressively more ^{13}C-depleted compositions. In a time-series set of experiments extending in duration from 30 seconds to 50 hours, Kueter et al. (2020) documented exactly the opposite trend; the $\delta^{13}C$-values of CH_4 continuously increased with reaction progress. They attributed this finding to a kinetic isotope fractionation in which the weaker ^{12}C–H bonds in CH_4 preferentially broke down to form elemental carbon, which caused the residual CH_4 to become more ^{13}C-enriched. The C-isotope data from the time-series experiments were modeled to extract a kinetic graphite–CH_4 fractionation factor of −6.2 ± 0.3‰, which contrasts sharply with the equilibrium fractionation factor at 800 °C ($\Delta^{13}C_{\text{graphite-}CH_4} = +2.2‰$). Kueter et al. (2020) speculated that an analogous, kinetically-controlled process involving methane oxidation may occur during natural diamond formation and could potentially explain why the great majority of diamond C-isotope profiles that do exhibit systematic variation show a trend to increasing $\delta^{13}C$-values even though the mantle is thought to be dominated by relatively reduced fluids.

To assess the relevance of their short timescale experiments to the much longer timescale growth of natural diamonds, Kueter et al. (2020) applied the growth-entrapment (GE) model of Watson and Liang (1995) and Watson (1996, 2004), which was originally developed to explain the origin of sector zoning in natural crystals. The GE model proposes that an interplay between the growth rate of the crystal (V) and the diffusion rate (D_i) of a narrow near-surface layer of half-width l determines if partitioning equilibrium will be established between the surface of the growing crystal and the growth medium. If the crystal growth rate is too fast or diffusivity in the surface layer too slow, a disequilibrium (i.e., kinetic) partitioning of elements or isotopes between the crystal and the growth medium can become entrapped at the crystal surface and later preserved within the growing crystal. The GE model defines a non-dimensional parameter, the growth Péclet number ($Pe = V*l/D_i$), which determines if a kinetic partitioning will likely be recorded in the crystal. Specifically, if the value of Pe is > ~1, entrapment of a kinetic partitioning is likely.

Using the growth rate of carbon globules in their experiments (~ 10^{-12} m/s) in combination with an estimated surface layer half-width (1 nm) and diffusivity (~ 10^{-32} m^2/s), Kueter et al. (2020) calculated a Pe number of ~ 10^{11}, which indicates that, as observed, kinetic rather than equilibrium fractionation of C-isotopes was inevitable in their experiments. Applying the GE model to the growth of natural diamonds is more challenging but Kueter et al. made the following bounding calculation. Using Koga et al.'s (2005) experimentally-determined estimate of carbon lattice diffusion rate in diamond ($D = 4.5 \times 10^{-30}$ m^2/s) at 1100 °C, a typical lithospheric diamond-formation temperature, they computed that a crystal growth rate of $V \leq \sim 10^{-21}$ m/s would be required to obtain a growth Péclet number low enough ($Pe \leq 1$) to attain C-isotope equilibration at the diamond rim. At that growth rate, it would take > 3 billion years to form a 250 μm-radius diamond crystal. Thus, if correct, these calculations would effectively preclude the establishment of C-isotope and by implication N-isotope equilibrium during the growth of natural diamonds.

We would offer the following critique of this model calculation. The calculation assumed that the diffusion rate of carbon in the surface layer would be the same as that determined by Koga et al. (2005) for carbon diffusion within a well-formed diamond crystal. However, as noted by Watson and Liang (1995) and elaborated by Watson (2004), the diffusion rate of the surface layer is likely faster and possibly many orders of magnitude faster than the diffusion rate within the crystal lattice. Without quantitative knowledge of the surface-layer diffusion rate, it is not possible to do a robust GE calculation constraining whether isotopic equilibrium is likely to be established during natural diamond growth.

Several observations may be helpful in framing the question. At the carbon globule growth rate of Kueter et al.'s experiments, a 1 mm-radius crystal could be grown in < 40 yr. The time required for lithospheric diamonds to grow is not known but timescales of 10^5 to 10^7 yr do not seem unreasonable. Indeed, sizable pockets of fluid are known to persist in stable continental crust for billions of years (Holland et al. 2013). If similar, long-lived fluid pockets exist in cratonic mantle, crystallization times of minerals in these pockets, driven by either simple cooling or redox reactions, could be quite protracted. For example, a growth rate of $V = 10^{-17}$ m/s allows a 1 mm-radius crystal to form in ~ 3.5 million years. At that growth rate, attaining $Pe \leq 1$ requires $D \geq \sim 10^{-26}$ m^2/s for the surface layer, which is about 3.5–4 orders of magnitude faster than the lattice diffusion rate for carbon in diamond at 1100 °C and roughly equivalent to the lattice diffusion rate at 1315 °C (Koga et al. 2005). It is not certain if such diffusivities for the surface layer are reasonable but the absence of clear examples of C- or N-isotope sector zoning in natural diamonds suggests that surface layer diffusivities must be significantly higher than lattice diffusivities; otherwise, isotopic sector zoning would be common in natural diamonds.

The isotopic composition of natural diamonds and their internal zoning patterns are helpful in assessing the general applicability of the Kueter et al. (2020) kinetic model. If the isotopic composition of gem diamonds is kinetically controlled, then for typical temperatures of diamond formation (~1150 °C), the kinetic CH_4-diamond C-isotope fractionation factor would be +4.7‰ (Kueter et al. 2020). Accordingly, the cores of gem diamonds crystallizing from fluids with a mantle-like C-isotope composition of −5‰ should have δ^{13}C-values of about −10 to −9‰. Such low δ^{13}C-values are atypical for peridotitic diamonds and zonation trends increasing outwards from distinctly ^{13}C-depleted cores are even rarer. These observations suggest that the kinetic fractionation factor of Kueter et al. is much too large to account for the average C-isotope compositions of peridotitic diamonds unless one supposes that the mantle fluids responsible for the formation of these diamonds have δ^{13}C-values of −1 to 0‰.

A comparison of isotopic zoning patterns in gem diamonds versus those in fibrous diamonds is also useful in evaluating Kueter et al.'s (2020) model and specifically their proposal that the isotopic compositions of many diamonds are determined by a kinetic fractionation process during crystallization from methane-rich fluids. As noted previously, kinetic isotope effects, if applicable, are more likely to be evident in rapidly-grown fibrous diamonds. For instance, if surface layer diffusion rates were fast enough to allow the establishment of isotopic equilibrium during formation of slowly-grown gem diamonds but not in rapidly-grown fibrous diamonds then diamond formation from CH_4-rich fluids would drive fractionation trends in opposite directions in the two cases; toward lower and higher δ^{13}C-values in gem and fibrous diamonds, respectively. In addition, assuming that gem and fibrous diamonds are, on average, formed from similar C-isotope composition fluids, the early crystallized diamond in the fractionation sequence would be markedly different in the case of diamond precipitation from methane-rich fluids because of the opposite signs of the kinetic and equilibrium diamond–fluid fractionation factors. Even in the case of diamond formation from oxidized fluids where the signs of kinetic and equilibrium fractionation factors should be the same, it seems likely from Kueter et al.'s (2020) results that the magnitude of the kinetic fractionation factor would be significantly larger, which should again lead to detectable differences in the $\delta^{13}C$-values of the first crystallized gem versus fibrous diamonds. The available data do not indicate systematic differences in the fractionation trends or isotopic compositions of gem and fibrous diamonds (Weiss et al. 2022, this volume). This observation suggests either that the isotopic compositions of *both* gem and fibrous diamonds are controlled by kinetic processes or *neither* are kinetically controlled. The evidence detailed in the paragraphs above suggests that the former scenario is unlikely.

In summary, it is our view that, at present, there is no compelling evidence to indicate that the C- or N-isotope compositions of natural diamonds are largely due to kinetic rather than equilibrium fractionation processes. Nevertheless, the exceedingly slow diffusion rates of carbon and nitrogen in diamonds make kinetic isotope effects a possibility, certainly during diamond growth on laboratory timescales but perhaps also on geological timescales. Further search for clear disequilibrium features, such as isotopic sector zoning in mixed habit diamonds, is recommended as these may help to determine if natural diamond growth rates are near a critical threshold such that second-order variations in these growth rates will determine whether diamonds grow in isotopic equilibrium with the growth medium.

Mixing of fluids

Mixing of fluids with one carrying oxidized (CO_2 or CO_3^{2-}) and the other reduced carbon species (CH_4, C_2H_6) and/or H_2 will cause redox reactions that produce diamond (or graphite) and water (e.g., reactions 8 and 10 in Deines 1980 and reactions 1 and 2 below). The same principle applies if a reduced fluid infiltrates a carbonate-bearing mantle substrate. This process can be associated with large variations in stable isotopic composition and nitrogen content of precipitated diamond if the fluids derive from isotopically and compositionally distinct sources (e.g., mixing of reduced mantle-derived and oxidized subduction-derived fluids). As discussed in the *Introduction* and further below, most natural diamonds show fairly limited internal variation in their carbon isotope composition across continuously grown sections (marked by an absence of hiatuses in CL images), suggesting that mixing between extreme end-member fluids does not usually occur at the site of diamond formation. Gradual mixing of distinct fluids or recharge through fresh pulses of isotopically similar (but not identical) fluid, however, likely cause at least some of the seemingly chaotic fluctuations observed even on the level of individual homogenous growth zones that involve switches from increasing to decreasing (or vice versa) $\delta^{13}C$ and/or N-content. Good examples can be found in a profile across a small (< 1 mm) diamond of unknown paragenesis from Aviat on the Rae Craton (AV6-111; Peats et al. 2012), in the George Creek (Colorado) diamond plates studied by Fitzsimons et al. (1999), the homogenous core zone of eclogitic diamond A46 from Argyle (analytical spots 1–5) and two homogenous zones of harzburgitic diamond DBP-460_7 from Kimberley/'*De Beers Pool*' (analytical spots 1–6 and 7–10; the latter two diamonds from Supplementary Table 1 of Howell et al. 2020). The small scale and complexity of these potential mixing relationships, however, has so far precluded the successful development of numerical mixing models. For a more detailed discussion of fluid mixing based on combined He–N–C isotope data, see Jacob and Mikhail (2022, this volume).

Isochemical precipitation of diamond and Rayleigh isotope fractionation in multi-component systems

Conventional models of diamond precipitation in the course of redox reactions between carbon-bearing fluids or melts and peridotitic or eclogitic diamond substrates are built on the premise that such mantle rocks contain sufficient iron that can be shifted between the divalent and trivalent states to buffer the redox state of infiltrating fluids or melts. In other words, in these models diamond precipitates because the oxygen fugacity of the fluid is buffered by the host rocks. Yet, for the most important diamond substrate, strongly depleted cratonic harzburgites (yielding ~ 56% of all mined diamonds; Stachel and Harris 2008), this premise does not apply. Even very minor amounts of infiltrating fluid or melt, capable of delivering or removing less than 50 ppm O_2, are sufficient to completely overwhelm the buffering capacity of such iron-, garnet- and pyroxene-poor rocks (Luth and Stachel 2014).

As a result of the high solidus temperature of cratonic harzburgites (Wyllie 1987), together with the lower temperatures of the lithospheric mantle relative to the asthenosphere, percolating melts should generally freeze in harzburgite whilst COH (carbon-hydrogen-oxygen) fluids will be able to pass through. For such COH fluids, redox neutral precipitation of diamond during either cooling or ascent (i.e., cooling and depressurization) was recently developed as an alternative mode of diamond formation (Luth and Stachel 2014; Stachel and Luth 2015) and is briefly reviewed here. In this model, diamond formation can occur as a consequence of two oxygen-conserving reactions:

$$CO_2 + CH_4 = 2C + 2H_2O \tag{1}$$

or

$$2C_2H_6 = 3CH_4 + C \tag{2}$$

The former occurs in response to cooling (isobaric or along a geotherm) and the latter in response to depressurization. Only reaction (1) is capable of creating significant diamond endowment (> 1 ppm or > 5 carats per ton) at constant depth (the requirement for a diamond mantle source) at low fluid–rock ratios. Reaction (1) also is more consistent with oxygen fugacity measurements on cratonic peridotites (e.g., Stagno et al. 2013), which predict that at typical diamond-forming pressure–temperature conditions, coexisting COH fluids would be at or near the water maximum (> 90 mol% H_2O) and contain minor amounts of both CO_2 and CH_4 (Luth and Stachel 2014; Stachel and Luth 2015).

During diamond precipitation following reaction (1), carbon isotopes fractionate between diamond and the two carbon species in the fluid (CO_2 and CH_4). The relative proportions of CO_2 and CH_4 in the fluid are expressed as XCO_2 (molar $CO_2/[CO_2+CH_4]$). Except for the case of equal proportions of CO_2 and CH_4 in the initial fluid ($XCO_2 = 0.5$), XCO_2 evolves during diamond crystallization to either higher or lower values. For initial fluids that are CO_2 dominated ($XCO_2 >$ 0.5), XCO_2 increases as diamond crystallization proceeds, leading to progressively more negative values for the diamond–fluid carbon isotope fractionation factor ($\Delta^{13}C_{diamond\text{–}fluid}$). Conversely, for methane-dominated fluids ($XCO_2 < 0.5$), XCO_2 decreases with diamond crystallization, which leads to progressively less negative and eventually positive $\Delta^{13}C_{diamond\text{–}fluid}$ values. The equations to model Rayleigh isotope fractionation in a multi-component system (RIFMS) were developed by Ray and Ramesh (2000). Assuming a mantle-like carbon isotope composition of the initial fluid ($\delta^{13}C$ of –5‰) and typical conditions for peridotitic diamond formation (1140 °C and 5 GPa), Stachel et al. (2017) used the RIFMS equations to model diamond precipitation following reaction (1) for XCO_2 from 0.89 (corresponding to the upper fO_2 limit of diamond stability at these *PT* conditions) to 0.10. In these calculations, the bulk diamond–fluid carbon isotope fractionation factor ($\Delta^{13}C_{diamond\text{–}fluid}$) scales with the relative proportions of CO_2 and CH_4 (i.e., XCO_2 and $1 - XCO_2$ respectively), each being multiplied by their respective fractionation factor, i.e., $\Delta^{13}C_{diamond\text{–}fluid} = XCO_2 \times \Delta^{13}C_{diamond\text{-}CO_2} + (1 - XCO_2) \times \Delta^{13}C_{diamond\text{-}CH_4}$. Except when $XCO_2 = XCH_4$, XCO_2 will vary as diamond precipitates and so will $\Delta^{13}C_{diamond\text{–}fluid}$. As illustrated in Figure 20, precipitated diamond invariably evolves along parallel trends, showing the same small increases in $\delta^{13}C$-values with fractionation up to the point at which either CO_2 or CH_4 becomes exhausted in the fluid. The absolute $\delta^{13}C$-values of diamond are, however, highly sensitive to XCO_2 and increase by 3.7‰ as XCO_2 decreases from 0.89 to 0.10 (Stachel et al. 2017).

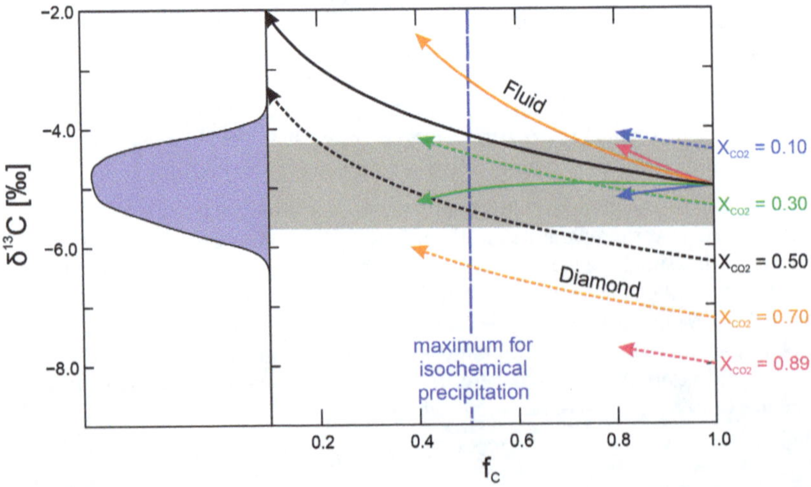

Figure 20. Evolution in carbon stable isotope composition of near water-maximum COH fluids and diamonds precipitated from such fluids during Rayleigh isotope fractionation in a multicomponent system (RIFMS; Ray and Ramesh 2000). Isotope fractionation during the diamond forming reaction $CO_2 + CH_4 = 2C + 2H_2O$ was modeled at a temperature-pressure condition of 1140 °C/5 GPa (Stachel et al. 2017). The evolution of the fluids is shown as **solid lines**, starting from an initial $\delta^{13}C$ of −5‰ (mantle value) at $f_C = 1$, with f_C being the remaining fraction of carbon-bearing species in the fluid. The evolution of precipitating diamonds is shown as **dashed lines**, matching the color of the associated fluid. XCO_2 (molar CO_2/[CO_2+CH_4]) indicates the relative proportions of the two diamond-forming carbon species in the fluid. Except for the case of $XCO_2 = 0.5$, the relative proportions of CO_2 and CH_4 evolve during diamond precipitation towards exhaustion of the minor species, at which point diamond formation stops. Isobaric cooling of a fluid, as the driver of isochemical diamond precipitation, is unlikely to exceed ~ 200 °C, limiting f_C to ≥ 0.5 (**vertical blue dashed line**). The **grey band** corresponds to the interquartile range in $\delta^{13}C$ for harzburgitic diamonds worldwide (Table 1). The distribution shown in **light blue** is a probability density function combining diamonds precipitated at four discrete XCO_2 values (0.4, 0.3, 0.2 and 0.1) and provides a good match for the distribution of $\delta^{13}C$-values in peridotitic diamonds worldwide (Fig. 2).

Isobaric cooling of fluid released, e.g., from hot intrusions of proto-kimberlite into ambient deep lithospheric mantle, likely will not exceed a ~ 200 °C temperature interval. As the amount of carbon precipitated isochemically from the fluid following reaction (1) scales with the temperature decrease, at most 50% ($f_C = 0.5$) of the carbon species in the fluid will precipitate as diamond (Stachel and Luth 2015). Based on this limit, diamond internal variations due to RIFMS processes will be smaller than 1‰. Also based on the limit $f_C \geq 0.5$, only fluids with XCO_2 between 0.4 to 0.1 precipitate diamond with carbon isotope compositions that fall between the 25th (−5.7‰) and 75th quartile (−4.2‰) for harzburgitic diamonds worldwide (shown as grey band in Fig. 20). Calculating a probability density function for all $\delta^{13}C$-values obtained from models for XCO_2 of 0.4, 0.3, 0.2 and 0.1 (calculated in f_C increments of 0.001, until either $f_C = 0.5$ or the point of CO_2 exhaustion was reached) indicates good agreement with the worldwide diamond distribution. As some fluctuation in the initial carbon isotope composition of diamond-forming fluids is likely, the variability in XCO_2 of ascending COH fluids may be even smaller than the modeled range from 0.4 to 0.1, in order to achieve the observed tight mode in the harzburgitic diamond distribution.

The isochemical diamond precipitation model of Luth and Stachel (2014) applies to harzburgitic diamond substrates below their solidus temperature. The model may, however, have applicability to lherzolitic and eclogitic diamond substrates as well (Luth 2017): in such rocks the influx of a hydrous fluid initiates melting, associated with strong partitioning of water from the fluid into the melt phase, which then should force diamond precipitation from the residual carbon-oversaturated COH fluid. This proposal awaits further detailed evaluation.

The calculations of Luth and Stachel (2014) are based on the thermodynamic model for pure COH fluids of Zhang and Duan (2009). Recent experimental work at pressures between 3 and 6 GPa showed, however, that at sub-solidus temperature conditions hydrous fluids contain up to ~40 wt% dissolved solids[6] (Adam et al. 2014; Kessel et al. 2015). The high solute content of the experimental fluids resembles that of high-density-fluids (HDFs) observed as inclusions in diamond (Navon 1991; Klein-BenDavid et al. 2009; Weiss et al. 2015, 2022, this volume). Whilst such high solute contents may affect the model of Luth and Stachel (2014) on a detailed level (potentially shifting fluid speciation and carbon solubility in *PT* space), the principle of isochemical diamond precipitation, during either isobaric cooling or ascent along a geotherm, is expected to remain valid. Based on the experimental work of Stagno and Frost (2010), HDFs derived from near-water maximum COH fluids in equilibrium with diamond (i.e., representing fO_2 conditions ~ ½ log unit more reducing than the enstatite + magnesite = olivine + diamond [EMOD] equilibrium) may carry a significant content of dissolved carbonate (equivalent to up to 20 mol% of CO_2), which then may allow for an additional oxygen conserving diamond-forming reaction between methane and dissolved carbonate to occur during cooling. As a variant of the isochemical precipitation model (Luth and Stachel 2014), a possible diamond-forming reaction involving a carbonated near-water maximum fluid passing through subsolidus peridotite is:

$$2MgCO_3 + 2CH_4 + Mg_2Si_2O_6 = 4C + 4H_2O + 2Mg_2SiO_4 \quad (3)$$

In this reaction, the sole role of the diamond substrate (the host silicate rock) is to accommodate liberated MgO through the conversion of orthopyroxene to olivine.

A question not explicitly addressed by the isochemical precipitation model is the impact of diamond precipitation from near-water maximum COH fluids on the water budget of the affected substrates. Diamondiferous peridotite xenoliths contain 25 ppm (125 ct/ton) of diamond on average, ranging up to 5500 ppm (Viljoen et al. 2004). During cooling by 200 °C (see above), a water-maximum COH fluid precipitates about 0.8 mol% diamond, which is equivalent to about 0.5 wt% diamond (see Fig. 15 of Stachel and Luth 2015). A diamond content of 25 ppm thus requires that the diamond substrates interacted with at least 0.5 wt% (5000 g/ton) fluid. Hydrous HDFs are documented as thin (≤ 1.5 μm) films around silicate and oxide inclusions (Nimis et al. 2016) and as primary fluid inclusions in both fibrous and smooth-surfaced monocrystalline diamonds (Jablon and Navon 2016). Mineral inclusions in diamond, however, are generally anhydrous (Matveev and Stachel 2009; Novella et al. 2015) and measured water contents in typical diamond source rocks (depleted cratonic peridotites) from the Kaapvaal Craton are low (< 80 ppm in olivine; Peslier et al. 2010). Metasomatism and associated diamond precipitation through essentially dry melts (<0.1 wt% H_2O in equilibrium with anhydrous olivine inclusions in diamond; Novella et al. 2015) is considered unlikely, as such dry melts cannot percolate through depleted cratonic peridotites that for normal geotherms are below their anhydrous solidus temperature. Thus, the observed low water contents can only relate to dehydrogenation of the peridotitic diamond substrates subsequent to diamond formation, likely caused by incipient melting (Matveev and Stachel 2009; Wang et al. 2021) or through more gradual upward migration of volatiles driven by continuous input of heat from the convecting mantle. Available data for the speed of hydrogen diffusion through diamond suggest re-equilibration of initially hydrous inclusions with anhydrous substrates may occur on the scale of days to thousands of years (Matveev and Stachel 2009; Cherniak et al. 2018).

[6] Dissolved solids are all non-COH (+ halogens + S) constituents of a fluid, irrespective if they are dissolved as ions or neutral species.

Diamond formation driven by pH changes in ionic fluids

Based on calculations conducted at 900 °C and 5 GPa using the Deep Earth Water (DEW) model, Sverjensky and Huang (2015) suggested that diamond may form in response to decreasing pH during interaction of an ionic hydrous fluid with coesite-bearing eclogite. Similar to the isochemical precipitation model outlined in the previous section, diamond formation via the destruction of formate ($HCOO^-$) and propionate ($C_2H_5COO^-$) occurs at nearly constant oxygen fugacity (Sverjensky and Huang 2015). Whilst constituting an appealing alternative to conventional models of diamond formation in settings with a geothermal gradient lower than the cratonic average (40 mW/m^2 model geotherm of Hasterok and Chapman 2011) and at depths near the graphite–diamond transition, geothermobarometric studies (e.g., Stachel and Luth 2015; Nimis 2022, this volume) indicate that pressure–temperature conditions of eclogitic diamond formation typically fall above the hydrous solidus of eclogite (Kessel et al. 2005), i.e., eclogitic diamond formation generally occurs in the presence of a melt, where the DEW model no longer applies. From a stable isotope perspective, although the diamond–fluid carbon isotope fractionation factors for most of the minor carbon species in the modeled fluids are unknown, CO_2, which is by far the most abundant carbon species in the fluid, should dominate the bulk fractionation factor and thus the isotope effects will be similar to diamond precipitation from an oxidizing fluid composed of neutral species (see above and Fig. 18).

Similar calculations using the DEW model to determine the speciation of fluids in equilibrium with peridotite (Sverjensky et al. 2014), indicated a non-ionic carbon speciation (CO_2, CH_4 and CO), in agreement with conventional models (e.g., Zhang and Duan 2009). This is of particular relevance as the isochemical precipitation model reviewed in the preceding section strictly applies only to strongly depleted peridotite (harzburgite).

Diamond through time—*in situ* carbon and nitrogen isotope data for diamonds with Paleoarchean to Meso/Neoproterozoic formation ages

Howell et al. (2020) conducted a large *in situ* study of the carbon isotope composition of diamond through time with the goal of examining possible temporal changes in mantle carbon speciation. In total, 908 carbon isotope measurements were obtained via multi-collector SIMS on fragments of 88 peridotitic and 56 eclogitic diamonds, with known formation ages ranging from the Paleoarchean (maximum age: 3.5 Ga) to the Mesoproterozoic (minimum age: 1.0 Ga). The key outcomes of the Howell et al. (2020) study are:

1. Cathodoluminescence imaging indicated that nearly half the diamonds formed during pulsed or episodic growth. Nevertheless, the observed variability in $\delta^{13}C$ (measured as absolute difference between highest and lowest value in single samples) was small in both peridotitic (mean: 0.8‰) and eclogitic diamonds (mean: 1.6‰). This limited variability across diamond fragments clearly implies that the heritage dataset of conventional bulk stable isotope analyses, obtained using diamond fragments ≤1 mg, is valid and its continued use is well justified.

2. No systematic trends in $\delta^{13}C$ or systematic co-variations of $\delta^{13}C$ and nitrogen content were observed in any of the 144 diamonds analyzed. This implies that kinetic fractionation effects, e.g., due to decreasing growth rates during outward growth (Reutsky et al. 2017), are uncommon, if they occur at all. Equally, Rayleigh fractionation during diamond growth is evidently uncommon. In view of the mild isotope effects of RIFMS (maximum variation in $\delta^{13}C$ during isobaric cooling by 200 °C is 0.9‰ but typically will be less; see Fig. 20), minor fractionations could remain unrecognized, but in general, the absence of coherent trends in the studied samples indicates that diamond formation rarely occurs under fluid-limited conditions where fractionation trends would be more apparent. An absence of fractionation trends in precipitated diamonds, and by inference in their parental fluids, however, does not imply that carbon isotopes are not fractionated between diamond and fluid.

3. No systematic variation is apparent in the mantle carbon isotope record since 3.5 Ga (Fig. 21). Local variability aside, from the Paleoarchean to the Mesoproterozoic the modes (not shown; see Howell et al. 2020 instead) for both peridotitic and eclogitic diamonds correspond to the mantle value of −5‰.

For nitrogen isotope measurements the worldwide database is less mature and, consequently, allows only for a preliminary assessment of possible variations through time. Combining bulk isotope analyses from literature with averages for SIMS traverses (in part calculated separately for distinct core and rim zones for clearly zoned diamonds) obtained on the same samples used for the Howell et al. (2020) study, $\delta^{15}N$-values associated with age information are available for 249 peridotitic and 97 eclogitic diamonds. For details on diamond formation ages, see Howell et al. (2020). From this data set, three broad age groups are established: (1) Archean (4.0–2.5 Ga; 144 peridotitic and 37 eclogitic), (2) Paleoproterozoic (<2.5–1.6 Ga; 71 peridotitic and 38 eclogitic) and (3) Meso- (<1.6–1.0 Ga) to Neoproterozoic diamonds (<1.0–0.54 Ga; 34 peridotitic and 23 eclogitic). Due to a strong bias of certain age groups to individual mines, the information derived from Figure 22 must be interpreted with some caution. For peridotitic diamonds, average and median values remain reasonably constant through time (e.g., median values of −0.8, +1.8 and −3.2‰ from oldest to youngest group), with the caveat that the Meso- to Neoproterozoic dataset only consists of diamonds from two locations, 1.4 Ga (Smit et al. 2010) diamonds from Ellendale 4 and 9, and 0.7 Ga (Aulbach et al. 2018) diamonds from Victor with quite disparate $\delta^{15}N$ characteristics (Ellendale being ^{15}N-enriched and Victor ^{15}N-depleted).

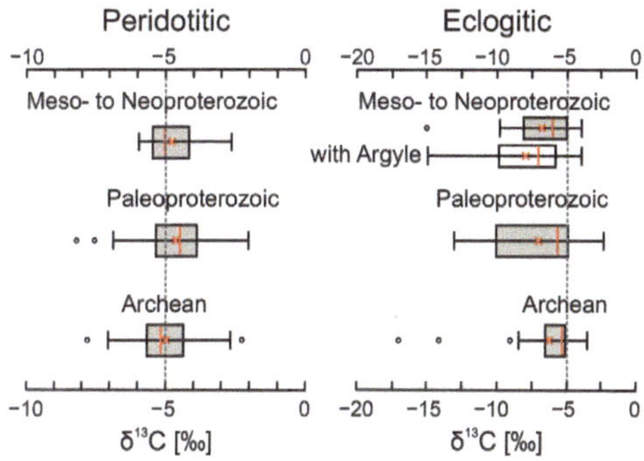

Figure 21. Carbon isotope composition ($\delta^{13}C$) of peridotitic (**left**) and eclogitic suite diamonds (**right**), grouped according to formation age. Median values (**red lines**), averages (**red crosses**), 25th to 75th percentiles (interquartile range; **boxes**) and ranges (**whiskers**) are shown. Outliers are indicated as **small circles**. The **blue dashed line** indicates the mantle value of about −5‰ (Deines 2002). Based on the carbon isotope data set of Howell et al. (2020), obtained on diamonds used in inclusion-dating studies or with otherwise constrained ages, and 22 additional analyses of Neoproterozoic diamonds from the Victor Mine. For Meso- to Neoproterozoic diamonds of the eclogitic suite two distributions are shown, one including diamonds from Argyle (labelled "with Argyle") and one (above) with eclogitic diamonds from Orapa, Jwaneng and Victor only. An apparent trend of decreasing mean and median $\delta^{13}C$ values with time disappears when the Argyle data set is excluded.

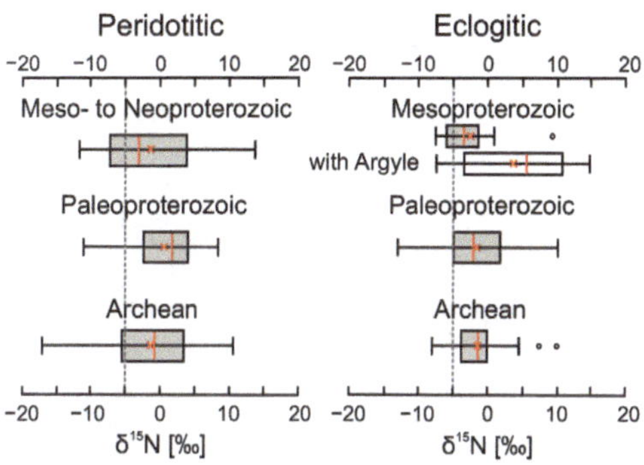

Figure 22. Nitrogen isotope composition ($\delta^{15}N$) of peridotitic (**left**) and eclogitic suite diamonds (**right**), grouped according to formation age. Median values (**red lines**), averages (**red crosses**), 25th to 75th percentiles (interquartile range; **boxes**) and ranges (**whiskers**) are shown. Outliers are indicated as **small circles**. The **blue dashed line** indicates the assumed mantle value of −5 (± 2)‰ (Cartigny and Marty 2013). For Mesoproterozoic diamonds of the eclogitic suite two distributions are shown, one including diamonds from Argyle (labelled "with Argyle") and one (above) with 1 Ga old eclogitic diamonds from Orapa and Jwaneng only.

For eclogitic diamonds, Figure 22 appears to indicate a strong increase in $\delta^{15}N$ in Mesoproterozoic diamonds, but this increase in the average (from −1.3 and −1.6 to +3.7‰, for Archean, Paleoproterozoic and Mesoproterozoic diamonds, respectively) and median $\delta^{15}N$-values (from −1.5 and −2.1 to +5.6‰) relates exclusively to the strongly ^{15}N-enriched eclogitic diamond suite from Argyle (mean of +9.5‰ and median of +10.7‰), whereas approximately 1 Ga diamonds from Orapa and Jwaneng document continuity in $\delta^{15}N$ through time (mean of −2.6‰ and median of −3.5‰).

Despite the limitations of the current data set, the above observations imply that the mantle source(s) for nitrogen in diamond did not undergo significant isotopic evolution between about 3.5 and 0.7 Ga.

ORIGIN OF LARGE RANGES IN $\delta^{13}C$ AND $\delta^{15}N$— THE RELATIVE ROLES OF MANTLE, SEDIMENT AND OCEANIC CRUST

As reviewed in the *Introduction* to this chapter, the explanation of ^{13}C-depleted carbon, observed principally in eclogitic, but also in websteritic and some peridotitic diamonds, as being sourced from subducted organic matter still remains a subject of scientific debate. To add further constraints, recent research increasingly included nitrogen isotope data for diamond, together with oxygen and sulfur isotope compositions determined for silicate and sulfide inclusions, respectively. When the first reliable isotope determinations of nitrogen in diamond became available for the different source parageneses (Boyd et al. 1987; Cartigny et al. 1997, 1998), a problem arose for the recycled sediment model in that the expected strictly positive $\delta^{15}N$-values (see Cartigny and Marty 2013, their Fig. 1 for the $\delta^{15}N$-distribution of both recent and ancient sediments) that characterize sediment-derived nitrogen were absent in the majority (~ 70%) of diamonds with low $\delta^{13}C$-values that were supposed to be indicative of sediment recycling (Cartigny 2005). Based on the expanded data set presented here, this relationship has changed given that ~ 70% (16 out of 22 samples) of diamonds with $\delta^{13}C < -15‰$ have positive $\delta^{15}N$ values (Fig. 8).

The recycled origin for C in eclogitic diamonds was bolstered by the observation of mass-independent sulfur isotope fractionation in sulfide inclusions of eclogitic diamonds (Farquhar et al. 2002; Thomassot et al. 2009; Smit et al. 2019b) and their near absence in peridotitic diamonds (Cartigny et al. 2009). Such fractionations only originated during Archean atmospheric processes and hence are restricted to Archean sediments, thereby providing a direct link between eclogitic inclusions in diamond and subduction of Archean sediments. The recycling model was further reinforced by the observation of prominent ^{18}O-enrichment ($\delta^{18}O > +6‰$, up to 16‰; $\delta^{18}O_{mantle}$ being taken as falling between +5 and +6‰) in eclogitic silicate inclusions, contained in diamonds with $\delta^{13}C$-values distinctly lower than the mantle range (Lowry et al. 1999; Schulze et al. 2003, 2013; Ickert et al. 2013; Zedgenizov et al. 2016). Questions arise from $\delta^{18}O$-values > +10‰ and up to +16‰, observed for coesite inclusions in diamond, as such high $\delta^{18}O$-values are exceedingly rare in both sediments and altered-oceanic crust. Cartigny et al. (2014) pointed out that the curvature of possible binary mixing trends in $\delta^{13}C$–$\delta^{18}O$ space between a mantle and a recycled component requires a lower carbon content in the recycled component, which is counter-intuitive in view of the low ratio of $C_{Mantle}/C_{Oceanic\ Crust}$ (~ 1/50, as deduced from C_{Mantle} of ~ 50 ppm of Marty 2012 and $C_{Oceanic\ Crust}$ of ~2000 ppm of Li et al. 2019). Simple binary mixing of a mantle and crustal component with fixed carbon concentrations, $\delta^{13}C$ and $\delta^{18}O$ is, however, not necessarily implied by the observed co-variations in $\delta^{13}C$ and $\delta^{18}O$. Instead, Ickert et al. (2013) proposed that only the most intensely altered, topmost parts of altered oceanic crust (AOC), with $\delta^{18}O \geq 6‰$, contain sufficiently high concentrations of primary organic carbon to allow for the growth of ^{13}C-depleted diamonds, whereas diamond formation in deeper AOC, associated with lower carbon contents and

$\delta^{18}O$-values, is dominated by infiltrating mantle fluids/melts, with carbon contents typically on the wt% level (e.g., Stachel and Luth 2015). Yet, this model is not universally accepted, as carbonates with high $\delta^{13}C$ are abundant in topmost parts of AOC, requiring carbonate removal by subsolidus decarbonation or partial melting reactions to precede diamond formation.

Models invoking the derivation of all C-isotope variability via within-mantle isotopic fractionation of asthenospherically-derived carbon-bearing fluids (Galimov 1991; Javoy et al. 1986) suffered from the difficulty in explaining the abundant circumstantial evidence of other signatures of crustal recycling in eclogites from both radiogenic and stable isotopes (Pearson et al. 2003; Jacob 2004). Attempts to reconcile subducted eclogitic substrates with mantle-derived carbon and nitrogen in diamonds called on the exclusive introduction of these components during metasomatic diamond growth (e.g., Thomassot et al. 2009) but suffered from the impracticability of generating the highly ^{13}C-depleted tail of the eclogitic diamond distribution through fractionation of mantle fluids (see section *Introduction*). Hence, while there is little doubt that carbon is subducted into the mantle, as evident from the mass balance and C-isotope composition of subduction zone volcanics (Mattey et al. 1984; Kelemen and Manning 2015; Plank and Manning 2019), disentangling the effects of subduction, mantle processes and fluid/melt evolution, and to develop a unifying theory to explain the C- and N-isotope data in particular, remained a major problem.

Early criticism of the requirement for subducted crustal carbon in diamonds generally focused on the inability of binary mixing to create the observed frequency distribution for $\delta^{13}C$-values in eclogitic diamonds (e.g., Javoy et al. 1986; Deines et al. 1991), while at the same time overlooking the likely mantle carbon input to the system. In addition, the inherent isotopic variability of any of the mixing endmembers involved makes the discovery of well-defined mixing trends, even between two endmembers, unlikely.

Building on an existing body of work on C- and N-isotope variability in AOC, including the observation of strongly ^{13}C-depleted values (Shilobreeva et al. 2011), Li et al. (2019) further investigated this reservoir as a source of the isotopically light carbon in diamonds and examined likely mixing relations in C–N isotope space. Given this model is among the most recent, here we review its strengths and weaknesses. In their model, Li et al. (2019) suggest that in addition to an asthenosphere-derived mantle component ($\delta^{13}C$ and $\delta^{15}N$ ~ −5‰), at least three AOC-hosted reservoirs are involved (Fig. 17) that together can explain the complex and varied C–N isotope variation in eclogitic diamonds. A key assumption in the model is the decoupling of the sources for carbon and nitrogen in these reservoirs, leading to the observed decoupling in their C- and N-isotope signatures. Li et al. (2019) implicated roles for metamorphic equivalents of: (1) "normal" marine carbonate, i.e., chemical sediment precipitated within AOC with a mean $\delta^{13}C$-value of ~ 0‰, together with clays formed by low-*T* alteration, with positive $\delta^{15}N$-values of up to +16‰; (2) biogenic carbonate (± organic matter) in AOC with a $\delta^{13}C$-value of ~ −30‰, again combined with low-*T* clays ($\delta^{15}N$ values of up to +16‰); and (3) AOC altered at relatively high temperatures, containing no carbonate or organic matter (i.e., negligible C) but ^{15}N-depleted clay ($\delta^{15}N$-values of −16 to −12‰), levering the N-isotope composition with little effect on the C-isotope composition. Mixing these three endmembers with a mantle component is a more satisfactory way of explaining the complex C–N isotope systematics of diamonds than considering sedimentary carbon alone (Fig. 17). All three AOC components will have some degree of isotopic heterogeneity making for a complex scatter of data—as observed. In addition, there may also be inhomogeneity in the isotope composition of mantle-derived nitrogen, with deep-rooted mantle plumes carrying a ^{15}N-enriched signature ($\delta^{15}N$ of +3 ± 2‰; Dauphas and Marty 1999; Marty and Dauphas 2003; Labidi et al. 2020) derived from ancient, deeply subducted material or possibly tapping primordial heterogeneities preserving strongly ^{15}N depleted signatures.

The model of Li et al. (2019) has significant advantages in being able to derive the C-, N- and O-isotope signatures of diamonds and their silicate inclusions in a spatially compact package, without requiring isotopically light carbon of exclusively organic-sediment origin to counterintuitively migrate stratigraphically downwards into the altered oceanic crust that constitutes the eclogitic diamond substrates. Instead, isotopically light and heavy C is an intrinsic component of altered oceanic crust that contains varied O-isotope signatures plus isotopically varied nitrogen being derived from secondary clays that may have light or heavy N-isotope compositions, depending on the temperature of AOC alteration.

The re-equilibration of isotopically distinct carbon in carbonate and reduced carbon, however, remained unaddressed. Also, the distribution and total range of δ^{13}C-values observed for diamonds and for AOC are not a perfect match, with high δ^{13}C-values (up to +5‰) being abundant in AOC and δ^{13}C-values below about −25‰ being very rare (e.g., Shilobreeva et al. 2011; Li et al. 2019). The suggested δ^{13}C-value of −30‰ for the biogenic carbonate—low-*T* clay end-member (Li et al. 2019; C2 in Fig. 17) already incorporates the inevitable metamorphic devolatilization of subducting slabs that drives the remaining carbonate to more negative δ^{13}C-values. For rare (2.4%) eclogitic diamonds with δ^{13}C-values below −30‰, AOC is an unlikely carbon source. Like biogenic carbonate, organic matter from the bio-alteration of igneous oceanic crust does not extend to such extreme levels of ^{13}C depletion (Banerjee et al. 2006; Kruber et al. 2008). Accordingly, while the model of Li et al. (2019) is attractive for explaining much of the apparently disparate C–N-isotope data for eclogitic diamonds, a role for isotopically light C within terrigenous or shelf sediments (average δ^{13}C of sedimentary organic carbon over the last 3.5 b.y. is −26 ± 7‰; Schidlowski 2001) is likely needed to explain diamond δ^{13}C-values below −30‰. For diamond δ^{13}C-values of about −40‰, as observed at the Jericho mine, involvement of methane in the formation of kerogen precursors for sedimentary organic carbon is required (Schidlowski 2001; Smart et al. 2011). Sedimentary inputs are also implied by the presence of anomalous Δ^{33}S variations in sulfides included within diamonds (Farquhar et al. 2002; Thomassot et al. 2009; Smit et al. 2019b). ODP drilling of forearc sediments shows that the down-going sediment inventory can be dominated by both terrigenous sediment and carbonate, whereas the carbon flux to arc magmas is dominated by the carbonate signal (Plank and Manning 2019). Subduction of some sediment along with oceanic crust into the deeper convecting mantle is demanded by modeling of the Hf–Nd isotope systematics of ocean island basalts (Chauvel et al. 2008).

A further open question that currently is very difficult to address is the role of isotope fractionation in creating some of the variance observed in Figure 17. Of the 335 combined C- and N-isotope analyses of eclogitic diamonds in our database, 239 (71%) fall into the C-isotope mantle range (δ^{13}C between −7 and −3‰). During diamond precipitation in a fluid-limited system, the N-isotope composition of diamond varies by an order of magnitude more than the carbon isotope composition (Mikhail et al. 2014; Hogberg et al. 2016; see also discussion of isotope fractionation above). Consequently, for the dominant population of eclogitic diamonds with a mantle-like C-isotope composition, elevated δ^{15}N-values (≥ 0‰) may be interpreted to represent either mixing of C and N derived from the mantle and from AOC (combinations of low temperature clay with either normal or biogenic carbonate, at $N/C_{Mantle} \ll N/C_{AOC}$; see Fig. 17) or the much stronger evolution of N-isotopes (due to a larger isotope fractionation factor or a lower residual fraction of nitrogen) during Rayleigh fractionation. We note, however, that the match between the δ^{15}N frequency distribution for eclogitic diamonds and the relative proportions of the expected AOC reservoirs is not perfect and that Rayleigh isotope fractionation, at least during diamond precipitation, appears to occur only rarely (see above). Furthermore, no coupled δ^{13}C–δ^{15}N–δ^{18}O dataset for AOC exist yet and the Li et al. (2019) model is based on a data set that was assembled from several studies. An important next step, therefore, is to collect a comprehensive isotopic dataset for AOC and to document its evolution during metamorphic re-equilibration and devolatilization.

Despite these currently unaddressed questions, Li et al. (2019) for the first time connected the broad ranges in observed C- and N-stable isotope signatures, their paragenetic association and their relationship to observed O-isotope variations and identified a combination of at least five major sources acting in various combinations:

1. Asthenosphere-derived typical mantle fluids carrying the "canonical" mantle C- and N-isotope signatures, possibly with a distinct N-isotope signature for deeper "plume mantle" sources (Dauphas and Marty 1999; Marty and Dauphas 2003).
2. Subducted shallow igneous oceanic crust that originally was altered at low temperatures, causing precipitation of biogenic and/or normal carbonate together with formation of clay.
3. Subducted deeper igneous oceanic crust that originally was altered at higher temperatures, including the formation of clay.
4. Subducted marine carbonate sediment with varying detrital clay contributions
5. Subducted terrigenous sediment containing kerogenous carbon.

The signature of this menu of widely varying C- and N-isotopes is enhanced by additional petrogenetic processes such as decarbonation reactions and isotopic fractionations.

FUTURE DIRECTIONS

To address the open questions discussed in this review, future research into the following topics would be beneficial:

- The speciation and partitioning behavior of nitrogen (K_N diamond–fluid) during diamond precipitation needs to be addressed experimentally over a range of oxygen fugacities.
- Nitrogen isotope fractionation factors between diamond and the various possible nitrogen species in diamond-forming fluids need to be investigated experimentally or with rigorous theoretical calculations.
- Elemental and isotopic partitioning data for nitrogen would allow to make full use of a rich dataset of nitrogen concentration measurements and a growing dataset of $\delta^{15}N$ analyses of diamond. For the relative rare cases where clear covariations between $\delta^{13}C$ and N-content occur, based on partitioning data the concentration of nitrogen in the fluid or the speciation of nitrogen could be constrained.
- Continued search for isotopic or elemental zoning patterns in individual diamonds (e.g., sector zoning) that can unequivocally be ascribed to kinetic isotope fractionation processes.
- Improved spatial resolution of stable isotope analyses could reveal hitherto unrecognized fractionation trends on the µm to sub-µm level.
- More extensive coupled C- and N-isotope studies need to be performed on altered oceanic crust and mantle, in the form of fracture-zone peridotites, to better evaluate the extent of coupling or decoupling of C- and N-isotopes in subducting slabs. This should be accompanied by experimental studies of the isotopic effects of devolatilization of these slabs.
- More extensive thermodynamic data relevant to fluid speciation at deep lithospheric mantle and asthenosphere / transition zone pressures would aid in understanding isotopic variations in diamonds formed in these environments. In particular,

the thermodynamic model for COH fluids in the mantle (Zhang and Duan 2009) needs to be expanded to include the role of dissolved solids.

- The observation of distinct features (higher N-content, lower N-aggregation state, more restricted range of $\delta^{13}C$-values) of sulfide-bearing diamonds relative to silicate-included diamonds from the same locality and of the same paragenesis needs to be studied in more detail to evaluate if sulfide-included diamonds have specific characteristics or if these differences just represent multiple generations of diamonds, with one being more sulfide-rich.

ACKNOWLEDGMENTS

We are grateful to the late Erik Hauri for being an ardent supporter of diamond research under the Deep Carbon Observatory banner. It was also Erik who pushed us hard to get the large database of stable isotope and nitrogen data for diamond that comes with this paper out into the public domain. Kelsey Graversen (University of Alberta) is sincerely thanked for conducting tedious and detailed error-checking of the database and for adding the DOIs for all references to the spreadsheet. Emilie Thomassot still found a few more wrong entries and we thank her for that. A special thanks to Richard Stern, managing director of the Canadian Centre for Isotopic Microanalysis, for supporting the research summarized in this review over the years through analytical developments and a trove of highest quality analyses. Juske Horita, Sami Mikhail and Zachary Sharp are thanked for their constructive input as reviewers. Karen Smit as our editor went above and beyond to get this chapter in good shape.

REFERENCES

Adam J, Locmelis M, Afonso JC, Rushmer T, Fiorentini ML (2014) The capacity of hydrous fluids to transport and fractionate incompatible elements and metals within the Earth's mantle. Geochem Geophys Geosystems 15:2241–2253

Akagi T, Masuda A (1988) Isotopic and elemental evidence for a relationship between kimberlite and Zaire cubic diamonds. Nature 336:665–667

Appora-Gnekindy I (1998) Etude expérimentale du fractionnement isotopique du carbone et de l'oxygène dans les systèmes CO_2–carbonate liquides: Application aux contextes carbonatiques. PhD thesis, Institut de Physique du Globe de Paris, 351p

Aubaud C, Pineau F, Hekinian R, Javoy M (2006) Carbon and hydrogen isotope constraints on degassing of CO_2 and H_2O in submarine lavas from the Pitcairn hotspot (South Pacific). Geophys Res Lett 33: L02308

Aulbach S, Creaser RA, Stachel T, Heaman LM, Chinn IL, Kong J (2018) Diamond ages from Victor (Superior Craton): Intra-mantle cycling of volatiles (C, N, S) during supercontinent reorganisation. Earth Planet Sci Lett 490:77–87

Badziag P, Verwoerd WS, Ellis WP, Greiner NR (1990) Nanometre-sized diamonds are more stable than graphite. Nature 343:244–245

Banerjee NR, Furnes H, Muehlenbachs K, Staudigel H, de Wit M (2006) Preservation of ~3.4–3.5 Ga microbial biomarkers in pillow lavas and hyaloclastites from the Barberton Greenstone Belt, South Africa. Earth Planet Sci Lett 241:707–722

Beyssac O, Chopin C (2003) Comment on "Diamond, former coesite and supersilicic garnet in metasedimentary rocks from the Greek Rhodope: a new ultrahigh-pressure metamorphic province established" by E.D. Mposkos and D.K. Kostopoulos–Earth Planet. Sci. Lett. 192 (2001) 497–506. Earth Planet Sci Lett 214:669–674

Bottinga Y (1968) Calculations of fractionation factors for carbon and oxygen isotopic exchange in the system calcite–carbon dioxide–water. J Phys Chem 72:800–808

Bottinga Y (1969a) Calculated fractionation factors for carbon and hydrogen isotope exchange in the system calcite–carbon dioxide–graphite–methane–hydrogen–water vapor. Geochim Cosmochim Acta 33:49–64

Bottinga Y (1969b) Carbon isotope fractionation between graphite, diamond and carbon dioxide. Earth Planet Sci Lett 5:301–307

Boyd FR, Pokhilenko NP, Pearson DG, Mertzman SA, Sobolev NV, Finger LW (1997) Composition of the Siberian cratonic mantle: evidence from Udachnaya peridotite xenoliths. Contrib Mineral Petrol 128:228–246

Boyd SR, Pillinger CT (1994) A preliminary-study of $^{15}N/^{14}N$ in octahedral growth form diamonds. Chem Geol 116:43–59

Boyd SR, Mattey DP, Pillinger CT, Milledge HJ, Mendelssohn M, Seal M (1987) Multiple growth events during diamond genesis: an integrated study of carbon and nitrogen isotopes and nitrogen aggregation state in coated stones. Earth Planet Sci Lett 86:341–353

Boyd SR, Pillinger CT, Milledge HJ, Mendelssohn MJ, Seal M (1988) Fractionation of nitrogen isotopes in a synthetic diamond of mixed-crystal habit. Nature 331:604–607

Boyd SR, Pillinger CT, Milledge HJ, Mendelssohn MJ, Seal M (1992) C and N isotopic composition and the infrared absorption spectra of coated diamonds: evidence for the regional uniformity of CO_2–H_2O rich fluids in lithospheric mantle. Earth Planet Sci Lett 109:633–644

Boyd SR, Pineau F, Javoy M (1994) Modelling the growth of natural diamonds. Chem Geol 116:29–42

Brey GP, Bulatov V, Girnis A, Harris JW, Stachel T (2004) Ferropericlase–a lower mantle phase in the upper mantle. Lithos 77:655–663

Bulanova GP, Pearson DG., Hauri EH, Griffin BJ (2002) Carbon and nitrogen isotope systematics within a sector-growth diamond from the Mir kimberlite, Yakutia. Chem Geol 188:105–123

Bureau H, Remusat L, Esteve I, Pinti DL, Cartigny P (2018) The growth of lithospheric diamonds. Sci Adv 4:eaat1602

Burgess R, Cartigny P, Harrison D, Hobson E, Harris J (2009) Volatile composition of microinclusions in diamonds from the Panda kimberlite, Canada: Implications for chemical and isotopic heterogeneity in the mantle. Geochim Cosmochim Acta 73:1779–1794

Burton JA, Prim RC, Slichter WP (1953) The distribution of solute in crystals grown from the melt. Part I. Theoretical. J Chem Phys 21:1987–1991

Cartigny P (2005) Stable isotopes and the origin of diamond. Elements 1:79–84

Cartigny P (2010) Mantle-related carbonados? Geochemical insights from diamonds from the Dachine komatiite (French Guiana). Earth Planet Sci Lett 296:329–339

Cartigny P, Marty B (2013) Nitrogen isotopes and mantle geodynamics: The emergence of life and the atmosphere-crust–mantle connection. Elements 9:359–366

Cartigny P, Boyd SR, Harris JW, Javoy M (1997) Nitrogen isotopes in peridotitic diamonds from Fuxian, China: the mantle signature. Terra Nova 9:175–179

Cartigny P, Harris JW, Javoy M (1998) Eclogitic diamond formation at Jwaneng: No room for a recycled component. Science 280:1421–1424

Cartigny P, Harris JW, Javoy M (2001) Diamond genesis, mantle fractionations and mantle nitrogen content: a study of $\delta^{13}C$–N concentrations in diamonds. Earth Planet Sci Lett 185:85–98

Cartigny P, Harris JW, Taylor A, Davies R, Javoy M (2003) On the possibility of a kinetic fractionation of nitrogen stable isotopes during natural diamond growth. Geochim Cosmochim Acta 67:1571–1576

Cartigny P, Stachel T, Harris JW, Javoy M (2004) Constraining diamond metasomatic growth using C- and N-stable isotopes: examples from Namibia. Lithos 77:359–373

Cartigny P, Farquhar J, Thomassot E, Harris JW, Wing B, Masterson A, McKeegan K, Stachel T (2009) A mantle origin for Paleoarchean peridotitic diamonds from the Panda kimberlite, Slave Craton: Evidence from ^{13}C-, ^{15}N- and $^{33,34}S$-stable isotope systematics. Lithos 112:852–864

Cartigny P, Palot M, Thomassot E, Harris JW (2014) Diamond formation: a stable isotope perspective. Annu Rev Earth Planet Sci 42:699–732

Casciotti KL (2009) Inverse kinetic isotope fractionation during bacterial nitrite oxidation. Geochim Cosmochim Acta 73:2061–2076

Chacko T, Mayeda TK, Clayton RN, Goldsmith JR (1991) Oxygen and carbon isotope fractionations between CO_2 and calcite. Geochim Cosmochim Acta 55:2867–2882

Chacko T, Cole DR, Horita J (2001) Equilibrium oxygen, carbon and hydrogen isotope fractionation factors applicable to geological systems. Rev Mineral Geochem 43:1–81

Chauvel C, Lewin E, Carpentier M, Arndt N, Marini JC (2008) The role of recycled oceanic basalt and sediment in generating the Hf-Nd mantle array. Nat Geosci 1:64–67

Cherniak DJ, Watson EB, Meunier V, Kharche N (2018) Diffusion of helium, hydrogen and deuterium in diamond: Experiment, theory and geochemical applications. Geochim Cosmochim Acta 232:206–224

Chinn IL, Perritt SH, Stiefenhofer J, Stern RA (2018) Diamonds from Orapa Mine show a clear subduction signature in SIMS stable isotope data. Mineral Petrol 112:197–207

Chopin C (2003) Ultrahigh-pressure metamorphism: tracing continental crust into the mantle. Earth Planet Sci Lett 212:1–14

Chrenko RM, Tuft RE, Strong HM (1977) Transformation of the state of nitrogen in diamond. Nature 270:141–144

Clayton RN, Goldsmith JR, Karel KJ, Mayeda TK, Newton RC (1975) Limits on the effect of pressure on isotopic fractionation. Geochim Cosmochim Acta 39:1197–1201

Craig H (1953) The geochemistry of the stable carbon isotopes. Geochim Cosmochim Acta 3:53–92

Craven JA, Harte B, Fisher D, Schulze DJ (2009) Diffusion in diamond. I. Carbon isotope mapping of natural diamond. Mineral Mag 73:193–200

Daulton TL, Ozima M (1996) Radiation-induced diamond formation in uranium-rich carbonaceous materials. Science 271:1260–1263

Dauphas N, Marty B (1999) Heavy nitrogen in carbonatites of the Kola Peninsula: A possible signature of the deep mantle. Science 286:2488–2490
Davies GR, Nixon PH, Pearson DG, Obata M (1993) Tectonic implications of graphitized diamonds from the Ronda, peridotite massif, southern Spain. Geology 21:471–474
Davies RM, Griffin WL, O'Reilly SY, Andrew AS (2003) Unusual mineral inclusions and carbon isotopes of alluvial diamonds from Bingara, eastern Australia. Lithos 69:51–66
De Corte K, Cartigny P, Shatsky VS, Sobolev NV, Javoy M (1998) Evidence of fluid inclusions in metamorphic microdiamonds from the Kokchetav massif, northern Kazakhstan. Geochim Cosmochim Acta 62:3765–3773
De Corte K, Cartigny P, Shatsky VS, Sobolev NV, Javoy M (1999) Characteristics of microdiamonds from UHPM rocks of the Kokchetav massif (Kazakhstan). *In:* Proc 7th Int Kimberlite Conf Vol. 1, Gurney JJ et al. (eds) Red Roof Design, Cape Town, p 174–182
De Stefano A, Kopylova MG, Cartigny P, Afanasiev V (2009) Diamonds and eclogites of the Jericho kimberlite (Northern Canada). Contrib Mineral Petrol 158:295–315
Deines P (1980) The carbon isotopic composition of diamonds: relationship to diamond shape, color, occurrence and vapor composition. Geochim Cosmochim Acta 44:943–961
Deines P (2002) The carbon isotope geochemistry of mantle xenoliths. Earth Sci Rev 58:247–278
Deines P (2004) Carbon isotope effects in carbonate systems. Geochim Cosmochim Acta 68:2659–2579
Deines P, Eggler DH (2009) Experimental determination of carbon isotope fractionation between $CaCO_3$ and graphite. Geochim Cosmochim Acta 73:7256–7274
Deines P, Haggerty SE (2000) Small-scale oxygen isotope variations and petrochemistry of ultradeep (> 300 km) and transition zone xenoliths. Geochim Cosmochim Acta 64:117–131
Deines P, Gurney JJ, Harris JW (1984) Associated chemical and carbon isotopic composition variations in diamonds from Finsch and Premier kimberlite, South-Africa. Geochim Cosmochim Acta 48:325–342
Deines P, Harris JW, Spear PM. Gurney JJ (1989) Nitrogen and ^{13}C content of Finsch and Premier diamonds and their implications. Geochim Cosmochim Acta 53:1367–1378
Deines P, Harris JW, Robinson DN, Gurney JJ, Shee SR (1991) Carbon and oxygen isotope variations in diamond and graphite eclogites from Orapa, Botswana, and the nitrogen-content of their diamonds. Geochim Cosmochim Acta 55:515–524
Deines P, Stachel T, Harris JW (2009) Systematic regional variations in diamond carbon isotopic composition and inclusion chemistry beneath the Orapa kimberlite cluster, in Botswana. Lithos 112:776–784
Donnelly CL, Stachel T, Creighton S, Muehlenbachs K, Whiteford S (2007) Diamonds and their mineral inclusions from the A154 South pipe, Diavik Diamond Mine, Northwest Territories, Canada. Lithos 98:160–176
Dobrzhinetskaya LF, O'Bannon EF III, Sumino H (2022) Non-cratonic diamonds from UHP metamorphic terranes, ophiolites and volcanic sources. Rev Mineral Geochem 88:191–256
Eggler DH, Baker DR (1982) Reduced volatiles in the system C–O–H: Implications for mantle melting, fluid formation and diamond genesis. In: High pressure research in geophysics. Akimoto S, Manghnani MH (eds) Center for Academic Publications, Tokyo, p 237–250
Evans T, Qi Z (1982) The kinetics of the aggregation of nitrogen atoms in diamond. Proc R Soc London A381:159–178
Evans T, Harris JW (1989) Nitrogen aggregation, inclusion equilibration temperatures and the age of diamonds. In Kimberlites and Related Rocks. Proc 4th Int Kimberlite Conf, GSA Spec Publ 14 Vol. 2, Ross J et al. (eds) Blackwell, Carlton, p. 1001–1006
Farquhar J, Wing BA, McKeegan KD, Harris JW, Cartigny P, Thiemens MH (2002) Mass-independent sulfur of inclusions in diamond and sulfur recycling on early Earth. Sci 298:2369–2372
Fitzsimons ICW, Harte B, Chinn IL, Gurney JJ, Taylor WR (1999) Extreme chemical variation in complex diamonds from George Creek, Colorado: a SIMS study of carbon isotope composition and nitrogen abundance. Mineral Mag 63:857–878
Fitzsimons ICW, Harte B, Clark RM (2000) SIMS stable isotope measurement: counting statistics and analytical precision. Mineral Mag 64:59–83
Fraley C, Raftery AE (2002) Model-based clustering, discriminant analysis, and density estimation. J Am Stat Assoc 97:611–631
Frost DJ, McCammon CA (2008) The redox state of Earth's mantle. Annu Rev Earth Planet Sci 36:389–420
Galbraith RF (2005) Statistics for Fission Track Analysis. Chapman and Hall/CRC, Boca Raton, Florida
Galimov EM (1991) Isotope fractionation related to kimberlite magmatism and diamond formation. Geochim Cosmochim Acta 55:1697–1708
Garai J, Haggerty SE, Rekhi S, Chance M (2006) Infrared absorption investigations confirm the extraterrestrial origin of carbonado diamonds. Astrophys J 653:L153–L156
Giuliani A, Kamenetsky VS, Phillips D, Kendrick MA, Wyatt BA, Goemann K (2012) Nature of alkali-carbonate fluids in the sub-continental lithospheric mantle. Geology 40:967–970
Green BL, Collins AT, Breeding CM (2022) Diamond spectroscopy, defect centers, color, and treatments. Rev Mineral Geochem 88:637–688

Gurney JJ, Harris JW, Rickard RS (1984) Silicate and oxide inclusions in diamonds from the Orapa Mine, Botswana. *In*: Kimberlites II: the mantle and crust-mantle relationships. Kornprobst J (ed) Elsevier, Amsterdam, p 3–9
Haggerty SE (2014) Carbonado: Physical and chemical properties, a critical evaluation of proposed origins, and a revised genetic model. Earth Sci Rev 130:49–72
Hanschmann G (1981) Berechnung von Isotopieeffekten auf quantenchemischer Grundlage am Beispiel stickstoffhaltiger Moleküle. ZFI-Mitteilungen 41:19–39
Harris JW, Hawthorne JB, Oosterveld MM (1986) A comparison of characteristics of diamonds from Orapa and Jwaneng kimberlite pipes in Botswana. Extended Abstr 4th Int Kimperlite Conf Geol Soc Australia Spec Publ 16:395–397
Harris JW, Smit KV, Fedortchouk Y, Moore M (2022) Morphology of monocrystalline diamond and its inclusions. Rev Mineral Geochem 88:119–166
Harte B (2010) Diamond formation in the deep mantle: the record of mineral inclusions and their distribution in relation to mantle dehydration zones. Mineral Mag 74:189–215
Harte B, Otter M, McKeegan K (1992) Carbon isotope measurements on diamonds. Chem Geol 101:177–183
Harte B, Fitzsimons ICW, Harris JW, Otter ML (1999) Carbon isotope ratios and nitrogen abundances in relation to cathodoluminescence characteristics for some diamonds from the Kaapvaal Province, S-Africa. Mineral Mag 63:829–856
Harte B, Taniguchi T, Chakraborty S (2009) Diffusion in diamond. II. High-pressure-temperature experiments. Mineral Mag 73:201–204
Hasterok D, Chapman DS (2011) Heat production and geotherms for the continental lithosphere. Earth Planet Sci Lett 307:59–70
Hauri EH, Pearson DG, Bulanova GP, Milledge HJ (1999) Microscale variations in C and N isotopes within mantle diamonds revealed by SIMS. *In:* Proc 7th Int Kimberlite Conf Vol 1. Gurney JJ et al. (eds) Red Roof Design, Cape Town, p 341–347
Hogberg K, Stachel T, Stern RA (2016) Carbon and nitrogen isotope systematics in diamond: Different sensitivities to isotopic fractionation or a decoupled origin? Lithos 265:16–30
Holland G, Lollar B, Li L et al. (2013) Deep fracture fluids isolated in the crust since the Precambrian era. Nature 497:357–360
Horita J (2001) Carbon isotope exchange in the system CO_2–CH_4 at elevated temperatures. Geochim Cosmochim Acta 65:1907–1919
Horita J, Polyakov VB (2015) Carbon-bearing iron phases and the carbon isotope composition of the deep Earth. PNAS 112:31–36
Horita J, Wesolowski DJ (1994) Liquid–vapor fractionation of oxygen and hydrogen isotopes of water from the freezing to the critical temperature. Geochim Cosmochim Acta 58:3425–3437
Hough RM, Gilmour I, Pillinger CT, Arden JW, Gilkes KWR, Yuan J, Milledge HJ (1995) Diamond and silicon-carbide in impact melt rock from the Ries impact crater. Nature 378:41–44
Howell D, O'Neill CJ, Grant KJ, Griffin WL, O'Reilly SY, Pearson NJ, Stern RA, Stachel T (2012) Platelet development in cuboid diamonds: insights from micro-FTIR mapping. Contrib Mineral Petrol 164:1011–1025
Howell D, Griffin WL, Yang J, Gain S, Stern RA, Huang JX, Jacob DE, Xu X, Stokes AJ, O'Reilly SY, Pearson NJ (2015a) Diamonds in ophiolites: Contamination or a new diamond growth environment? Earth Planet Sci Lett 430:284–295
Howell D, Stern RA, Griffin WL, Southworth R, Mikhail S, Stachel T (2015b) Nitrogen isotope systematics and origins of mixed-habit diamonds. Geochim Cosmochim Acta 157:1–12
Howell D, Stachel T, Stern RA, Pearson DG, Nestola F, Hardman MF, Harris JW, Jaques AL, Shirey SB, Cartigny P, Smit KV (2020) Deep carbon through time: Earth's diamond record and its implications for carbon cycling and fluid speciation in the mantle. Geochim Cosmochim Acta 275:99–122
Huss GR (2005) Meteoritic nanodiamonds: Messengers from the stars. Elements 1:97–100
Ickert RB, Stachel T, Stern RA, Harris JW (2013) Diamond from recycled crustal carbon documented by coupled $\delta^{18}O$–$\delta^{13}C$ measurements of diamonds and their inclusions. Earth Planet Sci Lett 364:85–97
Jacob DE (2004) Nature and origin of eclogite xenoliths from kimberlites. Lithos 77:295–316
Jacob DE, Mikhail S (2022) Polycrystalline diamonds from kimberlites: Snapshots of rapid and episodic diamond formation in the lithospheric mantle. Rev Mineral Geochem 88:167–190
Jablon BM, Navon O (2016) Most diamonds were created equal. Earth Planet Sci Lett 443:41–47
Javoy M, Pineau F, Iiyama I (1978) Experimental-determination of isotopic fractionation between gaseous CO_2 and carbon dissolved in tholeiitic magma–preliminary study. Contrib Mineral Petrol 67:35–39
Javoy M, Pineau F, Allegre CJ (1982) Carbon geodynamic cycle. Nature 300:171–173
Javoy M, Pineau F, Demaiffe D (1984) Nitrogen and carbon isotopic composition in the diamonds of Mbuji Mayi (Zaïre). Earth Planet Sci Lett 68:399–412
Javoy M, Pineau F, Delorme H (1986) Carbon and nitrogen isotopes in the mantle. Chem Geol 57:41–62
Johnson CN, Stachel T, Muehlenbachs K, Stern RA, Armstrong JP, EIMF (2012) The micro-/macro-diamond relationship: A case study from the Artemisia kimberlite (Northern Slave Craton, Canada). Lithos 148:86–97

Kagi H, Fukura S (2008) Infrared and Raman spectroscopic observations of Central African carbonado and implications for its origin. Euro J Mineral 20:387–393

Kaiser W, Bond L (1959) Nitrogen, a major impurity in common type I diamonds. Phys Rev 115:857–863

Kamiya Y, Lang AR (1964) On the structure of coated diamonds. Philos Mag 11:347–356

Kelemen PB, Manning CE (2015) Reevaluating carbon fluxes in subduction zones, what goes down, mostly comes up. PNAS 112:E3997–E4006

Kessel R, Ulmer P, Pettke T, Schmidt MW, Thompson AB (2005) The water-basalt system at 4 to 6 GPa: Phase relations and second critical endpoint in a K-free eclogite at 700 to 1400 degrees C. Earth Planet Sci Lett 237:873–892

Kessel R, Pettke T, Fumagalli P (2015) Melting of metasomatized peridotite at 4–6 GPa and up to 1200 °C: an experimental approach. Contrib Mineral Petrol 169:1–19

Kieffer SW (1982) Thermodynamics and lattice vibrations of minerals: 5. Application to phase equilibria, isotopic fractionation, and high-pressure thermodynamic properties. Rev Geophys Space Phys 20:827–849

Kjarsgaard BA, de Wit M, Heaman LM, Pearson DG, Stiefenhofer J, Janusczcak N, Shirey SB (2022) A review of the geology of global diamond mines and deposits. Rev Mineral Geochem 88:1–118

Klein-BenDavid O, Logvinova AM, Schrauder M, Spetius ZV, Weiss Y, Hauri EH, Kaminsky FV, Sobolev NV, Navon O (2009) High-Mg carbonatitic microinclusions in some Yakutian diamonds—a new type of diamond-forming fluid. Lithos 112 Suppl 2:648–659

Klein-BenDavid O, Pearson DG, Nowell GM, Ottley C, McNeill JCR, Cartigny P (2010) Mixed fluid sources involved in diamond growth constrained by Sr–Nd–Pb–C–N isotopes and trace elements. Earth Planet Sci Lett 289:123–133

Knoche R, Sweeney RJ, Luth RW (1999) Carbonation and decarbonation of eclogites: the role of garnet. Contrib Mineral Petrol 135:332–339

Koeberl C, Masaitis VL, Shafranovsky GI, Gilmour I, Langenhorst F, Schrauder M (1997) Diamonds from the Popigai impact structure, Russia. Geology 25:967–970

Koga KT, Van Orman JA, Walter MJ (2003) Diffusive relaxation of carbon and nitrogen isotope heterogeneity in diamond: a new thermochronometer. Phys Earth Planet Sci 139:35–43

Koga KT, Walter MJ, Nakamura E, Kobayashi K (2005) Carbon self-diffusion in a natural diamond. Phys Rev B 72:024108

Korsakov AV, Rezvukhina OV, Jaszczak JA, Rezvukhin DI, Mikhailenko DS (2019) Natural graphite cuboids. Minerals 9:110

Koval'skiy VV, Galimov EM, Prokhorov VS (1972) Isotopic composition of carbon from colored Yakutian diamonds. Trans (Doklady) USSR Acad Sci 203:118–119 (translated from Doklady Akademii Nauk SSSR, 1982, 203:1440–1442)

Krebs MY, Pearson DG, Stachel T, Stern RA, Nowicki T, Cairns S (2016) Using microdiamonds in kimberlite diamond grade prediction: a case study of the variability in diamond population characteristics across the size range 0.2 to 3.4 mm in Misery kimberlite, Ekati Mine, NWT, Canada. Econ Geol 111:503–525

Kruber C, Thorseth IH, Pedersen RB (2008) Seafloor alteration of basaltic glass: Textures, geochemistry, and endolithic microorganisms. Geochem Geophys Geosystems 9:12

Kueter N, Lilley MD, Schmidt MW, Bernasconi SM (2019a) Experimental carbonatite/graphite carbon isotope fractionation and carbonate/graphite geothermometry. Geochim Cosmochim Acta 253:290–306

Kueter N, Schmidt MW, Lilley MD, Bernasconi SM (2019b) Experimental determination of equilibrium CH_4–CO_2–CO carbon isotope fractionation factors (300–1200 °C). Earth Planet Sci Lett 506:64–75

Kueter N, Schmidt MW, Lilley MD, Bernasconi SM (2020) Kinetic carbon isotope fractionation links graphite and diamond precipitation to reduced sources. Earth Planet Sci Lett 529:115848

Labidi J, Barry PH, Bekaert DV, et al. (2020) Hydrothermal ^{15}N–^{15}N abundances constrain the origins of mantle nitrogen. Nature 580:367–371

Lavoisier A-L (1776) Second mémoire sur la destruction du diamant au grand verre brûlant de Tschirnausen, connu sous le nom de Lentille du Palais royal, Mémoires de l'Académie royale des sciences. 1772 II:591–616

Leech M, Ernst WG (1998) Graphite pseudomorphs after diamond? A carbon isotope and spectroscopic study of graphite cuboids from the Maksyutov Complex, south Ural Mountains, Russia. Geochim Cosmochim Acta 62:2143–2154

Li K, Li L, Pearson DG, Stachel T (2019) Diamond isotope compositions indicate altered igneous oceanic crust dominates deep carbon recycling. Earth Planet Sci Lett 516:190–201

Li Y, Keppler H (2014) Nitrogen speciation in mantle and crustal fluids. Geochim Cosmochim Acta 129:13–32

Liu J, Wang W, Yang H, Wu Z, Hu MY, Zhao J, Bi W, Alp EE, Dauphas N, Liang W, Chen B, Lin JF (2019) Carbon isotopic signatures of super-deep diamonds mediated by redox chemistry. Geochem Perspec Lett 10:51–55

Lowry D, Mattey DP, Harris JW (1999) Oxygen isotope composition of syngenetic inclusions in diamond from the Finsch Mine, RSA. Geochim Cosmochim Acta 63:1825–1836

Luth RW (1993) Diamonds, eclogites, and the oxidation state of the Earth's mantle. Science 261:66–68

Luth RW (2017) Diamond formation during parital melting in the Earth's mantle. Geol Soc Am 129th Annual Meeting, Seattle, abstract # 20–6

Luth RW, Stachel T (2014) The buffering capacity of lithospheric mantle: implications for diamond formation. Contrib Mineral Petrol 168:1–12

Luth RW, Palyanov YN, Bureau H (2022) Experimental petrology applied to natural diamond growth. Rev Mineral Geochem 88:755–808

Marty B (2012) The origins and concentrations of water, carbon, nitrogen and noble gases on Earth. Earth Planet Sci Lett 313–314:56–66
Marty B, Dauphas N (2003) The nitrogen record of crust–mantle interaction and mantle convection from Archean to present. Earth Planet Sci Lett 206:397–410
Matsuhisa Y, Goldsmith JR, Clayton RN (1978) Mechanisms of hydrothermal crystallization of quartz at 250 °C and 15 kbar. Geochim Cosmochim Acta 42:173–182
Mattey DP (1991) Carbon-dioxide solubility and carbon isotope fractionation in basaltic melt. Geochim Cosmochim Acta 55:3467–3473
Mattey DP, Carr RH, Wright IP, Pillinger CT (1984) Carbon isotopes in submarine basalts. Earth Planet Sci Lett 70:196–206
Mattey DP, Taylor WR, Green DH, Pillinger CT (1990) Carbon isotopic fractionation between CO_2 vapor, silicate and carbonate melts–an experimental-study to 30 kbar. Contrib Mineral Petrol 104:492–505
Matveev S, Stachel T (2009) Evaluation of kimberlite diamond potential using FTIR spectroscopy of xenocrystic olivine. Lithos 112:36–40
McCandless TE, Gurney JJ (1997) Diamond eclogites: Comparison with carbonaceous chondrites, carbonaceous shales, and microbial carbon-enriched MORB. Geol Geofiz 38:371–381
McDade P, Harris JW (1999) Syngenetic inclusions bearing diamonds from Letseng-La-Terai, Lesotho. *In:* Proc 7th Int Kimberlite Conf Vol 2. Gurney JJ et al. (eds) Red Roof Design, Cape Town, p 557–565
Meyer HOA (1987) Inclusions in diamond. *In:* Mantle Xenoliths. Nixon PH (ed) John Wiley and Sons Ltd., Chichester, p 501–522
Meyer HOA, Boyd FR (1972) Composition and origin of crystalline inclusions in natural diamonds. Geochim Cosmochim Acta 36:1255–1273
Milledge HJ, Mendelssohn MJ, Seal M, Rouse JE, Swart PK, Pillinger CT (1983) Carbon isotopic variation in spectral type II diamonds. Nature 303:791-792
Mikhail S, Howell D (2016) A petrological assessment of diamond as a recorder of the mantle nitrogen cycle. Am Mineral 101:780–787
Mikhail S, Sverjensky DA (2014) Nitrogen speciation in upper mantle fluids and the origin of Earth's nitrogen-rich atmosphere. Nat Geosci 7:816–819
Mikhail S, Verchovsky AB, Howell D, Hutchison MT, Southworth R, Thomson AR, Warburton P, Jones AP, Milledge HJ (2014) Constraining the internal variability of the stable isotopes of carbon and nitrogen within mantle diamonds. Chem Geol 366:14–23
Moore RO, Gurney JJ (1985) Pyroxene solid solution in garnets included in diamond. Nature 318:553–555
Mysen B (2018) Mass transfer in the Earth's interior: fluid–melt interaction in aluminosilicate–C–O–H–N systems at high pressure and temperature under oxidizing conditions. Prog Earth Planet Sci 5:6
Mysen B (2019) Nitrogen in the Earth: abundance and transport. Prog Earth Planet Sci 6:38
Navon O (1991) High internal-pressures in diamond fluid inclusions determined by infrared-absorption. Nature 353:746–748
Navon O, Hutcheon ID, Rossman GR, Wasserburg GJ (1988) Mantle-derived fluids in diamond micro-inclusions. Nature 355:784–789
Nimis P (2022) Pressure and temperature data for diamonds. Rev Mineral Geochem 88:533–566
Nimis P, Alvaro M, Nestola F, Angel RJ, Marquardt K, Rustioni G, Harris JW, Marone F (2016) First evidence of hydrous silicic fluid films around solid inclusions in gem-quality diamonds. Lithos 260:384–389
Nier AO (1947) A mass spectrometer for isotope and gas analysis. Rev Sci Instrum18:398–411
Nier AO, Gulbransen EA (1939) Variations in the relative abundance of the carbon isotopes. J Am Chem Soc 61:697–698
Nisbet EG, Mattey DP, Lowry D (1994) Can diamonds be dead bacteria. Nature 367:694–694
Novella D, Bolfan-Casanova N, Nestola F, Harris JW (2015) H_2O in olivine and garnet inclusions still trapped in diamonds from the Siberian craton: Implications for the water content of cratonic lithosphere peridotites. Lithos 230:180–183
O'Neil JR (1986) Theoretical and experimental aspects of isotopic fractionation. Rev Mineral 16:1–40
O'Neill HS, Rubie DC, Canil D, Geiger CA, Ross CR, Seifert F, Woodland AB (1993) Ferric iron in the upper mantle and in transition zone assemblages: Implications for relative oxygen fugacities in the mantle. *In:* Evolution of the Earth and Planets. Vol 74. Am Geophys Union Geophys Monograph Ser, p 73–88
Palot M, Cartigny P, Harris JW, Kaminsky FV, Stachel T (2012) Evidence for deep mantle convection and primordial heterogeneity from nitrogen and carbon stable isotopes in diamond. Earth Planet Sci Lett 357–358:179–193
Palot M, Pearson DG, Stern RA, Stachel T, Harris JW (2013) Multiple growth events, processes and fluid sources involved in diamond genesis: A micro-analytical study of sulphide-bearing diamonds from Finsch mine, RSA. Geochim Cosmochim Acta 106:51–70
Palot M, Pearson DG, Stern RA, Stachel T, Harris JW (2014) Isotopic constraints on the nature and circulation of deep mantle C–H–O–N fluids: Carbon and nitrogen systematics within ultra-deep diamonds from Kankan (Guinea). Geochim Cosmochim Acta 139:26–46
Palot M, Pearson DG, Stachel T, Stern RA, Le Pioufle A, Gurney JJ, Harris JW (2017) The transition zone as a host for recycled volatiles: Evidence from nitrogen and carbon isotopes in ultra-deep diamonds from Monastery and Jagersfontein (South Africa). Chem Geol 466:733–749

Pearson DG, Davies GR, Nixon PH, Milledge HJ (1989) Graphitized diamonds from a peridotite massif in Morocco and implications for anomalous diamond occurrences. Nature 338:60–62

Pearson DG, Davies GR, Nixon PH, Mattey DP (1991) A carbon isotope study of diamond facies pyroxenites and associated rocks from the Beni Bousera Peridotite, North Morocco. J Petrol Spec Vol 2:175–189

Pearson DG, Canil D, Shirey SB (2003) Mantle samples included in volcanic rocks: xenoliths and diamonds. Chapter 5 *In*: Treatise on Geochemistry vol. 2:171–275. Elsevier, Amsterdam

Pearson DG, Brenker FE, Nestola F, McNeill J, Nasdala L, Hutchison MT, Matveev S, Mather K, Silversmit G, Schmitz S, Vekemans B (2014) Hydrous mantle transition zone indicated by ringwoodite included within diamond. Nature 507:221–224

Peats J, Stachel T, Stern RA, Muehlenbachs K, Armstrong J (2012) Aviat diamonds: a window into the deep lithospheric mantle beneath the northern Churchill Province, Melville Peninsula, Canada. Can Mineral 50:611–624

Peslier AH, Woodland AB, Bell DR, Lazarov M (2010) Olivine water contents in the continental lithosphere and the longevity of cratons. Nature 467:78–81

Petts DC, Chacko T, Stachel T, Stern RA, Heaman LM (2015) A nitrogen isotope fractionation factor between diamond and its parental fluid derived from detailed SIMS analysis of a gem diamond and theoretical calculations. Chem Geol 410:188–200

Plank T, Manning CE (2019) Subducting carbon. Nature 574:343–352

Pokhilenko NP, Sobolev NV, Reutsky VN, Hall AE, Taylor LA (2004) Crystalline inclusions and C isotope ratios in diamonds from the Snap Lake/King Lake kimberlite dyke system: evidence of ultradeep and enriched lithospheric mantle. Lithos 77:57–67

Polyakov VB, Kharlashina NN (1994) Effect of pressure on equilibrium isotopic fractionation. Geochim Cosmochim Acta 58:4739–4750

Polyakov VB, Kharlashina NN (1995) The use of heat-capacity data to calculate carbon-isotope fractionation between graphite, diamond, and carbon-dioxide—a new approach. Geochim Cosmochim Acta 59:2561–2572

Ray JS, Ramesh R (2000) Rayleigh fractionation of stable isotopes from a multicomponent source. Geochim Cosmochim Acta 64:299–306

Regier ME, Pearson DG, Stachel T, Luth RW, Stern RA, Harris JW (2020) The lithospheric-to-lower-mantle carbon cycle recorded in superdeep diamonds. Nature 585:234–238

Reutsky VN, Harte B, EIMF, Borzdov YM, Palyanov YN (2008) Monitoring diamond crystal growth, a combined experimental and SIMS study. Eur J Mineral 20:365–374

Reutsky V, Borzdov Y, Palyanov Y, Sokol A, Izokh O (2015a) Carbon isotope fractionation during experimental crystallization of diamond from carbonate fluid at mantle conditions. Contrib Mineral Petrol 170:41

Reutsky V, Borzdov YM, Palyanov YN (2015b) Carbon isotope fractionation during high pressure and high temperature crystallization of Fe-C melt. Chem Geol 406:18–24

Reutsky VN, Shiryaev AA, Titkov SV, Wiedenbeck M, Zudina NN (2017) Evidence for large scale fractionation of carbon isotopes and of nitrogen impurity during crystallization of gem quality cubic diamonds from placers of North Yakutia. Geochem Int 55:988–999

Richardson SH, Gurney JJ, Erlank AJ, Harris JW (1984) Origin of diamonds in old enriched mantle. Nature 310:198–202

Richardson SH, Erlank AJ, Hart SR (1985) Kimberlite-borne garnet peridotite xenoliths from old enriched subcontinental lithosphere. Earth Planet Sci Lett 75:116–128

Richardson SH, Harris JW, Gurney JJ (1993) Three generations of diamonds from old continental mantle. Nature 366:256–258

Richet P, Bottinga Y, Javoy M (1977) A review of hydrogen, carbon, nitrogen, oxygen, sulphur, and chlorine stable isotope fractionation among gaseous molecules. Annu Rev Earth Planet Sci 5:65–110

Robert F, Rejou-Michel A, Javoy M (1992) Oxygen isotopic homogeneity of the Earth: new evidence. Earth Planet Sci Lett 108:1–9

Robertson R, Fox JJ, Martin AE (1934) Two types of diamond. Philos Trans R S (London), Ser A 232:463–535

Robinson DN, Scott JA, Van Niekerk A, Anderson VG (1989) The sequence of events reflected in the diamonds of some southern African kimberlites. *In:* Kimberlites and Related Rocks: Their Mantle/Crust Setting, Diamonds, and Diamond Exploration, GSA Spec Publ 14. Vol 2. Ross J et al. (eds) Blackwell, Carlton, p 990–1000

Rohrbach A, Schmidt MW (2011) Redox freezing and melting in the Earth's deep mantle resulting from carbon-iron redox coupling. Nature 472:209–212

Rohrbach A, Ghosh S, Schmidt MW, Wijbrans CH, Klemme S (2014) The stability of Fe–Ni carbides in the Earth's mantle: Evidence for a low Fe–Ni–C melt fraction in the deep mantle. Earth Planet Sci Lett 388:211–221

Rosenbaum JM (1994) Stable isotope exchange between carbon dioxide and calcite at 900 °C. Geochim Cosmochim Acta 58:3747–3753

Rudge JF (2008) Finding peaks in geochemical distributions: A re-examination of the helium-continental crust correlation. Earth Planet Sci Lett 274:179–188

Rumble D, Miller MF, Franchi IA, Greenwood RC (2007) Oxygen three-isotope fractionation lines in terrestrial silicate minerals: An inter-laboratory comparison of hydrothermal quartz and eclogitic garnet. Geochim Cosmochim Acta 71:3592–3600

Satish-Kumar M, So H, Yoshimo T, Mutsumi K, Hiroi Y (2011) Experimental determination of carbon isotope fractionation between iron carbide melt and carbon: ^{12}C-enriched carbon in the Earth's core? Earth Planet Sci Lett 310:340–348
Sautter V, Haggerty SE, Field S (1991) Ultradeep (>300 kilometers) ultramafic xenoliths—petrological evidence from the transition zone. Science 252:827–830
Schauble EA, Ghosh P, and Eiler JM (2006) Preferential formation of ^{13}C–^{18}O bonds in carbonate minerals, estimated using first-principles lattice dynamics. Geochim Cosmochim Acta 70:2510–2529
Scheele N, Hoefs J (1992) Carbon isotope fractionation between calcite, graphite and CO_2: An experimental study. Contrib Mineral Petrol 112:35–4
Schidlowski M (2001) Carbon isotopes as biogeochemical recorders of life over 3.8 Ga of Earth history: evolution of a concept. Precambrian Res 106:117–134
Schulze DJ, Harte B, Valley JW, Brenan JM, Channer DMD (2003) Extreme crustal oxygen isotope signatures preserved in coesite in diamond. Nature 423:68–70
Schulze DJ, Harte B, Edinburgh Ion Microprobe Facility staff, Page FZ, Valley JW, Channer DM, Jaques AL (2013) Anticorrelation between low $\delta^{13}C$ of eclogitic diamonds and high $\delta^{18}O$ of their coesite and garnet inclusions requires a subduction origin. Geology 41:455–458
Shatsky VS, Zedgenizov DA, Ragozin AL, Kalinina VV (2014) Carbon isotopes and nitrogen contents in placer diamonds from the NE Siberian craton: implications for diamond origins. Euro J Mineral 26:41–52
Shilobreeva S, Martinez I, Busigny V, Agrinier P, Laverne C (2011) Insights into C and H storage in the altered oceanic crust: Results from ODP/IODP Hole 1256D. Geochim Cosmochim Acta 75:2237–2255
Smit KV, Timmerman S, Aulbach S, Shirey SB, Richardson SH, Phillips D, Pearson DG (2022) Geochronology of diamonds. Rev Mineral Geochem 88:567–636
Shiryaev AA, Gaillard F (2014) Local redox buffering by carbon at low pressures and the formation of moissanite–natural SiC. Euro J Mineral 26:53–59
Simakov SK (2010) Metastable nanosized diamond formation from a C–H–O fluid system. J Mater Res 25:2336–2340
Simakov SK (2011) Nanodiamond formation in natural processes from fluid systems at low *P–T* parameters. Doklady Earth Sci 436:148–151
Slodkevich VV (1980) Polycrystalline aggregates of octahedral-shaped graphite. Doklady Akademii Nauk SSSR 253:697–700 (in Russian)
Slodkevich VV (1983) Graphite pseudomorphs after diamond. Int Geol Rev 25:497–514
Smart KA, Chacko T, Stachel T, Muehlenbachs K, Stern RA, Heaman LM (2011) Diamond growth from oxidized carbon sources beneath the Northern Slave Craton, Canada: A $\delta^{13}C$–N study of eclogite-hosted diamonds from the Jericho kimberlite. Geochim Cosmochim Acta 75:6027–6047
Smart KA, Tappe S, Stern RA, Webb SJ, Ashwal LD (2016) Early Archaean tectonics and mantle redox recorded in Witwatersrand diamonds. Nat Geosci 9:255–259
Smirnov GI, Mofolo MM, Lerotholi PM, Kaminsky FV, Galimov EM, Ivanovskaya IN (1979) Isotopically light carbon in diamonds from some kimberlite pipes in Lesotho. Nature 278:630–630
Smit KV, Shirey SB, Richardson SH, le Roex AP, Gurney JJ (2010) Re–Os isotopic composition of peridotitic sulphide inclusions in diamonds from Ellendale, Australia: Age constraints on Kimberley cratonic lithosphere. Geochim Cosmochim Acta 74:3292–3306
Smit KV, Shirey SB, Stern RA, Steele A, Wang W (2016) Diamond growth from C–H–N–O recycled fluids in the lithosphere: Evidence from CH_4 micro-inclusions and $\delta^{13}C$–$\delta^{15}N$–N content in Marange mixed-habit diamonds. Lithos 265:68–81
Smit KV, Stachel T, Luth RW, Stern RA (2019a) Evaluating mechanisms for eclogitic diamond growth: An example from Zimmi Neoproterozoic diamonds (West African craton). Chem Geol 520:21–32
Smit KV, Shirey SB, Hauri EH, Stern RA (2019b) Sulfur isotopes in diamonds reveal differences in continent construction. Science 364:383–385
Smith EM, Shirey SB, Nestola F, Bullock ES, Wang JH, Richardson SH, Wang WY (2016) Large gem diamonds from metallic liquid in Earth's deep mantle. Science 354:1403–1405
Sobolev NV (1977) Deep-seated inclusions in kimberlites and the problem of the composition of the upper mantle. (Translated from the Russian edition, 1974) AGU, Washington
Sobolev NV, Shatsky VS (1990) Diamond inclusions in garnets from metamorphic rocks–a new environment for diamond formation. Nature 343:742–746
Sobolev NV, Lavrent'ev YG, Pokhilenko NP, Usova LV (1973) Chrome-rich garnets from the kimberlites of Yakutia and their paragenesis. Contrib Mineral Petrol 40:39–52
Sobolev NV, Galimov EM, Ivanovskaya IN, Yefimova ES (1979) Isotopic composition of carbon from diamonds containing crystalline inclusions (in Russian). Doklady Akademii Nauk SSSR 249:1217–1220
Sokol AG, Palyanov YN, Tomilenki AA, Bul'bak TA, Palyanova GA (2017) Carbon and nitrogen speciation in nitrogen-rich C-O-H-N fluids at 5.5–7.8 GPa. Earth Planet Sci Lett 460:234–243
Stachel T (2014) Diamond. Mineral Assoc Canada Short Course Ser 44:1–28
Stachel T, Harris JW (1997) Syngenetic inclusions in diamond from the Birim field (Ghana)—A deep peridotitic profile with a history of depletion and re-enrichment. Contrib Mineral Petrol 127:336–352
Stachel T, Harris JW (2008) The origin of cratonic diamonds–Constraints from mineral inclusions. Ore Geol Rev 34:5–32

Stachel T, Luth RW (2015) Diamond formation—Where, when and how? Lithos 220–223:200–220

Stachel T, Brey GP, Harris JW (2000a) Kankan diamonds (Guinea) I: from the lithosphere down to the transition zone. Contrib Mineral Petrol 140:1–15

Stachel T, Harris JW, Brey GP, Joswig W (2000b) Kankan diamonds (Guinea) II: lower mantle inclusion parageneses. Contrib Mineral Petrol 140:16–27

Stachel T, Brey GP, Harris JW (2005) Inclusions in sublithospheric diamonds: Glimpses of deep Earth. Elements 1:73–78

Stachel T, Harris JW, Muehlenbachs K (2009) Sources of carbon in inclusion bearing diamonds. Lithos 112:625–637

Stachel T, Chacko T, Luth RW (2017) Carbon isotope fractionation during diamond growth in depleted peridotite: Counterintuitive insights from modelling water-maximum CHO fluids as multi-component systems. Earth Planet Sci Lett 473:44–51

Stachel T, Aulbach S, Harris JW (2022) Mineral inclusions in lithospheric diamonds. Rev Mineral Geochem 88:307–392

Stagno V, Frost DJ (2010) Carbon speciation in the asthenosphere: Experimental measurements of the redox conditions at which carbonate-bearing melts coexist with graphite or diamond in peridotite assemblages. Earth Planet Sci Lett 300:72–84

Stagno V, Ojwang DO, McCammon CA, Frost DJ (2013) The oxidation state of the mantle and the extraction of carbon from Earth's interior. Nature 493:84–88

Stern RA, Palot M, Howell D, Stachel T, Pearson DG, Cartigny P, Oh A (2014) Methods and reference materials for SIMS diamond C- and N-isotope analysis. University of Alberta, Education and Research Archive. Canadian Centre for Isotopic Microanalysis, Research Report 14–01:1–87

Stixrude L, Lithgow-Bertelloni C (2007) Influence of phase transformations on lateral heterogeneity and dynamics in Earth's mantle. Earth Planet Sci Lett 263:45–55

Sunagawa I (1990) Growth and morphology of diamond crystals under stable and metastable conditions. J Cryst Growth 99:1156–1161

Sverjensky DA, Huang F (2015) Diamond formation due to a pH drop during fluid–rock interactions. Nat Commun 6:8702

Sverjensky DA, Stagno V, Huang F (2014) Important role for organic carbon in subduction-zone fluids in the deep carbon cycle. Nature Geosci 7:909–913

Swart PK, Pillinger CT, Milledge HJ, Seal M (1983) Carbon isotopic variation within individual diamonds. Nature 303:793–795

Tappert R, Stachel T, Harris JW, Muehlenbachs K, Ludwig T, Brey GP (2005) Diamonds from Jagersfontein (South Africa): messengers from the sublithospheric mantle. Contrib Mineral Petrol 150:505–522

Taylor WR, Jaques AL, Ridd M (1990) Nitrogen-defect aggregation characteristics of some Australian diamonds: Time-temperature constraints on the source regions of pipe and alluvial diamonds. Am Mineral 75:1290–1310

Thomassot E, Cartigny P, Harris JW, Viljoen KS (2007) Methane-related diamond crystallization in the Earth's mantle: Stable isotope evidences from a single diamond-bearing xenolith. Earth Planet Sci Lett 257:362–371

Thomassot E, Cartigny P, Harris JW, Lorand JP, Rollion-Bard C, Chaussidon M (2009) Metasomatic diamond growth: A multi-isotope study 13C, 15N, 33S, 34S of sulphide inclusions and their host diamonds from Jwaneng (Botswana). Earth Planet Sci Lett 282:79–90

Thomson AR, Kohn SC, Bulanova GP, Smith CB, Araujo D, Walter MJ (2014) Origin of sub-lithospheric diamonds from the Juina-5 kimberlite (Brazil): constraints from carbon isotopes and inclusion compositions. Contrib Mineral Petrol 168:1–29

Thomson AR, Walter MJ, Kohn SC, Brooker RA (2016) Slab melting as a barrier to deep carbon subduction. Nature 529:76–79

Timmerman S, Yeow H, Honda M, Howell D, Jaques AL, Krebs MY, Woodland S, Pearson DG, Avila JN, Ireland TR (2019) U–Th/He systematics of fluid-rich 'fibrous' diamonds–Evidence for pre- and syn-kimberlite eruption ages. Chem Geol 515:22–36

Tolansky S, Komatsu H (1967) Abundance of type 2 diamonds. Science 157:1173–1175

Urey HC (1947) The thermodynamic properties of isotopic substances. J Chem Soc 562–581

Valley JW, O'Neil JR (1981) $^{13}C/^{12}C$ exchange between calcite and graphite–a possible thermometer in Grenville marbles. Geochim Cosmochim Acta 45:411–419

Vermeesch P (2012) On the visualisation of detrital age distributions. Chem Geol 312–313:190–194

Viljoen KS, Dobbe R, Smit B, Thomassot E, Cartigny P (2004) Petrology and geochemistry of a diamondiferous lherzolite from the Premier diamond mine, South Africa. Lithos 77:539–552

Vinogradov AP, Kropotova OI, Usitinov VI (1965) Possible source of carbon in diamonds as indicated by the $^{12}C/^{13}C$ ratios. Geochem Int 2:495–503 (translated from Geokhimiya, 1965, n°1966, 1643–1651)

Vinogradov AP, Kropotov OI, Orlov YL, Grinenko VA (1966) Isotopic composition of diamond and carbonado crystals. Geochem Int 3:1123–1125

Walter MJ, Bulanova GP, Armstrong LS, Keshav S, Blundy JD, Gudfinnsson G, Lord OT, Lennie AR, Clark SM, Smith CB, Gobbo L (2008) Primary carbonatite melt from deeply subducted oceanic crust. Nature 454:622–625

Walter MJ, Thomson AR, Smith EM (2022) Geochemistry of silicate and oxide inclusions in sublithospheric diamonds. Rev Mineral Geochem 88:393–450

Wang Y-F, Qin J-Y, Soustelle V, Zhang J-F, Xu H-J (2021) Pyroxene does not always preserve its source hydrogen concentration: Clues from some peridotite xenoliths. Geochim Cosmochim Acta 292:382–408
Watson EB (1996) Surface enrichment and trace-element uptake during crystal growth. Geochim Cosmochim Acta 60:5013–5020
Watson EB (2004) A conceptual model for near-surface kinetic controls on the trace-element and stable isotope composition of abiogenic calcite crystals. Geochim Cosmochim Acta 68:1473–1488
Watson EB, Liang Y (1995) A simple model for sector zoning in slowly grown crystals: Implications for growth rate and lattice diffusion, with emphasis on accessory minerals in crustal rocks. Am Mineral 80:1179–1187
Weiss Y, McNeill J, Pearson DG, Nowell GM, Ottley CJ (2015) Highly saline fluids from a subducting slab as the source for fluid-rich diamonds. Nature 524:339–342
Weiss Y, Cazs J, Navon O (2022) Fluid inclusions in fibrous diamonds. Rev Mineral Geochem 88:475–532
Wickman FE (1956) The cycle of carbon and the stable carbon isotopes. Geochim Cosmochim Acta 9:136–153
Wiggers de Vries DF, Bulanova GP, De Corte K, Pearson DG, Craven JA, Davies GR (2013) Micron-scale coupled carbon isotope and nitrogen abundance variations in diamonds: Evidence for episodic diamond formation beneath the Siberian Craton. Geochim Cosmochim Acta 100:176–199
Wilson MR, Kjarsgaard BA, Taylor B (2007) Stable isotope composition of magmatic and deuteric carbonate phases in hypabyssal kimberlite, Lac de Gras field, Northwest Territories, Canada. Chem Geol 242:435–454
Wirth R, Rocholl A (2003) Nanocrystalline diamond from the Earth's mantle underneath Hawaii. Earth Planet Sci Lett 211:357–369
Wyllie PJ (1987) Metasomatism and fluid generation in mantle xenoliths. *In*: Mantle Xenoliths. Nixon PH (ed) John Wiley and Sons Ltd., Chichester, p 609–621
Xu XZ, Cartigny P, Yang JS, Dilek Y, Xiong FH, Guo GL (2018) Fourier transform infrared spectroscopy data and carbon isotope characteristics of the ophiolite-hosted diamonds from the Luobusa ophiolite, Tibet, and Ray-Iz ophiolite, Polar Urals. Lithosphere 10:156–169
Yaxley GM, Ghosh S, Kiseeva ES, Mallik A, Spandler C, Thomson AR, Walter MJ (2019) CO_2-rich melts in Earth. *In*: Deep Carbon: Past to Present. Orcutt BN, Daniel I, Dasgupta R, (eds). Cambridge University Press, Cambridge, p 129–162
Zedgenizov DA, Harte B (2004) Microscale variations of $\delta^{13}C$ and N content within a natural diamond with mixed-habit growth. Chem Geol 205:169–175
Zedgenizov D, Rubatto D, Shatsky V, Ragozin A, Kalinina V (2016) Eclogitic diamonds from variable crustal protoliths in the northeastern Siberian craton: Trace elements and coupled $\delta^{13}C$–$\delta^{18}O$ signatures in diamonds and garnet inclusions. Chem Geol 422:46–59
Zhang C, Duan Z (2009) A model for C–O–H fluid in the Earth's mantle. Geochim Cosmochim Acta 73:2089–2102

FSC
www.fsc.org
MIX
Papier | Fördert
gute Waldnutzung
FSC® C083411

Kolibri 360 GmbH
Ferdinand-Jühlke-Straße 7
99095 Erfurt, Deutschland
produktsicherheit@kolibri360.de